Preface

The first reference text on surface mining was published by the American Institute of Mining, Metallurgical, and Petroleum Engineers, Inc., in 1968 and was appropriately titled "Surface Mining." That volume was the product of the dedicated labors of Professor Eugene P. Pfleider, his 3 coordinating editors, and 89 authors.

Now, 22 years later, the Society for Mining, Metallurgy, and Exploration, Inc. (formerly the Society of Mining Engineers of AIME) is publishing a completely new "Surface Mining" reference text. This book has involved a great deal of effort and perseverance by the authors, chapter editors, and coordinating editors. The intent of this revised edition was to completely rewrite, expand, and update the first edition to provide both a reference book for the working engineer and a textbook for the mining student. The book has taken far longer to prepare than the original ambitious plan and almost lost some of its authors, who had already written their contributions, when the Society had to shelve the project for two years due to financial constraints.

The Society and the profession are deeply indebted to the authors and editors who contributed both their knowledge and an enormous amount of time to this project. In addition, a very special acknowledgment is due to Marianne Snedeker, formerly Manager of Publications and currently Manager of Book Publishing, and her staff who have kept the project on course and organized during its development.

Special recognition is given the Seeley W. Mudd Memorial Fund Committee of AIME for financial support of the publication.

We hope that this publication will prove useful to many minerals industry professionals worldwide.

Dec. 8, 1989
Jakarta, Indonesia

Bruce A. Kennedy
Editor

Contributors

Editor

Bruce A. Kennedy
Managing Director
PT Pelsart Management Services
Jakarta, Indonesia

Associate Editors

Lee R. Rice
Lee R. Rice and Associates
Littleton, CO

Raj K. Singhal
Energy Mines and Resources Canada
Devon, AB, Canada

Editorial Board and Chapter Editors

Richard A. Bideaux
Consultant
Oro Valley, AZ
Chapter 3

Richard L. Brown
Vice President Exploration
Asarco Inc.
New York, NY
Chapter 2

Hugh W. Evans
Consultant
Union Terminal Assn., National Assn.
of 10th Mountain Division
Cincinnati, OH
Chapter 6

Guerdon E. Jackson
Management Consultant
Tucson, AZ
Chapter 4

Thys B. Johnson
Associate Director
Natural Resources Research Institute
University of Minnesota
Duluth, MN
Chapter 6

Bruce A. Kennedy
Managing Director
PT Pelsart Management Services
Jakarta, Indonesia
Chapters 1 and 9

Robert F. Winkle
Vice President
Pincock, Allen & Holt, Inc.
Tucson, AZ
Chapter 7

Robert Laurich
Principal Engineer
Lithium Corp. of America
Gastonia, NC
Chapter 5

Donald O. Rausch
Retired
Lakewood, CO
Chapter 8

Lee W. Saperstein
Chairman
Dept. of Mining Engineering
University of Kentucky
Lexington, KY

Authors

Aiken, George E.
Retired
Formerly with Kaiser Engineers, Inc.
Oakland, CA
Chap. 6, Sec. 6.3

Armstrong, David
Mintec, Inc.
Tucson, AZ
Chap. 5, Sec. 5.1

Ash, Richard L.
Retired
Formerly Professor
Dept. of Mining Engineering
University of Missouri-Rolla
Rolla, MO
Chap. 6, Sec. 6.2.2

Barker James M.
Industrial Minerals Geologist
New Mexico Bureau of Mines &
    Mineral Resources

Socorro, NM
Chap. 2, Sec. 2.10.3

Bauer, Alan*
Professor
Dept. of Mining Engineering
Queens' University
Kingston, ON, Canada
Chap. 6, Secs. 6.1.2 and 6.2.1

Bentzen, Edwin H., III
Senior Processing Engineer
Ore Sorters Inc.
Lakewood, CO
Chap. 2, Sec. 2.10

Berry, Charles W.
Professor
Mining Engineering Dept.
University of Utah
Salt Lake City, UT
Chap. 8, Sec. 8.2

Betzler, William F.
Retired
Minnesota Dept. of Revenue
Buhl, MN
Chap. 3, Sec. 3.4.4

Bideaux, Richard A.
Consultant
Oro Valley, AZ
Chap. 3, Sec. 3.1

Bolin, David S.*
Pincock, Allen & Holt, Inc.
Lakewood, CO
Chap. 3, Sec. 3.4.1

Bohnet, Ernest L.
President
Pincock, Allen & Holt, Inc.
Lakewood, CO
Chap. 5, Secs. 5. 4 and 5.6

Bolles, John L.

iv

Retired
Silver City, NM
Chap. 9, Sec. 9.4

Borquez, Guillermo V.
Bechtel, Inc.
San Francisco, CA
Chap. 4, Sec. 4.2

Boyd, James W.
President
John T. Boyd Co.
Pittsburgh, PA
Chap. 3, Sec. 3.4.3

Brazda, Lois L.
Amoco Corp.
Chicago, IL
Chap. 4, Sec. 4.3

Breeding, William H.
W.H. Breeding & Associates, Inc.
Summit, NJ
Chap. 2, Sec. 2.7

Brobst, Donald A.
Consulting Geologist
Reston, VA
Chap. 2, Sec. 2.10.2

Brown, Richard L.
Vice President Exploration
Asarco, Inc.
New York, NY
Chap. 2, Sec. 2.1

Burns, Charles L.
Vice President
Government Relations
Phelps Dodge Corp.
Washington, DC
Chap. 8, Sec. 8.9

Call, Richard D.
President
Call and Nicholas Inc.
Tucson, AZ
Chap. 6, Sec. 6.8

Canton, Perry A.
Minnesota Dept. of Natural Resources
St. Paul, MN
Chap. 3, Sec. 3.4.4

Carr, Donald D.
Branch Head
Mineral Resources & Data
    Management
Indiana Geological Survey
Bloomington, IN
Chap. 2, Sec. 2.10.11

Clar, Michael L.
Secretary-Treasurer
Engineering Technologies

Elliott City, MD
Chap. 6, Sec. 6.6.2

Cole, Clifford F.
Senior Project Manager
TRC Environmental Consultants, Inc.
Englewood, CO
Chap. 6, Sec. 6.7.2

Crosby, William A.
President
Mining Resource Engineering Ltd.
Kingston, ON, Canada
Chap. 6, Secs. 6.1.2 and 6.2.1

Dibble, M.F.
Consulting Engineer
Lakeland, FL
Chap. 2, Sec. 2.10.20

Donoso, Enrique Morel
President
Chuquicamata Div.
CODELCO-Chile
Santiago, Chile
Chap. 9, Sec. 9.6

Dorling, I.P.
Morrison-Knudsen Engineers, Inc
Boise, ID
Chap. 2, Sec. 2.9

Duncan, Larry D.
Stearns Catalytic
Denver, CO
Chap. 6, Sec. 6.5.3

Dupree, Annette C.
North Quincy, MA
Chap. 8, Sec. 8.4

Elayer, R.W.
Manager-Geology
Morrison-Knudsen Engineers, Inc.
Boise, ID
Chap. 2, Sec. 2.9

Elbrond, Jorgen
Professor
Dept. of Mineral Engineering
Ecole Polytechnique
Montreal, PQ, Canada
Chap. 6, Sec. 6.5.9

Files, Thomas I.
Project Manager
Marion Div., Dresser Industries Inc.
Marion, OH
Chap. 6, Sec. 6.4

Fitch, David C.
Manager, Southwest Exploration
Hecla Mining Co.
Reno, NV
Chap. 2, Sec. 2.4

Fletcher, David
Mineral Economics Dept.
Colorado School of Mines
Golden, CO
Chap. 8, Sec. 8.2

Gilliss, Archie M.
Falkirk Mining Co.
North American Coal Corp.
Bismarck, ND
Chap. 6, Sec. 6.12.4

Goldman, Harold B.
Geological Consultants, Inc.
San Francisco, CA
Chap. 2, Sec. 2.10.23

Goodner, William R.
Stearns Catalytic
Denver, CO
Chap. 6, Sec. 6.5.4

Goodrich, R.F.
Reclamation Engineer
IMC Fertilizer, Inc.
Bartow, FL
Chap. 6, Sec. 6.6.4

Grandt, Alten F.
Director of Reclamation
Peabody Coal Co.
St. Louis, MO
Chap. 6, Sec. 6.6.7

Grundstedt, Steven R.
Mining Analyst
Morgan, Paul & Co.
Sydney, NSW, Australia
Chap. 8, Sec. 8.6

Gunnett, John W.
Skelly and Loy
Harrisburg, PA
Chap. 6, Sec. 6.3

Hardie, Byron S.
Geological Consultant
Tucson, AZ
Chap. 2, Sec. 2.3

Hartman, Howard L.
Retired
University of Alabama
Consultant
Carmichael, CA
Chap. 6, Sec. 6.1.1

Hays, Ronald M.
Mining and Metallurgical Consultant
Ronald M. Hays and Associates
Edina, MN
Chap. 3, Sec. 3.4.4
Chap. 6, Secs. 6.5.2, 6.5.5, 6.5.6
    6. 5.7

Hemenway, David S.
Senior Counsel
Peabody Holding Co., Inc.
St. Louis, MO
Chap. 8, Sec. 8.7

Hennessy, J. Allen
Principal Mining Geologist
Mining & Metals Div.
Bechtel Civil & Minerals, Inc.
San Francisco, CA
Chap. 2, Sec. 2.8

Henning, Dieter
Reinbraun-Consulting GmbH
Rheinbraun Köln, Germany
Chap. 9, Sec. 9.2

Heyborne, L.S.
Manager
Mining Operations
Mine and Plant Operations Group
Bechtel Civil & Minerals Inc.
San Francisco, CA
Chap. 8, Sec. 8.5

Hoffman, Clarence W.
Performance Associates, Inc.
Danville, CA
Chap. 6, Sec. 6.11.2

Hoyle, Daniel Rodriguez
Mineral Industry Consultant
Formerly Vice President
Southern Peru Copper Corp.
Lima, Peru
Chap. 9, Sec. 9.5

Huber, Douglas W.
Mining Adviser
USAID/ISLAMABAD Agency for Inter-
   national Development
US Dept. of State
Washington, DC
Chap. 6, Sec. 6.12.1

Humphrey, James D.
Chief Engineer - Marketing Develop-
   ment
Marion Div., Dresser Industries, Inc.
Marion, OH
Chap. 6, Sec. 6.4

Hutnik, Russell J.
Professor of Forest Ecology
The Pennsylvania State University
University Park, PA
Chap. 6, Sec. 6.6.6

Jackson, Guerdon E.
Mining Consultant
Tucson, AZ
Chap. 4, Sec. 4.1

Johnson, Wesley G.

Mining Engineering Dept.
Colorado School of Mines
Golden, CO
Chap. 6, Sec. 6.12.3

Kadey, Frederic L., Jr.
Retired
Lantana, FL
Chap. 2, Secs. 2.10.6, 2.10.19

Kahle, Michael B.
Director of Marketing
Phillips Coal Co.
Richardson, TX
Chap. 6, Sec. 6.14

Kennedy, Bruce A.
Managing Director
PT Pelsart Management Services
Jakarta, Indonesia
Chap. 9, Sec. 9.1

Kerch, Richard L.
Consolidation Coal Co.
Pittsburgh, PA
Chap. 6, Sec. 6.7.2

Knudsen, Harvey P.
Associate Professor
Mining Engineering Dept.
Montana College of Mineral Science &
   Technology
Butte, MT
Chap. 3, Sec. 3.2

Kunze, Lutz
Vice President
Smith-Emery Co.
Los Angeles, CA
Chap. 5, Sec. 5.6

Leroy, A.J.
Palabora Mining Co., Ltd.
Phalaborwa, Rep. of South Africa
Chap. 9, Sec. 9.3

Levitt, Brian J.
Materials Handling Engineer
Stearns Catalytic
Denver, CO
Chap. 6, Sec. 6.5.3

Li, Ta M.
Chief Operating Officer
Pincock, Allen & Holt, Inc.
Lakewood, CO
Chap. 1, Sec. 1.2

Lill, John W.
American Barrick Resources Corp.
Toronto, ON, Canada
Chap. 9, Sec. 9.3

Loving, Gary A.
General Manager

Homestake Mining Co.
Lead, SD
Chap. 9, Sec. 9.4

Lowell, J. David
Geological Consultant
Lowell Mineral Exploration
Nogales, AZ
Chap. 2, Sec. 2.2

Madson, James L.
Manager, Morenci Branch
Phelps Dodge Corp.
Morenci, AZ
Chap. 9, Sec. 9.4

Malouf, E.E.
Consultant
Salt Lake City, UT
Chap. 6, Sec. 6.9.1

Marsden, Ralph W.*
Professor Emeritus
Dept. of Geology
University of Minnesota
Duluth, MN
Chap. 2, Sec. 2.5

Marshall, Darryl A.
Fluor Daniel
Redwood City, CA
Chap. 8, Sec. 8.6

Mathieson, Graham
General Manager
Griffin Coal Mining Co., Ltd.
Perth, WA, Australia
Chap. 3, Sec. 3.3

McCarter, M.K.
Professor
Mining Engineering Dept.
University of Utah
Salt Lake City, UT
Chap. 6, Sec. 6.9.2

McIndoo, Robert N.
Babbitt, MN
Chap. 6, Sec. 6.11.1

McKee, Guy W.
Professor of Agronomy
The Pennsylvania State University
University Park, PA
Chap. 6, Sec. 6.6.6

McKereghan, G.F.
International Minerals & Chemical Co.
Bartow, FL
Chap. 6, Sec. 6.6.4

McKie, P.W.
Director
Mine & Excavated Structures Design
Morrison-Knudsen Engineers, Inc.

Boise, ID
Chap. 2, Sec. 2.9

Meek, F.A., Jr.
Upshur Mine Complex
Enoxy Coal Inc.
Buckhannon, WV
Chap. 6, Sec. 6.7.1

Misagi, Leo
Head
MSHA Academy
US Dept. of Labor
Beckley, WV
Chap. 6, Sec. 6.12.2

Morey, Philip G.
President
Jval, Inc.
Redwood City, CA
Chap. 5, Sec. 5.7

Morris, Jerald S.
Vice President, Operations
Central Stone Co.
Hannibal, MO
Chap. 8, Sec. 8.6

Mortensen, Dennis K.
Vice President, General Manager
Cyprus Miami Mining Corp.
Claypool, AZ
Chap. 6, Sec. 6.10

Olson, Richard H.
President
Industrial Minerals Evaluations, Inc.
Golden CO
Chap. 2, Secs. 2.10, 2.10.24, 2.10.28

Parker, Harry M.
Vice President
Mineral Resource Development, Inc.
San Mateo, CA
Chap. 3, Sec. 3.4.2

Parkison, Gary
Westmont Mining Inc.
Tucson, AZ
Chap. 2, Sec. 2.10.27

Parr, Clayton J.
Kimball, Parr, Crockett & Waddoups
Salt Lake City, UT
Chap. 2, Sec. 2.11

Phelps, L.B.
Assistant Professor, Mining
    Engineering
The Pennsylvania State University
University Park, PA
Chap. 6, Sec. 6.6.3

Plute, Daniel P.
President

Procurement Services Associates
Stockton, CA
Chap. 8, Sec. 8.8

Pratt, Morton E.
Retired
Formerly Executive Engineer
Utah International
Moraga, CA
Chap. 9, Sec. 9.8

Presley, Gordon C.
Consultant
Englewood, CO
Chap. 2, Sec. 2.10

Ramani, Raja V.
Professor and Head
Mineral Engineering Dept.
The Pennsylvania State University
University Park, PA
Chap. 6, Secs. 6.5.1, 6.5.8, 6.6.2

Rausch, Donald O.
Retired
Lakewood, CO
Chap. 8, Sec. 8.1

Reed, Ronald G.
Pincock, Allen & Holt, Inc.
Lakewood, CO
Chap. 5, Sec. 5.5

Rendu, Jean-Michel
Newmont Gold Co.
Denver, CO
Chap. 3, Sec. 3.3

Ritchie, M.I.
Pathfinder Mines Corp.
San Francisco, CA
Chap. 9, Sec. 9.7

Rogowski, A.S.
Soil Scientist
US Dept. of Agriculture
Agricultural Research Service/North
    Atlantic Area
University Park, PA
Chap. 6, Sec. 6.6.5

Saperstein, Lee W.
Chairman
Dept. of Mining Engineering
University of Kentucky
Lexington, KY
Chap. 6., Sec. 6.6.1

Sargent, Fred B.
Vice President, Sales & Marketing
Marion Div., Dresser Industries, Inc.
Marion, OH
Chap. 6, Sec. 6.4

Savely, James P.

Chief Geological Engineer
Inspiration Consolidated Copper Co.
Claypool, AZ
Chap. 6, Sec. 6.8

Snedeker, Marianne
Manager of Book Publishing
Society for Mining, Metallurgy, and
    Exploration, Inc.
Littleton, CO
Chap. 1, Secs. 1.1 and 1.2

Sullivan, Ruth
Enoxy Coal, Inc.
Cincinnati, OH
Chap. 7, Sec. 7.2.2

Sweigard, Richard J.
Assistant Professor
Mining Engineering Dept.
University of Kentucky
Lexington, KY
Chap. 6, Sec. 6.6.2

Thomas, Stephen W.
Manager, Acquisitions and Development
Cyprus Gold Co.
Englewood, CO
Chap. 3, Sec. 3.4.5

Thompson, James V.
Consultant
Lafayette, CA
Chap. 4, Sec. 4.2

Thornburg, H. Andrew
Vice Chairman
Security Pacific Capital Markets Group
Los Angeles, CA
Chap. 4, Sec. 4.3

Ulatowski, Tomek
Senior Vice President
Crocker National Bank
San Francisco, CA
Chap. 4, Sec. 4.4

Vickers, Edward
Kailua Kona, HI
Chap. 4, Sec. 4.4

Voige, Raymond C.
Mine Safety Appliances Co.
Pittsburgh, PA
Chap. 6, Sec. 6.13

Waleski, William E.
Belle Ayre Mine, Western Div.
Amax Coal Co.
Gillette, WY
Chap. 7, Sec. 7.2.4

Waples, Bee R., Jr.*
Vice President

Mineral Div.
Kaiser Engineers, Inc.
Oakland, CO
Chap. 4, Sec. 4.1

Ward, Willard E., II
Geologist
Energy Resources Div.
Geological Survey of Alabama
Tuscaloosa, AL
Chap. 2, Sec. 2.6

Ward, Milton H.
President and Chief Operating Officer
Freeport McMoran
New Orleans, LA
Chap. 8, Sec. 8.3

Weber, Kenneth J.
Consultant
Iron Mountain, MI
Chap. 7, Sec. 7.2

Weinrich, B.E.
Assistant Professor of Mathematics

California University of Pennsylvania
California, PA
Chap. 6, Sec. 6.6.5

Weise, Hans
Reinbraun-Consulting GmbH
Reinbraun Köln, Germany
Chap. 9, Sec. 9.2

Wendt, Clancy J.
District Manager
Westmont Mining, Inc.
Sparks, NV
Chap. 3, Sec. 3.4.5

Whittle, Jeff
Director
Whittle Programming Pty. Ltd.
North Balwyn, Vic., Australia
Chap. 5, Sec. 5.3

Wicken, Oscar M.
SRM Associates
West Palm Beach, FL
Chap. 2, Sec. 2.10.13

Woodring, Kenneth
President
Hobet Mining Co.
Madison, WV
Chap. 7, Sec. 7.2.2

Yu, A.T.
Retired
Formerly President
ORBA Corp.
Clearwater, FL
Chap. 6, Sec. 6.10

Zdunczyk, Mark J.
Senior Geologist
Dunn Geoscience Corp.
Albany, NY
Chap. 2, Sec. 2.10.22

Zimmer, G.S.
President
GSZ Inc.
Tucson, AZ
Chap. 7, Sec. 7.2.5

*Deceased

Table of Contents

Chapter 1

Introduction

Bruce A. Kennedy, Editor

1.1. History of Mining

MARIANNE SNEDEKER

INTRODUCTION

A fascinating thread that runs through the history of mining is the continuing evolution of mining methods. Often, the initial exploitation of a deposit involved rudimentary scratching at outcrops and picking up pieces of ore from the surface. This surface method was then followed in many instances by the development of underground workings in the form of shafts and galleries. Finally, a surface operation, often on a large-scale, would take place. Two prime examples of this sequence of evolution are the Rio Tinto mines in Spain and the copper operations at Butte, Montana.

Mining was the second of man's endeavors—agriculture was the first. Since prehistoric times, mining has been integral and essential to man's existence (Hartman, 1987). Surface mining was certainly not a 20th century invention. The earliest relatively large-scale mining for outcropping native copper occurred between 5000 and 15,000 BC. Rock fragmentation was usually achieved by the cyclical application of fire and water; loading and haulage was performed by manual labor with stone, wooden, and bronze tools for excavation and animals and human beings were used for haulage (Michaelson, 1979).

EARLY ENDEAVORS

Mining began with Paleolithic man about 450,000 years ago. The first known mining was for nonmetallics (industrial minerals) where man recovered raw stone materials from surface excavations and shaped them by crude fabrication techniques. Flint instruments have been found with the bones of early man (Hartman, 1987) and early excavations for flint have been found in Obourg and Spiennes, Belgium. Utensils made from clay that date from 30,000 to 20,000 BC have been found in Czechoslovakia (Beall, 1973) and graphite, sometimes called "Plumbago" or "Black Lead," was used by primitive man to make drawings on the walls of caves and by the Egyptians to decorate pottery (Graffin, 1982).

The first use of metals was for decoration rather than for utility purposes because of their unusual character and rarity (Raymond, 1986). Metallic minerals particularly attracted early man and he usually used them in their native form, retrieved, probably, by washing river gravel in surface placers (Hartman, 1987). According to Agricola (1950) early exploration methods included trenches and "a divining rod shaped like a fork."

The cultural stages of the evolution of man are associated with minerals and are the Stone Age (prior to 4000 BC), Bronze Age (4000 to 1500 BC), Iron Age (1500 BC to 1780 AD), Steel Age (1780-1945), and the Nuclear Age (since 1945), according to Hartman (1987). A chronology of developments in mining technology is given in Table 1.1 (Hartman, 1987).

Mining was common in ancient times around the perimeter of the Mediterranean Sea. Greek writers, Herodytus and Aristotle, both mention mining (Beall, 1973) and caves in Spain that were occupied as early as the Paleolithic period (before 10,000 BC) and gravel deposits nearby have yielded artifacts of gold with a sun-religion significance (Dunning, 1970). Mining in the Rio Tinto district of Spain began far before the dawn of recorded history. The largest open pit in the area, the Atalaya, has old underground workings now cut by more recent surface diggings. The early inhabitants, the Iberians, had both gold and silver ornaments (Joralemon, 1973).

Bronze Age

Evidence of early copper mining exists in many parts of the world. For example, a recent archeometallurgical expedition has uncovered a prehistoric mining complex at Phu Lon ("Bald Mountain") on the Mekong River in Thailand, that may be dated as early as 2000 BC. Workers at this complex used massive river cobble mauls to break the friable skarn matrix that held quartz veins rich in malachite (Pigott, 1988). The world's oldest known copper smelting furnace, dating to 3500 BC, has been found near the modern Timna copper mine in Israel (Raymond, 1986).

The link between native copper and malachite might well have been suggested to Neolithic man by the common association of these two forms of the metal in outcrops. But the process by which he then learned how to extract copper from the malachite remains an historic mystery. One suggested answer is that both metal smelting and pottery making appeared to have evolved about the same time. The potter, the first technician in the management of heat, had under his control all the materials and conditions necessary for smelting copper (Raymond, 1986).

The advent of both the Bronze and Iron Ages was contingent upon man's discovery of smelting and learning to reduce ores to native metal or alloy form.

The art of rock breakage by fire setting was the first technological breakthrough in mining (Hartman, 1987). Archaeological evidence of copper smelting indicate that the technique may not have spread from the Near East to other areas, as often thought, but began independently at a number of sites. Some of these sites are (Raymond, 1986):

Rudna Glava, Yugoslavia, before 4000 BC
Italy, between 3000 and 2500 BC
Britain by 1900 BC
Scandinavia by 1500 BC
India about 3000 BC
Caucasus, southern Russia, by 2000 BC

The first true tin bronze appeared about 3000 BC with the earliest examples from the city states of Mesopotamia. In the following century, tin bronze was made in many areas of the Near East, including Egypt, Iran, Syria, Anatolia,

Table 1. Chronological Development of Mining Technology*

Date	Event
450,000 BC	First mining (at surface), by Paleolithic man for stone implements
40,000	Surface mining progresses underground, in Swaziland, Africa
30,000	Fired clay pots used in Czechoslovakia
18,000	Possible use of gold and copper in native form
5000	Fire setting, used by Egyptians to break rock
4000	Early use of fabricated metals; start of Bronze Age
3400	First recorded mining, of turquoise by Egyptians in Sinai
3000	Probable first smelting, of copper with coal by Chinese; first use of iron implements by Egyptians
2000	Earliest known gold artifacts in New World, in Peru
1000	Steel used by Greeks
AD 100	Thriving Roman mining industry
122	Coal used by Romans in Great Britain
1185	Edict by bishop of Trent gives rights to miners
1524	First recorded mining in New World, by Spaniards in Cuba
1550	First use of lift pump, at Joachirnstal, Czechoslovakia
1556	First mining technical work, *De Re Metallica*, published in Germany by Georgius Agricola
1585	Discovery of iron ore in North America, in North Carolina
1600s	Mining commences in eastern United States (iron, coal, lead, gold)
1627	Explosives first used in European mines, in Hungary (possible prior use in China)
1646	First blast furnace installed in North America, in Massachusetts
1716	First school of mines established, at Joachimstal, Czechoslovakia
1780	Beginning of Industrial Revolution; pumps first modern machines used in mines
1800s	Mining progresses in United States; gold rushes help open the West
1815	Sir Humphrey Davy invented miner's safety lamp in England
1855	Bessemer steel process first used, in England
1867	Dynamite invented by Nobel, applied to mining
1903	Era of mechanization and mass production open in U.S. mining with development of first low-grade copper porphyry, in Utah; while the first modern mine was an open pit, subsequent operations were underground as well

* Source: Hartman, H.L., 1987, *Introductory Mining Engineering*, John Wiley & Sons, New York, p. 6.

Cyprus. There was a society in Northern Thailand with bronze technology in the 3rd and 4th millenium BC. Bronze artifacts have been found in other locations in China (Raymond, 1986). Also in the Far East, Shang bronzes from Yin, China, settled about 1600 BC, are among the greatest artistic and technical achievements of early civilization.

Iron Age

The introduction of iron for making tools and weapons changed the life of early man in a vast number of ways. The earliest objects that have survived were made of meteoric iron, which contains a high percentage of nickel, and which were picked up from the ground (Raymond, 1986).

The Hittites, who settled in Anatolia, are credited with the invention of "good iron," and they flourished from 1400 to 1200 BC. After the Hittite kingdom was destroyed by European tribes, iron began to appear everywhere (Raymond, 1986).

The use of iron was made feasible through the development of three processes: "steeling," the addition of carbon to ore; "quenching," the sudden cooling of hot metal; and "tempering," the reheating of quenched metal to correct for brittleness. Steeling of iron seems to have developed in the Anatolia-Mesopotamia region of the Near East and then spread across Europe and Africa and into Asia (Raymond, 1986).

Greece and Rome

As the Iron Age progressed, new powers took over in the latter half of the 1st millenium BC as the older cultures of the Near East declined. The city-states of Greece were the first of these. One factor that made Greece—Athens, in particular—a great power was the rich silver-lead mines at Laurion. Once the silver mines in the hills of Attica became depleted, so too waned the Hellenistic world.

Rome replaced Greece as the dominant power and established a far-flung empire. Especially significant was the Roman conquest of immense mineral resources: gold in Gaul, Wales, the Balkans, and Persia; silver in the Pyrenees, Greece, and Anatolia; copper in Cyprus and the Sinai; and tin in Cornwall. A principal contributor to Roman mineral wealth was the Tartessus mine in Spain, now known as Rio Tinto, where the first exploitation was by trenches. The mine then went underground and today is a large open pit (Raymond, 1986).

Dark and Middle Ages

The Dark Ages descended after the Roman Empire fell and there was no advance toward a new level of civilization until almost the end of the 1st millenium AD. The interruption in the supply of metals was among many reasons for this standstill. The mining of metals ceased almost entirely in Europe, except for small quantities of easily accessible iron needed for weapons and tools (Raymond, 1986). A political development of significance to mining occurred in 1185 when the Bishop of Trent granted a charter to miners in his domain. It gave them legal as well as social rights, including the right to stake claims (Hartman, 1987).

Renaissance and Industrial Revolution

In Aachen, Germany, Charlemagne provided the leadership in the 9th century that began the emergence from the Dark Ages. He instituted drastic reforms in administration, finance, and education and he awakened a renewed interest in metals and metal working. Much of the wealth that made the Renaissance possible came from the Rammelsburg silver mine in the Harz Mountains. In addition, new deposits of silver and other metals were located in Saxony, Bohemia, and Moravia (Raymond, 1986).

The foundation of mining law, on which the laws in many parts of the world today are based, was laid in this period in Saxony (Raymond, 1986).

One of the world's greatest developments occurred in the 15th century when Guttenberg developed the printing process using moveable type cast from a mixture of lead and tin. Following this invention, there was a tremendous exchange of technology in almost every field of human activity. One of these early technological works was Georgius Agricola's *De Re Metallica* (Raymond, 1986).

The industrial revolution meant a soaring demand for minerals and spectacular improvements in mining technology (Hartman, 1987). By the Middle Ages, the British had become the world's largest users of coal. The first recorded accounts of coal as a fuel came from China during the Han dynasty (206 BC to 220 AD). Coal was first used in Britain during the Roman occupation when pieces were picked up on the seashores. Once the Romans left in the 5th century, the use of coal dwindled until the end of the 1600s, when Britain, an island rich in both iron ore and carbon fuel, was suffering an energy crisis and industrial stagnation from the depletion of wood resources, used in iron working to make charcoal (Raymond).

Then in 1709 Abraham Darby prepared some coke from Shropshire coal, mixed it with local iron ore, and charged his blast furnace at Coalbrookdale. Slowly his method spread throughout England. An immediate result of this widespread change to smelting iron with coke was the increase in the demand for coal. This development freed England from industrial stagnation (Raymond, 1986).

There were a number of significant developments during the Industrial Revolution that profoundly affected mining technology. The first of these was James Watt's steam engine, perfected in the 1760s. It provided the stimulus for the production of coal and iron, the evolution of machine tools, and the development of new forms of transport for people and goods (Raymond, 1986). The air compressor was another great innovation and resulted in the late 19th century in the development of the air-driven rock drill (Beall, 1973). An impetus for mechanization was the invention of electricity which made applications of machinery more flexible (Beall, 1973).

The introduction in Hungary in 1627 of black powder as a blasting agent was an important change in mining practice. Soon afterward black powder was also being used in the Cornish tin mines (Beall, 1973).

The dawn of global civilization began with Columbus' discovery of the New World in 1492. His voyage meant that direct sea contact between far distant continents was about to be established (Raymond, 1986).

CENTRAL AND SOUTH AMERICA

Native Civilizations

The Indians in South America used native gold, silver, and even platinum in the 1st millennium BC, but copper objects do not appear until about 500 AD (Raymond, 1986). There is evidence in the pre-Inca period of the use of mineral resources—salt, clay, and chalk (Boggio, 1983).

The Inca period, which began in the 12th to 13th century AD, reached its peak at about the time of the discovery of America. The majority of gold production during this pre-Hispanic period was by exploitation of placers or gold-bearing gravels in rivers. There was an abundance of tin in the plateau area (Bolivia) and this supply almost certainly influenced the appearance of Bronze (Boggio, 1983). In Peru the age of metals, corresponding with the Tiahuanaco cul-ture, began with the Christian era and lasted into the 14th century. There was lithic architecture of great dimensions; ceramics; and metalworking with gold, silver, and copper and tin, even attaining the level of bronze (Boggio, 1983).

Spanish Era

Beginning with the first voyage by Columbus, indications of gold were found in the newly discovered Carribbean islands (Prieto, 1973). During subsequent expeditions, Columbus established a mining camp at San Tomaso in San Domingo, from which he brought back more gold (Dunning, 1970). The Spanish explorers and conquistadores discovered and explored vast territories under the stimulus of a legend which told of a land abundant in gold (Prieto, 1973). Spain mined with the sword rather than the pick, as the conquistadores followed Columbus. In Peru they conquered the Incas, melted their mountains of gold decorative items, and cast the gold into ingots for shipment to Spain (Raymond, 1986).

Francisco Fernandez de Cordoba was exploring off the coast of the Americas in 1517 when he was blown off course in the Gulf of Mexico. He saw, on the Peninsula of Yucatan, evidence of a high civilization, the Mayans (Dunning, 1970). True mining by the Spaniards began in the lands discovered along the Gulf of Mexico in what was to be called New Spain. The first recorded copper mining by the Spaniards was in Cuba in 1524 (Beall, 1973). About 1525 the first silver mines exploited by the Spaniards were those of Morcillo in the present state of Jalisco, Mexico (Prieto, 1973). An event of great importance for the present-day Mexican iron and steel industry was the discovery in 1552 of the iron mines of Cerro de Mercado (Prieto, 1973).

Valdivia founded the cities of La Concepcion and Valdivia in Chile in 1550. At the same time rich gold mines were discovered in Confines and Quilacoya (Prieto, 1973).

Peru proved to be a country rich in mineral resources. Henrique Graces discovered mercury in 1558. The hill of Huantajaya, a rich silver deposit, was discovered in 1538 (Prieto, 1973). Cerro de Pasco had been a mining area for silver long before the arrival of the Spaniards. In colonial times great quantities of silver came from exploitation of surface deposits or pacos (Boggio, 1983).

In the district of Choco, Colombia, platinum was discovered in its natural state by Don Antonio de Ulloa and Don Jorge Juan in 1735 (Prieto, 1973).

In the Republican Period in Peru the first concession for exploiting guano deposits was signed in 1840. The 1830 nitrates began to be exploited at Tarapaca. The strong rise in copper price quotations in 1897 made this commodity one of the raw materials of great world demand and conditions were appropriate in Peru for a fast and growing participation in this market (Boggio, 1983).

Portugese America

A search for riches was also responsible for the Portugese exploration of the continent [South America] (Prieto, 1973). However, early attempts to find precious metals were not too successful. Some alluvial gold was discovered as early as 1541. The alluvial, gold-producing streams of Paranagua had been discovered by 1572, but production from the area was never large. The turning point came between 1693 and 1695 when Bandeirantes (Brazil's pioneers in the backlands) found rich gold placers in the Rio das Velhas, Rio das Mortes, and Rio Doce, north of Rio de Janeiro and west of Vitoria, in what is now the state of Minas Gerais. In 1721 new gold strikes were made at Cuiaba in Minas Gerais, followed five years later by discoveries in the present state of Goias (Prieto, 1973).

A Brazilian-born mineralogist, Manoel Ferreira de Camara Bethencourt e Sa (1765-1835) was the first to make pig iron at the Real Fabrica de Ferro do Morro do Pillar in Minas Gerais. The development of Brazil's iron industry dates from his time. Another development he fostered was the increased production of saltpeter, which induced him to set up a gunpowder factory (Prieto, 1973).

An event of profound importance was the discovery of diamonds in Minas Gerais by Bernardo da Fonseca in the mid-1720s. He found the diamonds in the gold washings of a place called Morrinhos in the Serro do Frio region (Prieto, 1973).

Recent History

Mention was made earlier of the importance of copper markets. The Incas first smelted copper from copper ores at Chuquicamata in the 1530s. Gold-quartz veins were found near the copper deposits but the Spanish were unable to do much with them because of the uninhabitable nature of the desolate and barren region of the Atacama desert. Modern exploitation did not begin until after 1881 when Chile established itself as the dominant power on the west coast of South America after the War of the Pacific. Initial investigation of Chuquicamata was made by churn drilling conducted by Edwin S. Berry and Walter A. Perkins in April 1912 (Parsons, 1933).

On Jan. 11, 1912, the Chile Exploration Co. had been organized by Guggenheim interests. The entire capital stock of the company was owned by the Chile Copper Co. of Delaware, organized 16 months later. Control of the company was acquired by the Anaconda Copper Mining Co. in 1923 (Parsons, 1933). William Braden formed Braden Copper Co. to develop the El Teniente mine in Chile. Control was subsequently bought by Guggenheim interests, later becoming Kennecott Copper Corp. Braden in 1913 founded Andes Copper at Potrerillos. The Chuquicamata mine followed in 1915.

The Cerro de Pasco Co. was formed by the Hearst-Haggin Syndicate with J. P. Morgan in 1902 and developed the mine at Oroya, Peru. After World War I, Cerro de Pasco developed the copper, lead, and silver mines at Morococha, Peru.

Since the early 1930s, nearly half the discoveries of great, low-grade copper ore bodies, usually dependent on large-scale surface mining, have been made in Peru and Chile (Joralemon, 1973).

In Peru Asarco won control of Toquepala and Quellaveco and Newmont acquired Cuajone. The two companies transferred their claims to Southern Peru Copper Co. Toquepala was opened in 1960 and Cuajone in 1964 (Joralemon, 1973).

In Mexico El Caridad was discovered between 1962 and 1968 and the Mexican government came to an agreement with Asarco to develop the property (Joralemon, 1973).

Changes in Peruvian mining laws in the period 1969-1978 led to the nationalization of many properties. In 1970 the state mining company, Empresa Minera del Perú, MINEROPERU, was formed. In the same year mining concessions that had been awarded to a number of companies were declared lapsed. Among those affected were Asarco, Cerro de Pasco, and Southern Peru Copper Corp. The concessions were then assigned to the new state company. On January 1, 1974, Cerro de Pasco was expropriated, it was stated, in the interests of the country and as a social necessity. The expropriated company became Empresa Minera del Centro del Perú, CENTROMIN-PERU (Boggio, 1983).

The Chilean government, in the same period, took similar action, forming Corporación Nacional del Cobre de Chile, CODELCO-Chile, the state company. The large mining operations at Chuquicamata, El Teniente, Andina, and El Salvador were expropriated.

NORTH AMERICA

Native Americans

Indians in Terra Incognita, the Spanish name for the vast southwest, part of which is now Arizona, mined salt, clay, pigment materials, stone, and turquoise by surface methods. There was quarrying of chert and obsidian for the manufacture of tools and weapons must have preceded most other forms of mining by several thousand years. The Apaches used vermillion cinnabar (mercury sulfide) as a body paint most probably taken from the mercury minerals found in La Paz county on the south flank of Cunningham Mountain in the Dome Rock mountain range. Here the Cinnabar mine was rediscovered by American prospectors in the 1880s (Canty and Greeley, 1987).

Probably as early as 3000 BC an elaborate copper culture began to grow up around the Great Lakes (Raymond, 1986). Copper articles were most common among the tribes that lived in what is now Michigan, Wisconsin, and the north shore of Lake Superior. Pits have been found that had existed long before the white man came. The original deposits were veins in the Keweenaw Penninsula on the south shore of Lake Superior (Joralemon, 1973).

Colonial Times

In the early days of settlement in North America, heavy bulk materials such as English brick, glass, and stone came from Europe as ship's ballast. In a very short time, extensive sand and clay deposits were utilized and the Boston, New Haven, New York, and Philadelphia area produced their own bricks. In the immediate areas around each of these commercial centers, lime and charcoal kilns were commonplace. Settlers moving back into the interior opened up deposits of marble, slate, soapstone, granite, limestone, and sandstone (Meade, 1988).

The first iron works was established at Saugus, MA, in 1646. It was imported from England and depended on the bog iron close by for feed. The Saugus works did not operate for many years, but the men who had worked there went elsewhere and helped set up forges and furnaces in the other Atlantic colonies (Raymond, 1986).

As the colonists gradually moved west into the interior, they exploited deposits as they found them. Two Frenchmen began mining lead at Mine la Motte in Missouri in 1718 (Beall, 1973).

Copper

There had been tales during the 17th and 18th centuries of extraordinary riches of copper along the shores of the Great Lakes in what is now Michigan. A geologist, Douglass Houghton, in 1840 went ashore by canoe on the Keweenaw Penninsula and found copper literally lying everywhere. After his report, there was a stampede of fortune hunters from the east who stumbled on the old Copper Culture of the Great Lakes. Who these early miners were is a mystery, since no burial grounds or other cultural evidence existed in the area—only mining tools and worked copper articles (Raymond, 1986).

The first modern-day mine in the area was the Cliff mine, discovered and opened by John Hays in 1843. This marked the first great mining boom in American mining history (Raymond, 1986).

Iron

Even as the first copper mines were being developed on the Keweenaw Penninsula, further west in the Michigan wilderness, a surveyor named Burt noticed erratic swings of his compass needle. With the help of a local Indian he had hired to help find outcrops of iron, in July 1845 he saw ahead of him a "mountain of solid iron ore 150 ft high. . . ." This first find turned out to be one of the world's largest deposits of iron, extending around the western shores of Lake Superior into Minnesota and Wisconsin. With the rich Mesabi Range at its core, the great iron province was to become one of the most productive ever known (Raymond, 1986).

The Mesabi Range was opened to development in the 1890s. Development of the steam shovel, which was to be vital in the development of the great copper porphyries, made surface mining the best possible exploitation method. Railroads went right into the pits. These procedures had clear economic advantages: they were cost efficient, resulted in larger production with fewer miners, and were simple to execute (Smith, 1987).

Gold Fever

Gold attracted the Spanish and Mexicans to the American Southwest, principally at first to New Mexico and Arizona and later California (Canty and Greeley, 1987).

Very little gold was found near the Atlantic coast, although a deposit in South Carolina, worked in the 1800s, is now being developed as a heap leach operation. Other deposits were located in Georgia and Virginia. In the west many deposits rich in gold and silver were found when the mountainous regions of Mexico were penetrated (Dunning, 1970).

California: In the 19th century gold was an inflammatory word in many parts of the world. In January 1848 James Marshall pulled a shiny pebble from the tailrace of Sutter's mill on the bank of the American River near Sacramento, CA. When news of the find became known, it started a stampede, as hordes of would-be miners poured into California to search for a fortune. In these workings, crude mining tools such as the wash bowl or pan, adapted from the paella pan of the Spanish settlers, were used. Early developments were the rocker, "long tom" cradle, sluices, and arrastras (crude, mule-powered crushers) used to separate gold-bearing sand from river gravel (Clamage, 1985; Raymond, 1986).

Colorado: In Colorado the search for gold had been going on for more than three centuries. In 1858 two prospecting parties were formed, one of which was made up of 19 Georgians. Most members of the two parties became discouraged and returned east; however, 12 Georgians headed by H. Green Russell stayed and established a camp near the mouth of Cherry Creek. Russell is credited with making the first important find about 80 miles above the mouth of the creek and other paying spots were found in the vicinity. Word reached the outside world and the Pike's Peak rush began (Dunning, 1970).

As a spinoff of this rush, prospectors climbed the mountains of Colorado and descended the canyons searching for gold and consequently a number of important mining districts were established in the period 1870-1900: the San Juans, Leadville, and Cripple Creek (Smith, 1988).

Nevada: When the rush to California subsided, many gold prospectors went in other directions. Among those returning via the Washoe were Peter O'Reilly and Patrick McLaughlin who did some placer mining around Washoe but were bothered by a heavy blue-black sand which clogged their sluice

boxes. They sent a sample to San Francisco for assay—the report showed that the heavy sand was almost pure silver. This was in about 1859. A third person, Henry P. T. Comstock, claimed to hold some interest in the land and although he only played a minor role in the discovery at Comstock/ Virginia City, the lode was named after him (Dunning, 1970).

Butte, Montana: In 1863 placer gold was discovered in Bannock in western Montana. The next year a party of emigrants found gold at the foot of a long yellowish hogback at the north edge of the Summit Valley just west of the Continental Divide. The new camp was called Butte and a miniature gold boom lasted for several years. While visiting Butte in 1872, William A. Clark was attracted by the big copper-stained outcrops on the hill and bought several claims along the biggest copper showing. He turned the district into a flourishing silver producer, with copper production as a byproduct (Joralemon, 1973).

In 1875 the Walker Brothers, leading bankers in Salt Lake City, sent Marcus Daly to Butte to investigate the silver possibilities. He and Mike Hickey located the Anaconda and Neversweat claims on a broad bank of yellow-stained, crushed rock that formed a bare streak across the slope of Butte Hill. This was the start of the Anaconda Co., developer of the "richest hill on earth" (Joralemon, 1973).

Alaska: A few early strikes, such as those at Juneau, had been made in Alaska. Prospectors found considerable gold along the 2000-mile Yukon River and its tributaries and by the fall of 1895 some spectacular finds had been made. A minor rush resulted at Forty Mile City in the Yukon. C. W. Carmack made a strike on Rabbit Creek in 1896 and it was this find that sparked the Klondike gold rush (Dunning, 1970).

The Porphyries

Two developments made possible the exploitation of the large, low-grade phorphry deposits in the US and South America. Progress in surface mining technology had been exceedingly slow with only minor changes in manually operated simple wheeled tools until finally steam shovels and rail haulage equipment were introduced in the early 1900s (Michaelson, 1979).

The second development was the flotation process. Largely pioneered in Australia, it soon spread worldwide and came to America just in time to compliment Daniel C. Jackling's innovation in open-cut mining at Bingham Canyon, Utah (Raymond, 1986).

Ely, Nevada, followed the Bingham Canyon operation. Jackling found the Ray, Arizona, and Santa Rita, New Mexico, properties. Miami and Inspiration came next. L. D. Ricketts and John Greenway developed the New Cornelia property at Ajo. Phelps Dodge developed Bisbee and Clifton-Morenci (Joralemon, 1973). Parsons (1933) lists the great power-shovel operations as Bingham Canyon, Chuquicamata, Nevada Consolidated (Copper Flat ore body), Chino, New Cornelia, and Sacramento Hill, Bisbee.

Recent History

In the US in the period 1930-1970 a number of large surface mining operations were developed. In the mid-1940s Phelps Dodge developed the Clay ore body, once considered too low grade to mine, and the Lavender pit in Bisbee in 1954. Around the turn of the century, Phelps Dodge had bought the Burro Mountain Copper Co., south of Silver City, NM, and in 1912 and 1915 bought two other properties in the area, now known as Tyrone. New developments in recovery techniques made it possible to treat the Tyrone oxidized ore and operations were started in 1969 (Joralemon, 1973).

Asarco began production at the Silver Bell ore body in Arizona in 1954 and started full production at the Mission mine in 1962 (Joralemon, 1973).

Pima in Arizona was brought on stream in 1959 by a joint venture of United Geophysical Co., Cyprus Mines Co., and Utah Construction and Mining Co. (Joralemon, 1973).

Duval opened two moderately successful properties at Copper Canyon and Battle Mountain in 1967. The Sierrita mine, one of the great ones, was opened in 1970 (Joralemon, 1973).

Anaconda bought the Twin Buttes and other properties from Banner Mining Co. and opened the property in 1971. Innovations were made in methods and equipment, such as high-speed belt conveyors for moving waste to the dumps (Joralemon, 1973).

Anaconda's mines at Butte had become deep and costs had increased to the point where many mines had been shut down. The company started a small open pit in the "Horsetail" area under the low ground southwest of Butte Hill and the great Berkeley pit developed (Joralemon, 1973).

Canada

In more recent times, large surface operations have developed in Canada. Bethlehem Copper Corp., under the direction of H. H. (Spud) Huestis, opened its surface mine near Merritt, BC, in 1962, paving the way for a whole flock of profitable, low-grade, open-pit copper mines. Bethlehem also acquired the Valley Copper-Lake area northwest of the original property (Joralemon, 1973).

Placer Development Co., led by John Simpson, found the Craigmont property a few miles south of the Bethlehem and started production in 1961. Noranda developed Brenda Mines Ltd., 96.6 km (60 miles) southeast of Craigmont in 1970 (Joralemon, 1973).

Texas Gulf Sulphur Co. in the mid-1960s did exploration work in the area north of the Noranda underground operations in the Mattagami Lake area. The result was the Ecstall mine, which in 1967, the first full year of production, produced 44.6 kt (49,200 tons) of copper, 203,8 Mt (224,640 tons) of zinc, and 112.5 Mg (13,968,000 oz) of silver (Joralemon, 1973).

AUSTRALIA

Gold

Australia's early prime industry was agriculture, although coal was mined underground in the Hunter Valley. In 1851 Edward Hargraves, returning to Sydney from California, found a stretch of land in Bathurst that reminded him of the California diggings. He found gold and started a rush (Park, 1988). Gold was first mined in the Timbarra region of New South Wales in 1853, initially as alluvials in rivers draining the Timbarra Tablelands, followed by mining aplite dikes and weathered adamellite, the latter by sluicing (Anon, 1988a). After the discovery of gold in New South Wales, discovery of gold later that year in Victoria started an even bigger rush (Raymond, 1986).

In 1851 the government in Sydney charged 30 shillings a month for a miner's license (laborers received 5 shillings a day). The fee and other grievances stirred rebellious feelings among the diggers and the situation finally came to a head in 1854 on a Ballarat hillside named Eureka where 150 diggers behind a hastily built stockade were charged by troopers and overwhelmed; 25 were killed and 30 wounded. Some

reforms resulted (Parks, 1988). The discovery of gold in Victoria established the state in its own right, forced the issue of democracy through the Eureka Stockade, and established Melbourne as Australia's financial capital (Anon, 1988).

Other gold discoveries took place in southern Queensland, tropical Queensland, and the Northern Territory in the 1860s; at Mount Morgans, Queensland, 1883; and in the early 1890s at Kalgoorlie and Coolgardie, Western Australia (Chadwick, 1987).

During the gold rushes in eastern Australia, arid Western Australia felt left out since conventional panning methods could not be used because of the lack of water. A dry blowing method was developed in which wind was used to carry away the lighter silt and sand. The "shaker" resembled a hopper on wheels with a blower attached. One man shovelled in the gold-bearing sand while his partner shook the hopper and worked the blower. These operations proved highly successful and a new gold rush started in Western Australia (Dunning, 1970).

Other Minerals

"Cousin Jacks," Cornishmen who emigrated to America and Australia, contributed greatly to mining history. The Cornishmen who came to Australia "found a landscape which they could never have imagined existed, with outcrops of metals so rich they could not have believed it" (Raymond, 1986). In 1841 two Cornish miners, within an hour's walk of Adelaide, saw a shiny metallic-looking rock jutting out of a grassy hillside and recognized it as galena. The mine they opened—the first metal mine in Australia—was called Wheal Gawler and exploited veins of rich silver and lead (Raymond, 1986).

Dutton, a pastoralist, late in 1842 on a sheep run 60 km north of Adelaide, was riding through a rain storm when he saw a patch of vivid green on a hillside. It was malachite. At Kapunda it was bursting out of the hillside, so soft and rich, assaying 23% copper, that in a few minutes a man could fill a wheelbarrow with it (Raymond, 1986).

Enormous iron deposits were discovered at Pilbara in Western Australia and are now being mined at the rate of hundreds of millions of tons per year (Raymond, 1986).

AFRICA AND ASIA

South African Gold and Diamonds

In 1886 in South Africa, George Walker, a stone mason, picked up a peculiar rock and noticed what appeared to be flecks of gold in it. His friend George Harrison helped him determine that it was really rich in gold. This was the forerunner of the biggest and richest gold field (The Witwatersrand) ever found (Dunning, 1970).

Diamonds were discovered in South Africa in the 1860s. The Jagersfontein and Dutoitspan pipes were discovered in 1870, followed by the discoveries at Bultfontein, De Beers, and Kimberly in 1871. The Premier pipe, the largest in the country, was discovered in 1902 and it yielded the 3106-ct Cullinan gem diamond in 1905 (Chadwick, 1988).

Mining expanded rapidly in the Kimberley area in the 1870s, based on the Dutoitspan, Bultfontein, De Beers, and New Rush mines. Cecil Rhodes arrived at the diamond fields late in 1871. He brought some degree of mechanization to the "Big Hole," the name by which the Kimberley mines were known (Chadwick, 1988).

Central Africa

When Stanley made his first journey across Africa in 1873-1875, he found the natives wearing leg bands of beaten copper wire. Later explorers found natives mining rich oxidized copper in open pits. In the Belgian Congo there are old workings in more than a hundred separate copper ore bodies (Joralemon, 1973).

Robert Williams, a deputy of Cecil Rhodes, formed Tanganyika Concessions Ltd. in 1899 to develop the Katanga copper deposits. A Belgian company, Union Miniere de Haut Katanga was formed in 1906, with Tanganyika Concessions as a large minority stockholder (Joralemon, 1973).

In Rhodesia Selection Trust, Ltd. and other companies developed the Roan Antelope, Mufulira, Nkana, and other mines of the northern Rhodesian Copper Belt (Joralemon, 1973).

Recent Developments

Palabora in South Africa was developed by Newmont Mining (manager), together with American Metal Climax (now Amax) and smaller South African companies. The mine began production in 1966 (Joralemon, 1973).

The O'okiep mine was opened in South Africa by Newmont in the 1930s (Joralemon, 1973).

In the Far East

Marcopper on Marinduque Island, New Guinea was developed by Placer Development and production started in 1969 (Joralemon, 1973).

Bougainville on one of the Solomon Islands started production in 1972 (Joralemon, 1973).

The Ertsberg mine, controlled by Freeport Minerals, is on an 11,500-ft tropical mountain in West Irian, Indonesia, and began production in 1972 (Joralemon, 1973).

The latest development in the area is Ok Tedi on Papua New Guinea.

References

Agricola, G., 1950 (1556), *De Re Metallica*, H.C. and L.H. Hoover, trans., Dover Publications, New York, pp. 36, 40.

Anon., 1988, "Victoria—the gold producer reopens fields and discovers others," *Australian Journal of Mining*, February, p. 84.

Anon., 1988a, "Gold deposit defies geological theory," *Australian Journal of Mining*, June, p. 32.

Beall, J.V., 1973, "Mining's Place and Contribution," *SME Mining Engineering Handbook*, Vol. 1, Sec. 1, A.B. Cummins and I.A. Given, eds., AIME, New York, pp. 1-1-1-13.

Boggio, M.S., 1983, *Peru: A Mining Country*, Vol. 1, *History*, Instituto Geologico Minero y Metalurgico, Lima, Peru, 239 pp.

Canty, J.M., and Greeley, M.N., eds., 1987, *History of Mining in Arizona*, Mining Club of the Southwest Foundation, Tucson, AZ, 279 pp.

Chadwick, J., 1987, "Kalgoorlie and more," *International Mining*, October, pp. 8-11.

Chadwick, J., 1988, "Diamonds—yesterday, today and forever," *International Mining*, March, pp. 10-15.

Clamage, S., 1985, "California's Mother Lode: The legend of '49," *Mining Engineering*, March, pp. 225-228.

Dunning, C.H., with Sadler, R., 1970, *Gold from Caveman to Cosmonaut*, Vantage Press, New York, 192 pp.

Graffin, G.D., 1982, "Graphite," *Industrial Minerals and Rocks*, 5th ed., Vol. 2, S.J. Lefond, ed., AIME, New York, p. 757.

Hartman, H.L., 1987, *Introductory Mining Engineering*, John Wiley & Sons, New York, pp. 1-6.

Joralemon, I.B., 1973, *Copper*, Howell-North Books, Berkeley, CA, 407 pp.

Meade, L., 1988, "Northeast USA and its minerals," *Industrial Minerals*, April, pp. 29-32.

Michaelson, S.D., 1979, "Open Pit Mining—Past, Present, and Future," *Open Pit Mine Planning and Design*, J.T. Crawford, III, and W.A. Hustrulid, eds., AIME, New York, p. 5.

Park, E., 1988, "In praise of my country-in-law on her 200th birthday," *Smithsonian*, January, pp. 128-135.

Parsons, A.B., 1933, *The Porphyry Coppers*, AIME, New York, 581 pp.

Piggott, V.C., 1988, "The Thailand Archaeometallurgy Project," *Journal of Metals*, January, pp. 36-37.

Prieto, C., 1973, *Mining in the New World*, McGraw-Hill, New York, 239 pp.

Raymond, R., 1986, *Out of the Fiery Furnace*, The Pennsylvania State University Press, University Park, PA, 274 pp.

Smith, D.A., 1987, *Mining America*, University Press of Kansas, Lawrence, KS, p. 86.

Smith, D.A., 1988, "How fleet the frontier: Colorado's San Juan mining district, 1870-1900," *Mining Engineering*, February, pp. 102-105.

1.2. Current Status and Future Trends

TA M. LI AND MARIANNE SNEDEKER

INTRODUCTION

Mining in the United States and elsewhere has undergone an evolution—or revolution—in the past few years, as an economic depression set in. In 1981 the iron ore operations were the first to be affected with the impact hitting the nonferrous industries in the following years. Mines closed, personnel were let go, and those operations still continuing began to tighten their belts.

The period 1983-1986 was characterized by low metal and mineral prices. Efforts to control production costs were characterized by improved mine production, better equipment utilization, and closer cooperation between labor and management. Concessions on wages were the key to the restartup of mines such as the Butte, MT, copper pit (Bohnet, et al., 1987).

In the same period (1982-1986) a bright spot was precious metals mining that has seen an upsurge in operations and new properties, particularly in the US and Australia. New mines and/or the expansion of existing operations kept coming on stream. Just recently, Newmont Mining, American Barrick, and Echo Bay announced an enlargement of their gold resource base and Asamera Minerals in Wenatchee, WA, disclosed an extension of the reserves adjacent to the Cannon mine.

As an inevitable consequence of the trends in the mining industry, the *bottom line* profitability of the major metal industries has eroded seriously. According to US Dept. of Commerce data, the steel industry has recorded a net loss in each of the last five years and total losses over the period amount to $12.5 billion (Sousa, 1988).

The corresponding losses of the nonferrous metal industries (e.g., aluminum, copper, and lead), while not as staggering as those of steel, have also been extensive. Collectively, the US nonferrous metal producers lost money in each of the four years between 1982 and 1985. Reflecting the major restructuring that has occurred in the nonferrous metal business over the last several years, however, this sector returned to the black in 1986 and registered a $760 million profit. Profit and loss trends in the metal industries in the period 1980-1986 for ferrous and nonferrous metal production are shown in Fig. 1 (Sousa, 1988).

By 1986 and more significantly in 1987, the economic condition of the nonferrous metals industry improved as commodity prices rose. The companies that survived had become "lean and mean," with drastic changes in the management style of their operations. Favorable negotiations with the mining unions, especially in Arizona and Utah, also contributed to this economic recovery.

PRODUCTION

The production index of the domestic metal industries when compared to the rest of the US economy show how these industries have underperformed the economy as a whole. Using 1977 as the base year, the Federal Reserve Board's (FRB) production index (Fig. 2) for the industrial sector of the economy as a whole increased by nearly 25% through 1986, while primary metal and metal mining indices declined by 24% and 28%, respectively, in the same period (Sousa, 1988).

Comparison of the FRB's production indices provides insight into another fundamental factor behind the recent decline of the metals industries: the US economy simply appears to be working smarter, making more by using less (Sousa, 1988).

Declining production and capacity utilization together with the compelling need to lower labor costs in order to become more competitive have had a devastating impact on the employment rolls of the metal industries. For example, total employment in the metal mining industries at the end of 1986 was approximately half what it had been less than a decade earlier, Fig. 3 (Sousa, 1988).

Metallic and Nonmetallic Production

Tables 1 and 2 show the material handled at surface mines in the United States by type and commodity. Table 3 lists the 25 leading metal and the 24 industrial mineral surface mines in the US in order of crude ore output (Tanner, 1988).

Coal

Of the 50 largest bituminous and lignite mines in the US, 38 are surface operations. Production data for these mines in 1986 are given in Table 4 (Anon, 1987).

In 1986 surface coal mines produced 496 Mt (536,444,000 st), 60.4% of the coal produced in the US. In 1985, 167,009 persons were employed in surface coal operations and worked an average number of 208 days. Production per man day was 3.9 t (4.32 st) (Anon., 1987).

Table 1. Material Handled at Surface Mines in the United States, by Type (Million Short Tons)

Type and Year	Crude Ore	Waste	Total*
Metals			
1981	592	1050	1650
1982	371	677	1050
1983	380	577	938
1984	429	614	1030
1985	411	499	911
Industrial Minerals			
1981†	1150	584	1740
1982‡	837	366	1200
1983†	1070	155	1230
1984‡	1060	286	1340
1985†	1260	450	1710
Total Metals and Industrial Minerals*			
1981	1750	1640	3390
1982	1210	1040	2250
1983	1450	712	2160
1984	1480	901	2380
1985	1670	950	2620

Adapted from Tanner, 1988, *US Bureau of Mines Yearbook*, Vol. 1.
* Data may not add to totals shown due to rounding.
† Includes industrial sand and gravel. Construction sand and gravel data were not available for 1981, 1983, and 1985 because of biennial canvassing.
‡ Crushed and broken and dimension stone data were not available for 1982 and 1984 because of biennial canvassing.
Metric equivalent: st × 0.907 = t.

8

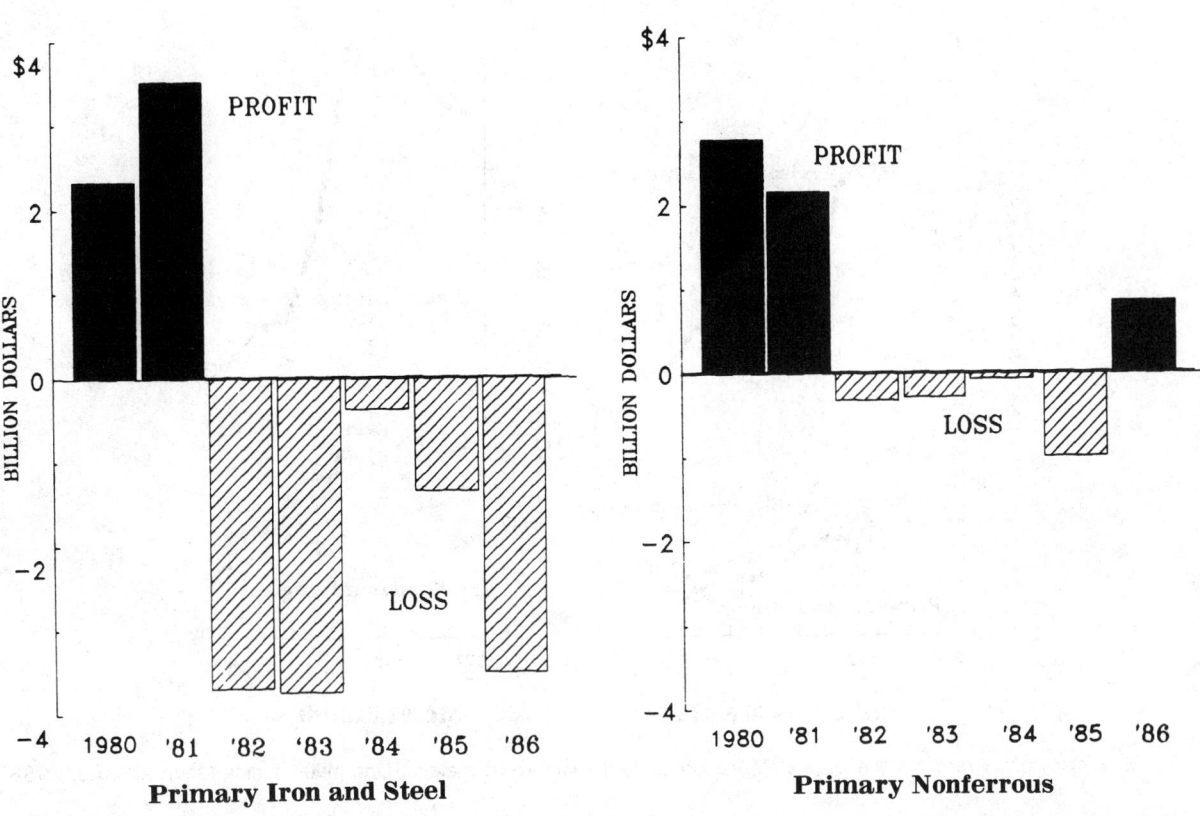

Fig. 1. Profit and loss trends in the metal industries. Left, primary iron and steel; right, primary nonferrous (source, Sousa, 1988).

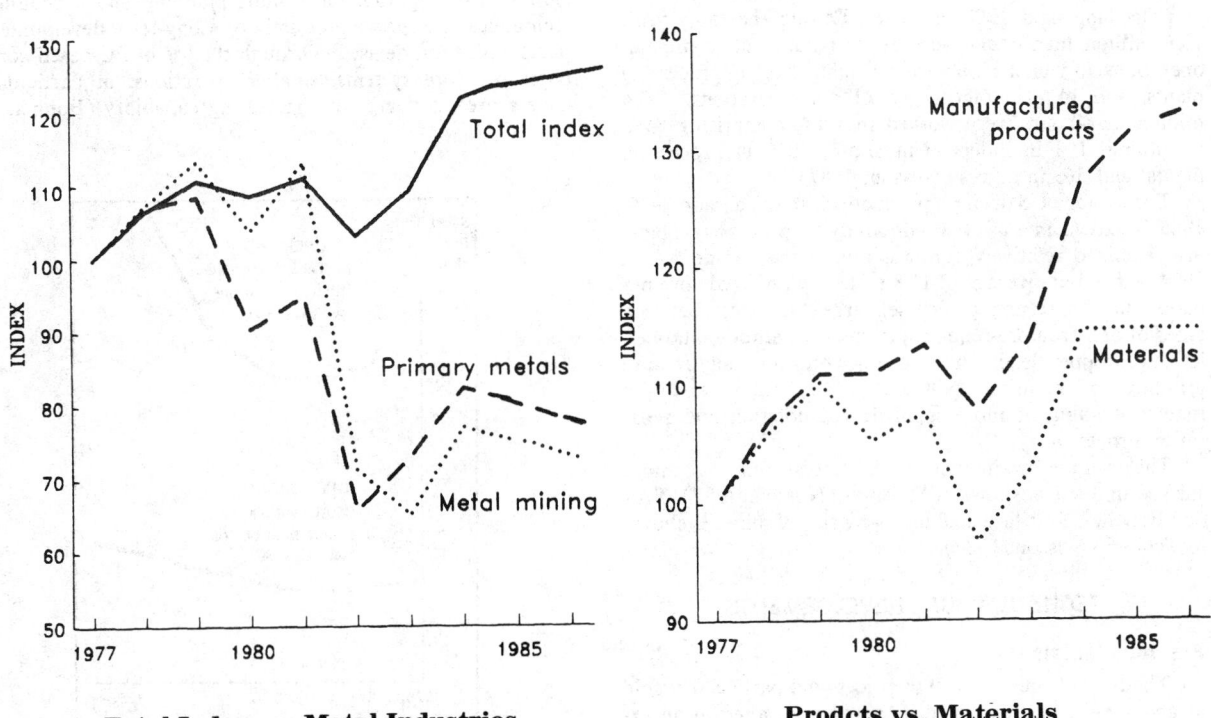

Fig. 2. Federal Reserve Board's industrial production index. Left, total index vs. metal industries; right, products vs. materials (source, Sousa, 1988).

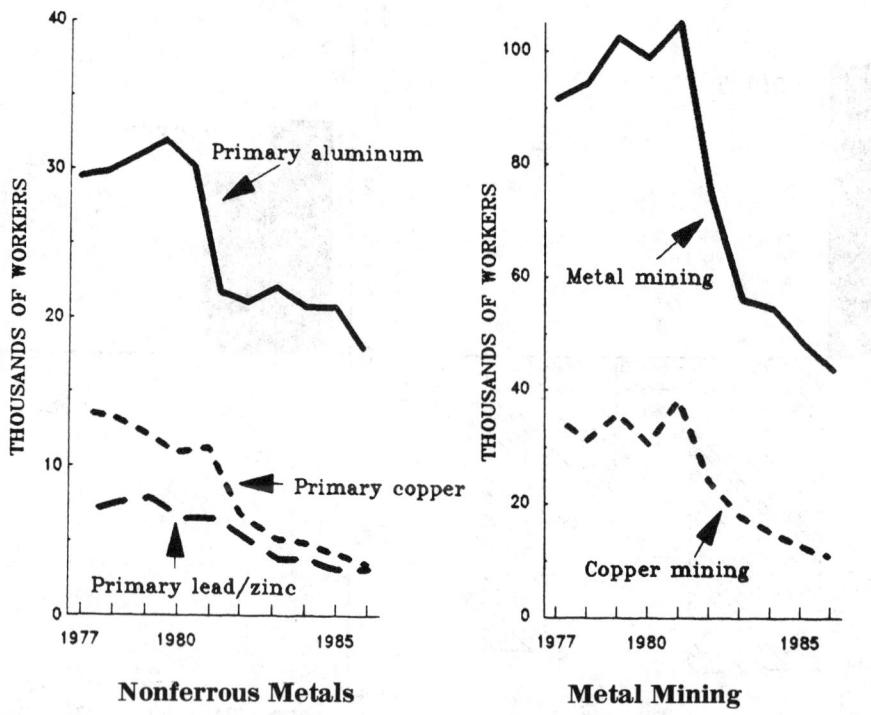

Fig. 3. Employment trends in the US metals industries. Left, nonferrous metals; right, metal mining (source, Sousa, 1988).

PRODUCTIVITY

According to the US Dept. of Labor statistics, a total of 93.7 million employee-hours (man-hours) were worked at metallic mineral operations in the United States in 1984. Of those, 47% were in mining, 2% in independent shops, 39% in processing, and 12% in offices. During the same year, 73.6 million man-hours were used at nonmetallic mineral operations, of which 35% were in mining, 51% in processing plants, and 14% in offices. At all coal operations, 363.4 million man-hours were worked in 1984, comprising 84% in mining, 1% in independent shops, 10% in processing plants, and 5% in offices (Nilsson, 1987).

The average US open pit productivity in the decade 1975-1985 is shown in Fig. 4. Productivity in processing plants has remained relatively constant during the decade but in 1984 reached an average of 11.8 t (13.0 st) of crude ore per man-hour. There are, of course, large differences between types of ore. Iron ore requires grinding, separation, and pelletizing. Copper, lead, zinc, and other ores normally require grinding and flotation (Nilsson, 1988). Coal may simply require washing or more sophisticated flotation and separation processes.

The average productivity in US metal and coal mines surface in 1984 is shown in Table 5 (Nilsson, 1987). Productivity in US surface coal mines by size of mines is shown in Table 6 (Nilsson, 1987).

TECHNOLOGICAL DEVELOPMENTS

Precious Metals

The level of precious metals prices, and particularly gold at about $400 an ounce and higher, has resulted in an explosion in exploration and surface development activities. Innovative technology has been the utilization of hydraulic shovels, continuous mining systems, computer-assisted pro-

duction control and scheduling, and improved bulk materials conveying and handling (Bohnet, et al., 1987).

Industrial Minerals

Development has been· at a brisk level. There have been efforts to incorporate more mine planning and scheduling techniques to improve productivity. Long-term development plans have been devised. Through the use of PC-based computer software systems, smaller operations, in particular, have a greater degree of engineering capability (Bohnet, et al., 1987).

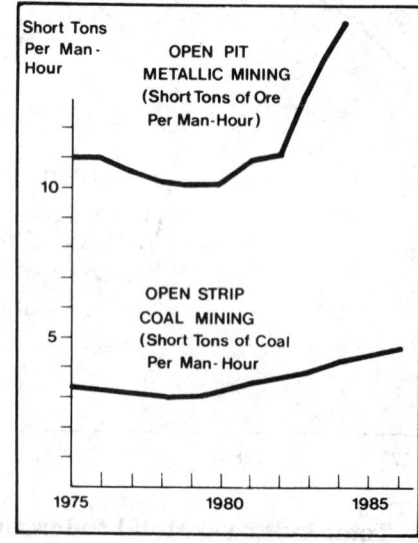

Fig. 4. Average productivity in US surface mines (Nilsson, 1988).

Table 2. Material Handled at Surface Mines* in the United States in 1985, By Commodity (Thousand Short Tons)

Commodity	Crude Ore	Waste	Total†
Metals			
Bauxite	795	W	795
Copper	170,000	260,000	429,000
Gold			
Lode	51,400	93,400	145,000
Placer	3,720	2,040	5,760
Iron Ore	165,000	73,300	238,000
Lead	—	W	W
Silver	2,590	8,780	11,400
Uranium	938	16,000	17,000
Zinc	W	—	W
Other‡	16,800	46,300	63,100
Total Metals†	411,000	499,000	911,000
Industrial Minerals			
Abrasives§	140	W	140
Asbestos	1,100	4,410	5,510
Barite	733	W	733
Clays	44,600	38,700ᵉ	83,300
Diatomite	635	—	635
Feldspar	1,510	W	1,510
Gypsum	11,700	7,120	18,800
Mica (scrap)	837	W	837
Perlite	672	W	672
Phosphate Rock	175,000	307,000	482,000
Pumice#	565	113	679
Salt	340	W	340
Sand and Gravel‖	28,800	—	28,800
Stone			
Crushed and Broken	987,000	80,900ᵉ	1,070,000
Dimension	2,440	1,290ᵉ	3,720
Talc, Soapstone, Pyrophyllite	1,240	8,430	9,670
Vermiculite	1,620	W	1,620
Other¶	2,150	2,300	4,450
Total Industrial Minerals†	1,260,000	450,000	1,710,000
Total All Commodities†	1,670,000	950,000	2,620,000

Source: Tanner, 1988, *US Bureau of Mines Mineral Yearbook*, Vol. 1.

ᵉ Estimated. W, withheld to avoid disclosing company proprietary data; included with "Other."

* Excludes material from wells, ponds, or pumping operations.

† Data may not add to totals shown because of independent rounding.

‡ Includes antimony, beryllium, manganiferous ore, mercury, molybdenum, nickel, platinum-group metals rare-earth metals, tin, titanium (ilmenite) tungsten, and metal items indicated by symbol W.

§ Includes abrasive stone, emery, garnet, millstones, and tripoli.

Excludes volcanic cinder and scoria.

‖ Includes industrial sand and gravel. Construction sand and gravel data were not available for 1985 because of biennial canvassing.

¶ Includes aplite, boron minerals, greensand marl, iron oxide pigments (crude), kyanite, magnesite, olivine, wollastonite, and industrial minerals indicated by symbol W.

Metric equivalent: st + × 0.907 = t.

Heap Leaching

A technological development introduced in the late 1960s, heap leaching, has made significant contributions to the viability of precious metals operations, both for low-grade deposits and the reworking of old properties.

Although the principles of heap leaching have a long history (mines in Hungary recycled copper-bearing solutions through waste heaps in the mid-16th century and Spanish miners percolated acid solutions through large heaps of oxide copper ore on the banks of the Rio Tinto about 1752), the first commercial application of the technology occurred in the late 1960s at Carlin Gold Mining Co. in Nevada (van Zyl, et al., 1988).

It is in the past ten years that heap leaching has developed into an efficient method of treating oxidized gold and silver ore. It has proven to be both an efficient way to extract precious metals from small, shallow deposits, as well as an attractive way to treat large, low-grade, disseminated deposits (van Zyl, et al., 1988). The technology is also being used to recover the metal values from waste dumps at old mining properties.

The results of the technological improvements which have occurred throughout the 1970s and 1980s can be seen in the dramatic production level increases. By 1986 production of gold from heap leaching had increased to over 30% of total US gold production from an estimated 6% in 1979 (van Zyl, et al., 1988).

Table 3. Leading Metal and Industrial Mineral* Surface Mines in the United States in 1985, in Order of Output of Crude Ore

Mine	State	Operator[†]	Commodity
Metals			
Morenci	Arizona	Phelps Dodge Corp.	Copper
Sierrita	Arizona	Duval Sierrita Corp.	Copper
Minntac	Minnesota	USX Corp.	Iron Ore
Empire	Michigan	Empire Iron Mining Co.	Iron Ore
Pinto Valley	Arizona	Newmont Mining Corp.	Copper
Hibbing Taconite	Minnesota	Pickands Mather & Co.	Iron Ore
Tyrone	New Mexico	Phelps Dodge Corp.	Copper
San Manuel	Arizona	Magma Copper Co.	Copper
Round Mountain	Nevada	Round Mountain Gold Corp.	Lode Gold
Tilden	Michigan	Tilden Mining Co.	Iron Ore
Erie Commercial	Minnesota	Pickands Mather & Co.	Iron Ore
Chino	New Mexico	Chino Mines Co.	Copper
Ray Pit	Arizona	Kennecott	Copper
Peter Mitchell	Minnesota	Reserve Mining Co.	Iron Ore
National Pellet Project-Itasca	Minnesota	M. A. Hanna Co.	Iron Ore
Thunderbird	Minnesota	Oglebay Norton Co.	Iron Ore
Inspiration	Arizona	Inspiration Consolidated Copper Co.	Copper
Eisenhower	Arizona	Asarco Inc.	Copper
Bagdad	Arizona	Cyprus Mines Corp.	Copper
Green Cove	Florida	Associated Minerals Corp.	Titanium
Golden Sunlight	Montana	Golden Sunlight Mines Inc.	Lode Gold
National Pellet Project-St. Louis	Minnesota	M. A. Hanna Co.	Iron Ore
Minorca	Minnesota	Inland Steel Mining Co.	Iron Ore
Zortman-Landusky	Montana	Pegasus Gold Inc.	Lode Gold
Esperanza	Arizona	Duval Sierrita Corp.	Copper
Industrial Minerals[‡]			
Noralyn	Florida	International Minerals & Chemical Corp.	Phosphate Rock
Swift Creek	Florida	Occidental Petroleum Corp.	Phosphate Rock
Kingsford	Florida	International Minerals & Chemical Corp.	Phosphate Rock
Suwanee	Florida	Occidental Petroleum Corp.	Phosphate Rock
Ft. Green	Florida	Agrico Chemical Co.	Phosphate Rock
Lee Creek	North Carolina	Texasgulf Inc.	Phosphate Rock
Haynsworth	Florida	American Cyanamid Co.	Phosphate Rock
Lonesome	Florida	American Cyanamid Co.	Phosphate Rock
Georgetown	Texas	Texas Crushed Stone Co.	Stone
Calcite	Michigan	USX Corp.	Stone
Clear Spring	Florida	International Minerals & Chemical Corp.	Phosphate Rock
Wingate	Florida	Beker Industries Corp.	Phosphate Rock
Payne Creek	Florida	Agrico Chemical Co.	Phosphate Rock
Hookers	Florida	W. R. Grace & Co.	Phosphate Rock
Ft. Meade	Florida	Mobil Oil Corp.	Phosphate Rock
FEC Hialea	Florida	Rinker Materials Corp.	Stone
Rockland	Florida	USS Agri-Chemicals	Phosphate Rock
Stoneport	Michigan	Presque Isle Corp.	Stone
Pennsuco	Florida	Tarmac Florida Inc.	Stone
Thornton	Illinois	General Dynamics Corp.	Stone
McCook	Illinois	Vulcan Materials Co.	Stone
Hardee	Florida	C. F. Mining Corp.	Phosphate Rock
Norcross	Georgia	Vulcan Materials Co.	Stone
St. Genevieve	Missouri	Tower Rock Stone Co.	Stone

Source: Tanner, 1988, *US Bureau of Mines Minerals Yearbook,* Vol. 1.
* Excludes brines and materials from wells.
[†] Reflects operator in 1985; does not indicate current (1988) operator in some cases.
[‡] Includes industrial sand and gravel. Construction sand and gravel were not available for 1985 because of biennial canvassing.

EQUIPMENT TRENDS

A most important development in surface mining in recent years has been the use of increasingly sophisticated on-board electronics and microcomputer systems for mining equipment. These systems range from those used to assist the personnel who is operating the hydraulic shovel or walking dragline to managing and monitoring the performance and productivity of the mine's mobile equipment (Tanner, 1988).

There has been over the past few years a very slow yet steady shift to electric power for mining vehicles. However, because of the high mobility and portability of diesel-powered vehicles, the diesel will probably continue to be preferred in the foreseeable future (Tanner, 1988).

Large shiftable mining equipment will be necessary in

Table 4. 38 Largest Bituminous Coal and Lignite Mines in the US, 1986

Company	Name of Mine	State	Production 1986	Production 1985	Date Opened
Thunder Basin Coal Co.	Black Thunder	WY	22,000,000	23,200,000	1975
Texas Utilities Co.	Martin Lake	TX	12,600,000	12,299,469	1950
The Carter Mining Co.	Rawhide	WY	12,394,359	12,237,000	1975
Decker Coal Co.	Decker East & West	MT	12,200,000	11,500,000	1972
AMAX Coal Co.	Belle Ayr	WY	12,145,900	12,829,379	1972
Kerr-McGee Coal Corp.	Jacobs Ranch	WY	12,100,000	13,000,000	1974
Texas Utilities Co.	Monticello	TX	12,100,000	11,990,354	1950
Western Energy Co.	Rosebud	MT	12,100,000	12,283,958	1968
AMAX Coal Co.	Eagle Butte	WY	12,000,280	11,808,014	1978
Cordero Mining Co.	Cordero	WY	11,300,000	10,100,000	1976
The Coteau Properties Co.	Freedom	ND	9,479,431	7,860,301	1983
The Carter Mining Co.	Caballo	WY	7,258,931	8,978,000	1977
Utah International Inc.	Navajo	NM	6,841,000	6,975,000	1963
Arch of Illinois	Captain	IL	6,717,000	5,400,000	1964
Peabody Coal Co.	Kayenta	AZ	6,600,000	7,274,000	1974
Bridger Coal Co.	Jim Bridger	WY	6,480,000	7,200,000	1963
Black Butte Coal Co.	Black Butte	WY	6,000,000	5,500,000	1979
The Falkirk Mining Co.	Falkirk	ND	5,766,308	5,874,649	1978
North Antelope Coal Co.	North Antelope	WY	5,700,000	5,713,000	1983
San Juan Coal Co.	San Juan	NM	5,216,000	5,328,000	1974
Texas Utilities Co.	Big Brown	TX	5,017,121	4,980,121	1950
Peabody Coal Co.	Black Mesa	AZ	4,800,000	2,351,000	1985
Pittsburgh & Midway Coal Mng.	McKinley	NM	4,717,000	4,940,000	1959
Spring Creek Coal Co.	Spring Creek	MT	4,664,000	2,800,000	1980
Washington Irrigation & Dev. Co.	Centralia	WA	4,609,000	4,425,000	1970
Northwestern Resources Co.	Jewett Mine	TX	4,300,000	31,000	1985
Triton Coal Co.	Buckskin	WY	3,990,400	3,958,600	1981
Mobil Coal Producing, Inc.	Caballo Rojo	WY	3,989,622	4,222,000	1983
AMAX Coal Co.	Ayrshire	IN	3,954,828	3,607,710	1971
Peabody Coal Co.	Lynnville Nos. 1 & 2	IN	3,600,000	3,259,000	1955
Rochelle Coal Co.	Rochelle	WY	3,572,000	211,000	1985
Central Ohio Coal Co.	Muskingum	OH	3,376,948	3,371,658	1952
Baukol Noonan Inc.	Center	ND	3,357,913	3,475,000	1973
Pyro Mining Co.	William Station	KY	3,292,121	2,713,466	1982
Colowyo Coal Co.	Colowyo	CO	3,143,919	3,129,327	1977
Pittsburg & Midway Coal Mng. Co.	Kemmerer	WY	3,128,000	3,418,000	1963
Texas Municipal Power Agency	Gibbons Creek	TX	3,077,774	2,829,600	1981
Consolidation Coal Co., Glenrock Coal Co.	Dave Johnston	WY	3,051,000	3,500,000	1958

Source: Anon., 1987, *1987/1988 Coal Mine Directory, United States.*
Metric equivalent: st $\times$ 0.907 = t.

the future as mining of deeper near-surface deposits becomes necessary. This will be particularly necessary where climatic conditions are extreme or overburden removal is difficult. An example is the Captain mine in Illinois where coal is transported by truck, but land is reclaimed by moving the overburden and topsoil by a conveyor system designed to handle 2.6 km^3/h (3366 cu yd per hr) (Tanner, 1988). The development and application of high-angle, elevating, and cross-pit conveyors result in considerable reductions in transportation costs. Conveyors are an alternative to trucks and other diesel-consuming transport. The cable belt conveyor system has undergone considerable design changes, improvements in performance and reliability, and operating costs (Singhal, et al., 1987).

The best features of trucks and conveyors can be combined by using an in-pit crusher. These can be fully mobile, semi-mobile, or fixed crushing plants. One advantage is to reduce material to the size limit transportable by conveyors (Singhal, et al., 1987).

Draglines that are crawler-mounted have resulted in reduced time and cost of erection. A major development has been the availability of long-boom draglines (Bohnet, et al., 1987; Singhal, et al., 1987).

Another development has been drill rigs with computer-based programmable controllers (Bohnet, et al., 1987).

Availability of suitable transmissions for large mechanical drive trucks has been an improvement in off-highway trucks. Other significant developments include renewed interest in trolley-assist, truck dispatch systems, higher horsepower, and greater capacity (Singhal, et al., 1987).

Continuous surface miners have been developed that incorporate a rotating cutter drum and conveyor discharge system (Bohnet, et al., 1987). Such surface miners are es-

Table 5. Productivity in US Surface Metal and Coal Mines, 1984

	Million metric ton	Million man-hours	Tons per man-hour
Metal Mines			
Waste and ore	937	23.6	39.6
Of which ore	381		16.1
Coal Mines			
Cleaned coal	496	120.0	4.1

Source: Nilsson, 1987.

Table 6. Productivity in US Surface Coal Mines in 1984 by Size

	Million metric tons of cleaned coal	Million man-hours	Tons per man-hour
Smaller than 0.9 Mt/a (1 million stpy)	176	71.4	2.5
Larger than 0.9 Mt/a (1 million stpy)	310	48.6	6.4
	486	120.0	4.1

Source: Nilsson, 1987.

pecially applicable to multiple seam mining where the seams are separated by thin bands of overburden or in cases where seams are split and where materials of different qualities must be separated. Surface miners can mine to very narrow limits, improving resource recovery and providing an uncontaminated product (Singhal, et al., 1987).

Hydraulic excavators are competitive with small to medium-size cable shovels and wheeled front-end loaders, especially in smaller open pits such as those in many of the new gold operations (Bohnet, et al., 1987).

Manually controlled bucket wheel excavators are relatively inefficient. A number of suitable microprocessor-based bucket wheel automation systems have become available. There is also a shift toward the all-hydraulic compact bucket wheel excavator (Singhal, et al., 1987).

Scrapers, front-end loaders, and electric cable shovels comprise the traditional loading equipment for surface mining. All classes of conventional loading equipment have undergone design changes, including improved electrics and incorporation of health monitoring and diagnostic systems. These changes are resulting in increased reliability, improved performance, and a lower unit cost of production (Singhal, et al., 1987).

Dredging is an attractive means of lowering mining costs, particularly in gold operations. Dredges are currently in use in Alaska, Ecuador, Brazil, Colombia, and Papua New Guinea (Bohnet, et al., 1987).

Especially in large surface mining operations, adequate communication has been a problem. On-board radios and computer equipment are focal points for improved communication and control and mine haulage efficiency (Bohnet, et al., 1987).

Other recent developments in surface mining technology include the wider use of emulsion-type explosives and developments already mentioned: increased use of hydraulic excavators, use of in-pit crushers, and shiftable belt conveyors (Tanner, 1988).

FUTURE TRENDS

Responding to increased competitivity, the future of surface mining has become a showpiece for technological innovations to meet sharply rising production costs. In the period of depressed commodity market prices, surface miners responded through the aggressive application of technology created by developments in mechanization and computers. In establishing a solid pattern of cost consciousness for mine productivity, future trends will see even greater mine productivities as a result of innovations in:

- Off-highway truck design and performance.
- Hydraulic shovel reliability and durability.
- Computer-aided controls in mine operations and design.
- Blasting agent utilization and detonation efficiencies.
- Management planning and manpower scheduling and utilization.
- Continuous mining and materials handling systems.

In closing, surface mining will continue in the forefront of innovation, mandated by the industry's commitment for excellence in cost competitivity and requirements for meeting world material demands.

References

Anon., 1987 *1987/1988 Coal Mine Directory, United States,* McGraw-Hill, New York, pp. 451, 452, 457.

Bohnet, E.L., Winkle, R.F., and Edmiston, K.J., 1987, "Surface Mining," *Mining Annual Review,* June, pp. 187-207.

Nilsson, D.S., "Productivity in Mining," *International Mining,* October, pp. 38-43.

Nilsson, D.S., 1988, "Open Pit Mining Productivity—An Update," *Mining Magazine,* June, pp. 506-511.

Singhal, R.K., Fytas, K., and Collins, J.L., 1987, "Open Pit Trends 1986," *International Mining,* August, pp. 41-50.

Sousa, L.J., 1987, *Problems and Opportunities in Metals and Materials: An Integrated Perspective,* Sec. 1, US Dept. of the Interior, Bureau of Mines.

Tanner, A.O., 1988, "Mining and Quarrying Trends in the Metal and Nonmetal Industries," *Minerals Yearbook 1986,* Vol. 1, US Bureau of Mines, pp. 7-45.

van Zyl, D., Hutchison, I., and Kiel, J., eds., 1988, *Introduction to Evaluation, Design and Operation of Precious Metal Heap Leaching Projects,* Society of Mining Engineers, Littleton, CO, pp. 3-4.

Chapter 2

Exploration and Geology Techniques

Richard L. Brown, Editor

2.1 Overview of Exploration

RICHARD L. BROWN

In this chapter a number of authors describe the kinds of geological thought and exploration techniques applicable to the original identification and subsequent mensuration, in terms of tonnage and grade, of mineral deposits judged by the geologist to be suitable for surface mining. Many of these techniques are also applicable to *grass roots* or systematic reconnaissance style exploration. It is appropriate that there be discussion of reconnaissance techniques in this handbook since that activity in established mining districts often continues long after initial production.

Exploration geology seems to have separated, as a discipline, from mining geology, not because the one group has a greater or less need than the other to know and understand all these techniques as much as that each group has different objectives. The mining geologist seeks new veins or other new extensions to ore bodies and expects to find these on a regular basis, whereas the exploration geologist must find new districts and knows that he will be lucky if he makes one or two such discoveries during a career. The mining geologist assists in the day-to-day problems of production, and reacts to the discipline which accrues to achieving daily and monthly goals. The discipline accruing to the discovery of a new mineral deposit in a new district is of a different sort. At any rate, the separate disciplines of exploration geology and of mining geology merge at that point in the history of a mineral deposit, after discovery while it is being explored and determinations are being made of the tonnage and grade—that is to say, the period in the history of an ore deposit when data necessary for a feasibility study are being prepared. Because so many of the techniques discussed in the following pages apply both to the general reconnaissance-type exploration and to the business of *drilling off* an ore body, the authors have been rather general in their treatment of the various forms of exploration geology which apply to the commodity they have discussed. In this introductory section, exploration techniques common to most commodities are addressed with the purpose of providing an overview of the duties of the exploration geologist as he takes a prospect from discovery to the final economic estimates.

A review of the technology directed at the search and discovery of economic mineral deposits illustrates once more that there is "nothing new under the sun." There has been a continuum from medieval times to the present day of man's knowledge of mineral deposits and of ways of finding them. A quick trip across Europe shows that the Phoenicians, Greeks, and Romans were all adept at reading gossans, at panning heavy minerals from stream sediments, and at a variety of other exploration techniques still used by today's geologists. It is also obvious that these ancients had more than a rudimentary grasp of many principles of economic geology.

The search for ore begins with the development of ideas as to where the search should be conducted. Application of the most modern geophysics, geochemistry, remote sensing, and other techniques cannot be made until the geologist has decided where the search should begin. The first things a geologist must decide are what the ore body he hopes to find looks like, what minerals are contained therein, and how it was formed. He must, in short, develop an empirical model before he leaves his office and goes to the field.

Most of the ideas, a few examples of which are described below, geologists use now and will use during the foreseeable future had been published by 1974. In March 1965, The Canadian Institute of Mining and Metallurgy held a symposium on volcanogenic deposits. The papers given at that symposium, published by the CIM later that year, form the bible used by most geologists as they plan their exploration for volcanogenic deposits. John Guilbert and David Lowell published their paper on "Mineral Zoning in Porphyry Copper Deposits" in *Economic Geology* during 1970. The so-called *Red Sea* book (*Hot Brines and Recent Heavy Metal Deposits from the Red Sea,* edited by Degans and Ross) was a 1969 publication. The term *plate tectonics* was firmly in place by 1970, and the *Journal of Geophysical Research* published its compendium of papers related to that subject in 1973. Kambalda in Western Australia was discovered in 1968, and the recognition that some nickel-copper deposits were derived from ultramafic volcanic rocks was made in print by a number of authors in the very early 1970s. Our knowledge of the so-called Mississippi Valley deposits lags far behind some of the other ore types mentioned above, but the sum of our knowledge of these deposits is pretty much contained in the August 1971 issue of *Economic Geology*.

The search for volcanogenic ores, many of which are mined from surface, has widened possibly further than any other type of exploration, and it may be well to describe the model which governs much of that exploration. In brief, the geologists who participated in the Canadian symposium in 1965 had noted that descriptions of synvolcanic and syngenetic ores described by German and Japanese workers corresponded closely with the results of our own mapping and observations of the Precambrian deposits in Northern Ontario and Quebec. As a result of this mapping, they were able to demonstrate that many of the Canadian deposits had been formed on ocean floors, apparently from brines derived from highly siliceous rhyolite domes. In addition, iron-rich silicate deposits often spread far from the volcanic dome and were deposited on the sea floor over wide regions. Thus the three elements of the volcanogenic model were: (1) the volcanic dome containing imbricate stringer zones; (2) the polymetallic sulfide deposits, formed on the ocean floor; and (3) widespread cherty pyrite bed. The complete system was usually covered by more recent volcanic rocks, often andesite or basalt. Thus the subsequent direction of massive exploration money at the andesite-rhyolite contacts in many shield areas throughout the world.

It would be difficult to pinpoint the decade during which geologists routinely began to map the distribution of clay alteration products around porphyry copper sulfide systems. A quick glance through bibliographies shows a number of papers on the subject were published in the 1930s. A number of company geologists were mapping such patterns routinely during the early 1950s. A number of papers authored by such people as S. C. Creasy, Richard and Courtright, and Paul Kerr demonstrated widespread interest in the subject during that decade. During the 1960s, Guilbert and Lowell, collectively and individually, published the results of their observations of alteration patterns in the southwest United States and in some other areas as well, and their 1970 paper cited earlier is now regarded by most North American geologists as the standard text on the subject. There is presently very little porphyry copper exploration near established mines or in new districts conducted which does not respond to the Guilbert and Lowell model.

Recently, it has become clear that predictable distribution of minerals containing fluorine, barium, and other elements, occurs around the previously known stacked intrusive complexes which host the molybdenum-porphyry systems. The discovery of molybdenum at Mt. Emmons in Colorado and at Pine Grove, Utah, can be attributed to recognition of this mineralogical distribution. Other important exploration programs, generated by recognition of similar features in other areas in the western United States, are in progress.

However, no such conclusive models have been developed for the Mississippi Valley deposit. While there is an excellent body of literature which describes most of the deposits in the six or seven type localities scattered around the United States and Canada, there is no single body of observations or of theory which is accepted by the majority of the workers in the field and which can be described as a common denominator underlying exploration activity. There is a need for additional work directed at these Mississsippi Valley-type deposits.

It would appear that geologists engaged in the exploration of these deposits are paying increased attention to the study of paleosurfaces and paleoecological environments which are dominated by carbonate-rich rocks. Of course, the internal characteristics such as collapse breccias, limestone-dolomite interfaces, and recrystallized dolomitic rocks are recognized and mapped, and trigger intense exploration when they are seen. Possibly there is some consensus that the margins of carbonate platforms are good places to look for these deposits, and in southeastern Missouri, criteria implicit in both the old and new leadbelts are carefully adhered to.

Also, there is no commonly accepted rationale governing exploration for replacement-type polymetallic sulfide occurrences hosted by carbonate rocks. Possibly researchers and explorationists had not been interested in this group of hydrothermal ore deposits because they are relatively rare, and the metal content, dominated as it usually is by lead and zinc, is apt to be relatively low and unremunerative. The Mexican deposits such as Plomsas, Santa Eulalia, Naica, Charcas, Providencia-Concepcion del Oro, Taxco, San Martin, La Encantada, Fresnillo, and Velardena are not well known, and the results of significant research, if any has in fact been conducted, are proprietary and locked up in mining company reports. Probably the most utilized lead in the search for these deposits is directed in the vicinity of veins and veinlet systems which can be classified either as feeder-type mineralization of the replacement bodies or as leakage from them.

The search for nickel-copper deposits during the 1950s and 1960s in general contemplated a Sudbury-type model in which it was supposed that a sulfide magma body had been injected from depth to a near surface position by any of a number of proposed mechanisms. The Sudbury, Ontario and the Thompson Lake, Manitoba districts both lie in the join between contiguous provinces of the Canadian Shield, and it has been widely assumed or hoped that additional deposits can be found along these sutures and supposed zones of weakness. Substantial exploration time and dollars have been expended in the search for deposits in these zones, and still continues, although on a much more limited scale. However, as noted earlier, subsequent to the discovery of nickel in the Kambalda district in Western Australia, it was recognized that some nickel-copper deposits are associated with ultramafic volcanic flows. The Travis and Wodell paper, published in the proceedings of the 12th Commonwealth Mining Congress, and A. N. Naldrett's paper entitled "Nickel Sulfides, Classification and Genesis," published by the CIM in 1973, each described this association. Nickel exploration has been at a low ebb during the last decade, due to unremunerative prices received by producers for that metal, but such nickel exploration work as does proceed is directed at both magmatic and the volcanically derived sulfides.

During the past 15 years, fair consensus at least regarding the morphology, if not the genesis, of uranium deposits hosted by sandstones and by quartz pebble conglomerates, has been achieved. Exploration, designed to test sulfide-rich as compared to oxide-rich portions of appropriate sandstone and quartz pebble conglomerate units, has developed. However, no such consensus has been reached in respect to the *vein-type* deposits of northern Saskatchewan or of northern Australia. These deposits appear to be characterized in both locations by high-grade pitchblende *veins* hosted in crystalline or in metamorphic rocks, covered by Proterazoic sediments in which are bedded carnotite or carnotite-type mineralization. While there is very little agreement amongst geologists as to how these deposits are formed, most organizations involved in the search for these deposits appear first (mainly by means of airborne electromagnetic and radiometric surveys) to search for signs of the bedded material in the overlying sandstones, and then to attempt to search for the veins.

Geologists who concern themselves with exploration, evaluation, and production of the various industrial minerals, as well as those involved in coal, oil shale, tar sands, and other similar materials, are obviously as interested in the genesis and geological environments of these deposits as are the hard mineral geologists. However, in most cases the deposits are huge and relatively easy to find and therefore the difficulty and cost of original discovery have not been as great as they have been in the case of the commodities discussed previously. The major challenge in the case of the industrial minerals, and in the case of the hydrocarbons, has been identifying major volumes of material which conform to various engineering and chemical standards. Therefore these geologists think not so much in terms of origin and empirical models, as they progress in their exploration work, as they do of quality control and engineering parameters.

The importance of these models, of course, is that they provide terms of reference and criteria to the geologist as he decides, for example, whether or not a given district or prospect warrants more expenditure. If the geologist maps a considerable number of features which conform to his model for a given deposit type, he may well decide to recommend drilling or some other form of physical exploration. If, on the other hand, his data shows that few of the elements of his model are present, he may decide that additional expenditure is not necessary or wise. Similarly, the geologist might use his model to tell him in which direction, either laterally or vertically, additional drilling should be planned. If he knows that the features he is mapping are usually found vertically above another feature of economic importance, he may decide to recommend deeper drill holes. The models, above all, give the geologists terms of reference and continuity of information which extend beyond the bore of the drill hole he is considering, and beyond perhaps, the geometery of the ore shoot being investigated.

The discovery by Newmont Mining Company of the Carlin gold deposit a decade or more ago generated a substantial amount of precious metals exploration in the basin and range province of the western United States. The primary exploration technique involved in this search has been the collection of samples, both geochemical and rock samples, which are assayed for gold, and a variety of other elements, such as mercury, which are thought by the geologists involved to be useful "indicator or pathfinder elements." This search often requires a collection of samples from wide areas, without much geological discrimination. Recently, emphasis has been placed on hot spring environments, the so-called jasperiod environment; the interest in this type of environment being generated by recent discoveries at Alligator Ridge, Nevada, and in the Caldera environment, similar to the one at McDermitt, Nevada, where mercury mineralization has been mined for some time.

In summary, it should be reiterated that as mineral exploration develops and grows more sophisticated, increasing care will be given to the development of the *empirical model*. The model is simply a generalized amalgam of features of known deposits, the enclosing host rocks, and all the various alteration patterns which are usually attendant to the mineralization. Once the model is produced and agreed upon by everyone involved, the next exploration step is to decide which geological province might provide all the various factors and features in the model. The third step is to identify, through literature searches and through inspection of old exploration records and other geological material, where within the district chosen the various features called for by the model might be found.

Prospectors as well as modern explorationists have always had models in mind. In former years the prospectors looked for signs of direct mineralization in outcrop, and proceeded then to test these outcrops by drilling or other means. The modern explorationist still hopes to find mineralization in outcrop, and on occasion will do so for many years hence. However, increasingly his work will consist of testing models, once he has found areas in the field which conform in most respects to the model.

Mining company managements, increasingly, will have to get used to the idea of drilling concepts or models rather than mineralized outcrops. One management which has already adopted this idea is that of Western Mining, the Aus-

tralian company. D. W. Haynes, its copper consultant, has explained in his paper entitled "Mining Technology in Mineral Resource Exploration" (published in *Proceedings,* Third Invitation Symposium on Mineral Resources in Australia, held in Adelaide in October, 1979 by the Australian Academy of Technological Sciences) how Western Mining geologists put together source rock theory with known information regarding the sedimentary basin on the Stuart Shelf to find the Olympic Dam deposit. The Western Mining geologists, Haynes explains, were looking for areas in the sedimentary basin in the state of Southern Australia where sediments similar to those hosting the Zambian Copper Belt might be found in proximity to basaltic rocks. In addition, a preconceived tectonic model was apparently postulated, and a lineament analysis, aided by data derived from Landsat images, was produced. Once the complete model had been settled upon, it was aggressively explored, and the important Olympic Dam copper-uranium discovery was made. Haynes points out that the mineral deposit which was discovered was not precisely the same type as was anticipated and this perhaps provides some food for thought. The Olympic Dam discovery is obviously not the first successful application of modeling. It is, however, one of the better documented cases of successful exploration generally designed to test a model.

Geological reasoning has improved and must improve even more if an adequate rate of discovery of mineral deposits is to be maintained. There has been parallel improvement of various geophysical, geochemical, and remote sensing techniques. Geophysics has progressed from Thomas Edison's dip needle, successfully used at Sudbury, Ontario, early in this century, to satellite-mounted magnetometers. Geochemistry has progressed from the practice of early Scandinavians, who during the Middle Ages chased mineralized boulders up streams of glacial debris to their source to determinations of 25 or more metals in soil samples. Remote sensing, in its strict sense, has been developed from the day prior to the Second World War when Canadian geologists made interpretations from oblique aerial photographs to today's interpretations made from enhanced remote sensed images from satellites.

Interpretation of leached outcrop and of gossan, which was started by Augustus Locke and Rowland Blanchard in the 1920s and carried on, among others, by Kenyon Richard and Harold Courtright in the 1940s and 1950s, appears to be a dying art, simply because surface mapping of large porphyry copper systems is not now an everyday activity. However, during the 1960s, enhanced evaluation of potential drill targets was made possible by the interpretations of clay alteration patterns combined with that of leached cappings. The gossans in Western Australia are far different from those of the western United States, but an awareness of the technique and a fair ability to apply it to gossans in that country occurred. Later the Australian expertise was transferred to South Africa, and a number of discoveries were made there as well.

Prototypes of much of the geophysical equipment in use today were in field prior to the Second World War, for example as previously referred to, Thomas Edison's dip needle, with which he discovered or, at least almost discovered, continuations of the nickel-copper ore at Falconbridge in the Sudbury District. Hans Lundberg, using an equipotential method which verged on electromagnetics, made a great discovery at Buchans Newfoundland, in 1926. Technology developed for military purposes during the war was put to work immediately thereafter, principally in the development of electromagentics and of airborne electromagnetics. Geiger counters, rarities before the war, become commonplace shortly thereafter. In the mid-fifties, relatively trouble-free ground electromagnetics systems were used routinely in the field and torsion spring magnetometers also were developed; these instruments dramatically increased the rate at which readings could be taken. Airborne electromagnetics and magnetics also became routine during the early 1950s. The combined use of these two techniques resulted in a very impressive string of discoveries, mainly in Canada, which continued at least until 1975. In recent years the rate of discoveries by AEM and AM has declined, for a variety of reasons. However, one can wonder if the application of a single technique will ever again result in a list of discoveries such as Thompson Lake, Manitoba; Heath Steele, New Brunswick; Mattagami and Joutel, Quebec; Timmins, Ontario; Sturgeon Lake, Ontario; and Crandon, Wisconsin. This list, while incomplete, represents an extraordinary record of discovery. Also, in the mid-fifties, Dr. Arthur Brandt completed development of the induced polarization method. The Geiger counter largely gave way in the 1960s to the scintillometer. Drill hole logging by regular radiometric methods became commonplace in the search for roll front uranium deposits in sandstones, particularly in the western United States.

The advent of various microelectronic devices has given geophysicists the capability of gathering enormous amounts of data. For example, modern airborne electromagnetic equipment provides as many as six audio frequency channels, two very low frequency electro EM channels, four gamma-ray spectrometer channels, and a magnetometer channel. All 13 channels are recorded on magnetic tape every half second.

The proton magnetometer can now achieve a sensitivity of about one gamma. High sensitivity magnetometers are used for identification of extremely subtle features in areas of very low magnetic relief. Airborne spectrometer surveys, utilizing 49 161 cm^3 (3000 cu in.) crystals are now routine. These crystals yield much higher count rates than has been the case; the differentiation between earth generated radiation and atmospheric radiation is measured separately and identified.

Of course all the additional information collected by the newer equipment has provided enormous challenges for the interpreter of the data. Sulfide sources are easily confused with nonsulfide sources as much these days as previously.

In spite of improvement of the equipment, it has not yet been possible to make significant improvements in the depth penetration of electromagnetics equipment. During the past five years, transient electromagnetic and audiomagnetic telluric systems, the latter utilizing either distant thunderstorms or nearby controlled sources, have come into use. This system reputedly can achieve depth penetration of several hundred meters. Presumably additional research as to the application of these systems to mineral exploration problems will continue.

Additional research and improvement of in-hole electromagnetic and induced polarization systems will continue. Most current interest appears to be directed at fixed source time domain units, fixed source continuous wave multifrequency units, and single frequency moving transmitter-receiver systems.

The broad range of portable gamma-ray detectors employed in airborne radiometric surveys is also adapted to in-hole surveys. It is anticipated that improvements in simple nondiscriminating scintillation counters for channel differential spectrometers will be made.

In summary, geophysicists have been able to make re-markable improvements in the portability and accuracy of their equipment. They have not been able to make substantial improvements in the depth penetration of the equipment because increased depth penetration means increased volumes of rock energized and therefore increased numbers of nonsulfide features which can cause responses and consequently signal noise with concurrent difficulties in interpretation. Because as many mineral deposits remain to be found at depths beyond the reach of present geophysical equipment as have been found in the near surface, we can confidently expect that additional research and development of equipment capable of seeing deeper into the earth will continue.

The basic principles of geochemical prospecting have been known for thousands of years. They were in fact successfully applied by early prospectors who traced visual indications of ore dispersion patterns in rocks, soils, and stream sediments back to bedrock sources. However, it was not until the 1930s that chemical emission spectrographic analytical methods began to permit trace element measurements. The Russians and Scandinavians began to use geochemical exploration techniques, as we know them, prior to the Second World War. Subsequent to the war, an explosion of geochemical exploration activity occurred, permitted by the development of inexpensive rapid colorimetric analytical techniques by the United States Geological Survey. Research activity spread to the United Kingdom and thence to other countries in western Europe during the 1950s. Students of Hawkes, Webb, Bloom, and Warren took the technique into almost every part of the world and many discoveries were made. As surveys were completed during the 1950s and 1960s, the various ways by which elements can become dispersed throughout the secondary environment became fairly well appreciated. During the 1970s startling advances, again made possible by the advent of microelectronics, were made in the analytical end of geochemical exploration. Atomic absorption instruments and techniques were refined and matrix corrections introduced as routine procedure for certain elements. X-ray fluoresence methods were greatly improved. Plasma spectrometry permitted significantly low detection limits for a number of elements. This improvement in analytical quality and reduction has resulted in a marked decrease in the dollar cost of exploration. Computer data handling capability has kept pace with analytical advances. Computerized data plotting as well as univariate and multivariate statistical procedures are widely accessible.

Many have pointed out that the improvement in equipment has not been matched by comparable improvement in the understanding of fundamental geochemical processes. Routine surveys have been laid out and interpreted without regard to the solid body of information which has been collected. Hopefully, future practice will catch up with the theory already in place.

The geologists also have improved hardware at their disposal and increased availability of this equipment should result in an increased rate of discovery. The microprobe is probably the most important of this new equipment. The ability to detect variations in metal content in individual minerals collected from various parts of districts and of mineral deposits will greatly increase the geologist's ability to predict projections and to site exploratory drilling and other exploration activity. Increased use of the fluid inclusion stage will result in increased knowledge of fluid inclusions and of hydrothermal fluid temperature, pressure, and composition which is not unknown in respect of many mineral deposits. Collection of this fundamental knowledge will result in better definition of mineral zoning patterns and will, as a result, guide exploration. Fluid inclusion research will become a routine part of modern mineral exploration.

The microprobe will make it possible to determine phase compositions of sulfide-silicate vein and wall rock assemblages. These studies also will assist in determinations of deposit zoning and increase reliable predictability of exploration parameters. Sulfur isotope studies will also become routine and will assist in the classification of ores and the placing of these ores in appropriate models.

Application of nearly all the techniques listed in the previous pages will result in the production of enormous amounts of data. At present, geologists use computers to store data and make various calculations as to tonnage, metal content, and economic viability of mineral deposits, but they have not found a way to utilize the deductive capabilities of computers to find deposits. Possibly they never will, as the human brain is still the best computer of all. However, it is obvious that increased research as to the application of computer techniques to exploration will continue.

In due course geologists will be able to make much better use of images produced by sensors placed in satellites than they now do. Much of the science and technology required to produce images which will identify various rock types now exists. These sensors have been flown from time to time in U-2-type aircraft, and experimental surveys such as famous ones over Saindoc, Pakistan, and Goldfields, Nevada, as well as the various case histories flown by a GEOSAT-NASA joint experiment, provide impressive data. It is clear that the potential impact on mineral exploration of remotely sensed data will be significant.

The ultimate exploration tool is still the diamond drill. We have seen substantial improvement in the reliability, portability and, most important, the percentage of core recovery achievable by drilling equipment in the past few years. The retrievable core barrel has been a great cost saver. However, costs of drilling in the past decade have increased drastically, and the mining business needs, and needs soon, additional improvement. Probably the next routine improvement will be the retrievable bit. If some way can be found to change drill bits without removing an entire string of drill rods, important savings will be achieved. It is certainly to be hoped that the drilling industry will continue its search for ways in which it can keep the costs of diamond drilling under better control than is now the case.

Both the exploration geologist and the mine geologist must at all times have a reasonably accurate perception of the economics of the project. It goes without saying that this perception of economics should apply at all stages of a project and will be applicable initially when a grassroots or reconnaissance program is devised. If a discovery is made in a new district, even during the very early stages of the assessment of the prospect, the geologist must make a back-of-the-envelope calculation designed to demonstrate that under prevailing and under forecast economic conditions, presuming that all assumptions as to tonnage and grade materialize, the mineral deposit he is modeling, will yield a suitable return on investment. During the early stages of such a project, there are few hard facts and many assumptions. As first order drilling, designed to determine the outlines of the deposit continues, these assumptions are replaced by facts, and estimates become more reliable.

A second occasion for an economic review of the project might reasonably occur when plans are made for the expensive closely spaced drilling required to determine final tonnage and grade figures. This estimate, of course, will utilize

many assumptions but substantial data will have been provided by the drilling completed. A crude estimate of tonnage and grade will be available, metallurgical data (supplied by bench tests on material from drill cores) will be on hand, assumptions concerning mining costs will be made possible, if only by comparison with other similar operations. Sufficiently accurate assumptions concerning the amount and cost of infrastructure can be made, and the cost of any necessary access roads and other similar items can be reasonably estimated. Presumably this estimate will be made by a team of construction and mining engineers, assisted by the geologist who should be on hand to provide data and interpretation of data concerning the nature and characteristics of the ore deposit, including among other things, continuity of grade, mineralogical zoning, and varying characteristics of host rock type.

There are, of course, various levels of economic estimates, leading from the first preliminary calculations to the full-scale detailed estimates as to the cost of a given project. In Fig. 1, various levels of economic estimates are defined. Probably, for reasons which will be explained later, the exploration geologist should be substantially involved during the early planning of a project, at least in those phases which involve mine design and planning. In the later design and planning stages, which involve mainly detailed estimates of construction, the geologist will be much less involved, but should in any case be available to the design and planning team until the mine opens.

If the results of the feasibility study described in the preceding paragraphs should indicate that a deposit will be financially attractive (should assumptions concerning both tonnage and grade be later confirmed), the next obvious step is to cost out the kind of detailed drilling program necessary to generate accurate ore reserve estimates. The most important consideration in this regard is drill hole spacing. This is a critical matter, as the cost of the drilling can be (even in, say, the case of only a medium-sized porphyry copper deposit) on the order of $10 million or even substantially more. However, as the preproduction expenditure in the case of such a deposit would run into hundreds of millions of dollars, it can easily be recognized that the cost of diamond drilling and sampling of drill results is a very poor place to try to save money. On the other hand, the geologist does have a responsibility to recommend a program which will adequately sample, but not overdrill, the prospect. We would all like to think that it should be possible through the application of some statistical formula or other to precisely establish optimum drilling spacings. Sadly, this is not the case. In many districts long practice may have established the proper drill hole interval, but many of the mineral deposits being discovered during the 1980s are in new districts, and some involve types of mineral deposits for which no precedents are available. Obviously the first matter to be considered in reaching such a decision is a variation of the metal content from hole to hole in the drilling already completed. If there is a small variation, then obviously the drill spacing can be much larger than in the case where there has been large variation of metal content.

Another matter to which the geologist must give his most careful consideration is sample size, or the length of core which will be included in each sample to be sent for assay. Assay costs in case of even a medium-sized deposit will be astonishingly large, and if, for example, the decision is reached to sample every 3 m (10 ft) run of core, rather than each 1.5 m (5 ft) run of core, these very large numbers can

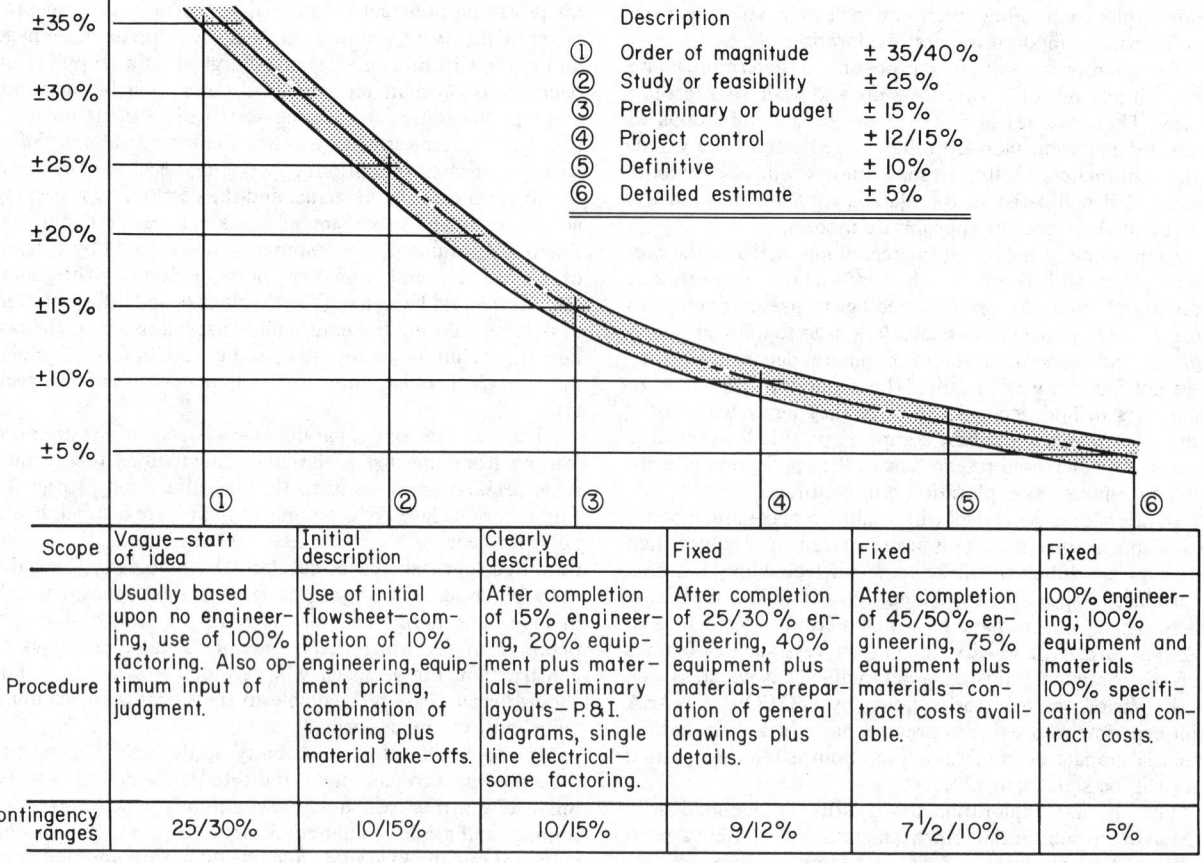

		Description	
	①	Order of magnitude	± 35/40%
	②	Study or feasibility	± 25%
	③	Preliminary or budget	± 15%
	④	Project control	± 12/15%
	⑤	Definitive	± 10%
	⑥	Detailed estimate	± 5%

	①	②	③	④	⑤	⑥
Scope	Vague–start of idea	Initial description	Clearly described	Fixed	Fixed	Fixed
Procedure	Usually based upon no engineering, use of 100% factoring. Also optimum input of judgment.	Use of initial flowsheet—completion of 10% engineering, equipment pricing, combination of factoring plus material take-offs.	After completion of 15% engineering, 20% equipment plus materials—preliminary layouts—P.&I. diagrams, single line electrical—some factoring.	After completion of 25/30% engineering, 40% equipment plus materials—preparation of general drawings plus details.	After completion of 45/50% engineering, 75% equipment plus materials—contract costs available.	100% engineering; 100% equipment and materials 100% specification and contract costs.
Contingency ranges	25/30%	10/15%	10/15%	9/12%	7½/10%	5%

Fig. 1. Definitions of various levels of economic estimates (prepared by A.O. Marsh, Jr., Central Engineering Dept., ASARCO, Inc.).

be halved. Here again, the optimum choice will result in the least expensive alternative which will completely achieve the required results.

Other decisions, such as staffing, choice of assay office facilities, core storage, provision for constant, high-caliber geological input, and interpretation of data on a current basis are all highly important matters to be costed into the final total budget. There is temptation to scrimp on all of the above-listed matters, but inadequate attention to any of them can in the long run be very costly.

Aside from making the proper arrangements as to drilling intervals, sampling intervals, standards of assaying and the like, there are many other matters which need to be considered. Large volumes of core will be coming in from the drill sites each day. Good facilities will be needed for logging, photographing, and storage of the core; splitting or sawing, and preparing the samples for assay is no small job. Literally hundreds of assays will be returned and of course these assays will need to be plotted and posted on the appropriate maps and sections on a current basis.

As a result of the necessity to perform all the tasks listed on a current basis, there is also need to provide for housing and eating facilities for as many as 30 or 40 people. Supervison must be provided not only for the geological end of the job but also for the logistical end. It is important to reiterate that the final authorization for expenditure should be constructed from costs carefully computed from realistic assumptions, and the recommended procedures should be completed as inexpensively as possible. However, reductions in costs that result in the production of poor samples and

poor interpretation can lead, at the very least, to severe cost overruns of the finally completed exploration job. And worst of all, if, as occasionally has happened, it is discovered after the mine has been opened that the ore "was not there," the entire preproduction cost has been wasted.

Obviously, the end result of the entire exercise is the production of an ore reserve. The geologist in charge of the project will ultimately have to certify the reserves since the success or failure of the entire mining project is, in the last analysis, dependent upon them. Obviously there are, in the case of any modern mining organization, a number of checks and the exploration geologist will not stand alone in this exercise. However, some one individual eventually is required to *sign-off* on the ore reserve. There are two kinds of numbers which will be generated and the meaning of these two following definitions should not be confused. The first is the mineral inventory. The mineral inventory simply is a listing of the pounds of metal confined within a given volumetric limit. This reserve may be stated without reference to the cost of the extraction. The second number, the ore reserve estimate, it best described by reflecting that ore is often defined as "naturally occurring materials which can be removed from the ground at a financial profit." This estimate must be made in connection with determinations of mining plans, the geometry of the ore itself and of internal waste, and obviously in reference to forecast metal prices.

Of course, the ore reserve estimate is the prime objective of detailed drilling. Secondary objectives include provision of material for metallurgical testing, rock mechanics information, and of course, the best possible determination of the geology. The determination of tonnage is a fairly routine arithmetic calculation. Practice varies from organization to organization, but it is common to make a measurement of the specific gravity of each run of core. If the metric system is in use, the volume of material expressed in cubic meters is simply multiplied by the specific gravity and the tonnage thereby determined. If the English system is in use, a *cubage factor* must be determined, and the volume of rock, as expressed in cubic feet, divided by this factor.

Determination of grade is a much more complicated matter. Corporate policy may dictate the method to be used, regardless of the type of mineral deposit under study. In the vast majority of operating mines in the United States, a simple polygon system is employed. Under this method of ore reserve calculation, the volume of ore closest to a sample point is assigned the grade of that sample. Efforts are made to design polygons so that the division point between adjoining sample points is more or less equidistant between them. The polygon technique has served many mines well over many years, but it sometimes fails to adequately predict trends within ore bodies which may, in fact, greatly influence grade. If such a condition should be suspected, it might be well to consider some form of moving average method similar to that devised by Dr. J. G. Krieg, the famous South African geologist and statistician. Under this technique, it is possible to assign a value of metal content in an area where there is no sample point, by considering the values of all the existing adjoining sample points. Sometimes, because all adjacent sample points are considered in calculating the grade of a block, some blocks are assigned higher values than are indicated by a sample point within the block. In extreme circumstances, ore can be extrapolated through a block in which there is a blank drill hole. Some engineers have extreme difficulty accepting a mathematical or statistical estimate which will draw ore grade contours through a blank drill hole. Regardless of their discomfort, Krieg's method or trend surface analyses may be the only way to accurately determine grade.

As previously noted, the geologist must remember in addition to the ore reserve data he must produce, he must also produce data which will aid in such matters as pit design and predictions of wall stability.

It is apparent from the foregoing that the exploration and mining geologist, to be really good at his job, must be expert in a wide number of fields. He must know geology and must be current in new geological understanding. He must be a good pragmatic prospector, must understand basic finance and economics, must be good at logistics in order to organize complete drilling and other kinds of exploration projects under his direction. In the following sections various authors describe in greater detail those factors which are important in the geological and exploration work concerned with deposits containing various important minerals and commodities.

2.2 Base Metal Exploration and Geology

J. David Lowell

EXPLORATION TECHNIQUES

To be suited for surface mining an ore deposit must occur at, or relatively near, the ground surface and must have horizontal dimensions which are relatively large in comparison with its vertical dimension. Stratabound, bedded, porphyry, or enrichment-blanket deposits are typically suitable for surface mining. Experience indicates that in bulk low grade base metal deposits underground block cave mining costs usually become equivalent to surface mining costs when the stripping ratio reaches somewhere in the range of 2:1 to 5:1. Relatively few lead or zinc deposits are suitable for surface mining, but many porphyry copper deposits can be mined by open pit methods, particularly those with supergene enrichment blankets.

A previous section in this chapter had described a number of exploration techniques and applications of techniques. Many of these techniques are applicable both to grass roots and to detailed exploration around known mineral deposits and in established mining districts.

In designing an exploration program to test a deposit in a given district, a number of factors should be considered. These include expected size of ore body, depth, percentage of premineral outcrop, presence of trace element dispersion halos, presence of silicate alteration halos, control of mineralization by structural features, association with certain types of intrusive bodies, association with certain paleo-basin environments, presence of soil or stream sediment anomalies or associated exotic or placer deposits, and physical properties of the deposit which would have a reasonable chance of producing a diagnostic geophysical response.

If the geophysical, geological, and geochemical surveys, which should be designed to accommodate the above-listed criteria, indicate that the showing or known mineral deposit might have commercial dimensions, then diamond drilling designed to determine the physical size of the deposit is probably indicated. If the results of this drilling, which is generally done with AX or BX sized core, indicate that a deposit of sufficient size to be of commercial importance exists, then detailed drilling and sampling should ensue.

SAMPLING AND TESTING

Drilling equipment of some kind must usually be employed to sample all three dimensions of a surface mined deposit. Drill samples usually consist of a split of percussion or rotary drill cutting or a split of drill core made using a core splitter or diamond saw. The whole drill core is sometimes used as a sample with a few pieces of core saved as a skeleton core record.

The drilling method or methods used will be selected considering drilling depths, rock conditions, and the sampling characteristics of the mineralized rocks. Common drilling methods arranged from least to most expensive are as follows:

1. Percussion drilling done from truck-mounted or Airtrac type crawler mounted rigs. Drilling depth is usually limited to about 91 m (300 ft) but can reach 152 m (500 ft) or more. Samples consist of relatively fine cuttings. Sample quality ranges from fair to good in good rock above the water table, to poor to fair in poor rock below the water table.

2. Truck-mounted rotary tricone and down-the-hole hammer drilling using as a drilling medium mud, compressed air, or high pressure-high volume air from a multiple stage air compressor. Drilling depths are usually limited to about 610 m (2000 ft), but with large oil field type equipment very deep holes are possible. Samples consist of fine-grained cuttings which range up to coarse chips when using the down-the-hole hammer in competent rock. Sample quality ranges from fair for mud rotary drilling to excellent for air rotary drilling in good rock at relatively shallow depth. Drill cores are often taken, either as spot cores to calibrate rock identification and sample accuracy, or as continuous wire line core if the rig is equipped for wire line coring.

3. Reverse circulation rotary or percussion drilling. In this system the drilling medium, usually compressed air, is circulated down the hole on the outside of the drill rods and the cuttings are blown back up the inside of the rods. Drill casing is usually carried at or near the bottom of the hole so that contamination of the cuttings sample is reduced to a minimum. Reverse circulation drilling is usually slightly more expensive than rotary drilling but produces a more accurate sample. It is widely used in sampling bulk low grade gold and oxide copper deposits.

4. Diamond core drilling using truck-mounted, skid-mounted, or hand-portable drill rigs powered by gasoline or diesel engines, or electric or hydraulic or compressed-air motors. Core recovery and sample accuracy has improved greatly in recent years and ranges from poor to good for *pack sack* type portable drilling, to good to excellent for large wire line rigs using carefully controlled drilling mud, swivel tube core barrels when drilling relatively large drill cores in good rock.

5. Large diameter core drilling for obtaining bulk samples for metallurgical testing. Core sizes are typically PQ [85.0 mm ($3\frac{11}{32}$ in.) diameter], or 152 mm (6 in.) diameter which requires core barrels fabricated for the project. Core recovery is usually excellent and sampling accuracy excellent. The heavy core is somewhat awkward to handle and costs are usually 50% higher than conventional core drilling, but the benefits far outweigh the added costs if this approach eliminates some underground headings and results in a better three-dimensional metallurgical picture of the ore body.

Sample preparation technique is often as important a problem as sampling and assaying accuracy. It should be planned and supervised by people with experience with the specific type of ore. Systematic and methodical checking of both sample preparation and assaying should be done using outside laboratories. A typical sample preparation procedure for a base metal development drilling project would be as follows:

NX size drill core is picked up daily at the drill rigs and hauled to a central core shed. Before splitting it is photographed on 35mm color film for later rock mechanics and structural interpretations. Core is then laid out on logging tables for geologic and geotechnical logging and testing. Rock Quality Designation (RQD) and core recovery measurements are made. Geologic logging and measurements such as point load testing are then done. Assay breaks are either on arbitrary lengths [2 or 4 m or (5 or 10 ft) are common sample intervals] or breaks picked by the geologist on the basis of lithology or ore mineralogy. This decision will be predicated to some extent on the expected grade control and selective mining problems. The geologic log sheet and rock mechanics log should include data entry format to facilitate key punch entry and easy readability. Computerization of geologic data should not be done at the sacrifice of good graphic geologic logs from which easily understandable geologic sections can be directly plotted.

Core samples are split using a hand or hydraulically powered core splitter and half of the core is preserved in the box for future reference or subsequent resampling. The assay split is crushed to 4.8 mm (3/16 in.) then split through Jones Splitters to one-eighth of the heads sample with the 4.8 mm (3/16 in.) reject stored for possible metallurgical testing or reassaying. The split is recrushed to −2.0 mm (minus 10 mesh), split to one-half weight and then pulverized using plate or ring and puck type pulverizers. Two assay pulps are split from the resulting sample. One of these is assayed and the second filed for check assays or composite assay samples.

Assaying procedures are highly variable in base metal deposits depending on the combination of elements which are of interest, effects of oxidation, presence of refractory constituents, and presence of native gold, silver, or copper, etc. Because of the many variables it is usually best to carry out a test program before deciding on an assay procedure for a given type of rock and mineralization. Most base metal assaying is now done by the atomic absorption method, but in some situations chemical colorimetric analyses are preferable, and a large volume of X-ray fluorescence and nuclear activation assaying is now being done where high assay precision is outweighed by the advantage of low cost and speed. It is very important in exploration programs to systematically send out for check assaying a split of every tenth or twentieth sample and to periodically compare the assaying results using statistical methods.

Other types of laboratory work are usually justified in base metal exploration programs. If hydrothermal alteration or precise identification of specific host rocks is important to the program, then a representative group of petrographic thin sections should be made. If sulfide textures or solid solution problems are important polished sections should also be made. X-ray analyses are useful in some mineralogical determinations, particularly in clay minerals, and the electron microprobe is often useful in identifying associations between elements and host minerals. Spectrographic analyses of composite samples are used to insure that no potential byproduct or coproduct in the ore has been overlooked. Absolute age determinations are being increasingly used to identify favorable ore-related intrusions since in most districts all hydrothermal mineralization is related to one intrusive event. These laboratory techniques in some organizations can be done *in-house* but in most cases lower cost and more accurate results can be obtained by using commercial labs or consultants with experience in the specific problem.

Bulk sampling for the purpose of obtaining representative metallurgical samples for bench or pilot mill tests usually involves a joint project between the metallurgist and the exploration geologist. The samples must have the volume and fragment sizes required for the metallurgical test and they must be sufficiently representative of the mass of the ore body in grade, mineralogy, grindability, etc. to demonstrate any lateral or vertical variations in the character of the ore body. Typical variations which may be present are a vertical change downward from strong oxidation to fresh sulfides, or a vertical or lateral metal zonation from, for

example, a copper-molybdenum core outward to a copper-zinc margin of a porphyry copper ore body. A metallurgical test program typically progesses from bench testing of composite assay reject drill core samples to pilot testing involving hundreds or thousands of tons of ore from underground declines, drifts, and raises or from large diameter core holes.

The following example from the Dizon project of the Benguet Corp. illustrates both efficient use of bulk sampling and the effective choice of development drill hole spacing. The Dizon ore body was originally tested by Nippon Mining and Mitsubishi with 42 BX and NX diamond core holes drilled on a grid of approximately 200 m (656 ft). Core recovery was relatively poor and a significant fraction of the gold originally present in the oxide portion of the ore body was lost from the core samples recovered. The Nippon project was also completed when the price of gold was low and the gold content was less significant, but at the present gold price, the 1.0 gmt gold credit has more value than the 0.5% copper credit.

One of the most important decisions in planning a development drilling project is selection of grid drill spacing. In porphyry copper projects this spacing usually falls in the range of 76 m (250 ft) (Mission mine, AZ development drill spacing) to 183 m (600 ft) (Kalamazoo mine, AZ drill spacing). Typical hole spacings are 100 m or 400 ft. Factors which influence this are the uniformity of ore grade and mineralogy, degree of structural dislocation, and the expected viability of the ore body: a deposit which is clearly viable can be drilled on wider spacing than a marginal ore body. A statistical study by Benguet of the Dizon drilling indicated a change of less than 0.01% Cu between the 200 m (656 ft) grid holes and the final 100 m (328 ft) grid holes.

Most of the 15 infill holes drilled in the Benguet exploration project were large diameter core holes [PQ 85.0 mm (3^{11}/$_{32}$ in.) core] which were laid out so as to complete a 100 m (328 ft) grid as shown on Fig. 1. Several holes also twinned previous Nippon holes to calibrate the accuracy of the previous sampling and assaying. Large diameter core samples were divided on 15 m (49 ft) bench intervals, and an individual bulk flotation bench test was completed for each bench interval in each drill hole through the entire thickness of the ore body. The result was a complete three-dimensional metallurgical picture of the ore body. In addition three adits were driven on two levels of the ore body from which a sufficient tonnage of ore was mined for pilot mill testing and development of a mill flow sheet. The large diameter core holes served three functions: calibrate the ore grade, fill in the drill grid to a 100 m (328 ft) spacing, and provide a representative set of bulk metallurgical samples.

Bulk samples from underground workings present a difficult sampling problem. A typical 18 t (20 st) muck sample must be reduced to a representative 4.5 kg (10 lb) assay sample. This normally requires construction of a sample mill. Channel and car samples may also be collected for comparison. An example of a well-designed sample mill is the Quintana Minerals mill installed during the evaluation of the 53.7 Mt 0.42% Cu, 0.012% Mo Copper Flat porphyry copper ore body at Hillsborough, New Mexico. (Dunn, 1981). The bulk metallurgical sample was derived from a 190 m (623 ft) minus 15° decline from which two drifts totaling 478 m (1568 ft) were driven into the ore body in a *Y* pattern. The sample mill was laid out as shown in Fig. 2.

GEOTECHNICAL STUDIES AND HYDROLOGY

Rock mechanics studies are done for essentially all new open pit base metal mines. The importance of this work is greater for deep pits and where wall rocks are relatively

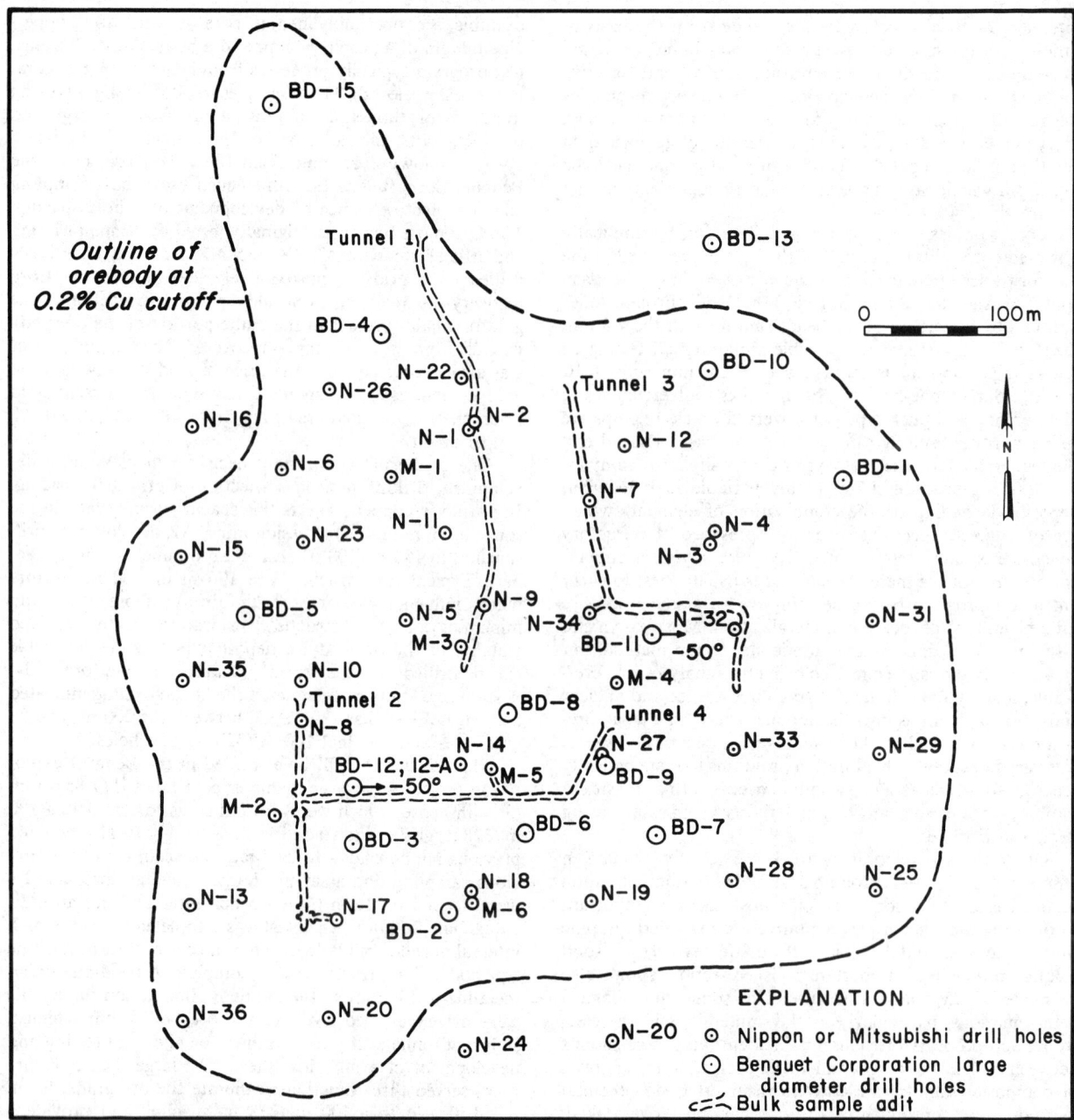

*Outline of
orebody at
0.2% Cu cutoff*

BD-15

Tunnel 1

BD-13

BD-4

BD-10

N-26   N-22

N-16

N-2

Tunnel 3

N-1

N-12

BD-1

N-6   M-1

N-7

N-11

N-4

N-23

N-3

N-15

N-5   N-9

N-34   N-32

BD-5

M-3

BD-11   -50°

N-31

N-35

N-10

M-4

Tunnel 2

BD-8   Tunnel 4

N-8

N-14

N-27

N-33

BD-12; 12-A

M-5

BD-9

N-29

-50°

M-2

BD-6   BD-7

BD-3

N-18

N-13

N-17

M-6

N-19

N-28

N-25

BD-2

EXPLANATION

N-36   N-20

N-20   ⊙ Nippon or Mitsubishi drill holes

N-24

N-20

◉ Benguet Corporation large
diameter drill holes

Bulk sample adit

0   100m

N

Fig. 1. Drill hole location map of Benguet Corp.'s Dizon copper ore body.

incompetent, but there are applications in all surface mining operations. RQD or other fracture density information should be collected during an exploration drilling project as soon as the possibility appears that the prospect might develop into a commercial ore body. Measurements usually made include point load testing, specific gravity, and sometimes compression testing.

These data are used to predict pit slopes, powder factors, fragment size, and water flow and give some information which can be integrated with metallurgical testing to determine overall grindability and power factors, etc.

Geostatistical analyses have been widely applied to mineral exploration in recent years, with the results ranging from good to poor. The best results have been obtained in mine planning studies in large homogeneous ore bodies lacking

structural dislocations; the poorest results have been obtained in studies intended to interpret the geology rather than the ore distribution in complex ore bodies. In general it is never safe to depend on a geostatistical ore reserve calculation during the exploration phase of a base metal project without carrying out a conventional ore reserve calculation based on polygons or sections and utilizing all available geologic maps and sections. Geostatistical techniques become increasingly useful as the overall shape and general grade distribution become clear and repetitive three-dimensional grade calculations are needed rather than geologic interpretations.

Water flows into the mine and a water source for the plant and camp are important problems which may be covered by the project geologist but which are usually given to a consulting groundwater geologist or hydrologist. If water

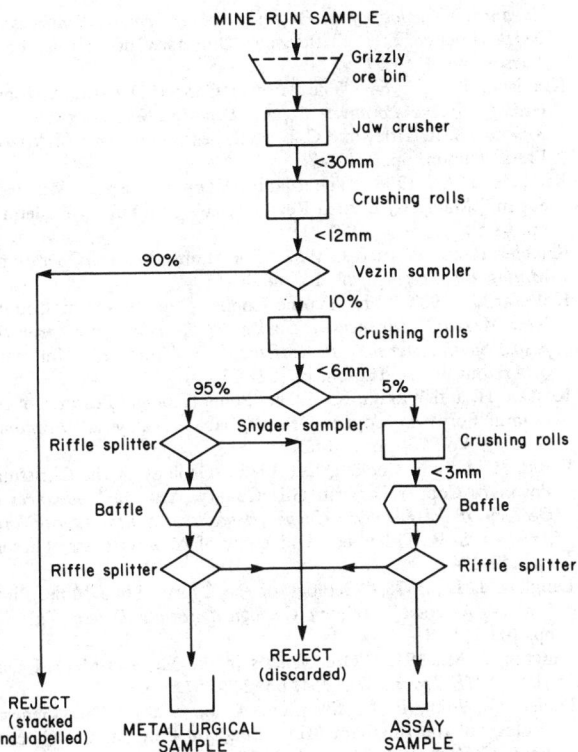

Fig. 2. Flowsheet showing preparation of underground samples.

problems are suspected, pump tests may be made on exploration drill holes, and water flow and static and drawdown water tables observed. However, since the average exploration drill hole is not well suited to hydrologic observations, except for measurement of static water table, groundwater information is usually collected from a well field constructed for that purpose. A recent paper by Montgomery and Harshbarger (1985) describes a very successful ground-water exploration project in the Andes range in Chile and Peru.

SELECTED REFERENCES

Aiken, D. M. and West, R. J., 1978, "Some Geologic Aspects of the Sierrita-Esperanza Copper-Molybdenum Deposit, Pima County, Arizona," *Arizona Geological Society Digest*, Vol. 11, pp. 117-128.

Ambrus, J., 1977, "Geology of the El Abra Porphyry Copper Deposit, Chile," *Economic Geology*, Vol. 72, No. 6, pp. 1062-1085.

Ambrus, J., 1978, "Chuquicamata Deposit," *International Molybdenum Encyclopedia*, A. Sutulov, ed., Alexander Sutulov Intermet Publications, Santiago, Chile, pp. 87-93.

Anderson, C. A., Scholz, E. A., and Strobell, J. D., Jr., 1955, "Geology and Ore Deposits of the Bagdad Area, Yavapai County, Arizona," Professional Paper 278, US Geological Survey, 103 pp.

Anon., 1977, "Lead Mining at Zeida in Morocco," *Mining Magazine*, Dec., pp. 608-617.

Baldwin, J. T., Swain, H. D., and Clark, G. H., 1978, "Geology and Grade Distribution of the Panguna Porphyry Copper Deposit, Bougainville, Papua New Guinea," *Economic Geology*, Vol. 73, pp. 690-702.

Barter, C. G. and Kelly, J. L., 1982, "Geology of the Twin Buttes Mineral Deposit, Pima Mining District, Pima County, Arizona," Chapter 20, *Advances in Geology of the Porphyry Copper Deposits, Southwestern North America*, S. R. Titley, ed., University of Arizona Press, Tucson.

Baumer, A. and Fraser, R. B., 1976, "Panguna Copper Deposit, Bougainville Island, Papua New Guinea," *Economic Geology of Australia and Papua New Guinea, 1. Metals*, Monograph 5, C. L.

Knight, ed., Australian Institute of Mining and Metallurgy, pp. 855-866.

Basin, D. and Hubner, H., 1964, "Cooper Deposits in Iran," Report No. 13, Geological Survey of Iran, 232 pp.

Bergey, W. R., Carr, J. M., and Reed, A. J., 1971, "The Highmont Copper-Molybdenum Deposits, Highland Valley British Columbia," *CIM Bulletin*, Vol. 64, No. 716, pp. 68-76.

Boyle, R. W., 1961, "Native Zinc at Keno Hill, Yukon," *Canadian Mineralogist*, Vol. 6, Part 5, pp. 692-694.

Bryner, L., 1968, "Notes on the Geology of the Porphyry Copper Deposits of the Philippines," *Mineral Engineering Magazine*, Vol. 19, pp. 12-23.

Bryner, L., 1969, "Ore Deposits of the Philippines—An Introduction to Their Geology," *Economic Geology*, Vol. 64, No. 6, pp. 644-666.

Campbell, N., 1967, "Tectonics, Reefs, and Stratiform Lead-Zinc Deposits of the Pine Point Area, Canada," *Genesis of Stratiform Lead-Zinc-Barite-Fluorite Deposits (Mississippi Valley Type Deposits)—A Symposium, New York, 1966*, Monograph 3, J. S. Brown, ed., *Economic Geology*, pp. 59-70.

Cargill, D. G., Lamb, J., Young, M. J., and Rugg, E. S., 1976, "Island Copper," *Porphyry Copper Deposits of the Canadian Cordillera*, Special Volume 15, A. Sutherland Brown, ed., CIM, pp. 206-218.

Carson, D. J. T., Jambor, J. L., Ogryzlo, P., and Richards, T. A., 1976, "Bell Copper: Geology, Geochemistry and Genesis of a Supergene-Enriched, Biotized Porphyry Copper Deposit with a Superimposed Phyllic Zone," *Porphyry Deposits of the Canadian Cordillera*, Special Volume 15, A. Sutherland Brown, ed., CIM, pp. 245-263.

Carvalho, P., Guimaraes, D., and Dequech, D., 1962, "Jazida Plumbo-Jincifera do Municipio de Vazante, Minas Gerais, Brazil," Bulletin 110, Dept. Nac. Producao Mineral, Div. de Fomento, Brasil, 119 pp.

Clark, A. H., 1979, "Potassium-Argon Age of the Cerro Colorado Porphyry Copper Deposit, Panama—A Reply," *Economic Geology*, Vol. 74, p. 695.

Clayton, R. L., 1978, "Alteration and Mineralization of the Cyprus Johnson Deposit, Cochise County, Arizona," *Arizona Geological Society Digest*, Vol. 11, pp. 17-24.

Cooper, J. R. and Silver, L. T., 1964, "Geology and Ore Deposits of the Dragoon Quadrangle, Cochise County, Arizona," Professional Paper 416, US Geological Survey, 196 pp.

Cornelius, K.D., 1969, "The Mount Morgan Mine, Queensland—A Massive Gold-Copper Pyritic Replacement Deposit," *Economic Geology*, Vol. 64, pp. 885-902.

Cornwall, H. R., 1982, "Petrology and Chemistry of Igneous Rocks: Ray Porphyry Copper District, Pinal County, Arizona," *Advances in Geology of the Porphyry Copper Deposits, Southwestern North America*, S. R. Titley, ed., University of Arizona Press, Tucson, pp. 259-274.

Cummings, R. B., 1982, "Geology of the Sacaton Porphyry Copper Deposit, Pinal County, Arizona," *Advances in Geology of the Porphyry Copper Deposits, Southwestern North America*, S. R. Titley, Ed., Univesity of Arizona Press, Tucson, pp. 507-522.

Dixon, D. W., 1966, "Geology of the New Cornelia Mine, Ajo, Arizona," *Geology of the Porphyry Copper Deposits, Southwestern North America*, S. R. Titley and C. L. Hicks, eds., University of Arizona Press, Tucson, pp. 123-142.

Dunn, P.G., 1984, "Geologic Studies During the Development of the Copper Flat Porphyry Deposit," *Mining Engineering*, Feb., pp. 151-159.

Eastlick, J. T., 1968, "Geology of the Christmas Mine and Vicinity, Banner Mining District, Arizona," *Ore Deposits of the United States, 1933-1967*, (Graton-Sales Volume), J. D. Ridge, ed., AIME, New York, pp. 1191-1210.

Eidel, J. J., Frost, J. E., and Clippinger, D. M., 1968, "Copper-Molybdenum Mineralization at Mineral Park, Mohave County, Arizona," *Ore Deposits of the United States, 1933-1967*, (Graton-Sales Volume), J. D. Ridge, ed., AIME, New York, pp. 1258-1281.

Einaudi, M. T., 1977, "Environment of Ore Deposition at Cerro de Pasco, Peru," *Economic Geology*, Vol. 72, No. 6, pp. 893-924.

Einaudi, M. T., Moore, W. J., and Wilson, J. C., eds., 1978, "An Issue Devoted to the Bingham Mining District," *Economic Geology,* Special Issue, Vol. 73, pp. 1215-1365.

Engineering and Mining Journal, 1982, *E/MJ International Directory of Mining,* McGraw-Hill, New York.

Fahrni, K. C., Macauley, T. N., and Preto, V. A., 1976, "Copper Mountain and Ingerbelle," *Porphyry Copper Deposits of the Canadian Cordillera,* Special Volume 15, A. Sutherland Brown, ed., CIM, pp. 368-375.

Fahrni, K. C., Kim, H., Klein, G. H., and Carter, N.C., 1976, "Granisle," *Porphyry Copper Deposits of the Canadian Cordillera,* Special Volume 15, A. Sutherland Brown, ed., CIM, pp. 239-244.

Fernandez, R. R., Brown, R. F., and Lencinas, A., 1973, "Pachon, Un Nuevo Porfiro Cuprifero Argentino, dto. Calingasta, Provincia de San Juan, Republica Argentina," *Jornadas Geol. Argentina,* Actas 5, Vol. 2, pp. 77-89.

Fleischer, V. D., Garlick, W. G., and Haldane, R., 1976, "Geology of the Zambian Copperbelt," *Handbook of Stratabound and Stratiform Ore Deposits,* Vol. 6, K. H. Wolf, ed., Elsevier, Amsterdam, pp. 223-352.

Flint, D. E., Nelson, F. J. and Stuart, R. J., 1975, "Gunung Bijih (Ertsberg) Copper Deposit, Irian Jaya, Indonesia," (Abstract), *Mining Engineering,* Vol. 27, No. 12, p. 71.

Fountain, R. J., 1972, "Geological Relationships in the Panguna Porphyry Copper Deposit, Bougainville Island, New Guinea," *Economic Geology,* Vol. 67, No. 8, pp. 1049-1064.

Frets, D. C. and Balde, R., 1975, "Mount Morgan Copper-Gold Deposit," *Economic Geology of Australia and Papua New Guinea. 1. Metals,* Monograph 5, C. L. Knight, ed., Australasian Institute Mining and Metallurgy, pp. 779-785.

Gilmour, P., 1982, "Grades and Tonnages of Porphyry Copper Deposits," *Advances in Geology of the Porphyry Copper Deposits, Southwestern North America,* S. R. Titley, ed., University of Arizona Press, Tucson, pp. 7-36.

Graybeal, F. T., 1982, "Geology of the El Tiro Ore Deposit, Silver Bell Mining District, Arizona," *Advances in Geology of the Porphyry Copper Deposits of Southwestern North America,* S. R. Titley, ed., University of Arizona Press, Tucson, pp. 487-506.

Greeley, M. N., 1978, "The Primary Copper Industry of Arizona in 1975 and 1976," Special Report No. 2, Arizona Department of Mineral Resources, 87 pp.

Gustafson, L. B. and Titley, S. R., 1978, eds., "Porphyry Copper Deposits of the Southwestern Pacific Islands and Australia," *Economic Geology,* Special Issue, Vol. 73, pp. 597-986.

Gustafson, J. K., Burrell, H. C., and Garretty, M. D., 1950, "Geology of the Broken Hill Ore Deposit, N.S.W., Australia," *Geological Society of America Bulletin,* Vol. 61, pp. 1369-1437.

Herbert, I. C., 1967, "Palabora," *Mining Magazine,* Vol. 116, London, pp. 4-25.

Hobbs, B. E., Ransom, D. M., Vernon, R. H., and Williams, P. F., 1968, "The Broken Hill Orebody, Australia: A Review of Recent Work," *Mineralium Deposita,* Vol. 3, pp. 293-316.

Hollister, V. F., Allen, J. M., Anzalone, S. A., and Seraphim, R. H., 1975, "Structural Evolution of Porphyry Mineralization at Highland Valley, B. C.," *Canadian Journal of Earth Sciences,* Vol. 12, pp. 807-820.

Hollister, V. F., 1978, *Geology of the Porphyry Copper Deposits of the Western Hemisphere,* AIME, New York, 219 pp.

Horton, D. J., 1978, "Porphyry-Type Copper-Molybdenum Mineralization Belts in Eastern Queensland, Australia," *Economic Geology,* Vol. 73, pp. 904-921.

Jansen, L. J., 1982, "Stratigraphy and Structure of the Mission Copper Deposit, Pima County, Arizona," *Advances in Geology of the Porphyry Copper Deposits, Southwestern North America,* S. R. Titley, ed., University of Arizona Press, Tucson, pp. 467-474.

Kents, P., 1975, "Geology and Mineralization of the Cerro Colorado Copper-Porphyry Deposit, Republic of Panama," Preprint No. 75-S-2, Society of Mining Engineers of AIME Annual Meeting.

Kihien, C. A. 1975, "Alteracion y su Relacion con la Mineralizacion en el Porfido de Cobre de Cerro Verde," Bol. 46, Tercer Congreso Peruano de Geologia, Parte II, Soc. Geol. Peru, pp. 103-126.

King, J. R., 1982, "Geology of the San Xavier North Porphyry Copper Deposit, Pima Mining District, Pima County, Arizona," *Advances in Geology of the Porphyry Copper Deposits, Southwestern North America,* S. R. Titley, ed., University of Arizona Press, Tucson, pp. 475-487.

Kinnison, J. E., 1966, "The Mission Copper Deposit, Arizona," *Geology of the Porphyry Copper Deposits, Southwestern North America,* S. R. Titley and C. L. Hicks, eds., University of Arizona Press, Tucson, pp. 281-287.

Kirk, H. J. C., 1966, "The Mamut Copper Prospect, Kinabalu, Sabah," Malaysia, Borneo Region, Geological Survey Bulletin 8, pp. 68-80.

Knobler, R. and Werner, J., 1962, "The Mantos Blancos Operation," *Mining Engineering,* vol. 14, pp. 40-45.

Kolessar, J., 1982, "The Tyrone Copper Deposit, Grant County, New Mexico," *Advances in Geology of the Porphyry Copper Deposits, Southwestern North America,* S. R. Titley, ed., University of Arizona Press, Tucson, pp. 327-334.

Kosaka, H. and Wakita, K., 1978, "Some Geologic Features of the Mamut Porphyry Copper Deposit, Sabah, Malaysia," *Economic Geology,* Vol. 73, pp. 618-627.

Koski, R. A., and Cook, D. S., 1982, "Geology of the Christmas Porphyry Copper Deposit, Gila County, Arizona," *Advances in Geology of the Porphyry Copper Deposits of Southwestern North America,* S. R. Titley, ed., University of Arizona Press, Tucson, pp. 353-374.

Langlois, J. D., 1978, "Geology of the Cyprus Pima Mine, Pima County, Arizona," *Arizona Geological Society Digest,* Vol. 11, pp. 103-113.

Langton, J. M., 1973, "Ore Genesis in the Morenci-Metcalf District," *SME Trans.,* Vol. 254, pp. 247-257.

Lanier, G., John, E. C., Swenson, A. J., et al., 1978, "General Geology of the Bingham Mine, Bingham Canyon, Utah," *Economic Geology,* Vol. 73, pp. 1228-1241.

Lombaard, A. F., Ward-Able, N. M., and Bruce, R. W., 1964, "The Exploration and Main Geological Features of the Copper Deposit in Carbonatites at Loolekop, Palabora Complex," *The Geology of Some Ore Deposits in Southern Africa,* S. H. Haughton, ed., Geological Society of South Africa.

Lopez, V. M., 1939, "The Primary Mineralization at Chuquicamata, Chile," *Economic Geology,* Vol. 34, pp. 674-711.

Loveman, M. H., 1917, "The Geology of the Bawdwin Mines, Burma, Asia," *Trans.,* Vol. 56, AIME, pp. 170-194.

Lowell, J. D., 1968, "Geology of the Kalamazoo Orebody, San Manuel District, Arizona," *Economic Geology,* Vol. 63, p. 645-654.

Lynch, D. W., 1966, "The Economic Geology of the Esperanza Mine and Vicinity," *Geology of the Porphyry Copper Deposits, Southwestern North America,* S. R. Titley and C. L. Hicks, eds., University of Arizona Press, Tucson, pp. 267-279.

Lyons, W. A., 1968, "The Geology of the Carahuacra Mine, Peru," *Economic Geology,* Vol. 63, No. 3, pp. 247-256.

MacKenzie, F. D., 1963, "Geological Interpretation of the Palo Verde Mine Based Upon Diamond Drill Core," *Arizona Geological Society Digest,* Vol. 6, pp. 41-48.

Magliola-Mundet, H., 1964, "Le Gisement de Cuivre de los Bronces de Disputada, Chile," *Chron. des Mines et de la Rech. Miniere,* Vol. 32, No. 330, pp. 120-127.

Manrique, C. J. and Plazolles, V. A., 1975, "Geologia de Cuajone," Bol. 46, Soc. Geol. Peru, Tercer Congreso Peruano de Geologia, Parte II, pp. 137-150.

McMillan, W. J., 1976, "Geology and Genesis of the Highland Valley Ore Deposits and the Guichon Creek Batholith," *Porphyry Deposits of the Canadian Cordillera,* Special Volume 15, A. Sutherland Brown, eds., CIM, pp. 85-104.

Metz, R. A. and Rose, A. W., 1966, "Geology of the Ray Copper Deposit, Ray, Arizona," *Geology of the Porphyry Copper Deposits, Southwestern North America,* S. R. Titley and C. L. Hicks, eds., University of Arizona Press, Tucson, pp. 177-188.

Montgomery, E. L., and Harshbarger, J. W., 1985, "Ground Water Development for Mineral Industry in Arid Zones of the Andean Highlands, South America," *Mining Engineering,* Jan., pp. 45-48.

Moolick, R. T. and Durek, J. J., 1966, "The Morenci District," *Geology of the Porphyry Copper Deposits, Southwestern North*

America, S. R. Titley and C. L. Hicks, eds., University of Arizona Press, Tucson, pp. 221-231.

Olmstead, H. W. and Johnson, D. W., 1966, "Inspiration Geology," *Geology of the Porphyry Copper Deposits, Southwestern North America,* S. R. Titley and C. L. Hicks, eds., University of Arizona Press, Tucson, pp. 143-156.

Pazour, D. A., 1980, "Cerro de Pasco; Centromin's Oldest and Largest Mine," *World Mining,* Vol. 33, No. 4, pp. 42-48.

Perry, V. D., 1935, "Copper Deposits of the Cananea District, Sonora, Mexico," *Copper Resources of the World,* Vol. 1, 16th International Geological Congress, pp. 413-418.

Perry, V. D., 1952, "Geology of the Chuquicamata Ore Body," *Mining Engineering,* Vol. 4, pp. 1166-1168.

Peters, W. C., James, A. H., and Field, C. W., 1966, "Geology of the Bingham Canyon Porphyry Copper Deposit, Utah," *Geology of the Porphyry Copper Deposits, Southwestern North America,* S. R. Titley and C. L. Hicks, eds., University of Arizona Press, Tucson, pp. 165-175.

Phillips, C. H., Gambell, N. A., and Fountain, D. S., 1974, "Hydrothermal Alteration, Mineralization, and Zoning in the Ray Deposit," *Economic Geology,* Vol. 69, pp. 1237-1250.

Pinto Linares, P. J., 1978, "Geologia y Mineralizacion de Tungsteno en la Veta San Cristobal, Peru," *Memoria,* C. Petzall, ed.; Segundo Congreso Latinoamericano de Geologia, Venez. Dir. Geol., Bol. Geol., Publ. Esp. 7, Tomo v, pp. 3839-3860.

Preece, R. K., III, 1979, "Paragenesis, Geochemistry, and Temperatures of Formation of Alteration Assemblages at the Sierrita Deposit, Pima County, Arizona," unpublished M.S. Thesis, University of Arizona, Tucson, 106 pp.

Prigogine, A., 1975, "Mamut Mine and Mill—Newest Southeast Asian Porphyry Development," *World Mining,* Vol. 37, No. 10, pp. 42-48.

Reed, A. J. and Jambor, J. L., 1976, "Highmont Linearly Zoned Copper-Molybdenum Porphyry Deposits and Their Significance in the Genesis of the Highland Valley Ores," *Porphyry Copper Deposits of the Canadian Cordillera,* Special Volume 15, A. Sutherland Brown, ed., CIM, pp. 163-181.

Richard, K. and Courtright, J. H., 1958, "Geology of Toquepala, Peru," *Mining Engineering,* Vol. 10, pp. 262-266.

Richard, K. and Courtright, J. H., 1966, "Structure of Mineralization at Silver Bell, Arizona," *Geology of the Porphyry Copper Deposits, Southwestern North America,* S. R. Titley and C. L. Hicks, eds., University of Arizona Press, Tucson, pp. 157-163.

Rose, A. W. and Baltosser, W. W., 1966, "The Porphyry Copper Deposit at Santa Rita, New Mexico," *Geology of the Porphyry Copper Deposits, Southwestern North America,* S. R. Titley and C. L. Hicks, eds., University of Arizona Press, Tucson, pp. 205-220.

Saegart, W. E. and Lewis, D. E., 1976, "Characteristics of Philippine Porphyry Copper Deposits and Summary of Current Production and Reserves," SME Preprint No. 76-I-79, 47 pp.

Shklanka, R., 1969, "Copper, Nickel, Lead, and Zinc Deposits of Ontario," Circular 12, Ontario Department of Mines and Mineral Resources, 394 pp.

Silberman, M. L. and Noble, D. C., 1977, "Age of Igneous Activity and Mineralization, Cerro de Pasco, Central Peru," *Economic Geology,* Vol. 72, No. 6, pp. 925-930.

Soregaroli, A. E., 1974, "Geology of the Brenda Copper-Molybdenum Deposit, British Columbia," *CIM Bulletin,* Vol. 67, No. 750, pp. 76-83.

Soregaroli, A. E. and Whitford, D. F., 1976, "Brenda," *Porphyry Copper Deposits in the Canadian Cordillera,* Special Volume 15, A. Sutherland Brown, ed., CIM, pp. 186-194.

Spatz, D. M., 1979, "Potassium-Argon Age of the Cerro Colorado Porphyry Copper Deposit, Panama—A Discussion," *Economic Geology,* Vol. 74, pp. 693-695.

Theodore, T. G., Silberman, M. L., and Blake, D. W., 1973, "Geochemistry and Potassium-Argon Ages of Plutonic Rocks in the Battle Mountain Mining District, Lander County, Nevada," Professional Paper 798-A, US Geological Survey, pp. 1-24.

Theodore, T. G. and deWit, M. P., 1976, "Porphyry-Type Metallization and Alteration at La Florida de Nacozari, Sonora, Mexico," Open-File Report 76-760, US Geological Survey, 28 pp.

Velasco, J. R., 1966, "Geology of the Cananea District," *Geology of the Porphyry Copper Deposits, Southwestern North America,* S. R. Titley and C. L. Hicks, eds., University of Arizona Press, Tucson, pp. 245-249.

Waddington, G. W., 1969, "Copper in Quebec," Special Paper 4, Quebec Department of Natural Resources, 395 pp.

Wall, J. R., Murray, G. E., and Diaz, T. G., 1961, "Geology of the Monterrey Area, Nuevo Leon, Mexico," Gulf Coast Association Geological Societies *Trans.,* Vol. 11, pp. 57-71.

Ward, H. J., 1978, "Environment of Ore Deposition at Cerro de Pasco, Peru—A Discussion," *Economic Geology,* Vol. 73, No. 6, pp. 1190-1194.

Waterman, G. C. and Hamilton, R. L., 1975, "The Sar Cheshmeh Porphyry Copper Deposit," *Economic Geology,* Vol. 70, pp. 568-576.

West, R. J. and Aiken, D. M., 1982, "Geology of the Sierrita-Esperanza Deposit, Pima Mining District, Pima County, Arizona," *Advances in Geology of the Porphyry Copper Deposits, Southwestern North America,* S. R. Titley, ed., University of Arizona Press, Tucson, pp. 433-466.

Wilkinson, W. H., Jr., Vega, L. A., and Titley, S. R., 1982, "The Geology and Ore Deposits at Mineral Park, Mohave County, Arizona," *Advances in Geology of the Porphyry Copper Deposits, Southwestern North America,* S. R. Titley, ed., University of Arizona Press, Tucson, pp. 523-542.

Wolfe, J. A., Manuzon, M. S., and Divis, A. F., 1978, "The Taysan Porphyry Copper Deposit, Southern Luzon Island, Philippines," *Economic Geology,* Vol. 73, pp. 608-617.

World Mining, 1963, "Palabora—Why It's the Most Exciting New Copper Mine in the 1960's," *World Mining,* Sept., pp. 30-36, 92.

World Mining, 1981, *1981-1982 World Mines Register,* Miller Freeman, San Francisco.

Young, M. J. and Rugg, E. S., 1971, "Geology and Mineralization of the Island Copper Deposit," *Western Miner,* Vol. 44, pp. 31-40.

Zweifel, H., 1972, "Geology of the Aitik Copper Deposit," 24th International Geological Congress, Section 4, Montreal, pp. 463-473.

2.3 Precious Metals Exploration and Geology

BYRON S. HARDIE

INTRODUCTION

Exploration and geological techniques have different meanings for different geologists. The definition of *technique* in Webster's Collegiate Dictionary varies somewhat with time, but science, ability, expertness, music, and writing are key words; implied is study, logical analysis, hard work, and experience. Each geologist and nongeologist is as unique as his fingerprints; likewise are the ore bodies for which he searches and sometimes finds.

The politics and economics of the 1960s and 1970s have turned gold into a desired metal again. The western Cordillera within the conterminous US is an important source of many metals and is increasingly important as a gold province. The classic gold vein districts are largely mined underground, but the recently discovered and now important disseminated gold deposits are nearly one hundred percent mined by surface methods. This is a result of geologic environment, modern earthmoving methods, metal price, metal content, and the present inability to locate nonoutcropping deposits.

In the western US, the current focus for disseminated, bulk-minable gold is in the Great Basin and largely in Nevada. None of the occurrences identified to date has been lacking in outcropping, assayable gold mineralization. Such mineralization is essentially impossible to identify by panning or other gravity methods. Gravity concentration does show associated minerals containing lead, zinc, copper, mercury, and other metals and nonmetals often associated with the gold mineralization. Chemical or fire-assay analysis is essential to identify the gold. All such gold deposits found to date have been within or adjacent to mineralized areas of record.

Nevada, centrally located in the Great Basin, contains the largest number of reported disseminated gold deposits. The discovery of the Carlin gold mine north of the town of Carlin in north central Nevada gave increased impetus to precious metal exploration by those who anticipated increasing gold prices in the early 1960s. The three and one-half million ounce gold content in the Carlin ore body, as originally defined in 1965, was ideally suited to surface mining and easy extraction by proven cyanide metallurgy.

The Carlin gold discovery was the result of a search by experienced geologists for a specific deposit type in a specific environment. The successful effort initiated by top management of the forty-year-old Newmount Mining Corp. was guided by pre-WWII gold mining experience (Ramsey, 1973).

The Carlin discovery was equally the result of several decades of intense, sophisticated geological study of structure, sedimentation, and mineralization by the US Geological Survey (USGS) (Roberts, et al., 1958; Roberts, 1960; Hardie, 1966; and Radtke, 1981).

Increasing gold and silver prices stimulated discovery of additional deposits in somewhat different environments than those present at the Carlin deposit.

CARLIN-TYPE DISSEMINATED-REPLACEMENT GOLD DEPOSIT

Various terms have been used to describe surface minable gold deposits and gold-silver deposits that contain minerals and native gold-silver disseminated in a similar manner to copper minerals in a porphyry copper deposit. The gold-silver and associated minerals are rarely detected by the unaided eye. The terms often used for this type gold and gold-silver deposit include: disseminated, disseminated-replacement, Carlin-type, micron, invisible, non-pannable, bulk-minable, and sometimes, porphyry-type.

The alteration of rocks occurring with Carlin-type gold and silver mineralization is seldom as spectacular as are the gossans occurring with outcropping copper deposits. The silica, clay, arsenic, antimony, mercury, pyrite, carbon, and carbonate alteration is often too subtle to recognize by casual visual inspection in outcrop, but often is recognized in new excavations that can be sampled in detail.

A brief description of the geologic and mineralogic setting of the Carlin-type disseminated-replacement gold deposit may explain some of the various descriptive terms used herein.

In 1971, the US Geological Survey modified earlier Nevada metal deposit classification by Ferguson (1929) to better accommodate the disseminated gold deposits of Getchell, Gold Acres, Carlin, and Cortez (Roberts et al., 1971). On the basis of form and host rock, gold deposits in north-central Nevada and southwestern Idaho were divided into three major classes: replacement deposits, disseminated deposits, and veins. The replacement deposits are divided into pyrometasomatic deposits, base-metal replacement deposits, and peripheral gold deposits. Carlin-type disseminated gold deposits (1971 terminology) are treated separately as a distinct class, and are described as a special kind of replacement deposit in that large amounts of carbonate are replaced by silica, but they contain mineral assemblages which more closely resemble those of the low-temperature veins than those of the replacement deposits. Unoxidized ores in the Getchell and Carlin deposits are characterized by pyrite and realgar; in the Carlin deposit by cinnabar, stibnite, and a little galena and sphalerite; and in the Cortez deposit by only pyrite and gold. In 1971 the known, disseminated gold deposits were considered to be spacially related to the Roberts Mountain thrust fault, located below the fault in carbonate rocks where the thrust had been domed. Fractured permeable ground and precipitants such as carbonate or organic carbon were considered to be requirements at that time.

Radtke and Dickson (1974) later dropped the Roberts Mountain thrust fault as a characteristically associated feature but included as host rocks thin bedded carbonaceous, silty carbonate, fine-grained silicified rocks and jasperoids, zones of oxidized rocks above unoxidized rocks, and argillized rocks.

DISSEMINATED GOLD AND GOLD-SILVER DEPOSIT EXAMPLES

Examples of disseminated gold and gold-silver deposits in different geologic environments as currently defined by active explorationists are listed here.

Carlin-Type Disseminated-Replacement Gold Deposit

The following deposits are considered Carlin-type (Radtke and Dickson, 1974a, 1974b; Hausen and Kerr, 1967)

Bell (Jerritt Canyon), Elko Co., NV   (Producing mine)
Carlin, Eureka Co., NV   (Producing mine)

Blue Star, Eureka Co., NV	(Inactive mine)
Bootstrap, Eureka Co., NV	(Producing mine)
Cortez, Lander Co., NV	(Producing mine)
Gold Acres, Lander Co., NV	(Producing mine)
Gold Quarry, Eureka Co., NV	(Inactive mine)
Mercur (Getty), Tooele Co., UT	(Producing mine)
Northumberland, Nye Co., NV	(Producing mine)
Pinson (Ogee), Humboldt Co., NV	(Producing mine)
Prebble, Humboldt Co., NV	(Prospect)
Standard Pershing Co., NV	(Inactive mine)
Tallman (Duval), Cassia Co., ID	(Inactive mine)
Whitecap (Manhattan), Nye Co., NV	(Inactive mine)
Maggie Creek, Eureka Co., NV*	(Producing mine)
Rain, Elko Co., NV*	(Prospect)

*Added 1983, by Hardie.

Bulk Disseminated Gold Deposits in Volcanic Environment

This class deposit is described as occurring in a variety of geological environments spanning most of geologic time (Worthington, 1981).

Delamar (Silver City), ID	(One-third income from gold)
Gilt Edge, SD	(Inactive)
Golden Sunlight, MT	(Producing)
Haile Gold Mine, SC	(Developing mine)
Ortiz, NM	(Producing)
Republic District, WA	(Producing)
Sunnyside, ID	(Inactive)
Tiger, AZ	(Inactive)

Disseminated Silver Deposits

Described by Graybeal (1981) as having a silver to gold ratio of greater than 33:1, and containing four to ninety million tons. These deposits are considered from the standpoint of distribution of values rather than the physical location of specific minerals, and they are characterized by assay cutoff values. Tabular mineral concentrations are minor and less selective mining is normally required compared to vein and replacement deposits. Examples are:

Candelaria, NV*	(Inactive)
Commonwealth, AZ	(Inactive)
Creede, CO	(Inactive)
Delamar, ID*	(Producing)
Flathead, MT	(Inactive)
Hardshell, AZ*	(Inactive)
Hercules, ID	(Inactive)
Landusky-Zortman, MT	(Producing)
Nevada Packard, NV	(Producing)
Rochester, NV	(Inactive)
Round Mountain, CO	(Inactive)
Santa Fe, NV	(Prospect)
Taylor, NV	(Producing)
Tombstone, AZ	(Producing)
Waterloo, CA	(Inactive)

*Watson (1977).

Bulk Tonnage, Low Grade Deposits

Described by Watson (1980) as a specific type that can sometimes be classed as a base metal deposit dependent on associated base metal prices. Examples listed earlier by Watson (1977) are included in the list above by Graybeal (1981). Other examples given by Watson (1977) are:

Hog Heaven, MT	(Inactive)
Langtry, CA	(Inactive)

Real de Angeles, Zacatecas, Mexico	(Producing)
Sam Goosley, BC	(Producing)

EXPLORATION

The objective of mining exploration is to find a new mine or identify an old one that can produce at a profit. This single function is the starting point required to extend the life of an active mining company. It matters not where or how the mine is found if it can produce a marketable metal that results in profit to both miner and consumer. The product can be any metal or nonmetal if compatible with corporate capability and consumer need.

Target-Identification

The listed gold and gold-silver deposits in preceding paragraphs are mineral occurrences that have either produced in the early part of this century or occur in active or formerly active mining districts. Not one represents an original discovery made more than a mile or two distant from known mineral or formerly producing mines—all outcropped and were partly eroded; not all are yet proven to be economic metal deposits. A few of these deposits are yet-to-produce and were partly talus-covered when discovered but showed visible surface alteration and small amounts (often less than 1 ppm) of detectable gold mineralization.

Only a half-dozen of the producing deposits and prospects listed would have been profitable in 1965 when the Carlin mine began production with the gold price fixed at $35 per troy ounce. Some of the mines and prospects listed will not be economic if prices of gold and silver do not continue to rise. The profit margin on many will be slim or nonexistent unless management continues to increase operating efficiencies. Management includes the intelligent assessment of geological-mineralogic controls of mineralization and accurate evaluation of metal content before large capital sums are committed. These latter items are deceptively basic but are often given too little attention.

Some of the known and economic or potentially economic deposits may be replicated within one or two miles from the original discovery. The geologist must learn by observation and experience to recognize visible alteration and a permissible environment that accompanies the sought-after metal. Red, brown, and light-colored gossan and silicification often rises above talus cover. Bleaching resulting from clay alteration that results in a leprous, unnatural blemish on the hillside can be a subtle guide to mineralization. Alteration effects that are not visible but identifiable by chemical means can be guides to economic mineralization. Recognition of this described environment will require careful and detailed field work. Less than total effort by the prospector-geologist will miss outcropping targets.

All environmental details that have proven to be guides to precious metals mineralization often occur without metal or economic metallization. When metal is present, it does not occur uniformly distributed through the host rock even though it may be disseminated in occurrence; it is possible to sample in or adjacent to any of the listed gold or gold-silver occurrences and conclude that they are silicified and bleached but barren of economic metallization. Even if arsenic and mercury sulfides are glaringly visible as in some developed gold and gold-silver deposits, their presence is no guarantee of metallization if identified on an otherwise barren-appearing hillside.

The preceding will sound familiar to all exploration geologists who have looked for mineral deposits using all available *state-of-the-art* geologic guides. The haunting thought is always present: "How do we find a gold-silver deposit that does not outcrop?" Success in finding outcropping disseminated gold deposits to date has been a function of some outcrop exposure and the ability to recognize the potential target.

To date geophysical methods have been of limited aid in the search for the disseminated gold and gold-silver deposit that does or does not outcrop. Copper deposits are relatively more responsive to geophysical probing, but they also are not always simple to detect.

The sophisticated chemical, x-ray, atomic and neutron analytical methods combined with space photography and sensing methods provide data on both the micro- and mega-scale (Ellett, 1981). The geologist has to work in between the extremes. Some geophysical technique application aids are helpful in the search for surface minable gold-silver deposits, but our instrumentation capabilities manage to stay ahead of our interpretive skills. The flood of numbers from computerized data must be properly handled to provide time required for field interpretation and application; most of today's deposits were found without today's technical backup aids. We must learn to convert our computerized numbers to recognizable three-dimensional concepts.

In the meantime, we continue with the methods that have become familiar and at least partially successful: face-to-face confrontation with rock, pick and sample sack, but leaning more on sophisticated analytical interpretation to collect raw samples with greater discrimination.

Investment Opportunity

The possibility of acquiring an identified mineral deposit or participating in its development and mining can contribute to the success of exploration. In a period of increasingly tight and expensive financing, exploration funds can often be more profitably applied by joint ventures involving two or more compatible partners. Such opportunities are often identified at the corporate level, but the exploration geologist should be able to recognize such potential as a result of daily exposure. The skills used in the discovery of new mineral deposits are needed in evaluating known mineral deposits.

Examples of mining companies, oil companies, utility companies, and other successful business organizations that have the financial capability to acquire developed and/or operating mines are recorded in the news releases and are well-documented in mining journals. Such acquisitions, if successful, are usually concluded following detailed examination of proven and/or potential ore reserves, mining, milling, refining, and marketing costs that will affect capital required and acceptable returns on investment; tax expense or credits are often critical factors in considering such acquisitions.

Examples of Successful Target Identification

In the past fifty years, mining, milling, and ore finding methods in the US have had variable success in meeting the increasing needs for many metals. The following examples illustrate the recent discovery approach used on some of the earlier listed gold and gold-silver deposits in which gold has been the predominant metal of interest:

Nevada Discoveries

Years 1930-1940

Getchell Mine, **Humboldt Co., NV:** Discovered by prospectors near old tungsten workings. Red and yellow arsenic in black carbonate accompanied gold in limestone.

Years 1940-1950

Gold Acres Mine, **Lander Co., NV:** Discovered by experienced mining company by examining a known producing gold district with turquois prospects in carbonate sediments.

Bluestar and Bootstrap Mines, **Eureka/Elko Counties, NV:** Discovered by prospectors mining turquois, barite, and antimony in a small placer gold district with known gold veins on dike contacts in sediments.

Years 1960-1970 [Gold decontrolled, 1968]

Carlin Gold Mine, **Eureka Co., NV:** Discovered in 1961 by experienced gold mining company geologists using a model patterned after mine districts at Manhattan, Getchell, and Gold Acres, all in Nevada. Outcrop sampling identified gold and antimony in the small Lynn mining district. Gold was found adjacent to the Roberts Mountain thrust fault mapped by the USGS in carbonate sediments.

Cortez Gold Mine, **Lander Co., NV:** Discovered by experienced mining company geologists exploring for silver and copper in an old silver-turquois mining district. USGS geochemical sampling identified gold in silicified limestone adjacent to the Roberts Mountain thrust fault.

Golden Sunlight, **Jefferson Co., MT:** Discovered by mining company geologists examining old gold mine on veins that showed disseminated gold in intrusive rock.

Years 1970-1980

Pinson Gold Mine, **Humboldt Co., NV:** Discovered by experienced geologists with a consortium of mining companies examining an old gold mine near the Getchell Gold Mine in carbonate sediments.

Bell (Jerritt Canyon) Gold Mine, **Elko Co., NV:** Discovered by experienced geologist and mining company exploring an old antimony district. Gold in carbonaceous sediments adjacent to Roberts Mountain thrust fault.

Smoky Valley Mine, **Round Mt., Nye Co., NV:** Discovered by experienced mining company geologist examining old lode and placer mining district. Gold in volcanics.

Alligator Ridge Gold Mine, **White Pine Co., NV:** Discovered by experienced geologist-prospector for experienced mining company. Gold in silicified limestone.

Sterling Gold Mine, **Nye Co., NV:** Discovered by experienced geologists for mining consortium in old gold-fluorite district in carbonate sediments.

California Discovery

Years 1970-1980

McLaughlin Gold Mine, **Napa Co., CA:** Discovered by experienced geologist for gold mining company at old mercury mine in known geothermal area near hot springs in volcanics, serpentine and clastic sediments.

Utah Discovery

Years 1970-1980

Mercur Gold Mine, **Tooele Co., UT:** Discovered by experienced geologists for oil company mining subsidiary in siliceous, carbonaceous, carbonate sediments in an old gold producing district.

Plan for Discovery

Nearly all of the listed examples were recognized or discovered by trained geologists having available sophisticated geologic and mineralogic backup supplied by capable mining companies.

The improving capability to interpret geological, geophysical, structural, and geochemical data will lead to discovery of the more subtle outcropping deposits that now elude easy detection, and eventually the nonoutcropping deposit. All of the known deposits are accompanied by a surrounding halo of identifiable if not easily recognizable alteration and anomalous indicator metals.

New deposits will continue to be found in areas known to contain gold and silver mineralization. Published literature gives some of the best leads to potential gold-producing areas. Examination of these areas will indicate potential exploration targets.

HYDROLOGIC MODEL

Ore deposition models are best constructed on the basis of mine openings and drill hole data developed as a result of commercial operations supplemented by detailed mineralogic studies of metallization, alteration, and fluid inclusions.

The hydrologic model for the Creede Mining District, Central San Juan mountains, Colorado (Wetlaufer et al. 1979), is viewed as a fossil geothermal system. The lower Miocene Ag-Pb-Zn-Cu-Au ore deposit has had extensive study (Steven and Eaton, 1975); Bethke et al., 1976; Barton et al., 1977; Bischoff et al., 1981). This deposit has many characteristics that are similar to those known for intermediate depths in active geothermal systems. The postulated hydrothermal system is seen as a nearly closed convection cell of deeply circulating saline fluids, depositing and leaching gangue and ore minerals. Alteration caps were probably caused by condensing of acid volatile compounds (H_2S and CO_2) that boiled off the hydrothermal solutions. The ore zone is thought to represent only the top of the system.

Isotopic studies have shown that recharge of active geothermal systems is dominantly supplied by meteoric (surface) waters. Many geothermal fields are located in volcanic regions and commonly are associated with silicic volcanic rocks thought to be characterized by shallow magma chambers. The Creede district is noted for middle-Tertiary volcanism that could easily furnish the energy needed to sustain a hydrothermal system.

The Carlin disseminated replacement gold deposit has been extensively studied by the US Geological Survey and others using drill hole and deep open pit exposures (Hausen and Kerr, 1967; Dickson et al., 1979; Radtke, 1981). The hydrothermal system has characteristics similar to those noted at Creede and at other deposits. The Carlin deposit is thought to have formed at shallow depths, about 1500 m (4921 ft) below surface. Hydrothermal fluids furnished by meteoric water heated by Tertiary volcanic energy are believed to have leached and deposited ore and gangue minerals from and within the host Paleozoic carbonate sediments. This model is more widely accepted when there is visible evidence of nearby intrusive activity.

More information from recently discovered deposits will undoubtedly modify current hydrologic concepts. Detailed studies of geothermal systems with and without presently economic amounts of metals have been led and guided by the worldwide background studies sponsored by the US Geological Survey and university research groups. These scientific contributions will continue to influence new economic metal discoveries.

Choosing a Target Area

A procedure often followed in choosing areas that may contain potential targets is outlined below. Choosing a place to dig or drill when the explorer is a prospector or employed by a large mining company should mean that there is no known available target area that has a better chance for

success in discovering the desired metal deposit (Broderick, 1948).

One suggested procedure:
1. Personal experience.
2. Published and open file reports by the US Geological Survey, US Bureau of Mines, US Bureau of Land Management.
3. State Geological Survey.
4. County records: assessor, county recorder.
5. Information data banks in state and federal agencies, universities, and public libraries.
6. Mining and geological society publications.
7. Consultants.

Geological Target Maps

A preliminary target search will often suggest several areas that should be given a first look in the field before committing to an initial sampling program.

All published maps covering the areas of interest should be acquired and prepared for field use. A map scale of 12,000:1- or 24,000:1-scale topographic map or recent aerial photographs make good initial geologic or reconnaissance maps. Maps of 2,400:1- or 6,000:1-scale will be required to show details of fracturing, mineralization, and bedding attitudes.

Major faulting, fracturing, and bedding attitudes are critical to control of mineralization and to eventual design of surface pits for mining. Characteristics of faulting and structurally disturbed areas should be modified and updated when drill testing provides information.

Water occurrence should be noted where observed in mapping and drilling. Water sampling is sometimes useful in locating mineralized structures as well as a possible supply for drilling and mining operations. Water occurrence can be a problem in drill sample collection and in pit design.

LAND STATUS

Location maps in the initial reconnaissance of potential target areas should include land classification designations to locate federal, state, or county restrictions on land use. Mining claim location and other private ownership information should be determined by field reconnaissance and record search of Bureau of Land Management and county assessor-recorder offices. A mineral survey of property boundaries is low-cost insurance against future loss.

SAMPLING

Sampling programs are designed to determine the metal content and the feasibility of economically recovering that metal from a volume of material being investigated. A quote from Griffiths (1962) is well worth emphasizing.

"Any scientific investigation is no better than its sampling plan; inadequate sampling cannot be subsequently offset by any procedure, experimental or statistical. The problem of sampling arises in the initial stages of an investigation when setting up the most efficient means of achieving the main objective of the experimental program, and it crops up again at various stages throughout the experiment in attaining required levels of precision of estimates from different measuring techniques. Because of its fundamental role in experimentation, the sampling pattern should be decided upon at the same time as the overall strategy of the program, i.e., at the beginning; generally in sedimentary petrography it is resolved when the experimenter becomes aware of it, a certainly inefficient and, possibly, disastrous practice."

Sampling is a major step in the discovery of a mineral deposit, and in the early exploration stage must receive close supervision by professional geologists. Sampling a mineral deposit after it has been identified is a critical step. Poor sampling procedure is common and has been for the past few decades of observation by the author. Careless sampling might not be damaging if there are no mineralogic characteristics that reduce metal recovery and no shortage of metal reserve. Remembering Griffiths' advice, careful sampling must be insisted on for every deposit at the outset of exploration, development, and mining; otherwise, a marginal deposit may fail after consuming unnecessary expenditures of time and dollars before problems are identified.

Disseminated gold deposits are limited in optional metallurgical extractive treatment and are often further plagued by interfering sulfides, carbon precipitants, and siliceous encapsulation of gold and/or silver. When low prices and high costs coincide, survival may hinge on effective initial sampling practice. Management efficiencies at this late stage could be nullified by late recognition of poor sampling procedures.

Disseminated replacement gold deposits appear to represent the definition of randomness and demand a systematic grid sampling plan that can be uniformly expanded or otherwise systematically altered as data accumulates. Gold deposits that contain less than two parts per million gold (0.05 ounces gold per ton) are economic at a $500 gold price and lower if metallurgical characteristics are favorable.

Biogeochemical sampling may be more effective in some areas than other surface sampling methods when near-surface conditions are unfavorable for otherwise representative sampling procedure. The fine-sized gold in disseminated gold deposits is often soluble in oxidizing conditions when organic acids and/or sulphur are present.

The following references are some that will serve as dependable guides when selecting a sampling program: Barnes, 1979; Curtin et al., 1974; Ferguson et al., 1977; Harris, 1981; McKinstry, 1948; Overstreet and Marsh, 1981; Peters, 1978; Rose, Hawkes, and Webb, 1979.

Concurrent with sampling for metal values, samples for alteration studies should be taken where indicated. Often the specific need for such samples is not indicated until metal anomalous areas are defined when assay results are available. Magnetic susceptibility of various rock units requires samples to determine the advisability of magentic surveys.

Sample Map

A topographic map on 6,000 :1- or 12,000 :1-scale can be used to establish a sampling grid that will develop as surface sampling and geologic mapping progress. Metal anomalies, alteration samples, metallurgical samples and major structure and rock contacts should be represented. This map will serve to plan a drill test if indicated by initial studies. Larger-scale maps (1,200 : 1- or 2,400 : 1-scale) will usually be required for detailed sample plotting and geological mapping.

Sample Preparation and Analysis

Effective collection of representative samples of any rock mass can only result from a carefully planned sampling program. There is always a tendency to favor the easily-chipped rock outcrop either on the surface or underground.

The detailed planning of sample collecting requires close supervision of sample helpers to avoid non-representative sampling. Inspection of sample book entries by the project geologist should be routine to help identify incorrect procedures used in sample collecting; drillers' reports should receive similar attention. Description of individual samples

in a sample book will often aid early recognition of mineralized rock types by comparing assay results.

Before sampling is begun, competent assay companies must be contacted and evaluated. Accurate sample preparation and assaying procedures are not to be taken for granted. Many mining companies now prepare their samples through the assay pulp stage to assure proper protection against salting and inaccurate size reduction procedure (Shaw, 1978).

Assay procedures are not usually difficult, but good housekeeping in commercial and private laboratories is an essential practice not always observed. Considering the critical nature of samples taken at the surface or from drill core and cuttings, it is imperative that the exploration manager assures himself that the most effective procedures are being followed. The samples may have less than one ppm gold and silver, and a decision may be made to commit millions of dollars on drilling an ore body that contains less than two ppm gold and forty ppm silver.

Assurance that correct sampling procedure is in place must be followed by continual checks on assaying results by comparative assays using coded duplicate samples and different assay firms. All competent assayers continually run checks on their own work.

DRILL TESTING

The decision to drill a mineralized area is ideally made as a result of several weeks or months of preparation of metal anomaly, geologic, property, water supply, and possibly, geophysical maps. A drill grid map and cross-section maps should be prepared to allow the project geologist to keep currently appraised of interpretations based on the results of drill sampling and geological logging. Access roads and drill sites must be prepared. Forest Service, Bureau of Land Management and/or local authorities must usually be contacted well in advance of drill site preparation. Negotiations with drill contractors follows a determination of those who are acceptable as judged by past performance. Assayers and sample preparation facilities should have already been arranged to provide sufficient service on assay results to avoid costly delays in drilling progress or inefficient drill planning. Sample collecting help should be already trained, and sample storage for future reference must be adequate to allow safe storage and retrieval of valuable samples.

Drafting help and office geological help should be adequate to plot and review drilling and interpret results. Interpretation of geophysical surveys or re-surveys should be kept current.

Several hundreds of tons of drill samples will be handled, i.e., collecting, splitting to assay duplicate and metallurgical test samples. Rockboards are often made from cuttings or chips from core or rotary-percussion samples to aid in geologic and mineralogic interpretation.

Drilling a ten million ton ore body can easily require 76 200 m (250,000 ft) of drilling equivalent to 625 holes 122 m (400 ft) deep. This equates to $2.5 million at ten dollars per 0.3 m (ft) for contract drilling. This type ore body will often require an equal number of holes outside the actual minable ore body and these are included in the numbers used in this example. The cost to the point of feasibility can quickly reach $5 million or $.50 per ton for a 10 million ton ore body.

The costs in dollars and manpower of a major drill program are of a magnitude to require constant review and frequent decision points as to whether or not a drill project should be continued. Such decisions can only be made with confidence when a well-organized plan allows daily review of all data. Currency of data available to make intelligent interpretations is aided by computer programs, assuming that all basic data are accurate.

Example: Ten Million Ton Disseminated Gold Ore Body

A few numbers relating to a drill test of a 9 Mt (10 million ton) disseminated gold or gold-silver deposit with rotary and percussion drilling will illustrate the logistics involved.

Assumptions:
Ore body size: 9 Mt (10 million tons)
Grade: 2.8 g/0.9 t (0.10 oz. average/ton)
Number of drill holes: 625
Average hole depth: 122 m (400 ft) $\times$ 124 mm (4 7/8 in.) diameter
Sample interval: 1.5 m (5 ft)
Sample weight per 1.5 m (five-ft) hole: 68 kg (150 lb)
Cost/foot drilling: $10 ($32.80/m)
Cost/sample assaying: $10

The drilling assumed in this example would result in 76 200 m (250,000 ft) of drill footage at an average of 14 kg (30 lb) of sample per 0.3 m (1 ft) [about 2.3 kg/0.3 m (5 lb/ft) for 51-mm (2-in.) diameter drill core]. The 3.4 kt (3750 st) of cuttings [or 567 t (625 st) of 51-mm (2-in.) diameter core] is often split four times using a Jones splitter to obtain two 4- or 4.5-kg (9- or 10-lb) samples [for an assumed 1.5 m (5 ft) drill sample interval] to provide for assay, rockboard, mineralogical and metallurgical test purposes.

The costs of this program for drilling and assaying alone would be $2.5 million for drilling and $500,000 for assaying.

Data Handling

Drilling data from sample assays, alteration studies and geological-mineralogical logging accumulates rapidly when good drilling progress is made. The number of drills operating on a project must not exceed, but should, at least, match the capability of the staff to effectively study, record, and interpret information acquired in a timely manner. A continuing problem on most drill programs today is preparation and analysis of samples and the plotting and interpretation of results. The availability of drill sample data controls the effectiveness of the drill program, and allows timely decisions.

Drill Sample Collection

The use of rotary and percussion drills in testing the disseminated metal deposits has resulted in better sampling in most instances than with diamond drill equipment. The cost of such drilling is often less than one-half that of diamond drilling. Samples from rotary-percussion drilling are two to six times as large as with drillcore, and sample recoveries are often more complete. Weighing samples can detect losses or caving in the hole that would otherwise go undetected. Screen analyses on samples will indicate the presence of coarse gold and the distribution of values. Losses of fines in rotary-percussion drilling can result in upgrading or downgrading of metal content in samples.

Drill Plan and Cross-Section Maps

Base maps should be constructed before drilling begins. Results of geological mapping and surface sampling should be plotted on plans and sections before drilling begins. An often critical step is proper orientation of vertical sections to allow effective drill hole orientation and geological interpretation.

All drilling data should be plotted currently to aid interpretation of mineral trends, alteration, attitudes of fault and dike structures with depth, and correlation of these interpreted features with those mapped at the surface. Often, close examination of drill sections can change the drill sampling plans and geologic interpretations based on surface study; also, additional surface study is often indicated by drilling results. The three dimension drill picture will often modify earlier concepts that may indicate termination of further drilling or may indicate additional drilling and possibly additional property acquisition.

Surface geologic plans and vertical cross-sections are used to confirm mineral projection and to give early indication of trends of total metal content for ore reserve estimation. Early in the geological mapping and drilling process, estimates of mining costs based on ore and waste distribution can be tested by hand and expanded by computer program.

Continuing estimates of geological reserves based on currently available drill results support judgment decisions and recommendations to management that should be made periodically.

Metallurgical samples should be collected and tested as soon as practicable to give mineralogic-alteration information and metal recovery numbers. This information is often valuable in making early estimates of ore classification that will indicate the need for different metallurgical treatment.

Assay-metallurgical information should be used by the geologist in conjunction with logging of drill core or cuttings. This approach will indicate the practicability of visual identification of metallurgical ore types.

Drill hole information should be used to evaluate the results of geophysical surveys or the need to recommend such surveys.

ORE RESERVES

The project geologist and his staff should maintain a continuing evaluation of geologic ore reserves. These evaluations are usually for an order of magnitude estimate and are not equivalent to the more detailed estimates prepared for final feasibility studies.

Geologic Reserves

Geologic reserve estimates are often made by outlining the ore exposed on vertical drill hole sections. Grade cutoff numbers and pit slopes are based on experience with current practice and economic guidelines as modified by metal prices, and mining and milling costs determined by operating experience. Metallurgical data developed during drill testing affect cutoff grades that may vary with different areas in the deposit. The influence of geologic structure, rock type, and alteration distribution must be accurately depicted on the geologic ore reserve sections. These data are essential to the final minable reserve estimates and pit design developed in feasibility studies. The original geologic reserve estimate will give a useful estimate of waste tonnage required to be stripped and/or mined internally during ore extraction. Dilution of ore with below cutoff grade and waste material will be estimated based on geological characteristics deduced from study of surface and drill hole information.

The geological reserve estimates usually should begin while drilling continues to define ore limits and extensions. This estimate will incorporate metallurgical data acquired by testing drill cuttings and samples from surface shafts, winzes or adits when indicated. In some instances, large diameter 152 or 203 mm (6 or 8 in.) drill core is effectively used at lower cost and better representative samples than obtained from other underground openings. Resident geologists at this stage often replace the exploration team geologists to coordinate with engineers in developing pit design and mine plans. Such matters as water supply, waste dump and tailings disposal plans should be assigned to the mine operation staff. The exploration team should be looking for another deposit.

Useful references on ore reserve estimation include: Brooks and Bray (1968), Hart and Sprague (1968), Hazen (1968), Peters (1978).

REFERENCES

Barnes, H. L., 1979, *Geochemistry of Hydrothermal Ore Deposits,* 2nd ed., John Wiley and Sons, New York.

Barton, P. B. Jr., Bethke, P. M., Roedder, E., 1977, "Environment of Ore Deposition in the Creede Mining District, San Juan Mts., CO," *Economic Geology,* Vol. 71, pp. 1-24.

Bethke, P. M., Barton, P. B., Jr., Lanphere, M. A., and Steven, T. A., 1976, "Environment of Ore Deposition of the Creede Mining District, San Juan Mts., CO," *Economic Geology,* Vol. 71, pp. 1006-1011.

Bischoff, J. L., Radtke, A. S. and Rosenbauer, R. J., 1981, "Hydrothermal Alteration of Graywacke by Brine and Seawater: Roles of Alteration and Chloride Complexing on Metal Solubilization at 200°C. and 350°C.," *Economic Geology,* Vol. 76.

Broderick, T. M., 1948, Discussion of "The Search for Concealed Deposits—A Reorientation of Philosophy," by S. G. Lasky, Trans. AIME, *Mining Geology,* Vol. 178, pp. 89-90.

Brooks, L. S. and Bray, R. C. E., 1968, "A Study of the Tonnage and Grade Calculations at the Geco Division of Noranda Mines," *Ore Reserve Estimation and Grade Control,* Special Vol. 9, Canadian Institute of Mining and Metallurgy, Montreal, p. 181.

Curtin, G. C., King, H. D., and Mosler, E. L., 1974, "Movement of Elements into the Atmosphere from Coniferous Trees in Subalpine Forests of Colorado and Idaho," *Journal of Geochemical Exploration,* Vol. 3, pp. 345-363.

Dickson, F. W., Rye, R. O., Radtke, A. S., 1979, "The Carlin Gold Deposit as a Product of Rock-Water Interactions," *Report 33 IAGOD,* J. D. Ridge, ed., Nevada Bureau of Mines and Geology.

Ellett, R. D., 1981, "Twenty Years of Exploration Technology—Have We Progressed," American Mining Congress, Denver, CO, Sept. 30.

Ferguson, R. B., et al., 1977, *Field Manual for Stream Sediment Reconnaissance,* D.O.E., Grand Junction, CO.

Ferguson, H. G., 1929, "The Mining Districts of Nevada," *Economic Geology,* Vol. 24, pp. 115-148.

Forrester, J. D., 1946, *Field and Mining Geology,* John Wiley and Sons, New York.

Graybeal, F. T. 1981, "Characteristics of Disseminated Silver Deposits in the Western United States," *Relations of Tectonics to Ore Deposits in the Southern Cordillera,* W. R. Dickinson and W. D. Payne, eds.

Griffiths, J. C., 1962, *Statistical Methods in Sedimentary Petrography,* Vol. I, Macmillan, New York, p. 609.

Hardie, B. S., 1966, "Carlin Gold Mine, Lynn District, Nevada," *Report 13,* Pt. A, Nevada Bureau of Mines, pp. 73-83.

Harris, J. F., 1981, "Sampling and Analytical Requirements for Effective Use of Geochemistry in Exploration for Gold," *Precious Metals in the Northern Cordillera,* A. A. Levinson, ed., Association of Exploration Geochemists Symposium, Vancouver.

Hart, R. C., and Sprague, D., 1968, "Methods of Calculating Ore Reserves in Elliot Lake Camp," *Ore Reserve Estimation and Grade Control,* Special Vol. 9, Canadian Institute of Mining and Metallurgy, Montreal, pp. 251-260.

Hausen, D. M., and Kerr, P. F., 1967, "Fine Gold Occurrence at Carlin, NV," *Ore Deposits of the United States, 1933-1967,* J. D. Ridge, ed., AIME, New York.

Hausen, D. M., Ekburg, C., Kula, F., 1982, "Geochemical and XRD—Computer Logging Method for Lithologic Ore Type Classification of Carlin-type Gold Ores," *International Gold/Silver Conference II,* Nevada Institute of Technology, Reno, May.

Hazen, S. W., 1968, "Ore Reserve Calculations," *Ore Reserve Estimation and Grade Control,* Special Vol. 9, Canadian Institute of Mining and Metallurgy, pp. 11-32.

Koch, G., and Link, R. F., 1972, "Sample Preparation Variability in Diamond Drill Core from the Homestake Mine, South Dakota," *Report of Investigation 7677,* US Bureau of Mines.

McKinstry, H. E., 1948, *Mining Geology,* Prentice Hall, Inc.

Overstreet, W. C. and Marsh, S. P., 1981, "Some Concepts and Techniques in Geochemical Exploration," *Economic Geology,* 75th Anniversary Volume.

Parks, R. D., 1957, *Examination and Evaluation of Mineral Property,* Addison-Wesley Press, Inc., Cambridge, MA.

Peters, W. C., 1978, *Exploration and Mining Geology,* John Wiley and Sons, New York.

Radtke, A. S., and Dickson, F. W., 1974a, "Controls on the Vertical Position of Fine-Grained Replacement Type Gold Deposits," Abstract, Fourth IAGOD Symposium, Varna, Bulgaria, pp. 68-69.

Radtke, A. S., 1974b, "Genesis and Vertical Position of Fine-Grained Disseminated Replacement-Type Gold Deposits in Nevada and Utah, U.S.A.," *Proceedings,* Vol. I, Fourth IAGOD Symposium, Varna, Bulgaria, pp. 71-78.

Radtke, A. S., 1981, "Geology of the Carlin Gold Deposit, Nevada," Open File Report 81-97, US Geological Survey.

Ramsey, R. H., 1973, *Men and Mines of Newmont,* Octagon Books, New York.

Roberts, R. J., Hotz, P. E., Gilluly, J., and Ferguson, H. G., 1958, "Paleozoic Rocks in North-Central Nevada," *Bulletin,* Vol. 42, American Association of Petroleum Geologists, pp. 2831-2857.

Roberts, R. J., 1960, "Alinements of Mining Districts in North-Central Nevada," *Geological Survey Research 1960,* P.P. 400-B, US Geological Survey, pp. B17-B19.

Roberts, R. J., Radtke, A. S., and Coats, R. R., 1971, "Gold-Bearing Deposits in North-Central Nevada and Southwestern Idaho," with a Section on Periods of Plutonism in North-Central Nevada by M. L. Silberman and E. H. McKee, *Economic Geology,* Vol. 66, No. 1, Jan.-Feb.

Rose, A. W., Hawkes, H. E., Webb, J. S., 1979, *Geochemistry in Mineral Exploration,* 2nd ed., Academic Press, New York.

Rye, R. O., and Wells, J. D., 1974, "Stable Isotope and Lead Isotope Study of the Cortez, Nevada, Gold Deposit and Surrounding Area," *Journal of Research,* Vol. 2, No. 1, US Geological Survey, pp. 13-23.

Shaw, M. J., 1978, "The Importance of Sampling and Sample Preparation in Assaying for Gold," Assayers Limited, Ontario, Canada.

Steven, T. S., and Eaton, G. D., 1975, "Environment of Ore Deposition in the Creede Mining District, San Juan Mts., CO," *Economic Geology,* Vol. 70, p. 1034.

Watson, B. N., 1977, "Large, Low-Grade Silver Deposits in North America," *World Mining,* Mar., pp. 44-49.

Watson, B. N., 1980, "Bulk Tonnage, Low-Grade Deposits" *Precious Metals Symposium,* Sparks, Nov., Nevada Bureau of Mines.

Wetlaufer, P. W., Bethke, P. M., Barton, P. B., Jr., Rye, R. O., 1979, "The Creede Ag-Pb-Zn-Cu-Au District, Central San Juan Mts., CO: A Fossil Geothermal System," *Report 33, IAGOD 1979,* J. D. Ridge, ed. Nevada Bureau of Mines and Geology.

Worthington, J. E., 1981, "Bulk Tonnage Deposits in Volcanic Environments" *Relations of Tectonics to Ore Deposits in the Southern Cordillera,* W. R. Dickinson and W. D. Payne, eds.

2.4 Uranium Exploration and Geology

DAVID C. FITCH

Uranium exploration is generally regarded as a specialized endeavor. Although some of the methods used are similar to those for exploration for other metals, and for oil and gas, there are certain techniques and approaches unique to the search for uranium. Unlike other metals, uranium typically occurs together with its radioactive daughter nuclides, and its presence is easily detected by a scintillation counter. Over the past 25 years, much of the land surface of a number of developed countries has been thoroughly prospected by airborne and ground scintillation surveys. Several of these efforts resulted in early discoveries of shallow uranium deposits. As a consequence of such activities, the chances are becoming increasingly remote for discovering a previously unknown uranium deposit that outcrops, especially in heavily prospected areas such as the western United States.

Therefore, a major part of most current exploration programs is directed toward searching for concealed deposits that have no direct surface expression. Such programs are based strongly on geologic factors, including a working knowledge of the size, shape, and grade, and a concept of the origin of deposits sought. Geologic criteria are used to select a favorable host rock and favorable areas that merit drilling. Company objectives, land availability, geologic setting of an area, and a continuing evaluation of field and drill data based on ore guides influences the overall exploration program and balances the exploration effort among several targets. Most uranium deposits occur as small targets within large areas of potential host rocks—a needle-in-the-haystack proposition. It is one of the more difficult metals to find in economic concentrations.

To be successful, an exploration program must begin with management that accepts the high risks involved and has the perseverance to see a program through perhaps as many as a thousand or more drill holes until discovery, an adequate budget, and a team of discovery oriented geologists.

TYPES OF URANIUM DEPOSITS

Known uranium deposits occur in a variety of host rocks and geologic environments, but significant deposits are commonly localized in specific hosts and environments. There are numerous papers describing the classification of uranium deposits and their origin. See, for example, Moench and Schlee, 1967; and Nash, et al., 1981, with their extensive bibliographies.

The following types are among the most significant:

Sandstone-type Host

The ore consists of uranium-enriched humate that coats sand grains and impregnates the sandstone, imparting a dark color to the rock. Deposits are commonly irregular in shape, elongate, and form distinct trends. They may be tabular or form rolls that cross bedding. A number of deposits show strong evidence of syngenetic or nearly syngenetic deposition in the host sandstone.

Examples of deposits are Colorado Plateau types, including the Morrison formation deposits of New Mexico; the Salt Wash deposits of Colorado and Utah; and the Chinle deposits, Lisbon Valley, Monument Valley, and White Canyon, Utah and Arizona; the Wyoming Tertiary deposits, Wyoming Cretaceous deposits, Texas deposits.

See Anon., 1980, 1980a, and 1980b for good summaries.

Precambrian Metamorphic Host

The ore consists of pitchblende, coffinite, and urano-complexes in certain favorable host rocks, commonly either 1.8 or 1.6 billion years old. Deposits are commonly controlled by structures or fractures.

Examples of deposits are Rabbit Lake, Key Lake, and Cluff Lake, Saskatchewan, Canada; Nabarlek and Jabiluka, Northern Territory, Australia; Roxby Downs, South Australia; Midnight mine, Washington; and Schwartzwalder mine, Colorado.

See Anon., 1978.

Precambrian Conglomerate Host

The ore consists of uraninite with thucolite (organic matter) and other heavy minerals, possibly as placer concentrates. Examples of deposits are Elliot Lake, Canada and Witwatersrand, South Africa.

See Pretorius, 1981.

Igneous Intrusive Hosts

The ore consists of finely disseminated uranium minerals in felsic igneous rocks. Examples of deposits are Rossing, Southwest Africa and Ilimaussaq, Greenland.

Other signficant types of deposits are in uraniferous lignite, black shale, and carbonate hosts such as the Yeelirrie, Western Australia (calcrete), and the Todilto (limestone) deposits in New Mexico (Carlisle, et al., 1978; Rawson, 1980; Schilling, 1975; Swanson, 1961; and Vine, 1962).

See Nishimeri et al., 1977.

GEOLOGIC LITERATURE STUDIES AND MAP COMPILATION

Based on literature and map review, experience, and ideas generated, a decision is made to search for a certain type or types of deposits that leads to the selection of a specific host rock known or thought to contain the types of deposits sought. Then a regional area of interest is selected. A regional map is compiled from available sources showing the outcrop distribution and known extent of the host rock together with all known uranium deposits and occurrences within the region. Deposits and occurrences that occur in the target host are marked with a bold map symbol to be easily seen. If drilling depths can be estimated, areas within shallow, open-pitable limits are stippled. The scale of mapping to be used depends on the size of a region. Initial maps are commonly prepared at several scales ranging from 1:500,000, 1:250,000, and 1:125,000. Base maps at these scales may be obtained from the US Geological Survey, state, province, or county sources. Geologic maps from which the target host may be traced are commonly available from federal or state sources.

Maps showing uranium occurrences for some regions are available from the US Department of Energy, the US Geological Survey, or the IAEA and federal governments in some foreign countries.

Extensive literature research is conducted during the initial map compilation in order to plot all known uranium occurrences in the region. In some cases, occurrences may be plotted from listings in federal or state publications. The US Department of Energy maintains public files of preliminary reconnaissance reports, arranged by county, that describe occurrences examined. Airborne scintillation and geochemical data of much of the United States was generated by the Department of Energy's National Uranium Resource Evaluation Program conducted several years ago. Other sources of data for the location of uranium occurrences are geology journal publications, scout reports, oil well gamma-logs, and in some cases, water well sample records on file with state agencies.

As the map compilation proceeds, the search is continued for all available data on geologic descriptions of deposits and occurrences within the region and the geology of the host rock. A tabulation of the occurrences is prepared with a description of the geology sufficient to make a preliminary ranking of their favorability. Based on data generated at this stage, favorable areas are selected for field study.

GEOLOGIC FIELD STUDIES

Before and during field evaluation of a favorable area, an effort is made to examine the geology of operating mines and known deposits similar to the targets sought. The geologic characteristics of mineralization in the host rock, alteration, and the geometry of the ore bodies are given special attention. Each observation of ore is examined with the following questions in mind: How did the ore form? Why is it located here? What are the halo characteristics related to ore that are not present in barren ground? When developing a concept of a deposit's origin to apply to the discovery of new deposits, it is important to keep in mind the idea of multiple working hypotheses. It is also necessary to modify the hypothesis as new facts are accumulated. An example of the chronologic development of genetic theories for sandstone-type uranium deposits in the United States is tabulated by Finch (1967). One should strive to develop, as soon as possible, a concept of origin that is accurate. Discussions of ideas with geologists who have discovered ore deposits are very helpful.

Ore guides are then developed, based on the above studies. These vary among the types of deposits and may range from general to specific. Ore guides common to nearly all uranium deposits are:

1. The presence of a proven host rock.
2. Anomalous uranium mineralization (two times background may be significant in some hosts; 20 times background may be required for other hosts).
3. Position of area with respect to mineralized trends, if any.

Additionally, the habit of uranium deposits is such that they commonly tend to occur in clusters, are elongate, and tend to form distinct trends.

A familiarization with the geologic setting and the experience gained from the above work may be applied to evaluating favorability of the study area. Generally, field examination of a potential area is conducted in stages. The first stage is a brief reconnaissance of the most significant occurrences and outcrops in the area. A hand-held scintillometer (Bailey and Childers, 1977) is commonly used to estimate the grade and extent of mineralization around previously plotted occurrences as well as detecting new ones. These are immediately examined and plotted on work maps. Careful attention to anomalies as low as two times background is prudent. Altered areas in the host rock are plotted, as well as the boundaries of the host.

Airborne scintillation surveys (Bailey and Childers, 1977) are commonly made at this stage, which may detect outcropping deposits, if any, or the surface expression of deposits concealed beneath shallow cover. The geologist directs the airborne work and generally flies as observer to insure the flight lines are as chosen. Flight patterns are selected to parallel and thoroughly cover outcrops of the potential host rock. Grid patterns are generally only used in areas where specific host rocks are unknown. Altitude above ground level should be as low as good safety practices permit, because radiation decreases with the inverse of the distance squared. Flight levels generally range from 15 to 76 m (50 to 250 ft) above the ground. Either an airplane or helicopter is chosen depending upon terrain. Hand-held or airborne scintillation equipment may be purchased or leased. Spectrometers that distinguish between uranium, thorium, or potassium radiation are commonly used, especially in areas of metamorphic or granitic host rocks which often contain thorium. Airborne equipment commonly provides for strip charts, digital print-outs, or computer tape storage of data.

It is good practice to calibrate a scintillation counter (Nininger, 1954) before and during use. Some are equipped with an internal calibration setting and an adjusting knob. Others use an external radioactive source and a calibration knob. It is also good practice to obtain hand samples of ore of known grade for a rough estimate of instrument response for various grades of mineralization. It is difficult to estimate grade in a large outcrop of mineralization due to a mass effect. For this purpose, face scanner attachments are available for several of the scintillation and geiger counters (Key et al., 1982).

Geochemical techniques have been used in a number of uranium exploration programs (Dall'Aglio, 1973; De Voto, 1978; Grimbert, 1972). These techniques include uranium analysis of sediments, stream sediments, stream water and ground water. Results of uranium anomalies in stream and ground waters are generally difficult to interpret. Much of the uranium in water may have been derived from disseminated sources that are uneconomic. Also, composition of the ground water typically determines the amount of uranium that may be held in solution.

Another technique is radon emanometry which consists of measuring radon gas concentration in soil gas generally on a grid pattern over potentially favorable areas. If used, this technique is usually applied to specific drill target areas rather than regional reconnaisance. One method (Miller and Ostle, 1973) uses a one-meter (3 ft) long steel tube driven into the ground. Soil gas is pumped from the tube into an alpha counter with a silver-activated zinc sulfide analyzer. The radioactivity of the sample may be measured in counts per second and the data plotted on a sample map. Background values for the area must be determined several times per day because barometric conditions will influence radon emanations. Another radon technique uses alpha-sensitive film that is placed inside a plastic cup in a shallow hole about 0.6 to 0.9 m (2 to 3 ft) deep (Fleischer et al., 1980). The cup is usually left in place for two to three weeks, then recovered and the film processed. Alpha tracks on the film are then counted. The alpha film method is available from a service company, which supplies the film and cups, processes the film, and furnishes results for a fee. A good summary of radon methods is given by Bailey and Childers (1977a).

FAVORABILITY MAPS

As the reconnaisance work progresses, all significant data is plotted on a regional map at a suitable scale. The data to be plotted includes, in order of importance: (1) grade of uranium mineralization and percent eU_3O_8 or radioactivity in multiples of background at all uranium occurrences, (2) map extent of the host rock or structure if applicable, and

(3) map extent of favorable areas based on alteration and mineralization. Only in this manner can a regional picture be obtained and any significant patterns recognized. Areas may be ranked on the map as favorable, semi-favorable, and unfavorable, or assigned a numerical favorability ranking. In some geologic environments, it is possible to assign a numerical rating to each of the ore guides developed, with the greatest weight to the most important guides, and derive a total number of each data point. From such data, it may be possible to contour a favorability map.

Because there is nearly always competition from other exploration groups, the field studies are generally completed as rapidly as possible in order to acquire lands and generate drill targets. One never has all the answers to an area before moving ahead. In most successful exploration programs, decisions are made with the best, although often incomplete, data available.

LAND ACQUISITION

Once a favorable area (or areas) is selected and it has been decided to move ahead, it is necessary to acquire exploration and mining rights, together with the right of access. In actual practice, a determination is often made prior to field studies that there is a reasonable expectation of acquiring lands in the area of interest under favorable economic terms. Uranium exploration generally requires large land holdings because it is not possible to predict the specific location of deposits before discovery. Land requirements range from a small program involving 26 to 52 km² (10 to 20 sq miles) to large programs containing 260 km² (100 sq miles) or more.

The cost of acquisition and annual holding costs are important factors that must balance within the overall exploration budget. Payments to the owner from future production as royalty or other interest is an important factor affecting the economics of any deposit subsequently mined.

Mineral ownership varies depending on the area. In the western US, for example, there are large areas of public domain upon which mining claims may be staked. There are a number of professional surveying firms that offer services of determining land status and properly staking and recording mining claims for a fee. In a number of cases, individuals or companies perform the work themselves. A summary of regulations governing claim staking may be obtained from the US Bureau of Land Management and the individual state geological survey or bureau of mines. The cost of staking and recording claims which average about 0.08 km² (20 acres) varies from a low of about $30/claim to $150/claim or more in rough, tree-covered country. A minimum of $100/claim annual assessment work is required to hold the claims. Mining claims held by others may be leased under terms that are negotiated. Private lands may be acquired under terms of a mining lease negotiated with the owners. Terms of a mining lease generally provide for payment of a royalty from production, annual lease rentals or advanced minimum royalties, and in some cases annual work requirements. Production royalty is regarded as one of the most important items as it affects the production costs and can make or break a deposit. Uranium royalties are typically in the range of 1 to 5% of the gross value of ores mined. Annual rentals or minimum royalties, if any, may range from 50 cents to $5 or more per acre.

Land acquisition in other countries depends on local laws and regulations. In some cases, exploration and mining concessions may be obtained, or joint ventures may be formed with government or private groups. A detailed discussion of land matters relating to uranium exploration and mining is given in Chapter 2, Section 11, of this publication. An excellent review of US mining law and leasing procedures with examples of typical lease agreements is given by Bailey and Childers (1977, Chapters 4 and 5).

DRILLING PROGRAM

Once a favorable area is selected and the lands are acquired, a drilling program is formulated and undertaken. Drilling is commonly conducted in three stages.

1. During the first stage, holes are drilled on a wide spacing to delineate favorable ground and to obtain geologic information. Holes may be spaced from 800 to 1 600 m (one-half to one mile) apart for sandstone-hosted uranium targets. Initial spacing may be closer for a Precambrian metamorphic or vein host where a more specific target area has been mapped.

2. During the second stage, holes are drilled on a closer spacing in the more favorable areas based on results of the first stage for the purpose of intersecting uranium mineralization. Holes may be spaced 300 to 900 m (1000 to 3000 ft) apart for a sandstone target, or closer for a metamorphic or vein target. Some programs use a grid pattern for drilling in the first two stages. However, if the targets sought are randomly located, a random drill pattern is preferred.

3. During the third stage, holes are closely spaced in mineralized areas to intercept ore-grade uranium mineralization. Once a discovery is made, four equidistant holes are drilled to offset the discovery hole (Fig. 1), and the resulting ore holes are again offset (Fig. 2). The purpose of offset drilling is to obtain a group of adjacent ore holes that is as large as possible, bordered by a rim of non-ore holes (Fig. 3). Barren or slightly mineralized holes are not offset. This is regarded as the most efficient method of determining an ore body outline that cannot be predicted before drilling.

Spacing between offset holes and the discovery hole varies depending on drill depths and continuity of mineralization. For shallow, open-pit uranium deposits, an initial offset pattern of 45 m (150 ft) is commonly used. If ore is difficult to intercept at this spacing, the pattern is moved in to 23 m (75 ft) or 15 m (50 ft). Commonly, one drill will be placed on offsets and a second drill is moved to widely spaced locations to seek additional deposits.

Choice of drilling equipment is based on the type of rock being tested and the accessibility of the area. For sandstone-hosted deposits, a truck-mounted rotary rig capable of drilling 12 or 12.7 cm (4 3/4 or 5 in.) diameter holes with a tricone rock bit is commonly used (Fig. 4) (Bailey and Childers, 1977b). A 1500 series rig is well suited for shallow drill holes and is capable of drilling to depths greater than 300 m (1,000 ft). The drill should have a dual circulation system for drilling with compressed air, or water and mud. The upper part of a hole is drilled by air as far as possible and the remainder of a hole is drilled by water and mud. An efficient method that is becoming increasingly common is stiff-foam air drilling. New types of detergent are mixed with high volume air to lift cuttings from a wet drill hole. Compressed air or water is pumped through a swivel connection at the top of the drill string, down the drill string, and through openings in the drill bit. The air or mud lifts drill cuttings up the hole to the surface where samples are caught. A portable steel mud pit is commonly used for efficiency and to eliminate the need for excavating earthen mud pits. A water truck with a capacity for 5 680 to 11 350 L (1500 to 3000 gal) is used to haul drilling water to the location. If a drill hole is completed entirely by air, it is generally filled with water to obtain a complete geophysical log.

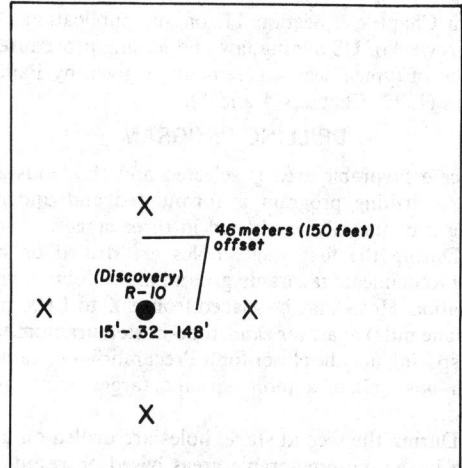

PLANNED DRILL HOLES

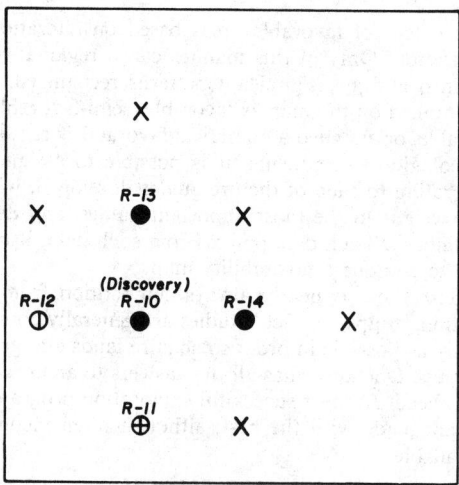

PLANNED DRILL HOLES

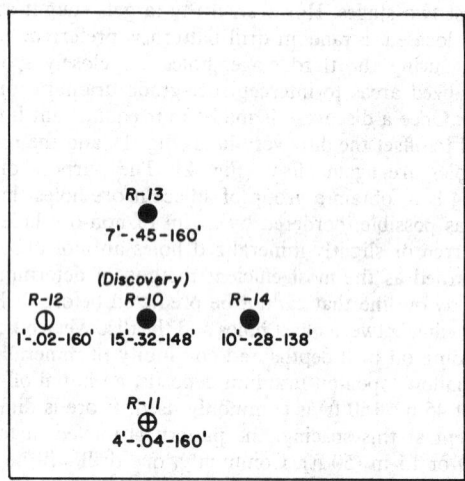

COMPLETED DRILL HOLES

- ● ORE GRADE & THICKNESS
- ◑ ORE GRADE, BUT BELOW MINIMUM THICKNESS.
- ⊕ STRONGLY MINERALIZED
- ⦶ WEAKLY MINERALIZED
- ○ BARREN TO TRACE

Fig. 1. Map showing drill sequence after discovery hole.

COMPLETED DRILL HOLES

- ● ORE GRADE & THICKNESS
- ◑ ORE GRADE, BUT BELOW MINIMUM THICKNESS.
- ⊕ STRONGLY MINERALIZED
- ⦶ WEAKLY MINERALIZED
- ○ BARREN TO TRACE

Fig. 2. Map showing drill sequence to delineate deposit.

Rotary drill samples are generally caught on 1.5 m (5 ft) intervals and placed on the ground in rows of 30 m (100 ft) (Fig. 5). A portion of each sample is placed in a sample bag and marked with the hole number and depth interval. Samples should be washed if collected from drilling mud. Considerable experience is required by the drillers to obtain representative samples.

In some areas, it may be decided to core (Bailey and Childers, 1977c), selected mineralized zones with a diamond core bit to obtain samples for stratagraphic information, chemical assay, or metallurgical testing. In most cases where core intervals are selected at shallow depths of less than

120 m (400 ft), it is economical to drill the hole by rotary methods to the core point, and then convert to a drilling mud suitable for coring. The tri-cone bit is then replaced with a core bit and core barrel, and coring performed through the mineralized zone and an additional 3 m (10 ft) for logging purposes. Choice of core diameter depends upon the volume of sample needed, and cost considerations as larger core is more expensive. Typically, NX sized core 4.45 cm (1.75 in.) diameter or NXWL 6.07 cm (2.39 in.) diameter is considered adequate.

Drill core is removed from the core barrel and placed in a wooden, V-shaped trough by the drillers. The core may be

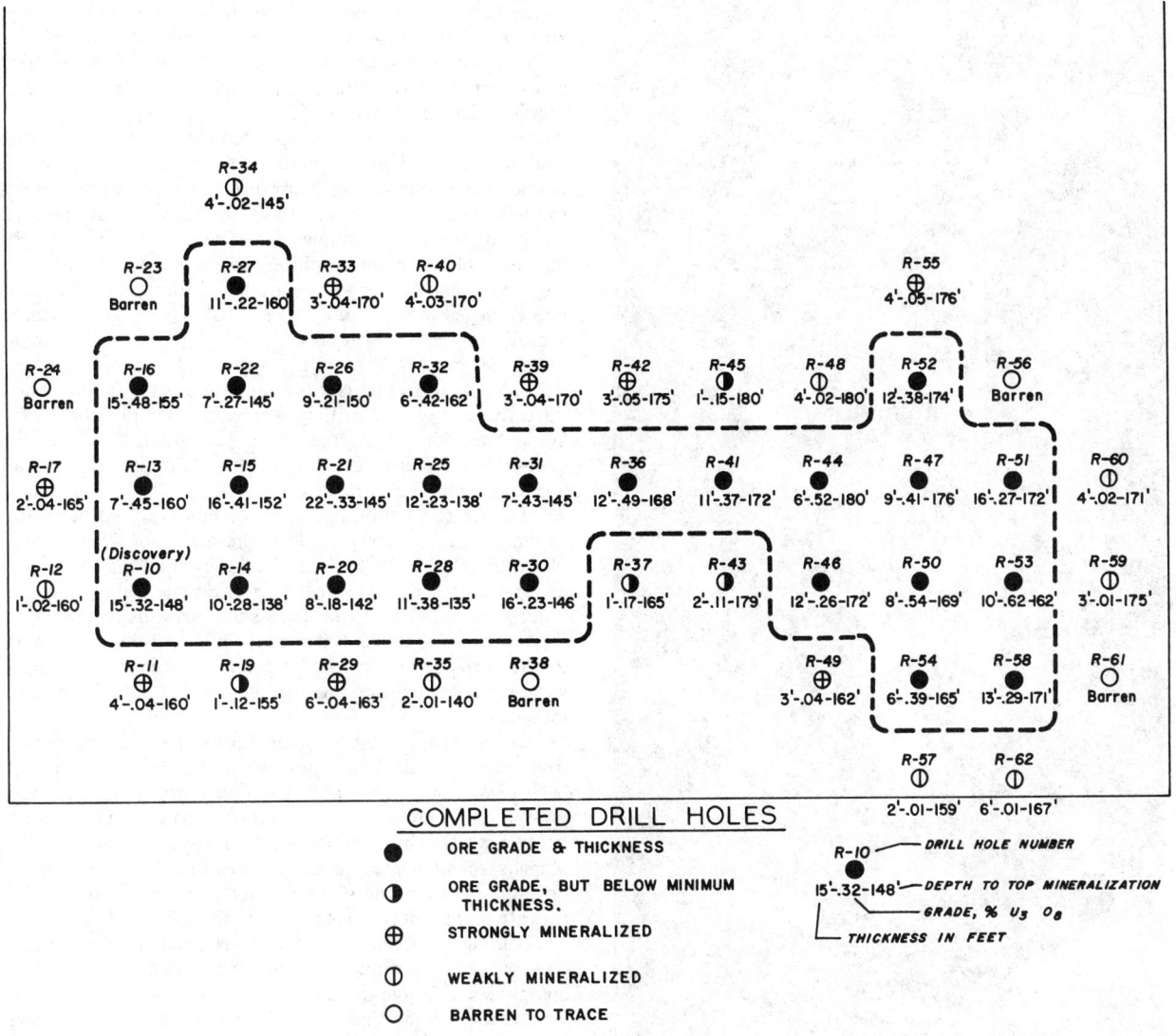

R-34
4'-.02-145'

R-23   R-27   R-33   R-40   R-55
○   ●   ⊕   ⦵   ⊕
Barren   11'-.22-160' 3'-.04-170' 4'-.03-170'   4'-.05-176'

R-24   R-16   R-22   R-26   R-32   R-39   R-42   R-45   R-48   R-52   R-56
○   ●   ●   ●   ●   ⊕   ⦵   ◐   ⦵   ●   ○
Barren 15'-.48-155' 7'-.27-145' 9'-.21-150' 6'-.42-162' 3'-.04-170' 3'-.05-175' 1'-.15-180' 4'-.02-180' 12'-.38-174' Barren

R-17   R-13   R-15   R-21   R-25   R-31   R-36   R-41   R-44   R-47   R-51   R-60
⊕   ●   ●   ●   ●   ●   ●   ●   ●   ●   ●   ⦵
2'-.04-165' 7'-.45-160' 16'-.41-152' 22'-.33-145' 12'-.23-138' 7'-.43-145' 12'-.49-168' 11'-.37-172' 6'-.52-180' 9'-.41-176' 16'-.27-172' 4'-.02-171'

R-12   (Discovery)   R-14   R-20   R-28   R-30   R-37   R-43   R-46   R-50   R-53   R-59
⦵   R-10   ●   ●   ●   ●   ⦵   ⦵   ●   ●   ●   ⦵
1'-.02-160' ●   10'-.28-138' 8'-.18-142' 11'-.38-135' 16'-.23-146' 1'-.17-165' 2'-.11-179' 12'-.26-172' 8'-.54-169' 10'-.62-162' 3'-.01-175'
        15'-.32-148'

R-11   R-19   R-29   R-35   R-38   R-49   R-54   R-58   R-61
⊕   ◐   ⊕   ⦵   ○   ⊕   ●   ●   ○
4'-.04-160' 1'-.12-155' 6'-.04-163' 2'-.01-140' Barren   3'-.04-162' 6'-.39-165' 13'-.29-171' Barren

R-57   R-62
⦵   ⦵

COMPLETED DRILL HOLES

2'-.01-159'   6'-.01-167'

●   ORE GRADE & THICKNESS

◐   ORE GRADE, BUT BELOW MINIMUM THICKNESS.

⊕   STRONGLY MINERALIZED

⦵   WEAKLY MINERALIZED

○   BARREN TO TRACE

R-10 ⟵ DRILL HOLE NUMBER
●
15'-.32-148' ⟵ DEPTH TO TOP MINERALIZATION
⟵ GRADE, % U_3O_8
⟵ THICKNESS IN FEET

Fig. 3. Drill map of ore deposit.

removed by gently tapping the core barrel, or if there is abundant clay, a pressure hose from the mud pump may be connected to one end of the barrel and the core pumped out with gentle pressure from the pump. The core is placed in a core box as shown by Fig. 6.

The core is examined and scanned with a geiger counter to select sample intervals. The sample intervals should be selected with boundaries based on changes in mineralization or lithology. Thin intervals may be sampled [as small as 0.15 m (0.5 ft) thick]. In thick mineralized zones, a combination of contiguous samples is taken whereby individual samples are generally held to a maximum of 1.5 m (5 ft) thick. The core for the sample is split with a core splitter and one-half bagged for assay.

The lithology of either rotary or core samples is logged by a geologist in the field using a hand lense or binocular microscope. The data recorded (Fig. 7) include rock type, color, and grain size. Special attention is given to describing alteration features and the presence of carbonaceous matter or pyrite in the host rock. The presence and amounts of hematite and limonite are also given special attention. Notes

are generally made of accessory minerals, however, detailed notes are not usually made of the percentage of the common minerals. Core recovery is also entered on the lithologic log.

For drilling in metamorphic or igneous terrain (Anon., 1968) either core drilling or percussion drilling is commonly employed. For steeply dipping targets, the drill must be capable of drilling at an angle so as to intercept the target at as high an angle as practicable. Drill holes that intercept a target at less than a 20 degree angle may either glance off the target or deflect into the target resulting in an exaggerated thickness.

For coring, truck-mounted drills are commonly used (Bailey and Childers,1977d). However, in areas of difficult access, skid-mounted drills are either moved in by winch and cable, or flown by helicopter. Helicopter-supported drill programs are generally expensive and occur in areas where future development costs may be expected to be high. Most core rigs now are equipped for wire line methods. A *messenger* is lowered by cable through the drill string to the inner core barrel at the bottom of the hole. Once a latch is completed, the cable retrieves the inner barrel with its core

Fig. 4. Typical drilling rig for uranium exploration.

to the surface. This method saves *tripping out* or removing all of the drill string to recover the lowermost core barrel.

Percussion drilling is also generally conducted with a truck-mounted drill, although portable units are available if required for rough terrain (Peters, 1978). A percussion or *hammer-tool* rig drills by compressed air fed to a pneumatic drill tool and bit at the bottom of the drill string. Hole diameter is commonly 15.2 cm (6 in.). Cuttings are carried to the surface and collected in a cyclone device that prevents escape of the fines. Samples are taken on a 1.5 m (5 ft) interval, split by a Jones splitter, and a manageable portion bagged and labeled for assay purposes. Lithology of the samples is logged in the field by a geologist. With a good driller, samples are generally representative to a depth of approximately 90 m (300 ft) or to the water table.

Percussion drilling is more rapid and is one-half to one-third the cost of core drilling. Percussion samples, however, are not as reliable as core. As a percussion hole reaches greater depth, there is usually slough and dilution by material above the zone being drilled.

Reverse circulation drilling is commonly used in exploration for other metals but less commonly for uranium exploration (Peters, 1978). The circulating fluid may be either air or water and mud. Drill pipe is double-walled and the fluid is pumped down the space between walls to the bit at the bottom of the hole. Cuttings are lifted up the center pipe to the surface. This method significantly reduces contamination in comparison to percussion or rotary drilling methods.

Drilling equipment with operators may be contracted from numerous drilling companies on a fixed price-per-foot basis or an hourly basis plus materials used. A less common practice is for an exploration company to own and operate its own drills. This practice nearly always requires that geologists spend a great deal of time directly supervising a drilling operation and obtaining parts and materials for the rig rather than devoting time to finding ore.

Drilling companies often have regional field offices in cities near exploration and mining activities. A listing of companies available in the area may be compiled from advertisements in the trade journals, listings in the appropriate telephone books, and phone calls to consulting geologic firms and geologists with other mining companies for recommendations. When the list is narrowed, it is advisable to call references for a recommendation of the drillers' abilities. Prices are often determined by inviting competitive bids from three or more contractors. Once a driller is selected, a drilling agreement spelling out prices and terms is prepared and signed by the driller and the mining company.

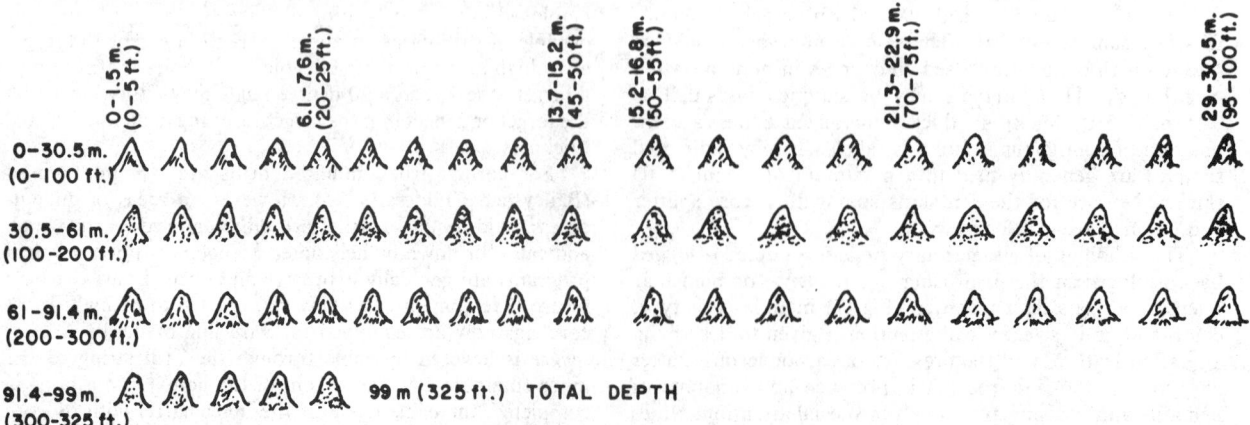

Fig. 5. Drill sample layout.

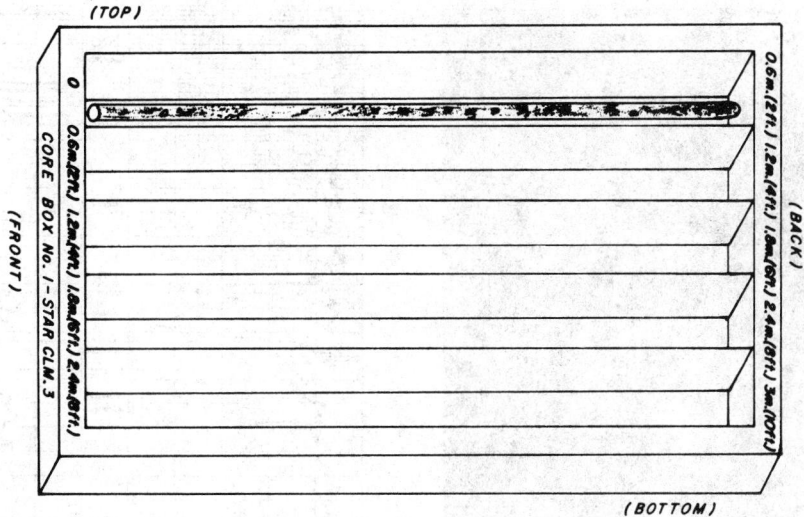

Fig. 6. Drill core box.

Drilling prices may vary widely depending on the area and rock types. Typical prices in 1983 in the western US are as follows:

Hole Depth	Type Host	Price per Ft (0.3 m)		
		Rotary	Percussion	Diamond Core
0-152 m	Sandstone uranium	$0.80-2	$3-6	$20-30
(0-500 ft)	Carbonate host	$1-4	$4-6	$20-35
	Igneous/metamorphic		$6-12	$15-32

Lost circulation	$50-$150 per hr plus material
Standby rates	$50-$80 per hr

SITE PREPARATION

Drill sites and access roads must be constructed where needed prior to drilling. Generally a bulldozer and operator are hired from a contractor on an hourly rate to prepare drill sites. Upon completion and abandonment of the drill hole, drill sites are reclaimed and seeded. Cost of drill sites varies greatly depending on terrain. Some accessible areas require a minimal expense of less than $30 per drill hole. In areas of difficult terrain, the cost of drill sites may range from $200 to $800 per hole, and more if extensive road building is required.

GEOPHYSICAL LOGGING

After a drill hole is completed, it is logged by a company-owned or logging service probe truck (Fig. 8) (Bailey and Childers, 1977e). The geophysical log generally in use was originally pioneered for the oil exploration industry. It consists of gamma-ray, resistivity, and self-potential curves plotted by depth (Fig. 9). The resistivity and self-potential (SP) curves provide bed boundaries and are mainly used for correlation. The resistivity curve, calibrated in ohms, is largely a measure of the formation water resistivity. Generally sandstones show a deflection to the right, or a greater resistivity

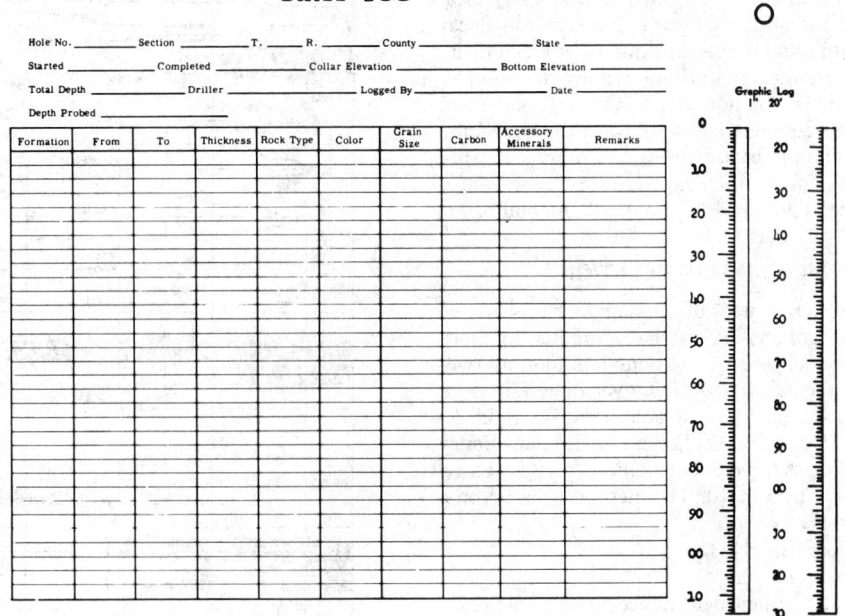

Fig. 7. Typical lithologic drill log form.

Fig. 8. Geophysical probe truck.

than claystones and shales. The spontaneous, or self-potential curve, indicates the natural potential differences in millivolts, between an electrode at the surface and an electrode in the probe in the drill hole fluid that is pulled up past different beds. This potential depends on a number of factors, but generally indicates the permeable zones, or sandstones, as a deflection left from the shale base line. Resistivity and SP logs are not as easy to interpret in igneous and metamorphic rocks. A summary of geophysical logging theory and practice is given in manuals prepared by Schlumberger (Anon., 1958) and Haun and LeRoy (1958). The gamma-ray log is used to interpret the amount of equivalent U_3O_8 in a zone by measuring the gamma radiation of radioactive uranium decay products. The scintillation probes in common use can delineate anomolous uranium mineralization down to approximately 7 or 15 parts per million eU_3O_8. Ore values derived by gamma-log interpretation are relatively accurate in a number of known uranium districts. However, in most programs, a decision is made before drilling expenditures become large to determine if disequilibrium is a problem. Disequilibrium problems are discussed later.

GAMMA-LOG INTERPRETATION

The equivalent U_3O_8 content of a zone may be calculated manually from the gamma-log, or by computer methods which record the counts per second at one-half foot intervals and then calculate the equivalent U_3O_8 over mineralized intervals. The manual method most commonly used is described in detail by Dodd and Eschliman (1972), and Bailey and Childers (1977e). A typical calculation form based on this method is shown by Fig. 10. The method is based on a relationship GT = KA where:

G = grade in percent eU_3O_8

T = ore thickness in feet

K = instrument constant determined in test holes

A = area of curve based on counts per second by 0.15 meter (0.5 ft) increments.

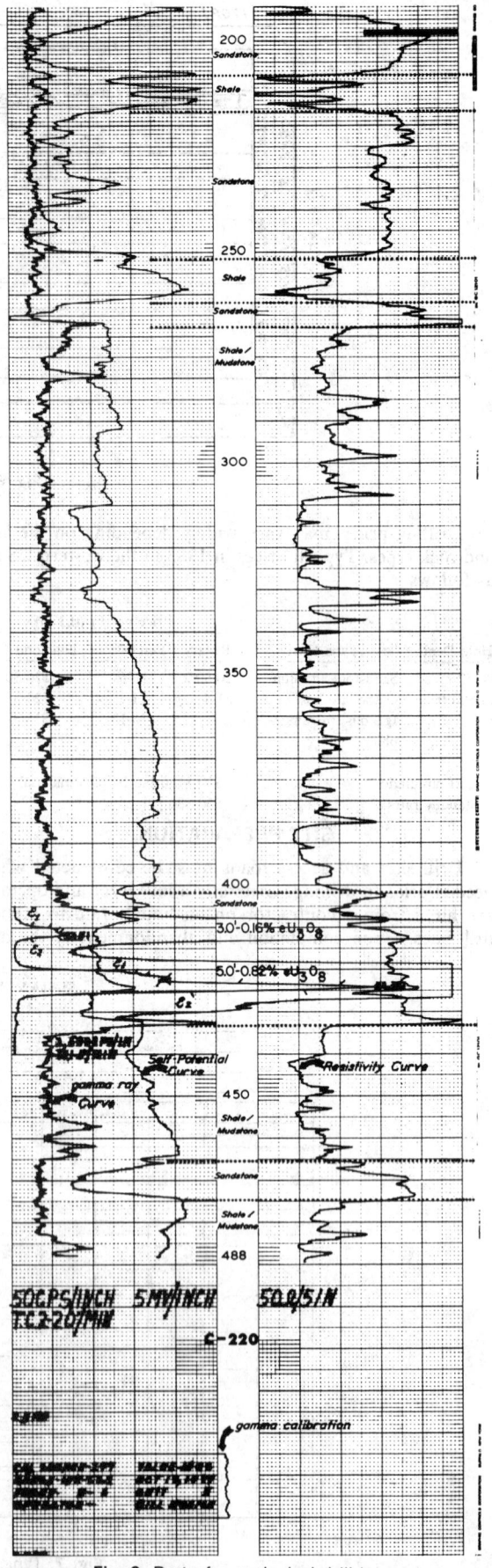

Fig. 9. Part of a geological drill log.

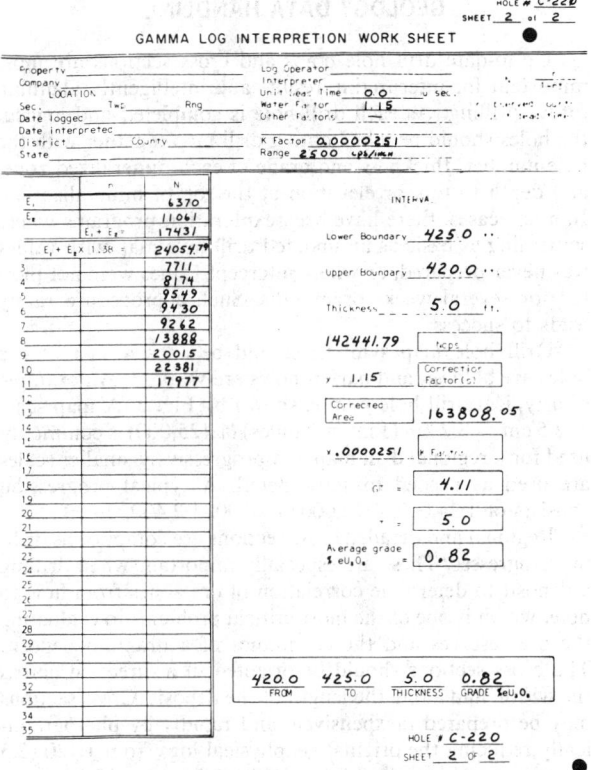

Fig. 10. Gamma log interpretation sheet.

First, an upper and a lower bed boundary are determined to the nearest 0.3 m (1 ft), at a point about halfway between background and the peak of the gamma-ray curve (Fig. 9). A reading of the counts per second (CPS) is made at each of these boundaries and termed E_1 and E_2. These are added and the total multiplied by 1.38 to provide for *tail effect* (shown by Fig. 10). Consecutive readings in CPS are then recorded for each 0.15 (0.5 ft) interval from the upper to the lower boundary. Each of these readings as well as the one for $E_1 + E_2 \times 1.38$ is corrected for instrument dead time by the formula $N = \dfrac{n}{1 - nt}$.

N = corrected counts per second
n = observed counts
t = unit dead time in seconds.

Most logging units now being used make this correction electronically and effectively have a zero dead time. The counts per second are totalled and multiplied by a water factor, if any. The result is termed the *corrected area* [which is CPS × 0.15 m (0.5 ft) increments]. The corrected area is then multiplied by the instrument K-factor. This result is divided by thickness to determine percent eU_3O_8. A printout of the CPS values by 0.15 m (0.5 ft) intervals is also provided by most logging units (Fig. 11). In addition, the digital data may be obtained on computer tape and the tape processed by the logging company or a company-owned computer to provide thickness and grade of ore intercepts at various cutoff grades.

It is important that the probe be calibrated with a known source before logging each hole. This is usually printed at the bottom of the geophysical log (Fig. 9). When ore zones are being encountered, it is also important that the logging unit periodically probe a standard test hole with known thickness and grade in percent U_3O_8. In the western US, several holes termed *test pits* are maintained by the Dept. of Energy for such purposes and to determine the instrument K-factor. It is good practice to request copies of test logs on a weekly or biweekly basis and file them within the sequence of ore holes drilled.

DISEQUILIBRIUM

Disequilibrium is a term for the disparity between uranium and its naturally occurring radioactive daughter products which are measured by the gamma log (Nininger, 1954a). Generally, checks are made for disequilibrium when drill-indicated reserves reach approximately 45 360 to 226 800 kg (100,000 to 500,000 lb) of contained U_3O_8. In new areas, disequilibrium is checked after the first few ore holes. The procedure consists of coring a mineralized intercept and also probing the hole upon completion. The core is split along intervals bounded by changes in mineralization, based on scanning the core with a geiger counter. As a practice, sample intervals are held to less than 1.5 m (5 ft). Contiguous mineralized samples may later be calculated to a composite by a weighted averaging method. Split core samples are then bagged, labeled, and forwarded to an assay laboratory. Each sample is crushed and a portion split for pulverizing to prepare a pulp which is then assayed. Two separate assays are requested: a scaler-radiometric or closed-can radiometric assay, and a wet chemical assay (Grimald, 1956). The two assay methods are performed on exactly the same pulp. The ratio between the two assays is the disequilibrium factor, which is generally reported in the form chemical:radiometric, and should be so stated when discussing disequilibrium. This factor, which is of obvious economic importance, may be in favor of either the chemical or the radiometric assay. The chemical assays are then compared

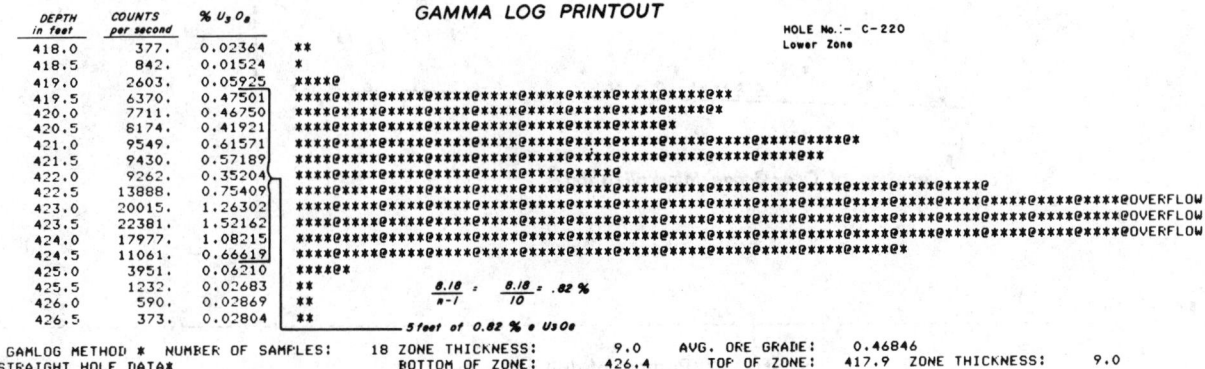

Fig. 11. Gamma log printout of mineralized zone.

to the gamma log results for another check. However, in no case should this comparison be used for determining disequilibrium. The gamma-probe detects radioactivity through a 0.76 m (2.5 ft) diameter cylindrical area centered about the drill hole (Dodd and Eschliman, 1972). The core samples a cylindrical area of 4.45 cm (1.75 in.) diameter.

Choice of an assay laboratory is very important because there is a wide variation in quality. It is a good practice to seek recommendations from several of the larger and well-known consulting engineering and geology firms with a highly regarded reputation for experience in evaluating uranium deposits.

After an ore-bearing drill hole is logged, it is a good practice to drift survey the hole to provide an accurate location of ore zone intercepts (Linton, 1963). Although it is not common, a drill hole 122 m (400 ft) deep may deviate from collar to bottom as much as 9 or 15 m (30 or 50 ft). Hole deviation is principally caused by variation in rock type and driller ability.

The gamma-ray resistivity and SP probe is removed from the cable and a drift tool attached. The drift tool takes consecutive readings of azimuth and inclination at intervals of 15 or 30 m (50 or 100 ft). The results are tabulated in X,Y,Z coordinates at recorded depths and most logging companies provide a computer-generated map which plots the drift (Fig. 12).

Geophysical logging units with operator may be contracted from a number of available logging companies. The units are commonly truck-mounted, but portable units are available for helicopter or backpack transport to remote areas. Typical costs for a logging truck on a monthly basis are in the range of $6,000 per month plus expenses and $.10 per ft (0.3 m) for a gamma-ray, resistivity, and SP log. Neutron density, caliper, dry-wall resistivity, and temperature logs typically cost an additional $.15 to $.20 per ft (0.3 m) each. In areas near their district offices, logging trucks may be obtained on a call-out basis. Costs are charged for mileage, setup, and the footage logged.

GEOLOGY DATA HANDLING

Up-to-date drill hole maps and cross sections are most important for interpreting results and intelligently planning future drilling. As each drill hole is completed and logged, the holes should be inked onto a drill map together with the hole number, thickness and grade of each mineralized zone, and depth to top, or elevation at the top of mineralization. In a few cases, there have been exploration programs where something as basic as an updated drill hole map with values was never prepared, or where intercept values were not plotted for several weeks or months. Such a procedure rarely leads to success.

Drill hole map symbols should be chosen so that ore holes are boldest, and barren holes are weakest. An example of a typical drill hole map is shown by Fig. 3. A map scale of 2.5 cm = 3.2 km (1 in. = 2 miles) (1:125,000) is commonly used for a regional drill map and progressively smaller scales are used as needed to show detail. A typical progression consists of 1:24,000, 1:12,000, 1:4,800, 1:2,400, 1:600.

Regional and detailed cross sections are compiled as drilling progresses. These are especially important when drilling a deposit to determine correlation of ore zones from hole to hole, which is one of the most critical problems in evaluating the ore reserves and the economics of a uranium deposit. The cross sections should be oriented in a direction across the width, and along the length of the deposit. Cross sections may be prepared inexpensively and rapidly by photographically reducing the original geophysical logs from 1:120 (2.5 cm = 3 m) (1 in. = 10 ft) to 1:600 (2.5 cm = 15 m) (1 in. = 50 ft). Prints of the *mini-logs* may be taped to cross-section paper and the correlation made on the resulting print. As an alternative, the gamma ray and resistivity curves may be traced from the mini-logs onto mylar for a cross section. In addition to mineralization, the lithology, alteration, and structure are plotted on the cross sections. Detailed lithologic data is especially important for igneous and metamorphic hosts to determine if ore values are related to a specific rock

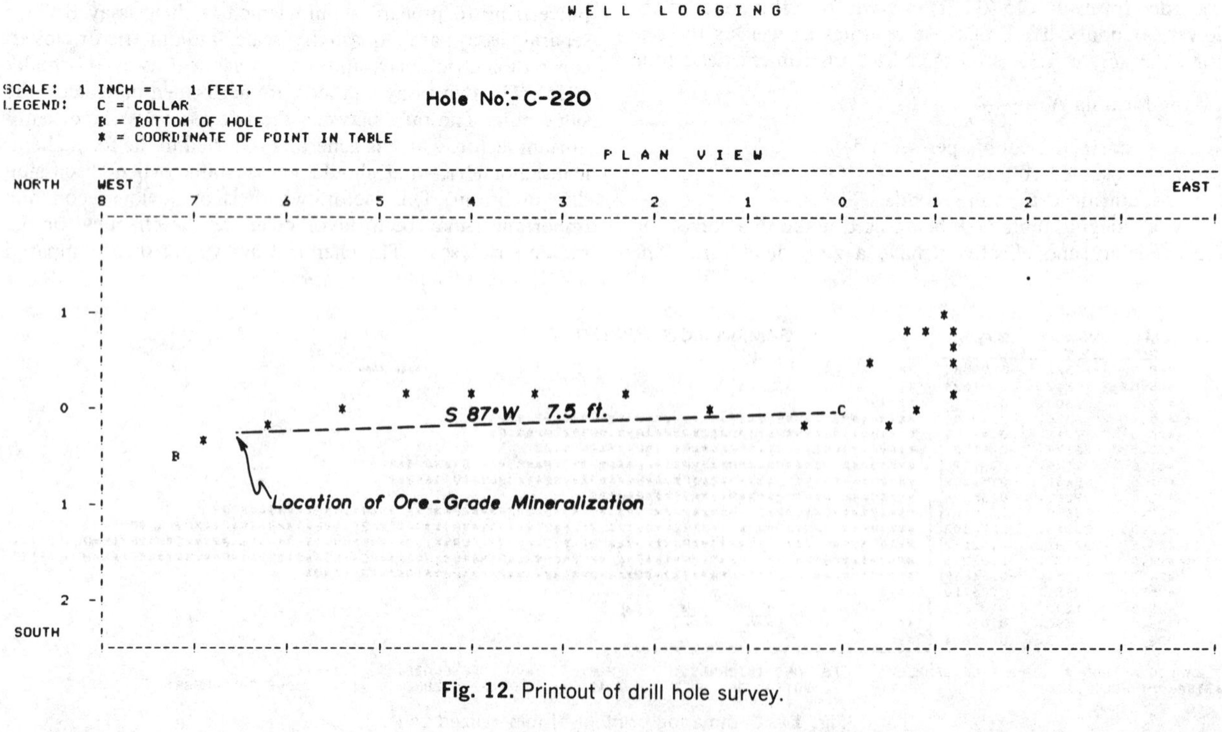

Fig. 12. Printout of drill hole survey.

type. Correlation of ore zones from hole to hole in uranium deposits often requires a considerable amount of judgment and experience. Multiple horizons, for example, are typical of sandstone uranium deposits (Fig. 13). As problems of correlation become evident, additional drill holes may be required to obtain the information needed to solve the problems.

Bench, or level, maps are plan maps prepared from the drill hole data to show the distribution of mineralization at various grade intervals on each consecutive bench or level. A typical bench map represents a 3 to 6 m (10 or 20 ft) vertical interval and is labeled by the basal elevation of that bench. Only the drill hole data within that 3 to 6 m (10 or 20 ft) interval is plotted on each drill hole. Ore reserves for each bench may then be determined by selecting a cutoff grade and using the general outline method or polygonal method as later described. Consecutive bench maps may then be superimposed, one above another, to determine continuity of ore from level to level.

Contour maps showing the product of grade in percent eU_3O_8 multiplied by thickness (GT or grade-thickness maps) are compiled for the total ore-bearing unit as well as individual subunits. Isopach maps of the host unit are also prepared. Favorability maps are compiled early in a drilling program and continually updated through the stage of drilling deposits. They are based mostly upon mineralization and alteration intensity together with additional, less certain ore guides that have been developed. It is worth the considerable amount of effort and imagination required to determine mappable geologic features related to ore.

ORE RESERVE ESTIMATION

After a number of contiguous ore holes have been completed and the correlation of ore zones from hole to hole has been determined for a deposit, ore reserves are generally calculated. The subject of uranium ore reserve estimation is covered in detail in Chapter 3, Section 5.3. Several other papers which describe ore reserve estimation are Bailey and Childers, 1979; Blanchfield, 1980; Grundy and Meehan, 1963; and Harris, 1977.

A brief description of the methods follows. Typically, ore reserves are determined by either the general outline or the polygonal method. The first step is to assign a cutoff value based on thickness and grade of mineralization. The cutoff is determined by the economic factors of market price less the anticipated mining, milling, and capital costs. As one example, the average grade of uranium ore mined in Wyoming (almost entirely open pit) in 1982 was 0.067% U_3O_8 (Anon., 1983) and the corresponding cutoff grade would be less than the average.

For the general outline method, a map boundary is drawn around contiguous ore holes within a single ore horizon, to

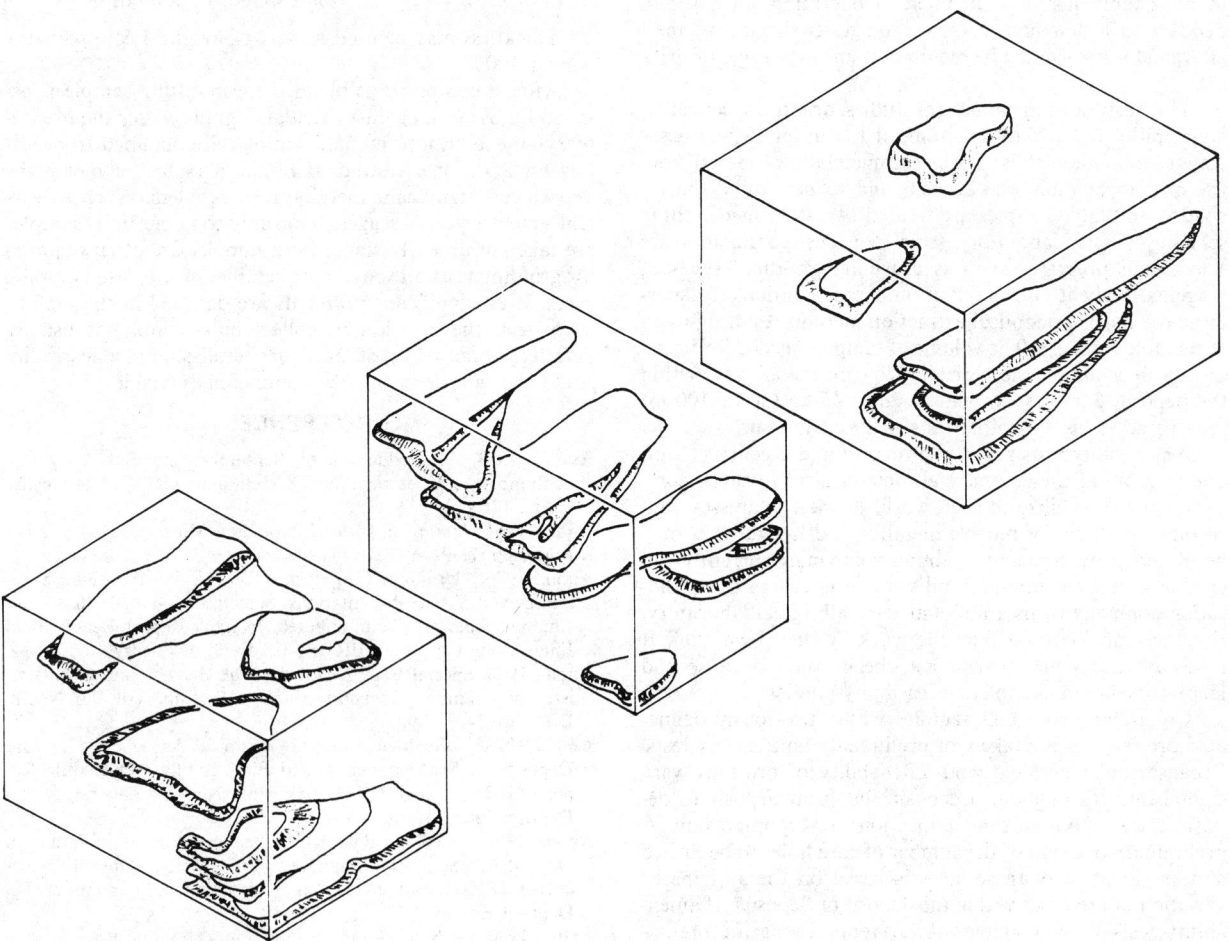

Fig. 13. Idealized exploded block diagram showing typical relationship of prefault ore bodies in the Ambrosia Lake district, New Mexico, Granger, et al., (1961).

provide a boundary to the deposit, assuming that drill holes are spaced 15 to 46 m (50 to 150 ft) apart. The outside boundary is positioned one-half the distance between an ore and non-ore hole. The area within the boundary is determined by a planimeter, and the average thickness of ore is determined by an arithmetic average of ore thickness in all of the drill holes within the boundary. The average grade is calculated by a weighted average, based on the sum of grade times thickness for the zones in each of the drill holes, divided by the sum of ore thickness of all of the zones. The volume of ore (area $\times$ average thickness) is converted to tons based on specific gravity of the ore.

The polygonal method yields similar results. First, polygons are drawn around each ore hole meeting cutoff. The polygons are constructed with boundaries that are typically one-half the distance from ore to non-ore holes. A tonnage and grade is calculated for each polygon. The tonnage is summed. An average grade is determined by a weighted average of tons times grade.

As mentioned previously, ore reserves may be calculated on each bench interval based on either the general outline or the polygonal method. A stripping ratio may then be determined based on tons of waste vs tons of ore for each level and for the total.

METALLURGICAL STUDIES

Samples for metallurgical studies are generally obtained in a deposit or several deposits when reserves appear sufficient to reasonably justify a commercial operation. In cases of deposits in a new area or new geologic environment, metallurgical samples may be required at an early stage of drilling.

The purpose of metallurgical studies, or more specifically, amenability tests, is to determine if the uranium and associated economic metals may be commercially extracted from the ores throughout the deposit, and to determine the recovery. Initially, samples are treated by established milling schemes at the laboratory scale. Various parameters are selected in processes such as crushing, grinding, types of reagents, reagent consumption, and precipitation, to determine the most economical extraction method resulting in an acceptable recovery. The volume of samples needed for initial testing depends on the variability of ore metallurgy within the deposit. Samples as small as 23 to 45 kg (50 or 100 lb) may be adequate for initial tests in some deposits.

Amenability tests may be performed under contract with one or more of the commercial metallurgical research companies that specialize in such work. Names, addresses, and recommendations of reliable metallurgical laboratories may be obtained by contacting mining companies currently operating successful uranium mills. In cases where the exploration company owns a mill and a metallurgical laboratory, they may prefer to conduct the work by their own staff. It is always advisable to obtain a check study by a second laboratory before completing feasibility studies.

Commonly, core hole samples within previously delineated ore reserves are taken for preliminary amenability tests. Because the mineralogy and amenability of ore may vary significantly throughout a deposit and from deposit to deposit, the selection of sample locations is very important. A preliminary decision of the number of core holes to be drilled and the location of these holes is based on the anticipated variation of ore types within the deposit or deposits. If amenability tests show a significant variation among the preliminary samples, additional coring may be required. In a case of a new deposit within a previously producing district with

a known history of commercial milling, the core holes may be as few as four or five per deposit. In cases of new deposits in unproven areas, numerous core holes may be required on a spacing as close as 61 or 152 m (200 to 500 ft). The number of holes and size of core depends also on the volume of samples required. Typically, NX sized core is chosen as a compromise between volume of sample needed and cost. Oversized cores ranging up to 15 or 20 cm (6 or 8 in.) in diameter are less commonly used because of excessive costs.

After the core is obtained, sample intervals are chosen based on changes in mineralization and lithology. The core is split or sawed in half and samples bagged for chemical assay. In cases where the entire core must be consumed for testing, the other half is bagged for assay along the same sample interval. After assay, the rejects and pulps are returned for preparing composite samples.

A decision of which samples to blend for a composite samples requires considerable attention to detail. The resulting composite should be representative and have a grade close to the average grade of the deposit. If there are two or more types of ore, samples from each type should be tested separately, followed by tests of the total composite of all types. Determination of anticipated grade of a composite sample may be calculated as a weighted average from the original assay data in the following manner:

Average grade

$$= \frac{\text{sum of (weight} \times \text{grade) for each sample}}{\text{total weight of composite}}$$

Thickness may be used instead of weight if core recovery is near 100%.

After a composite is blended the resulting sample is assayed for a check against calculated grade. After the preliminary amenability tests, bulk samples are obtained from test pits for additional testing. If preliminary tests indicate the deposit has significant metallurgical problems, such as variable recovery and reagent consumption, the bulk samples are taken at an early stage. Bulk samples are often required for grinding tests although core samples may be used in some cases. If no significant problems are detected in the preliminary test, the decision to collect bulk samples is usually deferred, because of cost, until an overall economic feasibility study shows a deposit to be commercially viable.

REFERENCES

Anon., 1958, "Introduction to Schlumberger Well Logging," Schlumberger Document No. 8, Schlumberger Well Surveying Corp., 176 pp.

Anon., 1968, *Mining Explained,* Northern Miner Press, Ltd., Toronto, pp. 73-92.

Anon., 1976, "Uranium Geophysical Technical Symposium: Summaries and Visual Presentations, September 14-16, 1976, Grand Junction, Colorado," unnumbered Open File Report, Bendix Field Engineering Corp. and ERDA, US Dept. of Energy.

Anon., 1978, Special Issue Devoted to the Geology and Geochemistry of Uranium, *Economic Geology Bulletin,* Vol. 73, No. 8, Dec., pp. 1401-1792.

Anon., 1978a, "Geological and Geochemical Aspects of Uranium Deposits: A Selected, Annotated Bibliography," Open File Reports GJBX-15, Vol. 1, 121, Oak Ridge National Laboratory, US Dept. of Energy.

Anon., 1979, "Geological and Geochemical Aspects of Uranium Deposits: A Selected, Annotated Bibliography," Open File Reports GJBX-15, Vol. 2, 132, Oak Ridge National Laboratory, US Dept. of Energy.

Anon., 1980, "A Symposium on the Grants Uranium Region, May 13-16, 1979, Alburquerque, New Mexico," *Memoir 38,* New Mexico Bureau of Mines and Mineral Resources, 401 pp.

Anon., 1980a, "Geological and Geochemical Aspects of Uranium Deposits: A Selected, Annotated Bibliography," Open File Reports GJBX-15, Vol. 3, 230, Oak Ridge National Laboratory, US Dept. of Energy.

Anon., 1980b, "Geological and Geochemical Aspects of Uranium Deposits: A Selected, Annotated Bibliography," Open File Reports GJBX-15, Vol. 4, Oak Ridge National Laboratory, US Dept. of Energy.

Anon., 1982, *Proceedings of the Symposium on Uranium Exploration Methods, June 1-4, 1982, Paris,* OECD Nuclear Energy Agency, 979 pp.

Anon., 1983, News Release, PR No. 83-22, April 1, US Dept. of Energy, Grand Junction, CO.

Bailey, R.V., and Childers, M.O., 1977, *Applied Mineral Exploration with Special Reference to Uranium,* Westview Press, Boulder, CO, 542 pp.

Bailey, R.V., and Childers, M.O., 1977a, *Applied Mineral Exploration with Special Reference to Uranium,* Westview Press, Boulder, CO, pp. 231-240.

Bailey, R.V., and Childers, M.O., 1977b, *Applied Mineral Exploration with Special Reference to Uranium,* Westview Press, Boulder, CO, pp. 345-430.

Bailey, R.V., and Childers, M.O., 1977c, *Applied Mineral Exploration with Special Reference to Uranium,* Westview Press, Boulder, CO, pp. 386-394.

Bailey, R.V., and Childers, M.O., 1977d, *Applied Mineral Exploration with Special Reference to Uranium,* Westview Press, Boulder, CO, p. 386.

Bailey, R.V., and Childers, M.O., 1977e, *Applied Mineral Exploration with Special Reference to Uranium,* Westview Press, Boulder, CO, pp. 394-415.

Blanchfield, D.M., 1980. "Methodology for Uranium Resource Estimates and Reliability," *Proceedings Uranium Industry Seminar,* October 22-23, 1980, Grand Junction, Colorado, ERDA GJO-108 (80), pp. 59-66.

Bowie, S.H.U., et al., eds., 1972, *Uranium Prospecting Handbook,* The Institution of Mining and Metallurgy, London, 346 pp.

Carlisle, D., et al., 1978, "The Distribution of Calcretes and Gypcretes in Southwestern United States and Their Uranium Favorability Based on a Study of Deposits in Western Australia and Southwest Africa (Namibia), "ERDA Open File Report GJBX-29 (78), 274 pp.

Dall'Aglio, M., 1973, "Geochemical Exploration for Uranium," *Uranium Exploration Methods, Proceedings of a Panel, Vienna, 10-14 April, 1972,* International Atomic Energy Agency, Vienna, pp. 189-208.

De Voto, R.H., 1978, *Uranium Geology and Exploration:* Colorado School of Mines, Golden, pp. 265-269.

Dodd, P.H. 1980, "Developments in Exploration Techniques," *Uranium Industry Seminar Proceedings,* October 22-23, 1980, Grand Junction, Colorado, U.S. Dept. of Energy, GJO-108 (80), pp. 105-121.

Dodd, P.H. and Eschliman, D.H., 1972, "Borehole Logging Techniques for Uranium Exploration and Evaluation," *Uranium Prospecting Handbook,* The Institution of Mining and Metallurgy, London, pp. 250-251, 244-276.

Emilia, D.A., and Kosanke, K.L., 1981, "Current Status of NURE Geophysical Research Projects," *Uranium Industry Seminar Proceedings,* October 21-22, 1981, Grand Junction, Colorado, U.S. Dept. of Energy, GJO-108 (81), p. 135-147.

Finch, W.I., 1967, in Geology of Epigenetic Uranium Deposits in Sandstone in the United States," Professional Paper 538, U.S. Geological Survey, pp. 86-90.

Fleischer, R.L., et al., 1980, "Radon Emanation Over an Orebody — Search for Long-Distance Transport of Radon," *Geology and Mineral Technology of the Grants Uranium Region, 1979, Memoir 38,* New Mexico Bureau of Mines and Mineral Resources, pp. 380-389.

Granger, H.C., et al., 1961, "Sandstone-type Uranium Deposits at Ambrosia Lake, New Mexico—An Interim Report," *Economic Geology,* Vol. 56, No. 7, pp. 1179-1210.

Grimald, F.S., 1956, "The Analytical Chemistry of Uranium and Thorium," *Contributions to the Geology of Uranium and Thorium by the United States Geological Survey and Atomic Energy Commission for the United Nations International Conference on Peaceful Uses of Atomic Energy, Geneva, Switzerland, 1955,* Professional Paper 300, US Geological Survey pp. 605-617.

Gimbert, A., 1972, "Uses of Geochemical Techniques in Uranium Prospecting," *Uranium Prospecting Handbook,* The Institution of Mining and Metallurgy, London, pp. 110-120.

Grundy, W.D., and Meeham, R.J., 1963," Estimation of Uranium Ore Reserves by Statistical Methods and a Digital Computer," *Geology and Technology of the Grants Uranium Region, Memoir 15,* New Mexico Bureau of Mines and Mineral Resources, pp. 234-243.

Harris, D.P., 1977, "Quantitative Methods for the Appraisal of Mineral Resources," ERDA Open File Report GJBX-14 (77).

Haun, J.D., and LeRoy, L.W., 1958, *Subsurface Geology in Petroleum Exploration,* Colorado School of Mines, Golden, CO. pp. 265-427.

Kelley, V.C., 1963, "Geology and Technology of the Grants Uranium Region," *Memoir 15,* New Mexico Bureau of Mines and Mineral Resources, 277 p.

Key, B.N., et al., 1982, "The Differential Face Scanner and Its Application for Grade Assessment," *Proceedings of the Symposium on Uranium Exploration Methods, Paris, 1st-4th June, 1982,* OEDC Nuclear Energy Agency, Paris, France, p. 623-638.

Linton, W.A., 1963, "Uranium Logging Techniques," *Geology and Technology of the Grants Uranium Region, Memoir 15,* New Mexico Bureau of Mines and Mineral Resources, pp. 222-233; 230-232.

Miller, J.M., and Ostle, D., 1973, "Radon Measurement in Uranium Prospecting," *Uranium Exploration Methods, Proceedings of a Panel, Vienna, 10-14 April, 1972,* International Atomic Energy Agency, Vienna, pp. 237-247.

Moench, R.H., and Schlee, J.S., 1967, "Geology and Uranium Deposits of the Laguna District, New Mexico, Professional Paper 519, US Geological Survey, 117 pp.

Nash, J.T., et al., 1981, "Geology and Concepts of Genesis of Important Types of Uranium Deposits," *Economic Geology Seventy-Fifth Anniversary Volume 1905-1980,* pp. 63-116.

Nininger, R.D., 1954, *Minerals for Atomic Energy,* D. Van Nostrand, New York, pp. 183-202.

Nininger, R.D., 1954a, *Minerals for Atomic Energy,* D. Van Nostrand, New York, pp. 172-176.

Nishimori, R.K., et al., 1977, "Uranium Deposits in Granitic Rocks," ERDA Open File Report GJBX-13 (77), 308 pp.

Peters, W.C., 1978, *Exploration Mining and Geology,* John Wiley, New York, pp. 438-440.

Pretorius, D.A., 1981, "Gold and Uranium in Quartz-Pebble Conglomerates," *Economic Geology Seventy-Fifth Anniversary Volume 1905-1980,* pp. 117-138.

Rawson, R.R., 1980, "Uranium in Todilto Limestone (Jurassic) of New Mexico—Example of a Sabkha-like Deposit," *Geology and Mineral Technology of the Grants Uranium Region 1979, Memoir 38,* New Mexico Bureau of Mines and Mineral Resources, pp. 304-312.

Schilling, F.A., Jr., 1975, "Annotated Bibliography of Grants Uranium Region, New Mexico 1950-1972," Bulletin 105, New Mexico Bureau of Mines and Mineral Resources, 69 pp.

Swanson, V.E., 1961, "Geology and Geochemistry of Uranium in Marine Black Shales, A Review," Professional Paper 356-C, U.S. Geological Survey, 112 pp.

Vine, J.D., 1962, "Geology of Uranium in Coaly, Carbonaceous Rocks," Professional Paper 356-D, US Geological Survey, 170 pp.

2.5 Iron Ore Exploration and Geology

RALPH W. MARSDEN

IRON ORES: TYPES AND CLASSIFICATION

An iron ore deposit is a mineral body of sufficient size, iron content, and chemical composition with physical and economic characteristics that will allow it to be a source of iron either immediately or potentially. Economic viability is essential. No definite limits can be set on the size, grade, or mineral composition as there is a considerable permissable range in the physical and chemical parameters for a commercially minable iron ore deposit. Geographic location, other competitive iron ore sources, the size and location of the market, minability, grade and/or concentratability, are critical in defining the economic viability of a particular iron-bearing deposit.

Iron forms an estimated 5% of the earth's crust; it is a common constituent in hundreds of minerals and rocks, and in small amounts has an almost universal distribution. Iron in ore commonly occurs as an oxide in the minerals hematite (Fe_2O_3), magnetite (Fe_3O_4), or goethite ($Fe_2O_3 \cdot H_2O$), and less often as maghemite (Fe_2O_3) and lepidocrocite ($Fe_2O_3 \cdot H_2O$). Ilmenite ($FeTiO_3$) may occur in some ores up to the allowable limits for titanium in an iron ore. When of sufficient size and grade, as in the Michipicoten district, Ontario, Canada, siderite ($FeCO_3$) may be mined as iron ore. Iron sulfides such as pyrite (FeS_2), marcasite (FeS_2), or pyrrhotite ($Fe_{1-x}S$) are sources of small amounts of byproduct iron ore after removal of base metals and sulfur, or they may be the primary source of iron oxide gossans in the outcrop area of sulfide deposits. Iron silicate minerals commonly contain in excess of 25% silica so are not generally considered iron ore minerals, although they may be constituent minerals in some iron ores such as in magnetite taconites and in the Minette ores of France. Iron-bearing silicate minerals are frequently the primary source of iron present in secondary oxides produced by weathering of iron-rich rocks that form residual type iron ores.

Iron ores occur in a variety of geological environments in deposits with widely differing shapes, sizes, origins, and with substantially different ore characteristics. The diverse nature of the geology of iron ores must be recognized in exploration, evaluation, and in the investigative methods used in their study so the best possible discovery and developmental system is followed. There is, however, a general similarity of geology, problems, and occurrence of iron ores within each ore type that allows a reasonable basis for exploration, evaluation, and development planning.

The classification of iron ores shown in Table 1, which is based on a combination of geological and commercial factors, separates the iron deposits into groups having similar relationships and characteristics. This grouping places emphasis on geological and process factors. A general description of each ore class follows that summarizes characteristics important in the exploration, development, production, and evaluation of these ores.

Sedimentary and Metasedimentary Deposits

Precambrian Cherty Iron-Formation Deposits: The largest and most widespread iron ore deposits in the world are associated with Precambrian cherty iron-formations. These formations include a family of layered silica-rich and iron-

Table 1. Classification of Iron Ore Deposits

I Sedimentary and Metasedimentary Deposits
  A) Precambrian cherty iron-formation ores
    1. Magnetite taconite ore
    2. Jaspilite ore
    3. Itabirite ore
    4. Siderite ore
  B) Ironstone ores
  C) Clastic iron ores
II Magmatic Deposits
  A) Kiruna type ores
  B) Titaniferous magnetite deposits
III Contact Metasomatic Ores
IV Massive Ore in Cherty Iron-formation
V Residual Deposits
  A) Lake Superior type ores
  B) OXIBIF ores
  C) Canga and River Terrace ores
  D) Brown ores
  E) Lateritic ores

rich sedimentary and metasedimentary rocks predominantly composed of chert or fine- to medium- to coarse-grained quartz and iron minerals as oxides, carbonates, or silicates. The origin and occurrence of cherty iron-formations is discussed in an extensive literature. Commercial deposits of world significance occur in iron-formations as direct shipping, merchantable ores, concentrating grade ores, and as concentrating grade iron-formation ores. These sedimentary or metasedimentary iron-bearing rocks have a common occurrence in Precambrian terranes in rocks of Archean and Proterozoic age older than 1.8 b.y., with some jaspilite type deposits found in rocks with ages as young as Late Precambrian/Cambrian. In Archean rocks, the cherty iron-formations are associated with volcanic, volcanogenic, and sedimentary belts as deposits from a few meters long and centimeters thick to bodies 100 m (328 ft) or more thick and several kilometers long. There is a distinct tendency for Archean deposits to be lenticular. In contrast, the iron-formations in the Proterozoic often have an extensive occurrence within rocks deposited in sedimentary basins as stratigraphic formations from 50 m (164 ft) to over 700 m (2297 ft) in thickness and several hundreds of kilometers long. Proterozoic iron formations are commonly associated with clastic, or chemical sedimentary, or metasedimentary rocks and may or may not have volcanic rocks associated with them.

Cherty iron-formations are composed dominantly of iron minerals and silica that together comprise over 90% of the rock. These rocks are banded and layered with alternating iron-rich and silica-rich layers, ranging from 1 mm (0.04 in.) to 10 cm (4 in.) or more in thickness. The iron content ranges from about 15 to 50%. The commonly accepted definition of cherty iron-formation requires a minimum 15% iron content. The original iron and silica-rich sediment was deposited in a variety of sedimentary environments ranging from very quiet, undisturbed conditions to areas with considerable wave and current action. Deposition of most iron-formations appears to have been in relatively shallow water. The Dales

Gorge member of the Brockman formation in Western Australia has a very fine, microlayering undisturbed over extensive areas indicating very quiet conditions of deposition. As a contrast, some stratigraphic zones in the Biwabik formation, Mesabi Range, Minnesota have a granular texture, are lenticular, and wavy bedded with pebbly to conglomeratic, fragmental, and cross-bedded layers. Locally iron-formations are intimately interstratified with volcanic rocks, or volcanogenic sediments, or fine to coarse clastic sediments and in some rather restricted, usually local areas, contain thin interbedded calcareous carbonate sediments. Knowledge of the potential variability of cherty iron-formation depositional environments is important in the exploration, development, production, and evaluation of these deposits.

All iron-formations have been changed from the original sediments by diagenesis and in most areas by metamorphism. There is a continuing debate on the nature of the primary minerals, the source of the iron and silica, and the conditions of deposition. The characteristics of the original sediment, its mineralogy, bedding features, and the degree of segregation of the iron and silica into laminae or layers appear to have had a considerable influence on the present mineralogy and the distribution of iron and silica.

The post-depositional history including both deformation and metamorphism introduces important variables that influence the potential economic value and commercial viability of cherty iron-formation deposits. The importance of structure is evident as the complexity of the rock structure may determine the tonnage available and minability of a particular deposit. A high-quality, concentratable iron formation 30 m (100 ft) in thickness with a relatively flat dip and exposed over a sizeable area may be economically valuable, whereas iron-formation of the same quality dipping steeply even with a strike length of several kilometers would have little if any value as an iron ore resource.

The metamorphic history of iron-formations also may be of major importance in determining the potential value of a cherty iron-formation as a source of iron ore. During metamorphism the mineralogy, texture, and grain size of the rock can be markedly changed so as to greatly enhance or decrease its possible commercial value. Commercial cherty iron-formation ores are composed of iron oxides with associated chert or quartz and iron minerals whereas many of the primary iron-formations as sediments may have been largely composed of iron carbonate or iron silicate minerals with a very fine grain size and an intimate mingling or interlayering of fine-grained cherty quartz. Through metamorphism, the primary carbonate and silicate may be changed to magnetite, quartz, and iron silicates and the grain size, including the resulting oxide minerals, increased to sizes that permit easy liberation of the iron minerals by grinding, thus changing a valueless rock into a potential ore. Iron formations are found in many metamorphic stages from the lower greenschist through the granulite facies with lithologies that range from carbonate-silicate-chert rock, to magnetite-rich cherty rock with small amounts of associated carbonates and iron silicate minerals, to largely magnetite-quartz rock, to magnetite-hematite quartz rock with small amounts of iron silicate, and to hematite-quartz rock. There is a wide range of metamorphic iron-formation types which are associated with a variety of country rocks. Microscopic studies of the iron-formation are particularly important in determining the mineralogy and the intermineral, textural relationships. This information is particularly needed for an evaluation of concentrating grades of cherty iron-formation deposits. These studies should use both thin and polished

sections and include geologists concerned with the field as well as with laboratory aspects of the work.

Cherty iron-formation ores include deposits of Precambrian iron-formations that can be mined and concentrated to yield an ore grade product. These concentrating grade iron-formation ores are of growing importance as they are mined and processed on a large scale in the United States, Canada, USSR, Sweden, Norway, Tasmania, and China. The essential feature of a commercial quality, concentrating grade, cherty iron-formation ore is the occurrence of iron oxide minerals that after grinding will yield enough high-grade concentrate to permit an economically viable operation. There are several lithologic types of iron-formation currently being mined and concentrated in commercial plants.

These iron-formation types may be conveniently grouped as magnetite taconite ores, jaspilite ores, or itabirite ores. They are being concentrated by three different concentration systems, so the ores are grouped on the basis of iron-formation lithology and the applicable concentration system. Each ore type will be described; factors important to their study, occurrence, evaluation, and processing discussed; and viable approaches suggested for their exploration, development, and evaluation.

Magnetite Taconite Ore—Magnetite taconite or taconite ore is a term applied to a concentrating grade of iron-formation that can be processed to yield a high-grade shipping product by magnetic separation methods after fine grinding. Magnetite taconite is the preferred term because the term *taconite* on the Mesabi Range of Minnesota, where the term originated, includes all types of iron-formation regardless of mineralogy or the possibility of concentration. Some ores may require upgrading by fine screening or flotation after magnetic concentration, but the main treatment method is by a magnetic system. The chief ore mineral is magnetite with most other iron and gangue minerals rejected into tailings.

Magnetite taconite ores are a banded, layered sedimentary or metasedimentary rock commonly composed of magnetite and chert or fine- to medium-grained quartz with varying amounts of iron silicate minerals, iron-bearing carbonates, or other iron oxides. These ores include a considerable range of iron-formation lithologies in which the principal iron mineral is magnetite. The range in mineral content of the magnetite taconite ores on the Mesabi Range is determined by the metamorphic rank of the iron-formation. The Biwabik iron-formation metamorphic rank ranges from the greenschist facies in the western and central part of the area to amphibole and to pyroxene facies in the eastern range area. The greenschist facies magnetite taconite is composed of magnetite, cherty and fine-grained quartz with varying amounts of greenalite, minnesotaite, stilpnomelane, siderite, ankerite, and minor chlorite and chamosite. The amphibole facies taconite ore is composed of magnetite, fine-grained quartz with varying amounts of cummingtonite-grunerite, hornblende, ankerite, and calcite. The pyroxene facies magnetite taconite ore is composed of magnetite, medium-grained [0.1 to 3.0 mm (0.004 to 0.12 in.)] quartz with varying amounts of hedenbergite, ferrohypersthene, cummingtonite-grunerite, hornblende, and fayalite with lesser amounts of calcite, almandine, and andradite garnet. There is a gradational relationship between the metamorphic facies. A complex interwoven series of lithologic zones of magnetite taconite ore types are found in the transition areas that may be 1 or 2 km (0.6 to 1.2 miles) or more in extent.

Even though the magnetite taconite ores on the Mesabi

Range bridge a considerable range of lithologic types they do not include all possible mineral phases of these ores. For example, a green biotite is common in magnetite taconite in the Kursk area, Russia, and the magnetite taconite ore at the Empire mine, Michigan, which is very fine grained, contains magnetite, hematite, quartz, siderite, ankerite, calcite, chamosite, riebeckite, ripidolite, stilpnomelane, and minnesotaite.

Common features observed in magnetite taconites that are important characteristics of these deposits include the following.

1. Most deposits occur as tabular bodies ranging from a thousand meters to tens of kilometers in length and tens or hundreds of meters in thickness. In areas of complex folding individual deposits may have various shapes and not be tabular.

2. All deposits are magnetic and may be investigated by magnetic methods, but the interpretation of the magnetic information is sometimes difficult because of remnant magnetization.

3. Under certain geological situations gravity methods may be used to give valuable information regarding the potential size and occurrence of the magnetite taconite deposit.

4. Weathering of a magnetite taconite body results in the oxidation of magnetite to hematite and sometimes to goethite. The zone of oxidation may extend to depths of over 100 m (328 ft) in certain areas. Oxidation may be an important factor to consider during the study of an ore body. Since the weathered zone commonly has an irregular pattern both horizontally and vertically, a careful evaluation of the surficial zone of a magnetite taconite deposit may be necessary to determine the character and quantities of the oxidized material and of the mixed magnetite-hematite-goethite zone material. Studies of the zone of oxidation usually require drilling, although test pits, shafts, adits, and tunnels may be used. These studies must be associated with a program of careful sampling and laboratory testing.

Exploration for magnetite taconite deposits is often started with an airborne magnetic survey using a 400 m (0.25 mile) line spacing and low terrain clearance. The magnetic survey may be preceded by geological mapping, but if the magnetic survey data are already available, the mapping is commonly more effective. Geological mapping is usually accompanied by a ground magnetic survey of potentially favorable areas to give accurate magnetic control for use in selecting drill hole locations. Remnant magnetization is common in Precambrian iron-formations, which often limits the evaluation of the magnetic survey results to physical perimeters of the magnetite taconite occurrence such as location, extent, and structure of the deposit. The magnetic surveys can give information needed to design a core drilling program that will sample the deposits, determine its detailed geology, grade, concentratability and minability, and permit an economic evaluation of the deposit.

Jaspilite Ore—Jaspilite ore includes cherty iron-formations that are composed of alternating fine to medium-fine grained, hematite-rich layers and cherty to fine-grained, jaspery quartz layers with minor associated magnetite and other iron minerals. These ores are commonly red to red-brown in color with steel grey hematite layering. They occur as sedimentary or metasedimentary formations associated with clastic sedimentary, and/or volcanic or volcanogenic sedimentary rocks. Jaspilite iron-formations have a range in geologic age from Archean to Late Precambrian/Cambrian.

Jaspilite ores, also termed Michigan Jasper ores, were first produced on a commercial basis at the Humboldt mine in 1954, and in 1956 at the Republic mine, Marquette Range, Michigan. Concentration is by froth flotation after fine grinding. The commercial Jaspilite ores in Michigan are crystalline with the hematite liberated from processing at about 0.21 mm (65 mesh) for some ore. The fine-grained ores are not being commercially concentrated but some deposits have been explored and much research on their benefication done.

Exploration methods that may be applied to jaspilite ores include geological mapping, airborne and ground magnetic surveys, and gravity surveys. Jaspilite iron-formations may contain sufficient magnetite to give a recognizable magnetic signature that will show the general size, shape, and extent of the deposit. Core drilling is the principal followup method to determine the nature of the deposit and to obtain samples for analysis and for laboratory concentration tests.

Itabirite Ores—Itabirite ores include commercial deposits of recrystallized, cherty iron-formation commonly composed of granular quartz, crystalline hematite, and martite with variable amounts of magnetite. The iron oxide-rich layers are interlayered with fine-to-medium and to coarse-grained quartz with more or less associated iron silicate minerals. The grain size varies from a fraction of a millimeter to several millimeters. These rocks characteristically are granular rocks and often are termed *quartzites* because of their texture. Their most common geological environment is with granulite facies rocks which often include granitic gneisses.

Itabirite ores are mined on a large scale in the Quebec-Labrador region of Canada at Fire Lake, Mt. Wright, Wabush Lake, and Labrador City. The first deposit mined in Quebec, at Jeannine Lake, has been exhausted. Itabirite ores range in composition from crystalline hematite and coarsely crystalline quartz with very minor amounts of iron silicate, to ores containing both magnetite and hematite interlayered with granular quartz with more or less associated iron silicate as amphibole or pyroxene, and minor calcite. There are extensive deposits of itabirite type ores in the Quebec-Labrador region of Canada, in Brazil, and in Venezuela. Similar deposits occur in Liberia and Sierra Leone, in West Africa, and possibly in other parts of this region north into Mauritania and south into Angola. Important characteristics of these ores are their simple mineralogy, moderate to coarse grain size, ease of liberation, and the simplicity of concentration.

Itabirite ores are mainly concentrated to yield a high-grade product by gravity methods followed by magnetic or electrical methods. In Quebec, the Fire Lake and Mt. Wright ores are concentrated by using spirals, and at Wabush by gravity methods followed by magnetic or high intensity electrical methods.

Itabirite deposits are typically tabular, occurring as stratigraphic formations associated with high-grade metamorphic or gneiss terranes. Even though deposits may contain only minor amounts of magnetite, magnetic survey methods can commonly be successfully used to supplement and assist geological mapping to show the location, extent, and the general shape and size of potential ore areas. In the Quebec-Labrador area magnetic anomaly areas with relatively low magnetic readings were associated with important hematite-quartz deposits such as at Jeannine Lake, Fire Lake, and Mt. Wright. In this region there are sufficient outcrops to permit a reasonable determination of the cause of magnetic anomalies. Airborne magnetic surveys may be used as a primary exploration method with followup ground surveys in areas of special interest. Followup ground magnetic surveys were helpful in showing the limits of the potential ore areas. Drilling is required to give positive knowledge of the extent of ore occurrence, the grade and concentratability of the iron-formation materials. Deposits with a relatively high magnetite content generally respond to magnetic surveys in ways that are similar to the response of magnetite taconite deposits.

Cherty iron-formation ores commonly occur as tabular, stratigraphic units within a sedimentary, bedded, iron-formation. In some cases the entire formation may be iron ore, but in other cases only certain stratigraphic units will have the necessary grade, mineral content, and liberation characteristics to be an iron ore. Since these ores were deposited as a sediment, they can commonly be studied by geological methods applicable to other bedded formations, including sedimentologic, stratigraphic, and structural methods assisted by geophysical surveys. Geological mapping appropriate to bedded deposits should receive important emphasis, particularly in early stages of an investigation. Airborne and ground magnetic surveys are also commonly made early in an exploration program. Magnetic methods are used as an almost universal method often as an early systems approach. With the high quality and sensitivity of modern airborne magnetic surveys, they may be expected to give valuable information regarding almost any type iron deposit. Surveys that give completely negative results are extremely rare. The surveys may not pinpoint ore, but give information on the location, occurrence, distribution, and structure of the iron-bearing rocks or associated formations. Magnetic surveys also give assurance of the continuity of formations between outcrop or drill holes.

The geological mapping and magnetic surveys are commonly followed by core drilling or other sampling methods. Drilling is the usual way to sample an iron deposit with samples sent to a research laboratory for analysis and process testing. Microscopic studies, both thin and polished section, should be done as soon as samples are available from outcrop or drill samples. It is important that variations in ore types, the mineralogy, and texture of iron-formation lithologies be recognized and their potential effect on concentratability determined. It is a geologist's responsibility to take the microscopic and research test information to the ore body and to show the distribution of ore grades and ore types within the deposit.

Gravity surveys should be considered as an exploration method if the iron deposit is nonmagnetic or where a magnetic survey gives insufficient or confusing results. There may be a sufficient density contrast between iron ore or iron-formation and wall rock to permit a reasonable correlation of the gravity readings with the size, shape, and structure of the deposit.

Down-the-hole magnetic susceptibility measurements may be a relatively inexpensive and effective way to obtain information on the magnetite content of the ore. Saturation magnetic measurements may be used in the laboratory to determine magnetite content of samples.

In dipping iron-formations, deviation of drill holes from the planned direction and inclination is common and may result in significant errors if the deviation is not recognized and corrections in location made on maps and sections. Gyroscopic controlled surveys of drill holes should be used as magnetite in the iron-formations prevents the use of other compass deviation methods. These surveys can be made on a contract basis with well service companies. Hole survey records show the inclination and the location of the drill hole from collar to bottom with a very small potential survey error.

Siderite Ores—Siderite iron ores of Archean age occur in a number of deposits in the Michipicoten district, Ontario, Canada. The siderite deposits are present as lenticular bodies that form the basal member of the Helen iron-formation, which appears to conformably overlie acidic volcanics and to be overlain by flows of intermediate composition. The Helen iron-formation is divided into three members that from the base upward are (1) siderite member, (2) pyrite member, and (3) banded silica member. The siderite member is composed of a fine-grained, gray-white to tan siderite with some disseminated pyrite and rare thin chert layers. Where of ore grade, the siderite contains from 33 to 37% Fe. It is sintered to remove CO_2 and S and to yield an oxide product as a shipping grade ore. The oxide sinter commonly contains about 50% Fe, 11% SiO_2, 1.7% Al_2O_3, 2.8% Mn, and 0.02% P. This ore is partly self-fluxing, with about 4% CaO and 6.8% MgO. Additional CaO may be added to give a super-fluxed ore with about 18.5% combined lime and magnesia, which reduces the iron content to about 48% Fe.

The major differences between the occurrence of the Helen type iron-formation ores and other Precambrian cherty iron-formation ores are (1) the segregation of the iron-formation into three distinct members with a significant part of the iron present in the siderite member, (2) the common occurrence of pyrite in the siderite ore and in the pyrite member, and (3) the occurrence of a large amount of bedded silica with a low iron content as the banded silica member. Most Precambrian cherty iron-formations contain very little pyrite and have the silica and iron closely linked as alternating silica-rich and iron-rich laminae or layers.

The Helen iron-formation and its members are lenticular on a local and a regional scale. The siderite member ranges from very thin to a maximum of 106 m (348 ft) at the Helen mine. Commonly the ore bodies are less than 60 m (200 ft) in maximum thickness. The pyrite member ranges from very thin to a thickness of 3 to 9 m (10 to 30 ft), but it is about 36 m (120 ft) thick locally in the northeastern part of the district. The banded silica member also ranges considerably in thickness from absent to 300 m (1000 ft) with a common range of from 30 to 60 m (100 to 200 ft).

The Michipicoten district is the only area where Precambrian siderite ore bodies are being mined. A number of deposits of small or rather low-grade sideritic carbonate bodies occur in other parts of Canada in Archean rocks, such as the Woman River area, the area south and east of Lake Abitibi, the Kirkland Lake, and Swayze areas.

Exploration for siderite ores follows a rather normal pattern that includes geological mapping, airborne magnetic surveys, detailed ground magnetic surveys, some local gravity surveys, and core drilling. Geological mapping can indicate the possible location of the siderite member, which commonly has limited outcrop, so this zone can be tested by other methods. An airborne magnetic survey commonly shows the location of the iron-formation often by the magnetic signature of the associated rocks. Ground magnetic surveys run in conjunction with detailed geological mapping give more specific magnetic information that may be related to the associated volcanic rocks. A few gravity surveys were run in past years with limited success. Problems associated with this method are the limited gravity contrast, terrain corrections needed, small size of the target areas, variable thickness of overburden, and rather high cost. While there is no record of the use of electromagnetic survey methods, they might be considered as the occurrence of disseminated and/or massive pyrite should give a recognizable response. Diamond core drilling commonly follows detailed geological and magnetic survey work to determine the thickness, distribution, and grade of siderite bodies.

Ironstone Ores: Ironstone ores are iron-rich sedimentary deposits composed of oolitic to granular hematite and/or goethite with varying amounts of carbonate and iron silicate minerals that are intercalated with sandstone, shale, carbonaceous shale, siltstone and limestone in rocks from Late Precambrian to Tertiary in age. Minable ironstone beds range from 2 to 7 m (7 to 23 ft) in thickness. The iron-bearing beds are often lenticular both along strike and down dip. The minerals in ironstone ores may be hematite, geothite, siderite, chamosite, iron-rich chlorite, and calcite, with a considerable range in content of the several minerals. They also contain a variable content of admixed sand, silt and clay, which together with the limy component gives a range from siliceous to self-fluxing ore. Ironstone ores have a rather high phosphorus content that ranges from 0.2 to 0.8% but may sometimes exceed 1.5%. Ironstone ores contain from 25 to 35% Fe, 4 to 15% SiO_2, 3 to 8% Al_2O_3, 2 to 15% CaO, and 0.2 to 0.8% P. Some deposits, as at Wabana, Newfoundland, Algeria, Morocco, and in the leached outcrop zones of other deposits, have an iron content that ranges from about 45 to 55% Fe. Even in the better grade ores, the SiO_2, Al_2O_3, CaO, and phosphorus contents are commonly rather high.

Exploration for ironstone ores usually has started from outcrops or iron-rich float ore visible at the surface. Work continues with geological mapping to show the stratigraphy and structure of the ore-bearing zone and of the associated formations. The ore zone is tested at the surface by pits and trenches and by drilling to show the distribution, thickness, and grade at depth.

Clastic Iron Ores: Iron-rich clastic sediments and sedimentary rocks deposited as sands and gravels have a worldwide distribution as modern and ancient beach, river, alluvial fan, and terrace deposits. Clastic iron deposits as unconsolidated sediments have a wide distribution, but only a few occurrences have sufficient size, grade, and availability to be of commercial interest. Beach and terrace deposits have been mined on a small scale in Japan for centuries with a minor production continuing. Deposits on the West Coast of North Island, New Zealand, are being mined and concentrated for domestic use and for export. The ore occurs on modern beaches and raised Pleistocene terraces. Iron is present as grains of titanomagnetite that contain about 7 to 9% TiO_2 and 0.36 to 0.40% V_2O_3. The titano-magnetite content in the sands ranges from small to 40 to 45%, with a lower commercial limit of about 8%. A very large alluvial fan type titaniferous magnetite-bearing deposit at Klukwan, Alaska, has been explored but not mined. Erosion of an extensive titaniferous magnetite-bearing mafic and ultramafic body has formed a large alluvial fan that is about 2.4 km (1.5 miles) long and about 0.8 km (0.5 mile) wide which contains about 12 to 13% iron as titaniferous magnetite grains in the fan and in iron-bearing rock fragments. Open pit mining and large scale concentration of the alluvial fan material after grinding and magnetic separation could yield an iron ore pellet that would contain from 64 to 65% Fe and 1.2 to 1.4% TiO_2. Thus the unconsolidated, modern clastic ores range from deposits in which the iron minerals are free from matrix to deposits where the iron minerals are still enclosed within rock fragments.

The paleo-clastic iron ores are commonly compact, more or less cemented, sedimentary rocks. Iron ore deposits of this type were formed by erosion of cherty iron-formation during the Precambrian and on the Mesabi Range, Minnesota, during the Cretaceous. The Mesabi Range Cretaceous ores were mined by open pit methods often as a part of a larger ore body within the iron-formation. Marquette Range deposits are small, with limited production. The Gamagara ores in the Sishen Region, South Africa, may be the largest known deposits.

Magnetite or titaniferous magnetite sand deposits may be explored in a preliminary way using an airborne magnetometer survey. A commercial deposit should give a weak but recognizable response. The physical exploration of unconsolidated, clastic iron ores generally is done using methods common to other placer deposits. The distribution of the potential ore areas is mapped, and then tested by drilling, test pitting, trenching, and possibly by shaft sinking. Drilling is done using a churn or a hammer drill. The sample validity is maintained by driving casing with the sample taken from within the cased hole. Care must be exercised to be certain a reliable sample is obtained. Exploration problems vary considerably and are related to the possible depth of the deposit, water conditions, and the size of the clastic materials.

Exploration for paleo-clastic deposits requires an understanding of the probable depositional environment, and the geological history after deposition. Geological interpretation of the likely sedimentological conditions is of primary importance. Since the most common ore mineral is hematite, with possibly some goethite and magnetite, magnetic surveys may or may not give useful information. As a general rule, magnetic surveys should be made using airborne and ground surveys to assist in broad geological interpretations and specific target selection. Core drilling is done to obtain samples for analyses and testing and to determine the character and the extent of the deposits.

Magmatic Iron Ores

Iron ores that form as a result of magmatic processes include Kiruna type ores and titaniferous magnetite deposits. The Kiruna type deposits are an important iron ore source whereas the titaniferous ores are of minor importance as a source of iron and are mined mainly as a source of titanium or vanadium.

Kiruna Type Ores: An interesting and diverse class of magmatic iron ore bodies is grouped together as Kiruna type deposits. They include deposits in the Kiruna district, Sweden, the Pea Ridge, Iron Mountain, and Pilot Knob deposits, Missouri, the Cerro Mercado deposit, Mexico, and a number of iron deposits such as the Algarrobo, Tofo, and the El Laco deposits in Chile. The Kiruna type ores are composed dominantly of magnetite and hematite commonly with an appreciable phosphorus content in apatite. The shape and size of the bodies varies widely from the very large tabular sill-like body at Kirunavaara, Sweden, which is 4400 m long (2.73 miles), averages 90 m (295 ft) wide, and is known to a depth of over 1500 m (4900 ft), to small veinlets, stockworks, and dikes, and to a deposit considered to be a surface flow at El Laco, Chile. These deposits are all associated with igneous rocks of intermediate to acidic composition, often show intrusive relationships with the associated rocks and show more or less alteration of the country rock. Magnetite is the most important iron ore mineral with varying amounts of hematite. Some deposits that contain mainly hematite are included in this class such as those at Iron Mountain, Missouri, which contain about 80 to 95% of the ore as massive hematite with some platy hematite and 5 to 20% magnetite (Murphy and Ohle, 1968). The Kirunavaara magnetite ore contains from 60 to 69% Fe, 0.01 to 2.3% P, 0.7 to 3.7%

SiO_2, 0.13 to 0.6% Al_2O_3, and 0.13 to 0.23% U_2O_5 (Parak, 1975). The Pea Ridge ore contains from about 45 to 60% Fe and a rather constant phosphorus content even though the magnetite-hematite ratio ranges from dominant magnetite to dominant hematite. Many of the Kiruna type iron ore deposits require concentration to remove phosphorous and to upgrade the ore to a high quality shipping product.

Usual exploration technique for Kiruna type deposits includes geological mapping, airborne and ground magnetic surveys, and gravity surveys. The ores in Sweden, Cerro Mercado, Mexico, at Iron Mountain and Pilot Knob, Missouri, and a number of Chilean deposits were discovered in outcrop and studied and explored by detailed followup geological and geophysical work. The Pea Ridge deposit and a number of other undeveloped iron ore deposits in Missouri were discovered by magnetic methods. The existence of the Pea Ridge deposit was indicated by an airborne magnetic survey flown in 1950 and found by drilling in 1953. The deposit occurs below about 396 m (1300 ft) of Paleozoic sedimentary rocks. Gravity surveys were used as an important exploration and evaluation method at the Iron Mountain and Pilot Knob deposits (Leney, 1956). The northwest ore body at Iron Mountain which is essentially a nonmagnetic, hematite ore was found by a gravity survey much later than the vertical, main ore body which was discovered by drilling under a strong magnetic anomaly. The northwest ore body occurred at a shallow depth and was mined by open pit methods whereas the vertical ore body was mined by underground methods. The Iron Mountain geophysical studies illustrate the importance of using both magnetic and gravity methods where nonmagnetic hematite ore may occur.

Titaniferous Magnetite Deposits: Titaniferous magnetite deposits have a worldwide distribution in basic igneous rocks, commonly gabbros or anorthosites, and in some metamorphic rocks of basic composition. These deposits occur as tabular to irregular shaped bodies or seams and veinlike areas or in layered mafic intrusives as tabular stratigraphic zones that range from massive to disseminated. The iron content ranges from less than 20 to over 60% with from 1.5 to 35% titanium and 0.1 to 2.0% vanadium. Titaniferous magnetite deposits have limited use as a source of iron ore because of the relatively high titanium content. The main purpose of mining these deposits is as a source of titanium or vanadium. There is more or less use of the magnetite byproduct. Iron is produced at the Sorel plant after processing the Allard Lake, Quebec ore. Some of the magnetite concentrate from the Sanford Lake deposit in New York is used as iron ore. Deposits in the Bushveld Complex, Republic of South Africa, are mined for their vanadium content. Titaniferous magnetite is a primary iron ore from beach and terrace deposits but not from lode deposits.

Contact Metasomatic Deposits (Magnitnaya Type)

An important class of iron ore deposits occurs as replacement bodies, commonly in limestones, at or near contacts with intrusive igneous rocks. These ores usually occur as massive, magnetite, or magnetite-hematite or hematite-magnetite bodies, often with common skarn minerals and small quantities of associated pyrite, pyrrhotite, and chalcopyrite. The skarn minerals include pyroxenes, amphiboles, garnet, epidote, chlorite, scapolite, apatite, and calcite. The ore bodies are often irregular to tabular in shape and range in size from deposits of a few thousand tons to several hundreds of millions of tons. Commercial deposits usually contain a minimum of a few million tons of ore grading from 35 to 55% Fe. These ores commonly require processing to upgrade the iron content and remove sulfur, copper, other base metals, and slag-forming elements to yield a high-grade, low-impurity product. Some deposits yield salable byproduct: copper and other metals.

Many of the known contact metasomatic iron ore deposits were found by direct observation of extensive areas of float ore and outcropping ore. Exploration of these deposits included geological mapping accompanied by magnetic surveys and core drilling.

Airborne magnetic surveys either made for iron ore exploration or as regional surveys have resulted in the discovery of a number of deposits. Exploration for blind contact metasomatic iron deposits may follow a rather classic pattern of (1) geological reconnaissance to select areas that appear to have a favorable geological environment with correct types of igneous intrusive and host rocks; (2) regional airborne magnetic surveys; (3) interpretation of the magnetic information with the selection of potential target areas; (4) detailed geological mapping and ground magnetic surveys possibly accompanied by gravity surveys; and (5) drilling. Replacement deposits commonly have a marked magnetic and density contrast with the associated country rock which permits an interpretation that includes a reasonable estimate of the size, shape, depth, probable reserve tonnage, and, with some drill core analysis data, the grade of the deposit.

Massive Iron Ores in Cherty Iron-formation

Massive, high-grade iron ore deposits occur in cherty iron-formations in geological situations that suggest an origin by a replacement process where silica has been replaced by hematite or magnetite and in deposits occurring at or near paleo-erosion surfaces that appear to be metamorphosed paleo-Lake Superior type ore bodies. The replacement type iron ores are commonly composed of compact, steel gray or crystalline, micaceous, or granular hematite or a mixture of hematite and magnetite or in a less common occurrence as granular magnetite. The common associated minerals are quartz and iron silicates. Ore bodies of replacement type have a considerable range in shape, size, structural relationships, and distribution within the cherty iron-formation. Deposits that occur at or near ancient erosion surfaces contain iron ores similar to replacement ores. There are conflicting opinions on their origin. The two ore occurrences will be considered together as both types of deposits have similar exploration and evaluation problems.

Iron ores at the Soudan mine, Minnesota, occur at the ends of lenticular belts of jaspilite iron-formation or are associated with folds. There is no recognized relationship of the ore to structures that would permit access of solutions involved in the replacement process although evidence of the replacement of chert layers by hematite is very clear. Deposits of granular, crystalline hematite occur in the axial areas of folds in parts of the Quebec iron ore district between the Matonipi Lake area and Labrador border and south of the Orinoco River between the delta area and the Caroni River. These iron ore deposits are commonly lenses and pods not large enough to be of commercial interest. The associated iron-formation is of the Itabirite type and country rocks are commonly gneiss.

Certain of the *Hard Ore* deposits in the Marquette Range, Michigan, are at or near a Precambrian paleo-erosion surface at the top of the Negaunee iron-formation. The Marquette Range *Hard Ores* were mined extensively in the Ishpeming-Negaunee area and are largely exhausted.

Exploration for hematite-rich replacement or Paleo-Lake Superior type iron ores may be difficult if they do not outcrop and do not contain sufficient magnetite to create a recognizable magnetic signature. Many deposits contain the same iron minerals as the associated iron-formation and are, difficult to locate using any geophysical method. A considerable density contrast may occur, so gravity surveys may be used to locate near surface deposits. However, gravity surveys are expensive, slow, and not easily adapted to the required accuracy as a reconnaissance method. After the initial discovery and the likely geological relationships to structures or to a paleo-surface are observed, drilling is the common followup exploration method. As an example, at the Soudan mine much underground drilling was done to test geologically favorable targets such as projected trends of known ore and favorable areas at the ends of iron-formation lenses. Exploration for Marquette Range *Hard Ores* was started from outcrop observations followed by geological mapping and drilling. Underground geological mapping, drilling, and a following of the ore continued during the long mining history. The principal method for successful exploration appears to be competent geological mapping and interpretation of the ore relationships followed by a carefully monitored drilling program.

Residual Iron Ores

Iron oxides are commonly formed at the earth's surface by the weathering of iron-bearing rocks. Where the oxide accumulations are of sufficient size and iron content they constitute sources or potential sources of iron. Five classes of iron ore deposits are formed by weathering processes: Lake Superior type ores, oxidized banded iron-formation-OXIBIF ores, Canga and River Terrace ores, Brown ores, and Lateritic ores. A large part of the iron ore moving in international trade is produced from naturally occurring, merchantable or concentrating grade, Lake Superior type iron ore deposits associated with Precambrian cherty iron-formation.

Lake Superior Type Deposits: Iron ores formed by the oxidation of iron minerals and the leaching of silica from cherty iron-formations with the residual accumulation of iron oxides have been extensively mined in Michigan, Minnesota, and Wisconsin, which resulted in these ores being termed *Lake Superior type* ores. They are now produced on a large scale in Australia, Brazil, India, Liberia, Mauritania, Republic of South Africa, USSR, and Venezuela. Large undeveloped deposits are known in Brazil and in Gabon and Guinea in Africa. Lake Superior type ores occur in deeply weathered areas at the present land surface and in favorable geological situations to considerable depth below the surface. In Michigan at the Mather mine, Lake Superior type soft ores were mined to depths of about 1 070 m (3500 ft) with ore known to occur to depths of over 1 525 m (5000 ft). On the Mesabi Range in Minnesota, slump structures are associated with the iron ore bodies. These structures formed because of the removal of about 30 to 45% of the original iron-formation volume during leaching of silica.

In the Lake Superior Region the base grade ore is assumed to contain 51.5% Fe with price adjustments made for ores that contain above or below this percentage of iron. Most merchantable ore and iron ore concentrates contain a higher iron content than the base grade. As an example, in 1970, 25 Mt of natural ore shipped from the Lake Superior region contained an average of 54.26% Fe, 8.87% SiO_2, 1.19% Al_2O_3, 0.61% Mn, 0.08% P, and 6.85% moisture.

Since about 1910, certain partly leached, natural iron ores have been beneficiated by simple methods such as washing, screening, and by gravity methods. These ore materials are termed *concentrating grade natural ores*. They are commonly included with the Lake Superior type ores as the geology of occurrence is similar and the two ore types often occur together with gradational relationships. The difference between the two ores is the efficiency of the removal of silica by leaching. The silica present in the concentrating grade ores occurs in a physical condition that will permit its removal to an acceptable concentrate grade by gravity benefication methods without extensive grinding.

Exploration for Lake Superior type, merchantable, and natural concentrating grade iron ores requires an understanding of the favorable geological conditions for ore occurrence. These deposits form as a result of oxidation and the leaching of silica from cherty iron-formation with the residual accumulation of iron oxides. Ore bodies commonly occur in areas that are most favorable for extensive ground water movement. Favorable exploration targets may be at the present land surface, along paleo-land surfaces, fault zones, or associated with troughlike structures that could channel ground water flow. Favorable geological structures include a variety of situations that will concentrate water movement such as synclinal folds, fault zones, troughlike zones at dike intersections and strongly jointed and fractured areas. During oxidation and leaching, primary iron silicate, iron carbonate, and magnetite are converted to hematite and goethite, with very minor maghemite and lepidocrocite and with associated varying, small amounts of residual clay, quartz, and manganese oxides. Ore development can be considered a part of the weathering process of a special rock, iron-formation during an extended, intense weathering cycle.

Exploration for Lake Superior type and concentrating grade natural ores is based on the accumulation of information on potentially favorable areas for intense, long-term leaching action by ground water. As in most exploration programs, geological mapping and direct field observations are used as fully as conditions permit. Field observations are commonly supplemented by magnetic surveys very early in the exploration program as some magnetic signature is usually identified. The usefulness of magnetic surveys varies considerably from merely locating the cherty iron-formation host to very directly indicating potential ore areas. Since the Lake Superior type ores and concentrating grade natural ores are formed by oxidation and leaching, the magnetic readings over ore bodies commonly are low relative to the associated iron-formation. Favorable exploration sites are located in low magnetic anomaly areas. A common exploration sequence following field studies is to conduct an airborne magnetic survey at a low terrain clearance of about 120 m (400 ft) and a line spacing of about 400 m (0.25 mile) with favorable areas shown by the airborne survey covered by ground magnetic surveys. The magnetic line and station spacing selected should give adequate detailed magnetic information to permit the geological interpretation of the magnetic readings. The ground magnetic survey area should be large enough to allow calculation of depth, width, and attitude of the potential ore-bearing structure.

Gravity surveys have a limited, though at times very useful, purpose in the exploration for Lake Superior type or concentrating grade natural ores. Since these ores are often rather porous, with a sizable moisture content, they may have a small, sometimes almost zero, gravity contrast with the enclosing rocks. This can severely limit the use of gravity methods. In some areas, such as in parts of the Cuyuna Range, Minnesota, regional gravity surveys have been used as an important geophysical method to supplement magnetic methods. The cherty iron-formation on the Cuyuna Range has a very weak magnetic response so gravity methods were used to trace the distribution of the iron-formation and to assist in the selection of potentially favorable areas.

Geological mapping is an important initial exploration approach used for iron ores that occur at the present surface in areas with a reasonable amount of outcrop. The mapping is often done after an airborne magnetic survey which commonly shows the location of the iron-formation. Airborne and ground magnetic surveys should be used in conjunction with geological mapping as they will give valuable information regarding the location, distribution, structure, and the character of the iron-formation. Field observations that can be made include variations in the character of the iron-formation, structural relationships, and the nature of the outcrop ore. The presence or absence of outcrops also may be helpful in the search for guides to the recognition of potential ore areas. The observed geological and magnetic data often indicate specific possible ore areas that should be tested by drilling.

Magnetic surveys are usually the principal exploration method used in areas with little or no outcrop. An exploration program in such areas commonly starts with a regional airborne magnetic survey of the selected area. This survey is usually run at about 400 m (0.25 mile) flight line spacing with low ground clearance. The magnetic data are then interpreted to show the location of the iron-formation, structural relationships, and potentially favorable areas. These areas are covered with ground magnetic surveys using appropriate line and station spacing. The possible ore areas are then tested by drilling.

An important aspect of any iron ore exploration program is the continuing reappraisal and evaluation of the available information as new and better data are received. In many cases interpretations made early in a program may need substantial adjustment as the work progresses. It is also important that as complete a drill record as possible be maintained in the form of drill core. Split drill core should be retained for restudy in the light of new facts generated as the investigations proceed. Core logging should involve the most capable geologists, as the best possible core interpretation is essential to a successful program.

OXIBIF Ores: Oxidized banded iron-formation ores include all concentrating grade cherty iron-formation material composed of earthy or very fine grained, compact, or steely hematite with more or less associated goethite interbedded with cherty quartz-rich layers. Characteristically these ores are the result of secondary oxidation with more or less leaching of cherty iron-formation. They could be termed "weathered cherty iron-formation." Marsden (1978) used the acronym OXIBIF for these ore materials as the term oxidized banded iron-formations is used to identify a variety of non-ore cherty iron-formation. As defined, the acronym is to be applied only to concentratable oxidized, banded, iron-formation materials. These ores require fine grinding to yield a commercial grade iron ore product.

OXIBIF ores are commonly red-brown in color and prominently layered with iron-rich layers dominantly hematite or goethite interbedded with brown to red-brown cherty quartz. The thickness of the layering varies from a millimeter to several centimeters with many visible layers observed in a mine bank or outcrop from 5 to 15 cm (2 to 6 in.) or more thick. These ores may be rather porous and friable to compact and massive depending on the degree of weathering and the original oxide, silicate, and carbonate content. The iron oxides and cherty quartz are often fine grained so a very fine grind may be required to give effective liberation of the iron

minerals. Commonly all primary iron silicate and carbonate minerals and magnetite have been oxidized, although a small amount of magnetite may be present. Soluble materials have been removed, leaving mainly Fe_2O_3 or $Fe_2O_3 \cdot H_2O$, SiO_2, and Al_2O_3. Extensive deposits of oxidized iron-formations occur in the Lake Superior region, in the Labrador trough, Canada, in the Thabazimbi region, South Africa, and may occur in other areas of the world. At this time OXIBIF ores have a limited commercial development as more economic iron ore deposits are available.

OXIBIF iron ore is currently being mined and processed at the Tilden mine, Marquette Range, Michigan. The ore is ground very fine, about 90% minus 500 mesh, to give effective liberation of the iron minerals; then the ore is concentrated using the selective flocculation-flotation system.

OXIBIF deposits are commonly weakly magnetic to essentially nonmagnetic, which often limits the use of magnetic methods in their exploration to surveys designed to show their general location and extent. Exploration for OXIBIF ores relies on geological mapping with accompanying magnetic surveys, both airborne and ground, followed by drilling and laboratory tests. It is worth the effort to attempt to determine whether detailed magnetic surveys are an aid in the determination of the distribution of ore quality iron-formation materials. The recognition of the concentratability of the oxidized banded iron-formation is based on laboratory testing. The extent and occurrence of OXIBIF materials requires a careful geological evaluation of the detailed stratigraphy and structure of possible ore areas. The exploration work must include a carefully planned, rather detailed sampling program, usually done by drilling followed by specific research testing of samples. The results of the sampling and testing is interpreted into the distribution of ore types within the deposit. A geological understanding of laboratory test results is most important in the evaluation of OXIBIF deposits.

Canga and River Terrace Deposits: Iron oxide crust deposits that contain variable amounts of cherty iron-formation and iron ore fragments cemented by goethite and hematite are common to tropical regions and in Western Australia where there are extensive areas of exposed cherty iron-formations. These deposits are termed *Canga* in Brazil. This term may also be applied to similar deposits that occur extensively as residual crusts of hematite and goethite on outcropping iron-formation in the Hamersley region of Western Australia. Canga is defined as a surficial crust commonly composed of iron-formation or iron ore fragments cemented by iron oxides. The content of fragments and secondary iron oxides that precipitated from solutions to form a cement varies. In both Brazil and Australia the Canga crust is believed to be of Tertiary age. In the Hamersley region the Tertiary Hamersley erosion surface is more or less eroded with the hard, oxide, ferricrete crust acting as a protective cap to hills.

In the Robe River, Duck Creek, Beasley River, and Fortescue River areas in the western part of the Hamersley Range there are extensive River Terrace iron deposits that are extensions of Canga accumulations. These River Terrace deposits occur as valley bottom iron oxide crusts composed of pisolitic and cellular goethite, with associated detrital materials. These deposits are described by Brandt (1973) as: "The derived type deposits have clearly originated from both mechanical and chemical weathering of jaspilites, whereby detrital accumulations were formed and were subject to supergene leaching of silica, hydration of original hematite and other iron minerals, and the deposition of much limonite cementing and replacing the original rock fragments. Their formation has involved the downslope transportation of enormous amounts of iron as detritus, in solution, or both. The deposits range from limonite-cemented screes on slopes which are sometimes continuous with in situ canga cappings on hilltops, to thick valley-floor accumulations of massive and pisolitic limonite without detrital material." The eroded pisolitic, river terrace deposits often form extensive mesa-like hills capped by an iron oxide crust often about 30 m (100 ft) to a maximum of about 60 m (200 ft) in thickness. The Canga and river terrace deposits commonly contain from 40 to 60% Fe, 4 to 10% silica, and have a 10 to 12% ignition loss. These ores are being mined in the Robe River area.

Since Canga and river terrace deposits occur as surficial crusts, they are exposed on hill slopes, and along ravines and streams so their occurrence and distribution can usually be mapped in the field or on aerial photographs. The thickness of the deposits and grade of the material is determined by drilling.

Brown Iron Ore: The term *brown ore* is applied to a group of iron ores that occur as residual accumulations from the weathering of a number of iron-bearing rocks. These include limestones and dolomites, greensands, sideritic bodies and rocks, and veins containing iron sulfides. Brown iron ores have a wide distribution and in the past were important sources of ore for local iron furnaces. They are now of limited importance for a variety of reasons, including their common small size, variable chemical composition, often rather high silica content, the exhaustion of many of the larger deposits, the concentration of iron and steel production into large integrated plants, and the ready availability of high-grade iron ore on a worldwide basis.

Discovery and exploration of brown iron ores is highly dependent on surface indications of possible ore occurrence which often consists of float pieces of iron oxide. The likely target areas are then sampled by test pits or trenches or by drilling. Leney (1966, p. 410) reports "Standard resistivity surveys with an arrangement of current and potential probes at the surface to measure resistivity of the ground have had some success in prospecting for shallow flat-lying ore bodies. They have been applied effectively in Minnesota's Spring Valley district to locate residual pockets of limonite-goethite ore occurring on a limestone surface under shallow cover, but so far have not been able to compete in cost with simple auger drilling. In Missouri, resistivity surveys are reported to be useful in measuring the thickness and lateral extent of ore-bearing clay residium in brown (goethite-limonite) ore deposits."

Lateritic Iron Ores: The lateritic weathering of ultra-mafic rocks including dunites, peridotites, and serpentinites in tropical and subtropical regions may develop residual iron oxide deposits that contain from 40 to 55% Fe. The iron-rich surficial deposits range in thickness from less than a meter to over 100 m (328 ft) with extensive areas underlain by deposits from 3 to 10 m (10 to 33 ft) thick. These laterite deposits are not a commercial iron ore at present because of the relatively high chromium and nickel content. Lateritic iron ores were mined in Cuba from the early 1900s until about 1950 and were used to make chromium-nickel steel. Lateritic iron deposits are composed of a rather massive, red-brown, hematite-goethite *crust ore* zone with a yellow, powdery limonitic zone below that rests on an irregular weathered bedrock surface. These ores commonly contain from 45 to 55% Fe, 8 to 12% Al_2O_3, 2 to 8% SiO_2, 0.2 to 2.5% Cr, 0.4 to 2.0% Ni, 10 to 30% moisture, and 6 to 14% combined

water. The powdery ores are very porous and high in both moisture and in combined water.

Since the lateritic deposits occur at the surface, their general distribution can often be determined by surface mapping with the depth determined by drilling, using a hand auger or power auger, trenching, and test pitting. The drill holes usually will stand as open holes, often for a number of years without caving, which makes sampling relatively easy and accurate.

MAPS AND MAPPING

Maps are essential to all exploration projects. The most basic group of maps is prepared and used for location control. These maps serve as base maps upon which other information is recorded either in the primary gathered form, or as compiled and coordinated data, to present results of the work in a readily usable form. The special maps are prepared to meet the needs of geologists, geophysicists, engineers, planners, evaluators, and mine operators at various stages during exploration, development, evaluation, and mining. Map scale selection is a fundamental decision of importance to the project. The map scale should be appropriate to the amount and type of information to be shown as well as to the purpose and expected use of the map. The map scale selection process should include the people developing the map and those who will use it. It should also fit into a rational system. A maverick scale is an unwelcome nuisance. The size of maps produced should also receive early attention so maps are convenient to use and easy to file. McKinstry (1948) suggests maps of 91 × 102 cm (36 × 40 in.) as a maximum and 46 × 61 cm (18 × 24 in.) as a minimum size. Another early decision made during an exploration program is the orientation of the survey control system. This is often the azimuth direction of the grid control for geological mapping and geophysical work. The implications of grid orientation with regard to future drilling, development, and mining should be considered as the first control work is done.

A review of available maps should be one of the first steps in an exploration program as location control is essential at all stages of the work. There are a variety of government-prepared maps and aerial photographs available that may be used during the initial phases of the work. Modifications of scale of regional and topographic maps and aerial photographs can be made to suit the geologist's needs during preliminary phases of a program. Early in an exploration program, maps made to suit the specific project needs will be required. These maps can be made as a part of the work program or under contract with one of a number of companies. The ready availability of map preparation services that can develop maps at a desired scale and accuracy is a modern reality. After selection of a contractor the most important decisions concern scale, accuracy, contour interval for contour maps, area to be included, and the time the work should be done. The contractee should exercise a reasonable level of supervision to insure the validity and accuracy of the completed work.

In iron ore exploration the likely occurrence of local magnetic anomalies that may influence field mapping should be recognized. The severity of the problem varies considerably from conditions where major compass deviations occur to almost no visible influence from the iron deposits. In strong magnetic deviation areas, a sun dial compass is useful in geological mapping for the accurate determination of strike and for direction control on survey lines. The sun dial compass is based on the principle of a sun dial to give the time of day, which is reversed to use the known time of day to give a true north direction at the reading point. Mapping iron ore deposits follows the usual methods applied by geologists and engineers. The geological work involves recording of special ore characteristics such as the mineralogy, texture, and the physical nature of merchantable ore and the mineralogy, texture, grain size, stratigraphic variations and zonation, weathering characteristics, potential minability, and concentratability for concentrating grade ores.

GEOPHYSICS

The application of geophysical methods to the exploration and evaluation of iron ore deposits is accepted as a standard procedure. Geophysical systems measure properties of minerals either in the ore body or in associated rocks that can be interpreted to give useful information regarding iron ore deposits. The successful application of a particular geophysical method depends on the presence of distinct, measurable properties interpreted in a way that assists in location and understanding of an iron ore body. The several methods used can often furnish recordable, objective information that complements and supplements other methods of study. Geophysical studies may be used effectively from the initial selection of possible exploration areas, throughout the exploration and development program, and sometimes during the mine operation. Geophysics should be considered as more than just an exploration tool as it may have other useful applications.

Studies of local variations of the earth's magnetic field using magnetic methods are the basis for much of the geophysical work related to exploration for iron ore. Other geophysical methods such as gravity, radiometric, electrical, or seismic have limited roles and when applied, usually have specific applications in contrast with the almost universal application of magnetic methods. A contributing factor to the wide use of magnetic methods is the fact that these surveys commonly give valuable information concerning regional geology such as the types of rock present, their distribution, continuity, and structure. A regional, broad purpose magnetic survey can often indicate the location of iron-bearing rocks and may in some cases show the location of iron ore bodies. The importance of the magnetic method in regional reconnaissance surveys cannot be overemphasized as it permits the use of airborne magnetic surveys as a primary screening system in the exploration for iron ore deposits. Airborne survey methods also have the advantage that the reconnaissance phase of the work can be done without the confining aspects of property control and access. Other airborne geophysical systems have almost no direct application to the exploration for iron ore, but electrical and radiometric methods are often run with regional magnetic surveys as the additional cost is relatively small and the exploration company may be able to use these data to look for other types of mineral deposits.

Magnetic Methods

Magnetic geophysical systems measure variations in the earth's magnetic field caused by differences in the magnetic characteristics of the underlying rocks. The earth's field is a spontaneous force that consists of the natural force field as modified by geological bodies. In the International System of Units (SI), the unit of measurement for a magnetic field is the tesla (T). The earth's magnetic field has a magnitude of about 0.6×10^{-4} T at the surface, so measurements of the field usually are reported in more convenient units of nanoteslas (nT); $1 \text{ nT} = 1.0 \times 10^{-9}$ T. Older magnetic surveys reported field measurements in gamma or Gauss units and may readily be converted to SI units by the relationships $1 \text{ nT} = 1$ gamma or $1 \text{ nT} = 0.1 \times 10^{-6}$ Gauss. The magnetic method may be used in a wide range of field circumstances in the air, on land or water, in drill holes, or on ore samples in the laboratory. There is a considerable range of available instrumentation, recording accuracy, and logistics of application, which eventually affects costs of conducting a magnetic survey. This flexibility of application is one of the factors that favors its use.

Rocks containing unusual amounts of minerals with ferromagnetic susceptibility are magnetized in the earth's field and minerals with a remanent magnetization maintain an independent magnetic field. Both distort the earth's normal field. Measurements of the magnetic field are compared with the normal field or relative to a selected magnetic field reference point. This comparative relationship is used to show local variations caused by anomalous rock bodies. Magnetic readings are taken over a grid pattern or along survey lines. The observed variations or anomalies are interpreted geologically to the likely material and body that may cause the observed magnetic response.

The magnetic method has been used in exploration for iron ore since about 1640 when it was first used in Sweden. The *dipping compass* or dip needle was used extensively in iron ore exploration until about 1950. Since that time a number of types of magnetometers have been developed that are more accurate and easier to operate in the field than the dip needle. The airborne magnetic method for use in exploration was introduced during the period 1946-1950 and has resulted in major changes in exploration for iron ore and other mineral deposits. There are four general types of magnetometers in use that include variometers, flux-gate magnetometers, proton magnetometers (nuclear resonance), and optically pumped or alkali vapor magnetometers. The variometer type instruments include the dip needle, super-dip, Schmidt type magnetometers, and torsion-head type magnetometers. They measure variations within the total magnetic field and can measure either the vertical or horizontal component; commonly, the vertical component is measured. These instruments can measure to a precision of about 10 gammas per scale division. The magnetometers are tripod mounted, which makes them cumbersome to use. The flux-gate magnetometer was developed for airborne magnetic surveys and later adapted to small, very portable units for ground surveys. Ground survey units measure the vertical component with a precision range of 0.5 to 12.5 gammas with a common accuracy of about 5 gammas. The Jolander and Sharpe magnetometers are flux-gate types. These instruments are direct reading, hand-held, do not require azimuth orientation, only coarse leveling, are lightweight, and convenient to use in the field. The proton magnetometers now have a wide use in airborne and ground magnetic surveys. These instruments have a high degree of accuracy and are used in airborne surveys recording at a precision of 0.5 to 1.0 gamma. The portable ground survey instruments are direct reading and measure the total magnetic field. The optically pumped magnetometers include the rubidium and cesium vapor magnetometers that measure the total earth's magnetic field with an accuracy to about 0.001 gamma. They have been developed for use in airborne magnetic gradiometers that measure and record gradients in the earth's magnetic field. This is a new airborne geophysical survey method expected to result in new applications in mineral exploration.

The magnetic properties of rocks, details of the several types of magnetic instruments, survey techniques, magnetic data processing, and the interpretation of magnetic observations are described in detail in a number of books and references, such as Hansen, et al., Heiland, Nettleton, Parasnis, Vacquier et al., and others. This discussion assumes a basic knowledge of the magnetic method and will focus on the applications to iron ore exploration, evaluation, and development.

Regional airborne surveys made as a part of an exploration for iron ore are usually run with a terrain clearance of about 50 to 150 m (165 to 500 ft) with a line spacing of from 50 to about 500 m (165 ft to about 0.3 mile). This type of coverage results in a good precision of magnetic observation along the flight line and a generally acceptable distance for lateral projection correlation between flight lines. The flight elevation is about minimum for safety and line location control in a fixed wing aircraft. A closer spacing rapidly increases costs without a significant increase in the accuracy needed for a detailed reconnaissance, regional survey. A wider spacing of flight lines is acceptable for some exploration situations but can materially decrease the confidence in lateral projection correlations and may miss magnetic anomaly areas associated with some types of valuable iron ore bodies. Competent geological knowledge is a prime requirement for the interpretation of airborne survey results. The survey data are plotted on isomagnetic contour maps prepared by the contracting survey company. The geophysicist and geologist making the interpretation may wish to examine the flight line data and check the contour control information to see if there may be other ways to connect the observed anomaly areas.

Ground magnetic surveys are run along profile lines, generally perpendicular to the strike of the anomaly area and at a line spacing determined to give adequate information for a study of the anomaly area. For best results, profile lines should be surveyed to give good location control. The magnetic readings are plotted as profile sections and contoured to show the magnetic variations on a map. The magnetic data can be interpreted by standard methods to determine the size, shape, dip, strike, depth below the surface, and probable volume of the anomalous body. Often the probable rock or iron ore type causing an anomaly is known, which reduces the range of likely geological situations that can result in the observed magnetic anomaly. Even when the likely nature of the anomalous body is known, there may be a sufficient number of unknown factors to prevent a unique identification of the cause of the anomaly. Core drilling is usually needed to show the nature of the magnetic material and to permit the interpretation of the magnetic data. Even after drilling, bodies that have a strong remanent magnetization may make the geometry of the anomalous body difficult to interpret.

After an iron ore deposit is located, and sometimes after an iron mine is in operation, detailed magnetic surveys are run to give more accurate information regarding the limits of ore occurrence. In a number of instances the delayed application of magnetic surveys is the result of a need for a better understanding of the contact zone of deposits that have gradational ore boundaries to better define the current economic ore limit. As an example, magnetic surveys were run in the 1950s and 1960s on a number of ore bodies on the Mesabi Range, Minnesota, to more accurately locate the margins of natural iron ore deposits. These surveys permitted a more accurate location of development drill holes needed to show the mine pit wall.

Magnetic measurements in drill holes may be made to define the shape of magnetic bodies cut by the drilling or to test the possibility that a drill hole may have narrowly missed an adjacent magnetic body. Magnetic observations also may be used to determine the content of the magnetic mineral present. A magnetic susceptibility probe is being used to determine the magnetite content of magnetite taconite intersected by blast holes at the Minntac mine. The results allow a close definition of the contact between waste rock and ore.

Gravity Methods

Gravity methods have a rather restricted application in the search for and evaluation of iron ore deposits. This is in part the result of the highly successful application of magnetic methods, the rather infrequent circumstances where gravity should be used, and in part due to the cost and problems related to gravity methods. Even though little used relative to magnetic methods, gravity surveys can give useful results when focused on specific applications. The successful use of gravity surveys requires a definite, measurable, density contrast between the iron ore body and country rock. It may give highly useful results on ore bodies with a strong remanent magnetization or on deposits that are essentially nonmagnetic. Since gravity measurements are directly related to rock density, gravity surveys can give, under favorable conditions, an approximate indication of the tonnage present in an iron ore body. Regional gravity mapping can also be a valuable addition to regional geological mapping, particularly when used in conjunction with an airborne magnetic survey. Both regional and specific gravity surveys were used successfully during exploration of the North Cuyuna Range, Minnesota, where only vague interpretations of the possible locations of the Trommald iron-formation could be made from magnetic surveys. At Iron Mountain, Missouri, Leney (1966) reports the use of a gravity survey that indicated an extension of iron ore to the south of the known ore body. Gravity surveys were also successfully used during exploration of the Steep Rock iron ore bodies, Ontario, Canada.

Other Geophysical Methods

A number of attempts to use electrical methods in iron ore exploration are reported by Leney (1966). The results from resistivity, self-potential, and other potential and electromagnetic surveys do not suggest an expansion of the use of these methods in the exploration and evaluation of iron ore deposits. They are seldom used although there may be special problems where one of the electrical methods may give helpful results. Resistivity was used with limited success in the exploration for limonitic iron ores that occur as pockets in limestones in the Spring Valley district, southern Minnesota. However, even here the exploration company concluded that geological studies accompanied by auger drilling were much cheaper, and gave equally satisfactory results in ore discovery.

Theoretically, electrical methods could be used as a mapping tool if a strongly conductive body, such as a graphitic slate, has a consistent relationship to a nonmagnetic iron-bearing formation. Also, if factors that affect iron ore minability, such as faults or leached zones, display unique electrical properties, electrical methods may be used to delimit these structures. Even in these situations, magnetic methods should be shown to be ineffective before electrical methods are applied.

The recent development of high resolution, shallow seismic survey equipment has added an important capacity for iron ore exploration. Due to velocity contrasts between most iron-bearing rocks and overburden materials it is possible to construct maps that show the depth of overburden. Some control drilling is required to insure that the correct velocity interface is selected so the ore-overburden contact is mapped. It seems possible that additional uses may be developed for high resolution seismic methods.

EXPLORATION DRILLING

Drilling is a critically important activity in the exploration and evaluation of an iron ore deposit. Drilling usually follows detailed surface geological and geophysical studies that have shown the possible occurrence of a valuable deposit. A drill program to investigate an ore target has three basic purposes and steps: (1) to show the occurrence of a potentially valuable deposit and give basic information on broad geological relationships and nature of the iron ore; (2) to determine the ore dimensions, chemical and physical nature of the ore, its concentratability, if the ore requires processing to yield a salable product, and to acquire other positive information that will permit a generalized evaluation of the tonnage potential and a preliminary estimate of its commercial feasibility, (3) to obtain sufficient detailed information, by a continuation of drilling and sampling, concerning the distribution of ore types and grades so a fully reliable feasibility study and economic evaluation can be made.

Planning a drill program requires judgments on the appropriate drill pattern, hole spacing, type of drilling and equipment, sampling procedure, elements to be determined by assay, and the research testing needs. A principal concern should be the accurate and efficient determination of the nature of the ore and of the ore body on a cost-effective basis.

The locations of the first few drill holes in a target area are critically important as a common perception within an exploration company is that these drill holes are situated in areas of maximum opportunity for ore discovery. Thus the success or failure of the first few drill holes may determine whether the project is continued or abandoned unless a very well-documented geological evaluation of all factors is presented. At this stage there should be a careful and reasonably complete geological appraisal of the ore potential before a decision is made regarding a partly drilled prospect. This evaluation should include a statement of the likely economic feasibility of the project. It is a waste of time and money to extend work on a deposit that is without commercial merit or to drop a property with a reasonable possibility of commercial value. The need for an early geological view of exploration potential is as valid for iron ore deposits as for other ores.

The importance of a well-considered start should be recognized in preplanning. The preplanning aspect includes the selection of drilling and sampling methods, equipment and drilling pattern, whether vertical or angle holes will be drilled, and if core drilling is done, the size of the core to be taken. The primary concern should be to insure that the drilling products give the proper information. Many geologists believe it is preferable to err on the side of too much detailed information rather than on the side of too little and possibly questionable results. Other geologists are very cost-conscious and emphasize the need to keep the costs low and to show rapid progress in the ore body study and target evaluation. A cost-effective approach to exploration is always needed but this must be coupled with reliable information. Lake Superior type iron ores are often soft and friable with mixed hard and soft layers which makes sampling and data gathering difficult. Thus, preplanning becomes very important so that accurate data may be obtained. In compact, rather uniform solid ores such as contact metasomatic ores

and magnetite taconite ores, a major problem may be the selection of the most cost-effective approach. Since some iron ores are very hard, the critical decision may be how to best obtain accurate core samples using diamond drills that give adequate amounts of material for analysis and testing.

Three general methods are used in iron ore exploration drilling and sampling: core, rotary, and percussion. For each method there is a considerable range in equipment and in the way it is applied. All three systems may be considered in shallow drilling, generally less than 90 m (300 ft) in depth. As a generalization, rotary and percussion drilling methods experience sampling problems in iron ores at relatively shallow depths because of the relatively low density of the gangue component and high density of the iron oxide minerals. The difference in density results in an easy separation and segregation of one or the other component during sample travel and recovery. There may also be a marked difference in the particle size between the gangue and ore minerals produced during rotary or percussion drilling. The drilling may yield a coarser and a finer component in either iron mineral or gangue so a sampling error may result as correct amounts of fine and coarser material may not be recovered. This can result in a considerable change in iron content in the samples obtained through the loss of either coarse or fine material. A loss of fine gangue material, often quartz, is common, which may result in a considerable apparent increase in iron content of the ore. Until modern, high core recovery diamond drill methods for sampling iron ores were developed in the 1950s and 1960s, elaborate and questionable formulas were used to combine the core and sludge samples received to give the ore assay. In past years, sampling of the Lake Superior type ore for open pit mining on the Mesabi Range included percussion drilling using a variety of drill machines such as churn drills and the use of a dry, punch coring method accomplished by driving a nonrotating, hollow bit into the ore to retrieve an unwashed, unsegregated sample. The current method used in exploration or development sampling of natural iron ores or concentrating grade natural iron ore on the Mesabi Range is by diamond drilling using N size, swivel type core barrels and bentonite mud as the drill fluid with the work done by skilled drill crews. This drill system commonly yields a sample with the needed amount and size of core for analysis and testing that has an acceptable accuracy. Even using this system there are interlayered hard and soft ores that may cause severe sampling problems.

Exploration drilling and sampling of compact, massive, nonfriable iron ores is usually done by core drilling. Diamond core drilling, in what may be termed solid ores, is cost effective. In most cases the work is accomplished with a very high core recovery, often between 95 and 100%. The wireline method is extensively used as it can give adequate sample material with excellent sample recovery at a reasonable cost. Surface set diamond bits and modern swivel type core barrels used with standard drill equipment and drilling methods commonly give satisfactory core samples, although the use of bentonite mud as the drilling fluid should be considered as it may improve drilling and sample recovery particularly in fractured zones. Detailed information on core drill equipment including wireline, core barrels, types of bits and advice on their use can be obtained from the drill equipment manufacturers and contract drilling companies.

Some iron mining companies, particularly those working in areas with substantial glacial cover, use a rotary drill to penetrate the glacial drift. This equipment must be of sufficient size and power to effectively drill in the glacial materials. The rotary drill is used to reach the bedrock surface where casing is set. Bedrock drilling is then done using diamond drill equipment. In bouldery surface materials it is sometimes best to use down-the-hole percussion equipment instead of a rotary drill. Overburden that contains large, hard boulders may be drilled with larger-sized diamond drill holes with casing set in bedrock. The use of bentonite in drilling has largely overcome the difficulties of drilling and setting casing in surface materials.

Drill Hole Surveys

Surveying drill holes should be considered an essential part of an exploration drilling program whenever accurate knowledge of the position of the hole from top to bottom is important to the evaluation of the deposit and to its eventual development. Deviation of drill holes from the planned direction and angle is common in holes that are over a few hundred feet in depth and should be expected. The severity of the vertical and horizontal changes ranges considerably. Deviations of from 5° to over 20° in inclination and up to 20° in horizontal direction have been observed in drill holes less than 245 m (800 ft) deep that were drilled in steeply dipping cherty iron-formation. Thus the change in bottom hole location can be substantial. Deviations in holes drilled in dipping, bedded, cherty iron-formation with alternating harder and softer layers can be remarkably rapid.

In iron ore exploration sufficient magnetite is commonly present to make magnetic borehole survey methods unreliable, so the use of a gyroscopic survey system is recommended. The first use of a gyroscopic system in a small diameter diamond drill hole was done at the Atlantic City taconite mine near Lander, WY, in 1966. Drilling was performed to obtain detailed information to permit the development of final open pit mine plans. This required that stripping be carried to the ultimate pit limit during mining. A substantial deviation in the inclination of drill holes was observed so horizontal deviation was suspected. Sperry Sun was encouraged to complete construction of a partly designed, small hole gyroscopic survey system that could enter a BX diameter hole. The resulting survey showed a substantial horizontal deviation in the exploration drill holes, requiring a relocation of the stripping limits. Mining has since shown that the gyroscopic survey results were accurate.

Gyroscopic drill hole surveys can be made on a contractual basis with well service companies. These surveys produce records that show the hole inclination and direction with a relatively small possible error. Well service company personnel are also available to assist with cost estimates and to assist in planning a drill hole survey program.

SAMPLING AND GRADES

Sampling is an important and difficult part of the exploration and evaluation of an iron ore deposit. Iron ore bodies of potential economic value commonly contain from a few million to several hundred million tons of ore material. The gathering of small amounts of material from properly distributed points in an ore body so the composite results of their analysis and research testing accurately represent, in a consistent way, the whole body and its variability can be a formidable undertaking. The accuracy and thoroughness of the sampling methods should be under constant review, with deliberate checks and evaluations to prevent gathering inaccurate samples that can lead to erroneous conclusions.

Iron ores can be conveniently divided into two major types: (1) natural iron ores, which include both merchantable ores and crude ore materials that can be beneficiated by simple methods, and (2) concentrating grade ores that require fine grinding before processing to yield an ore grade product. Each general type has its particular sampling methods and problems.

The sampling of a natural iron ore deposit must determine the distribution of various ore grade materials within the ore body in regard to physical and chemical characteristics and, where needed, the suitability and yield amounts of shipping product obtained from the nonmerchantable materials that require beneficiation. The usual and customary chemical requirements determined by sampling include analysis for Fe, SiO_2, Al_2O_3, P, Mn, CaO, MgO, S, loss on ignition, and moisture. In ores where other elements are commonly present, analysis for special elements may be required. As an example, an analysis for TiO_2 is required for titaniferous magnetite ores. Ores with a high iron content, low scavenger, slag-forming constituents (SiO_2, Al_2O_3, CaO and MgO) and very low amounts of deleterious elements are preferred. While scavenger, slag-forming constituents are needed in the blast furnace, it is preferable that these be added at the furnace in controlled additional amounts as required. The slag volume desired varies with blast furnace practice and the nature and amounts of elements such as phosphorus and sulfur that need to be removed in the slag. Most ores do not contain the desired ratio of material used for slag formation so low amounts of these impurities permit easy preparation of the furnace burden.

A considerable list of elements is considered deleterious. The more common, unwanted elements in iron ores include phosphorus, sulfur, and titantium, with a less common to uncommon occurrence of vanadium, copper, zinc, chromium, nickel, arsenic, lead, and tin. Phosphorus is common to most iron ore in amounts that may range from very low (about 0.005%) to very high (over 1% P). Because of the importance of phosphorus in steelmaking, iron ores have long been classified on their phosphorus content, as Bessemer— not over 0.045%; low-phos non-Bessemer between 0.045 and 0.180% P; and high-phos non-Bessemer over 0.180% P. There is a price differential that recognizes the phosphorus grades. With a decline in the production of merchantable ores and the major tonnage production of beneficiated ore and pellets from concentrating grade ores having a low phosphorus content, there has been a lessening of concern with the phosphorus content of ores in steelmaking practice in the United States. There are still major concerns with the phosphorus content in natural ores moving in international trade, as there seems to be a very limited market for high grade iron ore that contains more than 0.07% phosphorus. In the Lake Superior region, iron ore that contains over 2% manganese is classed as manganiferous ore with a premium paid for ore containing in excess of 5% Mn. There often is some small price recognition for ore with a manganese content below 5%. The iron ore industry also, in past years, has shipped a small tonnage of ore classed as siliceous ore. This ore should be considered a special situation ore and requires a special sales arrangement in order to be recognized as ore in the sampling and evaluation of an iron ore deposit. This fact is apparent from the analysis of the 1971 Tilden silica grade ore, which was 38.96% Fe and 40.41% SiO_2.

The physical characteristics of natural ore that influence the evaluation of these ores primarily concern the relative quantities of coarse, plus about 6.35 mm (0.25 in.), and fines. This distinction is important as it determines the amount of the shipping product that can be charged directly into a blast furnace after the ore has been crushed and sized in comparison with the ore that will require agglomeration by sintering or pelletizing before charging. The determination of the physical parameters is often poorly shown by samples obtained by drilling, so trenches, test pits, adits, or shafts are dug to obtain an adequate sample for crushing and screening tests.

The primary sampling of a natural ore deposit during exploration is commonly done by drilling. Drilling iron ore deposits that can be mined by open pit is done either in a grid pattern or along section lines, so cross sections can be prepared to show the distribution of grades of ore. Samples are commonly taken on 1.5 or 3 m (5 or 10 ft) intervals with a recognition in sampling of visible changes in the physical appearance of the ore and of geological boundaries. If core drilling is done, the core and/or sample is split, retaining one half for future study and reference. Samples, properly identified as to drill hole and footage, are sent for chemical analysis. Materials that are not of ore grade are sent for laboratory tests to determine suitability for beneficiation and the analysis of concentrated product. The laboratory personnel commonly determine the proper concentration tests to be used. The analyses and/or test results when received are plotted on maps and sections, and are identified as to ore grades and types for use in ore tonnage determinations and for mine planning. At many mines the drill and analysis results are entered into a computerized data system.

Sampling of concentrating grade, *hard rock* type ores, such as magnetite taconite, jaspilite, OXIBIF, and contact replacement ore, is commonly done by core drilling using diamond bits. In deposits such as Mesabi Range magnetite taconite having large lateral extent, the drilling is often done as a phased program with a relatively wide spacing of drill holes in the initial phase to quickly determine the ore occurrence and the size of the deposit with progressively closer drill hole and section spacing as the exploration drilling passes into development drilling. Often there is no marked demarcation between exploration work and development work in the sampling of an ore body, although there may be a modification of drill procedure. Splitting of the drill core is strongly recommended during the preliminary exploration stage, with one half of the core retained for later study during the exploration, development, and mining stages. After a competent general knowledge of the ore body has been obtained, and when the purpose of continued drilling is to give adequate knowledge of ore grade distribution for detailed mine planning and ore grading for mine production, the core size may be reduced with all ore grade core sent for analysis and testing. Drill core recovered should be logged by a competent geologist, geological boundaries marked, and sample units determined. The core footage included in a single sample for analysis or for laboratory tests should be varied as needed to show variations in the ore occurrence so maximum use can be made of the sample data in mine and production planning.

Bulk Sampling

Bulk sampling is an essential part of exploration and evaluation of iron ores. Large samples are required to check the accuracy of the drill samples and to obtain the needed information on the crushing and grinding characteristics of the ore as well as for use in detailed and large-scale metallurgical tests. In natural ores, bulk samples give information regarding the nature and quantities of various sized ore products that can be shipped. Herkenoff (1968) in a discussion of the need for large samples of natural iron ores points out the importance of information on how the ore fractures in blasting and breaks into fine sizes during mining, crushing, and in all subsequent handling steps, such as trucking, stock-

piling, ship loading, and ship unloading. He states, "On the Mesabi Range where jig and heavy-media ores occur, it is always critically important to determine what percent weights would be in the 64 mm (2.5 in.) plus 6.35 mm (0.25 in.), minus 6.35 mm (0.25 in.) plus 0.21 mm (65 mesh), and minus 0.21 mm (65 mesh) fractions because these splits directly affect capacity required for the respective heavy-media separation cyclone (or jig) and spiral plants which are to work in concert when processing the run-of-mine crude. Because HMS is more efficient than other processes, it is advantageous to keep as much material as possible in that size range. So the problem in sampling ores is to obtain material at least 102 mm (4 in.) in top size and to try to duplicate a typical plant feed structure."

"On the other hand, if the ore is found to require complete crushing and grinding to, say, minus 0.21 mm (65 mesh) for beneficiation by flotation, spirals, etc., there is not much use in striving to obtain material coarser than diamond drill core size . . ." As Herkenhoff indicates, even in iron ore districts as well known as the Mesabi Range, bulk samples of some ores may be required but for other ore bodies high quality drill samples may be sufficient to determine the beneficiation characteristics. In areas where the accumulated knowledge regarding the expected natural ore characteristics is not well known, bulk sampling may be required for all ore deposits.

Bulk sampling is recommended for concentrating grade iron ore deposits that require fine grinding to attain iron mineral liberation before concentration. Accurate data are needed concerning the various types and/or grades of ore present and all aspects of the mining and processing of the ore materials to the final agglomerated product. These data should include information on drilling, blasting, fragmentation, crushing, grinding, concentration, and agglomeration for use in mine pit design, mine production schedules and to permit the development of a flow pattern for the correct correlation of product mix and feed rates to each successive process unit. Needed information includes ore type classification, iron content, mineralogy, liberation characteristics, feed size distribution, grindability indices for each ore type, and the optimum feed rate relationships for each step of processing to final product. It is critically important that specific information be available concerning the various types of ore that must be blended to obtain the normal mill feed grade. This commonly requires the selection of bulk samples from each ore type so a blended mill feed can be prepared and tested in the laboratory and pilot plant. Much valuable data will be available as a result of analysis and bench scale testing of drill core samples, but reliable information for mine and plant design can only be adequately known from the mining and pilot scale processing of the various ore materials. As an example, before the Atlantic City taconite mine and mill were built, two 1360-t (1500-st) bulk samples were obtained from a 222-m (728-ft) long adit. The two samples represented major types of taconite ore that were recognized during test drilling. The bulk samples were shipped from Wyoming to the Pilotac plant located at Mountain Iron, MN, for plant tests. The bulk samples confirmed the drill results and laboratory tests and furnished needed information for final design plans (Cohlmeyer, et al., 1962). Whether bulk samples are sent to an existing plant for testing or a pilot plant is built at the ore deposit depends on the availability of a plant where the bulk sample can be sent and other factors. When the iron ore body at the Jeannine Lake deposit, Quebec, was tested, a pilot plant was erected on site.

Bulk samples can be obtained in a number of ways, such as from the surface area by test pits, trenches, or small open cut mine pits or from underground openings such as adits, shafts, tunnels, and drifts. In a number of iron ore deposits there may be a marked zone of oxidation and leaching at or near the surface, so surface samples may not be representative of the main body of ore. The validity of the bulk sample in regard to the purpose for which it is taken and the expected ore type is of paramount importance. The sample must be representative of the desired material. Because of the importance of the large-scale testing to several aspects of mine and plant design, there is small latitude for compromise or error. There often is a temptation to use easily available surface material even though it is somewhat altered. This approach should be firmly rejected. The ability to examine and to test bulk sample material from each ore type present in an iron deposit may be essential to the development of a final mine plan, plant flowsheet, and mill design. The several ore type bulk samples can be blended to a number of simulated mill feeds for process and flowsheet development planning.

EVALUATION OF IRON ORE DEPOSITS

Iron ore deposits are evaluated to appraise their worth for possible sale, or, most often, to determine the potential for a profitable mine operation. The results of an evaluation are commonly reported in monetary terms of the present worth of the property or as the cost and profitability per ton of shipped product. Evaluation methods used for an iron ore body are generally similar to the approach used for base or precious metal or industrial mineral deposits with the special recognition of factors characteristic of iron ore deposits and the iron ore industry. A characteristic of the iron ore industry having a major impact is the shipping of a product that is an unrefined, relatively low value, bulk material that moves in large amounts to rather restricted and specific market sites.

The evaluation of a property made to determine its value or profitability is often a team effort involving geologists, engineers, process metallurgists, and economists. In some cases, the evaluation should include the likely users of iron ore at the steel plant, as the profitability determination of an iron ore may reach to the eventual cost of hot iron metal made from the ore. Each team member brings special talents to the study. The geologist has the best knowledge of the physical occurrence and characteristics of the ore deposit; the engineers, of mine and production planning; process metallurgists, of plant requirements and design; iron makers, of feasibility aspects of metal production. The entire group working with economists links all elements into costs and potential profit. Fortunately or otherwise, the value and potential profitability fluctuates with time. An evaluation does not have a fixed validity even though it may be accurate when made. Also, each deposit is unique, so the evaluation of one ore body cannot be safely extrapolated to apply to its neighbor. This was well stated by Pardee (1957, p. 454): "All evidence available is considered for each mine and in this way each mine is treated as an individual problem." This individuality does not exclude the use of accumulated wisdom or the use of comparable information in preparing the evaluation, but requires specific treatment of each factor with regard to the deposit being evaluated.

An important question to consider during the exploration and development of an iron ore deposit is: when should an evaluation be made? There are conflicting ideas regarding this. The difference in opinion is, in part, a result of the perceived viability of a project. Some ore bodies are very obviously viable so the question to be answered is not whether

to continue the project but how profitable will it become. In other cases there may be serious questions regarding viability. Often a series of evaluations may be advisable, which may include: (1) a limited pre-exploration evaluation, (2) a limited mid-exploration evaluation, and (3) a post-exploration and predevelopment, full-scale feasibility evaluation. The pre-exploration evaluation is usually a rather limited effort as geological evidence may be sketchy or essentially lacking. The main purpose of this evaluation is to assess the likely geological conditions and the nongeological factors using a hoped-for, assumed ore body so an eventual discovery should be viable if it meets the expected ore body characteristics. The second, mid-exploration evaluation, would be made after ore has been found, the general geological aspects of the body known, with some analysis and research test information available. This evaluation would permit site specific, somewhat generalized data to be used with a reasonable knowledge of the ore types and grades, processing required, and the likely product analysis and characteristics. The post-exploration and predevelopment feasibility evaluation would be made when adequate information is available for a complete evaluation of the profitability potential.

The evaluation of an iron ore deposit involves a considerable range of factors from the physical, chemical, and geological to the political, governmental, social, and environmental. Ohle (1972) lists 22 factors, of which 9 are termed geological and 13 nongeological (see Table 2).

The list of factors in Table 2 is very condensed, as there are often a number of cogent aspects to each that should be included in the study. The amount and precision of the information required to permit an adequate evaluation will vary with the stage in exploration or development at which the evaluation is being made. An evaluation report should indicate its purpose and its time relationship in the exploration program as there may be a considerable range in data availability and in its accuracy. The validity of the conclusions may also vary with the quality of the evaluation data. The report should include a clear statement of information limitations. The pre-exploration evaluation report is often made using optimistic projections as a pessimistic outlook would launch few exploration projects.

Since the market for iron ore is at an iron and steelmaking facility, the availability of a market may be a critical factor in the evaluation of an iron ore body. A deposit that is viable in one location may be of little or no value or only be potentially viable at another. Property or plant ownership and/or long-term contract relationships of a steel plant to an established source of iron ore may materially influence the value of a new potential source of iron ore. A market survey and long-range study of iron ore requirements at all potentially available market sites may be essential to many iron ore exploration and development projects.

The physical facts needed include an estimate of capital costs, developmental requirements, product quality, etc. Ohle (1972) presents the possible evaluation relationships in four case studies. The four cases used by Ohle include one high grade, merchantable ore body, two magnetite taconite deposits, and a deposit of oxidized taconite (roast ore), which in this section is classified as OXIBIF ore. The same basic questions regarding the physical characteristics of a possible mine development such as the ore characteristics and grade, type of mine development, ore processing required, grade of the shipping product, and the estimated capital cost need to be answered for each ore type. The wide range in the physical aspects of the four cases is well shown by the estimated range in capital cost, from 21×10^6 to 240×10^6.

Table 2. Important Factors in the Evaluation of Iron Ore Deposits from Ohle (1972).

A. Geological Factors
1. Type and Grade: Direct shipping, wash, heavy media, taconite, etc.; Rice ratio,* impurities, and associated elements.
2. Tonnage: Crude and product, effect on capital cost and recuperation schedules; weight recovery.
3. Grain-Size: Grind size, texture, liberation of ore minerals, elimination of impurities.
4. Grindability: kW · hr per ton to reduce the ore to concentrating and agglomerating sizes.
5. Mineralogy: Magnetite, hematite, goethite, silicate, or carbonate. Impurity mineralogy—effect on the ability to separate the impurities in processing.
6. Distribution of ore types: Grades, textures, mineralogies—can selective mining be done?
7. Depth and nature of overburden: Sand and gravel, or rock. Open pit vs. underground mining.
8. Shape and attitude of the ore body: Tons per vertical foot. Effect on stripping ratio.
9. Location: Topographic effects, climate.

B. Nongeologic Factors
1. Market and Price: Individual company variations in requirements and locations of furnaces.
2. Politics and general business climate.
3. Transportation cost.
4. Labor and housing: Availability and cost.
5. Construction cost.
6. Power: Availability and cost: fuel.
7. Water supply.
8. Taxes.
9. Royalty rate.
10. Inflation factors.
11. Tailings disposal.
12. Environmental factors: Ecology.
13. Financing.

* Rice ratio $= \dfrac{Fe}{SiO_2 + Al_2O_3}$

When the results of a study of the physical facts concerning a deposit are available, the evaluation can move to an economic analysis that will present the cost and profit aspects of the evaluation. Table 3 gives a listing of the economic information needed to show the total cost of the product, after-tax profit, cash flow, payback time, and percent return on investment.

The summary economic analysis from Table 3 can easily be modified to suit any ore body and possible market. A determination of the market value of iron ore may be a problem as the price of iron ore is only published on a regular basis for Lake Erie ports. The Lake Erie price is regularly quoted in the market section of *Engineering and Mining Journal* and in some other journals such as *Skillings' Mining Review*. Much iron ore is sold under long-term contracts at negotiated prices. The ore price may not be publicly available. Spot iron ore purchases may also be negotiated so it may be difficult to document a current iron ore value based on actual ore sales except as adjusted to the Lake Erie price.

The evaluation of an iron ore deposit requires a special knowledge of the iron ore industry that often extends far beyond the locality of the ore occurrence. Today iron ore is a major commodity in international trade. Super-large bulk cargo vessels have reduced ocean transport costs to a level that permits a competitive relationship for a number of iron ore sources to be global in extent. This world view of the iron ore trade complicates the evaluation of any deposit as the physical facts of ore occurrence become a smaller part of the cost/profit analysis. The extension of the potential profitability of an ore to include its use at the steel plant further extends an in-depth study from the ore in place through the complete process to iron as metal.

REFERENCES

Anon., 1955, *Survey of World Iron Ore Resources,* Department of Economics and Social Affairs, United Nations, New York, p. 345.

Anon., 1970, *Survey of World Iron Ore Resources,* Department of Economics and Social Affairs, United Nations, New York, p. 479.

Table 3. Economic Analysis Summary Tabulation for Iron Ores*

Cost Item	Cost Per Ton	
	Crude Ore	Shipping Product
Mining, total cost per ton	———	
Processing, total cost per ton	———	
Total cost of crude ore	———	
Average concentration ratio - ———		
Average cost of product per ton		———
Agglomeration cost		———
Total cost shipping product		———
Developmental cost: exploration, pre-production		———
Employee benefits and related costs		———
General expenses		———
Total direct production cost		———
Royalty		———
Transportation costs: rail freight		———
lake or ocean freight		———
terminal and other		———
Depreciation		———
Interest on capital investment		———
Taxes		———
Total cost delivered ore		———
Market value at Lake Erie price or other		———
Profit		———
Payback time (years)		———
Percent return on investment		———

* Adapted from Ohle (1972) and Marsden (1978).

Anon., 1973, "Genesis of Precambrian Iron and Manganese Deposits," *Proceedings of the Kiev Symposium, August 1970,* Unesco, Paris.

Anon., 1974, "North American Iron Ore," *Engineering and Mining Journal,* Nov., pp. 83-114, 131-162; Dec., pp. 55-90.

Baxter, C. H. and Park, R. D., 1957, *Examination and Valuation of Mineral Property,* Addison-Wesley Publishing Co.

Blondel, F., and Marvier, L., eds., 1952, "Symposium Sur Les Gisements De Fer Du Monde," International Geological Congress.

Bleifuss, R. L., 1966, "The Origin of the Iron Ores of Southeastern Minnesota," Ph.D. Thesis, University of Minnesota, p. 125.

Boyum, B. H., 1975, "The Marquette Mineral District, Michigan," Cleveland Cliffs Iron Co., p. 59.

Brandt, R. T., 1973, "The Origins of the Jaspilitic Iron Ores of Australia," *Proceedings of the Kiev Symposium, 1970,* Unesco, pp. 59-68.

Carr, M. S. and Dutton, C. E., 1959, "Iron Ore Resources of the United States Including Alaska and Puerto Rico," *Bulletin 1082C,* US Geological Survey.

Chase, F. M., 1962, "Review—New Developments in Exploration and Investigation of Iron Ore Properties," *Proceedings*, Mining Symposium and Annual Meeting of AIME Minnesota Section, University of Minnesota, Minneapolis, p. 99.

Cohlmeyer, S. H., Henderson, A. S., and Morgan, R. C., 1962, "The Story of Atlantic City," *Proceedings*, Mining Symposium and Annual Meeting of AIME Minnesota Section, University of Minnesota, Minneapolis, pp. 133-137.

Collins, W. H., Quirke, T. T., and Thomson, E., 1926, "Michipicoten Iron Ranges," *Memoir 147,* Geological Survey of Canada.

Cummins, A. B. and Given, I. A., eds., 1973, *SME Mining Engineering Handbook,* AIME, New York, Sec. 5, pp. 5-1 to 5-105.

Cannon, W. F., 1976, "Hard Ore of the Marquette Range, Michigan," *Economic Geology,* Vol. 72, pp. 1012-1028.

Emery, J. A., 1968, "Geology of the Pea Ridge Iron Ore Body," *Ore Deposits of the United States,* AIME, New York, pp. 359-369.

Goodwin, A. M., 1962, "Structure, Stratigraphy and Origin of Iron-formation, Michipicoten Area, Algoma District, Ontario, Canada," *Bulletin,* Geological Society of America, Vol. 73, pp. 561-586.

Herkenhoff, E. C., 1968, *Surface Mining,* E. P. Pfleider, ed., AIME, New York, p. 108.

James, H. L. and Sims, P. K., eds., 1973, "Precambrian Iron-formations of the World," Special Issue of *Economic Geology,* Vol. 68, No. 7, pp. 913-1179.

Leney, G. W., 1966, "Field Studies in Iron Ore Geophysics," *Mining Geophysics,* Vol. 1, Case Histories, Society of Exploration Geophysicists, pp. 391-417.

Marsden, R. W., 1978, "Iron Ore Reserves of the Mesabi Range, Minnesota," *Proceedings,* 51st Annual Meeting, Minnesota Section, AIME and the 39th Annual Mining Symposium, University of Minnesota, Minneapolis, pp. 23-1 to 23-57.

Marsden, R. W., 1978, "Iron Ore Reserves of Wisconsin," *Proceedings,* 51st Annual Meeting, Minnesota Section, AIME and the 39th Annual Mining Symposium, University of Minnesota, Minneapolis, pp. 24-1 to 24-28.

McGannon, H. E. ed., 1964, "Iron Ores, The Making, Shaping and Treating of Steel," United States Steel Corp., pp. 148-209.

McKinstry, H. E., 1948, *Mining Geology,* Prentice Hall Inc.

Moberg, N. A., 1962, "New Developments in the Exploration and Investigation of Iron Ore Properties," *Proceedings,* 23rd Annual Mining Symposium and the Annual Meeting of the Minnesota Section, AIME, University of Minnesota, Minneapolis, 91-114.

Murphy, J. E. and Ohle, E. L., 1968, "The Iron Mountain Mine, Iron Mountain, Missouri," *Ore Deposits of the United States,* AIME, New York, pp. 287-302.

Ohle, E. L., 1972, "Evaluation of Iron Ore Deposits," *Economic Geology,* Vol. 67, No. 7, pp. 953-964.

Pardee, F. G., 1957, "The Michigan Mine Appraisal System, Appendix A," *Examination and Valuation of Mineral Property,* Addison-Wesley Publishing Co.

Parak, Tibar, 1975, "The Origin of Kiruna Iron Ores," *Sueriges Geologiska Undersokning,* Serie CNR 709.

Pfleider, E. P., ed., 1968, *Surface Mining,* AIME, New York.

Trendall, A. F., and Blocking, J. G., 1970, "Iron Formations of the Precambrian Hamersley Group, Western Australia," *Bulletin 119,* Geological Survey Western Australia, p. 366.

Willemse, J., 1969, "The Vanadiferous Magnetic Iron Ore of the Bushweld Igneous Complex," Monograph 4, *Economic Geology,* pp. 187-208.

2.6 Coal

Willard E. Ward, II

DEFINITION

Coal is a physically and chemically complex substance that has been defined in different ways over the years. Currently, the most widely accepted definition is that adopted by the American Society for Testing & Materials (ASTM) which is as follows:

"Coal is a readily combustible rock containing more than 50 percent by weight and more than 70 percent by volume of carbonaceous material including inherent moisture, formed from compaction and induration of variously altered plant remains similar to those in peat. Differences in the kinds of plant materials (type), in degree of metamorphism (rank), and in the range of impurity (grade) are characteristic of coal and are used in classification (ASTM, 1970, p. 70)."

CLASSIFICATION

Because of the complexity of its physical and chemical properties and its varied uses, the classification of coal is not a simple task. Many classification schemes for coal have been proposed over the years using a variety of parameters as criteria. Of the various approaches to classification, rank is one of the more important. Rank is a measure of a coal's thermal maturity, that is, its position in the coalification series. Coalification refers to the progressive transformation of peat through lignite, subbituminous, bituminous, and anthracite. The standard rank system used in North America is the ASTM system (Table 1). It is based primarily on fixed carbon, volatile matter, and calorific value and utilizes familiar rank terms such as lignite, bituminous, and anthracite. In the ASTM system, these terms have specific meaning with regard to the aforementioned parameters, but may conflict with meanings given to the same terms in another country's classification system.

Coals are also classified by type into two broad categories: (1) sapropelic or nonbanded coal, and (2) humic or banded coal. Nonbanded coals exhibit little or no apparent stratification, are frequently granular in texture, tend toward homogeneity, and are allochthonous in origin. Examples of nonbanded coals are boghead coal, composed primarily of algal remains, and cannel coal, which consists largely of spores.

Banded coals, by contrast, are composed of a series of layers which are parallel to the bedding and which can be distinguished on the basis of macroscopic characteristics such as luster, hardness, etc. These bands are known as lithotypes and are composed, in turn, of macerals, which are the microscopically identifiable components of coal. Macerals are defined on the basis of color, morphology, association, and fluorescence. Maceral analysis plays an important role in the coal evaluation process and yields valuable information concerning the nature of the paleoenvironment in which the coal was formed, the degree of thermal maturity of the coal (rank), and its suitability for particular uses.

Of the two types, banded coals are by far the more abundant and constitute the majority of the world's coal resources.

ORIGIN OF COAL

Coal is formed by the accumulation and preservation of organic material (primarily from plants) in swamp, marsh, or bog environments. This plant material is altered into peat by complex biochemical processes that are still poorly understood. Peat accumulates very slowly relative to the human lifespan. Accumulation rates in Florida and the Mississippi Delta are from 0.5 to 1 mm/a, whereas in Borneo rates of up to 4 mm/a have been recorded. Generally, accumulation rates are higher in tropical climates than in temperate to cool climates although the higher growth rates are partially offset by slower decomposition rates in cooler climates. Peat can accumulate whenever accumulation rates are higher than the rate of decomposition. Most ancient coals probably originated in temperate to tropical climates (Bustin, et al., 1983).

As geological conditions change in peat-forming areas, the peat deposits may become buried by subsequent influxes of sediment. Sedimentation may continue for very long periods of time and, when coupled with the subsidence of the depositional basin, can result in the burial of the peat deposit under thousands of meters (feet) of sediment. Heat and pressure generated by the weight of the sedimentary column, in addition to biochemical and geochemical processes, cause the coal to increase in rank. The level of coalification attained is primarily a product of temperature and length of time of heating. Because in most stratigraphic sequences temperature increases uniformly with depth, the more deeply buried coals are generally of higher rank. Generally, as rank increases, porosity, volume, volatile constituents, and water decrease, while fixed carbon, density, heating value, and reflectance increase.

The main byproducts of coalification are methane, carbon dioxide, and water. Water is lost early in the coalification process and the ratio of methane to carbon dioxide increases with rank (Bustin, 1983). Large volumes of methane may be generated during coalification and can be produced and marketed either in conjunction with underground mining or as an independent venture. In fact, in the Black Warrior Basin of Alabama, 25% of natural gas production on an annual basis currently comes from degasification of deeply buried coal beds.

DEPOSITIONAL ENVIRONMENTS

The thickness, lateral distribution, composition, and quality of a coal bed are determined to a great extent by the depositional environment. Moreover, Horne, et al. (1978) found that the aforementioned characteristics were determined by the depositional environments that preceded, were coeval with, and that immediately followed deposition of the peat. The preceding environment shapes the topography on which the peat is deposited and therefore affects the thickness and lateral extent of the deposit. Contemporaneous environments affect seam continuity and composition whereas later environments may affect the peat by partial or complete removal of the deposit by erosion or, if brackish or marine waters are introduced, alteration of peat chemistry and therefore of coal quality.

Coal-forming environments can be divided into two broad categories: (1) paralic, which refers to coastal or near-coastal marine settings, and (2) limnic, which refers to coals formed inland, usually in intermontane regions and under freshwater conditions. Generally, limnic coals are characterized by thick

Table 1. ASTM Classification of Coals by Rank*†

Class	Group	Fixed Carbon Limits, % Dry, Mineral-Matter-Free Basis		Volatile Matter Limits, % Dry, Mineral-Matter-Free Basis		Calorific Value Limits, Btu per lb (Moist,‡ Mineral-Matter-Free Basis)		Agglomerating Character
		Equal or Greater Than	Less Than	Greater Than	Equal or Less Than	Equal or Greater Than	Less Than	
I. Anthracitic	1) Meta-anthracite	98	—	—	2	—	—	Nonagglomerating
	2) Anthracite	92	98	2	8	—	—	
	3) Semianthracite#	86	92	8	14	—	—	
II. Bituminous	1) Low volatile bituminous coal	78	86	14	22	—	—	Commonly agglomerating§
	2) Medium volatile bituminous coal	69	78	22	31	—	—	
	3) High volatile A bituminous coal	—	69	31	—	14,000¶	—	
	4) High volatile B bituminous coal	—	—	—	—	13,000¶	14,000	
	5) High volatile C bituminous coal	—	—	—	—	11,500	13,000	Agglomerating
		—	—	—	—	10,500	11,500	
III. Subbituminous	1) Subbituminous A coal	—	—	—	—	10,500	11,500	Nonagglomerating
	2) Subbituminous B coal	—	—	—	—	9,500	10,500	
	3) Subbituminous C coal	—	—	—	—	8,300	9,500	
IV. Lignitic	1) Lignite A	—	—	—	—	6,300	8,300	Nonagglomerating
	2) Lignite B	—	—	—	—	—	6,300	

Source: ASTM Annual Book of Standards, 1981.   Metric equivalent: lb × 0.453 592 4 = kg.

* This classification does not include a few coals, principally nonbanded varieties, which have unusual physical and chemical properties and which come within the limits of fixed carbon or calorific value of the high-volatile bituminous and subbituminous ranks. All of these coals either contain less than 48% dry, mineral-matter-free fixed carbon or have more than 15,500 moist, mineral-matter-free Btu's per lb.

‡ Moist refers to coal containing its natural inherent moisture but not including visible water on the surface of the coal.

If agglomerating, classify in low-volatile group of the bituminous class.

¶ Coals having 69% or more fixed carbon on the dry, mineral-matter-free basis shall be classified according to fixed carbon, regardless of calorific value.

§ It is recognized that there may be nonagglomerating varieties in these groups of the bituminous class, and that there are notable exceptions in high volatile C bituminous group.

beds of limited lateral extent. Although some of the coals in the western United States are limnic in origin, most North American coal deposits appear to have formed in paralic environments.

Paralic environments can occur in back barrier, deltaic, or coastal and interdeltaic settings (Bustin, 1983). Back barrier coals develop landward of barrier islands, frequently in abandoned lagoonal basins that are formed between the barrier islands and the mainland. Back barrier coals are typically rather thin, laterally discontinuous deposits that are elongate parallel with depositional strike and that are usually high in sulfur and ash.

Coastal plain coals develop on low, relatively flat, subsiding coasts that have a high water table and little influx of sediment. Some of the more persistent coals in the Appalachians of the eastern United States may have been deposited in coastal plain settings. Modern coastal plain swamps that are active sites of peat accumulation include the Everglades of Florida and the Okefenokee Swamp of Georgia (Bustin, 1983).

Many ancient coals are interpreted to have formed in deltaic systems and thus depositional environments associated with deltas have been the subject of intensive investigation.

The following comments on coal-forming environments in deltaic systems are drawn from Horne, et al. (1978). Depositional modeling can be used to predict large-scale trends in coal deposits on a regional scale and are therefore useful in the initial phases of coal exploration. Further, small-scale variations in coal thickness, quality, and lateral continuity frequently can be predicted, providing data that can be extremely valuable in mine planning and development.

The following illustration (Fig. 1) was derived from a detailed data base developed from the coal-bearing carboniferous-age rocks of eastern Kentucky and southwestern Virginia and from similar environments in contemporary coastal areas. Fig. 1 illustrates the typical shape and lateral extent of coal deposits which form in the different environments within the deltaic setting.

Coals that form in lower delta plain environments are typically elongate parallel with depositional dip because the only environments suitable for peat accumulation are adjacent to relatively narrow levees on either side of distributary channels. Interdistributary bays occur between the distributary channels and are sites of accumulation of fine-grained bay-fill detrital sediments. Sites of peat accumulation on the lower delta plain are generally restricted to the elongate, relatively narrow areas between the levees and the interdistributary bays. Lower delta plain coals are usually relatively thin and contain splits caused by crevasse splays that breach the poorly developed levees along the distributary channels.

Upper delta plain-fluvial coals also tend to be elongate in the direction of depositional dip although they are not as continuous in that direction as the lower delta plain coals. Deposits typically formed as pod-shaped bodies on flood plains adjacent to coexisting meandering channels and exhibit significant thickness variations over short distances. Also, as in the case with lower delta plain coals, numerous splits can occur near the levees bordering active channels because of splays. Post-deposition shifting of channels can also complicate the sedimentary sequence by eroding the coal deposit and creating "washouts."

In some locales, a transitional zone exists between the lower and upper delta plain environments that exhibits characteristics of both lower and upper delta plain sequences. In the transition zone between the lower and upper delta plains, many of the large interdistributary bays (flood basins) that occur between distributary channels have filled with sediment and provide broad basins in which large coal swamps can develop. These broad, relatively uninterrupted basins provide a favorable environment for the formation of coal deposits that are typically more laterally extensive than those of the

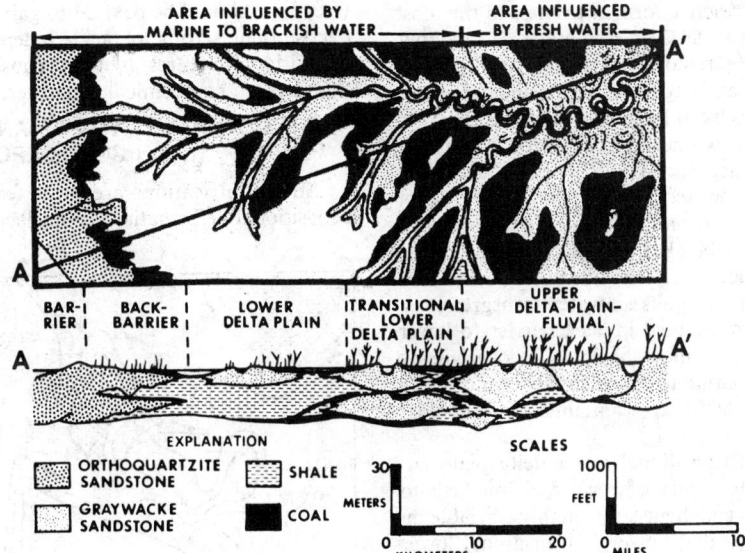

Fig. 1. Depositional model for peat-forming (coal) environments in coastal regions. Upper part of figure is plan view showing sites of peat formation in modern environments; lower part is cross section (AA') showing, in relative terms, thickness and extent of coal beds and their relations to sandstones and shales in different environments (Horne, et al., 1978, reprinted by permission of American Association of Petroleum Geologists).

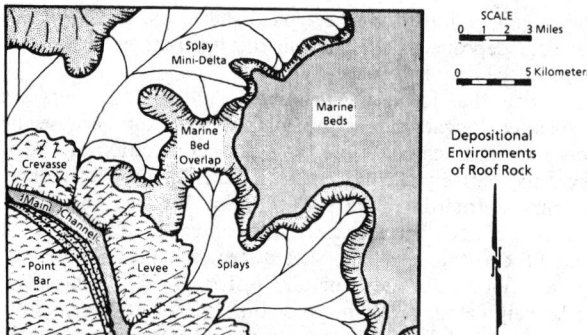

Fig. 2. Reconstruction of depositional setting immediately after formation of coal X. Diagram is based on data related to lithologic and sediment-thickness variations (Horne, et al., 1978, reprinted by permission of American Association of Petroleum Geologists).

lower and upper delta plain proper. Coals formed in this transitional zone share some characteristics with upper and lower delta plain coals such as splits that develop near levees and post-depositional washouts. Most of the more economically important coal beds in the Appalachian coal region are interpreted to have developed in this transitional zone between the lower and upper delta plains.

From the foregoing brief discussions, it is apparent that, in the initial phases of exploration, a knowledge of the depositional environments that control the shape and configuration of the coal body will enable explorationists to design a drilling program for maximum effectiveness and efficiency in defining the coal deposit. At the lease-tract or mine plan levels of exploration, more detailed drilling and evaluation may be desirable to predict areas of thick and/or high-quality coal.

Depositional environments also partially determine the sulfur content of coal deposits. Sulfur occurs in the form of iron sulfide (predominantly pyrite) in several ways in coal. A finely disseminated form sometimes referred to as framboidal pyrite is the most reactive form of pyrite and the most difficult to remove. It is so finely disseminated throughout the coal that it cannot be removed effectively in float-sink washability tests. Research suggests that framboidal pyrite originates from sulfur produced by microorganisms found in marine to brackish waters, but not in fresh water. It has been shown (Ferm, 1976; Caruccio, et al., 1977) that framboidal pyrite is most strongly associated with coals overlain by roof rocks deposited in marine to brackish-water environments. Exceptions occur when a blanket of sediment (such as a crevasse splay) is introduced early enough to shield the peat deposit from later marine to brackish-water transgressions. It follows that coals which formed in back barrier to lower delta plain environments are more likely to be overlain by sediments deposited by marine to brackish water and hence will be more likely to contain higher amounts of framboidal pyrite.

Coals that formed in transitional lower delta plain environments are subject to a mix of fresh and brackish to marine water influences and hence are highly variable in their sulfur content. Generally, however, transitional lower delta plain coals are considered to be lower in framboidal pyrite than coals deposited in lower delta plain and back barrier settings. This trend is thought to continue for coals formed higher in the delta plain in fluvial-upper delta plain settings where marine influence is uncommon. These coals are generally considered to be lower in finely disseminated

pyritic sulfur than coals formed in other delta plain depositional settings. An understanding of the depositional setting in which a coal bed formed can therefore be used to predict the amount and type of sulfur present and to guide the exploration for low-sulfur coals in areas where sulfur contents are usually high.

Investigations by Caruccio, et al. (1977) and Horne, et al. (1976), serve as examples to illustrate the potential usefulness to mine developers of understanding the depositional history of a coal bed. Using a data base of 450 core holes in a 518-km² (200-sq-mile area) located in the Appalachian coal region of the eastern United States, the investigators interpreted the target coal bed to have been deposited in a lower delta plain setting. Typically, coals interpreted as lower delta plain coals, where overlain by brackish to marine rocks, have sulfur contents of greater than 2% with 75% or more of the sulfur occurring in the form of framboidal pyrite (Caruccio, et al., 1977). Where deposits interpreted as freshwater splays were emplaced over the peat surface prior to the deposition of the marine rocks, the peat apparently was shielded from the sulfur-reducing bacteria, causing the sulfur content in the peat to remain low (Horne, et al., 1976).

Figs. 2 and 3 summarize the investigative results of Horne (1978). Fig. 2 is an interpretation of the depositional environments after deposition of a coal bed. The data suggest that the levees of a distributary channel in the southwestern part of the area were breached and splay deposits encroached to the north and east over the coal and into the marine-influenced interdistributary bay. Fig. 3 shows the distribution of disseminated sulfur in a target bed. A comparison of Figs. 2 and 3 illustrates the expected association between areas where the coal is overlain by marine beds (the eastern part of Fig. 2) and higher sulfur concentrations. In the western and southern parts of the diagrams where the wedge of nonmarine splay deposits covered the coal, sulfur contents are correspondingly lower.

The relationships shown in these diagrams between disseminated sulfur content and specific depositional environments suggest that exploration drilling programs at the lease-tract level should be devised to gain an understanding of the depositional setting of a coal deposit and to define such depositional features as might cause significant variation in the physical or chemical characteristics of the coal.

STRUCTURAL FEATURES AND THEIR EFFECTS ON COAL DEPOSITS

Structural features are those features provided by post-depositional deformation or displacement of the rocks. Such

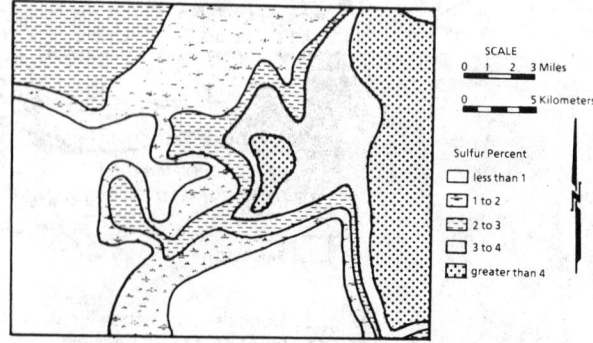

Fig. 3. Distribution of sulfur in coal X that cannot be removed in 1.50-density sink fraction of washability tests (Horne, et al., 1978, reprinted by permission of American Association of Petroleum Geologists).

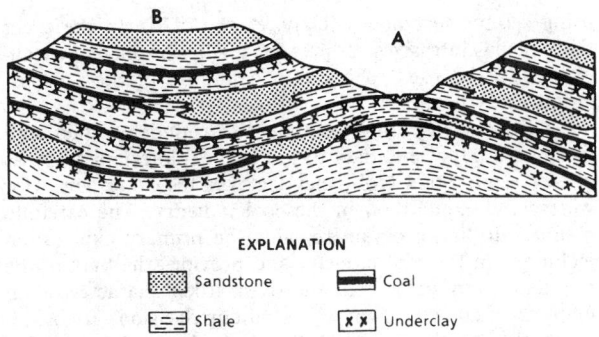

EXPLANATION

    Sandstone   Coal

    Shale   x x   Underclay

Fig. 4. Simplified cross section of coal-bearing sequence in the Appalachian coal region of the Eastern US where rocks have been gently folded into anticline (upfold) at A and syncline (downfold) at B (Simon and Hopkins, 1973).

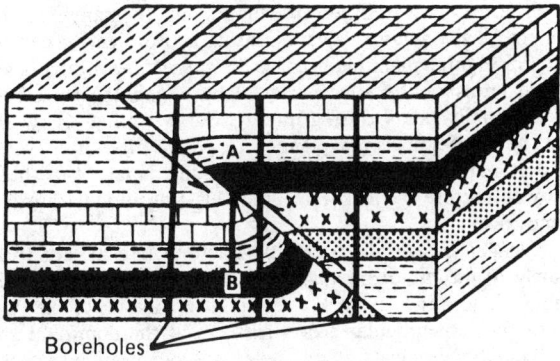

Fig. 6. Reverse fault showing displacement and direction of relative movement across fault plane. Hanging wall is upthrown relative to footwall. AB is vertical displacement (modified from Simon and Hopkins, 1973).

features can form concurrently with, or shortly after, deposition of the sediment, such as slumps or differential compaction of soft sediments having different densities. These soft-sediment structures can and do sometimes affect the continuity of coal deposits. More commonly, however, it is structural features such as folds and faults that formed later in the history of the rocks as a result of tectonic forces that determine the present attitudes of the coal beds.

Inclined or Folded Strata

Rock sequences deform plastically under conditions of high temperature and confining pressure and hence may be tilted or folded into a series of subparallel to parallel upwarps and downwarps termed anticlines and synclines, respectively (Fig. 4). Folding may be so intense as to lift the strata to the vertical or even to an overturned position. All uplifted strata are more susceptible to subareal erosion with areas of maximum uplift having the greatest degree of susceptibility. Therefore anticlinal crests are often severely denuded, creating a breached structure and interrupting the areal continuity of any coal beds. The tilting and folding of strata containing coal beds therefore complicates efforts toward correlation of beds from area to area and also imposes constraints on mining operations in areas of intense structural deformation. In most cases the overburden increases more rapidly away from the outcrop in downward pitching coal

beds than in flat-lying deposits, reducing the amount of coal that is economically recoverable.

Faults

A fault is a fracture or fracture zone along which displacement has occurred on one side of the fracture relative to the other. Faults are important considerations in coal exploration and mining and, depending on local conditions, can render an otherwise attractive area unsuitable for mining.

There are several types of faults defined by the direction of relative motion across the fault plane. The two types of faults most commonly encountered in coal exploration are normal faults and reverse faults. Normal faults occur where the block above the fault plane (termed the hanging wall) moves down relative to the lower block (the footwall) (Fig. 5). The effect of drilling through a normal fault is that of an apparent shortening of the rock section by the elimination of strata from the column of rock penetrated by the drill. This can be illustrated by visualizing vertical boreholes that penetrate the fault plane on the front panel of Fig. 5. The point where the wellbore enters the footwall is stratigraphically lower than the corresponding point on the hanging wall by an amount equal to the vertical displacement of the fault (AB).

In the case of a reverse fault, the hanging wall moves up relative to the footwall, and repeated sections of strata are encountered (Fig. 6). Vertical displacement is represented by line AB and, once again, each borehole allows a different interpretation of the nature and position of the target coal bed. Indeed, if the middle borehole was the only source of data, an observer might conclude that two coal beds were present if the intervening strata were not carefully evaluated. These examples emphasize the need for a carefully planned drilling program, especially in areas where existing data indicate the presence of faulting. Where faults are known to occur, the drilling program must be designed to yield sufficient data to allow adequate mapping of the type and extent of faulting present as well as the amounts of displacement so that the effects on the coal beds can be accurately determined.

Joints and Cleats

Joints are fractures in a rock mass across which no displacement has occurred. Joints are commonly planar, occur in groups of subparallel to parallel fractures called sets, and may extend, both vertically and laterally, for distances from

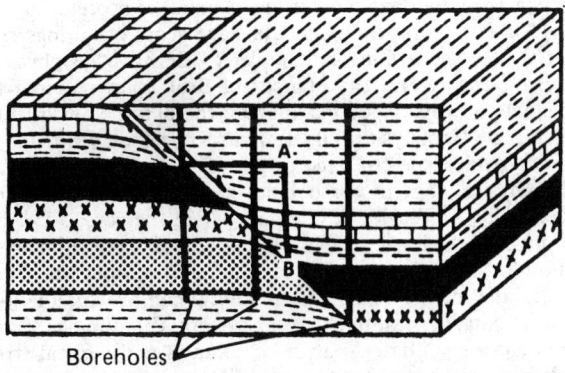

Fig. 5. Normal fault showing displacement and direction of relative movement across fault plane. Hanging wall is downthrown relative to footwall. AB is vertical displacement (modified from Simon and Hopkins, 1973).

as little as a few millimeters (inches) up to many tens of meters (feet) or more. Where jointing is prevalent, it can be a factor in mine planning because it represents existing planes of weakness in the overburden along which the rock will preferentially break during mining. Surface mine highwalls are therefore sometimes planned to parallel the orientation of dominant joint trends and hence take advantage of these natural fracture systems to facilitate blasting and overburden removal.

Cleats are naturally occurring fractures in coal beds (primarily in bituminous coals) that are morphologically analogous to jointing in rocks. Cleats typically occur in two mutually perpendicular sets. Fractures of the dominant set are called face cleats. Face cleats are penetrative, closely spaced fractures that serve as primary conduits for fluids such as methane gas, which is a byproduct of coalification, and ground water. Butt cleats form the complementary, less dominant cleat set and are typically irregular, nonpenetrative fractures that stop against a face cleat, occur over a broader range of orientations, and serve, to a lesser extent, as conduits for fluids. Because of their permeability, cleats in general, and especially face cleats, are often sites of mineralization and deposits of minerals such as pyrite, calcite, and others.

Cleat orientations can be important in mine planning for much the same reasons as joints, that is, they represent natural planes of directional weakness which can facilitate the cutting and loading of an exposed coal bed in a surface mine. Although probably of lesser importance generally than jointing in rocks, cleat orientations have determined, in certain cases, mine layout and the direction of mining.

Clastic and Igneous Intrusions

Perhaps of lesser importance in most locales than the previously discussed structural features is the intrusion of either clastic (sedimentary) material or igneous masses into a sequence of coal-bearing rocks. These intrusions may parallel bedding planes or cut across bedding. In the former case, the features are called sills, in the latter, dikes. These structures can range in thickness from a fraction of a millimeter (inch) up to many tens of meters (feet) and, in certain mining locales, can present significant problems. In the case of a clastic intrusion, the intruded material is waste material and must be separated and removed from the coal but does not alter the physical characteristics of the coal. In an igneous intrusion, the coal in the immediate vicinity of the intrusion is thermally altered. The alteration can result in an increase in rank or even the coking of immediately adjacent coal. An added problem with igneous intrusions is that the igneous rocks are much harder than coal and the associated sedimentary rocks, thereby increasing the difficulty of mining in these areas.

EVALUATION OF COAL DEPOSITS

Determination of the Amount of Coal in Place

Once the decision has been made to proceed with a detailed evaluation of the coal deposits on a particular tract with the purpose of opening a surface mine, a data base must be generated at a level of detail sufficient to characterize the coal and overburden. A number of geological or geophysical techniques can be used to provide data. In areas where a blanket of unconsolidated material was deposited on the erosional upper surface of the underlying bedrock, seismic refraction, seismic reflection, or, in some cases, gravity surveys can reveal the configuration of the bedrock surface. Also, faults with vertical displacements no smaller than 6.1 m (20 ft), or under ideal conditions 4.6 m (15 ft), can be identified using seismic techniques (Daly, et al., 1976). In the event that igneous intrusives are present, gravity or magnetic techniques can be used to assist in definition of the igneous-sedimentary boundary.

All of the foregoing techniques can, under certain conditions, supply useful data to the coal explorationist but, as general exploration tools, they lack the resolving power for widespread exploration in the coal industry. The carefully planned drilling program remains the primary exploration technique in the coal industry and provides the bulk of the raw data from which coal and overburden characterization maps are made and upon which mining decisions are based.

At the lease tract level, drilling is used primarily to define areas of thick coal and to determine coal quality. These data are then used to calculate measured reserves. Drill-hole density necessary to prove reserves varies with the complexity of the geology and the degree of consistency in coal bed thickness. In areas of structural complexity or where coal bed thickness is highly variable, drill-hole spacing may be as close as one hole every 1.6 ha (4 acres) (Reilly, 1968). Conversely, in geologically undisturbed areas where coal bed thickness is relatively constant, drill holes are sometimes spaced 0.4 km (0.25 mile) or more apart. Local variations in coal-quality parameters (such as sulfur content) constitute another reason to increase drilling density if those parameters are critical in determining the marketability of the coal. Accuracy of the reserve estimate should be within 20% and the drilling program should be geared to produce figures at this level of accuracy (Wier, 1976).

In planning a surface mine, coring of all the exploratory holes probably will not be required. A sufficient number of holes should be cored to allow the geologist to determine the depositional environment of the coal and thereby to make decisions for location of additional test holes and for mine planning. Data from the cored holes will also be useful in determining the type of blasthole equipment and bits as well as types of mining equipment that will be most appropriate. Otherwise, exploratory holes can be drilled by the less expensive air-rotary method. When a coal sample is needed in a particular area, a rotary hole is drilled to establish the elevation of the coal. A second hole is then put down immediately adjacent to the first, still using the rotary bit but stopping the hole just above the position of the coal bed as established by the previous hole. The core barrel and bit is then substituted for the rotary bit and the coal bed is cored.

Systematic sample or core descriptions and recording of coal thicknesses and depths are necessary to insure reliable integration of the data onto the various interpretive maps which are important tools in the evaluation process.

All exploratory holes should have geophysical logs run soon after the drilling is completed. Effective geophysical logging can reduce the number of drill holes required to evaluate a property by maximizing the data obtained from each hole. Geophysical logs serve as a check on written logs and provide a precise record of coal depth and thickness. Geophysical logs of cored holes also provide a means of identifying lithologies in intervals where the core is lost by comparing the logging tools' response to different lithologies in other cored intervals. The basic borehole geophysical suite (calibrated density, gamma, resistivity) can provide the following data: (1) coal thickness and depth; (2) lithologic data; (3) depositional data—nature of contacts and vertical stratigraphic sequences; (4) hydrologic data—aquifers, lost circulation zones, water levels; (5) identification of structural data and stratigraphic sequences; (6) recognition and correlation of specific coal intervals is augmented by their individual signatures; and (7) recognition of subtle mineralogic changes, such as alteration, that are difficult to discern from cuttings (modified from Crowder, 1986).

More sophisticated geophysical surveys can provide many more types of data such as the coal quality parameters of ash, carbon, volatile matter, heat content, moisture, mineral matter, and rank, at greater cost. Crowder (1986) estimates the cost of a basic geophysical logging suite at 10 to 20% of a total rotary drilling budget with more sophisticated logging suites increasing the cost to as much as 50%. A partial list of geophysical tools and their application to coal exploration is given in Table 2.

Once the data are assembled, the very basic task of correctly identifying and correlating the coal beds throughout the area of interest must be performed. Extra care must be taken in areas where multiple coal beds of varying thickness are present in close stratigraphic proximity to each other. Incorrect identification of the beds can result in a misleading evaluation of the property, which, in turn, can cause severe problems in the mining, preparation, or marketing aspects of the operation. Geologists commonly use physical characteristics of the overburden, physical and chemical characteristics of the coal bed, distinctive stratigraphic markers or sequences, signatures on geophysical logs, and any other pertinent data to assist in the correct identification and correlation of the coal beds. In more complex areas, additional data may have to be obtained in certain parts of the tract by the drilling of more closely spaced holes before correlations can be made with confidence.

When the coal bed stratigraphy has been worked out, it is useful to construct a series of maps using data obtained from field investigations, drilling, and laboratory work that will depict coal and overburden thickness, overburden to coal ratio, and significant analytical parameters. These maps may be prepared as a series of registered mylar overlays to allow simultaneous viewing of different combinations of data or they may be constructed separately. A structure contour map with the target coal bed as the datum horizon is also an extremely useful method to depict structural or stratigraphic features that affect the topography of the coal bed.

Table 2. Generalized Partial List of Some Drill-Hole Geophysical Measurements for Coal Exploration Applications

Log	Measurement	Applications	Limitations
Natural Gamma	Natural or gross gamma radiation	Lithology, and detection of uranium, potash, etc.	None
Gamma density Four-pi High resolution (HRD) Very High (VHRD) Dual-spaced	Scattered gamma intensity produced by a gamma source	Lithology, bulk density rock mechanics	Need caliper log for quantification. Four-pi is difficult to quantify. HRD-good densities for coal analysis. VHRD-good bed boundary definition-dual density overcomes caliper need somewhat
Resistance	Complex resistivity	Lithology, clay	Need open, fluid-filled holes. Focused gives best definition because of drill hole fluid resistivity
Resistivity-focused	Complex resistivity	Ash content	
Resistivity-normal	Complex resistivity		
I.P.	Complex resistivity	Sulfide, clay content	
Spontaneous potential, SP	Natural earth potentials	Lithology, mineral content	Open, fluid-filled holes, quantitative
Caliper	Drill hole diameter	Fracture, rock mechanics, change in diameter to correct other logs	Open hole
Temperature	Temperature	Fractures, borehole formation fluid flow	Open hole
Acoustic velocity	Acoustic pulse, travel time, and disbursement	Porosity, lithology, rock mechanics, fractures	Open, fluid-filled hole
Thermal neutron	Thermal neutron intensity produced by a neutron source	Porosity, lithology, content	Needs caliper corrections to be quantified
Dip log	Formation dip via resistivity measurements	Formation dip, fracture, structure dip, cross bedding	Open, fluid-filled hole
Flow meter	Fluid flow in drill hole	Fluid flow determinations	Open hole, lack of sensitivity to low flow rates
Spectral gamma	Natural gamma radiation at different energy levels	Radioactive coals, ash content, lithology	Very costly, slow
Deviation	Borehole inclination and bearing by magnetic or intertial systems	Borehole inclination and bearing-true thickness, sample location	Magnetic tools need open hole
Neutron activation	Neutrons resulting from fission of rock after it was activated by strong neutron source	Sulfur content direct mineral in-situ assay	Very costly, very slow, more development needed for coal applications
Others	More refined or compensated measurements for above logs	Better bed resolution and analysis	Development continuing

Source: Crowder, 1981.

Alternatively, the available data can be entered into a computer data base and appropriate software programs can be utilized to portray stratigraphic, mining, or economic conditions at desired scales.

Coal and overburden thickness maps (termed isopachs) are constructed by plotting the appropriate thickness values on a base map and constructing contour lines (isopachs) representing regularly increasing or decreasing thickness intervals using the plotted values as guides for positioning the contour lines. An example of a coal isopach map is given in Fig. 7. Similar isopach maps can be constructed showing overburden thickness. Isopleth maps can be constructed using the same principle as the isopach maps, but substituting various coal quality parameters for the thickness data (Fig. 8). This type of data presentation would have application where relatively small variations in coal quality would significantly impact the marketability of the coal.

Using the same coal and overburden thickness data, lines representing overburden to coal ratios or "mining ratios" (expressed as 5:1, 10:1, etc.) can be drawn. A ratio of 5:1 means that the line represents the limit at which 0.9 t (1 ton) of coal can be extracted by removing not more than 3.8 m³ (5 cu yd) of overburden. Strictly speaking, this is not a true ratio because it has dimensions of overburden volume per unit weight of extracted coal. The inclusion of these units gives rise to a difference between ratios computed in English units and ratios computed using metric units.

Ratios of thickness values alone cannot be used to deduce mine economics. These values must be converted to cubic meters (cubic yards) of overburden per tonne (ton) of coal to provide data to the analyst in units that can be more readily equated to mining costs. The conversion from feet (or meters) to cubic yards (or cubic meters) per ton (or metric ton) is outlined as follows from Wier (1976):

$$\text{Ratio} = \frac{OB(\text{ft})}{C(\text{ft}) \times SG \times 0.843}$$

or

$$\text{Ratio} = \frac{OB(\text{ms})}{C(\text{m}) \times SG}$$

where *OB* is the thickness of overburden, *C* is the thickness of the coal, and *SG* is the specific gravity of coal. Because

of the differences in the units of the English and metric systems, ratios calculated in the metric system are about 0.8 (0.842778) that of the English system. If specific gravity is not known, but the rank of the coal is known, the following specific gravity values are commonly used (Averitt, 1975):

Rank	Specific gravity
Anthracite	1.47
Bituminous	1.32
Subbituminous	1.30
Lignite	1.29

It should be noted, however, that it is important to use correct specific gravity values whenever possible because if the coal contains much mineral matter that has a higher specific gravity than coal, it will skew the results by decreasing the ratio and increasing calculated tonnages per unit area by a possibly significant amount.

Once the overburden lines have been established, the area can then be divided into mining units according to the mining plan and reserves can be calculated. The terms "measured," "indicated," and "inferred" are commonly used to define reserve categories with measured reserves having the highest level of reliability and inferred reserves the lowest. The distance between points of measurement distinguish the different reliability categories. Wood, et al. (1983) uses 0.8, 2.4, and 9.7 km (0.5, 1.5, and 6 miles) as the maximum distance between points of measurement for measured, indicated, and inferred coal, respectively. In mining, however, typical exploration drill-hole spacing is sufficiently close to classify all reserves as measured.

The first reserve determination is for total coal in place. The basic calculations in metric and English units, and using specific gravity for a given coal as determined in the laboratory, are as follows:

$$T = C\,(\text{m}) \times A\,(\text{m}^2) \times SG$$

where *SG* is specific gravity, *C* is thickness of coal in meters, and *A* is area in square meters.

In English units:

$$\text{Tons} = C\,(\text{ft}) \times A\,(\text{acres}) \times 1359.7 \times SG \quad (\text{Wier, 1976})$$

where *C*, *A*, and *SG* are the same terms as in the metric equation, and 1359.7 is a constant required to establish the correct tonnage factor to use with the English units.

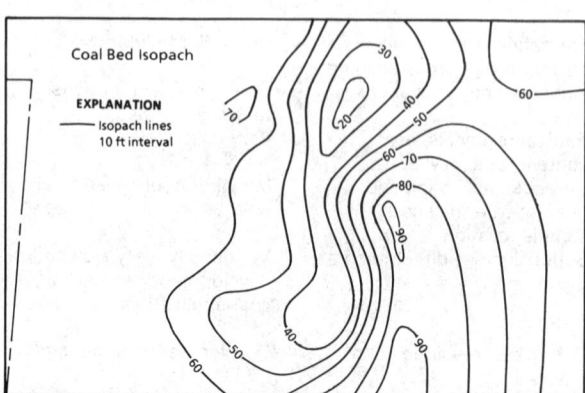

Fig. 7. Isopach map showing variation in thickness of a coal seam (modified from Wier, 1976).

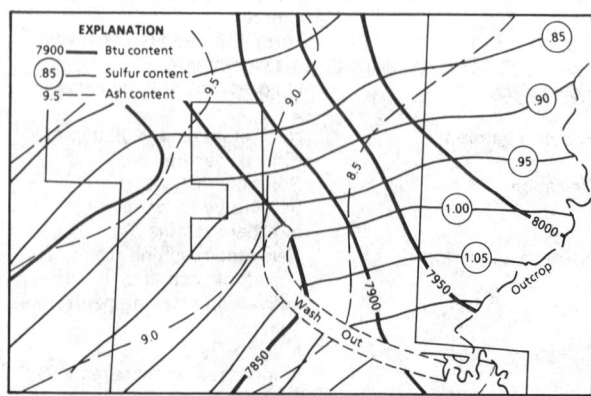

Fig. 8. Isopleth map showing variation in Btu, sulfur, and ash content (modified from Wier, 1976).

If average specific gravity values are used, the following values are frequently used as tonnage factors in a simplified form of the equation.

Rank	Tons of coal per acre per ft of thickness
Anthracite	2000
Bituminous	1800
Subbituminous	1770
Lignite	1750

In the case of bituminous coal the calculation would then be as follows:

$$\text{Tons} = C \text{ (ft)} \times A \text{ (acres)} \times 1800$$

Resource figures calculated from the preceding equations are estimates only and are not precise, but, with a thorough exploration drilling program, should be accurate to $\pm 10\%$ (Wier, 1976). This figure must then be adjusted downward to account for losses incurred during excavating, handling, processing, and transporting the coal. Anticipating the magnitude of these losses is usually accomplished by drawing on previous mining experience in the area, and in the case of processing or preparation loss, by reviewing washability studies. In some cases, cumulative loss can approach 50% of the total in situ reserve.

Coal Quality

The determination of coal quality is an integral part of the coal exploration process. It is as important as any of the other factors that are used to determine the mining potential of a given tract. Unlike many of the other factors, however, certain coal quality parameters can be changed to meet user specifications by coal cleaning technology and/or by blending with other coals. The following discussion touches on the more commonly used types of analytical data from the viewpoint of the information they convey to the explorationist about the suitability of his product for certain end uses.

Analytical procedures for testing coal have been continuously refined and updated for the past several decades by ASTM (American Society for Testing & Materials). Many major consumers and large coal-mining companies maintain well-equipped laboratories to analyze coal. The consumer does this to check the quality of the coal he purchases, whereas the producer must stay abreast of the quality of the product, especially as the mining operations advance into new territory. Analytical work is sometimes contracted out to independent laboratories either to run the primary analyses or to serve as quality control checks on the mining company's or consumer's results.

The analyses most frequently performed on coal include proximate analysis, calorific value, and sulfur. The proximate analysis consists of the determination of moisture, volatile matter, fixed carbon, and ash. Other types of analyses frequently performed on coal include ultimate analyses, the determination of free swelling index, and determination of trace element content.

Moisture: Moisture occurs naturally in coal beds in a number of different forms and has been determined by many methods. The most common method is through some procedural variant of the measurement of weight loss upon heating. These procedures measure the surface moisture and the inherent moisture, which is that moisture contained in the capillary system of the coal. Moisture contained in the molecular structure of the mineral matter present in coal (water of hydration) is not accounted for by this type of determination, nor does it include the moisture liberated by the thermal decomposition of organic matter in the coal. The water of hydration is commonly assigned a value of 8% of the ash value and the water of decomposition is not considered significant in most applications. For a more detailed discussion of analytical techniques used in this and other aspects of coal characterization, see Rees (1966). Total or "as-received" moisture, which is the value frequently given in coal analyses, is a combination of surface and inherent moisture and is used for calculating other parameters to the as-received basis. Total or as-received moisture values are critical because coal contracts are often based on as-received calorific values, usually measured in British thermal units per pound of coal, which are obtained by converting dry calorific values to as-received calorific values using total moisture content. Because of the tonnages involved in most contracts, and the fact that most contain penalty clauses for coal that does not meet specifications, even a small error in the moisture content value used to determine as-received calorific value can result in significant financial losses. Moisture content also plays an important role in handling and processing coal. As little as 0.5% surface moisture can cause coal to stick in a chute. Higher moisture contents also cause a decreased coke yield in coke ovens.

Volatile Matter: Volatile matter measurements do not reflect the actual amount of a given substance present in a coal sample but rather are measures of thermal decomposition products that form during the heating of a coal sample under rigidly specified conditions. Examples of volatile materials driven off in the heating process include water, hydrogen, carbon dioxide, carbon monoxide, hydrogen sulfide, chlorine, tar, ammonia, and a variety of organic compounds.

Volatile matter is a parameter used in some coal classification systems. It is used indirectly in the ASTM system for distinguishing between coals of medium volatile bituminous and higher rank. Volatile matter values provide useful information in matching specific coals to appropriate combustion equipment and are also of importance in selecting processes and conditions for the gasification and liquefaction of coal. As a general rule, the best metallurgical grade coking coals contain between 15 and 31% volatile matter and are ranked low- to medium-volatile bituminous in the ASTM system.

Fixed Carbon: Fixed carbon is the carbon that remains in the sample after determination of the volatile matter. The numerical value of fixed carbon is obtained by subtracting the sum of moisture, ash, and volatile matter from 100.

Fixed carbon values are used on a dry, mineral-matter-free basis as boundaries between coals ranked as medium-volatile bituminous and higher in the ASTM system. Because the amount of fixed carbon and ash is an approximation of the amount of coke produced, the fixed carbon value is used to estimate coke yield. It is also used in calculating the efficiencies of combustion equipment.

Ash: Ash is the noncombustible residue that is left when coal is burned. This ash residue derives from two basic sources within the coal bed including: (1) extraneous detrital particles of shale, clay, etc., and secondary mineral material such as calcite, pyrite, and marcasite; and (2) inorganic elements chemically bound in the organic compounds making up the coal. The detritus and secondary minerals make up the most significant part of the ash content. It should be noted that the terms ash and mineral matter are not synonymous. Ash, as stated previously, is the residue left after burning a quantity of coal in the presence of air. The amount of mineral matter present in a coal can be determined by a point count performed on a specially prepared coal sample using a petrographic microscope. A simpler method for determining mineral matter if the ash and total sulfur values are known is by using the Parr formula as follows:

$$\% MM = 1.08A + 0.55S$$

where MM is mineral matter, A is percentage of ash in the sample, and S is the percentage of total sulfur in the sample. The Parr formula is probably the method most widely used in the United States for determining mineral matter content.

The amount of ash contained in a coal, as well as the ash's composition, affect the coal's performance and therefore its success in the marketplace. Even in very clean coals the ash content may be 2 to 3% and ash contents of 10% or more are not uncommon in many productive coal beds. Carbonaceous material in the coal bed that contains excessive ash is frequently termed bone, bone coal, carbonaceous shale, black shale, rash, or any of a number of other locally used terms. The greater the ash content of a coal bed the lower is the heating value per unit weight of the coal, so that, in higher ash coal, more coal is required to produce a given amount of heat and disposal of the ash residue also becomes a problem. The ash content of a coal usually can be lessened by "washing" the coal in coal preparation plants. This usually entails grinding the coal to a specified size, then suspending it in a liquid with a specific gravity intermediate between that of coal and the mineral matter so that the mineral matter tends to sink in the solution and the coal tends to float. By repeating this procedure several times, a significant part of the mineral matter content can be removed from the coal.

Ash varies greatly in composition. It may contain varying amounts of silica and alumina derived from detrital minerals; iron oxides from siderite, pyrite, and marcasite; calcium oxides and carbonates from siderite; iron sulfide from pyrite and marcasite; and magnesium, sodium, potassium, phosphorus, and a wide range of trace elements (Tieman, 1973). Ash fusion temperatures, a measure of the temperatures at which coal ash begins to deform, softens, and becomes fluid, are important coal quality parameters that determine how the ash residue from a given coal will react when it is burned. Ash begins to deform at temperatures that range from 950° to 1700°C (1750° to 3100°F). Ash with fusion temperatures at the lower end of this spectrum is desirable in certain types of furnaces where ash is removed from the bottom in a liquid state but is undesirable in static fuel bed furnaces where removal of the residue is a difficult and costly process. Pyrite and marcasite (FeS_2), siderite ($FeCO_3$), calcite ($CaCO_3$), and other carbonate minerals are frequently responsible for low fusion temperatures in ash, whereas high silica or alumina contents are associated with higher fusion temperatures.

Calorific Value: In the ASTM system, calorific value is one of the primary rank-defining parameters for bituminous, subbituminous, and lignitic coals. Calorific value is usually reported in British thermal units per pound, or calories per gram, and can be easily converted from one system to the other.

Coal used for steam electric generation is sometimes sold at a fixed rate per million British thermal units with penalties for excess ash or sulfur (Wier, 1976). Because many contracts specify calorific value on an "as-received" basis and most analytical results are reported on a "dry" basis, dry values must be converted to "as-received" values by means of the following formula:

$$\text{Btu (as received)} = \text{Btu (dry)} \frac{100 - \text{moisture }\%}{100}$$

(Tieman, 1973)

The conversion formula contains percent moisture as a term, so the importance of accurate moisture values cannot be overstated.

Sulfur: Sulfur presents numerous problems in coal utilization. In combustion applications it can cause corrosion in the boiler or the buildup of heavy fouling in the boiler tubes. Large amounts of SO_2 are also generated upon combustion and may contribute to atmospheric pollution unless removed by limestone-based stack scrubbers. The same potential corrosion and pollution problems also apply to the liquefaction, gasification, and coking processes with the additional concern that unacceptably high levels of sulfur might be passed along through the coke to the iron and steel resulting in an inferior product (Ward, 1984).

Three commonly recognized forms of sulfur in coal are sulfate sulfur, pyritic sulfur, and organic sulfur. Of these, sulfate is the least important. In fact, sulfate sulfur contents are frequently on the order of 0.1%. Large sulfate values are sometimes indicative of a weathered sample. Relative amounts of pyritic and organic sulfur vary widely; in some coals total sulfur content is almost all organic whereas in other coals it is virtually all pyritic. It is important to analytically distinguish between organic and pyritic sulfur in coal because at least some of the pyritic sulfur can be removed by specific gravity separation methods. Pyritic sulfur occurs in the minerals pyrite and marcasite and, depending on the particle size of these minerals and the size to which the coal is crushed, half or more of this sulfur can be eliminated. That portion of the pyritic sulfur that occurs in finely disseminated form throughout the coal cannot be removed by specific gravity separation methods. The organic sulfur constituent is part of the hydrocarbon structure of the coal and cannot be removed by conventional coal cleaning technology. Although promising laboratory techniques designed to remove organic sulfur are under investigation, no commercially feasible process currently is available. Until washability tests are performed which can provide specific data, Wier (1976) advocates taking the sum of the organic sulfur content and one-half the pyritic sulfur content as a preliminary indicator of the final total sulfur content of the cleaned coal. Generally, only coals with low sulfur contents are used for steam electric generation. Average sulfur contents of coals received at US powerplants of 50 MW or greater generating capacity for the months October through December 1986 ranged from 0.16 to 5.6%. The overall average sulfur content of coal received at these same plants during the same period was 1.36% (US Dept. of Energy, 1987). Variations in maximum allowable sulfur contents in different areas are due primarily to differences in local regulations and to the presence of stack scrubbers at some facilities. The practice of blending different coals to achieve required specifications allows the use of some higher sulfur coals that would not otherwise be suitable for steam-electric generation.

Likewise, in coke production, the use of a high sulfur coal results in a decrease in the amount of coke that can be produced from a given amount of coal. Coal that cannot be cleaned to a sulfur content of less than 1.5% is not likely to be used, even as a blend, for coke production.

Free Swelling Index: Another commonly performed analytical procedure for coal is the determination of the free swelling index (FSI). The FSI is considered useful, although not definitive, in evaluating the coking properties of a coal.

It is a measure of the volume increase of a coal when it is heated under specific conditions and is reported in numbers from 0 to 9, with the higher values considered superior from a coking standpoint. FSI values generally increase with rank up to the anthracite rank but values within a given rank may vary widely. Generally speaking, coals with FSI values of 2 or less probably are not suitable for coke production and various users may require higher minimum FSI values for their specific equipment than others. Other tests that are used to predict the coking potential of a given coal include the Audibert-Arnu dilatometer, Gieseler plastometer, and Gray-King coke type, but the FSI is still the most commonly reported procedure of its type.

Ultimate Analysis: Ultimate analysis determines the percentages of the major constituent elements of coal. Determinations of hydrogen, carbon, nitrogen, oxygen, and total sulfur are reported. Typically, ultimate analyses are not performed on all coal samples but only on a representative number of samples. Data from ultimate analyses are used principally for research purposes and in certain classification systems, although there are commercial and industrial applications of the data. Specifically, ratios of carbon, hydrogen, and oxygen values are used to determine coal rank and as an aid in determining a coal's suitability for coke manufac-ture, gasification, or liquefaction. Data on oxygen content also are used in calculating boiler efficiencies.

Nitrogen present in the coal may react to form ammonium compounds when coal is carbonized in the coking process. These compounds can be extracted and marketed as fertilizer or for use in the manufacture of nitric acid. Ammonium compounds are also formed in the gasification and liquefaction processes. Their formation, however, utilizes some of the available hydrogen that would otherwise be used in the formation of the more valuable hydrocarbon end products. Also, during coal combustion, nitrogen forms oxides which become atmospheric pollutants when released. For these reasons, low nitrogen contents are usually preferred in coal (Ward, 1984).

Other elements for which analyses are commonly sought include chlorine and phosphorus. Chlorine contributes to corrosion and fouling problems and possibly to atmospheric pollution. A knowledge of the chlorine content is also essential in determining other parameters, including total sulfur. Phosphorus, which is concentrated primarily in the mineral matter, is undesirable in coking coals because, like sulfur, it can contaminate the steel end product.

Trace Elements: Most coals contain a wide range of trace elements, some of which tend to concentrate in the organic

Table 3. Trace Elements in Australian* and American[†] Coals

	Concentration (ppm air-dried coal)							
	USA						Australia, NSW and Qld	
	Illinois Basin		Appalachian		Western Coals			
Element	Range	Mean[‡]	Range	Mean[‡]	Range	Mean[‡]	Range	Mean[‡]
Ag	0.02-0.08	0.03	0.01-0.06	0.02	0.01-0.07	0.03	<0.2-1	<0.5
As	1.0-120	14	1.8-100	25	0.34-9.8	2.3	<1-55	3
B	12-230	110	5.0-120	42	16-140	56	1.5-300	60
Ba	5.0-750	100	72-420	200	160-1600	500	<40-1000	<100
Be	0.5-4	1.7	0.23-2.6	1.3	0.10-1.4	0.46	<0.4-8	1.5
Cd	0.1-65	2.2[§]	0.10-0.60	0.24[§]	0.10-0.60	0.18[§]	0.05-0.2	0.10
Co	2.0-34	7.3	1.5-33	9.8	0.6-7.0	1.8	<0.6-30	4
Cr	4.0-60	18	10-90	20	2.4-20	9.0	<1.5-30	6
Cu	5.0-44	14	5.1-30	18	3.1-23	10	2.5-40	15
F	29-140	67	50-150	89	19-140	62	15-500	80
Ga	0.8-10	3.2	2.9-11	5.7	0.8-6.5	2.5	1-20	4
Ge	1.0-43	6.9	0.10-6.0	1.6	0.10-3.0	0.91	<0.3-30	6
Hg	0.03-1.6	0.2	0.05-0.47	0.20	0.02-6.3	0.09	0.026-0.40	0.10
La	2.7-20	6.8	6.1-23	15	1.8-13	5.2	<4-50	10
Mn	6.0-210	53	2.4-61	18	1.4-220	49	2.5-900	150
Mo	0.3-29	8.1	0.10-22	4.6	0.10-30	2.1	<0.3-6	1.5
Ni	7.6-68	21	6.3-28	15	1.5-18	5.0	0.8-70	15
P	10-340	64	15-1500	150	10-510	130	30-4000	310
Pb	0.8-220	32	1.0-18	5.9	0.7-9.0	3.4	1.5-60	10
Sc	1.2-7.7	2.7	1.6-9.3	5.1	0.50-4.5	1.8	<0.3-30	3
Se	0.4-7.7	2.2	1.1-8.1	4.0	0.40-2.7	1.4	0.21-2.5	0.79
Sn	0.2-51	3.8	0.20-8.0	2.0	0.10-15	1.9	<0.9-15	<3
Sr	10-130	35	28-550	130	93-500	260	<20-~1000	100
Th	0.71-5.1	2.1	1.8-9.0	4.5	0.62-5.7	2.3	<0.2-8	2.7
U	0.31-4.6	1.5	0.40-2.9	1.5	0.30-2.5	1.2	0.4-5	2
V	11-90	32	14-73	38	4.8-43	14	4-90	20
W	0.04-4.2	0.82	0.22-1.2	0.69	0.13-3.3	0.75	<4-20	<10
Zn	10-5300	250	2.0-120	25	0.30-17	7.0	12-73	25
Zr	12-130	47	8.0-88	45	12-170	33	6-400	100

* Source: Swaine, 1977; updated 1980.
[†] Source: Gluskoter, et al., 1977 (after Ward, 1984).
[‡] Arithmetic mean.
[§] Numerous values below detection limit.

faction of the coal, while others have an inorganic affinity and are concentrated in the mineral matter. In some cases, trace element suites are distinctive enough to serve as aids in seam correlation or as indicators of the depositional environment. Boron, in particular, is more strongly associated with coals formed under marine influences. Some trace elements may act as catalysts or inhibitors during the complex reactions involved in coal conversion and may be transferred to the end products of those processes. Trace elements may also be released to the environment through combustion or through the weathering of the coal ash. Not all of the elements released into the environment are harmful but concentrations of toxic elements such as lead, arsenic, cadmium, or mercury might preclude the use of certain coals rich in those elements. Alternatively, other trace elements may be considered as potentially marketable byproducts of coal utilization. A list of trace elements and their concentrations in coals from different coal regions in the United States and Australia is given in Table 3.

Application of Coal Petrology: Coal petrology is the study by direct examination, usually microscopically, of the organic and inorganic components of coal. Petrologic studies form the basis for a broad range of relatively new techniques which have technologic applications of importance to those involved in coal exploration. For a thorough treatment of the subject, see Bustin, et al. (1985) from which the following comments are condensed.

Coal is a heterogeneous substance that is composed of components analogous to the minerals that are the constituents of inorganically derived rocks. These components are termed macerals and differ widely in physical and chemical properties and in their response to different technological processes. A knowledge of the petrographic composition of a given coal bed will allow the explorationist to predict its behavior in certain applications.

One area of technological application of coal petrology is in the area of coal cleaning by float-sink separation. Microscopic observations of the degree of intergrowth of the organic and inorganic constituents in both the float and sink fractions will indicate whether crushing to a finer size will increase the clean coal yield. Also, observing the type, distribution, and degree of intergrowth of the sulfur will give a preliminary indication of the probable methods of cleaning.

Another area where petrographic techniques have come to play a key role is in the production of coke. Through the employment of these techniques, predictions can be made concerning a coal's fluidity, FSI, and volatile matter content. Also, extensive research efforts through the years have shown that coal bed constituents can be classified as reactive or inert for given processes and that the information could be quantified to the point that the coal's behavior can be predicted with some accuracy. Two salient concepts resulting from these research efforts concerning coke quality prediction are: (1) an optimum mix of reactive to inert components of a given rank of coal will produce the best coke and (2) the percentages of this optimum mix will vary with rank. The importance of petrographic analysis to the steel industry is best illustrated by the fact that most steel producers now routinely conduct petrographic analyses to monitor blend quality and to evaluate new coals.

Although nearly 90% of the coal consumed in North America is used for combustion with most of that amount used for the generation of electricity, petrology has not played a significant role in the identification of desirable combustion characteristics. The primary reason for this is that factors most significant in defining the suitability of a coal for com-

bustion are either not directly measurable by petrographic techniques or they are more easily determined by other methods. Even so, some useful relationships have been identified through petrologic studies and it is an area of continuing research.

Finally, coal conversion technologies (primarily liquefaction) utilize petrographic data in identifying optimal coals for conversion. Rank and ratio of reactive to inert constituents are primary factors in determining a coal's suitability for conversion.

REFERENCES

American Society for Testing and Materials (ASTM), 1970, *Annual Book of ASTM Standards,* Pt. 33: Glossary of ASTM definitions and Index to ASTM Standards, p. 70.

American Society for Testing and Materials (ASTM), 1981, *Annual Book of ASTM Standards,* Pt. 26: American Society for Testing Materials, Sec. D-388, pp. 212-215.

Averitt, P., 1975, "Coal Resources of the United States, January 1, 1974," Bulletin 1412, US Geological Survey, 131 pp.

Bustin, R.M., et al., 1985, *Coal Petrology—Its Principles, Methods, and Applications,* 3rd ed., Geological Association of Canada, Short Course Notes, Vol. 3, Victoria, BC, 230 pp.

Caruccio, F.T., et al., 1977, "Paleoenvironment of coal and its Relation to Drainage Quality," Report No. EPA-600/7-77-067, Environmental Protection Agency, Interagency Energy—Environment Research and Development Program, 108 pp.

Crowder, R.E., 1986, "Cost-Effectiveness of Drill Hole Geophysical Logging for Coal Exploration," *Papers from the Third International Coal Symposium,* G.D. Argall, ed., Miller Freeman, San Francisco, pp. 195-207.

Daly, T.E., and Hagemann, R.F., 1976, "Seismic Methods for the Delineation of Coal Deposits," *Proceedings,* First International Coal Exploration Symposium, W.L.G. Muri, ed., Miller Freeman, San Francisco, pp. 192-226.

Ferm, J.C., 1976, "Depositional Models in Coal Exploration and Development, Sedimentary Environments and Hydrocarbons," R.S. Saxena, ed., AAPG/NOGS Short Course, New Orleans Geological Society, pp. 60-78.

Gluskoter, H.J., et al., 1977, "Trace Elements in Coal: Occurrence and Distribution," Circular 499, Illinois State Geological Survey.

Horne, J.C., Howell, D.J., and Baganz, B.P., 1976, "Splay Deposits as an Economic Factor in Coal Mining (abstract)," Geological Society of America Abstracts with Program, Vol. 8, p. 927.

Horne, J.C., et al., 1978, "Depositional Models in Coal Exploration and Mine Planning in the Appalachian Region," *Bulletin of the American Association of Petroleum Geologists,* Vol. 62, pp. 2379-2411.

Rees, O.W., 1966, "Chemistry, Uses, and Limitations of Coal Analyses," Report of Investigations 220, Illinois State Geological Survey, 55 pp.

Reilly, J., 1968, "Coal Mining," *Surface Mining,* E.P. Pfleider, ed., AIME, New York, pp. 821-848.

Simon, J.A., and Hopkins, M.E., 1973, "Geology of Coal," *Elements of Practical Coal Mining,* S.M. Cassidy, ed., AIME, New York, pp. 11-39.

Swaine, D.J., 1977, "Trace Elements in Coal," *6th Symposium on Trace Substances in Environmental Health,* D.D. Hemphill, ed., University of Missouri, Columbia, pp. 107-115.

Tieman, J., 1973, "Chemistry of Coal," *Elements of Practical Coal Mining,* S.M. Cassidy, ed., AIME, New York, pp. 40-48.

US Department of Energy, 1987, *Electric Power Quarterly,* US Department of Energy, Energy Information Administration, DOE/EIA-0397(86/4Q), 361 pp.

Ward, C.R., 1984, *Coal Geology and Coal Technology,* Blackwell Scientific Publications, 345 pp.

Wier, C.E., 1976, "Exploring Coal Deposits for Surface Mining," *Proceedings,* 1st International Coal Exploration Symposium, W.L.G. Muir, ed., Miller Freeman, San Francisco, pp. 540-561.

Wood, G.H., Jr., et al., 1983, "Coal Resource Classification System of the U.S. Geological Survey," Circular 891, US Geological Survey, 65 pp.

2.7 Placer

WILLIAM H. BREEDING

OCCURRENCE OF PLACERS

Placers are deposited by the action of water or wind. Aeolian or wind-deposited placers are generally not of commercial tenor, at least on any large scale, and will not receive further attention in this section.

Water-deposited placers include most placers mined commercially in the world today. Engineering data relates that the carrying power of water varies as the sixth power of its velocity. Whether carrying capacity varies as the fifth or sixth power, it will be appreciated that minor changes in velocity change the carrying capacity of a stream dramatically.

In planning land developments today, consideration is given to the effects of 25-year storms and 100-year storms, i.e., the severest storm to occur in those periods. When one thinks in terms of geological time, there are also 1,000-year storms and 10,000-year storms. Some of us have been privileged to see the results of very severe storms in our lifetimes, when brooks that one stepped over one day became raging torrents transporting 27-t (30-st) boulders the next. Severe storms result in major earth movements. Subsequent stream flows cause reconcentration of storm-deposited materials.

It is assumed that readers are familiar with the criteria that streams above a certain velocity will erode stream banks and bottoms, that streams at a certain velocity will transport suspended material, and that streams below a certain velocity will deposit transported material. Fine materials are, of course, deposited at lower velocities than coarse materials.

In accordance with the laws of hindered settling, coarse particles (gravel and coarse sand) will settle in general relationship to their specific gravities. Particle size and shape have effects but are subordinate to specific gravity. Moderately sized particles (sand) will settle similarly. Fine particles (slimes) have greater surface area per unit of mass than coarse particles, and the gravity component is limited by surface effects.

Heavy mineral concentrations may be expected to occur where streams lose velocity. Streams lose velocity when their gradients become flatter, where they change course, or deepen. More specifically, where a stream emerges from a canyon, on the inside of bends in a watercourse, or in deeper pools or "pot holes." These are the traditional places that placer miners look for concentrations.

Over the tens or hundreds of thousands or more years during which a placer is in formation, stream courses change and sites of heavy mineral deposition change. Glass models of placer deposits, based on extensive drilling results, have shown concentrations in similar patterns but at different sites. Certain levels of deposition are richer than others, and reflect the higher heavy mineral content transported during those periods of deposition.

In shallower placers the highest concentrations of heavy minerals are often on bedrock, which may be defined as the firm base on which alluvial material is deposited.

Ancient and Recent Placers

The division between ancient and less ancient and recent placers has never been established precisely to the author's knowledge. In some parts of the world ancient placers are called paleo-channels (Brazil); in others they may be called tertiary channels (California) reflecting the period of depositions. Many placers fall in the time frame between carbon-14 dating and potassium-argon dating.

It is believed that the tertiary placers of California were deposited 8 to 65 million years ago (Lindgren, The Tertiary Gravels of the Sierra Nevada of California, USGS Professionals Paper 73-1911) before and during the uplift of the current Sierra Nevada range. During this period, stream systems eroded auriferous sections of the western slope. The tertiary rivers were subsequently covered with volcanic ash, volcanic flows and later, alluvials. Today, the tertiary channels may be covered with a hundred to many hundreds of feet of later material. In general, channels in the headwater areas are higher in heavier mineral content. Channels are often narrower in headwater areas and coarse gravel and boulders prevail. Subsequent faulting and displacement may change original gradients, and add to the miner's and explorationist's problems.

Sites where recent drainage cut tertiary channels in California were first located and mined during the 1850–1884 period. The scale of the workings of that period is little appreciated today. Over 8046 km (5000 miles) of ditches carried water to hydraulic monitors that excavated 0.038 km³ (50 million cu yd) of material annually, and discharged tailings into the river systems. A court order suspended this practice in 1884.

The Ballarat/Bendigo area in Victoria, Australia, is another site of ancient placers which is well described in literature.

When areas of the world are first explored and developed by ambitious men, placer deposits, especially gold placer deposits, are among the first to receive attention. The gold fields of California in 1849, the Klondike of the Yukon in 1898, Nome, Alaska beaches in the early 1900s, the South Island of New Zealand in the later 1800s, and New South Wales in the late 1800s are a few examples. The placer tin fields of Southeast Asia followed in the 1900s. Lucrative gold fields in Colombia, South America were developed in the 1908–1920 period, expanded in the 1930s and late 1940s, and continue to produce today. The placer tin fields of the Brazilian Amazon were first recognized in the 1960s and production continues to expand annually.

What are known as heavy mineral placers, generally beach sand minerals not of the precious metals (Au, Pt) or high value specific gravities (SnO_2) types, have been developed in this century. They include rutile, ilmenite, and related minerals, zircon and monazite.

Large and rich placers may occur in areas where lode gold mines are few or nonexistent. Placer areas west of the Sierra Nevada range of California lie below serpentine belts where lode mines have been of the small vein type with rich localized ore shoots, often called *pocket mines*. In placer areas of Alaska, there have been few lode mines. In British Columbia, gold mines of size have been infrequent. In Colombia, substantial lode gold mines have been very few. In New Zealand, the best placer areas of the South Island have supported few lode mines. In Bolivia, the richest gold placers have been found in ancient conglomerates, or in more recent

77

placers resulting from their erosion and redeposition. Some placers have resulted from the erosion and direct deposition of vein materials, but they have been small. Most of the great placer gold fields of the world are second or third generation placers, i.e., the auriferous material has been eroded and redeposited two or three times.

Ancient placers may lie on varying grades due to subsequent tectonic movements. More recent placers may lie on grades of 0.5 to 3% or more. Smaller deposits with coarse gravel may be expected at the steeper gradients, and deposits with more sand and silt at the flatter gradients.

In areas of moderate to heavy rainfall, 76 to 760 cm (30 to 300 in.) per year, placers may be expected to contain rounded well-washed gravels and sands. Gravel with a shingled appearance on river bars provides an excellent environment for the collection of placer gold. In areas of lower rainfall, material may be subangular and contain more clay resulting from major movements during infrequent floods. In semiarid areas, alluvial fans, formed where intermittent streams emerge from hills and discharge onto flood plains, may contain heavy minerals. Fan deposits are found in Nevada, Arizona, and around lake basins in other states of the western United States. The monazite/zircon/ilmenite deposits of the Cascade basin of Idaho, which were mined with AEC/GSA support in the 1950s, were of this type.

In addition to placers deposited by the action of water, there are in-situ deposits which have been enriched by the erosion of lighter material accompanying the heavy minerals. When tin-bearing granites decompose and are subjected to erosion, the feldspars become clays and are washed away by rains and stream action. Quartz fragments remain behind along with heavier cassiterite. Unless quartz fragments are considerably coarser than the cassiterite, they too will be washed away, leaving rich concentrations of cassiterite behind.

Outcroppings of gold-bearing gossans subjected to weathering are other examples of in-situ enrichment. The enriched upper sectors are often mined profitably, but profitability generally declines rapidly as material becomes harder, and in-situ enrichment decreases. These surface enrichments have been called saprolitic in areas of the southeastern US.

Enrichments may also occur on the slopes between in-situ or *eluvial* deposits and stream bed *alluvial* deposits. These slope deposits are referred to as *colluvial* deposits.

Minerals found in placers in economic concentrations include gold, platinum, cassiterite, rutile, ilmenite, zircon, monazite, tantalite, columbite, diamonds (gem and industrial), gemstones (other than diamonds), abrasives (garnet and carborundum), chromite, and others in minor amounts. Silica and other light residuals may be salable.

POTENTIAL PLACERS

It is unlikely that high grade and significant placers of traditional minerals will be found at shallow depths along recent river systems with the exceptions of unexplored areas or remote areas without infrastructure.

Potential placers are more likely to be encountered: (1) in ancient or paleo channels, (2) at depth or with other problems not economically solved by earlier operators or equipment, and (3) in areas inaccessible to earlier engineers, or without infrastructure to support an economical operation. Minerals which previously had no commercial value add another factor.

To appraise the potential of a placer without expending major sums for equipment requires experienced judgment supported by convincing evidence of potential (usually samples). Deposits *above the water table* can be initially sampled for acceptable cost by methods described in Chapter 3 of this volume. For deposits *below the water table,* churn drilling is reliable, but time consuming, and adequate personnel and equipment may not be available for an initial appraisal. Caisson shafting is slow and generally confined to the gathering of confirmatory and supplementary data. Information is needed on the extent, depth, and nature of the placer material.

Seismic surveying is the most effective of the geophysical techniques for placer work; however, to be reliable, it should be referenced to certain known depths (from drilling, shafting, or tunneling). Properly undertaken and supervised, it will provide information on the depth and character of placer material. Seismic velocity is generally an adequate measure of material density, difficulty of excavation, and angle of repose; however, experience has shown that without adequate referencing and supervision, results can be misleading.

Double casing drills of both the direct and reverse circulation types, using water or air for recovery media, are much faster than churn drills. In sandy deposits, they have proven satisfactory for both qualitative and quantitative evaluations. In coarse placer gravel, most experienced placer engineers do not consider them satisfactory for quantitative evaluation; for qualitative results, they can be useful.

Rotary drills will provide information on the depths of placer materials, and will penetrate hard capping (such as lava flows) materials rapidly. Information on heavy mineral content is at best qualitative, i.e., it will show the existence of heavy minerals and if there is much or little, but not the precise quantities.

Vibrating drills are limited to the depths required to pack the core barrel. If a hole will stand open without caving, the coring drill may be reinserted. If the hole will not stand, the problems of casing the holes must be solved. These drills have been used to sample the first sections below the ocean floor.

Auger or post-hole type drills with highly portable, geared gasoline engine drives are effective above the water table, where holes will remain open. In tin-bearing sandy clays, they have been used to depths of 6 to 9 m (20 to 30 ft) effectively. Information on the depth and nature of material is obtained, and unless holes cave (when passing sand lenses, etc.), results are definitely qualitative and semi-quantitative.

It is not the purpose of this seciton to deal with operating costs, but in establishing what is a potential placer, attention must be given to size and the economics of scale. If all supervisory and supporting costs must be charged to one small producing unit, costs will be high. If there are several operating units, supervisory and supporting costs may be distributed. Experience has shown that adding a second (similar) operating unit will reduce costs of the second unit by about one-third, or one-sixth for each of the two units.

Two items that will severely increase operating costs are boulders and clay. Buried logs, acid water, and flooding are other considerations, but they can usually be handled by proper designs. In evaluating any placer, careful attention must be given to the quantity and size of boulders, and quantity and nature of clays (will they disintegrate under high pressure water jets; will they settle in decantation basins, etc.).

In assessing the potential of a placer, the evaluator is seeking to justify the cost of the next page of evaluation.

ECONOMIC PLACERS

When a potential placer is of great richness, considerable tolerance can be allowed in evaluating valuable mineral content. When a placer is very poor in valuable mineral content, it may be rejected without further work. Unfortunately, many potential placers are in the intermediate range of just good enough or almost good enough, and require very careful evaluation.

In placer evaluation, the evaluator should relate mineral content to volume. What is needed is mineral content per cubic yard or meter. Weighing bulk placer material is unrewarding and can be misleading. Weights of placer material will vary greatly with moisture content and with the nature of the material. Placer material normally weighs from 2500 lb to 3700 lb per cubic yard. For rough conversions, one cubic yard may be considered to weigh 1.5 short tons, and one cubic meter may be considered to weigh 2.0 short tons.

In evaluating placer material *above the water table* firm enough to stand unsupported, channel sampling is effective, requiring little equipment and more care than great expertise. As placers are deposited at low gradients, channels should be cut vertically. To limit the excavation of exposed banks, vertical cuts may be stepped, i.e., offset between elevations.

For evaluations made in traditional US or English units (milligrams of gold or pounds of tin per cubic yard), channels 6 in. x 12 in. have proven effective. If material is fine (limited coarse gravel), smaller channels may be used. If material is coarser, smaller channels are unadvisable and larger channels are worth considering. Sampled material is generally collected every 1.2 m (4 ft), or by strata, and concentrated by rocker and/or panning. Tailings from an entire channel should be collected and checked for values by repanning or tabling (if such equipment is available). Checking fine tailings is especially advisable when very fine heavy minerals (gold, cassiterite, rutile, etc.) are known to be present.

Where exposed banks are not available, but placer material is above the water table, shafts or trenches may be excavated. In shallower deposits [to about 6 m (20 ft)], trenches may be dug with a backhoe, or shafts may be excavated to that depth or deeper (with a larger hoe or by benching down). Bulldozer trenches are used but are generally more expensive and less acceptable environmentally.

Channel sampling of shafts or trenches is rapid, and provides information on the stratification of mineral values. If the responsibility of the samplers is not well established, different samplers may be used on opposite sides of a shaft or trench. Subsequent treatment of the entire volume excavated from a trench or shafts may be passed through sizing tests and a pilot plant (with a circuit similar to a commercial recovery circuit). The information obtained from bulk sample treatment will be helpful in designing a commercial plant. If material does not disintegrate readily in a small (pilot-plant-size) trommel, it may do so in a large one.

When sampling old hydraulic pits or other exposed banks, it is important to cut the banks to fresh surfaces. Heavier gold or cassiterite may fall from long-exposed surfaces, resulting in under-evaluations. Further, in old hydraulic pits, the former operators probably mined the best exposed material, and exposed faces may under-represent the remaining deposit. A few shafts excavated back from the face should provide useful supplementary data.

Sampling more than a few feet below the water table requires casing the hole or pit. Proven methods include Banka or Empire drilling, hand and mechanized churn drilling, and caisson shafting or drilling.

Banka drilling is an effective method in shallow alluvials, and where coarse and tight gravels are not prevalent. Casing is flush-threaded, and the drilling shoe at the lower end of the casing is sometimes serrated. Pressure is applied at the top of casing by men standing on a platform and by men dropping a weight on the casing head. The casing is rotated while pressure is applied. After a suitable casing advance, which is generally limited to 1 ft, or 0.25 or 0.50 m (depending upon the terrain and mineral sought) in economic ground, core removal is undertaken. In overburden or low-grade strata, longer drives may be used to increase drilling speed, but long drives generally result in poor core recovery and affect the accuracy of sampling. It is important to remember that the purpose of most placer drilling is to obtain an accurate sample, not to drill rapidly.

Core removal in gravel is generally by bailer, which is a piece of pipe with a flap or ball valve at the lower end and a plunger which pulls material into the barrel. In clayey ground, clay pumps, which are often pieces of pipe, open at the bottom, with side slots for core removal, may be used. Augers have been used at shallower depths, but are generally not the favored tool of experienced churn drillers.

Core removal or pumping usually proceeds until the casing has been cleared to the level of the drive shoe. Exceptions are: (1) when drilling in loose sand (often called rising ground) and a plug is left in the casing to reduce the inflow of alluvium, and (2) when drilling in very hard ground where it is impossible to drive the casing deeper within acceptable time. It is poor practice to pump or drill below the drive shoe; however, there are times when there is no other solution. When drilling below the drive shoe, appropriate notation must be made on the drill log, and positive corrections are not advisable.

In tight placer ground, there is an intermediate step between driving casing and pumping which consists of loosening material in the casing by dropping a chisel, star, or other shaped bit upon it. The bit breaks up placer material, including rocks, and also causes slimes by disintegrating clays. Because of the sliming effect and the possibility that stones may be driven downward, it is important to measure the height of material in the casing before and after pumping. Two or more measurements may be made to ascertain that pumping has reached the proper level before it is suspended.

The sequences in Banka or churn drilling are: (1) advance casing, (2) measure core rise in the casing, (3) loosen the core if necessary, (4) remove the core, (5) measure to ascertain that enough core has been removed, and (6) measure the core removed (volumetrically). Core removed from the casing is dumped into a trough and washed to a measuring box or pipe. For simplicity, some engineers use a pipe of drill casing for a measuring box. Overflow from the measuring box should be decanted into tubs and the slimes checked for mineral content. Slime volume may be added to the box or pipe measurement. Except in rising ground, core rise (the difference between before pumping and after pumping measurements) is generally greater than box or pipe measurement.

Hand and powered churn drills use the same basic procedures as the Banka drills, but casing is larger, especially for the powered drills, drilling shoes are not serrated, casing is rotated infrequently, casing couplings are used, and tools are lowered and removed by winch. Driving the casing is accomplished by dropping a heavy hammer on the drill (driving) head at the top of the casing. The heavy hammer is a pair of blocks bolted to the drill stem.

The standard placer exploration tool for over 50 years

has been the churn drill (Keystone, Bucyrus, Speed-Star, Cyclone, etc.) with 152 mm (6 in.) casing and a 191 mm (7.5 in.) drive shoe (see Fig. 1). Thirty years ago, it was common practice to use manila rope with drill tools, but recent practice uses left lay wire rope (to keep the tool joints tight). An experienced driller will bounce the drill bit on material in the casing, using the spring in the rope to keep from driving material downward.

On drills using 152 mm (6 in.) casing, cable is generally 19 mm (¾ in.) [though some use 16 mm (⅝ in.)] and cable on the bailers is 9.5 mm (⅜ in.). Crayon marks on the wire ropes indicate the distance to the end of the drill bit or bailer, and are used to measure core levels in the casing. Casing is marked at 1-ft (or appropriate meter*) intervals from the drill shoe before use.

Casing sizes vary slightly, but assuming a 6 in. interior diameter and a 7.5 in. drill shoe,

$$\left(\frac{7.5 \text{ in.}}{2}\right)2 = 44.18 \text{ sq in. of drill shoe interior}$$

$$\left(\frac{6.0 \text{ in.}}{2}\right)2 = 28.27 \text{ sq in. of drill pipe interior}$$

$$\frac{44.18 \text{ sq in.}}{144 \text{ sq in./sq ft}} = 0.3068 \text{ sq ft}$$

$$\frac{27 \text{ cu ft/cu yd}}{0.3068} = 88 \text{ ft of drive/cu yd}$$

Eighty-eight feet per cubic yard is considered the theoretical drive required to provide one cubic yard of core; however, drill shoes wear and experience has shown that the theoretical evaluation may be adjusted to provide greater accuracy. In the last century, an engineer named Radford began using *100 feet of drive to equal one cubic yard of material,* and it has been accepted by many in the industry.

Dividing drive shoe area by drill pipe area:

$$\frac{44.18 \text{ sq in.}}{28.27 \text{ sq in.}} = 1.56 \text{ theoretical core rise in the drill pipe per unit of advance}$$

Correcting this factor for shoe wear:

$$1.56 \times 0.88 = 1.37 \text{ adjusted core rise}$$

After studying production and drilling results of several companies for several years, the author proposed that 1.4 ft of core rise for 1 ft of drive be accepted in place of the several variations in use, and this practice was adopted and has been in use by several companies for 25 years. Different factors are obviously in use for casing of substantially different sizes, but the principle is valid.

An area where engineers and companies have differed is in the correction of drilling results. One major company did not use positive correction of drilling results for deficient cores and production results averaged 10 to 15% above drilling estimates. Another group of companies used positive corrections and obtained results (for five to seven dredges) of very close to 100% of drilling estimates. Monthly recovery/prospecting factors would vary, but annual and longer term factors were extremely close to 100%. The method of

treating exceptionally high holes (correcting for the nugget effect) also affects recovery factors.

Log Correction Techniques

Four methods used by drilling organizations are:
1. No corrections: seldom used at present.
2. Corrections for rises or volume in entire hole.
3. Corrections for rises or volume in pay streaks.
4. Corrections for rises or volume in individual drives.

Generally, the most conservative correction, whether it be from box measurement or from core rise measurement, is used, i.e., the correction giving the greatest negative correction is used.

Under certain conditions, methods *2* and *3* may be adequate, or nearly adequate, but method *4* is the only one believed applicable in all circumstances.

When evaluating a *gold property* in a district where recent mining or professional prospecting information is not available, representative gold samples must be sized and weighed; average weights are then assigned to No. 1, 2, 3, and 4 colors. These estimated weights are used for corrections to individual drives, and the algebraic total of the corrections adjusted as follows:

$$\text{mgs/cu yd} = \text{act. wt gold} + \text{corrections} \times \text{act wt} \div \text{est. wt}$$
$$\times \frac{\text{drive shoe factor}}{\text{dredging depth}}$$

Permissible Corrections

Maximum Positive Corrections in general use are given as follows:

Co. A	= 42%
Co. B (recent)	= 50%
Co. C (old)	= to about 200%
Co. C	= later 0%
Co. D	= 240% (lower river areas)
	100% (upper river areas)
Co. E	= 0%

If no boulders are present, the use of positive corrections to 100% may be acceptable. If boulders are present, the use of positive corrections must be reduced or eliminated. Experienced engineers regard excessive positive corrections wth caution.

Where hard, irregular bedrock is present, it is generally advisable to delete gold or tin recovered from the last foot or two of hole.

Similar corrections may be applied for *cassiterite (tin) evaluations,* but additional attention must be given to: (1) checking panners' tailings for fine tin, (2) checking coarse material for locked tin [passing + 9.5 mm (⅜ in.) material over heavy liquids is useful], and assaying samples. To avoid an excessive number of assays, material from various drives may be combined. If formations are relatively constant, errors should be minimal. If core recoveries and mineral content vary substantially, more assays are advisable.

If a 1-ft drive yields a core rise of 1.4 ft, there is no core correction. If a 1-ft drive yields a core rise of 1.7 ft, excess core has been recovered and a negative correction is applied of 1.4 ÷ 1.7 or 0.82 to the values recovered. If a 1-ft drive yields a core rise of 1.1 ft, deficient core has been recovered and a positive correction is applied of 1.4 ÷ 1.1 or 1.27 to the values recovered for that drive.

In gold prospecting, it is customary for the driller to estimate the quantity of gold recovered for each drive, and individual drives are corrected on that basis. Upon comple-

---

* Metric equivalents: in. × 25.4 = mm; sq in. × 645.16 = mm²; ft × 0.3048 = m; sq ft × 0.092 903 04 = m²; cu ft × 0.028 316 8 = m³; cu yd × 0.764 554 9 = m³.

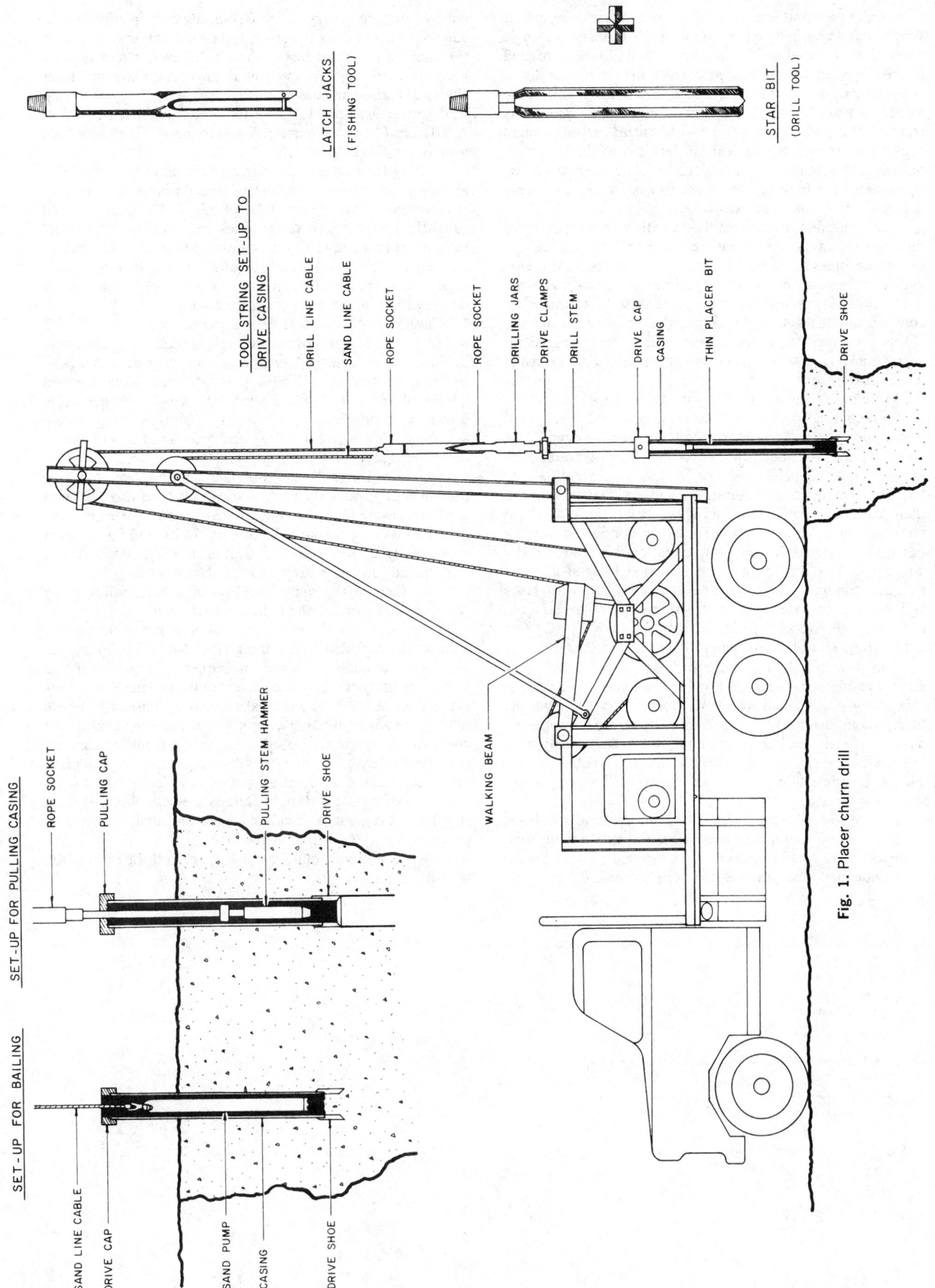

Fig. 1. Placer churn drill

tion of the hole and log, all gold recovered from the hole is weighed and checked against the panner's estimate. A single correction of actual weight of gold-estimated weight of gold is then applied to cumulative corrections. If the driller is consistent in his estimating (high or low), there will be no error. To assist the panner in his estimating, sample colors (size 1, 2, 3 and 4) are weighed and displayed on a small backboard. In a new gold field, drillers and panners may err in their first estimates, and extra care must be taken; however, if estimates are high or low, there will be no residual error if estimating is on a consistent basis.

In estimating tin values, and especially where other heavy minerals of similar appearance are present, it is difficult to eyeball estimate tin content for individual drives. In this case various drives may be combined and the concentrate assayed. Some prospectors use the zinc block test to estimate tin content, but the author has found this practice inaccurate. If cores are consistent, several drives may be combined for corrections. If they are inconsistent, caution and probably more assays are required.

Most placer engineers amalgamate gold samples. Common practice is to put the sample of gold and black sand into a porcelain or agate mortar with chemically pure (C.P.) mercury, and to grind with a porcelain or agate pestal until all visible and, hopefully, all commercially recoverable gold has been amalgamated. Amalgam and residual mercury is then placed in a small crucible with nitric acid and heated on a hot plate. This operation should be conducted under a hood with exhaust fan. The remaining gold is annealed (by additional heating on a burner), cooled, washed, and weighed. Samples may be collected from a group of holes and assayed to determine gold fineness. Most placer gold (bullion) will average from 850 (85%) to 950 (95%) fine gold. Alloying metals are generally silver and copper.

Churn drilling by experienced drillers has yielded satisfactory results for placer mining for many years, but drilling expertise is needed, and is often in short supply. In new fields there is a need for additional supporting information on the nature of the mineral and material to be mined. Also, despite the use of heavier drills, hydraulic assists, welded casing, and other modest improvements, churn drilling remains a slow process.

To check and supply supporting information to churn drilling, caisson shafts have been used, and they are effective to depths of 12 or 15 m (40 or 50 ft), especially if labor costs are reasonable. One practice has been to cut a 0.3-m (1 ft)

cube of material every 0.3-m (ft) of advance (using a knife-edged box), and later to screen and process all material from the shaft. Air pumps in the shaft will dewater and provide needed ventilation. Marsh gases can accumulate in placer shafts. If telescoping casing is used, a 51 mm (2 in.) reduction in diameter for each 1.2 m (4 ft) section has been effective. A 102 cm (40 in.) diameter is about the smallest in which a man can work effectively.

During the past decade, caisson drills have been available and they provide very reassuring data on both mineral content and the nature of material to be mined. The data obtained can help avoid expensive mistakes and will permit the development of an efficient mining plan and the use of efficient equipment. The only problem with caisson drilling is the high cost of the equipment and the relatively high cost of its operation, at least until depreciated.

Caisson drills come in various sizes from 0.500 m (19.7 in.) to 2.0 m (78.7 in.) or even larger, and are currently made in Europe and used there and elsewhere for pier, bridge, and building foundations. Properly sized, they have reached depths of 183 m (600 ft). They have been found acceptable in diamond prospecting, where large samples are required, and for gold prospecting. They have encountered and passed cemented gravels several feet thick.

One effective machine (BADE, made in Germany) uses oscillating hydraulic jacks of 136 t (150 st) to turn 900 mm (2953 ft) casing 22.5° (see Fig. 2). Raising and lowering jacks are 90-t (100-st). To date, drilling to 114 m (375 ft) has been successfully undertaken, and depths of 137 m (450 ft) are anticipated. In very deep ground, larger casing diameters may be used in the upper sections of a hole, followed by reduced diameter for the entire greater depth.

Cores are normally extracted with hammer grabs of both clam shell (two-blade) and orange peel (three-blade) design. Bailers are available, but used infrequently. Heavy star bits and other drop breakers are necessary when cemented layers are encountered. It is practical to pass all cores through a pilot plant approximating a commercial recovery circuit, but cores should be measured upon extraction from the caisson. (Duplicate measuring pipes that discharge into the pilot plant are effective, and avoid drilling delays.) Caisson joints are flush and of patented design. Details are best obtained from suppliers. Progress is equal to, or better than, churn drill progress for known equipment.

Fig. 3 is a sample of a form used to record the calculation of drilling results.

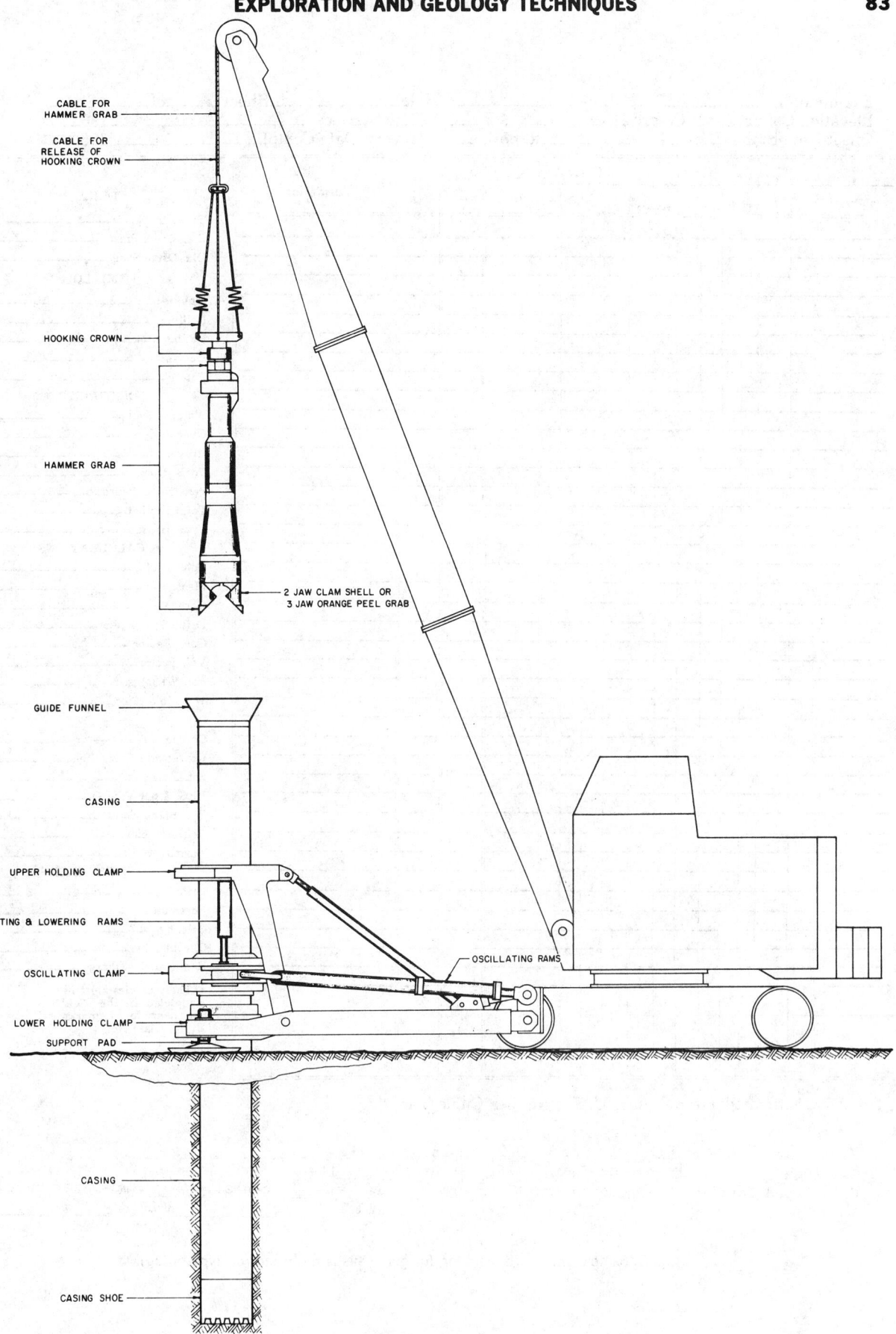

Fig. 2. Caisson drill (hydraulic casing oscillator).

Examination _____ Line _____ Hole _____ Sheet ____ of ____ Sheets
Elevation Collar _____ Co-ord N. _____ , E ____ Date Started _____ 196 ___
Offset from Stake, Bear _____ ft. hor., _____ ft. vert. Date Compl. _____ 196 ___

Pump Time		Depth Drilled		Core			Colors				Meas. Vol.	Correc- tions Mgs.	Est. Wt. Mgs.	Formation
Hr.	Min.	Ft.	1/10	Drive	Before Pumping	After Pumping	1	2	3	4				

DRILL •
Type & No. _____
Size Drive Pipe _____
Dia. Drive Shoe _____

TIME LOG •
Moving _____
Drilling _____
Pulling _____
Delays _____
Total _____

DEPTH, ETC. •
Water Level _____
Overburden _____
Gravel _____
To Bedrock _____
Penetrated Bedrock _____
Total Drilled _____
Type Bedrock _____

CALCULATIONS
Calc. Vol _____
Meas. Vol. _____
Core Vol. _____
Drive Shoe Factor _____
Core Factor _____
Vol. Factor _____
Est. Wt. Mgs. • _____
Wt. Gold Mgs. _____

Correction ✕ _____ = _____
Corrected Gold, Mgs. _____
Est. Fineness _____
U. S. $ per Fine Oz. _____
Wt. Black Sand Oz. _____
Bedrock elev. _____
Normal W. L. _____

PERSONNEL
Driller • _____
Foreman • _____
Calc. by _____
Checked by _____
Engineer in Charge _____

• These entries must be
completed in the Field
See over for Remarks
B. R. Class No.

Estimated Mean Value, U S. cents per Cubic Yard

Pay Stratum _____ ft. to ____ ft. = _____ cents.
Tailings _____ ft. Virgin Ground _____ ft. Bedrock _____ ft.
Calculated Total Dredging Depth (excl. water) _____ ft. = _____ cents. (= _____ mgs. per cu yd)
                                   LINE _____ HOLE _____

Fig. 3. Sample churn drill field log for gold (this is the preferred type of log).

DELAYS

Date	Cause of Delays	Hr.	Min.

TOPOGRAPHIC SKETCH

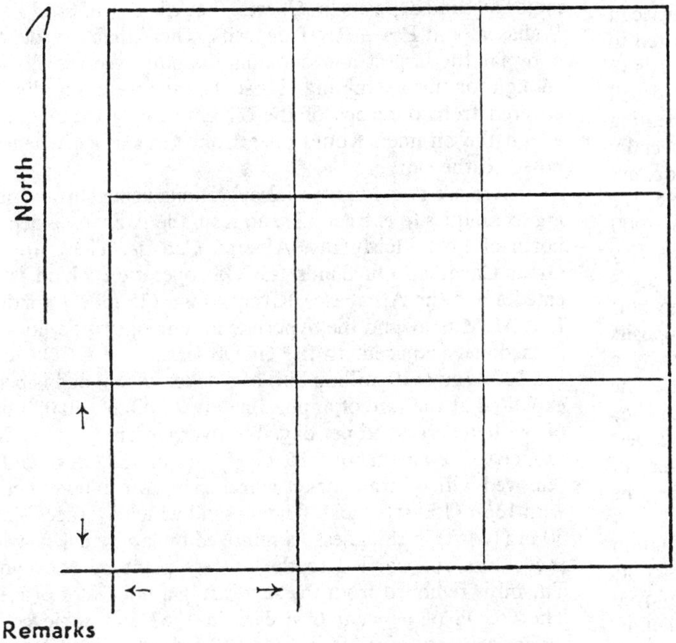

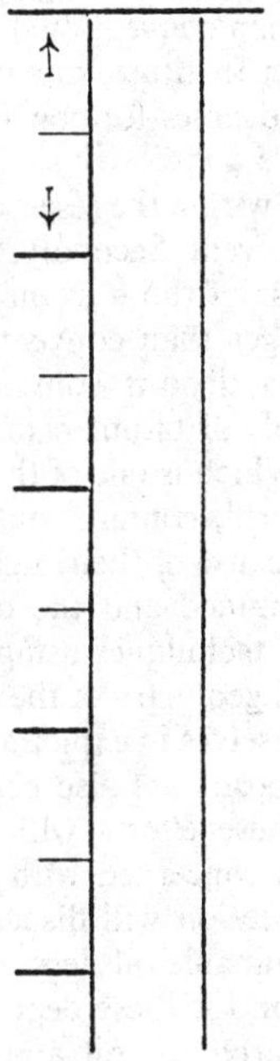

Remarks

Labor Charge: Survey Ref:

Fig. 3. Sample churn drill field log continued.

2.8 Tar Sands Exploration and Geology

INTRODUCTION

Tar sand is a term applied to sand impregnated with oil, or more precisely, with a viscous hydrocarbon called bitumen. A more technically correct term than tar sand is oil sand or bituminous sand. However, the term tar sand has been widely used and has been given official status by the Alberta Oil and Gas Conservation Act, which defines tar sand as "a sand containing a highly viscous crude hydrocarbon material not recoverable in its natural state through a well or by ordinary oil production methods." In this section, these deposits will be referred to as oil sand, the term most frequently used for such deposits in scientific literature and within the industry.

These deposits of oil are very similar to conventional oil pools except in two significant ways: First, they are more viscous than conventional crude oils with an API (American Petroleum Institute) gravity of 5 to 15 degrees, compared to 25 to 40 degrees for conventional crude oil. The deposits in place have a viscosity such that the oil is immobile and will not move within the reservoir unless it is mobilized by heating or by a solvent. Secondly, the deposits of oil sand, especially in the case of those in the Athabasca region of Canada, are much larger than conventional oil pools. For example, the Athabasca deposit contains approximately 143 Gm³ (900 billion bbl) of bitumen in place, whereas the Prudhoe Bay oil field, which is one of the ten largest conventional oil pools in the world, contains only about 2.4 Gm³ (15 billion bbl) of oil. Because of their high viscosity, these oil sand deposits must be mined and the oil separated from the sand by a variety of techniques using either hot water or solvents. The grade and geometry of the known deposits are such that they lend themselves to exploitation by surface mining techniques. These deposits are also being studied for in situ recovery of oil but these efforts will not be discussed here since this volume is concerned with surface mining.

This section will discuss the exploration and evaluation of these minable oil deposits. The techniques for grass-roots exploration for these deposits are similar to those employed in the search for oil and other energy resources; that is, regional mapping, stratigraphic studies, geophysical surveys, and exploration drilling. Very large deposits of oil sand have already been located and the major task is the delineation and evaluation of these deposits. Consequently, the early or grass-roots exploration will not be discussed in detail here, however, the techniques of drilling and sampling the deposits to determine their grade and extent, and to develop a data base required for surface mining planning, will be examined.

LOCATION OF THE MAJOR RESOURCES

Deposits of oil sand are known throughout the world, with the largest and most thoroughly studied located in Canada, the United States, and Venezuela. The world's most significant deposits are located in the northeastern portion of the Province of Alberta, Canada (Fig. 1). These Cretaceous oil sand deposits of Alberta have an estimated 143 Gm³ (900 billion bbl) of in-place bitumen, which is almost twice the recoverable conventional oil reserves in the world. However, only 10% of these reserves in Alberta are close enough to the surface to be considered recoverable by surface mining

techniques. The Orinoco petroleum belt in Venezuela is probably the second largest oil sand resource in the world estimated to contain 111 Gm³ (700 billion bbl) of oil.

Oil sand occurrences are known to exist in 24 states within the United States; however, only in six of these states are there large enough deposits to conceivably support a commercial operation. These are located in Alabama, California, Kentucky, New Mexico, Texas, and Utah. Deposits in these six states are estimated to contain approximately 4.8 Gm³ (30 billion bbl) of oil with the deposit in Utah containing approximately 90% of these US reserves.

DEVELOPMENT OF OIL SAND DEPOSITS

The only commercial development of oil sand has taken place in the Province of Alberta, Canada. There are four major oil sand deposits in Alberta: the Athabasca, Cold Lake, Wabasca, and Peace River deposits. The Athabasca deposit is by far the largest and contains the only reserves shallow enough for surface mining. These deposits were initially discovered from outcrops of the oil sand along drainages, in which the bitumen would ooze from the surface when exposed to the sun.

There are two commercial operations using surface mining techniques to extract oil sand from the Athabasca deposit north of Fort McMurray, Alberta, Canada. These are the Great Canadian Oil Sands (GCOS) operation which is located along the Athabasca River, 40 km (25 miles) north of Fort McMurray and the Syncrude mining operation, located immediately adjacent to the GCOS lease. The GCOS lease has 100 Mm³ (630 million bbl) of recoverable oil and is being exploited at the rate of approximately 7 900 m³ (50,000 bbl) of synthetic crude oil per day. The overburden on the GCOS property averages about 15 m (49 ft) in thickness and is removed with a large bucket wheel excavator in conjunction with 136-t (150-st) trucks. The oil sand which averages about 50 m (164 ft) in thickness, is removed by large bucket wheel excavators and carried to the process plant by conveyors. The oil is removed from the sand using a hot water process. The GCOS plant went onstream in 1967 and achieved its design capacity of 7 200 m³ (45,000 bbl) of synthetic crude per day by 1970.

The Syncrude mine went into operation in the late 1970s with a design capacity of 19 800 m³ (125,000 bbl) of synthetic crude oil per day. The oil sand averages approximately 11 to 12% bitumen by weight. Approximately 1.8 t (2 st) of oil sand is needed to produce one barrel of synthetic crude. Adding the handling of overburden and tailings material, 4.5 t (5 st) of material need to be moved per barrel of crude. At a designed capacity of 19 900 m³ (125,000 bbl) per day, the Syncrude operation will handle more than 544 kt (600,000 st) of material per day.

Currently there are no commercial oil sand operations in the United States. However, studies and pilot plant testing are underway on deposits in Utah, California, and Kentucky.

GEOLOGY

Oil sand deposits consist of accumulations of heavy oil within the pore space of permeable reservoir rocks, such as sandstones or unconsolidated sands. They are essentially the

Estimated In-place Resources*	
	billion bbl oil
Canada	900+
South America	700+
United States	29
Africa	1.7
Europe	0.3
Asia	NA
Total	1631+

*Modified from Walters, 1974.

Fig. 1. Location of major oil sands deposits.

same geologically as conventional oil reservoirs, except that the hydrocarbon is much heavier and cannot be recovered with conventional oil wells. These deposits are subject to the same type of geological analysis as conventional oil reservoirs, such as studies of source rocks, reservoir rocks, oil migration history, trapping mechanisms, and oil maturation and differentiation. The percent of oil in an oil sand deposit is a direct function of the porosity and permeability of the host rock. The distribution of permeable zones is controlled by the distribution of porous sand bodies and interbedded shales or other impermeable strata. Most oil sand deposits are thought to be formed when oil migrates from a source rock into a permeable reservoir rock and is exposed to water washing and a lessening of pressure with the consequential loss of the light hydrocarbons, leaving a heavy residue—the bitumen in the oil sands deposit. However, there are two schools of thought regarding the origin of the oil in these deposits. One theory is that the parent petroleum was a thermally mature crude similar to conventional crude oil that has migrated from some geographically remote source rock into the sand deposits and been subjected to biodegration and water washing to produce the oil sand deposit. The other theory postulates that the parent petroleum was an immature crude and has a source within the same formation or very close to the oil sand deposit. Whatever the source, it is clear that the oil was much less viscous at the time it entered the reservoir as it occupies all the porous and permeable zones within the reservoir.

One characteristic of oil sands that is important to the selection of processing techniques is whether the sand is water wet or oil wet. In the case of water wet oil sand, as in the Athabasca deposits, there is a thin film of water separating the bitumen from the sand grain (Fig. 2), which allows the bitumen to be removed by a hot extraction process. In the case of oil wet sands, the oil is in direct contact with the sand grains and is more difficult to remove, usually necessitating a solvent extraction process.

Most oil sand deposits are located in alluvial or deltaic sandstones near the edge of a large basin (Fig. 3).

These alluvial and deltaic environments usually contain a variety of sedimentary units with differing amounts of porosity and permeability. This variety results in a nonhomogeneus oil sand deposit. The oil content is directly proportional to the porosity of the host rock and its deposition was

controlled by the permeability. A study of the depositional environments can lead to an understanding of the variability in the deposit.

Any exploration or development effort should include a thorough study of the sedimentology and a reconstruction of the depositional history of the deposit area.

EXPLORATION TECHNIQUES

Oil sand deposits are essentially the same as conventional oil reserves except that they are more viscous because of the loss of light hydrocarbons. Exploration techniques used to locate these deposits are consequently the same as those used for conventional oil reservoirs. Regional or grass-root methods include identification of source-rocks, location of reservoir rocks with high porosity and permeability, and identification of traps in the reservoir rocks.

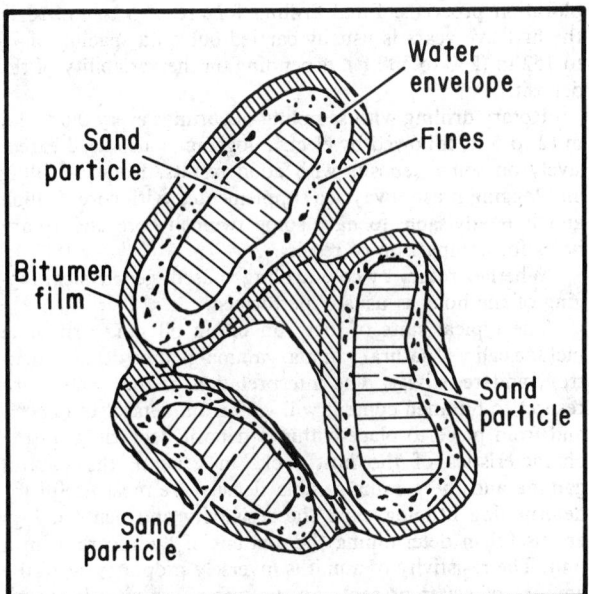

Fig. 2. Structure of typical oil sands particles (McConville, 1975).

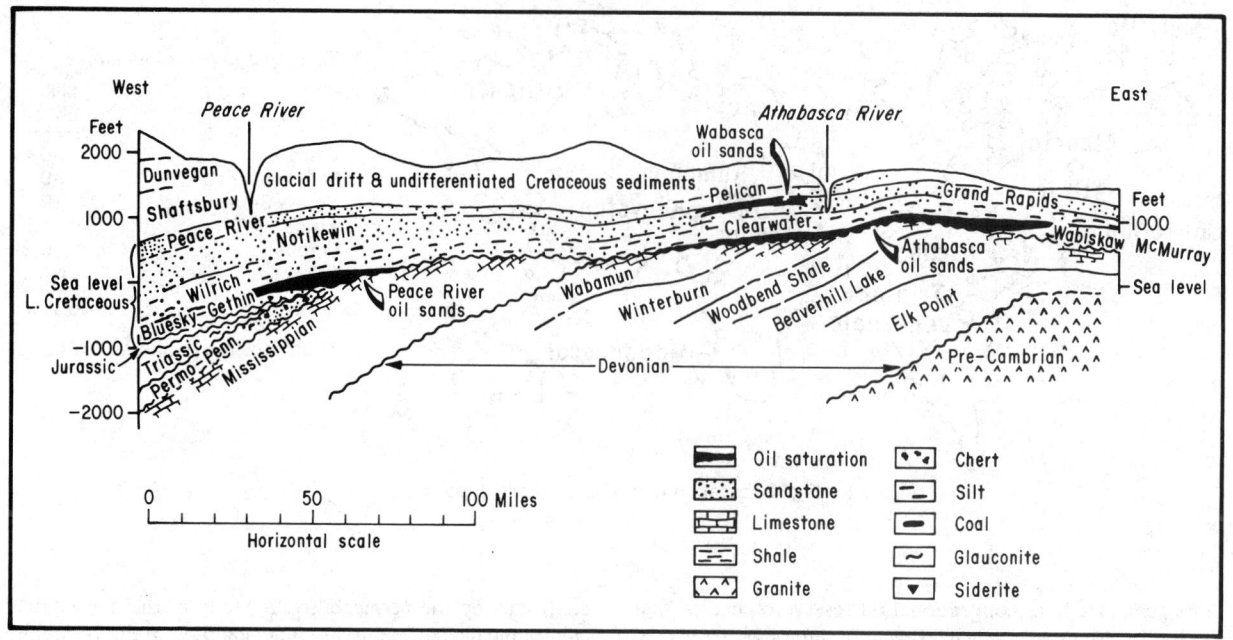

Fig. 3. Schematic geological east-west cross section of Northern Alberta oil sands (McConville, 1975).

DRILLING

Core drilling is the most widely accepted method for sampling oil sand deposits, although rotary drilling with electric logging and bulk sampling are also used.

The most common core size is 63.5 to 76.2 mm (2.5 to 3 in.) in diameter. The core should be described by a geologist and the intervals containing significant bitumen should be sampled by geologic unit or in 3 m (10 ft) intervals or less. Drill spacing varies from 1.6 km to 76 m (1 mile to 250 ft) depending on the deposit and the purpose of the drilling.

Reconnaissance drilling is often done on approximately 1.6 km (1 mile) centers with fill-in drilling at 0.8 km (0.5 mile) and 0.4 km (0.25 mile) centers as more detailed exploration proceeds. Final drilling for areas to be mined in the first five years is usually carried out on a spacing of 76 to 152 m (250 to 500 ft), depending on the variability of the deposit.

Rotary drilling with sampling of cuttings every 0.6 to 1.5 m (2 to 5 ft) followed by electric logging is also used extensively on some deposits with good results. Rotary drilling and logging must always be supplemented with core drilling and it is advisable to have some twinned core and rotary holes for comparison of results.

Whether rotary or core drilling is used, geophysical logging of the holes is usually performed.

The typical suite of logs run on an oil sand drill hole include caliper, natural gamma, gamma-gamma density, neutron, and resistivity. The interpretation of these logs with respect to bitumen content will vary from deposit to deposit and from place to place within a deposit depending on the characteristics of the host rock. In general, the natural gamma and the gamma-gamma density are most useful for determining lithology and the resistivity and neutron logs are useful in determining the amount of hydrocarbon in a unit. The resistivity of a unit is inversely proportional to the amount of water present. As the water decreases in an oil sand unit, the bitumen increases. Therefore the resistivity increases as the percent of bitumen increases. The neutron

log responds to the hydrogen and to a lesser extent to the amount of carbon in a formation. It can therefore be a useful direct measure of the bitumen (hydrocarbon) present. The neutron log is also useful in detecting zones of gas.

The natural gamma and density logs can also be correlated to the percent of bitumen in an oil sand unit. There is usually a decrease in bitumen content with increased natural gamma count. This is due to the fact that the natural gamma responds to radioactive elements such as potassium isotopes present in clays, and as the clay content increases, the porosity and permeability decrease, resulting in a lower bitumen content. The bitumen saturation tends to increase as the density of an oil sands unit decreases. The decrease in density is often due to an increase in porosity in the formation resulting in an increase in the amount of bitumen in the pore space.

These relationships are generalizations that will have exception in some deposits. The response of geophysical logs must be correlated with drill cores for each deposit and for various areas within a single deposit to insure reliable interpretation of the logs.

ASSAYING

One of the most important aspects of oil sands exploration is the determination of the amount of oil present in the rock.

There are three methods in common use for determining the amount of bitumen in an oil sand sample: the Dean-Stark method, the Soxlet method, and the modified Fisher assay. In some cases, a modification of these standard procedures is used for a particular deposit but the vast majority of samples are assayed using one of these standard techniques.

The Dean-Stark method is the most widely used analytical technique for determining the weight percent of oil in an oil sand sample. The technique consists of weighing the sample, then removing the bitumen and water with a hot solvent, such as toluene. The water evaporates and is trapped and weighed. The bitumen is lost in the solvent. The remaining sample is weighed and the weight percent of bitumen is determined by difference.

The Soxlet technique is similar to the Dean-Stark method except that the water is lost and the bitumen is recovered and measured directly. More specifically, the sample is weighed and hot solvent is percolated through the sample to remove the hydrocarbon. The solvent is then boiled off leaving the oil or bitumen which is then weighed directly. The method allows the oil to be tested for viscosity and other properties. It is advisable to run both the Dean-Stark and Soxlet on splits of the same sample so that both the water content and the properties of the oil can be measured on at least a few samples.

A third and less costly method for determining the amount of oil in a sample is to heat it in a retort and measure the amount of oil that comes out. This method is the modified Fisher assay used for oil shale. With this method there may be some small losses of the hydrocarbons or some hydrocarbon may remain in the sample. Therefore, duplicate samples should be tested using the Dean-Stark or Soxlet methods. Results from retorting have been quite successful on some deposits resulting in considerable savings in analytical costs.

Another indirect method of determining the amount of oil in a deposit is by inference from geophysical logs. This can only be accomplished after correlation between core analysis and geophysical logs has been established. This can be a cost effective and reliable method but must be used with caution because many geological variables can affect the geophysical response.

HYDROLOGICAL DATA

In addition to information on the geology of the oil sand deposit, information on the hydrology of the deposit and the surrounding area is essential to the design of a surface mining operation. There are several reasons for obtaining hydrological information, including the need for water in the process plant, the possible presence of unwanted water in the mining area, the possible effect of groundwater on slope stability, and the possible effect of mining and processing on the water resources of the area.

The first step in obtaining information on the hydrology is a data search of the surface and groundwater conditions in the area. This information may be available from government agencies or from oil and water wells in the area of the development. Another important source of information on hydrology is the geology of the deposit. At the same time that information on the general geology is gathered to characterize the oil sand resource, information on possible aquifers within the stratagraphic section can be obtained. Once information on the general hydrologic nature of the sediments and on potential aquifers is obtained, more detailed information can then be obtained during the exploration drilling phase of the project.

The same drill holes that are used to obtain information on oil sand grade and thickness can be used to obtain information on the groundwater hydrology. During the rotary or core drilling programs, areas where water loss occurs should be noted because this loss indicates a permeable zone. These zones should be correlated with the geologic information on aquifers. During the exploration planning phase, one should identify the number of holes that will be needed as groundwater observation holes. These usually range from 20 to 60% of the exploration holes. These holes should be drilled without significant drilling additives, or if one must use mud additives, they should be of the biodegradable type. This will allow information on water quality to be obtained that will not be biased by drilling mud additives. Once the holes are identified, the specific aquifer to be tested in each hole should also be identified. Generally only one aquifer per hole is tested. If problems occur during the exploration drilling that require the use of non-biodegradable mud additives, the observation wells can then be changed to alternative locations.

While the drill rig is still on the site of a test well permeability tests can be run to isolate the zones that are to be tested as potential aquifers. Also known as a packer test this involves sealing off a hole at a given test zone and pumping water into the zone to determine its permeability. Once the permeable zones are identified, one in each given test hole, a screen and plastic pipe can be inserted to allow water from this test zone to enter the pipe and be tested for quality. The test zone is sealed below and above to prevent contamination from any other aquifers.

If a significant aquifer is identified and that aquifer has to be dewatered before mining or is needed for water supply, a pump test should be planned. This involves drilling a new hole, usually larger than the exploration holes, to allow for a nominal 152.4-mm (6-in.) casing and pump capable of producing a few hundred gallons per minute. Observation wells should be established near this test well. If possible these wells should be the same wells established from the exploration drilling. The test well is then pumped to determine the yield of a given aquifer and the level of water in the surrounding observation wells is observed periodically during the tests to determine the drawdown of that particular aquifer. A test well should be put in for each significant aquifer from which either dewatering or water production is anticipated. The observation wells will be left in place and monitored at least once a year throughout the life of the mine and can also be used for testing the quality of the water before and during the mining operation.

GEOTECHNICAL DATA

Geotechnical information is essential to the planning of any surface mining operation. This information includes data on rock strength and characterization of the discontinuities such as faults, fractures, or bedding planes within the oil sand and surrounding rocks.

In open pit planning, geotechnical information is necessary to select the proper mining equipment and design the pit slopes. For instance, if a rock is very hard, it would not be suited for bucket-wheel excavators or draglines without first blasting the material.

A geotechnical study for an oil sand project is usually undertaken by a geotechnical group with specific expertise in this area. The first information that must be gathered is the geology of the deposit and its surrounding rocks. Separate units within the oil sand and overlying rocks should be identified for strength testing. Once the separate, identifiable units are recognized, samples for strength tests can be taken during the exploration drilling. These samples should be of unsplit core and be two to three times the core diameter in length. The tests to be performed on these samples should include a uniaxial compressive strength and triaxial shear test to determine the intact rock strength. The strength tests should be performed in at least two directions, parallel and perpendicular to the bedding.

Other important data characterizing the setting of an oil sand deposit include rock mass discontinuities such as the joints, faults, and bedding planes within a deposit. The techniques for obtaining data on these features include geologic mapping and fracture mapping where surface exposures are available and collection of Rock Quality Determination (RQD) data from drill core.

SUMMARY

Tar sands deposits constitute a hydrocarbon resource that far exceeds the world's resources of conventional oil. The most economical means of exploiting the near surface part of this resource is through surface mining. The exploration techniques described in this section should provide the information on the grade and geometry of the deposit as well as the hydrological and geotechnical conditions that are necessary to plan a successful tar sands surface mine.

REFERENCES

Baughman, G.L., 1978, *Synthetic Fuels Data Handbook,* Cameron Engineers, Inc.

McConville, L.B., 1975, "The Athabasca Tar Sands," *Mining Engineering,* Vol. 27, No. 1, pp. 19–38.

Waters, E.J., 1974, "Review of the World's Major Oil Sand Deposits," *Oil Sands—Fuel of the Future,* Memoir 3, Canadian Society of Petroleum Geologists.

2.9 Oil Shale Exploration and Geology

R. W. ELAYER
I. P. DORLING
P. W. McKIE

INTRODUCTION

Oil shale, long known as a potential source of energy, is found in many countries. The largest known deposits occur in the United States, primarily in Colorado, Utah, and Wyoming.

Even though numerous occurrences of oil shale are known, interest in their development has been, until recently, insufficient to support the thorough exploration and appraisal of known deposits or to encourage the search for additional deposits. However, during the past few years, a great deal of oil shale exploration and assessment work has been performed in the western United States, Australia, Morocco, and Brazil, with a view to commercial development.

In the past, oil shale has been commercially mined and used directly as a fuel, or processed into liquid fuels, in many parts of the world including Scotland, France, Sweden, South Africa, Australia, the Soviet Union, China, Brazil, and the United States. At the present time (1985), oil shale is being commercially mined and processed in only two countries. In the Soviet Union, (Estonia), oil shale is used directly as a fuel to generate electricity, and in the People's Republic of China (Fushun and Maoning), oil shale is used to produce liquid fuel and gas. Large open pit oil shale projects have been planned in Brazil, Morocco, and Australia while development of a commercial sized underground oil shale mining and processing operation is already underway in Colorado.

Definition of Oil Shale

Oil shale, in a strict sense, is a misnomer, since it does not contain oil and it is not necessarily a shale, as defined geologically. According to ASTM (1985), oil shale can be defined as a sedimentary rock containing insoluble organic matter that upon heating yields an oil similar to petroleum. Shale oil is the organic liquid obtained from the retorting of oil shale.

Oil shales are diverse in composition, lithologic association and genesis, and range in color from light shades of brown, green or red to dark brown, gray or black. Claystone, marlstone, siltstone, impure limestone, and black shale most often contain enough insoluble organic material to be classified as oil shale. Three general categories of oil shale are known: carbonate-rich shale; siliceous shale; and cannel or coaly shale.

Individual deposits of oil shale yield shale oil in greatly varying amounts. Some yield more than 38 L (10 gal) of shale oil per ton and others yield more than 380 L (100 gal) of shale oil per ton.

World Occurrence of Oil Shale

The world occurrences, geology, and resources of oil shale are well documented by Duncan and Swanson (1965). Oil shale deposits are found in every continent of the world, however, information on their quality, thickness, and extent is varied. The principal known deposits throughout the world are shown in Fig. 1.

The most important and largest known oil shale deposits of the world are found in the western United States. These deposits are located in Colorado, Utah, and Wyoming, and comprise some of the highest grade resources known. Other important deposits in the United States are found in Indiana, Kentucky, Illinois, Iowa, Michigan, Montana, Nevada, New York, Ohio, Oklahoma, Tennessee, Texas, and Alaska.

Many organic-rich shale deposits are known to exist in Canada, although little data are available concerning them. One of the most important Canadian resources is found in New Brunswick. Other deposits occur in Newfoundland, Nova Scotia, the Northwest Territories, Ontario, Quebec, and the Yukon.

Potentially large deposits of oil shale are known to exist in South America, particularly in Brazil. Smaller deposits are also known to occur in Argentina, Chile and Uruguay. Several large deposits of oil shale occur in Africa, notably in South Africa, Morocco, and Zaire. The Moroccans have been actively working on development of their Timahdit and Tarfaya deposits.

In Asia, the largest and most extensive known oil shale deposits are found in the People's Republic of China and the Soviet Union. About 180 "commercial" oil shale deposits occur in China, distributed in 21 provinces (Duncan and Swanson, 1965). The most important resources are located in Manchuria, Heilingkiang, Kwantung, and Sinkiang provinces. Many deposits of oil shale are reported in Siberia, particularly in Kazakhstan, Kuznetsk, and in northeast Siberia.

Elsewhere throughout Asia, rich but small deposits of oil shale are located in western Thailand, eastern Burma, and northwest Turkey. Extensive deposits also exist in Israel, Jordan, and Syria. Many small, high-grade oil shale deposits are found in Australia and New Zealand. In particular, oil shale is widely distributed in New South Wales, Queensland, and Tasmania.

The oil shale deposits of Europe are extensive and widely distributed. Some of the most important occur in Scotland and Estonia, France, Spain, Sweden, Austria, Bulgaria, Czechoslovakia, Italy, Germany, Switzerland, Yugoslavia, and Portugal. Detailed data are available concerning many European oil shale occurrences, particularly about those in Scotland, France, Estonia, Spain, and Sweden, which have been commercially developed during the past 130 years.

Oil Shale Geology

The world's known oil shale deposits range in age from Cambrian to Recent. The oldest deposits include those found in northern Europe, northern Asia, and east-central North America, while some of the youngest are found in the Green River formation of the western United States, and in China.

Three major depositional environments in which oil shale was formed have been identified. These are large lakes (lacustrine environment); shallow seas (continental platform or shelf environment); and small lakes, streams, and lagoons (coal-forming swamp environment). In these environments, oil shale was deposited from water containing a prolific quantity of flora and fauna, in conditions that precluded oxidation

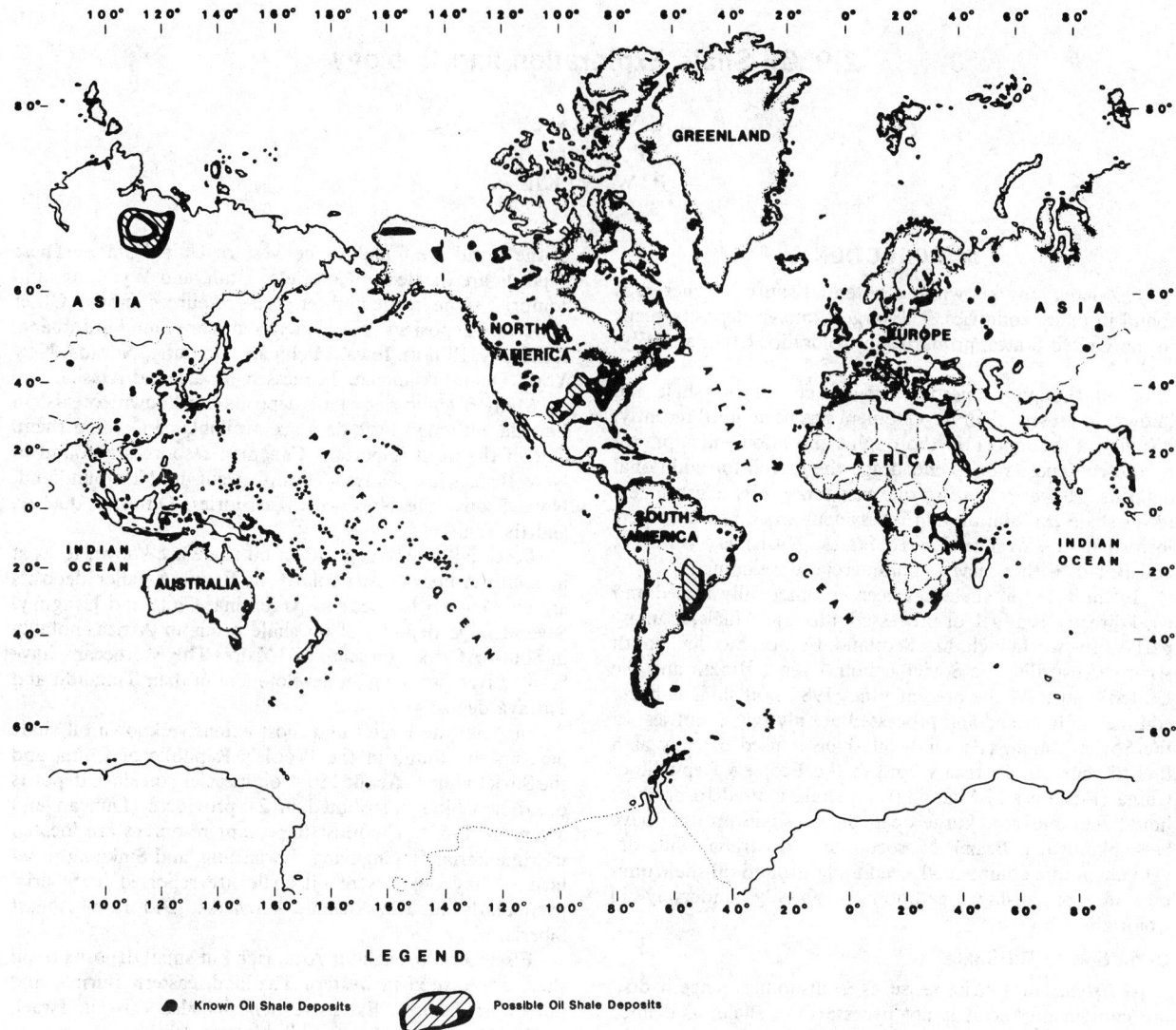

Fig. 1. Principal known deposits of oil shale throughout the world (Duncan and Swanson, 1965).

of the organic material which accumulated on the bottom of the depositional structures.

Oil shales formed in a large lacustrine environment include, most importantly, those of the Green River formation (Eocene in age) found in Colorado, Utah, and Wyoming; the Paraiba River Valley deposit (Pliocene) of southern Brazil; the Albert Shale (Mississippian) of New Brunswick, Canada; and the Aleksinac deposit in Yugoslavia. Such deposits form some of the highest grade, thickest, and most extensive oil shale resources in the world. They are predominantly calcareous and often interspersed with volcanic tuffs, clastic sediments, and carbonate rocks.

Those oil shale deposits which were formed in a shallow sea or continental platform/shelf environment, include the marine black shales of northern Siberia and northern Europe (Cambrian); the high-grade kukersite deposits (Ordovician) of Estonia; the Devonian black shales of the eastern and central United States; the Permian shales of Brazil, Uruguay and Argentina; and the Jurassic black shales of Europe and eastern Asia. These deposits are predominantly siliceous in

character and extend over large areas. They are associated with limestone, chert, sandstone, and phosphatic nodules.

Oil shales which were formed in small lakes and found associated with coal-bearing rocks include the thick, extensive, and high-grade deposits of Fushun (Tertiary) in Manchuria, China and those of late Carboniferous age in Kazakhstan in the Soviet Union. These deposits overlie, or are associated with, coal seams.

Oil Shale Resources

Duncan and Swanson (1965) compiled an order-of-magnitude estimate of the world's oil shale resources (see Table 1). Since this estimate was compiled, considerable work has been performed on oil shale deposits in many areas of the world. Known oil shale resources in Australia and Brazil, for example, have been considerably increased due to recent exploration.

EXPLORATION PHASES

Exploration for oil shale involves the initial "discovery," as well as the geologic work required to adequately char-

Table 1. Known World Resources of Shale Oil

| | Billion bbl | | | |
Continents	Proven	Probable	Possible	Hypothetical
Africa	10	90	ne*	534,000
Asia	20	84	3,702	697,400
Australia and New Zealand	small	1	ne*	121,000
Europe	30	46	300	177,200
North America	80	4,320	7,400	300,500
South America	50	750	7,200	244,000
Total	190	5,291	18,602	2,074,100

Source: Duncan and Swanson, 1965.
*Not estimated.

acterize a deposit for engineering planning. Once a viable deposit is confirmed and operation starts, development geology begins. The main thrust in development geology is geologic control for exploitation.

Exploration geology should not be approached as a cook book exercise. Every oil shale property, deposit, or region has its own particular geologic, physical, and economic characteristics that require one or more modifications to standard exploration practice. The exploration techniques presented here are most applicable to the oil shale deposits of the Green River formation located in Colorado, Utah, and Wyoming.

These techniques, with some possible modification, are applicable to oil shale deposits of the eastern United States, as well as those throughout the world.

The most important consideration in exploration is to identify and characterize a deposit at a minimum cost and in the shortest possible time. To accomplish this, a well organized and phased approach is necessary. This phased geological exploration effort, integrated with engineering studies and intermediate decision points, results in an efficient, well organized program. A flow diagram showing the three main phases of exploration is given in Fig. 2.

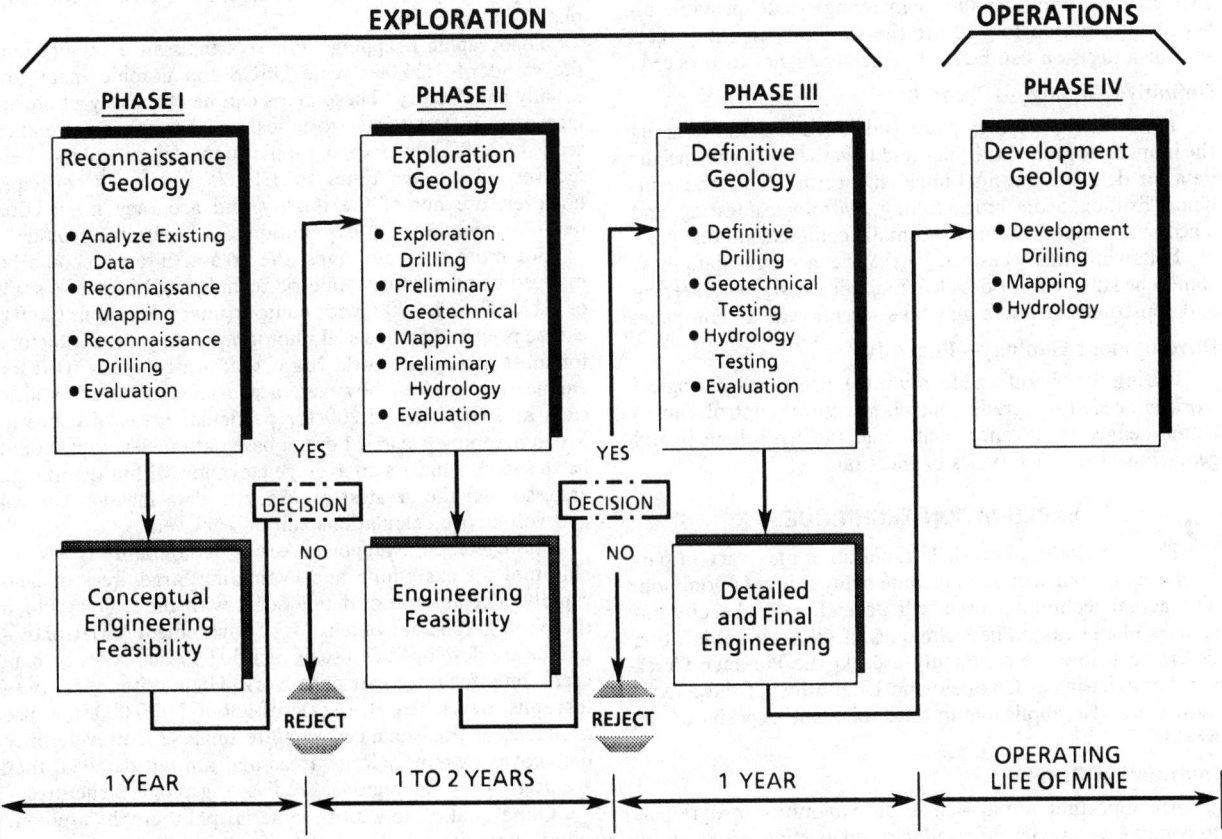

Fig. 2. Flow diagram showing the main phases of an oil shale exploration program, followed by development geology during operations

Reconnaissance Geology—Phase I

The reconnaissance geology phase represents the initial examination of a property or properties to identify tracts or areas having the best potential for acquisition and development. During this phase of work, existing data is collected and analyzed, aerial and ground reconnaissance geology mapping completed, and reconnaissance drilling performed. All information derived from this work is evaluated to determine the best potential areas for more detailed exploration.

The engineering effort which normally follows this phase of work is a conceptual engineering feasibility study. The quality of this study depends to a large extent upon the quality and completeness of the geologic data obtained during the reconnaissance phase. Following the initial exploration and engineering work, a decision can be made either to reject the area or property, or to move on to the next phase of exploration. The time usually required for this phase of work is one year, but less time may be needed if more initial data is available.

Exploration Geology—Phase II

During the exploration geology phase, more detailed and site specific data is acquired to better define the geology for feasibility engineering. This phase involves additional drilling, preliminary geotechnical testing, geologic mapping, preliminary hydrology testing, and geologic evaluation. This phase may last one or two years, depending upon the size of the property under investigation and the amount of drilling required.

The data obtained during this phase may be used in performing a feasibility study that would more accurately define the mining potential of the project. The exploration geology phase and feasibility engineering usually provide sufficient information to evaluate the feasibility of the project so that a decision can be made whether or not to proceed.

Definitive Geology—Phase III

The definitive geology phase is directed at better defining the immediate area of the mine and to gather very site specific data for detailed and final mine engineering. This phase includes drilling, geotechnical testing, hydrological testing, and a geologic evaluation, and is normally completed in one year.

Following this phase of work, the geologic data base should be sufficient for detailed final mine design engineering and construction, which may take several years to complete.

Development Geology—Phase IV

During the operating life of a mine, continued geological work is needed for grade control, structural control and to better define, in advance of mining, the hydrological and geotechnical characteristics of the strata.

EXPLORATION TECHNIQUES

The techniques of oil shale exploration may vary in particular cases, but will often include many things in common. The actual techniques used will depend upon the circumstances of the case. The techniques of oil shale exploration described below are commonly used in the Piceance Creek and Uinta basins of Colorado and Utah. Most of these techniques are also applicable to other oil shale deposits of the world.

Information Review

An important initial step in an exploration program is to compile and review all available and existing information. This information may be in the form of raw data, such as assay reports, or it may be reduced data which has been synthesized from the raw data. Reduced data, such as oil yield maps, are by far the easiest to work with, but for a reliable evaluation, all available information must be considered. It must be remembered, however, that with the passing of time, procedures and techniques are improved and popular theories are modified to reflect new technology. Existing information, therefore, must be viewed subjectively.

A bibliography is included in the reference section of this chapter that lists some of the many published sources of information on oil shale in the United States. This list is not all inclusive but is representative of the most commonly used sources. Since 1917, the US Geological Survey, US Bureau of Mines, Energy Research and Development Administration, Department of Energy, and several state geological survey offices have added to the available oil shale data base. Most information developed by private industry is confidential, and is not commonly available.

Information on international oil shale deposits usually can be obtained from the government geological agency in the country of interest. In former colonial countries, additional information may be available from government agencies in the mother country. Considerable geologic work is also performed in overseas countries by the US Geological Survey, the United Nations, and other world agencies.

Mapping

Good maps are necessary for performing any detailed geological or engineering work on a property. Topographic base maps are required for recording and presenting geologic information, and for engineering planning and design. Photogeologic maps should be prepared early in the reconnaissance phase. More detailed maps are required in the later phases of exploration.

Topographic Mapping: For reconnaissance exploration, the standard 1:24,000 scale USGS topographic maps are usually satisfactory. These maps can be effectively photoenlarged up to four times to a scale of 1:6,000. The smaller scale 1:62,500 USGS topographic maps can be enlarged effectively about five times to 1:12,000 scale. These maps, however, are not of the quality and accuracy needed for feasibility engineering, exploration, and definitive geology.

For more accurate maps, the area of interest should be mapped using photogrammetric techniques. Maps, at a scale of 1:12,000 or 1:6,000, with contour intervals of 6 m (20 ft), can be prepared from aerial photographs and are satisfactory for most geological work. Maps that are generally used for engineering studies, however, are usually of larger scales, such as 1:6,000 to 1:1,200 for particular areas of concern. When mapping is carried out, a base datum elevation should be specified, which is an average elevation within the mining area, so that the greatest accuracy in data location control and volumetric calculations can be achieved.

Photogeologic Mapping: Aerial photography is a valuable tool for examining and evaluating large areas of land rapidly. The amount of detail varies with the scale at which the photographs are taken. Most government aerial photographs are developed at a scale of 1:50,000 that renders them useful only for reconnaissance work. High altitude U-2 photographs, developed at a scale of about 1:120,000, are also available, as are enhanced satellite images. However, since they cover extremely large areas and are not detailed, they are useful only for regional studies of geologic structures.

Good quality, low altitude aerial photographs are commonly taken to cover the area of interest for photogeologic mapping. Photographs should be taken in the late fall or early spring when the deciduous trees are devoid of leaves and there is no snow on the ground. Overlapping vertical photographs should be taken for stereo viewing. A series of oblique photos should also be taken when there are cliff faces on the property since they do not appear clearly on vertical photographs. For geologic mapping, vertical photographs should be taken at a 1:12,000, or smaller, scale.

Government aerial photographs are produced primarily in black and white, which is adequate for most photogeologic mapping, since gray tonal contrasts and textures can be interpreted. Superior geologic mapping can be produced using color photography because color photographs show the actual color of the ground surface, making interpretations easier and more reliable.

Multispectral Imaging: Multispectral imaging, or remote sensing, from satellites, for example, Landsat, and aircraft has become a very valuable tool in recent years for regional studies as well as evaluations of large properties. The information derived from remote sensing is developed as an image, not as a photograph. An image of the earth's surface is developed by computer processing of digital sensor data. The sensors detect electromagnetic radiation in selected wavelength ranges, most of which are not visible to the human eye. The image is then enhanced by computer and presented in a variety of ways visible to humans. Satellite imagery, especially that from the new Thermatic Mapper (TM), has proven to be very useful in the evaluation of structures in oil shale exploration. Computer enhancement of the digital data can be used to show linear structural features. Radar imagery can be used to produce an image of the earth's surface that enhances linear features and is useful for performing structural studies.

Geologic Mapping: A preliminary photogeologic mapping study should be performed prior to any geologic mapping in the field. This will familiarize the field geologist with the project area and provide valuable geologic information that will help in field mapping. When thorough photogeologic mapping is available, the geologist's field time can be significantly reduced and used primarily for confirming the interpretations made from the photographs.

The most reliable method for compiling field data is to use frosted mylar overlays on aerial photographs so that data will not obscure features that may be important keys to underlying structures. Several overlays may be used to differentiate between various data such as structure and lithology. If the photographs are at a scale of about 1:12,000, the data recorded on them can be easily transferred to 1:12,000 scale base maps. Another common technique is to use aerial photographs, enlarged twice, on which mapping data can be directly recorded. Normal stereo pairs of the aerial photos can be used for reference.

Geotechnical Mapping: Another important objective of an exploration geology program is to determine the geotechnical characteristics of the strata. The feasibility of mining an oil shale deposit, by either underground or surface methods, is dependent upon the competency of the rock. Therefore, it is important to start gathering geotechnical data early on in the exploration geology phase.

Much of the information recorded during geologic mapping is also useful for geotechnical evaluation. Normally, however, geotechnical mapping records a great deal more detail about the characteristics of discontinuities, i.e., joints, partings, and faults.

Geotechnical mapping consists principally of area-wide geotechnical data gathering and site specific line mapping. The area-wide effort consists of developing a representation of the geotechnical characteristics of an area. The area is first surveyed to identify major structural features and regional joint patterns. Extensive data are collected concerning the strike and dip of the strata and of discontinuities such as joints, fractures, and faults. A detailed description is made of the beds, discontinuities, and bedding planes or partings. Data on strike and dip are normally analyzed using rose diagrams, histograms, and stereographic projections. Detailed line mapping presents site specific information on discontinuities. This mapping includes detailed observations of all discontinuities along a traverse section of about 30 m (100 ft) or more in length. Several sections are measured along several orientations to assure a complete investigation of all discontinuities. Detailed line mapping requires good unweathered rock exposures, which can be provided by driving a test adit or making an open cut face.

Drilling Program Planning

A detailed plan for the drilling program is made before any drilling is carried out on a property. Proper planning will ensure a coordinated program so that needed data is gathered in an efficient manner. The objective of the program must be agreed to by all interested parties. Once the program is approved, a geologist can begin program planning, which is closely coordinated with the project's engineering, geotechnical, and hydrological personnel. A well planned program will yield maximum information from each drill hole at a minimum cost.

The exploration drilling program should be supervised by an experienced geologist with a sound background in sedimentary and structural geology. The geologist responsible for the overall program will coordinate all geologic work and logistics. Large programs require several drill rigs, each of which has a geologist assigned to it. Each rig geologist is responsible for logging all the core recovered.

Drill Hole Design: One of the first steps in planning a drilling program is to determine the number of holes required and to show the location of each hole on a map. Only a few holes are required in a reconnaissance exploration geology program (Phase I), that is directed at identifying a potential property for more detailed exploration. However, in a definitive exploration geology program (Phase III) the objective is to accurately define the geology for final mine design. This may require many drill holes. The trend during the complete cycle of exploration geology (Phases I through III) and operations geology (Phase IV), is to increase definition of the geology through tighter drilling control, by increasing the number of drill holes and decreasing the spacing between them.

Drilling during an oil shale reconnaissance program (Phase I) is carried out to verify previous drilling data or gather information on the total stratigraphic oil shale section. During the reconnaissance phase, a minimum number of holes is required, generally with a spacing of up to many kilometers (Fig. 3a). During exploration drilling (Phase II), however, holes are usually spaced several kilometers apart (Fig. 3b), which should provide sufficient data for performing a mining feasibility study to determine the optimum mining area. During a Phase III definitive drilling program, holes are concentrated in the initial mine area, where the hole spacing varies between 1 and 2 km (0.6 to 1.2 miles). This spacing will usually provide the necessary detail for performing final mine design and grade control. Development drilling (Phase IV) is performed ahead of mining (Fig. 3d), during the mining operation where the holes are spaced at much smaller intervals.

Drilling and Coring Methods: Two drilling and coring

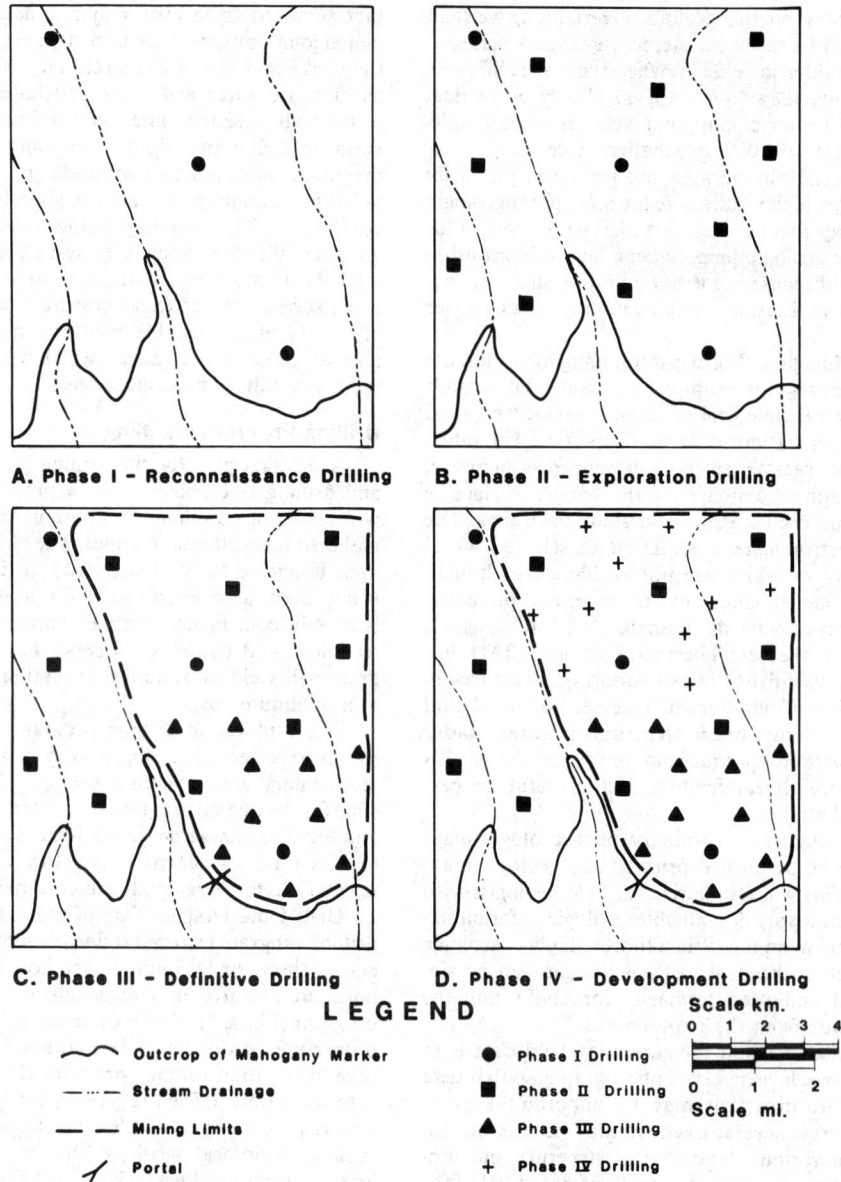

A. Phase I – Reconnaissance Drilling   B. Phase II – Exploration Drilling

C. Phase III – Definitive Drilling   D. Phase IV – Development Drilling

LEGEND

Scale km.
0 1 2 3 4

Scale mi.
0 1 2

- ⌢ Outcrop of Mahogany Marker
- ⋯ Stream Drainage
- – – Mining Limits
- ⤢ Portal

- ● Phase I Drilling
- ■ Phase II Drilling
- ▲ Phase III Drilling
- ✛ Phase IV Drilling

Fig. 3. Phases of drilling in iol shale exploration.

methods are basically applicable to oil shale. In one, a cylindrical rock sample is recovered from the drill hole during the drilling operation (coring), and in the other, cutting samples are recovered during rotary drilling. Usually, oil shale holes are cored from top to bottom, or at least through the main zone of interest. Rotary drilling is not satisfactory when samples are required for grade analysis, and should be used only as a cost efficient, rapid method for penetrating through rock overlying the target zone to be cored.

Diamond drilling is most commonly used for coring. Core drilling is expensive but it is the only method by which good, undisturbed samples can be obtained for assaying, structural geology studies, stratigraphical analysis, and geotechnical testing. Oil shale usually cores very well. Very good core samples of oil shale are normally recovered except through rubblized (due to burning or collapse) and fractured zones.

Two general types of coring tool are available: conventional and wireline. Conventional coring normally comprises a 6.1-m (20-ft) long core barrel that is placed at the bottom of the drill string. Once the barrel is full, the whole drill string is removed from the hole to recover the core from the barrel. The drill string is again lowered into the hole to core the next section of rock. This coring technique is simple but slow, particularly at depths greater than 150 m (500 ft). Progress is particularly slow when conventionally coring through rubblized zones, because the core barrel can become plugged by pieces of oil shale.

Wireline coring is most commonly used in oil shale, whereby core is recovered without pulling the drill string out of the hole after each run. The core is retrieved through the center of the drill string using a cable hoist with a special overshot latching device to attach to the inner core barrel.

Usually, two core barrels are used during coring so that while one is being emptied of core, the other is being filled. The use of two barrels saves time between runs and allows the core to be carefully extracted and prepared for logging.

The size of core to be recovered during a drilling program depends upon the purpose of the sampling program. Most coring operations in oil shale produce conventional NX or wireline NXWL core 5.4 cm (2-⅛ in.) and 4.76 cm (1-⅞ in.) in diameter, respectively. Core samples of either size are large enough for geotechnical testing and for assaying and are also fairly easy to handle and store neatly in core boxes designed to hold 3.05 m (10 ft) of core.

Conventional 7.35-cm (3-in.) or larger core is normally recovered when larger samples are required for bench-scale testing. Large diameter wireline core drilling, designated PQWL and 8.21 cm (3.35 in.) in diameter, has also been successful in oil shale.

Where a deep hole is to be drilled and continuously cored, and in all reconnaissance drilling, wireline core rigs should be used. These rigs are self-contained with a mud pump and all necessary equipment. Coring rates of over 75 m (250 ft) per 12-hr shift with a good driller can be attained. However, wireline rigs are slow and inefficient for rotary drilling.

Rotary drill rigs are used when short sections are to be cored at depth or when air drilling is desired. These rigs (Fig. 4) are also self-contained with mud pump, water injection pump, and air compressor to meet most requirements. In very deep holes, auxiliary air compressors and boosters are commonly used. Rotary drilling is accomplished using a tricone roller bit. A good driller can rotary drill over 152 m (500 ft) in 12 hours. Once the core point is reached, the tricone bit is replaced with a core barrel and core bit to recover the desired section of rock. Conventional coring rates with a rotary rig are usually about 35 to 50 m (100 to 150 ft) per 12-hr shift at hole depths of about 328 m (1,000 ft). Conventional coring often requires drill collars to add weight for increased penetration.

Core drilling is usually conducted using water as the drilling fluid, but drilling muds are added when needed to retain circulation. In many cases, however, the addition of drilling muds is insufficient to restore circulation when lost, and consequently drilling is carried out *blind*. Blind drilling can be safely accomplished providing sufficient water is pumped down the hole and penetration is slowed down to assure a clean, cool bit. This can be performed by a good driller without sacrificing core recovery or creating bad hole conditions.

Drilling with air is necessary if any drill holes are to be used later for hydrological monitoring or testing. The air compressor units found on most rotary drill rigs provide sufficient air when drilling shallow holes. However, supplementary air compressors and boosters are required for deeper holes, especially when large quantities of water are encountered. Air drilling is most suitable when coring conventionally because of the large volumes of air required. Wireline coring can be performed using air, but the anulus restrictions resulting from the larger-sized drill steel required for the method prevent a sufficient volume of air from reaching the bit in deep holes.

Permitting: The appropriate permits and approvals must be obtained before any drilling starts, whether the property is private, leased, or government-owned, to ensure that no

Fig. 4. Rotary drill rig wireline coring oil shale in Colorado.

violations of the law occur and that no needless damage is caused to the environment or natural resources. It is also advisable to establish early on in a program a rapport with the federal, state, and local government agencies and private land owners in the area of investigation. This will help assure a successful field program, and any future geological engineering, construction, and mining operations.

Logistics: Logistics include all those functions required to support a drilling operation. Adequate prior planning for logistics will assure a much smoother operation in the field and minimize lost time.

Accommodation is needed for the drilling and geology crews. It may be necessary to set up a field camp, depending upon the distance to a town with adequate accommodation. This maximizes the available daily work time. Normally, a drilling crew and geologists can work 12 hours per day if supported by a comfortable camp nearby. Work periods usually last for 10 days, followed by a four-day break.

A communications system should be established early on in a project. Radio-telephones are most appropriate. Through relay systems, radio-telephones have a tremendous range and can easily be integrated with the normal telephone network. When several drill rigs are operating to complete a program, it is advisable to maintain contact between rig geologists by the use of hand-held, two-way radios.

One important logistical consideration is the availability of water for drilling. Core drilling can consume significant amounts of water so adequate sources near the operation must be located and developed. Long water hauls are costly and require additional water trucks and holding tanks to avoid drilling delays.

Geological and Geotechnical Logging

During a drilling operation, samples recovered from each hole must be logged. This is usually performed by the geologist assigned to the drill rig. The samples may consist of cuttings recovered from a drill hole or core samples. Cuttings from a non-cored drill hole are normally collected periodically and segregated into piles representing 1.5 m (5 ft) of drill depth. Each pile is then examined by the geologist, who takes random samples, washes them to remove any drilling mud residue, and then describes their lithology. This logging is carried out at the drill site as the cuttings are recovered, so that the geologist remains fully aware of the drilling progress and can make timely decisions if the projected drilling depths need to be changed.

In a coring operation, the core barrel is first placed on sawhorses next to a core tray when it is recovered from the drill hole. Extracting the core from the barrel is accomplished in various ways depending upon the type of barrel used. A split-tube barrel is by far the easiest to clear. The top half of the inner barrel is removed, exposing the core. The lower half of the barrel is then laid in the core trough and carefully rotated so that the core gently rolls into it. The trough is usually V-shaped, made of steel or wood, is slightly longer than the core tube, and supported by sturdy, steel sawhorses. If a solid core barrel is used, the barrel must be elevated at one end so that the core will slide out of the bottom. However, care must be taken to assure that the core pieces are not dropped or misplaced as they emerge from the barrel. Another technique that is often used with much success requires air pressure to force the core out. A close-fitting piston is inserted into the top end of the core barrel which is sealed with a cap containing an air connection. Air pressure is slowly applied and the core is slowly pushed out of the barrel.

Invariably, the bottom of the core run is tightly held in the keeper spring and may need to be broken out with a hammer.

After the core has been cleared from the barrel, it is first cleaned with a scrub brush and water, then pieced together to determine its original length. Continuous lengths of oil shale core are often recovered, but frequently the core is rubblized so that it is very difficult to reassemble into its original form. In such a case, an estimate of its original length is made based on the quantity of material recovered. Lost core sections must be identified as best as possible and marked with a labeled slab of wood. The core is then marked with stripes along its entire length with broad, felt tip pens to permanently identify the orientation of the core. Normally, the right side of the core, looking at it as if it was standing in the hole, is shown in red and the left is indicated by blue. After *striping*, the depths are marked and labeled on the core with a broad black felt tip pen.

Core samples should be logged, wrapped, and boxed as soon as possible after recovery from the hole. This will minimize the chance for misplacing core sections. It is preferable to log the core while it is lying in the core tray before boxing it (Fig. 5). A typical form for logging is shown in Fig. 6. Each core run is logged on separate sheets. Space is provided for entering all data on the core including Rock Quality Designation (RQD) for geotechnical characterization. Also,

Fig. 5. Field geologist logging core before it is wrapped and boxed.

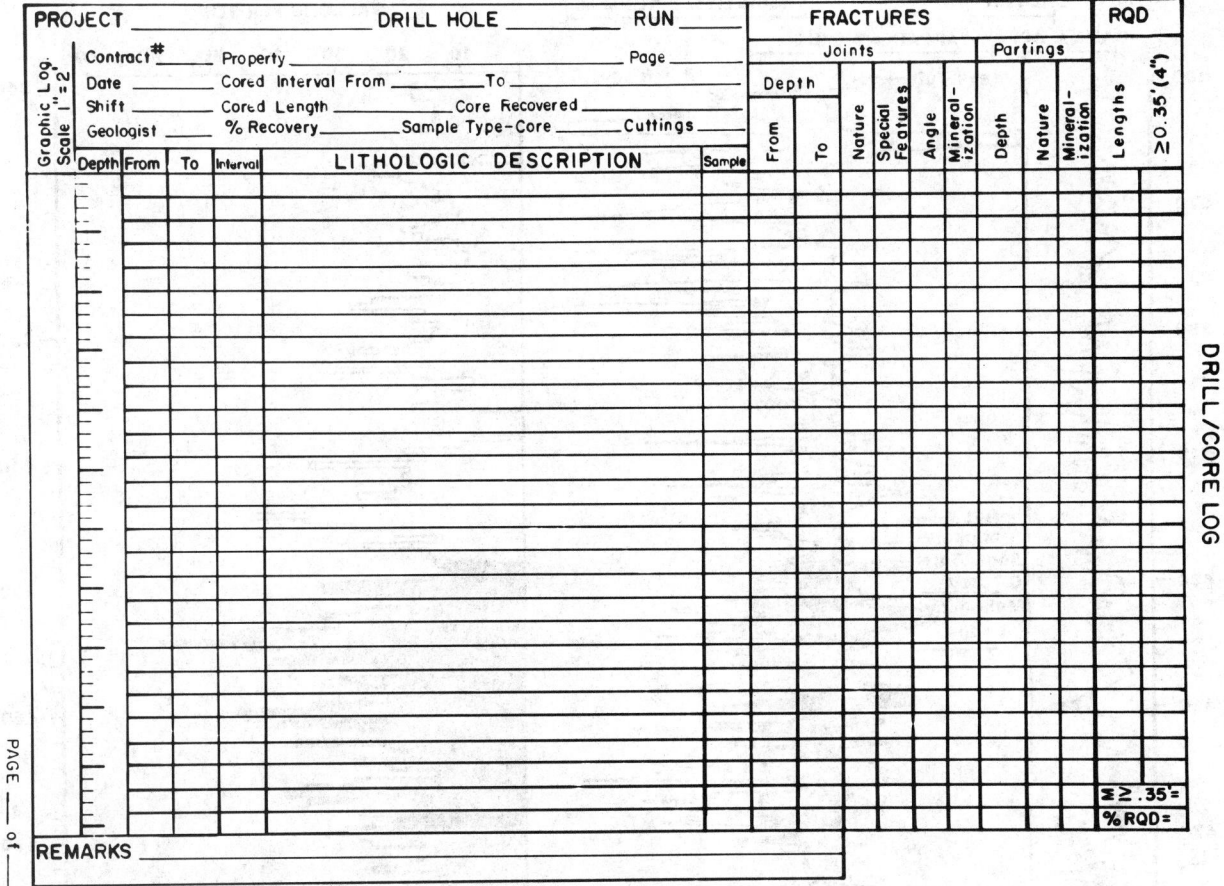

Fig. 6. Typical logging form used in oil shale exploration.

a cover sheet is completed showing the location, stratigraphic points, and other important information about each drill hole. It is important to note all information derived from logging so that there is a complete record of each hole.

After logging and before boxing the core, geotechnical samples are taken. These samples are usually wrapped in aluminum foil and dipped in paraffin wax to preserve the integrity of the sample. A wooden spacer is left in place of the geotechnical sample so that it can be returned to the core box after testing.

The method for placing the core in the storage box should be consistent. Normally, the core is placed so that the red stripe on the core is to the right, the top of the core section is in the upper left-hand corner of the box, and the core bottom is in the lower right-hand corner. With this method, the core boxes can be placed side by side to view large sections of the core.

Normally, Colorado oil shale cores are not wrapped for preservation during storage, but, to prevent any degradation from affecting test results, all samples and tests should be taken and completed soon after recovery. In cases where the oil shale oxidizes rapidly, the cores may be protected by wrapping in plastic core tubes or, in extreme cases, wrapped in aluminum foil and dipped in paraffin.

Geophysical Logging

Down-the-hole geophysical logging is commonly used in the oil, gas, and nonmetallic minerals industries, and has proven to be valuable in the geologic evaluation of oil shale. In geophysical logging a sonde is lowered into a bore hole by means of a cable. As the sonde is raised in the hole, it detects specific physical, electrical, and radioactive characteristics of the wall rock and bore hole. This information is transmitted to recording and processing equipment located on the surface. The resulting geophysical log provides a permanent, accurate, and detailed record of the bore hole.

The basic suite of geophysical logs normally used in an oil shale exploration program includes gamma-ray, caliper, gamma-ray density, and non-focused resistivity. These logs provide the geologist with additional valuable information for stratigraphic correlation, as well as an indication of lithology and oil shale grade, that cannot be derived by the logging of core or cutting samples alone.

A portion of a typical geophysical log, developed for an oil shale section in the Piceance Creek basin, Colorado, is shown in Fig. 7. On the far left is the natural gamma-ray curve, which exhibits the natural gamma radioactivity of the rock, measured in API units. This curve is useful for delineating sandstone and shale in the overburden, and tuff and barren marlstone beds in the oil shale. Barren marlstone and shale typically show values greater than 200 API. Oil shale, regardless of grade, exhibits values from about 80 to 120 API. Prominent tuff beds, such as the Wavy Tuff and Mahogany Marker beds in the Piceance Creek basin, will typically show values slightly higher than the surrounding oil shale. The sandstone in the overburden shows values from

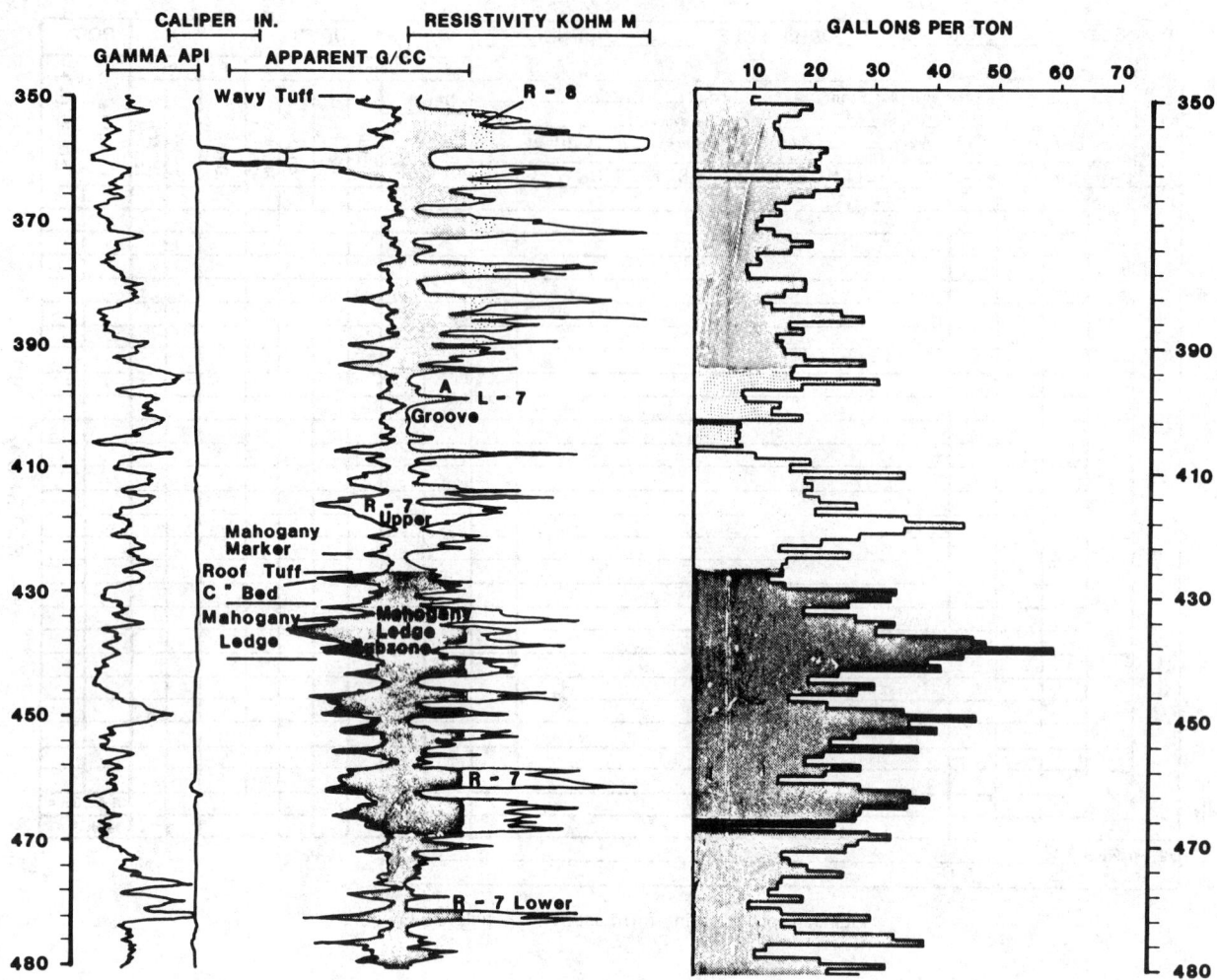

Fig. 7. Geophysical logs for an oil shale section from the Piceance Creek Basin, Colorado.

80 to 120 API with cleaner sands showing values below 40 API.

A caliper log is shown as the second curve from the left (Fig. 7). This curve shows variations in hole diameter and reveals important features such as vugs. It is useful for correlating vug zones and must be considered when evaluating other logs, such as density and natural gamma logs, that are dependent upon hole size. The caliper log is generated using a mechanical arm that detects irregularities in the hole wall.

The third curve from the left is a density curve, which has proven to be one of the most useful curves in the geophysical logging of oil shale. This log basically measures the electron density of the rocks encountered in a bore hole by bombarding them with gamma rays from a radiation source, and measuring the back scatter. The emitted gamma rays collide with the electrons in the rock, causing a back scatter and decrease of energy referred to as the Compton effect. The amount of back scatter is related to the electron density which is, in turn, related to the true bulk density of the rock (a function of grain density, porosity, and density of fluids filling the pore spaces).

The bulk density of oil shale is related to its grade. Generally, as the grade increases, its density decreases. The density curve, therefore, indirectly reflects the relative grade of the oil shale. This relationship has prompted many workers to explore the possibilities of accurately identifying the grade of oil shale from geophysical logs, thus eliminating much of the extensive, costly core drilling required to develop a property.

A field geologist can use the density log to accurately identify high-grade beds and low-grade markers. In the example presented in Figure 7 from the Piceance Creek Basin of Colorado, the high-grade Mahogany Ledge is clearly visible on the density log as a low-density, high-grade unit. Likewise, the lean A-Groove, B-Groove, and other lean zones can easily be recognized by their higher density.

The fourth curve shown in Fig. 7 is a resistivity log. It is a measure of the electrical resistance of the fluid contained within the rock unit. Water containing dissolved salts, for example, shows low resistance. Increased resistance reflects increased carbon and decreased pore space. Generally, higher grade oil shale shows high resistivity, which reflects the high carbon content. The leaner oil shales show lower resistivity.

Sampling

Sampling is a key part of an oil shale exploration program since it provides information on grade, quality, and rock characteristics. This information forms much of the foundation for future mine engineering studies and for determining the economic viability of a project.

Three techniques of oil shale sampling are commonly used to obtain representative samples: core, channel, and bulk. The techniques chosen in an actual application will depend upon the requirements of the tests to be made on the samples. Core samples are usually satisfactory for grade determination, geotechnical testing, and bench scale process testing. Channel samples are useful for grade determination and bench scale process testing. For pilot plant testing, bulk samples are required.

Core Sampling: One of the primary objectives of a drilling program is to obtain representative core samples of the oil shale section. A vertical, solid core is perhaps the most representative sample obtainable because it provides material for geotechnical testing and assaying and enables accurate geologic control.

After geologic logging has been completed and geotechnical samples have been removed, the core samples are normally sent to a laboratory where they are sawed in half. Sawing is preferred to splitting because of the better, more representative sample it gives. One half of the sample is returned to the core box for storage. The other half is retained for assaying. The core sections that were removed for geotechnical testing are returned for assaying after the testing is complete.

Channel Sampling: Channel samples are commonly collected from surface exposures at right angles to the bedding. Channel sampling provides material for grade determination only. Because of the extreme grade variability of oil shale in the vertical direction, it is extremely important to equally sample all portions of the zone of interest.

When surface exposures are to be sampled, a good site is chosen which is easily accessible and is free of the obvious effects of weathering or other forms of degradation. The surface of the exposure is cleaned with a brush to remove any foreign material. The location of the sample is marked and sampling begun. A channel is cut through the section of oil shale using a hammer and chisel. As pieces are broken out, they are placed in a sample bag. Care should be taken not to lose any pieces and to cut an even channel so that the whole section is equally sampled.

Channel sampling can also be carried out using a small hand-held rock saw. This technique is commonly used where electrical or air hookups are available. The saw is used to cut two parallel kerfs in the oil shale face. The resulting *key* between the kerfs is then broken out with a hammer and chisel. A careful operator can obtain very good, reproducible results using this technique.

Bulk Sampling: The feasibility of producing oil shale from a deposit must be determined ultimately by bulk sampling during the final phases of an exploration program. Bulk sampling is carried out primarily to validate grade analyses obtained from drill hole samples, to determine that these samples are representative of the deposit under investigation, and to provide a reliable basis for calculating reserves. Bulk sampling will also provide additional information on the geology of the deposit and geotechnical data necessary for developing a mining operation and for pilot plant testing. Bulk sampling must be carefully carried out to produce oil shale samples that are as representative of the to-be-mined characteristics as possible.

An oil shale bulk sampling program involves the extraction of quite large quantities of material, normally by underground mining but occasionally by surface mining. Typically, access is made by either a small-diameter shaft or

an adit, depending on the topography of the deposit under investigation, from which levels and/or crosscuts and ramps are driven into the oil shale.

Any variations in lithology, grade, or rock characteristics indicated by the drill holes will determine the extent of mining, both horizontally and vertically, required to obtain bulk samples.

Before any bulk sampling of an oil shale deposit is carried out, the procedures for collecting and analyzing samples from predetermined locations must be established. This will assure that acceptable correlations can be made with information derived from the drill holes. The procedures for collecting and preparing oil shale samples are based on the ASTM procedures for coal sampling and conform to established industry practices.

Testing

The object of any sampling program is to obtain representative samples on which tests can be made. Grade determination is the most common reason for sampling. Geotechnical and hydrological testing is also conducted on core samples to determine the strength and aquifer characteristics of the rocks in place. Tests for methane are also carried out, especially where an underground mine is envisaged.

Grade Determination: Oil shale grade determinations are conducted on samples using the ASTM standard procedures D3904-85 titled "Oil from Oil Shale (Resource Evaluation by the USBM Fischer Assay Procedure)." This procedure, commonly referred to as the modified Fischer assay procedure, involves the preparation of a representative sample approximately 100 g (3.5 oz) in weight by crushing and splitting the total sample section of core; heating the sample in a special retort according to a carefully controlled time temperature curve; and collecting and measuring the products of the destructive distillation process. Results are expressed in gallons per ton oil grade; water in gallons per ton; weight percent of oil, water, retorted shale, and gas plus loss; oil specific gravity; and coking tendency of retorted shale.

For accurate grade correlation and grade characterization of the mining zone, 0.3-m (1-ft) sections are normally analyzed for grade, at least through the zone of interest. Thicker sections can be analyzed in reconnaissance drilling and in zones other than the zone of interest, but the thicker sections smear the characteristic high- and low-grade markers, making accurate correlation impossible.

Geotechnical Testing: A geotechnical investigation considers not only the characteristics of intact and jointed rock, but also hydrological factors and the overall geological structure of the property under investigation. Factors such as hydrostatic pressure and zones of weakness (faults, shear zones, and dikes) that affect the long-term stability of a mine and related facilities, must be studied in detail so that they can be included in the mine design.

Geotechnical data derived from previous studies, published literature, and field mapping are used extensively in developing the basic geotechnical characteristics of a property. Additional geotechnical information concerning the property is acquired by shallow borings and deep hole drilling, and by testing of soils and rock samples. The purpose of these tests is to determine, more fully, the parameters necessary to carry out stability and foundation analyses for the design of a mine and related facilities.

Normally, a geologic exploration drilling program will be planned to include the collection of geotechnical and hydrological data to minimize drilling costs. Geotechnical logs are prepared for drill hole cores sampled by a geotechnical engineer. Selected samples are then sent to a laboratory for testing. Analyses of intact, jointed and broken rock samples are performed. Typical geotechnical laboratory procedures carried out for oil shale include the following tests: uniaxial compression, triaxial compression, indirect tension (Brazilian), direct shear, creep, and modulus of rupture.

Tests are also performed on retorted shale samples to determine water content, dry density, specific gravity, compaction, consolidation, and gradation characteristics. The data derived from these tests are used to establish mine design parameters, determine stability criteria, and foundation requirements.

Hydrological Testing: The data required for hydrologic investigations of a property originate from site specific field testing programs. Since the methodologies utilized to obtain subsurface hydrologic data are closely tied to geologic exploration, hydrogeologic studies are often conducted simultaneously with exploration drilling to maximize the use of each hole drilled and equipment available at the site. The intensity of the hydrologic field program is dependent upon the stage of geologic exploration.

The determination of preliminary hydrogeologic conditions is primarily based upon observations made during drilling. The geologist on site should be aware of and record indicators that exhibit potential hydrogeologic properties such as lost circulation, relative moisture content of cuttings (if drilling with air), geologic material suitable for aquifers (sands, gravels, silts), possible confining materials (clay lenses), increase in water production at various depths, and hole stability (caving). Since the majority of oil shale exploration drilling is accompanied by geophysical logging, several of the parameters measured in the logging process provide good preliminary indications of subsurface hydrologic conditions.

As the geologic exploratory program is expanded, a site specific hydrologic testing program can be designed and implemented to identify additional hydrologic properties required for mine planning. Several of the exploration holes may be cased to obtain potentiometric surfaces of individual aquifers. Groundwater gradients, flow rates, and potentiometric heads can then be determined on a preliminary basis.

In order to evaluate an aquifer unit for specific hydraulic properties (transmissivity and storativity) it must be artificially stressed and the subsequent changes measured. Several basic methods are available for this type of testing on a preliminary basis. Packer and Airlift Stem testing involves the use of an uncased hole while Slug and Airline testing is performed in small diameter, cased holes (Lowman, 1972 and Ferris, 1962).

In order to determine the hydraulic characteristics of a property to the degree of confidence needed to assess operational groundwater impacts (for dewatering, etc.), an aquifer testing technique, using a submersible pump (pump testing) and a series of observation wells, is routinely used.

During hydrological testing, samples of the aquifer water should be collected and preserved for laboratory analysis. The results of this analysis will further characterize the hydrologic regime of the property under investigation.

Methane Testing: The US Bureau of Mines has established, by laboratory absorption measurements on samples, that the amount of methane absorbed by oil shale is proportional to pressure and oil yield. The Bureau has also analyzed oil shale cores for their gas content by a direct measurement method. One recent study illustrating the use of this method is presented in Schatzel and others (1987). This publication presented the results of this study on the methane content of oil shale from the Cathedral Bluffs Mine, located in the Piceance Creek Basin. They found that cores taken from deep locations of an oil shale deposit, far from outcrops, yield more gas than cores taken at or near outcrops. This indicates, taking into account variations of oil yield and other data, that oil shale mines that are both deep and far from an outcrop will emit low levels of methane gas.

Where underground mining of oil shale is being considered, then, it is essential that some testing for methane is carried out, since there is a possibility that methane gas may be liberated from oil shale into the mine creating an explosion hazard.

Methane testing procedures for oil shale were developed originally from those used for determining the methane content of coal seams (Kissel et al., 1973). The procedure for oil shale involves recovering sections of core, sealing the samples in tubes, and monitoring the release of gas over a period of time. Analysis of the test results shows relative yields of methane gas in cubic feet per ton.

Oil shale core samples required for methane testing should be obtained from those holes that theoretically offer the best chance of encountering methane. As the core is recovered from the hole, it should be placed immediately in the tube, sealed, and protected from excessive heat. In the laboratory, the tube containing the core is connected to a gas collection and monitoring apparatus. The collected gas is sampled and analyzed to determine its content. The volume of collected gas will give an indication as to the amount of methane gas which might be expected in an underground mining situation at that site. After the test is complete, the core is removed, logged, and sampled for assay.

Geologic Evaluation

The purpose of a geological evaluation is to review and interpret all the available data concerning the property under investigation and to present a thorough description of the geologic model that is interpreted from the data. All sources of information and data should be documented.

An important part of the geologic evaluation is presentation of the data and interpretations of maps and cross sections. Maps commonly produced in a geologic evaluation of oil shale include surface geologic maps; surface structure and lineament maps; subsurface structure maps, isopach maps, isograde maps, and oil yield maps which show oil yields in barrels per thousand acres and are commonly used for resource estimation.

Computers are very effectively used for geologic evaluations, since they are excellent for manipulating data, identifying average grades for sections, and identifying zones meeting certain grade and thickness criteria. Computer graphics also aid in developing maps and subsequent volumetric calculations such as resource estimations.

REFERENCES

ASTM, 1985, "Standard Test Method for Oil Shale" (Resource Evaluation by the USBM Fischer Assay Procedure) Designation D3904-85, *Annual Book of ASTM Standards,* American Society for Testing and Materials, 13 pp.

Baughman, G. L., 1978, *Synthetic Fuels Data Handbook,* second ed., Cameron Engineers, Inc., 438 pp.

Beard, T. N., Tait, D. B., and Smith, J. W., 1974, "Nahcolite and Dawsonite Resources in the Green River Formation of Colorado," *Guidebook of the Energy Resources of the Piceance Creek Basin, Colorado,* Twenty-fifth Field Conference, Rocky Mountain Association of Geologists, pp. 101-109.

Belser, C., 1948, "Oil Shale Resources of Colorado, Utah and Wyoming," *A.I.M.E. Technological Publication No. 2358, Petroleum Technology,* Vol II, 11 pp.

Belser, C., 1951, "Green River Oil Shale Reserves of Northwestern Colorado," *Report of Investigations 4769, US Bureau of Mines,* 13 pp.

Bradley, W. H., 1964, "Geology of Green River Formation and Associated Eocene Rocks in Southwestern Wyoming and Adjacent Parts of Colorado and Utah," *Professional Paper 496-A,* US Geological Survey, pp. A1-A86.

Brown, J. W. and Repsher, R. C., 1972, "Detection of Rubble Zones in Oil Shale by the Electrical Resistivity Technique," *Report of Investigations, 7674,* US Bureau of Mines, 17 pp.

Cashion, W. B., 1967, "Geology and Fuel Resources of the Green River Formation, Southwestern Uinta Basin, Utah and Colorado," *Professional Paper 548,* US Geological Survey, 48 pp.

Cashion, W. B., 1959, "Geology and Oil Shale Resources of Naval Oil Shale Reserve No. 2: Uinta and Carbon Counties, Utah," *Bulletin 1072-0,* US Geological Survey, pp. 753-793.

Chong, K. P., et al., 1976, "Characterization of Oil Shale Under Uniaxial Compression," *Site Characterization Proceedings, 17th U.S. Symposium on Rock Mechanics,* Utah Engineering Exposition Station, Salt Lake City, pp. 5C5-1 to 5C5-8.

Dana, G. F. and Smith, J. W., 1972, "Oil Yields and Stratigraphy of the Green River Formations Tipton Member at Bureau of Mines Sites near Green River, Wyoming," *Report of Investigations 7681,* US Bureau of Mines, 46 pp.

Decora, A. W., et al., 1971, "A Rapid Method for Estimating Oil Yields of Oil Shale by Broad-Line NMR Spectrometry," *Fuel, Chemicals,* Vol. 15, No. 1, A.C.S. Preprints Division, pp. 38-46.

Decora, A. W., McDonald, F. R., and Cook, G. L., 1971, "Using Broad-Line Nuclear Resonance Spectrometry to Estimate Potential Oil Yields of Oil Shales," *Report of Investigations 7523,* US Bureau of Mines, 30 pp.

Donnell, J. R., 1957, "Preliminary Report on Oil Shale Resource of Piceance Creek Basin, Northwestern Colorado," *Bulletin 1042-H,* US Geological Survey, pp. 255-271.

Donnell, J. R., 1961, "Tertiary Geology and Oil Shale Resources of the Piceance Creek Basin between the Colorado and White Rivers, Northwestern Colorado," *Bulletin 1082-L,* US Geological Survey, pp. 835-891.

Duncan, D. C., 1976, "Geologic Setting of Oil-Shale Deposits and World Prospects," *Oil Shale,* T. F. Yen and G. V. Chilingarian, eds., Elsevier Scientific Publishing Co., Amsterdam, pp. 13-26.

Duncan, D. C. and Belser, C., 1950, "Geology of the Eastern Part of the Piceance Creek Basin, Rio Blanco and Garfield Counties, Colorado," *Oil and Gas Investigation Map OM-119,* US Geological Survey.

Duncan, D. C. and Swanson, V. E., 1965, "Organic-Rich Shale of the United States and World Land Areas," *Circular 523,* US Geological Survey, 30 pp.

Ferris, J. G., et al., 1962, "Theory of Aquifer Tests," *Water-Supply Paper 1536-E,* US Geological Survey, 173 pp.

Fertl, W. H., 1976, "Evaluation of Oil Shale Using Geophysical Well-Logging Techniques," *Oil Shale,* T. F. Yen, and G. V. Chilingarian, eds., Elsevier Scientific Publishing Co., Amsterdam, pp. 199-213.

Huggins, C. W., Green, T. E. and Turner, T. L., 1973, "Evaluation of Methods for Determining Nahcolite and Dawsonite in Oil Shales," *Report of Investigations 7781,* US Bureau of Mines, 21 pp.

Kissell, F. N., McCulloch, C. M., and Elder, C. H., 1973, "The Direct Method of Determining Methane Content of Coalbeds for Ventilation Design," *Report of Investigations 7767,* US Bureau of Mines, 17 pp.

Lowman, S. W., 1972, "Groundwater Hydraulics," *Professional Paper 708,* US Geological Survey, 70 pp.

Matta, J. E., LaScola, J. C. and Kissell, F. N., 1977, "Methane Absorption in Oil Shale and its Potential Mine Hazard," *Report of Investigations 8243,* US Bureau of Mines, 13 pp.

Robb, W. A. and Smith, J. W., 1974, "Mineral Profile of the Green River Formation Oil Shales at Colorado Corehole No. 1," *Guidebook to the Energy Resources of the Piceance Creek Basin, Colorado,* Twenty-fifth Field Conference, Rocky Mountain Association of Geologists, pp. 91-100.

Robb, W. A. and Smith, J. W., 1976, "Mineral Profile of Wyoming's Green River Formation-Oil Shales Sampled by Blacks Fork Core," *Earth Science Bulletin,* Vol. 9, No. 1, pp. 1-8.

Robinson, W. E., 1976, "Origin and Characteristics of Green River Oil Shale," *Oil Shale,* T. F. Yen and G. V. Chilingarian, eds., Elsevier Scientific Publishing Co., Amsterdam, pp. 61-79.

Schatzel, S. J., Hyman, D. M., Sainato, A., and LaScola, J. C., 1987, "Methane Contents of OilShale From the Piceance Basin, CO," *Report of Investigations 9063,* US Bureau of Mines, 32pp.

Smith, J. W., 1956, "Specific Gravity: Oil Yield Relationship of Two Colorado Oil Shale Cores," *Industrial and Engineering Chemistry,* Vol. 48, pp. 441-444.

Smith, J. W., 1958, "Applicability of a Specific Gravity Oil Yield Relationship to Green River Oil Shale," *Chemical & Engineering Data Series,* Col. 3, No. 2, pp. 306-310.

Smith, J. W., Trudell, L. G. and Stanfield, K. E., 1963, "Comparison of Oil Yields from Core and Drill-Cutting Sampling of the Green River Oil Shales," *Report of Investigations 6299,* US Bureau of Mines, 35 pp.

Smith, J. W., Trudell, L. G., and Stanfield, K. E., 1963, "Drill Cutting Sampling for Oil Yields of Green River Oil Shales," *Colorado School of Mines Quarterly,* Vol. 58, No. 4, pp. 113-127.

Smith, J. W. and Stanfield, K. E., 1964, "Oil Yields and Properties of the Green River Oil Shales in the Uinta Basin, Utah," *Guidebook to the Geology and Mineral Resources of the Uinta Basin,* Intermountain Association of Petroleum Geologists, pp. 213-221.

Smith, J. W. and Stanfield, K. E., 1964, "Oil Yields of Devonian New Albany Shales, Kentucky," *Bulletin of American Association of Petroleum & Geology,* Vol. 48, No. 5, pp. 712-714.

Smith, J. W. and Young, N. B., 1964, "Specific Gravity to Oil-Yield Relationships for Black Shales of Kentucky's New Albany Formation," *Report of Investigations 6531,* US Bureau of Mines, 13 pp.

Smith, J. W. and Stanfield, K. E., 1965, "Oil Shale of the Green River Formation in Wyoming," *Guidebook,* Nineteenth Field Conference, Wyoming Geological Association, pp. 167-170.

Smith, J. W., 1966, "Conversion Constants for Mahogany-Zone Oil Shale," *Bulletin, American Association Petroleum & Geology,* Vol. 50, No. 1, pp. 167-170.

Smith, J. W. and Harbaugh, J. W., 1966, "Stratigraphic and Geographic Variation of Shale-Oil Specific Gravity from Colorado's Green River Formation," *Report of Investigations 6883,* US Bureau of Mines 11 pp.

Smith, J. W. and Young, N. B., 1967, "Organic Composition of Kentucky's New Albany Shale: Determination and Uses," *Chemical Geology,* Vol. 2, pp. 157-170.

Smith, J. W., Thomas, H. E., and Trudell, L. G., 1968, "Geologic Factors Affecting Density Logs in Oil Shale," *Trans., Society of Professional Well Log Analysts,* 95th Anniversary, Logging Symposium, pp. 1-17.

Smith, J. W., Trudell, L. G., and Dana, G. F., 1968, "Oil Yields of Green River Oil Shale from Colorado Corehole Number 1," *Report of Investigations 7071,* US Bureau of Mines, 28 pp.

Smith, J. W., Trudell, L. G. and Stanfield, K. E., 1968, "Characteristics of Green River Formation Oil Shales at Bureau of Mines Wyoming Corehole Number 1," *Report of Investigations 7172,* US Bureau of Mines, 92 pp.

Smith, J. W., 1969, "Theoretical Relationship Between Density and Oil Yield for Oil Shales," *Report of Investigations 7248,* US Bureau of Mines, 14 pp.

Smith, J. W. and Young, N. B., 1969, "Determination of Dawsonite and Nahcolite in the Green River Formation Oil Shales," *Report of Investigations 7286,* US Bureau of Mines, 20 pp.

Smith, J. W., Beard, T. N. and Wade, P. M., 1972, "Estimating Nahcolite and Dawsonite Content of Colorado Oil Shale from Oil-Yield Assay Data," *Report of Investigations 7689,* US Bureau of Mines, 24 pp.

Smith, J. W., Trudell, L. G., and Robb, W. A., 1972, "Oil Yields and Characteristics of Green River Formation Oil Shales at Wasco Ex-1, Uintah County, Utah," *Report of Investigations 7693,* US Bureau of Mines, 102 pp.

Smith, J. W., 1976, "Relationship Between Rock Density and Volume of Organic Matter in Oil Shales," *Report of Investigation 76/6,* Laramie Energy Research Center, 11 pp.

Stanfield, K. E. and Frost, I. C., 1946, "Method of Assaying Oil Shale by a Modified Fischer Retort," *Report of Investigations 3977*, US Bureau of Mines, 11 pp.

Stanfield, K. E. and Frost, I. C., 1949, "Method of Assaying Oil Shale by a Modified Fischer Retort," *Report of Investigations 4477*, (Revision of US Bureau of Mines *Report of Investigations 3977*), US Bureau of Mines, 13 pp.

Stanfield, K. E., et al., 1954, "Oil Yields of Sections of Green River Oil Shale in Colorado, Utah and Wyoming, 1945-1952," *Report of Investigations 5081*, US Bureau of Mines, 153 pp.

Stanfield, K. E., et al., 1957, "Oil Yields of Sections of Green River Oil Shale in Colorado, 1952-1954," *Report of Investigations 5321*, US Bureau of Mines, 132 pp.

Stanfield, K. E., et al., 1960, "Oil Yields of Sections of Green River Oil Shale in Colorado, 1954-1957," *Report of Investigations 5614*, US Bureau of Mines, 186 pp.

Stanfield, K. E., Smith, J. W., and Trudell, L. G., 1964, "Oil Yields of Sections of Green River Oil Shale in Utah, 1952-1962," *Report of Investigations 6420*, US Bureau of Mines, 217 pp.

Stanfield, K. E., et al., 1965, "Bureau of Mines-Atomic Energy Commission Colorado Corehole Number 1, Rio Blanco County, Colorado," Open File at Laramie Petroleum Research Center Library.

Stanfield, K. E., Smith, J. W., and Trudell, L. G., 1967, "Oil Yields of Sections of Green River Oil Shale in Colorado, 1957-1963," *Report of Investigations 7051*, US Bureau of Mines, 284 pp.

Thomas, H. E. and Smith, J. W., 1970, "Caliper Location of Leached Zones in Colorado Shale," *Log Analyst*, Vol. II, No. 4, pp. 12-16.

Trudell, L. G., Beard, T. N., and Smith, J. W., 1970, "Green River Formation Lithology and Oil Shale Correlations in the Piceance Creek Basin, Colorado," *Report of Investigations 7357*, US Bureau of Mines, 212 pp.

Trudell, L. G., Roehler, H. W., and Smith, J. W., 1973, "Geology of Eocene Rocks and Oil Yields of Green River Oil Shales on Part of Kinney Rim, Washakie Basin, Wyoming," *Report of Investigations 7775*, US Bureau of Mines, 151 pp.

Trudell, L. G., Beard, T. N., and Smith, J. W., 1974, "Stratigraphic Framework of Green River Formation Oil Shales in the Piceance Creek Basin, Colorado," *Guidebook of the Energy Resources of the Piceance Creek Basin, Colorado*, Twenty-fifth Field Conferences, Rocky Mountain Association of Geologists, pp. 65-69.

Young, N. B. and Smith, J. W., 1970, "Dawsonite and Nahcolite Analyses of Green River Formation Oil Shale Section, Piceance Creek Basin, Colorado," *Report of Investigations 7445*, US Bureau of Mines, 22 pp.

2.10 Industrial Minerals

RICHARD H. OLSON, EDWIN H. BENTZEN, III, AND GORDON C. PRESLEY, EDITORS

1. Asbestos*

Asbestos is the generic name given to a group of fibrous mineral silicates found in nature. They are all incombustible and can be separated by mechanical means into fibers of various lengths and cross sections, but each differs in chemical composition from the others. The commercial grades of asbestos consist of a spectrum of lengths and sizes and are not definitive as are most organic fibers.

It is generally recognized that there are two main groups of asbestos. The first only contains the fibrous serpentine called chrysotile and comprises about 94% of the world production of asbestos. The second group contains five minerals in the amphibole series—crocidolite, amosite, anthophyllite, tremolite, and actinolite. The latter two have no commercial importance.

GEOLOGY

General

Chrysotile constitutes about 94% of the current world production of asbestos and, of this amount, all but a fraction of a percent is derived from deposits whose host rocks are ultrabasic in composition (Berger, 1963; Bates, 1969; Keith and Bain, 1932). This other fraction of the chrysotile production is derived from serpentinized dolomitic limestone. Among the other varieties of asbestos, amosite and crocidolite are found in certain metamorphosed ferruginous sedimentary formations and, together, account for some 5.3% of world production. Tremolite and anthophyllite make up the balance of the production and are generally found in association with highly metamorphosed ultrabasic rocks.

The bulk of world chrysotile production comes from the USSR and Canada. Southern Africa, including Zimbabwe, Swaziland, and South Africa, is the third largest producer with the balance coming from other parts of the world.

Of the Canadian production, by far the greatest amount comes from the Eastern Townships of Quebec with northern British Columbia, Ungava, Newfoundland, and Ontario contributing the balance. Russian production comes mainly from the Bazhenovo district in the central Urals, the Dzhetygara area of northwest Kazakhstan, the Kiembay area in the southern Urals and from Ak-Dovurak near the Yenesei River west of Lake Baikal.

The age of asbestos deposits varies greatly from earliest Precambrian in Zimbabwe and Swaziland to Upper Jurassic in California. The Ontario, Ungava, and Brazil deposits are all Precambrian, while those of the Eastern Townships, Vermont, and Newfoundland are all ascribed a Mid-Paleozoic age, associated with early folding in the Appalachian Mountain belt. The deposits of western Canada are connected with mountain building in the Late Paleozoic while the Russian deposits vary from Early Paleozoic to Late Paleozoic or Triassic in age.

*Excerpt from Mann, E.L., 1983, "Asbestos," *Industrial Minerals and Rocks,* 5th ed., Vol. 1, S.J. Lefond, ed., AIME, New York, pp. 435-484.

Structures

General deformation in the form of faulting, folding, or shearing evidently plays a major role in the localizing of asbestos deposits (Cooke, 1937). The introduction of swarms of small intrusive dikes of acidic to intermediate composition also contributes to the general fracturing and opening up of the rock which, in turn, leads to serpentinization and the formation of chrysotile fiber.

The majority of chrysotile ore deposits tend to occur in serpentinized peridotite rather than dunitic host material. This is probably due to the fact that serpentinized dunite bodies tend to flow when subjected to stress, thereby aiding in the development of fracturing in the surrounding more brittle peridotite.

Asbestos Veining

Chrysotile asbestos is referred to as cross fiber where the fibers lie transverse to the vein, and as slip fiber where the fiber lies in the plane of the vein (Gold, 1967; Cooke, 1937). In the case of cross fiber, the orientation may be anywhere from normal to sharply inclined to the vein walls, in which case it can be referred to as oblique fiber. The fiber in the veins may be straight, gently curved, or contorted. Fiber containing sharp flexures is apt to break easily.

The quality of the fiber ranges from soft and silky to harsh; from highly flexible to brittle; and from strong to weak.

Partings are often present in the veins, either as microscopic discontinuities or as irregular shaped inclusions composed of picrolite, serpentine, brucite, and magnetite, or combinations of these. The veins may be relatively persistent or short and lenticular, occurring in the form of stockworks or in parallel systems known as ribbon fiber. They may be fissure fillings, replacements, or stress-relief features, and are often the result of a combination of these processes.

Some very short fiber is derived from a variety referred to as "mass" fiber. Mass fiber consists of either the complete replacement of most of the original mineral constituents to fiber, each grain of which has a different fiber orientation, or a dense network of fine reticulating replacement veins. Both varieties result in a very high content of short fiber and the rock, which appears to be largely fibrous, is sometimes referred to as "fur" rock.

Other varieties of mass fiber include "platy" fiber, of which the Coalinga deposit is unique. A similar, but less important, variety is known as "mountain leather."

WORLD ASBESTOS DEPOSITS

North America

Canada: *Eastern Townships, Quebec*—Chrysotile asbestos was discovered in two localities in the Eastern Townships in 1878 and has been actively mined at Thetford Mines and Asbestos for over 100 years. The chrysotile deposits of Quebec's Eastern Townships occur intermittently along a major serpentine belt which arcs northeastward into the Gaspé

Peninsula in one direction and southward into the Appalachian Mountain belt of Vermont in the other.

The area yields over 80% of Canada's fiber production
and some 30% of the world total, thereby vying closely with
the Bazhenovo district of the Soviet Union for first place in
world production. Most of the major occurrences are located
along a sector 88.5 km (55 miles) long which lies parallel
to, and 80.5 km (50 miles) southeast of, the St. Lawrence
River extending from a point near East Broughton, 80.5 km
(50 miles) south of Quebec City, southwestward past Thetford Mines and Black Lake to Asbestos and Danville which
lie approximately 137 km (85 miles) east of Montreal.

This serpentine belt is a typical ophiolitic complex of
gabbroic and dioritic rocks, and of pyroxenite, peridotite and
dunite, serpentinized to various degrees with associated granite, rodingitic, and talc-carbonate rocks. This belt has been
described in the past as a partially serpentinized ultrabasic
intrusive emplaced in the crust along faulted zones of weakness. More recent study would indicate that this giant complex has been extruded onto or just under unconsolidated
aluminous and silicious sediments in a eugeosynclinal ocean
basin. Magmatic differentiation within the ultramafic magma
is thought to be responsible for the formation of the various
pyroxenite, peridotite, and dunite phases found in the complex.

The small granitic masses found within the ultramafic
complex have been considered intrusive but may be the result
of the ingestion and digestion of fragments of underlying sea
floor sediments. Subsequent faulting and shearing has resulted in numerous dislocations within and adjacent to the
ultramafic complex.

Consideration of this complex as a submarine extrusive
simplifies the problems of water for serpentinization, allowance for large volume changes, the occurrence of fiber as
open fracture fillings in stockworks, and banded fiber in
ribbon zones. It also explains the presence of "pillows,"
usually gradationally serpentinized from the surface inward.

The fiber occurs mainly in highly serpentinized peridotite
(harzburgite containing 10-15% enstatite) and is found in
numerous occurrences along the strike of the complex.

There are presently six deposits being mined in the Eastern Townships and numerous others are either dormant or
completely mined out. For purposes of description the deposits may be grouped on the basis of their distribution in
the belt, i.e., the Pennington dike deposits, those of Thetford
Mines area, the Black Lake deposits, and those of Asbestos.

Pennington Dike, Thetford Mines-East Broughton Area
(Anon., 1972; Merrill, 1957; Riordon, 1957; Rowbotham,
1970)—Separated from the main Thetford ultrabasic intrusive and extending northeastward for some 27 km (17 miles)
or more is the Pennington "dike" which has supported a
number of asbestos mining operations, both past and present,
over much of its length. This serpentinite body is actually
tabular and more sill-like in form, being closely comformable
with the country rocks which dip up to 60°SE. Shearing is
conspicuous throughout its length. This is interpreted as
being a major thrust fault, the plane of which occurs at the
base of the sill. Where the sill is narrow, shearing extends
from wall to wall and much of the rock is steatized, but the
wider portions which reach a thickness of up to 244 m (800
ft) are less affected. It is these wider bulges which contain
most of the major ore zones. Slip fiber predominates in all
the deposits, though some cross fiber is found in the wider,
less sheared sections, particularly toward the southwest.

The narrowness of the ore bodies and the attitude of the
sill are obstacles to mining any appreciable depth. Only two

mines, National Asbestos and Carey-Canadian, are presently
operating on the Pennington dike.

Thetford Mines Area (Cooke, 1937; Riordon, 1957,
1957a, b; Riordon and Laliberté, 1957)—A large mass of
periodotite is host to the Thetford Mines and Black Lake
deposits. The Thetford group lies at the northeast end of this
mass and contains four closely connected deposits, all of
which lie in a well-defined zone along the hanging-wall side
of a prominent fault structure marked by the presence of
intensive talc-carbonate alteration. The zone has a length of
1.8 km (6000 ft) and a width of 457 m (1500 ft) with the
largest and deepest deposits, the King-Bell-Johnson complex,
extending to a depth of 457 m (1500 ft) or more. These deep
zones are mined by underground methods. The other deposits
are the Bennett-Martin, the Beaver-Consolidated, and the
Beaver "C." The first two are now exhausted.

A number of subsidiary shears, most of which have dips
in the opposite direction to the main fault and strikes that
roughly parallel it, tend to break the ore zone into segments.
Most of these subsidiary shears are accompanied by numerous bodies of syenite, and evidently developed along the
original zones of weakness into which these acid rocks were
introduced. Some large bodies of serpentinized dunite situated within the ore zone contributed to the fracturing in the
surrounding peridotite by behaving as relatively incompetent
masses which tended to flow under stress.

Black Lake Area (Anon., 1972; Riordon, 1957a, c, d;
Riordon and Laliberté, 1957)—There are three major ore
bodies in the general vicinity of Black Lake. The British
Canadian-Megantic mine is over 1.6 km (1 mile) long and
0.8 km (0.5 mile) wide, and occupies a position adjacent to
a pronounced flexure in the peridotite contact immediately
east of the town of Black Lake. This ore body is cut up into
ore shoots by a series of roughly parallel shear zones and
syenite dikes striking tangentially to the folded contact. Large
bodies of serpentinized dunite are also found within the ore
body and produce low grade to barren sections.

Southwest of the town of Black Lake are the ore bodies
which comprise the Lake Asbestos mine, the lake having
been drained to permit mining of the deposit beneath. The
eastern-most of these ore zones is situated along the hanging-
wall side of a northwest-southeast transverse fault which dips
to the northeast. Like the major fault in Thetford this zone
has also undergone intensive steatization. Another ore body
to the west was located beneath a large, flat-lying acid dike
which evidently localized the fracturing in its vicinity and
no doubt was responsible for the formation of fiber in the
vicinity.

The Normandie mine is located about 1.6 km (1 mile)
to the west of Lake Asbestos and two miles southwest of the
town of Black Lake. This deposit consists of at least three
separate ore bodies lying along a northeasterly plunging fold
axis which manifests itself in a pronounced Z bend in the
peridotite-Caldwell schist contact in the vicinity. The old
Vimy Ridge deposit which is now completely worked out is
the most southerly of these and occurred close to surface,
with much of the upper portions of the ore body truncated
and removed by erosion.

The central zone lay completely buried beneath a capping
of drift and overburden, and was only discovered by regular
grid pattern diamond drilling in the peripheral areas surrounding the old Vimy Ridge mine. The Normandie open
pit mine, a very productive operation in its day, has now
been mined out.

Further deep drilling on the downdip extension of the
Normandie ore body led to the discovery of a deep ore zone

lying well beneath the overfolded capping of Caldwell quartzites. This deposit, the Penhale ore body, has been explored by underground methods and is now being considered for production.

The origin of these three ore bodies is closely related to the folding but may also be dependent on a fault zone trending southwest which parallels the fold axis to the east of the ore bodies.

Whereas the other deposits in the Eastern Townships are made up of a stockwork of asbestos veins, the Normandie, Vimy Ridge, and Penhale ore bodies are composed predominantly of a prominent system of parallel veins of ribbon fiber. These veins form a conjugate pattern but dip predominantly to the northeast with the rake of the ore zone.

The Vimy Ridge and Normandie ore bodies are separated by numerous large syenite dikes and some shearing, whereas the Normandie and Penhale deposits are separated by a zone of low grade to barren material.

Asbestos-Shipton Area (Allen et al., 1957; Bourassa, 1957; Rowbotham, 1970; Riordon and Laliberté, 1957)—The Jeffrey and the now-exhausted Nicolet ore bodies occur along the north and footwall side of a peridotite-dunite intrusive. The Jeffrey deposit, discovered over 100 years ago, has been producing since the early 1880s and is now the largest asbestos mine outside the USSR. The ore body itself has an ellipsoidal to cylindrical or roughly pipelike form, having an average diameter of about 610 m (2000 ft) and its major axis plunging about 55° to the southwest. The ore zone is bounded on both the footwall and hanging-wall contacts by major faults or zones of shearing which dip south-southeast at about 65° to 70°. The ore body is also cut by several irregular shears which trend roughly parallel to the footwall contact and divide the deposit into five major ore zones. Numerous acid dikes of dioritic to syenitic composition tend to follow the shear zones but a few exhibit later cross-cutting tendencies. The fiber is generally of good- to medium-grade cross fiber, though sections of slip fiber and mass fiber also occur.

The Nicolet ore body occurred as a narrow zone along the footwall contact, being bounded on its south side by a prominent zone of shearing. This deposit has now been exhausted.

Baie Verte, Newfoundland—Asbestos is currently mined near Baie Verte (Straw, 1961) on the north shore of Newfoundland. The Advocate ore body forms a continuous fiber zone within the highly serpentinized periphery of a massive elliptical pyroxenite body, 1.6 km (1 mile) long by 0.4 km (0.25) mile wide. The inward dipping fiber zone surrounds the massive pyroxenite core and is in turn surrounded by highly sheared serpentinite. The serpentinite is part of the 88.5-km (55-mile) belt of ultrabasic rocks that separate the schists of the Ordovician Fleur de Lys group from the volcanics to the east.

The massive pyroxenite core shows strong parallel, nearly vertical bands of pyroxene-rich and pyroxene-poor phases, suggesting that it is a fragment of differentiated oceanic crust caught up in a subsequent ultramafic flow. The fiber probably formed in fractures on the highly stressed, highly serpentinized periphery of this peridotite core.

The fiber occurs as a stockwork of good quality fiber veins averaging 6.35 to 7.9 mm (¼ to 5⁄16 in.) in length. After mining and milling, the fiber is loaded directly at the nearby terminal for shipment to overseas markets.

Reeves Township, Timmins, Ontario—Chrysotile asbestos occurs at Reeves (Douglas, 1970; Rowbotham, 1970; Hendry and Conn, 1957) 40 miles southwest of Timmins. This deposit lies near the northern tip of a large serpentine mass in contact with Precambrian metavolcanics. The south or footwall side of the ore body is bordered by a major east-west fault that dips 45°N in the serpentine. Extensive talc-carbonate alteration is common along this fault. Two northsouth diabase dikes cut the ore zone. These have caused pronounced alteration of the fiber along their contacts.

The deposit was mined for a number of years but finally closed in 1974 due to marginal grades and pervasive alteration of the fiber.

The deposit in Munro Township in eastern Ontario has long been exhausted by mining. This deposit occurred within a narrow ultrabasic dike bounded by faults containing pronounced talc-carbonate alteration. The ore zone was also offset by cross faulting.

Cassiar, Northern British Columbia—The Cassiar asbestos deposit (Douglas, 1970; Smitheringale, 1957) occurs at an elevation of over 1.8 km (6000 ft) near the peak of Mount McDame in the Cassiar Range. Fiber in the Cassiar deposit is long and of exceptional grade and quality and is associated with an ultrabasic sill intruded into a folded succession of volcanic flows and sediments of Devonian age. The asbestos occurs as cross fiber veins in dark green serpentine commonly 12.7 to 25.4 mm (½ to 1 in.) wide and up to 4.6 m (15 ft) long. This fiber is shipped in special containers by road to Whitehorse, then by narrow-gauge railway to Skagway and from there by ship to overseas markets.

Clinton Creek, Yukon Territory—Mining of asbestos at Clinton Creek (Douglas, 1970) near the Canada-Alaska boundary was short-lived and the relatively small deposit is now closed.

The deposit lies just south of the Yukon River within 241 km (150 miles) of the Arctic Circle. Here cross fiber averaging close to 6.35 mm (¼ in.) in length occurred in a small ultrabasic body enclosed by phyllites, argillites, and quartzites of Paleozoic age. A portion of the deposit has also undergone strong talc-carbonate alteration. This remote location necessitated costly road, rail, and shipping transportation similar to that of Cassiar asbestos which decreased the margin of profit considerably.

Asbestos Hill, Putunig, Ungava, Quebec—Situated some 64 km (40 miles) from tidewater on Quebec's Hudson Strait (Mann, 1962; Stewart, 1978) on the western margin of a large ultrabasic to basic intrusive, is one of Canada's newest and richest asbestos producers, Asbestos Hill. This deposit, about 762 mm (2500 ft) long and up to 61 m (200 ft) wide, occurs at a point where the dip of the serpentine changes from gently inward-dipping through vertical to steeply outward dipping from the intrusive complex. There is also evidence of a small drag fold where the dip changes most abruptly that appears to form the locus of the ore zone. Talc alteration is present in some zones. The ore is exceptionally high grade with the bulk of the fiber ranging from 4.8 to 11.1 mm (3⁄16 to 7⁄16 in.) in length and a minimum amount of short fiber. This produces a good group 4 asbestos-cement fiber. The ore near the surface was, until recently, mined by open pit but the deeper parts are now being considered for development by underground methods. The fiber is partially concentrated at the mine site before being transported to Deception Bay and thence by sea during a 3½ to 4-month shipping season to Nordenham, Germany, for final processing.

Other Potential Deposits—Several other deposits have been considered at various times for possible production. These include the Abitibi deposit situated 80.5 km (50 miles) north of Amos; the McAdam deposit 27 km (17 miles) east of Chibougamau, both in Quebec; the Lloyd Lake Midlothian Township deposit 64 km (40 miles) south of Timmins, Ont.; and the Garrison deposit 113 km (70 miles) east of Timmins. An attempt was made to develop the Midlothian deposit at Lloyd Lake but the project encountered serious financial problems and was halted.

United States of America: Chrysotile asbestos is mined near Eden, VT, and in two localities in California. Small quantities of chrysotile are also recovered from dolomitic limestones in Arizona.

Pacific Asbestos, Copperopolis, California—The Pacific Asbestos deposit (Leney and Loeb, 1972) is located between Copperopolis and Sonora in the foothills of the Sierras about 201 km (125 miles) east of San Francisco. This deposit occurs within a belt of ultrabasic rocks which trend northwest for 32 km (20 miles) and reach a maximum width of 6 km (4 miles). Strike faulting is common and the peridotites and dunites have been variously altered to serpentine. The ore body occupies a zone about 610 m (2000 ft) wide between two band of antigorite schist which are believed to represent zones of shearing within the intrusive. Fiber occurs as cross fiber up to 25.4 mm (1 in.) in length but most of the fiber is 6.35 mm ($\frac{1}{4}$ in.) or less. The ore body is cut by narrow dioritic dikes. Talc-carbonate alteration is also prevalent in some portions of the pit.

New Idria-Coalinga, California—The recognition of "platy chrysotile" in the New Idria ultrabasic intrusive led to the successful mining of this fiber in two or three operations northwest of Coalinga (Munro and Reim, 1962) in Fresno County. The ultrabasic body trends northwestward and covers an area some 24 km (15 miles) in length and 5 to 8 km (3 to 5 miles) wide, having been intruded into Jurassic sediments in Late Jurassic or Early Cretaceous times. The intrusive is composed of highly sheared serpentinite characterized by its extremely platy, slickensided nature. "Boulders" of massive serpentinized material occur scattered through the loose platy serpentine. There is also evidence of abundant landslide material, a factor which may have contributed to the extreme deformation of the serpentine.

The ore contains abundant short chrysotile fiber similar to the Canadian variety but low prices and severe safety and health regulations caused the closing of two of the three producers in the area.

Eden, Vermont—The southern extension of the Eastern Townships serpentine belt of Quebec continues intermittently southward down the Appalachian chain into northern Vermont (Rowbotham, 1970). Fiber occurrences have been known for many years over a considerable area in Lamoille and Orleans Counties. The deposit which has been producing for a number of years lies near the village of Eden. Here a substantial ore body, some 914 m (3000 ft) wide, is being mined at the foot of Mount Belvidere between elevations of 305 and 457 m (1000 and 1500 ft).

Slip fiber is the more predominant variety mined but some good quality cross fiber is also recovered. The host rock is a highly serpentinized peridotite with the fiber occurring along a contact between gneiss and amphibolite schist.

Mexico: *Cuicatlan, Oaxaca*—A large occurrence of slip fiber has been discovered at Concepcion Papalo east of Cuicatlan, and about 80.5 km (50 miles) north-northwest of Oaxaca in Mexico. The Pegaso occurrence is exposed on two sides of a ridge in highly dissected terrain and appears to be part of a tabular serpentinite body which forms the core of a highly folded and sheared sequence of rocks.

Ciudad Victoria, Tamaulipas—Occurrences of chrysotile fiber are known to occur in the Sierra Madre Oriental Mountains west of Ciudad Victoria (Rowbotham, 1970) in the State of Tamaulipas. These are generally small, discontinuous zones of ribbon fiber and are of little economic significance.

South America

Cana Brava, Goiás, Brazil: The Cana Brava asbestos deposit is located some 193 km (120 miles) north of Brasilia and 97 km (60 miles) northeast of Uruaçá in the State of Goiás. Little is known about the deposit except that it occurs in a large ultrabasic body which intrudes gneisses of the Brazilian Precambrian Shield. The ore is apparently a good quality cross fiber of medium length which occurs in two separate zones 549 m (1800 ft) long and roughly 61 m (200 ft) wide. The bulk of this fiber is utilized by local Brazilian manufacturers.

Las Brisas, Antioquia, Colombia: A deposit of chrysotile asbestos is at present being readied for production at Las Brisas, Antioquia, in central Colombia (Harris, 1973). This property lies about 129 km (80 miles) north of Medellin or roughly midway between Bogota and the Caribbean·port of Cartagena. The deposit is reported to comprise at least two separate zones containing fiber of distinctly different grade.

Europe

Balangero, Northern Italy: The bulk of the Italian chrysotile production comes from Balangero (Rowbotham, 1970), a large, low grade slip fiber deposit located near San Vittore about 32 km (20 miles) north-northwest of Turin. The ore zone occurs in a highly sheared serpentine body about 457 m (1500 ft) wide which trends north-south and is intrusive into paragneisses in the foothills of the Alps in Western Italy.

Zidani, Kozani, Northern Greece: Situated about 121 km (75 miles) southwest of Thessaloniki and 32 km (20 miles) due south of Kozani is a large, low grade slip fiber deposit. The deposit occurs in a highly sheared and fractured serpentinite, probably of dunitic composition which is flanked by crystalline marble to the west.

Troodos, Cyprus: Cyprus (Bear, 1963; Vokes, 1964) is an ancient source of asbestos, as it was known and exploited by the ancient Greeks and Romans. These deposits are some of the largest occurrences of chrysotile in the European area.

Asbestos had been mined for the past 70 years near Amiandos in the Troodos Mountains in the south-southwestern portion of the island. The chrysotile occurs in a body of bastite-serpentinite covering an area some 5.6 km ($3\frac{1}{2}$ miles) north-south and 4 km ($2\frac{1}{2}$ miles) wide, being roughly oval in plan. The best fiber is confined to the southern part of the area near the faulted eastern margin of the peridotite where deformation was most intense. The fiber occurs in small discontinuous lenticles or in larger composite veins of cross fiber and picrolite up to 12.7 mm ($\frac{1}{2}$ in.) wide and several feet long. The fiber seldom exceeds 6.35 mm ($\frac{1}{4}$ in.) in length. These deposits are probably Upper Cretaceous in age.

The Troodos mine is at an altitude of 1.5 km (5000 ft) and the crude ore is transported to Limassol by a 30.6 km (19-mile) long aerial ropeway for further processing.

Yugoslavia: Short fiber chrysotile has been produced from a number of mines in Yugoslavia (Rowbotham, 1970) for many years. Deposits are mined at Korlace in Serbia and near Tuzla, Bosnia. Details of the geology of these deposits are not available.

USSR: Russia (Nalivkan, 1960; Rowbotham, 1970; Smirnov, 1971; Stenho and Fronek, 1977; and Petrov and Znamensky, 1978) is now considered to be the largest producer of asbestos in the world. The USSR's known reserves are vast and are considered to surpass those of any other country.

There are four major areas of asbestos production in the USSR: the Bazhenovo deposits of the Central and Northern Urals, the Dzhetygara district in the Southern Urals of western Kazakhstan, the newly opened Kiembay district also in the southern Urals and the Ak-Dovurak deposits in Tuva ASSR some 966 km (600 miles) west of the southwest end of Lake Baikal.

Bazhenovo (Bajenovsk) District—According to its production and reserves, the Uralian deposits of Bazhenovo occupy first place in the USSR and rival the Eastern Townships of Quebec for the leading asbestos district of the world. These deposits are affiliated with the Bazhenovo-Alapajevsk or eastern serpentine belt which extends for a distance of about 177 km (110 miles) in the Central Urals. Most of the deposits are centered around the new town of Asbest which is 80.5 km (50 miles) northeast of Sverdlovsk and 1448 km (900 miles) east of Moscow.

The deposits are associated with fault zones in serpentinites which formed predominantly as the result of alteration of peridotites and, to a lesser extent, pyroxenites. Formation of the veins of chrysotile is thought to be by filling of fissures in a massif of ultrabasic rocks.

Mines in the Asbest area and others at Alapajevsk, Novaja and Lesnoje, together with the deposits at Rez (East Aninsk) Krasnouralsk, Lukosk and Krivisk have an annual capacity of at least 1.4 Mt (1.5 million tons) of fiber.

Besides the Bazhenovo group, a number of other deposits of asbestos are known.

Dzhetygara District, Northwest Kazakhstan—This constitutes the second largest area of production in the USSR and has been in production since 1965. The district has a planned annual production of about 726 kt (800,000 tons).

The Dzhetygara district lies some 225 km (140 miles) southwest of Kustenay. Little is known of the geology of the deposits except that they occur in serpentinized peridotites of Early Hercynian (Middle Devonian) age.

Kiembay (Kijembai), Southern Urals—This deposit, the third largest in the USSR, is located some 145 km (90 miles) east of Orsk in the Orenburg Administrative Region. The deposit occurs in a completely serpentinized ultrabasic intrusive. Severe weathering 46 to 61 m (150 to 200 ft) deep has caused the surface fiber to become weaker and relatively brittle.

Ak-Dovurak (Aktovrak), Tuva ASSR—Important deposits of asbestos have recently been developed near Ak-Dovurak, Tuva, near the Yenesei River to the west of Lake Baikal. These deposits are associated with a belt of serpentinized ultrabasic rocks of the Salair complex which forms part of the Lower Hercynian of the eastern Altai-Sayan Province. The ultrabasic intrusions occur intermittently in belts which extend along major faults.

At least six significant deposits have been discovered and the district is reported to contain vast reserves near the village of Kyzyl-Mazalyk and the Kedrovsk deposit 48 km (30 miles) to the south.

Other Deposits—Chrysotile asbestos deposits are known at Molodeznoje (Molodjezhnoje) some 209 km (130 miles) east of the northern end of Lake Baikal. The deposit is of excellent grade.

The Ilcirsk and Sajan deposits in the same general area also have good potential for future production.

A peculiar fiber deposit is located near Pechenge close to the Norwegian border. Although the deposit is large, the high iron content is very detrimental to the value and it is doubtful whether the deposit will ever prove to be economic.

Africa

Shabani, Zimbabwe: The Shabani mine (Laubscher, 1964; 1968; Oldham, 1968; Wilson, 1968), situated some 161 km (100 miles) east of Bulawayo, has the largest reserves of fiber in southern Africa. The deposit occurs near the base of a lenticular ultrabasic sill 14.5 km (9 miles) long and 2.4 to 5 km (1½ to 3 miles) wide which intrudes Early Precambrian gneisses. The sill strikes northwest and dips about 40°SW and is composed of dunite at the base, overlain by peridotite, pyroxenite, and gabbro. A swarm of younger diabase dikes cut the ore zone but these are themselves truncated by strong zones of shearing, both parallel and oblique to the basal contact. The intrusion and shearing has caused strong serpentinization of the dunite and heavy talc-carbonate alteration, particularly along the footwall shear zone. The fiber occurs as a stockwork of cross fiber veins described as "stress-controlled dilation seams." The quality is good and the length of the fiber up to 38 mm (1½ in.) in places.

Fiber is also mined at other localities in Zimbabwe, notably in the Mashaba, Belingwe, and Filabusi districts.

Msauli, South Africa: The Msauli mine (Hall, 1930; Pelletier, 1964; Van Biljon, 1964) is located 25¾ km (16 miles) south-southeast of Barberton in the eastern Transvaal and only 9.7 km (6 miles) southwest of the Havelock mine across the Swaziland border. The serpentine in which this ore body occurs is very similar to that of the Havelock deposit in what appears to be the faulted continuation of the same sill-like ultrabasic intrusive. The serpentine body averages 137 m (450 ft) wide and dips about 65° to the east. The best fiber is reported to occur to the north in association with major faulting. The fiber occurs as a stockwork of veins from 0.8 to 19 mm (1/32 to 3/4 in.) and is of the same general high quality as that of Havelock. However, some sections contain a higher percentage of magnetite.

Havelock, Swaziland: The Havelock mine (Pelletier, 1964; Van Biljon, 1964) is located within 0.8 km (½ mile) of Swaziland's western border with the Republic of South Africa, and lies some 19 km (12 miles) south-southeast of Barberton in the eastern Transvaal. Here, a sill-like body of serpentine of the ancient Jamestown Complex is intercalated in the Early Precambrian Fig Tree sedimentary sequence, all of which dip south at 55°. The ore zone has a strike length of 1.5 km (5000 ft) and widths from 18 to 107 m (60 to 350 ft) averaging about 33.5 m (110 ft). The ore body is traversed by a longitudinal fault zone with associated transverse faulting. Talc-carbonate alteration is prevalent in the western portion of the mine. The fiber occurs as a stockwork of high quality cross fiber seams, which average 12.7 mm (½ in.) or more in length and comprise 3% to 4% of the rock. Mining of the upper portion of the ore body was by open cast workings, the deeper portions now being mined by underground methods. The ore is transported by aerial cableway to Barberton for processing.

Other Deposits: Many small deposits of chrysotile asbestos occur at many locations in northern and eastern Transvaal but their combined production has never been appreciable. Most of these are elongated ribbon fiber zones which are in ultrabasic rocks or associated with dolomitic limestones.

Occurrences of chrysotile fiber are also reported from the Ingessana Hills and Qala en Nahl districts of eastern Sudan. Both areas have been explored extensively and several small deposits have been indicated, but the total potential remains relatively small.

Australia

Woodsreef (Transpacific Asbestos), New South Wales: The Woodsreef (Straw, 1964; Butt, 1978) asbestos deposit is located just east of Barraba approximately 547 km (340 miles) north of Sydney in northeastern New South Wales. The deposit occurs in a lens of serpentinite approximately 8 km (5 miles) long and 2.4 km (1½ miles) wide which forms part of the Great Serpentine Belt of New South Wales. The ore body occurs in a layered peridotite-dunite complex which has intruded sediments of Mid-Paleozoic age, and is approximately 701 m (2300 ft) long and 76 m (250 ft) wide. It is bounded on the western margin by prominent faulting and shearing. Strong shear zones are also prevalent within the fiber zone. The fiber occurs mainly as a stockwork of cross fiber seams and is generally short to medium in length. Although production commenced in 1972, the mine has encountered serious fiber recovery problems and the long-term viability of the deposit is still in question. Most of the fiber is exported to Japan but some is consumed locally.

New Zealand

Pyke Asbestos, South Island: A small occurrence of chrysotile fiber is reported near the headwaters of Pyke River (Babcock, 1978) high in the Southern Alps some 80.5 km (50 miles) northeast of Milford Sound on the South Island of New Zealand. Due to its isolated location and relatively low grade, it is not expected to be exploited in the near future.

Asia

China: Little is known about the geology of Chinese asbestos deposits apart from a few short reports written by various mining and trade delegations (Mamen, 1973). However, various production estimates indicate that China ranks fourth in world production.

The Lai-yüan district of Hopeh, situated approximately 161 km (100 miles) southwest of Beijing, was for many years the only source of chrysotile in China.

Emphasis has recently shifted to the Shihmien area which is located about 402 km (250 miles) west of Chungking in southwest Szechwan (the word *Shihmien* is translated as *rock cotton*). At least a dozen projects now contribute to the production from this area.

A small amount of fiber is also recovered from the Chinchou mine in Liaoning which lies some 426 km (265 miles) east-northeast of Beijing.

India: Indian asbestos production (private communication) is small and includes chrysotile of both ultrabasic and limestone origin, together with some amphibole varieties.

The small Roro deposit, located in southern Bihar State, about 282 km (175 miles) west of Calcutta, is an example of chrysotile production. Here the ribbon fiber occurs in five separate discontinuous zones along highly serpentinized peridotite bands and massive to sheared "clot" peridotite. One of these, the Upper Zone, is economic and contains from 7% to 15% fiber over minimum mining widths. The ore is hand-cobbed and sorted for milling.

Japan: Chrysotile asbestos is mined at two localities in the Yamabe district of Hokkaido, Japan (Rowbotham, 1970). The fiber is generally short and all of the production is utilized locally.

CHRYSOTILE IN DOLOMITE

In addition to the large deposits of chrysotile asbestos found in serpentinized peridotite there are comparatively small tonnages mined from serpentinized dolomitic limestones (duToit, 1946; Hall, 1930; Pelletier, 1964; Rowbotham, 1970; Van Biljon, 1964). Such fiber is often of high quality and is free of the magnetite found associated to a varying degree with deposits where the original host rock is of ultrabasic igneous origin.

Chrysotile of this type has been mined from the Carolina district in the Transvaal and in the Salt River and Sierra Ancha regions in Arizona where it is found in narrow bands associated with diabase sills. The only chrysotile of this variety known in the USSR occurs near Aspagash just south of Krasnovask on the Yenesei River.

CROCIDOLITE AND AMOSITE

Both crocidolite and amosite occur in South Africa (Cilliers and Genis, 1964; Cilliers et al., 1964; duToit, 1946; Frankel, 1953; Hall, 1930; Keep, 1961; Pelletier, 1964) in bedded metamorphosed sedimentary formations known as banded ironstones, that may be classed as ferruginous quartzite or probably more exactly as an iron-rich silicified argillite. Cross fiber veins are found as closely spaced ribbons roughly conformable with the bedding which in some localities is distorted and steeply dipping. Crocidolite veins are generally less than 50.8 to 76.2 mm (2 or 3 in.) thick with the bulk of the fiber between 6.35 to 19.05 mm (¼ to ¾ in.) in length. Amosite veins on the other hand often range up to 254 or 305 mm (10 or 12 in.) thick.

Crocidolite is found over a large area of Cape Province occurring in a belt of the Lower Griquatown series of the Transvaal system. One of the oldest producing areas is in the vicinity of Prieska on the Orange River with notable production coming from Westerberg and Koegas, located a few kilometers (miles) down the river. Another producing area of blue asbestos in the Cape Province is farther north in the Kuruman district, and at Pomfret, just south of the Botswana border.

Approximately 322 km (200 miles) southeast of Johannesburg, both crocidolite and amosite are found in similar formations in the vicinity of Pietersburg in northern Transvaal. In some places here, the two varieties have been known to occur side by side in the same vein. Such a vein, having fibers composed of crocidolite at one end and amosite at the other, is referred to as a "doublet." It is interesting to note that there is generally no weakening of the fibrils at the point of contact of the two varieties of fiber.

In the Pietersburg area a quite silky-fibered phase of amosite called *montasite* is found which is often a dusty blue in color and has sometimes been confused with crocidolite.

Widespread occurrences of amosite and crocidolite are known to occur in the Bababudan Hills north of Chikmagalur in Mysore, India. Both varieties of fiber occur in an assemblage of banded ironstones and shales similar to those of northeastern Transvaal but the amosite in particular is noticeably shorter. However, one area of amosite has been intensively investigated for possible production.

Deposits of crocidolite are found in the Hamersley range of Western Australia, as well as in Bolivia and elsewhere.

Amosite-bearing banded ironstone crops out for a distance of over 32 km (20 miles) near Penge in the Lydenburg district of the Transvaal. The fiber-bearing "reef" is being worked at three points along the strike, at Penge, and at the

Weltevreden and the Kromellenboog mines. A good cross section of the country rock has been exposed by diamond drilling to 305 m (1000 ft) in depth. The fiber-bearing banded ironstones and associated sedimentary members dip southwestward at about 20°. This sequence has been intruded by thin persistent sills of dolerite which are conformable with the bedding.

The following column is from a 366-m (1200-ft) diamond drill hole intersecting the fiber-bearing horizon at a depth of 305 m (1000 ft) below the surface at the "Amosa" mine which is part of the Penge property.

> Shales
> Dolerite sills
> Shales
> Bevetts conglomerate
> Banded ironstone (main fiber zone)
> Dolerite sill
> Banded ironstone (sometimes fiber-bearing)
> Dolomite

The Bevetts conglomerate is considered an important "marker" for the fiber-bearing formation.

In addition to the rock series in the foregoing, diabase cuts vertically across the whole formation at intervals. Evidence as to possible relationship of these dikes to the genesis of the amosite mineralization is inconclusive.

With the steady depletion of known amosite ore reserves, extensive exploration for additional ore was initiated a few years ago. Considerable success was achieved by utilizing a novel structural approach developed by the Cape Asbestos Group.

Careful mapping of the main marker horizons with respect to detailed topographic contours disclosed a system of broad, gentle north-south folds having amplitudes of about 15 m (50 ft) and wavelengths of 1.2 to 1.5 km (4000 to 5000 ft) between crests. A second, superimposed pattern of east-west cross folds was also detected having similar amplitude and a wavelength approaching 2 km (7000 ft). Careful structural mapping further indicated that each of the known or producing deposits was located on the loci of intersections of either anticlinal or synclinal structures. Consequently, by drilling similar loci, at least five new fiber occurrences were discovered and the sixth failed only because the precipitous terrain in the area prevented drilling directly on target. Nevertheless, a drill hole angled toward the target from the nearest accessible location did disclose traces of fiber on the periphery of the target.

This record of successful drilling gives considerable weight to the hypothesis of a structural control for the genesis of amosite fiber. Similar patterns of cross-folding have been noted in many crocidolite deposits in the Northwest Cape and elsewhere.

It should be remembered, however, that the main prerequisite for any fiber development is to have the necessary chemical components for a specific fiber variety. Without the required chemical constituents, no amount of folding, faulting, or alteration can produce any fiber.

It is interesting to consider this structural origin for the genesis of chrysotile fiber as well, because the effects of faulting, folding, and dilation by shearing or intrusion almost invariably occur in close proximity to fiber occurrences.

ANTHOPHYLLITE, TREMOLITE, AND ACTINOLITE

Production of amphibole asbestos (Rowbotham, 1970), other than crocidolite and amosite, includes significant amounts of anthophyllite, a small quantity of tremolite, and token amounts of fibrous actinolite.

Of the world's known anthophyllite deposits, Finland has been by far the most important producer. However, the famous deposits at Paakkila in the parish of Tuusniemi near Outokumpu in eastern Finland, as well as those of Maljasalmi in Kuusjarvi Parish some 16 km (10 miles) to the east have now been mined out and no further production is anticipated.

The anthophyllite occurred as a series of lenses of amphibolitized and serpentinized ultrabasic material originally thought to be dunitic in composition. These lenses appear to have become detached from larger bodies during an early period of severe deformation. The serpentinized amphibolite lenses occur in a biotite gneiss which is intruded by Late Karelian granites and pegmatite.

The size of these bodies was generally small, varying from 38 to 61 m (125 to 200 ft) in length and 9 to 24 m (30 to 80 ft) in width with a potential of 36 to 91 kt (40,000 to 100,000 tons) of anthophyllite asbestos in each pod.

Exploration for these isolated bodies is difficult and often costly as the area is drift-covered and most of the ore lenses are not exposed at surface.

Present mining of these ore bodies is by open pit with the ore being processed at a centrally located mill.

Other amphibole asbestos is reported from many countries including Yugoslavia, Bulgaria, Brazil, Japan, India, and Taiwan. Little is known of the mode of occurrence of these deposits or their present production status.

A small quantity of anthophyllite is mined near Greenmountain, Yancey County in North Carolina. Production has been small and erratic but is reported to be increasing slightly. Similar material is known to occur in the nearby state of Georgia.

Italy continues to produce a limited quantity of long-fibered tremolite from small deposits located at Val Malenco, in the Sondrio district some 97 km (60 miles) north of Milan. Reports of tremolite fiber are also known from the Aosta district north of Turin in the Italian Alps.

The Mavita anthophyllite mines of Mozambique were closed in 1975.

Although some occurrences of fibrous actinolite are reported, production is extremely limited and of negligible value.

PRINCIPAL DEPOSITS

In the foregoing sections, the major asbestos deposits of the world have been briefly described. Global distribution of these deposits appears to be rather haphazard, but closer examination reveals that their occurrence generally coincides with the earth's major chains of mountain building of all ages. This is particularly true of the Urals, the Appalachians, and Rocky Mountains, but the relationship is more difficult to recognize in Precambrian fold belts.

EXPLORATION AND EVALUATION

The following observations concerning exploration and evaluation apply mainly to chrysotile deposits and to those, in particular, which occur in ultrabasic rocks, and are essentially a review of current methods employed in Canada.

Geophysical

Owing to the lack of outcrop within areas of ultrabasic rocks, or within a belt in which these rocks are expected to occur, both aeromagnetic and ground magnetic surveys (Conn, 1967; Low, 1951) are often employed in the early stage of exploration for asbestos. Ground magnetic surveys may be used to check and define in more detail anomalies obtained by an airborne survey of a large area, or the ground survey alone may be used for the purpose of exploring a small area.

In the case of an airborne survey, flight lines are normally spaced at 0.4-km (¼-mile) intervals, and flown as close to 152-m (500-ft) elevation as possible. Where the terrain is rugged, helicopters are used which are capable of maintaining a constant altitude of 91 m (300 ft) above ground level. The spacing between profiles on a ground survey is usually 61 or 91 m (200 or 300 ft) and readings are taken at intervals of 15 or 31 m (50 or 100 ft).

Magnetic surveys are used to locate and define the areas of ultrabasic rocks and, within these, the areas which have been subjected to extensive serpentinization. This is possible because this type of alteration produces a higher content of secondary magnetite. Asbestos deposits in ultrabasic rocks are a result of intensive serpentinization, and for that reason asbestos veining is usually accompanied by a higher concentration of magnetite than is normally found in the barren serpentine. It follows, therefore, that magnetic anomalies obtained over an area of ultrabasic rocks are favorable places to explore for asbestos.

Modern instrumentation has made great strides in recent years, and it is now possible to conduct precise surveys with small lightweight magnetometers, in contrast to the cumbersome equipment used in the past.

Diamond Drilling

Diamond drilling is normally employed to probe beneath the overburden to assess and define the limits of an asbestos deposit. As asbestos ore bodies are usually large in volume, it is customary to drill vertical holes on a grid pattern. In the initial stages of exploration an interval of 122 m (400 ft) and sometimes more may be used, filling in to an interval of 31 m (100 ft) or even less, where an asbestos-bearing zone is encountered. In cases where a deposit is elongated in one direction, holes are generally spaced at closer intervals across the strike. Narrow, tabular targets are best explored by angle holes planned to give the attitude and true thickness of the body.

Care should be exercised to test the area with one or two preliminary drill holes to determine whether the fiber has a preferential vein angle or not. If the angles of the vein intersections appear random, no changes need be made to the program. If, on the other hand, the deposit exhibits prominent vein angles which diverge from the average 45°, then the attitude of all later drill holes should be changed to correct for this variation.

In the past, drill evaluation programs frequently used small diameter, AX [32.5 mm (1$\frac{9}{32}$ in.)] or EX [23.8 mm ($\frac{15}{16}$ in.)], core sizes for evaluating any deposits where shearing was minimal and where holes did not exceed 91 m (300 ft). Experience has shown, however, that large core sizes such as NX [54 mm (2$\frac{1}{8}$ in.)] or BX [41 mm (1$\frac{5}{8}$ in.)], are preferable for better depth penetration, more geologic data, and to aid in fiber logging and dry milling of the core.

Wire-line drill equipment and the use of nonrotating core barrels are also recommended to minimize fiber loss by grinding of the core during drilling.

Steps are taken to recover the sludge only where core recovery is poor, which is generally the case with slip fiber occurrences. Because of the tendency of the fiber to fluff up and remain in suspension, much greater settling tank capacity is required than is the case when recovering sludge from other minerals. Care must also be taken to avoid contamination by grease and vegetable matter as these cannot be burned off without damaging the fiber.

In regions where permafrost conditions are expected, every precaution should be taken to avoid freezing the string of drill rods down the hole. The use of suitable low-freezing drill lubricant media such as a concentrated brine solution is adequate to allow drilling in most areas of permafrost. However, due to the high percentage of fiber cuttings in the water return, the brine solution has to be renewed at frequent intervals—a factor which considerably increases the cost of drilling asbestos prospects in permafrost areas. A number of successful drill ventures under permafrost conditions have been completed, including Asbestos Hill which showed the frozen ground to persist to at least 305 m (1000 ft) below surface.

In areas where drilling is impractical, exposure by trenching or exploration beneath the surface by adit or shaft and lateral workings may offer the only means of assessing a deposit.

Evaluation

The evaluation (Conn and Mann, 1971; Dean and Mann, 1968; Oughtred, 1952) of any asbestos deposit entails the determination of its size, as well as the grade and quality of its fiber. Dimensions of a mineral zone are established by conventional methods, such as mapping, magnetic surveys, trenching, and diamond drilling. The value of the contained asbestos fiber is dependent on numerous physical properties such as fiber length, strength, flexibility, harshness, and color, besides the actual amount of fiber present.

The determination of grade cannot be based on a simple chemical analysis as both the fiber and the wall rock have essentially the same chemical composition. To avoid the complete crushing of the rock and physical separation of the fiber into different lengths, a method of visual evaluation has been developed which requires careful enumeration of the total number of fiber veins, together with the average length of fiber in each vein. These fiber lengths, if expressed in sixteenths of an inch, give an approximation of the grade, each 1.6 mm ($\frac{1}{16}$ in.) vein of fiber in a 1.5-m (5-ft) section being approximately 0.1% fiber.

As the price of fiber varies considerably depending on its length, the use of grade based solely on its percentage fiber content is of little significance. Instead, the product of these two variables, expressed in dollars per ton of rock, offers a far more meaningful value which can be used for direct comparison in the final evaluation.

Drill Core: The method of visual evaluation is ideally suited to the evaluation of drill core.

Each vein of cross fiber is logged and the length carefully measured and recorded in multiples of 1.6 mm ($\frac{1}{16}$ in.). As veins are often of irregular width, the average width of each vein should be estimated. Some veins are of a composite nature carrying partings, or the fiber may have kinks which cause it to break into shorter lengths. The fiber in some veins may be at right angles to the vein walls, in others sharply inclined. Allowance must be made for these conditions in arriving at the true lengths of fiber.

The more precise determination of fiber content calls for the measurement of each vein angle in the core to permit individual volume corrections. However, in practice it is usually assumed that the random angle is 45° and a factor of 1.414 (inverse of the sine of 45°) is used. However, this can be quite misleading if the average angle is rather small, so logging results should be checked against laboratory recovery and bulk sampling wherever possible.

As a rule, the visual reading gives an indicated lower yield of a higher value fiber than a corresponding mill test. This is to be expected as the visual readings disregard fiber lengths of less than 1.6 mm ($\frac{1}{16}$ in.) which are, of course, recovered in milling and are important in grading. Furthermore, a certain amount of pulverized host rock adheres to the fiber, further reducing the grade and increasing the yield by 20% or more. On the other hand, there may be some loss of veins in the core and the veins themselves contain foreign material such as nonfibrous serpentine, picrolite, and magnetite. There is also a tendency, on the one hand, for some breaking of fibers to occur in the mining and milling process while, on the other hand, the fluffed up milled product tends to remain on a screen which an unopened fiber would pass through. For most practical purposes, the discrepancies are compensating and, when a suitable volume correction factor is used, the ore values found by the two methods are generally quite close.

Slip fiber is often associated as a minor constituent in deposits made up mainly of cross fiber, and in others the slip fiber may be the predominant type. It is not easy to determine the slip fiber length by the normal visual methods of logging core; however, its presence may be recorded separately by vein widths and the percentage slip fiber determined in the same manner as used for cross fiber. Laboratory assistance is almost essential in evaluating slip fiber deposits.

Regardless of the method used in evaluating asbestos ore, the final answer should give the value of the fiber, the yield and, from this, the value of the ore in dollars per ton. A bulk sample properly milled and graded readily supplies the answer, whereas a laboratory test normally gives only the yield of an ungraded fiber and, therefore, of unknown value. The visual method, after due allowance for vein angles, gives the yield only and further calculations are required in order to approximate the fiber and ore value.

Face Readings

On the surface or in underground workings, channel sampling may be employed. On the other hand, it is also possible to log these surfaces in a similar manner to that used for drill core. One method is to take a linear reading along either wall of a drift or crosscut and another is to take cross sections at intervals of 1.5 m (5 ft) on the back and both walls to ensure that veins running parallel to the drive are not excluded.

An alternative method is to record all the veins in the face, walls, and back after each round. In order to arrive at a percentage, a factor based on the area involved is applied to the reading for each surface, and an average percentage for the round is determined in this manner.

The fact, however, that the rock tends to break along one or more of the numerous fault planes present rather than across the fault bounded blocks to expose the veins within them, makes it difficult to obtain representative results.

Bulk Sampling

The various methods of logging and sampling outlined seldom give entirely dependable results. The visual methods of logging will usually produce dependable results only where there is a relatively low content of shorts. Slip fiber presents a problem in this connection. Even laboratory results, which are dependent on complete extraction, usually give an appreciably higher fiber value than that obtained in a conventional mill. It is not easy to simulate in a laboratory the conditions to be found at an operating mine where the fiber from the time of blasting to the final product is subjected to

a good deal of handling, some of which is rather severe. The fibers, as a consequence, suffer some breakage in the process.

Bulk sampling is often resorted to as a means to check and to arrive at a suitable factor to be applied to drill core data. This may be done by diamond drilling a block of ground at close intervals prior to mining and milling. The core is then read visually and treated in the laboratory. Provided precautions are taken to avoid contamination and the sample is sufficiently large, results should be reasonably reliable despite the erratic distribution of fiber in the rock.

Tensile Strength

It is important to point out that the measurements made and the evaluations so obtained are based entirely on the length of the fiber which is today a secondary factor in fiber value. Most asbestos-cement fiber grades sold today are valued for the strength they lend to cement or other mixes and a standard scale of Strength Units has been established.

It is imperative then that any evaluation of a chrysotile deposit include Strength Unit evaluations which are laboratory tests involving the testing of an asbestos-cement tile made with the subject fiber. Once the inherent strength of the fiber from a particular ore body has been determined it is usually possible to equate strength to length distribution and dust content measurements.

Tonnage and Grade

To estimate the tonnage and grade of a deposit from diamond drill core data, individual drill holes may be weighted according to their interval using the polygonal method. An alternative procedure is to use cross sections, or groups of cross sections and, by weighting the individual holes in each section, determine the average grade for each section.

A third method employs contoured cross sections, wherein the contouring is based on a reasonable interpolation of the intervening area between drill holes. This method permits the estimator to make use of all available geological information in his interpolation. In open pit operations, contoured horizontal sections may be prepared in this manner to correspond with expected mining level intervals, and these serve as a useful guide to mining. Separate horizontal sections contoured for rock value and fiber value per ton permit the mine operator to produce a more balanced mill feed with respect to both fiber content and grades.

BIBLIOGRAPHY AND REFERENCES

Anon., 1966, *Testing Procedures for Chrysotile Asbestos Fiber*, 2nd ed., Quebec Asbestos Mining Assn.

Anon., 1972, *Geological Guide to the Asbestos-Mining Region of Southeast Quebec*, Quebec Asbestos Mining Assn., p. 32.

Allen, C.C., Gill, J.C., and Koski, J.S., 1957, "The Jeffrey Mine of Canadian Johns-Manville Company, Limited," *The Geology of Canadian Industrial Mineral Deposits*, 6th Commonwealth Mining and Metallurgical Congress, pp. 27-36.

Anderson, H.V., and Clark, G.L., 1929, "Application of X-Rays in the Classification of Fibrous Silicate Minerals Commonly Termed Asbestos," *Industrial & Engineering Chemistry*, No. 10, pp. 924-933.

Aruja, E., 1944, "Displacement of X-Ray Reflections," *Nature*, Vol. 154, p. 53.

Aruja, E., 1944a, "An X-Ray Study of the Crystal Structure of Antigorite," *Mineralogical Magazine*, Vol. 27, pp. 65-74.

Asbestos, monthly periodical, P.O Box 471, Willow Grove, PA 19090.

Avery, R.B., Conant, M.L., and Weissenborn, H.F., 1958, "Selected Annotated Bibliography of Asbestos Resources in the United States and Canada," Bulletin 1019-L, US Geological Survey, pp. 817-865.

Babcock, R.C., 1978, "The Pyke Asbestos Deposits, New Zealand," SME Preprint No. 78-H-84, AIME Annual Meeting, Denver.

Badollet, M.S., 1948, "Asbestos," *Encyclopedia of Chemical Technology,* Vol. 2, Interscience, New York, pp. 134-142.

Badollet, M.S., 1951, "Asbestos, A Mineral of Unparalleled Properties," *Transactions,* Canadian Institute of Mining & Metallurgy, Vol. 54, pp. 151-160.

Badollet, M.S., 1963, "Asbestos," *Encyclopedia of Chemical Technology,* Vol. 2, Kirk-Othmer, ed., Interscience, New York, pp.734-747.

Badollet, M.S., and Streib, W.C., 1955, "The Heat Treatment of Chrysotile Asbestos Fibers," *Transactions,* Canadian Institute of Mining & Metallurgy, Vol. 58, pp. 33-37.

Bates, R.L., 1969, "Metamorphic Minerals—Asbestos," *The Geology of Industrial Rocks and Minerals,* Dover Publications, New York, pp. 317-328.

Bates, T.F., et al., 1950, "Tubular Crystals of Chrysotile Asbestos," *Science,* Vol. 3, pp. 512-513.

Bear, L.M., 1963, "The Mineral Resources and Mining Industry of Cyprus," Bulletin No. 1, Geological Survey, Cyprus, 208 pp.

Berger, H., 1963, *Asbestos Fundamentals,* Chemical Publishing Co., New York, 171 pp.

Bourassa, P.J., 1957, "The Asbestos Mine of Nicolet Asbestos Mines Limited," *The Geology of Canadian Industrial Mineral Deposits,* 6th Commonwealth Mining and Metallurgical Congress, pp. 26-27.

Bowles, O., 1955, "The Asbestos Industry," Bulletin 552, US Bureau of Mines, 122 pp.

Bragg, W.L., 1937, "The Pyroxene and Amphibole Groups," Atomic Structure of Minerals, Cornell Univ. Press, p. 184.

Brindley, G.W., and Zussman, J., 1957, "A Structural Study of the Thermal Transformation of Serpentine Minerals to Forsterite," *American Mineralogist,* Vol. 42, No. 7-8, pp. 461-474.

Butt, B.C., 1978, "Exploration Forecasts and Exploitation Realities of the Woodsreef Mine, New South Wales, Australia," SME Preprint No. 78-H-71, AIME Annual Meeting, Denver.

Carroll-Porczynski, C.Z., 1956, *Asbestos,* The Textile Institute, Manchester, England.

Chidester, A.H., and Shride, A.F., 1962, "Asbestos in the U.S., Exclusive of Alaska and Hawaii," Mineral Investigation Research Map MP 17, US Geological Survey.

Cilliers, J.J.le R., 1964, "Amosite at the Penge Asbestos Mine," Vol. 2, *The Geology of Some Ore Deposits of Southern Africa,* Geological Society of South Africa, pp. 579-591.

Cilliers, J.J.le R., and Genis, J.H., 1964, "Crocidolite Asbestos in the Cape Province," *The Geology of Some Ore Deposits of Southern Africa,* Vol. 2, The Geological Society of South Africa, pp. 543-570.

Cilliers, J.J.le R., et al., 1961, "Crocidolite from the Koegas-Westerberg Area, South Africa," *Economic Geology,* Vol. 56, pp. 1421-1437.

Conn, H.M.K., 1967, "Geophysics and Asbestos Exploration," *Mining and Groundwater Geophysics,* Economics Geology Report No. 26, Geologic Survey of Canada, pp. 485-491.

Conn, H.K., and Mann, E.L., 1971, "Evaluation of Asbestos Deposits," SME Preprint No. 71-H-27, AIME Annual Meeting, New York, 9 pp.

Cooke, H.C., 1937, "Thetford, Disraeli and Eastern Half of Warwick Map Areas, Quebec," Memoir 211, Geological Survey of Canada, pp. 86-140.

Dean, A.W., and Mann, E.L., 1968, "The Evaluation of Chrysotile Asbestos Deposits," *Ore Reserve Estimation and Grade Control,* Special Vol. 9, Canadian Institute of Mining & Metallurgy, pp. 281-286.

Douglas, R.J.W., 1970, *Geology and Economic Minerals of Canada,* Economic Geology Report No. 1, Geological Survey of Canada, Dept. of Energy, Mines and Resources, 838 pp.

du Toit, A.L., 1946, "The Origin of the Amphibole Asbestos Deposits of South Africa," *Transactions,* Geological Society of South Africa, Vol. 48, pp. 161-206.

Fankuchen, I., and Schneider, M., 1944, "Low Angle X-Ray Scattering from Chrysotiles," *Journal of American Chemical Society,* Vol. 66, No. 3, Mar., pp. 500-501.

Frankel, J.J., 1953, "South African Asbestos Fibres," *Mining Magazine,* London, Nos. 2 and 3, pp. 89, 73-83; 142-149.

Genis, J.H., 1964, "The Formation of Crocidolite Asbestos," *The Geology of Some Ore Deposits in Southern Africa,* Vol. 2, Geological Society of South Africa, pp. 571-578.

Gold, D.P., 1967, "Local Deformation Structures in a Serpentinite," *Ultramafic and Related Rocks,* P.J. Wyllie, ed., John Wiley, New York, pp. 200-202.

Graham, R.P.D., 1944, "Serpentine Belt, Eastern Townships," *Geology of Quebec,* Geology Report 20, Dept. of Mines, Quebec, Vol. 2, pp. 439-443.

Hall, A.L., 1930, *Asbestos in the Union of South Africa,* Memoir 12, 2nd ed., Geological Survey of South Africa, p. 324.

Harris, H.I., 1973, "How Nicolet Proved and Evaluated Columbian Asbestos Deposit," *World Mining,* Vol. 26, No. 13, pp. 43-46.

Hendry, N.W., 1972, "The Outlook for Asbestos in Canada," *Bulletin,* Canadian Institute of Mining & Metallurgy, Vol. 65, No. 724, Aug., pp. 40-44.

Hendry, N.W., and Conn, H.K., 1957, "The Ontario Asbestos Properties of Canadian Johns-Manville Company, Limited," *The Geology of Canadian Industrial Mineral Deposits,* 6th Commonwealth Mining and Metallurgical Congress, pp. 36-44.

Hillier, J., and Turkevich, J., 1949, "Electron Microscopy of Colloidal Systems," *Analytical Chemistry,* Vol. 21, No. 4, Apr., pp. 475-485.

Hodgson, A.A., 1965, "Fibrous Silicates," Lecture Series No. 4, Royal Institute of Chemistry.

Keep, F.E., 1961, "Amphibole Asbestos in the Union of South Africa," *Transactions,* 7th Commonwealth Mining and Metallurgical Congress, Vol. 1, pp. 90-120.

Keith, S.B., and Bain, G.W., 1932, "Chrysotile Asbestos: 1. Chrysotile Veins," *Economic Geology,* Vol. 27, pp. 169-188.

Kula, J., and Wiser, J.P., 1970, "Msauli Asbestos Mill," *World Mining,* Sep., pp. 26-29.

Lamarche, R.Y., 1972, "Ophiolites of Southern Quebec," *Canadian Contributions 1-11 to the Geodynamics Project—A Symposium,* Earth Physical Branch, Dept. of Energy, Mines and Resources, Ottawa.

Laubscher, D.H., 1964, "The Occurrence and Origin of Chrysotile Asbestos and Associated Rocks, Shabani, Southern Rhodesia." *The Geology of Some Ore Deposits of Southern Africa,* Vol. 2, The Geological Society of South Africa, pp. 593-624.

Laubscher, D.H., 1968, "The Origin and Occurrence of Chrysotile Asbestos in the Shabani and Mashaba Areas, Rhodesia," Symposium on Rhodesian Basement Complex, *Transactions,* Geological Society of South Africa Annexure, Vol. 71, pp. 195-204.

Leney, G.W., and Loeb, E.E., 1972, "The Geology and Mining Operations at Pacific Asbestos Corporation," *Asbestos,* Vol. 54, No. 4, pp. 4-14.

Low, J.H., 1951, "Magnetic Prospecting Methods in Asbestos Exploration," *Transactions,* Canadian Institute of Mining & Metallurgy, Vol. 54, pp. 388-395.

Mamen, C., ed., 1973, "China's Mineral Industry," *Canadian Mining Journal,* Vol. 94, No. 1, pp. 21-31.

May, T.C., and Lewis, R.W., 1970, "Asbestos," *Mineral Facts and Problems,* Bulletin 650, US Bureau of Mines, pp. 851-863.

Merrill, R.J., 1957, "The Carey-Canadian Asbestos Deposit," *The Geology of Canadian Industrial Mineral Deposits,* 6th Commonwealth Mining and Metallurgical Congress, pp. 45-49.

Miles, K.R., 1942, "The Blue Asbestos-Bearing Banded Iron Formations of the Hammersley Range, Western Australia," Bulletin No. 100, Geological Survey of Western Australia, Pt. 1, pp. 5-37.

Mining Magazine, 1980, Vol. 143, No. 6, p. 537.

Munro, R.C., and Reim, K.M., 1962, "Coalinga Asbestos Fiber—A Newcomer to the Asbestos Industry," *Canadian Mining Journal,* Vol. 83, No. 8, Aug.; *Mining Engineering,* Vol. 14, No. 9, pp. 60-62.

Nalivkin, D.V., 1960, *The Geology of the U.S.S.R.—A Short Outline,* trans. by S.I. Tomkeiff, J.E. Richey, trans. ed., Pergamon Press, 170 pp.

Oldham, J.W., 1968, "A Short Note on the Recent Geological Mapping of the Shabani Area," Symposium on Rhodesian Basement Complex, *Transactions,* Geological Society of South Africa Annexure, Vol. 71, pp. 189-194.

Parry, J., 1980, "Zimbabwe Opens Its Doors," *Industrial Minerals,* No. 158, p. 55.

Pauling, L., 1930, "The Structure of the Chlorites," *Proceedings,* National Academy of Science, Vol. 16, p. 578.

Pelletier, R.A., 1964, *Mineral Resources of South-Central Africa,* Oxford University Press, 277 pp.

Petrov, V.P., and Znamensky, V.S., 1978, "Asbestos Deposits of the USSR," SME Preprint No. 78-H-106, AIME Annual Meeting, Denver.

Pundsack, F.L., 1955, "The Properties of Asbestos. I. The Colloidal and Surface Chemistry of Chrysotile," *Journal of Physical Chemistry,* Vol. 59, No. 9, Sep., pp. 892-895.

Pundsack, F.L., 1956, "The Properties of Asbestos. II. The Density and Structure of Chrysotile," *Journal of Physical Chemistry,* Vol. 60, No. 3, Mar., pp. 361-364.

Pundsack, F.L., and Reimschussel, G., 1956, "The Properties of Asbestos. III. Basicity of Chrysotile Suspensions," *Journal of Physical Chemistry,* Vol. 60, Sep., pp. 1218-1222.

Rabbit, J.C., 1948, "*A New Study of the Anthophyllite Series,*" *American Mineralogist,* Vol. 33, May-June, pp. 263-323.

Rice, S.J., 1963, "California Asbestos Industry," Mineral Information Service, California Div. of Mines, Vol. 16, No. 9, pp. 4-6.

Riordon, P.H., 1952, "Geology of the Thetford-Black Lake District of Quebec with Particular Reference to the Asbestos Deposits," Ph.D. Thesis, McGill University, unpublished.

Riordon, P.H., 1955, "The Genesis of Asbestos in Ultrabasic Rocks," *Economic Geology,* Vol. 50, No. 1, pp. 67-81.

Riordon, P.H., 1957, "The Structural Environment of the Thetford-Black Lake Asbestos Deposits," *Proceedings,* Geological Assn. of Canada, Vol. 9, pp. 83-93.

Riordon, P.H., 1957a, "The Asbestos Belt of Southeastern Quebec," *The Geology of Canadian Industrial Mineral Deposits,* 6th Commonwealth Mining and Metallurgical Congress, pp. 3-8.

Riordon, P.H., 1957b, "The Asbestos Deposits of Thetford Mines, Quebec," *The Geology of Canadian Industrial Mineral Deposits,* 6th Commonwealth Mining and Metallurgical Congress, pp. 9-17.

Riordon, P.H., 1957c, "The British Canadian Mine," *The Geology of Canadian Industrial Mineral Deposits,* 6th Commonwealth Mining and Metallurgical Congress, pp. 17-21.

Riordon, P.H., 1957d, "Normandie and Vimy Ridge Mines," *The Geology of Canadian Industrial Mineral Deposits,* 6th Commonwealth Mining and Metallurgical Congress, pp. 21-26.

Riordon, P.H., and Laliberté, R., 1957, "Asbestos Deposits of Southern Quebec," *Excursion B-08 Guidebook,* 24th International Geological Congress, Canada, pp. 1-21.

Robinson, K., and Shaw, E.R.S., 1952, "Summarized Proceeding of a Conference on Structures of Silicate Minerals (November 1951)," *British Journal of Applied Physics,* Vol. 3, Sept., pp. 277-282.

Rosato, D.V., 1959, *Asbestos, Its Industrial Applications,* Reinhold, New York, pp. 198-199.

Rowbotham, P.I., ed., 1970, "World Asbestos Industry," *Industrial Minerals,* No. 28, Jan., pp. 17-29.

Smitheringale, W.V., 1957, "The Mine of Cassiar Asbestos Corporation Limited, Cassiar, B.C.," *The Geology of Canadian Industrial Mineral Deposits,* 6th Commonwealth Mining and Metallurgical Congress, pp. 49-53.

Speil, S.S., and Leineweber, J.P., 1969, "Asbestos Minerals in Modern Technology," *Environmental Research,* Vol. 2, No. 3, Apr., pp. 166-208.

Stehno, J., and Fronek, R., 1977, "The Soviet Asbestos Deposits. A Guarantee for the Continuing Development of the Czechoslovakian Asbestos Cement Industry," *Stavivo* (Building Materials), No. 11, pp. 430-432. (Translation from Czechoslovakian technical journal.)

Stewart, R.V., 1978, "Geology and Evaluation of the Asbestos Hill Orebody," SME Preprint No. 78-H-69, AIME Annual Meeting, Denver.

St-Julien, P., 1967, "Tectonics of Part of the Appalachian Region of Southeastern Quebec," Special Publication 10, Royal Society of Canada, pp. 41-47.

Straw, D.J., 1955, "A World Survey of the Main Chrysotile Asbestos Deposits," *Canadian Mining & Metallurgical Bulletin,* Vol. 48, pp. 610-630.

Streib, W.C., 1978, "Asbestos," *Kirk-Othmer: Encyclopedia of Chemical Technology,* Vol. 3, 3rd ed., John Wiley & Sons, Inc., pp. 267-283.

US Bureau of Mines, *Minerals Yearbook,* annually.

Van Biljon, W.J., 1964, "The Chrysotile Deposits of the Eastern Transvaal and Swaziland," *Geology of Some Ore Deposits in Southern Africa,* Vol. 2, Geological Society of South Africa, pp. 625-669.

Vokes, F.M., 1964, "Asbestos Bearing Claims on Troodos," US Special Fund Project, Cyprus, United Nations unpublished report.

Warren, B.E., 1932, "Structure of Asbestos—An X-Ray Study," *Industrial & Engineering Chemistry,* Vol. 24, No. 4, pp. 419-422.

Warren, B.E., 1942, "X-Ray Study of Chrysotile Asbestos," *American Mineralogist,* No. 27, p. 235.

Warren, B.E., and Bragg, W.L., 1928, "The Structure of Diopside," *Zeitschrift fuer Krist,* Vol. 69, pp. 168-193.

Warren, B.E., and Hering, K.W., 1941, "The Random Structure of Chrysotile Asbestos," *Physical Reviews,* No. 59, p. 925.

Whittaker, E.J.W., 1952, "The Unit Cell of Chrysotile," *Acta Chrystalogica,* Vol. 5. pp. 143-144.

Whittaker, E.J.W., and Zussman, J., 1956, "The Characterization of Serpentine Minerals by X-Ray Diffraction," *Mineralogical Magazine,* No. 31, pp. 107-126.

Wilson, J.F., 1968, "The Mashaba Igneous Complex and Its Subsequent Deformation," *Symposium on Rhodesian Basement Complex,* Geological Society of South Africa Annexure, Vol. 71, pp. 175-188.

Yada, K., 1967, "Study of Chrysotile Asbestos by a High Resolution Microscope," *Acta Chrystalogica,* Vol. 23, pp. 704-710.

Yada, K., 1971, "Study of Microstructure of Chrysotile Asbestos by High Resolution Electron Microscopy," *Acta Chrystalogica,* Vol. A27, p. 659-664.

Zussman, J., Brindley, G.W., and Comer, J.J., 1957, "Electron Diffraction Studies of Serpentine Minerals," *American Mineralogist,* No. 42, pp. 133-153.

2. Barium and Strontium Minerals

Donald A. Brobst

ORES AND USES OF BARIUM

The chief ore minerals of barium are barite, the sulfate of barium, and witherite, the carbonate of barium, but barite is the much more common of the two. These ores are sources of barium and its compounds whose many uses are nearly hidden among the technical complexities of modern industrial products and processes.

Most of the barite produced is used in the petroleum industry as a weighting agent in the fluid circulated in drilling oil and gas wells. Barite is used in the manufacture of glass and ground barite is a common filler, extender, and weighting agent in many manufactured products. Barium ceramics have unusual electromechanical properties that make barium titanate useful in the electronics industry. Barium ferrate is used in permanent magnets. Barite is the common feedstock for the production of barium chemicals which have more than 2000 industrial applications in 17 major classifications. Barium metal is used as a "getter" to degasify TV and other vacuum tubes. Barium metal is alloyed with lead and calcium for use in bearings. The specifications of barium mineral products for these many uses generally are set by the user, but some negotiations between producer and user are common.

Barite, also called barytes, cawk, tiff, or heavy spar, has a calculated specific gravity of 4.5, but inclusions of other minerals may reduce this value considerably. Barite in most commercial deposits occurs as irregular masses, concretions, nodules, rosettelike aggregates, and in laminated to massive beds of fine crystallinity. The mineral occurs in many colors, although shades of white to gray and black are most common. The hardness is 2.5 to 3.5 on Mohs' scale (in which the hardness of a fingernail is about 2). Barite is virtually insoluble in water and acid and so is a useful chemically inert material. Thus, the weight, softness, and chemical inertness make it ideal for many of its uses.

Barite is commonly associated with quartz, chert, and jasperoid; calcite, dolomite, siderite, rhodochrosite, and celestite; fluorite; and various sulfide minerals, such as pyrite, chalcopyrite, galena, sphalerite, and their oxidation products. Barite is a common accessory mineral in many types of ore deposits, especially veins that are mined principally for other types of commodities including lead, zinc, gold, silver, fluorite, and rare-earth minerals. Barite is a potential byproduct of many of these and other metallic ore deposits. Barite in vein and residual deposits may contain as much as several percent strontium which substitutes for barium in the atomic lattice of the crystal because of the close similarity of the ionic size of the two metals in their bivalent state. Minor amounts of other metals commonly occur in barite (Brobst, 1958).

Witherite (barium carbonate) has a calculated specific gravity of 4.2 and a hardness of 3 to 3.5 on Mohs' scale. Crystals or complex masses are colorless to milky white or gray or tints of yellow, brown, or green. Witherite is a minor accessory in some barite deposits. In the United States, witherite in commercial amounts has been produced only from a mine at El Portal, Mariposa County, CA, and that operation ceased in 1950. Witherite is a highly desirable feedstock for barium chemical plants because of its solubility in acid.

It seems, however, that witherite constitutes only a small part of the world's barium resources.

Sanbornite, a simple barium silicate, is of potential interest as a source of barium for chemicals. Unlike many silicates, sanbornite is soluble in acid which permits the recovery of barium by further chemical processing. The mineral, long thought to be rare, has been found in abundance with other barium silicate minerals in rocks altered by the effects of heat and pressure from the invading rock materials of the Sierra Nevada batholith in Fresno County, California. The resource potential of sanbornite is virtually unassessed.

Barium occurs in many other minerals, but the concentrations are so low that none is likely to be a potential source of the element in the near future.

The deposits in which barite is abundant enough to recover as the primary commercial product generally are mined in open pits, although there are a few exceptions.

GEOLOGY OF BARITE DEPOSITS

The commercial deposits of barite may be geologically classified by their mode of occurrence into three major types: the vein and cavity-filling deposits, the residual deposits, and the bedded deposits (Brobst, 1958, 1983).

Vein and Cavity-Filling Deposits

Vein and cavity-filling deposits are those in which barite and associated minerals occur along various kinds of fractures, joints, bedding planes, and solution structures such as channels and sink holes that commonly occur in limestones. The vein and cavity-filling deposits commonly have sharp contacts with the enclosing rocks and large-scale replacements of the host rocks beyond the controlling structures are rare. The grade of the ore generally varies within the deposit and from deposit to deposit in a district. Within districts, the deposits are scattered and irregular and range in thickness from a few inches to a few feet and in length from tens to hundreds of feet. Many deposits of the vein-type have been mined by surface methods, but the nature of vein deposits generally means that at some time the mining operation might have to go underground.

Deposits of barite in veins are widely scattered throughout the United States. Many deposits in the western states are associated with igneous rocks of Tertiary age, although notable exceptions are the vein deposits of barite and rare-earth minerals associated with igneous rocks of Precambrian age at Mountain Pass, San Bernardino County, CA (Olson, et al., 1954) and in the Wet Mountains, Custer County, Colorado. In most of the midwestern and eastern states, the association of the deposits with igneous rocks is not obvious.

Barite deposits in solution structures are common in the collapse and sink structures of the midwestern and Appalachian states. These deposits are common in central Missouri where they are known as circle deposits. Many of these deposits have yielded rich ore, but the individual bodies tend to be small. These deposits have been mined by surface methods.

On weathering, vein and cavity-filling deposits may form valuable bodies of residual ore.

The barite and other minerals of the vein and cavity-filling deposits are typical of a suite precipitated from low-temperature hydrothermal solutions associated with igneous activity. Most geologists agree on such an origin for the vein deposits of the western states. A hydrothermal origin also is suggested for many deposits in the midwestern and eastern states where the mineral-bearing solutions seem to have traveled farther and were somewhat cooler before they formed the deposits. Sawkins (1966) has shown that evidence is accumulating to suggest that some barite and minerals of base metals may form vein deposits from circulating connate and ground waters.

Residual Deposits

Residual deposits occur in unconsolidated material and are formed by weathering from preexisting deposits. Many residual deposits of commercial value in the United States lie within the clayey residuum derived from limestone and dolomite of Cambrian and Ordovician age. Classic examples of these deposits occur in Washington and surrounding counties, Missouri (Brobst and Wagner, 1967; Wharton et al., 1981), and in the Appalachian region from Pennsylvania to Alabama (Brobst and Hobbs, 1968), but especially in the Sweetwater District, Tennessee (Maher, 1970) and the Cartersville district, Georgia (Kesler, 1950).

Most of the residual barite is white and translucent to opaque. It occurs commonly in tabular, fibrous, or dense fine-grained masses, but poorly to well-developed crystals are found. The barite in these deposits ranges from microscopic particles to irregular boulders weighing hundreds of pounds, but most of the masses are 2.5 to 15 cm (1 to 6 in.) in diameter. Chert, jasperoid, and drusy quartz are common in many deposits. Small amounts of pyrite, galena, and sphalerite occur on or in some of the barite, and locally lead and zinc minerals are recovered as a byproduct. Incompletely weathered rock fragments and red to yellow to brown residual clays typically make up the remainder of the deposit. The grade of these ores varies greatly, from less than 10% to 20% barite.

The size and shape of the deposits varies greatly. Some of the larger deposits in Georgia, Missouri, and Tennessee extend over hundreds of acres and are ideally suited to surface mining techniques. The shape of the residual deposit generally reflects the shape of the original deposit. The depth of the deposits depends on the local conditions. Some deposits in Georgia terminate at bedrock at depth of as much as 46 m (150 ft). Many of the deposits in Missouri rest on bedrock that lies only 3 to 5 m (10 to 15 ft) below the surface. These deposits are supremely amenable to mining by surface methods.

The source of the barite is generally thought to be vein material left after the host rock, generally limestone and dolomite, has been dissolved. Connate water and bacteria may play an important part in the redistribution of the barium in the carbonate environment, but their roles are not yet well understood.

Bedded Deposits

The bedded deposits include those in which barite occurs as a principal mineral or cementing agent in stratiform bodies in layered sequences of rocks. Many of these deposits are bedded concentrations of fetid, generally dark gray to black, fine-grained barite. The barite-rich beds have a thickness of less than 2.5 cm (1 in.) to as much as 15 m (50 ft) and contain as much as 50% to 90% barite. The barite beds are interlayered with dark chert, siliceous shale, and siltstone. Some barite-rich zones are more than 30 m (100 ft) thick. Some of these deposits contain millions of tons of barite easily beneficiated for use in drilling fluid and chemical plants. The structural configuration of the ore bodies and enclosing rocks make many of these deposits amenable to surface mining methods.

Since World War II, bedded deposits have yielded more commercial barite than any other type of deposit, notably in Arkansas, especially near Magnet Cove, Hot Spring County (Scull, 1958); Nevada, especially in the Battle Mountain region, in Eureka, Lander, and Elko counties (Papke, 1984); and near Castella, Shasta County, CA (Weber and Matthews, 1967). Valuable bedded deposits have been mined in West Germany, and more recently in Canada (Boyle and Jambor, 1966) and India (Meelakantam and Roy, 1979). In the last 20 years, extensive bedded deposits have been discovered in the Selwyn Basin, Yukon and Northwest Territories, Canada (Mako and Shanks III, 1984).

Early workers proposed that the barite beds were formed by the replacement of carbonate rocks by barium-bearing hydrothermal solutions from intrusive igneous activity. Many now believe that these deposits are of sedimentary origin and formed virtually at the same time as the enclosing rocks by organic and inorganic chemical processes of concentration and deposition (Miller, et al., 1977).

ORES AND USES OF STRONTIUM

Celestite, the sulfate of strontium, is the chief commercial source of the element strontium and its compounds. Strontianite, the carbonate of strontium, is the only other major strontium-bearing mineral, but it is so rare that it does not enter commercial markets. The principal uses of strontium are in the manufacture of strontium carbonate and nitrate. The carbonate is used principally in the manufacture of glass for TV picture tubes and ceramic ferrites for magnets. The nitrate is used for safety flares and other pyrotechnics where it imparts a brilliant red color to the flame.

The crystals of celestite commonly are tabular and closely resemble those of barite, but the mineral also commonly occurs in fibrous and granular forms. The hardness is 3-3.5 on Mohs' scale and the specific gravity of the purest forms is 3.95 to 3.97. The mineral is usually transparent to translucent, colorless to white, although some material is faintly red or blue.

GEOLOGY OF STRONTIUM DEPOSITS

Deposits of celestite are of the same general types as those of barite, namely, vein, residual, and bedded, with some differences in detail. Most commercial, or potentially commercial, deposits of celestite occur in or near sedimentary rocks as beds or lenses associated with gypsum, anhydrite, or rock salt. Under certain conditions, masses of celestite may be deposited from seawater with anhydrite. Some deposits occur in cavities or veins and disseminated in limestone and dolomite. Disseminations in shales, marls, and sandstones also occur. Celestite also occurs with some metallic ores, such as galena and sphalerite.

The known deposits of celestite in the United States are not currently mined because they are small and not economically competitive with deposits mined abroad. Deposits are described in at least 13 states including Arizona, Arkansas, California, Kentucky, Michigan, Missouri, New York, Ohio, Pennsylvania, Tennessee, Texas, Utah, and Washington (Schreck and Arundale, 1959).

The major deposits now supplying the world's markets are in Spain, where celestite is selectively mined from baritic, gypsiferous host rocks near Granada; Mexico, where the main celestite production comes from lens-shaped bodies concordant with the bedding of limestone host rock of Cretaceous age; England, where much celestite has been recovered since the 1880s from residual clays developed over marls of Triassic age in Gloucestershire; Iran, where the celestite occurs in marine sedimentary rocks of Miocene age; and Turkey, from deposits southeast of Sivas.

EXPLORATION TECHNIQUES FOR BARIUM AND STRONTIUM

Highly sophisticated technical methods for prospecting barite and celestite deposits have not been developed, mostly because new deposits have been fairly easy to find in known districts by applying the knowledge of the geologic association and the role of barium and strontium in the various geologic environments in which known deposits occur.

Useful geochemical techniques are available for exploration. A relatively simple and inexpensive turbidimetric test for barium was applied in a region of bedded deposits in western Arkansas (Brobst and Ward, 1965). The results indicated that barium anomalies could be detected that would offer targets for further detailed exploration. Geochemical studies of barium and strontium can be carried out with the use of emission spectrographic and X-ray fluorescence methods, but the equipment is costly. Some commercial laboratories can provide these analytical services. The use of a portable radioisotope fluorescence analyzer in geochemical prospecting for barite and celestite in the west of England has been described by Ball, et al. (1979).

Geophysical techniques of exploration for barite and celestite have not been widely used. Gravity surveys for residual barite deposits were undertaken in Missouri by Uhley and Scharon (1954). They reported differences of up to 35% between the tonnages of barite estimated in the studied tract and that subsequently mined. Klaus and Schroeder (1967) reported that the search by geophysical methods for hidden deposits of barite and fluorite in East Germany yielded less than ideal results. Bose and Vaidyanathan (1979), however, reported considerable success in tracing the large barite deposits in the Cuddapah District, India, by gravity surveys.

LITERATURE

Readers in search of information about barium and strontium, the occurrence and deposits of its ores, and other commercial and scientific aspects of the commodity and the industry will find an extensive literature available, especially in publications of the American Institute of Mining, Metallurgical, and Petroleum Engineers (AIME), the various state geological surveys, US Bureau of Mines, the US Geological Survey, and various technical journals devoted to the mining industry, such as *Industrial Minerals, Engineering & Mining Journal,* and *Pit and Quarry.* The literature cited in these various publications will lead readers to much more detailed information. Useful starting points include some general papers and reports by Ampian (1985, 1986, 1988), Dean and Brobst (1955), Brobst (1958, 1970, 1973, 1983), Ferrell (1985), Fulton (1983), Ober (1988), Schreck and Arundale (1959).

BIBLIOGRAPHY

Ampian, S.G., 1985, "Barite," *Mineral Facts and Problems,* US Bureau of Mines, Bulletin 675, pp. 65-74.

Ampian, S.G., 1986, "Barite," *Minerals Yearbook,* Vol. 1., "Commodity Reports," US Bureau of Mines.

Ampian, S.G., 1988, "Barite," *Mineral Commodity Summaries 1988,* US Bureau of Mines, pp. 18-19.

Ball, T.K., et al., 1979, "Geochemical Prospecting for Barite and Celestite Using a Portable Radioisotope Analyser," *Journal of Geochemical Exploration,* Vol. 11, pp. 277-284.

Bose, R.N., and Vaidyanathan, N.C., 1979, "Gravity Surveys for Barytes around Mangampeta, Cuddapah District, Andhra Pradesh (India)," *Journal of the Geological Society of India,* Vol. 20, No. 11, pp. 540-547.

Boyle, R.W., and Jambor, J.L., 1966, "Mineralogy, Geochemistry, and Origin of the Magnet Cove Barite-Sulphide Deposit, Walton, N.S.," *Canadian Mining and Metallurgical Bulletin,* Vol. 59, No. 654, pp. 1209-1228.

Brobst, D.A. 1958, "Barite Resources of the United States," Bulletin 1072-B, US Geological Survey, pp. 67-130.

Brobst, D.A., 1970, "Barite: World Production, Reserves, and Future Prospects," Bulletin 1321, US Geological Survey, 46 pp.

Brobst, D.A., 1973, "Barite," *United States Mineral Resources,* D.A. Brobst and W.P. Pratt, eds., Professional Paper 820, US Geological Survey, pp. 75-84.

Brobst, D.A., 1983, "Barium Minerals," *Industrial Minerals and Rocks,* 5th ed., Vol. 1, S. J. Lefond, ed., AIME, New York, pp. 485-501.

Brobst, D.A., and Hobbs, R.G., 1968, "Barite," *Mineral Resources of the Appalachian Region,* Professional Paper 580, US Geological Survey, pp. 270-277.

Brobst, D.A., and Wagner, 1967, "Barite," *Mineral and Water Resources of Missouri,* Vol. 43, Missouri Division of Geological Survey and Water Resources, pp. 99-106.

Brobst, D.A., and Ward, F.N., 1965, "A Turbidimetric Test for Barium and Its Geological Application in Arkansas," *Economic Geology,* Vol. 60, No. 5, pp. 99-106.

Dean, B.G., and Brobst, D.A., 1955, "Annotated Bibliography and Index Map of Barite Deposits in the United States," Bulletin 1019-C, US Geological Survey, pp. 145-186.

Ferrell, J.E., 1985, "Strontium," *Mineral Facts and Problems,* Bulletin 675, US Bureau of Mines, pp. 777-782.

Fulton, R.B., 1983, "Strontium," *Industrial Minerals and Rocks,* 5th ed., Vol. 2, S.J. Lefond, ed., AIME, New York, pp. 1229-1233.

Kesler, T.L., 1950, "Geology and Mineral Deposits of the Cartersville District, Georgia," Professional Paper 224, US Geological Survey, 97 pp.

Klaus, D., and Schoeder, N., 1967, "Erfahrungen beim Einsatz geophysicalischer Messfahrungen in der Flussspat- und Schwerspat Erkundung," *Zeitschrift fur Angewandte Geologie,* Vol. 13, No. 11-12, pp. 610-618.

Maher, S.W., 1970, "Barite Resources of Tennessee," Report of Investigations 28, Tennessee Div. of Geology, 40 pp.

Mako, D.A., and Shanks, W.C., III, 1984, "Stratiform Sulfide and Barite-Fluorite Mineralization of the Vulcan Prospect, Northwest Territories: Exhalation of Basinal Brines Along Faulted Continental Margin," *Canadian Journal of Earth Science,* Vol. 21, pp. 78-91.

Meelakantam, S., and Roy, S., 1979, "Barytes Deposits of the Cuddapah Basin," Records of the Geological Survey of India, Vol. 112, Pt. 5, pp. 51-64.

Miller, R.E., et al., 1977, "The Organic Geochemistry of Black Sedimentary Barite—Significance of Trapped Fatty Acids," *Organic Geochemistry,* Vol. 1, No. 1, pp. 11-26.

Ober, J.A., 1988, "Strontium," *Mineral Commodity Summaries 1988,* US Bureau of Mines, pp. 154-155.

Olson, J.C., et al., 1954, "Geology of the Rare-earth Deposits of the Mountain Pass District, San Bernardino County, California," Professional Paper 261, US Geological Survey, 75 pp.

Papke, K.G., 1984, "Barite in Nevada," Bulletin 98, Nevada Bureau of Mines and Geology, 125 pp.

Sawkins, F.J., 1966, "Ore Genesis in the Northern Apenine Orefield in the Light of Fluid Inclusion Studies," *Economic Geology,* Vol. 61, No. 2, pp. 385-401.

Schreck, A.E., and Arundale, J.C., 1959, "Strontium, a Materials Survey," Information Circular 7933, US Bureau of Mines, 45 pp.

Scull, B.J., 1958, "Origin and Occurrence of Barite in Arkansas," Information Circular 18, Arkansas Geology & Conservation Commission, 101 pp.

Uhley, R.P., and Scharon, LeR., 1954, "Gravity Surveys for Residual Barite Deposits in Missouri," *Mining Engineering,* Vol. 6, No. 1, pp. 52-56.

Weber, F.H., Jr., and Matthews, R.A., 1967, "Prospecting for Barite in Northern Shasta County," *Mineral Information Service,* Vol. 20, No. 9, California Div. of Mines and Geology, pp. 107-114.

Wharton, H.M., et al., 1981, "Barite in Residuum Derived from Vein and Replacement Deposits in Paleozoic Rocks," "Metallic Mineral Resource Potential of the Rolla 1 × 2 Quadrangle, Missouri as Appraised in September 1980," Open File Report 81-0518, US Geological Survey, pp. 23-25.

3. Borate Exploration

JAMES M. BARKER

INTRODUCTION

Borate minerals have been known and utilized since at least the 5th century when Byzantine artisans used Turkish borax in pottery glazes and as fluxes for precious metal. Beginning in the 13th century, borax from Kashmir, India, and Tibet was brought to Europe. About 1770, sassolite from Italy became the dominant source of borax in Europe and, Italy, along with Germany, produces borate today. Borax was first produced in western South America about 1850 and in the southwestern United States a little later.

Since the 1880s, the playa deposits of the southwestern United States, followed by massivly bedded borate and brine deposits in California, have dominated the world market. In addition, the largest known borate reserves were discovered about 1950 in Turkey which is now a close second to the United States in production and the leader in exports. Both Russia and China produce borates today. Chile, Bolivia, Argentina, and Peru intermittently produce borates. Recent major colemanite discoveries have been made in Sonora, Mexico, and borates continue to be found within existing districts in southeastern California.

Kistler and Smith (1983) review the world's borate producing regions and reserves. Barker and Lefond (1985) compiled 18 papers on borate geology, exploration, mining, processing, and end uses. Harben and Bates (1984) summarize the world borate situation.

The unique characteristics of borate derive from boron oxide (B_2O_3). Boron compounds are important in manufacturing applications, particularly in highly industrialized nations such as the United States, members of the European Economic Community, and Japan. The glass-forming properties of borate minerals and refined chemicals account for one-half US consumption of these products, specifically in the manufacture of specialty glasses such as pyrex, frits, and insulation-grade and textile-grade glass fibers. Inorganic chemicals with a borate anion in combination with either a metallic cation, usually sodium, or hydrogen, as in boric acid, display important chemical properties. These are applicable to industries ranging from agriculture to ceramics, to chemical and metallurgical processing to laundry products (Lyday, 1985, p. 91).

BORATE MINERALOGY

Boron-rich minerals vary widely in character due to boron's two coordination numbers, but most are hydrated, low in density, soft, either gray or white, monoclinic or triclinic crystals, and soluble in water and/or acid. Borate characteristics as listed in standard mineralogy texts often differ markedly from those of field occurrences owing to mode of formation, purity, age, or alteration.

At least 150 borate-bearing minerals exist (Table 1) but few are common or commercially important. Dominant industrial borate minerals are borax, kernite, colemanite, ulexite, probertite, priceite, and szaibelyite. Many others, useful as indicators in mineral exploration, occur as minor parts of borate ore bodies, or are mined on a small scale.

Borates are often produced with other commodities such as potash (KCl), soda ash (Na_2CO_3), salt cake (Na_2SO_4), and nahcolite ($NaHCO_3$) in lacustrine or brine deposits.

Other boron deposits may have varied coproducts or byproducts, depending on individual deposit characteristics.

BORON CHEMISTRY

Boron always occurs combined in nature. It has properties analogous to carbon, aluminum, and silicon and is concentrated in organic or silicate environments (Reynolds, 1972). Elements of similar character which may be found with boron include lithium, tantalum, cesium, beryllium, uranium, and thorium (Ostroschenko, 1967). Strontium-bearing (celestite) and arsenic-bearing (orpiment and realgar) minerals are common in borate deposits.

Boron is concentrated in residual magma or late magmatic fluids. Boron's large ionic diameter prevents it from easily entering the crystal structure of most silicates, thus making it highly mobile during igneous and metamorphic processes.

Many borate compounds are soluble and ground water and oilfield brines may contain high concentrations of boron. Undissociated boric acid and sodium borate in seawater form a pH-buffer system nearly as important as the carbonate one.

The geochemistry of boron has been documented in less detail than that of most other minor elements (Kemp, 1956). This is primarily due to analytical difficulties related to its low atomic number and close association with silicon and aluminum (Chorlton, 1973; Barsukov, 1961).

BORON DISTRIBUTION

Boron is widespread in nature but usually in trace amounts from which it cannot economically be recovered. Large-scale continental borate accumulation is restricted to evaporite sediments formed in the arid portions of the volcanic-tectonic belts of western North and South America, the eastern Mediterranean, or Asia. Marine accumulations are generally small.

Boron tends to be concentrated in marine sediments by adsorption on clay minerals or by substitution for silicon in clay minerals (Walker, 1975). Potassium-rich clays, such as illite or glauconite, usually contain from 100 to 300 ppm boron in contrast to the seawater concentration of about 4.6 ppm. The boron content of diagenetic or authigenic illite may be directly related to the boron content and salinity of the water in which the illite was altered or formed (Reynolds, 1965). The boron content may be detrital or temperature related. Clastic sediments, both marine and nonmarine, often have a high boron content from detrital tourmaline or illite (Walker, 1963; Shaw and Burgry, 1966).

BORATE MINING

Borates, the major type of boron minerals mined worldwide, are produced using both underground and surface mining techniques along with in situ ones. Surface mining of borates in the USA and Turkey produce the bulk of borates but significant deposits are mined using the other techniques. New underground and in situ mining projects have been implemented or contemplated in the 1980s. Underground mining was the most common form of mining into the 1950s including deposits now surface-mined such as Boron, CA (Obert and Long, 1962). Special situations, such as deposits under Death Valley National Monument, dictated under-

Table 1. Important Boron-Bearing Minerals

Mineral Name	Mineral Composition	Mohs Hardness	Specific Gravity	Crystal System	B_2O_3, Wt %	Boron, Wt %
Borax	$Na_2B_4O \cdot 10H_2O$	2-2 1/2	1.71	Mono.	36.5	11.3
Tincalconite	$Na_2B_4O_7 \cdot 5H_2O$	—	1.88	Hexa.	47.8	14.8
Kernite	$Na_2B_4O_7 \cdot 4H_2O$	2 1/2	1.91	Mono.	51.0	15.8
Inyoite	$Ca_2B_6O_{11} \cdot 13H_2O$	2	1.87	Mono.	37.6	11.7
Meyerhofferite	$Ca_2B_6O_{11} \cdot 7H_2O$	2	2.12	Tric.	46.7	14.5
Colemanite	$Ca_2B_6O_{11} \cdot 5H_2O$	4 1/2	2.42	Mono.	50.8	15.8
Ulexite	$NaCaB_5O_9 \cdot 8H_2O$	2 1/2	1.95	Tric.	43.0	13.4
Probertite	$NaCaB_5O_9 \cdot 5H_2O$	3 1/2	2.14	Mono.	49.6	15.4
Inderite	$Mg_2B_6O_{11} \cdot 15H_2O$	3	1.78	Tric.	37.3	11.6
Kurnakovite	$Mg_2B_6O_{11} \cdot 13H_2O$	3	1.85	Tric.	39.9	12.4
Szaibelyite	$MgBO_2(OH)$	3	2.62	Orth.	41.4	12.9
Sussexite	$MnBO_2(OH)$	3 1/2	3.30	Orth.?	30.4	9.4
Ludwigite	Mg_2FeBO_5	5	3.6	Orth.	17.8	5.5
Paigeite	Fe_2FeBO_5	5	4.7	Orth.	13.5	4.2
Howlite	$Ca_2B_5SiO_9(OH)_5$	3 1/2	2.53-2.59	Mono.	44.5	13.8
Priceite	$Ca_4B_{10}O_{19} \cdot 7H_2O$	3-3 1/2	2.42	Tric.?	49.8	15.5
Bakerite	$Ca_4B_4(SiO_4)_3(OH_3) \cdot H_2O$	4 1/2	2.7-2.9	Mono.	27.0	8.4
Hydroboracite	$CaMgB_6O_{11} \cdot 6H_2O$	2-3	2.17	Mono.	50.5	15.8
Searlesite	$NaBSi_2O_6 \cdot H_2O$	3 1/2	2.45	Mono.	17.1	5.3
Sassolite	H_3BO_3	1	1.46-1.50	Tric.	56.3	17.5
Boracite	$Mg_3B_7O_{13}CL$	7-7 1/2	2.91-3.10	Orth.	62.2	19.3
Datolite	$CaBSiO_4(OH)$	5-5 1/2	2.9-3.0	Mono.	21.8	6.8
K-Feldspar	$KBSi_3O_8$	6	2.9-3.0	Mono.	13.3	4.1
Tourmaline	$NaFe_3Al_6(BO_3)_3Si_6)_{18}(OH)_4$	7-7 1/2	3.10-3.25	Hexa.	9.7	3.0
Danburite	$CaB_2Si_2O_8$	7-7 1/4	2.97-3.02	Orth.	14.2	4.4
Kotoite (Jimboite)	$Mg_3B_2O_6(Mn_3B_2O_6)$	6 1/2	3.10	Orth.	18.3	5.7
Nordenskioldine	$CaSnB_2O_6$	5.5-6	4.2	Trig.	12.6	3.9
Luneburgite	$Mg_3B_2(OH)_6(PO_4) \cdot 6H_2O$	2+	2.05	Mono.?	7.0	2.2
Seamanite	$Mn_3(PO_4)(BO_3) \cdot 3H_2O$	4	3.08	Ortho.	9.3	2.9
Fluoborite	$Mg_3(BO_3)(OH,F)_3$	3 1/2+	2.85-2.98	Hexa.	18.7	5.8+
Hambergite	$Be_2(OH)(BO_3)$	7 1/2	2.359	Orth.	37.1	11.5

Source: Barker and Lefond, 1985.

ground borate mining in the 1970s and 1980s (Billie, Boraxo, and Sigma mines) using continuous miners adapted from coal mines. Great depth and low ore grade at the Hector deposit east of Barstow, CA, dictated in situ mining which was not carried past the testing phase.

Geology of Borate Deposits

Commercial borate deposits have formed in five main ways (Table 2): (1) by precipitation from brines in a permanent or semi-permanent shallow lake or a deep lake, (2) as crusts or crystals in mud of playas, (3) by direct precipitation near springs (warm or hot) or fumaroles, (4) by evaporation of marine water, and (5) by crystallization at or near granitic contacts or in veins. The deposits formed by precipitation in permanent shallow lakes are large and produce most of the world's borates. Other occurrences are comparatively small and most are now uneconomic.

Lacustrine Borates

Large-scale borate precipitation in shallow lakes requires that several conditions be met. A source of boron is necessary and is usually a nearby thermal spring, but may be a distant one. Weathering and erosion of preexisting borate deposits or bedrock is usually a secondary source. Rock weathering typically supplies relatively little boron but may be locally important.

Borates have relatively high solubilities so an arid climate helps produce boron supersaturation and retards borate dis-

solution. Nonborate saline minerals and sediment may predominate with most boron in an interstitial brine rather than in borate crystals. An interior drainage system with a relatively small drainage area concentrates boron and minimizes boron dilution by excessive inflow of water, other ions, or sediment.

The borate lake basin must either have at least occasional exterior drainage to remove accumulated brine containing competing ions, or must be stratified with complex brine-evolution mechanisms active. Subsequent diagenesis, remobilization, folding, faulting, and erosion normally alters and complicates the initially simple borate mineralogy and distribution.

The most common borates are those that form at surface or near-surface temperatures and pressures. One sequence in which these borates crystallize has been determined by field and laboratory work. Muessig (1959) summarized these data and found that the most likely primary mineral in each borate group was the highest hydrate (see top four groups, Table 1). The general applicability of these data are open to question because, depending on the activity of water and temperature, other borate mineral hydration levels can be primary (Barker and Barker, 1985). As borate minerals dehydrate during paragenesis, their density and hardness tend to increase along with their boron content. Replacement by calcium carbonate is common for colemanite and similar borates.

A stratified borate-rich lake can develop strong thermal

Table 2. Selected Borate Occurrences of the World

Borate Occurences	Type*	Age of Deposit	Minerals†
North America			
California			
Boron (Kramer)	L	Middle Miocene	B, K, U
Death Valley	L, P	Upper Miocene	U, C, P
Searles Lake	BR	Late Quaternary	B, K
Nevada			
Teels, Rhodes, &			
Columbus Marshes	P	Late Quaternary	U
Callville Wash &			
White Basin	L	Early Tertiary	C
Oregon	BR	Late Quaternary	B
Mexico			
Magdalena	L(P?)	Tertiary	C, H
Tubutama	L(P?)	Tertiary	C, H
Turkey			
Sultain Cayir	L	Late Tertiary	C, PR
Bigadic	L	Late Tertiary	PR, C, U
Emet	L	Late Tertiary	C
Kirka	L	Late Tertiary	B, C
USSR			
Inder	N	Permian	S
Lake Inder	BR	Late Quaternary	B
Europe			
Italy	H	Recent	SS
Germany			
Stassfurt	M	Upper Permian	BC
Hamburg	M	Upper Permian	BC
South America			
Chile	P	Quaternary	U
Bolivia	P	Quaternary	U
Argentina	P, H	Quaternary, Tertiary	U, B, K, C
Peru	P, H	Quaternary	U
Asia			
China	P, L	Quaternary	U, B, S
Tibet-Kashmir	BR, H	Recent	B

* L = lake deposit; P = playa deposit; BR = brine; H = thermal spring deposit; M = marine deposit.
† B = borax; C = colemanite; P = probertite; U = ulexite; K = kernite; S = szaibelyite; SS = sassolite; BC = boracite; PR = priceite; H = howlite.
Sources: Kistler and Smith (1983), Watt (1973), Wang (1974, 1975), Evans, et al. (1976), Barker and Lefond (1985).

stratification. A dilute surface layer over a hypersaline borate brine can trap solar energy so efficiently that temperatures exceeding 90°C are reached in these bodies of water called heliothermal lakes (Barker and Barker, 1985). This combination of elevated temperature and low activity of water alters the thermodynamics so that less hydrated borates, such as colemanite, are the stable mineral phase. Colemanite or other low hydrates can be a primary mineral under these conditions.

Playa Deposits

Borates in playas are formed as surface incrustations or crystals within near-surface sedimentary layers. The primary mode of formation is by repeated evaporation of incoming boron-bearing water and by capillary evaporation of ground water. The volume of sediment and other evaporite minerals is normally much greater than the borate volume so the borate crystals are matrix supported. Repeated solution-re-crystallization cycles result in massively bedded strata. Playas

may be related to an earlier lacustrine phase, so superimposed lacustrine/playa types are possible.

Thermal Spring and Fumarole Deposits

Thermal springs and fumaroles may emit boron-rich water or gases which precipitate boron-bearing minerals locally due largely to cooling. Such deposits may form proximal spring aprons or crusts of borate with the unprecipitated borate perhaps supplying distal lake or playa environments.

Marine Borate Deposits

Marine borate deposits are generally small. The borates are often a byproduct of mining some other evaporite mineral such as potash (Kuhn, 1962). The borate accumulation is generally an insoluble residue either differentially leached from local marine evaporites or formed diagenetically from them. The largest deposits (such as Inder) may be related to submarine thermal activity or evaporation of seawater with anomolously high boron content (Kistler and Smith, 1983).

Pegmatite Deposits

Boron is usually concentrated in the residual fluids associated with siliceous intrusions (Barsukov and Egorov, 1957). Boron-containing minerals (tourmaline, lepidolite, and others), may crystallize very late in a pegmatite (Oftedal, 1964) or boron may be mobilized into the country rock or into fluids feeding thermal springs. Boron may also be leached from country rock (particularly marine illite) by boron-poor residual magma or hydrothermal fluids (Ewers, 1977; Ellis and Mahon, 1967).

EXPLORATION FOR BORATES

Geophysical Exploration

Geophysical surveys utilizing gravity, magnetic, nuclear, electromagnetic, resistivity, and seismic methods have been employed for borate exploration. However, no borate deposit has been found using geophysics alone. Geophysical techniques are useful for delineating the shape, structure, or depth of known borate bodies or for solving local geological problems in borate terranes. Several methods discussed subsequently have significant but underutilized potential for locating unknown borate deposits.

Seismic Methods: Colemanite, but also limestone, tend to have seismic velocities greater than surrounding mudstone or shale. Borates often occur stratigraphically near basalt flows or other igneous bodies, both with seismic velocities which may be similar to the high colemanite velocities. These rocks supply, along with an occasional borate deposit, many high-velocity zones so other methods must be utilized to limit the number of exploration targets to a manageable total.

Density Surveys: The average density of borates is usually very near that of the surrounding rocks and sediment. Density contrasts approaching 1:1 are typical. Such small contrasts are not easily discernible using gravity techniques, although this method is useful in determining basement configuration.

Resistivity Surveys: Many borates have high resistivities as dry, single crystals, but field occurrences tend to have ion-rich interstitial water and associated nonborate minerals which mask this characteristic. Caliche deposits often inhibit energy penetration during resistivity studies in arid areas. However, resistivity surveys are a potentially useful, but as yet poorly understood, approach to borate exploration. Preliminary field studies suggest that large ore bodies may have sufficient contrast if the edges are traversed during measurement.

Nuclear Techniques: Boron is a very efficient thermal neutron absorber. It has a very large neutron capture cross section which is hundreds of times greater than other elements common to most borate deposits, especially sedimentary ones. Neutron methods are successful for indicating downhole presence of boron, up to a few hundred ppm, in oilfield brines and hydrocarbons using the neutron lifetime log, and tens of percent in bedded-lacustrine borates using the neutron-neutron log.

Magnetic and Electromagnetic Methods: Neither magnetic nor electromagnetic methods indicate the presence of borates directly. They do indicate details of subsurface geology such as faults, buried basalt, and other features that can lead to borate discoveries or deposit delineation. Paleomagnetic orientation and dating is useful in borate basin analysis and modeling.

Geochemical Exploration

Although it has been tried in Russia (Buyalov and Shuyryaylva, 1961), exploration for borates using geochemistry is not widespread. Boron has been used in Canada as a pathfinder element in exploration for metallic deposits of tin, lead, zinc, cadmium, gold, silver, tungsten, and others (Boyle, 1971).

Several approaches are possible using the following characteristics of boron: (1) Boron is often associated with elements such as strontium, lithium, tantalum, cesium, and beryllium; these elements have similar chemical properties. (2) Boron is very soluble and may be added to ground water by leaching of sedimentary borates (particularly sodium borates), during igneous and hydrothermal activity, or during metamorphism which characteristically shows a decrease in boron content in rocks with increasing metamorphic grades. (3) Plants require boron for healthy growth, but the concentration range is very critical (Levinson, 1974). Borates are added to both fertilizers and defoliants with slight differences in concentration required to affect the plants either way. Effects depend on the plant species as seen in Death Valley where serophytes grow on or near pure borate outcrops. Plants not adapted to boron stress are easily killed by a few ppm boron. Due to boron's high solubility, plants can be useful indicators of its presence in ground water. False-color infrared imagery in the proper wave length indicates subtle variations in plant chlorophyll, reflectance, etc., and could be used for regional studies. However, desert plants (xerophytes) associated with the arid or semi-arid climate of most borate terranes create a problem. These plants have small and sparse leaves, causing low image density and high-boron tolerance, causing no measurable effect. (4) Heavy mineral analysis, based primarily on detrital tourmaline, can be used in exploration for certain metallic deposits (Gleeson, 1968; Boyle, 1965), or borate-rich pegmatites. (5) Boron concentrations in desert varnish are much higher near borate deposits and in borate terrane in general (Lakin, et al., 1963).

Geological Exploration

Prior to 1970, borate exploration in the United States during this century was infrequent. Modern geological, geophysical, and drilling methods were rarely systematically applied, especially away from producing mines. In other regions, notably Russia in the 1930s and Turkey in the 1950s and 1960s, large borate deposits were found using geology and drilling in borate terranes and over suspected deposits.

Empirical Exploration Guides: A number of empirical exploration guides have proven useful in delineating borate terranes and for locating borates as follows: (1) tectonic belts; (2) dry and hot climate; (3) lacustrine rocks of Cenozoic age; (4) presence of igneous rock, thermal springs, and associated minerals and elements; (5) evaporite deposits, particularly containing borate minerals; (6) borate efflorescences and playas; (7) abundant limestone and brecciation; and (8) arsenic, lithium, and strontium anomalies.

Use of these parameters can rapidly and readily narrow areas of search and thus are valuable tools. Most are empirical guides, so the greater the number appearing in an area, the greater the potential for exploration success there.

Tectonic Belts—Borate ore deposits are usually associated with three main tectonic belts. The smallest deposits are formed at thermal springs and on playas in the Andes Mountains of South America. These are probably indirectly related to subduction along the Chile trench (Norman and Santini, 1985).

The largest borate deposits are in Turkey with smaller deposits in India, China, and Tibet. All are located along the tectonic belts extending from the eastern Mediterranean through the Himalaya Mountains into China.

The large deposits in the southwestern United States and north-central Mexico are related to the extension following overriding of the East Pacific rise by the North America plate. Separating these belts from tectonically similar ones without borates is a dry climate.

Dry and Hot Climate—A dry, hot climate, both at the time of deposition and later, is favorable for borates. High net evaporation helps saturate solutions relative to boron. Boron minerals, particularly sodium borates such as borax, are very soluble and so do not normally crop out under humid climates. The less soluble borates (ulexite, colemanite) are not highly resistant to weathering over geologic time.

Cenozoic Lacustrine Sediments—Most borate deposits are found in nonmarine sediments of Cenozoic age. These sedimentary rocks are of lacustrine origin for the largest deposits. Previously exploited playa or spring deposits are comparatively small while marine borate deposits are both rare and small.

Volcanic Rocks and Thermal Springs—The source of continental boron is generally attributed to volcanic emissions, particularly thermal springs. Areas with Tertiary volcanic rocks and existing thermal springs are favorable. This is so even if they do not now have a high boron content because boron may be supplied to a basin intermittently. However, very hot springs typically have a low boron content compared to warm ones (Barker and Barker, 1985).

Evaporite Minerals—Borates commonly occur without associated evaporite minerals. Large deposits (such as Searles Lake in California) can occur with them and certainly evaporites could be deposited in one part of a basin while borates are deposited in another part. The proximity of commercial borate deposits, most of which are evaporites, is the single most favorable condition in borate exploration. The presence of salt, potash, anhydrite, and gypsum is all favorable although these are often of marine, rather than continental, origin.

Borate Efflorescences and Playa Deposits—Crusts of borates on the surface, or disseminated crystals and aggregates near the surface, may indicate the existence of nearby massive borate deposits. Most of the world's playa deposits are low-grade surface accumulations and consist of ulexite, borax, tincalconite, and other minor borates. The playa borates in Death Valley were known for several years before exploration upstream exposed large, bedded, lacustrine borate outcrops. The Gerstley deposit (a long-active mine) near Shoshone, CA (Barker and Wilson, 1976; Noble, 1926b), was found by exploring behind outcrops which contained very minor radial growths of ulexite in mudstone.

Limestone and Brecciation—Limestone is often associated with borates. In the Death Valley region, the borates are embedded in calcareous shale with limestone abundance increasing near and within the borate body. During evaporation of a brine, both boron and carbonate saturation increases. Carbonates would begin to precipitate before borates and could continue to do so as borates precipitated. Such a mechanism would account for the large amount of limestone in and around lacustrine borate accumulations.

Sodium borates can alter to sodium-calcium or calcium borates by ion exchange or leaching. Leaching and alteration can continue until only limestone remains. The limestone may be pseudomorphous after borate and have a low B_2O_3 content of 1 to 3%. Such a situation in outcrop indicates possible borates at depth (downdip) below the near-surface weathering zone.

The alteration of sodium borates or other borates causes a volume decrease as water of hydration decreases. A result is the collapse brecciation common within borate deposits and in the surrounding rock. This brecciation is most pronounced above a borate ore zone and decreases away from it.

Arsenic, Lithium, and Strontium—Arsenic-bearing minerals, especially orpiment and realgar, suggest thermal spring activity. Their presence in lacustrine rock is a very favorable association. Both lithium and strontium are high at borate deposits. The lithium is typically in clay minerals such as hectorite or illite. The strontium is celestite, a sulfate, which commonly occurs as overgrowths on borate crystals such as colemanite. Regions with positive lithium or strontium anomalies are highly favorable for borate occurrences.

Field Tests: Four borate tests (Nemodruk and Karalova, 1965) are useful for field determination. These involve a flame test, reaction with turmeric paper, colorimetric determination, or taste.

Flame Test—Some boron minerals yield a yellowish-green flame when heated alone, but the majority require the application of H_2SO_4 or boric acid flux (3 parts $KHSO_4$, 1 part CaF_2). If decomposable by H_2SO_4, boron compounds burn with a yellowish-green flame [owing to the formation of acid methyl ester $B(OCH_3)_3$] when ignited in an evaporating dish with alcohol and concentrated H_2SO_4. Borates not decomposable by H_2SO_4 should be mixed with three parts of boracic acid flux and introduced into the flame on a clean platinum wire. A flash of green indicates the liberation of volatile boron fluoride (BF_3).

Turmeric-Paper Test—If turmeric paper is moistened with a dilute HCl solution of boron and dried, it assumes a reddish-brown color. It is then moistened with NH_4OH, and a bluish-black or grayish-blue color results, depending upon the amount of turmeric and boric acid present. It is advisable to run a blank test at all times. Solutions of zirconic, titanic, tantalic, columbic, and molybdic acids also color turmeric paper brown, so this test for boron can be employed only in their absence or with caution. Tantalum and columbium are especially prone to occur with boron due to geochemical similarities.

Colorimetric Test—Add several drops of polyvinyl alcohol (PVA) solution to a drop of iodine solution on a borate sample. The mixed solutions will turn blue if a borate is present. The change to blue is rapid for high boron concentration (over 300 ppm) but may take up to 30 min for 80 ppm. Nonborates may yield confusing colors: $CaCl_2$ is indistinguishable; $NaHCO_3$ is close; and Na_2CO_3 or K_2CO_3 have some particles appearing borate blue but most are purple. NaCl, $MgCl_2 \cdot nH_2O$, $BaCl_2$, and Na_2SO_4 (somewhat more blue to neutral), also yield a purple color.

The PVA solution is made by sprinkling 10 g of polyvinyl alcohol (DuPont "Elvanol" No. 7-50 or Braun FH 600) into 900 mL of cold distilled water while stirring vigorously. After a slurry is obtained, heat to about 75°C and continue stirring until the solution clears. Add 100 mL of concentrated HCl.

The iodine solution is made by dissolving 1.2 g of iodine and 3.0 g of KI in a small volume of distilled water, then diluting with 100 mL of water. Take 50 mL of this solution and further dilute to 250 mL to make an 0.02N solution. Store in dark glass bottles to extend shelf life. This reagent loses color after several months and is then ineffective.

Large amounts of organic materials such as acetone, N,N'-dimethyl-formamide, dioxane, ethanol, methanol, formic acid, and pyridine destroy the blue color, while others extract iodine, causing the color to fade.

Taste Test—Soluble borates have a sweetish, astringent taste.

Drilling: The drilling of borate deposits is similar in most respects to that of other evaporites. A drilling program for borates must be designed for a particular project with its unique conditions.

Water is typically acceptable as a drilling fluid for the less-soluble borates but hydrocarbon-based fluids may be necessary for more soluble ones. Care is needed to insure that excessive leaching of cores is avoided. Borate deposits are often vuggy causing lost circulation which can be exacerbated by solution of evaporites within the deposit.

Many borate deposits occur in lacustrine rocks with abundant expansive clay. This leads to excessive swelling by clay hydration. Hole problems result if water loss from the drilling fluid to the clay is not controlled by proper chemical treatment.

SUMMARY

Boron is widespread in nature. Large-scale economical accumulations are restricted to lacustrine sediments or rocks formed in arid portions of the volcanic-tectonic belts of western North America and South America, the eastern Mediterranean, or Asia. Dominant borate mineralization depends on temperature and activity of water. At low temperatures and high activity of water, higher hydrates are favored. At higher temperatures and lower activities of water, such as in a heliothermal lake, lower hydrates are the stable phase.

There are no sure methods for locating borate deposits but there are a number of empirical exploration guides that have proven useful in narrowing the search for borate deposits. These are: (1) tectonic belts; (2) a dry, hot climate; (3) lacustrine Cenozoic rocks; (4) the presence of igneous rocks, thermal springs, and associated minerals and elements; (5) evaporite deposits, particularly if borate minerals are present; (6) borate efflorescences and playas lakes; (7) limestone brecciation; and (8) arsenic, lithium, and strontium anomalies.

The search for new economic deposits of borate minerals requires diligent exploration and imaginative geologic thinking but is exciting and can be financially rewarding.

REFERENCES

Barker, C.E., and Barker, J.M., 1985, A re-evaluation of the origin and diagenesis of borate deposits, Death Valley region, California, *Borates: Economic Geology and Production,* J.M. Barker and S.J. Lefond, eds., AIME, New York, 274 pp.

Barker, J.M., and Lefond, S.J., eds., 1985, *Borates: Economic Geology and Production,* AIME, New York, 274 pp.

Barker, J.M., and Wilson, J.L., 1976, "Borate deposits in the Death Valley region," Report 26, Nevada Bureau of Mines and Geology, pp. 22-32.

Barsukov, V.L., 1961, "Some problems of the geochemistry of boron," *Geochemistry,* No. 7, pp. 596-608.

Barsukov, V.L., and Egorov, A.P., 1957, "Some geochemical characteristics of formation of hypogene borate deposits," *Geochemistry,* No. 8, p. 790-801.

Boyle, R.W., 1965, "Geology, geochemistry, and origin of lead-zinc-silver deposits of the Keno Hill-Galena Hill area, Yukon," Paper 111, Canadian Geological Survey, 302 pp.

Boyle, R.W., 1971, "Boron and boron minerals as indicators of mineral deposits," *Geochemical Exploration,* R.W. Boyle and J.I. McGerrigle, eds., Canadian Institute of Mining and Metallurgy Special Volume II, 112 pp.

Buyalov, N.I., and Shuyryaylva, A.M., 1961, "Geobotanical method in prospecting for salts of boron," *International Geology Review,* Vol. 3, No. 7, pp. 619-625.

Chorlton, L.B., 1973, "The effect of boron on phase relations in the granite-water system," unpubl. MS Thesis, McGill University, Canada, 95 pp.

Ellis, A.J., and Mahon, W.A.J., 1967, "Natural hydrothermal systems and experimental hot water/rock interactions (part II)," *Geochimica et Cosmochimica Acta,* Vol. 31, pp. 519-538.

Evans, J.R., Taylor, G.C., and Rapp, J.S., 1976, "Mines and mineral deposits in Death Valley National Monument, California," Special Report 125, California Division of Mines and Geology, pp. 21-32.

Ewers, G.R., 1977, "Experimental hot water-rock interactions and their significance to natural hydrothermal systems in New Zealand," *Geochimica et Cosmochimica Acta,* Vol. 41, pp. 143-150.

Harben, P.W., and Bates, R.L., 1984, *Geology of the Nonmetallics,* Metal Bulletin, New York, 392 pp.

Kemp, P.H., 1956, "The Chemistry of Borates, Part I," Borax Consolidated Ltd., London, 90 pp.

Kistler, R.B., and Smith, W.C., 1983, "Boron and Borates," *Industrial Minerals and Rocks,* 5th ed., S.J. Lefond, ed., AIME, New York, pp. 533-560.

Kuhn, R., 1962, "Geochemistry of the German potash deposits," Special Paper 88, Geological Society of America, pp. 460-466.

Lakin, H.W., et al., 1963, "Variation in minor element content of desert varnish," Professional Paper 475-B, US Geological Survey, pp. B28-B31.

Levinson, A.A., 1974, *Introduction to Exploration Geochemistry,* Applied Publ. Ltd., Maywood, pp. 388-389.

Lyday, P.A., 1985, "Boron," *Minerals Yearbook,* US Bureau of Mines, preprint, 12 pp.

Muessig, S., 1959, "Primary borates in playa deposits—minerals of high hydration," *Economic Geology,* Vol. 54, No. 3, pp. 495-501.

Noble, L.F., 1926b, "Note on a colemanite deposit near Shoshone, California, with a sketch of a part of the Amargosa Valley," Bulletin 785, US Geological Survey, pp. 63-73.

Norman, J.C., and Santini, K.N., 1985, "An overview of occurrences and origin of South American borate deposits with a description of the deposit at Laguna Salinas, Peru," *Borates: Economic Geology and Production,* J.M. Barker and S.J. Lefond, eds., AIME, New York, pp. 53-69.

Obert, L., and Long, A.E., 1962, "Underground borate mining, Kern County, California," Report of Investigations 6110, US Bureau of Mines, 67 pp.

Oftedal, I., 1964, "On the occurrences and distribution of boron in pegmatite," *Norsk. Geol. Tidsakr,* Vol. 44, pp. 217-225.

Ostroschenko, V.D., 1967, "Geochemistry of boron and cesium in the volcanic rocks of western Tien-Shan," *Geochemistry,* pp. 800-806.

Reynolds, R.C., 1965, "The concentration of boron in Precambrian seas," *Geochimica et Cosmochimica Acta,* Vol. 29, No. 1, pp. 1-16.

Reynolds, R.C., 1972, "Boron: element and geochemistry," *The Encyclopedia of Geochemistry and Environmental Science,* R.W. Fairbridge, ed., Van Nostrand and Reinhold, New York, pp. 88-90.

Shaw, D.M., and Bugry, R., 1966, "A review of boron sedimentary geochemistry in relation to new analyses of some North American shales," *Canadian Journal of Earth Science,* Vol. 3, pp. 49-63.

Walker, C.T., 1975, *Geochemistry of Boron,* Dowden, Hutchinson, and Ross, Inc., Stroudsburgh, 414 pp.

Walker, C.T., 1963, "Size fractionation applied to geochemical studies of boron in sedimentary rocks," *Journal of Sedimentary Petrology,* Vol. 33, No. 3, pp. 694-702.

Wang, K.P., 1975, "The People's Republic of China—a new industrial power with a strong mineral base," Special Publication, *Mines and Mineral Resources—China,* US Bureau of Mines, 96 pp.

Wang, K.P., 1974, "Boron: Commodity Data Summaries, Appendix 1, Mining and Mineral Policy," Third Annual Report, Secretary of the Interior, pp. 20-21.

Watt, K.A.L.G., 1973, "The borate industry," *World Minerals and Metals,* British Sulphur Corp., No. 12, pp. 5-12.

4. Chromite*

Chromite is the only ore mineral of metallic chromium and chromium compounds and chemicals. Because of this fact, *chromite* and *chrome ore* are used synonymously in trade literature. In commercial market quotations, chrome ore is the term commonly used. Chromite per se, because of properties imparted by its chromium content, is used in refractories and in special purpose molding sands for metal casting. There are many other minerals containing some, generally minor, amounts of chromium, but none are commercial sources of the element (Thayer, 1956).

Under present conditions, the US and North America, with the possible exception of Cuba, have no commercial ore reserves. There are chromite deposits, as will be described, but they would be used only under highly abnormal conditions if all outside sources were cut off for an extended period. All chromite used in the US is imported from the eastern hemisphere in which are contained the principal world reserves. The main suppliers in 1980 were the Republic of South Africa (43%), the USSR (21%), the Philippines (12%), Albania (11%), Turkey (8%), and others (5%). The strategic nature of chromite is obvious.

GEOLOGY AND MINERALOGY

Mineralogy

Chromite is a variety, or more properly stated, a composition range, of the spinel group of minerals (Stevens, 1944; Ulmer, 1970). This range can be expressed as $(Mg, Fe^{2+})O \cdot (Cr, Al, Fe^{3+})_2O_3$.

The composition also can be generalized as $R^{2+}O \cdot R^{3+}_2O_3$. Molecular ratio of divalent to trivalent parts theoretically should be, and generally is, close to one. A lower than one ratio usually indicates oxidation of some original FeO to Fe_2O_3 (Stevens, 1944).

Chromite compositions can be represented graphically by a prism of six theoretical end members as shown in Fig 1a. The curved pie-shaped volume within the prism is where the natural chromites occur. Extending this volume down to the $FeFe_2O_4$ corner may allow for chromium magnetites but chrome ore chromites are all near the $(Mg, Fe^{2+})(Cr, Al)_2O_4$ side as shown by the shaded area of the mid-plane projection, Fig. 1b. A further idea of composition range is gained from the projection of Fig. 1c. (All diagrams show molecular proportions.) The constituents shown by these diagrams account, in most chromites, for 99$^+$% of the compositions. But titanium, vanadium, manganese, and nickel are present in traces up to slightly more than 1% in most chromites.

TiO_2 is the most prominent of those accessory oxides and occurs from a few hundredths to about 1%. Reports of TiO_2 of 1.4 to 2.0% may indicate a separate titanium mineral, such as rutile or ilmenite. Unusual Ti-Cr spinels (not ores) have been reported (Cameron and Glover, 1973). Vanadium is seldom analyzed but accounts for around ½% (as V_2O_5) in Bushveld chromites (Cameron and Emerson, 1959). In most chromites vanadium amounts to less than 0.1%; MnO and NiO are very persistent constituents and generally each amounts to a few tenths of a percent. The latter three elements can be accommodated into the spinel structure.

Excerpt from Mikami, H.M., 1983, "Chromite," *Industrial Minerals and Rocks,* 5th ed., Vol. 1, S.J. Lefond, ed., AIME, New York, pp. 567-584.

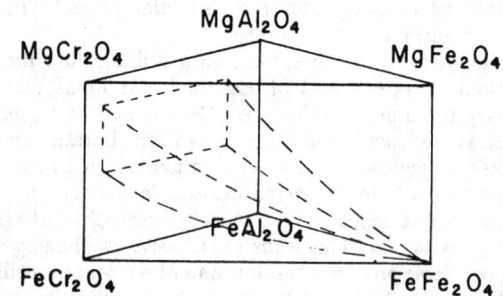

Fig. 1a. Prism representing molecular compositions of chromites in terms of six end members. Dashed lines outline volume containing most natural chomites.

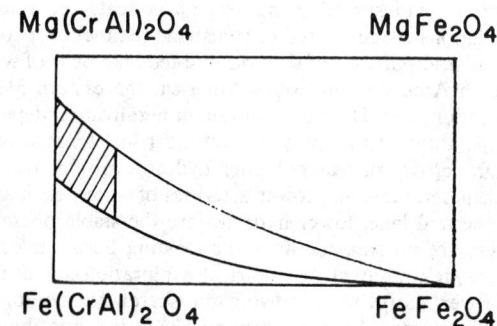

Fig 1b. Cross section through Fig. 3a. Shaded area contains the commercial chromites indicating relatively low amount of ferric iron.

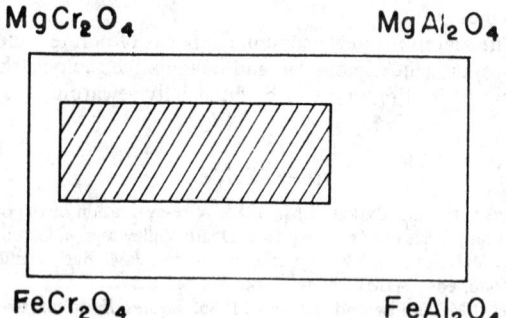

Fig. 1c. Projection of chromite compositions on (Mg, Fe^{2+}) $(Cr, Al)_2O_4$ face.

Chromites of economic interest have a Cr_2O_3 content of 25 to about 65%. Predominantly, the Cr_2O_3 of commercial ores ranges from 30 to about 60%. This refers to the chromite mineral free of gangue—which, of course, it never is. The always present interstitial gangue minerals are generally orthopyroxene, olivine, calcic plagioclase and their hydrous alteration products, serpentine, chlorites, and talc. In fact, the hydrous alteration magnesian silicates are probably more commonly present than the original minerals (although they may reveal, by microstructure, the original silicates from

Table 1. Analyses of Typical Chromites and Corresponding Chrome Ores

	No. 1		No. 2		No. 3		No. 4		No. 5		No. 6	
	C*	O*	C*	O*	C*	O*	C*	O*	C*	O*	C*	O*
Cr_2O_3	47.41	44.52	36.24	32.10	61.44	54.91	57.70	47.00	58.32	48.50	55.90	49.19
Al_2O_3	14.82	15.50	31.86	30.20	11.41	9.92	13.44	12.65	11.06	10.01	11.92	11.37
Fe_2O_3	9.21		2.97		nil		3.42		4.10		2.18	
FeO	16.86		11.32		12.53		11.66		11.10		18.80	
Tot. Fe as FeO†	(25.26)	24.72	(13.99)	12.72	—	12.41	(14.70)	11.93	(14.75)	13.28		18.27
MgO	11.40	10.10	17.10	18.06	13.66	14.92	13.29	15.46	14.23	18.83	10.36	12.35
CaO		0.30		0.44		0.70		1.77		0.40		0.48
SiO_2		2.24		5.00		5.02		5.71		6.94		6.58
TiO_2	0.41	0.43	0.38	0.30	0.17	0.20	0.39	0.32	0.06	0.05	0.76	0.67
MnO	0.09	0.07	0.11	0.08	0.16	0.14	0.24	0.06	0.15	0.12	0.30	0.26
NiO			0.12	0.10					0.20	0.18	0.09	0.08
V_2O_5	0.33	0.30							0.07	0.05	0.18	0.16
L.O.I.				0.35		0.92		3.95		1.20		
Total	100.53	98.17	100.10	99.35	99.37	98.64	100.24	98.85	99.29	99.56	100.49	100.44
Cr:Fe	1.67	1.58	2.30	2.22	4.31	3.90	3.45	3.47	3.45	3.25	2.37	2.37

*C = chromite, O = ore or concentrate.

†Separate FeO and Fe_2O_3 are generally not determined. Generally only Cr_2O_3, Total Fe, Al_2O_3, MgO, SiO_2, and CaO are available in out-turn analyses. These are given on a dried basis. CaO and SiO_2 ranged from nil to 0.17 in separated chromites and were omitted.

No. 1—Transvaal Steelport. 2—Philippine Masinloc. 3—Kempirsai (USSR). 4—Selukwe, Zimbabwe (talc-dolomite gangue). 5—Turkish. 6—Great Dyke.

which they were derived). Other less common to rare gangue minerals that have been found in various deposits include dolomite, magnesite, brucite, sepiolite, chrome tourmaline, uvarovite (chrome garnet), kämmererite, quartz, kaolin, pyrite, mica, and goethite. These minerals have been formed by metasomatism. A spectacular, though exceptional, example of this process is the large-scale alteration of original ultramafics to talc-quartz and talc-carbonate host rocks of the Selukwe, Zimbabwe, chromite deposits (Cotterill, 1969).

The geologist, engineer, or industry consumer will practically always be dealing with chrome ores or concentrates that are mixtures of pure chromite and essentially chrome-free silicate gangues. Enrichment of ore in chromium physically can only be accomplished by separating chromite from gangue, but the Cr_2O_3 content will never be higher than that of the chromite. Physical beneficiation methods are the only economical methods of upgrading chrome ore. Chemical methods for removing iron from chromite are known but not practical at present.

Table 1 lists some typical pure chromite analyses accompanied by analyses of the corresponding ores showing the diluent effects of the gangue minerals. SiO_2 and CaO are exclusively contained by the gangue and Cr_2O_3 by the chromite whereas MgO, Al_2O_3, and FeO are distributed in both minerals.

Physically, chromite is generally jet to dark grayish black with luster from shiny vitreous to resinous to dull. Hardness is 5.5 to 6.5 and specific gravity, 4.1 to 4.7. It is brittle, without cleavage, and fractures unevenly with conchoidal tendencies. Some chromites may react weakly to a hand magnet, but most will not. But all can be moved in a strong electromagnetic field.

Chrome ore generally occurs as massive aggregates of chromite grains 0.05 to 30 mm, more commonly 1 to 5 mm, in size. Interstitial gangue in shippable massive ore amounts from as little as 2 or 3% to as much as 25% for very high chromium chromite. The amount of gangue permissible before concentration becomes necessary depends, of course, on the chromite composition. As gangue increases, chrome ore grades from massive to disseminated ore. Gangue in crude ore in some mining operations can go well beyond 50% if beneficiation economics allow. Chromite grains range from near euhedral octahedrons to anhedral rounded to irregular blocky shapes. Cracks and fractures are common in some deposits. Microbrecciation to mylonitization is prominent in the Selukwe, Zimbabwe, ores (Cotterill, 1969). Lineation, foliation, and series of aligned partings, the so-called "pull-apart" textures, are found in some deposits (Thayer, 1964). Another common texture is nodular ore, also called grape, shot, pebbles, etc. This consists of spherical to ovoid aggregates mostly 5 to 40 mm in diameter. The nodules consist predominantly of chromite grains, generally in the 1 to 5 mm size range, in a silicate (commonly olivine-serpentine) matrix.

In the Transvaal stratiform deposits nodules of a texture and mineralogy somewhat different from those commonly seen in podiform deposits are found (Thayer, 1969). The Transvaal (Bushveld) nodules are spherical to ovoid bodies 5 to 50 mm in diameter. Cameron and Desborough (1969) state:

> Each nodule consists of one or more crystals of bronzite or plagioclase poikilitically enclosing abundant fine-grained chromite grains. Nodules are generally set in a matrix of coarser-grained chromite. They are high in silica compared to the matrix material, and portions of certain seams in which nodules are abnormally abundant are unworkable. At certain mines, light crushing of the crude ore disintegrates the matrix material, the nodules are then removed by simple screening, and the ore is correspondingly upgraded.

In podiform deposits, nodules generally are not excessively high in silica and may comprise good ore if sufficiently concentrated in the host rock matrix. For a further description of nodules and orbicules in podiform deposits, see Thayer (1969).

In thin section chromite ranges from dark yellow through orange reds to opaque. Polished sections in vertically reflected light are light to medium gray. Chromite is isometric;

hexoctahedral, in the space group Fd3m. Unit cell size *a* ranges continuously from about 8.200Å for high aluminum, low iron and chromium to ~8.330 for high chromium and iron chromites.

Geology

All primary chromites, and this means nearly all presently economic chrome ore deposits, occur in ultramafic rocks or their altered derivatives. The primary ultramafic rocks are peridotite, dunite, and pyroxenite and the principal derived rock is serpentine (properly called serpentinite). Serpentine is, in fact, the most common gangue mineral associated with chromite. Chlorite and talc schists are also alteration products of the primary silicate minerals. The ultramafic rocks are parts of peridotite or pyroxenite-gabbro complexes. Gabbros (including norite) and anorthosite are present near many chromite ore bodies in lesser amounts than the ultramafic rocks. In some chrome ore districts, e.g., Masinloc (Philippines) and Camagüey (Cuba), the ore bodies are found in peridotite near the peridotite-gabbro contact (Stoll, 1958; Davis et al., 1957). Gabbroic rocks occur in some ore bodies as cross-cutting dikes and inclusions. In some districts, e.g., the Bushveld, chrome ores may occur within norite or anorthosite, but mainly they occur in pyroxenite or peridotite.

Types of Deposits

Chrome ore deposits are divided by structural types into two major categories: stratiform and podiform. The *stratiform* deposits, exemplified by the chromite of the Bushveld complex of the Transvaal, are characterized by occurrence as parallel layers of great lateral continuity. In this stratiform class, the ore layers tend to be coextensive with the containing layered rock sequence much as are coal seams within sedimentary formations. The containing rock layers are themselves recognizably persistent, continuous, and extensive so that the occurrence and placement of the ore layers are amenable to conventional geologic mapping methods. Other examples of stratiform deposits are the Great Dyke of Zimbabwe (Worst, 1960); Kemi in Finland (Kahma, et al., 1962); Campo Formoso, Brazil (Thayer, personal communication); the Stillwater Complex, Montana (Jackson, 1964); Fiskenaesset, Greenland (Ghisler and Windley, 1967); and Bird River Complex, Manitoba (Jackson, 1964). The last three named are not economic ore deposits at the present time.

The other major type of chrome ore deposit, which includes all known deposits besides those already named, has been variously designated as sackform, piecemeal, discontinuous, or podiform. Following Wells, et al. (1946) and Thayer (1964), the term *podiform* will be used here. In contrast to the stratiform, the podiform deposits occur as pods, lenses, sackforms, slabs, and all sorts of irregular shapes. The ore bodies may also be tabular and occur in layers, but the important feature that distinguishes this type from the stratiform is that the ore bodies are not continuous, and in many districts follow no recognizable systematic distribution within the containing country rock. In fact, the distribution may be erratic, random, and unpredictable.

As mentioned previously, in some districts ore bodies seem to be concentrated in peridotite near its contact with gabbro, a fact that serves as an exploration guide, but since the designated areas are still very large in proportion to the size of the ore bodies, this criterion cannot locate individual ore bodies. Ore bodies may occur in strings or clusters, but these assemblages may give out without warning. The size of these podiform chrome ore bodies range from a few tons

(depending on economics of mining) to one that contained more than 14 million tons (Coto mine, Philippines). The majority of bodies range from a few tens of thousands to a few hundred thousand tons. Single ore bodies of even one million tons are quite rare, e.g., Gölolan mine, Turkey.

The contacts between massive chromite and country rock may range from very sharp to gradational in single ore bodies. By gradational is meant grading into rock by increasing amount of silicate and corresponding decrease of chromite. If the gradational zone is wide, the practical mining limit defines the boundary of the ore body. In some deposits, ore with high silicate content may be mined as disseminated ore and be subject to beneficiation. Float, or alluvial, chromite may contribute to production in some districts but generally is a minor factor. Such deposits derived from adjacent primary deposits are mined in Zimbabwe and Brazil. During 1980 a small chromite beach sand mine on Palawan Island in the Philippines was started up.

EXPLORATION, PROSPECTING, AND EVALUATION OF RESERVES

The most fundamental basis for searching for chromite is its virtually exclusive occurrence in ultramafic rocks that are, in turn, parts of ultramafic-gabbro complexes. But there are a number of large areas of ultramafic rocks in which no significant chromite ore bodies have ever been found, although there is no known basis for saying that they contain none. The second important point in identifying a chromite deposit as stratiform or podiform. The differences are so dramatic that there should be no confusion between the clearly recognizable stratiform deposits such as the Bushveld, Great Dyke, Stillwater, and Bird River on the one hand, and those presently designate podiform which encompasses most of the rest of the known commercial deposits. These include the Balkans-Asia Minor region, USSR, Zimbabwe exclusive of the Great Dyke, Caribbean, Philippines, India, New Caledonia, Pakistan, Iran, Sudan, and possibly others.

New deposits classified as stratiform in the past two decades are Kemi, Finland (Kahma, 1962); Campo Formoso in Bahia, Brazil (Thayer, personal communication); and Fiskenaesset, Greenland (Ghisler and Windley, 1967). The Brazilian deposit had been thought to be podiform but more detailed mapping has identified it as a sheared-out stratiform traceable for ten miles in heavily weathered terrane.

Prospecting and estimation of reserves in the stratiform complexes are subject to conventional geological field methods because there is a rational genetic basis for explaining and predicting the distribution of the chromite layers. The theory is that the chromite and containing ultramafites originated from a common liquid magma by differentiation and crystal settling or what has been termed *magmatic sedimentation*. Furthermore, these formations have been comparatively little disturbed from their original locations or attitudes. There has been some faulting and tilting, but no reemplacement. The structural displacements are still generally amenable to solutions by stratigraphic methods. The theory has difficulty in explaining every mineralogical and chemical detail, variation, and nuance found in the deposits, but overall it appears to be the best available and—more importantly for the mining geologist—it works.

The podiform deposits present a drastically different picture. Although the petrogenetic relations between chromite and host rocks appear to be similar, i.e., explainable by differential crystallization from a common magma, their size, shape, and, especially, their spacial arrangement and distribution are completely different from the stratiform. The ore bodies are separate, discrete, and generally appear to have been thoroughly scrambled. A currently favored theory of podiform origin proposed by Thayer (1969) is that these ore bodies were once layers in stratiform complexes but have since been broken up by tectonic forces within the earth's mantle and reemplaced tectonically, or by solid or quasi-solid flow into crustal rocks. The ultramafic complexes, of which podiform chrome districts are a part, are called alpine-type complexes because they are always characterized by intense regional deformation (Jackson and Thayer, 1972). It is postulated that the settled chromite layers (chromitite) were relatively well consolidated and cohesive as compared to the surrounding semisolid mixture of crystals and liquid, or mush, which was able to physically entrain and support the higher specific gravity chromite masses by virtue of motion and high effective viscosity. Broken segments of the chromite layers were jumbled and carried along in the moving stream and were frozen in place in varying attitudes and locations as the crystal mush finally solidified. In periods of relative quiescence, the chromite agglomerates would sink even in the mush in a manner simulating Stokes law according to which the velocity of descent is proportional to the square of the diameter of the body. This could explain the extreme paucity of very large podiform bodies (1 to 15 million tons) and the comparative abundance of small bodies.

The Selukwe district of Zimbabwe described by Cotterill (1969) possesses characteristics of both podiform and stratiform deposits. The district appears to consist of sliced-up remnants of an originally stratiform complex. Within some slices, ore zones (though not bodies) can be followed for the length of the slice. The actual ore bodies are most generally described as lenses or "pipelike" and are discontinuous or irregularly distributed. Zones of occurrence of ore bodies tend to extend the full strike distance of a slice but in the dip direction ore distribution is mostly restricted to much less than half the extent of the ore zone. Much of the chromite is strongly sheared, i.e., microbrecciated or mylonitized. Cotterill (1969) believes that the chromite lenses represent original cumulus chromite that settled in local furrows or depressions caused by buckling of silicate layers during tectonic movements. He does not believe the lenses were reemplaced through a semisolid silicate matrix but owe their present distribution to the segmentation of an original ultrabasic complex into slices. Jackson and Thayer (1972) classify the Selukwe district as a stratiform complex which it may be, viewed broadly petrogenetically. But from the practical viewpoint of finding ore bodies and calculating reserves, the Selukwe district is more closely akin to podiform districts than it is to classic stratiform deposits such as the Bushveld or Great Dyke. As more knowledge is gained, the transitional events between original chromite accumulation and final emplacement will become better known. In the Selukwe district these events seem somewhat more clear than, say, in Turkey or Masinloc but still far from obvious.

The implications of the petrogenetic hypotheses on exploration and prospecting for podiform ore bodies could be profound. If the podiform bodies are broken up remnants of originally stratiform deposits, only a tiny fraction of the original ore bodies have ever been found. How can the number and size of the original layers be estimated from the remnants? How much remains at economically exploitable levels, and do the fragments have a decipherable pattern of distribution? These are unanswered questions and the estimation of ultimate reserves of any podiform district is largely speculation. The fact that some of these districts have, in the past, managed to maintain stated reserves at, say, 10 times

annual production for 10 or 20 years indicates the estimators had no real idea of the reserves, but gave developed ore plus a reasonable inference based on experience, intuition, and an optimistic outlook.

Although there is no present scientific methodology for extrapolating podiform reserves, there are guides to prospecting that increase the probability of finding new ore bodies.

A great deal of interest has been shown recently in methods of exploration and prospecting for chrome ores. As mentioned previously, stratigraphic methods are best applied to stratiform deposits. But geophysical methods have had limited success as aids to exploring these deposits as at Kemi, Finland, where the extension of ore was traced under glacial deposits. The main interest lies in ore-finding methods for podiform deposits. The relationships of structural and textural features of ore bodies to those of the host rocks in some districts give clues to the probable locations of other ore bodies. These are empirical relationships and do not assure the occurrence of new ore bodies. Of geophysical methods, magnetic methods have been of little use. Gravimetric methods have had limited success in a few cases (Yüngül, 1956; Davis, et al., 1957). Hosking (1964) describes the possibilities of using geochemical methods for chrome ore prospecting mainly to narrow the areas for drilling. Drilling, drifting, or trenching near known ore bodies is the usual current method of discovering new bodies.

BIBLIOGRAPHY AND REFERENCES

Anon., 1961, Symposium on Chrome Ore, Central Treaty Organization, Ankara, Turkey, Sept. 1960, *Min. Res. Expl. Inst. Turkey*, 272 pp.

Anon., 1971, "Chromium," *Philippine Mining Yearbook*, Vol. 13, No. 7, pp. 26, 34.

Anon., 1971b, "Soviet Reveals Chromite Output," New York Times, April 17.

Anon., 1972a, *Chromium*, National Conference on Materials Policy, Interim Report, April, pp. 14-16.

Anon., 1977, "Chromium," *Mineral Commodity Profiles*, US Bureau of Mines, 14 pp.

Anon., 1980, "Chromium," *Mineral Commodity Summaries*, US Bureau of Mines, pp. 34-35.

Anon., 1980a, "Papua-New Guinea Chromite Prospects," *Industrial Minerals*, No. 152, May, p. 19.

Anon., 1980b, "Indonesia: Developments in Chrome Production," TIZ-Fachberichte Rohstoff-Engineering, Jahrgang, Indonesia p. 145.

Bala Sundaram, M.S., 1972, "A Review of Reserves and Demand Pattern of Iron, Chromium, Manganese Ores of India," *Eastern Metals Review*, Annual Number, Vol. 25, pp. 39-49.

Bosum, W., 1964, "Theoretical Limits of the Application of Geophysical Methods in Prospecting for Chromite," Chap. 8, *Methods of Prospection for Chromite*, R. Woodtli, ed., OECD, Paris, pp. 209-224.

Brantley, F.E., 1970, "Chromium," *Mineral Facts and Problems*, 1970 ed., US Bureau of Mines, Bulletin 650, pp. 247-262.

Broad, W.J., 1980, "Resource Wars: The Lure of South Africa," *Science*, Vol. 210, No. 4474, pp. 1099-1100.

Cameron, E.N., 1963, "Structure and Rock Sequences of the Critical Zone of the Eastern Bushveld Complex," Special Paper No. 1., Mineralogical Society of America, pp. 93-107.

Cameron, E.N., 1975, "Postcumulus and Subsolidus Equilibration of Chromite and Coexisting Silicates in the Eastern Bushveld Complex." *Geochimica et Cosmochimica Acta*, Vol. 39, pp. 1021-1033.

Cameron, E.N., 1980, "Evolution of the Lower Critical Zone, Central Sector, Eastern Bushveld Complex and Its Chromite Deposits," *Economic Geology*, Vol. 75, No. 6, pp. 845-871.

Cameron, E.N., and Desborough, G.A., 1969, "Occurrence and Characteristics of Chromite Deposits—Eastern Bushveld Complex," in "Magmatic Ore Deposits," H.D.B. Wilson, ed., Monograph 4, *Economic Geology*, pp. 23-40.

Cameron, E.N., and Emerson, M.E., 1959, "Origin of Certain Chromite Deposits of the Eastern Part of the Bushveld Complex," *Economic Geology*, Vol. 54, No. 7, pp. 1151-1213.

Cameron, E.N., and Glover, E.D., 1973, "Unusual Titanian-Chromium Spinels from the Eastern Bushveld Complex," *American Mineralogist*, Vol. 58, No. 3-4, Mar.-Apr., pp. 172-188.

Cotterill, P., 1969, "The Chromite Deposits of Selukwe, Rhodesia," in "Magmatic Ore Deposits," H.B.D. Wilson, ed., Monograph 4, *Economic Geology*, pp. 154-186.

Davis, W.E., et al., 1957, "Gravity Prospecting for Chromite Deposits in Camaguey Province, Cuba," *Geophysics*, Vol. 22, No. 4, pp. 848-869.

Dickson, T., 1980, "Chromite: Southern Africa Holds Sway," *Industrial Minerals*, No. 150, Mar., pp. 53-73.

Engin, T., and Aucott, J.W., 1971, "A Microprobe Study of Chromites from the Andizlik-Zimparalik Area, SW Turkey," *Mineralogical Magazine*, Vol. 38, Mar., pp. 76-82.

Engin, T., and Hirst, D.M., 1970, "The Alpine Chrome Ores of the Andizlik-Zimparalik Area, Fethiye, SW Turkey," *Trans./Sect. B*, Institution of Mining and Metallurgy, Vol. 79, pp. B15-B30.

Fernandez, N.S., 1960, "Notes on the Geology and Chromite Deposits of the Zambales Range," *The Philippine Geologist*, Vol. 14, No. 1, pp. 1-8.

Fourie, G.P., 1959, "The Chromite Deposits in the Rustenburg Area," Bulletin 27, Geological Survey of Union of South Africa, 45 pp.

Ghisler, M., and Windley, B.F., 1967, "The Chromite Deposits of the Fiskenaesset Region, West Greenland," Report No. 12, Geological Survey of Greenland, Copenhagen, Den., 39 pp.

Golding, H.G., and Johnson, K.R., 1971, "Variation in Gross Chemical Composition and Related Physical Properties of Podiform Chromite in the Coolac District, NSW Australia," *Economic Geology*, Vol. 66, No. 7, pp. 1017-1027.

Greathead, H.G., 1963, "Chromium," Minerals Bureau of the Government Mining Engineer's Div., Republic of South Africa, 18 pp.

Haéri, Y., 1961, "Geology of Iran's Chromite Deposits," *Symposium on Chrome Ores*, Central Treaty Organization, Ankara, Sep. 1960, Min. Res. Expl. Inst. of Turkey, pp. 21-26.

Heiligman, H.A., and Mikami, H.M., 1960, "Chromite," *Industrial Minerals and Rocks*, 3rd ed., J.L. Gillson, ed., AIME, New York, pp. 243-258.

Helke, A., 1962, "The Metallogeny of the Chromite Deposits of the Guleman District, Turkey," *Economic Geology*, Vol. 57, No. 6, pp. 954-962.

Hosking, K.F.G., 1964, "The Application of Applied Geochemical Methods to the Search from Chrome Ore Deposits," *Methods of Prospection for Chromite*, R. Woodtli, ed., OECD, Paris, pp. 149-174.

Jackson, E.D., 1964, "Primary Features of Stratiform Chromite Deposits," Chap. 8, *Methods of Prospection for Chromite*, R. Woodtli, ed., OECD, Paris, pp. 111-134.

Jackson, E.D., and Thayer, T.P., 1972, "Some Criteria for Distinguishing between Stratiform, Concentric and Alpine Peridotite-Gabbro Complexes," 24th International Geological Congress, Sec. 2, pp. 289-296.

Jankovic, A.S., 1964, "Prospecting for Chromite Deposits in Yugoslavia," *Methods of Prospection for Chromite*, Chap. 12, R. Woodtli, ed., OECD, Paris, pp. 203-208.

Kahma, A., et al., 1962, "On the Prospecting and Geology of the Kemi Chromite Deposit, Finland," *Bulletin de la Commission Geologique de Finlande*, No. 194, 91 pp.

Kurochin, M.G., 1971, "Rational Scheme for Beneficiating Kimpersai Chromite Ores," *Ogneupory (Refractories*, Eng. trans., Consultant Bureau. NY), Vol. 14, Sept.-Oct., pp. 504-506.

Mikami, H.M., 1975, "Chromite," *Industrial Minerals and Rocks*, 4th ed., S.J. Lefond, ed., AIME, New York, pp. 501-517.

Mikami, H.M., 1982, "Refractory Chromites from Southern Africa and Other Non-Masinloc Sources," *Proceedings*, Raw Materials for Refractories Conference, sponsored by US Bureau of Mines at the University of Alabama, Tuscaloosa, Feb. 8-9, pp. 286-308.

Morning, J.L., 1972, "Chromium," *Minerals Yearbook 1972*, US Bureau of Mines, pp. 289-299.

Morning, J.L., 1977, "Chromium," Preprint, *Minerals Yearbook 1977*, US Bureau of Mines, 11 pp.

Parasnis, D.S., 1964, "Some Aspects of Geophysical Prospecting for Chromite," *Methods of Prospection for Chromite*, Chap. 14, R. Woodtli, ed., OECD, Paris, pp. 225-233.

Pearre, N.C., and Heyl, A.V., 1960, "Chromite and Other Mineral Deposits in Serpentine Rocks of the Piedmont Upland in Maryland, Pennsylvania, and Delaware," 1082-K, US Geological Survey, pp. 707-833.

Peterson, E.C., 1981, "Chromium," *Minerals Yearbook 1980*, US Bureau of Mines, pp. 189-200.

Romagnoli, E., 1972, "Chrome Ore," *Engineering & Mining Journal*, Vol. 173, No. 3, Mar., pp. 116, 135-137.

Rossman, D.L., et al., 1959, "Chromite Deposits on Insular Chromite Reservation No. One, Zambales, Philippines," *Chromite*, Publication No. 19, Philippine Bureau of Mines, 12 pp.

Stanley, R., 1961, "Chromium in Southern Rhodesia," Mines Dept. of Southern Rhodesia, 21 pp.

Stevens, R.E., 1944, "Compositions of Some Chromites of the Western Hemisphere," *American Mineralogist*, Vol. 29, No. 1-2, pp. 1-34.

Stoll, W.C., 1958, "Geology and Petrology of the Masinloc Chromite Deposit, Zambales, Luzon, Philippine Islands," *Bulletin of the Geological Society of America*, Vol. 69, No. 4, pp. 419-448.

Sully, A.H., and Brandes, E.A., 1967, "Chromium," *Metallurgy of the Rarer Metals*, 2nd ed., No. 1, 373 pp.

Thayer, T.P., 1956, "Mineralogy and Geology of Chromium," *Chromium*, Chap. 2, M.J. Udy, ed., American Chemical Society Monograph 132, Reinhold, New York, Vol. 1, pp. 14-52.

Thayer, T.P., 1962, *World Resources, Chromium Minerals*, Chaps. 2, 3, Business and Defense Services Administration, US Dept. of Commerce, Washington, DC, pp. 3-47.

Thayer, T.P., 1964, "Principal Features and Origin of Podiform Chromite Deposits and Some Observations on the Guleman-Soridag District, Turkey," *Economic Geology*, Vol. 59, No. 8, pp. 1497-1524.

Thayer, T.P., 1969, "Gravity Differentiation and Magmatic Re-Emplacement of Podiform Chromite Deposits, "Magmatic Ore Deposits," H.D.B. Wilson, ed., Monograph 4, *Economic Geology*, pp. 132-146.

Thayer, T.P., and Lipin, B.R., 1978, "A Geological Analysis of World Chromite Production to the Year 2000 A.D.," *Proceedings*, Council of Economics of AIME.

Von Gruenwaldt, G., 1977, "The Mineral Resources of the Bushveld Complex," *Minerals Science and Engineering*, Vol. 9, No. 2, pp. 83-95.

Watson, G.A., 1980, "Chromium, Today and Tomorrow," *Iron and Steel Maker*, May, pp. 19-22.

Wells, F.G., et al., 1946, "Chromite Deposits of Del Norte County, California," California Div. of Mines, Pt. 1, Ch. 1, pp. 1-76.

Wilson, H.D.B., ed., 1969, "Magmatic Ore Deposits," Monograph 4, *Economic Geology*, 366 pp.

Woodtli, R., ed., 1963, *Methods of Prospection for Chromite*, OECD, Paris, 242 pp.

Worst, B.G., 1960, "The Great Dyke of Southern Rhodesia," Bulletin 47, Southern Rhodesian Geological Survey, 234 pp.

Yoshino, S., et al., 1981, "Materials from China," *Taikabutsu*, Vol. 33, No. 278, pp. 25-28.

Yüngül, S., 1956, "Prospecting for Chromite with Gravimeter and Magnetometer over Rugged Topography in East Turkey," *Geophysics*, Vol. 21, No. 2, pp. 433-454.

5. Clays*

INTRODUCTION

The term *clay* is somewhat ambiguous unless specifically defined, because it is used in three ways: (1) as a diverse group of fine-grained minerals, (2) as a rock term, and (3) as a particle-size term. Actually, most persons using the term clay realize that it has several meanings, and in most instances they define it. As a rock term, clay is difficult to define because of the wide variety of materials that compose it; therefore, the definition must be general. Clay is a natural earthy, fine-grained material composed largely of a group of crystalline minerals known as the clay minerals. These minerals are hydrous silicates composed mainly of silica, alumina, and water. Several of these minerals also contain appreciable quantities of iron, alkalies, and alkaline earths. Many definitions state that a clay is plastic when wet. Most clay materials do have this property, but some clays are not plastic; for example, halloysite and flint clay.

As a particle-size term, clay is used for the category that includes the smallest particles. The maximum-size particles in the clay-size grade are defined differently on various grade scales. Soil investigators and mineralogists generally use 2 micrometers as the maximum size, whereas the widely used scale by Wentworth (1922) defines clay as material finer than approximately 4 micrometers.

Some authorities find it convenient to use the term clay for any fine-grained, natural, earthy, argillaceous material (Grim, 1968). When used this way, the term includes clay, shale, or argillite, and some argillaceous soils.

Even though no standard definition of the term clay is accepted by geologists, agronomists, engineers, and others, the term is generally understood by those who use it. Clay is an abundant natural raw material, and it has an amazing variety of uses and properties which will be discussed.

As industrial minerals, clays are a complex group that consists of several mineral commodities, each having somewhat different mineralogy, geologic occurrence, technology, and uses. In this section, these commodities are organized as follows: (1) bentonite and fuller's earth; (2) kaolin, ball clay, halloysite, and refractory clays; and (3) miscellaneous clay and shale. Bentonite and fuller's earth are grouped together because they are closely interrelated. Though bentonite is a term based on mineral composition and fuller's earth is a term based on use, the two are virtually inseparable, because much, but by no means all, clay sold as fuller's earth is actually bentonite. The overlapping of the two terms is particularly evident where both bentonite and nonbentonite fuller's earth are used for the same purposes or products, such as in drilling muds, bleaching or clarifying fats and oils, and carriers for insecticides and fertilizers. Kaolin, ball clay, halloysite, and refractory clays are grouped together because they consist mainly of minerals of the kaolin group. Miscellaneous clay and shale is a grouping of several fine-grained materials of the type referred to as common clay in some reports.

BENTONITE AND FULLER'S EARTH

Definitions and Classifications

The term *bentonite* was first proposed in 1898 by Knight (1898), a year after he had named this clay taylorite; taylorite

*Excerpt from Patterson, S.H., and Murray, H.H. (revised by Murray, H.H.), 1983, "Clays," Industrial Minerals and Rocks, 5th ed., Vol. 1, S.J. Lefond, ed., AIME New York, pp. 585-651.

was found to be preoccupied. The name *taylorite* was after the Taylor ranch, the site of the first mine, near Rock River, WY, and the name *bentonite* is from the Benton Shale in which the clay was thought at that time to occur. The Benton Shale was, in turn, named after Fort Benton, MT, located more than 644 km (400 miles) north of Rock River.

Early in the 20th century, several geologists recognized that bentonite, mainly in beds in Cretaceous and Tertiary rocks, originated from transported volcanic materials. This recognition led to definitions based on origin, the one most widely quoted and generally accepted by geologists is the following definition by Ross and Shannon (1926):

> Bentonite is a *rock* composed essentially of a crystalline clay-like mineral formed by devitrification and the accompanying chemical alteration of a glassy igneous material, usually a tuff or volcanic ash; and it often contains variable proportions of accessory crystal grains that were originally phenocrysts in the volcanic glass. These are feldspar (commonly orthoclase and oligoclase), biotite, quartz, pyroxenes, zircon and various other minerals typical of volcanic rocks. The characteristic clay-like mineral has a micaceous habit and facile cleavage, high birefringence and a texture inherited from volcanic tuff or ash, and it is usually the mineral montmorillonite, but less often beidelite.

The difficulty in applying the foregoing definition to bentonite as an industrial mineral commodity is that it is based on origin and is restrictive to an ash, tuff, or volcanic glass parent material. Therefore, bedded deposits consisting of the clay minerals required by this definition but having uncertain origin or parent materials cannot properly be called bentonite. Furthermore, many deposits in the western United States and in other countries that have formed from rocks other than the types required by the definition are being mined and sold as bentonite.

Perhaps the best definition of bentonite as an industrial mineral is one given by R. E. Grim in a plenary lecture at the International Clay Conference (AIPEA) at Madrid, Spain, June 27, 1972. According to this redefinition, which will be used in this section, bentonite is a clay consisting essentially of smectite minerals (montmorillonite group of some usages), regardless of origin or occurrence. This definition solves the problem of the difference between the geologic and industrial usages of the term and overcomes the difficulty in assigning a name to smectite clay that formed from igneous rock other than ash, tuff, or glass, or those of sedimentary or uncertain origin. However, bentonite, when used with this meaning is still a rock term (consisting of more than one mineral), and it will not be possible to distinguish it from fuller's earth in many instances.

One way of classifying bentonite is based on its swelling capacities when wet or added to water. Bentonite having sodium (Na^+) as either the dominant or as an abundant exchangeable ion typically has very high swelling capacities and forms gel-like masses when added to water. Bentonite in which exchangeable calcium (Ca^{++}) is more abundant than other ions has much lower swelling capacities than sodium varieties. Some calcium types swell little more than common clay, and most crumble into granular masses in water. Intermediate calcium-sodium bentonites, the so-called mixed types, tend to swell moderately and to form gel of lesser volumes than equal masses of the sodium type. Bentonite, because of the general relationship of swelling and exchangeable ion characteristics, is commonly divided into the high-swelling or sodium, low-swelling or calcium, and moderate-swelling or intermediate types. The term *subben-

tonite (Davis, et al., 1940) is used inconsistently in industry for the low or moderate-swelling varieties. The authors believe the use of this term should be discouraged because of its implication of a low quality or low value and the lack of a mineralogical or use basis for it.

In the United States, bentonite is also classified by geographic location and the uses for which it is sold. Inasmuch as most of the low-swelling calcium type occurs in states bordering the Gulf of Mexico, this variety is commonly called *Southern bentonite.* The largest high-swelling sodium bentonite deposits and the major producing districts are in Wyoming and adjacent states. Therefore, this bentonite is commonly called *Wyoming* or *Western* type. Such terms as drilling mud bentonite, foundry bond bentonite, and taconite bond bentonite relating directly to use are applied in marketing. Other terms including high and low-yield bentonite, high and low-gel bentonite, and high and low-strength bentonite are also used to distinguish different grades.

The varied classifications notwithstanding, bentonite occurs in so many different varieties that some cannot logically be classified according to any of the foregoing groupings. One of these types is hectorite, which is a high-swelling lithium-bearing variety of smectite occurring mainly in California and adjacent states. It is, therefore, not a Wyoming type, and it occurs even farther west than the so-called western bentonites. Others are bentonites having magnesium (Mg^{++}) or hydrogen (H^-) as the most abundant or dominant exchangeable ion. These types are neither sodium nor calcium varieties, and apparently some are low swelling, whereas others have rather high-swelling capacities.

Still another type of bentonite, the so-called potassium type, K-bentonite, or metabentonite, occurs in Ordovician and other Paleozoic rocks at many places in Appalachian and Mississippi Valley regions and elsewhere. This bentonite, which is generally thought to have formed from volcanic ash, consists mainly of illite and mixed-layer minerals. Smectite minerals are ordinarily present in only minor quantities. Because of its mineral composition, this bentonite contains appreciably more potassium than most other types. The meta prefix originates from the idea that this bentonite was altered by low-grade metamorphism or diagenesis. Apparently, the only attempt to use K-bentonite in the United States was when five or six carloads were mined near Dayton, Rhea County, TN, about 1928; this bentonite was used for purifying lard (Gildersleeve, 1946), and no further mention of this type will be made.

The term *fuller's earth* is more or less a catchall for clay or other fine-grained earthy material suitable for bleaching and absorbent and certain other uses. It has no compositional or mineralogical meaning. The origin of the term dates back into antiquity; it was first applied to material used in cleansing and fulling wool, thereby removing the lanolin and dirt from it. When in the latter half of the last century it was found that some earths used for fulling would also serve in decolorizing and purifying mineral, vegetable, and animal oils, and the term *fuller's earth* was modified to include this usage.

Extensive use of fuller's earth in processing mineral oils in the first half of this century and the virtual end of its use in fulling fiber led to the general application of the term, fuller's earth, as meaning primarily earth used in petroleum processing. Further modification of the meaning came about as other uses developed for the earth and replaced oil processing as the major use. The term fuller's earth is retained, however, for fine-grained materials used for many purposes. Most of these uses require absorbent properties in one form or another, but the properties required for some uses, such as certain drilling muds and fillers, are other than absorbency and, therefore, result in further modification of the term.

The classification and understanding of fuller's earth are also complicated by several terms having more or less duplicate or overlapping meanings. When applied in the oil-processing sense, fuller's earth has the same meaning as *naturally active clay.* Fuller's earth and other clays which are treated with acid or otherwise altered to improve their desirable properties are called *activated clay* (Torok and Thompson, 1972). The terms *bleaching clay* and *bleaching earth* are applied mainly to both naturally active and activated clay, but they also include activated bauxite (Rich, 1960). Both naturally active and activated clays are included under the term *absorbent clays* (Nutting, 1943); therefore, this term has nearly the same meaning as bleaching clay. The term *absorbent clay* is applied to fuller's earth used for a wide variety of absorbing purposes which are different from that of processing oils.

History and Use

Bentonite: Bentonite mining began on the Taylor ranch near Rock River, WY, in 1888. This bentonite reportedly was shipped crude to Philadelphia where it sold for $25 a ton and was used in making cosmetics. A mine was opened in the Upton, WY, district on the western side of the Black Hills in 1903. The early method of drying was to rake the bentonite by hand over steel plates heated by wood or coal fires.

The date of the first mining of southern bentonite is somewhat conjectural because of the confusion between the terms bentonite and fuller's earth. One fuller's earth deposit near Gonzales, TX, that is almost certainly bentonite was mined as early as 1906, and another deposit in that same state mined during the Civil War (Lang, et al., 1940) may also have been bentonite. Large-scale mining of calcium-type bentonite began in Mississippi in the early 1930s (Bicker, 1970), and bentonite deposits were used as fuller's earth in several states about this time.

The value of bentonite as foundry-sand bond was recognized in the 1920s, and the iron and steel foundries have since been major consumers of bentonite. Bentonite was first used as drilling mud in the late 1920s or early 1930s, and bentonite is still one of the most efficient materials for drilling muds where the rocks penetrated contain only fresh water. The high-swelling Wyoming bentonite is the most efficient type for drilling mud, but the high-viscosity properties of hectorite make it useful in muds (Larsen, 1955). Some Texas bentonite is also used for drilling muds, particularly in shallow wells, and commonly it is treated with soda ash, polymers, or other chemicals to make it suitable for this use. Low-quality bentonite when sold for use in drilling mud is commonly classified merely as clay rather than bentonite (Anon., 1972d).

Bentonite production remained small during the first few decades of mining, and the first yearly total sufficiently large to be separated from miscellaneous clays in the US Bureau of Mines statistics was in 1930 when production was only a few thousand tons. Demands for bentonite increased during World War II, but a yearly total of 907 kt (1 million tons) was not reached until 1950. More than 1.8 Mt (2 million tons) was produced in 1966. Bleaching clay (oil refining, filtering, clarifying, and decolorizing), drilling mud, iron-ore pelletizing, and foundry-sand bonding have been the major uses of bentonite since World War II. Beginning about 1950, steel companies used Wyoming-type bentonite as a bond for pelletized taconite ore.

In addition to the major uses, bentonite is used in many miscellaneous products, and hundreds of patents for speciality uses have been issued or applied for. The speciality uses include filtering agents (one of which is a high-value product for clarifying wine, and another is a less costly one for treating waste water), water impedance (preventing seepage loss from reservoirs, irrigation ditches, and waste-disposal ponds, and seepage through basement walls, tunnel walls, and other structures), ingredients in cosmetics, animal feed, pharmaceuticals, colloidal fillers for certain types of paints, an additive to ceramic raw materials to increase plasticity, fire-retarding materials, and for many other purposes (Papin, 1964). The white bentonites occurring in Texas and Nevada and imported from Italy are particularly suitable for many of the speciality products, and a great deal of research evaluating them for many uses has been done. One of the unique uses of high-swelling bentonite that is likely to increase is in the "slurry trench" or "diaphragm wall" method of excavation in construction in areas of unconsolidated rock or soil (Blackman, 1969; Lang, 1971). Considerable quantities of bentonite were once used in making catalysts for petroleum refining, but the market for these types of catalysts in the United States declined after World War II. However, sizable quantities of catalysts have been produced from bentonite and similar clays in Japan, the United Kingdom, and elsewhere in recent years. Acid-activated bentonite is used for bleaching oils and in making multiple-copy paper requiring no carbon paper.

Fuller's Earth: Published reports give rather confusing accounts of the discovery of fuller's earth in the United States. Several state that fuller's earth was discovered in this country in 1893 near Quincy, FL, by the Owl Commercial Co. (predecessor of the present makers of White Owl cigars) during an attempt to burn brick from clays on tobacco property. The clay was not suitable for making brick, but an Alsatian immigrant employed as a farm worker recognized that it was similar to fuller's earth mined in Germany. His observation led to development of the first mine near Quincy two years later and to the use of this clay in processing mineral oils. Fuller's earth had been mined on a small scale, however, in Arkansas in 1891 (Miser, 1913), and tested for use in the refining of cottonseed oil.

Both of these reported discoveries are related to the use of fuller's earth in processing oils and fail to note its earlier use for other purposes. Clays and other earthy materials were undoubtedly used by the early settlers from Europe in cleansing wool and other materials. Little effort was made to search the historical records and to document this, but it was found that soldiers stationed near Perth Amboy, NJ, used Woodbridge fire clay for cleansing buckskins during the Revolutionary War (Cook and Smock, 1878). Fuller's earth associated with an iron-ore bed near Kent, CT, had been mined in the early 1800s (Silliman, 1820), and fuller's earth near Falls City, TX, had been used to bleach sugar during the Civil War (Lang, et al., 1940). Also, American Indians were reported to have used bentonite in cleaning blankets, and they probably dug it for other cleansing purposes before the Columbian period.

The use of fuller's earth in the refining of oils continued to be the major one for many years. The demand for fuller's earth for processing mineral oil increased rapidly in the first part of this century and reached a peak of approximately 288 kt (317,000 tons) in 1930; this was 91.1% of the total United States production that year. The production of fuller's earth for this purpose began to decrease thereafter, especially when activated bauxite was introduced in 1937 and mag-nesium silicate in 1940 as more efficient substitutes. The market for this use was also depressed by improvements in refining methods, which produce oils requiring little purification or bleaching. In recent years, the production of fuller's earth for use in refining mineral oils has maintained a rather uniform rate of 32 to 36 kt/a (35,000 to 40,000 tpy). The use of fuller's earth in processing animal and vegetable oils has never been a major one and has decreased to the point where it is included as a part of the miscellaneous uses in the US Bureau of Mines *Minerals Yearbooks*.

Palygorskite (attapulgite) fuller's earth was first sold for drilling mud in 1941. The market for this use expanded slowly and has maintained a level of 7 to 10% of the total United States production during the last few years. Most of the fuller's earth sold for drilling mud comes from the southern part of the Meigs-Attapulgus-Quincy district of Georgia and Florida. Palygorskite clays produced in this area are superior to most other fuller's earth for muds used in drilling salt formations, but because of high water loss, they are inferior to bentonite where the rocks drilled contain no saltwater.

Fuller's earth was used in significant quantities as a carrier for insecticides and fungicides by 1950, and the market for this use has grown at a rather uniform rate since that year.

The use of fuller's earth granules for absorbent purposes began during the 1930s, but this use did not expand significantly until the World War II period when fuller's earth was used as an absorbent for greases, oil, water, chemicals, and other undesirable substances on the floors of factories, filling stations, canning plants, aircraft hangers, decks and engine rooms of ships, and other installations. The absorbent granules are porous and ordinarily weigh less than 481 kg/m^3 (30 pcf). Because of their light weight, size, porosity, and absorbent properties, the granules are suitable for many uses, and since World War II many different markets for them have developed. Among other uses, they are now sold for litter and bedding for poultry, pets, and other animals, and as a soil conditioner in greenhouses and for golf courses,

Geology

Mineralogy: The principal clay mineral in bentonite is smectite, and many fuller's earth deposits also consist chiefly of this mineral. The term *smectite* is applied as a group name, and *montmorillonite* is a mineral species name. This usage conforms to the growing acceptance of the term smectite and does away with the confusing use of montmorillonite as both mineral species and group names. Montmorillonite, including both sodium and calcium varieties, is the most common member of the smectite group occurring in bentonite. However, saponite, a magnesian smectite, and hectorite, a lithium-bearing magnesian variety, are the major minerals in some bentonites.

Smectite minerals occur in extremely small particles; therefore, detailed information on them is difficult to obtain and is, in part, based on theoretical considerations. As the mineralogy of this group of clays is complex, the reader is referred to authoritative books and other references; only the following brief summary will be given here.

According to the most generally accepted structure, smectite consists of two silica tetrahedral sheets with a central octahedral sheet. The theoretical structural formula is $(OH)_4Si_8Al_4O_{20} \cdot nH_2O$, and the theoretical composition without interlayer material is SiO_2, 66.7%; Al_2O_3, 28.3%; and H_2O, 5%. However, smectite always differs from the foregoing theoretical formula and composition because of substitutions of various ions for silicon in the tetrahedral coordination and for aluminum in the octahedral sheet. Furthermore, the substituted ions are thought to be commonly of different valence than the theoretical ion replaced, which results in an unbalancing of the charge in a unit of smectite. This charge deficiency is balanced by exchangeable ions, which vary considerably and result in further differences in the compositions of smectite. Cations which are exchangeable in bentonite include sodium, calcium, potassium, magnesium, lithium, and hydrogen. The exchange capacity of most bentonite is within the range of 60 to 150 milliequivalents per 100 g.

Unit layers of smectite are thought to be stacked with the oxygen layer of one silica tetrahedral sheet adjacent to the similar layer in the neighboring unit. Only a very weak bond exists between neighboring units, and water or other polar molecules can enter between unit layers causing the lattice to expand in the c direction. This expansion by polar molecules and the collapse of expanded units with heat, which can be recognized by X-ray diffraction methods, is applied in studying smectite and distinguishing members of this group from other clay minerals.

The ion-exchange characteristics have an important role in controlling or influencing the physical properties of bentonite and the bentonite fuller's earths. In general, those bentonites that have Na^+ as the dominant exchangeable ion have very high swelling capacities and colloidal properties, and those in which Ca^{++} is the dominant ion tend to swell little more than other clays. Because of their high colloidal properties, high-swelling sodium bentonites are valuable as drilling muds and other uses requiring thixotropic suspensions. The calcium varieties usually have little value for such uses without beneficiation by the addition of soda ash or other chemicals, and even when so treated, are not ordinarily as efficient as the sodium types. The type of exchangeable ion also has a major role in controlling the bonding characteristics. As pointed out by Grim (1962), sodium bentonites have very strong foundry sand bonding properties after drying (dry strength), but only moderate strengths when moist or wet (green strength). Calcium bentonites, on the other hand, have high green strengths and low to moderate dry strengths. However, part of these differences in bond strengths may be due to particle size or other characteristics, and there is some evidence that the dry strength of some calcium bentonites can be increased by lengthening the mulling (mixing) time in the preparation of foundry sand.

All bentonites contain mineral impurities, which vary considerably in type and quantity present. A few are contaminated with minor quantities of clay minerals other than smectite, such as kaolinite and illite. The common nonclay minerals in bentonite include those listed in the older definition by Ross and Shannon (1926), and minor quantities of most of the accessory minerals in volcanic rock also may be present. Selenite is abundant in many deposits, and some beds contain carbonate veins and concretions. Zeolite minerals and opal, cristobalite, or other forms of poorly ordered silica are present in some bentonite deposits in western United States and at several other places in the world. Some deposits are contaminated with incompletely altered parts of the rock from which they formed. The vitric latitic tuff remaining in the bentonite at Cheto, AZ (Sloane and Guilbert, 1967), is an example of this type of material.

The principal fuller's earth deposits other than the bentonite type consist of palygorskite (attapulgite) and sepiolite. According to Bradley (1940), palygorskite has an ideal formula of $(OH_2)_4(OH)_2Mg_5Si_8O_{20} \cdot 4H_2O$, but substitution of Al^{3+} for either Mg^{++} or Si^{4+} takes place. It consists of two silica chains linked in amphibolelike structures and has both monoclinic and orthorhombic symmetry (Christ, et al, 1969). Sepiolite, $(Si_{12})(Mg_9)O_{30}(OH)_6(OH_2)_4 \cdot 6H_2O$, is thought to have three pyroxene-type chains (Grim, 1968), and therefore, the unit cell is somewhat larger than for palygorskite. Both palygorskite and sepiolite typically occur in fibrous and elongate lathlike particles. Sepiolite is also the mineral forming meerschaum.

Other nonbentonite fuller's earths include an uncommon type of rock called opal claystone (Heron, et al., 1965). This rock is more than three-fourths opal or another form of poorly ordered silica, and the remainder is montmorillonite and other mineral impurities. It was mined in South Carolina in the early part of this century and used for bleaching oil. Nonbentonite clays that have been used as fuller's earths in the past include the Anna kaolin in Illinois, halloysite in Utah (Schroter and Campbell, 1940), and glacial clays in Massachusetts. In addition to these nonbentonite clays, there are extensive deposits of fuller's earth consisting of very impure montmorillonite that are not ordinarily thought of as being bentonite. Two such deposits are the Porters Creek Clay (Paleocene) in the Mississippi Embayment region and the Twiggs Clay Member of the Barnwell Formation (Eocene) in Georgia.

Occurrence: Though bentonite deposits of Jurassic age are extensive in western United States and elsewhere in the world, virtually all deposits mined to date are of Cretaceous age or younger. All but a very minor tonnage of the high-swelling Wyoming-type bentonite produced has been mined from beds of Cretaceous age, and more than three-fourths of the total has come from a single formation—the Mowry Shale. The southern or calcium bentonite is mainly in formations of Cretaceous, Eocene, and Miocene age. The abundant miscellaneous bentonite deposits, other than the high-swelling type, in the western states are mainly of Tertiary age. Deposits scattered through Nevada, California, Oregon, and Idaho are mainly in upper Tertiary rocks, and many are thought to be Late Miocene or Pliocene in age. The widely known Cheto bentonite near Chambers, AZ, is in the Bidahochi Formation of Pliocene age, and deposits mined in western Oklahoma are also in Pliocene rocks. Deposits in the Ash Meadows district, NV, occur in rocks mapped by Denny and Drewes (1965) as Pleistocene in age, and if this age assignment is correct, they are among the youngest bentonites in the world.

Most bentonite occurs in beds or lenticular bodies aligned along a definite stratigraphic zone. Some beds extend for more than 322 km (200 miles), and other deposits, particularly the lenticular ones, are only a few hundred yards in diameter. Deposits mined range from about 0.3 to more than 9 m (1 to 30 ft) in thickness. Bedded deposits ordinarily have a characteristic sharp contact with underlying rocks and a gradational one with overlying strata. Deposits formed by hydrothermal processes tend to be irregularly shaped and to grade into the host rock in all directions.

Most bentonite, including both the high and low-swelling types, has a characteristic waxy or soaplike texture. Parts of deposits near the surface tend to be light-yellowish green or gray when the natural moisture is present and to become lighter in color when they dry. Deposits under considerable overburden tend to be bluish gray. The change to lighter shades near the surface is related to the oxidation of iron. Typical outcrops of high-swelling bentonite have a popcorn-like or frothy texture caused by alternate swelling and drying of the bentonite. Outcrops of low-swelling types commonly have a cracked appearance thought by some to resemble alligator hide.

Most bentonite in North America and elsewhere occurs in sedimentary rocks and was formed in place from volcanic ash or tuff. Some deposits have been formed by hydrothermal alteration of volcanic or other igneous rock, and other deposits of nearly pure smectite may be accumulations that were transported and deposited in marine or alkaline lake water with only minor postdepositional alteration.

The extensive palygorskite (attapulgite) fuller's earth deposits in the southern part of the Meigs-Attapulgus-Quincy district, Georgia and Florida, occur as tabular lenses and discontinuous beds in the Hawthorn Formation of Miocene age. This clay is green and bluish gray, and most of it has a waxy or soaplike texture.

In addition to the deposits in the United States, palygorskite fuller's earth is also mined in the following regions: (1) Pout district near Mbour, Senegal (Wirth, 1968); (2) Ukrainian district, USSR (Ovcharenko, et al., 1967); and (3) Mudh district, India (Siddiqui, 1968). The Senegal deposits are in marine beds of Early Eocene age. The Ukrainian palygorskite is interbedded with bentonite. Deposits in India occur between fossiliferous limestones and apparently are marine in origin.

Insofar as the authors are aware, the only sepiolite fuller's earth mined at the time this report was written is in the Vallecas and another district in Spain (Anon., 1972b). However, economic sepiolite deposits have been discovered recently in the Ash Meadows district, Nevada (private communication). The Spanish deposits occur in an evaporite sequence of Tertiary age. The Nevada deposits are closely associated with calcium, sodium, and magnesium bentonites of Pleistocene age.

Origin: The evidence for most bedded bentonite having formed from volcanic materials transported considerable distances in the atmosphere was presented by several geologists early in the century, including Hewett (1917), Wherry (1917), and recently by Slaughter and Earley (1965). This conclusion is based primarily on the purity of smectite, on the fact that the nonclay minerals in bentonite are angular and of the type occurring in volcanic rock, rather than mixtures of rounded grains characteristic of detrital sediments, and on the presence of relict shard textures in many bentonites (Ross and Shannon, 1926). Ample evidence supports the conclusion that the parent ash of most bentonite was deposited under marine conditions, but a few deposits apparently accumulated in alkaline lakes (Papke, 1969). As would be expected, different bentonites have formed from volcanic rock of somewhat different types, and the most common parent materials range from andesite to rhyolite in composition. Opinions differ concerning the process and time of alteration of the ash. Several authors have noted that change in ash would have begun with its contact with water, but when the alteration was completed is questionable. Papke (1969) believes that one type of bentonite deposit formed in alkaline lakes soon after deposition and another type formed more slowly as the result of alteration by ground water. Alteration by ground water requires burial by younger rocks and probably regional uplift and a considerable interval of geologic time. In the author's opinion, changes have taken place in bentonite deposits throughout virtually their entire geologic history. Some deposits may have been altered extensively before burial beneath younger rocks. However, the prominent cherty zone below several beds points to the downward transport of silica by ground water from altering deposits; this must have taken place after burial.

Bentonite formed by hydrothermal activity occurs in Nevada (Papke, 1969), California (Kelley, 1966), Spain (Anon., 1972b), and elsewhere. Deposits of this type commonly are associated with other minerals known to have formed by hydrothermal processes. They also tend to occur in irregularly shaped bodies rather than in beds, and they ordinarily grade into the host rock through partially altered zones. The distribution and shape of bentonite formed hydrothermally are commonly controlled by fault zones, joints, and other geologic features providing access for heated water.

Alteration of volcanic ash or tuff in restricted alkaline lakes heated by hot-spring activity is also thought to have formed bentonite (Ames, et al., 1958). The hectorite deposits in California probably formed this way, and this clay passed through an intermediate stage in which most of the ash was altered to zeolite. A magnesium bentonite called amargosite formerly mined near Shoshone, CA, also formed by the alteration of vitric volcanic material by hot springs (Sheppard and Gude, 1968).

Though the authors have doubts about the theory, some bentonite deposits are thought to have been deposited mainly in the form of smectite minerals. According to this explanation of origin, smectite minerals, having formed by weathering of volcanic or other igneous rocks, were transported, separated from impurities, and deposited. This theory has been applied to explain the origin of deposits in India (Siddique and Bahl, 1965), the USSR (Tazhibayeva and Galiyev, 1972), and Nevada (Papke, 1969). Bentonite fuller's earth deposits in the United Kingdom were thought by Robertson (1961) to have been transported from smectite-rich soils on the European landmass. Such soil would have formed mainly on nonigneous rocks.

As most fuller's earth deposits are bentonites, they have the same origins; however, some fuller's earths consist of minerals other than smectite and may be of quite different origin. The palygorskite (attapulgite) fuller's earth deposits in Georgia and Florida are one type that probably formed in a different way. These deposits have been investigated by several geologists and mineralogists, and no one has found convincing evidence of any volcanic materials having ever been present. The first author has investigated these deposits in detail and believes that the palygorskite in them formed in place from seawater evaporating in a tidal-flat environment. This theory is essentially the same as the "neoformation" idea of Millot (1964).

Distribution of Deposits

North America: *United States*—SOUTHERN BENTONITE AND FULLER'S EARTH. Southern or calcium bentonite is now mined in Texas, Mississippi, Alabama, Oklahoma, and Louisiana (Teague, 1972). Undeveloped bentonite deposits occur in South Carolina (Robinson, et al., 1961).

Fuller's earth is produced mainly in the southern states; and the Meigs-Attapulgus-Quincy district, Georgia and Florida is the leading fuller's earth-producing district in the United States.

Fuller's earth used mainly in bleaching animal and vegetable oils is now produced in Texas, and historically this state has ranked with the leading producers. The fuller's earth produced in Texas is mined from bentonite beds.

WYOMING OR HIGH-SWELLING BENTONITE DEPOSITS. The North Rocky Mountains-High Plains region has led the world in the production of Wyoming or high-swelling bentonite since the first mining of this clay. The major producing districts are as follows: (1) the northern and western Black Hills districts, Wyoming, Montana, and South Dakota; (2)

Kaycee-Midwest, Wyoming; (3) Greybull-Lovell, Wyoming and Montana; (4) Vananda, Montana; and (5) Chinook-Malta-Glasgow, Montana.

HECTORITE. Virtually all the hectorite mined recently has been near Hector, CA, the type locality. The hectorite occurs as a waxy soft nodular layer ranging from 1.8 to 2.4 m (6 to 8 ft) in thickness. It is associated with sandstone and clay beds of Tertiary age that accumulated in an alkaline lake environment. The beds below the hectorite contain travertine formed as the result of hot-spring activity (Ames, et al., 1958). The valuable hectorite deposits are overlain by Quaternary basalt.

A large deposit of hectorite has been discovered recently in the Amargosa Valley, California (private communication). Plans for mining and processing this clay were in an advanced stage of preparation when this report was written.

In addition to the Hector and the Amargosa districts, hectorite also occurs 26 km (16 miles) north of Amboy, CA, and in Yavapai County, AZ (Norton, 1965). Deposits not known to the authors probably occur at other localities, and there are many areas where the geologic conditions are favorable for more discoveries.

OTHER DEPOSITS IN WESTERN STATES. Bentonite (other than the Wyoming type and hectorite) and fuller's earth deposits are scattered throughout the western states. In general, bentonite production from these deposits has been small, as compared with that from the Wyoming and Southern types, because many of these clays are of lower quality and all of them are more distant from the major consumers. One district in which bentonite has been mined continuously for a long time is in the vicinity of Chambers [Cheto], AZ (Kiersch and Keller, 1955). This bentonite is a low-swelling clay used as bleaching earth and in making desiccants. Bentonite used for preparing livestock feed is mined in southwestern Idaho, southern California, and probably elsewhere. Deposits in central Oregon are used in the preparation of retardants used in fighting forest fires. Some of the white bentonite mined in Nevada is used in pharmaceuticals and beauty aids. A bentonite in the Ash Meadows district composed of saponite (Papke, 1969) was mined extensively several decades ago, and other deposits in the state have been mined for use in water impedance and for other purposes. Low-swelling or calcium-type bentonite also occurs and is produced in the region containing the Wyoming or high-swelling bentonite. Low-swelling bentonites are mined from deposits in the northern Black Hills and the flanks of the Big Horn Mountains. This bentonite is marketed primarily for foundry-sand bonding.

The most active fuller's earth-producing districts in the western states are in Kern County, CA. In addition to the deposits in Kern County, bentonite near Olancha and probably deposits at other localities in California have been processed for filtering and decolorizing agents in recent years.

RESOURCES. Resources of bentonite and fuller's earth in the United States are very large. The reserves of bentonite are at least one billion tons, and the total resources, including all bentonite that will be eventually suitable for one use or another, are considerably larger. This reserve estimate of one billion tons includes very large deposits that have a quality suitable for binding iron-ore pellets but that would be classed as submarginal for drilling mud and several speciality uses. The reserves of very high quality drilling-mud bentonite are limited and are much in demand. Resources of fuller's earth are tentatively estimated at two billion tons. However, almost any figure could be used here, if all the clays equal in quality to the lowest grade material used as bleaching clay in the

past were included. A large part of the fuller's earth resources is in the extensive Porters Creek Clay and the Twiggs Clay Member of the Barnwell Formation. These deposits are suitable for use in absorbent granules and certain other fuller's earth products, but they are not used in drilling mud and several of the speciality products.

Canada—Bentonite occurs in beds of Cretaceous and Tertiary ages at many places in western Canada (Ross, 1964; Spence, 1924). The major producing districts are Onoway (Baroid Canada, Ltd.) and Rosalind (Dresser Industries, Inc.) in Alberta and in the Prembina district (Prembina Mountain Clays, Ltd.) in Manitoba. Bentonite has also been mined near Princeton, BC, and deposits near Avonlea, Sask. (Anon., 1971), and along the Mackenzie River in the vicinity of Inuvik, NWT, have been investigated by industry.

According to Ross (1964), only a very few million tons of bentonite reserves have been proved in Canada. However, the potential resources of bentonite in the country must be very large, considering the large areas now known to contain scattered deposits and the general similarities of strata to Cretaceous and Tertiary rocks in the United States known to contain bentonite.

Mexico—The German firm Süd-Chemie AG with its Mexican affiliate Tonsil Mexicana SA has operated a bentonite plant in Puebla since 1967 (Anon., 1969k), and another plant is active intermittently at Monterrey. Bentonite and fuller's earth also occur in Querétaro, Michoacán, and Guanajuato, and a few of these deposits have been mined on a small scale (Esquivel and Zamora, 1958). Apparently, most Mexican bentonite deposits occur in sedimentary rocks, and some of these are of the nonmarine type.

South America: *Argentina*—Bentonite deposits are scattered throughout several provinces in Argentina (Bordas, 1947). The principal deposits mined are as follows:

Principal Deposits	Province
Rio Chico	Chubut
El Alamo y Las Hegueras	Mendoza
El Catalán	Neuquén
Del Lago Camino y Rob	Río Negro
La Emilia y Cristina	San Juan

A few thousand tons of fuller's earth are also produced each year. The fuller's earth deposits mined are in La Rioja and Río Negro provinces.

Brazil—A plant processing montmorillonite clays for use as foundry-sand bond was built at Sacramento, Minas Gerais, in 1962 (Anon., 1962). Bentonite has also been found at several other localities in the country.

Peru—Bentonite and fuller's earth occur in the provinces of Pisco, Canete, Paita, and Contralmirante Villar, and elsewhere (Caberra, 1963). Bentonite has been mined and used in iron-ore pelletizing in Peru for several years (Anon., 1967a).

Europe: *Cyprus*—A low-swelling calcium bentonite is mined in the Troulli district near Kambia (Anon., 1969l). A few thousand tons are produced annually. Part is processed in a small plant at Vassiliko, and part is shipped to Israel for beneficiation and use in several products (Anon., 1970e).

Czechoslovakia—Bentonite occurs in the northwestern, central, and eastern parts of Czechoslovakia (Gregor, 1967). The bentonite is chiefly of the calcium and magnesium types, and it formed from rhyolite tuffs. It is as much as 70% montmorillonite; the nonclay minerals present include cristobalite. Bentonite reserves in Czechoslovakia were estimated to be 10 million tons.

Apparently, the most productive bentonite district is in the northwestern part of the country, inasmuch as a bentonite processing plant having a yearly capacity of 91 kt (100,000 tons) was built at Zelenice (Anon., 1969l). One-third of the bentonite produced in this plant is reported to be available for export.

France—The company Société Française des Bentonites et Dérives SarL. markets several grades of bentonite, mined principally in central France (Anon., 1969j). The deposits mined are presumably those in the Department of Vienne and the Limousin region that have been worked for many years (Déribére and Esme, 1943). Other bentonite and fuller's earth deposits are in Vaucluse (Mormoiron, Sainte-Radegonde, and Apt) and Pyrénées. Most bentonite and fuller's earth deposits in France are chiefly smectite, but those at Mormoiron are mixtures of smectite and palygorskite (Millot, 1964).

Greece—Rather extensive deposits of bentonite occur in the eastern part of the island of Milos (Anon., 1969l). The deposits are as much as 30 m (100 ft) thick. Apparently they formed by the alteration of ash or tuffs of Pliocene age in a marine environment (Wetzenstein, 1972). One plant near deep water operated by the Greek company, Silver and Barytes Ores Mining Co., has a yearly capacity of 181 kt (200,000 tons). Three other companies are also mining bentonite on this island (Wayland, 1971). The bentonite producers on Milos have been particularly successful in selling bentonite for iron-ore pelletizing, mainly because of the nearness of deposits to deep-water shipping. Bentonite produced on this island is also sold for drilling mud and foundry-sand bond, and for other purposes.

In addition to the deposits on Milos, bentonite also occurs on Mikonos where three companies have been active in recent years. With production on both islands, Greece ranks among the leading producers of bentonite.

Hungary—Bentonite occurs at scattered localities in Hungary. Recently, bentonite-processing facilities were concentrated at Mod, under the supervision of personnel of the National Ore and Mineral Mining Enterprise (Sondermayer, 1967).

Italy—Italy is one of the leading European bentonite producers. The deposits mined are on the island of Ponza, off the western coast, and on Sardinia.

Poland—Bentonite deposits of Eocene and Miocene age occur in southern Poland; these are processed for use as bleaching clay (Nowacki and Ciechomska, 1964) and probably for other uses.

Romania—Romania has been one of the major East European bentonite-producing countries for several years. The bentonite is mined underground in the Alba-Iulia-Ocana Mures district (Neaçsu, 1969), and apparently it is produced elsewhere.

Spain—Bentonite and fuller's earth occur at several places in Spain. The most productive bentonite deposits are in the Cabo de Gata region in Almeria. Fuller's earth deposits consisting mainly of sepiolite occur at Vallecas and elsewhere in the Tagus basin (Anon., 1972b; Huertas, et al., 1971).

Switzerland—Bentonite is mined in Switzerland and used in bonding foundry sands (Grim, 1962).

United Kingdom—Calcium montmorillonite clays occur at many places in the United Kingdom, and fuller's earth has been mined since the Roman period nearly 2000 years ago (Anon., 1969f). Fuller's earth districts now active are at Redhill in Surrey, Combe Hay south of Bath in Somerset, and Woburn in Bedfordshire. In addition to these districts, large deposits of calcium montmorillonite have been discov-

ered in the Swindon-Abington district in Berkshire by the Institute of Geological Sciences (Poole and Kelk, 1971; Poole, et al., 1971), and other deposits occur near Clophill in Bedfordshire, and Maidstone in Kent.

USSR—The principal bentonite deposits in the USSR are in the Volga Region, western and central parts of the Ukraine, Transcaucasia Region, Kazakhstan, and Central Asia (Lebedinskii and Kirichenko, 1972; Anon., 1970a). Smaller bentonite resources also occur in eastern and western Siberia, Crimea, and in the northern Caucasus.

West Germany—Lower Bavaria has been a major producer of smectitic fuller's earth and bentonite for many years. The two major producers are Süd Chemie AG, which has a plant at Mooseberg (Anon., 1969k), and Erbslöh and Co., which operates a plant at Landshut (Anon., 1969e). The deposits are mined mainly by open pit methods, but some underground mining is still done. Both companies produce sodium-exchanged bentonite for use in foundry-sand bonding and for use in several products requiring colloidal properties. Smectite clays are also acid activated and exported to several countries for processing oils and waxes.

Yugoslavia—Calcium-type bentonite occurs in many places in Yugoslavia, and the total resources in the country are very large (Anon., 1969m). The bentonite is processed, including some sodium exchanging, at several places.

The best quality bentonite in Yugoslavia is in the Ginovci deposits, 81 km northeast of Skopje (Konta, et al., 1971).

Africa: *Algeria*—Calcium-type bentonite is mined in the Marnia and Mostaganem districts of Algeria (Anon., 1969l). Algerian bentonite production in recent years has ranged from 23 to 34 kt/a (25,000 to 38,000 tpy).

Kenya—Bentonite in the Athi River valley has been mined intermittently for use as foundry-sand bonding (Thompson, 1952).

Malagasy—Bentonite is produced on a small scale in this country, inasmuch as 90 t (99 tons) were reported to have been exported from that country in 1969 (Anon., 1969c).

Morocco—Extensive deposits of bentonite occur in the Camp Berteaux district and other localities near Taourirt in eastern Morocco.

Another type of clay consisting mainly of magnesium montmorillonite called *ghassoul* (rhassoul in French) is produced from deposits in the Tamdafelt district in the middle Moulaya valley (Jeannette, 1952; Eyssautier, 1952) and sold for several uses requiring absorbent clay.

Mozambique—A bentonite processing plant near Impamputo having a capacity of 22 to 24 t/d (24 to 26 tpd) began production in 1963. High-quality bentonite also has been produced at Lourenço Marques in recent years, and some of it is marketed in Europe (Anon., 1973a).

Senegal—Large fuller's earth deposits consisting principally of palygorskite (attapulgite) occur in the Pout region and elsewhere in Senegal (Wirth, 1968).

South Africa—Bentonite is mined in Transvaal and Orange Free State, and it is reported to have been discovered near Plettenberg Bay, Cape Province (Anon., 1972j; Woodmansee and Murchison, 1969). Production has increased in recent years and now virtually all of South Africa's bentonite requirements are fulfilled by the domestic output.

Tanganyika—The production of a few tons of bentonite in Tanganyika in 1958 was reported by Ross (1964).

Tanzania—Slatick (1969) noted the production of 184 t (203 tons) of bentonite in Tanzania and small quantities mined in other years.

Asia: *India*—Major fuller's earth deposits occur in at least five states in India, and bentonite occurs at several localities. Probably the best known deposits are in the Barmer district, Rajasthan (Siddique and Bahl, 1965). They are believed to be transported clays deposited in a marine embayment during late Tertiary time, and the clay minerals formed by the weathering of igneous rock. Other deposits occur in Tertiary beds in Gujarat and in Miocene rocks in Jammu and Kashmir; those in Madras and Bihar are of Jurassic age. Fuller's earth mined in the Mudh district is of Eocene age, and it is chiefly palygorskite (Siddique, 1968).

Japan—Two rather distinct types of smectite mineral deposits occur in Japan (Takeshi and Kato, 1969); one is called bentonite and the other, acid clay. The bentonite is composed chiefly of sodium montmorillonite, and the acid clay is montmorillonite in which H^- has substituted for exchangeable Mg^{++} and Ca^{++}. Deposits of both types occur at several places in Japan. Bentonite altered from bedded rocks called *Green Tuff* of Miocene age occur in Hokkaido and several places in Honshu (Takeshi and Kato, 1969; Urasima, et al., 1966). Large deposits formed by hydrothermal alteration of rhyolite and rhyolitic tuffs occur in Higashi-kambara-gun district, Niigata Prefecture (Takeshi, 1963). The Teikoku deposit, one of several in this district, is principally sodium montmorillonite. Bentonite formed by diagenic alteration of tuff beds occurs at Usui-gun, Gunma Prefecture, and several places in Yamagata Prefecture (Takeshi and Kato, 1969; Omori, et al., 1961). Several of these deposits contain appreciable quantities of zeolite, cristobalite, and detrital minerals, which contaminate the montmorillonite.

Large deposits of acid clay formed by diagenic processes occur at Tsuruoka-shi district, Yamagata Prefecture, at Kitakanbara-gun, Niigata Prefecture, and elsewhere. Acid clay formed hydrothermally also occurs at several places in Niigata.

Pakistan—Bentonite occurs at several places in Pakistan. Deposits in the Bhimber district of Azad Kashmir of probable middle Pliocene age have been mined since 1958 (Ali and Shahn, 1962). The bentonite is used for several purposes, including the preparation of surgical dressings.

South Korea—The Dong Bo Clay Industrial Co. operates a bentonite plant in Seoul (Anon., 1973a). The mines are located in North Kyongsang Province.

Turkey—Bentonite and fuller's earth occur at several places in Turkey. Bentonite has been produced on a small scale for domestic use as foundry-sand bond, wine purification, insecticide carriers, and other uses for several years. In 1973, plans were underway for expansion of mining and processing of sodium bentonite (Anon., 1973). Fuller's earth presumably has been used locally for centuries.

The older productive deposits apparently are at Killik, Eskişehir Province, and Kursunlu and Küçük, Çankiri Province (Anon., 1965). The sodium bentonite is near Resadiye, Tokat (Anon., 1973).

Oceania: *Australia*—A few hundred tons of bentonite have been produced in Australia in recent years (Kalix, 1971), and there has been intermittent mining of fuller's earth for a long time.

The bentonite was produced from Nanango Mineral Field and Ipswich in Queensland, and from lake deposits at Marahagee and Woodanilling in the southwestern part of Western Australia. Fuller's earth was mined on a small scale in the Dubbo district, New South Wales, as recently as 1969.

New Zealand—Bentonite mining began in New Zealand at Porangahau, Hawke's Bay, in 1940, and in recent years it has also been produced in the Harper Hills region of Canterbury.

Evaluation of Deposits

Exploration: The exploration for bentonite and fuller's earth deposits, as for virtually all minerals, requires first an understanding of the geologic occurrence of deposits, then sampling, testing, and appraising of results. An understanding of the geology of the deposits is essential in the early stages and is gained by first reviewing all published and other information on deposits under consideration or on similar rocks in the region. The second step is geologic field investigations, which may require mapping, but reconnaissance studies have been adequate for many deposits. The Wyoming bentonite beds are generally easy to find in the field, because of their tendency to swell when wet and to develop a "popcorn" or frothy texture caused by shrinkage when drying. The calcium or low-swelling bentonite and some fuller's earth deposits are not as easy to find as the swelling-type bentonite, and extensive drilling is commonly required.

Once located, deposits are explored by augering or drilling. The auger method has been widely used for bentonite, mainly because the overlying rocks can be penetrated by such equipment and the bentonite sampled satisfactorily, particularly if the hole is cleaned when the top of the bentonite is reached. Power augers are also used in some fuller's earth exploration, but they have generally been found to be unsatisfactory in searching for palygorskite (attapulgite) type deposits. These deposits are commonly explored with drilling equipment, and a fishtail bit is used to penetrate the soft sandy overburden, the cuttings being flushed with water. A core barrel with a toothed cutting head is used when the top of the clay is approached, and the clay is sampled with such equipment. A double-tubed barrel has been found to be effective in obtaining more complete recovery in some deposits. Core drilling is also used in exploring for hectorite, because of hard rock in the overburden.

KAOLIN, BALL CLAY, HALLOYSITE, AND REFRACTORY CLAY

Definitions and Classifications

Kaolin: The name kaolin is derived from the Chinese term *Kauling* meaning high ridge, the name for a hill near Jauchau Fu, China, where the clay was mined centuries ago. Apparently the first to use the name kaolinite for the mineral of kaolin were Johnson and Blake (1867). The term kaolin is now variously used as a clay-mineral group, a rock term (consisting of more than one mineral), an industrial mineral commodity, and interchangeably with the term china clay. The following mineralogical definition by Ross and Kerr (1931) is probably the most widely accepted one.

> By kaolin is understood the rock mass which is composed essentially of a clay material that is low in iron and usually white or nearly white in color. The kaolin-forming clays are hydrous aluminum silicates of approximately the composition $2H_2O \cdot Al_2O_3 \cdot 2SiO_2$, and it is believed that other bases if present represent impurities or adsorbed materials. Kaolinite is the mineral that characterizes most kaolins, but it and other kaolin minerals may also occur to a greater or lesser extent in clays and other rocks that are too heterogeneous to be called kaolin.

When applied as a term for an economic clay commodity, the foregoing definition must be modified to include some indication of use and to account for the fact that most kaolin now marketed is beneficiated to improve purity and whiteness. The following definition will apply to further discussions of kaolin in this chapter. Kaolin is a clay consisting of substantially pure kaolinite, or related clay minerals, that is naturally or can be beneficiated to be white or nearly white, will fire white or nearly white, and is amenable to beneficiation by known methods to make it suitable for use in whiteware, paper, rubber, paint, and similar uses. The term is applied without direct relation to purity of deposits. Many very large kaolin deposits are essentially pure, and they require little concentration during preparation for market. Most are slightly off-color and require bleaching, and others contain as little as 10% clay that must be washed and concentrated to recover marketable kaolin.

China clay is an ancient term originating from the use of clay, later found to be kaolinitic, in porcelain tableware and art objects in China. This term will be used rarely in this chapter.

Ball Clay: *Ball clay* is a term apparently originating in the United Kingdom many decades ago from the mining practice of prying out a lump of clay, rolling it into a crude ball, and loading it into a horse-drawn cart or wagon. Ball clay is defined by the American Society for Testing & Materials (Anon., 1971a) as a secondary clay commonly characterized by the presence of organic matter, high plasticity, high dry strength, long vitrification range, and a light color when fired. Kaolinite is the principal mineral constituent of ball clay, and it typically makes up more than 70% of this type of clay. The minor differences between the light fired color required by this definition and the white required for kaolin provide little practical basis for distinguishing these two types of clay. Therefore, ball clay is actually a variety of kaolin, but it is treated as a separate type of clay because of its wide application in marketing and as a use term in the ceramic industries.

The term ball clay is used in the United States, United Kingdom, India, Republic of South Africa, and a few other countries. Similar clay or clay used in the same products as ball clay is included with kaolin or china clay in most countries.

Halloysite: Halloysite occurs in two forms; one has the composition $(OH)_8Si_4Al_4O_{10} \cdot 4H_2O$ and the other $(OH)_8Si_4Al_4O_{10}$. The hydrated form is generally a dense porcelainlike hard clay that dehydrates at surface temperatures or slightly above to a white or light-colored porous, friable, or almost cottony-textured material. The dehydrated form is similar to kaolinite in composition and mineral structure, and the hydrated form has a c-axis spacing greater than that of kaolinite. The two forms of halloysite have caused considerable nomenclature problems. The system most commonly used now is to restrict the name *halloysite* to the hydrated form and to call the dehydrated form *metahalloysite*. However, some reports use the name *endellite* for the hydrated form and the name *halloysite* for the dehydrated form. As virtually all this clay below the surface occurs in the hydrated form, only the name *halloysite* will be used further in this section, except in the mineralogy discussion.

Halloysite, like ball clay, is a variety of kaolin, and it is included with kaolin in most production statistics. However, the term halloysite is accepted worldwide, and it is used in the clay industries, as well as by mineralogists, geologists, and soil scientists.

Refractory Clay: Refractory clay includes several varieties of kaolinitic clay that are used in the manufacture of products requiring resistance to high temperatures. The qualities and properties of refractory clays are expressed in terms of pyrometric cone equivalents (PCE) which is a method of

designating fusion points. Pyrometric cone equivalent is defined as the number of that standard cone whose tip would touch the supporting plaque simultaneously with that of a cone of the material being investigated, when tested in accordance with the Standard Method of Test for Pyrometric Cone Equivalent (PCE) of refractory materials (ASTM Designation C24) of the American Society for Testing & Materials.

The fusion points of products made from refractory clays range from just above PCE 19, the lower cutoff according to accepted definitions (Norton, 1968), to as high as cone 37. The refractory properties of kaolinitic clays are ordinarily classified, according to suitability for heat service, as low (PCE 19 to 26), moderate (PCE 26 to 31½), high (PCE 31½ to 33), and superduty (PCE 33 to 34). A few essentially pure kaolinite clays have fusion points as high as cone 35, and some refractory kaolins and fire clays containing diaspore, boehmite, or gibbsite have PCE's as high as cone 37.

The principal types of clay included in the refractory clay are fire clay, kaolin, and ball clay. Definitions of refractory kaolin and refractory ball clay are the same as given for these two types of foregoing definitions, except that the adjective "refractory" is added to indicate use. The definition of fire clay used in this chapter is essentially the one by Norton (1968), which assigns this term to all clays that are not white burning and have a PCE above 19. The term *fire clay,* therefore, excludes most kaolin and ball clay because they burn white, but it does include kaolin and ball clay that are colored when fired. This classification of refractory clay as a group name and the threefold subdivision are by no means universally applied. Other reports, including earlier ones by the authors themselves, have used the terms *refractory* and *fire clay* interchangeably.

Kaolin

History and Uses: The sedimentary kaolins of the Coastal Plain of Georgia and South Carolina and the residual kaolins of North Carolina have been known since Colonial times. Smith (1929) quotes from Sholes in his chronological history of Savannah, "1741—Porcelain Clay was discovered in or near Savannah by Mr. Duchet, and china cups made." In 1767, Josiah Wedgwood sent Thomas Griffiths to Macon County, North Carolina, to send a shipment of china clay to England (Goff, 1959). Minton (1922) recorded that, "As early as 1766 American clays from Georgia, Florida, and the Carolinas were being sent to England in considerable quantities." These clays were regularly imported and used by Wedgwood until the clays of England were available. In 1768 (Barton, 1966), the English kaolins in Cornwall were discovered, which virtually ended the mining of the Georgia kaolins for more than a century. In 1876 (Smith, 1929), the mining of the sedimentary kaolins of Georgia was started again by Riverside Mills of Augusta in Richmond County. Mining of the Georgia kaolins has been continuous since that time. In addition to the original pottery in Savannah, a second pottery was established near Bath, Aiken County, SC, at an early date. This plant was destroyed during Sherman's march to the sea (Lang, et al., 1940). However, kaolin mining was active again in Aiken County shortly after the Civil War and has continued with only minor interruptions until the present. South Carolina has ranked as the second leading kaolin-producing state for several decades. Kaolin mining in North Carolina started in 1888 in Jackson County (Parker, 1946), and in 1904, the earliest mines opened in the Spruce Pine district. The first kaolin operation in Vermont (Ogden, 1969) was in the early 1800s, and most of the product was sold in Troy and Albany, NY.

Kaolin has many industrial applications (Murray, 1963a), and new uses are still being discovered. It is a unique industrial mineral because it is chemically inert over a relatively wide pH range, is white, has good covering or hiding power when used as a pigment or extender in coated films and filling applications, is soft and nonabrasive, has low conductivity of heat and electricity, and costs less than most materials with which it competes. Some uses of kaolin require very rigid specifications including particle-size distribution, color and brightness, and viscosity, whereas other uses require practically no specifications; for example, in cement, where the chemical composition is most important. The better grades of kaolin make up most of the tonnage sold and, of course, have the highest value. Many grades of kaolin are specially designed for specific uses, in particular for paper, paint, rubber, plastics, and ceramics.

Geology: *Mineralogy*—By definitions, the clays discussed in this section consist chiefly of kaolin-group minerals. The members of this group are kaolinite, nacrite, dickite, and halloysite. With the exception of the hydrated form of halloysite, which has somewhat more water, all these minerals have essentially the same composition. Their overall crystal structures are also similar, but minor differences in arrangements of ions in octahedral positions, stacking of unit layers, crystal habit, etc., are sufficient to cause the use of separate names. Kaolinite is, of course, the most common member of the group, and halloysite obviously forms halloysite deposits. The other two members of the group rarely occur in economic deposits.

Structurally, kaolinite consists of an alumina octahedral sheet and a silica tetrahedral sheet. These sheets form triclinic crystals. The theoretical structure formula of kaolinite is $(OH)_8Si_4Al_4O_{10}$, and the theoretical composition is SiO_2 46.54%, Al_2O_3 39.5%, and H_2O 13.96%. There is relatively little ionic substitution in the mineral lattice, although there is evidence suggesting minor substitution of iron for aluminum in some kaolinite. The perfection or degree of ordering of kaolinite crystals varies considerably. By interpretation of X-ray diffraction data, Murray and Lyons (1956) have shown that the crystallinity of Georgia kaolin ranges from poorly ordered to very well ordered forms. Other investigators (Brindley and Robinson, 1946; Range, et al., 1969) have found similar variations in ordering in kaolinite from several localities. The more perfectly ordered forms of kaolinite commonly occur in hexagonal or subhexagonal crystals. Growths of crystals along the *c*-axis into accordionlike or vermicular forms are also common.

The clays described in this section contain, in their natural occurrence, a wide variety of mineral impurities. Georgia kaolin generally is 85 to 95% kaolinite, the remainder is mainly quartz, minor muscovite, biotite, smectite, ilmenite, anatase, rutile, leucoxene, goethite, and traces of zircon, tourmaline, kyanite, and graphite. Kaolin deposits in Latah County, Idaho, contain considerable ilmenite (Hosterman, et al., 1960). Kaolinite and halloysite make up from 10 to 40% of residual deposits in the Spruce Pine district, North Carolina. The bulk of these deposits consists of angular quartz, muscovite, microcline, and partly altered plagioclase (Parker, 1946). The well-known English kaolins, which are 10 to 40% kaolinite, consist mainly of quartz, mica, and potash feldspar; tourmaline is also present (Bristow, 1969).

Occurrence—Most kaolin is soft and plastic when natural moisture is present or where water has been added to dried material. One way of describing the natural texture is by the colloquial phrase—a good whittling clay—which refers to the ease with which the clay can be carved with a knife. Dry kaolin is ordinarily friable, and a common but not a diagnostic test for it is its stickiness when touched to the tongue. Some types of kaolin, which have apparently been under considerable overburden or which contain introduced silica, are hard and are sometimes referred to as flint kaolin. Though much more lithified than plastic varieties, this so-called flint kaolin is softer than most flint clay referred to in the section on fire clay, which also consists of kaolinite.

Kaolin and related clays occur in several different types of deposits. Many kaolin deposits throughout the world are in the form of tabular lenses and discontinuous beds in sedimentary rock. Most deposits of this type are of Cretaceous age or younger. The shapes and lateral extent of such deposits vary appreciably. Deposits as much as 18 m (60 ft) thick are common in some districts, and bodies of kaolin are known to extend laterally for more than a mile. Extensive sedimentary deposits of this type occur in the Georgia-South Carolina kaolin belt (Anon., 1970i), Arkansas bauxite region (Gordon, et al., 1958), and Ione district, California (Johnson and Ricker, 1948).

Residual kaolin deposits occur in weathered rock, and they are at or very near the surface, except where they formed during ancient weathering cycles and were later buried by younger rocks. Most residual kaolin deposits have formed from igneous rock. The age of kaolin formation is commonly controlled by the age of the present land surface rather than the age of the parent rock. Residual kaolin deposits are generally irregularly shaped and grade downward into the parent rock through partly altered weathered saprolite zones. They also commonly contain rounded boulders and irregular masses of fresh or only partly altered rock that were originally in the centers of joint blocks or otherwise protected from percolating water. Residual kaolin deposits occur in the Spruce Pine district, North Carolina (Parker, 1946); the Arkansas bauxite district (Gordon, et al., 1958); Latah County, Idaho (Ponder and Keller, 1960); Alberhill district, California (Cleveland, 1957); and elsewhere in the United States. They also occur in many other countries and are particularly common in Czechoslovakia (Kuzvart, 1969) and Brazil (de Souza Santos and de Souza Santos, 1972).

Some kaolin deposits throughout the world occur in rocks that have been hydrothermally altered. Such deposits are ordinarily in irregularly elongate pods or pipelike bodies aligned along joints, faults, or other permeable zones that allowed movement of warm or hot aqueous solutions. Some deposits of this type also occur in altered tuffs and are bedded. As with the residual type, the age of hydrothermal kaolin formation is not necessarily related to the age of the host rock. Most hydrothermal clay deposits are probably no older than Tertiary, though some occur in rocks that are much older. Insofar as the authors are aware, deposits in the Little Antelope Valley, California (Cleveland, 1957), contain the only hydrothermal kaolin now mined in the United States. It is used for refractory clay. However, kaolin formed by this process occurs at many places in the western states, and deposits were formerly mined on a small scale in several mining districts.

Origin—The two fundamental geologic processes that form most of the clay minerals of the kaolin group are weathering (Keller, 1964) and hydrothermal alteration (Sales and Meyer, 1949). These processes under certain conditions remove elements other than silicon, aluminum, oxygen, and hydrogen, the constituents of these clays. The clays form as the other elements are removed from preexisting minerals and possibly by the growth of new atomic arrangements from materials in solution. Though most kaolin minerals form in these ways, the origin of valuable deposits is quite another matter. Certainly several commercial clay deposits throughout the world have formed by hydrothermal processes, and others have formed in place by the weathering of other rocks. However, many of the large high-grade deposits of kaolin occur in sedimentary rocks and have either been (1) transported and deposited as kaolin; (2) altered from previously transported rocks by processes related to surficial weathering, submarine or subaqueous alteration, or postburial (diagenetic) changes; or (3) combinations of these possibilities.

Knowledge regarding the origin of kaolin deposits in the United States varies considerably. The origins of some of the deposits are reasonably well established, but geologists have widely differing opinions on how other deposits formed. Clearly, many of the scattered kaolins in western states formed by hydrothermal processes, but few such deposits have been significant sources of kaolin. In the Arkansas bauxite region (Gordon, et al., 1958) and in Latah County, Idaho (Hosterman, et al., 1960), are districts containing both transported clay and residual kaolin formed by weathering of igneous rock. In both districts, the kaolin grades downward through saprolitic zones into parent rock, which is nepheline syenite in Arkansas and granodiorite in Idaho. Also, in both districts, kaolin transported from the residual material occurs in sedimentary rocks—in marine shale in Arkansas and in lacustrine beds in Idaho. However, this explanation of origin is inadequate to account for all the kaolinitic clays in the regions containing these districts. The transported kaolin in Arkansas is of Wilcox age (early Eocene), and kaolinitic clays occurs in rocks of this age throughout a much more extensive region than can be related to the nepheline syenite in Arkansas. Similarly, the Miocene clays in Idaho cannot all be related to granodiorite. Another type of kaolin deposit that is reasonably well understood occurs as sinkhole fillings in the Valley and Ridge province in the eastern part of the United States. This type of kaolin apparently formed by the leaching of other aluminous minerals by downward-moving water in sinkholes. Probably the parent aluminous materials were mainly residuum from the solution of carbonate rock. This explanation is in part based on the presence of cores of gibbsitic bauxite in the centers of such deposits and the enclosing envelopes of cherty clay separating the kaolin from the carbonate rock in the walls of the sinkholes.

Strange as it may seem, the deposits whose origins are least understood include those in the extensive belt in Georgia and South Carolina which have been studied by many geologists and mineralogists and which lead the world as a source of commercial kaolin. Virtually all who have investigated these deposits agree that either the kaolin or the material from which it formed has been transported. Areas of disagreement center about explanations of how such very large deposits of fine-grained white clay could have formed in environments in which some sediments being deposited consisted of sand containing much more iron than does the clay. Furthermore an adequate explanation of origin of the deposits must account for the following facts: (1) some deposits of kaolin are finer grained than others; (2) some contain abundant accordionlike booklets and some do not; (3) gibbsite is present in some deposits and not in others; (4) lignitic layers or layers rich in organic matter are present in some deposits: and (5) smectite is present in some deposits in variable quantities.

One of the first geologic studies of the Georgia kaolin was by Ladd (1898); he described the deposits but did not discuss their origin. Veatch (1909) concluded that the low-

ermost member of the Cretaceous system, the Tuscaloosa, is the most important clay-bearing formation in Georgia. He indicated that the age of the formation in Georgia was determined only by its stratigraphic position. Veatch believed that the material making up the Tuscaloosa Formation was derived from disintegrated and decomposed rocks of the Piedmont Plateau. The greater part of the Piedmont region was a land surface for a very long period of geologic time. Just before the beginning of the Cretaceous, the Piedmont region was uplifted and tilted southeast. Streams rapidly carried the weathered residue to the Cretaceous sea. Veatch postulated that an enormous quantity of sediment was dumped rapidly at the stream mouths, forming extensive sand flats. Fine clay particles were deposited in nonpersistent and lenticular beds of white clay in the deeper and quieter waters near the stream mouths. He considered the Tuscaloosa in Georgia a nonmarine deposit.

Smith (1929) concluded that the basal Cretaceous sediments in Georgia were Upper Cretaceous and correlated them with the Middendorf Formation of eastern South Carolina. Smith believed that Veatch's explanation of origin was essentially correct, except that he agreed with Neumann (1927) that the kaolin was deposited in saltwater. Cooke (1943) assigned the basal Cretaceous sediments to the Upper Cretaceous under the name Tuscaloosa, which is still used. Kesler (1963) believes that in east-central Georgia there is no basis for subdividing the Upper Cretaceous, and he found that the kaolin lenses have no preferred stratigraphic position within the formation. Kesler theorized that feldspathic sands were deposited as delta sediments above sea level and were subsequently altered to kaolin. The kaolin was separated from the coarser sediment and deposited at points of lowest elevation. As the kaolin accumulated in the ponds, sand was washed from the margin into and over the edge of each deposit, forming gradational and interfingering contacts.

Bates (1964) reviewed the origin of the Georgia kaolins and summarized the mineralogic and geologic data, some of which support a marine origin and some a freshwater origin. Hinckley (1961) suggested that face-to-face flocculation of kaolin particles in a marine environment would result in the properties that distinguish a so-called "hard" kaolin, whereas edge-to-face flocculation in freshwater would produce a less dense, "soft" clay. Buie (1964) proposed the possibility that volcanic ash was first altered to montmorillonite and then to kaolinite. This is a radical departure from previously proposed theories of origin, and much more detailed field and laboratory work must be done before this hypothesis can be adequately tested. Jonas (1964) concluded from a detailed petrologic study that the originally deposited sediment was muscovite, feldspar, and quartz, and that postdepositional alteration and recrystallization produced the kaolin deposits now mined.

Recent studies of spores, pollen, and invertebrate fossils in and associated with kaolin indicate that many of the deposits thought to be Cretaceous are of Eocene age (Buie and Fountain, 1967; Cousminer and Terris, 1972; Scrudato, 1969), a possibility suggested by Eargle (1955). The first announcement of the Eocene age for any of the deposits was by Buie and Fountain, who on the basis of paleontological evidence established the presence of middle Eocene strata beneath some of the commercial kaolin deposits and above others. In one of the studies, Cousminer and Terris (1972) concluded through palynological work that some, if not all, of the sedimentary kaolin is middle Eocene in age. Results of palynological studies by R. H. Tschudy of the US Geological Survey on samples collected by Patterson have confirmed that some of the kaolin deposits are of Late Cretaceous age or older, and that some are no older than Claiborne (middle Eocene). Tschudy's work also revealed that sand beds of Paleocene age are present in the kaolin belt. This raises the possibility that some of the kaolin deposits may be of Paleocene age and that kaolin deposition took place during several different intervals of geologic time.

The recent findings on the range in age of Georgia kaolin deposits reinforce our opinion that the deposits have had a very complex history and origin, and the theories advanced to date do not provide adequate explanations. Probably the purity of the kaolin is the result of some sedimentary winnowing process, an idea favored by Kesler (1963) and others, as evidence now available suggests that the ultimate source was aluminum silicate minerals in the crystalline rocks to the northwest. The environment in which the deposition took place is yet to be established, but it was probably similar to the present coast of Georgia and South Carolina, where there are abundant offshore sand bars and swampy, muddy lagoons and winding estuaries. How iron was removed from the kaolin remains a problem; removal of iron may have been due mainly to chelating by organic acids from decaying plant materials, and oxidizing pyrite may have contributed to the process. Some of the differences in physical properties of the kaolin may be due to the fact that parts of the deposits, particularly the younger ones, have been reworked. Other differences in physical properties are probably due to varying degrees of diagenetic alteration, as most of the deposits now known to be Cretaceous in age contain more vermicular crystal growths and coarser particles than do the deposits known to be of Eocene age.

Major Kaolin-Producing Countries: *United States—* GEORGIA-SOUTH CAROLINA KAOLIN BELT. The principal reserves and the major kaolin-producing centers in the United States are in a belt extending from Twiggs County in the central part of Georgia to Lexington County in west-central South Carolina. This same belt extends southwest from Macon to Andersonville, GA (Zapp, 1965), and west to Eufaula, AL (Warren and Clark, 1965). The Andersonville and Eufaula districts contain appreciable kaolin resources, but both are centers of production of refractory kaolin and refractory bauxitic materials, and no kaolin, as defined in this report, is now produced in either district.

ALABAMA. Kaolin is mined from the Tuscaloosa Group in Marion County, Alabama (Clarke, 1964). It has been mined in this area for 40 years, and approximately 1 million tons has been shipped. This kaolin is relatively pure and is easily beneficiated by removing quartz and mica. Most of the production from this district has gone into various ceramic products.

CALIFORNIA. Filler-type kaolin is mined and processed near Ione and Bishop, CA. The Ione clay is sedimentary in origin (Bates, 1945), whereas the kaolin near Bishop is hydrothermal (Cleveland, 1957).

FLORIDA. The Florida kaolin is found in sand and gravel deposits of the Citronelle Formation (Calver, 1949).

IDAHO. Filler-type kaolin is mined and beneficiated in Latah County, Idaho.

NORTH CAROLINA. Near Spruce Pine, a primary kaolin (the term *primary kaolin* is used here as commonly applied in the kaolin industry for hydrothermal-type deposits and for residual deposits of kaolin formed by in-situ weathering) altered from pegmatite and aplite (Parker, 1946) is mined and beneficiated.

TEXAS. Near Kosse, TX, a kaolin deposit of early Eocene age in the Wilcox Group is mined and beneficiated as a filler

and ceramic clay (Fisher, et al., 1965). Pence (1954) described these deposits as nonmarine in origin. They contain 40 to 70% quartz, and the kaolin is a byproduct of a glass-sand operation.

OTHER STATES. Kaolin has been mined in the past or has been reported present in commercial quantities in Arkansas (Tracey, 1944), Illinois (Grim, 1934), Minnesota (Parham, 1970), Mississippi (Conant, 1965), New Mexico (Glassmire, 1957), Vermont (Ogden, 1969), and Virginia (Ries and Sommers, 1920).

RESOURCES. Resources of kaolin in the United States of a quality equal to or better than the lower grade clay now mined are at least 1 billion tons. The one grade of kaolin about which there is concern for long-range supply is the very high grade fine-particle-size white clay suitable as a paper coater and for certain other speciality uses (Buie, 1972). Reserves of this type of clay probably total 100 to 200 million tons. Potential resources of kaolin, including discolored and very sandy clay and that below more overburden than can be profitably stripped at present, are several times the estimated reserves.

*United Kingdom—*The United Kingdom is second only to the United States as a producer of kaolin or china clay and is the largest exporter of kaolin in the world (Anon., 1972g). China clay is the United Kingdom's chief export, and almost the entire output comes from the St. Austell area of Cornwall. Historically, the china clays in Cornwall were discovered by William Cookworthy (Barton, 1966) in the mid-1700s to be suitable for making porcelain. In Cornwall, an extensive region of kaolinized granite has been formed by hydrothermal alteration of feldspar (Bristow, 1969). The granite masses were emplaced in Permian time and are now exposed over large areas of Devon and Cornwall. The china clay deposits typically occur in funnel or troughlike bodies narrowing downward. Hot acidic solutions are thought to have migrated upward, guided by structural weaknesses such as faults and joints, and then to have been trapped under a roof of relatively impermeable rock. The hot solutions attacked the granite, altering the feldspar and mica to kaolinite. Quartz, tourmaline, and mica are the principal impurities. Drilling has shown kaolinization to exist at depths of more than 244 m (800 ft) (Bristow, 1969).

*USSR—*Kaolin is mined and processed in several districts in the USSR (Petrov, 1969). Residual, hydrothermal, and sedimentary deposits are worked. One kaolin deposit in the Transcarpathian region consists of virtually pure dickite (Rusko, 1973), and other deposits in this same district consist of mixtures of kaolinite, dickite, and halloysite. Gluhovetski district, southwest of Kiev, approximately midway between the cities of Zhitomir and Vinitsa, is one of the major centers of production. The best grades of kaolin are produced from deposits formed in weathered granite. Another major producing district is in the vicinity of Prosyanovski in the Dnepropetrovski region. The deposits in this area also occur in weathered granite and gneiss. Other large kaolin deposits occur in Kazakhstan. The largest sedimentary kaolin is the Pologa deposit on the Konok River near the Pologa Station. Many other kaolin deposits have been described by Petrov (1969). The Russian kaolin is used in the paper, ceramic, rubber, and chemical industries.

Minor Kaolin-Producing Countries: *North America—* MEXICO. Most of the kaolin in Mexico is in primary deposits originating by hydrothermal alteration of Tertiary rhyolitic rocks (Pasquera, et al., 1969). The deposits are found in almost every state, but the largest commercial deposits are in the central part of the country.

South America—ARGENTINA. The kaolin production in Argentina in 1969 was 73.4 kt (80,905 tons) (Anon., 1971b). Most of this kaolin was mined in the Blaya Dougnac, Don Carlos, and Maruja districts, Chubut. Districts in Río Negro and San Juan also had active mines in recent years.

BRAZIL. Kaolin deposits in weathered granites, pegmatities, and other crystalline rocks occur at many localities in Brazil, and recently very large transported deposits have been discovered. The deposits in weathered rocks have been mined for domestic use for many years. One region containing significant deposits of transported kaolin is along the Jari River (Colligan, 1973), a tributary of the Amazon. The deposits along this river are of Pliocene age (Klammer, 1971). Another region in which the kaolin deposits are being evaluated is along the Capim River approximately 200 km south of Bélem (White, 1973).

GUYANA. Very large transported kaolin deposits are associated with bauxite in Guyana and Surinam (Moses and Mitchell, 1963). These deposits are in sedimentary rocks underlying the Coastal Plain seaward from a shield area of ancient crystalline rocks.

OTHER COUNTRIES. Both residual and transported kaolin deposits occur at several places in Chile, Colombia, and Venezuela (Malkovský and Vachtl, 1969c). Hydrothermal deposits are also present in Chile, as indicated by the association of kaolin with alunite. Small tonnages of kaolin are also produced in Peru, Paraguay, and Ecuador (Ampian, 1973).

Europe—The kaolin operations in continental Europe are mostly small scale because of the scarcity of high-grade deposits. Most of the kaolin produced is still used by the ceramics industry and as filler for paper, and only minor quantities are sold for other applications. West Germany produces nearly 500,000 tpy kaolin, the main production being from Bavaria. The kaolin deposits in Brittany in France are being worked to a much greater extent, and production from this area will increase substantially. Czechoslovakia, where the production is controlled by the state, is becoming an important producer, and exports kaolin to other European countries.

AUSTRIA. Economically important kaolin deposits of primary origin occur on the *Bohemian Massif,* a deeply eroded complex of granitic and metamorphic rocks (Holzer and Wieden, 1969). Most of this kaolin is used in the ceramic industry and as filler in the paper and rubber industry.

BULGARIA. Sedimentary kaolin deposits of probable Pliocene age occur in northwestern Bulgaria (Karanov, et al., 1969). The individual deposits are small because the kaolin is associated with sand which fills sinkholes in a karst area underlain by Lower Cretaceous limestones.

CZECHOSLOVAKIA. Czechoslovakia is the principal producer of kaolin in Eastern Europe. The major deposits are in the vicinity of Karlovy-Vary, Plzeň (Pilsen), and Podborany in west Bohemia (Kuzvart, 1969). The kaolin is primary, is of high quality, and is used for the production of fine ceramics, paper, rubber, and refractories.

DENMARK. A small tonnage of primary kaolin is mined on the Island of Bornholm primarily for the Danish ceramic industry in Copenhagen (Almeborg, et al., 1969).

EAST GERMANY. Both primary and secondary kaolin deposits are worked in East Germany (Störr, et al., 1969). The kaolin is mined from an area north of Bautzen, Dresden, and Colditz.

FRANCE. Many small companies produce kaolin from districts in Brittany and around the Massif Central in central France. The deposits in Brittany are of approximately the same age as those in Devon and Cornwall in England (Anon., 1972f). In Brittany, the deposits are all primary in origin; and the deposits around the Massif Central are both primary and secondary (Damiani and Trautmann, 1969). About 80% of the French production is from Brittany and is near the coast.

GREECE. Low-grade deposits of kaolin suitable for use in cement, ceramics, refractories, and paper filler occur on the Aegean islands of Milos, Polyaegos, and Mytilene (Anon., 1972f). Reserves are small and production is limited.

HUNGARY. Most of the commercial kaolin deposits in Hungary are confined to an area in the northeast, in the Tokaj Mountains. These deposits are primary and have been derived from rhyolite (Varju, 1969). Secondary kaolin-bearing sandstones are mined and washed at Sarisap in the Visegrad Mountains west of Budapest.

ITALY. Both primary and secondary kaolin deposits are mined in Italy, primarily for the ceramic industry. Production is small and has never reached 90,000 tpy (Moretti and Pieruccini, 1969). The most important areas are in northwest Italy near Brella, on southern Sardinia, and near Rome.

POLAND. In the Lower Silesia area of Poland southwest of Wrocaw, primary kaolin derived from granite and gneiss is mined for local consumption mainly for the ceramic industry (Anon., 1970h; Gawronski and Kozydra, 1969).

SPAIN. Spain has the fourth largest production of kaolin in Western Europe (Anon., 1972f). Both primary and secondary deposits are mined in Spain. The deposits are widely scattered, but the commercially important production is concentrated in the northwest near Galicia-Asturias and in the east near Cuenca-Teruel-Valencia. The primary deposits of Galicia were formed by the hydrothermal alteration of mylonitized granite; reserves exceed 300 million tons. The secondary deposits in eastern Spain are Cretaceous in age and are principally kaolinitic.

WEST GERMANY. The largest kaolin mining district is in the townships of Hirschau and Schnaittenbach in Bavaria (Anon., 1972f). The deposits are sedimentary in origin, derived from decomposed granite in the Naab basin mountains, and are Triassic in age. These deposits consist of a mixture of quartz, feldspar, and kaolin. Other areas in West Germany where kaolin is worked are Oberneisen near Frankfurt, Hesse, and Westphalia (Lippert, el at., 1969).

OTHER COUNTRIES. Small tonnages of kaolin are produced in Romania, Sweden, Portugal, Yugoslavia, and Belgium (Ampian, 1973; Malkovský and Vachtl, 1969b).

Africa —REPUBLIC OF SOUTH AFRICA—The kaolin deposits of South Africa have been derived from the weathering of granite, shale, and tillite (Coetzee, 1969). Kaolin is mined near Capetown, Grahamstown, and Durban. With the exception of the large deposit of altered tillite near Grahamstown (Murray and Smith, 1973), deposits are small but of good color and quality. The Grahamstown deposit is very large, with reserves in excess of 50 million tons, but the quality ranges from excellent to very poor and it changes in very short intervals both horizontally and vertically.

SWAZILAND. Kaolin deposits of economic importance are found in the Malilangatsha Mountains of the Manzini district (Hunter and Urie, 1969).

OTHER COUNTRIES. Other African countries having minor kaolin production include Angola, Arab Republic of Egypt, Ethiopia, Kenya, Nigeria, Tanzania, Malagasy, Mozambique, and Morocco (Ampian, 1973; Malkovský and Vachtl, 1969c). The deposits in the Pugu Hills district near Dar es Salaam, Tanzania, are among those having economic potential.

Asia—SRI LANKA—The best known kaolin deposit on Sri Lanka is the Boralesgamuwa field, south of Colombo. The kaolin is probably Pliocene or Pleistocene in age (Pattiaratchi, 1969).

INDIA. Kaolin and related clay materials are mined in many locations in India (Arogyaswamy, 1968). The major producing area is in the vicinity of Chaibasa. The deposit mined in this district occurs in weathered granite. High-quality kaolin is also produced in the vicinity of Kundra, Chattanur, and Quilon, Kerala.

INDONESIA. Large residual kaolin deposits altered from granite are known to occur on the islands of Belitung and Banka (private communication). Although the deposits are large, little is known about the quality of the kaolin.

IRAN. The largest kaolin deposit in Iran is in the central part of the country near Simiron (Khadem, 1969).

JAPAN. Large primary and secondary kaolin deposits occur in Japan (Fuji, et al., 1969). The deposits are scattered throughout the islands. The primary deposits were formed by hydrothermal alteration of volcanic and granitic complexes and from in-situ weathering of granite, pegmatite, and volcanic rock. The secondary deposits are lacustrine, of lower Pliocene age.

KOREA. Primary and secondary kaolins occur in Korea (Heikes and Kim, 1965). The largest deposits, consisting of mixtures of kaolinite and halloysite, are in the Hadong-Sanchong area of southernmost Korea.

MALAYSIA. Residual kaolin altered from granite is being mined (Anon., 1970f) east of Kuala Lumpur.

PEOPLES REPUBLIC OF CHINA. Kaolin not only was used first in China, but advertisements in recent trade journals indicate that it is still being mined on a large scale and is available for export. Historically, deposits in the Ching-Tae-Chen district in Fuliang, Kiangsi Province, were the principal source of kaolin used in making fine ceramic ware (Pai, 1945). Other kaolin-producing regions in China include Singtze, Loping, Yukan, and other districts in Kaingsi Province, and Tzuhsien, Kaiping, Fengyn, and Chinghsing, Hopeh Province. Kaolin in deeply weathered granite occurs at several localities in Kwangtung Province which surrounds Hong Kong. Kaolinite clays associated with diasporic and bauxitic clays in Carboniferous, Permian, and Mesozoic coal measures occur in Szechuan, Honan, Kupeh, and Fukian Provinces and elsewhere. Residual kaolin formed by the weathering of arkosic sandstone occurs in Szechuan Province, and deposits formed from weathered feldspar in granite and pegmatite occur at several places in China.

PHILIPPINES. Philippine kaolin deposits are primary and are generally small and scattered. Most are hydrothermal in origin, although one deposit called Bukidnon is believed to have formed from in-situ weathering of a dacite (Comsti, 1969). The major deposits are on Luzon and Mindanao.

TURKEY. Both primary and secondary kaolin deposits are mined in Turkey, primarily for the Turkish ceramic industry.

OTHER COUNTRIES. The production of kaolin has been reported in recent years (Ampian, 1973) in Pakistan, Taiwan, Thailand, and South Vietnam.

Oceania—AUSTRALIA—Most of the kaolin deposits in Australia are primary and have been formed as a result of deep weathering of both silicic and mafic rocks (Gaskin, 1969). Secondary deposits in Western Australia are associated with the coal measures of Cretaceous age. Most of the deposits mined are within 200 km of the major cities of Melbourne, Sydney, Adelaide, and Perth. Extensive exploration for kaolin is being carried out in Australia by several

major companies, primarily because of the lucrative Japanese export market.

NEW ZEALAND. The only kaolin being produced in New Zealand at present is in the Northland some 120 km north of Auckland (Bowen, 1969).

Exploration, Evaluation

Exploration in the United States: The principal productive deposits of kaolin were found in the Carolinas and Georgia in the late 19th and early 20th centuries. Most discoveries were in outcrops in stream beds and road cuts. Today, exploration is carried out in geologically favorable areas, taking into account topography and elevation. Geologic and topographic maps, if available, and aerial photographs are used to lay out exploration drill-hole patterns. Most exploration holes are drilled to depths of as much as 61 m (200 ft) and on 122- to 244-m (400- to 800-ft) centers. Many deposits were missed in the past because only random spot drilling instead of pattern drilling was used. Also, because drilling was limited to 15 or 18 m (50 or 60 ft), many good deposits at greater depths were not discovered. Auger or fishtail drills are used to the top of the kaolin bed, and then a special core barrel is used to core the kaolin. Cores are difficult to recover, and special care by expert drillers is necessary to get good core recovery. In many areas, two or three kaolin beds separated by sand are penetrated where drilling is as deep as 61 m (200 ft).

Many geophysical methods of exploration for kaolin have been attempted, but to date none has been successful (Gross, 1960). Geochemical techniques also have been unsuccessful. The only positive method to locate and delineate kaolin deposits to date is by pattern drilling.

Evaluation of Deposits: After coring the kaolin and recording the thickness of the various strata above the kaolin and the thickness of the kaolin bed, a large-scale map is prepared. On this map, the drill holes are located and the thickness of the overburden and the kaolin are recorded. The cores are sent to the crude evaluation laboratory where the kaolin is prepared and tested for (1) percent grit or screen residue, (2) particle-size distribution, (3) low shear viscosity, (4) high shear viscosity, (5) brightness, and (6) leachability.

After all the tests are performed and evaluated and the overburden thicknesses and the thicknesses of the deposit have been determined, a decision is made whether the deposit is of a quality suitable for any of the present uses for kaolin or a potential aluminum resource, or whether the property should be dropped. The economic limits of the ratio of overburden to kaolin that can be mined economically range between 6 and 8 to 1, depending on the type of overburden, and the reclamation requirements imposed by federal and/or state law.

If the decision is made to keep the deposit for future mining, then additional drilling is done on 61- or 30.5-m (200- or 100-ft) centers. From this drilling information, additional maps and cross sections are prepared, from which a stripping and mining plan is devised.

The classification of a deposit as potential aluminum clay is of relatively recent practice as a result of the interest in using kaolin as a raw material to produce aluminum. The United States is dependent on foreign sources of bauxite supply approximately 90% of its aluminum requirements. The purchase of bauxite represents a significant outlay of dollars which contributes to our balance of payment deficits. Kaolin clays in Georgia and South Carolina contain 35 to 39% Al_2O_3 and represent a very sizable resource from which aluminum could be recovered if the economics become favorable.

Ball Clay

History and Uses: Probably the first deposits of the material now known as ball clay mined in the United States were those worked near Paris, TN, in 1860 for use in a local pottery (Hosterman, 1973). The first clay shipped to out-of-state consumers was that produced in west Tennessee in 1894 by a Mr. I. Mandle. Later, the industry expanded into Kentucky, and the ball clay districts in the western parts of these two states competed with districts in the United Kingdom (Anon., 1969d) for the world leadership in production. More than 80% of the ball clay tonnage produced is by companies active in Tennessee and Kentucky. The other states producing ball clay are Mississippi, Texas, and California. Ball clay is also produced intermittently in Maryland.

Ball clay is used principally in the manufacture of vitreous china sanitary ware, electrical porcelain, floor and wall tile, dinnerware, and artware (Phelps, 1972). It is also used in refractory products, ceramic glazes, and porcelain enamel slips.

Geology: *Mineralogy*—Kaolinite is the principal mineral in the ball clay deposits in Tennessee and Kentucky, but other clay minerals are present and nonclay minerals vary appreciably in abundance. The kaolinite in these deposits is fine-grained, and much of it is poorly ordered. The main clay minerals other than kaolin are illite, smectite, chlorite, and mixed-layer clay. Quartz is by far the most abundant nonclay mineral; in some deposits it makes up as little as 5% of the clay, but in others, as much as 30% (Olive and Finch, 1969). The other nonclay minerals ordinarily present occur in very minor or trace amounts and include plagioclase, potassium feldspar, and calcite. Organic matter in the form of black leaf imprints and disseminated lignitic and peaty materials are common in most ball clays.

Occurrence and Origin—Ball clay is extremely plastic. Deposits range from very light gray to nearly black. Most deposits are lenticular bodies varying considerably in size and shape. Some are more than 9 m (30 ft) thick and extend over areas 305 m (1000 ft) wide and 762 m (2500 ft) long, but most deposits are smaller. The physical properties of a typical deposit vary considerably, and layers and pockets mined from a given deposit are commonly separated in different stockpiles suitable for a particular use.

Most deposits in Tennessee and Kentucky are in beds of Claiborne age (middle Eocene), but a few, including one mined, are in Wilcox beds (lower Eocene) (Olive and Finch, 1969). The deposits mined in Texas near Henry's Chapel, northeastern Cherokee County (Fisher, et al., 1965) are also in Wilcox beds. The ball clay mined in Panola County, Mississippi (Bicker, 1970), is in a formation of Claiborne age, and the deposits are approximately the same age as younger ones in Tennessee and Kentucky. The deposits mined in Stanislaus County, California, are in beds of Paleocene or Eocene age (Cleveland, 1957). Those mined on a small scale in the Hart district in eastern San Bernardino County, California, are in hydrothermally altered Tertiary rhyolites (Kelly, 1966).

Most ball clay occurs in transported sedimentary deposits, but one minor producing district in California contains hydrothermally altered clay. The deposits in the Tennessee, Kentucky, and Mississippi districts are in sedimentary formations cropping out along the eastern edge of the Mississippi embayment. According to one theory of origin (Olive and Finch, 1969), they accumulated in an elongate broad

valley. The abundance of plant remains and organic matter associated with this clay indicates that the depositional basins probably were swampy.

Distribution of Deposits: *United States*—Any attempt to estimate the resources of ball clay is based on piecemeal information. The reserves, primarily in Tennessee and Kentucky, total probably as much as 100 million tons, and are owned by eight or ten different companies; they are thought adequate to sustain the industry for 100 years (Phelps, 1972). Additional reserves are in the producing districts in Mississippi, Texas, and California. Potential resources of low-grade ball clay, which probably would require beneficiation for market, occur in the vicinities of all of these districts. The prospects are also favorable for the discovery of more deposits of kaolinitic clay having the physical properties of ball clay in several southern states. Furthermore, if the term ball clay is applied to such deposits as the hydrothermally formed clay at Hart, CA, many areas in the western states are likely to contain large deposits.

Other Countries—The principal ball clay deposits in the United Kingdom occur in the Bovey Basin southeast of Dartmoor in Devonshire, the vicinity of Petrockstow in north Devonshire, and in the Wareham and Poole regions of Dorsetshire. These ball clays occur in sedimentary rocks of Tertiary age, consisting mainly of lenticular units of sand, clay, and lignite of probable lacustrine origin (Scott, 1929).

Ball clay occurs at several localities in India (Arogyaswamy, 1968), but production is limited, and in recent years India has been an importer of ball clay. The best quality ball clay produced is in the Than district of Gujarat. Other deposits are mined in two districts in Kerala.

Evaluation of Deposits: The prospecting techniques for ball clay are very similar to those outlined for kaolin. The most commonly used tests are very similar to those used for evaluating the ceramic properties of kaolin and heat-resistant properties of refractory clays. Chemical analysis for soluble sulfate in ball clays is commonly required because this material can be tolerated in only very minor amounts in most ceramic products.

Halloysite

History and Uses: Apparently the first halloysite used in the United States was that mined on a small scale near Rising Fawn, GA, in the late 1800s. This clay is reported to have been sold as a filler or extender in food (Veatch, 1909). Deposits near Gore, GA, were mined and used for making alum about 1912 (Broadhurst and Teague, 1954). A few hundred tons of halloysite was mined from deposits in Indiana and used for pottery and alum during and after World War I (Callaghan, 1948). The major source of halloysite in this country has been the approximately 1 million tons mined from the Dragon mine, Tintic district, Utah, by The Anaconda Co. It is processed into petroleum-cracking catalyst at the Filtrol Corp. plant in Salt Lake City, UT. Mining to supply this plant began in 1949. Minor tonnages of halloysite were produced from this same mine before 1949 for other uses. Halloysite occurring at several other places in Utah has also been used in making light-colored brick, firebrick, tile, and as a paper filler (Van Sant, 1964). Halloysite in Nevada has been used in portland cement in recent years (Papke, 1971). Deposits in Idaho consisting of mixed kaolinite and halloysite are used in refractories (Hosterman, et al., 1960), and deposits of similar clay-mineral composition in North Carolina (Parker, 1946) are mined for use in ceramics and other products.

Geology: *Mineralogy*—Halloysite is so similar to kaolinite that the two cannot be distinguished without careful
mineralogical work (Brindley, et al., 1963). Very fine tubular
structures commonly observable in electron micrographs of
dehydrated halloysite (metahalloysite) are ordinarily diagnostic of this mineral. The shortcoming of this criterion is
that not all metahalloysite occurs in the tubular form. A
reliable criterion is that the basal (001) spacing of halloysite
is at approximately 10Å; as halloysite converts irreversibly
to metahalloysite, the spacing collapses to about 7.2Å, which
is essentially the same spacing as for kaolinite. Halloysite,
therefore, can be identified by the greater spacing in X-ray
diffractograms of samples containing natural moisture.

Nonclay minerals and other impurities in halloysite vary
considerably in different deposits. Some deposits, such as
those in the Dragon mine, Utah, are exceptionally pure and
contain only minor quantities of contaminants, consisting
chiefly of pyrite, silica, dolomite, iron-bearing minerals and
noncrystalline materials, and manganese oxides (Morris,
1968). In most deposits, silica is the principal impurity; it
occurs in the form of quartz, cristobalite, chert, and poorly
crystalline forms. Many halloysite deposits are intermixed
with kaolinite, and several in western United States are
closely associated with alunite. Gibbsite and manganese minerals also occur in some halloysite deposits.

Occurrence and Origin—Small deposits of halloysite occur at many places in the United States, but known deposits
large enough to mine profitably are rare. The deposits in the
Dragon mine, Utah, the major producer, are two large pipelike bodies replacing lower Paleozoic limestone near the contact with monzonite porphyry (Kildale and Thomas, 1957).
The two bodies are separated by a zone of iron-oxide deposits
extending along the Dragon fissure (Morris, 1968). The occurrence of these deposits and the minerals associated with
them clearly indicate that they formed by hydrothermal activity. Other deposits thought by Papke (1971) to have
formed hydrothermally occur in the Terraced Hills, Washoe
County, Nevada. These deposits are bedded and apparently
altered mainly from tuff. The halloysite in the Gore district,
Georgia, is bedded and occurs in the Armuchee Chert of
Devonian age (Broadhurst and Teague, 1954). An adequate
explanation of these deposits has never been advanced, but
the extent of the beds suggests that they are of sedimentary
origin. Halloysite mixed with kaolinite in the Spruce Pine
district, North Carolina (Hunter and Hash, 1953; Parker,
1946), and the Bovill district, Idaho (Hosterman, et al.,
1960), is mainly residual and formed by the weathering of
igneous rocks. Several deposits in foreign countries are
thought to have formed hydrothermally, and some are the
result of surficial weathering.

Distribution of Deposits: *United States*—Virtually all
major productive halloysite deposits in the United States have
been referred to in the foregoing discussions. Most of the
many known scattered deposits are either too small or
too impure to be considered as significant sources of halloysite. Possible exceptions include the following: (1) the halloysite mixed with kaolin in weathered nepheline syenite
masses in the Arkansas bauxite region (Gordon, et al., 1958);
(2) the body of halloysite reported to have been exposed for
107 m (350 ft) in the adit of a mine in the Bullion (Railroad)
district, Nevada (Olson, 1964); and (3) rather extensive deposits reported near Bartow, GA (Anon., 1959). Possibilities
for the discovery of new large high-grade deposits are not
easy to evaluate. The likelihood of major discoveries does
not seem promising, but rocks in many large areas in western
United States have been altered hydrothermally, and several

of these could contain large deposits of halloysite. Also,
resources of halloysite in the old Gore mining district in
Georgia are estimated at more than a million tons (Broadhurst and Teague, 1954), and some use may again be found
for these deposits.

Other Countries—Halloysite occurs in many countries,
but, as in the United States, most deposits are small; the
authors are aware of small-scale mining only in Morocco,
Japan, Korea, Czechoslovakia, and New Zealand. The deposits mined in Morocco are in the Maaza district about 6
km west of Melilla (Hilali, et al., 1969; Martin Vivaldi and
Vilchez, 1959).

Halloysite mined in Japan includes deposits that formed
by hydrothermal alteration of granite and quartz porphyry
and others that were transported and deposited in sedimentary basins (Takeshi and Kato, 1969). Hydrothermal deposits occur at the Taishu mine, and white sedimentary
halloysite is in the Tajimi and Arikabe deposits.

The halloysite mined in Korea is presumably from the
San Chong and Tan Song deposits described by Kim and
Kim (1964).

The halloysite deposits mined in Czechoslovakia are in
the Michalovce district in the eastern part of the country
(Kuzvart, 1969).

The deposits in New Zealand occur in the Maungaperua
district on the North Island. This clay is mined by New
Zealand China Clays, Ltd., and sold for use in paper filler,
ceramics, paint extenders, and filler for adhesives and cosmetics. The halloysite is in massive deposits formed by hydrothermal alteration of andesite of middle Cenozoic age
(Bowen, 1969).

Several of the kaolin deposits in other countries consist
of mixtures of halloysite and kaolinite. One such mixed deposit occurs in altered felsitic volcanic rocks in the Burela
district, Spain (Anon., 1972b), the principal kaolin-producing center in that country. The kaolin produced contains
appreciably more halloysite than kaolinite, but the clay must
be considered to be a mixed type. Another mixed halloysite-
kaolinite deposit is in the Bukidon district in the Philippines
(Comsti, 1969).

Refractory Clays

History and Uses: The first plant in the United States
for making refractory products from clays began operation
at Salamander, NJ, in 1825. The refractory industry grew
rapidly and expanded from New Jersey to centers in Pennsylvania, the Ohio River valley, Missouri, and to western
states.

Refractory clays are used mainly in making firebrick and
block of many shapes, insulating brick, saggers, refractory
mortars and mixes, monolithic and castable materials, ramming and air gun mixes, and other products. A product called
chamottes in Europe (Anon., 1972, 1972i) and elsewhere is
made by calcining high-grade fire clays and other kaolinitic
clays. A similar material produced in the United States is
called merely calcined clay or calcined kaolin. A related
material called mullite refractory is made by calcining bauxitic clay or clayey bauxite. Further fabrication into finished
refractory products is required for chamottes and calcined
clay. Considerable tonnages of fire clay also have been used
in the past in the United States in the manufacture of light-
colored face brick, tile, stoneware, and other products.

Geology: The mineralogy, occurrence, and origin of refractory kaolin and refractory ball clay have been discussed
in the foregoing sections. Therefore, the following discussions

will be concerned with these subjects only as they apply to fire clay.

Mineralogy—The degree of ordering of the kaolinite, the principal mineral in fire clay, varies considerably. Most of the kaolinite in the harder and purer varieties of fire clay (flint clay) is very well ordered, and it is even more perfectly crystalline than the kaolinite in many kaolins. The kaolinite in the hard fire clay also tends to occur in coarser particles, and in the harder varieties the grains are interlocking. The kaolinite in the plastic varieties of fire clay, however, is ordinarily characterized by imperfectly crystalline structures. The more disordered form of this kaolinite has been widely referred to as fire clay mineral in an effort to distinguish it from the more perfectly crystalized forms.

Some of the better grades of refractory clays contain minerals that are richer in aluminum than kaolinite, and, therefore, higher alumina products can be made from them. Some refractory kaolin is a mixture of kaolinite and gibbsite, $Al_2O_3 \cdot 3H_2O$. The better grades of fire clay, that are exceptionally rich in aluminum, consist of kaolinite and diaspore, $Al_2O_3 \cdot H_2O$, and minor quantities of boehmite, $Al_2O_3 \cdot H_2O$.

Occurrence and Origin—The physical characteristics of fire clay vary considerably; the clays range from soft and plastic to flintlike. Fire clay, therefore, is divisible into plastic, semiplastic, semiflint, and flint types. Flint clay has unique properties for a clay; it lacks plasticity, breaks with a conchoidal parting into shardlike particles, and most of it is as hard as limestone. Typical diaspore-bearing flint clay is hard and has an oolitic, pisolitic, or nodular texture. Such clay is referred to as burley, birdseye, or nodular clay.

Most fire clay occurs in sedimentary rocks, and deposits range in age from Pennsylvanian to Tertiary. Fire clay is particularly common in rocks of Pennsylvanian age, and deposits of this age occur as underclays (seat earths) immediately below or closely associated with coalbeds. Fire clay of Cretaceous and Tertiary age occurs mainly in lenticular bodies. Some deposits of these ages are associated with lignite and are probably underclays, and others have apparently been transported and deposited in local basins in a nearshore, swamp, or flood-plain environment.

Opinions differ considerably on the origin of the extensive fire clays occurring below coalbeds. Some geologists (Keller, 1970; Patterson and Hosterman, 1963), including the authors, believe that most underclay deposits formed mainly by the alteration of aluminous sediments in a swampy environment. Some mineralogists, however, believe that these clays were transported and some sort of a sedimentary winnowing process caused the rather pure accumulations of kaolinite in fire clay. Some clays of Cretaceous and Tertiary age have been transported and formed by a process similar to those thought to have formed the transported kaolin deposits.

Distribution of Deposits: *United States*—Fire clay of Pennsylvanian age is widely scattered throughout the Appalachian region and parts of the Mississippi Valley (Anon., 1967). The very high quality diaspore-bearing clay, as well as other grades of fire clay, occur in the Clearfield district, Pennsylvania (Anon., 1964; Bolger and Weitz, 1952), and the Ozark region in Missouri (Keller, 1952; Keller, et al., 1954). High grade kaolinitic flint clay and semiflint clays are mined in the Olive Hill district, Kentucky (Patterson and Hosterman, 1963); Oak Hill district, Ohio (Stout, et al., 1923); and Somerset district, Pennsylvania (Hosterman, et al., 1968). Major districts producing fire clay suitable for low- and moderate-heat-duty refractory products include the Allegheny Valley and Beaver Valley districts, Pennsylvania;

Cordova district, Alabama; East Liverpool district, Ohio and West Virginia; and the Tuscarawas Valley and Hocking Valley districts, Ohio (Hosterman, et al., 1968); and a region including parts of Monroe, Audrain, Callaway, and Montgomery Counties, Missouri (McQueen, 1943).

The largest and best quality fire clay deposits in the Rocky Mountain region are in Fremont, Pueblo, Custer, Huerfano, Jefferson, and Las Animas Counties, Colorado. These deposits are in the Purgatoire Formation and Dakota Sandstone of Cretaceous age (Waagé, 1953). In addition to the fire clay mentioned in the foregoing discussions, other deposits are scattered throughout the western states (Anon., 1967; Mark, 1963; Van Sant, 1959, 1964). Districts in which these scattered deposits occur include several areas in King County and the Castle Rock area in Cowlitz County, Washington; the Molalla and Hobart Butte areas, Oregon; and the Alberhill area, California. Most of these deposits are of Cenozoic age.

Refractory kaolin is produced primarily in the following districts: (1) Georgia-South Carolina kaolin belt; (2) Andersonville, Georgia (Anon., 1972a); (3) Eufaula, Alabama; (4) Arkansas bauxite region; (5) scattered districts in Texas; (6) Latah County, Idaho; and (7) Ione, California (Anon., 1972e). Refractory ball clay is produced primarily in the western parts of Kentucky and Tennessee, but some of the ball clay produced in Texas may be used in refractory products.

Other Countries—Refractory clays are produced in many countries; however, the information available to the authors on worldwide production is sketchy and incomplete. This is partly because the distinction between fire clay and miscellaneous clay is not made in some countries, and clay used for refractory products is lumped with kaolin in others.

Those countries that produce more than 1 million tons of refractory (fire) clay annually include the United Kingdom, Federal Republic of Germany (West Germany), and Japan. The production in France exceeded a million tons a few years ago but has apparently dropped below this figure in recent years. Other countries producing 100,000 to 1 million tons annually include Argentina, Australia, India, Italy, Mexico, New Zealand, Sweden, the United Arab Republic, Uruguay, and Yugoslavia. Refractory clay is exported from the Peoples Republic of China, which is apparently a major producer. Fire clay is also mined in the USSR, Czechoslovakia, Hungary, Poland, and other East European countries.

Exploration and Evaluation: The methods used in exploring and evaluating fire clay deposits are in general similar to those used for bentonite and kaolin, except that different drilling and testing procedures are required. Drilling fire clay deposits, particularly those of Pennsylvanian age, ordinarily requires diamond bits and core barrels of the type used for minerals in hard rock. This is because the clay itself and the rocks overlying the deposits are hard. The most common test in the evaluation of fire clay is the determination of PCE (pyrometric cone equivalents).

In addition to the PCE requirement, fired test pieces also may be tested for such properties as high-temperature stability, hot strength, porosity, and spalling (Anon., 1972h; Norton, 1968). Chemical analyses are also required for some evaluations, and the contents of alkalies, alkaline earths, iron oxides, and a few less common elements are critical, because they act as fluxes when clay refractories are heated.

MISCELLANEOUS CLAY AND SHALE

Miscellaneous clay and shale includes a wide variety of clay and other fine-grained rocks that are used in many ways.

Most products made from them are fired and include such things as structural and face brick, drain tile, vitrified pipe, quarry tile, flue tile, conduit, pottery, stoneware, and roofing tile. Very large tonnages of these materials are bloated by firing to form lightweight aggregate, and large quantities are used in making portland cement. Uses not requiring firing of a finished product include filler for paint and other products and shale and clay used for packing dynamite in blast-holes and for plugging oil and gas wells that are no longer in use. Adobe building blocks are also used in unfired form in several southwestern states.

Properties

Clay and shale are used in so many different structural clay products that they necessarily have a wide range in physical properties. The properties are plasticity, green strength, dry strength, drying and firing shrinkage, vitrification range, and fired color. The properties desired vary with the structural clay product made. For example, a clay used in making conduit tile must be very plastic and have high green and dry strengths and uniform shrinkage, but for drain tile or common brick these properties do not have to be controlled so closely.

Occurrence

Miscellaneous clay and shale occur in many types of rocks ranging in age from Precambrian to Holocene. They include glacial clay, soils, alluvium, loess, shale, weathered and fresh schist, slate, and argillite. Some fire clay and kaolin are also included in the miscellaneous clay group, particularly when used in the manufacture of structural-clay products. This group consists of so many different types of rocks that only a few broad generalizations about their mineral makeup are possible. The most common mineral in many of them is one of the members of the mica group, and the dominant one in a given deposit may be illite, sericite, or one of the micas normally occurring in coarser grain sizes, such as muscovite and biotite. In addition to illite, the clay minerals present commonly are kaolinite, smectite, mixed-layer varieties, and chlorite. Some materials in this group actually contain more quartz and other detrital minerals than clay minerals.

Miscellaneous clay and shale are now mined in every state in the United States, except Alaska and Rhode Island, according to US Bureau of Mines *Minerals Yearbooks.* The states producing more than 1 million tons are Alabama, California, Georgia, Illinois, Indiana, Iowa, Louisiana, Maryland, Michigan, Missouri, New York, North Carolina, Ohio, Pennsylvania, South Carolina, Texas, and Virginia.

BIBLIOGRAPHY AND REFERENCES

Anon., 1959, "Halloysite in the Cartersville District," *Georgia Mineral Newsletter,* Georgia Geological Survey, Vol. 12, No. 2, pp. 42-43.

Anon., 1962, "Brazil," *Mining Journal,* Mar. 9, p. 244.

Anon., 1964, "Map of the Mercer Clay and Adjacent Units in Clearfield, Centre, and Clinton Counties, Pennsylvania," 4th Series, No. 12, Pennsylvania Geological Survey.

Anon., 1965, "Bentonite in Turkey," *Mineral Trade Notes,* US Bureau of Mines, Vol. 60, No. 4, pp. 12-14.

Anon., 1967, "Potential Sources of Aluminum," Information Circular 8335, US Bureau of Mines, 148 pp.

Anon., 1967a, "Area Reports: International," *Minerals Yearbook, 1965,* Vol. 4, US Bureau of Mines, p. 343.

Anon., 1969c, "Bentonite's Indispensible Role in Industry," *Industrial Minerals,* No. 25, Oct., pp. 9-14.

Anon., 1969d, "Buoyant Market for Ball Clays," *Industrial Minerals,* No. 17, Feb., pp. 9-15.

Anon., 1969e, "Erbsloh & Co.: Pioneer Producer of Sodium-Ex-changed Bentonite," *Industrial Minerals,* No. 27, Dec., pp. 33-34.

Anon., 1969f, "Laporte's Lead in Fuller's Earth, Continued Expansion at Redhill," *Industrial Minerals,* No. 24, Sep., pp. 15-16.

Anon., 1969g, "National Lead Company, Huge Reserves Held by Baroid Division," *Industrial Minerals,* No. 24, Sep., pp. 25-27.

Anon., 1969h, *Review and Forecast,* A Report Prepared for the Wyoming Natural Resources Board and the State Water Planning Program, Book I, "Summary," pp. 129-137; Book III, "Non-Fuel Minerals," pp. 1-68, A38-A60, Cameron Engineers, Denver, CO.

Anon., 1969i, "SAMIP: Mineria Isole Pontine Spa," *Industrial Minerals,* No. 24, Sep., pp. 21-23.

Anon., 1969j, "Ste. Francaise des Bentonites, A Range of Quality Clays," *Industrial Minerals,* No. 24, Sep., p. 45.

Anon., 1969k, "Sud-Chemie AG, Bleaching Earths from Bavaria," *Industrial Minerals,* No. 24, Sep., p. 44.

Anon., 1969l, "The Bentonite Industry Expands Capacity," *Industrial Minerals,* No. 24, Sep., pp. 10-14.

Anon., 1969m, "Yugoslavia's Bentonite Industry," *Industrial Minerals,* No. 27, Dec., pp. 35-37.

Anon., 1970, "An Emphasis on Minerals Marketing, Berk Ltd.'s Mineral Products Division," *Industrial Minerals,* No. 36, Sep., pp. 40-41.

Anon., 1970a, "Bentonite Clays of the Volga Regions," *Transactions,* Kazan Geological Institute, Ministry of Geology of the USSR, 1175 pp.

Anon., 1970b, "Bentonite from Sardinia, Laviosa a Leading Supplier of Processed Clays," *Industrial Minerals,* No. 31, Apr., pp. 47-48.

Anon., 1970c, "Fineness of Dispersion of Pigment-Vehicle Systems," *Annual Book of ASTM Standards, 1970,* American Society for Testing & Materials, D1210-64, pp. 210-215.

Anon., 1970d, "Hungary's Mineral Industry, Istvan Soha Details the Non-Metallics Produced," *Industrial Minerals,* No. 29, Feb., pp. 35-37.

Anon., 1970e, "Israel, New Process Up-Grades Cyprus Bentonite," *Industrial Minerals,* No. 39, Dec., 37-38.

Anon., 1970f, "Malaysia, Sanyo Backs Clay Project," *Industrial Minerals,* No. 35, Aug., p. 31.

Anon., 1970g, "Pacific Plans Kaolin Production," *Industrial Minerals,* No. 29, Feb., p. 39.

Anon., 1970h, "Poland, Kaolin Processing Commences," *Industrial Minerals,* No. 33, June, p. 46.

Anon., 1970i, "Processes for Extracting Alumina from Non-Bauxite Ores," Report 278, National Materials Advisory Board, 88 pp.

Anon., 1971, "Canada, Industmin Branches Out into Bentonite," *Industrial Minerals,* No. 45, June, p. 30.

Anon., 1971a, "Designation C-242-60," *Annual Book of ASTM Standards, 1971,* Pt. 13, American Society for Testing & Materials, 198 pp.

Anon., 1971b, *Estadictica Minera Año, 1969,* Argentina Direccion Nacional de Promocion Minera, 255 pp.

Anon., 1972, "An Introduction to Refractories," *Industrial Minerals,* No. 58, July, pp. 9-23.

Anon., 1972a, "CE Minerals: Georgian Bauxite and Kaolin Calcined for Refractory Grog," *Industrial Minerals,* No. 56, May, pp. 17-22.

Anon., 1972b, "Field Trip Guides," International Clay Conference, Madrid, Assoc. International pour l'Etude des Argiles (AIPEA), pp. II-1 to II-10, III-59.

Anon., 1972c, "Fullers Earth," *Industrial Minerals,* No. 62, Nov., p. 42.

Anon., 1972d, "Industrial Minerals in Oil Well Drilling," *Industrial Minerals,* No. 60, Sep., pp. 9-31.

Anon., 1972e, "Ione Calcined Kaolin, Interpace's Refractory Aggregate," *Industrial Minerals,* No. 59, Aug., pp. 37-41.

Anon., 1972f, "Kaolin in Europe and Japan," *Industrial Minerals,* No. 53, Feb., pp. 9-16.

Anon., 1972g, "Kaolin in the UK, English China Clay Defends Its Lead in World Paper," *Industrial Minerals,* No. 52, Jan., pp. 9-29.

Anon., 1972h, "Kaolin Operations of Gebrüder Dorfner at Hirschau," *Industrial Minerals,* No. 53, Feb., pp. 17-19.

Anon., 1972i, "Refractory Raw Materials, the Producers Reviewed," *Industrial Minerals,* No. 59, Aug., pp. 9-30.

Anon., 1972j, "South Africa: An Explosive Mining Potential," *Engineering & Mining Journal,* Vol. 173, No. 11, Nov., pp. 101-213.

Anon., 1973, "Bentonite in Europe and the Mediterranean," *Industrial Minerals,* No. 64, Jan., pp. 9-20.

Anon., 1973a, "Making Aluminum Without Electricity," *Business Week,* No. 2268, Feb. 24, pp. 58E-58H.

Ali., T.S., and Shahn, I., 1962, "The Bentonite Resources of Pakistan," *Symposium on Industrial Rocks and Minerals,* Central Treaty Organization, Lahore, Pakistan, Dec. 1962, The Mineral Research and Exploration Institute of Turkey, 1963, pp. 153-160.

Almeborg, J., et al., 1969, "Kaolin Deposits of Denmark," "Proceedings, Symposium I. Kaolin Deposits of the World. A. Europe," *23rd International Geological Congress, Prague, 1968, Report,* M. Malkovský, and J. Vachtl, eds., Vol. 15, pp. 75-84.

Ames, L.L., Jr., et al., 1958, "A Contribution on the Hector, California, Bentonite Deposit," *Economic Geology,* Vol. 53, No. 1, pp. 22-37.

Ampian, S.G., 1972, "Clays," *Minerals Yearbook 1971,* US Bureau of Mines, Vol. 1, pp. 301-327.

Ampian, S.G., 1980, "Clays," Preprint, 1978-1979, US Bureau of Mines, 41 pp.

Ampian, S.G., 1982, "Clays," *Mineral Commodity Summaries, 1982,* US Bureau of Mines, pp. 34-35.

Ampian, S.G., 1983, "Clays," *Mineral Commodity Summaries, 1983,* US Bureau of Mines, pp. 34-35.

Ampian, S.G., and Polk, D.W., 1981, "Clays," *Minerals Yearbook, 1980,* US Bureau of Mines, pp. 201-236.

Arogyaswamy, R.N.P., 1968, "Clays," *Economic Geology Bulletin 29,* India Geological Survey, Series A, 220 pp.

Barton, R.M., 1966, *A History of the Cornish China Clay Industry,* D. Bradford Barton, Ltd., Truro, 221 pp.

Bates, T.F., 1945, "Origin of the Edwin Clay, Ione, California," *Geological Society of America Bulletin,* Vol. 56, No. 1, pp. 1-38.

Bates, T.F., 1964, "Geology and Mineralogy of the Sedimentary Kaolins of the Southeastern United States—A Review," *Clays and Clay Minerals—Proceedings of the Twelfth National Conference . . . 1963,* W.F. Bradley, ed., Macmillan, New York, pp. 177-194.

Berg, R.B., 1969, "Bentonite in Montana," Bulletin 74, Montana Bureau of Mines and Geology, 34 pp.

Berg, R.B., 1970, "Bentonite Deposits of the Ingomar-Vananda Area, Treasure and Rosebud Counties, Montana," Special Publication 51, Montana Bureau of Mines and Geology, 5 pp.

Bicker, A.R., Jr., 1970, "Economic Minerals of Mississippi," Bulletin 112, Mississippi Geological, Economic, and Topographical Survey, 80 pp.

Blackman, A.G., 1969, "Bentonites Major Market: Two, Civil Engineering," *Industrial Minerals,* No. 25, pp. 23-25.

Bolger, R.C., and Weitz, J.H., 1952, "Mineralogy and Origin of the Mercer Fireclay of Northcentral Pennsylvania," *Problems of Clay and Laterite Genesis,* AIME, New York, pp. 81-93.

Bordas, A.F., 1947, "Contribucion al Concimento de las Bentonitas Argentinos," *Industria Minera, Cámera Argentina de Mineria,* Vol. 6, No. 68, pp. 43-48; No. 69, pp. 56-60; No. 70, pp. 96-100; No. 71, pp. 24-27; No. 72, pp. 51-53.

Bowen, F.E., 1969, "Kaolin Deposits of New Zealand," "Proceedings of Symposium I. Kaolin Deposits of the World. B. Overseas Countries," *23rd International Geological Congress, Prague, 1968, Report,* M. Malkovský and J. Vachtl, eds., Vol. 16, pp. 151-157.

Bradley, W.F., 1940, "The Structural Scheme of Attapulgite," *American Mineralogist,* Vol. 25, No. 6, pp. 405-410.

Brindley, G.W., 1957, "Fuller's Earth from near Dry Branch, Georgia, a Montmorillonite-Cristobalite Clay," *Clay Minerals Bulletin,* Vol. 3, pp. 167-169.

Brindley, G.W., et al., 1963, "Mineralogical Studies of Kaolinite-Halloysite Clays: Part I. Identification Problems," *American Mineralogist,* Vol. 48, Nos. 7, 8, pp. 897-910.

Brindley, G.W., and Robinson, K., 1946, "Randomness in Kaolin Clays," *Faraday Society Transactions,* Vol. 428, No. 4213, pp. 198-205.

Bristow, C.M., 1969, "Kaolin Deposits of the United Kingdom of Great Britain and Northern Ireland," "Proceedings of Symposium I. Kaolin Deposits of the World. A. Europe," *23rd International Geological Congress, Prague, 1968, Report,* M. Malkovský and J. Vachtl, eds., Vol. 15, pp. 275-288.

Broadhurst, S.D., and Teague, K.H., 1954, "Halloysite in Chattooga County, Georgia," *Georgia Mineral Newsletter,* Vol. 7, No. 2, pp. 56-61.

Buie, B.F., 1964, "Possibility of Volcanic Origin of the Cretaceous Sedimentary Kaolin of South Carolina and Georgia," *Clays and Clay Minerals—Proceedings of the Twelfth National Conference . . . 1963,* W.F. Bradley, ed., MacMillan, New York, p. 195.

Buie, B.F., 1972, "Future of the Kaolin Industry in Southeastern United States," H.S. Puri, ed., *Proceedings, 7th Forum on Geology of Industrial Minerals, Tampa, Fla., April 28-30, 1971,* Special Publication 17, Florida Bureau of Geology, pp. 103-107.

Buie, B.F., and Fountain, R.C., 1967, "Tertiary and Cretaceous Age of Kaolin Deposits in Georgia and South Carolina," Geological Society of America, Southeastern Section, 1967 Annual Meeting Program, p. 19 (G.S.A. Special Paper No. 115, p. 465).

Caberra, A.LaR., 1963, "I. Bentonitas, II. Diatomitas," *Peru, Ministerio de Fomento y Obras Publicas,* Minerales No—metalicos Serie, Memoir No. 7, 107 pp.

Callaghan, E., 1948, "Endellite Deposits in Gardner Mine Ridge, Lawrence County, Indiana," Bulletin 1, Indiana Division of Geology, 47 pp.

Calver, J.L., 1949, "Florida Kaolins and Clays," Information Circular 2, Florida Geological Survey, 59 pp.

Christ, C.L., et al., 1969, "Palygorskite: New X-Ray Data," *American Mineralogist,* Vol. 54, Nos. 1, 2, pp. 198-205.

Clarke, O.M., Jr., 1964, "Clay Deposits of the Tuscaloosa Group in Alabama," *Clays and Clay Minerals—Proceedings of the Twelfth National Conference in Clays and Clay Minerals . . . 1963,* W.F. Bradley, ed., Macmillan, New York, pp. 495-507.

Cleveland, G.B., 1957, "Clay," *Mineral Commodities in California,* L.A. Wright, ed., California Division of Mines, Bulletin 176, pp. 131-152.

Coetzee, C.B., 1969, "Kaolin Deposits of the Republic of South Africa," "Proceedings of Symposium I. Kaolin Deposits of the World. B. Overseas Countries," *23rd International Geological Congress, Prague, 1968, Report,* M. Malkovský and J. Vachtl, eds., Vol. 16, pp. 61-65.

Colligan, R.V., 1973, "Kaolin—U.S. Capacity Exceeds Use, but Sales to Grow in '73," *Engineering & Mining Journal,* Vol. 174, No. 3, pp. 158-159.

Comsti, F.A., 1969, "Kaolin Deposits of the Philippines," "Proceedings of Symposium I. Kaolin Deposits of the World. B. Overseas Countries," *23rd International Geological Congress, Prague, 1968, Report,* M. Malkovský and J. Vachtl, eds., Vol. 16 pp. 39-42.

Conant, L.C., 1965, "Bauxite and Kaolin Deposits of Mississippi, Exclusive of the Tippah-Benton District," Bulletin 1199-B, US Geological Survey, 70 pp.

Cook, G.H., and Smock, J.E., 1878, "Report of the Clay Deposits of the Woodbridge, South Amboy, and Other Places," New Jersey Geological Survey, Trenton, 381 pp.

Cooke, C.W., 1943, "Geology of the Coastal Plain of Georgia," Bulletin 941, US Geological Survey, 121 pp.

Cousminer, H.L., and Terris, L., 1972, "Palynology of Paleocene Clays from Georgia," Abstracts of Papers, 5th Annual Meeting, American Association of Stratigraphic Palynologists, 1 p.

Damiani, L., and Trautmann, F., 1969, "Les Dépots de Kaolins Français," "Proceedings of Symposium I. Kaolin Deposits of the World. A. Europe," *23rd International Geological Congress, Prague, 1968, Report,* M. Malkovský and J. Vachtl, eds., Vol. 15, pp. 141-178.

Davis, C.W., et al., 1940, "Bentonite, Its Properties, Mining, Preparation, and Utilization," Technical Paper 609, US Bureau of Mines, 83 pp.

Davis, J.C., 1965, "Bentonite Deposits of the Clay Spur District, Crook and Weston Counties, Wyoming," Preliminary Report No. 4, Wyoming Geological Survey, 17 pp.

Denny, C.S., and Drewes, H., 1965, "Geology of the Ash Meadows

Quadrangle, Nevada-California," Bulletin 1181-L, US Geological Survey, 56 pp.

Déribéré, M., and Esme, A., 1943, *La Bentonite, les Argiles Colloidales et Leurs Emplois,* Dunod, Paris, 175 pp.

de Souza Santos, P., and de Souza Santos, H., 1972, "Kaolin Clays of Brazil: Mineralogy and Properties," Program—1972 International Clay Conference (AIPEA), Madrid, June 25-30 (Paper presented without abstract or preprint).

Eargle, D.H., 1955, "Stratigraphy of the Outcropping Cretaceous Rocks of Georgia," Bulletin 1014, US Geological Survey, 101 pp.

Esquivel, J.M., and Zamora, S., 1958, "Informe Sorbre Minerales no Metalicos," Bulletin 44, Mexico Consejo de Recursos Naturales no Renovables, 152 pp.

Eyssautier, L., 1952 "L'Industrie Miniere du Moroc," Notes et Memoires, Morocco Service des Mines et de la Carte Géologique, 184 pp.

Fisher, W.L., et al., 1965, "Rock and Mineral Resources of East Texas," Report of Investigations 54, Texas Bureau of Economic Geology, 439 pp.

Fuji, N., et al., 1969, "Kaolin Deposits of Japan," "Proceedings of Symposium I. Kaolin Deposits of the World. B. Overseas Countries," *23rd International Geological Congress, Prague, 1968, Report,* M. Malkovský and J. Vachtl, eds., Vol. 16, pp. 29-37.

Gaskin, A.J., 1969, "Kaolin Deposits of Australia," "Proceedings of Symposium I. Kaolin Deposits of the World. B. Overseas Countries," *23rd International Geological Congress, Prague, 1968, Report,* M. Malkovský and J. Vachtl, eds., Vol. 16, pp. 115-150.

Gawronski, O., and Kozydra, Z., 1969, "Kaolin Deposits of Poland," "Proceedings of Symposium I. Kaolin Deposits of the World. A. Europe," *23rd International Geological Congress, Prague, 1968, Report,* M. Malkovský and J. Vachtl, eds., Vol. 15, pp. 217-223.

Gildersleeve, B., 1946, "Minerals and Structural Materials of East Tennessee," Report B., Tennessee Valley Authority Regional Products Research Division, 26 pp.

Glassmire, S.H., 1957, "Clay Mineral Potential of Northeastern New Mexico," New Mexico Economic Development Commission, 12 pp.

Goff, J.A., 1959, "Thomas Griffiths, A Journal of the Voyage to South Carolina, 1767, to Obtain Cherokee Clay for Josiah Wedgewood, with annotations," *Georgia Mineral Newsletter,* Vol. 12, No. 3, pp. 113-122.

Gordon, M., Jr., et al., 1958, "Geology of the Arkansas Bauxite Region," Professional Paper 299, US Geological Survey, 268 pp.

Gregor, M., 1967, "Industrielle Forschung und Verwertung von Bentoniten in der Tschechoslovakei," *International Clay Conference 1966, Jerusalem, Israel, June 20-24,* L. Heller, ed., Vol. 2, pp. 271-279.

Grim, R.E., 1934, "Petrology of the Kaolin Deposits near Anna, Illinois," *Economic Geology,* Vol. 24, pp. 659-670.

Grim, R.E., 1962, *Applied Clay Mineralogy,* McGraw-Hill, New York, 422 pp.

Grim, R.E., 1968, *Clay Mineralogy,* 2d ed., McGraw-Hill, New York, 596 pp.

Gross, G.W., 1960, "Location of Clay Deposits by Combined Self-Potential and Resistivity Surveys," *Trans. SME/AIME,* Vol. 217, pp. 124-130.

Heikes, G.C., and Kim, H.K., 1965, *The Kaolins of Korea,* issued by Mining Branch Industry—Engineering Division, US Operations Mission to Korea, Agency for International Development, 104 pp.

Heron, S.D., Jr., et al., 1965, "Clays and Opal-Bearing Claystones of the South Carolina Coastal Plain," Bulletin 31, South Carolina Division Geology, 65 pp.

Hewett, D.F., 1917, "The Origin of Bentonite," *Journal,* Washington Academy of Science, Vol. 9, pp. 77-96.

Hilali, E.A., et al., 1969, "Les gîtes de Kaolin du Maroc," "Proceedings of Symposium I. Kaolin Deposits of the World. B. Overseas Countries," *23rd International Geological Congress, Prague, 1968, Report,* M. Malkovský and J. Vachtl, eds., Vol. 16, pp. 55-59.

Hinckley, D.N., 1961, "Mineralogical and Chemical Variations in the Kaolin Deposits of the Coastal Plain of Georgia and South Carolina," Ph.D. Thesis, Pennsylvania State University, University Park., 194 pp.

Holzer, H.F., and Wieden, P., 1969, "Kaolin Deposits of Austria," "Proceedings of Symposium I. Kaolin Deposits of the World. A. Europe," *23rd International Geological Congress Prague, 1968, Report,* M. Malkovský and J. Vachtl, eds., Vol. 15, pp. 25-32.

Hosterman, J.W., 1973, "Clays," Professional Paper 820, US Geological Survey, pp. 123-131.

Hosterman, J.W., et al., 1960, "Investigations of Some Clay Deposits in Washington and Idaho," Bulletin 1091, US Geological Survey, 147 pp.

Hosterman, J.W., et al., 1968, "Clay," Professional Paper 580, US Geological Survey, pp. 167-188.

Huertas, F., et al., 1971, "Minerales Fibrosos de la Arcilla en Cuencas Sedimentarias Españolas. I. Cuenca del Tajo," *Spain Inst. Geológico y Minero Boletin,* Vol. 82, No. 6, pp. 28-36.

Hunter, C.E., and Hash, L.J., 1953, "Halloysite Deposits of Western North Carolina," Bulletin 58, North Carolina Division Mineral Resources, 32 pp.

Jeannette, A., 1952, "Argiles Smectiques et Rhassoul," *International Geological Congress 19th Algiers, 1952,* Chap. 20, Monograph Regionales Ser. 3, Maroc, No. 1, pp. 371-383.

Johnson, F.T., and Ricker, S., 1948, "Ione-Carbondale Clays, Amador County, Calif.," Report of Investigations 4213, US Bureau of Mines, 6 pp.

Johnson, S.W., and Blake, J.M., 1867, "On Kaolinite and Pholerite," *American Journal of Science,* 2d ser., Vol. 43, pp. 531-561.

Kalix, Z., 1971, "Clays Including Bentonite, Fuller's Earth and Damourite," *Australian Mineral Industry, 1970,* Review, pp. 82-87.

Keller, W.D., 1952, "Observations on the Origin of Missouri High-Alumina Clays," *Problems of Clay and Laterite Genesis,* AIME, New York, pp. 115-134.

Keller, W.D., 1964, "The Origin of High-Alumina Clay Minerals—A Review," *Clays and Clay Minerals—Proceedings of the Twelfth National Conference on Clays and Clay Minerals . . . 1963,* W.F. Bradley, ed., Macmillan, New York. pp. 129-151.

Keller, W.D., 1970, "Environmental Aspects of Clay Minerals," *Journal of Sedimentary Petrology,* Vol. 40, pp. 788-813.

Keller, W.D., et al., 1954, "The Origin of Missouri Fire Clays," *Clays and Clay Minerals—Proceedings of the Second National Conference on Clay Minerals . . . 1953.* A. Swineford, ed., US National Research Council Publication 327, pp. 7-46.

Kelley, F.R., 1966, "Clay," *Mineral and Water Resources of California, P. 1, Mineral Resources,* US 89th Congress, 2d Session, Senate Committee on Interior and Insular Affairs, Committee Print, pp. 126-134 (reprinted in Bulletin 191, California Division of Mines and Geology, pp. 126-134.)

Kesler, T.L., 1956, "Environment and Origin of the Cretaceous Kaolin Deposits of Georgia and South Carolina," *Economic Geology,* Vol. 51, No. 6, pp. 541-554.

Kesler, T.L., 1963, "Environment and Origin of the Cretaceous Kaolin Deposits of Georgia and South Carolina," *Georgia Mineral Newsletter,* Vol. 16, Nos. 1-2, pp. 2-11.

Khadem, N., 1969, "Kaolin Deposits of Iran," "Proceedings of Symposium I. Kaolin Deposits of the World. B. Overseas Countries," *23rd International Geological Congress, Prague, 1968, Report,* M. Malkovský and J. Vachtl, eds., Vol. 16, pp. 25-28.

Kiersch, G.A., and Keller, W.D., 1955, "Bleaching Clay Deposits, Sanders-Defiance Plateau District, Navajo County, Arizona," *Economic Geology,* Vol. 50, No. 5, pp. 469-494.

Kidale, M.B., and Thomas, R.C., 1957, "Geology of the Halloysite Deposits at the Dragon Mine," *Utah Geological Society Guidebook to the Geology of Utah,* No. 12, pp. 94-96.

Klammer, G., 1971, "Uber plio-pleistozäne und ihre Sedimente im unteren Amazonasgebiet." *Zeitschrift für Geomorphologie,* N.F., Vol. 15, No. 1, pp. 62-106.

Klinefelter, T.A., et al., 1943, "Hard and Soft Kaolins of Georgia," Report of Investigation 3682, US Bureau of Mines, pp. 1-20.

Klinger, F.L., 1965, "Mineral Industry of Finland," *Minerals Yearbook,* US Bureau of Mines, Vol. 4, pp. 229-238.

Knechtel, M.M., and Patterson, S.H., 1962, "Bentonite Deposits of the Northern Black Hills District, Wyoming, Montana, and South Dakota," Bulletin 1082-M, US Geological Survey, pp. 893-1030.

Knight, W.C., 1898, "Bentonite," *Engineering & Mining Journal,* Vol. 66, No. 17, p. 491.

Konta, J., et al., 1971, "Mineral and Chemical Composition of Four Varieties of Bentonite from Ginovci," *Acta Universitatis Carolinae-Geologica,* No. 3, pp. 175-187.

Kuzvart, M., 1969, "Kaolin Deposits of Czecholslovakia," "Proceedings of Symposium I. Kaolin Deposits of the World. A. Europe," *23rd International Geological Congress, Prague, 1968, Report,* M. Malkovský and J. Vachtl, eds., Vol. 15, pp. 47-73.

Ladd, G.E., 1898, "A Preliminary Report on a Part of the Clays of Georgia," Bulletin No. 6, Georgia Geological Survey, pp. 81-91.

Lang, W.B., et al., 1940, "Clay Investigations in the Southern States, 1934-35," Bulletin 901, US Geological Survey, 346 pp.

Lang, W.J., 1971, "Bentonite: The Demand and Markets of the Future," SME Preprint No. 71H29, AIME Centennial Annual Meeting, New York, Mar., 9 pp.

Larsen, D.H., 1955, "Use of Clay in Drilling Fluids," *Clays and Clay Technology,* J.A. Pask and M.D. Turner, eds., California Division of Mines Bulletin 169, pp. 269-281.

Lebedinskii, V.I., and Kirichenko, L.P., 1972, "Concerning the Collection 'Bentonitic Clays of the Volta Regions' (A Review and Discussion trans. From *Lithologiya i Poleznye Iskpaemye,* No. 3, 1972)," *Lithology and Mineral Resources,* Consultants Bureau, New York, Vol. 7, No. 3, pp. 393-394.

Lippert, H.J., et al., 1969, "Die Kaolinlagerstätten der Bundesrepublik Deutschland," "Proceedings of Symposium I. Kaolin Deposits of the World. A. Europe," *23rd International Geological Congress, Prague, 1968, Report,* M. Malkovský and J. Vacht, eds., Vol. 15, pp. 85-105.

Malkovský, M., and Vachtl, J., eds., 1969c, "Proceedings of Symposium I. Kaolin Deposits of the World. B. Overseas Countries," *23rd International Geological Congress, Prague, 1968, Report,* Vol. 16, 157 pp.

Mark, H., 1963, "High-Alumina Kaolinitic Clay in the United States," Mineral Resource Investigations Map MR-37, US Geological Survey.

Martin Vivaldi, J.L., and Girela Vilchez, F., 1959, "A Study of the Halloysite from Maazza (North Morocco)," *Silicates Industriels,* Vol. 24, No. 7-8, pp. 380-385.

Martin Vivaldi, J.L., and Linares, G.J., 1968, "Las Bentonitas de Cabo de Gata: II. Yacimiento de Palma del Muerto," *Spain Inst. Geológico y Minero Boletin,* Vol. 79, No. 6, pp. 65-71.

Martin Vivaldi, J.L., and Linares, G.J., 1969, "Las Bentonitas de Cabo de Gata, III, Consideracions sobre la Mineralógia y Genesis de los Yacimientos Estudiadon," *Spain Inst. Geológico y Minero Boletin,* Vol. 80, No. 1, pp. 74-80.

McQueen, H.S., 1943, "Geology of the Fire Clay Districts of East-central Missouri," *Missouri Geological Survey and Water Resources,* Vol. 28, 2d ser., 250 pp.

Millot, G., 1964, *Géologie des Argiles,* Masson, Paris, 320 pp. (trans. by W.R. Farrand and H. Paquet, published as *Geology of Clays,* Springer-Verlag, New York, 1970, 425 pp.).

Minton, L.H., 1922, "New Jersey's Part in the Ceramic History of America," *Ceramist,* Vol. 2, 271 pp.

Miser, H.D., 1913, "Developed Deposits of Fuller's Earth in Arkansas," Bulletin 530, US Geological Survey, pp. 207-220.

Moretti, A., and Pieruccini, V., 1969, "Italian Kaolin Deposits," "Proceedings of Symposium I. Kaolin Deposits of the World. A. Europe," *23rd International Geological Congress, Prague, 1968, Report,* M. Malkovský and J. Vachtl, eds., Vol. 15, pp. 201-209.

Morris, H.T., 1968, "The Main Tintic Mining District, Utah," *Ore Deposits of the United States, 1933-1967,* Vol. II, J.D. Ridge, ed., AIME, New York, pp. 1043-1073.

Moses, J., and Michell, W.D., 1963, "Bauxite Deposits of British Guiana and Surinam in Relation to Underlying Unconsolidated Sediments Suggesting Two-Step Origin," *Economic Geology,* Vol. 58, No. 2, Mar.-Apr., pp. 250-262.

Murray, H.H., 1963a, "Mining and Processing Industrial Kaolins," *Georgia Mineral Newsletter,* Vol. 16, pp. 3-11.

Murray, H.H., and Lyons, S.C., 1956, "Correlation of Paper Coating Quality with Degree of Crystal Perfection of Kaolinite," *Clays and Clay Minerals—Proceedings of the Fourth National Conference . . . 1955,* A. Swineford, ed., National Research Council, Pub. 456, pp. 31-40.

Murray, H.H., and Smith, J.M., 1973, "The Geology and Mineralogy of the Grahamstown, South Africa Kaolin Deposit," Program and Abstracts, 22nd Annual Clay Minerals Conference, Clay Minerals Society, p. 48.

Neacsu, 1969, "Bentonitele din Regiunea Alba-Iulia-Ocna-Muresului," *Editura Academiei Republicii Socialiste,* România, Bucaresti, 203 pp.

Neumann, F.R., 1927, "Origin of the Cretaceous White Clays of South Carolina," *Economic Geology,* Vol. 22, pp. 380-386.

Norton, F.H., 1968, *Refractories,* McGraw-Hill, New York, 228 pp.

Norton, J.J., 1965, "Lithium-Bearing Bentonite Deposits, Yavapai County, Arizona," Professional Paper 525-D, US Geological Survey, pp. D163-D166.

Nowacki, J. and Ciechomska, B., 1964 "Krajowe Surowce do Produkcji ziem Bielacgch (Domestic Raw Materials for Production of Bleaching Earths)," *Przeglad Geologiczny,* Vol. 12, No. 11, pp. 442-443.

Nutting, P.D., 1943, "Adsorbent Clays, Their Distribution Properties, Production, and Uses," Bulletin 928-C, US Geological Survey, pp. 127-221.

Ogden, D.G., 1969, "Geology and Origin of the Kaolin at East Monkton, Vermont," *Economic Geology,* Vermont Geological Survey, No. 3. 40 pp.

Olive, W.W., and Finch, W.I., 1969, "Stratigraphic and Mineralogic Relations and Ceramic Properties of Clay Deposits of Eocene Age in the Jackson Purchase Region, Kentucky, and in Adjacent Parts of Tennessee," Bulletin 1282, US Geological Survey, 35 pp.

Olsen, R.H., 1964, "Clays," *Mineral and Water Resources of Nevada,* US Geological Survey & Nevada Bureau of Mines, US Congress, 88th, 2d Session, Senate Document 87, pp. 185-189.

Omori, K., et al., 1961, "X-Ray Studies on Bentonite from Kaminoyama, Yamagata Prefecture," *Japan Association Mineralogists, Petrologists, and Economic Geologists Journal,* Vol. 45, No. 3, pp. 81-88.

Pai, C.C., 1945, "General Statement on the Mining Industry," Special Report No.7, China Geological Survey, 772 pp.

Papin, R., 1964, "Bentonite," *Encyclopedia of Chemical Technology,* Kirk-Othmer, ed., Vol. 3, pp. 339-360.

Papke, K.G., 1969, "Montmorillonite Deposits in Nevada," *Clays and Clay Minerals,* Vol. 17, No. 4, pp. 211-222.

Papke, K.G., 1971, "Halloysite Deposits in the Terraced Hills, Washoe County, Nevada," *Clays and Clay Minerals,* Vol. 19, No. 2, pp. 71-74.

Parham W.E., 1970, "Clay Mineralogy and Geology of Minnesota's Kaolin Clays," SP-10, Minnesota Geological Survey, 142 pp.

Parker, J.M., III, 1946, "Residual Kaolin Deposits of the Spruce Pine District, North Carolina," Bulletin 48, North Carolina Division Mineral Resources, 45 pp.

Pasquera, R., et al., 1969, "Kaolin Deposits of Mexico," "Proceedings of Symposium I. Kaolin Deposits of the World. B. Overseas Countries," *23rd International Geological Congress, Prague, 1968, Report,* M. Malkovský and J. Vachtl, ed., Vol. 16, pp. 105-110.

Patterson, S.H., 1972, "Fuller's Earth and Bentonite in the Southeastern States," *Proceedings Seventh Forum on Geology of Industrial Minerals,* H.S. Puri, ed., Special Publication 17, Florida Bureau of Geology, pp. 37-49.

Patterson, S.H., and Hosterman, J.W., 1963, "Geology and Refractory Clay Deposits of the Haldeman and Wrigley Quandrangles, Kentucky," Bulletin 1122-F, US Geological Survey, 113 pp.

Pattiaratchi, D.B., 1969, "Kaolin Deposits of Ceylon," "Proceedings of Symposium I. Kaolin Deposits of the World. B. Overseas Countries," *23rd International Geological Congress, Prague, 1968, Report,* M. Malkovský and J. Vachtl, eds., Vol. 16, pp. 17-24.

Pence, F.K., 1954, "Wilcox Sand-Kaolins of Northeastern Central Texas," Publication No. 5416, Texas University, 99 pp.

Petrov, V.P., 1969, "Kaolin Deposits of the USSR," "Proceedings of Symposium I. Kaolin Deposits of the World. A. Europe," *23rd*

International Geological Congress, Prague, 1968, Report, M. Malkovský and J. Vachtl, eds., Vol. 15, pp. 289-319.

Phelps, G.W., 1972, "The Ball Clays of Tennessee and Kentucky," SME Preprint No. 72H305, SME Fall Meeting, Birmingham, Oct., 11 pp.

Ponder, H., and Keller, W.D., 1960, "Geology, Mineralogy, and Genesis of Selected Fireclays from Latah County, Idaho," *Clays and Clay Minerals—Proceedings of the Eighth National Conference . . . 1959,* A. Swineford and P.C. Franks, eds., Pergamon Press, New York, pp. 44-62.

Poole, E.G., and Kelk, B., 1971, "Calcium Montmorillonite (Fuller's Earth) in the Lower Greensand of the Baulking Area, Berkshire," Report No. 71/4, Great Britain Institute of Geological Sciences, 56 pp.

Poole, E.G., et al., 1971, "Calcium Montmorillonite (Fuller's Earth) in the Lower Greensand of the Fernham Area, Berkshire," Report No. 71/12, Great Britain Institute of Geological Sciences, 65 pp.

Range, K.J., et al., 1969, "Fire-Clay Type Kaolinite or Fire-Clay Mineral? Experimental Classification of Kaolinite-Halloysite Minerals," *Proceedings of International Clay Conference, Tokyo, Japan, 1969,* Vol. 2, pp. 5-8.

Rich, A.D., 1960, "Bleaching Clay," *Industrial Minerals and Rocks,* 3d ed., J.L. Gillson, ed., AIME, New York, pp. 93-101.

Ries, H., and Sommers, R.E., 1920, "The Clays and Shales of Virginia West of the Blue Ridge," Bulletin 20, Virginia Geological Survey, pp. 20-21.

Robertson, R.H.S., 1961, "The Origin of English Fuller's Earth," *Clay Minerals Bulletin,* Vol. 4, No. 26, pp. 282-287.

Robinson, G.C., et al., 1961, "Common Clays of the Coastal Plain of South Carolina and Their Use in Structural Clay Products," Bulletin 21, South Carolina Division Geology, 71 pp.

Ross, C.S., and Kerr, P.F., 1931, "The Kaolin Minerals," Professional Paper 165-E, US Geological Survey, pp. 151-180.

Ross, C.S., and Shannon, E.V., 1926, "The Minerals of Bentonite and Related Clays and Their Physical Properties," *American Ceramic Society Journal,* Vol. 9, No. 2, pp. 77-96.

Ross, J.S., 1964, "Bentonite in Canada," Mines Branch Monograph 873, Canada Dept. of Mines and Technical Surveys, 61 pp.

Sales, R.H., and Meyer, C., 1949, "Results from Preliminary Studies of Vein Formation at Butte, Montana," *Economic Geology,* Vol. 44, pp. 465-484.

Schroter, G.A., and Campbell, I., 1940, "Geological Features of Some Deposits of Bleaching Clay," Technical Publication 1139, AIME, Vol. 4, No. 1, pp. 1-31.

Scrudato, R.J., 1969, "Origin of East-Central Georgia Kaolin Deposits [Abstract]," Abstracts with Programs, Geological Society of America, Vol. 1, Pt. 7, p. 203.

Sheppard, R.A., and Gude, A.J., 3d, 1968, "Distribution and Genesis of Authigenic Silicate Minerals in Tuffs of Pleistocene Lake Tecopa, Inyo County, California," Professional Paper 597, US Geological Survey, 38 pp.

Siddiqui, M.K.H., 1968, *Bleaching Earths,* Pergamon Press, New York, 86 pp.

Siddique, H.N., and Bahl, D.P., 1965, "Geology of the Bentonite Deposits of Barmer District, Rájasthan," Memoir, India Geological Survey, Vol. 96, 96 pp.

Silliman, B., 1820, "Sketches of a Tour in the Counties of New Haven and Litchfield in Connecticut, with Notices of the Geology, Mineralogy, and Scenery, etc.," *American Journal of Science,* Vol. 2, No. 2, pp. 201-235.

Slatick, E.R., 1969, "Mineral Industries of Kenya, Tanzania, and Uganda," *Minerals Yearbook, 1969,* US Bureau of Mines, Vol. 4, pp. 441-450.

Slaughter, M., and Earley, J.W., 1965, "Mineralogy and Geological Significance of the Mowry Bentonites, Wyoming," Special Paper 83, Geological Society of America, 91 pp.

Sloane, R.L., and Guilbert, J.M., 1967, "Electron-Optical Study of Alteration to Smectite in the Cheto Clay Deposit," *Clays and Clay Minerals—Proceedings of the Fifteenth Conference . . . 1966,* Pergamon Press, New York, pp. 35-44.

Smith, R.W., 1929, "Sedimentary Kaolins of the Coastal Plain of Georgia," Bulletin 44, Georgia Geological Survey, 482 pp.

Sondermayer, R.V., 1967, "Mineral Industry of Hungary," *Minerals Yearbook 1967,* US Bureau of Mines, Vol. 4, pp. 339-345.

Spence, H.S., 1924, "Bentonite," Report 626, Canada Mines Branch, 35 pp.

Störr, M., et al., 1969, "Kaolinlagerstätten der Deutschen Demokratischen Republik," "Proceedings of Symposium I. Kaolin Deposits of the World. A. Europe," *23rd International Geological Congress, Prague, 1968, Report,* M. Malkovský and J. Vachtl, eds., Vol. 15, pp. 107-140.

Stout, W.E., et al., 1923, "Coal-formation of Clays of Ohio," Bulletin 26, Ohio Geological Survey, 4th Series, 588 pp.

Takeshi, H., 1963, "On the Bentonite Deposits in Higashi Kambaragun, Niigata Prefecture," *Japan Geological Survey Bulletin,* Vol. 14., No. 1, pp. 29-38.

Takeshi, H., and Kato, C., 1969, "Minerals," *The Clays of Japan,* ed. by Editorial Subcommittee for 1969 International Clay Conference, Geological Survey of Japan, pp. 103-120.

Tazhibayeva, P.T., and Galiyev, M.S., 1972, "Genetic Features of South Kazakhstan Bentonites (Abstract)," *International Clay Conference, Madrid, June 25-30, 1972,* pp. 113-114.

Teague, K.H., 1972, "Southern Bentonite," SME Preprint No. 72H328, SME Fall Meeting, Birmingham, Oct., 7 pp.

Thompson, A.G., 1952, "Colonial Minerals Development—XII. Kenya," *Mining Journal,* London, Vol. 239, No. 6102, p. 120.

Torok, A., and Thompson, T.D., 1972, "Activated Bleaching Clay for the Future," *Trans. SME/AIME,* Vol. 252, pp. 15-17.

Tracey, J.I., Jr., 1944, "The High-Alumina Clays of Pulaski and Saline Counties, Arkansas," *American Ceramic Society Journal,* Vol. 27, No. 8, pp. 246-249.

Urasima, Y., et al., 1966, "Bentonite Deposits at Hattari Hokkaido (I)," *Japan Association Mineralogists, Petrologists, and Economic Geologists Journal,* Vol. 55, No. 1, pp. 17-24.

Van Sant, J.N., 1959, "Refractory-Clay Deposits of Colorado," Report of Investigations 5553, US Bureau of Mines, 156 pp.

Van Sant, J.N., 1964, "Refractory-Clay Deposits of Utah," Information Circular 8213, US Bureau of Mines, 176 pp.

Varju, G., 1969, "Kaolin Deposits of Hungary," "Proceedings of Symposium I. Kaolin Deposits of the World. A. Europe," *23rd International Geological Congress, Prague, 1968, Report,* M. Malkovský and J. Vachtl, eds., Vol. 15, pp. 179-200.

Veatch, O., 1909, "Second Report on the Clay Deposits of Georgia," Bulletin 18, Georgia Geological Survey, 53 pp.

Waagé, K.M., 1953, "Refractory Clay Deposits of South-central Colorado," Bulletin 993, US Geological Survey, 104 pp.

Warren, W.C., and Clark, L.D., 1965, "Bauxite Deposits of the Eufaula District, Alabama," Bulletin 1199-E, US Geological Survey, 31 pp.

Wayland, T.E., 1971, "Geologic Occurrence and Evaluation of Bentonite Deposits," *Trans. SME/AIME,* Vol. 250, pp. 120-132.

Wentworth, C.K., 1922, "A Scale of Grade and Class Terms for Clastic Sediments," *Journal of Geology,* Vol. 30, pp. 377-392.

Wetzenstein, W., 1972, "Die Bentonitlagerstätten im Ostteil der Insel Milos und Ihre Mineralogische Zusammensetung," *Geological Society of Greece Bulletin,* Vol. 9, No. 1, pp. 144-171.

Wherry, E.T., 1917, "Clay Derived from Volcanic Dust in the Pierre of South Dakota," *Journal,* Washington Academy of Science, Vol. 7, pp. 576-583.

White, M.G., 1973, "Probing the Unknown Amazon Basin, a Roundup of 21 Mineral Exploration Programs in Brazil," *Engineering & Mining Journal,* Vol. 174, No. 5, pp. 72-76.

Wirth, L., 1968, "Attapulgites du Sénégal Occidental," Rapport No. 26, Laboratorie de Geologie, Dakar, Université, Faculté des Sciences, 55 pp.

Woodmansee, W.C., and Muchison, R.C., 1969, "Mineral Industry of South Africa," *Minerals Yearbook, 1969,* US Bureau of Mines, Vol. 4, pp. 623-644.

Zapp, A.D., 1965, "Bauxite Deposits of the Andersonville District, Georgia," Bulletin 1199-G, US Geological Survey, 37 pp.

6. Diatomite

FREDERIC L. KADEY, JR.

Diatomite is a siliceous, sedimentary rock consisting principally of the fossilized skeletal remains of the diatom, a unicellular aquatic plant related to the algae. Thus, it has been formed by the induration of diatomaceous ooze and consists mainly of diatomaceous silica, a form or variety of opal which is first formed in the cell walls of the living diatom. Diatomaceous silica is not generally regarded as a synonym or the equivalent for diatomite, although it has been so used at various times. Accurately, diatomaceous silica is the preferred name for the principal mineral component of which the rock, diatomite, is composed. The terms diatomaceous earth and kieselguhr are used as synonymous with diatomite. The designations tripoli, tripolite, infusorial earth, etc., were used at one time but are now obsolete. With the changing nomenclature, these terms that were at one time correct when proposed and used for generations would be considered incorrect if used today in the light of current knowledge. The designation diatomite is reserved for those accumulations of diatomaceous silica that are of sufficient quality, size, and minability to be considered of potential commercial value.

GEOLOGY

Composition and Morphology

Diatomaceous silica qualifies as a mineral of organic origin in much the same way that aragonite and collophane do. The silica of the fossilized diatom skeleton closely resembles opal or hydrous silica in composition ($SiO_2 \cdot nH_2O$) (Cummins, 1960; Lohmann, 1960).

The silica is of acute biological significance, not only for the cell wall component, but also for the basic life process. Without silica, cell development ceases (Arehart, 1972).

In addition to bound water, varying between 3.5 and 8%, the siliceous skeleton may also contain, in solid solution or as part of the SiO_2 complex, small amounts of associated inorganic components—alumina, principally—and lesser amounts of iron, alkaline earths, alkali metals, and other minor constituents. Boron is reported to be an essential element for diatom growth (Lewin, 1966; Neals, 1967).

Since diatomaceous silica is not pure hydrous silica but contains other intimately associated elements, there is good reason to consider it a distinct type or variety. Associated with the diatomaceous silica, and integrated as part of the diatomite, may be variable amounts of organic matter, soluble salts, and particles of rock-forming minerals that were syngenetically deposited or precipitated with the diatom frustules. Sand, clay, carbonate, and volcanic ash are typical common contaminants. Other contaminating minerals may be present, such as feldspar, mica, amphiboles, pyroxenes, rutile, zircon—the result of weathering, then transporting, and subsequent redeposition of surrounding land masses. Commercial diatomite may also contain fragments and particles of other such organisms as silico-flagellates, radiolaria and siliceous sponges.

In a commercial diatomite, silica makes up the bulk of the chemical composition, usually over 86% and as high as 94%. Alumina and iron generally are at least 1.5 and 0.2%, respectively. This includes not only that believed to be incorporated as part of the skeleton but iron and alumina associated with many of the contaminants. Lesser amounts

of other elements, a small part of which may be secreted in the diatom skeleton, comprise the balance of the total chemical composition. The manner in which many of these elements are associated is not presently known. Table 1 illustrates the chemical composition of diatomites from various areas (Cummins, 1960).

Processed diatomite possesses an unusual particulate structure and chemical stability that lends itself to applications not filled by any other form of silica. Foremost among these applications is its use as a filter aid, which accounts for over half of its current consumption. Its unique diatom structure, low bulk density, high absorptive capacity, high surface area, and relatively low abrasion are attributes responsible for its utility as a functional filler and as an extender in paint, paper, rubber, and in plastics; and as an anti-caking agent; thermal insulating material; catalyst carrier; and chromatographic support; polish, abrasive, and pesticide extender to name a few representative applications.

The United States is the principal producing country, although diatomite is found in numerous other locations.

MINING METHODS

In the United States, diatomite is only extracted from quarries and open pits. Generally, a waste to ore (overburden) stripping ratio of up to 4:1 is ideal, although stripping ratios of up to about 7:1 have been economically feasible under particular conditions. Elsewhere in the world, although open pit mining is obviously preferable, underground mining methods have also been used successfully, where the stripping ratio would otherwise be economically prohibitive or the crude is overlain by dense material such as a basaltic cap.

In Iceland, commercial diatomite, deposited on the bottom of Lake Myvatn is dredged from beneath about 1 m of water and pumped in slurry to the processing plant 2 km away.

Diatomite of commercial significance in northern Brazil is being chopped into blocks by hand from shallow bogs and stacked in piles for field drying.

MODE OF OCCURRENCE AND ORIGIN

The frustule or siliceous skeleton of the diatom, a unicellular or noncellular microscopic algae, of the class Bacillariophyceae, and the order Bacillariaes, serves as the ultimate building block of which diatomite is composed. The group comprises over 300 genera and 12,000 to 16,000 species. In living form, accumulations of the diatom may be seen as the iridescent scum on ponds, the slippery gelatinous film on seaweed, on the bellies of certain species of whales, and other such varied habitats as oceanic ice floes, hot springs, moist soil, and particularly as masses of planktonic colonies on the open sea. Their natural function appears to be that of a food for other organisms of the sea. Furthermore, their role in controlling the geochemical balance of silica in marine and in lacustrine waters is scientifically recognized (Calvert, 1968; Kilham, 1971).

Environmental conditions for growth include at least the five following major requirements:

1) Large shallow basins (preferably 35 m or less in depth) for deposition, so that photosynthesis can occur. With regard to lacustrine deposits, a shallow lake provides suffi-

Table 1—Chemical Composition of Natural Diatomites from Various Localities (Oven-Dried Basis)

Constituent	Lompoc, CA	Maryland Calvert Formation	Nevada	Idaho	Kenya Soysambu	Japan Nilgata Earth	Russia Kamy-shlov Urals	Spain Alba-cete	Mexico Jalisco	Algeria (Primo Grade)
Silica (SiO_2), %	89.70	79.55	86.00	89.82	84.50	86.0	79.92	88.60	91.20	58.40
Alumina (Al_2O_3), %	3.72	8.18	5.27	1.82	3.06	5.8	6.58	0.62	3.20	1.66
Iron Oxide (Fe_2O_3), %	1.09	2.62	2.12	0.44	1.86	1.6	3.56	0.20	0.70	1.55
Titanium (TiO_2), %	0.10	0.70	0.21	0.07	0.17	0.22	0.48	0.05	0.16	0.10
Phosphate (P_2O_5), %	0.10	—	0.06	0.13	0.04	0.03	—	—	0.05	0.20
Lime (CaO), %	0.30	0.25	0.34	1.26	1.80	0.70	1.43	3.00	0.19	13.80
Magnesia (MgO), %	0.55	1.30	0.39	0.54	0.39	0.29	0.98	0.81	0.42	4.57
Sodium (Na_2O), %	0.31	1.31	0.24	1.03	1.19	0.48	0.65	0.50	0.13	0.96
Potassium (K_2O), %	0.41		0.29	0.22	0.91	0.53	0.72	0.39	0.24	0.50
Ignition loss, %	3.70	5.80	4.90	4.02	6.08	4.4	4.91	5.20	3.60	17.48*
Total	99.98	99.71	99.82	99.35	100.0	100.05	99.23	99.37	99.89	99.22

* Includes 13.9% CO_2.

cient actinic light for photosynthesis for not only pelagic diatoms, but also for benthonic forms attached to stones and to plants on the lake bottom. In the case of thick deposits of marine diatomites, there is evidence for a down warping of the basin of deposition, thus maintaining fairly shallow water for benthonic species. The open sea is reportedly the best environment for pelagic diatoms.

2) An abundant supply of soluble silica (Kilham, 1971; Lewin, 1959).

There is a worldwide correlation between the existence of thick diatomite deposits and proximity to volcanic ash occurrences. While volcanic ash does not necessarily have to accompany diatom deposition, some mechanism for increasing the silica content in marine and lacustrine bodies beyond the present day norm is necessary for the formation of commercially thick deposits. There are numerous examples wherein marine and nonmarine deposits meet this condition. Typical of these are at Lake Myvatn, Iceland; in the state of Jalisco, Mexico, where the deposits border the ancient shores of Lake Atotanilco; and those occurrences bordering Lake Rotorura, New Zealand; in Nevada the late Miocene Virgin Valley beds in Humbolt County, and the early Pliocene Esmeralda formation in Nye and Esmeralda Counties; the Payette formation in Idaho and eastern Oregon. These all are associated with volcanism. Deposits of marine diatomite exhibit a similar correlation; the late Miocene-early Pliocene Sisquoc formation which comprises the Lompoc, CA, diatomite; the middle and late Miocene Monterey formation of the Coast Ranges; and the middle Miocene Tremblor formation east of Coalinga, CA, are examples of diatomites associated with contemporary volcanism. Particles of volcanic ash are a common contaminant of some diatomites.

3) An abundant supply of nutrients. In most lakes that are nontoxic to diatom proliferation the supply of nutrients is often more available than is the supply of silica.

4) The absence of toxic or growth-inhibiting constituents in the water. Although few lakes contain toxic water in the usual sense, many in which the rate of evaporation exceeds inflow during long periods of the year build up concentrations of soluble salts to the point of inhibiting diatom growth (Jorgensen, 1969; Chin, et al., 1965).

5) A minimum supply of clastic sedimentary materials. While this, per se, is not a requirement for diatom growth, low nondiatomaceous contamination is paramount for the development of a commercially suitable deposit.

The effect of temperature, light, pressure, and other factors on diatom growth has been discussed (Berger, 1971; Felfoldy, 1961; Patrick, 1971). In addition to its unique siliceous skeleton, the living form has a nucleus, it produces certain protoplasmic substances by the process of photosynthesis, and, incidental to its metabolism, it manufactures oil and vitamins. The rate of reproduction of diatoms varies with the species from between two or three times a day to once a week; and one diatom may have 100 million descendants in 30 days. Deposition of the skeletal remains occurs after it has served its natural function. Thus, given the right conditions of environment and geologic location, tremendously thick deposits of diatomaceous ooze may build up on the floor of the containing body of water.

Potassium-argon dating of volcanic minerals and glass in North Pacific sediments has established Tertiary sedimentation rates of from less than 1 mm per 1000 years for deep sea red clay, to 1 cm per 1000 years for calcareous-siliceous ooze nearer the continent. In comparison, the rhythmic banding seen in Lompoc, CA, diatomites suggests a rate considerably faster than that—probably of the order of 1 mm or more per year. Gross has calculated a sedimentation rate of 4 mm per year for a 25% diatom-75% silt sediment, deposited in Sannich Inlet, BC. (Gross, et al., 1963)

After deposition, such subsequent geologic forces as consolidation, burial under what will later be overburden, regional uplift, and partial erosion come into play to expose, yet protect, the deposit for later discovery and exploitation.

Because of the delicate nature of the diatom skeleton, deposits of diatomite to be useful to industry cannot undergo any great degree of regional metamorphism or chemical alteration. For this reason, geologic conditions that have not resulted in an appreciable degree of consolidation or of cementation are preferable. When orogenic forces are excessive, the resulting metamorphosis produces opaline cherts, porcelanites, and similar more indurated materials of noncommercial interest.

In place, diatomite is soft and "punky" and has a chalk-like appearance. Color may vary from snow white in a pure, well bleached and dry deposit, to olive green or darker where substantial organic remains are still present and where moisture content is high. It may exhibit stratification, caused by either, or both, sedimentation of particularly flat beds or a preponderance of discoid diatoms, or by seasonally rhythmic deposition of clay and other impurities. On the other hand, it may be massive and show no stratification. It may be so loosely consolidated that, when handled, a field sample will readily break down to a powder, or it may be hard enough to crack "brittley" when struck with a hammer. In addition to induration through consolidation, precipitation of carbonate for example, or "baking" by volcanic flows can destroy an otherwise good deposit. The better quality diatomite is lightweight, usually possessing a block density between 320 to 545 kg/m³ (20 to 34 pcf).

CLASSIFICATION OF DEPOSITS

The various species of diatom thrive in either a marine or a lacustrine environment. Some forms live in brackish waters. Identification of the diatoms from an unknown deposit label it as having been laid down in either one environment or the other. It is, therefore, to be expected that a major criterion of deposit classification deals with whether it is of marine or of freshwater origin. Some investigators have added to the environments just mentioned, whether deposits are of modern lake, marsh, or bog origin (Durham, 1973).

These criteria are important because the diatom assemblages associated with marine environments are quite different from those that live only in freshwater habitats. The association of forms or the diatom assemblage, as seen by means of the microscope, not only serves to differentiate marine origin from freshwater, but also in many cases, to identify the deposit location from which an unknown sample may have come. Diatom assemblages, like fingerprints, are specific to individual locations. Because of the structural differences related to origin, as has been pointed out, diatomites have a range of properties and produce a range of effects in the numerous uses to which they have been applied. Numerically, most of the known deposits in the world are of lacustrine origin. Generally however, those of marine origin, although less numerous, tend to be larger.

DISTRIBUTION OF DEPOSITS

The occurrence of diatomaceous silica is widespread throughout the world. Although algae appeared quite early in geologic history, commercial deposits are generally restricted to sedimentary formations of Tertiary and of later age, and further limited ecologically by those conditions for formation that have been previous described. However, when one considers the numerous other limiting factors that must be taken into account before an occurrence qualifies as a commercial deposit—quality, minability, location, and size—then the numbers are few indeed.

North America

United States: Occurrences of diatomite have been reported from just about all of the east and the west coastal states, and indeed, from many bordering these states in the interior of the country. Commercial production, however, has been limited to a few of these. First American production was from Maryland, where from 1884 until 1930, marine diatomite was extracted from the Fairhaven member of the middle Miocene Calvert formation. This diatomite outcrops along the banks of the Patuxent, Rappahannock, and Potomac rivers, and in cliffs along Chesapeake Bay. It is contaminated with varying amounts of loosely held silica sand, most of which can be removed in processing. More intimately held montmorillonite, illite, and kaolinite are also present. Removal of the clay on an experimental basis has been attempted with little practical success, for other than low quality applications (Knechtel, and Hosterman, 1965).

The largest and most uniform deposits in the world are to be found in the vicinity of Lompoc, CA. The diatomite

sequence of commercial significance is of the order of 305 m (1000 ft) thick and is part of a thicker diatomite series of marine origin belonging to the Sisquoc formation of late Miocene or possibly early Pliocene age. Although the Lompoc diatomite may have been known since the time of the Spanish Conquistadors in the 1760s, it was not recognized as such until over a century later, and first mining was not started until about 1890. The strata at Lompoc, which present a good example of rhythmic bedding, are mined principally from a broad pitching syncline with related smaller anticlines and synclines.

Diatomite was mined from a deposit in the Palos Verdes Hills near Los Angeles starting in the 1930s, but that deposit is now depleted (Rowell, 1980). A massive embayment or commercial freshwater diatomite in Shasta County has been outlined recently. There are other minor operations in California (Clark, 1978).

Second to California in production is Nevada, where diatomite from freshwater deposits is mined in two principal areas from deposits near Lovelock and at Clark, near Reno. These are of late Tertiary or Pleistocene age. The beds at Clark are close to the surface and are relatively low in moisture.

Diatomite is also mined from a freshwater deposit near Basalt, about 97 km (60 miles) from Tonopah in southwestern Nevada. In the state of Washington, diatomite is produced from a lacustrine deposit near Quincy. At one time, there was production from a deposit at Kittitas, WA. Minor production has been reported from Oregon and from Arizona. Other states that have produced diatomite in the past include Idaho and Utah. A mixture of carbonate and freshwater diatomite in western Kansas was mined until 1977. Florida has bog and lake bottom deposits near Pensacola and in central Florida that have received exploratory attention in the past. Noncommercial bog and lake deposits are well known in Maine, New Hampshire, Massachusetts, and New York. Some of these were operated on a small scale in earlier years.

Canada: Production in Canada at present is limited to freshwater Miocene deposits near Quesnel, BC. Eastern Canada has well-known occurrences in Nova Scotia and New Brunswick that are not profitable to exploit at the present time.

Mexico: Diatomite occurs in several states in Mexico, and among these are Tlaxcala, Colima, Jalisco, Michoacan, and Mexico. Some production has been known since 1927. Good quality earth is mined near Catarina in Jalisco. Other commercial deposits are exploited at Zacapu and at Tuxpan by Kieselguhr de Mexico in Michoacan. Production has been reported at Magdalena and Cocula in Jalisco, at La Blanca in Tlaxcala, at San Martin Texmelucan in Puebla, and at Ixtlahuaca in the state of Mexico.

Europe

The most significant European sources of diatomite are freshwater Tertiary and Quaternary deposits located in the Massif Central area of southern France.

The Luneburger-Heide deposits of Western Germany were the first commercially mined deposits in the world, but have declined in importance in recent years. In Italy, commercial deposits are located at Arcidosso and Santa Flora and further south in the Viterbo area and at Castiglione in Teverina. Medium grade diatomite is mined in the area of Tombolia in northern Italy. Good quality Spanish diatomite is mined from lacustrine deposits between Hellin and Elche de la Sierra in the southeastern part of the country. Some of these deposits were mined from underground galleries for over 50 years, but in recent years have been converted to open pit operations. Some of the earth is a remarkably white color, but chert lenses and carbonate beds necessitate highly selective mining methods. There are other, low quality occurrences in Spain, including a marine occurrence near Almeria. In Iceland, a diatomaceous ooze of Holocene Age is dredged from Lake Myvatn and pumped in slurry to the processing plant where the volcanic ash is removed by hydroclones.

The diatomite of Skye, Scotland, has been known for some time. Some diatomite is mined from a wet lake basin deposit near Kendal in Westmoreland, UK. The earth is dark in color and is not suitable for filter aid manufacture. It is used mostly for thermal insulation.

An unusual, but commercially viable deposit, is the diatomite-clay mixture of Denmark called Moler. This earth is not suitable for filter aids because of the iron oxide and clay content, but it is exported elsewhere in Europe for fertilizer coating and for insulating. Moler is of marine origin and of Tertiary age and occurs on the islands of Fur and Mors in northern Jutland. Some diatomite of good quality has been produced in recent years in southern Denmark.

There are numerous deposits in the USSR, although little information is available. Freshwater diatomite deposits exist near Prilep, Jugoslavia, and in central Anatolia, Turkey. Romania has a lacustrine deposit near Adamclisi. Small unproductive deposits occur in many other European countries.

Africa

Impure diatomite, suitable only for insulating materials, is mined from a deposit in the Ermelco District, Transvaal, and from a deposit in the Prieska District, near Postmasburg, Cape Province.

East Africa produces both filter aid and filler grade products from deposits near Gilgil, Kenya. These were deposited in Pleistocene lakes in the Rift Valley. Several other smaller deposits are not of current economic interest.

Marine deposits of Miocene age occur near Sig (formerly St. Denis-du-Sig) in Algeria. These are mined from underground galleries. There is a marked similarity in appearance between the Algerian and Lompoc diatomites when viewed under the microscope. The diatomite from Algeria, however, is characterized by a noticeable amount of carbonate contamination and is deficient in many of the diatom types that, in the Lompoc earth, provide a better balance for filtration. Numerous, intensely folded, thin beds occur near Mosteganum, Algeria.

South America

Diatomite is mined from small, scattered deposits in Rio Negro Province in Argentina. Much of the earth occurs under a basalt cap, which necessitates mining from galleries and in one instance as an open pit operation after blasting away the basalt. There are also small impure occurrences in San Juan province near Calingasta, and in remote parts of Salta Province. There are numerous bog deposits in Brazil. The states of Ceará and Bahiá contain the most deposits, although eight other states are reported to have smaller impure occurrences. Peru has occurrences at Pisco, Piurá, Chiclayo, and Arequipa; and Chile has deposits of high brightness earth at Arica and Chiloe. In Colombia, diatomite occurs at Tunja and Antioquia. Diatomite occurs to a lesser degree in other Latin American countries.

Asia

There is considerable potential for development of deposits in the Far East. Japan has a growing industry based on its own marine and lacustrine deposits. Deposits, possibly suitable for local markets, exist in Indonesia and in Korea.

Diatomite deposits within China are widespread but localized and small. These are used mainly for production of construction and insulating materials.

There are several small deposits in Australia and in New Zealand, but for filter aid quality, these countries depend on imports, principally from the US. The pressure from increasing transportation costs, however, is lending motivation to the exploration and possible development of shallow mediocre quality bog deposits in Western Australia near Perth and Geraldton.

Elsewhere in the world are numerous other occurrences of diatomaceous silica which, for one reason or another, have not been exploited. Furthermore, accumulations of diatomaceous ooze are forming today that will constitute the deposits of "tomorrow."

EXPLORATION

Keeping in mind the previously described limiting criteria for formation, prospecting for diatomite entails reconnaissance of potentially suitable terrain. Because it is usually soft and easily eroded, white "showings" in stream banks or in road cuts should be investigated.

Since virtually all of the surface exposures of diatomite—at least in the United States—have been recognized, if not sampled, explorationists have shown increasing interest in geophysical methods as an opportunity to uncover as yet unknown diatoamceous sediments. Geophysical methods are increasingly contributing to the search for industrial minerals in general (Crice, 1980). Geophysical refraction seismic surveys have successfully outlined the depth of shallow freshwater diatomite basins. This works particularly well where the soft (low velocity) diatomite is underlain by higher velocity basalt (C.M. Smith, Grefco, Inc., personal communication). In addition to seismic methods, Manville Corp. has also used resistivity methods in diatomite exploration (Breese, Manville Corp., personal communication). Narrow pass-band infrared imagery (3-4 and 4.5-5.5 micrometers) has been used to recognize diatomite from aircraft by its thermal characteristics (Carter, 1971). Changes in vegetation have been noted over diatomite-bearing areas of bog deposits, and this might indicate the possible application of geobotanical techniques in prospecting.

When carried to the point of development, exploration of diatomite deposits is pursued in stages. After an occurrence has been recognized through prospecting, usually, the first step in exploration consists of a preliminary sampling of all visible outcrops. The nature of the material is noted and recorded by measuring the attitude of the beds and by observing all other visible structural and stratigraphic features. Sampling intervals are divided into visible increments if bedding, textural, and color changes or stratification are evident. If thick enough and no visible divisions are apparent, the stratigraphic interval spanned by each sample should be no more than 150 cm for each channel cut. In this way, nonvisible characteristics—diatom assemblage, chemical changes, etc.—may be noticed and characterized during testing of the samples. In subsequent exploration stages, the 150-cm interval may be reduced, if required.

The advantage of two- and three-stage exploratory programs is that exploration can begin with a relatively economically wide spacing of sampling locations. Upon testing of the samples from such a stage, a judgment can be made whether, because of unfavorable findings, the program should be aborted or if a subsequent program is needed in which the hole spacing is reduced.

Some preliminary field testing can be incorporated into the exploratory phase of deposit evaluation. Evidence of staining, degree of consolidation, judgment of color, bedding, and stratification, etc., are all field characteristics that are important, and which should be noted. Other obvious field aids include an HCl test for carbonate, a grit test by grinding the crude between teeth, and the noting of appreciable water solubles by taste.

The most useful tool in the field is a portable microscope. When performed by an experienced operator, microscopic examination in the field can be used to ascertain diatom constitution and contaminants, to direct the course of exploration, to provide a stratigraphic correlation, and to eliminate the shipping of useless samples to the testing center.

Pending favorable results from the laboratory evaluation of the samples, the next stage of exploration is planned to delineate the reserves within the area and to further assess the quality. With horizontal beds under relatively thin overburden and in areas of gently rolling topography, the digging or augering of vertical exploratory shafts to expose the entire stratigraphic column has proven highly satisfactory. In countries with low labor rates and in locations where mechanical equipment is difficult to maintain, hand-dug shafts up to 50 m in depth have many advantages. When these are of the order of 1 to $1\frac{1}{2}$ m in diam, ingenious "bird cage" types of sampling platforms, lowered by windlass into the hole from a tripod arrangement, have been used to support a geologist, who logs the hole and collects samples from the wall of the shaft. Whether hand-dug or machine-bored with a large auger, these openings in the deposit have the advantage that the structure and nature of the beds may be noted and correlated from hole to hole. Where overburden is minimal or nonexistent, holes dug by backhoe have been used. While these have the advantage of the speed of excavation that is provided by mechanical equipment, depth is limited to about 6 or 7 m at the most.

It should be emphasized at this point that an essential element in diatomite processing, hence also in diatomite sample preparation, is that the delicate diatom structures—as they exist in the deposit—be disturbed or broken down as little as possible. This diatom structure must therefore be carefully preserved to the extent possible during sampling as well. The previously described "nondrilling" sampling methods are most satisfactory from many viewpoints, not the least of which is structure preservation. Where drilling cannot be avoided, however, it is recommended, at the outset, that rotary drilling which pulverizes the cuttings to the extent that a powdered sample is flushed up the hole by reverse air circulation not be used. While this method of sample retrieval may not be as undesirable for more highly consolidated materials that are suitable for aggregate applications, it is virtually useless for filter aid and functional filler applications where the amount of whole diatom structure needs to be preserved for evaluation. Furthermore, the loss of dust plus the minimum critical size of particle that is flushed for a given vertical sampling interval renders representative sampling most unreliable (Krech, 1973).

The recommended sampling method is by core drilling. Core drilling in diatomite is specialized and requires special equipment and experienced drilling crews. Triple barrel, wire-line diamond core equipment is highly satisfactory for loosely consolidated material. The more loosely consolidated the diatomite is, however, the more difficult it is to obtain high percent recovery with as little as a 4- or 5-cm diam core. A 5-cm diam core of 150-cm vertical interval will not produce sufficient material in a low density crude for more than a few of the evaluatory tests that are required. Therefore,

careful and judicious compositing of suites of similar samples may be required.

Where any amount of topographic relief is present, trenching by bulldozer on hillsides will expose the bedding and will permit subsequent channel sampling. The opening of trenches by bulldozer across the bedding of dipping strata will remove overburden and expose the strata for sampling.

The positioning of drill holes, shafts, or trenches to adequately cover a deposit depends upon a number of factors. Sample positions are most commonly arranged systematically in a grid to cover the area to be explored. The distance between sample locations is dependent, among other things, on the lateral variation in important properties or characteristics of the diatomite. Where important properties are thought or known to change rapidly with lateral extent, holes must be placed more closely together than when a high degree of uniformity is experienced in preliminary study. (Sample holes, on as close as 30-m centers, have been used.)

Subsequent stages of exploration consist of additional specialized sampling and of the selection of bulk samples for plant scale trials. Whereas the foregoing methods are suitable for dry deposits, specialized techniques must be developed for bog deposits and for those occurring under lakes or ponds.

In the case of shallow lakes, sample locations are marked with survey poles driven into the ooze. Peat bog samplers that extract a 30- to 60-cm incremental sample have been designed with long shafts so that a sample can be extracted from as much as 8 m below the surface. These may be used from a boat or raft that is floated into position and anchored.

While much useful information can be collected in the field, and can lead at that point to a firm recommendation by the geologist to *not* consider the deposit further, the ultimate judgment of quality, however, is formed from the results of usually extensive testing in the laboratory.

For the interested reader, the Diatomite chapter in the 5th edition of *Industrial Minerals and Rocks* contains a comprehensive treatment of this commodity (Kadey, 1983).

BIBLIOGRAPHY AND REFERENCES

Arehart, J.L., 1972, "Diatoms and Silicon," *Sea Frontiers,* Vol. 18, No.2, Mar./Apr., pp. 90-94.

Berger, L.R., 1971, "Effects of Hydrostatic Pressure on Photosynthesis and Growth of Unicellular Marine Algae and Diatoms," National Technical Information Service, AD-720, Vol. 401, 11 pp.

Calvert, S.E., 1968, "Silica Balance in the Ocean and Diagenesis," *Nature,* Vol. 219, No. 5157, pp. 919-920.

Carter, W.D., 1971, "ERTSA—A New Apogee for Mineral Finding," *Mining Engineering,* May, pp. 51-53.

Chin, T.G., et al., 1965, "Influence of Temperature and Salinity on Growth of Three Species of Plankton Diatoms," *Ocenol. Limnol. Sinica.,* Vol. 7, No. 4, pp. 384-394.

Clark, W.B., 1978, "Diatomite Industry in California," *California Geologist,* R. Van Blaricom, ed., Northwest Mining Association, Spokane, WA.

Crice, D.B., 1980, "Seismic," *Practical Geophysics for the Exploration Geologist,* R. Van Blaricom, ed., Northwest Mining Assn., Spokane, WA.

Cummins, A.B., 1960, "Diatomite," *Industrial Minerals and Rocks,* 3rd ed., J.L. Gilson, ed., AIME, New York, pp. 303-314.

Durham, D.L., 1973, "Diatomite," in "United States Mineral Resources," Professional Paper 820, US Geological Survey, pp. 191-195.

Dunn, J.H., 1968, "Utilization of Silica by Diatoms," *Dissertation Abstracts B.,* Vol. 29, No. 4, p. 1447.

Felfoldy, L.J.M., 1961, "Effect of Temperature on the Photosynthesis of a Natural Diatom Population," *Am. Institute of Biology,* Vol. 28, pp. 95-98.

Gross, M., et al., 1963, "Varved Marine Sediments in a Stagnant Fjord," *Science,* Vol. 141, No. 3584, Sep. 6, pp. 918-991.

Hurley, J.P., et al., 1985, "Ground Water as a Silica Source for Diatom Production in a Precipitation-Dominated Lake," *Science,* Vol. 227, pp. 1576-1578.

Jorgensen, E.G., 1960, "The Effects of Salinity, Temperature, and Light Intensity on Growth and Chlorophyll Formation of Nitzschia orvalis," *Year Book,* Carnegie Institution, Washington, Vol. 59, pp. 348-349.

Kadey, F.L., 1983, "Diatomite" *Industrial Minerals and Rocks,* 5th ed., S.J. Lefond, ed., AIME, New York, pp. 677-708.

Kilham, P., 1971, "A Hypothesis Concerning Silica and the Fresh Water Planktonic Diatoms," *Limnol. Oceander (Livca),* Vol. 16, No. 1, pp. 10-18.

Khechtel, M.M., and Hosterman, J.W., 1965, "Outlook for Resumption of Diatomite Mining in Southern Maryland and Eastern Virginia," Professional Paper 525D, US Geological Survey, pp. 134-138.

Krech, W.W., 1973, "Fragmentation," *SME Mining Engineering Handbook,* A.B. Cummins and I.A. Given, eds., AIME, New York, pp. 11-1-11-123.

Lewin, J.C., 1959, "Silicon as an Essential Element for Diatom Cultures," *Recent Advances in Botany,* 9th International Botanical Congress, Vol. 1, pp. 253-254.

Lewin, J.C., 1966, "Boron as a Growth Requirement for Diatoms," *Journal of Phycologia,* Vol. 2, No. 4, pp. 160-163.

Lohmann, K.E., 1960, "The Ubiquitous Diatom—A Brief Survey of the Present State of Knowledge," *American Journal of Science,* No. 258-A, pp. 180-191.

Neals, T.F., 1967, "The Boron Nutrition of the Diatom, Cylindrothece fusiformis, Grown on Agar, and the Biological Activity of Some Substituted Phenylboronic Acids," *Australian Journal of Biological Science,* Vol. 20, No. 1, pp. 67-76.

Patrick, R., 1971, "The Effects of Increasing Light and Temperature on the Structure of Diatom Communities," *Limnol. Oceander,* Vol. 16, No. 2, pp. 405-421.

Pedersen, G.K., 1983, "The Fur Formation, a Late Paleocene Ash-bearing Diatomite from Northern Denmark," Vol. 32, pp. 43-65.

Rowell, H.C., 1980, "Diatom Biostratigraphy of the Monterey Formation, Palos Verdes Hills, California," MS Thesis, University of Southern California, 123 pp.

7. Feldspar*

Feldspars, the most abundant minerals of the igneous rocks, occur in numerous forms and mixtures. The feldspars of commercial significance are found in widely distributed pegmatites as large crystals usually free of iron-bearing impurities and thus suitable for hand-cobbing, and also in larger bodies where the ore contains various types of feldspar intermingled with quartz and relatively free of iron-bearing impurities, or at least readily unlocked from impurities present at relatively coarse mesh. The latter type of ore includes some pegmatites, and the "alaskite" bodies so common in the Spruce Pine district of North Carolina. On the west coast of the US and in Oklahoma there is—or has been—commercial recovery of natural feldspathic sand as a source of alumina in glass and ceramics. "Aplite" in commercial terminology is a light-colored igneous rock with a granitic composition and a fine sugary texture, often banded with readily removable iron impurities. Only one area in Virginia produces aplite for the glass industry; it is located near Montpelier in Hanover County.

According to Deer et al. (1966) the word feldspar derives from the Swedish *feldt* or *fält* (field) and "spath:" chunks of rock appearing in tilled ground overlying granite. Castle and Gillson (1960) indicate a Germanic origin, citing "spat," which is said to refer to any transparent or translucent material which is readily cleavable. The term "spar" has in the past been applied to minerals other than feldspar, such as barite, calcite, and fluorite. It is correct to call the latter "fluorspar." There is a true barium feldspar, but it is rare and of no present economic importance. What is recognized as feldspar or "spar" by present producers and consumers consists of three silicate minerals which, if pure, would have the formulas $KAlSi_3O_8$ (microcline or orthoclase), $NaAlSi_3O_8$ (albite), and $CaAl_2Si_2O_8$ (anorthite). These are almost never found in pure form in nature, but occur together in great abundance in a three-component system. A considerable number of different combined ratios of these three exists, with the exception that isomorphism between potassium spar and calcium spar is very limited. User specifications for feldspar products are largely built around the prevalent naturally occurring ratios. Passages following will refer to "K-spar," "Na-spar," or "Ca-spar."

NOMENCLATURE AND DESCRIPTIVE TERMS

Relating to both the mineralogy and the economics of feldspar are certain terms whose coverage it is useful to specify.

Commercially, *aplite* is a feldspathic rock mined and beneficiated in one location in Virginia, in which both titanium and feldspar minerals are present—the latter now being the major, if not the only, economic product. Technically, the term aplite when applied to this rock is of questionable accuracy. But, in light of accepted commercial usage, the term will be here used to designate it. Regarding the Virginia rock, Ross (1941) is cited.

Alaskite is another term requiring understanding if not precise use. A distinctive rock type customarily called by this name is found near Spruce Pine, Mitchell County, North

Carolina. An important major feldspar source, it contains somewhat higher levels of plagioclase feldspar and quartz than alaskite as defined in the *Glossary of Geology and Related Sciences,* which stated that "orthoclase, microcline, and subordinate quartz are the principal minerals" and "plagioclase may or may not be present." Brown (1962) notes a granitoid rock near Peakville in Bedford County, Virginia (a past commercial feldspar source) as being alaskite, but he also classifies the Spruce Pine rock as such. There is some professional disagreement as to whether the Spruce Pine rock is a granite or a pegmatite. In any event, the term "alaskite" will, in this chapter, pertain to the relatively coarse-grained granitelike Spruce Pine ore which is well known to the feldspar industry.

Graphic granite (also called corduroy spar) can be briefly described as a pegmatite rock predominating in K-spar, with the secondary mineral, quartz, forming a distinctive pattern in it. It is a source of feldspar where high K_2O assay is required.

Pegmatite is a prevalent, widely distributed, generally coarse-grained igneous rock from which are obtained predominantly potash feldspar and a certain number of other economic minerals. They occur frequently in association with granites, and usually have major minerals in common with these, but frequently a higher percentage of some (such as K-spar) in large crystals. And, in some instances, there are also additional, often more exotic, minerals present.

Perthite is a microscopic intergrowth of plagioclase in K-spar, having a high K-spar to Na-spar ratio. In graphic granite and in pegmatites, perthite is of common occurrence.

Commercially, *soda spar* is a mixture assaying 7% Na_2O or higher while *potash spar* contains 10% K_2O or higher. The term soda spar has recently changed. Formerly it denoted a higher Na_2O assay than at present. *Pottery spar* is a feldspar product which goes generally into a shaped, fired body where the spar exerts a fluxing action. Such feldspar is usually ground to −200 mesh or finer. The presence of a given level of K-spar is important here, although potash spar is not required in all ceramic applications. *Glass spar* is—with very few exceptions—a soda spar ground to −20 mesh or sometimes as fine as −40 mesh. It is expected to assay a given level of Al_2O_3—the most important chemical value which the spar furnishes to the melt.

Feldspathic sand is a naturally occurring or processed mixture of feldspar and quartz.

MINERALOGY AND GEOLOGY

Feldspar, or feldspar-plus-quartz, is now being obtained from a limited number of rock types. Most of these have been referred to previously.

The following, as present feldspar sources, appear to merit attention from a mineralogical standpoint:

1) pegmatite, including graphic granite;
2) "alaskite" as recognized commercially;
3) "aplite" in the same sense;
4) granite;
5) feldspathic sand;
6) feldspathic quartzite.

Pegmatite is discussed only briefly here, but several good references are cited in the bibliography.

Generally speaking, feldspathic pegmatite bodies have the following notable characteristics:

*Excerpt from Rogers, C.P., Jr., Neal, J.P., and Teague, K.H., 1983, "Feldspars," *Industrial Minerals and Rocks,* 5th ed., Vol. 1, S.J. Lefond, ed., AIME, New York, pp. 709-722.

1) derivation from residual magma solutions;

2) wide variety of accessory minerals with frequent evidence of magmatic alteration and replacement;

3) considerable variability in shape, size, and extent of body;

4) tendency toward a concentric sequence of mineral zones, with varying distributions and ratios of minerals; and

5) major minerals in common with associated granites.

Point 4 is commonly seen as the result of successive inward differential crystallization. This phenomenon follows a certain sequence of mineral assemblages (Cameron, et al., 1949).

With these characteristics, involving so many thermal and chemical variables, pegmatites contain feldspars and numerous other minerals in many combinations and a variety of crystal sizes. A list is cited of minor elements which can be found in various pegmatites (Castle and Gillson, 1960):

Sb	Sr	Cb	Zn	Sc	Sn
As	Be	Ta	Li	S	Ti
Bi	Cs	Cu	Mo	Th	Zr
Ba	Ru	Pb	W	U	Rare Earths

In addition, low concentrations of Cr, Co, Hg, Se, Te, V, and precious metals are occasionally found.

The following accessory minerals may be found in pegmatites in addition to the more usual ones (K-spar, Na-spar, quartz, and muscovite) (Williams, et al., 1954):

Lepidolite	Zircon
Spodumene	Tourmaline
Almandine	Topaz
Spessartite	Epidote

Allanite	Amblygonite	Columbite
Beryl	Fluorite	Tantalite
Apatite	Wolframite	Lithiophillite
Eudialite	Cassiterite	

To these could be added pollucite, petalite, and others.

Distribution of pegmatites in the United States, and their respective sizes and mineralogies, can be further studied by consulting Cameron, et al. (1949). The principal pegmatite regions of the US can be summarized as follows:

1) New England: southwestern Maine, New Hampshire, Massachusetts, and Connecticut.

2) Southeastern states from Virginia through the Carolinas to Georgia and Alabama.

3) Black Hills of South Dakota.

4) Rocky Mountain states, especially Colorado and New Mexico.

5) Small pegmatite districts in California, Arizona, Idaho, Texas, and Wyoming.

There are scattered pegmatites elsewhere, such as in southeastern Pennsylvania, southeastern New York, northern New Jersey, central Wisconsin, and central Washington.

Where pegmatite consists of large mineral crystals of K-spar, quartz, and not much else, the rock is—or has been—mined and handcobbed for potash feldspar (*block spar*). In other pegmatites, mineral substitutions and alterations have taken place, there are various minor and accessory minerals, and the rock is usually finer grained. Thus it requires beneficiation by relatively fine grinding (−20 mesh), and then by magnetic separation and/or froth flotation, to secure desired concentrates. In addition to feldspar, quartz, muscovite mica, spodumene, pollucite, and other economic minerals are obtained from the finer grained pegmatites. For commercial mining, the best zone or zones of any given pegmatite must be selected to yield the desired minerals in optimum avail-

ability and quality. Pegmatite containing spodumene occurs in contiguous bodies near Kings Mountain, North Carolina, and is an important source of feldspar and feldspathic sand—these being actually secondary to the spodumene content in economic importance. This pegmatite is the characteristic rock of the so-called tin-spodumene belt of the Carolinas. The belt can be briefly described as "a narrow, sinuous zone which strikes northeastward, paralleling in general the layering and foliation of the principal rock units of the area" (Broadhurst, 1956).

In considering theories regarding pegmatite formation, a concept involving mineral deposition from an aqueous or pneumatolytic environment may be of interest. This theory is intermittently referred to by Park, et al. (1964), wherein additional references are given, and by Turner, et al. (1960), dealing with related physical laws.

Graphic granite is cited by Cameron, et al. (1949) as one pegmatite manifestation. Williams, et al. (1954) describe graphic granite by the phrase "a cuneiform intergrowth of quartz and potash feldspar." This type of intergrowth is supposed to have come about through simultaneous or rapid crystallization from a viscous melt, along with a certain amount of replacement of massive K-spar by quartz. Like coarse-grained pegmatite, graphic granite will yield high-grade potash feldspar if beneficiated by grinding and flotation. If dry ground as is, the resulting feldspar product, being less pure, would be lower grade than fine-ground block spar.

The alaskite commercially mined in the area of Spruce Pine, NC, is defined by Parker (1952) as a Paleozoic igneous intrusive and also a stocklike pegmatite mass of Carboniferous age. Parker divides the intrusives of this area (those covered by his designation "pegmatite") into (1) some unevenly distributed aplites in the true mineralogical sense, (2) pegmatites of coarse-grained crystalline texture in locations mostly peripheral to the principal finer grained rock, and (3) the principal type, which is characterized by various writers as alaskite or granite. As elsewhere, the coarse-grained pegmatites have yielded commercial hand-mined block spar.

The alaskite is composed, roughly, of the following principal mineral ingredients.

Plagioclase:	45%
Quartz:	25%
Microcline:	20%
Muscovite:	10%

These proportions may vary; there may be less microcline or muscovite and sometimes none. Plagioclase and quartz are nearly always present. The commercially mined rock of the principal mass contains plagioclase-quartz-perthite-muscovite, usually in that order of abundance. In addition there are present minor amounts of garnet, biotite, and apatite plus very small quantities of beryl, tourmaline, epidote, and others (Parker, 1952).

Except for varying textures, the mostly peripheral coarse-grained bodies easily classified as pegmatites differ little in their collective mineralogy from that of the overall central body, leading to the conclusion of derivation from a common magma. Moreover, the prevailing mineralogical ratios of the central body are quite atypical of granite. Grain size of the alaskite is, on the average, between 0.25 and 0.5 in.: considerably coarser than the average granite or granodiorite (Parker, 1952).

The Spruce Pine alaskite has the characteristics of exceptional mineral purity, uniformity, coarse-grained texture, and large size of deposit, which are sought in commercial ore of feldspar.

The feldspathic aplite of Virginia, previously mentioned, is described by Castle and Gillson (1960) as essentially an intrusive mass having variable texture. It has presumably undergone magmatic alteration from coarse feldspar and is a host rock for titanium minerals not found in true aplite. According to Williams et al. (1954), aplite in the mineralogical sense developed from residual magmas and occurs, associated with pegmatite, as narrow, relatively homogeneous intrusives with such accessory minerals such as garnet, zircon, and tourmaline. These factors, plus differing content of K-, Na-, and Ca-feldspars, indicate that the Virginia rock should not be classified as true aplite. (For an additional reference on this rock, see Ross, 1941).

Granites are extremely abundant, a certain number of them being good sources of high-quality feldspar. Granite is coarse-grained acid plutonic igneous rock. Quartz generally makes up 25 to 30% of the rock; feldspar is the other major component. There is a wide range of minor constituents, among them muscovite, biotite, apatite, zircon, allanite, hornblende, magnetite, epidote, zoisite, and garnet. The feldspar component of granite is principally of the alkali members, which is to say crystalline feldspar containing some ratio of the potassium and sodium feldspars. The remaining feldspar is of the plagioclase series, i.e., a varying ratio of sodium and calcium feldspar is solid solution.

As long as such ores as the pegmatites and the Spruce Pine alaskite are readily available, granite nearby is unlikely to be mined for feldspar. Being generally finer grained than pegmatites, granite must, of course, be also beneficiated by froth flotation. Where granite has been beneficiated for feldspar in the United States, the operation has been in conjunction with quarrying for production of road or dimension stone: fines from such an operation can offer an economic advantage in terms of reduced mining and grinding costs. The granite must be sufficiently coarse-grained so that there is substantial mineral liberation at 40 mesh or coarser. Inclusions of minor, iron-bearing minerals must be at a minimum. Near Pacolet, Spartanburg County, South Carolina, one such granite quarrying operation has in the past furnished tailings which were beneficiated into feldspathic sand. This granite happens to be mineralogically close to the optimum for feldspar beneficiation. The operations are discussed by Eddy, et al. (1972).

Various beach and river sands containing economic amounts of feldspar have apparently acquired the feldspar principally from granitic bodies, and possibly also from feldspathic metamorphic rocks. Basis for this statement is in the assay of feldspar concentrates from these sands. Normal decomposition of feldspar involves first the breakdown and kaolinization of Na-spar, leaving behind a component higher in K-spar. This, if an assay is run on spar concentrate from feldspathic sand and the K_2O is high (perhaps 10% or more), that would indicate either extreme decomposition or that the feldspar originated in a pegmatite. Most concentrates from feldspathic sand, however, usually do not assay over 7% K_2O: considerably below usual pegmatite feldspars, and much more typical of feldspar from partially weathered granite, with part or all of its plagioclase component weathered away. The feldspar in these sands, then, probably is mainly from the original alkaline feldspar of granite.

Since each of the rock types discussed has its own distinctive mineralogy, reflected by the types and ratios of feldspars present, it follows that varying commercial chemical standards for assorted feldspar uses will be best met by different mineralogical combinations found in aplite, alaskite, granite, feldspathic sand, or pegmatite. (These are cited here in more or less rising order of their K-spar to Na-spar ratio.)

The presence or absence of Ca-spar is usually of slight importance to the feldspar user. Since pure Ca-spar (anorthite) contains a theoretical 36.7% Al_2O_3, it is welcome in most glass applications. In certain ceramic uses it may need to be limited.

The geology and mineralogy of feldspar are thoroughly discussed by Barth (1969) and the physical and chemical characteristics by Berry, et al. (1959) and Deer, et al. (1966).

DISTRIBUTION AND RESERVES

Feldspar is one of the most plentiful minerals in the earth's crust, occurring in various rock types within a short distance of almost anywhere. Thus, any absolute figure on reserve tonnage is academic. Any estimate of tonnage reserves would have to be relative to the economics of the mineral at the time. The feldspar market in the United States has been highly competitive for many years, and by now the surviving producers have located themselves on ore bodies of the best available quality, i.e., yielding a high percentage of feldspar easily concentrated as a quality product. Mineral purity, size of ore body, and coarse-grained liberation have been paramount considerations. Existing feldspar producers generally have available reserves which are calculated in quarter centuries, if not centuries. There may arise, of course, complications due to zoning or environmental limitations— or perhaps to shipping difficulties. Should such limitations preclude exploitation of many prime quality feldspar reserves, then the vast and widespread granite formations and the unworked alluvial sands may come into the picture. Cost and quality could well be variables of concern, but sheer physical reserves of marketable feldspar are on hand in copious quantity.

One rock type not previously discussed has recently emerged as an interesting potential reserve of potash spar. Located in San Bernardino County, California, this potential ore body is described by Sheppard and Gude (1965) as an altered tuff whose principal constituent is K-spar. It is part of the Barstow Formation and covers a wide area. Its chief disadvantage appears to be a high iron content in much of the feldspar.

FELDSPAR PRODUCTION

Producing Areas, US

In the United States, the state of North Carolina now holds a substantial lead in feldspar production. The state of Connecticut now appears to be second in feldspar production, followed by Georgia and California. Minor quantities of feldspar are shipped from Wyoming.

The Spruce Pine alaskite, measurable in terms of perhaps several square miles in area, irregularly shaped, with associated coarse pegmatites, is the principal feldspar ore of North Carolina. Around Bryson City in Swain County are pegmatites which have in the past yielded block feldspar. In recent years, North Carolina pegmatites in the area of Bessemer City (Gaston County), Lincolnton (Lincoln County), and Kings Mountain (Cleveland County) have been processed by flotation to yield feldspar concentrate. At these places, feldspar is a byproduct of spodumene production. One North Carolina company produces potash feldspar by flotation from a decomposed graphic granite.

South Dakota and Wyoming produce block spar from

pegmatites. This type was formerly produced in Connecticut, but in recent years a shift has been made to froth flotation, working finer grained pegmatites. A deeply weathered pegmatite in Georgia is processed by flotation in order to concentrate potash spar. The Georgia producer is developing a feldspar source in a weathered potash granite in Greene County to supplement present mill feed.

Feldspar Production Outside the US

A partial list of foreign producers would include the flotation plant of Bjorum Sibelco Norfloat A/S and Co. This plant has a capacity of approximately 65,000 tpy, and serves much of Europe and the British Isles. In West Germany, Franz Mandt and Sons operates three feldspar plants, using both imported hand-cobbed feldspar and local feldspathic pegmatites as sources of feed for fine grinding. West Germany also obtains considerable byproduct feldspar from local kaolin operations. In France, Denain-Anzin operates several dry grinding plants, using hand-cobbed spar from the Perpignan area. Maffei and Co. of Italy is one of the largest producers of feldspar in Europe, shipping over 200,000 tpy.

In Mexico, Materias Primas, with headquarters in Monterrey, operates a feldspar plant near Queretaro, using a feldspathic sand as feed; this raw material has been transported to the site by wind.

Many countries in South America produce limited amounts of feldspathic products, the same being true of Africa and Asia.

With the exception of "iron curtain" countries in Europe, it is believed all production not specifically described is by hand selection, rather than by more sophisticated beneficiation techniques.

Aplite Production

The one remaining aplite plant in the United States is located at Montpelier, VA, and is operated by The Feldspar Corp. It produces a low-iron aplite suitable for flint glass and some other glass applications and employs wet methods.

The other former producer at Piney River, VA produced aplite by dry methods suitable for amber and green glass and window glass. The Piney River operation was closed July 1, 1980.

EXPLORATION, MINING, AND PROCESSING

Exploration for feldspar and aplite deposits is fairly simple. A geologist with basic geologic experience in pegmatites, granites, and clays explores potential areas selected by a knowledgeable producer, seeking outcrops and small rock chips appearing as "float." Bulldozers and ditchers may be used to strip a relatively shallow overburden. Diamond drilling has been used to determine depth of a deposit, but is often costly with relation to the information obtained.

In those deposits appearing sufficiently promising, representative samples are collected and refined in a pilot plant—this often being preceded by bench work. The product is then evaluated by chemical assay and possibly by physical testing: button firing, sample batch runs, etc.

The feldspar industry has learned that, within rather broad limits, feldspathic minerals are suitable for glass and ceramics, provided iron-bearing impurities and the surrounding country rock can be separated from the finished product. Generally the deciding factor is whether, at approximately 20 mesh, or at most 30 mesh, the pure feldspathic mineral can be unlocked from its associated impurities. If it can, the product derived from flotation or magnetic treatment is evaluated as to suitability for available markets.

BIBLIOGRAPHY AND REFERENCES

Barth, T.F.W., 1969, *Feldspars,* John Wiley and Sons, New York, 261 pp.

Berry, L.G., and Mason, B., 1959, *Mineralogy—Concepts, Descriptions, Determinations,* Wm. H. Freeman, San Francisco, 612 pp.

Broadhurst, S.D., 1956, "Lithium Resources of North Carolina," *Information Circular 15,* North Carolina Div. of Mineral Resources, 37 pp.

Brown, W.R., 1962, "Mica and Feldspar Deposits of Virginia," *Mineral Resources Report 3,* Virginia Div. of Mineral Resources, 195 pp.

Cameron, E.N., et al., 1949, "Internal Structure of Granitic Pegmatites," *Economic Geology,* Monograph 2, 115 pp.

Castle, J.E., and Gillson, J.L., 1960, "Feldspar, Nepheline Syenite, and Aplite," *Industrial Minerals and Rocks,* 3rd ed., J. Gillson, ed., AIME, New York, pp. 339-362.

Clark, W.B., 1977, "Feldspar Deposit in the Ord Mountains, San Bernardino County, CA," *California Geology,* Vol. 29, pp. 81-85.

Deer, W.A., et al., 1966, *An Introduction to the Rock-Forming Minerals,* John Wiley and Sons, New York, 528 pp.

Eddy, W.H., et al., 1972, "Recovery of Glass Sand from South Carolina Waste Granite Fines," Report of Investigation 7651, US Bureau of Mines, 11 pp.

Kesler, T.L., 1961, "Exploration of the Kings Mountain Pegmatites," *Mining Engineering,* Vol. 13, pp. 1062-1068.

Park, C.E. Jr., and MacDiarmid, R.A., 1964, *Ore Deposits,* Wm. H. Freeman, San Francisco, 475 pp.

Parker, J.M., III, 1952, "Geology and Structure of Part of the Spruce Pine District, North Carolina," Bulletin 65, North Carolina Div. of Mineral Resources, 26 pp.

Potter, M.J., 1975-1980, "Feldspar," *Minerals Yearbook,* US Bureau of Mines.

Ross, C.S., 1941, "Titanium Deposits of Amherst and Nelson Counties, Virginia," Professional Paper 198, US Geological Survey, 56 pp.

Sheppard, R.A., and Gude, A.J., III, 1965, "Potash Feldspar of Possible Economic Value in the Barstow Formation, San Bernardino, California," Circular 500, US Geological Survey, 7 pp.

Turner, F.J., and Verhoogen, J., 1960, *Igneous and Metamorphic Petrology,* McGraw-Hill, New York, 602 pp.

Wells, J.R., 1971-74, "Feldspar," *Minerals Yearbook,* US Bureau of Mines.

Williams, H., Turner, F.J., and Gilbert, C.M., 1954, *Petrography—An Introduction to the Study of Rocks in Thin Sections,* Wm. H. Freeman, San Francisco, 406 pp.

8. Fluorspar*

Fluorspar is the commercial name for fluorite, a mineral that is calcium fluoride, CaF_2. The name, derived from the Latin word *fluere* (to flow), refers to its low melting point and its early use in metallurgy as a flux. It is the principal industrial source of the element fluorine.

Cryolite, sodium aluminum fluoride, Na_3AlF_6, is a rare mineral which has been found in commercial quantities only in Greenland. The natural material has been supplanted by synthetic cryolite for its principal industrial use in the manufacture of aluminum.

HISTORY

Fluorspar was used by the early Greeks and Romans for ornamental purposes as vases, drinking cups, and table tops. Various peoples, including the Chinese and the American Indians, carved ornaments and figurines from large crystals. Its usefulness as a flux was known to Agricola in 16th century Europe.

Fluorspar mining began in England about 1775 and at various places in the United States between 1820 and 1840. Production grew substantially following the development of basic open-hearth steelmaking, wherein it is used as a flux. Use was stimulated by growth of the steel, aluminum, chemical, and ceramic industries, particularly during World Wars I and II. Fluorocarbons entered the picture in 1931. The use of anhydrous hydrogen fluoride (HF) as a catalyst in the manufacture of alkylate for high octane fuel began in 1942.

Differential flotation for separating fluorspar from galena, sphalerite, and common gangue minerals in the 1930s, and the application of heavy-media concentrating methods to the treatment of low-grade ores in the 1940s were outstanding technological advances that facilitated increased production.

Recently, pelletizing and briquetting of flotation concentrates for use in steel furnaces and the development of flotation schemes for beneficiating ores containing abundant dolomite and barite have been major improvements in the industry.

USES OF FLUORSPAR

Fluorspar is used to make hydrogen fluoride, also called hydrofluoric acid, an intermediate for fluorocarbons, aluminum fluoride, and synthetic cryolite. It is used as a flux in the steel and ceramic industries, in iron foundry and ferroalloy practice, and has many minor specialized uses.

Hydrogen fluoride (HF) is produced by reacting acid grade (97% CaF_2) fluorspar with sulfuric acid in a heated kiln or retort to produce HF gas and calcium sulfate. After purification by scrubbing, condensing, and distillation, the HF is marketed as anhydrous HF, a colorless fuming liquid, or it may be absorbed in water to form the aqueous acid, usually 70% HF.

Synthetic cryolite, organic and inorganic fluoride chemicals, and elemental fluorine are made from hydrofluoric acid. The acid itself is important in catalysis in the manufacture of alkylate, which is an ingredient in high-octane fuel for aircraft and automobiles, in steel pickling, enamel stripping, glass etching and polishing, and in various electroplating

operations. The manufacture of one ton of virgin aluminum requires about 23 to 27 kg (50 to 60 lb) of fluorine content in synthetic cryolite and aluminum fluoride. This quantity, through improved technology and recovery practices, is being lowered significantly.

Elemental fluorine is prepared from anhydrous hydrofluoric acid by electrolysis. Gaseous at room temperature and pressure, fluorine is compressed to a liquid for shipment in cylinders or in tank trucks. Elemental fluorine is used to make uranium hexafluoride, sulfur hexafluoride, and halogen fluorides. Gaseous uranium hexafluoride is used in separating U^{235} from U^{238} by the diffusion process. Sulfur hexafluoride is a stable high dielectric gas used in coaxial cables, transformers and radar wave guides. Halogen fluorides have important applications, mostly as substitutes for elemental fluorine which is more difficult to handle.

Emulsified perfluorochemicals, organic compounds in which all hydrogen atoms have been replaced by fluorine, are undergoing investigation as promising blood substitutes. They transport oxygen and, in conjunction with a simulated blood serum, perform many functions of whole blood. With further development, ultimately they may be useful in saving lives of animals and humans in emergencies during periods of acute shortages of natural blood.

Inorganic fluorides are used as insecticides, preservatives, antiseptics, ceramic additives, in electroplating solutions, as fluxes, antioxidants, and many other ways. Boron trifluoride is an important catalyst.

Organic fluorides are volume leaders in the fluorine chemical industry. Fluorinated chlorocarbons and fluorocarbons are prepared by the interaction of anhydrous HF with chloroform, perchlorethylene and carbon tetrachloride, and are characterized by low toxicity and notable chemical stability. They perform outstandingly as refrigerants, aerosol propellants, solvents, and cleaning agents and as intermediates for polymers such as fluorocarbon resins and elastomers.

Fluorocarbon resins are inert compounds with an unusually low coefficient of friction which have found a number of applications for parts that cannot be oiled, such as the bearings inside of automobile doors, for window raising equipment, in small electronic equipment, and for the manufacture of chemical resistant gaskets and valve parts, pipe and tank linings, flexible tubing and containers, and cookware.

In the steel industry, fluorspar is used as a flux in basic open-hearth, basic oxygen, and electric furnaces where it is added to the heats in amounts ranging from 0.9 to 9 kg (2 to 20 lb) per ton of steel produced. The average in the US is currently about 2.7 kg (6 lb) per ton. It promotes fluidity of the slag and thus facilitates removal of sulfur and phosphorus from the steel into the slag. It serves the same purpose in iron foundries, where it is added to the cupola charge in the proportion of around 6.8 to 9 kg (15 to 20 lb) per ton of metal melted.

In the ceramic industries, fluorspar is used to make flint glass, white or colored opal glasses, and enamels. Flint glass mixtures commonly contain 3% fluorspar. Opal glasses, containing 10 to 20% fluorspar, are used in containers for foods, drugs and toiletries, and in ornamental glassware and lavatory and restaurant fixtures. Opaque enamels are used to cover steel stoves, refrigerators, cabinets, bathtubs, and cook-

*Excerpt from Fulton, R.B., and Montgomery, G., 1983, "Fluorite and Cryolite," *Industrial Minerals and Rocks,* 5th ed., Vol. 2, S.J. Lefond, ed., AIME, New York, pp. 723-744.

ing ware and for facings on brick and tile, and other structural materials. Fluorspar makes up to 3 to 10% of the weight of the enamel. Many types of welding rod coatings incorporate fluorspar or fluorspar mixtures. Ceramic grade fluorspar is used in the manufacture of magnesium and calcium metals and in the preparation of some manganese chemicals.

Lower grades of ceramic spar are used in making fiberglass insulation, in zinc smelting, as a clinkering aid in the manufacture of portland cement, as an inhibitor of vanadium green scumming in the manufacture of buff-faced brick, and as an abrasive on certain types of sandpapers. Various grades of fluorspar are used in electric furnace manufacture of calcium carbide, in making electrodes for arc lamps, and as a bonding material for abrasive wheels, among numerous minor uses.

GEOLOGY

Composition and Properties

Theoretically pure fluorite contains 51.1% calcium and 48.9% fluorine. Substitution of small percentages of cerium and yttrium for calcium has been noted. Inclusions of gases and fluids, such as petroleum and water, and of solid minerals such as pyrite, marcasite, and other sulfides are common. Free fluorine is present in some crystals. The fluorspar of commerce contains attached and admixed mineral impurities, such as calcite, quartz, barite, celestite, various sulfides, or phosphates.

Fluorite tends to occur in well-formed isometric crystals, forming cubes and octahedrons. It also occurs in massive and in earthy forms, and as crusts or globular aggregates with radial fibrous texture. Crystalline fluorspar exhibits a great range of colors, from colorless and waterclear to yellow, blue, purple, green, rose, red, bluish and purplish black, and brown. The colors may occur in alternating bands parallel to cube faces. They may be altered by exposure to X-rays, heat, ultraviolet light, and pressure. Colors are caused by a variety of factors including the presence of trace impurities and displaced ions in the lattice. Long exposure to sunlight, such as occurs on mine dumps, frequently results in the fading of the original coloration.

Some varieties fluoresce blue or violet under ultraviolet light or cathode rays. Some specimens phosphoresce when heated or after exposure to sunlight or ultraviolet light, and some exhibit triboluminescence.

The mineral has a hardness of 4, and is the type mineral of that hardness on the Mohs' scale. Its specific gravity is 3.18 when crystalline, but ranges from 3.01 to 3.6 in varous forms. The luster is vitreous. The mineral has perfect octahedral cleavage, and "diamonds" made by knocking the corners off of cubic crystals are seen in many collections.

When pulverized and treated with sulfuric acid, fluorite decomposes into gaseous hydrogen fluoride and calcium sulfate, which is the fundamental reaction in the production of hydrofluoric acid. When added to metallurgical slags, it imparts greater fluidity at lower temperatures making it valuable in steelmaking and ferrous and nonferrous foundry practice. In ceramic mixes it promotes crystallization around individual centers, and thus is useful in making opal glasses. Crystalline fluorite has a very low index of refraction (n = 1.4339), low dispersion, is isotropic, and has an unusual ability to transmit ultraviolet light. These are the properties which make it useful in optical systems as prisms and as components of high quality, special-purpose lenses. Synthetic fluorite has replaced the natural mineral for optical uses.

Modes of Occurrence

Fluorite occurs in a wide variety of geological environments, evidencing deposition under an extended range of physical and chemical conditions. At one extreme it is present as an accessory mineral in granites and related igneous rocks; while at the other extreme, it is found as crystals in geodes and as botryoidal linings in limestone caves.

From an economic standpoint, the most important modes of occurrence of the mineral are:

1) Fissure veins in igneous, metamorphic, and sedimentary rocks.

2) Stratiform replacement deposits in carbonate rocks.

3) Replacements in carbonate rocks along contacts with acid igneous intrusives.

4) Stockworks and fillings in shattered zones.

5) Deposits at the margins of carbonatite and alkalic rock complexes.

6) Residual concentrations resulting from the weathering of primary deposits.

7) Occurrences as recoverable gangue in base metal deposits.

Less common modes, in some places commercially important, include:

8) Fillings in breccia pipes.

9) Fillings in open spaces.

10) Pegmatites.

Least common but of potential importance are:

11) Deposits in lake sediments.

Fissure Veins: Fissure veins, usually along faults or shear zones, are the most readily recognized form in which fluorspar deposits occur the world over. Silica, calcite or other carbonates, iron, lead, and zinc sulfides, and barite are the typical associated minerals. In some vein deposits as in the Rosiclare district of southern Illinois, fluorite appears to have replaced a prior vein filling of calcite. Along some veins in carbonate rocks, fluorspar has replaced the wall rock at intersections with favorable beds, providing large minable tonnage mined from some deposits. Although vein structures are remarkably persistent, the fluorspar itself commonly occurs as lenses or ore shoots separated by barren or poorly mineralized portions of vein. Ore shoot widths of 0.6 to 9 m (2 to 30 ft) and lengths of 61 to 305 m (200 to 1000 ft) are common, but there is great variation from deposit to deposit. Vein systems may be up to several miles in length, and ore may be present to depths of 305 m (1000 ft) or more below the surface. The CaF_2 content of minable portions of veins normally ranges from 25 to 80%, although grades above 90% occur in limited areas.

Some of the world's great vein deposits include the Osor deposit in northeastern Spain, the Torgola deposit in northern Italy, the Muscadroxiu-Genna Tres Montis vein system in Sardinia, the Longstone Edge-Sallet Hole deposit in England, and the Rosiclare-Goodhope vein system in southern Illinois.

Stratiform Deposits: Stratiform, manto, or bedded deposits occur in carbonate rocks. Certain beds are replaced along or adjacent to structural breaks such as joints or faults. This relationship to structural features is very clear in some deposits, but obscure in others. Frequently there is a sandstone, shale, or clay capping. Typically there is evidence of loss of volume in the replaced zones with attendant development of gentle synclinical structures in overlying strata or of collapse structures, sometimes pipelike in shape. In some districts, as in southern Illinois, no connection is recognized between the mineralization and any igneous activity, whereas in others, such as the Encantada district in northern Coahuila, Mexico, the presence of rhyolite plugs and sills in the general vicinity of the spar deposits, and the association of spar with rhyolite injections along bedding planes, suggests an association. Stratiform deposits are known in many parts of the world and are particularly well developed in the Cave in Rock district of southern Illinois, in the northern part of the State of Coahuila in Mexico, and in the Ottoshoop (Zeerust) District of Transvaal in South Africa. In Illinois the deposits range from a few centimeters to 6 m (few inches to 20 ft) thick, 15 to 152 m (50 to 500 ft) wide, and are up to 7 km (4½ miles) long. In Coahuila the individual ore bodies are smaller but relatively more numerous and widespread. In the Ottoshoop District bedded deposits occur in an area 16 km (10 miles) long and 9.7 km (6 miles) wide in a dolmite facies underlying a prominent chert bed.

In stratiform deposits, textural features of the parent rock such as sedimentary banding are commonly preserved. Frequently associated with the banded ore is a massive crystalline type which appears to have filled open spaces left from the dissolving of limestone by the ore-bearing solutions or their precursors. Minerals accompanying fluorspar in stratiform deposits are calcite, dolomite, quartz, galena, sphalerite, pyrite, marcasite, barite, and celestite. CaF_2 content in minable deposits ranges from 15% upward. A few ore bodies in Illinois yield direct shipping metallurgical grade fluorspar containing 85% or more CaF_2.

Replacement Deposits: Replacement deposits in carbonate rocks along contacts with intrusive rhyolite bodies are well developed in the Rio Verde, San Luis Potosi, and Aguachile districts in Mexico. They include some of the largest and highest grade fluorspar deposits. The fluorspar is not thought to be contact metamorphic in origin, but introduced later, following the contact zone as a conduit replacing the limestone outward from the contact, either massively or selectively along certain beds. At Aguachile, cross sections show ore shoots resembling one side of a Christmas tree.

Stockworks: Fluorspar often occurs as stockworks and fillings in shear and breccia zones. Many occurrences in the American West are of the stockwork type and, though wide, usually have low overall CaF_2 content. Deposits in the Zuni Mountains of New Mexico and near Jamestown, CO, are examples. The Zwartkloof deposit in the Transvaal Province of South Africa consists of three vertical breccia zones in an east-west line in felsite containing stockworks of fluorite-carbonate veins. The largest zone is 61 by 183 m (200 by 600 ft) in plan and persists to 914 m (3000 ft) below the surface. The fluorspar grade is about 14%. The Buffalo deposit, near Naboomspruit in the Transvaal, consists of a network of fluorspar veinlets in sill-like bodies of fine-grained pink granite which are inclusions in coarse red granite of the Bushveld complex.

Carbonatite and Alkalic Rock Complexes: Fluorspar is a common mineral in carbonatite and alkalic rock complexes, rarely in sufficient abundance to be economic. The Okorusu deposit in South-West Africa (Namibia) of this type consists of a number of bodies of fluorspar in limestones, quartzites, and related rocks which have been intruded and metamorphosed by an alkaline igneous rock complex including a nepheline syenite stock. The fluorspar replaced bedded and brecciated limestone, marble, and quartzite, forming large lenticular masses. Apatite and quartz are abundant accessory minerals. At Amba Dongar, India, veins and replacement bodies of fluorspar occur in carbonate rocks bordering ankeritic carbonatite intrusives. Fluorite is also present in the carbonatite itself.

Residual Deposits: Concentrations of fluorspar in clayey and sandy residuum resulting from surficial weathering of fluorspar veins and replacement deposits in some places are sources principally of metallurgical spar. This category includes detrital deposits blanketing the apex of veins, as well as deeply weathered upper portions of the veins themselves extending to depths of 30.5 m (100 ft) or more. Such deposits were of major importance in Illinois and Kentucky. Similar deeply weathered ore has been mined in England, Thailand, and the Asturias district of northwestern Spain. Weathered residuum, called *kokoman,* is mined in the Marico district in South Africa, where it is the result of the weathering of gently dipping replacement bodies.

Gangue Mineral: Fluorspar occurs as a major gangue mineral in lead-zinc veins in many parts of the world, and in some, averaging 10 to 20%, is economically recoverable. Acid grade fluorspar is produced on a large scale from lead-zinc mill tailings near Parral in Mexico.

Breccia Pipes: Fluorite occurs in breccia pipes in the Thomas Range, Utah and near Beatty, NV. Pipes in the Thomas Range are circular to oval in plan, up to 46 m (150 ft) in diameter and over 61 m (200 ft) in depth, and are formed in dolomite by replacement along shattered zones associated with faults and intrusive breccias. Fluorspar occurs as soft friable masses and in boxworks of fine-grained more resistant veinlets, and is nearly unrecognizable. At the Daisy mine near Beatty, fluorspar replaces brecciated dolomite in pipelike bodies bounded by gouge zones along two sets of intersecting faults.

Fillings in Open Spaces: Fluorspar occasionally partially fills open spaces, both in veins and stratiform deposits. Spectacular examples of this type occur in the San Vicente district of northern Coahuila, where fluorspar occurs in veins and mantos as pure massive incrustations of mamillary, stalactitic, and stalagmitic forms. The Fluorspar-Gero-Penber vein system of the Northgate mine in Colorado is a similar occurrence where fluorspar occurs in botryoidal layers on the walls of open fissures and as concretionary coatings surrounding loose fragments of country rock. Lower parts of open areas of these fissures are partly filled with concretionary pebblelike masses, in places cemented into porous, rubbly aggregates.

Pegmatites: Many pegmatites contain minor amounts of fluorspar. Grade was high enough to support a mining operation at the Crystal Mountain occurrence in Montana, where three large tabular bodies of massive fluorspar containing minor amounts of biotite, quartz, feldspar, and other igneous-type accessory minerals, occur in coarse-grained biotite granite.

Lake Sediments: Fluorspar occurs in unconsolidated clayey and sandy pyroclastic sediments in the beds of former lakes near Castel Giuliano, about 40 km (25 miles) north of Rome, Italy. Fluorine of volcanic origin permeated the lake sediments resulting in deposition of minute disseminated crystals of fluorspar, which make up as much as 50 to 60% of the clayey parts and 15% of the sand parts of the deposits. It is accompanied by barite, apatite, calcite, dolomite, and opal.

DISTRIBUTION OF DEPOSITS

Fluorspar deposits occur on every continent. Major producers include the United States, Mexico, Spain, England, France, Italy, South Africa, China, Thailand, Mongolia, and the USSR. Lesser producers include Argentina, Brazil, Bulgaria, Czechoslovakia, Kenya, Tunisia, Morocco, West Germany, East Germany, Romania, India, and Korea.

North America

Canada: The only large production in Canada was from the St. Lawrence area of the Burin Peninsula in southern Newfoundland, which yielded over 2.7 Mt (3 million tons) of ore from 1940 until operation ceased in 1978. Other deposits occur in Ontario at Madoc and near Wilberforce and Cobden; at Rock Candy and Birch Island near Kamloops and near Quesnel Lake in British Columbia; along the Lower Laird River in northern British Columbia; and at Lake Ainslie on Cape Breton Island in Nova Scotia.

On the Burin Peninsula, fluorspar occurs in veins in granite and rhyolite porphyry along steeply dipping faults. Average thickness of higher grade veins is 0.9 to 1.5 m (3 to 5 ft) and of lower grade veins is 4.6 to 6 m (15 to 20 ft). Some veins have been traced for more than a mile.

Veins at Madoc are small and have been worked principally in wartime. The veins at Lake Ainslie are low grade fluorspar with associated abundant barite. A deposit at Birch Island remains undeveloped with reserves said to be 1.4 Mt (1.5 million tons) of 29% CaF_2 grade. Bedded deposits on the Lower Laird River near Mile 497 on the Alaska Highway in northern British Columbia have not proved to be of current commercial interest. Two low grade fluorite deposits near Likely, BC, were reported to be under exploration early in 1981.

Mexico: Mexico is the world's largest producer of fluorspar. Important deposits occur in San Luis Potosi, Coahuila, Chihuahua, and Guanajuato.

Many types of deposits are mined. In the Musquiz district of northern Coahuila, mantos in Cretaceous limestones are most important. Northwestern Coahuila has vein deposits and replacement deposits in limestone associated with rhyolite sills and plugs. Deposits in the Paila district of central Coahuila are veins. In San Luis Potosi and adjoining Guanajuato are several large replacement deposits in limestone associated with rhyolite intrusives. Numerous vein deposits occur in Chihuahua, but its principal production is acid grade concentrate recovered from sulfide mill tailings near Parral.

United States: *Alaska*—The Lost River deposit, located in the western portion of the Seward Peninsula 136.8 km (85) miles northwest of Nome, was worked intermittently for tin from 1904 to 1955 and has since been explored for fluorspar, beryllium, and tin. Fluorspar-tin-tungsten-beryllium mineralization occurs in acidic dikes, in skarn deposits along granite-limestone contacts, and brecciated limestone adjacent to thrust faults. Mineralization in one zone was reported to be 24.5 Mt (27 million tons) grading 16.3% CaF_2, 0.15% tin, and 0.03% tungsten. Fluorspar also occurs in six other zones in the district (Anon., 1972), but there has been no production.

Arizona—Although 56 occurrences of fluorspar were listed in one summary (Van Alstine, 1969) and 96 in another (Elevatorski, 1971), to date cumulative production has probably been less than 45 kt (50,000 tons). Deposits are widely scattered in the southern half of the state, principally as small epithermal veins in fissures and brecciated zones along faults in intrusive, metamorphic, and sedimentary rocks. No large replacement deposits have been found. Fluorite occurs mostly with calcite and quartz in non-metaliferous veins, although in the Castle Dome district it is a gangue mineral along with barite, quartz, and calcite in veins which were mined on a small scale for their lead-silver content.

The Castle Dome, Duncan, Sierrita Mountain, Tonto Basin, and Whetstone Mountain districts have been producers. The Lone Star mine in the Whetstone Mountain district yielded 9 kt (10,000 tons) of metallurgical spar in 1967, but then closed down.

California—Although there are many occurrences of fluorspar in California, only small amounts have been produced for ceramic and metallurgical purposes. Metspar was produced on a small scale from the Clark Mountain area of San Bernardino County in the 1950s and 1960s. California occurrences have been described by Chesterman (1966) and Elevatorski (1968).

Colorado—Colorado in some years ranked second in annual production, after Illinois. Aggregate production of crude ore to the present is estimated at about 2.3 Mt (2.5 million tons).

A total of 63 occurrences have been identified in the state (Van Alstine, 1964), including five major districts, all now idle. Northgate and Jamestown districts were the largest and most recently closed. The others are Browns Canyon, Poncha Springs, and Wagon Wheel Gap.

In the Jamestown district, fluorspar occurs in pipelike bodies and breccia zones in granite and granodiorite, and in associated veins. Output of the Burlington mine was treated at the former Valmont Mill near Boulder. Other mines include the Argo, Emmett, and Blue Jay, ore from which was concentrated to acid grade in a mill, located at Jamestown. Neither mill exists now.

At Northgate, fluorspar occurs in veins in shear and breccia zones in granite and schist, and has been surface mined and underground. During World War II the district produced metspar. Exploratory drilling continues from time to time and may result in reopening as prices rise.

The Browns Canyon district included a number of mines, developed on veins in rhyolite and in granite. The Wagon Wheel Gap mine was on a vein in a sheeted zone in rhyolite yielding metspar.

Idaho—The principal deposits of fluorspar in Idaho are in the Bayhorse-Keystone Mt. area of Custer County, southwest of Challis, and in the Meyers Cove, Big Squaw Creek, and Stanley areas (Anderson, 1964).

The Bayhorse dolomite replacement deposits, as outlined by core drilling and partially developed for mining, may exceed 2.9 Mt (3.2 million st) of 36% CaF_2, with a 25% grade cutoff. Other deposits, both vein and replacement, occur in the Bayhorse district and on adjoining Keystone Mt. The district has potential if production economics can offset freight disadvantages.

At Big Squaw Creek, a vein is exposed in the northeast canyon wall of the Salmon river west of Shoup. At Stanley, fluorspar occurs in shear zones in granite and volcanics.

Illinois-Kentucky—Current production is focused at two flotation mills situated in an area of 1813 km (700 square miles) in southern Illinois and adjacent Kentucky which for a number of years was the world's largest producer (Baxter, et al., 1973).

Vein deposits are the most numerous, but since 1950 the bedded deposits near Cave in Rock, IL, have been the source of most of the district's output, and outrank the vein deposits in terms of total all-time district production. Recently production from vein deposits has resumed importance as some of the older bedded deposits have been depleted.

In addition to the major district, a small production has been recorded from veins southwest of Lexington, in central Kentucky.

Vein deposits of the Illinois-Kentucky district occur along an extensive and intricate system of faults in sedimentary rocks of Mississippian Age. Most of these faults are the steeply dipping normal type, trending northeast, with displacement ranging from a few meters (feet) to over 305 m (1000 ft). The most important deposits are on faults of 15 to 152-m (50 to 500-ft) displacement. Veins range in width up to 9 m (30 ft) and to as much as 18 m (60 ft) in exceptional cases. They have been mined for 274 m (900 ft) vertically and for 3.2 km (2 miles) in length. Veins pinch and swell horizontally and vertically with the proportion of fluorspar to total vein material ranging from 0 to 100% within short distances. Near the surface and to depths as great as 76 m (250 ft), the veins contain much clay owing to the dissolving of vein calcite and limestone wall rock by circulating ground water leaving the argillaceous material in the limestone behind. Strong walls are found at depth in most places. A large inflow of water at high pressure was characteristic of the deeper mines, particularly at Rosiclare.

Major replacement deposits occur in Illinois northwest and north of the village of Cave in Rock; others occur in Kentucky south of Carrsville and near Hampton in Livingston County. The Cave in Rock group of mines occupies a zone of minor structural disturbance 1.6 km (1 mile) wide and over 9.6 km (6 miles) long, paralleling a major northeast trending fault which has maximum displacement of about 305 m (1000 ft). The deposits lie along one or both sides of minor fractures and small faults and are characteristically long and narrow. Major deposits occur along fracture-fault systems with less than 0.3 m (1 ft) of displacement. Widths of 15 to 46 m (50 to 150 ft) and thicknesses of 0.9 to 4.6 m (3 to 15 ft) are typical. Maximum widths of 99 to 107 m (325 to 350 ft) have been found. Lengths are mostly 61 to 457 m (200 to 1500 ft). One ore body is known to extend continuously for at least 4½ miles. The deposits occur through a vertical range of about 55 m (180 ft) in four preferred sets of limestone beds and follow these beds down dip from the outcrops (which were discovered originally in a prominent bluff) to depths well over 1000 ft.

Residual deposits have been less important producers, but have yielded high quality fluorspar. These "gravel spar" deposits are as much as 18 m (60 ft) wide and extend to depths of a hundred feet or more. The fluorspar occurs as weathered fragments from the size of boulders down to sand grains dispersed in a clayey matrix.

Montana—Most of the fluorspar production has come from the Crystal Mountain deposits near Darby in Ravalli County. The deposits of massive fluorite in granite and metasediments were discovered in 1951 and developed into one of the nation's major metallurgical grade fluorspar mines.

The mine was a side-hill benched open pit at the 2134-m (7000-ft) elevation on Sapphire Mt. Ore was mined, hand-sorted, and sized at the mine continuously from 1952 to 1973. This roughly concentrated ore was hauled 40 km (25 miles) by truck to Darby, where finished concentration was accomplished in a 54 t/h (60-ton per hr) heavy-media cone plant. The metspar was shipped by rail to steel plant consumers and government stockpiles.

Shipments from the accumulated fines at the heavy-media plant continued from 1973 to 1980 to briquetting plants serving the steel industry. Production from 1952 to 1980 totaled 513.5 kt (566,000 tons) of which 471.7 kt (520,000 tons) was metspar, 1.8 kt (2000 tons) was metspar-sized jig product 6.3 × 12.5 mm (¼ × ½ in.), and 39.9 kt (44,000 tons) was submetgrade fines. Mining and shipment of fines resumed in 1981.

Numerous other fluorspar occurrences have been recorded in Montana but have no production history. Fluor-

spar occurs as gangue in metalliferous veins in marginal portions of the Boulder batholith in west-central Montana, and is present as veins and replacement seams in intrusive syenite and in limestone in several areas characterized by alkalic intrusives in the north-central part of the state.

The presence of fluorspar has been used as a guide to gold deposits. In the Old Glory deposit in the Pryor Mountains, fluorspar occurs with the uranium mineral, tyuyamunite, replacing limestone in brecciated and cavernous portions of the Madison limestone.

Nevada—Fluorspar became commercially important in Nevada in 1919 when the Daisy mine near Beatty began its long productive life. Nevada deposits are described in a recent publication by Papke (1979). Most of the 62 deposits he describes lie in a broad, arcuate band across the southern part of the state, and northward through the central part. Total production has been about 522 kt (575,000 st) most from the Daisy (Crowell) mine in southern Nye County, the Baxter (Kaiser) mine in northeastern Mineral County, the Goldspar mine near the Daisy, and the Carp (Wells Cargo) mine in southeastern Lincoln County. About 19 other deposits have had some production, but only the Daisy mine is active. More than a fourth of the deposits are in the Quinn Canyon and Adaven district in the southeastern part of the state, but the small size of the deposits and remoteness of this area has hindered their development. Small metspar shipments have been made from Caliente.

Deposits are hydrothermal or pyrometasomatic. They occur as replacement bodies, veins, replacement and/or filling in jasperoid bodies, stockworks, and breccia pipes. The Daisy deposit is mined underground, with total past production of about 204 kt (225,000 st) It consists of irregular, generally steeply dipping, structurally complex replacement bodies in dolomite. The ore is soft, friable, and relatively low in silica. Most has been used as metspar. Some has been shipped to a southern California cement producer. The Baxter deposit, a vein in andesitic volcanic rocks, developed to a depth of 183 m (600 ft), had a total mine production of 165 kt (182,000 st), of which two-thirds was made into acid-grade concentrate in a mill operated from 1952 to 1957 by Kaiser Aluminum and Chemical Corporation at Fallon, NV. The Goldspar and nearby Mary deposits are breccia pipe and replacement bodies in dolomite. All the fluorspar from these surface mines was used in making cement. At the Carp mine an estimated 40.8 kt (45,000 st) of metspar was produced in four surface mines. The ore came from nearly flat, mantolike replacement bodies, generally concordant with bedding of the dolomite host rock.

New Mexico—There are numerous occurrences of fluorspar in New Mexico but production has been small. Van Alstine (1965) listed 65 areas in the central, north central, and southwestern parts of the state. The fluorspar map MR-60 of the USGS and accompanying tabulation in 1974, by Worl, Van Alstine, and Heyl, classified some 51 locations as to type, tonnage, and associated minerals.

Eight areas each produced at least 15 kt (20,000 tons) of crude ore and cumulative total statewide production has probably been about 635 kt (700,000 tons). Production peaked in World War II, when New Mexico was the fourth largest producing state after Illinois, Kentucky, and Colorado.

Most production came from the Zuni Mts., Sierra Caballos Mts., Gila district, Burro Mts., Cooks Peak, and from Fluorite Ridge and Tortugas near Las Cruces.

In the Zuni Mts., the "27" vein deposits were the most

productive, with the Mirabal deposit contributing. Until closed in 1953, these had produced totally about 203 kt (224,000 tons) of ore.

In the Cooks Peak district near Deming, White Eagle, Greenleaf and others in the Fluorite Ridge district, and Burro Chief and Shrine deposits in the Burro Mts. were important producers. All are vein deposits in brecciated or fractured zones along faults in igneous rocks. The Lyda K Mine, south of Truth or Consequences, is a siliceous vein in Precambrian granite. The Ruby mine north of Las Cruces contains veins in limestone near an igneous contact. At Chise small manto replacement bodies occur in limestone beds. The Bishops Cap deposit is a vein in limestone, with some adjacent replacement ore. In the Hansonburg district, east of San Antonio, fluorite occurs with lead and barite as vug filling in brecciated limestone.

The Salado deposit, southwest of Truth or Consequences, consists of tabular masses of fractured and brecciated jasperoid with the fluorspar forming a filling and cement. Between 2.3 and 2.7 Mt (2½ and 3 million tons) of low-grade mineralization have been estimated from drilling results. A thickness of over 12 m (40 ft) in places would permit open-pit mining, with little overburden. Difficult flotation metallurgy, low grade, and remoteness from market have inhibited development.

Oregon—Near Rome in the southeastern part of the state, fluorspar occurs as nearly spherical grains in volcanic tuff, tuffaceous mudstone, and mudstone of Tertiary lake deposits (Sheppard and Gude, 1969). The fluorspar content reaches 16% in some zones, but is generally much lower. The tonnage of contained fluorspar is estimated at 11 Mt. Because of very fine particle size, concentration to acceptable grade has not proved practical.

Tennessee—A small amount of metspar was produced from a small fissure vein in limestone in the central part of the state, where it occurs with barite and sulfides (Jewell, 1947).

Fluorite has been noted as a gangue mineral in zinc deposits near Carthage, locally comprising 12 to 15% of the ore but averaging much less.

A drilling program in the Sweetwater area indicated fluorspar-zinc-barite mineralization several hundred meters (hundred feet) below the surface in buried karst breccia zones similar to the Mascot district, TN zinc deposits. Exploratory drifting was done from a 183-m (600-ft) shaft sunk in 1979, now plugged.

Texas—Metallurgical grade fluorspar production began in 1971 in the Christmas Mts. south of Alpine from bedded limestone replacement mantos adjacent to intrusive rhyolite bodies. In late 1980, a 274-m (900-ft) decline was sunk beneath a laccolithic structure where exploration and development work continue.

In the Eagle Mts. southwest of Van Horn, fluorspar occurs as replacement of limestone and as fissure deposits in rhyolilte. In the period 1942 to 1950, about 13.6 kt (15,000 tons) of metallurgical gravel and acid grade were produced.

Utah—Most of the deposits are located in the west-central part of the state in the Spor Mountain, Indian Peak, Pine Grove, and Star districts. The Spor Mountain district in the Thomas Range northwest of Delta has been the most consistent producer. The deposits are classified as pipes, veins, and disseminated bodies in dolomite and volcanic rocks. Most of the production has come from the pipes. The most productive mines have been the Lost Sheep, Fluorine Queen, and Bell Hill.

In the Indian Peak district veins occur along faults in volcanic rocks. The Cougar Spar and Blue Bell mines were World War II producers of metspar. Small production has also been recorded from the Pine Grove and Star districts, and from the Rain Bow mine in Millard County and the Silver Queen mine in Tooele County.

Cumulative Utah production of crude ore to date is estimated to have been 226.8 kt (250,000 tons).

Washington—A small production of metspar has come from the Mitchem mine near Keller, in Ferry County.

Wyoming—There are reports of fluorspar in Crook and Albany counties.

Central America: Occurrences have been reported in Honduras and Guatemala, but details are not available.

South America: *Argentina*—Vein deposits have been mined intermittently in the Sierra Comechigones, 48 km (30 miles) northwest of Cordoba and others developed for production in the southern part of Rio Negro province and the adjoining northern part of Chubut. In recent years, metspar has been produced from a group of veins in the Sierra Grande district of Rio Negro, including the Delta vein, reportedly with 3.6 Mt (4.0 million tons) of drill-indicated ore of 51% CaF_2 grade. In Chubut a number of veins have been discovered in areas west of Puerto Madryn.

Brazil—Principal deposits are in the vicinity of Criciuma in the Santa Catarina province of southern Brazil. Small-scale production has been reported from deposits in the provinces of Rio Grande do Norte, Paraiba, and Bahia.

In the Criciuma district, known deposits are not large and continued output will depend on the discovery of new reserves and the resolution of water problems.

Chile—Small tonnage of fluorspar has been produced from veins in granite in the Province of Coquimbo. One deposit, the Mercedes mine, is near the village of Paihwano.

Europe: *France*—Important deposits occur in the departments of Haute-Loire, Pyrenees-Orientales, Var, Tarn, Saone-et-Loire, and Puy-de-Dome. Details have been described in a series of articles by Chermette (1972-1973, 1979).

Epithermal vein deposits occur in the Morvan, Auvergne, Limousin, and Albi districts of the Massif Central, the Maures and Esterel districts of the Mediterranean coastal area northeast of Toulon, in the eastern Pyrenees west of Perpignan, and in the Vosges. The vein deposits contain about 50% CaF_2, 20% silica, 5 to 50% barite, and 3 to 5% sulfides. Stratiform deposits with about 35% CaF_2 and 15% barite occur in the Morvan district. Deposits having stratiform characteristics are important producers in the Escaro district in the Pyrenees southwest of Perpignan. Deposits at Le Beix, Chavaniac, and Le Maine in the Massif Central are exhausted. La Charbonniere and Langeac have also closed. Fonte Sante (near Cannes in Var), Le Burc (near Albi in Tarn), Montroc (near Albi), Escaro and Chaillac continue in operation. Fonte Sante is noteworthy for the economic occurrence of sellaite (MgF_2) which is blended with the fluorspar product and is used at the Lyons hydrofluoric acid plant.

Germany, East—East Germany is estimated to produce about 100 kt/a of fluorspar. Deposits are in Saxony, Thuringia, and Anhalt, but details are lacking.

Germany, West—Production is derived from vein deposits in the Todtnau, Hesselbach, Oberwolfach, and Pforzheim districts in the Black Forest region in the southwest, and in the Naaburg and Sulzbach districts in Bavaria.

Greece—Fluorspar occurs as a gangue mineral in the silver ores of Laurium, but Greece imports its requirements.

Italy—Mines in Sardinia are the principal producers. Smaller outputs have come from Brescia, Trento, Bolzano, and Bergamo in northern Italy. The Genna Tres Montis-Muscadroxiu-S'Acqua Frida vein system in southern Sardinia near the village of Silius is credited with 8 Mt of ore reserve averaging 40 to 45% CaF_2, making it one of the world's largest deposits. Vein widths are 5 to 8 m. Barite and galena are recoverable accessory minerals. Numerous other vein deposits and a skarn deposit occur in southern Sardinia.

In northern Italy important deposits include Prestavel and Vallarsa near Bolzano, Torgola near Brescia, and Camerata Cornello near Bergamo.

The Pianciano deposit in the Castel Giuliano area 40 km northwest of Rome consists of fluorspar impregnating lake beds of volcanic ash (Spada, 1969). The fluorspar content, ranging from 20 to 55%, was derived from volcanic emanations along with barium, strontium, and phosphorous. The fluorspar is too fine-grained to be recovered by flotation, but tests indicate a possibility of producing metallurgical grade by hydrocyclone separation.

Norway—Veins are reported in the Buskerud, Telemark, Vest Agder, and Drammen regions west and southwest of Oslo. The Lassedal mine near Kongsberg was formerly a producer.

Spain—Principal deposits are in the Asturias region of northern Spain near the ports of Ribadesella and Aviles; in the southeast near the port of Almeria; and in south-central Spain near Seville and Cordoba. Deposits are both veins and replacement bodies. Acid-grade material is derived in the retreatment of old lead mine wastes in the Almeria-Berja area. Seven companies have produced acid grade, four in Asturias, two in the Seville-Cordoba region, and one near Berja. Grades of ore range from 40 to 45% for some of the vein-mines, down to 15 to 20% in the case of old lead mine waste. The large Osor vein in Gerona is exhausted.

Sweden—The sole producer in Sweden, the government-owned Yxsjoberg mine near Ludvika, 200 km northwest of Stockholm, where fluorspar occurs with tungsten and copper, closed in 1977. Other deposits have been reported between Branteriks and Onslunda in Skane in southern Sweden. The Osterlen district was formerly a small producer.

Switzerland—Occurrences are known, but none of commercial importance.

United Kingdom—Among numerous fluorite occurrences in the United Kingdom, present operations are located in the Southern Pennine Orefield of north Derbyshire and in the Northern Pennine Orefield of west Durham. Production comes from fissure veins and bedded replacement deposits, together with associated old lead mine waste and tailings dumps. The deposits occur in rocks of the Carboniferous Limestone Series (Mississippian). Ore is mined both underground and open pit. The deposits contain variable amounts of galena, sphalerite, calcite, barite, quartz, and iron sulfides. Some were originally mined for lead, even before written record.

USSR and Soviet Bloc—Deposits are reported in the Transbaikal, Tashkent, and Yaroslav regions. Czechoslovakia produces about 90 kt/a from deposits in Bohemia. Mongolian deposits are southwest of Ulan Bator in the Dorono Gobi district. Romania produces about 20 kt/a; Bulgaria about 25 kt/a, and Yugoslavia has a small production.

Africa: *Angola*—Deposits are reported in Angola, but details are lacking.

Kenya—A 100% government-owned acid grade mill, formerly jointly owned with IMC and Bamburi Portland Cement, produces concentrates relatively high in P_2O_5 from epithermal vein and replacement deposits in the Kerio Valley,

about 130 km northwest of Nairobi. Ore ranging from 30 to 45% CaF_2 is mined open pit.

Other similar deposits are reported along the Rift structure.

Malawi—Exploration for fluorspar deposits related to carbonatite structures has been reported; five such occurrences are noted by Van Alstine and Schruben (1980).

Morocco—The El Hammam deposit near Meknes is a nearly vertical vein on an east-west fault in sericite schist, grading about 45% CaF_2. Three other deposits are recognized in this region. Other veins occur in the Djebel Tirremi area, 10 km northeast of Taourirt, in eastern Morocco (Van Alstine and Schruben, 1980).

Mozambique—Large vein deposits occur approximately 150 km southeast of Tete. Remote from transportation and from sources of power, this area has been a minor producer of metspar. Further development may result as the area is opened up as a consequence of the construction of the Cabora Bassa dam on the Zambesi River. Six localities are listed by Van Alstine and Schruben (1980).

Nigeria—Deposits containing fluorspar with galena are reported in limestones in the Arufu district of Gongola State near the border with Benue State, in the Benue Trough about 80 km east of Makurdi, and fluorspar in greisens near Jos (Van Alstine and Schruben, 1980).

South Africa—The Republic of South Africa is ranked as third largest fluorspar producing country. Its large reserves indicate it will continue to be prominent in the world picture. Deposits occur in several areas, but production has come principally from mines in the western and northern parts of Transvaal.

Domestically, AECI consumes acid grade to make hydrofluoric acid and the steel industry consumes essentially all of the metgrade as well as about 20 kt/a of 80% CaF_2 briquettes.

The Buffalo open pit mine of Transvaal Mining & Finance, a subsidiary of General Mining and Finance, near Naboomspruit exploits 14 to 16% grade deposits consisting of a network of fluorspar veinlets in sills in a pink granite which is itself an inclusion in coarse Bushveld granite. Similar deposits occur nearby.

Near Warmbad, the Zwartkloof deposit consists of three breccia zones containing a stockwork of fluorspar veinlets in felsite.

At the open pit Vergenoeg mine of Chrome Chemicals, a subsidiary of Bayer, near Pienaarsrivier, fluorspar ore grading 40% occurs with abundant iron oxides (up to 50%) in a massive deposit in felsite.

In the Marico-Zeerust district, fluorspar occurs as replacements in dolomite and as associated residual deposits. Much of the ore in dolomite is disseminated along intersecting fractures. For many years metspar was produced by screening coarse fragments from residual surface deposits. Exploration has disclosed the existence of large reserves of primary ore probably on the order of 54.4 Mt (60 million tons) containing 15 to 30% CaF_2. Two open pit mines, the Chemspar mine of Phelps Dodge and the Marico mine of Barlow Rand, with capacity of about 120 kt/a, are located here. Armco Bronne, a subsidiary of Armco, Inc., has an active exploration program.

Namibia (South-West Africa)—The Okurusu deposit near Ojiwarongo consists of replacements or segregations in carbonatite and veins in quartzite. After yielding a few thousand tons of metspar, it has been idle. Reserves are indicated to be between 5.4 and 6.4 Mt (6 and 7 million tons) grading 40% CaF_2. Vein ore near Omaruru was mined open pit and concentrated to acid grade in 25 kt/a flotation mill from reserves in excess of 272 kt (600,000 tons). Vein and replacement deposits are known in other parts of the country. The lack of water and transportation facilities has hindered development.

Sudan—Fluorspar deposits occur at Jebel Semeih and Jebel Dumbeir approximately 322 km (200 miles) southwest of Khartoum and 1336 km (830 miles) by rail from Port Sudan. At Jebel Semeih a highly siliceous fluorspar vein occurs in a granitic host rock. At Jebel Dumbeir, fluorspar occurs in metasediments as replacements of bands of marble, as thin quartz-fluorite veins, and as bodies of fluorite associated with red feldspathic rock. No production is reported.

Tunisia—A flotation mill at the Hamman Zriba deposit near Zaghouan, 54 km south of Tunis, produces about 30 kt/a of acid grade concentrate from a bedded deposit in Jurassic limestone. Proven reserves are said to be 1.8 Mt (4 million tons) of ore containing 35% CaF_2 and considerable barite. In the same general area, replacement vein deposits occur at Djebel Staa, Hammam Djedida, Djebel Oust, and Djebel el Koho. Other deposits are at Bourchiba, on Cap Bon, and Bou Jaber.

Turkey—Fluorspar deposits in Turkey have been described (Anon., 1965); there is a very small annual production.

Asia: *China*—China is emerging as a major producer with an annual production of about kt (410,000 tons), mainly from deposits in the central and northern parts of Zhejiang Province, northeastern Hebei Province, northern Guangdong Province, northeastern Anhui Province, and eastern Shandong Province. Other production is at Taoling in Hunan Province, Ming Gang in Henan, at Kaiping in Liaoning Province, and at Yulin in Guangxi Province (a new discovery near rail). In all, fluorspar occurrences are reported in 17 provinces. China exported 95.3 kt (105,020 tons) to Japan in 1978 and 186.4 kt (205,489 tons) in 1979, mainly metallurgical grade.

India—At Amba Dongar, 563 km (350 miles) north of Bombay in Gujarat State, veins and replacement bodies occur in the carbonate wall rock around ankeritic carbonatite intrusives, similar to the occurrences at Okurusu in Namibia (Deans, et al., 1972). Reserves are reported to be 11.6 Mt of 30% fluorite content, of which proven reserves are said to be 4.7 Mt.

Other mineralized areas are reported at Chandri-Dungri in the Drug district of Madhya-Pradesh and at Mando-ki-Pal in the Dungarput district of Rajasthan.

Japan—Japan formerly produced less than 10 kt/a of fluorspar from the island of Hokkaido, but ceased production in 1972 and imports its requirements.

Korea, North—This country has an annual production of about 40 kt.

Korea, South—Fluorspar deposits occur in the northern, central, and southern parts of the country. Approximately 70 mines have been producers, formerly with annual production of about 50 kt, largely for export (Kim, et al., 1972), but production was down to 8 kt in 1979. Both vein and replacement deposits occur. Minable ore reserves have been estimated at 660 kt (Jee and Kye, 1971). Producers have included the Choonchon Shinpo, Dojon, Kumi, Busang, and Kumbo mines.

Pakistan—Minor production has been reported from the Kalat district 72 to 97 km (45 to 60 miles) south of Quetta in Baluchistan. There are also deposits south of Quetta.

Thailand—After the initiation of commercial fluorspar production in 1960, Thailand advanced to the status of one of the largest producers in the world.

Fluorspar deposits occur in a region from the northern border with Burma southward for over 1287 km (800 miles) to a point about 644 km (400 miles) southwest of Bangkok. Active mines are concentrated in three areas: Chiang Mai, Lumphun, and Mai Hongsun Provinced in the north; Kanchanaburi, Petchaburi, and Ratchaburi Provinces southwest of Bangkok; and Krabi Province in the south (Gardner and Smith, 1965; Anon., 1972; private communication, B. L. Hodge, 1981).

The northern deposits are mainly replacements in Paleozoic limestones which, so far, have been found suitable only for the production of metspar, derived by hand sorting with large resulting waste dumps containing low grade ore. Residual deposits are widespread. In the southern area, the deposits are veins in metamorphosed, fine-grained, fractured Paleozoic clastics and in granite.

Australia: Deposits of fluorspar have been found in all the Australian states except the Northern Territory (Liddy, 1971), but are undeveloped because they are small and remote. There have been investigations in the Chilagoe-Mungana-Almaden and Forsayth areas of northern Queensland, and in Western Australia in the Pilbara district 217 km (135 miles) southeast of Port Hedland, the Yinnietharra area 322 km (200 miles) from Carnarvon, and the Speewah Valley area southwest of Wyndham.

EXPLORATION

Because fluorspar resists chemical weathering, it can be traced in the soils overlying weathered veins. Cleavage fragments washed clean by rainwater or exposed in anthills or spoil piles from animal burrows are useful clues. Silicified veins resisting erosion may stand up like a "reef." Bedded deposits are less discernable, but in areas of sufficient topographic relief the outcropping edges of deposits or slumped fragments of ore may be found. Owing to its softness and cleavability, fluorspar does not survive in the beds of streams and ordinarily cannot be traced by panning.

The search for bedded deposits usually involves locating a mineralized horizon and following it with vertical drill holes drilled to intersect the favorable beds.

Geophysical prospecting methods are not applicable. Photogeology can be of assistance in locating and tracing structures and outcrops, as well as the areal distribution of stratigraphic units known to be favorable for veins or replacement deposits.

Geochemical methods have been employed with varying degrees of success. Fluorine anomalies have been found in ground waters and surface streams, in soil samples and in stream sediments, but with little practiced success in discovery of ore deposits. Various nonfluorine elements detectable geochemically can be guides to fluorspar deposits.

Once mineralization has been detected, prospect shafts, drifts, and core or churn drill holes are used to probe the structures.

BIBLIOGRAPHY AND REFERENCES

Anon., 1965, "Barytes and Fluorite Deposits of Turkey," *Maden Tetkik Arama Enstit. Yayinl,* No. 126, Ankara, 11 pp.

Anon., 1972, Annual Report, Lost River Mining Corp.

Anon., 1972a, "Progress in Modernizing Thailand's Fluorspar Industry," *Industrial Minerals,* No. 56, May, pp. 31-33.

Anon., 1973a, "Trade and Trends in Fluorspar," *Industrial Minerals,* No. 64, June, pp. 23-31.

Anon., 1978, "Samuk Enters the Major Fluorspar League," *Industrial Minerals,* No. 135, Dec., pp. 39-43.

Agricola, G., 1546, *Bermannus Sive de Re Metallica; De Re Metallica,* A. Hoover and H. Hoover, trans., Dover Publications, New York 466 pp.

Anderson, A.L., and Van Alstine, R.E., 1964, "Fluorspar," *Mineral and Water Resources of Idaho,* 88th Congress, 2nd Session, US Government Printing Office, Washington, DC, pp. 79-84.

Argall, G.O., Jr., 1949, "Fluorspar," in "Industrial Minerals of Colorado," *Quarterly of the Colorado School of Mines,* Vol. 44, No. 2, pp. 179-208.

Baxter, J.W., et al., 1973, "A Geologic Excursion to Fluorspar Mines in Hardin and Pope Counties, Illinois," Guidebook Series II, Illinois State Geological Survey, 28 pp.

Chermette, A., 1950, "L'Exploitation du Spath-Fluor en France de 1938 á 1946," *L'Echo des Mines et de la Metallurges,* Vol. 427, No. 3, pp. 547-549.

Chermette, A., 1964-65, "Les Ressources du Mexique en Spath-Fluor," *Mines et Metallurgie,* Oct., pp. 467-469; Nov., pp 519-521; Dec., pp. 571-574; Jan., pp. 25-27; Feb. pp. 77-79.

Chermette, A., 1972-1973, "Un Demi-Siècle de Spath-Fluor Francais (1922-1972)," *Mines et Metallurgie,* Oct., pp. 179-182; Nov./Dec., pp. 201-204, 213; Jan./Feb., pp. 9-12, 14; Mar., pp. 8, 11; Apr., pp. 10-13; May, pp. 9-16.

Chermette, A., 1979, "Le Spath Fluor est un Tournant," *Mines et Metallurgie,* No. 145, pp. 14-19 and No. 146, pp. 20-24.

Chesterman, C. W., 1966, "Fluorspar," *Mineral Resources of California,* Bulletin 191, California Div. of Mines and Geology, pp. 165-168.

Coope, B., 1978, "Fluorspar—Down But Not Out," *Industrial Minerals,* No. 125, Feb., pp. 39-66.

Cornwall, H. R., 1972, "Geology and Mineral Resources of Southern Nye County, Nevada," Bulletin 77, Nevada Bureau of Mines and Geology, 49 pp.

Dasch, M.D., 1964, "Fluorine," *Mineral and Water Resources of Utah,* Bulletin 74, Utah Geological and Mineralogical Survey, pp. 162-168.

Deans, T., et al., 1972, "Metasomatic Feldspar Rocks (Potash Fenites) Associated with the Fluoride Deposits and Carbonatites of Amba Dongar, Gujarat, India," *Transactions, Sec. B,* Institution of Mining & Metallurgy, Vol. 81, Bulletin 783, p. B6.

Dunham, K.C., 1952, "Fluorspar," Memoir, Special Report on Mineral Research, Geological Survey of Great Britain, Vol. 4, 143 pp.

Elevatorski, E.A., 1968, "California Fluorspar," *Mineral Information Service,* State of California, Vol. 21, No. 9, pp. 127-130.

Elevatorski, E.A., 1971, "Arizona Fluorspar," Arizona Dept. of Mineral Resources, 51 pp.

Ford, T.D., and Ineson, P.R., 1971, "The Fluorspar Potential of the Derbyshire Ore Field," *Transactions, Sec. B,* Institution of Mining and Metallurgy, Vol. 80, pp. B185-205.

Funnell, J.E., and Wolff, E.J., 1964, "Fluorspar (Fluorite)," *Compendium on Nonmetallic Minerals of Arizona,* prepared for Arizona Public Service Co. by Southwest Research Institute, pp. 105-114.

Gardner, L.S., and Smith, R.M., 1965, "Fluorspar Deposits of Thailand," Report of Investigation No. 10, Dept. of Minerals Resources, Bangkok, 42 pp.

Gillerman, E., 1953, "Fluorspar Deposits of the Eagle Mountains, Trans-Pecos Texas," Bulletin 987, US Geological Survey, 98 pp.

Goddard, E.N., 1966, "Geologic Map and Sections of the Zuni Mountains Fluorspar District, Valencia County, New Mexico," Map I-454, Miscellaneous Geologic Investigations, US Geological Survey.

Gossling, H.H., 1972, "A Review of the World's Fluorspar Industry, with Particular Reference to South Africa," Report No. 1424, National Institute of Metallurgy, Johannesburg, 30 pp.

Gossling, H.H., 1978, "Fluorspar, 1973-1980: A Commodity Profile," Report No. 3, Mineral Bureau, Dept. of Mines, Republic of South Africa, p. 36.

Grogan, R.M., and Bradbury, J.C., 1967, "Origin of the Stratiform Fluorite Deposits in Southern Illinois," *Genesis of Stratiform Lead-Zinc-Barite-Fluorite Deposits,* Monograph 3, The Economic Geology Publishing Co., pp. 40-51.

Grogan, R.M., and Bradbury, J.C., 1968, "Fluorite-Zinc-Lead Deposits of the Illinois-Kentucky Mining District," *Ore Deposits of the United States, 1933-1967,* Vol. 1, J.D. Ridge, ed., AIME, New York, pp. 370-399.

Guccione, E., 1972, "What's Going on in the Fluorspar Industry," *Engineering & Mining Journal,* Vol. 173, Dec., pp. 64-72.

Hodge, B.L., 1970, "The U K Fluorspar Industry and Its Basis," *Industrial Minerals,* No. 31, Apr., pp. 23-37.

Holdge, B.L., 1973, "World Fluorspar Development, Part 1," *Industrial Minerals,* No. 68, May, pp. 9-25.

Hodge, B.L., 1973a, "World Fluorspar Developments: Part 2," *Industrial Minerals,* No. 69, June, pp. 9-21.

Hodge, B. L., 1981, "Fluorspar," *Mining Annual Review, 1981,* pp. 116-117.

Horton, R.C., 1961, "An Inventory of Fluorspar Occurrences in Nevada," Report 1, Nevada Bureau of Mines, 31 pp.

Horton, R.C., 1962, "Fluorspar Occurrences in Nevada," Nevada Bureau of Mines, Map 3.

Horton, R.C., 1964, "Fluorspar," *Mineral and Water Resources of Nevada,* 88th Congress, 2nd Session, US Government Printing Office, Washington, DC, pp. 198-202.

Jee, J.M., and Kye, J., 1971, "Report on the Fluorite Deposits of Korea," Bulletin No. 13, Geological Survey of Korea, pp. 7-368 (English Abstract).

Jewell, W.B., 1947, "Barite, Fluorite, Galena, Sphalerite Veins of Middle Tennessee," Bulletin 51, Tennessee Div. of Geology, 114 pp.

Johnson, R.C., et al., 1973, "Economic Availability of Byproduct Fluorine in the United States," Information Circular 8566, US Bureau of Mines, 97 pp.

Jones, W.R., and Wolter, F.J., 1973, "Chemical Markets for Hydrofluoric Acid," Preprint No. 73-H-32, Society of Mining Engineers of AIME, SME Fall Meeting, Pittsburgh, Sep.

Kim, H., et al., 1972, "Mineral Requirements for Korea's Industrialization," Korean Institute of Mining, pp. 80-84.

Kostick, D.S., and De Filippo, R.J., 1980, "Fluorspar," Preprint, 1978-1979, US Bureau of Mines, 17 pp.

Liddy, J.C., 1971, "Fluorspar in Australia," *Australian Mining,* Jan., pp. 34-38.

MacMillan, R.T., 1970, "Fluorine," *Mineral Facts and Problems,* Bulletin 650, US Bureau of Mines, pp. 989-1000.

McAnulty, W.N., 1970, "Evaluation of Fluorspar Deposits," Preprint 70-S-63, Society of Mining Engineers, AIME Annual Meeting, Denver, Feb.

Northholt, A.J.G., and Highley, D.E., 1971, "Fluorspar," Mineral Dossier No. 1, Mineral Resources Consultative Committee, Institute of Geological Sciences, London, 31 pp.

Ortel, M.K., 1966, "Fluorspar," South African National Resources Development Council, Dept. of Planning, Vol. 5, 31 pp.

Papke, K.G., 1979, "Fluorspar in Nevada," Bulletin 93, Nevada Bureau of Mines and Geology, 77 pp.

Peters, W.C., 1958, "Geologic Characteristics of Fluorspar Deposits in the Western United States," *Economic Geology,* Vol 53, pp. 663-688.

Rothrock, H.E., et al., 1946, "Fluorspar Resources of New Mexico," Bulletin 21, New Mexico Bureau of Mines and Mineral Resources, 245 pp.

Sahinen, U.M., 1962, "Fluorspar Deposits in Montana," Bulletin 28, Montana Bureau of Mines and Geology, 38 pp.

Sheppard, R.A., and Gude, A.J., 1970, "Authigenic Fluorite in Pliocene Lacustrine Rocks Near Rome, Malheur County, Oregon." *Geological Survey Research 1969,* Professional Paper 650-D, US Geological Survey, pp D69-D74.

Snyder, K.D., 1978, "Geology of the Bayhorse Fluorite Deposit, Custer County, Idaho," *Economic Geology,* Vol. 73, No. 2, March-April, pp. 207-214.

Spada, A., 1969, "Il Giacimento de Fluorite e Baritina Esalativo-sedimentario in 'Facies' Lacustre, Intercalato nei Sedimenti Piroclastici della Zona di Castel Giuliano, in Province di Roma," *Industria Min.,* Vol. 20, pp. 501-518.

Staatz, M.H., and Osterwalld, F.W., 1959, "Geology of the Thomas Range Fluorspar District, Juab County, Utah," Bulletin 1069, US Geological Survey, 97 pp.

Tabor, J.W., 1953, "Montana's Crystal Mountain Fluorspar Deposit is Big and High Grade," *Mining World,* Vol. 15, No. 7, pp. 43-46.

Thurston, W.R., et al., 1954, "Fluorspar Deposits of Utah," Bulletin 1005, US Geological Survey, 53 pp.

Van Alstine, R.E., 1964, "Fluorspar," *Mineral and Water Resources of Colorado,* 88th Congress, 2nd Session, US Government Printing Office, Washington, DC, pp. 159-165.

Van Alstine, R.E., 1965, "Fluorspar," *Mineral and Water Resources of New Mexico,* Bulletin 87, New Mexico Bureau of Mines and Mineral Resources, pp. 260-267.

Van Alstine, R.E., 1969, "Fluorspar," *Mineral and Water Resources of Arizona,* Bulletin 180, Arizona Bureau of Mines, pp. 348-357.

Van Alstine, R.E., and Schruben, P.G., 1980, "Fluorspar Resources of Africa," Bulletin 1487, US Geological Survey, 25 pp.

Weller, J.M., et al., 1952, "Geology of the Fluorspar Deposits of Illinois," Bulletin 76, Illinois State Geological Survey, 147 pp.

Wood, H.B., 1971, "Fluorspar and Cryolite," *Minerals Yearbook,* US Bureau of Mines, pp. 517-530.

Wood, H.B., 1972, "Fluorspar," *Engineering & Mining Journal,* Vol. 173, Mar., p. 152.

Worl, R.G., et al., 1973, "Fluorine," *United States Mineral Resources,* Professional Paper 820, US Geological Survey, pp. 223-235.

Zijl, P.J. van, 1962, "The Geology, Structure, and Petrology of the Alkaline Intrusions of Kalkfeld and Okorusu and the Invaded Damara Rocks," *Aunale Univ. Stellenbosch,* Ser. A., Vol. 37, No. 4, pp. 237-346.

Zurowski, M., 1972, "Barite-Fluorite Deposits of Lake Ainslie—An Appraisal From an Economic Viewpoint," *Transactions,* Canadian Institute of Mining & Metallurgy, Vol. 75, pp. 318-321.

9. Gypsum and Anhydrite*

The two calcium sulfate minerals—gypsum and anhydrite—occur in many parts of the world, and gypsum has long been of economic importance in the family of industrial minerals. Gypsum, the dihydrate form of calcium sulfate ($CaSO_4 \cdot 2H_2O$) and anhydrite, the anhydrous form ($CaSO_4$) are frequently found in close association, and it is seldom that a calcium sulfate deposit will consist exclusively of one mineral or the other.

Although known gypsum deposits are extensive, anhydrite makes up the largest part of total calcium sulfate reserves. However, it has very minor economic use, and most of the following discussion will be devoted to gypsum.

Calcium sulfate is one of the principal constituents of evaporite deposits, and when pure, has the following composition:

	Lime, CaO, %	Sulfur Trioxide, SO_3, %	Combined Water H_2O, %
Gypsum	32.6	46.5	20.9
Anhydrite	41.2	58.8	—

Deposits of pure gypsum or of pure anhydrite which are large enough to be considered commercial have never been found because of both the metastable relationship between the two minerals and the presence of impurities such as calcium or magnesium carbonates, chlorides, other sulfate minerals, clay minerals, or silica. As a result most mine production of gypsum will range between 85 and 95% pure. Often it is used as mined, although in certain cases, one or more forms of mineral beneficiation are employed to upgrade the product.

END USES

The largest use for gypsum is based upon the unique property which calcium sulfate has of readily giving up, or taking on, water of crystallization. With the application of a moderate amount of heat in a process known as calcining, gypsum is converted to plaster of paris (the hemihydrate of calcium sulfate, $CaSO_4 \cdot \frac{1}{2} H_2O$) which, when mixed with water, will set or harden as the calcium sulfate returns to the dihydrate form. This semifinished product, usually called stucco, is then manufactured into a large variety of plasters, wallboard, and block for construction use, or into plasters for industrial applications. About 75% of the gypsum used in the United States is calcined for these purposes.

Uncalcined uses of gypsum are principally as a retarder for portland cement, as a soil conditioner, as a mineral filler, and other minor industrial applications. About 25% of the gypsum mined in the United States goes into these markets; however, in less developed countries where construction and other industrial uses may be quite limited, the use of uncalcined gypsum for portland cement retarder may well be the dominant—if not the only—market for the mineral.

It has often been noted that calcium sulfate constitutes

*Excerpt from Appleyard, F.C., "Gypsum and Anhydrite," *Industrial Minerals and Rocks,* 5th ed., Vol. 2, S.J. Lefond, ed., AIME, New York., pp. 775-792.

the world's largest reserve of sulfur, and minor use of both anhydrite and gypsum has been made to produce sulfuric acid or other sulfur compounds such as ammonium sulfate. However, because sulfur is readily available at lower cost from other sources, use of calcium sulfate has been limited to only a few locations where it could compete economically. No such use is reported in the United States at this time.

MINERALOGY

Anhydrite, $CaSO_4$, because of its geologically rapid conversion to gypsum, and relatively high solubility (about 0.2 g/100 g H_2O) is not often found outcropping in climates wet enough to support abundant vegetation except on steeply dipping slopes or other places where hydrated material is continuously removed. Anhydrite rock is most often light to bluish gray in color and under the microscope varies from granoblastic (ver Planck, 1952) to feltylath crystal aggregates.

Bassanite, $CaSO_4 \cdot \frac{1}{2}H_2O$, is a distinct phase intermediate between anhydrite and gypsum, but is identifiable only by X-ray diffraction or petrographic techniques using very carefully prepared samples. Bassanite is metastable under ordinary conditions (Wood and Wolfe, 1969); however, its occurrence in amounts under 1% is suspected to be widespread in calcium sulfate mineral deposits.

Gypsum, $CaSO_4 \cdot 2H_2O$, is the commonly observed calcium sulfate mineral in rock outcrops because of its diverse origins. It is easily distinguished from anhydrite by its inferior hardness (2.0 vs. 3 to 3.5) and lower specific gravity (2.2 to 2.4 vs. 2.7 to 3.0 for anhydrite). Most gypsum is white to grayish white although the impurities in any given deposit frequently determine the color of the rock and resulting products. Petrographically, most rock gypsum is granoblastic (Goodman et al., 1957). Commercial deposits show a great variety of crystal sizes and textures, both from deposit to deposit and within a single deposit. Relatively undisturbed deposits often contain texturally consistent stratigraphic units, but, in deposits that have been deformed, the gypsum is often recrystallized and shows a variety of textures. Many deposits exhibit porphyroblastic textures wherein two distinct sizes and ages of crystals occur (Holiday, 1970), and in some deposits fibrous gypsum similar to felty-lath anhydrite makes up the mass (Van't Hoff, et al., 1903).

Alabaster is a compact, very fine grained variety of rock gypsum prized by sculptors for its uniform workability under the chisel, and occasionally is found within commercial deposits. Fibrous gypsum composed of needle-shaped crystals in orientation parallel to the C-axis (Schmidt, 1914), is known as **satin spar** and is a type of stress mineral indicative of deformation. It is universally a secondary mineral and occurs widely as a fracture filling wherein the needle lengths are perpendicular to the fracture walls and less commonly in shear zones wherein the needles are parallel to the direction of movement. Large euhedral gypsum crystals and large cleavages commonly known as **selenite** form in fluid-filled spaces or in an easily deformable host material, and are sometimes mistakenly identified as mica by the uninitiated. Both satin spar and selenite are of little or no economic importance other than being accessories in rock gypsum, although one deposit consisting primarily of coarse selenite crystals is being worked near Apex, NV (Bergstrom, 1961).

ORIGIN AND OCCURRENCE

The calcium sulfate minerals are deposited by precipitation from aqueous solution when the concentration of the components and the physical conditions are suitable, with the majority of the deposits originating from evaporation and concentration of marine brines in a dry climate, i.e., evaporite conditions. That this is the basic mechanism of formation of economic calcium sulfate deposits is unquestioned; however, there is considerable discussion in the literature as to the physical environment in which precipitation takes place, and as to whether the originally precipitated mineral is gypsum or anhydrite.

The older, well established concept of deposit formation idealizes a brine-filled basin having restricted circulation (Branson, 1915; King, 1971), which permits limited replenishment of the brine as it evaporates. The resulting concentration causes precipitation of the contained salts in the inverse order of their solubility, and substantial thickness of essentially monomineralic deposits (e.g., calcium sulfate) are visualized as being derived from this type regime, an example of which is the A subzone of the Mississippian Windsor series of southeastern Canada.

A more recent concept of evaporite deposition has resulted from considerable work done in the Trucial Coast of Arabia (Butler, 1970; Kinsman, 1966, 1969) which describes the sabkha supratidal environment wherein calcium sulfate minerals are currently being deposited (sabkha, an Arabic word denoting salt-flat—Kinsman, 1966). These deposits are characterized by a distinctive sequence of sediments including lagoonal limestones, intertidal algal mat limestones, and precipitation of gypsum crystals and nodular anhydrite in the preexisting host sediments, which, if originally calcium carbonate, become dolomitized. Clastic sediments, if present, are unaffected chemically, but are contorted and displaced as the gypsum crystals and nodular anhydrite grow in them. In the sabkha regime, a basin is not required for deposition, although some concentration of the original seawater brine does take place in along-shore lagoons. Thick chloride accumulations would not be expected (Schroeder, 1970). The Mississippian evaporates of southwestern Indiana illustrate this type of deposit (Jorgensen and Carr, 1972).

The classical in-basin accumulations of sulfates and chlorides might be expected to have sabkha deposits along the margin of the basin, and the Silurian gypsum of Ohio, Ontario, and New York is possibly of this origin. The individual sulfate beds in sabkhas do not commonly exceed 6 to 12 m (20 to 40 ft) in thickness, although 61-m (200-ft) accumulations have been described in Jamaica, WI (Holliday, 1971).

Gypsum may also precipitate along fractures, bedding planes, or in other available spaces where ground water carrying sulfate ions from the oxidation of sulfides comes into contact with carbonate rocks. Such deposits, however, are of limited extent, and rarely—if ever—could be expected to have economic importance.

Because gypsum will precipitate wherever brine concentration is appropriate, and recrystallizes or dissolves readily as the metastable or temporary environments change, interpretation of geologic evidence to determine exact depositional environments is difficult at best. Several works (Dean and Schreiber, 1978; Kendall, 1978; and others) are helpful in determining depositional environments.

Gypsite deposits usually occur in semiarid and arid climates, and result from the solution of existing gypsum deposits by ground water, which in turn is drawn toward the surface by capillary action where it evaporates to deposit the dissolved sulfate as a porous aggregate of gypsum containing considerable impurities. Also, in a few cases, reworking of existing deposits by wind erosion has concentrated gypsum in sand deposits, as at White Sands, NM, or Cuatrocienegas, Coahuila, Mexico.

Of totally different origin are those calcium sulfate deposits which constitute part of the cap rock of salt domes in the Gulf Coast basin. The exact origin of these domes is not fully understood but most prevailing views involve flowage and upward movement of plastically deformed salt in response to overburden pressure from the overlying sediments (Martinez, 1974). Once the upward movement is initiated, it is possible for the top of the lighter salt mass to be injected into and through the overlying sediments.

The theory most widely accepted as to the origin of the cap rock is that it represents a residue of anhydrite and other relatively insoluble minerals which accumulate at the upper surface of the salt dome as it rises through water-bearing sediments and the salt is dissolved. The anhydrite crystals, which make up about 99% of the water-insoluble residue (Walker, 1974), are then recrystallized and compacted to form a massive rock which may be locally hydrated to form gypsum.

The total height of a salt dome can exceed 3048 m (10,000 ft), with the cap rock being well over 305 m (1000 ft). Most domes are roughly circular in plan view, and range in diameter from about 1000 m (3280 ft) to more than 6.4 km (4 miles) although there is great diversity in dome geometry and dome-to-cap-rock relationships. Most domes do not reach the present surface, although occasionally cap rock is found at or near the surface and might be of economic significance.

DISTRIBUTION OF DEPOSITS

Calcium sulfate minerals have been found within every geologic system from Silurian through Quaternary. The largest production in North America is from Mississippian rocks in southeastern Canada, Michigan, Indiana, and Virginia, followed by that from Permian rocks in Texas, Oklahoma, Kansas, Colorado, Wyoming, Nevada, and Arizona. Production from Tertiary rocks is obtained in California, Jamaica, and Arizona, from Jurassic rocks in Iowa, New Mexico, Colorado, Utah, Wyoming, and Montana, and from Silurian rocks in New York, Ontario, and Ohio. Elsewhere, Tertiary rocks in Australia, France, and Italy, Permian and Jurassic rocks in England, and the Triassic in France, Germany, and England contain large commercial deposits. Commercial deposits range in thickness from 101 cm (40 in.) to over 31 m (100 ft), and reserves of individual deposits are usually measured in millions of tons. Production rates commonly average 181 to 272 kt/a (200,000 to 300,000 tpy) from each operation, but range from a few thousand tons up to about two million tons.

Although gypsum is produced from rocks of many ages, many deposits are Quaternary in age due to the recent and continuing conversion of anhydrite to gypsum that occurs under certain conditions. This conversion of anhydrite to gypsum, as well as the reverse reaction, has been extensively studied because of its importance in the original sequence of deposition from brines, as well as in metamorphism that occurs in calcium sulfate deposits after deposition. Several reviews have been made (Braitsch, 1971; Hardie, 1967; MacDonald, 1953; Wood and Wolfe, 1969) of such variables as brine composition, temperature, water pressure, and overburden pressure. From these, it is evident that no one set of conditions applies to all deposits. On the contrary, all these variables, plus structural, lithological, and topographical considerations, which might affect water migration, must be examined in order to understand the degree of hydration of any given calcium sulfate deposit. To date, however, only limited attempts have been carried out in an effort to integrate theoretical models with field evidence of rock textures and structures.

Most commercial gypsum deposits are believed to have resulted from the action of surface and ground waters upon anhydrite, and hydration has been found at depths ranging from zero to more than 610 m (2000 ft) below the surface. In many deposits, the degree of hydration is readily predictable once the variables controlling it are understood; however, in others the degree of hydration can vary widely, with corresponding reduction in the amount of available reserves, and increase in the cost of mining. In all cases, the operator is well advised to thoroughly explore and understand the anhydrite-gypsum relationship prior to development of a mine.

Impurities in Calcium Sulfate Deposits

The number of minerals which can occur in evaporite deposits is quite large (Braitsch, 1971), and many can be found as impurities in calcium sulfate deposits. Their occurrence is often dependent upon the mode of formation of the deposit. In varying degrees, most gypsum and anhydrite deposits contain clastic sediments, commonly clay minerals and fine sands, as well as chemical sediments such as limestones and dolomites. In sabkha-type deposits, these preexisting sediments will fill intermodular spaces of the gypsum and anhydrite or form crude layers, and often appear to be replaced when they actually are displaced. Relatively insoluble evaporite minerals such as celestite, certain borates, some carbonate minerals, and silica may occur in calcium sulfate deposits as discrete crystals, crystal aggregates or nodules, and probably are in many cases original depositional features. However, strontium or boron, which may present in only trace amounts in anhydrite, may migrate during gypsification to result in strontium and boron mineral accumulations.

Soluble evaporite minerals, e.g., halite, sylvite, mirabilite, epsomite, and others are also frequently found in calcium sulfate deposits. These too are expelled from anhydrite during gypsification, and most commonly become associated with the clay minerals that may be present, probably being adsorbed on the clay mineral surfaces. They also are associated with the carbonate impurities where they may be absorbed into mineral surfaces, held as fluid inclusions, or included in a disordered lattice. In addition, these soluble minerals can occur as microfracture filling. The hydration halo of mixed gypsum-anhydrite commonly contains many times the soluble salt content of either unhydrated anhydrite or hydrated gypsym. Individual deposits show great variation in the quantity and type of soluble salts present, i.e., sulfates or chlorides, and it has been noted that, in general, chloride salts predominate in gypsum deposits of eastern North America, whereas sulfate salts exceed chlorides in western North American deposits.

In addition to the foregoing, evaporite mineral impurities whose origin is generally a feature of original calcium sulfate deposition, in any given mining operation may also be encountered as a result of structural features such as tight folding or faulting, or from recent erosional activity. These usually take the form of clays, sands, or gravels and occur most often in near surface deposits, although structural deformation can result in severe impurity problems at consid-

erable depth. More often than not, the greatest quantity of impurity in a gypsum deposit is anhydrite (and vice versa), a result of the metastable relationship of these two minerals, as previously discussed.

As utilized, most gypsum contains from 10 to 15% of impurities, although some deposits may be exceptionally pure (i.e., plus 95%) or somewhat impure (i.e., 80%). In general, the amount of impurity which can be tolerated depends upon: (1) the type of impurity, (2) the product being manufactured, and (3) the competitive situation.

Based upon the effect impurities may have upon the manufacturing process and finished products, impurities may be separated into three categories:

1) Insoluble or relatively insoluble minerals such as limestone, dolomite, anhydrite, anhydrous clays, silica minerals, etc.

2) Soluble chloride minerals such as halite, sylvite, etc.

3) Hydrous minerals such as the sulfate salts mirabilite and epsomite and the montmorillonite group of clays.

The first category, to the extent that the minerals replace gypsum, reduces the strength of rehydrated stucco and increases weights of the finished plaster or wallboard, i.e., more pounds of an impure stucco are required to obtain a given strength. Nevertheless many commercial gypsum deposits contain as much as 10 to 15% of these insoluble impurities. The presence of the second category, soluble chloride salts, affects calcining temperature and stucco slurry consistency and set time. These impurities are usually limited to no more than 0.02 to 0.03%. The principal impact of the third category is in moisture pickup of the finished product and on bonding characteristics of the gypsum stucco core of wallboard to its paper covering. Hydrous sulfate salts must be limited to 0.02 to 0.03% (too much causes the paper to peel off wallboard); however, hydrous clays up to 1.0 to 2.0% may be tolerated.

MAJOR NORTH AMERICAN PRODUCING AREAS

The principal gypsum producing areas in North America are shown in Fig. 1. In addition, there are many more known, but undeveloped gypsum deposits, especially in the midwestern and western states, southeastern Canada, and Mexico. Domestic reserves are geographically distributed in 23 states and have been estimated to contain 18.1 Gt (20 billion tons) or approximately 2000 years' supply at current production rates (Schroeder, 1970). Quantitative information is lacking on gypsum reserves in the rest of the world, but deposits are widespread, and inferred reserves are considered to be unlimited in relationship to requirements.

A brief description of the major North American producing districts follows.

Southeastern Canada

Marine evaporates of Mississippian (Visean) age underlie much of southern New Brunswick, northern Nova Scotia, and southwestern Newfoundland. The first production of gypsum in North America was recorded in Nova Scotia about 1735 and has grown to make this the largest gypsum mining region in the world, with an annual production exceeding 5.4 Mt (6 million tons).

Calcium sulfate is found in, or is correlative with, the Windsor Group which has been divided into five subzones: A, B, C, D, and E in ascending order. Although calcium sulfate is present in each of these subzones, the economic deposits of gypsum and anhydrite are confined to the Lower Windsor, i.e., subzones A and B. The most extensive is the A subzone, consisting of up to 305 m (1000 ft) of anhydrite

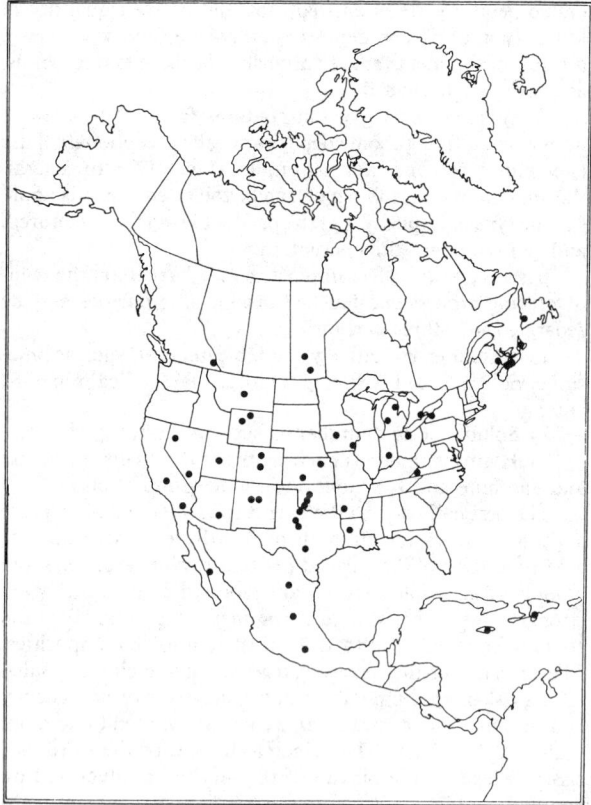

Fig. 1. Location of North American gypsum-producing districts.

with occasional thin dolomite beds. Where topographic conditions are favorable, surface hydration, which rarely exceeds 9 to 12 m (30 to 40 ft), has produced several deposits of dense, white, high purity gypsum.

The B subzone, which varies in thickness from 76 to 152 m (250 to 500 ft) consists of interbedded calcium sulfate, limestones, and siltstones. The calcium sulfate, representing about 80% of the section, has been hydrated to depths varying from 46 to 213 m (150 to 700 ft), although occasional remnant anhydrite lenses with associated soluble salts are encountered. Unlike most gypsum deposits which have only minor structural features, if any, the rocks of the B subzone are frequently deformed by complex plastic folding and by occasional faults of substantial displacement, which make for difficult mining conditions. Nevertheless, the B subzone accounts for approximately 75% of the region's total production.

The southeastern Canada gypsum deposits are usually covered by glacial till which may be as thick as 61 m (200 ft), but averages 12 to 15 m (40 to 50 ft). All mining is by open pit methods.

Iowa

The largest single gypsum producing district in the United States is at Fort Doge, IA, where mining began in 1872. Long considered to be an outlier of the Permian Basin, this deposit is now thought to be Jurassic in age (Cross, 1966). It consists of a single bed of gypsum up to a maximum of 9 m (30 ft) in thickness, but which probably averages nearer 3.7 to 4.3 m (12 to 14 ft). Its areal extent of some 39 km² (15 square miles) appears to be limited by post-depositional erosion, particularly during the period between

Late Cretaceous and the Pleistocene. Its continuity is interrupted by wide channels apparently formed prior to and/or during the glacial period, channels which cut through the gypsum into the underlying shales.

Overburden consisting of glacial till, varies from 6 to 18 m (20 to 60 ft), and the rock surface is commonly cut with mud-filled erosion crevices which must be cleaned out, thereby posing a mining problem; however, gypsum purity as used is usually plus 90%. Mining is by open pit methods, with four companies operating in the district.

Michigan

Another important gypsum source in the midwest is the Michigan Basin, with two producing districts: Grand Rapids in the southwestern portion of the basin, and Alabaster-National City near the northeasterly edge on Lake Huron. These deposits are in the Michigan formation of Mississippian age, and consist of multiple units of gypsum (5 to 40 ft) thick, separated by beds of shale varying from a few meters (feet) to 12 to 15 m (40 or 50 ft) thick.

Three operations are active in the Alabaster-National City area, using open pit mining methods to recover three gypsum beds [separated by 1.5 m (5 ft) of shale] at Alabaster, and a single bed at National City and Turner. Overburden is glacial till ranging in thickness from 12 to 21 m (40 to 70 ft), and at Alabaster, where the gypsum grades into anhydrite as the overburden thickness increases, a minor amount of anhydrite is also produced.

New York—Ontario

Gypsum is found in a belt of Upper Silurian rocks which extends eastward from the Niagara Peninsula in Ontario across New York to the vicinity of Utica. The first recorded mining of gypsum in the United States took place in this belt in 1808 near Syracuse. In recent years, however, all production has come from the Oakfield area near Buffalo in the western part of the state, where one operation currently is active. The economic gypsum here consists of a 91-122-cm (36-48-in.) thick bed in the Salina Group of Upper Silurian age, overlain by 3 to 12 m (10 to 40 ft) of thinly bedded shales and dolomites, covered by glacial till ranging up to 15 m (50 ft) in thickness.

Ohio

On the shores of Sandusky Bay of northern Ohio, gypsum has been continuously mined since 1849, with one open-pit mine being currently active. Several beds of calcium sulfate are found in the Tymochtee formation of Upper Silurian age, all of which are relatively impure due principally to inclusion of dolomite stringers and nodules and, to a lesser degree, anhydrite. As mined, this gypsum is probably the lowest grade of any commercial deposit in the United States (with the exception of gypsite in California) being in the 65 to 70% range; however, mine production is beneficiated by heavy media sink-float to produce an approximately 80 to 81% gypsum product.

Indiana

In 1952, gypsum was discovered in southern Indiana near the town of Shoals, and this has become one of the most important commercial gypsum producing districts in the United States. In part of the Illinois Basin, gypsum occurs in the St. Louis formation of Mississippian age, and lies 107 to 168 m (350 to 550 ft) below the surface of this rather hilly area. The bed averages 4.3 to 4.9 m (14 to 16 ft) thick.

The St. Louis limestones overlying the gypsum have been subjected to solution along bedding and joint planes to create a network of water-filled cavities of undetermined extent and location.

Virginia

In the western tip of Virginia near Saltville, gypsum has been produced continuously since 1835, first from shallow pits or quarries and currently from one underground mine.

The calcium sulfate originated in the McCready formation of Mississippian age, but the deposits are structurally very complex as a result of Appalachian mountain building, and in particular, the Saltville fault, a major northeast-south-west trending thrust fault which in the Plasterco area has a vertical displacement of over 4.3 km (14,000 ft). As mined, the calcium sulfate is found in part as a more or less continuous bed, approximately vertical and varying in width from 2 to 3 m (8 to 10 ft) up to 6 m (20 ft); and in part as remnants or fragments of an originally horizontal bed (or beds) known locally as "boulders."

Hydration of the calcium sulfate is highly irregular and, as with most deposits, is a function of the availability of ground waters. As a general trend, there is less hydration at depth than near the surface; however, the ultimate limit of sufficient hydration to support a viable operation is as yet unknown.

Permian Basin

One of the major geological features of the United States is the Permian Basin extending over parts of Texas, New Mexico, Oklahoma, and Kansas, and calcium sulfate is found extensively over this area. Where hydration has occurred, the resulting gypsum is usually of high (plus 92%) purity and, in fact, most of the specialty gypsum products which require high quality raw material are made from deposits of Permian age. The best purity and whitest color gypsum being commercially exploited in the United States is found in the Blaine formation of the Permian at Medicine Lodge, KS, and Southard, OK, and in the Easly Creek formation of the Permian at Blue Rapids, KS.

At the present time, there are seven operating mines in Texas, six in Oklahoma, and two in Kansas. All of the Texas and Oklahoma deposits are worked by open pit methods, but the two Kansas mines are underground.

Although considerable Permian calcium sulfate is found in southeastern New Mexico, none currently is being mined. There are, however, two operating mines(open pit) in the north-central part of the state working gypsum of the Todilto formation of Jurassic age.

Rocky Mountain and Intermountain States

This general area is particularly rich in calcium sulfate deposits, most of which are at, or just below, the surface and lend themselves to low cost open pit mining. As might be expected in this region, structural features play an important part in limiting the size of the deposits and in mine design. In general, purity is good, but the extent of hydration is frequently difficult to predict. A major deterrent to development is the cost of transportation due in part to the distance to roads and railroads, and, also, the long hauls to major markets.

Gypsum is currently produced in the Powder River Basin area of Wyoming from rocks of Upper Permian to Middle Jurassic age, and from the Front Range of Colorado in Larimer and Fremont Counties. Production has also been developed in Fergus County, MT (Piper formation, Jurassic age) and Sevier County, Utah (Arapien shale, Jurassic age).

In Nevada, two deposits are mined in Clark County and one in Pershing County, with most of the product from these operations being shipped to California markets. In Arizona, a deposit at Winkleman provides gypsum for a calcining and manufacturing operation in Phoenix, and three smaller operations in the state supply portland cement retarder and agricultural markets. All of these western operations are open pit mines with the exception of Montana.

California

Gypsum rock for five of the seven calcining and manufacturing plants in California is shipped into the state from Mexico (four plants) and Nevada (one plant). The other two utilize gypsum from a deposit in Imperial County, which in 1979 produced approximately 816 Mt (900,000 tons), making it the largest operating gypsum mine in the United States. This deposit, of Miocene age, occurs as a remnant of a calcium sulfate bed well over 30.5 m (100 ft) thick whose present extent is limited by structure and erosion. Nevertheless, it contains large reserves, in part because hydration has taken place to a greater degree than in many other cases. Little or no overburden is present, and mining is carried out by multi-bench open pit methods (ver Plank, 1952).

For many years, several low grade gypsite deposits have been worked in San Joaquin Valley to produce a 50 to 75% gypsum content material for use as a soil conditioner. These deposits are Holocene in occurrence and are of limited extent. Although of too low a purity for portland cement retarder or plaster products, they are important in treating the alkaline soils which predominate in the great agricultural valleys of the state.

Western Canada

Gypsum is produced from two open-pit mines, one in Manitoba, the other in British Columbia. These operations provide rock for eight calcining and manufacturing plants in the western provinces. Also, one plant in Vancouver is supplied with rock imported from Mexico.

Production from these deposits is used almost entirely in the manufacture of building products and as a portland cement retarder. In both cases, only mining and primary crushing operations are carried on at the mine site, with the crushed gypsum rock being shipped to plants located in or near the major cities of Winnipeg, Saskatoon, Calgary, Edmonton, and Vancouver.

Mexico

The primary use for gypsum in Mexico is as a retarder for portland cement and for building plaster, although in 1971 two small wallboard plants began operation in the city of Puebla in an attempt to introduce wallboard to the construction markets. Considerable gypsum is also exported to the West and Gulf Coasts of the United States, and on occasion, to the Far East.

Many large deposits are known to exist; however, at the present time the major producers are located on San Marcos Island in the Gulf of Baja California and in the states of Puebla, San Luis Potosí, and Nuevo Leon. In general, the gypsum is found in rocks of Mesozoic and Tertiary age.

The largest volume of production in Mexico is from San Marcos Island where an open pit mine and deep water ship loading facilities are operated to produce and ship a crushed gypsum rock. Another large highly mechanized open pit mine is operated at Estación La Borreguita in the state of San Luis Potosí to produce a crushed gypsum which is shipped by rail to local cement companies and to the port of Tampico for export.

In the state of Puebla, from the town of Izucar de Matamoros on the east to the village of Axochiapan on the west, a zone of gypsum bearing rocks of Miocene-Pliocene age occurs in which several small quarries are operated. These provide gypsum for some eight or nine cement plants in the Mexico City-Puebla area, for two new gypsum board plants at Puebla, and for several small calcining operations making gypsum plaster. The highest purity gypsum presently mined in Mexico comes from near Monterrey, Nuevo Leon, where at least two calcining plants operate to produce plaster for both construction and industrial uses.

Caribbean Area

Two deposits are being worked in the Caribbean Area: one at Bull Bay, Jamaica, about 16 km (10 miles) east of the capital of Kingston; and the other near Barahona, Dominican Republic, some 209 km (130 miles) west of the capital city of Santo Domingo. In both countries only small domestic gypsum markets have been developed primarily for portland cement retarder, and the major part of the production serves the export trade.

In Jamaica calcium sulfate is of Lower Eocene age, and occurs as several separate bodies of relatively small size in a northwest--southeast trending basin known as the Wagwater Trough (Holliday, 1971). Perhaps 60 to 70% of the calcium sulfate mass has been hydrated, and gypsum is currently being mined from three different deposits, utilizing multibench open pit methods.

The Dominican Republic gypsum is of probable Upper Miocene age, and the largest outcrop is found in the Cerro de Sal, a 19-km (12-mile) long hogback trending northwest-southeast. Calcium sulfate is interbedded with limestones, claystones, sandstones, and salt of the Cerro de Sal formation with a total thickness of some 610 to 762 m (2000 to 2500 ft). Bentonite clays and soluble salts are a potential problem associated with some stratigraphic members. The geological structure has been complicated due to several repetitions of the stratigraphic column caused by faulting; however, large deposits of economic gypsum may exist. Mining is by open pit methods. At least three other gypsum outcrops are known in this general area, but to date have not been developed for production.

GYPSUM RESOURCES OUTSIDE NORTH AMERICA

Reflecting its wide geologic occurrence and its basic use in the construction industry, gypsum is produced in some 72 countries around the world. No statistics are available as to the potential world reserves of calcium sulfate, but it can be safely predicted that they are enormous both in terms of total tons and of years at the present rates of consumption.

In many of the countries reporting production, gypsum was used only for portland cement retarder, this being by far the most common use of the mineral on a worldwide basis. As a general rule, the calcining of gypsum for use as plaster and wallboard building materials, or for industrial plasters, is limited to the United States, Canada, Europe (including Russia), and Japan, although a few wallboard plants have recently been built or are being considered in certain South American and Asian countries. In these countries, the general practice of open structural frames, usually made of wood, to form a building has historically required the use of some type of covering material for walls and ceilings, and gypsum products have gained wide acceptance for this use. That is, gypsum in the form of either plaster or precast wallboard is a versatile, fireproof covering material, and these products account for over two-thirds of the gypsum used in the United States and Canada.

Conversely, in those countries where the usual methods of construction rely heavily on masonry products, such as concrete, concrete block, stone, and brick (including adobe), the need for fireproof covering materials is much less, and the use of gypsum for the manufacture of building materials will be very small, or even nonexistent. The production and use of gypsum in a given country, therefore, is usually a result of its mode of use rather than the availability of gypsum deposits.

EXPLORATION METHODS

As for most minerals, a thorough understanding of the origin and occurrence of gypsum is essential before undertaking an exploration program. This may be obtained by a review of published literature, and a basic reference would be "Selected Annotated Bibliography of Gypsum and Anhydrite in the United States and Puerto Rico" by C.F. Withington and M.C. Jaster (1960).

By studying the stratigraphic column in regions where sedimentary rocks occur, an indication can usually be obtained from the lithology whether or not an evaporite environment may have existed, this being, of course, a requisite for the deposition of calcium sulfate. If evaporite rocks are indicated, further details can be obtained from examination of outcrops of the particular formations or, if outcrops are not available, from logs of oil or gas wells or from water wells which may have been drilled in the vicinity. Perhaps drill cuttings, or even core, might be available for inspection. As stratigraphic and lithologic details of an area are thus developed, structure should also be mapped so as to provide as detailed a picture of the area as the basic data will permit. From this, the possibility of calcium sulfate occurrence usually can be predicted and, also, it may be possible to judge the potential for hydration of the sulfate zones.

Sources of this type of information are published literature, unpublished university theses, geological surveys, geological departments of colleges and universities, oil companies, and water well drillers.

Geophysical work is only partially applicable to the search for calcium sulfate. The right combination of logs may indicate gypsum or anhydrite, but no log has yet been devised that will replace accurately a laboratory chemical or purity analysis. Geophysical logs are useful for determining structural details and can be used, in conjunction with other tests, to help determine mining environment conditions. Geochemical methods are rarely applicable to calcium sulfate, although heavy concentrations of sulfate in ground waters might serve as a useful clue, and in a few instances differences in plant species may indicate changes in rock types that can aid in surface mapping.

Even though the extent of gypsum can be mapped from surface outcrops, drilling is necessary to predict the amount and regularity of hydration, i.e., how much of the calcium sulfate is gypsum, and how much is either only partially hydrated or is anhydrite. For deeper deposits, additional drilling data is almost always required to map projections of lithologies and/or structures. Core drilling is usually preferred, taking BX (or larger) size core to provide adequate material for sampling and to permit good core recovery. In some cases where it is only necessary to determine the depth of overburden, or to locate the contact between gypsum and anhydrite, rotary drilling is sufficient and is much less expensive. Also geophysical methods can be successfully used under some conditions to determine the depth of overburden.

Although calcium sulfate beds may sometimes be continuous over tens—or even hundreds—of miles, the exploration geologist should keep in mind that they may also occur as discontinuous lenses. Also the extent of hydration is even less predictable, as is the occurrence of minor, but detrimental, amounts of impurities. Hence a successful exploration program requires knowledgeable consideration and interpretation of all the factors involved, and sums of money which can run into six figures.

BIBLIOGRAPHY AND REFERENCES

Adams, J.E., 1971, "Upper Permian Ochoa Series of Delaware Basin, West Texas and Southeastern New Mexico," *Origin of Evaporites,* American Association of Petroleum Geologists, Tulsa, OK, Reprint Series No. 2, pp. 60-89.

Appleyard, F.C., 1965, "The Locust Cove Mine," *Mining Engineering,* March, pp. 59-62.

Bergstrom, J.H., 1961 "Pabco's Gypsum Crystals Sparkle," *Rock Products,* April, pp. 90-95.

Borchert, H., and Muir, R.O., 1964, *Salt Deposits, The Origin, Metamorphism, and Deformation of Evaporites,* D. Van Nostrand, London, 338 pp.

Braitsch, O., 1971, *Salt Deposits, Their Origin and Composition,* Springer-Verlag, New York, 297 pp.

Branson, E.B., 1915, "Origin of Thick Gypsum and Salt Deposits," *Geological Society of America Bulletin,* Vol. 26, pp. 231-242.

Briggs, L.I., 1970, "Geology of Gypsum in the Lower Peninsula, Michigan," *Proceedings, Sixth Forum on Geology of Industrial Minerals,* Michigan Geological Survey, Misc. 1, pp. 66-76.

Butler, G.P., 1970, "Holocene Gypsum and Anhydrite of the Abu Dhabi Sabkha, Trucial Coast: An Alternative Explanation of Origin," *Proceedings, Third Symposium on Salt,* Northern Ohio Geological Society, Cleveland, Vol. 1, pp. 120-152.

Buzzalini, A.D., et al., 1969, ed. "Evaporites and Petroleum," *Bulletin,* American Association of Petroleum Geologists, Vol. 53, No. 4, April, pp. 775-1011.

Cole, L.H., 1930, *The Gypsum Industry of Canada,* Canadian Dept. of Mines, Publication 714, pp. 1-164.

Conley, R.F., and Bundy, W.M., 1958, "Mechanism of Gypsification," *Geochimica et Cosmochimica Acta,* Vol. 15, pp. 57-72.

Cross, A.I., 1966, "Palynologic Evidence of Mid-Mesozoic Age of Fort Dodge (Iowa) Gypsum," Abstract, San Francisco Meeting, Geological Society of America, Nov.

Dean, W.E., and Schreiber, B.C., eds., 1978, "Marine Evaporates," Mineral Short Course, *Notes,* No. 4, Society of Economic Paleontologists, 188 pp.

Friedman, G.M., and Sanders, J.E., 1978, *Principles of Sedimentation,* John E. Wiley & Sons, Inc., New York.

Gay, P., 1965, "Some Crystallographic Studies in the System $CaSO_4$—$CaSO_4 \cdot 2H_2O$, The Hydrous Forms," *Mineralogical Magazine,* Vol. 25 pp. 354-362.

Goodman, N.R., et al., 1957, "Gypsum," *The Geology of Canadian Industrial Mineral Deposits,* 6th Commonwealth Mining and Metallurgical Congress, pp. 111-137.

Ham, W.E., 1962, "Economic Geology and Petrology of Gypsum and Anhydrite in Blaine County," *Oklahoma Geological Survey Bulletin,* Vol. 89, pp. 100-151.

Ham, W.E., et al., 1961, "Borate Minerals in Permian Gypsum of West-Central Oklahoma," *Oklahoma Geological Survey Bulletin,* Vol. 92, 77 pp.

Hardie, L.A., 1967, "The Gypsum-Anhydrite Equilibrium at One Atmosphere Pressure," *American Mineralogist,* Vol. 52, Jan.-Feb., pp. 171-200.

Holliday, D.W., 1970, "The Petrology of Secondary Gypsum Rocks: A Review," *Journal of Sedimentary Petrology,* Vol. 40, No. 2, June, pp. 734-744.

Holliday, D.W., 1971, "Origin of Lower Eocene Gypsum-Anhydrite Rocks, Southeast St. Andrew, Jamaica." *Transaction Sec. B,* Institution of Mining and Metallurgy, Vol. 80, pp. B305-B315.

Jorgensen, D.B., and Carr, D.D., 1972, "Influence of Cyclic Deposition, Structural Features, and Hydrologic Controls on Evaporite Deposits in the St. Louis Limestone in Southwestern Indiana," *Proceedings,* 8th Forum on Geology of Industrial Minerals, Iowa Geological Survey, 195 pp.

Keith, S.B., 1969, "Gypsum and Anhydrite," *Arizona Bureau of Mines Bulletin 180,* pp. 371-382.

Kelly, K.K., et al., 1941, "Thermodynamic Properties of Gypsum and Its Dehydration Products," *Technical Paper 625,* US Bureau of Mines, 73 pp.

Kendall, A.C., 1978, "Calcium Sulfate Diagenesis, Mississippean of Eastern Williston Basin," Abstract, *American Association of Petroleum Geologists Bulletin,* Vol. 62, No. 3, p. 530.

Kerr, S.D., and Thompson, A., 1963, "Origin of Nodular and Bedded Anhydrite in Permian Shelf Sediments, Texas and New Mexico," *American Association of Petroleum Geologists Bulletin,* Vol. 47, pp. 1726-1732.

King, R.H., 1971, "Sedimentation in Permian Castile Sea," *Origin of Evaporites,* American Association of Petroleum Geologists, Tulsa, OK, Reprint Series No. 2, pp. 90-97.

Kinsman, D.J.J., 1966, "Gypsum and Anhydrite of Recent Age, Trucial Coast, Persian Gulf," *2nd Symposium on Salt,* Northern Ohio Geological Society, Cleveland, Vol. 1, pp. 302-326.

Kinsman, D.J.J., 1969, "Modes of Formation, Sedimentary Associations, and Diagenetic Features of Shallow Water and Supratidal Evaporites," *American Association of Petroleum Geologists Bulletin,* Vol. 53, No. 4, April, pp. 830-840.

MacDonald, G.F.J., 1953, "Anhydrite-Gypsum Equilibrium Relations," *American Journal of Science,* Vol. 251, pp. 883-898.

Martinez, J.D., 1974, "Tectonic Behavior of Evaporites," *Proceedings,* 4th Symposium on Salt, Northern Ohio Geological Society, Cleveland, Vol. 1, pp. 155-168.

McAide, H.G., 1964, "The Effect of Water Vapor Upon the Dehydration of $CaSO_4 \cdot 2H_2O$," *Canadian Journal of Chemistry,* Vol. 42, pp. 792-801.

Murray, R.C., 1964, "Origin and Diagenesis of Gypsum and Anhydrite," *Journal of Sedimentary Petrology,* Vol. 34, No. 3, Sep., pp. 512-523.

Newland, D.H., and Leighton, H., 1910, "Gypsum Deposits of New York," *New York State Museum Bulletin 143,* 94 pp.

Posnjak, E., 1938, "The System, $CaSO_4 \cdot H_2O$," *American Journal of Science,* Vol. 235A, pp. 247-272.

Posnjak, E., 1940, "Deposition of Calcium Sulfate From Sea Water," *American Journal of Science,* Vol. 238, pp. 559-568.

Pressler, J.W., 1979, Annual Advance Summary, "Gypsum in 1979," US Bureau of Mines, 17 pp.

Pressler, J.W., 1981, "Gypsum," *Minerals Yearbook 1980,* US Bureau of Mines, pp. 385-396.

Riley, C.M., and Byrne, J.V., 1961, "Genesis of Primary Structures in Anhydrite," *Journal of Sedimentary Petrology,* Vol. 31, pp. 553-559.

Schenk, P.E., 1969, "Carbonate-Sulfate-Redbed Facies and Cyclic Sedimentation of the Windsorian Stage (Middle Carboniferous), Maritime Provinces," *Canadian Journal of Earth Sciences,* Vol. 6, pp. 1037-1066.

Schmidt, R., 1914, "Uber die Beschaffenheit und Entstehung parallelfaseriger Aggregate von Steinsalz und Gips," *Kali.* 8, pp. 161, 218, 239.

Schroeder, Harold J., 1970, "Gypsum," *Mineral Facts and Problems,* Bulletin 650, US Bureau of Mines, pp. 1039-1048.

Shearman, D.J., 1966, "Origin of Marine Evaporites by Diagenesis," *Transactions Sec. B,* Institution of Mining and Metallurgy, Vol. 75, pp. B208 to B215.

Stone, R.W. et al., 1920, "Gypsum Deposits of the United States," Bulletin 697, US Geological Survey, 326 pp.

Stonehouse, D.H., 1970, "Gypsum and Anhydrite," *Canadian Minerals Yearbook,* Mineral Resources Branch, Dept. of Energy, Mines, and Resources, Ottawa, pp. 775-1011.

Stonehouse, D.H., 1972, "Gypsum and Anhydrite," *Canadian Minerals Yearbook,* Mineral Resources Branch, Dept. of Energy, Mines, and Resources, Ottawa, preprint, 7 pp.

Stonehouse, D.H., 1978, "Gypsum in Canadian Mineral Industry," Dept. of Energy, Mines and Resources, Ottawa, pp. 211-216.

Stonehouse, D.H., 1979, "Gypsum in Canadian Mineral Industry," Dept. of Energy, Mines and Resources, Ottawa, pp. 221-228.

Stonehouse, D.H., 1980, "Gypsum in Canadian Mineral Industry," Dept. of Energy, Mines and Resources, Ottawa, preprint, 8 pp.

Van't Hoff, J.H., et al., 1903, "Gips und Anhydrit," *Zeitschrift fuer Physik Chem.,* Vol. 45, pp. 257-306.

ver Planck, W.E., 1952, "Gypsum in California," Bulletin 163, California Div. of Mines, 151 pp.

Walker, C.W., 1974, "Nature and Origin of Cap Rock Overlying Gulf Coast Salt Domes," *Proceedings,* 4th Symposium on Salt, Northern Ohio Geological Society, Cleveland, Vol. 1, pp. 169-175.

Withington, C.F., 1962, "Gypsum and Anhydrite in the U.S.," Min. Inv. Res., Map MR-33 and text, US Geological Survey, 18 pp.

Withington, C.F., and Jaster, M.C., 1960, "Selected Annotated Bibliography of Gypsum and Anhydrite in the United States and Puerto Rico," Bulletin 1105, US Geological Survey, 126 pp.

Wood, G.V., and Wolfe, M.J., 1969, "Sabkha Cycles in the Arab-Darb Formation of Trucial Coast of Arabia," *Sedimentology,* Vol. 12, pp. 165-191.

Zen, E-An, 1965, "Solubility Measurements in the System $CaSO_4$-$NaCl$-H_2O at 35°, 50°, and 70° C and One Atmosphere Pressure," *Journal of Petrology,* Vol. 6, pp. 124-164.

10. Kyanite and Related Minerals*

INTRODUCTION

The sillimanite family of minerals, including kyanite, sillimanite, and andalusite, are anhydrous aluminum silicates with the formula $Al_2O_3 \cdot SiO_2$. Dumortierite and topaz are also included in this group because they are closely allied in composition and thermal behavior. However, neither are mined commercially today. They are typical metamorphic minerals which are found in metamorphic rocks on every continent. Sillimanite minerals are prized chiefly for their refractoriness and are important components in a broad range of acid refractory products, especially in mortars and castables. While these minerals have widespread occurrence, the consumption of sillimanite minerals is concentrated in the relatively few highly industrialized areas where refractories are manufactured, and which in turn are typically close to the major iron and steel producing regions of the world. Thus northern Europe, England, the United States, and Japan are the principal consumers of refractories and sillimanite minerals. Of these countries, only the US is a significant producer.

Kyanite

The largest use for domestic kyanite, both raw and calcined, is in the manufacture of refractory mortars, cements, castables, and plastic ramming mixes. In these applications kyanite constitutes from 10 to 40% of the mixture, the balance being refractory clays and coarser grog materials. A certain proportion of raw kyanite is used to offset the shrinking of the clay binder, whereas the calcined kyanite is used in the coarser sizes for body.

Sillimanite

Granular and massive forms of sillimanite are used in much the same way as granular and massive kyanite. Massive sillimanite from India and South Africa has been used for sawed block refractories principally in the glass industry in Europe. The massive varieties of kyanite and sillimanite have not enjoyed much use in the United States in recent years, but have been widely used by European refractory producers. Conversely, until recently, little granular domestic kyanite had found much application outside of North America. The use pattern is changing as a result of the depletion of the high grade massive deposits of the world, and of the increasing production of kyanite, andalusite, and sillimanite concentrates in India and Africa from deposits formerly thought to be unworkable. Andalusite concentrations are now being produced in France and are displacing or augmenting aluminum silicate sources from abroad.

The potential use of kyanite, sillimanite, and andalusite for the production of aluminum-silicon alloys has been investigated. To be an economic possibility, however, it appears that it would be necessary to produce aluminum silicate concentrates considerably purer than those currently produced and at a cost about half what the present market will bear. Short of some unforeseen and dire circumstances cutting off the free world trading and marketing of bauxite and silica, the likelihood of such production seems remote.

*Excerpt from Bennett, P.J., and Castle, J.E., 1983, "Kyanite and Related Minerals," *Industrial Minerals and Rocks,* 5th ed., Vol. 2, S.J. Lefond, ed., AIME, New York, pp. 799-807.

GEOLOGY

Mineralogy

The properties of the kyanite group of minerals are given in Table 1.

Classification of Deposits

Kyanite: *Kyanite Quartzite*—The kyanite produced in the United States comes from kyanite quartzites. Kyanite quartzites are rocks containing 15 to 40% kyanite and usually about 5% of other minerals such as pyrite, lazulite, rutile, and mica. The rock is characterized by an anomalously low content of potash and soda and the virtual absence of calcium and magnesium. The alumina content of kyanite quartzite ranges between 10 and 25% and averages about 18% Al_2O_3, which is generally similar to that of the schistose rocks which enclose the kyanite-bearing strata.

Kyanite quartzite deposits occur in the Piedmont region of the Appalachian Mountain system in a relatively narrow zone extending from northeastern Georgia through central South and North Carolina to southeastern Virginia. They are enclosed within a wider band of mildly metamorphosed acid metavolcanics and sediments known as the Volcanic Slate Belt in Virginia and the Little River Series in Georgia. The age of the rocks is uncertain, but several workers have suggested that they are lower Cambrian in age (Bennett, 1961).

More recent studies by Carpenter and Allard (1980) expand the mineralization and alteration theories of kyanite in the Georgia-South Carolina-McCormick district.

There are at least 13 distinct deposits of kyanite quartzite within this zone. All of them are known as "mountains" owing to their solitary prominence rather than their relatively modest height. Graves Mountain in Georgia, and Willis Mountain, Baker Mountain, and East Ridge all in Virginia are being mined at the present time. Until 1970, the Henry's Knob deposit in South Carolina was also mined (Espenshade and Potter, 1960).

In southeastern California and extending into southwestern Arizona there is a zone of metavolcanics whose composition and lithology resembles that of the Carolina Slate Belt in which kyanite quartzite deposits are also found. However, the kyanite in these deposits, while averaging 25 to 35% of the rock in apparently large and recoverable crystals, has proven to be contaminated with extremely fine inclusions of quartz. To date, it has not been possible to produce a competitive kyanite product from these western kyanite quartzite deposits.

Kyanite quartzite has been described in several places in the world, and usually seems to occur within a geologic framework similar to that of the deposits of the southeastern United States. Such deposits have been described and explored in Surinam, Norway, Kenya, and Austria (Varley, 1968).

Kyanite Schist and Gneisses—Kyanite is very common in the highly metamorphosed schists and gneisses of the metamorphic regions of the world. Typically, the kyanite occurs in quantities ranging from a percent or two to as much as 25% in a gangue of biotite, feldspar, muscovite, garnet, and occasionally hornblende and other common rock forming minerals. Rocks containing a few percent of kyanite are extremely abundant and widespread. They are exposed over hundreds of square miles in the eastern and western

Table 1. Properties of Kyanite Group Minerals

	Andalusite	Kyanite	Sillimanite	Mullite
Formula	$Al_2O_3 \cdot SiO_2$	$Al_2O_3 \cdot SiO_2$	$Al_2O_3 \cdot SiO_2$	$3Al_2O_3 \cdot 2SiO_2$
Crystal system	Orthorhombic	Triclinic	Orthorhombic	Orthorhombic
Cleavage	{110} good	{100} perf.	{010} good	{010} perf.
	{100} poor	{010} good		
		{001} parting		
Hardness	6-5-7	5.5-7*	6.5-7.5	6-7

*Varies with direction.

metamorphic areas of North America and in the metamorphic rocks of other continents.

Repeated attempts to recover the kyanite from such rocks have been made. The most recent attempt was in the Timiskaming district in western Quebec by North American Refractories Co. of Cleveland, OH. An earlier effort was made in the late 1930s near Burnsville, NC. At the present time no kyanite schists are being mined for kyanite.

In order for such kyanite to be economic it is necessary that the region be deeply weathered and of gentle relief so that a mantle or segregation of resistant kyanite nodules, cobbles, and boulders can be accumulated at the surface. It is further necessary for labor to be abundant and cheap.

A great deal of literature has been published by the US Bureau of Mines, the US Geological Survey, and agencies of other governments describing investigations of kyanite-bearing schists. An exhaustive study was conducted by the USBM in Idaho on the huge deposits of kyanite, sillimanite, and andalusite at Goat Mountain and the kyanite deposit on Woodrat Mountain near Kemiah (Van Noy, 1970). Hundreds of millions of tons of 25% aluminum silicate ore available for surface mining has been postulated. However, remoteness from the major markets and the difficulty of beneficiation have stalled development at this locality.

Massive Kyanite—Kyanite is found locally as nodules, knots, and huge boulder-sized segregations in very highly metamorphosed areas of aluminous sediments. This had been the principal source of kyanite from India for the past 40 years. Similar segregations were the basis for the kyanite production in Kenya which has been discontinued.

In several counties in Georgia, similar lumps of massive kyanite are found (Furcron and Teague, 1945). While the abundance and purity of the Georgia massive kyanite meet the requirements for commercial exploitation, high labor cost makes economic production by hand-sorting and gathering unfeasible and production has been limited to a few carloads during World War II.

Kyanite mineral segregations are probably the result of local pneumatolytic migration of silica and alumina during the late stages of regional metamorphism. Introduction of alumina does not seem to have been a factor since the overall composition of the segregations are similar to that of the country rock if the sample area considered is large enough to include the barren quartz segregations which invariably accompany the kyanite-corundum segregations. Aside from the size and abundance of the segregations in the Lapsa Bura deposits of India, they do not seem mineralogically dissimilar from kyanite segregations found occasionally in all kyanite schists.

Massive kyanite typically contains corundum and minor amounts of rutile. The kyanite is often felty and acicular and occurs in tightly interlocking aggregates. Kyanites from India are usually produced in lumps large enough to be hand-

sorted according to kyanite and corundum content, and three grades are offered. Massive Indian kyanite has properties quite unlike that of large coarse kyanite crystals. Massive Indian kyanite is essentially volume-stable and calcines to a dense white aggregate that is much prized by European refractory manufacturers. On the other hand, the coarse kyanite produced from Georgia placer deposits in the 1940s, and more recently in Kenya, crumbles and loses much of its density and physical strength upon calcining. Apparently the interlocking acicular crystal mode of "Indian" kyanite prevents such expansion and consequent breakdown.

Sillimanite: *Sillimanite Schists*—Sillimanite is a very common rock-forming mineral in metamorphic rocks of relatively high rank. It is common in a series of metamorphosed rocks to find sillimanite, kyanite, and andalusite interchanging occurrences in given strata as the local conditions of temperature vary, as in the proximity to intrusives. For this reason the aluminum silicate minerals are often used to identify parameters of metamorphic intensity. Sillimanite, while common, seldom occurs as potentially exploitable crystals. The typical mode is what is often called "fibrolite" which is a felty aggregate of extremely fine "whisker" of acicular sillimanite interlaced and interlocked with quartz, mica, and other minerals. Beneficiation is usually impossible. In some areas the sillimanite occurs as nodules and buttons which are marginally potential as in the Peltzer area of South Carolina (Espenshade and Potter, 1960).

In Hart County, Georgia, there is a northeasterly trending zone about 16 km (10 miles) long in which "matchstick" sillimanite occurs. Beneficiation tests have been conducted, from which it appears that a limited production of sillimanite in the 35 mesh range could be accomplished. However, the deposits are narrow and limited, and there is presently no existing market or incentive for production of sillimanite concentrates in the United States (Furcron and Teague, 1945).

Massive Sillimanite—Massive sillimanite has been produced for many years from the state of Assam, India. Production has been falling off in recent years as the cost of production has risen and as other materials are substituted for the sawed blocks which were formerly an important use of sillimanite. The deposits of Assam consist of huge segregations of sillimanite and corundum, often in intimate association, and weighing several tons. Considerable hand effort is employed to recover the boulders in a form suitable for sawing to refractory shapes, particularly for the English glass industry.

In the vicinity of Adelaide, South Australia, late-stage metasomatism resulted in a mixture of kaolin and included boulders and nodules of sillimanite. The sillimanite is recovered as a byproduct in the process of manufacturing refractory clays. Other deposits of residual boulders of massive sillimanite have been exploited in this region. Beneficiation has been tried with some success on the sillimanite itself and on the byproducts too low in grade to use directly. The Australian domestic market for such concentrates is limited, however, and thus far no large scale production has been reported.

The most important deposits of massive sillimanite-corundum occurs in the Republic of South Africa, in the Pella District near Pofadder, Namaqualand.

Andalusite: *Andalusite*—Andalusite is a frequent constituent of metamorphic rocks, although it is not as abundant or common as sillimanite or kyanite. It is found in argillaceous and micaceous slates, and in schists and gneisses resulting from the contact metamorphism of intrusive rock. Andalusite readily incorporates foreign matter in its crystals, and frequently grows around preexisting materials, including carbon. One variety, chiastolite, is so named for the crosslike inclusions of carbon oriented normal to the axis of the crystal.

In France, near Glomel in Brittany, an extensive, deeply weathered body of andalusite schist is being mined at the present time. Here the andalusite occurs as matchstick-sized crystals embedded in a fine-grained black groundmass composed of biotite, hornblende, muscovite, and feldspar. The andalusite is evenly disseminated and constitutes about 20% of the rock. The rocks in the area are very poorly exposed, and the geology of this occurrence is not well understood.

Near Canso, Nova Scotia, there is an extensive deposit of andalusite schist. Here the andalusite makes up about 15% of the rock and is evenly disseminated as large porphryoblasts averaging about 25.4 × 12.7 mm (1 × $\frac{1}{2}$ in.) in cross section. The groundmass is principally muscovite, garnet, and feldspar. The crystals of black andalusite incorporate about 10% of finely disseminated magnetite and muscovite, however, and effective beneficiation is not practical.

Near Hillsboro, NC, a monadnock of andalusite-pyrophyllite-sericite rock is being mined by the Piedmont Minerals Co. The ore consists principally of pyrophyllite and quartz and contains 15 to 20% of disseminated pink andalusite. The andalusite is mined along with the pyrophyllite, and the resulting mixture is almost entirely consumed by the parent company in the manufacture of refractory products.

In the Goat Mountain deposit in Idaho, andalusite coexists with kyanite and sillimanite. No attempt has been made to separate it from the other aluminum silicate minerals in the tests made so far. The andalusite in this deposit typically incorporates a great many deleterious impurities. The Goat Mountain deposit is extremely large, but to date no commercial exploitation of the deposit seems feasible owing to beneficiation difficulties (Abbott and Prater, 1954).

Residual Andalusite—Alluvial deposits of andalusite sand occur in South Africa. These deposits have been worked on a very large scale. The source of the andalusite is shales of the Pretoria Series that have been intruded and metamorphosed by the Bushveld Complex. The andalusite has been weathered free from the parent rock and subsequently concentrated by the action of wind and water. Apparently concentration is still going on. The andalusite sand commonly contains 50% of recoverable andalusite. The reserves are estimated to be in the neighborhood of 800,000 tons of +50% andalusite. In 1964 a heavy media separation plant was installed, and since then the production of high grade concentrates has increased markedly.

Reserves

Domestic: The potential supply of kyanite minerals vastly exceeds the potential market. Therefore, what constitutes a reserve cannot be defined without considering the probability of production. The two producing domestic kyanite deposits have enormous proven reserves. At least 65 million tons of kyanite quartzite containing at least 25% kyanite and amenable to surface mining were indicated by an extensive US Bureau of Mines drilling program undertaken in 1949 at Willis Mountain in Virginia (Jones and Eiletsen, 1954). The potential reserves in this locality are probably twice the indicated reserves. At Graves Mountain, geologic mapping and diamond drilling has indicated a reserve of 25% ore in excess of 30 million tons. From these figures, which translate into many decades of production at current rates, it is readily apparent that the existing producers have little interest in acquiring new reserves unless a considerable economic advantage would result. If the +10% kyanite-bearing schists and gneisses are included, the reserves in the United States and Canada are truly vast. At current prices these reserves are obviously submarginal.

World: Most of the comments made about the reserves of kyanite minerals in the United States apply to other areas of the world as well. However, the massive kyanite reserves of India, being relatively rare and unique in occurrence, are more susceptible to reserve considerations. The Lapsa Buru area is credited with 700,000 tons of massive kyanite to a depth of 3 m (10 ft), and the Madhya Predesh area with 250,000 tons of massive sillimanite.

In the Republic of South Africa, 800,000 tons of +50% andalusite sand are reported for the Transvaal and millions of tons of sillimanite reserves have been credited to Namaqualand in the Cape Province.

No figures have been reported from the Province of Brittany, France, where andalusite is produced. However, the body of andalusite schist is reported to be at least 1 km wide and about 10 km long, and to contain an average of about 15 to 25% andalusite. This translates into many tens of millions of tons of potential reserves.

It is apparent that the existing producers of aluminum silicate minerals have adequate reserves, with the possible exception of the unique and relatively rare deposits of massive kyanite-corundum and massive sillimanite-corundum. It should be noted, however, that the primary use for the massive varieties is being increasingly challenged by the more consistent synthetic mullite materials which are produced from more available high alumina clays and bauxite.

The heavy mineral beach sand deposits of the world generally contain a significant proportion of andalusite, sillimanite, or kyanite. The US Bureau of Mines has described a method of producing a kyanite coproduct from the ilmenite sand treatment plants in Florida. The potential production from heavy mineral deposits throughout the world is large.

Exploration and Evaluation

Kyanite and sillimanite are fairly common minerals, and are found in the metamorphosed rocks on every continent. Andalusite is less abundant but still common. However, kyanite is consumed principally by the highly developed industrialized nations and its consumption is therefore restricted to certain areas. For example, the bulk of the American production of kyanite is consumed in a relatively small region extending from the northeastern part of the United States to central Missouri. A similar pattern exists in Europe, where consumption is concentrated in those parts of England, France, and Germany where the producers of refractories are concentrated. Kyanite and related minerals must be delivered to the consuming areas at competitive prices. Therefore, a very important preliminary consideration in any exploration project is the prospective cost of delivering the kyanite to the market. In the United States, typical freight costs for delivering kyanite concentrates from traditional sources to consumers ranges between $30 and $40 per ton. A relatively small market exits on the west coast of the United States where the freight is as high as $90 per ton.

The second problem to be considered is the probability of producing a salable product. Kyanite consumers are notoriously finicky about the specifications of materials used by them in the manufacture of refractory products. Thus any deviation from the commonly accepted specifications imposes a considerable burden on products attempting to enter the market. Probably the most important consideration is this regard when evaluating a raw prospect is the liberation size of the kyanite crystals. In the western United States there are several otherwise attractive kyanite-quartzite deposits which cannot be beneficiated to the required specifications without grinding the ore to about −200 mesh, which is quite fine considering that most existing kyanite deposits can produce concentrates in a size range of at least −35 to 28 mesh.

If the initial investigation indicates that both freight and quality of product can be competitive, then the normal exploration techniques apply. These consist of surface sampling, trenching, geologic mapping, and some limited diamond drilling to test the material in depth. Kyanite deposits are in reality kyanite-bearing rocks produced by regional metamorphism. They have inherently consistent compositions, unlike the hydrothermal deposits which are typical of the precious and base metals. Geochemical and geophysical exploration techniques are generally not useful for evaluating kyanite and other aluminum silicate deposits.

After a deposit has been explored, the potential markets determined, and studies indicate that a competitive product could be produced and delivered to the market, the next step should be a pilot plant production of the material intended to be marketed. Bench scale and laboratory testing are not sufficient in this regard. As an example, in recent years a major attempt was made to produce kyanite from schists and kyanite-bearing gneisses in Canada. After the construction of a plant and the expenditure of a great deal of money it was found that a salable concentrate could not be produced economically. Similar failures have occurred in the United States and in other countries. Therefore, it is most important to produce, by reproducible pilot plant techniques, a quantity of concentrates from a representative selection of ores sufficient to permit prospective consumers to test the new material extensively in their plants. There is no laboratory substitute for this kind of field testing in the plants of the intended consumers. Quoted physical and chemical specifications are to be used only as approximations of the suitability of a given kyanite or aluminum silicate material. New producers of aluminum silicates have found to their sorrow that kyanite conforming to specifications supplied by the prospective consumers may not in fact be what those consumers are really willing to use when confronted with the decision to utilize a new material from a new area. Competitive testing and specific approval is the only sure way that the marketability of a given kyanite concentrate can be assured.

BIBLIOGRAPHY AND REFERENCES

Anon., 1972, "Refractory Raw Materials," *Industrial Minerals,* No. 59, Aug., pp. 13-17.

Abbott, A.T., and Prater, L.S., 1954, "The Geology of Kyanite-Andalusite Deposits, Goat Mountain, Idaho, and Preliminary Beneficiation Tests on the Ore," Pamphlet 100, Idaho Bureau of Mines and Geology, 27 pp.

Bennett, P.J., 1961, "The Economic Geology of Some Virginia Kyanite Deposits," unpublished Ph.D. Thesis, University of Arizona, Tucson.

Carpenter, R.H., and Allard, G.O., 1980, "Mineralization, Alteration and Volcanism in the Lincolnton-McCormick District, Georgia and South Carolina," Abstract, Bulletin of Geological Society of America, Nov., pp. 398-399.

Espenshade, G.H., and Potter, D.B., 1960, "Kyanite, Sillimanite and Andalusite Deposits of the Southeastern States," Professional Paper 336, US Geological Survey, 121 pp.

Furcron, A.S., and Teague, K.H., 1945, "Sillimanite and Massive Kyanite Deposits in Georgia," Bulletin 51, Georgia Dept. of Natural Resources, 76 pp.

Grametbaur, A.B., 1959, "Selected Bibliography of Andalusite, Kyanite, Sillimanite, Dumortierite, Topaz, and Pyrophyllite in the United States," Bulletin 1019-N, US Geological Survey, pp. 973-1046.

Jones, J.O., and Eiletsen, N.A., 1954, "Investigation of the Willis Mountain Kyanite Deposit, Buckingham County, Virginia," Report of Investigations No. 5075, US Bureau of Mines, 41 pp.

Klinefelter, T.A., and Cooper, J.D., 1961, "Kyanite—A Materials Survey," Information Circular 8040, US Bureau of Mines, 54 pp.

Potter, M.J., 1982, "Kyanite and Related Minerals," *Mineral Commodity Summaries 1982,* US Bureau of Mines, pp. 82-83.

Van Noy, R.M., 1970, "Kyanite Resources in the Northwestern United States," Report of Investigations 7426, US Bureau of Mines, 81 pp.

Varley, E.R., 1965, *Sillimanite, Andalusite, Kyanite,* Overseas Geological Surveys, Mineral Resources Div., London, 165 pp.

Varley, E.R., 1968, *Sillimanite,* Chemical Publishing Co., New York, 165 pp.

Wells, J.R., 1972, "Kyanite and Related Minerals," *Minerals Yearbook,* US Bureau of Mines, pp. 689-693.

Williamson, D.R., 1960, "The Sillimanite Group," *Mineral Industries Bulletin,* Colorado School of Mines, Vol. 3, No. 4, July, 12 pp.

11. Limestone and Dolomite*

DONALD D. CARR

INTRODUCTION

The physical and chemical properties and widespread distribution make limestone and dolomite so useful that in 1987 more than 764 mt (840 million tons) of these minerals were produced in the United States. They were produced in 46 of the 50 states and ranked second to sand and gravel in total tonnage.

About 75% of the land surface of the world is sedimentary rock, and limestone and dolomite compose about 20% of that area. Limestone and dolomite are exposed near the surface on all continents and range from Precambrian to Recent in age. Reserves are so large that the numbers become meaningless, but these reserves are not always where they are needed. For some uses, such as aggregates, other materials can substitute for limestone and dolomite, but, for uses requiring certain chemical properties, substitution may not be viable.

Although limestone and dolomite are distinctly different rocks with different mineralogical, chemical, and physical properties, they have many common characteristics. They are closely associated with each other in the geologic environment; some operators mine both in the same quarry. They are mined, processed, and transported in the same way. And they substitute for each other for many uses. Throughout this article the term "limestone" will be used for both limestone and dolomite unless a clear distinction is necessary.

DESCRIPTION

Mineralogy

Limestone is a sedimentary rock composed of more than 50% of the mineral calcite ($CaCO_3$); dolomite is composed of more than 50% of the mineral dolomite ($CaMg(CO_3)^2$). Aragonite has the same chemical composition as calcite, but it has a different crystal structure. It is important only in such modern deposits as oolites or oyster shell because aragonite is metastable and alters to calcite with time. Impurities are indicated by a qualifying adjective. For example, limestone with appreciable clay is designated as an argillaceous limestone. Crystal form, color, specific gravity, and other physical properties are important in the mineral identification of specific carbonate species (Table 1).

Chemical Properties

Although the chemical and physical attributes are independent properties, a carbonate rock is seldom considered only for its chemical properties. For most uses in which chemical attributes are specified, however, the degree of purity stands as the most important property. The chemical requirements for some uses do not easily fit into a pigeonhole classification. Although the minimum composition may indeed need to be specified, classification is useful for discussion

* In part from Carr and Rooney (1983). Curtis H. Ault and Brian D. Keith, Indiana Geological Survey, and Garland R. Dever, Jr., Kentucky Geological Survey, provided useful review and comment.

Table 1. Physical Properties of Calcite, Dolomite, and Aragonite.

Mineral	Chemical Composition	Physical Properties
Calcite	$CaCO_3$	Hexagonal crystal system, commonly good rhombohedral cleavage. Mohs hardness, 3. Specific gravity, 2.72. Commonly colorless or white but may be other colors because of impurities.
Dolomite	$CaMg(CO_3)_2$	Hexagonal crystal system, commonly good rhombohedral crystals with curved faces. Mohs hardness, 3.5 to 4. Specific gravity, 2.87, but common impurities such as iron can raise it to 2.95 or higher. Commonly white or pink.
Aragonite	$CaCO_3$	Orthorhombic crystal system. Mohs hardness, 3.5 to 4. Specific gravity, 2.93–2.95. Commonly colorless, white, or yellow but may be other colors because of impurities.

and evaluation. The terminology used in this article is as follows:

High-calcium limestone:	greater than 95% $CaCO_3$
High-magnesium dolomite:	greater than 42% $MgCO_3$
High-purity carbonate:	greater than 95% combined $CaCO_3$ and $MgCO_3$

Accurate chemical determinations in ranges greater than 95% $CaCO_3$ are difficult and should be entrusted only to a reliable laboratory. A useful list of testing companies capable of making chemical and physical determinations of carbonate rocks can be found in the *Directory of Testing Laboratories, Commercial-Institutional* (American Society for Testing and Materials, 1987).

Physical Properties

Although the suitability of a carbonate rock for industrial use may hinge on a small difference in chemical composition, the suitability based on physical properties is less demanding. Some physical testing is done by duplicating actual use conditions, but if actual use is not practical because of limited time or difficulty in duplicating the conditions, tests designed to correlate with actual use conditions are preformed. An example of the former is the burning of stone at calcination temperatures to measure decrepitation effects, and an example of the latter is the battery of physical tests for aggregates to measure the durability of stone during extended use and in certain weather conditions. Physical testing is described in detailed procedures from two main sources: the American Society of Testing and Materials (ASTM) and the American Association of State Highway Officials (ASSHO). Organizations that purchase aggregates generally follow the specifications given in AASHO and those that purchase dimension stone follow ASTM.

ORIGIN

Although carbonate sedimentation occurs in both freshwater and marine environments, the volume of freshwater carbonate is negligible; deposits of economic importance are almost entirely marine. Within the marine environment, carbonate deposition ranges from foraminiferal oozes in the deep sea floor to the fine-grained micritic algal mats of the supratidal environments. But probably the deposits of greatest economic value, because of their abundance and relative purity, are those that form on shallow-water banks of tropical areas in settings that are often referred to as carbonate factories. It is in these areas where optimum conditions exist for organic activity and organisms whose shells make up the deposits live in abundance. It is also where high-energy currents winnow out fine impure particles so they are not deposited with the carbonate.

In simplistic terms the environment of deposition and the organic community largely determine the geometry of the carbonate deposit. In some places lime-secreting organisms, such as corals, calcareous algae, and mollusks, are able to erect large wave-resistant structures, called reefs. As the reef grows, colonies of corals secrete calcium carbonate and attach their shells to the dead corals below. Where the waves break up the coral masses, debris is carried to the flanks and deposited in quieter water, building up steeply dipping beds of relatively pure carbonate grains. Many other calcareous-secreting organisms live in and around the reef and add to the mass of carbonate. Throughout the higher energy areas the skeletal parts may be broken up and moved by wave action until the delicate balance between grain size and current strength changes to allow deposition. It is generally in these high-energy zones where the carbonates are relatively pure and the finer impure clays and silts have been removed by currents to quieter areas for deposition.

Environment of deposition is significant to the economic geologist because it determines the size, shape, purity, and other economically significant characteristics of the carbonate-rock deposit. Limestones that form in high-energy zones generally contain little noncarbonate material and so may be the source of high-purity carbonate material. Micrite, which accumulates in zones of low energy, can be diluted by clay and silt-size noncarbonate material, although some micritic supratidal low-energy carbonates can be pure when no land sources of clay or silt contaminants are nearby.

Carbonate sediments are highly susceptible to postdepositional alteration and modification through chemical diagenesis. The origin of dolomite is especially significant to the economic geologist. Virtually all dolomite is a result of the alteration and recrystallization of calcium carbonate sediments or rocks by magnesium-bearing fluids, and very little dolomite may be precipitated directly from seawater. The mechanism of dolomite formation is not well understood, but, by whatever process, some magnesium ions can substitute for calcium ions in the calcite structure and convert it to dolomite. Good examples are the almost-pure dolomitic Silurian reefs in northern Illinois, Indiana, and Ohio and in southern Michigan.

DISTRIBUTION

Carbonate rocks suitable for most construction uses are widespread, but those of high chemical purity are more restricted in their distribution. Virtually all states except Delaware, Louisiana, New Hampshire, and Rhode Island (Table 2) and all Canadian provinces have potentially commercial deposits of high-calcium limestone and (or) high-magnesium dolomite. Twenty-one Mexican states have had production of high-calcium limestone or high-magnesium dolomite in recent years. Literature describing locations are given by Carr and Rooney (1983, Table 2).

Because deposits of carbonate rocks are present, however, does not necessarily mean that they can be exploited. In many areas competition is intense for potential mineral lands for construction sites, recreation areas, and highways; even nature has her own requirements for flowing streams, glacial cover, and soil development. People who have benefited by low-priced aggregates and building stone and have built homes in rural areas now look askance at quarries in their backyards. Society has imposed environmental controls, now firmly ensconced in state and federal statutes, that prohibit or restrict mineral production in areas where it might significantly affect the quality of the environment.

EXPLORATION

Exploration for limestone and dolomite in North America is largely the detailed examination of known deposits. Because limestone is a sedimentary rock, it occurs in strata generally of considerable areal extent, but the chemical and physical properties vary greatly in both lateral and vertical extent. Some data as to chemical composition and physical character are available for most such strata in the published reports or files of state, provincial, and national geological surveys. Most exploration for a new limestone deposit, therefore, begins with a search of these records to find the locations of deposits that satisfy the various economic factors and proceeds to a sampling program of favorable deposits. Geophysical techniques, if used at all, are used to determine the thickness of overburden. Geochemical techniques are not used.

All aspects of exploration are important, but the one most likely to be slighted is sampling. This is in spite of its importance in determining the validity of further study, for it may become the basis for hundreds of thousands and sometimes millions of dollars worth of development work. Its goal must be the accurate representation of the limestone deposit. Coring, rock bitting, and surface (ledge) sampling are the most common methods, and the choice among these depends on such matters as the geology of the deposit, the proposed use of the material, and the availability of equipment.

Special care should be taken when sampling weathered outcrops. In humid regions the surface layer of a carbonate rock may be leached of calcite and dolomite and therefore be less pure than the rest of the unit. On the other hand, in arid and semiarid regions, where evaporation exceeds precipitation for long periods of time, the surface layer may be enriched in calcite and dolomite. If surface samples must be taken under these conditions, the geologist should be aware of a potential bias.

Limestones used commercially are usually deposited in marine basins and across broad areas and have good lateral continuity. In the Pennsylvanian coal-bearing rocks of the Midcontinent, the marine limestones and shales as well as the coals are favored for correlation. But different depositional environments exist over these broad depositional areas and result in many different limestone types. Cressman (1973) has documented lateral variation and intertonguing of different types of limestone within a limestone formation.

Recognizing the depositional environment is a key in the exploration of limestone because it governs geometry, distribution, and composition of the various limestone types. Generally within a particular type of environment some consistency of properties can be predicted to enable the geologist

Table 2. **Principal Sources of High-Calcium Limestones and High-Purity Dolomites in the United States**

United States	Stratigraphic Unit	Limestone (L) or Dolomite (D)	Age
Alabama	Bangor Limestone	(L)	Mississippian
	Ste. Genevieve Limestone	(L)	Mississippian
	Girkin Limestone	(L)	Mississippian
	Warsaw Limestone	(L)	Mississippian
	Newala Limestone	(L)	Ordovician
	Ketona Dolomite	(D)	Cambrian
Alaska	Limestone of Kings River area	(L)	Jurassic
	Heceta Limestone	(L)	Silurian
Arizona	Mural Limestone	(L)	Cretaceous
	Escabrosa Limestone	(L)	Mississippian
	Redwall Limestone	(L)	Mississippian
Arkansas	Boone Formation	(L)	Mississippian
California	Shell beds	(L)	Quaternary
	Sierra Blanca Limestone	(L)	Tertiary
	Vaqueros Formation	(L)	Tertiary
	Martinez Formation	(L)	Tertiary
	Franciscan Formation	(L)	Tertiary to Jurassic
	Oro Grande Formation	(L)	Carboniferous
	Limestone of Pico Blanco	(L)	Paleozoic, undifferentiated
Colorado	Ingleside Formation	(L)	Permian
	Madison Group	(L)	Mississippian
Connecticut	*		
Delaware	†		
Florida	Ocala Limestone	(L)	Tertiary
Georgia	Bangor Limestone	(L)	Mississippian
Hawaii	*		
Idaho	Limestone of Snake River area	(L)	Triassic
	White Knob Limestone and	(L)	Permian to
	Wells Formation	(L)	Mississippian
	Madison Group	(L)	Mississippian
	Laketown Dolomite	(D)	Silurian
Illinois	St. Louis Limestone	(L)	Mississippian
	Galena Dolomite	(L)(D)	Ordovician
	Platteville Limestone	(L)	Ordovician
	Niagaran Series	(L)(D)	Silurian
Indiana	Ste. Genevieve Limestone	(L)	Mississippian
	Salem Limestone	(L)	Mississippian
	Jeffersonville Limestone	(L)(D)	Devonian
	Traverse Formation	(L)	Devonian
	Niagaran Series	(L)(D)	Silurian
Iowa	*		
Kansas	Plattsburg Limestone	(L)	Pennsylvanian
	Wyandotte Limestone	(L)	Pennsylvanian
	Keokuk Limestone	(L)	Mississippian
Kentucky	Monteagle Limestone	(L)	Mississippian
	Girkin Formation	(L)	Mississippian
	Ste. Genevieve Limestone	(L)	Mississippian
	Warsaw Limestone	(L)	Mississippian
	Camp Nelson Limestone	(L)	Ordovician
Lousiana	†		

Table 2.—Continued

United States	Stratigraphic Unit	Limestone (L) or Dolomite (D)	Age
Maine	*		
Maryland	Tomstown Formation	(D)	Cambrian
Massachusetts	*		
Michigan	Traverse Formation	(L)	Devonian
	Rogers City Formation	(L)	Devonian
	Dundee Formation	(L)	Devonian
	Niagaran Series	(L)(D)	Silurian
Minnesota	Cedar Valley Limestone	(L)	Devonian
	Galena Dolomite	(L)	Ordovician
	Platteville Limestone	(L)	Ordovician
Mississippi	*		
Missouri	St. Louis Limestone	(L)	Mississippian
Montana	Madison Group	(L)	Mississippian
	Meagher Limestone	(L)	Cambrian
	Hasmark Formation	(D)	Cambrian
Nebraska	*		
Nevada	Monte Cristo Dolomite	(D)	Mississippian
	Sultan Limestone	(L)	Devonian
New Hampshire	†		
New Jersey	*		
New Mexico	Fusselman Dolomite	(D)	Silurian
New York	Helderberg Group	(L)	Devonian
	Niagaran Series	(L)(D)	Silurian
	Trenton Limestone	(L)	Ordovician
North Carolina	*		
North Dakota	*		
Ohio	Vanport Limestone	(L)	Pennsylvanian
	Maxville Limestone	(L)	Mississippian
	Columbus Limestone	(L)	Devonian
	Niagaran Series	(L)(D)	Silurian
Oklahoma	St. Clair Limestone	(L)	Silurian
	Fernvale Limestone	(L)	Ordovician
	Royer Dolomite	(D)	Cambrian
Oregon	Franciscan Formation	(L)	Tertiary to Jurassic
	Isolated and unnamed	(L)	Triassic, Permian, and Carboniferous
Pennsylvania	Vanport Limestone	(L)	Pennsylvanian
	Niagaran Series	(L)(D)	Silurian
	Curtin Limestone	(L)	Ordovician
	Annville Limestone	(L)	Ordovician
Rhode Island	†		
South Carolina	*		
South Dakota	*		
Tennessee	Ste. Genevieve Limestone	(L)	Mississippian
	Holston Limestone	(L)	Ordovician
	Shady Dolomite	(D)	Cambrian

Table 2.—Continued

United States	Stratigraphic Unit	Limestone (L) or Dolomite (D)	Age
Texas	Shell beds	(L)	Quaternary
	Edwards Limestone	(L)	Cretaceous
Utah	Madison Group	(L)	Mississippian
	Water Canyon Formation	(D)	Devonian
	Laketown Dolomite	(D)	Silurian
	Fish Haven Dolomite	(D)	Ordovician
Vermont	*		
Virginia	Greenbrier Limestone	(L)	Mississippian
	Holston Limestone	(L)	Ordovician
	New Market Limestone	(L)	Ordovician
	Tomstown Formation	(D)	Cambrian
	Shady Dolomite	(D)	Cambrian
Washington	Metaline Limestone	(L)	Cambrian
West Virginia	Vanport Limestone	(L)	Pennsylvanian
	Greenbrier Limestone	(L)	Mississippian
	Tomstown Formation	(L)(D)	Cambrian
Wisconsin	Galena Dolomite	(L)	Ordovician
	Platteville Limestone	(L)	Ordovician
	Niagaran Series	(L)(D)	Silurian
Wyoming	Madison Group	(L)	Mississippian
	Bighorn Dolomite	(D)	Ordovician

Source: Hubbard and Ericksen (1973) and Carr and Rooney (1983).

* Potential commercial sources are present.
† Potential commercial sources are not present.

in planning the necessary sampling locations. In flat-lying rocks only a few drill holes may be needed, but in structurally altered areas many are required.

Carr (1973) found that individual high-calcium oolitic units in the Illinois Basin ranged from 0.8 to 3.2 km (0.5 to 2 miles) wide and as much as several miles long. The bodies were lenticular and as thick as 7.5 m (25 ft). Ault (1986) and Ault and Carr (1981) reported that in the Great Lakes area of northern Illinois, Indiana, and Ohio and in southern Michigan, reefal deposits, which are sources of high-purity carbonate rocks, ranged in size from 30 to 60 m (100 to 200 ft) in diameter and less than that in thickness to more than 15 km (9 miles) in diameter and 120 m (400 ft) in thickness. These reefs are commonly surrounded by interreef limestones and dolomites that have limited use as aggregates.

In structurally complex areas the variation in depositional environment as well as the structural component must be considered. Brown (1966) reported that in some Lower Devonian carbonate rocks used for cement raw material in the central Hudson River Valley, drill spacings of 60 by 120 m (200 by 400 ft) were necessary for quality control. Banino (1980) reported that in the Lehigh Valley, Pennsylvania, detailed mapping at a scale of 2.54 cm to 30.48 m (1 in. to 100 ft) was necessary in a complexly interfolded and faulted section of Cambro-Ordovician carbonate rock.

MINING

Limestone is mainly produced by open-pit and underground mining, but some modern deposits, such as shell and oolites, are produced by underwater dredging. The largest tonnage of stone is produced by open-pit mining because it is generally less expensive than underground mining, but reasons for underground mining can be compelling. Rooney and Carr (1971) summarized the advantages of underground mining as follows:

• Lack of near-surface deposits may make underground mining the only method.

• Unfavorable geologic structure may require that a surface mine move underground to follow a particular bed.

• More efficient use of reserves may be possible by underground mining in areas where overburden removal may make surface mining impractical.

• Underground mining may permit selective mining of a particular bed without removal of, or contamination by, the overlying materials or clay seams.

• Underground mining may permit all-weather mining and provide greater economies through year-around operation.

• Underground mining is more inconspicuous than open-pit mining, a factor that may lessen environmental problems. Underground mining may be feasible in urban areas where the surface is expensive or preempted for other uses.

• Underground mining may provide space for secondary uses, such as manufacturing, offices, storage, and waste disposal.

Open Pit

The techniques vary, depending on whether the product is crushed stone or dimension stone, but for both the first step is the removal of overburden to gain access to the stone to be quarried. Where overburden consists of soil or unconsolidated material, a bulldozer, a dragline, or a scraper may be used, but if the material is consolidated, blasting may be required to ease removal. For some dimension-stone operations, blasting is used little or not at all for fear of fracturing the stone below. Some care must be given to overburden removal; partial removal may lead to problems of contamination of the product, clogging of screens and other processing equipment, or unsafe working conditions for employees. In areas where solution may have produced grikes or solution features in the limestones, some hand dental work may be required to prepare the top ledge for removal.

In crushed-stone operations blasting is used not only to free the stone from the ledge but also to form the first stage of stone fragmentation. The material is then transported, generally by front-end loader, truck, and (or) belt conveyor, to a processing plant for final crushing, sizing, and stockpiling.

In dimension-stone operations the stone is set free from the ledge by some type of sawing or cutting device. Channeling machines, wire saws, and chain saws are the types most commonly used to make vertical cuts in the stone. If the stone does not have a natural bedding plane to allow separation of the base of the block, sleeved wedges may be driven into closely spaced horizontal drill holes. Once the stone is free on all sides, it can be tipped forward (turned down) and sectioned into mill blocks of desired size by using sleeved wedges in drill holes. The mill blocks are then transported to a staging area for drying (curing) before delivery to the mill for fabrication.

Underground

The techniques also differ for underground mining depending on whether the use is for crushed stone or dimension stone. For either use some access to the mining seam must be obtained; vertical shaft, slope, or horizontal drifts are used depending on conditions.

In crushed-stone operations some form of room-and-pillar mining is used, but whether the pillars are square, round, or elongate may depend on mining conditions and post-mine-use plans (Aughenbaugh, Christiansen, and Scott, 1974). Mine orientation and pillar design and configuration may depend on jointing and other structural considerations (Ault and Haumesser, 1989). The mine is generally developed by a heading-and-bench system, and the stone is brought to the surface for final processing and stockpiling.

Underground mining of dimension stone is less extensive in the United States than in Europe, but its popularity is sure to grow because of environmental and land-use considerations. Where underground mining is feasible a room-and-pillar method is used. The stone is cut from the mining seam by using a chain saw, called a gallery saw, capable of horizontal as well as vertical cutting. Underground mining holds promise of year-long operation, improved ability of removing only desirable blocks without removing excessive overburden, and reduced impact on the surface environment. Blocks quarried underground can be cured underground without fear of damage by freezing and then transported to the mill for processing.

USES

The uses of limestone and dolomite depend on the physical properties, chemical properties, or both. Physical properties are more important if stone is used "as is," such as for aggregate or building stone. Chemical properties are more important if stone undergoes changes from one form of matter to another, such as for manufacturing cement or lime. Chemical and physical properties are often interrelated; in whiteness, for example, the physical property of color is largely determined by the purity and chemical composition of the rock.

Limestone and dolomite account for about three-fourths of all crushed stone used in the United States; they account for less than one-third of the dimension stone (Fig. 1). Although limestone and dolomite have hundreds of uses, many of which are reported in Lamar (1961) and Siegel (1967), more than half of the tonnage produced each year is for construction aggregates (Table 3). The distribution of natural

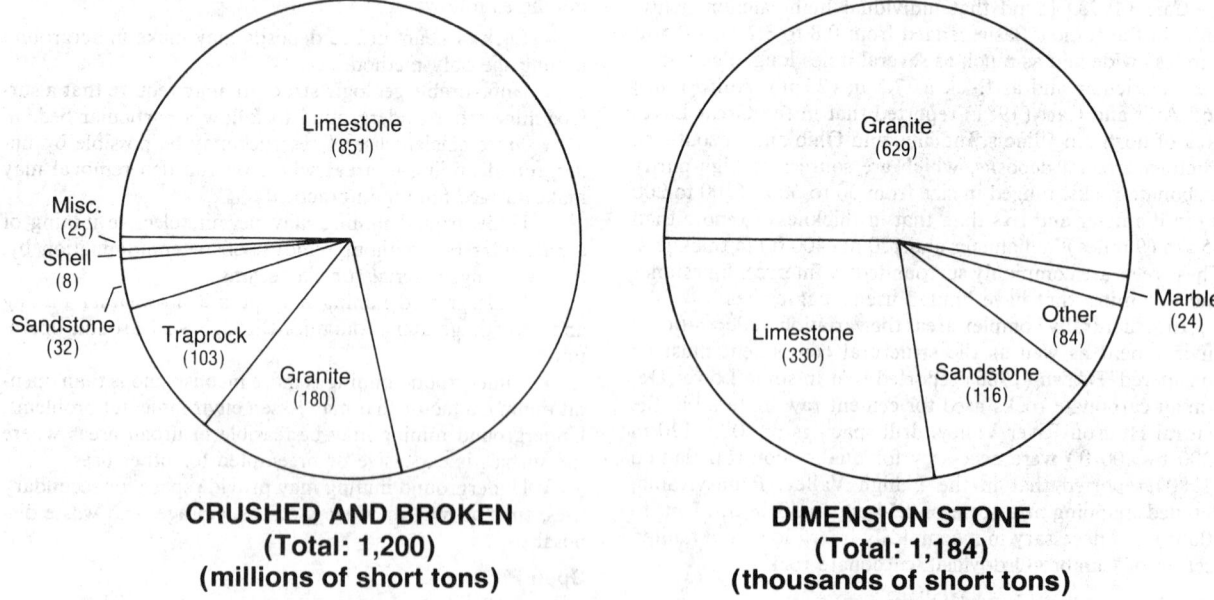

CRUSHED AND BROKEN
(Total: 1,200)
(millions of short tons)

DIMENSION STONE
(Total: 1,184)
(thousands of short tons)

Fig. 1. Diagrams showing types of rock used for crushed and dimension stone in the United States in 1987. From US Bureau of Mines *Minerals Yearbook.*

Table 3. Crushed Limestone and Dolomite Sold or Used by Producers in the United States in 1987

Use	Quantity, 1000 st	Value, $1000
Crushed and Broken		
Coarse aggregate, +1½ in.	30,227	114,495
Coarse aggregate, graded	152,250	655,474
Fine aggregate, −⅜ in.	43,428	184,970
Coarse and fine aggregate	200,104	691,622
Agricultural	22,807	115,514
Chemical and metallurgical	109,129	361,933
Special	283,156	1,332,612
Total*	841,101	3,456,620

	Quantity, st	Dimension Stone Cubic Feet, 1000	Value, $1000
Rough stone	218,542	2913	12,530
Dressed stone	111,301	1448	22,437
Total*	329,843	4361	34,968

Source: US Bureau of Mines *Minerals Yearbook.*
* Data may not add to totals shown because of independent rounding.
Metric conversion: in. × 2.54 = cm; ft × 0.305 = m. st × 0.92 = t.

aggregates in the United States was described by Langer (1988).

BYPRODUCTS AND WASTE MATERIALS

Crushed Stone

Removal and disposal of overburden are challenges facing operators of all crushed-stone operations. In most operations the overburden is wasted, but ideally, with a good mine plan, the material is placed where it will not need to be moved twice. In the startup of any open-pit operation the overburden must be removed to expose the mining seam. This material may be used for berms to create visual or sound barriers, which help with the attractiveness of the operation. As an alternative, the material may be stockpiled for future reclamation use by grading and seeding so as to prevent erosion and loss. In a few places overburden may have value for brick, cement raw materials, or fill.

Excess fines that may have potential use for other products are created in crushing limestone. The fines will generally have a composition different from that of the limestone being crushed because they contain the shale and soft material that pass through the screens. In agricultural areas fines generally work well for agricultural limestone. Although specifications vary from state to state, the requirements for agricultural limestone are sufficiently broad to allow the fines to be used without further processing. Fines may also be used for low-grade fillers and for some scrubber materials for stack-gas cleaning. In cold climates shipping of fine materials may be a problem because of the difficulty of unloading at the destination.

Dimension Stone

Overburden removal creates the same problem for dimension-stone quarries as for crushed-stone operations, but

an added problem is waste blocks. In most quarries the upper ledge must be wasted because of imperfections in the stone, and these blocks must be removed from the ledge in the same manner as good-quality stone. The dimension-limestone industry has long been characterized by a high ratio of waste materials to salable prime product, both in quarrying and in fabrication. In the Indiana building-limestone district, the principal visual evidence has been the grout piles of overburden and unused building stone. For years it has seemed to the public that blocks and coarse waste fragments, once quarried, should have some use, and many ideas have been advanced and attempts made to salvage the waste material for some purpose that would repay at least a part of its cost (Patton and Carr, 1982). The most common suggestion has been that it should be possible to crush the stone for one of the many purposes to which aggregates are put, but the waste is generally not well suited to most crushed-stone uses, and the cost of reworking the old grout piles is greater than the cost of obtaining new stone, freshly quarried by conventional crushed-stone practices. The size of the waste blocks induces handling and crushing expenses greater than the potential yield in dollars.

Some use of waste blocks has been made for breakwater and seawalls, but even for these uses flaws in the stone may make the stone unsuitable. Also, because the blocks are large and heavy, transportation costs are important. In areas near quarries, waste blocks and lapies have found use as lawn sculpture and ornamentation, and in one area near Oolitic, IN, an unsuccessful attempt was made to create a 29-meter-high (98-ft-high) pyramid, patterned at one-fifth scale after the Great Pyramid of Cheops. But all of these minor uses fail to dent the great grout piles that cover the landscape.

SOURCES OF INFORMATION

Publications of state geological surveys or their equivalents are the best references for information on the common industrial minerals and rocks. Most state publications are oriented toward areal geology, mineral resources, or the two combined rather than toward uses or methods. Some state publications purportedly on commodities contain valuable information on uses. Many states have geologic maps that show distribution of limestones, but the number that has published quadrangle maps (1:24,000 scale), the most useful for prospecting, is limited. Kentucky and Rhode Island are the only states to have complete coverage; some states have unpublished quadrangle maps on open file that are useful.

The US Bureau of Mines and US Geological Survey also publish information on limestone and dolomite, and their publications are useful for mineral statistics, mining characteristics, broad overviews, and, in some reports, site-specific information. Trade publications, such as *Rock Products, Pit and Quarry,* and *Industrial Minerals,* are good sources of information on specific mineral producers, plants, specifications, and trends in the industry.

Two professional organizations, which have meetings with consistently high-quality presentations on limestone, are the Industrial Minerals Division of the Society of Mining Engineers (SME) and the Forum on Geology of Industrial Minerals. Some papers presented at SME meetings are selected for publication in the Society's publications, MINING ENGINEERING and Transactions. The Forum is hosted each year by a different state geological survey. Most papers presented at the Forum appear in a proceedings volume, published by the host organization.

In the late 1950s and early 1960s oil companies began to concentrate research on carbonate rocks because of economic incentives. As a result, useful material was published on composition and texture (Ham, 1962; Scholle, 1978) and on depositional environments (Friedman, 1969; Scholle, Bebout, and Moore, 1983). Numerous articles and volumes that contain much useful information on limestone as a commodity and on the industry that produces it have been written. *Industrial Minerals and Rocks* (5th edition, 1983) has articles on "Limestone and Dolomite" (Carr and Rooney, 1983) and on "Lime" (Boynton, and others, 1983) as well as articles on crushed-stone uses that include limestone and dolomite, such as "Aggregates—Crushed Stone" (Schenck and Torries, 1983) and "Cement and Cement Raw Materials" (Ames and Cutcliffe, 1983). A useful handbook for quarry operators (Lamar, 1967) contains much information of interest to the general public.

Although designed as a textbook for the teaching of a course in industrial minerals, *Geology of the Industrial Rocks and Minerals* (Bates, 1969) contains useful information on selected limestone deposits. *Geology of the Nonmetallics* (Harben and Bates, 1984) has a chapter on "Limestone and Dolomite" that covers applied aspects of interest to industry, and *Industrial Minerals and Rocks* (Kuzvart, 1984), a handbook with a European perspective, has sections on limestone and dolomite. Broad coverage of limestone and dolomite in the United Kingdom is provided by Harris (1982), and a regional assessment is provided by Harrison and McL. Adlam (1985). The most comprehensive volume to cover limestone and its use as lime is *Chemistry and Technology of Lime and Limestone* (Boynton, 1980). The *Concise Encyclopedia of Mineral Resources* (Carr and Herz, 1989) gives comprehensive coverage for the various uses of industrial minerals of which limestone and dolomite are important commodities.

The properties of limestone that make a good crushed-stone aggregate are different from those for dimension stone; even the exploration, quarrying, processing, transportation, and marketing are different. Comprehensive coverage of these various aspects can be found in Barton (1968), Cox, Bridge, and Hull (1977), Currier (1960), Honeyborne (1982), Leary (1983), Patton and Carr (1982), Power (1983), and Rooney (1970).

REFERENCE LIST

American Society for Testing and Materials, 1987 "Directory of Testing Laboratories, Commercial-Institutional," ASTM Special Technical Publication 333F, 128 pp.

Ames, J.A., and Cutcliffe, W.E., 1983, "Cement and Cement Raw Materials," *Industrial Minerals and Rocks*, 5th ed., Vol. 1, S.J. Lefond, ed., AIME, New York, pp. 133-159.

Aughenbaugh, N.B., Christiansen, C.R., and Scott, J.J., 1974, "Conversion from Quarrying to Underground Mining," *Proceedings*, 10th Forum on Geology of Industrial Minerals, Ohio Geological Survey Misc. Rept. 1, pp. 17-24.

Ault, C.H., 1986, "Construction Materials: Crushed Stone," *Encyclopedia of Materials Science and Engineering*, M.B. Bever, ed., Pergamon Press, Oxford, pp. 813-820.

Ault, C.H., and Carr, D.D., 1981, "Search for High-Calcium Limestone in Silurian Reefs of Northern Indiana," *Bulletin*, Geological Society of America, Pt. 1, Vol. 92, pp. 641-647.

Ault, C.H., and Haumesser, A.F., 1989, "A Central Indiana Model for Predicting Jointing Characteristics in Underground Limestone Mines," *Proceedings*, 24th Annual Forum on Geology of Industrial Minerals, South Carolina Geological Survey.

Banino, G.M., 1980, "Cement limestone mining in a structurally complex setting," *Proceedings*, 14th Annual Forum on the Geology of Industrial Minerals," New York State Museum Bulletin 436, pp. 65-59.

Barton, W.R., 1968, "Dimension Stone," Information Circular 8391, US Bureau of Mines, 147 pp.

Bates, R.L., 1969, *Geology of the Industrial Rocks and Minerals*, Dover, New York, 459 pp.

Boynton, R.S., 1980, *Chemistry and Technology of Lime and Limestone*, 2nd ed., Wiley (Interscience), New York, 578 pp.

Boynton, R.S., et al., 1983, "Lime," *Industrial Minerals and Rocks*, 5th ed., Vol. 2, S.J. Lefond, ed., AIME, New York, pp. 809-831.

Brown, S.P., 1966, "New York Cement Producers Adjust to Geologic Complexities," *Ohio Journal of Science*, Vol. 66, pp. 123-130.

Carr, D.D., 1973, "Geometry and Origin of Oolite Bodies in the Ste. Genevieve Limestone (Mississippian) in the Illinois Basin," *Bulletin 48*, Indiana Geological Survey, 81 pp.

Carr, D.D., and Herz, N., eds., 1989, *Concise Encyclopedia of Mineral Resources*, Pergamon, Oxford, in preparation.

Carr, D.D., and Rooney, L.F., 1983, "Limestone and Dolomite," *Industrial Minerals and Rocks*, 5th ed., Vol. 2, S.J. Lefond, ed., AIME, New York, pp. 853-868.

Cressman, E.R., 1973, "Lithostratigraphy and Depositional Environments of The Lexington Limestone (Ordovician) of Central Kentucky," Professional Paper 768, US Geological Survey, 61 pp.

Cox, F.C., Bridge, D.McC., and Hull, J.H., 1977, "Procedure for the Assessment of Limestone Resources," Inst. Geological Sciences, Mineral Assessment Report 30, Her Majesty's Stationery Office, London, 14 pp.

Currier, L.W., 1960, "Geologic Appraisal of Dimension-Stone Deposits," Bulletin 1109, US Geological Survey, 78 p.

Friedman, G.M., ed., 1969, "Depositional Environments in Carbonate Rocks," Special Publication 14, Society of Economic Paleontologists & Mineralogists, 209 pp.

Ham, W.E., ed., 1962, "Classification of Carbonate Rocks," Memoir 1, American Association Petroleum Geologists, 279 pp.

Harben, P.W., and Bates, R.L., 1984, *Geology of the Nonmetallics*, Metal Bulletin, New York, 392 pp.

Harris, P.M., 1982, "Limestone and Dolomite," Inst. Geological Sciences, Mineral Assessment Report 23, Her Majesty's Stationery Office, London, 111 pp.

Harrison, D.V., and Adlam, K.A. McL., 1985, "Limestones of the Peak," Mineral Assessment Report 144, British Geological Survey, 40 pp.

Honeyborne, D.B., 1982, "The Building Limestones of France," Dept. of Environment, Building Research Establishment Report, Her Majesty's Stationery Office, 116 pp.

Hubbard, H.A., and Ericksen, G.E., 1973, "Limestone and Dolomite," *United States Mineral Resources*, D.A. Brobst and W.P. Pratt, eds., Professional Paper 820, US Geological Survey, pp. 357-364.

Kuzvart, M., 1984, "Industrial Minerals and Rocks," *Developments in Economic Geology*, Vol. 18, Elsevier, Amsterdam, 454 pp.

Lamar, J.E., 1961, "Uses of Limestone and Dolomite," Circular 321, Illinois Geological Survey, 44 pp.

Lamar, J.E., 1967, "Handbook on Limestone and Dolomite for Illinois Quarry Operators," Bulletin 91, Illinois Geological Survey, 119 pp.

Langer, W.H., 1988, "Natural Aggregates of the Conterminous United States," Bulletin 1594, US Geological Survey, 33 pp.

Leary, E., 1983, "The Building Limestones of the British Isles," Dept. of the Environment, Building Research Establishment Report, Her Majesty's Stationery Office, 91 pp.

Patton, J.B., and Carr, D.D., 1982, "The Salem Limestone in the Indiana Building-Stone District," Occasional Paper 38, Indiana Geological Survey, 31 pp.

Power, W.R., 1983, "Dimension and Cut Stone," *Industrial Minerals and Rocks*, 5th ed., Vol. 1, S.J. Lefond, ed., AIME, New York, pp. 161-181.

Rooney, L.F., 1970, "Dimension Limestone Resources of Indiana," Bulletin 42-C, Indiana Geological Survey, 29 pp.

Rooney, L.W., and Carr, D.D., 1971, "Applied Geology of Industrial Limestone and Dolomite," Bulletin 46, Indiana Geological Survey, 59 pp.

Schenck, G.H.K., and Torries, T.F., 1983, "Aggregates—Crushed Stone," *Industrial Minerals and Rocks*, 5th ed., Vol. 1, S.J. Lefond, ed., AIME, New York, pp. 60-80.

Scholle, P.A., 1978, "A Color Illustrated Guide to Carbonate Rock Constituents, Textures, Cements, and Porosities," Memoir 27, American Association Petroleum Geologists, 241 pp.

Scholle, P.A., Bebout, D.G., and Moore, C.H., 1983, "Carbonate Depositional Environments," Memoir 33, American Association Petroleum Geologists, 708 pp.

Siegel, F.R., 1967, "Properties and Uses of the Carbonates," *Developments in Sedimentology, Carbonate Rocks,* Vol. 9B, G.V. Chilingar, H.J. Bissell, and R.W. Fairbridge, eds., Elsevier, New York, pp. 343-393.

US Bureau of Mines, *Minerals Yearbooks,* annual, US Government Printing Office, Washington, DC.

12. Lithium Raw Materials*

INTRODUCTION

Lithium minerals occur predominantly in pegmatites which contain mineral assemblages derived from the crystallization of postmagmatic fluids or from the metasomatic action by residual pegmatitic fluids. They have been the traditional sources of raw materials for ceramic and chemical industries. With the discovery of Searles Lake, brines became new sources of lithium.

Indeed, the 1960s may well be referred to as the "brine decade" because several brine bodies of significant economic importance were discovered. In the United States, the brines of Clayton Valley, Nevada, are presently exploited by Foote Mineral Co. The brines of the Great Salt Lake of Utah, the Smackover formation, and the Imperial Valley geothermal field, have been defined as important resources of lithium. In South America, new occurrences have been identified in Bolivia and Argentina. At the Salar de Atacama, Chile, new production capacity was scheduled for 1984.

Following the termination of the AEC purchase program in 1959, several companies such as Maywood Chemical Co., American Lithium Chemicals, and Quebec Lithium Corp. were forced to close their operations. Nevertheless, with the development of new applications in glass ceramics, air conditioning systems, synthetic rubber, and metallurgy (aluminum potlines), the lithium industry grew steadily, and with the potential application of lithium metal in batteries and nuclear reactors, a healthy growth can be foreseen. New products and new applications are likely to stimulate exploration for expanded reserves as well as the search for new sources from which lithium could be extracted economically.

GEOCHEMISTRY OF LITHIUM

Lithium is the third element in the periodic table. It is the lightest of all the metals, having an atomic weight of 6.938, an ionic radius of 0.68Å, and a charge of +1.

The geochemistry of lithium has been extensively studied and has been summarized by Rankama and Sahama (1950), Goldschmidt (1937), and Hortsman (1957).

The distribution of lithium in igneous rocks is controlled by its size and its charge, and by the $(MgO+FeO)/Li_2O$ ratio. In the early stages of crystallization of a magma, that ratio is very large. Consequently, both magnesium and iron are removed by ferromagnesian minerals in preference to lithium which is then concentrated in the residual magma. This results in an enrichment of lithium in silicic rocks and pegmatites (Strock, 1936).

Pegmatites are coarse-grained igneous rocks formed by the crystallization of postmagmatic fluids. Minerals within pegmatites may also form by metasomatism (Jahns, 1955). Genetically, the pegmatites are associated with neighboring intrusives. Mineralogically, granitic pegmatites contain feldspar, quartz, and mica as the main constituents and a variety of exotic elements such as lithium, beryllium, tantalum, tin, and cesium, which may or may not occur in economically significant concentrations.

Detailed studies by numerous investigators (Cameron, et al., 1949, 1954; Hanley, et al., 1950; Jahns, 1953, 1955; Page,

et al., 1953) indicate that pegmatites often exhibit an internal zonal arrangement, with each zone containing a specific suite of minerals. The lithium minerals are usually found in the intermediate zones, and although as many as 13 zones have been recognized by Cameron, et al. (1949), a complete zonal arrangement is rarely found. Zoning of pegmatite bodies has also been observed on a regional basis. The regionally zoned pegmatite sequences exhibit mineral assemblages and complexity according to their respective distance from the granitic bodies to which they are genetically related.

Lithium is also found in small proportions in a variety of rocks. The average lithium content of igneous rocks is estimated at about 28 ppm Li. Sedimentary rocks contain an average of 53 ppm Li, the highest being recorded in shales (Hortsman, 1957).

Unusual amounts of lithium are found in the clay mineral, hectorite, which is expandable and belongs to the magnesian end member of the smectite group. Recent work by the US Geological Survey suggests that lithium-rich clays are ubiquitous in a number of clay environments.

Lithium is also present in significant amounts in waters associated with geothermal areas (White, 1957), in oil well brines (Mayhew and Heylmun, 1966), and certain brines of California (Searles Lake), Nevada (Clayton Valley), Utah (Great Salt Lake), and Chile (Salar de Atacama), Bolivia (Salar de Uyuni), and Argentina.

LITHIUM MINERALS

Although lithium occurs in some 145 minerals, only spodumene, lepidolite, petalite, amblygonite, and eucryptite have been commercial sources of lithium.

Spodumene, a lithium aluminum silicate ($LiAlSi_2O_6$), is a monoclinic member of the pyroxene group. It has a very pronounced cleavage plane (110) which results in typically lath-shaped particles upon breaking. The color of spodumene is variable, being nearly white in low iron variety and dark green in iron-rich crystals.

When clear, spodumene is considered a gemstone. Three varieties are known: *hiddenite,* the green variety from Alexander County, North Carolina, first discovered in Brazil; *triphane,* the yellow variety also from Alexander County; and the lilac-colored *kunzite* from the Pala district, California, Brazil, and Afghanistan.

Spodumene undergoes pseudomorphic alteration to a variety of minerals. Norton and Schlegel (1955) have described spodumene replacement by quartz, albite, perthite, muscovite, beryl, amblygonite, apatite, and tourmaline. Weathering commonly alters spodumene to kaolinite and to montmorillonite.

Spodumene constitutes the most abundant commercial source of lithium. Theoretically, it may contain up to 3.7% Li, but the actual lithium concentrations vary from 1.35% to 3.56%, probably as a result of sodium and potassium substitution for lithium. Spodumene concentrates typically contain 1.9% to 3.3% Li.

Lepidolite is a phyllosilicate with the general formula $K_2(Li,Al)_{5-6}\{Si_{6-7}Al_{2-1}O_{20}\}(OH,F)_4$. The chemical variability expressed in the formula stems from a structural complexity attributed to a mixture of different polymorphs which include muscovite, lithium muscovite, and polylithionite (Winchell, 1942). On the other hand, Foster (1960) and Deer, et al.

---

*Excerpt from Kunasz, I.A., 1983, "Lithium Raw Materials," *Industrial Minerals and Rocks,* 5th ed., Vol. 2, S.J. Lefond, ed., AIME, New York, pp. 869-880.

(1962) suggest that there is a continuous series between muscovite with a $2M_1$ structure to lepidolite, with $1M$, $2M_2$, and $3T$ structures. The structural transition takes place when the lithia content in the mica reaches 1.53%.

The lithium concentration in lepidolite varies between 1.53% to a possible theoretical maximum of 3.6%. In commercial deposits, the concentrations are more normally 1.4% to 1.9% Li. In addition to lithium, lepidolites also carry substantial concentrations of rubidium and cesium (Deer, et al., 1962).

The major commercial occurrences of lepidolite are located in Zimbabwe (Bikita), Namibia (Karibib), Canada (Bernic Lake, Manitoba), and Brazil.

Petalite, $LiAlSi_4O_{10}$, is a monoclinic mineral with a framework silicate structure. Its color is grayish white and more rarely pinkish. It has two cleavage directions which form an angle of 38.5°. The basal cleavage is perfect.

The theoretical lithium content of petalite is 2.27%. In actual commercial deposits, the concentration varies from 1.6% to 2.1% Li. Sizable deposits of petalite occur with lepidolite in Zimbabwe (Bikita), Namibia (Karibib), Brazil (Araçuai), Australia (Londonderry), in the USSR (eastern Transbaikalia), and Sweden (Utö).

In certain pegmatites, there is evidence that petalite alters to a mixture of spodumene and quartz. In the Bernic Lake pegmatites of Manitoba, Cerny, et al. (1971) have described pseudomorphs of spodumene and quartz after petalite.

Eucryptite is also a lithium aluminum silicate which is deficient in silica. It has a formula $LiAlSiO_4$ and may contain 5.53% Li. The only large deposit of eucryptite is found in Zimbabwe (Bikita) where its occurrence with quartz suggests a spodumene origin (Westenberger, 1963). The grade of the eucryptite is 2.34% Li.

Amblygonite, with the generalized formula $LiAl(PO_4)$ (F,OH), is the fluorine-rich end member of a phosphate series, while montebrasite represents the hydroxyl-rich end member. It occurs in white to gray masses. Basal cleavage planes are pearly, others are vitreous. Amblygonite weathers to earthy apatite, wavellite, and other lithium deficient phosphates. Although amblygonite may contain as much as 4.74% Li, commercial ores usually carry 3.5% to 4.2% Li. Amblygonite has been mined in Canada, Brazil, Surinam, Zimbabwe, Ruanda, Mozambique, Namibia, and the Republic of South Africa.

Brines. Lithium is found in commercial quantities in certain brine deposits. The brines are present in desert areas and occur in playas and saline lakes where solutions have been concentrated by solar evaporation. In Searles Lake where production of dilithium phosphate began in 1938, the lithium concentration is 70 ppm Li. In Clayton Valley, Nevada, lithium-bearing brines contain 200 ppm Li. Smaller concentrations of lithium (28 to 60 ppm Li) are found in the Great Salt Lake. Rich concentrations of lithium have been identified at the Salar de Uyuni in Bolivia (100-700 ppm) and the Salar de Atacama in Chile (1000-5000 ppm).

RESERVES

Projections of lithium metal requirements to satisfy the needs of the automotive and storage battery system as well as the potential use in thermal nuclear fusion energy generation resulted in concern regarding the availability of sufficient lithium. The US Geological Survey created the Lithium Exploration Group, which subsequently sponsored two symposia, one held in Denver, CO and another in Corning, NY. Concurrently, the Energy Research and Development Administration requested the National Academies of Sciences and Engineering to form a National Research Council Committee on Nuclear and Alternative Energy Systems (CONAES). One of the groups was the Lithium Subpanel consisting of experts from the industry and the government, formed to evaluate the lithium availability in the western world. The results of the Subpanel study were presented at the second lithium symposium held at Corning, NY in 1977. This document is considered the best reference regarding reserves and resources and the serious student of lithium should have this particular publication on hand (Penner, 1978). Many of the figures quoted in this review have been taken from the Subpanel study.

MAJOR PRODUCING DISTRICTS

At the present time, the United States and Zimbabwe supply the bulk of the free world's demand for lithium raw materials, and the United States is the leading producer. Within the communist bloc, the Soviet Union probably represents an important producer; production figures, however, are not available. The People's Republic of China is reported to produce lithium compounds from ores mined in the northwestern part of the country.

United States

Lithium ores and lithium chemicals in the United States are supplied by three producing areas. Two of the areas are in the tin-spodumene belt of North Carolina, while brines are exploited in Clayton Valley (Silver Peak), Nevada.

The tin-spodumene belt of North Carolina constitutes the largest developed reserve of lithium in the free world. The area is characterized by numerous pegmatites that intrude amphibolites and schists along the eastern margin of the Devonian Cherryville quartz monzonite (Kesler, 1961).

The pegmatites are typically unzoned, and spodumene is distributed throughout the bodies. The contact zones and aplitic layers contain substantially lower percentages of spodumene. On the average, the pegmatites consist of 20% spodumene, 32% quartz, 6% muscovite, 41% feldspar (including albite and microcline), and about 1% trace minerals, such as beryl, lithiophilite, columbite, tantalite, pyrite, sphalerite, apatite, and rhodochrosite (Kesler, 1961).

The tin-spodumene belt has a length of about 48 km (30 miles) and the exploration activity has been restricted to the upper 300 m (984 ft). The geological setting and distribution of pegmatites suggest that pegmatites may well occur at greater depth. Kesler, in a study of the potential lithium contained in the total belt down to a 1.5 km (4921 ft) level, has suggested the potential existence of a resource of 754 million tons at a grade typical for the area.

Other known pegmatite sources in the United States are not of great significance. These include the Black Hills of South Dakota, the Pala district of southwestern California, the White Picacho District of Arizona, the Harding and Pidlite mines in New Mexico, and several smaller areas in Colorado, Wyoming, Utah, and New England.

In Clayton Valley, located in Esmeralda County, Nevada, lithium-bearing brines occur in an undrained structural depression filled with Quaternary sediments composed mainly of clay minerals, including hectorite, volcanic sands, alluvial gravels, and saline minerals consisting of gypsum and halite (Kunasz, 1970).

In Clayton Valley, located in Esmeralda County, NV, lithium-bearing brines occur in an undrained structural depression filled with Quaternary sediments composed mainly of clay minerals, including hectorite, volcanic sands, alluvial gravels, and saline minerals consisting of gypsum and halite (Kunasz, 1970).

The brine which saturates the sediments is chemically simple. It occurs as a concentrated sodium chloride solution containing subordinate amounts of potassium and minor amounts of magnesium and calcium. The lithium concentration is variable; the composite average pumped into the ponds is 200 ppm Li.

Zimbabwe

The largest lithium-bearing area in Zimbabwe is that of the Bikita tinfields, located about 72 km (45 miles) east of Fort Victoria. Exploration has indicated the presence of an important mineralized zone within the Al Hayat, Bikita, and Southern sectors. The pegmatite is about 1.6 km (5100 ft) long and its width varies from 29 to 64 m (95 to 210 ft). It strikes north-northeast and dips from 14° to 45° east.

The pegmatite is asymmetrically zoned and contains a variety of commercially important lithium minerals as well as beryl and pollucite.

Drilling and development work have proved a reserve of 12 million tons of quarry ore grading 1.4% Li. Mining is presently done in the Al Hayat sector.

Other minor lithium-bearing occurrences are known in the Wankie, Salisbury, Umtali, Mtoko, Insizia, Matobo, and Mazoe districts (Toombs, 1962).

MINOR PRODUCING DISTRICTS

South America

Only two countries produce small amounts of lithium minerals in South America: Brazil and Argentina. The amblygonite deposit in Surinam has been essentially depleted.

In Brazil, lithium-bearing pegmatites are known in the districts of Minas Gerais and in the northeastern portion of the country which includes the states of Paraiba, Rio Grande do Norte, and Ceará (Heinrich, 1964).

The pegmatites which carry spodumene and amblygonite usually have been mined for cassiterite, tantalite, and beryl, while the lithium minerals have been sporadically recovered as byproducts.

An important occurrence of petalite has been reported in the tin-producing district near Araçuai, Brazil, located about 402 km (250 miles) northeast of Belo Horizonte in the state of Minas Gerais. The pegmatite is reported to contain about 100,000 tons of petalite grading 2.0% Li.

Spodumene reserves have been estimated at 300,000 tons, while lepidolite reserves are considered to be nearly exhausted.

In Argentina, lithium-bearing pegmatites occur in the western portion of the state in the Sierras Pampaneas, which include the productive districts of San Luis, Cordoba, and Catamarca. The pegmatites are zoned and contain spodumene. The reserves, considered to be small, total about 18,000 tons as spodumene (Angelleli and Rinaldi, 1963, 1965). Recently, the occurrence of brine has been reported in the salares of northwestern Argentina.

Namibia

Following a low production period, mining in the Karibib district was stimulated by the trade sanctions imposed on Southern Rhodesia (now Zimbabwe) in 1964. The district, located approximately 193 km (120 miles) from Walvis Bay on the Atlantic Ocean, contains several strongly zoned pegmatites which contain lepidolite, petalite, and small amounts of amblygonite.

Other African Deposits

In Africa, smaller productive areas occur in Mozambique, Uganda, and Ruanda.

In the Alto Ligonha area of Mozambique, lepidolite, amblygonite, and spodumene occur in zoned pegmatite bodies. Only lepidolite is exploited commercially. Amblygonite is rare, and spodumene is completely altered to kaolinite (Hutchinson and Claus, 1956).

Large amblygonite masses are found in pegmatite districts located west of the capital city of Kigal, in Ruanda (Valarmoff, 1954), and north of Kampala in Uganda (Roberts, 1948).

Other occurrences of lithium minerals include the Sudan, the Malagasy Republic, the Ivory Coast, and the Republic of South Africa (Nel, 1968).

Australia

The most important lithium-bearing districts occur in Western Australia. Mining is active only in the Coolgardie district where small quantities of petalite and minor amounts of spodumene and amblygonite are produced at Londonderry. In the same area near Mt. Marion, the Western Mining Corp. has discovered a spodumene deposit. Drilling has indicated reserves in excess of 1,000,000 tons of ore (Anon., 1968).

Other lithium mineral occurrences include spodumene near Ravensthorpe, lepidolite near Wodgina and Londonderry, and minor amblygonite near Euriowie, New South Wales (McLeod, 1965).

In addition to the producing areas described above, lithium minerals are known to occur in many European countries and in Ireland, India, Korea, and Japan. Sizable deposits exist in the Soviet Union but actual reserves data are not available.

BIBLIOGRAPHY AND REFERENCES

Anon., 1968, "Lithium Find for Western Mining," *Mining Journal,* Mar. 8.

Afghouni, K., 1978, "Lithium Ores in Brazil," *Energy,* S.S. Penner, ed., Vol. 3, No. 3, pp. 247-253.

Alexander, J.H., 1981, Lithium Annual Review, *Engineering and Mining Journal,* pp. 111-113.

Ames, L.L., et al., 1958, "A Contribution on the Hector, California Bentonite Deposit," *Economic Geology,* Vol. 53, No. 1, Jan.-Feb. pp. 22-37.

Angelleli, V., and Rinaldi, C.A., 1963, "Yacimientos de Minerales de Litio de las Provincias de San Luis y Cordoba," Info. No. 91, Argentina, Com. Nac. Energia Atomica, 79 pp.

Angelleli, V., and Rinaldi, C.A., 1965, "Resena Acerca de la Estructura, Mineralizacion, y Aprovechamento de Nuestras Pegmatita Portadoras de Minerales de Litio," *Acta Geologica Lilloana,* Vol. 5, pp. 1-18.

Cameron, E.N., et al., 1949, *Internal Structure of Granitic Pegmatites,* Monograph 2, Economic Geology Publishing Co., Urbana, IL.

Cameron, E.N., et al., 1954, "Pegmatite Investigations, 1942-1945, New England," Professional Paper 255, US Geological Survey, 352 pp.

Cerny, P., and Turnock, A.C., 1971, "Pegmatites of Southeastern Manitoba," Special Paper No. 9, Geological Association of Canada, pp. 119-127.

Deer, W.A., et al., 1962, "Lepidolite," *Rock-Forming Minerals,* Vol. 3, J. Wiley and Sons, New York, pp. 85-91.

Eardley, A.J., 1970, "Salt Economy of the Great Salt Lake," *Third Symposium on Salt,* J.L. Rau and L.F. Dellwig, eds., Vol. 1, The Northern Ohio Geological Society, Inc., Cleveland, pp. 78-105.

Foster, M.D., 1960, "Interpretation of the Composition of Lithium Micas," Professional Paper 354-B, US Geological Survey, 146 pp.

Goldschmidt, V.M., 1937, "The Principles of Distribution of Chemical Elements in Minerals and Rocks," *Journal of the Chemical Society,* pp. 655-673.

Hanley, J.B., et al., 1950, "Pegmatite Investigations in Colorado, Wyoming, and Utah, 1942-1944," Professional Paper 227, US Geological Survey, 125 pp.

Hegelson, H.C., 1968, "Geological and Thermodynamic Characteristics of the Salton Sea Geothermal System," *American Journal of Science,* Vol. 266, March, pp. 129-166.

Heinrich, E.W., 1964, "The Tin-Tantalum-Lithium Pegmatites of Sao Joao Del Rei District, Minas Gerais, Brazil," *Economic Geology,* Vol. 59, No. 6, Sep.-Oct., pp. 982-1002.

Hortsman, E.L., 1957, "The Distribution of Lithium, Rubidium, and Cesium in Igneous and Sedimentary Rocks," *Geochimica et Cosmochimica Acta,* Vol. 12, pp. 1-28.

Hutchinson, R.W., and Claus, R.J., 1956, "Pegmatite Deposits, Alto Ligonha, Portuguese East Africa," *Economic Geology,* Vol. 51, pp. 757-779.

Jahns, R.H., 1952, "Pegmatite Deposits of the White Picacho District, Maricopa and Yavapai Counties, Arizona," University of Arizona *Bulletin,* Vol. 23, No. 5, 105 pp.

Jahns, R.H., 1955, "The Study of Pegmatites, *Economic Geology,* 50th Anniversary Volume, Part II, pp. 1025-1130.

Kesler, T.L., 1961, "Exploration of the Kings Mountain Pegmatites," *Mining Engineering,* Vol. 13, No. 9, Sep., pp. 1062-1068.

Koenig, J.B., 1970, "Geological Setting of the Imperial Valley and its Geothermal Resources," Compendium of Papers Presented at the Imperial Valley—Salton Sea Area Geothermal Hearing, Oct. 22-23, Sacramento, CA, pp. E1-E5.

Kunasz, I.A., 1970, "Geology and Geochemistry of the Lithium Deposit in Clayton Valley, Esmeralda County, Nevada," Ph.D. Thesis, The Pennsylvania State University, 114 pp.

Kunasz, I.A., 1975, "Lithium Raw Materials," *Industrial Minerals and Rocks,* 4th ed., S.J. Lefond, ed., AIME, New York, pp. 791-803.

Kunasz, I.A., 1980, "Lithium, How Much?" *Foote Prints,* Vol. 48, No. 1, pp. 23-27.

Mayhew, E.J., and Heylmun, E.B., 1966, "Complex Salt and Brines of the Paradox Basin," *Second Symposium on Salt,* J.L. Rau, ed., The Northern Ohio Geological Society, Inc., Cleveland, pp. 221-235.

McLeod, I.R., 1965, *Atlas of Australian Resources Mineral Deposits,* 2nd ed., Department of National Development, Canberra.

Muessig, S., 1966. "Recent South American Borate Deposits," *Second Symposium on Salt,* J.L. Rau, ed., The Northern Ohio Geological Society, Inc., Cleveland, pp. 151-159.

Mulligan, R., 1965, "The Geology of Canadian Lithium Pegmatites," Economic Report No. 21, Geological Survey of Canada, 131 pp.

Nel, L.T., 1968, "Ore Deposits of Lithium in the Republic of South Africa," Atomic Energy Board, Republic of South Africa, PEL 23, 23 pp.

Norton, J.J., and Schlegel, D.M., 1955, "Lithium Resources of North America," Bulletin 1027-G, US Geological Survey, pp. 325-350.

Norton, J.J., 1965, "Lithium-Bearing Bentonite Deposit, Yavapai County, Arizona," Professional Paper 525-D, US Geological Survey, pp. D163-D166.

Page, L.R., et al., 1953, "Pegmatite Investigations, 1942-1945, Black Hills, South Dakota," Professional Paper 247, US Geological Survey, 228 pp.

Penner, S.S., 1978, "The Lithium Symposium 1977," *Energy,* Vol. 3, No. 3, 413 pp.

Rankama, K., and Sahama, T.G., 1950, "The Alkali Metals: Lithium, Sodium, Potassium, Rubidium, Cesium," *Geochemistry,* University of Chicago Press, Chicago, pp. 422-442.

Roberts, R.O., 1948, "Amblygonite and Associated Minerals from the Mbale Mine, Uganda," *Imperial Institute Bulletin,* Vol. 46, pp. 342-347.

Shawe, R.D., 1968, "Geology of the Spor Mountain Beryllium District," *Ore Deposits in the U.S., 1933/1967,* J. Ridge, ed., Vol. 2, AIME, New York, pp. 1143-1161.

Shawe, R.D., et al., 1964, "Lithium Associated with Beryllium in Rhyolitic Tuff at Spor Mountain, Western Juab County, Utah," Professional Paper 501-C, US Geological Survey, pp. 86-87.

Strock, L.W., 1936, "Zur Geochemie des Lithiums," Nachrichten von der Gesellschaft der Wissenschaften zu Goeltingen, Mathematisch-Physikalische Klasse, IV, N.F. 1, No. 15, 171 pp.

Toombs, R.B., 1962, "A Survey of the Mineral Industry of Southern Africa," Mineral Information Bulletin MR 58, Department of Mines and Technical Survey, Ottawa, 275 pp.

Valarmoff, N., 1954, "Matériaux pour l'Etude des Pegmatites du Congo Belge et du Ruanda-Urundi," *Annales de la Société Géologique Belgique,* Vol. 78, pp. 1-25.

Vine, J.D., 1976, "Lithium Resources and Requirements by the Year 2000," Professional Paper 1005, US Geological Survey, 162 pp.

Wegener, J.E., 1980, "Lithium Minerals in Zimbabwe with Special Reference to the Operations at Bikita Minerals (Pvt) Ltd.," International Conference on Zimbabwe, Sep. 1-5, Salisbury.

Wegener, J.E., 1981, "Profile on Bikita—Processed Petalite, The New Priority," *Industrial Minerals,* No. 165, June, pp. 51-53.

Westenberger, H., 1963, "The Lithium Minerals, Their Formation and Occurrence," *Review of Activities,* No. 6, Metallgesellschaft A.G., Frankfurt/Main.

White, D.E., 1957, "Thermal Waters of Volcanic Origin," *Geological Society of America Bulletin,* Vol. 68, pp. 1637-1658.

Winchell, A.N., 1942, "Further Studies of the Lepidolite System," *American Mineralogist,* Vol. 27, pp. 114-130.

13. Magnesite

OSCAR M. WICKEN

Magnesite ($MgCO_3$) when pure has a composition of 47.8% MgO and 52.2% CO_2. The pure mineral is sometimes, but rarely, found as transparent crystals resembling calcite, but most often it contains variable amounts of the carbonates, oxides, and silicates of iron; calcium; manganese; and aluminum in solid solution or as separate mineral entities.

Magnesite is one of the calcite group of rhombohedral carbonates that includes calcite ($CaCO_3$), siderite ($FeCO_3$), and rhodochrosite ($MnCO_3$) among others. The members of this group can enter into a wide range of substitutional solid solutions when the positive ions have similar radii. The radii of magnesium and iron ions are within 6% of each other; hence magnesite and siderite can form a complete series in which breunnerite (ferroan magnesite) is a well-known commercially valuable member. In contrast, the radius of the calcium ion is 36% larger than that of the magnesium ion, and only limited solution exists at each end of the $MgCO_3$-$CaCO_3$ series. Dolomite is not a member of the calcite group, but results when calcium and magnesium ions alternate in equal numbers in an ordered structure among carbonate ions. The result of these relationships is that calcite and dolomite often found intermixed with magnesite as identifiable crystal entities can be separated to varying degrees from the magnesite by beneficiation techniques.

Magnesite may be either macrocrystalline or cryptocrystalline (amorphous). The macrocrystalline form has a hardness of 3.5 to 4.0 on Mohs' scale. The specific gravity of the pure mineral is 3.02, but often the presence of iron carbonate in solid solution raises the value. The color may range from white to black with hues of yellow, blue, red, or gray. The color is not a significant indicator of composition, but in a given deposit an experienced person can often roughly grade magnesite by assessing color and crystallinity.

Cryptocrystalline magnesite is massive with no cleavage. The fracture is conchoidal; hardness is 3.5 to 5.0. The color is normally white but it can display tints of yellow, orange, of buff. Sometimes it is descriptively called "bone magnesite." The specific gravity ranges around 2.8 to 3.0 depending on accessory mineral inclusions.

The macrocrystalline form occurs in relatively few but large deposits, whereas the cryptocrystalline variety tends to occur in many but relatively small deposits.

Essentially all mined magnesite is dead-burned or calcined to make a variety of grades of refractory or caustic magnesias. Invariably provisions for some degree of beneficiation of the mined ore are required. The beneficiation steps may range from simple hand-sorting and scrubbing to more complex gravity, optical, and flotation methods to separate waste from the magnesite. However, occasionally some large deposits of macrocrystalline magnesite have blocks of sufficient purity to allow mining followed by direct firing of crushed and sized ore without other treatment. In any case, for most uses as a refractory material, the dead-burned product should contain at least 90% MgO; many commercial grades contain in excess of 95% MgO.

For uses as caustic magnesia in nonrefractory applications, the MgO content is often less important than other characteristics. Thus, magnesia contents of commercial grades can range downward to under 70% MgO and still be useful for certain applications.

ORIGIN, CHARACTERISTICS, AND EVALUATIONS OF DEPOSITS

Deposits of macro- and microcrystalline magnesite have practically no common features with each other as to origin, geologic structure, host rock, nor size. Each has its own unique features that dictate certain geologic study and exploration techniques. However, the mining and processing methods do have some common aspects in many cases.

Macrocrystalline Magnesite

The deposits are usually found in dolomitic terranes which can have areas measured in kilometers (square miles) and depths in hundreds of meters (feet). The magnesite itself occurs in the host structure as lenses and layers with horizontal dimensions that may be on the order of thousands of feet and depths of hundreds.

A variation in the magnesite-host rock relationship is seen in a deposit in Brazil where the magnesite is in a limestone measure (Bodenlos, 1954); however, even in this case the magnesite is not in direct contact with the limestone but is separated from it by a dolomitized zone.

Major deposits of macrocrystalline magnesite are found in Austria, Czechoslovakia, North Korea, China, Brazil, Canada, Nepal, and the United States. All are in dolomite host measures or zones. All are in regions of orogenic activity. Commonly such areas also show igneous activity, and an early and strongly supported theory postulated that magnesite resulted from the action of igneous intrusions and associated solutions on the dolomite (Wicken, Duncan, 1975). The dolomite measures themselves are widely viewed as sedimentary measures in which secondary placement of magnesite was brought about by hydrothermal action (Bain, 1924; Nishihara, 1956).

Deposits of this form of magnesite tend to be massive and physically homogenous. The magnesite formations are generally more resistant to weathering than associated formations; consequently, bold outcrops are characteristic of these deposits.

Exploration and Evaluation of Deposits: Outcrops are sampled by surface chipping and shallow trenching. Early evaluations of deposits are obtained by field mapping and from chemical and mineralogical analysis of samples to determine the quantity and distribution of accessory minerals containing silica, lime, iron oxide, and alumina. Tests to learn whether these undesirable minerals can be removed by ore dressing methods may be required as an aid to evaluation.

Following the preliminary surface investigations, a first-order drilling program would be carried out to assess the gross dimensions and the commercial potential of the deposit. Field mapping using standard techniques would be elaborated early by preparation of detailed maps showing surface configurations, geologic anomalies and features, and location of sampled sites and drill holes (Peters, 1978).

The drill-hole cuttings or cores are logged to show chemical and mineral features through the depth of the boreholes. The information from these first, widely spaced holes would be plotted to show cross sections of the deposit. Subsequent additional drilling would yield information to fill in the cross sections. At some point it may become desirable to log all

198

the drill-hole results into a computer data bank, particularly if the deposit shows substantial variation in thickness and composition at different levels. Graphic presentations would then be readily available to guide further exploration and subsequent mining operations (Carew, 1987).

Cryptocrystalline Magnesite

Deposits are found in contact with or close proximity to serpentized ultrabasic rock structures. The magnesite results as an alteration product of serpentine or allied magnesian rocks which have been subjected to the action of hydrothermal solutions and surface carbonate waters. The serpentine, which lies near or surrounds the magnesite, is itself an alteration product of ultrabasic rock. The relationship between temperature and carbon dioxide pressure is critical for the formation of the magnesite. The zones near or at the earth's surface are most favorably situated for the reactions. Hence, deposits of cryptocrystalline magnesite are found near the surface and have limited and decreasing extent in depth.

Major deposits of cryptocrystalline magnesite are found in Yugoslavia, Greece, Australia, India, South Africa, Zimbabwe, Turkey, Baja California, Colombia, and Central America.

Deposits of cryptocrystalline magnesite occur in three different modes. In one, the magnesite is found in large or small quantities in stockwork structures of serpentinized ultrabasic rock. The magnesite occurs with erratic dimensions from less than a centimeter (inch) to several meters (feet).

A second pattern is revealed, for example, by those deposits in Turkey and Yugoslavia. The deposits in Turkey are well-defined single or parallel veins more or less vertically dipping. The veins may be several meters (feet) thick with undulating horizontal continuity of hundreds of meters (feet). The deposits in Yugoslavia are similar but are more deeply buried. In both cases, the deposits are hydrothermal formations.

In a third pattern, magnesite is found as lacustrine deposits of nodules or concretions laid down in mud sediments in lakes and shallow embayments, or, as in the case of the Salda Lake deposit in Turkey, directly on the banks of the lake (Schmid, 1987). The nodules in all cases are secondary, hydraulically transported, and reworked mineral forms which have precipitated from solutions derived from the weathering of magnesian stockwork in nearby serpentinized ultrabasic rock formations. The nodules can range in size from fine gravel up to boulders 1m (3 ft) or more in diameter (Frost, 1987).

Exploration and Evaluation of Deposits: Stockwork structures of magnesite and ultrabasic rock are found characteristically as pronounced outcrops. The nearly white magnesite is easily seen against the dark gangue. In fact, beneficiation by optical methods is possible because of the color differences.

The outcrops are initially sampled by taking surface chip samples. These samples are useful for field and chemical tests to verify the presence of magnesite, establish the nature of associated rocks, and to give a first appraisal of the extent and dimensions of the deposit. However, in contrast to the information obtained from taking chip samples of macrocrystalline deposits, samples of stockwork structures yield little or no information as to the continuity and quantity of the magnesite. Also, in contrast to drilling for the evaluation of macrocrystalline magnesite deposits, a drilling program in stockwork has no early value. The important exploration of such exposures is by shallow trenching to observe the relative abundance, size, and continuity of veins and lenses.

Following favorable surface showings, adits may be driven to carry the exploration underground. At a certain point in the exploration program, large bulk samples should be taken for sorting tests to give information concerning the ratio of gangue to magnesite.

Magnesite deposits in vein structures are sampled by trenching along open exposures and by drilling following a program similar to that for macrocrystalline deposits.

In contrast to stockwork and vein deposits, both of which are found as outcrops, deposits of nodular magnesite often give no surface manifestation as the deposits may be covered with soil several meters (feet) thick. These lacustrine deposits—with or without soil cover—are evaluated in a program of excavating and drilling test pits and large diameter boreholes to determine the areal extent of the deposit, thicknesses of the nodule and overburden layers, and to provide samples for characterization of the nodules.

MINING PRACTICES

Depending on many circumstances such as anticipated reserves, kind of ore body, overburden, and ecological factors among others, mining can be by underground or by open pit methods or by a combination.

Underground Mining

The method has been used to some extent in the early development of many magnesite mines, particularly those that have had an operating history dating back several decades. Although the method has been largely abandoned elsewhere, it still survives in such places as, for example, Austria and Yugoslavia where the mountainous topography and thick rock overburden makes surface mining untenable. Where practiced, underground methods in large ore bodies are similar to those used in underground limestone operations; i.e., room and pillar patterns are developed, and the ore is removed by use of power loaders and trucks.

Open Pit Mining

This method is preponderately favored for ore removal. Stripping of overburden, the establishment of benches, and patterns of selective ore removal are guided by the information obtained from exploration and mapping programs. Stripping and ore removal are carried out with power shovels, trucks, scrapers, and front-end loaders. In the case of macrocrystalline deposits, the ratios of waste and overburden to ore are generally moderate as, for example, in Liaoning province of China some mines have ratios of 2.5:1 (Schmid, 1984); in the former operation of Northwest Magnesite Co. in Chewelah, WA, the ratio in the Red Marble deposit was 2.3:1. In the case of some stockwork deposits in Greece, the ratio sometimes exceeds 100:1. A modified form of open pit mining is found in Turkey where vertical vein deposits of cryptocrystalline magnesite are mined in open trench patterns on narrow benches.

The nodular forms of cryptocrystalline magnesite present entirely different recovery problems. As exemplified in the Australian deposits, the magnesite is not found as part of a rock formation, but as an unconsolidated layer of nodules under a shallow soil overburden a few meters (feet) thick. The strata of nodule concentration have variable thicknesses ranging from a few to several meters (feet). The concentrations and distribution of the nodules within the matrix of clay and miscellaneous silica, iron, calcium, and magnesium minerals are variable and unpredictable. Essentially, the overburden is stripped and the nodule layer is excavated using suitable earth-moving equipment. The magnesite is separated from the matrix by screening and scrubbing. Published information (Schmid, 1987) indicates a ratio of overburden and matrix to nodules, although highly variable, to be on the order of 3:1.

PRACTICAL ASPECTS OF EXPLORATION

Exploration today for magnesite focuses almost exclusively on finding extensions to known deposits. The extensions need not closely adjoin the existing deposit but can be separated from it by miles of barren ground. The key condition is that the extensions are in the same geologic horizon and structure as the known deposit.

Little or no planned exploration for magnesite is carried out in totally virgin fields. From time to time, an occurrence of magnesite may be noted in virgin fields by an observant individual, but such accidental discoveries rarely, or slowly at best, lead to the development of a profitable mine. Economic factors weigh against the activity.

Parties seeking other minerals have sometimes found magnesite in the course of exploration; for example, a search for tungsten mineralization in Nevada led to the discovery of the Gabbs deposit (Vitaliano, 1957).

The decision to explore a favorable site calls for a substantial amount of support information developed in an exploration program. Briefly put, the program must be designed to show objectives, to indicate estimated costs and benefits, to provide an appraisal of the regional situations, and to outline the scope and plans for reconnaissance.

The objectives of the exploration would be to locate a magnesite deposit at reasonable cost that could be developed into a profitable operation. The reasons for the search might be protective to assure adequate magnesite for continuing or future exploitation to meet market demands, to establish a position as insurance against interruption of supply from current suppliers, or to enter a new field of activity in industrial minerals. The costs and benefits would be analyzed and detailed for each option.

Appraisal of the regional situation would cover such matters as property ownership, exploration and mining rights, leasing, royalties, taxation, labor, political aspects, and environmental parameters.

The reconnaissance activity should deliver as much geologic information as is available from all sources; obtain local and areal maps showing topography, known deposits, and areas to be explored; and gather information about climate, rainfall, drainage, and vegetation.

Pertinent information about different areas in the world is available from a number of agencies. Government bureaus, foreign embassies, the United Nations, for example, are useful sources. Guides to information sources are available in many publications (Peters, 1978; Hoy, 1983). These and similar listings should be consulted as a prelude to serious exploration.

BIBLIOGRAPHY AND REFERENCES

Agterberg, F.P., 1989, "Computer Programs for Mineral Exploration," *Science*, Vol. 245, July 7, pp. 76-81.

Bain, G.W., 1924, "Types of Magnesite Deposits and Their Origin," *Economic Geology*, Vol. 19. No. 5, Aug., pp. 412-433.

Bodenlos, A.J., 1954, "Magnesite Deposits in Serra Das Equas, Brumado, Bahia, Brazil," Bulletin 975-C, US Geological Survey, pp. 87-170.

Carew, T.J., 1987, "Application of a Micro-computer to Exploration Geology: A Case Study," *Mineral Resource Management By Personal Computer*, Chap. 6, Ta M. Li, S.D. Handelsman, L. Korisars, eds., Society of Mining Engineers, Littleton, CO, pp. 47-52.

Fisk, R.L., 1953, "Changes in Primary Drilling at Northwest Magnesite," *Mining Congress Journal*, Vol. 39, Feb., pp. 48-54.

Frost, M.T., 1982, "The Magnesite Deposit at Main Creek, Savage River, Tasmania," *Economic Geology*, Vol. 77, No. 109, Dec., pp. 1901-1911.

Frost, M.T., et al., 1987, "The Kunwarara Magnesite Project—the development of a major new deposit in Queensland, Australia," *Proceedings*, 8th Industrial Minerals International Congress, Boston, MA, May, pp. 230-239.

Hoy, R.B., 1983, "Sources of Information for Industrial Minerals," *Industrial Minerals and Rocks*, 5th ed., S.J. Lefond, ed., AIME, New York, pp. 393-414.

Lawrence, L.J., 1965, "Exploration Procedures—Field Parameters of Mineral Exploration," *Transactions*, 8th Commonwealth Mining and Metallurgical Congress, Australia and New Zealand, Vol. 2, Chap. 4, pp. 55-59.

Lindgren, W., 1933, *Mineral Deposits*, 4th ed., McGraw Hill, New York, pp. 388-390.

Nishihara, H., 1956, "Origin of the Bedded Magnesite Deposits of Manchuria," *Economic Geology*, Vol. 51, No. 7, pp. 698-711.

Peters, W.C., 1978, "Mapping Surface Geology," *Exploration and Mining Geology*, John Wiley & Sons, New York, pp. 315-337.

Peters, W.C., 1978, "Sources of Preliminary Data and Information for Exploration Projects," *Exploration and Mining Geology*, John Wiley & Sons, New York, pp. 605-617.

Schmid, H., 1984, "China—the Magnesite Giant," *Industrial Minerals*, Aug., pp. 27-54.

Schmid, H., 1987, "Turkey's Salda Lake," *Industrial Minerals*, No. 239, Aug., pp. 19-31.

Schultes, H.B., 1986, "Baymag-High Purity Magnesium Oxide From Natural Magnesite," *CIM Bulletin*, Vol. 79, May, pp. 43-47.

Vitaliano, C.J. et al., 1957, "Geology of Gabbs and Vicinity, Nye County, Nevada," Mineral Investigations Field Studies Map, MF 52, US Geological Survey.

Wicken, O.M. and Duncan, L.R., 1975, "Magnesite and Related Minerals," *Industrial Minerals and Rocks*, 4th ed., S.J. Lefond ed., AIME, New York, pp. 881-896.

14. Manganese*

In 1774 a Swedish chemist, C. W. Schule, first recognized manganese as an element. That same year Schule's associate, J. G. Gahn, isolated the element manganese for the first time. In 1856 the Bessemer process of steelmaking gave birth to the economic importance of manganese. Later, in 1882, Robert Hadfield discovered high manganese steels.

Although the primary use of manganese is in the ferroalloy industry, two additional important uses for manganese are making chemicals and dry cell batteries. Manganese is also vital to plant and animal life. Various chemical compounds of manganese are used in fertilizers, feeds, glass manufacture, paints, varnishes, and for numerous medicinal and chemical purposes.

Although currently the world's largest consumer of manganese, the United States is producing only minor amounts of ore at the present time. This is due primarily to the fact that although huge deposits of manganese occur within US borders, all are of such low grade as to be economically unfeasible to mine.

For many years various government agencies and segments of private industry have devoted sizable sums of money and time in an attempt to upgrade ores from the various deposits, and also in an attempt to recover manganese from blast furnace slags. In view of current world prices, and the quantity of high grade material available, there is little likelihood that any of these approaches will be found to be economically feasible in the near future.

GEOLOGY AND MINERALOGY

Mineralogy

There are over one hundred minerals that contain manganese. These minerals vary from those with compositions that are predominantly manganese to those having only minor percentages:

Mineral	Chemical Composition	%Mn
Hausmannite	Mn_3O_4	72
Polianite	MnO_2	63.1
Pyrolusite	MnO_2	60–63
Cryptomelane	$KR_8O_{16}R$ (R=Mn)	Variable
Psilomelane	$Ba\ Mn\ Mn_8O_{16}\ (OH)_4$	45–60
Coronadite	$Pb\ R_8O_{16}R$ (R=Mn)	Variable
Hollandite	$Ba\ R_8O_{16}R$ (R=Mn)	Variable
Manganite	$Mn_2O_3 \cdot H_2O$	62
Braunite	$3\ Mn_2O_3 \cdot MnSiO_3$	62
Tephroite	$2\ MnO \cdot SiO_2$	54.3
Rhodochrosite	$MnCO_3$	47
Rhodonite	$MnSiO_3$	42
Spessartite	$3\ MnO \cdot Al_2O_3 \cdot 3SiO_2$	33.3
Wad	Hydrous Mn Oxides	Variable
Franklinite	$(Fe\ Mn\ Zn)O(Fe\ Mn)_2O_3$	Variable
Asbolan	Cobaltiferous wad	Variable
Alabandite	MnS	63.14

---

*Excerpt from Jacoby, C.H., 1983, "Manganese," *Industrial Minerals and Rocks*, 5th ed., Vol. 2, S.J. Lefond, ed., AIME, NY, pp. 897-908.

Geology

From a geological standpoint, most manganese deposits are complex. In general, it can be said that all primary deposits of manganese are carbonates or silicates. The most productive and profitable have been those of sedimentary origin and residual concentrations. Examples of these residual types of deposits are associated with the metamorphosed lodes of Madhya, Pradesh, India, and the lateritic deposits of Orissa and Bihar, India; nodules in the residual clays of the Philippine Islands, US southern Applachians and Arkansas, Ghana, and Brazil.

Sedimentary manganese deposits are best exemplified by the Nikopol district of Russia and the Tchiatouri area of Georgia, USSR.

Hydrothermal replacements are characterized by the rhodochrosite deposits of Butte and Phillipsburg, MT. In general, the hydrothermal-type deposit has not resulted in any large tonnages being produced, but may have given rise to the formation of sedimentary deposits throughout geological history.

Some sedimentary and residual-type deposits have been metamorphosed, giving rise to small high-grade ore bodies. These deposits are regionally metamorphosed, occurring in marbles, slates, quartzites, schists, and gneisses. Some of these metamorphosed deposits, such as the Franklin, NJ, deposit, are rich enough to be commercial without secondary enrichment. However, most of the exploitable deposits have been secondarily enriched.

Manganese is widely distributed and, to a greater or lesser degree, replaces two sets of elements; first, alkaline earths, calcium, barium, and magnesium and, secondly, aluminum and iron. Due to the diversity and complexity of manganese deposits, both with respect to deposition and chemistry, a wide range of impurities is almost invariably present in the ores.

DISTRIBUTION OF DEPOSITS

Distribution of manganese deposits can best be described by the term "scattered." They vary from occurrences such as the nodules on the bottom of Lake Michigan at Green Bay to the bog ores of Wickes, MT; the rhodochrosite of Phillipsburg, MT; the franklinite of Franklin, NJ; to the residual pyrolusite deposits of Cartersville, GA. More than 2100 manganese deposits are listed in 35 states. Currently the free world's largest deposits of manganese ore are in Gabon, Brazil, Republic of South Africa, Ghana, and India.

Foreign Deposits

Principal reserves in foreign countries are as follows:

	Million Tons of Mn
USSR	Unknown-in excess of 200
Republic of South Africa	300
Gabon	96
Brazil	46
Australia	44
India	22
World reserves, total	700

Russia possesses what are probably the world's two largest manganese deposits. The largest of these is the flat-lying Eocene deposit at Tchiatouri in Georgia, south of the Caucasus Mountains. These are beds of oolitic pyrolusite interbedded with layers of clays, sandstones, and marls. Upon beneficiation, primarily washing, these ores produce concentrates containing 50% MnO_2, and from 1 to 2% iron. The second great deposit of Russia lies in the Nikopol District on the Dnieper River in the Ukraine. Again, these are flat-lying beds close to the surface and easily accessible to strip mining. The beds of pyrolusite are interbedded with sand and glauconitic clays. These beds of Oligocene age rest on crystalline rocks covering approximately 415 km² (160 sq miles).

US Deposits

The largest deposit, or at least one of the largest, in the United States is located at Chamberlain, SD. This deposit is sedimentary in origin and consists of manganiferous concretions in the Pierre shale which is Upper Carboniferous in age. This formation varies in thickness from a few meters to 11.6 m (few feet to 38 ft) at Chamberlain.

The Batesville, AR, deposit contains a sporadic scattering of pyrolusite and wad-type residual materials in clay. Also buttons and thin layers of oxides are found in the Cason shale. This deposit rests on a fine-grained weathered limestone which gives rise to pinnacles, precluding the use of normal strip mining operations.

The Cuyuna iron range in Minnesota contains iron ore carrying varying quantities of manganese. The manganese content ranges from less than 1% to more than 17%. The manganese was originally sedimentary in origin but has undergone some secondary deposition in the filling of fractures and crevices.

Southern Appalachian deposits begin in the area of the Shenandoah Valley and extend south to the area of Cartersville, GA, and on into Alabama. While a wide variety of manganese minerals occur in lenses, beds, and residual pockets, only the oxides in residual pockets have been of any economic importance. In Virginia these occur as a series of small deposits extending for 129 km (80 miles) along the foot of the Blue Ridge Mountains. In general, most of the mines which have produced in excess of one million tons of manganese concentrates have been small operations conducted by local operators. Exceptions to this are the Crimora mine, working residual layers and pockets in the Erwin sandstone and the Kendall and Flick mines. These deposits, which have proven of economic importance, are the result of deep weathering. While the geological features of the primary deposits are undoubtedly dissimilar, the secondary residual deposits have many similarities.

The deposit at Artillery Mountains in Arizona has been operated from time to time during national emergencies. It has been estimated that the deposit contains approximately 159 Mt (175 million tons) of ore averaging from 3 to 4% manganese, with zones of material carrying values as high as 20% manganese. Beneficiation studies have made concentrates from which ferromanganese has been produced. The average iron content of these stratified oxides is 3% with 0.08% phosphorous.

The Three Kids property near Henderson, NV, was operated by Manganese, Inc. during World War II. At that time it was the largest producer of manganese in the United States. The primary ore, containing about 18% manganese, was beneficiated to produce a concentrate averaging 45% manganese.

Numerous other districts supply varying quantities of manganese during times of national emergency and stockpiling. The Anaconda Co. has produced approximately 1.8 Mkg (4 million lb) of manganese from the Butte District in Montana. Manganese, zinc, and silver have been deposited in the peripheral zone as rhodochrosite and manganosiderite. Almost all of the production has come from the mining of rhodochrosite. The Emma mine, located in the southernmost Anaconda vein, mined the "great bulge" at the east end. The Phillipsburg District gave rise to manganese, silver, lead, and zinc as replacement veins in limestones and shales. In part, the rhodochrosite of these deposits has been altered with the formation of high grade oxides. The primary source of the manganese was the Granite-Bimetallic vein, which produced silver and gold.

Aroostook County, ME, contains large reserves of manganese mixed with iron in a 1 to 2 ratio. The ore is a mixture of fine grained minerals including silicates and carbonates. These bedded deposits are steeply dipping slates which have been folded and faulted. Over 40 deposits are known to exist in an area 145 km (90 miles) long by 48 km (30 miles) wide.

Manganese Nodules on the Ocean Floors

The occurrence of large tonnages of manganese nodules in the three major oceans has been well established. Cardwell (1973) gives detailed chemical analyses for nodules taken from 54 different Pacific Ocean locations. The two major elements are manganese and iron. The minimal manganese analyses averaged 8.2% Mn, the maximum 50.1% Mn, and a median value of 24.2% Mn was determined. The percent nickel values on the same samples were 0.16%, 2.0%, and 0.99%, respectively.

The three manganese minerals found were todorokite, birnesite, and delta manganese dioxide. Todorokite varies in its chemical composition and can contain significant amounts of other elements substituted for manganese in the crystal lattice. It is basically a calcium, sodium, manganese, potassium, and magnesium oxide with three molecules of water. Birnesite contains less lattice substituted elements and has the typical formula:

$$(Na_7Ca_3) \ Mn_{70} \ O_{140} \ 28H_2O.$$

These three forms of manganese show the degrees of oxidation increasing from todorokite, through birnesite, to delta manganese dioxide. The only iron mineral which has been recognized in the nodules is goethite.

Archer (1973) in reviewing the progress and prospects of marine mining estimates that a commercial manganese nodule mining operation would operate at a rate between 0.9 and 2.7 Mt (1 and 3 million stpy) (dry weights). Allowing for processing losses, a 0.9-Mt (1 million-stpy) operation is likely to be capable of providing about 11.8 kt (13,000 tons) of nickel, 10 kt (11,000 tons) of copper, 2.3 kt (2500 tons) of cobalt, and 245 kt (270,000 tons) of manganese. It is highly probable that manganese and cobalt will be considered byproducts with the scale of mining depending mostly on the world demand for nickel.

EXPLORATION AND EVALUATION

Manganese deposits are one of the more difficult ores to evaluate. In general, the deposits are small and scattered and expensive geophysical means are not usually economically feasible. Also the geophysical exploration costs are often beyond the financial capacity of the small operator. However, since manganese has relatively high solubility, geochemical techniques can be an effective tool in an exploration program.

One of the main difficulties is securing representative samples of an ore body for proper evaluation and analysis. In the past, a large percentage of the holes drilled for samples were by cable or churn drill methods. These methods require extreme care in order to secure valid data. In most cases the data have been inaccurate.

Rotary drilling techniques are somewhat better, but again, extreme care must be exercised in order to secure representative samples. Where the deposit contains wad or soft oxides in clay, soil sampling methods can be used. Generally, where residual or detritus has occurred in a clay, the particle size of the manganese mineralization ranges from large boulders to -75 μm (-200 mesh) material. This makes soil sampling procedures difficult to employ.

In general the hard and deeper deposits are easier to sample with some degree of accuracy.

Although a great many manganese occurrences are known, the limited size of many of these deposits make them marginal to submarginal. Other large deposits are not economically exploitable due to the low concentration of manganese. The third consideration, and of prime importance, are the impurities associated with the manganese mineralization.

Due to the diversity and complexity of manganese formations, the impurities are many in number and complex in nature. Following is a broad generalization of the types of impurities:

1) Metallic impurities: iron, lead, zinc, copper, arsenic, and silver minerals.

2) Nonmetallic impurities: sulfur and phosphorous minerals.

3) Gangue: silica, alumina, lime, magnesia, and barium.

4) Volatiles: water, carbon dioxide, and organic matters.

Primary factors in the evaluation of deposits within the United States are the amenability of the material to beneficiation, and the price projection over the life of the necessary capital investment. Considerable effort has gone into the upgrading of ores from the various larger deposits within the continental limits of the United States.

Prior to World War II, few manganese operations had milling plants which were more complicated than washing, screening, jigs, tables, and, in a few instances, flotation. Now it is not only necessary to increase the manganese content, but also to decrease the percentage of impurities. Because each deposit is distinctly different from most other deposits, no single process is applicable to all.

BIBLIOGRAPHY AND REFERENCES

Archer, A.A., 1973, "Progress and Prospects of Marine Mining," *Mining Engineering*, Vol. 25, No. 12, Dec., pp. 31-32.

Boardman, L.G., 1961, "Manganese in the Union of South Africa," Paper presented at the 7th Commonwealth Mining and Metallurgical Congress Johannesburg, Republic of South Africa, Preprint, 12 pp.

Brooke, D.B., 1966, *Low-Grade and Nonconventional Sources of Manganese*, Resources for the Future, Johns Hopkins Press, Baltimore, MD, 123 pp.

Caldwell, A.B., 1971, "Deepsea Ventures Readying Its Attack on Pacific Nodules," *Mining Engineering*, Vol. 23, No. 10, Oct., pp. 54-55.

Cole, S.S., 1960, "Manganese Ore," *Industrial Minerals and Rocks*, 3rd ed., J.L. Gillson, ed. AIME, New York, pp. 545-549.

DeHuff, G.L., 1960, "Manganese," *Mineral Facts and Problems*, Bulletin 585, US Bureau of Mines, pp. 493-510.

DeHuff, G.L., 1965, "Manganese," *Mineral Facts and Problems*, Bulletin 630, US Bureau of Mines, pp. 553-571.

DeHuff, G.L., 1971, "Manganese," *Minerals Yearbook—1971*, Vol. 1, US Bureau of Mines, pp. 717-729.

DeHuff, G.L., 1972, "Manganese," *1972 Minerals Yearbook*, Vol. 1, US Bureau of Mines, pp. 757-769.

DeHuff, G.L., 1977, "Manganese," *Minerals Yearbook 1977*, US Bureau of Mines.

DeHuff, G.L., 1978, "Manganese," Preprint, 1978-1979, *Minerals Yearbook*, Bureau of Mines, 15 pp.

DeHuff, G.L., 1980, "Manganese in 1980," *Mineral Industry Surveys*, US Bureau of Mines.

DeHuff, G.L., and Jones, T.S., 1981, "Manganese," *Minerals Yearbook 1980*, US Bureau of Mines, pp. 543-553.

Dickson, T., 1980, "Manganese, Non-Metallic Market Development," *Industrial Minerals*, No. 155, Aug., pp. 21-29.

Dykstra, F.R., 1979, "Manganese—Its Strategic Implications," *Industrial Minerals*, No. 145, Oct., pp. 55-61.

Elkins, D.A., 1964, "Estimated Cost of Exploiting Enriched, Hard Manganese Ore from the Maggie Canyon Deposit, Artillery Mountains Region, Mohave County, Arizona," Report of Investigations 6438, US Bureau of Mines, 78 pp.

Farnham, L.L., and Stewart, L.A., 1958, "Manganese Deposits of Western Arizona," Information Circular 7843, US Bureau of Mines, 87 pp.

Fillo, P.V., 1963, "Manganese Mining and Milling Methods and Costs, Mohave Mining Company, Maricopa County, Arizona," Information Circular 8144, US Bureau of Mines, 29 pp.

Huttl, J.B., 1955, "How Manganese, Inc. Upgrades Complex Three Kids Ore," *Engineering & Mining Journal*, Vol. 156, No. 11, Nov., pp. 88-93, 116.

Jacoby, C.H., 1975, "Manganese," *Industrial Minerals and Rocks*, 4th ed., S.J. Lefond, ed., AIME, New York, pp. 821-836.

Johnson, A.C., and Trengove, R.R., 1956, "The Three Kids Manganese Deposit, Clark County Nevada: Exploration, Mining and Processing," Report of Investigations 5209, US Bureau of Mines, Apr., 31 pp.

Kline, H.D., 1958, "Methods and Costs of Mining and Washing Manganese Ore, Batesville District," Report of Investigations 5411, US Bureau of Mines, 46 pp.

Kline, H.D., 1962, "Methods and Costs of Mining and Washing Manganese Ore, Batesville District, Arkansas," Information Circular 8095, US Bureau of Mines, 22 pp.

Kraus, E.H., et al., 1959, *Mineralogy*, 5th ed., McGraw-Hill, New York, 686 pp.

Lewis, W.E., et al., 1958, "Investigation of Cuyuna Iron Range Manganese Deposits, Crow Wing County, Minn., Progress Report One," Report of Investigations 5400, US Bureau of Mines, 49 pp.

Mero, J.L., 1961, "Economics of Deep-Sea Mining," *Mining Congress Journal*, Vol. 47, No. 9, Sep., pp. 52-56, 68.

Mero, J.L., 1968, "Seafloor Minerals: A Chemical Engineering Challenge," *Chemical Engineering*, July 1, pp. 73-80.

Sears, C.E., 1957, "Manganese Deposits of the Appalachian Area of Virginia," *Mineral Industries Journal*, Virginia Polytechnic Institute, Vol. 4, No. 1, Mar., pp. 1-4.

Sheridan, E.T., 1970, "Manganese," *Mineral Facts and Problems*, Bulletin 650, US Bureau of Mines, pp. 315-331.

Sidwell, K.O.J., 1957, "The Woodstock, N.B. Iron-Manganese Deposits," *Canadian Mining & Metallurgical Bulletin*, Vol. 50, No. 7, July, pp. 411-416.

Tinsley, C.R., 1973, "In Search for Commercial Nodules, Odds Look Best in Miocene-age Pacific Tertiary System," *Engineering & Mining Journal*, Vol. 174, No. 6, June, pp. 114-116.

Tinsley, C.R., 1973, "Mining of Manganese Nodules: An Intriguing Legal Problem," *Engineering & Mining Journal*, Vol. 174, No. 10, Oct., pp. 84-87.

Tinsley, C.R., 1974, "Manganese-Prices Boosted by Record Steel Output in 1973," *Engineering & Mining Journal*, Vol. 175, No. 3, Mar., pp. 81-84.

Weiss, S.A., 1980, "Manganese, the Other Uses," *Metals Bulletin Book*, London, 360 pp.

Williamson, D.R., et al., 1959, "Future United States Manganese Sources," *Colorado School of Mines & Mineral Industries Bulletin*, Vol. 2, No. 4, July, 12 pp.

Young, W.E., 1966, "Manganese Occurrences in the Eureka-Animas Forks Area of the San Juan Mountains, San Juan County, Colorado," Information Circular 8303, US Bureau of Mines, 52 pp.

15. Mica*

Mica is a platy mineral occurring in a variety of complex hydrous aluminosilicate forms with differing chemical composition and physical properties. Principal minerals in the mica group include:

Muscovite—potassium mica (colorless or pale green/ruby)
Phlogopite—magnesium mica (dark brown or amber)
Biotite—magnesium-iron mica (black or dark green)
Lepidolite—lithium mica (lilac)

Selected properties are given in Table 1. Muscovite is most important commercially with phlogopite of lesser use. The laminated structure of mica enables it to be split into very thin films which are transparent and tough as well as having outstanding dielectric and insulating properties.

Commercially, mica is used in a number of forms (Chowdbury, 1941; Rajgarhia, 1951; Skow, 1962). *Sheet mica* consists of flat sheets mined from naturally occurring books of mica, free from defects and capable of being punched or stamped into required shapes. Sheet mica is separately classified into *blocks, films,* or *splittings* according to thickness.

Built-up mica, or *micanite,* is made by arranging overlapping splittings in layers cemented with a binder and pressed together at high temperature.

Reconstituted mica, or *mica paper,* is made by depositing fine flakes of high-quality scrap mica in a continuous film impregnated with binder.

Scrap and *flake mica* are normally unsuitable in quality or size for making sheet mica. Originally the term *scrap mica* referred to the waste byproduct of mining and fabricating sheet mica, and distinct uses were developed for this material. Subsequently it became economically feasible to mine lower quality mica for these uses, and the term *flake mica* was introduced. Although the two terms are still used somewhat synonymously, it is considered better, at least in the United

---

*Excerpt from Chapman, G.P., 1983, "Mica," Industrial Minerals and Rocks, 5th ed., Vol. 2, S.J. Lefond, ed., AIME, New York, pp. 915-929.

States, to use the term *flake mica* in referring to mica mined for uses other than as sheet mica; and to restrict the term *scrap mica* to mica mined for or with sheet mica, but which will not meet the specifications.

Synthetic mica, such as fluorophlogopite, is produced by slow crystallization of a melted blend of pure raw materials.

MINERALOGY

In the true sense the term *mica* does not relate to a particular mineral, but to a group or family of minerals of similar chemical composition and to some extent similar physical properties. These minerals are predominantly potassium aluminum silicates with varying amounts of magnesium, iron, and lithium. The precise formulas and isomorphic relationships of the various group members have been studied, but general agreement has not been reached (Grimshaw, 1971; Hurlbut, 1952; Skow, 1962). The general formula describing the chemical composition of micas is

$$X_2Y_{4-6}Z_8O_{20}(OH, F)_4,$$

where X is mainly K, Na, or Ca; Y is mainly Al, Mg, or Fe; and Z is mainly Si or Al (Deer, et al., 1962).

These minerals have an internal structure of the layered lattice type where the silicon atoms are in the center of a tetrahedral grouping of oxygen atoms. The groups are linked together in a single plane by three oxygen atoms that lie within the common plane. Each of these oxygen atoms is shared by two tetrahedra. These linked tetrahedral groups, when continually extended, produce a hexagonal network within the plane. This internal structure of the mica has been used to explain the external pseudohexagonal structure of a mica crystal. When a pair of silicon-oxygen sheets that have had about one-fourth of the silicon atoms replaced by aluminum to form a mica are oriented with the tetrahedra vertices pointing at each other, a firm double-layered structure is formed with hydroxyl groups and metallic atoms such as aluminum, magnesium, and lithium between them. These double-layered structures are joined together by potassium atoms. The cleavage plane of the mica is found between these double-layered structures.

Table 1—Selected Properties of Various Micas

Chemical Constituent	Muscovite	Phlogopite	Biotite
SiO_2	46	40	37
Al_2O_3	35	17	18
K_2O	10.5	10	9
MgO	0.5	26	8
Fe_2O_3	1	0.2	2
FeO	1	2.8	21
NA_2O	1	0.5	1
Minor	0.5	0.5	1
H_2O	4.5	3	3
Total	100	100	100
Specific gravity	2.77-2.88	2.76-2.90	2.70-3.30
Mohs' hardness	2½-3	2-2½	2½-4
Optic axial angle	30°-47°	0-15°	0-25°
Temp. of decomposition	400-500°C	850-1000°C	
Dielectric constant	6.5-9.0	5.0-6.0	
Specific heat, 25°C	0.206-0.209	0.206-0.209	

In addition to the better known members of the mica group—muscovite, phlogopite, biotite, lepidolite—there are zinnwaldite, a lithium-iron mica, fuchsite, a chromium mica, and roscoelite, a vanadium mica. Of the known micas, only muscovite and phlogopite exhibit any large commercial demand.

These micas crystallize in the monoclinic system with crystals that usually form in hexagonal or rhomb-shaped scales, prisms, or plates, with plane angles on the base of about 60° or 120°. The crystal faces are rarely smooth or well defined except for the basal plane. When a mica cleavage plate is struck sharply by a blunt needle, a six-rayed percussion figure is formed. The figure's most prominent line is parallel to the crystal's plane of symmetry; the remaining two lines are almost parallel to the crystal's prismatic edges. The percussion figure, in conjunction with the position of the optic axes as seen in an interference figure under the polarizing microscope, is useful in making a quick distinction between muscovite and phlogopite, which may closely resemble each other visually. Phlogopite, like biotite, has the plane of the optic axes parallel with the plane of symmetry of the crystal; muscovite has the plane of the optic axes perpendicular to the plane of symmetry. (When a blunt punch is pressed on a mica cleavage plate, a pressure figure is produced. The lines of the pressure figure are perpendicular to those of the percussion figure, but the pressure figure generally develops only partially.)

Minerals of the mica group have the distinctive physical property of being both flexible and elastic. On this basis they are readily distinguished from chlorite, talc, and vermiculite, flakes which will readily bend but will not snap back to their original position when released.

Muscovite has excellent basal cleavage that allows it to be split into very thin sheets exhibiting a high degree of flexibility, elasticity, and toughness. Very thin sheets of clear muscovite are transparent and colorless or almost colorless. In thicker sheets, it can be translucent and exhibit light shades of yellow, brown, green, or red. The mineral is strongly birefringent and exhibits a feeble pleochroism unless it is dark colored. The material is optically negative. The axial angle varies from 30° to 47° and is dependent on variation in the material's composition. The optic axes lie in a plane that is normal to the crystal's plane symmetry. Upon heating, muscovite commences to lose water at about 500°C. Muscovite is decomposed only by hydrofluoric acid.

The utility and value of sheet mica depends on the freedom from certain physical defects that occur as a result of the environment and events that occurred during and after crystallization of the mica. These defects are structural imperfections of the cleavage surfaces due to mineral inclusions.

Some of the structural imperfections associated with mica are reeves or cross grains which are lines, striations, or sharp folds in the plane of cleavage; "A" structure, which usually consists of two sets of reeves intersecting on a cleavage surface at about 60°; herringbone structure formed by two sets of reeves intersecting at about 120° along a central line of reeves; wedge mica which consists of mica runs that are thicker at one end than the other; warping which causes a mica characterized by shallow waves or ridges; ruled mica which has regular, sharply defined parting planes that intersect the basal cleavage plane at about a 67° angle; and tangle-sheet mica which is formed by an intergrowth of parts of one mica crystal with another. Most of these defects preclude use of the mica as sheet mica but do not interfere with its use as flake mica.

Mineral inclusions in mica can also limit the utility of sheet mica. These inclusions and intergrowths of other minerals are broadly classed as stains. Some mineral stains appear parallel with the cleavage, but others can penetrate the cleavage plane and cause pinholes. Examples of staining are those formed during mica crystallization and include mottling, inorganic stains, mineral inclusions, and intergrowths. Additional staining can be caused by the circulation of ground water. Mica deposits that have been stained will yield little usable sheet mica.

Phlogopite has a pearly to submetallic luster and varies from translucent to transparent in thin sheets. Phlogopite has been identified in colors ranging from brownish red through yellowish brown, greenish brown, and dark pearl grey to an almost colorless pale green. The color intensifies as the iron content of the mineral increases. Darker colored samples of this mineral exhibit greater hardness than those of lighter color. Phlogopite can contain fluorine which is usually most prominent in reddish brown samples and least in greenish specimens. Phlogopite generally exhibits a distinct pleochroism.

The mineral is optically negative and its optic axial angle varies from 0° to 15°, apparently increasing directly as a function of the iron content. The plane of the optic axes is parallel to the crystal's plane of symmetry. When heated, phlogopite will not dehydrate greatly until about 1000°C. The reaction to acids is variable. Hydrochloric acid will attack the mineral mildly. Hot concentrated sulfuric acid decomposes phlogopite completely.

HISTORICAL BACKGROUND

Sheet Mica

Evidence exists of the use of mica as far back as the earliest civilizations, especially in the Nile Valley and India prior to 2000 B.C.

One of the earliest uses of mica was as a medicine, and even today Hindu physicians use fused biotite. The lustre, iridescence and transparency of the larger sheets led to applications as windows, mirrors, adornments, and plates for painting mythological scenes and the like. The American Indians of the southern Appalachians are known to have used mica in ornamentation at grave sites in the 14th century.

The Romans were probably responsible for the name, from the Latin word *micare* meaning to shine or glitter. Pliny mentioned material scattered over the Circus Maximus to impart a shining whiteness, and window coverings of transparent stones; these may have been in reference to mica.

Muscovite probably derives from the Russian district of Moscovia where de Boodt found and identified potash mica in 1609. The Russians also used mica for medicine. Their men-of-war had mica portholes. Thin sheets of mica were used to protect the surface of icons from being kissed by worshippers.

Commercial mica mining was started in the United States in 1803 at the Ruggles Mines in Grafton County, NH. Primary uses of mica in the 1800s were for stove windows, shades for open-flame lights, and for furnace viewing glass. Although the United States originally was self-sufficient in its mica production, by 1885 India had become a major supplier of muscovite sheet.

The appearance of Thomas Edison's bipolar generator in 1878, his incandescent lamp in 1879, and his system of central-station power production in 1882 stimulated commercial development of electric power-generation.

Growth of the US electrical industry opened up many uses for sheet mica. Mica has proved to be the best insulation for use in commutator segments of large electric generators and motors. The US electrical industry grew so fast that the domestic reserves of large sheets were rapidly depleted.

A patent was issued in 1892 for the first built-up mica that maintained the dielectric and mechanical properties of natural mica. The use of splittings for built-up mica further depleted US reserves and led to increased imports of Indian mica.

Mica was used as spacers and insulators in the diode vacuum tube developed in 1904, and in the triode which followed in 1906. The vacuum tube industry grew in importance with the outbreak of World War I, and large quantities of mica were required to support the war effort. Mica retained its importance in the interwar period through use in the radio industry as well as in the growing electrical industry.

During World War II mica demand was heightened by the great increase in sophisticated electronic gear. The United States worked with the United Kingdom to stimulate production in India and other mica-producing countries. After the war, sheet mica production dropped off dramatically and the decline continues in the 1980s, caused by the rapid advancement of solid-state electronics.

Ground Mica

In 1890 J.S. Williams of the Richmond Mica Co. in Virginia produced ground mica as an ingredient for axle greases. Soon afterward the Vance brothers installed the first chaser mill in North Carolina producing wet ground mica for pigments, to be followed by wet-grinding by Williams.

From there the American industry grew. In 1908 Thomas English formed the English Mica Co. in Spruce Pine, NC. In the 1920s J.B. Preston Co. bought English Mica and merged it with the Biotite Mica Co. and Roofing Mica Co. Franklin Mica Products was started by John Davenport in Franklin, NC. In New Hampshire, what is now Concord Mica Co. was formed by Earl Moore. Diamond Mica Co. and Hayden Mica Co. were formed after World War II.

In the 1930s wet ground mica production started in Derby, England, and some dry grinding production began in Europe and India. Micronized mica emerged in Norway and England in the 1960s with further wet grinding plants in Norway, France, and South Africa.

The large growth in mica grinding came in dry ground mica from the 1940s onward, first with the uses in roofing felts and shingles in the US and, as that declined, with the emergence of drywall joint cements and oil-well lost circulation additives.

The 1970s saw the decline of water-grinding as rising energy costs made the products uneconomic and several plants in the US and Europe closed.

COUNTRIES OF ORIGIN

The most important countries producing mica are listed in Table 2, together with an estimate of their mica-scrap output. An attempt has been made to show how the total supply is distributed around the world for use in the manufacture of ground mica.

It will be seen that approximately 50% of the mica arises in the United States with four other countries—USSR, India, China, and Canada—making up the next 35%.

United States

Approximately 60% of US scrap is from North Carolina and consists of byproduct mica from the feldspar, kaolin, and lithium producers. Virtually all is ground in mills located near the mines. The remaining mica comes mainly from Alabama, Connecticut, Georgia, New Mexico, and South Carolina.

USSR

The pegmatite deposits of Karelia and Maura River regions appear to dominate production, but little information is published regarding mica-grinding. Apparently no exports are made outside Eastern Bloc countries.

India

High quality mica is mined in the United States of Bihar, Rajasthan, and Andhra Pradesh for sheet mica production, and the consequent scrap is exported for mica-paper manufacture or ground in India for home and overseas consumption.

Table 2 Countries of Origin and Consumption of Mica, t/a

Country of Origin as Raw Material	Eventual Consumption as Mica Powder or Flakes							
	US and Canada	USSR and Eastern Bloc	Europe	Far East and Australia	Middle East	Africa	South America and West Indies	Total
USA	105,000	500	2,500	1,000	200	300	500	110,000
USSR		35,000						35,000
India	1,000	7,000	3,000	5,000	4,000	1,000		21,000
China			8,000	4,000				12,000
Canada	12,000		1,500	1,500				15,000
Brazil	1,000		5,000				1,000	7,000
France			6,000					6,000
Spain			4,000					4,000
South Africa			1,500	500		1,000		3,000
Korea				3,000				3,000
Norway			2,000					2,000
Argentina			1,000				1,000	2,000
Italy			1,000					1,000
Others		1,000	1,000	1,000				3,000
European Re-exports	500	1,000	(4,000)	500	800	700	500	
Total	119,500	44,500	32,500	16,500	5,000	3,000	3,000	224,000

China

Mica is produced in a number of regions from pegmatite or as a feldspar byproduct. The most important area is Tientsin province.

Canada

Output consists of flake phlogopite mined in Quebec province and ground at a plant near Montreal. This is the only significant production of ground phlogopite on a world basis.

Madagascar

High quality phlogopite is mined for use as sheet mica to satisfy the electrical and mica paper markets of Europe and America.

GEOLOGIC FEATURES OF MICA DEPOSITS

Sheet Mica

Large crystals or books of muscovite and phlogopite mica are generally found in regionally metamorphosed rocks. Granitic pegmatites are the source of muscovite sheet. Phlogopite sheet is found in areas of metamorphosed sedimentary rocks into which pegmatite-rich granite rocks have intruded.

Pegmatites are generally the only source of usable quality sheet mica. Although there are many mica-bearing deposits throughout the world, only a few localities have deposits that are of economic importance. These pegmatites are light-colored, coarsely crystalline igneous rocks. They can be found as dikes or sills in metamorphic rocks and large granite intrusions. The mica crystals found in these deposits range from less than one inch to many feet in length. Variation of size within an individual deposit is not uncommon. Mica-bearing pegmatites have been known to exceed 61 m (200 ft) in thickness and 305 m (1000 ft) in length and have been worked to depths ranging from 61 to 152 m (200 to 500 ft). However, mica also has been produced from shallow surface deposits.

Pegmatites are composed primarily of feldspar, quartz, and mica. In many geologic situations accessory minerals such as garnet, tourmaline, and beryl occur with the primary pegmatite constituents. The distribution of minerals in pegmatites may be even, zoned, or segregated into layers.

Deposits of large books of phlogopite mica that can economically provide large quantities of sheet mica are available in only a few places. Skow (1962) has classified phlogopite deposits into vein, pocket, and contact deposits. Vein deposits, which are generally narrow and are enclosed in fine to medium grained pyroxenite, are the major source of phlogopite sheet mica. Pocket deposits have been found to be irregular in shape, size, course, and persistence. The surrounding pyroxenite of this type of deposit is usually more coarsely crystalline and open textured than that found with vein deposits. The phlogopite crystals may be irregularly distributed throughout the pocket along with crystals of other minerals such as pyroxene and calcite or may occur as very large solid masses of phlogopite.

Scrap and Flake Mica

Originally scrap mica was derived from the mining and processing of sheet mica and included poor quality sheet mica that did not meet the specifications for size, color, and quality. However, the industrial demand for scrap mica began to increase greatly in the 20th century, and mica deposits were mined specifically for the smaller size mica crystals or flake mica found in coarse-grained, weathered, granitic rocks such as alaskite and pegmatite, and in some schists. Flake mica is also recovered as a coproduct from the production of clay, feldspar, and spodumene.

Technology has been developed that permits the recovery of small particle size flake mica from various mica-bearing sources such as mica schist, graphitic mica schist, granite, silt, and recovery plant tailings. Good recovery rates for mica have been reported by the use of grinding and flotation techniques.

EXPLORATION

Sheet Mica

Exploration for sheet mica is not amenable to most of the methodology used to search for other minerals. The sporadic occurrence of large runs of books of mica crystals in the host rock makes it uneconomic to drill a deposit to any great extent to determine the mica content. Any drilling that is done can only delineate the existence of pegmatite and may reveal very little information about the mica content.

Prospecting for sheet mica has failed to advance beyond the trial and error method. Geological evidence is generally insufficient to supply enough information for the evaluation of a deposit. The most practical prospecting method for sheet mica is to sink a test pit in an effort to determine the percentage, size, and quality of trimmed sheet that might be obtained from a deposit. A potential deposit may be stripped or trenched by mechanical or hand methods to outline the size, shape, and attitude of the pegmatite vein for the purpose of selecting the sites for test pits or to supplement the information obtained from test pits. A common method of further exploration for sheet mica consists of sinking a shaft either down dip in the pegmatite body or vertically to cut the pegmatite at the desired depth. Drifts can then be driven along the strike of the deposit in both directions. Any large mica crystals that are found are visually examined by experienced mica workers who determine their freedom from structural imperfection and their suitability for sale. Steeply dipping pegmatites can be explored by the construction of adits. As yet, geophysical and geochemical techniques have not been applied to prospect and evaluate deposits containing sheet mica. The search for mica sheet remains basically a pick and shovel operation with some limited usage of mechanized equipment.

Flake Mica

Small particle size flake mica that is distributed in deposits of pegmatite, schist, and clay can be evaluated by conventional methods. The location of a deposit can be determined from surface geology. If the surface outcrops are of sufficient size, surface sampling may be justified, but it is more likely that the deposit would be augered or drilled to a shallow depth to furnish samples for testing and delineation of the deposit.

Ore samples obtained in this manner are crushed to a range of sizes to determine the optimum crushing size at which the ore will release the maximum quantity of mica. Chemical analysis is not effective for mica determination because of the wide variation of the chemical composition of the various micas. Petrographic analysis is not completely satisfactory to evaluate the mica content of a sample. The mica industry has not established a standard method of analysis to determine the mica content of an ore sample.

Sink-float techniques can be applied to the crushed samples of pegmatites and schists to assess the capability of separating mica from its accessory minerals by the use of a heavy liquid such as tetrabromoethane or mixtures of tetrabromoethane and kerosene or trichloroethylene. Experiments can be carried out to determine if mica can be extracted from the ore sample by some method of flotation.

As a result of preliminary studies, a flowsheet for the recovery of mica can be developed and a small pilot operation erected to determine the feasibility of the method of recovery.

The final test of any flake mica is reached when it is ground and/or classified to a particular particle size and accepted for an industrial end use.

BIBLIOGRAPHY AND REFERENCES

Chapman, G.P., 1980, "The World Mica Grinding Industry and Its Markets," *Proceedings,* 4th Industrial Minerals International Congress, Atlanta, 7 pp.

Chowdhury, R.R., 1941, *Handbook of Mica,* Chemical Publishing Co., Brooklyn, NY, 340 pp.

Deer, W.A., et al., 1962, "Sheet Silicates," *Rockforming Minerals,* Vol. 3, Longmans, London, 270 pp.

Grimshaw, R.W., 1971, *The Chemistry and Physics of Clays and Allied Related Ceramic Materials,* Wiley-Interscience, New York, 1024 pp.

Hawley, G.C., 1981, "Explaining Markets and Technologies for Mica from Suzor Township, Quebec," Preprint 81-121, SME-AIME Annual Meeting, Chicago, 10 pp.

Hill, T.E., Jr., et al., "Separation of Feldspar, Quartz and Mica from Granite," Report of Investigation 7245, US Bureau of Mines, 25 pp.

Hurlbut, C.S., Jr., 1952, *Dana's Manual of Mineralogy,* 16th ed., John Wiley, New York, 530 pp.

Ivey, K.N., and Haskiel, R.S., 1969, "Fluorine Micas," Bulletin 647, US Bureau of Mines, 291 pp.

Lesure, F.G., 1973, "Mica," "United States Mineral Resources," Professional Paper 820, US Geological Survey, pp. 415-425.

Petkof, B., 1975, "Mica," *Industrial Minerals and Rocks,* S.J. Lefond, ed., 4th ed., AIME, New York, pp. 837-850.

Preston, J.B., 1973, "Aluminum Potassium Silicate—Mica," *Pigment Handbook, Vol. 1, Properties and Economics,* John Wiley & Sons, pp. 249-263.

Rajgarhia, C.M., 1951, *Mining, Processing and Uses of Indian Mica,* McGraw-Hill, New York, 388 pp.

Skow, M.L., 1962, "Mica: a Material Survey," Information Circular 8125, US Bureau of Mines, 240 pp.

Tepordei, V.V., 1981, "Mica," *Minerals Yearbook,* US Bureau of Mines, pp. 563-572.

Zlobik, A.B., 1980, "Mica," *Mineral Facts and Problems,* Bulletin 671, US Bureau of Mines, 16 pp.

16. Natural Abrasives*

Abrasives include the substances that are used to grind, polish, abrade, scour, clean, or otherwise remove solid material, usually by rubbing action but also by impact (pressure blasting for example). They do not include abrasive tools, for instance, lathe tools and files—or polishing agents such as waxes, which act by filling pores.

Detergents and cleaners whose action is chemical rather than physical are omitted although some chemical-action polishes and cleaners may also contain solid abrasives, for example, many automobile and metal polishes.

GENERAL CONSIDERATIONS

The most important physical properties of materials that qualify them for use as abrasives are hardness, toughness (or brittleness), grain shape and size, character of fracture or cleavage, purity or uniformity. For making bonded abrasive products such as grinding wheels, additional important factors are stability under high heat and bonding characteristics of grain surfaces. The economic factors of cost and availability are always important.

No one single property is paramount for any use. For some uses extreme hardness and toughness are needed, as in diamonds for drill bits; for others, the factors of greatest importance are hardness and ability to break down slowly under use, to develop fresh cutting edges when grains become worn—for example: in garnet for sandpaper neither highly cleavable or friable grains nor extremely tough grains are wanted. For still other uses, great hardness is objectionable; for example, abrasives for dentifrices and for glass-cleaning soaps. For the most efficient use in the more critical applications, the different types of abrasives are rarely completely interchangeable; thus, while crushed quartz and garnet are both used in sandpaper, the papers are not at all interchangeable in their use applications.

In the last analysis, the choice of a high grade abrasive depends upon the quality and quantity of work done by the abrasive per unit of cost. Initial cost of an artificial abrasive may be much greater than that of a natural abrasive but the artificial abrasive may do so much better work than the natural one, and do it so much faster that the ultimate cost will be less. It is for this reason that artificial abrasives have largely replaced natural abrasives.

CLASSIFICATION

Abrasives may be divided into two general classes, natural and manufactured. The former includes all rocks and minerals used for abrasive purposes without chemical or physical change other than crushing, shaping, or bonding into suitable forms. Manufactured or artificial abrasives are made either by heat or chemical action from metals or mineral raw materials.

Table 1 lists most of the important abrasives, classified as to inherent types and the forms in which they are used industrially.

For most types of use there are manufactured products that can be substituted for the natural products, usually at higher initial cost but with higher efficiency. This is not always true; for example, there is no satisfactory manufactured substitute for garnet for making coated abrasive paper and cloth. For some abrasives whose use is gradually lessening—for instance, chaser stones—the making of manufactured substitutes has not been economically attractive, but for even such a low-priced commodity as pressure blasting sand there are substitutes such as steel shot, fused aluminum oxide, and silicon carbide grains.

The decline in the use of most natural abrasives and their replacement by manufactured abrasives has not been a net loss to the mineral industry, however, for virtually all manufactured abrasives are made from mineral raw materials.

NATURAL ABRASIVES

Corundum and Emery

Corundum and emery have become of relatively little importance both from the standpoint of domestic production and consumption and on a worldwide basis. No corundum is mined today in the United States or Canada and there are only two relatively small producers of emery, both in the Peekskill, NY, area. At this locality, a spinel emery occurs in veins in an igneous complex of hornblende and olivine pyroxenite. It is associated with mica schist in rocks containing sillimanite, cordierite, garnet, and quartz. Most of the domestic emery is used in tumbling barrels and for various types of nonslip floors and stair treads. Production in recent years has come from two producers and has averaged about 9 kt/a (10,000 stpy) valued at about $20 per ton.

Diamonds—Industrial

Their uses as abrasives are outlined here.

In 1981 the total world production of natural industrial diamonds of all types was about 31,100,000 carats, with US imports of 21,600,000 valued at $5.34 per carat.

There are three major types of natural industrial diamonds: (1) bort, which includes off-color, flawed, or broken fragments of diamonds unsuitable for gems, (2) carbonado, or black diamond, which is a very hard and extremely tough aggregate of very small diamond crystals, and (3) ballas, a very hard, tough, globular mass of diamond crystals radiating from a common center. Carbonadoes come only from Bahia, Brazil; ballas chiefly from Brazil but a few from South Africa. Bort comes from all diamond-producing centers. In addition, there is a considerable production of diamond dust and powder, waste from cutting gem diamonds.

The industrial diamond has become one of the most important and essential materials in modern industry. Diamond drilling, once used only for locating metallic ores, is now widely used also for exploring nonmetallic mineral deposits; for exploring geologic structures; for testing foundations for dams, buildings, and heavy machinery; for exploring internal condition in heavy concrete structures such as dams; for stope mining; explosive demolitions under special conditions; and other purposes.

Some of the most important uses are: diamond-drill bits for drilling rock and concrete; diamond dies for wire drawings; diamond-tipped tools for truing abrasive wheels and for turning and boring hard rubber, fiber, vulcanite, hard plastics, etc.; diamond-toothed (segmental) saws and rim-impregnated (continuous rim) saws for sawing stone, glass, quartz, metals, slicing expansion joints in concrete highways, etc.; wheels, both for grinding and for cutoff work, in which

*Excerpt from Hight, R.P., 1983, "Abrasives," *Industrial Minerals and Rocks,* 5th ed., Vol. 1, S.J. Lefond, ed., AIME, New York, pp. 11-32.

Table 1. Classification of Abrasives

Natural Abrasives

Superior Hardness (above 7 in Mohs' scale)
  Diamond H-10
  Corundum H-9
  Emery H-7 to 9
  Garnet H-6.5 to 7.5
  Staurolite H-7.0 to 7.5
Intermediate Hardness (H-5.5 to 7)
  Silica Abrasives
    Buhrstone
    Chalcedony
    Chert
    Flint
    Novaculite
    Quartz
    Quartzite
    Sandstone
    Silica sand
  Other Rocks and Minerals
    Argillaceous limestone
    Basalt
    Feldspar
    Granite
    Mica schist
    Perlite
    Pumice and pumicite
    Quartz conglomerate
Inferior Hardness (H-under 5.5)
  Apatite
  Calcite
  Chalk
  Clay
  Diatomite
  Dolomite
  Iron oxides
  Limestone
  Rottenstone
  Siliceous shale
  Silt
  Talc
  Tripoli
  Whiting

Manufactured Abrasives

Boron carbide
Boron nitride

Calcium carbonate (pptd.)
Calcium phosphate
Cerium oxide
Chromium oxide
Clay (hard burned)
Diamond
Fused alumina
Glass
Iron oxides
Lampblack
Lime
Magnesia (pptd.)
Manganese dioxide
Periclase (artif.)
Silicon carbide
Tantalum carbide
Tin oxide
Titanium carbide
Tungsten carbide
Zirconium oxide
Zirconium silicate

Metallic abrasives, including steel wool, steel shot, angular steel grit, brass wool, and copper wool
Porcelain blocks for mill liners and grinding pebbles

Types of Abrasive Products

Abrasive grains and powders, loose
Abrasive grains bonded into wheels, blocks, and special shapes
Coated abrasives; grains bonded to paper and cloth
Abrasive grains and powders; paste form; oil or water vehicles
Abrasive grains and powders; brick and stick form; grease, glue, and wax binders
Natural rocks shaped into grindstones, pulpstones, chaser stones, millstones, etc.
Natural rocks shaped into sharpening stones, such as oil stones, whetstones, scythe stones, razor hones, etc.
Natural stones shaped into rubbing and polishing stones such as holystones and pumice scouring blocks
Natural stones shaped into blocks for tube-mill and pebble-mill liners
Pebbles, natural and manufactured, for grinding mills

---

the working face consists of diamond grit bonded with a resinoid, metal, or ceramic product; diamond-tipped tools for cutting glass and for engraving gems; diamond powder for cutting gems. High-speed tool steels, cemented carbides, and other exceedingly hard, tough alloys can be cut and shaped efficiently with diamond tools, and diamond-tipped tools are essential for the rapid and accurate shaping, truing, and dressing of abrasive wheels.

Garnet

The name "garnet" is given to a group of iron-aluminum silicate minerals having similar physical properties, crystal forms, and general chemical formula.

Composition: In garnet, general formula $3RO \cdot R_2O_3 \cdot 3SiO_2$, the bivalent element may be calcium, magnesium, ferrous iron, or manganese, the trivalent element, aluminum, ferric iron, or chromium, rarely titanium; further, silicon is also replaced by titanium in some examples. There are three prominent groups and various subdivisions under

each, many of these blending into each other. These are as follows:

1) Aluminum Garnet
  A. Grossularite, calcium-aluminum garnet
    ($3CaO \cdot Al_2O_3 \cdot 3SiO_2$).
  B. Pyrope, magnesium-aluminum garnet
    ($3MgO \cdot Al_2O_3 \cdot 3SiO_2$).
  C. Almandite, iron-aluminum garnet
    ($3FeO \cdot Al_2O_3 \cdot 3SiO_2$).
  D. Spessartite, manganese-aluminum garnet
    ($3MnO \cdot Al_2O_3 \cdot 3SiO_2$).
2) Iron Garnet
  E. Andradite, calcium-iron garnet
    ($3CaO \cdot Fe_2O_3 \cdot 3SiO_2$).
3) Chromium Garnet
  F. Uvarovite, calcium chromium garnet
    ($3CaO \cdot Cr_2O_3 \cdot 3SiO_2$).

Properties

COLOR—Colors vary greatly but generally are as follows:
Grossularite—white, pale green, or yellow.
Pyrope—deep red to black.
Almandite—deep red, brownish red to black.
Spessartite—brown to red.
Andradite—black, green, and yellow green.
Uvarovite—emerald green.

Hardness ranges from 6 (grossularite) to 7.5 (almandite); some almandite has a hardness of between 8.0 and 9.0.

CRYSTAL SYSTEM—Cubic, commonly as rhombic do-decahedrons or tetragonal trisoctahedrons, or in combination of the two.

CLEAVAGE—Occasionally an indistinct dodecahedral cleavage is observed; some species of almandite possess a pronounced laminated structure, these are planes of weakness along which the mineral separates, this *parting* has no relation to the crystal form and is not a true cleavage.

FRACTURE—Garnets having a glassy structure usually have a marked conchoidal fracture but sometimes the mineral tends to break into thin flakes. In other varieties the fracture is sharp and uneven.

INDEX OF REFRACTION—The index of refraction of the garnet group ranges from 1.735 to 1.94.

TENACITY—Aggregates of crystal composed of many small individuals are brittle and shatter readily. Massive garnet and well formed crystals are remarkably tough and shatter with difficulty.

FUSIBILITY—Garnets having a high iron content, such as almandite, fuse at a temperature of about 1200°C. White garnets containing a considerable percentage of chromium are infusible.

OTHER PROPERTIES—Specific gravity of the garnet group ranges from 3.5 to 4.2; luster vitreous, resinous, or dull; transparent to opaque.

Occurrence: Garnet commonly occurs as accessory minerals in a large variety of rocks, more particularly in gneisses and schists but also in contact metamorphic deposits, in crystalline limestones, pegmatites, and in serpentines. It occurs as gangue in ore veins formed at high temperatures. As most varieties of garnet are resistant to chemical and mechanical erosion, they tend to be concentrated in the sands of present-day or preexisting beaches, streams, or other alluvial deposits.

Deposits of garnet are found in many foreign countries and in nearly every state. An estimated 95% of the world's production of technical abrasive garnet comes from a small area in the Adirondack Mountains, New York State. Eighty percent of current usage of garnet is accounted for by United States industry.

The superior abrasive quality of Adirondack garnet currently being mined for technical applications is due to a combination of crystal properties including hardness (which is at the top of the garnet hardness range). This garnet, basically a combination of almandite and pyrope, exhibits incipient lamellar parting planes which break under pressure into sharp chisel-edged plates. When crushed to a very fine size it retains its natural sharp irregular grain shape. These features, particularly important in coated abrasives, are also desirable for other abrasive applications. Since Adirondack garnet occurs in large crystals a complete range of sizes results from the crushing and grinding operations necessary for liberating and separating the garnet from matrix minerals.

United States—In recent years garnet production in the United States has been confined almost exclusively to two states, New York and Idaho.

NEW YORK. Although there were at one time several technical abrasive-grade garnet producers in New York, present production is confined to one company, the Barton Mines Corp., Gore Mountain, near North Creek. This company supplies garnet to coated abrasive, glass, and metal lapping industries throughout much of the world. It has been in continuous operation since 1878 and, through technical advances enabling it to utilize leaner ores, has constantly proved new ore reserves. This mine is believed to be the country's second oldest continuous operating mine under one management and, geologically, is one of the world's most interesting ore deposits.

The almandite-bearing rock, which is an igneous-metamorphic rock of uncertain origin, lies at the surface and is quarried. The ore body is over 2 km (1¼ miles) long and ranges in width from 15.2 to 91.4 m (50 to 300 ft). The principal gangue minerals are hornblende and plagioclase feldspar which constitute from 40 to 80% of the rock. Less abundant minerals include hypersthene, magnetite, biotite, apatite, and pyrite. The garnet content of the ore body ranges from 20 to less than 5% and averages slightly less than 10%. The garnet occurs as crystals, mostly imperfectly developed, known locally as "pockets." These range in size from a fraction of an inch to more than a foot in diameter. Occasionally crystals up to 914 mm (36 in.) in diameter have been seen; however, their average size is less than 102 mm (4 in.).

Nearly every crystal of garnet is surrounded by a rim of coarsely crystalline hornblende. The quarry faces present a striking appearance, showing crimson-red garnet crystals set in a coal black background. Specific gravity of the garnet averages about 3.95, while the hornblende ordinarily is 3.07 to 3.24, with some specimens of very dense hornblende having been found with a specific gravity as high as 3.40. The general matrix is diorite.

Elsewhere in New York, garnet is produced as a by-product of wollastonite mining at Willisboro by Interpace Corp. and is sold in graded form from sandblasting operations in the Northeast. This garnet occurs as loosely cemented fine crystals that appear to be remnants of weathering and water deposition within preexisting stream channels or along shorelines.

IDAHO. The only other important producer of abrasive garnet is the Sunshine Mining Co.'s subsidiary, Idaho Garnet Abrasive Co., whose deposit is on Emerald Creek and whose plant is at the creek's mouth near Fernwood, Benewah County, Idaho. This company works an alluvial deposit of almandite garnet, derived from the erosion of soft mica schists in which the garnets have a maximum grain size of about 4.8 mm (³⁄₁₆ in.).

The garnet-bearing gravel is mined by dragline.

Foreign Countries—Garnet is or has been commercially mined in Madagascar, Japan, Argentina, India, and Tanzania. Garnet mining in Russia has been reported but details regarding mining are not known.

Silica

Silica Sand: Silica sand is extensively used for pressure blasting, for the initial grinding or surfacing of plate glass, and as a cutting medium for gang saws on stone.

For pressure blasting, the main centers of production in the United States are Ottawa, IL, and Cape May, NJ. The Ottawa material occurs as a friable sandstone, which is broken down to its natural grain, washed, and screened. The grains are spherical in shape. The Cape May sand is subangular and uneven in shape but larger grain sizes are obtainable, more so than with the Ottawa sand.

In Canada, pressure blast sand is obtained from the decomposed rock containing friable quartz and china clay at Lac Remi, north of Montreal, Que.; from a similar type of material near Smoky Falls, north of Lake Nipissing, 40 km (25 miles) west of North Bay, Ont.; also from a friable quartzite at East Templeton, Que., a few miles northeast of Ottawa, Ont. The grain of the Canadian sands is sharp. The ranges of grain size are approximately: No. 1 between 20 and 35-mesh for light work; No. 2 between 10 and 28-mesh; No. 3 between 6 and 10, and No. 4 between 4 and 8-mesh. The last two are used for the heavy cast-iron work and steel work.

Pure, clean beach and river sands and Illinois sand are used for the preliminary or coarse surfacing of plate glass. The crushed sand is water-graded into a number of grades at the glass plants and fed to the surfacing machines. Approximately three tons of sand are required to surface one ton of plate glass.

Cutting sand, composed of sharp, solid quartz grains, is used as an abrasive for sawing stone. It is usually ungraded and about equivalent to a No. 1 pressure blasting sand.

Burnishing sand is a fine, rounded-grain silica sand of uniform size between 65 and 100-mesh, used in rolling down and burning gold decorations on porcelain.

Quartz: Crushed and graded quartz is used for the abrasive backing of "flint" sandpapers. Almost any deposit of massive white quartz is suitable. Being the cheapest of all the abrasive-coated paper, it still is sold in fair amount, mainly in hardware stores and by small jobbers. It is made only in the form of paper, not as cloth. True chalk flint from England and France is extensively employed for this purpose in Europe and has better cutting qualities and longer life than ordinary quartz.

Powdered quartz and silt are sometimes used for scouring compounds and for the harsher metal polishes.

Soft Siliceous Powder Abrasives: Many natural highly siliceous materials either occur as a powder or are used only in the powder form for mild abrasives. For the majority of these, use as an abrasive is of minor importance as compared with their principal applications.

Diatomite— Diatomite production in the United States comes primarily from California and Nevada with small outputs from Oregon and Arizona.

The amount used for abrasives is insignificant in comparison with its other applications—most important of which are filtration, fillers and insulation. Its abrasive uses include metal (silver) polishes, dental powders and pastes, and occasionally it is used as a friction agent in the manufacture of matches.

Pumice—Under the name of pumice are included lump pumice and pumicite or volcanic dust, the natural powder. Lump pumice is used by manufacturers of furniture and musical instruments; for dressing the wood and metal surfaces; by silver platers for preparing their metal surfaces; by lithographers for cleaning stone surfaces; for rubbing down and polishing fine tools and instruments; by restaurants for scouring grills and cooking utensils, and for domestic and toilet uses, such as hand cleaners. Pumicite or ground pumice is mainly used as a cleanser, the thin, sharp, and striated grains being particularly suitable.

Tripoli, Microcrystalline Silica, and Rottenstone—The fine-grained, porous materials, tripoli, microcrystalline silica, and rottenstone, are known to the trade as "sift silicas."

Tripoli, which in the United States comes from southwest Missouri and northeast Oklahoma, is mainly used in the form of made-up tripoli grease bricks or tripoli compositions for buffing and polishing. The compound is applied to a rapidly revolving belt or canvas wheel and used for the finishing or buffing of metals, plated products, and so forth. It is used also to a small extent in the manufacture of some scouring and cleaning powders and soaps; for the rubbing down of painted surfaces, such as automobile bodies. A similar but finer-grained material occurring in the northwest corner of Arkansas, about 80 km (50 miles) southeast of the Missouri deposits, is used mainly for oil-well drilling mud.

Microcrystalline (sometimes erroneously termed "amorphous") silica, which comes mainly from southwestern Illinois but to some extent from Wayne County, TN, also is used for buffing and polishing compounds. These compounds are termed "silica" by the trade and are much in demand for white "coloring" operations on high-class work. Chemically precipitated amorphous silicas also are used in polishing and buffing compositions.

Both tripoli and microcrystalline silica have been mined from deposits at Harrisburg, northwestern Georgia.

Rottenstone, a fine-grained gray-buff siliceous-argillaceous limestone, comes from Antes Forte, Lycoming County, PA, and is used as a polish base, for instance, for automobile polishes.

Nonsiliceous Soft Abrasives

Ground Feldspar is extensively used in scouring and cleaning compounds and for a window cleaner.

Chalk (calcium carbonate) is a soft, compact, fine-grained, white limestone composed of the calcareous remains of small marine shells. A small amount of this chalk—mainly from England and France—known as "whiting" is used as a very mild abrasive for hand polishing of nickel, gold, silver or plated ware, buttons, and similar materials.

China Clay (kaolin) and some pipe clays have been used successfully in polishing powders. Pipe clay at one time was the standard polish for naval and military tunic buttons.

BATH BRICK, used for scouring steel utensils, is made from a very fine-grained, quartzose clay found along the banks of the Parrot River in England.

STAUROLITE, a complex iron aluminum silicate, has been commercially recovered from various river placer deposits in the southeastern US. DuPont operates a deposit in Clay County, FL. It is used mostly for pressure blasting.

Sharpening Stones

Hand-used sharpening stones include scythestones, whetstones, oil stones, water stones, razor hones, holystones, and rubbing stones. While, in general, stones made from bonded artificial abrasive grains have largely replaced these natural abrasives there is still some small production of oilstones and whetstones in Arkansas, whetstones in Indiana, and scythestones in New Hampshire. Arkansas novaculite stones are still preferred for sharpening fine-edged tools for surgeons, carvers, and engravers. It is claimed that these stones give a smoother and longer-lasting edge.

Natural abrasive stones are made from a wide variety of material, including sandstone, novaculite, mica schist, siliceous argillite, shale, slate, and pumice. The superior cutting quality of some of these stones is due to the inclusion of well disseminated fine-grained inclusions of garnet or other minerals of superior hardness.

Grinding Pebbles and Tube-Mill Linings

Tube-mills, conical-mills, and cylindrical batch mills are used more often than any others for the fine grinding of hard ores, minerals, paints, chemicals, ceramic bodies and glazes and enamels, portland cement clinker, and similar materials.

Most efficient grinding can be done in mills with iron or steel liners and iron or steel balls, slugs, or rods, but where contamination of the color or chemical purity of the product by metals and metal oxides must be avoided, the grinding surfaces consist of blocks or bricks for the mill linings and natural pebbles or artificially prepared balls.

The most favored natural mill-lining material for most purposes is Belgian silex, which is a very hard, tough, more or less cellular quartzite resembling French buhrstone. This is imported in rectangular blocks more or less shaped to fit the curve of a mill. During World Wars I and II, when imports were cut off, and to a much lesser extent at other times, domestic substitutes have been used. These have consisted chiefly of quartzite from near Jasper, MN; Iron City, TN; and Baraboo, WI; and granites from Salisbury, Lilesville, and Faith, NC. Some of these materials are reported to give service equal to that of Belgian silex. Special hard, dense, porcelain blocks are used in some mills for grinding paint, ceramic, and chemical materials.

For grinding media, Danish flint pebbles, when available, have long been standard because of their superior hardness, toughness, and uniformity. These pebbles are found on the shores of Greenland but are marketed through Denmark. Other foreign sources of similar pebbles exist along the seacoasts of Belgium, France (between Le Havre and St. Valery-sur-Somme), Norway, and England. Seven sizes of Danish pebbles are marketed, ranging from 25.4 to 203 mm (1 to nearly 8 in.).

Domestic substitutes for grinding media are natural pebbles of flint, quartz, and quartzite, as well as artificially rounded (by tumbling small blocks in rotating cylinders) pebbles made from quartzite, granite, chalcedonized rhyolite, and other rocks. During World War II, pebbles were shipped from Encinitas Beach, CA (true flint); from Jasper, MN (quartzite); from the Austin chalk beds in south-central Texas (true flint); from Salisbury and Faith, NC (granite); from Baraboo, WI (quartzite), and several other points in the United States. Extensive deposits of quartzite pebbles in southwestern Saskatchewan have been reported. Beach pebbles were reported nonuniform in hardness and requiring careful sorting.

BIBLIOGRAPHY AND REFERENCES

Adams, W.T., 1980, "Abrasive Materials," *Minerals Yearbook, 1977,* US Bureau of Mines, Vol. 1, pp. 107-118.

Baskin, G.D., 1978-1979, "Abrasive Materials," Preprint, US Bureau of Mines, 1978-1979, 15 pp.

Baskin, G.D., 1980, Diamond (Industrial), *Mineral Commodity Summaries, 1980,* US Bureau of Mines, pp. 46-47.

Clarke, R.G., 1972, "Abrasive Materials," *Minerals Yearbook 1972,* US Bureau of Mines, pp. 123-134.

Evans, J.R., 1980, "Sand and Gravel," *Minerals Yearbook, 1977,* US Bureau of Mines, Vol. 1, pp. 797-819.

Harben, P., 1978, "Abrasives—Taking the Rough with the Smooth," *Industrial Minerals,* No. 134, Nov., pp. 49-73.

Heywood, J., 1942, *Grinding Wheels and Their Uses,* Penton Pub. Co., New York, 436 pp.

Hight, R.P., 1971, "Diamonds Industrial," *Encyclopedia of Industrial Chemical Analysis,* Vol. 11, John Wiley, New York, pp. 462-484.

Johnson, B.L., and Schauble, 1939, "Abrasive Materials," *Minerals Yearbook 1939,* US Bureau of Mines, pp. 1225-1240.

Meisinger, A.C., 1982, "Diatomite," *Mineral Commodity Summaries, 1981,* US Bureau of Mines, pp. 46-47.

Oskam, J., 1980, "Garnets, Abrasive and Gem," Preprint 80-382, SME-AIME Fall Meeting, Minneapolis, Oct., 5 pp.

Smook, J.F., 1982, "Abrasive Materials in 1981," *Mineral Industry Surveys,* US Bureau of Mines, 3 pp.

Tepordei, V.V., 1978-1979, "Sand and Gravel," Preprint, US Bureau of Mines, 1978-1979, 29 pp.

17. Nepheline Syenite*

Nepheline syenite is a silica deficient crystalline rock consisting of albite and microcline feldspars and nepheline, together with varying but small amounts of mafic silicates and other accessory minerals. Nepheline-bearing rocks are widely distributed around the world but only in Canada, Norway, the Union of Soviet Socialist Republics, and the United States are deposits worked commercially.

Because of its low melting point and fluxing ability, nepheline syenite was investigated in the early 1900s by many students of glass and ceramics. Although the nepheline syenite deposits in Methuen Township, Ont., Canada, were discovered as early as 1897, it wasn't until 1935 that the first mill was erected to produce nepheline syenite products.

In Canada, nepheline syenite deposits are worked by Indusmin Ltd. and International Minerals and Chemical Corp. (Canada) Ltd. who have combined production capacity in excess of 1800 stpd. In Norway, the Norsk Nefelin Div. of Elkem Spigerverket has produced nepheline syenite from a deposit on Stjernöy Island in western Finland since early 1960.

Nepheline syenite has been under investigation in the Union of Soviet Socialist Republics (USSR) since at least 1928. In 1951, the first commercial alumina works was brought on stream at Volkhov, near Leningrad, utilizing nepheline concentrates from the Kola peninsula as mill feed. Since then two others have been completed and more are planned. Large quantities of portland cement, sodium carbonate, and potassium carbonate are produced in conjunction with the alumina.

A material called nepheline syenite, but actually a pulaskite, is produced in Pulaski County, Arkansas, by the McGeorge Construction Co. for construction aggregates and by the 3M Co. for roofing granules. The Dome Fill Co. purchases the fines and uses them in the preparation of enclosed construction fill.

Chap. 16 on "Feldspar, Nepheline Syenite, and Aplite" in the 3rd ed. of *Industrial Minerals and Rocks* (Castle and Gillson, 1960) and *Nepheline Syenite Deposits of Southern Ontario* (Hewitt, 1960) contain excellent accounts of earlier commercial development of nepheline syenite as a glass and ceramic material. An English monograph on *Nepheline-Syenite and Phonolite* (Allen, et al., 1968) is a complete survey, at the time, of nepheline syenite which, with its very complete list of references, is an excellent source book.

END USES

Nepheline syenite from the Canadian and Norwegian producers finds use primarily in the manufacture of glass and ceramics. One producer also makes several grades of nepheline syenite that are useful as extender pigments and fillers in paint, plastics, and rubber. In the Soviet Union nepheline is used in increasing quantities in the manufacture of alumina, alkali carbonates, and portland cement and to a lesser degree in the manufacture of colored container glass.

Glass

By far the largest use of nepheline syenite is in the manufacture of glass products including container glass, fiber glass, opal glass, plate glass, sheet glass, and tableware glass. Nepheline syenite provides necessary additions of alumina and alkalis in the glass batch; it is low in silica and contains no free quartz; it has a favorable ratio of sodium oxide to potassium oxide of 2/1. Nepheline syenite bearing glass batches have lower viscosity and easier workability compared with those containing potash feldspar.

The lower fusion point of nepheline syenite lowers the melting temperature of the glass batch with attendant faster melting, high productivity, and fuel savings.

Ceramic Ware

Nepheline syenite is being used for the manufacture of a wide variety of whiteware products including dinnerware, sanitaryware, floor and wall tile, electrical porcelain, art pottery, chemical porcelain, dental porcelain, porcelain balls, and mill liners. In ceramics, nepheline syenite serves as a vitrifying agent, contributing to the glassy phase which binds other constituents together and gives strength to the ware. The ready fusibility and abundant fluxing capacity of nepheline syenite benefits manufacturers by permitting reduced body flux content, lower firing temperatures, or faster firing schedules than could be attained with other raw material combinations.

Extender Pigment and Fillers

Applications for nepheline syenite as extender pigments and inert fillers have been found and developed. These include exterior and interior latex and alkyd paint systems, traffic paints, metal primers, exterior wood stains, sealers, undercoats, and hardboard ground coats. Nepheline syenite for these products has desirable high dry brightness, high bulking value, low vehicle demand, extreme ease of wetting and dispersion, and a stabilizing pH value. In paint formulations, nepheline syenite has exceptionally good tint retention characteristics.

In plastics, nepheline syenite has found increasing use in rigid, flexible, and plastisol-type polyvinyl chloride, and in epoxy and polyester resin systems. It exhibits extremely low resin demand thus making high loadings possible. In PVC resins, it exhibits extremely low tinting strength, has a refractive index close to that of vinyl resin, and has a very low optical dispersion. It has a specific gravity lower than calcium carbonate and talc and the properties described for paint applications are beneficial in plastics applications as well.

Nepheline syenite is also used as an inert filler in the manufacture of foam carpet backing.

Alumina

In the USSR, nepheline syenite and nepheline are used to manufacture alumina for aluminum production, sodium and potassium carbonates, and portland cement. USSR production comes from Kola peninsula and Kiya Shaltyrsk deposits. The US Bureau of Mines (Anon., 1961-71) estimates that the present level of consumption of nepheline concentrates in the Soviet Union is 1 Mt (1 million ltpy). More recent data on nepheline syenite production in the USSR are not available. In addition, the Volkhov, Pikalevo, and Achinsk alumina plants require at least 3 Mt (3 million lt) of nepheline syenite annually. Nowhere at this time outside the Soviet Union is nepheline or nepheline syenite used as the basis for such industries.

*Excerpt from Minnes, D.G. (revised by Lefond, S.J., and Blair, R.E.), 1983, "Nepheline Syenite," *Industrial Minerals and Rocks,* 5th ed., Vol. 2, S.J. Lefond, ed., AIME, New York, pp. 931-960.

Product Specifications

Nepheline syenite used by the glass industry is characteristically a −30 mesh +200 mesh product having an alumina content in excess of 23%, alkali content in excess of 14%, and iron oxide content less than 0.1%. Other metallic contaminants must be very low. For the ceramic industry nepheline syenite is ground to typically −200 mesh and is characterized by its high fluxing and low level of iron and other impurities. For extender pigment and filler industries, nepheline syenite offers high dry brightness, uniformity, chemical inertness, and specific index of refraction and is supplied in various finenesses.

Market requirements vary and while the European markets accept a 0.1% Fe_2O_3 content, North American markets require 0.08% Fe_2O_3 or less. Norwegian nepheline syenite products are higher in alumina and alkalis than Canadian ones. Other contaminating metallic ions must be at acceptably low levels and refractory minerals must be absent.

GEOLOGY

Mineralogy

Nepheline syenite is a coarse to medium-grained crystalline *rock* generally considered of igneous origin in which the essential mineral constituents are microcline, orthoclase or albite feldspars, the feldspathoid nepheline, and ferro-magnesium minerals of which the principal are hornblende, pyroxene, and biotite. The most frequently occurring accessory minerals include magnetite, ilmenite, calcite, garnet, zircon, and corundum. Nepheline syenites are quartz-free. Commercial deposits of nepheline syenite contain at least 20% nepheline, at least 60% feldspar, and seldom more than 5% accessory minerals. For a deposit to be of interest, refractory minerals such as corundum and zircon must be absent or easily removed along with other accessory minerals and characterizing mafic silicates during processing.

Nepheline: Nepheline is the most common of the feldspathoid *minerals*. Its formula is $Na_3K Al_4 Si_4 O_{16}$. Potassium is always present in natural nepheline, most frequently in ratio amounts of Na/K: 3/1, although ratios of 4/1 and 6/1 are known. It crystallizes in the hexagonal system and frequently forms 6- or 12-sided prisms. It has a distinct and an imperfect cleavage and a subconchoidal fracture. It is brittle; the hardness is 5½ to 6 on Mohs' scale, the specific gravity is 2.5 to 2.7; the luster is vitreous to greasy, occasionally opalescent. It is colorless, white, or yellowish but when massive may be dark green, greenish, bluish-gray, brownish-red, and brick red. It is also known as elaeolite. It gelatinizes readily in acid and because it is more soluble than associated minerals can be readily spotted in outcrops by the pitted appearance of the rock surfaces. Nepheline alters to various zeolites, analcite, sodalite or cancrinite, and the gieseckite micaceous materials.

Feldspar: The major feldspars present are microcline and orthoclase with a theoretical composition SiO_2 64.7%, Al_2O_3 18.4%, and K_2O 16.9%; and albite with a theoretical composition SiO_2 68.7%, Al_2O_3 19.5%, and Na_2O 11.8%.

Pyroxene: The aegerine member of this series of minerals is customarily present in feldspathoid rocks but the theoretical composition ranges from aegerine SiO_2 52.0%, Fe_2O_3 34.6%, and Na_2O 13.4%; to that of diopside SiO_2 55.6%, CaO 25.9%, and MgO 18.5%.

Amphibole: A wide variety of amphiboles are present in nepheline syenites including hornblende, hastingsite, barkevikite, arfvedsonite, and riebeckite. These are calcium, magnesium, sodium, and iron aluminosilicates of various compositions.

Biotite: Two types of biotite are found in feldspathoidal rocks; the golden-brown variety, but more frequently the high-iron variety lepidomelane. It is a hydrated aluminosilicate of potassium, iron, and magnesium.

Accessory Minerals: Among the more frequently found accessory minerals are sodalite, noselite, hauynite, scapolite, cancrinite, calcite, apatite, magnetite, ilmenite, hematite, pyrite, zircon, sphene, pyrochlore, garnet, and corundum.

Nepheline syenites have a wide variety of textures which can be generally classified according to the origin of the rock; plutonic and hypabyssal nepheline syenites; nepheline syenite pegmatites, and the migmatic nepheline rocks or "nephelinized gneisses."

Classification of Deposits

Several methods can be used to classify nepheline syenite deposits—by the percentage of principal minerals present in the deposit, by their origin, and also by their color. Nepheline syenites have been studied intensively and works by Deer, et al. (1963) and Eitel (1965) provide an introduction to complete studies of nepheline and nepheline syenite. Little information about worldwide nepheline syenite occurrences and their commercial potential was available in concise form until publication of *Nepheline-Syenite and Phonolite* by Allen and Charsley in 1968. The book is an excellent summary of the present state of commercial interest in this and allied rocks.

Although a wide variety of names have been given to nepheline-bearing rocks and the term nepheline syenite is often given to any such rock containing 5% or more nepheline, it is unlikely that a rock would be of commercial interest unless the nepheline content is at least 20%.

Nepheline Syenite Deposits: Deposits are rarely of great areal extent, the largest being those in the Kola peninsula of the Soviet Union [1165 and 647 km² (450 and 250 sq miles)], Brazil [777 and 259 km² (300 and 100 sq miles)], and Greenland [207 km² (80 sq miles)]. Most others are typically much smaller, less than 52 km² (20 sq miles), and those in Blue Mountain, Ont. 7.8 km² (3 sq miles), and Stjernöy, Norway [0.5 km² (0.2 sq miles)] are even smaller. Classification may also be made based on the mode of occurrence and origin of the nepheline syenite. Moorhouse (1959) for one has divided them into five groups:

1) Feldspathoidal rocks associated with undersaturated volcanics.
2) Differentiated ring complexes, often associated with carbonatite and usually characterized by metasomatism around their borders.
3) Layered intrusives, possibly related to the ring complexes.
4) Borders or satellitic stocks associated with syenites or granites.
5) Nephelinized gneisses, usually associated with nepheline pegmatites.

The first type has no economic significance. Examples of the others would be: group 2, Nemegos, Ont.; Mont St. Hilaire, Que.; and Kola peninsula, Karelian ASSR; group 3, Ice River, B. C.; Julienhaab, Greenland; group 4, Blue Mountain, Ont.; group 5, Stjernöy, Norway.

A variety of origins are possible (Turner and Verhoogen, 1951): from low temperature residual magmas which have been conditioned by volatile components; by fractional crystallization of undersaturated olivine-basalt magma; by desilication of granitic magmas, that might have invaded the limestone with evolution of a CO_2-rich gas phase which may be important; nepheline syenite magmas, and metasomatic alteration of various preexisting rocks, i.e., "nephelinization."

Distribution of Deposits

Nepheline-bearing rocks are widespread throughout the world and deposits of nepheline syenite are common. Those in Canada, the United States, Norway, and the Union of Soviet Socialist Republics have been well described and much is known about those in other countries of Asia, Africa, and the South Pacific. Two of the largest known are the Khibiny [1165 km² (450 sq miles)] and Lovozero [647 km² (250 sq miles)] intrusives in the Karelian ASSR (Vlasov et al., 1966) of the USSR. Other large ones are located in Siberia. In Brazil, nepheline syenites occur in several extensive intrusive areas, the largest of which cover some 777 km² (300 sq miles) in Minas Gerais and São Paulo, and 259 km² (100 sq miles) in São Paulo. Other deposits are rarely large, seldom covering more than a few square miles or a few tens of square miles in area.

However size is not an overly important criterion in commercial considerations; more important are the purity and location of the deposits.

There are now seven known mining ventures in different parts of the world producing nepheline or nepheline syenite products; two in Canada, one in Norway, two in the USSR, and two in the United States.

In Canada, nepheline syenite deposits are common in several provinces. Occurrences in the vicinity of Lake Albanel, Gouin, Labelle, Mont St. Hilaire, and Oka, Que., are well known. In Ontario, Hewitt (1960) has described deposits in the vicinity of Blue Mountain, Haliburton-Bancroft, French River, Callander Bay, Nemegos, and Port Caldwell as well as other minor occurrences. In British Columbia, the Ice River and Kruger Mountain complexes are well known. Much of the interest in Canadian deposits has stemmed from the successful development of the Blue Mountain intrusive in Methuen Township but none other has proven pure enough to warrant exploitation.

In Norway, nepheline syenite is known in the vicinity of Oslo and on several islands in western Finland, although only that on the southwest side of Stjernöy Island has the necessary characteristics for commercial development.

Reserves: The only known deposits of nepheline syenite suitable for low-iron products are being mined at existing operations in Norway and Canada. Reserves are considered limitless. In Norway at Norsk Nefelin, the volume of proven nepheline syenite exceeds 135 million m³ representing nearly 400 million tons of rock of a nature the company considers suitable for mining. In Canada, Indusmin has published a reserve figure of 14 million tons of proven ore, sufficient for 35 years at present mining rates. The company owns 85% of the Blue Mountain intrusive and believes potential reserves of minable rock are satisfactory for an indefinite period.

International Minerals and Chemical Corp. (Canada) Ltd. owns 202 ha (500 acres) of nepheline syenite at Blue Mountain, Ont. Reserves are considered to exceed 30 million tons, ample for any production requirements.

In the Soviet Union, there are no firm indications of reserves of nepheline-bearing rocks suitable for chemical conversion. The estimated production of apatite from the Kola peninsula mines exceeds 27 Mt/a and consequently the operation generates more than 35 Mt of nepheline-bearing tailings annually from which an estimated 15 Mt of nepheline concentrate could be recovered. In addition many square kilometers of nepheline syenite are present so reserves can be considered adequate for projected chemical needs. Nepheline syenite reserves at Goryachegorsk in central Siberia have been estimated at over 100 Gt.

EXPLORATION

As with most industrial minerals the "place value" of a nepheline syenite deposit is critical. In addition to the following comments on field examination, the exploration geologist should consider the overriding importance of costs of production and the costs of transporting products to market.

Only by examining the interrelationship of all factors bearing on exploitation of a given nepheline deposit can one determine if development is warranted.

Field Techniques

The field geologist needs to search for a deposit that consists of a coarse to medium-grained rock. Crushing to minus US Standard 30 mesh should liberate all the impurities that need to be removed before a product is salable. Careful mapping of the deposit in detail will reveal textural, structural, and mineralogical changes that might affect mining. Samples need to be selected carefully to provide a representative suite on which initial microscopic, physical, chemical, and ceramic tests can be made.

If examination reveals that impurities are scattered through the feldspar crystals in a finely divided state, such a deposit would be unsuited for glass and ceramic uses. Only if impurities can be readily liberated by crushing should a deposit receive more detailed evaluation. One would normally expect to mine feldspathic materials by open pit methods and due consideration needs to be given to bench heights, berms, and mining regulations when determining if a given deposit would be satisfactory. Ore reserves should be large enough to sustain an operation for a long time because customers demand assurance of ample supply and, in industrial mineral ventures, ten or more years may be needed to repay the invested capital. Normally one would not consider a deposit having ore reserves for less than 20 years operation.

Consideration must be given to possible methods that might be used to beneficiate nepheline syenite from a deposit. Present feldspar and nepheline syenite benefication plants have relatively simple schemes for flotation and electromagnetic removal of impurities, and flotation separation of quartz and feldspar. Flotation, electrostatic, electromagnetic, and other techniques can all be successfully used on different mineral assemblages but if the mineral suite in a given deposit is too complex, the combination of techniques needed to remove impurities could well be too expensive and involve too many operating variables to permit economical operation.

In areas where chemical weathering is a factor, one must have fresh specimens with which to work. Nepheline dissolves from rock surfaces relatively easily, feldspars are altered to kaolin, iron oxides and sulfides may be leached away, all of which tends to distort results obtained when surface samples are examined.

Complex nepheline syenite deposits such as the ring-complex type and carbonatite type intrusives show little promise for utilization in glass and ceramics. They are too variable in normal feldspar and mafic constituents and furthermore may contain rare earths, halides, and numerous other trace elements that detract from the intended applications for the rock.

Nepheline syenite deposits of the sheet, gneissic, and border phase of granite and syenite intrusive types seem to offer the greatest chance of uniformity and quality needed.

Drilling

Deposits of merit need to be drilled to confirm the nature of the rocks and to provide samples for laboratory work. Depending on circumstances, AX, BX, and NX diameter core may be recovered. In laying out a drill program, care must be taken that holes intersect all rock types at appropriate angles of inclination. Bands of biotite gneiss, for instance, might be erratic in distribution and unless drilling is adequate, erroneous conclusions might be drawn about the amount of such impurities present. A deposit may be zoned and, although mapping indicates rock of one type, the core of the deposit may be another. There are always variations to be encountered in nepheline syenite deposits, as may be guessed from their mode of origin, and drilling a deposit provides assurance that surprises will be minimal.

Spacing of drill holes is determined by the uniformity of the rock and the stage of the investigation. Initial drilling is widespread, followed by grid drilling at perhaps 122-m (400-ft) spacing, followed by spacing of 30.5 m (100 ft) or less if conditions demand.

Geophysical and geochemical techniques are not a normal part of exploration for nepheline syenite although seismic surveys may be useful for helping to determine the depth of overburden.

Other Considerations

Consideration should be given to matters such as the size and character of the market, the mode and cost of transportation, availability of auxiliary services, labor supply, and laws that might be restrictive. Due concern for them at an early stage in exploration could provide a basis for judging which deposits have the greatest worth and save much time and wasted expense on further laboratory evaluation permitting one to study the rock for signs of product warrants.

BIBLIOGRAPHY AND REFERENCES

Anon., 1968, "Nepheline Syenite 30 Years Rapid Growth," *Industrial Minerals,* Metal Bulletin Ltd., London, England, No. 7, Apr., pp. 9-19.

Anon., 1961-1971, "Nepheline Syenite," *Canadian Minerals Yearbook,* Mineral Resources Div., Dept. of Energy, Mines and Resources, annual publication.

Anon., 1961-1971a, "Feldspar, Nepheline Syenite and Aplite," *Minerals Yearbook,* US Bureau of Mines, annual publication.

Allen, J.B., and Charsley, T.J., 1968, *Nepheline-Syenite and Phonolite,* Institute of Geological Sciences, Minerals Resources Div., London, England, 169 pp.

Baer, F.H., 1959, "Soviets Push Ambitious Aluminum Plans," *Engineering and Mining Journal,* Vol. 160, No.5, May, 1959, pp. 102-105.

Barth, T.F.W., 1963, "The Composition of Nepheline," *Schweiz, Min. Petr. Mitt.* 43/1, pp. 153-164.

Castle, J.E., and Gillson, J.L., 1960, "Feldspar, Nepheline Syenite and Aplite," *Industrial Minerals and Rocks,* J.L. Gillson, ed., 3rd ed., AIME, New York, pp. 339-362.

Cooper, J.D., and Wells, J.R., 1970, "Feldspar," *Mineral Facts and Problems,* Bulletin 650, US Bureau of Mines, pp. 977-988.

Deer, W.A., et al., 1963, *Rock-Forming Minerals,* Vol. 4, Longmans Green and Co. Ltd., London, England, pp. 231-269.

Dudkin, O.B., Kozyreva, L.V., and Pomerantseva, N.G., 1964, *Mineralogy of Apatite Occurrences of the Khibinsk Tundra,* Academy of Sciences, Moscow-Leningrad.

Eitel, W., 1965, *Silicate Science,* Vol. 3, Academic Press, New York, pp. 236-264.

Feitler, S.A., 1967, "Feldspar Resources and Marketing in Eastern United States," Information Circular 8310, US Bureau of Mines, 41 pp.

Gerasimovskii, V.I., 1956, "Geochemistry and Mineralogy of Nepheline Syenite Intrusions," *Geochemistry,* No. 5, pp. 494-510.

Grimshaw, R.W., 1971, *The Chemistry and Physics of Clays and Allied Ceramic Materials,* 4th ed., John Wiley and Sons Inc., New York, pp. 372-442.

Guillet, G.R., 1962, "A Chemical and Inclusion Study of Nepheline Syenite for Petrogenetic Criteria," unpublished M.A. Thesis, University of Toronto, Canada, 75 pp.

Harben, P.W., 1977, "Raw Materials for the Glass Industry," *Industrial Minerals,* Metal Bulletin, Ltd., London, pp. 6-14.

Heier, K.S., 1961, "Layered Gabbro, Hornblende, Carbonatite and Nepheline Syenite on Stjernöy, North Norway," *Norsk Geol. Tidsskr 41,* pp. 109-155.

Heier, K.S., 1964, "Geochemistry of the Nepheline Syenite on Stjernöy, North Norway," *Norsk Geol. Tidsskr 45,* pp. 205-215.

Heier, K.S., 1965, "Geochemical Comparison of the Blue Mountain (Ontario, Canada) and Stjernöy (Finnmark, North Norway) Nepheline Syenites," *Norsk Geol. Tidsskr 45,* pp. 41-52.

Heier, K.S., 1966, "Some Crystallo-Chemical Relations of Nepheline and Feldspars on Stjernöy, North Norway," *Journal of Petrology,* Vol. 7, Pt. 1, pp. 95-113.

Hewitt, D.F., 1960, "Nepheline Syenite Deposits of Southern Ontario," Ontario Dept. of Mines, Vol. 69, Pt. 8, 194 pp.

Huvos, J.B., 1980, "The Mineral Industry of Norway," Preprint, US Bureau of Mines, 14 pp.

Johnstone, S.J., and Johnstone, M.G., 1961, *Minerals for the Chemical and Allied Industries,* Chapman and Hall, London, England, pp. 188-194.

Minnes, D.G., 1975, "Nepheline Syenite," *Industrial Minerals and Rocks,* S.J. Lefond, ed., 4th ed., AIME, New York, pp. 861-894.

Moorhouse, W.W., 1959, *The Study of Rocks in Thin Section,* Harper and Brothers, New York, pp. 302-312.

Payne, J.G., 1966, "Geology and Geochemistry of the Blue Mountain Nepheline Syenite Body," Ph.D. Thesis, McMaster University, Canada, 183 pp.

Polkanov, A.A., ed., 1937, *The Northern Excursion, Kola Peninsula,* 17th International Geological Congress, Moscow, 119 pp.

Smothers, W.J., et al., 1952, "Ceramic Evaluation of Arkansas Nepheline Syenite," Arkansas Resources and Development Commission, Research Series No. 24, pp. 21.

Turner, F.J., and Verhoogen, J., 1951, *Igneous and Metamorphic Petrology,* McGraw Hill, New York, pp. 338-342.

Vlasov, K.A., et al., 1966, *The Lovozero Alkali Massif,* Oliver and Boyd, Edinburgh and London, 627 pp.

Watson, I., 1981, "Feldspathic Fluxes—The Rivalry Reviewed," *Industrial Minerals,* No. 163, Apr., pp. 21-45.

Watson, I., 1981a, "The Industrial Minerals of Scandinavia—Norway," *Industrial Minerals,* No. 171, Dec., pp. 34-45.

18. Olivine*

Olivine is a mineral containing a mixture of forsterite (Mg_2SiO_4) and fayalite (Fe_2SiO_4) in solid solution. The name *olivine* was first applied by Werner in 1790 (Hunter, 1941) because of the olive-green color of the mineral.

Olivine is the principal component of the rock dunite, which itself is a member of the peridotite group of ultrabasic igneous rocks. European uses of the terms *olivine* and *dunite* are somewhat different than US usage. In commerce European usage defines *dunite* as containing 36 to 42% MgO, 36 to 39% SiO_2, and loss on ignition of approximately 10%. The term *olivine* is used to designate a material containing approximately 85% forsterite, with a chemical composition of approximately 45 to 50% MgO, 40 to 43% SiO_2, 5 to 8% Fe_2O_3, and ignition loss of 1 to 2%.

USES

The principal use of olivine is in various applications involving hot metal. In excess of 90% of world uses involve hot metal, with approximately 75% as a slag conditioner in the blast furnace production of pig iron. Approximately 15% of total olivine usage is as a special foundry sand in mold making for the brass, aluminum, magnesium, and manganese steel foundries (Schaller, 1957, 1958; Snyder, 1957; Anon., 1977).

Olivine was first used as an industrial mineral in the early 1930s for a refractory material (Anon., 1970; Hunter, 1941). As a refractory raw material, olivine was first introduced in the United States as hand-cobbed, selected, shaped blocks of crude olivine. This use met with limited success. More recently, finely ground olivine blended with MgO and pressed into bricks, which are then fired, has found use in glass tank furnaces and open-hearth furnaces. Ramming or gunning mixes for basic furnace linings also utilize olivine. Olivine has been used in ladle linings with varied success. In Europe, substantial tonnages of olivine are utilized in refractory brick for night storage heaters (Anon., 1970).

A limited amount of olivine has been used in the past as a fertilizer (magnesium source) and has been used with rock phosphate to produce a magnesium phosphate as a plant food. The relatively high magnesia (MgO) content of olivine also attracted attention to it as a potential source of both magnesium compounds and as a source of metallic magnesium (Bengston, 1956; Hunter, 1941). The use of olivine as an additive in the blast furnace is relatively new, having been developed during the last two decades.

As a foundry sand olivine is used in mold making for the brass, aluminum, magnesium, and manganese steel foundries. It has a fusion point of approximately 1816°C (3300°F) and lower thermal expansion than silica; thus, it exhibits less defects in the castings such as "rat tails," "scabs," and "buckles." Little or no free silica occurs with olivine, so the silicosis hazard is reduced as compared to the use of normal silica sand in foundry applications.

Normally, coarse sand is used in manganese steel foundries and the finer sands in aluminum, magnesium, and brass foundries. Fine sand is used in the manufacture of "hot tops" where quartz sand is not permitted because of health hazards.

Industries using olivine have not standardized specifications. Normally, producers supply a product which meets individual customer requirements. Such requirements involve type and amount of impurities, grain size of product, refractory characteristics chemical analysis, etc.

GEOLOGY

Olivine occurs commonly as accessories in basic igneous and basic metamorphic rocks. Economic deposits are of magmatic origin and are restricted to essentially the dunite variety of peridotite. Dunites are medium to coarse-grained crystalline rocks, generally reddish brown on weathered outcrop, and are composed primarily of olivine.

Mineralogy

The name *olivine* is a generic term used to indicate a group of orthosilicate minerals in isomorphous series, with magnesium-rich forsterite (Mg_2SiO_4) and iron-rich fayalite (Fe_2SiO_4) as end points (Ramberg and DeVore, 1951; Reed, 1959). In commercial practice, use of the term olivine applies to deposits containing a mixture of forsterite and fayalite in solid solution with fayalite content usually restricted to less than 15%. Accessory constituents may include the primary minerals: ilmenite, magnetite, chromite, and garnet; and various secondary minerals as alteration products.

The chemical composition of typical olivine may be expressed either as $(Mg,Fe)_2 SiO_4$ or as $2(Mg,Fe)O.SiO_2$. The ratio of Mg:Fe varies from 16:1 to 2:1, passing from forsterite on the one hand to fayalite on the other.

Mode of Occurrence and Origin

Commercial olivine occurs as alpine-type and zoned dunite bodies. Controversy exists among various researchers concerning the origin and mode of emplacement (Astwood, et al., 1972; Bennett, 1940; Bowen and Schairer, 1935, 1936; Gaudette, 1963; Mossman, 1972; Ragan, 1959, 1961; Ross, et al., 1954; Yudin, 1959). There is general agreement, however, that most olivines have been emplaced in a partially crystallized condition (mush) and that in part, at least, foliation present in many unmetamorphosed dunites results from flowage during emplacement prior to complete crystallization.

All commercial bodies are of igneous origin and may be in part magmatic segregates from basic magmas. Domestic olivine bodies, of present or potential economic importance, range in age from pre-Ordovician for the North Carolina deposits, to post early-Tertiary age for the Washington dunites.

Distribution of Deposits

Large deposits of olivine-bearing dunites crop out in Norway, Sweden, USSR, Austria, Japan, New Zealand, Zimbabwe, South Africa, United States, New Caledonia, Italy, Greece, Spain, India, Brazil, and Canada (Anon., 1970; Bennett, 1940; Brothers, 1960; Du Rietz, 1935; Francis, 1956; Gwinn, 1943; Hunter, 1941; Ragan, 1961; Roberts, 1947; Yudin, 1959).

In the United States, numerous lenslike bodies extend in a belt from northeastern Georgia, northeastward across western North Carolina. Some 25 of these deposits are relatively fresh, with each containing up to five million tons of recoverable olivine.

---

*Excerpt from Teague, K.H., 1983, "Olive," *Industrial Minerals and Rocks,* 5th ed., Vol. 2, S.J. Lefond, ed., AIME, New York, pp. 989-996.

Many small to large dunite bodies occur in northwestern Washington, with the Twin Sisters deposit being the largest.

Taylor (1967) describes eight zoned ultramafic complexes in southeastern Alaska containing cores of dunite. Some of these olivine cores are as great as one mile in diameter.

Reserves

The largest body of olivine in the United States is the Twin Sisters deposit in Whatcom and Skagit Counties, Washington. This dunite has an outcrop area of approximately 93 km^2 (36 sq miles) and has a relief of about 1.5 km (5000 ft).

Other countries containing large reserves of olivine include Norway, where the Aaheim deposit is reported to contain two billion tons. Several other Norwegian deposits contain up to five million tons. In Sweden the largest deposit is at Arutats and is reported to be comparable in size to the Twin Sisters deposit. In Austria large deposits of serpentinized-olivine occur in the Province of Styria. Scotland, Spain, and Italy contain large deposits of serpentinized olivine. In New Zealand, the Dun Mountain dunite is reported to be approximately 2.4 km (1½ miles) in diameter. This deposit is located near Nelson, New Zealand (Gwinn, 1943). A dunite deposit in the coastal area of southern New Zealand has been described by Mossman (1972) "as exceeding 600 meters in thickness." This occurrence apparently is a magmatic differentiate in the Greenhills Ultramafic Complex.

Unaltered dunite crops out as a lenslike body approximately 40 km (25 miles) south of Ste. Anne des Monts on the Gaspé Peninsula of Quebec, Canada. Dunite is also reported in Newfoundland.

EXPLORATION

Most of the dunites containing economic olivine outcrop as prominent ridges or domes. Soils developed from the weathering of dunites support limited vegetation, thus many deposits crop out as "balds."

As a first step in exploration, an olivine deposit should be thoroughly investigated by reconnaissance techniques followed by detailed topographic and geologic mapping. Grab samples, representing various varieties of dunite which occur in the deposit, should be collected for petrographic and chemical evaluation. Grain size, alteration products, and distribution of impurities should be noted, both in samples and throughout the deposit. Portions of the deposit covered with soil should be trenched to determine whether or not the bedrock is olivine. For example, some of the North Carolina olivines, originally regarded as being relatively continuous, have shown by trenching and drilling that areas covered by soil are underlain by non-olivine material that has resulted from alteration of the olivine.

After the surface extent of the olivine deposit is established, details of vertical variation should be determined. As most olivine bodies in the Southeastern deposits are rather closely fractured and jointed and may contain alteration products considerably softer than olivine, core drilling is not recommended for determining vertical continuity. Some variety of percussion drilling, such as Air-Trac drilling or down-the-hole drilling is normally superior to core drilling in olivine deposits. The driller can note variations in drilling rates as well as character of cuttings. If the drill is equipped with a vacuum pump, continuous samples of cuttings can be collected and evaluated.

EVALUATION OF DEPOSIT

A commercial olivine should contain in excess of 40% MgO, before any beneficiation. Many olivine deposits, such as Twin Sisters in Washington, and those in Norway and Sweden, are relatively pure and are amenable to evaluation by dry crushing, followed by screen analysis of the crushed product and chemical and petrographic analysis of the various size fractions. Refractory characteristics also should be determined on the various screen sizes.

Olivine samples which contain objectionable impurities can be evaluated by crushing to a size that will permit removing these impurities and evaluating the clean olivine. Perhaps the most efficient laboratory method to accomplish removal of impurities from crushed or broken samples is to use heavy liquids. Most objectionable impurities which occur with olivine "float" on a liquid of specific gravity of about 3.0. Accessory chromite and olivine report as "sinks." If complete liberation is obtained prior to separation with heavy liquid, the resultant olivine product can be evaluated as outlined previously. Olivine deposits, which require beneficiation in order to produce a usable olivine concentrate, should yield at least 65% product to be considered commercial.

Most accessory minerals associated with olivine lower its refractory characteristics. For example, serpentine, chlorite, and vermiculite contain 12 to 17% water of crystallization, and unless these materials are removed from the olivine, they will contribute to a high loss-on-ignition of the product. Also, the platy minerals, chlorite and vermiculite, react explosively when heat shocked. Olivine sands which are to be coated should contain a minimum amount of platy minerals because of the surface area-weight ratio.

BIBLIOGRAPHY AND REFERENCES

Anon., 1970, "Opportunities for Increasing Olivine Output," *Industrial Minerals,* No. 29, Feb., pp. 11-21.

Anon., 1976, "The Industrial Minerals of Spain," *Industrial Minerals,* Apr., pp. 15-51.

Anon., 1977, "Olivine and Dunite—Blast Furnace Usage Adds New Dimension," *Industrial Minerals,* May, pp. 39-50.

Ashby, G., 1976, "Minerals in the Foundry Industry," Second Industrial Minerals International Congress, Munich.

Astwood, P.M., et al., 1972, "A Petrofabric Study of the Dark Ridge and Balsam Gap Dunites, Jackson County, North Carolina," *Southeastern Geology,* Vol. 14, No. 3, Sep., pp. 183-194.

Beckius, K., 1970, "Olivine: Its Properties and Uses," *Industrial Minerals,* No. 29, Feb., pp. 22-26.

Bengston, K.B., 1956, "Magnesium from Olivine *via* Chlorination: A Possibility, Trend in England," University of Washington, Vol. 8, Jan., pp. 21-26, 35-36.

Bennett, W.A.G., 1940, "Ultrabasic Rocks of the Twin Sisters Mountains, Washington," *Bulletin,* Geological Society of America, Vol. 51, p. 2019.

Bowen, N.L., and Schairer, J.F., 1935, "The System MgO, FeO, SiO$_2$," *American Journal of Science,* 5th Ser., Vol. 29, pp. 151-217.

Bowen, N.L., and Schairer, J.F., 1936, "The Problem of the Intrusion of Dunite in the Light of the Olivine Diagram," *Proceedings,* 16th International Geological Congress, Vol. 1, pp. 391-396.

Brothers, R.N., 1960, "Olivine Nodules from New Zealand," *Proceedings,* 21st International Geological Congress, Pt. 13, pp. 68-81.

Dana, E.S., and Ford, W.E., 1940, *Textbook of Mineralogy,* John Wiley, New York, 851 pp.

Du Rietz, T., 1935, "Peridotites, Serpentines, and Soapstones of Northern Sweden," *Geologiska Foreningen i Stockholm,* Forh. 401, Vol. 57, pp. 133-260.

Francis, G.H., 1956, "The Serpentine Mass of Glen Urquhart, Iverness-Shire, Scotland," *American Journal of Science,* Vol. 254, pp. 201-226.

Gaudette, H.E., 1963, "Geochemistry of the Twin Sisters Ultramafic Body, Washington," unpub. Ph.D. Thesis, University of Illinois, Urbana, 104 pp.

Goldschmidt, V.M., 1938, "Olivine and Forsterite Refractories in Europe," *Industrial & Engineering Chemistry,* Vol. 30, pp. 32-33.

Gwinn, G.R., 1943, "Olivine," Information Circular 7239, US Bureau of Mines, 11 pp.

Hunter, C.E., 1941, "Forsterite Olivine Deposits of North Carolina and Georgia," Bulletin 41, North Carolina Dept. of Conservation & Development, Div. of Mineral Resources, 117 pp.

Kaufman, A.J., 1952, "Industrial Minerals of the Pacific Northwest," Information Circular 7641, US Bureau of Mines, p. 37.

Misra, K.C., and Keller, F.B., 1978, "Ultramafic Bodies in the Southern Appalachians: A Review," *American Journal of Science,* Vol. 278, pp. 389-418.

Mossman, D.J., 1972, "The Geology of the Greenhills Ultramafic Complex, Bluff Peninsula, Southern New Zealand," *Abstracts with Programs,* Vol. 4, No. 7, Oct., p. 605.

Ragan, D.M., 1959, "The Mode of Emplacement of the Twin Sisters Dunite, Washington," *Bulletin,* Geological Society of America, Vol. 70, pp. 1742-1743.

Ragan, D.M., 1961, "Geology of the Twin Sisters Dunite in the Northern Cascades, Washington," unpub. Ph.D. Thesis, University of Washington, 88 pp.

Ramberg, H., and DeVore, G.W., 1951, "The Distribution of Fe and Mg in Coexisting Olivines and Pyroxenes," *Journal of Geology,* Vol. 59, pp. 193-210.

Reed, J.J., 1959, "Chemical and Modal Composition of Dunite from Dun Mountain, Nelson," *New Zealand Journal of Geology and Geography,* Vol. 2, No. 5, pp. 916-919.

Roberts, M., 1947, "Washington's Vast Olivine Deposit," *Mining Congress Journal,* Vol. 33, No. 6, pp. 29-32.

Ross, C.S., et al., 1954, "Origin of Dunites and of Olivine-rich Inclusions in Basaltic Rocks," *American Mining,* Vol. 39, pp. 693-737.

Schaller, G.S., 1957, "New Foundry Sand," *Modern Metals,* Vol. 13, No. 9, Oct., pp. 82-86.

Schaller, G.S., and Snyder, W.A., 1958, "Industrial Applications of Olivine Aggregate," *Modern Castings,* Vol. 33, No. 6, June, pp. 99-104.

Snyder, W.A., 1957, "How to Use Olivine Sand," *Foundry,* Vol. 85, No. 9, pp. 100-105.

Taylor, H.P., Jr., 1967, "The Zones Ultramafic Complexes of Southeastern Alaska," *Ultramafic and Related Rocks.*

Yudin, M.I., 1959, "Dunite of the Boris Mountain Range and Their Origin," *Academy of Science Bulletin,* USSR, Geol. Ser. 2., pp. 47-62.

19. Perlite

FREDERIC L. KADEY, JR.

Perlite, as a volcanic glass, has been recognized since the 3rd century, BC (Langford, 1978).

Today the name perlite is applied to both the hydrated volcanic glass, generally of rhyolitic composition, and to the lightweight aggregate that is produced from the expansion of the glass after it has been crushed and sized.

For the purpose of this discussion, however, only the unexpanded ore will be considered. Petrologically, it was originally described as a glassy rhyolite that had a pearly luster and concentric, onionskin parting (Johannsen, 1939). Since the end of World War II when its commercial significance as a lightweight aggregate became important, other textures associated with the classical texture have become included within the definition of perlite and will be described herein.

GEOLOGY

Occurrences of perlite are restricted to several Tertiary to Quaternary age rhyolitic belts that trend in a generally north-south direction around the world. Commercially suitable deposits generally occur as domes of several hundred feet in height, although glassy zones in welded ash-flow tuffs and others associated with dikes and sills also have been reported. Mining is by ripping and blasting from open pits. Nowhere in the world has it been found economical to extract perlite as an underground operation.

In the United States, New Mexico leads in production with Arizona, California, Nevada, Idaho, and Colorado following in approximately that order. The principal use for expanded perlite is as a lightweight insulating aggregate in cryogenics, in plaster, concrete, and in loose fill insulation. Expanded perlite is also used in horticultural applications, and, after subsequent milling and classification, as a filter aid.

The United States is the world's largest producer and consumer of perlite. In its naturally occurring form, perlite is a rhyolitic glass that contains from 2 to 5% combined water. While perlite also can occur as andesitic or dacitic glass, these latter types are of negligible commercial significance. Table 1 lists the chemical composition of a few typical perlites (Anderson, et al., 1956; Langford, 1979).

What sets perlite of commercial significance apart from other volcanic glasses is the fact that under the proper conditions of preparation—crushing and sizing—it will, when rapidly introduced into a flame of sufficient temperature, expand or "pop." All of the elements of composition contribute to the expansibility of the rock. The role of the combined water, however, is the most significant because it is believed not only to produce a fluxing effect in the softening of the highly siliceous glass prior to expansion, but it is also responsible for the explosive force of expansion through volatilization during heating. The current theory of the origin of water in perlite is now less controversial than it was a decade earlier. The 2 to 5% range for total combined water in perlite was originally thought to have been chilled or frozen into the glass as it was injected under an ice sheet or into a lake (Huntting, 1949). Now, there is universal acceptance of the theory that perlite was formed by the secondary hydration of obsidian after its emplacement (Ross and Smith, 1955; Friedman, et al., 1966; Jesek and Noble, 1978).

Studies have shown that the combined water in perlite exists in at least two forms (Lehmann and Knauf, 1973; Lehmann and Rossler, 1974): one as molecular water and the rest as hydroxyl water. The ratio of one kind of water to the other is different for perlites of different origins. The composition of the perlite, particularly the amount of MgO and CaO, is believed to influence the rate of hydration of the obsidian.

An emplacement model for perlitic domes suggests that the outermost material will be pumiceous, grading inward into increasingly more compact textures and finally into a felsitic core. The following textures that fit the model have been observed at No Aqua Mountain, New Mexico, and elsewhere (Whitson, 1982). While these textures also may be known by other names, the following seems to be a convenient and logical nonmenclature.

Pumiceous

Near surface, this is a lightweight and frothy perlite where vesiculation is less confined by lithostatic and hydrostatic pressures. The vesicularity and the degree of distortion of the vesicles is a function of local confining pressures and of the amount of the supercooled liquid. Flattening of vesicle walls increases with depth while elongation occurs during flow. Hence, in depth, the vesicles of pumiceous perlite when viewed with an X10 hand lens appear elongated and taffylike.

At the No Aqua deposit, this taffylike texture—shardy— while actually a gradational subunit of the pumiceous textured material has different and commercially advantageous properties compared with the more vesicular main unit. It has, therefore, been mapped separately from the pumiceous texture. Pumiceous perlite is usually light gray but may be buff.

During exploration, near surface pumiceous perlite is difficult to core drill. In the field, the more compacted pumiceous varieties of perlite—shardy—are characteristically difficult to sample because they tend to powder when struck by a geologist's pick and much of the energy is absorbed rather than expended in breaking a chip off the outcrop.

Granular

This texture is found adjacent to and deeper than pumiceous perlite in the emplacement model. It is more dense than the overlying textures, has a "sugary" or saccharoidal appearance, and commercially is highly satisfactory from the milling, classification, and expansibility standpoints. It cores well during drilling and blasts easily during mining. Color usually is buff but varies from gray to brown. It often displays flow banding.

Classical

This is the typical pearl gray material with "onionskin" concentric parting. Classical texture also may be dark gray to black. The name is derived from the classical description of perlite in the literature (Johannsen, 1939). Classical perlite is found stratigraphically below the granular perlite in the dome. Obsidian, whenever present, is always encased in classical perlite, lending substance to the theory that perlite is formed from the hydration of obsidian. The concentric rings around each obsidian nodule were probably formed as the result of volume increase due to the hydration which advanced only as far as the perlite-obsidian interface. Toward the interior of the flow, the amount of obsidian in classical perlite tends to increase. Also found in perlite deposits are a rhyolitic or felsitic core, flow breccia, and other materials and structures associated with volcanic flows and domes.

In addition to obsidian which may contain what have been identified as tridymite inclusions, often such nonexpansibles as quartz, feldspar, biotite, magnetite, and other accessory minerals as well as products of devitrification, may be present.

Using the No Aqua deposit as a model, Whitson (1982) has expressed, in cross section, the lithologic zonation of these described textures.

DISTRIBUTION OF DEPOSITS

Perlite deposits are restricted to Tertiary or younger volcanics of rhyolitic composition. Because rhyolitic glasses are unstable, they devitrify with age. Very young acidic volcanics have not had sufficient time to hydrate; hence they are still essentially obsidian and not expansible within the temperature range of perlite. There is a belt of Tertiary and Quaternary rhyolites that begins in Iceland, extends south into Ireland, Scotland, through the Massif Central of France into some of the Aegean Islands, through Sardinia and mainland Italy and farther south into Morocco, Algeria, and into South Africa. A Pacific belt splits, with a western limb extending through Japan, the Philippines, China, New Zealand, and Australia. The eastern part of the circum-Pacific belt extends from Alaska (Plafker, 1963) through western Canada, down through the western United States into Mexico, and emerges in parts of the Andes in Chile and Argentina.

North America

While perlitic occurrences have been noted in Alaska and western Canada, the continental United States has the largest known reserves and commercial production in the world (Jaster, 1956; Meisinger, 1985).

In the United States, perlite deposits are confined to the western states. New Mexico accounts for the major share of US production, with substantial deposits at No Agua, Socorro, Grants, and smaller occurrences in the southwest part of the state. Arizona, Nevada, and California have substantial deposits with Idaho, Colorado, Utah, Oregon, and Washington reporting lesser occurrences.

Mexican perlite is centered in the state of Sonora.

South America

Argentina is the major source of South American perlite with most deposits in Salta Provence. The only other significant South American deposits are in the Laguna de Maula area of Chile.

Europe

Greece is the largest western European producer of perlite. Major production comes from the island of Milos. The Greek islands of Kos and Lesbos in the Aegean Sea also have commercial grades of perlite from which there has been some production. Numerous other Greek islands contain perlite.

Turkey has substantial reserves of high quality perlite in the western part of the country. There are also occurrences in central and eastern Turkey as well.

Italy is an important source of perlite with the most important deposits located on the island of Sardinia near the town of Uras (Carta, 1976). Perlite is mined on the island of Ponza.

In eastern Europe. Hungary (Perlaki and Szoor, 1973) and the Soviet Union are significant producers of perlite (Rudnyanszky, 1978). Bulgaria, Czechoslovakia, and Yugoslavia have smaller but growing industries (Vladimirov, 1975).

Iceland has a commercially challenging but as yet unproductive perlite deposit at Priest Mountain (Prestahnukur), located about 100 km northeast of Reykjavik. The remoteness of the deposit, the relatively high concentrations of obsidian, and the short four-month mining season offset the otherwise good quality of perlite in many parts of the mountain. A smaller occurrence at Lodmundarfjordur has been investigated, but the location and quality do not justify commercial consideration.

Northern Ireland did at one time have limited production from deposits in the Sandy Braes area of County Antrim.

Small deposits in France and Germany are devitrified or have not been economical to mine.

Africa

Africa has commercial occurrences of perlite at Nxwala Estate, Zululand, and in the Lemombo Mountains, both in northern Natal, South Africa (Coetzee, 1976).

Asia

Japan is the most significant producer of perlite in Asia, which because of the volcanic nature of the islands has numerous perlite deposits. Dominant among these are deposits at Kitakata, Ippongisita, and at Arita. Other deposits are located at the Kushiro prefecture of Hokkaido; at Akita; in Yamagata and Osaka; in Yamaguchi; and in Kumamoto and Fukuoka.

New Zealand has a substantial deposit at Kenlieth and several other deposits have been described (Thompson and Reed, 1954). A large deposit is located on Great Barrier Island, located a few miles off the east coast, northeast of Aukland. Perlite also occurs in Queensland, Australia, and in the Philippines on Luzon Island.

In summary, deposit occurrence worldwide corresponds with island arc and continental volcanism related to plate boundary movement or oceanic spreading center volcanism (Breese and Piper, 1985).

EXPLORATION

Exploration for perlite is restricted to those areas of rhyolitic volcanism between Tertiary and Quaternary age. Any glassy rhyolites of earlier than the Tertiary age have long since devitrified and are not of commercial interest. Younger acidic volcanics will not as yet have hydrated from obsidian. Perlite is recognized in the field by its characteristic appearance, whether it be a rather light gray pumiceous to classical onionskin or the gray to buff granular texture. The presence of marekanite or obsidian nodules (*Apache teardrops*), while detrimental in commercial perlite, are a good field indicator that the prospector may not be far from "pay dirt." Once grab samples have been taken from outcrops or shallow trenches, and upon receipt of encouraging test results, the next step includes the careful mapping of the dome or flow. Careful field work should ascertain the nature and distribution of rock textures and identify other such features as flow banding, joint systems, faults, etc.

The next step, diamond core drilling, is one in which considerable difficulty may be encountered. The friable nature of some perlitic textures, together with fracturing and jointing, may result in poor core recovery. Often only sludge samples can be obtained. In many cases, such mechanical difficulties as hung bits, lost drill steel, and loss of circulation of drilling fluid may be encountered. Such problems can be minimized with the use of a triple tube wire line core barrel and NQ size core. On some occasions and in less accessible terrain, portable drilling equipment will produce adequate results and often can furnish information not otherwise attainable. The spacing between drill holes and the intervals into which the core is to be split for testing depends on the variability of quality. In many occurrences, variability is so great that the economics of the matter determines the closeness of the drilling and sampling pattern. In general, multistage drilling programs, in which drilling is begun on wide spacings, then tightened up in subsequent stages, are most economical in time and costs. A preliminary indication of quality, when determined in the field, is useful in the exploration for perlite. This can eliminate the shipping of worthless samples back to the laboratory. Whether a crude perlite sample is worth laboratory testing can be ascertained by a petrographic examination for percent contamination, refractive index of the glass, and other criteria (Kadey, 1963).

Geophysical methods had not heretofore been used to any extent in the exploration for perlite deposits. Recently, however, there have been some encouraging results using seismic methods (R. Breese, Manville Corp., personal communication). For the interested reader, the Perlite chapter in the 5th edition of *Industrial Minerals and Rocks* contains a comprehensive treatment of this commodity (Kadey, 1983).

BIBLIOGRAPHY AND REFERENCES

Anderson, F.G., et al., 1956, "Composition of Perlite," Report of Investigations 5199, US Bureau of Mines, 13 pp.
Breese, R.O.Y., and Piper, J.R., 1985, "Occurrence and Origin of Perlite," SME preprint 85-359.
Carta, M., et al., 1976, "The Industrial Minerals of Sardinia: Present Situation and Future Prospects," 2nd Industrial Minerals International Congress, Munich, pp. 41-55.
Coetzee, C.B., 1976, "Perlite in Lenses and Intermittent Layers," Perlite—South Africa Geological Survey Handbook, pp. 397-398.
Friedman, I., et al., 1966, "Hydration of Natural Glass and Formation of Perlite," *Bulletin*, Geological Society of America, Vol. 77, Mar., pp. 323-328.
Huntting, M.T., 1949, "Perlite and Other Volcanic Glass Occurrences in Washington," Report of Investigation No. 17, Division of Mines and Geology, State of Washington, 77 pp.
Jaster, M.C., 1956, "Perlite Resources of the United States," Bulletin 1027-1, US Geological Survey, pp. 375-403.
Jesek, P.A., and Noble, D.C., 1978, "Natural Hydration and Ion Exchange of Obsidian: An Electron Microprobe Study," *American Mineralogist,* Vol. 63, pp. 266-273.
Johannsen, A., 1939, *A Descriptive Petrography of Igneous Rocks,* University of Chicago Press, 318 pp.
Kadey, F.L., 1963, "Petrographic Techniques in Perlite Evaluation," *Trans. SME-AIME,* Vol. 229, pp. 332-336.
Kadey, F.L., 1983, "Perlite," *Industrial Minerals and Rocks,* 5th ed., Vol. 2, S.J. Lefond, ed., AIME, New York, pp. 997-1015.
Langford, R.L., 1979, *Mineral Dossier No. 21,* Institute of Geological Sciences, London.
Lehmann, H., and Knauf, A., 1973 "Entwicklung Einer Pruemethode 2UR Beuteilung Der Blahfahigkeit Von Perlitgestein," *Tomind-Ztg.,* Vol. 97, No. 3, pp. 65-66.
Lehmann, H., and Rossler, M., 1974, "A Contribution of the Nature of Water-Binding in Perlites," *Thermal Analysis,* Vol. 2, *Proceedings,* 4th ICTA, Budapest, pp. 619-628.
Meisinger, A.C., 1985, "Perlite," *Mineral Facts and Problems,* Bulletin 675, US Bureau of Mines, pp. 1-7.
Perlaki, E.I., and Szoor, Gy., 1973, "The Perlites of the Tokaj Mountains," *Acta Geologica Academiae Scientiorum Hungaricae,* Vol. 17 (1-3), pp. 85-106.
Plafker, G., et al., 1963, "Investigations for Perlite in the Alaska Range," Bulletin 1155, US Geological Survey, 49-66.
Ross, C.S., and Smith, R.L., 1955, "Water and Other Volatiles in Volcanic Glasses," *American Mineralogist,* Vol. 40, Nos. 11, 12, Nov.-Dec., pp. 1076-1089.
Rudnyanszky, P., 1978, "Development and Uses of Perlite in Hungary," Perlite Institute Annual Meeting, Dubrovnik, Yugoslavia, May, pp. 135-138.
Thompson, B.N., and Reed, J.J., 1954 "Perlite Deposits in New Zealand," *New Zealand Journal of Science and Technology,* Series B., Vol. 36, No. 3, pp. 208-218.
Vladimirov, I., 1975, "The Perlite Industry in Bulgaria," Perlite Institute Annual Meeting, Athens, Greece, pp. 39-46.
Whitson, D.N., 1982 "Geology of the Perlite Deposit at No Agua Peaks, New Mexico," *Industrial Rocks and Minerals of the South West,* Circular 182, New Mexico Bureau of Mines and Mineral Resources, pp. 89-95.

20. Phosphate Rock

M. F. DIBBLE

Phosphate rock refers to commercial concentrations of apatite which occur in both igneous and sedimentary deposits. The quality of marketable phosphates has historically been based on the content of tricalcium phosphate, $Ca_3(PO_4)_2$ or "bone phosphate of lime" (BPL). The concentration in igneous rock deposits and chemical processing facilities is more generally referred to as percent phosphorus pentoxide (P_2O_5), which is equal to BPL times 0.4567.

The principal mineral found in igneous deposits is fluorapatite, $[Ca_{10}(PO_4)6F_2]$ which in its pure form contains about 42.2% P_2O_5 (92.2 BPL) and 3.8% fluorine. Phosphate rock mined in the United States is found exclusively in sediments. The specific mineral found in these sedimentary deposits is francolite, a carbonate fluorapatite, of the general formula $[Ca_{10-a-b}Na_aMg_b-(PO_4)_{6-x}(CO_3)\ F_{0.4x}F_2]$ (McClellan) that can contain up to 2.0% molecular CO_2.

Total world production of phosphate rock products was approximately 158 Mt in 1987. Estimated US production for 1988 is 46 Mt, of which Florida will provide 34 Mt or 74% (Stowasser, 1988). Over 85% of the total phosphate rock mined is used for fertilizer products, either as direct application or the manufacture of upgraded products. Of the phosphate mines in the US, 27 of the 28 are open pit.

SEDIMENTARY PHOSPHATE RESOURCES

Nearly 80% of the total phosphate rock mined in the world is pelletal in form and is found in sedimentary deposits. These francolite pellets are of marine origin and have been found in sedimentary rocks ranging in age from Proterozoic (Precambrian) to Holocene (Cathcart). Many theories have been postulated as to the origin of these types of deposits. It appears that the areas of deposition, because of their scope, had to be open to oceanic water circulation for long periods of time. In general, "phosphorite precipitation took place as the cold, chemically supercharged and somewhat toxic upwellings moved across the shallow platforms and into coastal environments. The biologically stressed shallow-water environments received the bacterially precipitated microcrystalline phosphorite mud or microsphorite, as well as the other biologically produced phosphate grains" (Riggs, 1978). Depending on the energy level locally, some of these muds were aggregated into pelletal forms as they were reworked by submarine currents, rivers/streams, wind, or a combination of any or all of these. The quantity of francolite in these type deposits can vary from less than 3.0 to an excess of 38% P_2O_5.

Marine formations that are phosphatized can cover very large areas such as the Phosphoria Formation in the western United States. This formation is of Permian age and covers an area in excess of 349 623 km² (135,000 sq miles). Thickness varies from a few millimeters (inches) to many meters (feet) and phosphate content ranges from 16 to 36% P_2O_5. This formation has been mined for many years from outcrops along a 724-km (450-mile) long belt.

COMMERCIAL SEDIMENTARY DEPOSITS

Central Florida has been a major source of pelletal phosphates since the discovery of the "Land Pebble Deposits" in the 1880s. This rock was originally deposited during the mid to late Miocene. The present commercial phosphate production of the Central Florida District is the result of the reworking of this rock during the Pleistocene age. This consisted of weathering of unconsolidated pellet phosphorite beds, leaving residual concentrations of francolite pellets. Subsequent high energy sedimentary action resulting from fluctuating sea levels cyclically reworked and redistributed these deposits. Particle enrichment also occurred because of supergene alteration. This constant reworking caused the intermingling of enriched phosphate pellets with silica sand, skeletal land and sea animal remains, and clays.

The Central Florida District covers an area of about 1812 km² (700 sq miles). The commercial mineral zone, locally called matrix, ranging from 0.9 to 7.6 m (3 to 25 ft) in thickness, is covered with 3 to 14 m (10 to 45 ft) of overburden. The upper part of the Central Florida District occasionally contains a lateritic leached zone with a high concentration of uranium-bearing aluminum phosphates. Historically, the pelletal pebble fraction coarser than 1.0 mm has been of high enough grade to be screened, washed, and sold as product. The -1.0 mm pellets, which occurred intermixed with silica sand and clays, were formerly discarded as debris.

The development of two-stage anionic/cationic froth flotation in the late 1930s led to the recovery of the -1.0 mm $+100\ \mu$m phosphatic pellets by removing the silica (insol) waste. Concentrates occasionally exceeded 80% BPL (37% P_2O_5) from these high grade deposits. (BPL is more commonly used by miners in this district.) The development of the two-stage flotation process made possible the successful reprocessing of the old debris dumps resulting from earlier pebble mining.

During the glory years of mining in this district, a general rule of thumb for this matrix was that it was composed of one-third pelletal phosphate, one-third silica sand or insol, and one-third clay. As mining has depleted higher grade ores and moved southward toward the fringes of the Central Florida District, pebble and flotation-sized pellets are less enriched and contain higher concentrations of carbonates. Typical pebble grades of the 1950-1960s of 68% to 74% BPL (31.1 to 33.9% P_2O_5) are now more likely to be 62% to 68% BPL (28.4 to 31.1% P_2O_5). Flotation concentrates now range from 66 to 72% BPL (30.2 to 32.9% P_2O_5) from the high 70%'s ($+35\%$ P_2O_5) in earlier years.

FUTURE CENTRAL/SOUTH FLORIDA RESOURCES

The Hawthorn Group of Miocene age covers most of peninsular Florida with the exceptions of the Ocala Arch and Sanford High. Underlying the Bone Valley Member, Pease River Formation of the Hawthorn Group (Scott, 1985), it consists of impure carbonates with varying concentrations of well rounded phosphate pellets.

Representing a significant, more costly, future phosphate rock resource is a weathered extension of the Hawthorn Group. This formation underlies and extends southward from the Central Florida District. This southern extension can be subdivided into three distinct units. The lowest is predominantly carbonate with pelletal francolite particles intermingled. The middle unit is a transitional zone with pelletal phosphate particles intermixed with dolosilts and clays. The upper unit is primarily clastics with phosphate pellets, quartz sand, clays, dolosilts, and distinct dolomitic particles.

The upper clastic unit is further subdivided into two zones. The upper zone somewhat resembles typical Central Florida District rock with some exceptions. The matrix grade is lower, the enrichment of the pellets is less, and the ratio of CaO:P_2O_5 is higher, approximately 1.55 (the higher number usually reflects greater sulfuric acid consumption when the rock is acidulated). The -100 μm clay slimes are higher in quantity and the pebble fraction poorer in both quality and quantity. The MgO content of normal concentrate is significantly higher than what is found in the Central Florida District. This upper zone is not found throughout the southern extension and varies significantly in thickness.

The lower zone is quite different from the upper. The presence of dolomite produces an unacceptably high level of MgO when conventional phosphate concentration methods are used. The coarser individual pellets also contain inclusions of dolomite as well as silica. Carbonates are found as distinct beds and as separate grains intermingled with the pellets. Clays and dolosilts can be found as individual beds and interspersed throughout the matrix section. Pebble and concentrate from most of this lower zone is generally unacceptable for use in a typical wet process acid plant unless there is reduction of the MgO content.

The southern extension is generally considered the logical direction of movement for the existing mining companies and operations when the Central Florida District is depleted in the fairly near future. Development will need to consider both zones as part of a mining sequence to maximize resource recovery. Significant progress has been reported by various investigators (Lawver, et al., 1984) to reduce the dolomite (MgO) content of product from both of these zones.

Table 1 was developed to define the fundamental differences between the traditional Central Florida deposits and the southern extension of the weathered Hawthorn.

The southern extension of the Central Florida District represents a large potential resource, more costly to produce and lower in quality than the present district. Table 1 shows that potential mining costs (per ton of product) should be significantly higher than the Central Florida District because of the sheer magnitude of material that needs to be stripped,

mined, and then replaced during required reclamation. The removal of the dolomite in the concentration step would also add to the product costs. The waste clays, which are roughly 2.5 times the quantity of the Central Florida District per ton of product, will present challenging obstacles in obtaining necessary permits for mining as well as actual disposal.

NORTH CENTRAL FLORIDA PHOSPHATE RESOURCES

Extensive shallow Hawthorn phosphate resources exist east of the Ocala Arch on the North Florida Platform. One company operates two mines near Lake City. This rock has also been reworked but less extensively than the Central District. The alteration and supergene enrichment has been less which has limited the availability of marketable pebble products.

Extensive regional prospecting in north Florida and south Georgia was conducted during the 1960s in the general area of these existing operations. In most cases the quantity of pelletal phosphates was too low, at that time, or was located in areas of great environmental sensitivity.

DEEP MIOCENE DEPOSITS, NORTH CAROLINA

The Miocene was also the most important period in the development of the extensive phosphorites formed throughout the southeastern United States continental margin. These deposits are found from Chesapeake Bay in Virginia to south of Cape Canaveral (Cape Kennedy) in Florida.

One of the more important and proven phosphogenic provinces of these Miocene sediments is the Pungo River Formation. The phosphate-enriched section occurs as highly truncated sequences of multiple depositional units in eastern North Carolina (Riggs, 1986). Cape Lookout High subdivides a large Neogene basin into two regional embayments. The northern or Aurora embayment is the site of one of the largest operating phosphate mines in the United States.

The Aurora phosphate district is one of the shallower, more accessible, occurrences of commercial grade rock in this coastal margin and is mined by open pit. Overburden is partially removed by a dredge stripping the upper 11 m (35 ft). After dredging a specific area, the phosphate-rich Pungo River Formation is dewatered by depressurizing the underlying Castle Haynes Eocene Limestone Formation. The overburden remaining after dredging, from 11 m (35 ft) down

Table 1. Differences Between Central Florida Deposits and the Southern Extension

	Central District	Southern Extension
Matrix thickness	Average 25 ft	Up to 65 ft
Matrix grade, P_2O_5 %	11.4 to 13.7%	5.5 to 8.2%
Matrix yards per ton product	3.0 to 4.0	4.0 to 6.5
Matrix & overburden, tons moved per tons prod.	7.5 to 12.0	12.0 to 14.0
Tons prod. per acre-foot mined	400 to 600	200 to 400
Pebble product, P_2O_5 %	10% of matrix 28.4 to 31.1 direct salable	5% of matrix 18.3, upgrading required
Flotation product, P_2O_5 %	30.1 to 32.9 using standard double flotation	29.2 to 31.5, req. Additional dolomite removal stage req.
Product CaO:P_2O_5	1.50 to 1.52	1.55 to 1.57
Product MgO% content	0.2 to 0.5%	0.7 to 1.3%
Combined iron and aluminum oxide	1.4 to 2.6%	2.0 to 3.8%
Waste material	Quartz sand and clay	Sand and clay plus dolomite particles, marl
Tons waste clay per ton of product.	0.9 to 1.1	2.0 to 2.5

Metric equivalents: yard $\times$ 0.914 = meter; ton $\times$ 0.907 = tonne; foot $\times$ 0.305 = meter; acre-foot $\times$ 1233 = cubic meter.

to about 28 m (93 ft), is stripped by large dragline and the spoil dumped into the adjacent mined-out pit. This same dragline also mines the matrix section, which can range from 11 to 12 m (35 to 40 ft) in thickness, and stockpiles it on top of the dredged surface. A smaller dragline then reclaims the stockpile and dumps the matrix at a portable pumping station to be slurried by high pressure water. The slurried matrix is then pumped to the process facility (Hird, 1988).

A program is underway to replace the dredge and remove the upper overburden by use of a bucket wheel excavator. This improves the operational efficiency by moving upper waste material at natural percent solids and opens up the area for the dragline more effectively. High solids content allows more efficient transport of this overburden with a conveyor belt system (Breza, 1987).

The francolite pellets at this deposit show evidence of only localized reworking. The product tons per acre is much higher than in the Central Florida District and can approach 32 to 36 Mt (35,000 to 40,000 tons) per acre. The flotation feed grade is also much higher and finer in mesh size than from rock presently mined in Central Florida. No commercial pebble product is produced.

The final product grade obtainable is about 29.7 to 30.7% P_2O_5. The $CaO:P_2O_5$ ratio is close to 1.60 due to the lack of supergene enrichment. Concentrates from this rock have a lower combined iron and aluminum oxide content compared to Central Florida concentrate. The rock also contains a high concentration of organic carbon that can be removed by calcination if necessary.

ALTERNATIVE DEEP MIOCENE RESOURCES

South of the Cape Lookout High is the Onslow embayment, which underlies the continental shelf in Onslow Bay. Two significant phosphate rock occurrences have been discovered by shallow drilling and surface sampling. Seismic data has indicated that substantial additional resources may be present and extend well beyond the areas sampled.

Moving further south is the Beaufort High (Riggs, 1986) that apparently provided the same conditions for phosphate rock deposition as the Cape Lookout High. Low concentrations of phosphate minerals have been observed west of this High in the Beaufort, SC, deposits (Woolsey, 1977). South of this High occurs the Chatham, GA, embayment where seismic data has indicated a substantial Miocene section (Riggs, 1986).

DEEP MIOCENE DEPOSITS, NORTHEAST FLORIDA

A major Miocene resource has been partially defined within the Jacksonville, FL, Basin. The deposition of phosphorites appears to be controlled by the presence of the Sanford High. The magnitude of this resource is unknown since only the western and southern end of the deposit has been roughly defined. The JOIDES deep drill core hole, J-1, drilled on the continental shelf 45 km (28 miles) east of Jacksonville, indicated a Miocene thickness of at least 79 m (260 ft), beneath a cover of 18 m (60 ft) of Plio-Pleistocene sediment (Riggs, 1986). Onshore drilling in this basin indicated commercial thicknesses and grades of Miocene phosphates at least 80 km (50 miles) to the southwest of J-1.

Extensive core drilling conducted during the late 1960s indicated that a sizable primary phosphate resource exists north of St. Augustine, FL. Subsequent drilling further defined this resource and indicated that it should project eastward under the continental shelf for some distance (Brown, 1979). If continuity were established between the landside

phosphates, the J-1 hole, and an eastward extension, this could represent a substantial phosphatic resource.

The landside deposit consists of five to six different lithologic units. The lowest two units are referred to as "D" and "E" with the lowest, E, representing the largest potential source of phosphate rock. This E zone appears to be always present throughout the phosphogenic environment and can vary from 1.5 to 11 m (5 to 36 ft) in thickness. The top of the D zone, when present, occurs at depths from 55 to over 91 m (180 to over 300 ft) below the surface. The two zones are separated by a barren clay member and the E zone is usually capped with dolomite of differing degrees of hardness. This barren zone can vary in thickness from 0 to 4.6 m (0 to 15 ft). Overall the E zone feed sized material ($-600 +74$ μm) will usually exceed 18.3% P_2O_5. There is no pelletal pebble-sized particles present. The upper D zone is lower grade than E, and ranges from 9.1 to 16% P_2O_5. Some reworking is evident in the D zone as indicated by water-rounded silica sand particles and less clay than the E zone.

The product grade indicated from experimental borehole mining tests shows many similarities to the product obtained in North Carolina (Davis, 1982). The Jacksonville Basin concentrate has a slightly higher P_2O_5, lower $CaO:P_2O_5$ ratio, lower organic carbon content, lower combined iron and aluminum oxide content, and equal to slightly higher MgO content. The combination of the minor elements present (sometimes referred to as the minor element ratio or MER), including iron and aluminum oxide plus the MgO divided by the P_2O_5 content, is from 0.055 to 0.060.

DEEP MIOCENE PHOSPHATE ROCK POTENTIAL EXTRACTION TECHNIQUES

Most of the deep Miocene deposits, other than the relatively shallow North Carolina Pungo River deposits, are too deep to be mined by conventional open pit methods. The Jacksonville Basin rock, for example, is covered by recent to Plio-Pleistocene sand and shell hash from 18 to 24 m (60 to 80 ft) in depth. This is followed by clay, sandy limestone, and interbedded clay and shells. Below this and consistent throughout that mineral province is a massive dark green rubbery clay section that is at least 18 m (60 ft) in thickness. Underlying this clay section is a dolomitic rock that can range from 0 to 2.4 m (0 to 8 ft) in thickness, with varying degrees of hardness, that caps the D zone phosphate rock (Brown, 1979).

Experimental borehole mining tests to slurry mine phosphate rock from this and North Carolina deposits were made during the 1960s (McKelvey, 1985). In 1980, a limited program with improved technology, was conducted in conjunction with the US Bureau of Mines in the Jacksonville Basin (Savanick, 1985). In 1984-1985 additional tests conducted by private companies near that site were encouraging and demonstrated that this rock could be mined successfully if a flooded cavity environment was maintained (Dibble, 1985). The waste clay resulting from mining was reinjected back into the cavities. The nature of the overburden, at that location, eliminated the need for casing below the upper contact of the dense clay member. The casings used in the latest tests were pulled and could have been recycled repeatedly.

Borehole mining technology is still in its embryonic stage. Information gained during the most recent borehole mining tests indicated several areas where improvements were possible, making this concept more economically attractive (Dibble, 1987). Deep Miocene phosphate resources found both on land and under the continental shelf represent the largest potential major phosphate rock source. Borehole mining technology will need to be developed to recover a major portion of this resource. By proving this technology to be economically competitive with conventional deep open pit mining or dredging for the offshore rock, the United States agricultural sector could remain free from dependence on foreign phosphate rock sources.

FERTILIZER PRODUCTION—FUTURE SUPPLIES

The Central Florida District rock have been the standard of comparison for new sedimentary phosphate deposits. Chemical processing plants for fertilizer manufacture have traditionally been geared to use this rock. Adjustment to plant operations has been necessary as mining has moved southward and rock quality has diminished. Rock supplied to these same plants from the deep Miocene deposits would probably require additional plant modifications. Because the deeper Miocene deposits have not been extensively reworked, the rock quality throughout a given deposit is very consistent. A process facility geared to use this type of rock would have very stable operations because of this inherent consistency.

The Central Florida deposits are being rapidly depleted. The location of the very large western Phosphoria deposits, with their high mining, processing, and transportation costs to the consuming market, does not indicate that they can be used to replace much of this depleted rock. The United States, in order to maintain its present fertilizer production capacity with domestic rock into the 21st century, needs to make substantial changes from traditional sources. Either production must be greatly expanded in North Carolina, the processing technology must be commercialized to mine the Central Florida southern extension of the Hawthorn, or the mining technology must be developed to recover the other deep on- and offshore Miocene deposits.

To determine the direction the US fertilizer industry must take, these two major future sources of rock need to be compared. This comparison must assume that eventually the metallurgical problems will be solved to commercially process the southern extension of the Hawthorn Formation. It will also be assumed, for the sake of comparison, that the technology necessary for deep Miocene deposit mining can also be developed. Many other factors besides technical ones will influence the direction taken, so Table 2 can only serve to illustrate fundamental differences between these two sources.

FUTURE SOURCE COMPARISONS

The deeper Miocene phosphate rock has many apparent advantages over the southern extension Hawthorn rock as illustrated in Table 2. In addition, the borehole mining technology causes a minimal disturbance of the surface and almost immediately upon completion of mining, the surface would be available for other purposes. This is more environmentally acceptable than open pit mining followed by required reclamation. In the event that the offshore deposits could also be mined, less environmental damage would occur when compared to dredging. The operational concept calls for mining to start from the bottom of the mineral section and to work upwards. Dredging works the opposite since the overburden has to be physically relocated, and then the phosphate. Stirring up the fine sediments with a dredge could cause some turbidity that may not be environmentally acceptable (Woolsey, 1987).

The deep Miocene phosphate deposits do have some disadvantages when compared with the southern extension rock. Most of the major fertilizer plants are located convenient to the Central Florida deposits. The open pit mining technology using draglines is well proven and has been relatively cost effective. Mining by large draglines recovers a significantly higher percentage of the available phosphate rock than what can be expected with a borehole system. Mining tradition in the region already exists because of present mining in the Central Florida District. Although environmental pressures against open pit operations are already high and intensifying, phosphate mining has historically provided substantial financial contributions to the region.

There is no phosphate mining experience in areas where the deep Miocene deposits occur except in North Carolina. Near shore lands are becoming more valuable, and it will becoming increasingly more difficult, as large land holders such as timberland interests are broken up, to put together large enough tracts to support mining ventures. This may force a greater emphasis, in the long-term, on the offshore Miocene deposits, which may be more costly to produce than the deeper onshore deposits.

EXPLORATION METHODS

The limits of the Central Florida District have been well defined for many years. A flurry of exploration activity in the 1960s also helped define the general limits of the southern Hawthorn extension. Exploration usually consisted of reconnaissance drilling using a pattern of one to four core holes per square mile. Core samples were then taken from the matrix section encountered and divided into lithologic units or splits and carefully logged by the geologist.

Samples are obtained for moisture and bulk density calculations. Each split is washed in the preparation laboratory which removes the clay fraction, usually $-100~\mu m$, through cycloning or screening. The coarse fraction is also separated at approximately 1.0 mm size to determine the nature or quantity of the pelletal pebble fraction. The remaining material is the feed-sized fraction which is processed by flotation in the laboratory cell. Generally these splits are processed in accordance with the same flowsheet used in existing plants of that specific area. In the case of cores from the southern extension, additional concentration steps can be made to upgrade the product quality.

Calculations are made from the results obtained in processing core samples to determine bulk density, product distribution, product quality, clay content, product tons per acre, and tons per acre-foot. An acceptable reconnaissance hole is usually followed by a denser pattern that may eventually end up with as many as 16 holes per 40 acres. Mine production plans are usually developed with this hole density in the Central Florida District.

Most of the large tracts of land existing in the southern extension have already been acquired by mining companies or land speculators knowledgeable of the phosphate mining industry. Large tracts are usually optioned and then drilled. The early definition of the district was made from reconnaissance drilling from roadways and other rights-of-way.

Exploration for the deeper landside Miocene deposits is much more complicated than prospecting a known, relatively shallow deposit. The geologist in charge of an exploration project must have a good working knowledge of the stratigraphic sections to be encountered. He should also know the interrelationships of the various southeastern highs and basins and the conditions leading to phosphate rock deposition.

One of the best screening methods to determine the presence of these Miocene phosphates is the gamma-ray logger. Much of the early work in defining the North Carolina

Table 2. Differences Between the Southern Extension and Miocene Deposits

	Southern Extension	Miocene (north FL-NC)
Matrix & overburden tons moved per tons prod.	12.0 to 14.0	1.75 to 2.5 (o'b not moved, drilled through with B.H.)
Matrix grade, P_2O_5 %	5.5 to 8.2	13.7 to 22.8
Tons prod. per acre-foot mined	200 to 400	600 to 1,000
Pebble product	5% of matrix, 18.3 P_2O_5 %. Upgrade?	Not present.
Flotation product, P_2O_5 %	29.2 to 31.1 req. Dolomite removal	30.6 to 31.1 Single-stage flotation possible.
Product CaO:P_2O_5	1.55 to 1.57	1.50 to 1.55
Combined iron and aluminum oxide	2.0 to 3.8%	0.9 to 1.1%
Product MgO	0.7 to 1.3%	0.65 to 0.85%
Product organic carbon	Low	Up to 2.0%
Minor element ratio	0.10 to 0.13	0.055 to 0.065
Tons waste clay per ton product	2.0 to 2.5	0.4 to 0.6
Relcamation required	Total surface of mined area needs to be reclaimed	No surface disturb.; waste clay and sand injected back into mined out cavity

Metric equivalents: ton $\times$ 0.907 = tonne; acre-foot $\times$ 1233 = cubic foot.

resource was in gamma logging water wells (White, 1984). The Jacksonville Basin exploration work found that the initial dolomite section overlying the D zone phosphate rock had a very strong gamma-ray kick. The phosphate rock zones below this cap closely reflected the concentration of francolite pellets present by the magnitude of the gamma count. The dolomite always found under the phosphate sections also had a sharp kick at the top and bottom. Once below the Miocene dolomite into the Eocene limestone of the Floridian acquifer, the radiation was very minimal.

The landside drilling of these deeper deposits necessitated more careful monitoring of the drilling progress by a competent geologist. The higher costs of drilling these deeper holes mandated that the amount of information collected be maximized. If the depth precluded the possibility of open pit mining, the overlying sections have to be carefully logged to determine their future impact on alternative mining methods such as borehole mining. The individual phosphate sections need to be evaluated to determine capping stability, barren clay sections, dolomite stringers, and the nature of formations separating individual phosphate sections. In all cases, completed drill holes should be gamma-logged to coordinate depths and define the interfaces of the marker horizons.

The processing of cores obtained in deeper Miocene deposits is basically similar to that practiced in the Central Florida Hawthorn extension. The geologist should carefully log the lithologic features and define the presence of barren clay stringers or beds. If borehole mining proves to be the only applicable mining method, the nature of the clay zones could have a profound effect on the mining recovery and on the eventual surface processing facilities.

For example, in open pit mining of phosphate rock with a dragline essentially all of the matrix section is mined and slurried for transport to the washer/concentrator. All the clay contained in the mined section is eventually broken up and subsequently sent to a clay (slimes) storage area at a very low percent solids. The results would be somewhat similar in dredging but possibly to a somewhat lesser degree. The effect of borehole mining would be much different as the clay sections are much harder to slurry than the higher grade phosphate sections. A fairly high percentage of this resistive clay will not be slurried but will remain in the cavity as large lumps. The barren clay lumps that do report to the surface could easily be screened out and separated at the mining site. The borehole test programs in both North Car-

olina (Morrow, 1980) and north Florida reported less clay in the slurry than was indicated from the analysis of the drill cores. In the most recent tests at least one-third of the total clay present did not slurry and remained as coarse clay lumps (Dibble, 1985).

Since the deeper Miocene deposits are essentially primary in origin and have not been reworked, they are relatively consistent over a fairly large area. The drilling density of the existing North Carolina deposit does not need to exceed one hole per every 60 acres. The lack of reworking in the Jacksonville Basin, with the indicated consistency over large areas, should not require a density of more than one hole per 80 acres.

The offshore Miocene deposits will be much more expensive to prospect than the onshore occurrences. Seismic work conducted in Onslow Bay, NC (Riggs, 1987), indicated that this tool can be used to broadly define the presence of Miocene age formations at depth. To date there has not been any deep drill holes in that area to provide correlation. Shallow vibra-core holes drilled in this area indicated the possible top of phosphate-containing Miocene beds. The phosphate materials found in these samples had been diluted by biogenic debris and rock fragments resulting from ongoing erosion typical of that environment.

The ability to drill deep core holes in these basins will require a stable drilling platform. The high cost of conducting such a program eliminates most private companies from sponsoring such a project. Additionally, the lack of assurances to the potential sponsor of eventual ownership of the mineral found makes such exploration efforts difficult to rationalize. The development of these offshore resources will not be done in the near future unless such exploration efforts are funded by federal or state agencies.

Active efforts are needed to develop borehole mining technology and define these large offshore phosphorites. Without this effort, the southern extension of the Hawthorn may be the only future source of domestic phosphate rock developed. Unfortunately, it is questionable whether this higher cost rock will be able to compete economically with the abundant sedimentary phosphate rock from Morocco (Stowasser, 1985). The depletion of the Central Florida deposits could turn the United States into a net importer of phosphate rock to meet domestic requirements while the enormous, deep, potentially lower cost Miocene resources remained relatively undeveloped.

BIBLIOGRAPHIC REFERENCES

Breza, H.M., 1987, Private Communication.

Brown, C.K., 1979, Private Communication.

Cathcart, J.B., "Sedimentary Phosphate Deposits Of The World—Present Status And Outlook For Future," US Geological Survey, Denver, CO.

Davis, B.G., Llewellyn, T.O., and Sullivan, G.V., 1982, "Beneficiation Of A Phosphate Ore Produced By Borehole Mining," RI 8681, US Bureau of Mines, Tuscaloosa, AL.

Dibble, M.F., and Long, H., 1985, "Agrico Mining Company North Florida Project Borehole Mining Report," presented to State of Florida Dept. of Environmental Regulation, Northeast District, Jacksonville, FL, November.

Dibble, M.F., 1987, "Borehole Slurry Extraction Of Phosphate," Engineering Foundation Conference, In Situ Recovery of Minerals, Santa Barbara, CA, October.

Hird, J.M., 1988, Private Communication.

Lawver, J.E., et al., 1984, "New Techniques In Beneficiation of The Florida Phosphates Of The Future," *Minerals & Metallurgical Processing, Trans. SME-AIME,* Vol. 276, pp. 89-106.

McClellan, G.H., "Quality Factors Of Phosphate Raw Materials," International Fertilizer Development Center, Muscle Shoals, AL.

McKelvey, V.E., 1985, "New Technology And Exploration May Much Extend The Life Of The Southeastern Phosphate Industry," Open-File Report 85-580, US Geological Survey, St. Cloud, FL.

Morrow, D.W. 1980, Private Communication.

Riggs, S.R., 1978, "Phosphate Sedimentation In Florida—A Model Phosphogenic System," East Carolina University, Greenville, NC, March 21.

Riggs, S.R., 1986, "Future of U.S. Phosphate Resources: A New Perspective," *Economics of Internationally Traded Minerals,* Society of Mining Engineers, Littleton, CO.

Savanick, G.A., 1985, "Borehole Mining Of Deep Phosphate Ore In St. Johns County Florida," *Transactions,* Society of Mining Engineers, Vol. 278, pp. 144-148.

Scott, T., 1985, "The Geology Of The Florida Peninsula And Its Relationship To Economic Phosphate Deposits," International Geological Correlation Program, May.

Stowasser, W.F., 1988, "Mineral Survey—Phosphate Rock," US Bureau of Mines, June.

Stowasser, W.F., 1985, "The Outlook For The United States Phosphate Rock Industry And Its Place In The World," preprint 85-116, Society of Mining Engineers, AIME, Annual Meeting, Feb. 24-28, New York, NY.

White, B., 1984, Private Communication.

Woolsey, J.R., 1977, "Neogene Stratigraphy Of The Georgia Coast And Inner Continental Shelf," Doctoral Dissertation, University of Georgia, Athens, GA.

Woolsey, J.R., and Dibble, M.F., 1987, "Systems Applicable To Phosphorite Mining On Continental Shelves," Offshore Technical Conference, April, 27-30, Houston, TX.

21. Pyrophyllite*

Most technical and statistical data published on pyrophyllite relating to production figures, uses, markets and sales, have in the past traditionally linked the mineral with the talc and soapstone. This is due to several common superficial physical properties resulting in substitute uses and applications. However, these minerals are not similar in chemical composition and do not generally occur in a similar geological environment.

Pyrophyllite now stands on its own in the marketplace—particularly in ceramic applications—due to the combination of excellent heat shock and creep resistance properties of the mineral. The increase in pyrophyllite use in the field of ceramics indicates the continuing realization of these marked unique properties. Several new fields of ceramic applications, previously the preserve of high-alumina materials, have been developed on the intrinsic properties of pyrophyllite.

Pyrophyllite is an aluminum silicate with the molecular formula of $Al_2O_34SiO_2H_2O$. It would be more appropriate to group pyrophyllite with such materials as kyanite, diaspore, andalusite, and certain high-alumina clays in relation to its general applications (Stuckey, 1928). These materials all have good heat shock resistance properties at high temperatures and are widely used for the manufacture of refractories. For those reasons, they are interchangeable in some applications. Pyrophyllite also has many uses in nonceramic applications, the most important to date being as a substitute for talc in fillers.

The major pyrophyllite mining operations are located in the US, Japan, Korea, Canada, and Australia, where the mineral is widely used as a raw material for general ceramic, mineral filler, and dilutant applications.

COMMERCIAL APPLICATIONS

Pyrophyllite was first used in the last century for lining stoves and fireplaces, and also in the carving and polishing of ornamental objects. The rise in industrial demand for new products and applications resulted in the introduction of pyrophyllite as an alternative, mainly for talc and soapstone. Industrial filler applications followed for paint, paper, cosmetics, rubber, pesticides, crayons and pencils, and bleaching soap.

The first use of pyrophyllite in ceramic applications was prior to World War II in the manufacture of ceramic cladding tiles in the US. This was followed by the production of kiln furniture refractories and whiteware when pyrophyllite was offered by suppliers in North Carolina as an alternative to some local clays.

The largest tonnages of pyrophyllite used today are in refractory linings in steel ladles in Japan. The consumption of pyrophyllite in Japan is second only to clay and chamotte among refractory raw materials used. Recently this technology as been adopted in Australia. Japanese and Australian practice has resulted in appreciable technical and economic benefits, particularly when pyrophyllite is used in combination with zircon. The potential growth for this application is substantial if US and European steelmakers follow.

Investigation and research into the properties of pyrophyllite determined the following main advantages of pyrophyllite-based ceramics: high corrosion resistance for molten iron, steel, and slags; good heat shock resistance; good indices of deformation under load and hot creep resistance (refractories); and increase in mechanical strength of whiteware products promoted by a better distribution of mullite in finished products (also obtained at lower firing temperatures than normally required in production of triaxial ceramics). White cement clinker formulation based on pyrophyllite generally has an advantage of a low iron content when compared with batches made up from silica and clay mixtures. A lower temperature of clinker sintering of batches is possible due to the higher reactivity of the pyrophyllite. White road aggregate made from calcined pyrophyllite rock exhibits high polished stone values (PSV), high aggregate impact values (AIV), and high luminance factors (LF).

These intrinsic properties of pyrophyllite disclosed potential ceramics uses which subsequently have been commercially exploited in industrial countries around the world in contact refractories for lining steel ladles; kiln furniture for tunnel kilns; refractory mortars; ingredients for vitreous china, stoneware, chinaware and wall tile bodies; manufacture of white cement; calcined pyrophyllite for use in white road-making aggregate; and refractories for use in furnaces, melting ceramic and enamel fritts.

GEOLOGY

Mineralogy

Pyrophyllite is an hydrated aluminosilicate which has a phyllosicate sheet or layer structure . This structure consists of two (Si_2O_5) nets between which are interposed a layer of groups of octahedral aluminum cations surrounded by oxygen and hydroxyl anions. The crystalline system of pyrophyllite is monoclinic.

The structural formula of pyrophyllite, $Si_4O_{10}Al_2(OH)_2$, also could be presented in the oxide form $4SiO_2 \cdot Al_2O_3H_2O$. The centidecimal theoretical pure composition of pyrophyllite is SiO_2, 66.7%; Al_2O_3, 28.3%; and H_2O, 5%.

Pyrophyllite ore bodies often contain the following characteristic mineral assemblage: diaspore, corundum, pyrophyllite, kaolinite, alunite, silica (quartz), sericite, and montmorillonite (Iwao, 1953). A diagrammatic representation of Iwao's concept of mineral segregation within hydrothermal deposits derived from volcanic rocks in Japan is given under "Classification of Deposits." The mineral assemblage at Pambula, NSW, Australia is somewhat similar: diaspore, chlorite-cookeite, pyrophyllite, kaolinite, chalcedony (quartz), and sericite.

Diaspore occurs commonly as a minor constituent in most pyrophyllite deposits. Less than 20% of the samples contain diaspore, with minor amounts of kaolinite in the massive pyrophyllite zones.

Chlorite-cookeite occurs near the central shear zone of the ore body where quartz and sericite are virtually absent. Appreciable quantities of chlorite and, to a lesser extent, diaspore are associated with the pyrophyllite. The chlorite has a dioctahedral X-ray diffraction pattern and is characterized by abundant aluminum and a small significant lithium content, and has been identified as the mineral cookeite ($Al_{3.93}$ $Fe_{0.02}Mg_{0.02}Li_{0.81}Al_{0.74}Si_{3.26}O_{10}(OH)_8$).

*Excerpt from Cornish, B.E., 1983, "Pyrophyllite," *Industrial Minerals and Rocks,* 5th ed., Vol. 2, S.J. Lefond, ed., AIME, New York, pp. 1085-1108.

Pyrophyllite occurs as a solid mass varying in color from green to yellow to white. It has physical properties similar to talc. The mineral has a low hardness and a good mechanical strength.

Kaolinite often occurs in associated shear zones and is probably developed during a retrogressive stage, possibly by the silicification of diaspore but more likely from cookeite through the loss of lithium.

Chalcedony appears throughout pyrophyllite ore bodies as in Australian and Korean deposits, with the exception of the pure pyrophyllite lenses ranging in size from large inclusions of quartz to small grains.

Sericite occurs in solid lensoid masses similar to pyrophyllite and can generally be separated by visual inspection. The percentage of alkalis in sericite lenses is in the range of 0.05 to 5%. Sericite is primarily a potash alkali.

Classification of Deposits

A twofold classification of pyrophyllite deposits is evident based on geological mode of occurrence.

Hydrothermal Deposits: These are the most common and are associated with acid-volcanic complexes. Hydrothermal solutions originating within the complex passed along channelways formed by faults or shears and transformed the country rock from rhyolite or feldspar-quartz to pyrophyllite-quartz. Hydrothermal alteration is indicated by dilution of alkali and iron oxide values and abnormally high lithium oxide values. The Pambula pyrophyllite deposit near Eden, NSW, is such an example.

Figs. 1 and 2 show diagrammatic representation of Iwao's (1953) concept of mineral segregation within hydrothermal deposits derived from hydrothermal rocks in Japan showing a hypothetical erosion surface and subsequent surface-exposed mineralogical relationship.

Metamorphic Deposits: These are less common and are found to be associated with metamorphosed volcanic ashes. Pyrophyllite-schist is developed and typically grades irregularly into other schistose rocks. In this case dilution of K_2O values is difficult to explain in terms of isochemical metamorphism. The recording of pyrophyllite-schist near Peak Hill, NSW, is an example of this metamorphic origin.

Principal Producing Countries

Japan: Pyrophyllite was first discovered in Japan in 1797 on Mount Omotoyama, at the site of the Mitsuishi mine of the Shinagawa Mining Co. In the early 1800s, sawn blocks of pyrophyllite were used for carving images and in the manufacture of slate pencils, signature seals, and firebricks.

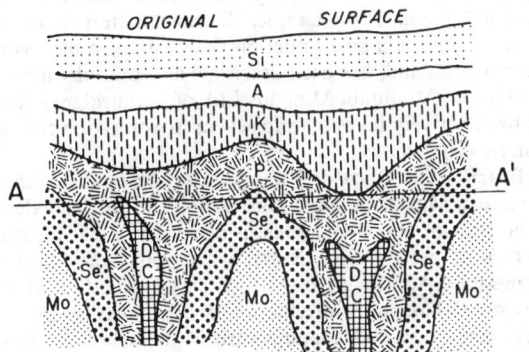

FIG. 1 Diagrammatic representation of Iwao's concept of mineral segregation within hydrothermal deposits derived from volcanic rocks in Japan. Source: Modified from Iwao (1953).

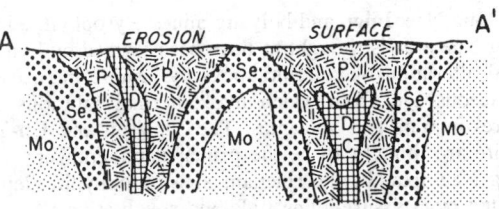

FIG. 2 Diagrammatic representation of Iwao's concept of mineral segregation. D = diaspore, C = corundum, P = pyrophyllite, K = kaolin, A = alunite, Si = silica, Se = sericite, Mo = montmorillonite. A-A' is a hypothetical erosion surface.

In 1885, deposits in the Mitsuishi district were surveyed by the Imperial Geological Survey, and this work established the existence of abundant reserves of usable quality ore.

The most important pyrophyllite production is in Okayama. This prefecture is near the south coast of southwest Honshu Island, approximately 145 km (90 miles) west of Osaka. Fifty-seven of the 70 mines in Japan are known to have produced various quantities of pyrophyllite. Of these 57 principal mines, 27 are located in Okayama prefecture.

The largest production from Okayama has been from the Shinagawa-Mitsuishi Ohira, Kawabe, and Kato mines. Significant production has also been recorded from the Shinagawa-Makata, Fukuyama, and Kiyotaki mines in Hyogo prefecture. The Shokozan mine in Hiroshima prefecture has also produced significant quantities of pyrophyllite. In Hiroshima and Hyogo prefectures, massive pyrophyllite deposits in rhyolite and porphyrite include the accessory minerals diaspore, kaolinite, corundum, pyrite, and alunite.

The second important pyrophyllite-producing area in Japan is in Kanzaki-gun, Hyogo prefecture. Six mines are known in this area, which is located 113 km (70 miles) northeast of Okayama.

Three important mines also exist in the Kinosaki-gun, Hyogo prefecture. These mines occur between the Okayama and Kanzaki districts, in which the largest reserves of pyrophyllite in the world occur.

The Japanese pyrophyllite deposits were formed by the hydrothermal alteration of porphyry and liparite. In Okayama prefecture, pyrophyllite occurs as veins and irregular masses within sedimentary and volcanic rocks. Kaolin accompanies pyrophyllite within the upper parts of the ore bodies, with nodules of diaspore occurring in the central part of the ore mass, and pyrite and marcasite in the lower parts. The deposits were probably formed by the replacement of andesites, shales, agglomerates, rhyolites, and porphyries which were permeated by hot hydrothermal solutions associated with volcanism.

South Korea: Pyrophyllite is mined in two major areas of South Korea, both near the coast in the southern part of the peninsula. The largest production is from the Wan-Do mine at Haenam. Major mines in the Haenam area are the Sungsan, Hwansan, Okmesan, Dae-Do, and Chin-Do mines. The deposits mined are tabular, and are associated in places with diaspore.

The second major pyrophyllite-producing area in South Korea is at Dong Nae where the Nilyang, Yangsan, Kimhae, Pusan, and Kyong-nam mines are worked. The deposits in the Dong Nae area were formed by the alteration of intermediate volcanic rocks including andesite, trachyte, trachytic andesite, and quartz porphyry. The host rocks of the Pusan and Kyongnam mines consist of quartz-porphyry and feldspar porphyry, while trachytic andesite is the host rock in the Dong Nae, Imgi, and Nilyang mines. Pyrophyllite in the Dong Nae area is thought to have been formed by hydrothermal alteration of volcanic rocks by granite intrusives of Cretaceous age, from which hydrothermal solutions filled fissures in tuffs and volcanics. The Yangsan and Un-Yang granite are thought to be the principal mineralizers in the Dong Nae area. In the Haenam area pyrophyllite deposits resulted from fissure-filling tuffs and rhyolites.

The major reserves of pyrophyllite in South Korea are primarily located at Wan-Do, where inferred ore reserves total approximately 30 Mt.

United States: The production of pyrophyllite in the US is limited to the states of North Carolina and California. Annual US production is approximately 100 kt, and reserves of various grades of pyrophyllite are estimated at 1.5 Mt.

Principal production in the US is in Moore and Randolph Counties in North Carolina. Other occurrences in North Carolina are in Granville, Orange, Alamance, and Montgomery Counties. The North Carolina belt actually starts near the Virginia line and extends across the state through South Carolina to Graves Mountain, Georgia. Other occurrences of pyrophyllite in the US have been reported in the South Mountain area of Pennsylvania and in California.

Pyrophyllite occurs in irregular, lenticular, or bedded deposits. The pyrophyllite deposits of North Carolina occur within rhyolite volcanic rocks and form oval or lenticular bodies up to 152 m (500 ft) wide and several hundred meters (feet) long, located within fracture zones. According to Stuckey (1928) the North Carolina deposits were formed through metasomatic replacement of Precambrian acid tuffs and breccias of both dacitic and rhyolitic composition. Pyrophyllite occurs with quartz and sericite, forming the central part of these lenses and grading outward into impure zones. The bulk of production comes from a few large deposits mined by open-cut and underground methods and used in ceramic filler dilutant and carrier applications.

The other district in the US where pyrophyllite has been produced commercially is in California. Sericitic pyrophyllite is mined in Inyo and San Diego Countries for ceramic applications. The San Diego County pyrophyllite deposit was described by Jahns and Lance (1950). An important deposit called the Pioneer mine occurs in the San Dieguito area northeast of La Jolla and 12 km (7½) miles from Escondido, the shipping point by rail. The deposits occur in the Santiago Peak volcanic series, which comprises miscellaneous rocks of Jurassic age. The pyrophyllite deposits, which occur within an area of about 2.6 km² (1 sq mile), represent progressive stages in the alteration of the volcanic rocks. Nearly all the pyrophyllite mined has been obtained from open pits in a single mass of high-grade schist 46 m (150 ft) long and 4.6 m (15 ft) wide.

Australia: The main pyrophyllite deposit in Australia is located at Pambula near the deep water port of Eden on the southeast coast of New South Wales. An intrusive and extrusive volcanic assemblage formed in an active tectonic environment during the Upper Devonian period provided the basis for extensive hydrothermal activity and burial metamorphism. This hydrothermal activity associated with the volcanism formed irregular and lenticular mineralization, rich in aluminous phases, by the remobilization and the dissipation of the alkali components of the host rock (Taylor, 1978). Suitability of this host rock and structural competence of the volcanic sheet control the geometry and distribution of the deposits, with optimal conditions favoring large, strongly cleaved, broadly altered lensoid bodies.

It is probable that the mineralization at Pambula origi-

nated by a similar process to the zonal distribution of mineralization in the Japanese deposits as explained by Iwao (1953). However, the segregation process of diaspore-pyrophyllite-kaolinite-alunite-silica was achieved through the alteration of lavas and pyroclastic rocks at shallow depth by ascending acid solutions of volcanic and subvolcanic derivation. During ascent the temperature decreased from 600° to 150°C and the solution became progressively more acidic, selectively mobilizing the alkalis which hydrated and laterally giving rise to zones enriched in sericite and montmorillonite. Simultaneously the residue (mainly alumina and silica) reacted with the components of the solution.

Recoverable reserves of all grades of pyrophyllite total approximately 10 Mt, with an estimated additional 20 Mt in place.

Canada: The only current commercial source of pyrophyllite in Canada is on the Avalon Peninsula of Newfoundland, near the small town of Toxtrap. Pyrophyllite was first extracted in the area in 1904 when a shipment of 1750 tons was sent to the US. Thereafter production continued intermittently, the most consistent period being 1938-1948 when Industrial Minerals Co. of Newfoundland mined and milled 6500 tons. However, Newfoundland Minerals Ltd., a wholly owned subsidiary of American Olean Tile Co. Inc. of the US, acquired the operation in 1956. Since then production has been continuous.

The pyrophyllite occurs in a narrow belt running south of Conception Bay in eastern Newfoundland and along the east side of the Precambrian Holyrood granite batholith. It is thought that hydrothermal solutions, moving up via faults and shear zones from the granite, may have recrystallized rhyolitic rocks into the fine-grained assemblage of quartz, sericite, and pyrophyllite. Rhyolitic rocks remote from the granite showed no signs of pyrophyllite mineralization.

A future Canadian source of pyrophyllite may well be the Senneterre deposit located in Carpentier Township, Abitibi region, Quebec, which has been actively prospected since the 1930s. It was not until 1962, however, that a mapping project conducted by the Quebec Dept. of Natural Resources identified the rock in the shear zone as pyrophyllite. Domtar Ltd. optioned the property between 1964 and 1974. Since then it has been in the hands of Descarreaux & Mousseau Tremblay Inc. of Quebec. The property, which is within the mineral-rich Precambrian Shield area of Canada, has pyrophyllite mineralization in a series of zones associated with sheared metamorphic volcanics and pyroclastics ranging in composition from andesites to rhyolites. Again, a metasomatic origin has been assigned to the deposit.

Some occurrences of pyrophyllite have been reported in western Canada. For example, the northeastern portion of Kyuquot Sound, a small inlet on the west coast of northern Vancouver Island, has been described as a pyrophyllite area. In addition, Mountain Minerals Ltd. of Lethbridge, Alberta, has been actively investigating several pyrophyllite claims in western Canada.

Brazil: The Onca de Pitangui and Pitangui deposits in the state of Minas Gerais and the Santano de Parnaiba deposit in the state of São Paulo contain more than 1 Mt measured reserves of pyrophyllite. Brazil's vast refractory producer, Magnesita SA, has an output of about 60 kt/a at Belo Horizonte. The entire production is consumed by the company, largely for the manufacture of ladle linings. Finapa Assessoria Commercial Industrial Ltda. of São Paulo claims to have a 100 kt/a capacity pyrophyllite mine which supplies the ceramics industry. Cia. Minera Agregados Calcareous SA of Lima, Peru, has been mining and processing a mineral similar in chemical composition and uses to pyrophyllite, although it is mineralogically different. The company has also isolated a large deposit of pyrophyllite—the only one on the Pacific coast of South America.

Minor Producing Countries

India: Three deposits at Madhya, Aradesh, Rajasthan, and Uttar Pradesh produce a total of approximately 15 kt of pyrophyllite annually.

Argentina: Argentinan pyrophyllite is mined principally from the Juanita del Puerto mine in La Rioja province and is used in the manufacture of electrical porcelain and vitreous enamel. It produces less than 10 kt/a.

South Africa: Pyrophyllite, locally referred to as *wonderstone,* is mined near Ottosdal in the Lichtenburg district. The deposit is of the hydrothermal type and crops out in a series of elongate lenses associated with two lineaments which transgress rhyolite and acid types of the Dominion Reef series of the Witwatersrand System. The material mined in fact is a metasedimentary rock made up of pyrophyllite (89%), chloritoid, epidote, and rutile. The product, with an average chemical composition of 55.5% SiO_2, 35.24% Al_2O_3, 0.31% MgO, 2.8% TiO_2, and 6.15% loss on ignition, is different from a standard pyrophyllite in having a Mohs' hardness of 8. Naturally, the end uses are therefore different, the main ones being bushes for submerged pumps, inserts for sluiceway nozzles, spigots for coal washing plants, and abrasion- and acid-resistant linings for pipes and coal chutes.

Swaziland has been producing small tonnages of pyrophyllite for some years now, particularly from the Usetu area between the Mkhondvo and Great Usutu Rivers. However, production is only on the order of a few hundred tons a year.

Thailand: Minor production of pyrophyllite occurs in central Thailand for general ceramic uses in Southeast Asia. The mining capacity of the Thailand operation is on the order of 12 kt/a. Mineralogically the Thailand material is a mixture of pyrophyllite, kaolinite, and diaspore according to X-ray diffraction analysis.

Turkey: Turkey has been a commercial producer of pyrophyllite since 1976 when Endustrie Mineralleri ve Taslari Sanayi (Industrial Minerals of Turkey Ltd.) started up its production in the Malatya Province of eastern Turkey. Approximately 6 Mt of pyrophyllite are contained in hydrothermally altered dacite tuffs found in lenses scattered over a wide area. Three main grades are mined: a green colored massive block variety; a white soft clay material; and a sericite-contaminated dark green pyrophyllite.

Current production is around 3 kt/a. The entire production is currently being consumed by Canakkale Ceramic Refractories, part owners of Industrial Minerals of Turkey, in its production of wall tiles and cordierite products, and as a natural grog for kiln furniture such as setters, slabs, pins, and posts. There are plans to produce pyrophyllite for the production of pyrophyllite/zircon ladle lining bricks. Later the market for pyrophyllite as a paper filler will be examined.

EXPLORATION

Exploration programs for pyrophyllite are usually conducted in areas where geological evidence indicates that underlying rocks belong to acid-volcanic complexes. Pyrophyllite is found in hydrothermal alteration zones associated with faults or shears which transect these felsic-igneous rocks.

Field Techniques

Aerial photographs are most useful in reducing the area to be explored to a manageable size. This is achieved by plotting on the photographs linear features which may represent faults or shear zones. The geologist then examines each of these lineaments on the ground and undertakes reconnaissance work along favorable zones.

Pyrophyllite is generally identified in the field by its translucent green color and waxy feel. In addition, outcrops of chalcedonic pyrophyllite tend to be bolder than the enclosing rocks and display a characteristic weathering pattern, whereas outcrops of chloritic pyrophyllite are typically subdued and feature jagged boulders.

When a discovery has been made, drilling or trenching is performed to provide fresh samples for preliminary laboratory evaluation.

Geochemical

The pattern of alteration in pyrophyllite deposits is usually zoned, with a central core of chloritic pyrophyllite surrounded by imperfectly segregated zones of chalcedonic pyrophyllite and sericitic pyrophyllite which grade into unaltered host rocks. Geochemical techniques have been found to be successful in delineating the zonal pattern and defining the margins of the ore body.

Surface samples could be collected on a 10-m^2 grid basis. In areas of good outcrop conditions, clean representative samples can be obtained without difficulty. However, in areas of poor outcrop conditions shallow drilling or trenching may be necessary.

Contoured plots of alumina and total-alkali values are prepared and superimposed on a geological plan. The results not only define the zonal pattern but also provide a good basis for designing detailed drilling programs. As an example, in the evaluation of the Pambula deposit 3000 samples, both surface and subsurface, were analyzed for potash alkalis and alumina values. These were used to prepare histogram plots and cumulative frequency distribution curves. These plots confirmed the bimodal distribution of both Al_2O_3 and K_2O populations, probably representing both a primary and a secondary alteration phase and two periods of mineralization. The primary mode has a definite positive skew with the greater proportion of the mineralization showing alkali values at less than 0.5% K_2O. The mean values of K_2O and Al_2O_3 suggest a composition for the pyrophyllite mineralization occurring during the primary alteration phase of 12.7% Al_2O_3 and 0.57% K_2O.

Drilling

Two kinds of drilling operations generally apply to pyrophyllite exploitation, diamond drilling and percussion drilling.

Diamond drilling provides core samples to enable the geologist to predict the subsurface behavior of the ore body and to find and correlate the zonal pattern determined from surface observations. Diamond drill holes are usually inclined at low angles and designed to penetrate the entire width of the steep dipping ore bodies.

Percussion drilling is carried out as a fore-runner to production operations for detailed grade evaluation. Vertical holes are usually drilled on a square grid spacing of between 1 and 3 m and chip samples are collected at 1- or 2-m intervals to depths of up to 10 m. Compressed air raises drill chips to the surface and into a cyclone where they are removed from the airstream and discharged into a sample collecting tube.

EVALUATION OF DEPOSITS

Specifications

A variety of grades is available from each deposit. Grade classification takes account of market requirements on the one hand and mineralogy of the ore on the other. Ore is graded according to its pyrophyllite mineral content and the nature of abundance of other minerals in the ore. Mineralogy is determined indirectly by chemical analysis, principally for Al_2O_3, K_2O, and Fe_2O_3 contents.

Evaluation

Grades and reserves are calculated by conventional geological techniques and the results used to optimize quarry design.

BIBLIOGRAPHY

Bowman, R., and Tauber, E., 1979, "Potential Applications of Two High Grade Kaolinitic Clays and a Pyrophyllite Deposit," *Journal Australian Ceramic Society,* Vol. 15, No. 1, pp. 5-8.

Chappell, F., 1960, "Pyrophyllite," *Industrial Minerals and Rocks,* 3rd ed., J.L. Gillson, ed., AIME, NY, pp. 681-686.

Cornish, B.E., 1980, "Australian Pyrophyllite and Its Growing Influence in the World Markets," 4th Industrial Minerals International Congress, Atlanta, Georgia, pp. 179-184.

Cornish, B.E., Tauber, E., and Nichol, D., 1980, "Pambula Pyrophyllite: Production and Applications," *Proceedings,* 9th Australian Ceramic Conference, pp. 132-134.

Fergusson, C.L., et al., 1979, "The Late Devonian Boyd Volcanic Complex, Eden, N. S. W.: A Reinterpretation." *Journal Geological Society Australia,* Vol. 26, pp. 87-105.

Harben, P., 1978, "Pyrophyllite Stands on Its Own," *Industrial Minerals,* No. 129, June, pp. 19-25.

Iwao, S., 1953, "Ceramic Minerals of Japan," Society Mineral Geology, Japan, Vol. 1, No. 2, pp. 105-124.

Jahns, R.H., and Lance, J.F., 1950, "Geology of the San Dieguito Pyrophyllite Area, San Diego County, Calif.," Special Report 4, California Div. Mines, 32 pp.

Jolly, W.T., and Smith, R.E., 1971, "Degradation and Metamorphic Differentiation of the Keweenawan Tholeiitic Lavas of Northern Michigan, U.S.A.," *Journal of Petrology,* Vol. 13, Part 2, pp. 273-309.

Lipman, P.W., et al., 1972, "Cenozoic Volcanism and Plate Tectonic Evolution of the Western United States," Phil. Trans. Royal Society, London, A., pp. 217-248, 271.

Loughnan, F.C., and Steggels, K.R., 1976, "Cookeite and Diaspore in the Back Creek Pyrophyllite Deposit near Pambula, N.S.W." *Mineralogical Magazine,* Vol. 40, pp. 765-772.

Parsons, W.H., 1969, "Criteria for the Recognition of Volcanic Breccias: Review," *Memoir 115,* L. Larson et al., ed., Geological Society of America, pp. 263-304.

Pauline, M., and Jones, R., 1979, "Zircon-Pyrophyllite—An Economical Steel Ladle Refractory," AIME Conference, Detroit, March.

Phillips, B.L., 1976, "Synthetic Aggregates for Road Surfaces," *Proceedings,* 8th ARRB Conference, Vol. 8, No. 1, p. 1.

Smith, R.E., 1968, "Redistribution of the Major Elements in the Alteration of Some Basic Lavas During Burial Metamorphism," *Journal of Petrology,* Vol. 9, Part 2, pp. 195-219.

Stuckey, J.L., 1928, "The Pyrophyllite Deposits of North Carolina," North Carolina Department of Conservation and Development Bulletin, Vol. 37, 62 pp.

Tauber, E., et al., 1973, "Stoneware Bodies Based on Pyrophyllite," *Journal Australian Ceramic Society,* Vol. 9, No. 2, pp. 47-51.

Tauber, E., et al., 1974, "Earthenware Tile Body Based on Pyrophyllite and Orthophosphoric Acid," *Journal Australian Ceramic Society,* Vol. 19, pp. 46-49.

Tauber, E., et al., 1976a, "Development of a China (Porcelain) Body from Australian Minerals," *Interceram,* Vol. 26, p. 211.

Tauber, E., et al., 1976b, "Replacement of Silica by Pyrophyllite in Vitreous China Products," *Interceram,* Vol. 25, p. 195.

Tauber, E., and Pepplinkhouse, H.J., 1972, "Ceramic Properties of Pyrophyllite from Pambula, N.S.W.," *Journal Australian Ceramic Society,* Vol. 8, No. 3, pp. 62-64.

Taylor, J.R.P., 1978, "Pambula Eden Regional Geology," Unpub. Thesis, Monash University, Melbourne, Australia.

Wall, V.J., 1976, "Gold and Pyrophyllite Mineralization in the Devonian Acid Volcanics of the Yalwal-Eden Belt," *Bulletin,* Australian Society Exploration Geophysics, Vol. 7, No. 1.

Wall, V.J., and Kesson, S., 1969, "Pyrophyllite-Bearing Rocks in a Regionally Altered Volcanic Sequence," *Abstracts,* Part 7, Geological Society Australia, 232 pp.

22. Quartz/Silica Sand

MARK J. ZDUNCZYK

INTRODUCTION

Commodity Description

Industrial grade silica is usually sand and is essentially pure quartz (SiO_2).

Location

Deposits of industrial silica sand are relatively common and occur in many geographic regions. In the United States, these deposits are mined in the following states: Arkansas, Alabama, California, Colorado, Connecticut, Florida, Georgia, Idaho, Illinois, Indiana, Louisiana, Michigan, Massachusetts, Minnesota, Missouri, Nevada, New Jersey, New York, North Carolina, South Dakota, Tennessee, Texas, Virginia, Washington, West Virginia, and Wisconsin.

Products

The main silica products are glass and foundry sand. Other products include filtration sand, blast sand, proppant sand, traction (engine) sand, as well as other specialty sand products. The chemical and physical characteristics of the two main products follow.

For glass or melting sand, the SiO_2 content of the processed product must exceed 99.5%, the Al_2O_3 should consistently be less than 0.30%, and the Fe_2O_3 should not exceed 0.030% by weight. Limitations are also placed on TiO_2, Cr_2O_3, CaO, MgO, ZnO, Na_2O, and K_2O, the amount depending on the glass product which is manufactured. For example, the bottle (glass container) industry can tolerate varying amounts of minor oxides, but float glass (plate glass) requires a more consistent oxide content. To some degree, glass manufacturers can adjust their processing to accommodate sand that does not meet rigorous specifications.

Grain size for glass sand should have less than 1% retained on the 550 μm (30 mesh) sieve and have 8 to 10% retained on the 375 μm (40 mesh) sieve. No more than 5% should pass the 140 mesh sieve.

Heavy minerals, such as ilmenite and leucoxene, are limited by weight and number of particles. These limitations are placed on the material retained on the 550-375 μm (30 to 40 mesh) sieves. Generally, the glass manufacturer provides these specifications.

For foundry sand the SiO_2 content must exceed 98%, but less emphasis is placed on the other oxides except the base oxides, such as CaO and MgO. The greater amount of base oxides a sand contains, the more synthetic binder is necessary to make the mold or cast. The amount of synthetic binder is called acid demand.

Grain size for foundry sand varies depending on the products a foundry makes. Generally, the coarse grain sand is used for large iron and steel castings while a finer grain sand is used for smaller brass, aluminum, and other alloy castings. Grain shape also plays a role in the need for chemical binders. Round grains free of base oxides will have less surface area than angular grains and, therefore, use less binder. Heavy minerals generally have little effect on foundry sands.

Types of Occurrence

Industrial silica is mined from sedimentary sandstones, both loosely and well cemented, metamorphic quartzites and orthoquartzites, igneous pegmatites, and unconsolidated sand deposits derived from the aforementioned rock types. Some quartz-rich kaolin clay deposits and quartzo-feldspathic deposits may occasionally be of economic interest.

MINING METHODS

Silica sand producers surface mine by various methods. Some operators mine unconsolidated sand by means of a front-end loader, diesel shovel, or equivalent. The sand is loaded into trucks which transport the material to a nearby hopper that feeds the processing plant. When the water table is high, dredging may be used. Dredging removes sand below the water table usually with a 20- to 30-cm (8- to 12-in.) suction pipe. Sometimes "cutterheads" are used in conjunction with the suction pipe and break the material free from the bank. The sand is then transported from the dredge in slurry through a pipeline to a stockpile. Hydraulic mining may be used in unconsolidated and loosely consolidated sand. In this method, water is jetted under high pressure against a bank, washing the sand into a sump where the sand is transported through a pipeline to a stockpile and later processed. A dragline clamshell is often used in unconsolidated sediments above and below the water table and in some instances an hydraulic excavator may be used in shallow deposits.

In rock, such as quartzite, well-cemented sandstone, or pegmatite, drill and blast techniques are used. The fragmented rock is mined by shovel or front-end loader and loaded into trucks which haul to a nearby hopper that feeds a crusher. After crushing, it is conveyed to the processing plant. Underground operations are limited to a few cases where the deposit is consolidated and the overburden is too thick to strip economically. The room-and-pillar method of mining is used in these cases.

EXPLORATION

The different geologic occurrences of silica sand lend to different types of exploration techniques. However, a literature search is usually the first step in exploration. Along with this search, defining the market area is very important, with the market a major factor determining whether a site is explored. Most state geological surveys are well aware of their state's silica sand deposits and can be very helpful in providing information. Government agencies are generally less helpful unless a significant study was undertaken in that area. Nevertheless, most of the silica sand deposits in the United States are known and some have been investigated several times. Once a site is selected, the geologist must perform a site reconnaissance and subsequent geologic mapping to determine the number of drill holes and/or test pits locations and patterns.

In unconsolidated deposits such as those similar to New Jersey or South Carolina where the water table is high, using a rotary drill rig with a 7.6-cm (3-in.) OD split spoon sampler is recommended. A split spoon sampler is a pipe which is split down the middle where it can be opened to observe and collect the sample. The spoon is driven into the deposit by a 136-kg (300-lb) weight. A "drag bit" is used to deepen the hole to the desired depth where the next sample is to be taken. Bentonite mud is usually employed during drilling to case the hole and prevent collapse. Mud drilling rigs should be equipped with a large mud pump [10 × 12.7 cm (4 × 5 in.)] 136-kg (300-lb) weight to drive the sampler and an experienced mud driller. Failure to have these essential components may result in poor recovery of samples, untimely delays, and wasted efforts. During this type of drilling, continuous sampling is very important. After logging the samples, they are examined petrographically and splits are sent to analytical laboratories for determination of the critical oxides.

In loosely consolidated or consolidated sandstone, quartzites or pegmatites core drilling is usually required. A drill rig, equipped with NX size wire line (54.9-mm) coring capabilities is generally used. When the rock is loosely cemented, bentonite mud may be used to help in recovery of a full sample. The core is logged by a geologist who is particularly interested in both type and amount of cementation in addition to the degree of iron staining on the quartz grains.

Two types of auger drilling can be used in unconsolidated deposits. Hollow stem augers with 7.6-cm (3-in.) OD split spoon sampler is limited to depth and water table. A split spoon sampler is periodically lowered through the hole in the augers. Solid stem augers have been used, but the samples which flow up the sides of the augers are mixed and not usually representative. This type of drilling is inexpensive, but it is not a recommended method to explore a deposit.

Water jet drilling can drill quickly, but produces a mixed sample that may be difficult to interpret. Test pitting with backhoe or excavator shovel can be useful in shallow deposits. Geophysical techniques, such as seismic or resistivity, are not very useful when quality of the material must be tested.

Percussion air drill rigs, such as those used to drill holes for blasting rock, can be used for quickly determining depth. However, the samples are mixed from various horizons and, therefore, difficult to use to delineate quality.

Rugged, swampy, or mountainous terrain is often encountered in exploration drilling. All terrain drill rigs are available; however, proper drilling and sampling equipment should be placed on the rig.

The number and location of drill holes generally rely on economics. Most geologists and producers would prefer to delineate the deposit as accurately as possible because the final silica product must meet stringent specifications. Therefore, tight delineation of the variations within the deposit is desirable. The geologist, with the information available from drill holes, literature, mapping, and geologic intuition, must decide whether the deposit is worth exploiting.

23. Sand and Gravel

Harold B. Goldman

In commercial usage "sand" applies to rock or mineral fragments ranging in size from three-thousandths of an inch to a quarter of an inch. "Gravel" consists of rock or mineral fragments larger than 6 mm (¼ in.) ranging up to 89 mm (3½ in.) maximum size.

The construction industry consumes 90% of the sand and gravel produced; the remainder is sand used for specialized products such as glass.

UTILIZATION

The building industry uses sand and gravel chiefly as aggregate in portland cement concrete, mortar, and plaster; the paving industry uses sand and gravel in both asphaltic mixtures and portland cement concrete. Aggregate is commonly designated as the inert fragmental material which is bound into a conglomerate mass by a cementing material such as portland cement, asphalt, or gypsum plaster.

Portland cement concrete consists of sand and gravel surrounded and held together by hardened portland cement paste. Concrete mixes commonly contain 15 to 20% water, 7 to 14% cement, and 66 to 78% aggregate. Sand and gravel used as concrete aggregate have to meet many requirements (Goldman and Reining, 1983). Premature deterioration of concrete has been traced in many instances to the use of unsuitable aggregates.

Asphaltic mixtures used predominantly for paving consist of combinations of sand, gravel, and mineral filler [material finer than 0.07 mm (0.003 in.)], uniformly coated and mixed with asphalt produced in the refining of petroleum. Except for the addition of mineral filler, sand and gravel used as asphaltic aggregate must meet the same general physical requirements as materials used for portland cement concrete aggregate.

General Requirements of Aggregates

Construction aggregate has many requirements that are difficult to meet if only unprocessed material from natural deposits is used. Suitable material is composed of clean, uncoated, properly shaped particles which are sound and durable. Soundness and durability are terms used to denote the ability of aggregates to retain a uniform physical and chemical state over a long period of time so as not to cause disruption of the concrete when exposed to weathering and other destructive processes. To have these attributes, individual particles must be tough and firm, possessing the strength to resist stresses and chemical and physical changes, which may cause swelling, cracking, softening, and leaching. The aggregate should not be contaminated by much clayey material, silt, mica, organic matter, chemical salts, and surface coatings.

The quality of aggregate depends upon its physical and chemical properties. These, in turn, may be inherent mineralogical and textural features of the rock or may be the effects of later changes such as tectonic fracturing, mechanical or chemical weathering, or incrustations.

The physical properties most significant with regard to concrete are: (1) abundance and nature of fractures and pores, (2) particle shape and surface texture, and (3) presence of material which may cause volume change. An aggregate is considered to be physically sound if it is adequately strong and capable of resisting the agencies of weathering without disruption or decomposition. Minerals or rock particles that are physically weak, extremely absorptive, and easily cleavable are susceptible to breakdown by weathering. The use of such materials in concrete reduces strength or leads to early deterioration by promoting weak bond between cement and aggregate, or by inducing cracking, spalling, or popouts. Severely weathered, soft, micaceous, or porous materials may cause localized stresses to develop in concrete by swelling and shrinking during wetting and drying or freezing and thawing cycles.

Suitability of the Various Rock Types: Sedimentary rocks show a wide range in physical qualities and suitability. Sandstones and limestones, if hard and dense, are ordinarily satisfactory, but many sandstones are friable and excessively porous and commonly are clay-bearing. Shales generally make poor aggregate material, being soft, light, weak, and absorptive. Most igneous rocks are satisfactory, being normally hard, tough, and dense. Tuffs and certain flow rocks may be extremely porous and have high absorption and low strength. Metamorphic rocks differ in character. Most marbles and quartzites are usually massive, tough, and dense. Gneisses are ordinarily very tough and durable. Some schists contain micaceous minerals which are undesirable because they are soft, laminated, and absorptive. Micaceous minerals are susceptible to weathering along cleavage planes and thereby impair strength and durability. Some schists and slates in particular are thinly laminated and tend to assume flat slabby shapes which lack strength.

Any or all of these rock types may be rendered undesirable because of harmful exterior coatings. Weathering processes, particularly the action of ground waters, deposit these coatings. The most common coatings are calcium carbonate, clay, silt, opal, chalcedony, iron oxide, manganese oxide, and gypsum. Particles with these coatings are generally undesirable as aggregates. The bond between particle and coating may be weak and decrease the strength of the aggregate-cement bond.

Chemical Properties: The chemical properties which may affect service life of concrete are (1) reaction of certain rocks and minerals with high-alkali cement (alkali-aggregate reactivity); (2) leaching of water soluble substances; (3) solution of certain secondary minerals, such as the zeolites, to release sodium and potassium which aids in attacking susceptible aggregate particles; and (4) oxidation by weathering to produce compounds that may retard cement hydration.

Alkali-aggregate reactivity has been discussed at length in many publications. A reactive aggregate is any rock, gravel, or sand that contains one or more constituents that react chemically with the alkalies (sodium and potassium) in some types of portland cement. This reaction causes expansion, cracking, and deterioration of concrete and arises from osmotic pressures produced by the formation and hydration of alkali-silica gels. The gels are formed through interaction between reactive silica in the mineral aggregate and the alkalies which are liberated by the cement during hydration. Opal (amorphous hydrous silica) is the most conspicuous aggregate material reacting in this manner. Other rocks and minerals known to be reactive are: glassy volcanic rocks of medium to high silica content (andesite and rhyolite), chalcedonic rocks, certain phyllites which contain a hydromica, and the minerals tridymite and heulandite. Any rock containing a significant proportion of reactive substances may be deleteriously reactive; thus normally non-reactive sandstone, shale, basalt, granite, and other rock types may be harmful if impregnated or coated with opal, chalcedony, or other reactive substance.

Certain sulfide minerals, such as the iron sulfides, pyrite, and marcasite, oxidize and cause unsightly rust stains or generate acidic compounds injurious to the concrete matrix and cause popouts. Chemical salts such as sulfates, chlorides, carbonates, and phosphates also occur in some aggregates. Some of these substances dissolve easily or react to impede the setting of the cement.

A chemical alteration contributing to physical unsoundness is hydration. Shales, clays, and some rock nodules are examples of materials which expand when they absorb water and shrink as they dry. This absorptive character increases the rock's susceptibility to disruption by weathering.

Most of the foregoing features can be observed in the field, and on the basis of these observations, laboratory test procedures can be set up to further evaluate the properties of the aggregate.

GEOLOGIC OCCURRENCE

In the United States, sand and gravel is obtained commercially from rock units of many types and ages in open pit operations. The principal sources are along existing or ancient river channels and in glaciated areas; marine and lake deposits and older geologic formations are significant but of lesser importance.

Classification of Deposits

Stream Deposits.

Stream Channel Deposits—Stream channel deposits consist of sand and gravel deposited in stream beds along present or former stream courses. Most channel deposits are generally easily accessible and easily mined.

Materials in these deposits are desirable as aggregate for many reasons. The natural abrasive action of stream transport has removed most of the soft weak rocks, leaving only the harder and firmer particles. These latter particles have undergone some degree of rounding and are subrounded to well-rounded, a desirable aspect for use in concrete.

Most channel sand and gravel deposits are replenished by material carried by seasonal floods, except in portions of the streams downstream from dams. Overburden is rarely present, but high flood waters may leave silt, clay, and wood debris covering parts of some channel deposits. The size of gravel gradually decreases downstream in the streams with long reaches. Commercial production is concentrated in the deposits where a proper blend of sand and gravel can be obtained. Fortunately, the favorable portions of many streams occur in flat-lying areas near population centers.

Channel deposits normally are free of excessive amounts of silt and clay and most deposits contain sand and gravel in the size gradations necessary for concrete design. Mining operations are often relatively simple, consisting of no more than washing and screening to obtain suitable aggregate. In spite of these qualities some deposits are unsuitable because they contain harmful ingredients such as physically unsound or chemically reactive rocks. The nature of the material in the stream channel is determined in large part by the nature of the source rocks within its drainage area. Different geological formations in a drainage basin contribute a variety of rock types which show a wide range in chemical composition and degree of weathering.

Flood Plain Deposits—True flood plain deposits consist of material deposited on plains bordering streams by periodic overflow of the streams from their channels. The sediments deposited are normally composed of silt and sand grains. However, fine materials may mantle usable deposits of sand and gravel, particularly in areas where, in the geologic past, the streams were more vigorous and transported greater volumes of coarser material. The sand and gravel in these older, deeper flood plain deposits is similar to that in channel deposits and is suitable for use after overlying flood plain silt layers are removed.

Terrace Deposits—Stream terrace deposits are benchlike deposits of sand and gravel which border a stream but lie above the level of the present flood plain. These deposits are remnants of older flood plains through which the stream has cut. Terrace deposits, because they are above the stream level, may be more desirable than stream channel deposits if water tables are shallow and abundant ground water makes stream channel and flood plain operations difficult. However, there is little possibility of replenishment of terrace deposits. The materials in these deposits have the general properties of stream channel materials, but weathering processes may have diminished the quality of some of their constituents by converting certain minerals to clay. Not all terrace deposits can be used for aggregate. Some deposits are only thin veneers of sand and gravel deposited on a stream-cut bedrock surface. The materials in the deposit may not be suitable if they have undergone post-depositional weathering or if the grains have become coated with harmful substances due to the action of ground waters.

Alluvial Fans: An alluvial fan is formed when streams carrying large volumes of sand and gravel down a steep mountain slope enter an adjacent valley or plain. The abrupt change in slope causes a decrease in the speed of the stream. This change causes the stream to deposit the sand and gravel it has been carrying. The deposited matter spreads in a gently sloping fan-shaped mass from the mouth of the canyon onto the valley floor. The heavier, coarser material is deposited near the mouth of the valley, while the finer material is carried out toward the edges of the fan. Alluvial fan deposits have been found to reach depths up to and occasionally more than about 61 m (200 ft).

Fan deposits, because of the frequent shifting of the stream channel, ordinarily contain lenticular beds or tongues of poorly sorted sand and gravel interbedded with varying proportions of silt and clay. The particles are subangular to angular in the fans built by streams with small drainage areas. The particles in large fans, covering several square kilometers (miles), are generally subangular to subrounded. Suitable aggregate is obtained from alluvial fans which are free from thick clay lenses.

In the past, the deposits were replenished during periods of flooding until the time that flood control dams were built to control the rivers.

Glacial and Fluvial-Glacial Deposits: Sand and gravel is produced commercially from open pit operations in glacial deposits in the midwestern and northeastern parts of the United States. The bulk of the sand and gravel was deposited directly or indirectly from continental ice sheets which originated in Canada and spread south into the United States. Glacial deposits that have been subjected to stream action are called fluvial-glacial deposits.

The glaciation period consisted of advances and recessions of the continental ice sheets which were several thousand feet thick, and moved roughly north to south.

Fragments of the rocks in the areas over which the glaciers passed, such as the igneous rocks in Canada, were picked up and incorporated in the glacier. When the glacial ice melted, the debris of clay, silt, sand, and pebbles and boulders were deposited as glacial till. When the advance and melting of the ice were in the balance, the debris deposited along the margin of the glacier formed irregular ridges termed moraines. As the water flowed away from the melting glaciers, the glacial debris was transported and deposited in outwash plains or elongate valley trains which are now major sources of sand and gravel (Ekblaw and Lamar, 1964). In addition, kames (hills) and eskers (ridges) of sand and gravel were deposited by melt waters within the ice sheet.

The fluvial-glacial deposits exhibit a better degree of sorting and contain less clay than the true glacial deposits and are of high enough quality to yield satisfactory aggregate comparable to true stream sand and gravel. Generally, the fluvial-glacial deposits are thinner than stream deposits in regions of considerable topographic relief.

Outwash Deposits—Outwash debris varies in size with the distance from the former glacial ice. The coarsest materials, pebbles, cobbles, and boulders, were deposited near the glacier, and the finer materials, sand, silt, and clay, were carried successively farther away from it. Since the water velocities fluctuated, the maximum size of transported particle varied and the deposits are characterized by great variation in particle sizes. Instances are cited where layers of gravel abruptly terminate, to be replaced by thick accumulations of silt or clay.

Kames and Eskers—Kames are hills composed of material deposited by glacial melt water in vertical crevasses or holes in the glacier, generally at or near its front. Eskers consist of material deposited in the beds of streams that flowed under or in the glacier. When the ice melted, the kames remained as rounded hills, and the eskers were left as ridges. The materials in these deposits, which show a wide variation in size and physical characteristics, are a common source of sand and gravel.

Beach and Dune Deposits—During the last stage of glaciation, the Lake Michigan basin was occupied by a larger lake whose level varied with the advances and recessions of the glacier. Each stage is marked by beach ridges of gravelly sand. In places, the beach sand has been blown into dunes. These deposits are suitable as sources of sand, although urbanization and preservation as natural and recreational areas have removed many of them from consideration as minable reserves.

Moraines—True glacial deposits such as moraines have been transported solely by glacial ice with little or no reworking by streams. Morainal debris contains material of diverse size, shape, and quality ranging from pulverized rock flour to erratic boulders and generally is usable only after expensive processing.

Dredge Tailings: As a result of bucket dredge operations for gold, huge furrowlike ridges of gravels have been accumulated as tailings along many of the western state streams. A normal gold-dredging operation involves washing and screening the stream deposit. The oversize gravel, which ordinarily consists of everything over 13 or 19 mm (½ or ¾ in.), is washed and sent to tailings piles behind or alongside the dredge. The undersize fine gravel and sand are then processed to remove the gold and returned to the dredge pond and subsequently covered by other oversize tailings. This results in segregated deposits of fine gravel and sand which underlie coarser gravels, requiring that considerable material be processed by aggregate producers to obtain the desired gradings of sand and gravel. Most commercial operations employ equipment to crush the oversize tailings. Tailings piles which have not been subjected to excessive

weathering contain material as suitable as the related stream deposit.

Older Geologic Formations: Pre-Quaternary formations, particularly partly consolidated sedimentary beds of sandstone and conglomerate, afford usable sources of aggregate when post-depositional weathering has not been too complete. Ordinarily, these formations have been subjected to long periods of weathering and are either too well cemented or contain too much clayey material to be processed economically. Accessibility of the deposit and depth of overburden also present problems.

The pre-Quarternary formations which are composed of well-indurated sedimentary rocks or hard crystalline rocks require expensive quarrying operations to obtain suitable material for aggregate.

Beach Deposits: Sand and gravel formed by the winnowing action of currents and waves on a beach make excellent aggregate. The gravel and coarse sand size particles are generally well-rounded, hard, and firm. The sands are commonly composed predominantly of resistant quartz and feldspar. Most beach deposits, however, are thin and lack proper size gradation.

Distribution of Deposits

The occurrence of sand and gravel in the United States is related to geologic processes. On a geographic-geologic basis, in the northern states, the principal sand and gravel resources are various types of glacial and outwash glacial deposits. Marine terraces, ancient and recent geologically, are the major sand and gravel sources in the Atlantic and Gulf Coastal Plains. River deposits are the prime sources in several of the southeastern and south-central states. In the Great Plains, sand and gravel are mainly stream deposits. On the West Coast, deposits are principally alluvial fans, stream deposits, beach and dune sand, and fluvial-glacial deposits.

Offshore deposits, although of minor importance tonnagewise, are mined along the east coast from Washington to Boston and in the Great Lakes.

EXPLORATION

Exploration for sand and gravel is generally guided by the needs of the potential producer. There are certain criteria that a deposit should meet in order to be of economic interest for a commercial producer, such as proximity to market, ease of obtaining the necessary mining permit, and a deposit of sufficient quantity to amortize the cost of a modern processing plant.

As sand and gravel is a low-cost, high-volume commodity, it cannot be transported far before the cost of transportation equals or exceeds the cost of the processed material at the plant. Thus, in most urbanizing areas the proximity to market would determine the target area for exploration. In aggregate-deficient parts of the country where materials are transported hundreds of miles, the target area naturally would be much larger.

Another important criteria to be met before field exploration begins is the socioeconomic restrictions on a potential mining operation. If the governing bodies have adopted a policy of refusing to grant operating permits, there would be no purpose for exploring in that jurisdiction. Also, there should be an awareness of the opposition that would be generated by conservation and ecology minded groups to granting of mining permits.

The size of the deposit is critical. In some parts of the country, where large volumes of sand and gravel are consumed, new sand and gravel plants cost millions of dollars.

In order to amortize the cost of such a plant, the deposit should contain sufficient reserves for at least 20 years.

Naturally, in prospecting for sand and gravel for non-commercial production, such as for use in dams, the foregoing criteria rarely apply. In this instance, the quantity and quality are the prime factors to consider.

Field Technique

Preliminary field evaluation by a geologist or by an engineer with some geologic training may often suffice to evaluate a deposit as a possible source of aggregate material. The characteristics that bear upon the usefulness of a deposit as a source of aggregate material have been determined by the geologic processes by which the deposit was formed and subsequently modified. A basic understanding of the processes, therefore, is helpful to the person evaluating it. The important characteristics are types and physical condition of the rock, grading, rounding and degree of uniformity, particle size and shape, location, thickness and type of overburden, and ground water level.

By the judicious use of geologic maps, a geologist can determine the type of materials to be encountered in a certain drainage area and predict the type of alluvial material to be found in a given stream channel. He thereby can recognize likely sources of aggregate and often can rule out areas unlikely to contain suitable materials. The geologist uses commercial sand and gravel deposits as a yardstick to guide him in evaluating new deposits. By comparing service records and laboratory test results with the petrographic character of commercially proven aggregates, the geologist builds up a background of data from which he can extrapolate to predict the behavior of untested deposits of similar petrographic character. This technique has been used successfully by the US Bureau of Reclamation in preliminary planning studies for large structures that require large amounts of suitable aggregates. In many instances field evaluation has obviated the need of expensive and time-consuming laboratory tests.

Target Areas—As outlined in the section on geology, those areas to be prospected for sand and gravel are predominantly those associated with either the present stream drainage or ancient drainages. Clues to buried or older deposits may be gleaned from well drilling logs, geophysical data, and from geologic reports.

Aerial photographs, coupled with topographic maps, are essential tools. Any geologic information in the form of maps or reports should be researched. In most instances, the state geological surveys can provide the necessary maps.

The initial exploration can be greatly expedited with the use of airplanes, preferably a helicopter, to observe promising geologic features. Preliminary investigations, which include determining such features as crude estimates of volume of available material, thickness and type of overburden, and water and power availability and access, are based primarily upon visual examination, either by a geologist or someone with experience in aggregate production.

Detailed Exploration: The method of exploration will depend upon the particular set of conditions; the objective being to obtain representative samples of the entire deposit and to delineate the dimensions of the deposit.

No matter what method of exploration, ultimately representative samples must be obtained so that standard laboratory acceptance tests can be performed on the samples.

Where the deposit is shallower than 9 m (30 ft), test pits may be excavated with bulldozers, backhoes, clamshells, or dragline. The location of the water table often will determine the type of test pit.

For deeper deposits, truck-mounted drill rigs should be used. Ordinarily, truck-mounted bucket augers, which use buckets from 46 to 91 cm (18 to 36 in.) in diameter, perform satisfactorily when large boulders are not present in the deposit. When drilling beneath the water table, steel casing is used to prevent the hole from caving.

Other drilling rigs, such as clamshells and reverse circulation drills, perform satisfactorily. Choice of drilling method will depend upon the availability of competent drillers in the area.

Geophysical methods can be used in conjunction with a drilling program. Geophysics may be used to check continuity of geologic units between drill sites or beyond a known deposit. For example, electrical resistivity techniques may be used to locate thick zones of clay interbedded in the sand and gravel. The depth of a sand and gravel deposit resting on a hard bedrock surface can be determined by means of refraction seismic methods.

The cost of geophysical methods equals the cost of drilling, on an hourly basis. However, much more information is obtained within the same time interval through geophysics.

Obviously, geophysical methods are useful exploration tools but cannot substitute for the other methods when samples must be obtained for laboratory testing. Standard laboratory acceptance tests are used to determine the suitability of the deposit. Equally important as laboratory testing is the sampling in the field. Where possible, large bulk samples should be obtained and quartered in the field. If it is possible, enough material should be stockpiled and trucked or railed to an existing plant to determine size gradation and quality of the finished product.

As an integral part of the exploration program, cross sections should be drawn of the deposit and estimates made of the quantity of material available. This figure should be reduced by the percentage of waste material, such as beds or lenses of clay and silt, and the loss of material to setbacks and finished slopes required by mining permits.

General Specifications

The study of the results of laboratory tests of aggregates that have good service records in concrete has led to the establishment of certain minimum requirements of specifications to which aggregates are expected to confirm. These specifications are designed so that completely serviceable concrete may be made, using any aggregate that meets the requirements. Most specifications are written by government agencies, engineering societies, and concrete technologists and attempt to conform to one standard set of specifications, those set up by the American Society for Testing and Materials. Modifications of these standards for certain types of concrete work make it difficult to compare individual requirements of the various organizations. Therefore, to evaluate the suitability of a deposit by judging the test results of selected samples is a difficult task. Some deposits which may not meet certain required specifications may have to be utilized because of other outside factors, such as the greater expense of hauling a more suitable aggregate.

In general, aggregate from an untried deposit will be satisfactory for most uses if it meets the minimum standards of the US Army Corps of Engineers, and the US Bureau of Reclamation.

BIBLIOGRAPHY AND REFERENCES

Anon., 1963, *Concrete Manual,* 7th ed., US Bureau of Reclamation, Denver, CO, 642 pp.

Ekblaw, G.E., and Lamar, J.E., 1964, "Sand and Gravel Resources of Northeastern Illinois," Circular 359, Illinois State Geological Survey, 8 pp.

Goldman, H.B., 1956, "Sand and Gravel for Concrete Aggregate," *California Journal of Mines and Geology,* Vol. 52, No. 1, pp. 79-104.

Goldman, H.B., 1959, "Franciscan Chert in California Concrete Aggregate," Special Report 55, California Div. of Mines and Geology, 28 pp.

Goldman, H.B., 1959a, "Urbanization and the Mineral Industry," *Mineral Information Service,* California Div. of Mines and Geology, Vol. 12, No. 12, Dec., pp. 1-5.

Goldman, H.B., 1961, "Sand and Gravel in Northern California," Bulletin 180A, California Div. of Mines and Geology, 38 pp.

Goldman, H.B., 1961a, "Urbanization—Impetus and Detriment to Mineral Industry," *Mining Engineering,* Vol. 13, No. 7, July, p. 717.

Goldman, H.B., 1962, "Aggregate from 'Fossils,'" *Rock Products,* Vol. 65, No. 11, pp. 65-68.

Goldman, H.B., 1964, "Sand and Gravel in Central California," Bulletin 180B, California Div. of Mines and Geology, 58 pp.

Goldman, H.B., 1968, "Sand and Gravel," *Mineral and Water Resources of California,* Bulletin 191, California Div. of Mines and Geology, pp. 361-369.

Goldman, H.B., 1969, "Sand and Gravel in Southern California," Bulletin 180C, California Div. of Mines and Geology, 56 pp.

Goldman, H.B., and Klein, I.E., 1958, "Sand and Gravel Resources of Cache Creek in Lake, Colusa, and Yolo Counties, California," *California Journal of Mines and Geology,* Vol. 54, No. 2, Apr., pp. 237-296.

Goldman, H.B., and Klein, I.E., 1961, "Sand and Gravel Resources of the Kern River, near Bakersfield, California," Special Report No. 70, California Div. of Mines and Geology, 33 pp.

Goldman, H.B., and Reining, D., 1983, "Sand and Gravel," *Industrial Minerals and Rocks,* S.J. Lefond, ed. AIME, New York, pp. 1151-1166.

24. Talc

RICHARD H. OLSON

COMMODITY DESCRIPTION

Talc is a mineral with fixed chemical composition. In industry, however, the name is applied to a number of mixtures of talc and talclike minerals, e.g., anthophyllite, antigorite, chlorite, dolomite, magnesite, mica, pyrophyllite, serpentine, and tremolite. The term "talc," therefore, to a mineralogist has a distinctly different connotation than to a miner or ore-purchaser. The remainder of this article is largely directed toward the latter two groups.

The term "steatite," originally a mineralogist's term applied to pure talc, in today's industrial usage generally refers to a massive variety of talc suitable for machining into electrical insulators.

The term "soapstone" is a rock name, not a mineral name. It refers to a massive coherent rock which is generally composed predominantly of impure talc containing relatively minor amounts of chlorite, mica, and various amphibole and pyroxene minerals. Its marked tendency to retain block form when broken out of its outcrop resulted in its abundant use by prehistoric Indians for the manufacture of utensils and ornaments.

As with most other ores, local terminologies have been developed for talc ores and related rocks by both the miners and producers. Examples of this plus a discussion of technological terms applied to talc products are given in Chidester, Engel, and Wright (1964, pp. 8-10).

Bodies of more or less pure talc are referred to as "talc schist." In the mining of talc ores it is not uncommon to have in near proximity separate and distinct bodies of pure talc schist and tremolitic ores. Therefore, tremolite can be a coproduct of talc mining or talc can be a coproduct of the mining of tremolitic talc ores.

Recent federal government investigations into health and safety matters have resulted in tremolite being likened to asbestos. A summary of the results of one of the early government-sponsored symposia on this subject was compiled by Goodwin (1974). Such investigations have led many consumers to switch away from tremolite-containing products over the past decade or so. No such firm stands have yet been taken on this matter re tremolite, as is the case with asbestos, and this reluctance to decide one way or the other continues to complicate the lives of those who manufacture products from tremolitic ores.

MINING METHODS

Although talc in the past has been mined to a large extent underground, the tendency is now more and more toward surface mining. Districts such as the Allamoore mining district in western Texas and that in the southwestern corner of Montana are virtually completely dedicated to open-pit mining. Vermont, which has in the past been dominated by underground mining [some of it to depths greater than 305 m (1,000 ft)], is now tending toward being almost completely open pit. The Gouverneur mining district in north-central New York, which until some 20 years ago was virtually totally an underground district, now has a significant proportion of open-pit mining. In the near future when special ores are found which have relatively minor competition from open-pit ores it will not be uncommon for them to be mined

underground; but when such a competitive edge is lacking, it is likely that open-pit mining will be considered.

GEOLOGY

Most talc deposits are stratiform and concordant with the enclosing rocks. Although they clearly show such tendencies, it is still nevertheless common for them to be called "veins." Although several types of origin are known, virtually all of the talc ores currently being mined are one of two types:

1) Deposits associated with sedimentary rocks.
2) Deposits associated with ultramafic igneous rocks.

Talc deposits are also associated with mafic igneous rocks such as gabbro, but this type of association is far more important re soapstone than talc and therefore will not be dealt with in this discussion.

The formation of talc is invariably epigenetic; talc ores are always of secondary origin.

Talc deposits formed by the alteration of sedimentary rocks generally have a carbonate precursor (siliceous dolomite being the most common), although to a much lesser degree magnesite and even limestone may also serve as the parent rock. The best example of this type of talc formation is that of the large district in the southwestern corner of Montana where all of the known commercial talc has been formed by the alteration of extremely old Precambrian dolomite (Olson, 1976).

Siliceous rocks may also be the precursor of talc ores, although this is far less common. Argillite, phyllite, quartzite, and schist are the major lithologic types in this category of sedimentary precursor.

Talc deposits associated with serpentinite and related ultramafic rocks such as dunite, peridotite, and pyroxenite are typified by the large talc mining district of central Vermont. Such deposits characteristically consist of a serpentinite core surrounded by a shell of talc-carbonate ore which in turn may be surrounded by a relatively thin shell of pure talc schist ore.

In the past, and even now to some extent, talc deposits derived from sedimentary rocks have dominated the US talc mining industry. This has been due in no small part to the fact that such ores may commonly be mined directly with no wallrock or included contaminants (i.e., "horses") and be milled directly on a dry basis. Those deposits associated with ultramafic igneous rocks, however, invariably must be beneficiated by wet methods (generally flotation) in order to obtain the high purity commonly achieved in products derived from the ores of sedimentary host rocks. The dry brightnesses of filler and extender products derived from ores of ultramafic igneous host rocks are generally considerably less than those of the products derived from sedimentary host rocks, particularly dolomites. Nevertheless, the tendency in recent years and probably for the immediate future is toward an ever-increasing importance of the position of the former ore types over those of the latter.

Most of the talc ores in the US have been formed in Precambrian rocks. This is invariably true in the Montana, New York, and west Texas districts and is generally true in the California-Nevada ("Death Valley") district. The talc ores of the Appalachians, however, have generally been formed in Early Paleozoic rocks. In either case we are dealing with extremely old rocks which seldom crop out because such an enormous amount of tectonic uplift and subsequent erosion is necessary in order to bring them either to the surface or to place them at relatively shallow depths. Therefore, very few areas of limited geographical extent are capable of having commercial talc deposits, as opposed to most other industrial minerals ores such as clay, gypsum, limestone, phosphate, etc.

The geology of talc deposits in the US is described in great detail by Chidester, Engel, and Wright (1964).

The US currently has six talc mining districts. The geology of the five most important (production-wise) of these is briefly summarized.

California-Nevada

The California-Nevada talc mining district (also known as "Death Valley") is to all intents and purposes now dormant, due to a variety of reasons including cost factors and environmental constraints. It is about 322 km (200 miles) long and has an average width of 48 km (30 miles), although it locally may be as much as 121 km (75 miles) wide.

The district is too large and its geology too complicated to be satisfactorily summarized here, but essentially all of its deposits have been formed by metamorphism and hydrothermal replacement of siliceous and silicated magnesian marbles and limestones (Engel and Wright, 1960).

One deposit, the "Western-Acme" mine complex, extends for more than 1524 m (5000 ft) along strike, is as much as 24 m (80 ft) wide, and has been mined downdip for at least 107 m (350 ft).

Two main types of ore are present in this district: the "hard" ore of the miners which is markedly tremolitic, and the "soft" ore of the miners, which is talc schist. Many deposits contain both types, but the tremolitic ore is generally by far the more abundant.

In the early days most of the mining was performed underground, but in recent years there has been a pronounced tendency more toward surface operations.

Wright (1969) gives an excellent and detailed description of numerous mines and deposits in the central portion of the district.

Montana

The most important talc mining district in the US is that in the southwestern corner of Montana. This is currently the most important domestic source of tremolite-free talc ores from which pure and highly desirable products can be produced directly from crude ores without the need for beneficiation techniques, other than sorting.

The earliest significant geological investigation of the Montana talc mining district is that of Perry (1948), an excellent field study which was performed in the burgeoning days of the district. Olson (1976) describes the mines and deposits of the Ruby Range in more detail and Berg (1979) does so for the remainder of the district.

High-purity talc deposits in southwestern Montana, as presently known, are totally restricted in their occurrence to dolomitic marble in the Precambrian Cherry Creek Series. This generally white to pale green or yellowish-green talc commonly occurs as conformable lenses and stringers within the marble. Although structural controls in the strictest sense have not been recognized, the best talc bodies appear to occur spatially close to large bodies of granite gneiss.

The host rocks have been subjected to upper amphibolite grades of regional metamorphism which included the development of intense isoclinal deformation. Apparently the formation of the talc occurred during the retrograde phase (greenschist grade) of the regional metamorphic event and it was almost certainly formed in Precambrian time.

Ross, et al. (1955) map the geological unit in which the talc has formed as covering an area about 290 km (180 miles) long east-west and 137 km (85 miles) wide (north-south), but the dolomitic marble which invariably hosts the talc does not occur within it over this entire extent.

The mechanism of the formation of Montana talc ore bodies was probably similar to that postulated by Van Horn (1948) for those similar talc ores in the Murphy, NC, area.

New York

The New York talc mining district was the site of the first talc mining in the US, which commenced about 1878. This district in the north-central portion of the state is the smallest in area of all the talc mining districts of the US, yet it appears to be dominant in its proven and probable ore reserves and it dominated domestic talc production for much of the early part of this century.

The talc-containing unit is a dolomitic marble which probably averages only 76 m (250 ft) wide or so in its least contorted areas; much greater widths, which may be locally observed to be as much as 610 m (2000 ft) thick, have probably been formed by structural flowage. Brown and Engel (1956) divided the rock sequence into 15 units, of which only unit 13 is known to contain ores. It is peculiar that this unit, which contains numerous deposits of talc and tremolite, is also the host for all of the well-known high-grade zinc deposits of the district. The talc and zinc ores, although they occur within the same unit, do not occur together.

Tremolitic ores have been proven by underground workings to downdip depths of at least 335 m (1100 ft) and by drill holes (sunk for zinc exploration) to downdip depths of more than 914 m (3000 ft).

Although pure talc schist has been found locally along the hanging wall contact between the tremolitic ore bodies and the overlying dolomite, these bodies have proven to be too erratic in their occurrence and too intensely secondarily stained to be exploited as ore.

Field mapping by workers of the US Geological Survey indicates that the tremolite, talc, and associated minerals were formed more than 10^9 years ago.

Although the talc ore zones locally pinch and swell, they are on the whole conformable with the enclosing dolomitic marble layers. The regional dip averages about 45° to the northwest; departures from this relatively simple structural framework are only locally significant. All of the rocks in this locality have been complexly folded and generally subjected to regional metamorphism of amphibolite grade.

Texas

Texas has two talc mining districts, the Llano district in the central part of the state and the Allamoore district in the western part of the state. The latter is the one most recently put into production and is now the only one of importance.

The Allamoore talc mining district extends along strike for about 32 km (20 miles) east-west and is up to 8 km (5 miles) wide. All of its ore deposits occur in the Allamoore Formation, which consists of thousands of feet of carbonate rocks, volcanic rocks, and phyllite (King and Flawn, 1953).

Not only is this district "one of a kind" within the US, it is also the one about which there is the least agreement as to its geology and the genesis of its ores. Some workers feel that phyllite is the host rock while others feel that the talc was formed by the replacement of primary sedimentary magnesite. There appears to be little disagreement, however, that the talc formed in Precambrian time.

No ore body is known to have been "bottomed-out" by either drilling or mining. Talc is commonly found from the grass roots down and the soil cover is seldom greater than 0.3 m (1 ft) thick.

The ores for the most part are an extremely fine-grained admixture of dolomite and talc; beneficiation techniques for the recovery of pure talc have not yet been developed.

With the exception of a few small light-colored ore bodies, Texas crude ore is dark-colored, generally gray to black. With the sole exception of the volumetrically insignificant light-colored bodies, Texas talc ores have no possible uses as fillers and extenders, yet they serve as some of the finest ceramic crude ores known anywhere in the world, with superb pressing qualities and excellent firing characteristics.

Virtually all of the production has been from large open pits and the mining costs are certainly the lowest currently known in the US.

Vermont

Vermont has no carbonate-derived talc ores. Its ores have been formed from ultramafic bodies which have been locally serpentinized and then converted into talc and/or talc-carbonate mixtures. The talc deposits are commonly associated with dunite, peridotite, and serpentinite, which were probably the precursors of the talc itself.

The talc ores are generally dark-colored and are relatively high in iron content, particularly near the surface where weathering processes have concentrated the iron. Although the mineral talc is commonly a major constituent, the grades of Vermont talc ores are generally considerably lower than those of most of the other US talc mining districts.

In northern Vermont entire ultramafic bodies have been converted into talc-carbonate ore mixtures which may have widths on the order of 30.5 m (100 ft) or so. In southern Vermont a similar body has a width of about 53 m (175 ft). At least 145 separate talc occurrences are known in central Vermont within an area having a strike length of about 241 km (150 miles) (north-south) and an east-west width ranging from 8 to 40 km (5 to 25 miles).

To manufacture the most profitable product line, i.e., high-purity tremolite-free talc products, it is always necessary when using Vermont talc ores to employ froth-flotation techniques. Although the resulting products do not approach those from Montana or New York on a dry brightness basis, the manufacture of such products is quite profitable. In the past, several Vermont talc producers, however, have commonly ground the talc ore as mined and then simply subjected that ground product to size-classification techniques, thereby producing low-value dusting compounds and fillers for roofing products and for the rubber industry.

GENESIS

Talc is always of secondary origin. Those ores derived from sedimentary host rocks have generally been formed by metamorphic or metasomatic processes associated with regional high-grade metamorphic alteration. In several instances the formation of the talc ores has occurred during the retrograde metamorphic events associated with such metamorphism which attained grades as high as amphibolite rank. Those ores derived from ultramafic igneous rocks have generally been formed by the reaction between serpentinite and introduced CO_2.

STRUCTURAL CONTROLS

Most talc ore bodies are stratiform in habit, being more or less concordant with their enclosing tabular rock units.

Therefore they are in essence tabular ore bodies and commonly relatively easy to mine. In many instances, however, structural deformation may have been intense, resulting in the development of isoclinal folding and a distinct tendency for the ore body to noticeably pinch and swell, both along strike and dip. Later block faulting may also locally complicate the task of the talc miner. Developmental and production-exploration drilling, which is seldom performed due to a variety of reasons, is nevertheless desirable with virtually all talc deposits, more for the need to know whether folding or faulting may have displaced or otherwise modified the distribution of the talc body than toward the need for knowledge of grade control. In the past most talc miners have just shot off the next round and hoped that they stay in ore; this technique (or lack of same) is changing, albeit very slowly, toward a more enlightened attitude.

EXPLORATION

In the exploration phase it is most important to recognize that talc is extremely limited as to which ages of geologic formations it may occur within as potentially commercial bodies. The next stage of such an effort should be a thorough literature search coupled with visits to governmental and academic institutions where past or present workers may have been pursuing investigations into talc and related mineral occurrences. The use of unpublished open-file materials is important here. Lastly, known deposits should be visited and studied in order to more intelligently guide the search for their currently unknown analogues.

Once the actual physical sites deemed to be most favorable for physical exploration have been chosen, it must be realized that the life of a talc exploration geologist is tremendously complicated and made extremely difficult due to the tendency of talc to almost never crop out! It is not at all unusual for such explorationists to literally have to get down on their hands and knees and then study float fragments commonly considerably smaller in size than one's fingernail. The mapping of the areal patterns of the most abundant type of float rock may then be extrapolated into the compilation of bedrock geologic maps. These maps may then be used to guide exploration trenching with the backhoe or bulldozer across the mapped zones of talc-related rock types.

Drilling may then be performed after the presence of interesting quantities of talc or talc-related alteration has been observed and mapped in the trenches. Core drilling is always preferred on the first pass, because with it the nature and distribution of both talc and its associated impurities can be detected without question. On subsequent drilling phases, reverse circulation methods on closer spacings may be employed.

Geophysical and geochemical methods have not yet proven to be very successful in the exploration for talc. Soil-sampling methods, however, have proven their worth in both Alabama and Montana. Such techniques may vary widely in the details of the actual procedures followed, but essentially the soil is sampled, generally with a hand-operated soil auger, and the resulting samples are then processed and studied in the laboratory, where they are carefully sized and then subjected to X-ray diffraction.

REFERENCES

Berg, R.B., 1979, "Talc and Chlorite Deposits in Montana," Memoir 45, Montana Bureau of Mines and Geology, 65 pp.

Blount, A.M., and Vassiliou, A.H., 1980, "The Mineralogy and Origin of the Talc Deposits near Winterboro, Alabama," *Economic Geology*, Vol. 75, No. 1, pp. 107-116.

Brown, C.E., 1973, "Talc," *United States Mineral Resources,* Professional Paper 820, US Geological Survey, pp. 619-626.

Brown, J.S., and Engel, A.E.J., 1956, "Revision of Grenville Stratigraphy and Structure in the Balmat-Edwards District, Northwest Adirondacks, New York," Bulletin, Geological Society of America, Vol. 67, No. 12, pp. 1599-1622.

Chidester, A.H., Billings, M.P., and Cady, W.M., 1951, "Talc Investigations in Vermont," Preliminary Report, Circular 95, US Geological Survey, 33 pp.

Chidester, A.H., Engel, A.E.J., and Wright, L.A., 1964, "Talc Resources of the United States," Bulletin 1167, US Geological Survey, 61 pp.

Clifton, R.A., 1985, "Talc and Pyrophyllite," *Mineral Facts and Problems,* Bulletin 675, US Bureau of Mines, pp. 799-810.

Engel, A.E.J., and Wright, L.A., 1960, "Talc and Soapstone," *Industrial Minerals and Rocks,* 3rd ed., J.L. Gillson, ed., AIME, New York, pp. 835-850.

Furcron, A.S., Teague, K.F., and Calver, J.L., 1947, "Talc Deposits of Murray County, Georgia," Bulletin 53, Georgia Geological Survey, 75 pp.

Goodwin, A., 1974, "Proceedings of the Symposium on Talc, Washington, D.C., May 8, 1973," Information Circular 8639, US Bureau of Mines, 102 pp.

King, P.B., and Flawn, P.T., 1953, "Geology and Mineral Deposits of Precambrian Rocks of the Van Horn Area, Texas," No. 5301, Texas University Publications, 218 pp.

McMurray, L., and Bowles, E., 1941, "The Talc Deposits of Talledega County, Alabama," Circular 16, Geological Survey of Alabama, 31 pp.

Neathery, T.L., 1968, "Talc and Anthophyllite Asbestos in Tallapoosa and Chambers Counties, Alabama," Bulletin 90, Geological Survey of Alabama, 98 pp.

Neathery, T.L., et al., 1967, "Talc and Asbestos at Dadeville, Alabama," Report of Investigations 7045, US Bureau of Mines, 57 pp.

Olson, R.H., 1976, "The Geology of Montana Talc Deposits," chapter in Special Publication 74, Montana Bureau of Mines and Geology, pp. 99-143.

Papke, K.G., 1975, "Talcose Minerals in Nevada—Talc, Chlorite, and Pyrophyllite," Bulletin 84, Nevada Bureau of Mines and Geology, 62 pp.

Perry, E.S., 1948, "Talc, Graphite, Vermiculite and Asbestos in Montana," Memoir 27, Montana Bureau of Mines and Geology, 44 pp.

Roe, L.A., 1975, "Talc and Pyrophyllite," *Industrial Minerals and Rocks,* 4th ed., S.J. Lefond, ed., AIME, New York, pp. 1127-1147.

Roe, L.A., and Olson, R.H., 1983, "Talc," *Industrial Minerals and Rocks,* 5th ed., S.J. Lefond, ed., AIME, New York, pp. 1275-1301.

Ross, C.P., Andrews, D.A., and Witkind, I.J., 1955, "Geologic Map of Montana," US Geological Survey.

Spence, H.S., 1940, "Talc, Steatite, and Soapstone; Pyrophyllite," Memoir 803, Mines and Geology Branch, Canada Dept. of Mines and Resources, 146 pp.

Van Horn, E.C., 1948, "Talc Deposits of the Murphy Marble Belt," Bulletin 56, North Carolina Dept. of Conservation and Development, Div. of Mineral Resources, 54 pp.

Winkler, H.G.F., 1974, *Petrogenesis of Metamorphic Rocks,* Springer-Verlag, New York, 320 pp.

Wright, L.A., 1968, "Talc Deposits of the Southern Death Valley-Kingston Range Region, California," Special Report 95, California Div. of Mines and Geology, 79 pp.

25. Titanium*

Elemental titanium has become famous as a space age metal, because of its high strength/weight ratio and resistance to corrosion. However, the major use is in the form of titanium dioxide pigment, which because of its whiteness, high refractive index, and resulting light-scattering ability, is unequaled for whitening paints, paper, rubber, plastics, and other materials. A relatively minor use is in welding rod coatings, in the form of the mineral rutile. The only commercially important titanium ore minerals at the present time are ilmenite and its alteration products, and rutile.

Titanium was discovered by Gregor in 1790, as a white oxide which he discovered from menaccanite, a variety of ilmenite occurring as a black sand near Falmouth, Cornwall. Barksdale (1966) stated that the fundamental chemical reactions on which the present-day titanium industry is based were known before 1800, although it was not until 1918 that these pigments were available commercially on the American market.

Pings (1972) outlined the early history of the titanium industry in the United States, referring to the work of Guise (1964) and others:

Mining of titanium minerals in the United States began sometime between 1880 and 1900, in Chester County, Pennsylvania. Small quantities of rutile were also produced during that time in North Carolina and Georgia. In 1901 rutile was mined from a deposit near Roseland, VA, and was used in making titanium chemicals and for coloring ceramics. Ilmenite in the deposit was produced as a separate item in 1913, and rutile and ilmenite were obtained from this deposit through 1921.

The mining of titanium bearing beach sands began in 1916 near Mineral City and Pablo Beach, FL, for the purpose of making titanium tetrachloride to be used in tracer bullets, flares, and smokescreens. The titanium pigment industry was founded in 1918. By 1928 a large part of the domestic production of rutile, ilmenite, and zircon came from Florida. However production ceased in 1929 in Florida with the mining of newly discovered deposits in Virginia which were mined until 1968. The discovery of new deposits in Florida, and the development of better mining and concentrating methods for lowgrade sands, led to a return of activities in this area. The large deposits near Tahawus, NY, were brought into production in 1942 by National Lead Co. (now NL Industries, Inc.), and by 1949 this company was the leading producer of ilmenite in the world. In 1948 the extensive deposits of ilmenite and ilmenite-bearing iron ores were discovered in eastern Quebec.

The large Trail Ridge sand deposit in Florida was discovered by E.I. du Pont de Nemours & Co., Inc. geologists in cooperation with the US Bureau of Mines and the Florida Geological Survey, and was developed by Du Pont in the late 1940s (Garnar, 1980).

The beginning of the modern titanium metal industry was in 1948, when Du Pont produced the first metal. US Bureau of Mines reports, which gave details of the Kroll process, together with the attractive properties of the metal for military aircraft, led to a concerted effort by industry and government to develop a large-scale titanium metal industry, which reached a peak capacity of over 36,000 stpy from six producers by 1958 (Pings, 1972a).

The major developments affecting the titanium metal industry since the late 1950s have been fluctuations in the

production rate of military aircraft, and the increasing use of the metal in commercial aircraft and for industrial applications. A major change in the titanium dioxide pigment industry starting in 1957 was the development and expansion of the chloride process, which generally uses higher grade TiO_2 raw materials such as rutile in contrast to the relatively low TiO_2 ilmenite feed used in the older sulfate process. The resulting increased demand for rutile led to higher prices, which in turn stimulated the development of processes for making rutile substitutes by beneficiation of ilmenite.

Since 1960, new deposits of sand ilmenite were developed in New Jersey, Georgia, and Florida.

GEOLOGY

Mineralogy

Although titanium is the ninth most abundant element of the lithosphere, comprising an estimated 0.62% of the earth's crust, there are only a few minerals in which it occurs in major amounts: rutile, anatase, and brookite (which are polymorphs of TiO_2), ilmenite and its alteration products, including leucoxene, perovskite ($CaTiO_3$), and sphene ($CaTiSiO_5$). Anatase may be emerging as a significant ore mineral of the future, but ilmenite, altered ilmenite, leucoxene, and rutile have been the only large volume ore minerals through 1980.

Ilmenite: The chemical formula of theoretically pure ilmenite is $FeO \cdot TiO_2$. It was shown by Ramdohr (1950) that up to 6% Fe_2O_3 may be dissolved in solid solution, and at 1050°C a continuous solid solution series exists between ilmenite and hematite (Nicholls, 1955). Hematite may, and often does, occur with ilmenite as minute exsolution lamellae. Magnesium and manganese may substitute for the ferrous iron in ilmenite, which can produce the rare end-members $MgTiO_3$ (geikelite) and $MnTiO_3$ (pyrophanite), but usually these two elements are present as minor impurities. Magnetite is a common associate of ilmenite in igneous and metamorphic rocks, and in such coexisting pairs chromium, nickel, and vanadium tend to concentrate in magnetite while manganese concentrates in ilmenite.

In basic igneous rocks, notably anorthosites, gabbros, and basic lavas, ilmenite frequently occurs in intimate intergrowths with magnetite. The ilmenite forms lenses following octahedral parting planes in the magnetite host, and magnetite may, in turn, form crystallographically oriented inclusions within the ilmenite lenses.

Altered Ilmenite and Leucoxene: In sand deposits ilmenite frequently exhibits a degree of alteration caused by oxidation and removal of iron. The end product is essentially TiO_2. The process was described by Temple (1966).

Alteration is an extremely slow process that is aided by elevated temperature, so that older sand deposits in temperate and tropical regions of the world generally contain high-TiO_2 ilmenite. Younger deposits, for example, those found on modern beaches, and those in the higher latitudes usually contain unaltered ilmenite with a TiO_2 content around 50%, near the theoretical level for pure ilmenite.

Rutile: Rutile, the high pressure, high temperature polymorph of TiO_2, is the commonest form in nature and is a widespread accessory mineral in high grade metamorphic gneisses and schists and in igneous rocks. It is also a common detrital mineral.

*Excerpt from Lynd, L.E., and Lefond, S.J., 1983, "Titanium Minerals," *Industrial Minerals and Rocks,* 5th ed., Vol. 2, S.J. Lefond, ed., AIME, New York, pp. 1303-1362.

Commercial rutile concentrates run 95% TiO_2 or more, with SiO_2, Cr_2O_3, V_2O_3, Al_2O_3, and iron oxides comprising the remainder. Analyses of rutile from other occurrences may show major amounts of tantalum and columbium, which can enter titanium minerals because of the close similarity in ionic radius between Ti^{+4} and both Cb^{+5} and Ta^{+5}. There is also a high iron variety termed ferroan rutile.

Rutile may form by alteration from ilmenite or anatase, and while it is very stable over a broad range of geologic conditions, occasionally processes may be reversed so that rutile alters to sphene, possibly ilmenite, and more rarely anatase.

The characteristic color of rutile is reddish brown, but it may be black, violet, yellow, or green.

Classification of Deposits

Titanium minerals have been mined from both rock and sand deposits. Until about 1942, nearly all of the ilmenite and rutile produced commercially came from sand deposits, but about one third of the world's ilmenite in 1982 came from rock deposits. Rutile, however, is now produced exclusively from sand deposits.

Rock Deposits: *Anorthositic Deposits*—Nearly all of the known commercially important rock deposits of titanium minerals are associated with anorthositic or gabbroic rocks, and are of three main types: ilmenite-magnetite (titaniferous magnetite), ilmenite-hematite, and ilmenite-rutile. Ilmenite-magnetite deposits usually contain ilmenite and magnetite as granular intergrowths that can be separated rather readily to yield concentrates of ilmenite and magnetite which may be essentially homogeneous minerals, or may consist of intimate intergrowths of one mineral in the other. Ilmenite-hematite deposits usually contain these minerals as intimate intergrowths and yield an ilmenite-hematite, or hemo-ilmenite concentrate rather than a separate concentrate of each mineral. Ilmenite-rutile deposits contain rutile and ilmenite either as separate concentrations or occurring together.

Miscellaneous Deposits—Other types of rock deposits in the United States that have been mined, or seriously considered as sources of titanium, include a deposit of ilmenite disseminated in schist in Yadkin Valley, North Carolina (Broadhurst, 1955), and a complex deposit of rutile, anatase, and brookite in a pegmatitic phase of alkalic rocks surrounding sediments at Magnet Cove, AR (Fryklund and Holbrook, 1950; Fryklund, et al., 1954).

A perovskite deposit in southwestern Colorado owned by Buttes Gas & Oil Co. was estimated to contain about 50 million tons of TiO_2 (Thompson, 1977). Processes to convert perovskite to titanium dioxide were being investigated through 1982.

The US Bureau of Mines (Llewellyn, et al., 1980) and the US Geological Survey (Force, 1980) investigated the porphyry copper ores and mill tailings as a possible source of rutile. It was concluded that this material, which contains about 0.3% of potentially recoverable fine-grained rutile, could constitute a sizable domestic resource, but more work is needed to develop an economic recovery process.

Major occurrences of anatase and ilmenite in weathered carbonatite bodies at Tapirá, Salitre, and Catalão in Minas Gerais, Brazil, are under investigation as new raw material sources. Feasibility and pilot plant studies to determine grades and recoveries of possible future commercial titanium mineral concentrates have been carried out. Modifications in pigment manufacturing processes may be necessary to accommodate potential production as a feed, or beneficiation or slagging may be used to make concentrates acceptable to

existing pigment plants, but successful exploitation by any avenue will cause these and possibly other carbonatite occurrences to be classed as an important rock ore type.

Occurrences of titanium minerals in other types of rocks are known, but are not considered to be of commercial significance (Force, et al., 1976).

Sand Deposits: Beaches, bars, dunes, and stream sands in many parts of the world are enriched by gravity segregation of the heavy minerals that are chemically resistant to weathering and physically hard enough to withstand considerable abrasive action. The energy of currents, waves, and to a lesser extent, winds, mobilizes the sand grains and permits them to behave as individuals within a fluid medium. Where the energy decreases, the heavier particles fall out while the lighter ones are carried farther. The composition of the heavy mineral concentrate depends on the nature of the geologic terrane being subjected to weathering and erosion and supplying materials to transporting streams. If available at the source, titanium-bearing minerals, zircon, magnetite, chromite, rare earth minerals such as monazite, staurolite, kyanite, sillimanite, garnet, xenotime, precious metals, and diamonds may be concentrated in this way. The resultant gravity segregations are usually dark colored and are called "black sands." Not all black sands are ore mineral occurrences; those found on the beaches of volcanic islands, for example, may be mostly amphibole and pyroxene.

There is a wide variation in titaniferous black sands in terms of heavy mineral concentration, percentage of titanium minerals within the heavy mineral suite, and TiO_2 content of the titanium mineral concentrate. All of these factors interact in rating deposits according to quality and as to whether they are today's reserves or possible resources of the future. Furthermore, the presence or absence of valuable coproduct or byproduct minerals such as zircon is a complicating factor.

Sand deposits in which rutile is the only economically important titanium mineral occur along the eastern shore of Australia. Ilmenite, altered ilmenite, and rutile form inland elevated strand-line deposits in Western Australia and in older sands of the Atlantic Coastal Plain of the United States. Ilmenite and altered ilmenite are the principal titanium ore minerals in other Western Australian districts; in Kerala, India; in deposits north of the Black Sea in the USSR; and in Florida and Georgia. Relatively unaltered ilmenite is found in large beach and dune occurrences along the northeastern coast of South Africa, in the Nile Delta of Egypt, and in still other Western Australian deposits, those closest to the present coast. Sand deposits of titaniferous iron ores occur as dune and beach deposits in many volcanic areas, of which those in New Zealand are the outstanding examples.

Mode of Occurrence and Origin

Rock Deposits: The anorthositic deposits contain titanium minerals either as massive ore, or disseminated in rock ranging from anorthosite to gabbro in composition, and varying in grade from solid ore to almost barren rock. While there are metamorphic and hydrothermal features in some parts of these deposits, for the most part, they seem to have formed at magmatic temperatures by magmatic processes.

The only other type of ilmenite rock deposit to be exploited commercially was the Yadkin Valley deposit in Caldwell County, North Carolina (Broadhurst, 1955), which was operated by a Glidden Co. subsidiary from 1942 to 1952. The ore consists of small masses of ilmenite disseminated throughout a talcose body which lies conformably with the enclosing quartzite and mica schist.

The Hot Spring County, Arkansas, deposits have been studied extensively for possible commercial development, and some rutile was recovered from the Magnet Cove deposit from 1932 to 1944. The deposits occur in a complex mixture of alkalic igneous rocks intruding folded sedimentary and metamorphic rocks. Much of the titanium is in the form of rutile or brookite, but many other titanium minerals have been identified. The ore minerals are intimately associated with gangue (Fryklund and Holbrook, 1950; Fryklund, et al., 1954; Toewe, et al., 1971).

Sand Deposits: Titanium-bearing black sands are found mainly in ancient or modern ocean and sea beaches around and occasionally within continental land masses. They frequently form highly visible surficial layers between the high and low water marks which may extend intermittently along coasts for miles, but such concentrations, containing perhaps 80% heavy minerals, are not mined on a large scale because they are usually too shallow and narrow to represent major reserves. Minable bodies are multilayered occurrences of a similar nature left behind by retreating seas, or coastal dunes formed when heavy minerals from black sand beaches were being transported inland by wind action. Heavy minerals tend to be disseminated within such dunes rather than layered as in beach-type deposits.

The history of a black sand ore body may be simple or complex. The essential elements are: (1) a "hinterland" of crystalline rocks in which the heavy minerals were accessory constituents, (2) a period of deep weathering, (3) uplift with rapid erosion and quick dumping into the sea of the products of stream erosion, and (4) emergence of the coastline with longshore drift and high-energy waves acting during the process of shoreline straightening. There may be intermediate stages such as partial concentration of the heavy minerals in a coastal plain sediment and subsequent elevation, erosion, and reconcentration. The sand brought to the sea by rivers is picked up and carried away from their mouths by longshore currents, forming offshore bars and filling in bays between headlands, particularly during storms. Where bars are formed, the sand-carrying waves drag bottom and lose their energy so that the heavy minerals fall on the seaward side while the light minerals are cast over the bar and into the quieter water beyond. Layer upon layer of varying concentrations of heavy minerals accumulates on the growing bar in this way. Where bays are being filled with sand, both heavy and light minerals are churned from the bottom by landward-rushing waves and are hurled up the beach slope. The smoother, slower retreat of each wave mobilizes the uppermost layer of sand deposited there, and draws away the light minerals, to be picked up again and again by waves as currents move them along the coast, while leaving the heavy minerals behind. Alternating periods of stormy and calm weather leave alternating layers of high and low concentrations of heavy minerals in the beach sand as it advances toward the sea.

DISTRIBUTION OF DEPOSITS

Reserves

Ilmenite resources in rock and sand deposits in the United States, recoverable under the economic and technological conditions of 1981, were estimated to contain 18 million tons of titanium dioxide. Rock deposits in New York account for 29% of the total. The remainder is in beach and river sands in Florida (Garnar, 1980), Georgia, New Jersey (Markewicz, 1969), and Tennessee. An additional 94 million tons of titanium dioxide in ilmenite and 48 million tons in perovskite

(Thompson, 1977), not recoverable under present conditions, occur in identified resources, widely dispersed throughout the United States. Major occurrences lie in Colorado, Minnesota, New Jersey, New York, and Wyoming.

Reserves of rutile in the United States are in sand deposits in Florida, Georgia, and Tennessee. The rutile in these deposits contains about 1.7 million tons of titanium dioxide. Identified but presently subeconomic deposits in Arizona, Arkansas, California, Florida, North Carolina, South Carolina, Tennessee, Utah, and Virginia contain an estimated 13 million tons of titanium dioxide in the form of rutile. The rutile in the Arizona and Utah resources occurs mainly as an accessory mineral in porphyry copper ores and mill tailings (Force, 1980; Llewellyn and Sullivan, 1980).

Estimates for world titanium reserves and resources are shown in Table 1. World reserves of ilmenite are estimated to contain 266 million tons of titanium dioxide, of which US reserves account for 7%. The titanium dioxide content of world rutile reserves is estimated at 97 million tons; US reserves represent 1.6% (Lynd, 1983).

Rock Deposits—Principal Producing Countries

United States: *Sanford Lake District, New York*—The Sanford Lake deposits (Gross, 1968), are located in the heart of the Adirondack Mountains. Discovery of these titaniferous magnetite deposits dates back to 1826. Several attempts to exploit the ores for iron prior to 1942 proved uneconomical because of difficulties with the associated titanium, and the isolated location (Bachman, 1914; Masten, 1923). The district is within the large anorthosite massif making up the central high peak area of the Adirondacks. All of the various low silica rock types associated with the massif are found within the boundaries of the district. These consist of both the Marcy and Whiteface types of anorthosite and of gabbroic anorthosite, gabbro, and different grades of titaniferous magnetite ores. All of the rocks contain the same minerals and differ only in the percentages of these constituents.

There are four mineralized areas where an economic grade of ore has been found. Three of these have both gabbroic-type ore and anorthositic-type ore. The fourth has only gabbroic ore. The TiO_2 contents of the ores range from 9.5 to over 30.0% TiO_2. Ore bodies of both types are related to gabbro and conform to the configuration of the gabbro bodies within the anorthosite.

Table 1. World Reserves of Ilmenite and Rutile
Thousands of Short Tons of TiO_2 Content

Country	Ilmenite	Rutile
Arab Republic of Egypt	—	—
Australia	27,000	10,000
Brazil	2,000	67,000*
Canada	26,000	—
China	34,000	—
Finland	3,400	—
India	22,000	3,000
Norway	52,000	—
Sierra Leone	—	3,000
South Africa, Republic of	57,000	5,500
Sri Lanka	21,000	5,000
United States	18,000	1,700
USSR	7,000	3,000
Total	270,000	140,000

*Mainly anatase. Metric equivalent: 1st × 0.907 184 7 = 1t.

Except for a few thousand tons mined before 1900, all ore production has come from the Sanford Hill-South Extension ore body. Mining began on Sanford Hill in 1942, and as a result of further exploration and development work, was transferred in the early 1960s to the South Extension part of the ore body, which was overlain by glacial till and Sanford Lake.

GENERAL GEOLOGY. Rocks of the Sanford Lake district are regarded as members of a genetically related anorthositic series, the whole being part of the large Adirondack anorthosite massif. Locally, anorthosite grades into gabbro by an increase in the content of mafic minerals. Buddington (1939) divided this sequence into four rock types depending upon the amount of ferromagnesian minerals present: anorthosite with 0 to 10%, gabbroic anorthosite with 10 to 22.5%, anorthositic gabbro with 22.5 to 35%, and gabbro with over 35% mafic minerals.

For mine mapping purposes, rock types are designated as anorthosite, gabbroic-anorthosite, gabbro, and the various grades of ore.

ECONOMIC GEOLOGY. The titaniferous magnetite ore bodies of the district are of two types. One type, referred to locally as "anorthositic ore," is associated with anorthosite waste rock as coarse-grained massive lenses of irregular shape, with little structure and having sharp contacts with the anorthosite. The second ore type, "gabbroic ore," occurs as fine to medium grained oxide-enriched bands within the gabbro, with well defined structures similar to those in the gabbro. Contacts with the gabbro range from distinct to gradational.

In the Sanford Hill-South Extension ore body, both types of ore occur. The anorthositic type forms a footwall ore body and the gabbroic ore forms a hanging wall ore body. These are separated by various widths of anorthosite and/or gabbro rock. In some instances the two types of ore are in direct contact.

MINERALOGY OF THE ORE BODIES. The ores of the Sanford Lake district contain both ilmenite and magnetite as granular aggregates and disseminated grains. The ratio of total Fe: TiO_2 is generally 2:1 or greater in the anorthositic ore, and is less than 2:1 in the gabbroic ore. Gabbroic ore is finer-grained than anorthositic ore.

The magnetite grains are seldom homogeneous, and normally contain some ilmenite in solid solution and up to 35% of ilmenite as exsolution intergrowths. The intergrown ilmenite occurs as tabular plates oriented parallel to the octahedral planes of the magnetite or along the boundary between magnetite grains.

A second highly magnetic phase, ulvospinel (Fe_2TiO_4), has been identified as a fine network within the magnetite of the Sanford Hill-South Extension ores. This ulvospinel was first identified by Ramdohr (1956), using high magnification metallography. Kays (1965) used X-ray techniques along with chemical assay data to estimate that the magnetite grains may have an ulvospinel content of 34%.

About 0.5% vanadium occurs in solid solution within the magnetite. No separate vanadium mineral has yet been identified.

Ilmenite occurs as a matrix around magnetite and is observed under the microscope to be corroded by magnetite. It is finer-grained than magnetite and can be distinguished megascopically in coarse-grained anorthositic ore by its high luster and conchoidal fracture, compared to the dull luster and parting planes in magnetite. In finer-grained ore, the ilmenite and magnetite are more equigranular and cannot be recognized individually in hand specimens.

Other minor minerals identified in the Sanford Hill-South Extension ore body include chalcopyrite, sphalerite, molybdenite, prehnite, barite, leucoxene, scapolite, epidote, orthoclase, and quartz.

ORE GENESIS. Investigators have generally considered the Sanford Lake ores to have formed by magmatic segregation, as a result of such processes as simple segregation, filter-pressing of residual liquid with injection into the wall rock, and gravitational purification of the oxide melt by floating out of crystallized silicates with later injection, with replacement playing a minor role (Balsley, 1943; Bateman, et al., 1951; Buddington, et al., 1955; Evrard, 1949; Osborne, 1928; Stephenson, 1945). Gillson (1956), however, related the variations in the anorthosites and the origin of the gabbro and ore to a series of pneumatolytic replacements involving andesination of Marcy anorthosite. Later solutions were assumed to be the source of the ferromagnesian and ore minerals.

Heyburn (1960) made reference to two stages of gabbro emplacement, one of them associated with the anorthosites, and the second a gabbroic intrusive rich in iron and titanium with sufficient volatiles to allow partial replacement of anorthosite by ore minerals.

Kays (1965) proposed that during granulation and shearing of the anorthosite, calcium and aluminum were released from the original laboradorite plagioclase, resulting in andesinization, but by a different means than proposed by Gillson (1956). Creation of pressure gradients by rock fracturing was then postulated to cause iron, magnesium, and titanium to migrate to the low pressure fractured areas, where they reacted with the calcium and aluminum released by andesinization to form ferromagnesian silicates, garnet, ilmenite, and magnetite. The resulting replacement could be partial, as in gabbro, or complete over large volumes as in anorthosite.

Sun (1971) concluded that the Cheney Pond deposit is more likely to be of magmatic rather than replacement or metamorphic origin, stressing evidence of magmatic crystallization temperatures and the layered nature of the ore bands, which resembles the structure of ultramafic stratiform sheets.

Virginia Titanium Deposits—The titanium ores of Virginia are also associated with anorthosite, and are of two types: (1) rutile and ilmenite disseminated in the border facies of the anorthosite and bordering gneiss, and (2) rutile and/or ilmenite with apatite in the form of dikelike masses which Watson and Taber (1913) named nelsonite after the county in which they are most common. Deposits of both types have been mined commercially, utilizing mainly the soft, saprolite portions of the ore bodies which form an upper layer about 9 to 37 m (30 to 120 ft) thick.

Rutile was first mined commercially at Roseland, VA, in 1900 by the American Rutile Co. The ore was obtained by open cut mining of both saprolite and hard rock. The mine was operated intermittently until 1949 when increased labor costs, low grade ore, and low market prices for rutile led to a shutdown of operations. The ore consists almost exclusively of disseminated rutile in the anorthosite (Fish, 1962; Hillhouse, 1960).

Commercial surface mining for ilmenite began in 1930 from a large saprolite ore body along the Piney River on the old Warwick Tract. This mine was first operated by the Vanadium Corp. of America, and later by the Southern Mineral Products Corp. (Hillhouse, 1960) until it was acquired by the American Cyanamid Co. in 1944. In 1958 the company transferred operations to the deposit on the S.V. Wood property near Lowesville about 5 km (3 miles) to the west (Fish, 1962), supplying ilmenite to its nearby sulfate process pigment plant until 1971, when both mine and pigment plant were closed down, mainly because the supply of readily mined soft ore at this location was running out.

Fish (1962) estimated the saprolite deposits in the area to contain reserves of titanium-bearing material in excess of 20 million tons, averaging 7.0% TiO_2. Unweathered rock beneath the saprolite contains comparable titanium values, but mining and processing this ore would be more expensive. The ore is derived from two types of rocks; one, containing ilmenite as the dominant titanium mineral, is a diorite, and the other, containing ilmenite and rutile, is an anorthosite.

The nelsonite deposits are higher grade, but of limited size. Many of them are lenticular and pinch out at shallow depths (Fish and Swanson, 1964).

Davidson, et al., (1946) studied the Piney River deposit and concluded that it was emplaced as a large dike fingering into its walls. Replacement was agreed to have occurred, but this was not regarded as proof that the main development of ore was by replacement.

Theories proposed regarding the origin of the Virginia titanium ores in general include magmatic segregation from the anorthosite, favored by Watson and Taber (1913); differentiation from a granodiorite magma, suggested by Moore (1940); and replacement by invading solutions, proposed by Ross (1941, 1947) and Hillhouse (1960).

Yadkin Valley Deposit, North Carolina—The Yadkin Valley deposit is about 21 km (13 miles) north of Lenoir, in Caldwell County. The ore occurs as a series of narrow, close-spaced lenses which form a nearly continuous vein about 305 m (1000 ft) long. The ore consists of small masses of ilmenite disseminated throughout a talcose body which lies conformably with the enclosing quartzite and mica schist (Broadhurst, 1955). Some rutile occurs as fine inclusions in the mica, but was not recovered.

The ore zone is parallel to the gneissic structure, is about 9 m (30 ft) thick, and was mined by quarrying, running about 30 to 35% TiO_2. The mine was operated from 1942 to 1952 by the Yadkin Mica and Ilmenite Co., a subsidiary of Glidden Co. A flotation ilmenite concentrate was produced which contained 49 to 52% TiO_2 (McMurray, 1944), unusually high grade for a rock ilmenite.

Hot Spring County, Arkansas Deposits, (Magnet Cove, Christy, Hardy-Walsh)—These deposits occur in a complex mixture of alkalic igneous rocks intruding folded sedimentary and metamorphic rocks, many of which contain various quantities of titanium-bearing minerals (Fryklund and Holbrook, 1950). Much of the titanium is in the form of rutile or brookite, but many other titanium minerals have been identified. Rutile was recovered commercially from the Magnet Cove (Reed, 1949a) deposit from 1932 to 1944. A considerable amount of work has been done on ore dressing techniques for these ores (Fine and Frommer, 1952; Fine, et al., 1949). The features contributing most to the difficulty of utilizing these deposits are their comparatively low grade, the larger deposits averaging 3 to 6% recoverable TiO_2 (Fryklund, et al., 1954), the formation of large amounts of slimes during grinding, and the intimate association of ore minerals and gangue, resulting in low recoveries and comparatively low grade concentrates (Toewe, et al., 1971). No commercial output of titanium minerals has been reported from the Christy (Reed, 1949) or Hardy-Walsh deposits.

Other US Deposits—Other deposits of potential commercial importance include the titaniferous magnetite deposits in the Laramie Range, Wyoming (Diemer, 1941; Frey, 1946; Pinnell and Marsh, 1954). Metamorphic rocks were intruded by anorthosite, which in turn was cut by gabbro and titaniferous magnetite dikes. The main dike at Iron Mountain consists of a granular aggregate of homogeneous ilmenite and magnetite containing very small intergrown ilmenite inclusions. In the Taylor deposit, as much as 60% apatite is locally present. Resources of high grade ore containing about 45% Fe and 20% TiO_2 may amount to about 30 million tons. Further metallurgical research would be needed, and the geographical location is at present unattractive for commercial development.

Other large deposits occur in the San Gabriel Mountains in Los Angeles County, California, as disseminations in a gabbroic facies of an anorthosite (Moorhouse, 1938). Very large tonnages may be available, but the average grade is only about 4.5% TiO_2.

Smaller deposits occur in nearly every one of the western states, for example in Montana (Wimmler, 1946); Minnesota (Broderick, 1917); Wichita Mountains of Oklahoma (Merritt, 1939); Boulder County, Colorado (Jennings, 1913); and the San Juan district, Colorado (Singewald, 1913).

Canada: *Quebec*—The geology and general characteristics of the Canadian titanium deposits were well described in an extensive and thorough report by Rose (1969). He concluded that the titaniferous magnetite and ilmenite deposits in eastern Canada were formed by magmatic differentiation and injection, mainly in Precambrian time, presumably from 850 million to 1500 million years ago. There is an unmistakable genetic relationship between these deposits and anorthositic rocks. The anorthosite bodies, as exemplified by the Morin, St. Urbain, Lac St. Jean, Sept-Iles, and Lac Allard anorthosites, appear in general to be composite, multiple intrusions composed mainly of anorthositic and gabbroic (noritic) rock.

The main deposits include: the sill-like Lac Tio ilmenite-hematite deposits being mined by QIT-Fer et Titane, Inc. (formerly Quebec Iron and Titanium Corp.), estimated to hold about 100 million tons of high grade open pit material; the dikelike Magpie Mountain medium grade titaniferous magnetite deposits, averaging 43% iron and 6% titanium, probably containing over 250 million tons of open pit material; the St. Urbain deposits with over 20 million tons of high grade ilmenite-hematite; and several million tons of slightly lower grade material near Ivry, and in the Lac du Pin-Rouge area, near St. Hippolyte-de-Kilkenny. Substantial amounts of low-grade titaniferous magnetite and ilmenite occur at many other localities in the Morin, Lac St. Jean, Sept-Iles, Lac Allard, and St. Urbain anorthosite areas (Rose, 1969).

Kish (1972) studied the chemical composition of the Quebec deposits, particularly with regard to the vanadium content, and reported that vanadium concentrations of economic importance have so far been found only in the titaniferous magnetite deposits. The composition and origin of selected titaniferous deposits were discussed by Lister (1966).

ALLARD LAKE (LAC TIO) DEPOSITS. The geology of the area in which the Allard Lake deposits occur was investigated by Retty (1944), who noted numerous concentrations of ilmenite within a mass of anorthosite, along the shores of several lakes. These ilmenite showings, while not large enough to warrant commercial development, led to detailed exploration of the anorthosite area in 1946 by Kennco Explorations, Ltd., which resulted in the discovery of the large Lac Tio deposit. The deposit lies between Allard Lake and

Puyjalon Lake, about 40 km (25 miles) north of Havre St. Pierre on the north shore of the St. Lawrence River.

The geology of the area and the nature of the ilmenite deposits have been well described by Hammond (1949, 1952), and by Hargraves (1959). The most important lithological unit in the area is the Allard Lake anorthosite. It is one of several anorthosite masses occurring at intervals in the southeastern part of the Precambrian shield, in a line trending northeast from the Ontario-Quebec boundary to the Labrador coast. The Allard Lake anorthosite mass is about 145 km (90 miles) long, and 32 to 48 km (20 to 30 miles) wide, its length paralleling the Gulf of St. Lawrence. The several facies of the anorthosite range from almost pure feldspar rock through anorthositic gabbro, ilmenite-rich anorthosite, and norite. The norite occurs as steeply dipping sheets, as much as 6 km (4 miles) long and 914 m (3000 ft) thick, intruded into the anorthosite, and is rich in hemoilmenite and magnetite.

The Lac Tio deposit is a flat-lying, tabular body about 1097 m (3600 ft) long, and 1036 m (3400 ft) wide. With an estimated tonnage of 125,000,000 st of ilmenite averaging 32% TiO_2 and 36% Fe, it is the largest known ore body of its type in the world.

The ore consists of exsolution intergrowths of ilmenite and hematite, with coarse-grained ilmenite containing numerous blades and lenses of hematite up to 0.3 mm wide, which in turn may contain similarly shaped but smaller inclusions of ilmenite, in parallel orientation. The typical high grade ore contains about 75% ilmenite, 20% hematite, and 5% gangue minerals consisting of pyroxene, feldspar, and minor amounts of pyrite, pyrrhotite, and chalcopyrite.

The ore occurs in anorthosite and anorthositic gabbro and is identical in character with that in other deposits in the area. Inclusions of anorthosite are found in the ore, and do not show evidence of replacement by ilmenite. The ilmenite is very coarse-grained along its contacts with anorthosite, and the contacts are very sharp. Tiny dikelets of coarse granular ilmenite commonly cut the anorthosite at contacts with the ore bodies. On a basis of such direct field evidence, Hammond (1949, 1952) concluded that the Allard Lake ores are late magmatic, probably emplaced by a process of late gravitational liquid accumulation, with injection of the oxides into fractures within the anorthosite.

The ilmenite-hematite ore produced from the Lac Tio operation contains about 34% TiO_2 and 40% Fe. Because of its high iron content it is not used directly for titanium dioxide pigment manufacture, but is subjected to further concentration and electric furnace smelting at Sorel, Que., to produce iron metal, and a high-TiO_2 slag which is widely used as a feed for sulfate process pigment plants.

China: The largest titanium deposit in China is reportedly the 1.1 billion ton Panzihua titaniferous magnetite deposit containing about 7% titanium in the form of ilmenite, near Dukou, Sechuan province (Brady, 1981).

Norway: The iron-titanium provinces of Norway have been described by Geis (1971), Vokes (1968), Carstens (1957), Michot (1956), Hubaux (1956), and others. There are three types of ilmenite deposits: (1) ilmenite, (2) vanadium-bearing magnetite-ilmenite, and (3) apatite-bearing magnetite-ilmenite. Production of ilmenite has come only from Type 1 deposits (Storgangen and Tellnes), except for a small amount produced from a Type 2 deposit (Rodsand), as a byproduct from magnetite production. These three commercial deposits all occur in the Egersund anorthosite in southwestern Norway.

Finland: *Otanmäki Deposit*—The ilmenite-magnetite deposit at Otanmäki is located almost at the geographical center of Finland. The Otanmäki deposit and a neighboring deposit at Vuorokas were discovered in 1938 as a result of magnetic surveys by Veikko Okko of the Geological Survey of Finland (Harki, et al., 1956).

Rock Deposits—Potential Sources

Brazil: *Tapira and Salitre Titanium Deposits*—These deposits are unique in that their major titanium mineral is the anatase (or octahedrite) polymorph of TiO_2, rather than rutile or ilmenite. The deposits occur in an alkaline pipe, 6 km (4 miles) in diameter, about 48 km (30 miles) southeast of Araxá in Minas Gerais. According to the National Department of Mineral Production, proved and indicated reserves contain 108 million st of TiO_2 (Anon., 1980a). In places, anatase may make up as much as 70% of the ore. In preliminary tests concentrates containing up to 86% TiO_2 were produced with very low chromium and vanadium content, said to be suitable for producing TiO_2 pigments by the chloride process. These concentrates have some solubility in sulfuric acid (Anon., 1972).

Through 1979, about 12 million tons of anatase-bearing ore had been mined and stock-piled in conjunction with phosphate production.

Mexico: *Pluma Hidalgo Deposit, Oaxaca*—The Pluma Hidalgo Deposit occurs in an area known for some time to have scattered occurrences of rutile, and is located about 125 km (78 miles) south of the City of Oaxaca, the state capital. Starting in 1953, Republic Steel Corp. carried out an extensive program of exploration, but did not find high grade ore bodies large enough to justify a mining operation. In 1957, when the price of rutile dropped substantially, the project was discontinued.

Paulson (1964) described the dominant country rock as a quartz feldspar gneiss, or granulite, which may be classed as an anorthosite by analogy with the anorthosite associated with the similar Virginia titanium deposits. The common mineral association in ore zones is ilmenite, rutile, and apatite in a green rock that is mostly chlorite.

Sand Deposits—Principal Producing Countries

United States: All commercially important titanium mineral sand deposits in the United States are within the Atlantic and Gulf Coastal Plain geologic provinces. States which have established or potentially exploitable heavy-mineral-bearing sands are Florida, Georgia, New Jersey, and Tennessee.

Concentrations of heavy minerals in the coastal areas of the southeast are related to both recent and ancient marine shorelines, the latter of Pleistocene age. Up to seven old shorelines have been identified by different investigators (Cooke, 1941, 1945; Flint, 1940, 1942, 1947; MacNeil, 1949; Parker and Cooke, 1944), but the most widely recognized are: (1) Okefenokee, 46 m (150 ft), (2) Wicomico, 31 m (100 ft), (3) Pamlico, 8 to 11 m (25-35 ft), and (4) Silver Bluff, 2 to 3 m (8-10 ft) above sea level. All are regarded as lines of farthest marine transgression during interglacial and postglacial periods, and all have produced commercial heavy mineral deposits. Mining began where black sands were exposed on modern beaches, and when these deposits were exhausted, operations moved inland to the larger and harder-to-find deposits of the older elevated strands. Florida has three heavy mineral mines operating with possibilities for more in the future; Georgia's only deposit exploited to date recently has been worked out but, again, other occurrences of possibly economic size and grade are known.

In central New Jersey two operations have recovered titanium minerals and byproducts from Pliocene sands of the Cohansey formation, about 24 km (15 miles) inland from the present shore.

The host for heavy mineral deposits in western Tennessee which have been considered for mining is the Cretaceous McNairy sand out-cropping within the Mississippi embayment of the Gulf Coastal Plain. Past, present, and possible future ore bodies in each of the three areas of occurrence are considered in turn below.

Florida and Georgia—Storm-line concentrations of heavy minerals in modern beach sands have been mined for ilmenite at two locations on Florida's eastern beaches. Mining began at Pablo Beach near Ponta Vedra in 1916 (Martens, 1928), and Riz Mineral exploited a deposit on the ocean front near Vero Beach in later years. Similar modern beach, bar, and barrier island black sands are found on the northeast and west coasts of the peninsula and along the shore of the western panhandle. The ratios among species in heavy mineral suites in the western Florida occurrences are different from those in eastern deposits, probably because of dissimilar ultimate sources on opposite sides of the southern Appalachian Mountains. The potential ore reserves in the known near-shore concentrations are small, with the possible exception of Amelia Island in extreme northeastern Florida, Cumberland Island just to the north in Georgia, and several others of the Sea Island chain formed as barrier islands during the development of the Silver Bluff shoreline. Land values in the islands have increased because of their growing popularity as resort areas over approximately three decades since the heavy mineral deposits were first drilled, so that mining may not be economically competitive with tourism and recreation.

Elevated sand bars were mined by Riz Mineral south of Vero Beach and by Humphreys Gold Corp., on behalf of Titanium Alloy Manufacturing Co., in a Pamlico shoreline feature just west of Jacksonville Beach. These deposits are now mined out.

Just south of the St. Marys River, which forms Florida's northern boundary, and northeast of the community of Yulee, ITT-Rayonier owns land containing an unmined heavy mineral deposit which has been drilled by several organizations, most recently by Pennsylvania Glass Sand Corp., the ITT-Rayonier subsidiary. Ore grade lenses form low north-south ridges, interspersed with shallow swamps, and rest on clay and shell beds. The Yulee deposit is also a Pamlico shoreline development.

In Georgia other black sand concentrations, all related to the Pamlico shoreline, are located north and south of Brunswick and south of Savannah. Potential ore reserves are modest in all known cases, as they are at Yulee, but exploitation might be possible if the same mining equipment could be used at several deposits sequentially, and if a heavy mineral concentrate could be economically transported to a central separating plant.

TRAIL RIDGE, FL. The broad sand ridge that extends from the southern parts of Clay and Bradford counties in north-central peninsular Florida for 201 km (125 miles) northward into southeastern Georgia, called Trail Ridge, is thought to have formed by one or more of several processes in late Miocene or Pliocene time.

Du Pont's Trail Ridge ore body occupies the southern 29 km (18 miles) of the western part of the feature, and is from 2 to 3 km (1 to 2 miles) wide. The base of the ore body is at an elevation of 44 m (145 ft) to more than 61 m (200 ft) above sea level, and the average thickness is 11 m (35 ft). It overlaps barren coarse sand on the east. An indurated layer of concentrated organic material comprised of decomposed remnants of roots, branches, and trunks of trees underlies the central portion, and the western margin lies on pre-Pleistocene clayey sand in many places (Grogan, et al., 1964).

The heavy minerals are thinly layered and disseminated in brown, oxidized, cross-bedded sand. In random lenticular areas at various depths organic materials and clayey minerals cement the sand grains to form a poorly consolidated sandstone locally called "hardpan." The average grade in heavy minerals is low, only about 4%, but there are about 45% high-TiO_2 minerals in the heavy mineral suite (Pirkle and Yoho, 1970).

The titanium ore minerals are altered ilmenite, leucoxene, and a very little rutile. Staurolite, zircon, tourmaline, spinel, kyanite, sillimanite, monazite, corundum, and topaz are also present. Authigenic pyrite occurs in small amounts associated with present-day swamps. The absence of monazite, garnet, and epidote, which are common in the underlying formations and in other southeastern heavy mineral deposits, implies different source rocks and a different age for the Trail Ridge minerals. The absence of these three minerals makes the separation and marketing of staurolite possible.

Alteration of the titanium minerals by oxidation and leaching of iron has been extensive.

GREEN COVE SPRINGS, FL. About 32 km (20 miles) east-southeast of the south end of Trail Ridge an ilmenite-leucoxene-zircon-monazite heavy mineral deposit was mined by Titanium Enterprises, a joint venture of American Cyanamid and Union Camp Corp., until 1978 when mining ceased for economic reasons. In 1980 the property was purchased and mining resumed by Associated Minerals (U.S.A.) Ltd., Inc., a subsidiary of the Australian firm, Associated Minerals Consolidated Ltd. The ore body is near the eastern margin of the Duval Upland, which is thought to be a regressional beach ridge plain, and is probably related to an ancient shoreline at an elevation of 27 to 31 m (90 to 100 ft) above present sea level or to the Wicomico shoreline of MacNeil (1949).

The ore zone at Green Cove Springs is 16 to 19 km (10 to 12 miles) long, 1 km (¾ mile) wide, and it averages about 6 m (20 ft) thick. The heavy minerals comprise 3 or 4% of loose to slightly consolidated quartz sands which are underlain by subgrade quartz sands, by brown and gray sands containing some clay, by shell beds, and finally, at a depth of 31 m (100 ft) or more below surface, by limestone and dolomite.

The grain size of both quartz and heavy minerals is smaller than at Trail Ridge. Also, the heavy mineral suite, containing a significant percentage of monazite together with minor amounts of epidote and garnet, is different. It has been suggested (Pirkle, et al., 1974) that while the Trail Ridge concentration of heavy minerals was derived from the Northern Highlands of the northwestern part of peninsular Florida, the Green Cove Springs deposit was formed from the sands of the Duval Upland to the east.

BOULOUGNE, FL. Humphreys Mining Co. from 1974 to 1979 mined a heavy mineral deposit about 3 km (2 miles) south of the town of Boulougne on the St. Marys River, which forms the boundary between extreme northern Florida and Georgia. It was located near the eastern edge of the Duval Upland at an elevation above sea level the same as that at Green Cove Springs, and had similarly fine sands containing monazite, garnet, and epidote. The two deposits probably had a common genesis.

FOLKSTON, GA. Du Pont, in a program of systematic examination of elevated bars and old high level shorelines along the Atlantic Coast, discovered the Folkston heavy mineral deposit in 1952. It was situated in a flat, broad elevated area a few miles north of the Boulougne, FL, ore occurrence and the two may at one time have been one, later cut by the St. Marys River. The flat area is better described as a marine terrace rather than a sand bar or old shoreline, as it is up to 6 km (4 miles) wide and over 32 km (20 miles) long. The zone in which commercial deposits of heavy minerals were found, 8 km (5 miles) east of Trail Ridge, is not geomorphically different from the remainder of the terrace. It is a relatively thin layer of sand resting on clay. Three lenses were outlined in development drilling: the Main Area, the West Extension, and the North Extension. Each of the lenses was bounded by swamps, underlain by barren sand and clay, which were probably original depressions rather than erosion channels. Humphreys Mining Co. completed mining of the Main Area for Du Pont in 1974 and moved mining and primary concentrating equipment to Boulougne, FL.

New Jersey—The New Jersey Geological Survey, in an exploration program which grew out of observations of minor titanium mineral concentrations during investigation of monazite placers as sources of radioactive materials, identified the upper Tertiary sediments in the northern part of New Jersey's Coastal Plain as an ilmenite province in 1956 (Markewicz, 1969). The main heavy mineral formations were found to be the Miocene Kirkwood marine micaceous sand, silt, and clay, the Pliocene (?) Cohansey fluvial poorly sorted quartz sand, and the Pleistocene Cape May sands and gravels derived from the Kirkwood and Cohansey and from sediments farther inland. The Cohansey formation, which contains the greatest concentrations among these three, is ilmenite-rich in the northern third of its extent in New Jersey, in the vicinity of Lakehurst, about 48 km (30 miles) southeast of Trenton. Exploratory drilling by private companies resulted in the discovery of four ore bodies, and two heavy mineral mines were established.

LAKEHURST, NJ. The deposit of Glidden Pigments Group of SCM Corp. was located at Legler, about 3 km (2 miles) due north of Lakehurst, and was mined from 1962 to 1978. A mantle of Pensauken sand and gravel from 0.3 to 3.0 m (1 to 10 ft) thick overlay the Cohansey ore body, which was from 6 m (20 ft) to more than 12 m (40 ft) thick, with the average being 8 m (25 ft). A barren red sandstone 6 to 12 m (20 to 40 ft) thick underlay the ore, and beneath it was a zone in the Kirkwood formation, more than 12 m (40 ft) thick in places, averaging 4.5% heavy minerals. Only the Cohansey ore was mined.

MANCHESTER MINE. The Manchester mine of Asarco, Inc., also in the Cohansey formation, was about 97 km (60 miles) south of New York City, near Lakehurst, NJ. Geological features are similar to those at the Glidden operation at Lakehurst. Reserves were reported to be 180 million tons of sand averaging about 4% heavy minerals (Anon., 1974a) and 1.95% TiO_2 (Li, 1973). The operation was expected to have a 20- to 22-year life from 1973 when production began, based on a planned initial rate of 155,000 tons and a possible

ultimate rate of 185,000 tpy of ilmenite, averaging 63% TiO_2. All ilmenite produced was under a sales contract to Du Pont. The mine was closed in March 1982 because of escalating costs and the prospect of a long-term oversupply situation.

Tennessee—A reconnaissance exploration program conducted by Du Pont to investigate reports of surface heavy minerals in 1957-1958 led to discoveries of ore grade occurrences of titanium minerals in the Cretaceous McNairy formation. McNairy is a poorly consolidated, clay-bearing fine-grained sandstone, about 91 m (300 ft) thick, which outcrops in a north-south belt just west of Kentucky Lake. The dip is very gentle to the west. Heavy minerals, including ilmenite, leucoxene, rutile, zircon, monazite, and minor amounts of kyanite, staurolite, tourmaline, and xenotime, are concentrated in the lower portion of the McNairy above the underlying Coon Creek formation. The ilmenite is somewhat altered, and concentrates have shown an average of about 62% TiO_2.

Ethyl Corp. and Kerr-McGee carried out exploratory drilling programs during the period 1970-1972, and both companies outlined several ore bodies. Production plans were deferred because of environmental problems and changing markets.

Australia: Australia is the world's most important heavy mineral sand mining country, producing 64% of the rutile mined in 1982, and 38% of the ilmenite.

Heavy mineral concentrations occur widely around the Australian coast, and small black sand deposits have been reported in Tasmania, Victoria, South Australia, and the Northern Territory. Those at Nepean Bay on Kangaroo Island have been mined in the past. In addition to onshore beach-type deposits, offshore occurrences have been discovered and explored in shallow water along portions of the New South Wales-Queensland coasts. In 1982, however, mining was restricted to three areas: (1) the coast of New South Wales and Queensland from Newcastle to Gladstone, (2) near Bunbury, Capel, and Busselton on the west coast of Western Australia south of Perth, and (3) near Eneabba 29 km (18 miles) inland and about 225 km (140 miles) north of Perth.

As elsewhere, the heavy mineral grains were derived from the erosion of granites, intrusives, quartz reefs, and sandstones, and transported to the coast and concentrated by water and wind action. They occur in present-day beaches, fossil beaches, buried strand lines, and coastal dunes, and may be up to nearly 32 km (20 miles) inland as in Western Australia.

Ratios among individuals in Australian heavy mineral suites vary geographically. For example, between Sydney and Newcastle, rutile and zircon comprise about 90% of the concentrates, while farther north in New South Wales the rutile plus zircon content drops to 60 to 70%. Continuing northward, in southern Queensland the ilmenite level increases to about 60% of the heavy minerals. In the area between Busselton and Bunbury in Western Australia ilmenite makes up about 90% of the heavy mineral fraction, and near Eneabba concentrations contain 40 to 60% ilmenite, and about 10% rutile.

There is also a geographical variation in the chemical composition of ilmenite. Along the east coast ilmenite is generally too high in chromium to be suitable as a raw material for pigments manufacture by the sulfate process, and has found only a limited market in the Japanese steel-making industry. The chromium content does, however, decrease from south to north, and the ilmenite is apparently suitable for upgrading to rutile grade material by some beneficiation processes. In Western Australia ilmenites from the Bunbury-Busselton area average 54 to 57% TiO_2, while Eneabba ilmenite grades around 61% TiO_2.

Titanium mineral mining began in eastern Australia in 1934. Ore bodies were modern beach placers containing up to 50% heavy minerals, mostly rutile and zircon. All these high grade deposits have been exhausted over the past 40 years, and the average grade in the east by 1980 was about 3% heavy minerals. The increased cost of working lower grade deposits farther inland has been offset by improved techniques for mining and processing, and some ore bodies containing less than 0.25% TiO_2 as rutile (plus at least as much zircon) are now mined successfully on the east coast. In Western Australian ore bodies titanium minerals must be more highly concentrated because of the relative scarcity of rutile, and although there are wide variations within individual deposits, the average is between 5 and 10% heavy minerals. Mining of heavy mineral sands began in Western Australia in 1956.

The development of the large and important Eneabba ore field began in 1970. Rutile occurs in the heavy mineral suite there with ilmenite and leucoxene, so that with full-scale production commencing in 1974 and 1975, Western Australia became an important source of rutile for the first time.

Eastern Australia—Ore bodies are in ancient beaches and bars in New South Wales and parts of Queensland. On North Stradbroke and Fraser Islands, they are in high coastal dunes.

The opposition of conservationists to sand mining on Fraser Island led to the withdrawal of export licenses for Fraser Island concentrates at the end of 1976, and to restrictions on sand mining in other areas in Queensland and New South Wales.

Western Australia—Heavy mineral concentrations occur in three separate strand lines in the Bunbury-Busselton area: the present beach, the Capel line, and the Yoganup line, with the Yoganup line being oldest and 16 km (10 miles) inland. The TiO_2 content of the titanium mineral fraction of the heavy minerals increases with distance from the present coast.

Longshore drift in the area is southward and westward toward Cape Naturaliste, and it appears that longshore currents acting in the past sorted sands brought to the coast by north-westward-flowing rivers, creating heavy mineral deposits along beaches near their mouths. The existence of the old Yoganup and Capel lines indicates two past episodes of shoreline standstill.

Commercial concentrates of ilmenite, leucoxene, zircon, and monazite are produced from this region of Western Australia.

Exploration northward toward Perth and beyond resulted in the discovery of heavy mineral occurrences in old strand lines at Boyanup, Waroona, Bull's Brook, and Gin Gin. Rutile appears significantly in the heavy mineral suites at Bull's Brook and Gin Gin. The ilmenite is reported to be high in chromium content and it could be more suitable as a raw material for beneficiation than for direct sale.

ENEABBA. It was suggested by Lissiman and Oxenford (1973) that the heavy mineral deposits at Eneabba were formed by agencies acting similarly to those which formed the ore bodies in the Bunbury-Busselton area. Rapid erosion of a Mesozoic sedimentary rock high-land contributed sand to a westward-flowing stream which emptied into a bay partially protected on the southwest from south-to-north longshore currents by a headland. Longshore drift produced countercurrents within the bay which concentrated black sands on the beaches south of the mouth of the stream. There are several distinct strand lines marked by north-trending bands of high heavy mineral concentration at the bottom of the ore zone, and they are thought to have been formed during relatively brief periods of sea level standstill. The strand lines are at successively lower levels from east to west. The geologic history of the ore field is made complex by evidence of reworking of the higher strand lines by waves which must have been associated with a sea which rose again, perhaps repeatedly, and by wind which created disseminated heavy mineral sands between and over the strand line deposits. Ore bodies in the northern portion of the district, which is about 16 km (10 miles) long and over 2 km (1½ miles) wide, are thought to have been formed by wind concentration of heavy minerals from black sand dunes then exposed in areas to the south.

Total reserves of the Eneabba district are in excess of 20 million tons of heavy minerals. Rutile, ilmenite, leucoxene, zircon, and monazite are present in variable proportions from one area to another, but the average ratios may be 1:5:0.2:2:0.2, respectively.

KING ISLAND. Beach placers containing heavy minerals are located along the central part of the east coast from Naracoopa some 11 km (7 miles) along Sea Elephant Bay north to Cowper Point. The heavy mineral content is high, up to about 60%. Rutile and zircon occur in about equal proportions. Accessory minerals are leucoxene, ilmenite, magnetite, garnet, and cassiterite.

China: Coastal sand deposits in Guandong and Guangxi provinces contain ilmenite, zircon, and other heavy minerals (Brady, 1981).

India: At one time India was a leading producer of ilmenite from the state of Kerala (formerly Travancore-Cochin). The beach sands were mined in the Manavalakurichi (M.K.) area and later the Quilon deposit of ilmenite near Chavra was put into production. These deposits supplied the bulk of the titanium ore used by the US prior to World War II.

The two deposits have more differences than similarities. The ilmenite in the M.K. deposit analysed only 54% TiO_2 and the sand was rich in garnet and monazite. The ilmenite in the Quilon deposit analyzes about 60% TiO_2. The sand carried almost no garnet and is high in monazite in only two places.

Malaysia: Substantial tonnages of ilmenite are produced in Malaysia as a byproduct of tin mining either as rough concentrates in mills using magnetic separation techniques or in the form of *Amang*, a crude mixture of heavy minerals which has to be further treated to recover the ilmenite content. The main centers of production are at Perak and Ipoh. The deposits are alluvial in nature.

Sierra Leone: A deposit along the Sherbro River was originally developed as a joint venture of Pittsburgh Plate Glass Co. and British Titan Products. Although relatively high grade (±1.5% TiO_2) the sands were extremely fine and disseminated in lumpy lateritic clay. Reserves are estimated to be between 3 and 30 million tons of contained rutile. The ore problems plus mechanical and political difficulties caused Sherbro Minerals to shut down in 1971 after only two years of operation.

Bethlehem Steel acquired 85% ownership of the property in 1976 (15% Nord Resources) and conducted an extensive prospecting and geological investigation that extended into 1977, finding higher grade deposits (±2.5% TiO_2) than that mined by Sherbro. Improvements were made in both the wet and dry flowsheets and as of 1981 the operation now called

Sierra Rutile Limited was approaching specifications and design capacity of 110,000 stpy.

Rutile grades are highest in the topsoil, averaging 2.5% TiO_2, and in the basal sands and gravels, with up to 3.0% TiO_2. The average grade of the Mogbwemo deposit where Sierra Rutile began mining in 1979 is over 2.0% TiO_2 (Anon., 1981).

South Africa, Republic of: Shipments of rutile and zircon were started in 1977 from the Richards Bay mineral sands operation in Natal. Richards Bay Minerals manages the combined affairs of Tisand Pty. Ltd. and Richards Bay Iron and Titanium Pty. Ltd., the companies responsible for the operation of the mine and the smelter, respectively. A high-TiO_2, low-magnesium-plus-calcium slag (RB slag) containing 85% TiO_2 has been produced from ilmenite since 1978, and is reportedly suitable for use in either the sulfate or chloride titanium dioxide pigment processes. Since 1980, production rates have approached planned annual capacities of 440,000 tons of RB slag, 220,000 tons of low-manganese iron, 53,000 tons of rutile, and 110,000 tons of zircon. Proven reserves were estimated at 770 million tons grading 5% ilmenite, 0.3% rutile, and 0.65% zircon. Principal shareholders at the end of 1980 were QIT-Fer et Titane, Inc. (32%), General Mining Union Corp. (30%), and the Industrial Development Corp. of South Africa Ltd. (16%).

Sri Lanka: Sri Lanka contains extensive beach deposits of titanium-bearing sands at Pulmoddai, Tirukkovil, Kelani River, Kalu River, Modoragam River, Kudremalai Point, Negombo, and Induruwa.

The Pulmoddai area contains 5.6 million st of titaniferous material with 2.451 million st of contained TiO_2. The deposit extends for a distance of 7 km (4½ miles), has a maximum width of about 91 m (300 ft), and a thickness of about 2.4 m (8 ft). There is no overburden. The deposit contains about 80% ilmenite and rutile.

The separation of rutile has been adversely affected by the presence of excessive amounts of residual ilmenite and quartz in the tailings. The separation of zircon has been hampered by inadequate water and insufficient wet tabling equipment to handle the extremely fine-grained Pulmoddai ore.

Sand Deposits—Potential Sources

Brazil: Brazilian deposits of heavy minerals on ocean beaches occur in a zone about 161 km (100 miles) long, extending north from the northeast corner of the state of Rio de Janeiro up into the state of Espirito Santo as far north as the Rio Doce. There is also a zone in southern Bahia. Deposits are known to exist near Natal in the state of Rio Grande do Norte.

Egypt: Rich deposits of black sands occur along the northern beaches of the Nile Delta for about 241 km (150 miles). There are two types: one is dark in color and contains about 70 to 90% heavy minerals and a second which is grayish-yellow to dark gray and contains about 40% heavy minerals. The deposits cover areas 10 km (6 miles) long and a few meters (yards) wide and up to 0.5 m (1½ ft) in thickness.

New Zealand: Ilmenite and associated heavy minerals are found in New Zealand on the west coast of South Island from Jackson Bay to Karamea and on the west coast of North Island at the mouth of Waikato River, Muriwai, and at Manukan Heads. Total reserves near Westport are estimated to be between 17 to 31 million tons. Tests of the sand from Westport indicate the titanium content of the ilmenite averages about 47%

Approximately one billion tons of sand are available between Karamea and Jackson Bay from which ilmenite could be recovered. These deposits on the west coast of South Island are believed to be one of the world's major reserves of low chromium ilmenite.

South Africa: Ilmenite-bearing sands occur in scattered deposits along the east coast of the Republic of South Africa from the area of East London to the Mozambique border and northward. Ilmenite sands also occur along the west coast particularly in the Vannhyrodorp district.

Titanium minerals are found on the east coast in three types of deposits: (1) older red and brown coastal sands, (2) recent dunes and beach sands, and (3) alluvium in coastal lagoons.

Older red and brown coastal sands occur in a belt of undulating fixed dunes which is roughly parallel to the coast. The belt ranges from a few meters (yards) to 6 km (4 miles) or more in width. Some dunes are as much as 152 m (500 ft) high but the average height is much less.

Ilmenite concentrates occur in the alluvium in coastal lagoons, i.e., lagoon sands and in the bars at river mouths.

On the west coast, ilmenite-bearing sands exist in the Vannhyrodorp District between Strandfontein and the mouth of the Zout River. One deposit, a recent dune, skirts the coast for 6 km (4 miles) and another on the Geelwae Karoo is found in a narrow modern beach about 8 km (5 miles) long. The sands are unconsolidated and fine to medium grained.

Titaniferous sandstone containing up to about 50% ilmenite with lesser amounts of rutile and zircon occur in the Ecca series of the Karoo system. They outcrop in several areas, principally near Bothaville about 483 km (300 miles) from the east coast in the Orange Free State. Total reserves are estimated at 85 million tons of material, some of which might be recovered with future technological improvements.

Uruguay: Black sands are found northeast of Montevideo at Agua Dulces, in the state of Rocha. The sands extend along the coast for a distance of 11 km (7 miles) with an average depth of 5.5 m (18 ft). Reserves are estimated at 3.3 million tons of heavy minerals, with an average 2.5% heavy mineral concentrate.

China: Titanium sand deposits are known at Luanping and Chente, and the coast of Jiangsu province.

Korea: Resources of 2,535,000 tons with 507,000 tons of contained TiO_2 have been identified.

Mozambique: Rutile and ilmenite are known to exist on the coastal beaches of Mozambique north of 16°30' latitude and near Vila Luisa, about 40 km north of Lourenco Marques.

USSR: Deposits have been identified along the Black Sea, Aldan River, Boludka River, Pit River in Siberia, Sylvista River, Chusovaya River, and Vizhai River in the Urals.

EXPLORATION AND EVALUATION

Hardrock Ilmenite Deposits

Hardrock ilmenite deposits, because of their inherent magnetic properties, are readily amenable to the application of aero and ground magnetic geophysical surveys. With few exceptions, these deposits respond to such an application by reflecting abnormally positive magnetic intensities (gammas).

Examples can be cited, however, where negative magnetic anomalies are associated with such deposits. In such cases, the ore occurs either as a titaniferous-hematite concentration or a titaniferous-magnetite occurrence when the titanium content is greater than the iron content. Theoretically, it would seem possible that a commercial deposit may exist where no magnetic response would be obtained. Such a deposit has not yet been found.

Once such anomalies are mapped, further exploration may take place in the form of detailed surface geological observations and ultimate drilling to first test the anomalies and hopefully to delineate a viable ilmenite deposit.

MacIntyre Development: The original discovery of magnetite-ilmenite deposits at MacIntyre Development, Tahawus, NY, was made from surface outcrops. Extension of these outcrop locations has been made through use of geophysical methods and diamond drilling.

Geophysical—Since the ore is magnetic, the geophysical methods used have been for the purpose of outlining magnetic anomalies. Early work was by dip needle readings over a grid pattern extending outward from surface outcrops. With the advent of the aerial magnetometer the entire property was flown. This produced anomalies well beyond any previous work extending to areas of deep overburden and underlake area.

Aerial work was followed with detailed ground magnetometer readings taken on 15-m (50-ft) spacings along section lines 30 m (100 ft) apart. This served to guide the location of diamond drill holes for exploratory drilling on one or two sections with drill holes spaced 122 m (400 ft) apart on section.

The initial ground magnetometer work on any anomaly is a section across the center of the anomaly at right angles to the long axis indicated from aerial work. The magnitude and gradient of the ground anomaly compared to the magnitude and gradient of the aerial anomaly, along the same section, gives an indication of depth and configuration of the source of the anomaly.

Drilling—If initial diamond drilling indicates an economic deposit, development drilling proceeds on a set grid pattern of 61 or 91 m (200 or 300 ft). Intermediate holes are drilled where necessary to fill in major gaps in geologic interpretation of structure, or ore continuations.

Core Sampling—Drill cores are visually logged and split for chemical assay. In ore areas, samples are taken of the entire core and composited in 1.5-m (5-ft) intervals for assay. In waste areas samples may be taken every 1.5 m (5 ft) and composited for a maximum of 8 m (25 ft) of core. Routine assaying is done to determine TiO_2 and total iron content. Based upon past experience, the nominal cut-off for ore is 13.5% TiO_2.

Mill Testing—Metallurgical laboratory work is done using core assay rejects to determine milling characteristics. Concentrates from the laboratory work are used for testing end use performance. Complete analyses are run on concentrates to determine presence and quantity of detrimental elements.

Tellnes Deposit, Norway: The Tellnes ilmenite deposit near Hauge-I-Dalane in Norway was discovered in 1954 as a result of an airborne magnetometer survey carried out for Titania A/S, with the object of locating new sources of ilmenite to provide an alternative to developing additional underground ore at Storgangen. Anomalies were also found at four other locations, so that ground geological study and diamond drilling had to be done at all of these sites to be certain that the best deposit was selected for development. A total of 10,058 m (33,000 ft) of drill holes was completed in 1956, and showed the total amount of ore in the Tellnes deposit to be about 300 million tons, averaging 18% TiO_2 (Brun, 1957; Anon., 1978a).

Allard Lake (Lac Tio) Deposit, Quebec: Preliminary investigation of the original ilmenite discoveries at Allard Lake (Hammond, 1949), indicated that all of these were too small to be of commercial importance. A detailed exploration program was carried out in 1946 to more thoroughly explore the area of 647 km² (250 square miles) where ilmenite had been found along certain of the lakeshores.

Otanmäki Deposit: The Otanmäki ilmenite-magnetite deposit and a neighboring deposit at Vuorokas were discovered in 1938 as a result of magnetic surveys by the Geological Survey of Finland (Harki, et al., 1956).

Sand Deposits

Exploration: There are only a few large areas of the world where the granite-clan rocks and high-grade metamorphic gneisses which are likely to contain ilmenite (not titaniferous-magnetite) and rutile are close enough to continental margins to have contributed their erosion products to the sediments of coastal plains. Well-sorted sands are much more likely hosts than unsorted sands. These are the areas on which exploration efforts should be focused. Since the alteration of ilmenite to remove iron is aided by humic acid developed by the decomposition of organic material near the water table in hot and humid climates, it follows that the highest TiO_2 ilmenites are more likely to be found in the tropical and temperate regions of the world.

Titanium minerals are dark-colored and their concentration, as in black beach sands, tends to be fairly readily noticeable against the light brown or white quartz. Many sand ore bodies, therefore, have been discovered through surface observation of high-grade placer zones formed on beaches and along the courses of streams, and by following their traces into the larger, lower grade concentrations which constitute economic ore bodies.

There are areas in which potential heavy mineral concentrations in ancient beach sands may be masked by younger sand, gravel, or soil. Exploration under these circumstances then involves interpretation of geomorphic and subsurface geologic data to define areas which could have been beaches or dunes in the past, and then drilling to obtain samples.

Field Techniques—In areas where heavy mineral concentrations are suspected but where the concentration level is difficult to estimate because of dissemination, as in dunes, hand panning of samples is an excellent method of rapidly producing a clean concentrate. A hand magnet gives a quick identification between ilmenite and magnetite or titaniferous-magnetite. The presence of potentially valuable byproduct zircon, monazite, kyanite-sillimanite, etc., can be determined with a hand lens. A Geiger counter or scintillometer also helps in identifying radioactive minerals such as monazite and zircon. Wind and current action can develop very high-grade layers of heavy minerals on the surface which have no economic significance; thus samples for panning and examination should be taken from the maximum convenient depth. Often an idea of the vertical distribution of heavy minerals can be gained from observation of wave-cut cliffs in sands behind the beaches, cutbanks in stream margins, road cuts, construction excavations, and material encountered in sinking or drilling wells. Dark, heavy minerals may even be observed in the sand brought to surface by burrowing insects and small animals.

Hand augering with a two-man crew is possible to a depth of 6 m (20 ft) or more with a jointed auger stem and is reasonably rapid and inexpensive. Some heavy-mineral bearing sands contain so much clay that they are difficult to hand auger and a mechanical auger mounted on a light off-the-road vehicle (like a post-hole digger) is preferable. The samples so obtained are not precise because of the contamination of the hole by sand falling into it. The hole cannot be advanced nor reliable samples obtained below the water table when the material is free-flowing and easily disturbed by the auger.

Drilling—For the first phase of development, drilling equipment should have a depth capability of at least 15 m (50 ft) and preferably 30.5 m (100 ft). The selection of the type of drill will depend upon the "stiffness" of the ground as influenced by clay, hardpan, caliche, indurated and iron-oxide cemented layers; the presence of roots, stumps, and other organic material; as well as upon the elevation of the water table.

The truck-mounted jet drill is a popular exploration and development tool of the southeast US. One-inch flush-jointed steel pipe, with a chisel lower end or bit and perforations near the bit to permit water to jet forward and downward, is attached to a hammer of about 136 kg (300 lb) which is activated in 0.3-m (1-ft) strokes by a hand-held rope about a capstan winch. Water is supplied to the bit by a swivel connection above the hammer and by a duct through it. As the hammer rises, falls, and turns, it causes the bit to chop and churn into the ground and drives a 59-mm (2-in.) casing downward. The water pipe is within the 59-mm (2-in.) casing. The bit of the water pipe is about 25.4-mm (1-in.) or less ahead of the casing when casing and hammer are in contact. The sand sample is flushed out of the hole continuously by water rising in the annulus between water pipe and casing, and is collected at the surface in vessels or tubs. A jet drill under favorable conditions with short moves between holes can drill about 61 m/d (200 fpd) of hole. Its disadvantages lie in its inability to penetrate very hard layers, and the casing can stick in tenacious clay.

The auger-within-an-auger type of drill has good depth and penetration capabilities, particularly in the less free-flowing sands. The internal auger is advanced about 1.5 m (5 ft) to collect a sample on its flights, then the external auger is put down to the same depth to form a casing, and the internal auger is withdrawn for sample collection. There is some possibility of particularly fluid sands below the water table invading the casing when the internal auger is raised.

In Australia extremely light, gasoline-powered augers, portable and operable by two-man crews, have been developed and used for rapid, low-cost depth investigations. They are particularly successful in high, clean dunes.

A truck-mounted rotary drill driven by rapid strokes of a hydraulic hammer is also very useful. The drill string is a double pipe, and compressed air forced down the inner pipe returns the sample in the annulus between the drill-pipe and casing. A drag bit with hardened cutting edges may be used.

It is important in all types of machine drilling to avoid contamination of samples by material originating higher in the walls of the hole. The hole and casing should be flushed free of loose sand after collecting each sample, and before the bit is advanced farther.

The nature of the sand and other materials penetrated by the drill is usually logged by a geologist, who makes a rough estimate of heavy mineral content and variation by panning a small portion of each sample.

Depending on hole diameter, a 0.9 or 1.5-m (3- or 5-ft) sample may be too large for laboratory requirements. If so, a simple heavy duty riffle may be used to reduce the weight to about 2 kg (5 lb). The bulk of the samples may then be bagged or drummed and left at the hole site for future reference.

Geophysical Techniques—Some titanium mineral ore bodies contain sufficient iron to be detectable by magnetometers in ground or airborne surveys, but many are too dilute and shallow to respond.

If monazite or radioactive zircon is present, low-level radiometric surveys may help in exploration. Scintillometers and Geiger counters are sometimes useful on the ground to define horizontal limits of ore and to detect internal variations in heavy mineral concentrations close to the surface.

The problem of downward contamination during drilling can lead to a misrepresentation of the total depth of ore because the heavy minerals, being higher in specific gravity, are hardest to remove by flushing with air or water and may be reported even when the hole has been advanced into barren underlying sand. A scintillometer probe with a surface recording instrument can show a sudden decrease in radioactivity from the walls of the hole at the true bottom of the ore. Radiometric surveying of all or at least some of the holes drilled during development of an ore body is a highly desirable check on the fidelity of drilling and sampling.

Evaluation of Deposits: An economic titanium mineral deposit must have reserves large enough to support depreciation over a period of at least 10 to 20 or more years. The capital investment in 1980 was in the range of $75 to $80 million in the US for a mine and mill plant with an output of 100 to 200 thousand stpy of ilmenite (or equivalent rutile) with given "normal" geologic parameters. Significant contributions can be made by zircon and other byproducts. Another general rule is that a new and separate ore body, if its production is to be all ilmenite which cannot be treated in an existing mill, should have a minimum reserve of about 1 million tons of recoverable TiO_2 in the titanium minerals. Small, high-grade concentrations are uneconomic under the present conditions.

The definition of economic reserves depends, of course, upon many factors, among them:

1) Cost of mining and milling, as influenced by depth of overburden (if any); cost of surface and mineral rights; and availability of water, power, labor, and transportation facilities for bulk shipments.

2) Recoverability in mining and milling.

3) Cost of treatment and disposal of waste slimes.

4) Cost of waste water treatment and land reclamation.

5) Distance to markets and cost of transport.

6) Ability of markets to absorb the type of titanium minerals to be produced, and prevailing prices for titanium minerals and byproducts.

BIBLIOGRAPHY AND REFERENCES

Anon., 1972, "Brazilian Titanium," *Mining Journal,* Vol. 278, No. 7121, Feb. 11, pp. 118-119.
Anon., 1974, "Pulmoddai's Mineral Sands," *Industrial Minerals,* No. 77, Feb., p. 27.
Anon., 1974a, "U.S. TiO₂ Mine on Stream," *Mining Magazine,* Vol. 130, No. 1, Jan., p. 7.
Anon., 1977, "RBM Progress Report," Sep., Richards Bay Minerals, 4 pp.
Anon., 1978a, "Titania: The Largest Producer of Titanium Minerals in Europe," *Mining Magazine,* Vol. 139, No. 4, Oct., pp. 365-371.
Anon., 1978b, "Rautauruukki—A Major Force in World Vanadium Supplies Is Still Expanding," *World Mining,* Mar., pp. 44-46.
Anon., 1980a, "Titanio, Anuário Mineral Brasileiro," Brasilia, Vol. IX, p. 358.
Anon., 1980b, "Australia's Mineral Resources: Mineral Sands," Australian Department of Trade and Resources, 10 pp.
Anon., 1980c, "Australian Mineral Sands Processing Industry— Potential for Expansion," Commonwealth/State Joint Study

Group on Raw Materials Processing, Australian Government Publishing Service, Canberra, pp. 17-18.

Anon., 1980d, "South Africa—Mining at Richards Bay," *Mining Journal,* Vol. 295, No. 7579, Nov. 21, pp. 411-413.

Anon., 1981, "Sierra Rutile," *Mining Magazine,* Vol. 144, No. 6, June, pp. 458-465.

Bachman, F.E., 1914, "The Use of Titaniferous Ores in the Blast Furnace," *Iron and Steel Industry Yearbook,* pp. 370-419.

Balsley, J.R., Jr., 1943, "Vanadium-Bearing Magnetite-Ilmenite Deposits Near Lake Sanford, Essex County, New York," Bulletin 940-D, US Geological Survey, pp. 99-123.

Barksdale, J., 1966, *Titanium, Its Occurrence, Chemistry, and Technology,* 2nd ed., Ronald Press, New York, 691 pp.

Bateman, A.M., et al., 1951, "Formation of Late Magmatic Oxide Ores," *Economic Geology,* Vol. 46, No. 4, June-July, pp. 404-426.

Beals, M.D., and Merker, L., 1960, "Three New Single Crystal Materials," *Materials in Design Engineering,* Jan., pp. 12-13.

Bishop, E.W., 1956, "Geology and Ground-Water Resources of Highlands County, Florida," Report of Investigation 15, Florida Geological Survey, 115 pp.

Brady, E.S., 1981, "China's Strategic Minerals and Metals—Titanium," *The China Business Review,* Vol. 8, No. 5, Sep.-Oct., pp. 62-65.

Broadhurst, S.D., 1955, "The Mining Industry in North Carolina from 1946 through 1953," Economic Paper No. 66, North Carolina Dept. of Conservation and Development, Div. of Min. Resources, pp. 26-27.

Broderick, T.M., 1917, "The Relation of the Titaniferous Magnetites of Northeastern Minnesota to the Duluth Gabbro," *Economic Geology,* Vol. 12, No. 8, Dec., pp. 663-696.

Brooks, H.K., 1966, "Geological History of the Suwanee River," *Geology of the Miocene and Pliocene Series in the North Florida-South Georgia Area,* N.K. Olson, ed., Guidebook for Atlantic Coastal Plain Geological Assn., 7th Field Trip and Southeastern Geological Society, 12th Field Trip, pp. 37-45.

Brun, R.M., 1957, "The Tellnes Story," *Ilmeniten* TITANIA, A/S, Norway, Summer issue.

Buddington, A.F., 1939, "Adirondack Igneous Rocks and Their Metamorphism," *Geological Society of America Memoir* 7, pp. 19-48.

Carstens, H., 1957, "Investigations of Titaniferous Iron Ore Deposits, Part I Gabbros and Associated Titaniferous Iron Ore in West-Norwegian Gneisses," *K. Norske Vidensk. Selsk. Skr.,* No. 3, 67 pp.

Cooke, C.W., 1941, "Two Shore Lines or Seven?," *American Journal of Science,* Vol. 239, No. 6, pp. 457-458.

Cooke, C.W., 1945, "Geology of Florida," Bulletin 29, Florida Geological Survey, 339 pp.

Davidson, D.M., et al., 1946, "Notes on the Ilmenite Deposit at Piney River, Virginia," *Economic Geology,* Vol. 41, No. 7, Nov., pp. 738-748.

Diemer, R.A., 1941, "Titaniferous Magnetite Deposits of the Laramie Range, Wyoming," Bulletin No. 31, Geological Survey of Wyoming, 23 pp.

Evrard, P., 1949, "Differentiation of Titaniferous Magmas," *Economic Geology,* Vol. 44, No. 3, May, pp. 210-232.

Fine, M.M., and Frommer, D.W., 1952, "Mineral Dressing Investigation of Titanium Ore from the Christy Property, Hot Spring County, Arkansas," Report of Investigations 4851, US Bureau of Mines, 7 pp.

Fine, M.M., et al., 1949, "Titanium Investigations . . . The Laboratory Development of Mineral Dressing Methods for Arkansas Rutile," *Mining Engineering,* Vol. 1, No. 12, pp. 447-452.

Fish, G.E., Jr., 1962, "Titanium Resources of Nelson and Amherst Counties, Virginia (In Two Parts) 1. Saprolite Ores," Report of Investigations 6094, US Bureau of Mines, 44 pp.

Fish, G.E., Jr., and Swanson, V.F., 1964, "Titanium Resources of Nelson and Amherst Counties, Virginia (In Two Parts) 2. Nelsonite," Report of Investigations 6429, US Bureau of Mines, 25 pp.

Flint, R.F., 1940, "Pleistocene Features of the Atlantic Coastal Plain," *American Journal of Science,* Vol. 238, No. 11, pp. 757-787.

Flint, R.F., 1942, "Atlantic Coastal 'Terraces'," *Washington Academy of Sciences Journal,* Vol. 32, No. 8, pp. 235-237.

Flint, R.F., 1947, *Glacial Geology and the Pleistocene Epoch,* John Wiley, New York, 589 pp.

Force, E.R., 1980, "Is the United States Geologically Dependent on Imported Rutile?" Presented at 4th Industrial Minerals International Congress, Atlanta, GA, 4 pp.

Force, E.R., et al., 1976, "Geology and Resources of Titanium," Professional Paper 959-A through F, US Geological Survey.

Frey, E., 1946, "Exploration of Iron Mountain Titaniferous Magnetite Deposits, Albany County, Wyoming," Report of Investigations 3968, US Bureau of Mines, 37 pp.

Fryklund, V.C., Jr., and Holbrook, D.F., 1950, "Titanium Ore Deposits of Hot Spring County, Arkansas," Bulletin No. 16, Arkansas Research and Development Comm., Arkansas Div. Geology, 173 pp.

Fryklund, V.C., Jr., et al., 1954, "Niobium and Titanium at Magnet Cove and Potash Sulphur Springs, Arkansas," Bulletin 1015-B, US Geological Survey, pp. 23-57.

Garnar, T.E., Jr., 1980, "Heavy Minerals Industry of North America," Presented at 4th Industrial Minerals International Congress, Atlanta, GA, 13 pp.

Geis, H.P., 1971, "A Short Description of the Iron-Titanium Provinces of Norway, with Special Reference to Those in Production," *Minerals Science Engineering,* Vol. 3, No. 3, pp. 13-24.

Gillson, J.L., 1959, "Sand Deposits of Titanium Minerals," *Trans. SME-AIME,* Vol. 214, pp. 421-429; *Mining Engineering,* Vol. 11, No. 4.

Grogan, R.M., et al., 1964, "Milling at Du Pont's Heavy Mineral Mines in Florida," *Milling Methods in the Americas,* N. Arbiter, ed., Gordon and Breach, New York, pp. 205-229.

Gross, S.O., 1968, "Titaniferous Ores of the Sanford Lake District, New York," *Ore Deposits in the United States, 1963/1967,* John D. Ridge, ed., AIME, New York, Vol. 1, pp. 140-153.

Guimond, R., 1964, "Quebec Iron and Titanium Corporation, A Study in Growth," *Canadian Mining Journal,* Vol. 85, No. 11, pp. 47-53.

Guise, F.P., et al., 1964, "Titanium in the Southeastern United States," Information Circular 8223, US Bureau of Mines, 30 pp.

Hammond, P., 1949, "Allard Lake Ilmenite Deposits," *Canadian Mining & Metallurgical Bulletin,* Vol. 42, pp. 117-121.

Hammond, P., 1952, "Allard Lake Ilmenite Deposits," *Economic Geology,* Vol. 47, No. 6, Sep.-Oct., pp. 634-649.

Hargraves, R.B., 1959, "Petrology of the Allard Lake Anorthosite Suite and Paleomagnetism of the Ilmenite Deposits (Quebec)," Ph.D. Thesis, Princeton University, Princeton, NJ, May, 193 pp.

Harki, I., et al., 1956, "Discovery and Mining Methods at Finland's Largest Fe-Ti-V Mine," *Mining World,* Vol. 18, Aug., p. 62.

Heyburn, M.M., 1960, "Geological and Geophysical Investigation of the Sanford Hill Ore Body Extension, Tahawus, New York," Unpublished M.S. Thesis, Syracuse University, Syracuse, NY, 48 pp.

Hillhouse, D.M., 1960, "Geology of the Piney River-Roseland Titanium Area, Nelson and Amherst Counties, Virginia," Unpublished Ph.D. Thesis, Virginia Polytechnic Institute, Blacksburg, VA, 169 pp.

Hoyt, J.H., 1967, "Pleistocene Shore Lines: Guide to Tectonic Movements, Northern Florida and Southern Georgia," *Abstracts, 1967 Annual Meeting,* Geological Society of America, New Orleans, LA, p. 104.

Hubaux, A., 1956, "Various Types of Black Ores of the Egersund Norway Region," Bulletin 79, Ann. Soc. Geol. Belg., pp. 203-215.

Jennings, E.P., 1913, "A Titaniferous Iron Ore Deposit in Boulder County, Colorado," *AIME Trans.,* Vol. 44, pp. 14-25.

Kays, M.A., 1965, "Petrographic and Modal Relations, Sanford Hill Titaniferous Magnetite Deposit," *Economic Geology,* Vol. 60, No. 6, Sep.-Oct., pp. 1261-1297.

Kish, L., 1972, "Vanadium in the Titaniferous Deposits of Quebec," *CIM Bulletin,* Mar., pp. 117-123.

Li, T.M., 1973, "Startup of Manchester Mine and Mill Boosts U.S. Production of Primary Ilmenite," *Engineering & Mining Journal,* Dec., pp. 71-75.

Lissiman, J.C., and Oxenford, R.J., 1973, "The Allied Mineral N.L. Heavy Mineral Deposit in Eneabba, W.A.," *Conference Volume,* Australasian Institute of Mining & Metallurgy, pp. 153-161.

Lister, F.G., 1966, "The Composition and Origin of Selected Iron-Titanium Deposits," *Economic Geology,* Vol. 61, No. 2, Mar.-Apr., pp. 275-310.

Llewellyn T.O., and Sullivan, G.V., 1980, "Recovery of Rutile from a Porphyry Copper Tailings Sample," Report of Investigations 8462, US Bureau of Mines, 18 pp.

Lynd, L.E., 1983, "Titanium," *Mineral Commodity Profile,* US Bureau of Mines, 17 pp.

MacNeil, F.S., 1949, "Pleistocene Shore Lines in Florida and Georgia," *Shorter Contributions to General Geology,* Professional Paper 221-F, US Geological Survey, pp. 93-106.

Markewicz, F.J., 1969, "Ilmenite Deposits of the New Jersey Coastal Plain," *Geology of Selected Areas of New Jersey and Eastern Pennsylvania and Guidebook of Excursions,* S. Subitzky, ed., Rutgers University Press, New Brunswick, NJ, pp. 363-382.

Martens, J.C.H., 1928, "Beach Deposits of Ilmenite, Zircon, and Rutile in Florida," 19th Annual Report, Florida Geological Survey, pp. 124-154.

Masten, A.H., 1923, *The Story of Adirondac,* Princeton Press, Princeton, NJ, 199 pp.

McMurray, L.L., 1944, "Froth Flotation of North Carolina Ilmenite," *Trans. AIME,* Vol. 173, 1947; *Mining Technology,* Jan. 1944.

Merritt, C.A., 1939, "Iron Ores of the Wichita Mountains, Oklahoma," *Economic Geology,* Vol. 34, No. 3, May, pp. 268-286.

Michot, P., 1956, "The Deposits of Black Ores of the Egersund Region," Bulletin 79, Ann. Soc. Geol. Belg., pp. 183-201.

Moore, C.H., Jr., 1940, "Origin of the Nelsonite Dikes of Amherst County, Virginia," *Economic Geology,* Vol. 35, No. 5, Aug., pp. 629-645.

Nicholls, G.D., 1955, "The Mineralogy of Rock Magnetism," *Advances in Physics* (Supplement to *Philosophical Magazine*), Vol. 4, p. 113.

Nilsen, A.E., 1972, "Extraction of Iron from Titaniferous Ores," US Patent 3,647,414, Mar. 7.

Osborne, F.F., 1928, "Certain Magmatic Titaniferous Ores and Their Origin," *Economic Geology,* Pt. 1, Vol. 23, No. 7, Nov., pp. 724-761; Pt. 2, Vol. 23, No. 8, Dec., pp. 895-922.

Parker, G.G., and Cooke, C.W., 1944, "Late Cenozoic Geology of Southern Florida," Bulletin 27, Florida Geological Survey, 119 pp.

Paulson, E.G., 1964, "Mineralogy and Origin of the Titaniferous Deposit at Pluma Hidalgo, Oaxaca, Mexico," *Economic Geology,* Vol. 59, No. 5, Aug., pp. 753-767.

Pings, W.B., 1972, "Titanium, Pt. 1," *Colorado School of Mines Industries Bulletin,* Vol. 15, No. 4, July, 13 p.

Pings, W.B., 1972a, "Titanium, Pt. 2," *Colorado School of Mines Industries Bulletin,* Vol. 15, No. 5, Sep., 17 pp.

Pinnell, D.B., and Marsh, J.A., 1954, "Summary Geological Report on the Titaniferous Iron Ore Deposits of the Laramie Range, Albany County, Wyoming," *Mines Magazine,* Vol. 44, No. 5, p. 30.

Pirkle, E.C., and Yoho, W.H., 1970, "The Heavy Mineral Ore Body of Trail Ridge, Florida," *Economic Geology,* Vol. 65, No. 1, Jan.-Feb., pp. 17-30.

Pirkle, E.C., et al., 1974, "The Green Cove Springs and Boulougne Heavy Mineral Sand Deposits of Florida," *Economic Geology,* Vol. 69, No. 7, Nov., pp. 1129-1137.

Pirkle, F.L., 1975, "Evaluation of Possible Source Regions of Trail Ridge Sands," *Southeastern Geology,* Vol. 17, No. 2, Dec., pp. 93-114.

Ramdohr, P., 1956, "Die Beziehungen von Fe-Ti Erzen und Magmatischen Gesteinen," Bulletin No. 173, Comm. Geol. Finlande, pp. 1-18.

Reed, D.F., 1949, "Investigation of Christy Titanium Deposits, Hot Spring County, Arkansas," Report of Investigations 4592, US Bureau of Mines, 10 pp.

Reed, D.F., 1949a, "Investigation of Magnet Cove Rutile Deposits, Hot Spring County, Arkansas," Report of Investigations 4593, US Bureau of Mines, 9 pp.

Retty, J.A., 1944, "Lower Romaine River Area, Saguenay County, Quebec," Report 19, Quebec Dept. of Mines & Geology, pp. 3-29.

Rose, E.R., 1969, "Geology of Titanium and Titaniferous Deposits of Canada," Economic Geology Report No. 25, Geological Survey of Canada, 177 pp.

Ross, C.S., 1941, "Occurrence and Origin of the Titanium Deposits of Nelson and Amherst Counties, Virginia," Professional Paper No. 198, US Geological Survey, 59 pp.

Ross, C.S., 1947, "Virginia Titanium Deposits," *Economic Geology,* Vol. 42, No. 2, Mar.-Apr., pp. 194-198.

Singewald, J.T., Jr., 1913, "Titaniferous Iron Ores of the United States, Their Composition and Economic Value," Bulletin 64, US Bureau of Mines, 145 pp.

Stephenson, R.C., 1945, "Titaniferous Magnetite Deposits of the Lake Sanford Area, New York," Bulletin No. 340, NY State Museum, 95 pp.

Sun, S.S., 1971, "Fission Track Study of the Cheney Pond Titaniferous Iron Ore Deposit, Tahawus, N.Y.," Ph.D. Thesis, Washington University, St. Louis, MO, June, 134 pp.

Temple, A.K., 1966, "Alteration of Ilmenite," *Economic Geology,* Vol. 61, No. 4, June-July pp. 695-714.

Thompson, J.V., 1977, "Appraising Large Diameter Core and Percussion Drilling for Bulk Samples," *Engineering and Mining Journal,* Vol. 178, No. 8, Aug., pp. 80-82.

Toewe, E.C., et al., 1971, "Evaluation of Columbium-Bearing Rutile Deposits, Magnet Cove, Arkansas," Prepared for US Department of the Interior, Office of Minerals and Solid Fuels, Contract No. 14-01-0001-1738, Battelle Memorial Institute, Columbus, OH, Sep., 134 pp.

Vokes, F.M., 1968, "Forelesninger i Malmgeologi I," *Noregs Malmgeologi,* Hösten, Vol. 99, p. 488.

Watson, T.L., and Taber, S., 1913, "Geology of the Titanium and Apatite Deposits of Virginia," Bulletin 3A, Virginia Geological Survey, 308 pp.

Wimmler, N.M., 1946, "Titaniferous Magnetite Deposits in Montana," Report of Investigations 3981, US Bureau of Mines, 12 pp.

26. Vermiculite*

Vermiculite is the name generally applied to the group of hydrated ferromagnesian aluminum silicates that are characterized by the ability to expand when heated. This process, called exfoliation, results in a lightweight product of commercial value. Most uses of vermiculite are for the expanded material. The chief markets are in construction, agriculture, and horticulture, with lesser uses in general industry.

COMPOSITION AND PROPERTIES

Vermiculite, in its natural state, has the characteristic micaceous habit, a perfect basal cleavage which causes splitting into thin laminae that are soft, pliable, and inelastic. The structure of vermiculite is basically that of a talc. The prominent monoclinic crystal faces are often marked by lines at 60° and 120°. Hardness varies from 1.5 to 2 or more; specific gravities are between 2.1 and 2.8; color varies from almost clear to amber, bronze, brown, green, or black. Vermiculite feels like talc, especially when wet.

Although much research has been performed on the chemical and structural composition, there is not yet complete agreement on the exact formula. This is to be expected when different workers have examined the many different varieties. Vermiculite is not considered to be a single mineral species but a family of related minerals. The structural formula for a trioctahedral vermiculite may be written:

$$(H_2O)—(Mg,Ca,K)—(Al_2,Fe,Mg)—$$
$$(Si,Al,Fe)_4O_{10}(OH)_2$$

Hydrobiotite also occurs with vermiculite and is usually considered a vermiculite for commercial uses.

When heated quickly to elevated temperatures, vermiculite expands by exfoliating at right angles to the cleavage into wormlike particles. The name vermiculite is derived from the Latin *vermiculare,* to breed worms. This characteristic of expansion is the result of the mechanical separation of the layers by the rapid conversion of contained water to steam. The decrease in bulk density of commercial grades is usually approximately 10 times, from 801 to 80 kg/m² (50 to 5 pcf), but varies depending on the quality, size, and furnace efficiency. Individual flakes may expand up to 30 times. Vermiculite may also be expanded by soaking in chemicals such as hydrogen peroxide, weak acids, and other electrolytes. Color change during expansion is dependent upon the type of vermiculite and furnace conditions. Heating in an oxidizing atmosphere produces a dull gray or tan color, whereas a reducing atmosphere can produce a bronze or gold color.

The expansion of the vermiculite crystal results in large pores being formed between the platelets. Thus, exfoliation makes available a large increase in void volume which is important in the application of vermiculite as a chemical carrier.

Mathieson and Walker (1954) state vermiculite must be regarded as a true clay mineral. The characteristic properties of the mineral, such as high cation exchange capacity, organic complexing ability, and variable interlamellar distance are very similar to those of montmorillonite. The cation exchange capacity of vermiculite is one of the highest of all the clay minerals. The interplatelet space is accessible to penetration

by some electrostatically neutral molecules. Water and glycerine are two common substances whose molecules may be so imbibed.

In the natural state and under normal atmospheric conditions, water occupies the spaces between the silicate layers. The crystal *d*-spacing is near 14.2Å. By differential thermal analysis, it has been determined that the water is released at three temperature ranges. The "unbound water" is released near 149°C (300°F). This water is reversible and comes to equilibrium with the environment. The second water, designated "bound water," is removed at about 260°C (500°F). This is the water necessary for exfoliation. The third water is released at approximately 871°C (1600°F) and is probably hydroxyl. Little of this water is released by commercial exfoliation and when it is, a very noticeable change occurs in the color and physical characteristics of the product.

Vermiculite is closely related to biotite and phlogopite. The essential difference is that the unit cell of vermiculite contains a layer of water and the biotite contains a layer of potassium. Biotite or phlogopite are almost always associated with vermiculite in commercial deposits and are sometimes intermixed within crystals or across a single crystal face. The term *hydrobiotite* has been used for those varieties where the analysis indicates that there is a layered mixture of the two minerals in some definite proportion.

ORIGIN

Detailed investigations have been made of the three largest commercial deposits of vermiculite. According to Boettcher (1966), the Rainy Creek alkaline-ultramafic complex near Libby, MT, represents a composite of successive intrusions of igneous rocks emplaced into the Precambrian metasedimentary rocks. Most of the biotite in the inner body of pyroxenite has been altered to hydrobiotite and vermiculite. From his studies, Boettcher suggests the vermiculite is a product of leaching of the biotite by ground waters, whereas the hydrobiotite may represent a higher temperature alteration product.

Stewart (1949) in his report on the vermiculite deposits at Tigerville, SC, states that hydrothermal activity was necessary in the formation of the high grade deposits only to the extent of furnishing the biotite, which was later altered to vermiculite by meteoric waters. Buie and Stewart (1954) determined the paragenetic sequence as:

pyroxene → amphibole → biotite → vermiculite
(hypogene)   (supergene)

Libby (1975) suggests that the potassic ultramafic plutons of the Enoree, SC, vermiculite district probably were intruded into Late Precambrian to Cambrian sedimentary rocks and subsequently were metamorphosed. Studies indicate that both the vermiculite and hydrobiotite form under weathering conditions.

The Palabora deposit in South Africa at Loolekop is located in a carbonatite complex. Both the hydrothermal and weathering theory have been proposed for the formation of the vermiculite. There is a gradation of vermiculite to phlogopite or biotite with increasing depth.

The Palabora carbonatite complex (Dekun, 1965) is about 6.4 × 2.4 km (4 × 1½ miles) in size. The dolomite core is surrounded by a thin inner ring of altered phoscorite (serpentine-apatite-magnetite) and an outer ring of diopside

---

*Excerpt from Strand, P.R. (revised by Stewart, O.F.), 1983, "Vermiculite," *Industrial Minerals and Rocks,* 5th ed., Vol. 2, S.J. Lefond, ed., AIME, New York, pp. 1375-1381.

pyroxenite. Outward from this is a discontinuous 2.4-km (1½-mile) wide ring of fenitized gneisses. Numerous alvikite and orthoclasite veins have been injected into the pyroxenite ring. Serpentine, apatite, and magnetite occur in both inner zones, and a pegmatitic pyroxenite zone occupies the center of the northern ultrabasic ring. Vermiculite occurs only in the ultrabasics, especially close to serpentinized patches. Vermiculite also occurs disseminated with apatite. The principal vermiculite areas are located (1) in the north-central pegmatic pyroxenite 0.8 km (0.5 mile) from the limit of the carbonatite plug; (2) 1.2 km (0.75 mile) south of the core; and (3) in a zone near the southeastern rim of the pyroxenite.

A vermiculite deposit near Louisa, VA, occurs in a body of basic pyroxenite, about 6.4 to 8 km (4 to 5 miles) in diameter, surrounded by gneisses and granitoid type rocks typical of the Piedmont. The vermiculite occurs in varying flake size but nearly all are less than 10 mesh. The grade of vermiculite varies considerably within a few meters (feet) both laterally and with depth. The vermiculite appears to have been altered from biotite by surface weathering. However, little, if any, geological investigation has been conducted on this deposit.

DISTRIBUTION OF DEPOSITS

Vermiculite occurrences have been reported from many countries of the world. As the mineral has become more widely recognized, new occurrences have been reported. Market conditions, location, size, grade, quality, and economics are some of the factors affecting the commercial value and development of a deposit.

In the United States, vermiculite deposits or occurrences are found in Alabama, Arizona, Arkansas, California, Colorado, Georgia, Idaho, Kansas, Montana, Nevada, New Mexico, North Carolina, Pennsylvania, South Carolina, Texas, Virginia, and Wyoming. Production has been reported from California, Colorado, Georgia, Montana, Nevada, North Carolina, South Carolina, Texas, Virginia, and Wyoming. However, production is reported today only from mines in Montana, South Carolina, and Virginia.

Other countries where deposits have been reported are Argentina, Australia, Brazil, Canada, China, Egypt, Finland, India, Japan, Kenya, Korea, Mexico, Morocco, Pakistan, Zimbabwe, Republic of South Africa, Spain, Tanzania, Uganda, and the USSR. Minor production has been reported from Argentina, Brazil, China, Canada, Egypt, India, Kenya, Korea, Mexico, Spain, and Tanzania. Some vermiculite is produced in the USSR, but quantitative data are not available.

PROSPECTING AND EXPLORATION

In the United States vermiculite deposits of consequence are located in two principal areas: the Piedmont region from Alabama to Pennsylvania and in the Rocky Mountain Range from Montana southward into New Mexico, including the central mineral region of Texas. A few deposits occur in eastern Canada in the Precambrian shield.

Vermiculite deposits usually are covered by vegetation because the minerals are soft and when weathered have a considerable water-holding ability. Outcrops showing vermiculite are rare, but in some areas the mineral can be recognized in road cuts. Outcrops of associated rocks such as biotite, alkalic pyroxenites, and dikes of carbonatites or syenite, generally are more noticeable. In most instances, the only field evidence will be the presence of vermiculite flakes in the soil or in stream beds.

The larger commercial size deposits usually are found associated with ultrabasic rock, commonly pyroxenite. The pyroxenite may be intruded by numerous dikes such as pegmatites, syenites, carbonatites. Numerous veins containing vermiculite are found in ultramafic intrusives such as dunites, peridotites, and pyroxenites, but these do not often represent commercial size deposits.

In a vermiculite-bearing formation, the rock will vary from barren to a high content of vermiculite. The bulk of the material containing vermiculite will fall in the 20 to 30% content range. Drilling or trenching is necessary to prove the presence of ore in economic quantities and grade. Since the preservation of flake size is most important, auger or large rotary hole drilling is necessary with special attention to bit design. Diamond core drilling is most impractical.

The primary requirement of a commercial deposit is that the vermiculite be of acceptable quality. The concentrate must expand to a high degree without decrepitating, and the expanded particles must be strong enough to withstand handling. The biotite or phlogopite content must be low, or occur in such a form that it can be kept from mixing with the mill feed. The deposit predominantly must contain flake size material larger than 65 mesh, as there is little demand for the −65 mesh size.

Another characteristic of a commercial deposit is that the ore body be large enough to be mined with modern earthmoving equipment. Vermiculite production is very energy-intensive, as a high percentage of the commercial cost is for mining, reagentizing, drying, expanding, and transportation.

Deposits are considered to be high grade if their content of +65 mesh vermiculite is over 30%, and uneconomical or at least borderline if the content is 20% or below.

BIBLIOGRAPHY AND REFERENCES

Bassett, W.A., 1959, "The Origin of the Vermiculite Deposits of Libby, Montana," *American Mineralogist*, Vol. 44, No. 3-4, Mar.-Apr., pp. 282-299.

Boettcher, A.L., 1966, "The Rainy Creek Igneous Complex Near Libby, Montana," Ph.D. Thesis, Pennsylvania State University, 70 pp.

Buie, B.F., and Stewart, O.F., 1954, "Origin of Vermiculite at Tigerville, South Carolina" (abstract), *Geological Society of America Bulletin*, Vol. 65, No. 12, Pt. 2, pp. 1356-1357.

Dekun, N., 1965, *Mineral Resources of Africa*, Elsevier, Amsterdam, pp. 440-441.

Haines, S.K., 1978, "Vermiculite," *Mineral Commodity Profiles*, US Bureau of Mines, 10 pp.

Heinrich, E.W., 1966, *The Geology of Carbonatites*, Rand McNally & Co., Chicago, 555 pp.

Libby, S.C., 1975, "Origin of Potassic Ultramafic Rocks in the Enoree Vermiculite District, South Carolina," Ph.D. Thesis, Pennsylvania State University.

Mathieson, A.L., and Walker, G.F., 1954, "Crystal Structure of Magnesium Vermiculite," *American Mineralogist*, Vol. 39, pp. 231-255.

Meisinger, A.C., 1979, "Vermiculite," *Minerals Yearbook, 1979,* US Bureau of Mines.

Meisinger, A.C., 1981, "Vermiculite," Preliminary Annual Report, US Bureau of Mines.

Meisinger, A.C., 1982, "Vermiculite," *Mineral Commodities Summaries 1982*, US Bureau of Mines, p. 170.

Stewart, O.F., 1949, "Origin and Occurrence of Vermiculite at Tigerville, South Carolina," M.S. Thesis, University of South Carolina.

Strand, P.R., 1975, "Vermiculite," *Industrial Minerals and Rocks*, 4th ed., S.J. Lefond, ed., AIME, New York, pp. 1219-1226.

27. Wollastonite

GARY PARKISON

INTRODUCTION

Wollastonite is an industrial mineral which is commonly grouped with other filler-extender minerals such as kaolin, calcium carbonate, and talc. Its consumption pattern and usage depend primarily on its unique physical properties. Wollastonite is the only commercially available white mineral that is wholly acicular with typical length to diameter ratios ranging from 3:1 to 20:1 with an average diameter of 3.5 μm.

Most current production is derived from open-pit mines with the world's two largest producers both located in upstate New York. Reserves associated with currently producing operations are generally quite large and sufficient for many years of production given current consumption. There are a number of deposits on a worldwide basis which could be viable potential producers given current product prices and an expanded market.

MINERALOGY, USES AND SPECIFICATIONS

Wollastonite, or calcium silicate ($CaSiO_3$), has an ideal composition of 48.3% CaO and 51.7% SiO_2. Substitution of the calcium by iron, magnesium, and manganese is quite common. It is most commonly white or cream in color, has a hardness of $4\frac{1}{2}$ to 5 on Moh's scale, a specific gravity of 2.8 to 3.0, and has a melting point of about 1540°C. The most diagnostic property of wollastonite is, however, its common appearance as bladed, radiating, crystal masses. The crystals are generally acicular with aspect ratios (length:width) ranging from 3:1 to >20:1. Owing to perfect cleavage properties, this acicular habit is preserved in crushed and ground wollastonite. A broader discussion of mineralogy can be found in Andrews (1970).

It is the acicular nature of wollastonite which determines its utility in the marketplace as a functional filler-extender mineral. The acicular nature in conjunction with a relatively low melting temperature have contributed to the use of wollastonite in ceramic tile body and glaze formulations. The use of wollastonite in both architectural and special use paint is due primarily to its white color, acicular nature, and low oil absorption. Plastics applications are currently the largest domestic consumer of wollastonite where its acicularity contributes to the impact and flexural strengths in both thermoplastic and thermosetting resin systems. Other applications are as an asbestos substitute in brake linings, backing material for linoleum flooring, and in marine wallboard. A more in-depth discussion of applications is contained in Power (1986).

Specifications for wollastonite used in the various ceramic applications are ball- or pebble-mill ground material with aspect ratios generally less than 5:1 and particle sizes from -72 to -44 μm (-200 to -325 mesh). More finely ground material, typically 99% less than 44 μm (-325 mesh), is used in paints. High aspect ratio materials (10:1 to >20:1) with particle sizes ranging from -72 to -36 μm (-200 to -400 mesh), are produced via attrition mills and are typically used for plastics and other applications. Coated or chemically surface-treated wollastonite usage is becoming increasingly popular in high performance plastics applications.

A more thorough description of the uses and specifications for wollastonite can be found in Power (1986), Elevatorski and Roe (1983), and Smith (1981).

Synthetic wollastonite has been produced for some years both in the United States and Europe. These plants are generally not economic relative to the costs of natural wollastonite production but do offer more consistent chemistry and product supply. The synthetic products are produced by mixing finely ground SiO_2 and $CaCO_3$ in a rotary kiln to produce synthetic wollastonite via the sintering process (Rosner and Kurczyk, 1980). On a worldwide basis the production of synthetic wollastonite does not significantly displace natural wollastonite, with the former being restricted to the low aspect ratio "powder" uses primarily in ceramic and metallurgical applications.

GEOLOGY AND MINING

Formation

Wollastonite is most typically a product of contact and regional metamorphism, although minor amounts have been found associated with carbonatites. In a typical contact metamorphic environment with a granitic magma intruding impure calcareous rocks (i.e., limestone or dolomite), the farthest out metamorphosed zone will be the tremolite zone, where dolomite and quartz will react to form tremolite and calcite. The tremolite zone is succeeded by the zone closer to the intrusion where tremolite and quartz react to form diopside. Adjacent to the intrusion in the highest temperature zone calcite and quartz react to form wollastonite while diopside is stable (Winkler, 1974). The quartz can be introduced from the granitic magma, be mobilized from the surrounding sediments, or present within the host rock itself. Similar paragenetic reactions can also form garnet which is also stable with wollastonite, diopside, and calcite and together comprise a rock type commonly referred to as skarn. Within the skarn zone any unreacted calcite would typically be of sufficiently large grain size as to constitute marble.

The reaction

calcite + quartz = wollastonite + carbon dioxide
$$CaCO_3 + SiO_2 = CaSiO_3 + CO_2$$

typically takes place in the higher temperature ranges (>500°C.) of shallow contact metamorphism only where the CO_2-rich fluid phase is sufficiently dispersed to allow the reaction to proceed (Winkler, 1974). While small amounts of wollastonite are relatively common in contact metamorphic aureoles, the conditions to create sufficiently large deposits to be of commercial interest are quite rare and seemingly would be dependent on appropriate host rock bulk chemistry, chemistry of the igneous intrusive, and associated pressure and temperature. In western North America wollastonite is often found as rims and masses around and between remnant chert horizons in what is now calcite marble (Joeston and Fisher, 1988). This type occurrence is often associated with porphyry copper-related skarn deposits (Einaudi, 1981a,b). Even more specific conditions are required for the generation of highly acicular masses of wollastonite.

Significant Deposits—North America

The two largest producing wollastonite deposits in the world are both located in upstate New York flanking the Adirondack Mountains. Wollastonite is mined by NYCO near the community of Willsboro and by R.T. Vanderbilt some 160 km (100 miles) to the west in the Gouveneur region. The NYCO deposits have been described by Andrews (1970) and de Rudder (1962), and include the Deerhead, Fox Knoll and Lewis deposits. The current open-pit mining operation at the Lewis mine is described by Herod (1984) and involves mining and crushing the ore at the mine site then transporting the ore in trucks some 25 km (15 miles) to a mill at Willsboro. At the mill garnet and diopside are separated from wollastonite via dry magnetic separation following fine crushing. Much of the NYCO production is of high aspect-ratio material. In the area of the deposits Precambrian Grenville marble has been intruded and partly replaced by rocks of anorthositic composition to produce mixed gneisses containing wollastonite with garnet and diopside. At the Lewis deposit, original reserves were estimated at 5.4 Mt (6 million st) grading about 60% wollastonite, 30% garnet, and 10% diopside. The wollastonite-rich zone occurs as a single shallow-dipping sheet some tens of feet thick which is intercalated with thinner wollastonitic and gneissic bands. Total ore reserves held by NYCO in the three properties likely exceed 10 Mt (Elevatorski and Roe, 1983). This figure does not include the recently discovered reserves at the Oak Hill deposit adjacent to the Lewis mine.

The deposit being exploited by R.T. Vanderbilt is similar to that of NYCO but lacks garnet and hence is lower in iron content, which make this wollastonite better suited to ceramic applications (Harben and Bates, 1984). The wollastonite zone is mined by both open-pit and underground methods, is steeply dipping, and contains an estimated 0.9 Mt (1 million st) of ore (Power, 1986). Wollastonite production is of low aspect ratio.

Some 160 km (100 miles) to the northwest of the R.T. Vanderbilt deposit in southeastern Ontario, Canada, is a more recent discovery of wollastonite in a geologic setting similar to that of the New York deposits. Drilling by the Cominco/Platinova joint venture has defined total ore reserves exceeding 0.9 Mt (1 million st) in two separate deposits grading from 39 to 47% wollastonite (Anonymous, 1988).

The only other deposits which have been mined for wollastonite in the United States are in California. A small amount has been mined from a deposit in Kern County where Paleozoic limy sediments have been intruded by Mesozoic granitic rocks where the wollastonite is associated with diopside and garnet (Troxel and Morton, 1962). A larger but unknown amount of wollastonite has been mined from several deposits in the Big and Little Maria Mountains near Blythe in southern California. Wollastonite-bearing rock occurs sporadically throughout a thick succession of Paleozoic siliceous and calcareous sediments. Local pods of high purity massive wollastonite can be traced for up to 600 m (1800 ft) in length (Troxel, 1957). Hoisch (1987) has determined that the wollastonite formed where large volumes of high temperature fluids were channeled along permeable or fractured zones during a Late Cretaceous regional metamorphic event.

Large undeveloped wollastonite deposits are located in both the southern and northern extremes of Death Valley National Monument. The Warm Spring Canyon deposit hosts a lens of almost pure wollastonite some 12 m (35 ft) thick and 250 m (750 ft) long within moderately dipping carbonate rocks (Troxel, 1957). Extensive wollastonite mineralization in the Hunter Mountain area has been described by Clark (1980) and Andrews (1970). In this area, wollastonite with lesser amounts of garnet, epidote, and amphiboles have preferentially replaced favorable beds in Paleozoic limestone adjacent to a Cretaceous quartz monzonite intrusion (McAllister, 1965). Most of the wollastonite is quite compact and short-fibered and occurs along a band up to 100 m (300 ft) wide and more than 8 km (5 miles) long. Reserves are thought to exceed 22.5 Mt (25 million st) grading 65 to 85% wollastonite. Other wollastonite occurrences in California have been documented by Murdock and Webb (1966) while occurrences in Arizona, Nevada, and Utah are described by Anthony, et al. (1977) and Andrews (1970).

Some of the largest known wollastonite deposits in the world are found about 65 km (40 miles) northwest of Hermosillo, Sonora, Mexico. The Pilares and San Hector deposits both outcrop prominently in isolated hills rising from the desert pediment. Both areas are underlain by Paleozoic age, thickly interbedded, moderately north-dipping quartzite and calcitic marble in close proximity to Cretaceous granitic rocks. Wollastonite mineralization at grades over 70% and many tens of meters thick can be traced for thousands of meters along strike and alternate with marble zones with from 30 to 40% or less wollastonite. Most of the wollastonite is white or buff colored and found as radiating splays or clusters with individual crystals from 1 to 3 cm in length. Commonly associated with the wollastonite are finer-grained quartz, calcite, diopside, and garnet. Reserves of material grading more than 50% wollastonite at both the Pilares and San Hector deposits exceed 30 Mt (33 million st).

Other wollastonite deposits in Mexico are described by Andrews (1970). The only current wollastonite production in Mexico is by General de Minerals, SA, which operates four small open pit and underground mines in the La Blanca district about 50 km east of Zacatecas (Smith, 1981). Relatively thin and erratic steeply dipping zones grading 50% or less wollastonite are developed in thin-bedded limestone. The gray to white wollastonite is associated with considerable calcite, quartz, garnet, and diopside. The wollastonite zones are within one-half mile of a granitic stock. Hand-cobbing upgrades the run-of-mine ore prior to trucking to a grinding mill located near Zacatecas.

Significant Deposits—Europe and Asia

While wollastonite production from North America dominates the world market, a significant and increasing share of production comes from Europe and Asia, specifically Finland, India, and China. Oy Partek AB produces wollastonite from an open pit at the Lappenranta deposit in southeastern Finland in conjunction with limestone production for cement manufacture (Power, 1986; Harben and Bates, 1984). Wollastonite occurs within an elliptical Archean limestone mass in a zone some 650 m (2000 ft) long by about 50 m (150 ft) wide. The zone is fairly low grade, averaging about 20% wollastonite along with considerable calcite, dolomite, and quartz but no garnet (Andrews, 1970). The deposit is thought to have been formed through contact metamorphism by the enclosing younger granite.

Wollastonite production in India is currently from the Belkapahar deposit in the state of Rajasthan by Wolkem Pvt. Ltd. (Power, 1986). Wollastonite is found within limestone horizons which have been replaced by pyroxenites and gneiss in association with contact metamorphism (Sinha, 1986). After hand sorting, the wollastonite has few impurities and is of high aspect ratio. Reserves are very large and production currently is via open-pit mining.

China has been an intermittent exporter of wollastonite for the last several years. Wollastonite is currently produced from numerous small open pit mines in several northeastern Chinese provinces (Fountain, 1986). Most of the operations are labor intensive and make extensive use of hand-cobbing to produce salable lump ore primarily for domestic use. High-grade wollastonite layers are typically enclosed within calcitic marble and most deposits are associated with iron and copper mines.

The USSR is thought to produce a significant amount of wollastonite from several deposits for domestic consumption only. Kuzwart (1984) describes a deposit in Tadzhikistan which hosts high grade coarse-grained wollastonite within Paleozoic limestones at their contact with syenodiorites. Other deposits in the USSR are described by Andrews (1970). Past producing deposits in Kenya, Namibia, and Sudan are further described by Andrews (1970) and Power (1986). Power (1986) also describes potentially commercial wollastonite deposits in Turkey and Greece. Minor production from New Zealand is reported by Powers (1986) and several occurrences in Australia are briefly summarized by Johns (1986).

EXPLORATION

All commercial wollastonite deposits have been found to be associated with calcareous metamorphic rocks, and most of these deposits are near or along zones of contact with granitic intrusive rocks. Chances are good that wherever granitic rock intrudes a sequence of calcareous sediments, at least a minor amount of wollastonite will be developed. Of importance in determining whether a large amount of wollastonite will result is the pressure and temperature conditions associated with the metamorphic event and the bulk chemistry of the calcareous rock. In this latter regard, a siliceous limestone would be preferred as in most instances an isochemical process for the origin of wollastonite within the host rock is suggested. The specific conditions which lead to the generation of long fiber or high aspect ratio wollastonite vs. short fiber wollastonite both in outcrop and in milled product have not been described.

Potentially helpful information which can be used to focus the search for commercial wollastonite deposits could include published descriptions of skarn or carbonate-hosted base metal deposits, specifically porphyry copper type deposits in addition to descriptions of known wollastonite deposits and occurrences. In this instance, a review of state or province publications on mineral resources and/or systematic reviews of minerals can be most helpful. Once a prospective area is determined or located, on-site investigations should initially be concentrated along or near the contact between the intrusive rock, if present, and favorable calcareous rocks. Particular attention might also be paid to the probable metamorphic zonation with wollastonite closer to the intrusive with diopside and tremolite at increasing distance. In arid environments, wollastonite-rich zones are typically more resistant to weathering than the enclosing rocks while in humid areas the opposite is often the case.

EVALUATION

Evaluation of the commercial significance of a wollastonite occurrence involves a number of physical, chemical, and metallurgical tests in addition to the development of an accurate marketing study. The initial testing should involve the collection of outcrop and/or subsurface samples via core or rotary drilling, chip and channel sampling, trench sampling, etc. Testing should be supported and directed by de-

Table 1. Some Typical Specifications of Commercial Wollastonite

Component		Properties
CaO	46-47%	Brightness (Hunter) 85-95
SiO$_2$	49-51%	at −200 mesh sizes
Fe$_2$O$_3$	0.1-1.0%	Aspect ratio 3:1 to >20:1
Al$_2$O$_3$	0.1-1.0%	for high aspect uses
MgO	0.3-2.0%	Oil absorption 20-40
LOI	0.5-2.0%	mg/100 g

tailed geologic mapping to determine the width, length, and attitude of the wollastonite mineralization and its host rock associations. Particular attention should be paid to variations in wollastonite and gangue mineral contents and physical aspects of the wollastonite-rich zone. X-ray diffraction testing and petrographic examinations of representative and select samples should be used to determine modal mineralogy in conjunction with routine X-ray fluorescence analyses for major and minor elements. In many cases, the modal and chemical analyses are correlative and provide a good basis for further grinding and/or beneficiation studies. Grinding of unbeneficiated and beneficiated feed should utilize one of a number of grinding options including a jet and/or attrition mill to maximize aspect ratio. Beneficiation can employ hand-sorting, wet or dry magnetic separation, flotation, or a combination of methods. Products from all testing phases should be reanalyzed for chemistry and mineral percentages and evaluated for both reflectance or brightness in addition to aspect ratio. Some representative specifications for commercial wollastonite products are shown in Table 1.

PRODUCTION

Wollastonite demand is currently growing both in the United States as well as worldwide, with production capacity in the four major producing countries now about equal to demand (Choate, 1988). Demand is forecast to continue to grow at a rate of about 5% per annum, with the most dramatic growth in high aspect ratio products used in plastics applications. Somewhat slower growth is forecast for paint, ceramics, and other applications.

The United States dominates the world wollastonite market both in terms of production and consumption. About 40% of the domestic production is currently exported, primarily to Europe and Japan. Current domestic consumption is estimated at about 45 kt (50,000 st) valued at roughly $13 million. Worldwide production in 1987 was about 135 kt (150,000 st). NYCO produced about 45 kt (50,000 st); R.T. Vanderbilt, 27 kt (30,000 st); Wolkem, 23 kt (25,000 st); Partek, 23 kt (25,000 st); and others, 18 kt (20,000 st) (Choate, 1988).

Production typically involves open-pit and/or underground mining using conventional drill/blast techniques. Mining is often of a selective nature and the mined ore is commonly upgraded via hand sorting. Further beneficiation and grinding are often performed at a more centralized location which requires shipping of the lump ore some distance. In some instances, the lump ore is sold as is with further processing performed by the consumer.

While many wollastonite operations are found in proximity to base-metal mines, wollastonite is typically the only economic material produced. However, a possibly salable garnet byproduct is produced at NYCO's New York operations and filler-extender grade calcium carbonate is produced via flotation by Oy Partek. Other potential byproducts associated with certain wollastonite operations could include tungsten and tin minerals, fluorite, magnetite, base-metal sulfides, etc.

Because of the acicular nature of wollastonite, it is an irritant to the skin and is regulated by OSHA as a nuisance dust. Additional medical evaluations of wollastonite by Sollman and Berger (1988) and the World Health Organization (1987) suggest that there is only limited evidence of carcinogenicity in experimental animals and inadequate evidence of carcinogenicity for humans when exposed to wollastonite. However, there was some decreased pulmonary capacity when comparing wollastonite workers to an unexposed control group. Based on these findings, new regulations governing the production of wollastonite could be forthcoming.

REFERENCES

Anonymous, 1988, "Platinova Tests Wollastonite," *Industrial Minerals,* No. 253, p. 12.

Andrews, R.W., 1970, *Wollastonite,* Institute of Geological Sciences, Her Majesty's Stationery Office, London, 114 pp.

Anthony, J.W., Williams, S.A., and Bideaux, R.A., 1977, *Mineralogy of Arizona,* University of Arizona Press, Tucson, pp. 204-205.

Choate, L.W., 1988, "Wollastonite," *Mining Engineering,* Vol. 40, No. 6, p. 437.

Clark, D.W., 1980, "Hunter Mountain Wollastonite, Northern Death Valley Area, California," *Geology and Mineral Wealth of the California Desert,* D.L. Fife and A.R. Brown, eds., South Coast Geological Society, Santa Ana, CA, pp. 294-298.

de Rudder, R.D., 1962, "Mineralogy, Petrology and Genesis of the Willsboro Wollastonite Deposit, Willsboro Quadrangle, New York," PhD Thesis, Indiana University, 136 pp.

Elevatorski, E.A., and Roe, L.A., 1983, "Wollastonite," *Industrial Minerals and Rocks,* 5th ed., S.J. Lefond, ed., AIME, New York, pp. 1383-1390.

Einaudi, M.T., 1981a, "Description of Skarns Associated with Porphyry Copper Plutons," *Advances in Geology of the Porphyry Copper Deposits,* S.R. Titley, ed., University of Arizona Press, Tucson, pp. 139-183.

Einaudi, M.T., 1981b, "General Features and Origin of Skarns Associated with Porphyry Copper Plutons," *Advances in Geology of the Porphyry Copper Deposits,* S.R. Titley, ed., University of Arizona Press, Tucson, pp. 185-209.

Fountain, K., 1986, "Chinese Wollastonite, Industry and Commerce," *Proceedings,* 7th Industrial Minerals International Congress, G.M. Clarke and J.B. Griffiths, eds., Metal Bulletin PLC, London, pp. 117-125.

Harben, P.W., and Bates, R.L., 1984, *Geology of the Nonmetallics,* Metal Bulletin, Inc., New York, 392 pp.

Herod, S., 1984, "NYCO Strengthens Operation with New Mine, Crushing Plant," *Pit & Quarry,* June, pp. 36-40.

Hoisch, T.D., 1987, "Heat Transport by Fluids during Late Cretaceous Regional Metamorphism in the Big Maria Mountains, Southeastern California," *Geological Society of America Bulletin,* Vol. 98, pp. 549-553.

Joesten, R., and Fisher, G., 1988, "Kinetics of Diffusion-Controlled Mineral Growth in the Christmas Mountains (Texas) Contact Aureole," *Geological Society of America Bulletin,* Vol. 100, pp. 714-732.

Johns, R.K., 1986, "Wollastonite in Australia," in "Letters to the Editor," *Industrial Minerals,* No. 226, p. 19.

Kuzvart, M., 1984, *Industrial Rocks and Minerals,* Elsevier, Amsterdam, 287 pp.

McAllister, J.F., 1965, "Geologic Map of the Ubehebe Peak 15 Minute Quadrangle," US Geological Survey Map GQ-95, scale 1:62,500.

Murdoch, J., and Webb, R.W., 1966, "Wollastonite," *Minerals of California,* California Div. of Mines and Geol. Bulletin 189, pp. 388-396.

Power, T., 1986, "Wollastonite-Performance Filler Potential," *Industrial Minerals,* No. 220, pp. 19-34.

Rosner, K., and Kurczyk, G., 1980, "Synthetic Alkaline Earth Silicates—New Materials for Ceramics, Paint and Papers," *Proceedings,* 4th Industrial Minerals International Congress, G.M. Clarke, ed., Metal Bulletin Plc., London, pp. 165-173.

Sinha, R.K., 1986, *Industrial Minerals,* 2nd ed., A.A. Balkema, Rotterdam, 379 pp.

Smith, M., 1981, "Wollastonite—Production and Consumption to Climb," *Industrial Minerals,* No. 167, pp. 25-33.

Sollman, K.J., and Berger, S.E., 1988, "Medical Evaluations of Wollastonite—A Review," Preprint 88-138, SME, 7 p.

Troxel, B.W., 1957, Wollastonite, *Mineral Commodities of California,* California Div. Mines and Geology Bulletin, Vol. 176, pp. 693-697.

Troxel, B.W., and Morton, P.K., 1962, *Mines and Mineral Deposits of Kern County, California,* California Div. of Mines and Geol. County Rept. 1, pp. 344-345.

Winkler, H.G.F., 1974, *Petrogenesis of Metamorphic Rocks,* 3rd ed., Springer-Verlag, New York, 320 pp.

World Health Organization, 1987, *IARC Monographs on the Evaluation of the Carcinogenic Risk of Chemicals to Humans, Silica and some Silicates,* Vol. 42, Intl. Agency for Res. on Cancer, World Health Organization, New York, pp. 145-158.

28. Zeolites

RICHARD H. OLSON

COMMODITY DESCRIPTION

Zeolites are highly crystalline hydrated aluminosilicates which have a framework structure enclosing interconnected cavities occupied by relatively large cations and water molecules. More than 30 distinct species of natural zeolite minerals are known and in addition several synthetic species have been manufactured for which there are presently no known counterparts.

The development of synthetic zeolite minerals in the late 1940s and early 1950s (products which were then patent-protected with extremely high profit margins) resulted in the subsequent search for any possible natural analogues. This initial search for possible natural zeolite ores of monomineralic composition was aggressively pursued in the late 1950s and early 1960s and eventually resulted in the knowledge that such natural ores could probably not compete with most of the sophisticated high-value uses of the synthetic products. Similar aggressive exploration efforts in the late 1970s and early 1980s were geared more toward lower-value uses than toward those of the synthetic products and were not generally aimed at replacing them or even partially substituting for them.

GEOLOGY

In the early days of exploration for natural zeolite ores full-time attention was devoted to searches for the space-filling euhedral type of occurrences which are of such keen interest to mineralogists and rockhounds. The best examples of such occurrences are those in the Triassic diabase sills spectacularly exposed as steep to vertical cliffs along the shores of the Bay of Fundy in Nova Scotia. The most detrimental features of such deposits are that they are generally not monomineralic (a compositional feature that would be necessary in order to compete with the synthetics) and would have extremely high mining and beneficiation costs.

In 1958, thanks largely to the discovery of a landmark paper by Bramlette and Posnjak (1933), it was realized that economic natural zeolite ore bodies were much more likely to be found as "single-species" occurrences in diagenetically altered volcanic ash deposits which could be mined comparatively cheaply and would not generally require expensive beneficiation. Virtually all of the attention devoted to the exploration for natural zeolite deposits since early 1958 has been paid to this type of occurrence. Sheppard (1971, 1973, 1983), Sheppard and Gude (1968, 1969), and Munson and Sheppard (1974) describe this type of occurrence in great detail.

Clinoptilolite is far and away the most abundant of the natural zeolite species currently known as occurring in continental deposits. In deep sea deposits, phillipsite is similarly far and away the most common of the known natural zeolite species in such an environment.

Although at least 20 natural zeolite species have been reported as occurring in sedimentary rocks, only 9 of these are known to be at all common. These nine are, listed in the approximate order of decreasing abundance: clinoptilolite (including heulandite), analcime, mordenite, erionite, phillipsite, chabazite, laumontite, and ferrierite.

No byproducts or coproducts are presently known to be associated with natural zeolite ore bodies. Such are not impossible, however, and might be achieved in the future.

MINING METHODS

Natural zeolite mining is presently performed completely by open-pit methods. It is conceivable that underground mining methods might be employed were a potential substitute for a synthetic product to be discovered in a country where underground mining could be done relatively cheaply, but this would be a relatively rare instance indeed.

Selective mining methods are commonly employed, but several natural zeolite deposits do not lend themselves readily to mechanized mining methods. One deposit in southeastern Arizona has an ore bed on the order of only 15 to 30 cm (6 to 12 in.) thick and is totally hand-mined after the top of the ore bed is encountered and swept clean, as one might do with a warehouse floor. Another deposit in northeastern Nevada has an ore bed only 46 cm (18 in.) or so thick. These are admittedly extremes of thinness, but they do show how thin these exploration targets may be; the greater the value of the product (which may be in dollars per pound), however, the more one can tolerate such extremely thin thicknesses.

GEOLOGY

Most commercial natural zeolite deposits have been formed by the diagenetic alteration of volcanic ash deposits formed in playa lake environments and are therefore generally tabular and concordant in nature. Their geographical distribution, even though generally restricted to desert environments, is nevertheless widespread as compared with most other industrial mineral ores. It is no exaggeration to state that the extent of many, if not most of them, may be expressed in square miles rather than in acres.

GENESIS

The largest and potentially most valuable natural zeolite deposits belong to the "open-system" and "closed-system" types as described by Sheppard (1983). These terms are hydrologic terms: the "open-system" type of deposit is formed by the alteration of volcanic glass by meteoric waters, while the "closed-system" type is formed by the reaction of trapped connate water with the volcanic glass. The future economic potential of natural zeolite deposits appears to be largely restricted to those of these two systems; therefore, this discussion will only consider these two.

Both of these types of natural zeolite deposits are zoned and one who attempts to mine them should be aware of this feature. In the "open-system" type, a vertical zonation of authigenic silicate minerals is common. With the "closed-system" type, there is commonly a basinward lateral zonation from unaltered volcanic glass to zeolites and finally to potassium feldspar in the central portion of the basin.

The genesis of most potentially commercial natural zeolite ores is relatively simple, i.e., a pseudomorphic replacement of amorphous volcanic glass by highly crystalline zeolite minerals. It is likely that most such deposits were at their purest when they were formed and that virtually all of the deleterious included material in such ores was formed secondarily over geologic time. A common method of formation of such deposits results in a zonal arrangement related to the original configuration of the depositional basin; consequently, care must be taken not to mine outside of the zeolite zone and therefore include extraneous material in the mined ore.

Natural zeolite deposits are invariably quite young geologically, for such minerals are not time-stable. It is likely that no commercial natural zeolite deposit older than Cretaceous will ever be found.

STRUCTURAL CONTROLS

The tabular and concordant nature of natural zeolite ore bodies dictates that their mining is relatively simple. Many of them can be traced along strike for several miles and steep dips are uncommon except where block-faulting, associated with nearby mountain-building, has taken place. Folding of the strata seldom complicates the life of the zeolite miner, but block-faulting commonly does. Grade control is not easily accomplished because visual inspection alone generally can not aid in the determination of whether ore minerals are present or absent. Virtually all natural zeolite samples must be subjected to X-ray diffraction analysis in order to determine whether or not they may represent potential ore.

EXPLORATION

Airborne reconnaissance is useful in the exploration for natural zeolite deposits, particularly in areas where the desert basins have little topographic relief and difficult road access. Areas of interest, which can commonly be readily identified from the air by their light color and overall geological habit, are spotted upon maps or aerial photographs. Follow-up visits are then made on the ground.

Even though visual identification of just which zeolite species may be present is difficult, if not impossible, to achieve in the field with either the hand lens or even with the petrographic microscope, the geologist with a little field experience will soon be able to discern with a high degree of success whether or not the precursor rock has been zeolitized at all. This is a subjective art and can not really be taught, but it can be developed in the interested neophyte by visiting many known deposits and manually working with such ores.

Compared with most other ores, the exploration for natural zeolite ores is relatively simple. A thorough literature search is desirable, followed up by visits to governmental and academic agencies and institutions in order to consult with past and present workers in either natural zeolites or in geological systems associated with the formation of natural zeolites.

REFERENCES

Barrer, R.M., 1978, *Zeolites and Clay Minerals,* Academic Press, New York, 497 pp.

Bramlette, M.N., and Posnjak, E., 1933, "Zeolitic Alteration of Pyroclastics," *American Mineralogist,* Vol. 18, No. 4, Apr., pp. 167-171.

Breck, D.W., 1974, *Zeolite Molecular Sieves: Structure, Chemistry and Use,* John Wiley, New York, 771 pp.

Clifton, R.A., 1987, "Natural and Synthetic Zeolites," Information Circular 9140, US Bureau of Mines, 21 pp.

Deffeyes, K.S., 1959, "Zeolites in Sedimentary Rocks," *Journal of Sedimentary Petrology,* Vol. 29, pp. 602-609.

Deffeyes, K.S., 1968, "Natural Zeolite Deposits of Potential Commercial Use," *Molecular Sieves,* Society of Chemical Industry, London, pp. 7-9.

Hay, R.L., 1963, "Stratigraphy and Zeolitic Diagenesis of the John Day Formation of Oregon," *Geological Science,* California University Publications, Vol. 42, No. 5, pp. 199-262.

Hay, R.L., 1966, "Zeolites and Zeolitic Reactions in Sedimentary Rocks," Special Paper 85, Geological Society of America, 130 pp.

McBain, J.W., 1932, *The Sorption of Gases and Vapors by Solids,* Chap. 5, George Rutledge and Sons Ltd., London.

Meier, W.M., and Olson, D.H., 1978, *Atlas of Zeolitic Structure Types,* Structure Commission of the International Zeolite Assn., 99 pp.

Mumpton, F.A., 1973, "Worldwide Deposits and Utilisation of Natural Zeolites," *Industrial Minerals,* No. 73, pp. 30-45.

Mumpton, F.A., ed., 1977, "Mineralogy and Geology of Natural Zeolites," *Short Course Notes,* Vol. 4, Mineralogical Society of America, 233 pp.

Mumpton, F.A., 1983, "Commercial Utilization of Natural Zeolites," *Industrial Mineral and Rocks,* 5th ed., Vol. 2, S.J. Lefond, ed., AIME, New York, pp. 1418-1426.

Munson, R.A., and Sheppard, R.A., 1974, "Natural Zeolites: Their Properties, Occurrences and Uses," *Mineral Science and Engineering,* Vol. 6, No. 1, pp. 19-34.

Papke, K.G., 1972, "Erionite and Other Associated Zeolites in Nevada," Bulletin 79, Nevada Bureau of Mines & Geology, 32 pp.

Sheppard, R.A., 1971, "Zeolites in Sedimentary Deposits of the United States—A Review," *Molecular Sieve Zeolites-I,* R.F. Gould, ed., Advances in Chemistry Series 101, American Chemical Society, pp. 279-310.

Sheppard, R.A., 1973, "Zeolites in Sedimentary Rocks," Professional Paper 820, US Geological Survey, pp. 689-695.

Sheppard, R.A., 1983, "Zeolites in Sedimentary Rocks," *Industrial Minerals and Rocks,* 5th ed., Vol. 2, S.J. Lefond, ed., AIME, New York, pp. 1413-1418.

Sheppard, R.A., and Gude, A.J., III, 1968, "Distribution and Genesis of Authigenic Silicate Minerals in Tuffs of Pleistocene Lake Tecopa, Inyo County, California," Professional Paper 597, US Geological Survey, 38 pp.

Sheppard, R.A., and Gude, A.J., III, 1969, "Diagenesis of Tuffs in the Barstow Formation, Mud Hills, San Bernardino County, California," Professional Paper 634, US Geological Survey, 35 pp.

29. Zirconium / Hafnium*

Zirconium and hafnium are curious elements because they are almost always found together in nature. Zirconium was discovered by Klaproth in 1789 and isolated 35 years later by Berzelius. Because hafnium's chemical properties are so close to those of zirconium, it was not discovered until 1923 when Coster and Von-Heresey detected it using X-ray spectrographic analysis. These elements occur most commonly in nature as the mineral zircon ($ZrSiO_4$) and less commonly as the oxide baddeleyite (ZrO_2). They also are found as a variety of other silicates. Zircon was identified as a component in alluvial and beach sands in 1895, but it was not produced in any quantity until 20 years later. Zircon is always a coproduct from TiO_2 mining and processing. During World War I it was produced as a coproduct of beach sand mining for titanium minerals just south of Jacksonville Beach, FL, and was patented as a refractory. It was not until the 1930s when Zircon Rutile, Ltd. began mining at Byron Bay on the east coast of Australia that zircon was first used as a foundry sand. Later in the 1940s, NL Industries, Humphreys Gold Corp., and E. I. du Pont de Nemours & Co., Inc. began production of zircon sands from fossil beaches in northeast and northcentral Florida. Baddeleyite first became available as a commercial product in 1916, but never in the quantity of zircon.

Zircon holds a unique position as an industrial mineral because it is used for both its physical and chemical properties as well as an ore of zirconium and hafnium metals. Australia is the major zircon producer followed by South Africa and the US. Zircon is also produced in India, Sierra Leone, Sri Lanka, Malaysia, China, Thailand, and Brazil. Undeveloped heavy mineral beach sand reserves containing zircon are known to occur in Egypt, Malawi, Senegal, and Tanzania. Locations are shown in Fig. 1.

MINERALOGY

Zirconium and hafnium are always present together in naturally occurring compounds. Most commonly they form as the silicate sometimes containing iron, calcium, sodium, manganese, and other elements. Less commonly they are found as oxides in combination with titanium, thorium, calcium, and iron. Zircon is the most common form. It is extremely resistant to weathering; therefore, it is often present in ancient beach sands and placer deposits. Although resistant to alteration from external sources, it is vulnerable to internal alteration as the result of thorium and uranium substituting for zirconium either in the zircon lattice or in solid solution. Alteration to the metamict state takes place as radioactive emanations from these elements disorder the crystal lattice, accompanied by hydration, reduction in specific gravity, and changes in color. The mineralogy of zircon and baddeleyite and variation of physical and chemical properties are shown in Table 1.

The effects of zircon's internal alteration and darkened color can be corrected by heating to 1000°C for 30 min or so. A certain time-temperature relationship makes it possible for decolorization and reordering of the crystal lattice to take place when heating at higher temperatures for shorter periods or at lower temperatures for longer periods. After heating, X-ray diffraction patterns are sharper, the grains may be harder, and, except for particles of malacon, the grains will be white or colorless in the absence of grain surface coatings. The color change is only temporary and the particles slowly darken with time.

Zircon fluoresces light yellow in shortwave ultraviolet light. If exposed to ultraviolet light for extended periods of time, white (calcined) zircon turns purple. With time the induced color begins to fade and eventually disappears (e.g., calcined Florida zircon exposed to ultraviolet light for 24 hr still retains faint, but discernible, purple color after six months).

OCCURRENCE OF ZIRCON AND BADDELEYITE

Commercial zircon is available only in the form of sand mined, for the most part, from ancient beach sand deposits. In the beginning, zircon occurs as an accessory mineral in a variety of igneous and metamorphic rocks, especially those containing sodic feldspars such as granite, syenite, diorite, etc. It is one of the earliest minerals to crystallize from a cooling magma and frequently incorporates inclusions (e.g., apatite, magnetite). The zircon in these rocks often crystallizes out in tetragonal prisms with pyramidal terminations but it also commonly occurs as rounded grains in igneous rocks. It is sometimes found in metamorphic rocks such as gneiss and schists. One occurrence has been reported in a meteorite.

Before the zircon particles can become part of a beach sand deposit, their host rocks must undergo a series of events that will ultimately liberate the zircon for transportation to a seacoast. This begins with the exposure of the host rocks to subaerial weathering. The rocks are broken down into smaller fragments and transported downhill by rainwater and gravity. The rocks further decompose to the point at which the smaller zircon grains are liberated from the enclosing feldspars and quartz. The mineral particles are transported by streams and ultimately end up along a marine shoreline. Here the action of waves, tidal currents, and wind may remove lighter quartz, forming a heavy mineral deposit rich in zircon. Many of the older beach sand deposits have been consolidated and the grains cemented together to form sandstones containing concentrations of titanium minerals, zircon, monazite, and other heavy minerals. Deposits such as these occur in the four-state area of Utah, Wyoming, New Mexico, and Colorado. Since this type of deposit requires more costly mining methods, crushing, and grinding, they are less attractive for mining and exploitation than are the more recent less consolidated deposits. All of the commercial zircon products today are mined and separated from relatively young beach sand deposits occurring on or near active coast lines.

Australian Zircon Deposits

Australia is the major world producer and exporter of zircon sand. Deposits are found on both the east and west coasts of Australia. In eastern Australia major heavy mineral sand deposits lie between Broken Bay, New South Wales, to the south and Cape Clinton north to Rockhampton, Queensland. The richest occurrences are concentrated in the area between Yamba and North Stradbroke Island. Although

---

*Excerpt from Garnar, T.E., Jr., 1983, "Zirconium and Hafnium Minerals," *Industrial Minerals and Rocks,* 5th ed., Vol. 2, S.J. Lefond, ed., AIME, New York, pp. 1433-1446.

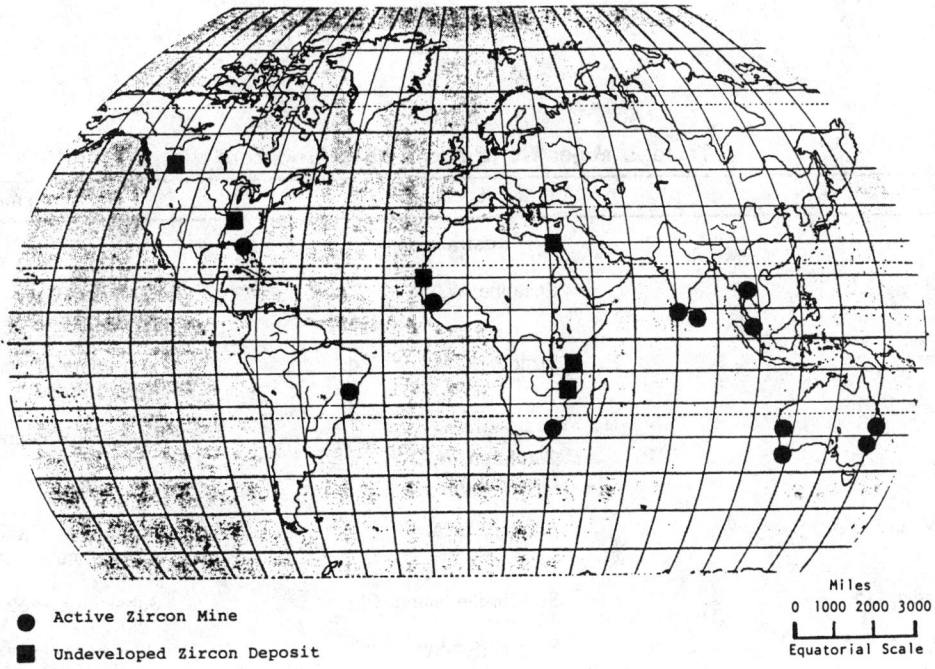

● Active Zircon Mine

■ Undeveloped Zircon Deposit

Miles
0 1000 2000 3000
Equatorial Scale

Fig. 1. Location of world zircon deposits

there had been production of zircon from a mine at Ponte Vedra, FL, between 1918 and 1929, the east Australian mines were the first major producers of commercial zircon sands beginning in the 1930s.

The east Australian heavy mineral beach sands deposits are of both Pleistocene and Holocene age. Zircon and its coproduct rutile have been derived from the breakdown and erosion of well sorted silica sandstone in the late Paleozoic and Mesozoic basins of eastern Australia. Zircon-bearing sands as mined contain less than 0.5% zircon. Zircon from the east coast of Australia ranges in color from dark tan to pale lavender. Because of iron-bearing grain surface coatings, these zircon sands become brownish orange to orange on heating to 1000°C. Grain size and shape vary from deposit to deposit. The eastern Australia zircon is characterized by subangular to rounded particles with many elongated prismatic grains. Particle size ranges from 90 to 100 μm depending upon which mine produced the zircon. Until 1983, five companies produced zircon from beach sands along the east coast of Australia. Two have since shut down.

Mineral sand deposits of Western Australia are found along ancient beaches which lie at different elevations ranging from 10 to 200 m above present sea level. The strand line deposits are part of a series of Pleistocene and Recent sediments which exist as a thin veneer throughout the Perth basin. These overlie much older marine and continental deposits. Shoreline deposits formed during stable Pleistocene sea level stands. The age of these deposits ranges from fairly young to perhaps over a hundred thousand years. Heavy mineral sand deposits along the southern coast of Western Australia near Cable and Bunbury are very rich in the mineral ilmenite ($Fe_2O_3 \cdot 3TiO_2$) but contain some zircon.

Zircon has become a major export of Western Australia since the discovery of the deposits near the town of Eneabba. The Eneabba heavy minerals occur at different strand lines ranging from 82 to 120 m above sea level. The presence of clay, induration, and grain surface coatings created difficult mining and mineral separation problems; however, these have largely been corrected by the producers. The mineral grains are extensively coated with ferrugenous aluminum silicates

Table 1. Mineralogy of Zircon and Baddeleyite

	Normal Zircon	Altered Zircon (Hyacinth)	Much Altered Zircon (Malacon)	Baddeleyite
Crystal	Ditetragonal Dipyramidal	Ditetragonal Dipyramidal	Amorphous	Monoclinic Prismatic
Specific gravity	4.6-4.7	4.2-4.6	3.9-4.2	5.4-5.7
Mohs hardness	7.5	7	6-7	6.5
Color	White	Purplish	Dark	Variable
Index of refraction	1.92-1.96	1.90-1.92	1.76-1.90	2.19
Cleavage	110 poor	110 poor	None	001 Perfect
Loss on igition	0.1%	0.5-1%	10%	

Table 2. Major World Zircon Sand Producers

Producer	Location	Remarks
Australia		
Allied Eneabba Pty., Ltd.	Eneabba, W.A.	60% Du Pont 40% Allied Minerals
Associated Minerals Consolidated, Ltd.	Stradbroke Island, Qld.	Subsidiary of Consolidated Goldfields
	Capel, W.A. Geraldton, W.A. Eneabba, W.A.	Formerly Western Titanium
Cable Sands Pty., Ltd.	Bunbury, W.A.	Subsidiary of Kathleen Investments, Ltd.
Consolidated Rutile, Ltd.	Stradbroke Island, Qld.	Subsidiary of Cudgen RZ, Ltd.
Cudgen RZ, Ltd.	Kingscliff, NSW	General Mining (49%)
Rutile and Zircon Mines	Newcastle, NSW	Peko-Wallsend, Ltd. Kathleen Investments
Westralian Sands, Ltd.	Capel, W.A.	
Brazil		
Nuclemon	Bahia and Espirito Santo States	Subsidiary of Nuclebras
India		
Indian Rare Earths Ltd.	Manavalakurichi, Tamil Nadu, Chavara, Quilon	
Kerala Minerals & Metals	Chavara, Quilon	Owned by the Government of Kerala
South Africa		
Richards Bay Minerals (Tisand Pty, Ltd.)	Richards Bay	Quebec Iron & Titanium, General Mining, Industrial Development Corp., S.A. Mutual Life Assurance Soc., Southern Life Association
Sri Lanka		
Ceylon Mineral Sands Corp.	Pulmoddai China Bay	
United States		
E. I. du Pont de Nemours & Co., Inc.	Starke, FL Lawtey, FL	
Associated Minerals Consolidated, Inc.	Green Cove Springs, FL	Subsidiary of Consolidated Goldfields

as a result of laterization. The grain surface coatings cause the zircon to take on a light orange color when heated to 1000°C.

South African Deposits

Heavy minerals are produced from a high dune paralleling the Indian Ocean about 160 km north of Durban. The ore body being mined is 17 km long and 2 km wide and is reported to contain 700 million tons of ore. Life of the present ore body is estimated at 30 years.

The zircon grains are rounded to irregular or prismatic in shape and are slightly finer than Du Pont Florida zircon but coarser than Associated Minerals, US. Like Australian zircon the grains have surface coatings which turn a dark orange color on heating.

There are reported to be similar concentrations of heavy minerals in dunes to the north of Richards Bay.

Baddeleyite is recovered from mill tailings at the Palabora copper mine in the Transvaal. The ZrO_2 and HfO_2 range from 97 to 99%.

Baddeleyite may occur either as a primary or secondary mineral. It is sometimes found in alluvial deposits associated with ilmenite, zirkelite (mixture of zircon and baddeleyite), apatite, and perovskite. It is also found in Brazil, Sri Lanka, Italy, and the United States.

US Deposits

A small amount of zircon was produced from a beach sand mine located near Ponte Vedra, FL, between 1918 and 1929. There were no companies producing heavy minerals in Florida from 1929 to 1939. In 1940 a small quantity of heavy minerals including zircon was produced from surface concentrates found in the beach sand near Melbourne, FL; this operation was later moved to a dune area near Vero Beach and operated until about 1963. Between 1943 and 1963 zircon was produced from a deposit at Arlington, a suburb of Jacksonville, FL. In 1963 production began from a heavy mineral deposit near Folkston, GA. When this deposit was mined out in 1974, the mining and concentrating equipment was moved a few miles south to a deposit near Boulougne, FL, where mining continued until the ore was mined out in 1979. Zircon production began from an old shoreline deposit near Green Cove Springs, FL in 1973. Zircon from this deposit averages about 98 μm in size. This is similar to the zircon produced from the Jacksonville deposit during World War II.

The largest domestic supply of zircon sand comes from the Trail Ridge deposit in north central Florida. This large deposit was discovered by Du Pont in 1946 and has been in continuous production of zircon since 1950. The deposit was formed at the height of Florida's submergence during Pleistocene times. Reworked sands transported from older sand features were redeposited forming the ridge. Waves, currents, and wind action removed part of the lighter silica sand, leaving an enrichment of heavy minerals in sand dunes on the western flank of this prominent geographic feature. After deposition of the heavy mineral-bearing sands, the ridge was covered with wind blown sand, forming an overburden.

Zircon from the Trail Ridge deposit is extremely uniform with respect to particle size and grain morphology as well as chemical purity. The grains are free of surface coatings and have been slightly etched as a result of wind action during formation of the deposit. The grains have no surface coatings, therefore, the zircon turns white on heating.

Undeveloped zircon deposits are also known to exist in western Tennessee.

Major world zircon sand producers are listed in Table 2.

Hafnium Metal

Hafnium, though relatively abundant in nature, is sparsely disseminated and very costly to extract. Zircon sand is the only commercial source of this metal. Hafnium's melting point is 4000°F, which is higher than the melting point of zirconium. Heat-resisting parts for special purposes have been made by compacting hafnium powder to a density of 98%. The metal has a close-packed hexagonal structure. It has excellent resistance to a wide range of corrosive environments. Because of its high thermal neutron-capture cross section and excellent strength up to 1000°F, hafnium is useful in unalloyed form in nuclear reactors.

WORLD ZIRCON RESERVES

The US Bureau of Mines reports identified world resources of zircon exceed 60 million tons; however, only a small fraction is economic to mine. Of this 60 million tons, the US has about 14 million tons. Studies of zircon associated with phosphate, sand, and gravel have been carried out by several US governmental agencies and private interests. In all cases the tonnage of zircon and other heavy minerals produced in those industries is insufficient to be considered economic at this time. An important future source may be Canada tar sands. Although the percent heavy minerals in the tar sands is less than 1, about 41,000 st of zircon could be produced from the 95 million st of tar sands mined annually.

THE FUTURE OF ZIRCON IN NORTH AMERICA

The demand for zircon will be set by the needs of the various consuming industries and by the availability of equal or lower cost alternative products.

The supply of zircon will be determined by demand for titanium minerals and availability of deposits. It is doubtful that new heavy mineral mines will open until the current oversupply of rutile and zircon passes and prices stabilize at higher levels. Several known deposits have already been put off limits. Others may be eliminated if governmental attitudes toward mining and the environment follow the trend of the 1970s.

Additional North American reserves are available for development to meet domestic needs. These will be developed if the delivered cost of the zircon and other heavy minerals is competitive with that from foreign sources and yet high enough to provide producers a reasonable return on their investment.

BIBLIOGRAPHY AND REFERENCES

Anon., 1969, "Pattern of Zircon Use May Change," *Industrial Minerals*, No. 16, Jan., pp. 9-13, 15-17.

Anon., 1969a, "Titanium Minerals—Uruguay," *Mineral Trade Notes*, US Bureau of Mines, Vol. 66, No. 6, pp. 25-26.

Anon., 1969b, "World Zircon Producers—Present Production and Future Potential," *Industrial Minerals*, No. 16, Jan., pp. 19, 21, 23-27.

Anon., 1970, "On the Trail of Zirconium Oxides," *South African Mining and Engineering Journal*, Vol. 81, Pt. 1, No. 4034, pp. 1085, 1087, 1089, 1091.

Anon., 1970a, "South Africa—Foskor's Huge Phosphate Reserves," *Industrial Minerals*, No. 28, Jan., p. 36.

Anon., 1971, "Titanium Minerals—Part 1, The Producers Reviewed," *Industrial Minerals*, No. 43, pp. 9-13, 15-23.

Anon., 1973, "Zirconium and Hafnium," *Commodity Data Summaries*, US Bureau of Mines, pp. 166-167.

Brooks, C.K., 1969, "On the Distribution of Zirconium and Hafnium in the Skaergaard Intrusion, East Greenland," *Geochimica et Cosmochimica Acta*, Vol. 33, No. 3, Mar., pp. 357-374.

Brooks, C.K., 1970, "The Concentrations of Zirconium and Hafnium in Some Igneous and Metamorphic Rocks and Minerals," *Geochimica et Cosmochimica Acta*, Vol. 34, No. 3, Mar., pp. 411-416.

Calver, J.L., 1957, "Mining and Mineral Resources," Geological Bulletin No. 39, Florida Geological Survey, pp. 15-31.

Carpenter, J.H., and Griffith, R.H., 1960, "Production of Monazite from Alluvial Concentrates," Carpco Technical Bulletin, Carpco, Jacksonville, FL, 10 pp.

Chao, E.C.T., and Fleischer, M., 1960, "Abundance of Zirconium in Igneous Rocks," *Report*, 21st International Geological Congress, Copenhagen, Part 1, pp. 106-131.

Clarke, R.G., 1970, "Zirconium and Hafnium," *Minerals Yearbook 1970*, US Bureau of Mines, Vol. 1, pp. 1205-1211.

Coope, B., 1982, "Titanium Minerals—Focus on Production," *Industrial Minerals*, No. 178, July, pp. 27-35.

Dana, E.S., 1932, *A Textbook of Mineralogy*, W.E. Ford, ed., 4th ed., John Wiley & Sons, New York, 851 pp.

Dow, V.T., and Batty, J.V., 1961, "Reconnaissance of Titaniferous Sandstone Deposits of Utah, Wyoming, New Mexico, and Colorado," Report of Investigations 5860, US Bureau of Mines, 52 pp.

Fleischer, M., 1955, "Hafnium Content and Hafnium-Zirconium Ratios in Minerals and Rocks," Bulletin 1021-A, US Geological Survey, pp. 1-13.

Franco, R.R., and Loewenstein, W., 1948, "Zirconium from the Region of Poços de Caldas," *American Mineralogist*, Vol. 33, No. 3-4, pp. 142-151.

Frondel, C., 1957, "Zirconium—Mineralogy and Geochemistry," *Advances in Nuclear Engineering*, Vol. 2, *Proceedings*, 2nd Nuclear Engineering and Science Congress, Philadelphia, Pt. 2, pp. 305-312.

Gadsden, J., 1972, "Zirconium and Hafnium," in "Mining Annual Review," *Mining Journal*, London, p. 94.

Garnar, T.E., Jr., 1978, "Geologic Classification and Evaluation of Heavy Mineral Deposits," 12th Forum on the Geology of Industrial Minerals, Information Circular No. 49, Georgia Geological Survey, Atlanta, pp. 25-36.

Garnar, T.E., Jr., 1980, "Heavy Minerals Industry of North America," Presented at 4th Industrial Minerals International Congress, Atlanta, GA, 13 pp.

Gerasimovskii, V.L., 1956, "Geochemistry and Mineralogy of Nepheline Syenite Intrusions," *Geochemistry*, No. 5, pp. 494-510.

Gottfried, D., and Waring, C.L., 1964, "Hafnium Content and Hf/Zr Ratio in Zircon from the Southern California Batholith," Professional Paper 501-B, US Geological Survey, pp. B88-B91.

Guimarães, D., 1948, "The Zirconium Ore Deposits of the Poços de Caldas Plateau, Brazil, and Zirconium Geochemistry," Boletim 6, Minas Gerais Instituto de Tecnologia Industrial, pp. 1-40 (Portuguese), 41-79 (English).

Hansen, J., 1968, "Niobium Mineralization in the Ilimaussaq Alkaline Complex, South-West Greenland," *Proceedings*, 23rd International Geological Congress, Prague, Sec. 7, "Endogenous Ore Deposits," pp. 263-273; reprinted as Miscellaneous Papers No. 60, Grönlands Geologiske Undersögelse.

Hess, H.D., 1962, "Hafnium Content of Domestic and Foreign Zirconium Minerals." Report of Investigations 5856, US Bureau of Mines, 62 pp.

Horn, M.K., and Adams, J.A.S., 1966, "Computer-Derived Geochemical Balances and Element Abundances," *Geochimica et Cosmochimica Acta*, Vol. 30, No. 3, Mar., pp. 279-297.

Kauffman, A.J., Jr., and Holt, D.C., 1965, "Zircon—A Review, with Emphasis on West Coast Resources and Markets," Information Circular 8268, US Bureau of Mines, 69 pp.

Klemic, H., et al., 1973, "Zirconium and Hafnium," US Mineral Resources, Professional Paper No. 820, US Geological Survey, pp. 718-722.

Kramer, J.W., et al., 1976, "Survey of Heavy Minerals in Surface Mineable Area of Athabasca Oil Sand Deposit" *Canadian Mining and Metallurgical Bulletin*, Vol. 69, No. 776.

Lewis, R.M., 1978, "Possible Recovery of Heavy Minerals from Phosphate Tailings," Preprint No. 78B300, SME-AIME Fall Meeting, Orlando, FL, 7 pp.

Lissiman, J.C., and Oxenford, R.J., 1973, "The Allied Mineral N.L. Heavy Mineral Deposit in Eneabba, W.A.," Conference Volume, Australian Institute of Mining and Metallurgy, pp. 153-161.

Lyakhovich, V.V., and Shevaleyevskii, I.D., 1962, "Zr:Hf Ratio in the Accessory Zircon of Granitoids," *Geochemistry*, No. 5, pp. 508-524.

Lynd, L.E., 1978, "US Dependence on Foreign Sources of Heavy Mineral Concentrates," Preprint No. 78H366, SME-AIME Fall Meeting, Orlando, FL, 17 pp.

Lynd, L.E., 1980, "Zirconium and Hafnium," *Minerals Industry Surveys*, US Bureau of Mines, 1 p.

Lynd, L.E., 1981, "Zirconium," *Mineral Commodity Summaries*, US Bureau of Mines, pp. 182-183.

Martens, J.H.C., 1928, "Beach Deposits of Ilmenite, Zircon, and Rutile," 19th Annual Report of the Florida State Geological Survey, pp. 124-125.

Palache, C., Berman, H., Frondel, C., 1951, "Baddeleyite," *Dana's System of Mineralogy*, John Wiley & Sons, New York, pp. 608-610.

Pirkle, E.C., Pirkle, W.A., and Yoho, W.H., 1974, "The Green Cove Springs and Boulogne Heavy Mineral Sand Deposits of Florida," *Economic Geology*, Vol. 69, pp. 1129-1137.

Pirkle, F.L., 1975, "Evaluation of Possible Source Regions of Trail Ridge Sands," *Southeastern Geology*, Vol. 17, No. 2, Duke University, pp. 93-114.

Sørensen, H., 1970, "Low-Grade Uranium Deposits in Agpaitic Nepheline Syenites, South Greenland," *Uranium Exploration Geology, Proceedings*, International Atomic Energy Agency Panel on Uranium Exploration Geology, Vienna, pp. 151-159.

Stamper, J.W., and Chin, E., 1970, "Hafnium," *Mineral Facts and Problems*, Bulletin 650, US Bureau of Mines, pp. 587-594.

Stamper, J.W., and Chin, E., 1970a, "Zirconium," *Mineral Facts and Problems*, Bulletin 650, US Bureau of Mines, pp. 825-835.

Stow, S.H., 1968, "The Heavy Minerals of the Bone Valley Formation and Their Potential Value," *Economic Geology*, Vol. 63, No. 8, pp. 973-975.

Sullivan, G.V., Browning, J.S., 1970, "Recovery of Heavy Minerals from Alabama Sand and Gravel Operations," Technical Progress Report No. 22, US Bureau of Mines.

Taylor, S.R., 1966, "The Application of Trace Element Data to Problems in Petrology," *Physics and Chemistry of the Earth*, Vol. 6, Pergamon Press, Oxford, England, pp. 133-213.

Tolbert, G.E., 1966, "The Uraniferous Zirconium Deposits of the Poços de Caldas Plateau, Brazil," Bulletin 1185-C, US Geological Survey, 28 pp.

Turekian, K.K., and Wedepohl, K.H., 1961, "Distribution of the Elements in Some Major Units of the Earth's Crust," *Bulletin*, Geological Society of America, Vol. 72, No. 2, pp. 175-191.

Vainshtein, E.E., et al., 1959, "The Hf/Zr Ratio in Zircons from Granite Pegmatites," *Geochemistry*, No. 2, pp. 151-157.

Vlasov, K.A., 1966, *Geochemistry and Mineralogy of Rare Elements and Genetic Types of Their Deposits—Vol. 1, Geochemistry of Rare Elements*, IPST No. 2123, Israel Program for Scientific Translations, Jerusalem, 945 pp.

Wedow, H., Jr., 1967, "The Morro do Ferro Thorium and Rare-Earth Ore Deposit, Poços de Caldas District, Brazil," Bulletin 1185-D, US Geological Survey, 34 pp.

Wessel, F.W., 1958, "Zirconium Raw Materials Supply," *Survey of Raw Material Resources, Proceedings*, 2nd International Conference on Peaceful Uses of Atomic Energy, Geneva, Vol. 2, pp. 17-20.

Williams, L., 1964, "Titanium Deposits in North Carolina," Information Circular 19, Div. of Mineral Resources, North Carolina Dept. of Conservation and Development, 51 pp.

Winchell, A.N., 1951, "Zircon," *Elements of Optical Mineralogy. Part II—Description of Minerals*, John Wiley & Sons, New York, pp. 494-495.

2.11 Acquisition and Exercise of Rights to Use Land in Connection with Surface Mining Operations

CLAYTON J. PARR

INTRODUCTION

This section deals with the acquisition, maintenance, and exercise of rights to make use of lands in connection with surface mining operations. It is presumed that the mineral rights necessary to exploit an ore occurrence have been obtained and that the focus has shifted to considerations of land use.

At a surface mine, land is needed for uses related both to mineral extraction activities and to mining related activities. Property rights are required for stripping beyond the ore zone, and for waste dumps, leach piles, ore storage, ore transportation systems, haul roads, crushers, and other mine related purposes. Space is also needed for mine support facilities such as maintenance and repair shops, warehouses, mine offices, change houses, equipment storage areas, power substations, and water storage and distribution systems. A mill will require a separate set of property rights, the most critical of which are those needed for tailings disposal.

Often, the operator already holds most of the needed lands through mining claims or mineral leases. If they do not provide the necessary rights to conduct mining and mine related activities, or if additional land is needed, then independent methods of land acquisition must be utilized.

For a discussion of methods for acquisition of mining rights, reference can be made to "Mining Law," Section 2, in the *Mining Engineering Handbook,* published in 1973 by the Society of Mining Engineers of AIME.

FEDERAL LANDS

Unpatented Mining Claims

If unpatented mining claims are included in the mine project area, they might provide convenient sites for mine related surface activities. The original federal mining law provides that the owner of an unpatented mining claim has exclusive possessory rights of all the surface within the boundaries of the claim.[1] Notwithstanding several judicial decisions invalidating claims used solely for nonmine related uses,[2] claim owners commonly used the surface of claims for other purposes, especially for cabin sites. That possessory rights which accompany a mining claim do not include the right to conduct nonmine related activities was clearly established by the Multiple Surface Use Act of 1955.[3] This Act provides that any mining claim located after July 23, 1955 "shall not be used, prior to issuance of patent, for any purposes other than prospecting, mining, or processing operations and uses reasonably incident thereto." The Act further provides that rights under any mining claim thereafter located will be subject, prior to the issuance of patent therefor, "to the right of the United States to manage and dispose of the vegetative surface resources thereof and to manage other surface resources thereof (except mineral deposits subject to location under the mining laws of the United States)." Thus, the uses that may be made of the surface of an unpatented mining claim are limited to those reasonably related to the development and extraction of minerals.[4]

The right to make use of the surface of an unpatented mining claim in aid of mining, however, is not without its limits. Surface use activities for mining purposes must be performed in good faith.[5] Thus, if a claim were located solely for the purpose of surface use, it might be said not to have been located in good faith, and therefore be subject to challenge by the United States.[6] In addition, according to recent authority, mining related activities on a claim may be proscribed if they are determined to be unnecessary and unreasonably destructive to surface resources.[7]

Some uncertainty exists relative to the use of an unpatented claim in connection with mining on other properties. Even where a claim is located in good faith, if it is subsequently used for mine related purposes incident to the extraction of minerals outside of the claim, then, although the law is not entirely clear on the subject, the operator takes the risk that the activity might be enjoined and damages in trespass awarded.[8] Generally, however, if a claim has been located in good faith for its mineral potential and is part of a mineralized block that is covered by a valid mining plan, there is little risk of a challenge to reasonable mine related surface use. That risk could be sufficient, however, to necessitate consideration of alternative land acquisition procedures.

An inherent risk in using any unpatented mining claim for mine ancillary purposes is that ownership may be lost relatively easily as compared to other forms of land ownership. For instance, the claim may be invalidated by reason of noncompliance with federal location and claim maintenance requirements or the claim may be subject to forfeiture to a junior locator for failure to perform annual assessment work.

Every unpatented claim must be independently supported by a valid discovery of minerals.[9] A claim not supported by a legally sufficient discovery is subject to challenge by a subsequent locator who is better able to demonstrate a discovery,[10] by a conflicting user through a private contest,[11] or by the United States through a government initiated contest.[12] The basic legal standard for discovery is the so-called *prudent man* test which examines whether the minerals are of such a character that a person of ordinary prudence would be justified in the further expenditure of his labor and means, with a reasonable prospect of success in developing a valuable mine.[13] In a government initiated contest, the claimant must also be able to show that by reason of accessibility, bona fides in development, proximity to market, existence of present demand, and other factors, the deposit is of such value that it can be mined, removed, and disposed of at a profit.[14] These tests require that there be an actual physical exposure of mineral-bearing rock in place within the limits of a claim.[15] An important maxim to remember is that information sufficient to justify additional exploration is not enough to establish a discovery, at least as against the United States.[16] Clearly then, use of an unpatented mining claim for mine related purposes carries with it an inherent risk that the claim might not be able to withstand careful scrutiny as to whether it is supported by a legally sufficient discovery.

An unpatented mining claim can also be lost if legally

sufficient annual assessment work is not performed.[17] Although failure to perform valid annual assessment work has been thought only to render a claim subject to forfeiture to the locator of an overlapping junior claim, the United States has asserted a separate right to declare a claim cancelled when assessment work has not been performed.[18]

It is common practice in a mine area to utilize mine work on a few claims as assessment work for a much larger group of contiguous claims. Group assessment work is valid, however, only if the work performed off of any particular claim benefits or tends to benefit that claim and is done for the purpose of developing each claim in the claim group.[19] This direct benefit requirement is sometimes difficult to establish where the claim is being maintained or being utilized primarily for mine related purposes without being projected for mining pursuant to an overall mining plan.

Loss of a claim can also result from failure to make annual filings of evidence of performance of assessment work with the local county recorder and the Bureau of Land Management.[20]

A more serious problem relative to rights of surface use is presented where the mining claim has been located on land which has been patented by the United States subject to a reservation of minerals. Since the subject relates to both mining claims and federal mining leases, it is discussed separately later in this section.

Millsite Claims

Millsites provide a simple, straightforward method for acquiring federal land for use in connection with mining operations. Because the location of a millsite is a self-initiated right, however, it is, like a mining claim, of tenuous validity and therefore often lacking in the security of title necessary for a major investment.

Three types of millsite claims are recognized in the federal mining law. They are:

1. Nonmineral land not contiguous to a vein or lode used or occupied by the proprietor of a lode mining claim for mining or milling purposes;

2. Nonmineral land not contiguous to a vein or lode appropriated by the owner of a quartz mill or reduction works; and

3. Nonmineral land used or occupied by the proprietor of a placer claim for mining, milling, processing, beneficiation, or other operation in connection with the placer claim.[21]

Thus, a millsite must be tied to a lode or placer mining claim, or be used specifically for a quartz mill or reduction works.

The terms *quartz mill* and *reduction works* have been interpreted strictly by the Department of the Interior to refer literally to plants that crush quartz rock for the extraction of gold, or to plants that reduce metals from their ores, such as smelting works or cyanide plants.[22] It is questionable, therefore, whether a conventional base or precious metal concentrator is a *reduction works*. Thus, *independent* millsite claims are not commonly used because of the fear that the desired use will not fit the narrow definitions. Accordingly, most millsites are used in connection with lode or placer mining claims.

Millsites can be located in plots up to 2.02 ha (5 acres)[23] utilizing location procedures similar to those for lode or placer claims.[24] The land must be unappropriated public domain open to location under the mining laws.[25] Reference must be made to the laws of the state involved in order to find out the precise requirements for location. Generally, there is no specific dimensional configuration required for a millsite, but usually they are in square or compact rectangular configurations.

The land upon which a millsite is located must be nonmineral in character. This can pose problems with respect to millsites located in close proximity to a mine site since the likelihood of mineralization is greater in such areas. The risk of a determination that the land is mineralized is that the millsite claim can be declared to be void through a contest proceeding. Mineralization on millsite lands also creates a risk that the land might be appropriated through a lode or placer mining claim located by another party. A millsite can be located on lands known to be valuable for leasing act minerals.[26]

Although the rule is not free from doubt, as against junior locators, the mineral or nonmineral character of the land is determined at the date when rights to a millsite attach, which would be the date of location of the millsite if the millsite is put to immediate use, or if not, the date when use or occupation for use begins.[27] With respect to the United States, however, a millsite claim can be defeated if mineralization is found at any time prior to the issuance of a patent.[28] The presence of mineralization can be established with less information than is required to demonstrate a discovery of valuable minerals.[29]

Some persons have attempted to reduce the risk of invalidation by locating lode or placer claims over millsite claims.[30] That practice is not encouraged, however, because the location of the mining claim can be construed as an admission that the land covered by the millsite is mineral in character.[31]

An additional consideration that might influence the use of a millsite is that in order for a millsite claim to be valid the lode or placer claim serving as the basis for the millsite must be valid. Accordingly, if an unpatented lode or placer claim is determined to be invalid, a dependent millsite attached to that claim also will be invalid.

The requirement of noncontiguity has been interpreted fairly loosely and usually will not pose a problem where there is some slight contact between the millsite and the mining claim which is not outright boundary-to-boundary overlap.[32]

There is no maximum number of millsites that can be held, but the number can be limited by considerations of good faith and reasonable need.[33] Only as much ground may be included in an individual millsite as is actually needed for the particular use involved, but a paring down of a millsite for such reason would ordinarily occur only in connection with a patent proceeding.[34]

The contemplated use of the millsite must be a "mining or milling purpose" under the statute.[35] Generally all that is required is that the use be some step in, or directly connected with, the process of mining or milling.[36] Millsites can be used for facilities related to the acquisition and exercise of a water right, but they cannot be used for purposes which would ordinarily be permitted under an easement, such as roads, transmission lines, and pipelines.[37] There is some uncertainty as to whether a millsite can be obtained and used exclusively for the deposit of waste.[38]

Read literally, the statute requires that a millsite's validity be sustained by an actual continuing use.[39] Mere intention or purpose of performing acts of use or occupancy will not satisfy the statute.[40] A reasonable time to make use of a millsite claim will be permitted,[41] but in practice, a millsite will almost have to be located well in advance of the date when the mine related activity is commenced. Although there is a question as to the validity of an idle millsite, a prospective use designated specifically in a mine plan might be sufficient to protect the millsite. The validity of a millsite will nevertheless remain tenuous until an actual use is commenced, and then it is only secure as long as that use is continuous.

It is not necessary that annual assessment work be performed on a millsite; however, a notice of intent to hold must be filed each year both in the county records and with the Bureau of Land Management pursuant to regulations of the Bureau of Land Management.[42]

A millsite may be patented, but only after or contemporaneously with the issuance of a patent on the associated mining claim.[43] The validity tests described above will be applied most strictly during the course of a patent proceeding.

Federal Mineral Leases

As is the case with any other situation where mining rights are held pursuant to a lease, reference must be made to the language of the lease itself in the first instance to determine what rights exist relative to use of the surface in aid of mining operations. Typically, the granting clause of a federal lease gives the lessee the right to construct and maintain on the leased lands such structures or other facilities as may be necessary for the mining, preparation, and removal of the leased minerals. This right may be expanded to include the right to construct and maintain works, buildings, plants, structures, and appliances necessary to the mining, processing, and removal of the deposits. The express grant of the second right is important, because reference to processing as a permitted use removes doubt as to whether concentrators or other mineral processing facilities can be constructed on the premises.

Although federal leases are generally issued on standard forms, specific provisions, often imposing onerous burdens, are added in the form of special stipulations. The stipulations can impose detailed requirements, particularly with respect to surface protection and reclamation and especially where national forest lands are involved. Newer leases involving minerals whose mining may cause widespread land disturbance, such as oil shale and tar sands, are likely to have extensive stipulations.

The real governing factor in defining the right of surface use is the regulatory approval that must be obtained for any proposed mining or mine related activities. Applicable regulations require an operator to submit a detailed mining plan for approval before conducting any operations under a federal mineral lease.[44]

Under BLM rules published in Part 23 of Title 43, Code of Federal Regulations, the mining plan is submitted to the mining supervisor, but he must consult with the other agencies involved and the BLM district manager with respect to surface protection and reclamation requirements before approving the plan.[45] The plan must describe proposed methods of operating, including a description of proposed roads or vehicular trails and the size and location of structures and facilities to be built, and it must provide information about water use and contamination, erosion control procedures, and environmental impact mitigation control measures.[46] Details must be provided relative to the timing and methods of reclamation and revegetation.[47]

Additional BLM regulations that relate primarily to mining considerations are set forth in Part 3570 of Title 43, Code of Federal Regulations. They require that the plan include information relative to mining design and operation, and also data concerning runoff and water pollution control, mitigation measures for the protection of fish and wildlife, and a plan of reclamation.[48]

Before approving a mine plan, the authorized officer must consult with the agency having administrative jurisdiction over the lands concerning surface protection and reclamation.[49] This process could create additional requirements for information and additional burdens in the way of mitigating adverse environmental impacts.

If the mining plan is approved, the operator must provide a performance bond before commencing operations, to ensure that commitments made in the mining plan are carried out.[50]

Although grants of surface use rights in individual leases seem to refer to mining on the leased tract itself, approval for use of the leased lands in whole or in part in aid of mining on other federal leases can be approved on a site specific basis. If, however, the surface of such federal leased lands is held under private ownership, then a new problem arises. Surface owners whose rights derive from patents issued under the Coal Acts of 1909 and 1910, the Agricultural Entry Act of 1914, and the Stock-Raising Homestead Act of 1916 can prevent use of the surface of the tract involved to support simultaneous or independent production on other lands.[51]

Special purpose leases providing for the use of surface rights can be obtained in connection with mining on federal phosphate and sodium leases.[52] A phosphate lease holder can obtain up to 32.4 ha (80 acres) of land if the Secretary of Interior determines that the lands are necessary or convenient for the extraction, treatment, and removal of phosphate deposits.[53] Tracts can be in separate plots so long as they do not exceed 32.4 ha (80 acres) in the aggregate. The lands cannot be included in a National Forest.

A sodium lease holder may obtain rights to use a single tract of unoccupied nonmineral public land for campsites, refining works, and other purposes connected with and necessary to the proper development and use of deposits covered by the permit or lease.[54]

Surface Use Where the United States Has Issued a Patent Subject to a Reservation of Minerals

The federal government has issued land patents to private owners with a reservation of minerals to the United States under the coal acts of 1909[55] and 1910,[56] which apply to coal only, and the Agricultural Entry Act of 1914,[57] the Stock-Raising Homestead Act of 1916,[58] and the Taylor Grazing Act of 1934.[59] Special procedures must be followed with respect to locating and operating mining claims on such lands, and in conducting operations thereon pursuant to federal mineral leases. This discussion focuses on the Agricultural Entry Act of 1914 and the Stock-Raising Homestead Act of 1916.

Patents issued under the Stock-Raising Homestead Act of 1916 contain a reservation to the United States of all minerals in the lands and the right to prospect for, mine, and remove the same.[60] A party who has located mining claims or obtained a federal mineral lease on lands patented under the Stock-Raising Homestead Act can reenter the tract involved and occupy as much of it as may be required for mining. Before the reentry, however, the claimant must (1) secure written consent or waiver of the surface owner, (2) make an agreement with the surface owner for payment of damages to crops or other tangible improvements, or (3) execute a bond or undertaking to the United States for the benefit of the owner.[61]

The Agricultural Entry Act of 1914 allowed entry of lands withdrawn as phosphate, nitrate, potash, oil, gas, sodium, sulphur, or asphaltic minerals lands or classified as valuable for those deposits, with the proviso that a subsequent patent would be subject to a reservation to the United States of the deposits on account of which the lands were withdrawn or classified or reported as valuable, together with the right to prospect for, mine, and remove the same.[62] Where there has been a reservation under the Act, the mine operator operating under a federal lease for the reserved minerals must make payment of damages to the owner or give a good and sufficient bond or undertaking therefor in an action instituted in any competent court to ascertain and fix damages.[63]

Much uncertainty has existed under these two acts as to the extent of use that may be made of the patented land by a mineral interest holder and the amount of compensation that must be paid for damages. Generally, the mineral developer can use the entire surface so long as such use is necessary and reasonable for the extraction of minerals from the associated tract of land.[64] Although the question is not entirely free of doubt, surface mining is considered to be permissible if it is reasonable and necessary for the recovery of the reserved minerals.[65] The method adopted in exercising rights of surface use, however, must be necessary and cannot result in use of more of the surface than is reasonably required.[66] Whether the use is excessive or in fact permitted at all under the law is subject to determination on a case-by-case basis. For example, the right to "mine and remove" minerals has recently been interpreted as encompassing the right to construct a geothermal power plant,[67] but the extent to which the decision of the court will be applicable to processing of other kinds of minerals remains to be seen.

A mining claim holder or mineral lessee does not have a right to make use of the surface of patented land to support simultaneous or independent production on other lands.[68]

Patentees under the various acts are entitled to compensation for damages caused by the mineral claimant or mineral lessee to "crops and improvements" under the Agricultural Entry Act[69] and to "crops and tangible improvements"[70] under the Stock-Raising Homestead Act. Because the "improvements" were clearly contemplated to be agricultural improvements, the Supreme Court of the United States held in 1927 that compensation for nonagricultural improvements need not be made.[71] It is not clear, however, that the rule in that case would be extended to all nonagricultural improvements or that it would be applied so rigidly in a modern setting.[72] Since 1949, compensation is also payable in respect of damage that may be caused to the value of the land for grazing purposes by removal of the reserved minerals by strip or open pit mining methods.[73]

Because of the tenuous nature and uncertain extent of the rights of a mining claimant or federal mineral lessee to make use of the surface of lands patented with a reservation of minerals in the United States, it is standard practice to enter into a surface use agreement with the surface owner in such situations, such as is described later in this section. A forced entry procedure becomes necessary only when attempts to arrive at an agreement with the surface owner have been unsuccessful.

Special Procedures for Federal Lands

Land Exchanges: Authority has been granted to the Secretary of the Interior with regard to Bureau of Land Management administered lands and to the Secretary of Agriculture with regard to lands in the National Forest System, to dispose of public land or interests therein by exchange.[74] The exchanged lands must be in the same state, and there must not be a difference in appraised value of the exchanged tracts greater than 25%.[75] That difference can be made up in cash equalization payments running either way.[76] Thus, a mine operator can purchase lands that are desired by the BLM or the US Forest Service and exchange them for needed federal lands in the same state of near equivalent value in the vicinity of a mining operation.

BLM exchange procedures are established in Part 2200 of Title 43, Code of Federal Regulations; those of the Forest Service are in Section 254 of Title 36, Code of Federal Regulations.

Exchange procedures, although workable, are lengthy and complex, thereby limiting their usefulness. Even though the Secretaries of Interior and Agriculture must consider the public interest in acting upon exchange applications,[77] the procedure is subject to the vagaries of a complex administrative process. Because of the long time periods required between initiation and completion of the transaction, the approach is feasible only if the contemplated need can be deferred for several years, or if interim rights to the land can be acquired under millsite claims or temporary permits.

Before any decision is made about seeking a land exchange, contact should be made with local representatives of the US Forest Service or the District Office of the BLM. The initial inquiry will reveal whether certain lands are desired by these agencies that might be suitable for purchase and exchange for federal lands near the mine site. The interest acquired for exchange need not necessarily be the entire fee title but could be a lesser interest such as water rights, or easements for roads and trails. Informal negotiations with the District Ranger can be culminated in a more formal Statement of Intent that spells out the basic understanding reached by the parties.[78]

If these informal negotiations are successful, a formal land exchange proposal complying with the regulations is then prepared. Of the numerous steps that follow, the most important is the appraisal process which determines the current fair market value for the lands or interests in the lands subject to the exchange.

Surface Use of BLM Lands by Lease, Permit, or Easement: The Secretary of Interior has authority to issue easements, permits, leases, and licenses for "the development of small trade or manufacturing concerns."[79] Rights can be obtained under this authority for many miscellaneous mine-related uses. Those that require rights-of-way for linear type uses would ordinarily be obtained under rights-of-way provisions described later in this section.

The nature of the right granted is tailored to the use for which the land is needed. If there is to be substantial construction and the investment of a large amount of capital, a *lease* will be issued for a specific term consistent with the time required to amortize the capital investment. If the use is for a short term not to exceed three years, and involves little or no construction or investment, a *permit* will be issued. A permit may be renewed or revoked. Nonpossessory and nonexclusive *easements* can be granted to provide rights for limited linear type uses that are not covered under other regulatory provisions. Short-term noncommercial activities that do not appreciably damage the lands are considered to be *casual uses* for which no formal approval is required.[80]

Rights granted under these provisions are essentially negotiated contracts consistent with the regulations in Part 2920, of Title 43, Code of Federal Regulations. Processing is begun by informal discussions with the Bureau of Land Management District Office having jurisdiction over the public lands in question. It is possible at this stage for the authorized officer to issue a permit for a land use if the proposed use is in conformance with Bureau of Land Management plans, policies, and programs, local zoning ordinances, and any other pertinent requirements, and will not cause appreciable damage or disturbance to the public lands, their resources, or improvements.[81]

If a more permanent right is required, formal processing is initiated by the filing of a proposal by the person desiring to make the land use.[82] The proposal is then thoroughly reviewed to evaluate the appropriateness of the proposed use, particularly with respect to BLM land use plans, which must have been completed for the area in question.[83]

Barring any problems, the initial approval step is then taken by the BLM through publication of a Notice of Realty Action indicating the availability of public lands for nonfederal uses through lease, permit, or easement.[84] Formal applications can then be submitted by the person who initiated the process and by other interested parties.[85]

If interest is expressed by several parties, the authorization may be offered on a competitive basis, and if not, on a negotiated, noncompetitive basis.[86] This decision is mainly a matter of judgment on the part of the authorized officer.[87] In any event, the proposal by the person selected must not be for less than fair market value.[88]

Terms and conditions are then prepared consistent with the regulations.[89] Rental fees are to be paid annually in advance based on fair market value or as determined by competitive bidding.[90]

An important factor in deciding whether to go through this process is that the land use applicant must reimburse the United States for reasonable administrative and other costs incurred in processing the application and in monitoring construction, operation, maintenance, and rehabilitation of authorized facilities including, if required, costs of preparing an environmental impact statement.[91] These costs can be substantial and should be evaluated before an attempt is made to use procedures of this type.

In general, even though the procedures may be onerous, acquisition of lands through this method may be the most appropriate for many uses required in connection with a mining operation. Because it is frequently desirable to obtain title ultimately to lands that will be the subject of a large investment, emphasis may be placed on other methods that lead to the acquisition of title such as land exchanges. Even in such cases, temporary rights might be obtained under these regulations that will permit the desired use pending the ultimate acquisition of title.

Rights-of-Way on BLM Lands: The Secretary of the Interior is authorized to grant rights-of-way for roads, ditches, pipelines, tunnels, and other transportation facilities on public lands under Title V of The Federal Land Policy and Management Act of 1976 (FLPMA).[92] Under the same authority, rights-of-way may be granted for reservoirs; storage and terminal facilities in connection with systems for the transportation or distribution of liquids and gases (other than water and oil and natural gas); facilities for the storage of materials in connection with pipelines, slurry, and emulsion systems; conveyor belts for transportation and distribution of solid materials; power distribution facilities; and radio communications facilities.[93] Of particular importance to mining operations are rights-of-way for reservoir facilities, power distribution facilities, and facilities for the storage of materials in connection with conveyor belts.

BLM regulations under this authority are set forth in Group 2800, of Title 43, Code of Federal Regulations. The regulations refer to a right-of-way grant, which is an instrument formally authorizing the use of a right-of-way over, upon, under or through public lands for construction, operation, maintenance, and termination of a project, and to a temporary use permit, which is an irrevocable, nonposses-

sory, nonexclusive privilege, authorizing temporary use of public lands in connection with construction, operation, maintenance, or termination of the project.[94] The terms of such grants or permits may range from a month to a year or a term of years to perpetuity, except that the term for a temporary use permit shall not exceed three years.[95] Terms and conditions are to be established by the authorized officer consistent with the regulations.[96]

Initial contact should be made with the Bureau of Land Management office responsible for management of the affected public lands prior to submitting a formal application. Certain preapplication activities are authorized.[97] After determining the practical considerations involved in pursuing an application, the applicant files a formal application with the Area Manager, the District Manager, or the State Director having jurisdiction over the affected public land.[98] Application processing will require an environmental analysis in accordance with the National Environmental Policy Act of 1969.[99]

An important consideration is that the applicant must reimburse the United States for administrative and other costs incurred in processing applications, including preparation of reports and statements pursuant to the National Environmental Policy Act of 1969, before the right-of-way grant or temporary use permit will be issued.[100] These costs must be borne even if the application is denied. Interim payments may be required.

Rental fees are to be paid annually in advance based on the fair market value of the right being granted.[101] A bond may be required to assure performance of the permittee's obligations.[102]

Rights-of-Way and Special Use Permits on National Forest Service Lands: The Secretary of Agriculture has elected to establish one set of regulations for the issuance of rights-of-way under Title V of FLPMA, special use permits and leases, and other special use authorizations.[103] Regulations are contained in Section 251 of Title 36, Code of Federal Regulations. The regulations provide for the granting of easements, leases, permits, and term permits, all of which are described as "special use authorizations."[104]

The regulations suggest that informal discussions be held prior to the filing of a formal application in order to identify problems and to advise the potential applicant of what lies ahead if an application is pursued.[105]

Applications are filed with the District Ranger or Forest Supervisor having jurisdiction over affected land except in certain circumstances, such as where the lands are under the jurisdiction of more than one agency or more than one unit of the Forest Service.[106] In addition to describing the project in substantial detail, the application must include proposed measures and plans for the protection and rehabilitation of the environment during construction, operation, maintenance, and termination of the project.[107] Additional information may be required by the authorized officer.

Whether an application will be approved, and if so, under what conditions, is largely left to the discretion of the authorized officer. For example, an application may be denied if the authorized officer determines that the proposed use would be incompatible with the purposes for which the lands are managed or if the proposed use would "not be in the public interest."[108]

The regulations establish standard terms and conditions that are applicable to any special use authorization, most of which are elaborations of the general requirement that measures will be taken to protect the environment and promote safety.[109].

If approved, a term will be established which is to be no longer than is necessary to accomplish the purpose of the authorization and is reasonable in light of all circumstances important to the federal land manager. Authorizations exceeding 30 years will provide for revisions of terms and conditions at specified intervals.[110]

Prior to construction of facilities under the special use authorization, the holder must submit plans of all developments within the authorized area for review and approval.[111] Authorization for construction may be conditioned on furnishing a bond or other security to assure that the holder's commitments will be honored.[112]

Rental fees are established by the authorized officer based on the fair market value of the rights and privileges authorized as determined by appraisal or other sound business management principles.[113]

The Forest Service Special Use Authorization regulations currently contain no provisions for cost reimbursement.

Purchase of BLM Lands: Purchases may be made of public lands (except lands in the National Wilderness Preservation System, National Wild and Scenic River System, and National System of Trails), if the land is not suitable for federal management, if it was acquired for a specific purpose that is no longer valid, or if disposal will serve important public objectives including expansion of communities and economic development which cannot be achieved prudently or feasibly on other public land and which outweigh other public objectives and values.[114] An opportunity for congressional action is required if tracts are sold in excess of 1012 ha (2,500 acres).[115] Regulations under the statute are in Part 2710 of Title 43, Code of Federal Regulations.

Although this procedure may seem attractive initially as a method for acquiring rights to federal lands, it is actually of limited value. The thrust of the statute and regulations is to require that the lands for sale be identified primarily through the agency land planning process rather than through requests of applicants. Although expressions of interest can be made, public lands are offered only on the initiative of the Bureau of Land Management after they are determined to be suitable for sale.[116]

If a sale is to be made, it is through competitive bidding procedures. The process is initiated by the publication of a Notice of Realty Action offering the land for sale.[117]

Minerals on any lands sold pursuant to these procedures are to be reserved to the United States.[118]

STATE LANDS

State lands are much like federal lands with regard to acquisition and exercise of mine-related land use rights except that with few exceptions mineral rights are available only through the leasing method. Rights to use the surface of leased lands will depend on rights granted under the lease instrument, applicable regulations, and advance review and approval of mine plans by the state land agency.

State lands, like federal lands, are often patented with reservations of minerals to the state. Reference must be made to the applicable statutes and regulations, and to the case law of the particular state in question, to determine the rights of the mineral lessee from the state vis-a-vis the surface owners.

All states have procedures for the acquisition of rights to state lands, usually in the form of leases, permits or rights-of-way. Some state land agencies have authority to effect land exchanges and most have power to sell state lands under certain circumstances.

FEE LANDS

Introduction

This part discusses the rights of a mineral owner or lessee to make use of the surface of a tract of land for the conduct of mine-related activities where neither the surface nor minerals are owned by a governmental entity. Such *fee lands* include patented mining claims. The discussion focuses on the most common situation where the mine operator is a lessee, either of a lessor who owns both surface and minerals, or of a lessor who owns only the minerals subject to surface ownership by a third party. Surface uses may be related to mining just on the tract in question, or to mining being conducted wholly or partially on other lands.

Surface Owned by Lessor

Where a lease is involved, reference is made in the first instance to the contractual document to see what agreement, if any, was made by the parties relative to use of the leased lands for the conduct of mining operations. Although a mineral lessee has broad implied rights to make use of the surface, as is discussed in more detail below, the parties can limit, expand, or condition those rights by contract.

Even where broad surface use rights are granted, the lessee may still be subject to a limitation imposed by law to exercise those rights in a manner that is not excessive. The reasonableness of the proposed use can be determined in part on the basis of whether the lessee has alternative methods to accomplish the intended purpose which will have a more limited adverse effect on the lessor's surface operations.[119]

In general, where the operations are conducted in a reasonable fashion without negligence, and without contravention of the terms of the lease, there will be no liability on the part of a mineral lessee for surface damages.[120] Parties to a mineral lease can, and often do, however, provide for the payment of surface damages in accordance with procedures that would be appropriate to a separate surface use agreement such as is discussed later in this section.

A more difficult situation is presented where the lease is silent as to the particular use that the mine operator desires to make of the leased land. A lessee is generally recognized as having the right to make all use of the leased land that is reasonably incident to the extraction of minerals therefrom, which rights extend to making reasonable use of the surface or dumping waste.[121] Whether certain uses not inherently a part of the mining operation, such as employee housing, are reasonably necessary to the conduct of the operation, is a matter not necessarily free of doubt and will have to be determined on a case-by-case basis.[122]

Of particular concern to a surface mine operator, however, is whether, in the absence of a specific grant, the implied leasehold rights include the right to destroy the surface through the utilization of surface mining methods. As a general rule, a grant of mineral rights will also carry with it the right to extract the minerals through the use of such mining methods as are reasonably necessary. Specific circumstances around a given lease, however, may raise a question as to whether the parties contemplated possible destruction of the surface through the use of surface mining methods.[123]

Surface Not Owned by Lessor

A far more difficult situation exists where the surface of the land to be mined is not owned by the mineral lessor or mineral owner. Although the same general rules discussed above apply with regard to express and implied rights to make use of the surface, they are applied more strictly where a third party who does not have an economic interest in the mining operation is involved.[124]

In general, reference must again be made to the instrument wherein the original severance of the surface and mineral estates was accomplished to see to what extent the surface use rights were granted, limited, or conditioned. Since the parties were often not contemplating a mining operation when the severance occurred, there is far less likelihood that such specificity will exist as that found in a typical mining lease. Accordingly, the implied rights of surface use become vital.

In general, the mineral estate is considered to be dominant over the surface estate.[125] Thus, the mineral owner or lessee has the right to use as much surface as is reasonably necessary to have access to and to carry out the extraction of minerals. This includes rights of egress and ingress. The mineral owner or lessee continues to have the right to realize the benefits from the development of the mineral estate so long as the activities do not unreasonably interfere with the rights of the surface owner.[126]

The principal difference between this situation and the mining lease situation relates to whether the implied rights include the right to destroy the surface in the conduct of the mining operation. The best that can be said in this regard is that existence of the right to destroy the surface will depend on the intention of the parties when the severance was accomplished. Since generally that intent cannot be determined by an examination of the operative legal document, resolution of the question is subject to the vagaries of the law of the jurisdiction involved and the circumstances surrounding a particular situation.[127]

The implied rights of a mineral owner or lessee to make use of the surface owned by another may also be limited by statute. Several states have enacted surface owner protection legislation that requires consent of the surface owner to mining activities.[128]

Use in Connection with Mining on Other Lands

In general, neither a mineral lessee or mineral owner may make use of the surface of a tract of land in aid of mining operations on other lands.[129] Rarely does an instrument that severs the surface estate from the mineral estate provide for use of the severed surface in aid of mining operations on other lands. It is not unusual, however, to find that a grant of such rights has been made in a mining lease.[130] The granting clause may, for example, extend the right to make use of the surface in connection with mining on the leased or "other lands." Alternatively, the rights may be granted separately in a specific provision. Although such provisions are enforceable, they are much more likely to stand up against judicial scrutiny if the landowner was aware that he was granting such broad rights in the lease and if he is to be paid separately for uses unrelated to mining from the lands in question.

SURFACE USE AGREEMENTS

A separate surface use agreement may be necessary to acquire rights to use fee lands outside of the immediate area where mining will take place. Also, where there has been a severance of surface and minerals in a tract to be mined, the potential of disputes with the surface owners will generally justify a separate surface use agreement notwithstanding the right of the mineral owner to make reasonable use of the surface. In some states the implied rights of surface use have been severely limited by surface owner consent requirements. Moreover, even where surface use rights have been referenced

in a mining lease, it may still be necessary to obtain an additional agreement to avoid controversy where the surface use rights are not spelled out with sufficient clarity or are not supported by reasonable consideration. This is especially true where mining is going to take place on a tract of land that is being used for other purposes.

Obviously, the length and content of a surface use agreement will be determined largely by the existing circumstances at the site. Nevertheless, there are certain general guidelines that can be followed to ensure the acquisition of adequate rights and to promote a harmonious relationship between the landowner and the mineral operator. In general, in areas where the land is being used for other purposes by the surface owner, an effort should be made to allow those uses to continue to the extent they are compatible with the mine related uses, and there should be full and fair compensation to the landowner on the basis of the degree of mine related use that is made. The mine operator must have the right to make use of the land as needed, but as always, the byword for the successful assertion of those rights is reasonableness.

Following are brief discussions of basic provisions in a surface use and access agreement.

1. Grant of rights: The grant of rights should be extremely broad, but as specific as possible with respect to uses that are contemplated. Use of the tract of land in question for mining purposes should be permitted in connection with mining on that land and on other lands developed by the operator in an area within a specified distance from the tract in question. If appropriate, the agreement should recite that it constitutes consent of the surface owner in lieu of a bond or undertaking required under the provisions of the Stock-Raising Homestead Act.

The rights granted generally should be those "reasonably required or convenient" to the mining operator for the conduct of its operations with a listing of what specific uses might be included. For example, as to a surface mine the rights granted should include the right to construct, install, maintain, and operate stockpiles, waste dumps, tailings impoundments, leach piles, ponds, pipelines, and related facilities. There should be a right to destroy the surface if the land might be included within the dimensions of an openpit or strip mine or might be covered by dumps.

There should also be a provision granting the operator the right to use the land for ingress and egress required or convenient to the grantee for construction, maintenance, and use of linear facilities such as roads, conveyor belts, power lines, and pipelines.

2. Term: The rights granted should be for a fixed period consistent with the proposed or possible life of the mine, and so long thereafter as the operations continue.

3. Multiple Use: So as to minimize adverse effects on the landowner's property and to permit multiple use to the degree possible, there should be fairly detailed provisions concerning the giving of advance notice of proposed operations, maintaining fences, gates, and cattleguards, preventing access by livestock into operational areas or to contaminated water, and limitations of recreational use by employees. Access by the landowner to areas of the property that are not being used in connection with the mining operation should be insured.

4. Reclamation: Where reclamation is possible, the mine operator should have the obligation to do so. The operator may reasonably be required to post a reclamation bond to insure that the reclamation is performed.

5. Indemnification: There should be an obligation on the part of the operator to indemnify and hold the landowner harmless in respect of any liability that is incurred by reason of the operator's activities. Conversely, if the landowner is to conduct active operations on the affected ground, it is appropriate to request that the indemnification obligation be reciprocal.

6. Compensation: Compensation can take a variety of forms. Generally, there is an initial payment in the nature of a bonus or first year's rental. Simply by reason of the burden placed on the land by the grantor's rights, it may be appropriate, especially if the land is not part of the mineralized area, to pay an annual fee in the nature of a rental. Where a pipeline, roadway, or other linear use is made by the operator, payment can be established on the basis of the overall length of that right-of-way subject to maximum width limitations.

Where the nature of the mine related use causes the land to be unavailable for use by the landowner, a higher rental value per acre affected is appropriate. If the land is not reclaimable, then full value should be paid at a preestablished price or at a price determined by appraisal. Repair or replacement value is appropriate for any damage or destruction to livestock, water rights, or physical features on the property.

The mine operator always would like to have the right to acquire title and exclusive control over portions of the land that would be the subject of substantial investment. This is usually accomplished through an option to purchase title to such areas. The landowner might require that the entire tract be purchased in such event because of the diminution of value that may be experienced as to the remainder of the land. If a price is not negotiated, it can be established by appraisal either at fair market value or at some preagreed increment in excess of fair market value.

In some instances, either because of the insistence of a landowner with negotiating leverage, or because it is in the best interest of the parties, a royalty will be paid to the surface owner. Such royalties can be in the form of a wheelage fee based on volume of material removed from or across the tract in question, or if mining is performed on the tract in question, in the form of a net return from sales royalty. As a rule, royalties are to be avoided because they are more in the nature of tribute rather than fair compensation for reasonable surface use.

7. Taxes: Most agreements provide that the landowner will continue to pay ad valorem property taxes except those attributable to the operations of the grantee.

8. Default: There should always be a default provision allowing notice of default and an opportunity to cure.

9. Assignment: Care should be taken to insure that the rights granted under the agreement remain in effect even though the landowner makes an assignment of his ownership rights. Since the whole purpose of the agreement was to avoid disputes by providing compensation for surface damage, it is appropriate to require that the right to receive payment run with ownership of the land. Because of the importance to the owner of the reputation and financial substance of the mine operator, the landowner might require that he have the right to consent to any assignment by the operator subject to limitations of reasonableness.

10. Surrender: The operator will generally want the opportunity to surrender the property at any time either in whole or in part. The landowner may, however, require that a release be an all or nothing situation. Upon termination, the operator should have the obligation to remove all facilities and reclaim the ground where reclamation is feasible.

ACQUISITION OF RIGHTS BY CONDEMNATION

As a method of last resort when negotiations fail, a mine operator can consider acquiring rights to privately held lands by condemnation. Lands held by a governmental entity generally cannot be acquired through condemnation.

The theory of eminent domain, or condemnation, is that the public interest is served by allowing a procedure whereby interests in land, or title itself to the land, can be taken by the state where necessary so long as due process procedural rights are followed and just compensation is paid. It has been recognized that the rights of the sovereign states to exercise condemnation rights can be provided as well to private institutions whose activities are determined to equate to the public interest.[131] Mining, which occupied such a central part in the initial economic development of various states, has been determined in some instances, particularly in the Rocky Mountain states, to be an activity that is in the public interest.[132]

Condemnation, however, is not a panacea. Certain realities must be weighed before a determination is made to go that route. Even in the best of circumstances, one can face a lengthy and expensive legal proceeding, the adverse impacts of which might outweigh the benefit that is derived. Because of the vagaries of the law, final success might not be assured.

In deciding whether condemnation rights are available, state statutes must be examined to determine whether the state involved has granted eminent domain powers to mining companies to acquire property for mine related uses. In this regard, the statute might contemplate that any exercise of eminent domain procedures for such things as access rights to a parcel of land be made by the owner of that parcel of land, rather than by a lessee.[133] Since in many instances the mine operator will be a lessee, that uncertainty alone might limit the usefulness of the condemnation procedure.

In states where a private right of eminent domain is recognized but specific public uses are not enumerated, it will be necessary to determine whether the use to be made of the acquired land will be a public use.[134] This may be determined by such considerations as whether the public at large will be permitted to use the land, such as an access road, or whether, as several courts in the Rocky Mountain states have determined, that public use is shown if the public will benefit from the use indirectly by reason of the economic contribution of the mining operation involved.[135]

If the use is authorized, then a showing must be made that there is a *necessity* for acquisition of the property rights being sought.[136] These rights will usually be for easements for such things as access roads, tailings disposal sites, tunnel rights, or dumpsites for mining wastes. Where alternatives are available, the showing of necessity may be more difficult, but deference is given to the judgment of the mining operator as to the importance of a particular selected plan. Further, the showing of absolute necessity is not generally required, as it is sufficient to show that there is a reasonable necessity for the proposed use.[137]

Procedurally, it can be critical to the mine operator that immediate occupancy is acquired even though the question of damages is left for determination through a more lengthy proceeding. Some states have a procedure for such immediate occupancy.[138] Obviously, before that occupancy can be achieved, a showing must be made that the criteria discussed above are clearly satisfied. Equally obvious is that there could be a lengthy trial-type proceeding just to establish the right of immediate occupancy.

Once the right to condemn is determined, then a determination of damages must be made. Most states allow an award based on the value of the land or the interest acquired for its "highest and best use."[139] From the standpoint of the mining company, that value might not have been reflected in the earlier negotiations which were more keyed to the mining company perspective. The determination of the *fair market value* can be a lengthy process with no assurance that the company will end up in a better position than it would have had it simply raised the original price.

PERMITS

Introduction

Numerous permits relative to environmental protection are required in connection with a surface mining operation. Procedures for acquiring such permits are described in Chapter 1 of Environmental Considerations, entitled "Environmental Laws and Regulations Applicable to Mining Operations," in the *Underground Mining Methods Handbook,* published in 1982 by the Society of Mining Engineers of AIME.

Permitting requirements of most significance to a surface mining operation are those that relate to surface protection and reclamation. This part describes mined land protection regulations applicable to mining operations conducted on mining claims on BLM and Forest Service administered lands. These regulations, although unique to the particular category of lands involved, are typical in their structure to surface protection legislation adopted by many states.

BLM Surface Management Regulations

Regulations were adopted by the BLM in 1980 that provide for the protection of surface resources on certain unpatented mining claims from the adverse effects of mining operations. The regulations are contained in Subpart 3809 of Title 43, Code of Federal Regulations. They are not applicable to lands in the National Park System, National Forest System, National Wildlife Refuge System, acquired lands, Stock-Raising Homestead lands, land where the surface has been patented by the United States, or lands under wilderness review and administered by the Bureau of Land Management.[140] Thus, as a general rule, there will be a question whether the regulations are applicable in any situation where the lands are subject to intense management regulations. In such instances, there will be special requirements and special rules governing the conduct of mining operations.

The essential purpose of the regulations is to prevent "unnecessary or undue degradation of federal lands which may result from operations authorized by the mining laws."[141] Since the regulations effect a new policy of regulating the conduct of mining operations on unpatented claims in contrast to over 100 years of a basic hands-off policy, they lack the stringency of other federal and state regulatory requirements and they reflect recognition of some of the practical aspects of the conduct of exploration and mining operations on unpatented claims.

In general, the regulations place a self-policing responsibility on the operator when disturbance is minor but they insure conscientiousness as to surface protection by requiring notification to the BLM of such activities. The agency is directly involved in approving plans for major disturbances.

The regulations impose a litany of requirements that should be part of any responsible mining operation, and they provide an example of procedures that must be followed under almost all surface protection and reclamation laws.

It is recognized that the burdens of administering a permit procedure for activities of minor impact would be burdensome and generally unnecessary. Accordingly, a category of activities referred to as *casual use* has been defined. Persons engaging in such activities are responsible only for insuring that the operations will not create unnecessary or undue degradation of the federal lands and for compliance with all pertinent federal and state laws.[142] This means that no notice need be given to the BLM of casual use operations, but the operator must nevertheless conduct the operation in such a manner that surface disturbance shall not be greater than that which would normally result from a prudent operation of similar character, and the operator must initiate and complete reasonable mitigation measures including reclamation of disturbed areas.[143] Also, there must be compliance with federal laws pertaining to air quality, water quality, solid waste disposal, protection of fish, wildlife, and plant habitat, protection of cultural and paleontological resources, and protection of survey monuments.[144] With regard to a surface mining operation, the casual use category would be of most significance during conduct of exploration or development drilling operations.

If the proposed operation will cause a cumulative surface disturbance of 2.02 ha (5 acres) or less during any calendar year, the operator must notify the authorized officer at the District Office of the Bureau of Land Management at least 15 days before commencing operations.[145] Operations include all activities carried out in connection with prospecting, discovery and assessment work, development, extraction, and prospecting of mineral deposits locatable under the mining laws and all uses reasonably incident thereto whether on a mining claim or not.[146] This includes access roads, transmission lines, pipelines, and other means of access for support facilities across federal lands in the vicinity of the unpatented claims.

The notice must contain information about the identity of the operator, a statement describing the activities that will be conducted, and a statement that reclamation will be completed and reasonable measures taken to prevent unnecessary or undue degradation of federal land as required by the regulations.[147] There are also certain standards governing the conduct of activities carried out under a notice relating to the width of access routes, disposal of tailings, dumps, and other waste produced by the operation and the conduct of reclamation.[148] Disturbed land must be reclaimed to the extent necessary to prevent erosion, landslides, and contaminated runoff.[149]

No approval is required for a notice filed pursuant to this requirement; however, the operation is subject to inspection by BLM representatives to insure compliance with regulatory requirements.[150] Further, the BLM must be notified when reclamation has been completed so that an inspection can be made to determine its adequacy.[151]

Of most interest to the operator of a surface mine is the requirement that operations be preceded by submission and approval of a plan of operations where the disturbance level will exceed 2.02 ha (5 acres) per annum.[152] Such a plan will also be required in special areas such as areas designated as a part of the National Wilderness Preservation System and administered by the Bureau of Land Management, areas withdrawn from operation of the mining laws in which valid existing rights are being exercised, and designated areas of critical environmental concern.[153]

The contents of the plan of operation, which dwell on the nature of the operation and the surface protection and reclamation procedures that will be followed, are established in the regulations.[154] Required time periods for review are established, which allow only 30 days for the initial review subject to discretionary extensions of up to 60 days upon notice to the applicant, but the time involved can be extended if the application is returned for additional information or if an environmental impact statement is required under the National Environmental Policy Act.[155] The agency must also make required assessments of cultural resources and of factors important under the National Historic Preservation Act or the Endangered Species Act.[156]

In any major operation to be conducted on federal lands there is a very high likelihood that an environmental impact statement will be required, greatly extending the time necesary for approval of a plan. In any event, an environmental assessment must be performed by the agency.[157] Approval of a plan can be conditioned upon incorporation by the operator of specific requirements imposed by the agency to prevent degradation and achieve reclamation.[158]

Of vital importance to the operator is the requirement that a bond be submitted in an amount specified by the authorized officer to cover the estimated cost of reasonable stabilization and reclamation of the area disturbed.[159]

As would be expected, any significant modification of an approved plan triggers a requirement for the submission and approval of a revision to the plan.[160]

Both a notice of operations and a plan of operations must specify the location of access routes for operations, and such access roads are subject to requirements for prevention of unnecessary or undue degradation.[161] Although access cannot be denied under the regulations, it can be managed through special requirements tailored to the circumstances.

There will be instances when the operation is subject to regulation both under the BLM surface protection laws and state surface mine reclamation laws. The regulations attempt to deal with this by saying that the BLM regulations do not preempt the state laws, and they call for consultation between BLM and state representatives to establish a joint federal/state program for administration and enforcement.[162] Whether such a program is in effect in a particular state or whether there will be possible confusion and duplication are matters that must be addressed on a case-by-case basis.

Since the regulations are not based on any particular statutory mandate, the enforcement mechanisms are limited. When an operator does not comply with the regulations, he can be served with a notice of noncompliance or enjoined from the continuation of the operation by a court order. In addition, if unlawful acts are conducted, the operator may be liable for damages.[163]

Similar regulations have been adopted by the BLM in respect of mining on lands which have been identified for possible inclusion in the Public Wilderness System. Review and approval of a plan of operations is required for any proposed operations which will cause a significant disturbance to the environment.[164]

Forest Service Surface Protection Regulations: Any surface mining operations involving unpatented mining claims conducted on National Forest System lands are subject to regulations set forth in Part 252 of Title 36, Code of Federal Regulations.[165] In general, these regulations follow the same format as the BLM regulations discussed earlier.

Basically, any person proposing to conduct operations which might cause disturbance of surface resources must give a notice of intention describing the proposed operation to the District Ranger having jurisdiction over the area in which the operation will be conducted.[166] Operations include all functions, work, and activities in connection with development, mining, or processing of mineral resources and all uses reasonably incident thereto, including roads and other means of access on land subject to the regulations regardless of whether the operations take place on or off mining claims.[167] If the District Ranger determines that the operations will likely cause significant disturbance to surface resources, the operator must submit a proposed plan of operations.[168] The proposed plan of operations need not be preceded by a notice of intent if it is obvious that significant disturbance will occur. There are limitations to the requirement for submission of a notice of intent or plan of operations, but they will not be relevant in connection with a proposed surface mining operation.

The plan of operations must describe the type of operations proposed and how they would be conducted, the type and standard of existing and proposed roads or access routes, the means of transportation used or to be used, the period during which the proposed activity will take place, and measures to be taken to meet requirements for environmental protection as set forth in the regulations.[169] The environmental protection regulations generally require that adverse environmental impacts on National Forest surface resources be minimized and that federal laws and regulations pertaining to air quality, water quality, and solid waste disposal be complied with.[170] Special requirements are set forth relative to the protection of scenic values, fisheries and wildlife habitat, and roads.[171]

Central to the function of the regulations is a requirement that upon exhaustion of the mineral deposit or the earliest practical time during operations, or within one year of the conclusion of operations, unless a longer time is allowed by the authorized officer, reclamation of surface disturbed by the operations be performed.[172]

There is no set requirement as to time permitted for a plan of operations to be reviewed. An environmental analysis will be performed in connection with each proposed operating plan and if necessary an environmental impact statement will be prepared.[173]

Once approval is obtained, the operations must proceed in accordance with the plan. Any substantial change in conditions will require a formal modification of the plan.[174]

The right of access granted to the holder of mining claims is recognized, but the regulations impose requirements for prior approval and minimization of environmental disturbance resulting from construction and use of access routes.[175]

A bond or cash deposit will ordinarily be required in connection with a surface mining operation to insure that reclamation will be performed. The amount of the bond is based on the estimated cost of stabilizing, rehabilitating, and reclaiming the area of operation.[176]

Periodic inspections are conducted to insure that the operator is complying with the regulations and plan of operations. Upon failure to comply, a notice of noncompliance is given which if not complied with within 30 days will expose the operator to sanctions.[177]

The provisions of the regulations apply as well to wilderness areas within the National Forest System.[178] Operations on wilderness lands are, however, subject to more stringent requirements.

State Mined Land Reclamation Laws

Reference should be made to the laws and regulations of the state involved to see if a mined land protection and reclamation law exists that will be applicable to the contemplated surface mining operation. Many states have such laws.

State mined land protection and reclamation laws usually will follow the same general format. When mine related activity reaches a certain threshold level, usually based on the type of activities or degree of disturbance, a mine plan must be submitted to the regulatory agency for review and approval. The plan must show how environmental protection will be achieved during mining, and how reclamation will be achieved afterward. Standards of performance designed to achieve protection of surface resources and to protect water quality will often exist.

Central to the effective application of any such law is a bonding requirement to secure reclamation obligations. Bonding requirements, although necessary to ensure that the burden of reclamation does not fall on the state, can pose severe practical problems for the mine operator.

If mining is to take place on federal or state lands, concurrent jurisdiction of federal and state regulatory agencies may result in duplication of effort and occasionally inconsistent requirements. Usually such a problem can be worked out through a state federal cooperative agreement or informally through coordinating between state and federal agencies.

REFERENCES

1. 30 U.S.C. § 26 (1976).
2. Teller v. United States, 113 F. 273 (8th Cir., 1901); United States v. Rizzinelli, 182 F. 675 (D. Id. 1910).
3. 30 U.S.C. § 612 (1976).
4. For a thorough discussion of this subject, see Miller, J., 1983, "Surface Use Rights Under the General Mining Law," *Rocky Mountain Mineral Law Institute*, Vol. 28.
5. *See* United States v. Nogueria, 403 F.2d 816 (9th Cir. 1968); *cf.,* Bagg v. New Jersey Loan Co., 88 Ariz. 182, 354 P.2d 40, 45 (1960).
6. *Ibid.* at 824. *See also* United States v. Zweifel, 508 F.2d 1150, 1156 (10th Cir. 1975); United States v. Rizzinelli, 182 F. 675 (D. Id. 1910).
7. United States v. Richardson, 599 F.2d 290 (10th Cir. 1979), *cert. denied* 444 U.S. 1014 (1980).
8. Parr, C.J. and Kimball, D.A., 1971, "Acquisition of Non-Mineral Land for Mine Related Purposes," *Rocky Mountain Mineral Law Institute*, Vol. 23, pp. 595, 637.
9. 30 U.S.C. § 23 (1976).
10. *See,* for example, Ranchers Exploration and Development Co. v. Anaconda Co., 248 F. Supp. 708 (D. Utah 1965).
11. 43 C.F.R. § 4.450 (1981).
12. 43 C.F.R. § 4.451 (1981); *See* Parr, C.J., 1968, "Government Initiated Contests Against Mining Claims," *Utah Law Review*, p. 102, reprinted in *Rocky Mountain Mineral Law Review*, Vol. 6, p. 1 (1968); Tilden, MW., 1979, "Defending a Mining Claim Contest," *Rocky Mountain Mineral Law Institute*, Vol. 25, p. 12-1.
13. Castle v. Womble, 19 Interior Dec. 455, 457 (1984).
14. Coleman v. United States, 390 U.S. 599 (1968).
15. Rocky Mountain Mineral Law Foundation, 1976, *American Law of Mining*, Vol. 1, § 4.21, Matthew Bender, New York City; Henault Mining Co. v. Tysk, 419 F.2d 766 (9th Cir. 1969).
16. Converse v. Udall, 399 F.2d 616 (9th Cir. 1968), *cert. denied*, 393 U.S. 1025.
17. For a discussion of assessment work requirements, see *SME Mining Engineering Handbook*, Vol. 1, AIME, New York, p. 2-14 (1973).
18. Hickel v. Oil Shale Corp., 400 U.S. 48 (1970); 43 C.F.R. § 3851.3a (1984); Nelson, W.H., 1971, "Assessment Work Before and After the Oil Shale Case," *Rocky Mountain Mineral Law Institute*, Vol. 17, p. 435.
19. *See* Knutson, R.D., and Morris, J.H., Jr., 1980, "Locating Maintaining, and Patenting Groups or Large Blocks of Mining Claims," *Rocky Mountain Mineral Law Institute*, Vol. 26, pp. 517, 578.
20. 43 C.F.R. Subpart 3833 (1984).
21. 30 U.S.C. § 42 (1976).
22. *See* Harris, R.W., 1976, "The Law of Millsites: History and Application," *Natural Resources Lawyer*, Vol. 9, pp. 103, 126-28.
23. 30 U.S.C. § 42(a) (1976).
24. Rocky Mountain Mineral Law Foundation, 1976, *American Law of Mining*, Vol. 1, § 5.31, Matthew Bender, New York; (hereinafter cited as *American Law of Mining*).
25. Harris, *supra,* note 2, at 115.
26. *American Law of Mining*, Vol. 1, § 5.32.
27. *See* Cleary v. Skiffitch, 28 Colo. 362, 65 P. 59, 62; *American Law of Mining*, Vol. 1, § 5.32.
28. *See* Reed v. Bowron, 32 Land Dec. 383, 386 (1904).
29. United States v. Silver Chief Mining Co., 40 Int. Bd. of Land Appeals 244, 248 (1979).
30. Harris, *supra,* note 22, at 118-19.
31. *See* United States v. Moorehead, 59 Interior Dec. 192, 198 (1946).
32. Harris, *supra,* note 22, at 81.
33. United States v. Swanson, 14 Int. Bd. of Land Appeals 158, 173-74 (1974); Alaska Copper Co., 32 Land Dec. 128, 131 (1903).
34. Hard Cash and Other Mill Site Claims, 34 Land Dec. 325, 327-28 (1905); *see* Parr, C.J. and Kimball, D.A., 1977, "Acquisition of Non-Mineral Land for Mine Related Purposes," *Rocky Mountain Mineral Law Institute*, Vol 23, p. 595.
35. 30 U.S.C. § 42 (1976).
36. *See* Alaska Copper Co., 32 Land Dec. 128, 129-31 (1903); Greer, G.L., 1967, "Millsites: Nonmineral Mining Claims," *Rocky Mountain Mineral Law Instiute*, Vol. 13, p. 143.
37. *Ibid.* at 157.
38. *Ibid.* at 158-60.
39. United States v. Dora M. Werry, 14 IBLA 242, 251-52 (1974); United States v. S.M.P. Mining Co., 67 Interior Dec. 141 (1960).
40. United States v. Silver Chief Mining Co., 40 Int. Bd. of Land Appeals, 244, 247 (1979).
41. Reed v. Bowron, 32 Land Dec. 383 (1904).
42. 43 C.F.R. § 3833.2-1(d) (1984).
43. Union Phosphate Co., 43 Land Dec. 548, 551 (1915).
44. 43 C.F.R. § 23.8 (1984).
45. *Ibid.*
46. *Ibid.,* § 23.8(b).
47. *Ibid.,* § 23.8(c).
48. 43 C.F.R. § 3572.1(c)(6) (1984).
49. *Ibid.,* § 3570.0-2(c).
50. 43 C.F.R. § 23.9 (1984).
51. Bourdieu v. Seaboard Oil Corp., 38 Cal. App. 2d 11, 100 P.2d 528, 544 (1940); Mountain Fuel Supply Co. v. Smith, 471 F.2d 594, 596 (10th Cir. 1973).
52. Procedures for granting of such rights to sodium or phosphate lessees are set forth in 43 C.F.R. Part 3540 (1984).
53. 30 U.S.C. § 214 (1976).
54. 30 U.S.C. § 263 (1976). Procedures for the granting of such rights to sodium or phosphate lessees are set forth in 43 C.F.R. Part 3540 (1984).
55. Act of March 3, 1909, ch. 270, 35 Stat. 844.
56. Act of June 22, 1910, ch. 318, 36 Stat. 584.
57. Act of July 17, 1914, ch. 142, 38 Stat. 509.
58. Act of December 29, 1916, ch. 9, 39 Stat. 862.
59. Act of June 28, 1934, ch. 865, 48 Stat. 1269.
60. *See* 43 C.F.R. § 3814.1(a) (1984).
61. *Ibid.,* § 3814.1(c).
62. Act of July 17, 1914, ch. 142, § 1, 38 Stat. 509.
63. 43 C.F.R. § 3813.1 (1984).
64. Lacey, J.C., 1976, "Conflicting Surface Interests: Shotgun Diplomacy Revisited," *Rocky Mountain Mineral Law Institute*, Vol. 22, pp. 731, 753.
65. *Ibid.* at 758. *But see* Note, 1973, "Surface Damages From Strip Mining," *Denver Law Journal*, Vol. 50, pp. 369, 371-73.

66. Twitty, H.A., 1961, "Law of Subjacent Support and the Right to Totally Destroy Surface in Mining Operations," *Rocky Mountain Mineral Law Institute,* Vol. 6, p. 497, 519-21.
67. Occidental Geothermal, Inc. v. Simmons, 543 F. Supp. 870 (N.D. Cal. 1982).
68. Bouirdieu v. Seaboard Oil Corp., 38 Cal. App. 2d 11, 100 P.2d 528, 534 (1940); Mountain Fuel Supply Co. v. Smith, 471 F.2d 594, 596 (10th Cir. 1973).
69. *See* 43 C.F.R. § 3813.1 (1984).
70. *See* 43 C.F.R. § 3814.1(c) (1984).
71. Kinney-Coastal Oil Co. v. Kiefer, 277 U.S. 488 (1927).
72. Lacey, *supra,* note 64, at 764-65.
73. 30 U.S.C. § 54 (1976); *see* 43 C.F.R. § 3814.1(b) (1984).
74. 43 U.S.C.A. § 1716(a) (Supp. 1981); 16 U.S.C. §§ 485-86 (1976); 7 U.S.C. § 428a(a) (1976); 16 U.S.C. § 555a (1976); 7 U.S.C. § 1010-12 (1976).
75. 43 U.S.C.A. § 1716(b) (Supp. 1981).
76. *Ibid.*
77. *Ibid.,* § 1716(a).
78. Forest Service Manual § 5432.8.
79. 43 U.S.C.A. § 1732(b) (Supp. 1981).
80. These terms are all defined in 43 C.F.R. § 2920.0-5 (1984).
81. 43 C.F.R. § 2920.2-2 (1984).
82. *Ibid.,* § 2920.2-3.
83. *Ibid.,* § 2920.2-5.
84. *Ibid.,* § 2920.4(a).
85. *Ibid.,* § 2920.5-1.
86. *Ibid.,* § 2920.5-4.
87. *Ibid.,* § 2920.5-4(b).
88. *Ibid.*
89. *Ibid.,* § 2920.7.
90. *Ibid.,* § 2920.8.
91. *Ibid.,* § 2920.6.
92. 43 U.S.C.A. §§ 1761-71 (Supp. 1981).
93. *Ibid.,* § 1761(a).
94. 43 C.F.R. § 2800.0-5(h) (1984).
95. *Ibid.,* § 2801.1-1(h).
96. *Ibid.,* § 2801.2.
97. *Ibid.,* § 2802.1.
98. *Ibid.,* § 2802.2-1.
99. *Ibid.,* § 2802.4(d)(1).
100. *Ibid.,* § 2803.1-1.
101. *Ibid.,* § 2803.1-2.
102. *Ibid.,* § 2803.1-3.
103. Specific authority for the issuance of special use permits not over 30 years and for not over 80 years is provided under the Act of March 4, 1915; 16 U.S.C. § 497 (1976).
104. 36 C.F.R. § 251.51 (1984).
105. *Ibid.,* § 251.54(a).
106. *Ibid.,* § 251.54(b).
107. *Ibid.,* § 251.54(e)(v)(4).
108. *Ibid.,* § 251.54(h).
109. *Ibid.,* § 251.56.
110. *Ibid.,* § 251.56(b).
111. *Ibid.,* § 251.56(c).
112. *Ibid.,* § 251.56(e).
113. *Ibid.,* § 251.57.
114. 43 U.S.C.A. § 1713(a) (Supp. 1981).
115. *Ibid.,* § 1713(c).
116. 43 C.F.R. § 2710.0-6(b) (1984).
117. *Ibid.,* § 2711.1-2.
118. *Ibid.,* § 2711.5-1.
119. *See* Getty Oil Co. v. Jones, 470 S.W.2d 618 (Tex. 1971); Flying Diamond Corp. v. Rust, 551 P.2d 509 (Utah 1976); Lopez, O.M., 1980, "Upstairs/Downstairs: Conflicts Between Surface and Mineral Owners," *Rocky Mountain Mineral Law Institute,* Vol. 26, pp. 995, 1007.
120. Annot., 53 A.L.R.3d 16, 31-32 (1973).
121. Reeves, G.E., and Alfers, S.D., 1977, "Dumps and Tailings," *Rocky Mountain Mineral Law Institute,* Vol. 23, pp. 419, 426.
122. Annot., 53 A.L.R.3d 16, 41 (1973).
123. *See* Annot., 53 A.L.R.3d 16 (1973).
124. For a general discussion of this subject see Lacey, J.C., 1976,

"Conflicting Surface Interests: Shotgun Diplomacy Revisited," *Rocky Mountain Mineral Law Institute,* Vol. 22, p. 731.
125. Lacey, *supra,* note 123, at 735-37; Lopez, *supra,* note 119, at 1003.
126. *See* Lopez, *supra,* note 119, at 1003.
127. *See* Annot., 70 A.L.R.3d 383 (1976); Ferguson, F.E., Jr., 1974, "Severed Surface and Mineral Estates—Right to Use, Damage or Destroy the Surface to Recover Minerals," *Rocky Mountain Mineral Law Institute,* Vol. 19, p. 411; Lopez, *supra,* note 119, at 1003-06.
128. Hultin, P.F., 1983, "Recent Developments in Statutory and Judicial Accommodations Between Surface and Mineral Owners," *Rocky Mountain Mineral Law Institute,* Vol. 28.
129. Hi Hat Elkhorn Coal Co. v. Kelly, 205 F. Supp. 764, 766 (D., Ky. 1962); Annot., 83 A.L.R.2d 665 (1962); Allen, R.G., 1980, "Utilization of Adjacent Properties, Cross-Mining, and Commingling," *Rocky Mountain Mineral Law Institute,* Vol. 26, pp. 419, 420-21.
130. Parr, C.J., and Kimball, D.A., 1977, "Acquisition of Non-Mineral Land for Mine Related Purposes," *Rocky Mountain Mineral Law Institute,* Vol. 23, pp. 595, 648-51.
131. Martz, C.O., Love, R., and Kaiser, C.L., 1983, "Access to Mineral Interests by Right, Permit, Condemnation or Purchase," *Rocky Mountain Mineral Law Institute,* Vol. 28, pp 1075, 1104-24.
132. Parr, C.J., and Kimball, D.A., 1977, "Acquisition of Non-Mineral Land for Mine Related Purposes," *Rocky Mountain Mineral Law Institute,* Vol. 23, pp. 595, 652.
133. Martz, Love and Kaiser, *supra,* note 131, at 1111-16.
134. Roth, U.L., 1982, "'To Take or Not To Take, That Is the Question': Acquisition of Mining and Mine-Related Rights Through Eminent Domain," *Rocky Mountain Mineral Law Institute,* Vol. 27A, pp. 739, 751-59.
135. *Ibid, see, e.g.,* Dayton Mining Co. v. Seawell, 11 Nev. 394 (1876); State ex rel. Butte—Los Angeles Mining Co. v. District Ct., 103 Mont. 30, 60 P.2d 380 (1936); Highland Boy Gold Mining Co. v. Strickley, 38 Utah 215, 78 P. 296 (1904), *aff'd* 20 U.S. 527 (1906).
136. *See* Roth, *supra,* note 134, at 759-62.
137. *See* Schara v. Anaconda Co., 610 P.2d 132, 137 (Mont. 1980), *cert. denied,* 449 US. 920 (1981); Overman Silver Mining Co. v. Corcoran, 15 Nev. 147, 156 (1880); Marsh Mining Co. v. Inland Empire Mining and Milling Co., 30 Idaho 1, 165 P. 1128 (1917).
138. *See* Roth, *supra,* note 134, at 766-70.
139. See generally Campbell, R., 1969, "Condemnation of Mineral Properties—Related Aspects of Just Compensation," *Rocky Mountain Mineral Law Institute,* Vol. 15, p. 305.
140. 43 C.F.R. § 3809.0-5(c) (1984).
141. *Ibid.,* § 3809.0-1.
142. *Ibid.,* § 3809.1-2.
143. *Ibid.,* § 3809.0-5(k).
144. *Ibid.,* § 3809.2-2.
145. *Ibid.,* § 3809.1-3(a).
146. *Ibid.,* § 3809.0-6(f).
147. *Ibid.,* § 3809.1-3(c)(2).
148. *Ibid.,* § 3809.1-3(d).
149. *Ibid.,* § 3809.1-3(d)(4).
150. *Ibid.,* § 3809.1-3(e).
151. *Ibid.,* § 3809.1-3(d)(5).
152. *Ibid.,* § 3809.1-4.
153. *Ibid.*
154. *Ibid.,* § 3809.1-5.
155. *Ibid.,* § 3809.1-6(a).
156. *Ibid.,* § 3809.1-6(a)(5).
157. *Ibid.,* § 3809.2-1(a).
158. *Ibid.,* § 3809.1-5(c)(5).
159. *Ibid.,* § 3809.1-9.
160. *Ibid.,* § 3809.1-7.
161. *Ibid.,* § 3809.3-3.
162. *Ibid.,* § 3809.3-1.
163. *Ibid.,* § 3809.3-2.
164. 43 C.F.R. Part 3802 (1981).

165. For a discussion of these regulations, see Dempsey, S., 1975, "Forest Service Regulations Concerning the Effect of Mining Operations on Surface Resources," *Natural Resources Lawyer,* Vol. VIII, p. 481.

166. 36 C.F.R. § 252.4 (1981).

167. *Ibid.,* § 252.3(a).

168. *Ibid.,* § 252.4(a).

169. *Ibid.,* § 252.4(c).

170. *Ibid.,* § 252.8.

171. *Ibid.*

172. *Ibid.,* § 252.8.

173. *Ibid.,* § 252.4(f).

174. *Ibid.,* § 252.4(e).

175. *Ibid.,* §§ 252.12, 252.8(f).

176. *Ibid.,* § 252.13.

177. *Ibid.,* § 252.7.

178. *Ibid.,* § 252.15.

Chapter 3

Ore Reserve Estimation

Richard A. Bideaux, Editor

3.1 Introduction

RICHARD A. BIDEAUX

HISTORICAL OVERVIEW

The art of ore reserve estimation has been in its period of greatest advance over the last 20 years. The purpose of an ore reserve estimate is to first assist in determining if a property is worth mining, and, if so, to guide its later development. Thus, ore deposit models are the underlying foundation for numerous consequent economic decisions, and the correctness of those decisions will be directly dependent on the accuracy of the ore reserve estimates. It is now possible to produce, for the smallest to largest of deposits, spatially detailed mineral inventories of tonnage and grade or quality, including statistically derived estimates of these values' probable accuracy and precision.

The ability to provide such desirable ore reserve estimates rests on the modern development of improved geological, geometrical, and mathematical techniques of mineral deposit model construction. Most of these improved methods depend in turn on the pervasive use of digital computers that have provided dramatically larger storage capacities and higher speeds at an ever-decreasing cost over this same time period.

Ore reserve estimation is now a suitable topic for a doctoral dissertation in mining engineering. Nevertheless, it remains an art, requiring practice and judgment in its application. Even with the power of present techniques, or those likely to be developed in the future, this must always remain so, due to the inherent geological complexity of mineral deposits. From very limited data, well-considered geological inferences must be drawn, subject to frequent review as new information becomes available.

Any ore reserve estimate must begin with the collection and treatment of the geological samples drawn from and defining the ore body. From the earliest stage at which a preliminary ore reserve estimate can be made, these available samples may represent as little as 10^{-6} to 10^{-8} of the bulk of the mineral deposit penetrated. Even just prior to a production decision only 10^{-5} to 10^{-6} of the deposit may be available for inspection and analysis. The samples available to control ongoing mining may bulk as much as 10^{-3} to 10^{-5} of the material to be mined. Considering the cost of collection and importance of this sample data, on which all further computations rest, accurate information about each sample must be obtained and made readily available for use in any of these cases. Information on their accurate spatial locations and as much of their geological character as might affect later mining decisions should be collected. It is now possible to economically provide virtually unlimited computer storage and processing for this basic data.

The first steps in deposit model construction involve geometrical and geological considerations. For a surface mine, representation of both the preexisting topography and working faces must be incorporated. The deposit's geological features must then be interpreted to the degree necessary both to guide efficient mining and to assist future geological exploration. With computer assistance, this topographical and geological data can presently be captured and stored to almost any degree of detail desirable.

The greatest recent advances in ore reserve estimation have been made in the area of grade or quality computation. First there has been a greater appreciation and usage of ordinary statistics. These methods depend on several assumptions about the nature of the geological samples collected. The ore body is considered to be made up of a total population of all such possible samples that could be taken. The available set of sample data is then representative of the whole deposit to the degree that it has been drawn on a random and unbiased basis from the total population. The individual sample values obtained are considered to be statistically independent or spatially uncorrelated with one another. These sample values from deposits of many metallic and industrial minerals, as well as the calorific content of coal, will often exhibit a symmetrical frequency distribution or grade histogram. Ordinary statistics provide efficient estimators and descriptors of the mean value and dispersion about the mean of such distributions.

Deposits of some other commodities, especially where the material of economic interest is present in quantities little more than that of trace elements, often exhibit a different underlying distribution of values. As examples, these include notably lode gold, silver, and diamond deposits, molybdenum porphyries, roll-front uranium deposits, and also contaminants in coal such as sulfur. Rather than being symmetrical, their grade frequency distributions are assymetrical, skewed towards a long high-grade tail, and often truncated at low grades. These distributions can usually be transformed to a more symmetrical form by taking logarithms of the individual sample values, leading to their designation as lognormal. Much of the worth of such deposits may be represented by values falling in the high-grade tail of these distributions, so lognormal distributions have received considerable attention in recent years. A body of statistics to deal effectively with such cases has been developed.

The next necessary step in mineral deposit model construction required some method of extension of sample grades throughout the volume of the model. Virtually every method ever used for manual computations in the past has been programmed for the computer and extensively analyzed as to its characteristics. At their simplest, these include relatively crude methods that regard the deposit essentially as a point value or distribution, such as the general outline method. Somewhat more complex but still straightforward methods build polygonal and triangular models, either in plan or section, where each drill-hole sample value is considered to entirely represent the whole of a geometrical fraction of the deposit, or is simply averaged with at the most a few other samples.

In earlier years, more complex models were built using linear regression or trend-surfaces, sometimes in three dimensions. These have fallen into disfavor, partially because of the complexity of their construction and display, and also because of an inability to objectively determine the number of determining coefficients to be used in the model. With the wide-spread availability of computers necessary for their handling, even more detailed model geometries have been developed, beginning with regular or variable-block three-dimensional models. These were followed by development of models more specialized in their geometries to the types of deposit under examination, such as regularly gridded seam models for coal and other stratabound commodities. The most general type of digital model possible, composed of polyhedra of arbitrary shape and size, has seen little use as yet due to the details of its construction and complexity in its manipulation. However, these models must find use eventually for accurate and efficient evaluation of the most geologically and structurally complex types of deposits. These models might better be able to explicitly represent such complex geological deposits as steeply dipping, anamorphosing metallic veins, the recumbent folds found in some coal deposits, and the geometry and zoning of roll-front uranium deposits.

Along with the computer's ability to handle more detailed geology and geometry in deposit models came more complicated methods of weighting together sample grade values. In general, these consider all samples available within some distance of the center of the block of ground being estimated, with the weighting schemes being an empirical function of distance and sample geometry, usually incorporating observed anisotropisms in the mineralization.

A complete mathematical theory, the methods of which go collectively under the name of geostatistics, has been derived, largely perfected, and accepted over these last two decades. These geostatistical methods are intended to meet many of the problems of mathematical assessment of ore reserves. Compared to the assumptions of ordinary statistics, geostatistics can assess the degree to which nearby sample values are in fact found to be spatially correlated. This information is used in constructing a model of the grades throughout the deposit that is of minimum deviation or a best possible estimate compared to the sample data. The block grades so derived have been found to deviate from as-mined values on average by the minimum amount possible.

Further, geostatistics allows the computation of error bounds on the precision of these grade estimates, taking into account the sample variability and geometry, as well as the size of the individual blocks of ground being estimated. The accuracy of the resulting model can be assessed through successive deletion and estimation of the missing value of the actual samples.

A principal tool of geostatistics, the variogram, allows definition of the spatial correlation or continuity of mineralization. Variograms thus quantify the concept of distance of influence of individual samples and permit study of the irreducible error in sampling of a deposit. They provide derivation of an optimum function with which to combine sample values to obtain estimates of the grade of intervening blocks of ground, and the associated error of estimate. Knowledge of the variogram also permits prediction of the location and additional number of samples necessary to reduce the error bounds on an ore reserve estimate to any predefined level.

By geostatistically deriving the expected frequency distributions of material expected to be encountered in a mining block, more accurate estimates of overall recoverable ore can result. Grade-tonnage curves obtained from this data have been found to be quite accurate, and explain why many operating mines had to derive empirical adjustment factors in the past.

COMPUTING EQUIPMENT

A consensus on the complement of presently available computing equipment desirable for ore reserve analysis seems to have been reached. Components that should be directly available to a user include a terminal, usually with graphics capability, a coordinate digitizer, and hard-copy plotter. Many choices are possible for each of these functional components. Digitizers are used to capture drill-hole locations, surface topography, and geological outlines. The *(x, y)* position of the digitizer, while being moved over property maps, is converted to the deposit model coordinate system either on a point-by-point or continuous basis. Accuracy of the digitized data can be quickly and cheaply checked by display on the terminal, or converted to hard copy by the plotter. Computed data, often shown by means of contoured maps, can also be displayed and plotted.

Some remarks on the computer and its data storage capabilities can also be made, presuming there is some choice in its selection—often quite large and capable corporate machines are already available, with which the data terminal is used to communicate. Care must be taken to insure that the computer has sufficient precision in its internal representation of decimal numbers to not create a significant loss of accuracy in extended calculations. The representation of the coordinate system chosen for the deposit can sometimes exceed this limit; a possible solution is to translate the deposit by subtraction of constants from all coordinates, to maintain numerical significance of the least significant digits. Alternatively, it may be necessary to make some computations in double-precision mode.

Using block models as a reference, an average model size for a deposit to be mined by large-scale surface methods is perhaps $100 \times 100 \times 50$ or 500,000 blocks. Depending on detail or the amount of data associated with each block, such a model may then occupy 2 to 10 megabytes of storage (or as little as 0.2 to 2 megabytes if data storage compression techniques are used). Such models are best handled in sequentially accessed parts, either a horizontal or vertical section brought into the computer's main memory at one time. A section through such a model may require 20 to 200 kilobytes, with perhaps double that central memory necessary for the computer program and its ancillary data. The largest models may be in the range of several million blocks; here data compression is a necessity to keep the model storage within available capacity.

Close attention must be paid to backups of the sample data and models at various stages of construction. Double backups are advised for critical and expensive-to-replace data. Backed up files should be printed, mapped, or otherwise operated on to verify that they are correctly made and usable.

IMPLICIT MODELS

Manual computation of ore reserves is still quite appropriate for smaller properties and for larger properties that are either geologically quite simple or too complex for computer modeling! Reserves for an ore body that is quite homogeneous and with reliable continuity can be fairly easily maintained by entirely manual methods, even if the property is of large size—an example might be a limestone quarrying operation for a cement plant. A computerized ore body model is not entirely necessary since mine plans relying on selective mining are not necessary to be implemented. Powerful desk top calculators are of great assistance, and digitizers with microcomputer processors can be used solely as digital planimeters.

At the other end of the spectrum, the most complex ore bodies can be so discontinuous and unpredictable in their mineralization that an explicit computer model can be very difficult to construct and of little reliability. Highly faulted phosphate rock or barite deposits and lensed kaolin deposits might be some examples here. The partial geological interpretation given by the mine geologist to drill-hole results shown on cross section are the only reasonable basis for mine planning.

It should be noted, however, that a computer-based statistical model of such types of deposits can be of considerable help in predicting production from deposits of discontinuous or difficult-to-localize mineralization. While an explicit model cannot be constructed, averages, distributions, and trends of mineralization can be analyzed and used as a basis for meeting production requirements. The desired quantity of mineral may not be exactly located and able to be produced in the very short-term, but in the medium- to long-term, expected production can often be predicted quite successfully. Other examples of deposits to which such models have been applied are placer tin, gold, and alluvial diamonds and also large and geologically complex uranium deposits.

DRILL-HOLE DATABASES

The computer is of great assistance in helping to organize and maintain large drill-hole data files. It can be called on when there are as few as several tens of holes, and becomes a virtual necessity as the number of drill holes reaches into the thousands, or in a few cases, tens of thousands, and the sample intervals into perhaps hundreds of thousands. (Grab or bulk samples can be treated as drill holes at a point, while channel samples can be stored as short drill holes.)

Construction of such a drill-hole data base, even prior to its use, can be a major geological and data processing task. Some man years of effort can go into collecting the data and entering it into the computer. Required is the ability to store, for each drill-hole, a unique identifier, perhaps information on the drilling contractor, collar location coordinates, and down-the-hole survey data. Collection of this last is often underemphasized, yet has proven to be of substantial value in most cases. Obviously, lack of knowledge of reasonable exact locations of samples taken from the drill-hole can be a contributing source of error in later ore reserve estimates. There are cases known where unsurveyed holes, thought to be vertical, have returned to the surface; and ore bodies thought to be drilled out, have been entirely missed by the intended main haulage shaft.

The results of geological and chemical examinations of the samples are then added to the files containing the geometrical parameters of each drill hole. Generally, at least until a preliminary understanding of the most important data

is gained, any decisions should be on the side of storing too much data. Too early a censoring or prejudgment of what may be important can, unfortunately, lead to relogging of the samples. It is relatively easy to add all captured information initially, but quite expensive to later add further data to the files. One of the most common errors concerning bedded deposits, such as coal, is for the geologist to decide an interval is not of economic thickness and thus omit logging it. While this judgment as to minability may be entirely correct, the overall correlation of that bed may be essential in controlling the construction of a model of the deposit, particularly for beds that pinch and swell over their extent. Missing data values should be explicitly coded, recognized, and appropriately treated in further processing.

All engineering decisions should be deferred to as late a stage in data collection and model construction as possible. The main utility in using computer-resident data is in the ability to quickly reevaluate with different engineering parameters, so the data should exhibit the full range of values seen in the real-world deposit.

After entry, the sample data base must be verified. The files should be scanned by computer programs designed to catch simple errors and full printed listings visually examined. The drill-hole locations should be plotted in plan view if made from the surface, and ideally a set of cross-section plots made on whichever drill-hole shows at least once. Assay or quality values are conveniently displayed alongside the drill-hole traces.

SAMPLING PROBLEMS

All parties, including geologists, mining engineers, and data analysts, who are assisting or later will be using an ore reserve analysis derived from drill-hole results, bear responsibility to try to detect anomalous or erroneous sampling problems. While various methods to help detect such problems can be programmed for the computer, the variety and subtle nature of some such errors can at present only be recognized by human analysis. Operator biases and equipment miscalibration can often be demonstrated by statistical treatment of replicated analyses, by displaying their differing relative accuracy and precision. More subtle problems may be due to the methods of chemical analysis employed, or sample preparation. An interesting total metal grade may be of no value if it is later found to be locked into a mineral not amenable to recovery in the plant. The analytical methods should be similar to those used in a recovery plant, with determinations based on total sample decomposition rarely appropriate.

It is virtually impossible to take subsamples that are representative of the whole, whether they be the original drill-hole samples or splits thereof. This is due to differential hardness and specific gravity effects; any mechanical process has varying recoveries of the different mineral components depending upon their physical properties. For example, down-the-hole salting effects can sometimes be observed, when a rich but thin metalliferous zone has been penetrated. Such high-grade fines can contaminate deeper samples, or the bottom of the hole.

One of the most difficult effects to detect is bias in the location of the drill holes, violating statistical assumptions of randomness or equiprobable sample locations. These may be due to the drilling crew's unwillingness to collar holes at the intended, but perhaps difficult to occupy, stations when shallower slopes are somehow related to mineralization; or a geologist's unwitting "nose for ore" or even subconscious desire to "look good"—thereby selectively overdrilling high grade zones.

A somewhat subjective judgment concerns that of continuity and stationarity of the economic components of potential ore-grade material. Geostatistical variograms can provide necessary but not sufficient data bearing on geological continuity of mineralization. There are all too many cases where deposits thought to be well defined by surface drilling have been found to actually consist of discontinuous and hence uneconomic lenses.

A much rarer but very misleading situation can arise in analysis of drilling results from folded bedded deposits. It is necessary to have drill-hole penetrations at more than twice the frequency of the periodicity of folding, otherwise an effect known as aliasing occurs. Underdrilling results in an interpretation of a much lower periodicity and amplitude, or much smoother folding. While the actual tonnage later encountered is greater than anticipated, so will be the mining problems due to unanticipated tighter folding. In one case of a valuable coal deposit that proved eventually to consist of an extraordinarily large number of parallel, flat-lying discrete beds the first company with an option on the property rejected it. This was based on an interpretation that recumbent folds must be present, when in fact that was not the case. An ability to discriminate individual beds on the basis of their trace element chemistry might have assisted a correct determination in this case.

Stationarity of mineralization refers to its similarity over various parts of the deposit. Detection of significant differences in the continuity, anisotropy, or grade distribution over different parts of a deposit should be sought and, if detected, submodels must be constructed.

Perhaps the highest level of ore reserve analysis is to recognize that the available data is insufficient to make an ore reserve estimate that will have a high enough degree of confidence to justify further investment in the deposit. In such cases, usually a statement can be made as to the additional amount of information that it appears necessary to acquire, whether by using ordinary statistics or geostatistics. The drilling results obtained to the present can be analyzed, and the best apparent locations for obtaining further data can be indicated. Their locations can be optimized to give a maximum reduction of confidence bounds on consequent ore reserve estimates. If the expenditure necessary to acquire this information is greater than the return on its value, the property should be abandoned. Deposits of some minerals can present almost insurmountable estimation problems. Their inherent variability is so great that they would almost have to be mined by sampling to bring the accuracy of ore reserve estimates within the bounds to prudently justify investment. Nevertheless, many such deposits are in fact profitable.

Density determinations present another set of problems. Accurate and precise densities are a necessity, since their use in tonnage factors is one of the largest components of error in overall tonnage estimates. If density determinations are made on drill core samples, multiple determinations are necessary. Removal and weighing of a bulk sample, with its density determined by measurement of the volume of the excavation, is an excellent method.

If densities are available, for example, to only three digits of numerical significance, that is also the maximum number of significant figures to which any reserve tonnages can be reported. Gross errors in density determinations, such as using values obtained from the other mines in a district, have led to seriously overestimating actually obtained recoveries. For deposits of materials of especially high or low density, such as iron ores or borates, it will be necessary to derive and use a function of assay grades to compute expected densities and tonnages for the individual parts of a model.

SURFACE TOPOGRAPHY

Surface topography is another type of information that it is necessary to obtain quantitatively for an ore reserve estimation of a property to be mined from the surface, almost regardless of the type of model or the methods used in its construction. This is formally a digital terrain model, of interest in many areas besides mining, such as geography, geomorphology, forestry, military, etc. Again, the adequate definition of surface topography can be somewhat of an art, as interpolation of the data is usually required. Some presently available methods can produce such surface models entirely automatically from stereographic aerial photographs, or by digital stereoplotter operator interpretation. The more usual method is by preparation of an intermediate contour map at a vertical resolution appropriate to the surface roughness; then these contours can be digitized in continuous mode and the resulting (x,y,z) triplets interpolated to assign elevations on the desired planar grid. In terms of reliability and ease of producing the desired data, manual digitization of elevations at the intersections or centers of an appropriate grid overlay, followed by manual computer entry, is quite feasible for properties of up to intermediate size. The selection of digitization resolution can require some judgment; steep surface topography requires relatively finer resolution, and a balance must be made between adequate representation of the surface features vs. the corresponding resolution imposed on the underlying volume, since too fine a surface resolution can result in an unwieldy ore deposit model.

The representation of surface topography by such digitization can be a later source of controversy. Accuracy of the quantized representation should be checked first by computer programs designed to find gross errors (that are always present), then by a plotted contour map for final verification by overlay on the source map. Due to the quantization of the topography, this check plot will never exactly overlay the (essentially analog) source map, with quantization effects most apparent as departures from smoothness along ridge lines and valleys. However, the errors introduced are more apparent than real; plotted cross sections of the source and quantized data, without vertical exaggeration, will usually show a negligible departure between the two sets of data. For most types of computer models, it should be realized that the elevations at grid points will be treated further as horizontal flat planes with step boundaries between adjacent plates. Again, comparatively little error in subsequent volume calculations is usually involved, with counterbalancing errors, so that errors introduced into later tonnage figures can be ignored for practical purposes.

GEOLOGY

A geometrical model of the subsurface geology should first be constructed and checked, before grades or qualities are derived. Only after it has been verified and accepted should modeling of grade or quality attributes begin. Based on those geological units that are mappable, the deposit geology should be interpreted and transferred onto plan and/or section maps. The units chosen are usually able to be visually distinguished by a geologist, but for some highly altered deposits, more detailed petrographic work may be needed. If a wide range of recognizable geological units has been logged from the core and carried in to the drill-hole data base, the economic mineralization should be statistically studied in each coded unit. This can aid the geologist and ore reserve analyst in determining which units have statistically similar mineralization, which can then be considered for combination in the model, and those units that show bi- or multi-modal distributions, indicating that there may be petrological subunits requiring further discrimination.

For most deposits, even those with complex geology, it is rare that there are more than half-a-dozen rock types requiring distinction in their computer models. The overall effort is to make the geological model sufficiently complex to represent reality, but not overly so, requiring extra effort throughout its construction.

Special attention must be paid to the nature and location of contacts between rock types, as these will often be ore-waste boundaries. If faults are present, their relationship to the mineralization must be determined. The geometry of folding in bedded deposits is equally important. The interpretation of deposit geology will ultimately be reflected in tonnages of ore; note that it necessarily involves subjective interpolation and extrapolation from the geology logged on the penetrating drill holes.

Geological boundaries that are interpolated within the limits of drilling will be subject to some errors of location, but these will also be largely counterbalancing, especially when considered in the large. Overextrapolation of units beyond the boundaries of drilling is potentially a source of much more serious problems. Compared to the interior of a model, material around the margin can assume significant volumes within very short extrapolation distances. The most serious ore reserve misestimates have arisen either from geological projection of mineralized material too far outside the bounds of known data, or across fault structures not recognized or thought to be mineralized on both sides. An apparent doubling or more of tonnage can then easily result. In some cases, the subsequent mining feasibility study has been equally in error, forcing the mine to close with near-total loss of the capital investment as the true situation came to be recognized.

Geological outlines can be captured either by digitization, or manual coding onto a grid, in the case of block models. For gridded seam models, the locations of outcrops can be estimated and incorporated, along with burnouts and wash-outs in coal deposits. These digitized outlines should be plotted back to scale and visually checked for accuracy before their inclusion in a model. For block or gridded seam models, it is possible then to interact the geological outlines with the blocks so as to carry fractions of more than one geological unit into each block. For manual coding, these percentages are estimated and loaded directly into the blocks.

While some models with blocks able to be composed of as many as four different rock types have been built, it is rarely actually desirable to hold such detail. It can perhaps be justified if there are unmineralized dikes or sills cross-cutting more important mineralized units, when the dikes are below the limit of resolution of their thickness by whole blocks in the model. While these dikes might be mappable on the surface, their location below the surface is rarely well established.

It is possible today to economically work with quite large models, so that consideration should be given to halving the grid size and staying with only one rock type per block. Experiments made on multiple rock type per block models, converted to only the majority rock type in the same size blocks, indicates only about 1% difference in overall tonnages resulting from the lower spatial resolution. The increased resolution afforded by split blocks may be justified when more

data is available and ore reserves to be based on selective mining are considered.

Once the model geology is loaded, it should be plotted back and visually examined. Both plans, and preferably vertical sections in orthogonal directions, should be made. If the geology has been loaded from sections in only one direction, it will often give a curious and unintended appearance when seen in plans and sections at right angles, perhaps requiring corrections before the geological model is judged acceptable.

For geology that has been explicitly coded into a model, rather than being derived automatically from drilling results, a further cross-check is desirable. The block in which every drill-hole sample is located should be computed, and the match of the geology coded for the sample should be compared with that in the block. Any mismatches must then be noted and reconciled.

Occasionally no geologist can be found willing to try interpretation of complex drilling results, yet an ore reserve estimate must be made. Here an empirical, statistically based model must be resorted to, with the deposit geology interpolated directly from the drill-hole geology. Each block is assigned the rock type of the nearest drill-hole sample. This is the discrete digital equivalent of polygon assignment in two-dimensional plan sections, or a polyhedral assignment in three dimensions. Models made in such fashion should ideally be used only for global reserve assessment, although considerable reliability can be placed on the overall results.

GRADE ASSIGNMENT

Once the geological model is prepared, extension of the grades throughout the model can follow. In the following sections, every method in current use is discussed in considerable detail, so only general remarks will be made here. The selection of a suitable method is controlled by the geological nature of the deposit, as well as being a balance struck between the time, money, expertise, and facilities available to make a model.

As with manual methods, polygonal or triangular models are the most easily constructed, if suitable programs are available. The set of polygons constructed by determining perpendicular bisectors of the lines joining adjacent drill holes is unique. A triangularization of the same data, where triangle edges exist if the polygon sides to which they are perpendicular also exist, is likewise unique, except in cases of exactly square drilling patterns.

These types of models are not able to hold either the geological or grade data with much resolution, however, so their use should be restricted to small deposits, or those that are densely drilled, or fairly uniform in their mineralization. Biases of overdrilling in high-grade material must be particularly watched for in their use, and they are especially susceptible to overestimation of tonnages by unrealistic extrapolation outside the boundaries of known data.

Any type of deposit can theoretically be modelled in a three-dimensional block model, and this type is also the most easily constructed and manipulated on a digital computer. It should always be the first choice of model to be considered, to be rejected only if there are other overriding considerations. Bedded deposits can often be efficiently handled by gridded seam models. By using a partial block strategy, geology can be incorporated into any of these model types if the effort is warranted.

The rock types in the samples must be matched to the same rock type or rock types found to have similar grade distributions in the model for grade estimation. This cannot be overemphasized—again, uncritical model construction not honoring this principle has in the past resulted in highly inaccurate models, with disastrous consequences to the property owner. In the case of faults cutting off mineralization, it may be necessary to doubly code a petrologically identical rock type on either side of the fault to allow distinction of mineralized and unmineralized equivalents.

It should always be borne in mind that all of the methods of grade assignment throughout a model involve mathematical interpolation and extrapolation that are inherently smoothing processes. As such, they will be found to be more or less accurate in their prediction of grades when the estimated material is actually mined. Polygonal and triangular grade interpolations do not realistically model the probably smoother distributions that will actually be found, but may still prove not too far in error in favorable cases of slowly varying, continuous mineralization. Various entirely empirical interpolation methods such as those using inverse distance will more accurately model rapidly varying grade distributions with a still relatively modest expenditure of computer time. The geostatistical methods undoubtedly can be expected to provide the best estimate of what will ultimately be found, but require considerable knowledge in their application, as well as much larger expenditures of computer time. Nevertheless, they have been growing in popularity and will deservedly continue. With ever faster and cheaper computers and improved computer programs, the consequences of the greater computing time requirements for geostatistical analyses will have decreased impact.

However a grade model is constructed, it must also be extensively verified before much credence is placed on results obtained from it. A thorough and necessary check involves comparison of frequency distributions of mineralization of the various rock types for overall correspondence with the sample distributions from which they were derived. Another sufficient check should also be made of the input sample values individually against the interpolated values. This will help insure that gross errors have not been incorporated into the model, unfortunately all too easy to do even with long-used and well-validated computer programs.

CLASSIFICATION OF ORE RESERVES

In the interest of consistency of presentation, the following section is reproduced verbatim from Readdy, Bolin, and Mathieson, 1982. Their article in SME's companion *Underground Mining Methods Handbook* is also recommended as an overview of ore reserve calculation; while directed to deposits suited to underground mining, most of their discussion applies equally well to deposits minable by surface methods, or a combination of each.

The definition and classification of ore reserves has varied over the period of development of the modern mining industry. *Ore* is generally understood to be any naturally occurring, in-place, mineral aggregate containing one or more valuable constituents that may be recovered at a profit under existing economic conditions. This definition ignores special situations, such as wartime production, or those cases when an otherwise unprofitable deposit may be exploited for political or social reasons.

Ore reserves are classified with respect to the confidence level of the estimate. Traditionally, ore reserves have been classified as proven (measured), probable (indicated), possible, and inferred. Historically, proven ore has been regarded as that which is "blocked out," i.e., measured, sampled, and assayed on four sides; probable ore as blocked on three sides; possible ore as blocked on two sides; and inferred ore as ore grade material that is known on only one side.

More recently the US Bureau of Mines (USBM) introduced the following ore reserve classifications:

Measured Ore

"Measured ore is ore for which tonnage is computed from dimensions revealed in outcrops, trenches, workings, and drill holes, and for which the grade is computed from the results of detailed sampling. The sites for inspection, sampling, and measurement are so closely spaced and the geologic character is so well defined that the size, shape, and mineral content are well established. The computed tonnage and grade are judged to be accurate within limits that are stated, and no such limit is judged to differ from the computed tonnage or grade by more than 20%."

Indicated Ore

"Indicated ore is ore for which tonnage and grade are computed partly from specific measurements, samples, or production data and partly from projection for a reasonable distance on geologic evidence. The sites available for inspection, sampling, and measurement are too widely or otherwise inappropriately spaced to outline the ore completely or to establish its grade throughout."

Inferred Ore

"Inferred ore is ore for which quantitative estimates are based largely on a broad knowledge of the geologic character of the deposit and for which there are few, if any, samples or measurements. These estimates are based on an assumed continuity or repetition for which there is geologic evidence; this evidence may include comparison with deposits of similar type. Mineral bodies that are completely concealed may be included if there is specific geologic evidence of their presence. Specific estimates of inferred ore usually include a statement of the special limits within which the inferred ore may occur."

If geostatistical methods are used, providing estimation variances for each unit of the model, a probabilistic classification of reserves on a unit basis is then possible. The combined estimation variances of the portions of the reserves larger than individual units still cannot be defined, however. Because of intercorrelations of the data used to derive the grade estimates, a strength of the geostatistical method, the variances of adjacent units are not independent, so cannot be combined as in ordinary statistics. Nevertheless, a good idea of the overall relative precision of the reserves can be gained. In any event, the continuity of the reserves in any category should be examined on maps covering the deposit, as an initial guard against the infeasibility of actual mining.

REFERENCE

Readdy, L.A., Bolin, D.S., and Mathieson, G.A., 1982, "Ore Reserve Calculation," *Underground Mining Methods Handbook*, W.A. Hustrulid, ed., AIME, New York, pp. 17-38.

3.2 Computerized Conventional Ore Reserve Methods

The first use of computers in ore reserve estimation was simply to automate the traditional manual methods of ore reserve estimation. Due to the power and speed of the computer, new estimation techniques became available for the ore reserve analysis (O'Brian, Weiss, 1968). Among these new techniques were automation of the polygonal method, various interpolation techniques based upon inverse distance weighting, and the geostatistical techniques developed in the last 20 years.

This section describes the polygonal method and the distance weighting interpolation methods that are in current use by the mining industry. Geostatistical methods of ore reserve estimation are discussed in the next section. The description of each method in this section includes the method of calculation, the type of model produced, and a discussion of the properties and characteristics of each method.

DESIRABLE PROPERTIES OF ORE RESERVE METHODS

An ore reserve method can be judged on several different criteria, but certainly the main criterion is simply how correct are the estimates. Correctness refers not only to the overall reserve figures, but also to the local block grade estimates that will form the basis of the mine plans. Let $Z_1, Z_2, .. Z_n$ be the actual grades of blocks within an ore deposit, and let $Z_1^*, Z_2^*, .. Z_n^*$ be the corresponding estimates of the grades made by a particular ore reserve method. Three particularly informative measures of prediction accuracy can be determined by comparing the true grades with the estimated grades. First, we can determine whether the estimator is unbiased by calculating the average error, $\overline{E}$ as given in:

$$\text{Average error} = \overline{E} = \frac{1}{n} \sum_{i=1}^{n} (Z_i - Z_i^*)$$

The average error should be very small, and is usually found to be when the comparison is made on all the blocks in the deposit, regardless of cutoff grade. For most mining projects it is very important to determine whether the estimates are unbiased for grade estimates above the cutoff grade. A good estimator should be unbiased for all cutoffs. This second measure is often referred to as the conditional unbiasedness property and can be measured by examining the errors made for block estimates above specified cutoffs.

The third measure of accuracy is the variance of the errors as follows:

$$\sigma_e^2 = \frac{\sum_{i=1}^{n} [(Z_i - Z_i^*) - \overline{E}]^2}{n - 1}$$

The lower the variance, the better the estimator.

The foregoing three measures of accuracy can only be obtained by direct comparison of predicted grades to actual mined-out grades. In the absence of actual mined-out block grades, a comparison can still be made by comparing estimates of individual drill-hole composites with the actual value of the drill-hole composites. Such comparisons are widely used to validate the ore reserve method chosen for a particular project (Knudsen and Kim, 1978; Readdy, et al.,

1982; Kane, et al., 1982). The methodology is simple: a composite is removed from the data set and an estimate of the removed composite is made from the surrounding data using the particular ore reserve method. The process is repeated many times, usually over all available composites. The measure of accuracy is then calculated for the resulting set of errors.

POLYGONAL METHOD

Two types of ore reserve models can be constructed using computerized versions of the polygonal method. The first model consists of a set of computer calculated polygons with appropriate calculations for tonnage and grade. Fig. 1 shows an example of a computer generated polygonal model of a uranium mine. The second model is a standard block model where the grade assigned to each block in the model is the grade of the nearest drill-hole composite. As can be seen in Fig. 2, such a model is a close approximation of a standard polygonal model. Both models are in current use, although the block model is easier to utilize in subsequent computerized mine planning.

Computer-Drawn Polygon—Traditional Method

A computer program for the polygonal method should do three basic tasks. First, it should define polygons around each drill-hole composite. Second, it should compute the area of the polygon from which the tonnage can be calculated. Third, the program should store the polygon data for later mine planning and ore reserve reporting by either computerized or manual methods.

The first task of defining a polygon around drill holes is not easy to program, but, fortunately, several algorithms and programs have been published. One that works well is the algorithm published by Green and Sibson (1978). Commercial software for the polygonal method is also available.

An Approximate Polygonal Method

Most of the computerized mine planning systems developed for open pits have been based upon the standard block model for the mineralization inventory. Using the polygonal model to assign grades to this block model results in each block being assigned the grade of the nearest assay composite.

Whereas a computer program for the traditional polygon method is difficult to write, the approximate polygonal method is quite easy. The main algorithm of the program consists of calculating the distance from the center of a block to each surrounding assay composite. The grade of the closest assay composite is then assigned to the block. This is illustrated in Fig. 3, where hole AM 148 is found to be closest to the center of the outlined block. This grade assignment is repeated for each block in the model.

Choice of the Maximum Polygon Size

The size of the polygons is directly related to the drill-hole spacing. In areas where the holes are closely spaced, the polygons will be small. Likewise, in areas where the drill holes are far apart, the polygons will be proportionately larger. Usually, however, a limit will be set on the maximum size of the polygons. This limit has been commonly referred to as the maximum range of influence of an assay.

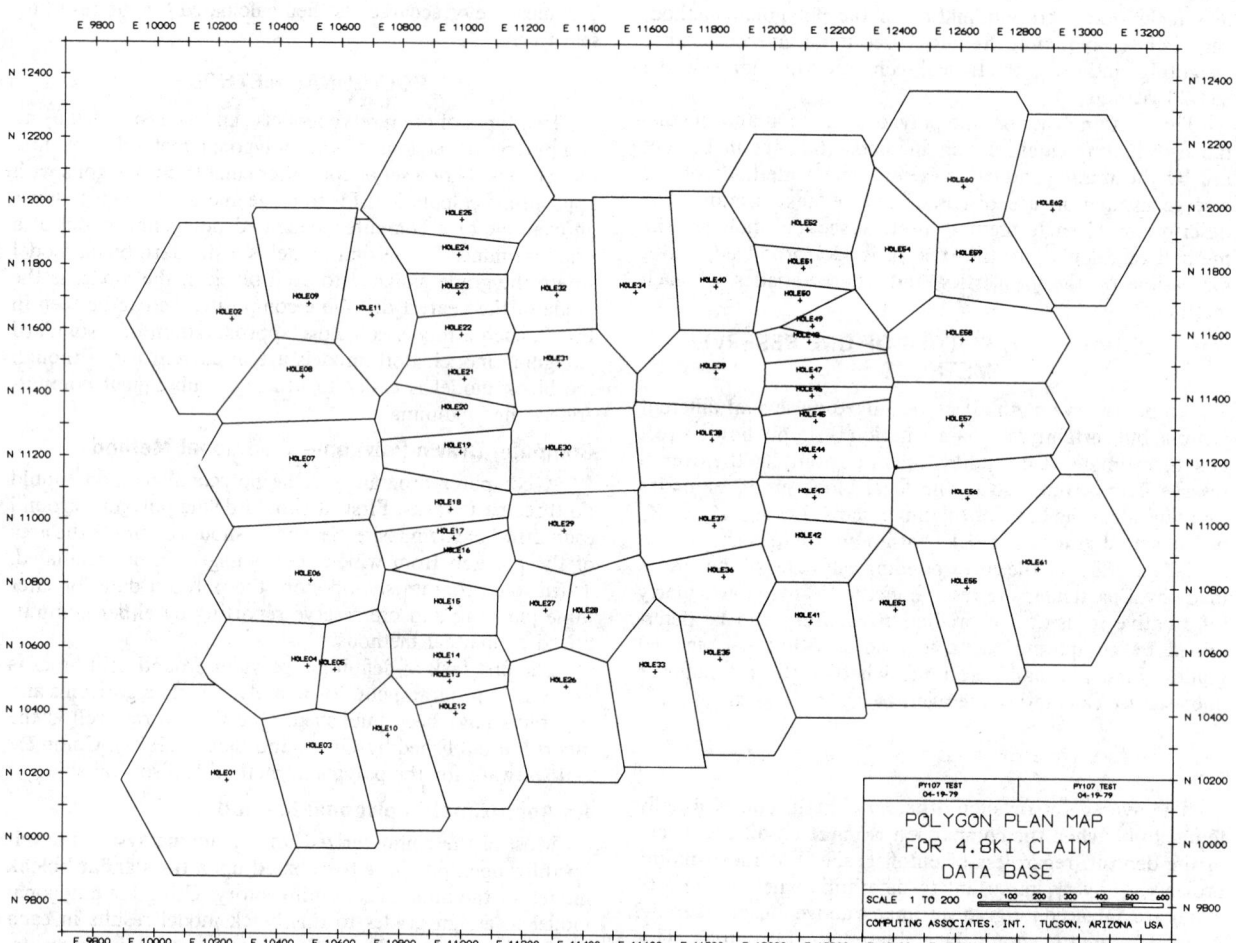

Fig. 1. The polygonal method (*courtesy of* Control Data Corp.).

0.27	0.27	0.06	0.06	0.06	0.06	0.32	0.32	0.51	0.76	0.76	0.76	1.14	1.14	1.14	1.14	1.74	1.74	0.84	0.84
0.27	0.27	0.06	0.06	0.06	0.32	0.32	0.32	0.32	0.76	0.76	0.76	1.14	1.14	1.14	1.14	1.14	0.84	0.84	0.84
0.27	0.13	0.06	0.06	0.06	0.32	0.32	0.32	0.32	0.76	0.76	0.76	1.14	1.14	1.14	1.14	1.08	0.84	0.84	0.84
0.13	0.13	0.13	0.13	0.06	0.32	0.32	0.32	0.32	0.76	0.76	0.76	1.14	1.14	1.14	1.08	1.08	1.08	0.84	0.84
0.13	0.13	0.13	0.13	0.13	0.04	0.04	0.04	0.93	0.93	0.93	0.93	0.93	1.08	1.08	1.08	1.08	1.08	1.08	0.84
0.13	0.13	0.13	0.13	0.13	0.04	0.04	0.04	0.93	0.93	0.93	0.93	0.93	1.77	1.08	1.08	1.08	1.08	1.08	1.08
0.13	0.13	0.13	0.13	0.06	0.04	0.04	0.04	0.93	0.93	0.93	0.93	0.93	1.77	1.77	1.77	1.77	1.08	1.08	0.67
0.45	0.13	0.06	0.06	0.06	0.04	0.04	0.04	0.04	0.93	0.93	0.93	0.85	1.77	1.77	1.77	1.77	1.77	0.67	0.67
0.64	0.55	0.55	0.06	0.06	0.06	1.00	1.00	1.00	1.00	0.85	0.85	0.85	0.85	1.77	1.77	1.77	0.67	0.67	0.67
0.64	0.55	0.55	0.06	0.06	1.00	1.00	1.00	1.00	1.00	0.85	0.85	0.85	0.85	0.85	1.77	1.77	0.67	0.67	0.67
0.64	0.55	0.55	0.55	0.06	1.00	1.00	1.00	1.00	1.00	0.85	0.85	0.85	0.85	0.85	1.36	1.36	0.67	0.67	0.67
0.50	0.50	0.55	0.55	0.64	0.64	1.00	1.00	1.00	0.67	0.67	0.67	0.85	0.85	0.85	1.36	1.36	1.36	0.67	0.03
0.50	0.50	0.50	0.48	0.64	0.64	0.64	0.64	0.67	0.67	0.67	0.67	0.44	0.44	1.36	1.36	1.36	1.36	1.36	0.03
0.50	0.50	0.48	0.48	0.64	0.64	0.64	0.69	0.69	0.67	0.67	0.44	0.44	0.44	1.36	1.36	1.36	1.36	1.36	0.03
0.50	0.50	0.48	0.48	0.48	0.64	0.64	0.69	0.69	0.69	0.67	0.44	0.44	0.44	1.36	1.36	1.36	1.36	0.99	0.99
0.43	0.43	0.43	0.48	0.48	0.64	0.69	0.69	0.69	0.69	0.69	0.44	0.44	0.44	0.44	1.36	0.99	0.99	0.99	0.99
0.43	0.43	0.43	0.75	0.75	0.75	0.53	0.53	0.69	0.69	0.69	0.75	0.75	0.75	0.75	0.99	0.99	0.99	0.99	0.99
0.43	0.43	0.43	0.75	0.75	0.75	0.53	0.53	0.53	1.28	1.28	0.75	0.75	0.75	0.75	0.99	0.99	0.99	0.99	0.99
0.43	0.43	0.75	0.75	0.75	0.75	0.53	0.53	1.28	1.28	1.28	0.75	0.75	0.75	0.75	0.99	0.99	0.99	0.99	0.99
0.22	0.22	0.75	0.75	0.75	0.75	0.53	0.53	1.28	1.28	1.28	0.66	0.75	0.75	0.75	0.75	0.99	0.99	0.99	0.99

38100   38350

Fig. 2. Example of a standard block model with grade assignment by the polygonal methods.

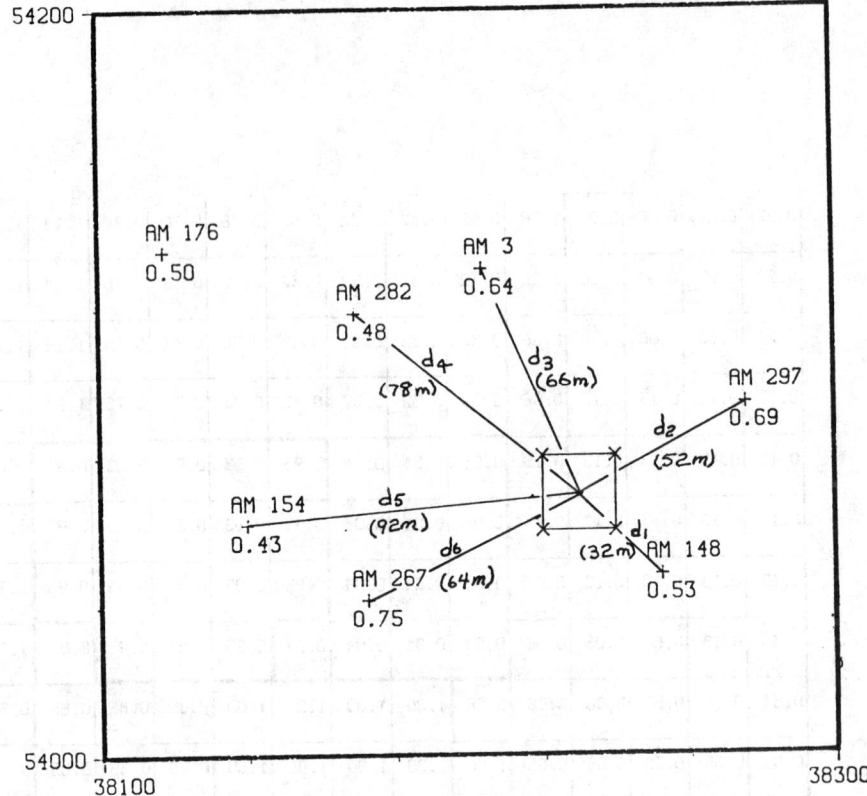

Fig. 3. Approximate polygonal method. The grade of the closest drill hole, AM 148, is assigned to the block.

The choice of the maximum polygon size has always been quite subjective and based on the experience of the ore reserve practitioner. However, the maximum range of influence can be defined as the distance at which grade values are no longer correlated. This distance can be determined from a variogram study (see Section 3 for more information).

Characteristics of the Polygonal Method

The polygonal method of ore reserve estimation has been in use since the early 1900s (Popoff, 1966), and is still used by some companies, especially when manually preparing ore reserve estimates.

The polygonal method is unique among the estimators discussed in this section, because the polygons drawn around each drill hole define both the volume of ore and its grade. This aspect of the polygonal method means that it can be used both to estimate grades and to estimate the boundary of the mineral deposit. As an estimator of the boundary of a deposit, the polygon is unbiased and gives sharp boundaries rather than the smooth gradational boundaries an interpolation method such as inverse distance method would give.

As an estimator of grade values, the polygon does a poor job compared with the other estimators discussed in this section. In several case studies (Knudsen and Kim, 1978; Readdy et al., 1982; Baafi and Kim, 1981), the polygonal method has been found to give individual block estimates that have significantly greater error than inverse distance weighting or kriging. This is illustrated in Fig. 4. Of greater

danger than large errors in individual block estimates, however, is the fact that the estimates of blocks above cutoff will be biased. This is illustrated by Fig. 5, taken from Baafi, where the average error in estimated coal thickness increases as the thickness increases.

DISTANCE WEIGHTING INTERPOLATION METHODS

Several technological advances in ore reserve estimation and mine planning are directly attributable to the use of the computer. No longer was the ore reserve analyst limited to simplistic methods, such as the polygonal method, but could instead utilize various smoothing and interpolation methods that were simply too time-consuming without the computer. The development of the first computerized mine planning systems that were based on the standard block model also gave impetus for using interpolation methods to estimate the grade of ore in each block of the model. The development of computerized contouring programs also spurred the use of a variety of interpolation methods, because the contouring procedures require values on a regularly spaced grid.

Although a large variety of interpolation methods have been developed, this section will be limited to discussing several nonstatistical techniques in common use in the mining industry. The methods discussed are applicable both to estimating grades in a standard 3-D block model, or to estimating grades (or thickness, etc.) at each point in a 2-D grid such as is used to model coal seams.

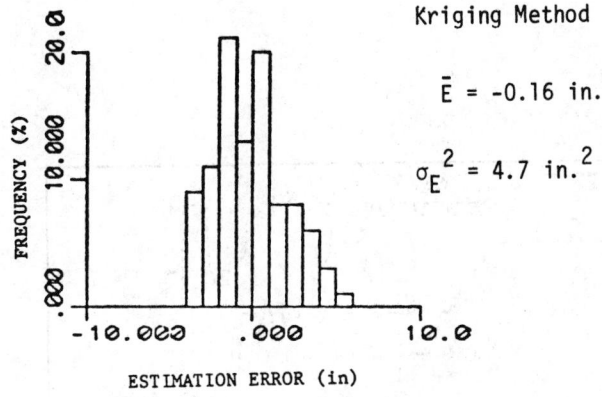

Kriging Method

$\bar{E} = -0.16$ in.

$\sigma_E^2 = 4.7$ in.^{2}

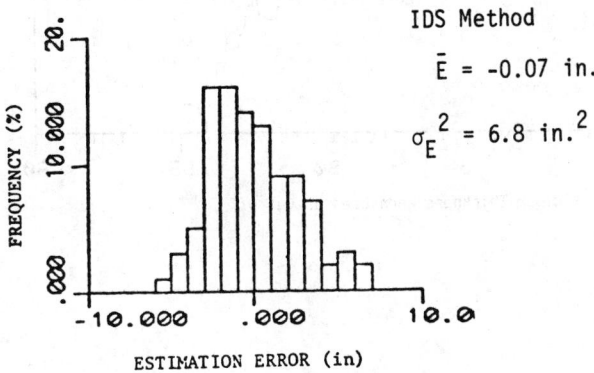

IDS Method

$\bar{E} = -0.07$ in.

$\sigma_E^2 = 6.8$ in.^{2}

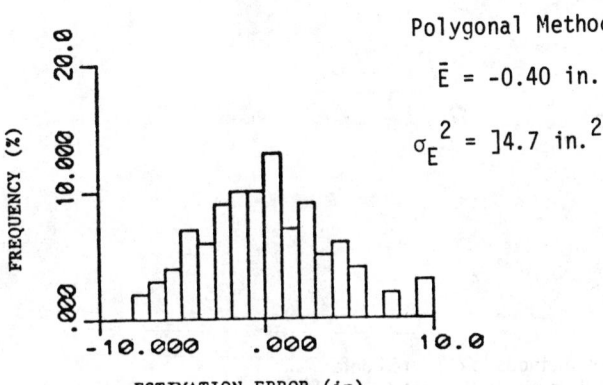

Polygonal Method

$\bar{E} = -0.40$ in.

$\sigma_E^2 =]4.7$ in.^{2}

Fig. 4. Histogram of estimation errors for various ore reserve methods (Baafi and Kim, 1982).

General Formulas

The object of interpolation is to predict the grade of a point or block by a weighted average of the nearby data points $Z(x_i)$. The general formula for distance weighting is

$$Z^*(v) = \sum_{i=1}^{n} \lambda_i Z(x_i)$$

and

$$\sum_{i=1}^{n} \lambda_i = 1$$

The weights, λ_i, are chosen as some function of the distance between the sample and the block. The most common form of weighting is to weight by the inverse of the distance raised to some power, r, as in:

$$\lambda_i = \frac{d_i^{-r}}{\sum_{i=1}^{n} d_i^{-r}}$$

Common choices of the power are 1, 1.5, 1.75, and 2. As the power gets larger, the faster the weights decay as d gets larger. The weight λ_i is not only a function of the power r, but also a function of the number of samples included in the interpolation. As more samples are included, the weight λ_i gets smaller; thus, the samples included in the interpolation are limited to those lying within a specified search radius.

Interpolation methods are susceptible to clustered data points, such as frequently occur in mining, where the clusters often occur in the high grade portions of the deposit (Ripley, 1981). This is shown in Fig. 6 where the cluster will dominate the other samples in estimating the value of point x.

One way to minimize this effect is to estimate the value of x using only a specified number of the nearest samples in each octant around x. This procedure is described by Sampson (1978) in the manual for *Surface II*.

Although the equation for distance weighting shows that the sample weighting is a function of the distance only, the weighting method can easily be altered to be a function of both distance and direction (Knudsen, 1975). This allows the effects of anisotropy in the mineralization to be included in the weighting.

An example of the calculations for a distance weighting interpolation method is shown in Fig. 7.

Validation of Interpolation Parameters

Specific values of the interpolation parameters used in a given interpolation should be chosen to give the optimal estimation. The only practical way to insure the best choice of the weighting exponent, the search radius, and the anisotropy factors is to validate the choices by using the comparison technique discussed earlier in this section. Using a specified set of parameter values, a comparison is made between the estimated grades of individual drill-hole composites and the true grades of the composites. Using this validation technique we can decide whether the exponent should be 1.75, 2.0, or some other value. The parameters giving the best estimation results are chosen for interpolation.

An example of this technique is shown in the following table. A comparison was made for six values of the weighting exponent.

Exponent	1.00	1.25	1.50	1.75	2.00	2.25	2.50	2.75
Error variance	1.89	1.87	1.88	1.89	1.91	1.94	1.97	2.01

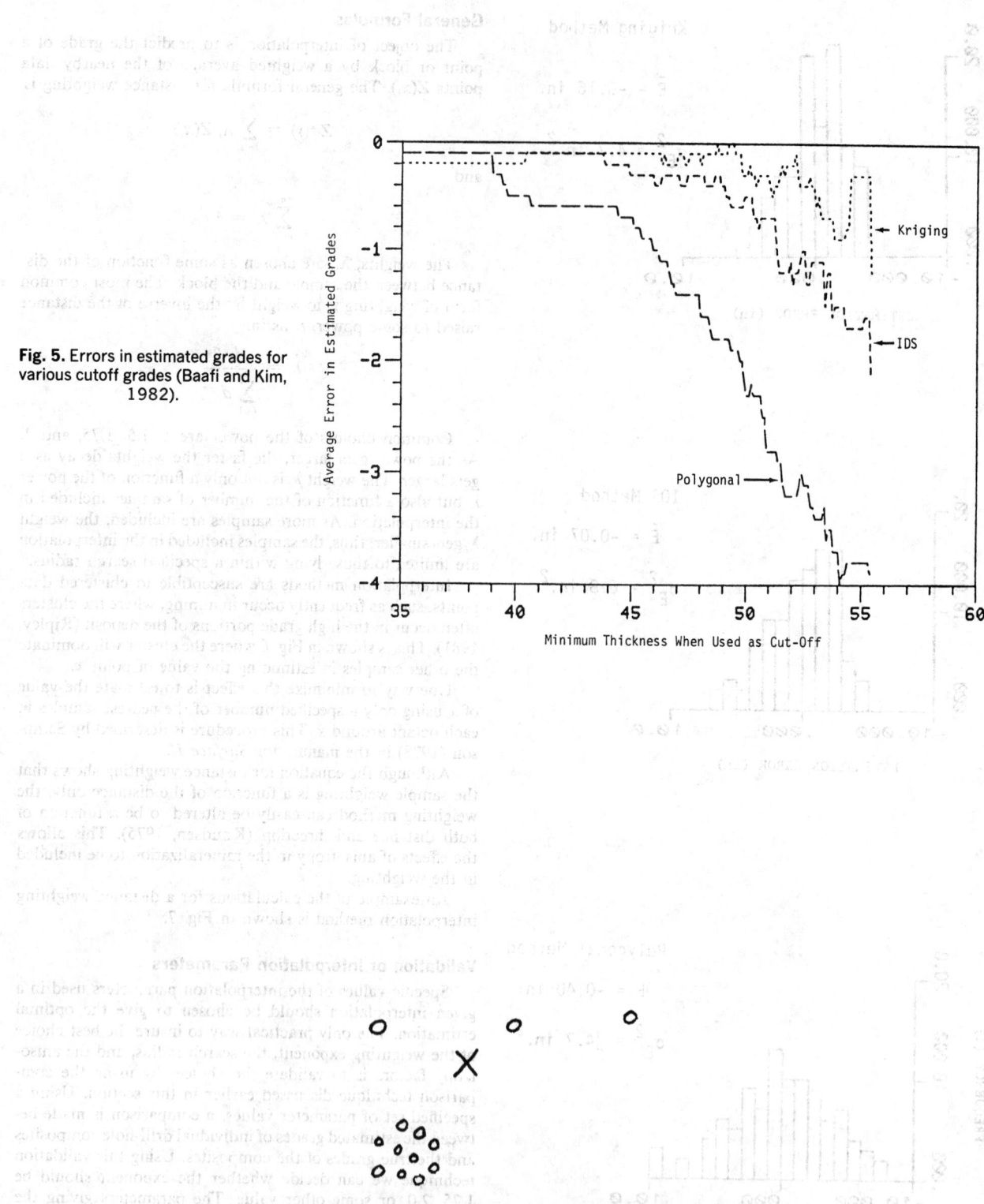

Fig. 5. Errors in estimated grades for various cutoff grades (Baafi and Kim, 1982).

Fig. 6. Susceptibility of interpolation methods to clustered data points. The ten clustered points will dominate the rest for estimation at X (from Ripley, 1981).

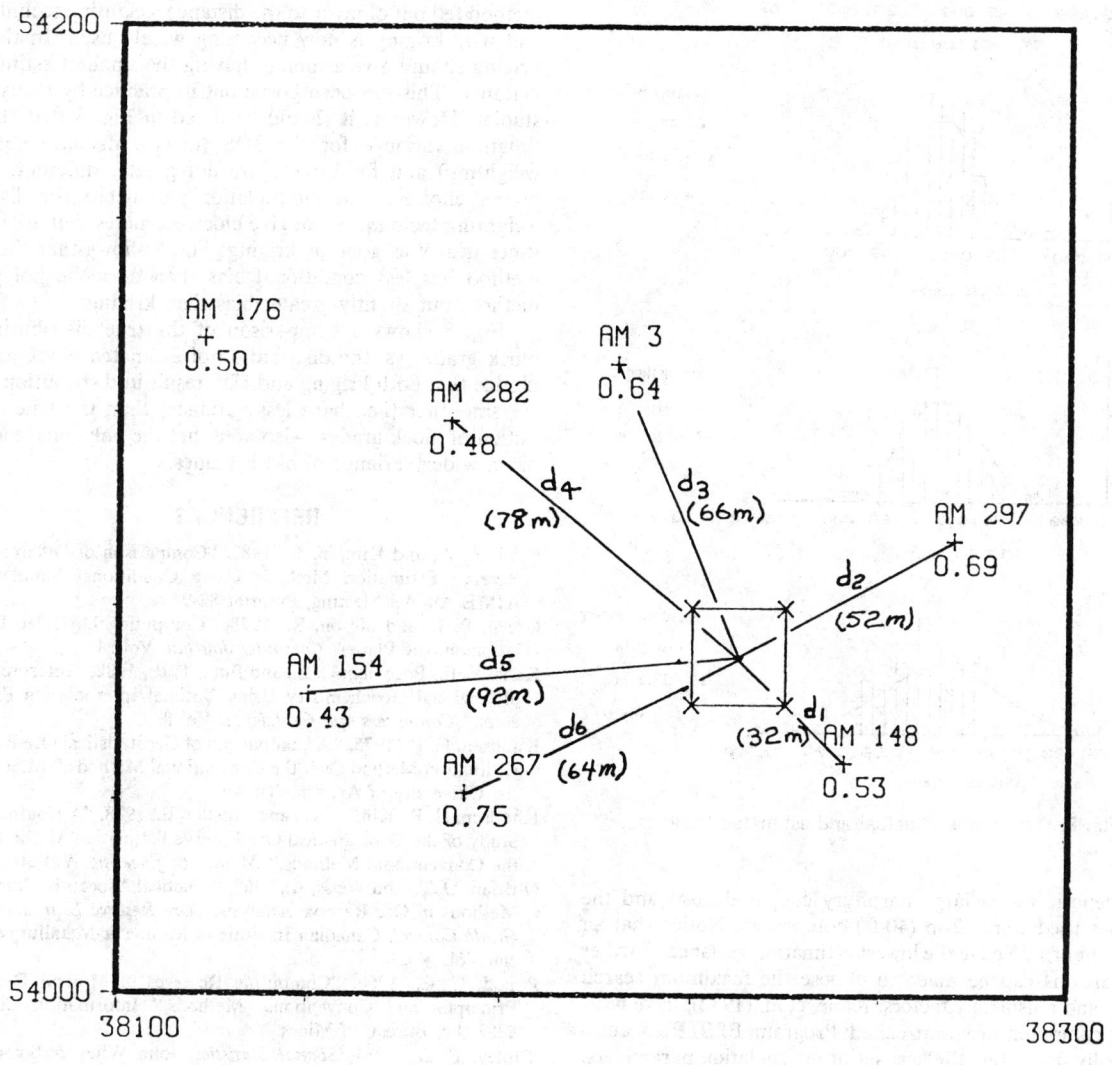

Fig. 7. Example calculations for inverse distance squared weighting.

$$\lambda_1 = \frac{\dfrac{1}{32^2}}{\displaystyle\sum_{i=1}^{6}\dfrac{1}{d_i^2}} = \frac{\dfrac{1}{32^2}}{\dfrac{1}{32^2}+\dfrac{1}{52^2}+\dfrac{1}{66^2}+\dfrac{1}{78^2}+\dfrac{1}{92^2}+\dfrac{1}{64^2}} = 0.464$$

likewise $\lambda_2 = 0.176$, $\lambda_3 = 0.109$, $\lambda_4 = 0.078$, $\lambda_5 = 0.056$, $\lambda_6 = 0.117$

$$Z^*(v) = \sum_{i=1}^{6}\lambda_i\,Z(x_i)$$

$$= 0.464 \times 0.53 + 0.176 \times 0.69 + 0.109 \times 0.64 + 0.078 \times 0.48$$
$$+\ 0.056 \times 0.43 + 0.117 \times 0.75 \ \%Cu = 0.59 \ \%Cu$$

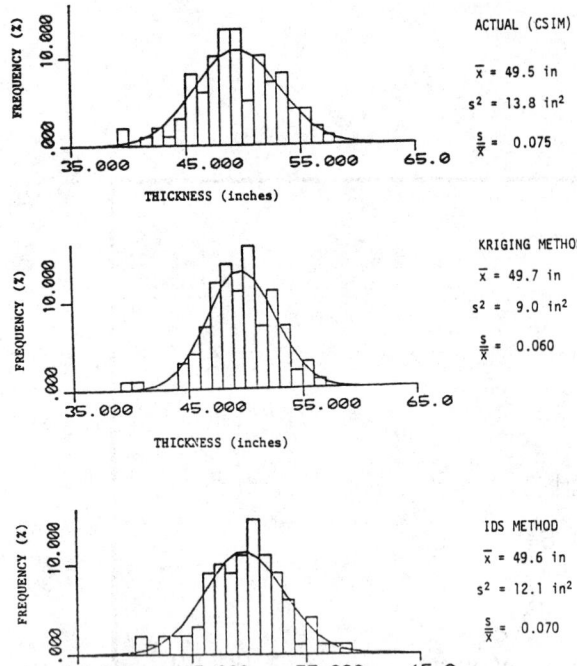

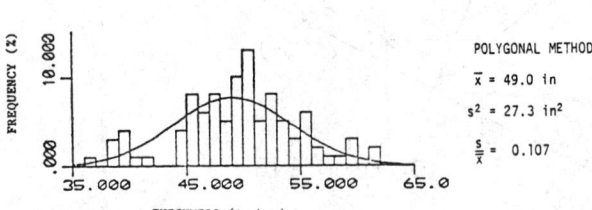

Fig. 8. Histograms of actual and estimated thickness.

modeling of open pit mines. Whereas the distance weighting interpolators largely replaced the traditional methods of estimation in the late 1960s, the distance weighting interpolators are now being rapidly replaced by the geostatistical technique of kriging.

Figs. 4 and 5 illustrate the reasons why the polygonal method fell out of favor to the distance weighting techniques, and why kriging is now becoming widely used. In theory, kriging should give estimates having the smallest estimation variance. This has been borne out in practice by many case studies. However, it should be noted in Fig. 4 that the estimation variance for the IDS (inverse distance squared weighting) and for kriging are not greatly different. With proper choice of the interpolation parameters the distance weighting techniques can give block estimates that are sometimes nearly as good as kriging. Fig. 5 shows that the IDS method has less conditional bias than does the polygonal method, but slightly greater bias than kriging.

Fig. 8 shows a comparison of the true distribution of block grades vs. the distribution of estimated block grades. Notice that both kriging and IDS result in distributions that are smoother (i.e., have less variance) than the true distribution of block grades. Also note that the polygonal method has a wider variance of block grades.

REFERENCES

Baafi, E. Y., and Kim, Y. C., 1982, "Comparison of Different Ore Reserve Estimation Methods Using Conditional Simulation," AIME Annual Meeting, Preprint 82-94.

Green, P. J., and Sibson, R., 1978, "Computing Dirichlet Tessellations in the Plane," *Computer Journal*, Vol 21.

Kane, V. E., Begovich, C. L., and Butz, T. R., 1982, "Interpretation of Regional Geochemistry Using Optimal Interpolation Parameters," *Computers and Geoscience*, Vol 8.

Knudsen, H. P., 1975, "A Comparison of Geostatistical Ore Reserve Estimation Method Over the Conventional Methods," M.Sc. Thesis, University of Arizona, Tucson.

Knudsen, H. P., Kim, Y. C., and Mueller, E., 1978, "A Comparative Study of the Geostatistical Ore Reserve Estimation Method Over the Conventional Methods," *Mining Engineering*, Vol 30.

O'Brian, D. T., and Weiss, A., 1968, "Practical Aspects of Computer Methods in Ore Reserve Analysis," *Ore Reserve Estimation and Grade Control*, Canadian Institute of Mining & Metallurgy, Special Vol. No. 9.

Popoff, C. C., 1966, "Computing Reserves of Mineral Deposits: Principles and Conventional Methods," Information Circular 8283, US Bureau of Mines.

Ripley, B. D., 1981, *Spatial Statistics*, John Wiley & Sons, New York.

Readdy, L. A., Bolin, D. S., and Mathieson, G. A., 1982, "Ore Reserve Calculation," *Underground Mining Methods Handbook*, W. A. Hustrulid, ed., AIME, New York, pp. 17-38.

The deposit was a large porphyry copper deposit and the samples used were 12-m (40-ft) composites. Notice that an exponent of 1.25 gave the lowest estimation variance. Further comparisons can be made to choose the maximum search radius and anisotropy factors. Kane, et al. (1982), have written a computer program called Program BESTP to automatically determine the best set of interpolation parameters.

Characteristics of Distance Weighting Method

Distance weighting interpolation methods have been widely used by the mining industry, especially for ore-body

3.3 Statistical and Geostatistical Methods

JEAN-MICHEL RENDU AND GRAHAM MATHIESON

INTRODUCTION

The economic value of a mineral deposit is first a function of its ore reserves. During the exploration and feasibility phases of a mining project, the need for timely and reliable ore reserve estimates is of critical importance particularly as a basis for mine planning in the early production years. This is a first step in assessing the economic merit of a deposit and must be done to permit monitoring of the appropriateness of continued capital investment in the project.

Underestimation of reserves may result in rejection of a viable project, significant opportunity losses, and possible embarrassment if the deposit is later successfully mined by a competing company. Overestimation may result in the inappropriate development of a subeconomic deposit, disappointing results, and significant financial losses. Poor evaluation may also lead to an undesirable mine development sequence, improper mill design, and inappropriate mining equipment selection.

Once in production, reserves and local block grades are reevaluated periodically for medium- and long-range planning, perhaps annually, down to weekly or even daily for short-term planning and grade control. Poor reserve estimation will result in waste blocks being predicted as ore, while ore blocks will be treated as waste, with direct impact on the grade of anticipated mill feed and cash flow. The precision that can be attained in each evaluation will vary significantly with the information available, including sampling and mapping data as well as the current understanding of the deposit geology. Similarly the method to be used for reserve evaluation must be adapted to take into account the quantity and quality of information to be processed, as well as the purpose for which the reserve estimate is to be made.

Whatever the amount of effort put into the calculation of the reserves, some degree of uncertainty will remain concerning the true characteristics of the deposit. This is a direct consequence of the need to estimate the properties of thousands of tons of material from samples whose total weight may only be a few kilograms. The overall precision with which the reserves are estimated is a function of numerous factors, including the type of mineralization, the geologic complexity of the deposit, the quality and quantity of sample information available, as well as the methodology used to process this information.

Considering the complexity of the parameters that must be taken into account and the interdisciplinary nature of the studies required, reserve estimation must be a team effort, involving geologists, mining engineers, mineral process engineers, statisticians, and, usually, computer specialists. Furthermore, in all instances where the quality of the reserve estimates is likely to have a significant influence on the financial and economic health of a mining company, high level executive attention should be given to the study to ensure not only that the correct answer is obtained, but that its meaning and its basis is also fully understood.

The main purpose of this chapter is to present the principal statistical and geostatistical techniques of ore reserve estimation as they apply to surface mining operations. To fully appreciate both the power and the limitations of these techniques, the preceding remarks should be kept in mind throughout. They were very aptly summarized by the authors of an excellent guide to the understanding of ore reserve estimation published by The Australasian Institute of Mining and Metallurgy (King, et al., 1982):

"Ore reserve estimation is not a matter of mere calculation but a procedure which involves, explicitly or implicitly, judgment and assumption about geological, operational and investigational factors. The calculations therefore form only part, and not necessarily the most important part, of the overall procedures.

An ore reserve statement should, where appropriate, be not merely an estimate of what is in the ground but a prediction, involving a further stage of judgment and assumption, of what will be fed to the mill or recovered.

For practical and statistical reasons related to the limitations of sampling and the kind and character of the ore, accuracy of prediction, especially of grade, will rarely exceed and will commonly not reach two significant figures."

Statistical methods of reserve evaluation do not take into account the physical position of the samples with respect to each other. They are extremely useful in the early analysis of a deposit and can be used to verify and interpret the sample values available, as well as to obtain preliminary reserve estimates. Geostatistical methods, on the other hand, are based on the common observation in mineral deposits that samples taken close to each other are more likely to have similar values than if they are located far apart. For this reason they are particularly well adapted to detailed reserve evaluations, when relatively dense sample information is available.

Statistical and geostatistical methods take into account the variability in an ore body's mineralization, and can be used to quantify the uncertainty associated with the reserve estimates. As with any other reserve estimation procedure, considerable judgment is needed in the use of geostatistics but the methods are easily adapted to take into account the specific properties of each deposit, the nature of the sample information available, the degree of continuity present in the mineralization, as well as the geologic controls that may influence the ore distribution. For this reason, geostatistics is increasingly becoming accepted as an adaptable method, likely to give improved estimates in most situations (Clark, 1979; David, 1977; Journel, et al., 1978; Rendu, 1981; Royle, 1980).

STATISTICAL METHODS

Introduction

Traditional statistical methods are based on the assumption that all sample values are equally representative of the properties of the deposit under study. The physical position of the samples with respect to each other is not taken into account. For this reason, application of these methods to reserve evaluation is usually limited to the early stages of ore body analysis, when a global estimate is sufficient, or to the study of highly variable mineralizations in which the similarity between sample values is negligible, even at short distances. Statistical methods are also extremely useful to test the reasonableness of estimates obtained using more sophisticated approaches, such as kriging. Many of the visual

301

and mathematical tools described here can also assist in the analysis, verification, and interpretation of sample data and in this respect they may be found useful throughout the life of a mining project.

Perhaps the easiest and fastest method of estimating the average grade of a deposit, or part of a deposit, consists of calculating the average value of the samples in the area of interest, ignoring their relative position. As an example, if a coal seam is intersected by ten drill holes and the seam thickness is measured in each hole, the average thickness of the seam can be estimated by the mean value of the ten drill-hole intercepts. If the drill holes are approximately uniformly distributed, this estimate is likely to be perfectly acceptable. Simple statistical formulae can then be used that give reasonable preliminary estimates of the coal tonnage in the deposit and the precision with which it is known.

If on the other hand, we consider an uranium vein which has also been intersected by ten drill holes, and want to estimate the average grade times thickness (uranium accumulation measured in meter $\times$ %U308), the average value of the ten drill holes is likely to give a very poor estimate of the vein properties. Indeed, the uranium accumulation may vary from a minimum of say 0.01 %U308 over 0.5 m (1.6 ft) [0.005 m (0.02 ft) $\times$ %U308] to a maximum of say 1.00% U308 over 1.5 m [1.5 m (5 ft) $\times$ %U308], and the mean value of the ten drill holes will be extremely sensitive to the presence of extreme values, especially extremely high values. In such instances where a very high level of variability is observed, more complex statistical methods may be required.

Visual Methods of Statistical Analysis

The first step in a statistical analysis of a mineral deposit involves plotting the *histogram* of the sample values. To calculate a histogram, the values are first sorted in increasing order. They are then grouped into class intervals and the number of samples falling within each class is calculated. Examples of histograms are given in Fig. 1 for a coal seam. The heat contents of the corresponding samples are listed in Table 1. These values have been grouped in increments of 0.23 MJ/kg (100 Btu per lb) as shown in Table 2 and the histogram plot shows, for example, that five samples have heat contents between 23 (9900) and 23.3 MJ/kg (10,000 Btu per lb).

Histograms can be extremely instructive. They show the range of values that can be expected, and they can be used to detect extreme, possibly erroneous values. The value that is occuring most often can be visually determined. It is known as the histogram *mode*. The midpoint value below which 50% of the samples are likely to be located is the histogram *median*.

In the foregoing example, the heat content varies from 21.2 to 23.8 MJ/kg (9000 to 10,100 Btu per lb), the most likely heat content is between 22.7 and 22.9 MJ/kg (9600 and 9700 Btu per lb) and the median is 22.88 MJ/kg (9697 Btu per lb). Histograms can also help to detect mixtures of geologic environments (Fig. 2).

From the histogram, the *frequency distribution* can be calculated where the number of samples falling in a given class interval is replaced by a proportion given as a percentage of the total number of samples. For example, Table 2 shows that 18.37% of the samples have a value between 22.7 and 22.9 MJ/kg (9600 and 9700 Btu per lb). If the samples shown in this table are representative of the actual coal seam, this distribution also indicates that 18.37% of the entire deposit has a value between 22.7 and 22.9 MJ/kg (9600 and

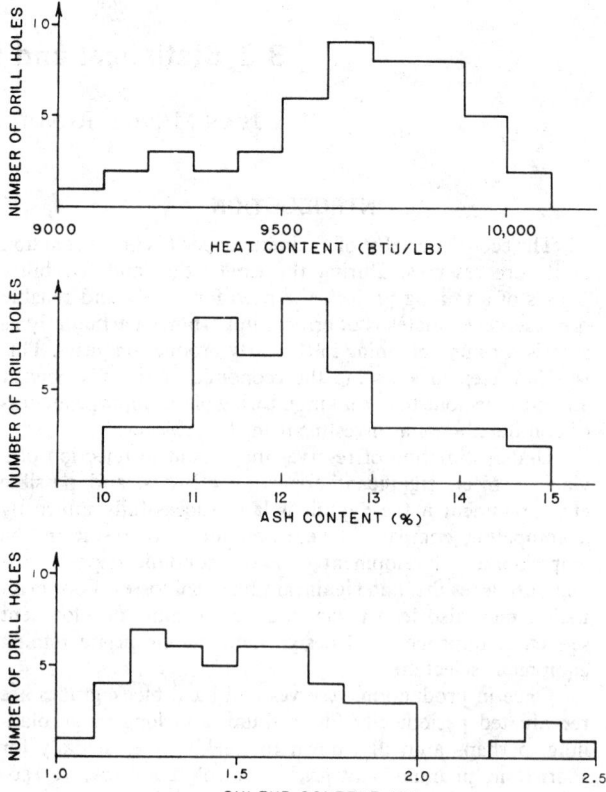

Fig. 1. Examples of histograms from a coal seam.

9700 Btu per lb). Extreme care must, however, be applied when using sample statistics to determine the economic properties of deposits that can only be mined using panels or blocks of size significantly larger than the samples. The influence of block size and selectivity on the grade-tonnage relationship is the subject of extensive discussions later in this section. Sample statistics will give reasonable estimates when applied to deposits with relatively low variability such as coal or industrial minerals. Only an order of magnitude estimate will be obtained if highly variable values are studied, such as gold or uranium grades.

Another tool that is useful in the statistical study of sample values is the *cumulative frequency distribution,* that indicates the proportion of the samples with value less than a given upper limit. Looking again at Table 2, one sees that 34.69% of the samples have a heat content of less than 22.7 MJ/kg (9600 Btu per lb). This also means that 65.31% of the sample values exceed 22.7 MJ/kg (9600 Btu per lb). Again, if the samples are representative of the deposit, one

Table 1. List of Coal Sample Heat Contents (Btu per lb*)

9000	9125	9723	9542	9772	9682	9852
9501	9311	9180	9450	9540	9880	9690
9580	9405	9605	9210	9410	9780	9788
9611	9706	9851	10083	9295	9630	9870
9820	9950	9972	10100	9572	9215	9865
9623	9932	9671	9745	9585	9900	9382
9712	9841	9730	9945	9800	9691	9697

* To convert to MJ/kg multiply by 2.326^{-3}.

Table 2. Distribution of Heat Contents in a Coal Seam

Heat Content, Btu per lb*		Histogram (Number of Samples in Class)	Frequency Distribution (Proportion of Samples in Class, %)	Cumulative Frequency Distribution, %
Lower	Upper			
9000	9100	1	2.04	2.04
9100	9200	2	4.08	6.12
9200	9300	3	6.12	12.24
9300	9400	2	4.08	16.33
9400	9500	3	6.12	22.45
9500	9600	6	12.24	34.69
9600	9700	9	18.37	53.06
9700	9800	8	16.33	69.29
9800	9900	8	16.33	85.71
9900	10000	5	10.20	95.92
10000	10100	2	4.08	100.00
9000	10100	49	100%	100%

* To convert to MJ/kg multiply by 2.326^{-3}.

may be able to conclude that 65.31% of the entire deposit will exceed 22.7 MJ/kg (9600 Btu per lb).

Cumulative frequency distributions are usually plotted using a probability scale along the horizontal axis. The probability scale is such that if the sample values are approximately normally distributed, the cumulative frequency distribution will be well represented by a straight line. The cumulative frequency distribution of the sample values listed in Table 2 is plotted in Fig. 3.

Another graphical representation of the distribution of sample values which is extremely helpful in understanding the properties of a deposit is the *scatter diagram*. A scatter diagram can only be plotted if at least two values are available for each sample. Scatter diagrams for a coal deposit are shown in Fig. 4. These diagrams show a positive but weak correlation between ash content and sulfur content, no correlation between seam thickness and heat content, a strong negative correlation between ash content and heat content, and a weak negative correlation between sulfur content and heat content. These plots can be used to detect anomalous values, as well as mixture of geologic populations (Fig. 5).

The use of such graphical methods for the preliminary analysis of sample values is usually extremely instructive. One should, however, remember that such methods can give meaningless results unless the geology of the deposit is first taken into consideration to classify the sample values. The spatial distribution of the samples and drill holes must be kept in mind when interpreting the results. For example, frequency distributions may be meaningless if the drill holes analyzed are located on a 50-m (164-ft) grid in the high grade part of the deposit and a 200-m (656-ft) grid in the lower grade parts of the deposit. Finally, all statistical analyses require that the values analyzed correspond to samples, composites or blocks of constant size. Using geostatistical terminology, it will be said that the values must correspond to *supports* of constant size.

The concept of support size is extremely important in all statistical studies of geological data. This is clearly seen in Fig. 6 that shows the grade of samples of variable size on the right-hand side, while the corresponding average grade of composites of 3-m (10-ft) length is plotted on the left-hand side. The sample size varies from 3 cm (0.1 ft) to 9 m (29.5 ft). The highest sample assay is 2.0% over 3 cm (0.1 ft), while the composites do not exceed 1.0%.

A numerical example of the influence of support size is shown in Fig. 7. Channel samples of 1-m (3.3 ft) length have been taken along a drift in a high-grade copper deposit. The values of the 1-m samples vary from 1% to 15%, while 8-m (30.6-ft) composites remain between 5.25% and 8.25%. Also shown in this figure is the effect of the support size on the frequency distribution. Typically, when the support size increases, the values get closer to each other and, therefore, closer to the mean of the ore body. Similarly, if longer or larger samples are used to analyze a deposit, the variability of the corresponding values is likely to be reduced, unless an increase in assaying errors counters the effect of the decreased geologic variability.

Fig. 8 illustrates the hypothetical distribution of all 3-m (10-ft) samples and 15-m (49-ft) composites in a porphyry copper deposit. If a 0.5% Cu cutoff grade is applied to the 3-m (10-ft) samples, 55% of the deposit is above cutoff, averaging 1.5% Cu. On the other hand, if the same cutoff grade is applied to 15-m (49-ft) composites, 65% of the deposit is above cutoff, averaging 1% Cu. Clearly, mining will occur at a scale greater than 3-m (10-ft) samples or even 15-m (49-ft) composites. Blocks of size 15 × 15 × 15 m (49 × 49 × 49 ft) might be considered as selectively minable, whose distribution will show even less variability than the one exhibited by the 15-m (49-ft) composites and will typi-

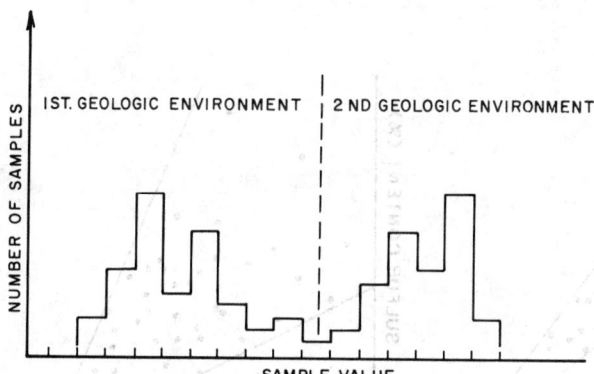

Fig. 2. Histogram indicating a possible mixture of geologic environments.

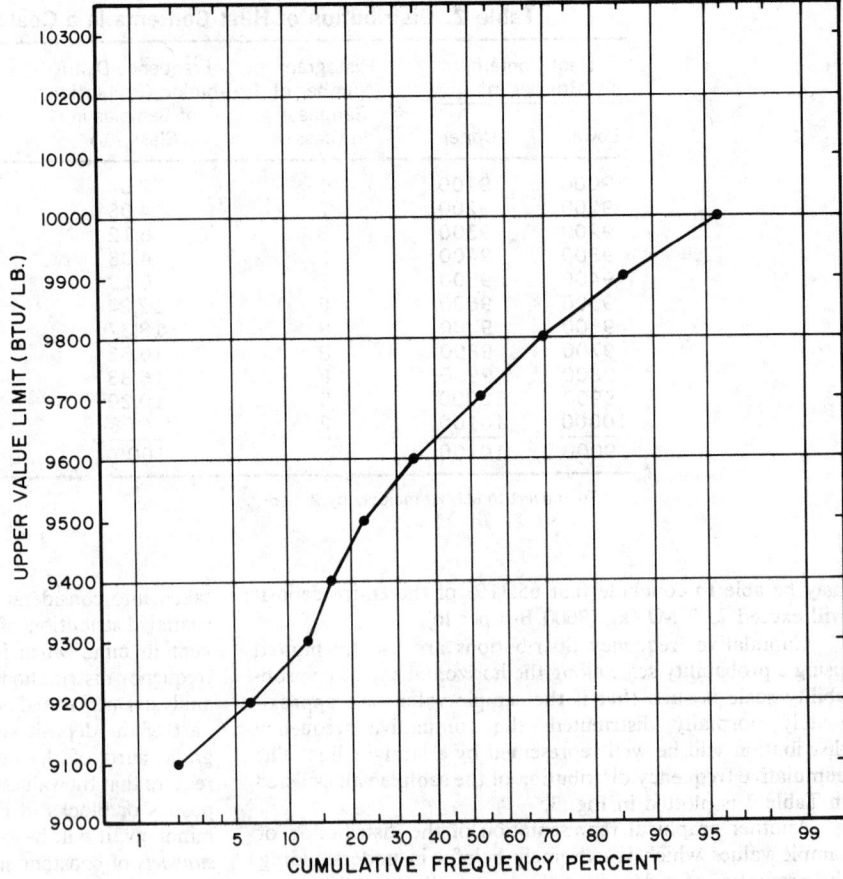

Fig. 3. Cumulative frequency distribution of heat content in coal seam.

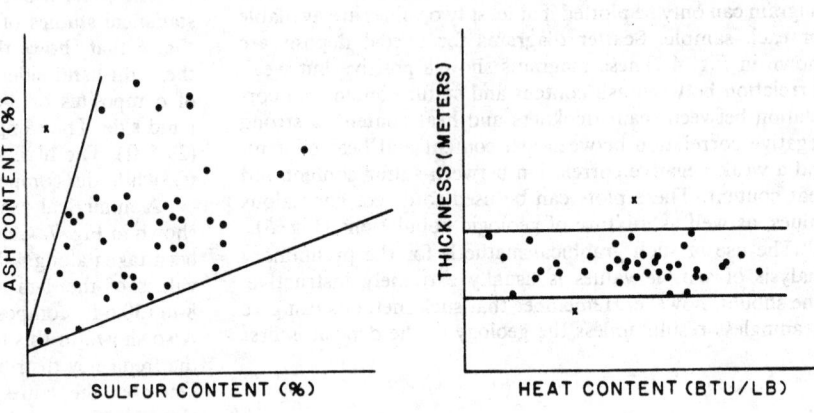

Fig. 4. Examples of scatter diagrams.

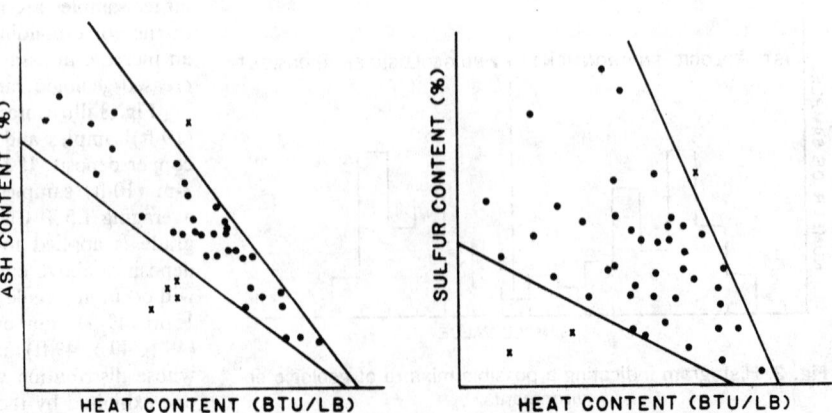

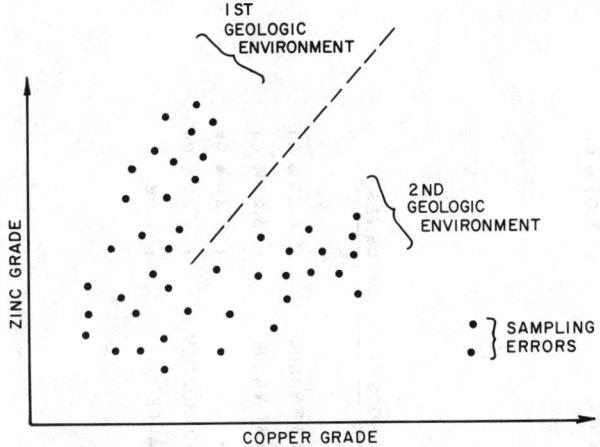

Fig. 5. Scatter diagrams indicating errors and mixture of geologic environments.

cally result in higher tonnage and lower grade estimation. Geostatistical methods exist that can be used to estimate the distribution of block grades from the sample distribution. These methods will be discussed later.

Normal Distributions

The methods described here apply to normally distributed sample values, i.e., values whose frequency distribution is symmetrical around the mean and can be plotted as a straight line on probability paper (Fig. 3). These methods are also commonly used when the sample values are only approximately symmetrically, but not necessarily normally distributed (Fig. 1). They can also be applied to large number of samples, independent of their distribution, under conditions specified later. Examples of deposits to which these methods can be applied include iron, coal, potash, phosphate, and industrial minerals as well as some high grade base metal ore bodies. Seam or vein thicknesses can also often be treated in the same fashion.

If n samples have been taken from a mineral deposit, with values x_i, $i = 1, 2, \ldots n$, an estimate of the average value of the deposit is the sample *mean* $\bar{x}$:

$$\bar{x} = \frac{1}{n}(x_1 + x_2 + \ldots + x_n)$$

For example, if five drill holes intersect a coal seam indicating coal thicknesses of 5, 2, 7, 4, and 3 m, (16, 6.6, 23, 13, and 10 ft), respectively, the average thickness of the deposit can be estimated at 4.20 m (13.8 ft). This number is, however, only an estimate of the true average coal thickness which could be thicker [say 5 m (16 ft)] or narrower [say 3 m (10 ft)]. The degree of certainty with which the deposit characteristics are known is a function of the number of samples available, the variability in the sample values, and of whether the sample values are truly representative of the deposit. If μ is the unknown average thickness of the deposit, $\bar{x}$ is an estimate of μ. The difference between $\bar{x}$ and μ represents the

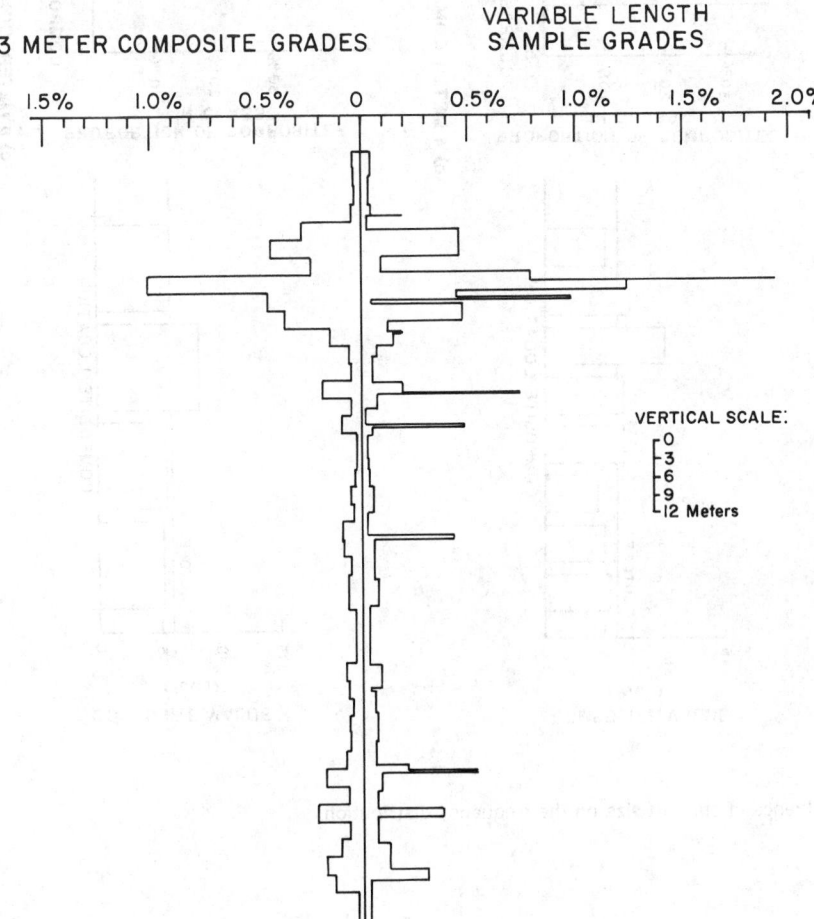

Fig. 6. Comparison of samples of variable size, with fixed length composites.

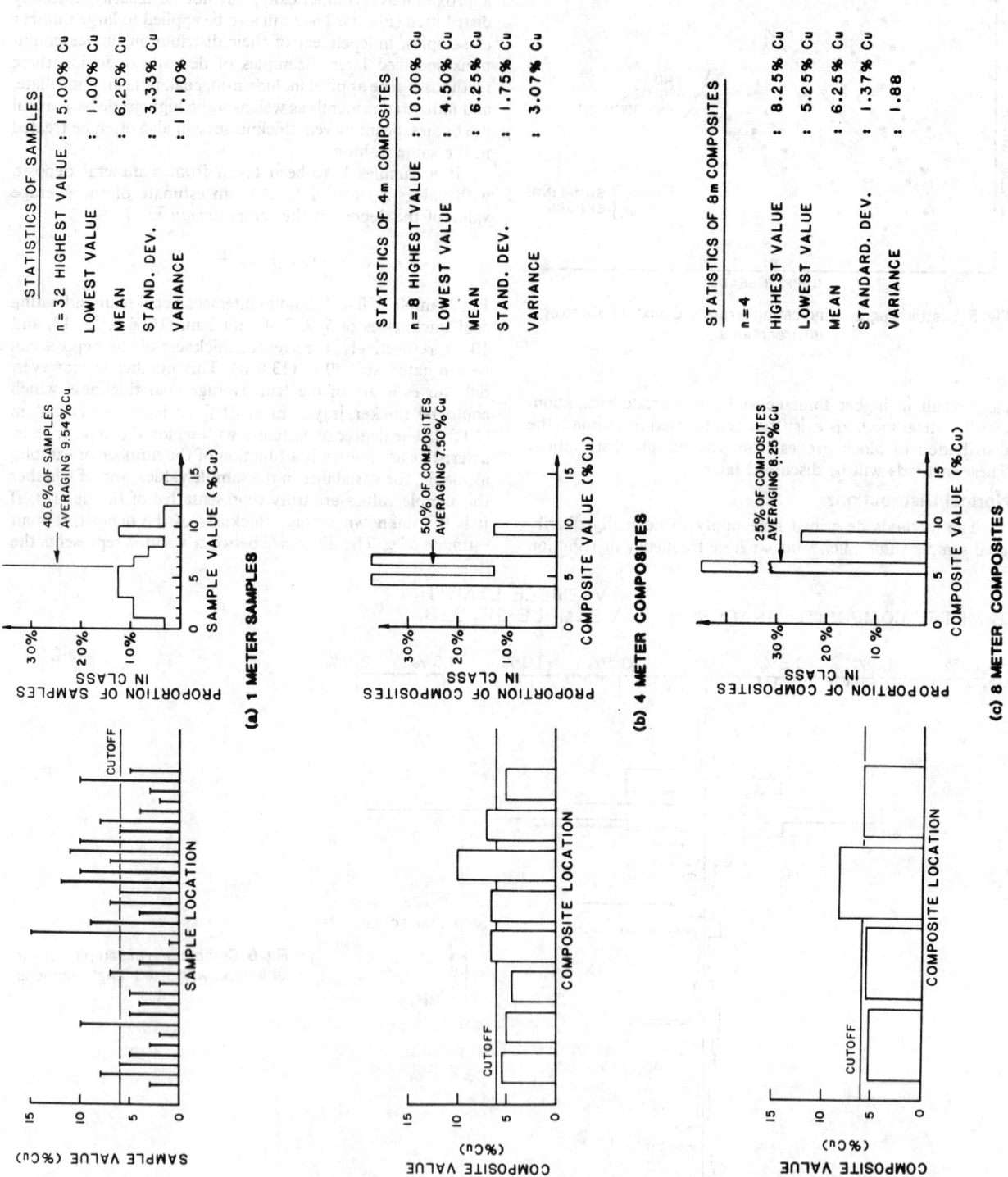

Fig. 7. Influence of support size on the frequency distribution.

	3 METER SAMPLES	15 METER COMPOSITES
PROPORTION ABOVE CUT OFF GRADE	55%	65%
AVERAGE GRADE ABOVE CUT OFF GRADE	1.5% Cu	1.0% Cu

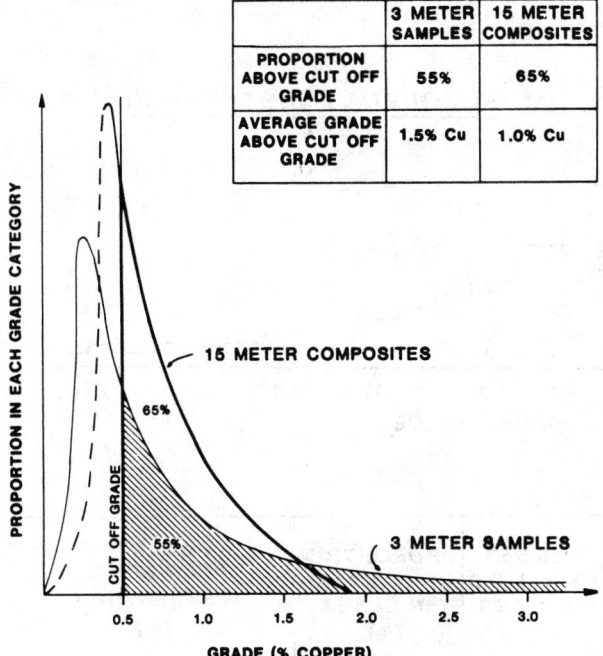

Fig. 8. Influence of support size on the tonnage and grade above cutoff grade.

error made in estimating the deposit parameter in question. This difference is unknown, but statistical methods can be used to quantify its likely magnitude. The expected value ($E[\ ]$) of the squared difference between $\bar{x}$ and μ is known as the *error variance* σ_E^2:

$$\sigma_E^2 = E\,[(\bar{x} - \mu)^2]$$

and its square root σ_E is the (standard) error of estimation of μ. The magnitude of σ_E gives an indication of the precision with which the mean is estimated. In many circumstances it is reasonable to state that there are two chances in three (68% probability) that the true value μ is greater than $\bar{x} - \sigma_E$ and less than $\bar{x} + \sigma_E$. There are also approximately 19 chances in 20 (or 95% probability) that the true value of the deposit exceeds $\bar{x} - 2\sigma_E$ and is less than $\bar{x} + 2\sigma_E$. The error variance is calculated from the estimated sample variance s^2 as follows:

$$s^2 = \frac{1}{n-1}\,[(x_1 - \bar{x})^2 + \ldots + (x_n - \bar{x})^2]$$

$$= \frac{1}{n-1}\left(\sum_{i=1}^{n} x_i^2 - n\bar{x}^2\right)$$

$$\sigma_E^2 = s^2/n$$

Using the coal seam example given previously, we calculate $n = 5$, $s^2 = 3.70$, and $\sigma_E^2 = 0.74$ or $\sigma_E = 0.86$ m. Therefore if μ is the unknown average thickness of the deposit,

Prob (3.34 m $< \mu <$ 5.06 m) = 68%

Prob (2.48 m $< \mu <$ 5.92 m) = 95%

If the sample values reasonably well approximate a normal distribution, the confidence intervals can be calculated with increased precision, using the Student's t statistics, a method described in detail in most texts on statistics or geostatistics (David, 1977; Rendu, 1981).

The foregoing formulae can generally be applied to most deposits, independent of whether or not the sample values are normally distributed, provided the number of samples is large enough, such that the error of estimation σ_E does not exceed 20% of the mean.

Lognormal Distributions

In many more deposits the sample distribution is not symmetrical, but is characterized by a longer tail towards the high values (Fig. 9): this is known as a positively skewed distribution. This type of distribution is extremely common among lower grade deposits, characterized by a high variability in sample values. It is commonly observed not only in gold and precious metal deposits, but also in uranium, tin, molybdenum, copper, and zinc deposits, as well as in the sulfur content of coal deposits and in the thickness of extremely variable veins. If sample values have a very skewed distribution, the simple statistical methods discussed earlier in the context of the normal theory may give meaningless results.

In many instances, as illustrated in Fig. 10, the frequency distribution of the sample values will plot as a straight line on logarithmic probability paper. The distribution is then said to be lognormal and the logarithms of the sample values are normally distributed. Logarithmic probability paper is characterized by a probability scale on the horizontal axis and a logarithmic scale on the vertical axis. The data plotted in Fig. 10 are listed in Fig. 9. Also shown in Fig. 10 are graphical methods of analysis of lognormal distributions that are discussed by Krige (1978) and Rendu (1981). The lognormal distribution of sample values was first recognized by Dr. Sichel in the South African gold fields (Sichel, 1952).

In some instances, the lognormal assumption is not exactly satisfied and the cumulative frequency distribution shows a downward curvature when plotted on lognormal probability paper (Fig. 11). D. G. Krige showed (1960) that a lognormal distribution can often be obtained if a constant is added to the sample values. The distribution is then said to be three-parameter lognormal. The three parameters are the mean, the logarithmic variance, and the additive constant. The additive constant is best determined graphically by trial and error, using a plot of the cumulative frequency distribution.

A simple example will be used to indicate some of the properties of the lognormal approach. Consider a block of ore from a gold deposit in which seven blastholes are located, having assays of 10, 11, 12, 15, 18, 20, and 200 milliounces* per ton (moz/ton) (Block 3 in Fig. 12). If the normal statistical methods discussed earlier are used to analyze this block, its average grade is estimated at 40.9 moz/ton, with 90% probability that the true average grade of the block is at least 0 moz/ton (the calculated value is −10.7 moz/ton) but not more than 92.5 moz/ton. The confidence limits were calculated using the Student's t statistics, and the lower limit is obviously meaningless. If, on the other hand, lognormal statistical methods are used, the block average grade is estimated at 31.5 moz/ton, with 90% probability that the true average grade is at least 16.7 moz/ton but not more than 163.6 moz/ton. These numbers appear more reasonable and reflect one of the properties of the lognormal theory; namely that it is less sensitive to extreme values. The downside risk, as represented by the lower confidence limit, is also better bounded. The two other examples given in Fig. 12 (Blocks 1 and 2) show only a small difference in results when applying

---

* 1 milliounce = 0.001 oz.

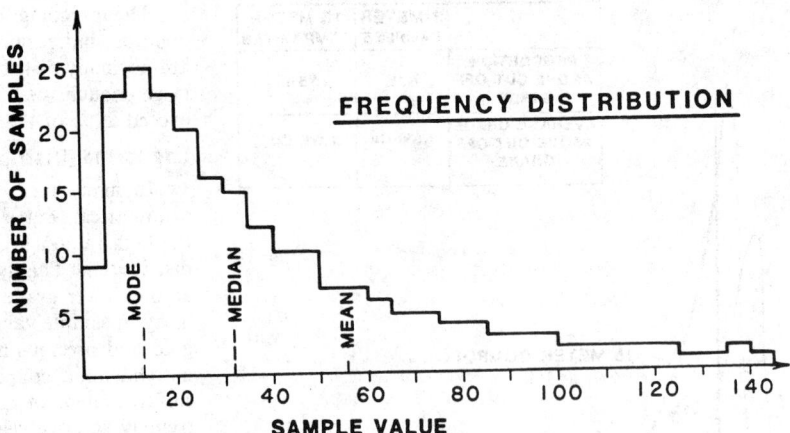

Fig. 9. Example of lognormal distribution.

CLASS LIMITS		NUMBER OF VALUES IN CLASS	PROPORTION OF SAMPLES IN CLASS (%)	CUMULATIVE FREQUENCY DISTRIBUTION (%)
LOWER LIMIT	UPPER LIMIT			
—	4.8	9	3.6	3.6
4.9	10.0	24	9.6	13.2
10.1	20.9	51	20.5	33.7
21.0	43.7	68	27.3	61.0
43.8	91.5	56	22.5	83.5
91.6	191.3	30	12.0	95.6
191.4	400.0	9	3.6	99.2
400.1	—	2	0.8	100
TOTAL		249	100.0	100

normal and lognormal theory to data with relatively low variability. Unfortunately, this example should not be taken to imply that a lognormal approach is always preferable, as there are cases where it will also give meaningless results.

Consider n sample values x_i, $i = 1, 2, \ldots n$ taken from a lognormal distribution with additive constant β. If we define $y_i = \ln(x_i + \beta)$, the logarithmic mean and logarithmic variance are estimated as follows:

$$\bar{y} = \frac{1}{n}(y_1 + y_2 + \ldots + y_n)$$

$$V = \frac{1}{n}[(y_1 - \bar{y})^2 + \ldots + (y_n - \bar{y})^2]$$

The median of a lognormal distribution is estimated by $m = \exp(\bar{y}) - \beta$. In first approximation if n is larger than 20 and V is less than 2.0, the mean value of the deposit can be estimated by T calculated as follows:

$$T = \exp(\bar{y} + V/2) - \beta$$

Upper and lower confidence limits for this mean value can be estimated using the following formula:

$$T_p = (T + \beta)\psi_p(V; n) - \beta$$

with:

$$\psi_p(V; n) = \exp\left(\frac{\sigma^2}{2} + t_p\sigma\right)$$

$$\sigma^2 = \frac{V}{n}\left(1 + \frac{V}{2}\right)$$

and where t_p is the Student's t statistic.

The values $t_p = -1$ and $t_p = +1$ can be used to calculate the lower and upper 68% confidence limits. Values $t_p = -2$ and $t_p = +2$ can be used to calculate the lower and upper 95% confidence limits. For example, consider a molybdenum deposit from which 30 samples have been taken with a three-parameter lognormal distribution, the additive constant having value $\beta = 0.02$ Mo%. The logarithmic mean and logarithmic variance are first calculated, say $\bar{y} = -2.30$ and V

= 1.2. The average grade of the deposit is estimated at T = exp $(-2.30 + 1.2/2) - 0.02 = 0.163$ Mo%. Furthermore, $\sigma^2 = (1.2/30 (1 + 1.2/2) = 0.064$; hence there is approximately 95% probability that the average grade of the deposit is at least $(0.163 + 0.02) \times$ exp $(0.064/2 - 2\sqrt{0.064}) - 0.02 = 0.094$ Mo% but not more than $(0.163 + 0.02) \times$ exp $(0.064/2 + 2\sqrt{0.064}) - 0.02 = 0.293$ Mo%.

The foregoing formulae are only approximations and are not necessarily acceptable when the number of samples becomes small. For these conditions, Dr. Sichel developed exact statistics which require use of tables, as can be found in Sichel (1966), Rendu (1981), Krige (1978), David (1977) or Wainstein (1975). The average value μ is then estimated by the Sichel's T estimator defined as follows:

$$T = \exp{(\bar{y})} \gamma_n(V) - \beta$$

where $\gamma_n(V)$ is obtained from tables. Confidence limits for the value μ are calculated using the formula:

$$T_p = (T + \beta)\psi_p(V; n) - \beta$$

where $\psi_p(V; n)$ is also obtained from tables. The mathematical formulae that can be used to calculate $\gamma_n(V)$ and $\psi_p(V; n)$ are discussed by Sichel (1966) and Wainstein (1975).

Lognormal theory was first successfully used in the 1950s in the evaluation of South African gold mines, where it was commonly necessary to achieve accurate estimation of new deposits from a small number of drill-hole intercepts. It has been used extensively since then in many types of deposits, including disseminated and vein gold, molybdenum, tin, sulfur in coal deposits, bismuth in zinc deposits, lead, copper, zinc, and uranium.

In many instances, the distribution of the sample values well approximates the two- or three-parameter lognormal model. In other instances, significant departure from these models may be observed. Such departures may be indicative of a number of factors, including the superposition of a number of mineralizing events, or simply that the grade distribution is indeed not lognormal. In all instances, simple explanations should first be sought for complex value distributions. An irregular sampling density will often result in an excessive representation of the high-grade portions of the deposit. A mixture of geologic environments may also produce apparently abnormal distributions. The presence of clay zones in the middle of a disseminated gold deposit may result in an excess of low-grade values. Secondary enrichment may, on the other hand, result in an excess of high-grade values. Changes in the sampling density can be taken into account by using declustering techniques that consist in weighting the sample values in inverse proportion to the drill-hole density. If mixed geologic zones are suspected, additional geologic analysis and partitioning of the deposit may be required before meaningful statistical results can be obtained.

Once again it should be emphasized that conventional statistical methods, even though limited in their application, still have an important role to play in checking and sorting out data, as well as in the controlled estimation of reserves under specific conditions.

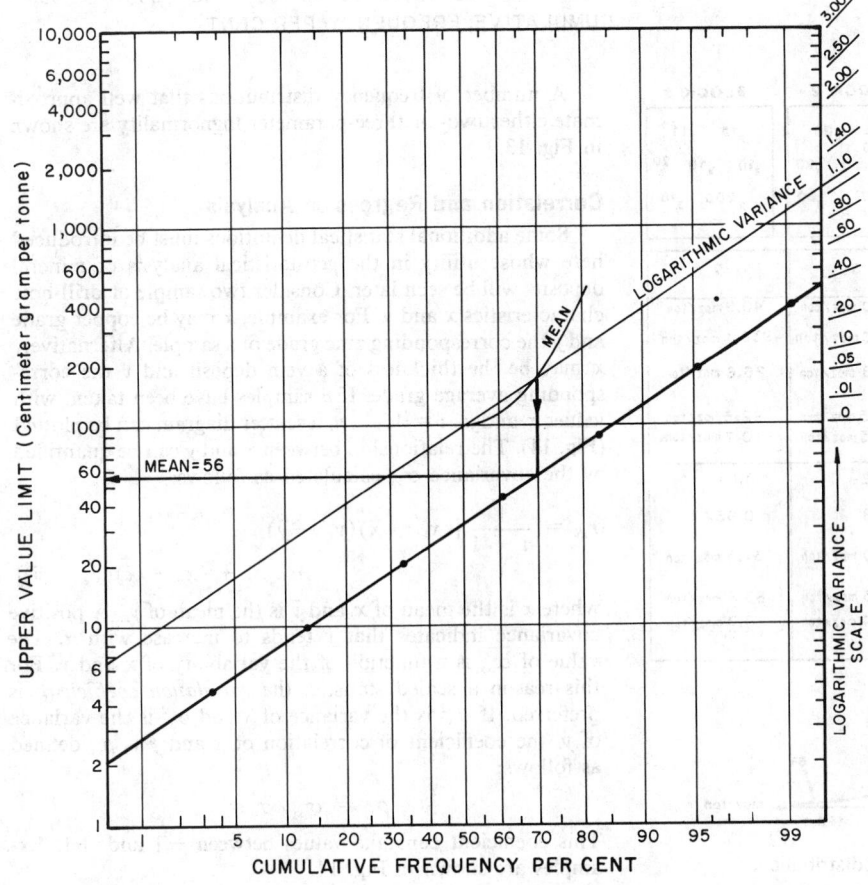

Fig. 10. Graphical study of lognormal distribution.

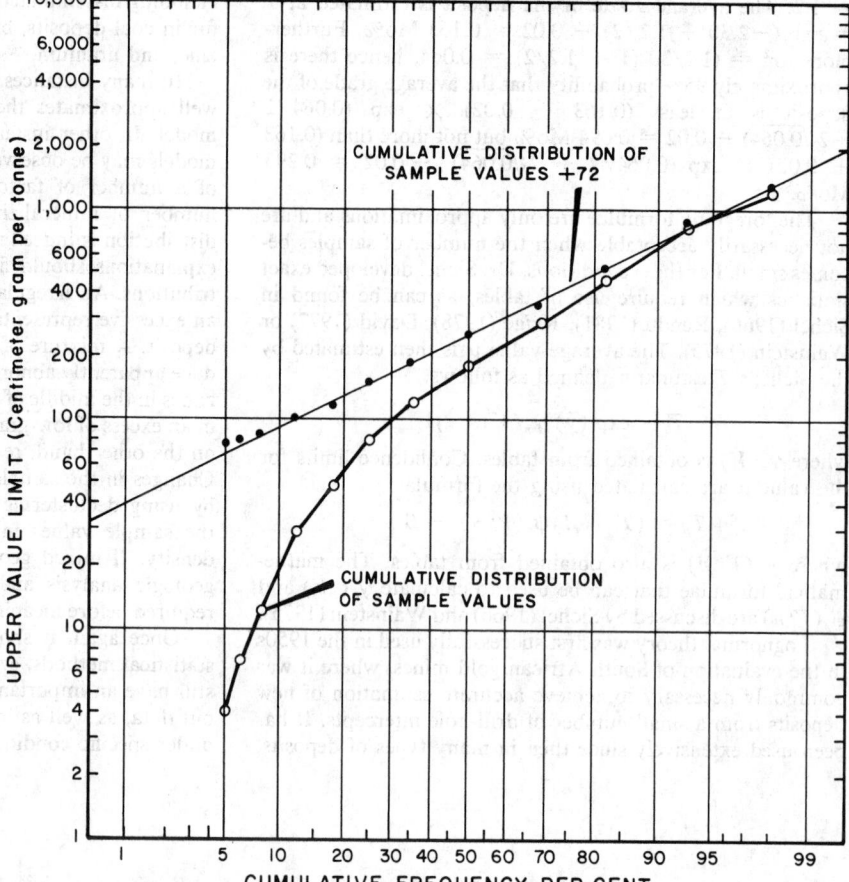

Fig. 11. Example of three parameter lognormal distribution.

CUMULATIVE DISTRIBUTION OF SAMPLE VALUES +72

CUMULATIVE DISTRIBUTION OF SAMPLE VALUES

UPPER VALUE LIMIT (Centimeter gram per tonne)

CUMULATIVE FREQUENCY PER CENT

	BLOCK 1	BLOCK 2	BLOCK 3
	x^1 x^3 x^2 x^2 x^5	x^{10} x^{30} x^{20} x^{20} x^{50}	x^{12} x^{11} x^{10} x^{18} x^{20} x^{200} x^{15}

	NUMBER OF COMPOSITES	5	5	7
NORMAL	**MEAN**	2.60 moz/ton	26.0 moz/ton	40.9 moz/ton
	VARIANCE OF SAMPLES	2.30 moz/ton	230.0 moz/ton	4938.1 moz/ton
	STANDARD DEVIATION OF MEAN	0.68 moz/ton	6.8 moz/ton	26.6 moz/ton
	90% CENTRAL CONFIDENCE LIMITS (using Student's t)	{ 4.05 moz/ton 1.15 moz/ton	{ 40.5 moz/ton 11.5 moz/ton	{ 92.5 moz/ton -10.7 moz/ton
LOG NORMAL	**LOG MEAN**	0.82	3.12	3.01
	LOG VARIANCE OF SAMPLES	0.28	0.28	0.93
	SICHEL'S t	2.60 moz/ton	26.0 moz/ton	31.5 moz/ton
	90% CENTRAL CONFIDENCE LIMITS (using Sichel's ψ)	{ 6.35 moz/ton 1.74 moz/ton	{ 63.5 moz/ton 17.4 moz/ton	{ 163.6 moz/ton 16.7 moz/ton

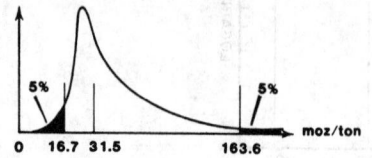

5%   5%

0   16.7 31.5   163.6   moz/ton

Fig. 12. Statistics: Lognormal distribution.

A number of frequency distributions that well approximate either two- or three-parameter lognormality are shown in Fig. 13.

Correlation and Regression Analysis

Some additional statistical definitions must be introduced here whose utility in the geostatistical analysis of mineral deposits will be seen later. Consider two sample or drill-hole characteristics x and y. For example, x may be copper grade and y the corresponding zinc grade of a sample. Alternatively x may be the thickness of a vein deposit and y the corresponding average grade. If n samples have been taken, with values x_i and y_i, $i = 1, \ldots n$, a scatter diagram can be plotted (Fig. 14). The relationship between x and y can be quantified by the covariance σ_{xy} calculated as follows:

$$\sigma_{xy} = \frac{1}{n-1} [(x_1 - \bar{x})(y_1 - \bar{y})$$
$$+ \ldots + (x_n - \bar{x})(y_n - \bar{y})]$$

where $\bar{x}$ is the mean of x_i and $\bar{y}$ is the mean of y_i. A positive covariance indicates that y tends to increase with x. The value of σ_{xy} is a function of the variability of x and y. For this reason a scaled statistic, the *correlation coefficient*, is preferred. If σ_x^2 is the variance of x and σ_y^2 is the variance of y, the coefficient of correlation of x and y is ρ_{xy} defined as follows:

$$\rho_{xy} = \sigma_{xy}/\sigma_x \sigma_y$$

This coefficient can take values between -1 and $+1$. Examples are shown in Fig. 14.

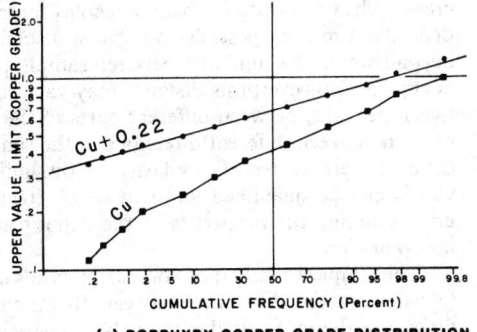

(a) PORPHYRY COPPER GRADE DISTRIBUTION

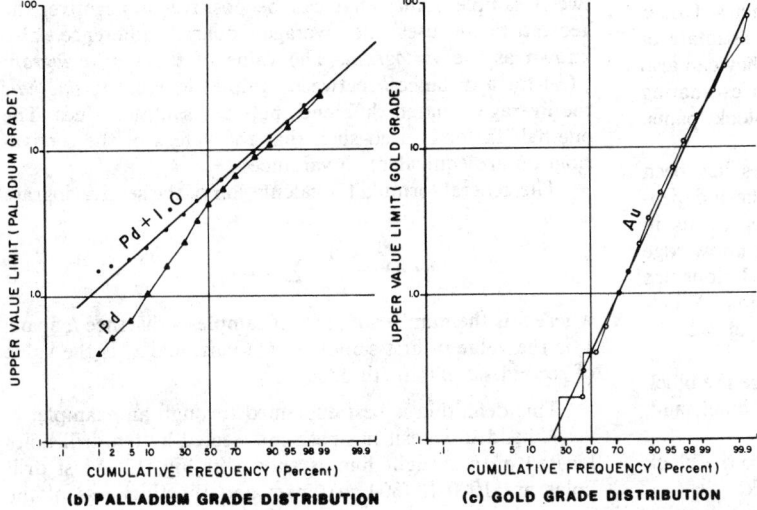

(b) PALLADIUM GRADE DISTRIBUTION   (c) GOLD GRADE DISTRIBUTION

Fig. 13. Examples of cumulative frequency distributions.

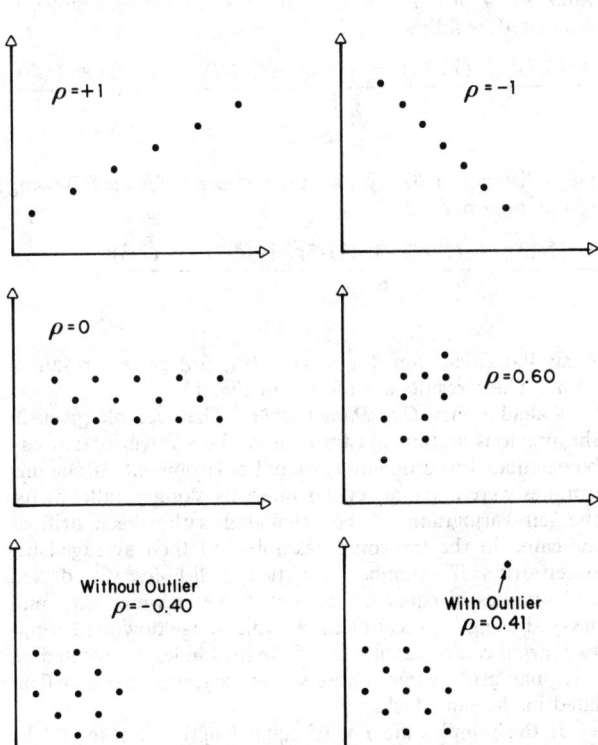

Fig. 14. Examples of correlation coefficients.

One must be careful when interpreting coefficients of correlation. Values close to 1 or -1 usually indicate a strong correlation, while values close to zero indicate a poor or no correlation. However, outliers, extreme values, or incorrect values may result in a meaningless correlation coefficient. For this reason it is recommended that scatter diagrams be plotted before calculating the correlation coefficient. This is illustrated in the lower part of Fig. 14.

In some instances it is desirable to estimate the value of a variable y from the value of x. For example, in a massive sulfide deposit, x might represent the copper grade and y the specific gravity. In simple cases, linear regression can be used for this purpose. The expected value of y given x is $\hat{y}$ calculated as follows:

$$\hat{y} = ax + b$$
$$a = \sigma_{xy}/\sigma_x^2 = \rho_{xy}\sigma_y/\sigma_x$$
$$b = \bar{y} - a\bar{x}$$

The error made when estimating y by $\hat{y}$ is given by the following formula:

$$\sigma_E^2 = E\left[(y - \hat{y})^2\right] = \sigma_y^2(1 - \rho_{xy}^2)$$

If the sample values x and y are lognormally distributed, the covariance and variances should be calculated using the logarithm of the sample values. The following regression equation must then be used (David, 1977):

$$\hat{y} = \exp\left[a \ln(x) + b + \frac{1}{2}\sigma_y^2(1 - \rho_{xy}^2)\right]$$

When using regression analysis, it is important to realize that the regression line is not the same depending on which variable is treated as the y (*dependent*) variable, and which is treated as the x (*independent*) variable.

THE FOUNDATIONS OF GEOSTATISTICS

The Geostatistical Approach

The statistical methods described earlier permit the estimation of the average value of a deposit parameter and the calculation of confidence limits for this value. They are based on the very important assumption that sample values are independent of one another and are equally representative. It has been demonstrated that these methods yield unreliable results when the drill-hole spacing is highly variable but they also suffer from at least two other major deficiencies. Under many circumstances, they lead to an incorrect estimate of the certainty with which the deposit is known. They can also be extremely unreliable for the essential task of estimating the value of a subset of an ore body such as a block, panel, mine section or bench.

As soon as a reasonable number of samples has been obtained from the deposit, the assumption of the independence of these samples becomes invalid and represents in fact an ignoring of an important portion of the knowledge gained during the sampling program, i.e., the relationships between the sample values and their relative positions. Whatever method is used for the valuation of a block of ore, the following assumptions are always made:

1) The values of samples located near or inside the block being estimated are related to the value of the block and, most importantly,

2) The values of the samples located closest to the block are most closely related to the value of the block.

These assumptions will hold true only if the following condition is also satisfied, as it is in virtually every known type of mineral deposit:

$$\gamma(1000) = \frac{1}{2} \cdot \frac{(5\text{-}5)^2 + (5\text{-}7)^2 + (12\text{-}11)^2 + (11\text{-}8)^2 + (8\text{-}7)^2 + (7\text{-}2)^2 + (3\text{-}3)^2}{7}$$

$$= 2.86$$

3) There exists a relationship between sample values which is a function of their distance apart.

$$\gamma(2000) = \frac{1}{2} \cdot \frac{(5\text{-}7)^2 + (7\text{-}12)^2 + (12\text{-}8)^2 + (11\text{-}7)^2 + (8\text{-}2)^2 + (2\text{-}3)^2}{6}$$

$$= 8.17$$

The geostatistical method of ore reserve estimation is distinguished from other methods available by its inclusion of this factor. At the outset, it involves estimating the spatial relationship that exists between the known sample values and from that estimation deriving a model of this relationship for the ore body or some zone within it. This model is then used to calculate the relationships between the known sample values and the unknown block values and thus to produce the best possible estimate of this block value.

The Semivariogram

Definition: Once a geologic understanding of the deposit has been obtained and preliminary nonspatial statistics have been completed, more complex statistical methods, that take into account the relative position of the sample values, must be used.

It is obvious that samples located a few meters (feet) apart are likely to indicate similar properties of the mineralization, and that this similarity would be expected to de-

crease when the distance between holes increases. In some deposits, it may be possible to define a distance of influence beyond which the similarity between sample values becomes negligible. However, this distance may vary significantly between deposits, between different parts of the same deposit, or even between different directions at the same point in the deposit. This degree of similarity or dissimilarity between values can be quantified and can be of great assistance in understanding the properties of the deposit and optimizing its evaluation.

The simplest method of comparing two sample values is to calculate the difference between them: the smaller the difference the more similar the values. Generally, when the distance between samples increases, the difference between values increases. Rather than working with differences between sample values that can be positive or negative, the geostatistician uses the average squared difference, also known as the *variogram*. The value of the *semivariogram* $\gamma(h)$ for a distance h between samples is equal to *one-half* the average squared difference between sample values. The one-half factor is used such that the values of the semivariogram are equivalent to variances.

The general formula for calculation of the semivariogram is:

$$\gamma(h) = \frac{1}{2n} \sum_{i=1}^{n} (x_i - x'_i)^2$$

where n is the number of pairs of samples a distance h apart, x_i is the value of first sample in ith pair, and x'_i is the value of second sample in ith pair.

This definition is best explained through an example:

A bedded deposit has been intersected by ten drill holes located on a straight line as shown in Fig. 15. Most drill holes are 1000 ft (304 m) apart and the thickness of the deposit has been measured at each point. The semivariogram value for a distance of 1000 ft (304 m) between holes is calculated as follows:

For a distance of 2000 ft (608 m) between holes the following results are obtained:

A similar calculation for 3000 ft (912 m) gives a result of 15.67. These results are plotted in Fig. 15.

Calculations: *One Dimensional*—The example given in the previous section illustrates how the semivariogram can be calculated in a one-dimensional environment. If channel samples were taken at regular intervals along parallel drifts, the semivariogram will be calculated within each drift as indicated in the foregoing example and then averaged between drifts. If a number of vertical drill holes were drilled and samples of equal length and diameter were taken and assayed using identical methods, an average downhole semivariogram can be calculated. If the drill holes are located on a regular grid, between-hole semivariograms can be calculated in the same fashion.

If the samples are not of equal length, they should be first grouped or divided to form composites of equal length. As indicated earlier, statistics calculated on values corresponding to *supports* of variable size are usually meaningless.

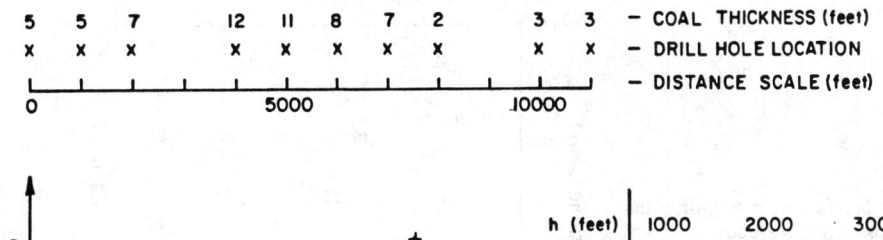

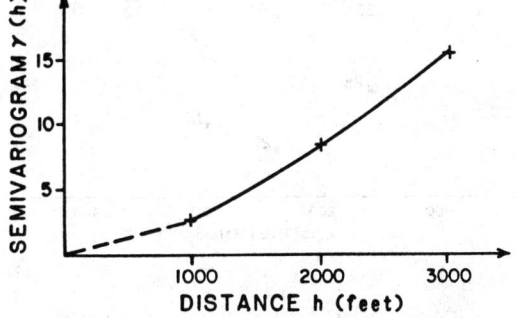

Fig. 15. Calculation of semivariogram.

h (feet)	1000	2000	3000
γ (h)	2.86	8.17	15.67

If a number of drill-hole sizes, sampling methods, or assaying methods were used, each data set should be statistically analyzed separately. If the statistics indicate no significant difference, it may be possible to combine different types of data in the semivariogram calculation.

In many instances, it may be necessary to use sample values that are not strictly comparable. This is commonly the case when a deposit has been analyzed over a number of periods by different operating companies using different types of equipment. Minor differences such as in drill-hole diameters can often be ignored. But chip samples taken along a drift should not be combined with core samples taken ahead of drifting.

Two-Dimensional—When a vein or bedded deposit is analyzed, the values of interest can often be treated two-dimensionally. The information supplied by each drill-hole intercept is reduced to an ore thickness and an average grade between hanging wall and footwall. The grade values can be analyzed directly or alternatively the product of grade times thickness (the accumulation) can be calculated. The thickness analyzed may be either the true thickness or the thickness measured normal to a reference plane, this being either a horizontal or vertical plane or a dipping plane parallel to the deposit. If, in the reference plane case, the drill holes are approximately located on a regular grid the semivariogram can be calculated using a grid search as shown in Fig. 16a.

The direction and distance between samples whose value is to be compared, is specified by a number of blocks n_i in the east direction and n_j in the north direction. Fig. 16a shows the pairs of samples to be considered in the following directions: east-west, north-south, northeast-southwest, and north 33.7° east ($n_i/n_j = 2/3$).

This grid search method is also applicable when a very large number of sample values is available for analysis, as may be obtained by blasthole sampling or underground channel sampling for grade control. The sample values can then be averaged within blocks of equal size and the semivariogram is calculated between the block averages. Compared with other methods of semivariogram calculations, the grid search method is remarkably inexpensive.

The method commonly used to calculate the semivariogram in a two-dimensional environment is illustrated in Fig. 16b and uses an angular search. The value of the semivariogram is to be estimated in a direction α for a distance h between samples. If the reference plane is horizontal, α will represent the azimuth. Two samples with value x_1 and x_2 will be considered in this estimation, provided (1) the direction defined by these two samples is greater than $\alpha - d\alpha$ and less than $\alpha + d\alpha$, and (2) the distance between samples is greater than $h - dh$ and less than $h + dh$. $d\alpha$ and dh are the tolerance angle and distance that should be kept small whenever possible. The result obtained is an estimate

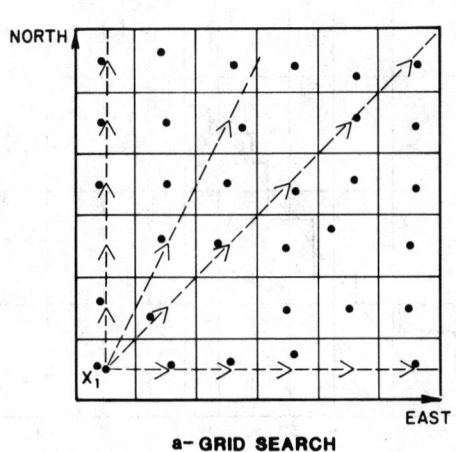

a- GRID SEARCH

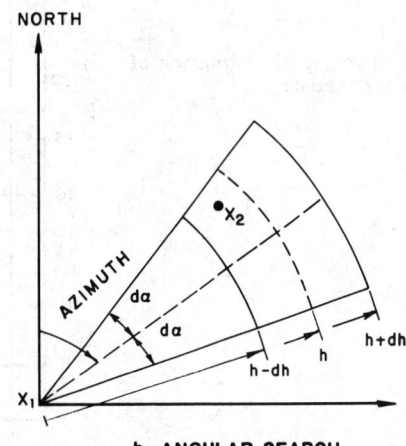

b- ANGULAR SEARCH

Fig. 16. Two-dimensional search.

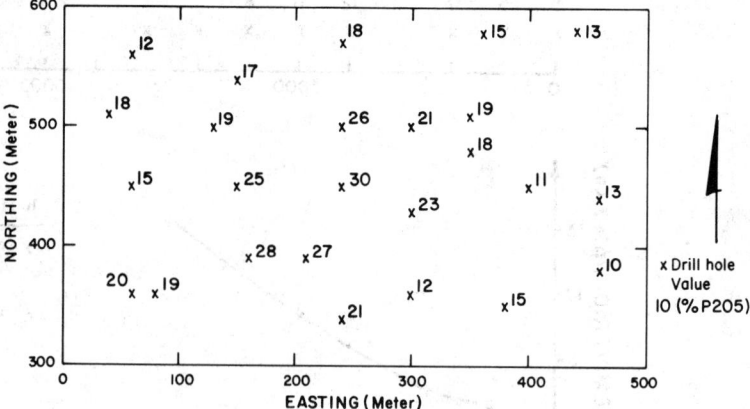

Fig. 17. Drill hole location map and drill hole values, test case ore body.

of the average value of the semivariogram when the distance between samples varies from $h - dh$ to $h + dh$ and the direction varies from $\alpha - d\alpha$ to $\alpha + d\alpha$. The average distance between the samples used to estimate the semivariogram should always be calculated, as it may be significantly different from the nominal distance h if the sample pattern is highly irregular. The average direction corresponding to the pairs of samples retained, can also be used in graphical representations.

Consider the 25 sample values given in Fig. 17. Their cumulative frequency distribution is shown in Fig. 18. It differs only slightly from a normal distribution. The semivariogram has been calculated in the following directions: north-south, northeast-southwest, east-west, northwest-

southeast. An angular tolerance of 22.5° was used. The distances considered were grouped in classes of 75 m (246 ft). The results obtained are given in Table 3. For example, in the north-south direction, for a nominal distance of 37.50 m (123 ft), 11 pairs of samples were retained. The average distance between these samples was 59.32 m (194.69 ft) in direction north 1.8° west. The calculated value of the semivariogram was 14.59. These results are plotted in Figs. 19 and 20. Fig. 19 is a plot of $\gamma(h)$ as a function of the distance h. It shows that the lowest level of variability is in the north direction, while the highest level of variability is in the east or northwest direction. In all directions, the semivariogram shows first an increase in value followed by a sharp decrease. This could reflect periodicities or symmetry in the distri-

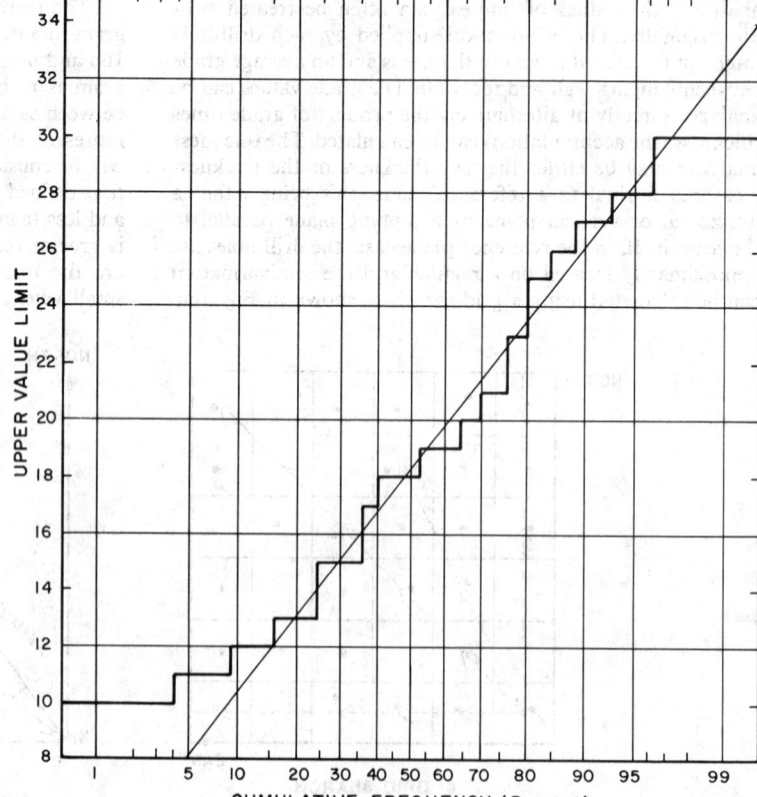

Fig. 18. Cumulative frequency distribution of test case data.

Table 3. Experimental Semivariogram of Test Case Data

Direction (Azimuth)		Distance Class, m*					
		0 to 75	75 to 150	150 to 225	225 to 300	300 to 375	375 to 450
North-south	Number of pairs	11	16	15	4	—	—
	Average distance, meter	59.32	117.39	175.19	231.65	—	—
(0° + 22.5°)	Average direction, °	−1.8	0.2	0.4	6.2	—	—
	Semivariogram	14.59	17.37	20.77	2.75	—	—
Northeast-southwest	Numbers of pairs	3	21	20	14	8	2
	Average distance, meter	60.84	116.58	188.96	262.52	332.38	430.50
	Average direction, °	32.7	46.3	48.0	49.2	57.2	59.3
(45° + 22.5°)	Semivariogram	6.33	14.86	32.90	37.36	29.69	21.25
East-west	Number of pairs	8	23	37	22	22	12
	Average distance, meter	52.77	106.55	188.43	257.40	324.14	404.56
(90° + 22.5°)	Average direction, °	92.5	86.8	91.7	90.8	91.9	89.5
	Semivariogram	10.87	31.00	44.32	46.77	31.77	17.29
Northwest-southeast	Number of pairs	2	22	20	12	3	3
	Average distance, meter	58.31	114.29	190.47	264.92	320.48	399.05
	Average direction, °	135.0	130.7	137.5	129.2	122.4	117.6
(135° + 22.5°)	Semivariogram	21.25	36.64	48.85	38.96	14.17	3.67

* To convert meters to feet, multiply by 3.281.

bution of the sample values. Symmetry is clearly visible in Fig. 17 and the semivariogram reflects the shape of the high grade central zone apparent on this figure.

To determine the direction of best continuity, it is often helpful to plot the semivariogram in the form of isovariogram curves (Fig. 20). These curves show the distance at which the semivariogram is expected to reach a given value in a given direction. The isovariogram curves are elongated in the direction showing the best continuity in mineralization.

In the test case under study, the best direction of continuity is approximately north 25° east, and the worst direction north 65° west. Additional semivariograms have therefore been calculated in these directions, also with a tolerance angle of 22.5° and distance classes of 75 m (246 ft). The results obtained are given in Fig. 21. This figure also

shows a model that was used to represent the semivariogram, derivation of which will be discussed later. The least variability again appears in the north 25° east direction that corresponds to the axis of the ore shoot. The highest level of variability is in the direction north 65° west. The bell shape shown by the semivariogram in the latter direction reflects the presence of an ore zone approximately 200 m (656 ft) wide with low-grade material on both sides.

Three-Dimensional—Consider now a large porphyry copper ore body sampled by 200 vertical drill holes. Typically, if the deposit is to be mined using 12-m (40-ft) benches, the samples will be grouped in 12-m (40-ft) long composites. Within reasonably homogenous geological environments, an average downhole semivariogram will be calculated using the one-dimensional approach discussed earlier. Also, horizontal

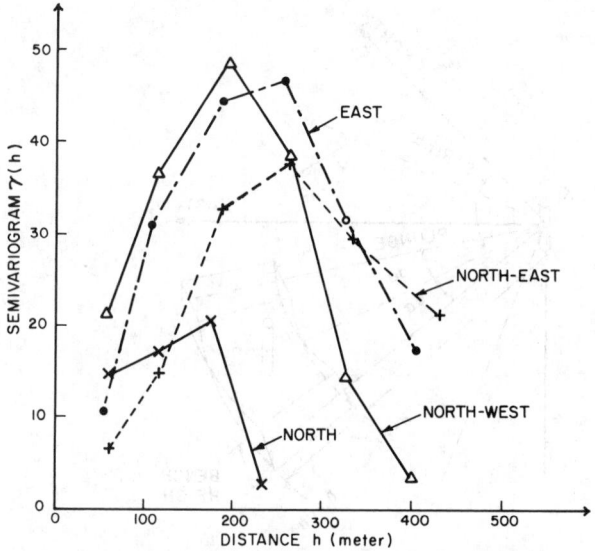

Fig. 19. Directional semivariograms in test case ore body—one-dimensional representation.

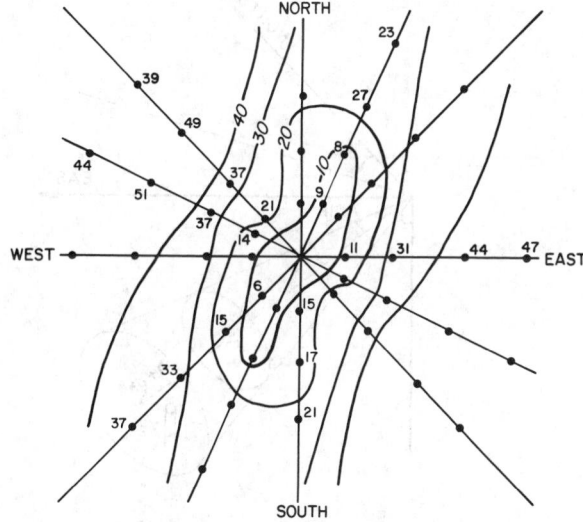

Fig. 20. Directional semivariogram in test case ore body—two-dimensional representation.

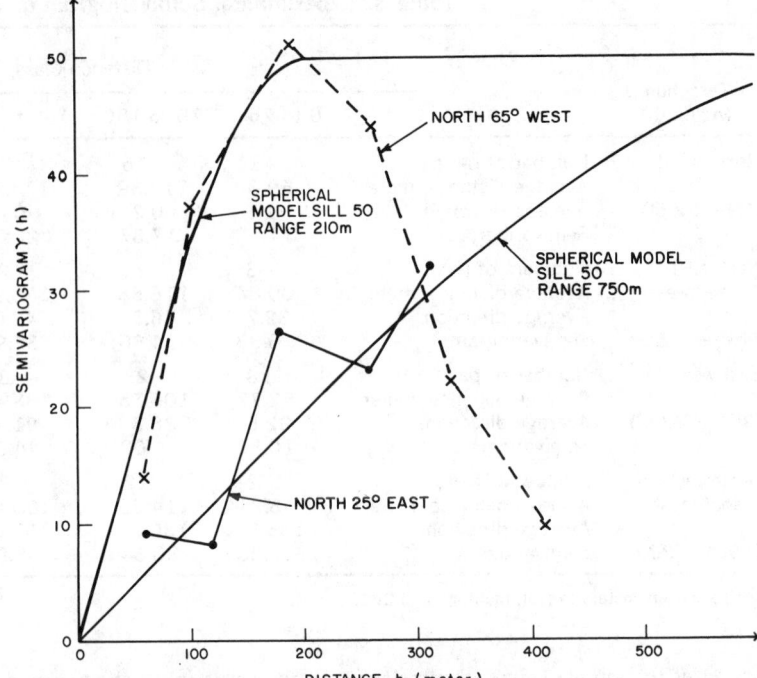

Fig. 21. Calculated semivariogram and fitted model in principal directions—test case ore body.

semivariograms will be calculated within each bench and averaged between benches. In such a case this preliminary semivariogram study will be used to determine the direction of best continuity in the horizontal plane and to compare the horizontal and vertical directions. The results obtained may be useful in the development of exploration drilling, mine design, and grade control strategies. Unfortunately, a semivariogram study that considers only horizontal and vertical directions is not based on any geological foundation and can lead to incorrect conclusions. Except for a small number of significant exceptions, geological controls have no reason to be preferentially directed in a horizontal fashion.

The method most commonly used to calculate semivariograms in a three-dimensional environment is based on a conical search (Fig. 22a). The direction of interest is defined by an azimuth and dip. A tolerance angle is specified around this direction and the pairs of samples are grouped according to distance. The directions to be considered are chosen taking known geologic controls on the mineralization into account, wherever possible. The conical search is particularly helpful in large disseminated deposits, such as porphyries. Its main drawback, however, is that it usually requires considerable computer resources. It is also poorly adapted to the analysis of stratified, bedded, or fault-related deposits, or any type of deposit in which the controls of mineralization are essentially planar. For example, consider an ore body whose economic mineralization is fracture controlled and the major fracture zones are striking east and dipping 60° north. The directions

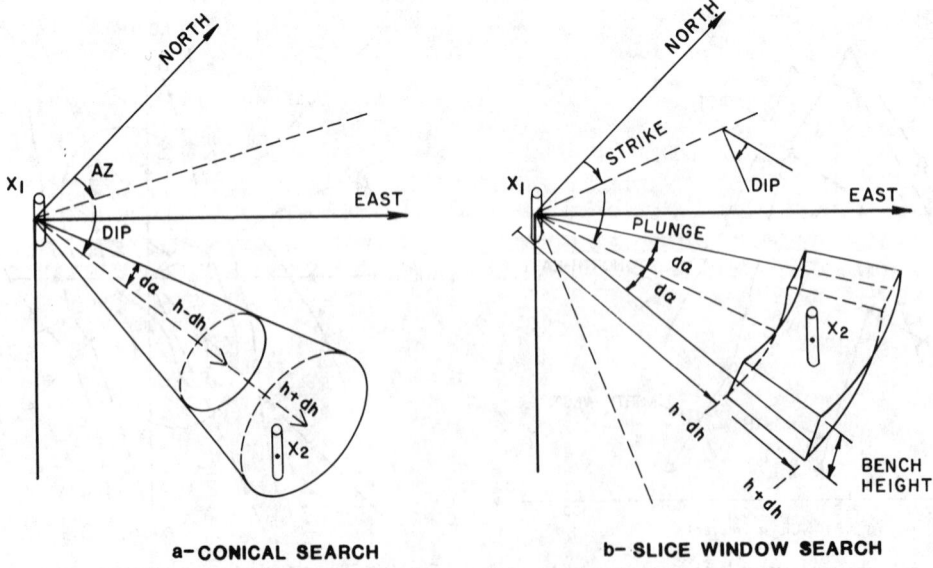

a- CONICAL SEARCH   b- SLICE WINDOW SEARCH

Fig. 22. Three-dimensional search.

to be considered will be (1) one direction normal to the fracture plane orientation, i.e., dipping 30° south, and (2) a number of directions within the fracture plane.

The anisotropy of the semivariogram will then be defined by three principal directions:

1) The direction of best continuity that would normally correspond to the plunge direction within the fracture plane orientation.

2) The direction normal to plunge within the fracture plane.

3) The direction of least continuity which is typically normal to the fracture plane.

A method of semivariogram calculation particularly well adapted to the analysis of mineralization with planar controls is the "slice" window search illustrated in Fig. 22b. The method initially involves preparation of the data so as to assign the bench composites to parallel artificial slices through the deposit, whose strike and dip is user-controlled and is chosen as a function of the known geology. The semivariogram is then calculated within these inclined slices. The technique has the advantage of involving minimal computer time compared with the conical search.

This approach is only applicable to mineralization with planar control. If there is only a limited geologic understanding of the deposit and the direction of the ore shoots is unknown, one can calculate the bench semivariogram in three perpendicular planes (horizontal, vertical east-west, and vertical north-south). A model of a semivariogram is then fitted in each plane. These planar models are then combined to obtain a three-dimensional mathematical representation that can be used to determine the likely principal directions of continuity (Rendu, 1983). Clearly, this approach should be used in last recourse, and only if geologic understanding is minimal.

An example of three-dimensional anisotropic semivariogram obtained in a massive sulfide deposit with well defined geologic controls is shown in Fig. 23. Semivariogram models are shown in this figure whose derivation is discussed in the next section.

Semivariogram Modeling: Once a semivariogram has been calculated, it must be interpreted by fitting to it a mathematical formula or "model" that will help to identify the characteristics of the deposit and yield numerical parameters that describe the deposit's continuity. Examples of such models are shown in Fig. 24. To the practiced eye, each model has a particular significance.

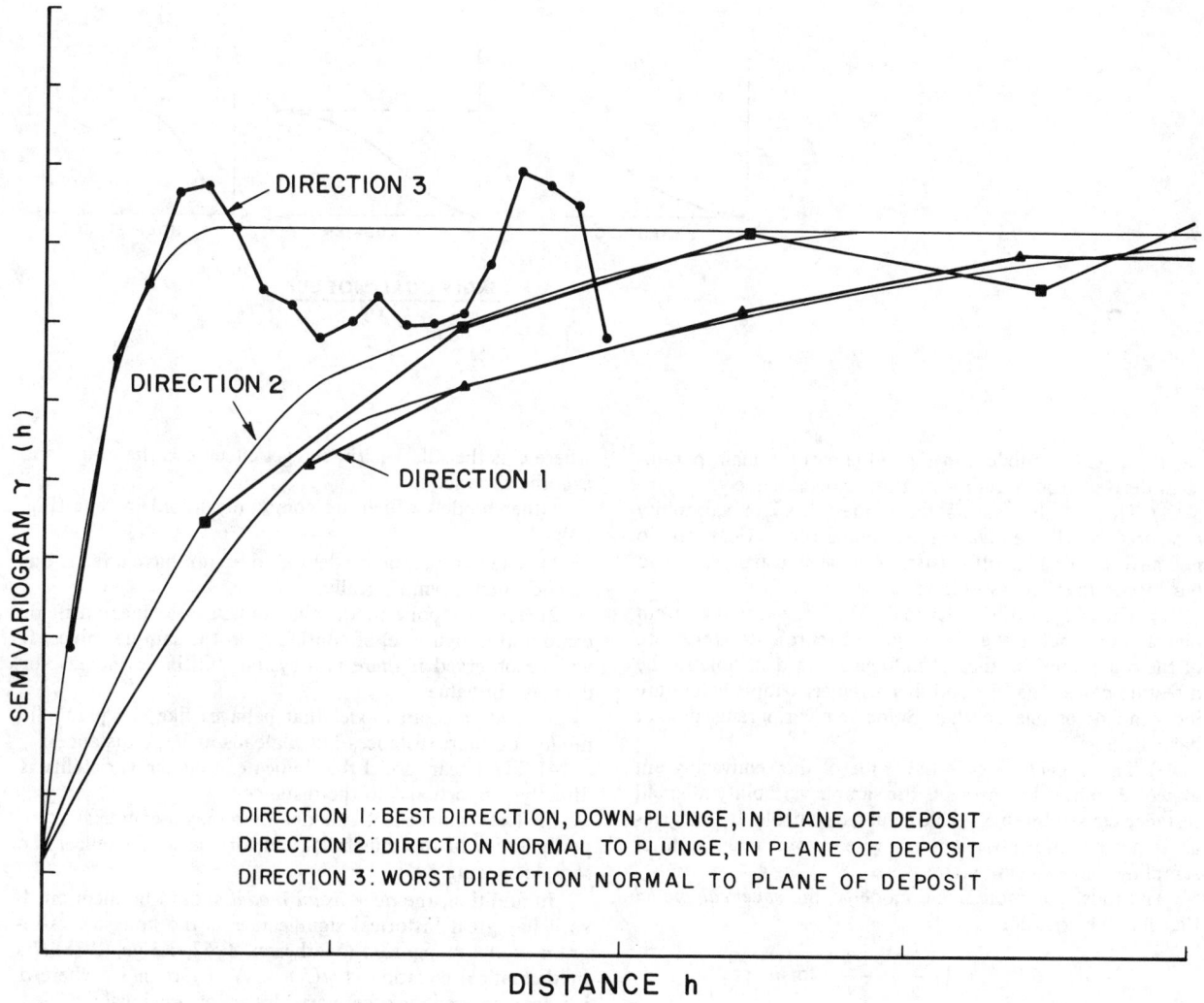

DIRECTION 1: BEST DIRECTION, DOWN PLUNGE, IN PLANE OF DEPOSIT
DIRECTION 2: DIRECTION NORMAL TO PLUNGE, IN PLANE OF DEPOSIT
DIRECTION 3: WORST DIRECTION NORMAL TO PLANE OF DEPOSIT

Fig. 23. Three-dimensional anisotropy in massive sulfide deposit.

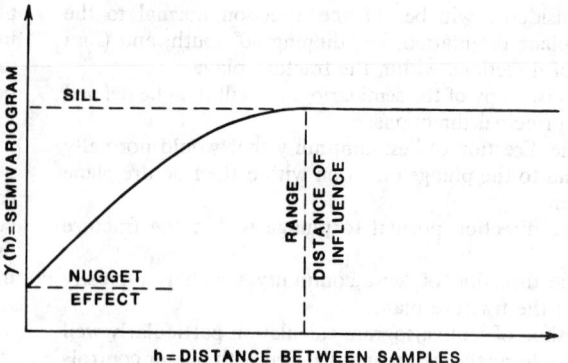

GENERAL MODEL

Fig. 24. Models of semivariograms.

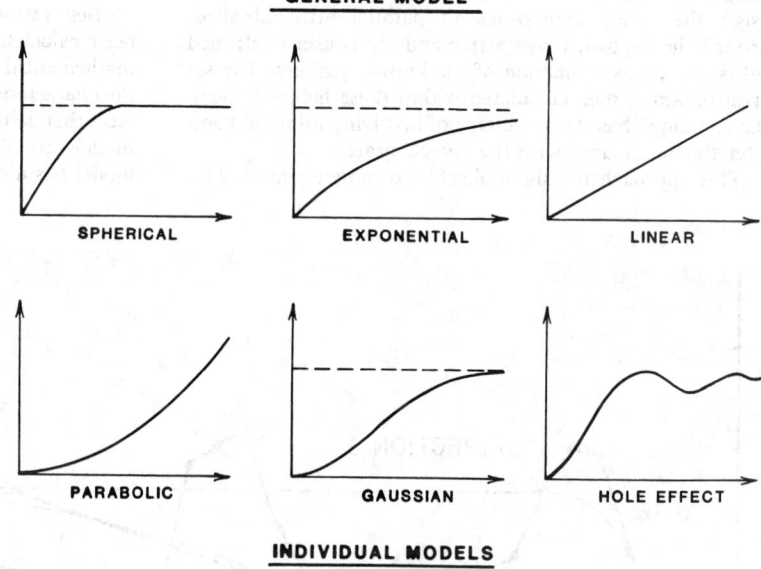

INDIVIDUAL MODELS

The "general model" in Fig. 24 shows the main parameters derived from a semivariogram model, namely:

1) The *sill* that shows the highest level of variability measured by the semivariogram. Some semivariograms do not have a sill. If a sill exists, its value is usually close to but higher than the sample variance.

2) The *range* is the distance at which the semivariogram plateaus or reaches the sill value and represents a measure of the maximum distance of influence of a drill hole in the direction concerned. Beyond this distance, sample values are independent of one another. Some semivariograms do not have a range.

3) The *nugget effect* is the value of the semivariogram at zero distance. It represents the sample variability at small distance caused by small-scale geologic controls. It also gives an important indication of the presence and magnitude of sampling and assaying errors.

The most common single model is the *spherical model* that has the equation:

$$\gamma(h) = N + C\left(\frac{3}{2}\frac{h}{a} - \frac{1}{2}\frac{h^3}{a^3}\right) \text{ for } h \le a$$

$$\gamma(h) = N + C \qquad\qquad\qquad \text{ for } h > a$$

where C is the sill, N is the nugget effect, a is the range, and h is the distance lag.

Other models which are commonly found include (Fig. 24):

1) The exponential model that does not have a range but reaches a sill asymptotically.

2) The parabolic model that indicates a linear drift or trend and a high level of continuity in the sample values. It will be observed if there is a systematic linear increase or decrease in values.

3) The gaussian model that behaves like the parabolic model for short distances but plateaus at large distances.

4) The linear model that indicates that the variability is directly proportional to the distance.

5) The "hole effect" model that may be indicative of periodicities in the mineralization, or may only reflect the sample spacing.

In addition, the *de Wijsian model* should be mentioned, as it has great historical significance as the first semivariogram model recognized (Matheron, 1962; Krige, 1978). Its mathematical equation is $\gamma(h) = N + 3\alpha \ln(h)$ where α is a constant and the other variables are as previously defined.

Most semivariograms can be represented by the sum of

a number of submodels, including a nugget effect, a short-range submodel, and a long-range submodel. With the exception of the parabolic, gaussian and hole-effect models, a combination of spherical submodels can be used to represent most other calculated or so called experimental semivariograms.

The *parabolic model* has equation:

$$\gamma(h) = N + \frac{d^2}{2} h^2$$

If the distance between samples is measured in meters, this equation indicates a linear drift in the direction under study; the sample values increasing or decreasing by an amount equal to d units per meter. Parabolic models are commonly observed when highly continuous variables are studied, such as vein or seam thicknesses. They also result when systematic increases or decreases in grade occur, as illustrated by the downhole semivariogram in the supergene sulfide mineralization of a porphyry copper deposit (Fig. 25).

The *gaussian model* is also indicative of a local drift, but over a shorter distance than the one indicated by the parabolic semivariogram. The equation of the gaussian model is:

$$\gamma(h) = N + C [1 - \exp(-h^2/a^2)]$$

For short distances less than $a/2$, the model indicates that the sample values tend to increase or decrease by an amount equal to d units per meter, where:

$$d^2 = 2C/a^2$$

A gaussian model with nugget effect was used to represent the semivariogram of a coal seam shown in Fig. 26. The

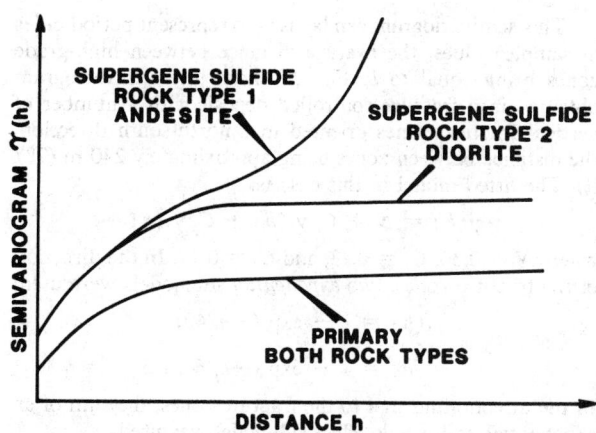

Fig. 25. Down hole semivariograms in porphyry copper deposit.

nugget effect has value of 0.8 and in the north-south direction, the gaussian model has sill $C = 8.0$ and range $a = 1800$ m (5906 ft). The model indicates a regular change in thickness in this direction of $d = 2.22$ m per 1000 m (7.3 ft per 3281 ft).

A number of equations have been proposed for semivariograms which present a *hole effect*. One possible equation is:

$$\gamma(h) = N + C \left[1 - \exp\left(\frac{-h}{a}\right) \cdot \cos\left(2\pi \cdot \frac{h}{b}\right) \right]$$

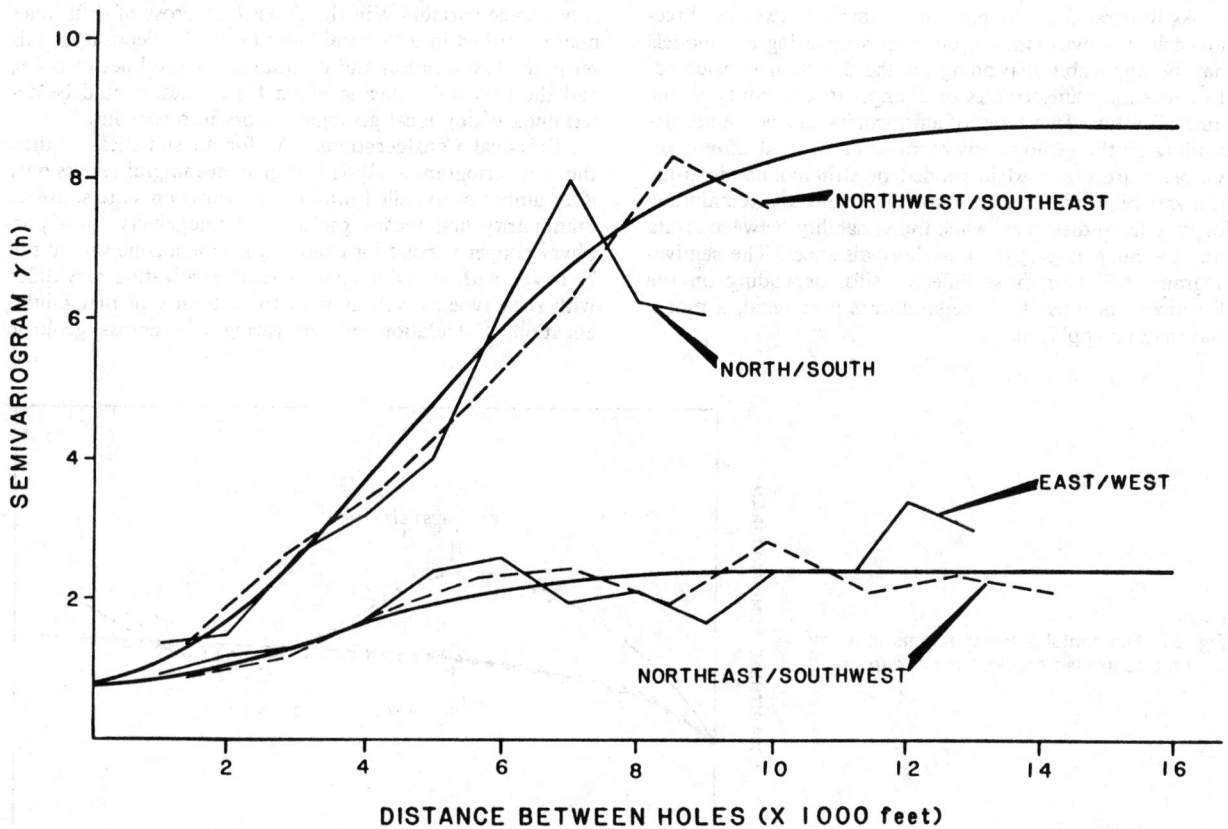

Fig. 26. Semivariogram of coal thickness.

This semivariogram can be used to represent periodicities in sample values, the average distance between high-grade zones being equal to b. Fig. 27 shows the semivariogram obtained in a fracture-controlled deposit with a number of vertical fracture zones oriented in a north-south direction, the distance between zones being approximately 240 m (787 ft). The fitted model in this case is:

$$\gamma(h) = N + C_1 \gamma_1(h) + C_2 \gamma_2(h)$$

where $N = 0.30$, $C_1 = 0.23$, and $C_2 = 0.17$. In the direction of the fracture zones, two *exponential* submodels were used:

$$\gamma_1(h) = 1 - \exp(-h/40)$$

$$\gamma_2(h) = 1 - \exp(-h/600)$$

In the direction normal to the fracture zones, the sum of an exponential and a hole effect submodel was used:

$$\gamma_1(h) = 1 - \exp(-h/15)$$

$$\gamma_2(h) = 1 - \exp(-h/1150) \cos 2\pi (h/240)$$

The first submodel $\gamma_1(h)$ represents the short-range continuity that is clearly much better in the direction of the fracture zones. The second submodel $\gamma_2(h)$ represents the long-range variability and indicates that the average distance between fracture zones is 240 m (787 ft). If drilling is intended to discover new fracture zones for exploration purposes, the optimal drill-hole spacing in the east-west direction is 240 m (787 ft). An equal or larger spacing can be used in the north-south direction. On the other hand, if drilling is intended for grade control and mine planning, the spacing between holes should reflect the short-range variability, with a possible spacing of 40 m (131 ft) in the direction of the ore zones and 15 m (49 ft) in the perpendicular direction.

As illustrated in the previous example in two- or three-dimensional environments, different semivariogram models may be applicable depending on the direction considered, thus reflecting anisotropies or changes in continuity in the mineralization. Two types of anisotropies can be found, depending on the geologic environment considered. *Zonal anisotropies* are observed in bedded or stratabound deposits. The variability within a stratum may be low and remain low for very large distances, while the variability between strata may be much larger, even at short distances. The semivariograms will then have different sills, depending on the directions considered. If the strata are horizontal, a model that may be applicable is:

$$\gamma(h) = \gamma_1(x, y) + \gamma_2(z)$$

where x, y, and z represent the easting, northing, and vertical components of the distance h between samples.

A more common form of anisotrophy is the *geometric anisotropy*. If such an anisotropy is present, the sill remains the same in all directions, but the range is variable. Often the range in any direction can be represented by the radius of an ellipse or ellipsoid. For example, consider a deposit with ore zones elongated in the north 60° east direction (azimuth $\alpha_0 = 60°$); the semivariogram being spherical with range $a = 600$ m (1969 ft) in this direction and $b = 400$ m (1312 ft) in the perpendicular north 30° west direction. The range d of the semivariogram in any direction with azimuth α will be calculated as follows:

$$d = \sqrt{1 \Big/ \left(\frac{\cos^2 (\alpha - \alpha_0)}{a^2} + \frac{\sin^2 (\alpha - \alpha_0)}{b^2} \right)}$$

If nested structures are present, neither the direction nor the ratio of anisotropy need be the same for each structure. An example is illustrated in Fig. 28. Structures "en echelon" are clearly visible, the ore pods being oriented in a roughly north 20° south direction while the mineralized areas containing these pods are elongated in a north-south direction. If only a one dimensional representation of the semivariogram had been used (Fig. 29) the different structures would not have been detected.

In most deposits, the sampling density will be such that the short-range variability, including the nugget effect, will be difficult to estimate. In three-dimensional bodies, such as porphyry deposits, the nugget effect can be determined using the downhole semivariogram. In many instances, some limited close spaced drilling will be required to determine the short-range variability in the deposit. A cross of drill holes may be drilled in a "typical" section of the deposit for this purpose. The location and orientation of the lines of holes, and the drill hole spacing along these lines, should be determined taking local geologic factors into account.

Practical Considerations: As for all statistical studies, the semivariogram analysis will give meaningful results only if a number of so called *stationarity* constraints are satisfied. Stationarity first means geological homogeneity. In a porphyry copper deposit for example, a sulfide zone should not be mixed with an oxide zone. The mineralization may differ with rock type as well as with the intensity of that mineralization. Calculation of semivariograms across geologic

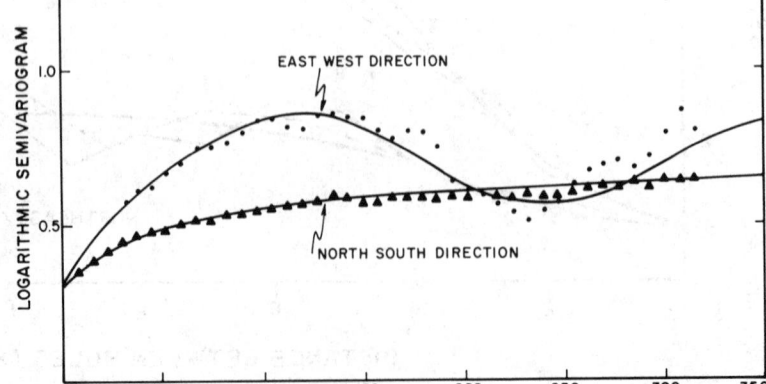

Fig. 27. Horizontal semivariograms in a fracture controlled precious metal deposit.

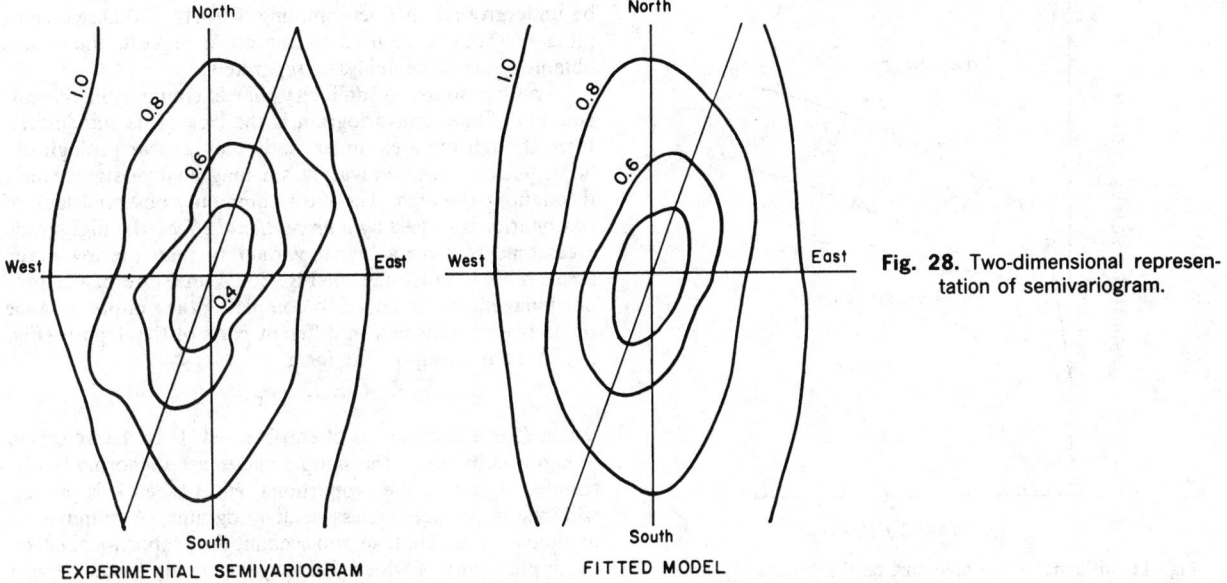

Fig. 28. Two-dimensional representation of semivariogram.

boundaries will give meaningless results. The significance of geologic discontinuities such as faults on the semivariogram should be obvious. Fig. 25 shows the semivariograms obtained in three different geologic environments in a porphyry copper deposit. Fig. 30 illustrates the influence of changes in average grade on the semivariogram in different sections of a precious metal deposit.

A geostatistical study should first start with a thorough review of all geological information concerning the deposit, including rock types, folding, faulting, zones of primary and secondary enrichment, and a model of the genesis of the deposit if available. In most situations, mixing of geologic environments will result in meaningless or erroneous geostatistical results.

The influence of the sample or support size on the semivariogram is similar to the influence of sample size on the sample variance: the larger the sample size, the lower the sample variability. This is illustrated in Fig. 31 where the semivariograms of 2- and 4-m (6.7- and 13-ft) composites are compared. The decrease in variability is known under the name of *regularization* and can be calculated from the semivariogram model. If $\gamma(h)$ is the semivariogram for very small samples and $\gamma_w(h)$ the semivariogram for samples of size $w = 4$ m (13 ft), it can be shown that $\gamma_w(h) = \overline{\gamma}(w; w_{+h})$

$- \overline{\gamma}(w; w)$ where the quantities $\overline{\gamma}(w; w_{+h})$ and $\overline{\gamma}(w; w)$ are a function of the sample size w, the distance h, and the semivariogram $\gamma(h)$ (Rendu, 1981). This mathematical relationship can be extremely useful to verify a semivariogram model. The difference between the semivariogram $\gamma_w(h)$ for small samples of size w and the semivariogram $\gamma_W(h)$ for large samples of size W, is approximately equal to:

$$\gamma_w(h) - \gamma_W(h) = \overline{\gamma}(W; W) - \overline{\gamma}(w; w)$$

This relationship is also used to determine the likely effect of different selective mining methods on the grade tonnage relationship. If σ_w^2 is the sample variance and σ_W^2 is the block variance, the difference between variances can be calculated by the same formula:

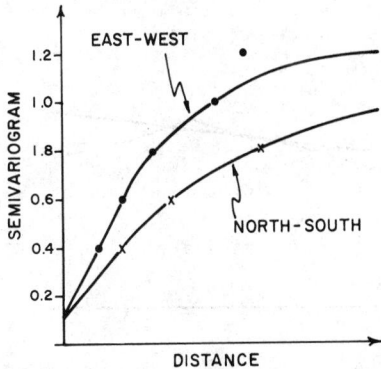

Fig. 29. One-dimensional representation of semivariogram.

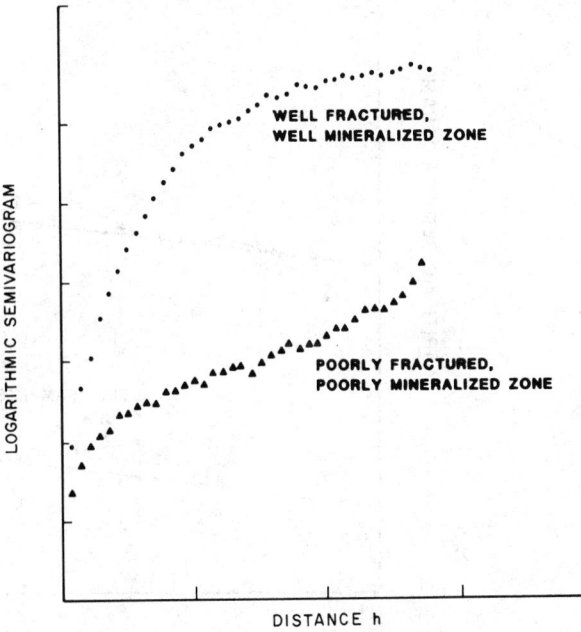

Fig. 30. Logarithmic semivariograms in two zones of a precious metal deposit.

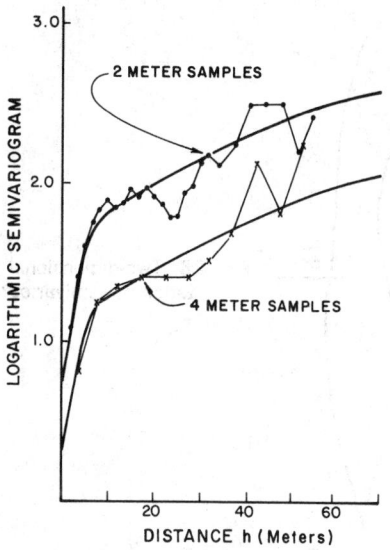

Fig. 31. Influence of sample size on the semivariogram.

$$\sigma_w^2 - \sigma_W^2 = \overline{\gamma}(W;W) - \overline{\gamma}(w;w)$$

The use of this formula will be discussed later.

Differences in sampling methods will also result in differences in the semivariogram. In Fig. 32 the results obtained in a tin deposit are illustrated. Two types of samples were analyzed, core samples and bulk samples taken from successive rounds along horizontal drifts. The semivariograms shown in Fig. 33 represent the properties of the thickness of a precious metal vein deposit, as measured by drilling and

by underground channel sampling. Clearly, if different sampling methods were used to analyze a deposit, the values obtained should be analyzed separately.

Another source of difficulty that is common in the calculation of the semivariogram is the lack of its *stationarity*. Even though the area under study may appear geologically homogeneous, the semivariogram may change significantly throughout the area. The most commonly observed lack of stationarity is caused by a *proportional effect:* the high-grade areas tend to have a higher variability than the low-grade areas. This is illustrated in Fig. 30. A measure of the proportional effect is obtained by comparing the sample variance σ^2 and sample mean $\overline{x}$ in different parts of the deposit (Fig. 34). A relationship of the form:

$$\sigma^2 = K \overline{x}^2$$

where K is a constant, is often observed. It is characteristic of deposits in which the sample values are lognormally distributed. Ignoring the proportional effect when it is present will result in meaningless semivariograms. A number of methods are used to take into account the proportional effect, the applicability of which will vary depending on the deposit under study. These methods include:

1) Division of the deposit into "grade zones" or zones of fairly uniform grade, within which the proportional effect can be ignored.

2) Calculation of a relative semivariogram. The relative semivariogram is $\gamma(h)/\overline{x}^2$ where $\overline{x}$ is the mean value of the samples used to calculate each point of the semivariogram $\gamma(h)$.

3) Calculation of the logarithmic semivariogram. If the sample values are approximately lognormally distributed, the semivariogram of the natural logarithms of the sample values will be independent of the sample grade.

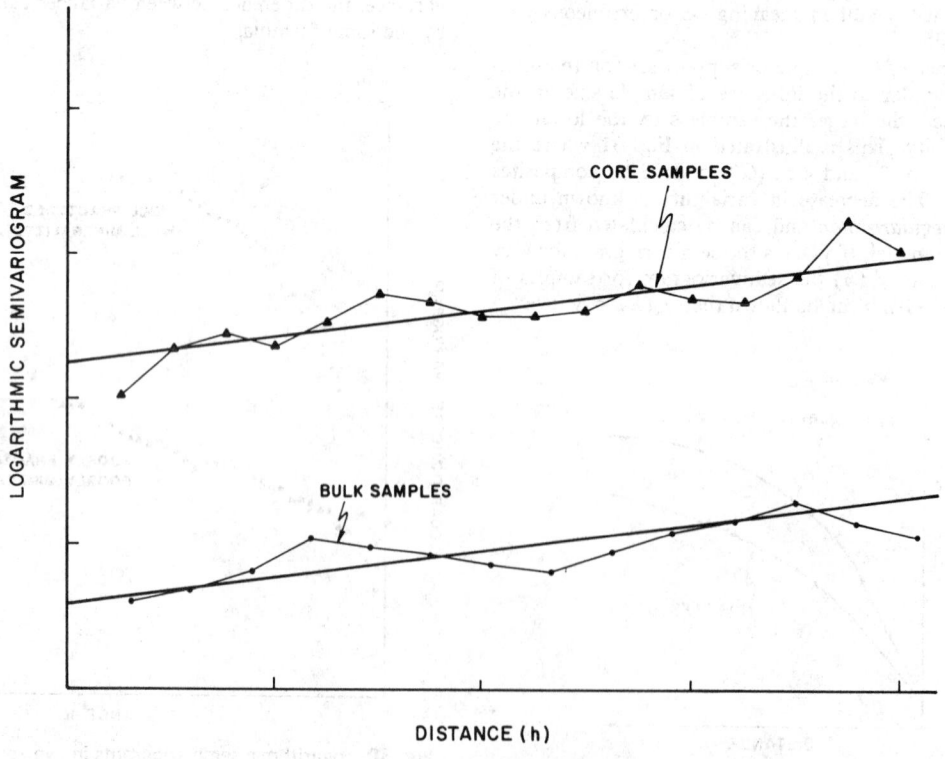

Fig. 32. Comparison of semivariograms in a tin deposit.

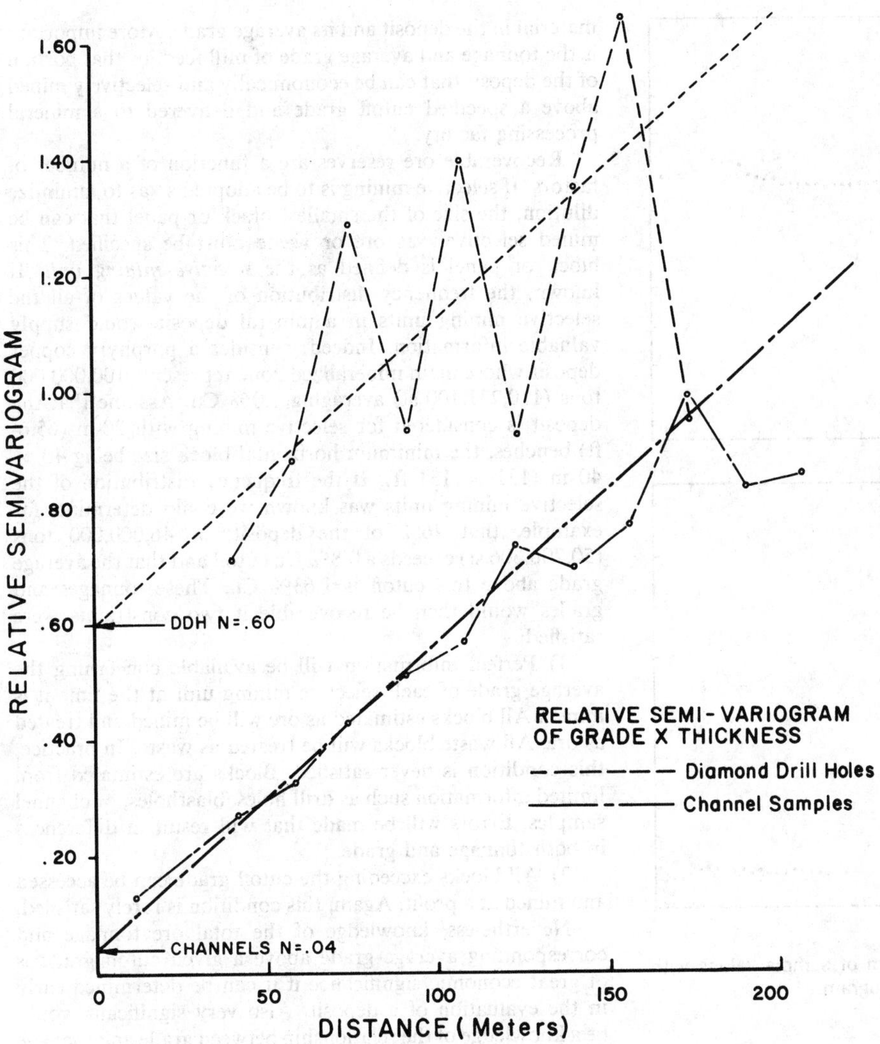

RELATIVE SEMIVARIOGRAM

DDH N=.60

RELATIVE SEMI-VARIOGRAM
OF GRADE X THICKNESS

– – – – – Diamond Drill Holes
— - — - — Channel Samples

CHANNELS N=.04

DISTANCE (Meters)

Fig. 33. Comparison of diamond drill hole and channel sample semivariograms in a silver vein deposit.

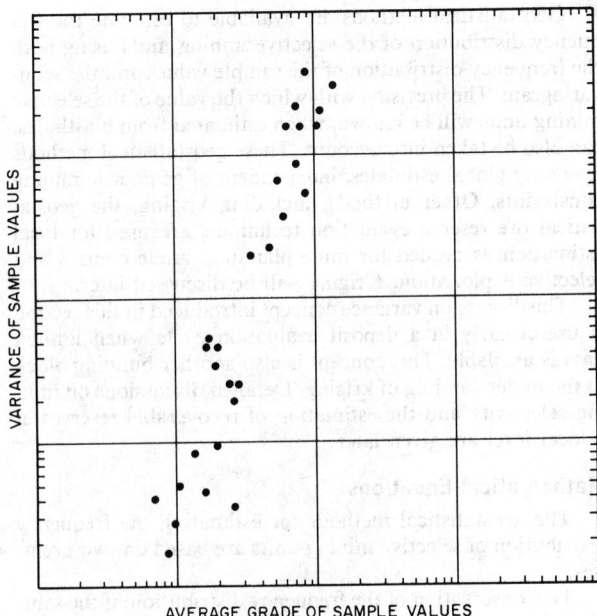

VARIANCE OF SAMPLE VALUES

AVERAGE GRADE OF SAMPLE VALUES

Fig. 34. Proportional effect in a uranium deposit.

Fig. 35 shows the average downhole semivariogram calculated in a precious metal deposit, first using untransformed sample values, then using the logarithm of the sample values. This figure dramatically illustrates the impact of the logarithmic transformation of the data prior to semivariogram estimation.

Some Applications of the Semivariogram: The semivariogram study is one of the most important parts of any geostatistical study. The semivariogram is a prerequisite to the use of the kriging method of ore reserve evaluation. In addition, the semivariogram may supply information concerning:

1) The direction of best continuity in the mineralization.
2) The orientation and size of high-grade pods.
3) The average spacing between ore zones and periodicities in the mineralization.
4) The strike and dip of fracture zones and the plunge of the mineralization within these zones.
5) The distance of influence of a sample or the distance beyond which the similarity between values is negligible.
6) The precision with which the mineralization is known at the point where it is sampled.

Using this information, answers can be given to some critical questions such as:

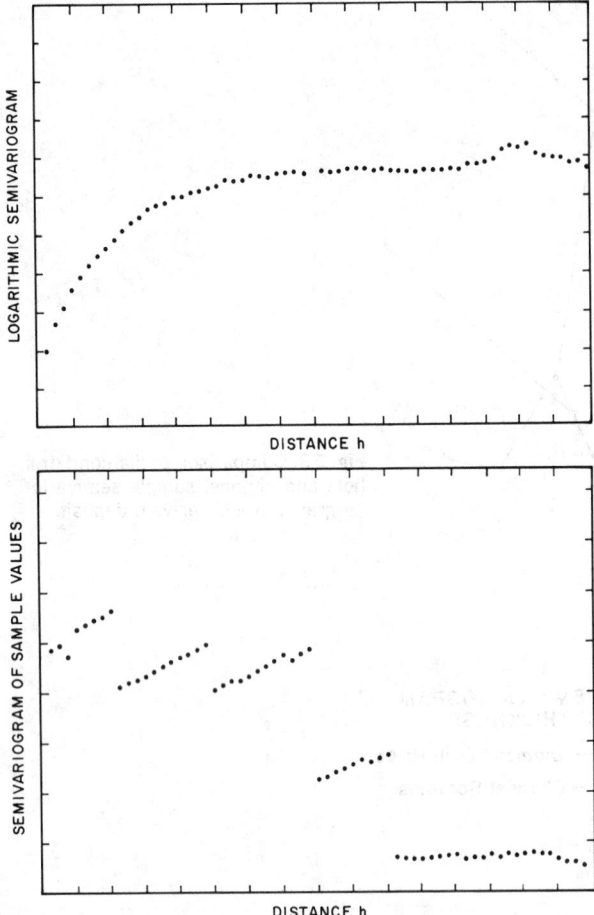

Fig. 35. Comparison of semivariogram of sample values with logarithmic semivariogram.

1) What is the optimal drill hole orientation and drill hole spacing required to complete a drilling program?

2) What will be the effect of a highly selective mining method, as opposed to a large scale bulk mining method, on the tonnages and grades mined?

3) What sampling method, including sample spacing, should be used for grade control during production?

Hypotheses concerning the genesis of the ore body can be tested and conclusions can be drawn concerning future drilling to find new deposits in the same district or new ore zones in the same deposit.

In summary, there are very few deposits where the process of exploration and evaluation cannot be enhanced through semivariogram analyses. In addition to an improved understanding of the geology of the deposit such analysis provides a quantitative model of the variability of the mineralization that can be directly used as a basis for subsequent reserve calculations.

DISPERSION VARIANCE AND GRADE TONNAGE CURVE

Introduction

Some degree of mining selectivity is adopted in the exploitation of most mineral deposits. The economic value of a deposit is only in part a function of the total tonnage of

material in the deposit and its average grade. More important is the tonnage and average grade of mill feed, or that portion of the deposit that can be economically and selectively mined above a specified cutoff grade and delivered to a mineral processing facility.

Recoverable ore reserves are a function of a number of factors. If selective mining is to be adopted so as to minimize dilution, the size of the smallest block or panel that can be mined selectively as ore or waste must be specified. This block or panel is defined as the *selective mining unit*. If known, the frequency distribution of the values of all the selective mining units in a mineral deposit would supply valuable information. Indeed, consider a porphyry copper deposit whose main mineralized zone represents 100,000,000 tons (110,231,100 st) averaging 1.0% Cu. Assume that the deposit is considered for selective mining with 20-m (65.6-ft) benches, the minimum horizontal block size being 40 × 40 m (131 × 131 ft). If the frequency distribution of the selective mining units was known we could determine, for example, that 46% of the deposit, or 46,000,000 tons (50,706,306 st) exceeds a 0.8% Cu cutoff and that the average grade above this cutoff is 1.63% Cu. These tonnages and grades would then be recoverable if two constraints were satisfied:

1) Perfect information will be available concerning the average grade of each selective mining unit at the time it is mined. All blocks estimated as ore will be mined and treated as ore. All waste blocks will be treated as waste. In practice, this condition is never satisfied. Blocks are estimated from limited information such as drill holes, blastholes, or channel samples. Errors will be made that will result in differences in both tonnage and grade.

2) All blocks exceeding the cutoff grade can be accessed and mined at a profit. Again, this condition is rarely satisfied.

Nevertheless, knowledge of the total ore tonnage and corresponding average grade above a given cutoff grade is of great economic significance if it can be determined early in the evaluation of a deposit. Also very significant would be a knowledge of the relationship between grade and tonnage as a function of cutoff grade and of the degree of selectivity with which the deposit will be mined.

Geostatistical methods are available to estimate the frequency distribution of the selective mining units using both the frequency distribution of the sample values and the semivariogram. The precision with which the value of the selective mining units will be known, when estimated from blastholes, can also be taken into account. These geostatistical methods give only global estimates, independent of geometric mining constraints. Other methods, including kriging, the geostatistical ore reserve evaluation technique, are used for local estimation as needed for mine planning, grade control and selective exploitation. Kriging will be discussed later.

The dispersion variance concept introduced in this section is useful early in a deposit evaluation cycle when limited data is available. This concept is also another building block in the understanding of kriging. Detailed discussions on mining selectivity and the estimation of recoverable reserves at a local level are given later.

Mathematical Equations

The geostatistical methods for estimating the frequency distribution of selective mining units are based on two premises:

1) Conservation of the frequency distribution: if the sample values are normally distributed, the block values will also be normally distributed (conservation of normality). If the

sample values are lognormally distributed, so will be the block values (conservation of lognormality).

2) The variance of the block values in the deposit can be calculated from the variance of the sample values and the semivariogram, or from the variance-volume relationship to be discussed later.

The validity of these premises has been proven theoretically for the normal distribution and has been tested extensively using results obtained in various deposits.

The best known relationship between grade and tonnage is probably Lasky's relationship (Lasky, 1950):

$$g_{+c} = a - b \ln (T_{+c})$$

where g_{+c} is the average grade above cutoff grade, T_{+c} is the tonnage above cutoff grade, and a and b are coefficients that vary with both the deposit and the mining method.

This relationship can be derived from the lognormal theory and has been observed in many deposits, from porphyry copper to disseminated gold. The values of the coefficients a and b are a function of the variability of the ore in the deposit and of the degree of selectivity with which it will be mined. They usually have to be determined empirically, using historical data. If geostatistics is used, on the other hand, the relationship between tonnage and grade can be determined early in the study of a deposit, using the sample values. The average grade μ of the deposit can be calculated from the sample values. The dispersion variance of the block values σ_w^2 can also be determined from the variance or the semivariogram of the sample values.

Normal Distribution: If the sample values are normally distributed, then the block values are also normally distributed with mean μ and variance σ_w^2. The tonnage of ore above a given cutoff x_c is given as a function of the total tonnage T, and is estimated by the following formula:

$$\frac{T_{+c}}{T} = F(y)$$

where

$$y = (x_c - \mu)/\sigma_w$$

and

$$F(y) = \frac{1}{\sqrt{2\pi}} \int_y^{+\infty} \exp(-t^2/2) \, dt$$

The average grade of this ore material is calculated as follows:

$$\frac{\mu_{+c}}{\mu} = 1 + \frac{1}{\sqrt{2\pi}} \cdot \frac{\sigma_w}{\mu} \exp(-y^2/2)/F(y)$$

The value of the F function can be obtained from statistical tables of the cumulative normal distribution. Graphical methods can be used to calculate the tonnages and grades above cutoff grade (Fig. 36). The foregoing equations are also easily programmed on a pocket calculator. As an example, consider a bedded deposit, with average thickness of 5 m (16 ft). The variance of the thickness of 50 × 50-m (164 × 164-ft) panels has been estimated at 4 m² (43 sq ft). Assuming a normal distribution, it can be calculated that 84% of the deposit will have a thickness in excess of 3 m (9.8 ft) and will average 5.60 m (18 ft). How the variance of these 50 × 50-m (164 × 164-ft) panels can be calculated will be discussed in the next section.

These results could have been obtained graphically, using Fig. 36. We calculate: block mean to be 5 m (16.4 ft) and block standard deviation to be $\sqrt{4}$ equals 2 m (6.6 ft).

$$X = (\text{cutoff-mean})/(\text{standard deviation}) = (3-5)/2 = -1$$

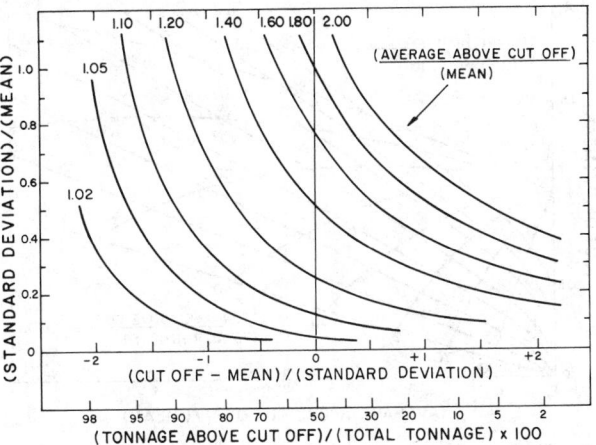

Fig. 36. Determination of tonnage-grade relationship, normal distribution.

Using the horizontal axis we see that indeed approximately 84% of the total tonnage is expected to exceed the cutoff grade. Furthermore:

$$Y = (\text{standard deviation})/(\text{mean}) = 2/5 = 0.4$$

Using these values of X and Y the following ratio is estimated graphically:

$$(\text{average above cutoff})/(\text{mean}) = 1.12$$

The average thickness above cutoff is therefore expected to be 12% above the mean, or 1.12 × 5 = 5.60 m (18 ft).

Lognormal Distribution: For all practical purposes, if the sample values are lognormally distributed, the distribution of the block values will also be lognormal. Even though not mathematically proven, this theorem, known as conservation of lognormality, has been fully verified in numerous deposits.

If the mean value of the deposit is μ and the logarithmic variance of the block values is σ_{ew}^2, then the tonnage and average grade above cutoff grade will be calculated as follows:

$$\frac{T_{+c}}{T} = F\left[\frac{1}{\sigma_{ew}} \ln\left(\frac{x_c}{\mu}\right) + \frac{\sigma_{ew}}{2}\right]$$

$$\frac{\mu_{+c}}{\mu} = F\left[\frac{1}{\sigma_{ew}} \ln\left(\frac{x_c}{\mu}\right) - \frac{\sigma_{ew}}{2}\right] \Big/ F\left[\frac{1}{\sigma_{ew}} \ln\left(\frac{x_c}{\mu}\right) + \frac{\sigma_{ew}}{2}\right]$$

Again, a graphical method can be used to calculate these tonnages and grades (Fig. 37). These formulae are also easily programmable on a pocket calculator. If a copper deposit has average grade 1.0% Cu and the logarithmic variance of 40 × 40 × 20-m (131 × 131 × 65.6-ft) blocks is 0.6, the tonnage of ore above a 0.8% Cu cutoff grade is 46% of the total tonnage in the deposit and it averages 1.63% Cu. If a reduced level of mining selectivity is applied with blocks of size 50 × 50 × 25 m (164 × 164 × 82 ft), the logarithmic variance will be reduced, possibly to a value of 0.5. Then 48.5% of the total tonnage will exceed 0.8% Cu, with an average grade of 1.54% Cu.

To obtain these results graphically, the ratio (cutoff)/(mean) = 0.8 is read along the vertical axis in Fig. 37. Given a logarithmic variance of 0.6, the following ratios can be estimated:

$$(\text{tonnage above cutoff})/(\text{total tonnage}) = 46\%$$

$$(\text{average grade above cutoff})/(\text{mean}) = 1.63$$

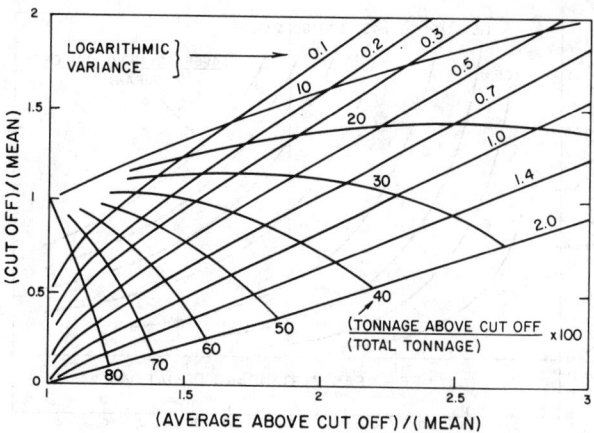

Fig. 37. Determination of tonnage-grade relationship, lognormal distribution.

The influence of selectivity on the grade-tonnage relationship is illustrated in Fig. 38 for a gold deposit. When the selectivity decreases and the smallest block size that can be mined selectively increases, the average grade above cutoff grade decreases. At the same time, except possibly for very high cutoff grades, the tonnage above the cutoff grade increases.

Comparison between a tin and a molybdenum deposit is shown in Fig. 39. In the tin deposit, most of the variability

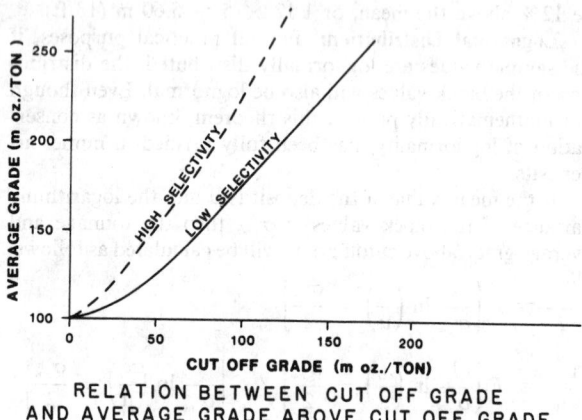

RELATION BETWEEN CUT OFF GRADE AND AVERAGE GRADE ABOVE CUT OFF GRADE

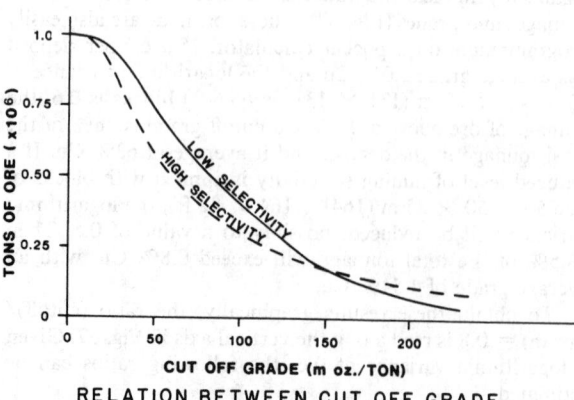

RELATION BETWEEN CUT OFF GRADE AND TONNAGE ABOVE CUT OFF GRADE

Fig. 38. Influence of selectivity on grade tonnage relationship.

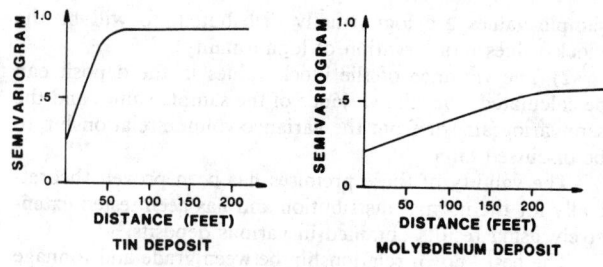

LOGARITHMIC SEMIVARIOGRAMS

BLOCK SIZE (FEET)	TIN DEPOSIT		MOLYBDENUM DEPOSIT	
	1,000,000 TONS AT 0.1 % Sn		1,000,000 TONS AT 0.1% MoS₂	
	CUT OFF GRADE 0.07% Sn		CUT OFF GRADE 0.07% MoS₂	
	TONNAGE ABOVE CUT OFF (TONS)	AVERAGE GRADE ABOVE CUT OFF (% Sn)	TONNAGE ABOVE CUT OFF (TONS)	AVERAGE GRADE ABOVE CUT OFF (% MoS₂)
10x10x10	470,000	0.17	670,000	0.13
50x50x50	560,000	0.14	700,000	0.12

TONNAGE – GRADE RELATIONSHIP

Fig. 39. Influence of semivariogram on grade tonnage relationship.

in grade occurs over short distances, approximately 20 m (65.6 ft), while the grade is less variable and relatively continuous over much larger distances in the molybdenum deposit. Increasing the selectivity by reducing the block size from 20-m (65.6-ft) cubes to 4-m (13-ft) cubes results in a 21% increase in grade and an 18% decrease in tonnage in the tin deposit. The same increase in selectivity has a much smaller effect on the molybdenum deposit, the grade increasing by only 8% and the tonnage decreasing by 4%. These results are directly derived from the semivariogram and illustrate the fact that the impact of mining selectivity is more pronounced as the variability of the economic mineral in the deposit increases.

Dispersion Variance, Variance-Volume, and Variance Additivity Relationship

Earlier, we looked at the distribution of the sample values and calculated the mean value and the variance of these values. By definition, the sample variance calculated in a given area is the *dispersion variance* of the samples in the area. There are two variances of interest to the geostatistician, the dispersion variance, to be discussed here, and the estimation or error variance that will be analyzed later.

If, instead of considering samples, we consider blocks of a size that can be mined selectively, the mean value of all the blocks within a given area is equal to the mean value of the samples. However, the variability of the block values is less than the variability of the sample values. The block dispersion variance is a decreasing function of the block size. Furthermore the dispersion variance corresponding to a given block size is also a function of the size of the area within which it is calculated. If only a small section of a deposit is considered, either high-grade or low-grade, the block values will have a low variability. If a large section of the deposit is considered, both low-grade and high-grade zones will be represented and the variance would be expected to increase.

The main relationship between variance and size of area is known as the *variance additivity relationship* or Krige's relationship. Consider an ore body O within which samples of size w have been taken. The variance of the sample values in the ore body is equal to the variance of the samples in selective mining units of size W, plus the variance of the selective mining units in the ore body:

$$\sigma_D^2 \, (w \text{ in } O) = \sigma_D^2 \, (w \text{ in } W) + \sigma_D^2 \, (W \text{ in } O)$$

The relationship between the sample (dispersion) variance $\sigma_D^2 \, (w \text{ in } W)$ and the block size W is known as the variance-area or variance-volume relationship. The larger the block size the larger the variance of the samples within the block. Given the variance-volume relationship, it is possible to calculate the dispersion variance of the block values in the ore body for any block size:

$$\sigma_D^2 \, (W \text{ in } O) = \sigma_D^2 \, (w \text{ in } O) - \sigma_D^2 \, (w \text{ in } W)$$

The variance volume relationship is also sometimes defined as the relationship between the dispersion variance of the block values in the orebody $\sigma_D^2 \, (W \text{ in } O)$ and the block size W. This variance is a decreasing function of the block size.

The variance of samples w in blocks W can be calculated directly from the sample values. For example, the variance of the samples in the ore body is $\sigma_D^2 \, (w \text{ in } O)$. Alternatively, the dispersion variance can be calculated from the semivariogram, in which case the following formula will be used:

$$\sigma_D^2 \, (w \text{ in } W) = \bar{\gamma} \, (W;W) + N$$

where N is the nugget effect of the semivariogram and $\bar{\gamma} \, (W;W)$ is the average value of the semivariogram within the block W.

The $\bar{\gamma}$ function can be calculated using either auxiliary functions (known as F functions) and graphs or by numerical integration. The graph reproduced in Fig. 40 can be used to calculate the average value of a spherical semivariogram in a block of size $L \times D$.

To illustrate the use of this graph, consider the test case analyzed earlier (Fig. 17). Twenty-five samples are available in the ore body with mean 18.60 and variance

$$\sigma_D^2 \, (w \text{ in } O) = 30.92$$

For distances less than the range (Fig. 21), the semivariogram is spherical without nugget effect, with a sill $C = 50$ and a range

$$a_1 = 750 \text{ m (2460 ft) in the direction north } 25° \text{ east}$$

$$a_2 = 210 \text{ m (689 ft) in the direction north } 65° \text{ west.}$$

If we want to estimate the dispersion variance of 100×100-m (328×328-ft) blocks, we can use Fig. 40. Since $L/a_1 = 0.133$ and $D/a_2 = 0.476$ we read

$$F(L;D)/C = 0.24$$

hence

$$\sigma_D^2 \, (w \text{ in } W) = \bar{\gamma} \, (W;W) = F(L;D)$$
$$= 50 \times 0.24 = 12.00$$

The dispersion variance of the blocks W in the deposit is therefore expected to be:

$$\sigma_D^2 \, (W \text{ in } O) = 30.92 - 12.00 = 18.92$$

The sample values being approximately normally distributed (Fig. 18), we can assume that the block values are also

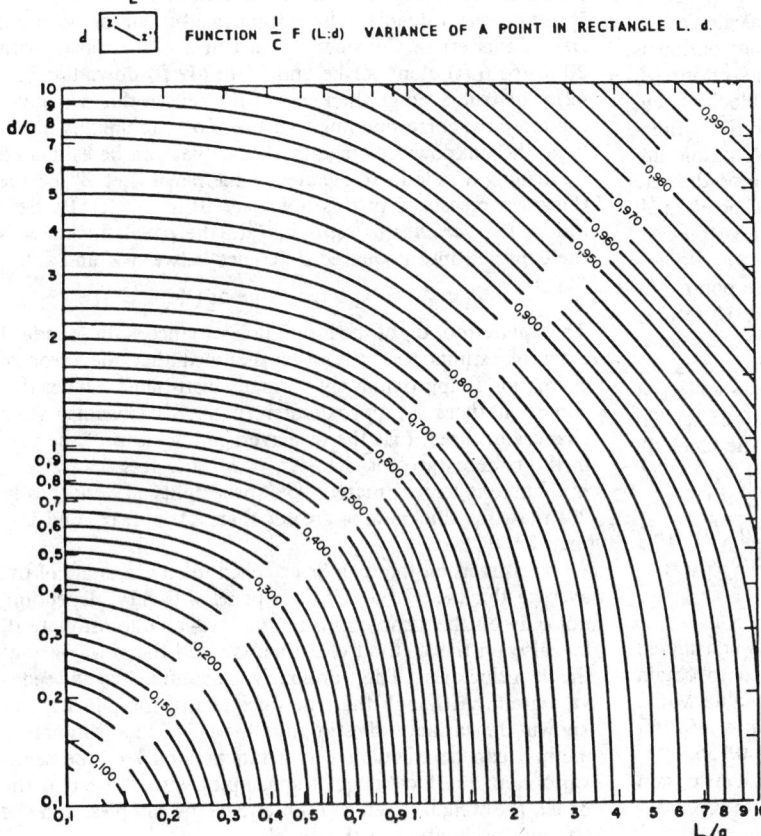

Fig. 40. Two-dimensional F function for the spherical semivariogram.

normally distributed, with mean 18.60 and variance 18.92. Using the formulae given previously for the normal distribution, we estimate that 37% of the deposit exceeds a 20.00 cutoff with average grade 23.00. Visual analysis of the sample values in Fig. 17 indicates that this result is reasonable.

In many instances, and more specifically when estimating block values by kriging, the *covariogram* $\sigma(h)$ is used rather than the semivariogram. The covariogram is defined as:

$$\sigma(h) = \sigma^2 - \gamma(h)$$

where σ^2 is the variance of the samples in the deposit. Then the block variance is calculated directly:

$$\sigma_D^2 (W \text{ in } O) = \overline{\sigma}(W;W)$$

where $\overline{\sigma}(W;W)$ is the average value of $\sigma(h)$ within the block W. This approach is required when the sample values are lognormally distributed and only the logarithmic semivariogram $\gamma_e(h)$ is available. The logarithmic covariogram is:

$$\sigma_e(h) = \sigma_e^2 - \gamma_e(h)$$

where σ_e^2 is the logarithmic variance of the samples in the deposit. The relative semivariogram is $\sigma(h) = \exp[\sigma_e(h)] - 1$ and the relative dispersion variance of the block is equal to $\overline{\sigma}(W;W)$. The logarithmic dispersion variance of the blocks is obtained as:

$$\sigma^2_{De}(W \text{ in } O) = \ln[\overline{\sigma}(W;W) + 1]$$

This is the logarithmic variance that must be used to calculate the grade tonnage relationship for lognormally distributed block values (Fig. 37).

Numerical Calculation of $\overline{\gamma}(W;W)$ or $\overline{\sigma}(W;W)$

By definition $\overline{\gamma}(W;W)$ is equal to the average value of the semivariogram $\gamma(h)$ when all possible pairs of points in the block W are taken into consideration. In this calculation, the nugget effect N must be ignored as its contribution is treated separately in all equations given in this section.

There is an infinite number of points in the block W and an exact calculation of $\overline{\gamma}(W;W)$ would be prohibitively time-consuming even with the fastest computers. A simple approximation method is commonly used, known as discrete representation. The block is divided into a number of small volumes of equal size and a single point is taken within each volume. The value of $\overline{\gamma}(W;W)$ is then estimated as the average value of the semivariogram between these points. As an example, consider a block of size $40 \times 40 \times 15$ m ($131 \times 131 \times 49$ ft). The semivariogram has been calculated using 15-m (49-ft) composites. The model accepted is the sum of a nugget effect $N = 0.2$ and an isotropic spherical model with sill $C = 0.8$ and range $a = 160$ m (525 ft). To calculate $\overline{\gamma}(W;W)$, one could choose to divide the block W into four equal cubes of $40 \times 40 \times 15$ m ($131 \times 131 \times 49$ ft). A first approximation would then be (Fig. 41):

$$\overline{\gamma}(W;W) = [4\gamma(0) + 8\gamma(20) + 4\gamma(20\sqrt{2})]/16$$
$$= (4 \times 0 + 8 \times 0.149 + 4 \times 0.210)/16$$
$$= 0.127$$

This approximation is poor and in practice it is considered that a 16-point discrete representation is needed to obtain acceptable results (Fig. 41). The corresponding value would be 0.149. If Fig. 40 is used, given $H/a = D/a = 40/160 = 0.25$, we obtain $\overline{\gamma}(W;W) = 0.8 \times 0.195 = 0.156$.

$\overline{\sigma}(W;W)$ can be calculated directly from the covariogram function $\sigma(h)$, using the same method of discrete representation of the block W.

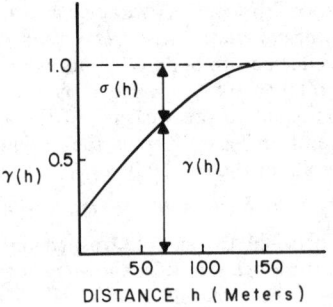

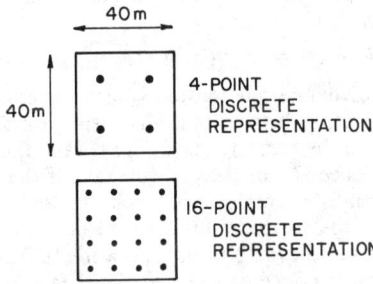

Fig. 41. Estimation of $\overline{\gamma}$ or $\overline{\sigma}$ by discrete representation.

GEOSTATISTICAL RESERVE EVALUATION

Error of Estimation

Definition: When estimating an ore body or part thereof from a number of sample values, an error is inevitably made. It can be very valuable to have some measure of the potential size of this error. Consider a panel in a vein deposit, with 20 m (65.6 ft) along strike and 15 m (49 ft) downdip (Fig. 42). The true average thickness of the vein in this panel and therefore the corresponding tonnage of ore are unknown. Let Z be the unknown average thickness that can be estimated by drilling a hole in the center of the panel. Let Z^* be the thickness measured in this hole, say 10 m (33 ft). By definition, the *variance of estimation* of the panel thickness is the expected mean squared difference between Z and Z^*:

$$\sigma_E^2 = \text{expected value of } (Z - Z^*)^2$$

The square root σ_E of the estimation variance is the standard error of estimation. It is often assumed that the error of estimation is approximately normally distributed. Under this condition there is approximately a 68% probability (two chances in three) that the true average thickness of the vein in the panel is greater than $Z^* - \sigma_E$ and less than $Z^* + \sigma_E$. There is approximately 95% probability (19 chances in 20) that this thickness is greater than $Z^* - 2\sigma_E$ and less than $Z^* + 2\sigma_E$.

The precision that can be attached to any estimate of the average thickness of a vein in a panel will vary, depending not only on the sample information available to estimate it, but also on the method of estimation and the variability of the vein thickness. This variability is quantified by the semivariogram. Although the true thickness of the panel is unknown, the variance of estimation for a particular estimation method can be calculated as a function only of the semivariogram, the location of the samples with respect to the panel, the weight given to each one of the samples, and the size and orientation of the panel.

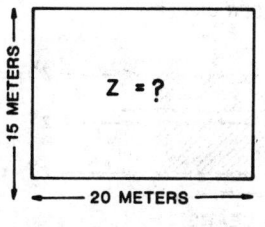

PANEL TO BE ESTIMATED

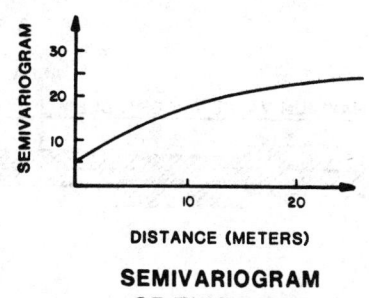

**SEMIVARIOGRAM
OF THICKNESS**

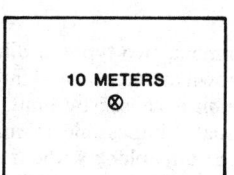

ONE CENTRAL SAMPLE

$Z^* = 10$ meters

$\sigma_E = 3.5$ meters

Fig. 42. Error of estimation.

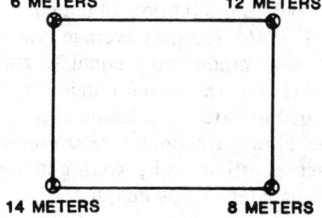

FOUR CORNER SAMPLES

$$Z^* = \frac{6 + 12 + 14 + 8}{4} = 10 \text{ meters}$$

$\sigma_E = 2.3$ meters

Given the isotropic semivariogram shown in Fig. 40, the precision with which a panel of 20 × 15 m (65.6 × 49 ft) would be estimated from a single central sample can be calculated. A standard error of 3.5 m (11.5 ft) is obtained. If the central sample indicated a 10-m (33-ft) thickness, there is 68% probability (two chances in three) that the average thickness of the panel is less than 13.5 m (44 ft) and greater than 6.5 m (21.3 ft).

In the same vein deposit, consider a panel of the same size, 20 × 15 m (65.6 × 49 ft), estimated by the arithmetic average of four corner samples. The corresponding standard error is 2.3 m (7.5 ft). If the thicknesses observed at each corner are 6, 12, 14, and 8 m (19.7, 39, 46, and 26 ft), respectively, the average thickness of the vein will be estimated at 10 m (33 ft) with 68% central confidence limits 12.3 and 7.7 m (40.4 and 25.3 ft).

Error of Estimation and Risk Analysis: The error of estimation is useful for estimating the risk involved in opening a new mine. The need for additional drilling can be quantified by comparing the cost of drilling with the expected resulting decrease in the estimation error. The optimal locations of additional drill holes can also be determined as they must be chosen to minimize the resulting error of estimation.

The error of estimation can also be used to determine, on a monthly or yearly basis, the precision with which the properties of the mineral to be produced are known. Confidence limits for critical parameters such as pounds of sulfur per million Btu in a coal seam or bismuth content in a massive sulfide zinc deposit can be obtained. The potential for unacceptable variations in these parameters can be detected and might lead to a decision for additional drilling, a change in

a mining strategy, and/or an increase in storage and blending capacity (Fig. 43).

Error of Estimation and Selective Mining: In most mining operations, some degree of selectivity is applied during mining. Blocks or panels whose estimated value exceeds a specified cutoff grade are processed as ore blocks while blocks estimated below cutoff grade are treated as waste. The decision to classify a block as an ore or waste block is made on the basis of the detailed but incomplete information available at the time mining takes place.

This information, which may include channel samples, exploration drill holes, or assayed blastholes, is usually not sufficient to allow exact calculation of the block values. Consequently some blocks will be estimated as being below cutoff grade whose true grade exceeds the cutoff grade. They will be incorrectly treated as waste blocks. Other blocks will be estimated as being above cutoff grade whose true grade is below cutoff grade. They will be incorrectly treated as ore blocks. This situation is illustrated in Fig. 44. A scatter diagram is shown in which each block is represented by a point, with horizontal coordinate the (known) block estimated value, Z^*, and vertical coordinate the (unknown) block value, Z. Ideally, one would mine only the ore blocks with true value Z exceeding the cutoff grade. In fact, one will mine the blocks whose estimated value Z^* exceeds the cutoff grade.

The scatter diagram is divided in four sectors. The first sector contains ore blocks that will be treated as waste. The second sector contains ore blocks that will be treated as ore. The third sector contains waste blocks that will be treated as ore. The fourth sector contains waste blocks that will be

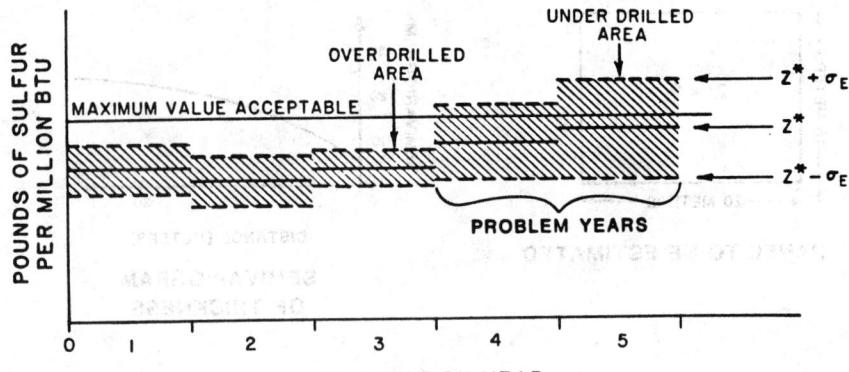

Fig. 43. Estimation error and quality control.

treated as waste. The number of blocks falling in the first and third sectors should be minimized.

Consider a gold deposit averaging 0.10 oz per ton (Fig. 45). If the polygon method is used to estimate the selective mining units from a single central sample, the following results are likely to be observed (Krige, 1962):

1) If the economic cutoff grade is less than the average grade of the deposit (say 0.05 oz per ton), using this cutoff grade for selective mining will result in a significant tonnage of ore material being treated as waste. Indeed, the tonnage of ore material being mined can be increased by decreasing the effective cutoff grade to a level below the economic cutoff grade. Alternatively, and this is the approach often taken, a "dilution factor" is taken into account, preferably based on past experience, to bring up the value of the low grade blocks that are recognized as being underestimated.

2) If the economic cutoff grade is higher than the average grade of the deposit (say 0.20 oz per ton), most and possibly all the material sent to the mill will average less than the economic cutoff grade. The average grade of the blocks estimated at cutoff grade is significantly less than the cutoff grade. Again, the overestimation of the high grade blocks is usually recognized and a dilution factor is introduced to reduce these block values.

In a selective mining environment, two types of dilution should be taken into account, known as external and internal dilution. The first *external dilution* is caused by limitations in the mining method. It is usually impossible to mine a $15 \times 15 \times 20$-m ($49 \times 49 \times 65.6$-ft) block without some dilution from surrounding material. The second, *internal dilution,* is of a completely different origin, as it reflects a weakness in the method of reserve evaluation.

Only if a smoothing method of block estimation such as the kriging method is used will the situation obtained be closer to an ideal situation (Fig. 46), the true average grade of the blocks estimated at cutoff grade being equal to the cutoff grade. If such a method is used, the need for an internal dilution factor may disappear, but external dilution still remains to be accounted for. How to estimate recoverable reserves from the in-situ reserves estimated by kriging is the subject of extensive discussions later in this chapter.

A Case Study Using Cross Validation

To illustrate these problems, a method of analysis known as "cross validation" has been applied to a porphyry copper deposit. A total of 200 drill-hole composites of equal length were available for analysis. Each composite was successively removed from the data set, and the same composite point in

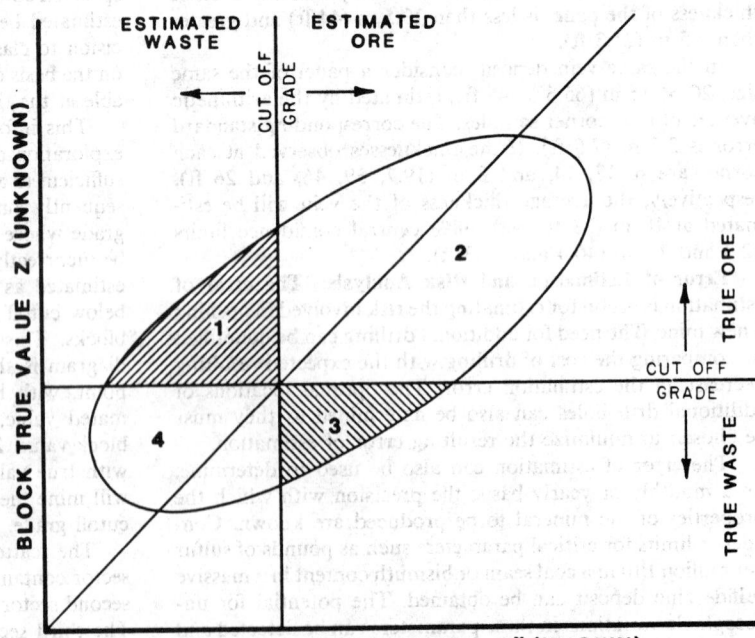

Fig. 44. Error of estimation and selective mining.

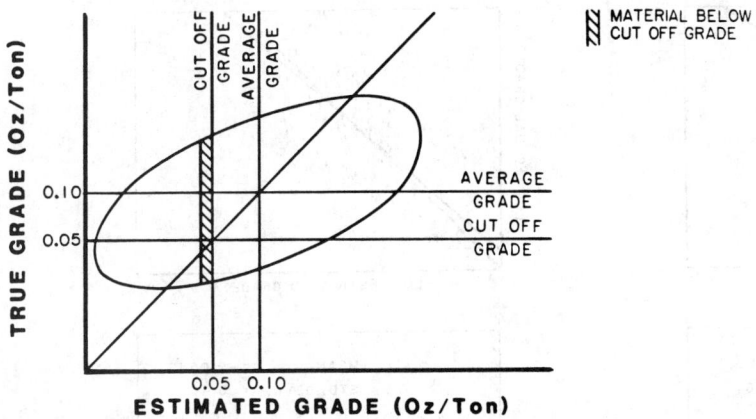

LOW ECONOMIC CUT OFF GRADE:
THE EFFECTIVE CUT OFF GRADE SHOULD BE LOWER.

Fig. 45. The polygon method and selective mining.

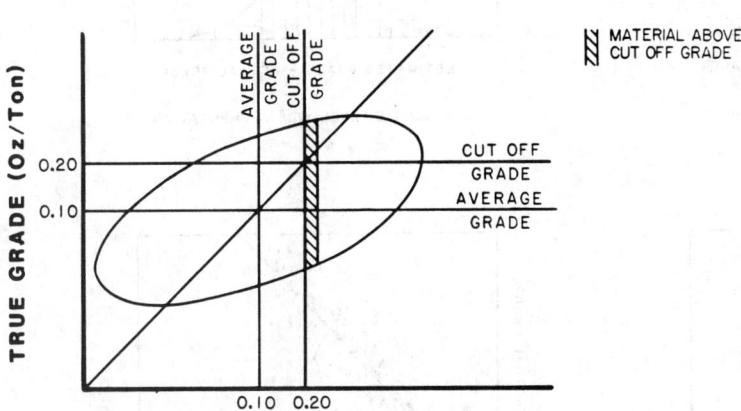

HIGH ECONOMIC CUT OFF GRADE:
THE EFFECTIVE CUT OFF GRADE SHOULD BE HIGHER.

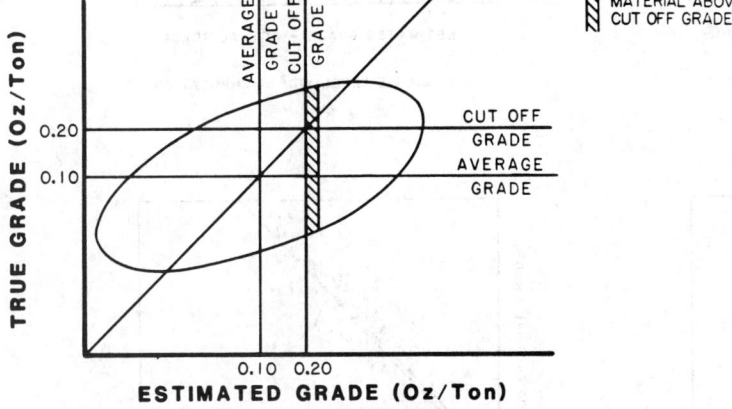

Fig. 46. The Kriging method and selective mining.

the ore body was estimated from surrounding assay values using a variety of methods: nearest sample grade assignment (polygonal approximation), inverse distance square interpolation using both two-dimensional bench search and three-dimensional ellipsoidal search, and finally kriging (Readdy, et al., 1982).

The estimated composite values were then compared with the true assays through the use of linear regression. The most accurate estimation method would be that which yielded a regression line passing closest to the origin with a slope closest to 45° and with minimum variability or scatter about the best fit line. Using geostatistical terminology, this method would best approximate the desirable conditions of minimum error variance along with conditional and local unbiasedness.

For each estimation method, the differences between true assay values and estimated grades were also computed and plotted in the form of a histogram of errors. Once again, the best method would be the one that would have a mean error of zero (global unbiasedness) and the least variance in this error distribution (minimum error variance). Fig. 47 gives the results of these studies for each of the methods examined.

The polygonal method (Fig. 47a), that consists of giving to a block the value of the nearest composite located in the same bench, shows the largest error variance. Considerable local biases are also shown by the significant departure be-

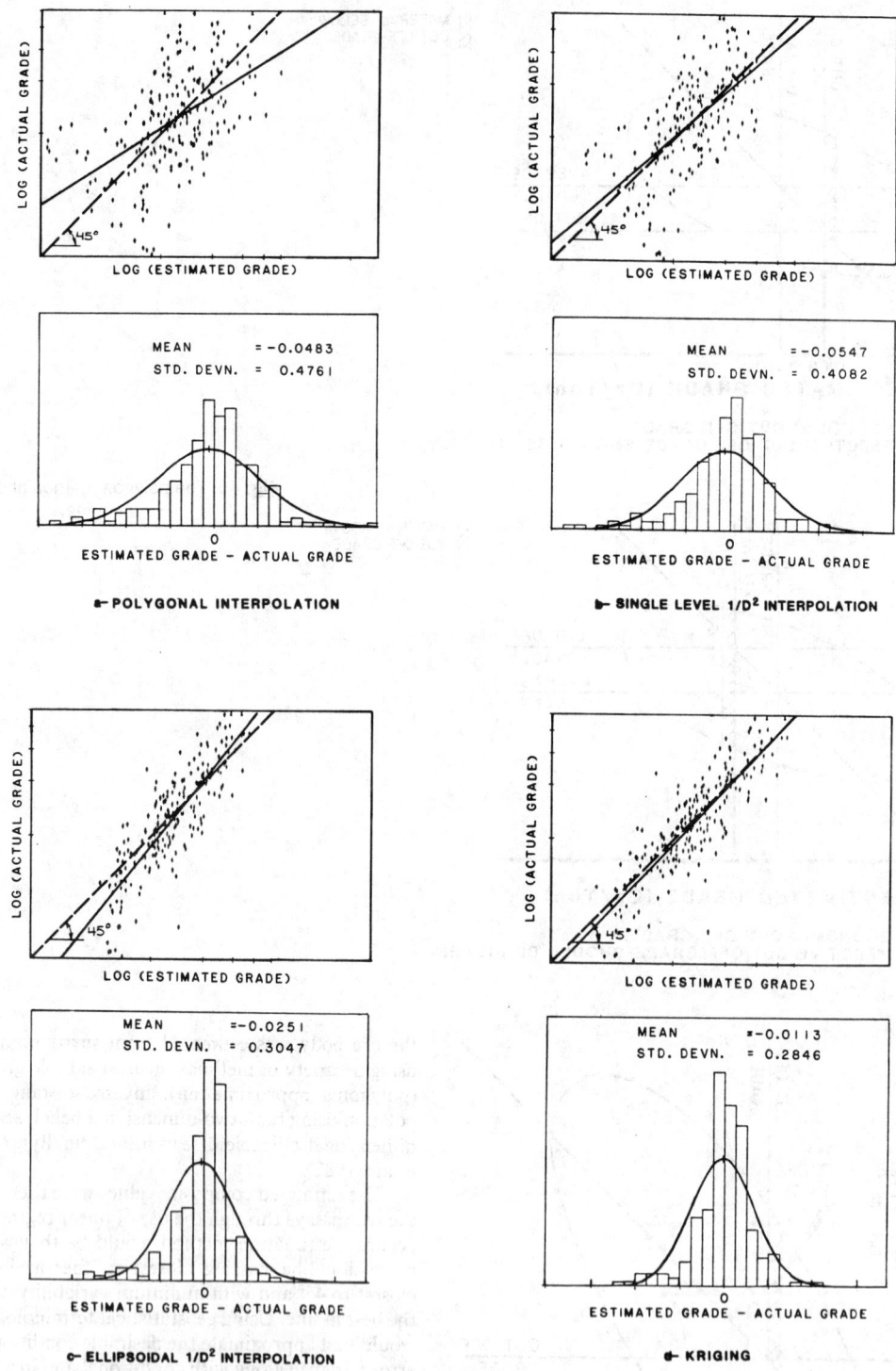

47. Analysis of cross validation results.

tween the regression line and the 45° line. Low values are underestimated, while high values are overestimated.

Both the error variance and the local biases are significantly reduced if the blocks are estimated using a multi-sample smoothing procedure such as the inverse square of the distance method, even when considering only composites within a single bench (Fig. 47b). Further reduction in error variance is obtained by taking into account the composites

in the neighboring benches (Fig. 47c). It is interesting to note that in this case, low-grade blocks are overestimated and high-grade blocks are underestimated. This inversion in local bias results from the inability of the inverse square of the distance method to deal properly with autocorrelations between composite values, the result being excessive smoothing.

As indicated in Fig. 47d, the best results are obtained

when the kriging method is used, for which global bias, local biases, and the error variance are reduced to a minimum. All the results that are expected according to the geostatistical theory have been verified by this cross-validation example. This technique, also known as "jacknifing," can be very useful in comparing the results of using various semivariogram models and kriging schemes to estimate a deposit. Such analysis is often desirable particularly in the predevelopment stage of a property when close-spaced blasthole data is not available for use in validation of the estimation methodology.

Calculation of Estimation Variance

All traditional methods of reserve evaluation consist in estimating the value of a panel or a block W as a linear combination of surrounding sample values. If n samples are used to estimate the block, its estimated value will be:

$$Z^* = \sum_{i=1}^{n} b_i x_i$$

where b_i is the weight given to the ith sample. The sum of the weights is equal to 1.0. If the inverse square of the distance method is used for example, and d_i is the distance between the ith sample and the center of the block being estimated, then

$$b_i = K/d_i^2$$

where

$$K = 1/\sum_{i=1}^{n} (1/d_i^2)$$

The error of estimation of the true block value Z is given in the following equation:

$$\sigma_E^2 = \sigma_W^2 + \sum_{i=1}^{n}\sum_{j=1}^{n} b_i b_j \sigma_{ij} - 2\sum_{i=1}^{n} b_i \sigma_{iW}$$

where: σ_W^2 is the block variance; when j is not equal to i, σ_{ij} is the covariance between the sample i and the sample j; when $j = i$, $\sigma_{ij} = \sigma_{ii}$ is the variance of the sample i, and σ_{iW} is the covariance between the sample i and the block W.

The value of these variances and covariances can be calculated directly from the autocovariance function $\sigma(h) = \sigma^2 - \gamma(h)$ where σ^2 is the sample variance and is usually taken equal to the sill of the semivariogram. The block variance σ_W^2 can be calculated using auxiliary functions or by numerical integration as discussed earlier:

$$\sigma_W^2 = \overline{\sigma}(W;W) = \sigma^2 - \overline{\gamma}(W;W)$$

σ_{ij} is equal to the sample variance σ^2 for $i = j$, and to the covariance $\sigma(h_{ij})$ for $i \neq j$, where h_{ij} is the distance between the two samples. σ_{iw}, also denoted $\overline{\sigma}(w_i;W)$ is equal to the average value of the covariance $\sigma(h)$ when h is the distance between the sample and all possible points in the block W. Auxiliary functions can be used in simple cases. More commonly, numerical integration will be used, that is based on discrete representation of the block W.

As an example, consider a block of size 40 x 40 m (131 x 131 ft) and two samples at a distance of 35 m (115 ft), as shown in Fig. 48. Assume that the semivariogram can be represented by a nugget effect N = 1.2 and a spherical model with sill C = 2.4 and range a = 100 m (328 ft). Using the F auxiliary function (Fig. 38) we calculate $\overline{\gamma}(W;W) = 0.30C$ = 0.72. Hence $\sigma_W^2 = 2.40 - 0.72 = 1.68$. Furthermore: $\sigma_{11} = \sigma_{22} = C + N = 3.60$ and $\sigma_{12} = \sigma(70) = 3.60 - \gamma(70) = 0.29$.

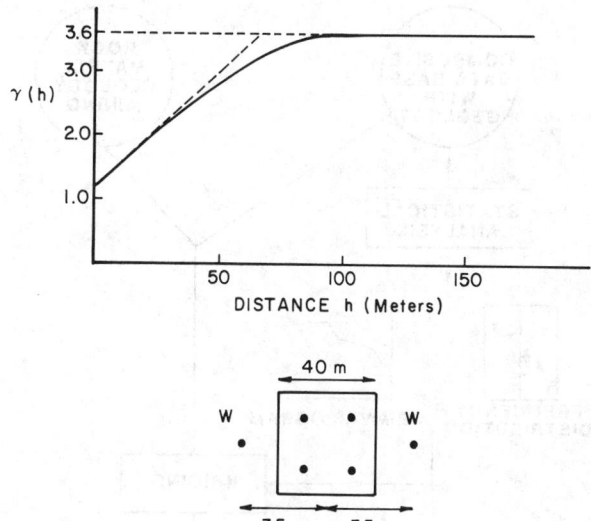

Fig. 48. Calculation of the error of estimation.

The covariances σ_{1W} and σ_{2W} are equal. If a four-point discrete representation is used, we can write:

$$\sigma_{1W} = \frac{1}{4}[2\sigma(\sqrt{(25)^2 + (10)^2})$$
$$+ 2\sigma(\sqrt{(45)^2 + (10)^2})] = 1.16$$

Typically, a 16-point discrete representation is preferred, from which a value $\sigma_{1W} = 1.15$ is obtained.

If the block was to be estimated by the mean value of the two samples, then $b_1 = b_2 = 1/2$ and:

$$\sigma_E^2 = 1.68 + \frac{1}{4}[2 \times 3.60 + 2$$
$$\times 0.29] - 2 \times 1.15 = 1.33$$

Note that, provided the stationarity conditions are satisfied, this error of estimation is dependent on the location of the samples w_1 and w_2 but *not* on their value.

Kriging

General: Intuitively, when estimating a panel or block of ore, the closer samples should be given more weight than the ones farther away. The directions showing the best continuity should also be considered in assigning a weight to a sample. In addition, the relative position of the drill holes with respect to each other should be taken into account. The size of the block to be estimated is also an important factor. Most important is the fact that the optimal method of reserve estimation should be a function of the specific properties of the deposit or part of the deposit within which the block is located. The geostatistical method of reserve estimation takes all these factors into account (Fig. 49).

As indicated earlier, the precision with which a block is estimated can be quantified using the semivariogram. The error of estimation is a function of the estimation method or, more specifically, of the weight given to each sample in estimating the block. Instead of calculating the error of estimation that corresponds to a given set of sample weights, it is possible to calculate the weights that will result in the minimum error by solving a system of linear equations, known as the *kriging system,* that makes extensive use of the semivariogram. The resultant weights then form the basis for estimation of thickness, qualities, grade, etc.

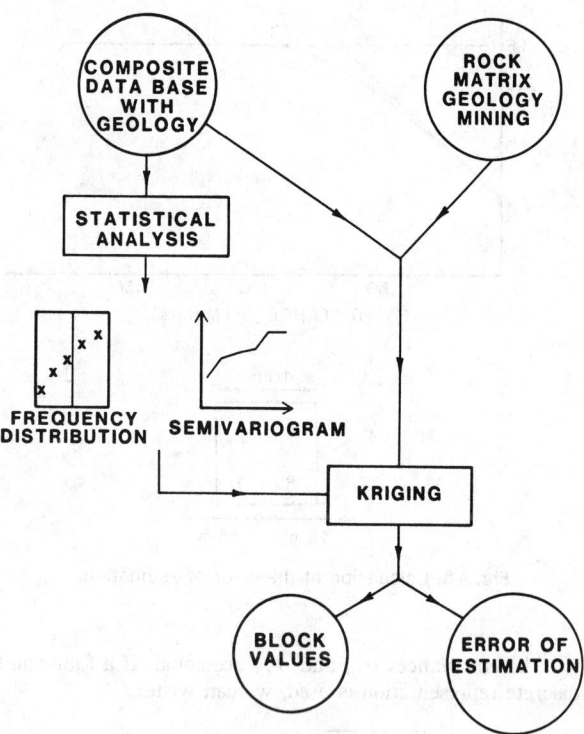

Fig. 49. Geostatistical reserve estimation.

An illustration of a simple kriging problem is given in Fig. 50.

Nine drill-hole intercepts are available within a bench to estimate a block of 122 × 122 m (400 × 400 ft). The average grades measured at each hole are shown in Fig. 50 as well as the semivariogram calculated from all the drill holes in the deposit. If the polygon method is used, the block is estimated by the central sample, with value 4% Zn and an error of 1.20% Zn is made. If the mean of the nine samples is used, an average of 6.11% Zn is calculated with an error of 0.72% Zn. If kriging is used, the central sample is given most of the weight and the outside samples are given weights that reflect the anisotropy shown by the semivariogram. The block estimated average grade is 5.10% Zn with an error of estimation of 0.52% Zn.

Kriging can be used to estimate the reserve within an entire deposit, to estimate the reserves scheduled for production on a yearly basis, or to estimate a regular grid of blocks. If a regular grid is estimated, the corresponding errors of estimation can be contoured, as well as the estimated block values. Areas with relatively high errors can be studied with respect to the proposed mining sequence to determine whether additional drilling is required. The block values estimated by kriging can also be used for computerized or manual mine design.

The kriging method of evaluation is the only method that uses to its full potential the statistical and geologic properties of the deposit as represented by the semivariogram. The kriging weights and the error of estimation depend on the block size, block location and sample locations, and on the semivariogram. They are independent of the sample values. This is true only if the necessary stationarity conditions are satisfied.

Another application of the kriging method to estimation of a porphyry copper deposit is illustrated in Fig. 51. In this figure, a block of 15 × 15 m (50 × 50 ft) has been estimated from 23 surrounding drill holes. A de Wijsian semivariogram model has been used. The weight given to each sample is shown that can be used to classify the samples according to their influence on the block.

Two rings of samples are observed. In the first ring, comprising four samples, all weights exceed 0.10; the total weight assigned to these four samples is 0.71. In the second ring, comprising nine samples, all weights are below 0.05. Samples outside the second ring have no influence on the block. The results obtained are a function of the distance between the sample and the block, of the sample spacing and of the semivariogram. The samples closest to the block have an influence on the weight assigned to the farther samples. This is known as the *screen effect*. It can be shown that the screen effect is most significant when the semivariogram shows a small nugget effect, while it disappears completely in the presence of a pure nugget effect.

The kriging method of block estimation has been applied to the test data set given in Fig. 17, using the semivariogram model shown in Fig. 21. Blocks of 100 × 100 m (328 × 328 ft) have been estimated using the ten samples closest to each block. The results are shown in Fig. 52. All calculations were made on a programmable pocket calculator.

For most ore bodies, block estimation will entail kriging of separate zones (lithologic, structural, etc.) within the deposit, each with its corresponding set of drill-hole data and semivariogram model. It is then necessary to assign a zone identifier to both the drill-hole assay composite data and the three-dimensional block model. A model that portrays the volumetric boundaries between various zones is normally created in advance of kriging and is used to control the block estimation process. Such a model is often referred to as the *geologic model or rock matrix*.

Kriging Calculations: If a block W is estimated from the sample values x_i using a formula of the form:

$$Z^* = \sum_{i=1}^{n} b_i x_i$$

the error made in estimating the true value Z is a quadratic function of the sample weights b_i. Kriging consists of calculating the weights that should be given to the sample values, so as to minimize the error of estimation. It can be shown (Rendu, 1981) that these weights are obtained by solving the following system of linear equations:

$$\sigma_{11}b_1 + \sigma_{12}b_2 + \cdots\cdots + \sigma_{1n}b_n + \lambda = \sigma_{1w}$$
$$\sigma_{12}b_1 + \sigma_{22}b_2 + \cdots\cdots + \sigma_{2n}b_n + \lambda = \sigma_{2w}$$
$$\cdot$$
$$\cdot$$
$$\cdot$$
$$\sigma_{1n}b_1 + \sigma_{2n}b_2 + \cdots\cdots + \sigma_{nn}b_n + \lambda = \sigma_{nw}$$
$$b_1 + b_2 + \cdots\cdots + b_n = 1$$

There are $n + 1$ equations with $n + 1$ unknowns, the sample weights b_i, and the lagrange multiplier λ that must be introduced to impose the constraint that the sum of the weights adds up to 1.0.

This system of equations is known as the *kriging system*. We saw earlier how all the variances and covariances in this system can be calculated from the semivariogram. The optimal weights that are solutions to this system of equations

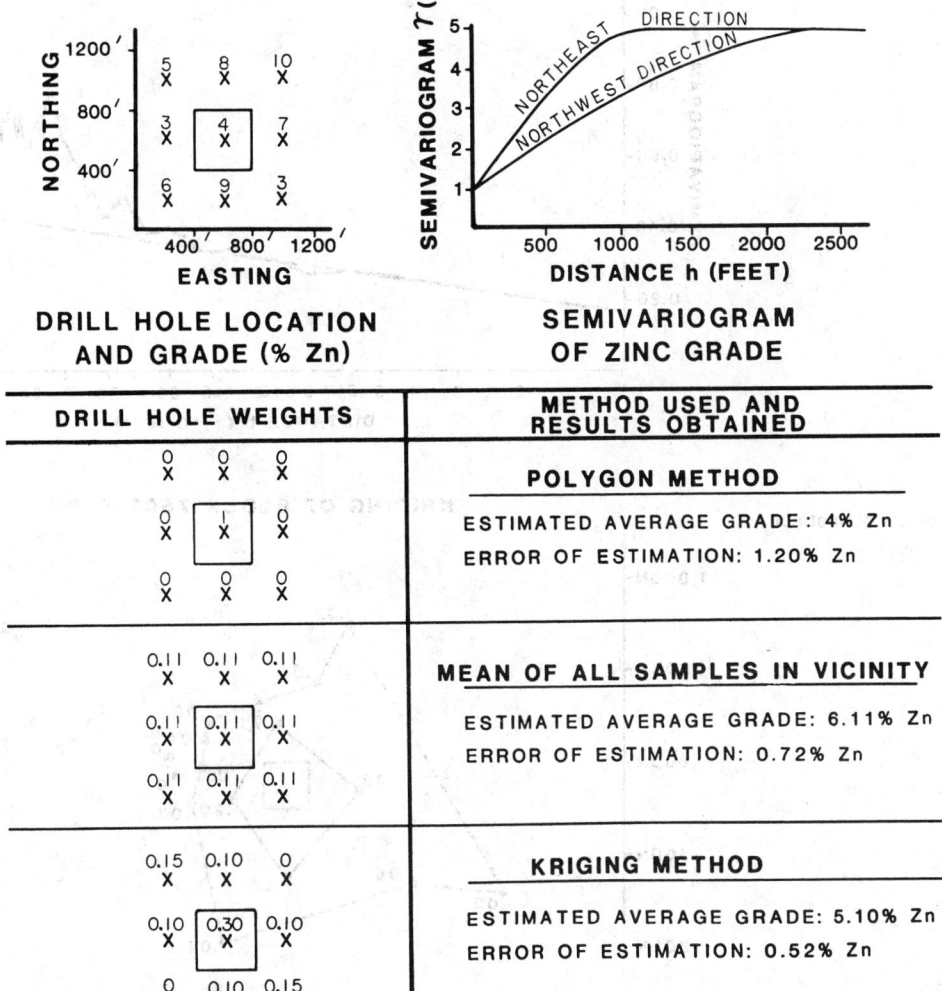

Fig. 50. Comparison of estimation methods.

DRILL HOLE WEIGHTS	METHOD USED AND RESULTS OBTAINED
$\begin{matrix}\underset{X}{0} & \underset{X}{0} & \underset{X}{0} \\ \underset{X}{0} & \boxed{\underset{X}{1}} & \underset{X}{0} \\ \underset{X}{0} & \underset{X}{0} & \underset{X}{0}\end{matrix}$	**POLYGON METHOD** ESTIMATED AVERAGE GRADE: 4% Zn ERROR OF ESTIMATION: 1.20% Zn
$\begin{matrix}\underset{X}{0.11} & \underset{X}{0.11} & \underset{X}{0.11} \\ \underset{X}{0.11} & \boxed{\underset{X}{0.11}} & \underset{X}{0.11} \\ \underset{X}{0.11} & \underset{X}{0.11} & \underset{X}{0.11}\end{matrix}$	**MEAN OF ALL SAMPLES IN VICINITY** ESTIMATED AVERAGE GRADE: 6.11% Zn ERROR OF ESTIMATION: 0.72% Zn
$\begin{matrix}\underset{X}{0.15} & \underset{X}{0.10} & \underset{X}{0} \\ \underset{X}{0.10} & \boxed{\underset{X}{0.30}} & \underset{X}{0.10} \\ \underset{X}{0} & \underset{X}{0.10} & \underset{X}{0.15}\end{matrix}$	**KRIGING METHOD** ESTIMATED AVERAGE GRADE: 5.10% Zn ERROR OF ESTIMATION: 0.52% Zn

are the *kriging weights*. The corresponding minimum error of estimation is the *kriging error* and can be written in the following form:

$$\sigma_k^2 = \sigma_w^2 - \sum_{i=1}^{n} b_i \sigma_{iw} - \lambda$$

As an example, consider the case discussed in "Calculation of Estimation Variance" and illustrated in Fig. 48. Clearly by reason of symmetry, estimation of the block by kriging using only the two samples shown in Fig. 48 can only give the solution $b_1 = b_2 = 0.5$, and an error of estimation as calculated in that section is $\sigma_k^2 = 1.33$. If on the other hand a third sample is considered, in the center of the block, the weight to be given to this and the other two samples must be calculated. Using a 16-point discrete representation of the block when needed, the following variances and covariances are obtained:

$$\sigma_w^2 = 1.68$$
$$\sigma_{1w} = \sigma_{2w} = 1.15$$
$$\sigma_{3w} = 1.87$$
$$\sigma_{11} = \sigma_{22} = \sigma_{33} = 3.60$$

$$\sigma_{12} = 0.29$$
$$\sigma_{13} = \sigma_{23} = 1.19$$

Hence the kriging system:

$$3.60\, b_1 + 0.29\, b_2 + 1.19\, b_3 + \lambda = 1.15$$
$$0.29\, b_1 + 3.60\, b_2 + 1.19\, b_3 + \lambda = 1.15$$
$$1.19\, b_1 + 1.19\, b_2 + 3.60\, b_3 + \lambda = 1.87$$
$$b_1 + b_2 + b_3 = 1$$

The solution to this system is $b_1 = b_2 = 0.267$, $b_3 = 0.466$, and $\lambda = -0.443$. The corresponding estimation variance is $\sigma_K^2 = 0.638$. If the polygon method was used, only the central sample would be considered and the error of estimation would be $\sigma_E^2 = 1.68 + 3.60 - 2 \times 1.87 = 1.54$.

Some Remarks About the Smoothing Effect of Kriging

According to geostatistical theory and practical experience, if a deposit is mined selectively, the optimal method of evaluation of individual blocks is by kriging. However even if kriging is used, an error will remain in the estimation of the blocks. Consequently the tonnage mined and the cor-

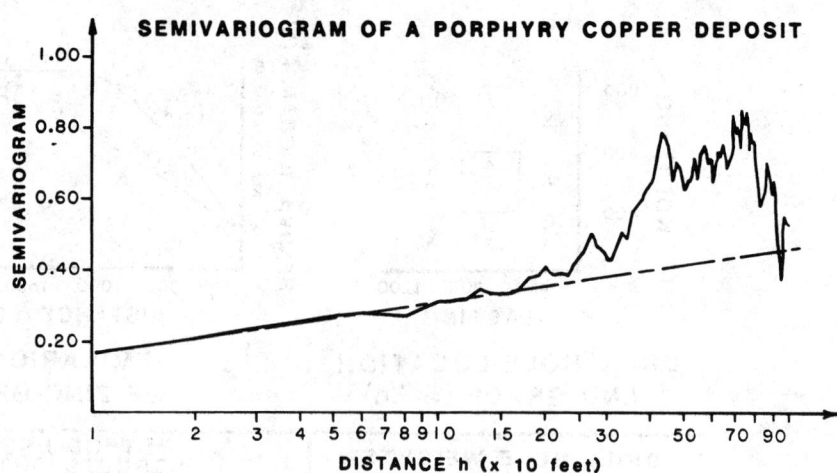

Fig. 51. Kriging and screen effect.

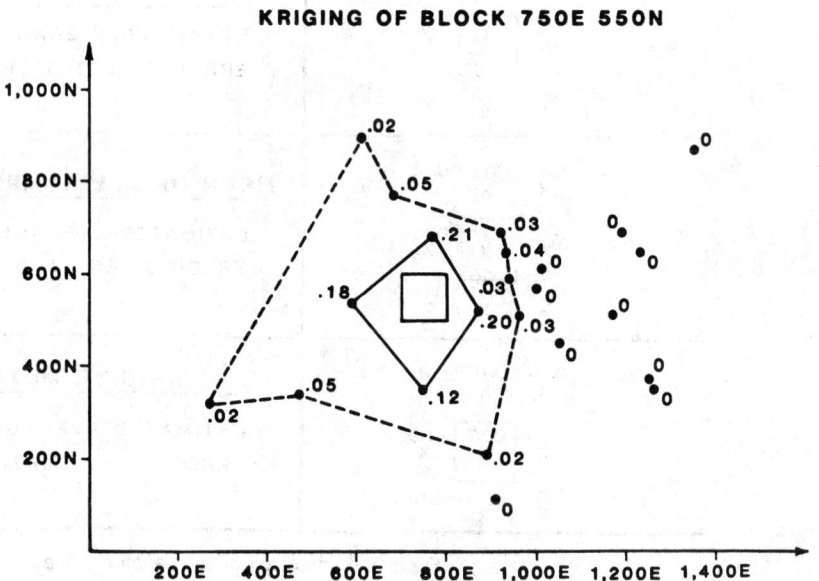

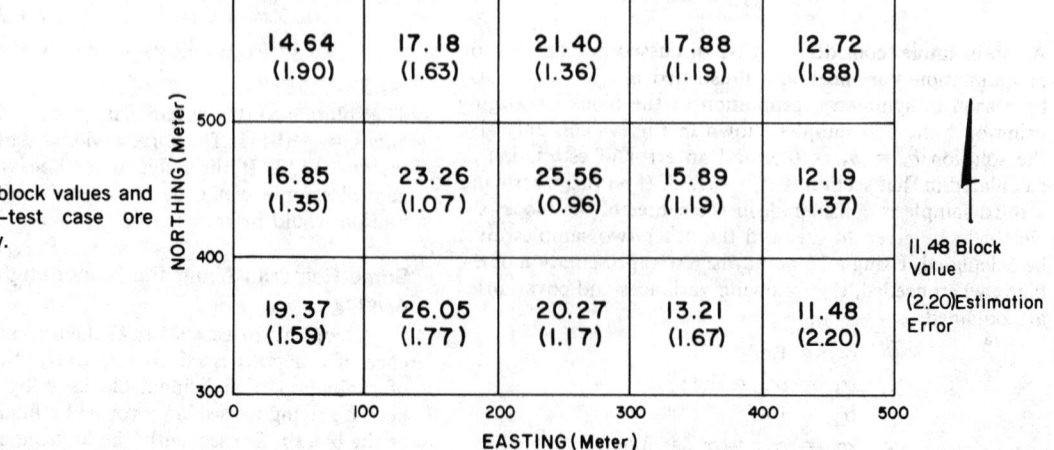

Fig. 52. Estimated block values and estimation errors—test case ore body.

responding average grade will differ from the tonnages and grades that would be mined if we had exact information concerning the true grade distribution.

In the study of the dispersion variance, it was shown that the likely effect of selectivity and cutoff grade can be estimated by assuming a normal or lognormal distribution of the block values with known mean and variance. If we had exact information, the variance to be used would be the variance of the true block values Z. Since we have only limited information, the variance that must be used is the variance of the estimated block values Z^*. This dispersion variance is

$$\sigma_D^2 = \sigma_W^2 - \sigma_K^2 - 2\lambda$$

In most practical applications, λ is negative and small when compared with σ_K^2. The dispersion variance of the estimated block values is then less than the variance of the true block values. This variance reduction is known as the *smoothing effect* of the kriging procedure.

For example, consider a gold deposit with average grade of 5 g/t (0.18 oz per ton) to be mined selectively by blocks of 20 × 20 × 10 m (65.6 × 65.6 × 32.8 ft). Using the semivariogram it is possible to estimate the block relative dispersion variance. Assuming that a value $\sigma_W^2 = 1.4$ was obtained and given a cutoff grade of 4 g/t (0.14 oz per ton), we would expect 41% of the deposit to be above cutoff grade, averaging 9.3 g/t (0.33 oz per ton) if we had perfect information. These results were obtained assuming a lognormal distribution of the true block values, with mean 5 g/t (0.18 oz per ton) and logarithmic variance $\ln(1 + 1.4)$. In practice, the block values are estimated from blasthole assays. Assuming that the relative error of estimation of the blocks from these blastholes is $\sigma_K^2 = 0.4$ and ignoring the lagrange multiplier, the relative dispersion variance of the block estimates will be $\sigma_D^2 = 1.4 - 0.4 = 1.0$. Using this dispersion variance we can now expect that 44% of the deposit will be estimated above cutoff, averaging 8.5 g/t (0.3 oz per ton). If the blasthole sampling can be improved to the point where the relative kriging error is reduced to 0.2, the estimated ore tonnage will be 42% of the deposit, at an average grade of 8.9 g/t (0.31 oz per ton), significantly closer to the optimum of 41% at 9.3 g/t (0.33 oz per ton).

Other Geostatistical Methods

In theory, the linear kriging method of reserve estimation discussed in the foregoing is a minimum variance, conditionally unbiased estimator only if minimum stationarity conditions are satisfied and the sample values are normally distributed. If departure from this ideal but unfortunately rather uncommon model occurs, linear kriging still remains the best linear method of estimation but can give conditionally biased results.

Other geostatistical methods have been developed that are applicable under specific circumstances. Logarithmic kriging (Journel and Huijbregts, 1978; Rendu, 1979) was developed to estimate deposits in which the sample values are lognormally distributed. Disjunctive kriging (Matheron, 1976; Rendu, 1980) can be used to estimate deposits within which the sample frequency distribution is well known but cannot be represented by simple models such as the normal or lognormal model. Universal kriging (Journel and Huijbregts, 1978) was developed to estimate block values when drifts are present and the necessary stationarity conditions are no longer satisfied.

Finally, geostatistical methods have been developed to assist in the evaluation of possible grade control, blending, and short-term mining and processing problems. These methods, known as "ore-body simulation" (Journel and Huijbregts, 1978) are aimed at creating a computer model of the ore body that shows the same short-range variability as is present in the deposit itself. This computer model can then be used to detect problem areas, before exploitation actually begins, possibly resulting in changes in the mining, blending, and/or processing of the ore. Ore-body simulation must not be used for reserve estimation, however.

EVALUATION OF RECOVERABLE RESERVES

Objective

Since the ore reserves of a mining company are its primary asset and represent its potential for economic viability, any mineral property evaluation should include the best possible estimate of ore reserves that are regarded as *economically recoverable*. It is therefore very important to review the implications of the word "recoverable" in the context of this section on statistical and geostatistical methods.

Earlier we have dealt with the available statistical and geostatistical methods for estimating in situ global or geologic reserves. Some discussion has also been given on the estimation of overall deposit tonnages and grades that might be considered recoverable given a certain cutoff grade and mining selectivity. The intention here is to introduce concepts involved in the conversion of an in situ deposit block model and corresponding reserve estimate into one which is considered recoverable in response to a selected mining and mineral processing production stream. The questions of cutoff grade, dilution, mining selectivity, etc., will be covered in more detail than was possible in earlier sections.

Implicit in the recoverable reserve notion is the fact that what is of interest to the mining company is the quantity of metal or valuable constituent that can be recovered and economically sold. It is not the quantity that may exist as an in situ resource or even that which is regarded as minable, i.e., merely contained within certain defined pit limits. We should be interested in properly estimating not only the ore tonnage and grade that will be recovered during mining, but in addition, the metal content that will be recoverable during processing.

To properly arrive at a recoverable reserve estimate the study team, comprising geologists, mining engineers, and metallurgists, should address each of the following interwoven factors:

1) *Cutoff grade,* referring to grade above which material is regarded as ore and dispatched to the mill.

2) *Mining recovery,* referring to that tonnage and grade on a block by block basis that is expected to be realized in practice given a certain cutoff grade and selectivity in mining the deposit (*recoverable ore reserves* in this discussion). This may be based on a classification of in situ block grade estimates after making some allowance for dilution and ore loss or on estimates of recovered fractions of the blocks that will be mined above a chosen cutoff grade and of the corresponding grades.

3) *Minability,* referring to that tonnage and grade that is estimated to be recoverable or achievable through mining on a local block by block basis, but which is also contained within economic ultimate pit limits and is demonstrably minable through the life of the project (*minable ore reserves* in this discussion).

4) *Metallurgical,* referring to that fraction of metal content in run-of-mine mill feed that will be recovered by the processing plant and will therefore report within the product

to be sold, e.g., the concentrate (*recoverable metal* in this discussion).

Discussion of each of these factors follows. While each is important in a thorough evaluation of recoverable ore reserves, all but the question of mining dilution are treated very briefly in this section because these aspects are addressed in greater detail elsewhere in this handbook. Mining dilution in different ore-body environments, however, is discussed at some length; first because its significance has often been underestimated historically, and second because geostatistics provides a means by which the effects of mining selectivity and ore dilution can be studied.

Cutoff Grade

To geostatistically estimate reserves for deposits that have gradational boundaries, a cutoff grade must be selected, above which material will be classified as ore and below which it will be considered as waste. Obviously, the ore reserve figures calculated will depend on the cutoff grade chosen.

Ore reserves, whether they be geologic or minable, are normally estimated and reported based on a so called *internal* cutoff grade. This is that grade at which the value of recoverable metal contained in each ton of ore just exceeds the cost of recovering that metal. Within defined pit limits, the near-cutoff grade material will typically be mined either as ore or waste in order to uncover higher grade material below. Therefore, this internal cutoff grade calculation generally only includes the cost difference between ore and waste mining and not the total mining cost. The internal cutoff grade does not include an allowance to carry the cost of any waste stripping that will inevitably be necessary to expose the ore. To be economic, the pit increments at the ultimate pit wall will therefore contain ore of higher grade than the internal cutoff grade.

In long-term mine planning, an analysis of cutoff grade should not be limited to a simple break-even calculation. For a given mill capacity, an increase in cutoff grade provides a means of increasing mill feed grade and concentrate production, and hence annual profitability. Coincident with such an increase in cutoff grade, however, will be an increase in the mining capacity required to satisfy the plant at full feed capacity and a decrease in the ore body's life.

Generally, a set of cutoff grades can be identified for each production period that simultaneously balances each production stage (i.e., mining, concentrating, refining) at its maximum capacity. A further set of "optimum" cutoff grades can also be determined that maximize the present value of future cash flows based on given assumptions regarding product price, costs, and capacities (Lane, 1964; Blackwell, 1970; Taylor, 1972).

To evaluate recoverable reserves, cutoff grades must therefore be chosen. Depending on the company's objectives these grades may or may not include some minimum profit element and may or may not represent an optimum choice. A rigorous analysis of the economic effects of varying cutoff grades throughout the life of the mine cannot be properly undertaken until an ultimate pit and a corresponding sequence of pit development phases has been determined. This permits an evaluation of tonnage-grade inventories for each logical and minable pit expansion or phase, that in turn provides the necessary input data for variable cutoff grade analysis. The estimate of recoverable ore reserves that is finally provided might be based on a variable cutoff grade strategy through time or a fixed break-even cutoff grade, depending on the management philosophy and the amount of detail with which the property is being evaluated.

In some mineral deposits cutoff grades are chosen such that maximum (or minimum) product grades or qualities are achieved. Examples can be found in iron ore, chromite, coal, limestone, bauxite, phosphate, etc. These deposits have a very high proportion of ore mineral (compared to copper, gold deposits, etc.) and generally high homogeneity and continuity. Mining therefore tends to require blending from various parts of the deposit to achieve a run-of-mine product that meets a set of required constituent grades, i.e., minimum calorific value, minimum iron, maximum ash, minimum silica, etc. The emphasis is on control of product quality and impurities rather than on marginal economic value.

Typically, a geostatistical ore-body model is obtained by dividing a deposit into blocks of a given size for which in situ grade estimates can be calculated by kriging. Blocks are then classified as ore or waste and accumulated, depending on the cutoff grade.

Where it is proposed that mining should or will be more selective in terms of ore/waste discrimination than the in situ model block size, geostatistics provides a means of estimating that portion of each block that will be recovered as ore along with its corresponding grade. Such estimates of so-called "recovered fractions" and "recovered grades" (by mining) are made, assuming a certain selective mining unit size and orientation and a certain cutoff grade. In this case recoverable ore reserves within a given zone are evaluated by accumulating estimates of recoverable fractions or volumes and recoverable grades for all blocks or cells within the volume of interest.

The resultant reserve estimate is valid for the selected cutoff grade only. Any classification of material above various grade limits in this reserve merely represents an estimate of the distribution of ore tonnages within grade intervals that could be expected if the deposit was mined with the chosen degree of mining selectivity and the chosen cutoff grade. To obtain reserves at some other cutoff grade, it is necessary to reestimate the corresponding recoverable fractions and grades block by block. Alternatively, adjustments may be made on a global basis.

Further discussion on the calculation of cutoff grades is given in Chapter 5, "Planning and Design of Surface Mines," while the geostatistical estimation of recoverable fractions and grades given a cutoff grade is addressed in the following.

Mining Recovery

Having generated a computerized geologic model for a mineral deposit, an ore reserve estimation team should make estimates of dilution in ore grade and loss of metal content to the waste dump. This assessment should recognize external dilution and losses that may occur at the ore zone boundary (i.e., inclusion of peripheral barren or low-grade material) or internally (i.e., inclusion of internal waste that may be incompletely discarded during mining) or perhaps unavoidably through the ore reserve modeling process.

The significance of these factors will depend on the ore body's geometry and distribution of mineralization, along with the degree of mining selectivity and production control that should be practiced or that can be achieved operationally. It will also depend on the methodology used in generating the in situ block grade model.

Dilution displaces higher grade ore from mill feed, reduces head grades, and hence directly reduces profitability. In addition, because process tailings often have a constant assay for a given mineral processing facility, dilution also tends to reduce metallurgical recovery irrespective of whether the waste minerals have deleterious chemical effects on the process or not. Consequently it is important, particularly for low-grade and more erratic deposits. It is very rare that a mining operation realizes the mill-head grades that were predicted from exploration drilling and feasibility level ore reserve estimation. Ore tonnage is typically higher than expected and the grade lower—a situation that leads to lower than expected returns. This may necessitate a mill expansion early in the mine's life so as to meet the scheduled concentrate production levels that are necessary to service the mine's debt repayment schedule. In some cases, oversight of the significance of this factor has contributed to financial failure of the mining operation.

Often "apparent dilution" arises indirectly in response to the reserve estimation methodology adopted. For example, in a highly nonhomogeneous deposit, polygonal methods will tend to overestimate ore grade and underestimate ore tonnage partially because these methods fail to smooth the grade distribution of samples. They effectively assume that mining will differentiate ore from waste on the basis of volumes equivalent to the size of the samples. In other words, they fail to recognize the grade variance-volume relationship wherein the grades of the selective mining units are known to have a reduced variability compared with samples, and therefore when a cutoff grade is applied, mined tonnages and grades do not match those that were predicted.

Problems can occur when the grades of large in situ blocks are estimated via some sort of multi-sample smoothing procedure (be it geostatistical, inverse distance weighting, etc.). Although this process may reliably estimate the overall average grade of these in situ blocks, a reserve generated simply by classifying these blocks as ore or waste depending on their grade may overestimate mined ore tonnage and underestimate mined grade. This is likely to be true whenever detailed grade information will be obtained, e.g., from blasthole assaying that is not available when calculating in situ reserves. The problem will be more pronounced where the homogeneity in the mineralization is poor.

In summary, dilution and ore loss must be recognized in ore reserve estimation in not only the traditional sense of improper ore zone boundary selectivity but also inherently within the estimation methodology itself (i.e., oversmoothing) and in the fact that one is in effect estimating the boundary between ore and waste based on sparse exploration drilling rather than the close spaced blasthole assays that will be available in production.

An examination of some specific deposit environments and their geometries in terms of dilution and modeling impacts follows. For simplicity, only two general deposit categories—regular stratiform and irregular deposits—are discussed.

Regular Stratiform Deposits: Under this heading, we have the flat-lying, stratabound, single and multi-seam type of deposits as can be found in coal, bauxite, tar sands, phosphates, oil shale, etc. Such deposits, because of their small dip angles, are normally mined so as to respect the hanging wall and footwall surfaces as far as possible. Mining conforms to these surfaces rather than to horizontal benches.

It is common in such cases to construct a computerized gridded model of the deposit by estimating the elevations of hanging wall and footwall surfaces for each grid cell as shown in Fig. 53. Geostatistically, it is often better to estimate the footwall surface and seam thickness directly since the latter variable typically demonstrates good continuity in variogram analysis. The hanging wall surface is then calculated indirectly.

In this type of deposit there is normally a very sharp contact between the valuable mineral and the waste overburden and interburden strata. These interfaces have good continuity and can be estimated with high precision. However, depending on their local irregularity and on the mining equipment used, some dilution and loss will generally occur. The mining method will normally call for cleaning off all waste material from the hanging wall and consequently in this process some of the coal (or ore mineral as the case may be) is unavoidably mined and lost as waste. As the coal is extracted, it is similarly difficult to clean precisely to the footwall with high production mining equipment and consequently some waste dilution will typically occur.

Computerized seam-type modeling systems normally allow for the user to specify two thickness values to be used

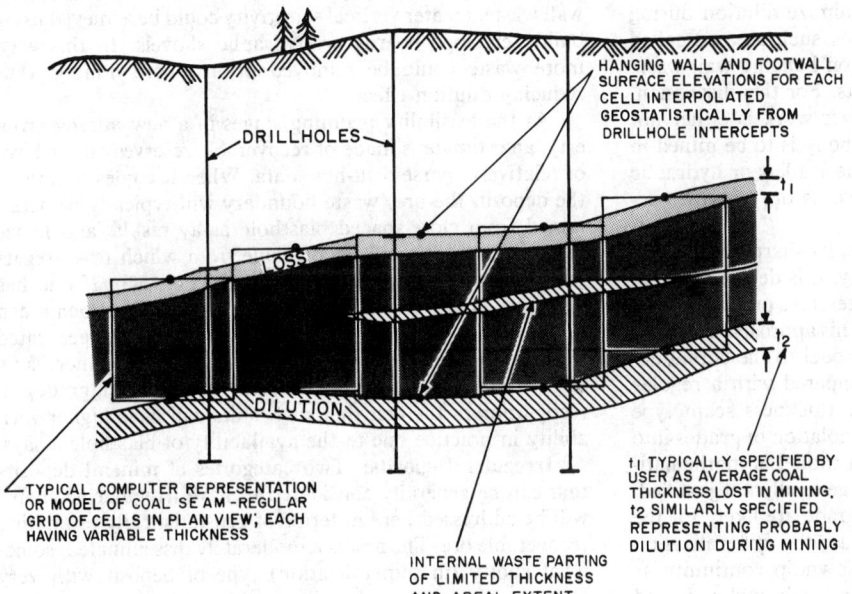

Fig. 53. Regular flat-lying stratiform deposit.

DRILLHOLES

HANGING WALL AND FOOTWALL SURFACE ELEVATIONS FOR EACH CELL INTERPOLATED GEOSTATISTICALLY FROM DRILLHOLE INTERCEPTS

LOSS

DILUTION

t_1

t_2

TYPICAL COMPUTER REPRESENTATION OR MODEL OF COAL SEAM-REGULAR GRID OF CELLS IN PLAN VIEW; EACH HAVING VARIABLE THICKNESS

INTERNAL WASTE PARTING OF LIMITED THICKNESS AND AREAL EXTENT

t_1 TYPICALLY SPECIFIED BY USER AS AVERAGE COAL THICKNESS LOST IN MINING. t_2 SIMILARLY SPECIFIED REPRESENTING PROBABLY DILUTION DURING MINING for hanging wall losses and footwall dilution. Losses and dilution are then computed by multiplying these thicknesses by the cell area in plan view and the number of cells in the area of interest. Reserves are reduced by the loss quantity and increased by the dilution quantity at its estimated quality or grade (i.e., that of footwall waste). The thicknesses selected by the user are very much a matter of judgment based on past experience with other similar deposits; on the degree of irregularity in the seam surfaces; on the mining equipment selected (e.g., scrapers, dozer/loader, etc.); and on whether there is a difference in material properties (e.g., color, radioactivity, etc.) between the valuable mineral and the adjoining waste. In modeling a tar sands deposit, for example, it may be better to respect lithological or strata boundaries (e.g., sandstone containing the bitumen vs. barren shales) so as to simplify production control procedures and hence to minimize boundary dilution, rather than to define vertical ore and waste limits based on grade alone.

In addition to the footwall dilution and hanging wall loss problems in estimating recoverable reserves in stratabound deposits, one also must deal with internal waste bands or partings. These may or may not be of sufficient thickness to selectively discard as waste and are typically of limited areal extent. The footwalls and hanging walls of such partings are not normally modeled as surfaces but are interpolated into cells as an average parting thickness or perhaps as a waste fraction. The computer system then reports internal waste parting volumes separate from the mineral by accumulating these fractions or thicknesses over all cells within the area of interest.

To arrive at recoverable reserves, the reserve estimation team must then decide on what percentage of the parting volume will probably be discarded during mining and what will be washed out in a preparation or processing plant (if any). Having done so, quality or grade estimates of the recoverable mineral can be derived.

Let us now examine a regular stratiform deposit that again has sharp boundaries between the massive ore body and waste but which no longer can be easily mined so as to follow these boundaries exactly because of its steeply dipping attitude. Such a deposit is depicted in simple form in Fig. 54a. If the ore body was relatively soft and dipping at less than about 20°, dozers could be used to clean off hanging wall waste material and therefore minimize dilution during loading operations. Dozer attachments such as a so-called slope board can also be used to clear off hanging wall waste for more steeply dipping environments. For this discussion, however, let us assume that such dozer work is either not justified or possible, and that the ore body is to be mined in regular horizontal benches by front-end loaders or hydraulic excavators. Material will be dispatched as ore or waste depending on blasthole grades.

Because of its relatively high grade, its sharp boundaries, and its very strong downdip continuity, it is decided to geologically model this deposit for in situ reserves purposes using an inclined grid as shown in Fig. 54a. This approach normally permits maximum alignment of the model to the actual deposit footwalls and hanging walls compared with a regular horizontal block model or a variable thickness seam type model. In addition, geostatistical interpolation of grades into a multi-level, variable thickness seam model is mathematically very complex, cumbersome, and generally not justified.

More reliable kriged estimates of grade are likely for the inclined cuboid blocks proposed because the mineralization often demonstrates more dominant downdip continuity as opposed to horizontal continuity. In such inclined grid

models, the estimation is normally undertaken with only mineralized drill-hole samples rather than with a statistical mixture of barren and mineralized samples that can lead to theoretical and practical difficulties. This mixture, of course, would be the case if a regular horizontal block model was created from the outset.

Having created an inclined gridded model by kriging and estimated geologic reserves, it is now necessary to convert this model into one suitable for mine planning purposes because the deposit is to be surface mined on horizontal benches. A reliable assessment of dilution and loss impacts is therefore necessary.

At least one ore body modeling system known to the authors provides an effective means of handling this problem from a mining standpoint via a so-called grid rotation procedure. The system firstly subdivides all cells (ore and waste) of the inclined input matrix into a regular array of points as shown in Fig. 54a. A grade and a volume is assigned to each point (the volume is important if several inclined and overlapping grids of different cell size are to be combined). The user then specifies a horizontal grid of cells whose size is equivalent to a selective mining unit. The system merely conducts a weighted averaging of the grades of all points that fall within each new horizontally oriented small cell. In this way, the grades of cells along the deposit boundaries are diluted with adjacent waste or low grade material. In some cases, ore will be lost because when it is combined with the waste, the selective mining unit or new cell in the model will not make the cutoff grade. The resultant model of small selective mining units can be used for mine planning and minable reserves evaluation. Reserves are obtained through an accumulation of cells having grades above a selected cutoff value. In practice selective mining units will, of course, not be mined either as complete ore or waste cuboids. The ore/waste boundary will be determined on the basis of blastholes and hence, will be irregular. This concept, however, does provide a suitable and more reliable means of estimating ore for planning purposes.

This technique provides a means of geometrically handling dilution and ore loss and should give reasonable results as long as there is a sharp boundary and definite grade difference between the ore and waste zones. If the ore mineral is, for example, distinctly different in color than the hanging wall waste, greater vertical selectivity could be achieved using bulldozer slope boards or hydraulic shovels. In this way, more waste could be removed prior to ore mining, thus reducing dilution effects.

At the feasibility planning stages of a new mining property, an estimate is made of recoverable reserves on the basis of relatively sparse drill-hole data. When it comes to mining the deposit, the ore/waste boundary will typically be determined from close-spaced blasthole assay results and hence more information will be available from which to segregate waste from ore. This is illustrated in Fig. 54b. If one has blasthole grades available, local or global adjustments can be made to the deposit model and reserve estimates generated from exploration holes only. This can be done by increasing the variance of estimated selective mining unit grades in recognition of the fact that such units will have higher variability in practice due to the availability of blasthole assays.

Irregular Deposits: Two categories of mineral deposits that can be generally considered to have irregular geometry will be addressed here in terms of geostatistically estimating recoverable ore. The first is a moderately disseminated, somewhat spotty (in mineralization) type of deposit with very gradational boundaries, while the second is more vein-dom-

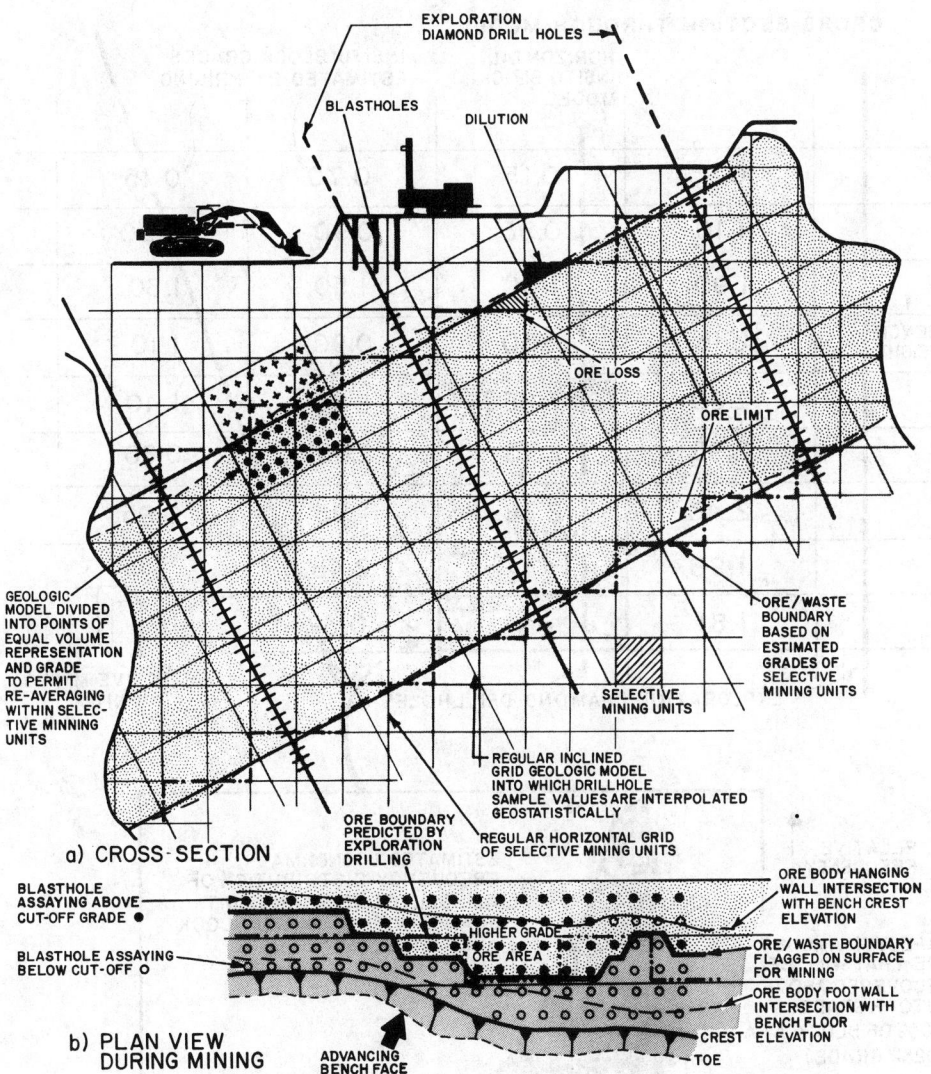

a) CROSS-SECTION

b) PLAN VIEW DURING MINING

Fig. 54. Steeply dipping stratiform deposit.

inated and consists of intermixed barren and mineralized zones.

Consider the deposit in Fig. 55. A horizontal in situ block model was defined with lateral dimensions roughly one-third of the exploration drill-hole spacing. The frequency distribution of drill-hole samples was found to behave lognormally. Following a semivariogram analysis, the blocks were lognormally kriged resulting in the indicated grades. At this point, it would be possible to simply classify these in situ blocks as falling above or below the selected cutoff grade and report reserves accordingly. However, in this deposit it is recognized from an examination of drill core and from semivariogram studies that the mineralization has a high degree of short range variability (high nugget effect) even though the semivariogram range is relatively long and the sill variance relatively low.

It is further recognized that due to the nature of this deposit, potential exists for minimizing the mill size and hence the capital investment, and for maximizing mill head grades even though this may mean a higher cost operation involving a selective mining operation and engineered grade

control. It is proposed, therefore, to mine the deposit with greater ore to waste discernability or selectivity than the in situ block size. Recoverable reserve estimation must therefore attempt to take this into account.

As previously discussed, geostatistics recognizes that the value of a geologic variable is not only associated with its location but its support volume as well. Variability in grade for example decreases with increasing support size. If the impact of selective mining is to be estimated, then the first step is to develop a frequency distribution of selective mining unit grades within the in situ blocks. This process usually assumes that if the sample grades are distributed lognormally, then so will be the selective mining unit grades. To assume lognormality of small, mineralized selective mining units is often not unreasonable. However, this assumption becomes less reliable for larger units.

We have seen earlier ("Dispersion Variance and Grade Tonnage Curve") how, given semivariogram models for the deposit in question, one can determine the dispersion variance $\sigma_D^2(w \text{ in } W)$ of the small blocks w within the in situ blocks W. In fact, we do not know the true value of the in situ

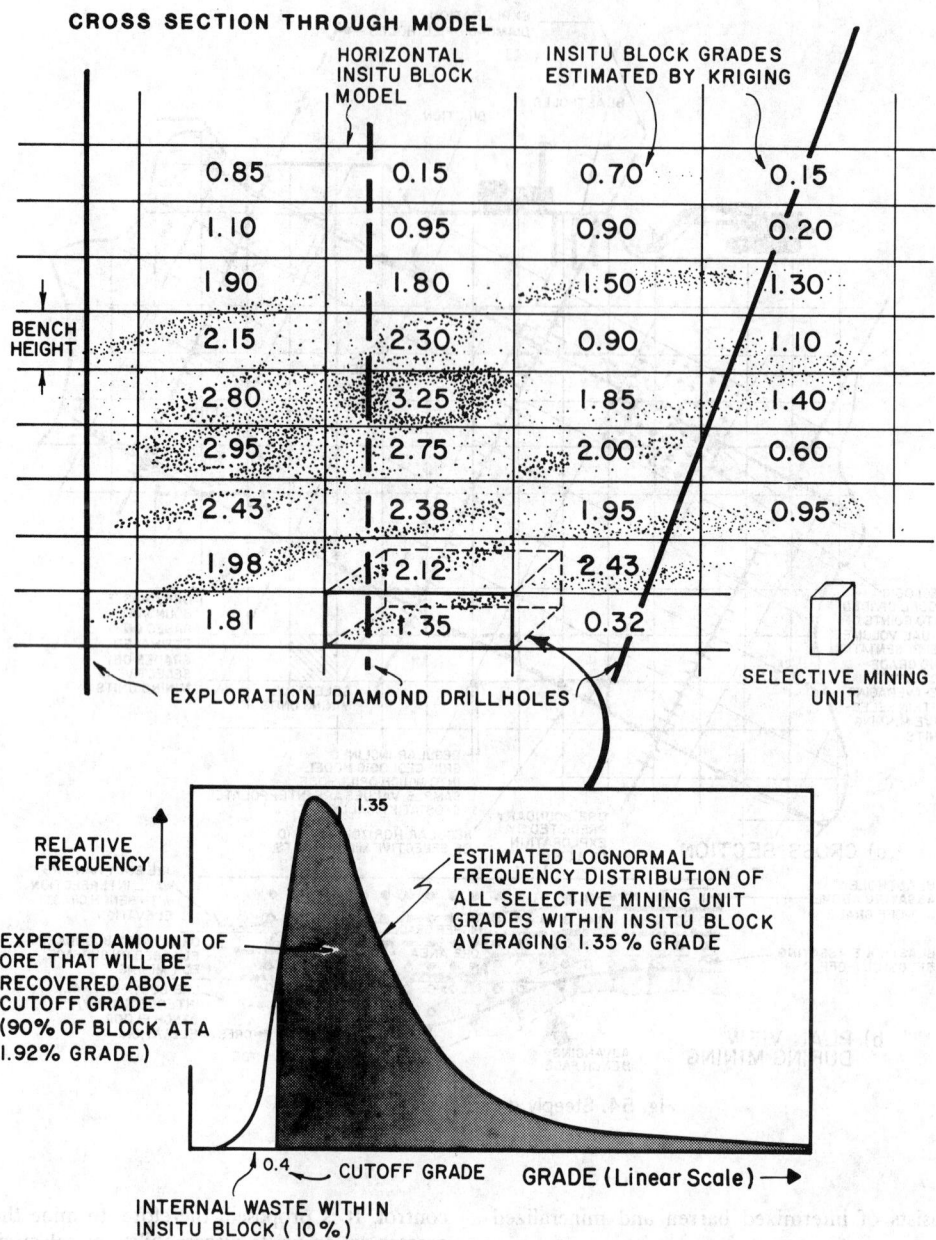

Fig. 55. Irregular transitional and disseminated ore zone.

blocks, but only their estimated value μ_K which was calculated by kriging, with an estimation variance σ_K^2. What is needed is the variance of the small blocks with respect to their expected value μ_K. It is estimated as follows:

$$\sigma^2 = E[(\mu_w - \mu_K)^2] = \sigma_D^2(w \text{ in } W) + \sigma_K^2$$

If the sample values are lognormally distributed, the semivariogram is either logarithmic or relative and the variance σ^2 is a relative variance. The corresponding logarithmic variance is calculated as follows:

$$\sigma_E^2 = \ln(1 + \sigma^2)$$

The assumption is then made that the selective mining units are either normally or lognormally distributed with a mean equal to the estimated block mean μ_K and variance σ^2 or σ_K^2 as calculated previously. The formulae given in "Dis-

persion Variance and Grade-Tonnage Curve" are used to estimate the fraction of the block that will be recovered as ore during mining, along with its average grade.

When applied to lognormally distributed values, this method of analysis is often referred to as the "lognormal shortcut." More complex methods can be used and have been successfully applied to selected deposits, such as the multivariate lognormal approach or disjunctive kriging. More detail on the theory and results achieved with these methods are given in Parker (1979), David (1977), and Journel and Huijbregts (1978).

An illustration of the method is shown in Fig. 55 where 90% of an in situ block is expected to be recoverable during mining at a grade of 1.92%. Recoverable fractions and grades are then accumulated over all blocks within the area of influence to arrive at a reserve estimate. Thus, in theory, we are able to recognize the potential effects of selective mining using geostatistical methods.

This procedure is more accurate than one involving direct kriging of selective mining units themselves unless the drill-hole spacing is small enough to make such estimation meaningful. Typically, samples taken for ore body evaluation are taken far apart and justify estimation of in situ reserves for relatively large blocks only. Only when detailed grade information is available, usually through blasthole sampling, will it be possible and justified to estimate small selective mining units directly.

The procedure previously described does not locate within the block where the ore will be mined, but merely estimates that there will be a certain percentage of each block recovered above cutoff grade with the envisaged degree of mining selectivity. This apparent limitation is not really a problem for long term mine planning in that we are dealing with large volumes or combinations of such blocks for a given year's production estimate. What is more important is that we have recognized the potential effect of mining selectivity and attempted to estimate recoverable ore accordingly.

Let us now consider a deposit (Fig. 56) that comprises sharply bounded but extremely irregular ore lenses interspersed by barren rock. In this case, the sample grade distribution demonstrates a significant proportion of barren samples as well as a population of mineralized sample grades that can be regarded as being lognormally distributed. As one would expect, the frequency of barren subzones increases toward the deposit margins. It is again possible to use basically the same estimation approach as previously described but with some modification.

The normal procedure is to first composite the drill-hole grade data into regular lengths [say 2 m (6.6 ft)] and determine if the composites are mineralized (indicator 1) or barren (indicator 0). Variograms on the indicator are then evaluated and an estimate of the percentage of barren material is obtained by kriging for each in situ block from the surrounding drill holes. Next, the average grade of the portion mineralized is estimated based on neighboring mineralized samples only and the corresponding semivariogram. Problems may be encountered with this approach in poorly mineralized areas. The overall average grade of the in situ block is then obtained

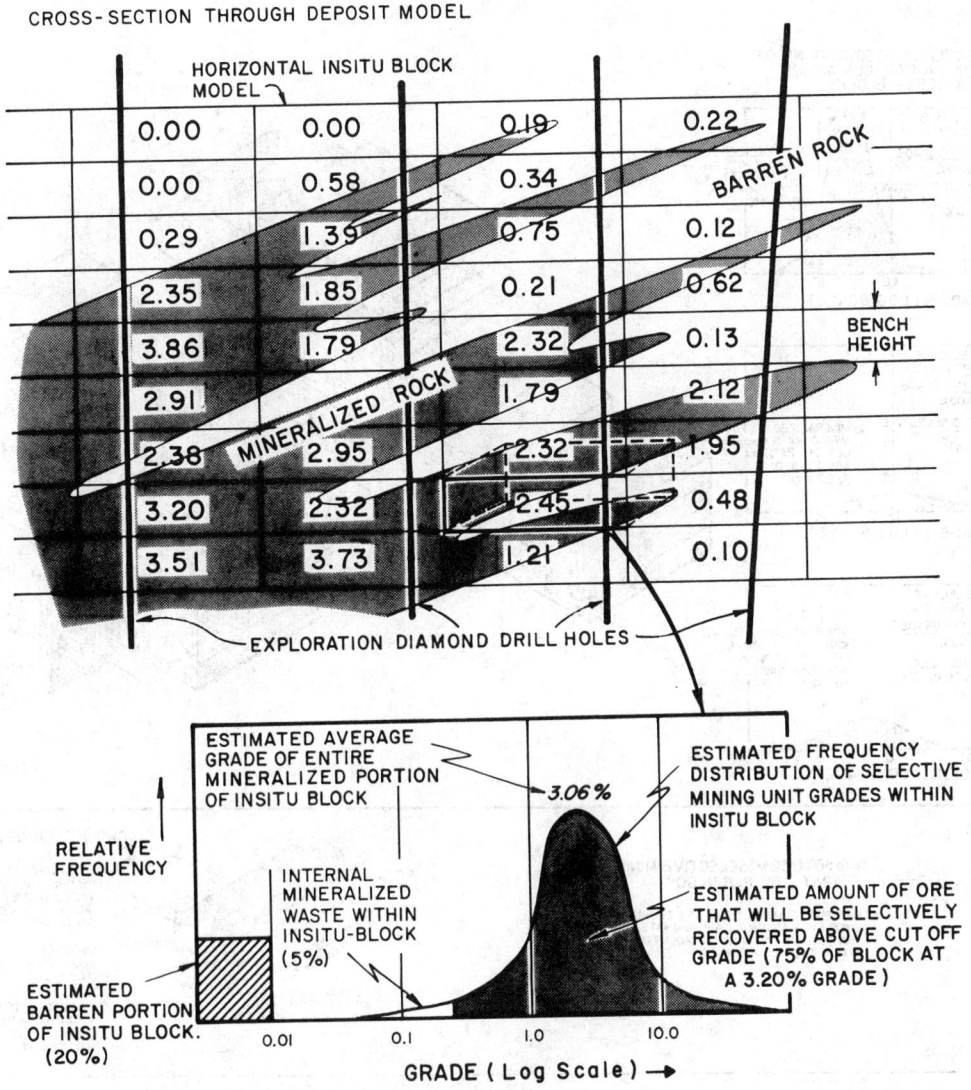

CROSS-SECTION THROUGH DEPOSIT MODEL

Fig. 56. Irregular sharply bounded ore zone.

by combining barren and mineralized fractions. Block by block recovered fractions and grades are then computed as in the previous example, taking into account the mineralized portion only. Adjustment for the barren portion is required for tonnage estimation. Finally, reserves can be computed as before. This methodology has been used on numerous occasions particularly for high grade, nonhomogeneous ore bodies where a high degree of selectivity is necessary. A case study is given in Parker and Switzer (1975).

A key assumption in this procedure is that the entire barren proportion of each block can be discarded as waste during mining, irrespective of the degree of mining selectivity adopted. This is a reasonable assumption if the selective mining unit is very small and/or if the mineralized zones are separated by sufficient distances. Given this assumption, however, along with the inherent uncertainty in the estimation methodology itself, its susceptibility to error in estimating grades where recovered fractions are small (i.e., in the distribution tails), the fact that very limited data is available on such a small scale, and the inevitability of less than perfect selectivity in practice (production emphasis is normally on tonnage mined, not *ore* extracted), the ore reserve estimation team should make an additional subjective allow-

ance for dilution and metal loss. Such adjustments may represent a problem to some who regard geostatistics as the final and comprehensive tool, but realistically they are essential to the reliability and credibility of the recoverable reserves which are ultimately stated. This point underscores the importance of balanced judgment throughout the ore reserve estimation process to thoroughly review, question, and interpret the geostatistically derived data.

The concepts given in the previous example are perhaps more dramatically presented through the use of Fig. 57 that illustrates actual geostatistical estimates derived on a high grade base metal deposit. The figure shows a cross section of ore reserve blocks in this deposit for which individual estimates of recoverable fractions, recoverable grades, in situ block grades, and precision are given. From this figure, the following observations can be made:

1) That the uncertainty (lower left value) on average block grades is in general quite high and tends to increase toward the edges of the ore zone as one would expect (i.e., generally where the density of drill-hole data diminishes).

2) That the estimated percentage of recoverable ore (upper right value) increases toward the center of the ore zone reflecting greater ore concentration and continuity there.

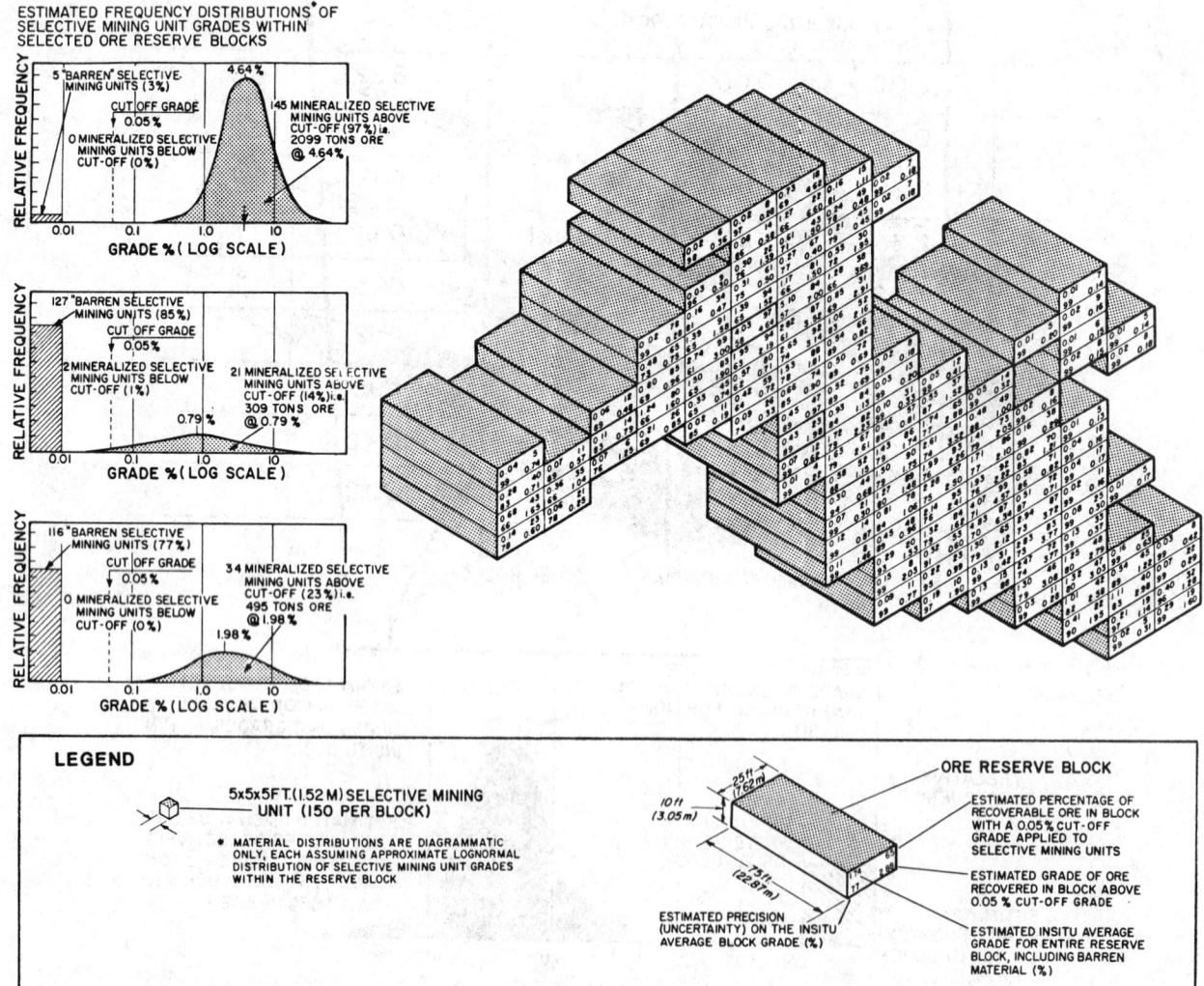

Fig. 57. Cross-section through high grade base metal deposit showing impact of selective mining on ore reserve estimates.

a) NUMERICALLY

MATERIAL	NUMBER SELECTIVE MINING UNITS	TONNAGE	MATERIAL DISTRIBUTION (%)	GRADE (%)	LBS. METAL
BARREN	10,920	159,100	52	0	—
MINERALIZED WASTE BELOW 0.05/	1,260	18,540	6	0.01	4,000
ORE >0.05%	8,820	127,160	42	1.88	5,259,000
TOTAL SECTION	21,000	304,800	100	0.78	5,263,000
INSITU BLOCKS— WITH TOTAL INSITU AVE. GRADE>0.05%		238,550	79	0.99	5,196,000

Fig. 58. Material distribution within ore body cross-section shown in Fig. 57.

b) SCHEMATICALLY

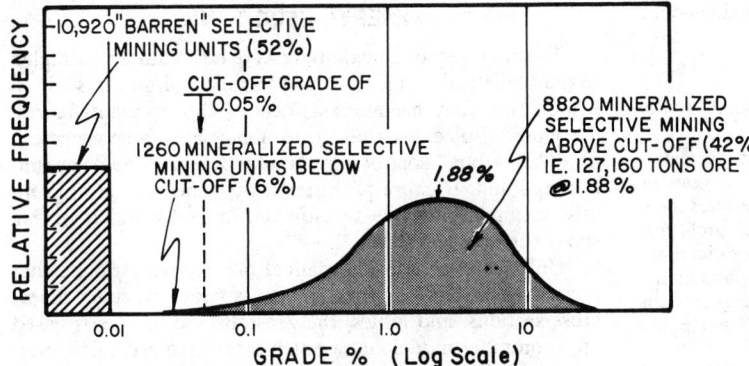

3) That in cases where the average or in-situ block grade is low, low proportions of very high grade ore are sometimes predicted for the ore recovered at the chosen selectivity. These estimates typically have poor reliability. The proportion recovered in these cases is low, however, and therefore the cumulative effect is not particularly significant.

4) That a picture of extreme variability is evident across the ore body. This variability supports the case for selective ore mining, at least as a basis for feasibility study planning.

5) Most of the metal content tends to be concentrated in the central core, or richest portion of the ore zone, that is some 18 to 36 m (60 to 120 ft) wide.

For illustrative purposes Fig. 57 also shows estimated grade distributions within a few distinctly different ore reserve blocks (insets on left). The frequency distributions of grade within these blocks show that barren material generally constitutes most of the waste. In other words, selective mining units, if they are mineralized, are probably going to be above cutoff grade. As previously described, the percentage barren in each case was estimated via an indicator kriging procedure that does *not* take into account the degree of mining selectivity intended. Taking mining selectivity into account is only possible in the procedure used to estimate recovered fractions of the mineralized portion using standard distribution theory.

Except in the case of a few very central blocks, this model suggests that most of the volume of a given ore reserve block will probably be discarded as waste. In fact, in this deposit only 28% of the total ore zone volume (ore reserve blocks whose in situ grade is greater than or equal to 0.01%) is estimated to be recoverable ore.

It is instructive to study the distribution of material within all of the ore-body blocks illustrated in Fig. 57. This distribution is shown schematically and numerically in Fig. 58. We find that for the envisioned 5 × 5 × 5-ft cube (1.52 × 1.52 × 1.52-m) mining selectivity, 42% of this cross section's total in situ block volume would be mined as ore at a grade of 1.88%. If, however, the deposit was mined with no greater selectivity than the 75 × 25 × 10-ft (22.87 × 7.62 × 3.05-m) in situ blocks, then we could expect significant dilution. With the same cutoff grade, 88% more tonnage would make ore grade. The average mined ore grade, however, would fall by almost one-half. The impact on mill size, capital costs, and process costs per kilogram (pound) of metal produced would obviously be dramatic. In such a high grade deposit, meticulous ore mining practices are warranted, at least for planning and costing purposes. In operation, such practices can always be easily relaxed but it can be embarassing and costly to a mine faced with the need to implement the reverse.

A final but nevertheless very important subject under this heading of mining recovery is that of bulk density. Often mining companies obtain extensive specific gravities on core samples and use averages of these, perhaps by rock type, to convert their volumetric deposit model into a tonnage based reserve. Many ore deposits by their very nature are highly fractured and permeable. They therefore have a true in situ bulk density that is perhaps up to 10% less than the core sample specific gravity tests suggest. This directly affects the ore reserves and estimates of expected metal recovery. To gain a proper fix on this variable, it is wise to undertake true in situ bulk density determinations through the typical excavation and sand filling technique if ore exposures are available at outcrop or within underground bulk sampling drifts. Alternatively, such tests should be done on large diameter core samples rather than on small cores such that the volume occupied in fracture voids will be taken into account. Particular care must be taken in *vuggy* ores where normal specific gravity tests may be unreliable.

Minability

To provide an estimate of minable ore reserves, one must undertake a determination of potential ultimate or economic pit limits and accumulate estimates of block by block recovered fractions (or full blocks as the case may be) and grades within these limits. In addition, it should be demonstrated that the reserves contained within this ultimate pit are practically accessible and minable throughout the pit development sequence, by way of sufficient mine planning (Mathieson, 1982). Pit plans for the early production years should in particular be designed and evaluated in detail. Procedures for doing this are covered elsewhere in this book.

Metallurgical Recovery

As has been previously recognized in this section, the term recoverable reserves:

"assumes and should imply that the valuable constituent is economically recoverable and, second, in relation to the first few years of an operation, that recoverability is assured, since in these years there may not be time to cope with unexpected metallurgical problems. Acceptance of this future widens the reach of ore reserve estimation. Instead of being, as it so often is, a matter of dimensions and grade, it becomes an assessment of the whole production process from in situ definition to separation of a saleable product." (King, et al., 1982)

In relation to metallurgical recovery, ore grindability will affect mill throughput and hence the timing of cash flows. Sufficient testing should be undertaken on representative samples taken from all ore rock types, particularly if autogeneous or semiautogeneous milling is being contemplated. Such tests should also be made on composite samples representative of yearly mine production from different ore rock types.

It is becoming increasingly important to adequately address such "geo-metallurgical" factors within feasibility and early engineering studies. Significantly different metallurgical response and hence metal recoverability is often encountered between different lithologic units within the same ore body. This might be due to varying degrees of mineral dissemination and association between mineral assemblages (affecting the grind and process stages required for liberation); the degree of alteration, mineral tarnishing, etc., and hence the ore hardness and natural flotation potential; or perhaps an association with certain gangue minerals that may promote or inhibit concentration of the economic minerals of interest.

All of these factors strongly suggest that geologists and mining engineers must play an active part in collecting drill core and bulk samples for metallurgical testing to ensure that they are representative of each ore rock type to be mined. Particular care should be taken in blending these samples in a manner that properly models the proposed mine plan. It should be recognized in this sample selection stage that ore mined and delivered to the mill will normally constitute a mixture of other waste or ore rock types between or adjacent to those that primarily constitute the ore. Metallurgical tests on such dilution impacts may also be warranted.

In addition to the metallurgical team, geologists and mining engineers should also thoroughly review metallurgical test data on the samples they have collected. Bench-scale tests on regular exploration core samples should be compared with tests done on larger diameter cores and/or bulk samples. Such reviews may suggest a need to recognize varying metallurgical response within the deposit modeling and reserve estimation methodology. This may simply entail the assignment of metallurgical parameters to each rock type or a more rigorous block-by-block estimation of these parameters.

In deposits that are highly variable metallurgically, it may be desirable to geostatistically evaluate the spatial correlation between metallurgical recovery and grindability with rock type and incorporate this into the recoverable reserve estimation procedures. Although not commonly done in the past, such studies, followed by kriging of metallurgical parameters, may prove in some highly variable deposits to be essential, if one is to gain sufficient confidence in the recoverability of metal from within a given mine plan, particularly in the early production years.

RESERVE REPORTING

To some, geostatistical ore reserve estimation is shrouded in mathematical complexity and technical jargon to such an extent, that they become skeptical of the resultant figures. They prefer to believe that the estimates have been generated in a "black box" sort of way, totally devoid of geologic and mining input. Because of this, and the general need to properly document any reserve estimate, some discussion on this subject is appropriate.

Unlike a conventional manual ore reserve estimate that is accompanied by largely self explanatory working maps, cross sections, and tables, the geostatistical reserve is based on a computer model and a much more involved calculation process. Adequate documentation of the methodology and the results is therefore essential if the reserve estimate is to have any credibility. The minable reserves estimation report should include as many of the following items as possible:

1) A complete discussion on the geologic setting of the ore body and the perceived geologic factors that influence and control the distribution of mineralization. Any conclusions that were reached, for example, with regard to relationships between mineral grades and structure and/or rock type, etc., should be addressed. Representative geologic maps and sections should be included. This should clearly establish to the reader that the "geo" in geostatistics has not been overlooked.

2) A brief exploration history for the deposit including details on any geophysics, drilling programs, core recovery, back reaming studies, bulk sampling, mapping, etc.

3) A complete description of the sample preparation and assaying procedures that were adopted. Results of all check assay analyses on split core, small and large diameter core, and between laboratories, etc., should be summarized along with any petrographic studies demonstrating distribution and gradation of mineral assemblages.

4) A discussion of results obtained from various specific gravity and in situ bulk density tests, justifying figures used within the reserve evaluation.

5) A description of the rock matrix or similar geologic input to the computer used to control geostatistical analysis and block interpolation.

6) Discussion on drill-hole coordinate survey reliability. A drill-hole location plan along with representative cross sections showing drill holes and grades should be provided.

7) Plots of actual semivariogram results within respective geologic zones of the deposit, and in different directions along

with semivariogram models chosen. This should include discussion on the geologic significance of the ore-body zonation that was used (i.e., mineralogic, alteration, rock type, structural, etc.) and any grade anisotropies found. The procedures and results of any cross validation or "jacknifing" tests conducted to verify the semivariogram models and the kriging method should also be given.

8) Grade frequency histograms of drill-hole bench composites within each grade zone, along with their cumulative probability plots and pertinent statistics.

9) A discussion of the modeling and kriging scheme used to estimate the reserves. Model dimensions, cell size, base-point coordinates and model orientation, etc., should be provided and illustrated. This should also permit the reader to understand the process used to estimate the various geologic zones in the deposit. The treatment of "fuzzy" boundaries, corresponding to gradational changes, and "sharp" boundaries, such as faults with major displacement, should be discussed here.

10) A series of bench plans and cross sections, showing not only the estimated block values and the precision with which they were estimated, but also the actual composite values used for estimation and the relevant geologic controls. These plans and sections should be reviewed in detail by the project geologists to assess the reasonableness of the results.

11) A summary of the block estimation errors or precisions by year for the first five to ten years of production and a discussion on how these errors might relate to a traditional ore reserve classification. This precision or uncertainty should incorporate not only the quantitatively derived estimation errors from kriging, but also a subjective estimate of the errors associated with grade zone definition, sample preparation, estimation methodology, and any known factors that, for whatever reason, could not be modeled, e.g., the effect of thin barren dikes cutting across an ore body that cannot be properly located.

12) A discussion on mining dilution, proposed mining methods, and inherent selectivity and procedures used within the ore reserve estimation process to account for these factors.

13) A plan of the proposed ultimate pit, along with the criteria used to derive it and plans demonstrating the minability of the ore reserves. It is important that the proposed mine plan and schedule incorporate sufficient insurance to cover a possible reduction in mined ore tonnage above the cutoff grade due to poor block estimates or perhaps greater mining selectivity due to blasthole data availability (i.e., discarding as waste some of the material predicted as ore). Obviously, if the mine plan does not recognize this possibility, however remote it is believed to be, there is a risk that the mine might either have to cut back on planned mill throughput or embark on a "crash" stripping program early in its life. The mine plan should provide sufficient preproduction waste stripping to obviate this concern.

14) A discussion on the chosen cutoff grade strategy and metallurgical characteristics related to geology insofar as they may effect annual process recoveries and metal production.

15) The reserve report should be accompanied by relevant geologic maps, data listings, bench plans, and cross sections.

In short, thorough documentation along the lines outlined increases a reviewer's "comfort index" by providing him with a full range of information from which he can understand the procedures used and the assumptions made. The reserve estimation team, through the discipline of preparing a detailed report, inevitably improves the reserve reliability and dramatically enhances the credibility of the result.

Statements on ore reserves in annual company reports should be well qualified such that they are truly meaningful to potential investors and shareholders. It should be remembered that the statement is only an estimate with uncertainty, and is far from being precise no matter how much statistical and geostatistical analysis is done. Statements should specify whether tonnage and grade estimates refer to in situ or recoverable reserves, should state the cutoff grade used, and what amount is considered to be in the measured (proven) and indicated (probable) categories. Comments should also be made regarding the level of risk in estimating reserves in the deposit in question in terms of its continuity and/or homogeneity.

Often, ore reserve estimation is treated as a mechanical process, of a purely mathematical nature, whose purpose is solely to manipulate a number of sample values to obtain tonnage and grade estimates. The result of this computational exercise is then summarized in statements of the form: "The ABC deposit contains 250,352,000 t (275.9 million st) of ore averaging 0.52%Cu." The numbers obtained are used by the mining company for internal financial and technical planning, and are published in stockholders reports, in a form that too often implies a high degree of certainty concerning their precision. Even though there is no question that reserve estimates must be made and disclosed, there is also a need for appropriate qualifications to be attached to these estimates, as to their meaning and the unavoidable imprecision involved.

Calculated reserves can represent global estimates characteristic of an entire deposit, independent of location, continuity, or exploitability. They can represent in-situ geologic reserves, where geologic controls and location are taken into account, but mining constraints have been ignored, except possibly in a very broad sense such as being subject to minimum thickness or grade constraints. If the necessary information is available, recoverable reserves could be calculated, taking into account mining method, mining selectivity, influence of cutoff grade, dilution, recovery possibly including metallurgical recovery, and other economic and technical constraints. Finally, minable reserves might be estimated as that ore tonnage and grade that is contained by and demonstrably minable within economic ultimate pit limits.

Clearly, the economic value of the ABC deposit will not be the same whether the 250,352,000 t (275.9 million st) of ore averaging 0.52% Cu represent global reserves, recoverable, or minable reserves. These reserve estimates will also have different meanings depending on the precision with which they are known. Estimation errors in excess of 50% are not uncommon, but in favorable conditions, the precision of the estimates may be as high as 10%. All of these factors must be taken into account when reporting reserve estimates internally and externally, as well as when using these estimates for decision purposes. We must remember that despite modern computer-based ore reserve estimation technologies, we are still typically dealing with relatively sparse data and often a highly variable ore body to be estimated. In consequence it is important that the estimating team appropriately convey their inherent uncertainty in the deposit model created or reserve estimate obtained.

REFERENCES

Blackwell, M.R.L., 1970, "Some Aspects of the Evaluation and Planning of the Bougainville Copper Project," 9th International Symposium for Decision Making in the Mineral Industry, Canadian Institute of Mining & Metallurgy, Montreal, Special Vol. 12, June.

Clark, I., 1979, *Practical Geostatistics,* Applied Sciences Publishers, London.

David, M., 1977, *Geostatistical Ore Reserve Estimation,* Elsevier, Amsterdam.

Journel, A.G., and Huijbregts, C.J., 1978, *Mining Geostatistics,* Academic Press, London.

King, H.F., McMahon, D.W., and Bujtor, G.J., 1982, "A Guide to the Understanding of Ore Reserve Estimation," Supplement to *Proceedings No. 281,* Australasian Institute of Mining and Metallurgy, Melbourne, March.

Krige, D.G., 1960, "On the Departure of Ore Value Distribution from the Lognormal Model in South African Gold Mines," *Journal,* South African Institute of Mining and Metallurgy, Nov., pp. 231-244.

Krige, D.G., 1962, "Effective Pay Limits for Selective Mining: Economic Aspects of Stoping Through Unpayable Ore," *Journal,* South African Institute of Mining and Metallurgy, Vol. 62, pp. 345-364.

Krige, D.G., 1978, *Lognormal de Wijsian Geostatistics for Ore Evaluation,* South African Institute of Mining and Metallurgy Monograph Series, Johannesburg.

Lane, K.F., 1964, "Choosing the Optimum Cutoff Grade," *Quarterly,* Colorado School of Mines, Vol. 59, No. 4, Oct.

Lasky, S.G., 1950, "How Tonnage and Grade Relation Help Predict Ore Reserves," *Engineering & Mining Journal,* pp. 81-85.

Matheron, G., 1962, "Traite de Geostatistique Appliquee," *Memoires,* Bureau de Recherches Geologiques et Minieres.

Matheron, G., 1976, "A Simple Substitute for Conditional Expectation: The Disjunctive Kriging," *Advanced Geostatistics in the Mining Industry,* Guarascio, et al., eds., Reidel, Boston, pp. 221-236.

Mathieson, G.A., 1982, "Open Pit Sequencing and Scheduling," SME Preprint 82-368, 1st International SME-AIME Fall Meeting, Honolulu, Hawaii, Sept.

Parker, H.M., and Switzer, P., 1975, "Use of Conditional Probability Distributions in Ore Reserve Estimation—A Case Study," *Proceedings,* 13th International APCOM Symposium, Clausthal, West Germany, Oct.

Parker, H.M., 1979, "The Volume-Variance Relationship: A Useful Tool for Mine Planning," *Engineering & Mining Journal,* Oct.

Readdy, L.A., Bolin, D.S., and Mathieson, G.A., 1982, "Ore Reserve Calculation," *Underground Mining Methods Handbook,* W.A. Hustrulid ed., AIME, New York, pp. 17-38.

Rendu, J.M., 1979, "Normal and Lognormal Estimation," *Mathematical Geology,* Vol. 11, No. 4.

Rendu, J.M., 1980, "Disjunctive Kriging: Comparison Theory with Actual Results," *Mathematical Geology,* Vol. 12, No. 4.

Rendu, J.M., 1981, *An Introduction to Geostatistical Methods of Mineral Evaluation,* South African Institute of Mining and Metallurgy Monograph Series, Johannesburg.

Rendu, J.M., 1984, "Interactive Graphics for Semivariogram Modelling," *Mining Engineering,* Sept., pp. 1332-1340; SME-AIME *Trans.,* Vol. 276.

Royle, A.G., 1980, "Why Geostatistics?," *Geostatistics,* McGraw-Hill, New York.

Sichel, H.S., 1952, "New Methods in the Statistical Evaluation of Mine Sampling Data," *Transactions,* Institution of Mining & Metallurgy, Vol. 61, pp. 261-288.

Sichel, H.S., 1966, "The Estimation of Means and Associated Confidence Limits for Small Samples from Lognormal Populations," *Proceedings,* Symposium on Mathematical Statistics and Computer Applications in Ore Valuation, South African Institute of Mining & Metallurgy, Johannesburg, pp. 106-122.

Taylor, H.K., 1972, "General Background Theory of Cutoff Grades," *Transactions A,* Institution of Mining & Metallurgy, Vol. 81, July.

Wainstein, B.M., 1975, "An Extension of Lognormal Theory and Its Application to Risk Analysis Models for New Mining Ventures," *Journal,* South African Institute of Mining & Metallurgy, Vol. 75,. pp. 221-238.

3.4 Application of Methods

3.4.1 Precious Metal Deposits

DAVID S. BOLIN

INTRODUCTION

Precious metal deposits, for the purpose of the following discussion, are considered to be those in which gold and/or silver are the principal commodities, both in terms of metal content and value. Not included are those deposits in which either or both metals are recovered as a byproduct, even though such byproducts may constitute a significant percentage of the total dollar value of the ore, and, in some cases the deposit might not be economic to mine without the precious metal content. Also not included in this discussion are platinum group metal deposits, although this group of deposits is sometimes classed with the precious metals. This presentation is confined to lode deposits; placer deposits are discussed in a separate section of this volume.

The sampling and evaluation of precious metal deposits presents some unique and complex problems in ore reserve analysis. Many of these problems are related to the widely variable geological environments in which these deposits occur and the often highly erratic distribution of values within a deposit. The problems tend to be magnified due to the high unit value of the precious metals in which a relatively small error in the estimation of grade can produce a large variation in value.

Assaying of precious metal deposits has become an increasingly critical factor. This is particularly the case in recent years, with the decrease in economic cutoff grades in response to the dramatic increase in gold and silver prices. Average economic metal concentrations are now often in the range of only a fraction of a part per million, a level that requires great care in sample preparation and analytical procedures to produce reliable figures for reserve estimation.

While there have been many developments in geostatistical techniques and computer modeling for ore reserve computation, the fact remains that the most critical phase in the evaluation of a precious metal deposit is that of sampling. This is the phase most susceptible to the generation of errors and biases, and the numerical results of the sampling program provide the basic data for all subsequent calculations.

The design of a sampling program should be given careful consideration. The sampling program should be designed for each individual project and should take into account the geological environment and mineral distribution characteristics of the deposit. The sampling program must be executed with care and attention to detail if reliable results are to be obtained without excessive expenditures.

TYPES OF DEPOSITS

Gold and silver deposits occur in a wide range of geological environments and have a variety of physical and mineralogical characteristics. In the following discussions of deposit types, these various characteristics will be emphasized in as much as they impact on the design of a sampling program. The classification system utilized should not be considered to have any genetic connotations in the geological

sense. The details of the various deposit types are summarized on Table 1.

This table gives an indication of the average mineral distribution characteristics of the various types of deposits. The tonnage column gives the range of tonnages for deposits known or in production in 1982. The average ore grade given is the grade being produced in 1982; however, many of the mines have operated at higher grades in the past. The grade value given for gold deposits is gold grade, without respect to silver credits, and for silver deposits, the value given is silver grade without regard to gold credits. The range of ore values is the range of values expected to be encountered between the 5 and 95% limits of the frequency distribution of the deposit type.

The average grain size given is the expected range in particle size of the valuable mineral for each type of deposit. This is an estimated value, and in some deposits, the range can be considerably greater.

The last column, the expected coefficient of variability, gives an indication of the variability of ore grade, and therefore the degree of precision to which a given ore grade estimate can be carried out, even assuming the sampling and assaying to be perfectly executed. The coefficient of variability is defined as:

$$C = \frac{s}{w}$$

where C is the coefficient of variability, s is the sample standard deviation, and w is the sample mean. This coefficient is dependent upon sample size, and the value given here is for a sample weight of approximately 40 kg (88 lb), the amount in a 1.5 m (5.0 ft) exploration rotary drill intercept.

The higher the coefficient of variation, the more difficult will be the estimation of ore grades within narrow limits. In general, values of the coefficient between 1.0 and 1.5 will produce reasonably good precision on a grade estimate. Values between 1.5 and 2.5 will have only fair precision, and should be treated with caution. Values greater than 2.5 are likely to produce grade estimates with a very high degree of imprecision, a factor which must be considered in the final evaluation of a deposit.

Gold Deposits: Those deposits for which gold is the major constituent, both in content and value include the following:

1) Stratabound conglomerate deposits of the Witwatersrand type.

2) Hydrothermal veins and breccia zones of mesothermal and epithermal association.

3) Disseminated deposits, usually in volcanic or sedimentary rocks.

Stratabound Conglomerate Deposits: The Witwatersrand type gold deposits are the most prolific known in the free world. The type locality is the Witwatersrand basin of

Table 1. Sampling Characteristics of Precious Metal Deposits

Type of Deposit	Tonnage t × 10⁶	Average Ore Grade, g/t	Range of Ore Values, g/t	Average Grain Size, mm	Gold/ Silver Ratio	Coefficient of Variability
Gold/Silver						
Stratiform Conglomerate	0.5 - >10	8 - 20	1 - 50	0.05 - 0.15	30	0.8 - 1.2
Veins and Breccias	0.1 - 5	10 - 40	1 - >1000	0.1 - 10	5 - 15	+ 2.5
Disseminated Deposits						
Stratiform	0.5 - >10	4 - 10	0.1 - 30	<.01	1 - 10	1.25 - 1.75
Disseminated-Hydrothermal	0.5 - >100	1 - 6	0.1 - 30	<.01 - 0.10	1 - 5	1.5 - 2.5
Shear Zone	0.1 - >10	1 - 10	0.1 - 50	0.05 - 0.1	1 - 5	2 - 4
Silver/Gold						
Veins and Breccias	0.1 - >50	100 - 500	10 - >1000	1 - 10	0.1 - 1	1.5 - >3
Disseminated	0.5 - >50	60 - 200	10 - >1000	0.1 - 1	0.1 - 0.5	1.5 - 2.5

English equivalents: 1t × 1.102 311 = 1 st; 1 g/t = 0.029 03 oz per st; 1 mm × 0.03937 = 1 in.

South Africa, although similar deposits are known in Canada, Brazil, and elsewhere. The deposits are characterized by usually thin conglomeratic beds of great lateral persistence. Mineralization and gold values tend to be rather uniform, sometimes, although not invariably, over distances of several hundred meters. The gold occurs as free gold with an average size of approximately 80 μm. The gold/silver ratio averages about 30.

Due to the general regularity of the distribution of values, and the fine grain size at the gold, there is usually little difficulty in sampling deposits of this type. The upper and lower boundaries of the seams or "reefs" are usually quite sharp, although lateral boundaries can be poorly defined.

Hydrothermal Veins and Breccia Zones: Deposits of this type are responsible for the majority of the sampling problems associated with precious metal deposits. This category includes a wide variety of deposits including:

1) Gold-quartz veins such as those of the Canadian Shield and the Mother Lode region of California.

2) Gold-sulfide veins, with varying amounts of quartz, calcite or other gangue such as the Homestake deposit at Lead, SD, and Central City, CO.

3) Epithermal veins and breccia pipes such as those in the San Juan Mountains of Colorado and Goldfield, NV.

The gold in the first group usually occurs as the native element; however, in the other groups the gold may be free, bound in sulfide minerals, or, more rarely, as various gold tellurides. Vein boundaries are usually very sharp, and the gold mineralization tends to occur as discreet ore shoots within the veins. Values are normally very erratic, but grades may occasionally be in the range of 5-10 kg/t (12-24 lb per st). Grain size of the gold varies from 0.1 to over 10 mm (0.004 to 0.39 in.), a characteristic that further complicates sampling of deposits of this type.

Disseminated Deposits: Deposits in this category have received increasing attention with the steady increase in the price of gold in recent years and are the deposits usually mined by open pit methods. These deposits occur in a variety of geologic environments, but are most commonly encountered in volcanic and sedimentary rocks.

The deposits are characterized by relatively large tonnage with average grades that vary from 0.1 to about 6.0 g/t (0.003 to 0.175 oz per st). The ore bodies occur in stockworks and shear zones, as lenticular stratiform ore zones, and as columnar chimney-shaped bodies. These deposits, with a few exceptions, have not been extensively reported in the literature, but seem to fall into the following general types:

1) Stratiform deposits in carbonaceous sedimentary rocks, such as Carlin, NV.

2) Disseminated, often pipelike bodies in sedimentary or volcanic rocks such as Pueblo Viejo, Dominican Republic, or Round Mountain, NV.

3) Wide fracture or shear zones in various host rocks such as the Zortman-Landusky deposits of Montana.

Gold distribution in these deposits tends to be somewhat more regular than vein deposits. The boundaries of economic mineralization tend to be poorly defined and limits are often based on assay cutoffs. Within a single disseminated deposit, the range of values is usually somewhat more restricted than the vein deposits with values rarely exceeding 15 g/t (0.44 oz per st). In some deposits, economic mineralization is confined to the oxidized material due to metallurgical considerations.

Silver/Gold Deposits: Silver and silver/gold deposits occur primarily as veins and occasionally as disseminated deposits. Vein deposits, with the possible exception of the Cobalt, Ont., type, are most often epithermal deposits. Typical examples include Guanajuato, Mexico; Tonopah, NV; Comstock, NV; and the epithermal silver deposits of the Sierre Madre Occidental of Mexico. These deposits are most often encountered in volcanic terrains in the form of veins and breccia bodies. These deposits are usually of somewhat more complex mineralogy than gold veins and silver occurring as the native element, simple sulfides, silver sulfosalts, and tellurides. Base metal minerals are common, and most deposits contain at least minor gold values.

The vein deposits suffer the same problems as the vein gold deposits. The values vary from a few parts per million to several kilograms per ton, and distribution of values tends to be highly erratic. The vein walls are usually sharply defined, and economic values are often confined to discreet ore shoots with distinct tops and bottoms.

Known disseminated silver deposits are mostly stockwork or shear zone controlled, with the grade of mineralization closely related to degree of fracturing. The values tend to have a rather restricted range, usually varying between 60 to 300 g/t. Mineralization is more regular in distribution than the vein deposits, and the limits of economic values are usually defined by assay boundaries. Typical examples of this type include Delamar, ID; Rochester, NV; and, Calico, CA.

SAMPLING

The design of a sampling program for a precious metal project must be tailored to the type of mineralization known or expected to be present, and to the amount of previous exploration or development of the property. Traditionally, precious metal deposits have been sampled by means of prospect pits, trenches, and underground development. Within more recent years, drilling, both core and noncore, has assumed a more prominent role in the evaluation of gold and silver deposits, particularly the disseminated types.

Channel Sampling

Channel sampling of underground or surface workings is the method most often applied to lenticular ore bodies of various types. In the classical application of the method, a channel of constant width and depth is cut across the vein or ore zone. If the vein is of less than mining width, the walls are usually sampled separately to provide information for the calculation of dilution.

In practice, the cutting of precise channel samples is a laborious and often very expensive process, and the collection of "chip channel" samples has become more common. If reasonable care is exercised in collecting the sample, the chip samples are often statistically indistinguishable from traditional channel samples.

Although usually applied to vein type deposits, channel or chip channel sampling of crosscuts and trenches is utilized in the evaluation of disseminated precious metal deposits. These samples are usually cut in a continuous line along the rib of a crosscut or trench with the length of individual samples varying from 1 to 3 m (3 to 10 ft). Veins or breccia zones are usually sampled separately.

Samples are cut with a hammer and moil or, if available, an air or electric powered chisel. The cut sample is collected on a canvas tarp for homogenization and splitting if required. The mechanics of channel sampling have been described in numerous publications, for example, McKinstry (1948).

Drilling

Evaluation of precious metal deposits by drilling has become more common in recent years due to the increase in exploration for disseminated type deposits. Although drilling is applicable to vein, disseminated, and strataform deposits, the method has often proven to be of doubtful value in many vein occurrences. While drilling is often used to confirm the continuity of vein structure and mineralization, the assay values obtained are often unreliable. This is due to the usually highly variable distribution of values in vein systems and often poor sample recovery in fractured and brecciated veins.

Core Drilling: This drilling method provides lithological and structural information as well as providing a sample for assay; however it is usually the most expensive of the methods available for general use.

In instances where substantial drilling is required to evaluate a deposit, it is usually considered to be good practice to core a certain percentage of the holes in the drill program. The actual percentage will vary with the available budget and the degree of complexity of the geology, but usually averages about 10%.

Core samples offer the advantage that sample location can be closely defined, and there is little possibility of contamination or loss of values as long as recovery is good.

On the negative side, core drilling is expensive and the usual sample obtained (NX or BX) is relatively small. In cases where the ore boundaries are fairly well known, the cost can sometimes be lowered by rotary drilling to near the ore boundary and then continuing with core.

Noncore drilling: Noncore drilling techniques have become increasingly popular in recent years due to the low cost relative to diamond drilling combined with the large number of holes required for the evaluation of high tonnage, low grade deposits. The noncore drilling methods in common use include: (1) conventional rotary drilling, (2) downhole hammer drilling, (3) reverse circulation drilling, and (4) percussion drilling. Each of these methods offers certain advantages and disadvantages. Percussion drilling, either from a small air-track machine or underground longhole drill, is usually very cheap and simple to operate. The disadvantages are the restricted depth range, usually less than 30 m (100 ft), the small diameter drill hole, and often poor sample recovery.

Conventional rotary or hammer drilling is relatively cheap, but there are often problems with sample recovery and possible contamination from uphole high grade zones. The equipment is usually truck-mounted and road construction may be required for access.

Some of the problems with sample recovery may be overcome by reverse circulation drilling. This technique utilizes double walled drill pipe with the circulating medium, air or water, passed down through the outer space and recovered, along with the sample, through the center pipe. The principal disadvantages are the relatively high cost and the size of the equipment.

The circulating medium for sample recovery may be either air, water, or mud, however, air is preferred if possible. Sample handling is considerably easier and the problems of obtaining water for drilling are avoided if air drilling is utilized.

A potential problem is the high pressure air required for reverse circulation drilling. It has sometimes been suspected that values may be lost by being blown into fractures, although this has not been documented. For this reason, some companies have gone to low pressure-high volume air systems in order to mitigate this possibility.

Sample recovery is usually by means of a cyclone attached to the air return system. These systems work quite well, and few problems have been reported. The recovery is calculated by weighing the sample recovered in each interval, commonly 2 or 3 m (6.5 to 10 ft). It is also a good practice to occasionally sample the cyclone dust overflow, particularly when oxidized silver mineralization is known or suspected to be present.

The drilling density determined for a given deposit is a function of the geologic complexity, continuity of mineralization and economic constraints. In disseminated deposits, spacing of drill holes is often 15 to 30 m (50 to 100 ft) in somewhat irregular mineralization, to as much as 100 m (330 ft) in more regular mineralization. The usual procedure is to begin the evaluation drilling at the widest spacing possible to outline the deposit, then fill in the drill pattern to a point that grade precision is considered to be adequately defined. Often variogram studies (see Section 4) may help in defining optimum drill spacing.

Bulk Sampling

At some point in the evaluation of a precious metal deposit, the collection of bulk samples for confirmation of grade estimates and metallurgical testing will be required. These samples are usually collected from existing underground workings, driving new headings or large diameter drilling.

In existing workings samples up to several tonnes may be obtained by slabbing along the rib of a crosscut or by cutting a channel down the back utilizing short drill holes and explosives. When driving new headings for sampling, samples may be taken on car lots.

SAMPLE PREPARATION

Sample Size

The size of the sample required to provide a reliable grade estimate is dependent upon the absolute content of the precious metals, the grain size of the valuable minerals, and various other factors. The general rule of thumb has always been "the bigger the better."

In practice, of course, sample size is limited by physical and economic constraints. Most samples taken for the evaluation of an ore deposit range from 10 to 50 kg (22 to 110 lb), the range of weight of a 1.5-m (5.0-ft) chip sample or drill interval. Once a sample has been broken, the determination of a representative sample weight becomes amenable to a more rigorous examination.

The problems of determining minimum sample weights have been investigated by many authors (for example, Gy, 1979; David, 1977, pp. 331-345). The work of Pierre Gy has resulted in the presentation of a formula for the calculation of the fundamental sample variance given various parameters of the ore in question. The simplified version of this formula is:

$$s^2(FE) = \frac{Cd^3}{Ms}$$

where $s^2(FE)$ is the fundamental sampling variance; d is the maximum particle diameter in centimeters; Ms is the sample weight in grams; and C is a sampling constant where $C = clfg$, c being the mineralogical composition factor, l the liberation factor, f the shape factor, and g the particle size distribution. This formula may be applied in several manners. The two most common are: (1) given a maximum acceptable sample variance, to calculate the sample weight necessary to achieve this precision; (2) given a sample size, to compute the expected precision of the estimate. This formula may be utilized for order-of-magnitude estimates of minimum sample size; however there are several factors that may produce substantial errors.

The mineralogical composition factor, c, is very simple to compute if the ore mineral is free gold in a siliceous matrix. In this case the factor is primarily a function of the gold assay grade. In complex silver ores in which silver may be present in several different minerals, the factor is not a simple function of grade and calculation should be made based on known or estimated mineral composition.

The liberation factor, l, may be difficult to estimate, particularly in gold deposits, due to the variation in size that may range from 30 μm (500 mesh) to 1680 μm (10 mesh) or larger in a single sample. It is always good practice to pan a crushed sample of the ore to obtain an idea of the size distribution of the gold particles.

The shape factor for most calculations is assumed at 0.5 that represents a spherical particle shape. Both gold and silver, particularly in the native form, often occur as wires, flakes, and other forms that may require an adjustment to the value of this factor.

While there are problems with the application of this formula to sampling of precious metal deposits, it has been used in several cases and has produced generally good results (Vallee, et al., 1976). The formula should not be applied blindly; the various components of the formula should be carefully evaluated in the light of the geological and mineralogical characteristics of the ore.

As an example, consider a 1.52-m (5.0-ft) intercept in an 114 mm (4 ½-in.) rotary drill hole. This sample will weigh approximately 40 kg (88 lb), and the maximum chip size should be about 6.4 mm (0.25 in.) and a gold particle size of 10 μm. What might be a minimum sample weight at this sample particle size to provide a 95% probability that a 5.0 g/t (0.15 oz per st) gold would be known within $\pm$ 1.0 g/t (0.03 oz per st)? Using a modified form of Gy's equations:

$$Ms = \frac{4\, cflg\, d^3 A^2}{a^2}$$

where $c = 3.8 \times 10^6$, $f = 0.5$, $l = 0.001 - 0.65 = 0.04$, $g = 0.25$, $A = 5$, $a = 1$; $Ms = 494$ kg. This weight considerably exceeds the amount of sample available, which indicates that, under the stated parameters, the entire sample should be reduced prior to splitting in order to achieve a precision of 1.0 g/t (0.03 oz per st).

It is important to remember that application of this formula (or any other formula) will result in only an order-of-magnitude estimate for sample size. It is also significant that the theory applies only to broken material, that is that the sample weight is assumed to come from a mass of the same particle size, a condition that is not fulfilled in the direct sampling of ore in the ground. For further information and examples, refer to Gy (1979).

Sample Reduction

A series of comminution and sampling steps are required in order to arrive at a sample of one assay ton for analysis. Each of these steps involves potential errors and opportunity for the generation of biases. Each sampling step must be analyzed to ensure that sufficient sample weight of the proper size is being taken to produce reliable results. For example, a split half of a drill core weighing approximately 8 kg (17.6 lb) must be crushed prior to splitting. If the gold averages about 100 μm (150 mesh) in size and 5 g/t (0.15 oz per st) in grade, and the entire 8 kg (17.6 lb) is crushed to 1.0 mm (0.04 in.), and a 2-kg (4.4 lb) split is taken, then the sampling variance would be:

$$s^2(FE) = \frac{cflg\, d^3}{2000}$$

$$s^2(FE) = 0.076$$

where $c = 3.8 \times 10^6$, $f = 0.5$, $l = \sqrt{0.01 \div 0.1} = 0.32$, $g = 0.25$, and $d = 0.1$ cm (0.04 in.) that is a reasonable figure. On the other hand, if the core is crushed to only 6.4 mm (0.25 in.) prior to splitting, the calculated variance would be:

$$s^2(FE) = \frac{cflg\, d^3}{2000}$$

$$s^2(FE) = 8.09$$

where $c = 3.8 \times 10^6$, $f = 0.5$, $l = \sqrt{0.01 \div 0.64} = 0.12$, $g = 0.25$, and $d = 0.64$ cm (0.25 in.). A variance of this magnitude would indicate a very low precision on the estimate of grade.

The point of the previous examples is to emphasize that in order to sample low grade gold deposits with a reasonable degree of precision, the samples should be reduced to 1.7

mm (10 mesh) or smaller prior to making any split. In this example as well as the preceding, an ore grade has been assumed. In the initial stage of exploration of a property, the grade distribution is unknown. However, as exploration proceeds, there will be some indication of what range of grades are to be expected in a given deposit.

ASSAYING

The traditional technique of assaying for gold and silver is the fire assay, and the method is still considered to be the reference standard of the industry. Atomic absorption spectrometry methods have been developed in recent years, most of which produce good results when proper care is exercised in sample digestion and preparation of standards. Some companies rely on atomic absorption for routine assays due to the somewhat lower cost, particularly in areas where there is sufficient experience to confirm the reliability of the method. It is considered good practice, however, to fire assay up to 25% of any sample group to maintain control.

As in all sampling programs, it is necessary to establish a check sampling procedure by which a certain percentage, usually about 10% of the sample pulps, are renumbered and resubmitted to the same laboratory. Samples should also be sent to outside laboratories for check analyses. It should be mentioned that there will usually be a wide variation in check assays on an individual sample basis. The mean for specific lots, however, should check fairly closely.

At an early stage of the evaluation program of a precious metal deposit, if cyanidation [either for heap leaching or conventional countercurrent-decantation (CCD) circuit methods] appears to be a potential recovery method, bottle roll tests often are carried out on drill samples. These tests give an idea of cyanide extractability of the precious metals in the ore.

STATISTICAL CONSIDERATIONS

Precious metal deposits are usually characterized by high variability. Shown on Table 1 are the coefficients of variation of approximately equivalent size samples for various types of precious metal deposits. By comparison, the coefficients of variation for various other deposit types are shown in Table 2. The high variability of the precious metal deposits is readily apparent in comparison with those other deposits.

Most precious metal deposits have a lognormal frequency distribution, examples of which are shown in Fig. 1. When a deposit has this type of distribution, the calculation of the lognormal mean and variance often gives more accurate results than a straight arithmetic average. The tables prepared by Sichel may be utilized for the calculation of lognormal

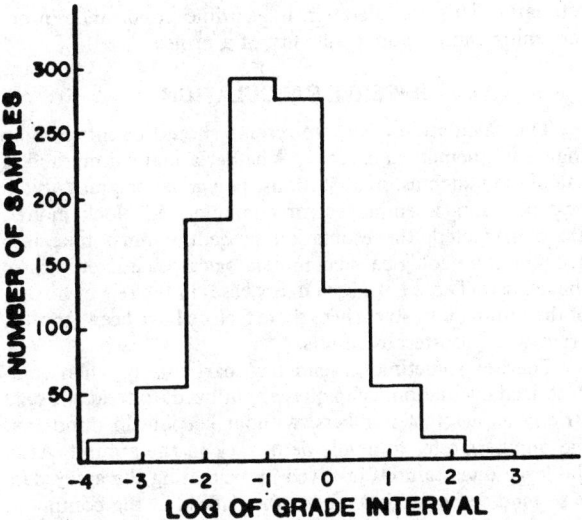

Fig. 1a. Histogram of disseminated silver sample values, stockwork silver deposit, Rochester District, Nevada.

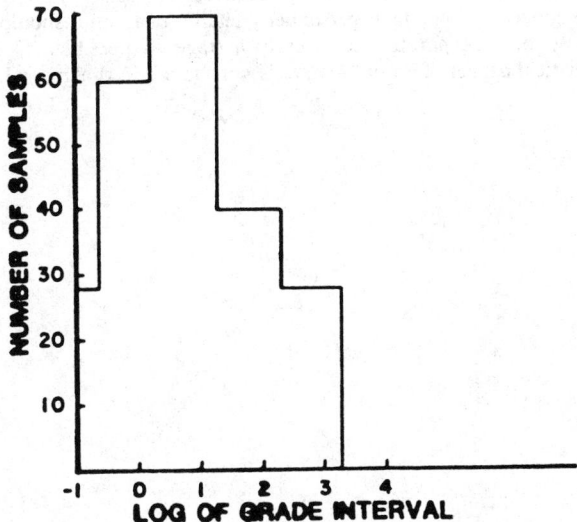

Fig. 1b. Histogram of disseminated gold sample values, stockwork (hydrothermal hot springs) gold deposit, Nevada.

means. The methodology of the calculations is detailed in Section 4 of this chapter.

The use of lognormal averages also provides a means of accounting for the erratic high values that usually occur in any set of assay data from a precious metal deposit. Another method which has been applied in some mines is to cut all high values to a value equal to the mean plus twice the standard deviation of the original data set. This method often has a tendency to undervalue the grade of many deposits.

The critical aspect of the grade estimate of a precious metal deposit is the realization that the grade represents a point on a probability distribution, not an absolute value. If the sampling and assaying has been carried out correctly and without bias, the grade estimate should be reasonably precise, but the high natural variability of this type of deposit will always inject a high degree of uncertainty in most grade

Table 2. Average Coefficients of Variation from Various Mineral Deposits

Type of Deposit	Coefficient of Variation
*Silver (veins)	1.218
*Uranium (sandstone)	1.359
*Tungsten (vein)	2.21
Tin (vein)	2.11
Copper (porphyry)	0.95
Gold (vein)	4.33
Gold (disseminated)	1.63
*Lead	1.427
*Zinc	1.19

*Data from Hazen and Meyer, 1966.

estimates. This consideration is of prime importance in determining the economic viability of a project.

RESERVE CALCULATION

The calculation of an ore reserve should be much more than a mathematical exercise. Whether a manual method of calculation such as cross sections, polygons, or grade-thickness contours is employed, or computerized block models are constructed, the estimation procedure must take into account the geological and mineralogical characteristics of the deposit. There have been many cases in the recent history of the mining industry where these factors have been ignored, often with disasterous results.

There is sometimes a tendency, particularly when computerized evaluation techniques are utilized, to treat the data strictly as a set of numbers, without keeping in mind that the numbers refer to a volume of rock in the ground. Also, the large uncertainties involved in generating the assay data base tend to be forgotten once the data is in the computer. When the computer models are constructed, geological boundaries such as structural features, lithologic contacts, and mineralizational zones must be respected.

For these reasons, the preparation of an ore reserve should be a cooperative effort between geologists, mining engineers, and systems personnel. Metallurgical input should also be incorporated at as early a stage as possible. The critical aspect of an ore reserve is not the ore in the ground,

but the amount of final product that can be profitably extracted. If the final metal production is significantly less than anticipated, the entire exercise of reserve estimation has been in vain.

REFERENCES

Cochran, W.G., 1963, *Sampling Techniques,* John Wiley, New York, 413 pp.

David, M., 1976, "What Happens If—Some Remarks on Useful Geostatistical Concepts in the Design of Sampling Patterns," *Sampling Practices in the Mineral Industries,* Australasian Institute of Mining and Metallurgy, Melbourne Branch, pp. 1-16.

David, M., 1977, *Geostatistical Ore Reserve Estimation,* Elsevier, Amsterdam, 364 pp.

Gy, P., 1979, *Sampling of Particulate Materials,* Elsevier, Amsterdam, 431 pp.

Hazen, S.W., and Meyer, M.L., 1966, "Using Probability Models as a Basis for Making Decisions During Mineral Deposit Exploration," Report of Investigations 678, US Bureau of Mines, 83 pp.

Koch, G.S., and Link, R.F., 1970, *Statistical Analysis of Geological Data,* Vol. 1, John Wiley, New York, 375 pp.

Koch, G.S., and Link, R.F., 1971, *Statistical Analysis of Geological Data,* Vol. 2, John Wiley, New York, 438 pp.

McKinstry, H.E., 1948, *Mining Geology,* Prentice-Hall, Englewood Cliffs, NJ, 680 pp.

Vallee, M., Filion, M., and David, M., 1976, "Of Assay-Tons and Dollars, or Can You Trust Gold Assay Values," Canadian Institute of Mining & Metallurgy, 78th Annual Meeting, Quebec, 1976.

3.4.2 Reserve Estimation of Uranium Deposits

HARRY M. PARKER

INTRODUCTION

Uranium deposits are nearly universally characterized by erratically distributed mineralization that is commingled with barren material. Because of this commingling and the relatively high costs of beneficiation of the ores to produce a salable product (twice to three times the cost per ton of treatment of copper ores), uranium mines tend to be small operations that allow the maximum practicable segregation of ore from waste.

The very close-spaced sampling, on which decisions to select ore from waste are made, is generally done just prior to or during mining, and hence this information is not available for making reserve estimates. Instead, wider-spaced exploration or development sample data must be used to predict the tonnage of ore and grade of ore within an area of the deposit or even within the deposit taken as a whole. Precise blocking of ore and waste, that is possible in making reserve estimates for the more massive and/or continuous deposits of iron ore, coal, or many base metal ores, is not possible for uranium. This factor is recognized in the planning of uranium operations, and sufficient stockpiles and/or exposed mineralized zone (typically three to six months mill feed) are maintained so that fluctuations between predicted reserves and actual production are smoothed out.

Because of the erratic character of the mineralization, it is often difficult to determine geologic controls on ore/waste contacts or even the outlines of mineralized areas with a degree of certainty sufficient to increase the accuracy of reserve estimates. Because of this difficulty in locating precise ore/waste boundaries, reserve methods in common use are either deterministic (polygons) or statistical (probabalistic). Both methods have been successfully applied in the evaluation of reserves for a large number of prospects and operating mines.

The appropriate use of these methods is described in subsequent sections. However, before considering the methodology, it must be emphasized that the accuracy of any reserve estimate depends ultimately upon the accuracy of the samples used. In most sandstone-hosted uranium deposits and in some "hard-rock" deposits, the assay is not a chemical or other direct analysis, but a radiometric equivalent based on counts of gamma radiation received per time interval at a detector. Proper probe calibration is important to guarantee that a systematic bias is not introduced in the "assays."

It is also important to recognize that most of the gamma radiation is emitted by daughter-product elements. The uranium and/or its daughter products may be locally selectively leached and/or redeposited leading to "disequilibrium" and more or less gamma radiation arriving at the detector than there would be expected for a given grade of uranium if this migration had not occurred. Investigation of disequilibrium and local correction for it is critically important to produce unbiased reserve estimates. This is particularly true in the case of roll-front deposits where uranium and its daughter products have been successively oxidized, leached, reduced, and redeposited. Disequilibrium correction factors of between 0.5 and 2 are not uncommon in these environments.

METHODS OF ASSAYING

An extended discussion of various assaying techniques is beyond the scope of this chapter. In general two methods of assaying are in common use, X-ray fluorescence (XRF) and natural gamma logging.

X-Ray Fluorescence (Core or Chip Samples)

This method has proven to be a rapid inexpensive replacement of wet chemical analyses. The sample is irradiated by X-rays that cause elements within the sample to give off or fluoresce X-rays in turn. Characteristic spectra are emitted for each element, and the number of emitted X-rays at characteristic wave lengths are proportional to the amount of the element present. A spectrometer is used to measure the count rate of emitted radiation at various wave lengths. Because elements other than uranium will often emit X-rays of nearly the same wave lengths as those emitted by uranium, adjustments termed matrix corrections are often made by computer to give a refined estimate of the amount of uranium present. A helpful discussion of the subject is given in Levinson (1974).

Natural Gamma Logging (Within Drill Holes)

Gamma rays are emitted as uranium radioactively decays to lead. As given in Dodd, et al., (1967), uranium itself has a rather long (4.5×10^9 years) half-life compared to some of its less stable daughter products such as radium (1622 years), radon (3.82 days), lead-214 (26.8 min), and bismuth-214 (19.7 min). Of these isotopes the latter two are the significant gamma ray emitters. The implications of this are explained by Dodd, et al., (1967):

"If uranium is in secular equilibrium with its gamma-emitting daughters, and if there is an insignificant contribution of gamma rays from other sources, then the concentration of uranium in the rocks immediately surrounding a bore hole can be determined from a gross-count gamma-ray log with high sensitivity and practical accuracy. Because of the geochemical characteristics of the various radionuclides, these conditions are seldom, if ever completely fulfilled. Fortunately, for many economically valuable concentrations of uranium these conditions are sufficiently met, or data may be obtained to supply adequate corrections, to determine the thickness and mean equivalent grade of the mineralized zone from the gross gamma-ray log and by calculation, to closely approximate the grade of incremental layers within the zone."

Scott, et al., (1961), have shown that the grade-thickness product is proportional to the area under the peaks of anomalies (plotted in counts per second) on radiometric logs. The counts per second recorded on the log are a biased estimate of the counts per second emitted by the rock surrounding the drill hole. Casing, if present, water, and mud in the hole absorb some of the radiation before it can reach the detector, and hence the reading is generally increased by a factor often termed the *water correction factor*. The correction for casing is sometimes made separately (S.P. Morzenti, personal communication). Also, when the detector receives an X-ray photon, a pulse of electric current is generated causing a "count" to register. While this "registration" is occurring, other incoming photons cannot be counted. The time it takes this to occur, usually only a few microseconds, is termed the *dead time* of the detector. The following formula generally is used to estimate the counts per second actually emitted by the rock:

$$N = \frac{(\text{water correction factor})(\text{measured counts per sec})}{1 - (\text{measured counts per sec})(\text{dead time in sec})} \quad (1)$$

Thus if the water correction factor were 1.07 and the dead time were 5.5×10^{-6} sec, an observed count rate of 6000 would be adjusted to 6639 counts per sec.

Prior to the widespread use of computers, the grade and thickness of ore for each peak were calculated by hand, as described by Linton (1963). Figs. 1 and 2, reproduced from Stoll (1972), illustrate the technique. Basically the thickness is taken to be the distance between the upper and lower half-peak amplitudes measured to the nearest $\frac{1}{2}$ ft (15 cm). As shown in Figs. 1 and 2, the accumulation (grade $\times$ thickness or GT) in % U_3O_8 in feet is found to be:

$$GT = K \left(\sum_{i=1}^{\ell} N_i + 1.38 (E_1 + E_2) \right) \quad (2)$$

Here the factor K permits conversion of counts per second into grade-thickness units. It is determined by calibrating the probe in a hole drilled through a "test pit" containing known grade. Details of the procedure are provided in Scott, et al., (1961), and Dodd, et al., (1967).

The grade (measured in percent U_3O_8 in the United States and percent uranium in much of the rest of the world) is obtained by dividing GT by the estimated thickness in feet. This is so even though the number of measurements between the half-peak amplitude marks E_1 and E_2 is $2T$. If a grade were to be desired for a $\frac{1}{2}$ ft (15 cm) zone, ignoring the ends of the anomaly, then:

$$G (0.5) = KN$$

$$G = 2 KN \quad (3)$$

Scott (1963) developed a more sophisticated method to predict the grade using a computer. As shown in Fig. 3, a mineralized zone is made up of a series of layers, each of which has anomalous gamma radiation, and which superimpose to give the observed anomaly. Scott hypothesized that a series of $\frac{1}{2}$ ft (15 cm) layers could be found to which synthetic grades could be assigned that would reproduce the anomaly. These synthetic grades could then be used in reserve estimation. He next calculated the amplitude of an anomaly for a $\frac{1}{2}$ ft layer at various distances from the center:

i	Distance from Center, (ft)	Relative Amplitude	W_i
$i-3$	-1.5	2%	0.01
$i-2$	-1.0	8%	0.04
$i-1$	-0.5	40%	0.20
i	0.0	100%	0.50
$i+1$	0.5	40%	0.20
$i+2$	1.0	8%	0.04
$i+3$	1.5	2%	0.01

The amplitude at position i on a log may be thought to be made of contributions from adjacent positions using the stated weight factor W_i. To obtain the grade at i, as a first approximation the grades are assigned to all positions using the formula:

$$g_i^{\circ} = 2 KN_i \quad (4)$$

Grades for the three positions above and below the logged zone are assigned $g_i^{\circ} = 0$.

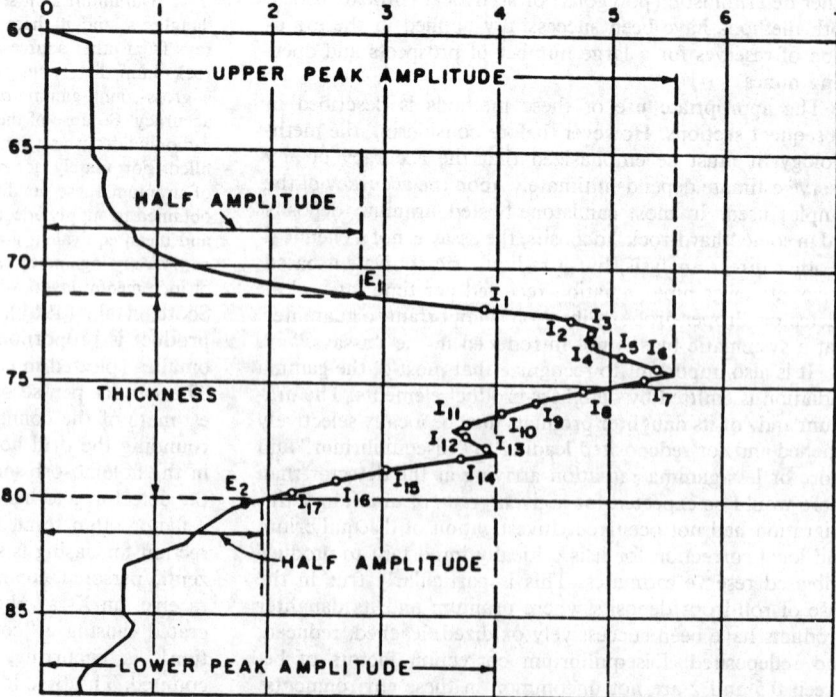

Fig. 1. Thickness determination, numerical integration points, determination of tail areas and central area (Stoll, 1972).

UNITED STATES ATOMIC ENERGY COMMISSION
GRAND JUNCTION, COLORADO
GAMMA RAY LOG INTERPRETATION WORK SHEET

Claim LEASE NO. 100 - JONES
Company LPI
LOCATION
Sec. 1 Twp. 1N Rng. 101 W
Date logged 3/10/70
Date interpreted 3/20/70
District GULF COASTAL PLAINS State TEXAS

Log Operator L.E. PHARR
Interpreter M.G. STOLL
Unit Dead Time 5.5+ u sec.
Water Factor 1.10
Other Factors NONE
K Factor 0.000004632
Range 100K

Probe No. 102
Ratemeter No. 1
Unit No. 2
Tail Factor 1.38
Standard Reading NONE
Disequilibrium Ratio NONE

Inches	n	N	Inches	n	N
E_1 2.80	28,000	33,097			
E_2 1.80	18,000	19,978			
		$E+E_1$ = 53,075			
$E+E_1$ X 1.38		73,244			
3.90	39,000	49,650			
4.65	46,500	63,479			
4.84	48,400	65,950			
4.85	48,500	66,144			
5.10	51,000	70,883			
5.40	54,000	76,814			
5.20	52,000	74,806			
4.77	47,700	64,665			
4.35	43,500	57,180			
4.05	40,500	52,107			
3.75	37,500	47,244			
3.745	37,450	47,165			
3.925	39,250	50,056			
3.70	37,000	46,463			
3.04	30,400	36,503			
3.00	26,000	30,338			
2.20	22,000	25,028			

INTERVAL

Lower Boundary __ 80.0 ft.
Upper Boundary __ 71.5 ft.
Thickness _____ 8.5 ft.

996,717 — Σ Ncps
x 1.10 — Correction Factor(s)
Corrected Area — 1,096,389
x 0.000004632 — K Factor
GT = 5.078
T = 8.5
Average grade % eU₃O₈ = 0.597

GT X Disequilibrium Ratio = NA
+ Thickness =
Corrected grade % U₃O₈ = NA

55 X 10^{-6}

EXAMPLE OF STEP PROCEDURE NECESSARY FOR EACH "N" VALUE.

$$N = \frac{n}{1 - n t}$$

WHERE

n = Observed Count Rate   N = 28,000
t = Unit Dead Time   1-(28,000 x 5.5 x 10⁻⁷)
N = Corrected Count Rate   N = 33,097

Fig. 2. Gamma-ray log interpretation worksheet (Stoll, 1972).

The grades g_i° are assigned to G_i, the observed composite anomaly. Then the synthetic anomaly grades are assigned for all i except the three positions on either end of the anomaly:

$$G_i' = \sum_{i=i+3}^{i=i+3} W_i g_i \qquad (5)$$

For the first time $g_i = g_i^\circ$. The difference between G_i and G_i' is found and added to the g_i as a correction:

$$g_i \text{ (new)} = g_i + (G_i - G_i') \qquad (6)$$

For the first time $g_i = g_i^\circ$. Then the process is repeated with a new set of G_i' being computed from the g_i (new). The differences are taken again and a new set of g_i (new) created, etc. Should a g_i (new) be found negative, it is set to zero.

When the maximum difference is found to be less than significant, typically 0.0025% U_3O_8, the synthetic grades g_i (new) are accepted and the process is terminated. Generally convergence is obtained within 10 to 15 iterations. It is wise to set a maximum of 25 to prevent endless looping for exceptional cases.

Programs using this algorithm have been developed by the Grand Junction Office of the Department of Energy (formerly the Energy Research and Development Administration, ERDA, and before that the Atomic Energy Commission, AEC). The programs have been referred to variously as GAMLOG, MDDAT, or MDAT and have been used by many companies in the US mining industry.

S.P. Morzenti of Exxon Minerals (personal communication) has commented that some probes in current use give different shaped anomalies from those studied by Scott, and therefore different weighing factors should be used. The reader should consult the probe manufacturer for advice.

The successful use of gamma logs to predict the uranium content is dependent on the assumptions presented in the papers referenced being met. Key among these is the assumption of horizontal stratification of the mineralization, that is met in most undeformed sandstone deposits and a

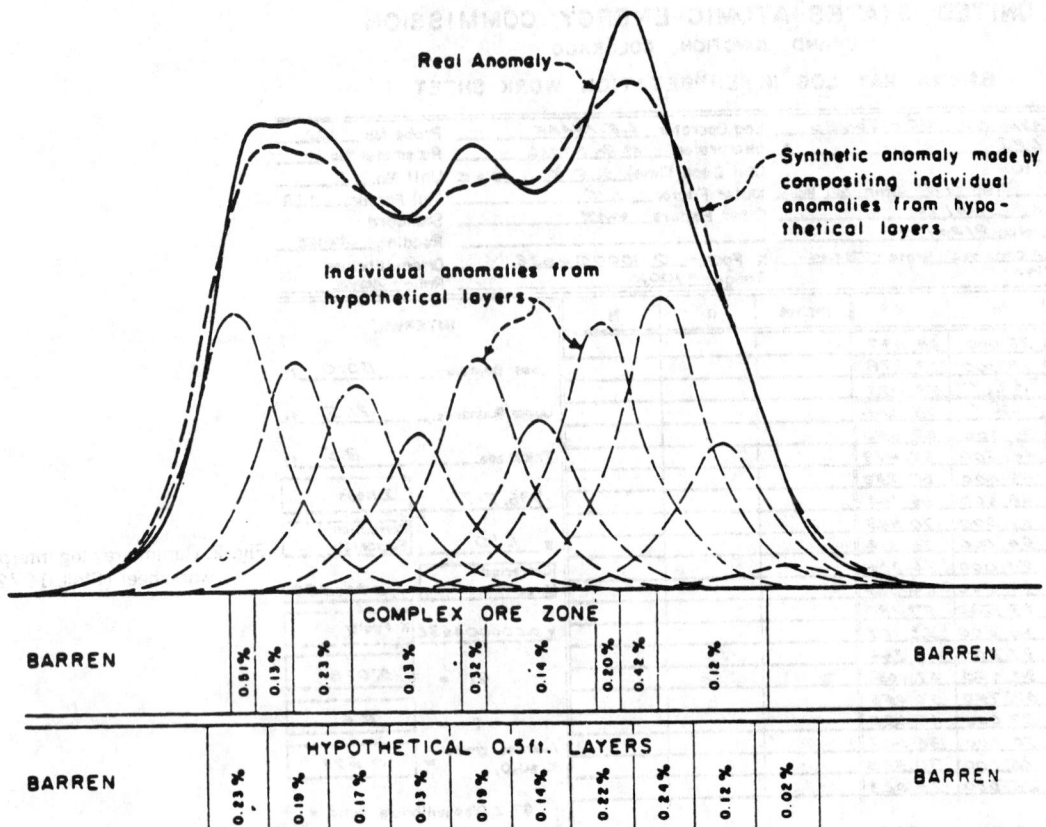

Fig. 3. Complex uranium ore zone and equivalent hypothetical ½ ft (15 cm) thick layers with associated gamma-ray anomalies (Scott, 1963).

constant *K*-factor that is independent of grade. Robert Enwall of Rocky Mountain Energy Company (personal communication) has found that significant overestimation of the grade occurred when these assumptions were made for a low-grade (less than 0.05% U_3O_8) deposit with mineralization contained along steeply dipping fractures.

Other Techniques

Several logging techniques employing neutron fission are beginning to come into commercial use. These techniques, described by Czubek (1972), involve pulsed generation of neutrons from a source in the drill hole and measurement of the epithermal neutron time distribution (prompt fission neutron method, PFN) or the delayed neutron distribution from fast neutron fission or uranium-238 or thermal neutron fission of uranium-235. These techniques are not dependent on equilibrium between uranium and its daughter products and offer promise of rapid, accurate direct measurement of the uranium content of ores.

More recently, as is described in *Engineering & Mining Journal* (1980), a probe has been developed that measures only the gamma radiation from protactinium-234. Protactinium is the first daughter product of uranium and has a half-life of only 26 days. Hence it will be in equilibrium with the uranium from which it is produced. This tool has been used in routine exploration and drilling programs and is considered operational and reliable (Ken Weissenburger, CONOCO Inc., personal communication). The method has the advantage over the neutron fission methods in that the latter use a "hot" source that requires special handling and

could be lost in the process of logging with environmental consequences.

DISEQUILIBRIUM CORRECTIONS

The discrepancy between radiometric gamma log equivalent assays and chemical or XRF assays is often loosely termed *disequilibrium*. As shown by Hansink (1976) if uranium is leached from the tail zone of a roll front, less soluble daughter products such as radium, radon, lead, and bismuth that emit 98% of the gamma rays remain, and the grade, if not corrected, will be overestimated. Conversely in the protore zone ahead of the nose of the roll front, uranium is being or has been redeposited without the time necessary to build up concentrations of gamma-ray-emitting daughter products.

In fact the discrepancy between gamma logs and assays of core may be due in various degrees to the following factors:

1) Geometry of mineralization other than horizontal.

2) Use of linear *K* factor when nonlinear should apply.

3) Instrument drift, inaccurate water correction factor, dead time correction, etc.

4) Volume-variance relationship of sample volumes (core volumes are smaller than volumes swept by the probe). The ratio of chemical to gamma log value is sometimes called the sampling factor.

5) Inaccuracies or systematic biases in collection of cores such as poor core recovery or sample preparation.

6) Inaccurate match between cores and gamma logs.

7) Presence of anomalous potassium-40 in feldspars, micas, and clay minerals, or thorium in hard-rock deposits.

8) True disequilibrium between uranium and its daughter products. This includes migration of uranium or its daughter products (such as radon which may move into highly porous or fractured zones).

Measurement of disequilibrium has been described by Rosholt, et al., (1965), Santos (1975), Killeen and Carmichael (1976), and Ostrihansky (1976). As indicated previously simple study of radiometric vs. chemical analyses of the same sample usually is not enough. Disequilibrium may be overshadowed by discrepancies due to other factors previously listed. A good rule (Sam Smyth, personal communication) is to have at least 10% of the drill holes be cored to establish geographic and hopefully geologic control on the discrepancies.

Correction for discrepancies has been found to be relatively straightforward for deposits where remobilization has not occurred. Fig. 4 shows a linear regression correction which was made for a stratabound uranophane deposit. This should be contrasted to Fig. 5 that is taken from a roll-front deposit in Wyoming. Here lognormal regression has been used to make the correction. The procedure for doing this is as follows:

1) Plot both chemical (chem) and radiometric (gamma) assays separately on lognormal probability paper. For each distribution find a constant C to add to the assays Z such that $Z + C$ is lognormal (plots as a straight line, see Fig. 6). Details of the procedure are described in Rendu (1978).

2) Let:

$$Y_{chem} = \ln (Z_{chem} + C_{chem}) \qquad (7a)$$

$$Y_{gamma} = \ln (Z_{gamma} + C_{gamma}) \qquad (7b)$$

Perform a linear regression of Y_{chem} on Y_{gamma} and find a predicting equation:

$$Y^*_{chem} = A (Y_{gamma}) + B \qquad (8)$$

3) Because Y^*_{chem}, the expected value of Y_{chem} given Y_{gamma}, is not the logarithm of the mean (expected value of ($Z_{chem} + C_{chem}$) given ($Z_{gamma} + C_{gamma}$) that is desired, a final correction is required:

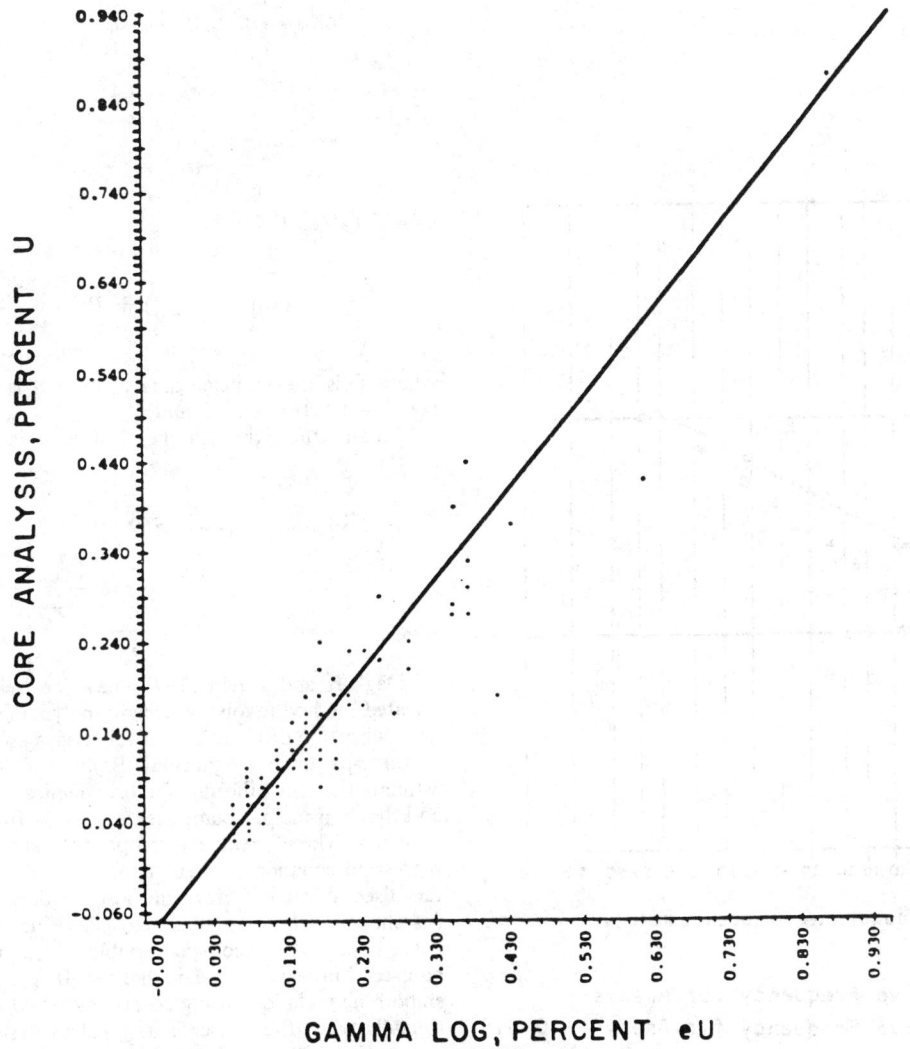

Fig. 4. Scatterplot of core analyses vs. gamma log and linear regression line.

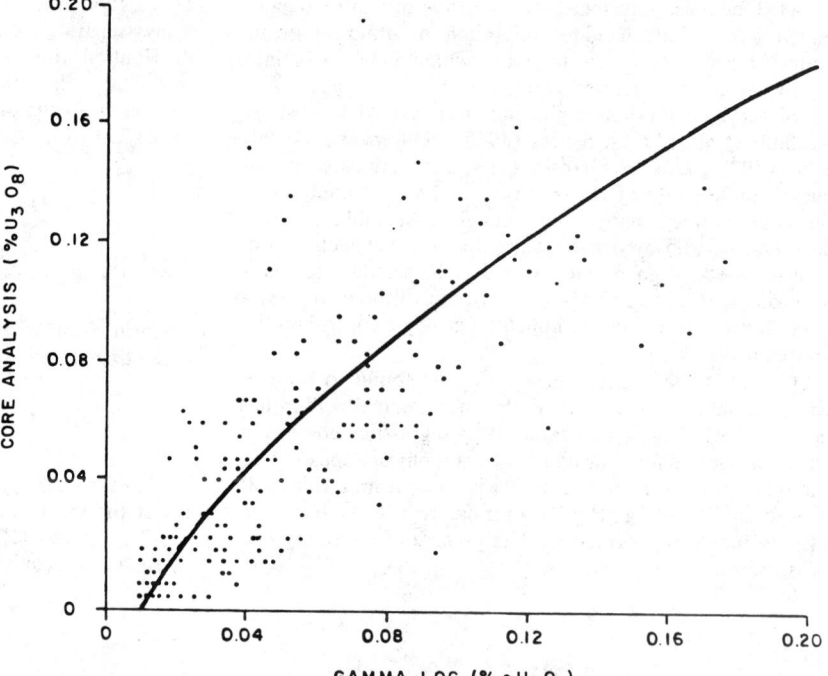

Fig. 5. Scatterplot of core vs. gamma log analyses and nonlinear regression line obtained by lognormal regression.

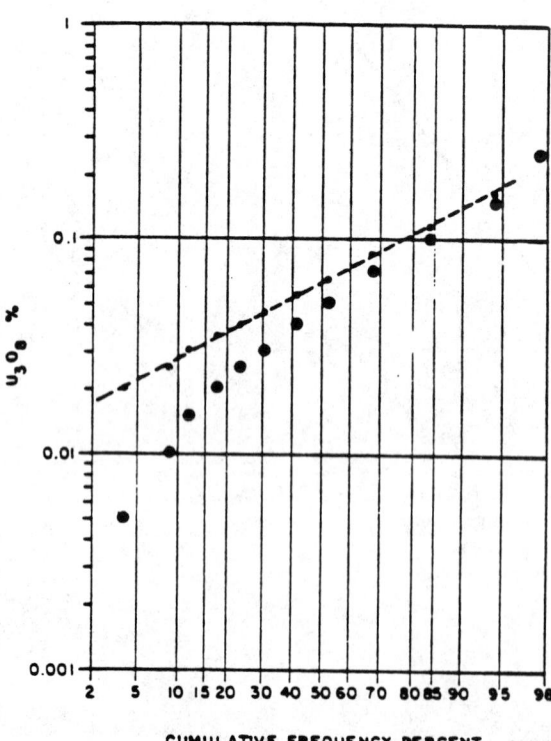

- ● Cumulative Frequency for Assays
- ● Cumulative Frequency for Assays + 0.015

Fig. 6. Plot of uranium assays on lognormal probability paper.

$$(Z_{chem} + C_{chem})^*$$

$$= \exp\left[A\,(Y_{gamma}) + B + \frac{s^2\,(1 - \rho^2)}{2}\right] \quad (9)$$

where s^2 is the variance of Y_{chem} and ρ is the correlation coefficient between Y_{chem} and Y_{gamma}.

Finally, the estimated chemical value is

$$Z^{\cdot}_{chem} = (Z_{gamma} + C_{gamma})^A \exp\left[B\right.$$

$$\left. + \frac{s^2\,(1 - \rho^2)}{2}\right] - C_{chem} \quad (10)$$

Dagbert and David (1979) have devised a more complicated method involving hermite polynomials that may be used when the distributions of the assays are not lognormal or three-parameter lognormal. Bryan and Roghani (1982) estimate the disequilibrium factor, chemical/radiometric, by co-kriging principal components formed from 12 geologic variables. These predicted components are then used in a regression equation to predict the disequilibrium factor. As the disequilibrium factor may also be dependent on grade (as shown in Fig. 5), it is suggested that the gamma log value be added to geologic variables if this procedure is to be used. Marbeau and Marechal (1980) provide an illusory smoothing technique using co-kriging on composites; unfortunately when the scatter is as great as is shown in Fig. 5, no method will provide a very accurate correction. In the face of this, and making the assumption that uranium leached from the tails of rolls is redeposited in the protore zone ahead of the noses, some companies do not attempt to correct for disequilibrium in roll-front environments. Reserve estimates based on these assumptions may or may not be biased globally, but most certainly will be locally, particularly where the deposit is above the water table. Here oxidation and leaching have generally occurred. As a consequence, only the daughter products remain.

Other companies, such as Exxon's Highland, WY, operation, calculate disequilibrium factors for geologic environments, such as roll near wing, wing, far wing, etc., and monitor the actual disequilibrium encountered during production. (S.P. Morzenti, personal communication).

RESERVE ESTIMATION USING DETERMINISTIC METHODS

Polygonal Approach

As stated in the introduction, most uranium mines are small-scale, highly selective operations. Typically the ore stringers in a roll-front deposit may be less than 1.5 m (5 ft) thick and 4.6 m (15 ft) wide perpendicular to strike (Sandefur and Grant, 1976). As indicated by these authors, drilling on 3-m (10-ft) centers would be required to outline the ore bodies. The cost of this sampling would be prohibitively expensive.

Sandefur and Grant show that the reserves may be accurately determined by weighting the grade-thickness product of ore in each hole by its polygonal area of influence. The following steps are followed:

1) Ore intercept definition. A minimum minable thickness is defined, typically 0.7 to 1 m (2 to 3 ft). Very often 15 cm (0.5 ft) may be added to the top and bottom at grade or zero grade to represent dilution. A minimum grade is defined that the intercept must meet to be accepted. Reserves are often categorized according to the intercept criteria, e.g., "2 ft (0.7 m) of 0.04," "3 ft (0.9 m) of 0.06," etc.

The assays, whether corrected radiometric equivalent or chemical, are examined to determine the intercepts for each hole. This process is often computerized. The intercepts are generally chosen without regard to a common datum or bench level, as mining typically follows the ore/waste contacts. The thickness and grade-thickness product of each intercept are noted. Sometimes the intercept will be tagged with a level or geologic horizon indicator. This permits tabulation of reserves by geographic position or geologic horizon within the deposit.

2) A polygonal area of influence is computed for each hole. Polygonal boundaries are perpendicular bisectors of lines connecting adjacent holes. (The area of influence of isolated holes is often restricted.) Fig. 7 shows a typical map.

3) The reserves in an area of a deposit are simply calculated:

$$\text{Accumulation} \atop \text{(metal content)} = Q^* = K_Q \sum_{i=1}^{n} A_i \, q_i \qquad (11)$$

$$\text{Tonnage} = T^* = K_T \sum_{i=1}^{n} A_i \, t_i \qquad (12)$$

$$\text{Grade} = G^* = \sum_{i=1}^{n} A_i \, q_i \bigg/ \sum_{i=1}^{n} A_i \, t_i \qquad (13)$$

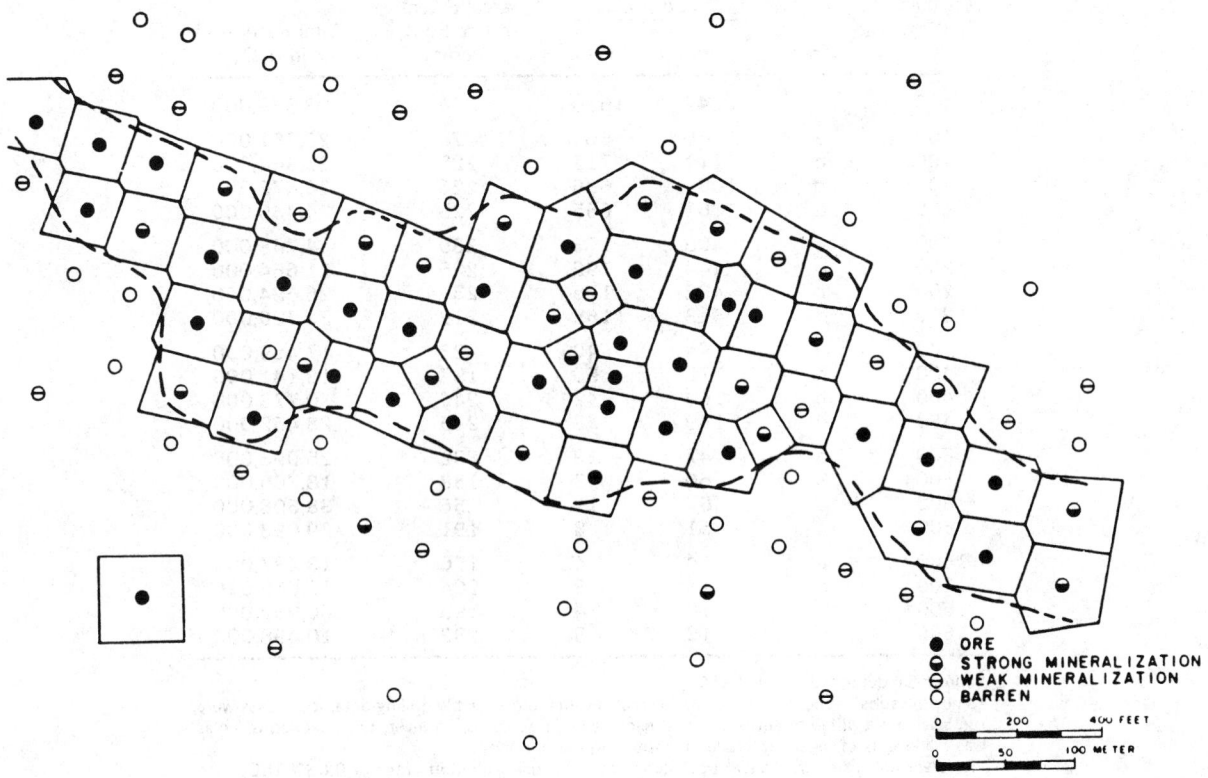

● ORE
◒ STRONG MINERALIZATION
⊖ WEAK MINERALIZATION
○ BARREN

Fig. 7. Polygonal areas of influence for a stratabound uranium deposit.

where A_i is the area of polygon of influence for hole i; t_i is the thickness of ore intercept(s) for hole i; q_i is the grade-thickness product of ore intercept(s) for hole i; K_Q is the conversion factor to obtain kilograms, pounds, etc., of uranium; and K_T is the density or tonnage factor.

It is common practice in many mines to drill higher grade or accumulation areas on a tighter spacing. Because these intercepts often have limited areal extent, it is appropriate to weight these holes less in a reserve estimate, as is done by the polygonal method. A lower weighting of closer holes also reflects the possible spatial correlation and duplication of information provided.

Fig. 7 shows that it is possible, particularly if a specific geologic horizon is considered, to have non-ore holes commingled with ore holes. Given the remarks referenced earlier by Sandefur and Grant, the polygons must be thought of as tools to assign areas of influence (weights) to the assays to obtain an unbiased estimate of reserves, and *not* tools to block out ore on a local basis for mine planning. During mining, ore will be found to occur in non-ore polygons and vice versa.

Table 1, reproduced from Sandefur and Grant (1976), shows the variation in polygonal reserves for various hole spacings. There is little change once the number of holes in ore becomes statistically large [at the 61-m (200-ft) spacing]. As explained subsequently, geostatistics may be used to define the appropriate hole spacing.

Journel and Huijbregts (1978, p.464) and Parker (1979) have shown polygonal estimates to be biased in many mines

because of internal dilution toward overestimation of grade. Where the cutoff grade is above the mean, the tonnage is also overestimated. Where the cutoff grade is below the mean, the tonnage is underestimated. This is because the grade and tonnage estimates are based on a declustered (area of influence weighted) distribution of sample volumes, and the assumption is implicitly made that selection of ore from waste will be made on similar sized volumes. In many (other than uranium) mines this does not occur. The selection is based on estimates of larger volumes, up to several thousand tons in extent, that are comprised of many sample-sized volumes, some of ore and some of waste grade material, leading to the internal dilution mentioned previously.

In highly selective uranium operations the difference between polygonal estimates and recovered reserves has *not* been found to be a problem. This is because of the degree of selectivity. Probing of ore in the pit on spacings as close as 0.6 m (2 ft), mining with as small as 0.8 m³ (1 cu yd) backhoes, and scanning of truck loads, typically containing 23 to 45 t (25 to 50 st), all contribute to accurate selection of ore from waste using selective mining units of similar volume to that detected by a gamma logging probe.

Cross-Sectional Approach

The cross-sectional approach has been used to outline hard-rock uranium deposits, such as those found in the Athabasca region of Saskatchewan or Northern Territory of Australia. The same erratic character of grade is seen in vertical section as in the horizontal direction for the roll-

Table 1. Polygonal Ore Reserves Based on Various Drill Hole Spacings

Minimum Spacing, ft	Case*	No. of Holes Total	In Ore	Areal Extent of Ore Body, acres	Total Reserves,† lb U₃O₈
0		2142	1630	224	23,338,000
100	a	1050	665	224	23,751,000
100	b	1111	712	225	21,864,000
100	c	1088	686	226	26,472,000
100	d	1061	665	225	24,460,000
200	a	403	173	230	24,748,000
200	b	444	192	225	21,664,000
200	c	432	175	234	26,684,000
200	d	413	165	219	23,720,000
400	a	144	40	223	17,469,000
400	b	172	50	197	18,141,000
400	c	169	42	242	30,773,000
400	d	149	38	215	28,452,000
800	a	47	12	202	25,996,000
800	b	60	13	188	18,765,000
800	c	57	11	256	38,608,000
800	d	51	8	191	29,093,000
1600	a	14	4	176	13,937,000
1600	b	19	3	204	19,615,000
1600	c	18	4	195	60,952,000
1600	d	15	3	232	10,498,000

Source: Sandefur and Grant, 1976.
*Since the subset process chooses certain holes within the available drilling data, four cases were run selecting a different set of holes in each case. In all cases, however, the selection of holes was made without regard to grade or position in the ore body.
† Reserves calculated by standard polygonal methods, using a cutoff grade of 0.03% U₃O₈.
Metric conversions: 1 ft × 0.3048 = 1 m; 1 acre × 0.4047 = 1 hectare; 1 lb × 0.4536 = 1 kg.

front deposits. Ore intersections are projected into blocks surrounding drill holes. Because ore is localized in veins or high-grade pods, accurate assessment of reserves is difficult. The geostatistical approaches outlined in the next section are only beginning to be implemented and offer hope of improvement in the accuracy of reserve estimates.

Inverse-Distance Approach

With the advent of computers it was recognized by Weiss and O'Brian (1971) that locally weighted averages of bench composites could provide better estimates than polygons of the reserves on a local basis. This is because the use of all samples that are spatially correlated with the block should provide a better estimate than using the closest sample. Invoking the theory that closer samples should be better predictors than ones further away, the inverse-distance (often inverse distance squared) method of assigning weights to samples was tried. Unfortunately in most cases computer users did not understand the highly selective nature of uranium mining and the limited extent of the ore bodies. They mistakenly performed inverse-distance weighting on ore *and* waste composites alike. This resulted in a markedly lower estimate of overall grade than would be obtainable with polygons and geologically unrealistically smoothed ore/waste contacts. As a result, the uranium industry has largely rejected use of this method.

Were inverse-distance weighting applied to grade-thickness and thickness of ore intercepts to estimate the tonnages and accumulation of ore in mine planning units, the method would be useful in providing both unbiased estimates of overall reserves and better and unbiased estimates of local reserves than that provided by polygonal estimators. However, as shown in the next section, more accurate geostatistical methods have been developed that perform the same function.

Remark

Much has been said throughout this paper of the high degree of selectivity of uranium operations and the need to be cognizant of this in developing reserve estimates. This is true for most but *not all* operations in existence. There are in existence and were until the recent decline in uranium sales price several contemplated low-grade deposits at which large-scale less-selective mining methods are used or were envisaged. Reserve estimation for these deposits have thus far been mainly treated using geostatistics. As explained in the next section, a deterministic approach to reserve estimation for these deposits is not recommended.

RESERVE ESTIMATION USING STATISTICAL METHODS

Because of the erratic distribution of the grade and tonnage of uranium ore bodies within a deposit, statistical methods that make inferences about a population (the ore bodies) from samples are useful, and as shown by Sandefur and Grant (1980) may provide a superior accuracy of prediction. There is a trade-off, however. The methods and their applicability must be thoroughly understood to achieve success. A detailed explanation of the theory would be of necessity too lengthy to be provided here. The reader is referred to the section of this chapter by J.M. Rendu for an introduction. Excellent texts on basic statistics as applied to geological and mining problems have been written by Koch and Link (1970) and Davis (1973). Textbooks on geostatistics by David (1977), Journel and Huijbregts (1978), Rendu (1978), and Clark (1980) are also helpful. Discussion papers on uranium reserve estimation using geostatistics have been written by Parker and Chavez (1979), David and Dagbert (1980), and Marbeau and Marechal (1980). The selection of the appropriate method depends on the density of data, geology, and selectivity of mining method. Described in the following are methods that have been shown to work and for which published case studies are available.

Geostatistical Setting

As mentioned previously, uranium deposits tend to be comprised of distinct ore bodies within a zone of weak or even barren mineralization (see Fig. 7). Thus two populations exist that can be referred to as the barren and the mineralized population. As shown in Fig. 8 taken from Parker, et al. (1979), the grade of the mineralized population is often approximately lognormally distributed (and by adjustment of the barren/mineralized cutoff can sometimes be made to be lognormally distributed). Sandefur and Grant (1980) have also shown that the grade-thickness product of ore intercepts may also be lognormally distributed.

Where the deposit is thick and comprised of many interlayered ore bodies, that may happen if the cutoff criteria are low, nearly every hole may have a positive grade-thickness product, as shown in Fig. 9. On the other hand, if the grade-thickness product is being computed over a single stratigraphic horizon, with high cutoff criteria, many holes that intersect only the barren or weakly mineralized population may occur, as indicated in Fig. 8. In these cases it can be advantageous to estimate the area underlain by ore by coding each hole with an indicator 1 if ore bearing and 0 if barren, and kriging (refer to the section by J.M. Rendu) this variable to estimate the proportion of the area underlain by ore.

Obtaining variograms for uranium deposits is a difficult task given the strong local proportional effect (Fig. 10) between mean and standard deviation within pods of mineralization and the erratic distribution of grade and change in directions of anisotropy throughout the deposit. The proportional effect may be handled by calculating relative variograms (variograms that are scaled by dividing by a function of the mean $f(m)$). Very often $f(m) = m^2$ has been found to apply and can be theoretically proven in the lognormal case. A full explanation is found in Journel and Huijbregts (1978, pp.187-191, 250-251). Alternatively, variograms on the logarithms may be calculated (Fig. 11), that may be converted to a relative variogram provided the underlying distribution is multivariate lognormal using the following formula:

$$\gamma_{rel} = \exp\left[C_{\log}\right]\left[1 - \exp\left(-\gamma_{\log}(h)\right)\right] \quad (14)$$

where $C_{\log}$ is the sill of the variogram of logarithms.

Knudsen and Kim (1979) found that by calculating the variogram parallel and perpendicular to the roll-front axes, significant anisotropy in the spatial variability (ratio of the ranges of 2.5) could be observed. Parker, et al., (1980), coded the strike direction of the roll fronts within the deposit and locally oriented the variogram search cone (Fig. 12) to obtain the along- and across-strike variograms shown in Fig. 13. If the roll-front orientation varies even more than indicated by exploration drilling, the successfulness of this technique maybe limited.

One characteristic of nearly all variograms is a relatively short range [often less than 50 m (163 ft)] and a high nugget effect. This implies that accurate local reserve estimates will be difficult to obtain without close-spaced drilling. It also means that selective mining units that are *significantly* larger

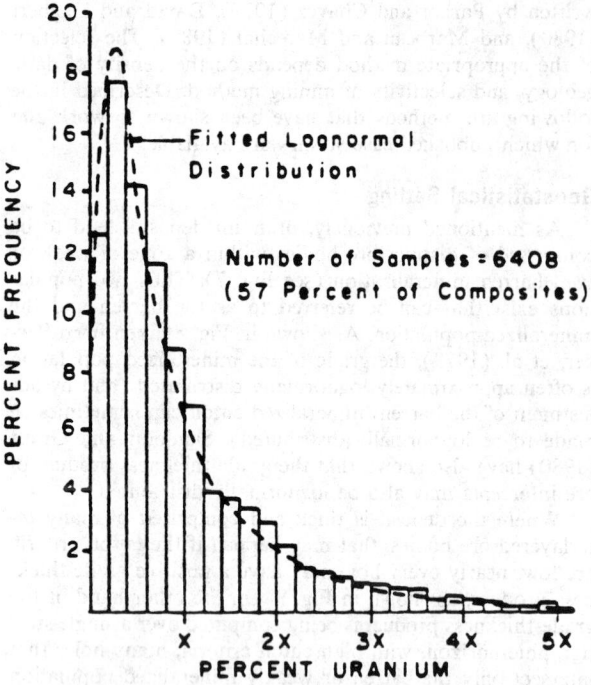

Fig. 8a. Histogram of mineralized population.

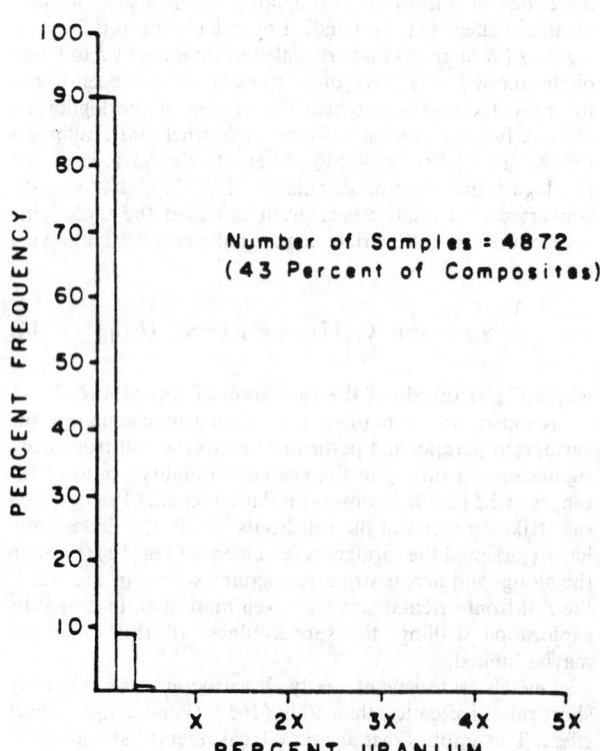

Fig. 8b. Histogram of barren population.

than sample-sized volumes will have markedly different grade tonnage curves from that of samples (polygonal estimates).

Reserve Estimation with Sparse Data

This section applies to the early exploration stage, when for example 50 or less holes are available, of which some may be barren and some mineralized, and the spacing is beyond the typical range of influence of a variogram. In this case estimation of mean ore grade-thickness by taking the arithmetic mean of positive grade-thickness intercepts can provide a misleading estimate of the mean grade-thickness of ore. As demonstrated by Koch and Link (1970), when samples are drawn from a skewed distribution, the maximum likelihood is that the arithmetic mean of samples will be below the deposit mean. This is because much of the uranium content is contained in rare rich zones that will most likely not be intersected in the first pass of drilling. For example, in one project on which the author monitored the reserves, the mean grade-thickness rose by 20% over the course of going from 50 to 400 holes. On the other hand, occasionally the first pass may intersect one of the rich pods in which case the arithmetic average of the data would be biased high, and extreme caution is warranted in making reserve estimates.

The grade-thickness is often approximately lognormally distributed. This can be checked by plotting the available data on log probability paper. In this case the Sichel t estimator may be used to find an unbiased estimate of the mean. A description of the technique is given in Sichel (1966) and in David (1977). This method uses the logarithms of the sample data to estimate the mean. It must be used with discretion since it only yields an unbiased estimate when the underlying distribution is lognormal and when the data are distant enough from one another to allow assumption of their independence. This is difficult to verify with a small amount of data. The Sichel method also provides an estimate of confidence limits on the estimate. These are heavily dependent on inferences made about the distribution and calculation of the variance and statistical independence of samples, and must be weighed with caution.

The distribution of thickness of ore-bearing holes tends to be less skewed and is best estimated using the arithmetic average of the data. Computation of tonnage obviously depends on extrapolation around and interpolation between ore-bearing holes. In the absence of sufficient information upon which to base a mine plan, classifying this tonnage as reserves is inappropriate. Using geologic judgment to outline favorable areas and *provided the drilling density is uniform,* the area underlain by ore may be taken to be equivalent to the proportion of ore holes multiplied by the total area judged to be favorable.

Given the estimated average grade-thickness is q^*, the estimated average thickness of ore-bearing holes is t^*, and the proportion of area underlain by ore is estimated to be p^*, the resources may be estimated to be:

$$\text{Accumulation (metal content)} = Q^* = K_Q \, A \, p^* \, q^* \qquad (15)$$

$$\text{Tonnage} = T^*P = K_T \, A \, p^* \, t^* \qquad (16)$$

$$\text{Grade} = G^* = q^*/t^* \qquad (17)$$

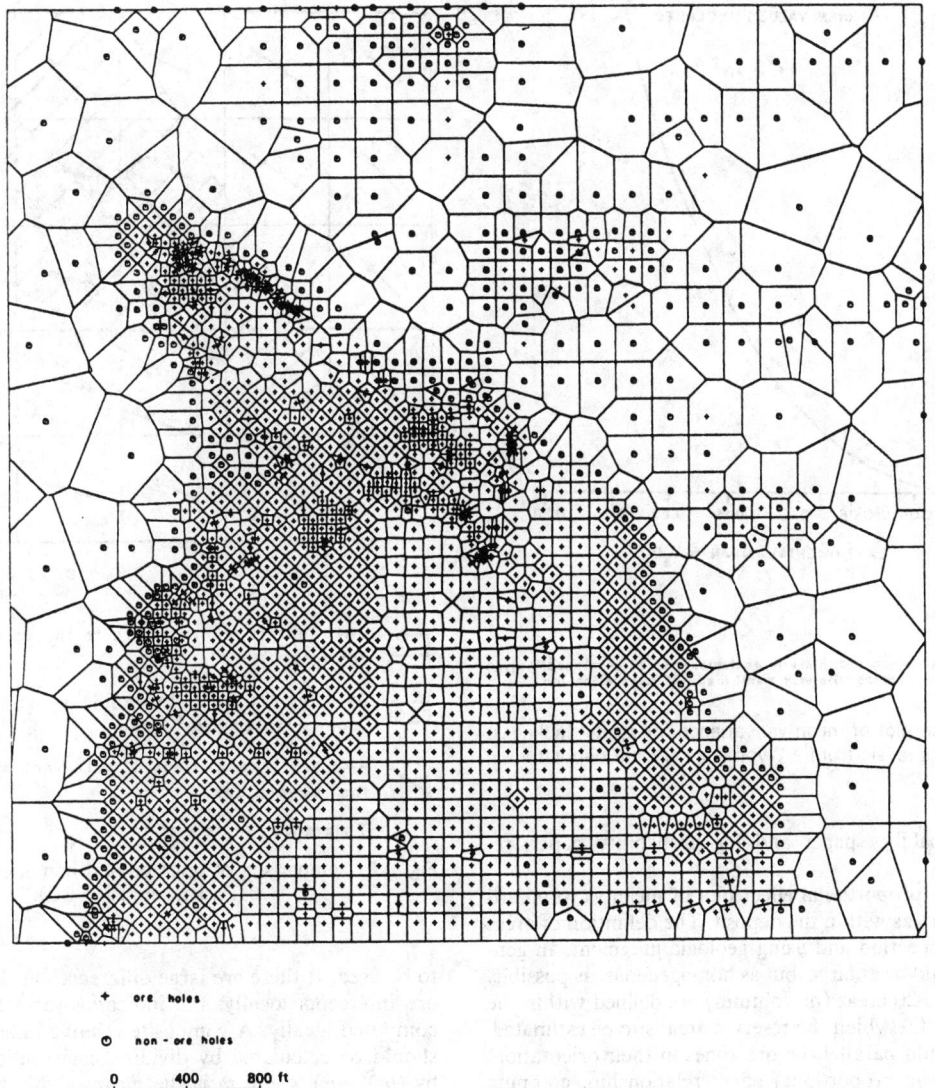

+   ore holes

⊙   non - ore holes

0   400   800 ft

Fig. 9. Ore and barren holes penetrating several mineralized horizons using a low cutoff grade (Sandefur and Grant, 1976).

where A is the total area (ore bearing and barren); K_Q is the conversion factor to obtain kilograms, pounds, etc., of uranium; and K_T is the density or tonnage factor.

This approach was found to work well for the author in drilling out several township sized areas near Crownpoint, NM. The approach also provided unbiased estimates for a simulated deposit based on a mined out area in the Grants, NM, mineral belt. In general, as the drill hole spacing is reduced by infill drilling, so is the favorable area A. This is offset by an increase in p^* and the area underlain by ore remains relatively constant. Confidence limits on the estimates may be calculated, as is described in the next section.

Reserve Estimation For Very Selectively Mined Deposits

This section describes techniques for reserve estimation of deposits for which the frequency distribution of selective mining units is essentially the same as that of the samples. The procedures must be separately applied for each set of cutoff criteria. In setting these criteria, the method of mining must be taken into account (e.g., lift height for excavator,

ore loss due to overlying waste removal, etc.). The following steps should be followed:

1) Correct the radiometric assays for disequilibrium and other factors and obtain the thickness and grade times thickness product for ore-grade intercepts. These intercepts can be calculated on a diluted basis using the comments made previously in the section on polygonal estimators as guides. Intercepts may be restricted to geologic horizons or slices as desired. As an example of the latter, composites were made every 0.9 m (3 ft) for a Wyoming roll-front deposit. The thickness and grade-thickness product of ore-grade composites were accumulated every 2.7 m (9 ft). These resultant accumulated thickness and grade-thickness variables were used for kriging.

2) Plot the thickness and grade-thickness variables and define areas of interest for reserve estimation. Compute the histogram of thickness and grade-thickness variables within this area. If there are more than 15% barren (zero thickness) intercepts, it is suggested an indicator variable be defined as 1 if the thickness is nonzero and 0 if the thickness is zero.

3) Make a plot of thickness and grade-thickness on cu-

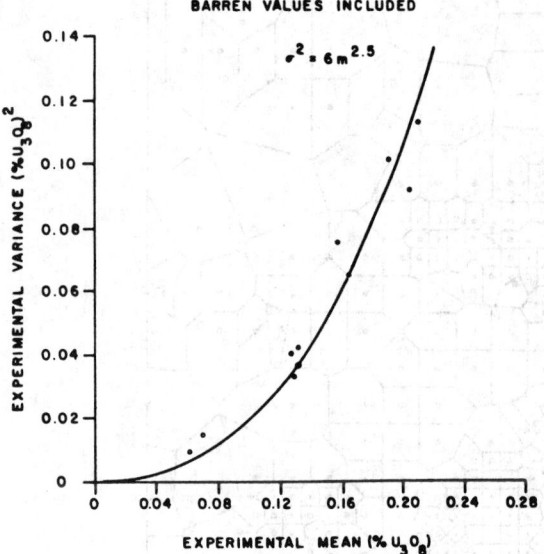

- Data for benches with nearly equivalent areas of mineralization
- Data for benches with very small areas of mineralization

Fig. 10. Scatterplot of mean vs. variance for grade-thickness for various levels from a Wyoming roll-front deposit.

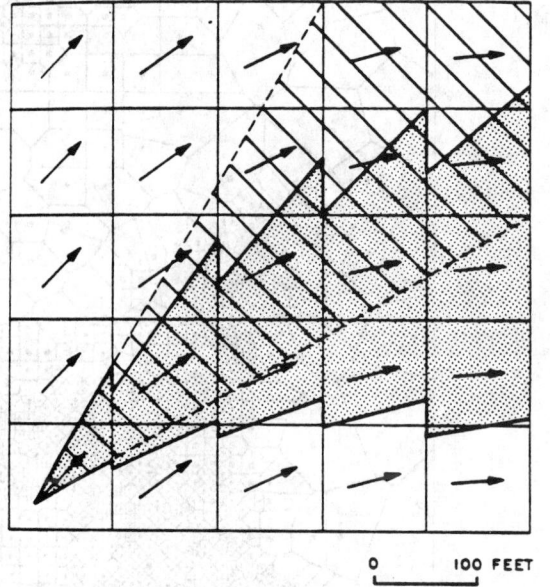

0   100 FEET

Data points contributing to the oriented variogram

Data points contributing to a conventional (non-oriented) variogram

Along strike attitude of roll front within the superblock

Tolerance angle = 15°

Fig. 12. Comparison of oriented and non-oriented variograms (Parker, et al., 1980).

mulative probability paper and log probability paper, respectively.

4) Make a proportional effect plot of mean vs. standard deviation for areas within the deposit. The definition of areas is made by inspection and using geologic judgment. In general, they should be as large, but as homogeneous, as possible. Typically 10 to 20 areas (or volumes) are defined within the area (volume) for which the reserves area is to be estimated. The areas should parallel the ore zones in their orientation.

5) Using the proportional effect relationship, compute composite relative variograms (see Journel and Huijbregts, 1978). The variograms should be oriented as necessary to follow geologic trends. Logarithmic variograms may be used and converted to relative variograms if the underlying distribution is shown to be lognormal.

6) Compute variograms on the indicator variable if it is

to be used. If there are large differences in the proportion of ore intercepts locally, the indicator variograms should be computed locally. A composite relative indicator variogram should be calculated by dividing each indicator variogram by $(m)(1-m)$ where m is the mean of the indicator data in an area and compositing the resultant variograms.

7) Compute the correlation coefficients between thickness and grade-thickness product. In general, thickness and grade-thickness product will be correlated between 0.3 and 0.7.

8) Perform ordinary linear kriging for the grade-thickness product q^*, thickness t^*, and indicator p^*, if used. Kriging should be done using the relative variograms. Sizes of blocks kriged should be smaller than the areas used to compute proportional effects (i.e., the samples and block should be close enough together so that stationarity of the variogram may be assumed). Blocks need not be rectangular and may be irregular areas corresponding to mineralized trends.

9) Compute the estimation variances for the thickness, grade-thickness product, and indicator variable, if used. These are computed by multiplying the relative estimation variance by $f(m)$, where m is usually taken to be the estimate.

10) Compute the estimated metal content, tonnage, and grade for each block:

$$\text{Accumulation (metal content)} = Q_i^* = K_Q A_i p_i^* q_i^* \qquad (21)$$

$$\text{Tonnage} = T_i^* = K_T A_i p_i^* t_i^* \qquad (22)$$

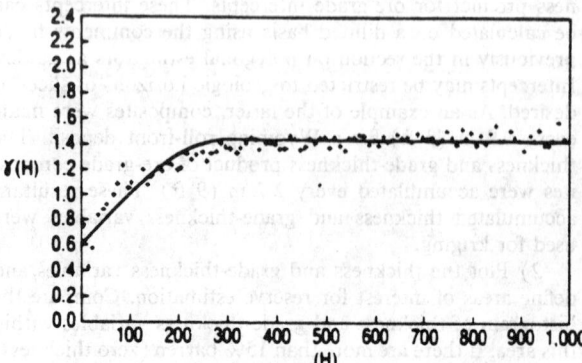

Fig. 11. Logarithmic variogram of a total grade thickness, distance between samples in feet (H), (Sandefur and Grant, 1980).

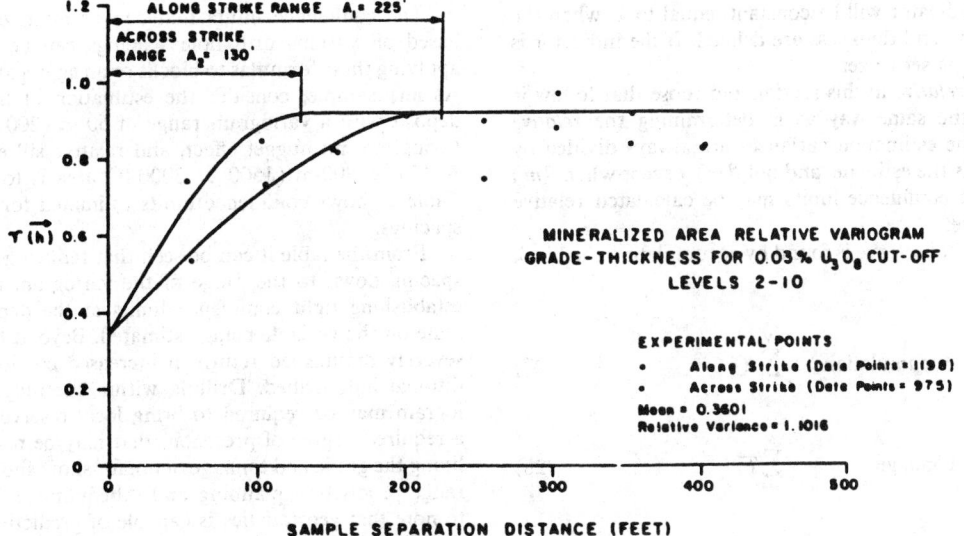

Fig. 13a. Oriented variogram for grade-thickness (Parker, et al., 1980).

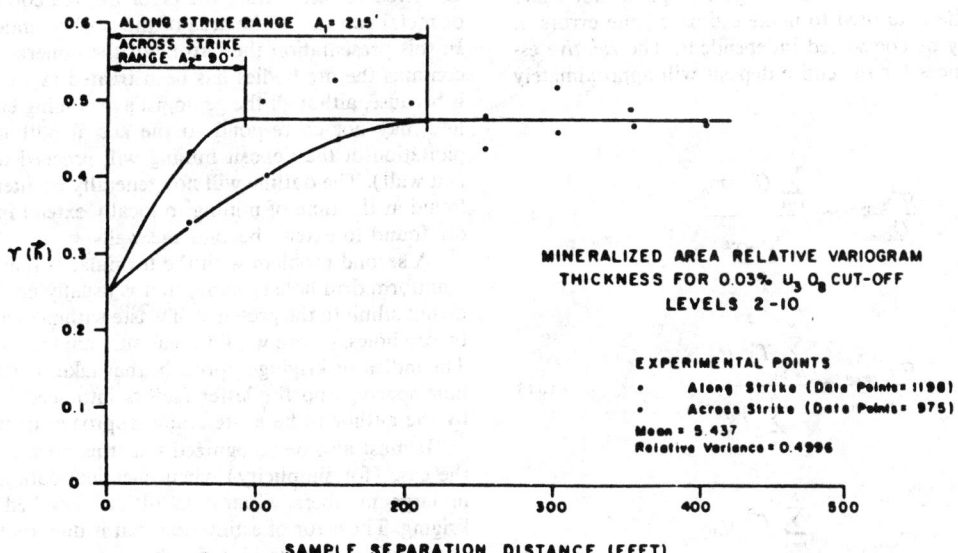

Fig. 13b. Oriented variogram for thickness (Parker, et al., 1980).

$$\text{Grade} = G_i^* = q_i^*/t_i^* \qquad (23)$$

where A_i is the total area of the block; K_Q is the conversion factor to obtain kilograms, pounds, etc., of uranium; and K_T is the density or tonnage factor. If the indicator is not used, set $p^* = 1$.

11) Compute the *relative* estimation variance for accumulation, tonnage, and grade as follows:

$$\frac{\sigma^2_{Q_i}}{Q_i^2} \approx \frac{\sigma^2_{p_i}}{p_i^2} + \frac{\sigma^2_{q_i}}{q_i^2} \qquad (24)$$

$$\frac{\sigma^2_{T_i}}{T_i^2} \approx \frac{\sigma^2_{p_i}}{p_i^2} + \frac{\sigma^2_{t_i}}{t_i^2} \qquad (25)$$

$$\frac{\sigma^2_{G_i}}{G_i^2} \approx \frac{\sigma^2_{q_i}}{q_i^2} + \frac{\sigma^2_{t_i}}{t_i^2} - 2\,\rho_{qt}\frac{\sigma_{q_i}\,\sigma_{t_i}}{q_i t_i} \qquad (26)$$

where $\sigma^2_{p_i}$ is the estimation variance (not relative estimation variance for the indicator), $\sigma^2_{q_i}$ is the estimation variance (not relative estimation variance for the accumulation), $\sigma^2_{t_i}$ is the estimation variance (not relative estimation variance for the thickness), and ρ_{qt} is the correlation coefficient between accumulation and thickness.

These formulas are standard applications for computing the variances for the product and quotient of random variables (see Journel and Huijbregts, 1978, pp. 424-426). In practice, p, q, and t are replaced by their estimates p^*, q^*, and t^*. The formulas assume that there is no correlation between indicator and grade-thickness or thickness. This is because the indicator will be constant, equal to 1, when the grade-thickness and thickness are defined. If the indicator is not used, $\sigma^2_{p_i}$ is set to zero.

The term *relative* in this section and those that follow is not used in the same way as in determining the *relative* variogram. The estimation variances are always divided by m^2, where m is the estimate, and not $f(m)$, except when $f(m) = m^2$ so that confidence limits may be calculated relative to the estimate.

12) The total reserve is found by accumulating the block values:

$$\text{Accumulation} = \sum_{i=1}^{N} Q_i^* \tag{27}$$

$$\text{Tonnage} = \sum_{i=1}^{N} T_i^* \tag{28}$$

$$\text{Grade} = \sum_{i=1}^{N} Q_i^* \Big/ \sum_{i=1}^{N} T_i^* \tag{29}$$

13) Provided the blocks are large enough to not share too much of the data used to make estimates, the errors of estimation may be considered independent. The *relative* estimation variances for the entire deposit will approximately be:

$$\frac{\sigma^2_{Q_{\text{Deposit}}}}{Q^2_{\text{Deposit}}} \approx \frac{\sum\limits_{i=1}^{N} Q_i^{*2} \, \sigma^2_{Q_i}}{\sum\limits_{i=1}^{N} Q_i^{*2}} \tag{30}$$

$$\frac{\sigma^2_{T_{\text{Deposit}}}}{T^2_{\text{Deposit}}} \approx \frac{\sum\limits_{i=1}^{N} T_i^{*2} \, \sigma^2_{T_i}}{\sum\limits_{i=1}^{N} T_i^{*2}} \tag{31}$$

$$\frac{\sigma^2_{G_{\text{Deposit}}}}{G^2_{\text{Deposit}}} \approx \frac{\sum\limits_{i=1}^{N} T_i^{*2} \, \sigma^2_{G_i}}{\sum\limits_{i=1}^{N} T_i^{*2}} \tag{32}$$

14) Relative confidence limits may be put on quantity of metal, tonnage, grade, etc., by assuming the distribution of errors is normal. This assumption is reasonable if the relative standard deviation of the estimation errors is less than ±25%. For example:

Relative
90% Confidence
Limits

$$= \pm 1.645 \sqrt{\text{relative estimation variance}} \tag{33}$$

Thus if the relative estimation variance for tonnage were 0.01, the relative 90% confidence limits would be 0.1645; that is, with 90% confidence the true tonnage should be within ±16% of the estimated tonnage.

The confidence limits that can be attached to reserves based on various drill hole spacings can be estimated by applying these formulas to blocks or an aggregation of blocks. As an example, consider the estimation of thickness in a deposit with a variogram range of 60 m (200 ft), spherical variogram, no nugget effect, and relative sill equal to 0.50. A 450 × 900-m (1500 × 3000-ft) area is to be explored. Table 2 shows confidence limits estimated for various hole spacings.

From the table it can be seen that reducing the drill hole spacing down to the range of the variogram is effective in establishing tight confidence limits at the deposit (global) scale on the variable being estimated. Beyond this there is a severely diminished return in increased confidence per additional hole drilled. Drilling within the range of the variogram may be required to bring local reserve estimates to a required degree of precision, that may be restated as outlining the grade and tonnage of ore for short and intermediate range production planning and scheduling. It is important to note that geostatistics is capable of predicting the results established empirically by Sandefur and Grant (1976), without resorting to fill-in drilling.

In many geostatistics texts, for example, David (1978) and Journel and Huijbregts (1978, pp. 428-438), formulae are given for estimating the error in prediction for tonnage or metal content in the area of the surface underlain by ore. In this presentation the outline of the mineralized area that contains the ore bodies has been treated as a constant. This is because, although the geologist's or mining engineer's outline may not correspond to the actual outline, in the exploitation of the deposit mining will proceed to the outline (pit wall). The outline will not generally be altered for waste found at the time of mining to locally extend into the pit or ore found to extend beyond the walls.

A second problem with the formulae is that they assume a uniform drill hole spacing, that is usually not the case, and do not admit to the presence of waste within areas surrounded by ore holes, or ore within areas surrounded by waste holes. The indicator kriging approach, that takes both nonuniform hole spacing and the latter factors into account, is believed by the author to be a preferable approach to the problem.

It must also be recognized that this presentation ignores the case (for simplicity) when chemical data are presented in large numbers. Guarascio (1976) handled this by co-kriging. The error of estimation that is due to disequilibrium and other factors is also disregarded (for simplicity). If a downhole variogram of residuals (actual chemical value-corrected value) shows a short range and/or a high nugget effect the error due to local incorrect adjustment for disequilibrium may be neglected. If not, better corrections are needed, or more data enabling better corrections must be taken. Although maps showing plots of residuals can be helpful, for most deposits insufficient data are collected to

Table 2. Confidence Limits vs. Hole Spacings

Hole Spacing, meters (ft)	No. of Holes	±90% Relative Confidence Limits %	Increase in No. Holes	Decrease in Confidence Limits
125 (400)	37	18.7	N.A.	N.A.
90 (300)	57	12.9	25	5.9
60 (200)	128	5.9	71	7.0
30 (100)	512	0.9	384	5.0
15 (50)	2048	Nil	1536	0.9

allow for study of the horizontal regionalization of disequilibrium or its correction. Therefore assessment of the additional error due to disequilibrium in a reserve estimate is difficult, and statements as to the precision of reserve estimates should be treated with caution.

Similarly, there is no allowance for error in the computation of the density K_T in the formulas. For most deposits this is standard practice. If a variable density is suspected, it should be investigated and densities assigned spatially to rock units using lithologic logs.

Reserve Estimation for Less Selectively Mined Deposits

This section describes techniques for reserve estimation of deposits for which the frequency distribution of selective mining units has significantly (at least 10%) lower variance than that of samples or composites of samples.

Assessment of Grade-Tonnage Curves for Selective Mining Units: The first task is to establish the selective mining unit. This will normally be the volume surrounded by four blastholes, but may be as small as a truck load if scanners are used to classify the loads. In many cases computing the reserves for several selective mining unit sizes can be useful. The variance of selective mining unit grades will be inversely proportional to the volume (Parker 1979). Using a selective mining unit height as the composite length, the declustered frequency distribution within the mineralized portion of the deposit should be calculated. The *mineralized portion* should be that area in which the ore occurs. It should include barren areas that may be encountered during mining, but exclude barren areas around the fringes of mineralization that will probably not be included within the pit. Polygonal areas of influence should be used to weight (decluster) the composites in constructing the frequency distribution. If a polygon construction program is not available, the deposit may be divided into blocks. The number of composites in each block is noted and the weight assigned to a composite is the volume of the block divided by the number of composites. The mean and variance of the frequency distribution are computed.

Horizontal variograms are then made. If variograms of composite grades are erratic, logarithmic or relative variograms may have to be used. In this case these should be transformed to give an estimate of the average variogram of composites for the deposit. If necessary the sill of this transformed variogram should be rescaled to the variance of the frequency distribution, $\sigma^2_{composites}$.

The mean of the frequency distribution of selective mining units will be the same as the mean of the frequency distribution of samples, m. The variance of the selective mining units will be smaller:

$$\sigma^2_{smu} = \sigma^2_{composites} - \overline{\gamma}_{smu} \qquad (34)$$

σ^2 composite is best estimated by the sill of the variogram. $\overline{\gamma}_{smu}$ can be calculated from graphs in Journel and Huijbregts (1978) or found by taking the average value of the variogram between points within the selective mining unit (usually 25 to 100 points).

There are a number of ways to adjust the frequency distribution of composites to obtain the frequency distribution of selective mining units. The two most common are the affine correction and the indirect lognormal correction (Journel and Huijbregts, 1978, pp. 468–471).

In the first instance (affine correction) the grades of the *distribution* of selective mining units are estimated by:

$$Z_{smu} = \frac{\sigma_{smu}}{\sigma_{composites}} (Z_{composite} - m) + m \qquad (35)$$

To form the frequency distribution, the influence (tonnage) assigned to a Z_{smu} is the tonnage fraction assigned to the corresponding $Z_{composite}$. The tonnage above cutoff is found by multiplying the total tonnage of the deposit times the tonnage fraction above cutoff. The grade above cutoff is found by a tonnage fraction weighted average of the Z_{smu} above cutoff.

To use the second method (indirect lognormal), the distribution of composites should be verified as approximately lognormal. The assumption is made that the distribution of *smu* will be approximately lognormal. Under these assumptions the mean of the two distributions will be the same, m. The variance of the logarithms will be approximately:

$$\beta^2_{composites} = \ln \left(1 + \frac{\sigma^2_{composites}}{m^2} \right) \qquad (35a)$$

$$\beta^2_{smu} = \ln \left(1 + \frac{\sigma^2_{smu}}{m^2} \right) \qquad (35b)$$

Then the proportion of material above cutoff Z_c assuming strict lognormality will be:

$$P^c_{composites} = \phi \left[\frac{1}{\beta_{composites}} \ln \left(\frac{m}{Z_c} \right) - \frac{\beta_{composites}}{2} \right] \qquad (36A)$$

$$P^c_{smu} = \phi \left[\frac{1}{\beta_{smu}} \ln \left(\frac{m}{Z_c} \right) - \frac{\beta_{smu}}{2} \right] \qquad (36b)$$

where ϕ is the standard normal cumulative density function that is tabulated in introductory statistics texts.

The grade of material above cutoff, assuming strict lognormality, will be:

$$G^c_{composites} = \frac{m \, \phi \left[\frac{1}{\beta_{composites}} \ln \left(\frac{m}{Z_c} \right) + \frac{\beta_{composites}}{2} \right]}{P^c_{composites}} \qquad (37a)$$

$$G^c_{smu} = \frac{m \, \phi \left[\frac{1}{\beta_{smu}} \ln \left(\frac{m}{Z_c} \right) + \frac{\beta_{smu}}{2} \right]}{P^c_{smu}} \qquad (37b)$$

Then the frequency distribution for selective mining units may be found by:

$$p_{smu} = \frac{P^c_{smu}}{P^c_{composites}} p^c_{composites} \qquad (38)$$

where p_{smu} is the estimated actual proportion of selective mining units above cutoff Z_o and $p^c_{composites}$ is the actual proportion of composites above cutoff. If p_{smu} is multiplied by the total tonnage in the deposit, the tonnage above cutoff may be estimated.

If the grade above cutoff g_{smu} is desired, the formula is:

$$g_{smu} = \frac{G^c_{smu}}{G_{composites}} g^c_{composites}$$

where $g^c_{composites}$ is the tonnage-weighted average of grades of composites above the cutoff.

Once the frequency distribution of selective mining units is obtained, the estimated recoverable metal content, tonnage, and grade from selective mining units above a given cutoff can be computed. An example of the application is given in Table 3. In this example the effect of selective mining unit height on recoverable reserves was investigated.

For this deposit the 3-m (10-ft) case was selected because the 1-m (3-ft) case was not deemed practical considering the high ore and waste production rate specified in the mine design criteria.

A comment is made that recoverable reserves computed using this method must be taken as rough global estimates. Often there is insufficient close-spaced drilling to accurately define the variogram, particularly the nugget effect, at spacings required to calculate γ_{smu}. The assumption is made that selective mining units will be perfectly selected at the time of mining. Considering that blasthole probing and face and truck scanning, if properly corrected for local disequilibrium and other factors, can accurately predict the grade, further slight adjustment of the frequency distribution for grade control errors is usually not warranted.

In any case, the relative changes in recoverable reserves with change in selective mining unit volume may be assessed using this approach.

Assessment of Local Reserves: Local reserve estimation may be done by ordinary linear kriging if the estimation variances are not so high as to cause appreciable smoothing effects. This may be investigated by kriging a panel of sufficient size to be useful for mine planning with a typical configuration of data:

Variance of kriging estimates $\simeq$ variance of   (39)
panel grades — estimation variance

For some deposits rich and poor zones can be delimited by geologic contacts. In these cases kriging should be restricted to the ore-bearing area. The selective mining unit/panel size may vary within the deposit.

For large-scale mining operations a panel may be larger than a selective mining unit (see Fig. 13). It is important to compare the variance of the kriged panel estimates to σ^2_{smu}. If the difference is small (as a rule-of-thumb less than 20%),

the grade-tonnage curve and mine planning, based on kriged estimates, will be representative of the recoverable reserves. An adjustment for the additional sampling available and better selection possible at the time of mining can be made to the grade-tonnage curve of kriged estimates using either the affine or indirect lognormal method explained previously.

Unfortunately, except in operating mines with a wealth of blasthole and other development drilling data, there will be a large difference between the variance of the kriging estimates of panels and the variance of the selective mining units. This difference occurs because the data are widespaced and sample-panel correlation is small, leading to large estimation variances. In addition selective mining units are usually much smaller than panels (that may be used for mine planning), which further increases the discrepancy between the variance of kriged estimates and that of recoverable reserves.

In these cases what is desired is not the prediction of each selective mining unit's grade within the deposit or even within a panel. What is required is an estimate of the frequency distribution of the grades of selective mining units within a panel. Using the frequency distribution, the amount of waste, amount of ore-grade selective mining units, and grade of ore-grade selective mining units can be predicted for each panel.

Nonlinear geostatistical methods are used to make the predictions. These methods, though they have been tried on several deposits, have not been tested to the extent that linear kriging estimators have been tested. They nearly all depend on strong hypotheses of stationarity of the frequency distribution of grade from place to place within the deposit, and make inferences as to the shape of the distribution of selective mining units using the distribution of composite samples.

Lognormal Shortcut—The easiest method to apply is the lognormal shortcut method of David, et al. (1977). Although the method is described to apply to blastholes in a copper deposit, it has been adapted to prediction of selective mining units. It has been successfully used (S.P. Morzenti, Exxon Minerals, personal communication) in an operating mine, with the benefit of production data to correct variances and remove biases. The conditional distribution of selective mining unit grades within the panel is presumed to have a lognormal distribution with mean and variance:

Mean = ordinary linear kriged estimate, Z^* (40)

Variance = Kriging estimation variance +   (41)
dispersion variance of selective
units within a panel

The dispersion variance of selective mining units within a panel is estimated to be:

$$\sigma^2_{smu/panel} = \bar{\gamma}_{panel} - \bar{\gamma}_{smu} \tag{42}$$

Given the mean and variance of the lognormal distribution, the proportion of ore and grade of ore can be estimated using Eqs. 35b, 36b, and 37b.

Correct use of this technique relies heavily, especially where the cutoff grade is above the mean, on the assumption that the distribution of selective mining units will be lognormal and that the formula for the variance is correct. In deposits with strong proportional effects, Eqs. 41 and 42 should be evaluated using relative variograms and variances and then adjusted by multiplying by $f(Z^*)$. Usually $f(Z^*)$

Table 3. Effect of Selective Mining Unit Height on Recoverable Reserves*

	Selective Mining Heights			
	1 m (3.3 ft)	3 m (9.8 ft)	6 m (19.7 ft)	12 m (39.4 ft)
Contained Uranium	1.00	0.88	0.78	0.58
Tonnage	1.00	0.95	0.90	0.79
Grade	1.00	0.92	0.86	0.73

* 1-m height taken as base case and figures quoted are relative to it.

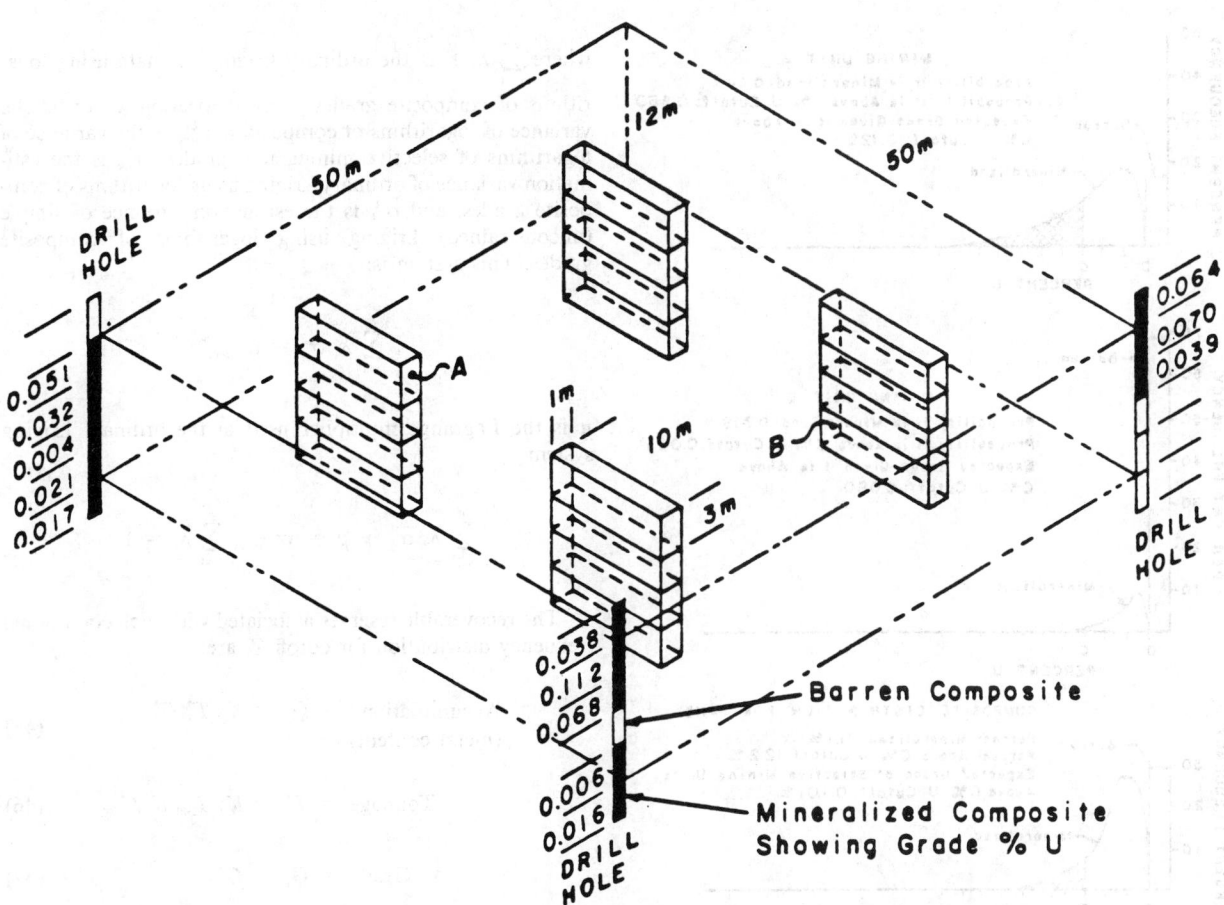

Fig. 14. Relative configuration of selective mining units, panel for mine planning, and drill holes (Parker, Journel, and Dixon, 1979). There are 1000 selective mining units in a panel, but only 16 of them distributed on 4 evenly spaced stacks are considered for the estimation of the minable reserves within the panel.

$= Z^{*2}$. In addition, the formula should be recognized to be an approximation of the more complex method described subsequently. The extent to which the assumption of lognormality and the approximation holds cannot be verified prior to mining. The author prefers the more complicated method to be described next that provides a more rigorous estimate of the conditional frequency distribution of grade within a panel.

Conditional Lognormal Distributions—Parker, et al., (1979) made an extensive geostatistical study of the Imouraren uranium deposit, located in Niger. Two populations are present, a barren population comprising 43% of composites and a mineralized population comprising 57% of composites (Fig. 8). Ore and waste are commingled. Mine planning panels were chosen to be 50 × 50 × 12 m (164 × 164 × 39 ft) or approximately 64,000 t (70,500 st). Within each of these panels are 1000 selective mining units [each a 64-t (70-st) truck load] as shown in Fig. 14. The spacing of the holes given in the figure is representative. The probability distribution, conditional on the data, for each of the selective mining units shown in the figure was calculated. If this probability distribution is extended into the surrounding 4000 t (4400 st), it may be considered a conditional frequency distribution of selective mining units in the vicinity. If all 16 conditional probability distributions are combined, the con-

ditional frequency distribution of all the selective mining units within the panel is obtained (see Fig. 15).

To obtain a conditional probability distribution for a selective mining unit, the following assumptions were made:

1) The barren population is independent of the mineralized population. This assumption was verified by cross-variogram analysis.

2) Selective mining units are either all barren or all mineralized. This assumption was verified by examining pairs of samples at the same level in close-spaced twinned holes.

3) The mineralized population is multivariate lognormally distributed. This assumption could not be checked; however the successful local cross-validation indicates the assumption is not inconsistent with the data.

4) The mineralized selective mining units are lognormally distributed and multivariate lognormally distributed with the composite data. From a mathematical point of view, these hypotheses are inconsistent since the sum of lognormal distributions (aggregation of composite supports within selective mining units) are not lognormal. When the difference in variance is less than 15%, the assumption has been shown to work well in practice. This assumption could not be checked. On a global basis the results agreed well (within 5% of an independently globally adjusted frequency distribution of selective mining units).

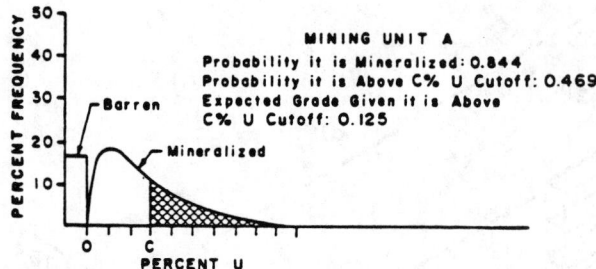

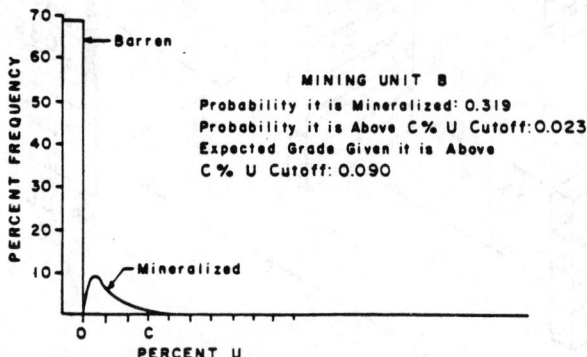

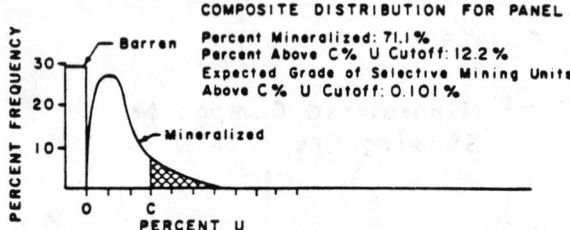

Fig. 15. Conditional probability distributions for selective mining unit grades in the panels shown in Fig. 14 (Parker, Journel, and Dixon, 1979).

The theory used is developed in Journel (1980). Its application is described in Parker, et al., (1979). In brief, a conditional probability distribution is characterized by:

p^* = the ordinary kriged estimate of the indicator; 1 if composite is mineralized and zero if not.

z^*_{minz} = the mean of the conditional lognormal probability distribution of the selective mining unit, given it is mineralized.

σ^2_{minz} = the variance of the conditional lognormal probability distribution of the selective mining unit, given it is mineralized.

The formulas for z^*_{minz} and σ^2_{minz} are given by:

$$z^*_{minz} = \exp\left[\sum_{i=1}^{n} \lambda_i\, y\,(x_i)\right.$$

$$\left. + \frac{\sigma^2 - \sigma^2_{smu} + \sigma^2_{ok}}{2} + \mu\right] \quad (43)$$

$$\sigma^2_{minz} = (z^*_{minz})^2 \exp\left[\sigma^2_{sk} - 1\right] \quad (44)$$

where $\sum_{i=1}^{n} \lambda_i\, Y_i$ is the ordinary kriging estimate using logarithms of composite grades $y(x_i)$ at location x_i, σ^2 is the variance of logarithms of composites, σ^2_{smu} is the variance of logarithms of selective mining unit grades, σ^2_{ok} is the estimation variance of ordinary kriging using logarithms of composite grades, and σ^2_{sk} is the estimation variance of simple (unconstrained) kriging using logarithms of composite grades. This system is:

$$\sum_{i=1}^{n} \lambda_i\, \sigma_{ij} = \sigma_{j,\,smu}$$

μ is the Lagrange multiplier used in the ordinary kriging system

$$\sum_{i=1}^{n} \lambda_i\, \sigma_{ij} + \mu = \sigma_{j,\,smu}; \quad \sum_{i=1}^{n} \lambda_i = 1$$

The recoverable reserves associated with each conditional frequency distribution for cutoff Z_c are:

$$\text{Accumulation} \atop \text{(metal content)} = Q^*_i = K_Q\, T^*_i\, G^*_i \quad (45)$$

$$\text{Tonnage} = T^*_i = K_T\, T_{infl}\, p^*_i\, P^*_{i,\,z_c} \quad (46)$$

$$\text{Grade} = G^*_i = G^*_{i,\,z_c} \quad (47)$$

where $P^*_{i,\,z_c}$ and $G^*_{i,\,z_c}$ are given by Eqs. 36b and 37b, and T_{infl} is the tonnage influenced by the conditional probability distribution. K_Q and K_T are conversion factors stated earlier. For the whole panel:

$$A_{panel} = \sum_{i=1}^{n} Q^*_i \quad (48)$$

$$T_{panel} = \sum_{i=1}^{n} T^*_i \quad (49)$$

$$G_{panel} = \sum_{i=1}^{n} T^*_i\, G^*_i / \sum_{i=1}^{n} T_i \quad (50)$$

As described by Parker, et al. (1979), it is important to cross-validate the variance of logarithms σ^2 locally. This cross-validation can be done by kriging logarithms of composite grades, that have been temporarily removed from the dataset, using logarithms of surrounding composites as data. Within an area of the deposit the reduced errors of logarithms are calculated:

$$e_i = \frac{y_i - y^*_i}{\sigma_{i,\,ok}} \quad (51)$$

The deposit variogram with sill σ_o is used to do this kriging, and predict $\sigma^2_{ok,\,i}$. The average value of e^2_i is computed:

$$w^2 = \frac{\sum_{i=1}^{n} e_i^2}{m} \qquad (52)$$

Then the local variance of the logarithms of composite grades is estimated to be $w^2\sigma^2_o$. The sill of the variogram of logarithms of composite grades is rescaled locally by w^2.

It is also important to cross-validate the conditional probability distributions on a local basis. While this cannot be done on selective mining units prior to mining, it can be done on composites. It is the experience of the author that if the underlying data are approximately lognormally distributed on a local basis, the technique works well. If there is a departure from lognormality, biases of up to 20% in recoverable grade and tonnage may be produced, and the method must be rejected. To some extent the indicator-mineralized cutoff may be moved to make the mineralized distribution as lognormal as possible, but the important thing to remember is that cross-validation must be performed exhaustively to check that the method does not produce biased results. Fig. 16 shows the cross-validation for the Imouraren (Niger) uranium deposit.

As a final check, the global frequency distribution of selective mining units can be compared with the frequency distribution obtained by compositing the local conditional frequency distributions over the whole deposit. Prior to mining, this is the only cross-check that can be made to validate the reasonableness of the hypothesis of multivariate lognormality.

Disjunctive Kriging—Matheron (1976) proposed a solution to the problem of predicting local recoverable reserves that does not depend on any underlying assumptions of lognormality. This method, known as disjunctive kriging, was described by Marechal (1975) and was first applied to uranium deposits by Jackson and Marechal (1979). The reader is also referred to a related study by Sans and Martin (1983). The technique is involved mathematically and computationally. The frequency distribution of the data is transformed into a normal distribution. The variogram of the transformed distribution is calculated and modeled. Using kriging, combinations of hermite polynomials describing a pseudo probability density function of grade are produced for selective mining units. It has been observed that these pseudo density functions may attach negative probability for some grade intervals. In addition, where the distribution of grade is nonstationary, the method, as others that do not correct for this, may give biased results. To be useful it must, like the lognormal methods, be cross-validated and the transformation into a normal distribution repeated on a local basis if necessary. Journel (1982) and Verly (1983) have investigated simpler methods that do not attach negative probability to grade intervals using multiple indicator variables and multigaussian transformation methods, respectively. These methods differ from those previously mentioned in that they predict the frequency distribution of composite ("point support") values within a panel. An adjustment is then made to this distribution, similar to the affine or indirect lognormal method, to provide an estimate of recoverable reserves based on selective mining units. At present, these methods have not been tested on uranium deposits, but show promise of being more robust than the lognormal or disjunctive kriging methods. The formulae required and assumptions made are as simple or simpler. Judging from experience, an exhaustive cross-validation of results is still recommended.

Remarks—In the absence of detailed geology from very

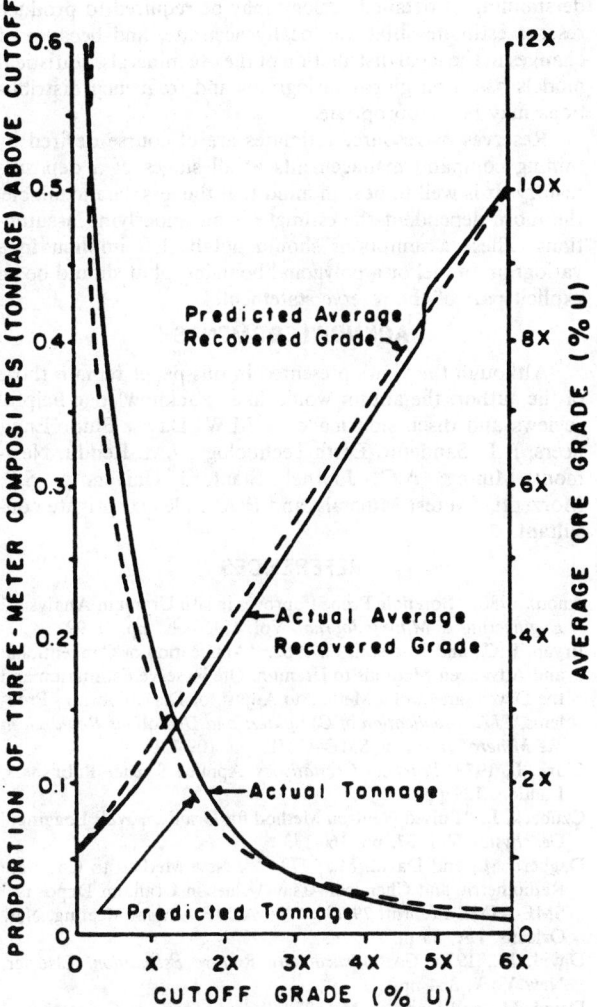

Fig. 16. Results of cross-validation of conditional frequency distributions on 3-m (10-ft) support (Parker, Journel, and Dixon, 1979) "Actual" curves are based on distribution of over 7000 composites.

close-spaced samples, the cost of which would be prohibitive, the statistical approach to ore reserve estimation is preferable to the deterministic approach in that it mathematically characterizes the continuity of the mineralization. Despite their mathematical complexity, it must be recognized that the statistical models of reserves that are derived using these methods are geologically simple. In many cases these models must ignore changes in the distribution of grade, lithology, mineralogy, or structure, sometimes for lack of data, and in others because the mathematical model that would take them into account would be intractable.

There are as yet too few guideposts to be provided by reserve estimates calculated at various stages of exploration, development, operation, and exhaustion to make this section of the handbook complete. The character of the undeformed stratabound deposits is fairly well known, because of extensive mining experience. The metamorphosed and igneous deposits of the Athabasca (Canada), Northern Territory (Australia), and Rossing (Namibia) genre are far fewer, and newer. Early work on some of these deposits (J. O'Leary, personal communication) has shown that a thorough understanding of detailed geology may be required to produce reserve estimates that are locally accurate, and because of changes in the local distribution of the ore minerals, statistical models based on global variograms and frequency distributions may be inappropriate.

Reserves or resource estimates are of course desired by mining company managements at all stages of a deposit's history. It is well to bear in mind that the less data available, the more dependent the estimate is on underlying assumptions. These assumptions should not be left implicit in a variogram model or a polygonal boundary, but should be an explicit part of the reserve statement.

ACKNOWLEDGMENTS

Although the views presented in this paper remain those of the author, the author would like to acknowledge helpful reviews and discussion made by M.W. Davis, Fluor Engineers; R.L. Sandefur, Earth Technology; J.M. Rendu, Newmont Mining; A.G. Journel, Stanford University; S.P. Morzenti, Everest Minerals; and R.A. Bideaux, private consultant.

REFERENCES

Anon., 1980, "Borehole Probe Improves in situ Uranium Analysis," *Engineering & Mining Journal,* Vol. 181, Feb., pp. 98-99.

Bryan, R.C., and Roghani, F., 1982, "Application of Conventional and Advanced Methods to Uranium Ore Reserve Estimation and the Development of a Method to Adjust for Disequilibrium Problems," *17th Application of Computers and Operations Research in the Mineral Industry,* SME-AIME, pp. 109-130.

Clark, I., 1979, *Practical Geostatistics,* Applied Science Publishers, London, 129 pp.

Czubeck, J., "Pulsed Neutron Method for Uranium Well Logging," *Geophysics,* Vol. 37, pp. 160-173.

Dagbert, M., and David, M., 1979, "A New Method to Correlate Radiometric and Chemical Assay Values in Uranium Deposits," SME-AIME Preprint 79-136, 1979 AIME Annual Meeting, New Orleans, LA, 23 pp.

David, M., 1977, *Geostatistical Ore Reserve Estimation,* Elsevier, New York, 364 pp.

David, M., and Dagbert, M., 1980, "Case Studies in Geostatistical Ore Reserve Estimation of Uranium Deposits," *Uranium Exploration and Mining Techniques,* International Atomic Energy Agency, Vienna, pp. 111-137.

David, M., Dagbert, M., and Belisle, J.M., 1977, "The Practice of Porphyry Copper Deposit Estimation of Grade and Ore-Waste Tonnages Demonstrated by Several Case Studies," *15th Application of Computers and Operations Research in the Mineral Industries,* Australasian Institute of Mining & Metallurgy, Parkville, Vic., Australia, pp. 243-254.

Davis, J.C., 1973, *Statistics and Data Analysis in Geology,* John Wiley, New York, 550 pp.

Dodd, P.H., Droullard, R.F., and Lathan, C.P., 1967, "Borehole Logging Methods for Exploration and Evaluation of Uranium Deposits," *Mining and Groundwater Geophysics,* Canadian Geological Survey, pp. 401-415.

Guarsascio, M., 1976, "Improving the Uranium Deposits Estimations (The Novazza Case)," *Advanced Geostatistics in the Mining Industry,* D. Reidel, ed., Dordrect, Holland, pp. 351-368.

Hansink, J.D., 1976, "Equilibrium Analysis of a Sandstone Roll Front Uranium Deposit," *Exploration for Uranium Deposits,* International Atomic Energy Agency, Vienna, pp. 683-693.

Jackson, M., and Marechal, A., 1979, "Recoverable Reserves Estimated by Disjunctive Kriging: A Case Study," *16th Application of Computers and Operations Research in the Mineral Industry,* SME-AIME, pp. 240-249.

Journel, A.G., 1980, "The Lognormal Approach to Predicting Local Distributions of Selective Mining Unit Grades," *Mathematical Geology,* Vol. 12, pp. 285-304.

Journel, A.G., 1982, "The Indicator Approach to the Estimation of Spatial Distributions," *17th Application of Computers and Op-*

erations Research in the Mineral Industry, SME-AIME, pp. 793-806.

Journel, A.G., and Huijbregts, C.J., 1978, *Mining Geostatistics,* Academic Press, New York, 600 pp.

Killeen, P.G., and Carmichael, C.M., 1976, "Determination of Radioactive Disequilibrium in Uranium Ores by Alpha-Spectrometry," Canadian Geological Survey Paper 75-38, pp. 1-17.

Knudsen, H.P., and Kim, Y.C., 1979, "Development and Verification of Variogram Models in Roll Front Type Uranium Deposits," *Mining Engineering,* Vol. 31, No. 8, pp. 1215-1219.

Koch, G.S., and Link, R.F., 1970, *Statistical Analysis of Geologic Data,* Wiley, New York, 2 Vols.

Levinson, A.A., 1974, *Introduction to Exploration Geochemistry,* 2nd ed., Applied Publishing, Ltd., Calgary, 924 pp.

Linton, W.A., 1963, "Uranium Logging Techniques," *Geology and Technology of the Grants Uranium Region,* V.C. Kelley, ed., New Mexico Bureau of Mines and Mineral Resources Memoir 15, Socorro, NM, pp. 222-233.

Marbeau, J.P., and Marechal, A., 1980, "Geostatistical Estimation of Uranium Ore Reserves," *Uranium Evaluation and Mining Techniques,* International Atomic Energy Agency, Vienna, pp. 89-110.

Marechal, A., 1976, "The Practice of Transfer Functions: Numerical Methods and Their Application," *Advanced Geostatistics in the Mining Industry,* D. Reidel, ed., Dordrect, Holland, pp. 253-276.

Matheron, G., 1976a, "A Simple Substitute for Conditional Expectation: The Disjunctive Kriging," *Advanced Geostatistics in the Mining Industry,* D. Reidel, ed., Dordrect, Holland, pp. 221-236.

Matheron, G., 1976b, "Forecasting Block Grade Distributions: The Transfer Functions," *Advanced Geostatistics in the Mining Industry,* D. Reidel, ed., Dordrect, Holland, pp. 237-252.

Ostrihansky, L., 1976, *Radioactive Disequilibrium Investigations, Elliot Lake Area, Ontario,* Canadian Geological Survey Paper 75-38, pp. 19-48.

Parker, H.M., 1979, "The Volume-Variance Relationship: A Useful Tool in Mine Planning," *Engineering & Mining Journal,* Vol. 180, Oct., pp. 106-123.

Parker, H.M., and Chavez, W.X., 1979, "The Use of Geostatistics in the Delineation of Grade and Tonnage of Underground Sandstone-type Uranium Deposits," SME-AIME Preprint 79-120, 1979 AIME Annual Meeting, New Orleans, LA, 33 pp.

Parker, H.M., Farrell, C.W., and Srivastava, R.M., 1980, "Confidential Ore Reserve Study, Wyoming," Report, Fluor Mining and Metals, Redwood City, CA.

Parker, H.M., Journel, A.G., and Dixon, W.C., 1979, "The Use of Conditional Lognormal Probability Distributions for the Estimation of Ore Reserves in Stratabound Uranium Deposits—A Case Study," *16th Application of Computers and Operations Research in the Mineral Industry,* SME-AIME, pp. 133-148.

Rendu, J.M., 1978, *An Introduction to Geostatistics Methods of Mineral Evaluation,* Monograph, South African Institute of Mining & Metallurgy, Johannesburg, 84 pp.

Rosholt, J.N., Tatsumoto, M., and Dooley, J.R., 1965, "Radioactive Disequilibrium Studies in Sandstone, Powder River Basin, Wyoming and Slick Rock District, Colorado," *Economic Geology,* Vol. 60, pp. 477-484.

Sandefur, R.L., and Grant, D.C., 1976, "Preliminary Evaluation of Uranium Deposits. A Geostatistical Study of Drilling Density in Wyoming Solution Fronts," *Exploration for Uranium Ore Deposits,* International Atomic Energy Agency, Vienna, pp. 695-714.

Sandefur, R.L., and Grant, D.C., 1980, "Applying Geostatistics to Roll Front Uranium in Wyoming," *Geostatistics,* McGraw-Hill, New York, pp. 127-143.

Santos, E.S., 1975, "A Characteristic Pattern of Disequilibrium in Some Uranium Ore Deposits," *Journal of Research,* US Geological Survey, Vol. 3, pp. 363-368.

Sans, H., and Martin, V., 1983, "Technical Parameterization of Uranium Reserves to be Mined by Open-Pit Method," *Proceedings,* 2nd Nato Geostatistics Congress, Lake Tahoe, NV, 17 pp.

Scott, J.H., 1963, "Computer Analysis of Gamma-ray Logs," *Geophysics,* Vol. 28, pp. 457-465.

Scott, J.H., et al., 1961, "Quantitative Interpretation of Gamma-Ray Logs," *Geophysics,* Vol. 26, pp. 182-191.

Sichel, H., 1966, "The Estimation of Means and Associated Confidence Limits for Small Samples from Lognormal Populations," *Symposium on Mathematical Statistics and Computer Applications in Ore Valuation,* South African Institute of Mining & Metallurgy, Johannesburg, pp. 106-123.

Stoll, M.G., 1972, "Field Methods of Determining Uranium Ore Reserves," Report 12-A, Atomic Energy Commission, Grand Junction, Colorado, 14 pp.

Verly, G., 1983, "The Multigaussian Approach and its Applications to the Estimation of Local Reserves," *Mathematical Geology,* Vol. 15, No. 1, In Press.

Weiss, A., and O'Brian, D.T., 1971, "Practical Aspects of Computer Methods in Ore Reserve Analysis," *Decision-Making in the Mineral Industry,* Canadian Institute of Mining & Metallurgy, Spec. Vol. 9, pp. 109-113.

3.4.3 Coal Ore Reserve Estimation*

JAMES W. BOYD

INTRODUCTION

An accurate determination of the quantity, quality, and minability of a reserve is essential for any coal or mineral property valuation. The foundation of such a valuation is the reserve estimate; the roots are the exploration and geological programs. The exploration program incorporates a study of the surface geology, subsurface characteristics, sampling of the deposit for quality analysis, and previously mined areas. Reliability of the reserve is determined by the thoroughness of the exploration program and variability of the deposit. Considerations such as transportation availability, markets for the products, and environmental regulations also affect the value of a reserve.

In any valuation study, an engineer develops an estimate of the present-day reserve base, isolating the minable reserve from the gross resource. Present-day reserves are capable of being marketed and are minable within a reasonable time period by available technological methods. A clear distinction exists between a minable reserve and a gross resource. The United States Geological Survey (USGS) has developed standard definitions for differentiation between the two terms within the coal industry.

Reserve: Virgin and/or accessed parts of a coal reserve base which can be economically extracted or produced at the time of determination considering environmental, legal, and technological constraints. The term reserve need not signify that extraction facilities are in-place or operative. Reserves include only recoverable coal, i.e., coal that is or can be extracted from a bed during mining. Reserves are derived from the reserve base, which includes bituminous coal and anthracite 28 in. [71 cm] or more thick occurring at depths to 1000 ft [304.8 m], subbituminous coal 5 ft [1.5 m] or more thick that occurs at depths to 1000 ft [304.8 m], and lignite 5 ft [1.5 m] or more thick that occurs at depths to 500 ft [152.4 m]. Reserves also include thinner and/or more deeply occurring beds of coal of these types currently being mined (Wood, et al., 1983).
Resource: A naturally occurring concentration or deposit in the earth's crust, and in such forms and amounts that economic extraction is currently or potentially feasible. Tonnage estimates are determined by summing the estimates for identified deposits of coal 14 in. [35.56 cm] or more thick for bituminous and anthracite coal with less than 6000 ft [1829 m] of overburden and 30 in. [76.2 cm] or more thick for lignite and subbituminous coal with less than 6000 ft [1829 m] of overburden (Wood, et al., 1983).

From the USGS definition, it can be concluded that all coal reserves are considered resources, but not all coal resources are considered reserves. If coal prices increase and/or new technology is developed, deposits classified as resources could become reserves. Governmental action, in the form of new legislation, and the development of additional markets also influence classification changes. The USGS system of reserve and resource classification identifies present-day minable coal reserves. The definitions are general in nature and require refinement based on the particular property being examined.

A reserve estimate is not simply an exercise in geometry to determine the total resource deposit, subsequently disregarding such factors as seam thickness and depth of overburden. Reserve calculations are oriented toward minability, which requires engineering experience and judgment.

Unless specifically requested, mining valuation studies are restricted to reserves. Resources are not included because some may be considered unminable and, if included in a valuation study, would distort the most important item to be addressed: namely, determination of the present-day reserve base. Resource estimation also creates confusion, particularly in nonmining professionals, and results are often misinterpreted and consequently misused.

Many factors are considered by a mining engineer when determining the present-day reserve base of a property. Primary emphasis is given to: geology, exploration, development of reserve parameters, transportation and markets, environmental considerations, and reserve calculation methods. The balance of this section will examine these topics.

GEOLOGY

Geologic factors influence the parameters selected for the final reserve estimate; therefore, a formal investigation of available published geologic data is mandatory for a reserve study. The USGS, USBM, and various state mining and geologic departments collect and assemble a wealth of information concerning geology, thickness of deposits, and mined-over areas.

The geology of the deposit and surrounding strata is studied in detail prior to formulating a reserve estimate. Determination of the thickness and consistency of the deposit is essential.

Geologic data from the subject property and others in the general vicinity are examined to determine the existence of geologic disturbances or anomalies. Published sources may reveal whether seam thinning is regular or abrupt in nature. Extreme geologic disturbances may eliminate the deposit from consideration or the reserve value may be heavily discounted.

Topographic relief influences reserve calculations. A gently rolling plateau or relatively flat horizon offers less difficulty for surface mining than a high relief escarpment, which can negate cost-effective surface mining. Surface slopes limit the maximum bench width utilized during surface mining. Topographic relief also influences selection of a transportation method and siting of process facilities.

Studies of aquifers, flood plains, and drainage areas are essential to reserve determination. The presence of a large aquifer may hinder future mine development or cause additional pumping units to be installed.

* Reprinted with permission from *Fundamentals of Coal and Mineral Valuations* by James W. Boyd, © 1986, John T. Boyd Co., Pittsburgh, PA.

EXPLORATION

Collection of published geologic data is the initial step in reserve determination. An exploration program (including core and/or rotary drilling, seam outcrop and mine measurements, and geophysical logging) further defines the geologic characteristics of the deposit and provides sufficient information for the mining engineer to formulate a reserve estimate.

When evaluating a reserve, the mining engineer is often provided with numerous seam measurements from the subject property. All drill hole and other pertinent seam data provided should have been observed and recorded by responsible drillers, geologists, and mining engineers. Before purchasing a reserve, a prudent buyer performs carefully designed check drilling to verify results. Absence of a well-defined exploration program decreases reliability of the reserve estimate, while a well-designed exploration drilling and sampling program supplies information on reserve quantity, reserve quality, overburden depth, optimal mining method to be utilized, outcrop/suboutcrop determination, seam structure, and mined-over areas.

Reserve Quantity

A predetermined core drilling and seam measurement program permits the engineer to develop seam thickness trends and minable reserve areas. Based on available information, minable seam heights for the deposit are determined and subsequently utilized in estimating reserve quantity. Additional discussion of reserve quantity is included in this section under the heading *Development of Reserve Parameters*.

Reserve Quality

Core drilling and bulk sampling permit the engineer to make a determination of coal quality.

Overburden Depth

Overburden depth is an important parameter considered in the final reserve estimate. The depth of overburden strata influences such items as selection of mining method, type of mining equipment to be used, strip ratios, blasting procedures, and minable limit of the reserve area.

Optimal Mining Method to be Used

Mining methods include underground, surface (contour or mountaintop), and auger operations. The exploration program provides the engineer with information which permits a selection of the most appropriate and cost-effective mining method for extracting the reserve.

Outcrop/Suboutcrop Determination

Seam outcrop is that portion of a deposit which appears at or near the surface of the ground. The outcrop is located using the results of the exploration program. Location (or absence) of an outcrop influences selection of appropriate mining methods. Outcrop location influences type of surface mining, i.e., whether the reserve is minable by the contour surface method or total removal can be utilized.

A suboutcrop was at one time an outcrop that, over the years, became covered with unconsolidated material. A "clinker zone" was also an outcrop but the coal became altered by ignition and burning.

Seam Structure

Seam structure refers to the structural dip (incline) of the deposit. Seam structure is used in conjunction with topographic relief to determine seam outcrop locations. Surface pit layout is influenced by seam structure.

Mined-Over Areas

Core drilling and field exploration may confirm existing mined-over areas or locate unknown mined areas, but these methods alone are not precise determinations. Other sources, such as aerial photographs and mine maps, are used in conjunction with exploration data to identify previously mined areas.

DEVELOPMENT OF RESERVE PARAMETERS

Parameters selected for calculation of a reserve estimate are site-specific. The following factors are important in developing an estimate of minable reserves: ownership or control, reserve classification, reserve category, seam thickness, overburden classification, in-place reserves, raw coal recovery, dilution, and processing and salable products.

Ownership or Control

Establishment of basic ownership or control rights is of primary concern in the calculation of reserves. The final reserve estimate categorizes minable acres and tons of the deposit by ownership classification.

A detailed title search should be performed by a qualified legal representative to verify ownership. If such a search is not feasible, the engineer states that representations of ownership have been provided and accepted without benefit of a title search. Ownership categories include:

Fee	Possession of complete mineral and surface rights
Surface Lease	Control by lease of surface rights
Mineral Lease	Control by lease of mineral rights
Surface Only	Ownership of surface rights
Mineral Only	Ownership of mineral rights
Federal Lease	Lease of mineral deposit and/or surface rights from the federal government
Adverse	Properties not owned or leased; future acquisition may be justified

Reserve Classification

Coal reserves are classified into three major categories (measured, indicated, and inferred) based on the reliable occurrence of the deposit.

The definitions which follow are oriented toward mine application and development and are generally in conformance with published government standards. Particular attention should be given to the criteria of having seam data points in a systematic grid pattern. Areas of measured and indicated reserves should not be shown around isolated drill holes.

Some latitude is warranted in particular cases, such as the use of a partially measured classification, which is applied where rotary drilling partially estimates the reserve. Use of all rotary holes to classify a property excludes it from the measured category as no information relating to seam quality is obtained. Deviations from the classification guidelines need to be clearly justified based on available data.

In cases where available seam information is comprised of a large number of thin or no coal data points, it is advisable to limit the reserve estimate to potential coal horizon acreage and classify the category as speculative, or simply not estimate any reserve due to insufficient data.

Fig. 1 depicts the classification system based on core holes, active mines, and other seam data points. The following provide a description of the three classification categories.

Measured reserves are tonnages computed from seam measurements as observed in diamond drill core holes, mine workings, and/or seam outcrop prospect openings. The

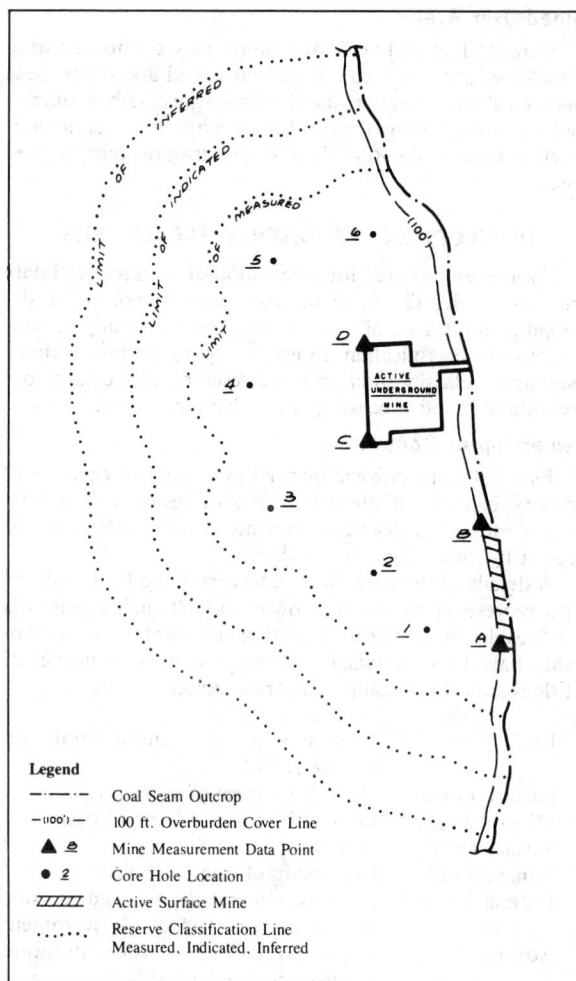

Fig. 1. Example of coal reserve classification system (not to scale).

points of observation are spaced to verify uniformity in seam thickness, physical character and quality, continuity of the seam horizon, and overall minability. The maximum acceptable distance between seam data points varies with the geologic nature of the seam being studied. Generally a distance of 0.5 mile (0.8 km) between points is recognized as the allowable standard for coal. If seam anomalies are present, this distance is reduced accordingly. For seams of known geologic continuity, where thickness and minability have been clearly defined over large areas by available seam data and historical mining experience, the maximum distance between points can be increased to 1 mile (1.6 km).

It should be noted that this definition is predicated on a systematic arrangement of holes in a grid pattern and does not allocate an area of measured reserve around isolated or wide centered holes. The computed tonnage is judged to be accurate within 20% of the actual tonnage.

Where data is not available (other than regularly spaced measurements along the outcrop) and the continuity of the outcrop is measured in miles (indicating the uniformity of the seam), an area of measured coal may be assigned extending a distance of 0.5 mile (0.8 km) (or the assigned maximum distance between seam data points) inby the outcrop. The same criteria may be applied to existing mine

workings provided these workings are extensive, regularly developed, and reveal the seam to be uniform and consistent.

Indicated coal tonnage is computed partly from inference of published geologic sources and use of available seam data points. Under normal conditions, the points of observation are a maximum 1 mile (1.6 km) apart, or approximately twice the distance established for the measured category.

Inferred coal tonnage is located beyond the limits of the identified classification. Reserves in this category should be identified by published geologic sources and sporadic seam data points. Inferred reserves are highly speculative and normally 0.75 mile (1.2 km) beyond known data points.

Reserve Category

Group 1: Reserves presently being mined or similar to those being extracted. The reserve has favorable potential for near-term development.

Group 2: Reserves which appear to be economically minable under favorable market conditions. Economic study and/or additional exploration is required to determine current minability. The reserve has marginal to good potential for near-term development.

Group 3: Reserves which appear to have uncertain mining potential because of less favorable seam height or mining conditions. A detailed mining study and market analysis are required. The reserve has doubtful near-term mining potential.

Seam Thickness

Seam data points should demonstrate uniformity in thickness. If extreme variations occur in total seam thickness or within the physical structure of the seam, reserve classification definitions are modified because of nonuniformity.

Based on seam thickness, a mining limit is assigned to the reserve. Seam data points having a thickness below the assigned mining limit are not used to classify the reserve. These measurements are actually demonstrating the unminable portion of the reserve.

A study of all available sources is made to determine whether seam thinning is regular and uniform or abrupt in nature. When abrupt thinning occurs, it is more reliable to project the limit of the minable or consistent coal area and assume normal seam thickness to the limit line. Reserves adjacent to the projected limit line are not considered measured unless data points are closely centered.

Want areas around seam data points recording no thickness measurement are projected midway between the no thickness point and adjacent minable thickness point.

Overburden Classification

Because the depth of overburden influences selection of mining methods and stripping ratios, reserves may be categorized by overburden depth. Use of overburden categories for surface minable reserves is complicated due to seam thickness and economic restrictions. Three frequently used classification groups are:

0 to 500 ft (0 to 152 m)

0 to 100 ft (0 to 30.5 m)
100 to 150 ft (30.5 to 45.7 m)
150 to 500 ft (45.7 to 152 m)

0 to 200 ft (0 to 61 m)
200 to 500 ft (61 to 152 m)

In-Place (In Situ) Reserves

In-place tonnage is the amount of coal in the ground prior to deductions for mining and processing losses. In-place reserves are calculated by determining the material specific gravity. Based upon the specific gravity, a ton per acre-foot (tonne/hectare/meter) or ton per acre-inch (tonne/hectare/centimeter) factor is developed. Multiplying that factor by the product of minable reserve acres and seam thickness yields in-place tons.

Raw Coal Recovery

A portion of the in-place coal reserve is not recoverable and is considered to be lost coal. The recovery percentage applied to an in-place reserve is affected by current mining methods and techniques, seam characteristics, classification of the reserve and depth of cover. If geologic anomalies are suspected, a reduction in coal acres or mining recovery may be justified. Mining recovery is also influenced by experience in neighboring mines operating in the same seam under similar mining conditions. A recovery percentage applied to in-place tons yields a raw recoverable product.

Dilution

Dilution is contamination of the deposit by foreign rock material or water. Rock partings are found either inside or outside the seam. In underground mining, thin in-seam partings are rarely separated during extraction. In-seam partings can often be removed in the pit during surface mining. Out-of-seam dilution is extraneous rock material, mined during the extraction process, that is not considered part of the minable seam.

Processing and Salable Product

Cleaning coal at a processing facility removes a portion of the impurities in order to produce an optimum salable product. A detailed sampling and analysis program provides the information necessary for calculating the clean coal recovery percentage which, when applied to the run-of-mine (ROM) tonnage, yields total clean coal product tons.

TRANSPORTATION AND MARKETS

Reserve marketability and delivered cost are influenced by the existing or future transportation network. An isolated reserve area may require construction of transportation facilities thus increasing costs.

Transportation from a mining operation can be by truck, belt conveyor, railroad, barge, pipeline, or an intermodal movement involving two or more of these methods. Competitive transportation options can increase the value of a property because transportation costs can be the highest component of the total delivered product price. Over half the delivered cost for sand and gravel and western coal is in the form of transportation costs.

In order for a reserve to be viable, markets must exist for the product. With no market, the deposit has only a residual value. Whether a market is unavailable because of the quality of the deposit or solely because of market conditions must be determined by the engineer.

ENVIRONMENTAL CONSIDERATIONS

Environmental topics are addressed during the valuation process. Type of mining influences environmental requirements.

Surface mining has come under the greatest amount of environmental scrutiny. Noise, air and water pollution, waste disposal, and land reclamation are some of the critical areas affected by environmental regulations. The Federal Surface Mining Control and Reclamation Act of 1977 (SMCRA)

defines guidelines for environmental control in the United States for coal mining operations. Many states have formulated mining regulations beyond the requirements of the federal government.

The burning of high sulfur coals by electric utility generating stations has led to the development of various proposals for acid rain legislation. As of this writing (1986), the federal government has been unable to formulate a comprehensive acid rain program. However, individual states such as New York and Wisconsin have passed acid rain legislation in an attempt to reduce sulfur dioxide emissions.

RESERVE CALCULATION METHODS

Major methods of reserve calculations include:
- Construction of isopach maps (lineal interpolation)
- Polygon construction (irregular grid construction)
- Computer-generated reserve (regular grid construction)

Construction of Isopach Maps (Lineal Interpolation)

This is the traditional method of calculating reserves. An isopach map defines, by means of contour lines, the varying thicknesses of a designed deposit. Based on structure contours, coal seams are correlated, minable seam thicknesses determined, and outcrop locations plotted. Seam measurement points are plotted on a map, and seam thickness is interpolated between existing data points. Fig. 2 is an example of a reserve map showing contour isopach lines. The isopachs are classified in increments, such as 36 to 42 in. (91 to 107 cm) and 48 to 54 in. (122 to 137 cm), and reserves are calculated based on these increments. Areas between seam thickness isopachs and other categories (i.e., classification, overburden cover lines, etc.) are calculated by planimetric measurement, and acres (hectares) of minable reserves are determined.

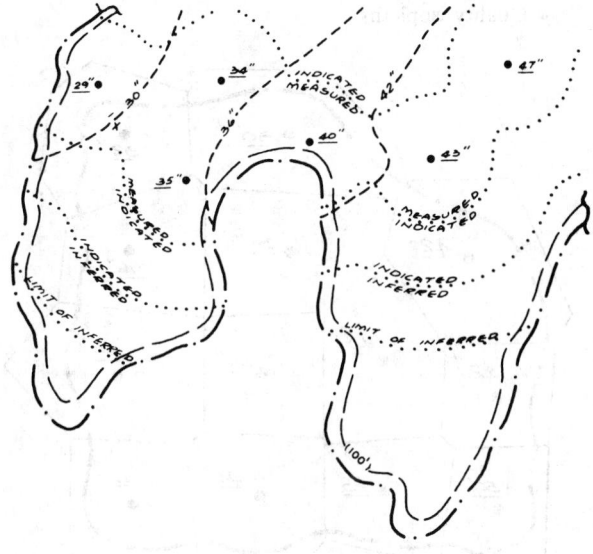

Legend

—·—·— Coal Seam Outcrop

—(100')— 100 ft. Overburden Cover Line

● 35" Core Hole Location with Total Mineable Seam Thickness (inches)

········ Reserve Classification Line Measured, Indicated, Inferred

---36"-- Seam Isopach (inches)

Fig. 2. Example of coal seam reserve map (not to scale).

The advantage of isopach lines is that they provide a reliable base for assigning seam thickness and permit the engineer to visually locate various thickness zones, a capability essential for mine design.

Polygon Construction (Irregular Grid Construction)

The polygon method of calculating reserves uses perpendicular bisectors to define areas of influence for each seam thickness data point. Each data point is assigned an area of seam thickness influence, and this area is planimetered and assigned a seam height based on the thickness of the seam data point. The polygon method is an accurate means of calculating reserves; however, thickness trends are difficult to discern. For mine planning purposes, the construction of seam thickness isopachs is recommended. Fig. 3 is an example of polygon construction.

Computer-Generated Reserves (Regular Grid Construction)

For this method, drill hole data and other seam measurements are collected and a data base is formed. The raw drill hole data include information regarding location, elevation, lithology, seam correlation, and quality. All collected data are entered into the computer; then a regular grid system is generated, with each grid composed of seam thickness, quality, elevation, topography, and other pertinent information. Maps showing reserve and seam quality data are constructed from the drill hole data base and displayed using cross sections and isopach contours. After map construction, the elevation extends to reserve calculation and mine planning. The following are developed from the data base:

- Coal/parting/seam thickness isopachs
- Seam structure
- Interburden isopachs
- Overburden contours
- Tons per acre (hectare)
- Quality isopleths

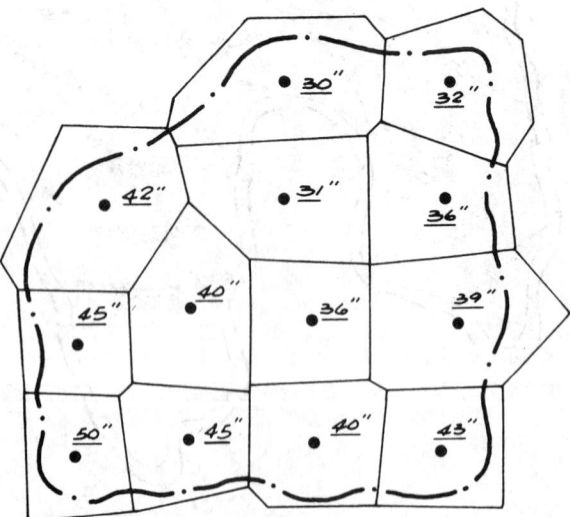

Legend

● *42"*   Core Hole Location with Total Mineable Seam Thickness (in.)

— · —   Coal Seam Outcrop

Note: All reserves in this example considered measured.

Fig. 3. Example of polygon construction.

- Strip ratio
- In-seam waste
- Topography

Applying the traditional hand calculating approach limits the number of changes that can be made quickly during the course of reserve estimation. Computer-generated reserves permit flexibility and can reduce the amount of time required to reestimate a reserve.

For increased detail, geostatistics are applied toward reserve estimation to account for the spatial relationship between data points and provide mathematically determined confidence levels. As opposed to classical statistics, in which samples are assumed to be random and independent of each other, geostatistics recognize geologic variables as having some degree of continuity within an area. This is observed by closely spaced samples tending to be more similar than data points separated by larger distances.

These values, which can display both a random nature yet exhibit limited continuity are termed "regionalized variables." The underlying assumption of geostatistics is that distribution of the differences of regionalized variables is consistent throughout a deposit, being dependent only upon position of data points. The distribution variance is primarily used in geostatistics to measure the interdependence of sample values and their influence over adjacent areas within a deposit. A low variance of distribution differences indicates the data points to be highly related.

The variogram, or more commonly semi-variogram, is used to model the deposit and study the influence of samples by graphing variance or half-variance vs. distance between sample points.

Through use of the semi-variogram, it is possible to calculate the variance of error distribution resulting from assigning values of a sample point to a larger block of the deposit. It is obviously desirable to minimize this estimation variance, decreasing the probable error, thus producing a true or more accurate estimate. Errors are often assumed to be evenly distributed in order to apply standard confidence intervals.

For estimation of reserve, the deposit is divided into blocks with weighting coefficients assigned to nearby samples producing a value for each block. As with all estimation techniques, the sum of the weighting coefficients should be unity, thus providing an unbiased estimate. The geostatistical procedure that provides the set of coefficients that produces an unbiased estimate and minimizes the estimation variance is kriging.

The mathematics involved in construction of the semi-variogram and computation of the kriging system are tedious and well-suited for computer application. Several commercial geostatistical packages, with mapping and plotting programs to display data and results, are currently available.

SUMMARY

The foundation of accurate property valuation is the reserve estimate, a multi-faceted study which has been discussed in this section. Accepted standards guide the implementation of a reserve study, but site-specific characteristics determine the final applicable parameters. Favorable transportation options, available markets, and environmental considerations are necessary; if they are absent, the reserve may be of marginal value.

REFERENCE

Wood, G.H., Jr., et al., 1983. *Coal Resource Classification System of the United States Geological Survey,* Circular 891, US Geological Survey.

3.4.4 Iron Ore

Ronald M. Hays, William F. Betzler, and Perry A. Canton

INTRODUCTION

Both conventional manual and computerized methods are used to estimate iron ore reserves. Examples of application of these methods are presented for two taconite deposits on the Mesabi Iron Range: Biwabik Reserve and Prindle Reserve.

MESABI IRON RANGE

The Mesabi Iron Range, in northeastern Minnesota, is the primary source of iron ore for US steel mills.

The Biwabik iron formation of the Mesabi Iron Range outcrops for approximately 175 km (109 miles). This formation has a simple structure and is 100 to 225 m (330 to 740 ft) thick, striking east-northeast, and dipping 5° to 12° SE. A Z-shaped bend or cross fold, the Virginia Horn, occurs near the center of the Range. A major fault, the Biwabik fault, occurs in the eastern part of the Range. Minor local structural features are common. The Biwabik iron formation is divided into four stratigraphic units. From top to bottom, these are Upper Slaty, Upper Cherty, Lower Slaty, and Lower Cherty. The Virginia argillite formation and glacial material overlay the iron formation and the Pokegama quartzite underlies the iron formation.

The Biwabik formation has been altered and enriched in many areas, resulting in various ore types. Of principal importance are magnetic taconites, large low grade magnetite ore bodies. Important factors in estimating taconite reserves are geology, stratigraphy, concentrate grade, ratio of concentration, grindability, and stripping ratio. Blending also is important because of ore variability. Estimating reserves is complicated because ore and waste are intermixed in various horizons, with mining restricted because of equipment limitations.

Biwabik Reserve

The Biwabik Reserve, located between the cities of Gilbert and Biwabik on the eastern Mesabi Range, is a large magnetic taconite reserve controlled by Jones & Laughlin Steel Corp. The portion of this reserve considered in this study is designated as the "East Pit" and is approximately 3.0 × 1.5 km (1.9 × 0.9 miles) and has 64 drill holes. This East Pit has been estimated by both conventional manual and grid computerized methods.

Prindle Reserve

The Prindle Reserve, located north of the community of Parkville (city of Mountain Iron), is part of the "East Pit" of US Steel Corp.'s Minntac operation. There are 180 drill holes that extend over approximately 1.78 × 10⁶ m² (440 acres) in section 36 except for the southeast quarter. Reserves have been estimated for 0.32 × 10⁶ m² (80 acres) of the Prindle Reserve by both the grid and block computerized methods.

CONVENTIONAL MANUAL METHOD

The conventional manual method is the precomputer method of estimating reserves in flat or gentle dipping ore bodies such as the iron ores of the Mesabi Range. Topographic maps are made from ground or aerial surveys. Cross sections then are drawn to accommodate the existing drill holes and incorporate the planned drill-hole grid. Geologic information and test data from drill-hole cores are added to cross sections and the configuration and stratigraphic layering of the magnetic taconite ore body are determined. For each drill hole, the minable taconite layer, or layers, are determined and the ore vs. waste ratio calculated and compared to the cutoff stripping ratio.

Using mining, metallurgical, and economic data available from taconite operations in the area, a mining plan is developed: a "first approximation" of the pit outline based on drill holes meeting the stripping ratio limitation is laid out on the map and cross sections. Cross-sectional areas of surface (glacial overburden), waste (nonmagnetic or lean magnetic formation), natural ore (if any), and magnetic taconite ore are determined by planimeter. Distances of influence are determined for each cross section. Based on cross-sectional areas, distances of influence, and material densities, the volumes and tonnages, including grade where appropriate, are calculated and tabulated. An overall stripping ratio for the pit is determined. The pit outline then is adjusted inward or outward to optimize ore recovery within the overall stripping ratio and also to minimize mining costs. Manpower requirements to determine reserves by this method limit the number of alternatives to be analyzed.

GRID COMPUTERIZED METHOD

The grid computerized method uses a series of computer programs to transform topographic elevations, drill-hole data, and geologic knowledge into a three-dimensional model of the deposit and then generates graphic displays and estimates surface (glacial overburden), waste (nonmagnetic formation), resources (minable and nonminable iron-bearing materials), and reserves (minable iron-bearing materials). This computerized system has great versatility, but requires considerable user interaction.

Topographic elevations normally are obtained by digitizing surface contour maps, but can be developed from survey data or drill-hole collar elevations. These topographic elevations are used to generate a grid matrix consisting of rectangular grids with surface elevation estimated for each grid node. For topographic surfaces, the grid matrix usually is generated by a distance-weighted function. The grid dimensions used for the surface topography grid matrix must be used for all subsequent grid generation. The grid matrix is a numerical representation of the surface topography. Topographic data must be generated for the entire area to be evaluated because a model can be developed only under grids having topographic elevations.

Raw drill-hole data are placed on a data base file and checked extensively for errors. Drill-hole data include identification, mining company, mine name, location, collar elevation, and incremental data. Incremental data include geologic horizon, geologic structure, chemical analyses, and beneficiation results and analyses.

The computerized system uses raw drill-hole data to prepare refined data base files for (1) mapping of stratigraphic units; (2) estimating and mapping surface, waste, and resources; and (3) estimating and mapping overburden (surface and waste), reserves, and stripping ratio. The system distinguishes between resources and reserves based on user-supplied economic related criteria.

The cutoff values for reserve estimates can be based on average or absolute values. Using average cutoff values maximizes reserve tonnage by converting resources to reserves if the combined values comply with the reserve classification criteria. Using absolute cutoff values minimizes reserve tonnage since resources cannot be converted to reserves. Using average cutoff values tends to increase the property's present value while using absolute cutoff values tends to increase the property's rate of return.

Based on raw drill-hole data and user-supplied criteria, the computerized system performs the following steps to generate a refined data base file for estimating and mapping minable reserves: (1) organizes data; (2) determines material classification (surface, waste, resource, reserve, direct shipping ore, or unknown) of drill-hole increments; (3) determines stratigraphic unit of drill-hole increments; (4) compiles drill-hole data for same material classification and stratigraphic unit; (5) compiles drill-hole data for surface, waste, resources, and reserves based on minimum minable thickness and maximum stripping ratio (considers waste both above and intermixed with resources and reserves); and (6) compiles drill-hole data for overburden, reserves, and bottom of pit.

Grid matrices then are generated based on the spatially distributed data in the refined data base files. Grid matrices can be generated by a variety of techniques, including (1) various distance-weighted functions with a variety of search procedures, (2) universal kriging, and (3) polynomial regression. Each grid matrix that is a numerical representation of a drill-hole parameter can be displayed as a contour map. Grid matrices are developed in a specified sequence and compared with previously generated matrices to assure compatibility of values in all grid matrices. The system also performs a statistical analysis comparing the calculated and actual values for drill-hole data parameters.

Using grid matrices, contour maps and cross sections can be generated for drill-hole parameters. Grid matrices also are manipulated to estimate volumes and/or tonnages of various materials.

For estimating reserves, volumes and/or tonnages of overburden and reserves are estimated for each grid and totaled for a user-designated area. The usual graphic displays are surface topography, bottom of pit, thickness of overburden, thickness of reserve, stripping ratio, and various cross sections.

Similar procedures are used to map stratigraphic units and to estimate and map surface, waste, and resources.

The grid computerized method provides numerical data, maps, and cross sections related to many parameters necessary for estimating reserves and subsequent mine planning.

BLOCK COMPUTERIZED METHOD

The block computerized method was developed for natural iron ores. An extension of that method has been adapted for use with taconite ores. This method combines some activities of the conventional manual method of estimating iron ore reserves with computerized activities.

A topographic map is prepared based on an aerial survey and field work. Cross sections then are developed with consideration given to geology of the formation and the planned drill-hole grid. Drill holes are plotted on cross sections and the iron formation is divided into layers based on geologic information, drill-hole chemical analyses, and beneficiation testing data. Next, "blocks" are delineated on the cross section by the mining engineer using equipment criteria and physical and geological characteristics of the pit area.

Blocks divide the layers into "slices" that are the basic unit of the block computerized method of estimating. Each slice is planimetered to determine the area and this information, along with all other pertinent data (drill-hole footage, geologic horizon, chemical analysis, etc.) related to this slice, is entered into the raw slice data base file. This slice data file is combined with a raw drill-hole data base file in a slice-block program to produce refined slice and block data base files for subsequent computer programs.

A computerized program could be developed that would give a reserve estimate tonnage based on blocks and appropriate criteria for the blocks. A computerized mining program has been developed that brings in additional file data on "lift and haul," equipment production rates, standard costs, etc. This mining program is a series of computer programs in which each block is first considered as waste (glacial overburden or iron formation) to be disposed of on a waste stockpile and, second, in the case of iron formation, as crude ore to the plant for processing. This allows for the determination by costs and earnings in each alternate as a means of block classification. The summation of blocks provides a reserve estimate tonnage with analyses plus certain cost and earnings data.

In taconite estimation, methods have been further computerized as compared to natural ore procedures. Basic data are taken from a data base file to plot required cross sections and maps. Computerized programs then assist in accomplishing the following steps:

1) Classifying each drill hole-increment as crude ore or waste based on parameters established by plant practice.

2) Economically evaluating data for each drill hole by combining waste and crude incrementally from the collar to the bottom and determining if the hole is economic and, if so, the location of the economic bottom of the hole; then assigning an economic index to the drill hole.

3) Contouring and plotting the economic indices to delineate the ore body limits, both as to lateral extent and bottom configuration.

4) Determining mining methods and designated bench elevations upon review by planning engineers using the previously determined ore body configuration; introducing the block concept of reserve estimation.

5) Creating and measuring slices with metallurgical data by analyzing drill-hole data from top to bottom; combining the slices to form blocks (this computer program eliminates the manual measurement of slices).

6) Combining blocks of about 100,000 t (98,400 lt) to form blast units of about 600,000 t (590,500 lt); combining blast units to form shovel units of about 4×10^6 t (3.9×10^6 lt) each.

These computerized programs give a versatile tool for mine planning. On a daily basis, however, closer control is desired. Mine engineers locate shovels daily as to coordinates and elevation. The computer then scans a 152-m (500-ft) radius of the location using all available data. Such data can consist of one to five drill holes, plus the magnetic iron information obtained from contoured susceptimeter readings. Using this information, the computer program calculates the best values to assign to that shovel location. Data from shovel locations are then blended to obtain the most desirable crude feed to the plant.

COMPARISON OF METHODS

Iron ore reserve estimates have been made for the Biwabik Reserve using conventional manual and grid computerized methods and for the Prindle Reserve using grid and block

Table 1. Comparison of Iron Ore Reserve Estimates

Method	Biwabik Reserve, t	Prindle Reserve, t
Conventional Manual	337×10^6	
Grid Computerized	344×10^6	48.0×10^6
Block Computerized		48.7×10^6

English equivalent: 1 t $\times$ 0.984 206 4 = 1 lt

computerized methods. The three methods have different approaches to defining parameters. However, to the extent possible, comparative reserve estimates are based on similar parameters, resulting in similar economic related criteria for minable reserves.

Table 1 presents iron ore reserve estimates for the Biwabik and Prindle Reserves using the three methods. The conventional manual method has been used for many years on the Mesabi Range and a good correlation exists between estimated resources using this method and actual mined crude ore. Thus, reserve estimates made using the conventional manual method serve as the standard for determining reliability of estimates made by other methods. Reserve estimates for the Biwabik Reserve are based on average cutoff values and show an excellent correlation between estimates made by conventional manual and grid computerized methods. For the Prindle Reserve, absolute cutoff values are used, with an excellent correlation between estimates made by grid and block computerized methods. Thus, all three methods appear to provide reliable estimates of crude iron ore.

Table 2 presents iron ore reserve estimates for the Biwabik Reserve made by the computerized grid method with various parameters and procedures. Estimate 1 of 344×10^6 t (339×10^6 lt) uses the same parameters and procedures as the Biwabik Reserve estimate presented in Table 1. These reserve estimates show that the indicated changes in reserve classification criteria, minimum minable thickness, and grid dimensions did not substantially affect the reserve estimate. However, reducing maximum or cutoff stripping ratio from 3.25 to 1.50 reduced the reserve estimate from 344×10^6 t to 281×10^6 t (277×10^6 lt) as illustrated by Estimate 4. The first six estimates are based on average cutoff values while estimate 7 is made on absolute cutoff values and reserve classification criteria used for the Prindle Reserve. Thus, the estimate 7 with 288×10^6 t (283×10^6 lt) is less than other estimates except for estimate 4 with a 1.50 cutoff stripping ratio.

Table 3 presents iron ore reserve estimates for 0.32×10^6 m² (80 acres) of the Prindle Reserve made by the computerized grid method. Parameters and methods changed in these estimates are stripping ratio and reserve limits or cutoff basis. As expected, reserve tonnage was greater with higher maximum or cutoff stripping ratio. Also, average cutoff values resulted in greater reserve tonnage than absolute cutoff values.

The conventional manual method has the advantages of (1) proven correlation between estimated reserves and actual

Table 2. Iron Ore Reserve Estimates Using Computerized Grid Method with Different Parameters and Procedures for Biwabik Reserve

Estimate	Parameters and Procedures	t
1	$A_1, B_1, C_1, D_1, E_1, F_1$	344×10^6
2	$A_2, B_1, C_1, D_1, E_1, F_1$	338×10^6
3	$A_1, B_2, C_1, D_1, E_1, F_1$	345×10^6
4	$A_1, B_1, C_2, D_1, E_1, F_1$	281×10^6
5	$A_1, B_1, C_1, D_2, E_1, F_1$	349×10^6
6	$A_1, B_1, C_1, D_1, E_2, F_1$	339×10^6
7	$A_3, B_3, C_3, D_1, E_1, F_2$	288×10^6

A Reserve classification criteria
  1) Concentrate grind, −200 mesh; minimum concentrate weight recovery, 20.00%; minimum concentrate grade, 63.00% Fe; maximum concentrate silica content, 10.00% SiO_2
  2) Concentrate grind, −200 mesh; minimum concentrate weight recovery, 15.00%; minimum concentrate grade, 64.50% Fe; maximum concentrate silica content, 7.00% SiO_2
  3) Concentrate grind, −200 mesh; minimum concentrate weight recovery, 21.00%; minimum concentrate grade, 65.00% Fe; maximum concentrate silica content, 7.00% SiO_2
B Maximum minable thickness
  1) 6.10 m
  2) 4.57 m
  3) 9.14 m
C Maximum stripping ratio (overburden/reserve)
  1) 3.25
  2) 1.50
  3) 3.00
D Grid dimensions
  1) 51.67 x 50.90 m
  2) 105.02 x 104.01 m
E Grid generation technique
  1) Constrained distance-squared weighting function
  2) Universal kriging
F Reserve limits
  1) Cutoff based on average values
  2) Cutoff based on absolute values

English equivalents: 1 t x 0.984 206 4 = 1 lt; 1m x 3.280 84 = 1 ft.

Table 3. Iron Ore Reserve Estimates Using Computerized Grid Method with Different Parameters and Procedures for Prindle Reserve

Estimate	Parameters and Procedures	t
1	$A_1, B_1, C_1, D_1, E_1, F_1$	65.0×10^6
2	$A_1, B_1, C_2, D_1, E_1, F_1$	61.0×10^6
3	$A_1, B_1, C_1, D_1, E_1, F_2$	48.0×10^6
4	$A_1, B_1, C_2, D_1, E_1, F_2$	40.6×10^6

A Reserve material classification
  1) Concentrate grind, −270 mesh; minimum concentrate weight recovery, 21.00%; minimum concentrate grade, 65.00% Fe; maximum concentrate silica content, 7.00% SiO_2
B Maximum minable thickness
  1) 9.14 m
C Maximum stripping ratio (overburden/reserve)
  1) 3.00
  2) 1.50
D Grid dimensions
  1) 51.35 $\times$ 51.43 m
E Grid generation technique
  1) Constrained distance-squared weighting function
F Reserve limits
  1) Cutoff based on average values
  2) Cutoff based on absolute values

English equivalents: 1 t $\times$ 0.984 206 4 = 1 lt; 1 m $\times$ 3.280 84 = 1 ft mined ores, (2) flexible decision-making criteria for special situations, (3) minimum equipment requirements, (4) no retraining of geological and mining personnel, and (5) maps and cross sections usable for mine planning. However, this method has the disadvantages of (1) high labor requirements, (2) high costs, (3) long time to complete (several months), and (4) high error potential.

The grid computerized method has the advantages of (1) low person-power requirements, (2) low cost, (3) minimum time to complete (a few hours after raw drill-hole data base file is created), and (4) low error potential. Indications are that reserve estimates by this method also correlate with actual mined ore. This is the quickest, easiest, and least expensive method of estimating reserves using a variety of parameters and procedures. The maps, cross sections, and data files generated by this method can be used for general mine planning, but are inadequate for detailed mine sched-

uling. However, the method can be modified for detailed mine scheduling. Disadvantages of this method are (1) lack of decision-making flexibility since the same criteria are used for the entire reserve, (2) need for computer, printer, plotter, and/or related equipment, and (3) need for specially trained personnel.

The block computerized method is between the other two methods in (1) labor requirements, (2) cost, (3) time to complete, and (4) error potential. Experience indicates that reserve estimates by this method are highly reliable and provide the vital decision-making tools to maximize the economic return of an operation. In addition to this advantage, this method also provides maps, cross sections, and data files to be used for detailed mine planning and scheduling. As with the grid computerized method, this method also needs a computer and related equipment and specially trained personnel.

3.4.5 Placer Sampling and Reserve Estimation

CLANCY J. WENDT AND STEPHEN W. THOMAS

Placer Sampling

One of the most difficult tasks associated with placer mining is the sampling of the deposit. More placer projects have failed due to inaccurate assessment of the reserves than to any other reason. Within the realm of placers, those containing valuable minerals with a high unit value are more difficult to sample than those with larger bulk, lower unit value minerals. Some items to consider when sampling a placer deposit are:

1) A relatively large size sample is needed for accurate valuation of the ground being tested. Placers are composed of many sizes of gravel that make a representative sample difficult to obtain.

2) When sampling placers for high unit value minerals such as gold, any error in mineral content of the sample will be magnified in the calculation of reserves.

3) Values usually are erratically distributed within the gravel mass. Therefore, some placers with a more uniform value distribution may be adequately assessed with a minimum number of samples, while a deposit with a high erratic distribution of values may not be adequately sampled regardless of how many samples are taken.

4) The investigation of a placer deposit should be made by or be under the direction of a person experienced in the art of placer sampling.

5) During a sampling program, items that must be observed and noted in addition to the sample size and valuable mineral content should include boulder size and number, clay content, bedrock conditions, water, frozen ground, false bedrock, and any other physical characteristics that would affect mining of the deposit.

Sampling Programs

The steps to be followed in approaching a placer sampling program are outlined below:

Reconnaissance: (1) check status of land ownership, (2) physical characteristics of area, and (3) research mining history of the area.

Choosing a Sampling Method: The main methods to consider are (1) existing exposures, (2) hand-dug pits or shafts, (3) backhoe trenches, (4) bulldozer trenches, (5) other machine-dug pits or shafts, (6) churn drill holes, (7) other drilling methods, or (8) bulk samples.

Special Problems Associated with Placer Sampling: These are (1) large rocks and boulders, (2) erratic high values, (3) uncased holes, (4) small diameter holes, and (5) salting.

Sample Processing or Washing: The considerations are the actual sample washing and the sample washing equipment.

Data Processing: Data processing consists of record keeping, reporting values, and assay procedures.

Reserve Estimation and Placer Valuation: Methods that can be used for reserve estimation and placer valuation are (1) block method, (2) triangle method, (3) polygonal method, (4) cross-section method, or (5) method of diamonds.

Conclusion: The placer sampling program report ends with a conclusion.

SAMPLING

The sampling methods outlined previously vary from simple grab samples on existing exposures to sophisticated drilling methods such as churn drilling. In this section, the various methods are discussed briefly and the applicability of each is indicated:

Existing Exposures

If existing exposures are available, they can be tested for potentially valuable minerals by taking a grab sample and panning the sample. Advantages of taking samples from surface exposures are the low cost and the speed at which the samples can be taken. The disadvantages are that you can only sample what is on the surface and no quantitative information can be produced.

Hand-Dug Excavations

These may be pits, trenches, or shafts and are suited to dry, shallow ground. This method of sampling is not used in the United States much today due to the high cost of labor but it can be effectively used in remote parts of the world where trained labor is not available or general labor is relatively cheap. The method provides a good bulk sample and is often used to verify drilling results by sinking a shaft over a drill hole. Bedrock values and characteristics can also be accurately determined when the excavations are sunk to bedrock.

Backhoe Trenches

Backhoes are a very versatile piece of equipment for sampling relatively shallow, up to about 6-m (20-ft) deep deposits. Backhoes are mobile, fast, can dig fairly hard ground, and are inexpensive compared to hand-dug excavations. Once a trench is opened up, channel samples are taken by hand or by using the backhoe or a bulk sample can be made with all of the material from the excavation. For backhoe sampling programs, the ground must be fairly dry and stable. Care must be taken when using a backhoe or any mechanical equipment to keep all fuel and lubricants away from the sample material as sample contamination can result, causing the fine gold to float and thereby reducing the value of the sample.

Bulldozer Trenches

Bulldozers are best suited to work in ground where trenches are to be dug 3 m (10 ft) deep or less and are in dry, stable ground. The greatest advantage of prospecting placers with a large bulldozer trench is that the trenches permit good visual inspection of the ground. Other advantages and disadvantages are similar to those associated with backhoe trenches.

Other Machine-Dug Excavations

Machine-dug excavations are shafts or pits which are dug using powered equipment such as large augers, bucket drills, or clamshell-type excavators. Digging shafts with the aid of powered equipment requires experienced operators and suitable machine access. Dry, stable ground is also needed for successful sampling operations when digging shafts and pits.

Augers: Augers for sampling placers can vary from small hand-held posthole size machines up to very large truck-mounted machines. Augers are relatively inexpensive to operate and can provide large volume samples. The disadvantages of using augers are their inability to penetrate ground with boulders, gold sorting may occur with spiral-type augers, and their inability to perform in water-saturated ground.

Bucket or Clamshell Type Excavators: These machines are usually quite large in size and allow taking a fairly large bulk sample. The advantages of using this equipment are its ability to allow visual inspection of bedrock, use of caissons to hold the hole open, and the capability of obtaining a fairly accurate sample volume. The disadvantages are the need for good access for the large equipment and a fairly slow digging speed.

Churn Drills: Churn drilling is used in deep or wet ground where sampling by pits, trenches, or shafts is not feasible. The churn drill (Fig. 1) utilizes a heavy casing with a drive shoe at the bottom, a chisel-shaped bit, and a vacuum-type sand pump for removing the sample from the hole. There are three main types of churn drills that are differentiated by their size: the hand-held "Ward," the light "Hillman" or "Airplane," and the heavy "Bucyrus-Erie" or "Keystone" drills.

The advantages of using churn drills for sampling placer deposits are (1) the sample is very reliable; (2) equipment is fairly portable; (3) few mechanical problems are encountered; and (4) technical data and interpretative information are available.

The disadvantages of using churn drills are (1) very slow penetration rate, (2) large boulders create many problems,

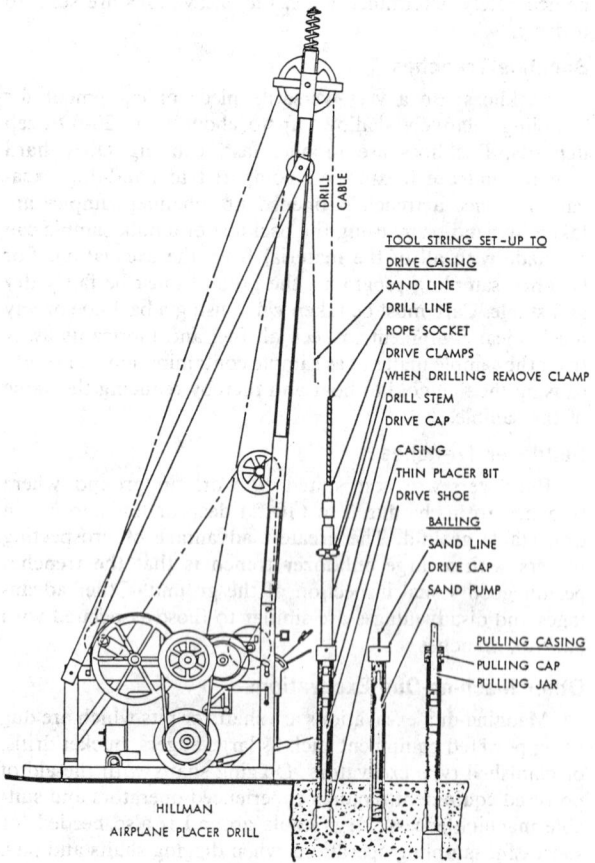

Fig. 1. Churn drill for placer sampling.

(3) problems from "running" ground, (4) requires skilled operator and careful drilling techniques, (5) fairly small sample size, and (6) the larger equipment requires good access.

The churn drill is a tried and proven process for sampling placer deposits. Abundant information exists on both the process and method of sampling. Due to the churn drill's reputation for producing reliable sampling results, it is still the most popular method of sampling deep placer ground in use today.

Other Drilling Methods

Due to the slowness and high cost associated with churn drills, other drilling methods are being tested for placer sampling. Some of the other types of drills are the reverse-circulation, hammer, and rotary; the vibratory; and the resonant or sonic drill. Details of each drill type follow.

Reverse-Circulation: With a reverse-circulation drill, the sample cuttings are blown up the inner section of a double-walled pipe. This method of cuttings removal reduces the chances of sample contamination as compared to conventional drilling techniques that force the drill cuttings out between the drill pipe and the surrounding wall rock. The advantages of using the reverse-circulation drill are high penetration rates at a fairly low cost per foot, its ability to penetrate boulders, and the ability to drill fairly deep deposits.

The main disadvantages are the need for good equipment access and the sample may not be representative of the actual material being sampled due to sorting. The use of a combination of churn and reverse-circulation equipment would give the operator the advantage of both speed and reliability. A correction factor will be necessary in most cases for the reverse-circulation drilling results.

Vibratory Drills: Vibratory drills have been tried recently for sampling placer deposits. The vibratory drills developed to date are quite portable, very fast in fine grained materials, and a fairly undisturbed core is obtained. The disadvantages are the small sample size, blockage in the core barrel, inability to penetrate boulders, and reduced penetration rates in clay.

Resonant or Sonic Drills: These drills have also been developed fairly recently and the technology associated with this drilling method is still being tested. Advantages of the resonant drill are its ability to penetrate boulders, it is relatively fast, the sample is generally undisturbed, and the method has good depth penetration. The disadvantages are the need for good equipment access, fairly high operating costs, core may be lost if vibration is used to extract the rods, and the method can only be used in fairly competent ground.

Bulk Samples

Wherever possible, bulk samples should be taken in placer deposits having an extremely erratic value distribution such as in deposits where the valuable mineral is found as widely spaced nuggets. Bulk samples are also needed for pilot plant test work.

SPECIAL PROBLEMS ASSOCIATED WITH PLACER SAMPLING

Due to the physical nature of placer deposits with a mixture of material ranging from large boulders to sand, and the nature of the valuable mineral contained in the deposit with medium to high unit value minerals erractically distributed within the deposit, placers are extremely difficult to sample. As mentioned previously, more placer mines have failed due to improper sampling than due to any other cause.

Some of the special problems associated with placer sampling follow.

Large Rocks and Boulders

When sampling placers, there is a tendency to bypass areas containing many boulders or to sample the easily collected finer material around the boulders. In many cases, this is essentially salting the sample since the valuable minerals in a placer deposit usually occur in the finer material.

The most direct solution to sampling areas containing boulders would be to take samples large enough to contain a representative portion of the boulders to give accurate value estimates. Since it is not physically possible to take large enough samples to include the boulders in most sampling situations, how does one consider the effect of boulders without including them in the sample? The most common solution is to visually estimate the volume and insert a correction factor into the end sample volume calculations.

Erratic High Values

The methods used for estimating the value of placer ground rely on the assumption that the value found in a particular sample extends halfway to the next sample. While evaluation of ground having a generally low or fairly uniform average of values can be done using the standard reserve estimation methods, erratic high value samples in a deposit cause problems.

Methods for adjusting erratic high values so an overevaluation of the ground does not occur includes resampling erratic areas and using the lower value determined or determining what the highest reasonable value should be and then keeping all sample values within that limit.

Uncased Holes

The use of uncased drill holes should be discouraged in placer sampling, especially in sampling ground with a high unit value mineral. If uncased holes are used, there is a tendency to unintentionally salt the sample and get overvalued sample results. This problem arises due to the fact that without casing, an excess of material can get into a sample without the evaluator knowing where, within the hole, the material came from. An exception to always casing drill holes is when drilling frozen ground. If the ground being sampled is well frozen, casing is usually not used.

Small Diameter Holes

Due to the large "nugget effect" associated with sampling ground containing high unit value minerals, the use of small diameter drill holes is not recommended for sampling deposits for gold, platinum, etc. When sampling deposits that have fine-grained material and contain low unit value minerals, small [50 mm (2-in.) diameter] holes may be used.

Salting

Salting of samples can occur intentionally or unintentionally. Intentional salting is the deliberate addition of valuable mineral to a sample. Unintentional or innocent salting is usually the result of careless or improper working procedures. No matter how salting occurs, the results can mean the failure of a project after many thousands or even millions of dollars have been spent bringing a property into production.

SAMPLE PROCESSING OR WASHING

Sample Washing

Once a sample has been taken from a placer deposit, it must be washed to separate the valuable mineral from the waste material. The valuable mineral thus separated is then weighed to determine the value of the ground being tested. Most washing devices use some type of riffled surface to retain the heavy minerals as the lighter waste material is washed away. In dry washers, a current of air floats away the lighter material while leaving the heavy minerals behind.

No production equipment designed for the recovery of heavy minerals actually recovers 100% of the mineral. Because of this, it is important to select a sample washing system that will indicate the commercially recoverable mineral content of a sample. Other essential features of a sample washing system are low initial cost, easy maintenance, easy transport and setup, acceptability of a wide range of material sizes, efficient washing of the sample, efficient use of available water, ability to process large and small samples, good mineral recovery, ease and speed of cleanup, and reliability.

Sample Washing Equipment

Miner's Pan: The old-fashioned gold pan is still the most widely used device for washing small placer samples. The pan is well suited for washing small samples but is not well suited for handling samples over 13.6 kg (30 lb) or if samples are taken very frequently.

Sluice Box: A sluice box (Fig. 2) is generally defined as an elongated wood or metal trough equipped with traverse riffles through which alluvial material is washed to recover the heavy minerals. Sluice boxes are sometimes, but erroneously, called "long toms."

The function of riffles in a sluice box is to retard the movement of heavy minerals allowing them to settle while the lighter material flows on through the sluice. Carpeting is often used under the riffles to increase the recovery of fine gold. In some operations, mercury is placed in the riffles to enhance gold recovery.

Rockers: A rocker (Fig. 3) is a short, sluicelike trough with curved traverse supports that permit it to be rocked from side to side. A flow of water, aided by the rocking motion, carries the material down the trough where the heavy minerals are trapped by riffles. Rockers have been used almost unanimously for washing samples by the dredging industry. A rocker requires less water than a sluice and is capable of recovering very fine gold.

Dry Washers: Dry washers are used in arid regions where water is scarce and a "dry" recovery plant is proposed for production. The dry washer uses a bellows to blow air up through a sluice box with a porous bottom, causing the lighter material to progress down the sluice while the heavy minerals are trapped behind the riffles. In order for a dry washer to work efficiently, the feed must be very dry. Damp, sticky material cannot be processed until it is dried. A typical dry washer is shown in Fig. 4.

Other Washing Equipment: In addition to the previously discussed washing equipment that has been used since the early years of placer mining, there are two other types of washing machines that have come into use relatively recently.

The "Denver Gold Saver" and other comparable machines utilize a revolving trommel screen to separate the coarse and fine materials with the fines going to a shaking riffle for separation of the heavy minerals. The advantages of the "Gold Saver" type of system are its ability to process relatively large samples and to recover fine gold.

A second machine that has been used for about the last 40 years is the mechanical pan. Once the coarse material in a sample is screened from the fines a mechanical panning device can collect the heavy minerals. As the name implies, the equipment is designed to duplicate hand panning.

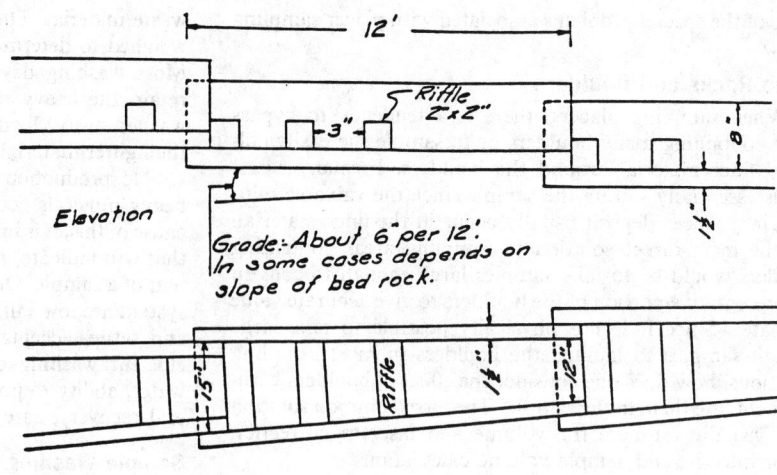

Fig. 2. Sluice box design (upper panel) and general sluice arrangement (lower panel).

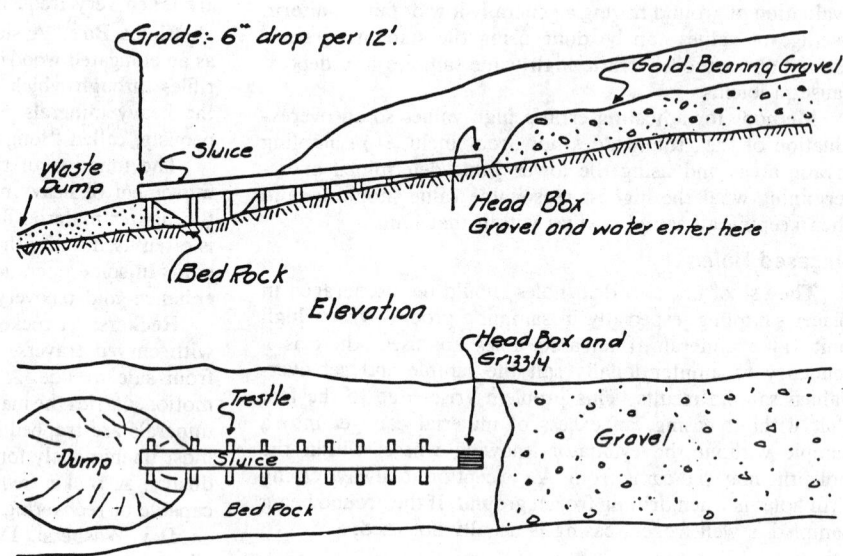

DATA PROCESSING

Record Keeping

Accurate, systematic records must be kept for proper placer evaluation. As an example, data that must be recorded on a log sheet when churn drilling a placer gold deposit includes:

1) Name of property
2) Location
3) Date
4) Drill-hole line number
5) Hole number
6) Hole collar elevation
7) Time (time of day is given for each bailing. A summary of time consumed in drilling, pulling, moving, repairs, and delays is also given)
8) Depth of the drive shoe for each sample interval
9) Depth of pumping for each drive or sample interval
10) Total hole depth
11) Core rise in the pipe for each drive
12) Core left in the pipe after pumping, i.e., the plug thickness
13) The length of core removed
14) Volume bucket measurement
15) Classification of colors (count the number of No. 1, 2, and 3 colors)
16) Estimate the gold weight based on the color count
17) Formation (note the visible physical characteristics of the formation being drilled)
18) Depth and nature of overburden
19) Labor used
20) Depth of the pay gravel
21) Depth to bedrock
22) Nature of bedrock
23) Thickness of pay zone
24) Diameter of the drive shoe
25) Theoretical volume of core removed
26) Measured volume of core removed

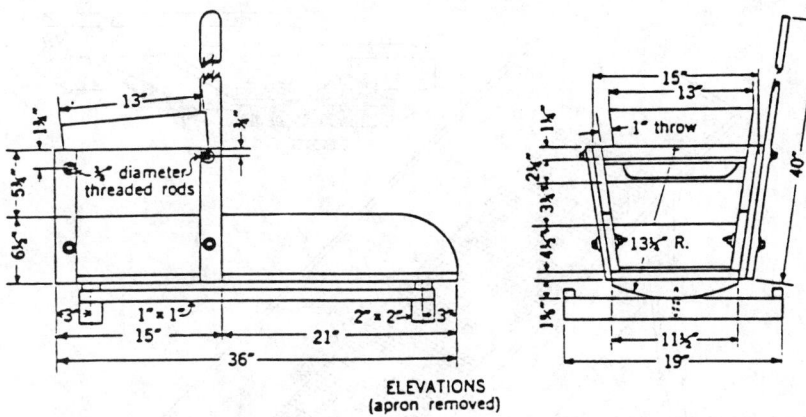

ELEVATIONS
(apron removed)

Fig. 3. Rocker for placer sample washing.

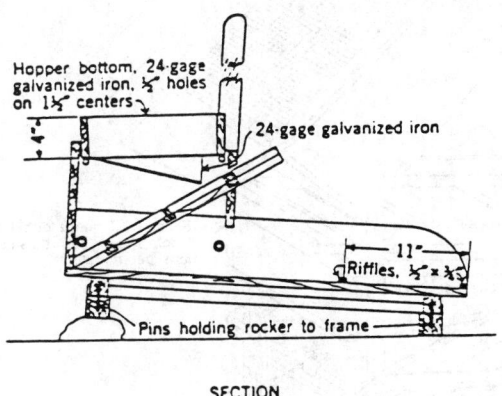

SECTION

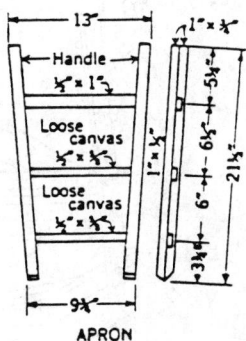

APRON

27) Weight of gold recovered in milligrams
28) Fineness of gold
29) Drive shoe constant used in calculations
30) Value in cents per cubic meter (cubic yard)
31) Price of gold used in calculations
32) Signature of the driller, panner, and helper

It is essential that detailed records are kept as they will be used to determine the value of the deposit as well as to determine its minability. Information included in the drill logs will also be used in selecting a mining method.

Reporting Values

In the United States, the volume of placer material is always reported as bank cubic yards. The value of ground in gold placers is reported in cents per cubic yard as well as in milligrams of gold per cubic yard. The fineness of gold and the price used in the value determination are also reported. Minerals other than precious metals are reported in kilograms per cubic meter (pounds per cubic yard), percent, or the particular unit customarily used for the commodity in question.

Splitting Samples

When a gold or other high unit value mineral sample is taken, it should never be split or reduced prior to assay determination. Any attempt to divide a sample to reduce its volume will yield erratic assay results. When dealing with fine size placer minerals having a low unit value, a reduction of the sample size by mixing and splitting is an acceptable procedure.

Assay Procedures

At this point, the procedures for determining the value of a sample need to be discussed. The only correct way to determine the amount of valuable mineral in a placer sample is by weighing. After the volume of the total sample taken has been determined, the amount of valuable mineral in the sample is measured by weighing. After the sample volume and valuable mineral quantity have been determined, a value can then be placed on the ground represented by that sample. It should be noted that placer samples should never be fire or AA assayed. The only time fire assaying is acceptable is to determine the fineness of the gold.

RESERVE ESTIMATION AND PLACER VALUATION

After samples have been collected, washed, and assayed, reserves for a deposit can be estimated. There are many placer reserve estimation methods available. Some of these are the block, triangle, polygonal, cross sectional, and diamond methods. The figures with each description graphically demonstrate these methods.

Block Method

The value using the block method (Fig. 5) is calculated as follows:

1) Find the volume of each block, length times width times depth.

2) Multiply the volume by the value per cubic meter (cubic yard) in each block.

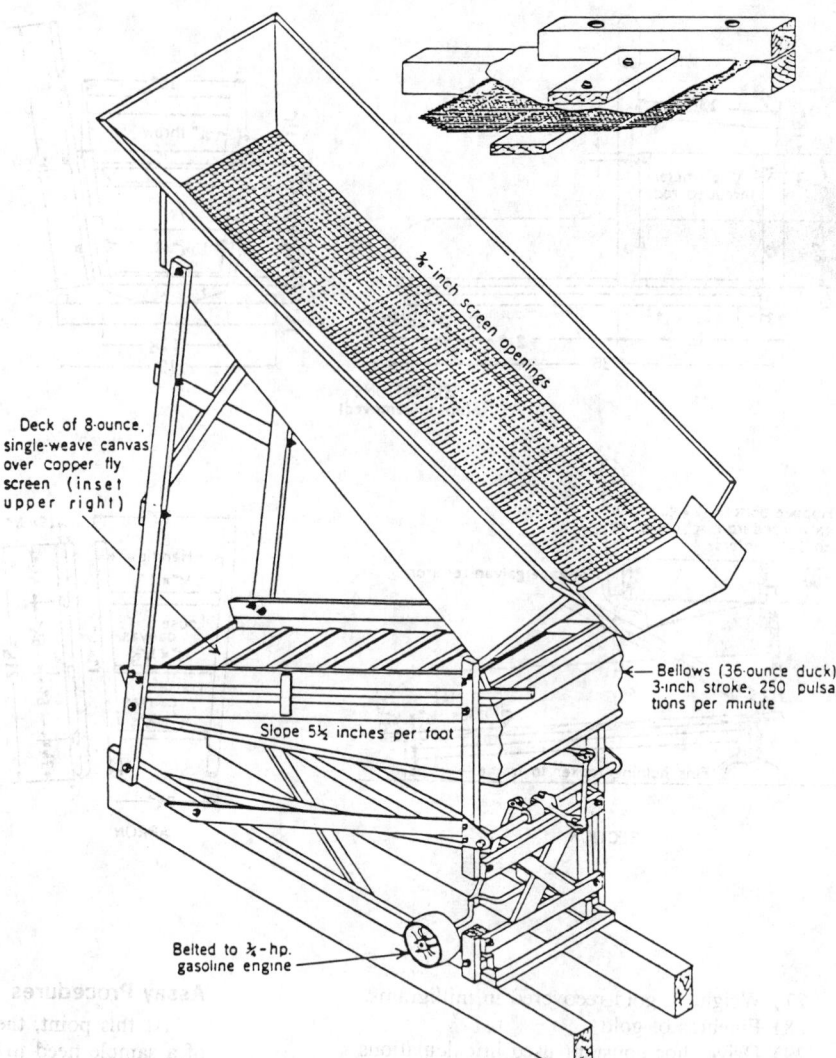

Fig. 4. Dry washer.

Deck of 8-ounce,
single-weave canvas
over copper fly
screen (inset
upper right)

¼-inch screen openings

Bellows (36-ounce duck),
3-inch stroke, 250 pulsa-
tions per minute

Slope 5½ inches per foot

Belted to ¾-hp.
gasoline engine

3) Find the sum of all the weighted values to obtain the total value.

4) Find the average grade by dividing the total in step 3 by the total volume to obtain the value per cubic meter (cubic yard).

Triangle Method

If Fig. 6A is used, the the method of calculating the value of the section is:

1) The volume equals the average of depth of the three drill holes times the area of the triangle.

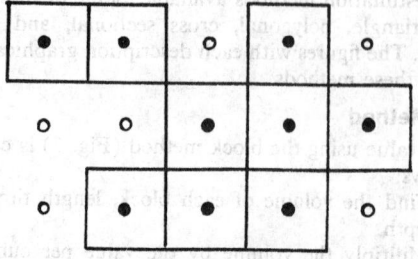

Fig. 5. Block method.

2) The average gold content is equal to the gold value of each of the three holes times the depth of each of the holes divided by the sum of the depths of the three holes.

3) Total volume equals the summation of values in step 1.

4) Total gold content equals the total volume in step 3 times the average from step 2.

5) The average grade equals the value in step 4 divided by the value in step 3.

Fig. 6B is used when the pay streak is parallel to the long dimension of the triangles. The method of finding the value is the same as Fig. 6A.

Polygonal Method

Either method may be used in finding the value using polygons. There may be need either for regular or irregular drill spacing. Fig. 7A is for regular polygons and is calculated as follows:

1) The total volume equals the sum of the volumes of the individual polygons. Find the volume of the polygons by multiplying the area of the polygon times the depth of the drill hole.

2) Total gold content equals the sum of the grades of each hole times the volume of each polygon.

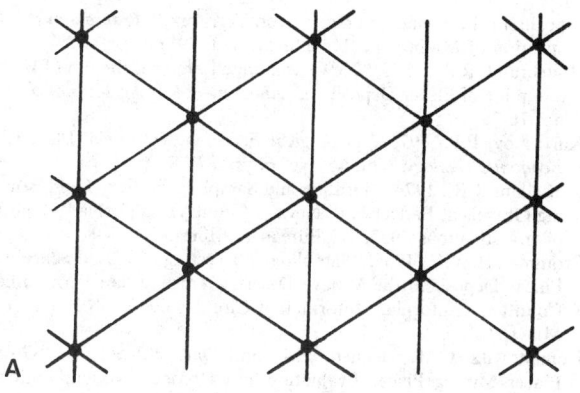

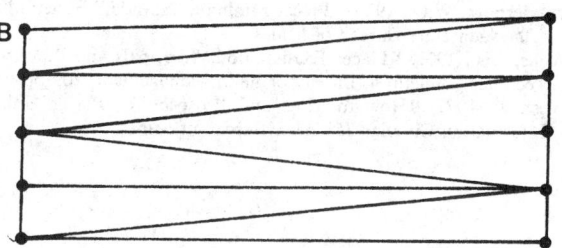

Fig. 6. Triangle method.

3) Average grade equals the value in step 2 divided by the value in step 1.

The volume (area) of each irregular polygon in Fig. 7B may be found by using a planimeter or by dividing the polygon into regular triangles and rectangles.

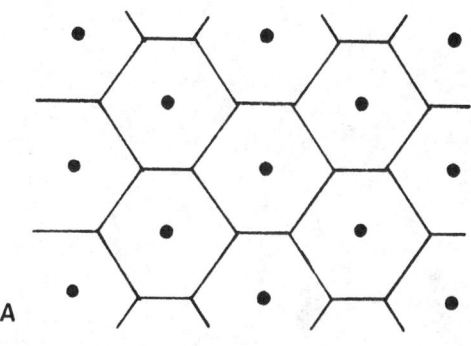

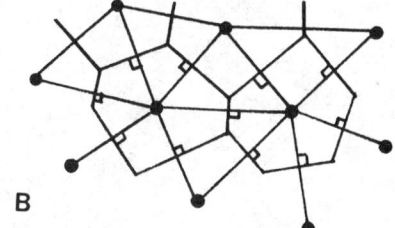

Fig. 7. Polygon method.

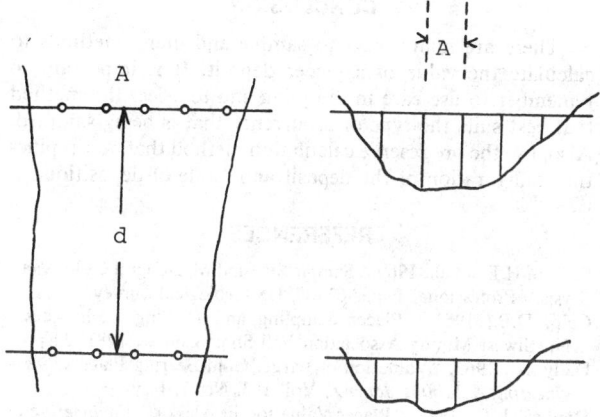

Fig. 8. Cross section method.

Cross Section Method

This method is very similar to that of the triangle and may be used as a check:

1) Area of a cross section (A in Fig. 8) is the average of the depths of the boreholes $(di + dj)/2$ times the distance between the boreholes (W).

2) The total volume is one-half the sum of the areas of all individual sections A and B times the distance between the traverse (L).

3) The total gold content is the sum of the gold contents of each section. The gold content of a section is the volume of the section times the value of the drill hole.

4) The average grade is the value in step 3 divided by the value in step 2.

Method of Diamonds

The method of diamonds is much the same as that of triangles. The drill hole is located at the center of the diamond and apexes midway between drill holes on adjacent lines. This method is best used for regularly spaced holes. The total area is equal to the sum of all diamonds that may be treated as right triangles for all practical purposes:

1) Total volume equals the area of each diamond (A, Fig. 9) times the depth of each hole through the pay zone.

2) Total gold value equals the sum of the gold value in each hole times the volume of each diamond.

3) To find the value per cubic meter (cubic yard), divide the value in step 2 by the value in step 1.

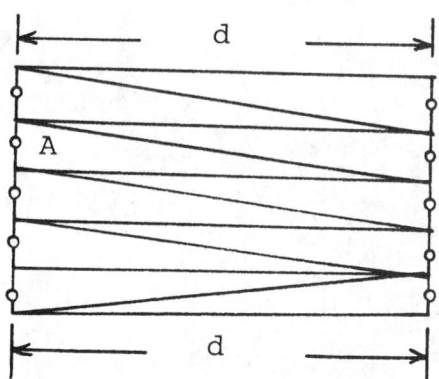

Fig. 9. Method of diamonds.

CONCLUSION

There are many ways to sample and many methods to calculate the value of a placer deposit. It is important to remember to use care in sampling and to select the method that best suits the type of occurrence that is being sampled. Also, use the ore reserve calculation method that best typifies the configuration of the deposit and mode of deposition.

REFERENCES

Clifton, H.E., et al., 1969, "Sample Size and Meaningful Gold Analysis," Professional Paper 625-C, US Geological Survey.

Colp, D.B., 1981, "Placer Sampling and Drilling Techniques," Northwest Mining Association Fall Short Course, 1981, 34 pp.

Daily, A., 1962, "Valuation of Large, Gold-Bearing Placers," *Engineering & Mining Journal,* Vol. 163, No. 7, July.

Donhey, L.C., 1942, "Placer Valuation in Alaska," *Engineering & Mining Journal,* Vol. 143, No. 1, Jan.

Fricker, A.G., 1976a, "Placer Gold - Measurement and Recovery," *Proceedings,* Australasian Institute of Mining & Metallurgy Sampling Symposium, Melbourne, Sept.

Gardner, W.H., "Drilling for Placer Gold," Beaver Falls, PA, Keystone Driller Co., unpublished report.

Hester, B.W., 1970, "Geology and Evaluation of Placer Gold Deposits in the Klondike area, Yukon Territory," *Transactions,* Institution of Mining and Metallurgy, Vol. 79, pp. B60-67.

Hildebrand, R.A., 1976, "New Technique Enhances the Art of Bulk Sampling of Placer Deposits, *Engineering & Mining Journal,* July, pp. 76-78.

Kartashov, P.P., 1971, "Geological Features of Alluvial Placers," *Economic Geology,* Vol. 66, No. 6, pp. 879-85.

McLellan, R.R., 1974, "Drilling and Sampling Tertiary Gold Bearing Gravels at Badger Hill, Nevada Country, California," Report of Investigations 7935, US Bureau of Mines.

Prommel, H.W.C., 1935, "Sampling and Testing of a Gold-Scheelite Placer Deposit in the Mojave Desert, Kern and San Bernardino Counties, California," Information Circular 6960, US Bureau of Mines, 18 pp.

Romanowitz, C.M., Bennett, H.J., and Dar, W.L., 1970, "Gold Placer Mining Placer Evaluation and Dredge Election," Information Circular 8462, US Bureau of Mines.

Staley, W.W., 1931, "Elementary Methods of Placer Mining," Pamphlet 35, Idaho Bureau of Mines and Geology.

Vanderburg, W.O., 1936, "Placer Mining in Nevada," Bulletin No. 4, Nevada State Bureau of Mines.

Wells, J.H., 1973, "Placer Examination, Principals and Practice," Technical Bulletin 4, Bureau of Land Management, 209 pp.

West, J., 1971, "How to Mine and Prospect for Placer Gold," Information Circular 8517, US Bureau of Mines.

Chapter 4

Feasibility Studies and Project Financing

Guerdon E. Jackson, Editor

4.1 Introduction

Guerdon E. Jackson
Bee R. Waples, Jr.

Feasibility studies, financial analysis, and project financing are necessary to bring together the large amount of basic data that will have been generated about a mining prospect over a period of time. As projects have grown in size, the amount of money required to bring a mine from the exploration stage to an integrated operation has increased, and it has become necessary to present a detailed plan to financial institutions for financing of projects. For many years mining companies provided funds from their reserves or profits to explore and bring new mines into production without turning to outside financing. This has generally not been the case in recent times.

It is the intent of this chapter to provide a basis for gathering and organizing required data in a systematic way and to provide guidelines and methods for varying depths of studies and analyses to meet a variety of conditions.

The feasibility study takes on many general meanings, from the commonly used term of *prefeasibility* to a full integrated feasibility study. In the first sense, prefeasibility can include order-of-magnitude estimates, whose only purpose is to provide a quick answer as to whether additional work and more detailed data should be investigated. In a multimillion dollar development, prefeasibility studies could be interpreted as milestone guideposts to assure that the project is still a viable entity.

The minerals industry has evolved a series of types of feasibility study estimates with varying degrees of accuracy that are used for different purposes. The prefeasibility study is often done early for comparative analysis in order to make decisions as to the continuation or rejection of a property, whereas a full feasibility study is normally produced for submittal to financial institutions to obtain the money for implementation of the project. When a feasibility study for financial purposes is required, then more than just a mine evaluation is involved. A complete analysis of mine, process plant, and infrastructure must be made in order to have a complete document that is acceptable to both the owner applicant and the receiving financial group.

If the capital required for a project is small in comparison to the assets of the owner's company, then a feasibility study accomplished within the owner's organization may be acceptable by the financial organization. On the other hand, if the capital expenditure is large and the total ability to repay the expenditure is predicted on the project being viable, then a recognized third-party company is usually contracted to make the feasibility study. Interestingly, most large banks or financial groups have evaluation departments that employ mining engineers, process engineers, and economists who review the feasibility studies and recommend actions to their financial officers for acceptance or rejection.

Statistics indicate that a major ore body requires ten to fifteen years from discovery to production. The feasibility study is a major milestone along the way.

First and foremost among the feasibility study estimates is the ore reserve estimate. Without an ore body with a sound calculated ore reserve, there is no mine. Details on ore reserve calculations are addressed later in the chapter.

Development and exploitation of a mine are the base anchor and require imagination and experience. However, most ore bodies are complex; and processing of the ore is a necessary part of any project to produce a salable product. Many excellent texts exist for processing information: including the *Handbook of Mineral Dressing* by Taggart and many symposia publications by AIME, such as *Flotation* and *The International Symposium on Hydrometallurgy*. These are comprehensive metallurgical works that can be used to plan various flowsheets for processing the ore.

Laboratory testwork can confirm the proper method for processing the ore. It is also an advantage to the feasibility study if a pilot plant has been operated and the process data obtained from the laboratory has been confirmed with pilot plant backup.

Once the ore can produce a product, a market must be established. No product is acceptable unless there is a market for it, and many times, it is required that a market study be included in the feasibility study.

When it has been demonstrated that the ore can produce a marketable product, the infrastructure requires complete investigation. Table 1 is a checklist of the infrastructure items that must be investigated to assure successful operations. Depending upon the location of the ore body, the infrastructure problems may be severe. Domestically, many of the necessary services may be in place. On the other hand, in some foreign locations, the ore body must be able to pay for entire new communities, as well as all services. The burden generated by the infrastructure may be the pivotal influence on the feasibility study.

When all capital costs are estimated and indirect costs are included, a financial analysis is made. A section on the various elements to be considered is found later in the chapter.

New mining ventures frequently are characterized by rel-

Table 1. Infrastructure in Mining Projects

I. Definitions
   A. Resources
     1. Natural (water, sand, gravel, soil, labor)
     2. Improved (roads, power, railroads, steel, lumber, shops, trucks, skills)
   B. Supplies
     Fuel, food, parts, reagents
   C. Infrastructure
     All facilities not a part of the process facilities.
II. Surveys
   A. Country
     Available essentials
   B. Area
     Pinpoint necessities
   C. Plant Location
     Careful analysis
III. Checklist
   A. Power
     1. Available
       a. Utility
       b. Others will build
     2. Project must build
   B. Water
     1. Available
       a. Close by
       b. Long route
     2. Poor source or quality
       a. Sea water only
       b. Brackish
     3. Sewage disposal
   C. Transportation
     1. Roads
       a. Exist
       b. Build by others
       c. Build by project
     2. Railroads
       a. Exist
       b. By others
       c. By project
     3. Airport
       a. Exists
       b. Needed
         1. Yes, by others
         2. Yes, by project
         3. No
     4. Water
       a. Type
         1. Lake
         2. River
         3. Ocean
         4. Wells
       b. Facilities exist
       c. Facilities must be built
   D. Communications
     1. Telephone
       a. Exists
       b. Project cost
       c. Not possible
     2. Teletype
       a. Exists

       b. Project cost
       c. Not possible
     3. Radio network
       a. Exists
       b. Project cost
     4. Mail service
       a. Exists
       b. Requires some project cost
   E. Reagents
     Chemicals
       a. Available
       b. Requires imports
   F. Supplies
     1. Food
     2. Parts
     3. Materials
   G. Townsite
     1. Housing
     2. Stores—garages, cinema, cafes, shops, markets, hotel, guest houses, or motel
     3. Hospital
     4. Police
     5. Clubs
     6. Recreation—parks, gyms, swimming pools, athletic fields
     7. Banks
     8. Churches
     9. Post office
     10. Government departments
     11. Schools
     12. Fire stations
     13. Rail or bus station
   H. In-Plant
     1. Offices
     2. Storage rooms
     3. Maintenance shops
     4. Warehouses
     5. Security
     6. First aid
     7. Fire protection
IV. Negotiations
   A. Governmental
     1. National
     2. State
     3. City
     4. Village
   B. Agencies
     1. Public
     2. Private
   C. Others
     1. Social law requirements
     2. Tax concessions
     3. Duty concessions
     4. Language
V. Comparisons
   A. Projects
     1. Locations
     2. Markets
     3. Products
   B. Percentage of project cost

atively large capital expenditures, long development times, and high commercial risk. Financial analysis thus becomes one of the key elements in the resource development process.

The financial analysis section in the chapter describes methods for establishing the necessary financial criteria. The time value of money, discount rates, accounting for inflation, and cash flows are discussed.

Methods of evaluating investments, including measurements of performance, sensitivity analysis, optimization, and risk analysis are discussed in detail.

To complete the section on financial analysis, taxation and depletion and their accounting are discussed with emphasis on practice in the United States. This section closes with a review of foreign project requirements, and other financial and economic considerations.

This chapter concludes with a discussion of project financing. A detailed review of general financing concepts reveals the sometimes controlling role of project financing in resource development. Among the factors discussed is the role of the sponsoring mining company, financing costs, cash flow, credit evaluation, cash flow lending, and the development of the financing plan. Project viability factors and tax consideration are also reviewed.

Project financing is discussed in detail from the project planning and financing decision viewpoint. Credit terms and the general availability of creditors is reviewed. Significance of the debt coverage ratio, the impact of leveraging, and debt structures are discussed. A review of the impact of unanticipated capital cost overruns and the timing of borrowing conclude the discussion.

REFERENCES

Evans, D. J. I. and Shoemaker, R. S., eds., 1973, *The International Symposium on Hydrometallurgy*, AIME, New York.

Fuerstenau, M. C., ed., 1976, *Flotation*, AIME, New York.

Taggart, A. F., 1945, *Handbook of Mineral Dressing*, John Wiley & Sons, New York.

4.2 Feasibility Studies

Guillermo V. Borquez
James V. Thompson

INTRODUCTION

The purpose of a feasibility study is to demonstrate, on paper, the technical and/or economic practicability of a project prior to execution. The initial effort in this section is to establish a feasibility study model and to show how to develop its capital and operating cost.

Generally, surface mining feasibility studies are only a part of a larger all-inclusive feasibility study that might include beneficiation and processing as well as infrastructure facilities. An exception to this might be a mine where the ore is shipped to a nearby custom processing facility. There are several examples of this type of operation in the uranium industry in the western United States.

The objective of the feasibility study is usually economic and the methods employed are somewhat independent of the mineral commodity. However, in a country which might have massive unemployment, ample inexpensive hydroelectric power, and no domestic petroleum production, the objective might be the utilization of domestic resources for the benefit of the country rather than hard line feasibility in the traditional sense. For example, trolley operated haulage units might be selected regardless of the more favorable apparent economics of employing diesel trucks and imported oil.

BASIC DATA REQUIRED

General

Obviously the location, elevation, climatic environment, and local infrastructure should be known. The local infrastructure, including such items as housing, power and water availability, communications and transportation systems, labor availability and quality, is important advance information.

Nature of the Ore Body

Once the local setting of the ore body has been ascertained, its general nature should be studied to determine: 1) placer deposit, open or confined area; 2) rippable waste and/or ore; 3) hard rock waste and/or ore; and 4) geological vs. minable ore body.

Geological vs. Minable Ore Body

Geological Ore Body: The geological ore body generally includes all the body of material that contains the valuable mineral to be mined. The geological ore body may contain high and low grade ore zones that are not physically or economically minable, because of depth, isolation, or amount of included or covering waste.

Minable Ore Body: The minable ore body is that portion of the geological ore body that can be extracted at a profit. The limitations to the minable ore body may be stripping ratio, grade of ore, alteration of ore, depth, excessive water, or environmental considerations.

At the beginning of an operation, the minable ore body may not be completely delineated and, as operations proceed, more of the geological ore body becomes minable because of the learning curve, improved methods and equipment, and increases in the prices of the mineral product being mined. Overall changes in technology and economics may also cause the removal of ore from the minable category. A typical example of this was the vast reserves of wash and jigging ore on the US iron ranges in the North Central States. The advent of pellet technology and the resulting economic advantages in blast furnace operation brought about the abandonment of iron ore reserves that could only be beneficiatied to an iron content in the low 50% Fe range.

Ore Reserves and Waste to Ore Ratios: In order to make a meaningful feasibility study of a site specific mining operation, it is necessary to have as much data as possible concerning ore reserves and the waste to ore ratio. While ore reserves are often based on surface trenching and pitting and at times indirect geophysical surveys, the most common exploration methods usually involve core drilling or air hole drilling.

A surface mining operation may produce any combination of the following products: ore, stripped waste, included waste that occurs between distinct bodies of ore, low grade material to be stockpiled for future treatment or treatment by a separate process.

It is, of course, necessary to quantify the amounts of each material to be mined and handled. Most copper mines in the southwest US will produce most of the above products and the same is true for some uranium mines and iron ore mines.

Selection of Mining Methods: While this subject is discussed in greater depth elsewhere in this volume, it is of course important to emphasize that the mining methods must be defined before a feasibility study can be undertaken. In some cases, for example in many placer operations, it is difficult to separate mining from concentration. A bucket line dredge, cutterhead dredge, bucket wheel dredge, or simple dragline operation usually dredges the material directly to the processing plant, which is frequently on board the dredge or floating alongside in the dredge pond.

Stripping: Stripping is discussed in more detail elsewhere in this volume. Most ore bodies require the removal of a waste covering of various depth; frequently the stripping operation involves different methods and equipment. Many coal and phosphate deposits are stripped by a large dragline or shovel which casts the overburden in windrows to the side of the exposed ore. Many uranium deposits in the West, particularly in Wyoming, can be stripped by ripper and scraper. The general objective of stripping is to expose the ore with minimum transport of stripping material, if transportation is required. Removal can be accomplished by conveyor systems, diesel, diesel-electric or trolley trucks, or even scrapers at times. Pit slurrying and pipeline systems can be used where conditions are favorable. It is generally undesirable to crush and wet overburden material because of the slime problem. In the early days of California placer mining, hydraulic methods were employed for the removal of overburden but the resulting slime problem in river waters ultimately closed down these mines.

Ore Mining: The excavation of ore and its transportation to processing plant facilities may involve quite different methods and equipment than those employed in stripping. In the feasibility study, the ore mining portion may be considered separately if the amount of stripping is considerable and the nature of stripping waste is quite different than the ore. For example; soft ores such as the phosphate ores of Idaho can be removed by ripper and scraper, and in some western US uranium mines, the ore is removed from small lenses by very selective backhoe excavation.

Production Scheduling: The extent to which mining is planned and scheduled will depend upon the end use of the feasibility study. If the study is destined to aid in the decision to accept or reject a property, a conceptual plan can be developed to arrive at quantities of ore and waste for costing purposes. On the other hand, if the study will be used to obtain funding, detailed plans should be prepared. For feasibility studies, this may involve computer-assisted calculation of ore reserves and computer-prepared plans and sections of the pit at various stages in the life of the surface mine. Computer modeling might be used to develop production schedules on a yearly basis for the first five years of the mine and on a five-year basis thereafter.

Definitions: A feasibility study must adhere to some carefully defined terms and these should be set forth at the beginning of any feasibility study report. The following are some important definitions.

Frame of Time—A feasibility study should be based on data and estimates for a one-year frame of time. Economic factors developed in the study will ultimately be used in some kind of economic analysis and these are always on an annual basis. The study may involve projected production rates over a number of years, but often the preliminary effort is to develop a full production model first. The total frame of time will depend on the life of the ore body and the economic criteria selected for the evaluation.

Operations Scheduling—It is important to define the annual hours of scheduled operation. Most very large mines are scheduled to operate 365 days per year, three shifts per day. However, certain sections such as the administrative group in the mine office may only work a 40 hour week on day shift only. Drilling and blasting is frequently scheduled for only five days per week, and often for only one or two shifts per day. Certain maintenance functions are more heavily manned on day shift and some may be on a 40 hour week basis. Some large mines in the United States will plan a total shutdown for ten legal holidays, frequently to avoid the premium pay required for operation during these periods. Labor is usually scheduled on a rotating shift basis and an effort is made to hold each employee to a 40 hour week. If the mine is to operate three shifts a day, seven days a week, it may be advantageous for the primary crusher to operate on the same schedule. From the standpoint of worker satisfaction, the best schedule is probably five days a week, two shifts per day. However, this can cause downstream difficulties. For very large mines, the storage necessary for 48 hours of mill consumption of primary crushed ore could be quite costly, particularly in severe winter climates.

It is not uncommon to schedule Monday day shift, both in the mine, primary crusher and concentrator, for scheduled maintenance, but this period is still part of overall scheduled time. A feasibility study should contain a table that outlines the mine schedule similar to the example shown in Table 1. The table should contain the following: scheduled days per year, scheduled shifts per day, scheduled legal holidays when the mine is shut down, average tons per day of stripping waste, average tons per day of ore, peak tonnage delivered to the primary crusher, and any other data pertinent to scheduling and production.

Factors Which Affect Productivity—Usually a mining feasibility study is concerned with a nonexistent operation. No productivity criteria may be available, and frequently there is no similar experience in the area. The engineer works with judgment factors and, where possible, with experience from similar mines. The following are productivity factors and the definitions of these factors as they apply to feasibility studies. Operating mines may adopt different definitions.

Overall Job Efficiency—This is an hourly factor and it refers to the average number of minutes per hour that a machine or group of closely linked systems will operate while in service in the mine. Downtime during the hour is caused by the following: fueling and servicing of equipment; recess and lunch time, if lunch must come out of an eight hour shift; poor coordination of shovels, haulage vehicles; and crowding at the dump point.

Machinery manufacturers often talk about the 50-minute hour which results in an overall job efficiency of 83.5%. However, in many operations, a 45-minute hour and 75% overall job efficiency is more realistic.

Mechanical Availability—This is a term that may cause some confusion. In a feasibility study the concern is the mechanical availability of a machine assigned to the job. A machine can be 100% available simply because it is not scheduled to perform work. Loss of mechanical availability refers to time when the machine is substantially out of operation for repairs during the period of time when it would normally be scheduled for production. Machinery manufacturers tend to be somewhat optimistic about this number, but if an overall number could be picked for all machines in an average surface mine, it would probably be about 85%.

Generally, mines which are scheduled for a high percentage of total annual hours will have lower mechanical availability of scheduled units because there was simply less unscheduled time for maintenance. It should be emphasized that mechanical availability is related to time lost for maintenance when the machine was scheduled for operation and does not consider maintenance done on unscheduled time.

One Wyoming uranium open pit mine can document the fact that when a seven-day-week operation is used, scraper availability drops to about 65%. This means that a larger scraper fleet must be scheduled. If a five-day-week is used, scraper fleet availability increases and fewer machines are scheduled.

Annual Outage Factor—Most mines are subject to some kind of loss of production that can only be measured on an annual basis. Examples are: 1) electrical storms and snow storms which knock out transmission lines and substations and block roads; 2) flash floods producing uncontrollable water in the pit, haulage road damage, and slides in the mine; 3) moving large units of equipment, such as draglines and shovels, which may have to be *walked* by trailing auxiliary power; 4) external causes, such as breakdowns in the transportation systems, strikes in some other segments of the industry, and local labor disturbances.

Without a backlog of experience, it is recommended that some figure be used for annual outage factor if for no other reason than to indicate that the items have not been overlooked. If a 95% factor is used, meaning that 5% of the scheduled time is lost, this would be about 18 days in a mine scheduled for operation 365 days per year. The 95% factor is probably too low for a mild dry climate with an adequate public power system such as in Arizona; however, in a region of severe winter climate and heavy snowfall, 95% is probably reasonable.

Production Utilization—This is the figure often confused with *availability*. The concern here is with the amount of time on an annual basis that the machine is actually productive. A machine may be 100% *available,* but it has no work to do.

For the purpose of feasibility studies, production utilization can be considered as the product of all of the foregoing. For example, if a mine is scheduled for six days per week, three shifts per day, minus ten legal holidays, 7,248 hours are scheduled for operation. The production utilization would be 0.75 (job efficiency) × 0.85 (mechanical availability) × 0.95 (annual outage factor) = 0.61.

The production utilization for the preceding example is 61% of scheduled time, or 7,248 × 0.61 = 4,421 hours for most machines. This is a conservative figure, and if actual operating experience closely related to the project under study is available, such information should be used in preference.

Costing Hours—Operating and maintenance labor should be carried on a separate table as an annual cost and not assigned to each machine on an hourly basis. This will be discussed in more detail later. Ultimately, the annual use hours must be costed. The time lost in the overall job efficiency factor is not considered as free time and direct operating cost is calculated for the full hour. However, direct operating cost is not accruing during the downtime related to the mechanical availability factor. Also the machine is not accruing direct operating cost during periods of annual outage. The annual costing hours, for the example already discussed, are determined as follows:

Total scheduled hours	7,248
Mechanical downtime	1,087
(7,248 × 0.15)	
Annual outage (7,248 × 0.05)	362
Costing hours (rounded up)	5,800

The costing hours for each machine are multiplied by the number of machines or fraction thereof to arrive at total costing hours for the particular type of machine.

MACHINE SIZING, UTILIZATION AND SELECTION

This can be a major engineering effort involving computer programs and detailed studies, but information for such a study is seldom available in the preliminary feasibility phases. Much time can be wasted in detailed studies of equipment selection which affect the reliability of estimates to only a

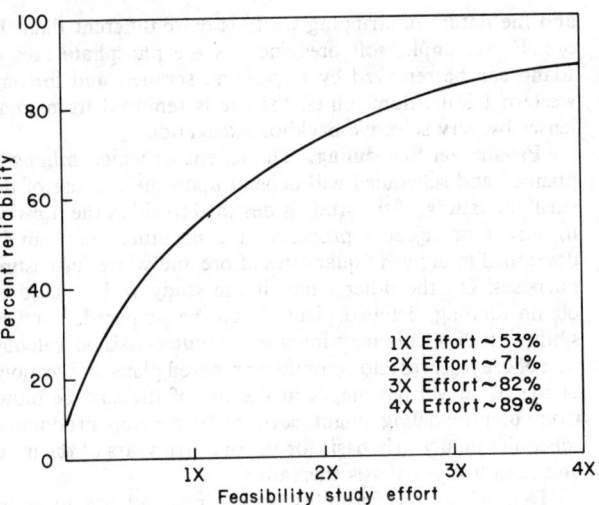

Fig. 1. Reliability vs. effort (not an absolute analogy).

small degree (see Fig. 1, which is a hypothetical comparison of reliability vs. effort). It should be noted that many older open pit mines tend to get over-equipped and caution should be used when using equipment performance data from the large older mine as an example.

For the purposes of the discussion that follows, the reader is referred to Table 1. Using this information as a basis, equipment will be sized for a hypothetical operation in order to demonstrate the principles involved in preparing a feasibility study.

Drilling and Blasting

Some engineers prefer to design elaborate drilling patterns for feasibility studies, but generally this effort is unnecessary. There is usually not enough information on hand to design such a drilling pattern unless a pilot mine with full scale benches has been operated. Considering the contingencies employed in a feasibility study, it would be necessary to know all of the factors which affect drilling and blasting.

The most important factors to know or to assume, in drilling and blasting, are a reasonable penetration rate for the size of drill hole selected and the powder factor. The

Table 1. Basic Criteria, Hypothetical Open Pit Mining

Annual tons of ore and waste combined*	12,700,000
Annual legal holidays of total shutdown	10
Scheduled operating days per week†	6
Annual scheduled operating days§	302
Scheduled shifts per day	3
Scheduled hours per year, 302 × 24	7,248
Average daily tonnage 24 hr day (ore and waste)	42,053
Average hourly tonnage	1,752
Peak delivery to dumping points (1752 ÷ 0.75)	2,336
Overall job efficiency (45 min. hour)‡	75%
Average mechanical availability of scheduled time‡	85%
Annual outage factor	95%

* It is assumed that ore and waste are of the same physical nature and that the haulage distance for both is the same.

§ 52 × 6 = 312 −10 = 302, however, 365 − 10 − = 03. In this case, 302 days have been assumed.

† Blasthole drilling scheduled 5 days, 2 shifts.

‡ Blasthole drills 60% and 80%.

Table 2. Selecting Number of Drills Required

Hole size	22.86 cm
Bench height	12.20 m
Hole depth	13.72 m
Total hole volume	0.56 m³
Percent of hole depth filled with explosive	60%
Volume of explosives	0.34 m³
Bulk density of explosives average	803 kg/m³
Weight of explosives in hole	271 kg
Explosives factor, kg/t rock blasted	0.15
Tons broken per hole	1 807 t
Total tons ore and waste per year	12.7 Mt
Total holes per year, 12,700,000 ÷ 1,807	7,028
Total length of hole, 7,028 × 13.72	96 424 m
Drilling rate while drilling the hole	25.9 m/hr
Actual drilling time required, 96,424 ÷ 25.9	3,723 hr
Scheduled annual hours, 5 days, 2 shifts, 10 holidays	4,000 hr
Overall job efficiency	60%
Mechanical availability	80%
Annual outage factor	95%
Production utilization 0.60 × 0.80 × 0.95	45.6%
Actual productive hours: 4,000 × .456	1,824 hr
Drills required: 3,723 ÷ 1,824	2.05
Drills in use or available	2.0
Drills owned	3.0
Costing hours: [4,000 − (4,000 × 0.20) − (4,000 × 0.05) × 2.05]	6,150 hr

Note: Time lost from overall job efficiency is not free time because most of it is for drill movement nonoperating time due to mechanical availability and annual outage factors are not changed for direct operating cost.

penetration rate is the meters drilled per hour while actually drilling, and the powder factor is the quantity of explosives required per ton of rock broken. With these two factors, a reasonable estimate can be made of the number of drills required.

Table 2 gives a typical feasibility analysis of drill requirements. Many of the line items are assumptions based on experience and in some cases field observation. Referring to Table 2, the hole size selected was 22.86 cm. Generally, larger holes require less drilling but may not yield as good fragmentation as a larger number of smaller holes. The hole depth is 13.72 m, 12.20 m is in the bench and 1.52 m is in the toe. The hole volume is 0.56 m³ and it will be filled to 60% of hole volume with explosives resulting in a total volume of 0.34 m³. These are arbitrary assumptions for a preliminary analysis.

The bulk density of the explosives is 803 kg/m³ and the weight of the explosive in the hole is 271 kg. This is a critical figure. The explosive factor is 0.15 kg of explosive per metric ton of material blasted and, therefore, each hole breaks 271 ÷ 0.15 = 1 807 t. The total tons of ore and waste to be blasted each year is 12 700 000. The holes per year are 12 700 000 ÷ 1 807 = 7 028 holes per year. The total length of holes is 7 028 × 13.72 = 96 424 m.

Next, it is necessary to develop production utilization. This is the product of assuming an overall job efficiency of 60%, a mechanical availability of 80%, and an annual outage factor of 95%, yielding a production utilization of 45.6%. The low overall job efficiency accounts for moving the drill from hole to hole and the low mechanical availability accounts for the rough usage that blast hole drills may encounter.

The drilling in this example has been scheduled for two shifts per day, five days per week, minus ten legal holidays which results in 250 scheduled days or 4 000 scheduled hours.

With production utilization of 45.6%, the actual drilling hours equal 1 824. In this amount of actual drilling time,

96 424 m of drilling must be accomplished in a year. As shown in Table 2, the drilling could be accomplished by 2.05 drills which means that two drills would be in use and that the operation would own three drills. The third drill could be a deferred purchase after the first year. The costing hours for the drills would be 6 150 hours as shown by the calculation at the bottom of Table 2.

Secondary Size Reduction

This can be a costly activity. Every reasonable effort should be made to avoid secondary blasting methods by perfecting better primary blasting. However, it is often necessary to do secondary lump breaking of some kind. The modern trend is away from explosive methods. Drop balls and similar devices tie up an expensive machine, which may be slow to move.

Consideration should be given to hydraulic hammers mounted on a controllable boom and on rubber tires with self-contained on-board power sources. However, the hydraulic hammer may be expensive and underutilized. A secondhand or unused small dragline may be available which could be equipped with a drop ball, or a portable compressor and wagon drill might be used to drill a single hole in a large boulder for explosive breaking.

For the purpose of this example, a 0.40 m³/s compressor and a 1.18 cm track-mounted hammer drill has been selected. Such a unit can do other drilling jobs around the mine as well as secondary breaking. It is arbitrarily scheduled for 800 hours per year and there will be days when it is not used at all. Mechanical availability during scheduled time should be near 100%.

Loading Machines

A typical open pit loading machine is the standard cable operated dipper and boom shovel that has been in use for many years. A more modern piece of equipment with bucket capacities up to 15.3 m³ is the hydraulic front shovel. Since

1970, front-end loaders have been perfected to the point where they can be used in coarse blasted rock and can be either rubber tired or crawler mounted. For the purposes of this discussion, Table 3 gives the data required to select the number of 7.65 m³ conventional shovels. The factors used in Table 3 are judgment factors, in part based on manufacturer's information and in part on field observations.

Fill Factor: Fill factor is the percent of total capacity of the bucket that is actually filled during each pass into the muckpile. Coarse, poorly blasted ore usually gives a low fill factor. Finer ore and a relatively smooth bottom will give a high fill factor.

Actual Bucket Capacity: This is obtained by multiplying the fill factor by the bucket size.

Swing Time: Swing time is the time in seconds that it takes for the operator to crowd the bucket into the muckpile, pull out, turn, and dump into the haulage vehicle and return to the muckpile. The swing time for shovels is usually less than the similar cycle time for front-end loaders. However, this depends on the nature of the muckpile and the skill of the operator. Generally, conventional shovels will perform better in coarsely blasted rock than front-end loaders. Front shovels were first introduced in the mid 1960s and were mostly of European design. By the early 1980s most US manufacturers of excavating machinery were offering competitive designs. Sizes have ranged up to 19 m³, but the more common large size is about 15 m³. The advantages and disadvantages of front shovels compared to conventional cable operated shovels is hardly significant in a preliminary feasibility study. Many manufacturers make both types, and they have accumulated considerable operating data.

Continuous Capacity: The continuous passes per hour are determined from swing time and this multiplied by bucket capacity gives the hourly continuous production that would be expected. Continuous capacity is determined first and then discounted for production utilization later. See Table 3 for an example of this calculation.

Swell Factor and Bulk Density of Blasted Material: This is difficult to obtain even from operating mines. In the example on Table 3, a swell factor of 0.67 has been employed, and this means that a bank m³ after blasting will have a volume of 1.49 m³. The weight of a bank m³ is 3.0 t, which means that the weight of a loose m³ would be 3.0 ÷ 1.49 = 2.0 t. The tons per hour capacity of the loading machine at 80% fill factor on a continuous basis is then 6.12, giving a total of 1 571 t/h continuous capacity.

Production Utilization and Annual Shovel Capacity: Employing a 75% overall job efficiency, 85% mechanical availability, and 95% annual outage factor, the production utilization is 60.6%. The mining shovels are scheduled for three shifts per day, six days per week, minus 10 legal holidays which equals 302 days or 7,248 scheduled hours. When this is multiplied by the production utilization, the result is 4 392 hours of continuous production. Multiplying this number by the tons per hour capacity of the shovel, the annual capacity becomes 6 899 832 t. The annual production of ore and included waste is 12.7 Mt/a. Dividing the annual requirements by shovel capacity indicates that 1.84 shovels are required.

Two shovels would be in use, and rather than owning a third shovel, the operation could be equipped with a 9.18 m³ front-end loader, rubber tire mounted, which could back up the shovels and do other utility work. It can be noted in this calculation that the annual tonnage can safely be made when one shovel is shut down for short periods of time and production could be easy to sustain on an annual basis, if the front-end loader can be used as backup. To provide absolute assurance that loading will go on at all times, it would be necessary to equip the mine with more equipment than is justified when production is considered on an annual or even a weekly basis. Certainly, two shovels and a large front-end loader are adequate for the annual tonnage required.

Haulage Units

For the purposes of this example, 77.1 t (85 st) rear dump trucks will be used. The selection of an electric wheel or

Table 3. Selecting Boom and Dipper Type Shovel

Bucket size	7.65 m³
Fill factor, well blasted rock	80%
Average bucket capacity 7.65 × 0.80	6.12 m³
Swing time	28 sec
Passes per minute, continuous operation 60 ÷ 28	2.14
m³ per hour continuous 6.12 × $\frac{60}{28}$ × 60	787 m³
Swell factor	0.67
Weight of bank m³ in place	3 t
Loose m³ = 1 ÷ 0.67	1.49 m³
Weight of loose ore 3 ÷ 1.49	2.0 t/m³
Tons per pass, continuous 6.12 × 2	12.24 t
Tons per minute, continuous 12.24 × 2.14	26.19 t
Tons per hour, continuous 26.19 × 60	1 571 t/h
Annual tons to be loaded	12,700,000
Annual shovel hours 12,700,000 ÷ 1,571	8,084 hr
Annual scheduled hours, 3 shifts, 6 days, 10 holidays	7,248 hr
Overall job efficiency	75%
Mechanical availability	85%
Annual outage factor	95%
Production utilization 75% × 85% × 95%	60.6%
Productive hours 7,248 × 0.606 =	4,392 hr
Shovels required 8,084 shovel hours ÷ 4,392	1.84
Shovels in use	2.0
Costing hours [7,248 − (7,248 × .15) − (7,248 × .05)] × 1.84 =	10,669 hr

Table 4. Haulage Truck Selection, 85-t (94st) Trucks

Cycle time continuous	12.5 min
Trips per hour, continuous 60 ÷ 12.5	4.8
Tons per hour continuous 4.8 × 85	408 t/h
Overall job efficiency	75%
Mechanical availability	85%
Annual outage factor	95%
Production utilization	60.6%
Scheduled hours per year (365 − 52 − 11) × 24	7,248 hrs
Productive hours 7,248 × 0.606	4,392 hrs
Annual production per truck year 4,392 × 408	1,791,936 t
Annual production required	12,700,000
Trucks required 12,700,000 ÷ 1,791,936	7.09
Trucks in use	8.0
Trucks in fleet	9.0
Costing hours [7,248 −(7,248 × .15) −(7,248 × .05)] × 7.09 =	41,111

Note: If trucks were capable of operating 100% of scheduled time only 4.29 trucks would be required. A fleet of eight trucks could in theory handle all the requirements without any spare, but if more than three trucks are out of service, the production schedule could not be met and a spare truck is provided to insure production if more than three trucks are out of service.

mechanical drive should be given careful study during the feasibility analysis.

Cycle Time: For the purpose of this example, it will be assumed that ore and waste are hauled the same distance (1 067 m one way) and that 12.7 Mt of combined ore and waste will have to be hauled each year.

The cycle time for trucks can be determined with greater reliability if an accurate profile is available over the haulage route. In the early phase of feasibility studies, this must be calculated from a hypothetical mining plan or assumed from limited information.

Manufacturer's catalogs contain data from which the speed for loaded and empty trucks can be calculated for actual distances and grades.

A typical truck cycle might be:

Maneuvering for position at the shovel	1.0 min.
Loading	3.0 min.
Accelerating the loaded truck	1.0 min.
Haulage to dumping point, 1 607 m away	2.5 min.
Decelerate and dump	1.5 min.
Return empty	1.5 min.
Total	12.5 min.

A critical portion of the cycle time is the time required to load the truck. In this example, the 77.1 t (85 st) truck might be somewhat large for a 7.65 m³ shovel because six passes might be considered excessive to some operators. However, the 77.1 t (85 st) truck is an efficient unit and fewer truck drivers would be required.

The use of computer programs to optimize truck and shovel combinations is recommended for feasibility studies when adequate data is available.

Determining the Number of Haulage Units: Table 4 gives the calculations for determining the number of haulage units without necessarily optimizing the shovel-truck combination and without details of the haulage road profile.

Support and Auxiliary Equipment

For the example under consideration, drills, shovels and trucks are the front line, regularly scheduled production equipment. Other support and auxiliary equipment is required as follows:

Rubber Tired Front-End Loader: In this example, a large front-end loader is proposed for emergency use when a reg-

ularly scheduled loading shovel is shut down for an extended period. This machine has a capacity of 9.18 m³ and its performance could be calculated in the same manner as the shovel in Table 3. It would probably have about the same fill factor, but its swing time or cycle time would be somewhat more. Depending upon conditions and the skill of the operator, it might have about 90% of the annual capacity of a 7.65 m³ shovel. As an alternative, the operation could be equipped with two smaller front-end loaders of lesser capacity.

The front-end loader is useful for removing isolated segregations of included waste or low grade ore. In any event, costing hours must be assigned to the front-end loader. Arbitrary scheduling might be one shift per day, six days per week, and with such light duty scheduling, its mechanical availability could be assumed to be 100% and the annual outage factor insignificant. When accounting for ten legal holidays and a six day work week, the scheduled annual days amount to 302, and the annual costing hours on a single shift basis will be 2,416 hours.

Rubber Tired Bulldozers: Where traction is no problem and there are no excessive lumps of very large coarse rock, a rubber tired bulldozer is very useful around an open pit mine. Compared to a track mounted dozer, the rubber tired bulldozer can be moved rapidly from one work site to the other.

Some of the duties which can be assigned to the rubber tired bulldozers are: 1) sweeping up fly rock around the shovel to keep a clear path for haulage trucks; 2) crowding the muckpile so that the dipper on the shovel may achieve a higher fill factor; 3) keeping a generally smooth haulage road in the immediate vicinity of the shovel, performing such duties as filling chuck holes that may accumulate water; and 4) leveling wind rows on waste dump.

Before the development of the rubber tired bulldozer, a track mounted bulldozer would be assigned to a large shovel. Much of the time this machine was idling without much work to do. Since the development of rubber tired bulldozers, if the shovels are not too far apart, one rubber tired dozer may service two shovels. For the purposes of this example, two machines of the 231 kW class are assumed. Such a machine may be oversized for some of the duties but it would have the capacity for more rigorous duty when required. There should be one rubber tired dozer assigned to the two shovels and one other assigned to the waste dump. It may not be absolutely imperative that the waste dump station be constantly attended by one of the bulldozers. Backup is provided by a track mounted bulldozer and a motor grader, discussed later. The costing hours for the two bulldozers combined is about the same as the costing hours for the two shovels, which amounts to 10,612 per-year.

Track Mounted Bulldozers: The dozing power of a track mounted bulldozer exceeds that of the rubber tired machines and its principal duty would be in the pioneering of cuts for new benches and pit roads. In an emergency, it could be used on the waste dumps and around the shovels. For this example a machine in the 343 kW class is used, and if it were scheduled for use on day shift only, it should have almost a 100% mechanical availability during scheduled time and the annual outage factor would not be significant. The costing hours would be about 2,400 per year.

Motor Grader: A motor grader is indispensable for most open pit mines to properly maintain haulage roads. If the motor grader is scheduled for day shift only, six days per week, its mechanical availability should be almost 100% of scheduled time and the annual outage factor would be insignificant. The costing hours would be about 2,416. There are no tasks assigned to the motor grader that are so imperative that they cannot be delayed for minor maintenance purposes. In addition, both the rubber tired and track mounted bulldozer, and to a lesser extent, the front-end loader, can provide emergency backup for the motor grader.

Smaller Dump Trucks: There are frequent duties around a large open pit mine where smaller utility dump trucks can be used. In this example it is assumed that 32 t (35 st) trucks would be available for miscellaneous use. Among these uses are emergency use for ore haulage in case of unexpected multiple breakdowns of the main truck fleet and the haulage of road dressing material for the haulage roads. The total costing hours for the three trucks could be assumed to be about 2,400 hours.

Service Vehicles: Most surface mines find it necessary to sprinkle the haulage roads to control dust and, therefore, a water sprinkling truck is necessary. In a dry arid climate such as the southwestern United States, the truck may be used all year around; but in more humid northern climates, it might not be used for more than five or six months per year. Indeed, mud may be a greater problem than dust. For the purpose of this example, an arid dry climate is assumed and the sprinkling truck would be used for a few hours every day. For study purposes, assume 1,000 hours per year.

Most surface mines will have a field fueling and lubrication truck and a field repair truck with welding, hard facing, and cutting equipment on board. Some large mines may also have a field tire changing truck, but in most locations in the United States, the major tire manufacturers have service organizations very close by that can dispatch a truck from their shop to the mine to perform this service.

Supervisors and engineering staff are usually provided with pickup trucks. Haulage of explosives and the loading of blastholes can often be contracted in United States mines.

All of the aforementioned vehicles, with the exception of pickup trucks, are generally underutilized. Even in a dry climate the sprinkling truck would probably be in use no more than about 1,200 hours per year. The same applies to fuel and repair trucks, and for both of these vehicles, a combined allowance of 2,000 hours per year would be adequate for this example. It will be assumed that the mine in the example being discussed here will be equipped with an explosives truck whose scheduled time would hardly exceed

600 hours per year. For this example, no tire changing truck would be provided.

For a mine that moves a total of 12 700 000 t per year, there would probably be about eight pickup trucks for supervisors and engineers. Use of such vehicles will vary widely, but for the purposes of this example, it is assumed that the total usage would be about 10,000 hours per year.

Pit Drainage and Lighting: These items may add significant cost to the mining operation. Many surface mines which do not have natural drainage may have to resort to pumping water from low points in the pit. For this example, it will be assumed that the mine is developed on a self-draining hillside location. If an open pit mine is to be developed on relatively flat terrain, a great deal of drilling and hole pumping may be required to obtain an estimate of the quantity of water to be removed. Accurate information can be difficult to obtain in the preliminary phases. With sufficient information, projections of water quantities can be made by competent geohydrologists. Complex problems can be encountered where lakes must be drained, streams rerouted, and well points established around the active mining pit.

Most mines scheduling a night shift would have pit lighting. All of the equipment has attached headlights or floodlights. Most shovels and large blast hole drills are electrically operated, therefore, power is generally available in the pit and lights may be strung along the principal haulage loads. Portable floodlights can be placed in the shovel and dumping areas. Developing the equipment and cost for pit lighting is a straightforward engineering task and need not to be discussed in detail here. For the purpose of this example, it will be assumed that pit lighting is provided from the power distribution system in the mine.

Power Distribution: When shovels and blast hole drills are electrically powered, power distribution facilities from incoming power lines involve portable switch houses and trailing cables. While these units do not involve large operating costs, they are a significant item of capital cost. For the example being considered, three 1,000 ampere switch houses and four trailing cables of at least 457 m (1,500 ft) capable of handling 8,000 volts are assumed. Besides the secondary distribution system, the feasibility study must consider the requirements for incoming power and a loop around the pit, if necessary.

Radio Communication: Every mining operation today should make maximum use of two-way radio communications which is inexpensive in both capital cost and operating cost. A repeater station would be required which is located either on a high elevation of land or a tower so that maximum range may be obtained in the system. All supervisors' trucks, haulage trucks, field service, repair trucks, and shovels should be equipped with a two-way radio unit. Other units such as the rubber tired bulldozer, motor grader, front-end loader, and water truck might be included.

MAINTENANCE FACILITIES

General Maintenance Concept for Feasibility Studies

The ultimate objective of feasibility studies is usually economic and involves the development of capital and operating cost. Many large mining operations will have central shops that perform all the maintenance for the mine, concentrator, processing plant, railroad, port facilities, and town site, as applicable, on a work order basis. The difficulty with this concept in a feasibility study is that it becomes necessary to prorate maintenance costs in order to determine a maintenance cost for the mining department.

The preferred method for feasibility studies is to assume that the mining department operates its own maintenance facility for the use of the mine. In some instances this might include the primary crusher. For this example, it will be assumed that the maintenance facility is solely for the mine and under the supervision of the mining department.

Nature of the Facilities

Mine maintenance shops can vary from elaborate, fully equipped, enclosed, and heated facilities to simple open shelter in a mild climate. The location of the mine, the surrounding infrastructure, and transportation facilities to and from the supply centers affect the nature of the shop. In a remote area, in the subarctic for example, the mine shop would have to be enclosed and heated with the equipment and parts inventory necessary to sustain the operation with the supply centers located thousands of miles away. In a remote tropical area, the structure might be in open shade but the facility would have to be fairly complete.

In remote areas the mine shop will have to carry a large inventory of spare parts, be capable of rebuilding engines, transmissions, and electrical components. A tire recapping shop may also be required.

In mining districts such as the northern US iron ranges and the southwestern copper mining districts, suppliers maintain service centers in medium to large communities within the district. Tire service is often provided on a contract basis. Major machinery companies that manufacture open pit mining machinery keep stocks of spare parts nearby and can usually provide engines and transmissions on an exchange basis.

From the foregoing discussion, it can be seen that maintenance facilities are affected by the local infrastructure and climate conditions. As stated in the beginning of this section, information concerning infrastructure and climatic conditions is essential in the preparation of a feasibility study. For the purposes of the example referred to herein, it is assumed that the maintenance shop is in a mild climate requiring some minimum enclosure and provisions for restricted space heating.

Sizing the Maintenance Facilities

The maintenance facility usually involves a rectangular building with numerous bays along one long side. These bays are usually equipped with roll-up doors that remain open most of the time, weather permitting. The building should be high enough so that it can accommodate a haulage truck with the bed raised to the maximum vertical position. It is desirable for the building to be equipped with an overrunning bridge crane with the capacity to remove the largest engine. Heating can be provided on a *spot basis* with several types of heating devices.

In preparing a feasibility study it is important that supporting maintenance facilities not be overlooked. These include spaces for welding, electrical and instrumentation repair, and washdown. In remote areas requirements for oil reclaiming and components repair may be needed.

One of the important concepts to be developed during the preparation of the feasibility study is the amount of repair and maintenance work to be done at the mine. Because the cost of space and equipment can be significant and because the cost of downtime for equipment improperly maintained can become prohibitive to an operation, the repair and maintenance facilities should be developed in some detail during the feasibility process. See Table 5.

CAPITAL COSTS

Definition of Capital Costs

A general definition of capital costs would be those items of project cost that will be depreciated for capital recovery and tax purposes. Whether or not certain initial costs are capitalized or expensed is frequently influenced by management policy and income tax requirements. For example, the preproduction cost required to bring an open pit mine into production might be capitalized or expensed depending upon a particular company's policy. As a general rule most mining companies would prefer to expense these items if it resulted in a tax advantage that was allowed by law.

Items Specifically Included in Capital Costs: Generally, capital costs include: 1) exploration and other preproduction cost depending on management policy; 2) mining machinery including freight, erection, and initial spare parts inventory; 3) permanent structures such as maintenance shops, mine offices, and warehouse; 4) all of the equipment and initial supplies in permanent structures; and 5) operating capital.

For a complete mining facility which might involve concentrators, smelters, refineries, and other infrastructure, the list and definitions of capital costs would be more extensive.

Items Specifically Excluded from Capital Costs: This category also depends upon management policy and tax matters but, in general, facilities located outside the immediate area of the mine are normally excluded from feasibility studies.

Types of Feasibility Study Estimates

For the purpose of feasibility studies, four general types of estimates have been devised for both capital and operating costs. A summary of these definitions is included in Table 6. Generally, feasibility studies restricted to mining only seldom become overly involved with these definitions. With the exception of more detailed design for permanent facilities, the definitions vary only slightly from Type I to Type IV. There is no difference in a quotation for equipment obtained for Type I or Type II. Type III and Type IV might involve written equipment specifications. The feasibility study types are somewhat more applicable for complete feasibility studies that involve an entire project, not just the mining phase alone.

Mobile Mining Equipment.

In the preparation of a mining feasibility study, one of the most important items of capital cost is the list of mobile mining equipment. This list should include the following: 1) FOB factory cost of the machine fully equipped with all of the accessories required for the job; 2) dry weight of the machine; 3) export packing, if required; 4) installed diesel or electrical power, although not important for small vehicles; 5) freight from factory to job site, including actual deliveries to the job site; 6) import duties and special taxes; 7) job site erection cost; and 8) spare parts inventory.

Many of the above items are essential for operating cost calculations. The weight of the equipment may be required for freight calculations. In addition, if the weight and cost of a machine are known, a rough estimate can be made of the cost of another machine if its weight is known and if it is of the same order of complexity.

Most mining machinery in the United States originates in the middle west or the upper middle west, and inland freight charges in the United States can be estimated by contacting railroads and truck lines.

Table 5. Major Items of Shop Equipment

Item	Size	Qty	Unit Wt, kg
Overhead crane		1 ea	
Shop supply air compressors	17.3 m³/min	1 ea	2,950
Steam cleaner	—	1 ea	410
Forklift truck	1.8 t	1 ea	2,720
Welders, shop	400 amp	2 ea	365
Welders, field	600 amp	2 ea	455
Pipe/bolt threader	15 cm pipe, 5 cm bolt	1 ea	2,270
Band saw	—	1 ea	860
Hydraulic drill press	25 cm	1 ea	455
Oil reclaiming unit	190 L/hr	1 ea	455
Blacksmith anvil	—	1 ea	230
Pedestal grinder	—	1 ea	270
Work benches	0.76 × 1.52 × 0.91 m high	5 ea	270
Cleaning tanks	0.91 × 3.05 × 0.91 m deep	2 ea	680
Tool lockers	36 compartments	1 ea	1,360
Welding booth	3 room	1 ea	180
Tire press	Light vehicle	1 ea	180
Tire press	Heavy vehicle	1 ea	1,815
Jib crane	0.9 t	1 ea	1,000
Engine positioner	—	1 ea	410
Transmission positioner	—	1 ea	410
Hydraulic puller	—	1 ea	1,815
Differential stand	—	1 ea	365
Hose reels, lube oil, air, water, grease	—	4 ea	270
Barrel pumps	—	2 ea	270
Electrical test equipment	—	lot	90
Battery charge	Light vehicles	1 ea	45
Injector pump tester	—	1 ea	20
Injector nozzle tester	—	1 ea	20
Sump pumps	—	2 ea	45
Grit blaster w/enclosure	—	1 ea	320
Chain hoists	1.8 t	5 ea	140
Miscellaneous tools			
Subtotal before sales tax			
Sales tax @			
Subtotal before contingency			
Contingency @ 10%			
Total			

Import duties and other government assessments on overseas projects may require careful investigations. Even though foreign projects may be undertaken by mining organizations owned by the government, duties on imported equipment may be assessed.

Most large mining machinery arrives on the job in a knock-down condition. A large dragline, for example, can require many months to erect. Erection and start up of large mining machinery is usually under the supervision of a field engineer provided by the vendor. His service, if not included in the initial price of the machinery, will be an extra cost charged by the vendor.

Spare parts inventories can represent a significant capital cost; the cost of this inventory is an interest-bearing item. In the United States, where vendors maintain field service organizations, inventory may be kept to a minimum. On overseas projects, spare parts inventories are usually larger. Often an overseas project is being financed by some source of international funding and the tendency is to provide a large initial spare parts inventory since it may be difficult to obtain government licenses to import spare parts once the project is operational.

For a preliminary feasibility study, 5% of the FOB factory cost of the machinery is adequate for estimating an initial stock of spare parts at US locations, but overseas projects may require as much as 10 to 20%.

Permanent Structures

Usually the permanent structures for an open pit mine are relatively simple architectural and structural designs. Unless the mine includes a crushing plant, the permanent facilities are usually a maintenance shop, mine office, warehouse, outdoor storage, fuel storage facilities, powder magazine and blasting material storage, and a changehouse. For a preliminary study, these facilities can be estimated by the square foot employing locally obtained unit costs. If the maintenance shop is to be equipped with an overhead crane, the building must be designed to include columns and crane rails for the support of the crane. Table 7 gives an example of a preliminary estimate for permanent structures for an open pit mine. More detailed estimates, of course, require detailed design with quantity take offs.

OPERATING COST

Operating costs for a feasibility study should be developed for each element of cost. Caution should be exercised when using the operating costs of existing mines. Criteria may be obtained from existing mines, but even a primary feasibility study should document each item and group of operating costs.

If the feasibility study involves a surface mine only, the operating cost should include the highest level of resident management. If the feasibility study is for an integrated mining, concentrating, and smelting operation, the operating cost estimate should include the highest level of management

directly related to the mine. This would generally be the office of the mine manager. Generally, *front office* cost should not be prorated against the mining department.

Definitions

Generally, operating costs include: 1) supervision and labor, salaries and wages; 2) labor burden; 3) all expendable mining supplies; 4) operation of major mining equipment including maintenance parts; 5) electrical power; 6) allowance for mine department undistributed overhead such as office supplies, engineering supplies, and general maintenance supplies; and 7) local property taxes and insurances, in the case where the mine is producing ore for direct shipment to other utilization points.

Items which are generally not included in a mining fea-

Table 6. Types of Feasibility Study Estimates§

Item	Type I	Type II	Type III	Type IV
Site				
Plant capacity	Assumed	Preliminary	Optimized	Finalized
Geographical location	Assumed	General	Approximate	Specific
Maps and surveys	None	If available	Available	Detailed
Soil and foundations tests	None	None	Preliminary	Final
Site visits by project team	Possibly	Recommended	Essential	Essential
Process				
Process flowsheets	Assumed	Preliminary	Optimized	Finalized
Bench-scale tests	If available	Recommended	Essential	Essential
Pilot plant tests	Not needed	Recommended	Recommended	Essential
Energy and material balances	Not essential	Preliminary	Optimized	Finalized
Facilities Design				
Nature of facilities	Conceptual	Possible	Probable	Actual
Equipment selection	Hypothetical	Preliminary	Optimized	Finalized
General arrangements, mechanical	None	Minimum	Preliminary	Complete
General arrangements, structural	None	Outline	Outline	Preliminary
General arrangements, other	None	Minimum	Outline	Preliminary
Piping drawings	None	None	One-line	Some detail
Electrical drawings	None	None	One-line	Some detail
Specifications	None	Performance	General	Detailed
Basis for Capital Cost Estimating				
Estimates prepared by	Project Engr	Sr Estimators	Sr Estimators	Est Dept
Vendor quotations	Previous	Single source	Multiple	Competitive
Civil work	Rough sketch	Drawing estimate	Drawing estimate	Take-offs
Mechanical work	% of machinery	% of machinery	Man-hr/ton	Man-hr/ton*
Structural work	Rough sketch	Prelim drawings	Take-off/ton	Take-off/ton*
Piping and instrumentation	% of machinery	% of machinery	Take-off	Take-off*
Electrical work	$ per kW	$ per kW	Take-off	Take-off*
Indirect costs	% of total	% of total	Calculated	Calculated
Contingency†	20-20%†	15-20%†	15%†	10%†
Operating Cost Determination				
Labor rates	Assumed	Investigate	Get contracts	Get contracts‡
Labor burden	Assumed	Calculated	Calculated	Calculated‡
Power costs	Assumed	Actual	Actual	Contract‡
Fuel costs	Assumed	Verbal quote	Letter quote	Contract‡
Expendable supplies	Assumed	Verbal quote	Letter quote	Contract‡
Reagents	Assumed	Verbal quote	Letter quote	Contract‡
Parts	Assumed	Verbal quote	Letter quote	Letter quote
Economic Analysis D.C.F.	Not meaningful	If requested	If requested	If requested
Use of Estimates	Comparison rejection	Feasibility	Budget	Funding

* Often subject to subcontract bids.
† In this definition the percentage assigned to contingencies is a judgment factor and is not to be interpreted as meaning that estimates are necessarily accurate within this percentage range, nor is there an implied reference to any order of accuracy.
‡ Contracts can be solicited if project is near-term.
§ Table courtesy of Kaiser Engineers.

Table 7. Estimated Capital Cost, Mine Maintenance Shop

	Item	Total Cost US$ × 1000
1	Structure 10,368 m^3 @ $62.21 m^3	$ 645
2	Shop equipment	913
3	Fuel handling equipment	33
4	Millwright labor and material, 20% of $946,000	189
5	Piping labor and material, 10% of $946,000	95
6	Electrical labor and material, 10% of $946,000	94
7	Subtotal—Equipment and installation, Lines 2 thru 6	$1,324
8	Subtotal—Direct field cost, Line 1 + 7	$1,969
9	Contractor's Field O.H., Camp, Plant and Profit, 30% of line 8	591
10	Subtotal—Field constructed, Line 8 + 9	$2,560
11	Engineering, Procurement, Construction Management, 12% of line 10	307
12	Total—Before contingency, Line 10 + 11	$2,867
13	Contingency, 20% of Line 12	573
14	Total	$3,440,

sibility study are: 1) nonresident management and sales costs; 2) income related or other taxes, because these are matters to be addressed in the economic analysis; 3) depreciation, interest, and royalties, which also are to be addressed in the economic analysis; and 4) transportation of ore beyond the local dumping points.

Basis for Direct Operating Cost Calculations

Every attempt should be made to develop operating cost. Unit cost from other mines can be used as a credibility check, but the engineer should develop costs specifically related to the project involved in the feasibility study. Many items are judgment factors applied by the engineer to indicate that the item was not overlooked, rather than to demonstrate absolute accuracy.

Supervision and Labor: Labor rates should be the current rates employed in the area. If there is no current mining activity in the area, the labor rates should reflect the local wage rate structure.

In a unionized mining area, the local labor unions will publish labor rates. In the United States, union wage scales may show very little percentage difference in rates between various categories. A high degree of skill may not show a large difference over unskilled employees. This is not true in some of the developing countries where skills are rewarded handsomely as compared to unskilled wages, and this differential can be a matter of several hundred percent.

Supervisory and engineering personnel are usually paid a monthly salary, and such rates are not difficult to obtain. Local precedents should be followed where possible. In a surface mining operation, most of the supervisors with some exceptions are engineering graduates. The wage scale for supervisors may be higher than for engineers. It is not unusual for the engineering staff to be rotated through supervisory positions, and often the distinction between supervisors and engineers is difficult to ascertain.

The most important item in developing the labor cost is the manning table. It is generally divided into three categories: 1) supervision, engineering, and clerical; 2) equipment operators; and 3) maintenance people.

The manning table should always be on an annual basis. In the example which has been used in this section, the total annual scheduled operating hours are 7,248. It is assumed that all labor is paid for 52 weeks per year, 40 hours per week, or 2,080 hours. However, an employee, if allowed two weeks vacation and five days sick leave, in addition to the 10 legal holidays, only works 1,872 straight time hours. If a position such as a shovel operator must be manned for all scheduled hours, it will require 3.87 employees for two shovels. If the operation were fully scheduled at all times, or 8,760 hours per year, the position would require 4.7 employees. For this example, 4.5 employees should be used as an average because it may be assumed that there are excused absences and employees with long periods of service that may qualify them for more than two weeks vacation. For example, there are eight trucks in the active fleet and these trucks must be manned for 57,984 hours per year; dividing these hours by 1,872 indicates a requirement of 3.87 drivers per truck, or a theoretical requirement of about 31 drivers. To cover the aforementioned extended seniority vacations and excused absences, the total truck drivers who would be paid for 2,080 hours per year would be at least 35 in number.

Similar reasoning can be applied to other operations. Inasmuch as drilling and blasting is only scheduled for a five-day week, two shifts per day, the manning calculations would of course take this reduced schedule into account.

Unavoidable Overtime: Usually an effort is made to avoid all overtime by scheduling individuals in such a manner that no person works more than 40 hours per week. However, it is almost impossible to avoid some overtime. For a small mine scheduled for six days per week operation, the labor would probably be paid for six days, and in the United States, this would amount to at least time and a half for the sixth day or weekly pay for 52 hours. The manning table should be based on sufficient labor to avoid overtime, but for the purposes of a feasibility study, 10% can be added to direct hourly wages as a line item for large mines and 15% or more for small mines.

Labor Burden: Labor burden means *fringe benefits*. Caution should be exercised when obtaining a factor for labor burden from existing operations. Such figures may contain items which, for a feasibility study, are covered elsewhere. In the example under consideration, sick leave, vacation, and holiday pay should not be a part of the labor burden because the employees are paid for holidays when the mine is not scheduled to operate and sick leave and vacations have been included in the calculation of the 1,872 hours per year that the average employee works. Overtime need not be in labor burden because it has already been accounted for as a line item on the manning table.

Labor burden usually consists of the following: 1) *Statutory Burden:* This includes items mandated by law, such as the employer's contribution to Social Security, Workmen's Compensation Insurance, Unemployment Insurance, and other costs that result from government action; 2) *Benevolent Labor Burden:* These are the items that an employer must pay to be competitive in the labor market to keep capable people. These include health insurance, group life insurance, pension plans, and other items directly related to wages and employment; 3) *Union Enforced Burden:* This includes items that may be the result of direct union negotiation.

Labor burden can be obtained from other operating companies and government agencies, but care should be exercised in identifying the included costs. As an approximation, true cash labor burden in the United States will be 25 to 35% of direct wages. However, it may be higher in some employee categories, and it can be higher in older operations that have many employees with many years of service. Overseas, it is not uncommon to find burdens of 50 to 100% or more. The best way to determine true cash labor burden is to make an actual investigation by contacting governmental authorities, insurance companies, and whatever labor unions may be involved.

Manning the Operation: In Table 8 for the example under consideration, it will be noted that there are 12 supervisors for 126 employees and 12 clerical and engineering employees. There is one supervisor for each 9.5 employees. The ratio of operators to maintenance personnel is 1.68 to 1.0. This would be considered a good ratio; however in some operations, the ratio is one to one. A rough rule for feasibility studies is two operators to one maintenance employee.

More maintenance people will be required if on-site engine and transmission rebuilding is done and if tire shops capable of recapping tires are maintained. Older fleets of equipment may also require more maintenance personnel and the same is true if operation is scheduled for 100% of all time.

It is important when investigating operating mines to find out which people are classified as maintenance and which as operators.

In a feasibility study, the manning table need not be organized in a manner that distributes personnel by shift.

Table 8. Direct Operating Cost Supervision and Labor for a 14 Mt/a Open Pit Mine

	Number of People	Hours Paid	Hourly or Annual Wage $	Annual Cost $
Mine superintendent	1	Salary	50,000	$ 50,000
General mine foreman	3	Salary	40,000	120,000
Drilling and blasting foreman	1	Salary	35,000	35,000
Shift foreman	3	Salary	35,000	105,000
Maintenance foreman	1	Salary	40,000	40,000
Maintenance shift foreman	3	Salary	30,000	90,000
Mining engineer—geologist	1	Salary	35,000	35,000
Planning engineer-surveyor	1	Salary	30,000	30,000
Draftsman-rodman	2	Salary	25,000	50,000
Exploration driller	1	2080	12.00	24,960
Drill helper	1	2080	9.00	18,720
Tim, Mtc, Supply clerks	3	2080	8.00	49,920
Secretary	1	2080	7.00	14,560
Messenger, sampler, truck driver	1	2080	7.00	14,560
Janitor	1	2080	6.00	12,480
Subtotal: Supervision	24			$ 690,200
Blasthole drill operators	2	2080	11.00	45,760
Drill helpers	3	2080	9.00	56,160
Lead man blasting	1	2080	9.00	18,720
Blasters helpers	3	2080	8.00	49,920
Shovel operators	8	2080	12.00	199,680
Heavy equipment operators	12	2080	10.00	249,600
Truck drivers	35	2080	8.00	582,400
Subtotal: Operators	64			$1,202,240
Electrician A	3	2080	12.00	74,880
Electrician B	3	2080	11.00	68,640
Mechanics A	8	2080	10.00	166,400
Mechanics B	8	2080	9.00	149,760
Field service men	4	2080	8.00	66,560
Maintenance labor	8	2080	7.00	116,480
Parts and tool room men	4	2080	8.00	66,560
Subtotal: Maintenance	38			709,280
Subtotal: Direct wages	126			$2,601,280
Unavoidable overtime, 10% of hourly labor (operators & maintenance)				191,152
Burden, 30% of all wages				838,862
Total				$3,631,734
Round to nearest $1000				$3,632,000

This is a matter which depends on experience and actual operations. It can generally be assumed that engineering and clerical help will work a 40 hour week on day shift. It can also be assumed that there will be more maintenance people on day shift than on the other two shifts. Drilling, blast hole loading, and blasting can probably be scheduled two shifts per day, five days a week. It is only important that the manning table have sufficient people. Only people operating production machinery should be distributed approximately equally on 3 shifts.

Machinery Operation: In Tables 2, 3, and 4 and the text that follows these tables, annual *costing* hours have been assigned. These are the hours of actual machinery operation that must be paid for. Table 9 provides calculations for determining the annual cost of mining machinery. Costing hours are multiplied by the hourly cost of the machinery exclusive of operating and maintenance labor.

It may be desirable to present a table that gives the hourly operating cost of each piece of mobile mining equipment by categories such as fuel, lubrication, engine supplies, tires, repair parts, and electrical power. In a mining feasibility study, such items as taxes, insurances, depreciation, and interest should not appear on this table because these items are more appropriately covered elsewhere, particularly in the section on economic analysis.

Details of operating costs are not always easy to obtain. The best source of information would be the records of an operating open pit mine that is similar to the mine being studied. Machinery manufacturers can frequently provide cost breakdowns. A word of caution is necessary concerning maintenance costs obtained from equipment manufacturers. These costs invariably include maintenance labor and in the example under consideration here this item is carried on the mining department manning table. It is therefore necessary to factor out maintenance labor cost from the maintenance figure provided by vendors. In the United States, labor is generally one-half to two-thirds of the maintenance cost and the remainder would be parts; but in overseas situations, particularly where duty is charged on imported parts, the labor portion may be only 10 to 25%.

The table showing the breakdown of hourly cost for each machine, particularly on overseas jobs, can be quite valuable. It is a guide for logistics calculations because it provides information on parts and supply quantities to be transported and stored.

The hourly cost of machines in an open pit mine varies widely. Older fleets may cost more, and new fleets considerably less. The figures obtained from vendors are usually averages and should be adjusted, if necessary, employing judgment factors.

The annual cost of operating mining machinery is the result of costing hours multiplied by hourly operating cost with the aforementioned exclusions.

For small mines that do not have redundant equipment and for large mines that prefer to keep maintenance forces at a low level, a line item should be added to the machinery cost table for rental of equipment during major breakdowns and for off-site maintenance contracts. This is a judgment factor; but for small mines it could be as high as 25% of the total machinery operating cost, and for large mines it will be in the range of 10 to 15%. To a certain extent, this is a contingency factor.

Drilling Supplies: The hourly operating cost for blast hole drills is not intended to include drill bits. These are included in Table 10. The life of drill bits depends on numerous factors, most of which might not be known in a preliminary feasibility study. If air hole drilling with rotary drills or down-the-hole hammers have been used in the exploration program, some idea might be obtained about the possible bit consumption. Also, operating mines with similar ore and waste could be a source of such information. Vendors of drill bits can be a valuable source of information if first hand knowledge is not available from on-site drilling of the deposit or from similar mining operations.

Table 9. Annual Cost of Mobile Mining Machinery

Major Items of Machinery	Annual Costing Hours*	Cost per Hour† $	Total Annual Cost‡ $ × 1000
Blasthole drills, exclusive of bits	6150	75.00	461
Secondary drill rig	800	21.00	17
7.6 m³ shovels	9915	115.00	1,140
77 t haulage trucks	45290	42.00	1,902
9 m³ rubber tire front-end loader	2416	48.00	116
230 kw rubber tire bulldozer	10612	25.00	265
345 kw truck-mounted bulldozer	2400	41.00	98
Motor grader	2400	17.00	46
32 t small dump trucks	2400	21.00	50
Service vehicles, ANFO, fuel, repair	3800	12.00	46
Supervisors pickup trucks	10000	8.00	80
30 m³ front-end loader	800	12.00	10
Subtotal	96958		$4,226
Allowance for rental equipment 10%			422
Allowance for less than expected productivity, 10%			423
Total			$5,071

* See Tables 2, 3, and 4 and text.
† Exclusive of operating and maintenance labor.
　Direct operating cost, no ownership cost included
　Information obtained mostly from vendors and represents averages. Costs should be updated.
‡ Rounded to nearest $1000.

Table 10. Annual Cost of Drilling and Blasting Supplies

	Annual Cost $ × 1000
Drill bits 97000 m ÷ 300 m/bit × $4000/bit	1,293
Slurry 0.025 kg/t × $1.43/kg × 12,700,000 t	454
ANFO 0.165 kg/t × $0.804/kg × 12,700,000 t	1,684
Blasting cord 451000 m @ $0.25/m	113
Boosters 1.0 kg/hole × 7028 hole × $1.43/kg	10
Secondary blasting @ 8% of above total including bits	283
Total	$3,837

Explosives: In Table 10, the cost of explosives is tabulated. This is a straightforward calculation and the most important information is the powder factor. From the powder factor and the annual tons to be blasted, the explosives cost can be determined. What the explosive mix will be is a matter of judgment and, again, an operating mine with similar ore and waste would be the best source of information. Manufacturers of explosives and the literature they publish are also good sources.

Undistributed Mining Department Overhead: This item is frequently overlooked. It is recommended that this be developed in some detail as shown in Table 11. The item is referred to as *undistributed* because it represents those costs not directly assignable to labor burden, machine operation, or explosives. This item is most often forgotten by engineers who make operating cost estimates by the *unit cost* method. Usually unit costs are available for drilling and blasting, loading, hauling, etc., but none of these costs contain department overhead. A percentage factor can be applied for overhead, but it is risky and at times not very convincing.

Reliability of Estimates and Credibility Checks

Most mining engineers with some experience have certain credibility checks that they can apply to operating cost calculations. For example, in the United States, wages and burden should be about 40% (plus or minus 10%) of the

Table 11. Undistributed Mining Department Overhead

	Annual Cost $
Mine office heat, 200 m² @ $30/m²/yr	6,000
Electric power, 200 m² @ $30/m²/yr	6,000
Telephone, 12 outlets @ $300/yr/outlet, basic	3,600
Water and sewage, inside employees 12 @ $50	600
Office supplies, $300 yr, all staff employees, 24 × $300	7,200
Engineering supplies, $600 × 4 employees	2,400
Repair office and engineering equipment 24 employees × $150	3,600
Safety and Training supplies 126 employees × $50	6,300
Equipment usage during training 500 hr @ $40 hr	20,000
Radio service 35 units @ $150 yr/unit	5,300
Crew bus* 230 days × 105 km × $1.24/km	30,000
Office building maintenance allowance 200 m² × $15.00/m²	3,000
LD phone and telex allowance $300/month × 12 mos	3,600
Professional cost†, 12 professionals @ $1000/yr each	12,000
Exploration drilling, 1400 hr @ $75/rig hour	105,000
Assaying $1.0 per hole 7000 blastholes + 3000 Exploration boreholes @ $10.00 each	100,000
Subtotal, Mine and Engineering offices	$314,600
Heat, 1000 m² @ $22/m²/yr‡	22,000
Electric power 1000 m² @ $35/m²§	35,000
Telephone, 4 outlets @ $300/yr/outlet	1,200
Water and sewage, 38 employees @ $50/yr each	1,900
Small tool replacement allowance, 34 employees × $300 yr	10,200
Undistributed maintenance supplies $3.00 machine hour	291,000
Repair shop equipment $300/employee, 34 employees	10,200
Shop building maintenance 1000 m² @ $10.00/m²	10,000
Subtotal, Maintenance shop	381,500
Total	$696,100

* Assume a subsidized crew bus and mileage cost covers depreciation, tax, insurance, etc.
† Publications and professional activity such as AIME.
‡ Lower heat level than office building.
§ Higher power use than office building.

Table 12. Summary of Direct Cost Mining Department

	Annual Cost $ × 1000	Cost of Material Moved $/t
Supervision, labor and burden	0.29	3,632
Mobile machinery	0.40	5,071
Drilling and blasting supplies	0.30	3,837
Undistributed overhead	0.05	696
Local taxes (non-income related) and insurance*	0.06	720
Total before contingency	$1.10	13,956
Contingency†, 20%	0.22	2,791
Total	$1.32	16,747
Annual Tons Material Moved: 12,700,000		

* Unless firm data is available, use 3% of total mining department capital cost which, in this example, would be about $24,000,000 × 0.03 = $720,000.

† In this definition the percentage assigned to contingencies is a judgment factor and is not to be interpreted as meaning that estimates are necessarily accurate within this percentage range, nor is there an implied reference to any order of accuracy.

direct operating cost of an open pit mine. There are always special circumstances that would change this percentage. The direct operating cost of mining machinery will depend on such things as fuel, remoteness of the operation, etc., but it will amount to about 37% (plus or minus 10%) of the direct operating cost. Blasting supplies can vary widely. For explosives alone the percentage is 12 to 15% in most instances. Explosives can be very expensive overseas. Undistributed mining department overhead will vary widely, but it is about 10%. See Table 12.

Production in terms of total tons of material per man shift in the mining department is a good credibility check. Unless there are unusual circumstances or a very small mine is involved, less than 200 t of total material moved per man shift is unacceptable. The study should be reviewed for underestimation of machine capacity and too many employees. Two hundred to three hundred tons per man shift is in the acceptable range, 300 to 400 t is good, and 400 to 600 t is excellent and achievable in many operations. It is important to note that this figure is based on total tons of material moved and this includes ore and waste, and the total number of employees in the mining department.

ECONOMIC ANALYSIS

Economic analysis is not very meaningful if the mine is a part of a larger integrated concentrating and smelting complex. If the mine sells raw ore with no beneficiation except primary crushing, then the mine becomes an economic unit generating revenue.

The details of discounted cash flow analysis, sensitivity analysis, and payout time are discussed later in this work. However, a simple *spot* cash flow analysis is frequently helpful. An example is given in Table 13. This procedure is used only to indicate whether or not a project is in the range of profitability. The analysis is for some nth year in the future when the project has reached design production and overcome initial start-up difficulties. Such an analysis assumes constant dollars. It can be used for preliminary comparison

Table 13. *Spot* Cash Flow Analysis for Preliminary Feasibility Studies*

	Annual $ × 1000
Gross revenue 12.7 Mt @ $2.64/t	33,528
Less direct operating cost 12.7 mt × $1.32/t	16,764
Operating profit	21,336
Depreciation $24,000,000 ÷ 8 years	−3,000
Interest @ 15% on 75% of investment, 0.15 × .75 × $24,000,000	−2,700
Depletion 15% of gross revenue† 0.15 × $33,528,000	−5,029
Before tax profit	10,610
All income-related taxes, assume 50%	5,305
After tax profit	5,305
Add back depletion	5,029
Net profit	10,334
Add back depreciation	3,000
Cash flow‡	13,334

Payout time, $24,000,000 ÷ $13,334,000 =	1.8 years	
ROI $10,334,000 ÷ 24,000,000	43%	
ROE $10,334,000 ÷ 6,000,000	172%	

* Assume a mining contract delivering ore to the buyer's crusher from a deposit owned by the mining contractor. Assume all material moved is ore.

† Some foreign countries have no provision for depletion. See US tax laws on depletion.

‡ Assume no investment credit or tax holiday.

of alternates such as shovels and trucks vs. bucket wheel excavators and conveyors. More sophisticated programs are adapted to a computer program.

ECONOMIC VIABILITY-CASH FLOW ANALYSIS

Feasibility studies may have many different purposes. However, in many cases the final objective will be to serve as a reliable document to be presented to a lender. Therefore, the economic viability of the mining venture must be demonstrated in the feasibility study.

The main test of the overall economic viability, from a lender's point of view, is through cash flow analysis. "A cash flow forecast that extends at least through the life of the proposed loans—and preferably for some additional years—should be prepared under the costing and marketing assumptions justified elsewhere in the report." (Gibbs and Sroka, 1978.)

The final decision on alternate choices, either of different mining projects or changes in parameters, such as production rates, methods, etc., is based on a rational continuous process, namely economic analysis.

Principles of an Economic Model

In general, an economic model for investment decisions should consider four important principles as given by Haynes and Massie, 1969.

The Incremental Principle: A decision is sound if it increases revenue more than costs, or if it reduces costs more than revenue. This seems obvious; however, its application is not obvious at all, and using average costs as the basis for the decision model could lead to error, or at least it does not present a complete picture of the investment alternatives.

The Principle of Time Perspective: A decision should take into account both the short and long term effects on revenues and costs, giving appropriate weight to the most relevant time periods.

The Opportunity Cost Principle: Decision making involves a careful measurement of costs. The company must evaluate the opportunity cost of investing in the proposed project as compared to other investment possibilities (Joy, 1980).

The Discounting Principle: If a decision affects costs and revenues at future dates, it is necessary to discount these costs and revenues to present values before a valid comparison of alternatives is possible.

The four principles, previously mentioned, are the framework of economic analysis which will interrelate the areas of mineral resources, technology for recovering the resources, the annual capacity of the proposed mine operation, the capital investment required, the estimated operating costs, and the profitability criteria for the final investment decision.

In the final analysis, the evaluation of the profitability of the investment is measured by the difference between the summation of the present value of the expected proceeds over future years vs. the capital invested today. This could be simply expressed by *P. Value of Cash flow = P. Value of Investment + P. Value of Profit.*

In summary the economic analysis in the feasibility study is performed by formulating a model which should include the following elements: 1) evaluation of the main variables; mineral reserves, production rates, recoveries, cost estimations, prices of the commodity, regulatory and environmental factors; 2) profitability criteria; 3) cash flow projection model; and 4) test of the cash flow model.

Evaluation of the Main Variables

Several major variables are involved in the economics models. These variables are classified either as Industry Stimulant Group, Economic Stimulant Group, or Regulatory Stimulant Group (Beasley and Pfleider, 1972; Pfleider, 1980).

The Industry Stimulant Group model contains: 1) mineral reserves, tonnage and grades (cut-off and average); 2) production rate and grade, corrected for dilution; 3) alternative mining, mineral processing, and metals recovery methods; 4) anticipated product recoveries in the mining, processing and metals recovery steps. And, for each of the above set of variables: 1) operating and overhead costs; 2) transportation and sales expenses; 3) capital costs for mine, plant and infrastructure; and 4) working capital requirements.

The Economic Stimulant Group model consists of: 1) market prices of products; 2) equity—loan ratios; and 3) interest rates and payback times of loans.

The Regulatory Stimulant Group model variables are: 1) depreciation and amortization rates; 2) depletion allowances, if any; 3) royalties and/or profit splits; 4) tax rates for ad valorem, property, production, profits, custom duties, transfer of dividends; and 5) tax moratoriums, investment credits.

Some of these variables have already been described in detail elsewhere in this chapter; however, they are repeated here with the objective of presenting an overall view of all the factors involved in the economic analysis.

While the Industry Stimulant Group has its impact in the design and estimating phase, the Economic Stimulant Group (market prices and loan arrangements) and the Regulatory Stimulant Group (representing taxing structures) are of equal or perhaps greater importance because the firm has only limited or no control over these variables.

In general, the main objective of the Economic Model is to estimate the annual return. However, in this process the projected return is dependent upon two types of parameters identified by Wells (1978) as "nondiscretionary and discretionary parameters." Discretionary parameters include mining and beneficiation methods, annual production capacity, sequence of mining, cut-off grade, design of the mine, processing plant, and facilities required to obtain an efficient production. Nondiscretionary parameters are those which the firm cannot control directly, such as selling price of the product, basic costs of labor and supplies, taxes, regulations, etc.

Profitability Criteria Used in the Evaluation

The criteria to be selected for the evaluation will depend on the investment objective of the parties concerned in the mining venture. Several profitability criteria are widely used by the mining industry, financial institutions, and government agencies to evaluate individual projects and to compare alternates between mutually exclusive mining ventures. There are two general groups of profitability criteria: 1) *Non-discounted cash flow*-payback period and average internal rate of return; 2) *Discounted cash flow*-Hoskold Formula, Net Present Value (NPV), Internal Rate of Return (IRR or ROR), Profitability Index (PI), Payback Period.

For evaluation purposes, the discounted cash flows are the only profitability criteria used, since the time value of money is the most important principle in the evaluation process. The following is a brief review of the profitability criterion of this group.

The Hoskold Formula has been used since 1877; however, its use is limited today. It calculates present value by discounting future annual income at a risk rate while building a sinking fund to repay investment at a safe rate. The present value *(PV)* is calculated by the following formula:

$$PV = \frac{A}{\dfrac{r}{(R^n - 1)} + r^1}$$

where *PV* is present value, *A* is annuity (annual cash flow), *r* is safe discount rate, r^1 is risk rate, *n* is number of years, *R* is $(1 + r)$.

Net Present Value (NPV) has been widely used since the 1960s. It calculates net present value by discounting estimated annual cash flows to a common point of time at a selected discount rate considering the risk of the investment.

$$\text{Net present value } (NPV) = \sum_{t=0}^{N} \frac{\text{Cash Flow } (CF)}{(1 + k)^t}$$

where *k* is discount rate, *t* is number of years.

Internal Rate of Return (IRR) or (ROR) is the calculated rate that makes the present value of cash inflow equal to that of cash outlays, or

$$\text{Internal rate of return } (IRR) = \sum_{t=0}^{N} \frac{\text{Cash Flow } (CF)}{(1 + 1RR)^t} = 0$$

where *t* is number of years.

Profitability Index (PI) is defined as the present value of future cash flow divided by the initial investment. This criterion is also called the benefit-cost ratio and is used in the capital rationing situation. This is the situation where a budget constraint is imposed, and the firm may not invest in all acceptable projects.

$$\text{Profitability Index } (PI) = \frac{\displaystyle\sum_{t=1}^{N} \dfrac{\text{Cash Flow } (CF)}{(1 + k)^t}}{\text{Original Investment}}$$

where *k* is discount rate, *t* is number of years.

Payback Period represents the number of years required in the cash flow analysis for the accumulated cash flow to equal investments. This is measured either from the initiation of the project or from the start of the operations. This criterion is especially useful for mining ventures with high risk or short life. Normally the payback period is expected to be one to three years for high risk projects, and six to ten years for low risk projects.

The selection of the criterion for the evaluation of the mining property will depend upon the criteria best suited for the investment objective. However, there is a trend to use the present value and rate of return techniques for economic analysis. These techniques, in conjunction with sensitivity and risk analysis, are good analytical tools for simulating the impact of the uncertainty on the profitability measurement, either present value or rate of return. These simulation techniques will be treated in more detail later in this section.

Rate of Return Vs. Present Value

The Internal Rate of Return (IRR or ROR) is easy to understand because it has intuitive economic meaning. It works well on simple accept/reject problems but its main drawback could be that it may not give the best selection criteria for complex accept/reject problems (multiple rates), mutually exclusive of choices and capital rationing.

The Net Present Value (NPV) is relatively easy to calculate. It is considered by many people as the best method for mutually exclusive ranking problems and for accept/reject decisions. Its main disadvantages could be that some-

times it is not easy to understand the concept and it may not work well for capital rationing situations. In the latter case, the profitability index (PI) works better than net present value (NPV).

Probably the most troublesome part of the net present value calculation is the estimation of the discount rate, which will convert the cash flows projected to the net present value at a given point in time.

Various financial treatises give details for estimating the discount rate; however, in this section, the authors provide an overview of one method for estimating the discount rate using the equation known as *Capital Asset Pricing Model, CAPM* (Joy, 1980).

The interest rate or discount rate (K_x) may be estimated by the following linear equation.

$$K_x = K_{rf} + (K_m - K_{rf}) \text{ Beta}_x$$

where K_x is the interest rate required for *X* investment, K_{rf} is the risk-free interest rate, K_m is the interest rate at market value, Beta_x is the investment risk for *X* investment. "In principle, K_{rf} is observable in the market place and K_m can be estimated. The interest rate for US Bonds is usually taken as a basis to estimate the risk-free rate (K_{rf}) for long term investments."

"The most common method of estimating Beta_x is by regression analysis. Rates of return on the firm's stock, *Re,* and rates of return on a market index, *Rm* values are statistically related by the straight-line equation called market model." (Joy, 1980.)

$$Re = a + \text{Beta}_x Rm$$

About five years of monthly rate of return are used to estimate Beta_x. Further discussion of this model is beyond the scope of this work. Nevertheless, to complete our overview, we add the following values: 1) For a risk-free investment, $\text{Beta}_x = 0$; 2) For $K_x = K_m$, $\text{Beta}_x = 1$. 3) For $K_x = 1.4 K_m$, $\text{Beta}_x = 2$.

Cash Flow Model

Financial treatises give several reasons for using cash flow in the evaluation of investment alternatives. The three main reasons for using cash flow in the evaluation of mining investment are summarized as follows: 1) Cash flow yields a better measurement of the net economic benefit associated with the mining venture; 2) Cash flow avoids accounting ambiguities; 3) Mining companies are permitted depletion allowances in addition to the depreciation allowances. These noncash expenditures are tax deduction items that are added back into the cash flow stream.

The cash flow after tax stream consists of four main parts:

$$\text{Net Cash Flow} = A - B + C - D$$

where *A* is gross income, *B* is cost deductions, *C* is noncash expenditure, *D* is capital expenditure. A typical format for estimating annual cash flow is presented in Table 13a.

Test of the Cash Flow Model Evaluation

In the evaluation of a mining project, we consider three basic elements for discriminating between alternatives. These elements are: 1) project cash flow estimations; 2) appropriate discount rate; and 3) profitability criteria for the decision making process consistent with the objectives of the firm.

Regardless of the methodology used to develop the estimate of the profitability of the project, the fact is that the result is subject to a level of uncertainty. This level will depend upon the degree of reliability of each of the variables involved in the calculation. The uncertainty is due to the

Table 13a. General Format for Estimating Annual Cash Flow

Gross Income (A)
        Gross value of production
      − Value of royalties payment*
  $ = Gross income from mine property

Cost Deductions (B)
      − Mine operating costs
      − Milling operating costs
      − Environmental and reclamation costs
      − Deferred exploration
      − Ore development
      − Deferred pre-production development
      − Interest on loan*
      − Net taxable loss carried forward*
      − Depreciation
      − Other undistributable costs*

Net Income
  $ = Net income before taxes and depletion
      − State and local taxes
      = Net income before depletion
      − Depletion
      − Federal income taxes
  $ = Net income after taxes and depletion

Non-Cash Expenditures (C)
(Add back) + Depreciation
      + Depletion
      + Deferred exploration
      + Deferred pre-production development
      + Net tax loss carried forward*
  $ = Net Cash Inflow

Capital expenditures (D)
      − Exploration costs
      − Pre-Production costs
      − Plant and equipment
        − initial
        − replacement
      − Payment on loan principal*

Cash Flow $ = Net Cash Flow
  * If any deduction is applicable.

fact that the projections of future cash flows as predicted by the model, are only at best, *good estimates*. These estimates are obtained from data and conditions as known at the time of the model formulation.

In order to evaluate the effect of changes in values of variables incorporated in the model on the profitability criteria, two techniques are generally used. One is the simulation using the *Monte Carlo* technique or risk analysis simulation, and the other is *sensitivity analysis*. With the advent of the computer and rapid simulation programs, there is an increasing application of these techniques.

Sensitivity Analysis: Sensitivity analysis varies the range of a single variable at a time. The procedure is repeated for the other variables. In this fashion the most sensitive parameters of the model are detected. The process for testing the sensitivity of the model consists of changing one variable and holding constant the others at their estimated most likely value. The variation for the single value will range from its minimum to its maximum estimated value. The effect, in either present value or rate of return, is recorded in a matrix form with the main parameters and their respective changes in values.

Simulation Using Risk Analysis: Risk analysis is a systematic analysis of the risk associated with various dependent variables to develop a final expected distribution of the profitability measurement. This distribution is used for the decision making process of the investment alternative.

This technique is also known as *Probabilistic Analysis*. It simulates numerous variations in the range of all the main variables at the same time. The simulation is carried out by computer programs using randomly selected values from given statistical distributions.

This approach is widely used today. Its general application in evaluation of projects was originally proposed by D.B. Hertz in 1964. The main feature of the technique is that it gives the decision makers a schedule indicating the frequency distribution of the profitability criteria and their respective relative probability of occurrence. At the same time, it will indicate if the investment has any chance of being a total loss.

When the feasibility study is performed with the objective of selecting a mining venture between various alternate investments, risk analysis is a powerful analytical tool for identifying in a realistic way the alternate that offers the greater economic comparative advantage in line with the objectives of the firm.

Often, the alternates differ in location, tonnage and grade distribution, size of equipment, operating costs, life of operation and equipment, etc. The main factors for discriminating between these types of alternatives are the estimation of capital and operating costs. These costs are deterministic events by the nature of their development.

The cost estimation of a new mining venture, regardless of the detail contained in the estimates, is subject to a level of uncertainty. This uncertainty is, in fact, a pervasive feature of the cost estimating process because of the lack of complete knowledge about the conditions or efficiency of the future operation. A search in the technical literature shows that there are several good descriptions of the risk analysis method.

The main concept and elements are summarized here to present a comprehensive overview of the analysis. It is hoped that the description will give the reader some insight for practical applications of this technique for the evaluation of mineral deposits or other selection of alternatives in the mineral industry.

Type of Risk in Mine Evaluations

Risk is inherent in the decision to invest capital for converting mineral resources into future expected profit or, simply stated, in mining jargon, there is always a risk in the process of converting mineral resources into ore.

There are two types of risk in mine evaluations: 1) risk inherent to the physical characteristics of the mineral body, and 2) risk associated with future events.

Risk Inherent to the Physical Characteristics of the Mineral Body: The mineral reserves obtained from the geologic model contain a level of uncertainty, depending upon the amount of exploration data used as a basis for the model. This is due to the fact that the reserves estimation is based upon a small, physically known fraction of the mineral body. Therefore, the tonnage and grade of the entire deposit are inferred from a relatively small sample. Consequently, the specific mineral content, based on the information available at the time of the study, is limited by the sample population and the geometry of the mineral body.

The advent of geostatistics permits us to greatly reduce this uncertainty. Geostatistics reveals a great deal about the idiosyncrasies of mineral distribution and its spatial location. Consequently, the reliability of the in situ reserves estimations are considerably improved. It should be emphasized that by far the most important factors in the reliability of the reserves estimates are good core recoveries, good sampling, and good assaying procedures. It would be a fallacy to use a sophisticated geostatistical model with questionable data base. Therefore, in a complete feasibility study, it is essential to review the sampling procedures and confirm the reported assays.

The expected mill head grade will depend upon the grade distribution of the deposit, the economic constraints imposed to define the limit of the ore body, and the proposed mining plan. However, this grade will be subject to variation due to dilution as a consequence of the following factors: geometry of the ore body, mining practice and unknown geological factors at the time of the study.

Risk Associated with Future Events: In general, a mineral deposit takes three to seven years to bring into production after the firm has made the investment decision. Depending upon reserves, it will take another 10 to 30 years to exploit the ore body considered in the evaluation. During this period many events could happen that will affect the following parameters: price of the commodity, initial and replacement capital costs, operating cost, taxation policies, and change in government policies regarding partial or total

ownership of the mineral properties. In the life of the properties, including the development period, there is the possibility that all the main parameters involved in the evaluation could deviate from the original estimates. Several combinations of variations in parameters could occur and these variations will affect the profitability estimate of the property under evaluation.

Procedure for Simulation by Risk Analysis

The process of performing a systematic risk analysis for the measurement of profit associated with changes of the main dependent variables consists of three main elements: 1) estimation of the basic values of the main variables; 2) estimation of the probability of occurrence of each estimated value; and 3) selection, at random, of one value from the probability distribution for each variable. The general methodology to be used is illustrated in Fig. 2.

Estimates of the Basic Values of the Main Variables: Up to this point, the basic values of the main variables have been established with the formulation of the geological model (including cut-off grade, average grade, and tonnage), revenue forecast, and the development of capital and operating costs. A typical schedule of the basic values of the main variables is as follows:

1. Total tonnage at given cutoff grade, number of tons.
2. Average in situ grade, percent.
3. Dilution, percent.
4. Average mill head grade, percent.
5. Annual mining rate, number of tons.
6. Metallurgical recovery, percent.
7. Estimated annual revenue, dollars.
8. Estimated capital costs, dollars.
9. Estimated operating costs, dollars.

Estimates of the Probability of Occurrence of Each Basic Value: The estimation of the probability of occurrence of each basic value is the step that is most likely to trouble the mine evaluators. However, there are three general ways of approaching the estimation: 1) mathematical probability, 2) inferred probability, and 3) subjective probability.

Mathematical Probability—is the generation of stochastic variates for several distributions, such as normal, uniform, binominal, and is easily manipulated by a computer program. Naylor, et al., (1966) provide formulas and programs for generating the probability distribution.

Inferred Probability—by fitting curves has been suggested by Harris (1970) when there is historical or other data available and when the probability distribution law is unknown. He suggests that, "The estimate be made by the fitting of a polynomial of some unspecified degree to the data or the grouping of data into class intervals, and then calculating the relative frequencies for each class."

Subjective Probabilities—are typically cases where there are components for which there is no historical information applicable for the fitting of a probability distribution. The estimation of capital and operating costs are some of these components. A general procedure for forming a subjective probability distribution is given in the following section.

The estimation of the average value of each component is expanded to represent the possible range of variation of the basic values and the likelihood that these values will be reached. The range could be estimated with some degree of confidence by experienced engineers involved in developing the basic values of the cost components of the feasibility study. The degree of confidence is directly related to the

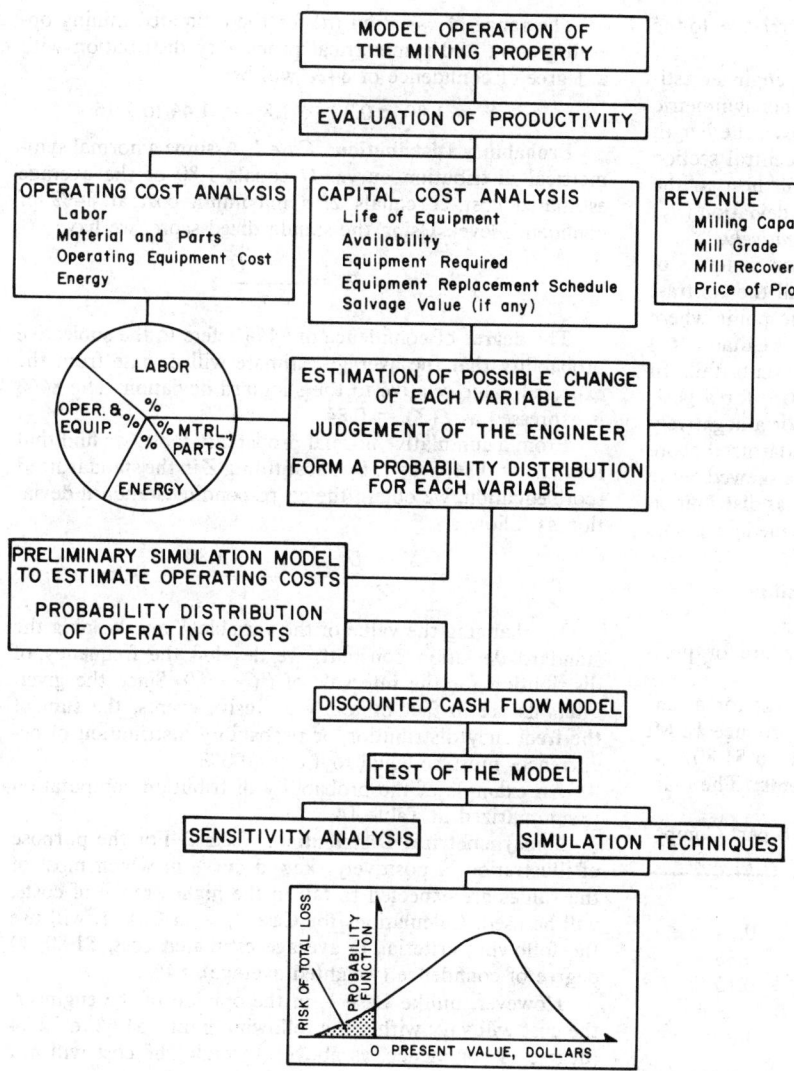

Fig. 2. Logic diagram for a risk analysis applied to a mine feasibility study.

amount of detail contained in the estimates and judgment from the experts.

Procedure for Forming a Subjective Probability Distribution Curve

Probability distribution, in general, is an idealization of the way certain events, or intervals, might occur. Subjective probability, in particular, quantifies the judgments from experts with respect to the particular range and pattern that their predictions may form.

In the estimation of the costs of new mines, experienced engineers involved in developing these estimates have a good idea of the degree of confidence they have in their estimate. This confidence is directly related to the amount of detail involved in developing the cost of the average estimated values. The greater the confidence in the estimate, the less will be the possible value range about the given estimate number. Conversely, the less the confidence, the greater the possible value range.

The sequence to follow in the process of forming a subjective probability distribution of possible cost values is: 1) identify the main cost components; 2) express the degree of confidence for the range of values of each estimated com-

ponent cost; 3) compute the weighted average degree of confidence for the total of all estimated costs; and 4) compute the range of values and their respective degree of confidence.

Two general cases are possible for forming the subjective probability distribution of values: 1) symmetrical distribution, and 2) skewed distribution.

Symmetrical distribution occurs when the engineer believes that the mean of the estimated value (U) might vary in an amount (d) on either side of the true value. The possible range of values for this case would be:

$$X = U + d$$
$$d = X - U.$$

In this case, the distribution curve is symmetrical and the *standardized score* approach is applicable in forming the probability distribution. If X is considered a value of a random variable, then the corresponding standardized score, relative to the probability distribution is $Z = \dfrac{X - U}{\sigma}$. "The Z score tells how many standard deviations (σ) away from the mean (U) is X." (Hays 1963.) In other words, the standardized score is a deviation from expectation relative to the standard deviation.

Skewed distribution is the case when the engineer estimates that the mean value (U) will vary in a nonsymmetric way, generally described as skewed. In this case, the length of the tails of the distribution relative to the central section are not equal. When the deviation (d_1) is furthest to the right, the curve is *skewed positively*. When the deviation (d_2) is furthest to the left, the curve is *skewed negatively*.

In describing the skewness of a distribution in terms of measures of central tendency, we should recall the contrast between mean and median. Mean (M) is the point where the sum of deviations about the mean is zero. Median (Mo) is the point at which exactly 50% of the events fall. In general, for a symmetrical curve, M equals Mo; for a positively skewed curve M is less than Mo; and, for a negatively skewed curve M is larger than Mo. The standardized score with some modification can also be applied to a skewed curve of probability distribution. However, a triangular distribution seems to be more adequate to infer the intermedian points of this type of distribution.

Example of Inferring a Subjective Probability Distribution

The following example illustrates the procedure for quantifying a subjective probability distribution.

The estimated annual mining operating cost for a new open pit mine is $21,600,000. The mine will produce 12 Mt of ore per year. The mine operating cost is then $1.80/t.

Identification of the Main Cost Components: The main cost components are:

	% of Total	Cost per Components, $
Labor	43	0.77
Equipment	30	0.54
Material	20	0.36
Energy	7	0.13
Total	100	1.80

Degree of Confidence: Based on the details involved in the performance evaluation of the equipment, the labor productivity and other economic factors, the engineers feel the basic amount of each cost component will not vary more than plus or minus 20%. However, they feel that due to the more known factors in some components than others, the degree of confidence for each cost component is as follows:

	Degree of Confidence %
Labor	90
Material	85
Equipment	85
Energy	65

Weighted Average Degree of Confidence.: An illustration of the computation of the related cost components is given as follows:

The range of variation (d) for the estimated mining operating cost, and symmetrical probability distribution with a degree of confidence of 84% will be:

$$X = 1.80 \pm 0.20 \times 1.80 = 1.44 \text{ to } 2.16$$

Probability Distribution: *Case 1.* Assume a normal symmetrical distribution curve. U equals 1.80 of the average estimated cost, X equals 2.16 maximum cost at 84% of confidence level. Using the standardized score, we have

$$Z = \frac{X - U}{\sigma}$$

The degree of confidence of 84% refers to the subjective probability that the average estimate will deviate from the expected value relative to the standard deviation. The 84% is expressed as $F(Z) = 0.84$.

From a cumulative normal probability table, we find that for $F(Z) = 0.84$; $Z = 1$. By substituting Z in the standardized score equation, we obtain the corresponding standard deviation as follows:

$$\sigma = \frac{X - U}{Z} = \frac{2.16 - 1.80}{1} = 0.36$$

By changing the value of the variable X and keeping the standard deviation constant, we develop the frequency of distribution for the intervals of ($X - U$). Since the given intervals are in fact mutually exclusive events, the sum of the frequency distribution or probability distribution of occurrences must be equal to 1.0 or 100%.

An example of the probability distribution computation is summarized in Table 14.

Nonsymmetrical Distribution: *Case 2.* For the purpose of illustration, a positively skewed curve in which most of the values are expected to fall in the higher range of costs, will be used. Calculations for Case 2, as in Case 1, will use the following criteria: 1) average estimated cost, $1.80; 2) degree of confidence (weighted average), 84%.

However, unlike Case 1, in the opinion of the engineer, the cost will vary within the following limits: $1.62 to $2.34 (degree of confidence, as above). Overall, the cost will not fall under $1.50 or exceed $3.30.

The foregoing criteria yield a triangular distribution that simply reflects the judgment of the engineer in that a value close to the estimated average ($1.80/t) is assigned a higher probability of occurrence than a value further from the estimated average. For practical purposes, it is assumed that the probability of occurrence of the intermediate values vary linearly, with respect to the probability of occurrence of the best estimated average. By solving for the area percent relationship of the triangles shown in Fig. 3, and estimating the probability of occurrences of the intermedian values, the subjective probability distribution is quantified as shown in Table 15. This type of distribution can also be obtained by using mathematical expressions for both positively and negatively skewed curves as given by O'Hara (1982). However, since the original data base reflects only the subjective decisions of the engineer, a simple approach is adequate at this point.

Components	% of Total	Cost Component, $	Confidence Level	Product
Labor	43	0.77	90	69.30
Material	30	0.54	80	43.20
Equipment	20	0.36	85	30.60
Energy	7	0.13	65	8.45
Total	100	1.80	84	151.55

Table 14. Computation of Probability Distribution

Mining Operating Costs Intervals $	Midpoints (X) $	σ	Average Estimated Costs (U) $	Z Values $Z = \dfrac{X-U}{\sigma}$	Interval Probability Distribution %	Cumulative Probability F(Z) %
0.37–0.77	0.57	0.36	1.80	−3.42	0	0
0.78–0.98	0.88	0.36	1.80	−2.56	0.5	0.5
0.99–1.17	1.08	0.36	1.80	−2.00	1.8	2.3
1.18–1.28	1.23	0.36	1.80	−1.58	3.4	5.7
1.29–1.35	1.32	0.36	1.80	−1.33	3.5	9.2
1.36–1.52	1.44	0.36	1.80	−1.00	6.8	16.0
1.53–1.60	1.56	0.36	1.80	−0.67	9.1	25.1
1.61–1.70	1.66	0.36	1.80	−0.39	9.7	34.8
1.71–1.89	1.80	0.36	1.80	0	15.2	50.0
1.90–1.98	1.94	0.36	1.80	0.39	15.2	65.2
1.99–2.09	2.04	0.36	1.80	0.67	9.7	74.9
2.10–2.22	2.16	0.36	1.80	1.00	9.2	84.1
2.23–2.33	2.28	0.36	1.80	1.33	6.7	90.8
2.34–2.40	2.37	0.36	1.80	1.58	3.5	94.3
2.41–2.63	2.52	0.36	1.80	2.00	3.4	97.7
2.64–2.80	2.72	0.36	1.80	2.56	1.8	99.5
2.81–3.25	3.03	0.36	1.80	3.42	0.5	100.0
3.26–3.30	3.28	0.36	1.80	—	0	—

*Negative Z Mean Frequency = 1 − F(Z).

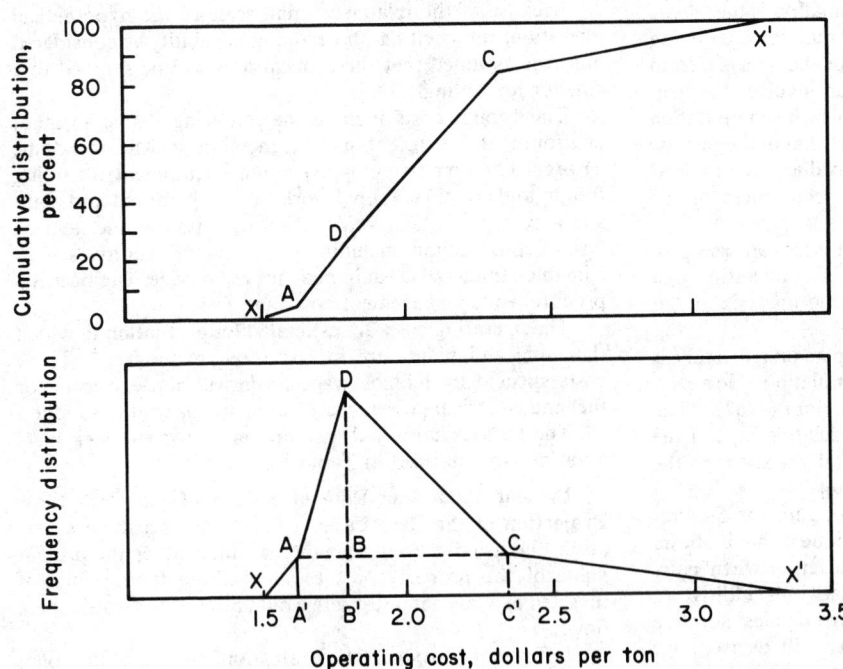

Fig. 3. Triangular probability distribution (positively skewed).

Area

AB = 0.18   ΔABD = $\dfrac{0.18}{0.72} \times 84\%$ = 21

BC = 0.54   ΔXA'A = (25 − 21) = 4

AB+BC = 0.72   ΔCBD = $\dfrac{0.54}{0.72} \times 84\%$ = 63

ΔX'C'C = (75 − 63) = 12

Total = 100%

Table 15. Probability Distribution Positively Skewed Curve

Mining Operating Costs Intervals $	Midpoint X $	Interval Probability Distribution %	Cumulative Probability %
1.40–1.60	1.50	0	0
1.61–1.68	1.64	4	4
1.69–1.71	1.70	8	12
1.72–1.88	1.80	13	25
1.89–1.99	1.94	17	42
2.00–2.09	2.05	11	53
2.10–2.19	2.15	11	64
2.20–2.29	2.25	10	74
2.26–2.42	2.34	10	84
2.43–2.53	2.48	3	87
2.54–2.64	2.59	2	89
2.65–2.75	2.70	2	91
2.76–2.86	2.81	2	93
2.87–2.97	2.92	2	95
2.98–3.08	3.03	2	97
3.09–3.19	3.14	2	99
3.20–3.30	3.25	1	100
3.31–3.33	3.32	0	—

Simulation of Values from Probability Distributions

This simulation is the last step in the process of performing a risk analysis evaluation. This is the easiest and fastest operation in the analysis because it can be performed by a computer program.

The most common procedure for simulation is the *Monte Carlo* technique which is the process of randomly sampling values of the main variables contained in the analysis from a probability distribution. The simulation involves the generation of random values for each variable, the computation of the profitability measures, and the repetition of the process many times (100 to 1,000) until enough values are obtained to construct the overall probability of occurrences of the profitability measures.

The simulation produced by the computer can also provide the distribution curve. Therefore, it is an operation that frees the engineer from having to run the analysis and it takes only a few minutes to perform.

Almost all the published technical papers on risk analysis are predicated on the *Monte Carlo* simulation. However, there is a new technique developed by O'Hara (1982), "The procedure is based on the mathematical relationships of the skewed curve to determine the incremental variations in the estimated value of each element parameter."

The O'Hara approach is called *Root Sum of Squares (R.S.S.)*. The distinct feature of this technique is that it allows for the interrelationship of parameter variation when pairs of parameters are judged to be compensative or additive in their joint effect on the profitability measure. Typical additive situations are ore reserves vs. mill grade, mill recovery vs. grade of concentrates.

The evaluation of the advantages and disadvantages of the *Monte Carlo* vs. *Root Sum of Squares* techniques with regard to their application to risk analysis is beyond the scope of this chapter.

Case Study of A Simulation Using Risk Analysis

Basic Criteria and Assumptions: A hypothetical 12,700,000 overall tons per year mining operation in the western United States shall be used as a case study. The following assumptions are made: 1) The deposit is within reasonable distance of a small population center; 2) Out of the total tonnage mined, a production level of 10 Mt/a of copper ore is expected; 3) Average copper grade is 0.87% with no byproducts; 4) Metallurgical recovery is 87%. 5) The waste-to-ore stripping ratio is 0.27:1.

Because of the relatively small scale of the hypothetical operation, no smelting and refining capability is considered and it is assumed that the concentrates will be shipped to a smelter for tolling.

The capital costs include the following items: 1) land, environmental, exploration, and metallurgical testing costs; 2) preproduction stripping; 3) mining equipment including freight and erection; 4) civil work, site preparation, and plant site access, including construction of a power line; and 5) cost of construction, including wages, payroll additives, consumables, materials, equipment usage, service, engineering, procurement, management costs and fees.

The operating costs are estimated for a situation in which the mine and plants are in operation in mid-1982. These costs specifically include wages, salaries, burden, costs for fuel and electrical power, spare parts, and operating supplies.

The basic criteria and assumptions for the model operation are summarized in Table 16.

Present Value and Rate of Return of the Cash Flow Projection of the Base Case: In performing a mine evaluation, the usual procedure is to determine either the present value of the property or the value of the funds which if invested at a given interest rate would return the same income flow.

In order to produce the evaluation, the cash flow plan should be developed in line with the criteria and assumptions summarized in Table 16. The cash flow projection for the estimated average basic conditions is illustrated in Table 17.

From the net cash flow plan given in Table 17, the present value and rate of return of the investment are calculated. This calculation shows a net present value of $71,750,590 and an internal rate of return of 20.9%.

Sensitivity Test: As discussed in the section under Sensitivity Analysis, the model may be sensitive to changes in the variables. This model, for instance, is very sensitive to

Table 16. Basic Criteria and Assumption for the Economic Analysis Case Study

1. Initial Investment
   a) Capital already invested—land, exploration, studies, met. tests $17,000,000
   b) Plant construction and equipment to be expended in 3 years as follows:
Year 1 (ending 1984)	42,500,000
Year 2 (ending 1985)	54,700,000
Year 3 (ending 1986)	54,800,000
	$152,000,000

2. Replacements Cost
   Year 7 $6,000,000

3. Life of the Operation   12 years

4. Depreciation
   Straight line method over an average life of 12 years   S/L

5. Depletion Allowance
   50% of net income

6. Operating Costs
Mining including primary crushing	18,000,000
Concentration	36,000,000
Transportation of concentrates	6,400,000
Total	$60,400,000

7. Working Capital
   Estimated @ 3 months of operation
   (0.25 × 60,400,000) $15,100,000

8. Annual Production Capacity
   a) *Design capacity*   10,000,000 tons of ore
      @ Av grade of 0.87% Cu
      production of concentrate of 28% Cu   = 270,320 tons of concentrate

   b) *Start-up*
      First year (end 1984): 50% of capacity   = 135,160 tons of concentrate
      Thereafter   270,320 tons of concentrate

9. Revenue
   Copper concentrates @ $480/t.
   At design capacity, annual revenue $129,754,000

10. Financing
    The purpose of the evaluation is to discriminate between alternative investments; therefore, all investment is assumed to be from equity for all practical purposes.

11. Escalation
    No escalation is applied

12. Profitability Measure
    Since it has been assumed that the evaluation is aimed toward selection between mutually exclusive alternatives, the present value of the discounted cash flow will be used as the profitability criterion.

13. Discount Rate
    The discount rate or interst rate is estimated as follows:
    $K_x = K_{rf} + (K_m - K_{rf})$ Beta$_x$
    K_{rf} = risk-free interest rate, 10%
    K_m = interest rate at market value, 15%
    Beta$_x$ = investment risk = 0.80
    $K_x = 10 + (15-10)0.8 = 14\%$

14. Salvage Value
Equipment	$13,700,000
Working capital	15,100,000
Total	$28,800,000

Table 17. Net Cash Flow Projection of the Base Case Operation ($ × 1000)

	Preproduction period			Year of operation						
	1984*	1985	1986	1†	2	3-5	6	7	8-11	12
Gross income				64877	129754	129754	129754	129754	129754	129754
Cost deductions:										
operating costs				30400	60400	60400	60400	60400	60400	60400
depreciation				10766	10766	10766	10766	11766	11766	11766
deferred exploration and development				3400	3400	3400	—	—	—	—
Net income before taxes				20311	55188	55188	58588	57588	57588	57588
State and local taxes @ 6%				1219	3311	3311	3515	3455	3455	3455
Net income before depletion				19092	51877	51877	55073	54133	54133	54133
Depletion (50% of net income)				9546	25938	25938	27536	27066	27066	27066
Net income after depletion				9546	25938	25938	27536	27066	27066	27066
Federal tax @ 50%				4773	12969	12969	13768	13533	13533	13533
Net income after taxes and depletion				4773	12969	12969	13768	13533	13533	13533
Non-cash expenditures (add back)										
depreciation				10766	10766	10766	10766	11766	11766	11766
depletion				9546	25938	25938	27536	27066	27066	27066
deferred exploration and development				3400	3400	3400	—	—	—	—
salvage										28800‡
Capital expenditures										
pre-production costs										
initial plant and equipment	−42500	−54700	−54800							
replacement								6000		
working capital				7550	7550					
Net cash flow ($ × 1000/yr)	−42500	−54700	−54800	20935	45523	53073	52070	46365	52365	81165

* Construction start in 1984
† Production start in 1987
‡ Including working capital

change in revenue brought about by a change in either the grade of ore or the price of the product.

The profitability measures illustrated in Table 19 are obtained by simulating various mill head grades and keeping constant all other parameters. Similarly, if only the price of the ton of concentrate changes in the same percentage, the values of the profitability measures will change as shown in Table 18.

The grade and price are typical parameter pairs that are compensative or additive in their joint effect on the profitability measure. Therefore, a change in revenue could be due to the joint effect of the variability of certain parameter pairs, for example, grade to the mill and price of the product.

Test of the Model Under Uncertainty: There are many possible combinations of the main variables that would yield a present value of $71.75 million or 20.9% rate of return. However, at this point, we do not know what the probability is that these profitability measures or another value, either higher or lower, can occur. Therefore, it will be necessary to simulate various combinations of changes in the values of the main variables. In this way, it is possible to evaluate the effect of these changes on the profitability measure and determine the frequency of occurrence of either the required present value or rate of return.

The simulation is performed following the general procedure described in the section under "Procedure for the Simulation by Risk Analysis," which consists of the following three main elements: 1) estimates of the basic values of the

Table 18. Sensitivity Test, Mill Grade vs. Profitability Measures

Ore Mill Grade		Profitability Measures	
Grade Ore % Cu	Percent Change of Grade	Present Value $ × 1000's	Rate of return %
0.73	−16	−360	0
0.75	−14	11,180	1.0
0.81	− 7	41,910	10.0
0.87*	0	71,750	20.9
0.93	+ 7	105,330	23.0
1.00	+14.9	143,229	27.2
1.05	+21.0	168,753	33.0

* Most likely average grade. This value is used as the reference base to compute the ± percent change of grade.

Table 19. Probability Distribution of Estimated Annual Revenue (Positively Skewed Curve)

Revenues Intervals $ × 1000's	Midpoint $ × 1000's	Interval Probability Distribution %	Cumulative Probability Distribution %
—			
—			
100,562–107,050	103,806	0	0
107,051–113,536	110,293	3	3
113,537–120,024	116,780	3	6
120,025–126,510	123,267	14	20
126,511–132,997	129,754	20	40
132,998–139,484	136,241	17	57
139,485–145,971	142,728	17	74
145,972–152,458	149,215	17	91
152,459–158,945	155,702	3	94
158,946–165,432	162,189	3	97
165,433–171,919	168,676	3	100
171,920–178,406	175,163	0	—

main variables; 2) estimates of the probability of occurrence of each assumed value; and 3) simulation technique involving selection at random of one value from the probability distribution for each variable.

Values of the Main Variables: The main variables are established in the basic criteria and assumptions for the case study given in Table 1. However, it is emphasized that some variables have joint effect (such as annual production rate, grade to the mill or price of the commodity) on the annual revenue which could result from either additive or compensatory effect. In this context there are only three main variables and their values are: 1) revenue at full production, $129,754,000; 2) capital expenditure present value at the beginning of year one of operation, $196,526,000; 3) operating cost at full production, $60,400,000.

Probability Distribution of the Main Variables: Following the procedure described in the sections under "Procedure for Forming a Subjective Probability Distribution Curve" and "Example of Inferring a Subjective Probability Distribution," the probability distributions for the three main variables were formulated. Tables 19, 20, and 21 give the probability distributions of the revenue, capital cost, and operating cost intervals, respectively.

Simulation Technique: The simulation involves selecting at random, by computer, one value from the probability distribution of each variable. The combination of these values is then used to calculate the present value of the cash flow of the operation. This process is repeated many times to form a probability distribution curve of the present values.

Evaluation of the Simulation Model: The present values of the cash flow and their respective frequencies of occurrences obtained by the simulation process are tabulated in Table 22. The values shown on this table are used to form the risk profile of the feasibility study depicted in Fig. 4.

From Fig. 4 the following conclusions can be drawn:

1) The most likely range of the present value is $62.6 to $77.4 million at a confidence level of 95%. There is a 50% probability that the value will fall below this range, and a 37% probability that the value will exceed this range.

2) The property has a 35% probability of not attaining the minimum present value required to cover at least a rate of return of 14%.

3) The property has a 5% probability of being a total loss.

It is hoped that this systematic approach to the evaluation of a surface mining property will provide support to those readers involved in mine feasibility studies.

Table 20. Probability Distribution of Estimated Capital Costs (Positively Skewed Curve)

Capital Costs Intervals $1000's	Midpoint $1000's	Interval Probability Distribution %	Cumulative Probability Distribution %
142,522–154,522	148,522	0	0
154,523–166,523	160,523	3	3
166,524–178,524	172,524	3	6
178,525–190,525	184,525	13	19
190,526–202,526	196,526	21	40
202,527–214,527	208,527	18	58
214,528–226,528	220,528	17	75
226,529–238,529	232,529	15	90
238,530–250,530	244,530	4	94
250,531–262,531	256,531	3	97
262,532–274,532	268,532	3	100
274,533–286,533	280,533	0	—

Table 21. Probability Distribution of Estimated Operating Costs (Positively Skewed Curve)

Operating Costs Intervals $1000's	Midpoints $1000's	Interval Probability Distribution %	Cumulative Probability Distribution %
47,797–51,397	49,597	0	0
51,398–54,998	53,198	5	5
54,999–58,599	56,799	10	15
58,600–62,200	60,400	25	40
62,201–65,801	64,001	18	58
65,802–69,402	67,602	16	74
69,403–73,003	71,203	10	84
73,004–76,604	74,804	6	90
76,605–80,205	78,405	5	95
80,206–83,806	82,006	3	98
83,807–87,407	85,607	2	100
87,408–91,008	89,208	0	—

Table 22. Present Value Probability Distribution (from Simulation)

Present Value $ Million From	Up to	Frequency	Cumulative Frequency	Percent
−40	−20.1	0	0	0
−20	0	5	5	5.0
0.1	+20	6	11	11.0
20.1	40	10	21	21.0
40.1	60	15	36	36.0
60.1	80	20	56	56.0
80.1	100	22	78	78.0
100.1	120	16	94	94.0
120.1	140	5	99	99.0
140.1	160	1	100	100.0
160.1	180	0	—	—

Mean $ 70
Standard deviation 36.85
Standard deviation of mean 3.7
Fiducial interval ± 7.4
Interval limit @ 95% of confidence 62.60–77.4

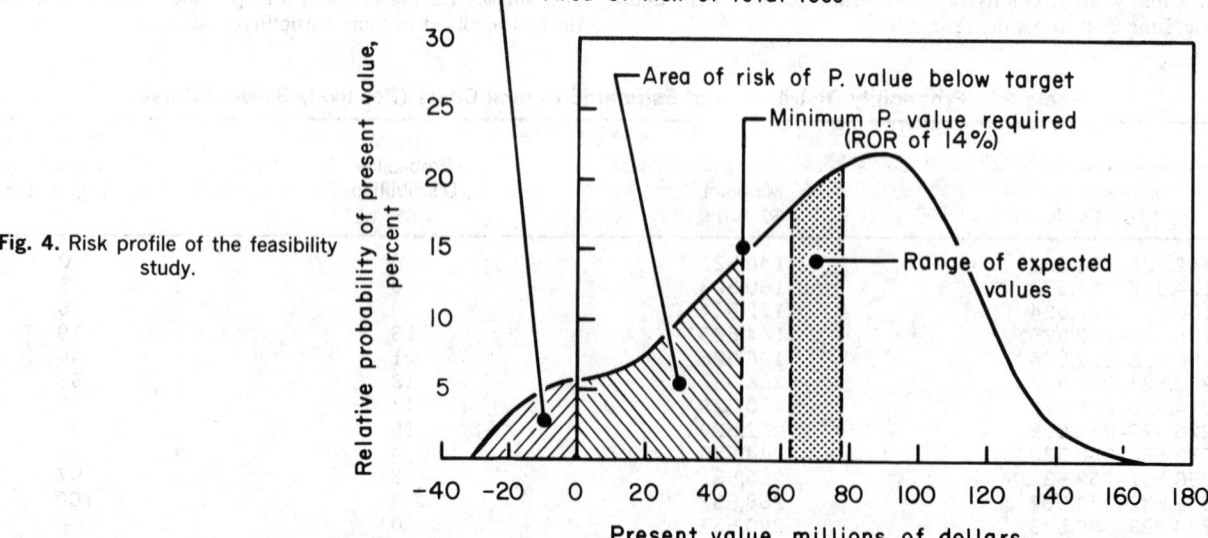

Fig. 4. Risk profile of the feasibility study.

REFERENCES

Beasley, C. A. and Pfleider, E. P., 1972, "Profitability Sensitivity Analysis of a Mining Venture," Mine Evaluation Seminar at the University of Minnesota.

Gibbs, J. N. and Sroka, J., 1978, "What Bankers Look for in Project Loan Applications," *Mining Engineering*, Dec., pp. 1646-1648.

Harris, D. P., 1970, "Risk Analysis in Mineral Investment Decisions," *Trans.* SME-AIME, Vol. 247, Sept., pp. 193-201.

Haynes, W. W. and Massie, J. L., 1969, *Management Analysis Concepts and Cases*, second ed., Prentice-Hall, Inc., NJ, pp. 412-413.

Hays, W. L., 1963, *Statistics*, Holt, Rinehart and Winston, Inc., New York.

Hertz, D. L., 1964, "Risk Analysis in Capital Investment," *Harvard Business Review*, Jan.-Feb., pp. 95-106.

Joy, M. O., 1980, *Introduction to Financial Management*, Richard D. Irwin, Inc.

Naylor, T. H., et al., 1966, *Computer Simulation Techniques*, John Wiley, New York, p. 352.

O'Hara, T. A., 1982, "Analysis of Risk in Mining Projects," *CIM Bulletin*, Vol. 75, No. 843, July, pp. 84-90.

Pfleider, E. P., 1980, "Mineral Valuation and Development in a World of Rapid Change," *Anales del Congreso Cincuentenario, Mineria de Cobres Porfidicos*, Vol. I, Nov., Santiago, Chile, pp. 93-109.

Wells, H. M., 1978, "Optimization of Mining Engineering Design in Mineral Valuation," *Mining Engineering*, Dec., pp. 1676-1684.

4.3 Financial Analysis

Lois L. Brazda
H. Andrew Thornburg

INTRODUCTION

The high risk and large capital expenditures that are characteristic of the mineral industry have made financial analysis one of the crucial elements in the resource development process. As with any business endeavor, the object is to provide a satisfactory return to the owners of the enterprise, whether they be public or private, consistent with the objectives of the enterprise. This section describes how criteria for financial analysis are established, presents methods of evaluating investment, and discusses areas of particular concern such as risk analysis, taxes, and corporate objectives. An attempt has been made to make the presentation universal, but in specific areas such as accounting and taxes more emphasis has been placed on United States practice. The Appendices contain examples of both.

Preceding sections of this chapter have described the various stages of feasibility studies and planning that are needed to gather the data required for evaluation of a project. From a financial analysis standpoint, the principal characteristics that set the mineral industry apart from others are the depleting nature of the resource itself and the very long project cycles. The financial cycle of a typical mining project is shown in Fig. 1.

To prepare an evaluation, criteria must first be established to assess the flow of funds over the project cycles and then a rigorous framework for measuring the performance of a project must be established.

Time Value Concepts

In a free market economy the time value of money or interest rate concept lies at the heart of all financial transactions. To properly evaluate the mineral investments with their long project lives, it is essential to have a methodology for evaluating past, present, and future cash flows.

Interest is the equivalent of rent paid for the use of money. Therefore, money borrowed today must be compounded at the appropriate interest rate to repay the principal and the rent due at some future date. The mathematical relationship for a Future Value *(FV)* given a Present Value *(PV)* or principal amount borrowed or invested for N periods at an interest rate *(i)* is:

$$FV = PV (1 + i)^N$$

For example: $PV = \$1$, Interest $= 20\%$ per period $(i = 0.2)$, $N = 3$ periods

$$FV = (\$1) (1 + 0.2) (1 + 0.2) (1 + 0.2) = \$1.728$$

Conversely a future value that includes rent must be discounted to find the present value. The corresponding formula is:

$$PV = \frac{FV}{(1 + i)^N}$$

It is important to distinguish how interest is compounded in the project evaluation so results are comparable and un-

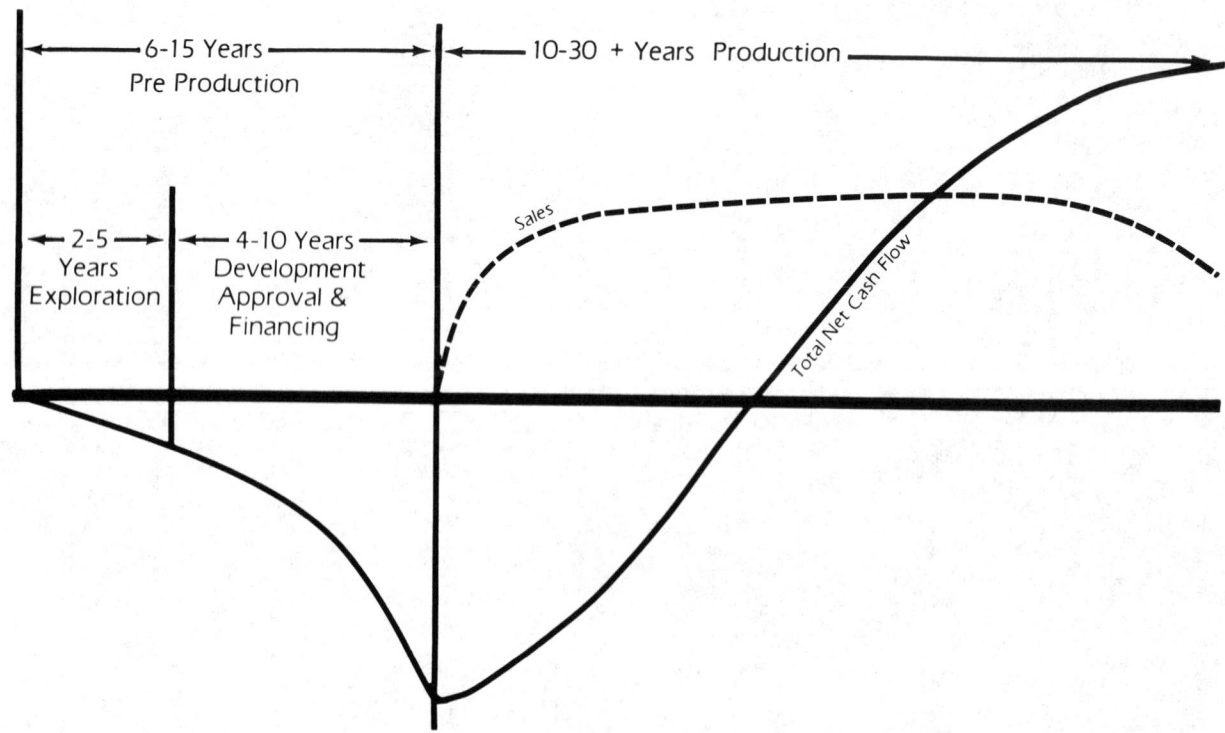

Fig. 1. Typical mining project cash flow.

424

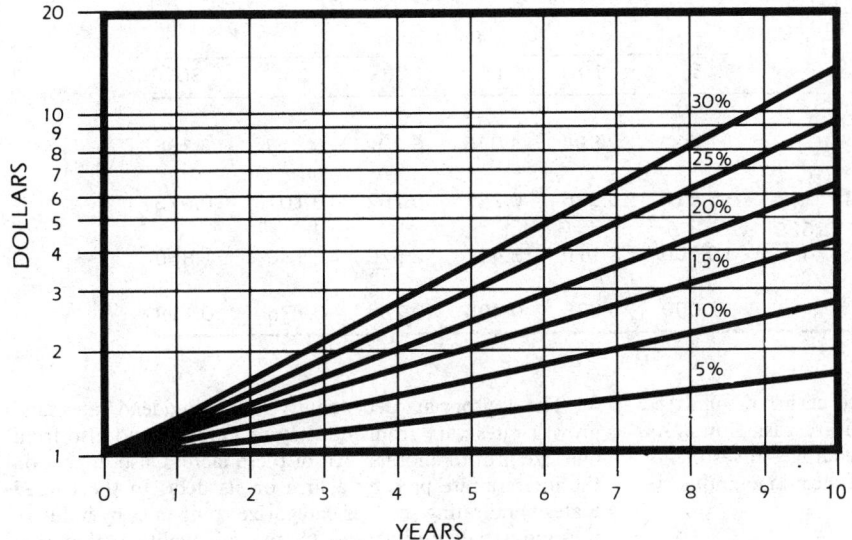

Fig. 2. Compound growth.

derstandable by all concerned. Simple compounding is expressed in terms of the annual percentage rate of interest (APR for those familiar with US lending terminology). In the above example the APR was 20%. If interest was compounded quarterly, instead of annually, as is often the case with models utilizing corporate financial data, the quarterly rate that would yield the same result is not 20/4 or 5% but 4.6635% as shown from the following:

$$FV = (\$1) (1 + 0.046635)^{12} = 1.728$$

The effective rate or APR is still 20%, but in terms of a quarterly evaluation model the annual rate or nominal rate (number periods in a year times the interest rate for the period) would be referred to as 18.654%. The mathematical limit to the continuous compounding power series of $(1 + i) (1 + i) \ldots$ can be expressed as:

$$FV = PVe^{iN}$$

The continuous compounding interest rate that satisfies the example is a nominal rate of 18.232%, a difference from the effective (APR) rate of 1.768%, yet the results over the three-year period are identical. It is clear, especially at higher interest rates, that the financial analyst should be specific as to the basis used for an evaluation. It is suggested the results be expressed in terms of the effective annual rate no matter what nominal rate is used.

The financial implications of both compounding (Fig. 2) and discounting (Fig. 3) become increasingly pronounced as rates increase due to the geometric nature of the compounding process. Table 1 illustrates the striking effect after ten years of annual compounding. Over 10 years the difference in future values between 5% and 30% is not six times but 8.46 times. Starting with the same investment at year 0, it only takes 1.86 years at a 30% rate to reach the future value attained over 10 years at a 5% rate (5.37 times as long).

When considering the evaluation methods presented later in this section, it is important to remember the reverse implications of discounting. A dollar earned ten years from now is worth $0.614 today at a 5% rate, while at 30% it is only worth $0.118 (19.2% of the value at a discount rate of

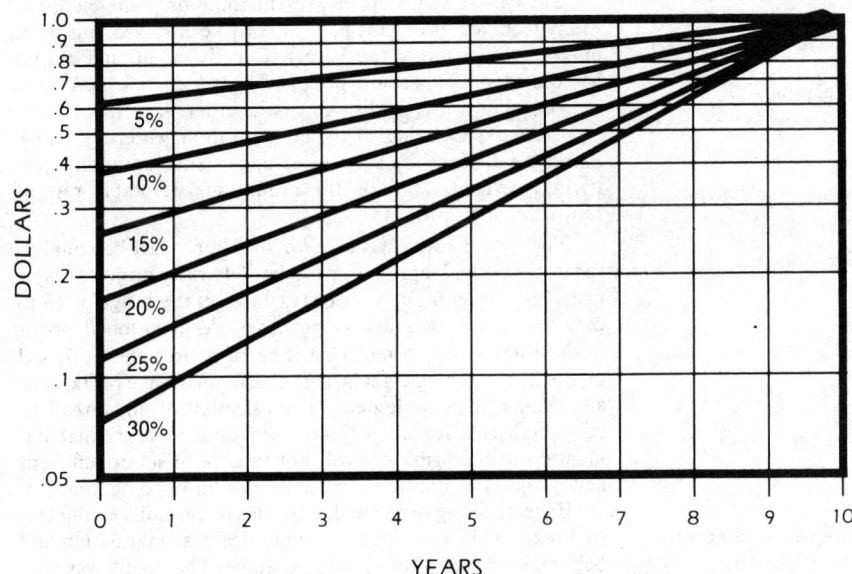

Fig. 3. Discounted values.

Table 1. Examples of Compounding and Discounting

	% Per Annum					
	5	10	15	20	25	30
Future Value in 10 Years ($1 invested today)	1.629	2.594	4.046	6.192	9.313	13.786
Present Value ($1 earned in year 10)	0.614	0.386	0.247	0.162	0.107	0.073
Years to Future Value of $1.629 ($1 invested today)	10.000	5.119	3.491	2.676	2.186	1.860
Present Value of $1.629 (earned in year 10)	1.000	0.628	0.403	0.263	0.175	0.118

5%). Higher interest rates in discounted cash flow analyses place rapidly diminishing values on future cash flows. As will be shown later, this generally discourages investments with the long lead times characteristic of the mining industry.

Establishing a Discount Rate

There is a general agreement in the mining industry regarding the need to recognize the time value of money in financial evaluations, but the selection of the appropriate rate to use is the subject of much disagreement, and with good reason. A government entity considering an investment might select the interest rate paid on government bonds, under noninflationary conditions, with a maturity matched to that of the project. To this rate, which in the United States has historically been in the range of 2%, a premium may be added if the project's risk is considered to be substantially different from projects normally undertaken by the government, and it must also be adjusted for inflation.

The analysis for a shareholder-owned firm with publicly traded stock contains the same basic elements but is much more complex. The primary objective of the firm is the maximization of the wealth of the shareholders over time. The shareholders benefit from the success of the firm through the receipt of dividends and the appreciation of the market price of the shares of common stock.

It is not possible to fully explore all the financial theory underlying the determination of the cost of capital to a firm. However, let us assume the cost of capital to a firm is primarily a function of the following:

$$\text{cost of capital} = (E) \times (D + G) + (1 - E) \times (1 - T) \times (I)$$

where E = equity as fraction of total capital
= equity/(debt + equity)
D = dividend rate
G = growth rate in equity
T = corporate tax rate
I = before-tax interest rate on debt

An example of the application of this formula is:

Equity	= 70%
Debt	= 30%
Dividend rate	= 5%
Equity growth rate	= 7%
Interest rate	= 12%
Tax rate	= 50%
Cost of capital	= 0.7(5 + 7) + 0.3(1 − 0.5)(12)
	= 8.4 + 1.8
	= 10.2%

For those requiring a more rigorous discussion, a number of books cover the topic more thoroughly (Gordon, 1974).

The appropriate debt/equity ratios, dividend rates, and growth rates vary from industry to industry and also from one size firm to another. All of these factors also impact on the interest rate paid by a firm on its debt. In the United States bond rating services categorize companies by industry (i.e. industrial, utility) and by overall quality within that industry. The long term debt rates charged by the market to firms of varying qualities (Aaa being the highest) and US Treasury Bonds during the 1970s are shown in Fig. 4. For analysis purposes, many firms define their cost of capital as that long term rate that leaves unchanged the market price of the common stock (Gentry, 1977). As a project increases in size in relation to the capital base of the firm, this assumption becomes difficult to maintain. The large swings in interest rates illustrated in Fig. 4 also indicate the problems of assigning a single cost of capital rate to projects with long lives, especially if the firm must borrow at short-term rates to finance long-term projects. Thus, it is evident, a project that might be quite acceptable to one company would be rejected by another who finds its cost of capital is too high in relation to project returns.

It has also been suggested that a firm risk adjust the rate of return by class of investment project as follows (O'Neil, 1979):

Class 1—Replacement of equipment in an ongoing operation. Market is known, technology is proven, so that risk is fairly low. Maybe a discount rate of 10% is acceptable here.

Class 2—Expansion of present mine or plant facilities. Many technical problems have already been solved, but there may be a question in marketing. Can the additional output be sold at pre-expansion prices? The added risk here may indicate that a higher discount is in order, say, 15%.

Class 3—Opening of a new operation, entering a new market, etc. Here the sources of uncertainty are many, justifying, perhaps, 20% as the minimum acceptable rate of return on such projects.

The lowest rate classification should at least be equal to the firm's cost of capital. For moderately risky projects major firms have recently used after-tax rates in the range of 15 to 22% for project evaluation purposes. As mentioned previously, there is no correct rate. The rates are selected based on the firm's strategic goals, past costs, inflation expectations, and management judgment. The selection of an unrealistically high rate for all projects can result in a gradual liquidation of the firm as it will not be able to select sufficient new projects to continue to operate as an ongoing concern.

Expected long term yield must also be carefully evaluated. In 1982 the US experienced a rapid decrease in inflation and only slow moderation in interest rates. The result was that

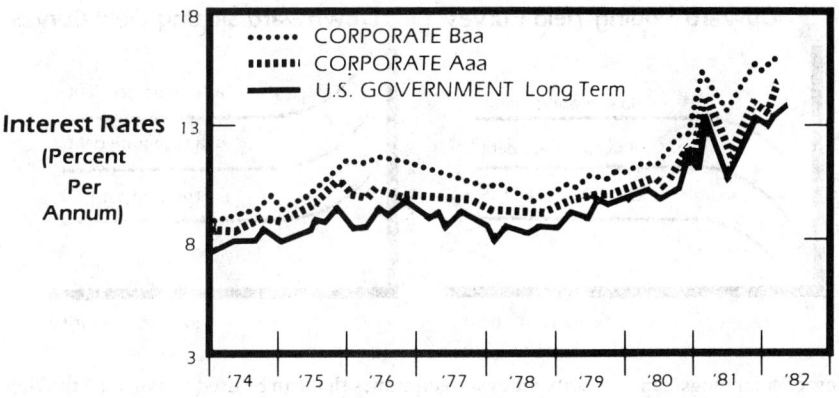

Fig. 4. United States long-term debt rates.

real interest rates, the difference between the market rate and the inflation rate, were in excess of 6% instead of the 2 to 3% that prevailed in the 1970s. Using current bond rates as a proxy for the discount rate under these conditions would distort the long term valuation of a project if one did not believe this relationship would be maintained over the life of the project.

Accounting for Inflation

During the 1950s and 1960s when prices rose at rates of only 1.5 to 2%, business schools turned their attention to fine tuning of the economy and business planning. With very low rates of inflation and increased access to computer-based analysis techniques, elaborate models could be constructed using indices to adjust prices and costs from one time period to another. Then came the 1970s and early 1980s, which turned both national economic and business planning into a shambles. Basic forecasts for the most volatile years could be in error by more than 20 to 30% per annum. Worse yet, in the mineral industry, conventional wisdom was that prices of a depleting mineral asset should rise to at least equal to or exceed inflation in general to offset the cost of bringing in new production. But, this was not the case for many minerals. A new word was coined, *stagflation*, to indicate that economic growth could stagnate while cost inflation could continue at high rates.

The experience of the Hecla Mining Co. in their attempt to diversify from silver into copper by developing the Lake-

shore Copper Project is but one example of the problems facing the industry. Fig. 5 plots the inflation adjusted cost of copper against the cost estimate for bringing this Arizona mine into operation (O'Neil, 1982). Hecla started work on their underground mining project in 1969 when both the US economy and the price of copper were strong. Ground control and mining problems delayed preproduction development. The scheduled startup date slipped from 1973 to 1975. Cost estimates increased from $100 million in 1969 to $185 million in 1975 as capital costs were impacted by the delay and by the first OPEC-induced cost increases. Real copper prices continued to fall and in 1978 Hecla abandoned the project which by then had a total investment of $200 million. What may have been a sound decision in 1969 almost forced Hecla into bankruptcy in 1978 when their net assets dropped from $199 million to $12 million. Only spectacular rises in silver prices in 1979 saved the company.

Inflation is a real problem that has major impacts on the following factors:

1. Interest rates directly reflect the financial community's opinion and current forecast of future rates (which does not necessarily mean it is an accurate forecast). In the broadly based US financial market on any given day interest yield curves can be plotted for securities with varying quality ratings and maturities. Fig. 6 shows both upward sloping curves typical of the pattern as the economy begins to recover from a recession and downward sloping curves that indicate the market perceives long term demand for money will drop.

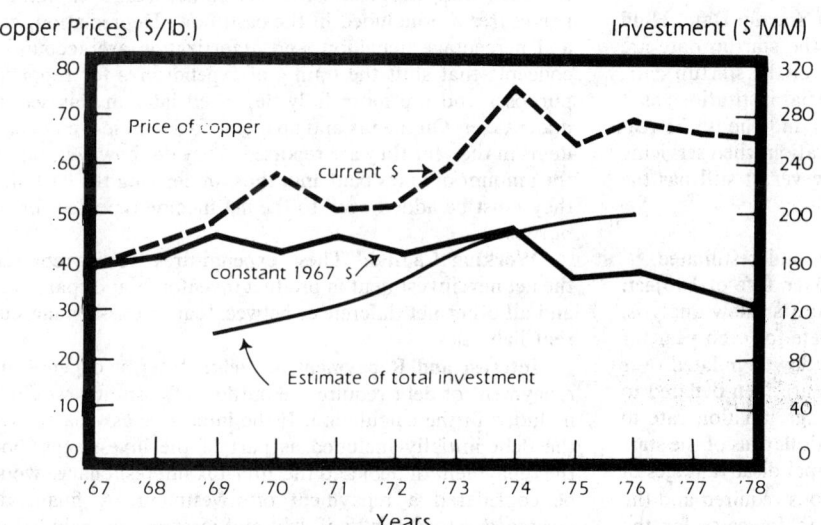

Fig. 5. Economic parameters during development of the Lakeshore mine.

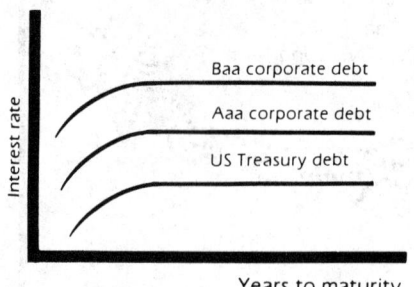
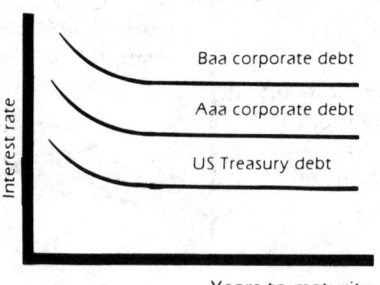

Fig. 6. Long-term bond yield curves.

Access to long term funds and the timing of financings can be critical to the success of large projects. Fig. 4 illustrates the variability of the US cost of funds during the 1970s.

2. Access to capital in rapidly changing times can be limited. In 1974 the utility industry faced a severe capital shortage as investors became wary of the utility's ability to cope with the changing economy. Likewise, lenders try to lower their risk by shortening repayment terms. This can actually make mining projects more risky as cash flow will be decreased in the usually critical early years.

3. Accounting problems can be significant as reported profits are distorted during inflation. The sale of low cost inventories results in high current profits, but producers must figure new inventory at higher costs, which depresses cash flow. Depreciation charges do not properly reflect the cost of replacing capital equipment. In the US major tax and accounting changes were required to cope with the turmoil caused by the economy of the 1970s. Later in this section taxes and depreciation accounting are discussed.

4. Price and cost forecasting so necessary to proper investment planning becomes difficult as historical trends are greatly distorted and perhaps altered permanently.

There are three basic methods to consider inflation in project cash flow analysis:

Method 1—Constant Dollars, No Inflation: This method values all costs and revenues as of the date of the estimate. It implicitly assumes future costs and revenues will inflate at the same rate. The disadvantages of the method are:

1. Capital costs are not identified at their actual ultimate cost, which makes financial planning difficult.

2. Taxes are underestimated because the value of depreciation is overstated.

Method 2—Inflate All Variables to Startup Date, Hold Constant Thereafter: Cash flows after the startup date are expressed in constant dollars valued as of the startup date. This method is frequently used by financial institutions as it does account for increased capital costs and should be reasonably accurate for the first year of operation when servicing the debt will be the most difficult. However, it still has the following disadvantages:

1. Taxes are still underestimated.

2. Equipment replacement costs are underestimated.

Method 3—Inflate All Variables Over Life of Project: This method provides the most realistic cash flow analysis. One must assume an explicit inflation rate for each year for each variable. Each component of cash flow is inflated over the life of the project. These cash flows are then deflated to the beginning of the project at an average inflation rate to give present values expressed in current dollars as of the start of the evaluation time frame. The principal disadvantages of this method are the number of calculations required and the problem of coming to a consensus as to forecasts for the appropriate values for the component inflation rates. Sensi-

tivity analysis techniques that can be used to assist in studying the impact of changes in assumptions are described later in this section.

Cash Flow Defined

The time value concept previously described relies on the identification and analysis of cash flows that consist of the investment and the subsequent return on and recovery of the investment. During the preproduction stage the investment consists of current expenditures plus capital expenditures (capital being defined for accounting and tax purposes as an investment with an economic life greater than one year). During production, revenues are received from the sale of product, and the cash flow is equal to the revenues less current capital expenditures and current cash expenditures, including taxes. During the post-production phases remaining assets are sold providing a return of investment, and cash is expended for items such as reclamation.

Deriving cash flow from tax and financial accounting statements can be complicated and requires an understanding of accounting. Table 2 shows how the same project would be reported differently for tax, cash flow, and financial book reporting purposes. Net income represents the return on the investment while depreciation and depletion represent a recovery of the investment. Four items requiring special explanation in the determination of cash flow are capital recovery charges, taxes, working capital, interest, and sunk costs.

Capital Recovery Charges and Taxes: Capital expenditures require special handling for financial and tax reporting. Comparison of the statements in Table 2 demonstrates that for cash flow purposes only the actual year's capital expenditures are included in the cash flow. Depreciation, cost and percentage depletion, and amortization are accounting concepts that shift the timing of expenditures for reporting purposes and are more fully described later in this section under taxes. On the tax and financial books these are noncash items in the year they are reported. They do, however, reduce the amount of taxes paid and, thus, in deriving the cash flow they must be added back to the net income reported for tax purposes.

Working Capital: These expenditures would represent the net new investment in product inventories and spare parts and all other net differences between current assets and current liabilities.

Interest and Repayment of Debt: Interest expense and repayment of debt require the outflow of cash and would be included in the calculation. If the interest is associated with the debt initially included as part of the investment, both the repayment of debt and the after-tax interest charge would be considered a repayment of investment. A financially leveraged return results if debt and interest are included (in this case both are excluded so the analysis only represents

Table 2. Comparison of Tax, Cash Flow and Shareholder Statements

Tax Basis
Profit and Loss Statement

		$(000)
Net Sales (Revenue)		10,000
Direct costs, labor and supplies (operating costs)		6,000
Gross operating profit		4,000
Less: General and administration expense	1,500	
Depreciation and amortization	1,600	
Percentage depletion	450	3,550
Taxable income		450
Less: Federal and state income tax, 50%		225
Add: Investment tax credit		75
Net income		300

Cash Flow Statement

Net income		300
Add: Depreciation	1,600	
Percentage depletion	450	2,050
Less: Increase in working capital	500	
Capital expenditure in plant	200	700
Net cash flow		1,650

Financial Book Basis
Profit and Loss Statement

Net sales		10,000
Direct costs, labor and supplies		6,000
Gross operating profit		4,000
Less: General and administration expense	1,500	
Depreciation and amortization	1,400	2,900
Taxable income		1,100
Less: Federal and state income tax, 50%		225
Deferred taxes		100
Net income		775

the return on the investor's capital).

Sunk Costs: Sunk costs are expenditures that have already been made and may or may not be recoverable. They should be ignored when making investment comparisons. Exploration and nonrecoverable development costs incurred prior to the date of the analysis are examples of the costs that are excluded.

As it may not be clear why these charges should be ignored, consider the following example: Assume $3 million has been spent to develop ore body A, and ore body B has been found without cost. Ore body B needs $2 million for exploration after which it can be sold for $3 million. Ore body A requires $1 million before it can be sold for $3 million. If one were to proceed with both projects, the total cash flow for ore body A would be a $1 million loss, while ore body B would yield $1 million profit. Yet, ignoring the previous investment which is not recoverable shows a 3-for-1 undiscounted future return for A and only 3-for-2 undiscounted return for B. Thus, continued investment in A is the preferred alternative.

Having now defined time value methodology, discount rates, and cash flow, attention can now be turned to measuring the performance of projects.

MEASUREMENT OF PERFORMANCE

The primary goal of measuring project performance is to establish criteria against which one can distinguish acceptable and unacceptable projects. The evaluation criteria chosen should (1) be consistent with the firm's primary objective; (2) recognize that bigger benefits are preferable to smaller benefits; (3) recognize that early benefits are preferable to later benefits; (4) be conformable and realistic, and (5) rank the proposals in the order of their desirability (Gentry, 1979).

Over the years numerous project evaluation techniques have been developed to provide guidance in making investment decisions. Both nondiscounted and discounted methods (which are generally considered most reliable because they recognize the time value of money) are currently in use. The most commonly used methods are presented here.

Payback Method

Because of its simplicity, rather than its merit, the payback method historically has been one of the most frequently used measures of performance. Payback is the amount of time required for the project to generate sufficient cash flow to repay the firm's initial investment in the project.

$$\sum_{n=0}^{N} CF_n = 0$$

This is a nondiscounted evaluation method that does not take into consideration the overall life of the investment nor the timing or size of the cash flows beyond recapture of the

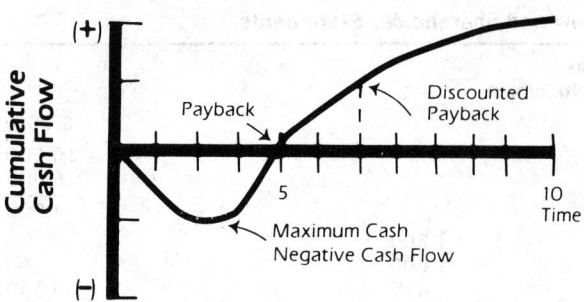

Fig. 7. Payback method.

initial investment. It is most commonly used only as a guideline, particularly in evaluating projects that are extremely risky or where the accuracy of long term data is very uncertain.

If the cash flows are discounted in the following way, the time value of money during the payback period is recognized:

$$\sum_{n=0}^{N} \frac{CF_n}{(1 + Disc)^n} = 0$$

The resulting discounted payback period is a more accurate measure of these initial cash flows, but it still does not measure the benefits of the project beyond the payback period. At best, it should be used as a constraint or hurdle against which a project is measured and not as a value to be optimized. Fig. 7 shows the discounted payback period must always be longer, but varies with the discount rate.

Net Present Value Method

The net present value (NPV), also known as present worth (PW), of a project is defined as follows:

$$NPV = PV \text{ Benefits} - PV \text{ Costs} = \sum_{n=0}^{N} \frac{CF_n}{(1 + Disc)^n}$$

The present values of both benefits and costs are calculated at the company's required rate of return. Projects with a

NPV greater than zero will at least earn more than their costs at the minimum acceptable return.

It should be noted when this method is used to compare mutually exclusive projects, the project with the highest NPV may depend on the discount rate used to evaluate the project, as illustrated by the following example.

The two projects each require an investment of $10,000 and the cash flows from the projects are expected to total $4,380 per year for three years for Project 1 and $2,433 per year for six years for Project 2. As can be seen in the following table and in Fig. 8, for a discount rate in excess of 7.7% Project 1 yields a greater NPV, while below this rate Project 2 would be preferred.

Discount Rate, %	Present Value—Annual Cash Flows, $		Project Preferred
	Project 1	Project 2	
0	13,140	14,598	2
3	12,389	13,180	2
6	11,708	11,964	2
9	11,087	10,914	1
12	10,520	10,000	1
15	10,000	9,204	1

Internal Rate of Return (Discounted Cash Flow) Method

Expressed in terms of the NPV method just described, the Internal Rate of Return (IRR) is the discount rate at which the NPV at time zero of all cash flows is equal to zero (the NPV of all benefits is equal to NPV of all costs). For the previous example the IRR's for Projects 1 and 2 are 15% and 12%, respectively. The formula for calculating the IRR is as follows:

$$0 = \sum_{n=0}^{N} \frac{CF_n}{(1 + IRR)_n}$$

If the project cash flows are uneven, the solution of this relationship involves an iterative trial-and-error process to converge on the exact IRR that sets the NPV equal to 0. The use of the method has been dependent on the availability of computers and advanced electronic calculators. If one

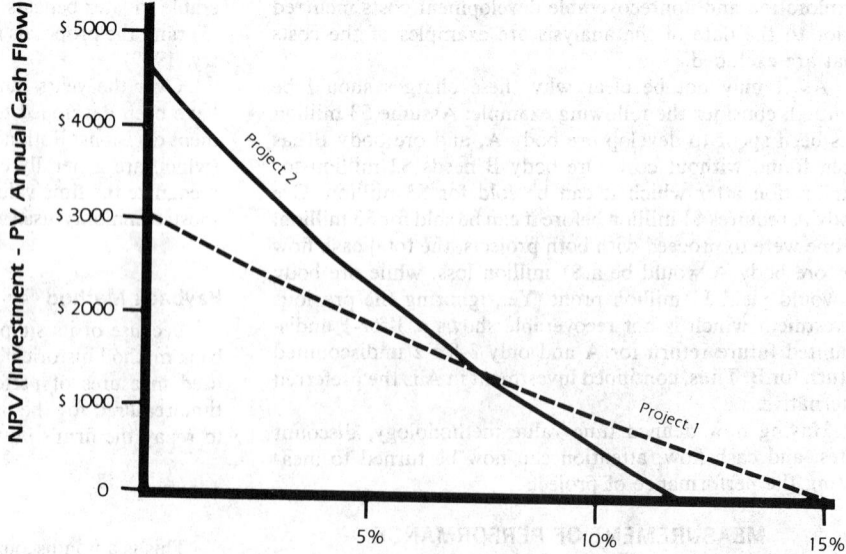

Fig. 8. NPV of two projects vs. discount rate.

must calculate the value manually, then a rough approximation of the IRR that should be used as a starting point in the iterative process is (Whitney, 1979):

$$\frac{0.07}{(\text{Investment Expenditures}/\text{Average Cash Flow})}$$

If the cash flow stream is an even series and the initial investment occurs at time 0, then the calculations can be greatly simplified by using interest tables. The initial investment is divided by the annual cash flow and the resulting factor is compared to the factors given in tables for the present value of an annuity.

The IRR method, also known as the discounted cash flow method (DCF), is today the most widely used by industry to measure the attractiveness of a project. When comparing evaluation criteria between firms, it is important to establish a common definition of the cash flows being discounted. The most common practice is to use total cash flow after tax to compute a return on the total capital invested (ROI) in the project. The result is often referred to as the DCF-ROI.

In some financial evaluations investors may wish to analyze just the return on their equity investment (ROE) net of any debt and debt service requirements. The ROE will then be a financially leveraged return (see the financing section in this chapter). Also, one should be sure that the method of determining the discount rate is consistent (i.e., continuous discounting or discrete). This is often a source of confusion and does vary by firm.

The IRR method has disadvantages, particularly when it is used to compare projects. It does not consider the relative size of the investments nor the annual cash flows. Thus, making comparisons between projects with very different lives or cash flow patterns is difficult. With regard to individual projects, it is also possible to have multiple rates of return that are economically meaningless but mathematically correct. Each time the annual net cash flow changes sign, it will yield an additional rate of return that will satisfy the resulting polynomial IRR equation (see Fig. 9). Fortunately, most projects start with an initial investment period followed by positive cash flows. The problem most typically arises if there are very large reclamation costs after a mine ceases operation, but they can also arise with projects built in overlapping stages during which the additional investments exceed the cash flows from previously constructed portions of the project. In the latter case, it is possible to avoid the problem by doing an incremental IRR on each stage of the project.

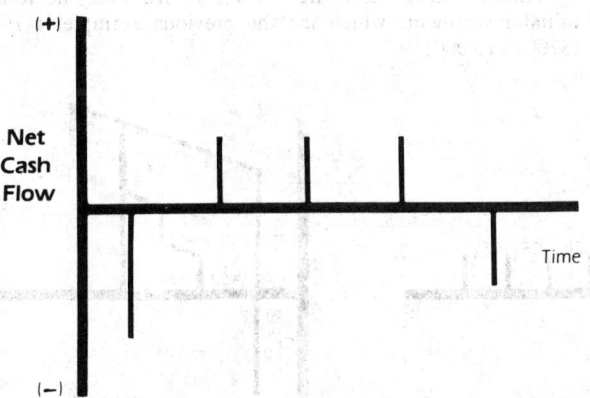

Fig. 9. Example of cash flow which will produce two IRR rates.

Profitability Index Method

The Profitability Index (PI), also known as the discounted benefit/cost ratio, is defined as follows:

$$PI = \frac{\text{PV of future benefits}}{\text{PV of future costs}}$$

The present value of benefits and costs are both calculated at the same discount rate. A PI of 1.00 or better indicates the project will at least cover its cost based on the company's required rate of return. The advantage of the PI is that it indicates relative magnitudes and thus enables the comparison of competing projects. As shown in the following example, Project 2 would be preferred even though the NPV's are identical:

	Project 1	Project 2
PV Benefits, $	1,000,000	300,000
PV Costs, $	900,000	200,000
Net Present Value (NPV), $	100,000	100,000
PI	1.11	1.50

Wealth Growth Rate Method

As mentioned previously, the NPV and IRR techniques present problems when comparing projects with different lives because of the implied assumptions regarding reinvestment of cash flows over the remaining life of the project. The Wealth Growth Rate (WGR) method developed by C. W. Berry assumes the positive cash flows are invested at the firm's average reinvestment rate of return (usually defined as the firm's cost of capital) and are compounded at that rate until the end of the longest lived project alternative. (See Fig. 10.) The Wealth Growth Rate is defined as the interest rate that will equate the future value of the capital investment with the future values of the reinvested cash flow from the project (Berry, 1972). Only projects with a WGR greater than the reinvestment rate should be considered.

The technique is best suited for large enterprises where the individual project will not substantially alter the long term reinvestment rate assumptions. Its advantages are that it uses cash flow rather than profit, recognizes the time value of money, and provides for direct comparison of projects.

Growth Rate of Return Method

The growth rate of return (GRR) method is very similar to the wealth growth rate method as it uses the firm's investment rate to discount or compound all project cash flows to a specific point in time. Thus, it is especially useful in evaluating competing projects. Positive cash flows occurring before the selected point in time *(t)* are compounded at the reinvestment rate while those occurring after are discounted by this rate to time *t*. The growth rate of return (GRR) is that rate at which the compounded net present value of the investment at t_o will equal the future value of the positive cash flows at *t*. (See Fig. 11.) If the project's GRR is greater than the reinvestment rate and also the GRR of competing projects, then it is selected.

This method does not yield multiple solutions nor require the trial-and-error solutions of the IRR method while it provides selection criteria consistent with the NPV and PI methods. The rankings of projects do not change regardless of the time horizon selected. It suffers from the same disadvantages as does the wealth growth rate method.

Competitive Ranking Method

The inability to forecast drastic economic changes one year hence, much less many years into the future as required by the more traditional DCF techniques, has led many ex-

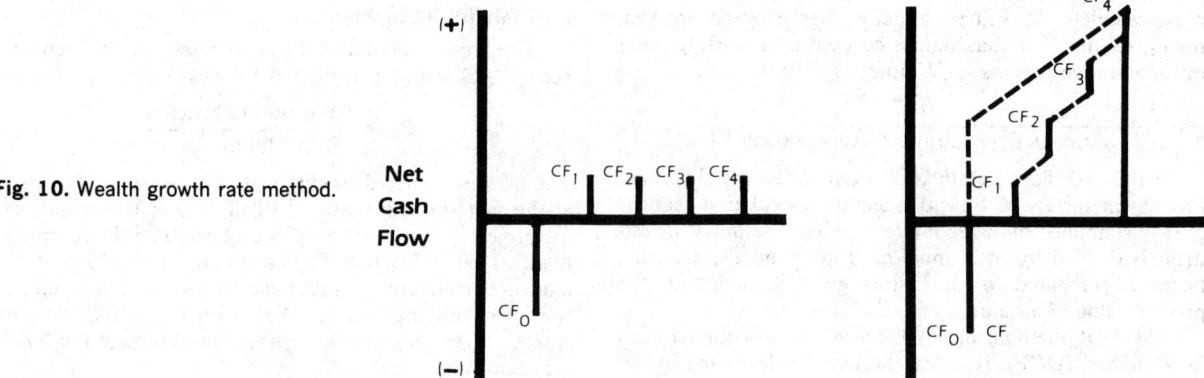

Fig. 10. Wealth growth rate method.

ecutives to also require projects to be ranked in relation to their competitive production cost position in the industry. New projects are then screened against an upper cost limit. Using this criterion at least assures the firm that if prices fall precipitously for an extended period, higher cost producers will be required to reduce production sooner. Fig. 12 illustrates a copper supply curve developed by a major metals industry consulting firm (O'Neil, 1982). Similar curves for 1982 would show virtually all US producers operating at a loss. It is recommended that this method be used for evaluation of all significant projects.

One should also be concerned with the corollary to the question of investment—what happens if one does not invest? Other industries such as the US auto industry could have benefited greatly if they had not limited their analysis of the immediate profitability of producing a small car to challenge the first imported Japanese cars. By not accepting smaller initial returns, they ultimately lost market share as the Japanese built volume and lowered unit costs.

The competitive nature of most markets also makes it difficult in practice to actually reach DCF projections, especially with regard to cost saving projects. Other firms are also initiating their own programs that could lower overall industry costs and, thus, cause prices to drop due to competitive pressure. The greatest problem associated with competitive ranking is of course finding reliable data on one's competitors. Fortunately for most major mineral commodities, a significant amount of published data exists, and several consulting firms can offer assistance in constructing competitive cost curves.

Accounting Rate of Return Method

The Accounting Rate of Return (ARR) method is used to measure a project's profitability as an annual percentage of total capital investment. Some firms calculate the return on an annual basis while others calculate it on the basis of an average over a specified period, generally five years.

$$ARR = \frac{\text{Average Book Income}}{\text{Average Book Capital Investment}}$$

OR

$$= \frac{\text{Average Book Income}}{\text{Total Invested Capital}}$$

The method is generally used by firms whose stock is publicly traded. Analysts who follow these stocks do not have access to the results of specific projects, but they compare company performance by measures such as return on equity and return on total assets. Large projects that produce large variations in reported earnings could introduce price fluctuations in the firm's stock. The ARR is an attempt to indicate this impact.

To illustrate the use of the method, assume a project has been capitalized on the books at the end of year one at $10,000. For book purposes the investment is written off on a straight line basis over ten years and the book profits are as shown in the following:

Year	Book Investment, $	Book Profits, $	Annual ARR, %
1	10,000	600	6.00
2	9,000	600	6.67
3	8,000	800	10.00
4	7,000	900	12.86
5	6,000	600	10.00
5 year average	8,000	700	8.75

Another measure is the average return over the total initial investment, which for the previous example is 7% ($700/$10,000).

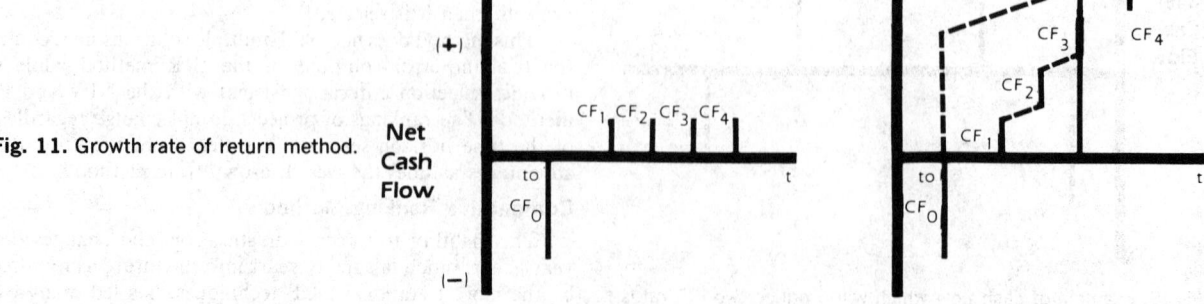

Fig. 11. Growth rate of return method.

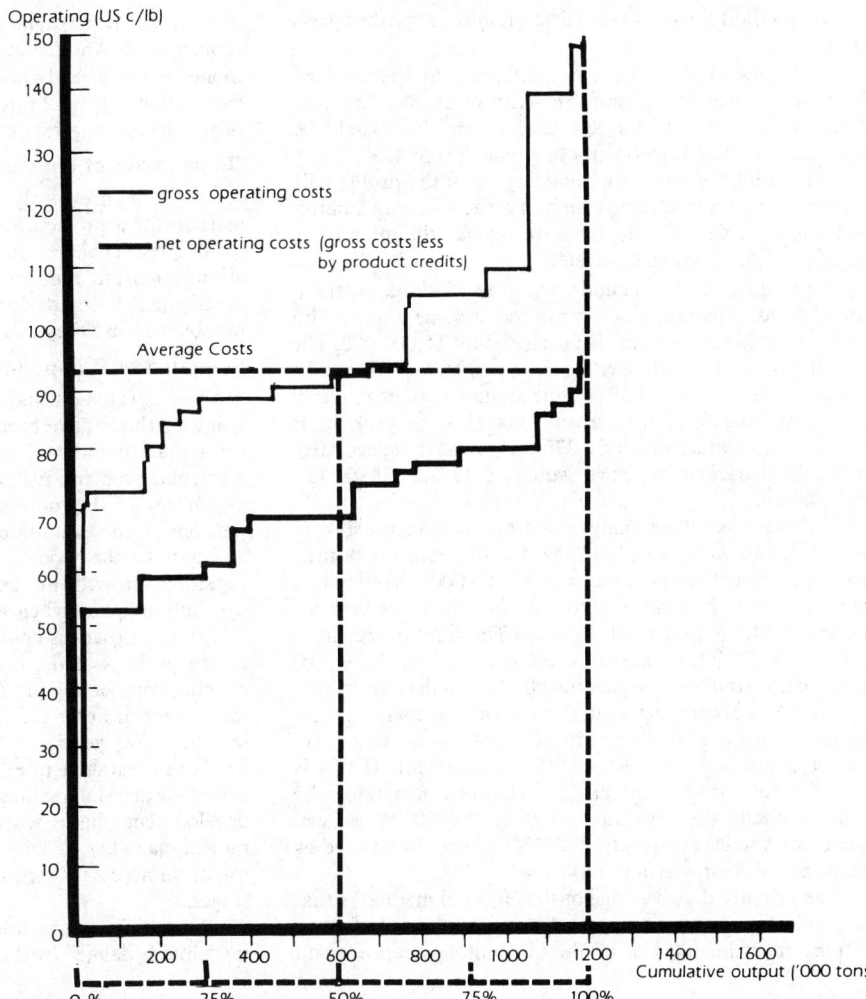

Fig. 12. Competitive cost comparison cumulative USA output against USA operating costs, 1980.

The principal disadvantage of the ARR method is it is a nondiscounted evaluation measure. It does not take into account project lives, the time value of money, or benefits such as percentage depletion and accelerated depreciation. The tax and accounting implications of a project are more fully described later in this section.

Profit Contribution Method

The term profit contribution is intended to encompass all of the net earnings impacts of a project as they are forecast to affect the consolidated financial accounting statements. In a full-cycle economic evaluation, a project is carried from a point where it first affects the balance sheet and net earnings through the point where it last affects them. By assuming liquidation of the assets, equipment, and business of the project, all their earning, saving, or benefit-producing power is extinguished and the books cleared. The algebraic summation of all profit contribution elements is equal to the cumulative net cash flow. No discounting is involved over the life of the project. Thus profit contribution is of limited value except that it places a project in perspective as to its possible impact on overall corporate earnings prospects. Its calculation also serves as a check that the balance sheet

treatment of the project and the net cash flow calculations used to prepare NPV or DCF analyses conform.

Maximum Negative Cash Position Method

While not a performance measure as such, maximum negative cash position is an indicator of the magnitude of a project. It does not provide an indication of how long funds are at risk, only the magnitude of the risk. Stated together with profit contribution, it provides some measure of quality. It is useful because it places a project in perspective as to its impact on overall corporate financing requirements and the firm's total financial exposure.

Hoskold Method

In 1877, M.D. Hoskold proposed a mine valuation method that was an attempt to recognize both the high risk of mining and the depleting nature of ore bodies. Today the method is not used by mine financial analysts because it is inconsistent with current accounting, budgeting, and income tax management practices. Unfortunately, in some states of the United States, the method was used to formulate mine property taxation systems, and thus, until the laws are changed, the legacy will remain.

The method assumes two basic premises regarding cash flows:

1. The initial investment is considered to have a *high risk* that will generate a uniform series of profits. The rate of return is known as the speculative rate that would be required to attract investments to similar properties.

2. Because the mine is a depleting asset, the profits will be reinvested at a safe return such that the resulting sinking fund will provide sufficient funds to replace the mine once the original ore body is exhausted.

Fig.13 graphically demonstrates the Hoskold methodology. As an illustration of the method, assume a mine with a remaining life of ten years is purchased for $5,000,000. The annual profits are estimated to be $1,200,000 and can be invested at a safe rate of 10%. The annual amount required for the sinking fund to recover $5,000,000 in year 10 is $313,727 per year leaving $886,273 per year as the speculative profit. The speculative return would be 17.7% ($886,273/$5,000,000).

If this project were analyzed on the basis of no sinking fund, i.e., an annuity of $1,200,000 for 10 years, the returns from the project would equal the $5,000,000 investment if interest were compounded at 20.2%. The difference between these two rates of return is important. The point to recognize is that the 17.7% speculative return is the result of two separate investments, one yielding 20.2% and the other yielding 10.0%. This implies that the decision to make the investment with the 20.2% return does not require that there also be a decision to make the 10% investment. If this is true, the 10% investment has no relevance in making the decision about the investment having the 20.2% return. Therefore, the 20.2% return, not 17.7%, should be used as the measure of project attractiveness.

The primary disadvantage of the Hoskold method is that companies commonly do not set up sinking funds for recouping investments, just as they do not put depreciation allowances in a special sinking fund for future equipment replacement. Another disadvantage is that the time value of money is not directly involved in the risk or speculative rate calculation. The risk rate calculation actually takes the form of the accounting rate of return calculation.

Comparison of Financial Measurements

As stated previously, there is no single correct evaluation method for a project. Most firms will use a combination of techniques. Table 3 compares the major methods of evaluation against a number of financial criteria. It should be noted the ARR and Hoskold method are not directly comparable because they rely on profits rather than on cash flow.

Selection of Competing Projects

The advantages and disadvantages of each of the foregoing methods have been briefly described as have the cautions that should be heeded in applying the methods to a particular problem. It should also be noted that the successful use of any of the more sophisticated techniques is very dependent on the assumptions made regarding the data serving as input to the model. The strategic goals of a company regarding growth, markets, and risk capacity must also be carefully weighed when selecting the performance measures.

Two approaches commonly used when comparing competing projects at the preliminary stage are the hurdle and ranking approach. Hurdles are minimum goals that a project must meet in order for it to be continued (e.g., payback in less than five years, or NPV equal to or greater than the firm's reinvestment rate). This approach is particularly well suited for decision making in the exploratory and preliminary development phases where uncertainty regarding the accuracy of many key variables is high. A more rigorous approach might unnecessarily penalize a project in the very early stages.

A ranking approach lists all projects in order of decreasing value as defined by the particular method of measurement

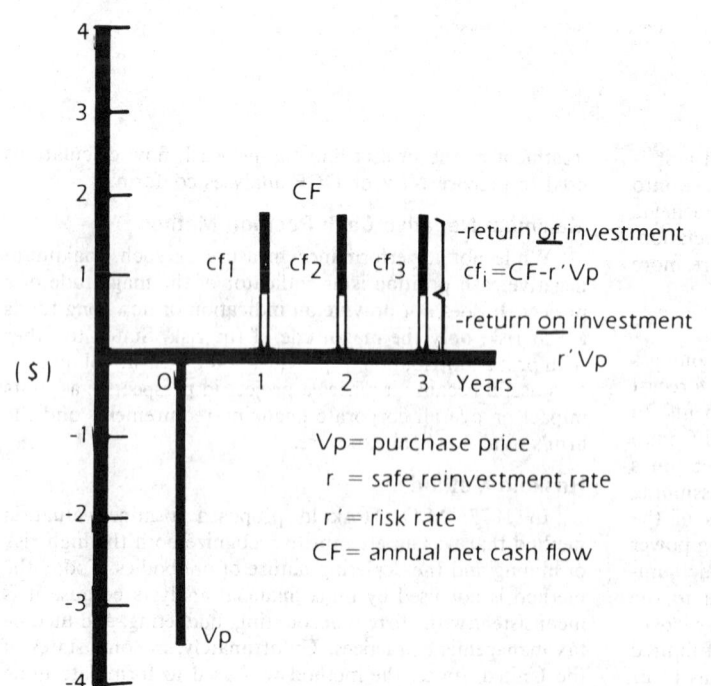

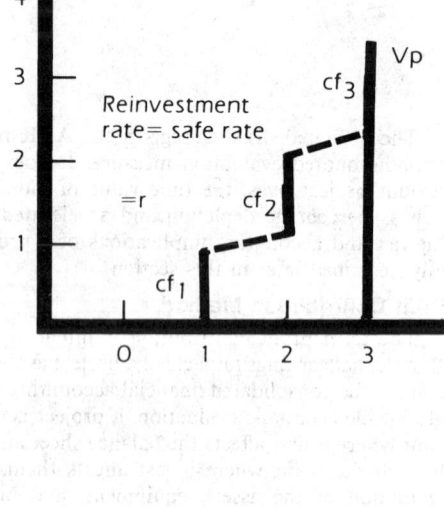

Fig. 13. Hoskold method.

Table 3. Comparison of Financial Measurements

Items for Financial Measurement	Accounting Rate of Return	Payback Period	Discounted Payback Period	NPV	PI	IRR	WGR	Hoskold
Uses profit or cash flow?	Profit	Either	Either	Cash Flow	Cash Flow	Cash Flow	Cash Flow	Profit
Recognizes time value of money?	No	No	Yes	Yes	Yes	Yes	Yes	Yes
Requires reinvestment rate in calculation?	No	No	No	Yes	Yes	No	Yes	Yes
Assumes a sinking fund?	No	No	No	No	No	No	No	Yes
Results in a form of a rate of return?	Yes	No	No	No	No	Yes	Yes	No
Can yield multiple solutions?	No	No	No	No	No	Yes	No	No
Compares different investment requirements?	Yes	Maybe	Maybe	No	Yes	Yes	Yes	No
Accounts for benefits after payback period?	Yes	No	No	Yes	Yes	Yes	Yes	Yes
Appraises market value of project?	No	No	No	Yes	No	No	Yes	Yes
Ranking may vary with different reinvestment rates?	No	No	Yes	Yes	Yes	No	Yes	Yes
Recognizes explicitly life of the project?	No	No	No	No	No	No	No	No

Source: Berry, 1972.

(e.g., PI). Again, care should be exercised at the early stages of a project so that slight differences in values do not eliminate a project that may have a lower profitability but a high strategic value to the firm.

SENSITIVITY ANALYSIS AND OPTIMIZATION

All the methods described so far try to interpret, in a consistent manner, a firm's projections regarding future values. The real problem with financial analysis is that no matter how elaborate the procedures, one is still trying to predict the outcome of a vast array of key independent and dependent parameters, many of which, such as overall economic activity, inflation, and cost of funds, are far more complicated than the project itself.

Sensitivity Analysis

A deterministic approach to project variables is the most common method of evaluating future possible outcomes. Each parameter deemed critical, such as ore grade, recoveries, capital costs, operating costs, sales prices, and inflation is varied to determine its effect on the measure of performance. Fig. 14 illustrates the simple comparison of price vs. return on equity. If the percentage change in values of the

parameters with respect to the base or expected case is normalized, a *spider diagram* can be constructed as illustrated by Fig. 15 in order to directly compare several parameters.

The common practice is to select the minimum, most likely, and maximum values expected for each parameter and to calculate the corresponding value of the performance measure. Thus, if one were to consider price, capital costs, and operating costs as the critical parameters, there are 27 specific *what if* outcomes (i.e., 3 × 3 × 3). The financial analyst and management combine their subjective judgments to select the range of values considered most likely to occur and rate the projects accordingly. If all the cases produce performance values in excess of the hurdle values, the decision process is greatly simplified. Unfortunately, many times the studies will produce values above and below the minimum acceptable values. Probabilistic methods described later may be the best method to quantify the likelihood of their outcomes.

Optimization

For any given set of assumptions it is generally possible to optimize the results. Linear programming, an operations research technique that systematically solves simultaneous equations, has been used effectively by integrated mining

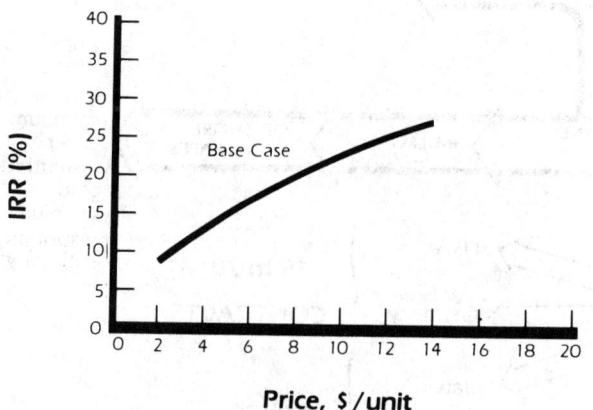

Fig. 14. Typical sensitivity graph.

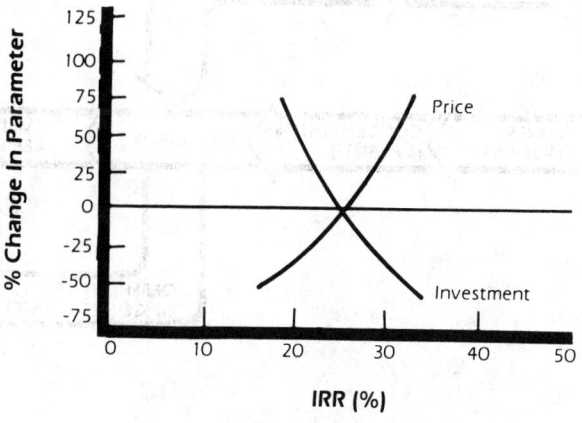

Fig. 15. Typical spider diagram.

companies such as Cerro Corp. to optimize the results of current operations (Anon., 1972). Many variables including metals prices and operating costs such as labor, transportation, raw materials, and capital costs are incorporated in the model together with constraints such as ore grades, quantities, and the availability of purchased concentrates and metals. The Cerro zinc model contains some 370 variables and more than 150 basic equations that solve the problems posed in Fig. 16. *What if* solutions can be found for both current and projected changes in values.

Others have applied the linear programming technique to entire industries in an attempt to determine the complex interrelationship of variables. The nickel industry has been studied (Copithorne, 1973) to determine optimum pricing and marketing strategies. Properly constructed models can be used to determine the competitive ranking of projects and the outcome of future changes in the variables. The principal limitation in using these techniques is the collection of accurate information, especially with respect to one's competitors.

RISK ANALYSIS

In contrast to the deterministic approaches just described, probabilistic methods attempt to relate the expected variance of outcomes in a systematic manner. This permits the analyst and management to more fully express their degree of uncertainty regarding future events. This section does not attempt to cover all the mathematics of probability theory; however, several examples illustrate how the techniques can be utilized.

Discrete Probability Distributions

The assumption of a finite number of independent outcomes implies the ability to assign a unique probability to the occurrence of each variable. As an example, consider a sensitivity study for a new project which consists of capital costs, ore grade, and market prices, each of which could reasonably be expected to be independent. It is assumed a copper mining project will be operated over a ten-year period and management estimates the probabilities of occurrence for each variable as follows:

Capital Costs	Ore Grade	Market Price
$10MM; P = 0.10	0.45%; P = 0.50	$0.70/lb; P = 0.05
12MM; P = 0.60		0.90/lb; P = 0.90
15MM; P = 0.30	0.50%; P = 0.50	1.10/lb; P = 0.05

The probability of a specific combination occurring is simply the multiplication of the probability for each variable (e.g., $10MM cost, 0.45% grade, and $0.90/lb price is (0.1) (0.5) (0.9) = (0.045). The probability tree, Fig. 17, shows the possibilities given a mining rate of 500,000 tons per year, direct operating costs of $700,000 per year, and a tax rate of 50%. There are 18 possible outcomes (3 capital costs × 2 ore grades × 3 market prices).

The discounted cash flows range from 5.4% to a high of 26.2% with a mean expected value of 14.1%. Figs. 18 and 19 are histogram plots of the DCF-ROI probability distribution and its cumulative probability. They show an 89% probability the DCF-ROI will be between 9% and 18%. This type of presentation gives management a much clearer idea of the variance of the outcomes.

Conditional Probability

The preceding example was greatly simplified. Frequently, the variables are conditionally dependent on previous events. This is typical of exploration programs as the success of each stage is dependent on the success of the prior stage. A decision tree is shown in Fig. 20 to illustrate the possible outcome of a two-stage exploration program that will result in a property value of $8,000,000 if successful.

The expected value of this program is less than zero and, thus, would be rejected. The minimum probability of success that would make the company indifferent to the outcome of 2a and 2b is given by the risk capacity equation:

$$P = \frac{1}{\frac{\text{NPV success}}{\text{NPV failure}}} + 1 = \frac{1}{\frac{2,900,000}{5,100,000}} + 1 = 0.6376$$

where P = the risk capacity or minimum acceptable probability of success.

Conditional probabilities will exist between many variables such as ore grade vs. tonnage and ore grade vs. mill recoveries. Thus, care must be exercised in building economic models. A variety of statistical methods exist to determine the degree of dependence or correlation between variables. As a practical matter, a balance must be struck between complexity and the cost of building the model.

Getty Oil Co. is one of the major oil companies that has attempted to quantify the very subjective variables involved

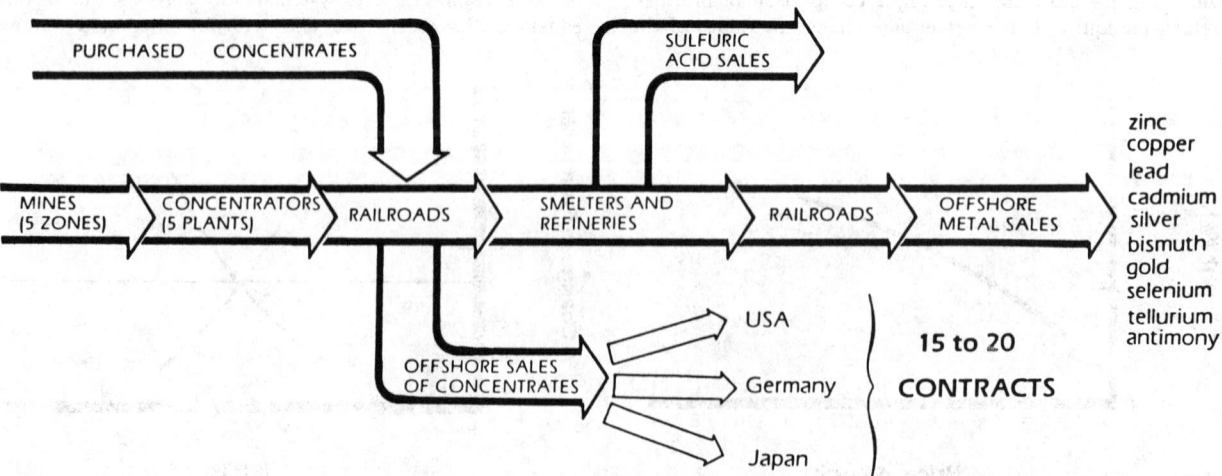

Fig. 16. Cerro De Pasco's worldwide operations.

Capital Cost	Ore Grade	Price	Cumulative Probability	Annual Net Cash Flow	DCFROI (%)
	.45%p=.5	$0.7p=.05	.0025	1,725,000	11.4
		$0.9p=.90	.0450	2,175,000	17.4
	cum.p=.05	$1.1p=.05	.0025	2,625,000	23.0
$10 MM p=.1					
cum.p=.1	.5%p=.5	$0.7p=.05	.0025	1,900,000	13.8
		$0.9p=.90	.0450	2,400,000	20.2
	cum.p=.05	$1.0p=.05	.0025	2,900,000	26.2
	.45%p=.5	$0.7p=.05	.0150	1,825,000	8.5
		$0.9p=.90	.2700	2,275,000	13.8
	cum.p=.3	$1.1p=.05	.0150	2,725,000	18.6
$12 MM p=.6					
cum.p=.6	.5%p=.5	$0.7p=.05	.0150	2,000,000	10.6
		$0.9p=.90	.2700	2,500,000	16.2
	cum.p=.3	$1.1p=.05	.0150	3,000,000	21.5
	.45%p=.5	$0.7p=.05	.0075	1,975,000	5.4
		$0.9p=.90	.1350	2,425,000	9.9
	cum.p=.15	$1.1p=.05	.0075	2,875,000	14.0
$15 MM p=.3					
cum.p=.3	.5%p=.5	$0.7p=.05	.0075	2,150,000	7.2
		$0.9p=.90	.1350	2,650,000	12.0
	cum.p=.15	$1.1p=.05	.0075	3,150,000	16.5
Cumulative Probability	10,000	10,000	1.0000		

Annual Cash Flow Calculation

Capital Investment	$10,000,000
Ore Mined (tons/yr)	500,000
Grade	.45%
Copper (lbs./yr)	4,500,000
Price	$.90/lb
Sales	$4,050,000
less:	
Operating Costs	700,00
Depreciation (over 10 years)	1,000,000
Taxable Income	2,350,000
less:	
Taxes (50%)	1,175,000
Net after Taxes	1,175,000
plus:	
Depreciation	1,000,000
Net Cash Flow	$2,175,000

Fig. 17. Discrete probability analysis.

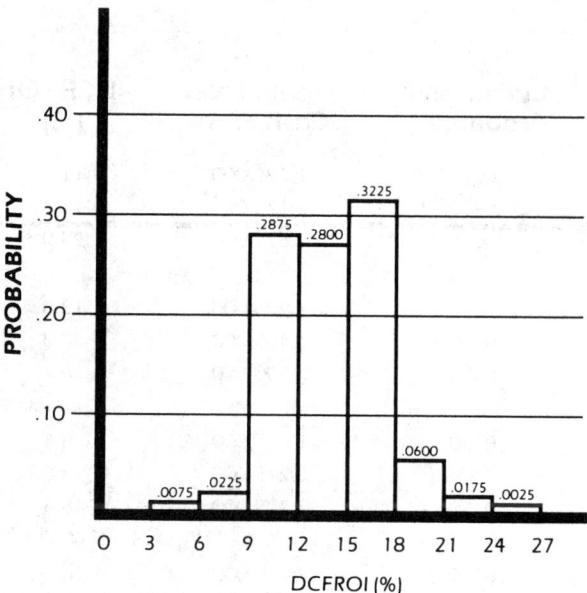

Fig. 18. DCFROI probability histogram.

in political risk analysis for foreign investment. The following risk factors are often included in their evaluations:

1. Civil disorder
2. War
3. Sudden expropriation
4. Creeping expropriation
5. Domestic price controls
6. Adverse tax changes
7. Production restrictions
8. Product export restrictions
9. Repatriation limitations
10. Devaluation risk

These factors are related through a series of equations to risk adjust projects for a variety of these occurrences (Bunn and Mustafaouglu, 1978). The company relies on a panel of experts to establish the probability distributions and relationships necessary to solve this extremely complex model.

Continuous Probability Distributions

For discrete probability distributions, the number of possible outcomes is finite. By assuming continuous distributions, one can use statistical sampling methods to generate the probability distribution of the performance measure described previously. This is the basis of Monte Carlo simulation analysis. Common probability distributions that might be considered of each variable are shown in Fig. 21. Referring to the discrete probability example of Fig. 17, the rectangular distribution might be used for ore grade, the normal distribution for price, and the triangular distribution for capital cost.

Using a random number generator, the Monte Carlo simulation method selects a unique value for each variable and sequentially calculates the measure of performance (e.g., DCF-ROI, NPV). The process is repeated a sufficient number of times to produce a stable probability distribution for measure of performance. Complex models may require more than 1,000 simulations. Thus, the method has gained acceptance only as large-scale computers have become available (O'Neil, 1979).

The most important benefit of this analysis is the ability

to compare the risk variance of the possible outcomes for competing projects. Two projects may have the same expected values, but the project with a greater variance may be much less attractive because it is much riskier.

TAXATION

No firm can operate in the US without paying a share of its earnings to the federal and state governments. Tax calculations in the mining industry are complex, and thus careful tax planning is essential to utilize all the available methods of minimizing this cost. Laws are constantly changing, and although this section will outline the pre-1984 basis, some aspects already have changed. The major concepts for calculation of current tax liabilities are expenses, depreciation, amortization, depletion, and investment tax credits (ITCs). Later in the section, deferred taxes, which is a financial concept, will be discussed. Only the current tax liability is relevant for the cash flow calculation, but deferred taxes enter into a firm's reported net income figures. Examples of tax calculations for a project are shown in the Appendix.

The first necessary distinction is between expensed and depreciable project expenditures. Expensed items are those that can be deducted totally from revenue in the year of expenditure. These are generally intangible items, such as drilling costs. Depreciation is a deduction allowed for tax calculations to reflect wear and tear, usage, and obsolescence of equipment and facilities used in a trade or business. Depreciation is over a period of time set by law and can vary from 3 to 15 years, based on the current US tax provisions. Amortization is similar to depreciation, i.e., a periodic writeoff, but applies to intangible assets. Amortization usually applies to relatively small sums in mining projects and will not be discussed further.

Operating costs at an ongoing mine, e.g., labor, energy, and supply costs for mining, milling, administration, and treatment charges, can be expensed for tax purposes. Prior to 1982, exploration and development for a potential new

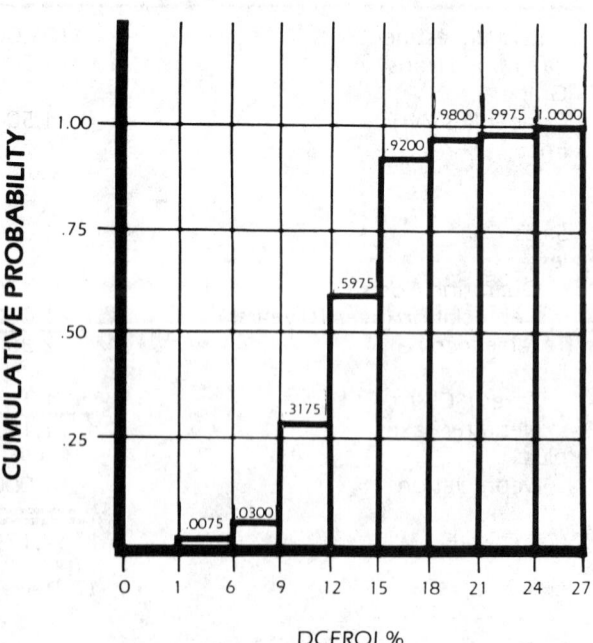

Fig. 19. DCFROI cumulative probability histogram.

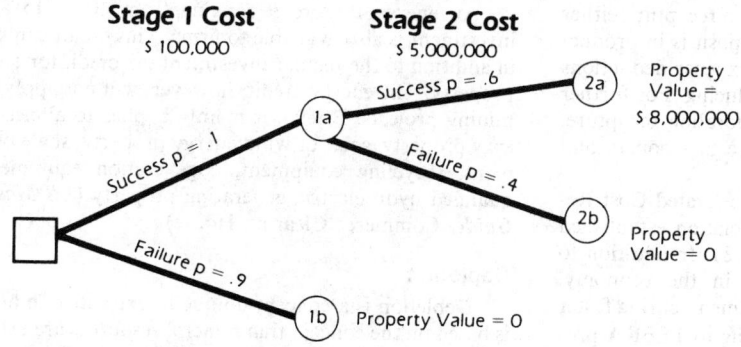

Fig. 20. Decision tree analysis.

The Expected Benefits are Calculated as Follows:

OUTCOME	EXPECTED BENEFIT
1b = (.9) (-100,000)	= - 90,000
2b = (.1) (.4) (-100,000 - 5,000,000)	= -204,000
2a = (.1) (.6) (8,000,000 - 5,100,000)	= +174,000
	-120,000

mine had also been fully deductible. Exploration expenditures are defined for tax purposes as expenses incurred in finding and proving the quality of a mineral deposit that had not previously been commercially exploited. The development stage starts once the deposit is defined so that the quality and quantity could justify commercial exploitation. In 1982,

the Tax Equity Fiscal Responsibility Act of 1982 (TEFRA) was enacted. TEFRA allows only 85% of exploration and development expenditures to be deducted in the year spent. The remaining 15%, as well as most equipment and facilities, must be depreciated based on the life of the asset. One other aspect is relevant to exploration expenditures. Any amount

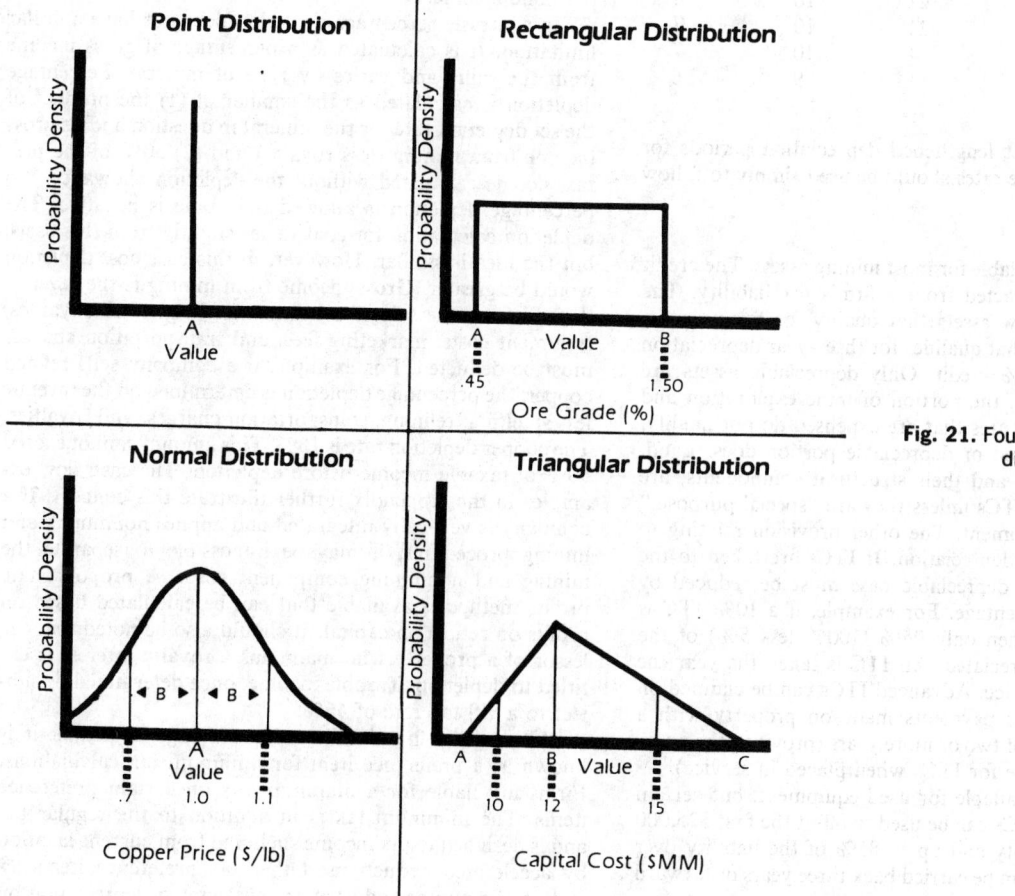

Fig. 21. Four common probability distributions.

that is expensed when incurred is subject to recapture either through depletion or income when the deposit is in production. Both exploration and development expense deductions can also be deferred until the mine is producing. For further explanation of this approach and/or exploration recapture, one of these tax references is valuable (e.g., Coopers and Lybrand, Burke and Bowhay).

TEFRA basically reconfirmed the Accelerated Cost Recovery System (ACRS) method of depreciation established in the major tax legislation, enacted in 1981. In addition to recognizing the inflationary conditions in the economy, ACRS was established to encourage investment, curing faster writeoffs than the previous laws. According to TEFRA provisions, most mining and milling equipment and facilities can be depreciated over five years. Some trucks and other short-lived assets are eligible for three-year depreciation. Buildings generally require a 15-year depreciation period, but if specifically associated with a mining asset such as the mill housing, they qualify for five-year depreciation. Land cannot be depreciated. Depreciation begins once the asset is placed in service, not when the funds are expended. Both new and used equipment can be depreciated on the ACRS system. Salvage value, the value of the item when its usefulness to the firm has ended, is ignored; i.e., the full asset value is depreciated. The depreciation rates are as follows:

Asset Life	%/Year			
	3 Year	5 Year	10 Year	15 Year
Year 1	25	15	9	5
2	38	22	14	10
3	37	21	12	9
4		21	10	8
5		21	10	7
6			10	7
7-10			9	6
11-15				

The 1984 Tax Act lengthened depreciation periods for certain assets, so these rates should be used simply to follow the examples.

Tax Credits

ITCs are also available for most mining assets. The credit amount can be subtracted from a firm's tax liability. The credit is 10% for new assets that qualify for five-year depreciation. An asset that qualifies for three-year depreciation is only allowed a 6% credit. Only depreciable assets are eligible for ITCs, i.e., the portion of mine exploration and development expenditures that are expensed do not qualify, whereas the capitalized or depreciable portion does. Land, warehouse buildings, and their structural components, are also not eligible for ITCs unless they are "special purpose," i.e., they house equipment. The other provision relating to ITCs is its effect on depreciation. If ITCs are taken to the extent available, the depreciable base must be reduced by half of the ITC percentage. For example, if a 10% ITC is taken on the mill, then only 95% (100% less 5%) of the mill cost can be depreciated. An ITC is taken the year the asset is placed in service. Advanced ITCs can be claimed on construction progress payments made on property with a construction period of two or more years (provided the property would be eligible for ITCs when placed in service).

ITCs are also available for used equipment, but certain limitations apply. ITCs can be used to offset the first $25,000 of a firm's tax liability and up to 85% of the liability over that amount. ITCs can be carried back three years or forward fifteen years.

An energy tax credit ranging from 10 to 15% of the investment is also available to firms. This credit can be taken in addition to the regular investment tax credit for qualifying property. The energy credit, however, will not apply to most mining projects. Briefly, it mainly applies to alternative energy property, solar or wind energy property, shale oil equipment, recycling equipment, cogeneration equipment, and qualified hydroelectric generating property (*US Master Tax Guide,* Commerce Clearing House).

Depletion

Depletion is a concept unique to extractive industries. It is based on the concept that mineral resources are exhaustible as well as difficult to replace. Depletion is an additional deduction available for tax calculations. Depletion must be calculated on a mine-by-mine basis, i.e., a firm cannot consolidate its mines. There are two methods to calculate depletion, cost, and percentage (also known as statutory). Depletion must be calculated both ways, and the larger sum claimed. Generally the percentage method is larger. Cost depletion is determined by the basis in the property applicable to the minerals. Thus, if a firm acquires the mineral rights to an undeveloped property for $2 million, this is the cost depletable basis. Exploration expenditures that are not expensed can be included in this base. This amount can be written off as the minerals are mined on a unit-of-production basis. For example, if the acquired reserves above are mined over ten years in equal amounts, the annual cost depletion is $200,000. Once the basis has been exhausted or reached zero, no further deductions are allowed. The cost depletable basis is also reduced by any percentage depletion taken on the mine income.

In contrast, percentage depletion does not have a dollar limitation. It is calculated as a percentage of gross income from the mine and varies by type of mineral. Percentage depletion is calculated as the smaller of (1) the product of the set depletion rate for the mineral in question and its gross income from mining (less royalty) and (2) 50% of the pretax income calculated without the depletion allowance. No percentage depletion is allowed if income is negative. The depletion calculation for coal varies slightly from this basis, but the idea is similar. However, in this case cost depletion would be greater. Gross income from mining is the revenue derived from the sale of concentrates. In general, royalties, treatment costs, marketing fees, and transportation charges must be deducted. For example, if a company sells refined copper, the percentage depletion is determined on the revenue less smelting, refining, transportation charges, and royalties. The copper depletion rate is 15%. This amount cannot exceed 50% of taxable income before depletion. The cash flow examples in the Appendix further illustrate this concept. If a company is vertically integrated and applies nonmining and mining procedures, it may be impossible to separate the mining and nonmining component costs. A proportionate profits method is available that can be calculated based on profits on return on capital. It should also be noted that the lessor of a property who maintains a royalty interest is entitled to depletion. Taxable income, once determined, is subject to a US tax rate of 46%.

Although a firm is allowed percentage depletion, it is known as a preference item for minimum tax calculations. Firms are liable for a minimum tax on certain preference items. The minimum tax is in addition to the regular tax and is designed to tax income sheltered from current taxation by accelerated deductions. Other tax preference items affecting the mining industry are accelerated depreciation on real property and amortization of certified pollution control equipment. For further discussion of this, consult a tax reference guide (e.g., *US Master Tax Guide,* Coopers and Lybrand). However, if a firm is taking percentage depletion and taxes are reduced to zero or near-zero, the minimum tax will most likely apply.

State Taxes

State taxes are another cost to a project and can often be significant. Five main types of state taxes exist: (1) ad valorem property tax, (2) severance tax, (3) income tax, (4) sales tax, and (5) excise tax. Further detail of these can be obtained in the reference material or from the appropriate state agency. State taxes now can be deducted as an expense for federal taxes. (This may be changed under revisions to the tax code.) Some states calculate their income tax using the unitary concept. The exact formula varies from state to state, but the tax is calculated based on a percentage of a firm's property, payroll, and sales located within its borders. Each state computes the apportionate formula differently making it possible for a firm operating in several states to be taxed on more than 100% of its income. This tax often results in a higher actual tax than the project itself would generate on a stand-alone basis. Again, consult a tax adviser to determine a basis for this tax.

Capital Gains

Another type of tax is capital gains, which is triggered if an asset is sold for more than its original value. Asset sales are also subject to taxes if the sale is at more than the tax book value. Assume an asset was purchased for $10,000 and depreciation of $3,700 has occurred on the tax books, resulting in a tax book basis of $6,300 (a five-year ACRS asset with two years depreciation). If the asset is sold for $7,000, then depreciation of $700 must be recaptured and included as income. This is subject to the 46% tax rate. If the asset was sold for $11,000, then all the depreciation must be recaptured as income, and capital gains must be paid on $1,000, the amount over the asset cost. The current capital gains tax rate for firms is 28%. The asset may also be subject to partial ITC recapture depending on the length the firm owned it. An unusual provision exists for coal royalties. These are subject to taxation at the 28% capital gains rate, rather than the 46% ordinary income rate.

Structure and Strategy

The structure of a project can be very important for tax consequences, in addition to legal consequences. Among the areas that are different are at what level tax elections are made, whether tax benefits/losses flow through to other corporate entities, if dividend payout to the corporation is taxable, and if disproportionate allocation of income, expenses, and depreciation is possible. For example, in a partnership, tax elections are made at the partnership level. Likewise, if a corporation is less than 80% owned, then 15% of the dividends are taxable. A tax adviser could define the advantages and disadvantages of each structure.

Tax strategies are available to minimize a firm's liability if carefully considered. One method is to set up a two company structure, a land and a mining company. The land company acquires the property and subleases it (retaining a royalty interest) to the mining company. The mining company develops the mine and pays the royalty to the land company upon production. The land company can take cost or percentage depletion on the royalty income, and the mining company gets percentage depletion. This results in greater combined depletion deductions vs. a one-company structure

where cost depletion is generally lost. Another method to increase income in mining firms is to remove all debt from the mining company, increasing depletion when the 50% of pre-tax income limitation applies. A third method is to remove fast depreciation equipment from the mining company and lease it back using longer lives. These are sophisticated concepts, and before being evaluated and utilized, should be checked thoroughly with a tax adviser as to the precise mechanisms.

Deferred Taxes

The previous discussion on taxes centered around the firm's actual current tax liability. The concept of deferred taxes also exists and is used in financial accounting or to report income figures. Before the discussion of deferred taxes, it is necessary to explain how they arise. The use of ACRS depreciation does not accurately reflect the actual usage of the asset over its life and distorts the tax effects. Thus, for a firm's financial books the asset is usually depreciated on a straight line or units of production basis. These methods allocate the life of the asset more accurately based on its actual usage. The straight line method allocates the cost of the asset equally by years over the actual life of the asset. The units of production method allocates the cost of the asset as a percentage of the production in each year it is used. The table that follows shows the differences in depreciation methods using ACRS, straight line, and the units of production method. The asset is assumed to cost $1.0 million and has a ten-year life. These examples assume no ITC has been taken; otherwise, the depreciable base would be reduced.

		Annual Production, mm Lbs	$thousand		
			ACRS	Straight Line	Units of Production
Year	1	5	150	100	50
	2	5	220	100	50
	3-5	10	210	100	100
Year	6-10	12	0	100	120
Total		100	1,000	1,000	1,000

The financial book reporting system considers this difference in depreciation. It is reflected in the income by showing a deferred tax entry. In early years when tax book depreciation is greater, deferred taxes are positive, whereas in later years, when the reverse is true, deferred taxes are negative. Table 4 shows years 1 and 10 for the mine in the example.

As can be noted, the pre-tax income is $50,000 higher on the financial book basis in year 1. However, the project is incurring an additional tax cost that must be noted. In years 6 through 10 the asset will no longer enjoy any tax benefits due to tax depreciation. This is reflected in the deferred tax component of $23,000 which is the difference between the financial book and tax book depreciation adjusted for the tax effects. Likewise, in year 10, deferred taxes are a negative $46,000. Deferred taxes are a noncash charge, and for cash flow purposes, are an additional source of funds. This simple example ignores depletion on financial books and the effect on deferred taxes due to ITC implications. Only cost depletion, if there is any, is reported on financial books.

FOREIGN PROJECTS

This section is written assuming a US-based firm as the investor. The analysis, however, could apply to a non-US-based firm investing in its country, the United States, or another country. Although the same economic analysis tech-

Table 4

| | Year 1, $thousand | |
	Tax Books	Financial Books
Revenue	1,000	1,000
Operating costs	500	500
Depreciation	150	100
Pre-tax income	350	400
Depletion*	100	—
Pre-tax income	250	400
Taxes (46%)—current	112	112
—deferred	—	23
After-tax income	138	265

| | Year 10, $thousand | |
	Tax Books	Financial Books
Revenue	1,000	1,000
Operating costs	500	500
Depreciation	—	100
Pre-tax income	500	400
Depletion*	100	—
Pre-tax income	400	400
Taxes (46%)—current	184	184
—deferred	—	(46)
After-tax income	216	262

* Treatment of depletion for tax book income may vary by firm.

niques already outlined apply regardless of a project's location, the return may change considerably for several reasons. Even in the United States, taxes vary by state and royalties vary by project. While state taxes and royalties generally have minor impacts on the DCF-ROI or other measures of return, the potential for variation is considerably greater for a foreign project. Taxes, both US and foreign, currency values, currency convertibility, repatriation of cash flow, and expropriation can all have significant effects on a project's return. Perhaps the key point of this discussion is investment outside a firm's country of incorporation involves substantially more complexities and uncertainties increasing the range of potential outcomes and project risk.

Foreign Taxation

Breaking the components that need to be quantified by topic, the most easily quantifiable is foreign taxes. Few generalizations can be made about taxes among countries worldwide since the codes vary dramatically. The first step is to retain an experienced tax adviser to ascertain all the potential tax implications. Depreciation methods can range from straight line to 100% writeoffs. For example, Canada allows certain assets to be depreciated to the extent of a project's income. Foreign tax codes for which items can be expensed also vary. As discussed previously, most stripping and certain other mine development costs can be expensed in the year incurred in the US, but tax treatment of these items varies among countries. In certain countries, additional expenditures may even be expensed. Tax holidays, periods in which no taxes are assessed, are also written in tax codes of certain countries and may often be negotiated where they are not. Mining is often regarded as a strategically desirable industry, i.e., development of natural resources has positive implications for growth, security, power, and independence. One extreme example of a tax holiday is gold production in Australia. Australia currently has no tax on income from gold mines. The same principle of stand-alone vs. integrated may also apply in foreign countries and should be checked. Most

countries allow mining income to be consolidated, i.e., profits from one mining project can offset losses from another. Many countries also allow mining and nonmining projects to be consolidated. (This is dependent on the structure of the firm's operations in that country.) Loss carryforward and carryback provisions should also be ascertained. The allowed time periods are different throughout the world.

The concepts of depletion and investment tax credits (ITCs) should also be investigated. Certain countries have similar concepts; for example, Canada has earned depletion and resource allowance provisions and also ITCs. Another tax that is unique to foreign projects is the dividend withholding tax. Before funds can be repatriated to the US or another country, most countries impose a dividend withholding tax. This tax is not due until the funds are taken out of the country, rather than when earned. The impact of this additional assessment must be considered in the project's profitability.

One important aspect remains to be discussed relating to taxes on foreign projects, namely, the US tax liability associated with the project. If the taxes in the foreign country are lower than the tax would be in the US, the earnings will be subject to US taxes at some time. The US tax liability is calculated according to US tax laws, not foreign laws. Again, this is an area where tax advice is extremely valuable. Depending on the project structure, the tax implications vary widely. Although foreign tax rates are often higher than US and no additional liability will be incurred, certain structures, such as one with a US corporation as the joint venturer in the foreign project, allow percentage depletion when determining the US tax liability. The timing of the US tax liability is also affected by the structure. Taxes can be due as soon as the income is earned or not until it is repatriated. Since numerous structures are available, this area offers high potential for cash flow improvements.

Generally, assets outside the US do not qualify for ACRS depreciation rates per the 1982 TEFRA legislation. Thus, when computing the US liability, slower depreciation rates must be applied. Double declining balance methods are generally allowed.

A further consideration is the concept of creditable vs. deductible foreign taxes. Creditable taxes are those taxes paid to foreign governments that can be used to offset a firm's US tax liability dollar for dollar. Many foreign taxes, such as Canadian federal and provincial income taxes, qualify. The general rule is that the tax must be an income, war profits, excess profits, or dividend withholding tax. However, these definitions are broad, and interpretation from one's adviser and/or the IRS is suggested. Deductible taxes are those that only can be deducted as a project expense for the US tax liability calculations, i.e., they are treated the same as an operating expense. Thus this tax does not offer as significant an advantage as a creditable tax. The Ontario, Canada, mining tax is an example of a deductible tax.

Another consideration for a US-based multinational firm with several foreign projects is the concept of tax pools. If the firm has projects in several countries, worldwide income can be consolidated into tax pools, i.e., a firm can combine earnings from all foreign countries. Further this is not limited to mining projects, but applies to all foreign income. Thus, if in one country, the company is paying foreign taxes greater than its associated US tax liability, and in another country the foreign taxes paid are less than the US tax liability, the one project will generate excess foreign tax credits which can be utilized to offset the US tax liability from the other project. Carryforwards and carrybacks are also allowed on these credits. For reference, the current US regulations allow a carryback of two years and carryforward of five years. The following table shows a simple example of this concept. In reality, the concept is more complex, and the implications of the project on a firm's income should be checked by the tax adviser. This example also assumes the deductions for US and foreign tax calculations are identical, which is seldom the case.

Year 1

	Country A Tax Rate 30%	Country B Tax Rate 60%	Total Foreign	US Basis Tax Rate 46%
Revenue	100	130	230	230
Costs (including noncash)	70	60	130	130
Pre-tax income	30	70	100	100
taxes	9	42	51	46
After-tax income	21	28	49	

A credit of $5 ($51 vs. $46) is generated to be carried forward or back in this example. Even though the tax rate in Country A is only 30%, the higher tax rate of 60% in Country B prevents this firm from owing any additional liability on Country A's income. Further, a credit is generated which could be partially used in the next year. An example of one way this could occur follows:

Year 2

	Country A	Country B	Total Foreign	US Basis
Revenue	250	130	380	380
Costs (including noncash)	170	60	230	230
Pre-tax income	80	70	150	150
taxes	24	42	66	69
After-tax income	56	28	84	

The US tax liability is $69 whereas foreign income taxes paid are only $66. Thus, the $3 due to the US can be offset by the previous year's $5 credit, and a credit of $2 would still be available.

Other Considerations

Currency values and convertibility of currency are very important parameters. However, it is impossible to predict future currency relationships. Government changes, shifts in demand patterns, oil discoveries, and numerous other factors enter into this component. Noticeable depreciation and appreciation of currencies are very common during mine life periods. This can be particularly important when a firm sells its products in one currency and makes purchases in another. Most readers are familiar with the hyperinflation that has occurred in several South American countries in the 1970s and 1980s. However, exchange rates change widely even among politically stable countries. Consider the Japanese yen vs. the US dollar. The rate has varied from about 170 to 360 yen per dollar from 1970 to the present.

Inflation is a related yet additional aspect. Differential inflation rates among countries can offset changes in currency exchange rates. However, fixed rates preclude this adjustment, and other factors also enter into the relationship. Despite the uncertain nature of foreign currency values and

inflation, some assumptions on both exchange rates and inflation must be made for evaluation purposes. Historical data is helpful, especially for noticing trends. However, this is an extremely uncertain area for evaluation purposes.

Convertibility of foreign currency is another consideration. This relates to repatriation of cash flow to the US and also the currency value at which the funds are repatriated. Often, a country will not allow or only allow a portion of a project's cash flow to be taken out of the country or taken out at a fixed unrealistic rate. If this is the case, the project's cash flow is vastly different from the theoretical cash flow shown by the calculation. Again, this is an important provision in the structuring of the agreement. Contacts with other firms who have dealt with the specific country can be valuable.

Expropriation is another risk in foreign projects, i.e., a firm may be forced out of a country. One way to evaluate this event is to determine the project return assuming some fraction of the project (e.g., 10 years cash flow on a 20-year project) or use the same technique but add a buy-out sum.

To clarify the foreign project discussion, an example was created for the hypothetical project and is shown in the Appendix. No US taxes are ever due since the foreign tax liability always exceeds the US liability. The assumptions for depreciation, tax rates, exchange rates, and other salient factors are shown. This case also assumes the company has no other foreign projects with which to consolidate the project, either in that country or for US tax purposes.

OTHER FINANCIAL/ECONOMICS CONSIDERATIONS

In addition to equity investment decision analyses, a firm often has numerous other financial decisions. Examples include asset sale or disinvestment opportunities, mergers and acquisitions, farm-in/farm-out of properties, shutdowns, rate acceleration projects, post-investment evaluations, and leasing decisions. Discounted cash flow methods are applicable to all of these decisions, but additional analysis is also necessary. This section will highlight the major issues and provide guidance as to important parameters.

Disinvestment/Acquisition

First consider disinvestment, also known as redeployment, or the firm's disposal of a project. What criteria should be used? The first is to evaluate the property on a point-forward basis. The future cash flow of the project should be determined and discounted at some hurdle rate. This rate is often set lower than the investment hurdle rate. If the project has any debt associated with it, that amount should be subtracted from the value. Likewise, working capital at the time of disposal should be considered. If positive, working capital could add value. Major items on the accounts should also be evaluated. For example, some receivables may need to be written off as bad debt. Likewise the value of spare parts may have increased due to inflation. Short-term debt should not be double-counted in the debt components. Tax consequences are also very important. Sale of an asset usually generates taxes, and the amount generated can vary depending on the nature of the sale transaction. Among items to be considered are depreciation recapture, ITC recapture, and capital gains taxes. Once the after-tax value of the asset is determined, a firm has an idea of an acceptable price. However, strategic considerations are often important and should not be overlooked. If this project has nonquantifiable synergistic effects with other projects, marketing programs, or another area of the firm's business, these factors should be weighed in the decision.

Mergers or acquisitions are similar to disinvestment except the firm is the buyer rather than the seller. Again, a discounted cash flow method is appropriate to determine the value of the asset. Other performance measures, such as competitive position and market share, can be ascertained. Factors such as strategic goals and cash availability are also important. A firm has some flexibility here on how to acquire an asset. Different methods are available such as stock for stock, stock for assets, etc., and can result in different depreciable bases for tax books (a real dollar effect). Debt and working capital should again be quantified from the purchaser's side. Potential liabilities, e.g., pensions, lawsuits, environmental problems, should also be assessed.

Farm-in/farm-out analysis is a form of acquisition/divestment. Payments for purchase or sale occur over time in the form of capital investments, options, royalties, lump sum amounts, percentages of income or cash flow, or other varied methods. Here again, the cash flow streams for one's firm have to be separated from the total project cash flow. The present value at desired discount rates and/or DCF-ROI can then be calculated. If these are revenue or income dependent, price sensitivity analysis is highly desirable.

Shutdown

Shutdown decisions are often only near-term decisions but may be long term if a mine's reserves are almost depleted or the mine occupies a very high cost position in a glutted market. Shutdown, care and maintenance, and startup costs should be determined and compared to the operating cash flow for the period of time in question. Several sets of shutdown time frames can be evaluated to consider shorter and longer economic recovery periods. All costs that will be ongoing, such as property taxes, fixed power contract base amounts, or other fixed costs should be included in the shutdown cash flow. If a permanent shutdown is being considered, tax and salvage effects can be very important. Assets can be sold or written off, but reclamation may also be required. A determination of these amounts and their tax effects is needed.

Rate Acceleration

Rate acceleration analysis involves whether to increase the size or move forward the timing of a project. The opportunity involves more cash in early years at the expense of foregoing cash in later years. Thus, two cash flows can be determined and often two DCF-ROIs result. One method of handling this is to compare the present value for acceleration vs. nonacceleration. This should meet the minimum corporate present value hurdle rate. More sophisticated analyses use the double DCF-ROI concept. The upper return should be compared with the firm's DCF-ROI hurdle rate (i.e., the project can compete with other company investment opportunities). The lower return is the rate the firm would be willing to pay an outside party to fund that project. As this is a very brief discussion of rate acceleration, further reading on this subject is recommended if utilization of this analysis is desired.

Post-Investment

Post-investment analysis, the evaluation of an ongoing project vs. its expected performance at the time of acceptance, is often overlooked despite its significant value as a tool for future decision making. This involves a search through a firm's books to determine the actual performance. The future for the project should also be assessed in the current environment. Often this analysis provides management with disinvestment candidates, shows areas in which strategic mistakes were made, or points out a changing environment for future project analysis and direction for the firm.

Leasing Decisions

Another area of economic decision-making a firm may face is the lease vs. buy or lease vs. do-nothing analysis. If an asset in an investment proposal (or even the total proposal) may be either purchased or leased, the investment evaluation must first be performed to see if the asset should be acquired at all. This decision is based on the firm's hurdle rate and other competing investment opportunities. If this analysis indicates that the asset should be acquired, then the next step is to perform the lease vs. buy evaluation to determine *how* the asset should be acquired, i.e., lease or purchase. The after-tax cash flow associated with the lease should be determined and subtracted from the purchase cash flow and the discount rate determined. (For simplicity, costs identical to each case need not be included since the net effect is zero.) This will equate to the implicit interest rate associated with the lease. The relevant criterion to judge this return is based on the after-tax cost of corporate long-term debt. (Leases can be separated into categories, e.g., financing and operating, with separate debt rates set for each.) Table 5 presents an example of this analysis. This example only includes the costs/savings that are different for each case.

Table 5

	Year	Buy	
Purchase	0	(1000)	
Depreciation savings/ITC	1	171	
Depreciation savings	2	104	
Depreciation savings	3	100	
Depreciation savings	4	100	
Depreciation savings	5	100	
Total		(425)	

	Year	Lease	
Lease cost	0	—	(1000)
Lease cost	1	(130)	301
Lease cost	2	(130)	234
Lease cost	3	(130)	230
Lease cost	4	(130)	230
Lease cost	5	(130)	230
Total		(650)	225

Buy less lease costs	Year	
	0	(1000)
	1	301
	2	234
	3	230
	4	230
	5	230
Total		225

Assumptions:
1. 5-year asset life/5-year lease
2. ACRS depreciation (95%)
3. 10% ITC on purchased asset in year 1
4. $1000 purchase price
5. Annual lease cost of $260 (before tax)
6. 50% income tax rate
7. All other revenues and costs same for both lease and purchase

Result: DCF-ROI = 7% (Mid-year continuous discounting)

The example yields a DCF-ROI of 7%. If the after-tax cost of debt for the firm is greater than this amount, the asset should be leased. The project life used in both the lease and purchase cases must be equivalent for the analysis to be correct. If lease payments extend beyond the project life, they should be included or replaced by an appropriate buy-out penalty. Similarly, if salvage or depreciation tax credits extend beyond the project life, they should also be included. If sensitivities are investigated, they should be performed on both the lease and buy evaluation on equivalent bases.

This calculation does not consider strategic factors. If a firm is capital short, leasing may be the only option. Leasing may also be desirable if the asset may soon become obsolete or if upgrading is possible.

The lease vs. do-nothing evaluation is more intangible. Two steps are required. First, a hypothetical purchase option that is equivalent to the lease option at the allowable cost of leasing should be developed. This can be accomplished using the lease vs. buy evaluation procedure. Second, an equity investment evaluation should be performed using the hypothetical purchase option. If the hypothetical purchase option is acceptable, then by definition the lease is acceptable since the lease is equivalent to the purchase.

CONCLUSIONS

The foregoing discussion was intended to acquaint readers with economic and financial analysis. An appendix is included with examples that should aid in performing the cash flow or other investment techniques discussed. Again, we would like to emphasize the quality of the input data and the importance of sensitivity analysis. A DCF-ROI based on scoping data may be highly inaccurate. Price assumptions are often the most important. Low metals prices have caused the downfall of many projects and greatly affected the financial position of many firms.

APPENDIX
ASSUMPTIONS
CONSTANT DOLLAR CASE

Production

10 year life

	Year 4	Years 5-13
Copper (million lb)	75	100
Gold (thousand oz.)	30	40

First year production figures recognize startup inefficiencies—production at 75% of capacity.

Prices

Copper: $1.00 per pound
Gold: $500 per ounce

Operating Costs

	$ million/($/lb)			
	Year 4		Years 5-13	
Mining	20.0	(.27)	20.0	(.20)
Milling	15.0	(.20)	15.0	(.15)
Treatment	22.5	(.30)	30.0	(.30)
Administration	5.0	(.06)	5.0	(.05)
	62.5	(.83)	70.0	(.70)

Mining, milling, and administration costs were assumed to be the same for the startup year (year 4) to reflect inefficiencies. Treatment charges based on actual pounds treated.

Capital Costs

	$ million			
	Year 1	Year 2	Year 3	Total
Stripping and other expensed items	5	20	5	30
Mining equipment	10	10	10	30
Mill, crusher, other facilities	20	30	40	90
Total	35	60	55	150

All prior exploration and development costs are assumed to be sunk costs and thus ignored. Sustaining investment of $2 million in years 5 and 6, and $4 million in years 6, 7, and 8.

Tax/Depreciation

Development:	85% expensed in year of expenditure 15% capitalized — 5 year ACRS depreciation on 95% upon production.
Mine, Mill, Facilities:	5 year ACRS depreciation on 95% of expenditures. Depreciation begins when assets are placed in service, i.e., mining equipment depreciation starts in year 1. This is a simplifying assumption since some assets enjoy faster depreciation and others (e.g., buildings) have slower rates. 95% rate is used since 10% ITC is taken.
Tax Rate:	50% tax rate upon startup 46% tax rate in development years 50% rate (years 1-3) assumes both federal and state taxes. Lower tax rate assumed in development years since many state taxes are based on revenues and thus credits are not available in development years.
Investment Tax Credit:	10% in year of expenditure Assets not placed in service immediately are assumed to have two year or longer construction periods and thus qualify.
Depletion:	15% of revenue less treatment charge up to a maximum of 50% of pretax income.

The tax assumptions used in this example assume the tax losses can be consolidated with the firm's other income for US tax purposes, i.e., the firm is profitable (this is not a stand-alone case).

Working Capital

Assumed to be two months of revenues. This is a simplifying assumption and reflects spare parts inventory plus other effects, such as timing differences in payables and

Table A.1. Handbook Copper Company
Constant Dollar Case
(All figures in millions of dollars unless otherwise indicated.)

Year	1	2	3	4	5	6	7	8	9	10	11	12	13	Total
Production														
Copper, mm lb	—	—	—	75.0	100.0	100.0	100.0	100.0	100.0	100.0	100.0	100.0	100.0	975.0
Gold, m oz.	—	—	—	30.0	40.0	40.0	40.0	40.0	40.0	40.0	40.0	40.0	40.0	390.0
Price														
Copper, $/lb	—	—	—	1.00	1.00	1.00	1.00	1.00	1.00	1.00	1.00	1.00	1.00	—
Gold, $/oz.	—	—	—	500.00	500.00	500.00	500.00	500.00	500.00	500.00	500.00	500.00	500.00	—
Revenue														
Copper	—	—	—	75.0	100.0	100.0	100.0	100.0	100.0	100.0	100.0	100.0	100.0	975.0
Gold	—	—	—	15.0	20.0	20.0	20.0	20.0	20.0	20.0	20.0	20.0	20.0	195.0
Total	—	—	—	90.0	120.0	120.0	120.0	120.0	120.0	120.0	120.0	120.0	120.0	1170.0
Treatment charges	—	—	—	22.5	30.0	30.0	30.0	30.0	30.0	30.0	30.0	30.0	30.0	292.5
Net smelter return	—	—	—	67.5	90.0	90.0	90.0	90.0	90.0	90.0	90.0	90.0	90.0	877.5
Mining	—	—	—	20.0	20.0	20.0	20.0	20.0	20.0	20.0	20.0	20.0	20.0	200.0
Milling	—	—	—	15.0	15.0	15.0	15.0	15.0	15.0	15.0	15.0	15.0	15.0	150.0
Administration	—	—	—	5.0	5.0	5.0	5.0	5.0	5.0	5.0	5.0	5.0	5.0	50.0
Expensed capital	4.3	17.0	4.2	—	—	—	—	—	—	—	—	—	—	25.5
Tax book depreciation	1.4	3.5	5.5	19.6	26.1	23.5	22.2	21.1	3.0	2.8	2.4	1.6	0.8	133.5
Depletion base income	(5.7)	(20.5)	(9.7)	7.9	23.9	26.5	27.8	28.9	47.0	47.2	47.6	48.4	49.2	318.5
Tax book depletion	—	—	—	4.0	11.9	13.2	13.5	13.5	13.5	13.5	13.5	13.5	13.5	123.6
Taxable income	(5.7)	(20.5)	(9.7)	3.9	12.0	13.3	14.3	15.4	33.5	33.7	34.1	34.9	35.7	194.9
Taxes	(2.6)	(9.4)	(4.5)	2.0	6.0	6.6	7.2	7.7	16.7	16.8	17.0	17.5	17.8	98.8
Investment tax credit	3.1	4.3	5.1	—	0.2	0.2	0.4	0.4	0.4	—	—	—	—	14.1
After-tax income	0.0	(6.8)	(0.1)	1.9	6.2	6.9	7.5	8.1	17.2	16.9	17.1	17.4	17.9	110.2
Plus: Depreciation expensed capital	5.7	20.5	9.7	19.6	26.1	23.5	22.2	21.1	3.0	2.8	2.4	1.6	0.8	159.0
Depletion	—	—	—	4.0	11.9	13.2	13.5	13.5	13.5	13.5	13.5	13.5	13.5	123.6
Less: Capital expenditures	35.0	60.0	55.0	—	2.0	2.0	4.0	4.0	4.0	—	—	—	—	166.0
Working capital	—	—	—	15.0	5.0	—	—	—	—	—	—	—	(20.0)	—
Net cash flow	(29.3)	(46.3)	(45.4)	10.5	37.2	41.6	39.2	38.7	29.7	33.2	33.0	32.5	52.2	226.8
Cumulative cash flow	(29.3)	(75.6)	(121.0)	(110.5)	(73.3)	(31.7)	7.5	46.2	75.9	109.1	142.1	174.6	226.8	226.8
DCF-ROI, %	—	—	—	—	—	—	1.6	7.7	10.8	13.1	14.7	15.8	17.1	17.1

receivables. Note working capital increases from year 4 to 5 as revenues increase. All is recovered at end of project.

Royalties

None.

Inflation

None since this is a constant dollar example.

Economic Results

DCF-ROI, %:   17.1 (mid-year continuous discounting)

Present value, $ million
@ 13%:   26.3
@ 15%:   12.3

Payback period: year 7

Maximum cash out,
$ million: 121.0 in year 3

See Table A.1.

ASSUMPTIONS
FINANCIAL BOOKS — ESCALATED DOLLAR CASE

Production, Prices, Operating Costs, Capital Costs, Working Capital, Royalties, Inflation, Economic Results

Equivalent to escalated dollar case.

Tax/Depreciation

Development, mine, mill, facilities: 10 year straight line depreciation on 100% of asset value.

Sustaining investment: 5 year straight line depreciation on 100% of asset value.

Current taxes: Same as tax books. 46% rate prior to startup. 50% rate upon startup. Some investment tax credit and depletion assumptions.

Deferred taxes: Calculated as the difference between tax book and financial book depreciation times the tax rate in that year. The only years that deviate are years 1 and 13. Since only 95% of assets are depreciated on the tax books, the remaining 5% is added to the first year tax book depreciation for deferred tax calculations. Likewise, the difference between the 46% and 50% tax rates generates a difference which is balanced out at the end of the project.

Return on capital: Ratio of financial book income to average capital employed. Average capital employed is the average of the beginning of year capital employed and the end of year capital employed. Capital employed at the beginning of the project is zero (nothing has been invested to date). End of year capital employed is net income less net cash flow. For example, year 1 has a beginning of year capital employed of zero and an end of year capital employed of 3.3 − (−31.0) = 34.3 for an average capital employed of $17.5 million. Also capital employed at the end of the project is also zero.

See Table A.2.

ASSUMPTIONS
FOREIGN PROJECT ESCALATED DOLLAR CASE

Production/Royalties

Same as constant dollar example.

Prices, Operating Costs, Capital Costs

Base level same as constant dollar example.

Inflation

Same as escalated dollar case. This assumes inflation is the same in the foreign country as in the United States. Six percent per year for all capital costs, operating costs, and prices.

Working Capital

Same as escalated dollar case. Assumed to be two months of revenues.

Exchange Rate

Figures shown in US dollars. Exchange rate held constant throughout, i.e., dollar and foreign currency value relationship is the same throughout the project.

Tax/Depreciation

Development, mine, mill, facilities: Eight year doubling declining balance depreciation converting to eight year straight line depreciation when latter exceeds the former.
Depreciation begins upon production.

Sustaining investment: Five year doubling declining balance depreciation converting to five year straight line depreciation when latter exceeds the former.

Foreign taxes: 60% upon startup. Tax losses carried forward and utilized against profits.
No investment tax credits.
15% withholding tax on dividends.
Assumes dividends repatriated as earned.

Depletion: None.

US Taxes: None due since foreign liability exceeds US liability.

Table A.2. Handbook Copper Company
Escalated Dollar Case
(All figures in millions of dollars unless otherwise indicated.)

Year	1	2	3	4	5	6	7	8	9	10	11	12	13	Total
Production														
Copper, mm lb	—	—	—	75.0	100.0	100.0	100.0	100.0	100.0	100.0	100.0	100.0	100.0	975.0
Gold, m oz.	—	—	—	30.0	40.0	40.0	40.0	40.0	40.0	40.0	40.0	40.0	40.0	390.0
Price														
Copper, $/lb	1.06	1.12	1.19	1.26	1.34	1.42	1.50	1.59	1.69	1.79	1.90	2.01	2.13	—
Gold, $/oz.	530.00	562.00	596.00	631.00	669.00	709.00	752.00	797.00	845.00	895.00	949.00	1006.00	1066.00	—
Revenue														
Copper	—	—	—	94.5	134.0	142.0	150.0	159.0	169.0	179.0	190.0	201.0	213.0	1631.5
Gold	—	—	—	18.9	26.8	28.4	30.1	31.9	33.8	35.8	38.0	40.2	42.6	326.5
Total	—	—	—	113.4	160.8	170.4	180.1	190.9	202.8	214.8	228.0	241.2	255.6	1958.0
Treatment charges	—	—	—	28.4	40.1	42.6	45.1	47.8	50.7	53.7	56.9	60.4	64.0	489.7
Net smelter return	—	—	—	85.0	120.7	127.8	135.0	143.1	152.1	161.1	171.1	180.8	191.6	1468.3
Mining	—	—	—	25.2	26.8	28.4	30.1	31.9	33.8	35.8	38.0	40.2	42.7	332.9
Milling	—	—	—	18.9	20.1	21.3	22.6	23.9	25.3	26.9	28.5	30.2	32.0	249.7
Administration	—	—	—	6.3	6.7	7.1	7.5	8.0	8.4	9.0	9.5	10.1	10.7	83.3
Expensed capital	4.5	19.1	5.1	—	—	—	—	—	—	—	—	—	—	28.7
Tax book depreciation	1.5	3.8	6.2	22.2	29.6	27.0	25.8	24.7	4.6	4.5	3.8	2.6	1.3	157.6
Depletion base income	(6.0)	(22.9)	(11.3)	12.4	37.5	44.0	49.0	54.6	80.0	84.9	91.3	97.7	104.9	616.1
Tax book depletion	(6.0)	(22.9)	—	6.2	18.1	19.2	20.3	21.5	22.8	24.2	25.7	27.1	28.7	213.8
Taxable income	(6.0)	(22.9)	(11.3)	6.2	19.4	24.8	28.7	33.1	57.2	60.7	65.6	70.6	76.2	402.3
Taxes	(2.8)	(10.5)	(5.2)	3.1	9.7	12.4	14.3	16.6	28.6	30.3	32.8	35.3	38.1	202.7
Investment tax credit	3.3	4.8	6.0	—	0.3	0.3	0.6	0.6	0.7	—	—	—	—	16.6
After-tax income	0.1	(7.6)	(0.1)	3.1	10.0	12.7	15.0	17.1	29.3	30.4	32.8	35.3	38.1	216.2
Plus: Depreciation expensed capital	6.0	22.9	11.3	22.2	29.6	27.0	25.8	24.7	4.6	4.5	3.8	2.6	1.3	186.3
Depletion	—	—	—	6.2	18.1	19.2	20.3	21.5	22.8	24.2	25.7	27.1	28.7	213.8
Less: Capital expenditures	37.1	67.4	65.5	18.9	2.7	2.8	6.0	6.4	6.8	—	—	—	(40.2)	194.7
Working capital	—	—	—	—	7.9	1.6	1.6	1.8	2.0	2.0	2.2	2.2	—	—
Net cash flow	(31.0)	(52.1)	(54.3)	12.6	47.1	54.5	53.5	55.1	47.9	57.1	60.1	62.8	108.3	421.6
Cumulative cash flow	(31.0)	(83.1)	(137.4)	(124.8)	(77.7)	(23.2)	30.3	85.4	133.3	190.4	250.5	313.3	421.6	—
DCF-ROI, %	—	—	—	—	—	—	5.4	11.5	14.8	17.3	19.1	20.3	21.8	—

Table A.3. Handbook Copper Company
Foreign Project—Escalated Dollar Case
(All figures in millions of dollars unless otherwise indicated.)

Year	1	2	3	4	5	6	7	8	9	10	11	12	13	Total
Production														
Copper, mm lb	—	—	—	75.0	100.0	100.0	100.0	100.0	100.0	100.0	100.0	100.0	100.0	975.0
Gold, m oz.	—	—	—	30.0	40.0	40.0	40.0	40.0	40.0	40.0	40.0	40.0	40.0	390.0
Price														
Copper, $/lb	1.06	1.12	1.19	1.26	1.34	1.42	1.50	1.59	1.69	1.79	1.90	2.01	2.13	—
Gold, $/oz.	530.00	562.00	596.00	631.00	669.00	709.00	752.00	797.00	845.00	895.00	949.00	1006.00	1066.00	—
Revenue														
Copper	—	—	—	94.5	134.0	142.0	150.0	159.0	169.0	179.0	190.0	201.0	213.0	1631.5
Gold	—	—	—	18.9	26.8	28.4	30.1	31.9	33.8	35.8	38.0	40.2	42.6	326.5
Total	—	—	—	113.4	160.8	170.4	180.1	190.9	202.8	214.8	228.0	241.2	255.6	1958.0
Treatment charges	—	—	—	28.4	40.1	42.6	45.1	47.8	50.7	53.7	56.9	60.4	64.0	489.7
Net smelter return	—	—	—	85.0	120.7	127.8	135.0	143.1	152.1	161.1	171.1	180.8	191.6	1468.3
Mining	—	—	—	25.2	26.8	28.4	30.1	31.9	33.8	35.8	38.0	40.2	42.7	332.9
Milling	—	—	—	18.9	20.1	21.3	22.6	23.9	25.3	26.9	28.5	30.2	32.0	249.7
Administration	—	—	—	6.3	6.7	7.1	7.5	8.0	8.4	9.0	9.5	10.1	10.7	83.3
Tax book depreciation	—	—	—	42.5	33.0	25.6	21.4	18.2	19.2	16.9	15.7	1.5	0.7	194.7
Taxable income	—	—	—	(7.9)	34.1	45.4	53.4	61.1	65.4	72.5	79.4	98.8	105.5	607.7
Taxes	—	—	—	—	15.7	27.2	32.0	36.7	39.2	43.5	47.6	59.3	63.3	364.5
After-tax income	—	—	—	(7.9)	18.4	18.2	21.4	24.4	26.2	29.0	31.8	39.5	42.2	243.2
Plus: Depreciation	—	—	—	42.5	33.0	25.6	21.4	18.2	19.2	16.9	15.7	1.5	0.7	194.7
Less: Capital expenditures	37.1	67.4	65.5	18.9	2.7	2.8	6.0	6.4	6.8	—	—	—	—	213.6
Working capital	—	—	—	—	7.9	1.6	1.6	1.8	2.0	2.0	2.2	2.2	(40.2)	(18.9)
Net cash flow in country	(37.1)	(67.4)	(65.5)	15.7	40.8	39.4	35.2	34.4	36.6	43.9	45.3	38.8	83.1	243.2
Dividend withholding tax	—	—	—	2.4	6.1	5.9	5.3	5.2	5.5	6.6	6.8	5.8	12.5	62.1
Net cash flow to US	(37.1)	(67.4)	(65.5)	13.3	34.7	33.5	29.9	29.2	31.1	37.3	38.5	33.0	70.6	181.1
Cumulative cash flow	(37.1)	(104.5)	(170.0)	(156.7)	(122.0)	(88.5)	(58.6)	(29.4)	1.7	39.0	77.5	110.5	181.1	—
DCF-ROI, %	—	—	—	—	—	—	—	—	0.2	4.1	6.8	8.4	10.8	—

SURFACE MINING

Table A.4. Handbook Copper Company
Financial Books—Escalated Dollar Case
(All figures in millions of dollars unless otherwise indicated.)

Year	1	2	3	4	5	6	7	8	9	10	11	12	13	Total
Production														
Copper, mm lb	—	—	—	75.0	100.0	100.0	100.0	100.0	100.0	100.0	100.0	100.0	100.0	975.0
Gold, m oz.	—	—	—	30.0	40.0	40.0	40.0	40.0	40.0	40.0	40.0	40.0	40.0	390.0
Price														
Copper, $/lb	1.06	1.12	1.19	1.26	1.34	1.42	1.50	1.59	1.69	1.79	1.90	2.01	2.13	—
Gold, $/oz.	530.00	562.00	596.00	631.00	669.00	709.00	752.00	797.00	845.00	895.00	949.00	1006.00	1066.00	—
Revenue														
Copper	—	—	—	94.5	134.0	142.0	150.0	159.0	169.0	179.0	190.0	201.0	213.0	1631.5
Gold	—	—	—	18.9	26.8	28.4	30.1	31.9	33.8	35.8	38.0	40.2	42.6	326.5
Total	—	—	—	113.4	160.8	170.4	180.1	190.9	202.8	214.8	228.0	241.2	255.6	1958.0
Treatment charges	—	—	—	28.4	40.1	42.6	45.1	47.8	50.7	53.7	56.9	60.4	64.0	489.7
Net smelter return	—	—	—	85.0	120.7	127.8	135.0	143.1	152.1	161.1	171.1	180.8	191.6	1468.3
Mining	—	—	—	25.2	26.8	28.4	30.1	31.9	33.8	35.8	38.0	40.2	42.7	332.9
Milling	—	—	—	18.9	20.1	21.3	22.6	23.9	25.3	26.9	28.5	30.2	32.0	249.7
Administration	—	—	—	6.3	6.7	7.1	7.5	8.0	8.4	9.0	9.5	10.1	10.7	83.3
Financial book depreciation	—	—	—	17.0	17.5	18.1	19.3	20.6	21.9	21.4	20.9	19.6	18.4	194.7
Before-tax income	—	—	—	17.6	49.6	52.9	55.5	58.7	62.7	68.0	74.2	80.7	87.8	607.7
Taxes: Current	(2.8)	(10.5)	(5.2)	3.1	9.7	12.4	14.3	16.6	28.6	30.3	32.8	35.3	38.1	202.7
Deferred	2.8	10.5	5.2	6.8	6.0	4.5	3.3	2.0	(8.7)	(8.4)	(8.6)	(8.5)	(6.9)	—
Investment tax credit	3.3	4.8	6.0	—	0.3	0.3	0.6	0.6	0.7	—	—	—	—	16.6
After-tax income	3.3	4.8	6.0	7.7	34.2	36.3	38.5	40.7	43.5	46.1	50.0	53.9	56.6	421.6
Plus: Depreciation	—	—	—	17.0	17.5	18.1	19.3	20.6	21.9	21.4	20.9	19.6	18.4	194.7
Deferred taxes	2.8	10.5	5.2	6.8	6.0	4.5	3.3	2.0	(8.7)	(8.4)	(8.6)	(8.5)	(6.9)	—
Less: Capital expenditures	37.1	67.4	65.5	18.9	2.7	2.8	6.0	6.4	6.8	—	—	—	—	194.7
Working capital	—	—	—	—	7.9	1.6	1.6	1.8	2.0	2.0	2.2	2.2	(40.2)	
Net cash flow	(31.0)	(52.1)	(54.3)	12.6	47.1	54.5	53.5	55.1	47.9	57.1	60.1	62.8	108.3	421.6
Cumulative cash flow	(31.0)	(83.1)	(137.4)	(124.8)	(77.7)	(23.2)	30.3	85.4	133.3	190.4	250.5	313.3	421.6	421.6
DCF-ROI, %	—	—	—	—	—	—	5.4	11.5	14.8	17.3	19.1	20.3	21.8	—
Return on capital, %	19.2	7.6	4.9	5.2	24.1	28.7	35.6	43.6	51.8	60.5	76.1	95.9	218.5	—

The tax assumptions used in this example assume the firm either has no operations in that country to utilize tax losses and/or that tax losses cannot be consolidated, i.e., this is a stand-alone example.

Economic Results

DCF-ROI, %:   10.8 (mid-year continuous discounting)

Present value, $ million
@ 13%: (17.3)
@ 15%: (30.0)

Payback period: year 9

Maximum cash out,
$ million: 170.0

Comments

The higher tax rates (both foreign and dividend withholding), no pre-production tax benefits, and lack of ITC and depletion cause the return to drop noticeably from equivalent US project (10.8% vs. 21.8%).

See Table A.3.

ASSUMPTIONS
ESCALATED DOLLAR CASE

Production, Tax/Depreciation, Royalties

Same as constant dollar example.

Prices, Operating Costs, Capital Costs

Base level same as constant dollar example.

Working Capital

Assumed to be two months of revenue as in constant dollar example. Working capital increases every year due to inflation of prices.

Inflation

Six percent per year for all capital costs, operating costs, and prices. As an example, year 1 copper prices are $1.06 per pound, gold costs are $540 per ounce, and capital costs are $37.1 million. The equivalent figures in the constant dollar example were $1.00 per pound, $500 per ounce, and $35 million.

Economic Results

DCF-ROI, %:   21.8 (mid-year continuous discounting)

Present value, $ million
@ 13%: 76.8
@ 15%: 53.1

Payback period: year 7

Maximum cash out,
$ million: 137.4 in year 3

Comments

The inflation causes the DCF-ROI to rise since all parameters are increasing at the same rate. This may not hold true if different factors escalate at different rates. Note also that the total cash flow is higher as is the maximum cash out.

See Table A.4.

BIBLIOGRAPHY

Anon., 1972, "A Computer Model to Upgrade Zinc Profits," *Business Week,* Aug. 26.

Bagnall, R. and Copithorne, L., 1973, "An Economic Analysis of the Nickel Industry (An Application of Linear Programming)," *Proceedings of the Council of Economics,* Feb., AIME, New York, pp. 113-121.

Berry, C.W., 1972, "A Wealth Growth Rate for Capital Investment Planning," Ph.D. Thesis, The Pennsylvania State University, University Park, PA.

Bunn, D.W. and Mustafaoglu, M.M., 1978, "Forecasting Political Risk," *Management Science,* Nov.

Copithorne, L., 1973, "A Linear Programming Model of the Australian Nickel Industry," Dept. of Economics, The Australian National University, Jan.

Gentry, D.W. and Hrebar, M.J., 1977, "Economic Principles for Coal Property Valuation," SME-AIME, Oct.

Gordon, M.J., 1974, *The Cost of Capital to a Public Utility,* Michigan State University, East Lansing.

Mustafaoglu, M.M., 1979, "Integration of Sudden Expropriation Risk Into Project Economics," *Trans. SPE-AIME,* Sept.

O'Neil, T.J., 1979, "Mine Development and Valuation," *Computer Methods for the 80's in the Mineral Industry,* A. Weiss, ed., AIME, New York.

O'Neil, T.J., 1982, "Mine Evaluation in a Changing Investment Climate: Part 1," *Mining Engineering,* Vol. 34, No. 11, Nov., pp. 1563-1566.

O'Neil, T.J., 1982, "Mine Evaluation in a Changing Investment Climate: Part 2," *Mining Engineering,* Vol. 34, No. 12, Dec., pp. 1669-1672.

Stermole, F.J., 1974, *Economic Evaluation and Investment Decision Methods,* 2nd ed., Investment Evaluations Corp., Golden, CO.

Whitney, J.W. and R.E., 1979, *Investment and Risk Analysis in the Minerals Industry,* Revision No. 2, Whitney & Whitney, Inc., Reno, NV.

The following tax guides are helpful:
1983 US Master Tax Guide, Commerce Clearing House, Inc., updated annually.
1982 Income Taxation of Natural Resources, Burke and Bowhay, updated annually.
Financial Reporting and Tax Practices in Nonferrous Mining, Coopers and Lybrand, updated periodically.
Price Waterhouse Tax Guides. Price Waterhouse prepares booklets on almost every country. Other major accounting firms have similar information.

4.4 Project Financing

GENERAL CONCEPTS

EDWARD VICKERS

Financing has always been an important consideration in resource development, but today it has reached the point where it may well be the controlling element in bringing an ore body into production. The proper application of financing to a project may be the difference between success or failure of the venture. Financing and technical considerations are not independent, but are inexorably linked. The engineer today must be aware of financing implications every step of the way, from the beginning to the completion of a feasibility study. This is not to suggest that the engineer assume the role of the corporate treasurer, but rather that the two arms of management must work in close concert to achieve the desired result.

The role of the mining company is shifting from its classic position of an organizer and provider of funds to that of a sponsor, providing the upfront risk money and managers of the project. In this environment, new partnerships are forged. These partnerships take on many forms, but certainly the lender is one of the most important partners. Other partnerships are also forged by the joining together of more than one mining company, offering host country participation, and, due to the heavy reliance on sales contracts, even involving the purchaser of the product. As a partnership takes form, a new project emerges as a separate entity, a new mining company in other words, relying heavily on its intrinsic worth to attract the capital needed to bring it into production.

Financing costs of a new venture are often equal to, or exceed, all of the other operating costs put together. These costs are real and demanding on the cash flow of the mine. It is no longer satisfactory for the engineer to address himself only to the technical issues that give rise to an optimum mine design. He must focus equally on the implications of this design on the financing of the ore body. If the cost of financing is not considered, the engineer may feel free to employ his skills constrained only by the characteristics of the ore body in the development of an optimal exploitation system. These design characteristics include scale of operation, annual output of metal, timing of both the construction and production. In the real world, however, each of these factors significantly affects the financing requirements, and the credit strength that the deposit can convey to the suppliers of capital. The highest possible profit from an ore body may well become subservient to the financing constraints. A mining venture of any particular size is only financially attractive if it reflects a desired profit and at the same time is able to service its debt obligations.

Let's begin with fundamentals. What is a banker going to look for? The process of credit evaluation begins with a detailed analysis of the project. Lenders look specifically to the realism of stated assumptions. This often requires independent validation of the major aspects of the venture. In particular the lenders will be concerned with the level of capital cost, contingency provisions, provisions for overruns, ore reserves, operating cost estimates, scope of the project, availability of labor and materials, lead time for delivering critical items of equipment, and possible infrastructure requirements. A great deal of time is spent by the lenders in evaluating price assumptions. The purpose of this analysis is to quantify the risk associated with the project's development. Lenders want to be assured firstly, that sufficient funds are available for the cost of the project, including any potential overruns, and secondly, that future cash flow generation will be sufficient to meet the assumed debt obligations.

Mine finance, whether it be general corporate borrowing, or project financing, is ultimately characterized as cash flow lending. Security instruments for the lender will be taken on the fixed assets of the property, including a lien on the reserves. This may be viewed as having great value from the mining company's perspective, but the lenders will place little reliance on it in terms of its liquidated value. Their focus will be on the expected cash flow from the property, and the reliability of such cash flow.

Against this backdrop, let us focus for a moment on what is often described as project financing. Many definitions exist, but primarily, it is the financing of a particular mine, in which the lender is looking principally at the cash flow and earnings from the mine as the source of funds from which the loan will be repaid, and secondarily, at the assets created by such financing as collateral for the loan. This differs from normal corporate borrowing, in that the funds are specifically linked to a project with the loan secured by the venture itself but not necessarily guaranteed by the mining company which is sponsoring the project.

The first step in the development of a financing plan for a particular project is the realistic determination of the amount of borrowing such a project can support. Traditionally debt equity ratios of 20 to 25% might well define the level of borrowing. To the degree that the lender is asked to take a higher risk, this ratio is more likely to be in the one third-two thirds range. The degree of leverage the project can withstand depends basically on two factors. First is the ability of the project to meet the scheduled debt repayment in a satisfactory manner. The second is to maintain a level of commitment on the part of the borrower that the lenders on a subjective basis feel is necessary to assure a serious commitment. The normal project loan will generally permit an initial grace period of from two to four years, during which time interest only will be payable on the debt. The total length of a loan will include this grace period. There is no exact rule as to the maturity of any given financing. It is a function of capital markets at the time, the risk inherent in the project, and a number of other considerations. The commercial banks will normally be providing funds for 8 to 12 years, including the grace period. Institutional lenders very often may be persuaded to go up to 15 years. These lenders, however, are much more security conscious and it thus becomes a tradeoff from the sponsor's point of view between maturity and credit support. The credit repayment terms will normally be a level amortization of the principal with a declining interest component as the debt is reduced.

The building block of project financing is basically the process of the allocation of credit or risk in a project. The most fundamental element of security is, of course, the ore body itself. Beyond this basic concept, a number of credit instruments and credit support considerations can be applied to give rise to a financeable project. Completion guarantees, working capital and deficiency agreements, requirements with respect to the minimal level of production, sales contracts, insurance coverage, all form a package that creates the financial strength necessary to meet the lending criteria.

In assessing the viability of the project, the lenders must first be convinced that the project will indeed be completed and performed according to the general specifications set forth in the feasibility study. This kind of assurance is normally provided, by the sponsor of the project, in the form of a completion and performance agreement. It is at this juncture that many projects without senior sponsors have their most severe problems. A junior mining company may well have a very attractive project with outstanding capabilities of producing a highly reliable cash flow, but not have sufficient financial wherewithal to assure the lenders that the project will be completed at the anticipated cost. Realism on this issue is fundamental if the project is to be viable.

Without creditworthy sponsors, some form of an association with an institution capable of supporting these undertakings will be essential. It is only after performance and completion has been achieved that the lenders will be relying on the project cash flow for repayment. Until that time, regardless of the problem, the lenders want others to shoulder those risks.

With performance and completion having been achieved, the lenders are now relying on the cash flow of the venture to provide the debt service requirements. This, however, will not be done on the basis of spot marketing of the product. With the growth of project financing, the role of sales contracts has increased substantially. A lender may be willing to take certain commercial risks but the ability to sell the product is not one of these. Long-term contracts are unquestionably one of the most important aspects of project lending. These contracts must have a number of special characteristics to be satisfactory to the lenders. By and large, they have to extend beyond the term of the lending, provide either a floor price or margins sufficient to insure debt service, and, of course, levels of production necessary to achieve such ends.

This arrangement takes the miner away from the more traditional spot sales and forces him to live with a consumer over a long time frame. As such, the character of long-term contracts has undergone significant changes. They must incorporate escalation provisions to insure that margins can be maintained. These generally are not a single price escalator but cover a whole range of cost factors with an attempt to keep them on a current cost basis. However carefully this is done, contemplating changes in the economic environment of a long time period is a difficult task. For final protection, many contracts have some form of a gross inequities provision. This has the effect at times of opening up a contract to the detriment of either the buyer or the seller. Similar provisions with respect to regulatory changes can result in an interruption in the flow of the product. All such changes or cancellation provisions become very worrisome to the lenders. This results in something of a tug-of-war over the various objectives sought in long-term contracts. The lenders will immediately seek out anything that can cancel the contract. To insure against cancellation, however, can often have negative ramifications on the economic attractiveness of a venture.

In the optimal design of a financing plan, a significant but relatively new force has come into play. This is the heavy emphasis on tax considerations. Perhaps more than in any other industry, the federal government is truly a partner in any mine development. Tax considerations can significantly affect the economics of a project. Tax implications are found in preproduction expense, depletion, depreciation, investment tax credits, and in the case of coal, royalty income. These considerations affect financing in that reducing taxes provides a greater cash flow for debt service. To take maximum advantage of tax benefits can have the net effect of significantly reducing financing costs.

A traditional form of financing that relied principally on tax timing considerations was lease financing. Leasing had its birth more in the transportation and computer field, but it is finding more and more markets in the mining structure, particularly coal. Many coal developers are simply not in a sufficiently high tax-paying position to make maximum use of investment tax credits and accelerated depreciation. Ideally suited for lease financing are large ticket items in open pit operations. Underground equipment can likewise often be leased effectively but its short life takes away some of the advantages. In the same way that lease financing is attractive because of unused investment tax credits and depreciation, other tax considerations can be employed in mine financing. It is beneficial if preproduction cost and depletion shields are in the hands of taxpaying entities, but to achieve this often involves transfer of mineral ownership. Such transferences certainly introduce a significant complication to financing, but the benefits in reducing financing costs can justify the time and effort involved. Conceptually, one can go as far as to consider structures similar to a sale and lease back to a total mine. Optimal use of tax considerations in mine financing is a somewhat specialized tax field with only a few players at this point in time. Tax-oriented financing structures will no doubt become increasingly important in mine developments.

A feasibility study is a complex undertaking requiring careful analysis of many variables to insure the most economic development of an ore body. It may take six to twelve months to complete and cost anywhere up to a million dollars. Financing considerations are often left to be accomplished, unfortunately, in a few short weeks, and often times by those unfamiliar with the implications of technical decisions already cast in stone. In light of its tremendous significance on the variability of the project, financing considerations should be highlighted and carefully considered throughout all aspects of the feasibility study. In the final analysis, it is the proper marriage of both technical and financial considerations that gives rise to a viable project.

PROJECT PLANNING AND THE FINANCING DECISION

Tomek Ulatowski

Introduction

The financing of any major capital-intensive project, is normally predicated on three factors: first, on the economic viability of such an undertaking; second, on credit terms and its structure; and third, on the general availability of credit. The examples used in this article demonstrate the relevance of concurrently considering these factors in designing an appropriate financing structure for a project.

To illustrate how sensitive new investments are to changes in financial variables, an interactive computer model has been built using hypothetical operating and capital cost values.*

* Content is based in part on "Importance of Financing in Project Planning," by T. Ulatowski, *Mining Engineering,* June 1978, and "Mine Financing: When Is the Best Time to Borrow?" by T. Ulatowski, E. Frohling, F.M. Lewis, *Engineering and Mining Journal,* May 1977.

In all simulations, operating costs were projected in constant terms. The significance of a particular financial variable change was equated in terms of resulting breakeven price per ton of product. It was assumed that the simulated project was self-standing, and it had to demonstrate economic viability on its own.

Significance of the Debt Coverage Ratio

One of the first steps typically taken by a lender in analyzing the financial viability of projects is to test the projected cash flow stream before debt service to the anticipated debt charges (interest and scheduled debt principal repayments). This ratio, calculated on an annual basis, is referred to as cash flow debt coverage ratio (DCR).

The lender must be satisfied that the projected revenue stream is not too optimistic and that the operating and capital costs are conservative. Since the intent of this article is to illustrate the importance of financial parameters on a project, sensitivities of nonfinancial variables are left out. Without doubt, however, they are equally important to any potential lender.

In most cases, lenders require that the debt coverage ratio in any given year be considerably above 1.0. The actual expected ratio will depend on a variety of factors such as, creditworthiness of the sponsors, strength and type of completion undertakings, product purchase/sale agreements, degree of leverage, country credit standing, level of confidence in the projections, sociopolitical considerations and others.

The impact of varying this criterion on the required price per ton of product is illustrated in Fig. 1. If the lender were satisfied with a DCR of 1.20 in any given year he would lend on the basis of a projected price of $73.5/ton in the first year of operations. On the other hand, should the lender

require the DCR to be 1.80, the price would have to be $115.3/ton. In most projects, lenders expect to have a projected debt coverage fall within the range of 1.20 to 1.60. Since the product price is independent and cannot differ significantly from the market, such increases in the required DCR could make the project unrealistic as the resulting breakeven price would vary too much from the lender's conservative forecast. Therefore, it is essential for project sponsors to develop high quality information that decreases risk and may lower required DCR. In summary, debt coverage ratio may be viewed as a quantitative instrument employed by creditors in measuring a project risk.

Impact of Leveraging on Project Economics

One way to improve the creditworthiness of a project is to lower the leverage, i.e., the ratio of total debt to equity. The results of the simulations in Fig. 1 were based on a debt/equity structure of 75 to 25. Fig. 2 demonstrates the dramatic improvement in financeability of a venture achieved by reducing the leverage ratio first to 50/50 and then to 25/75.

The minimum breakeven price in the latter case could be as low as $48.9/ton, assuming a DCR of 1.3, whereas it would have to be increased to $89.2/ton if the project were leveraged to 75/25.

There is a direct correlation between lender's risk, debt coverage, and leverage. It must be recognized that in extending credit, a typical lending institution is not in a position to accept a great deal of risk. Therefore, to attract financing, sponsors must structure projects in a manner that minimizes the lender's risk exposure. One way to accomplish this is to increase the effective amount of equity in the project through the injection of common equity and also, if necessary, of subordinated debt.

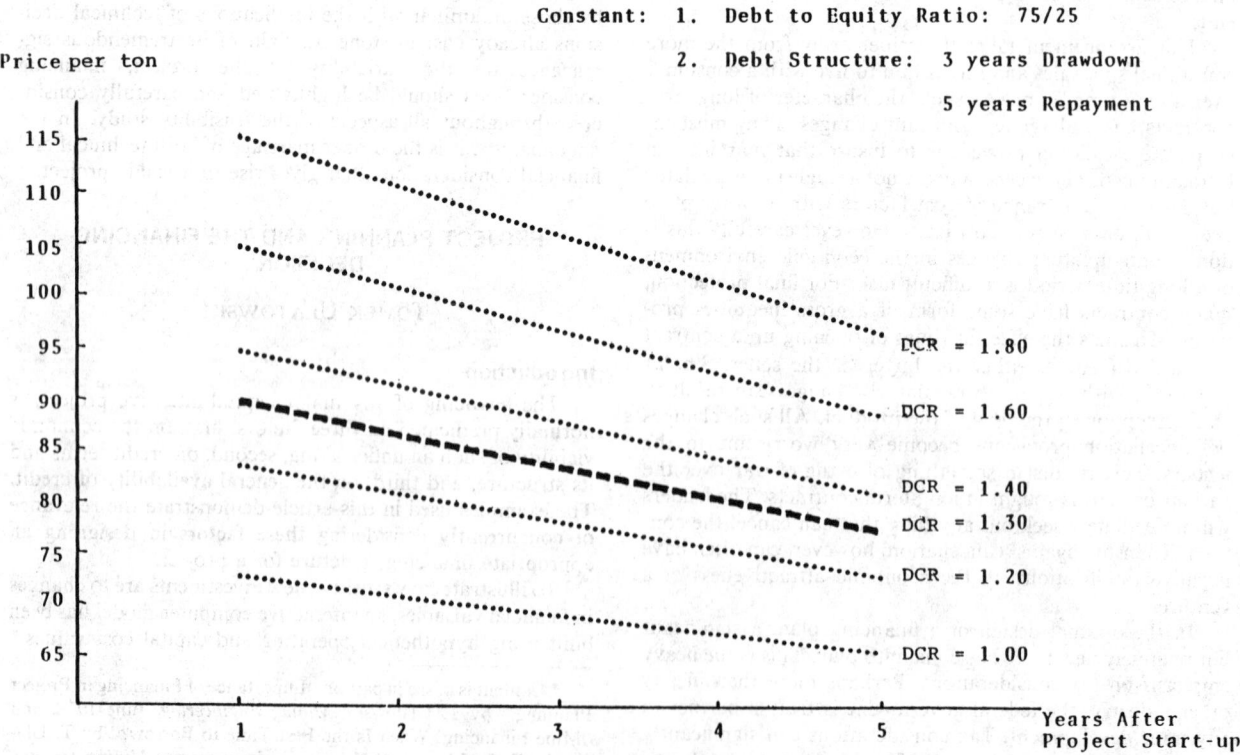

Fig. 1. Impact of debt coverage ratio (DCR) on price per ton.

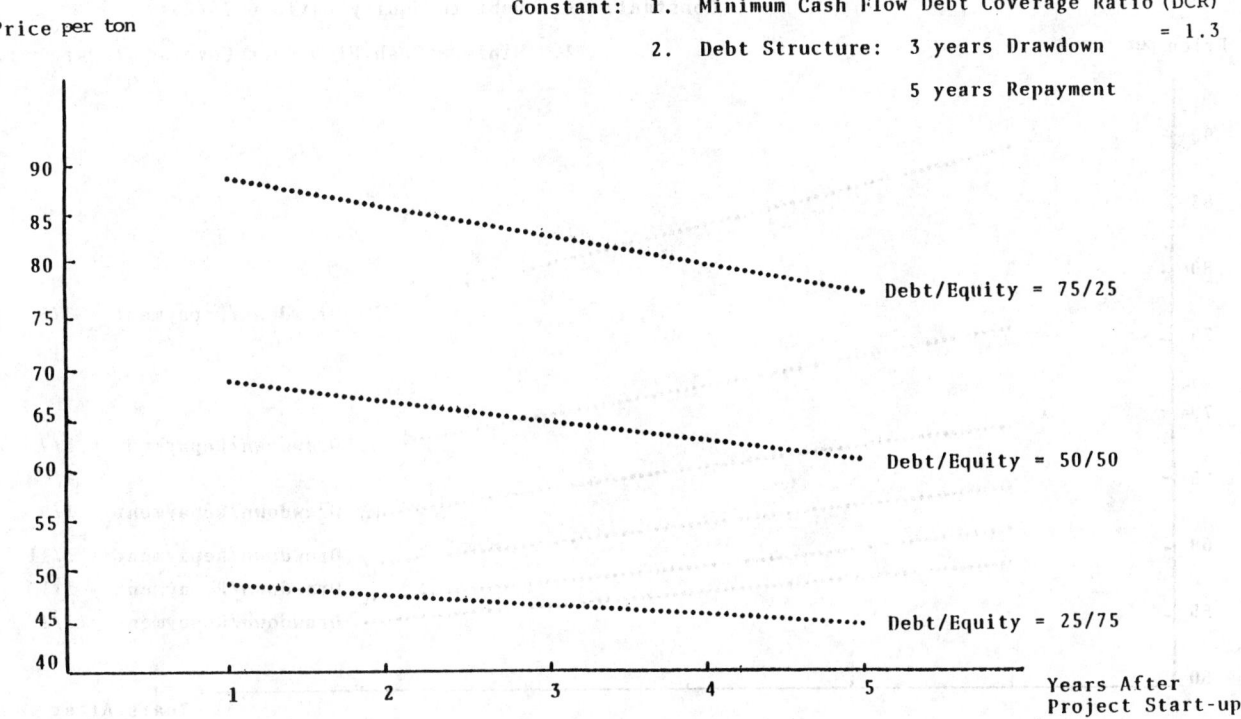

Fig. 2. Impact of leveraging on price per ton.

Debt Structure and Project Economics

Other things equal, projects in countries with healthy economies and access to extensive capital markets will attract the most desirable financing terms. For instance, a prime borrower in the United States today could place bonds with a maturity of up to 30 years. A similar borrower in a developing country would most likely have to be satisfied with commercial loans carrying overall tenor of no longer than 8 to 12 years from the time of obtaining commitment to the date of last installment repayment.

The implications of this for project sponsors are illustrated in Fig. 3. The longer the overall loan tenor and repayment period, the more attractive are the project economics and the more financeable is the venture. For example, by holding debt/equity constant at 75/25 and assuming the required DCR of 1.3, and then varying the period of debt repayment from 5 to 15 years, the minimum price required by lenders drops from $89.2/ton to $59.9/ton.

It is imperative, therefore, that the project sponsors evaluate all possible financing sources with great care, keeping in mind that the tenor of credit is one of the most crucial variables in establishing the viability of a new project.

Unanticipated Capital Cost Increases

One of the greatest areas of concern to a potential lender in financing a new project is the risk of unanticipated capital cost increases. The risk the lender faces is basically twofold: noncompletion, caused by insufficient funding or delay, and deteriorated economics resulting from unanticipated capital cost overruns. The first concern is normally dealt with via requirements that sponsors issue irrevocable and unconditional completion guarantees releasable only upon satisfaction of predetermined completion criteria. The second concern results from the fact that higher capital costs, even if funded by the guarantors, usually produce lower cash flows. From the standpoint of a lender if the unanticipated cost overruns are funded with senior loans, the actual DCR would deteriorate in relation to the projections and lenders' risk would, therefore, increase.

In Fig. 4, the basic capital cost (level at which the decision to proceed with the investments is made) was increased in 15% increments up to 60% above the original estimate. The impact of such cost increases is most pronounced on the price per ton lenders would expect in order to lend to the project. Using 75/25 leverage, eight-year tenor of the credit and a DCR of 1.30 the project could be financeable with the price of $89.2/ton, if there were no additional capital costs. On the other hand, if capital cost were to rise by 60%, the price required in the first year of operations would have to be $121.1/ton.

Timing of Borrowing

Fig. 5 reflects typical characteristics of a major project development: low expenditures in early phases with poor confidence level as to final project outcome. It was indicated earlier that loan tenor has a major impact on project economics. Therefore, if an eight-year loan tenor available in the market for the project is assumed, timing of borrowing becomes critical. In most cases, sponsors attempt to secure financing as early as possible in project development. This may be a good practice when credit is easy to obtain and on relatively good terms. In the example, however, eight years from the date of commitment to the final repayment is a maximum, meaning had the sponsors borrowed at the inception of the project development they would have had only four years to amortize the debt. By delaying formal credit commitments and by financing initial construction costs entirely with equity, it is possible not only to substantially

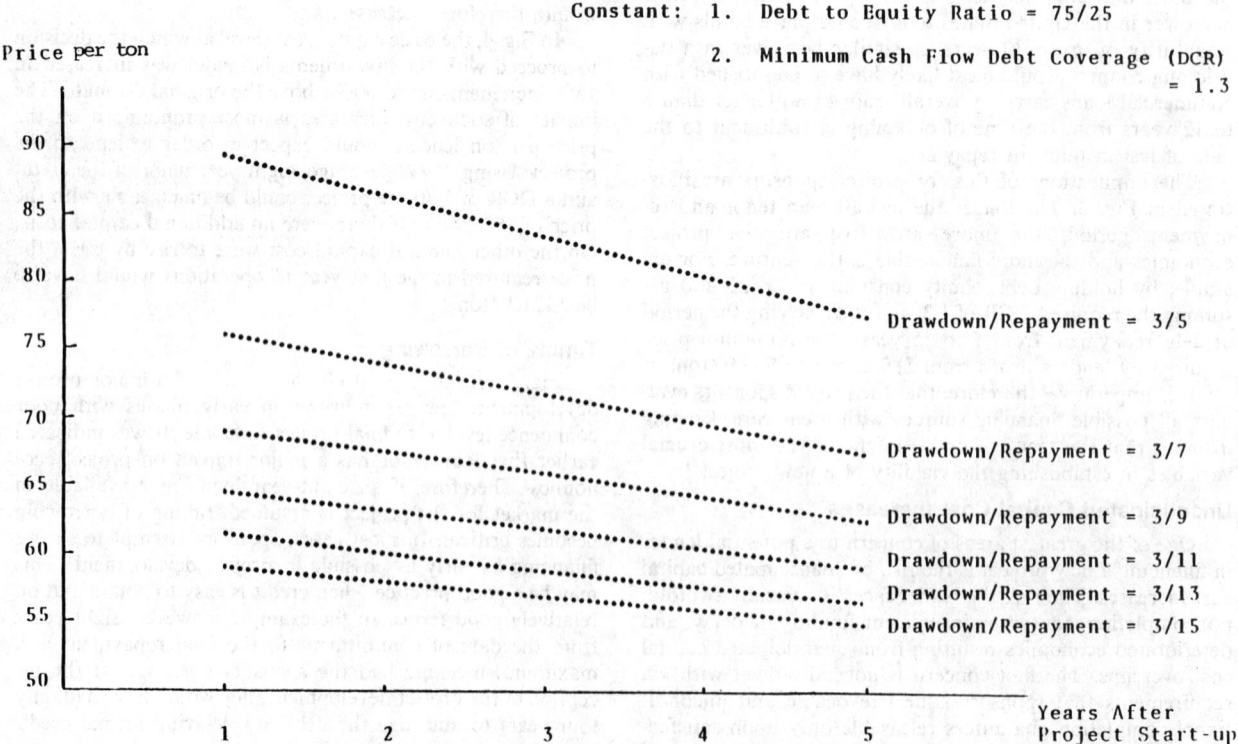

Fig. 3. Impact of different debt structures on price per ton.

Fig. 4. Impact of capital cost increases on price per ton.

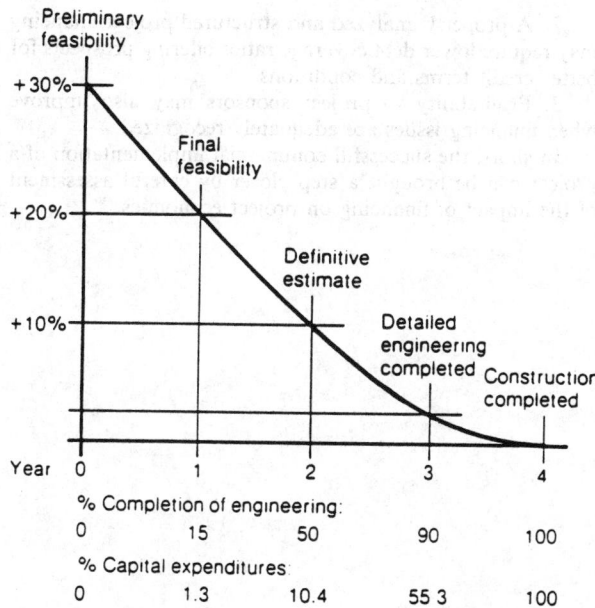

Fig. 5. Accuracy of cost estimates vs. engineering progress.

improve cash flow and debt coverages during the critical initial years of operations, but also to improve the economic returns to the project sponsors.

Fig. 6 reflects dramatic improvements in debt coverages using three different debt/equity structures of 25/75, 50/50, and 75/25. The delay of two years in executing bank credit in case of 50/50 leverage allows for two years longer amortization period, which improves debt coverage from 0.94 to 1.24 in the first year of operations. This produces a difference between a project that clearly is not financeable to the one that begins to meet lenders' criteria.

To one's surprise, perhaps, internal rate of return on equity (IRE) (Fig. 7) also increases, although initial expenditures are funded solely by equity. If there is a 75/25 debt/equity structure, IRE improves from 13.57% to 14.23% by deferring commitment and increasing repayment period by two years.

Referring again to Fig. 5, one may surmise that in the early phases of project development there is a great deal of risk. This risk can be reduced by additional engineering studies which, in relation to total project cost, are not that substantial. Obtaining financing too early can penalize the project since lenders will implicitly demand higher DCR, which may result in much tougher credit terms or even denial of credit. However, proper balancing of project developmental work with timing of financing commitments may improve lenders confidence in the project, lower debt coverages, and enhance owners' profitability.

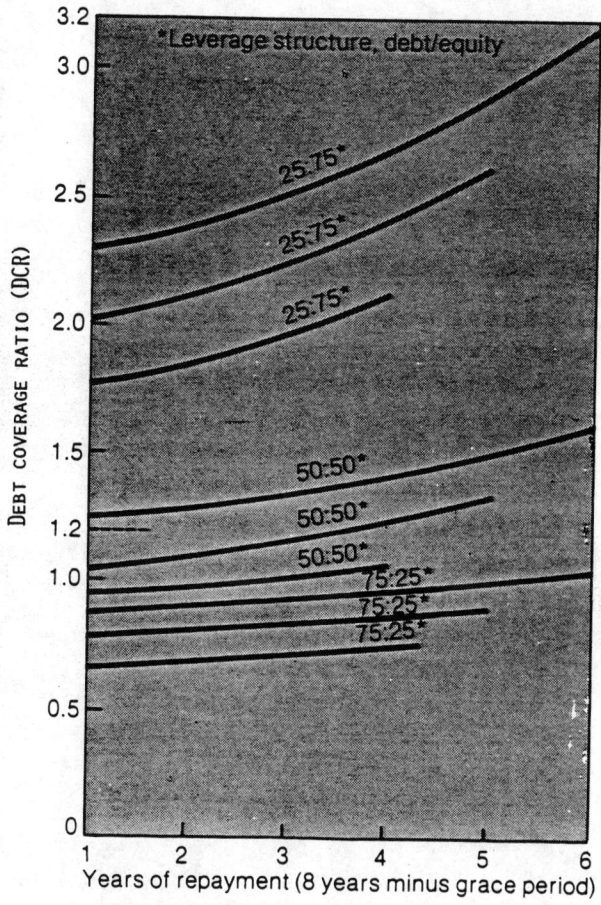

Fig. 6. Impact of postponing debt commitments on DCR.

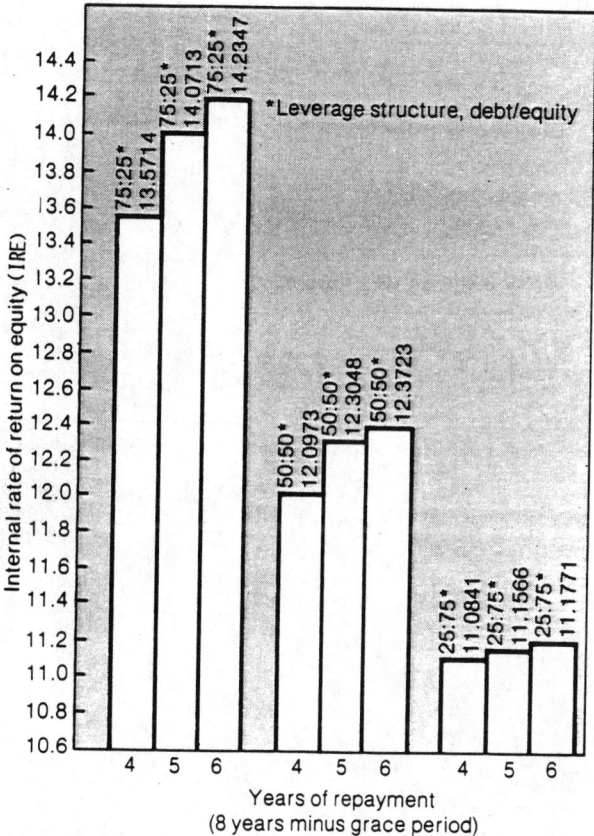

Fig. 7. Impact of postponing debt commitments on IRE.

Conclusions

The preceding has attempted to highlight how critical financial variables are in planning new projects. The main conclusions are summarized as follows:

1. The sensitivity of project cash flows in relation to fluctuations in leverage, debt structure, or timing of borrowing is typically of major importance.

2. A properly analyzed and structured project financing may require lower debt coverage ratios offering prospects for better credit terms and conditions.

3. Profitability to project sponsors may also improve when financing issues are adequately recognized.

In short, the successful commercial implementation of a project can be brought a step closer by careful assessment of the impact of financing on project economics.

Chapter 5

Planning and Design of Surface Mines

Robert Laurich, Editor

5.1 Definition of Mining Parameters

DAVID ARMSTRONG

INTRODUCTION

Many factors govern the size and shape of an open pit. These must be properly understood and used in the planning of any open pit operation. The importance of each will depend on the particular project, but the following are the key items affecting the pit design: geology, grade and localization of the mineralization, extent of the deposit, topography, property boundaries, production rates, bench height, pit slopes, road grades, mining costs, processing costs, metal recovery, marketing considerations, strip ratios, and cutoff grades. This section will discuss several of these factors.

BENCH HEIGHT

The bench height is the vertical distance between each horizontal level of the pit. The elements of a bench are illustrated in Fig. 1. Unless geologic conditions dictate otherwise, all benches should have the same height. The height will depend on the physical characteristics of the deposit; the degree of selectivity required in separating the ore and waste with the loading equipment; the rate of production; the size and type of equipment to meet the production requirements; and the climatic conditions.

The bench height should be set as high as possible within the limits of the size and type of equipment selected for the desired production. The bench should not be so high that it will present safety problems of towering banks of blasted or unblasted material or of frost slabs in winter. The bench height in open pit mines will normally range from 15 m (49 ft) in large copper mines to as little as 1 m (3.3 ft) in uranium mines.

PIT SLOPES

The slope of the pit wall is one of the major elements affecting the size and shape of the pit. The pit slope helps determine the amount of waste that must be moved to mine the ore. The pit slope is usually expressed in degrees from the horizontal plane.

A pit wall needs to remain stable as long as mining activity is in that area. The stability of the pit walls should be analyzed as carefully as possible. Rock strength, faults, joints, presence of water, and other geologic information are key factors in the evaluation of the proper slope angle. The slope may be stated as a simple, overall average for the pit (e.g., 45°), but a more detailed study may show that the physical characteristics of the deposit cause the pit slope to change with rock type, sector location, elevation, or orientation within the pit. Fig. 2 illustrates how the pit slopes may vary in the deposit.

A proper slope evaluation will give the slopes that allow the pit walls to remain stable. The pit walls should be set as steep as possible to minimize the strip ratio. The pit slope analysis determines the angle to be used between the roads

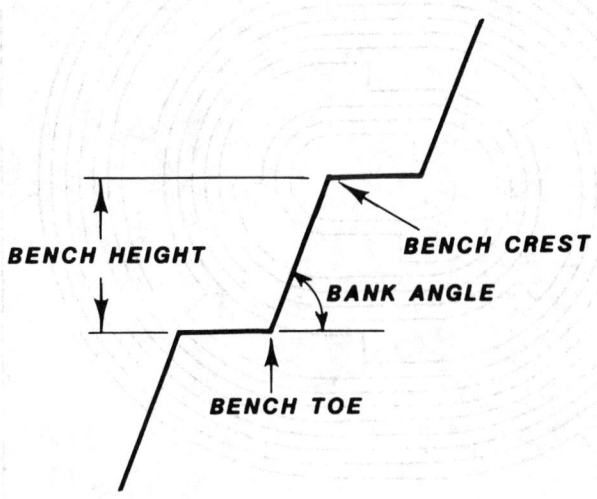

Fig. 1. Bench cross section.

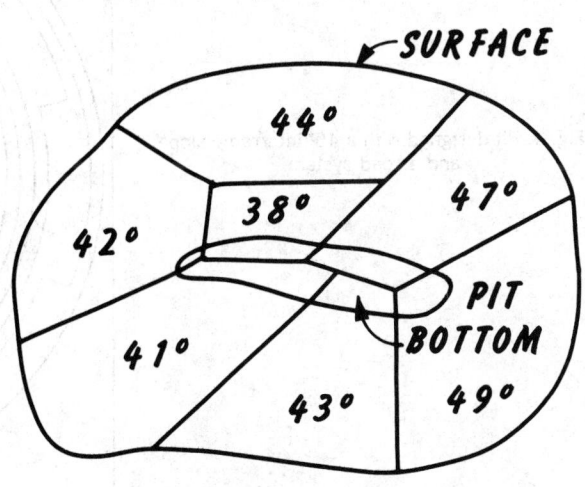

Fig. 2. Example of pit slopes varying in a deposit.

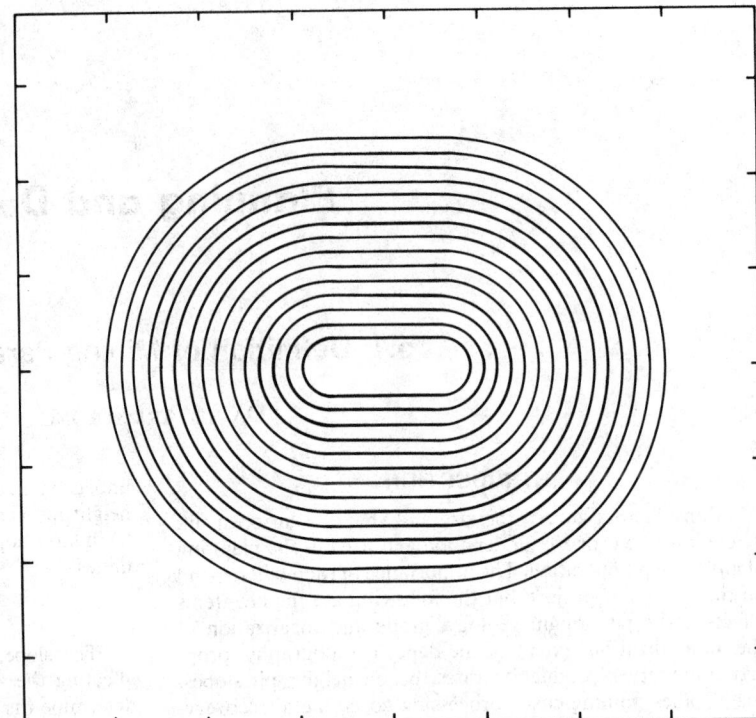

Fig. 3. Pit designed with a 45° pit slope.

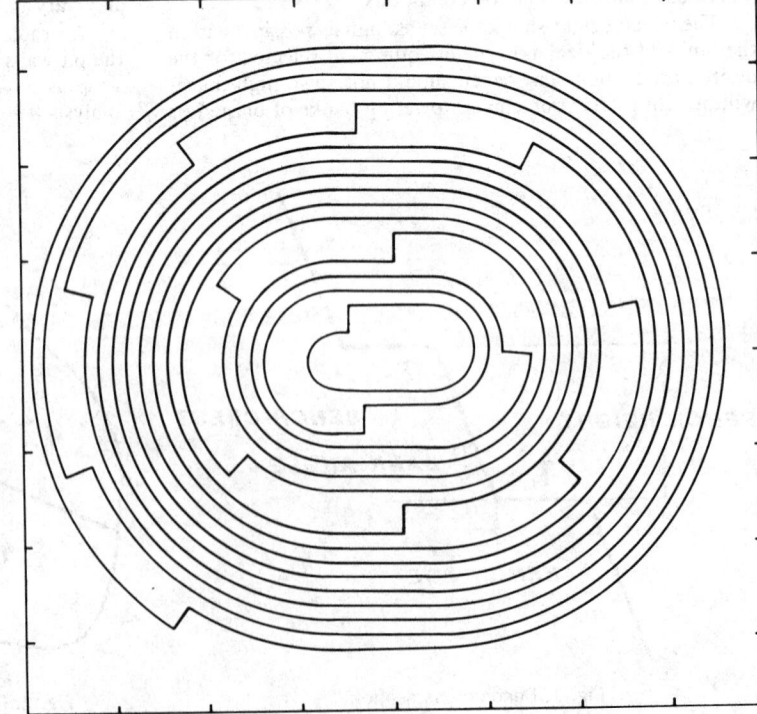

Fig. 4. Pit designed with a 45° interramp slope
and a road system.

in the pit. The overall pit slope used for design must be flatter to allow for the road system in the ultimate pit.

Figs. 3 and 4 show the need to design the pit with a lesser slope to allow for roads. The pit in Fig. 3 has been designed with a 45° angle for the pit walls. The pit in Fig. 4 uses the same pit bottom and the 45° interramp slope between the roads, but, a road has been added. Note the

larger pit that results. In the example, almost 50% more tonnage must be moved to mine the same pit bottom.

In the early design of a pit a lesser pit slope can be used to allow for the road system. The pit in Fig. 5 was designed with an overall slope of 38°. The overall slope to use will depend on the width, grade, and anticipated placement of the road.

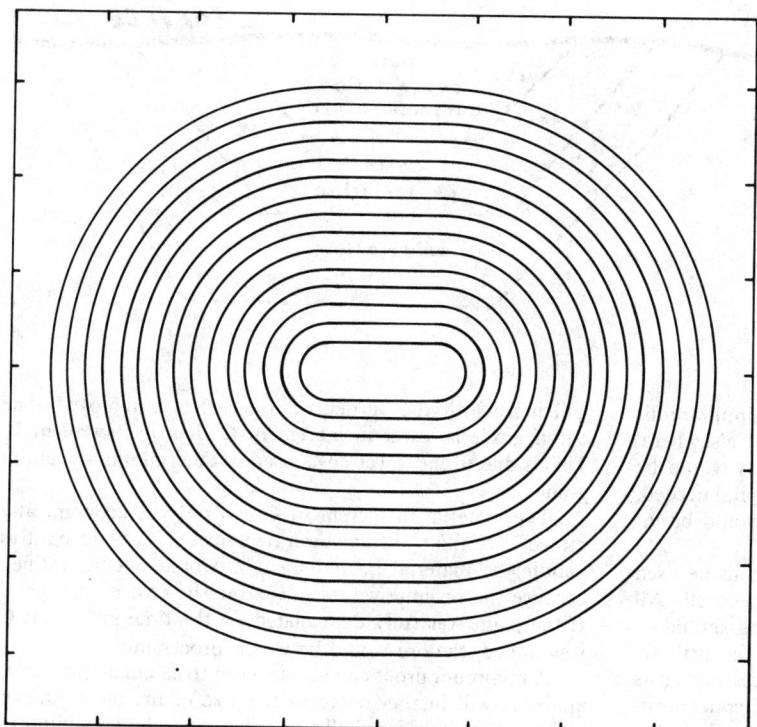

Fig. 5. Pit designed with a 38° overall slope to allow for a 45° interramp slope and a road system.

Fig. 6 shows a vertical section of a pit wall from Fig. 4. The interramp angle is projected from the pit bottom upward to the original ground surface at point B. The overall pit slope angle is the angle from the toe of the bottom bench to the crest of the top bench. Point A shows the intercept of the overall pit slope angle with the original ground surface.

CUTOFF GRADE

As stated by Taylor (1972), a "cutoff grade is any grade that for any specified reason is used to separate any two courses of action." The reason used in setting a cutoff grade usually incorporates the economic characteristics of the project.

When mining, the operator must make a decision as to whether the next block of material should be mined and processed; mined and stockpiled; mined (to expose ore) and sent to the waste dump; or not mined at all. The grade of the block is used to make this decision.

For any block to be deliberately mined, it must pay for the costs of mining, processing, and marketing. The grade of material that can pay for this but for no stripping is the breakeven mining cutoff grade.

A second cutoff grade can be used for blocks that are below the mining cutoff grade and would not be mined for their own value. These blocks may be mined as waste by deeper ore blocks. The cost of mining these blocks is paid for by the deeper ore. The final destination of these blocks is then only influenced by costs for the blocks once they have been mined. The blocks can be processed at this point if they can pay for just the processing and marketing costs. Because the revenue for the block does not need to cover the mining cost, the milling cutoff grade is lower than the mining cutoff grade.

The cutoff calculation depends on the point of the cutoff decision in the life of the mine. In deciding whether to mine one more block at the end of the mine life, the only costs

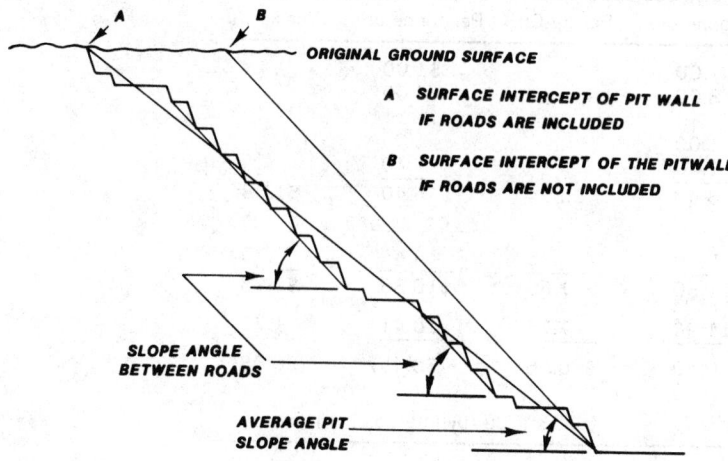

ORIGINAL GROUND SURFACE

A   SURFACE INTERCEPT OF PIT WALL
    IF ROADS ARE INCLUDED

B   SURFACE INTERCEPT OF THE PITWALL
    IF ROADS ARE NOT INCLUDED

SLOPE ANGLE
BETWEEN ROADS

AVERAGE PIT
SLOPE ANGLE

Fig. 6. Vertical section through a pit wall.

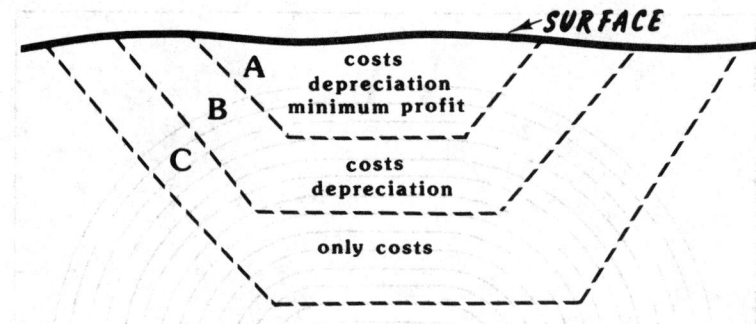

Fig. 7. Relative pit sizes using different levels of costs.

used would be the cash operating costs and a minimum profit to reflect the opportunity costs of using the money elsewhere. For a decision to mine one more year, the costs would be the cash operating costs, plus the replacement capital needed, plus all general and administrative costs that would be incurred.

For a mine in the planning stage, the costs to be used are more complex and must be carefully considered. All direct costs of mining, processing, and marketing should be used. In the mining phase this would include the drilling, blasting, loading, and hauling costs. The processing costs would cover crushing, conveying, grinding, and concentrating costs. Depending on the final form of the product, the marketing costs could include concentrate handling, smelting, refining, and transportation. Additional direct costs for royalties and taxes would also be included.

Overhead costs should also be added to the calculation. The general and administrative costs for the mine, mill, and administrative office staff should be included. Until the size of the pit has been determined and the associated overhead costs developed, the costs to be used for the calculation can only be estimated and are therefore subject to later refinement.

Depreciation is used in the calculation for the purpose of setting the pit size. As shown in Fig. 7, the size of the pit

will increase if the burden of some costs is removed. The cutoff grade is lower in increment C than in increment B. This is due to the lower costs used in determining the cutoff grade.

The material in increment C can only be economically mined after the plant has been depreciated. A plant built to handle the material in increment C would not be justified because the revenue would not cover the cost of the plant. If the plant was fully depreciated by the time increment C was mined, the ore would be worth processing.

A minimum profit can also be used to calculate the cutoff grade. It will further decrease the size of the pit as shown by increment A in Fig. 7. The purpose of adding a minimum profit is twofold: (1) it confirms that a block is ore only if it can be mined and processed at a profit; and (2) it sets an economic limit below which a company would find an alternate investment more attractive.

The amount of minimum profit to be used is a difficult decision. A true profit calculation would include the role of depreciation, depletion, and taxes. At the design stage, these are not known. An approximation can be made by increasing the costs.

Other costs and changes in revenue can be included if they are known. These would include recoveries that vary with the ore grade, mining costs that vary with the distance

Table 1. Calculation of Breakeven Cutoff Grade

	Per tonne ore	Per kg Cu	Per tonne ore	Per kg Cu
Head Grade	0.80% Cu		0.70% Cu	
Recovery	85% Cu		85% Cu	
Recoverable Copper Per Tonne	6.80 Kg		5.95 Kg	
Costs	Per tonne ore	Per kg Cu	Per tonne ore	Per kg Cu
Mining	$1.00		$1.00	
Processing	3.00		3.00	
General & Administrative	1.00		1.00	
Depreciation	1.40		1.40	
Total	$ 6.40	$0.94	$ 6.40	$1.08
Freight, Smelting Refining	5.10	.75	4.46	.75
Total	$11.50	$ 1.69	$10.86	$1.83
Value @ $1.75/Kg	$11.90	1.75	10.41	1.75
Net value	$ 0.40	$ 0.06	($0.45)	($0.08)
Cutoff grade	0.753% Cu (by interpolation)			

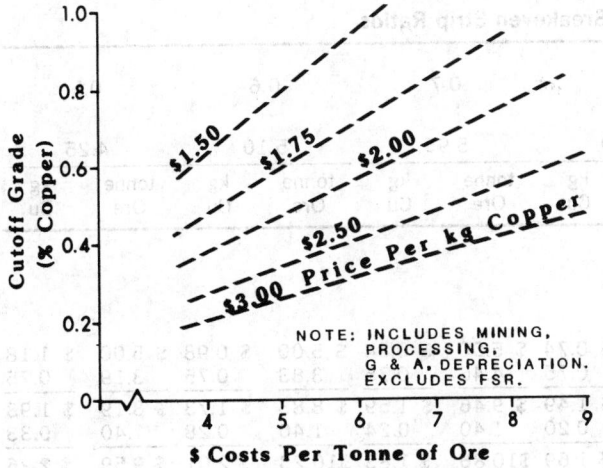

Fig. 8. Cutoff grades for different costs and metal prices.

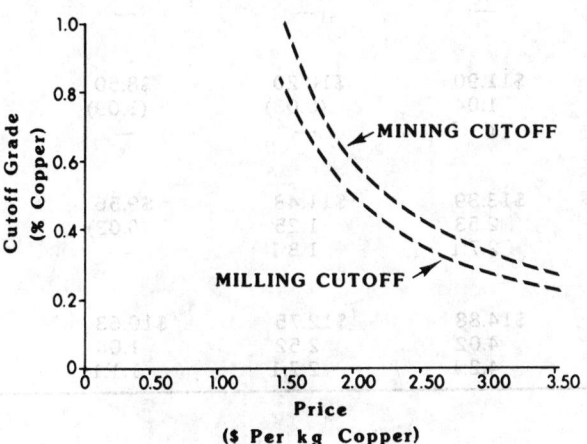

Fig. 9. Relationship of mining and milling cutoff grades.

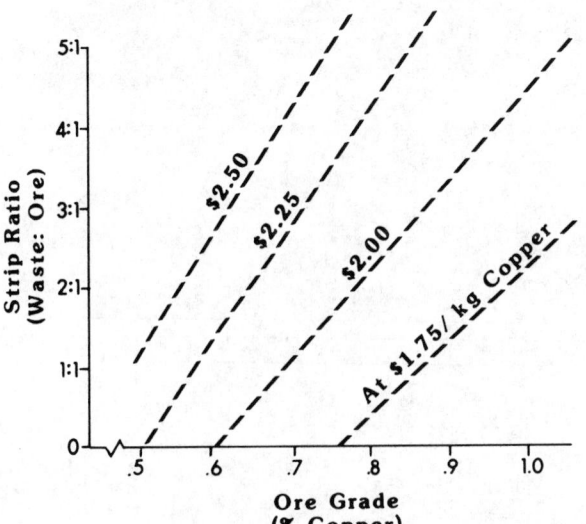

Fig. 10. Strip ratios for different ore grades and metal prices.

or elevation of haulage, and the time lag between stripping the waste from a block of ore and the mining of the ore. These values should only be added if they are well known and the added degree of sophistication is warranted.

Table 1 is the calculation of the mining cutoff grade for a copper project with the following parameters:

30	kt/d (33,000 st pd) of ore mined for 20 years
$300,000,000	capital cost (including replacement capital)
$1.00	mining cost per tonne of ore
$0.95	mining cost per tonne of waste
$3.00	processing cost per tonne of ore
$1.00	general and administrative (G&A) cost per tonne of ore
$0.75	freight, smelter, and refining (FSR) cost per kilogram of copper
85%	overall copper recovery

The results are shown graphically in Fig. 8. Note that the cutoff grade will increase as the costs increase. The difference between the mining cutoff grade and the milling cutoff grade is shown in Fig. 9.

STRIP RATIO

The strip ratio is the ratio of the number of tonnes of waste that must be moved for one tonne of ore to be mined. The results of a pit design will determine the tonnes of waste and ore that the pit contains. The ratio of waste and ore for the design will give the average strip ratio for that pit. This differs from the breakeven strip ratio used to design the pit. The breakeven strip ratio refers only to the last increment mined along the pit wall. The strip ratio is calculated for the point at which break even occurs and the necessary stripping is paid for by the net value of the ore removed.

The calculation for the breakeven strip ratio (*BESR*) is:

$$BESR = (A - B)/C$$

where:

A = revenue per tonne of ore
B = production cost per tonne of ore (including all costs to the point of sale, excluding stripping)
C = stripping cost per tonne of waste

In certain studies a minimum profit requirement is included in the formula.

$$BESR = [A - (B + D)]/C$$

where:

D = minimum profit per tonne of ore.

Table 2 contains the information for calculating the strip ratio for the example used in calculating the cutoff grade previously. The results are shown graphically in Fig. 10.

REFERENCE

Taylor, H.K., 1972, "General Background Theory of Cutoff Grades," *Transactions* (Section A: Mining Industry), Institution of Mining and Metallurgy, Vol. 81, pp. A160–A179.

Table 2. Calculation of Breakeven Strip Ratios

Head Grade (% Cu)	1.0		0.9		0.8		0.7		0.6		0.5	
Kg Cu recovered per tonne ore	8.50		7.65		6.80		5.95		5.10		4.25	
	tonne Ore	kg Cu	tonne Ore	kg Cu	tonne Ore	kg Cu	tonne Ore	kg Cu	tonne Ore	kg Cu	tonne Ore	kg Cu
COSTS:												
Mining*	$ 1.00											
Milling	3.00											
G&A	1.00											
	$ 5.00	$ 0.59	$ 5.00	$ 0.65	$ 5.00	$ 0.74	$ 5.00	$ 0.84	$ 5.00	$ 0.98	$ 5.00	$ 1.18
FSR	6.38	0.75	5.74	0.75	5.10	0.75	4.46	0.75	3.83	0.75	3.19	0.75
	$11.38	$ 1.34	$10.74	$ 1.40	$10.10	$ 1.49	$ 9.46	$ 1.59	$ 8.83	$ 1.73	$ 8.19	$ 1.93
Depreciation	1.40	0.16	1.40	0.19	1.40	0.20	1.40	0.24	1.40	0.28	1.40	0.33
Total Cost	$12.78	$ 1.50	$12.14	$ 1.59	$11.50	$ 1.69	$10.86	$ 1.83	$10.23	$ 2.01	$ 9.59	$ 2.26
BREAKEVEN STRIPPING RATIO:												
@ $1.75/kg Cu												
Value	$14.88		$13.39		$11.90		$10.41		$8.93		$7.44	
Net	2.10		1.25		.40		(0.45)		(1.30)		(2.15)	
Ratio†	2.2:1		1.3:1		0.4:1		—		—		—	
@ $2.00/kg Cu												
Value	$17.00		$15.30		$13.60		$11.90		$10.20		$8.50	
Net	4.22		3.16		2.10		1.04		(0.03)		(1.09)	
Ratio†	4.4:1		3.3:1		2.2:1		1.1:1		—		—	
@ $2.25/kg Cu												
Value	$19.13		$17.21		$15.30		$13.39		$11.48		$9.56	
Net	6.35		5.07		3.80		2.53		1.25		(0.03)	
Ratio†	6.7:1		5.3:1		4.0:1		2.7:1		1.3:1		—	
@ $2.50/kg Cu												
Value	$21.25		$19.13		$17.00		$14.88		$12.75		$10.63	
Net	8.47		6.99		5.50		4.02		2.52		1.04	
Ratio†	8.9:1		7.4:1		5.8:1		4.2:1		2.7:1		1.1:1	

* Excludes stripping cost.
† At stripping cost of $0.95 per tonne of waste.
( ) Indicates negative value.

5.2 Ultimate Pit Definition

INTRODUCTION

There are probably as many ways of designing an ultimate open pit as there are engineers doing the design work. The methods differ by the size of the deposit, the quantity and quality of the data, the availability of computer assistance, and the assumptions of the engineer.

As the first step for long or short-range planning, the limits of the open pit must be set. The limits define the amount of ore minable, the metal content, and the associated amount of waste to be moved during the life of the operation. The size, geometry, and location of the ultimate pit are important in planning tailings areas, waste dumps, access roads, concentrating plants, and all other surface facilities. Knowledge gained from designing the ultimate pit also aids in guiding future exploration work.

In designing the ultimate pit, the engineer will assign values to the physical and economic parameters discussed in the previous section. The ultimate pit limit will represent the maximum boundary of all material meeting these criteria. The material contained in the pit will meet two objectives.

1. A block will not be mined unless it can pay all costs for its mining, processing, and marketing and for stripping the waste above the block.

2. For conservation of resources, any block meeting the first objective will be included in the pit.

The result of these objectives is the design that will maximize the total profit of the pit based on the physical and economic parameters used. As these parameters change in the future, the pit design may also change. Because the values of the parameters are not uniquely known at the time of design, the engineer may wish to design the pit for a range of values to determine the most important factors and their effect on the ultimate pit limit.

MANUAL DESIGN

The manual method of designing pits involves considerable time and judgment on the part of the engineer. The usual method of manual design starts with the three types of vertical sections shown in Fig. 1:

1. Cross sections spaced at regular intervals parallel to each other and normal to the long axis of the ore body. These will provide most of the pit definition and may number from 10 to perhaps 30, depending on the size and shape of the deposit and on the information available.

2. A longitudinal section along the long axis of the ore body to help define the pit limits at the ends of the ore body.

3. Radial sections to help define the pit limits at the ends of the ore body.

Each section should show ore grades, surface topography, geology (if needed to set the pit limits), structural controls (if needed to set the pit limits), and any other information that will limit the pit (e.g., ownership boundaries).

The stripping ratio is used to set the pit limits on each section. The pit limits are placed on each section independently using the proper pit slope angle.

The pit limits are placed on the section at a point where the grade of ore can pay for mining the waste above it. When a line for the pit limit has been drawn on the section, the grade of the ore along the line is calculated and the lengths of the ore and waste are measured. The ratio of the waste and ore is calculated and compared to the breakeven stripping ratio for the grade of ore along the pit limit. If the calculated stripping ratio is less than the allowable stripping ratio, the pit limit is expanded. If the calculated stripping ratio is greater, the pit limit is contracted. This process continues on the section until the pit limit is set at a point where the calculated and breakeven stripping ratios are equal.

In Fig. 2, the grade on the right side of the pit was estimated to be 0.6% Cu. At a price of $2.25 per kg of copper, the breakeven stripping ratio from Fig. 3 is 1.3:1. The line for the pit limit was found using the required pit slope and located at the point that gave a waste:ore ratio of 1.3:1. At the limit

$$\frac{\text{Length of waste } (XY)}{\text{Length of ore } (YZ)} = \frac{1.3}{1}$$

On the left side of the section, the pit limit for the 0.7% Cu grade was similarly determined using a breakeven stripping ratio of 1.7:1. If the grade of the ore changed as the pit limit line was moved, the breakeven stripping ratio to use would also change.

The pit limits are established on the longitudinal section in the same manner with the same stripping ratio curves. The pit limits for the radial section are handled with a different stripping ratio curve, however. As shown in Fig. 4, the cross sections and the longitudinal section represent a slice along the pit wall with the base the same length as the

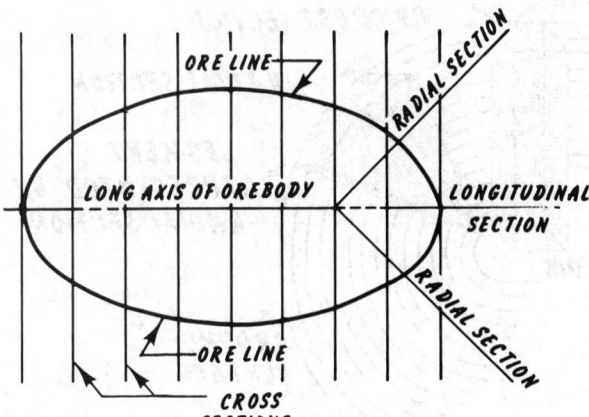

Fig. 1. Types of vertical sections used for a manual pit design.

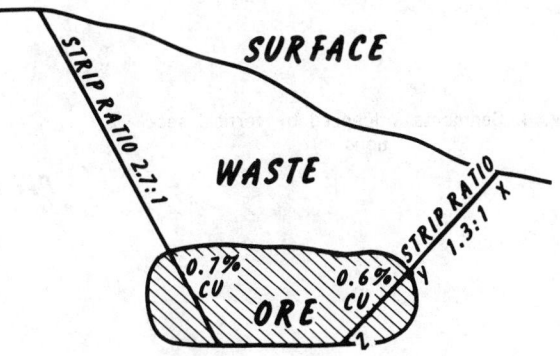

Fig. 2. Pit limits shown on section.

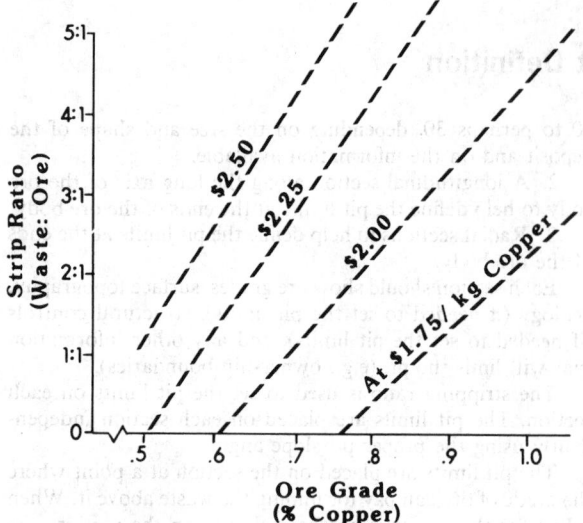

Fig. 3. Strip ratios for different ore grades and metal prices.

surface intercept. The radial section represents a narrow portion of the pit at the base and a much wider portion at the surface intercept. The allowable stripping ratios must be adjusted downward for the radial sections before the pit limit can be set.

The next step in the manual design is to transfer the pit limits from each section to a single plan map of the deposit. The elevation and location of the pit bottom and the surface intercepts from each section are transferred. If a pit slope change occurred on a section, its position is also transferred.

The resultant plan map will show a very irregular pattern of the elevation and outline of the pit bottom and of the surface intercepts. The bottom must be manually smoothed to conform to the section information.

Starting with the smoothed pit bottom, the engineer will develop the outline for each bench at the point midway between the bench toe and crest. The engineer manually expands the pit from the bottom with the following criteria:

1. The breakeven stripping ratios for adjacent sections may need to be averaged.

2. The allowable pit slopes must be obeyed. If the road system is designed at the same time, the interramp angle is used. If the preliminary design does not show the roads, the outline for the bench midpoints will be based on the flatter overall pit slope that allows for roads.

3. Possible unstable patterns in the pit should be avoided. These would include any bulges into the pit.

4. Simple geometric patterns on each bench make the designing easier.

When the pit plan has been developed, the results should be reviewed to determine if the breakeven stripping ratios have been satisfied. The pit can be divided into sectors on the pit plan and each sector checked for the waste:ore ratio. Two ways the stripping ratios for each sector can be checked are:

1. The pit limits from the pit plan maps can be transferred back to the sections and the stripping ratio can then be calculated from the sections.

2. The bench outlines can be transferred to each individual bench map. The ore and waste lengths are measured along the bench outline for each sector. The results for each bench are combined to calculate the stripping ratio for that sector. The ore grade for the sector is the weighted average (by length) of the grade of the ore along the pit limit for each bench.

The total reserves for the pit and the average stripping ratio are determined by accumulating the values from each bench. On each bench the ore tonnes above the breakeven cutoff grade are measured and the average grade of the ore is calculated. The tonnes of waste are also measured. The total of the tonnes of ore and the total of the tonnes of waste on each bench give the average stripping ratio for the pit.

COMPUTER METHODS

As should be appreciated, the manual design of a pit gets the planning engineer closely involved with the design and increases the engineer's knowledge of the deposit. The procedure is cumbersome, though, and is difficult to use on large or complex deposits. Because of the lengthiness of the procedure, the number of alternatives that can be examined is limited. As more information is gathered or if any of the design parameters change, the entire process may have to be repeated. Another drawback to the method of manual design is that the pit may be well designed on each section, but,

Fig. 4. Segments influenced by vertical sections.

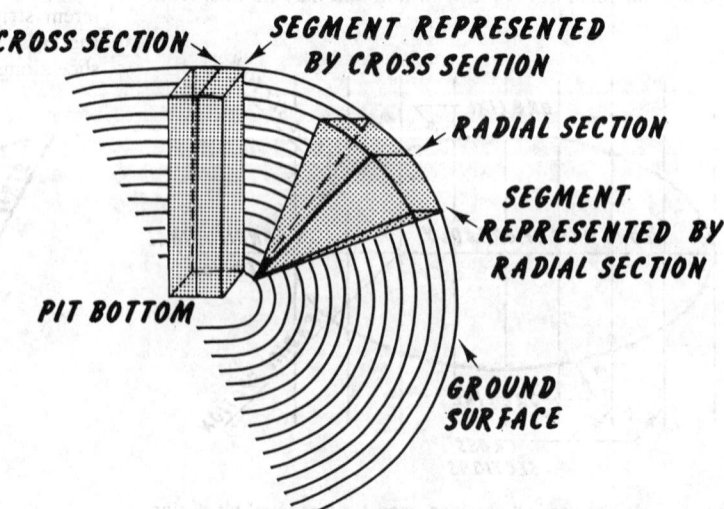

when the sections are joined and the pit is smoothed, the result may not yield the best overall pit.

The growth of computer usage has allowed engineers to handle greater amounts of data and to examine more pit alternatives than with manual methods. The computer has proved to be an excellent tool for storing, retrieving, processing, and displaying data from mining projects. Computer applications have been developed to take much of the burden of pit design from the engineer.

The computer efforts can be divided into two groupings:

1. Computer-assisted methods. The calculations are done by the computer under the direct guidance of the engineer. The computer does not do the entire design but only does the brunt of the calculation work with the engineer controlling the process. Examples would be the two-dimensional Lerchs-Grossman technique and the three-dimensional design using an incremental pit expansion method.

2. Automated methods. These are capable of designing the ultimate pit for a given set of economic and physical constraints without intervention by the engineer. One category of automated methods contains the mathematically optimal techniques using linear programming, dynamic programming, or network flows. A second category has the heuristic methods, such as the floating cone method that produces an acceptable pit, but not necessarily an optimal one. As the cost of computer processing decreases, better automated methods will be forthcoming.

Another characteristic differentiating the types of computerized methods is the use of either a whole or partial block for mining. In a whole block method, each block is mined either as a unit or left intact; in a partial block method, a portion of each block can be mined. Each type has certain advantages:

1. *Accuracy*—With the use of partial blocks, the tonnage of small volumes can be calculated quite accurately. The overall tonnage of the pit may be accurate using a whole block method, but, the accuracy is less for smaller volumes.

2. *Physical constraints*—The desired pit slopes and pit boundaries are approximated by the mined blocks. The use of whole blocks may result in pit walls that are unacceptable in terms of operations and slope stability. Some whole block techniques may assume the block size is a function of the pit slope and some may not allow the slope to vary in the pit. Smoothing is usually required for an ultimate pit designed using whole blocks.

3. *Cost*—When properly used, whole block methods have generally proven to be less costly in terms of computer costs than partial block methods. As a result, several pit configurations can be quickly analyzed with a whole block method to give a good basis for a more detailed partial block analysis.

LERCHS-GROSSMAN METHOD

The two-dimensional Lerchs-Grossman method will design on a vertical section the pit outline giving the maximum net profit. The method is appealing because it eliminates the trial-and-error process of manually designing the pit on each section. The method is also convenient for computer processing.

Like the manual method, the Lerchs-Grossman method designs the pit on vertical sections. The results must still be transferred to a pit plan map and manually smoothed and checked. Even though the pit is optimal on each section, the ultimate pit resulting from the smoothing is probably not optimal.

The example in Fig. 5 represents a vertical section through a block model of the deposit. Each square represents the net value of a block if it were independently mined and processed. Blocks with a positive net value have been shaded in the figure. The block size has been set in the example so that the pit profile will move up or down only one block at most as it moves sideways.

Step 1

Add the values down each column of blocks and enter these numbers into the corresponding blocks in Fig. 6. This is the upper value in each block of Fig. 6 and represents the cumulative value of the material from each block to the surface.

Step 2

Start with the top block in the left column and work down each column. Put an arrow in the block pointing to the highest value in:
1. the block one to the left and one above,
2. the block one to the left,
3. the block one to the left and one below.

Calculate the bottom value for the block by adding the top value to the bottom value of the block the arrow points to. The bottom value in each block represents the total net value of the material in the block, the blocks in the column, and the blocks in the pit profile to the left of the block. Blocks marked with an *X* cannot be mined unless more columns are added.

Step 3

Scan the top row for the maximum total value. This is the total net return of the optimal pit. For the example, the optimal pit would have a value of $13. Trace the arrows back to get the outline of the pit. Fig. 7 shows the pit outlined on the section. Note that even though the block on row 6 at column 6 has the highest net value in the deposit it is not in the pit. To mine it would lower the value of the pit.

	1	2	3	4	5	6	7	8	9	10	11
1	-$2	-$2	-$4	-$2	-$2	-$1	-$2	-$3	-$4	-$4	-$3
2	-$5	-$4	-$6	-$3	-$2	-$2	-$3	-$2	-$4	-$5	-$5
3	-$6	-$5	-$7	$6	$13	-$2	-$5	-$4	-$7	-$4	-$6
4	-$6	-$6	-$8	-$8	$17	$8	$5	-$6	-$8	-$9	-$7
5	-$7	-$7	-$8	-$8	$6	$21	$5	-$8	-$8	-$9	-$7
6	-$7	-$9	-$9	-$8	-$5	$22	-$8	-$8	-$8	-$9	-$8
7	-$8	-$9	-$9	-$9	-$8	$10	-$9	-$9	-$9	-$9	-$9

Fig. 5. Vertical section showing the net value of each block.

Fig. 6. Section after the search procedure.

	1	2	3	4	5	6	7	8	9	10	11
1	-2 / -2	-2 / -2	-4 / -4	-2 / -2	-2 / -2	-1 / -1	-2 / -2	-3 / 2	-4 / -1	-4 / 13	-3 / 10
2	-7 / X	-6 / -8	-10 / -12	-5 / -9	-4 / -6	-3 / -3	-5 / 5	-5 / 3	-8 / 17	-9 / 8	-8 / X
3	-13 / X	-11 / X	-17 / -25	1 / -11	9 / 0	-5 / 10	-10 / 8	-9 / 25	-15 / 16	-13 / X	-14 / X
4	-19 / X	-17 / X	-25 / X	-7 / -32	26 / 15	3 / 18	-5 / 34	-15 / 31	-23 / X	-22 / X	-21 / X
5	-26 / X	-24 / X	-33 / X	-15 / X	32 / 0	24 / 39	0 / 46	-23 / X	-31 / X	-31 / X	-28 / X
6	-33 / X	-33 / X	-42 / X	-23 / X	27 / X	46 / 46	-8 / X	-31 / X	-39 / X	-40 / X	-36 / X
7	-41 / X	-42 / X	-51 / X	-32 / X	19 / X	56 / X	-17 / X	-40 / X	-48 / X	-49 / X	-45 / X

Fig. 7. Optimal pit outline.

	1	2	3	4	5	6	7	8	9	10	11
1	-$2	-$2	-$4	-$2	-$2	-$1	-$2	-$3	-$4	-$4	-$3
2	-$5	-$4	-$6	-$3	-$2	-$2	-$3	-$2	-$4	-$5	-$5
3	-$6	-$5	-$7	$6	$13	-$2	-$5	-$4	-$7	-$4	-$6
4	-$6	-$6	-$8	-$8	$17	$8	-$5	-$6	-$8	-$9	-$7
5	-$7	-$7	-$8	-$8	$6	$21	$5	-$8	-$8	-$9	-$7
6	-$7	-$9	-$9	-$8	-$5	$22	-$8	-$8	-$8	-$9	-$8
7	-$8	-$9	-$9	-$9	-$8	$10	-$9	-$9	-$9	-$9	-$9

INCREMENTAL PIT EXPANSION

The incremental pit expansion technique is a trial-and-error process guided by the engineer. Although this method will not necessarily produce an optimal pit, in the hands of a skillful engineer it is a very powerful tool. Either whole or partial blocks can be used.

The engineer will digitize the outline of a new pit bottom or an expansion to a pit wall. The computer projects this shape upwards in conformance with the pit slopes to be used. The resulting expansion should be graphically shown to the engineer for confirmation that the increment is as expected.

If the expansion is agreeable to the engineer, a tabulation is done for the material in the increment. The shape of the expansion at the midpoint of each bench is used with the block values for the bench to calculate the grade, tonnes of ore, tonnes of waste, revenues, and costs of the increment. If the increment meets the criteria of the engineer, it is kept in the pit and another outline is digitized. In this manner, the size of the pit gradually grows as the engineer outlines each increment and decides if it meets the design criteria.

To be most effective, the design should progress from the upper benches downward and from the higher grade areas outward on each bench. This is to ensure that only those blocks that can pay for themselves will be included in the pit.

FLOATING CONE METHOD

The most popular automated method has been the floating cone method. The concept is similar to the incremental pit expansion but the manual intervention can be minimized or eliminated.

Instead of a digitized bottom, one block or a group of blocks forms the base of the expansion. If the grade of the base is above the mining cutoff grade, the expansion is projected upward to the top level of the model as in Fig. 8. The resulting cone is formed using the appropriate pit slope angles.

All blocks that are encompassed by the cone (and are not considered previously mined) are tabulated for the costs of mining and processing and for the revenues derived from the ore. If the total revenues are greater than the total costs for the blocks in the cone, the cone has a positive net value and is economic to mine. The surface topography is then altered to reflect the simulated mining of the cone. The topography is left unchanged unless the cone value is positive.

A second block is then examined, as shown in Fig. 9. Assuming the first cone had a positive value and was included in the pit, only the blocks in the shaded portion need be tabulated.

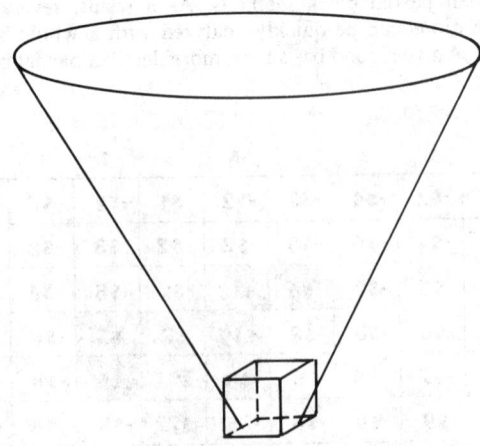

Fig. 8. Cone centered on a base block.

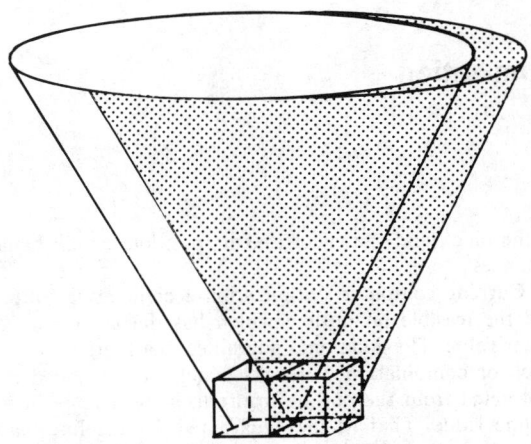

Fig. 9. Cone formed by a second base block.

Each block in the deposit is examined in turn as a base block of a cone. For a large model, this can be a costly process. The resulting pit is also dependent on the pattern in which the next base block is chosen. For example, a base block on an upper level may not have been economic when initially examined. If part of the waste covering it is stripped by mining a cone from a lower level, the block should again be checked before another block from a lower level is used as a base block. This is necessary to make each cone pay for itself.

Because of this potential problem, an engineer can intervene in the process. The engineer can define a smaller volume in which all base blocks will be checked by the computer. From the results of the cones in this smaller volume, the engineer can specify another volume to check. With this added control, the selection sequence of base blocks is less of a problem.

REFERENCES

Barnes, M. P., 1980, *Computer-Assisted Mineral Appraisal and Feasibility,* AIME, New York.

Kim, Y. C., 1978, "Ultimate Pit Limit Design Methodologies Using Computer Models—The State of the Art," *Mining Engineering,* Vol. 30, No. 10, pp. 1454-1459.

Koskiniemi, B. C., 1977, "Hand Methods in Open-Pit Mine Planning and Design," *Open Pit Mine Planning and Design,* J. T. Crawford and W. A. Hustrulid, eds., AIME, New York, pp. 187-194.

Lerchs, H., and Grossman, I. F., 1965, "Optimum Design of Open-Pit Mines," *Transactions,* Canadian Institute of Mining and Metallurgy, Vol. 68, pp. 17-24.

Miller, V. J., and Hoe, H. L., 1982, "Mineralization Modeling and Ore Reserve Estimation," *Engineering and Mining Journal,* Vol. 183, No. 6, pp.66-74.

Soderberg, A., and Rausch, D. O., 1968, "Pit Planning and Layout," *Surface Mining,* E. P. Pfleider, ed., AIME, New York, pp. 141-165.

Pana, M.T., and Davey, R. K., 1973, "Pit Planning and Design," *SME Mining Engineering Handbook,* A. B. Cummins and I. A. Given, ed., AIME, New York, pp. 17.1-17.19.

Pana, M.T., and Davey, R. K., 1973a, "Open-Pit Mine Design," *SME Mining Engineering Handbook,* A. B. Cummins and I. A. Given, ed., AIME, New York, pp. 30.7-30.19.

Taylor, H.K., 1972, "General Background Theory of Cutoff Grades," *Transactions* (Section A: Mining Industry), Institution of Mining and Metallurgy, Vol. 81, pp. A160-A179.

5.3 Open Pit Optimization

JEFF WHITTLE

INTRODUCTION

Computer hardware, and to a lesser extent software, has for the last 20 years consistently advanced at a rate which has exceeded all expectations. As a result, calculations which were difficult or impossible to do only a few years ago can now easily be completed on a computer small enough to fit on a desk and costing only a few months' salary. What is more, the calculations can be done by users with very little knowledge of computers.

Pit optimization is a field which has benefitted greatly from this process in recent years, and we can now go far beyond simple optimization of a pit outline. Thorough sensitivity work, which has often only received lip service in the past, can now be carried out routinely on every ore body that is examined. Management can be offered the real possibility of trading profit for reduced corporate risk in an explicit manner.

Pit optimization was touched upon briefly in the previous section, but we will now go into it in much more detail and describe what can be done at the time of writing (early 1990). There will undoubtedly be further developments.

THE MEANING OF PIT OPTIMIZATION

The first thing to realize is that any feasible pit outline has a dollar value which can, in theory, be calculated.

By feasible, here, we mean that no wall slope is steeper than the rock can support after allowing for the insertion of haul roads and safety berms. That is, we are talking about overall pit slopes.

To calculate the dollar value we must decide on a mining sequence and then conceptually mine out the pit, progressively accumulating the revenues and costs as we go. If we wish to allow for the time value of money—that is the fact that a dollar we receive today is more valuable than one that we (might) receive next year—then we must discount the revenues and costs by a factor which increases with time.

The second thing to realize is that in doing this calculation we have, in effect, allocated a value to every cubic meter or to every block of rock. What is more, we have allocated these values without taking any account of the mining which has gone before, except that the value may depend on the position of the block and the effect that its position has on haulage distances.

Current computer optimization techniques attempt to find the feasible pit outline which has the maximum total dollar value. The good ones guarantee that there is no single block or combination of blocks which can be added to or subtracted from the outline to produce an increase in total outline value. That is, they guarantee the absolute mathematical maximum. They also exclude any block combinations which have a zero value.

Once we have fixed the block values and the slopes, we have fixed the optimal outline, and it is important to make the point that there is only one optimal outline. If we assume that there are two outlines of the same value, then it is easy to show that the two taken together would produce an outline of higher value. Consequently the assumption of the existence of two different optimal outlines of equal value is false.

If the block values increase then, in general, the optimal pit gets bigger. If the slopes increase then, in general, the optimal pit gets deeper.

Of course, we have to know the pit outline in order to calculate the values of the blocks, particularly if the time value of money is important. Conversely, we have to know the block values in order to find the optimal outline. We therefore have a chicken and egg situation, and we will return to this.

A SIMPLE EXAMPLE

Let us assume that we have a flat topography and a vertical rectangular ore body of constant grade as is shown in Fig. 1. Let us further assume that the ore body is sufficiently long in strike for end effects to be ignored. Under these circumstances, we only have to concern ourselves with a section.

In this simplified case there are eight possible pit outlines that we can consider, and the tonnages for these outlines are given in Table 1.

If we assume that ore is worth $2.00 per tonne after all mining and processing costs have been paid, and that waste costs $1.00 per tonne to remove, then we obtain the values shown in Table 2 for the possible pit outlines.

When plotted against pit tonnage, these values produce the graph in Fig. 2. With these very simple assumptions the outline with the highest value is number five.

There are other things that we can learn from this curve.

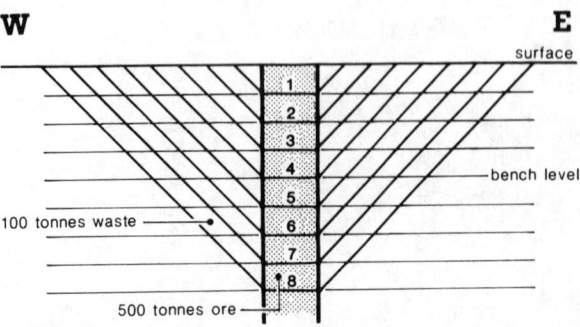

Fig. 1. Simplified ore body.

Table 1. Tonnages for the Possible Pit Outlines

Pit	Ore	Waste	Total
1	500	100	600
2	1000	400	1400
3	1500	900	2400
4	2000	1600	3600
5	2500	2500	5000
6	3000	3600	6600
7	3500	4900	8400
8	4000	6400	10400

470

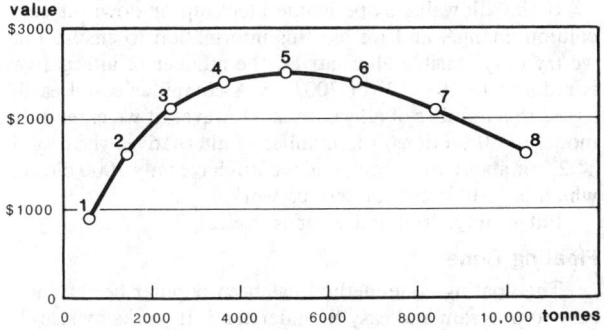

Fig. 2. Relationship between pit tonnage and value.

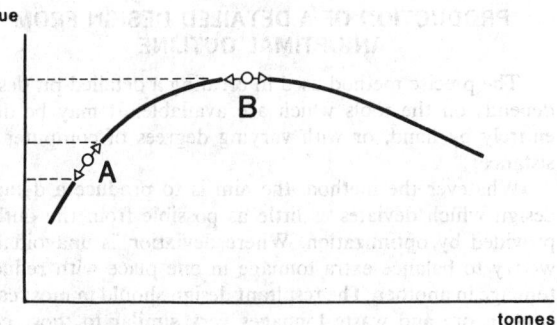

Fig. 3. The effect of small design changes at different points on the tonnage/value curve.

Firstly, outlines four and six have values which are close to that of outline five, and this is not just an artefact of this particular ore body. For any continuous ore body, as the pit is expanded towards optimality, the last shell which is added will have only a small positive value. If it had a large one, there would probably be another positive shell to follow. This means that in this case, and in the vast majority of real ore bodies, the curve of value against tonnage is smooth and surprisingly flat at the peak. It is common to find that a 10% range of pit tonnage covers only a 1% range of pit value. The trick is to find the peak, and good optimizers guarantee to do this.

Secondly, consider Fig. 3. If we are working without an optimizer and doing a detailed design for a realistically complex ore body, then we might be working away from the peak at 'A', where changes in pit tonnage can have a significant effect on the value of the pit. In fact, generations of mining engineers have learned that a series of small adjustments, involving a great deal of work, can significantly affect the profitability of the mine. Contrast this with starting from an optimized outline at 'B'. From this point, providing that ore and waste are kept in step with each other, it is difficult to go wrong. Certainly there is no need to experiment with small adjustments. Since, with modern software, we can plot this graph for real ore bodies, we can actually find out how much freedom of movement we have before we start the detailed design. In other words, designs based on optimized outlines are very much easier to do.

THE EFFECTS OF SCHEDULING ON THE OPTIMAL OUTLINE

When we schedule a pit, we plan the sequence in which various parts of it will be mined and the time interval in which each is to be mined. This affects the value of the mine

Table 2. Values of the pits if ore is worth $2 per tonne and if waste costs $1 per tonne.

Pit	Value
1	900
2	1600
3	2100
4	2400
5	2500
6	2400
7	2100
8	1600

because it determines when various items of revenue and expenditure will occur. This is important because the dollar we have today is more valuable to us than the dollar that we are going to receive or spend in a year's time. There are various reasons for this:

• Delayed revenue may increase our need to borrow funds and pay interest, thus reducing the effective revenue;

• Delayed revenue may not eventuate—one of the risk factors;

• Delayed expenditure may reduce our need to borrow funds and pay interest, thus reducing the effective expenditure;

• Something unexpected may go wrong with the operation—another risk factor; etc.

The standard way to allow for this is to discount next year's dollar by a certain percentage and to apply that idea cumulatively into the future. Thus we discount future revenues and costs by a particular discount rate and reduce them all to a net present value.

There are two discount rates. The **notional** discount rate is applied to actual revenues and costs which are likely to occur. That is, revenues and costs which follow the inflation rate. Thus the notional rate (typically 20%) includes an allowance for inflation. It is correct to use this, provided that we inflate our revenues and costs for future years. However, we are then in the position of guessing at the future inflation rate and then guessing at a figure to correct for it! It is easier to work out revenues and costs in today's dollars and then to use the **real** discount rate (typically 10%), which does not allow for inflation.

In what we will call **worst case** mining, each bench is mined completely before the next bench is started. Waste at the top of the outer shells is mined early, and the cost is discounted less than the revenue from the corresponding ore which is mined much later. This can make the outer shells uneconomic. The optimal pit for worst case mining is thus generally smaller than is indicated by simple optimization using today's costs and revenues. This can easily be seen by referring to Fig. 1.

In what we will call **best case** mining, each shell is mined in turn and thus the related ore and waste is mined in approximately the same time period. In this case, the optimal pit is usually close to the one obtained by simple optimization. Unfortunately, if we try to mine each shell separately, mining costs usually increase and cancel out some of the gains.

In small pits, worst case mining may be the only possibility. The larger the pit, the more opportunity there is for creative sequencing, and the closer it is possible to get to best case mining.

PRODUCTION OF A DETAILED DESIGN FROM AN OPTIMAL OUTLINE

The precise method used in creating a detailed pit design depends on the tools which are available. It may be done entirely by hand, or with varying degrees of computer assistance.

Whatever the method, the aim is to produce a detailed design which deviates as little as possible from the outline provided by optimization. Where deviation is unavoidable, we try to balance extra tonnage in one place with reduced tonnage in another. The resultant design should in most cases contain ore and waste tonnages very similar to those contained by the optimal outline. If it is not possible to achieve this, then it may be that the slopes were not set correctly for the optimization. For example, insufficient allowance may have been made for the effect of haul roads.

While all reasonable steps should be made to follow the optimal outline, the shape of the graph shown in Fig. 2 should be borne in mind. Provided that waste is not included without the ore which it uncovers, small deviations from the outline have little or no effect on the pit value. A useful concept is to say that the spirit of the outline should be followed rather than the detail. Certainly the square edges of the blocks on the outer surface of the outline are irrelevant. As a starting point, a smooth line should be drawn through them as is shown in Fig. 4. Remember that the block edges are artefacts, they do not represent geological or grade boundaries.

The achievement of the necessary minimum mining widths at the bottom of the pit is often cited as a problem with pit optimization. This problem is more apparent than real in that, for large disseminated or near horizontal ore bodies, the necessary adjustments at the bottom of the pit are usually easy, whereas, for steeply dipping reef structures, it may be possible to put extra constraints into the optimization so as to ensure the necessary width. In the remaining cases, some loss of pit value will be involved in adjusting the bottom of the pit, but it should never exceed 1 or 2%.

THE AVAILABLE OPTIMIZATION METHODS

All currently available methods of optimization attempt to find the optimal outline in terms of a block model. That is, they try to find the list of blocks which has the maximum total value while still obeying the slope constraints.

The enormity of this problem is seldom appreciated.

Trial and Error

Consider a trivial model with only one section and 10 benches of 10 blocks. If we take a very simple-minded approach, each of the 100 blocks can either be mined or not, so there are 2^{100} or 10^{30} alternatives, many of them not feasible. Even if a computer could assess a million alternatives a second, it would still take three million times the current age of the universe to find the best one!

If the allowable slope is one block up or down at each column change, and we use this information to ensure that we try only feasible alternatives, the number of alternatives is reduced to 10×3^9 or 200,000. A computer could easily assess this number of alternatives. However, if we extend the model to 10 sections, the number of alternatives rises to 10×2^{99} or about 10^{30} again, and we still have only 1000 blocks, which is insufficient for serious work.

Put simply, trial and error is useless.

Floating Cone

The floating cone method has been popular because it is easy to program and easy to understand. It works by searching through the block model for ore blocks and then assessing the value of the inverted cones which have to be mined to expose them. If the value of a cone is positive, it is mined out and all the blocks it includes are changed to air blocks. The search then continues.

Unfortunately, this simple-minded approach rarely finds the optimal pit because of two distinct problems; one causes it to omit profitable ore from the pit and the other causes it to include non-profitable ore.

The first occurs because it cannot try all possible combinations of ore blocks, as that would be a trial and error process, and we have seen that that is computationally unreasonable. Most pits are viable in part at least because numbers of ore blocks combine to pay for the stripping of waste above them, when no individual block or even close group of blocks can do so. The floating cone method cannot detect this co-operation between different parts of the ore body if neither part is viable in its own right.

The second occurs for slightly more technical reasons. In Fig. 5 there are three small ore bodies and their corresponding waste volumes, with their values and costs shown. A floating cone program will examine A and will find that the corresponding cone has a total value of $(40-20-30)$ $= -10$, and so is not worth mining. It will then examine B, will find a cone of value $(200-80-30) = +90$ and will convert it to air, leaving the values shown in Fig. 6.

If a floating cone program is to work correctly, whenever it converts a cone to air, it should start searching again at the top of the model. However, this is computationally very expensive so that most programs continue their search downwards and would consider C next.

At this time the cone for C has a total value of $(40-50+40-20) = +10$, so that the program mines it. This should not happen, because some of the value of ore body A is being used to help pay for the mining of waste

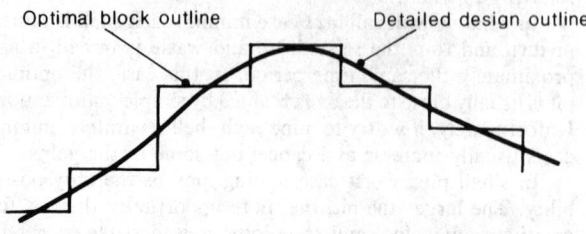

Fig. 4. Smoothing out the block outline.

Optimal block outline   Detailed design outline

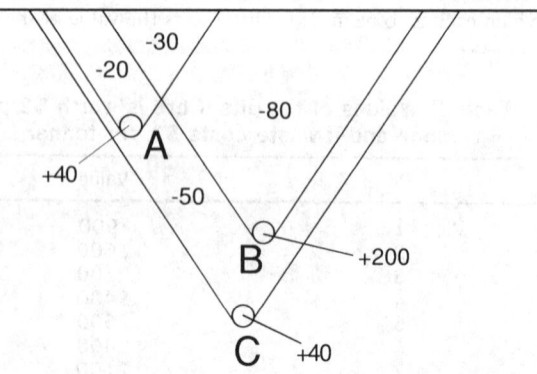

Fig. 5. Ore and waste values before floating cone run.

-30
-20
-80
+40
-50
A
B   +200
C   +40

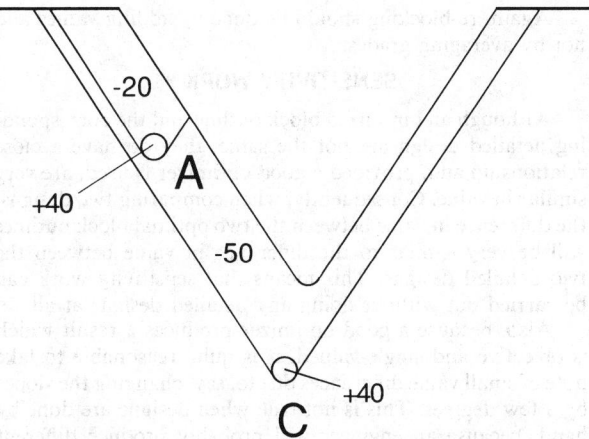

Fig. 6. Ore and waste values after the removal of ore body B and its corresponding waste.

(the −50 region) which is below it. The true optimal pit in this case includes A and B, but not C.

Apart from being easy to understand and program, the one advantage that the floating cone method has over other methods is that, if instead of using just one block the program uses a disk of blocks as its starting point, then this can ensure a particular minimum mining width at the bottom of the pit.

Two-Dimensional Lerchs-Grossmann Method

In 1965 Lerchs and Grossmann gave two different methods for open pit optimization in the same paper. One works on a single section at a time. It only handles slopes which are one block up or down and one across, so that the block proportions have to be chosen so as to create the required slopes. This method is easy to program and is reliable in what it does, but, since sections are optimized independently, there is no guarantee that successive sections can be joined up in a feasible manner. Consequently a good deal of manual adjustment is usually required to produce a detailed design. The end result is erratic and unlikely to be truly optimal.

Two later variants of this method exist. One (Johnson, Sharp, 1971) uses the two-dimensional method both along sections and across them, in an attempt to join them up. The other (Koenigsberg, 1982) uses a similar idea but works in both directions at once. Both are restricted to slopes which are defined by the block proportions and neither honors even these slopes at 45° to section. This last point is best illustrated by running the programs on a model which contains only one (very valuable) ore block. The resulting pit is diamond shaped rather than circular, with slopes correct in the E-W and N-S directions, but much too steep in between.

Three-Dimensional Lerchs-Grossmann and Network Flow

The second method given by Lerchs and Grossmann (1965) was based on a graph theory method, and Johnson (1968) published a network flow method of optimizing a pit. Both guarantee to find the optimum in three dimensions regardless of block proportions. Both, naturally, give the same result.

Both are difficult to program for a production environment where there are large numbers of blocks. Nevertheless this has been achieved and programs are now available which

can run on any computer from a PC upwards. Most of these use the Lerchs-Grossmann method.

Because these programs guarantee to find the sub-set of blocks with the absolute maximum value consistent with the slope constraints, the alterations to the pit outline caused by small slope or block value changes are reliable indicators of the effect of such changes. This has opened up the field of real sensitivity analysis, where the effects of slope, price and cost changes can be measured accurately. With other methods, only the crudest sensitivity work is possible.

This has led to the development of programs which automate some aspects of sensitivity analysis to the point where graphs of net present value against, say, total pit tonnage, can easily be plotted. Further mention of this will be made later.

CALCULATING BLOCK VALUES

The correct calculation of block values is essential for any optimization. If the block values are wrong, the optimized pit outline will also be wrong.

For optimization purposes, there are two basic rules which must be followed when calculating the value of a block.

The First Rule

Calculate the block value on the assumption that it **HAS** been uncovered and that it **WILL** be mined.

No allowance for assumed stripping ratios should be made, because stripping is precisely what pit optimization works out. If a stripping ratio is assumed when calculating the block values, the result of the optimization is being prejudged.

Similarly, take no notice of any pre-conceived breakeven cutoff. The use of a breakeven cutoff can be helpful in manual pit design; it is inappropriate for optimized pit design. A consequence of this is that a block model in which only rock containing grades above a breakeven cutoff is designated as ore, is also inappropriate for pit optimization.

The only relevant cutoff in this context is that grade at which the revenue from recovered product will just pay for the cost of processing and any extra mining cost which is only applicable to ore.

Second Rule

Include any on-going cost which would stop if mining were stopped.

This is because, when the optimization program is adding a block to the pit outline, it is effectively extending the life of the mine. It must therefore pay for all the costs involved in extending the life of the mine.

Incremental costs such as fuel costs, wages, etc. must obviously be included in the cost of mining or processing, whichever is involved.

Overhead costs **WHICH WILL STOP IF MINING STOPS** must also be included. If the mine throughput is to be limited by the overall mining capacity, then these overheads should be included in the mining costs. If the throughput is to be limited by the processing capacity, then these overheads should be included in the processing cost, because only the addition of an ore block extends the life of the mine.

Nonrecoverable upfront costs, such as the cost of building access roads, should not be included in the costs used in optimization. Although these may be paid for with a loan which is to be repaid over a number of years, these repayments will be required whether mining continues or not. If the value of the optimized pit is less than the nonrecoverable upfront costs, then the mine should not be proceeded with.

SURFACE MINING

474

BLOCK SIZES

There are four block sizes which are relevant in this work.

For Outlining the Ore Body

The size of the block that is needed for outlining the ore body depends on the shape and size of the ore body and on the particular computer modelling package that is being used. It may be quite small, which can lead to a model consisting of millions of blocks.

For Calculating Block Values

The value of blocks should be calculated with a block size which is similar to the selective mining size. That is, a parcel of rock should not be so small that it could not be mined separately, nor so large that grades are artificially smoothed. This block is sometimes bigger than that needed for outlining the ore body, requiring blocks to be combined and their grades averaged.

For Designing a Pit

There is now considerable experience in pit design using optimization techniques and, assuming that the pit occupies most of the width and length of the model and that the outline is not too convoluted, then a full model of 100,000 to 200,000 blocks is usually more than sufficient for pit design purposes. This leads to a block size which may be bigger than that for calculating values.

If it is necessary to re-block the value model, then it should be done by adding component block values and **NOT** by averaging grades.

For Sensitivity Work

If we want to do a series of optimizations using, say, different product prices so as to plot a graph of pit value against price, a model of 20,000 to 50,000 blocks will give just the same shape of graph with a very small shift of absolute value. Thus, most optimizations for sensitivity work can be done very quickly and this approach generally leads to a much more thorough sensitivity analysis.

Again, re-blocking should be done by adding values and not by averaging grades.

SENSITIVITY WORK

Although an optimized block outline and the corresponding detailed design are not the same, they do have a close relationship and, provided a good optimizer is used, are very similar in value. Consequently, when comparing two designs, the difference in value between the two optimal block outlines will be very similar to the difference in value between the two detailed designs. This means that sensitivity work can be carried out without doing any detailed designs at all.

Also, because a good optimizer produces a result which is objective and single-valued, it is quite reasonable to take note of small value differences due to, say, changing the slopes by a few degrees. This is not true when designs are done by hand, because an engineer will probably produce different designs on different days, without any change of slope.

During sensitivity work, we explore the economic and slope sensitivity of the mine. We sort out the general scale of mining and hence the operating costs. We decide approximately where the haul roads are to go and adjust the slopes in these regions to the average slope.

This requires a large number of quick optimization runs. However, it is probably the most valuable part of the whole design exercise because it inevitably leads to a much better understanding of the ore body and its economics. Graphs can be prepared which show how various characteristics of the mine, such as value or tonnage, are related to product price, costs, etc.

Probably the most significant graph is the one shown in Fig. 7. This relates net present value (NPV) to total pit tonnage for a given throughput and product price.

First, a set of optimal outlines is prepared, where each is optimal for a different product price. For some fixed product price, each of the outlines is then scheduled as though it was to be the limiting pit. If an automated practical scheduling scheme is available, it should be used. In producing Fig. 7, two limiting schedules have been used. Best case

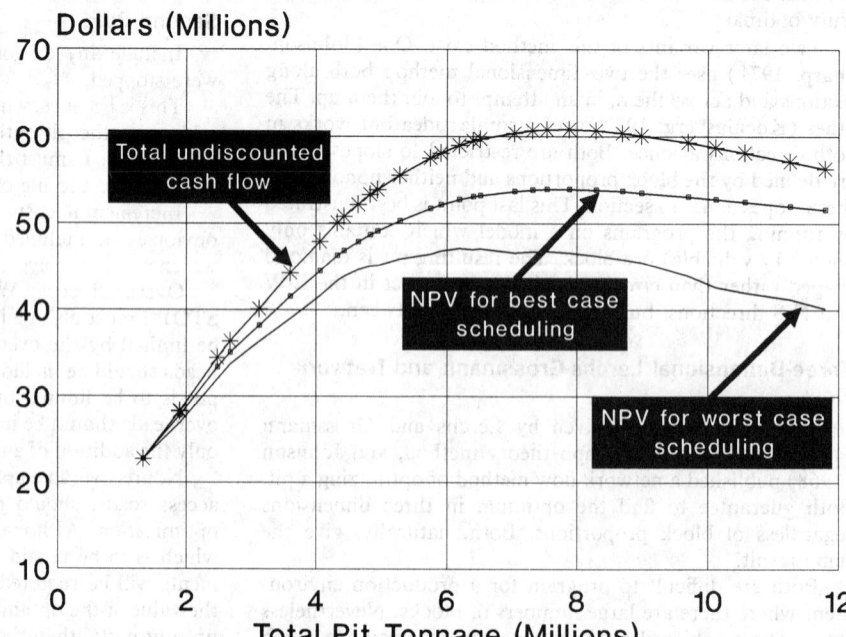

Fig. 7. Undiscounted, best case, and worst case NPV plotted against pit tonnage.

scheduling involves mining with many small pushbacks or cutbacks. Although in no sense a practical schedule, it indicates the highest possible NPV. Worst case scheduling involves completing the mining of each bench before starting the next. This is usually practical, but produces the lowest possible NPV.

The NPV for any practical mining schedule must lie somewhere between the two lower curves, with smaller pits tending towards the bottom curve and larger pits providing opportunities to get nearer to the middle curve.

This graph, which can be plotted for different product prices, is the single-most useful presentation known to the writer. It is meaningful to engineers, accountants, and management alike and can usefully be discussed in committee. It allows profit and corporate risk, in the form of mine life (pit tonnage), to be related and traded explicitly. Once a pit size has been chosen, it is easy to use the corresponding pit outline as a starting point for the detailed design.

This graph can be prepared by using any good optimizer and by doing a lot of work. However, software now exists which will produce the data for it automatically and quickly.

CONCLUSION

We have seen how good pit optimizers can be used not only to help design ultimate pit outlines, but also to carry out sensitivity analysis to an extent which is not possible without them.

Pit optimization is a tool which, used properly, can greatly speed and ease the process of pit design and can significantly increase the value of most pits. It can also be used to reduce the corporate risk involved in mining.

REFERENCE LIST

Johnson, T.B., 1968, "Optimum Open Pit Mine Scheduling," Ph.D. Diss. University of California, Berkeley, CA, 120 pp.

Johnson, T.B., and Sharpe, R.W., 1971, "Three Dimensional Dynamic Programming Method for Optimal Ultimate Pit Design," Report of Investigation 7553, US Bureau of Mines.

Koenigsberg, E., 1982, "The Optimum Contours of an Open Pit Mine: An Application of Dynamic Programming," *Proceedings*, 17th APCOM Symposium, AIME, New York, pp. 274-287.

Lerchs, H., and Grossmann, I.F., 1965, "Optimum Design of Open Pit Mines," *CIM Bulletin*, Canadian Institute of Mining and Metallurgy, Vol. 58, January.

5.4 Optimum Production Scheduling

Ernest L. Bohnet

The objective of production scheduling is to maximize the net present value and return on investment that can be derived from the extraction, concentration, and sale of some commodity from an ore deposit. The method and sequence of extraction, and the cutoff grade and production strategy will be affected by the following primary factors:

1. Location and distribution of the ore in respect to topography and elevation;
2. Mineral types, physical characteristics, and grade/tonnage distribution;
3. Direct operating expenses associated with mining, processing, and converting the commodity into a salable form;
4. Initial and replacement capital costs needed to commence and maintain the operation;
5. Indirect costs such as taxes and royalties;
6. Commodity recovery factors and value;
7. Market and capital constraints;
8. Political and environmental considerations.

The procedure used to establish the optimal mining schedule can be divided into three stages. The first defines the extraction order or mining sequence, the second defines a cutoff grade strategy that varies through time and will be optimal for a given set of production parameters, and the third defines which combination of production rates of the mine, mill, and refinery will be optimal, within the limits placed by logistical, financial, marketing, and other constraints.

In order to develop an optimum production schedule, a sequence or extraction order inside of the so-called ultimate pit must first be determined. The extraction sequence depends on two subsets of parameters. The first deals with the strip ratio associated with recovering the ore, the grade of that ore, and the physical location of that ore in respect to availability through time.

The second subset of parameters consists of costs associated with starting and maintaining the whole operation. Direct operating costs can be used to define a breakeven cutoff grade and strip ratio, but the objective of mine planning is to devise a strategy that will optimize the total investment. Operating at breakeven cutoffs and strip ratios is only optimal for the final phase at the end of the mine life.

Before the mine production planning commences, a great deal of work has already been completed in exploration and modeling the deposit. From this work, a number of tentative assumptions have been made, including the most probable mining method and hence, the bench height, type, and approximate size of the loading equipment and the mining selectivity. Other test work and assumptions will also have been made regarding the type of process needed to recover the commodity. These parameters will be used to estimate the most probable range of mining and processing costs.

The design of the mining phases can be accomplished by rough manual approximation after review of the bench plans and cross sections, or analytically by computer techniques. Each method has advantages and disadvantages as applied by an experienced engineer, but the method chosen is determined by the accuracy requirements and available funding. If the study objective is very preliminary, with little basic data available, then manual methods can be justified. If the study is to be a sound basis for investment and development of the mine, and a great deal of information has been collected, then a very thorough computer analysis is warranted.

Computer designed phases can be determined by feeding the data developed and stored in a computer block model into a set of programs that can be used to calculate an economic phase limit. The objective is to develop three dimensional equal profit potential surfaces throughout the mineral deposit. Each surface has to be sufficiently spaced apart to allow adequate room for mining the slices between the surfaces. Since the distance between equal profit potential surfaces will vary, some manual adjustments will be required, as well as the addition of haul roads out of a phase and if required, access left for the next phase. See Fig. 1.

Manual methods depend on having an experienced engineer review the bench plans and cross sections through the deposit to visually pick out the higher grade targets that have reasonable strip ratios. For example, it would be incorrect to first target high grade areas for mining having very high strip ratios that reduce the net value of the recovered ore below the net value of medium grade ore in another area with much less stripping. The manual method is only a first step estimate and, therefore, it will not be as accurate as a computerized technique.

Computerized pit limit determinations can be made using the 2-D Lerchs-Grossman method, or three-dimensional techniques, such as the floating cone or the 3-D Borgmann pit design method. The results of the last two systems are nearly the same, the difference being that the floating cone is a computerized trial-and-error method while the Borgmann is an analytical method.

The strategy used in developing the computerized pit phases is to use higher costs or lower commodity prices in the initial phase and then, for each successive phase, lower costs or higher commodity prices are used. The net effect of this strategy is that the initial phase will have a high breakeven cutoff grade and high net value per ton of ore mined, and each following phase will have a lower cutoff grade and net value per ton of ore.

The accuracy of the estimated costs, recoveries, and commodity price parameters need only be reasonably precise, since they are only used to determine the location of the ores with the relative highest to lowest net values.

Cost estimates needed for determining the extraction sequence can be broken down into three distinct major categories: (1) costs per ton of material mined; (2) costs per ton of ore treated; and (3) costs per pound of commodity produced.

The costs per ton of material mined include the direct mining costs per ton for the drilling, blasting, loading, hauling, ancillary equipment, and mine general and administrative functions. The cost per ton of the capital and replacement expenditures for mine mobile equipment related to the total material mined is also included because the major mobile mining equipment is consumed in approximate direct proportion to the amount of material handled by the equipment, unlike the major initial capital cost components of the plant and infrastructure. The plant components that are replaced

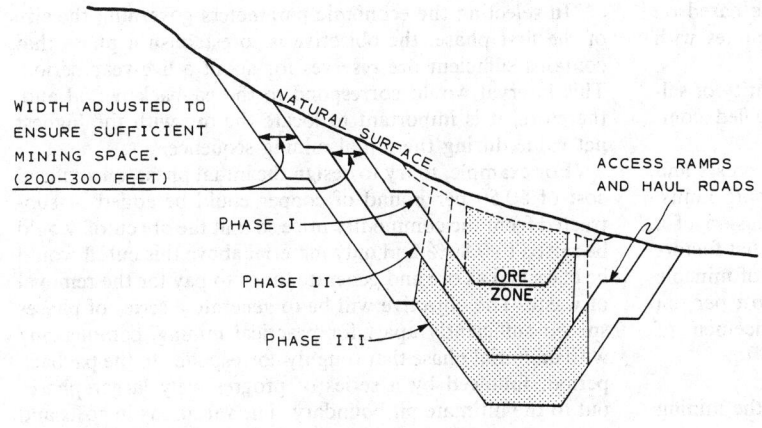

WIDTH ADJUSTED TO
ENSURE SUFFICIENT
MINING SPACE.
(200-300 FEET)

NATURAL SURFACE

ACCESS RAMPS
AND HAUL ROADS

PHASE I

PHASE II

PHASE III

ORE
ZONE

Fig. 1. Internal pit phases: typical pit cross section.

PHASE I DESIGNED USING A 0.80% COPPER CUTOFF
CONTAINS ORE OF HIGHEST NET VALUE.

PHASE II DESIGNED USING A 0.60% COPPER CUTOFF
CONTAINS ORE OF MEDIUM NET VALUE.

PHASE III DESIGNED USING A 0.30% COPPER CUTOFF
CONTAINS ORE OF THE LOWEST NET VALUE,
AND ORE IN LAST FINITE SLICE IN PIT WALL
CONTAINS ORE OF ZERO NET VALUE.

or repaired are usually included in the operating maintenance costs per ton of ore, and are not related to the total tonnage moved in the mine.

The depreciation cost to cover mobile surface mine equipment purchase and replacement will usually be in the range of $0.15 to $0.25 per ton of material mined. The magnitude of the cost will depend on the size, type, and anticipated life of the equipment, the mine production level, the haulage distances, and the work schedule.

The most important reason for including mine mobile equipment depreciation is that the method of producing three-dimensional equal profit surfaces in the ore deposit must consider all relevant costs per ton of material mined. If the equipment depreciation were not included, then the cost per ton of material mined will be understated and the phase design will tend to move out into higher stripping ratio areas. This would not matter if the relative strip ratios were equal in all directions around the pit perimeter, but this is rarely so. In most mineral deposits, there will be areas of high and low strip ratios, so moving too far out into a high strip ratio area will lower the net value of the phase, and the 3-D surface generated will no longer be of equal value.

The significance of this costing philosophy can be realized in mining operations where the direct operating costs per ton of material are low relative to the mine equipment depreciation costs per ton of material.

As an example, compare two surface mining operations, one located in the Philippines, the other in Alaska. The net value of a mining increment in each of the pit walls is compared.

	Philippines	**Alaska**
Direct mining cost/ton of material	$0.25	$0.80
Gross value/ton of ore f.o.b. mine	$3.00	$3.00
Indicated breakeven strip ratio	12-1＝11:1	3.75-1＝2.75:1

If $0.20/ton of material is added for mobile mine equipment depreciation: breakeven strip ratio		
ratio	5.67:1	2.00:1
% change	48%	27%

In summary, the net value of the ore in an incremental slice has to be sufficient to carry all direct operating costs and the initial and replacement expenditures for mobile mine equipment. If the mine equipment depreciation is not included, areas with a much higher breakeven strip ratio will be incorporated into the mine phase, resulting in an overstatement of the net value of ore derived from those high strip ratio areas in the phase.

The second cost collection area is the cost per ton of ore treated and includes expenditures applied to the ore once it has left the mine area. These costs are not related to the total quantity of material removed from the surface mine, but only applicable to the ore tonnage to be treated. Direct costs applied would be:

1. Extra costs associated with transporting the ore to treatment facilities;
2. Crushing and grinding costs;
3. Concentrating cost; and
4. Overhead costs to cover site and head office administrative and general expenditures as well as marketing, sales, and property management costs.

The third cost category is the expenditures incurred per unit of salable commodity(ies) produced. This would cover the sums spent for concentrate handling and transportation, smelting, and refining, and any royalties or taxes that relate to gross revenues rather than profits.

In addition, a certain amount could be inserted into this category to ensure a minimum profit per pound of salable product.

In order to determine the quantity of salable product, recoveries have to be estimated for the concentrating, smelting, and refining processes. Recoveries should be based on pilot plant results or on recoveries obtained at mines with similar ores and processes.

Gross revenues are determined from the quantity of salable commodity produced multiplied by a specified commodity price.

There are specific reasons for including some costs and excluding others in the determination of the mining limits. The best manner to justify the inclusion or exclusion of a cost parameter is to first answer the question of what factors are reasonably known or unknown. The quantity of minable ore reserves, the strip ratio, and the associated cost per ton of ore for capital are not known at the commencement of the design. Reasonable estimates can be made as to:

1. Mining cost per ton of material;
2. Mining equipment depreciation or cost of the mining equipment consumed per ton of material mined;
3. Ore treatment costs;
4. General overhead costs per ton of ore;
5. Anticipated recoveries;
6. Direct charges per pound of salable product;
7. Commodity price; and
8. Minimum profit expected per pound of commodity produced.

Using these estimated factors, a breakeven cutoff grade can be determined and the final pit limits and total minable ore reserves determined for the breakeven cutoff grade. From these basic parameters, optimization routines can be applied in order to determine the best extraction sequence.

Table 1 illustrates an example of typical base economic parameters used to determine phase increments in a copper mine. If the values were used without modification, the ultimate pit limits would be determined. Since the objective is to define internal phases of higher net value, either an artificial cost is added to the cost per unit of salable product or the commodity value is lowered.

In selecting the economic parameters governing the size of the first phase, the objective is to establish a phase that contains sufficient ore reserves for about a five-year period. This interval would correspond to the payback period and, therefore, it is important to locate the ore with the highest net value during the initial mining sequence.

For example, to try to design the initial phase, an artificial cost of $0.50 per pound of copper could be added or subtracted from the commodity price so that the ore cutoff would be raised to 0.80% and only material above this cutoff would be classified as ore and generate funds to pay for the removal of waste. The objective will be to generate a series of phases spaced sufficiently apart for practical mining, commencing with an initial phase that roughly corresponds to the payback period, followed by a series of progressively larger phases out to the ultimate pit boundary. The variations in costs and the number of phases are determined by combining judgment with a trial-and-error method. One computerized technique that can provide guidance in selecting the various phases and economic parameters is the 2-D Lerchs-Grossman method. By selecting a few typical sections through the ore deposit, relatively inexpensive 2-D pit limits can be determined for various economic parameters. These results can then be used to set the variable costs needed to determine the pit limits using 3-D computer techniques.

The preceding discussion has described the method used to define the internal pit phases. This is the first stage in defining the optimum production and cutoff grade strategy. The second step is the determination of the optimum cutoff grade strategy to be used from one phase to the next, for a defined trial production rate. Only cutoff grades equal to, or less than, the cutoff grade for a particular phase can be used for determining the optimum cutoff grade. If an attempt is made to use a higher cutoff grade, the physical shape of the phase will no longer be valid since the pit wall location depended on revenues from a specified amount of ore that

Table 1. Basic Economic Parameters Used to Determine a Sequence of Phases (Copper Mine Example)

a. Direct mining costs per ton of material:	$0.80
Mobile mining equipment depreciation per ton of material:	$0.20
Total costs for category (a), per ton of material mined:	$1.00
b. Ore treatment costs per ton of ore:	$3.00
General and overhead costs per ton of ore:	$0.75
Total costs for category (b), per ton of ore processed:	$3.75
c. Smelting, refining, and transportation costs per lb of copper:	$0.35
Insurance, property taxes, and royalties per lb of copper:	$0.05
Total costs for category (c), per lb of copper recovered:	$0.40
(A minimum profit or variable cost can be added to this total)	
d. Plant, smelting, and refining recoveries:	85% or 0.85
e. Commodity price per lb of copper recovered:	$1.25
(This value can be varied to expand or contract phases)	

$$\text{Breakeven cutoff equals: } \frac{(a) + (b)}{\frac{2000(d)[(e) - (c)]}{100}} = 0.33\%$$

$$\text{Internal cutoff grade equals: } \frac{(b)}{\frac{2000(d)[(e) - (c)]}{100}} = 0.26\%$$

will no longer be available. That is, previous low grade ore blocks will now be waste with negative revenues.

In the situation where low grade rock has to be removed from the pit to expose ore, a lower cutoff grade can be used to determine if that low grade material should be processed. This lower cutoff grade is called the internal cutoff grade and is determined by ignoring the mining cost in the breakeven cutoff grade calculation.

The optimum cutoff grade will usually start at a somewhat higher level than the breakeven cutoff grade and will be reduced in time to equal the internal breakeven cutoff grade. The higher the production level, or for a marginal deposit, the less difference there will be between the optimum and breakeven cutoff grades.

The optimum production level can be determined on a strictly economic basis, but with large ore deposits, other constraints such as mining logistics, marketing, and financing will provide limits.

The best strategy can be determined graphically by varying the production rate and cutoff grade for a number of combinations. Fig. 2 illustrates the results from twelve strategies: three production rates and four cutoff grade alternatives. For example, if three pit phases were determined using breakeven cutoff grades of 0.80, 0.60, and 0.30% copper, then four alternative cutoff grade strategies could be:

| | Cutoff grades used inside of each phase | | |
Alternative	0.80 phase	0.60 phase	0.30 phase
1	0.8	0.6	0.30
2	0.6	0.6	0.30
3	0.45	0.45	0.30
4	0.30	0.30	0.30

These cutoff strategies can then be applied to each of the three alternative production rates.

A second, more rigorous method to determine the best mine/mill/refinery production rate and cutoff strategy is to use the approach presented by K. F. Lane. Lane's method considers the constraints placed on the operation by the mine, mill, and refinery (or market). Utilizing the grade/tonnage curves for each of the phases developed in the previous stage and combining this with the three categories of costs, plus a fixed cost per year, the optimum cutoff grade strategy on a net present value basis can then be determined, for a given set of production parameters of the mine, mill, and refinery (market). This analysis is more accurate than a graphic solution since the program will fluctuate the cutoff grades through time to match the unique physical distribution of the ore in the various increments.

In order to develop a practical extraction sequence using this method, a scheduling program has to be first applied to the phases to define the progressive mining sequence. This allows the program to recognize the internal stripping variations and permits the progressive removal of material from more than one phase simultaneously; it permits prestripping of an outer phase as ore is being drawn from an internal phase.

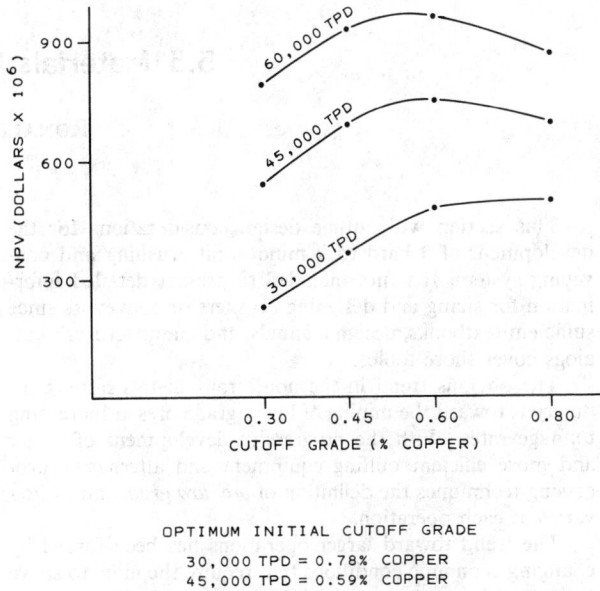

OPTIMUM INITIAL CUTOFF GRADE

30,000 TPD = 0.78% COPPER
45,000 TPD = 0.59% COPPER
60,000 TPD = 0.56% COPPER

Fig. 2. Graphic solution to maximize NPV (NPV vs. production rate and cutoff grade alternatives).

In order to determine the optimum production capacities of the mine, mill, and refinery that will maximize the net present value, a trial-and-error method is used. Initial capital costs estimates are subtracted from the NPV to allow ranking of the various production rate alternatives.

Lane's method allows the mine planner to readily try more alternatives and complete sensitivity analyses on commodity price, recoveries, and other cost parameters.

The preceding discussion has outlined the procedures that can be used to determine the optimum production schedule and cutoff grade strategy. The completed analysis will serve to guide the engineer in detailing the production schedule for both short and long range plans. In cases where the operation is in existence, the same analysis can be completed so that fluctuations in commodity values and costs can be quantified and the plan altered accordingly to optimize the extraction strategy.

It should be noted that in following through the procedures outlined, the basic pit design should be completed first with few or no constraints. The design can then be modified and the costs resulting from constraints can then be quantified. Constraints may be the number of working faces needed for a particular production rate, waste dump locations, drainage routes, property boundaries, ore delivery points, market capacity, environmental constraints, and the availability of personnel, equipment, and financing.

Designing the best and most practical production schedule for each unique ore deposit is a complex task, and only by using a logical procedure that isolates and provides a solution for one set of variables at a time can a satisfactory and optimum solution be determined.

5.5 Materials Handling Ex-Mine

RONALD G. REED

This section will outline design considerations for the development of a hard rock mine in-pit crushing and conveying system. It is not intended to present detailed information for sizing and designing crushers or conveyors since sufficient textbooks, design manuals, and manufacturer's catalogs cover those topics.

The obvious trend in the nonferrous metals mining industry is toward the mining of lower grade ores at increasing tonnage rates. With the progressive development of larger and more efficient milling equipment and alternative processing techniques the definition of *ore, low grade,* and *waste* varies at each operation.

The trend toward larger operations has been forced by changing economic conditions that require the mine to strive to lower costs per unit of product recovered. The primary cost elements that determine the profitability of a project are commodity price, capital cost, and operating expenses such as energy (i.e., electricity, fuel), labor, equipment, maintenance, taxes, and environmental protection measures.

Some of the main characteristics that make each mining project distinctive are the following:

1. Location
2. Topography
3. Size and type of deposit
4. Mineral grades/distribution
5. Weather conditions
6. Excess or lack of water
7. Type and cost of energy available
8. Cost of obtaining trained maintenance and operation personnel
9. Management/company objectives

Therefore, any optimum pit design must take into account both the character of the mine and the economic factors as reflected in the costs to recover each unit of salable product.

In the past, the primary crushing, fine crushing, and mill complex tended to be located in relative close proximity and the majority of the horizontal and vertical travel distances from the ore source to the crushing station was handled by truck haulage. Due to rising fuel and maintenance costs, economic conditions have forced the pit designer to minimize the distance the trucks have to transport the ore, and to bring the primary crusher closer to the source and thus utilize conveyors to perform a much larger proportion of the ore transport requirements.

Data generated from actual installations have shown that properly designed crushing and conveying systems compare to truck haulage systems as follows:

1. Significantly lower operating and maintenance costs.
2. Higher initial capital costs, but with lower present value costs when compared to the life of the operation.
3. Improved foul weather operating conditions.
4. Can provide comparable operating flexibility in certain circumstances.

As a comparative example, a study was performed to determine the capital and operating costs of crushing and conveying of waste material from a bench of elevation 1 220 m (4,000 ft). This location was chosen because of the space needed around a crushing station that would allow four 155-t (170-st) trucks to dump simultaneously to a two-crusher station.

A weighted average haul distance of 2 880 m (9,450 ft) on an 8% grade was used for comparing the crusher-conveyor cost with the truck haulage expenditures. The use of trucks allows an incremental buildup of the dump and the average dump elevation height was 1 376 m (4,516 ft) in contrast to the final dump height of 1 405 m (4,610 ft), which was used in the crusher-conveyor study.

Using an annual tonnage of 41.73 Mt (46,000,000 st), truck haulage operating costs were calculated:

Haul distance (crusher to dump)	Haul time (minutes)
2 880 m (9,450 ft) of that 1 966 m (6,450 ft) is at 8% with 915 m (3,000 ft) level	15.30

This 2 880 m (9,450-ft) haul distance required a minimum of 14 trucks for haulage. Maintenance and fuel costs amounted to $0.1694/t ($0.1537/st). The costs per tonne for the crushing and conveying system were calculated to be $0.0872/t ($0.0791/st). The installation charge for the fixed crushing and conveying system composed of two 1.37 by 1.88 m (54- by 74-in.) gyratory crushers, 2 060 m (6,760 ft) of 1 500 mm (60-in.) conveyor, and a tripper and stacker was estimated to cost $28,000,000.

For this comparison example, the following assumptions were made:

1. Minimum 10-yr plant life with crusher station area accessible during the full plant life.

2. The waste system is part of an expansion to the existing plant, therefore, making a phased start-up of the waste system possible.

3. Trucks purchased by year:

Period	Truck purchases
Preproduction	7
Year 1	6
Year 2	2
Year 3	1
Year 4	3
Total	19

4. Eight-year truck life.

5. Conveying system usable for full plant life. Replacement parts for the conveyors are carried in the operating costs.

6. Straight-line depreciation for five years.

This study is purely hypothetical and may seem somewhat simplistic. Many other factors must be considered for any one operation as mentioned earlier in this section. Table 1 illustrates the economic benefits of an in-pit crushing-conveying system.

This crushing-conveying system as compared to truck haulage has a payback period of a little over three years, and

Table 1. Economic Benefit Comparison In-Pit Crushing vs. Truck Haulage
(All Dollars × 10⁶)

Year	PP	1	2	3	4	5	6	7	8	9	10
Production (tonnes × 10⁶)	15.4	28.1	32.7	34.5	41.7	41.7	41.7	41.7	41.7	41.7	41.7
Trucks											
Op Costs	−2.60	−4.76	−5.53	−5.84	−7.07	−7.07	−7.07	−7.07	−7.07	−7.07	−7.07
Capital	−4.84	−4.15	−1.38	−0.69	−2.07	—	—	—	−2.76	−6.22	−1.38
Depreciation	0.45	0.83	0.96	1.02	1.21	0.76	0.38	0.25	0.44	0.82	0.95
Cash flow	−6.99	−8.08	−5.95	−5.51	−7.93	−6.31	−6.69	−6.82	−9.39	−12.47	−7.50
Conveyors											
Op Costs	−1.34	−2.45	−2.85	−3.00	−3.64	−3.64	−3.64	−3.64	−3.64	−3.64	−3.64
Capital	−28.00	—	—	—	—	—	—	—	—	—	—
Depreciation	2.58	2.58	2.58	2.58	2.58	—	—	—	—	—	—
Cash flow	−26.76	0.13	−0.27	−0.42	−1.06	−3.64	−3.64	−3.64	−3.64	−3.64	−3.64
Cash Flow (Conv-Trucks)	−19.77	8.21	5.68	5.09	6.87	2.67	3.05	3.18	5.75	8.83	3.86

The net present value at a 15% discount rate is $8,124,000.

Table 2. Crusher System Characteristics Desired by Mine Personnel

Gyratory type

Large capacity
(2.27-3.63 kt/h, or 2,500-4,000 stph)

Freedom from clogging

Maximum frequency of moves (one/year)

Average frequency of moves (one every two years)

Large feed openings
(1.37 m or 54-in. minimum)

Ability to crush rock with compressive strength
up to 32.2 kt/m (50,000 psi)

High reliability of system (85%)

Freedom from bridging

Low maintenance cost

All-weather operation
(−40 to +50 C or −40 to +120 F, rain, snow)

22 hours/day operation, 350 days/year

Moderate noise

Dust control at transfer points

Operate in a pit with 12-15 m (40-50 ft) bench heights

Ability to move up 10% maximum grades
(15 m or 50 ft wide)

Maximum relocation time of two weeks

A surge capacity of 360 t (400 st)

Ideally compatible with trucks or trains

Have a minimum of conveyors
(but also contain redundancy where necessary)

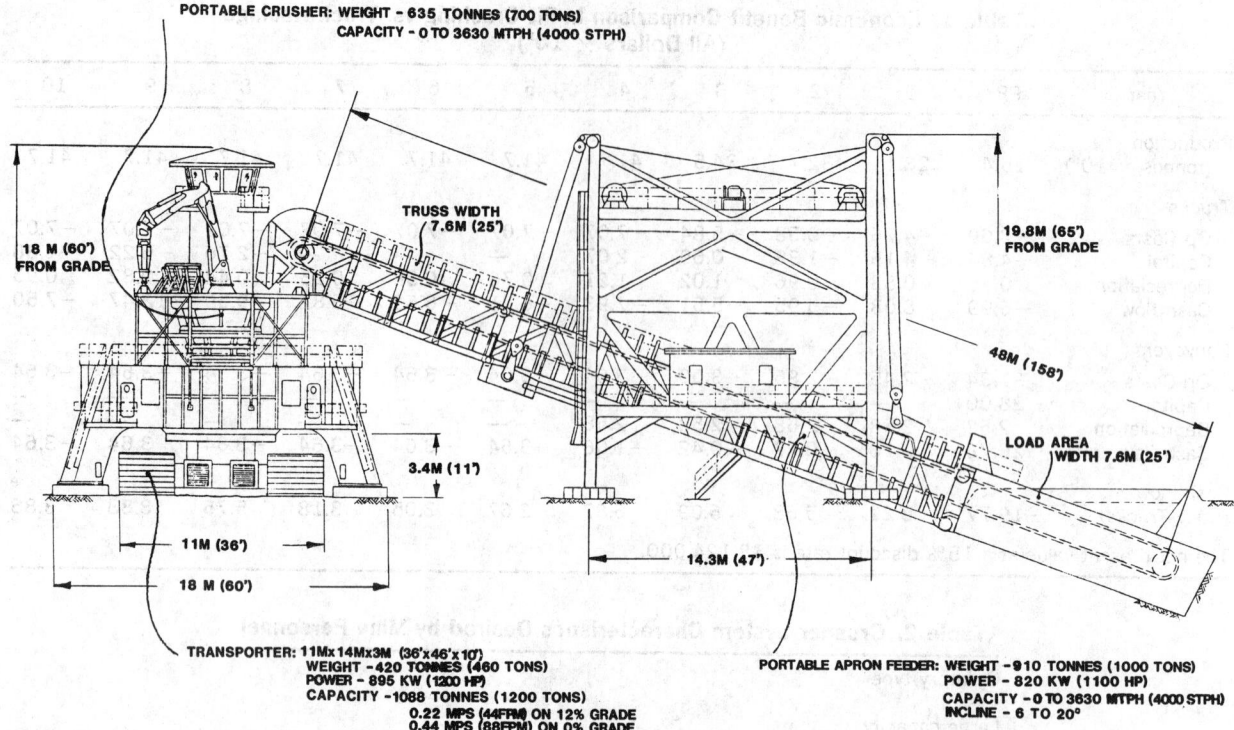

PORTABLE CRUSHER: WEIGHT - 635 TONNES (700 TONS)
CAPACITY - 0 TO 3630 MTPH (4000 STPH)

TRUSS WIDTH
7.6M (25')

18 M (60')
FROM GRADE

19.8M (65')
FROM GRADE

48M (158')

3.4M (11')

LOAD AREA
WIDTH 7.6M (25')

11M (36')

18 M (60')

14.3M (47')

TRANSPORTER: 11Mx14Mx3M (36'x46'x10')
WEIGHT - 420 TONNES (460 TONS)
POWER - 895 KW (1200 HP)
CAPACITY - 1088 TONNES (1200 TONS)
0.22 MPS (44FPM) ON 12% GRADE
0.44 MPS (88FPM) ON 0% GRADE

PORTABLE APRON FEEDER: WEIGHT - 910 TONNES (1000 TONS)
POWER - 820 KW (1100 HP)
CAPACITY - 0 TO 3630 MTPH (4000 STPH)
INCLINE - 6 TO 20°

Fig. 1. Portable crusher and feeder at Sierrita mine, Duval Corp.

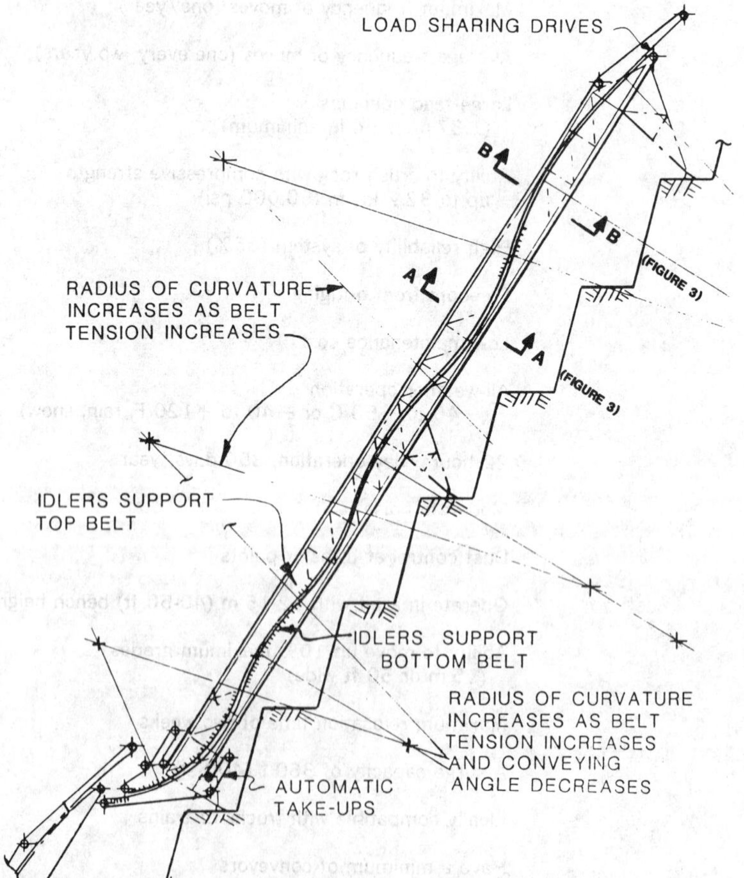

LOAD SHARING DRIVES

B

B

(FIGURE 3)

A

A

(FIGURE 3)

RADIUS OF CURVATURE
INCREASES AS BELT
TENSION INCREASES

Fig. 2. Schematic of snake sandwich conveyor.

IDLERS SUPPORT
TOP BELT

IDLERS SUPPORT
BOTTOM BELT

RADIUS OF CURVATURE
INCREASES AS BELT
TENSION INCREASES
AND CONVEYING
ANGLE DECREASES

AUTOMATIC
TAKE-UPS

over the 10-yr life has a rate of return of 26.42%. It should be noted that ore usually has to be crushed as preparation for further milling operations. Therefore, the payback for an ore crushing system would be shorter.

With the crushing and conveying system offering potential cost savings, the main drawback is the lack of mobility of the fixed crushing-conveying system.

In certain circumstances with the large tonnages being mined, it can become increasingly difficult to find suitable locations for a fixed crusher/conveying system that will have sufficient longevity to pay for itself.

Mining industry surveys sponsored by the US Bureau of Mines have determined it is difficult to find in-pit crusher locations that are both economically advantageous and that have a useful consecutive life of more than two years. It is obvious that with the dismantling and erection of *permanent systems*, the costs become prohibitive. The studies have also determined that what is needed is a reliable, reasonably priced, *portable* crushing and conveying system. The findings of this survey are listed in Table 2.

As further reported, this system consisting of a crusher, rock breaker, crane, and hopper-feeder would cost between $6 and 8 million, plus erection costs. This is roughly equivalent to the installed costs of one permanent single crusher.

A system using a 1.52 by 2.26 m (60-in. by 89-in.) gyratory crusher (Fig. 1) has been installed at the Duval Corp.'s Sierrita-Esperanza open pit mine south of Tucson, Arizona. Studies conducted by Sierrita indicate a cost savings of $0.29 per ton of rock processed by the portable crushing system can be achieved.

The in-pit portable crusher is definitely an alternative worth considering for any large mining operation that has to handle more than 9.1 Mt (10 million st) per year. An in-pit crushing system does have many design considerations that could affect pit production, such as (1) site preparation, (2) alternative ore/waste handling system during crusher downtime, and (3) downtime during system dismantling and re-erection.

Mr. Calvert D. Iles, of Duval Corp., stated, "As a mine becomes more dependent on conveyor haulage, the effects of downtime become increasingly severe." Crusher sites have to be located to minimize the lifts of truck haulage and positioned in the mine plan to minimize moves and possible shutdowns.

The Sierrita in-pit crushing system is made portable by a Dutec (Duval Technology Engineering and Construction Co.) designed transporter. This transporter is capable of moving the apron feeder, crusher, and belt conveyor drive modules. The transporter, measuring 11 m (36 ft) in width and 14 m (46 ft) in length, can handle loads up to 1 088 t (1,200 st) on a 12% grade. The load is kept level when on a ramp by four hydraulic cylinders that adjust automatically.

A mobile 1.37 by 1.89 m (54-in. by 74-in.) gyratory crusher was installed in 1980 at Phosphate Development Corp. Ltd., South Africa (FOSKOR) to crush up to 2 700 t/h (2,950 stph) of run-of-mine ore to 250 mm (-10 in.). (See Table 3.)

Although definite strides are being taken in the technology of in-pit crushing, in-pit conveying is still a problem. With maximum design angles of about 15°, the conveyors still have to be routed around the pit, across roads, and through embankments.

Studies are currently underway to determine what can be done to improve the technology of in-pit belt conveying. Consideration is being given to the loop belt concept by Stephens-Adamson. This method works on the principle that

Table 3. General Specifications of Mobile Crusher (FOSKOR)

Feed hopper capacity	30 m³
Primary apron feeder	
Width	2.4 m
Length (c-c head and tail sprockets)	23.6 m
Inclination	27°
Conveying speed	0.08-2.4 m/sec through variable speed hydraulic drive
Drive motors	2 × 180 kW hydraulic
Crusher	A.C. 54/74 gyratory with 450-kW drive from 3,300-V power; rated 2 700 t/hr
Crusher Discharge Apron Feed	
Width	2.6 m
Length (c-c head and tail sprockets)	9.2 m
Inclination	15°
Conveying speed	0.3 m/sec
Drive motor	1 × 55 kW
Travel rollers, chain, and drive sprocket are CAT D9. Entire apron feeder can be moved to rear for crusher maintenance.	
Slewing belt conveyor	
Width	1.6 m
Length (c-c head and tail pulleys)	15.0 m
Belt speed	1.5 m/sec
Luffing inclination	0-15°
Max. discharge height	4.5 m
Slewing range	90°
Walking mechanism	
Ground pressure when crushing	200 kPa
walking	240 kPa
Max. gradient (walking)	1:10
Walking speed	1.0 m/min
Length of stride	0.6 m
Lifting cylinders	3 × 400 mm dia
Horizontal stride cylinders	3 × 400 mm dia
Hydraulic pumps	3 × 180 kW
Operating pressure	About 180 bar
Electrical data	
Supply voltage	3,300 V
Crusher motor	3,300 V
Other motors	500 V
Auxiliary transformer	1,000 kVA, 3,300/500 V
Installed power for crushing operation	Approx. 920 kW
walking operation	Approx. 580 kW

depends on the radial forces generated by putting S curves in the profile of a *sandwiched* conveyor. By continually varying the curvature, a lift is achieved within "geometrical constraints," conforming to specified mine slopes (see Figs. 2 and 3).

Many high angle conveying concepts have been studied and the sandwich belt conveyor seems the most promising.

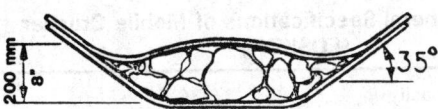

CARRYING BELT SUPPORTED
ON 35° TROUGHING IDLERS
SECTION A

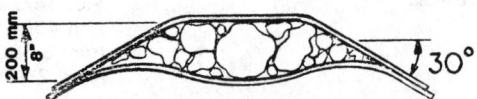

COVER BELT SUPPORTED
ON 30° TROUGHING IDLERS
SECTION B

Fig. 3. Scaled representation of belt sandwich and material [1500 mm (60-in.) conveyor].

Tonnages above 907 t/h (1,000 stph) and high lifts of up to 107 m (350 ft) per module should be achieved. A prototype using this method has not yet been made.

It is obvious that the technology of in-pit crushing and conveying is changing. The rapidity of this change depends mainly on economics. These methods may not provide solutions to every open pit mine's current economic problems, but in-pit crushing and conveying methods are becoming significantly more practical and will become more economically attractive in both existing and future surface mining operations.

REFERENCES

Anon., 1982, "FOSKOR Pioneers a Mobile-Crusher," *Engineering and Mining Journal,* Nov.

Bowman, E. D., Jr., and Ilves, C. D., 1983, "Duval's Portable In-Pit Crushing System," paper presented at 44th Annual Mining Symposium, Duluth, Minnesota, Jan. 13.

Frizzell, E. M. and Utley, R. W., 1983, "U.S.B.M. Designs In-Pit Movable Crusher Based on Mine Personnel Survey," *Mining Engineering,* Apr.

Ilves, C. D., 1983, "Costs of In-Pit Crushing," *Mining Engineering,* Apr.

Santos, J. A. and Frizzell, E. M., "Evolution of Sandwich Belt High Angle Conveyors," paper presented at Fourth Annual Meeting of The Canadian Institute of Mining and Metallurgy.

5.6. Waste Disposal—Planning and Environmental Protection Aspects

ERNEST L. BOHNET AND LUTZ KUNZE

DUMP DESIGN

A waste dump is an area in which a surface mining operation can dispose of low grade and/or barren material that has to be removed from the pit to expose higher grade material. In some instances, material has to be removed for other indirect reasons, such as pit wall stabilization and for haul road construction.

The first step in designing a dump is the selection of a site or sites that will be suitable to handle the volume of waste rock to be removed during the mine's life. Site selection will depend on a number of factors, the most important of which are:

1) Pit location and size through time.
2) Topography.
3) Waste rock volumes by time and source.
4) Property boundaries.
5) Existing drainage routes.
6) Reclamation requirements.
7) Foundation conditions.
8) Material handling equipment.

All of these parameters will be considered during the site selection process. Once a site or number of alternative locations have been selected, the designing of the dumps can commence, using the same points utilized in defining the best potential dump locations.

The objective of dump planning is to design a series of waste disposal phases that will minimize the horizontal and vertical distances between the source and the disposal area. Since material handling costs are usually the largest single component of the mining cost, well designed dumps play a very important and critical role, affecting the expense of the total operation.

The pit mining sequence and production schedule will be completed prior to dump design with the objective of maximizing the return on the investment. Therefore, two of the most important parameters affecting dump design have been set before any design effort commences: the pit location and size through time and the waste production schedule and source location. These two parameters define where the dumps can start, how fast they will advance, and the ultimate volume that the disposal area must contain. The location where dumping can commence may not necessarily be outside of the pit limits. In some instances, internal dumping may be the most economical and practical method of establishing haul roads to the disposal area or to later pit phases. Also, as an alternative, it may be wiser to dump short and rehandle the material in the future if the economic benefits of this can be demonstrated. This can affect the pit design in the sense that later phases adjacent to the dumps could have higher ratios than the original design. Therefore, these areas should be examined in more detail and the haulage cost savings gained by dumping short compared to the potential ore reserve loss.

The pit mining sequence will define the rate and source of the waste rock. Generally, waste material from upper areas should be hauled to dumps located at higher elevations and lower waste dumped in lower locations. This is common sense if haulage costs are to be minimized. Although this is the ideal objective, topography, property boundaries, drainage routes, dump stability, environmental considerations, and other constraints may make this objective difficult or impossible to achieve.

Topography will limit the available areas and usually defines the type or shape of the waste dump. More common dump configurations are valley fills (complete or partial), hillside wedges, fan and terraced dumps, and combinations of these. If the pit mining sequence permits backfilling one area that has been depleted while another adjacent area is still active, then this alternative can be preferable instead of extending dumps over virgin areas, depending on haulage and reclamation costs.

Dump areas can also be limited by existing drainage routes and property boundaries. In both of these cases, an economic comparison should be completed to weigh the relative cost and potential savings that would result from removing the constraint, e.g., drainage diversion or property purchase.

Before commencing a dump design, two additional parameters must be determined. The material swell factor and angle of repose are very important factors in determining the dump volume needed and the overall dump slopes. In situ material, when mined, will swell from 10 to 60%, depending on the type of material and fracture frequency. In hard rock operations, the swell factor is commonly between 30 to 45%, meaning that one in situ unit will swell to a volume of 1.30 to 1.45 units.

Loose density tests should be performed to determine the anticipated swell. These figures will first be used to size loading equipment buckets and haul truck box sizes. The second use is to quantify the volume of dumping room that will be necessary to dispose of the material from the mine. Loose material will compact to some degree after placement on the dump. This will depend on the type of material, size distribution, moisture content, disposal method, and the height of the dump. Common compaction numbers will range from 5 to 15%. Crushed and conveyed waste will not have a compaction factor as great as that of waste placed in low lifts by 154-t (170-st) haul trucks.

A second parameter that must be determined is the angle of repose of the loose dump material. Dry run-of-mine rock will usually stand between 34 to 37°. The lower the dump height, the more rapid the advancement and the more irregular the rock pieces, the higher the angle. For design purposes, a conservative slope of 1.5:1 (34°) is recommended in order to safely project the anticipated toe position. Measurements of existing talus slopes will also give a good indication of the expected long-term dump face angle.

The dump configuration will also be affected by the haulage method and by stability and reclamation considerations. The three methods of material handling in order of use are: truck, conveyor, and rail. Truck haulage is used in more instances because of its flexibility and lower capital cost. In particular cases, conveyors are more economical to use for waste disposal with their lower operating costs and where large tonnages have to be transported over either large horizontal or vertical distances. As in-pit mobile crusher development progresses, the tonnages handled by conveying systems will increase substantially. Rail haulage is in use at only a few of the older surface mining operations and is not considered as an alternative for many future operations.

Stability considerations will affect the design of the dumps either by lowering the ultimate height or reducing the overall slope. The slope can be reduced either by building the dump in lifts or by dozing. Sometimes, a combination of these two methods is necessary for reclamation purposes.

The intermediate phases of a dump may vary to a large extent from the planned final dump configuration. For stability reasons, lower lifts or toe dumps may have to be established during the earlier stages. As the mine life progresses, additional lifts can be placed above the lower dumps, subject to future design criteria. By this, it is meant that the berm left on a lower dump must have a design width to facilitate future reclamation, overall slope reduction for reasons of stability, or leaving sufficient width for an access haul road to a future disposal area.

Mining operations are conducted in many different topographic and climatic conditions. These conditions will require changing the techniques used to safely start and maintain a dump. A high wedge-type dump may be safe in a dry climate if it progresses over a rocky and competent base. The same dump would most likely fail if it progresses over wet hillside soils or permafrost. For this reason, geotechnical studies are very important in predicting the stability of both intermediate and final dump phases. Pertaining to the same situation, dump stability monitoring is also very important in cases where failure has a high probability. The degree of monitoring will depend on the consequences and risk of failure.

Continual monitoring will reduce the risk of injury and equipment damage. Failures are acceptable if this risk can be minimized and the failure will not affect downstream facilities, equipment, and personnel. Some northern operations even use failures as a method of material transport and reclamation, since the failures shorten haul distances and lower the overall dump slope to facilitate reclamation.

Particular emphasis should be placed on drainage in designing both intermediate and final dump phases. Dumps constructed using haul trucks have nearly an impermeable surface so that rainfall or melting snow will pond on the top of the dump or cascade over the face if care is not exercised in the dump design and construction. Dumps should, therefore, be built at a slightly adverse gradient for three reasons:

1) Carry runoff away from the crest.

2) A positive gradient means that haul trucks will have to power back to the dump crest rather than rolling back. As a safety feature, this will also reduce the chance of parked equipment accidentally rolling toward and over the crest.

3) Most mining operations set a speed limit below what a loaded haul truck is capable of achieving. For this reason, a 1 to 2% uphill gradient will not slow haulage, but will increase dump capacity and shorten haul distances.

Waste dumps that progress over or fill up drainage routes must also have special design considerations. If run-of-mine rock is end-dumped from the tip head, then given sufficient dump height, gravity will segregate the larger and smaller fragments. The larger material will roll to the bottom of the dump and will normally form a very permeable base. The finer material gathering in the upper portions of the dump will tend to form a nearly impermeable surface, especially with heavy haulage traffic. Waste dumps built with this natural segregation are free draining and offer little chance of saturation unless the base material weathers rapidly and will have reducing permeabilities through time. The high base permeability will allow the dumps to progress over small drainage routes and not block the flow. For larger streams, the shifting stresses placed on the base of the dump as it advances can jeopardize any drainage structure, such as a culvert. Therefore, a diversion tunnel is preferable where long-term drainage is critical.

In the case where dump failures occur, a number of corrective procedures can be implemented. These may be as simple as rerouting surface drainage or slowing the rate of advance, or as expensive as modifying the profile and design of the dump. One of the most common methods of stabilizing dump failures and allowing use of the dump to continue is to place more material at or on the toe of the failure. If haulage access to the toe of the dump is not feasible due to elevation differences, then dozers may have to be employed to push material downslope onto the toe. This may be helpful if reclamation regulations require a 2:1 slope for topsoil and revegetation placement.

As a new dump is started in a virgin area, small failures can be anticipated if the dump commences as a wedge type on relatively steep terrain. For this reason, it is better to advance a new dump slowly and not count on all the waste being disposed of at one tip point.

In order to enhance the stability of the initial dump, lower benches may have to be notched into the hillside to key the base of the dump into the slope. It may also be necessary to clear off vegetation, such as trees and brush, and, in some instances, to remove soils and other unconsolidated materials that would not provide a stable dump base if the risk or magnitude of failure was unacceptable.

Another component of dump design deals with operating considerations. If a side hill or contour dump is under construction and a tracked dozer is assigned to the dump, then the tracked dozer can be used to establish a pioneer road in advance of the dump. This road, established at a slightly lower elevation than the dump crest, can be used to collect drainage, act as a level control, give additional dump width, and serve such purposes as a small vehicle and lighting plant parking site. Care must be exercised so the cutting of the pioneer road does not undercut the hillside.

Access to the dumps should be aligned to provide good visibility of the congested area around the dump head. In many instances, the access road will have to be wider than normal to allow it to be used for other purposes, such as a park-up area for mine equipment at the end of a shift, a pull-out area for fueling, a truck weighing station, and for dump lighting. A general rule of thumb is that haul roads should normally be five times the width of the trucks using them. This width would include ditches and berms and allow sufficient room for road maintenance vehicles to work safely while trucks are using the road. Preferably, graders may be able to blade the roads while the haul trucks are using another route, but this is not always possible.

A permanent lighting system can be installed along the route because of the relative long life of most dump access roads and for safety reasons.

The dump width at the tip head should be sufficient to allow for a moderately sized turning circle of the haul trucks. For large trucks, this should be between 61 to 91 m (200 to 300 ft). The length of the active dumping face depends on the number of truck fleets hauling to the area. Commonly, a distance of 30 m (100 ft) should be allowed per loading unit's truck fleet operating to that tip head. High berms should always be maintained along the total dump crest length, except at the tip head where the berm height should be equal to the radius of a haul truck's tire.

A tracked dozer at the dumping point is preferred over a rubber-tired dozer for a number of reasons, including:

1) Greater traction that allows it to push more material when the ground is wet or icy.

2) The tracks crush larger rock fragments, thereby reducing truck tire damage that occurs at the dumping area.

3) A tracked dozer with a winch can free stuck equipment readily.

4) A tracked dozer can push material farther over the bank in safety, since traction is spread over a larger area and not at just four points.

In most instances, dump material will not have the same supportive strength as the same material in situ, especially in wet climates. Rolling resistances may increase with traffic to a point of impassibility. Additional thinner lifts of more competent rock may have to be placed on the dump surface to maintain haul roads.

The operating differences between intermediate and final dump configurations can be quite large. For example, the material handling methods may change from truck haulage to crushing and conveying as distances increase. Since the prime objective of dump design is economics, the initial dump should have the shortest haul distance. As the mine progresses, haul distances will become longer and vertical haulage more excessive. Reducing the rate of future material handling cost increases using well-thought-out alternative methods and designs is the objective of good dump planning. Accomplishing this task may mean leaving lower routes open as a future access to potential dump areas if the future discounted savings balances today's cost sacrifice. In rugged terrain, this may mean that lower lifts at the base of a high dump will have to be established early, since later access may be impossible or too costly to construct. Several dumping points of various haulage distances should be available on a daily basis so that when the operation is short of trucks, closer dump points can be utilized to maintain production and when truck availabilities are high, longer hauls can be used.

Climatic conditions coupled with mine locations outside of the United States also will have some bearing on the dump design and operation. Less stringent safety and environmental regulations will allow more economic mine operations.

Politics may also intervene and demand short-term economic savings that will be costly for the operation in the long run.

In comparisons of one design to another, a method should be used that first establishes a base case. Then other designs can be completed and the economic and other advantages and disadvantages weighed.

If a choice exists as to the elevation of the dump, then the preferable order of haulage gradients is level, downhill, and uphill last. If haulage costs are equated to level, -8% downhill and $+8\%$ uphill, then the cost differential for a unit of distance is approximately 1.0, 1.46, and 2.38 for a 154-t (170-st) haul truck. This means that waste dumps should be designed level from the start point and only after the dump has progressed a certain horizontal distance will an upper lift become more economic. As an example, only when the horizontal haul distance exceeds 457 m (1500 ft) will it be more economical to lift the material 15 m (50 ft) and start another lift closer to the pit (see Fig. 1). If a dump were mistakenly designed so that all the volume was dumped from an elevation 15 m (50 ft) higher than necessary, and if 90.7 Mt (100 million st) could have been dumped at the lower elevation first, then the direct cost increase would be approximately $0.019/t ($0.017 per st) or a total of $1,700,000. Additional capital and replacement costs would also be incurred due to the increase in the number of haul units required.

Therefore, it is very important to recognize the best economic dumping plan and material handling method and to weigh the cost and effect of constraints such as stability, reclamation, drainage, and property boundaries.

Waste dump planning is usually not as critical or as detailed as mine design. This is due to the fact that the mine is the source of the ore and revenues. However, good waste dump design can be critical in minimizing costs and increasing the value of the ore produced. Improper waste dump planning can mean the difference between profit and loss and often should receive more attention and detail.

STABILITY OF MINE WASTE DUMPS

The overall stability of mine waste dumps is dependent on a number of factors such as:

1) Topography of the dump site.

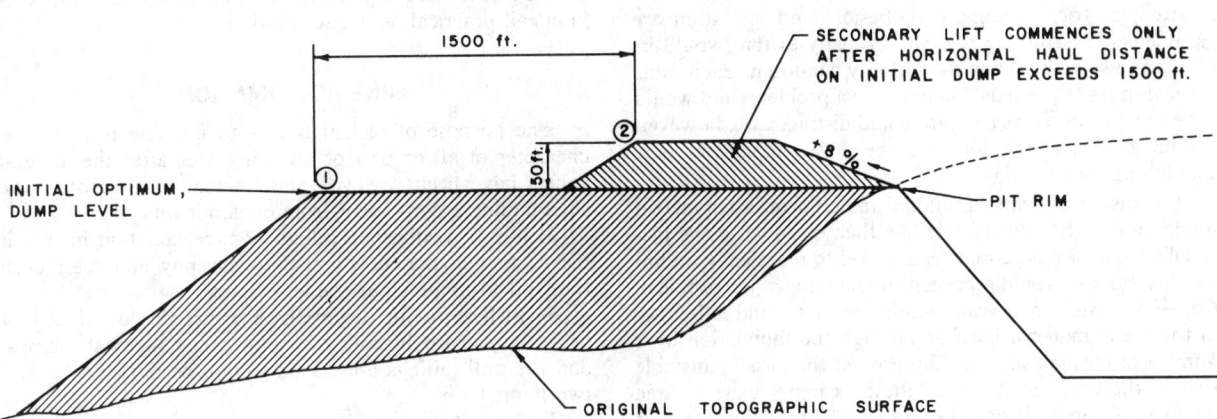

NOTE: LOCATIONS ① AND ② WOULD HAVE EQUIVALENT HAULAGE COSTS

Fig. 1. Equivalent truck haulage distances.

2) Method of construction.
3) Geotechnical parameters of the mine waste.
4) Geotechnical parameters of the foundation materials.
5) External forces acting on the dump.
6) Rate of advance of the dump face.

All of these factors combine in various ways during the life of a mine waste dump to aid in the stability of the dump or to contribute to its instability. The various technical analyses used to assess the stability of dumps are well documented in literature (Winterkorn and Fang, 1975; Anon., 1977) and will not be covered in this section. The factors affecting stability mentioned earlier will instead be discussed.

The choice of dump sites and their topography usually is limited to within an economic distance from the mine and, since rearranging of site topography is rare, the topography usually becomes a fixed condition. The crucial aspect of topography is the existing slope of the natural ground upon which the dump is to be constructed. Analyses show that factors of safety begin to drop significantly above a ground surface inclination of 20°, regardless of the strength parameters of either the waste or foundation material.

Mine waste dumps are usually constructed by one of two common methods: in lifts or layers or by end-dumping. End-dumping is a controlled failure process where the waste material is deposited forming a slope at or close to its angle of repose and the factor of safety is accordingly close to one. Since the front face is always advancing during the life of the dump, the slopes are not stabilized by flattening with conventional earthmoving equipment until closure of the dump. Monitoring of the *live* dump face is recommended to anticipate and deal with slope failures. The mine waste dumps constructed using an end-dumping technique are sometimes referred to as "built from the top," whereas, layered dumps are said to be constructed "from the bottom up." Layered dumps can be controlled, which adds significantly to their overall stability; however, they require a relatively gently sloping topography and usually entail a longer haul distance in the early years of the mine. Layered dumps are preferred where weak foundation conditions exist, since the load application can be controlled to allow for strength gains by consolidation and pore pressure dissipation.

The geotechnical properties of the mine waste and the foundation material are major factors in determining the overall stability of a mine waste dump. Such characteristics as strength, friction angle or cohesion, and gradation are parameters determining the type of analyses that would be selected to solve or define the stability condition. Each mine waste dump site presents a unique set of problems and would have to be analyzed as a separate and distinct case; however, certain general conclusions may be drawn based on some simplifying assumptions.

For instance, coarse frictional material on a competent foundation with a slope angle less than or equal to the material's angle of repose may be dumped to practically unlimited heights and would represent the safe side of the problem. Cohesive wastes on a weak foundation that could fail either in the waste material itself or through the foundation by a number of mechanisms would represent the unsafe, unstable side of the problem. Between these extremes exist a large number of combinations that have different failure modes and, therefore, must be carefully analyzed.

The most commonly occurring combinations are coarse frictional materials on weak shallow foundations or weak foundations extending to considerable depth. The failure modes associated with these are horizontal translation of the waste, deep seated rotational shear failure through the foundation, or a combination of the two. For cohesive wastes on shallow or deep, weak foundations, the failure modes become difficult to distinguish and the type of analyses used becomes a matter of experience and judgment. It should be remembered that regardless of the sophisticated analysis and the capacity of computers to produce numbers to several decimal places, the reliability of the calculated factors of safety is dependent primarily on the degree to which the input parameters or assumptions made are representative of the actual conditions existing in the waste material and of the foundation of the dump. This is where an engineer's experience is vital to assess the stability of a mine waste dump by determining whether the assumptions and choice of analyses are reasonable.

External forces, such as water and earthquakes, often play a decisive role in the stability of a mine waste dump and, therefore, should be carefully considered in the analysis. Of the two, earthquake or seismic forces are a relatively straightforward factor that can be readily accommodated in most stability analyses by determining the location of the mine waste dump in relation to seismic zones and inputting the proper seismic coefficient into the analyses as an additional horizontal force.

The effect of water on the stability of mine waste dumps is more difficult to evaluate and, as a general rule of thumb, measures should be taken to prevent water from entering the dump. Water pressure buildup in a dump will always lower the factor of safety and, therefore, should be prevented, if possible.

It is not always possible to avoid building mine waste dumps across drainage courses and, therefore, provisions should be made for unimpeded passage of flows either around the dump by ditching to divert surface waters or by providing a coarse, filter-protected drainage layer beneath the dump. The top surface of the dump should be sloped away from the leading edge of the dump face to eliminate ponding during periods of rainfall and snowmelt. In some instances, it may be necessary to include inclined drainage layers at the perimeter of the dump to maintain drained conditions.

In summary, the stability of mine waste dumps is dependent on a combination of factors and, although sophisticated numerical analyses are available to solve stability problems, practical experience and engineering judgment are still the dominant ingredients needed to arrive at an economical, practical, and safe solution.

MINE RECLAMATION

The purpose of reclamation is to upgrade the physical character of all or part of a mining area after the mineral values have been removed and, thereafter, to protect the surrounding environment from contaminants.

However, restoration differs from reclamation in that it means to recreate the original topography and reestablish the land to its previous use and character.

In surface mining operations, the three largest areas that are reclaimed are the mine excavation, the mine waste dumps, and the mill tailings area. This section pertains to the first two items.

Different types of surface mining operations will tend to facilitate or hamper programs that reclaim both the pit and waste dump area. If the commodity extracted is a bedded deposit of large extent and of relatively shallow depth such as in coal mines or if a number of separate pods of ore are mined from individual pits as in uranium deposits, then backfilling of worked-out areas is a common method of waste disposal and reclamation. Waste material removed from the initial box cut or pit can either be stockpiled and later transported to fill the final excavation or the stockpile could be reclaimed and not moved and the last pit left with little reclamation effort applied.

However, in most surface operations in commodities other than coal, the amount of backfilling is restricted or totally impractical. Therefore, most of the reclamation effort is directed toward the waste disposal area.

If the mining area must be restored to original topography, then very few operations can afford to accomplish this task. In many instances, new topography can be constructed to be superior to the original. During excavation, the waste material swells from 10 to 60%, depending on the type of material and its size and shape distribution. Therefore, in a coal operation, for example, the amount of coal removed will usually not compensate for the larger volume of waste due to swelling. The net effect is that a greater volume exists after excavation and if used as pit backfill the new level of topography has to be higher than the original, e.g., with strip ratios greater than 3.3:1 and a swell factor of 30%. This can work in reverse when the volume of the minerals removed is greater than the increase in volume obtained from waste rock swell, e.g., strip ratio lower than 3.3:1 and a swell factor of 30%. In coal mine reclamation, it is sometimes mandatory that the backfilled waste be redeposited in the same stratigraphic order. This can further add to the mining complexity and reclamation cost. When a reclamation plan is devised, normally all topsoil from both the pit and dump area has to be removed and stockpiled, and erosion protection methods applied at the beginning of the mining activity. This could be done in phases as the pit and dump areas expand through time. In order to minimize costs, a good plan will reduce double handling so that soils removed from one area can be directly laid down on another area where the mining activity has been completed.

The degree of effort in mine reclamation or restoration will be sensitive to a number of factors including:

1) Government regulations.
2) Prior land use.
3) Proximity to population centers.
4) Desired aesthetic value and visibility.
5) Quantity and types of airborne or waterborne contaminants that could be released from the excavation and dumps.
6) Local vegetation types and densities.
7) Climatic conditions.
8) Location of mine in respect to state and country.
9) Mining company policy.
10) Cost and effect on mine development.
11) Prior reclamation efforts in the area.
12) Dump configuration.
13) Type and size distribution of dump material and available soil quantities and quality.
14) Natural and introduced fauna species of area.
15) Proposed end use of land.
16) Availability of water and water-table level.
17) Topography.
18) Existing and rerouted drainage flows with respect to quantities, flood levels, and frequency.
19) Downstream water uses.
20) Future potential of currently uneconomic minerals in the waste dumps and mineralization adjacent to the pit excavation.

All of these factors will have to be considered in designing a suitable reclamation plan that will meet with approval from the mining company, the public, and the designated governing authorities. Each property will be unique in the sense that the factors will have different effects and weights on the cost of the reclamation plan.

In most surface mining operations, the waste material removed from the pit is deposited on an adjacent area. The area required for waste disposal is usually equal to or greater than the pit area because the disturbed waste material has a greater volume than in situ, a lower slope angle than the pit walls, and rarely can the material be stacked as high as the pit is deep. Normally, the first area available for reclamation is a part of the waste dump. If the pit area cannot be economically backfilled, then the reclamation efforts applied to this area will only take place toward the end of the mine's life.

In designing waste dumps, particular consideration has to be given to reclamation needs if the cost is to be minimized. If the overall slope of the dump face has to be reduced to prevent erosion and to allow placement of top soils and vegetation, then the dump design should consider terracing to minimize the amount of material rehandling. As illustrated in Fig. 2, the cost of rehandling decreases in proportion to the square of the reciprocal of the number of terraces into which a dump can be broken. Therefore, a dump constructed using three terraces will have only one-ninth the rehandle dozing costs of a dump of similar height with no terracing. In order to facilitate reclamation efforts, a berm should be left on each terrace level. This will lower costs by providing easier access to the faces for equipment spreading topsoil and for revegetation efforts. The berms can also serve as erosion protection and drainage diversions, if necessary. In order to select the number of terraces, increases in material handling costs due to the increase in the number of lifts cannot be allowed to exceed that saved in reducing reclamation dozing costs.

The main hazards to a reclamation project will be erosion and leakage of contaminated waters that will hamper revegetation or be hazardous to life. Both of these problems usually can be corrected through proper drainage control and treatment. Drainage channels will need to be rock-lined if the channels are to remain in the same location without excessive bank erosion.

The depth of topsoil spread on the waste disposal areas will depend to a large extent on the amount that is available. A desirable thickness would be 0.6 m (2 ft) or greater and, to a large extent, will depend on reclamation regulations pertaining to the area. In no case should it exceed the current depth of soil cover.

Reclamation of large, deep open pits is very limited if the pits cannot be backfilled. Typical reclamation efforts in these cases would be to allow the pit to fill with water. This area then could be used for recreational purposes such as fishing or as a water reservoir. The pit crest can be blasted and dozed into the excavation to create a more gentle slope along the edge of the water. This will facilitate water access and will provide a more suitable area for vegetation regrowth and waterfowl habitat. Care has to be exercised using this technique if the chance of bank failure is increased. Pit walls that extend a large distance above the water level can be recontoured, in some cases, through blasting and dozing to blend in with the surrounding relief.

Pit excavations can also be used for disposal of other surplus materials, but strict controls have to be applied to avoid ground-water contamination and gaseous emissions.

In cases where the pit is a shallow excavation of large lateral extent above the water table and not backfilled due

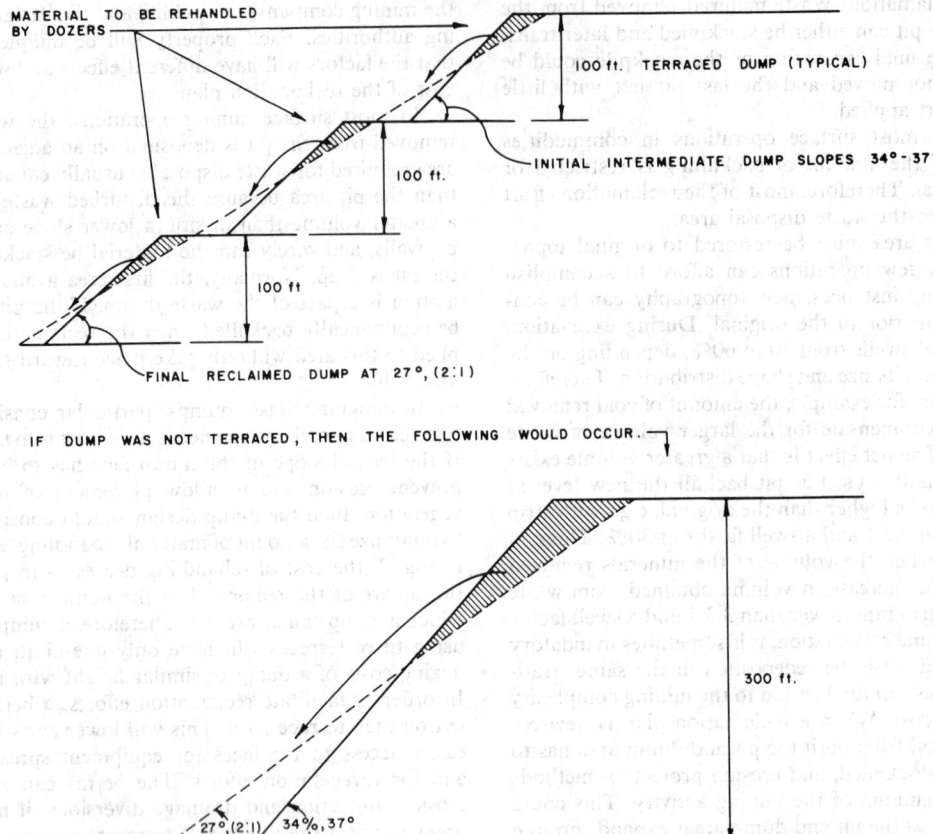

Fig. 2. Reclamation of terraced mine dumps. In the example of a dump that is not terraced, the net result is a threefold increase in material dozed and a threefold increase in dozing distance, leading to a total cost increase of nine times that of the terraced dumps.

to a lack of material, the pit can be reclaimed by spreading topsoil and planting suitable vegetation on the pit floor.

GROUND WATER AND CONTAMINATION

Economics dictate that the mine waste be disposed of in the proximity of the mine. This, in turn, limits the selection of disposal sites free from any water course and/or closeness to the ground-water sources. Mine waste disposal can be done economically and be environmentally safe if hydrologic factors are carefully considered. With larger and better material handling equipment, surface mining operations have increased in number and size in hard rock, uranium, coal, and in other minerals. Not all mine wastes are detrimental to the water resources; however, a potential for surface and ground-water pollution does exist due to some of the contaminants found in mine wastes. The parameters that play an important role in determining the extent of possible pollution and the relative effort needed to counter the effects are: (1) type of mine waste, (2) location of disposal site, (3) climatic conditions, (4) method of placement, (5) foundation condition, and (6) regulatory requirements.

In general, mine waste materials consist of overburden, soils, and rocks and may or may not contain minerals and/or contaminants. Metal mine wastes, and many nonmetal mine wastes, in general, exhibit low contamination potential. In order to make mine waste dump operations a success, in terms of economics, stability, and environmental acceptabil-

ity, each site must be evaluated on a case-by-case basis. An overall feasibility and planning study should include the potential of water contamination. The study does not have to be extensive, but must identify the constituents that may be regarded as potential sources of contamination. Based on a mineralogical study, the waste material may be grouped into low, moderate, or high potential categories.

Location of the dump site should also be taken into consideration, since a dump in an isolated area may require different consideration than one located in proximity to a populated area. Other location considerations are:

1) Dump placement in a stream, as a valley fill.
2) Dump placement on a mountain ridge or a side hill.
3) Upstream watershed.
4) Location of ground-water table and seasonal variation.
5) Type of downstream water utilization.

Climatic conditions also are an important factor in an environmental study. As an example, if certain minerals are detached due to frequent water action, the dump, which might have been regarded as noncontaminating in an arid climate, may become a major source of water pollution in an area that receives more precipitation. Also, weathering of the waste material would contribute toward physical contamination of the water, i.e., erodable material would transport as sediment in the surface runoff. This physical contamination is mainly restricted to surface water and can be dealt with easily by providing peripheral diversion ditches and sediment collection ponds.

The design of sediment ponds and diversion ditches depends upon the site's specific conditions and upon regulatory requirements.

How the waste material is to be deposited would also influence the potential release of contaminants from the dump. Of particular importance from the water pollution standpoint is the permeability of the dump. A waste dump created by end dumping normally tends to be loose and more porous, exhibiting high permeability. On the other hand, a dump constructed in layers exhibits low permeability and has a lower rate of infiltration which, in turn, reduces the risk of contaminating water resources.

From the ground-water impact standpoint, important factors are the permeability characteristics of the soils and bedrock upon which the dump is constructed and the topography of the area. As an example, if the dump is placed on relatively level ground, ample time is provided for infiltration since a perched water level would be created within the dump, and this would allow the water to react with dump material. In contrast, a dump placed on steeply sloping ground under similar precipitation conditions would allow less time for infiltration.

The surface water and ground-water measures taken should, of course, comply with applicable regulatory requirements.

Surface Water

Physical and chemical contaminants can be treated easily through the use of a sedimentation pond. A pond constructed downstream of the dump will retain eroded material transported in the water and collected runoff then is released through an outlet mechanism. For chemical contaminants, it would be essential to determine the amount and type of chemical constituents prior to specifying the necessary treatment. Generally, metal mine wastes will not contaminate the surface water, and, therefore, treatment may not be necessary. However, it would be prudent to collect water samples and determine the water quality. To this end, it is recommended that surface water quality analyses be conducted prior to the commencement of mining activities. The design of sedimentation ponds and their appurtenant works should be based on prudent engineering practices and should meet the appropriate regulatory requirements. At present, the surface coal mining regulations for the United States require that the sedimentation yield be calculated using either the universal soil loss equation or 304.8 m^3/ha (0.1 acre-ft per acre) of disturbed area within the upstream drainage area.

Although the universal soil loss equation was developed to predict erosion from agricultural land, it is useful in estimating the sediment load from the disturbed area. Recent research activities have been directed toward estimating sediment load from surface mining activity. In addition to the estimation of sediment load, the following hydrologic criteria should be considered in designing a sedimentation pond as set forth in 30 CFR 816.45-.47 and .49 (Code of Federal Regulations, Title 30, Mineral Resources; current issue revised as of July 1, 1984, and reissued each July 1).

1) *Detention Time*—The sedimentation pond should have a capacity sufficient to provide a theoretical detention time of 24 hr for the 10-year, 24-hr precipitation event. It must provide adequate detention time to allow the effluent from the ponds to meet state and federal effluent limitations [30 CFR 816.46(c)(iii)(B)].

2) *Dewatering*—The storm storage should be evacuated through a nonclogging device such as a conduit spillway. The inlet elevation of the discharging mechanism should be at or higher in elevation than the maximum sedimentation storage volume.

3) *Emergency Spillway*—The emergency spillway in combination with the principal spillway must be capable of safely passing a 100-year, 6-hr precipitation event [30 CFR 816.46(c)(iii)(I)(2)(a)] or, if of smaller size as defined in ¶ 77.216(a), must be capable of safely discharging a 25-year, 6-hr precipitation event [30 CFR 816.46(c)(iii)(I)(2)(c)].

Embankment Design

The design of the embankment should be performed in accordance with prudent geotechnical engineering principles and practice. Mine waste, if suitable, can be used in the construction of the embankment. The following criteria, as set forth in 30 CFR 816.49, should be met in the design of the embankment:

1) Embankments shall have a minimum static safety factor of 1.5 for the normal pool with steady seepage saturation conditions and a seismic safety factor of at least 1.2.

2) Embankments shall have adequate freeboard to resist overtopping by waves and by sudden increases in storage volume.

3) Foundation and abutments for the impounding structure shall be designed to be stable under all conditions of construction and operation of the impoundment. Sufficient foundation investigations and laboratory testing shall be performed in order to determine the design requirements for foundation stability.

4) All vegetative and organic materials shall be removed and foundations excavated and prepared to resist failure. Cutoff trenches shall be installed if necessary to ensure stability.

5) Slope protection shall be provided to protect against surface erosion at the site and protect against sudden drawdown.

6) Faces of embankments and surrounding areas shall be vegetated, except that faces where water is impounded may be riprapped or otherwise stabilized in accordance with accepted design practices.

7) Impoundments shall include a combination of principal and emergency spillways which shall be designed and constructed to safely pass the design precipitation event (see previously).

8) The vertical portion of any remaining highwall shall be located far enough below the low-water line along the full extent of high wall to provide adequate safety and access for the proposed water users.

The design criteria for the sedimentation ponds (impoundments) described, although they appear to be stringent, are, in most cases, not overly difficult to incorporate. The design and construction of such facilities can be achieved economically and practically by including the system in the overall mine and waste dump planning efforts. It should be noted that these reflect current regulations and, as such, can change in the future.

Removal of Sediment, Maintenance, and Monitoring

Accumulated sediment should be removed from the pond when 60% of the design capacity is utilized. The removed sediment can be hauled to and deposited on the dump. The embankment and appurtenant works should be monitored and maintained so that the structures will continue to function as intended. The settling pond monitoring program should include:

1) Sediment level.
2) Water level and quality.
3) Seepage through embankment and foundation.
4) Sediment accumulation in outlet pipes.
5) Functioning of gates and other mechanical systems.
6) Observation of embankment for any signs of deterioration.

Ground-Water Contamination and Monitoring

Generally, metal and nonmetal mine waste will not contribute to the ground-water contamination. However, in surface coal mines the materials immediately above and below the coal are likely to contain pyritic shale which, in time, may impact on the ground water. In addition to the waste materials, climate, topography, geologic features, and the ground-water level below the ground surface can affect the levels of ground-water contamination. Careful planning and proper placement of materials can avoid the risk of contamination.

A ground-water monitoring program should be established to evaluate the impact on ground water. This program should include:

1) Installation of observation wells upstream, downstream, and in the vicinity of the waste dump.
2) Water level readings on a regular basis covering wet and dry seasons.
3) Water quality analysis by collecting water samples from the observation wells.

Results of water level fluctuations and water quality dated prior to construction of a waste dump and during the operation should be compared to determine the impact on ground-water contamination.

PERMITTING

Permitting for the disposal of mine waste can be considered an integral part of the permit requirements related to the mining operation, reclamation, operating plans and procedures, and water pollution measures. With the recent developments in regulatory requirements and procedures, it is common for many federal and state regulatory authorities to become involved. Federal acts and authorities that may be currently applicable but may change in the future are:

1) National Environmental Policy Act (NEPA) of 1969, NEPA 42, United States Code (USC) 4321, et seq.

2) Federal Water Pollution Control Act Amendments of 1972, amended as Clean Water Act of 1977 and further amended by Clean Water Act of 1978, 33 USC 1251 et seq.

3) Resource Conservation and Recovery Act (RCRA) of 1976, 42 USC 6901 et seq.

4) Federal Mine Safety and Health Act of 1977, 30 USC 951 et seq.

5) Surface Mining Control and Reclamation Act of 1977 (for coal), 30 USC 1201 et seq.

Timely commencement of the mining operation will contribute considerably in monetary gain and, therefore, it is essential to determine, file, and obtain approval for all the necessary regulatory permits. Overburden and leach dumps are subject to Environmental Protection Agency Regulations, 40 CFR 240 et. seq. Currently, however, specific requirements exist only for hazardous solid wastes. The Nov. 19, 1980, amendment excludes solid wastes from the regulations pertaining to the extraction, beneficiation, and processing of ores and minerals, except for coal, phosphate rock, and overburden removed for the mining of uranium ore. Most of the regulatory provisions of 30 CFR are directed toward mining, operational safety, and health controls; however, the Office of Surface Mining Regulations 30 CFR, Pt. 710 through 950, cover surface coal mining and surface activities for underground coal mining. Of particular importance for surface coal mining activities is Pt. 816 of 30 CFR.

Regulations, permit requirements, and permitting agencies often change, so it is recommended that in order to acquire an up-to-date detailed description of these requirements, the appropriate state and federal agencies be contacted at the outset of the mine planning program.

Listings of federal agencies and agencies involved in permitting and/or administering in selected western states are presented in Tables 1 and 2, respectively.

REFERENCES

Anon., 1977, "Waste Embankments," *Pit Slope Manual,* Chap. 9, Report 77-01, CANMET.

Office of the Federal Register, 1981, National Archives and Record Services, General Services Administration, Code of Federal Regulations, 30, revised July.

Vandre, B.C., 1985, "Scoping Regulatory Requirements," *Design of Non-Impounding Mine Waste Structures,* M.K. McCarter, ed., AIME, New York, pp. 79-88.

Winterkorn, H.F., Fang, H.Y., 1975, *Foundation Engineering Handbook,* Chap. 10, Van Nostrand, New York.

Table 1—Federal Agency(ies) Jurisdiction and Responsibilities, Mining Operations, and Reclamation Control

Minerals Covered	Surface Management Agency	Agency(ies) Responsibilities or Controls
Locatables (30 USC 22) gold, silver, copper, lead, zinc, molybdenum, uranium, etc.	Forest Service (FS)	Operation plan approval—FS consultation—Department of Interior agencies (36 CFR 252.5: plan of operations) permit for waste embankment off claim (36 CFR pt. 251-land uses, special use permit)
	Bureau of Land Management (BLM)	Operation plan approval—BLM (43 CFR 3809.1-6: plan approval, 43 CFR 3802.1-5: wilderness study area plan approval)
Leasables (30 USC 181 et seq) phosphate, sulfur, oil shale, tar sands, potash, etc. (coal excluded)	Bureau of Land Management, Forest Service, Bureau of Indian Affairs (BIA)	Mining plan approval—U.S. Geological Survey (USGS) consultation—FS, BLM, BIA (30 CFR 231.10: operating plans)
Minerals on Indian lands (25 USC 397)		Wastes to be disposed of in accordance with terms of lease and USGS directions (general mining orders) (30 CFR 231.51: disposal of waste) lease issuance—BLM lease issuance and stipulation consent—FS, BIA
Coal	Non-federal, non-Indian lands	Permits issued and plans approved by OSM or state with federally approved program (30 CFR pt. 730-736—federal program for state)
	Federal lands	Permits issued and plans approved—Office of Surface Mining (OSM) concurrence-states with approval programs (30 CFR 741.17: criteria for permit approval or denial)
		Consent—USFS (30 CFR 741.20: permit review processing for operations on national forest system lands)
	Indian lands	25 CFR pt. 177—surface exploration, mining, reclamation of lands subpart B—coal operations compliance—OSM

Table 2.   Programs for Mining Operation and Reclamation Control, Selected Western States

State	Mineral or Commodity Covered	Acts	Rules and Regulations	Technical Guidelines	Administering Agency(ies)
			Scope of Program		
Alaska	No direct state control				Department of Natural Resources 323 E. Fourth Ave. Anchorage, AK 99501 (907) 279-5577
Arizona	Reclamation requirements apply to state lands as a condition of mineral leases. Local government land-use controls and permit activities may be applicable to mining and reclamation				Arizona State Land Department 1624 W. Adams St. Phoenix, AZ 85007 (602) 271-4621
California	All minerals (1975)	X	X	X (Division of Mines and Geology)	Department of Conservation 1416 Ninth St., Rm. 1341 Sacramento, CA 95841 (916) 445-0514
Colorado	All minerals excluding oil, gas, and geothermal (1973, as amended 1976)	X	X	. . .	Department of Natural Resource 1845 Sherman St. Denver, CO 80203 (303) 892-3401
Idaho	All minerals (1972) Dredging (1955, amended 1957, 1969, 1970, 1977)	X	. . . (Dredging)	. . .	Department of Lands State Capitol Bldg. Boise, ID 83720 (208) 334-3569
Montana	(1) Coal and uranium (1973) (2) bentonite, clay, phosphate rock, scoria, and sand and gravel (1973) (3) other minerals (1971, as amended 1974-75)	X	X	X (Partial)	Department of State Lands Capitol Sta. Helena, MT 59601 (406) 587-4560
Nevada	Local governmental land-use controls and permit activities may be applicable to mining and reclamation				Nevada Bureau of Mines and Geology University of Nevada Reno, NV 89507 (702) 885-4369
New Mexico	Coal (1972)	X	X		Office of Surface Mining Bureau of Mines and Mineral Resources Campus Sta. Socorro, NM 87801 (505) 827-5451
Utah	All minerals excluding geothermal, oil, and gas	X	X	. . .	Department of Natural Resources 1588 W. N. Temple Salt Lake City, UT 84116 (801) 533-5771
Washington	All minerals (1970)	X	X	. . .	Department of Natural Resources Public Lands Bldg. Olympia, WA 98504 (206) 753-6183
Wyoming	All minerals (Environmental Quality Act. 1973. amended 1974-75)	X	X	X (Partial)	Department of Environmental Quality State Office Bldg. W Cheyenne, WY 82002 (307) 777-7756

5.7 SURFACE COAL MINES

PHILIP G. MOREY

There are two different, but valid, approaches in surface mine design. One approach is to consider the problem to be defining the sequential removal of blocks of material to satisfy the criteria of production. However, an equally valid approach—and one shared with the majority of operations-oriented mining engineers—is that material movement is inextricably bound with the equipment used. To these engineers, mine design that does not specify the equipment and how it is employed is incomplete. In such an environment, the selection of equipment will affect the sequence and the dimensions of the material blocks to be moved, even as the material will affect the selection of equipment. The two considerations are two aspects of a single problem. This latter approach is the basis of the following discussion on surface coal mine design.

Surface coal mining generally involves removal of large amounts of overburden or waste for every ton of coal produced. Higher productivity per unit invested and per man-shift is achieved in surface coal mining than in underground coal mining. For example, underground productivity per man-shift varies from approximately 14 to 73 t (15 to 80 st), depending on the mining method. Surface coal mining productivity per man-shift generally ranges between 18 to 181 t (20 to 200 st), depending on the mining method and the amount of stripping required to uncover coal. At most surface coal mines, removal of overburden accounts for a high percentage of total mine investment and operating costs.

Because the working area is relatively unrestricted, compared to an underground working area, large excavating and haulage equipment can be used. Because such large equipment units are involved in the mining process, individual equipment units can be interdependent. For maximum productivity and efficiency, it is necessary to match equipment units to each other and to the ratio of materials that must be removed. In turn, the mine design must match mine equipment to geologic constraints of the deposit.

Selection of a mining system must be based on careful consideration of several factors, including geographic and geologic information, size of the deposit and the distribution of coal seams, availability of equipment and its compatibility with other equipment. The expected mine life, rate of production, and production buildup also must be given consideration in order to select a mining system that provides the most favorable engineering economics.

GEOGRAPHIC FACTORS

The geographic location of the surface coal mine may govern the adaptability of mining equipment. Periodic and annual rainfall, temperature extremes, unusual weather conditions, and altitude may affect equipment productivity and mine design. For example, surface mine design and equipment selection for an east Texas lignite mine must consider the impact of rainfall in the area. Wet overburden is difficult to handle, pit and waste slopes are difficult to maintain, pumping of water from the pit will be required, and haul roads must be specially constructed and maintained in order to be passable during inclement weather. In northern climates, extreme winter temperatures may affect the scheduled working days and equipment productivity. Frozen soil may have to be drilled and blasted before it can be excavated.

Equipment must be constructed with special materials to withstand temperature extremes. The impact of climate must be considered carefully before proceeding with the design and implementation of a mining plan.

GEOLOGIC FACTORS

Area topography and geology generally are the most important factors affecting the selection and design of a surface coal mining system. Surface attitude, number of minable coal seams and their general dip may limit the mining system alternatives.

Desired quality of the run-of-mine (ROM) coal produced may complicate mine design by increasing the number of production areas that must be made available. Thickness and characteristics of overburden and interburdens will influence system design and the selection of major and secondary excavation equipment.

The mine design engineer is not expected to be a knowledgeable coal geologist. However, it is important for the engineer to understand the analytical methods and tools used by the geologist so that geologic results can be utilized with confidence for mine design. During an exploration program, the experienced coal geologist collects data from which depositional environment and stratigraphy can be interpreted. Its accurate interpretation, coupled with good coal seam correlation, establishes the framework for mine design, environmental analysis, and feasibility studies.

The geologist uses his experience to identify basic lithology from the exploration drill-hole cuttings and cores. The information results in the identification or recognition of different materials in the stratigraphic column. Since most of this identification is made by the inspection of drill cuttings, the vertical position of strata is only approximate. A further problem exists in the identification of various coal layers within the stratigraphic column. Seams vary in dip and thickness, frequently part (divide), disappear, and reappear.

The coal geologist utilizes geophysical drill-hole logging techniques combined with cores and drill-hole cuttings to correctly interpret and locate lithologic sequences (Anon., 1981). These logging tools are:

Gamma ray reacts to the potassium isotope K40 that is present in most shales. The *shale line* is the estimate of where 100% shale would be represented on the log. Sandstones and sand typically have values much less than shale. The *sand line* is the estimate of 100% sand (0% shale). Readings between the sand line and shale line usually indicate sandstones and shale. Readings above the shale line generally indicate the presence of marine bands such as coal or uranium mineralization. Inferior coal may also give readings within the sand and shale lines.

Density log is a measurement of the bulk density of material through which the hole has been drilled. The density log of coal is unique because of coal's low density and may clearly distinguish between materials of near-equal gamma ray deflection. The *long-spaced density probe* permits deeper penetration of induced gamma rays into material, thus providing a better estimate of rock density, but decreased sensitivity to identifying contacts between beds of material of different density. The *bed resolution density probe* permits accurate identification of bedding contacts with decreased accuracy of bulk density.

Caliper log measures variability in the diameter of the drill hole. The caliper log is used to compensate for density readings in caved areas of the hole.

Neutron log responds to hydrogen, and to a lesser extent, to carbon, and generally confirms the gamma ray log. Readings for coal are high because of the hydrogen and carbon content; shale readings are high because of the OH content present in mica and because of free water content in the shale. Sandstone and limestone differ due to variations in porosity and the effect of contained water.

Resistivity log responds to variation in electrical resistivity of material. Usually, shale readings are low and all other materials, including coal, are higher than shale. However, local variations can occur that will cause coal and sandstone to respond with low readings. Therefore, the resistivity log should be used primarily as a confirmation tool in conjunction with other logs. Exclusive reliance on the resistivity log can cause serious errors in seam correlation.

Sonic log gives a measure of compressibility and responds similarly to the density log. The sonic log can be utilized as a coal rank indicator, especially in defining inferior coal portions that may lay within a major coal seam. As with the resistivity log, the sonic log should be used more as a confirmation or ranking log.

The sonic log, in combination with the density log, can provide the basis for comparing relative strength, *S,* of material. The general formula is expressed as:

$$S = P/a - \Delta t^2$$

where *P* is the density of material, mass per unit volume; *t* is the sonic transit time per unit length; and *a* is a gravitational constant in appropriate units.

The gravitational constant, *a*, equals:

1.0 $\dfrac{\text{kg/m}}{\text{N/sec}^2}$ in International units

32.2 $\dfrac{\text{1bm/ft}}{\text{lbf/sec}^2}$ in English units

Gamma ray and neutron logs are especially useful in cased holes; the casing has no effect on their response.

Fig. 1 compares each log response to the drill-hole lithology. Based on these logs and identification of the materials, the geologist correlates various burdens and coal seams. From this correlation, considerable data can be compiled and presented in the form of geologic maps to assist the engineer in mine design.

In addition to the geophysical logs and drill cuttings, cores are taken from coal seams, including the seam contacts, for coal and ash quality analysis. Cores of overburden, interburden, and underburden also are taken for material strength analysis and environmental studies.

Each of the processes of data development serves a specific purpose. The potential accuracy of surface coal mine design is directly related to the quantity and quality of data developed. The geophysical logs, in particular, should be considered as a set of tools for data development. No single log should be used for correlation.

Fig. 2 illustrates a typical topographic map. Topographic maps, generally at a scale of 1 in = 100 ft (1:1200) or less, are preferred for detailed mine planning. Frequently, preliminary mine planning utilizes published US Geological Survey 7.5' quadrangles of scale 1:24,000. In the absence of suitable topo maps, aerial photographs or stadia surveys with ade-

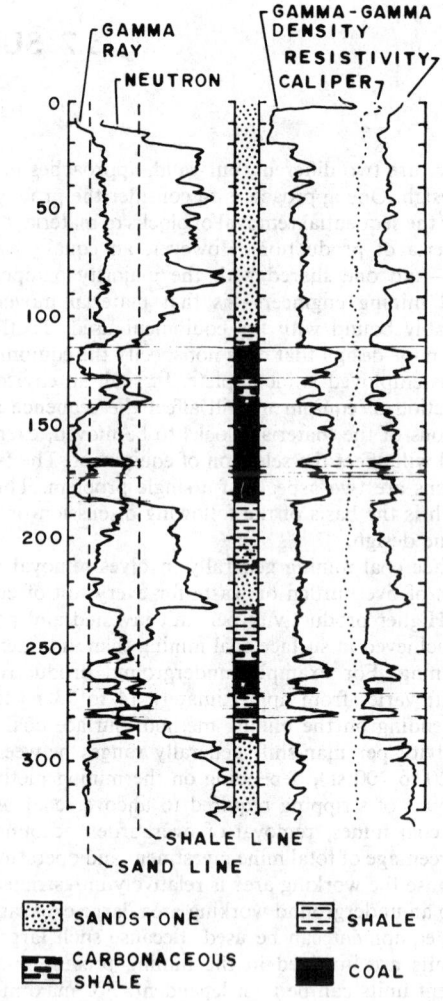

Fig. 1. Geophysical logs as compared to drill-hole lithology.

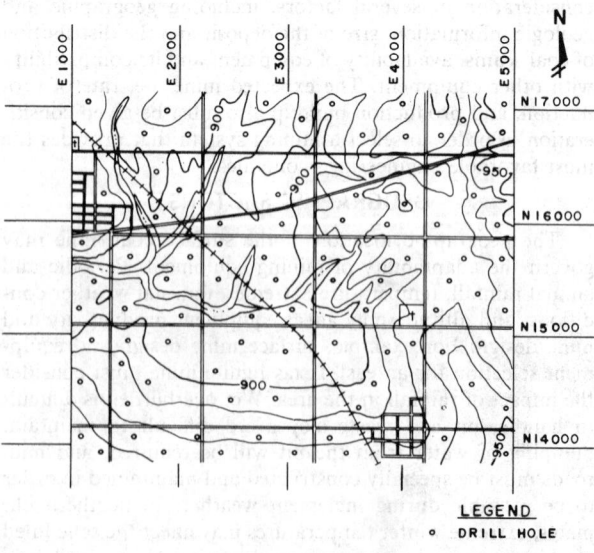

Fig. 2. Topographic map.

quate ground control may be used as a mapping base. Occasionally, topo maps—originally prepared from aerial photos—are in error. Topo maps provide information about surface features necessary for mine design. In addition to providing contours, regional drainage patterns can be identified. Any known environmental impact that may restrict mine design will be shown (townsites, isolated buildings, roads, trails, cemeteries, etc.). Generally, topo maps used during coal exploration become the base map from which the mine is designed. On these maps pertinent exploration data, such as drill-hole locations, are documented.

In addition to topo maps several geologic maps are essential for mine design:

Structure Contour Maps. Maps showing lines of equal elevation of a horizon of interest are known as structure contour maps. These maps are prepared to depict the structure of seams, and to show folds, faults, and other anomalies whose correct interpretation is critical for mine planning.

Isopachs. Maps indicating contour lines of equal thickness are known as isopachs. Overburden, interburden(s), and coal seam(s) isopachs contribute to understanding the three-dimensional geologic model of the coal deposit. Fig. 3 is an example of a coal seam isopach.

Isopleths. Maps indicating contour lines of equal value are known as isopleths. Ash, sulfur, moisture content, and coal heating value are commonly depicted on isopleths. Fig. 4 is an example of an ash content (percent) isopleth.

Resource Stripping Ratio Contours. Maps indicating contour lines of equal resource (vertical) stripping ratio contribute to understanding the economic or physical limitations of mining.

Stripping ratio is defined as the in-place volume unit of burden that must be removed to obtain one weight unit of recoverable coal as defined in cross section. To estimate recoverable coal, all inferior coal, top and bottom seam contacts, and partings must be removed from the total coal thickness measurements. When estimating the loss of coal at seam contacts, excavating or loading equipment that will make the separation at the contact must be considered. Also quality restraints may dictate the amount of coal that must be wasted at contacts. Generally, losses of 10 to 15 cm (4 to 6 in.) at top and bottom of each seam are allowed as equipment-generated losses.

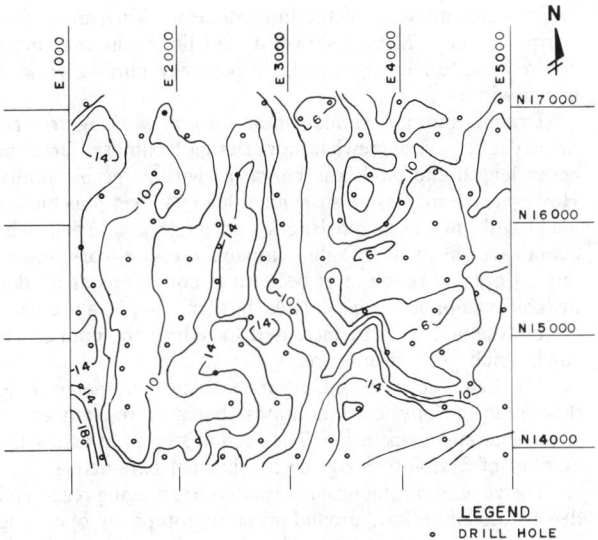

Fig. 4. Isopleth, percent ash.

The resource stripping ratio should not be construed as the true stripping ratio resulting from the mine plan. The resource stripping ratio is a first attempt to identify economically desirable deposit boundaries. The ratio does not include the impact of designed pit slopes and preproduction stripping.

All of these maps can be generated to scale by computer. Computer programs interpolate x, y, and z values from the geologic logs of bore holes to produce the desired map. It is most important to ensure that geologic logs are accurate, that computerized data is correct, and data processing performs correctly. Even though computerized data appears correct, inconsistencies in data gathering still may result in erroneous computer interpolation. To correct these errors, the geologist or engineer may be required to make judgmental corrections to ensure continuity of interpolation.

MINE DESIGN CRITERIA

Given the geologic data for a specific coal property, the next step is to define those factors that will influence or restrain mine design. In the initial planning stages of a new mine, the effect and interaction of these factors must be examined carefully in order to select the appropriate system among several available. Key parameters influencing mine design can be divided into three distinct groups: geologic, mechanical (equipment specifications), and operational parameters.

Geologic parameters dictate general pit features of depth, width, and slopes. Surface terrain, depth, and dip of coal seam(s) may limit mine design options. Geohydrologic and geotechnic parameters may restrict or eliminate equipment applicability. General geologic parameters governing mine design are overburden and interburden depth, thickness of coal seams and partings, density of burden and coal, burden swell factor, bench angles, and spoil angle of repose.

Mechanical parameters of equipment must meet the constraints of geologic parameters. Working radii of equipment units—dragline, backhoe, bucket wheel excavator (BWE) or shovel swing radius, truck-turning radius, scraper loading distance and turning radius—are related to the geologic parameters that impact the immediate working area.

Mine equipment design generally is based on standard specifications with limited flexibility. Therefore, physical lim-

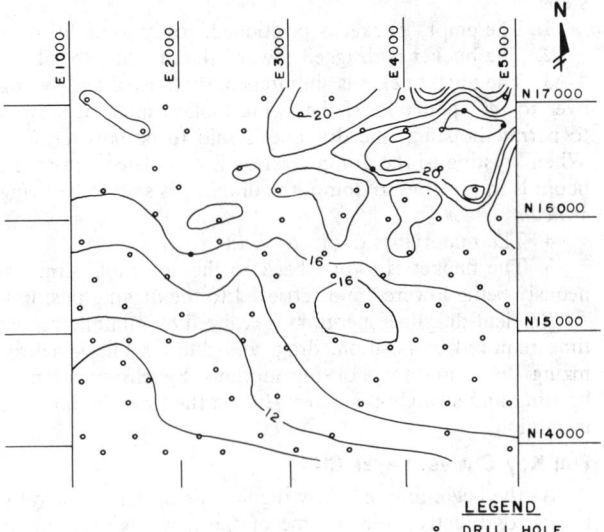

Fig. 3. Isopach of coal seam.

itations are imposed on the immediate working area of an equipment unit. Mining shovel design limits the maximum controllable bench height while imposing minimum working bench widths.

Draglines impose similar limitations on working area geometry, but have somewhat more design flexibility. Dragline boom length and angle can be varied within specific limits. However, the maximum suspended load (bucket plus bucket load) will vary as boom length and angle are varied. The overall weight of the dragline impacts the safe working distance from the bench crest because of conditions of burden instability. For less consolidated burden, the dragline base, or tub, diameter can be increased to reduce the pound-per-square-inch ground pressure.

The first step in examining mechanical parameters is to determine the physical limitations that may restrain or restrict equipment operation. The second step is to assess the amount of flexibility for given mechanical parameters.

The general mechanical parameters are digging reach and depth, maximum load, ground pressure, equipment operating and clearance radii, dumping height, and work cycle time.

Operational parameters, in part, are dictated by the specifications of equipment capable of working under restraints imposed by the geology of the deposit. For example, the depth of burden is the major factor influencing the selection of a dragline. Because of the dragline's flexible mechanical parameters, several draglines and/or production rates may be considered. However, draglines are the most difficult to match to specific coal production rates. Varying depth of overburden and the options of varying panel width, digout length, boom length and angle, and bucket size must be examined jointly to estimate the most favorable mine design. In comparison, shovel excavators can be selected easily by designing more or less working benches with excavator units.

General operational parameters affecting mine design are scheduled annual working hours, operating efficiency factors, bucket fill factor, availability and utilization of equipment, mobility of equipment, downstream constraints on coal quality, and the desired coal production rate.

While the objective of general mine design is to develop a mine plan to produce a product at a desired and sustained level under the most favorable economic conditions, individual surface coal mine design frequently is affected more by matching geologic, mechanical, and operational parameters other than the desired production rate. Geologic parameters often limit the size of individual pit operations. Occasionally, geologic parameters dictate a greater, more favorable, production rate. For example, depth of overburden above a coal seam may require a very large dragline with sufficient digging depth and dumping reach to uncover the seam. Since burden removal is likely the greatest unit cost of this operation, the most favorable production rate will be the rate at which coal can be uncovered. To produce coal at a lesser rate would require the dragline to occasionally stand by to wait for coal production to catch up. Dragline standby would penalize overall economics of the operation.

DRAGLINE STRIPPING METHODS

A simplified dragline operation is illustrated in Fig. 5. The stripping cycle begins with the dragline at position 1, cutting a trench, referred to as the *key cut,* along the newly formed highwall. The distance from the previous key cut position to the new position is referred to as the *digout* length. The key cut is made to maintain the panel width and uniform highwall. Without a key cut, the panel width would narrow with each subsequent digout, because the dragline could not

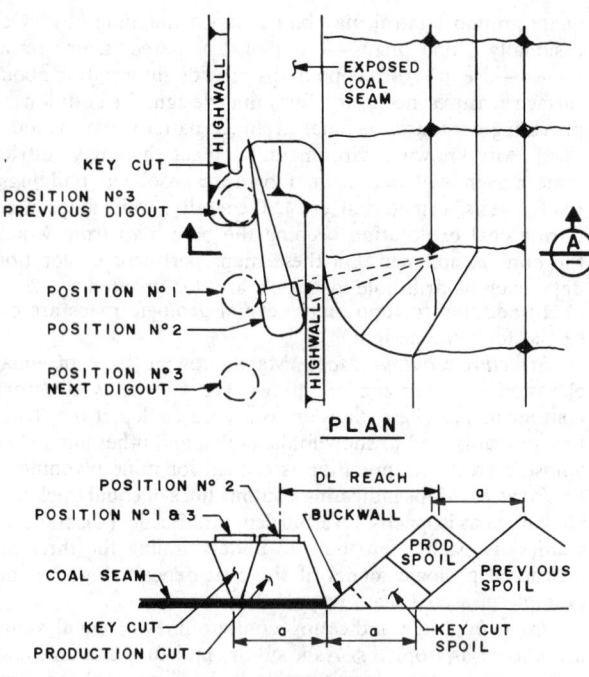

Fig. 5. Typical dragline operation.

control the bucket digging against an open face. The dragline deposits the key cut material in the bottom of the mined-out pit off the coal and against the previous spoil pile. More stable spoil from the key cut may be placed in the very bottom next to previous spoil to form the buckwall which provides a more stable spoil slope that can be steepened if deemed necessary.

When the key cut has been completed, the dragline is moved to position 2 to complete excavation of the digout. This is known as the *production cut,* and the material is cast on top of the key cut spoil. When the digout has been completed, the dragline is moved to position 3, the beginning of the next stripping cycle (next digout).

The operating cycle of the dragline consists of five basic steps:

1. The empty bucket is positioned, ready to be filled.
2. The bucket is dragged toward the dragline to fill it.
3. The filled bucket is simultaneously hoisted and swung over to the spoil pile. If the swing motion must be slowed to permit hoisting, the dragline is said to be *hoist critical.* When hoisting to the dump position is completed before the boom is in position to dump, the dragline is said to be *swing critical.*
4. The material is dumped on the spoil.
5. The bucket is swung back to the cut while simultaneously being lowered and retrieved to the digging position.

Efficient dragline operation is realized by minimizing the time required to position, drag, and dump while synchronizing the swing and hoisting motions. Synchronization of hoisting and swinging is dependent on the time the boom is in motion.

Full Key Cut vs. Layer Cut

At the beginning of a new digout, the dragline generally is placed directly over the toe of the new highwall to be formed. From this position, the dragline can establish a uniform and safe highwall if the burden is sufficiently stable.

In this position, the dragline excavates the key cut which is more than the width of the bucket at the bottom of the cut. When the cut has been completed, the dragline moves over to make the production cut. The two positions generally are required because of the limited reach of the dragline in relation to the panel width being stripped. Large draglines, operating under ideal conditions, may be able to excavate the total digout from the one position over the highwall. Such situations are the exception, not the rule.

When operating conditions permit excavation of the digout from one position over the highwall, the dragline generally excavates the digout in layers. The key cut is formed, one layer at a time, by excavating along the highwall before the completion of each layer. Cutting in layers can be performed from the production cut position; however, the highwall slope will require dressing by dozer while the dragline is digging. Under such circumstances, some mines also have adequately dressed the highwall by dangling a heavy section of chain from the bucket and dragging the chain along the wall. Other mines, because spoil area is critical, have progressively stepped the dragline toward the spoil while excavating in the layer cut method. This procedure has the tendency to pack spoil as tightly as possible on the spoil slope.

Layer cutting generally increases dragline productivity with a corresponding decrease in operating cost. Increased productivity is realized by progressively decreasing the average swing angle as the dragline walks in the direction of the spoil pile.

Dragline Panel Width: Panel width is defined as the width of the cut taken by the dragline, as it progresses from digout to digout, along the highwall from one end of the pit to the other. Panel width, one of the most important parameters affecting dragline productivity, is influenced by depth of overburden, dragline boom length, hoist and swing time, and available spoil area. Since panel width becomes the available operating area in the pit bottom, coal loading operations are also affected.

Several operational factors must be considered in the selection of panel width. A wide pit generally is favorable for coal loadout and permits greater safety for men and equipment. The minimum practical pit width is dictated by the maneuverability of coal loading and hauling equipment.

If the available space for placement of spoil is critical, such as might occur when crowding spoil to open haul roads through the spoil, narrow panels permit greater flexibility to deal with such problems. In general, the wider the panel, the less dragline walking time is required.

Productivity variations, because of panel width, are directly related to whether or not the dragline is swing critical. Small draglines can become swing critical at panel widths less than the width required for practical coal operations in the pit. Their cycle time also increases dramatically. Larger draglines may not become swing critical until the panel width exceeds 50 m (150 ft). Fig. 6 compares cycle time and productivity vs. panel width for large and small draglines.

Bench Height: The height above the coal seam at which the dragline is positioned is defined as the bench height. Selection of the bench height is based on numerous operational factors and topographic restraints.

The complex relationship of bench height (which could be equal to overburden depth), panel width, dragline dumping reach and dumping height, as well as material characteristics such as swell and angle of repose, influence greatly the dragline's capability to dispose of burden off the coal.

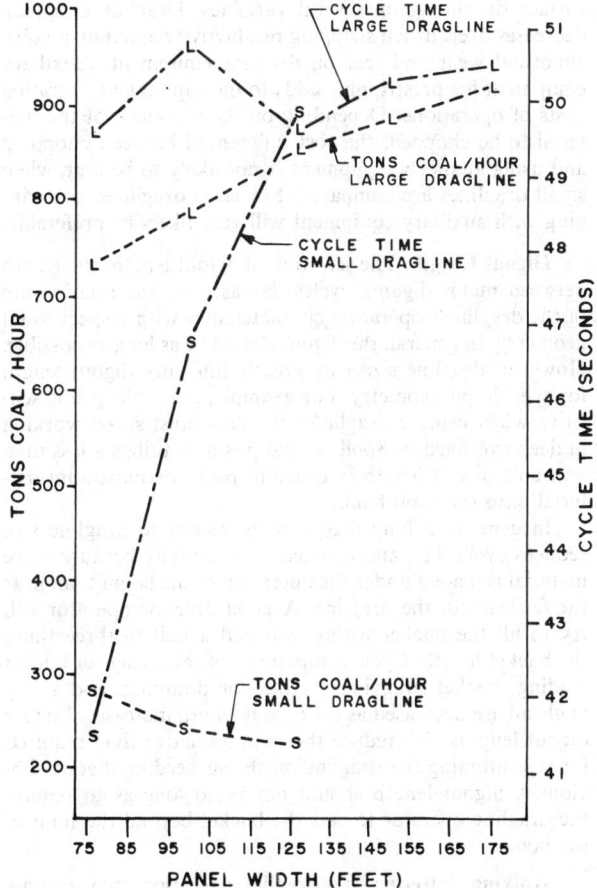

Fig. 6. Cycle time and productivity vs. panel width for large and small draglines (Anon., 1978).

The dragline's digging depth, while related to burden depth, rarely becomes a factor in dragline performance.

The bench height must be selected primarily on the basis of fitting the dragline's specific characteristics to the required pit geometry. In general, the bench height should be as high as possible within the limit of required dragline reach.

Undulating topography may complicate a simplified selection of bench height. Two alternatives are available to alleviate the problem:

1) The dragline can be used to cut and fill to develop a common bench elevation. Cutting, termed *chopping* or *overhand digging,* increases the cycle time and reduces the bucket fill factor, thereby reducing effective productivity. Fill material must be rehandled, thus reducing overall production. Chopping has very special advantages: the dragline reach required may be shortened, rehandling of burden may be avoided, fill may not be required to create a level working surface, a level return path for deadheading can be provided, and subsoil can be placed back in its relative position on top of the spoil.

2) Auxiliary equipment can be used to perform the cut and fill operation. Care must be exercised to ensure that filled areas are stable. Utilizing auxiliary equipment offers the benefit of freeing the dragline for its primary function of stripping burden from the coal.

Whether the dragline cuts and fills its working pad or auxiliary equipment is utilized to prepare a level working surface depending on several variables. Dragline chopping decreases overburden stripping productivity and may involve abnormal wear and tear on dragline equipment. Auxiliary equipment for prestripping adds to the capital and operating costs of operations. Depending on the thickness of the material to be chopped, the cost differential between chopping and using auxiliary equipment is not likely to be high when small draglines are compared. For large draglines, prestripping with auxiliary equipment will very likely be preferable.

Digout Length: The selection of digout length, the length between major digging cycles, is based on the relationship of the dragline's operating characteristics with respect to pit geometry. In general, the digout should be as long as possible. However, dragline size may greatly influence digout length for specific pit geometry. For example, digout length is sensitive when using a dragline with slow hoist speed working in deep overburden. Spoil critical pits may utilize a less than desirable digout length in order to pack the maximum material onto the spoil bank.

In general, a long digout with respect to dragline size reduces cycle time and increases productivity because more material is loaded under the outer end of the boom than near the fairleads of the dragline. A good dragline operator will try to fill the bucket within two and a half to three times the bucket length. Cycle components of retrieving for bucket loading, bucket dragging, payout for dumping, and swing angle all are decreased as the digout length increases. Longer digout lengths also reduce the nonproductive time required for repositioning the dragline on the succeeding digout. Obviously, digout length should not be so long as to require the dragline operator to cast the bucket beyond the limit of the boom.

Walking Patterns: In a dragline operation, two separate walking cycles are involved: *deadheading* and *walking within the digout* (see Fig. 5, positions 1, 2, and 3). When a panel has been completed, there are two options available for the dragline. One, the dragline can wait for the coal to be mined to the end of the pit, then turn around and begin the next panel in the opposite direction. This procedure is termed *laying over* at the end of a panel. Two, the dragline can turn around and travel part or all of the way down the panel to begin the next cut. This procedure is termed *deadheading*. If the dragline travels part way down the panel, it cuts into the next panel and strips in the opposite direction. If the dragline travels to the other end of the pit, it cuts into the next panel and strips in the same direction.

Whether a dragline lays over or deadheads depends primarily on the production time lost. Contractual requirements, such as exposed coal inventory, may eliminate the layover option and force deadheading. Some mines opt to lay over at the end of a panel because ground conditions are not favorable for deadheading. Other mines limit layover to a maximum of two shifts. If waiting for coal production involves two or more lost dragline shifts, the dragline will be deadheaded.

When deadheading is feasible, the decision should be made on the basis of minimum lost dragline production. Fig. 7 illustrates the approximate time required for deadheading draglines. Deadhead time is based on 33% of the specified dragline walking speed. Such a large discount factor must be used to account for various delays in deadheading, such as maneuvering, cable handling, ground preparation, and minor breakdowns.

In Fig. 5, the standard dragline positions within the digout were illustrated. From position No. 1., the key cut is

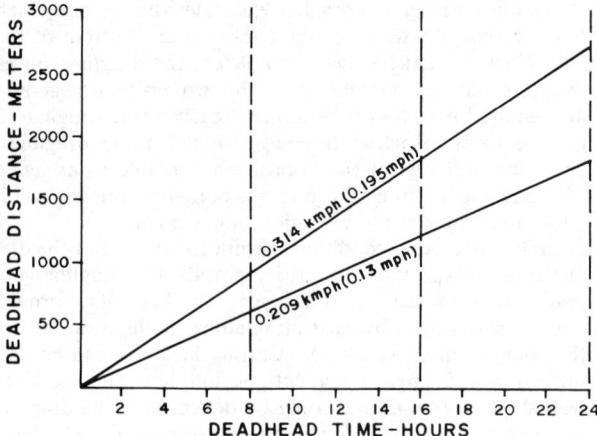

Fig. 7. Deadhead time vs. distance (Anon., 1978).

excavated; from position No. 2, the production cut. Realistically, additional positions are utilized by the dragline as layers are stripped from the digout. Depending on the location of the fairleads, it may be necessary to walk the dragline to successive positions systematically closer to the digout face to permit clearance of the bucket chains from the fairleads.

The greater the digout length, the less walking time will be required per panel. Repositioning in the digout can affect cycle time. Therefore, walking patterns must be considered when selecting digout length. Time spent in repositioning the dragline can be estimated by discounting the walking speed of the dragline. Generally the discount factor is much less than the factor utilized for deadhead estimates. Based on observations by the author, a discount to 15 to 20% is appropriate.

Pit Shape: The new dragline pit begins with the initial cut, termed the *box cut,* made along the outcrop, subcrop, or property boundary. To open the box cut, excavated material is spoiled to one or both sides of the cut. The material lying on the newly created highwall must be moved or spread out evenly by auxiliary equipment. The material lying on the cut wall that will become the spoil side may, or may not, have to be moved depending on reclamation requirements.

Because of rolling topography, the mine engineer may be inclined to design the box cut along a uniform contour. Generally, succeeding cuts are designed parallel to the box cut. As a result, this type of pit develops a meandering design. In Fig. 8, the effects of pit curvature are illustrated. The figure is an idealized compound curve with equal radii of curvature to illustrate the difference in spoil area between an inside and outside curve. Obviously, outside curves provide more spoil area. Depending on depth of overburden, panel width, radius of curvature, and operating parameters, severe operating problems may occur on inside curves. Dragline cycle time will increase, spoil crowding will occur, and coal may be lost by being covered with spoil.

To remedy the problems caused by inside curves, several options can be considered. Panel width may be decreased, material may be cast short and rehandled by extending the bench, a small auxiliary dragline may be utilized on the spoil to pull back excess spoil, the spoil pile may be steepened, or the pit may be straightened by stripping a series of short panels. Generally, the most favorable solution is to straighten

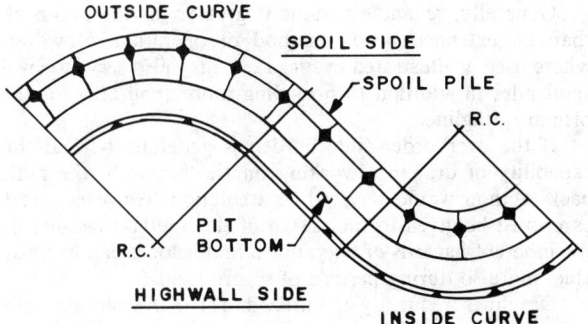

Fig. 8. Pit curvature.

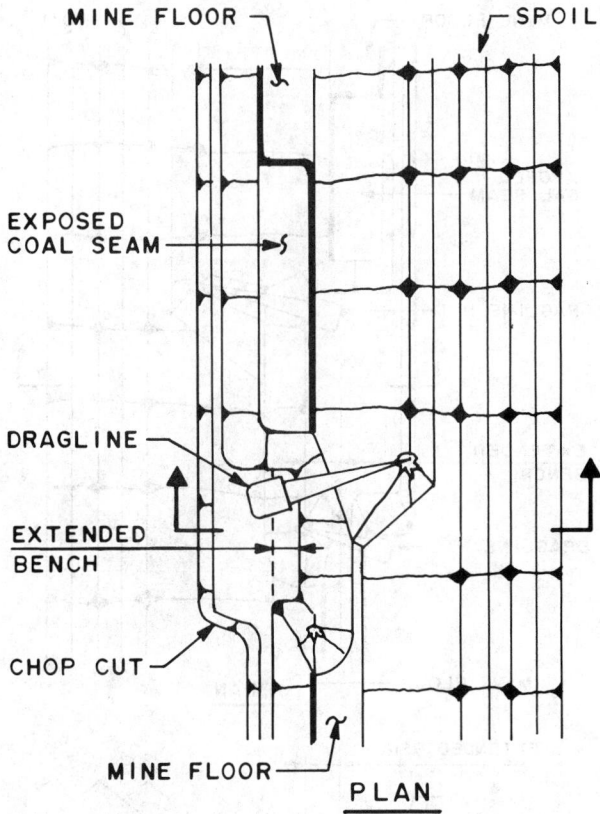

PLAN

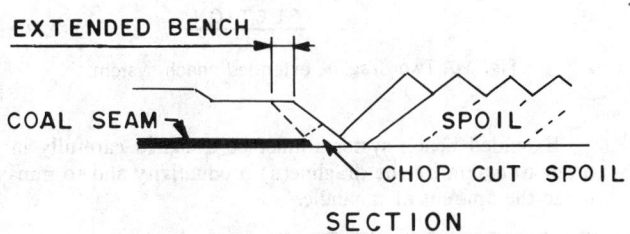

SECTION

Fig. 9. Extended bench and crop cut single dragline.

the pit. Spoil steepening is also an effective method for disposing of relatively small amounts of excess spoil. The dragline bucket is positioned on the spoil slope where steepening is desired and dragged down and across the top of coal. The bottom part of the spoil pile is steepened and coal is cleaned in the process. Digging efficiency during this process is reduced, cycle time increased, and rehandling reduces effective productivity. Steepened spoil slopes may present special hazards to equipment and personnel because they are more prone to failure.

Spoil Patterns: There are three basic methods of spoiling. When using short digouts and casting at a near 90° angle, a uniform ridge line can be created. This configuration makes maximum use of the available spoil room. As the digout length is increased, uniformity of the ridge line is lost and individual peaks of spoil are created.

With sufficient spoiling area, the dragline operator may cast material from both the key cut and production positions at angles less than 90°. While the dragline stripping cycle will improve, spoil piles appear to be ragged and irregular. An aerial view of the operation will show a definitive pattern to the irregularity. In reality, spoil peak grading will be reduced by this method of spoiling.

Dragline cycle time can be reduced by dumping the loaded bucket on the fly, that is, before the dragline swings to the ultimate dumping position. This procedure, termed *radial casting,* gives the spoil a cross bedded appearance. Provided that there is sufficient spoil room, radial casting tends to spread the spoil more effectively, reducing spoil grading costs.

Since distance between spoil ridges is equivalent to the panel width, narrower panels will reduce spoil grading costs. However, such reductions in cost generally will be offset by increases in dragline operating cost if the dragline is not swing critical.

Dragline Extended Bench Systems

Where overburden depth or the panel width exceeds the limit at which the dragline can sidecast the burden from the coal, a bridge of burden can be formed between the bank and the spoil which effectively extends the reach of the dragline. The bridge extends the bench on which the dragline is operating. The bridge is formed by material falling down the spoil bank or by direct placement with the dragline. To remove the bridge material from the top of coal, it must be rehandled.

Extended bench systems are adaptable to many configurations of pit geometry. Fig. 9 demonstrates a dragline forming its working bench by chopping material from above

the bench and forming the bridge, then moving onto the bridge to remove it from top of coal.

Fig. 10 demonstrates a two-dragline extended bench stripping sequence. The primary dragline strips overburden and spoils it into the previously excavated panel. This material is leveled, either by tractor-dozers or the secondary dragline, to form the bench for the secondary dragline. The secondary dragline first strips material near the highwall, then moves on to the bridge to move the rehandle material. In a two-dragline system, one machine must operate at the pace set by the other. Therefore, mine design must consider their respective capacities when assigning respective digging depths.

Fig. 11 illustrates an extended bench operation in which three coal seams are to be recovered. The primary dragline strips overburden to the top of the first seam. Coal is removed, then a small parting dozed into the pit and the second coal seam removed. The secondary dragline strips the large interburden to the third and final seam.

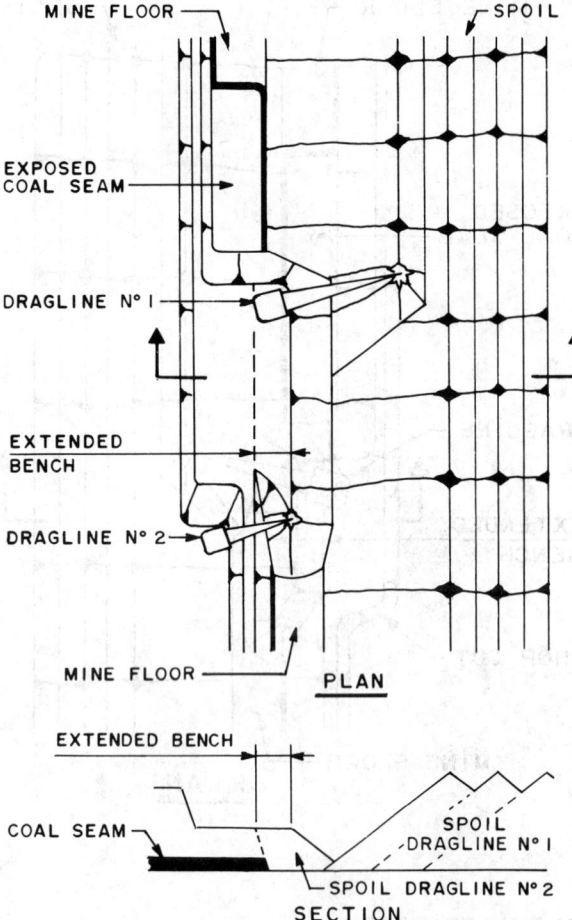

Fig. 10. Two dragline extended bench system.

Extended bench systems must be designed carefully in order to maximize the dragline(s) productivity and to minimize the amount of rehandle.

Contour Stripping with Draglines

In this method, dragline stripping proceeds along the coal outcrop or subcrop, with each successive panel following the line of the original panel. In rolling terrain, the pit twists and turns, developing a series of *S* curves. If the dragline is operating at near maximum geometric limits, it will become *spoil-bound* (unable to deposit stripped burden) when working an inside curve. Under such conditions a smaller secondary dragline may be used to pull back the spoil sufficiently to permit the primary dragline to complete stripping of the coal. Haul roads generally are designed into an outside curve where some spoil crowding is possible because of the wider spoil bank arc.

Dragline Pull-Back Method

Occasionally, overburden to be stripped will be beyond the capacity of the dragline to spoil off the coal by any of the previous methods described. In this case, a secondary dragline can be placed on the spoil bank to pull back sufficient spoil to make room for complete removal of overburden.

The pull-back method is illustrated in Fig. 12 in which a small dragline is pulling back overburden at inside curves of a contour mine.

Generally, rehandle volume is greater for the pull-back than an extended bench method of operation. However, where used as illustrated in Fig. 12, it may also serve to level spoil piles in addition to providing more spoil area for the primary dragline.

If the overburden/interburden is generally beyond the capability of draglines working on the highwall, the pull-back method would seem to be a solution. However, great care must be given to the design of this method because of the inherent hazards of operations. Spoil slopes can be unstable, more so during periods of severe rainfall.

Draglines frequently are utilized to strip overburden from deeper coal seams than originally intended. Occasionally, spoil slopes cannot be maintained at designed angles. Various methods have evolved to stack more material into the spoil bank to alleviate these problems. The more common methods are described briefly:

Buckwalls involve building the base of the spoil adjacent to the pit with competent material so that a steeper spoil slope near the base can be maintained.

Coal fenders require leaving a small wedge of coal untouched in the pit so that more spoil can be packed on the spoil slope.

Outside pit involves modifying the pit shape in order to develop the outside curve concept which increases the spoil area relative to the stripping area.

Fig. 13 demonstrates a modified outside curved pit plan developed at a southern Illinois mine. This plan permits stacking of extra spoil on each side of the designated pit haul roads and also dramatically decreases the time required to

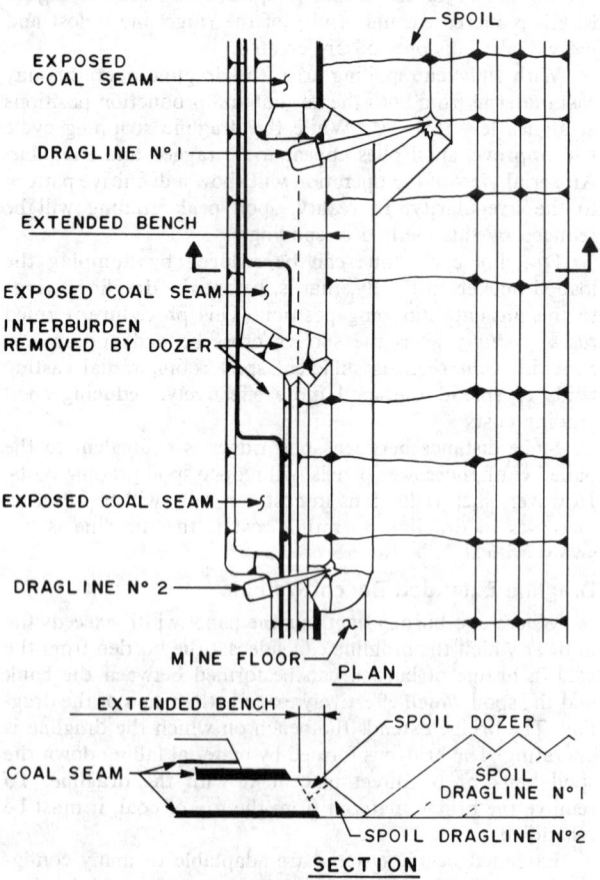

Fig. 11. Two dragline extended bench for three coal seams.

deadhead the dragline from one panel to the beginning of the next panel. This plan will work well if doglegs are maintained at 60°. Lesser angles will work if overburden is shallow and the dragline is oversized in relation to its digging depth.

Drilling and Blasting of Overburden for Draglines

Overburden drilling and blasting is more critical for dragline stripping than for shovel digging. Shovels have the ability to crowd the dipper into the bank, providing leverage to dig difficult or poorly blasted material. Draglines have leverage only by dragging the bucket over the material. Such leverage is translated to severe strain on the bucket lip and teeth. In poorly blasted material, dragline productivity can drop more rapidly than that of shovels working in similar material.

Selective placement of explosives and blasting agents may be critical to the surface coal mine operation. Many coal seams are overlain with sedimentary beds of varying hardness and thickness. Improper placement of the charge in the blast-

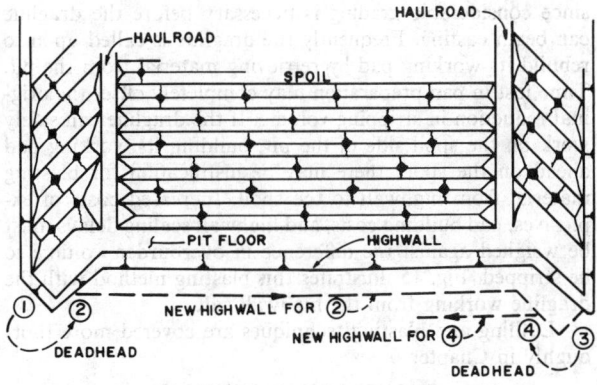

① STRIPPING MACHINE DIGS OUT AT PIT END.

② DEADHEADS BACK TO THIS POINT AND SHEARS IN AT SHARP OUTSIDE CURVE.

③ DIG OUT.

④ DEADHEAD BACK AND SHEAR IN.

Fig. 13. Modified outside curving dragline pit.

hole can cause blast energy to travel along planes of greater weakness and through softer material. Under such conditions, harder beds of material will tend to break in large blocks or fragments. To ensure adequate placement of the blast charge, it is necessary that drill operators log differences in material or drill penetration rates and provide this information to the blasting foreman.

For dragline stripping, there are two general methods of blasting overburden in common use. One method utilizes a blasthole pattern with a buffer zone to contain the blasted material against the highwall. Fig. 14 illustrates this pattern. The advantage of this pattern is to contain the blasted material within the dragline working area and avoid large broken material that must be handled with difficulty.

The width of the buffer zone, combined with the powder factor, are critical elements in efficient utilization of this method. This method is useful, especially if the dragline is performing a chop cut prior to the key and production cuts.

The other method of blasting overburden is similar to the standard open pit blasting procedure. Its purpose is to blast as much material into the spoil area as possible, thereby reducing the amount that must be stripped by the dragline. The resultant advantage is debatable in the author's opinion,

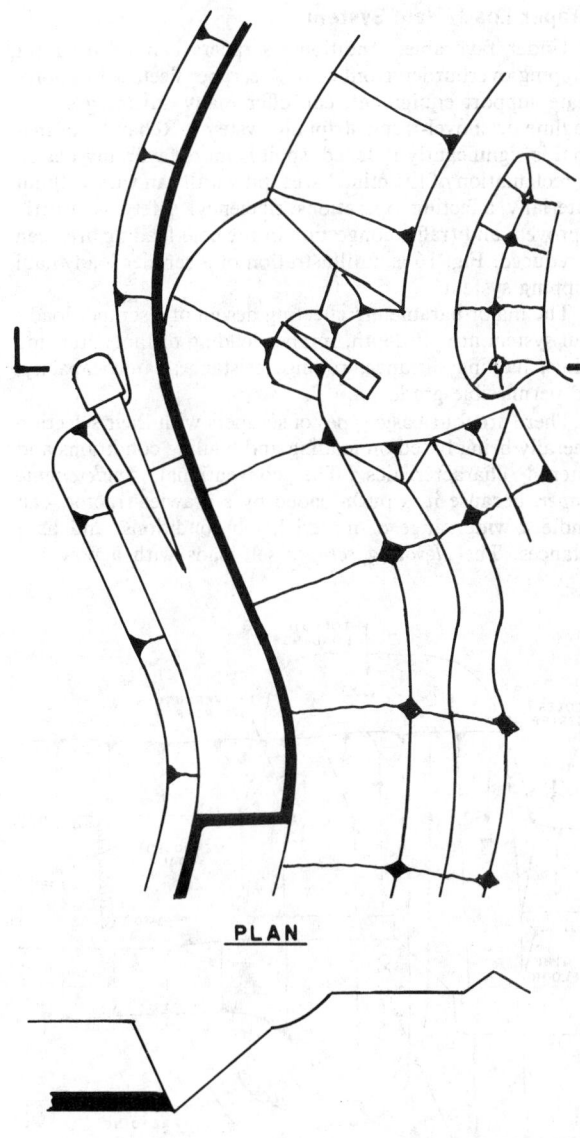

PLAN

SECTION

Fig. 12. Dragline pull-back method.

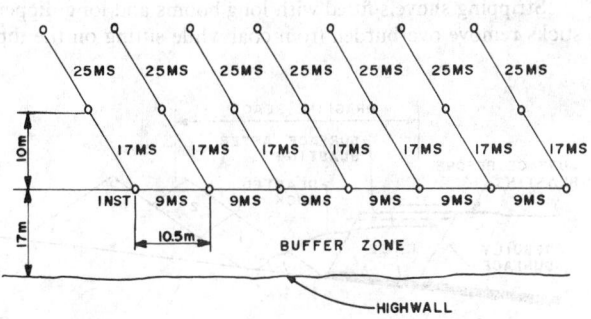

Fig. 14. Blasting pattern to provide maximum fragmentation with a buffer zone (Anon., 1981).

since considerable grading is necessary before the dragline can begin casting. Frequently the dragline is called upon to rebuild its working pad by retrieving material from the pit. Time lost in pad preparation may completely offset the original reduction in stripping volume. If the dragline can safely work on the spoil side of the pit, building its working pad ahead on the spoil, there may be justification for blasting material from highwall to the spoil. Increased costs of explosives, pad building costs, and highwall scaling delays must be weighed against the difference in overburden volume to be stripped. Fig. 15 illustrates this blasting method with the dragline working from the highwall side.

Drilling and blasting techniques are covered more thoroughly in Chapter 6.

SHOVEL-TRUCK SYSTEMS

There are a multiplicity of geologic situations adaptable to a shovel-haul truck system for removing burden from coal. The shovel-truck system could be substituted for all of the foregoing systems. Generally, the shovel-truck system is considered for one of the following reasons:

1) The coal is sufficiently covered by burden to be beyond the digging depth of draglines.

2) The burden is sufficiently unstable that dragline operations would be hazardous.

3) Toxic material within the burden requires special emplacement in the spoil.

4) The coal seam is thick enough to prevent adequate dragline spoiling off the coal.

5) The burden must be drilled and blasted. When blasted, the broken material is blocky and does not swell sufficiently to facilitate easy digging. (Since shovel/truck removal of overburden generally costs three times as much as dragline stripping, or more, the costs of closer drill-hole spacing and increased powder factor for the dragline system should be examined as a possible alternative.)

Because of a wide range of equipment sizes, shovel-truck systems have a decided advantage when attempting to match specific production goals. The equipment fleet easily can be increased in proportion to increased coal demand.

The design of shovel-truck systems for overburden removal follows similar standards of operation as in general open pit design as described in Section 6.3. Perhaps the only significant difference is in the treatment of economic limits. In surface metal mines, variability of ore grade and current market price greatly influence the stripping ratio. In surface coal mining, however, the coal product generally has consistent quality over large areas. Therefore, stripping ratio and stable long-term market price contracts more directly control minable limits.

Stripping Shovel System

Stripping shovels fitted with long booms and long dipper sticks remove overburden from coal while sitting on the top

of coal. Stripping shovels up to 138 m³ (180 cu yd) have been used for overburden removal in the midwest. Interest in stripping shovels has waned in the past decade because of their limited operating depth and relatively higher operating cost than the newer, longer boom, draglines. Where currently in use, stripping shovels are limited to relatively flat-lying seams and shallow burden. Because of their large crawlers, it is difficult to change their working bench level to follow undulating or pitching coal seams. As burden depths have increased, the mine engineer has been required to utilize auxiliary stripping equipment to remove near surface material in order to keep the stripping shovel working.

The pit configuration for a stripping shovel is similar to the basic dragline pit. The major parameters controlling its operation are the operating and dumping radii and dumping height.

The new, larger draglines with long boom and large bucket capacity have tended to render the stripping shovel obsolete.

Scraper Load/Haul System

Under favorable conditions, scrapers can be used for stripping overburden from coal. A scraper fleet, with appropriate support equipment, can offer many advantages over dragline or shovel-truck stripping systems. Rehandle of material is significantly reduced, spoil is more favorably placed for reclamation of the mined area, pit width can vary without materially affecting operations efficiency, safety is greatly improved, and traffic congestion in the coal loading area can be reduced. Fig. 16 is an illustration of a scraper load/haul stripping system.

The major parameters affecting design of a scraper load/haul system are pit depth, scraper loading distance, turning radius, hauling distance, rolling resistance, rated capacity, and permissible grade.

There are four basic types of scrapers with their selection generally being based on loading and hauling conditions and material characteristics. The conventional *single-engine scraper,* because it is push-loaded by a crawler tractor, can handle a wide range of material, job conditions, and haul distances. The *elevating scraper* self-loads with a powered

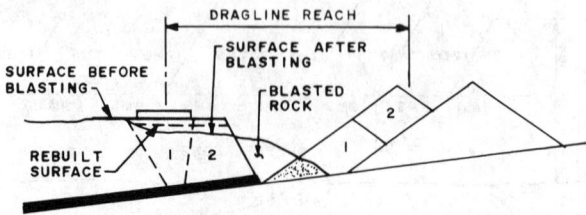

Fig. 15. Blasting burden into pit. (Note small area of muck requiring no dragline handling.)

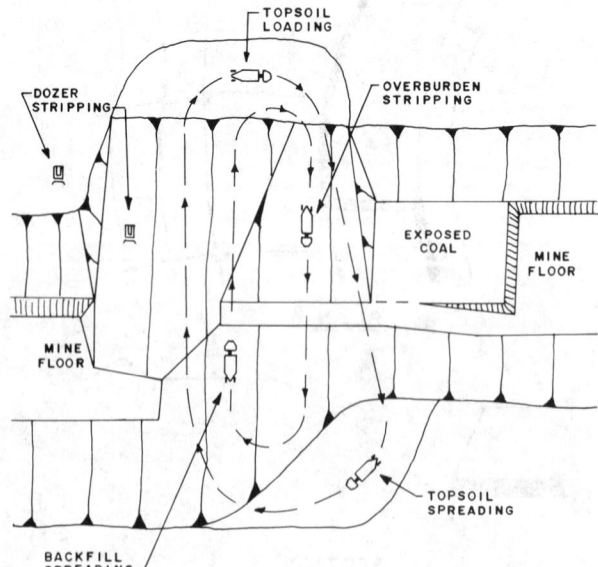

Fig. 16. Scraper haul mining system.

elevator that lifts material from the cutting edge up into the scraper bowl. It is limited to handling material up to gravel size because of the mechanical reliability of the elevator. The *dual-engine scraper* can handle the same materials as a conventional machine and is able to work in wet ground conditions where a conventional scraper might be limited by poor traction. *Push-pull scrapers* are dual-engine scrapers that work in tandem as a unit, the lead scraper first being push-loaded by the second scraper, then the second scraper being pull-loaded by the first. Once loaded, they separate and haul as individual dual-engine scrapers. Push-pull scrapers eliminate the need for push-dozers and generally can work in soft burden. For all four scraper types, very wet ground conditions reduce their productivity in varying amounts.

Various studies comparing scrapers to draglines have been performed by the author and others (Chironis, 1981; Anon., 1978). These studies have shown that scrapers should be considered in the removal of soft overburden less than 20 m (66 ft) in depth. While operating costs may be higher for the scraper system, initial capital investment is lower, thus creating a more favorable return.

Important in comparing the scraper system to other systems is the elimination of spoil grading equipment. Also, the self-loading scraper is frequently used for removing, stockpiling, and replacement of topsoil. Mine design utilizing a scraper haul system could eliminate the stockpiling phase of the topsoil removal and replacement cycle.

Most surface coal mines utilize scrapers for numerous reclamation activities. Some dragline operations put them to additional use for skimming off overburden in rolling terrain.

BUCKET WHEEL EXCAVATOR SYSTEMS

Given suitable geologic conditions, the bucket wheel excavator (BWE) can compete with dragline and shovel/truck systems. The BWE, a continuous mining system with high productivity, can be used as a primary or secondary excavator and is adaptable to all hauling and materials handling systems.

BWEs vary in size from an approximate theoretical output of 200 to 20 000 m^3/h (262 to 706,300 cu yd per hr). The larger BWEs have been used extensively in German and Australian brown coals. The smaller BWEs have found wide acceptance in many East European countries, but only sporadically in the US and other parts of the world where their size is proportionate to material thickness or production requirements.

The BWE generally is limited to excavating material having a cutting resistance less than 70 kg/cm (4700 lb per ft). Table 1 lists specific resistances for various types of overburden and coal. Laboratory methods for measuring cutting resistance are indicative at best because theoretical measurements vary with the size of the sample tested. Therefore, they should be used only as a guideline for specific material.

The methods of BWE operation are described according to the position of the BWE when excavating the face: *frontal* or *face mining, full block mining,* and *half block mining.* In each method, there are two alternative cutting techniques: horizontal, termed *terrace cutting,* and vertical, termed *drop cutting.* Each operational method and cutting technique has specific uses and affects mine design differently.

The frontal or face mining method is especially useful in separating soil or sublayers requiring special placement in the backfill. In this method, the BWE moves along the face using either the terrace or drop-cut technique of slicing. For handling soil or sublayers, terrace cutting is preferable. The face mining method does not require a slewable boom, but

Table 1. Specific Cutting Forces of Virgin Material for Bucket Wheel Excavator Excavation

Material Type	Specific Cutting Forces, kg/cm
Earth	10–30
Loess	20–40
Sand (fine, coarse, wet, or dry)	10–40
Clayey sand	10–50
Gravel, fine	20–50
Gravel, coarse	20–80
Sandy loam and wet loam	20–60
Dry loam	20–80
Clay, wet	30–65
Clay, dry	50–120
Clay, schistose	35–120
Sandy clay	20–65
Clayey slate	50–160
Slate	70–200
Sandstone (easy digging)	70–160
Sandstone (hard digging)	160–280
Gypsum	50–130
Lime	30–120
Phosphate	80–200
Marl	60–140
Limestone	100–180
Weathered granite	50–100
Alluvial, light consolidation	30–60
Alluvial, heavy consolidation	70–150
Alluvial, medium consolidation	50–80
Hard coal, normal	50–100
Hard coal, frozen	100–160
Lignite	20–70
Brown iron ore	190–210

Source: Anon., 1973. (Conversion factor: 1kg/cm = 67.197 lb per ft)

does require a long boom crowd. Face mining may be desirable if stable bench slopes can be maintained and if material is cast directly to the spoil side of the pit. Fig. 17 illustrates face mining with direct overcast.

Full block mining is the most common method utilized for removal of large, thick deposits of loosely to semiconsolidated material. The BWE continuously slews across the face block while the boom is crowded into the face. Terrace cutting the block is more common than drop cutting. If the boom cannot be extended (crowded), the depth of the terrace cut will be limited by the distance the BWE can advance. Use of crowdless machines must take into consideration the soil-bearing characteristics over which the machines must travel. Also, their crawlers are subjected to more mechanical wear and tear. Full block mining with shiftable bench conveyor is illustrated in Fig. 18.

Face block mining is particularly useful when it is desirable to remove extensive layers of overburden by terracing. The BWE, traveling parallel to the face, continuously slews across the face block in making the terrace cut. This method requires machines with long bucket wheel booms. Face block mining is especially useful in the selective excavation of topsoil or toxic layers requiring preferential placement. Fig. 19 illustrates the face block mining method.

Material removed by a BWE is overcast directly onto the spoil pile, transferred to large haul units, or to belt conveyors that transport material around the pit to the spoil.

A typical BWE mining system utilizing conveyors to transfer material to the spoil side of the pit will consist of

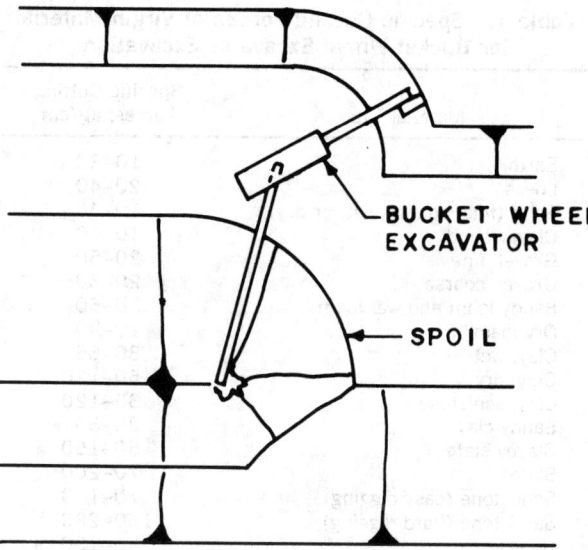

Fig. 17. BWE face mining with direct overcast.

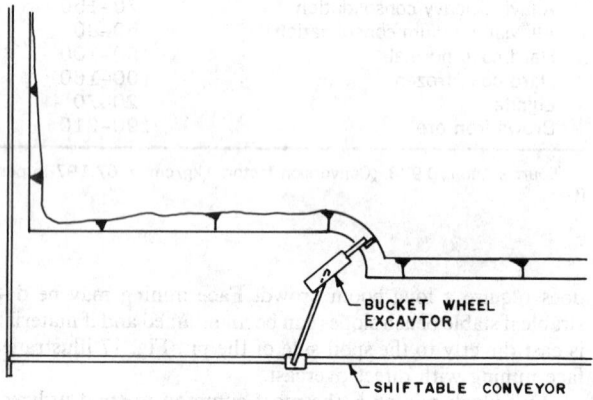

Fig. 18. BWE full block mining technique.

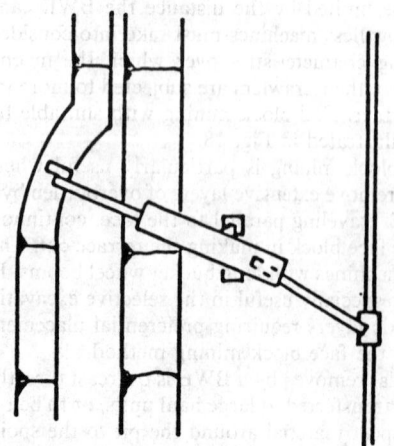

Fig. 19. BWE face block mining technique.

several equipment units that comprise the system. These units may consist of:

BWE excavates in situ material and discharges it either directly into the traveling hopper on the face conveyor or into the load hopper of a band wagon or belt wagon.

Band wagon or *MTC* (mobile transfer conveyor) receives material from a BWE and discharges it into the traveling hopper on the face conveyor. Its primary function is to add horizontal and vertical range to the digging capabilities of the BWE, thus limiting the frequency of face conveyor moves.

Belt wagon is similar to a band wagon, but has a longer discharge boom and is also used as a stacker.

Conveyors

Face conveyor, located on an excavation bench parallel to the advancing bench face, extends the full length of the face and is equipped with a traveling hopper car that allows the conveyor to be loaded at any point along its length.

Side conveyor, located on the pit side at right angles to the face conveyor, receives overburden material from the face conveyor and discharges it onto the spoil conveyor.

Spoil conveyor, located on a backfill bench, receives overburden material from the side conveyor and discharges it into the load hopper of the belt stacker. This conveyor extends the full length of the backfill bench and is equipped with a traveling tripper car that allows the conveyed material to be discharged at any point along its length.

Belt stacker receives overburden material from the spoil conveyor and stacks it onto a backfill bench.

Ancillary equipment includes tractor dozers for bench preparation, leveling and spreading, and conveyor shifting.

As mining progresses, the BWE mining system must be moved. Because of lost production when shifting the conveyor components, mines incorporating BWE systems favor long mining panels. In addition to increasing output of the BWE, rehandle volume relative to the total deposit is reduced. Long panels, however, may not be the desirable mine design. Increased capital costs of conveyors and development of the box cut may offset the benefits of long mine panels.

Productivity of BWEs

Manufacturers generally quote BWE production in theoretical output. The theoretical rate must be discounted to account for lost productive hours due to equipment and system unavailability, lost time for nonproductive maneuvers, cleanup, etc., and operator inefficiency. Fig. 20 illustrates a typical distribution of hour components from which utilization, availability, and productivity are calculated. It is important to note in Fig. 20 that equipment operating costs are based on operating hours and productivity is based on effective productive hours.

As shown in Fig. 20, operating delays can represent a large percentage of scheduled hours. Therefore, care should be given in their estimation to ensure reasonable production estimates. Equipment rearrangement involves movement of BWEs from bench to bench in the pit. Face, side, and spoil conveyors must be moved after completion of each panel cut. Encountering large rocks will slow production while the BWE digs around and dislodges them. Adverse digging conditions may be encountered as in unstable face slopes. Digging conditions may require frequent stoppages for bucket cleaning. For example, in Australian brown coal mines, bucket plugging is so severe that Australian engineers have spent considerable effort in designing buckets that tend to be self-cleaning. This is accomplished by installing a loosely fitted chain matting in the cutout back of the bucket.

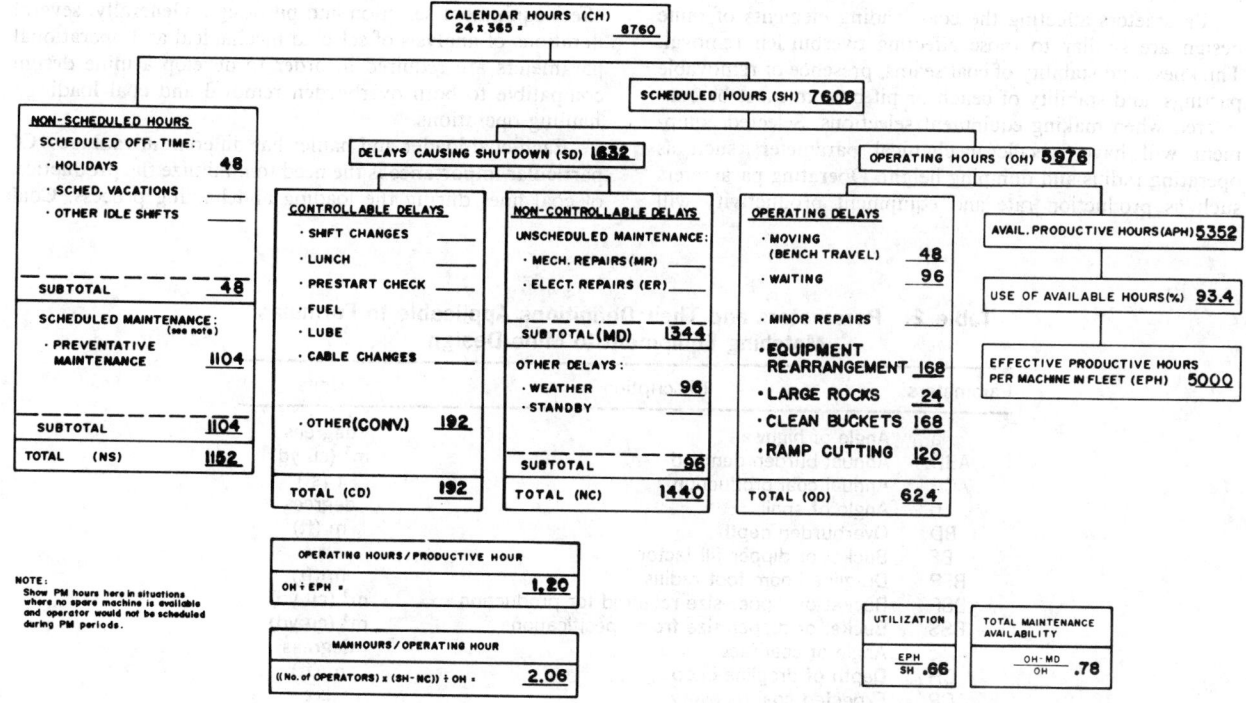

Fig. 20. Annual schedule of equipment hours for bucket wheel excavator.

Conveyor shifting is a time-consuming operation. To ensure rapid repositioning and realignment, conveyor shifting must be well coordinated. Shifting is accomplished by dozers fitted with a special clamp that grabs the rail of the conveyor base. At a travel rate of 5 km/h (3 mph), the belt can be shifted approximately 80 cm (31 in.) per dozer pass. When repositioning is complete, from one to three alignment passes are made with the dozer traveling at about 2 km/h (1.2 mph). Preshifting and setup preparation generally consume about 2 hr each. This schedule assumes well maintained working areas, equipment, and a well-coordinated shifting crew. Generally, the BWE and associated equipment undergo preventative maintenance while the conveyors are being shifted.

To compare a mine design consisting of several small or compact BWE systems with other primary excavators, consideration must be given to phased buildup of production, and under certain conditions, phasing in BWE systems may show more favorable economics than other primary excavators. On the other hand, large unit production requirements may dictate a BWE of such size that economy in scale may be offset by interest payments on capital investment.

Given appropriate digging conditions, the BWE should be given consideration in mine design because of its high production capacity and materials handling capabilities. Since the BWE is a highly specialized excavator, the engineer is referred to Anon. (1973) and Rasper (1975).

COAL REMOVAL METHODS

Conditions encountered in coal removal vary as much as in overburden removal. Coal removal must be matched to overburden removal in order to maintain a properly balanced production cycle. Since the stripping cycle generally is more cost sensitive than the coal loading and hauling cycle, coal loaders and coal haul fleets generally are oversized to actual production requirements to ensure sufficient capacity to recover coal production lost by nonrelated stripping delays.

Coal loading from seams in narrow pits requires loaders with large operating reach and/or dumping height. Loading in narrow pits generally begins with the haul truck located on the top of coal and the loader in the pit bottom. The loading sequence ends with both units on the pit bottom. Fig. 21 illustrates a shovel with extended boom loading trucks in this manner. In unusual cases of bad pit bottom conditions, all coal removal may take place from the top of coal.

For normal loading operations, the engineer has the option of cable or hydraulic shovels or front end loaders. When pit bottom conditions are unsuitable, backhoes may be used for loading from top of coal. Scrapers also may be used for loading and hauling. In recent years, a new type of coal loading machine has been tested and, under specific conditions, has been found to perform satisfactorily. This loader, *Easi-Miner®*, consists of a tractor-driven rotating drum with ripper teeth offset on its circumference. The drum, hydraulically operated, rotates and chews up coal from top downward as the unit is slowly propelled forward. The unit can be controlled to very selectively mine layers of seam (or partings) from about 10 cm (4 in.) up to 60 cm (2 ft). While still considered to be somewhat a prototype machine, its workability appears very promising, especially in coals where production quality is demanding.

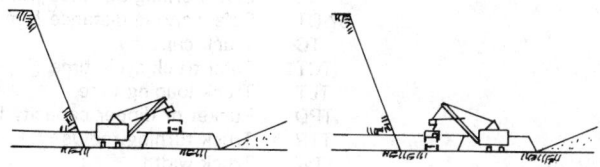

Fig. 21. Coal loading sequence in narrow pits.

Parameters affecting the coal loading elements of mine design are similar to those affecting overburden removal. Thickness and stability of coal seams, presence of removable partings, and stability of bench or pit bottom must be considered when making equipment selections. Selected equipment will have specific mechanical parameters such as operating radius and dumping height. Operating parameters such as production rate and equipment productivity will affect equipment selection and pit design. Generally, several iterations of analysis of selected mechanical and operational parameters are required in order to develop a mine design compatible to both overburden removal and coal loading/hauling operations.

Each coal loader and hauler has different advantages. Of particular importance is the need to minimize the production of coal fines during the loading and hauling process. Coal

Table 2. Parameters and Their Definitions Applicable to Formulas Matching Equipment to Mine Design

Parameters	Description	Units
a	Angle of highwall	degrees
ABD	Annual burden demand	m³ (cu yd)
ACP	Annual coal production	t (st)
b	Angle of spoil	degrees
BD	Overburden depth	m (ft)
BF	Bucket or dipper fill factor	.xx
BFR	Dragline boom foot radius	m (ft)
BSP	Bucket or dipper size required for production	m³ (cu yd)
BSS	Bucket or dipper size from specifications	m³ (cu yd)
c	Angle of coal face	degrees
CH	Depth of dragline chop	m (ft)
CR	Expected coal recovery	.xx
CT	Cycle time	sec or min
DB	Density of burden	t/m³ (st/cu yd)
DC	Density of coal	t/m³ (st/cu yd)
DDH	Dragline dumping height	m (ft)
DOL	Dragline digout length	m (ft)
DR	Dragline reach	m (ft)
DOR	Dragline, shovel, backhoe operating radius	m (ft)
DSY	Dragline base (tub) diameter	m (ft)
DT	Delay time (truck)	min
EBD	Equivalent reduction in burden thickness gained by dragline chop	m (ft)
ES	Average speed of truck, empty	m/min (fpm)
FT	Fixed time (truck)	min
HC	Coal seam thickness	m (ft)
HD	Height of dragline above coal seam	m (ft)
HDT	Haul distance (truck)	m (ft)
HS	Height of spoil peak above pit floor	m (ft)
IU	Intermediate calculation for simplification	—
LS	Average speed of truck, loaded	m/min (fpm)
MSL	Maximum suspended load of dragline	kg (lb)
MT	Dumping and spotting time (truck)	min
MWS	Minimum working width based on DOR	m (ft)
MWT	Minimum working width based on TTR	m (ft)
NTR	Number of trucks required	—
OE	Operator efficiency	.xx
PW	Panel width	m (ft)
RCP	Hourly coal production required	t (st)
RCR	Clearance radius of shovel or dragline	m (ft)
s	Swell factor for burden	.xx
SF	Swell factor for coal	.xx
SH	Scheduled equipment hours per year	hr
SR	Stripping ratio, burden to recovered coal	m³/t (cu yd/st)
SWDH	Safe working distance from highwall for dragline tub	m (ft)
SWDS	Safe working distance from toe of spoil pile	m (ft)
SWDT	Safe working distance from toe of highwall	m (ft)
TC	Truck capacity	t (st)
TCT	Total truck cycle time	min
TLT	Truck loading time	min
TPD	Bucket or dipper capacity for loading	t (st)
TTR	Truck turning radius	m (ft)
TW	Truck width	m (ft)
"	Annual equipment utilization factor	.xx

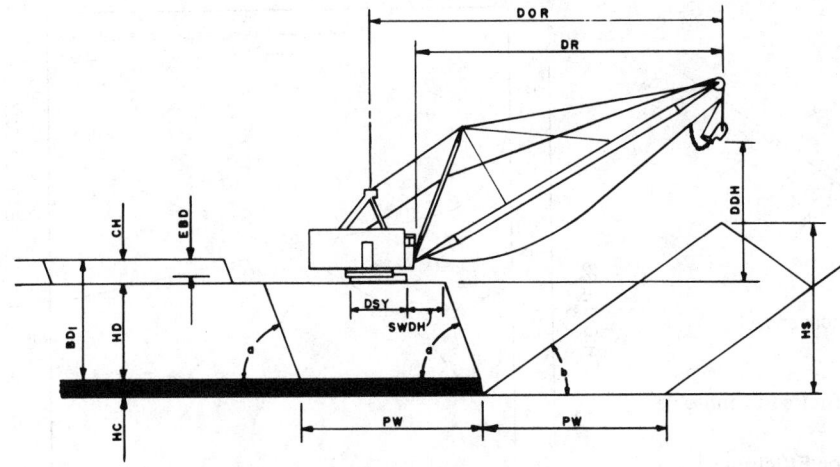

Fig. 22. Geometry of dragline pit, basic stripping method.

loading equipment frequently is used to remove coal partings. The parting is either directly casted off coal or loaded into end-dump trucks for disposal in mined-out pit areas.

ENGINEERING DESIGN EXAMPLES

Equipment for surface coal mines is selected based on the numerous parameters discussed in this section. Various formulas that match equipment to mine design are given. The list of parameters and their definitions, applicable to these formulas, are given in Table 2. The parameters are illustrated in Fig. 22 and 23. Except for exponents, standard computer notation is used in formulas for ease of printing.

Use of Swell Factors

Frequently, mines are designed and equipment selected based on burden swell factors experienced at neighboring mines. For example, a lignite mine in the Texas Wilcox formation reports an average swell factor of 35%. Based on 35% swell, mine design indicates that the burden can be removed with the operating dragline without rehandle. In actual operations, however, an extended bench method with 20% rehandle of burden is required to uncover coal. Why? The initial swelling of burden as it is dumped on the spoil by the dragline is much greater than the measured swell which probably is determined on an annual basis for management.

Knowledge of the settlement characteristics of spoil is necessary to ensure that mine design correctly defines conditions at the moment of spoiling burden. Settlement characteristics vary according to the material classification and hydrologic conditions. An analysis of the settlement characteristics of the burden at a Texas lignite mine estimated that 75% of all settlement occurs within one year after mining. Fig. 24 illustrates the settlement characteristics for the burden (Schneider, 1977). Based on an annual measurement of 35%, initial swell is more likely 45% or more. This difference in swell could represent approximately 3 m (10 ft)

Dragline Pit Design Example*
Basic Stripping Method for Single Seam Mining

$$DOR = IV + DOL \, {}^2\!/_4 \, IV$$

where

$$IV = (BD - EBD)/\tan a + HC/\tan c + (BD - EBD)\cdot(1 + s)/\tan b + PW/4 + DSY/2 + SWDH$$

and

$$EBD = (CH \tan b/\tan a)/1 + s)\dagger$$

Then

$$DR = DOR = BFR$$

and

$$DDH = BD(1 + s) + (PW/4)(\tan b) - HD - DC$$

Also,

$$HS = [(DOR^2 - DOL^2)^{\frac{1}{2}} - DSY/2 - SWDH \, BD/\tan a - HC/\tan c]\cdot\tan b$$

With *DR* and *DDH*, dragline specifications may be examined for appropriate dragline relative to the following specifications:

$$BSS = MSL/[2000 + (2000 \, DB/(1 + s)]$$
$$BSP = ABD(1 + s)/(BF\cdot SH\cdot u\cdot OE\cdot 3600/CT)\cdot(HD/BD) + [(CH/BD)\cdot(BF\cdot U)/CT]\dagger$$

where

$$ABD = ACP\cdot SR/CR$$

*Derivation of formulas given and geometry of other pit designs found in Dagdelen, 1979; Morey and Lee, 1979; and Fund, 1981.

†For basic stripping without chop down: *EBD* = 0, *HD/BD* = 1, *HC* = 0.

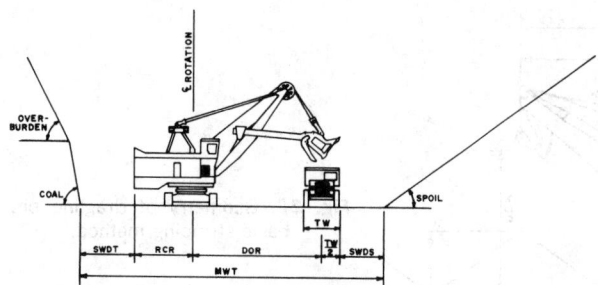

Fig 23. Minimum work width for coal shovel.

of additional dragline dumping height required in mine design.

Calculation of Utilization and Operator Efficiency Factors

In determining equipment productivity, it is common practice to discount operating time by applying an operating efficiency factor termed, "the 50-minute hour." This factor, equivalent to 0.833, is intended to include discounts for operator inefficiencies and minor operating delays. If operating conditions are abnormal, more discount is taken in terms of fewer "minutes per hour."

In the author's opinion, there are 60 min in every hour and the "minute-hour" factor has been used to account for inadequate analysis of the equipment operation. As equipment units have increased in size and unit capacity and as operating conditions vary, overdesign or underdesign of mining systems has resulted from this inadequacy.

With the advent of more refined engineering analysis made possible by computer application, it is possible to calculate many of the operating delays that affect productivity. In addition, time-motion studies can more appropriately assess the operator's efficiency without penalizing him for other operating delays.

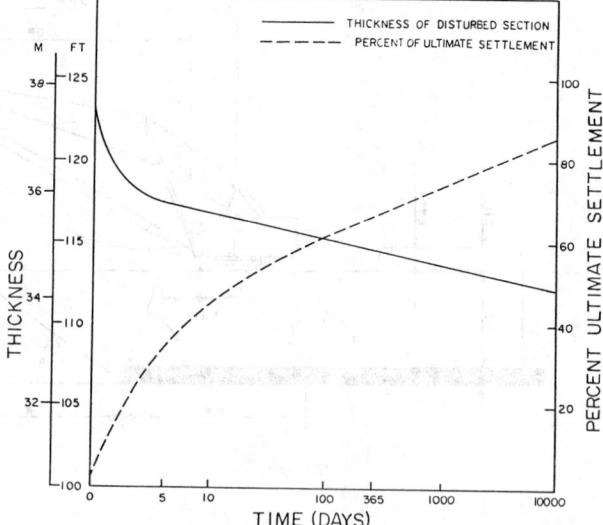

Fig. 24. Thickness of a disturbed section and the percent of ultimate settlement recovered as a function of time since disturbance of an initial 30-m (100-ft) undisturbed section (Schneider, 1977).

Referring back to Fig. 20, Annual Schedule of Equipment Hours, operator efficiency assigns accountability to the operator only for those hours that can be productive.

While utilization is simply a measure of the ratio of time utilized to time scheduled, accountability for all delays that affect utilization is assigned to respective areas of responsibility. The utilization factor should be used in engineering design calculations and for equipment fleet sizing.

REFERENCES

Anon., 1973, "Materials Handling Research: The Bucket Wheel Excavator," Information Circular 8580, US Bureau of Mines.

Coal Shovel Selection

$$BSP = [RCP/(3600/CT)](1 + SF)/(DC)(BF)$$

where

$$RCP = (ACP/CR)/(SH)(u)(OE)$$

Burden shovel can be selected by substituting *ABD* for *ACP*, *s* for *SF*, *DB* for *DC*, with *CR* = 1.

Bucket size, *BSP*, is calculated in loose volume equivalent to in-bank (in-place) volume.

Minimum Truck Requirements for Coal Shovel

$$TLT = (TC/TPD)(CT)$$

where

$$TPD = (BSP)(BF)(DC)/(1 + SF)$$
$$FT = DT + MT + TLT$$

Therefore,

$$TCT = FT + (HDT/LS) + (HDT/ES)$$

and

$$NTR = [(TCT/TLT) \cdot u \cdot OE] + 1$$

Factors *u* and *OE* are assumed to be fleet averages. One truck is added, assuming all other trucks are in motion.

Minimum Working Width for Coal Shovel

$$MWS = SWDT + RCR + DOR + TW/2 + SWDS$$

or,

$$MWT = SWDT + 2TTR + SWDS$$

If *MWS* is equal, or less than the dragline panel width, it is selected. If the maximum designed dragline panel width is less than *MWT,* an extended bench method must be considered.

Anon., 1978, "Economics of Shallow Overburden Stripping," Caterpillar Tractor Co., Peoria, IL.

Anon., 1981, "Coal Interpretation Manual," BPB Instruments Ltd., Loughborough, England.

Anon., 1982, "Optimal Dragline Operating Techniques," Final Report, Contract No. DE-AC01-77ET12195, US Dept. of Energy.

Chironis, N.P., 1981, "Scraper Use Picks Up Steam," *Coal Age,* March.

Cummins, A.B., and Given, I.A., eds., 1973, *SME Mining Engineering Handbook,* Vol. 2, AIME, New York.

Dagdelen, K., 1979, "Dragline Simulation and Selection for Single Flat-Lying Coal Seams," MS Thesis, Colorado School of Mines, Golden.

Fund, R., 1981, *Surface Coal Mining Technology,* Noyes Data Corp., Park Ridge, NJ.

Morey, P., and Lee, C., 1979, "Optimization of Dragline Operation," *Mining Congress Journal,* June.

Rasper, L., 1975, *The Bucket Wheel Excavator,* Trans Tech Publications, Clausthal, Germany.

Schneider, W.J., 1977, "Analysis of the Densification of Reclaimed Surface Mined Land," MS Thesis, Texas A&M University, College Station.

Chapter 6

Mine Operations

Hugh W. Evans, Editor
Thys B. Johnson, Assistant Editor

6.1 Drilling

6.1.1 Drilling Principles

HOWARD L. HARTMAN

INTRODUCTION

In virtually all forms of mining, rock is broken by drilling and blasting. Except in dimension stone quarrying, drilling and blasting are required in most surface mining. Only the weakest rock, if loosely consolidated or weathered, can be broken without explosives, using mechanical excavators (rippers, wheel excavators, shovels, etc.) or occasionally a more novel device, such as a hydraulic jet.

In the mining cycle, drilling performed for the placement of explosives is termed *production drilling*. Drilling is also used in surface mining for purposes other than providing blastholes. It finds application during exploration for obtaining drill hole samples and during development for drainage, slope stability, and foundation-testing purposes. Only in the production phase of surface mining, however, are unique or specialized drilling methods employed; the discussion in this section is directed primarily to that application of drilling.

There are minor applications of rock penetration in surface mining other than drilling. In quarrying, dimension stone is freed by cutting, channeling, or sawing. Usually mechanical means or sometimes a thermal jet is employed to produce a cut or kerf, outlining the desired size and shape of stone block.

A variety of geologic materials may be encountered in drilling. But whether they are ore or waste is usually of less consequence in selecting a drilling method than how resistant they are to penetration and how they occur geologically. The same drill may be used in overburden and in ore—but different drilling methods may be required in the same mine for markedly different ore or overburden formations.

Likewise, both consolidated and unconsolidated materials may have to be drilled. While soils and other loose materials do not require blasting, on occasion they may have to be penetrated by a drill when they overlie rock, or when they can be economically cast by explosives. The latter practice is termed *explosives stripping*, and is covered in the subsequent section on "Overburden Removal."

This section is concerned primarily with the principles and engineering selection of drilling equipment to meet specified job requirements in surface mining, and secondarily with the drilling methods themselves. Following a presentation of the general principles of rock penetration, the remainder of the section considers the engineering application of the major types of production drilling equipment in surface mines.

CLASSIFICATION OF METHODS

A classification of drilling methods can be made on several bases. These include size of hole, method of mounting, and type of power. The scheme that seems the most logical to employ, a generic one, is based on the form of rock attack or mode of energy application leading to penetration.

Since drilling occupies only one category in the classification, the more general term *rock penetration* is preferred for all methods of forming a directional hole in rock. Hence it is preferable to speak of jet piercing as a method of thermal penetration rather than thermal drilling. *Drilling* is reserved for the mechanical attack systems.

The classification advocated here is a general one, applicable to all kinds of mining and encompassing all forms of rock penetration. Thus machines used for cutting as well as drilling are included. This classification bears some resemblance to one for rock fragmentation methods (such as blasting and other wholesale breakage techniques), since the principles are identical, and rock breakage is the common objective.

The known methods, based on rock attack, are listed in Table 1 in general order of importance today, with those having widest application appearing first. When a commercial or operational machine exists that employs a particular principle, it is so identified. An alternative classification, one employed by Maurer (1968, 1980), utilizes the rock disintegration mechanism (see "Analyzing Drill Performance" and Table 3).

Mechanical Attack

The application of mechanical energy to rock can be performed basically in only one of two ways: by percussive or rotary action. Combining the two results in hybrid methods termed roller-bit rotary and rotary-percussion drilling. The mechanical category, of course, encompasses by far the majority (probably 98%) of rock penetration applications today. In surface mining, roller-bit rotaries and large percussion drills are the machines in widest current use, with rotaries heavily favored.

Thermal Attack

Although penetration principles other than drilling are known, only two have been utilized commercially in surface mining. One is thermal penetration (the other is fluid). The only thermal method having practical application today is

Table 1. Classification of Rock Penetration Methods, Based on Form of Attack*

Form of Energy Application	Method	Machine
Mechanical (Drilling)	Percussion	
	Drop tool	Churn or cable-tool drill
	Hammer	Rock drill, channeler
	Rotary, drag bit	
	Blade	Auger or rotary drill
	Stone-set	Diamond drill
	Sawing	Wire-rope, chain, or rotary saw
	Rotary, roller-bit	Rolling-cutter drill
	Rotary-percussion	
	Hammer	Rock drill (independent rotation)
	Rotary	Rolling-cutter drill (superimposed percussion)
Thermal	Flame	Jet piercer, jet channeler
	Plasma	Plasma torch
	Hot Fluid	Rocket
	Fusion	Subterrene
	Freezing	(Conceptual)
Fluid	Jet	Hydraulic jet, monitor, cannon
	Erosion	Pellet-impact or abrasion drill
	Bursting	Implosion drill
	Cavitation	Cavitating drill
Sonic	Vibration	High-frequency transducer
Chemical	Explosion	Shaped charge, capsule, projectile
	Reaction	Rock softener, dissolution
Electrical	Electric arc or current	Electrofrac drill
	Electron beam	Electron gun
	Electromagnetic induction	Spark drill
Light	Laser	Electromagnetic radiation beam
Nuclear	Fission	(Conceptual)
	Fusion	(Conceptual)

* Energy applications listed in approximate order of present practicality. For classification by mechanism of rock disintegration, see Maurer (1980).

flame attack with the jet piercer or channeler. It penetrates the rock by spalling, an action associated with hard rocks of high free-silica content. Because of its ready capability of forming various shapes of openings, oxygen or air jet burners are used not only to produce blastholes but to chamber them as well and to cut dimension stone. Jet piercing of blastholes, however, has decreased in popularity in recent years as mechanical drills have improved in versatility and penetrability.

Fluid Attack

While disintegration of rock by fluid injection is an attractive concept, the end result is more likely fragmentation than penetration. To produce a directed hole with pressurized fluid from an external source, jet action or erosion appears to be more feasible, but commercial application to date is limited. Hydraulic monitors have been used for over a century to mine placer deposits and to strip frozen overburden; and more recently, high-pressure hydraulic jets have been applied successfully to the mining of coal, gilsonite, and other consolidated materials of relatively low strength. In one penetration/fragmentation device, hydraulic and mechanical attack mechanisms assist and complement one another. For large holes, the hydraulic jet alone may be competitive with drilling.

Sonic Attack

Sometimes referred to as vibratory drilling, this method as presently conceived is a form of ultra-high-frequency percussion. Attractive but not presently commercial, actuation of sonic devices by hydraulic, electric, or pneumatic means is possible.

Chemical Attack

Chemical reaction, because of the time element, may be more attractive as an accessory rather than a primary means of penetration. The use of explosives is a distinct possibility, however, and several alternative systems are under investigation. Additives to the drilling fluid, termed softeners, have shown some improvement in penetration rate in conventional drilling.

Other Methods of Attack

While some attempts to employ other forms of energy (electrical, light, or nuclear) have been made, the remaining methods in Table 1 must be classified in the experimental or hypothetical category at present.

Future Applications

Maurer (1980) among others is optimistic about the future of novel penetration devices, citing successful laboratory and field tests where they have outperformed conventional drilling methods. It is probable, however, that their value for rock penetration in the near future will be limited to (1) supplementing mechanical energy (drilling) systems for special circumstances, and (2) creating very large or deep holes. Their application to general blasthole drilling in surface mining seems less attractive.

THEORY OF PENETRATION

Since the vast majority of rock penetration in surface mining is carried out by mechanical attack systems, the remainder of this section is devoted almost entirely to drilling.

Operating Components of System

There are four main functional components of a drilling system (and of most other penetration systems): (1) drill (energy source); (2) rod (energy transmitter); (3) bit (energy applicator); and (4) circulation fluid. These components are related to the utilization of energy by the drilling system in attacking rock in the following ways:

1. The *drill* is the prime mover, converting energy from its original form (fluid, electrical, pneumatic, or combustion-engine drive) into mechanical energy to actuate the system.

2. The *rod* (or drill steel, stem or pipe) transmits energy from the prime mover or source to the bit or applicator.

3. The *bit* is the applier of energy in the system, attacking rock mechanically to achieve penetration.

4. The *fluid* cleans the hole, controls dust, cools the bit, and at times stabilizes the hole.

In commercial drilling machines, attention has focused to some extent on reduction of energy losses in transmission. This has led to the introduction of down-hole (in-the-hole) drills, both of the large percussion variety and the roller-bit rotary (electro- and turbodrill) type, although the latter has found application mainly in oil well boring. They replace mechanical energy transmission with fluid or electrical transmission, which usually results in more energy reaching the bit and faster drilling.

Functions of Rock Drilling

A drilling (or any penetration) system must perform two separate functions in order to achieve advance into rock: (1) fracture and break material from the solid, and (2) eject the debris formed. The first phase is, of course, actual penetration, while the second is cuttings removal. Both affect drilling and drill performance but are distinct and separate functions of the process.

Mechanics of Penetration

As indicated earlier, there are only two basic ways to attack rock mechanically—by percussion and by rotation—and the four classes of commercial drilling methods to be discussed utilize these principles or combinations of them. It is the bit/rock interaction that governs the efficiency of energy transfer and the nature of the breakage process.

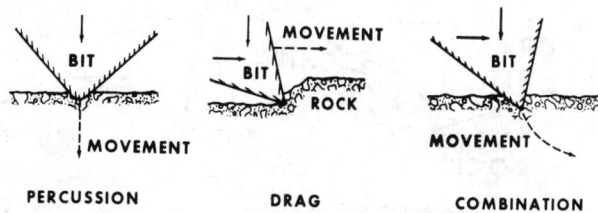

Fig. 1. Types of drilling action in mechanical attack on rock (after Maurer, 1967).

Causing rock to break during drilling is a matter of applying sufficient force with a tool to exceed the strength of the rock. This resistance to penetration of rock is termed its *drilling strength;* it is not equivalent to any of the well-known strength parameters. Further, the stress field created by the tool must be so directed as to produce penetration in the form of a hole of the desired shape and size. These stresses are quasi-static in nature, because forces are applied slowly in the drilling process. The inertia force, induced stress wave, and rate-of-loading effects in rock drilling have been demonstrated to be negligible (Maurer, 1962).

The different ways in which percussion, rotary, and combination drills attack rock are compared in Fig. 1.

Percussion Drilling: The *applicator* in a percussion drill is a chisel-shaped or button-studded tool that impacts the rock with a hammer-like blow. The stress effective in breaking the rock acts essentially in an axial direction and in a pulsating manner. Rotation enables the bit to strike the rock in a different spot on consecutive blows, a mechanism called *blow indexing,* which forms contiguous craters and ultimately a directed hole in the rock. The rotational torque applied, however, is usually not responsible for any penetration of the rock, since it is small in magnitude and, with rifle-bar rotation, is operative between blows only. Likewise, the sole function of the applied thrust is to keep the bit in contact with the rock.

The sequence in crater formation is as follows (Fig. 2): (1) the rock is inelastically deformed, with crushing of surface irregularities; (2) subsurface microcracks form from stress concentrations and confinement at the bit/rock interface enclosing a wedge of material, which is crushed; (3) secondary cracks propagate along shear trajectories to the surface, forming large fragments or chips; and (4) broken particles are ejected by the rebound of the bit and the cleaning action of any circulation fluid, resulting in the formation of a crater. The sequence is repeated with succeeding blows, except that indexing tends to provide additional "free faces" that may aid rock breakage and increase crater size. Indexing is not a sensitive variable, however, nor does it lend itself to precise control in drilling machines (Hartman, 1966).

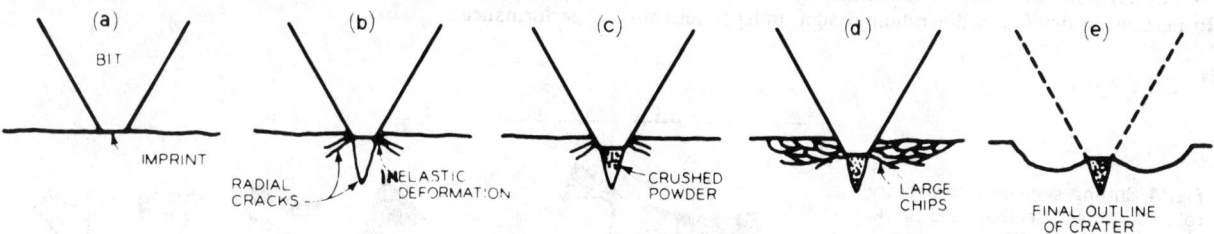

Fig. 2. Cutting sequence of a percussion drill bit (after Hartman, 1959).

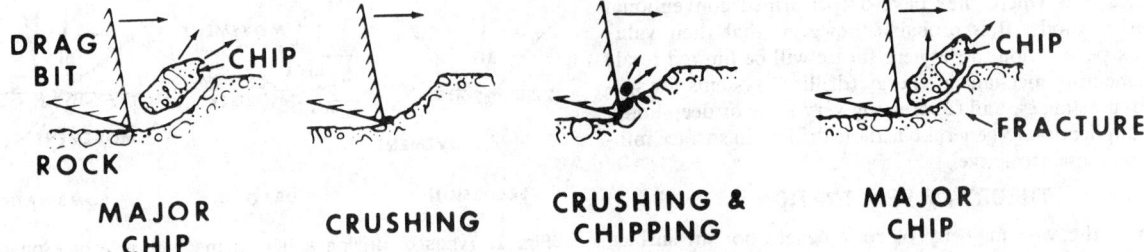

Fig. 3. Cutting sequence of a rotary drag bit (after Goodrich, 1957, and Maurer, 1967; by permission of The Curators of the University of Missouri).

In essence, insofar as rock fracture is concerned, the two predominant mechanisms in percussion drilling are crushing and chipping.

Rotary Drag-Bit Drilling: The planing or plowing action of a drag-type rotary drill bit is performed by a variety of tools, including blade and diamond drills as well as rope, chain, and rotary saws. Regardless of the geometry of the device, drag action at the cutting surface is supplied by two forces: thrust, a static load acting normally, and torque, the tangential force component of a rotational moment acting on the rock surface.

The mechanism of penetration in drag-bit drilling is as follows (Fig. 3): (1) as the cutting edge of the bit comes in contact with the rock, elastic deformation occurs; (2) the rock is crushed in the high-stress zone adjacent to the bit; (3) cracks propagate along shear trajectories to the surface, forming chips; and (4) the bit moves forward to contact solid rock again, displacing the broken fragments. One may conceive of the thrust as being responsible for indentation and the tangential force for plowing.

The similarity in cutting action of these two basic drilling systems is striking. One concludes that, under mechanical attack, rock fails alternatively by crushing and chipping, whether the energy is applied by percussion or rotation.

Rotary Roller-Bit Drilling: Essentially, the same form of drill rig may be used with rolling-cutter bits as with drag bits, employing the same forces to achieve penetration (although higher levels of thrust and torque are utilized, and heavier machines are customary). However, the geometry of the roller bit is such that a hybrid cutting action, a combination of percussion and rotary, results (Simon, 1956). As the bit turns, cutting teeth mounted on each rotating cone alternately engage the rock, impacting, indenting, and (with "soft-rock" bits) planing it (Fig. 4). The same crushing and chipping occur, however, as in the two basic systems; only the proportions differ.

Rotary-Percussion Drilling: This is also a hybrid form of drilling, combining independent percussive and rotational actions (Bullock, 1976). Generally, percussion bits (with buttons or asymmetric wings) or sometimes roller bits are used. In percussion drills of a down-hole design, independent ro-

tation is utilized, and the cutting action can be adjusted from straight percussion to rotary percussion. The superimposing of percussion on a rotary system means that higher impact forces are realized than in straight rotary drilling, but thrust and torque-induced forces are still operative. In rotary-percussion drilling, rock failure occurs by crushing and chipping, the proportion being a function of the drilling action.

Factors Influencing Drilling

A number of factors affect rock penetration or cuttings removal in the drilling process. These in turn largely determine the performance of a given drilling machine.

The various factors may be grouped in six categories: (1) drill, (2) rod, (3) bit, (4) circulation fluid, (5) drill hole, and (6) rock.

Those design factors in categories 1 to 4, components of the drilling system itself, are referred to as *operating variables*. They are controllable within limits, interrelated in some instances, and must be selected to match the environmental conditions reflected by category 6, rock type. Those variables of prime importance in the various drilling systems are listed in Table 2.

The *drill hole factors* of category 5, size, depth, and inclination, are dictated primarily by outside requirements and are independent variables in the drilling process.

The *rock factors* of category 6 are environmentally derived. They are also independent variables in the drilling process and include the following (Tandanand, 1973):

1. Material properties (resistance to penetration, porosity, moisture content, density, Shore hardness, compressive strength, coefficient of rock strength, etc.)

2. Geologic conditions (petrologic and structural-bedding, fractures, folds, faults, joints, etc.)

3. State of stress (in situ pressure and pore pressure—unimportant in shallow holes).

Another group of factors is external to the drilling process itself and may be referred to as *job* or *service factors*. These include operational variables related to labor, supervision, job site, scale of operations, power availability, and weather. While job factors are not involved in the mechanics of rock penetration, they may exert considerable influence on drill performance.

Fig. 4 Cutting sequence of a rotary roller bit (after Cheatham and Gnirk, 1967).

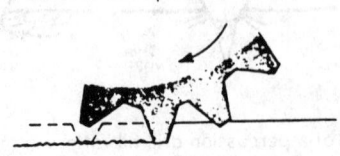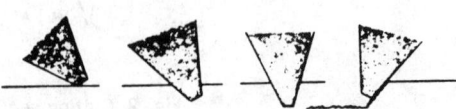

Table 2. Operating Variables in Drilling and Their Effects on Different Methods

	Percussion	Rotary	Rotary-Percussion
1. Drill			
Drill power	x	x	x
Drill thrust	x	x	x
Drill torque		x	x
Drill rotary speed		x	x
Blow energy	x		x
Blow frequency	x		x
2. Rod			
Rod dimensions	x	x	x
Rod geometry	x	x	x
Material properties	x	x	x
3. Bit			
Bit diameter	x	x	x
Bit geometry	x	x	x
Material properties	x	x	x
4. Circulation Fluid			
Fluid flow rate	x	x	x
Fluid properties	x	x	x

ANALYZING DRILL PERFORMANCE

For the operating details of application with a particular drilling method, the reader is referred to the latter part of this section. The discussion here is restricted to the basic aspects of drilling applications.

Performance Parameters

While more sophisticated criteria have been proposed, the following are adequate and employed almost exclusively in evaluating the performance of a given drilling system or in comparing different systems:

1. Energy or power
2. Rate of penetration
3. Bit wear
4. Cost

Under particular field circumstances, any one of these parameters may govern. In surface mining, energy or power consumption is becoming of increasing concern; but energies or powers, if compared, are generally of more concern because of their effect on penetration rate. Both penetration rate and bit wear are popular criteria, with rate more general in usage and wear more common for deep drilling where bit changes must be minimized.

But preeminent as a yardstick in any drilling situation is cost, for it collectively reflects all the other factors and is the ultimate measure of feasibility. A drill can have high availability and be novel, fast, and environmentally acceptable; but if it is not cost-effective, then an alternative system should be sought. (It is well to remember, however, that the goal in mining is the minimization of *all* rock breakage costs, and that drilling cannot be analyzed independently of blasting and comminution.)

Nevertheless, the first three parameters (energy, rate, and wear) enter into determination of the cost and largely control it. For this reason, it is desirable to know the quantitative, individual effect of pertinent operating variables (listed in Table 2) on energy, penetration rate, and bit wear, because they in turn determine the drilling cost.

Energy and Power: Formulas for determination of energy and power output in drilling systems are tabulated below, together with reference sources.

1. *Percussion* (Wells, 1950; Pfleider and Lacabanne, 1961):

$$\text{blow energy } E = \tfrac{1}{2} mv^2 = c\,WL^2B^2 \tag{1}$$

$$\text{power } P = BE = c\,WL^2B^3 \tag{2}$$

where m is piston mass $= W/g$, v is piston impact velocity, c is a constant $= 0.3$ to 0.5×10^{-6} (in English units), W is piston weight, L is piston stroke, and B is blow frequency. If the drill is pneumatic powered, then its energy and power output are most readily changed by varying the air pressure. Mathematically, the effect of gage air pressure p is

$$\begin{aligned} E \propto p \text{ and } P \propto p^{1.5}, &\text{ since } B \propto p^{0.5} \\ \text{Thus, } P \propto p^{1.5}\,a^{1.5}\,L^{0.5}&/W^{0.5} \end{aligned} \tag{3}$$

where a is piston area.

2. *Rotary* (Teale, 1965):

$$\text{energy (work/revolution) } E = E_f + E_r$$

$$= Fh + 2\pi T \tag{4}$$

$$\text{power } P = P_f + P_r = FR + 2\pi NT \tag{5}$$

where E_f is thrust energy, E_r is rotational energy, P_f is thrust power, P_r is rotational power, h is depth of penetration per revolution $= R/N$, R is rate of penetration, F is thrust, T is torque, and N is rotary speed (revolutions per unit time). Because the thrust component of the total energy and power in most rotary drills and boring machines is small (usually only 1% or less), the work done by the thrust is generally neglected (Roxborough and Rispin, 1972; Rowlands, 1974).

3. *Rotary percussion* (Inett, 1956; Bullock 1974):

energy and power:
percussion and rotary values additive.

Mining rotary-percussion drills of the hammer type are now usually hydraulically actuated, which permits higher efficiencies and increased power levels delivered to the bit.

One useful measure of the drilling efficiency of a particular machine is the *specific energy*, or the energy consumed per unit volume of rock broken (Teale, 1965; Bailey and Dean, 1967):

$$e = \frac{E}{V} = \frac{P}{AR} \qquad (6)$$

where V is volume of rock broken and A is hole area. Specific energy varies with the drilling method, operating variables, drill hole geometry, and rock properties. As such, it is a sensitive measure of the effectiveness of a particular drilling system. It has also been shown by Tandanand (1973) to correlate directly with the coefficient of rock strength, compressive strength, and Shore hardness, all rock properties identified in "Theory of Penetration." A comparison of specific energy requirements for different drilling methods (plotted against hole size) appears in Fig. 5; in these tests, rotary percussion was most efficient.

Rate of Penetration: The rate of penetration for any drilling process is customarily expressed as a linear advance rate and may be defined by the relation:

$$R = \frac{\Delta V}{\Delta t} / A \qquad (7)$$

where t is time and $\Delta V/\Delta t$ is the average volume rate of rock removal with respect to an interval of time. In drilling, the volume of rock broken has long been known to be directly proportional to the energy applied (Simon, 1956; Hartman, 1959):

$$V \propto E \qquad (8)$$

neglecting the threshold energy to initiate penetration. Eq.

8 further yields the very important and frequently confirmed relation between rate of penetration and drill power (for a given bit size),

$$R \propto P \qquad (9)$$

This is the basis for the repeated observation that, in a given system, *the only way to drill rock faster is to supply more energy to the bit.*

General relations for the various drilling systems can be developed from the specific energy (drilling strength) relation, Eq. 6 (Maurer, 1967):

$$R = \frac{P}{Ae} \qquad (10)$$

At optimum conditions (peak efficiency), the minimum value of specific energy attained approaches the compressive strength of the rock being drilled (Teale, 1965).

Values comparing the specific energy, maximum power, and maximum potential drilling rate have been estimated for all the known methods of rock penetration and are given in Table 3, together with the applicable rock disintegration mechanism. While not precise, the data raise the interesting question of whether any other method of rock attack can penetrate rock more efficiently than mechanical attack. In fairness to the novel methods, however, Maurer (1980) points out that, in large holes where a kerf can be cut, a comparison of penetration efficiency and rate should be based on *specific kerfing energy.*

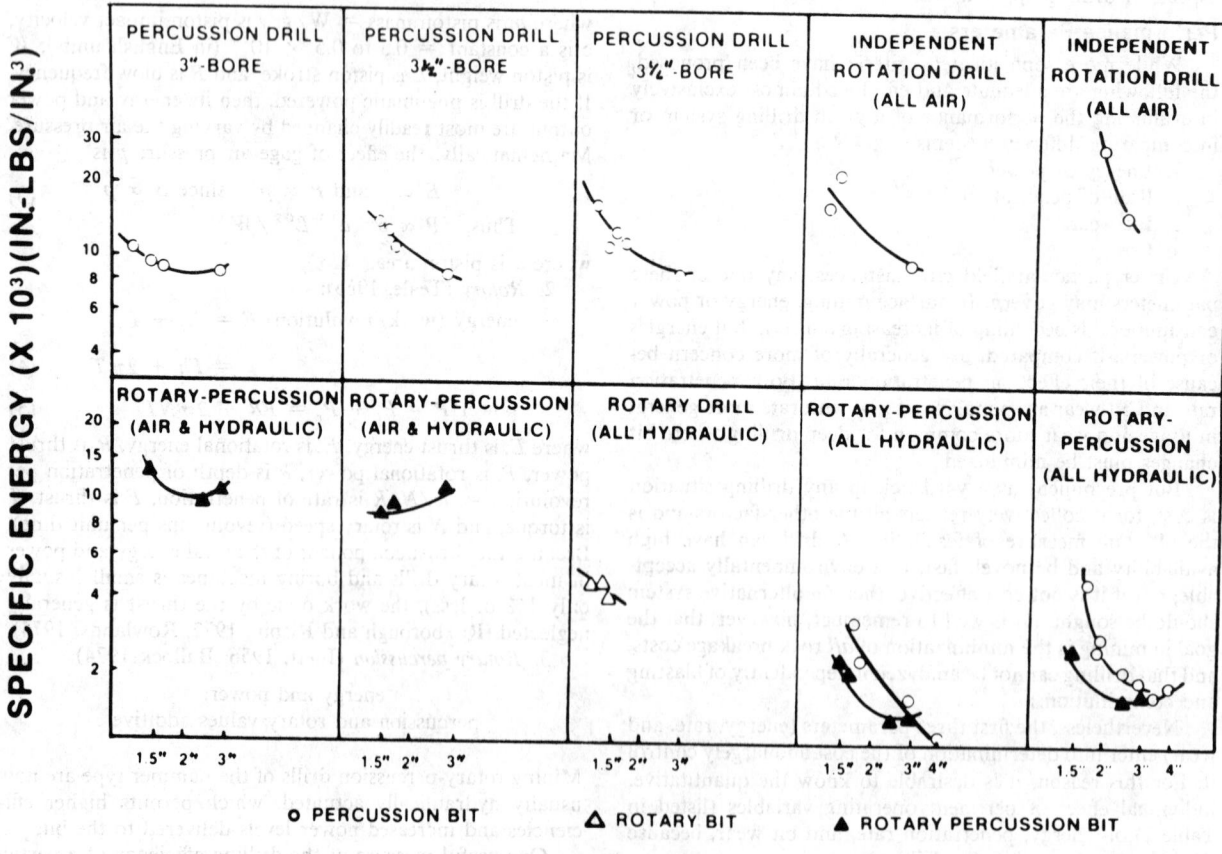

Fig. 5. Comparison of specific energy for different drilling systems; rock type: Bonneterre dolomite (after Bullock, 1976).

Table 3. Comparison of Specific Energy, Power, and Rate of Penetration for Different Methods of Rock Penetration

Penetration Device	Application	Rock Removal Mechanism	Specific Energy J/mm³	Maximum Power to Rock, kW	Maximum Potential Penetration Rate mm/min
Rotary (roller bit)*	Field drill	Mechanical	0.2-0.5	15-20	140-850
Spark*	Laboratory drill	Mechanical	0.2-0.4	75-150	350-1400
Erosion*	Laboratory drill	Mechanical	2.0-4.0	750-1500	350-1400
Explosive*	Field drill	Mechanical	0.2-0.4	55-75	260-700
Forced-flame	Field drill	Spalling†	1.5	225-450	280-560
Jet piercing	Field drill	Spalling†	1.5	75-150	90-180
Electric disintegration	Laboratory drill	Spalling†	1.5	75-110	90-140
Pellet*	Laboratory drill	Mechanical	0.2-0.4	7-15	40-140
Turbine (roller bit)*	Field drill	Mechanical	0.4-1.3	20-30	30-140
Plasma	Laboratory test	Spalling†	1.5	60-90	80-110
Electric arc	Laboratory test	Spalling†	1.5	35-65	40-80
High-frequency	Laboratory test	Spalling†	1.5	20-45	30-60
Plasma	Laboratory test	Fusion	5.0	60-90	20-30
Electric heater	Laboratory drill	Fusion	5.0	35-75	10-30
Electric arc	Laboratory test	Fusion	5.0	30-65	10-30
Nuclear	(Conceptual)	Fusion#	5.0	930-1860	10-30
Laser	Small-hole test	Spalling	1.5	9-18	10-20
Electron beam	Small-hole test	Spalling†	1.5	7-15	10-20
Microwave	Laboratory test	Spalling†	1.5	7-15	10-20
Induction	Laboratory test	Spalling§	1.5	4-7	5-10
Laser	Small-hole test	Fusion	5.0	7-15	3-6
Electron beam	Small-hole test	Fusion	5.0	7-15	3-6
Electron beam	Small-hole test	Vaporization	12.0	7-15	1-2
Laser	Small-hole test	Vaporization	12.0	5-10	1-2
Ultrasonic*	Laboratory drill	Mechanical	20.0	4-7	0.4-0.7

Sources: Maurer (1968) and Clark (1971).
Based on 200 mm boreholes.
* Water-filled borehole.
† Limited to highly spallable rock such as taconite.
‡ Limited to highly spallable rock with high electrical conductivity.
§ Limited to highly spallable rock with high magnetic susceptibility.
\# 1000 mm diameter drill.

Formulas to determine rate of penetration for the three common rock drilling methods, incorporating theoretical or experimental parameters, follow.

1. *Percussion* (Hartman, 1959; Maurer, 1967):

$$R = \frac{BE}{Ae} \tag{11a}$$

$$R = \frac{V_c Bn}{A} \tag{11b}$$

where V_c is volume broken per cutting edge and n is number of bit cutting edges. Note that V_c, like e, must be determined experimentally. A more fundamental relation may be developed from Eq. 3 for pneumatic powered drills (Pfleider and Lacabanne, 1961):

$$R = \frac{k\,(pa)^{1.5}\,(L/W)^{0.5}}{Ae} \tag{11c}$$

where k is a proportionality constant. Again, values of k and e must be arrived at empirically.

Another factor that affects the performance of a percussion drill is thrust. Studies (Hustrulid and Fairhust, 1972) have demonstrated that the value of thrust must be optimized to yield the maximum rate of penetration for a given set of operating conditions. The effect of thrust on penetration rate for varying air pressure is shown in Fig. 6; notice the optimal thrust levels, which can be estimated theoretically but are best determined experimentally.

2. *Rotary* (Teale, 1965; Maurer, 1967):

$$R = \frac{2\pi NT}{Ae - F} \tag{12a}$$

This equation follows from Eqs. 5 and 6. For most drag-bit drills, the thrust is small compared with other terms and may be neglected. With roller-bit drills, F is large and must be considered. Further, thrust and torque are interrelated in a rotary drill, and their effects cannot be considered independently.

a. *Drag-bit rotary* (Nishimatsu, 1972; Tandanand, 1973):

$$R = \frac{2\pi NrF \cot(\phi - \alpha)}{Ae - F} \tag{12b}$$

where r is effective bit radius, ϕ is angle of cutting friction, and α is rake angle of bit. This relation considers the effect of more variables and resolves the forces acting on a drag bit into vertical and horizontal components; however, it is more complicated to employ than Eq. 12a. The effect of thrust on penetration rate for different rock types is shown in Fig. 7.

b. *Roller-bit rotary* (Maurer, 1967):

$$R = \frac{V_c n_t N}{A} \tag{12c}$$

where n_t is number of tooth impacts per revolution. This formula follows from Eq. 7, but V_c varies for individual bit teeth. An empirical approach was taken by Simon (1956,

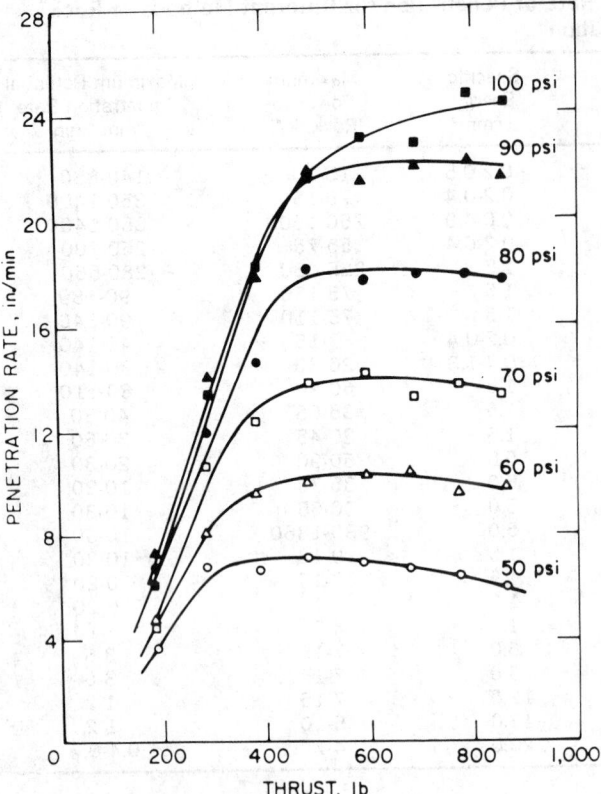

Fig. 6. Relationship between penetration rate and thrust as a function of operating air pressure; rock type: taconite (after Paone, Madson, and Bruce, 1969).

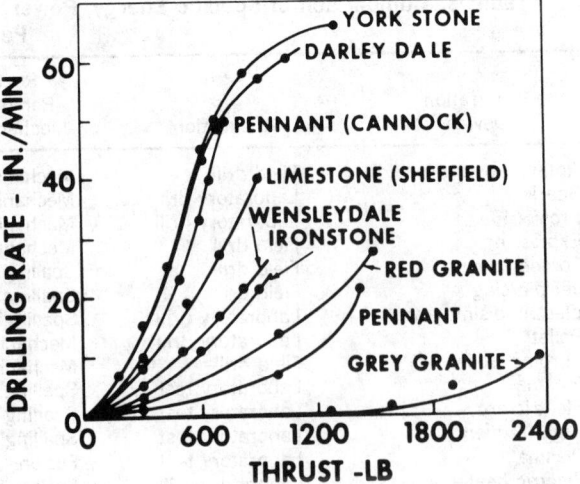

Fig. 7. Effect of thrust on drag-bit drilling rate for various rock types; bit diameter: $1^{11}/_{16}$ in., rotary speed 180 rpm (after Alpan, 1951-1952, and Maurer, 1967; by permission from the Institution of Mining Engineers).

Bit Wear: To date, there are no quantitative ways to relate bit wear or bit life to the operating variables in drilling. The term *wear* (actually, rate of wear) here refers to the loss of dimension, weight, or shape of cutting element in the bit unit time or length of hole. It is an inverse measure of the bit life, usually expressed in footage of hole, during which a tool is effectively and economically penetrating rock at an acceptable rate.

The mechanism of bit wear itself is not well understood. In percussion drilling, it occurs mainly as *impact wear,* and

1967), who developed a series to relate the variables of roller-bit drilling:

$$R = k_1 \frac{NF}{De} \pm k_2 \frac{NF^2}{D^3 e^2} \pm k_3 \frac{NF^3}{D^5 e^3} \pm \dots k_n \frac{NF^n}{D^{2n-1} e^n}$$

(12d)

where $k_1, k_2, \dots k_n$ are constants. The relationship simplifies to the following for shallow holes and good cuttings removal (Maurer, 1962):

$$R = k \frac{NF^2}{D^2 e^2}$$

(12e)

Fig. 8 demonstrates the relation of penetration rate to rotary speed and thrust; note that R is proportional to N only in the low range where good cleaning occurs.

3. *Rotary percussion.* Because it is a combination method in which variables interact, rotary-percussion drilling (like roller-bit drilling) does not lend itself to simple mathematical analysis. Rate of penetration is a function of drilling power and all the other variables operative in rotary and percussion drilling (Bullock, 1974, 1976). Drill performance is best determined by experiment.

4. *Jet piercing.* Thermal penetration with commercial jet piercers achieves advance chiefly by spalling. Rate of penetration is thus proportional to spallability. Although spalling is complex and not entirely predictable, a relationship can be used to estimate spallability (Calaman and Rolseth, 1968):

$$\text{spallability} \propto \frac{(\text{thermal diffusivity}) (\text{thermal expansion}) (\text{grain size})}{(\text{compressive strength})}$$

(13)

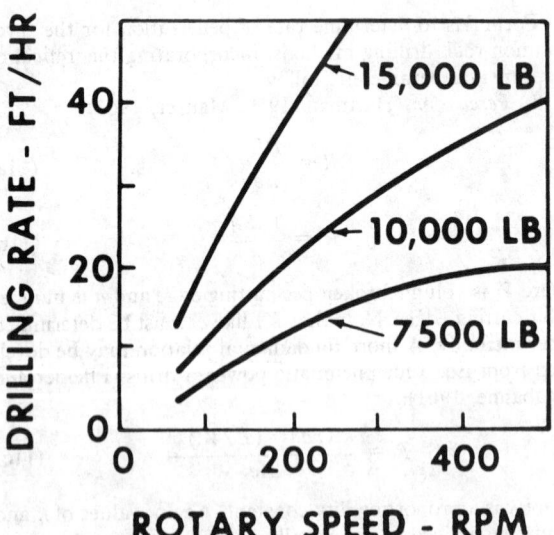

Fig. 8. Effect of rotary speed and thrust on roller-bit drilling rate; rock: Beekmantown dolomite; downhole pressure: atmospheric (after Rowley, Howe, and Deily, 1961).

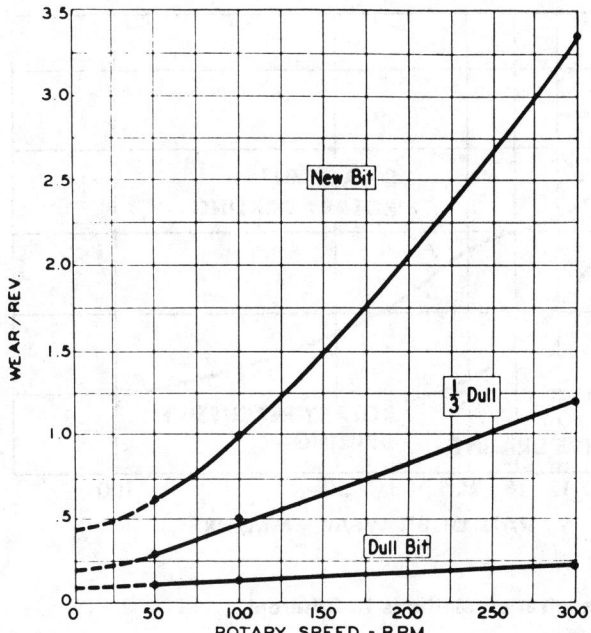

Fig. 9. Tooth wear vs. rotary speed for a roller bit (after Williamson, 1968).

1. Determine and specify the conditions under which the machine will be used, such as the job factors (labor, site, weather, etc.), with safety the ultimate consideration.

2. State the objectives for the rock-breakage phases of the production cycle of operations—considering excavation and haulage restrictions, pit-slope stability, crushing capacity, production quota, pit geometry—in terms of tonnage, fragmentation, throw, vibrations, etc.

3. Based on blasting requirements, design the drill hole pattern (hole size and depth, inclination, burden, spacing, etc.).

4. Determine the drillability factors, and, for the kind of rock anticipated, identify the drilling-method candidates that appear feasible (manufacturers can perform rock drillability tests and recommend drills and bits).

5. Specify the operating variables for each system under consideration, including drill, rod, bit, and circulation fluid factors.

6. Estimate performance parameters, including machine availability and costs, and compare. Consider the power source and select specifications. Major cost items are bits, drill depreciation, labor, maintenance, power, and fluids. Bit wear and costs are critical but difficult to project.

7. Select the drilling system that, in best satisfying all requirements, has the lowest overall cost, commensurate with safe operation.

Items 4 and 6 are the most difficult steps to accomplish in the entire design procedure, primarily because of the present unreliability of drillability determination and drill performance prediction.

Table 4 is a qualitative attempt to aid in drill selection, relating application to rock type. The rating employs a relative scale of rock drillability (4 = highest), with corresponding examples. There is a prevailing tendency for the field of application of each method to expand toward the right, into more resistant materials.

The end point of the process is cost estimation (see the following section). While published data are not transferrable and quickly become obsolete, trends and ranges are significant. *In general, with off-the-shelf technology, the lowest costs are obtainable in soft rock with rotary drag-bit drilling, in medium and hard rock with rotary roller-bit and rotary-percussion drilling, and in very hard rock with percussion drilling.* In some rocks that are spallable, especially hard siliceous

in drag-bit rotary drilling, mainly as *abrasion wear,* with some mixed effect evident (Clark, 1982b). Combination effects occur in roller-bit rotary drilling and rotary percussion drilling, as would be expected. Percussion-bit wear has been described by Montgomery (1968) and Engel (1976) as microspalling. Drag-bit wear has been analyzed by Tandanand (1973) and Czichos (1978) as abrasion, adhesion, and surface fatigue. While concerned mainly with bearing life rather than tooth life, Burr and Marshek (1982) have attempted to quantify roller-bit wear. The effect of rotary speed on bit wear for sharp and dull roller bits is shown in Fig. 9; predictably, a sharp bit dulls more rapidly as N is increased.

The following wear relations are based on the major drilling variables and are empirical and qualitative only, with w some measure of bit wear.

1. *Percussion* (Inett, 1956; Larsen-Basse, 1973):

$$w \propto \frac{\text{(blow energy) (fluid viscosity) (rock hardness)}}{\text{(cutting edge angle) (number of edges) (bit hardness) (fluid flow rate)}} \tag{14}$$

2. *Rotary* (Goodrich, 1957; Fish, 1961; Tsoutrelis, 1969):

$$w \propto \frac{\text{(thrust) (rotary speed) (rock hardness)}}{\text{(bit cutting angle) (bit hardness) (fluid flow rate)}} \tag{15}$$

The primary effect of bit wear is to limit the application of a drilling method to rocks of acceptable hardness. In Fig. 10, it is very evident that drag-bit drills are operationally and economically feasible to operate only in soft and medium rock, while rotary-percussion and roller-bit drills are applicable to medium and hard rock. Percussion drills alone, however, are effective in very hard rocks.

Selection Procedure for Drilling Method

The selection of a particular machine for production drilling in a surface mine is the most critical kind of drill evaluation that the pit engineer is called upon to make. It is a true engineering design problem, requiring value judgments. Generally, the procedure follows these steps (Capp, 1962):

rock, jet piercing may be economic. Hydraulic penetration appears promising for soft to medium rock. Rotary and percussion are sufficiently competitive in the medium-to-hard range to make the choice depend on particular circumstances, although the trend in surface mining is to roller-bit drilling for practically all rock conditions.

ACKNOWLEDGMENT

The author is especially grateful to his colleague and former student, Sathit Tandanand, US Bureau of Mines, Minneapolis, who contributed generously to the writing and editing of this section.

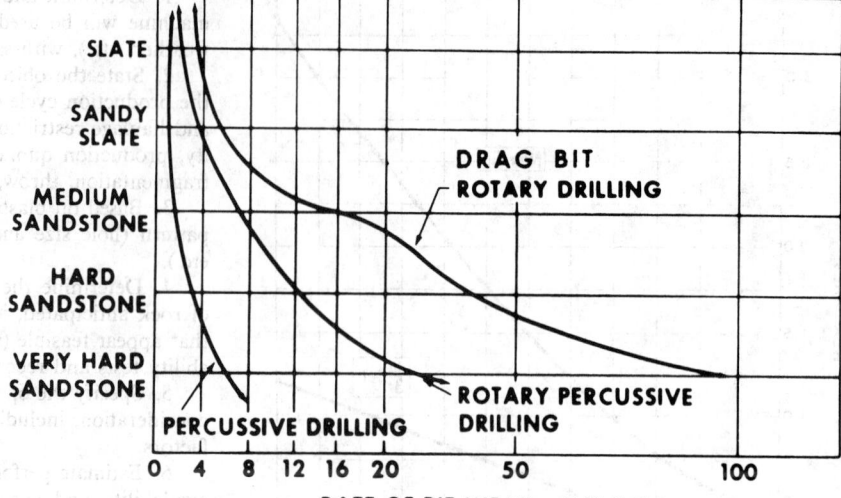

Fig. 10. Rate of bit wear (after Fairhurst and Lacabanne, 1957, and Maurer, 1967, by permission of *Mine & Quarry Engineering*).

Table 4. Application of Drilling and Penetration Methods to Different Types of Rock

Drilling Method	Rock Type/Drillability			
	4 Soft (shale, weathered limestone)	3 Medium Hard (limestone, weathered sandstone)	2 Hard (granite, chert)	1 Very Hard (quartzite, taconite)
Hydraulic jet	x	x		
Rotary, drag-bit	x	x		
Rotary, roller-bit	x	x	x	
Rotary percussion	x	x	x	
Percussion	x	x	x	x
Thermal jet piercing			x	x

REFERENCES

Alpan, H. S., 1951-1952, "Factors Affecting the Speed of Penetration of Bits in Electric Rotary Drilling," *Trans. Institution of Mining Engineers,* Vol. 111, Part 2, p. 374.

Bailey, J. J., and Dean, R. C., Jr., 1967, "Rock Mechanics and the Evolution of Improved Rock Cutting Methods," *Proceedings, 8th Symposium on Rock Mechanics,* AIME, New York, pp. 396-409.

Bullock, R. L., 1974, "Industry-Wide Trends Toward All-Hydraulically Powered Rock-Drills," *Mining Congress Journal,* Vol. 60, No. 10, Oct., pp. 54-65.

Bullock, R. L., 1976, "An Update of Hydraulic Drilling Performance," *3rd Proceedings, Rapid Excavation and Tunneling Conference,* AIME, New York, pp. 627-648.

Burr, B. H., and Marshek, K. M., 1982, "An Equation for the Abrasive Wear of Elastomeric O-Ring Materials," *Wear,* Vol. 81, pp. 347-356.

Calaman, J. J., and Rolseth, H. C., 1968, "Jet Piercing," Chapter 6.4, *Surface Mining,* E. P. Pfleider, ed., AIME, New York, pp. 325-337.

Capp, F. M., 1962, "Factors in Rotary Drill Evaluation," *Mining Congress Journal,* Vol. 48, No. 12, Dec., pp. 20-23.

Cheatham, J. B., Jr., and Gnirk, P. F., 1967, "The Mechanics of Rock Failure Associated with Drilling at Depth," *Proceedings, 8th Symposium on Rock Mechanics,* AIME, New York, pp. 410-439.

Clark, G. B., 1971, "Rock Disintegration—The Key to Mining Progress," *Mining Engineering,* Vol. 23, No. 3, Mar., pp. 47-51.

Clark, G. B., 1982a, "Principles of Rock Drilling and Bit Wear. Part 1," *Quarterly Colorado School of Mines,* Vol. 77, No. 1, Jan., 118 pp.

Clark, G. B., 1982b, "Principles of Rock Drilling and Bit Wear. Part 2," *Quarterly Colorado School of Mines,* Vol. 77, No. 2, Apr., 42 pp.

Czichos, H., 1978, *Tribology—A Systems Approach to the Science and Technology of Friction, Lubrication and Wear,* Elsevier Scientific Publishing, Amsterdam, pp. 103-130.

Engel, P. A., 1976, *Impact Wear of Materials,* Elsevier Scientific Publishing, Amsterdam, 339 pp.

Fairhurst, C., and Lacabanne, W. D., 1957, "Hard Rock Drilling Techniques," *Mine & Quarry Engineering,* Vol. 23, Apr., p. 157; May, p. 194.

Fish, B. G., 1961, "The Basic Variables in Rotary Drilling," *Mine & Quarry Engineering,* Vol. 27, Jan.-Feb., pp. 29-34, 74-81.

Goodrich, R. H., 1957, "High-Pressure Rotary Drilling Machines," *Bulletin Missouri School of Mines,* Technical Series, Vol. 94, pp. 25-45.

Hartman, H. L., 1959, "Basic Studies of Percussion Drilling," *Trans. AIME,* Vol. 214, pp. 68-75.

Hartman, H. L., 1966, "The Effectiveness of Indexing in Percussion and Rotary Drilling," *International Journal of Rock Mechanics & Mining Sciences,* Vol. 3, pp. 265-278.

Hustrulid, W. A., and Fairhurst, C., 1972, "A Theoretical and Experimental Study of the Percussive Drilling of Rock," *International Journal of Rock Mechanics & Mining Sciences,* Vol. 9, pp. 431-449.

Inett, E. W., 1956, "A Study of Drill Bit Wear in Percussive Drilling," *Mine & Quarry Engineering,* Vol. 22, July, pp. 275-280.

Larsen-Basse, J., 1973, "Wear of Hard Metals in Rock Drilling: A Survey of the Literature," *Powder Metallurgy,* Vol. 16, pp. 1-32.

Ledgerwood, L. W., 1960, "Efforts to Develop Improved Oilwell Drilling Methods," *Journal of Petroleum Technology,* Vol. 18, No. 4, Apr., pp. 61-74.

Maurer, W. C., 1962, "The 'Perfect-Cleaning' Theory of Rotary Drilling," *Journal of Petroleum Technology,* Vol. 14, No. 11, Nov., pp. 1270-1274.

Maurer, W. C., 1967, "The State of Rock Mechanics Knowledge in Drilling," *Proceedings, 8th Symposium on Rock Mechanics,* AIME, New York, p. 355.

Maurer, W. C., 1968, *Novel Drilling Techniques,* Pergamon Press, New York, 114 pp.

Maurer, W. C., 1980, *Advanced Drilling Techniques,* Petroleum Publishing, Tulsa, 698 pp.

Montgomery, R. S., 1968, "The Mechanism of Percussion Wear of Tungsten-Carbide Composites," *Wear,* Vol. 12, pp. 309-329.

Nishimatsu, Y., 1972, "The Mechanics of Rock Cutting," *International Journal of Rock Mechanics & Mining Sciences,* Vol. 9, pp. 261-270.

Paone, J., Madson, D., and Bruce, W. E., 1969, "Drillability Studies—Laboratory Percussive Drilling," *Report of Investigation, US Bureau of Mines 7300,* 22 pp.

Pfleider, E. P., and Lacabanne, W. D., 1961, "Higher Air Pressures for Down-the-Hole Percussive Drills," *Mine & Quarry Engineering,* Vol. 27, Oct.-Nov., pp. 464-468, 496-501.

Rowlands, D., 1974, "Diamond Drilling with Soluble Oils," *Trans. Institution of Mining & Metallurgy,* Vol. 83, Series A, pp. 127-132.

Rowley, D. S., Howe, R. J., and Deily, F. H., 1961, "Laboratory Drilling Performance of the Full Scale Rock Bit," *Journal of Petroleum Technology,* Vol. 13, No. 1, Jan., p. 71.

Roxborough, F. F., and Rispin, A., 1972, *The Mechanical Cutting Characteristics of the Lower Chalk Report,* Transport and Road Research Laboratory, University of Newscastle Upon Tyne, pp. 163-186.

Simon, R., 1956, "Theory of Rock Drilling," *Proceedings, 6th Annual Drilling Symposium,* University of Minnesota, Minneapolis, Oct., pp. 1-14.

Simon, R., 1967, "Rock Fragmentation by Concentrated Loading," *Proceedings, 8th Symposium on Rock Mechanics,* AIME, New York, pp. 440-454.

Tandanand, S., 1973, "Principles of Drilling," Sec. 11.3, *SME Mining Engineering Handbook,* AIME, New York, Vol. 1, pp. 11-5 to 11-24.

Teale, R., 1965, "The Concept of Specific Energy in Rock Drilling," *International Journal of Rock Mechanics & Mining Sciences,* Vol. 2, pp. 57-73.

Tsoutrelis, C. E., 1969, "Determination of the Compressive Strength of Rock In Situ or in Test Blocks Using a Diamond Drill," *International Journal of Rock Mechanics & Mining Sciences,* Vol. 6, pp. 311-321.

Wells, E. J., 1950, "Penetration Speed References for the Drillability of Rocks," *Proceedings, Australian Institute of Mining & Metallurgy,* Vol. 158-159, Sept.-Dec., pp. 453-464.

Williamson, T. N., 1968, "Rotary Drilling," Chapter 6.3, *Surface Mining,* E. P. Pfleider, ed., AIME, New York, pp. 300-324.

6.1.2 Drilling Application

ALAN BAUER

WILLIAM A. CROSBY

HISTORICAL DEVELOPMENT

Up until relatively recent times the principal method of blasthole drilling used percussive techniques. Early developments in the 19th century progressed from hammering on hand-held drill steel to steam-operated machines and then to piston-type drills. In 1897 J.G. Leyner introduced hollow drill steel to allow compressed air and water to be delivered to the drill bit to be used to flush the drill chippings clear of the hole. Two main types of percussion drills were developed: the piston drill and the hammer drill. In the piston drill, the drill steel was attached to the piston and both reciprocated and rotated. Piston drills now have largely been replaced by the hammer drill for percussion drilling. Drills currently in use employ pneumatic hammers either above ground or down-the-hole as well as the more recently introduced above ground hydraulic hammers. Percussion drill blastholes vary up to a normal operating size of 190.5 mm (7 ½ in.) with some machines being used at 228.5 mm (9 in.) diameter.

After the Second World War, efforts to improve drilling productivity, particularly in hard drilling applications such as in taconite in the Mesabi Range, prompted the development of the rotary jet-piercing drill. At that time, rotary blasthole drills were not well developed and could not economically drill hard materials. The jet piercers enjoyed good success in hard spallable rock drilling for many years. With the development of the rotary drill, however, present-day use of the jet piercer drill has become extremely limited.

The techniques and principles of rotary drilling were to a large degree developed by the petroleum industry for the purpose of drilling oil wells to depths in excess of 6096 m (20,000 ft). Prior to about 1950, rotary drills were being used in mining but drilling was performed with drag bits confining the operations to soft materials such as those found in coal, porphyry copper, and soft natural iron ore mines. The replacement of water by air circulation to clean the blasthole and the introduction of the rotary tricone drill bit paved the way for major improvements in rotary drill performance. At the same time, the introduction of bulk, low cost explosives and the increasing demand for unit productivity did much to create a demand for larger blastholes. The result has been that, by far, the majority of all primary blasting in surface mines today is now being performed using rotary blasthole drills producing holes in the 165 to 445 mm (6½ to 17½ in.) size range.

PERCUSSION DRILLS

Percussion drills generally play a minor role as compared with rotary machines in surface mining operations. Their application is limited to production drilling for small mines, secondary drilling, development work, and wall control blasting.

There are two main types of drill mounting. The smaller machines utilize drifter-type drills placed on self-propelled mountings designed to tow the required air compressor. Typical hole sizes are in the 63 to 150 mm (2½ to 6 in.) range (Fig. 1). The larger machines are crawler-mounted and self-

contained (Fig. 2). Drill towers permit single pass drilling from 7.6 to 15.2 m (25 to 50 ft) with hole sizes in the range of 120 to 229 mm (4¾ to 9 in.) in diameter. These larger machines are almost exclusively operated using down-the-hole hammers.

For many years these machines were exclusively operated using pneumatic hammers. In the last 15 years, hydraulic machines have been introduced in the smaller size range. The higher capital cost of these hydraulic drills is offset by lower operating costs and increased productivity compared with pneumatic machines. Another aspect that is becoming increasingly more important is the reduced noise produced by the hydraulic drills. The advantages of the hydraulic machines may be summarized as follows:

1. They are self-contained, diesel powered, and do not require an auxillary compressor for drill operation.

2. The energy delivered in each stroke is markedly higher than for the pneumatic counterpart resulting in faster penetration rates.

3. The absence of exhaust air results in lower noise levels and reduced freezing problems as compared with pneumatic drills.

4. Energy consumption is reduced by up to 66%.

5. There is reduced machine and drill bit wear.

The surface-mounted percussive drill offers a number of advantages over down-the-hole drills. Surface-mounted drills often have higher penetration rates than down-the-hole drills in shallow holes at the same operating air pressure since the hammer piston area can be made larger than the blasthole area. For down-the-hole hammers, the piston area must necessarily be smaller than the blasthole area. Surface-mounted drills are attractive for secondary drilling work because they can drill smaller holes than down-the-hole machines, which are limited to a minimum size of approximately 100 mm (4 in.). Finally, the drill is never in the hole so there is no fear of it being lost in a caving hole.

On the other hand, for holes larger than 150 mm (6 in.), a down-the-hole machine is necessary. Energy losses in the drill string, particularly for all but the shallowest holes, contribute to decreased penetration rate with depth for the surface-mounted drills. The location of the hammer directly behind the drill bit in a down-the-hole machine means much lower energy losses regardless of the depth of hole. Furthermore, air used to operate the hammer is directly available to aid drill chip removal from the blasthole. Two other advantages of the down-the-hole machine as compared with the surface-mounted drill are the noise dampening effect of having the hammer in the blasthole and an increased life for drill rods and couplings as they are not required to transmit energy from the hammer to the drill bit.

Percussion Drill Productivity

Fig. 3 presents drill penetration rates plotted against hole diameter for percussive drills operating at low air pressures of up to 0.7 MPa (100 psi), drilling in materials of two different compressive strengths. It will be observed that penetration rate decreases with both increasing hole size and increasing rock strength.

Fig. 1. Self-propelled, hydraulic crawler-mounted drifter drill.

Blasthole drill productivity, regardless of the type of drill, is dependent on drill penetration rate and blasthole pattern size, which itself is dependent on hole diameter. As percussive drills are limited in hole size to a practical maximum of no more than 228 mm (9 in.) and few if any are used with greater than 190 mm (7½ in.) productivity improvements for these types of drill have concentrated on an increase in the penetration rate. Eq. (3) in the "Principles" part of this section presents the relationship between penetration rate and piston area, weight and stroke length as well as operating air pressure. As the piston area and stroke length are fairly inflexible for a given hole size, one main thrust to help improve percussion drill productivity has been to improve the drill penetration rate by increasing the hammer operating pressure. A typical increase in penetration rate experienced when changing from 0.7 MPa (100 psi) to 1.7 MPa (250 psi) would be of the order of 200% with an approximate doubling in the actual hole production rate.

Some attempts have also been made to improve drill penetration rate by using high frequency blows. However, it has been found difficult to hold the hammer together using this approach, especially at the higher air pressures.

The other option is to use hydraulic machines to improve the energy available at the drill bit. Table 1 presents a detailed test comparison between one hydraulic drill and eight pneumatic machines drilling a fine-grained, dense hornfels having a compressive strength of the order of 200 MPa (30,000 psi). The penetration rate for the hydraulic drill is shown to be from 20 to 100% higher than the pneumatic machines.

Percussion Drill Costs

Fig. 4 presents cost vs. blasthole size for percussive drills operating up to 0.7 MPa (100 psi) air pressure in the same materials as compared in Fig. 3. As for the penetration rate, the costs are dependent on both the blasthole size and the strength of the rock.

Fig. 5 compares the cost for two sizes of hammer drill with a medium-sized rotary machine. This comparison is made for hammer drills operating with 0.7 MPa (100 psi) air pressure. The high cost for hammer drilling is partly a result of lower penetration rates obtained as compared with rotary machines (Fig. 6). While an increase in operating air pressure has greatly improved blasthole penetration rates, the drilling cost improvements have been less impressive because of higher maintenance costs, lower machine availabilities, etc. The result is that high pressure 178 mm (7 in.) hammer drilling cost is approximately 100% higher than 250 mm (9⅞ in.) rotary drill per unit volume of material blasted.

A comparison between pneumatic and hydraulic surface-mounted drill costs follows:

	Pneumatic	Hydraulic
Investment cost	1.0	1.27
Energy cost	1.0	0.24
Drill steel cost	1.0	0.86
Overall operating cost	1.0	0.78

A complete operating cost comparison between the two drill types is presented in Table 2.

Fig. 2. Self-contained, crawler-mounted, down-the-hole hammer drill.

JET-PIERCING DRILLS

The jet-piercing process relies upon a characteristic of rock known as spallability. The rock is broken down, or spalled, as a result of differential expansion of the rock crystals by thermally induced stresses. The jet-piercer drill essentially consists of a burner fixed to a blowpipe that produces a high temperature flame (of the order of 4300°F) with a high velocity of approximately 5000 fps by burning fuel oil in oxygen. The burner is directed into the ground in the same manner as a rotary drill bit using the blowpipe, with drills having a typical drilling depth capability of 15.2 m (50 ft). In addition to fuel oil and oxygen, water is also used to cool the burner and, in the form of steam, it helps eject the spalled rock cuttings from the blasthole. Hole diameters range from a minimum of approximately 229 mm (9 in.) up to 457 mm (18 in.) using chambering burners.

As stated in the introduction, very few jet-piercer drills are now in operation. One of the main problems with the system has been the high cost of oxygen and fuel oil, which has helped to make the drill uneconomic. Other disadvantages to the drill include the difficulty of supplying the process fluids (particularly in cold climates), and the irregular blasthole profile that makes consistent blast design difficult.

For detailed discussion on jet-piercing, the reader is referred to the first edition of *Surface Mining*, Chapter 6.4 by Calaman and Rolseth (1968).

ROTARY DRILLS

In rotary drilling, the drill bit attacks the rock with energy supplied to it by a rotating drill stem. The drill stem is rotated while a thrust is applied to it by a pulldown mechanism using up to 65% of the weight of the machine, forcing the bit into the rock. The drill bit breaks and removes the rock by either a ploughing-scraping action in soft rock, or a crushing-chipping action in hard rock, or by a combination of the two.

Compressed air is supplied to the bit via the drill stem. The air both cools the bit and provides a medium for flushing the cuttings from the hole. Water may be used in addition

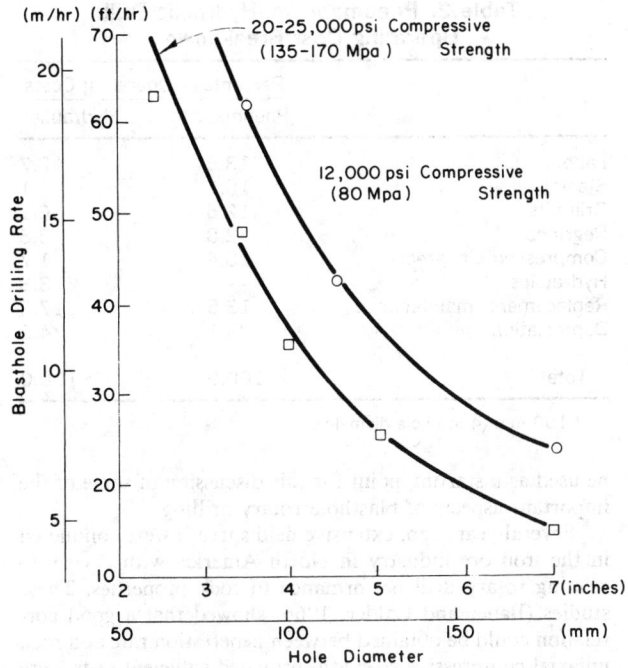

Fig. 3. Penetration rate vs. hole diameter for percussive drills operating at low pressure.

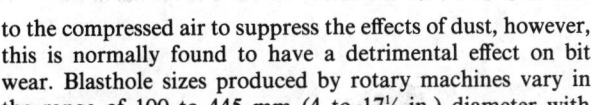

Fig. 4. Cost vs. hole size for percussive drills operating at low pressure.

to the compressed air to suppress the effects of dust, however, this is normally found to have a detrimental effect on bit wear. Blasthole sizes produced by rotary machines vary in the range of 100 to 445 mm (4 to 17½ in.) diameter with the most common sizes being 200, 250, 311, and 381 mm (6⅞, 7⅞, 9⅞, 12¼, and 15 in.). These drills usually operate

in the vertical position (Figs. 7 and 8), although many types can drill up to 25 or 30° off the vertical. Drills are manufactured that can drill horizontal holes used in overburden stripping where hard bands of material are located low in the highwall face. This technique eliminates wasteful and difficult vertical drill footage through soft materials.

Table 1. Comparison of a Hydraulic Drill vs. Conventional Pneumatic Drills

Manufacture		A*	A*	B*	C*	D*	E*	F*	G*	H†
Crawler model		Roc-601	Roc-701	ATD3700	ECM-350	RAM	CDR-12E	STD-350	WR-11	Hydrofore
Compressor size	cfm	700	1050	750	750	800	600	1050	750	160
Steel type		T-38	T-38	1600	1600	1-½ in.	1600	T-38	1600	T-38
Steel length	ft	12	12	10	10	10	10	12	12	10
Bit diameter	in	2-½	2-½	2-½	2-½	2-½	2-½	2-½ 2-¾	2-½	2-½
Bit type (S)		X,B	X,B	X	X,B	X,B	X,B	X,B	X,B	X,B
Number of holes drilled		10	9	4	7	10	5	5	9	17-½
Hole depth	ft	60	60	60	60	60	60	60	60	60
Total drilling	ft	600	540	240	420	600	300	300	540	1052
Total drilling time	h.m.s.	5, 50, 18	2, 32, 20	2, 32, 20	3, 51, 49	4, 31, 43	3, 46, 56	2, 58, 39	5, 15, 36	6, 34, 29
Average drilling time for 60 ft	m.s.	35.02	27.26	38.05	33.07	27.10	45.23	35.43	35.04	22.35
Average add rods time for 60 ft	m.s.	3.00	1.15	4.34	5.12	2.38	3.00	1.37	2.53	1.51
Average pull rods time for 60 ft	m.s.	5.51	3.50	5.42	5.16	6.02	6.00	5.52	3.30	4.34
Average total time 60 ft hole	m.s.	43.53	32.31	48.21	43.35	35.50	54.23	43.12	41.27	29.00
Penetration, drilling only	in/min	20.55	26.24	18.91	21.74	26.50	15.86	20.16	20.53	31.88
Penetration, complete cycle	in/min	16.41	22.14	15.21	16.52	20.09	13.24	16.67	17.37	24.83
Fuel consumption	i.g.	64.0	54.9	27.6	55.4	59.8	38.8	54.8	66.6	43.0
Gallons per hour of drilling		10.97	12.82	10.90	14.39	13.23	10.26	18.47	12.68	6.53
Feet per gallon		9.38	9.83	8.70	7.58	10.03	7.73	5.47	8.11	24.46
Drilling time, best hole	m.s.	32.25	23.08	32.25	29.01	23.28	40.12	30.09	29.00	20.55
Drilling time, worst hole	m.s.	40.44	33.06	43.00	37.53	32.20	52.30	42.05	48.05	24.20
Average drilling time Rod #1	m.s.	4.45	3.46	4.30	3.16	2.49	5.04	2.55	3.40	2.27
Rod #2	m.s.	5.54	5.03	5.17	4.12	4.27	6.04	6.28	5.56	3.59
Rod #3	m.s.	7.04	5.37	5.57	5.30	4.28	7.04	7.42	7.39	4.05
Rod #4	m.s.	8.13	6.18	6.28	6.07	4.48	8.04	9.21	8.51	3.43
Rod #5	m.s.	9.06	6.42	7.38	6.46	5.19	9.04	9.17	8.58	3.58
Rod #6	m.s.	N.A.	N.A.	8.15	7.14	5.19	10.03	N.A.	N.A.	4.23

Source: Anon., 1975.
* Pneumatic Drills.
† Hydraulic drill.

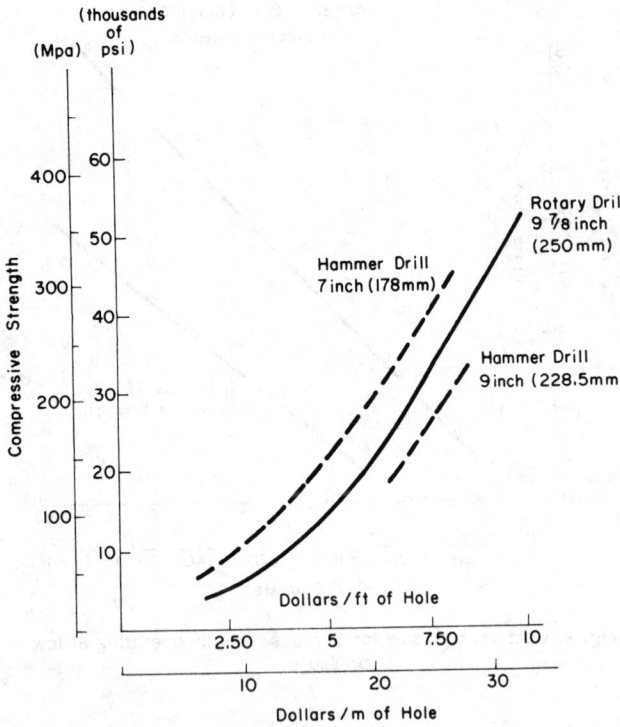

Fig. 5. Comparison between hammer and rotary drill costs.

Table 2. Pneumatic vs. Hydraulic Drill Operating Cost Breakdown

	Percentage Operating Costs	
	Pneumatic*	Hydraulic*
Labor	13.5	12.7
Steels	10.1	11.1
Drill bits	12.6	16.1
Regrinds	2.8	3.6
Compressed air / grease	18.4	1.7
Hydraulics	—	3.3
Replacement / maintenance	13.5	17.3
Depreciation	29.1	34.2
Total	100.0	100.0

* 100 mm (4 in.) hole diameter.

be used as a starting point for this discussion of some of the important aspects of blasthole rotary drilling.

Several years ago, extensive field surveys were conducted in the iron ore industry in North America with a view to relating rotary drill performance to rock properties. These studies (Bauer and Calder, 1966) showed that a good correlation could be obtained between penetration rate and rock uniaxial compressive strength, provided sufficient tests were conducted to obtain a sufficiently meaningful rock strength. Instrumented field tests also indicated that the penetration rate could be correlated linearly with the weight/inch of bit diameter and with the rotary speed. The results of this work

Rotary Drill Penetration Rate

One of the most important factors in drilling is how fast can drill hole be produced while the machine is actually drilling. This factor almost entirely influences productivity and has a strong influence on unit costs. It will, therefore,

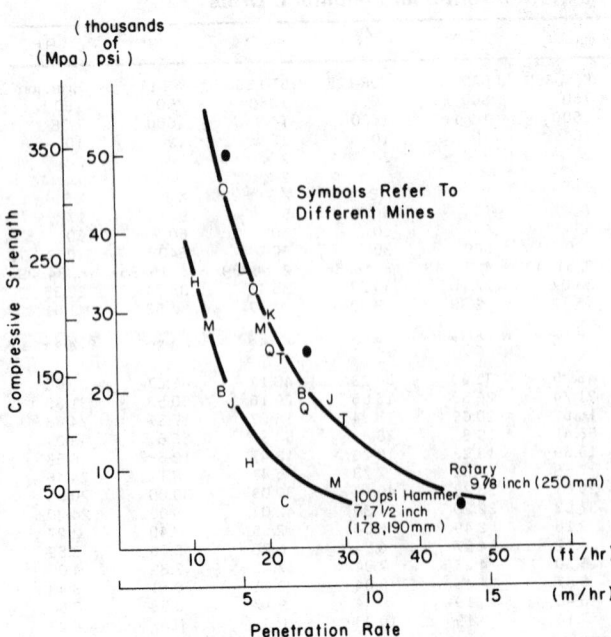

Fig. 6. Penetration rate vs. rock compressive strength for rotary and hammer drills.

Fig. 7. Large electrically powered rotary drill.

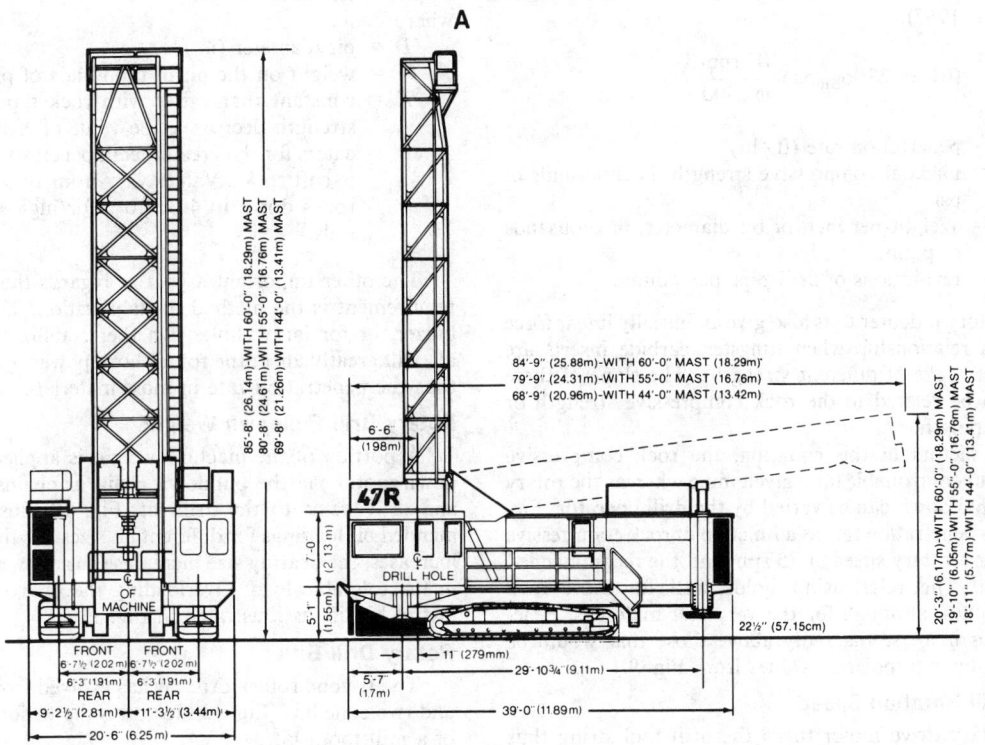

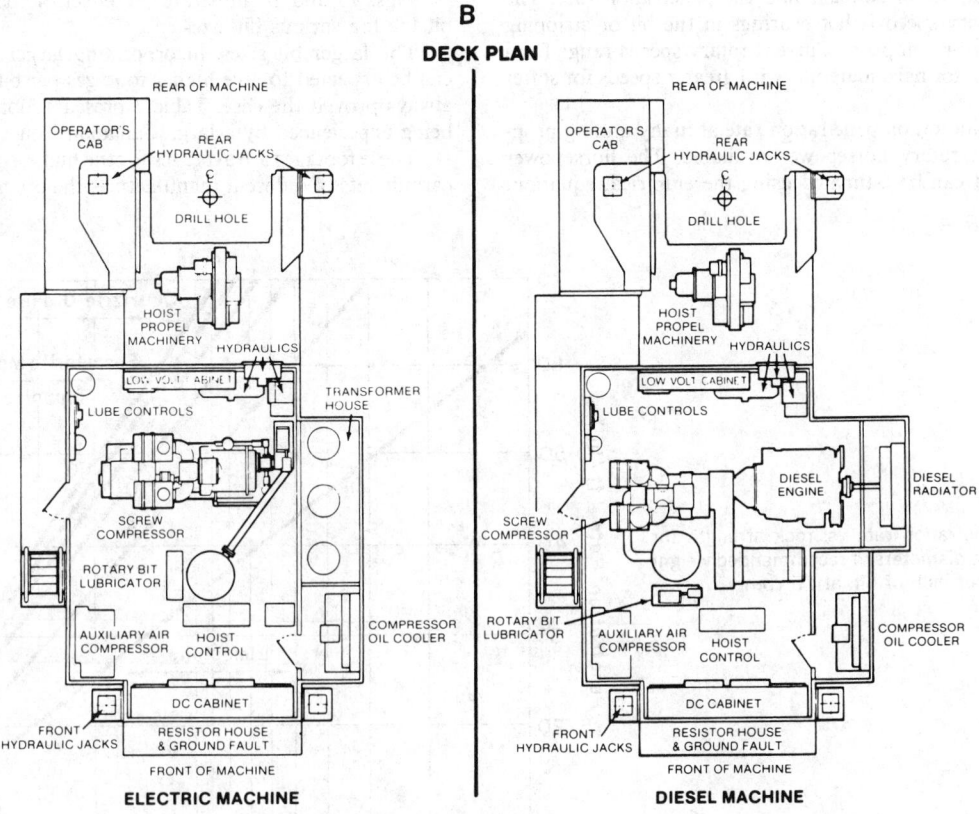

Fig. 8. Large rotary drill elevations (A) and deck plans (B) for an electric and diesel machine.

can be expressed by the following empirical equation (Bauer and Calder, 1967),

$$P = (61 - 28 \log_{10} Sc) \frac{W}{\phi} \cdot \frac{rpm}{300}$$

where

P = penetration rate (ft/hr)

Sc = uniaxial compressive strength, in thousands of psi

W/ϕ = Weight per inch of bit diameter, in thousands of pounds

rpm = revolutions of drill pipe per minute.

Laboratory indenter tests also give an initially linear force penetration relationship when tungsten carbide inserts are pushed into rocks of different strengths. The slope of these graphs can be related to the rock compressive strength or crushing strength.

Of the factors in this equation, the rock compressive strength is uncontrollable for a given mine whereas the rotary speed and pulldown can be varied by the drill operator. Fig. 9 illustrates penetration rate as a function of rock compressive strength for a rotary speed of 75 rpm and the recommended bit loading. The relationship holds well for most rock strength values, although for the very soft materials, penetration rates increase markedly above those that would be anticipated by extrapolating values from Fig. 9.

Rotary Drill Rotation Speed

The rotary drive motor turns the drill tool string thus turning the drill bit at the bottom of the hole. This action brings the successive lines of drill bit compacts into contact with the base of the hole. As the rotary speed increases, so does the number of contacts and the penetration rate. The limit to rotary speed is hot bearings in the bit or stripping of the heel row compacts. Current rotary speeds range from 60 to 90 rpm for hard materials with greater speeds for softer rocks.

The limitation on penetration rate at many mining properties is the rotary horsepower available. The horsepower requirement can be estimated using the empirical equation:

$$hp = K \cdot rpm \cdot D^{2.5} \cdot W^{1.5}$$

where

D = bit diameter (in.)

W = weight on the bit in thousands of pounds

K = constant that varies with rock type. As material strength decreases, the value of K increases. This caters for the greater teeth penetration experienced in soft rocks. Values vary from 14×10^{-5} for soft rocks down to 4×10^{-5} for high-strength materials.

The other important aspect as regards the rotary power requirement is the method of stabilization. The type of stabilizer, or for larger holes, whether stabilization is used at all, will greatly affect the rotary horsepower requirement and thus the penetration rate in most materials.

Rotary Drill Pulldown Weight

A portion of the machine weight is applied by the pulldown motor via the pulldown chain or chains, rotary head and drill stems to the drill bit. Fig. 10 illustrates recommended bit loadings for different bit sizes. As the bit diameter increases, the bearing size increases thus allowing an increase in the tolerable load. Overloading the bit results in severe loss of bit life as illustrated in Fig. 11.

Rotary Drill Bits

The tricone rotary drill bit has evolved from the drag bit and two cone bits. Fig. 12 illustrates the various components of a mill tooth bit.

Mill, or steel tooth bits are used for soft rock with the cutoff being a medium-strength limestone. Tungsten carbide bits are used for all harder rocks.

Figs. 13 and 14 illustrate the effect of rock strength on bit life for various bit sizes.

The larger bit sizes, incorporating larger bearings, etc., can be expected to give higher footages per bit. This has not always proved the case. Table 3 presents footages currently being experienced by a large Canadian iron ore operation.

These footages are averages for the highest-grade tungsten carbide bits of different manufacture, the ore having variable

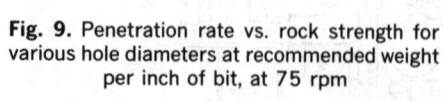

Fig. 9. Penetration rate vs. rock strength for various hole diameters at recommended weight per inch of bit, at 75 rpm

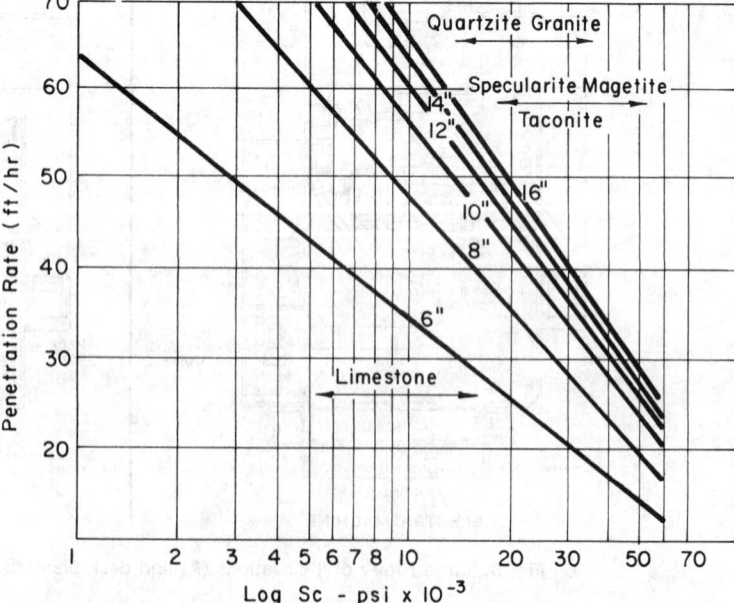

(thousands of lbs)
(Kgm)
Weight per mm of Bit Diameter
Weight per inch of Bit Diameter
160
140
120
100
9
8
7
6
5
6 8 10 12 14 16 (inches)
150 200 250 300 350 400 (mm)
Bit Diameter

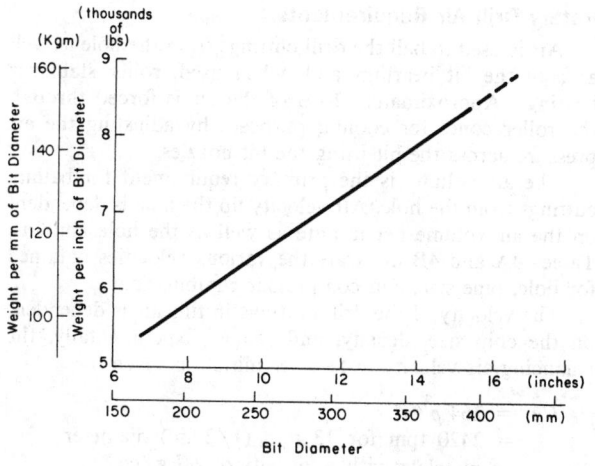

Fig. 10. Recommended pulldown weight per inch of bit diameter vs. bit diameter.

(m) (ft)
Bit Life
400
360
320
280
240
200
1400
1200
1000
800
600
2000 4000 6000 8000 10000
Weight/inch of Bit Diameter (lbs)
40 80 120 160
Weight/mm of Bit Diameter (Kgm)

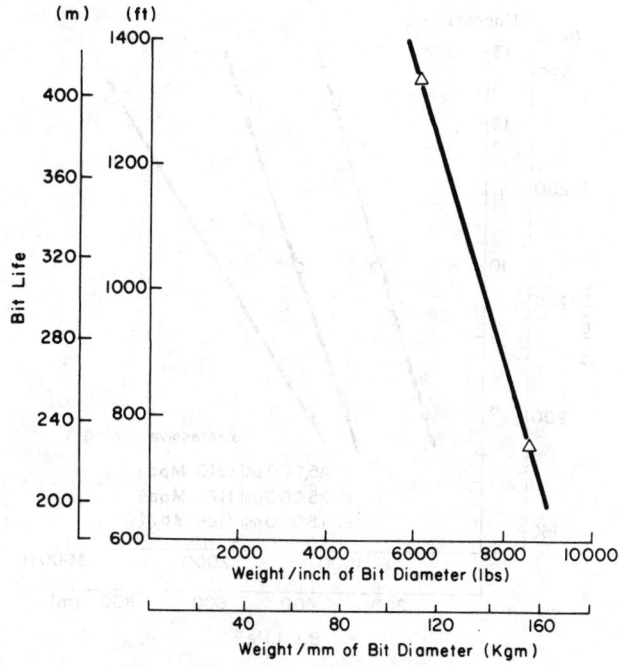

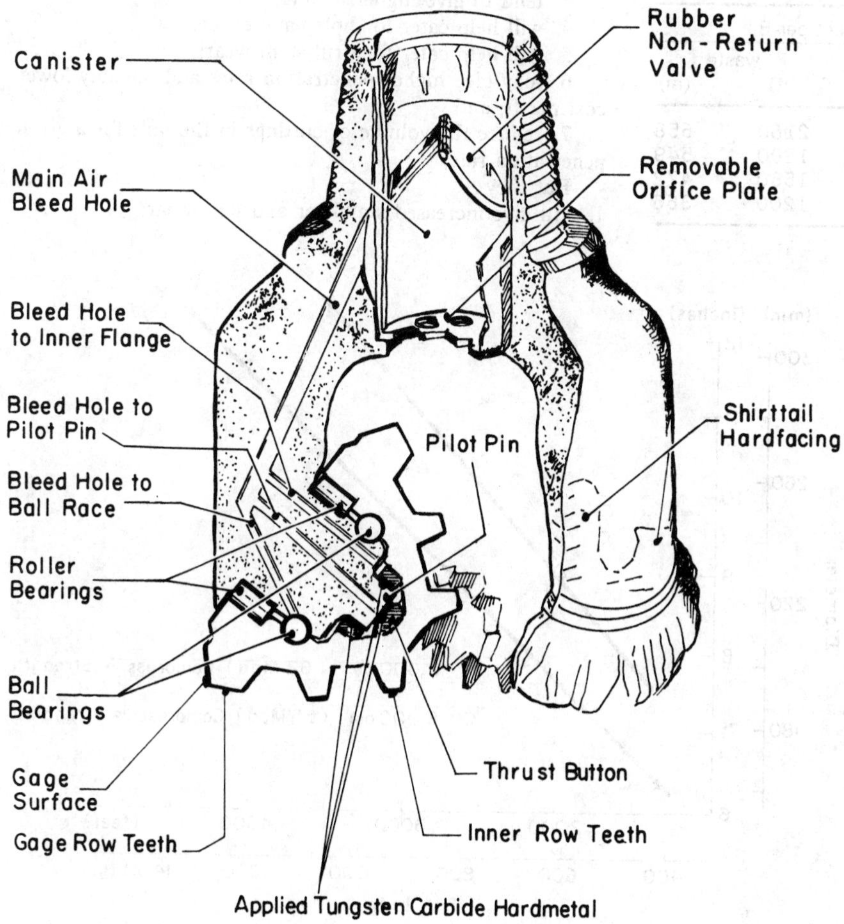

Fig. 11. Bit life vs. pulldown weight for 250 mm (9⅞ in.) diameter tungsten carbide rotary bits in a hard formation.

compressive strengths averaging approximately 204 MPa (30,000 psi). As can be seen the values are completely reversed from what may be expected, introducing another important factor in the drill operation (which also affects penetration rate) the air supply and bailing velocity.

Canister
Rubber Non-Return Valve
Main Air Bleed Hole
Removable Orifice Plate
Bleed Hole to Inner Flange
Bleed Hole to Pilot Pin
Pilot Pin
Shirttail Hardfacing
Bleed Hole to Ball Race
Roller Bearings
Ball Bearings
Gage Surface
Gage Row Teeth
Thrust Button
Inner Row Teeth
Applied Tungsten Carbide Hardmetal

Fig. 12. Diagrammatic illustration of a mill tooth tricone rotary drill bit.

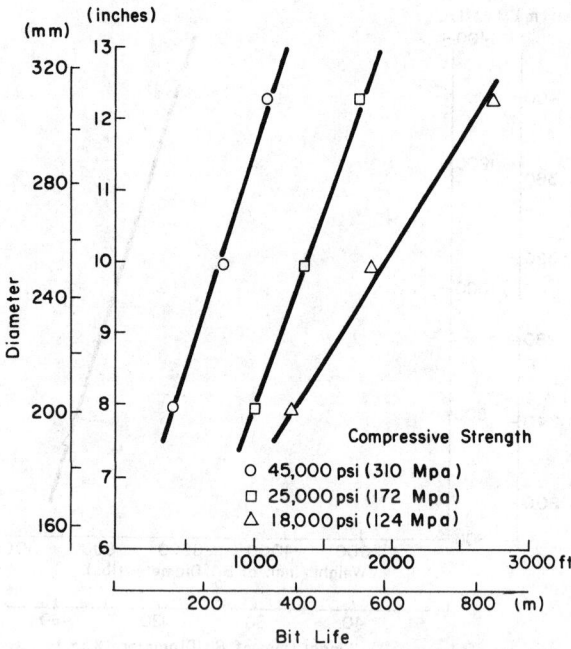

Fig. 13. Tungsten carbide rotary bit life vs. diameter for different rock strengths.

Table 3. Average Drill Bit Footages for a Large Iron Ore Mine in Northern Canada

Drill Bit Diameter		Average Feet per Bit			
		ore		waste	
(in.)	(mm)	(ft)	(m)	(ft)	(m)
9⅞	250	1800	549	2160	658
12¼	311	1450	442	1800	549
15	381	1400	427	1550	472
17½	445	1200	366	1200	366

Rotary Drill Air Requirements

Air is used to bail the drill cuttings from the hole as well as cool the bit bearings and, when used, roller stabilizer bearings. Approximately 20% of the air is forced through the roller cones for cooling purposes by adjusting the air pressure across the bit using the bit nozzles.

The air volume is the primary requirement for bailing cuttings from the hole. Air velocity up the hole is dependent on the air volume per minute as well as the hole annulus. Tables 4A and 4B illustrate the various velocities obtained for hole, pipe size, and compressor combinations.

The velocity of the drill cuttings in this air is dependent on the chip size, density, and shape. Experimentally, the balancing air velocity in feet per minute is given by:

$$Um = 264\, p^{1/2}\, d^{1/2}$$
$$= 2420 \text{ fpm for 13 mm (1/2 in.) diameter}$$
platelets with a density of 2.7 g/cc.
d = diameter of the chip in inches
ρ = density of chip in lb/ft³

At air velocities above this balancing value, the chips begin to move, their velocity being approximately one half the excess air velocity above the balancing value. A bailing velocity of 1800 mpm/s (6000 fpm) is usually adequate to bail 13 mm (1/2 in.) chips. Fig. 15 illustrates a typical air requirements chart.

Factors involved with choosing the air velocity are that higher velocities:

1. give higher bailing velocities;
2. will bail larger chips;
3. tend to give higher bit life;
4. will help cater for hole cavities, etc.;
5. will help cater for drill stem wear;
6. may give higher penetration rates and possibly lower cost per ft; and
7. reduce the volume of cuttings in the hole for a given penetration rate;

But they
1. will give increased stabilizer and pipe wear;

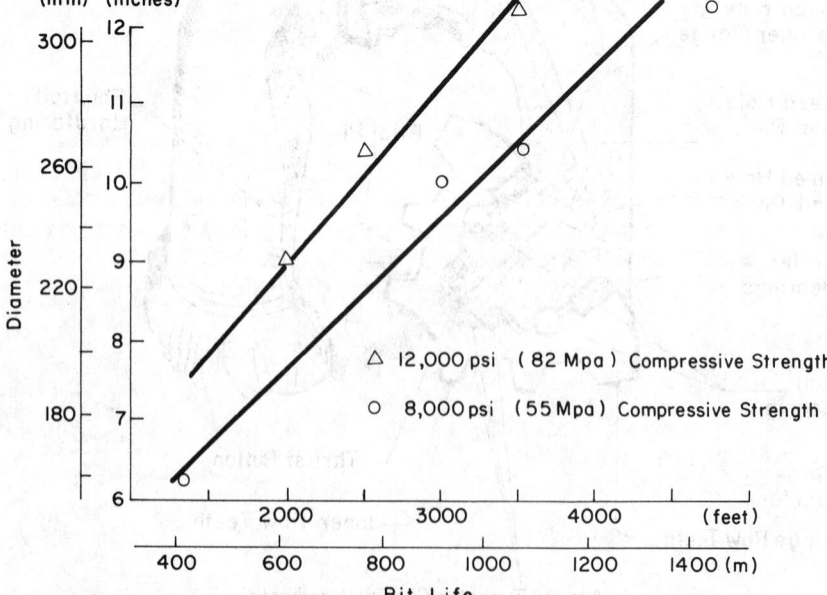

Fig. 14. Steel toothed rotary bit life vs. diameter for two rock strengths.

Table 4. Bailing Velocities for Different Drill Pipe and Bit Diameters.
A. Imperial Units.

Drill Pipe Diameter	Bit Diameter	Annular Area	Allis-Chalmers 11-L 773 c.f.m.	12-L 982 c.f.m.	17-L 1310 c.f.m.	19-S 1550 c.f.m.	Sullivan Screw 1300 c.f.m.	Gardner-Denver 900 Screw 925 c.f.m.	1200 Screw 1170 c.f.m.	LeRoi 100-S-DS 450 c.f.m.	Gardner-Denver WFN 880 c.f.m.	WFO 1075 c.f.m.
Inches	Inches	Sq. in.	40 p.s.i.	40 p.s.i.	40 p.s.i.	40 p.s.i.	100 p.s.i.	50 p.s.i.	50 p.s.i.	125 p.s.i.	100 p.s.i.	100 p.s.i.
4½	5⅝	8.95								7,240		
	6	12.37								5,240		
	6¼	14.78								4,385		
5	6	8.64								7,500	14,665	17,915
	6¼	11.05								5,865	11,465	14,010
	6¾	16.15		8,755	11,680					4,015	7,845	9,585
5½	6¾	12.03	9,252	11,754	15,680			11,072	14,000	5,385	10,535	12,865
	7⅜	18.95	5,870	7,458	9,949		9,873	7,025	8,885	3,420	6,680	8,165
	7⅞	24.95	4,461	5,668	7,560	8,945	7,503	5,340	6,750	2,600	5,080	6,205
6¼	7⅜	12.04	9,245	11,744	15,667	18,538	15,548	11,060	13,990	5,380	10,525	12,855
	7⅞	18.03	6,173	7,843	10,462	12,379	10,383	7,385	9,345	3,595	7,025	8,585
6¾	7⅞	12.93	8,609	10,936	14,589	17,262	14,478	10,300	13,030		9,800	11,970
	8½	20.97	5,308	6,743	8,996	10,643	8,927	6,350	8,035		6,040	7,380
	9	27.84	3,998	5,079	6,775	8,017	6,724	4,785	6,050		4,550	5,560
7	8½	18.26	6,096	7,744	10,330	12,223	10,252	7,295	9,225		6,940	8,475
	9	25.14	4,427	5,624	7,504	8,878	7,446	5,295	6,700		5,040	6,160
	9⅞	38.11	2,920	3,710	4,950	5,957	4,912	3,495	4,420		3,325	4,060
7¼	9	22.34	4,983	6,330	8,444	9,991	8,380	5,960	7,540		5,670	6,930
	9⅞	35.31	3,152	4,005	5,342	6,321	5,302	3,770	4,770		3,585	4,380
7¾	9	16.45	6,767	8,596	11,467	13,568	11,380	8,100	10,240		7,705	9,410
	9⅞	29.42	3,783	4,806	6,412	7,586	6,363	4,525	5,725		4,305	5,260
	10⅝	41.49	2,682	3,408	4,547	5,380	4,512	3,210	4,060		3,055	3,730
8⅝	9⅞	18.16	6,130	7,787	10,388	12,291	10,308	7,335	9,275		6,975	8,525
	10⅝	30.23	3,682	4,678	6,240	7,383	6,193	4,010	5,575		4,190	5,120
	12¼	59.43	1,873	2,379	3,174	3,756	3,150	2,240	2,835		2,130	2,605
9¼	10⅝	21.46	5,187	6,589	8,790	10,400	8,723	6,205	7,850		5,905	7,215
	12¼	50.66	2,197	2,791	3,624	4,406	3,695	2,630	3,325		2,500	3,055
10¾	12¼	27.10	4,107	5,218	6,960	8,236	6,908	4,915	6,215		4,775	5,715
	15	85.95		2,195	2,596		2,178	1,550	1,960		1,475	1,800

2. will give increased dust deflector and deck bushing wear; and

3. may damage borehole walls in soft drilling.

A further consideration when selecting the air volume is the altitude at which the drill will be working. Table 5 presents the compressor altitude multipliers that should be used to compensate for increased mine elevation above sea level.

Most new drills are now purchased with rotating screw compressors that tend to give lower drill bit life as a result of poor lubrication as compared with the vane compressor. Many mines have achieved much increased bit life by adding lubrication oil to the drilling water. One large copper mine in Africa has obtained a doubling of their 311 mm (12¼ in.) drill bit life by using 7 L/hr oil in 9 L/min water with air at 23 m³/min.

Rotary Drill Stems

Other factors involved when selecting drill stems (apart from the bailing velocity consideration) include: (1) fabricated or integral drill steel; (2) thread size and type; (3) wall thickness; and (4) types of connection.

High wear occurs behind the bit as this is a turbulent air zone. Also the section of rod just behind the bit is in the hole the longest. Once wear begins in this area, it will be further accentuated by the increased annulus further disturbing the air flow. The use of two box ended rods helps to reduce this type of wear by offering the opportunity to up-end the rods in addition to the usual exchange practice.

Rotary Drill Stabilizers

There are two main types of stabilizer on the market, the blade and roller.

Table 4. Bailing Velocities for Different Drill Pipe and Bit Diameters.
B. Metric Units.

Drill Pipe Diameter (mm)	Bit Diameter (mm)	Annular Area (sq. cm)	Allis-Chalmers 11-L 21.9 c.m.m. 0.27 Mpa	Allis-Chalmers 12-L 27.6 c.m.m. 0.27 Mpa	Allis-Chalmers 17-L 37.1 c.m.m. 0.27 Mpa	Allis-Chalmers 19-S 43.9 c.m.m. 0.27 Mpa	Sullivan Screw 36.8 c.m.m. 0.68 Mpa	Gardner-Denver 900 Screw 26.2 c.m.m. 0.34 Mpa	Gardner-Denver 1200 Screw 33.1 c.m.m. 0.34 Mpa	LeRoi 100-S-DS 12.7 c.m.m. 0.85 Mpa	Gardner-Denver WFN 24.9 c.m.m. 0.68 Mpa	Gardner-Denver WFO 30.4 c.m.m. 0.68 Mpa
114	143	57.7								2207		
	152	79.8								1597		
	159	95.4								1337		
127	152	55.7								2286	4470	5460
	159	71.3								1788	3495	4270
	172	104.2		2669	3560					1224	2391	2922
140	172	77.6	2820	3583	1779			3375	4267	1641	3211	3921
	187	122.3	1789	2273	3032		3009	2141	2708	1042	2036	2489
	200	161.0	1360	1728	2304	2726	2287	1628	2057	792	1548	1891
159	187	77.7	2818	3580	4775	5650	4739	3371	4264	1640	3208	3918
	200	116.3	1882	2391	3189	3773	3165	2251	2848	1096	2141	2617
171	200	83.4	2624	3333	4447	5262	4413	3139	3972		2987	3648
	216	135.3	1618	2055	2743	3244	2721	1935	2449		1841	2249
	229	179.6	1219	1548	2065	2444	2049	1459	1844		1387	1695
178	216	117.8	1858	2360	3149	3726	3125	2224	2812		2115	2583
	229	162.2	1349	1714	2287	2706	2270	1614	2042		1536	1878
	251	245.9	890	1131	1509	1816	1497	1065	1347		1013	1237
184	229	144.1	1519	1929	2574	3045	2554	1817	2298		1728	2112
	251	227.8	961	1221	1628	1927	1616	1149	1454		1093	1335
197	229	106.1	2063	2620	3495	4136	3469	2469	3121		2348	2868
	251	189.8	1153	1465	1954	2312	1939	1379	1745		1312	1603
	270	267.7	817	1039	1386	1640	1375	978	1237		931	1137
219	251	117.2	1868	2373	3166	3746	3142	2236	2827		2126	2598
	270	195.0	1122	1426	1902	2250	1888	1222	1699		1277	1561
	311	383.4	571	725	967	1145	960	683	864		649	794
235	270	138.5	1581	2008	2679	3170	2659	1891	2393		1780	2199
	311	326.8	670	851	1105	1343	1126	802	1013		762	931
273	311	174.8	1252	1590	2121	2510	2106	1498	1894		1455	1742
	381	554.5			669	791	664	472	597		450	549

Blade type stabilizers are generally cheaper in terms of initial purchase price but require rebuilding for each drill bit. In hard ground they give very poor stabilization after one or two holes have been drilled. A relatively high rotary torque is required in comparison with the roller stabilizer. Replaceable wear bars and sleeves can be used to speed up rebuild and help to reduce stabilization costs.

The roller stabilizer can be a throwaway or replaceable roller type. Lower rotary head torque is required for operation and stabilization is generally better than that provided by blades. The throwaway type have larger rollers, (no rebuild pin assembly, etc.) and usually give more economical life.

Experimental stabilizers include: (1) roller stabilizer with nonrotating outer casing; (2) chamber reamer; and (3) hole reamer stabilizer using a smaller pilot bit.

For larger drill bit sizes, 311 mm (12¼ in.) and up, some operations dispense with stabilization, thus eliminating the loss of rotary horsepower.

Rotary Head Shock Subs

In recent years rotary head shock subs have been developed by a number of manufacturers in order to reduce drill vibration, cut maintenance costs, and increase drill availability and bit life. The shock sub is essentially a large shock absorber which is fitted between the drill and drill pipe on top-drive machines. Shock subs are designed as follows:

1. Type A (Swivel Mounted). This consists of an adapter mounted element incorporating eighteen segments, each consisting of two heat treated alloy steel drive lugs molded together with rubber. These segments are retained by end plates to which the adapters fasten.

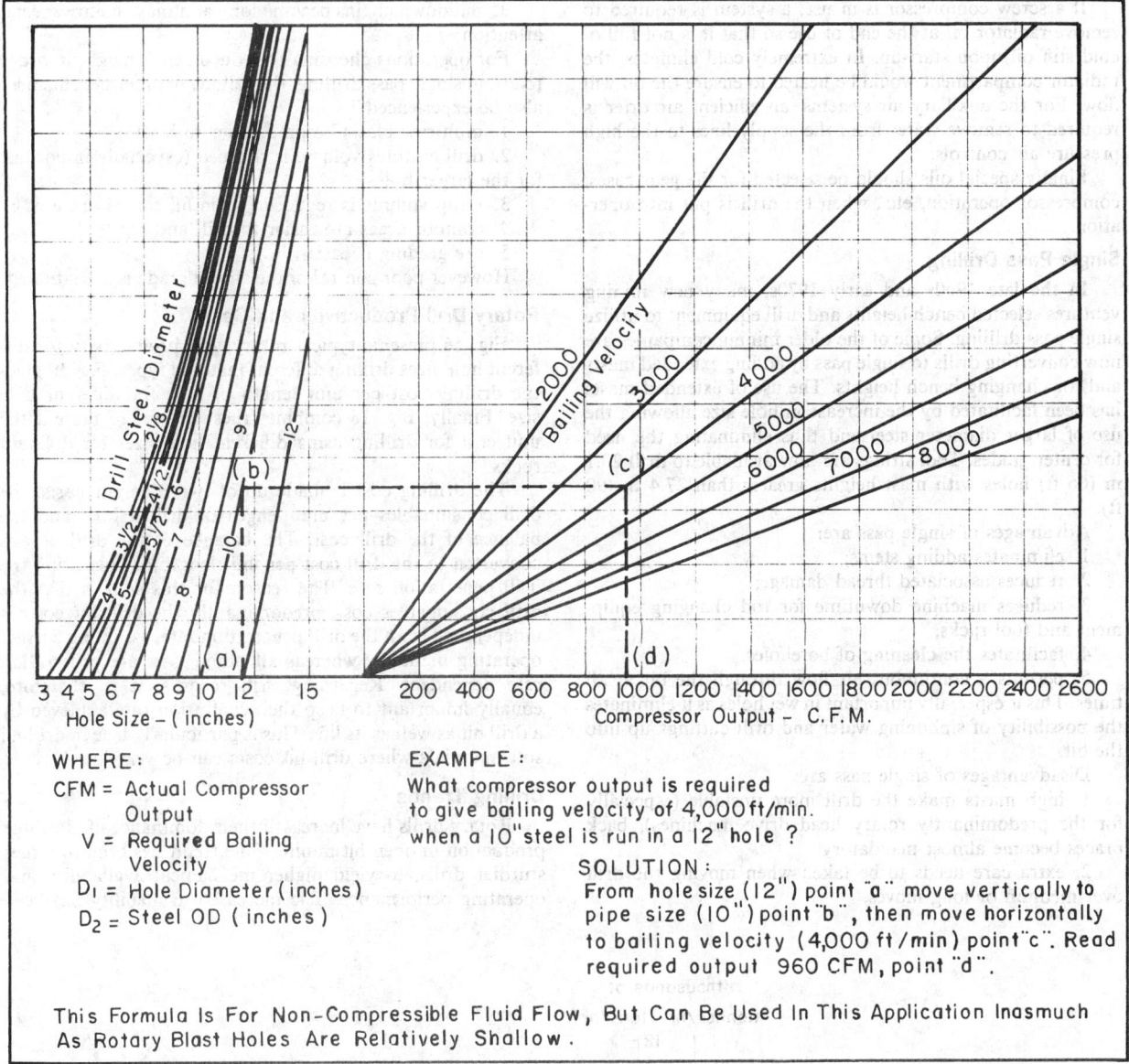

WHERE:

CFM = Actual Compressor
          Output
V = Required Bailing
       Velocity
D_1 = Hole Diameter (inches)
D_2 = Steel OD (inches)

EXAMPLE:

What compressor output is required
to give bailing velocity of 4,000 ft/min.
when a 10" steel is run in a 12" hole?

SOLUTION:
From hole size (12") point "a" move vertically to
pipe size (10") point "b", then move horizontally
to bailing velocity (4,000 ft/min) point "c". Read
required output 960 CFM, point "d".

This Formula Is For Non-Compressible Fluid Flow, But Can Be Used In This Application Inasmuch
As Rotary Blast Holes Are Relatively Shallow.

Fig. 15. Air requirements chart.

2. Type B (Gas Spring). This works on a gas spring principle using nitrogen under pressure to achieve a soft spring.

Several operators report that machine maintenance costs
have been reduced up to 18%, drill availability increased by
up to 7½%, noise levels have been reduced (in one reported
case from 120 to 84 db), and penetration rates and bit life
increased considerably.

**Table 5. Multipliers for Air Consumption of Drills
at Various Altitudes**

Altitude (ft)	0	1,000	2,000	3,000	4,000	5,000	6,000
(m)	0	305	610	915	1,220	1,525	1,830
Multiplier	0	1.03	1.07	1.10	1.14	1.17	1.21

Cold Weather Operation

Drills to be used in cold climatic conditions should be
purchased with a cold weather package.

In construction, high stress or impact areas should be
manufactured using normalized steel having good notch
toughness at cold temperatures. For operator comfort, double
or triple glazing and additional insulation in the cab are
useful along with additional heating capability as compared
with regular machines.

Additional heating is also required for the machinery
house to assure adequate start-up. Heated water tanks and
lines are required for the water injection system if used. A
special blowdown system is necessary to purge water lines
for stand-down periods. An auxiliary diesel generator is commonly employed to run heaters should the drill come off
power.

If a screw compressor is in use, a system is required to remove radiator oil at the end of use so that it is not full of cold stiff oil upon start-up. In extremely cold climates, the radiator compartment would be heated to ensure the oil will flow. For the auxiliary air systems, an efficient air drier is required to remove water from the supply lines to the high pressure air controls.

Finally special oils should be selected for the gear cases, compressor operation, etc., when the drill is put into operation.

Single Pass Drilling

In the late 1960s and early 1970s, many new mining ventures selected bench heights and drill equipment to utilize single pass drilling. Some of the older mining companies are now converting drills to single pass by adding extended masts and/or changing bench heights. The use of extended masts has been facilitated by the increase in hole size allowing the use of larger diameter steel and thus eliminating the need for center guides. Tool strings are now available to drill 20.1 m (66 ft) holes with mast heights greater than 27.4 m (90 ft).

Advantages of single pass are:
1. eliminates adding stems;
2. reduces associated thread damage;
3. reduces machine downtime for rod changing equipment and tool racks;
4. facilitates the cleaning of boreholes;
5. permits a continuous air flow through the bit at all times. This is especially important in wet holes as it eliminates the possibility of siphoning water and drill cuttings up into the bit.

Disadvantages of single pass are:
1. high masts make the drill more unstable (especially for the predominantly rotary head drive machines); back braces become almost mandatory;
2. extra care needs to be taken when moving the drill over medium or long moves,

3. pulldown chains become long and may require special attention.

For operations choosing to reduce bench height in order to attain single pass drilling, the following other benefits may also be experienced:
1. multirow blasts become easier to blast;
2. drill cuttings volume is reduced (especially important for the larger holes);
3. ramp volume is reduced (when fill ramps are used);
4. contour areas are easier to drill and blast;
5. ore grading is easier.
However poor control of the shovel grade is accentuated.

Rotary Drill Productivity and Costs

Fig. 16 presents typical rotary drill productivity for different hole sizes drilling different material types. Fig. 17 gives the drilling cost per unit length of hole for different hole size. Finally, Fig. 18 combines these graphs to present the unit cost for drilling using different hole sizes for different rocks.

The drilling cost is made up of two different parts, the drill consumables per unit length of drilled hole and the balance of the drill cost. The balance of the drill cost is converted to the drill cost per unit length of hole using the drill penetration rate. The reason for this split is that the drill consumables cost, predominantly the drill bit cost, is independent from the drill penetration rate, assuming correct operating methods, whereas all other costs are penetration rate dependent. Regarding drill records, it is, therefore, equally important to keep the penetration rate achieved by a drill bit as well as its life. This is particularly true in drilling soft materials where drill bit costs can be very low.

Drilling Trends

Rotary drills have increased their dominance of blasthole production in open pit mining. The trend has been to larger sturdier drills, to yield higher mechanical availability and operating performance. The increased availability has been

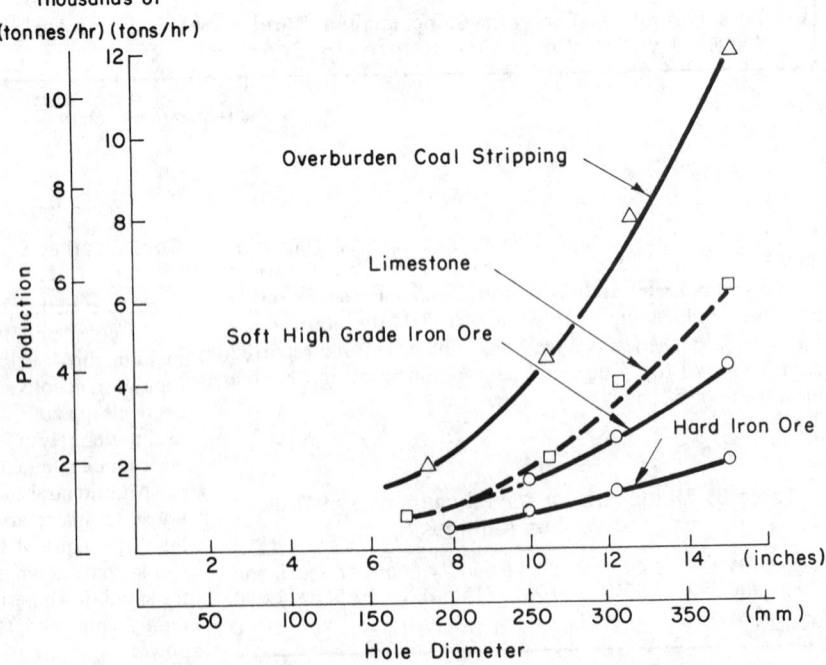

Fig. 16. Tons drilled per operating hour in different materials using rotary drills of different hole diameter capability.

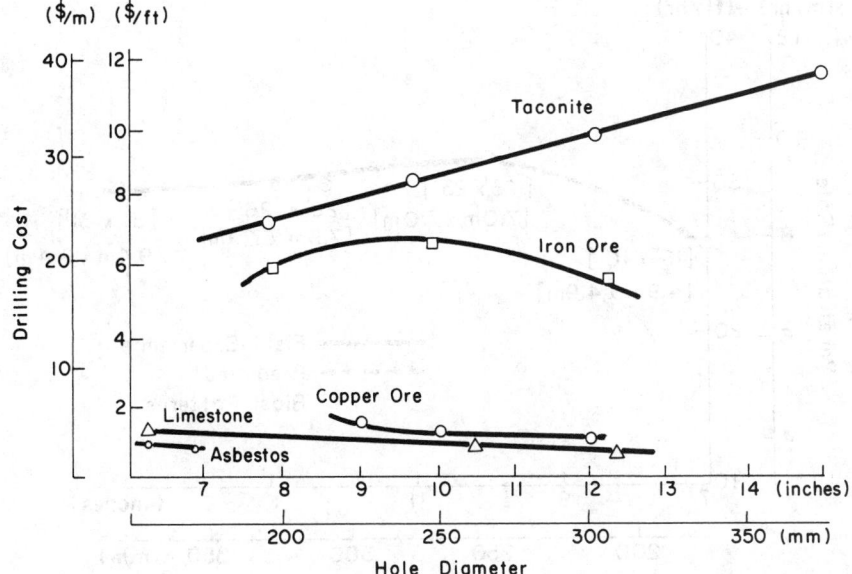

Fig. 17. Drilling cost for rotary drills drilling holes of different size.

achieved by improvement of crawler track frames, masts, propel chains, pulldown mechanisms, rotary head drives complete with automatic lubrication and greasing, on the new models of machine. Drill pipe feed rates have been increased so that pulldown weight can also be applied in very soft formations that fail easily and the penetration rates increased in such applications. Increased rotary drive horsepower and improved mast designs have permitted higher rotational speeds, which have in turn permitted higher penetration rates and lower costs to be achieved in many instances. The improved mast designs permit the use of higher speed drilling by reducing vibration on the drill.

Productivity improvements have been tried by using chambering or reaming drill bits. Techniques that have been tested include drill bits having extendable arms that could

increase the hole diameter at depth. None of these techniques to date has proved successful on a practical and economic basis.

Improvements have also been attempted in the weight distribution on the drills so that a higher percentage of the gross machine weight can be used for generating pulldown; the improved mast design on some of the models also helps to serve this purpose by putting more weight over the drilling tools.

It is noteworthy that the trend has been to larger and larger drills with increased pulldown weight capability. In very hard rock drilling, the larger machines are being used at a hole size that is well within their capability and therefore the machine is not taxed; this results in less breakdown and increased performance. Since the penetration rate is propor-

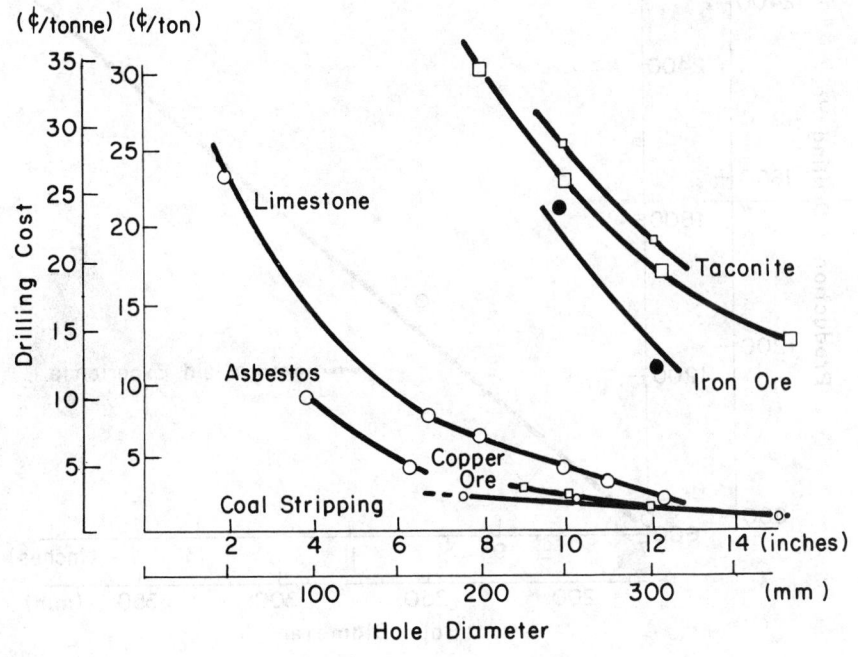

Fig. 18. Drilling cost per ton vs. hole size in different rocks.

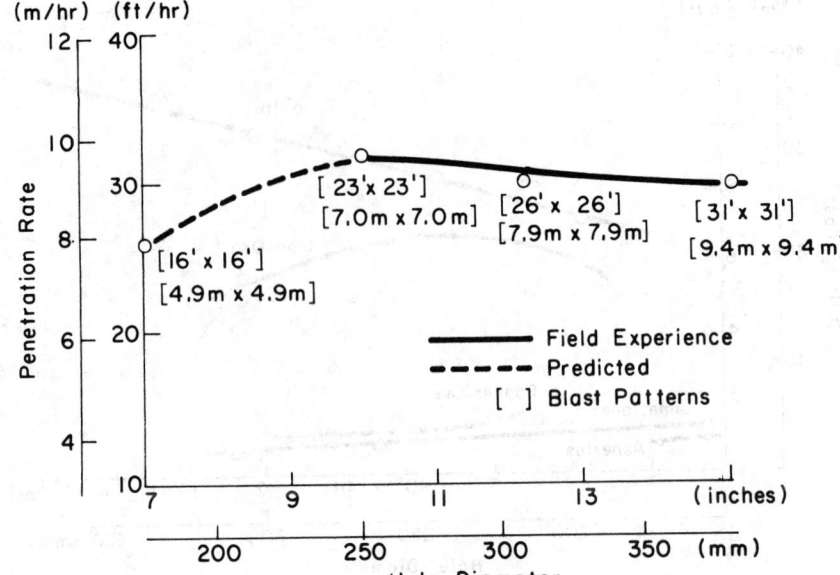

Fig. 19. Penetration rate vs. hole diameter for rotary drilling in hard iron ore [compressive strength, 200 MPa (30,000 psi)].

tional to the rpm and the weight/length of bit diameter, the bit technology has been to improve bearings, cone structure, and carbide quality for hard rock drilling. This has allowed the W/ϕ and rpm to be increased to increase performance. At the same time, bit life has not decreased, or, at least, not to an extent that the gains due to the penetration rate increase have been negated. For example, in North America in hard drilling, bit costs can represent up to 60 to 65% of the total operating cost on the drill so that if increases in penetration rates are accompanied by decreases in bit life, it is not nec-

essary for the bit life to decrease too far before any gains due to productivity increase are cancelled. In such operations considerable attention must be paid to penetration rate and bit life to produce the minimum cost/foot of hole. In soft formations, the bit cost only represents approximately 10% of the total drilling cost per unit length so that one can tolerate a significant decrease in bit life for increases in productivity and still gain financially.

Depending on the material type and hole size, rotary horsepower is proportional to (rpm)·(weight/in. of bit di-

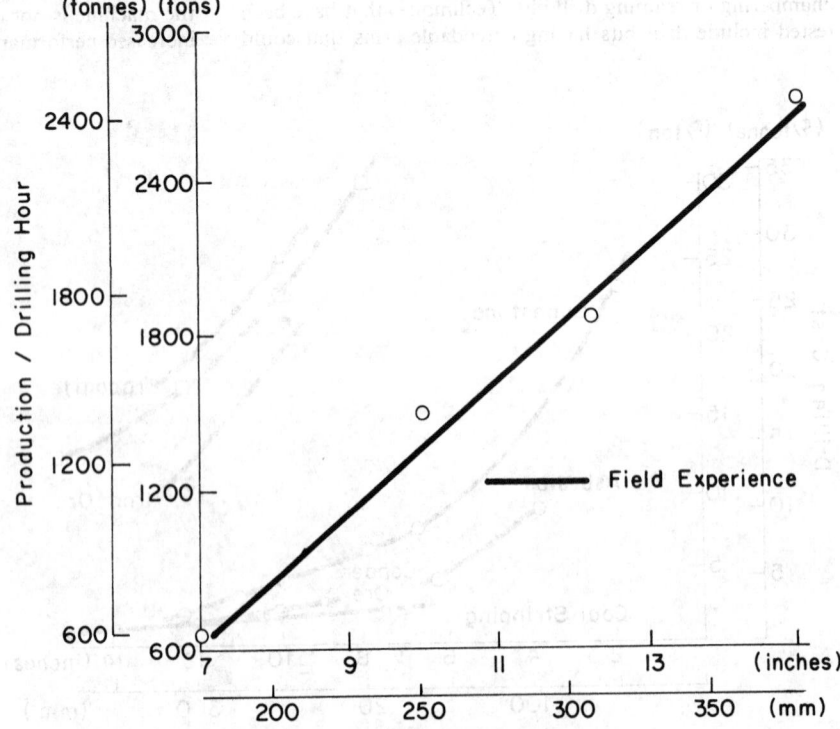

Fig. 20. Production rate vs. hole diameter in the same hard iron ore.

ameter)[1.5]. To double the rpm capability of the machine under the same load conditions, it is necessary to double the rotary horsepower. Future developments should come in this area since it seems entirely feasible with suitable rotary head modifications to double rpm capability, which would double production capability.

In iron ore mining 250, 311, and 381 mm (9⅞, 12¼, and 15 in.) diameters are standard sizes of hole, with the two smaller sizes being in the majority. The percentage of 381 mm (15 in.) holes has been continually increasing. This larger hole size was initially used in the softer formations and in higher tonnage operations where ore grading requirements are not a problem. Today this has changed and 381 mm (15 in.) holes are being used in the hardest formations.

Productivity has also been increased significantly at most operations with hole size increases. Larger bits have larger bearings and can accept higher loadings and still give satisfactory life. In going from 175 to 250 mm (6⅞ to 9⅞ in.) bit sizes, the operating weights increase from 89 to 116 kg/mm (5000 to 6500 lb/in.) of bit diameter for hard formation drilling. With similar rotary speeds, this produces a penetration rate increase of approximately 30%. Bit sizes larger than 250 mm (9⅞ in.) are being run at 107 kg/mm (6000 lb/in.) of bit diameter in order to maintain adequate bit life, which often represents 60 to 65% of the total operation cost of the drill in hard formations [compressive strength 200 MPa (30,000 psi) or greater]. Holes of 381 mm (15 in.) diameter can be drilled almost as quickly as 250 mm (9⅞ in.) holes and, when scaled for the tonnage increase per hole, this represents a significant productivity increase.

In softer formations or formations in which the fragmentation is preformed, patterns have generally increased in proportion to the diameter of the hole so that in going from 311 to 381 mm (12¼ to 15 in.) holes, productivity has often increased by approximately 40% and there has been a corresponding decrease in cost/ton. In hard massive formations, the productivity has been increased but not to the same degree. Recently 445 mm (17½ in.) diameter holes have been tested at two locations, as a desire to extend the productivity increases and economic gains. The results were relatively poor. More recently, work on the Mesabi Range has shown some productivity advantage of this large hole drilling and blasting in hard rocks with preformed fragmentation, but the economic advantage has yet to be seen due to bit technology lag. In Labrador, 381 mm (15 in.) diameter blastholes are being used successfully in hard [270 MPa (40,000 psi), compressive strength] massive quartz, magnetite, and specular hematite.

Fig. 19 presents the drill penetration rate vs. hole size in a hard massive low-grade iron ore, and Fig. 20 shows the actual productivity obtained at one Canadian mine in going from 250 to 311 mm (9⅞ to 12¼ in.) to 381 mm (15 in.)

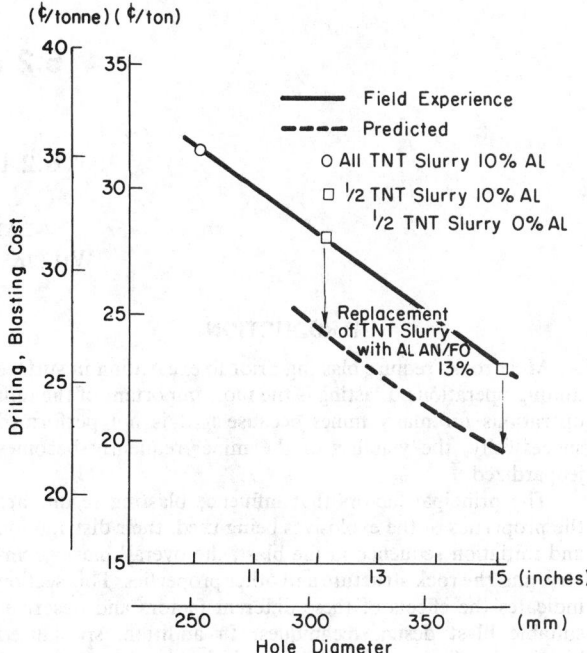

Fig. 21. Total drilling and blasting costs for the same fragmentation using holes of different diameter in the same hard iron ore.

diameter blastholes on the same size of drill. This represents a productivity increase of 73%. Fig. 19 also contains the equivalent blast patterns for excellent fragmentation using a 10% aluminized TNT slurry, and 0% aluminized TNT slurry, and based on 63.5 Mt (70 million st) mined at the larger hole sizings. Pit floor digging of the large hole blasts is eased significantly due to the higher concentration of energy in the toe of the blast. This has resulted in higher shovel productivity. Fig. 21 shows the total direct drilling and blasting costs, which were reduced from 32¢ to 27.5¢/ton. Machine capital cost savings also accrue due to the productivity increase.

REFERENCES

Anon., 1975, *Technical Newsletter,* Jarvis Clark Co. Ltd., Oct. 6.

Bauer, A. and Calder, P. N., 1966, "Drilling in Open Pit Iron Mines," *American Mining Congress,* Sept.

Bauer, A. and Calder, P. N., 1967, "Open Pit Drilling Factors Influencing Drilling Rates," Fourth Canadian Symposium on Rock Mechanics, Ottawa, Mar.

Calaman, J. J. and Rolseth, H. C., 1968, "Jet Piercing," *Surface Mining,* E. P. Pfleider, ed., AIME, New York, pp. 325-337.

6.2 Blasting

6.2.1 Blasting

ALAN BAUER
WILLIAM A. CROSBY

INTRODUCTION

Most rocks require blasting prior to excavation in surface mining operations. Blasting is the most important of the unit operations for many mines because if it is not performed successfully, the viability of the mine frequently becomes jeopardized.

The principal factors that influence blasting results are the properties of the explosives being used, their distribution and initiation sequence in the blast, the overall blast geometry, and the rock structure and other properties. This section indicates the effects of these different factors and describes suitable blast design techniques. In addition, specialized blasting methods such as throw blasting and wall control are discussed along with some of the detrimental aspects of blasting, vibration, air blast, and noise.

EXPLOSIVES

There are four main types of explosive used in surface mining: slurries, dry mixes, emulsions, and the hybrid heavy ANFO. These products will be considered separately.

Blasting Agents Development

Historically, prior to the *lease-lend* fertilizer disasters at Texas City and Hamburg in the 1940s, ammonium nitrate (AN) and fuel was not being used in bulk form as a blasting agent. Its use up to that time primarily was in canned explosives which used blends of AN to increase the density, cheap fuels such as Bunker C, and densifying agents such as ferrosilicon. Because of the viscosity of the Bunker C fuel, these mixes were often made hot. The cost was high due to the use of the cans and the canning process. Although the density was about 1.1 g/cm,³ the bulk strength was low due to the large annular space around the can when it was loaded into the borehole. Pelleted TNT was often put into the holes to surround the lower cans and to reduce the effect of the decoupling in the toe. The other explosive in use was dynamite.

During the early 1950s AN mixed with fuel oil was introduced to the blast site as a bulk explosive. In many instances it proved to be ideal, but it did have its drawbacks. These mainly revolved about its lack of water resistance and low bulk density, which in turn produced a low bulk strength.

By the mid 1950s, the drilling and blasting of very hard iron formation at Schefferville in north-eastern Quebec, with its attendant very high drilling costs and wet conditions, prompted Cook and Farnam to set out to invent explosives having good water resistance and high bulk strength. Initially two systems were developed, both using AN and water, with one system based on TNT or other high explosives as the sensitizer, and the other which used fine suitably coated aluminum for the same purpose. The basic high explosive (HE) sensitized slurry consisted of ammonium nitrate, water, thickeners, and the HE (TNT) sensitizer. The quantity of TNT used varied, depending on the critical diameter desired

and the temperature likely to be encountered. However, for a critical diameter of 75 mm (3 in.) at 0° C, a typical composition would be approximately 25% TNT, 15% water, 59% AN, and 0.5 to 1.0% of gum thickener. The gum thickener served two purposes: first, it held the coarse high density TNT in suspension and second, it formed a jellylike continuous matrix which imparted good water resistance to the slurry if the gum was suitably crosslinked. Energetic metal fuels such as aluminum were soon added to these slurried explosives to increase their bulk strength further. The basic TNT slurry had a density of approximately 1.45 g/cm³ and a weight strength relative to ANFO of 85%, so that its bulk strength was 1.48 relative to free poured porous prills and fuel oil. This meant much greater toe pulling capability and in many instances larger drilling and blasting patterns. The explosive cost was sometimes increased, but the reduction in drilling cost often made up for it and the net result was an overall reduction in drilling and blasting costs with better results.

Nitro-carbon nitrate (NCN) aluminum-based slurries had been developed but these were not rushed into production immediately in Canada on account of the climatic conditions in the northern part of the country. Borehole temperatures run as low as −2° C and temperatures on the benches periodically as low as −45° C with −20° to −30° C for extended periods of up to several weeks.

More recently fuel oil air bubble sensitized slurries were introduced to replace ANFO where holes cannot be dewatered. Phosphates and other coating agents were introduced to stop the aluminum from gassing, and also to prevent the pH changing, thereby breaking down the gel. Fig. 1 shows the general class of blasting slurries.

By the early 1960s slurry had been well tested in many areas and there was a considerable demand by mine operators, many of which were loading at levels of approximately 25 000 kg/d (55,000 lb/per day), which made packaging expensive. Transportation was also a costly item since the explosive plants were often remote from the mines. This meant that all of the ingredients in the packaged TNT slurry travelled at the high explosive freight rate. A minimum of 15% water was being hauled large distances.

These factors provided the impetus for the development of bulk slurry mix trucks. Fig. 2 shows a Canadian Industries Ltd. Gelmaster slurry truck.

Consideration of the solubility curve vs. temperature for AN clearly shows that by going to elevated temperatures all of the AN can be put into solution using the 15% water that the formula requires, and this then acts as the vehicle to carry the AN in the slurry mix truck.

For many mixes, the liquor is often partially pregelled. This has the advantage of putting a substantial quantity of air bubbles into the liquor. With NCN mixes the problem is to hold air bubbles on the aluminum surface. Premixing of the aluminum and other fuels by some producers enables

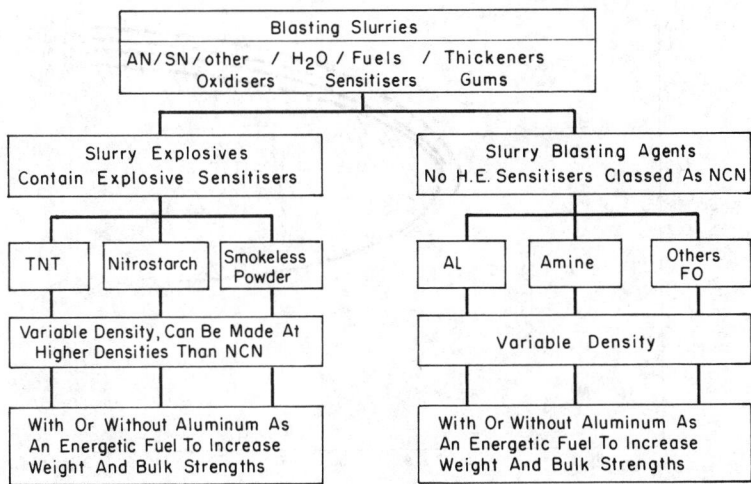

Blasting Slurries
AN/SN/other / H₂O / Fuels / Thickeners
Oxidisers Sensitisers   Gums

Fig. 1. Classes of blasting slurries.

this to be done properly. Others do it by *Slurrexing*, by adding gassing agents to the mixture. *Bicarbonate* produces carbon dioxide. The quantity is varied to vary the density.

Compositions based on aluminum or air bubble sensitization have critical densities beyond which they will not shoot. These are considerably lower than HE sensitized slurries and have to be controlled carefully. In wet holes, pumping air can cause problems and so can pumping a low density slurry. Many densities in the hole are 1.2 to 1.3 g/cm^3 and pumping density is usually lower than this at approximately 1.1 g/cm^3.

Truck pumping rates are variable with the larger units now being able to deliver at 700 kg/min (1500 lb per min). Weightometers are commonly fitted to the bins, offering the opportunity to automate the explosive delivery record keeping. This facility, along with automatic drill recording units, has spurred the introduction of computer programs designed to give complete control of the drilling and blasting costing and evaluation.

Slurries

Although fuel oil/gas bubble sensitized water gels are still very popular and relatively cheap, they do suffer from the following disadvantages. First, because they are gas bubble sensitized, they are subjected to desensitization under hydrostatic heads. This is overcome to some degree by adding

aluminum to the basic slurry or by incorporating additional gas into the product. For wet hole application, the minimum fully gassed density that can be employed is 0.95 g/cm^3. With loading beneath water and with a slurry column above the end of the hose, the hydrostatic head on the slurry is sufficient to bring the slurry density to greater than 1.05 g/cm^3 and thereby avoid floating. However the critical density of the basic fuel gas bubble slurry is approximately 1.15 g/cm^3 for a 200-mm (8-in.) critical diameter. This means the maximum hydrostatic head in a wet borehole that can be tolerated for this critical diameter is 0.1-MPa (14-psi) overpressure. For dry holes this is not such a problem since the slurry can be gassed down to densities much lower than 0.95 g/cm^3, so that under the existing hydrostatic head conditions it will not be precompressed beyond its critical density.

Fig. 3 shows the effect of density on the critical diameter and the detonability limits of a typical air bubble fuel oil slurry containing different percentages of aluminum. Because of the cost of aluminum and the large amounts needed in deep holes such as encountered in dragline stripping operations, if the extra energy is not really required, there has been a trend to mechanical means of sensitization for lower cost. This can be achieved by the use of microballoons or the use of solid void-filled ingredients in the compositions.

The microballoons in combination with a lower aluminum content permit retention of the sensitivity at lower cost than the high aluminum content mixes.

Table 1 shows the critical density and maximum hydrostatic heads that chemically gassed compositions can be used under in wet holes. For comparison purposes TNT-based slurries are included. Some of the amine-sensitized slurries have similar properties to the TNT ones, but they also suffer from the disadvantages of higher cost than the low grade fuel oil slurries. Other slurries are sensitized using methylamine nitrate and ethylene glycol mononitrate. The shelf life difference between the two classes of slurry in Table 1 is also noteworthy and clearly the use of the low grade fuel oil slurries in wet holes is limited to a short in-hole exposure time.

Emulsions

Because of the foregoing, blasting emulsions, in which the oxidizer is dispersed throughout the continuous fuel oil emulsion phase, are receiving increasing attention.

The manufacture of blasting agent emulsion is usually performed using an ammonium nitrate water solution or a

Fig. 2. Gelmaster slurry truck.

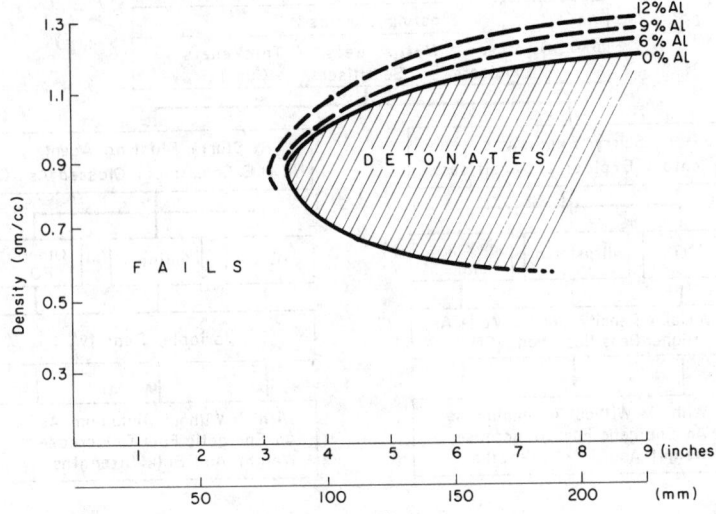

Fig. 3. Effect of density on the critical diameter of aluminized air bubble slurries containing different percentages of aluminum.

calcium nitrate, ammonium nitrate water solution emulsified in fuel oil using 1 to 2% of a suitable emulsifier such as sodium oleate or sodium stearate. The calcium nitrate is sometimes used to replace some 50% of the ammonium nitrate to help reduce cost, although the calcium produces solid oxides as a product of detonation so that useful energy output is somewhat reduced. Other salts sometimes used include sodium nitrate, to perhaps 33%, to reduce solution temperature and early mixes used ammonium perchlorate to improve sensitivity. The water content in the solution varies in the 14 to 20% range and solution temperatures range from 60 to 70° C, depending on the salts used. The fuel oil content is commonly approximately 4% but varies to produce the correct oxygen balance. For consistency, varying amounts of wax may also be employed as part of the fuel and to improve shelf life. For a field mix system perhaps 1% wax would be employed.

As with the NCN slurry system, aluminum may be added to increase the weight and bulk strengths of the emulsion as well as to improve the sensitivity. In North America some manufacturers incorporate 1 to 2% glass microballoons in all of their emulsions to ensure sensitivity and to help control

density. Microballoons would be added if the emulsion were to be used by itself or if added to ANFO to make heavy ANFO. The cost of the microballoons is approximately twice that of aluminum and so a 1.5% addition produces an emulsion of cost similar to a 3% Al fuel oil bubble sensitized slurry.

While emulsions have good water resistance to standing water, they do not have good resistance to water jetting into the borehole since they do not have the physical strength of a gel.

Blasting emulsions have very flat velocity of detonation charge diameter curves. They rise rapidly from the critical diameter to the thermohydrodynamic value, in contrast to the non-high explosive slurries which have a much more gradual rise associated with them. This probably is a reflection of a more intimate degree of mixing. With the significantly higher detonation velocities at intermediate diameters of charge 75 to 175 mm (3 to 7 in.), it would appear that the emulsion products would have a possible performance edge in this size range.

Emulsions can be gassed in a similar manner to the NCN slurries and sometimes microballoons are added to help con-

Table 1. Critical Densities, Hydrostatic Head Limits and Shelf Life for TNT Slurries and Fuel Oil Air Bubble Sensitized Slurries When Used in Wet Holes

Explosive	Critical Density (g/cm³) For a 150-mm (6-in.) Unconfined Critical Diameter at 5°C	Maximum Hydrostatic Head Mpa	psi	Shelf Life Wet Holes
TNT Slurries				
20% TNT Slurry (T-3)	1.55	0.78	115	Many Months
+ 10% Al (M-210)	1.57	0.85	125	Many Months
Gassed T-3	1.15 (min.)	0.78	115	Many Months
Fuel Oil/Gas Bubble Slurry				
Basic Slurry	1.15	0.10	14	2 Days
+ 5% Al	1.25	0.29	42	2 Weeks
+ 10% Al	1.30	0.41	60	3 Weeks
+ 12% Al	1.32	0.44	65	3 Weeks
+ 17% Al	1.37	0.51	75	3 Weeks

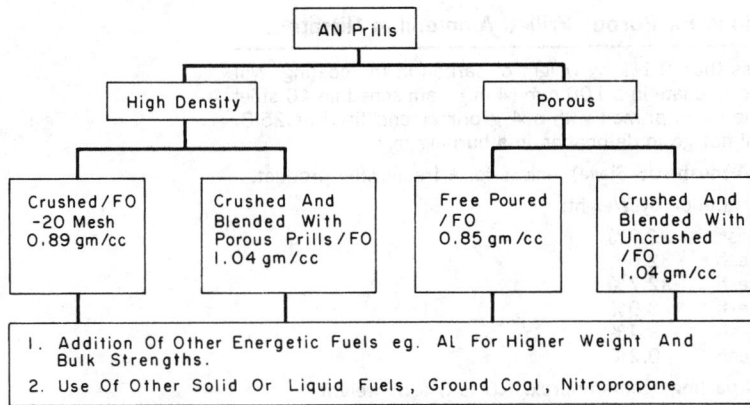

Fig. 4. Classes of ammonium nitrate dry mixes.

trol density in both emulsion and heavy ANFO. High velocities producing high brisance is the result, which can be very beneficial when blasting the higher strength rocks.

Ammonium Nitrate Dry Mixes

Fig. 4 illustrates the classes of ammonium nitrate dry mix explosives.

There are a number of different forms of ammonium nitrate, all of which have been used in explosives at one time or another. The porous prills are most suitable for use with fuel oil directly as a blasting agent. The anti-caking agents usually used are Kieselguhr up to 2 to 3% or surface active agents such as some of the sulfonates, a fraction of 1%, or combinations of the two. Ammonium nitrate is an oxidizer and for each molecule, (NH_4NO_3), there is an excess of one atom of oxygen.

A variety of fuels could be used with the ammonium nitrate. Fuel oil is cheap and easy to apply. For maximum energy output, the oxygen balanced mix of approximately 94.4% AN to 5.6% FO gives the optimum mix.

The energy output can be calculated for different ratios of ANFO. Maximum energy output occurs when the excess oxygen is used by the fuel to give only carbon dioxide, steam, and nitrogen. This is called the zero oxygen balanced mixture. If the nitrogen combines with oxygen, as occurs in an

underfueled mix, an endothermic reaction occurs producing nitrogen oxide, reducing energy output. Further underfueling or mixing with air produces the telltale brown fumes of nitrogen dioxide. Brown fumes from an ANFO blast indicate that either there was insufficient fuel present or the ANFO was attacked by water.

Fig. 5 shows the effect on the energy output of different percentages of fuel oil added to AN. On both sides of the oxygen balanced mixture, there is a fall off in energy output, but it is more pronounced on the fuel lean side. Because of this, most operators run the mixture slightly fuel rich.

Fig. 6 is a schematic of the feeding and mixing arrangement of an ANFO mix truck. Calibration is afforded by interlocking the speed of the AN delivery auger to the back of the truck with the fuel oil pump. In view of the sensitivity of the energy output to variations in fuel oil content, great care must be taken in maintaining accurate calibration. For this reason some operators, notably in Africa and Australia, are now going to the "concrete style" ANFO mix truck concept.

Table 2 presents the specifications for porous prilled ammonium nitrate suitable for the manufacture of ANFO. Porous prills usually hold the fuel oil well but dense prills do

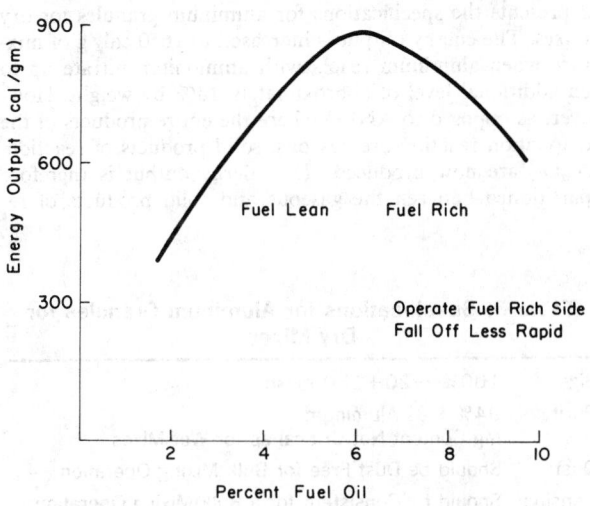

Fig. 5. Energy output vs. percent fuel oil added to ammonium nitrate.

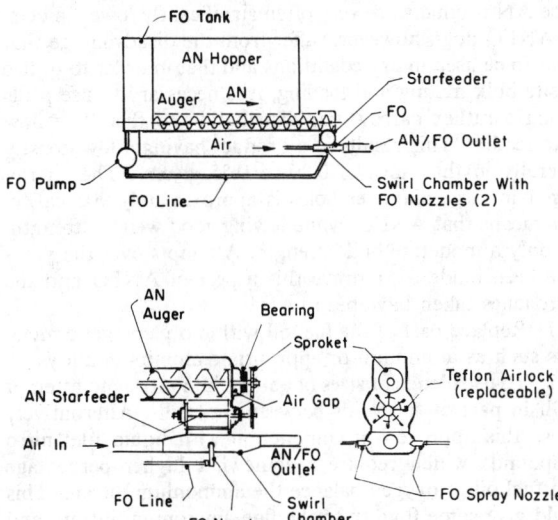

Fig. 6. Schematic of the feeding and mixing arrangement of an ANFO mix truck.

Table 2. Specifications for Porous Prilled Ammonium Nitrate

Oxidizer Specifications:	Less than 0.1% by weight of carbon in the coating. Will not detonate in a 100-mm (4 in.) diam schedule 40 steel pipe when primed with a 4 g primer and fired at 25°C. Will not go to detonation in a burning test.
Size:	General −6+14 mesh (US Sieve), unless for a freeflowing product.

Typical Size Distribution by Weight:

− 6 +10 mesh	22.8%
−10 +12 mesh	32.0%
−12 +14 mesh	32.7%
−14 +16 mesh	8.9%
−16 +20 mesh	3.4%
−20 +35 mesh	0.2%

Coating:	Clay and special parting agents approx. 1.6 to 3% by weight
Density:	Different products have bulk density 0.82 to 0.90 g/cm³ and should not vary widely for a particular product.
	Some low density products at 0.68 to 0.70 g/cm³ tend to be too sensitive.
Porosity:	Minimum fuel oil retention 6% by weight, and should retain this indefinitely.
Free Flowing:	
Critical Diameter:	When mixed with 6% by weight of fuel oil and free poured will shoot unconfined in a 125-mm (5-in.) diam cardboard 1 m (3 ft) long when primed with 60 g of pentolite at one end. Should not be cap or fuse sensitive.
Hardness:	Should be such that there will not be excessive prill breakdown in handling or in the normal temperature cycling that can occur during a reasonable time period (2–3 months). This would clog auger, cause caking, etc.

not (less than 2% when mixed). It is necessary to increase the surface area of this form of AN to get adequate FO retention. This can be done by incorporating hammer mills on the back of bulk trucks and, with the correct size of screen, procure a product that has the necessary surface area. As a rule if the dense prills are crushed to 96% −850 μm (−20 mesh), then the fuel oil retention characteristics are as good as porous prills. To do this, a 6-mm (0.25-in.) screen is used and the hammer mill is run at about 1800 rpm. The reason for going to such a system is purely economic in that these AN products are very often significantly lower in cost.

ANFO does, however, suffer from the disadvantage that it has to be used in dry conditions and that in order to utilize on-site bulk mixing and loading, it is necessary to use prills having a rather narrow size distribution so that they flow quite readily. This results in a product having a low density generally in the range of 0.80 to 0.85 g/cm³. The energy output in large diameter holes is approximately 903 cal/g. This means that ANFO, while having good weight strength, has only a moderate bulk strength. Attempts over the years have been made to improve this aspect of ANFO and the approaches taken have been to:

1. Replace part of the fuel oil with more energetic metal fuels such as aluminum or appropriate aluminum alloys.

2. To use blends of sizes of ammonium nitrate to attempt to fill in part of the voids between the prills. Alternatively to use this approach in conjunction with liquid fuel nitro compounds which require a somewhat higher percentage than fuel oil to oxygen balance the ammonium nitrate. This would give some fluidity to the fine ammonium nitrate and ease the void filling operation.

3. To use a blasting emulsion with the appropriate emulsifying agents and to use this to fill or partially fill the voids

between prills, depending upon the level of addition. This is the so-called heavy ANFO. The emulsion in some instances is sensitized with microballoons or other hollow gas containing material. In other cases it is not.

With regard to the addition of energetic fuels to ammonium nitrate and fuel oil, it is well recognized that as aluminum is added, the fuel oil is reduced in order to produce the optimum energy output. The particle size of the aluminum is also selected so that it will be fully reactive in the blasthole size in use. It should also be dust free to avoid the possibility of hazards associated with dust in handling. Table 3 presents the specifications for aluminum granules for dry mixes. The energy output is increased to 1650 cal/g of mixture when aluminum reacts with ammonium nitrate up to an additional level of approximately 18% by weight. However, as opposed to ANFO where the entire products of the detonation reaction are gaseous, solid products of reaction, Al_2O_3, are now produced. The energy output is therefore partitioned between the gaseous and solid products of re-

Table 3. Specifications for Aluminum Granules for Dry Mixes

Size:	100% −20+150 mesh
Purity:	94% Plus Aluminum
	Mg Content Not Critical as for Wet Mixes
Dust:	Should be Dust Free for Bulk Mixing Operation
Density:	Should be Consistent for a Bulk Mixing Operation so that Calibration is not Being Changed Continually
Flow:	Should be Free Flowing for a Bulk Truck Application

action. In practice and in test evaluations, half of the energy associated with solid products of reaction cannot be obtained usefully from the system. Evidently this is because during the work process in which the explosive gases are expanding, the solid products of reaction cannot transfer all of their contained heat rapidly enough to do useful work. At the same time, the gas volume per unit weight of explosive also has been reduced as aluminum is added. Despite these short-comings, there is a significant increase in the energy output as aluminum is added to ANFO. Fig. 7 is a plot of the strength relative to ANFO of Al/ANFO containing various percentages of aluminum. Since the addition of aluminum to ANFO also increases the density slightly, this will influence the relative bulk strength. Table 4 gives the bulk strength compared to ANFO for Al/ANFO containing various percentages of aluminum up to 15% by weight. The optimum formulations for these various percentage aluminum additions are indicated. Fig. 7 indicates that the increase in energy output with aluminum addition is approximately linear up to 7% by weight. After this the incremental increase in energy output with further aluminum addition decreases. For this reason unless the drilling cost is excessive, it is usual to find a maximum addition of aluminum to ANFO of 15% by weight and more commonly a maximum percentage addition of 7 to 10%. For dry hold blasting, the use of Al/ANFO in combination with ANFO is often an economic proposition.

Heavy ANFO

More recently, attempts have been made to densify ANFO using emulsions of ammonium nitrate, calcium nitrate water, and fuel oil with the appropriate emulsifiers to form a water in oil emulsion. Such compositions containing up to 45 to 50% by weight of emulsion are now being used for wet holes. Bulk trucks have been developed to blend the emulsion with AN or ANFO and to auger feed the product through water leads to considerable water inclusion in the column, which seriously reduces the effectiveness of the charge. For holes containing relatively small amounts of water, the shock on impact is sufficient to take the emulsion from the solid ammonium nitrate surface and allow part of the ammonium nitrate to go into solution. Again, this reduces the effectiveness. Blasthole polyethylene liners with slits in the wall are now being used to successfully load heavy ANFO in water-filled holes.

Comparison of Slurries, Emulsions, Heavy ANFO, and Dry Mixes

Fig. 8 shows the effect of the addition of aluminum on the weight strength of ANFO compared with values for typical slurries of both NCN and the TNT sensitized types, emulsions, and heavy ANFO. From this graph, it can be seen that the increase in energy output for each blasting

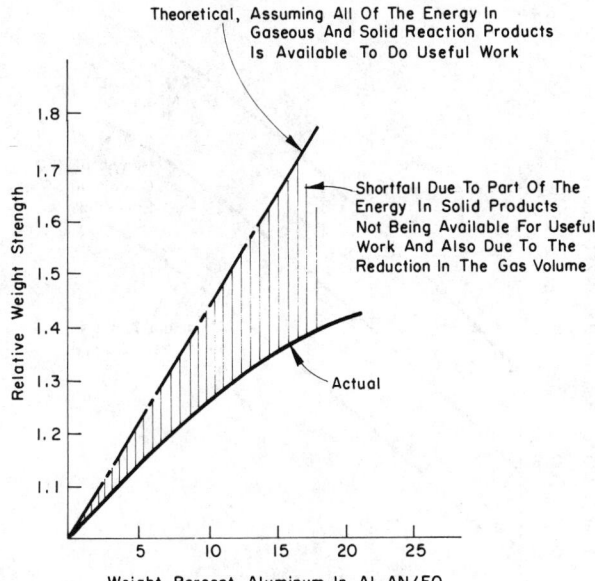

Fig. 7. Relative weight strength of Al/ANFO containing various percentages of aluminum.

agent is comparable for increased quantities of aluminum addition.

Fig. 9 presents the corresponding blasting agent bulk strengths relative to ANFO for different aluminum additions. Now the advantage of increased loading density becomes apparent. The TNT-sensitized slurry and heavy ANFO give the highest relative bulk strengths closely followed by the bagged TNT-sensitized product, which is slightly lower because of the voids introduced into the loaded explosive column. Heavy ANFO bulk strength does vary depending on the proportion of emulsion to ANFO used and the final loading density. Basic NCN slurry and emulsion have similar strengths if they are manufactured using the same oxidizers and have the same water contents.

Table 5 shows how the weight and bulk strength concepts may be used. Relative powder factors can be obtained from the relative weight strengths by dividing the powder factor for ANFO by the weight-strength relative to ANFO.

Toe pulling capability or collar height is dependent on the action of the ends of the charge. This is approximately hemispherical and is therefore related to the one-third power of the useful energy in the bottom or top part of the charge (approximately eight diameters). For a given fixed diameter of blasthole, the toe pulling ability is proportional to (the bulk strength)$^{1/3}$. If the toe distance divided by the (bulk

Table 4. Relative Bulk Strength for AL/ANFO

Explosive	Density, g/cm³	Relative Weight Strength	Relative Bulk Strength (Relative to ANFO at a Density of 0.83 g/cm³)
ANFO	0.83	1.00	1.00
Al/ANFO 5/90.5/4.5	0.87	1.13	1.18
Al/ANFO 7/89/4.0	0.88	1.18	1.25
Al/ANFO 10/87/3	0.91	1.24	1.36
Al/ANFO 15/83.6/1.4	0.94	1.35	1.53

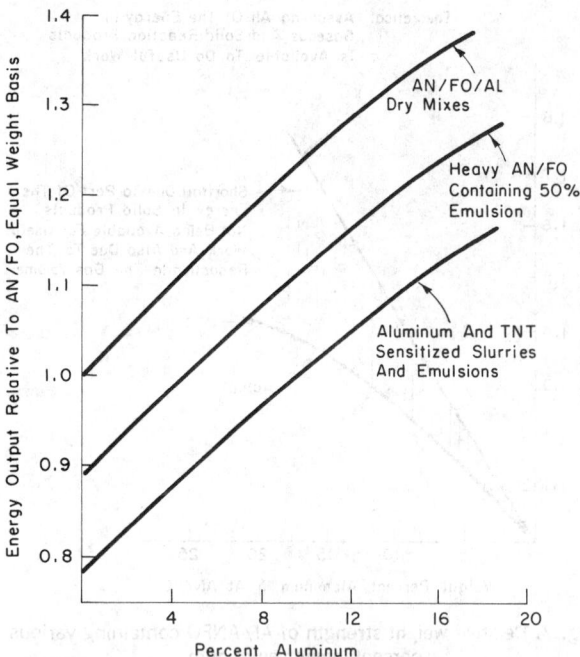

Fig. 8. Weight strengths relative to ANFO for Al/ANFO dry mixes and for aluminum and TNT-sensitized slurries and emulsions containing different percentages of aluminum.

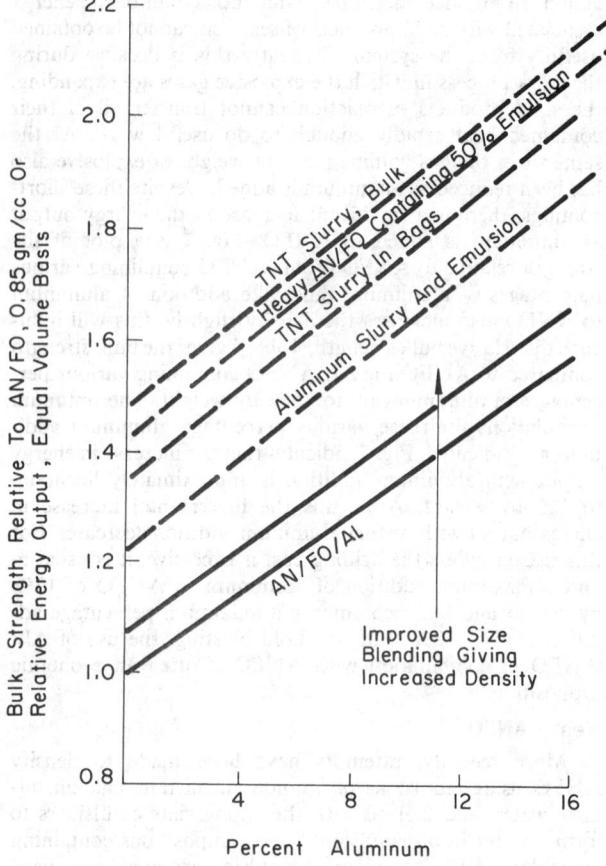

Fig. 9. Bulk strengths of dry mixes, emulsions, heavy ANFO, and slurries with different aluminum contents.

strength)$^{1/3}$ is kept constant, then the similar results will be produced.

Collar heights are scaled in a similar manner, the burden is now the depth from the top of the bench to half way down the top eight diameters of charge. If this scaled distance is

Table 5. Relative "Powder Factors" and Toe Pulling Capability for Different Blasting Agents

Explosive	Density, g/cm³	Weight Strength, ANFO=100	"Powder Factor" Relative to ANFO	Bulk Strength Relative to ANFO at a Density of 0.83 g/cm³	Toe Pulling Capability Compared to ANFO (Rel. Bulk Strength)$^{1/3}$
ANFO	0.83	100	1.00	1.00	1.00
ANFO/Al-87/3/10	0.91	124	0.81	1.36	1.11
ANFO/Al-84/3/13	0.93	131	0.76	1.46	1.13
Metallized Slurry, 1% Al, 12% H₂O	1.25	86	1.16	1.26	1.08
FO Slurry, 16% H₂O	1.20	77	1.30	1.11	1.04
Metallized Slurry, 10% Al, 14% H₂O	1.30	106	0.94	1.66	1.18
TNT Slurry, 25% TNT, 14% H₂O	1.45	87	1.15	1.48	1.14
TNT Slurry, 20% TNT, 10% Al, 13% H₂O	1.50	111	0.90	2.01	1.26
Emulsion, 16% H₂O	1.25	77	1.30	1.16	1.05
Heavy ANFO, 50% ANFO, 50% Emulsion	1.35	89	1.16	1.45	1.13
Aluminized Heavy ANFO, 10% Al	1.37	114	0.91	1.88	1.23

kept constant, then similar results will be produced at the top of the bench.

Explosives Selection Factors

The factors that go into the selection procedure are as follows:

1. Critical diameter and the factors that affect it, such as hydrostatic pressure and temperature.

2. Minimum primer weight and the factors which affect it.

3. Density.
4. Weight strength.
5. Bulk strength.
6. Gap sensitivity.
7. Water resistance, loading procedures, gelling rates, etc.
8. Coupling or decoupled properties if used for wall control work.
9. Shelf life.
10. Sensitivity primacord downlines.
11. Reliability and quality control for bulk operations.
12. Overall drilling and blasting economics.

Careful observation indicates that blasting results can be correlated with the total energy input per ton of rock, providing there is adequate subgrade, bottom charge, stemming, delays, etc. The most important characteristics of explosives for blasting are their useful energy outputs per unit weight and per unit volume, that is, their weight and bulk strengths. Typical values have been presented in Table 5.

PRIMING AND INITIATION SYSTEMS

The onset of detonation is without doubt thermal in origin and since reaction rates are exponential functions of temperature, then those explosives that generate the highest pressures at the primer-explosive interface will be the most efficient primers, in other words, those explosives having the highest detonation pressure. The considerable work by Cook, et al. (1968), shown in Fig. 10, clearly demonstrates that the size or weight of the primer necessary to prime ANFO or blasting slurries is a function of the detonation pressure of the primer. Those primers having high detonation pressure such as cast pentolite (detonation pressure 210 kbars) require relatively small weights compared to the less efficient lower detonation slurry primers (detonation pressures 110 kbars

or less). Because of this, it has been usual to use high pressure cast primers for priming of blasting agents. Weights usually used are from four to six times the minimum primer weight, determined at an unconfined charge diameter of somewhat greater size than the critical diameter of the explosive to be primed.

Fig. 11 shows typical priming test results for crushed high density prills/FO in steel pipe. It shows a minimum primer weight of 105 g (0.23 lb) of pentolite. For large diameter blastholes, 450 g (1 lb) primers are adequate.

Slurries for large diameter blasthole application under normal conditions have minimum primer weights of approximately 80 to 100 g (0.18 to 0.22 lb). One pound primers once again are adequate. With NCN slurries the minimum primer weight is strongly dependent on the temperature of the slurry. If it is hot then it can become cap sensitive.

Emulsions and heavy ANFO have similar priming requirements to regular ANFO and slurries. For most blasting 450-g (1-lb) cast pentolite or procore primers are standard.

The predominant method of priming, which is easy to do, as well as being extremely safe, is to use detonating fuse downlines to the cast primers from the surface primacord truck lines with nonelectric detonating delays between holes. In some parts of the world, there has been trouble with misfires due to poor tensile strength and PETN packing in the core of the detonating fuse. Heavier blasting to produce better top fragmentation shows up these deficiencies and, in hard rock blasting with relatively short collars, it is necessary to use high quality detonating fuse downlines. Surface trunk lines can be of lower quality, 10 g/m (40 grains per ft) and lower tensile strength. Downlines are tied to the trunk lines. Care should be taken with primacord angles to ensure they are as near right angles as possible.

In areas subject to massive ground motion along contacts, which can cause cutoffs close to the collars of the holes prior to their firing in the normal sequence, use is being made of nonelectric down-the-hole delays such as the toe-det system developed by Canadian Industries Ltd. A standard long delay (100-125 msec) on a low energy detonating cord (LEDC) is put down each hole. LEDC is used so that the primer, into which the delay is placed, will not be shot ahead of the delay by the detonating cord. The blast is then tied up in the normal manner using a normal surface delay sequence.

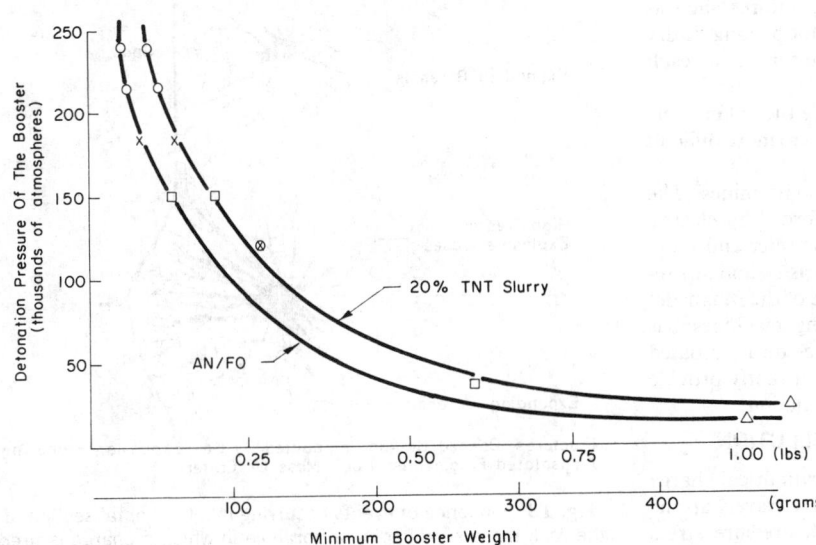

Fig. 10. Minimum booster weight as a function of booster detonation pressure (Cook, 1968).

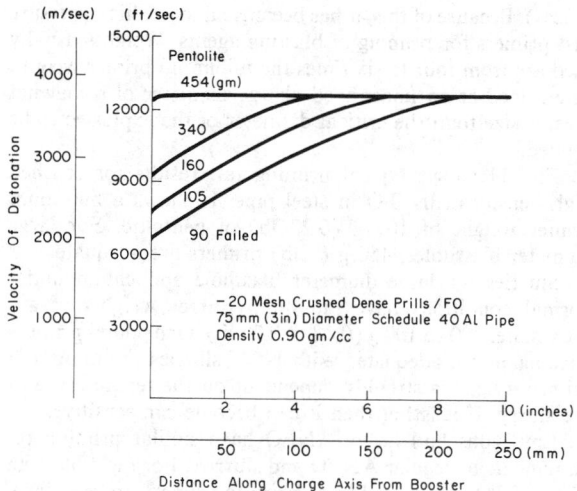

Fig. 11. Velocity run-up distances for −20 mesh crushed dense prills/FO fired in 75-mm (3-in.) diam pipes with different booster weights.

The effect of this on firing is to have several rows of surface detonating cord and downlines detonated before the first row of holes start to fire. Alternatively different delay numbers could be used in subsequent rows of holes in the blast, but this does not give any flexibility in the tie up. If the face is not dug out prior to loading the blast, this can give difficulty.

Other nonelectric delay systems include the Deckmaster produced by Atlas Powder Co. This unit uses a regular downline and has the facility to utilize any of a number of different delays between the downline and the primer.

The shock tube system, commonly called Nonel, is proving very successful. These units consist of plastic tubes containing a very light coating of explosive adhering to the inside. The detonation is of such low energy that the tube is not broken during firing. Because of the low activation energy, delays manufactured using Nonel are potentially much more accurate than cordline systems. Other advantages include virtually no blast noise and relative safety as Nonel is not susceptible to stray electric currents, etc. Surface delaying is common using Nonel. Down-the-hole use requires special tubing for explosives loaded at high temperatures such as slurries and emulsion; sleep times should not be long in dry mixes to guard against deterioration of the tube as a result of contact with fuel oil.

Gas systems have had limited use. While they offer some of the advantages of the Nonel system, they are more difficult and less flexible to use.

Electric blasting is still employed by some mines. The low noise and relatively accurate delay afforded by electric blasting caps is essentially duplicated in the safer and more flexible Nonel system. Safety aspects and easier hookup requirements have prompted the development of the Magnedet system. Also the use of sequential blasting machines has expanded the number of charges that may be delay initiated in a single blast. However electric systems currently provide only a small proportion of surface blast initiation.

BLAST PATTERN AND DELAY SELECTION

Fig. 12 indicates what happens when a cylindrical charge is fired in a borehole. As the detonation wave travels up the explosive column from the primer, a high pressure stress

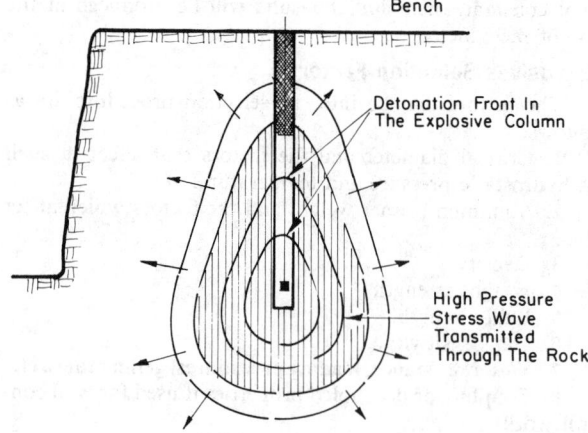

Fig. 12. Section through the face when the charge detonates.

wave travels out into the rock mass. Fig. 12 indicates the position of the detonation front and the stress wave at different time intervals. For a bottom primed charge, the stress wave envelope is pear shaped. Close to the charge, the rock fails in compression due to the high borehole pressures produced and beyond this tensile fractures are produced due to the transverse hoop stress. Fig. 13 is a horizontal section through the charged borehole. It indicates a pulverized zone close to the charge with tensile radial fractures proceeding

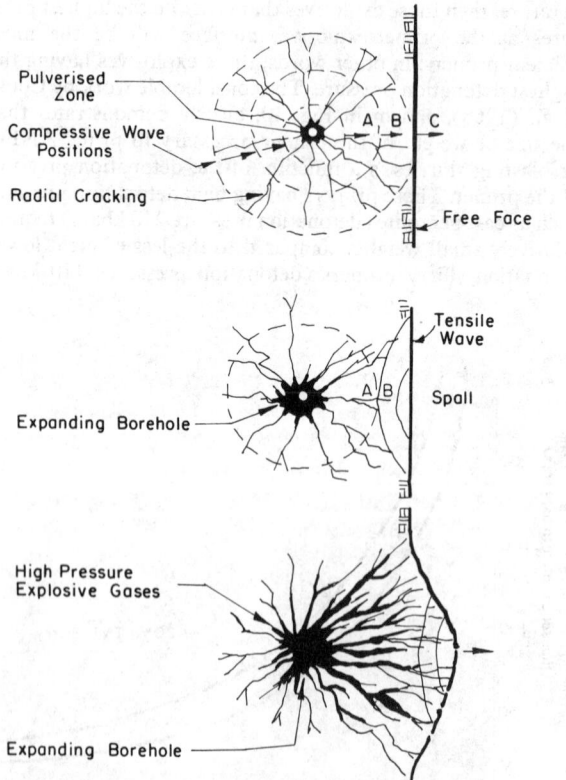

Fig. 13. Sequence of events occurring in a horizontal section of the rock mass surrounding a borehole in which a charge is fired.

beyond this. These are caused by the expanding compressive wave trying to tear the rock apart. Fig. 13 shows three successive positions of the expanding compressive wave front at different times indicated by A, B, and C, with the radial cracking proceeding behind the compressive front at a slower speed. At a free face the radial expanding compressive wave is reflected as a tensile wave and travels back towards the borehole causing a series of reflected wave tensile failures. This occurs in massive brittle rocks and, if the reflected tensile failures or slabs meet up with the subsurface failures, an isolated fragmented rock mass or crater is produced. In most bench blasts, however, the rock formations have been blasted over on the previous lift, resulting in inherent fractures being opened up to the degree that the surface reflected tensile failure is minimal and the bulk of the failure proceeds from the borehole outwards. As the radial compressive wave proceeds outwards, it produces tensile failure at right angles to it. Since the wave decays as it travels outward, this occurs close to the borehole where the tangential stress is high enough. The high pressure gases originating at the borehole wall rush into these cracks and attempt to wedge them open. If the burden on the charge is such that the compressive wave is still strong enough to produce tensile failure when it reaches the face, then failure will occur throughout the burden, and as the rock is unloaded due to the reflection of the compressive wave, it is now possible for the expanding gases to wedge open the cracks and start to expel the rock mass. In opening up the cracks, wedging action takes place, and considerably more fracturing results because of this. Since the gases have to sense the release of load prior to any appreciable wedging occurring, then the fracturing behind the charge will be less than in front of the charge; additionally, they will be tight.

For a given material type and explosive charge per unit length of hole, there is a maximum size of burden that can be used and still produce a full crater. Fig. 14 is a schematic illustration of the effect of varying the burden on a constant charge in the same formation. This is a horizontal section through the borehole to the face. In blasting rock it is desirable to detach the burden and produce a full crater. An optimum crater condition is produced in which the rock is completely isolated back to the charge. At depths deeper than this, a shelf of unbroken material remains between the

charge and the free face. For a multiple hole shot, it is possible that this material could be isolated by poor fragmentation. At shallower depths than optimum, the crater is reduced in depth, increased in radius, volume is reduced, fragmentation is finer, the crater assumes more of a bowl than a trumpet shape, and flyrock and noise start to become appreciable. The sequence of events is the same for craters from both spherical as well as off the sides of cylindrical charges. For craters from spheres in brittle rocks, the optimum true crater occurs often at scaled depths of burial, $d/W^{1/3}$ equal to 0.8 m/kg$^{1/3}$ or 2.0 ft/lb$^{1/3}$, approximately. The factor, $d/W^{1/3}$, is defined as the distance from the rock surface to the charge center (d), divided by the charge weight (W). The contained depth is at a scaled distance of 1.6 m/kg$^{1/3}$ or 4.0 ft/lb$^{1/3}$.

As the material gets softer, the optimum scaled depth increases. Fig. 15 shows the scaled crater depth vs. scaled depth of burial curve for craters produced off the sides of cylindrical charges fired in massive quartzite. The scaled optimum depth is in the range of 1.2 to 1.45 m/kg$^{1/2}$ (2.7 to 3.2 ft/lb$^{1/2}$) and containment occurs at a scaled depth in the range of 2.25 to 2.7 m/kg$^{1/2}$ (5.0 to 6.0 ft/lb$^{1/2}$). Scaled depth in this instance is the depth to the center of the charge divided by the half power of the explosive weight per unit length of hole. For softer rocks the optimum would move to increased values of scaled depth. By way of interest, frozen materials have curves that go from optimum crater sizes to no crater much more rapidly than a brittle rock. The optima are similar.

From Fig. 16 it can be seen that the scaled optimum crater radius occurs at a scaled depth of burial which is close to that for the optimum in depth. Fig. 17 has the same data plotted so that the change in the crater radius depth ratio can be seen. At the optimum crater condition this has an average value in this formation of approximately 0.75. However, as the burden is reduced, this continues to increase. For this reason in hard rocks there is considerable advantage in shooting blasts en echelon if they are laid out in a square pattern or along the long axis if the pattern is staggered. This produces a reduced burden, more displacement, and better fragmentation. There does not appear to be any advantage in going to spacing to burden ratios greater than those encountered with such blasts. Such schemes would lead to more complicated blasthole layouts and, if laid out in a row by

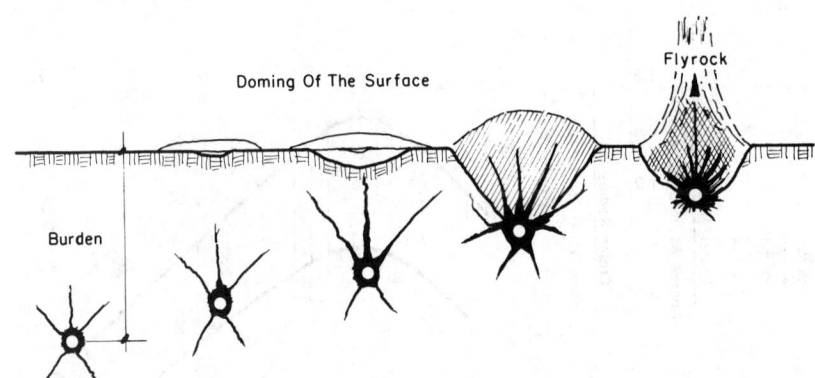

Doming Of The Surface

Flyrock

Burden

Completely Contained, Only Failure Is Pulverisation Near The Charge And Radial Tensile Failure Running Out From It.

Start Of Surface Failure. Burden Not Broken. Some Doming Of The Surface.

Surface And Subsurface Failure Almost Meet There Will Be A Shelf Of Unbroken Rock Between The Two. Doming Or Surface Bulging.

Full Crater, Burden Completely Broken Out. Surface And Subsurface Failures Run Through To The Surface.

Full Crater, Lower Volume Than Optium Fine Fragmentation. Noise, Flyrock, Bowl Shaped Crater.

Fig. 14. Schematic of the effect of decreasing the burden on charges fired in rock.

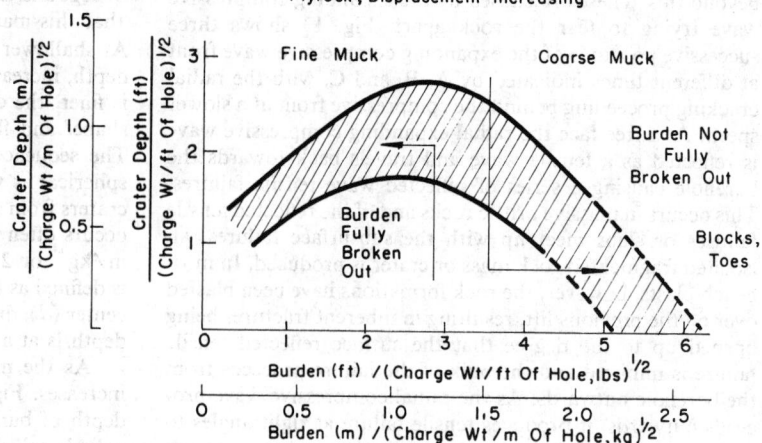

Fig. 15. Change in crater depth produced by different burdens on cylindrical charges fired in quartzite.

row manner, the excessive hole spacing along the front of the blast would make for difficulties, also along the back and sides. This then calls for a considerable amount of extra drilling if the tonnage per blasthole is the same with just changes being produced in the burden to spacing ratio. The en echelon firing has the advantage of keeping the blast more confined along the front, causing less scatter of the muckpile. With 250-mm (9⅞-in.) holes on a 7.6 × 7.6-m (25 × 25-ft) pattern and an explosive loading of 60 kg/m (40 lb per ft) of hole, the scaled burden for the body of the charge would be 1.6 m/kg^½ (3.6 ft/lb^½) if fired en echelon. The curves indicate a noticeable difference in result and this is seen in practice. In soft materials the difference is not as pronounced.

In order for the crater to be produced and for the material to start to be affected, there is a necessary lapse time. If subsequent delay periods are fired prior to this occurring, then the shot starts to become choked. The pit floor conditions fall off, the muck is tight, and in many instances an inadequate delay time interval will result in a significant increase in flyrock from the top of the blast and back throw behind the blast. For some years this minimum delay time

had been receiving considerable attention by Chung, et al. (1975). In the absence of an adequate theory, they began photographing using a dual camera system, numerous blasts to observe the time at which face motion started in bench blasting and also the time and velocity with which the top of the bench started to move. It will be seen that framing rates of the order of several hundred per second are adequate for this work. Fig. 18 shows a plot of a bench face profile from such movies and Fig. 19 a distance time plot for different burdens. Note the time lags before any motion starts to take place. In Fig. 20 similar data are plotted for small diameter blastholes and therefore small burdens. The interesting outcome of this is shown in a plot of time before the face starts to move vs. the burden for holes of different size, Fig. 21. This indicates that the minimum delay time for burden detachment in rock should be no less than 3 millisec/m (1 millisec/ft) of burden. In many instances, there is an advantage in extending the delay period to greater values to produce better lateral relief. These conclusions are in agreement with rules of thumb that have generally been employed.

The upper limit to the delay time that can be used is imposed by the possibility of cutoff holes. Factors influencing

Fig. 16. Change in crater radius produced by different burdens on cylindrical charges fired in quartzite.

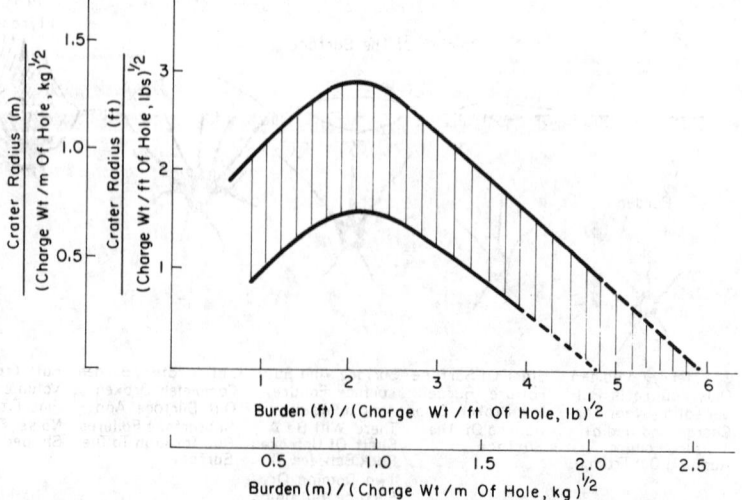

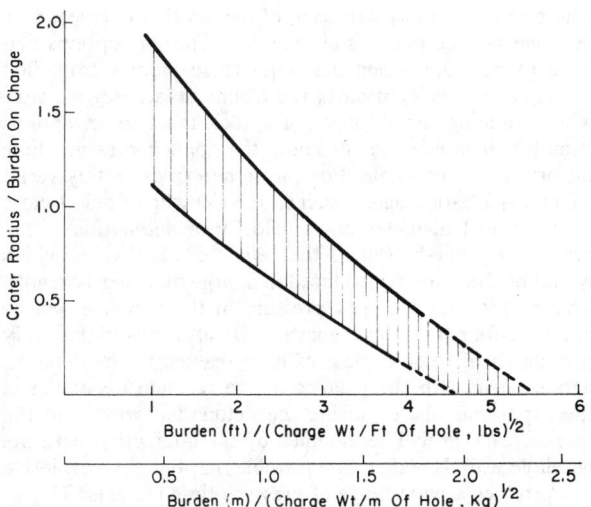

Fig. 17. Effect produced on the crater radius/burden ratio with different burdens on cylindrical charges fired in quartzite.

this are the collars on the holes, the spacing of the holes, the charge per unit length of hole, and the material type. Once again, high-speed photography can be used to determine the time delay that occurs in blasting before there is appreciable top movement. Fig. 22 shows typical field data. From these movies, it is also apparent that one can consider the top eight diameters of the cylindrical charge to act like a sphere located four diameters down from the collar. In Fig. 23, the delay time before the top of the shot starts to move is plotted vs. scaled depth of burial of the equivalent spherical charge in rock. One can use this to determine what the scaled distance is from the collar of the hole on the next delay period to see if the maximum permitted delay time is greater than that required to produce the separation. If not, the collar or initiation system has to be adjusted.

Other methods used to determine the maximum permitted delay interval are concerned with the particle velocity generated at the collar of the next row of holes and the amount of elongation that detonating cord can take prior to its failure. This requires assumptions in the length of downline which is stretched, since if one assumes that a 4%

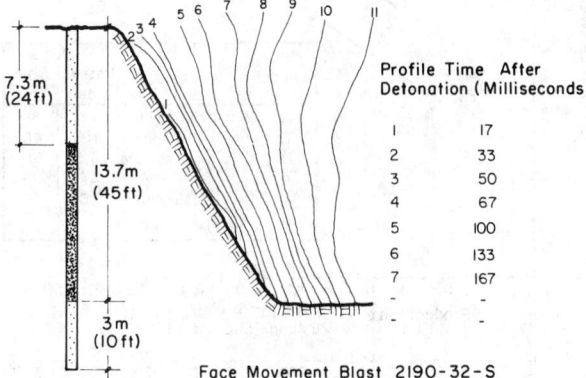

Fig. 18. Profiles seen with high-speed cameras used in the determination of required delay times.

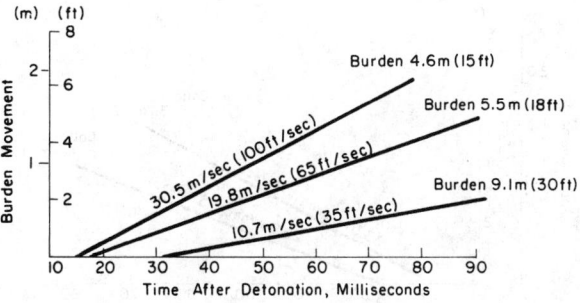

Fig. 19. Burden movement in large-hole blasting 250 mm (9 ⅞ in.) in hard rock.

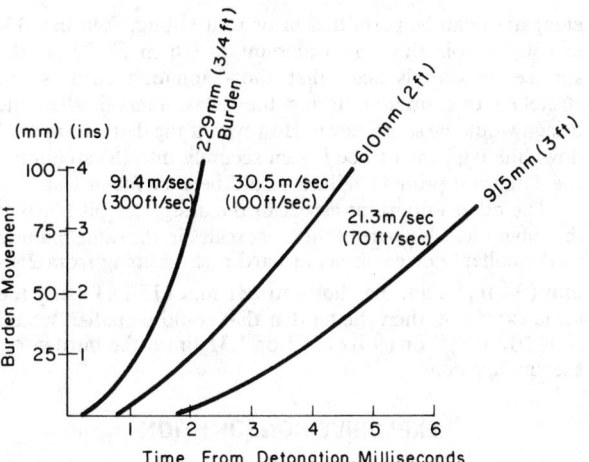

Fig. 20. Burden movement in small-hole, 32 mm (1 ¼ in.) blasting.

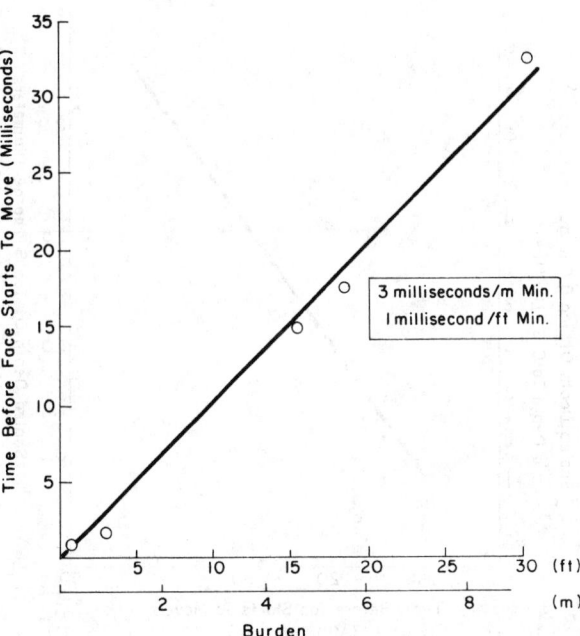

Fig. 21. Onset of movement for burdens of different size.

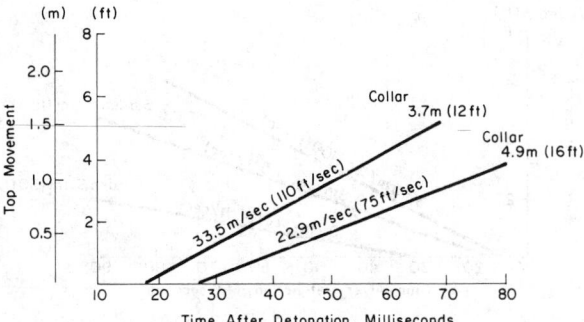

Fig. 22. Top movement in large-hole blasting, 250 mm (9⅞ in.) using high density slurry.

elongation can be permitted prior to it failing, then in a 15-m (50-ft) hole this could amount to 0.6 m (2 ft). If the surface velocity is such that the detonating cord is not stretched this amount during the delay interval, then the design would be satisfactory. However, if the detonating cord downline happened to be frozen securely into the stemming the extension prior to failure would be very much less.

The other important aspect in the design for pit blasts is the subgrade. This part of the blast scales in the same manner as the collar. For example, in hard rock in going from 250-mm (9⅞-in.) diam blastholes to 281 mm (15 in.) using the same explosive, then the burden that could be pulled would be $(250/381)^{2/3}$ or $(9⅞/15)^{2/3}$ or 1.31 times the burden on the smaller hole.

EXPLOSIVE CONSUMPTION

In recent years attempts have been made to model blast designs, in some cases right from first principles. The authors of some of these models have made great claims to their capability and in some instances they have even decried the common present day approach of the blasting engineer who uses energy factors in his blast design. The assumptions that have to be made when designing these models from first principles in reality negates the claims made. For example when studying the response of a rock mass to an applied stimulus, it is essential to know the applied pressure time history in the borehole. For the same explosive this varies widely. A blasting agent such as ANFO when fired at close to its critical diameter has a velocity of detonation of approximately one-half of its theoretical or *ideal* value which would be the velocity produced in a large diameter borehole. This means that the peak pressure in the borehole will be only one-fourth of the theoretical. It also means that only one-fourth of these explosives have reacted in the determination wave with the balance of the reaction occurring in the expansion phase outside the detonation wave and the rate of reaction will be modified by the interaction with the borehole wall. It is therefore possible that the same explosive can produce a wide range of pressure time impacts. This in turn will produce a wide variation in hoop stress around the borehole which will influence the degree and content of the radial cracking. Models built from first principles have ignored the explosive property variability. The next difficulty with the models that are claimed to produce fragment size distributions is that of the number of cracks in the major fracture sets (one such set would be those radiating from the borehole wall), their orientation, and extent. Difficulties will obviously arise if the assumption is made that the material is homogeneous and isotropic.

One way out of this difficulty is to use an empirical type of approach in which size distributions are determined in the field using photographic methods or photographic examples of known size distributions as a comparison base. Providing all of the important items such as the geometric, geological, delay time and sequencing, operational control,

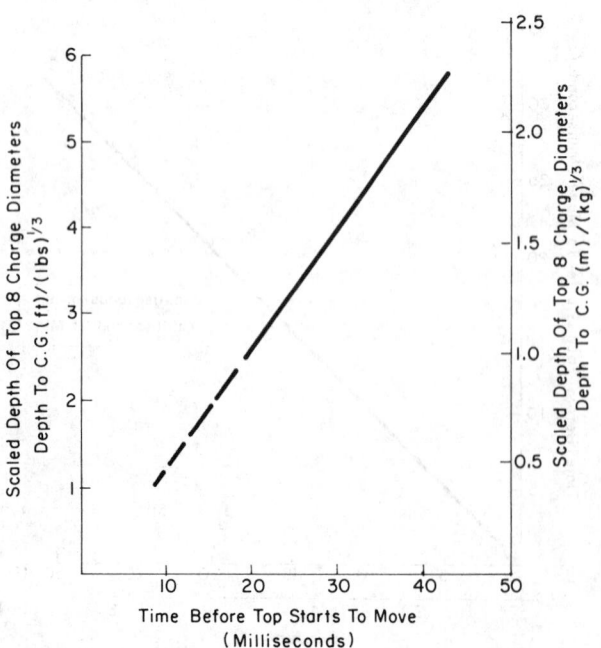

Fig. 23. Delay time prior to bench top motion.

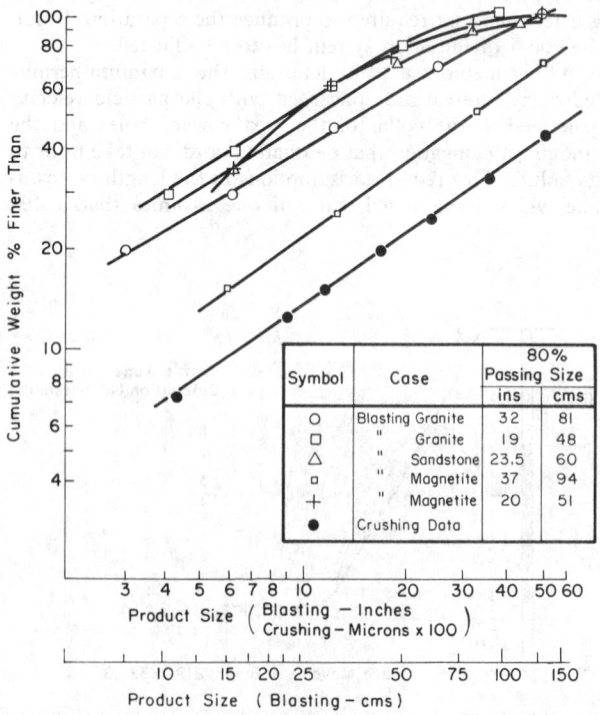

Fig. 24. Size distributions from blasting and crushing operations.

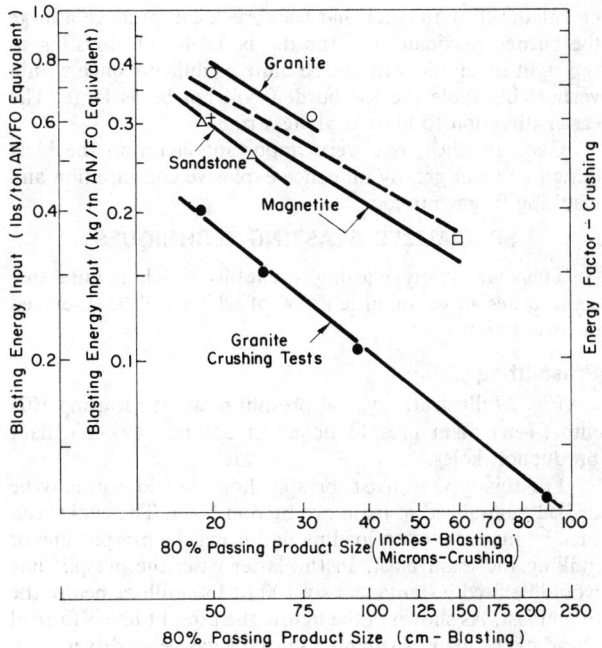

Fig. 25. Energy-size plot of the data from Fig. 24.

where E is the energy input in lbs ANFO or equivalent ANFO per ton of rock, K is the 80% passing fragment size, A is the blasting work index, and $K_1/K_2 = \{E_2/E_1\}^2$ in the same material.

The ratio of the 80% passing size is equal to the reciprocal of the energy input/ton squared.

The use of the 80% passing size is typical of crushing operations. For blasting operations, it is necessary to use a much higher percentage than this, particularly as it refers to the undersized material in a large tonnage operation. Fig. 26 shows some of the difficulty with this in that the curve at the high percentages in many cases is nonlinear. It also indicates the difficulty in interpolating close to 100%. Nevertheless, curves such as these do provide useful information and an estimate can be made of the 100% passing size for a given explosive consumption. This can be used with the preceding equation to estimate the 100% passing size at any other energy input. This has been plotted for sandstone and also for a soft siltstone in Fig. 27, which is a plot of the maximum fragment size vs. explosive consumption. By picking off the fragment size and its corresponding explosive consumption from the curve, the energy required to produce any other size can be calculated. The results from this approach are not perfect but are reasonable if it is recognized that chunks can be obtained off the face, sides, and back of a blast. It can be a good and valuable guide.

Cunningham (1982, 1983) has shown that other size distribution relationships can be used with success to predict or modify size distributions produced by blasting. To do this they have made use of the Rosin-Rammler and Kuznetsov (1973) equations in the development of their KUZ-RAM model for predicting size distributions.

As illustrations of the types of geological effect on fragmentation, Fig. 28 is a sketch of a horizontal slice from the charged borehole to the rock face. In case 1, the charging and burden are such that a full crater would be produced. The rock structure is such that it has no influence on the crater. In case 2, there is a parallel set of fractures running at a small angle to the direction of blasting. The resulting crater is influenced by this because of the energy being reflected by these planes thereby being trapped in a narrow region. The crater which is produced is reduced in radius and the burden that can be broken out fully right back to

etc., are observed in the blasting operation, then the resulting size distribution of the muck can be related to the energy input.

Fig. 24 is a plot of the product size vs. cumulative weight percent finer than. The results are compared to those of typical crushing operations. The similarity of shape and slope is readily apparent.

Fig. 25 is a plot of the 80% passing size vs. energy input for blasting operations. The dotted line in Fig. 25 defines the relationship between the energy input and the 80% passing product size, and can be determined for this portion of the line from

$$E = \frac{A}{K^{1/2}}$$

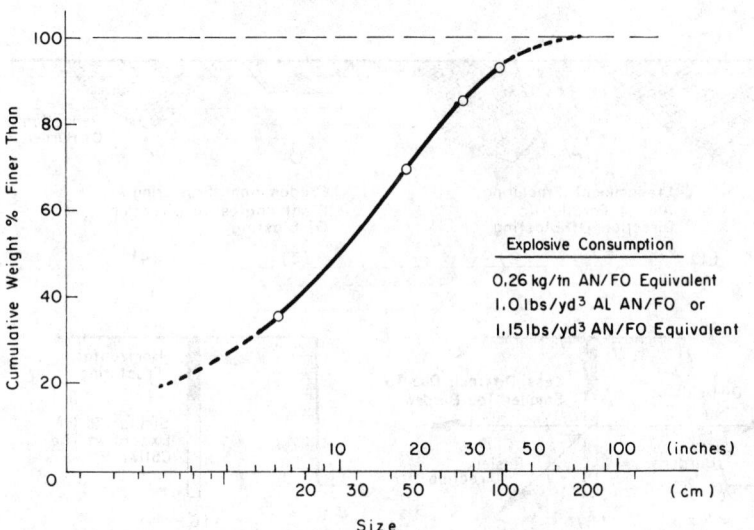

Fig. 26. Estimated size distribution from a sandstone blasting operation.

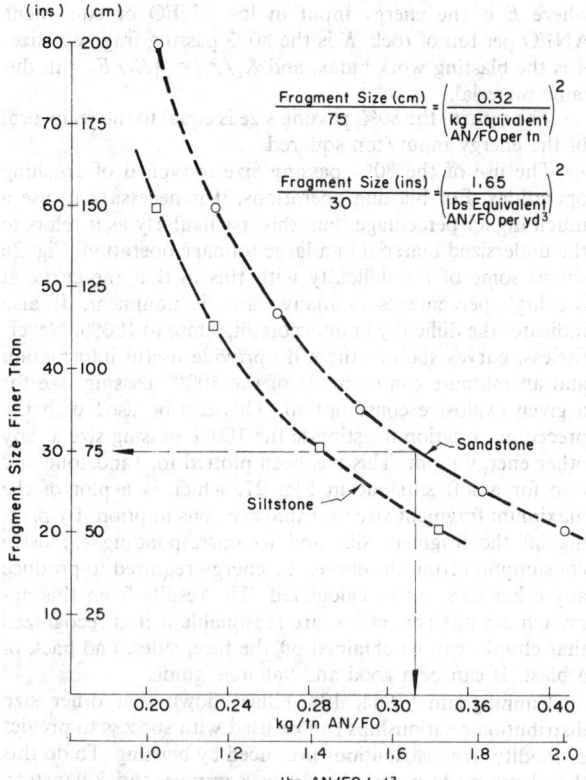

$$\frac{\text{Fragment Size (cm)}}{75} = \left(\frac{0.32}{\frac{\text{kg Equivalent}}{\text{AN/FO per tn}}}\right)^2$$

$$\frac{\text{Fragment Size (ins)}}{30} = \left(\frac{1.65}{\frac{\text{lbs Equivalent}}{\text{AN/FO per yd}^3}}\right)^2$$

Fig. 27. Fragment size vs. explosive consumption for a sandstone and siltstone.

the charge can be increased slightly. In case 3, the predominant rock fracturing is at right angles to the direction of blasting and this makes for difficulty in hard rocks with wide fracture spacing. The fracture planes reflect the energy and make it more difficult to blast in this direction. The burden has to be reduced. If the same burden is used as in case 1, then it will not be fully detached and there will be sections of unblasted material. In case 5, if a horizontal section is run through the borehole, the blasting to the left or down dip will be difficult and similar to case 4. The difficulty can

be enhanced by the fact that backbreak can produce a large toe burden particularly if the dip is 35° to 50°. Blasting to the right or up dip will not be quite as difficult since with a vertical blasthole the toe burden will not be as large. The easier direction to blast is along strike.

Geology, then, is a very important factor in the blast design and can greatly influence explosive consumption and resulting fragmentation.

SPECIALIZED BLASTING TECHNIQUES

There are many blasting situations which require specialized design techniques, a few of which will be discussed in this section.

Presplitting

Fig. 29 illustrates typical presplit blast layout using 102-mm (4-in) diam presplit holes for 250-mm (9⅞-in) diam production holes.

For this type of blast, presplit holes would normally be drilled first, ahead of main production holes. The choice can then be made between loading and firing the presplit line or infilling the main blast. In the latter case, the presplit line would be fired instantaneously 100 to 150 millisec before the main blast. As shown in the figure, the presplit line is formed ahead of the main blast and allows the gas being driven back from the buffer row through the radial cracks to terminate at the presplit line.

The presplit row in Fig. 29 has a spacing of 2 m for a 102-mm (4-in.) diam hole and is inclined at 15° to the vertical. The presplit angle is somewhat dictated by rock structure although a slight angle is preferred regardless of structure for long-term stability as well as for best initial results with large production holes.

The figure illustrates the upper bench where two benches will finally run together to form the final face between berms. Presplit drill requirements become clear when presplit holes needed for the next bench are considered. The drill must be capable of drilling close to the previously produced bench face at an angle of 15° beneath itself so the face can be continued to depth. Currently, this means some form of drifter drill is required limiting the hole size to 102 to 127 mm (4 to 5 in.) diam.

The back row of the main production blast, termed the buffer row, must also be carefully designed with respect to standoff distance from the presplit row and spacing as well

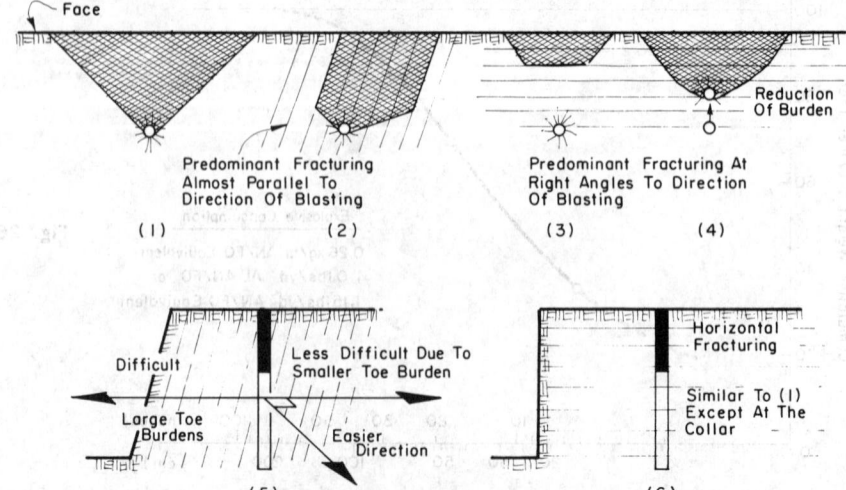

Fig. 28. Illustrations of the effect of rock structure on crater formation.

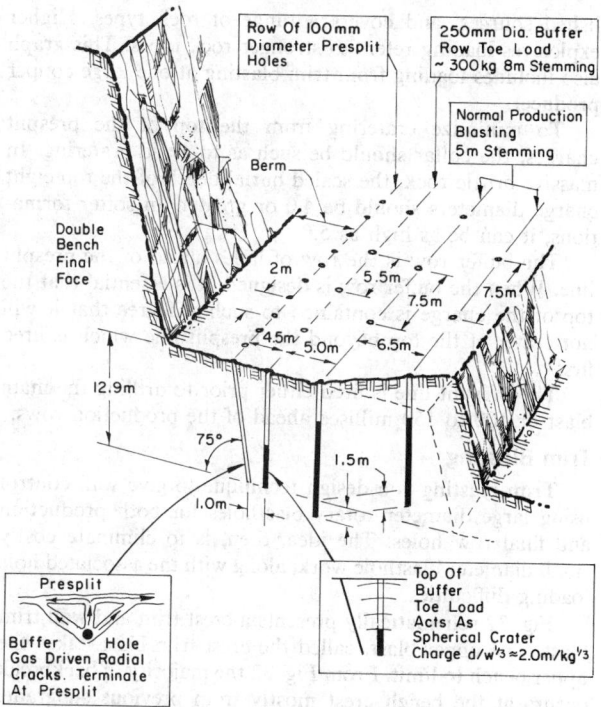

Fig. 29. Presplit blast coupled to a 250-mm (9 ⅞-in.) diam production blast.

as explosives load. The inset sketch on the right side of Fig. 29 shows how the top portion of the buffer row hole charge acts as a spherical crater charge breaking to the bench surface. Subsequently, main blastholes after the buffer row are designed at regular spacing, burden, and loading for the type of material blasted.

One further point to note from Fig. 29 is the subgrade or, more accurately, lack of subgrade used on the presplit and buffer row holes. This is to prevent damage to the bench below or to the wall at that point.

Fig. 30 is a graph of recommended hold spacings used in presplitting as a function of hole diameter. These data cover various types of material, but are not fully specified in the open literature.

According to the *Pit Slope Manual* (1977), hole spacing is defined as:

$$S = \frac{\phi (Pw + T)}{T}$$

where ϕ is hole diameter, Pw is pressure at the borehole wall, and T is rock tensile strength.

For a decoupling ratio producing a constant borehole wall pressure in a given rock type, hole spacing is proportional to the blasthole diameter.

Line of constant spacing/hole diameter, ranging from 8:1 to 14:1, have been included in Fig. 30. Generally, published references also show a linear relationship between hole diameter and hole spacing.

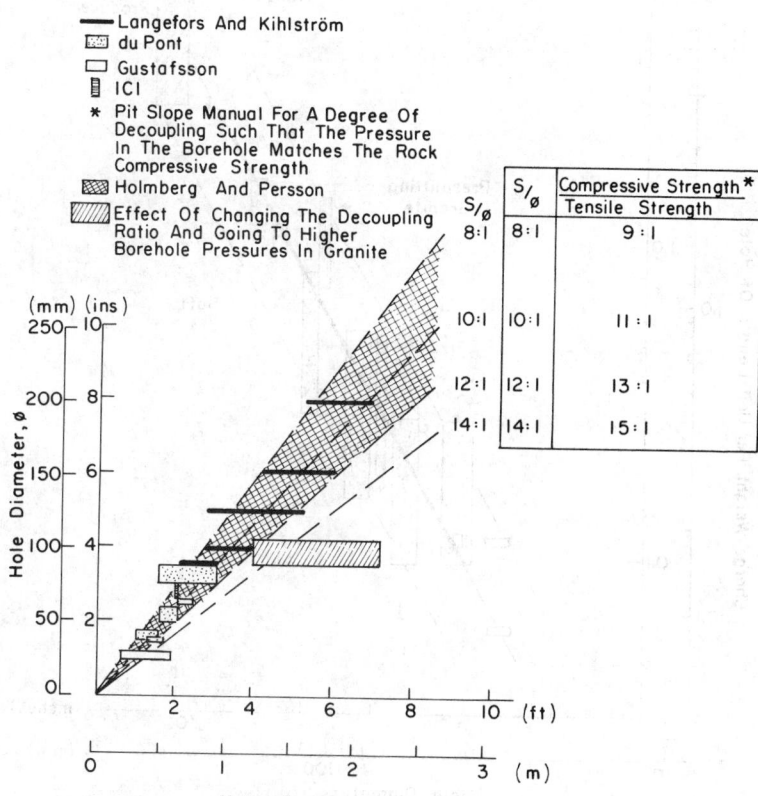

S/ø	Compressive Strength*
	Tensile Strength
8:1	9:1
10:1	11:1
12:1	13:1
14:1	15:1

Fig. 30. Published relationships between hole diameter and spacing for presplitting.

From the spacing equation, if pressure at the borehole wall is made equal to the dynamic compressive strength of the rock mass, then the spacing must be less than or equal to the ratio of the rock's compressive-to-tensile strength plus 1. For many strong rocks, this ratio ranges from 12 to 15:1.

The first requirement for presplit line loading is adequate decoupling in presplit holes to eliminate crushing. An airspace around a charge is the best way of achieving this. The explosive gases expand out from the charge adiabatically, and the factor of reduced pressure can be expressed as the ratio of the charge diameter to the hole diameter to the 2.6 power, being twice the ratio of the specific heats of the explosive gases, about 1.3. The pressure in bars at the side of the explosive charge can be calculated from the velocity of detonation, D in m/s and the charge density, in g/cm^3, by the following equation:

$$P_{\text{side charge}} = \frac{(0.00987) \, \rho \, D^2}{8}$$

and the pressure at the wall of the borehole in bars is then:

$$(0.00987) \, \rho \frac{D^2}{8} \left\{ \frac{(\phi c)}{(\phi h)} \right\}^{2.6}$$

where ϕc is charge diameter and ϕh is hole diameter.

Fig. 31 is a graph of the charge weights per unit length of borehole vs. the presplit hole diameter from various published sources, and covers a range of rock types. Higher explosives loading refers to stronger rock types. This graph also includes loading from trim blasting at one large copper producer.

To minimize cratering from the top of the presplit charges, the collar should be such as to avoid cratering. In massive brittle rock, the scaled burial depth of the top eight charge diameters should be 4.0 or greater. In softer formations, it can be as high as 5.0.

The buffer row is the row of holes ahead of the presplit line. When the buffer row is designed, it is essential that the top of the charge is contained to such a degree that it will not crater at the top beyond the presplit line which is fired first.

The presplit line is fired either prior to drilling the main blast or 100 to 150 millisec ahead of the production rows.

Trim Blasting

Trim blasting is a design technique to give wall control using large diameter rotary blastholes for both production and final row holes. The idea, then, is to eliminate costly small diameter blasthole work, along with the associated hole loading difficulties.

Fig. 32 schematically presents a crest trim and wall trim blast. The upper blast, called the crest trim blast, takes the upper bench to limit. From Fig. 32 the majority of backbreak occurs at the bench crest mostly from previous subgrade. The blast has three distinct components, similar to the presplit blast. A trim row is used to produce the final wall in

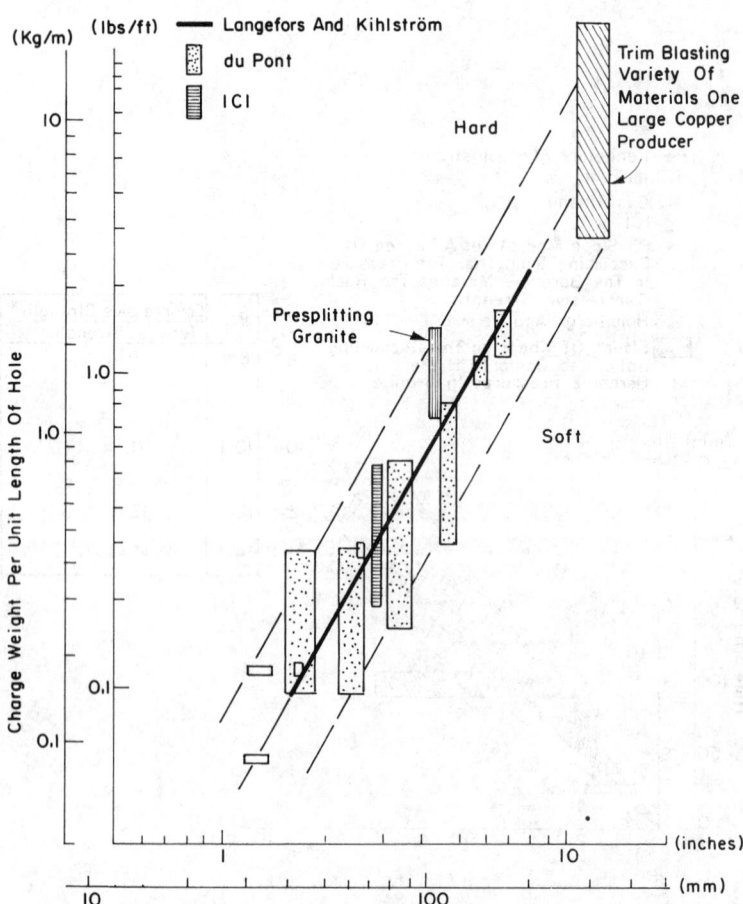

Fig. 31. Published charge weight per unit length of borehole for wall blasting.

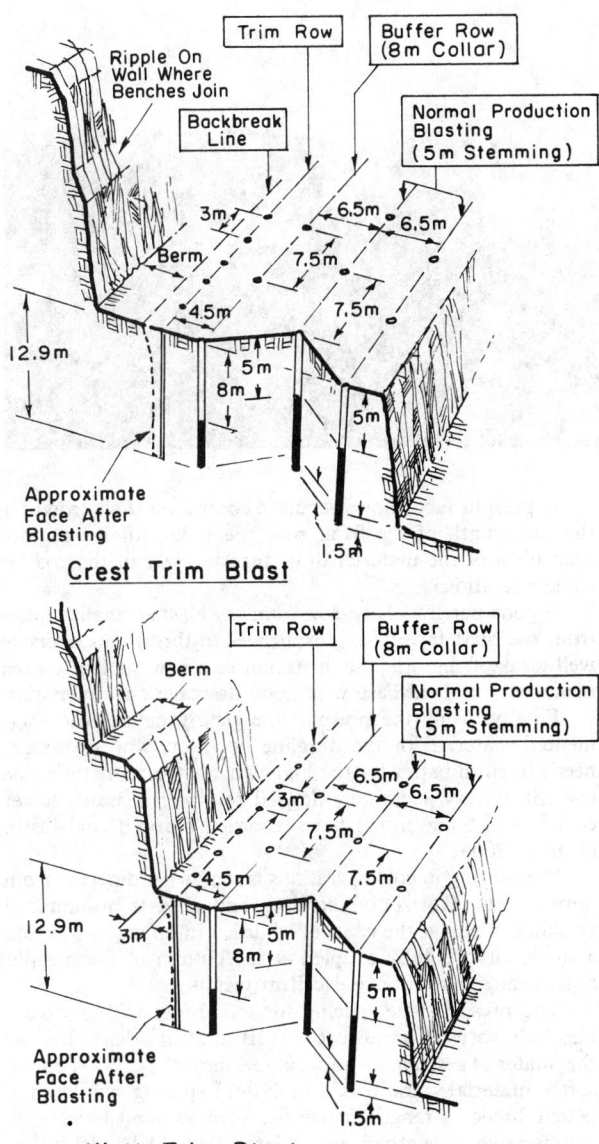

Fig. 32. Crest and wall trim blast using 250-mm (9⅞-in.) diam holes in hard brittle rock.

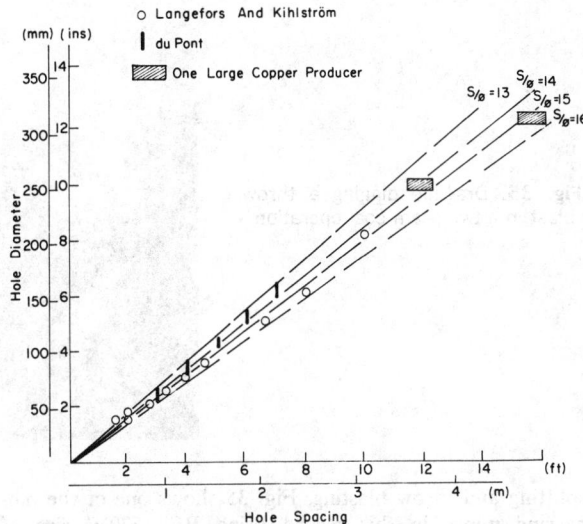

Fig. 33. Relationships between hole diameter and hole spacing for trim blasting.

a similar manner to the presplit row. The trim row is decoupled as shown in Fig. 31. If stemming is used, then the decoupling calculation is performed using 40% to 50% voids. Again, a buffer row is designed as the last row of the main blast, with increased stemming to prevent cratering back at the surface through the trim row. Normally, two other regular rows of holes would be used in front of the buffer row to complete the trim blast. These two rows would be at normal spacing and burden and would be loaded using the standard procedure for the appropriate material type. All holes would be vertical and would be of production size. For example, in the case of Fig. 32, holes are 250 mm (9⅞-in.) diam.

Fig. 33 is a graph of hole spacing used in trim blasting as a function of hole diameter from literature references. Spacing generally ranges from 12 to 16 times the hole diameter. These values correspond to a range of rock hardness

from hard to moderately soft and are for adequately decoupled charges.

The second bench blast used to complete the two bench face, termed the wall trim blast in Fig. 32, is shown on the lower portion of the figure. The blast is similar to the crest trim blast, except the trim row is now 3 m (10 ft) from the toe position. This distance is dictated by drilling equipment size. Some backbreak from the wall trim blast makes the final wall almost continuous, with a slight ripple remaining where the two benches join. It should be noted that this trim blasting technique does not prove effective where the rock structure dips at a shallow angle into the pit.

Presplitting and Casting in Dragline Operations

Fig. 34 illustrates the relationship between drilling and blasting costs, dragline productivity and overall unit costs. A number of mines are now attempting to achieve minimum overall cost by adjusting their drilling and blasting cost to the optimum. In many cases the optimization includes pre-

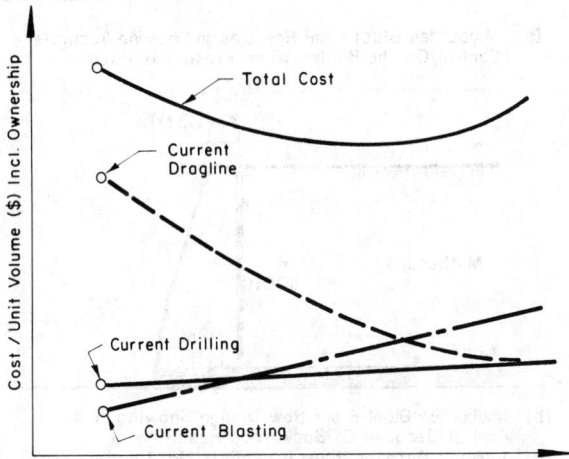

Fig. 34. Overall unit cost vs. dragline productivity projection for different blasting costs.

Fig. 35. Dragline digging a throw blast in a two-seam coal operation.

splitting and throw blasting. Fig. 35 shows one of the pioneering mines in this regard using BE 1570W size of draglines.

One major reason tor presplitting, apart from improved safety, occurs at operations with water problems. The presplit commonly dewaters the blastholes, which enables easy use of dry mix explosives rather than the more expensive slurries. When presplitting a 50-m (165-ft) wide pit for a 300-m (1000-ft) length, the entire presplit block commonly moves up to 1 m (3 ft) towards the mined cut. Main blastholes drilled after the presplit has been fired are usually dry, even in areas expected to be filled with water after drilling.

The second major reason for presplitting is to enhance the throw blasting technique. Fig. 36 shows the reason.

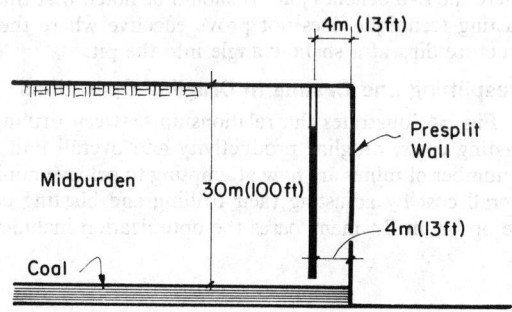

(a) Midburden Blast Front Row Design Showing Accurate Control On The Burden When Presplit Is Used.

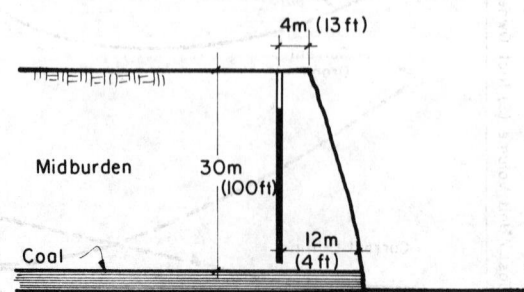

(b) Midburden Blast Front Row Design Showing Poor Control Because Of Backbreak Resulting In Varying Burdens When Presplit Is Not Used.

Fig. 36. Comparison between front row burden with and without a presplit.

A presplit face allows accurate control on the burden for the full length of the front row blastholes, allowing more than 60% of the material to be thrown clear of the coal at some operations.

A good portion of oversize from any blast normally comes from the blast face. Using explosives to throw this oversize well clear of the dragline operation to a new spoil pile area makes an excellent base with good drainage characteristics.

Finally, using the presplit line clearly defines the fragmented material for the dragline operator, who no longer needs to cut a batter for the highwall. At the same time, the coal rib is very accurately defined resulting in much lower coal losses. Some mines have recorded reduced coal losses of up to 90%.

Presplitting in coal operations is somewhat different from normal pits. Horizontal bedding and a very pronounced weakness plane at the coal/overburden interface means that a single charge, fully coupled at the bottom of the presplit hole, is adequate to give excellent results.

The presplit hole spacing for a 250-mm (9⅞-in.) diam blasthole varies from 3 to 6 m (10 to 20 ft) depending on the material's strength and competency. Highly fractured softer materials require a 3-m (10-ft) spacing, while competent higher strength materials, such as sandstones with compressive strength from 82.7 to 103.4 MPa (10,500 to 15,000 psi) are satisfactorily presplit using a 5- to 6-m spacing.

Presplit holes are drilled into the coal and loaded with 60 kg (130 lb) of explosives (usually slurry in wet conditions). The charge is placed on the hole bottom or suspended just above the bottom by 1 to 1.5 m (3 to 4 ft) using a borehole liner. No stemming is used, and it is preferable to dewater if possible. The presplit row is then fired instantaneously, prior to any main blasthole pattern.

The presplit line is drilled around the complete proposed blast area. The cutoff to the face is made at the blast cutoff angle, which is usually either 45° or 60° with the face. Once an area has been presplit, the entire block of material must then be blasted to avoid poor fragmentation at blast interfaces. This main overburden blast is shot in the normal manner with a 3 to 4 m back row standoff from the presplit line for a 250-mm (9⅞-in.) diam hole. Fig. 37 presents an overall plan view of the presplit and main blast showing the method of delaying and relative positions of the blasts. If operating and scheduling constraints warrant, the presplit blast may be drilled and fired by itself.

An important control on the quantity of throw is the blast depth/width ratio (McDonald, et al., 1982). Fig. 38

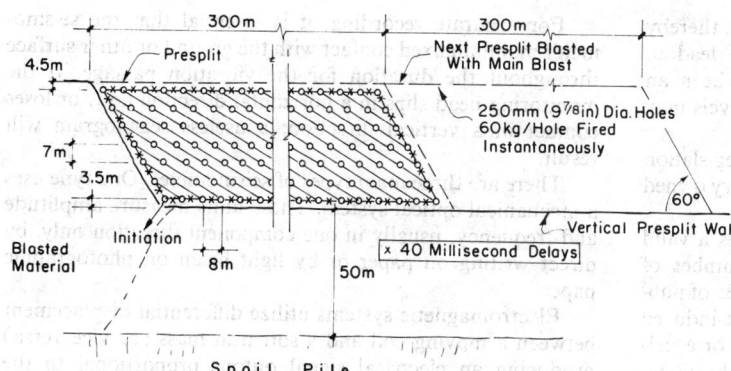

presents the relationship between percentage blast throw vs. blast depth/width ratio experienced at Rietspruit Opencast Services.

Crater Blasting

Cratering using spherical charges finds wide application in surface mining operations. For example, crater blasting test results are commonly used to design the correct blasthole collars either for fragmentation or flyrock control.

Other more specialized applications include blasting tar sand to prevent frost penetration during winter operations, coyote blasting, and the disruption of planes of weakness in slopes to aid stability.

BLAST VIBRATION AND AIR BLAST

The principal factors that affect vibration levels at a given point of interest are the weight of explosive fired, the distance from the blast, the delay period used, if any, and the blast geometry. The type of explosive, although weight strengths can differ by 40% or more, has not been shown to significantly affect vibration levels other than for very close measurements [6 m (1.8 ft) or less]. Geology does materially

affect levels of vibration, but due to each measuring site being unique, it must be evaluated on a site by site basis.

In general, a scaling factor based on distance is used, being derived from the effects of geometrical dispersion of the outbound ground motion from the explosion. The total energy introduced into the ground by the charge detonation varies directly with the charge weight. The vibration propagates outward from the blast such that the volume of rock subjected to the compressional wave increases. Since the energy in the ground is distributed throughout successively larger volumes of rock, the peak ground motion levels must decrease.

Considering bench or overburden blasts having relatively deep blastholes, the explosive column approximates the shape of a cylinder having a much greater than 6:1 length to diameter ratio. The expanding wave front from charges such as these adopt an expanding cylindrical shape. The volume of this compression cylinder varies as the square of its radius. Thus the peak particle motion at any point from this type of shot will be inversely proportional to the square of the distance from the blast.

The scaled distance, therefore, combines the effects of charge weight (W) on the geometrical dispersion of the vibration at distance (d) in the form $d/W^{1/2}$. An empirical equation of the form

$$V_{max} = K \left(\frac{d}{W^{1/2}} \right)^m$$

relating the peak particle velocity with scaled distance must therefore be developed. Local site factors for each vibration component, K and m, allow for the influence of geological characteristics on the peak particle velocity and can be determined from a logarithmic plot of peak particle velocity versus scaled distance.

This scaling relationship must be further modified when delayed blasts are to be considered. The explosive weight becomes the explosive weight per delay. This is the total explosive weight fired instantaneously. Delays of 15 millisec or greater have been found sufficient length to isolate individual detonations.

Vibration Effects

The problem of vibration from blasting is one of the most pressing facing mine operators in populated areas, both from a potential structural damage claim situation and from the standpoint of the annoyance effect on people.

Generally approaches to the problem have been made in two ways. First, attempts have been made to predict vibration

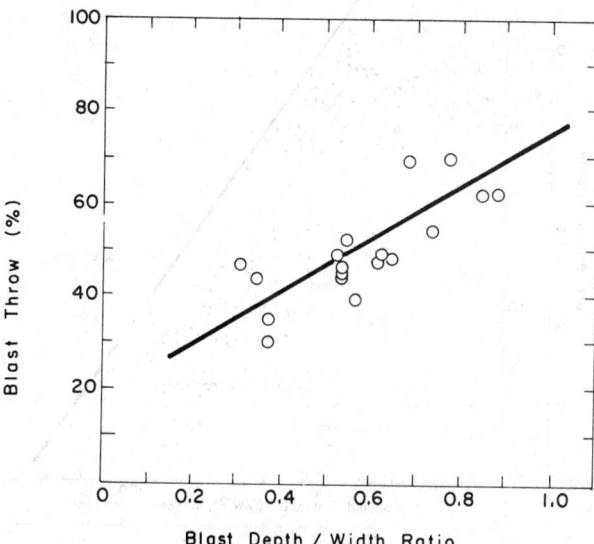

Fig. 38. Relationship between percentage blast throw vs. depth/width ratio experienced at Rietspruit Opencast Services (McDonald, et al., 1982).

levels from blasts and to correlate these with damage, thereby permitting the avoidance of conditions that may lead to damage or annoyance. The second approach has been an investigation of methods whereby blast vibration levels may be reduced.

This in turn has led in many instances to legislation concerning blasting in proximity to nonmine or quarry owned property such as dwellings or plants.

In recent years the question of what constitutes a valid criterion for damage has been discussed by a number of authors. Extensive statistical reviews have been made of published vibration data to determine which of the blast-induced parameters such as particle displacement, velocity, or acceleration best measure the damaging effects of a vibrational wave on a structure.

The current conclusion is that damage can best be correlated with peak particle velocity of the wave regardless of the direction of the motion; longitudinal, vertical, or transverse (Edwards and Northwood, 1960; Langfors, et al., 1958; Nichols, et al., 1971). Because of this, some investigations use the peak resultant particle velocity. It has also been concluded that due to the complex nature of the waves generated during blasting, such as P, S, Rayleigh, Love waves, etc., that it is impossible to correlate damage theoretically with a specific wave form. Because of this, it is usual to lump all of the waves together by measuring the composite wave form and to correlate the results empirically.

Some investigations have given consideration to the frequency response spectra of structures which is sometimes being used. For general application this is not necessary.

Blast Vibration Measurement

In the measurement of vibration, the objective is to detect and record the vibratory motion of the ground. The quantities measured must result in a full description for the vibratory event. This requires three orthogonal components of either particle displacement (x), velocity (v), or acceleration (a), to be recorded as a function of time (t). The three components are termed:

1. Longitudinal, normal to the direction of the blast in the horizontal plane.

2. Transverse, perpendicular to the direction of the blast in the horizontal plane.

3. Vertical, perpendicular to the direction of the blast in the vertical plane.

The basic problem associated with the measurement of blasting vibrations is the establishment of a fixed point in space from which to measure. During the passage of seismic energy, the entire environment is in motion which includes the measurer and the measuring instrument. The seismometer (or vibrograph) overcomes this problem by establishing a reference point within itself that tends to remain fixed during the movement of its housing.

For typical blast vibrations it is necessary that the seismometer should be able to record the following ranges:

Frequency, 1 to 100 Hz.
Displacement, 0.00254 to 12.7 mm.
Velocity, 0.254 to 254 mm/s.
Acceleration, 0.005 to 2 g.

For overburden blasts and other large blasts measured at distances greater than 305 m (1000 ft), the blast frequency is generally in the 1 to 20 Hz range (Attewell and Farmer, 1964). Higher frequencies are experienced closer to the shot point and frequency responses of 500 Hz and more are required for instruments measuring close to blasts in high strength rocks such as granites.

For accurate recording, it is essential that the seismometer maintain fixed contact with the ground or other surface throughout the duration for the vibration passage. If the monitoring head slips in a horizontal direction and/or loses contact in a vertical direction, a useless seismogram will result.

There are three main types of seismometer. One type uses a mechanical optical system. These units measure amplitude and frequency, usually in one component direction only, by direct writing on paper or by light beam on photographic paper.

Electromagnetic systems utilize differential displacement between a moving coil and a soft iron mass (or vice versa) producing an electrical signal output proportional to the seismic event. This unit is referred to as a transducer. The signal so produced is then amplified by a stepped gain setting and fed into a recorder to give velocity. Integrating and differentiating circuits may be included to read displacement or acceleration. Output is usually via a timed, variable speed, direct print paper, the three mutually perpendicular components being measured simultaneously.

Both of the previous systems suffer to some extent from sympathetic resonance of the instrument at certain frequencies. This is overcome using calibration charts. In an attempt to correct minor interpretational difficulties due to resonance, a third type of seismometer has been developed based on solid-state accelerometers feeding into integrating circuits.

Large-Scale Open Pit Blasts

Fig. 39 presents peak particle velocity plotted vs. scaled distance for open pit and rock strip mine multiperiod blasts which contains more than 10,000 data points recorded worldwide (representative data only being shown on the graph itself). As such, Fig. 39 represents a very conservative design criteria when the upper limit line is used for design purposes.

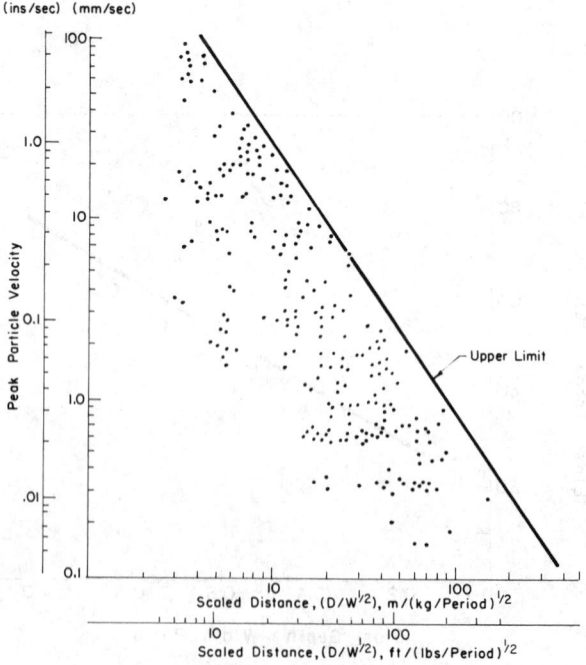

Fig. 39. Typical vibration data from multiperiod delay blasts in open pits and strip mines.

Table 6. Types of Damage Related to the Peak Particle Velocity in the Ground Waves from Blasts

Type of Structure	Type of Damage	Peak Particle Velocity Threshold at Which Damage Starts	
		mm s	in. per sec
Rigidly Mounted Mercury Switches	Trip Out	12.25	0.5
Houses	Plaster Cracking	50	2
			Set initial Limit of 125 mm/s (5 in. per sec) Maximum at the Crusher
Concrete Block as in a New House	Cracks in Blocks	200	8
Cased Drill Holes Retaining Walls, Loose Ground Mechanical Equipment Pumps	Horizontal Offset	375	15
Compressors	Shafts Misaligned	1000	40
			Beyond 250 mm/s (10 in. per sec) Major Damage Starts, Such as Possible Cracking of Cement Block
Prefabricated Metal Building on Concrete Pads	Cracked Pads Building Twisted and Distorted	1500	60

Table 6 presents the type of damage that can be expected in relation to the threshold values of peak particle velocity experienced in the ground waves from blasts. As can be seen, the onset of plaster cracking in houses occurs at a threshold peak particle velocity value of 50 mm/s (2 in./sec). This is an almost universally accepted damage control criteria in North America. In Fig. 40, the peak particle velocity damage threshold values have been superimposed on the upper limit line. In addition, vibration levels for trucks and buses and damage to water wells are shown. This, then, offers a useful guide with regard to safe scaled distances that may be employed.

Human Response to Blast Vibration

Fig. 41 shows the results of the experimental work of Goldman (1948) in which tests were conducted using sinusoidal mechanical vibrations of different frequencies and particle velocity on human test subjects. Since the frequency of pit blasts are often in the range of 10 to 30 cycles per second, then this data can be used to determine the range of scaled distance from the pit blast peak particle velocity upper limitations at which certain human reactions take place. This has been done in Fig. 42 from which it can be seen that blast vibration effects became intolerable to humans at levels appreciably lower than levels at which structural damage takes place. The result is that often complaints can be received due to human response and not due to a situation producing damage, and in these instances a reduction in the peak particle velocity limit to take care of this situation is a good philosophy, particularly if such a change does not produce operational difficulties. Limits used in the US of peak

particle velocity of 12.5 mm/s (0.5 in. per sec) reduce the number of complaints by a factor of three compared to 50 mm/s (2 in. per sec). It has been noted by the USBM that the percentage of people complaining about blast vibration on one construction site was as high as 30% at 50 mm/s (2 in. per sec), 10% at 12.5 mm/s (0.5 in. per sec), and 1% at 2 mm/s (0.08 in. per sec), which is just in the perceptible range. In some instances, therefore, the human response becomes the major element in the blast design.

The current blasting code for Ontario, Canada, calls for a maximum peak particle velocity of 10 mm/s (0.4 in. per sec).

Noise and Air Blast

Whenever a pressure wave travels through the air faster than a sound wave, it produces a shock wave. The airborne pressure wave emanated from an explosion, called air blast, is a shock wave. At first it travels at supersonic speed, but, depending on the magnitude of the energy released by the explosion, it will decay in time to an ordinary sound wave.

Noise or sound is a pressure wave traveling through air at approximately 335 m/s (1100 fps). A sound wave consists of a train of high pressure regions following one another in rapid succession with the frequency of the sound and separated by regions of lower pressure. The frequency of sound varies from approximately 20 to 20,000 Hz, which is the range that can normally be detected by the human ear.

Air blast acts in a similar manner to noise or sound as it is also a pressure-type wave. However, it has a greater speed than that of sound and its frequency is less than 20 Hz, so consequently it cannot be heard. Its effect, though,

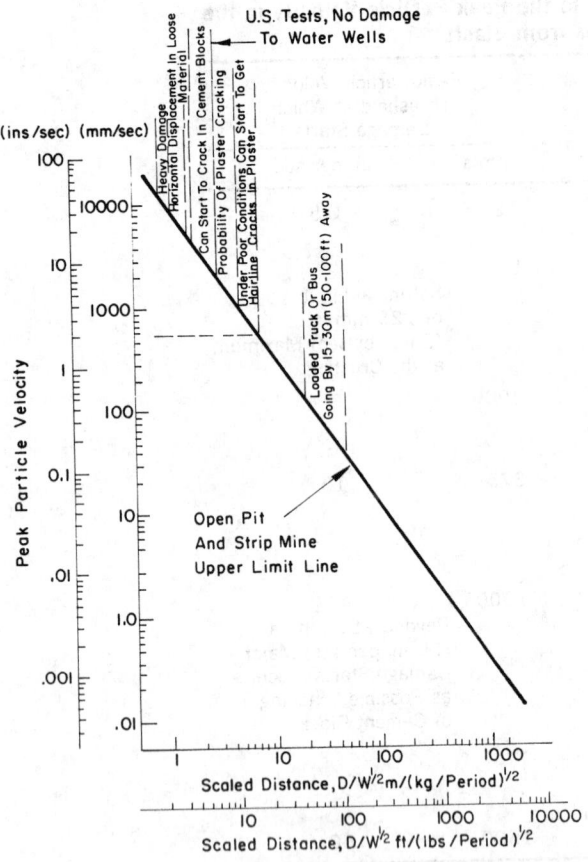

Fig. 40. Structural response at different scaled distances from pit blasts.

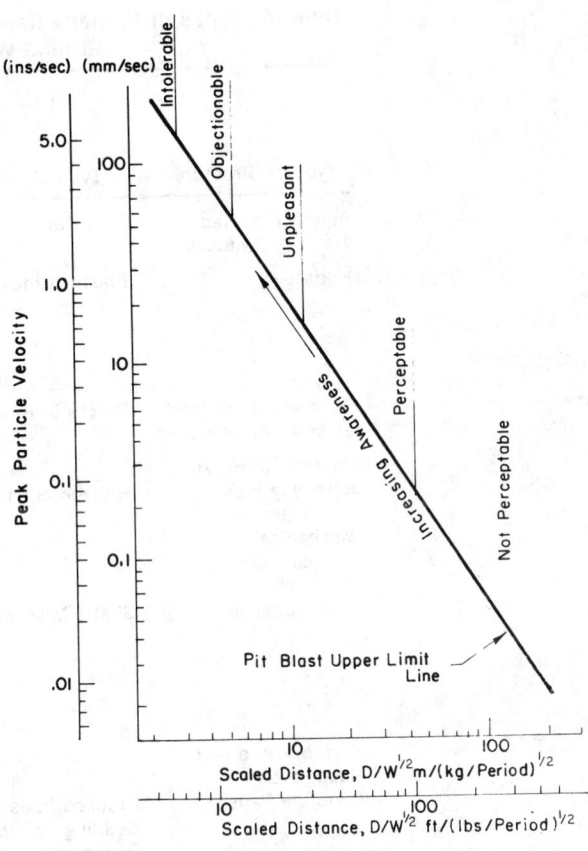

Fig. 42. Anticipated human response to blast vibrations at different scaled distances from pit blasts.

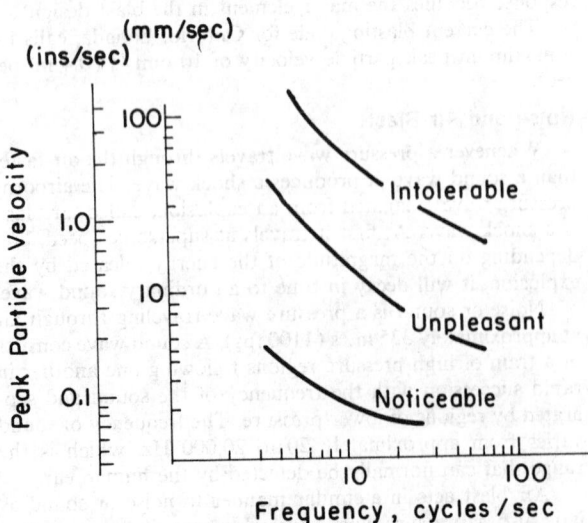

Fig. 41. Response of the human body to mechanical vibration (Goldman, 1948).

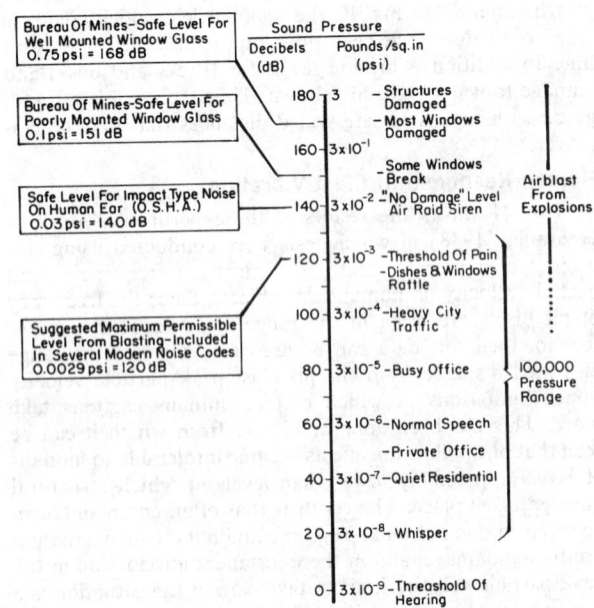

Fig. 43. Human and structural response to sound pressure level.

is apparent in the rattling of windows, etc. The positive phase of the air blast wave is much greater than a sound wave and the following lower pressure region actually changes to a negative value.

To get a feel for the relative sound levels, Fig. 43 presents typical human and structural responses to different scale level.

Air Blast Control

In pit blasting, noise or air blast can be due to any of the following items:

1. Primacord trunk lines.
2. Lack of proper stemming material.
3. Inadequate stemming height.
4. Overdug or overloaded front row holes, collars or burden near the crest too small due to backbreak.
5. Delay sequence.
6. Atmospheric conditions such as temperature inversions or wind in the direction of concern.
7. Secondary blasting.

The first item can often be eliminated as a hazard by going to lower grain count primacord trunk lines or Nonel, covering with drill cuttings, or using electrical blasting. Overdigging or overloading of the front row of holes can be eliminated by flagging for the shovel operator and the stemming heights in the body of the blast increased, but this has to be balanced vs. the top fragmentation. The blasts can be laid out so that a smaller number of holes are fired per delay period. Plaster or mudcap shooting is particularly hazardous in this regard with most of the explosive energy going into air blast.

Delays can be used to isolate the effects of one charge from the next in secondary blasting. Arrange for an approximate minimum of 50 millisec to lapse between the arrival of the noise from the successive shots. If the closest one is fired first, then the delay to be used can be calculated by dividing the distance difference between the two shots by the sound speed 335 m/s (1100 fps), and subtracting this from 50. Perhaps a good rule of thumb might be to shoot the closest shot first with all others having delays between them of 50 millisec. Care should be taken to ensure that the blast wave from one shot does not dislodge the next.

When a blast is well stemmed with 12-mm (0.5-in.) crushed rock material and is initiated electrically or with a low energy system such as Nonel, then the front row of the blast is the major area of concern. For example, if the face moves out, it drives the air ahead of it like a piston. If the blast is initiated to produce a row by row blast, then this piston effect is more pronounced and the air blast is enhanced compared to shooting the blast en echelon. However if the en echelon blast is initiated at the furthest point from the observer, then the effect observed depends upon the spacing and delays chosen.

The peak overpressure at a given point is influenced by atmospheric conditions particularly at long distances away from the blast. Wind and temperature altitude profiles have the major influence. The speed of sound in air

$$C = \sqrt{\frac{\gamma P}{\rho}} = \sqrt{\gamma nRT}$$

where T is temperature, γ is the ratio of specific heats, P is pressure, ρ is air density, R is a gas constant, and n is the number of moles so that increases in temperature cause increases in sound speed.

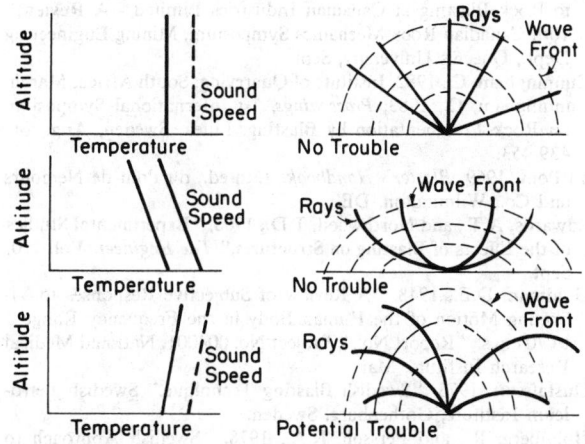

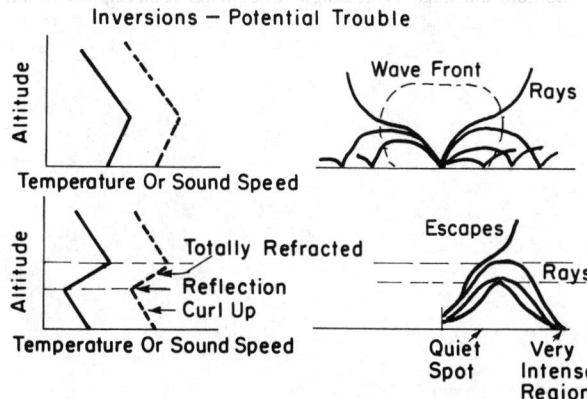

Fig. 44. Effect of altitude temperature profiles on air blast propagation (Edwards and Northwood, 1960).

Fig. 44 shows typical atmospheric temperatures (Cook, 1958), conditions which can occur, and the potential problems associated with temperature inversions or with an increasing temperature with altitude profile. These conditions if they exist can increase the peak overpressure by as much as a factor of five.

In situations in which air blast from production shooting can be a problem, it is often customary to fire a small surface shot and measure the peak overpressure at the point of interest. If the reading is normal, then the main production shot is fired. If it is excessively high, then the blast will be delayed. However, unless the mine equipment has been scheduled for other areas, then this is not a very attractive scheme since the time lost could be considerable. If airports are close by, then the local meteorological information can be used in determining the hazard. If winds are in evidence then inversions will be absent, and one can also obtain information from how the smoke rises from smokestacks, whether it rises initially and then spreads out.

BIBLIOGRAPHY

Attewell, P.B., and Farmer, I.W., 1964, "Attenuation of Ground Vibrations from Blasting," *The Quarry Manager's Journal*, June.
Cook, M.A., 1958, "Science of High Explosives," American Chemical Society Monograph No. 139, Reinhold Publishing Corp.
Cook, M.A., 1968, "Explosives, A Survey of Technical Advances," *Industrial and Engineering Chemistry*, July, pp 44-55.
Chung, S., et al., 1975, "Applications of High-Speed Photography

to Rock Blasting at Canadian Industries Limited - A Review," 10th Canadian Rock Mechanics Symposium, Mining Engineering Dept., Queen's University, Sept.

Cunningham, C., 1982, Institute of Quarrying, South Africa, March.

Cunningham, C., 1983, *Proceedings,* 1st International Symposium on Rock Fragmentation by Blasting, Lulea, Sweden, Aug., pp. 439-453.

du Pont, 1969, *Blaster's Handbook,* 15th ed., du Pont de Nemours and Co., Wilmington, DE.

Edwards, A.T., and Northwood, T.D., 1960, "Experimental Studies of the Effects of Blasting on Structures," *The Engineer,* Vol. 210, Sept.

Goldman, D.E., 1948, "A Review of Subjective Responses to Vibrating Motion of the Human Body in the Frequency Range 1 to 70 c.p.s.," Report No. 1, Project No. 004001, National Medical Research Institute, Mar.

Gustafsson, 1973, "Swedish Blasting Technique," Swedish Petroleum Institute, Gothenburg, Sweden.

Holmberg, R., and Persson, R.A., 1978, "Swedish Approach to Contour Blasting," *Proceedings,* 4th Conference on Explosives and Blasting Techniques, Annual Meeting, Society of Explosives Engineers, New Orleans, LA.

ICI, "Excavating with Explosives," Stevenson, Scotland.

Kuznetsov, V.M., 1973, *Soviet Mining Science,* Vol. 9, No. 2, pp. 144-148.

Langfors, U., Kihlstrom, B., and Westerberg, H., 1958, "Ground Vibrations in Blasting," *Water Power,* Feb., pp. 335-338; 390-395; 421-424.

Langfors, U., and Kihlstrom, B., 1963, *The Modern Techniques of Rock Blasting,* John Wiley and Sons Inc., New York.

McDonald, K.L., Smith, W.K., and Crosby, W.A., 1982, "Productivity Improvements for Dragline Operations Using Controlled Blasting in a Single and Multiple Seam Opencast Coal Operation at Rietspruit, South Africa," Annual General Meeting, Canadian Institute of Mining and Metallurgy, Quebec City, April.

Nichols, H.R., Johnson, C.R., and Duvall, W.I., 1971, "Blasting Vibrations and Their Effects on Structures," Bulletin 656, US Bureau of Mines.

Pit Slope Manual, 1977, "Perimeter Blasting," Chap. 7, CANMET, May.

6.2.2 Design of Blasting Rounds

RICHARD L. ASH

INTRODUCTION

For mining competent materials, the production cycle begins with the blasting, the results of which affect all subsequent operations. Improper blasting causes economic and operational difficulties and can create serious safety and environmental problems, particularly when conducted near inhabited areas. It is important, therefore, that all blasts be carefully designed at the onset so they are compatible with the mining conditions and surroundings. It is no longer acceptable to blast without careful advance planning. For all practical purposes it can be assumed all blasts will use explosives confined within a number of boreholes (Fig. 1).

Primary requisites for a successful blasting round from an operational standpoint are that it provides a sufficient quantity of suitably fragmented and properly placed material at the lowest practical cost, possesses adequate flexibility in its basic design to accommodate changing mining conditions, be relatively simple to apply, and generates as few objectionable side effects as possible. A well designed blast is a controlled blast, with the blasthole pattern so arranged that best advantage is obtained from the explosives' energies, the properties of the material being blasted, and all environmental and operational factors present at the mine site. It is the purpose of this discussion to review the significant variables and relevant engineering principles and to present helpful guidelines by which to design blasts for surface mining conditions.

GENERAL CONSIDERATIONS

Blasting is basically a three-dimensional work process, involving the conversion of chemical energy from explosive compounds into gas pressures applied to perform mechanical work. An explosive confined within a borehole functions much like the air and fuel mixture in the internal combustion engine. The work is accomplished during blasting by the generated gas pressures loading materials in such a way that they are displaced, or moved, causing them to fracture. To design a blast, therefore, requires a working knowledge of the capabilities of explosives, how they function in boreholes, and how materials respond to explosive loadings. Because it is quite likely that most explosives used will be the relatively insensitive blasting agents, close attention must be given to the priming, confinement control, and selection of charge diameters that will ensure complete explosive reactions. For purposes of this discussion all products employed for blasting will be termed as explosives whether or not they are defined technically as such.

Characteristics of Explosives

Industrial explosives are a relatively inexpensive energy source but a very powerful tool, having mechanical work-energy equivalents that may range in value from some 167 kj/kg (5×10^5 ft-lb/lb) to in excess of four times that amount. The explosives are essentially mixtures composed of two or more chemical ingredients that are combined in such a way they will react exothermically after being ignited to quickly form gaseous pressures and sudden heat releases. Pressures developed may vary from as low as 1.4 up to about 13.8 GPa (0.2 to 2.0×10^6 psi) depending on the particular explosive mixture. There must be adequate ignition energy provided and suitable temperature and pressure conditions present to properly initiate, then sustain an explosive reaction.

Important Properties: Explosives are characterized by several interrelated properties that affect their performances in the field and are always considered when selecting a product for use. An explosive's density, sensitivity, velocity, water resistance, and fumes are the most important properties; although strength is considered by some people also to be significant. The shock energy and expanding gas energy ratings obtained from underwater tests provide the best measure for comparing explosive strengths according to most explo-

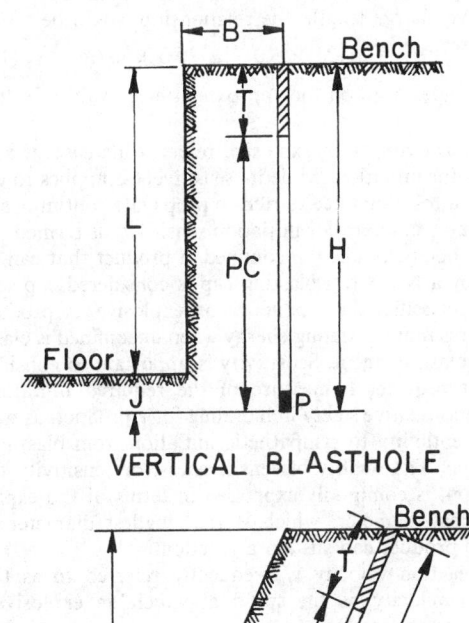

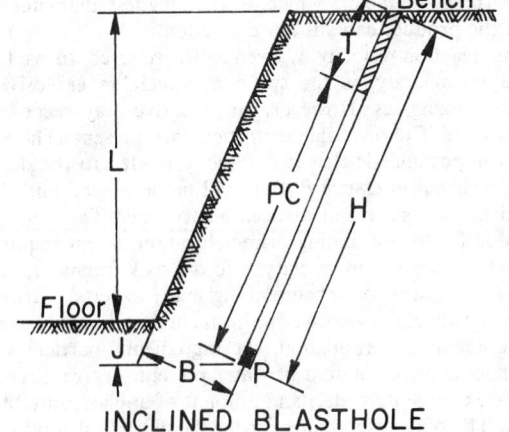

B – BURDEN   L – LEDGE HEIGHT
J – SUBDRILL   H – HOLE DEPTH
T – STEMMING   P – PRIMER
PC – CHARGE LENGTH

Fig. 1. Blasthole nomenclature.

sive engineers and scientists. The average blaster, however, normally cannot directly relate such strength ratings to field practice in a practical way.

An explosive's density determines the quantity on a weight basis that can be loaded into a given volume. As a general rule, denser explosives provide more energy per unit volume and are usually preferred where drilling is expensive. The three commonly used density ratings are specific gravity SG_e, loading density d_e, and mass density ρ_e. For approximation purposes, one can consider that most explosive products will fall within the range of 0.8 and 1.6 sp gr, the ρ_e values being the same but expressed in g/cc. Loading density, which depends on an explosive's specific gravity, provides the explosive's weight that would be contained within a cylindrical volume (as for a borehole) with a given charge diameter and unit length. A useful expression for calculating loading density is as follows:

$$d_e = 7.854 \times 10^{-4} D_e^{2} (SG_e), \text{ kg/m} \qquad (1a)$$

in which D_e is the explosive's diameter in mm. If the explosive's charge diameter is in inches and d_e is desired in lb/ft of explosive charge length, the relationship would be

$$d_e = 3.405 \times 10^{-1} D_e^{2} (SG_e), \text{ lb/ft} \qquad (1b)$$

Fig. 2 provides a graph for approximating d_e values in lb/ft.

The sensitivity of an explosive refers to its ease or susceptibility for initiation, while its sensitiveness applies to the ability for a reaction once started to propagate continuously. The common standard for initiation sensitivity is termed the minimum booster. When unconfined, a product that can be initiated by a No. 8 test blasting cap is considered cap sensitive and classifies the product as an explosive. A product that requires more initiating energy when unconfined is classified as a blasting agent. Sensitivity is important to consider because it provides a measure of the required initiator's strength and relative safety in handling for a product, as well as its susceptibility to sympathetic initiation from blasthole to blasthole. The quality of sensitiveness, or sensitivity for propagation, is commonly expressed in terms of the explosive's critical diameter, which is the smallest diameter at which the product can sustain a reaction.

The reaction velocity v_e, frequently referred to as the detonation velocity, is the speed at which an explosive's reaction propagates. However, an explosive may react but not detonate if unfavorable conditions are present. The velocity's importance is that it is directly related to the detonation or shock pressure P_d that will be developed and the explosive's acoustical impedance, defined later. The velocity is also useful for estimating primer location design requirements. How significant velocity is to the rock fragmentation process is a source of debate among many experts. Various factors can affect the velocity's value including the explosive's specific chemical formulation, the ingredients' particle sizings, mass density, amount of water present, degree of confinement, charge diameter, and amount of applied initiating energy. The range of v_e values for most commercial products is from 1700 to 7600 m/s (5500 to 25,000 ft/s) as illustrated in Fig. 3.

An explosive's water resistance is its ability to withstand exposure to water without losing efficiency. The two primary effects from water that can result are cooling of a reaction and possible leaching out of soluble ingredients. Both effects can alter the reaction process, resulting in a lower energy release and the generation of toxic gases, or the explosive may misfire. The visible emission during a blast of reddish-

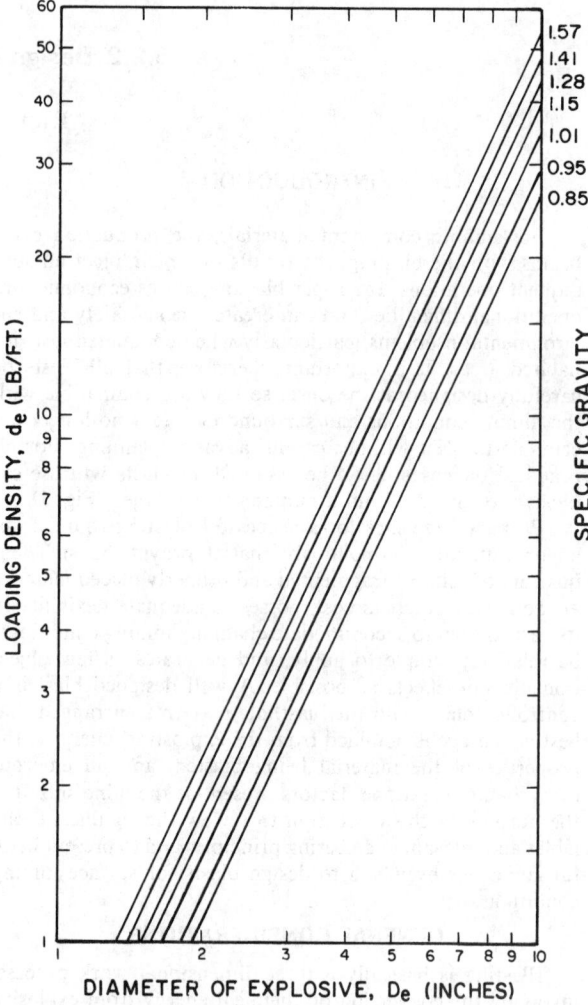

Fig. 2. Explosive densities as a function of charge diameter.

brown nitrogen oxides often indicates an inefficient reaction caused by water deterioration of mixtures containing nitrate compounds.

The fume rating for an explosive provides an indication of the amount of toxic gases generated by the product's particular chemical mixture when tested under certain prescribed conditions. The importance of potential fume generation by a given product is as applicable to surface blasting as it would be in underground usage. This is because certain fumes are heavier than normal air and can be trapped in muck piles and lay in pit bottoms. The oxides of nitrogen are particularly hazardous gases possessing such characteristics. Carbon monoxide is another toxic gas that could be produced but is slightly lighter than dry air; but where humidity is high it may be intermixed in the local atmosphere.

Initiation and Propagation Mechanisms: An explosive's initiation is thermal in origin, beginning as combustion at minute explosion centers known as hot spots. Voids in the explosive are necessary for hot spots to form. If too few voids are present, the explosive cannot react satisfactorily and is called dead pressed. The initial combustion, or oxidation chemical reaction from which heat and light are evolved, begins comparatively slowly due to the requirement to build

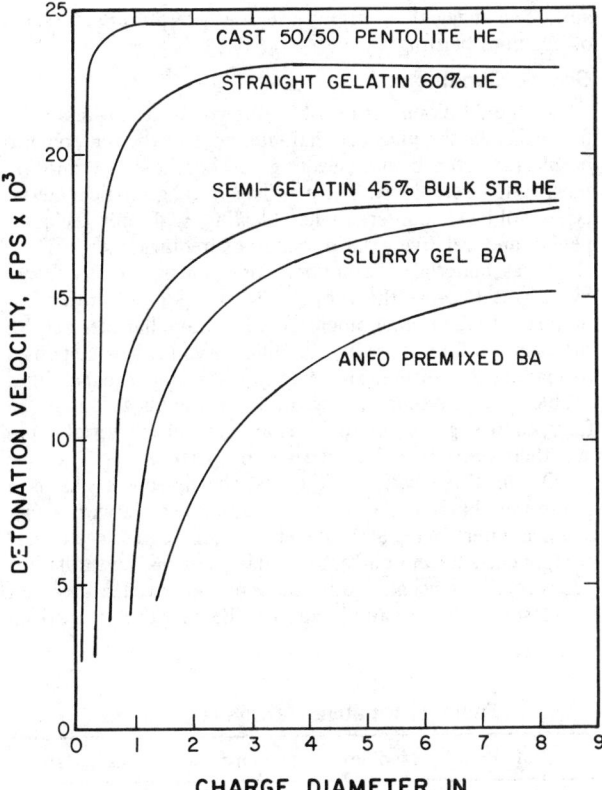

Fig. 3. Reaction velocity variations with charge diameter for selected explosives.

up a critical concentration of reactants. Whenever the heat generation rate exceeds that lost due to convection and conduction, the gas pressures continue to grow, accelerating the heat generation and attendant pressure rise. As the number of hot spots increases and spreads throughout an explosive, they consolidate into a pressure front. When the combustion transcends into a rapidly progressing, relatively self-sustained, and near-spontaneous oxidation with high heat and pressure formation accompanying the reaction, it is called deflagration. If the rapidity of energy release continues to increase so that compressional waves are propagated through the unreacted portion of the explosive in excess of its characteristic sound velocity, a low-order detonation results. When conditions are favorable, temperature, pressure, and velocity gradients may continually grow steeper until a stable, fully developed detonation results, with its supersonic shock wave preceding the combustion zone. Fig. 4 illustrates an explosive column that was initiated at one end with a full detonation positioned halfway through the column.

Adequate confinement for explosives is necessary because it minimizes temperature and pressure losses. Also, since reaction products expand laterally as well as longitudinally, both the charge diameter and length must be large enough to not hamper movement of the reaction. Those dimensions are related geometrically to the explosive's chracteristic reaction length. Most high explosives have short reaction zones and smaller critical diameters than do the blasting agents.

To ensure rapid formation of hot spots and prompt development of high-order detonation in a main explosive charge, a good primer must be employed to provide adequate heat and pressure quickly over as large an area as possible.

A primer is a cap-sensitive explosive unit containing an initiating device; a booster is an explosive used to intensify an explosive reaction but it does not contain an initiator and may or may not be cap sensitive. A primer is considered satisfactory if it has four minimal qualities: cap sensitivity, an SG_e of 1.2 or higher, a v_e of at least 4600 m/s (15,000 ft/s), and immunity to the environment. An explosive satisfying the density and velocity minimum requirements will develop a detonation pressure of approximately 80 κb (1.2 $\times$ 10^6 psi). Immunity to the environment requires that the explosive be unaffected by water, diesel fuel oil, normal ambient temperatures, lack of confinement, and poor storage conditions. Ideally, the primer has a diameter that matches as closely as possible the diameter of the explosive it is to initiate for maximum efficiency, or is sufficiently greater than the critical diameter of the explosive used as a blasthole's main charge.

The ideal gaseous products formed by a properly oxygen-balanced mixture consist essentially of some combination of nontoxic carbon dioxide, inert nitrogen, and steam. If aluminum, calcium, or sodium, for example, is included in an ingredient and a reaction is complete, a metallic oxide such as Al_2O_3, CaO, or Na_2O will be produced as a solid residue. The difference between the sum of the heats of formation of the products generated and the sum of those heats for the initial ingredients determines the maximum amount of heat that can be potentially released by the reaction of an explosive. The values thus obtained for various mixtures, termed their heats of explosion Q_e in κcal/g, are considered by some persons as their respective blasting capabilities. Use of Q_e to rate explosives, however, is misleading because explosives are employed to provide pressures and not just heat. A more useful rating than κcal/g is the mechanical equivalent of work expressed in terms of joules per gm or ft-lb/lb, which relates both the potential heat content and gas volume produced.

Explosives' Pressures: As illustrated by Fig. 4, there are two distinct and separate pressures generated by explosives during blasting: that from detonation P_d and that from the explosion products P_e. Each has its own unique characteristics. Since P_d results from physical movement, its value is

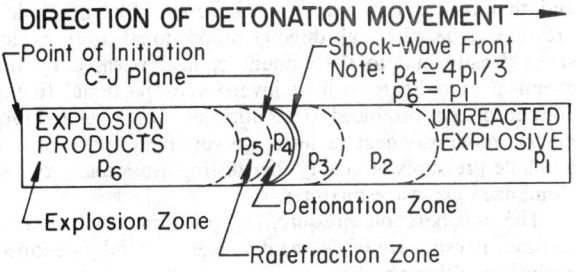

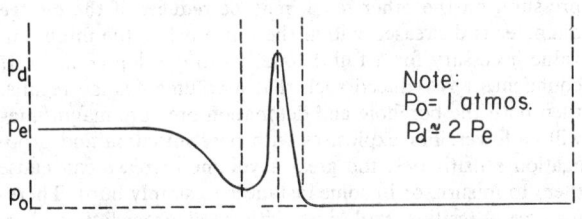

Fig. 4. Characteristics of a detonation reaction.

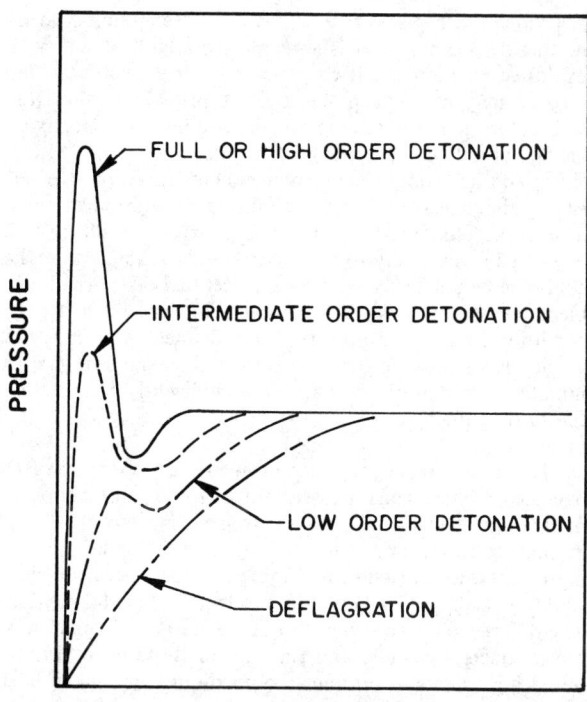

Fig. 5. Pressure-time profiles for explosive reactions.

and when engaged in practices such as presplitting and trim, or cushion, blasting.

Ground Conditions

In the development of a blasting round it is essential to first consider the physical characteristics of the materials to be blasted, even before selecting the borehole size and explosives to be employed. Frequently, several geologic material types will be encountered when blasting, with only the competent and relatively hard materials requiring explosives. Rocks as materials are neither homogeneous nor isotropic by nature. Because the various rock types are more or less unique as to their mineralogical composition, hardness, structure, and texture, there will be differences in their responses to loading, fracturing, and energy transmission capabilities (Table 1). It is essential, therefore, that the blast designer be fully knowledgeable of the various geologic material types and their respective characteristic properties.

Of the three basic rock types, the igneous rocks as a group are characterized by a crystalline texture with a mechanical interlocking arrangement of their respective mineral components. Exceptions are the glasses, lavas, and tuffs. Igneous rocks are usually quite coherent and have textures that are closely similar in all directions. The finer the grain sizing

a function of the kinetic energy generated and is proportional to the product of the first power of the explosive's mass density and the second, or squared, power of the reaction velocity. Therefore, if an explosive reacts at one-half its ideal rate, the detonation pressure will be only one-quarter of its maximum theoretical steady-state value (Fig. 5). On the other hand, the pressure from the reaction products, more commonly called the borehole pressure, is independent of the reaction velocity and is produced regardless of whether or not the explosive detonates. The magnitude of P_e will be a function of only an explosive's chemical composition, density, and that its reaction went to completion. The borehole's pressure value, although directly proportional to an explosive's density and to the amount of heat released by the chemical ingredients, will be inversely proportional to the volume of gas produced. Therefore, explosives generating higher amounts of heat but lesser gas volumes provide greater borehole pressures, a quality that distinguishes many of the aluminized explosive mixtures.

The unit borehole pressure, i.e., psi, ideally will remain constant irrespective of charge diameter for a fully confined explosive; although the total force produced and charge weight per unit length that is used will vary in direct proportion to the ratio of the diameters squared. The detonation pressure, on the other hand, may be reduced if the charge diameter is decreased within the range below the minimum value necessary for a full detonation to develop (Fig. 3). If confinement is reduced such that a volume change results, then both the borehole and detonation pressure magnitudes will be lower. For explosives with poor initiation and propagation sensitivities, too great a volume increase can cause them to misfire, or in some instances to simply burn. Therefore, cap-sensitive explosives with good propagating characteristics must be used under poor confinement conditions

Table 1. Relative Hardness of Rocks

Soft	Medium	Hard	Very Hard
Asbestos	Limestone	Granite	Iron ore
Gypsum	Dolomite	Quartzite	Taconite (some)
Slate	Sandstone	Iron (some)	Granite (some)
Shale		Ores	Trap rocks
Talc		Trap rock	
Soft limestone		(basalt)	

Sonic Velocities of Rock Types from Various Locations

Rock Type	Specific Gravity	Velocity ft/sec
Granite gneiss	2.65	18,350
Granite gnesis	2.62	12,150
Granite	2.80	18,800
Granite	2.70	15,900
Granite	2.67	19,500
Basalt (trap rock)	2.88	17,150
Basalt (trap rock)	2.96	21,700
Gabbro	3.10	23,100
Sandstone	1.87	6,900
Quartzite	2.63	16,000
Taconite	2.67	12,600
Taconite	2.72	18,800
Taconite	2.95	20,000
Limestone	2.74	20,900
Limestone	2.70	16,300
Limestone	2.54	14,200
Limestone	2.34	11,000
Marble	3.04	22,000
Dolomite	2.76	15,100
Dolomite	2.51	16,400
Slate	2.71	18,100
Water	1.00	4,600 to 4,900
Air	—	1,000

Source: Burkle, W.C., 1980, "Geology and Its Effect on Blasting," *The Explosive Engineer*, Supplement, No. 3, Apr., p. 1.

the greater the strength, and many of the rocks will have comparatively high sonic velocities. As materials, they are strong and break in a brittle fashion, being quite elastic and dense with low or negligible porosities.

Sedimentary rocks have widely varying strengths, definite and pronounced structural and textural orientations of the constituents, and highly variable porosities and energy propagating velocities. The mineral aggregates are bonded together either chemically or have a cementing agent as a matrix. Limestones and dolomites, for example, are usually quite strong with medium-rate and relatively stable sonic velocities. The sandstones, however, exhibit light densities, low strengths comparatively, and propagation velocities that often decrease with travel distance away from the impacting source until a lower level stable rate is attained. Sandstones also may have porosities that are particularly susceptible to saturation with water. Shales, on the other hand, are fairly dense and nonporous but are very weak due to their highly platey or laminated structure.

The metamorphic rocks such as marble, quartzite, and slate, for example, possess properties of both the igneous and sedimentary types. Because they were formed from the other two primary types of rock as a consequence of added heat or pressure, or a combination of those agents, the metamorphics are usually quite dense and strong. Their energy propagating characters are such that they frequently exhibit high sonic velocities.

Density and Swell: The significance of density in blasting is that as a rule one could expect more energy and thus more work to be required for displacing and fragmenting the heavier materials. Most dense rocks are stronger and tougher. Density of in-place rocks may be expressed in three ways: as ordinary density, d_r in kg/m³ (lb/ft³), as specific gravity, SG_r that is unitless, and as mass density ρ_r in g/cc (lb-s²/ft⁴). The ρ_r is usually employed in theoretical calculations such as when considering a material's dynamic energy transfer property or acoustical impedance. The latter is the product of a material's mass density and its compressional wave velocity.

Except for the very heavy iron ores with densities of about 5 kg/m³ (310 lb/ft³), most ores and rocks requiring blasting have in-place densities in the 2 to 3 kg/m³ range (125 to 187 lb/ft³). For first approximations, an average SG_r of 2.65, or d_r of 2.12 kg/m³ (165 lb/ft³) can be used for most materials without introducing excessive error.

Although not strictly a property like density, the swell, or expansion, of material when broken from the intact or solid condition must always be considered for blast design (Fig. 6). The swell factor S_f will be unique for a given material and fragmental sizing mix. It is defined as the ratio of the volume of a unit weight of intact material to that when broken. The factor's value will usually be greater than 0.5 and less than 1. Table 2 lists approximate S_f values for various earth materials considering average sizing mixes. A good mean value of S_f for estimating purposes is 0.7, or as a rough approximation, it is common to assume that there will be a normal expansion of 50%, or half again, the original in-place volume when the material is broken out of the solid bank.

As shown by Fig. 6, minimal volumetric expansion during ordinary bench blasting occurs in two directions for a box cut, while for the corner cut the expansion will be in three directions. If swell is assumed to be uniform, each solid dimension toward an open face could be expected to expand to an amount at least equal in value to the initial solid dimension divided by the material's characteristic S_f value to a root power equal to the number of open surfaces. For the box cut the factor would be $S_f^{1/2}$, while $S_f^{1/3}$ would be appropriate for a corner cut.

The effects of swell must be accounted for in designing not only a blast's entire volume but also for the respective volumes of each row of holes and those for each individual blasthole fired independently or set of holes initiated with close interval firing times. Otherwise, the displacement necessary for proper fracturing and displacement of the broken material as holes are fired will be impeded. Failure to leave adequate space for expansion is a major contributing cause for airblast, excessive flyrock, intensified ground vibrations, overbreak, toe, uneven pit floors, and difficult digging caused by packing.

Geologic Structure: Competent blasters all recognize that rock structure exerts a most important, if not dominant, influence on the results achieved from blasting. Discontinuities such as jointing, stratification, foliation, schistosity, faults, and unconformities all serve as natural planes of weakness along which a rock mass will separate the easiest. The planes control not only the locations where fractures will most likely start but also the directions or paths along which breaks in a material once initiated can be easily extended. Most discontinuities can be detected and recognized by their

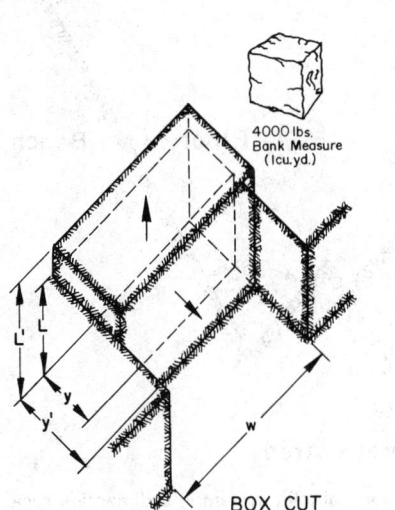

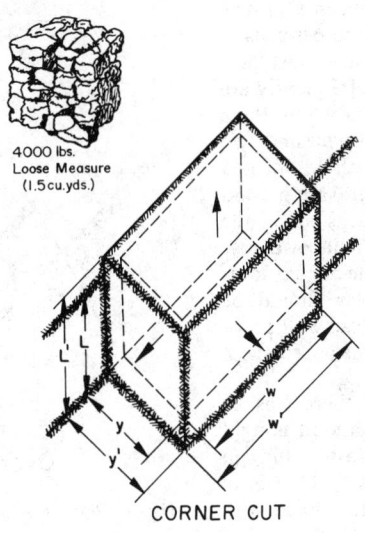

4000 lbs. Bank Measure (1 cu.yd.)

4000 lbs. Loose Measure (1.5 cu.yds.)

BOX CUT

CORNER CUT

Fig. 6. Swell during bench blasting.

Table 2. Approximate In-Bank Weights and Swell Factors

Material	Approximate Swell Factor,				In-Bank
	S_f	$S_f\frac{1}{2}$	$S_f\frac{1}{3}$	Avg.	lb/cf
Coal	0.65	0.81	0.86	0.84	85-100
Bauxite Ore	0.75	0.87	0.91	0.89	120
Salt Rock	0.65	0.81	0.86	0.84	145
Copper Ore	0.74	0.86	0.90	0.88	140
Caliche	0.65	0.81	0.86	0.84	140
Granite	0.62	0.79	0.85	0.82	170
Quartzite	0.66	0.81	0.86	0.84	160
Iron Ore (Hematite)	0.85	0.92	0.95	0.94	180
Iron Ore (Magnetite)	0.85	0.92	0.95	0.94	200
Iron Ore (Taconite)	0.57	0.75	0.83	0.79	315
Limestone	0.59	0.77	0.84	0.81	165
Gypsum	0.57	0.75	0.83	0.79	175
Shale	0.75	0.87	0.91	0.89	160
Dolomite	0.64	0.80	0.85	0.83	180
Slate	0.77	0.88	0.92	0.90	170
Sandstone	0.63	0.80	0.85	0.83	140
Trap Rock	0.67	0.82	0.88	0.84	175
Basalt	0.66	0.81	0.86	0.84	190
Clay, Light	0.82	0.91	0.94	0.93	104
Earth, Dry	0.85	0.92	0.95	0.93	104
Gravel, Dry	0.89	0.94	0.96	0.95	122
Sand, Dry	0.89	0.94	0.96	0.95	122
Avg. for Rock	0.70	0.83	0.88	0.85	165
Percent Increase	1.43	1.20	1.14	1.18	

more or less continuous flat surfaces when exposed. Caves, cliffs, mud seams, and the general trend of river channels through rocky ground show a preference to follow such structural orientations.

Jointing is present in all rock types. Joint planes always occur in sets, one of which is parallel or nearly so with a rock formation's dip and strike with two or more in number oriented approximately perpendicular to the first set of planes. Joints may be tight, open, or filled, thus exhibiting different strength and stress-energy propagating properties. As the number of differently oriented planes increases within a rock mass, the less competent it becomes. For materials having joint planes of uniform strength and with nearly equal included angles, blasting patterns are relatively simple to design. The igneous rocks as a group will approximate this condition, with angles between joint planes being near 60°.

Sedimentary and certain metamorphic materials, however, present a different condition, and balancing an explosive energy is more difficult to achieve. In this case there is usually one particular direction along which joints are most pronounced, and angles between the various sets frequently are not equal. Such a condition presents situations where there will be two opposite tight or acute-angled corners and two wide-angled parts within an area (Fig. 7). Blastholes located in the tightly jointed areas always tend to aggravate ground vibrations and cause overbreak, opening cracks along the jointing planes back into the solid ground. To reduce adverse effects, blasts are best directed out of the wide-angled locations whenever possible, and perimeter holes should be aligned with the dominant joint set. Serious backwall problems can arise from overbreak when alignment is crosswise with joint planes (Fig. 8).

It is important to note that for jointing there was no visible movement parallel with the planes. If a joint is tight and contains no filling, stress wave energy passes through the joint as if it were not there since there is no change of material's properties on either side. However, if filled with

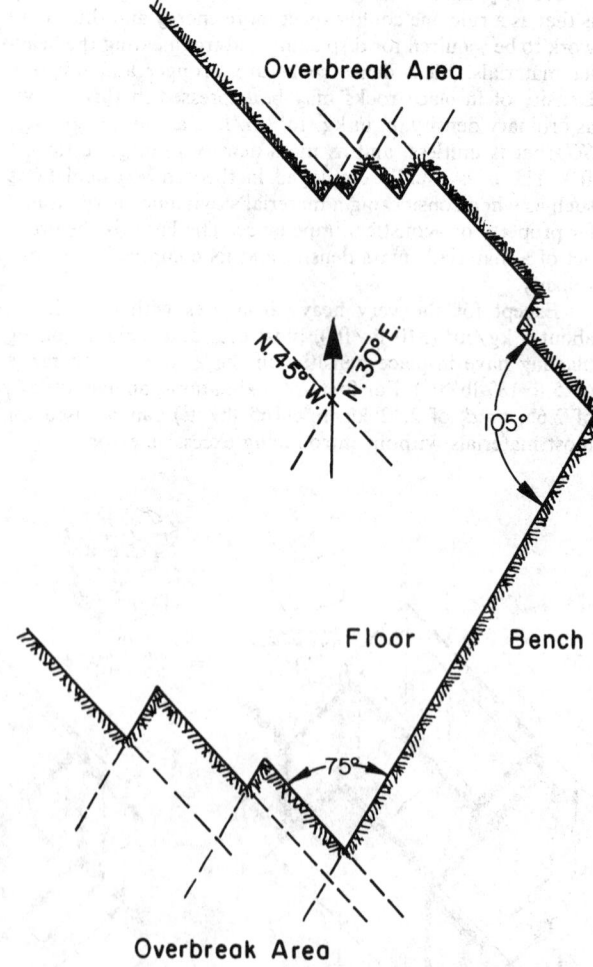

Fig. 7. Plan view of vertical joint system in sedimentary rock.

WEDGE DIPPING INTO
PIT, UNDERCUT, NOT
STABLE SCALLOPING
ALONG PIT WALL

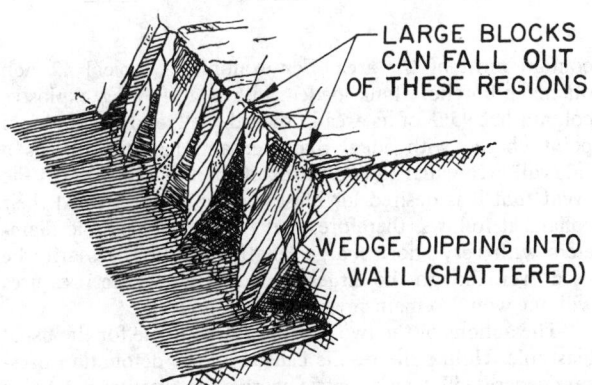

LARGE BLOCKS
CAN FALL OUT
OF THESE REGIONS

WEDGE DIPPING INTO
WALL (SHATTERED)

Fig. 8. Example overbreak damages from holes not aligned with joint planes (Bauer, 1978).

different material or open, the joint will function as an impedance discontinuity, providing a location for scattering of stress waves. A fault, on the other hand, involves an offset of material on either side of a failure plane and usually coincides with the direction of jointing. More often than not there also will be some gouging or scouring of material within the fault zone. Thus, a fault provides a failure plane at which separation is easy as well as where stress wave energy can be reflected and dispersed.

Stratification exists where sediments are deposited in beds or layers and results from interruptions in the deposition process. The material on one side of a plane has a somewhat different lithology than that on the other side. The planes separating the various strata normally offer virtually no resistance to movement and are commonly referred to as partings or slip planes. If a bed is thick and massive except for its transverse jointing, uncontrolled slippage can lead to the production of large slabs or blocky rock sections, high bottoms, and possible blasthole cutoffs. Where beds are inclined, additional problems may be presented. For example, beds dipping downward into a bench or highwall provide a condition for possible toe development and uneven pit floors but with a relatively stable bench wall. On the other hand, beds that dip upward into a highwall can leave a gouged-out pit floor as well as an unstable final wall (Fig. 8).

Banding such as foliation and schistosity as found in certain metamorphic rocks responds to blasting much like

bedding. The coherence of this class of rocks will depend on the minerals present. For example, as the amount of constituent mica increases, the competency decreases. Failure under stressing is always favored along the mica orientations when sufficiently concentrated. Foliation is the general term applied to the laminated structure resulting from segregation of different minerals into layers in metamorphic rocks while schistosity is foliation occurring in the coarser grained rocks containing micaceous minerals.

Unconformities, as opposed to bedding planes, are generally irregular and have materials on either side with quite different properties. They were formed where there was a long period of time elapsing between the deposition of the younger material on older-age rock that had been subjected to erosion or long weathering.

Energy Transmission and Ground Water: As pressures are produced within a blasthole, the surrounding rocks respond by developing stresses and their related strains, or internal deformations. The initial rapid loading effects of the pressures are that only a portion of the rock mass is affected at any one time, an action by which particles within the mass are set into vibrational motions. Thus, compressional or sonic waves from in-line movements and distortional or shear waves from sideways displacements will be generated, which traverse through the rock mass at different rates. A particle is in motion for only a short time, and once set in motion has a maximum velocity when it passes through a position of zero displacement. The rate of travel of particle movements through the ground by the compressional action v_p is faster than the rate of lateral motion travel v_s. Thus, at a given point in the ground away from a blasthole, the longitudinal or compressional motions arrive first with sideways movements occurring shortly after. The surface wave-type motions are generated wherever the compressional and shear body waves encounter open ground surfaces. The energies of the surface waves travel at lower rates and decay more slowly with distance than those of body waves. The formation and propagation of surface waves are favored by pronounced geologic layering.

In the design of blast, the ground's characteristic sonic velocity v_p should always be taken into account (Table 1). First of all, the greater the match between the ground's acoustical impedance $\rho_r v_p$ with the explosive's impedance $\rho_e v_e$, the greater will be the amount of dynamic or shock energy transferred into the ground when the explosive first reacts. However, in a practical sense, the amounts will be low in most instances because of the large difference normally present in the respective ρ_r and ρ_e values. Only where dense explosives with high reaction rates are used are impedance mismatches kept to a minimum. The practice of decoupling such as used in cushion blasting, for example, uses the principle of providing the greatest possible mismatch of impedance for the expressed purpose of minimizing dynamic shock transfer.

The combination of the ground's v_p and explosive's v_e, charge length PC, and primer location is important also because it determines the shape of the composite stress front generated within the ground from each successive point along an explosive column. As shown by Fig. 9 for end-initiated blastholes, a conical stress form develops when v_e exceeds v_p. If the velocity ratio v_e/v_p is equal to one or less in value, the spherical form is developed. In this latter case, the resultant effect is one by which each incremental stress generated by successive points along a column reinforces those induced earlier. For columns that are center-primed, or initiated simultaneously at both ends, the stress forms where v_e exceeds

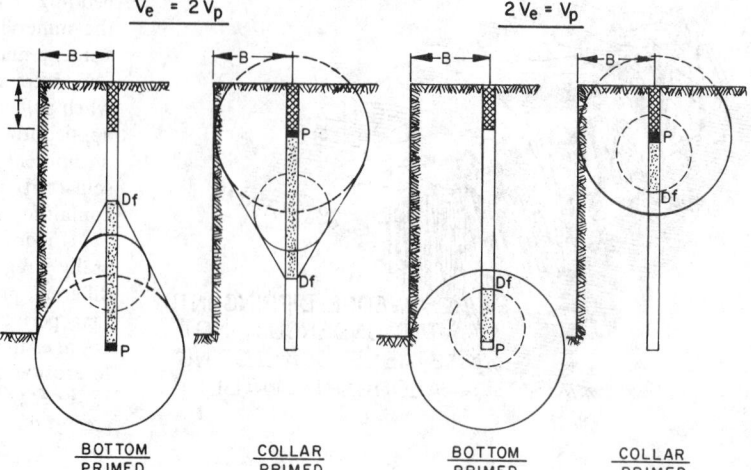

Fig. 9. Stress wave forms from end
initiated blastholes.

v_p will be merely double images of those for the single end-primed conditions.

Depending on the degree of saturation, fracturing of wet ground during blasting can be more extensive than when it is dry. Under wet conditions more than normal collar stemming amounts are recommended to control possible airblast and flyrock. The presence of water in ground has the effect of reducing the ground's compressibility, causing it to be less capable of absorbing strain energy. Because of wet ground's increased rigidity, the generated lateral strains are greater for the same applied longitudinal compressive strains, decreasing the ground's compressive and shear strengths. A blasting design, therefore, should include provisions to account for changed moisture conditions, particularly for the control of possible intensified detrimental side effects.

BLASTING MECHANICS

The purpose of blast design is to distribute explosive energy in such a way that certain fragmentation and muck pile displacement requirements will be satisfied. The primary means by which this is accomplished is by selecting the proper borehole size and locating the respective boreholes making up a round in their proper positions. To develop a suitable pattern, however, requires a working knowledge of a borehole's blasting characteristics when acting independently and together with other blastholes.

Because of its cylindrical shape, the typical blasthole directs most of its energy laterally, or perpendicular to its axis, not equally outward in all directions as does a spherical idealized charge. For that reason the burden dimension B is generally measured perpendicular to a borehole's length. The directional effect occurs from the fact that most of a hole's surface area against which explosive pressures act lies in a blasthole's sidewall. The condition becomes apparent when one thinks of a hole as a pipe or tube whose respective areas of surface could be described by the expression

$$A = 1.57D^2(2L/D + 1), \qquad (1)$$

in which A is the total area and D and L are the diameter and length, respectively. In the expression the $2L/D$ ratio is that share of the total area contributed by the sidewall while the 1 represents the portion due to the sum of the two ends. Thus, for many blastholes whose lengths are commonly sixty or more times their diameters their sidewalls contain over 99% of the surface area. Even when lengths are shorter and diameters proportionately increased, the sidewalls still contain most of the area. For example, a typical 12-inch diameter borehole bulk-loaded with a 20-ft long explosive column has 95% of its area in the sidewall, and a cylindrical point charge with equal diameter and length still has a sidewall area equal to two thirds of the total surface. In the event that it is desired for blastholes to crater toward their collars, it follows, therefore, that the largest possible diameters with very short charge lengths should necessarily be used even though the largest area for which the pressures will act would remain present on the hole side.

The actions of the two pressures are unique for the usual blasthole. Unlike the results caused by the detonation pressure generated by a concentric charge, the pressure produced by an explosive confined within a borehole is usually not effective. For a borehole the detonation pressure's energy is directed primarily axially through the explosive's column center. As a consequence of the detonation front's rapid movement through the column, the pressure is directed at an angle into the sidewall, reducing its thrust's effectiveness against the surrounding rock. Furthermore, the pressure's contact area is such that only a small portion of the borehole's rock surface is affected at any time instant, which moves as a very narrow circle of pressure that proceeds from the primer in progressive order along the wall to the charge's opposite end at which point the pressure is terminated (Fig. 4). The borehole pressure, on the other hand, undergoes a rapid pressure buildup at the explosive's reaction rate at the end of which it stabilizes at its full value to pressurize the entire surface area of the explosive chamber. It follows, therefore, that the borehole pressure is the dominant cause for rock breakage in field blasting, and it is the near-sudden buildup of that pressure in a borehole that is the major cause for generating ground vibrations from blasting.

Rock Breakage Mechanisms

Three distinct processes can break rock during blasting: slabbing from stress wave reflection, radial fracturing, and flexural rupture. True slabbing by reflection of stress waves, however, is considered absent or at least negligible under conditions normally present in field blasting. For slabbing to result, a free surface must be present at a burden distance within twenty borehole diameters to be even detectable.

Radial fracturing, or the formation of cracks oriented in lines with a borehole, results from tangential or hoop stresses generated in the rock surrounding a blasthole. The cracks can be initiated in a borehole's sidewall by both the detonation and borehole pressures if their respective energies are sufficient to overcome the rock's tensile strength. Fracturing in the sidewall results when pressurizing of the borehole cavity causes the wall surface to expand or stretch. Crack initiation by the detonation pressure is short-lived however, because of the momentary applied nature of that pressure. Cracks can also be started in the ground away from a borehole by stress waves as they encounter voids and other impedance discontinuities, providing particle motion stress levels at such locations produce strains sufficient in strength to cause tensile dislocation. Crack initiation by stress waves also is brief because stress waves quickly outdistance the cracks, whose peak velocities can approach only from 0.15 to 0.40 times a ground's sonic velocity.

Radial cracks form in a borehole wall initially at a primer, developing fracture planes that parallel the borehole axis as the explosive column reacts. The cracks form, however, only where explosive energy is located. Extension of cracks once started is accomplished by the borehole pressure as it loads the surrounding ground. As long as adequate strain energy is present at crack tips, the fractures will continue to extend, following paths that are normal to the maximum principal tensile stresses. More energy is needed to start cracks than to extend them, and as cracks become longer less and less pressure is needed for further extension. The extension process occurs in a similar manner to a wedging or splitting action.

Where there is only one free face in which a borehole collar is located, the radial cracks form a symmetrical pattern about the hole. In that the ground can displace only toward the single free surface, an uplifting of the collar rock results, causing cracking to extend to the surface. The result is a loosening of the collar rock and hole stemming and this always causes some airblast and flyrock. Use of excessive burden distances to the sides of holes produces a similar effect.

If there is a second free face located parallel or nearly so to a borehole, radial cracks will begin to distort toward the second face as it is brought closer to the hole. Because of the open surface to the side the high pressure forces the ground in between to displace toward that face, causing the borehole to assume an elliptical shape oriented perpendicular to the side surface. The extended sides of the hole, thus, are placed in greater tension than are the hole portions nearest and farthest from the side face. As the burden is reduced even more to the side face, the cracks reduce in number as well as extend more from the hole's sides until they intersect the side face. Where a borehole is properly located, there is little or no radial cracking in back of the hole and collar overbreak is generally absent. With the addition of a third free surface, or two side faces as on a bench corner, the effect is to offer further reduction in resistance for the radial cracking process as long as the two side faces are located at closely similar distances from a blasthole.

The outer edges of the radial crack system define the limits of the blasthole crater formed (Fig. 10). Two factors determine the crater's apex angle: the orientations of the geologic structural planes and the relative distances from a borehole to all open faces. A narrow crater apex angle with back shatter always indicates an excessive side burden. A borehole with but two open faces will have a crater apex angle that may vary from as little as 60° up to as broad an angle as double that amount. As the burden distance decreases the apex angle widens up to its maximum value. The reverse effect occurs as the burden distance increases until a crater angle of about 60° is reached; below this point the

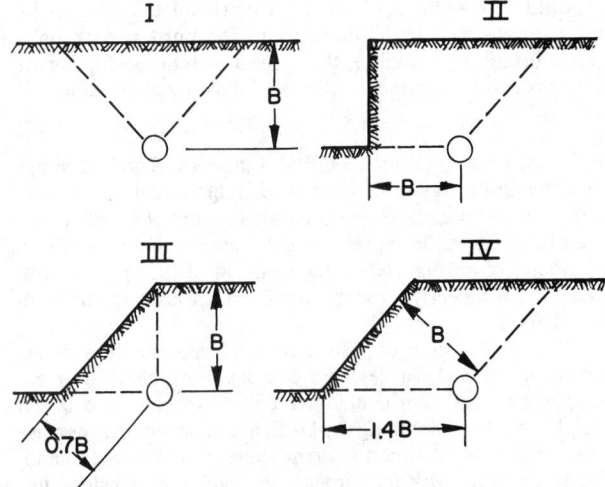

Fig. 10. Basic crater forms.

intervening rock becomes difficult, if not impossible, to be displaced outward. If three free faces are present, the crater formed has an apex angle from about 135 up to 180°. In this situation the preexisting corner angle as well as distances to the two side surfaces will determine the resulting crater angle following blasting. For estimating and general design purposes, however, it is customary to assume a 90° apex angle for craters obtained from blastholes having two free faces and 135° for the three free-faced condition.

The segments of rock bounded by radial crack planes are fixed in place where stemming is used and at borehole bottoms if no pronounced discontinuity planes are present along which easy slippage can occur. The breakup of the intact rock segments, however, is accomplished by flexural rupture through bending of the segments as they are displaced outward by gas pressure expansion action. The bending action occurs in both directions, i.e., in the hole's length plane and its diameter plane, the process being clearly visible during any blast. Rock movement and the accompanying bending action occur shortly after radial fracturing is completed, the initial movement beginning after a time lapse following a blasthole's firing of about 3.3 ms/m (1 ms/ft) of burden distance. Without the flexural rupturing complete breakup of burden rock would not be possible.

Burden Rock Stiffness

Breakup of rock from flexural rupture by a blasthole is controlled by the stiffness condition of the burden rock. The principle of stiffness can be easily understood if it is recognized that a thick short block is always more difficult to break by bending than would be a long, slender, more flexible segment of material. In applying the stiffness concept to blasting, the burden rock on a single blasthole when radially cracked can be considered as a modified rock beam with a fixed hole bottom and collar zone that is restrained but allowed to rotate. In the case of several blastholes fired together, or fired at close-interval timing so their burden rocks respond more or less as a single unit, the combined rock mass would function like a wide rock beam or plate.

As with any beam structure, the stiffness property K of a blasthole's burden rock defines the amount of loading force or pressure needed to produce a unit external displacement. According to ordinary beam theory, the value of K will vary directly with the product of the beam material's Young's

Modulus, its width, and the third power of the ratio of its thickness to its length. In terms of the burden rock on a single blasthole, therefore, the applicable relationship for its stiffness can be represented by the following expression:

$$K \sim E_r S(B/L)^3, \text{ N/m (lb/ft)} \qquad (2)$$

in which E_r is the elastic modulus of the rock, S is the average crater width or borehole spacing, B is the burden dimension, and L is the open face length paralleling the blasthole (Figs. 1 and 10). From the expression it can be seen that the stiffness character of burden rock is independent of the applied pressure but is linearly proportional to the rock's modulus of elasticity.

The B/L ratio is of particular significance to the stiffness of burden rock because of its compounding effect. For example, with restraint conditions, charge spacing, and bench height constant, doubling the burden dimension will increase the stiffness by as much as eight times. On the other hand, for a blasthole with its burden and spacing dimensions unchanged, increasing the bench height by just 25% reduces the burden rock's stiffness to one half that existing previously. In other words, relatively minor changes in either the burden or bench height with the other dimension remaining constant will cause large adjustments to the stiffness condition of burden rock.

Reductions in stiffness are greatest when B/L ratios are decreased within the 1 to 0.5 value range, the reductions becoming less and less significant with each incremental additional decrease in ratio. Because of the stiffness effect the use of larger diameter holes with their greater burdens for relatively short bench heights invariably produces coarser fragmentation and a potential for developing problems of stronger ground vibrations, overbreak, and the like. To avoid the problems resulting from excessive stiffness, therefore, it is recommended that as a rule blasthole burdens should be limited at a maximum to one half or less the bench height dimensions. Fig. 11 is a set of plots obtained by finite element analysis showing the initial displacements resulting during the bending process for three different bench height conditions with a common burden dimension.

In much of blasting, the hole collar and bottom regions often cause the most problems. At both locations the flexural tensile stresses are maximum at the explosive charge ends, causing the fractures to start at the borehole and then extend

BACK-BREAK VS. BACK-SHATTER

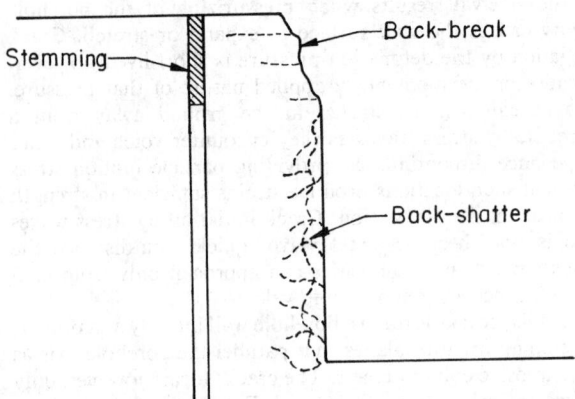

Fig. 12. Effects from excessive burden rock stiffness (Anon., 1975).

toward the open faces. The movement action of the collar zone is normally one of rotational uplift produced by the bending action of the burden rock below. Backbreak at the collar results where the rock mass in that region is too stiff to break completely through at the borehole (Fig. 12). In this case the amount of unloaded or stemmed hole collar relative to the burden dimension, or T/B ratio, controls the stiffness of collar rock in the same way the B/L ratio affects the bending capability of the lower burden rock (Fig. 1). To reduce the stiffness, radial fracturing must be raised higher toward the hole collar so as to reduce the rock's thickness between the explosive charge and open collar face. Figs. 13 and 14 illustrate three techniques for solving the problem as may be appropriate for the conditions.

Bending at the hole bottom is similar to that at one end of a beam fixed at both ends and is necessary to eliminate toe if there is no free parting at that location (Fig. 15). Normally drilling below grade level prevents toes. The effect of the subdrilling is that it lowers the position of the max-

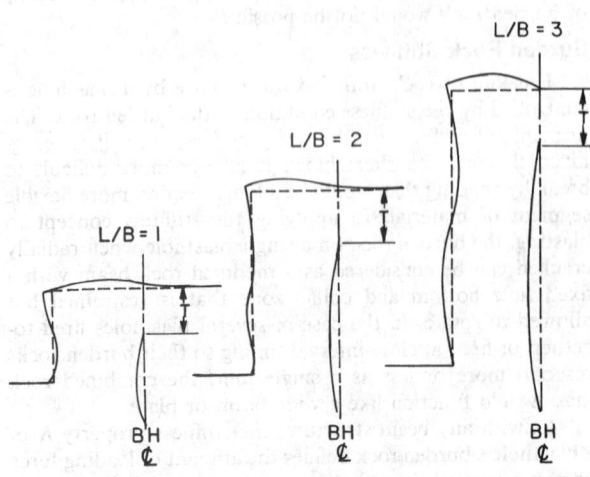

Fig. 11. Bending conditions during bench blasting (Smith, 1976).

TOP BREAKAGE IMPROVEMENT

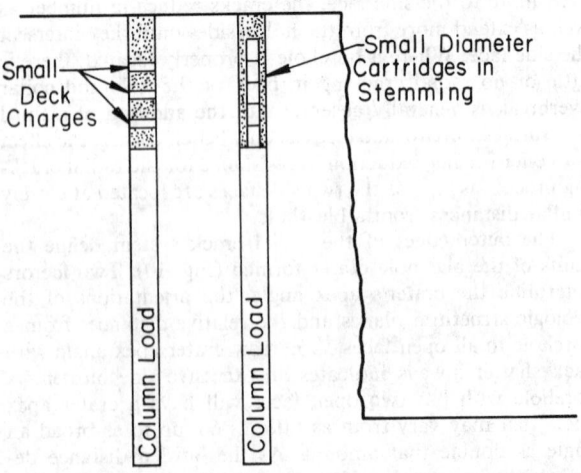

Fig. 13. Example techniques for breaking collar rock (Anon., 1975).

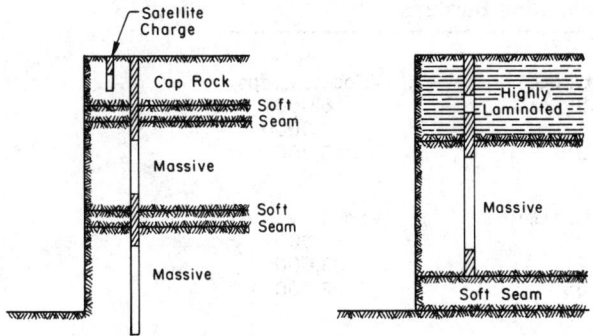

Fig. 14. Satellite charge and stemming applications.

imum moment during bending so fractures can proceed perpendicularly from the borehole to the open face at floor level. However, subdrilling will be of little help if the upper burden rock cannot be moved outward in advance by bending because it is too stiff.

BLAST DESIGN GUIDELINES

There are nine design variables normally considered as important and over which the blaster has some measure of control when designing a blasting round. Six of the factors are dimensions: the explosive's charge diameter D_e, the blasthole burden B, collar stemming T, subdrilling J, and hole length H, and the spacing distance S between adjacent holes. The remaining three variables are the explosive's specific gravity SG_e, the primer locations P, and the respective firing times of the blastholes. The various factors are all interdependent with values for the dimensional ones having been determined empirically. Frequently certain of the variables are combined in the form of dimensionless ratios such as the burden in terms of the charge diameter, stemming as a function of the burden, and spacing as some multiple of the burden, for example.

The problem with developing a design for a blast is that a blaster must select the combination of variables best suited for the conditions existing at a given site and operation. The recommended procedure is to first consider the single blast-

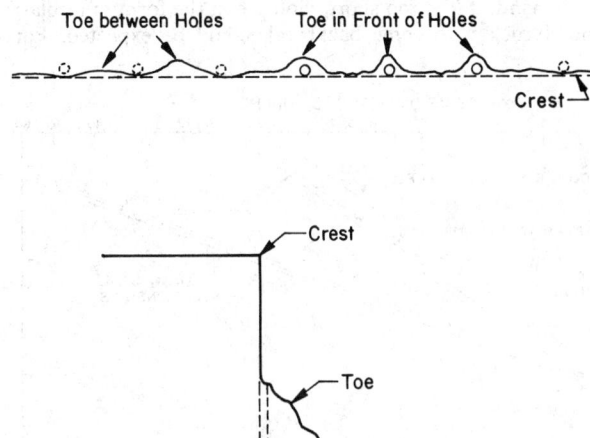

Fig. 15. Two types of toes resulting from bench blasting (Anon., 1975).

hole and its probable radial cracking and flexural breaking effects, taking into account the ground conditions and being certain its individual design specifications are correct. Secondly, all possible interactions between neighboring blastholes comprising the round must be taken into account so that the desired overall results will be obtained. In each case tight conditions should be avoided in so far as possible.

Single Blasthole Relationships

The charge diameter D_e is the most critical dimension of the design factors because it controls the explosive quantity at any one location within a blasthole for any given explosive type (Fig. 2). It may also affect the explosive's reaction velocity (Fig. 3), thus influencing the dynamic characters of the borehole and detonation pressures, the shape of the composite stress form generated in the ground, and to some degree the relative frequency of cracks in the blasthole's immediate vicinity. Smaller-diameter holes develop more fractures in their walls than do larger ones because of their greater curvatures. Selection of a specific charge diameter may be restricted, however, because the available borehole sizes will depend on the drill equipment on hand. But wherever possible it is recommended that the D_e be no larger than 0.015 times the bench height L, the limiting point at which the burden dimension will not impose excessive stiffness on the burden rock.

The burden dimension, or B, is defined as the distance from a blasthole to the nearest open face at the time the hole is fired. Normally the burden would be considered in a direction perpendicular to a borehole's axis (Fig. 1), while for the crater type blast it is measured from the upper charge end to the open face at a blasthole's collar. In the latter situation, it is often called the depth of burial. Whether for bench or crater blasting, its optimum value will be that which produces the desired fragmentation and displacement of the broken ground that provides the most efficient subsequent removal. The burden's primary function is that it provides the major control over the directions and extents of radial cracks and is one of the main factors affecting the stiffness character of burden rock. There is no common burden for all explosives and materials; the best dimension will depend on each given situation and should be determined after initial field tests.

For first approximating the burden's value there are a number of relationships. Many formulas utilize charge volume, weight, or diameter as the basic independent variable with the burden, or burial depth, being a function of the cube or square root of the variable. Other expressions consider the burden as dependent on the product of the charge diameter and the square root of the ratio of the explosion pressure to the rock's tensile strength or directly on the square root of the product of the charge diameter and hole length. Still another proposed relationship determines instead the required charge weight for a given burden dimension. In most formulas various empirical constants are included to account for the explosive's and material's properties. Some of the relationships, on the other hand, require initial tests to determine the necessary values for the needed constants. Regardless of the relationship, most provide burden values that are generally comparable within certain limits.

Of the many formulas, a relationship developed from the analyses of numerous field blasts that is relatively simple for blasters to use and applicable for most conditions for estimating the burden is as follows:

$$B = 0.036 \, D_e \, (SG_e/SG_r)^{1/3}, \text{ m} \qquad (3a)$$

Table 3. Representative Values for Estimating Burdens

Properties of Explosives

	(Density, SG_e)	(Velocity, v_e, fps)
Minimum	0.8	6,000
Maximum	1.6	24,000
Average	1.3	15,000

Properties of Rocks

	(Density, SG_r)	(Velocity, v_p, fps)
Minimum	1.92	6,000
Maximum	3.21	23,000
Average	2.65	12,000

$(SG_e/SG_r)^{1/3}$ Ratios

	(Light Density Explosive)	(Heavy Density Explosive)
Light Rock	0.75	0.94
Heavy Rock	0.63	0.79
Average	0.79	

in which B is in meters and D_e is in millimeters. If D_e is in inches and B is given in feet, the relationship is

$$B = 3 D_e (SG_e/SG_r)^{1/3}, \text{ ft.} \qquad (3b)$$

Table 3 lists representative $(SG_e/SG_r)^{1/3}$ ratio values covering most explosive-rock combinations.

The burden may also be estimated on the basis of an explosives loading density d_e as follows:

$$B = 1.087 d_e^{1/2}, \text{ m} \qquad (4a)$$

in which B is in meters and d_e is in kg/m of charge length. For the burden dimension in feet and loading denity in lb/ft, the expression is

$$B = 3.562 d_e^{1/2}, \text{ ft.} \qquad (4b)$$

Fig. 16 illustrates the ranges and trends of B values in ft as a function of D_e in inches and d_e in lb/ft.

To confine explosive gases until they have had time to adequately fracture and start ground movement, stemming material is placed in the collar region of a blasthole (Fig. 1). Crushed rocks of medium to coarse sizings are considered best as stemming; they tend to lock or wedge in the hole when it is pressurized. Since the stemming material must be first compressed before it can be ejected, the confinement occurs as a brief delayed action. Stemming material that is wet or has a low compressibility, such as finely sized cuttings, provides little or only minimal confinement. If the burden

rock below is sufficiently flexible, its bending outward and the accompanying collar rock rotational uplifting action cause the borehole to partially close at its collar, helping to further lock the stemming material in place.

Experience has shown that relatively dry collar stemming for most bench blasting conditions should fill a hole collar a distance T of from 0.5 to 0.7 the burden dimension. A convenient approximation relationship frequently used is as follows:

$$T \sim 2B/3. \qquad (5)$$

Because the stemming volume occupies about 1.5 times that of solid rock due to swell when compressed by the explosive gases, the T dimension, or the collar rock's thickness, therefore, will become about 0.5 times the B dimension, i.e., the collar rock's length. The result is to decrease the stiffness of the collar rock, aiding in its breakup by flexure when uplifted, similar to the B/L ratio limit of 0.5 at which burden rock becomes excessively stiff as discussed earlier.

Collar priming requires more stemming than bottom priming as does also a relatively stiff burden rock that inhibits the necessary collar rock uplift. However, too large a T value results in the formation of boulders or blocks from the collar zone with a potential for severe backbreak or undercutting when rock structural conditions are unfavorable. On the other hand, if T is too short, violence in the form of airblast and flyrock with some backbreak could be expected. For

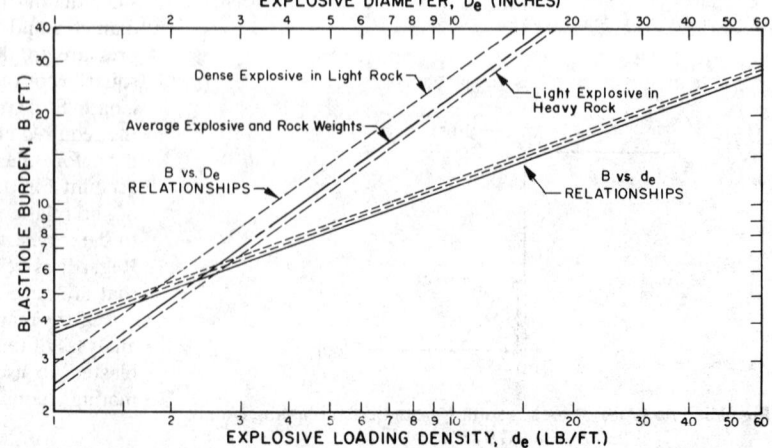

Fig. 16. Chart for estimating blast-hole burdens.

crater blasting in which breakage is intended toward the collar, the T dimension should equal the estimated B value.

Stemming is applied not only in the borehole collar but also wherever it is necessary to confine explosive gases that might otherwise vent such as soft mud seams, open fractures, and the like, or where it is desired not to damage material as with coal (Fig. 14). The two-thirds rule is not normally applied in such cases, or for deck loading, in which, instead of a continuous explosive column, several separated charges are contained in the same borehole to bypass the areas to be avoided. In those situations, the rule is to employ stemming in the amount of about one-third the burden dimension on each side of the hole portion to be left void of explosive.

As a consequence of the particular free face conditions in blasting with a fixed borehole bottom, blastholes must be drilled below floor or grade level in massive ground. The actual amount of subdrilling needed depends on the burden distance, orientation of the blasthole relative to all open faces, and the geological structural conditions at its bottom. If an easy shear plane exists at that location, little or no subdrilling normally would be required. Otherwise, as a first approximation the rule is that the subdrill J should be about equal to one-third the B dimension for hard ground, a relationship used for estimating the subdrill being

$$J \sim B/3 \qquad (6)$$

In the event some high bottom remains following blasting or burden rock above is quite stiff, i.e., a low B/L ratio, more subdrill is required but should not exceed one half the burden's value. Excessive subdrilling wastes drilling and explosives, causes intensified ground vibrations, and often leaves fractured and gouged pit floors. To satisfactorily displace bottom rock at a blasthole's toe region, it is always necessary for the burden rock above to be moved first. Faces that are perpendicular to floors are placed in high compression so are more prone to develop toes; angled faces relieve the tightness, thereby usually requiring less subdrill. Bottom-primed holes also generally need less subdrilling than those employing collar priming.

Where the bench height L is a fixed amount the hole length H similarly will be predetermined, the H dimension being the sum of L and J. For an opening sinking cut the value of H is considered to be a function of the cut depth that can be successfully pulled. In that regard a common principle frequently followed is that only about half the narrowest width of the opening cut could be expected to be properly broken. Irrespective of the type of cut, the hole depth must be compatible with the burden dimension if excessive burden rock stiffness is to be avoided with a length of at least twice the burden.

Long hole depths are generally undesirable, particularly when they are of small diameter. Hole lengths greater than four or more times their burdens are subject to potential cutoffs of their explosive columns in layered ground. For the smaller diameters, the error in maintaining proper hole alignment is compounded as length is increased. In addition, having to work at the edges of high bench faces while drilling and charging front rows of holes with their small burdens presents a hazard. Blasts with extra long holes can be difficult to control and should always use multiple primers. To avoid difficulties it is recommended, therefore, that insofar as practical, blasthole lengths should be kept within a range of two to four times their burden dimensions. A relationship for estimating the appropriate hole length H in meters with subdrilling in terms of explosive charge diameter D_e in millimeters is as follows:

$$H \sim 0.097 D_e, \text{ m.} \qquad (6a)$$

For a D_e expressed in inches the approximate H in feet would be

$$H \sim 8.1 D_e, \text{ ft.} \qquad (6b)$$

It should be noted from the above relationships that for any given D_e, the bench height L would be from two and one-half up to four times the estimated range of B dimensions for most explosive-rock combinations, with an L value of approximately three times B for the average 2.65 specific gravity rock in which a 1.3 specific gravity average explosive is used for the blasting.

The primer position for an explosive column charge determines that portion of the burden rock which is first radially cracked and begins movement (Fig. 9). As hole depths increase, differences in blast effects become more pronounced. Collar priming, for example, usually promotes a waterfall effect, causing the broken material from a blast to be stacked in high piles adjacent to the vertical face. On the other hand, bottom priming tends to throw and spread the fragmented rock over a pit floor. Except where a high muck pile is necessary for efficient loading, bottom priming is the preferred practice from the standpoint of safety and explosive energy utilization. Locating the primer in the hole bottom ensures there will be no unexploded explosive remaining in the bottom after firing and provides maximum confinement, assuming adequate collar stemming is present and the explosive column is not too long. Regardless of primer location, the initiator, or detonator, in a primer should be pointed toward the bulk of the blasthole total charge. Initiators as well as all explosive column charges are very directional in their execution.

Collar, or direct, priming produces an ideal situation for premature collar rock fracturing and movement. Where the upper part of the burden rock is displaced too early, the reduced confinement can result in severe airblast, excessive flyrock, and the formation of toes. Top-of-blasthole priming also favors the creation of backbreak, cutoffs to adjacent blastholes by top rock movement, and potential misfires due to excessive pressure drops for explosives still in the process of reacting. Collar priming should be employed only where instantaneous or very closely timed initiation is present, blastholes are relatively short, vertical uplifting or cratering is required, or where multiple priming is needed.

For small-diameter holes, i.e., those less than 100 mm (4 in.), the primer should be the first charge loaded into a hole. In the larger-diameter blastholes, the primer can be positioned where the most difficult rock to break is located within a bench. Normally where single priming is used, locating the primer at floor level to help reduce ground vibrations will be satisfactory rather than in the subdrill, providing the primer is of adequate size and the explosive in the subdrill is reasonably sensitive. Since it is the opening in the bottom created by subdrilling that provides the stress conditions necessary for the removal of toe, some operators, to reduce ground vibration levels, load no explosive in the subdrill but instead fill it with coarse rock fragments or similar type material. However, the practice requires either a pronounced seam at floor level or a bench height in excess of three times their blasthole burden dimensions.

Multiple priming of a blasthole is required where there are charges separated by stemming, termed deck loading; seamy ground conditions presenting a possibility for cutoffs; and the hole is quite long. In some states, at least two primers are required. The use of deck charges may be needed to reduce the charge quantity fired at any one time so as to minimize ground vibrations, for economy by decreasing the overall powder factor, and where it is necessary to bypass weak zones such as mud seams (Fig. 14). Each deck charge will require its own primer. In the first case the deck charges employ different initiation time-delay periods. For the other two instances all decks may be fired at or near the same time if the total charge amount is not excessive and initiation is begun at the hole bottom. The exception to initiating the bottom charge first would be where it is necessary for top rock to be lifted or moved first to provide relief for lower decks. If deck loading is used to reduce powder factor only, it is important that deck loads in adjacent blastholes be alternated, or staggered, relative to their respective positions in hole lengths to ensure that explosive distribution is uniform throughout a bench's burden rock.

A long small-diameter hole generally requires multiple primers because it is particularly vulnerable to premature ground movement while its explosive column is still reacting. There are two reasons for this: explosives typically react at slower velocities in small diameters and a hole's burden is small relative to its borehole length, creating a very flexible burden-rock condition. To be certain all the explosive charge will react completely before ground movement occurs, therefore, requires multiple primers to be placed along the charge column, being fired at nearly the same time or at least in a sufficiently rapid sequence before radially cracking is complete and ground movement commences.

To determine the need for primers in addition to the one at a hole's bottom, an estimate can be made by ascertaining the earliest time a crack could propagate from the hole to the nearest open face and comparing that time with that required for the explosive column to completely react. Under most circumstances it can be assumed that the shortest possible distance for any crack to propagate would be the burden as measured from the bottom primer. Since cracks can propagate at the very maximum at a rate of about one-third a rock's sonic velocity, the shortest crack propagation time for a blasthole then would be $3B/v_p$. On the other hand, the minimum time for an explosive column to completely fire would be equal to the powder column length PC divided by the explosive's velocity, or simply PC/v_e. For most situations, therefore, a blasthole's explosive column should be no longer than three times the burden dimension for a single primer located at its hole bottom. If high velocity explosives are used in low sonic-velocity rock, longer powder columns could be used without additional primers located higher in the explosive column, the reversed condition requiring shorter column lengths. As a means for estimating the maximum length of explosive charge PC_{max} for any single primer, the following expression will be appropriate:

$$PC_{max} = 3B(v_e/v_p). \qquad (7)$$

Where the actual explosive column exceeds the PC_{max} value, the additional required primers should be spaced equidistant along the column as may be appropriate. For example, considering a 75-mm (3-in.) diameter blasthole with a 1.83-m (6-ft) burden, an 11.6-m (38-ft) powder column, and a v_e/v_p ratio of 0.6 so that PC_{max} would be 3.24 m (10.6 ft), a total of 3.58 or 4 primers would be recommended, spaced 2.9 m (9.5 ft) apart beginning with the bottom primer.

Multiple Hole Design Principles

The typical blasting round consists of a number of blastholes arranged in one or more rows of holes in such a way that certain specific requirements will be satisfied. As a rule, the arrangement will be designed to produce a given mini-

mum tonnage or volume of broken ore or rock, the in-place bench height and area of which will have their respective dimensions more or less limited because of restrictions imposed by the equipment used to load or strip the broken material. To produce the necessary volume of rock, the blasts will be one of three possible forms: the sinking cut for initially opening an area, the box or trenching cut for extending an excavation lengthwise, and the corner cut (Fig. 6). Once a bench has been established most cuts will be of the corner type until such time as a new opening will have to be developed. Regardless of the cut type, however, the same principles upon which to design an appropriate drill round will always apply.

There are three rules for locating drill holes that should be followed in every case when designing a blast: (1) tight conditions should be avoided whenever possible with adequate space provided for each hole to account for swell, (2) the proper burden should exist for each and every blasthole at the time it fires, and (3) holes should be positioned so the radial cracking and flexural rupturing processes will not be impeded. Insofar as practical, holes should also be aligned with the rock's joint planes and particularly so the perimeter holes of the blasted area ensure clean breaklines at the edges. In addition, a round's cut holes, i.e., the first blastholes to fire, must be properly located so they will provide the necessary relief for all subsequently fired blastholes.

Equally important to the burden and hole depth dimensions is the charge spacing in a blast's design. The spacing dimension S is defined as the distance between any two adjacent blastholes, measured perpendicular to their burdens. The spacing in conjunction with the firing times of adjacent charges and the rock's structural discontinuities provide the primary means for controlling whether or not the holes will interact during a blast. Where the spacing dimension is not compatible with the timing and rock character, the fragmentation and the control over detrimental side effects will be affected. For example, too wide a spacing produces poor breakage, toes, and humps on faces between holes. Too close a spacing generates excessive fracturing and collar cratering between holes, blocky burden rock, and toes. Airblast and flyrock frequently occur also under such a condition. In selecting a spacing it is most important, therefore, that one first consider the probable directions radial cracks will follow, any possible interactions that may occur between holes in any direction, and the general overall stiffness conditions of the burden and collar rocks of the neighboring blastholes at the respective times they are intended to be fired.

For design purposes, it is common to consider the spacing in terms of the burden on a blasthole in the form of a ratio. For most situations the S/B ratio's value will be from 1 to 2; although in certain instances where conditions are favorable, larger ratios have been used with satisfactory results. In general, blastholes fired independent of one another will require a spacing ratio between 1 and 1.5, a 1.41 value being the ideal geometric balance for breakage of massive material. For rocks with joint planes located nearly perpendicular to one another, the following first approximation for independently fired holes will be applicable:

$$S = 1.41B. \qquad (8a)$$

Where a rock's joint planes are oriented at close to 60° with one another, blastholes with long-delay interval firing times generally require closer spacings as given by the following relationship:

$$S = 1.15B. \qquad (8b)$$

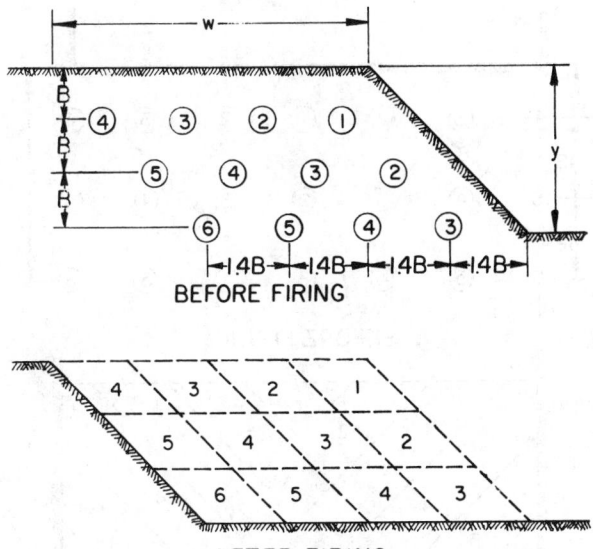

Fig. 17. Plan view of angled corner cut with staggered pattern for echelon independent delay firing.

Minor adjustments to the spacing should be made if the overall fragmentation achieved and general appearances of the bench faces remaining after blasting are not satisfactory. However, modifications in the spacing cannot be expected to compensate for an extremely stiff burden rock condition, e.g., a B/L ratio that exceeds 0.5 in value. Figs. 17 and 18 provide a comparison of two three-row corner cut designs in which holes are timed to fire separately using the spacing relationships given previously. In the diagrams it should be noted that the individual areas of rock broken out in their respective diameter planes are well balanced with uniform burdens to their side faces at the time each fires. Also, in conformance with crater form IV given on Fig. 10, the re-

spective burden rocks can be moved out easily as a consequence of their characteristic wide crater-apex angles.

For blastholes fired at close interval timings or simultaneously so interactions can occur between adjacent holes, the spacings must be extended to produce uniform breakage with a minimum of detrimental effects. Otherwise, radial cracks from borehole sides will join prematurely before breakage by bending out of the solid can develop. For a sufficiently large enough bench height, i.e., the B/L ratio is less than 0.5, ideal geometric balancing between charges will exist where the spacing dimension is double that of the common burden. However, field experience and laboratory tests on models show that as burdens are increased for a given bench height, spacings must be reduced to compensate for the stiffening of the burden rock where burdens are larger than one-third the bench height. The effect is apparently geometrically controlled and independent of the rock type. The similarity of trends is illustrated by Fig. 19, which shows the maximum permitted S/B values at various L/B ratios for complete shearing between holes obtained from tests conducted on models of three completely different materials. On the basis of field data and laboratory test results, the relationship for approximating charge spacing for blastholes fired closely together with hole depths three or less times their burdens would be the following:

$$S = (BH)^{1/2}. \tag{9}$$

In those situations where H is more than three times the value of B for blastholes fired at close time intervals, or more or less together, the S dimension of $2B$ usually will be satisfactory. For rock conditions under which it is difficult to break cleanly between holes, a slightly reduced spacing to about $1.8B$ generally is quite sufficient.

Two basic blast designs in which holes in the same row are fired at the same time are illustrated in Figs. 20 and 21, one pattern being arranged with displacement of the broken material directed at any angle into the pit area opposite the bench corner, the other being designed to spread the broken rock uniformly out in the pit parallel with the bench's length. It should be noted, however, that there is slightly more bench area broken by the latter of the two patterns in spite of the

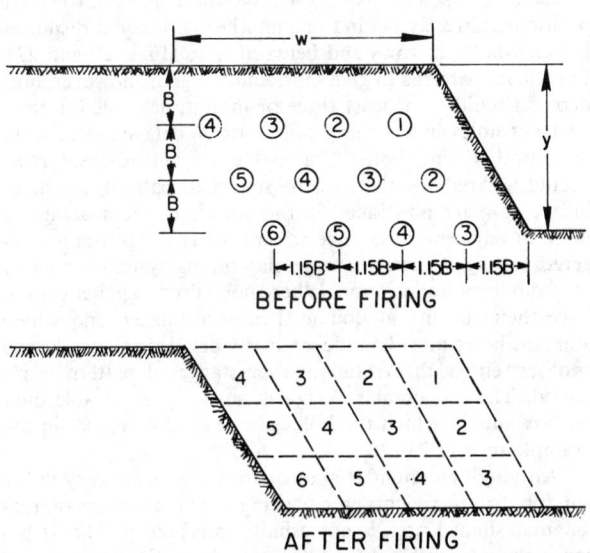

Fig. 18. Plan view of angled corner cut with staggered pattern for echelon independent delay firing modified for 60° degree rock jointing.

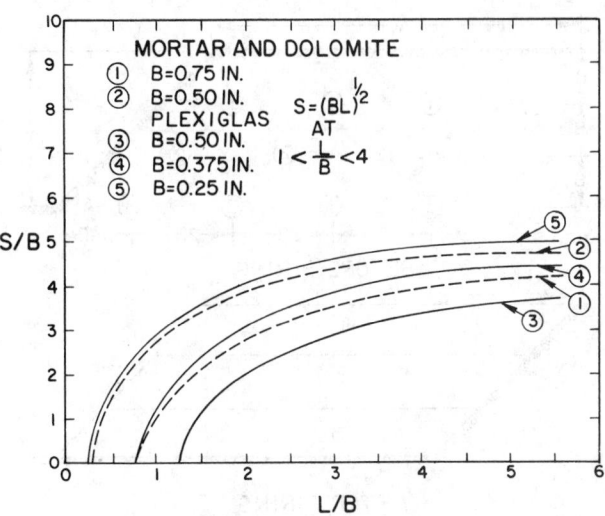

Fig. 19. Relationships between S/B and L/B design ratios for close-interval delay firing.

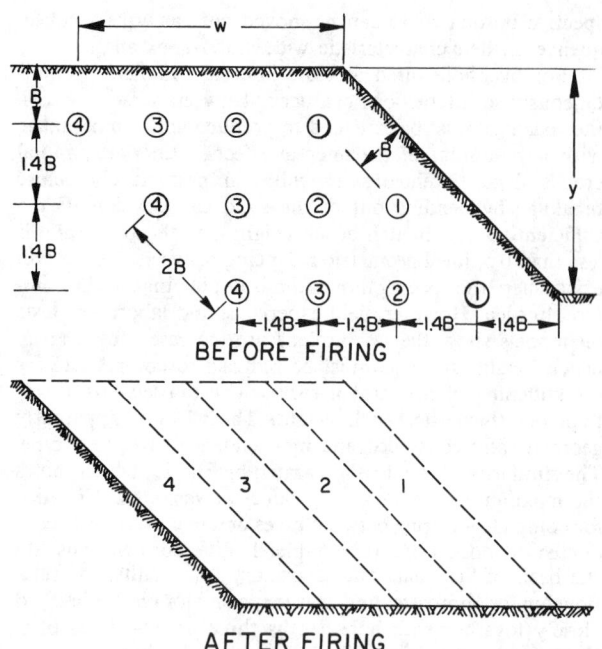

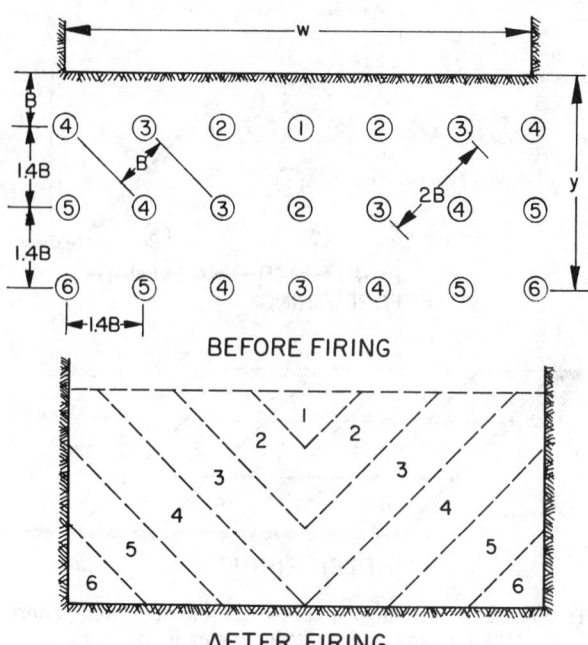

Fig. 20. Plan view of angled corner cut with square pattern for echelon close-interval delay firing.

Fig. 22. Plan view of V-cut with square pattern for close-interval delay firing.

fact both arrangements contain twelve holes each, burdens are constant for all holes when they fire, and the respective spacings are correct for the initiation timing system employed. On the other hand, the latter pattern is subject to causing overbreak at the end of each hole row while the design using delayed firing times in rows paralleling the bench length or *w* direction will be devoid of endbreak. The reason for this difference is that the extended length of open face beyond an end hole's location of a series of holes fired together creates a condition by which the composite burden rock for a row will function in a manner similar to a cantilever wide-beam or plate. The action is often clearly visible when such a blast is in progress. To avoid the problem, it is always

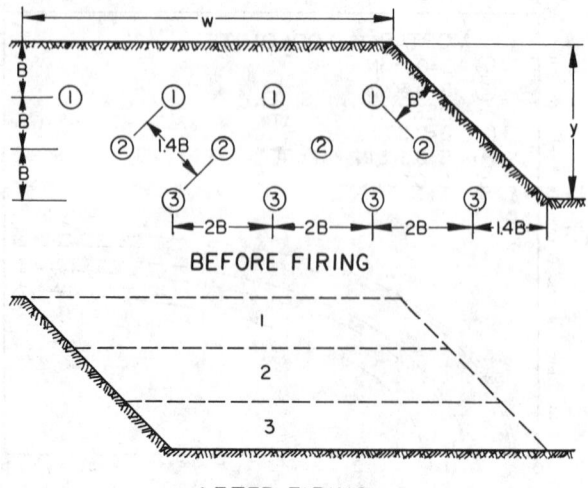

Fig. 21. Plan view of angled corner cut with staggered pattern for frontal close-interval delay firing.

best to delay the firing of the last hole in a row until the neighboring hole in that row has been fired and most of its burden rock has been moved out of its place. The spacing for the last hole should also be reduced as for any independently fired blasthole.

There are only two basic types of drill patterns: the square and the staggered arrangements. The distinction between the two depends on how the orientation of hole rows is defined. For corner cuts, the alignments of holes paralleling a bench's length, or *w* dimension, are considered as the rows (Figs. 20 and 21). For the sinking and box cuts, the rows normally would be the lines of holes paralleling the cut's longest dimension. The square pattern will be where holes in rows are positioned directly behind one another with equal distances between holes in rows and between rows (Figs. 20 and 22). For square patterns to give satisfactory results, however, hole depths should be at least three or more times hole burdens, adjacent holes in the same row must be delayed relative to one another, and holes located laterally in adjacent rows should be fired together. In the staggered pattern, the holes in one row are positioned in the middle of the spacings of holes in adjacent rows. The square pattern becomes a staggered pattern if the initiation delay-timing system is rotated 45° from its initial position. Where holes fired together cannot have their spacing at double their true burden and where adjacent holes must be independently delayed relative to one another, either the rectangular or staggered pattern is required. The rectangular design is normally applicable only for box cuts in which the holes are delayed as shown in the example in Fig. 23.

Although rectangular and square patterns are easy to lay out for drilling, a corner cut using either a square or rectangular shaped area is not usually satisfactory. This is because the square block of burden rock on the open corner is virtually impossible to break properly due to its shape (Fig. 24), and the last hole in the front row of holes with such patterns quite often tends to overbreak. To avoid the prob-

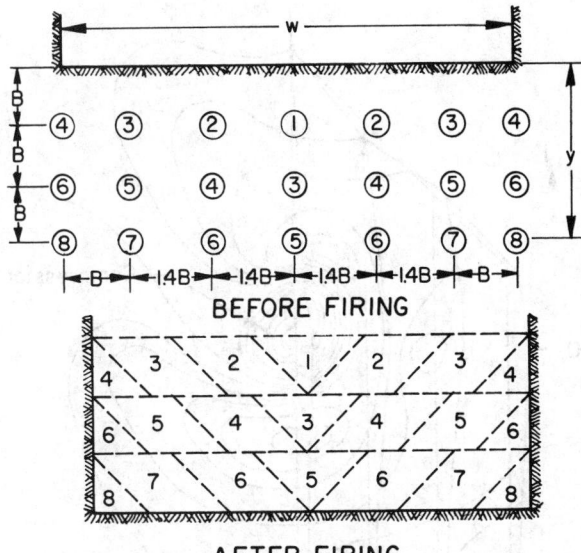

Fig. 23. Plan view of V-cut with rectangular pattern for independent delay firing.

lems, the two side faces for corner cuts are best situated when angled with one another. For the opening box cut to develop angled bench faces, the square pattern can be employed if hole depths are sufficiently long enough compared to their burdens and only those holes fired together laterally from the front of the cut to its back are used. Otherwise, a stag-

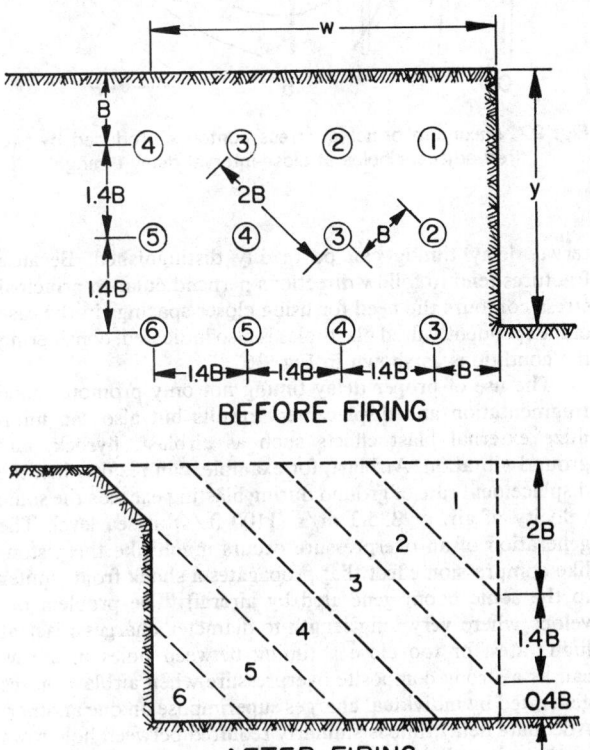

Fig. 24. Plan view of square corner cut with square pattern for echelon close-interval delay firing.

gered pattern with holes on row ends delayed most often is applicable, as illustrated in Fig. 25.

Equally important with the use of proper burdens, spacings, and hole depths is the selection of a compatible initiation delay-time pattern for the various explosive charges. Three time elements should be considered for each blasthole: (1) the time t_m for the burden rock to first begin its initial movement at the primer location, (2) the time t_e required for the explosive in the hole to react completely, and (3) the time t_b for all of the broken material to move out at least one full burden distance. For any two holes to be fired independently so no interactions are possible between them, then the total minimum delay time in their respective firings should be no less than the sum of the three time elements, or $t_d = t_m + t_e + t_b$. The move-out time is particularly critical for certain kinds of blasts, especially from the standpoints of controlling toe formation, overbreak, and potential violence caused by tight conditions.

A common rule for t_m is to assume that rock movement begins approximately 3.28 milliseconds for each meter of burden (1 ms/ft) after the primer's initiation, and broken rock can be considered to move out from faces at an average rate of about 23 m/s (75 ft/s). The time required for an explosive column to react t_e would be its length PC divided by the explosive's v_e. Therefore, to determine the minimum delay-time interval in seconds for independent firing of blastholes, the following approximation can be used:

$$t_d = B/305 + PC/v_e + B/23, \text{ sec,} \qquad (10a)$$

in which B and PC are in meters and v_e is in m/s. If B and PC are measured in feet and v_e is in ft/s, the relationship would be

$$t_d = B/1000 + PC/v_e + B/75, \text{ sec.} \qquad (10b)$$

For example, to avoid all possible interactions between two neighboring bottom-primed blastholes with 3-m (10-ft) burdens, 9.14-m (30-ft) powder columns, and containing explosives that react at 3050 m/s (10,000 ft/s), the estimated required minimum delay-time interval in their respective firings would be 143 ms. On the other hand, the full face of burden rock would start its initial movement outward in an estimated time lapse from initial firing of approximately 13 ms, or the sum of only t_m and t_e. Any time interval of delay firing used for initiating the second blasthole greater than 13 ms but less than 143 ms following firing of the first hole,

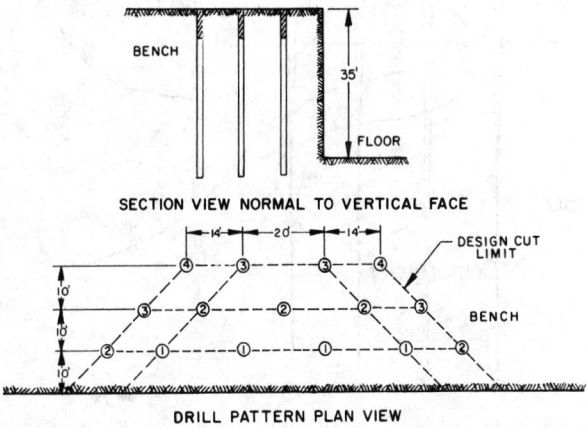

Fig. 25. Plan view of opening box cut using independent and close-interval delay firings arranged to avoid overbreak.

therefore, could be expected to produce interactions in the ground from the two holes.

Rock breakage enhancement effects from the use of millisecond delay initiation has been observed for many years. The effects result from strong shearing stresses developed by the differential displacements generated in the ground where borehole pressures from adjacent blastholes are applied while the respective burden rocks are still under pressure loadings. Mutual stress-wave interactions seldom if ever would be possible in normal field blasting due to the delay time-lapse intervals involved and the characteristic rapid propagation rates of such waves. For example, under the situation given above, stress waves would have developed and propagated away from the first hole in ground having a sonic velocity of 3000 m/s (9840 ft/s), a distance equal to about ten burdens in the three 13-ms time interval preceding any firing of the second blasthole. Figs. 26 and 27, produced by computer simulation using the finite element analysis technique, illustrate the differences in ground stressing conditions in the charge diameter plane for two delay-fired blastholes spaced at double their burden values and aligned with a dominant joint plane orientation. Comparing the two plots, the shearing interaction effects generated by the holes fired at close-in-

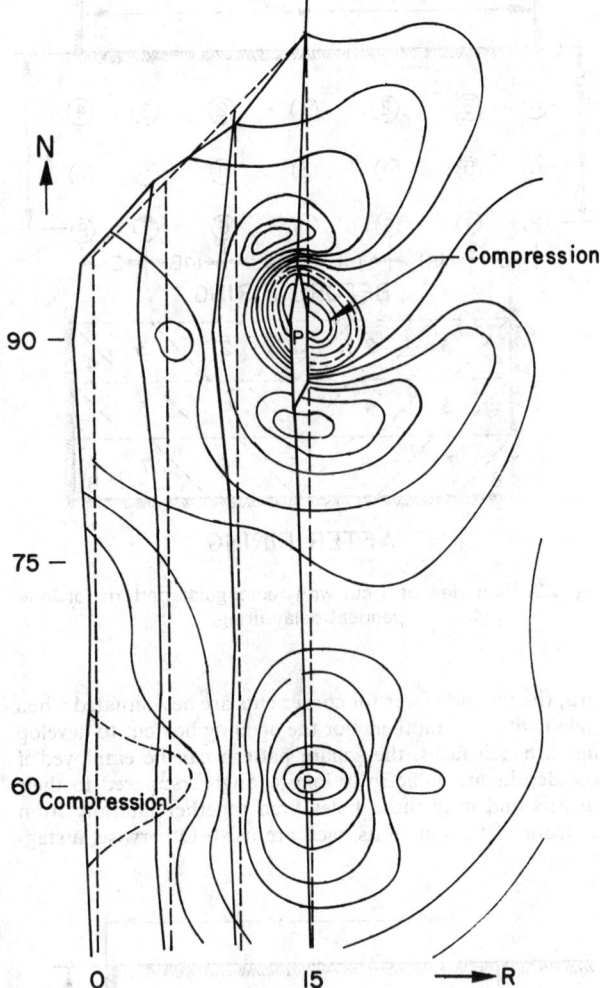

Fig. 27. Maximum principal stress contours produced by two fired adjacent holes at close-interval delay timing.

terval delay timing can be readily distinguished. Because fractures tend to follow directions perpendicular to principal stress contours the need for using closer spacings in the case of independently fired blastholes is also indicated, considering the conditions as shown in Fig. 26.

The use of proper delay timing not only promotes good fragmentation and displacement results but also can minimize external blast effects such as airblast, flyrock, and ground vibration. Airblast, for example, can result when the displacement rate of ground during blasting exceeds the sonic velocity of air, or 335.3 m/s (1100 ft/s) at sea level. The generation of an overpressure occurs much like the piston-like compression effect that propagates a shock front similar to the sonic boom generated by aircraft. The problem develops where very long length-to-diameter charges react at high rates, or too close a timing between holes in a row causes a strong composite overpressure when airblast pulses generated by individual charges superimpose on one another. Adequate delay time is similarly required between hole rows so there is sufficient relief to avoid reinforcement effects of airblast and ground vibration.

Several guidelines have been developed to aid blasters in the selection of delay time intervals that will provide suitable

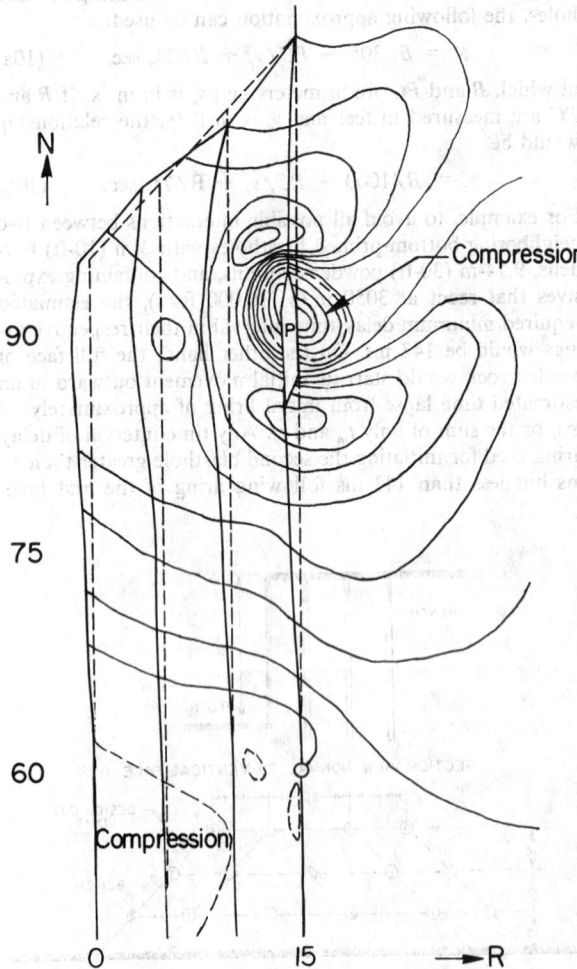

Fig. 26. Maximum principal stress contours produced by fired corner hole with pressurized crack developed along joint plane but adjacent second hole not fired.

fragmentation while keeping airblast and ground vibration at controllable levels. The rules are as follows:

1. The minimum delay time interval between holes in the same row should exceed 3.28 ms/m of burden (1 ms/ft) with an upper limit of 16.4 ms/m (5 ms/ft) to achieve good fragmentation in massive rock;

2. The delay time interval between rows of holes should be two to three times the delay used between holes in a row;

3. The rate of blast propagation along an open face should be less than the sonic velocity for air, i.e., 335.3 m/s (1100 ft/s) and preferably half that value where possible but at least 7 ms/m of spacing (2 ms/ft) to prevent airblast from superposition; and

4. The minimum delay time interval between adjacent charges, whether between holes in a row or between rows, or between deck charges in a single hole, should be at least equal to the nominal delay intervals of the delays used to control ground vibration.

To calculate the rate of blast progression along a free face to control airblast, the spacing distance between holes simply needs to be divided by the delay time interval used. For example, if the spacing between holes in a row for the blastholes described earlier were 5.5 m (18 ft) and the delay time interval were 50 ms, the rate of blast progression would be 110 m/s (361 ft/s) or only about one-third the sonic velocity for air, thus presenting no problem. With that particular delay interval, also, there would be rock breakage enhancement between adjacent holes because for completely independent firing a delay time of some 143 ms or more would be required between the firings.

Most regulatory authorities consider two charges fired at delay time intervals of 9 ms or more as two separate firings as far as controlling ground vibration is concerned. Few blasting rounds, regardless of the number of charges and their locations, present a delay timing problem since the introduction of the sequential timing blasting machine. Where surface delays and sequential timers are employed, however, in-hole delays are also recommended so as to reduce the possibility of cutoffs that might occur due to early ground movements.

Powder factor is not normally a sound index upon which to design blasts. The powder factor relates the yield of material blasted to the quantity of explosive used and will be extremely variable for the different types of rounds, explosive-rock combinations, and burden, spacing, and bench height conditions. Higher energy explosives, for example, can break more rock than can lower energy products, and box cuts provide lower amounts of broken material than do corner cuts using the same blasting design relationships. Comparing Figs. 17, 18, and 20 through 23, one will observe that not only can powder factors vary between cut types but also will differ for the same basic kind of cut depending on the particular delay-timing arrangement used. For surface blasting powder factors can vary from as low as 0.15 to as much as 1.5 kg/m^3 (0.25 to 2.5 lb/yd^3), with 0.3 to 0.6 kg/m^3 (0.5 to 1.0 lb/yd^3) being the most typical. On a yield weight to explosive used weight basis powder factors could vary from 3.4 t/kg to 13.5 t/kg (1.7 to 6.8 st/lb), depending on the materials blasted, explosives used, and kind of fragmentation desired. In general, for most blasting in construction and mining, the factors will range between 4.8 and 7.6 t/kg (2.4 to 3.8 st/lb). However, other than determining the powder factor for cost purposes, its use as a basis upon which to design blasts is not recommended.

6.3 Overburden Removal

GEORGE E. AIKEN
JOHN W. GUNNETT

GENERAL OVERVIEW

Any evaluation of a specific need for overburden removal requires that the engineer first be aware of the various designations of overburden, the state-of-the-art removal systems, and methods to transport these materials to permanent deposition areas.

When concerned with the development of a mine, the overburden removal system should be in harmony with the mine planning of the future. Unless this and the other factors influenced by overburden removal schemes are carefully considered, operations may suffer in efficiency and costs.

This section examines the definition of overburden, the tools to remove it, and the transport methods available for moving it to disposal sites. Table 1 is a generalized guide to excavating and transport methods that may be used for overburden removal, and Table 2 is a similar table more oriented to surface coal mining.

Overburden Defined

Depending on the user, the term overburden can be defined in several senses. For the miner, overburden is used in the context of any material, loose or consolidated, lying over a mineral deposit of ore or coal. This applies especially to surface mine operations.

To the builder, civil engineer, or soils technician, the term can refer to only loose soil, sand, or gravel that lies above the bedrock. The term should be well defined then as to the type of material(s) that are being considered to design for the most effective removal method. Having defined the overburden type, the excavation tools may be assessed.

Excavation Techniques

These are, for all practical purposes, limited to the presently produced production mining equipment. A discussion of each of the potential machines can be found in a later section. The most common machines in everyday use are: draglines, power shovels, dozer/front-end loaders, dredges, motor scrapers, and bucket wheel excavators (BWEs).

Occurrences: Excavator selection should be well suited to the type of overburden occurrence. This will involve consideration of depth of material, hardness, abrasiveness, moisture, suboverburden topography, waste dump sites, ore overburden interfaces, and surface topography.

Engineering studies to select optimum methods must ensure that the machines can efficiently cope with the conditions to be encountered. Where alternate choices may be possible, it may be necessary to carry out detailed comparative cost analyses studies.

General Techniques: In most instances the overburden is removed only far enough ahead of mining to ensure an adequate ore supply. However, the entire operation of overburden removal must be examined from the standpoint of project requirements. Where backfill for construction may be in short supply, the overburden stripping operation may need to be accelerated. Other factors that may affect the overburden removal techniques are the need for road construction, tailings, and water retention dams.

Consideration should be given to the multiple use of construction earthmoving equipment in the overburden removal scheme. Providing economically acceptable and compatible equipment may be selected, the overall costs can be reduced and double handling the material may be avoided. It is not uncommon to find that this equipment is fully depreciated at the start-up of production. This may allow the mine the chance to select different equipment that could be more advantageous for overburden only.

Another technique which some owners may find eco-

Table 1. General Material Types

Type of Equipment	Topsoil*	Vegetation	Soft and Dry*	Soft and Wet*	Medium Hard or Highly Fractured*	Dense Hard Overburden*
Excavator						
Dozer/Front-end Loader	X	X	X		X	X(?)
Scraper	X		X		X(?)	
Shovels			X		X	X
Draglines			X	X	X	X
BWE	X		X		X(?)	
Dredging				X		
Transport						
Scraper	X		X		X(?)	
Front end Loader	X		X		X	
Trucks	X	X	X	X(?)	X	X
Belt Conveyor	X		X		X	X(?)
Hydraulic	X(?)		X(?)	X		
Railroad		X			X	X
Other						
Crusher						X

*(?) Depends on specific conditions.

Table 2. Equipment Rating—Overburden Removal

		Dragline	Shovel	Shovel and Truck Comb.	Front-End Loaders	Dozers	Front-End Loader and Truck Comb.	Bucket Wheel Excavator	Scrapers		
									Elevating	Full Power	With Push Tractor
Thickness	0-30 ft	2	1	1	1	1	1	1	1	1	1
	30-60 ft	1	1	1	2	2	2	1	1	1	1
	60-100 ft	1	1	1	3	3	3	2	2	2	2
	> 100 ft	1	2	2	—	4	—	3	3	3	3
Characteristics	Poor Fragmentation (Blocky-Large Breakout Force)	2	1	1	3	1	3	—	—	—	—
	Moderately Blocky	1	1	1	2	1	2	—	2	2	2
	Good Fragmentation (Low Breakout Force)	1	1	1	1	1	1	3	1	1	1
	Unconsolidated	1	1	1	1	1	1	1	1	1	1
Transport Distance	50-150 ft	1	3	—	1	1	—	3	—	—	—
	150-300 ft	1	—	2	1	1	3	1	3	3	3
	300-500 ft	2	—	1	2	2	1	2	1	1	1
	500-1000 ft	—	—	1	—	—	1	—	1	1	1
	> 1000 ft	—	—	1	—	—	—	—	1	1	1
Coal Seam Support Characteristics	Good (Hard)	1	1	1	1	1	1	1	1	1	1
	Moderate	1	2	2	1	1	1	1	1	1	1
	Poor (Soft)	1	4	4	2	1	2	2	1	1	1
Segregation Capability	—	C	A	A	A	A	A	A	B	B	B
Production Capability	—	A	A	A	A	A	A	A	A	A	A
Flexibility Under Varied Field Conditions	Good	A	A	A	A	A	A	B	A	A	A
	Fair	A	B	B	B	A	A	B	A	A	A
	Poor	A	C	C	C	B	B	C	B	B	B
Mobility	—	C	A	A	A	A	A	C	A	A	A

Legend
1. Should be considered
2. May be considered
3. May be considered under certain conditions
4. May be considered special situation
A. High
B. Moderate
C. Low

nomically convenient is the use of contract stripping. This relieves the owner of any need for capital expenditures over the preproduction/construction period. Contract stripping can even be continued into the production years if economically rewarding for both owner and contractor.

With widely dispersed, smaller deposits, the excavator units need to be of small capacity and highly mobile. One or two large machines may be inappropriate due to their lack of selectivity between ore and overburden or their need for frequent moving about to work multiple faces.

Topographic conditions must be suitable for the selected excavator. Very steep, mountainous areas usually call for benching using shovel or front-end loader type excavators. Gentle to moderately sloping conditions often are more favorable to the dragline or bucket wheel excavator (BWE). Dozing and front-end loader combinations may be applicable over a wide range of topographic conditions.

Where the overburden is hard and must be blasted, the excavator should be able to efficiently and economically load the blasted rock chunks. In very soft overburden, the excavator must be supported as it moves across the area to be stripped. This will require a machine having low unit bearing pressures. These and other special problems of the excavator are discussed in the section on Overburden Excavation.

Overburden Relationships: Closely related to the quantity of overburden to be excavated is the relationship to pit development. Enough overburden has to be stripped back to allow the planned sequential development of the mine. Depending on depth of overburden, mine scheduling, disposal space availability, weather conditions, material stability, and cost relationships, the rate and method of waste excavation will be factors in sizing equipment.

Since overburden has already been defined as the capping on the mineral reserve, the excavation rate must be related to a designed stripping ratio. This is a parameter universally used in surface mining operations. The interpretation is not always the same and the term should be carefully defined to apply to a specific mine or condition. For example, it is quite different in the coal industry where it refers to the cubic yards (bank) of overburden removed per ton of coal mined.

However, in hard rock mining to define the stripping ratio in the example of Fig. 1, consider mining of the ore ABC at depth h. Having defined the slope angle at this mine section, the waste (overburden) section AICD must be removed. The ratio of this overburden stripped to mine the ore is known as the average stripping ratio (SR).

$$SR = \frac{\text{Volume waste at depth } h}{\text{Volume of ore at depth } h} = \frac{\text{AICD}}{\text{ABC}}$$

Thus SR can be shown to be a geometrical relationship. It can be noted in Fig. 1 that initially the overburden stripping ratio will have a theoretical value equal to infinity, e.g., no ore is recovered. In the case illustrated, when the first ore is mined the value of the average stripping ratio will have its highest value. This will decrease as the ore recovered increases with depth and will reach a minimum at some point. The average stripping ratio will begin to increase again as the open pit volume increases and the pit becomes deeper.

A ratio of the incremental amount of overburden to an incremental amount of ore is defined as the *incremental stripping ratio*, or SR. This is illustrated in Fig. 1 where the incremental volume of overburden CDEF must be removed to recover the increment of ore at depth h_1, or area BCFG.

$$SR = \frac{\text{Volume of incremental waste at depth } h_1}{\text{Volume of incremental ore at depth } h_1} = \frac{\text{CDEF}}{\text{CBFG}}$$

This relationship is also geometric, but using economic parameters the maximum incremental overburden removal ratio can be calculated. This maximum ratio will define the ultimate overburden to ore ratio or pit limit. At this point the value of the recovered mineral pays the costs of mining, overburden removal, beneficiation, and a minimum acceptable profit.

Fig. 1 assumes a homogeneous ore body, but if waste strata or large horsts occur that must be selectively removed during mining, these volumes will be additive to the overburden increasing the SR. Therefore, the amount of overburden that may be economically removable to reach ore may be decreased and reduce the amount of recoverable ore. Thus when the incremental stripping ratio exceeds the economic stripping ratio, open pit or surface mining ceases to be profitable.

Transport Systems

The excavated overburden must be transported to a deposition site. Economics dictate that the transport distance be as short as possible, consistent with the mine plan, and any other special requirements, such as the eventual rehabilitation of the mined area or other environmental concerns.

Selection of the transport method should be analyzed. This will ensure a system of optimum cost and one most suitable to the characteristics of the material to be moved.

Material Characteristics: Depending on the nature of the material to be transported, there may be either a very limited choice of equipment or a wide selection. See the chart of equipment choices matched to equipment types, Fig. 1. There are more alternates to select from for handling consolidated materials than for soft, water-saturated materials.

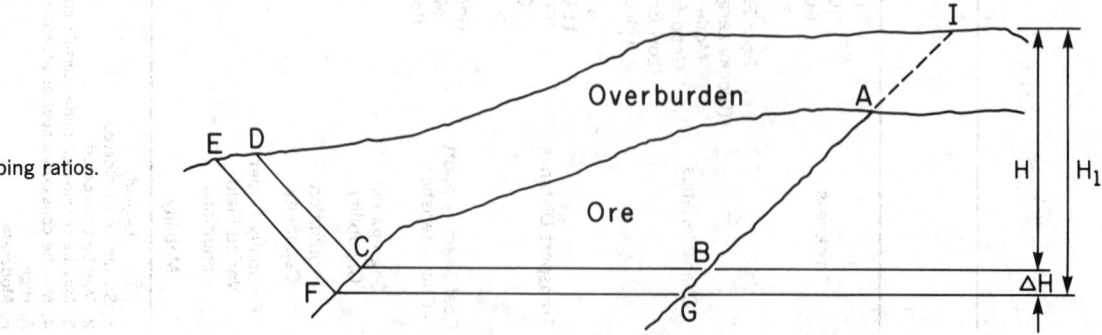

Fig. 1 Stripping ratios.

This transport system should also be selected to suitably take other aspects into consideration such as material size, degradation, hardness, and abrasiveness.

Equipment Review: Once the significant parameters of material characteristics are defined, equipment for transporting the material may be reviewed. Generally, several methods of transportation will be found to be feasible. Technical evaluation will limit these several feasible methods, as a rule, to two or three most likely to be appropriate.

Some of the technical aspects that may be limiting factors are the production level demanded, the equipment capital cost vs. operation life, the length and elevation difference of the transport route, any special requirements in placing material in the dump, and others.

As an example, a small operation with a short life will not likely be able to support either a railroad or expensive overland belt conveyor. The small operator will select low capital cost mobile transport of trucks or scrapers. Economy of size becomes apparent when dealing with very large production tonnages. These are the operations that will support long depreciation periods for expensive high technology systems with their potential for minimum operating costs.

Mine Compatibility: Most operators agree that interchangeability of pit equipment is good. Therefore, if possible, consider similar transport equipment for both overburden and ore. This offers the advantage of a smaller parts inventory, a narrower range of maintenance requirements, and a resulting lower operating cost.

Environmental Concerns: Disposal of overburden may often involve issues of environmental concern. Some of these are visual and others may be related to the discharge of noxious or hazardous materials.

Where overburden must be dumped in sites at some distance from the mine, the visual aspect of the dumps may have to be considered. To present dumps as visually attractive to viewers may require contouring, and seeding or foresting to harmonize with the surrounding country.

The mineral content of the dumps should be well known. This will provide information as to the steps that must be taken to contain leached solutions. When these are allowed to escape freely without neutralization, severe damage may occur to the immediate quality of life including stream pollution and damage to plant or animal life.

Runoff waters from leached overburden dumps may be a possible contaminant to subsurface aquifers. These frequently feed to more distant sources of human water supplies. To avoid this potential hazard requires a good knowledge of the local geology, e.g., fissures, faults, and local hydrogeology.

An increasing concern for protection of the environment during mining and recent enactment of stringent mining regulations mandate that the mine operator give extra thought to the environmental planning process. However, by incorporating basic principles into any mining technique, high productivity with little environmental disturbance can be achieved. Principles include:

1) Practice spoil segregation and immediate burial of toxic material. (Topsoil comes in A, B, and C layers, each of which must be segregated under certain regulatory requirements.)

2) Where topsoil is of sufficient quantity and reasonable quality, remove and reapply on regraded spoils in one operation, if possible.

3) Incorporate as many production operations as possible into reclamation procedures, thus integrating unit operations.

4) Have a preplanned reclamation scheme which can easily be integrated with a feasible mining plan.

OVERBURDEN MATERIALS

Overburden, as discussed here, will refer to the first referenced definition as stated in the section, Overburden Defined. Thus any material lying over an ore or coal deposit is defined as overburden. The approach to its removal will depend on the material characteristics. Commonly encountered types of material are: topsoil, soft overburden, medium hard rock, and hard rock.

Topsoil

This is a general term applied to the near surface portion of the material lying above the ore. Topsoil has sufficient agricultural nutrients to support varying degrees of vegetation growth. This layer or layers of soil may only be a few inches deep or extend to several feet in depth.

In recent times its segregation in the removal stage may have prime importance. This has occurred from the impact of governmental regulations for replacing topsoil as it originally existed on mined areas. To do this requires that each layer be carefully excavated and placed in an area where relatively easy recovery can be made. When replacement is required, topsoil should not be mixed with the agriculturally poor underlying materials.

Topsoil Characteristics: In addition to the agricultural attributes of topsoil, it is an unconsolidated material. It is frequently loose and plastic in nature. The water content may vary, and mineralogically it may contain clays, sand, fully decomposed rock, loess, marine remnants or combinations of these.

Where topsoil exists and must be removed, it may be necessary to call on qualified soils experts to accurately define the topsoil. This could be the case where topsoil grades gradually into a relatively poor agricultural material. The latter may be a soft or decomposed rock from which the topsoil originated.

Soft Overburden

This may either be gradational downward from the topsoil interface or be the primary material on top of the ore. Typically it might be a glacial moraine of unsorted material, a rock advanced in decomposition but with original structure, agriculturally poor sands, barren or marginal grade laterites, clays, or other soft material that is not classed as topsoil.

Under certain circumstances part or all of these materials may be stockpiled separately for later land rehabilitation. Some soft overburdens not considered as topsoil, with an addition of nutrients, may make a soil capable of supporting vegetation cover. This can be done to retain recontoured land, minimize erosion, and ultimately improve the original land condition.

Excavation of soft overburden can be done for a minimum cost; separate dump locations may not be required. Frequently the soft overburden goes to the main waste dumps without segregation. Soft overburden may also be useful as road ballast.

Soft overburden normally can be excavated without drilling and blasting. Equipment typically used for the soft in situ material, either dry or with a high water content, is indicated in Tables 1 and 2.

Medium Hard

This classification covers a wide range of rock types. It includes most semi-decomposed rock, highly fractured or well jointed rock systems, shales and schists with easily ruptured parting planes, many cemented conglomerates, and other rocks that can be broken for loading with moderate forces.

Rock breakage may be done using drilling and light blasting, ripping by a dozer, fracturing and breaking with a dozer blade, or in a few cases by the milling action of a bucket wheel excavator (BWE).

Overburden of medium hardness is normally of no value for agricultural purposes, but it may be used as a base for returning mined areas to the original topographic contours. Where rock overburden has been stripped from an open pit operation, it is unlikely that it would be used to fill the pit after mining is finished. The most likely environmental solution would be to reshape and contour waste dumps, adding an adequate cover of topsoil if required, and planting some form of vegetation adaptable to the conditions. The abandoned open pit would most likely fill with water becoming a lake in a potential recreation site. The final disposition of harder rock overburden is similar to the medium hard.

Hard: Drill and Blast

This is overburden that must be drilled and blasted before it can be dug and loaded. It may occur from the surface to the ore interface or lie at some depth below the surface. In most cases the hard overburden is the barren country rock surrounding the ore body. Fig. 17 shows the relationship of a typical hard rock overburden to the ore zone.

Equipment for excavating and transporting the hard rock must be robust to survive the impacts of loading and abrasion wear. The principal machines used are the power shovel, dragline, and hydraulic shovel. Smaller operations will often use the rubber tired front-end loader; but for sustained, long-term loading of hard rock, the shovel and dragline are preferred.

Where draglines or direct casting shovels are used, the blasted rock is usually side cast to disposal and end dump trucks will work with power shovels. In some cases where the broken rock is used for civil construction, the transport system is automatically defined. In recent years, with fuel- and cost-conscious planning influencing mine layouts, interest is widespread in crushing rock waste in the pit and moving it by belt conveyor. Having established the character of the overburden material, the choice of the excavator can be made.

OVERBURDEN EXCAVATION

Most of the major methods for excavating the overburden are cyclical in nature. This means that they experience interruptions in the digging and material-moving elements of the operation. The important exceptions to this are dredges and the bucket wheel excavator. Each of the major overburden excavators is discussed here from the standpoints of method of application, suitable material for handling, machine characteristics, support functions required, and selection criteria.

Dozers / Front-End Loaders

Dozers or pusher-type tractors are either mounted on crawlers or on rubber tired wheels. Rubber-tired dozers have little application to overburden removal when working alone. Crawler-type tractor dozers are frequently found working alone pushing loose or blasted overburden in preproduction benching operations.

Tractor push blades are available in a variety of designs for different applications. The range of the hardness of the material handled by the crawler dozer may be very large. A crawler tractor's pushing ability is only limited by the machine weight and horsepower. This is also true for the rubber-

Table 3. Characteristics of Rubber Tired and Track Type Dozers

Dozer Type	Horsepower	Approx. Weight (Dozer Only) St	Capacity Loose cu yd per hr
Track	180	18.2	458
	270	29	628
	385	40	889
Rubber-Tired	300	32	366
	400	41	562

Metric equivalents: hp × 0.746 = kW; st × 0.91 = t; cu yd × 0.765 = m³.

tired dozer, but under most conditions of overburden removal where a machine works unassisted, the crawler dozer is preferred. It generally takes a heavier rubber-tired dozer to exert a given push force than it does a crawler tractor, due to the greater coefficient of friction for the crawler-type dozer (Shand, 1970). Table 3 illustrates this relationship within a 30 m (100 ft) push distance. The rubber-tired dozer is more often found as a secondary cleanup machine working with a loader or other overburden excavators especially where mobility is necessary.

Front-end loaders are either crawler- or wheel-mounted. The latter is used almost exclusively in surface mining because of its mobility. The front-end loader is primarily used for loading trucks, however, where the overburden deposition distance is relatively short, a load-and-haul type operation may be employed. It is common to find the track dozer working with the wheeled front-end loader. Where the hauls are short, the wheeled dozer works well from a stockpile as

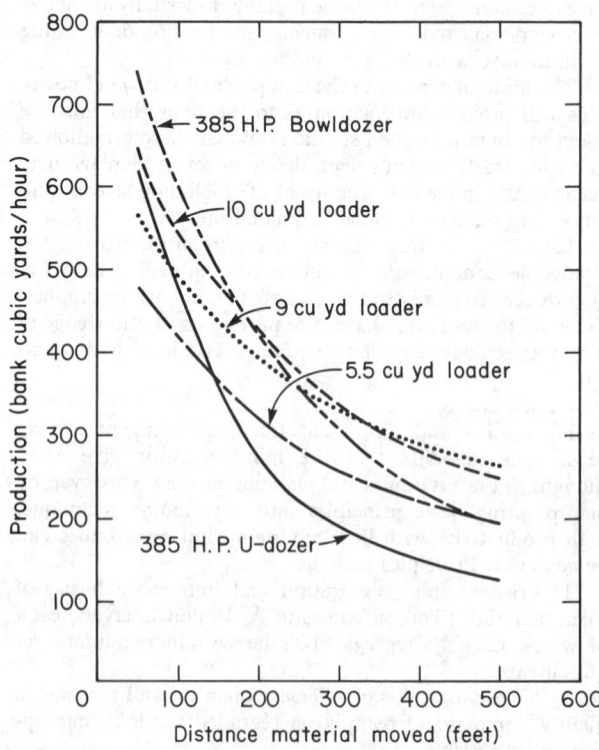

Fig. 2. Generalized production comparison for front-end loaders vs. bulldozers.

a load-haul-dump machine, the stockpile material being excavated and pushed into a pile by the track dozer.

Fig. 2 shows the effect of distance on dozer and front-end loader production. As a general rule of thumb, dozer push distances should be maintained at less than 61 to 76 m (200 to 250 ft) while front-end loader tram distance should not exceed 152 m (500 ft) with 91 m (300 ft) being a more desirable limitation for economic production.

The types of overburden that can be dozed are loose and/or friable. Typically a dozer may be used to move a topsoil layer downslope from a hillside benching operation. Crawler dozers are frequently used during mine development for moving shot rock overburden. In preproduction this may involve opening the initial mine benches by dozing the blasted rock over the edge of a steep decline. Gravity carries the dozed rock down the hillside and under optimum conditions no further handling is required.

Overburden removal by ripping and dozing has increased in surface mining with the advent of the larger machines. Tandem ripping, adding the weight and power of a second tractor, will further expand the rippability range (Jackson, 1979). However, the costs of ripping must be evaluated against the resulting higher repair costs. Costs of 120 to 130% of the depreciation for repairs are not unusual (Jackson, 1979). Production ripping, where feasible, can loosen material for one-third to one-half the cost of drilling and blasting. Light blasting with ripping can also be, in some cases, cost effective. Seismic refraction tests help to determine the degree of rock rippability. Fig. 3 shows the potential performance of a Caterpillar D10 tractor.

Dozers/Trap Loaders (Belt Loaders): Trap loading operations are traditionally associated with earthmoving contractors. Since the mining industry is generally unfamiliar with the equipment, its application in suitable material has been limited. A few mining operations have used trap loaders or belt loaders with great success. One of the most remarkable applications was the use of the trap loader and shiftable conveyor belt transport system for soft ore from a South American iron ore deposit (Anon., 1975). A sand and gravel operation also uses the belt loader/shiftable conveyor system in California.

Belt or trap loaders are heavy duty units made up of a material-receiving box or trap and a discharge conveyor belt. Design variations are possible that permit loading by dozers, front-end loaders, or even rubber-tired scrapers. Material can be dozed to the loading station from 61 to 91 m (200 to 300 ft) away. Two or more stations may be operating with stations being leapfrogged to maintain production.

Scrapers

Although the scraper has been around since the early 1930s, the machine is relatively new in surface mining. In the 1970s it became a proven tool for stripping overburden. Because of its excellent mobility, low initial cost, and sturdy construction, the scraper will continue to have widespread application.

Scrapers can be classified into four general types: (1) tandem-powered; (2) push-pull; (3) single engine; and (4) elevating.

The tandem-powered machine is preferred for steep grades and poorer underfoot conditions with the front and rear engines providing the traction needed. Push-pull scrapers are combined in a "buddy" system that can self-load each scraper. On the travel leg, they separate from their special loading hitches to move independently. The single engine machine requires a pusher dozer and will perform well for removing thin lifts, mixing soils, or special grading.

Synchronized teams of scrapers and dozers can econom-

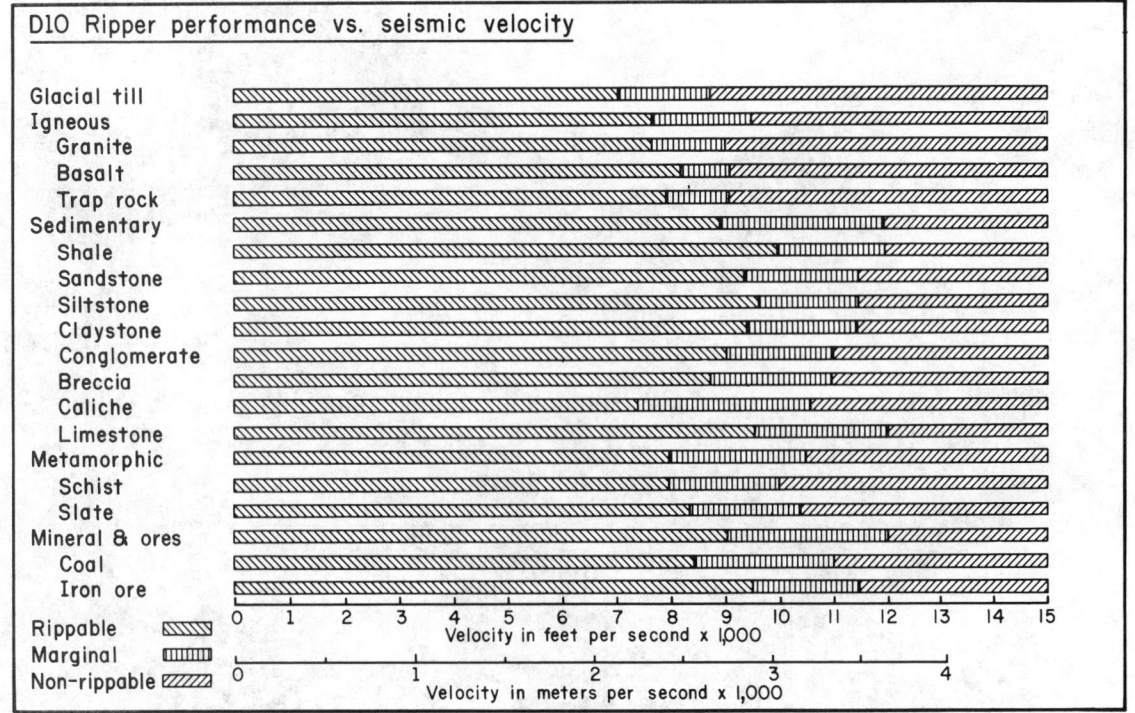

Fig. 3. Caterpillar D10 ripper performance vs. seismic velocity. Performance estimated by seismic wave velocities. Multiple or single shank No. 10 ripper. Production shown for 14 rippers.

ically remove overburden 20 m (65 ft) or more in depth (Chironis, 1979).

Frequently, the scraper may be used in combination with other mining machines. As an example, the early use of scrapers, during the delivery period of the main excavators, may allow the mine to reach full production at an earlier date. Scrapers may also remove an upper level of overburden, eliminating the need for any dragline rehandle.

The Lucky Mc mine in Wyoming is believed to be the first mine to use scrapers for overburden removal (Anon., 1969). Since then scrapers have been used to move a wide range of overburden types. In Arizona at the Twin Buttes mine, 140 m (460 ft) of overburden was moved by a scraper belt conveyor system. Other mines using scrapers for overburden include phosphate, uranium, aggregate, coal, iron, and copper.

Draglines

There are three types of draglines: (1) truck-mounted, (2) crawler-mounted, and (3) walking.

For overburden removal, it is necessary to consider only the crawler and walking machines since the truck-mounted machines are small in capacity and are not designed for overburden removal work.

Some years ago a dragline was considered to be a trans-porter rather than a production tool. It moved material over fairly long distances while machine moving speed and production output were not much considered. The dragline's prime purpose was to transport material out of the way of mining. Today the dragline is an important production tool designed to move large volumes requiring increased power and speed of bucket movements. Longer dragline booms have increased the possible transport distance and the working range of the machine.

Dragline usage has outpaced shovel applications in a large percentage of the overburden removal work. Where conditions of pit design can take advantage of the reach and dumping radius of the dragline, much deeper overburden can be removed for less cost than with shovels and trucks.

The dragline is versatile and easily maneuvered from a top-digging position as opposed to a shovel in a pit situation. Bank or bench failures that may be catastrophic for other pit equipment seldom impede dragline operations.

Topographic variations at the overburden-ore interface are easily adjusted for in dragline operations. Overall planning can correct for digging variations of a localized nature by a change in cut width. In some cases draglines must remove material from a bench above the working floor to allow a single machine to handle the full depth of overburden. This is usually not as efficient as the standard operation, but

Fig. 4. Extended bench modified area mining with rehandle.

may be more cost effective than additional machines for the top bench material. The operation of working a face above the machine support level is called chopping.

Draglines will handle a wide range of materials, both in situ and shot rock. Fragmentation of rock must be suitable in character for the dragline bucket size selected. Shovels will be less affected in productivity by large blocky material, and a shovel tends to have a higher loading factor than a similar size drag bucket.

Draglines will permit a higher percentage of coal recovery than a shovel, although the dragline will require closer supervision. The dragline has a lower maintenance cost than does a shovel.

Draglines are ideally suited to the deep overburden so frequently encountered today. However, applications studies should be carried out to determine the optimum scheme. In the case of upper overburden removal (chopping), combinations of equipment working with the dragline must be considered where rehandle becomes a significant factor in productivity.

Some of the possible alternative combinations with dragline operations are dozers, scrapers/dozers, shovels/trucks, front-end loaders/trucks, BWE/conveyors, shovels/cross pit conveyors, and shovels/around-pit conveyors.

One possible way to reduce the dragline size is to bench ahead with dozers pushing a lift of overburden into the pit. The dragline then might work from an extended bench and rehandle the prestripped material, as illustrated by the artist's drawing in Fig. 4 and the schematic drawing in Fig. 5. The other alternatives also involve removing the upper overburden, but transporting it independently of the dragline work.

Before deciding to employ the extended bench mining method, the operator should calculate the rehandle volume and determine the added cost and production decline. Comparative evaluations with alternative mining systems and/or equipment purchases lead to the most cost effective decision. Many operators are forced to make the best possible use of existing equipment, and the extended bench method may afford a technically viable means of extending the dragline digging depth. Several factors are considered in determining the volume of rehandle. First is the angle of repose in which material will naturally set if not compacted or disturbed. In an area mine where no rehandle is normally experienced, this angle is formed by the material after the dragline disposes of it in the adjacent cut.

The second factor is the highwall angle or slope of the highwall exposed in excavation. The third is the cutback angle not normally recognized in calculations. In Fig. 6 this is the angle the material will form when the dragline is excavating the extended bench. With conventional spoil material consisting of shales and sandstone, the cutback angle usually runs between 50° to 60°. This is much shallower than

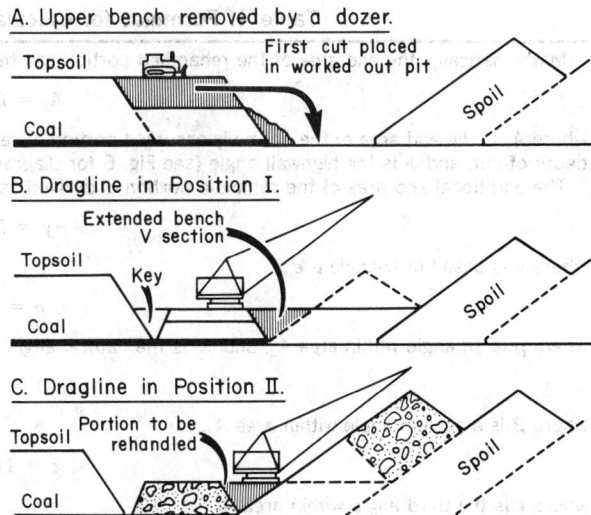

Fig. 5. Schematic of extended bench method.

a highwall angle of 75° to 85°. Table 4 indicates the basic method of recalculating this area of rehandle.

Other methods can be used in the field to determine the rehandle percentage more accurately than by using the formula depicted in Table 4. These methods may be some type of surveying either by conventional surveys in the pit floor or by photogrammetrics. Use of these methods will determine bank overburden quantities necessary to uncover the coal; by determining actual dragline cycle times, exact rehandle percentages can be derived.

In addition to the prominent position of draglines in overburden removal from coal, they have been similarly applied to other minerals. Some of these include iron, phosphate, bauxite, alluvial gold, aggregate rock, and laterite.

Extended Bench Operation: The extended benching technique allows the dragline to handle overburden covers that are beyond its normally rated range. Thus, in situations where isolated pockets or consistently deep overburden covers are encountered, the dragline, by extending its bench, is able to mine areas beyond its normal reach capability.

This technique is based on extending the dragline support level out from the old highwall and closer to the spoil. This is accomplished by casting material over the edge of the old highwall and building the extended bench. The extended bench material may be moved either by the dragline and/ or by dozers and normally would come from one or a combination of two places depending on the situation. In a simple side casting situation, the material would come from the key

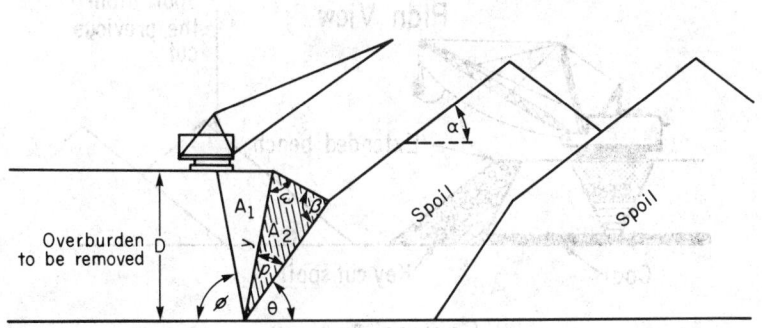

Legend

D = Depth of cut
ϕ = Highwall angle ($\simeq 80°$)
θ = Cutback angle ($\simeq 56°$)
α = Angle of repose ($\simeq 38°$)
ϵ, β, ρ = Angles in the additional rehandle portion
A_1 = End area of normally assumed rehandle portion
A_2 = Area of additional rehandle portion
y = Base of $\Delta \rho, \epsilon, \beta$

Fig. 6. Detailed geometry of revised rehandled portion.

Table 4. Formulas for Calculation of Revised Rehandle Wedge

Mathematically, the end area of the rehandled portion can be calculated by the formula:

$$A_1 = D^2 \div \tan \phi \tag{1}$$

where A_1 is the end area of the normally assumed portion of rehandle bounded on one side by the previously cut highwall, D is the depth of cut, and ϕ is the highwall angle (see Fig. 6 for diagram).

The additional end area of the rehandle portion is derived as follows:

$$y = D \div \sin \phi \tag{2}$$

where y is based of triangle $\rho \, \epsilon \, \beta$

$$< \rho = \phi - 1.3 \, \alpha \tag{3}$$

where ρ is an angle within area A_2, and α is the natural angle of repose of the loose spoil.

$$< \beta = \alpha + 1.3 \, \alpha \tag{4}$$

where β is a second angle within area A_2.

$$< \epsilon = 180° - \rho + \beta \tag{5}$$

where ϵ is the third angle within area A_2

The end area of the additional rehandle portion can then be calculated by the formula:

$$A_2 = \frac{y^2 \, (\sin \epsilon) \, (\sin \rho)}{2 \sin \beta} \tag{6}$$

(see Fig. 6 for diagrammatic representation).

Total end area (A_T) of this is now simply computed by the formula:

$$A_T = A_1 + A_2 \tag{7}$$

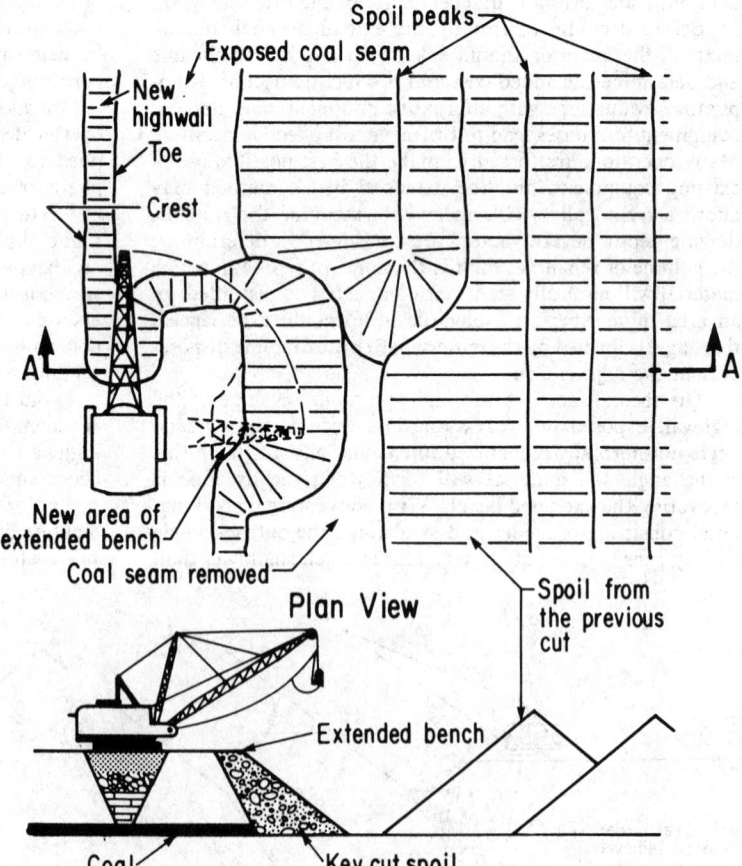

Fig. 7. In simple sidecasting situation, material comes from the key cut.

cut. In an advanced bench situation, the material would come from the key cut and the advanced bench chop-down area. Figs. 7 and 8 show plan and cross-sectional views for both situations. The material from the advanced bench area can be used only if it is stable and can support the dragline. Once the material is leveled and a base is prepared for the dragline, the machine can reposition itself on this extended bench and remove material from the final cut and the previous extended bench. See Figs. 9 and 10.

By using this technique the dragline is able to dig overburden covers that create soil piles beyond the reach of the machine. However, rehandle situations are created with rehandle volume dependent upon overburden depth. Careful study and close supervision are essential with extended bench operation.

Power Shovels

Two types of mining shovels currently in use are stripping and quarry. The two machines differ mainly in the duty they perform. A stripping shovel is associated almost exclusively with coal mining for overburden removal. The stripping shovel places the overburden beside itself in the mined out area. Mechanically the two shovel types are similar, but the stripper is overall a much larger machine with greater overall digging range.

A quarry shovel is used principally for open pit mining loading into trucks or rail cars. The quarry shovel is the workhorse of difficult mining and rock waste removal. In recent years the hydraulic shovel has challenged the traditional power shovel in some of the harder digging materials, but it will not supplant it in the overall excavation of hard rock in large open pits. Where overburden and ore are hard rock, requiring drilling and blasting, a shovel section will provide maximum loading flexibility. Where a transport unit must be loaded by the excavator, the shovel "spots" a target more accurately and less spillage results.

Shovels require good underfoot conditions, which dictate a hard rock overburden condition. In the larger sizes, lower groundbearing pressures for the excavator favor the dragline. In planning any overburden stripping program, consideration of the soil and/or rock physical characteristics must be made. Acceptable slope stability conditions may require extensive enlargement of an open pit to ensure good bench safety. Highwall conditions in coal mining may dictate the amount of overburden unloading required for pit safety.

Capital cost of power shovels is expensive compared to

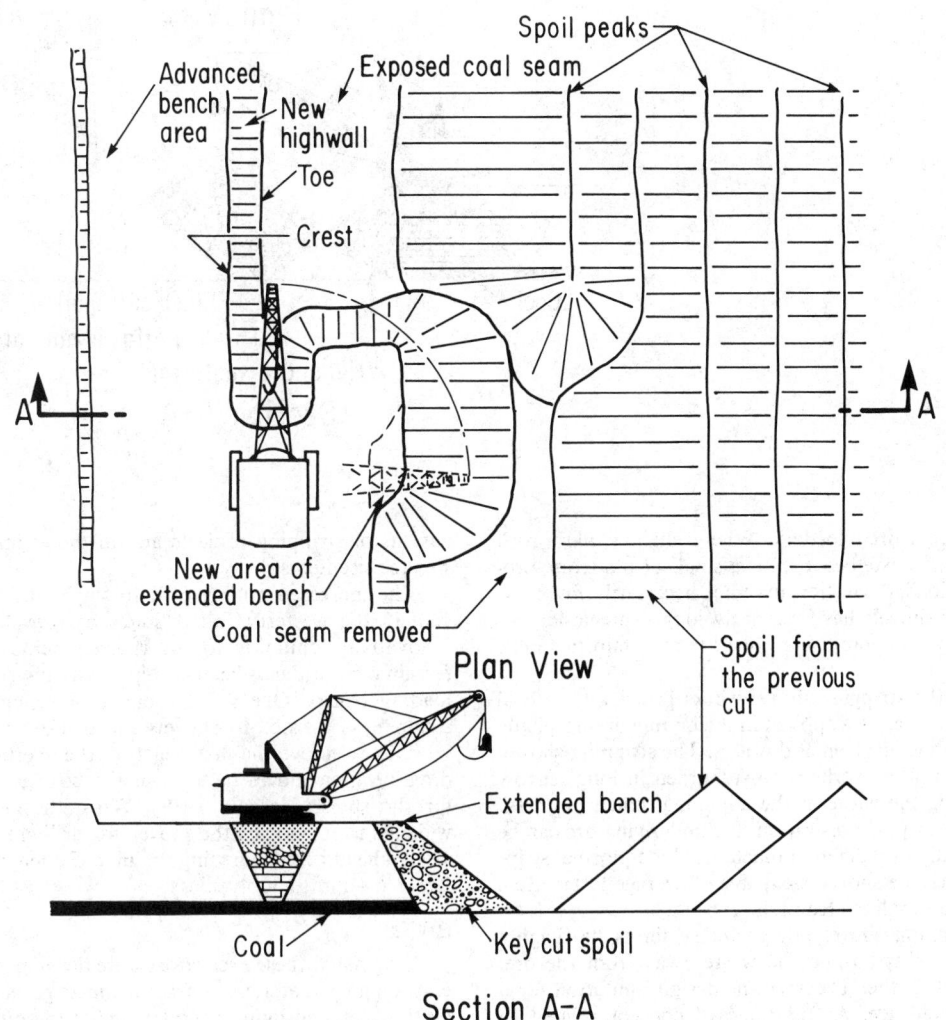

Plan View

Section A-A

Fig. 8. In advanced bench situation, material comes from the key cut and advanced bench chop-down area.

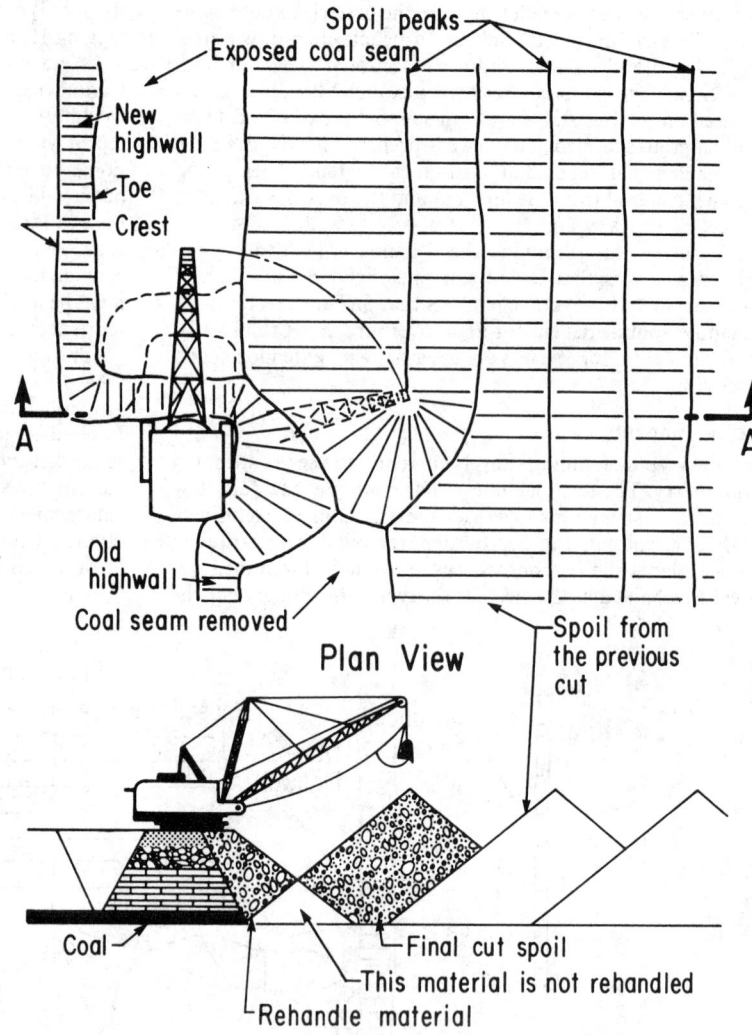

Fig. 9. Dragline repositioned on extended bench.

dozers or rubber-tired loaders. Where shallow hard rock overburden and waste occur, plus a small or moderate production rate, loader selection may not necessarily favor the shovel. Power shovels have relatively long depreciation periods requiring an appropriate mine life to obtain favorable ownership costs.

Although the stripping shovel works principally in coal overburden removal, it is applied in the mining of phosphate, bauxite, ironstone, gypsum and others. The stripping shovel, as the name implies, works the overburden in long narrow strips, dumping the waste in the adjoining mined out area. Following the stripping operation, the underlying ore can be removed, usually by a different machine. Under normal stripping conditions, the shovel is capable of average faster cycle times than the dragline (Rumfelt, 1968).

For a stripping shovel pit, geometry limits its digging depth by the ability to pile the waste away from the ore. Waste spoil piling then becomes the design limitation for a particular shovel size. A "usefulness" concept related to gross shovel weight has been developed: usefulness is defined as the dumping reach multiplied by the nominal size of the dipper (Rumfelt, 1968). This concept can help in early eval-

uations of stripping projects and in the estimates of shovel size requirements.

The increased mining of both single and multiple seam coal in the western United States has developed new and innovative techniques for overburden removal. In rugged terrain overburden is being stripped with quarry-type shovels loading trucks. One western operation using draglines for stripping is assisted by shovels and trucks.

These remove the first bench of the overburden and the dragline then moves in to open the box cut making room for the spoil (Jackson, 1979). When the bench has been widened to six cuts by the shovel, the drill crews can operate safely ahead of the pursuing dragline. Pit length ranges from 1.6 to 6.4 km (1 to 4 miles).

BWEs

Originally, these excavators were developed for relatively easy digging materials. These included gravel, sand, loam, marl, clays, and lignites. Today one of the outstanding successes for BWEs is in overburden removal in the easy digging lignite brown coalfields of the Federal Republic of Germany. However, the newer generations of BWEs have incorporated

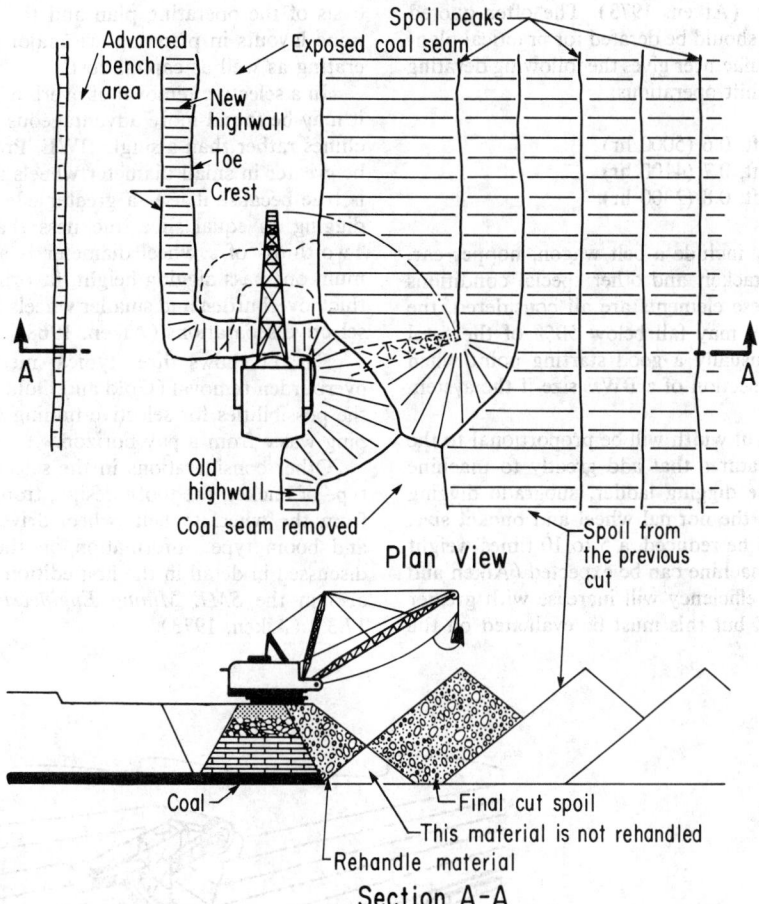

Fig. 10. Repositioned dragline removing material from the previous extended bench.

modification designs giving them better operating characteristics for more difficult digging materials. Their use is now extended to highly compact sediments, shales, black coal, some limestones, tar sands, and others. Although blasting of some of these materials can assist wheel digging, normally rock-type materials needing blasting should not be considered for BWEs.

Bucket wheel/belt conveyor transport systems can achieve high production rates. Recently a new class of giant excavator went to work in Germany handling up to 240 000 m³ (8,475,474 cu ft) per day. This is equivalent to handling 5.3 m³/s (187 cfs).

This BWE has frequently exceeded the rated capacity with rates to 254 000 m³ (8,969,877 cu ft) per day in good digging (Anon., 1976). At the same time, the BWE can offer the operator in shallow ore horizons the advantage of precise cutting of overburden, using small buckets, and near 100% ore recovery.

The BWE is used to load trucks but few examples exist, since the belt conveyor is the preferred method. BWEs are continuous excavators that produce a constant flow of material. To maintain truck loading requires a continual queue-up to keep a unit always under the loading conveyor of the BWE.

Because the BWE has less operating flexibility than shovel trucks, a careful study must be executed before selecting the machine. This includes detailed pit layouts based on explo-

ration results and geology of the deposit. The object of this is to keep the relocation of excavators and conveyor belts to a minimum. To avoid relocations, the size of the BWE in terms of output and maximum vertical digging reach must be chosen in line with the maximum depth and output of the pit (Beneche, 1982). As a rule of thumb, the vertical reach of the BWE, when added for all mine units, should equal the pit maximum depth.

Since large BWEs are only suitable to long-term operations because of their high capital cost, they can compare favorably with draglines and power shovels in the following: (1) instantaneous power demand, (2) weight to output ratio, (3) loading shocks, and (4) power consumption (Aiken, 1968). BWEs have the additional advantages of having (1) wider operating benches and more stable pit slopes, (2) improving a shovel efficiency in underlying hard rock when soft overburden is removed with a BWE, (3) close control in selective mining, (4) easier and cheaper rehabilitation of overburden spoil, (5) the exposure of more reserves with wider benches, (6) an ability to deliver material both above and below the working level, and (7) the ability to make high and/or deep cuts (Aiken, 1973).

Selection of a BWE for a specific production rate can be more complex than for other excavators. Shovel output and digging factors are well established for power shovels. This is not true for the BWE, since within a basic size range, machines of equal capacity can vary in profile dimensions, weight, horsepower, etc. (Aiken, 1973). The often-quoted theoretical BWE output should be derated for practical planning purposes. One manufacturer gives the following derating factors for 3, 2, and 1 shift operations:

3 shift, 0.6 (5000 hr)
2 shift, 0.7 (4100 hr)
1 shift, 0.8 (2300 hr).

A complete system may include a belt wagon, hopper car, cable reel car, mobile stacker, and other special conditions of operations. When these elements are all considered, the overall BWE availability may fall below 50% of the total calendar time. This is usually a good starting point when making a preliminary selection of a BWE size if the system is complex.

Digging height and cut width will be proportional to the boom length. Design features that add greatly to machine cost are increases in the digging ladder, subgrade digging ability, and increases in the normal wheel and bucket size. When these weights can be reduced, a 5 to 10 times weight reduction for the total machine can be expected (Aiken and Wohlbier, 1968). BWE efficiency will increase with greater digging and cut heights, but this must be evaluated on the basis of the operating plan and the total economics. Good mine layouts in planning are major factors in reducing operating as well as capital costs.

In a selective removal of overburden from a pay horizon, it may be found more advantageous to use two smaller machines rather than a single BWE. Production efficiency may be greater in small diameter wheels than in large ones. This is true because it uses a greater effective wheel diameter in digging an equal slice thickness than the larger machine. Two-thirds of a wheel diameter is approximately the optimum one pass digging height, favoring the larger BWEs for thick overburden and smaller wheels for thin overburden and selective separations (Aiken, 1968).

Fig. 11 shows three typical methods of BWE use for overburden removal (Gold and Gold, 1974). These visualize the possibilities for selective mining of overburden or stripping waste from a pay horizon.

Other considerations in the selection of a BWE are the type of bucket, the tooth design, transfer feeder for material from the wheel to belt, wheel drives, suspension systems, and boom type. Information on these elements has been discussed in detail in the first edition of *Surface Mining* and also in the *SME Mining Engineering Handbook,* Section 17.3.2 (Aiken, 1973).

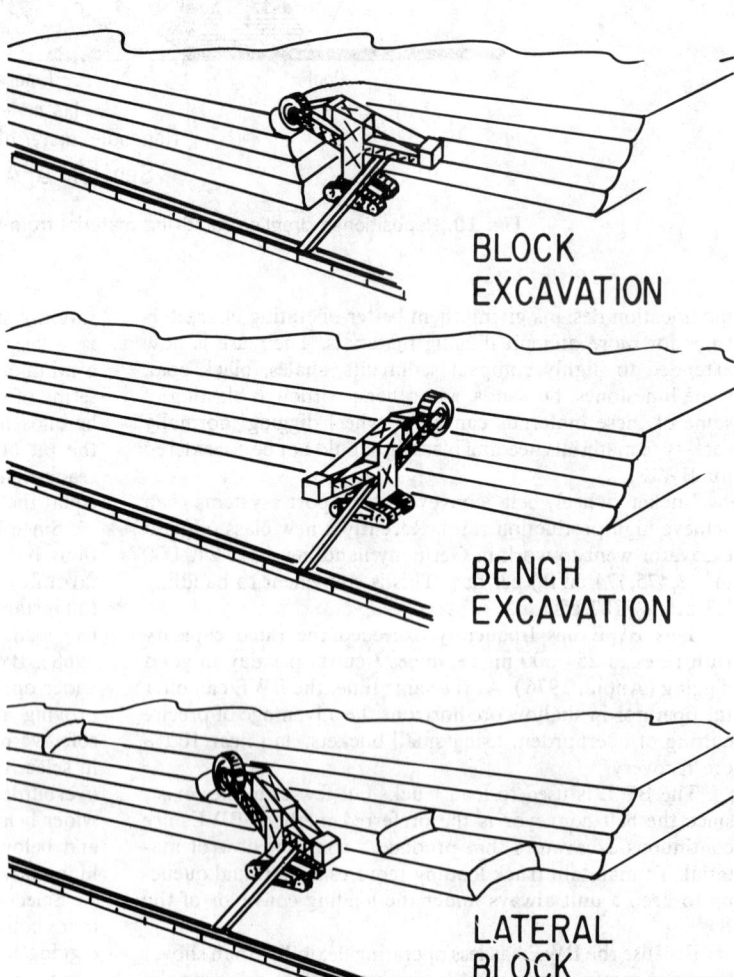

BLOCK EXCAVATION

BENCH EXCAVATION

LATERAL BLOCK EXCAVATION

Fig. 11. Three typical methods of BWE use for overburden removal.

Dredging/Monitors

These two wet methods for excavating overburden may be treated together. Both may involve pump selections, pressure heads, and production of a slurry. Monitors are sometimes mounted on the bow of dredges to cut down dangerously high digging faces. Hydraulic monitors are more frequently used in land-based stripping operations. Soft, easily eroded materials are typically handled by monitors, whereas dredges are equipped to handle a wider range of material hardness.

Dredges are either classed as bucket line or hydraulic. The removal of water-saturated materials may also be done with a barge-mounted clamshell or drag bucket. These have minor use for application to overburden removal and are not considered here. Both hydraulic and bucket-line dredges have been used for overburden removal.

Because most water-saturated overburden is not very difficult digging, it would be unusual to find an economically justified case for dredging by a bucket-line dredge. These may be as much as 1.4, or even more, in capital cost over a large hydraulic dredge of equal capacity and digging depth. Almost the only circumstance justifying overburden removal by bucket-line dredge is a situation in which the underlying pay mineral is dug by the same machine. This presents obvious problems in providing a continuous flow of mineralized material. To avoid this problem, a two-ladder bucket-line dredge was once considered for a tin property in Malaysia (McKay and Davis). For all practical purposes, the hydraulic cutterhead dredge for overburden is the only machine to be considered.

A recent innovation in dredge design has been the combining of a bucket wheel fitted ladder and a bucket-line ladder dredge into a single machine, the idea being that the BWE will dig overburden ahead of the bucket-line ladder digging the underlying mineral-bearing horizon.

The most simple hydraulic dredge is a pump, barge-mounted, connected to a hinged suction pipe that can be submerged at an angle of 45° below the horizontal. Material recovered is pumped via a floating pipeline to a discharge area. In harder materials, the suction pipe may carry an independently driven toothed cutterhead. With an increase in dredge size, power, and digging depth, more sophisticated controls and equipment add greatly to the capital cost.

Large cutterhead hydraulic dredges are capable of digging difficult material. In-place coral and blasted rock have been handled successfully. The hydraulic dredge is at its optimum efficiency when used to excavate and transport overburden from lode or alluvial deposits at sites distant from the mine area (Barker and McKay, 1972). A dredge can provide continuous excavation with comparable availability, or better, to other machines including the BWE (Barker and McKay, 1972).

Both bucket-line and hydraulic dredges always operate in prepared ponds that allow for water control. This includes water quantity, turbidity, and pond level. The dredge can operate at either higher or lower elevations by control of the pond level. Above water dredging faces are normally limited to about 1.6 m (5 ft) for hydraulic dredges and 4.6 m (15 ft) for bucket lines. Dredge faces that slough to the pond may be carried higher if caution is exercised. Under these conditions one Australian sands operation carried a face approaching 70 m (230 ft). To safeguard the dredge, bow monitors knock down the high sand face to a safer angle.

Designing and building hydraulic cutterhead dredges to handle up to 3800 m³/hr (134,195 cfm) or bucket lines to dig up to 2000 m³/hr (70,629 cu ft per hr) in harder materials is not a problem. Jet lift or jet assist as a device for increasing the upward flow of solids toward the suction pipe has been used successfully in loose material.

This has been pioneered by the Elliott Machine Corp. and their machine is called the *Dragon Wheel*. The Dragon Wheel can operate with equal efficiency in either direction of swing. This has resulted in a 2:1 improvement over the conventional cutterhead and even up to 3:1 increase in output. The underwater BWE dredge is reported to have eliminated former objections by placer miners to hydraulic dredges. See Fig. 12.

Dragon Wheels are presently (December 1982) operating at 23 sites throughout the world. Where these machines are applicable, Dragon Wheels will have a lower first cost than a comparable bucket-line dredge. Units built to date have been in the range of 75 to 1119 kW (100 to 1500 hp) (McDowell, 1980).

In easy digging overburden, a 406 mm (16 in.) discharge rated dredge would handle about 45 400 L/min of slurry at 15 to 20% solids (12,000 gpm). This would be pumped at 6.1 m/s (20 fps). Pumping power would be about 825 kW (1100 hp). Where pumping head conditions may vary, a nominal kilowatt dredge pump rating is selected and a booster pump may be used for head increases.

Hydraulicking monitors, in addition to their use with dredges for reducing hazardous bank heights, have been successfully used as the prime excavator for overburden removal. Material, ideally, should be of a sandy nature with only a minimum of clay and not compacted. A sandy slurry produced at densities up to 50% of volume may be achieved, but during operations will vary constantly. Where the clay content is high, free hydraulicking is less successful. In the sandy-clay material of Florida phosphates, monitors are always assisted by draglines. These do the excavation and the monitors pulp the loosened clay materials. At one bauxite operation, when clay was encountered, the hydraulicking efficiency was reported to fall off rapidly.

Test work has indicated that sandy slurries will flow on negative 2% gradients. However, contained rock and boulders are not moved easily from the face until downhill slopes exceed 5%. Optimum mixing of solids and water is best achieved on very steep gradients of +20%. In all instances of free face monitoring, the operation efficiency is enormously increased by slot dozing material to a slurry mixing point. This method is helpful in reducing overall pusher distances. Trials should be conducted to determine the best range of distances for monitors from the sluicing face.

A case history is described in a later part of this section where 32 Mt (35 million st) of overburden was removed from one of the world's largest copper ore bodies. Another unique monitor application for overburden removal was applied by Alcoa in Suriname. This is described later more fully, including updated cost estimates (Koch, 1975).

OVERBURDEN TRANSPORT

Excavator Transport

The choice of the transport method for the excavated overburden must be in harmony with the type of excavator. There are some combinations that will not be acceptable and others will have a broader range of acceptability. It is relatively easy to reject the noncompatible methods, but those fitting multiple combinations must be considered carefully to reach an optimum solution. This may also include a combining of more than a single method of transport. All feasibly

Fig. 12. Dragon Wheel BWE hydraulic dredge.

applied transport methods are discussed as to compatibility with the excavator options.

Trucks

These can be considered the workhorse of overburden transport machines. Trucks offer maximum flexibility of operation in equipment size, trafficability, engine power, material adaptability, and the matching of the excavator. Trucks perform well in all types of climates and operating conditions. This has made them very attractive as transport around the globe.

Truck units come in a variety of configurations to fit almost any type of overburden removal operation. The choice of a truck depends on the conditions of each case. Overburden is mainly moved by the standard, two-axle end dump truck. The body of this truck can be loaded by shovels, front-end loaders, draglines, BWEs, and even dozers working with belt loaders. In special conditions, the three-axle truck may be employed. End dump design is the best method when dumping over the edge of an advancing fill. Most overburden is not free flowing, but rather a loose material mixed with nonuniform rock or blasted rock of varying sizes. This requires the open design of end dump units, but a few special cases may permit the use of the bottom dump truck. The unit can be very useful for moving a soil overburden free of large rocks, since it can spread and compact the material in one operation.

Trucks should be size related to the intended excavator. On average haul distances, the excavator should be able to fill the truck body in three to five passes. Filling of the truck body must be in agreement with the truck weight-carrying ability. Overloaded trucks will be subject to excessive tire wear and abnormal operating costs. Therefore, bank and loose material unit weights must be considered when body sizes are selected for a specific truck.

In operations where overburden removal will continue for the economic truck life period, the body size choice may be based on hauling a single material. Where truck interchangeability is planned with the ore hauling operation, body size should be carefully considered. Very often the ore may weigh considerably more than the overburden, creating a potential for overloading trucks designed to carry the latter material. Either the owner must rely on underloading of ore to overburden trucks or select a single truck capacity for carrying both materials. Fuel and labor costs have forced owners to periodically review the advisability of switching from trucks to other systems, such as conveyors, in-pit crushing with conveyors, or electrification of long, fixed uphill hauls with trolley-assisted trucks.

Belt Conveyors

Large increases in fuel costs over the last 15 years, and the fact that truck haulage may amount to 50% of the mine direct costs, have created strong interest in belt conveyor transport. Large-scale conveying is not new. Several metal mines, including the alluvium overburden removal phase at Twin Buttes, Arizona, have successfully used conveyors. Overburden at Twin Buttes was handled by scrapers without crushing in the pit. Dumping was to a grizzly hopper feeding a belt system. At another operation, a glacial moraine overburden was stripped from an iron ore deposit and conveyor transported about 4.8 km (3 miles). A dragline excavated the material and discharged to a portable hopper. Oversize was scalped at the hopper before feeding the belt conveyor.

A common application of conveyors is in a shiftable system where the excavator is a BWE. In most cases of a single bench, the system operates on the same level as the excavator. This feeds to a face conveyor that discharges to an extensible conveyor removing the overburden from the pit. The extensible conveyor allows the face conveyor to be shifted forward with minimum lost time. Overburden is either moved out of the mine area or deposited in the mined out areas of the pit.

The belt-conveyor, compared with truck haulage, is more energy efficient, less labor-intensive, and more inflation-proof. To expand the use of belt conveyor systems to more difficult materials has required the introduction of mobile crushers. Although they may be fed with either a power shovel or dragline, the latter may be more cost effective as a volume producer. The choice may be dependent on the mine plan and work the excavator must do. Combining the dragline with a belt conveyor system and mobile crusher utilizes proven cost-effective tools, increases productivity, and allows

for greater overburden depth mining (Frizzell and Martin, 1977).

A combined dragline-crusher-conveyor system requires a surge hopper to ensure uniform belt loadings. The extra surge capacity covers intermittent digging cycles and minor operating delays. Past experience suggests that a top opening on the hopper of 13 × 13 m (43 × 43 ft) may be minimum for a dragline of 30 to 53 m³ (1059 to 1871 cu ft) bucket volume. The operator must hit this dump opening with a minimum of delays and spillage.

The US Bureau of Mines (USBM) sponsored a program concerning the development of in-pit crushing and belt conveying methods. Incentives for such a program stemmed from the rising costs of truck fuel and the high cost of developing alternative fuel sources. A multimine survey was conducted that clearly indicated a need for in-pit crushing and conveying (Johnson, Frizzell, and Utley, 1981). Their study showed large movable gyratory crushers to be the only practical crushing solution and such a system has been designed. Several commercial installations exist at this time and the concepts appear practical and economically effective. In those areas of the world where fuel costs have been substantially higher than in the United States, direct feed to overburden handling conveyors has reached an advanced stage of technical development and application. Table 5 is a summary of mine operating conditions as determined in the USBM-sponsored study.

USBM examined several designs for movable in-pit crushers. They all included the same elements of design shown in

Table 5. Summary of Mine Survey Data

	Composite Data from Iron Mines Surveyed	Composite Data from Copper Mines Surveyed
Pit Characteristics		
Maximum haul grade, %	10%	10%
Average bench width	215 m (700 ft)	75 m (250 ft)
Bench height	12-14 m (40-45 ft)	12-14 m (40-45 ft)
Expected change in depth (per 5 years)	60 m (200 ft)	45-150 m (150-500 ft)
Truck Haulage		
Number of ore trucks	47	20-60
Number of waste trucks	47	20-60
Capacity:		
Unit ore capacity	90-154 t (100-170 st)	90-180 t (100-200 st)
Unit waste capacity	90-154 t (100-170 st)	90-180 t (100-200 st)
Average number of shovels	14	9
Shovels, size	11 m³ (14 cu yd)	7.5-15.3 m³ (10-20 cu yd)
Rock/day:		
Ore	58.3 kt/d (60,000 stpd)	36.3-110 kt/d (40,000-120,000 stpd)
Waste	40.8 kt/d (45,000 stpd)	72.6-81.5 kt/d (80,000-200,000 stpd)
Hauls		
Ore, length	4.8 km (3 miles)	2.4 km (1.5 miles)
Waste, length	3.2 km (2 miles)	3.2 km (2 miles)
Ore, climb	73 m (240 ft)	84 m (280 ft)
Waste, climb	76 m (250 ft)	150 m (500 ft)
Ore, time (min/load)	17	25
Waste, time (min/load)	23	25
Maintenance		
Average haulage (men/unit)	7	5
Average fuel consumption	1550 L/d (410 gal/d)	1890 L/d (500 gal/d)
Crusher		
Number	2-4	1-2
Size	1525 mm (60 in.)	1525 mm (60 in.)
Capacity	3630 t/h (4000 stpd)	2270-3630 t/h (3000-4000 stph)

Fig. 13. Movable crushers with crawler transporters.

Fig. 13 but with individual adaptations to a unique pit problem. USBM survey results indicated more variations in conveying systems than in the crusher. Fig. 14 shows a very workable system that could be adapted to thick overburden. A crusher of this type may result in a lesser cost than a fixed crusher housed in a conventional concrete structure (Johnson, Frizzell, and Utley, 1981).

The first major, high capacity, in-pit ore crushing system was installed at Duval's Sierrita copper-molybdenum open pit mine in Arizona. Crusher capacity is up to 3.6 kt/hr (4000 stph) using a 60 × 89 gyratory. Duval estimates the system will save $8 to $10 million a year in operating costs (Anon., 1982). Pit trucks dump to a portable apron feeder linked to the crusher. The entire system is moved by a crawler-type transporter. Duval will license the system to others under a technology agreement (Anon., 1982).

Railroad

When a high production rate can be maintained over a long period of time, the railroad will provide the lowest cost per ton of any system (Shand, 1970). The railroad has a high initial cost that severely limits its use for overburden. When rail haulage is used for overburden, the track system has been designed principally for ore movement. In most cases, the overburden if transported by rail is done because the system is installed to handle ore.

Most waste is disposed of by dumping from a higher elevation with an outward building pile being formed. This can only be done by railroad if side dump cars are used. Due to the side dumping mechanism, the cars are heavier and

more costly than flat bottom ore cars used in a rotary dumper. Train length is usually limited by the extra weight of side dump cars. Waste dumps created by side dump cars must have the track shifted outward as the dump grows and may be more costly to maintain than those made by trucks. On the other hand, a very long dump section of track on the edge of a deep canyon may be very practical. This may create visibility problems for the train engineer, but it may not be important if a two-way radio is used to keep the locomotive engineer informed of the train movement and dumping progress (Aiken, 1970).

Rail movement of waste is probably only justified where the main mine ore transport is by rail and waste and ore car interchangeability is practical. Waste volume should also be large to experience the low costs of rail movement. Some of the advantages of rail waste hauls are: (1) relatively small labor force for potential tonnage moved; (2) can handle wide range of materials and big volumes; (3) low cost for long hauls; and (4) can be remote-controlled. The major disadvantages of a rail system are: (1) high initial cost; (2) grade design limitations; and (3) inflexibility of the system.

Hydraulicking

This can be termed the excavation of unconsolidated or loosely consolidated materials by the impact of high pressure water, or in areas where frozen material is being removed after thawing. Both mining and transportation are by the same medium. Special nozzles, giants or monitors, are used to break down the materials with pressures of 5273 g/cm² (75 psi) and higher.

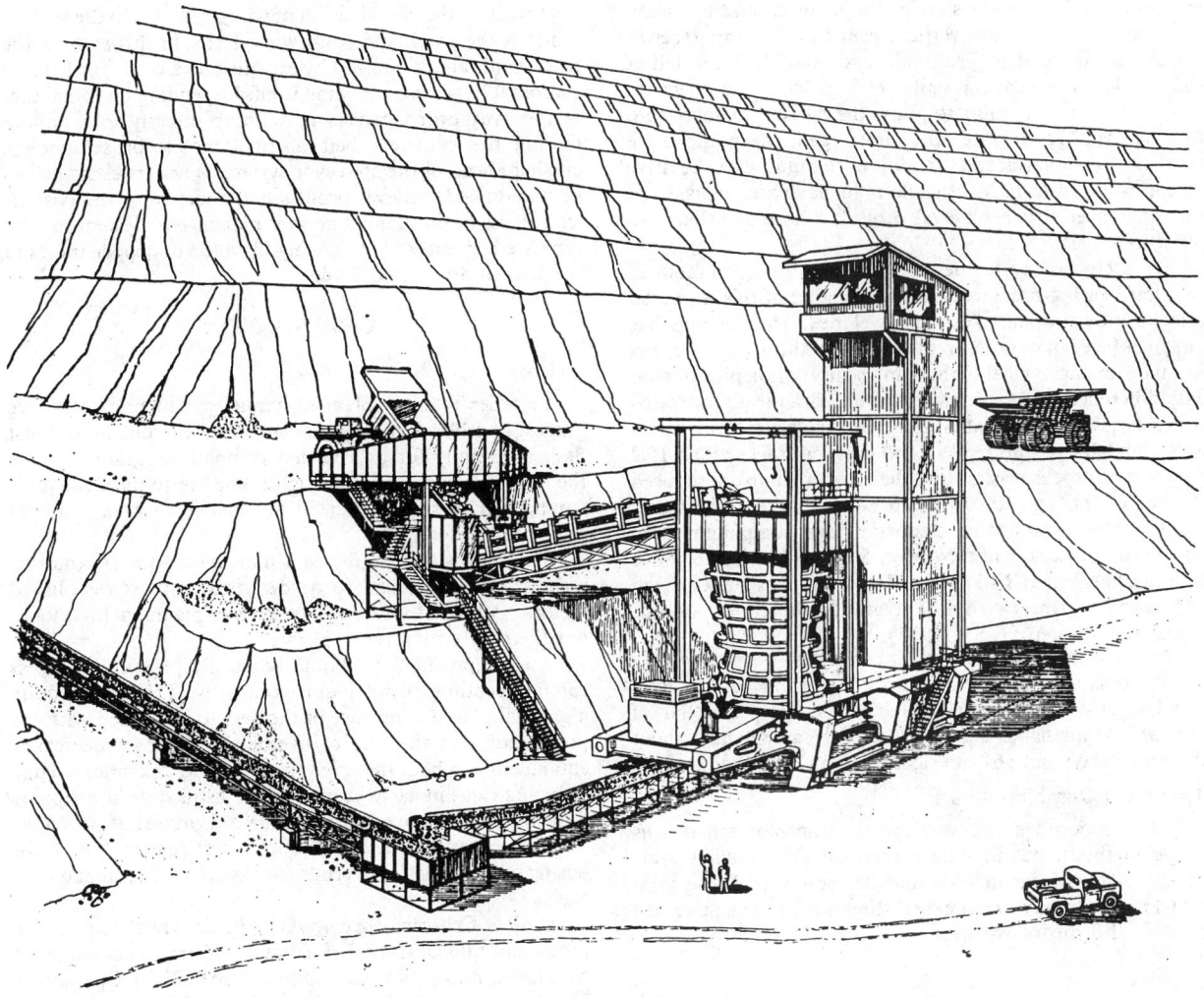

Fig. 14. Movable crusher concept with constant feed.

Modern-day hydraulicking tools are more efficient than the old equipment of years past. The equipment is more mobile and controls are easier to operate and may be automated. Pipe size to weight is better, making site moves faster. Monitors can be remotely operated for better water control. No longer are clumsy jockeyboxes, heavy skids, and deadmen needed as well as the hard-to-control straight pipe monitors. The present-day monitor uses hydraulic thrust for easy precision positioning control.

Old time hydraulicking was mainly concerned with moving alluvial gravels through a sluice to recover minerals. Today's monitor can handle twice as much yardage when used for stripping or washing material. This can be verified by inspecting old monitor production tables.

Conditions that must be met for successful hydraulicking are: (1) sufficient water; (2) spoil disposal area; (3) sufficient runoff gradient; and (4) amenable material.

These parameters apply for unassisted *free* hydraulicking. For materials of a more compact nature, hydraulicking techniques today employ a bulldozer. It can rip and doze semicompact material to a slot trench. Monitors are directed into the trench to further disintegrate and transport the material.

Dredging/Pipeline

The hydraulic dredge previously has been described as an excavating machine. Its pumping aspect is often incorporated as a transport tool as well, aside from discharging excavated material back to the pond. One of the best examples of a dredge and slurry pipeline transport is the overburden stripping at the Caland mine. The ore body was under 30 m (100 ft) of water and overlain with an average of 90 m (300 ft) of overburden. This 124 million m³ (3.5 million cu ft) of overburden was transported 6 km (4 miles) to a disposal site (Li, 1976).

Two 7900-kW (10,600-hp) dredges did the pumping of the slurried overburden. Material was handled at the rate of 2700 m³/min (3500 cu yd per min). The entire job was completed in about seven months from start of pumping. A more recent (1976) slurry pipeline project at Caland was the removal of some 203 000 m³ (265,000 cu yd) of erosional silt. This material resulted from erosional deposition of silt into the mine area.

The 1976 project was carried out by slurrying the silt with a Marconaflo system, a patented device that uses a gooseneck swivel nozzle to produce a high pressure water jet. Impact of the water slurries the material and a caisson unit with a top suction vertical centrifugal pump receives the gravity fed slurry. The pump and nozzle are located at the bottom of the caisson unit. See Fig. 15. Sinking jets are provided below the units to allow the caisson to settle into the material. To operate fully submerged, the pump drive mechanism is interfaced at the top of the unit with the drive motor at the surface. Macronaflo units have been considered for other overburden removal projects where a siltlike material may exist (Schabas, 1976).

Of interest is another device used for overburden removal of a low bearing-pressure material. Because of this, a special crawler-mounted slurrifier was developed. The machine has an open-faced jaw crusher mounted on the front of a hydraulic gun car, typical of those in use in Florida phosphates. The difference is the addition of the hydraulically operated movable jaw grizzly that is designed for vertical movement above and below the level of the crawler tracks. Plus 15.2 cm (+6 in.) rocks washed into the crusher sump are reduced by the 182 000 kg (400,000 lb) pressure exerted. The slurry at 25% solids is pumped to a constant level sump, monitored by sensing devices, and through a 50.8-cm (20-in.) pipeline to a discharge area 10 km (6 miles) away. Instrumentation and monitor panels ensure that any malfunctions are immediately detected (Koch, 1975).

Some of the advantages claimed for this system are: (1) increased safety, can operate at distance from vertical face; (2) low cost transport; (3) higher tonnage per dollar of capital; (4) low labor costs; (5) can get equipment without delivery delay; and (6) can operate in any weather.

Transport Combinations

This section has discussed specific transport schemes as single methods, but for major economies in moving overburden, some of the methods may be combined. These combined methods may have interest stimulated by the increased cost of all forms of energy.

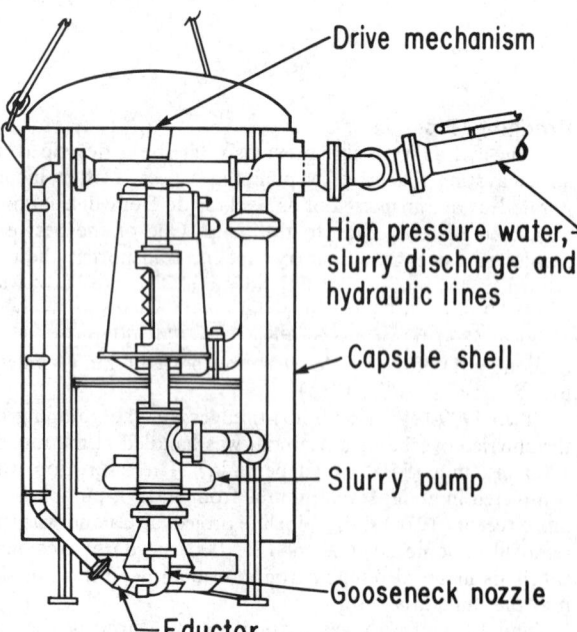

Fig. 15. Marconaflo system for slurrying the silt at Caland mine.

Probably the combination of widest interest to the surface miner is the truck and conveyor system. In this system, the truck hauls are minimized by establishing a conveyor loading station at or within the mine itself. Depending on the nature of the overburden, trucks may dump directly to a hopper feeding the conveyor belt or to a mobile or semimobile crusher ahead of the conveyor system. When truck hauls can be maintained on level profiles and short runs, the cost effectiveness of this combination is maximized. Systems of this type are currently (1987) being installed in copper mines in the United States and Chile.

CASE STUDIES

Iron

In order to meet planned annual production rates, ore grades, and blending requirements, iron ore operators must design carefully for overburden stripping programs. Unless the stripping can expose enough crude ore to meet the mine commitments, shipments will fall behind and/or material grade will suffer.

This case study examines a mine where the overburden stripping program had to be designed to meet established factors. Among those considered was planning to smooth out the overburden removal rate.

An equipment fleet should be ideally planned so excess capital spending is not required initially. This may require a gradual planned increase of the fleet to be phased with ore production. As the rate of overburden removal decreases, equipment can be transferred to ore mining operations. High stripping rates in early years may be justified if the stripping time span is similar to the equipment depreciation schedule.

This case deals with the problems of removing and disposing of two types of overburden material: alluvial and rock waste.

Project Criteria: *Topography:* Alluvial valley with ore in prominent hills of low relief. *Climate:* Warm to hot summers with heavy rain at times. *Elevation:* 40 m (131 ft). Operations: 330 days per year, three shifts per day, 8 hr per shift, and 75% operating efficiency. *Material:* Alluvium, 1.7 t/m³ (0.53 st per cu ft); rock, 2.7 t/m³ (0.084 st per cu ft).

Selection Parameters: The primary requirement to be met in the overburden stripping was the removal of sufficient barren material to allow for a blended plant feed. This was a constraint imposed by the project costs and metallurgical requirements. Two types of ore had to be blended in equal proportions until the capping ore was exhausted.

To satisfy the 50/50 blend to the mill, 48 Mt (53 million st) of alluvium had to be removed in 24 months. About 89 Mt (98 million st) of rock waste had to be moved concurrently. The quantities of overburden were calculated by a computer program based on exploration input. This is shown in Table 6, a sample printout giving the ore and waste/alluvium amounts in the initial pit. The term overburden here refers only to the alluvium with waste representing the rock overburden.

Level-by-level printouts were made for the start-up mine pit. The results of this were graphically plotted, as seen in Fig. 16. This indicates the need to remove nearly 120 Mt (132 million st) of overburden, to 120 m (394 ft) deep, to expose the required 50/50 mixture of two ores.

Alluvium Alternates: Soil borings and test pits indicated the soft, unconsolidated nature of the alluvium. It ranged between clays, sandy clay, and loess. Fig. 17 shows the relationship of the overburden to the ore zone. The rock waste must be classed as overburden since it had to be removed

Table 6. Quantities of Ore and Waste in Iron Mine, North Pit, Pit Tonnage Evaluation

	Ore Body #1			Ore Body #2			Ore Body #3			Ore Body #4			Total Ore		
	Mt	Ave % TFE	Ave % FEO	Mt	Ave % TFE	Ave % FEO	Mt	Ave % TFE	Ave % FEO	Mt	Ave % TFE	Ave % FEO	Mt	Ave % TFE	Ave % FEO
Ore A	0.0	0.0	0.0	0.0	0.0	0.0	0.508	28.04	5.61	0.136	28.83	5.77	0.644	28.21	5.64
Ore B	0.0	0.0	0.0	0.0	0.0	0.0	0.026	29.13	11.65	0.006	26.73	10.69	0.032	28.67	11.47

	Mt	Ave % TFE	Ave % FEO
Total Ore	0.676	28.23	5.92
Overburden	23.688		
Waste	89.209		

Cost to remove overburden	$ 15,396,939.0
Cost to remove waste	$ 89,209,264.0
Cost to remove ore	$ 675,511.62
Concentrating cost	$ 3,006,026.00
Pelletizing cost	$ 1,992,759.00
Total	$110,280,464.62
Ore value	$ 7,479,224.00
Value to cost ratio	0.0678
Pellets, t	233,725.
Stripping ratio	167.13

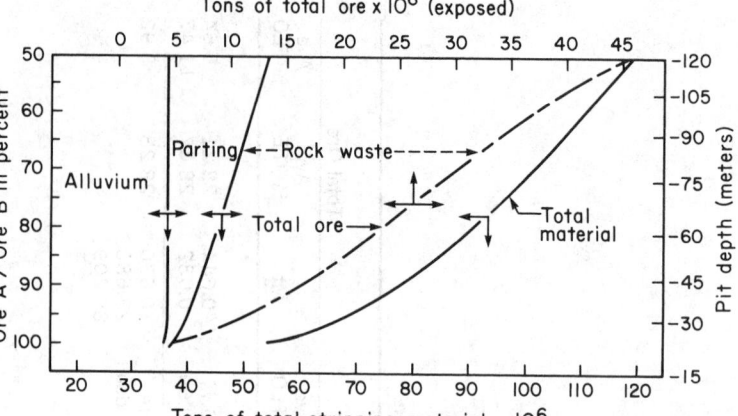

Fig. 16. Overburden stripping required for various proportions of ore exposure.

to reach the ore. The time frame of the overburden removal scheme was paramount to the method selected.

Two alternate schemes were applied and estimated for capital and operating costs. Based on the experience of a large open pit in the southwest United States, it appeared rubber-tired motorized scrapers would work well. These machines are particularly suited to soft, sometimes moist, loose overburden. Combined with the scrapers was a belt conveyor loading station and overland belt.

The second alternate considered was an all scraper fleet delivering the alluvium to the dump site. This is the same location in both alternates. Although other feasible methods were technically acceptable, the ones discussed here were selected as best fitting a restricted time frame imposed on the operations start-up.

Comparison Methods: An all scraper fleet was compared with a scraper fleet delivering to a drive-over hopper feeding a belt conveyor system. Delivery of material from both systems was to the same disposal site. This site was adjacent to the open pit that would be eventually refilled with the overburden. Refilling was a requirement to free up agricultural land that was temporarily holding the waste dumps.

The optimum scraper size was determined by computer haulage simulation. This indicated the largest size scraper available to handle over 22 kt (24,250 st) per shift. A belt conveyor 1.22 m (4 ft) wide by 3.4 km (2 miles) long was used in the combined scheme. At the discharge end of this method, bottom dump trucks were used to distribute the alluvial overburden to the waste area. Fig. 18 illustrates the system components. Table 7 is a typical haulage simulation computer output.

A second computer program, using scrapers, assigned operating costs for the trucks and conveyor system and calculated the optimum number of units to work with the conveyor. From these data, the system capital and operating costs for a unit of material were estimated.

Similar procedures were followed for the alternate alluvial overburden method. This was the application of an all scraper fleet to load and transport the material. The ability of scrapers to dig, load, and transport and discharge, with travel speeds to 50 km/h (31 mph), made this study of interest. After analyzing the total costs of the system, the comparative study results favored the scraper/truck scheme with a conveyor belt by 13%.

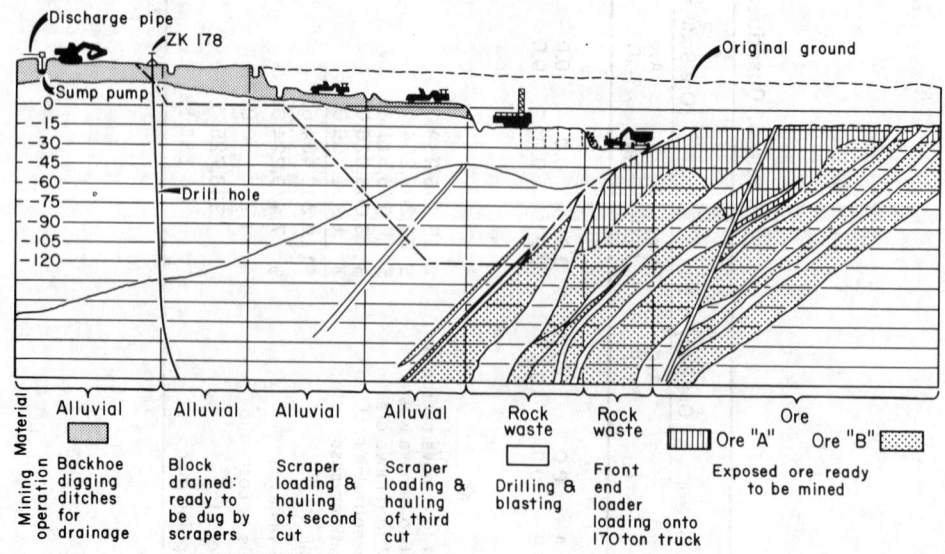

Fig. 17. Typical mining cycle for overburden removal.

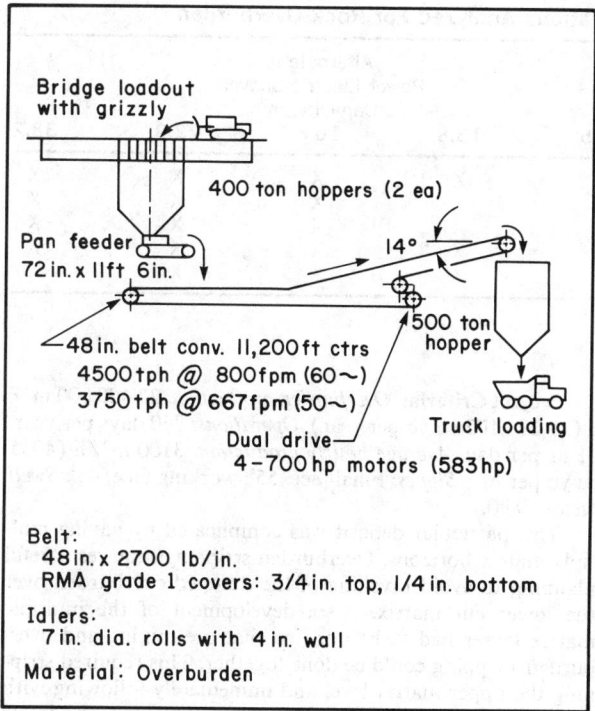

Belt:
48 in. x 2700 lb/in.
RMA grade 1 covers: 3/4 in. top, 1/4 in. bottom
Idlers:
7 in. dia. rolls with 4 in. wall
Material: Overburden

Fig. 18. Scraper/belt system.

Rock Overburden: To begin mining and have an advance supply of ore available required removal of waste rock at 35.7 Mt/a (39,352,500 stpy). In the overburden removal program, this rate would continue for 30 months. The rock material, with an average compressive strength of 1360 kg/cm² required drilling and blasting.

The criterion of having to expose two types of ore for the initial mining required taking the removal rock overburden to the −120 m (−394 ft) pit level. Graphically this is demonstrated in Fig. 17, showing the complete cycle of overburden removal for alluvium and rock. The program would cover a total development time of 30 months. Total material removed, alluvium and rock, would amount to 137.25 Mt (151.29 million st).

Two alternates were studied for overburden rock removal. Alternate one was front-end loaders working with 109 and 154 t (120 and 170 st) trucks. Alternate two examined many combinations of truck sizes, 109 to 317 t (120 and 350 st) and electric power shovels, 11.5 to 38.2 m³ (406 to 1349 cu ft). See Table 8. Performance output of the equipment was determined by computer simulation. The same programs were used as those applied to the alluvium.

The comparative results indicated that the front-end loader and truck combination was the most economical. This was particularly true when trucks were loaded from each side in the model. Recommendations for the rock overburden removal equipment combined the front-end loader and truck fleet with power shovels. The decision to include power shovels was based on the need for loading from high benches, interchangeability of loaders for ore work, and an owner preference for power shovels over front-end loaders.

Phosphate

In this case, the overburden consisted of well-consolidated but relatively soft siltstone or fine-grained sandstone. The formation was thinly bedded and contained some very hard chert beds. This meant that some blasting would be required prior to removing the harder material.

Topography of the area was flat to gently rolling. Dissecting river and creeks, generally dry, cross the mining area. The generally flat terrain lent itself to dragline operation with no need for extensive cutting to maintain operating grades for the dragline tub. Locally, the formation is gently

Table 7. Stripping Belt/Scraper

Tandem Tractor Scraper 708.7 kW (950 hp)
Shift Time = 0.0 Sec
PL Mult = 1.00

Rated Payload = 47 174 kg   Empty Weight = 66 725 kg (147,100 lb)
                                                   Payload = 47 174 kg (104,000 lb)

Haul Road   Initial Vehicle Speed, 0.0 mph*

Sec No.	Dist (ft)	Roll Res	Grade (%)	Vel Limit (mph)	Max ss Vel (mph)	Top Vel (mph)	Last Vel (mph)	Accum Time (min)
1	656.	5.0	0.0	15.00	22.25	15.00	15.00	0.54
2	2296.	3.0	2.30	25.00	22.09	21.79	15.00	1.85
3	328.	3.0	2.30	15.00	22.09	15.00	0.0	2.13

Return Road   Initial Vehicle Speed, 0.0 mph

Sec No.	Dist (ft)	Roll Res	Grade (%)	Vel Limit (mph)	Max ss Vel (mph)	Top Vel (mph)	Last Vel (mph)	Accum Time (min)
1	328.	3.0	−2.30	15.00	80.56	15.00	15.00	0.26
2	2296.	3.0	−2.30	25.00	80.56	25.00	15.00	1.33
3	656.	5.0	0.0	15.00	32.23	15.00	0.0	1.86

Haul Time   2.13 min
Return Time   1.86 min
Fixed Time   1.90 min
Cycle Time   5.89 min
Production 480.16 t/h (529.28 stph)
Haul and Return Distance, 2000 m (6560 ft)
Avg Speed, 12.6 mph†

* Metric equivalent: mph × 1.609 344 = km/h.
† Does not consider job efficiency or cut-and-fill distances.

Table 8. Summary of Loader/Truck Combinations Analyzed For Rock Overburden

Rear Dump Trucks Capacity, t	Alternate I: Front-end Loaders Capacity, m³		Alternate II: Power Electric Shovels Capacity, m³				
	9.94	19.88*	11.5	13.8	16.8	22.9	38.2
109 t (120 st)	X	X	X	X	X	X	X
154 t (170 st)	X	X	X	X	X	X	X
181 t (200 st)			X	X	X	X	X
213 t (235 st)			X	X	X	X	X
317 t (350 st)			X	X	X	X	X

* Two front-end loaders of 9.94 m³ (total: 19.88 m³) loading one truck.

folded tending to let the overburden depth vary tens of feet over a horizontal distance of several hundred feet. This uneven occurrence of the overburden interface with the matrix contributed to the choice of the dragline. Fig. 19 shows this uneven overburden-ore interface.

Alternative overburden removal schemes to the dragline were studied. All draglines for both overburden and ore were considered. The ore matrix would be delivered to conveyor loading points for transport to the washer.

Motor scrapers were estimated as too expensive due to their high consumption of diesel fuel. In this case, available natural gas for power generation offered very low unit kilowatt per hour costs. On the other hand, trucking costs were $0.085 per ton mile in the year of the study vs. electric power at $0.0083 per ton mile.

In exploration the chert had been shown to break into boulders up to 610 mm (24 in.). The slopes of the four-year-old exploration trenches maintained verticality, but the material in the ravelled state reposed at 26° to 30°. This required the draglines to deposit overburden well off the matrix to be mined.

All the overburden was densely packed and would offer firm resistance to a dragline bucket. This was especially so since the drag direction was parallel with the hard surfaces of the cherty horizon. Experience in working the relatively light material, 1.71/t/m³ (0.05 st per cu ft), indicated sufficient competency to justify light blasting before loading. A powder factor of 0.17 g/t (0.0003 lb per st) was sufficient to loosen overburden without mixing with the phosphate matrix.

Project Criteria: *Overburden production:* 22 400 000 m³/a (29,298,102 cu yd per year). *Operations:* 340 days per year, 21 hr per day. *Average hourly production:* 3100 m³/h (4055 cu yd per hr). *Slopes:* Final face, 55°; working face, 45°. *Swell factor:* 0.80.

This particular deposit was complicated by having multiple matrix horizons. Overburden stripping requires careful planning to avoid ravelling of the dumped overburden over the lower cut matrix. Also, development of the multiple matrix lenses had to be done so that ore mining and overburden stripping could be done together. This required stripping the upper matrix level and immediately following with the mining to permit the dragline access to strip the lower horizon. Fig. 20 illustrates the proposed mining sequence with the offset bench method. Ore matrix was taken by shovel and truck operations.

Table 9 shows the calculation of the dragline output. Self-loading scrapers were selected for bench preparation ahead of tub moves, to do some stripping in very deep overburden areas, and for land rehabilitation work. In addition, the other major auxiliary equipment provided were bulldozers and a motor grader.

Laterites

These materials are the end product of long tropical weathering of serpentinized ultramafic rocks and the derived clastics. Laterites are developed in situ. Where they are nickeliferous, there are often well defined physical and mineral grade differences on a sharply defined horizontal basis. It is not uncommon to have to remove an upper horizon of lean

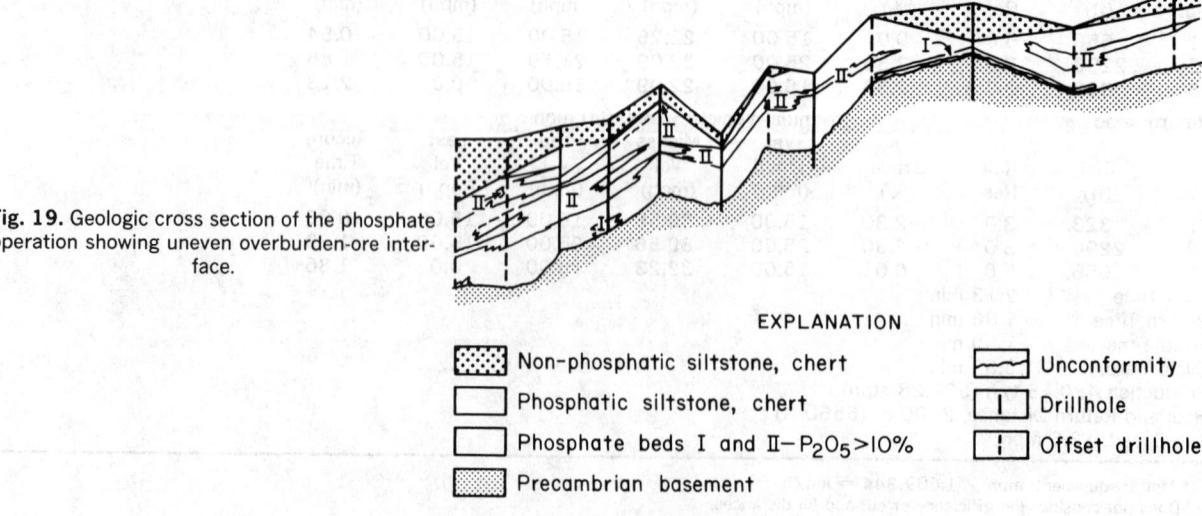

Fig. 19. Geologic cross section of the phosphate operation showing uneven overburden-ore interface.

EXPLANATION

▨ Non-phosphatic siltstone, chert

☐ Phosphatic siltstone, chert

☐ Phosphate beds I and II–P₂O₅>10%

▨ Precambrian basement

∿ Unconformity

☐ Drillhole

☐ Offset drillhole

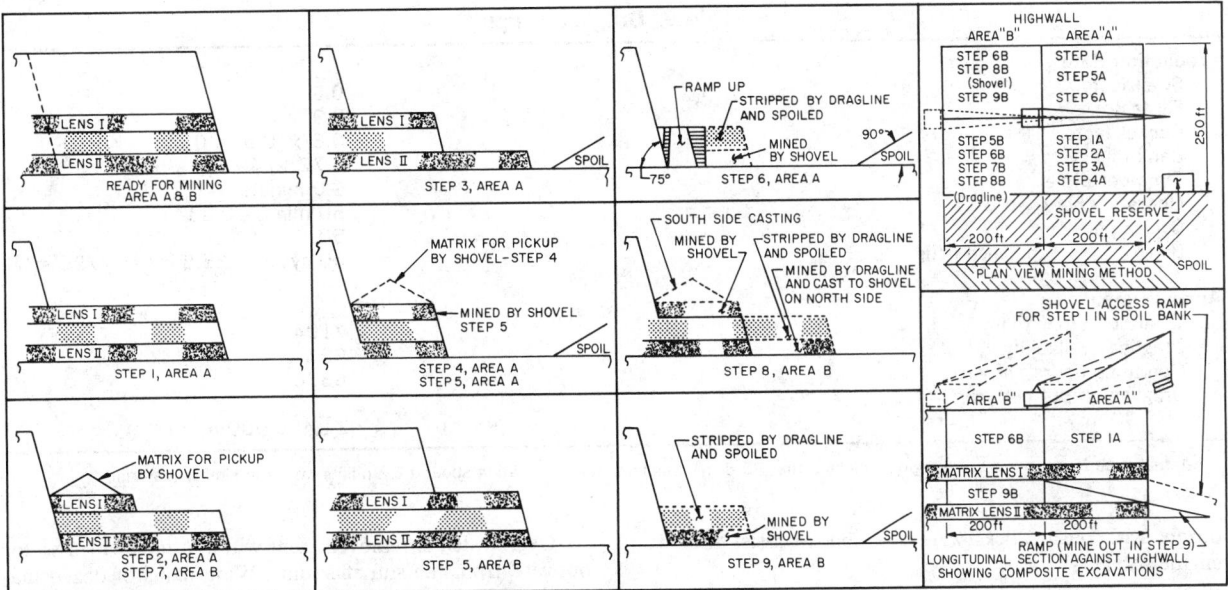

Fig. 20. Proposed mining sequence with the offset bench method.

Step. 1. The dragline excavates the full width of the cut to lens I, developing sufficient matrix to occupy the shovel in the same area until reaching the location of two lenses. At this point, the shovel schedule should be such that a reserve area of matrix is left behind while the shovel starts mining matrix in the immediate area of the dragline.

Step 2. The dragline mines matrix in the southern half of the area, casting back to the northern half for loadout by the shovel.

Step 3. The dragline strips waste in area A to lends II.

Step 4. The dragline mines matrix from lens II, area A, and casts back to the north for loadout by the shovel.

Step 5. The dragline strips overburden above lens I, area B, on the south side. The shovel mines out lens I, area A, north side, and retreats to the reserve loading area to continue mining matrix.

Step 6. Waste is excavated by the dragline above lens II on the north side of area A, leaving a ramp into area B. The dragline continues stripping in area B above lens I. The shovel returns to area A and mines lens II on the north side, leaving a ramp to the top of lens I in area B.

Step. 7. The dragline mines lens I on the south side of area B, casting back to the north side for shovel loadout. Shovel ramps up to lens I, north side, area B.

Step 8. The dragline strips overburden above lens II, area B, south side, and then mines the matrix, casting back to the north for shovel loadout. After loading all of the dragline production from the south side of area B, the shovel mines out lens I, north side, area B, and ramps back down to the floor of the pit, mining out the ramp as it retreats.

Step 9. The dragline strips the overburden above lens II, north side, area B, and the shovel mines the matrix in lens II. The sequence is ready to be repeated as the shovel ramps back up on waste to the top of lens II.

laterites to expose the lower horizons of mining grade nickel ores. The separations of the horizontal layers are not always sharp and careful grade control must be exercised. This includes the removal of the lean ore or overburden occurring on the upper horizon. It may also be necessary to provide for a degree of pit perimeter layback to ensure good slope stability and safe ore mining conditions.

The overburden depth to be removed in this case will vary according to an assay cutoff for nickel. To make this selective type of operation effective, the use of motorized scrapers was recommended. The deposit has a tropical location with a well defined wet and dry season. Because of the inherent moisture of laterites, the materials cause operating problems with any type of mining equipment. When undisturbed laterites maintain stable steep slopes, but as soon as they are disturbed, they slump into a very sticky and super slippery clay mass.

Excavation and mechanical handling of wet laterite creates more than ordinary problems due to the viscous and sticky conditions. This required keeping overburden removal confined to the drier months of the year. This does not fully eliminate the handling problems, but small 10.7 m³ (379 cu ft) motorized scrapers, with front and rear drive engines,

were considered to have the necessary maneuverability to negotiate the poor underfoot conditions. Some of the deposit topography is steep, 30% or more, and the high-traction twin-powered units were considered adaptable to the conditions. The separate engine drives of the tractor and scraper make them a good tool for the wet sloppy conditions of the deposit.

Since the best laterites tend to be found on the ridges and highs, downslope pushing of the scrapers was to be done whenever possible. Erosion tends to carry away the laterite on the slopes and in the valleys. This meant that the average scraper haul for the entire life of the mine would be about 350 m (1148 ft). In the early years the haulage by scraper was maintained at one-half the average 350 m (1148 ft). Overburden waste was spread on nonore areas adjacent to the mining areas.

Scraper flexibility also permitted close grade control. The stripping depth was indicated to the operator when a calcium oxide (lime) charge, previously placed at cutoff depth, was smeared along the cut. To ensure good visibility of the cutoff mark point, overburden removal was scheduled for 12 hr per day, on two 6-hr shifts.

In addition to the eight scrapers shown in Table 10 three

Table 9. Dragline Output

Production Rate	
Swell factor	0.8
Fill factor	0.9
Bucket factor	$0.8 \times 0.9 = 0.72$
Bank m³/cycle	$0.72 \times 49.7 = 35.8$ m³
Time per cycle	1.0 minute
Operating hour	50 min. = 0.83
Cycles per hour	50
Bank m³ per scheduled hour	50 cycle $\times$ 35.8 m³ = 1790 m³/h
Annual Output	
Total hours per year	7104
Dragline availability	90%
Operating hours per year	6376
* Dragline output per year:	
1790 m³/h $\times$ 6376 hrs	11,400,000 m³/a

* Actual experience in the same region indicated that 35.8 m³ draglines were excavating about 12.2 million m³/a in similar material.

dozers with push blocks were available to assist the twin engine scrapers.

Bauxite

A modified hydraulicking method is used by the Aluminum Co. of America (Alcoa) for removing 18 to 21 m (60 to 70 ft) of overburden from its Suriname deposit. Previous stripping methods for removing a 12 to 15 m (40 to 50 ft) highwall of material of varying consistency at up to 3.4 million m³ (4.5 million cu yd) per year had been uneconomical. Material also had to be spoiled about 11 km (7 miles) away. The stripping and transport problem led to the development of the "Double-Hinged Crawler-Mounted High-Wall Slurrifier." This machine is an open-faced jaw crusher mounted with monitors and pumping equipment on a crawler mounted platform (Koch, 1975).

Mining is done with the monitors that wash the material to the slurry sump; the crusher handles the oversize; and the slurry pumps transport the mixture to spoil using six booster pumps. Water is supplied from a lake by three 448-kW (600-hp) vertical turbine pumps. At the slurrifier unit, this monitor water is boosted to 17.6 kg/cm² (250 psi).

The pumping rate for the 25% solids slurry is 910t/h (1000 tph), powered by a 746-kW (1000-hp), 40.6-cm (16-in.) pump. Flexible 50.8-cm (20-in.) rubber lines are used to connect with 50.8-cm (20-in.) steel pipe at the slurry exit and water entry points in the pit. Normal operating kilowatts are 1790 (2400 hp) for the on-board equipment.

Overburden is removed from within a fan shaped sector outward from the slurrifier unit. Wing dams of dozed material guide the slurried bank material to the pickup point. When the radius of the fan reaches 107 m (350 ft), the setup is relocated parallel with the face.

Costs for this unit in 1982 would be $3,000,000. This includes all on-board and remote location boosters. It is reported that in the 1981-1982 fiscal year labor was $212,000, plus an additional $182,000 in labor to relocate the unit and prepare the sump site. Maintenance costs amounted to $346,000 for replacement spares and wear parts. If power is charged at $0.05/kW/h and the operation is on a 12-hr, six days per week schedule, it is estimated that the cost for 3.4 million m³ (4.5 million cu yd) per year is $982,000 or $0.29/m³ ($0.22 per cu yd). The unit is made by Hardcastle Industries, Tampa, FL.

Copper

A contemporary example of the application of hydraulic excavation is the case study of a large copper mine in the South Pacific area. Overlying much of the mine site was a layer of mainly volcanic ash, clastics material, and weathered rock. This ranged from a thin mantle to up to 61 m (200 ft) in depth, with the average being 15 m (50 ft). The entire area was covered with thick jungle vegetation. An additional advantage for a hydraulicking scheme was the steep terrain and heavy year-round rainfall.

Table 10. Scraper Design Selection

Average one-way haul	350 m (1150 ft)
Overburden output, bank	6000 m³/d (7850 cu yd per day)
Simulation production rate of a scraper, 60 min/hr	148 m³ (194 cu yd) loose
Effective hour	45 minutes
Equipment availability	75%
Machine output	500 m³/shift (645 cu yd per load)
Material swell factor	0.8
Production/day loose	7500 m³ (5734 cu yd)
No. of scrapers, 2 shifts	7.5 units
$\dfrac{7500 \text{ m}^3}{2 \text{ shifts}} \div 500$ m³/shift	
Operate	8.0 units
To buy, including spares	11.0 units*

* Based on probability of having at least 8 units available 75% of the time, with up to 10 units available 20% of the time.

The contractor and owner carried out extensive hydraulicking tests to confirm viability. This work was supported by the manufacturer of the hydraulic monitor equipment. Testing was to develop design data on the following: (1) types of material that can be hydraulicked without mechanical aid, (2) mechanical assistance required, (3) areas to be included in the hydraulicking program, (4) slurry transportability vs. lump size, (5) water and monitor duty, (6) water supply data, and (7) slimes containment.

Materials to be tested by the monitor were slope wash soil and tuff and completely weathered, highly weathered, and moderately weathered rock. These initial field tests confirmed the feasibility of hydraulicking.

From the beginning of hydraulicking in April to September of the next year, production hydraulicking had removed some 15.5 million m³ (20.3 million cu yd) of mixed material types. This was at an average rate of 10 920 000 m³/a (14.3 million cu yd per year). See Table 11 for equipment for the production period.

In addition, a three-section 22.26 × 13.72 × 3.05 m (75 × 45 × 10 ft) intake structure was constructed to feed clarified water to the main pump intake. The intake stations had to be moved further upstream to avoid heavy sedimentation. This caused excessive pump and monitor wear. Normally, only three to four monitors [2.54 cm (1 in.)] operated in the ore area due to the water supply limitations.

Throughout the program, the shifts and number of people varied due to circumstances. During the first five months, the normal operation was three 8-hr shifts per day, six days per week; and then for most of the program, two 10-hr shifts per day, six days per week.

Average production in the latter period was 160 000 m³ (209 272 cu yd) per week at a rate of 138.5 m³ (181 cu yd) per dozer hour, continuously. The average water consumption was 9.1 L/s (143.4 gpm) per 1.0 m³ (1.308/cu yd). The highest rate achieved for any one hour was 213.4 m³ (279.1 cu yd) per dozer hour using 3.22 L/s (51 gpm) of

water. A low monitor water consumption was helped by the up to 50% contained water in the tuff.

Some of the conclusions reached from the production results were: (1) 40 to 50% by volume of solids optimum; (2) thicker slurries did not move well on flatter gradients; (3) grades over 4% kept the water-solids content well mixed; (4) optimum mixing of water and solids was on a −20% or more gradient, after initial water contact; (5) on gradients flatter than −5%, optimum mixing did not occur and rock buildup occurred; (6) rock buildup splits the channel flow reducing efficiency; (7) high monitor pressure in bulldozer cuts may force a slurry forward as a wave that can carry boulders to 1 m (3.3 ft); (8) average tractor push dozing was about 60 m (200 ft); (9) ground sluices at ±10% helped decrease dozer push distances; (10) slot dozing was extensively used along with ripping where required to loosen boulders; (11) optimum monitor stream distance was 9 to 15 m (30 to 50 ft) in ground sluices.

Personnel concluded that hydraulicking was a viable proposition for removal of unconsolidated overburden.

Coal Case Studies

Of all the minerals extracted by surface mining methods, coal mining represents not only the largest industry but also the broadest range of equipment applications. Geographic variations in geology and topography have presented mine engineers with a variety of design considerations for selection of the most technically and economically sound approach to removing overburden and the exposed coal seam. While large direct casting units such as draglines and shovels represent potentially the most inexpensive means of overburden removal, mine operators confronted with rolling to steep terrain and thick multiple or pitching coal seams must resort to more mobile equipment. In these conditions loading shovels and front-end loaders working in conjunction with trucks provide viable alternatives for overburden and coal removal.

For the most part, mining practices can be characterized

Table 11. Production Equipment, Copper Case Study

No. of Items	Description
7	Centrifugal pumps, Thompson-Castlemain, 254 cm (10 in.), 157.7 L/s (2500 gpm), 198.2 m (650 ft) head at 1430 rpm. With engines pumps rated at 126.2 L/s (2000 gpm), 198.2 m (650 ft) head at 766 rpm.
5	8YJC Paxman engines.
2	V12-700-IP Cummins engines.
1	SCO 5W Kelly and Lewis pump, 126.2 L/s (2000 gpm), 198.2 m (650 ft) head at 1800 rpm. Used with Cummins V12-700-IP.
17	D8H Caterpillar dozers.
2	TS 14 Terex scrapers.
6	15.24-cm (6-in.) Stang Intelligiant Monitors.
4	10.16-cm (4-in.) Stang Intelligiant Monitors.
6	2.54-cm (1-in.) Stang Intelligiant Monitors.
2	CAT 561 Pipelayer.
Lot	Floodlights.
3	Pipe carriers (Muskateer).
Lot	40.64-cm (16-in.) main-line Naylor pipe.
Lot	30.48-cm (12-in.) main-line Naylor pipe.
Lot	30.48-cm (12-in.) and 20.32-cm (8-in.) Naylor branchline pipe to monitors.

regionally as a result of topographic and geologic conditions. The Appalachian coalfields including the states of Pennsylvania, West Virginia, and eastern Kentucky are characterized by rolling to steep terrain. Thus, the most common equipment applications are front-end loader/truck haulback, dozer-loader modified block cut mining, shovel/truck mountaintop removal, and small dragline modified area mining. In contrast, the topography of the midwest coalfields is flat overlying relatively thin coal seams. Surface mining in Illinois, Indiana, and western Kentucky consists of area mining with large draglines, direct casting shovels, and in one case a tandem bucket wheel excavator/shovel operation.

The western coalfields including Arizona, Colorado, Montana, New Mexico, North Dakota, and Wyoming contain the largest potential coal reserves in the nation. Topography of the entire region varies from flat plains to more rugged Rocky Mountain terrain. Where the surface is flat enough and the coal seams are thin to moderately thick, dragline area mining operations are utilized. However, the coal seams in Wyoming and Montana are often 9 to 30 m (30 to 100 ft) thick resulting in overburden coal ratios of 3:1 or less in many instances. Concentration must be placed more on the proper coal loading, hauling, and storage facilities than on overburden removal procedures. In addition, the thick seams present major spoiling problems for direct casting equipment resulting in use of open pit mining techniques for removal of overburden and coal with shovel-truck teams.

Case Study No 1: This case study describes a typical quarry type operation in a northern plains mine. The mine is situated south of Gillette, WY. This area is characterized by plains and moderately sloping hills, with land use being primarily range and cattle grazing. Rain and snow average about 30.5 to 127 cm (12 to 50 in.). The temperature will be below freezing on some 190 days and above 32.2°C (90°F) on about 20 days per year.

Coal deposits are a low sulfur subbituminous material averaging 18 608 kJ/kg (8000 Btu/lb). Coal seams range between 18.3 to 24.4 m (60 to 80 ft) in thickness and dip at about 2°. In the south, topsoil is 0 to 1.5 m (0 to 5 ft) thick and overburden 12.2 to 15.2 m (40 to 50 ft) thick. In the north, the topsoil is 0 to 1.5 m (0 to 5 ft) and overburden is 12.2 m (40 ft) of subsoil and 30.5 m (100 ft) of clays and sands. The streams and aquifers running through the area are given consideration in premining plans.

The work schedule for all operations is 3 shifts per day, 6 days per week, 52 weeks per year. Excluding 11 holidays, 903 shifts are scheduled per year. One or two shifts per year are usually lost to weather conditions.

Normally exposed coal inventories are 4.6 to 5.5 Mt (5 to 6 million st) or about 4 to 5 months production. The total overburden to coal ratio will run about 1.5:1.

Topsoil and subsoil are removed with contractor rubber tire scrapers of the 13.7- to 16.8-m³ (18- to 22-cu yd) class. The annual quantity of contractor overburden will vary but typically will be between 1.15 to 1.53 million m³ (1.5 to 2.0 million cu yd). This level of production will require a fleet of 6 to 12 scrapers with an average of 8 working. Contracts are let on a one year basis and include topsoil removal, reclamation, and contouring after mining, and placement of sands to reconstruct a subterranean aquifer removed during mining. Topsoil and subsoil are also stockpiled for long-term storage.

Overburden from the rectangular pit is hauled by 109-t (120-st) capacity, two-axle, end-dump trucks. Haulage is around the pit ends to spoil piles on the opposite highwall.

The overburden varies in thickness along the advancing highwall and stripping is normally from one to three working benches. Terraces to accommodate overburden haulage are designed around the pit ends at the level of each active bench. See Fig. 21 for an overview of mine activities. Other equipment required in the overburden removal scheme are two 22.9-cm (9-in.) blasthole drills. These drills have a 7.3 to 9.2 m (24 × 30 ft) pattern with a 1-m (3.3-ft) subdrill below grade. Drilling stops at the overburden to coal interface with all material being blasted except the topsoil and alluvium. Blasting is with ANFO at a powder factor of 300 g/m³ (18.7 pcf) of overburden. Novel trunkline and noiseless trunkline delays are used in detonation.

Excavation is by five 17.6-m³ (23-cu yd) dipper shovels. Dozers and scrapers are contractor provided with the mine providing four rubber-tired dozers for cleanup. The blasted overburden will vary per year, but will run about 15.3 million m³ (20 million cu yd) per year. Mine motor graders maintain the haul roads. The terraced overburden dump on the opposite side of the pit is dozer regraded according to the reclamation plan.

Surface stream beds that were disturbed by mining are being reconstructed with beds of sand and underlying impermeable clays. The use of sands and clays for aquifer and stream bed reconstruction necessitates some selective mining of the overburden to assure the availability of materials with the most favorable physical characteristics at the proper time. This does not pose any scheduling problems, since sands and clays are present in sizable quantities throughout the overburden.

Scrapers strip topsoil and brown subsoil from the advancing highwall, transport, and deposit them according to the reclamation plan. A 1.5-m (5-ft) thick layer of unconsolidated brown subsoil is placed uniformly over the regraded overburden followed by a uniformly thick 0.6-m (2-ft) layer of topsoil. For field use, topsoil is defined as surface material containing root medium. Topsoil and subsoil having the most favorable plant growth medium are selected on site by the reclamation department for use in land reclamation. The mining department then directs the scrapers to the areas of selected materials.

Management of revegetation operations is performed by an operating unit of the mine. It directs the placement of topsoil and the seeding with wheat of reclaimed areas. Reseeding has not been necessary even though irrigation is not done. Artificial maintenance of reclaimed land is not cost acceptable.

Contour Mining—Truck Haulback Case Study: The mine is located in north central West Virginia in the Allegheny Plateau. Precipitation averages 1143 mm (45 in.) annually with temperatures ranging from −5° to 100°F. Terrain is moderately steep (greater than 25°) and covered with grasses, brush, and eastern hardwoods. This particular operation is reaffecting 33 ha (82 acres) containing old strip cuts along the Pittsburgh coal seam outcrop that ranges from 1.7 to 1.9 m (5.5 to 6.5 ft) thick. Overburden units consist of calcareous shales and soapstones with intermittent sandstone units.

Of primary consideration in selecting the basic mining approach was the existing regulations preventing downslope spoiling of overburden material and requiring restoration of the land to approximate original contour. Since the entire hilltop could not be economically mined due to excessive stripping ratios, the operator could not utilize small draglines and was forced to consider controlled placement either by dozer/loader or truck haulback operations. Preliminary

Fig. 21. Mine overview, coal case study No. 1.

evaluations indicated that a dozer/loader mine would experience difficulty with placement of the initial cut material and could not meet desired production goals. In addition, the confined working conditions and short reserve life were more conducive to mobile front-end loaders vs. quarry shovels.

The basic mining method selected for implementation (Fig. 22) is a classic example of truck haulback contour mining. Use of the mobile haulage units allowed the operator to dispose of the initial cut material on an adjoining lenticular ridge segment that had been mined prior to enactment of current stringent reclamation laws. Fig. 23 shows the disposal area with respect to the total mine site. This facilitated excavation of a 213-m (700-ft) long cut prior to initiation of concurrent backfilling activities.

As illustrated by Fig. 24, the mining plan consists of topsoil removal and storage followed by overburden drilling and blasting on 3.6 × 4.5-m (12 × 15-ft) centers with ANFO used as the blasting agent.

A variable powder factor is used according to the amount of sandstone encountered. Overburden removal progresses in an alternating contour block-cut method. The 10.7-m³ (14-cu yd) rubber-tired front-end loader accompanied by a 287 kW (385 hp) dozer loads two 45-m (50-st) off-road rock trucks. Each cut is 61 m (200 ft) long with a 18 to 21 m (60 to 70 ft) highwall. Pit widths range from 29 to 37 m (95 to 120 ft). These pit dimensions reflect the past experience of the operator concerning efficient operating conditions and a maximum average stripping ratio of 12:1 [bank cubic yards (bcy) per ton].

When each cut is completed, overburden operations move to the opposite side of the initial cut to facilitate coal removal in the exposed pit. Coal is blasted with dynamite prior to extraction. One 5-m³ (6.5-cu yd) front-end loader loads coals into 23-t (25-st) contracted coal haulers. Operations will progress in this alternating method until the hill is completely encircled.

As each mined out section is backfilled, dozers regrade and prepare areas for reseeding. Seed mixtures include Birdsfoot Trefoil and Kentucky −31 fescue. Knockulate is used for a hardy first growth season cover.

With overburden removal unit operations working two 8-hr shifts per day and all other operations being performed during one 8-hr shift, the mine has produced in excess of 181 440 t (200,000 st) per year. In selecting the equipment to perform the overburden removal and achieve the desired production goals, several combinations of front-end loaders and trucks were evaluated. The loader cycle time was established at 37.9 sec from time studies at other active operations. As called for by the plan, the average round trip haul distance was 366 m (1200 ft). Utilizing these parameters, computer simulations were performed to determine the production potential, delay time, and number of trucks required for a 6.9-m³ (9-cu yd) loader and a 10.7-m³ (14-cu yd) loader working independently with 45- and 68-t (50- and 75-st) trucks. Narrow pit limits were deemed undesirable for larger loaders and trucks. Fig. 25, 26, 27, and 28 graphically depict the theoretical production for each combination. The 6.9-m³ (9-cu yd) loader and two 45-t (50-st) trucks could potentially produce 5171 t (5700 st) of overburden per shift while the

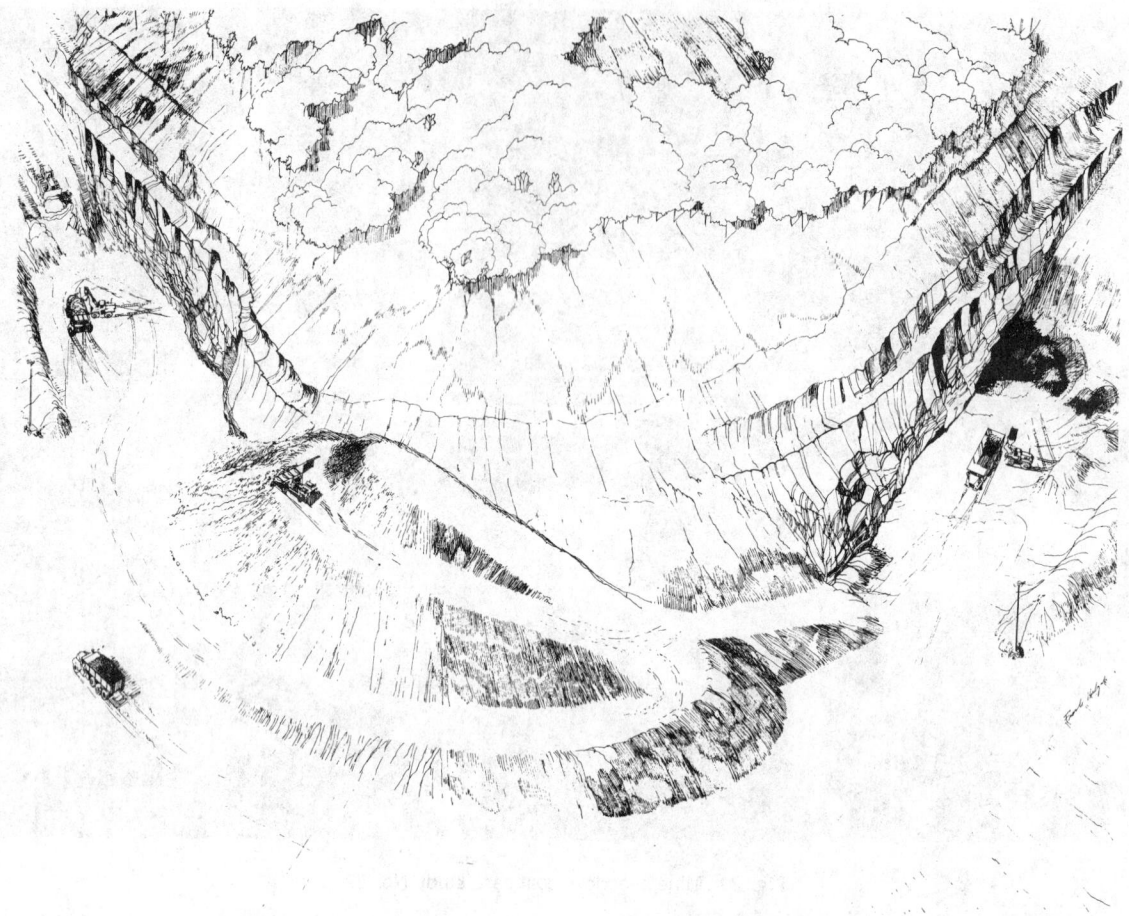

Fig. 22. Case study of a truck haulback mine.

same trucks and a 10.7-m³ (14-cu yd) loader could produce 7167 t (7900 st) per shift. Resultant economic analyses confirmed that the 10.7-m³ (14-cu yd) front-end loader and two 45-t (50-st) trucks provided the lowest operating costs due to maximum utilization of the loader with minimum truck delay. While one 45-t (50-st) and one 68-t (75-st) truck may have improved the economics, the operator preferred to maintain a standard fleet to minimize his spare parts inventory.

Mountaintop Removal—Shovel Truck Case Study: The problems presented by this multiple seam reserve were typical of those confronting many surface mine operators considering development of a mountaintop removal operation.

Typically, mountaintop removal methods had been designed as the logical extension of contour stripping the most common practice in Appalachia. Thus, an entire mountain was mined through a continuous series of contour cuts. While this technique proved to be a viable surface mining method, there were some deficiencies:

1. Limited bench space forced haulage of a large percentage of overburden to valley fill disposal sites, imposing high haulage costs.

2. Operating costs per ton of coal mined increased as outcrop coals were depleted and advancing operations encountered consistently higher overburden to coal ratios.

3. Progressive contour lifts and requirements for large

valley fills to compensate for limited backstack space forced exposure of large surface areas for extended periods of time.

Recognizing the need to alleviate these deficiencies and maintain the total resource recovery benefits of mountaintop removal mining, cross-ridge mining was chosen by the operating company and the site was selected as a government demonstration project. Cross ridge, unlike contour methods, is designed to effect total recovery through a series of cuts perpendicular to the long axis of a ridge. This type of operation affords several potential advantages:

1. Maintains relatively consistent operating costs and production by excavating both high and low cover coal in the same cut.

2. Minimizes active mine and valley fill disturbed surface areas by creating sufficient space for overburden backstacking on the mine bench following removal of an initial cut.

3. Lowers overall mining costs through a reduction of valley fill and associated haulage requirements.

The area's topography is typical of most of southern West Virginia with steep mountains and narrow valleys or hollows. This rugged nature is evidenced by differences in elevation of over 610 m (2000 ft). Consequently, coal seams outcrop on the reserve and allow relatively easy access for mining.

Approximately 74 ha (184 acres) of minable area is associated with the cross-ridge demonstration site, containing over 2.7 Mt (3 million st) of potentially clean recoverable coal at an average strip ratio of 6.7:1 bank cu yd (bcy) per ton. This tonnage is contained in the 5-Block and Stockton-Lewiston (site specific nomenclature) seams that average 76 mm (30 in.) and 3 m (10 ft) of clean recoverable coal, respectively. Total production from these seams had risen from 99 792 t (110,000 st) in 1975 to 428 198 t (472,000 st) in 1980. The cross-ridge demonstration contributed 171 461 t (189,000 st) from April to December 1980. However, production in 1981 experienced a sharp decline primarily due to the United Mine Workers (UMWA) strike from April to mid-June. Cross-ridge production during this 12-month period was only 70 762 t (78,000 st), while total production was 301 190 t (332,000 st).

Mining operations are separated into two working areas, each having a power shovel as its main overburden excavation unit. A 13.7-m³ (15-cu yd) P&H 2100-BL has been assigned to the demonstration project, while a 16.8-m³ (22-cu yd) BE 295-B is working in another area not associated with the demonstration. Both pits function in a similar manner with the exception of an end-dumped fill being constructed on the demonstration site.

The first step in the mining process involves clearing and grubbing vegetation. Then two Caterpillar 657B scrapers remove the upper horizon soil material and stockpile it on a previously constructed backstack lift. Once the topcover is removed, Caterpillar D-9 and Komatsu 355 dozers prepare benches for blasthole drilling.

Blastholes are drilled on a 3.6 × 3.6-m (12 × 12-ft) spacing by four Drilltech D-40K 171-mm (6¾-in.) rigs and one Ingersoll-Rand ECM-350 crawler drill. Each hole is loaded with ANFO explosive and stemmed the final 3.3-m (11 ft). When overburden above the 5-Block seam is blasted, dozers push material over the highwall by a slot method. On a bench below, two Dart D600C front-end loaders load the spoil into Euclid R-35 and R-50 off-highway trucks for transport to on-bench backstack areas. Caterpillar D-8 dozers spread and compact the material into the 15-m (50-ft) lifts.

To ensure a clean marketable raw product, the final 0.6

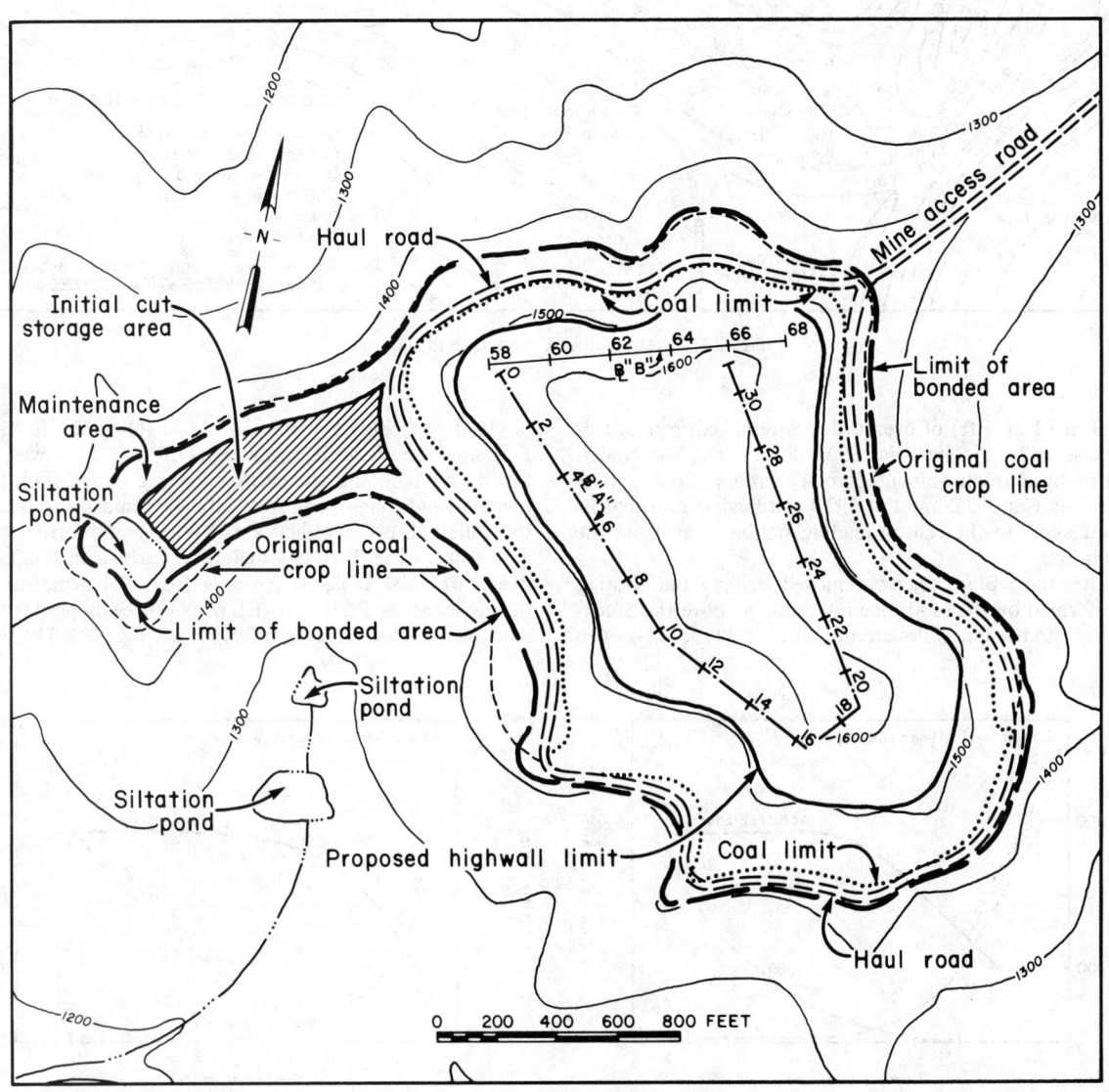

Fig. 23. General location map of mining area, truck haulback mine, showing disposal area in relation to the total mine site.

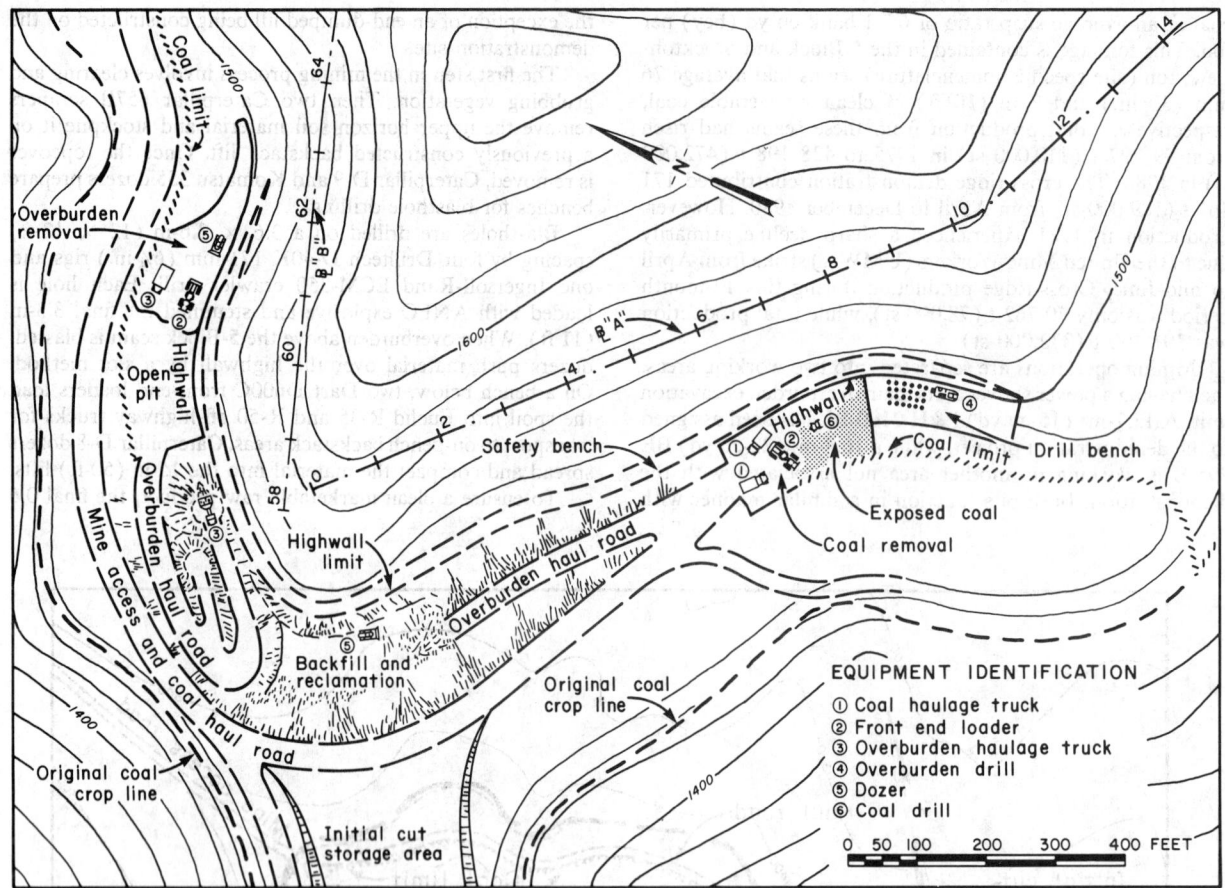

Fig. 24. Truck haulback mine layout plan map.

to 0.9 m (2 or 3 ft) of overburden is removed from the 5-Block seam by a Caterpillar 992A loader, and the coal is cleaned by a tractor-mounted rotary broom. Coal is then loaded via Cat 992C and Cat 992B end loaders onto trucks for transport to the central loading station away from the permit area.

After the 5-Block seam is removed, drilling and blasting is performed on the sandstone interburden above the Stockton-Lewiston seam. Holes are drilled on a 4.9 × 4.9-m (16

× 16-ft) spacing to within 6 m (20 ft) of the Stockton-Lewiston. The hole depth averages 23 m (75 ft) with 6 m (20 ft) of stemming added after loading the ANFO. Fragmented sandstone overburden is loaded and transported to the valley fill or backstack area by the two electric shovels and a fleet of rock trucks including ten Euclid R-105 and five Dart 3100 vehicles. Records from the demonstration indicate that the P&H 2100-BL places three dipper loads per truck at a bucket fill factor of 90% (see Fig. 29). The shovel

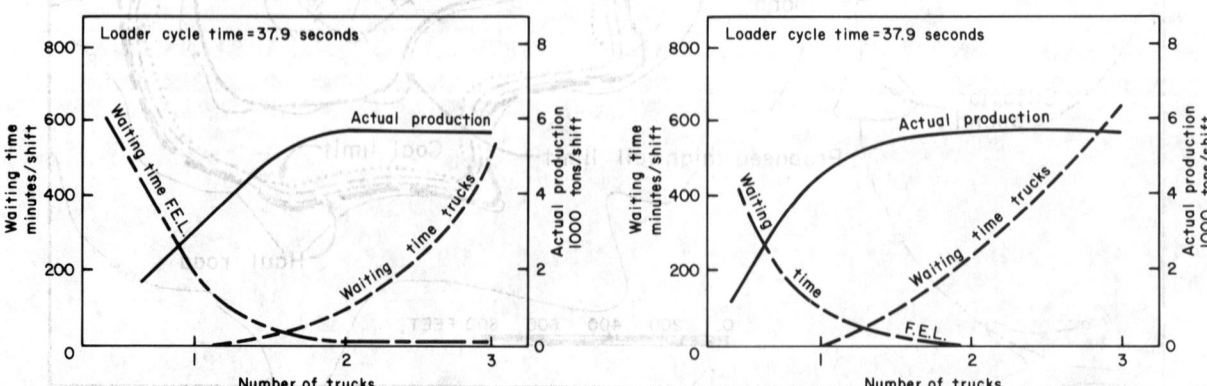

Fig. 25. Equipment delay time/production graph for a 45-t (50-ton) truck and one 6.9-m³ (9-cu yd) loader system.

Fig. 26. Equipment delay time/production graph for a 68-t (75-ton) and one 6.9-m³ (9-cu yd) loader system.

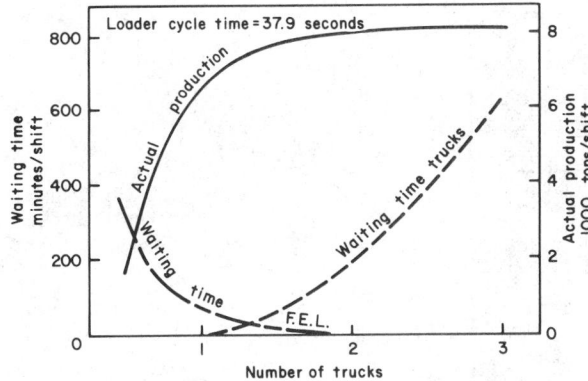

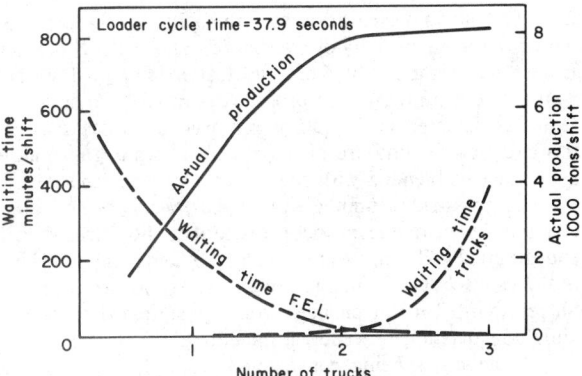

Fig. 27. Equipment delay time / production graph for a 68-t (75-ton) truck and one 10.7-m³ (14-cu yd) loader system.

Fig. 28. Equipment delay time / production graph for a 45-t (50-ton) truck and one 10.7-m³ (14-cu yd) loader system.

has experienced an 85% mechanical availability and an average of 38 sec per cycle. With trucks hauling an average one-way distance of 518 m (1700 ft), the shovel produces 190,000 bcy per month during normal operations.

The 6 m (20 ft) of overburden remaining above the

Stockton-Lewiston seam is usually mined by the shovels as a second lift. This material is a mixture of sandstone and shale and requires blasting prior to loading. The location of spoil placement depends upon material composition. Only durable sandstone is placed in the demonstration fill.

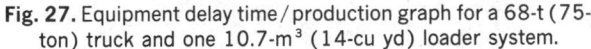

Fig. 29. Shovel truck loading at a mountaintop removal operation.

As in the final excavation over the 5-Block seam, the last 0.6 or 0.9 m (2 or 3 ft) of overburden above the Stockton-Lewiston is removed by front-end loader. The coal surface is also swept by the rotary broom. Because of partings, the seam, as depicted in Fig. 30, is removed in three lifts. The top and bottom lifts are of high quality, while the middle split must be blended with the better coal prior to sale.

The cross-ridge mining system has received an added economic benefit due to incorporation of the durable rock end-dumped valley fill. Approved by the Department of Natural Resources (DNR) under an experimental variance, this fill represented major changes from the standard West Virginia construction procedures, including:

1. The fill is being constructed from the head of the hollow as compared to beginning at the toe (see Fig. 31).

2. All material is being dumped from the Stockton-Lewiston bench vs. hauldown by rock trucks.

3. The final fill face will be terraced on 15-m (50-ft) benches rather than construction of 15-m (50-ft) benches with compaction in 1.2-m (4-ft) lifts throughout the fill area.

4. External drainage is diverted to side ditches vs. the West Virginia procedure requiring direction of all drainage to a central rock chimney drain.

Since the fill must contain 80% durable rock (the Standard Slaking Test is acceptable to the DNR) by volume, most of the 1.1 million m³ (1.5 million cu yd) of material placed in the fill to date has come from the interburden between the 5-Block and Stockton-Lewiston seams. This durable sandstone unit has facilitated achievement of a stable fill and, through natural segregation, development of an 2.4-m (8-ft) underblanket for drainage. When completed the fill will have allowed permanent disposal of 5.9 million m³ (7.8 million yd³) of excess spoil material.

Costs and productivity have been documented thoroughly for the P&H 2100BL shovel and Dart 3100 trucks. Considering an average haul distance of 518 m (1700 ft), the shovel/truck team has averaged approximately 560 bcy per operating hour. Three trucks have provided an efficient match, keeping the 11.5-m³ (15-cu yd) shovel productive. Examination of only basic direct costs (Table 12) associated with the shovel/truck team reveals a cost of $0.74 per bcy.

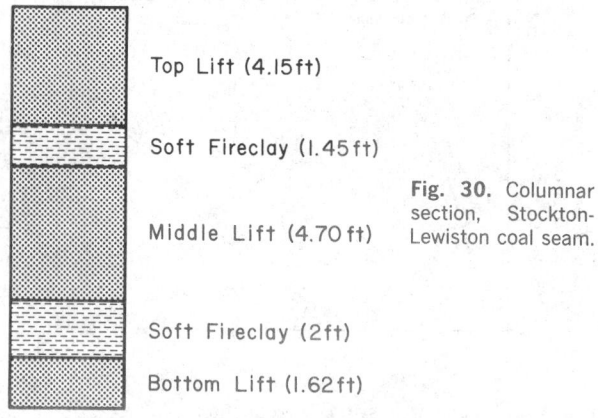

Top Lift (4.15 ft)

Soft Fireclay (1.45 ft)

Middle Lift (4.70 ft)

Soft Fireclay (2 ft)

Bottom Lift (1.62 ft)

Fig. 30. Columnar section, Stockton-Lewiston coal seam.

For comparison, construction of a West Virginia style fill would have resulted in an additional 1585-m (5200-ft) hauldown, increasing the average round trip cycle time including load/dump from 8 to 16 min. Operational plans would have required three additional trucks to maintain the P&H 20s present hourly production level. The resultant cost per bank cubic yard would have been $1.13. With an average stripping ratio of 6.7:1, the end-dumped valley fill has produced a minimum savings of $2.60 per ton. Factoring in overhead, benefits, and other operational costs, including added wear and higher maintenance costs on the trucks, even greater savings have been realized. While these savings will decrease as the fill approaches completion, the impact in terms of present-day dollars is extremely important to the life of mine profitability.

Area Mining — Dragline Case Study: The mining site under study is located in the midwestern coalfields. Topography is characteristically flat to gently rolling, with cropland cover and agricultural land use. The overall mine area is approximately 648 ha (1600 acres).

Two coal seams are present and considered of minable thickness and marketable quality. Overburden and interbur-

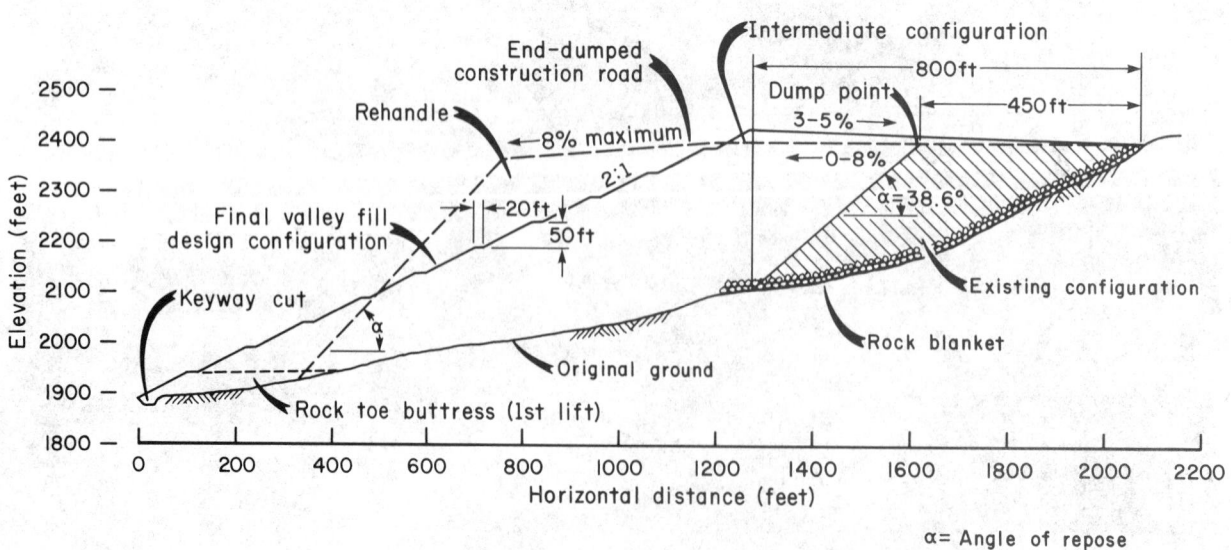

Fig. 31. Profile of end-dumped valley fill.

Table 12. Basic Direct Cost Associated with Shovel/Truck Team

Unit	Depreciation	Fuel/Power	Lube	Tires	Parts/Supplies	Labor	Total
11 m³ (15 cu yd) shovel	$84.36	$31.40	$20.14	$ —	$35.61	$22.00	$193.51*
90-t (100-st) trucks	$22.14	$ 9.07	$ 4.33	$16.88	$11.12	$10.31	$ 73.85*

* Direct cost per bank cubic yard per operating hour.

den at the site total approximately 38 m (126 ft). Cover above the upper seam consists of 1.2 m (4 ft) of topsoil, 3.7 m (12 ft) of unconsolidated subsoil, 12 m (40 ft) of shale, 5 m (18 ft) of clay and sandy shale, and 9 m (31 ft) of sandstone directly above the 1.2-m (4-ft) thick upper seam. Interburden down to the lower 3-m (10-ft) thick seam consists of 6.4 m (21 ft) of sandy shale. Both seams are flat lying and maintain a consistent thickness throughout the mining property. The raw recoverable reserves are approximately 28.6 Mt (31.5 million st) for the composite seams.

Based on the general site information, it was determined that the reserve site was particularly suitable for development as an area mine. Both overburden and interburden are mined with a large dragline, utilizing a simple sidecast technique on the upper seam and a less efficient chop-down technique from the spoil side on the lower seam. However, to comply with federal requirements for topsoil handling in prime farmland, a separate system was needed to remove and segregate 1.2 m (4 ft) of A horizon and 3.7 m (12 ft) of combined B and C soil horizons. Due to the need to transport and stockpile a minimum of three cuts of topsoil and subsoil prior to replacement, scrapers were selected for this operation.

The virgin stripping ratio for the mine is 9:1 (bcy/ton). Annual coal production was projected for approximately 1.4 Mt (1.5 million st) per year, which would result in achievement of a 20-year mine life. This production level was consistent with market requirements and provided for a 20-year depreciation of the dragline purchase.

With a knowledge of the production rates required and all other site information, the equipment selection was made to best meet these parameters. Assuming that the top 4.9 m (16 ft) of cover would be removed as a separate operation, the choice of a dragline was made on the following criteria:

1. Minimum cut width of 32 m (105 ft), 30% swell factor, 75° highwall angle and 34° angle of spoil repose.

2. Based on its range diagram, the dragline had to have an operating radius of at least 90 m (295 ft).

3. It had to have a capacity of digging to a depth of 34.7 m (114 ft).

4. Considering the stripping ratio and approximately 10% rehandle, the bucket size had to be of sufficient capacity to meet the production requirements in both the simple sidecast and chop-down techniques.

Several manufacturers' specifications for draglines were examined to meet these conditions. This review revealed that a 2570-W dragline with 72.6-m³ (95-cu yd) bucket and a 94-m (310-ft) boom at 30° would meet the requirements. Following are the production factors utilized in selecting the dragline as well as the scrapers and drills.

Bucyrus-Erie 2570-W Dragline
1) Bucket size 72.6 m³ (95 cu yd)
2) Fill factor, 90%
3) Cycle Time, + 63 sec
4) Job efficiency, 89%
5) Capacity per hour: (1) simple sidecast seconds 3106

bcy per hr @ 100% availability; (2) chop-down method = 1398 bcy per hr @ 100% availability

6) Mechanical availability = 85%

Bucyrus-Erie 60-R Drill 31 cm (12¼ in.)
1) Drill pattern = 7.9 × 7.9 m (26 × 26 ft) (average)
2) Drilling rate = 0.5 m (1.7 fpm) (average)
3) Job efficiency = 82%
4) Capacity per hour = 2094 bcy per hr @ 100% availability
5) Mechanical availability = 80%

Caterpillar 651-B Scraper - 32 cu yd (struck); 44 cu yd (heaped)
1. Haul segments = 30 m (100 ft) @ +4%; 1250m (4100 ft) @ 0%; 45.7 m (150 ft) @ −8%; 152 m (500 ft) @ 0%; 38 m (125 ft) @ +10%; 1250 m (4100 ft) @ 0%
2. Bench rolling resistance = 6%
3. Load, maneuver, and dump time = 1.4 min
4. Travel time = 14.9 min
5. Total cycle time = 16.3 min
6. Fill factor = 90%
7. Job efficiency = 87%
8. Capacity per hour = 91 bcy per hr @ 100% availability
9. Mechanical availability = 75%

Following removal of the topsoil and subsoils, the selected mining method was implemented by excavating an initial cut across the entire width of the mine. This cut provided an open face for all future cuts perpendicular to it and provided access for coal removal. Upon completion of the first cut, the dragline opened a second box cut perpendicular to the first. After exposing the first seam by sidecasting, as shown in Fig. 32, the dragline walks down the spoil side of the cut on a bench that has been prepared by leveling or filling and exposes the second seam, as shown in Fig. 33 by using a chop-down technique. The dragline alternates between these two operating modes until the 2438-m (8000-ft) cut is completed and the subsequent parallel cut is ready for excavation. Working three shifts per day, 341 days per year, the dragline produces in excess of 14 million bcy per year.

Prior to dragline excavation, the overburden is drilled and blasted utilizing two B.E. 60-R drills. The 31-cm (12¼-in.) holes are spaced at 7.9 × 7.9 m (26 × 26 ft) with 0.6 kg (1.3 lb) of ANFO employed per bank cubic yard of overburden.

Coal loading operations are conducted on a one shift per day basis at a safe working distance behind the dragline operation. A 12-m³ (16-cu yd) shovel loads the coal into five 145-t (160-st) bottom dump coal haulers for transport to the tipple.

Regrading of the area is accomplished by bulldozers that cap and fill dragline spoil piles. Following storage of topsoil from the initial cut and the first three regular cuts, topsoil is directly replaced on graded spoil by the scrapers. Thus, concurrent regrading and reseeding is accomplished enhancing reclamation potential.

Fig. 32. Dragline side cast mining.

BIBLIOGRAPHY

Aiken, G.E., 1973, "Open-Pit and Strip-Mining Systems and Equipment," *SME Mining Engineering Handbook,* AIME, New York, Vol. 2, Sec. 17, p. 47.

Aiken, G.E. and Wholbier, R.H., 1968, "Continuous Excavators," *Surface Mining,* 1st ed., AIME, New York, p. 478.

Anon., 1969, "A Kick Is Really a Boost. . . .," *Engineering and Mining Journal,* p. 74.

Anon., 1975, "Samarco: Major New Material Handling Concepts for Iron Ore," *Engineering and Mining Journal,* Nov., p. 125.

Anon., 1976, *World Mining,* Aug. p. 60.

Anon., 1982, "Duval Corp. Develops New Portable Ore Crushing System," *Skilling's Mining Review,* Nov. 6, p. 12.

Argall, G.O., Jr., 1972, "Rip, Strip, Mine Copper Ore with Cats," *World Mining,* Aug., p. 85.

Barker, G., and McKay, C., 1972, "Some New Concepts in Dredge Design," *Mining Magazine,* July, p. 25.

Beneche, C.J., 1982, "BWE's Offer Flexibility in Production Goals," *Mining Equipment International,* July, p. 15.

Chironis, N.P., 1975, *Coal Age,* Oct., p. 156.

Frizzell, E.M. and Martin, J.W., 1977, "Cross-Pit Conveyor Systems," *Mining Congress Journal,* Aug. p. 67.

Gold, R.J., and Gold, O., 1974, "Wheel Excavators and Conveyors For Use in Surface Coal Mining," *Mining Congress Journal,* Nov., p. 40.

Jackson, D., 1979, *Coal Age,* Dec., p. 74.

Johnson, R.N., Frizzell, E.M., and Utley, R.W., 1981, "Movable In-Pit Primary Crushers," Presentation at American Mining Congress, Denver.

Koch, E.T., Jr., 1975, "Double-Hinged Crawler-Mounted High-Wall Slurrifier," *Trans. SME-AIME,* Vol. 258, pp. 229-232.

McDowell, A.W.K., 1980, "The Underwater BWE-A Review of 15 Years of Ellicott Experience," *World Dredging and Marine Construction,* May.

McKay, C., and Davis, C., "Dredging as a Mining Method," a symposium presentation.

Rumfelt, H., 1968, "Cyclical Methods-Shovels and Backhoes," *Surface Mining,* 1st ed., AIME, New York, p. 435.

Schabas, W., 1976, "System Sprays Silt into Slurry," *Canadian Mining Journal,* Sept., p. 39.

Shand, A.N., 1970.

Fig. 33. Chop-down mining.

6.4 Loading

FRED R. SARGENT

HISTORY AND DESCRIPTION OF EXCAVATORS

Henry Rumfelt in the first edition of *Surface Mining* presented the general history of cyclical machines used in excavating in surface mines. It will be helpful to review part of that material before proceeding.

The history of excavating machinery is not clearly defined but the machines seem to have been originally developed for dredging activities, principally involving rivers and harbors. The first, as far as can be determined, was a steam "spoon dredge" developed by Grimshaw of Boulton and Watt of Sunderland, England, in 1796. The land machine forerunner of the present-day power shovel was not developed until 1835. It was invented by William Smith Otis, a young partner of a Philadelphia contracting firm by the name of Carmichael and Fairbanks. The moving force behind this invention was the heavy activity in railroad construction. (See Fig. 1.)

From this early start, the power shovel has developed from steam powering to internal-combustion-engine powering and to electric powering. In both directions, it has developed to its present state of a finely engineered and precision-built tool. Through the interim, the railroad-type shovel, about a half-swing arrangement designed especially for making railroad cuts and mounted on rails, was developed. The rail-mounted shovel gradually gave way to the full revolving shovel. It was high on the list in 1912 and was practically extinct by 1927. From the beginning, a few of the small revolving shovels designed primarily for the general contractors' market had found their way into quarry and mining operations.

During this same period, the *stripping* shovel had been introduced—a 1.9 m³ (2.5 cu yd), 90-t (100-st), full revolving stripping machine with a 12.5 m (41 ft) boom. Although this shovel was designed especially for stripping, it was also used for heavy-duty, open pit mining service. In 1925, the heavy-duty full revolving quarry and mine machine was developed.

Strip mining shovels evolved through a somewhat separate path. Hand-digging plows and mule-drawn slip scrapers were first reported used by Kirkland, Blakeney and Groves near Danville, IL, for strip mining coal. In 1877, Hodges and Armil used an Otis-type steam shovel in stripping a property near Pittsburg, KS. In 1855, Wright and Wallace used a wooden dredge with wheels applied for land mobility for coal strip mining near Danville, IL. The machine was equipped with a 15.2-m (50-ft) boom.

The first true *stripping* shovel began work in 1911 in the Mission Field just out of Danville, IL. The machine had a 19.8-m (65-ft) boom, 12.2-m (40-ft) dipper handle, and 2.73 m³ (3.5 cu yd) dipper. It was steam-powered and propelled on four four-wheel bogies, which in turn ran on rail tracks. It operated in 6 to 9 m (20 to 30 ft) of shale and gravel overburden uncovering 2 m (7 ft) of coal. Electric powering came into the picture about 1916. Ward Leonard Electric powering came later and has continued to be the basic system of powering through the present.

Hydraulic leveling jacks were introduced early and have become standard on all *stripping* shovels. Special features such as counterbalanced hoist, bridge strand-type boom suspension, wire rope crowd, and "knee-action" front have all been introduced, and some of these are used on the modern *stripping* shovel (Fig. 2).

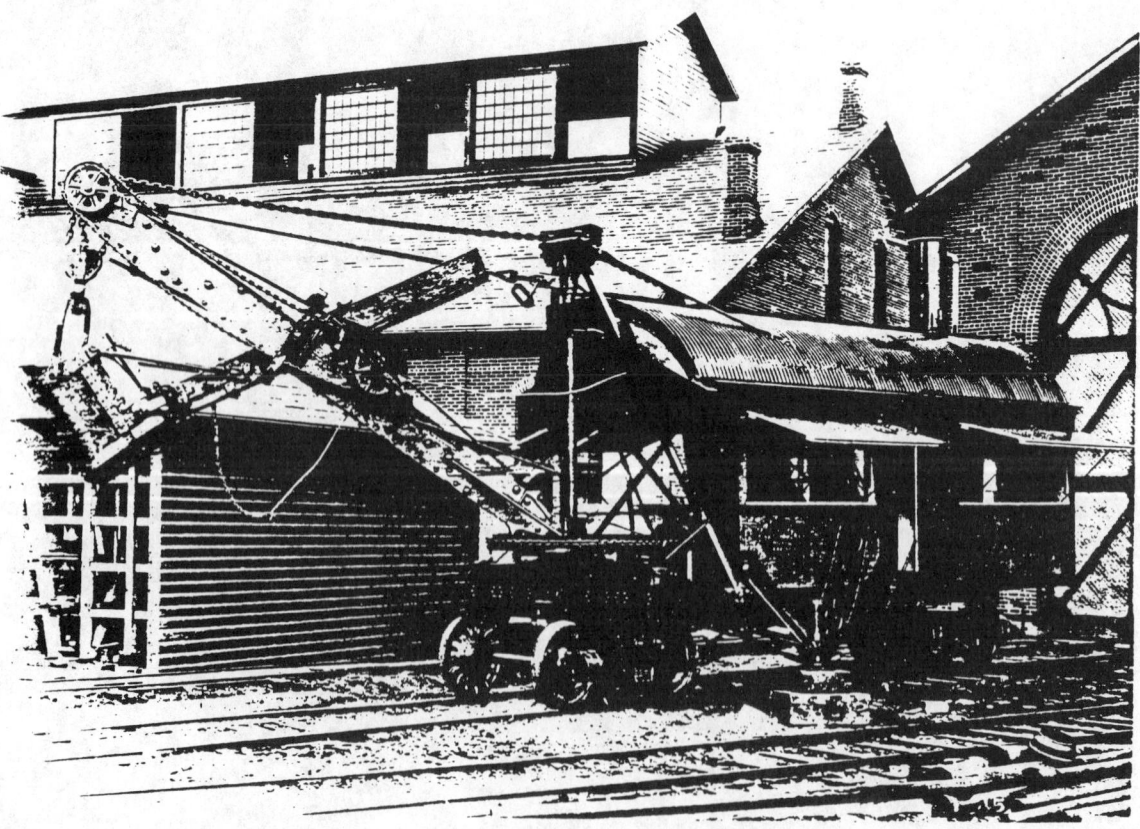

Fig. 1. An early excavator, Barnhart's steam shovel.

Fig. 2. World's largest stripping shovel Marion Model 6360 with 138-m³ (180-cu-yd) dipper.

Evolution has produced nomenclature that indicates the *mining shovel* to be a machine mounted on only two crawlers, while the *stripping shovel* is mounted on four sets of crawlers each having two crawler belts placed side by side, thus the term eight crawler machine used to denote a stripping shovel.

The digging element of a power shovel is a dipper attached to a handle. The dipper lip, the principal cutting edge, is attached to the dipper front and is equipped with replaceable teeth. The handle is connected to the back side of the dipper. The dipper bottom is a hinged door and has a latch for dumping the contents.

In operation, the dipper is pulled through the bank by hoist cables and, on the modern type, is held against the bank by the crowd motion which extends or retracts the handle lengthwise. The handle is pivoted about a fixed point, and it is positioned lengthwise by crowd motion; and the dipper is pulled through the cut. When the dipper is full, the machine swings sidewise and the dipper load is dumped through the door.

The entire machine is usually mounted on a system of crawlers. The crawlers are endless belts of links hinged or pinned together. They permit mobility and the positioning of the machine with respect to the bank.

Electric energy is furnished to the machine through a trailing cable and usually powers a large motor-generator set with multiple generators. In the full Ward Leonard system, each generator furnishes d-c power to separate motors which drive the principal motions of the shovel: hoist, crowd, and swing. By the use of such devices, shovels of remarkable performance, speed, and smoothness have been developed primarily for these surface mining activities.

Similarly the history of the hoe is not clearly defined. Patents as early as July 17, 1855 (E.O. Rood) describe a dredging apparatus which uses a digging motion involving a boom and handle that is directed toward the machine. Later on there were land-mounted machines. However it was not until about 1925 that the device became quite popular.

The name describing the device has gone through an evolution. It has been described by many names such as "backhoe," "dragshovel," "back action," and "pull-shovel." Now, however, the trend is to call it a "hoe."

The main advantage of the machine is its ability to dig below itself with a positive action. It has been used in trenching and basement excavation for domestic dwellings.

While the history of *dragline* excavators is not nearly so defined as that of shovel equipment, early dragline models used dragline booms on excavators in the early 1900s. Dragline booms were applied to crawler excavators to extend the excavator's reach and to dig into wet areas or areas that would not support the shovel's tracks.

The first walking dragline as it is presently known was patented by Oscar J. Martinson of the Monaghan Manufacturing Corp. This company eventually became part of the Bucyrus-Erie Co. Another early manufacturer of walking draglines was the Page Engineering Co. of Chicago, IL. Marion Power Shovel entered the walking dragline business in the 1930s. The first walking draglines were built with essentially the same equipment as the shovel but without the lower frame and crawlers of shovel equipment. Instead, the crawlers were replaced with a tub upon which the machine sits. The shoes or pontoons at the side of the machine are attached to the revolving portion of the dragline, and when the machine walks, the shoes lower themselves to the ground and the eccentric device attached to the shoes elevates the machine and slides it backwards. This process is then repeated after the shoes leave the ground again and the machine thus walks its way to its new digging location.

Early dragline excavators generally were less powerful machines than the shovel excavator, both from a connected horsepower standpoint and the fact that the dragline relies on pulling the bucket through the ground, whereas the shovel excavator is directly connected to the dipper and is both pulled and forced through the earth by a rigid connection to the dipper.

Dragline excavators originally were utilized in moist or wet conditions where the low-bearing pressure of the tub would allow the machine to propel on top of the ground while digging below its tub and in stripping conditions where blasting was not required or was minimal. Early dragline excavators became known as excavating tools for digging "easy" material, whereas the shovel excavator was generally used in the harder or blasted materials. The evolution to the use of the dragline in rough, broken rock took many years and was brought about generally by better blasting practices and materials, as well as better tools for drilling of the ground itself and higher horsepowers in the dragline. The most important feature of draglines, however, was their ability to dig deeper overburden depths than the stripping shovel could effectively handle. Often these draglines were used in conjunction with a shovel wherein the dragline excavator would take the top portion, or soil above the rock portion which would have to be blasted. The blasted portion of the overburden would then be removed with the shovel excavator.

From the first draglines utilized in the mining industry until the early 1960s, the largest walking dragline was a machine with 68.6 m (225 ft) of boom coupled with a 26.8-m³ (35-cu yd) bucket; this size machine had been continually produced from the early 1940s. Thus from 1900 to 1940, the industry went from 1.5 or 2.3 to 11.5 m³ (2 or 3 to 15 cu yd). From 1940 to 1960, the size of draglines increased from 11.5 to 26.8 m³ (15 to 35 cu yd). During the 1960s the era of the super dragline was to take rapid hold in the mining industry. When it was sold in late 1960, the Marion 7900 with 83.8 m (275 ft) of boom and a 30.6-m³ (40-cu yd) bucket was the world's largest dragline. By 1969, when the Bucyrus-Erie 4250W went to work with its 94.5-m (310-ft) boom and 168.2-m³ (220-cu yd) bucket, the industry had increased dragline size fivefold in less than ten years (Fig. 3).

While only one of these 168.2-m³ (220-cu yd) dragline excavators has been produced, approximately 50 walking draglines of the 76.5 m³ (100 cu yd) size and larger are currently working in mining operations through the free world. The average size of draglines sold has doubled during the last decade.

REVIEW OF DEVELOPMENTS IN CYCLICAL EXCAVATING EQUIPMENT, 1968-1983

Rumfelt and Boulter in the first edition of *Surface Mining* (1968) gave various developments that had occurred up until that year in the excavating machinery business. Developments in the ensuing 21 years have been numerous and can best be described by commenting on these by the type of machinery.

Mining or Quarry-Type Shovels

The trend to larger two-crawler machines has accelerated with the advent of haulage trucks driven by electric wheel motors. In 1968, 11.5-m³ (15-cu yd) shovels and 90-t (100-st) trucks predominated. Marion Power Shovel in the early 1960s built two 291M two-crawler machines (Fig. 4) capable of utilizing 19 to 26.8-m³ (25 to 35-cu yd) dippers; however, truck sizes were too small and the machines were used only for stripping overburden in coal. Harnischfeger Corp. in 1982 built a Model 5700 (Fig. 5) with a 45.9-m³ (60-cu yd) dipper for loading trucks. Now, as in 1962, economical truck size has not kept pace with shovels.

Harnischfeger was first to market static power conversion for electric mining shovels, thus eliminating the rotating motor generator sets and replacing the generator with thyristors. Initial problems of all manufacturers with statics were with low power factors, fuse protection, cooling of the thyristors, and troubleshooting. These problems have now been eliminated.

All manufacturers of electric mining shovels now design propelling systems that utilize motors and gearing on the lower works in their larger shovels from 9.2 m³ (12 cu yd) and upward in size. This system eliminates maintenance on

Fig. 3. World's largest dragline used for overburden removal in a coal operation. B-E model 4250W with 168-m³ (220-cu-yd) bucket.

Fig. 4. Marion 291M mining shovel.

the gearing and shafting from the revolving frame down to the lower works.

Other innovations in mining shovels include a trend toward increased cutting efforts by the combination of higher hoist and crowd horsepowers and front-end geometry, dual motor propel systems, and design of modular components for the reduction of shovel downtime.

One innovation on the market is the Marion 204M SuperFront mining shovel which utilizes a radical departure in front-end geometry. A pitch system allows a flat pass and greater cutting effort vs. the conventional shovel front end. The machine utilizes a nominal 24.5-m³ (32-cu yd) dipper (Fig. 6).

Hydraulic Excavators

If there has been any revolution in excavators, it has been in the introduction of large, heavy-duty hydraulic excavators into mining operations. First developed by German manu-

Fig. 5. 46-m³ (60-cu-yd) class coal loading shovel. P & H Model 5700.

Fig. 6. The 24-m³ (32-cu-yd) SuperFront mining shovel with its unique front-end geometry.

facturers as small construction machines, the larger hydraulic shovels and backhoes have created a market which can be classified between electric shovels and front-end loaders.

During the booming mid-1970s, hydraulic excavators were readily available, whereas electric shovels were not, due to backlogged plants, and this shorter delivery coupled with faster erection time brought the hydraulic excavator into mining. Thus the hydraulic excavator made inroads into mines and new mining ventures. The largest usage of hydraulic excavators is in quarries and coal mining where digging conditions are easier than in metal mines. They are being utilized as both stripping excavators loading trucks and also as coal loaders, both in front shovel and backhoe configurations.

While hydraulic units up to 23 m³ (30 cu yd) have been manufactured, the 7.6 to 17 m³ (10 to 22 cu yd) are the most popular sizes (Fig. 7).

Since initial hydraulic excavators in mining were adaptations of construction machines, three-shift, seven-day-per-week applications caused low availabilities and high costs. Newer units are more rugged and have shown higher availabilities even with high utilization common in mining excavators.

Both diesel and electric driven machines are available and have essentially replaced smaller diesel and electric cable shovels.

Due to the lack of large hydraulic motors, rugged mine conditions in large volume metal mines, and lack of mine technicians trained in high pressure hydraulics, it is doubtful that in the near future hydraulic machines will entirely replace the dependable electric cable shovel with its low unit cost capabilities.

Stripping Shovels

The last stripping shovel (eight-crawler machine) sold was a Marion 5900 in 1969, and while these workhorses of the US coal mining industry continue to operate removing overburden in the midwestern US coalfields, the trend away from these large shovels continues. Walking draglines have replaced the shovel due to deeper coal overburden depths, multiple coal seams in the same field, and greater flexibility with the dragline. Because the stripping shovel operates in the pit, it has limited depth capability.

No new developments from either Bucyrus-Erie or Marion Power Shovel indicate that, unless some new market should develop, the era of the stripping shovel is dead and

the machines now working will eventually be scrapped as their dedicated coalfields work out.

Walking Draglines

From 1968 to 1989, large walking draglines have largely been modernized by changing various features of the machines to provide greater mechanical availability of the machine. Walking systems and booms have been improved, and better methods of monitoring both production and problems have been established. Improved weld criteria, metallurgy, and design features have been instituted by the manufacturers to help operators increase uptime. Size ranges in 1968, while primarily up to 45.9 m³ (60 cu yd), have jumped up to 76.5 to 84 m³ (100 to 110 cu yd) in 1989 with a preponderance of 45.9, 57.3, and 84 m³ (60, 75, and 110 cu yd) machines working. Only six draglines in the free world have bucket capacities over 88 m³ (115 cu yd). Yet approximately 250 draglines are working or scheduled with bucket capacities of 42 to 88 m³ (55 to 115 cu yd) (Fig. 8).

Draglines are now working multiple seam operations vs. primarily single seams 15 years ago. Extended bench, spoil side stripping, and other dragline techniques have produced ways for operators to utilize this lowest-unit-cost machine in areas previously not indicated as being applicable.

In the smaller walking draglines, manufacturers have designed modular quick erection and teardown machines for use in smaller deposits of mineral. In addition, static power conversion systems are utilized as in electric mining shovels, thus eliminating motor generator sets for conversion of incoming a-c power to d-c for motion drive motors.

Crawler Draglines

Neither design nor utilization of crawler draglines has changed appreciably since 1968. Smaller coal mining operations are the primary users, along with bauxite, lateritic nickel, and small phosphate operations. Construction materials quarries, such as coral rock and sand and gravel, also utilize crawler machines.

The larger units for mining now reach boom lengths of 78 m (255 ft) and bucket capacities of up to 16.8 m³ (22 cu yd), Fig. 9.

Backhoes and Clamshells

Backhoe excavators in 1968 were primarily cable type, with limited usage in the mining industry. Recent machines are entirely hydraulic with respect to the front-end structure and have found more acceptance in coal loading where the

Fig. 7. A typical 11.5-m^3 (15-cu-yd) class hydraulic front shovel in a coal operation.

backhoe can sit on top of the coal and load the truck on this same level, thus keeping both machines out of wet or mucky conditions. In some applications, the backhoe sits on top of the mineral and loads the truck which sits on the bench below it at the bottom of the mineral (Fig. 10). Cycle times are less in this instance.

Clamshell excavators are essentially draglines fitted with clamshell buckets and different rigging of the hoist and drag ropes to perform opening and closing of the clam bucket. Other than below waterline digging of coral rock, rehandling and dredging construction projects associated with mining, there is little use of large clamshells in mining (Fig. 11).

Fig. 8. An 8.7-m^3 (110-cu-yd), Marion Model 8750 walking dragline.

MINING AND QUARRY SHOVELS

FRED R. SARGENT

Machine Features—Mechanical

The basic mining shovel and its components are illustrated in Fig. 12. A deck layout of a shovel is shown in Fig. 13. The main mechanical parts of the shovel are: crawler side frames which bolt or weld to the lower frame or carbody; the revolving frame upon which are mounted the swing and hoist machinery and major electric drive components; the boom, which generally has the crowd machinery attached; and the dipper handle and dipper. A gantry and suspension ropes to hold the boom in position and a covering or house over the machinery and electrics comprise the remainder of the machine.

The mining shovel is a rugged, heavy-duty machine designed to take the abuse imposed on it from continual digging forces and shock applied by the motor torques which drive the dipper into and upward through the digging face. With proper maintenance, these machines can work 24 hr per day, 7 days per week.

Propelling machinery, including motor(s), is either mounted on the carbody or driven by the hoist motor through shafting and gearing from the revolving frame. The final gear reduction boxes are connected to the crawler side frames to provide propelling of the machine.

Each motion of hoist, swing, crowd, and propel is driven by a heavy-duty high-torque, low-speed motor or motors which are required for high production at minimum cost.

Fig. 9. Large crawler dragline P & H Model 2355.

Fig. 10. A typical backhoe.

Fig. 11. Barge-mounted clamshell.

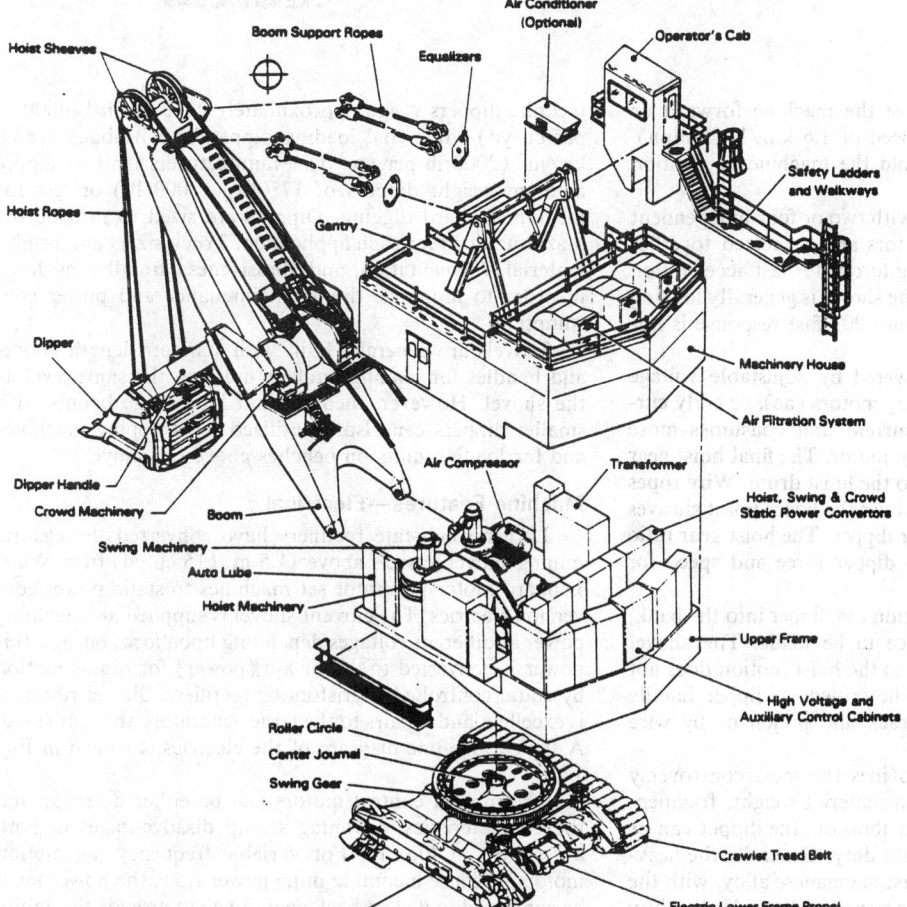

Hoist Sheeves

Boom Support Ropes

Equalizers

Air Conditioner (Optional)

Operator's Cab

Hoist Ropes

Gantry

Safety Ladders and Walkways

Dipper

Dipper Handle

Crowd Machinery

Boom

Air Compressor

Transformer

Machinery House

Air Filtration System

Swing Machinery

Auto Lube

Hoist Machinery

Hoist, Swing & Crowd Static Power Convertors

Upper Frame

High Voltage and Auxiliary Control Cabinets

Roller Circle

Center Journal

Swing Gear

Crawler Tread Belt

Electric Lower Frame Propel

Lower Frame and Crawler Assembly

Fig. 12. Exploded view drawing of a typical mining shovel.

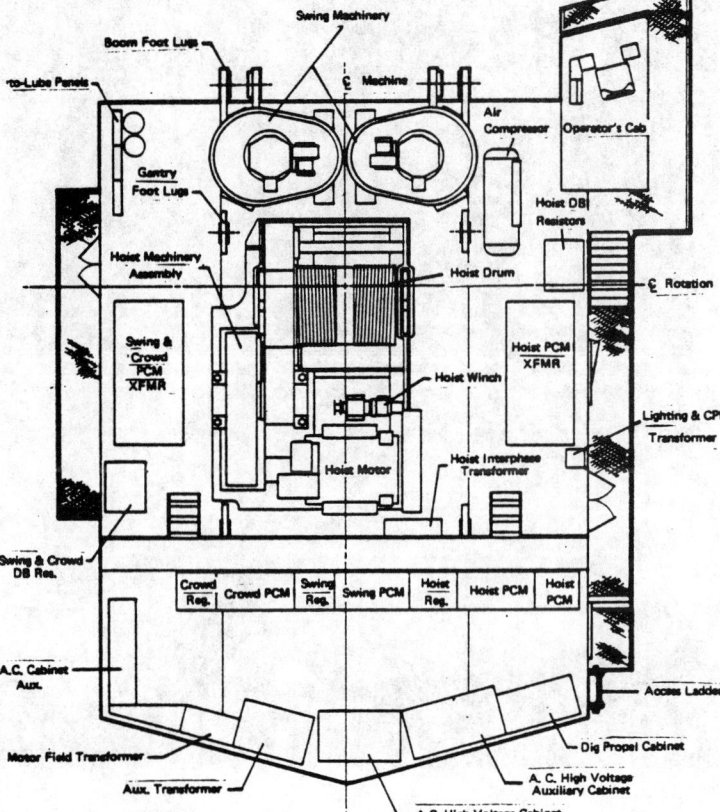

Fig. 13. Deck layout of a typical mining shovel.

The propel motion will drive the machine forward or backward at an approximate speed of 1.6 km/h (1 mph). Large shaft or motor brakes hold the machine in position when digging.

The swing motion is driven with two or four independent sets of gear reductions; the motors are connected together electrically to provide high swing torque for fast acceleration and deceleration. Since the mining shovel is generally loading trucks or railcars, and only swings 90°, fast response is important for good cycle time.

The hoist machinery is powered by adjustable voltage motors (dc), adjustable frequency motors (ac), or eddy current slip coupling. The eddy current unit consumes more power than an adjustable voltage motor. The final hoist gear reduction connects the motors to the hoist drum. Wire ropes attached to the drum are reeved over the boom point sheaves and connect to the dipper bail or dipper. The hoist gear ratio is selected to provide optimum dipper force and speed for most digging conditions.

The crowd motion is to position the dipper into the bank, allowing a narrow or deep slice to be made. The shovel operator can control the crowd so the hoist motion does not stall and yet fills the dipper. The crowd or dipper handle can be driven by means of a rack and pinion or by wire ropes.

The shovel dipper itself provides the most controversy among operators. Depending on material weight, fragmentation, and toughness, as well as abrasion, the dipper can be heavy duty, medium duty, or light duty. Generally the heavier dippers are constructed of cast manganese alloy, with the lighter dippers of fabricated wear resistant alloys. Heavy duty

taconite dippers weigh approximately 2373 kg/m³ (4000 lb per cu yd) while coal loading dippers weigh about 1186.6 kg/m³ (2000 lb per cu yd). Manufacturers tend to supply medium weight dippers of 1780 kg (3000 lb) or less for medium to hard digging. Dippers are sized to provide the maximum load for each application. Truck size, bank height, material fragmentation, and abrasiveness are all considered in order to minimize dipper maintenance and power consumption.

Shovels are generally built with standard length booms and handles for loading haulage units on the same level as the shovel. However, medium and long-range booms with smaller dippers can also be utilized for stripping machines and for loading units on benches above the shovel.

Machine Features—Electrical

Larger solid-state rectifiers have converted the electric mining shovel in sizes above 11.5 m³ (15 cu yd) from Ward Leonard motor generator set machines to static power conversion electrics. The current shovel is supplied a-c incoming power at different voltages depending upon location, and this power is converted to d-c or a-c (power) for motor motion by static controlled thyristors or rectifiers. Shovel response is excellent and repairs to the large generators are eliminated. A typical one-line diagram of the electrics is shown in Fig. 14.

The motion control motors can be either dc or ac and the characteristics, advantages, and disadvantages of both are shown in Fig. 15. For variable frequency a-c motion motors, because incoming mine power is ac, the power must be converted to dc and back again to ac to provide the ability

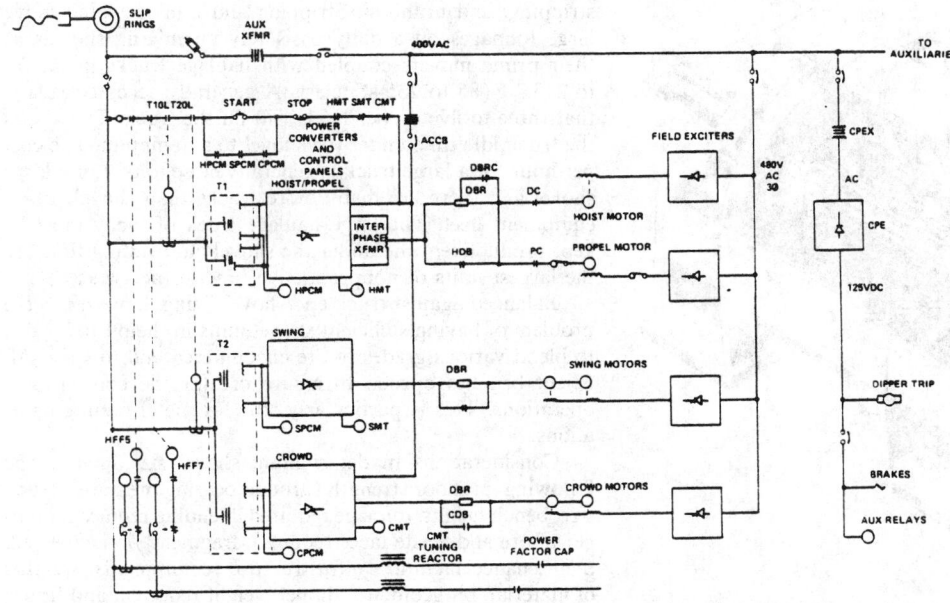

Fig. 14. A typical one-line diagram of electrical system for a mining shovel.

to vary frequencies. This additional rectification causes a certain power loss and adds to the complexity of the system. Certain motor advantages are claimed, however, such as lack of brushes and commutator wear in the a-c motor system.

A standard construction d-c mill-type motor used on excavators is shown in Fig. 16. The motors are adjustable voltage (speed) motors. They are built for fast reversing on a constant basis and have a split frame for easy removal of the armature. These motors are available in 44.8 to 969.8 kW (60 to 1300 hp) at 475 V, the normal rating voltage, with peak power of 67.1 to 1417.4 kW (90 to 1900 hp) at peak voltage of 575 V.

Incoming a-c power to mining shovels is normally 4160 or 7200 V at 60 Hz in the US and 6600 V at 50 Hz in most

Fig. 15. A comparison of ac vs. dc.

TECHNOLOGY			
Converter	Converter Inverter	Converter Inverter	Diode Bridge Inverter
DC	**AC** (VVI)	**AC** (CCI)	**AC** (PWM)

FEATURE	DC	AC (VVI)	AC (CCI)	AC (PWM)
Efficiency Motor Controller	High High	High Medium	High Medium	High Medium
Reduced Speed Operation	Wide	Medium	Medium	Medium
Space Efficiency	High	Medium	Low	Low
Dynamic Braking	Yes	Yes	Yes	Yes
Constant HP Speed Range	Wide	Medium	Medium	Medium
Regeneration	Yes	No	Yes	No
Complexity and Maintenance	6 Thyristors	18 Thyristors 6 Diode Filter Capacitor Filter Reactor	12 Thyristors 6 Diode Large Filter Reactor Commutation Capacitor	12 Thyristors 12 Diode Filter Capacitor Filter Reactor
Multi-Motor Operation	Yes	Yes	No	Yes
First Cost	Low	High	High	High

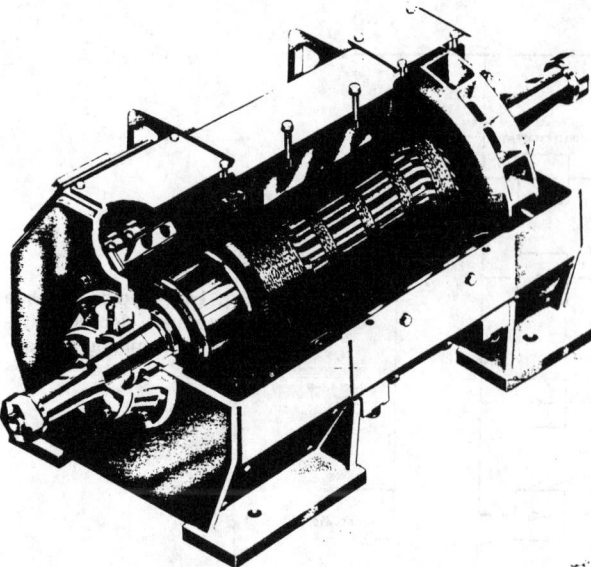

Fig. 16. A standard construction d-c mill-type motor.

foreign countries. Power supply is critical to good shovel operation and shovel manufacturers prefer voltage fluctuations to be limited to ±10% of the nominal voltage. Trailing power cable to the shovel is three conductor SHD type with a ground conductor.

Static power conversion shovels using conventional converters have inherently poor power factors and must be corrected by using switching capacitors or a newly designed four-legged bridge where phase to neutral connections can be utilized to improve the power factor. Static power converters generate harmonics that must be filtered out.

Newer static systems utilize microprocessors and digital systems for both control of the system and diagnostics of problems in the system itself. Analog systems are in use by some manufacturers.

Application and Use of Mining Shovels

Mining shovels are utilized in many different types of operations, but the largest users are in copper, iron ore, coal stripping, and uranium. Stripping and mining which move large tonnages on a daily basis rely on mining shovels as their prime movers coupled with haulage trucks in the 77 to 213.2-t (85 to 235-st) class. A generally accepted rule is that three to five dipper loads will fill the truck. This limits the truck idle time under the shovel to a minimum. The cost per hour for a large truck is generally close to that of a large shovel and there are many more trucks than shovels in an equipment fleet. The large haulage truck of today must be kept productive; time under the shovel and waiting time are the largest units of nonproductive time in the cycle.

Balanced against truck and shovel sizing, however, is the problem of having sufficient shovel units in the pit to be able to blend various grades of ore encountered and to keep sufficient benches opened to assure ore for the concentrator operations. This is particularly true for metal mining operations.

Considerations in determining shovel size include the following: pit floor strength (ground bearing pressure), truck size, bench height, tonnage required, blending required, number of ore and waste faces required, fragmentation expected, maintenance facilities, infrastructure requirements, weights of material, pit geometry, dipper weight required, and future possibilities in all these items.

Figs. 17 and 18 show typical shovel loading operations at coal stripping and metal mining operations. Fig. 19 shows a range drawing for a typical 25-yard-class shovel.

In recent years the trend has been not only to larger shovels in the 13 to 26.8 m³ (17 to 35 cu yd) range, but also in utilizing shovel/truck fleets to prebench ahead of large stripping draglines as deposits of coal become deeper. In many multiple-seam coal operations, draglines cannot be easily utilized and shovel/truck combinations become the major equipment application.

Due to increasing costs of oil and diesel fuel, the trend to full electric shovels continues. Full-diesel shovels and diesel-electric versions are declining and are virtually unused except in small operations where power line construction is uneconomical and where mine life is comparatively short.

Productivity and Costs

The element most consistent in surface mining is inconsistency. No two mineral deposits are the same and this can be said for production rates and costs of shovel equipment. When estimating, however, it is important to utilize a con-

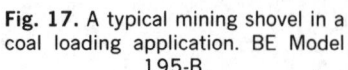
Fig. 17. A typical mining shovel in a coal loading application. BE Model 195-B.

Fig. 18. A typical mining shovel loading operation in a metal mine. Marion Model 191 m.

sistent approach to the factors used. For more details on definitions for hours utilized, bank vs. loose yards, fill and dipper factors, etc., refer to the section on draglines.

Material in situ has a certain specific gravity and is utilized as the starting point to obtain shovel production rates. Next a swell or swell factor is applied to give a loose weight after blasting or loading. The dipper fill factor (percent of struck capacity) gives the volumetric equivalent of the material as compared with the volume of the dipper.

Fill factors of dippers range from 40 to 50% for heavy blasted rock and are up to 120% for loose lighter weight alluvial materials. Generally, 85% fill factor is normal for overburden materials that are well blasted.

Mining shovels have cycle times that vary from 22 to 35

sec, depending on swing angle, material weight and preparation, and operator skill. The longest percentage of time is spent swinging—about 60% of the cycle. The average time for swing both ways in a typical 90° swing is 18 to 20 sec. Hoisting or digging time will vary greatly depending on material being loaded, but 6 to 8 sec is normal. Dumping time is approximately 2 sec. A typical production calculation is shown in Table 1.

While the theoretical ability of the shovel to produce can never be utilized in actual practice, mine operators can keep productivity high by reducing maintenance time on shift and improving truck scheduling to the shovel. Shovel production rates for various sizes of machines are shown in Table 2.

Actual cost figures for electric mining shovels depend on many factors such as the accounting system, labor rates, mine conditions, etc. Some general rules commonly used to estimate cost are:

Power costs will run from 0.15 to 0.25 kWh per st moved (0.3 to 0.5 kWh per bank cubic yard).

Maintenance labor, supplies, lube, oil, parts, etc., will run 10 to 15% of the installed cost of the machine per year, depending on digging conditions, quality of maintenance, and operating hours per year.

Ownership costs and maintenance costs should be approximately equal per year.

Table 3 illustrates typical costs for a 13 m³ (17 cu yd) shovel, and shows averages over a 10-year period, 3-shift, 7 day-per-week operation. Each year will be higher or lower, depending on repairs and replacement parts during the year. The figures represent a hard rock metal mine.

Ownership costs are given in Table 4 for a typical 13 m³ (17 cu yd) shovel operating in the US.

Machine Selection

Shovel selection, like other large dollar expenditures, attracts top management attention and often price and personal preference become the selection criteria. In past years other criteria have been:

1) *Weight*—with today's use of high alloy steels and improved design, weight is often confused with strength.

2) *Horsepower*—rated motor horsepower alone does not take into account the thyristor or generator capacity required to "drive" the motor.

Considerations for correct shovel selection from a technical standpoint will be weight (in certain locations) and

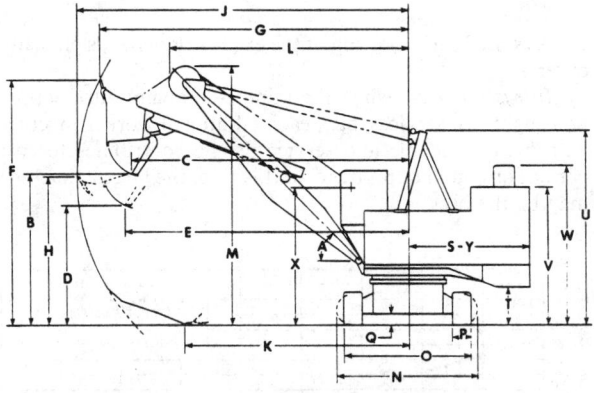

WORKING RANGES (Based on 25 Cubic Yard Mining Dipper)

A — Boom Angle . . . . 45°
B — Dumping Height — Maximum . . . . 33'-6"
C — Dumping Radius @ Maximum Height . . . . 56'-0"
D — Dumping Height @ Maximum Radius . . . . 23'-10"
E — Dumping Radius — Maximum . . . . 57'-8"
F — Cutting Height — Maximum . . . . 52'-0"
G — Cutting Radius @ Maximum Height . . . . 60'-6"
H — Cutting Height @ Maximum Radius . . . . 28'-8"
J — Cutting Radius — Maximum . . . . 67'-6"
K — Radius of Clean-up . . . . 46'-10"
L — Clearance Radius — Outside Boom Point Sheave . . . . . 49'-9"
M — Clearance Height — Over Boom Point Sheave . . . . . . 52'-3"

Fig. 19. Range drawing for a typical 19-m³ (25-cu yd) mining shovel.

Table 1. Example of Production Calculation for a 13-m³ 17-(yd³) Shovel

Material weight (in situ) cu yd	4000 lb
Swell factor, 1/1.3 (assume 30% swell)	0.77
Material weight (loose cu yd)	3077 lb
Dipper size	17 cu yd
Fill factor (estimated)	0.85
Cycle time (average, sec)	28
Stph (theoretical)	2858
$\dfrac{3600}{28} \times 17 \times 0.85 \times \dfrac{3077}{2000} =$	
Availability, %	83
Stph (practical)	2372
Stph loading trucks (assume 80% truck/shovel efficiency)	1898

Note: This is shovel production only. To calculate mine production, the truck efficiency must also be considered. Metric equivalents: lb $\times$ 0.453 592 4 = kg; st $\times$ 0.907 184 7 = t; cu yd $\times$ 0.764 554 9 = m³.

Table 2. Mining Shovel Productivity, per scheduled hour

Capacity, cu yd	Loading, Rock, bcy	Loading, Rock, st
8	330-460	650-900
12	500-690	1000-1350
17	700-1000	1400-1900
25	1050-1450	2050-2800
32	1300-1850	2600-3600

* Assumes average conditions, loading trucks, minimum swing angle, 0.85 fill factor, and material weight 3000 lb per loose cubic yard. Metric equivalents: cu yd $\times$ 0.764 554 9 = m³; st $\times$ 0.907 184 7 = t.

Table 4. Example of Ownership Costs for a 13-m³ (17-cu yd) Shovel

Price f.o.b. factory including options	$3,400,000
Freight, ballast, erection	300,000
Cost of investment	$3,700,000
Depreciation straight line per year (assume 20-yr life)	$ 185,000
Taxes, insurance, interest: 20%/yr	$ 388,500
$\dfrac{20+1}{2 \times 20} \times \$3{,}700{,}000 \times 0.20$	
Annual total	$ 573,500
Assume 6000 hr per year	
Ownership cost per hour	$ 95.58
Assume 1900 stph	
Ownership cost per ton	0.050

horsepower, but a better analysis would be to compare performance curves of each motion. Swing torque (ability to accelerate and decelerate from a still position) is important. Swing rpm is not important as the shovel swings are usually 90° or less. Swing is the largest component of cycle time. Hoist effort is important, so as not to stall in the bank, as stalling the motor increases cycle time and contributes to motor burnout.

Crowd effort is least important in terms of cycle time; however, crowd effort and speed must be matched to the hoist effort and speed for ease of operation and maximum production. A general rule often used is that crowd effort should be one-half of hoist effort at stall. Refer to Figs. 20-25 for performance curves of a typical shovel.

Table 3. Example of Operating Costs for a 13-m³ (17-cu yd) Shovel

	Per Hour
Operating labor (operator $16, oiler $14)	$ 30.00
Power (360 kW/h @ $0.05/kWh)	18.00
Operating supplies (lube, wire, rope, teeth)	15.00
Mechanical maintenance labor	20.00
Mechanical maintenance parts	20.00
Electrical maintenance labor	8.00
Electrical maintenance parts	10.00
Operating cost per hour	$121.00
Operating cost per ton (assume 1900 stph)	$ 0.064

As stated earlier, sizing of the excavator depends on many criteria:

Infrastructure—when the mining company must supply housing, town services, etc., each additional worker together with his dependants and service people contributes to cost per ton or pound of product. In this case, the largest machine may be the best.

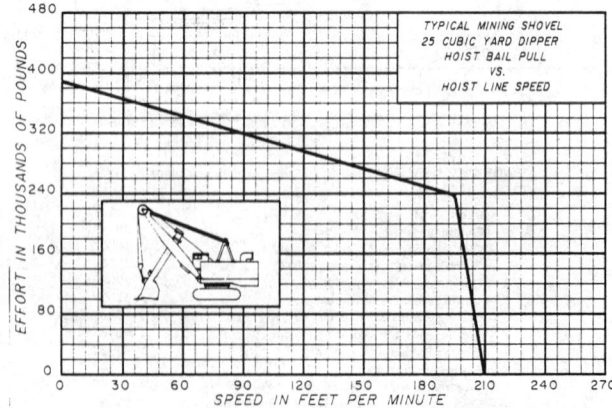

Fig. 20. Electrical performance curve for a typical 19-m³ (25-cu yd) mining shovel showing hoist pull vs. hoist line speed.

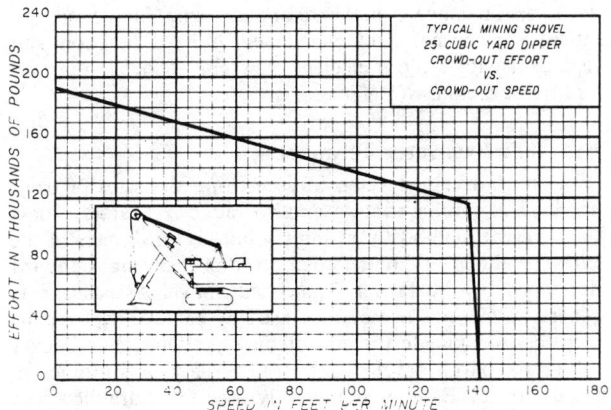

Fig. 21. Electrical performance curve for a typical 19-m³ (25-cu yd) mining shovel showing crowd-out effort vs. crowd-out speed.

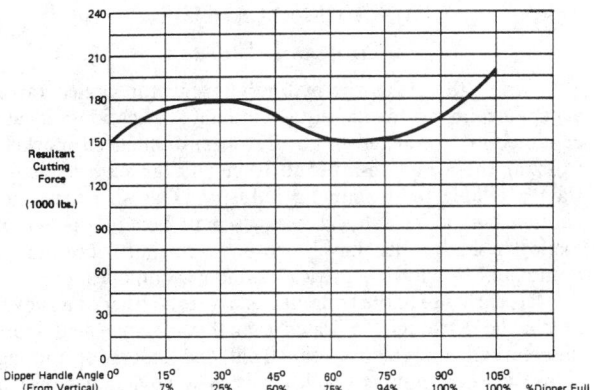

Fig. 24. A cutting force curve for a typical 19-m³ (25-cu yd) mining shovel.

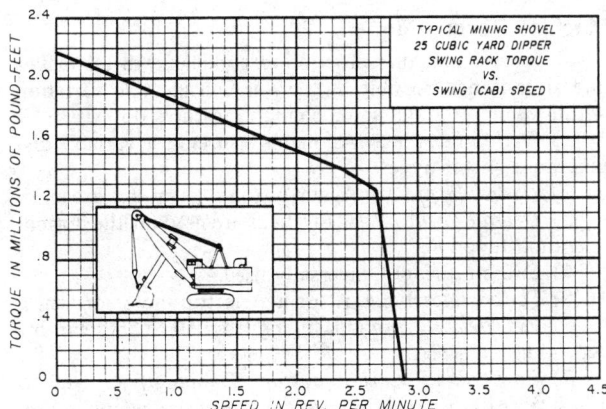

Fig. 22. Electrical performance for a typical 19-m³ (25-cu yd) mining shovel showing swing rack torque vs. swing (cab) speed.

Bench height—the shovel teeth should be able to scale the crest of the bank. Safety laws may govern.

Tonnage required—shovel size should be the largest available to produce the tonnage within the constraints of blending, number of ore and waste faces needed, and ore grades available. However, consideration must be given to the mining system as a whole. For instance, one large shovel, when down, will leave the truck fleet idle and the possibility of other phases of the operation, such as crushers, etc., dependent on this situation.

Maintenance facilities—proper crane capacity, shop overhaul facilities, and warehouse stock levels may dictate the size needed.

Truck size—the needed production by the truck fleet along with ore deposit will dictate shovel sizing. Thought should be given to future truck sizes as the haulage fleet may be economic only for ten years, whereas the shovel will be economic for twice that period.

Dipper size—each size shovel is capable of many dipper sizes, depending on dipper and material weights. Once the general size range of shovel is decided upon, discussions with manufacturers should firm up the exact specifications required, dependent upon fragmentation, weights of materials, maintenance systems, union contracts, etc.

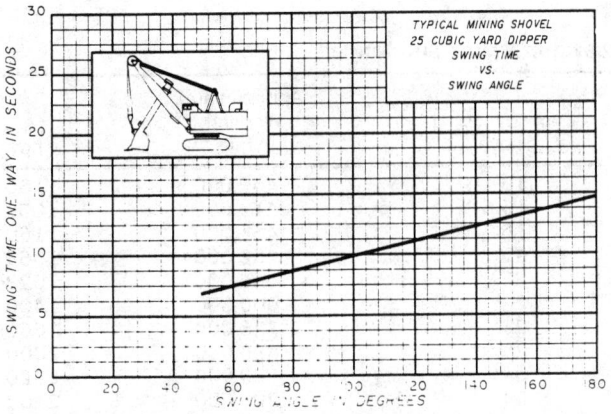

Fig. 23. Electrical performance curve for a typical 19-m³ (25-cu yd) mining shovel showing swing time vs. swing angle.

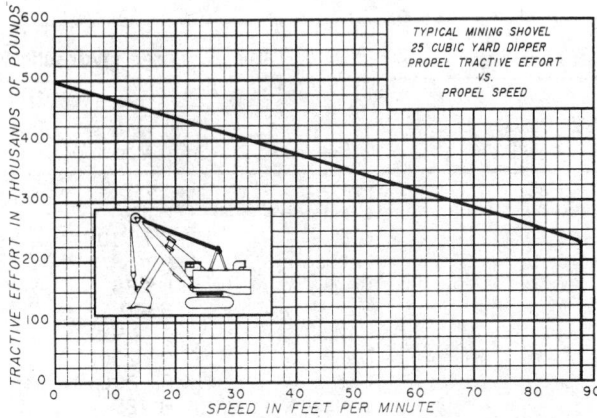

Fig. 25. Electrical performance curve for a typical 19-m³ (25-cu yd) mining shovel showing propel tractive effort vs. propel speed.

HYDRAULIC EXCAVATORS

Thomas I. Files

During the 1970s the hydraulic excavator gained rapid acceptance in worldwide surface mining applications. Its acceptance first occurred in the European-influenced markets utilizing the 3.8 to 7.6-m³ (5 to 10-cu yd) capacity machines then available to the mining industry. This size machine, equipped as a face shovel, was used in lieu of the wheel loader or small cable shovel then prevalent in surface mines, dominated by North America's mining techniques.

The early use of the hydraulic excavator, either as a shovel or backhoe, strongly indicated that this machine had some inherent advantages over the front end loader or mining shovel in some mining applications. Key advantages over a front end loader or mining shovel are:

Front End Loader

Lower hydraulic excavator operating costs due to:

1) Elimination of tire maintenance cost.

2) Less horsepower consumed to load material, therefore, fuel cost per yard of material moved is lower.

3) More rugged structural and drive train components may be used, since the front tire and axle loading are not the predominant machine design parameters.

4) Greater digging forces (crowd and breakout) may be applied to the material bank.

Mining Shovel

1) Greater mobility (if diesel driven), higher travel speed, and more important, weight and improved steerability are available with the hydraulic excavator.

2) A digging envelope permitting selectivity in penetrating the bank.

3) Higher cutting forces.

With the acceptance of the larger size of hydraulic excavator, there is a demand for hydrostatic components (pumps, motors, valves, cylinders, filters, and hoses) of a higher rated horsepower for larger sizes of machines. As these higher horsepower components are developed, larger sizes of machines are becoming available. Ten makes and models of hydraulic excavators are currently available from European and US manufacturers. For convenience, the makes and models listed in Table 5 are listed by 11.5, 15.3, and 23-m³ (15, 20, and 30-cu yd) shovel buckets.

Unlike front end loaders or mining shovels, hydraulic machines are in a period of dynamic product development. New models offer lower production cost through basic design improvement that increases the machine efficiency and re-

duces maintenance cost. Therefore, an owner may choose to retire or place a machine in a less utilized function and replace it with a more efficient, new model sooner than rebuild cost dictates. (Refer to Fig. 7.)

Hydraulic Excavator Terminology

Confusion can develop in comparing the sizes of front end loaders, mining shovels, and hydraulic excavators, unless the different methods of rating the buckets or dippers of the three products are understood. A dipper rating standard developed by the Power Crane and Shovel Association is generally, but not necessarily, used by cable shovel manufacturers to indicate the size of their products. The bucket rating standards developed by the Society of Automotive Engineers (SAE) is predominantly used by hydraulic excavator manufacturers to note the capacity of their products. When evaluating one type of machine vs. another or when comparing the same type of machine, the manufacturer's rated payload in pounds of capacity should be used as the most accurate indication of capacity.

The SAE has also published a list of standard terminology to be used when discussing hydraulic excavators. For instance, a hydraulic excavator equipped with a shovel front is officially called a *hydraulic shovel* when used in mining.

Machine Description

A general machine description applicable to all makes and sizes is not possible as there is considerable variation within each, although some common features do exist:

1) *Hydraulic pumps:* All manufacturers use variable displacement piston types.

2) *Hydraulic motors:* Again, piston-type motors are used.

3) *Operator controllers:* Joysticks are used for the normal operation function.

4) *Swing bearing:* Universally used.

5) *Power:* Diesel engines or a-c electric motor driven.

6) *Fast field assembly:* Ranging from 50 to 250 man-hr, depending on size.

7) *Low ground bearing pressure:* Ranging from 68.9 to 172.4 kPa (10 to 25 psi), depending on size and manufacture.

8) Dual path hydrostatic propel drive.

9) Dozer-type crawler component through the 11.5 m³ (15 cu yd) class.

10) Use of planetary reduction gearing.

Illustrated in Fig. 26 is the common nomenclature features that would be applicable to each major component of a typical hydraulic shovel.

Table 5. Large Hydraulic Excavator Makes and Models

Size Class, cu yd	Make	Model	Capacity, cu yd Shovel	Approximate Working Wt, lb	hp
+15	P&H	1550	15	450,000	1100
	Demag	H 185	15.6	419,000	1000
	O&K	RH 120C	16	466,000	1160
	Liebherr	994	15	453,000	1065
+20	Demag	H 285	22	666,000	1523
	Marion	3560	22	600,000	1400
	Hitachi	Ex 3500	23	723,000	1660
	O&K	RH 200C	25	880,000	2000
+30	O&K	RH 300	34	1,130,000	2320
	Demag	H 485	34	1,200,000	2103

Metric equivalents: cu yd × 0.764 554 9 = m³; lb × 0.453 592 4 = kg; hp × 0.746 = kW.

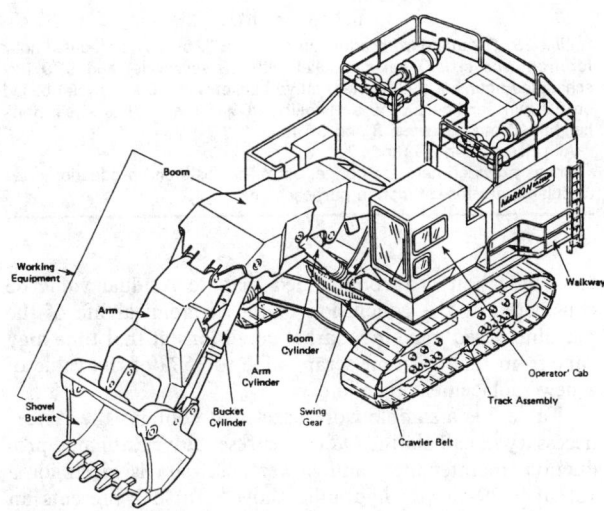

Fig. 26A. Hydraulic excavator nomenclature.

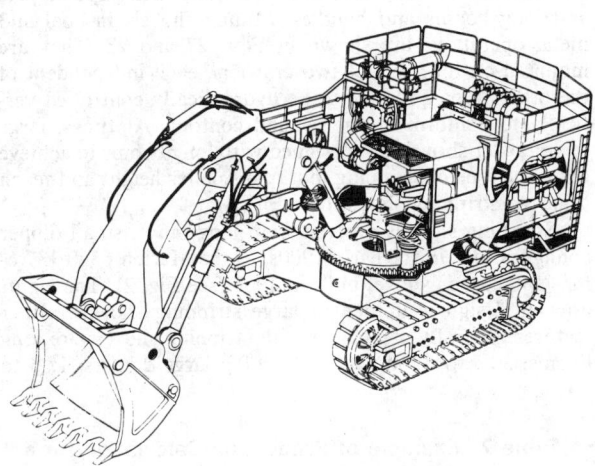

Fig. 26B. Hydraulic excavator cutaway view, front shovel version.

Component	Operating Hours
Cylinders	3,500- 5,000
Main pumps	5,000- 8,000
Swing motors	4,000- 7,000
Propel motors	12,000-15,000
Valves	Over 15,000
Operator controllers	3,000- 5,000
Diesel engine: Top	6,000- 8,000
Major	10,000-12,000
Clutch	Over 15,000
Torsional dampener	Over 15,000
Hoses	Over 15,000
Propel brakes	Over 15,000
Swing brakes	4,000- 7,000

Application

Comparison of Hydraulic Shovel with Front End Loaders and Mining Shovels: Before developing application studies using hydraulic shovels and hoses, the unique abilities of the machine as compared with front end loaders, mining shovels, or draglines should be considered. The key unique features of the hydraulic shovel are:

1) Higher penetrating forces per bucket capacity that may be applied to the bank and that affect bucket fillability.

2) Ability to penetrate the bank at any elevation above the pit floor permitting *top down* and selective loading.

3) Flat pass capability which is particularly effective in thin seam material loading.

4) High mobility when compared with an electric powered mining shovel.

Bucket Factor: The hydraulic shovel bucket may be pivoted at any point within the operating envelope to permit driving the bucket teeth into the material face, allowing removal and loading from the top of the material face downward. This feature, not available on mining shovels, assists in preventing any material overhang that may occur in poorly blasted material. This ability also allows the operator to segregate and load different material in the face.

Truck Sizing: The majority of the hydraulic shovels used in surface mining are equipped with the *clam* type bucket. The clam action bucket allows the machine operator to control the dump speed. Reduced material dumping speed, particularly on the first load, will reduce the shock load on the haul truck bed, chassis, suspension, and tires. This allows the hydraulic shovel to operate with smaller capacity trucks than practical with mining shovels. The dumping target or bed size becomes more of a criterion than maximum bucket load per pass.

Productivity Estimates

The factors that affect the productivity of a hydraulic shovel or backhoe are the same as those used in calculating production for front end loaders, mining shovels, and draglines. Some variation could occur between brands of machines due to differences in bucket crowd and breakout forces, swing speed, and, more important, swing acceleration and braking time. For additional information on productivity estimating, refer to *Productivity* in the Walking Dragline section.

Ownership and Operating Costs (O&O)

Actual ownership costs will probably be unique to each owner and dependent on his individual depreciation rate, interest, taxes, and insurance.

For a comparison of the O&O cost of one type of machine with another (hydraulic shovel vs. mining shovel or front

Maintenance and Estimated Component Life

Hydraulic shovels require maintenance procedures that are unique to this product. Maintenance of the relatively simple hydraulic circuits on haul trucks, dozers, front end loaders, and similar machines does not necessarily qualify the mine's maintenance department to maintain or repair the hydrostatic systems used on hydraulic shovels or backhoes.

The life of the major components used in the power train of hydraulic shovels and backhoes is undergoing rapid improvement by manufacturers in development of systems as well as the individual components. Most manufacturers offer the major components on a rebuild exchange basis to minimize repair cost and downtime.

Replacement time will vary for individual manufacturers, as well as the conditions and maintenance at a specific mine. As a general rule of thumb, replacements should be considered when components reach the following hours:

Table 6. Hydraulic Excavator Ownership and Operating Costs

Name _____   Model _____
Project _____   Date _____

Ownership

1.   Purchase price $ _____
2.   Extras   _____
3.   Freight   _____
4.   Trail cable   _____
4A.   Subtotal   _____
5.   Ballast (counterweight)   _____
6.   Erection   _____
7.   Total price   _____
8.   Life of machine, years   _____
8A.   Total scheduled hours   _____
8B.   Hours/year   _____
9.   Depreciation cost/hour   _____
      $\dfrac{\text{(Item 7)}}{\text{(Item 8A)}}$
9A.   Average investment   _____
      $\dfrac{\text{(Item 7} \times \text{(Item 8} + 1)}{2 \times \text{Item 8}}$
10.   Interest   _____ %
11.   Taxes   _____ %
12.   Insurance   _____ %
13.   Total   _____ % × Line 9A
14.   Interest, taxes, and insurance cost/
      scheduled hour   _____
      $\dfrac{\text{(Line 13)}}{\text{(Line 8B)}}$
15.   Total scheduled hourly ownership cost $ _____

Operating

                                              Electric
16.   Hours operation per year (scheduled)   _____
      $\dfrac{\text{(Item 8A)}}{\text{(Item 8)}}$ or 8B
17.   Maintenance and supply cost/hour   _____
      $\dfrac{\text{(10\% to 20\% Item 4A)}}{\text{8B}}$
18.   Cost electric power/hour   _____
      (@ Avg. power demand of 662 kwh)
18A. Fuel oil cost/hour (@ $1.00/US gal)   _____
19.   Operator rate/hour   _____
20.   Oiler rate/hour   _____
21.   Total operating cost per hour   _____
      (Sum Item 17 through 20)
22.   Total ownership and operating costs per
      scheduled hour $ _____
      (Item 15 plus Item 21)
23.   Total ownership and operating costs per
      cubic yard $ _____
      $\dfrac{\text{Line 22}}{\text{bcy per scheduled hour}}$

NOTES:
Line 5: Standard price includes ballast.
Line 6: Erection costs include cost for tools, ties, etc.
Line 17: Includes repair labor and material. Use 10 to 20% of 4A depending on scheduled hours (2700 to 8000) and digging conditions (sand, taconite) less 2% for electric machine.

Table cont.

Line 23: For standard production rate, use 1230 bcy/scheduled hour for front shovel (20 cu yd bucket with 28 sec/cycle) and 920 bcy scheduled hour for backhoe (16 cu yd bucket with 30 sec/cycle) based on 50 min per hour, 80% availability, 90% fill, and 25% swell. Bank height assumed between 3 and 35 ft.
All prices are in US funds.
Prices provided here are budget only. The costs and production rates calculated are for estimation purposes only.

end loader), it is recommended that no residual value be considered at the conclusion of the economical life of the machine. Used machine market conditions at that time may vary from their current scrap value to 25% of the value of a new replacement machine.

Table 6 is a sample work sheet that contains the factors necessary in calculating O&O. Representative estimated production, maintenance, and power cost data is given for a 15.3-m³ (20-cu yd) hydraulic shovel. Table 7 presents an example of a production calculation.

STRIPPING SHOVELS

Stripping shovels, as opposed to mining or quarry type shovels, are normally very large machines with large dippers and long booms and handles. Mining shovels in coal and metal operations are shown in Figs. 27 and 28. They are mounted on four sets of two crawlers, each independent of the other three and able to be hydraulically controlled vertically to conform to the ground contour. At times, large two-crawler shovels are designed with long booms to achieve the same type of stripping, but to a smaller height and reach than the stripping shovel (See Fig. 4).

Stripping shovels evolved from long-range, small dipper configurations in the early 1900s, to the Marion 6360 137.6-m³ (180-cu yd) shovel built in 1966 (see Fig. 2). The eight-crawler design is unique to large stripping shovels. Other features generally associated with stripping shovels are long booms: 30.5 to 67 m (100 to 220 ft); large dippers: 15.3 to

Table 7. Example of Production Calculation for a Hydraulic Shovel

1. Bucket capacity, SAE heaped	20 cu yd
2. Material weight, lb/lcy	3000
3. Swell factor	0.77
4. Bucket fill factor	0.95
5. Bank yard factor (Item 3 × Item 4)	0.73
6. Cycle time, estimated 90° swing	30 sec
7. Bcy/cycle (Item 1 × Item 5)	14.6
8. St/cycle $\left(\dfrac{\text{Item 1} \times \text{Item 2} \times \text{Item 4}}{2000}\right)$	28.5
9. Propel time factor	0.90
10. Cycles/hr, theoretical $\left(\dfrac{3600}{\text{Item 6}}\right)$	120
11. Bcy/hr @ 100% Avail. (Item 7 × Item 9 × Item 10)	1577
12. Availability	0.75
13. Bcy/hr, practical (Item 11 × Item 12)	1183
14. St/hr, practical (Item 8 × Item 10 × Item 12)	2565

These calculations are estimates of the hydraulic shovel possible production; to calculate the actual shovel and mine production, both the mine management rating and the truck efficiency must also be considered.

Fig. 27. A typical mining shovel in a coal loading application. BE Model 295-B.

137.6 m³ (20 to 180 cu yd); mammoth weight: 907 185 to 1 224 699 kg (2 to 27 million lb); hinged dipper handle; and propel motors on crawlers.

A deck layout of a stripping shovel is shown in Fig. 29. This shows the multiple motors utilized to provide hoist and swing effort and the multiple motor generator sets utilized to provide power to the d-c motion motors of hoist, crowd, swing, and propel.

All stripping shovels use electric Ward Leonard control of the motion motors. In this system, an individual generator directly connected to the a-c driving motor supplies power electrically to its mating d-c motor.

Use and Application

While there are a few stripping shovels at work outside the US, most of these large machines currently work in the midwestern US, stripping overburden from seams of coal. Generally, these coal seams are covered by less than 30.5 m (100 ft) of overburden and are reasonably flat lying.

The last stripping shovel was sold in 1969 and the trend is to utilize large walking draglines instead of shovels. The shovel must work in the pit, as opposed to above it and, with the shallower overburden depth seams being exhausted, coal mines must go to equipment able to handle these deeper seams. A rule of thumb commonly used is that the shovel can handle overburden depth roughly one-half the boom length; thus, a 61-m (200-ft) boom is required on a shovel to be able to spoil 30.5 m (100 ft) of overburden.

The mammoth structure of base and crawlers adds more weight to a shovel than a tub and walking mechanism does to a dragline; thus more weight is required per yard of capacity in the shovel. Another item that adds weight is the crowd handle, which is not utilized on a walking dragline.

Fig. 28. A typical mining shovel loading operation in a metal mine. Marion Model 201 m

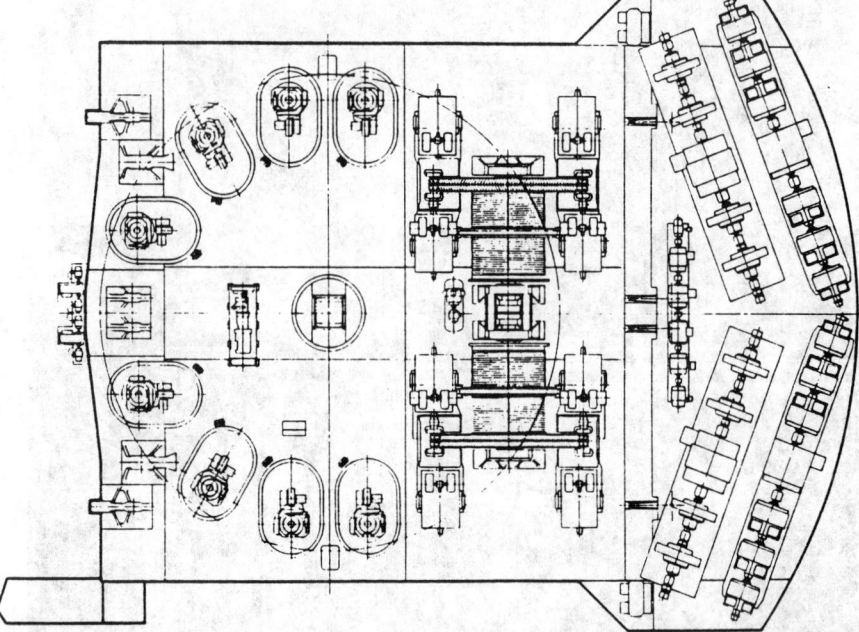

Fig. 29. Stripping shovel deck layout.

Generally, with regard to spoil reach, the cost per invested dollar is less with a dragline than a shovel.

Shovels do work in tandem with draglines, with the dragline taking the top cut and depositing it far enough away from the lower shovel cut so the shovel has room to spoil the material. Alternately, a smaller *pullback* dragline may be used to take excess spoil which the shovel cannot handle and deposit it in the previous cut, giving the shovel additional spoil area.

A typical spoil diagram for a stripping shovel in a two-seam operation is shown in Fig. 30. The shovel is shown in its pit digging the cut ahead of itself and depositing the spoil material 90° from the cut direction. The shovel sits on top of the coal or other mineral. Slope of the spoil and highwall and depth and width of cut determine the dumping radius of the shovel required. This is for a straight cut. Stripping with either inside or outside curves will allow lesser or greater spoil room for the shovel and give it more reach or radius.

WALKING DRAGLINES

JAMES D. HUMPHREY

The dragline mining system is a relatively simple, versatile, low cost mining method. A single dragline has the capability of operating over a wide range of overburden depths and material characteristics. It is particularly adept in dealing with overburdens that have fairly low cohesion and/or high water content. In addition, the dragline is quite capable of working very competent rock that has been blasted. Fundamentally, the walking (tub-mounted) dragline (see Fig. 31) differs from other types of draglines, e.g., crawler- or truck-mounted, only by its working base and the means by which it propels.

The walking dragline is slower, hence less mobile, but is more maneuverable and, because the weight is distributed to a tub, it has lower ground-bearing pressures. Because of the capital associated with the larger size of walking draglines, they are generally used in long-life (10 to 40 years) operations.

Walking draglines encompass an extensive range of bucket capacities from approximately 7 to 168 m³ (10 to 220 cu yd); all draglines with bucket capacities above 15 m³ (20 cu yd) typically fall into the walking dragline classification. Boom lengths of walking draglines vary from approximately 37 to 128 m (120 to 420 ft) long.

Crawler-mounted draglines (see Fig. 58) are generally limited to bucket capacities of less than 19 m³ (25 cu yd). The crawler dragline is generally utilized in operations that require more mobility (propel speed). In addition, it is applied in one location for a shorter duration than a tub-mounted dragline. Crawler draglines, due mostly to their size, lend themselves to bolt and pin assemblies and diesel power, both desirable features for short-term operations.

Fig. 30. Stripping shovel two-seam cut sequence.

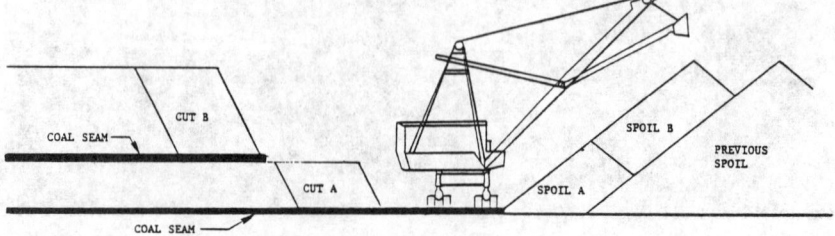

Fig. 31. 58-m³ (75-cu yd) class dragline, Marion Model 8200.

Machine Features—Mechanical

The mechanical features of a walking dragline can be divided into seven basic components: the tub or base, rotating frame, swing machinery, hoist and drag machinery, propel machinery, the bucket, its ropes and rigging, and the front-end components (see Fig. 32).

Tub: The tub is a rigid circular base that provides complete support for the dragline at all times except during propel mode (see Fig. 33). The weight of the upper components of the dragline transferred to the tub by the roller circle, an independent ring of interconnected rollers that acts as a large bearing for slewing of the rotating frame. The center journal transfers the lateral forces (drag forces) to the tub and provides the transfer point for the electrical power from the trail cable via the collector rings to the electrical equipment on the rotating frame. The rotational forces (during swing) are transferred by main rotating gear segments. These gear segments are bolted to the tub and form a ring gear, which on some machines can be as large as 16.5 m (54 ft) in diameter.

Rotating Frame: The rotating frame or machinery deck is the main platform of the dragline through which the reactive forces of swinging, digging, and propelling are distributed (see Fig. 34A). The rotating frame rests on the roller circle and rotates about the center journal during the swing motion. *All* machinery and superstructure components are mounted to the rotating frame. (See Fig. 34B).

Swing Machinery: The swing machinery consists of a number of individual swing units mounted to the rotating frame (see Fig. 35 and 36). Each of these swing units, of which there may be from one to eight, is comprised of a vertically mounted d-c motor which, through a series of gear reductions, drives a main swing shaft. The main swing shafts, to which the swing pinions are attached, extend vertically below the rotating frame and mate with the main rotating gear segments which are mounted to the tub and thereby slewing the rotating frame (see Fig. 35). More recently, the concept of using smaller but more numerous motors (up to 16) and planetary gear cases has been implemented.

Hoist and Drag Machinery: The hoist and drag machinery consists primarily of motors, gear reductions, and wire rope drums. These can be found in two basic arrangements: the independent or the synchronous hoist and drag system. The independent arrangement is utilized on all sizes of walking draglines, while the synchronous arrangement is utilized only in draglines with less than 15 m³ (20 cu yd) of rated bucket capacity.

The independent hoist and drag system is comprised of two sets of separate and independent d-c motors and gear reductions, each of which drives a rope drum (see Fig. 37). As few as one or as many as 12 motors, depending on machine size, may drive either gear reduction. The gear ratios and motor horsepower will generally be the same for the hoist and drag. The placement of the drag drum in front of the hoist drum or vice versa and whether the ropes are overwrapped or underwrapped on the drums varies between manufacturers and/or machine models.

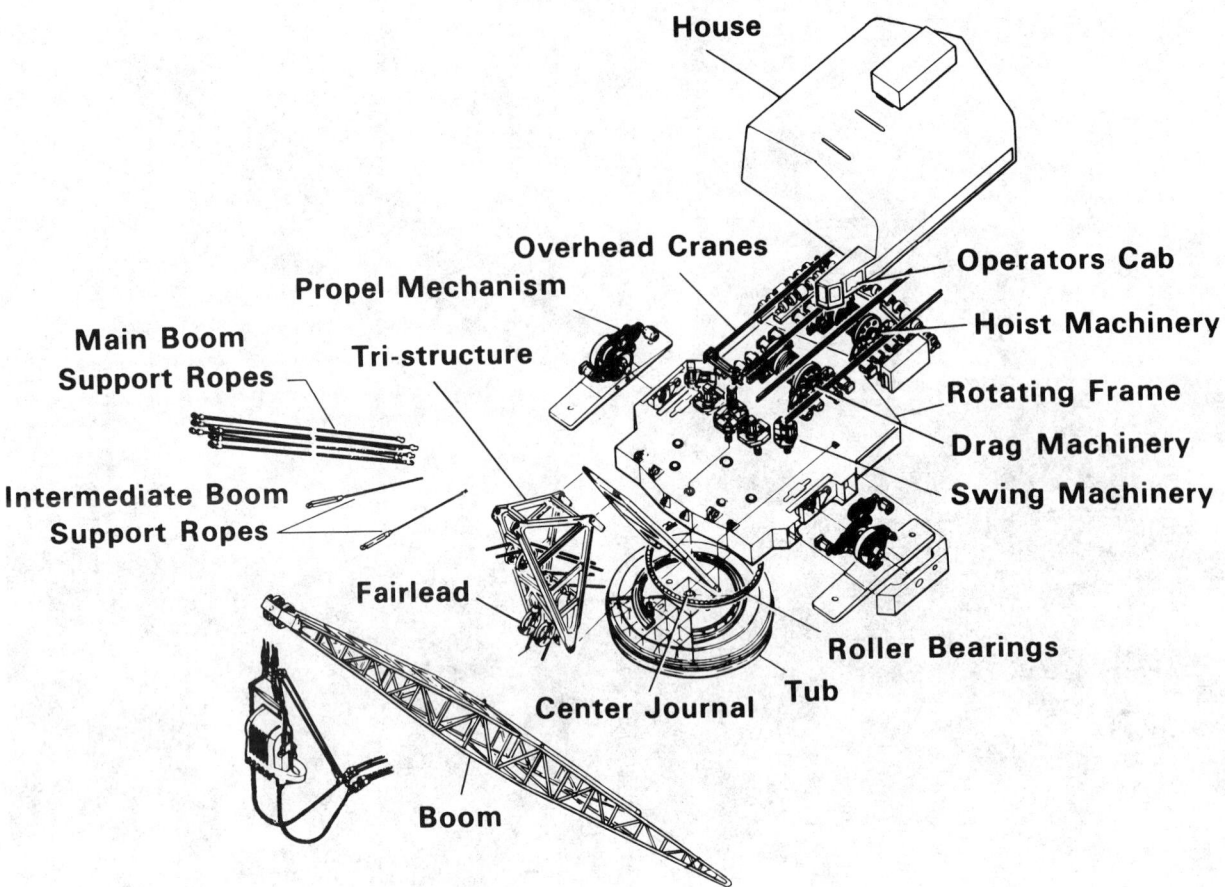

House

Overhead Cranes

Propel Mechanism

Operators Cab

Main Boom Support Ropes

Tri-structure

Hoist Machinery

Rotating Frame

Drag Machinery

Intermediate Boom Support Ropes

Swing Machinery

Fairlead

Roller Bearings

Center Journal

Tub

Boom

Fig. 32. Exploded view of the 8200 walking dragline showing the basic components.

A synchronous hoist and drag system differs from the independent system because its hoist and drag gear reductions are driven by the same motor (see Fig. 38). The gear reduction units are interconnected and therefore rotate together. The rope drums are clutched to the reduction units, so that the drums may be disengaged either individually or coincidentally, if necessary. An undriven drum, when disengaged from the drive, may be held stationary with an independent brake. These clutching and braking actions are independent of each other and are controlled by the operator. The ropes are normally directed around their respective drums so that when one drum reeves in its rope, the other drum is paying out its rope or vice versa. This is the normal pattern of rope movement during the largest part of a typical digging cycle.

Propel Machinery: Much like the hoist and drag machinery, the propel machinery is commonly found in two configurations: independent and interconnected. On the independently propelled dragline, there are two propel units. Each unit is comprised of one or two d-c motors, a gear reduction, a walking eccentric, and a shoe (see Fig. 39). The propel units are mounted on the sides of the rotating frame (see Figs. 34A and 34B). There is considerable variety in the design of the units, especially the walking eccentric component. However, they all accomplish a similar walking action, which is to move the shoes simultaneously through a nearly elliptical pattern. As the shoes are driven through the lower part of the propel motion, they contact the ground

and begin to take the weight of the machine. The propel units are placed on the rotating frame to the rear of the center of gravity, so as the shoes accept the load, the dragline is tipped a few degrees forward and the rear of the tub is lifted clear of the ground. As the propel motion continues, the propel units slide the machine 1.8 to 2.4 m (6 to 8 ft) to the rear and set it down. The front edge of the tub remains in contact with the ground and will carry about 20% of the weight of machine. The shoes will carry the remaining 80%. The final movement in the propel cycle lifts the shoes clear of the ground into the carry position at the top of the eccentric. The machine is now prepared to take another step or switch to normal swinging and digging operations. Since the propel unit is built into the rotating frame, the dragline may rotate and propel virtually in any direction; the only major restriction is boom clearance.

In some older or smaller models, the propel units are interconnected with a propel shaft driven from a single point. Often the propel drive in this type of system is taken from the drag motors through a jaw clutch. Propel systems that utilize hydraulic power and/or have three or four propel units have been developed for a limited number of special machines.

It is vital that the left and right propel loads occur as nearly simultaneously and equally as possible to minimize stress to the propel components. This requires some method of coordination between the units. On the interconnected type of propel system, the propel shaft provides the timing.

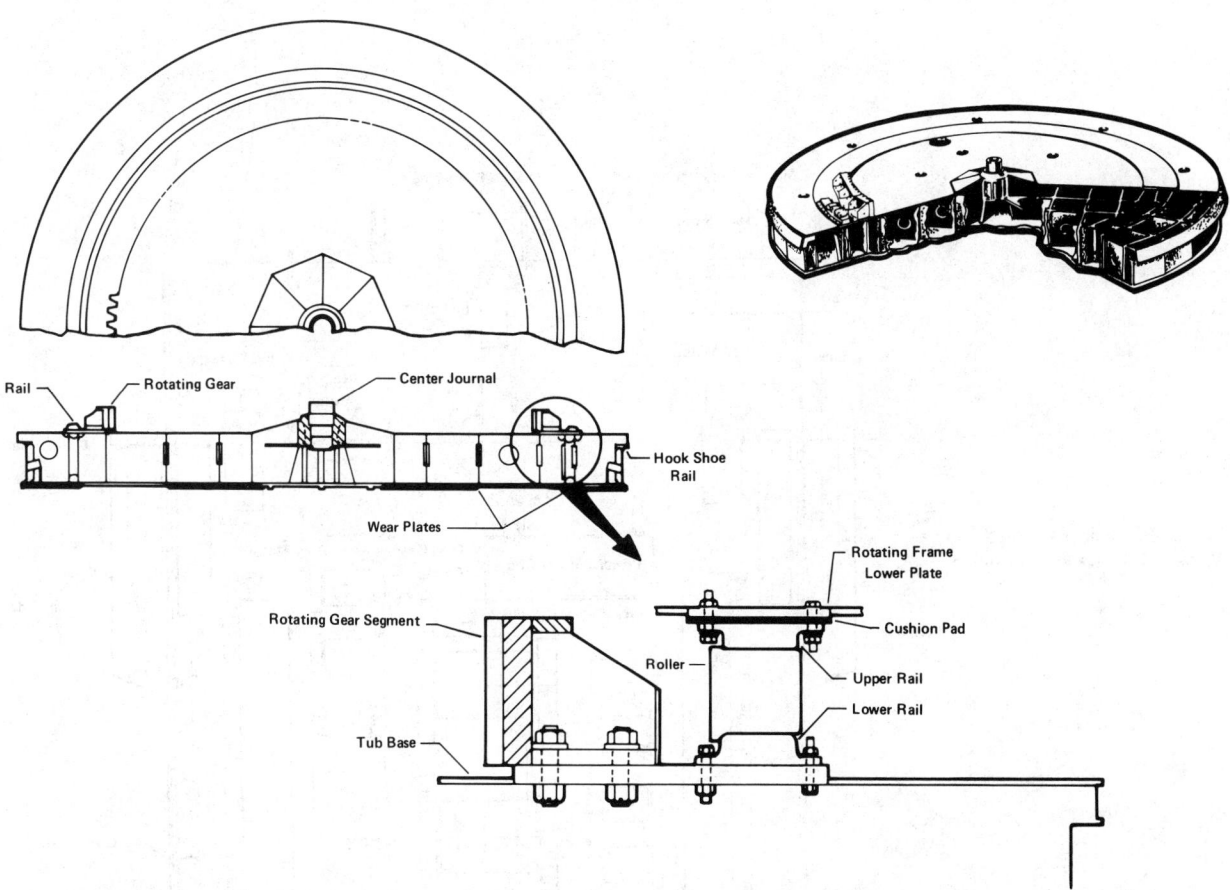

Fig. 33. Diagram of the tub for a walking dragline with roller circle rotating gear and center journal.

In modern independent propel systems, computers control motor loads and eccentric travel differences.

Another common propel system is the crawler frame and mechanism (see Fig 58). This is a common means of locomotion in a mine. The crawlers for draglines are most noticeably different because they are designed for high deadweight loads rather than for high speeds or tractive effort.

Bucket, Ropes, and Rigging: The purposes of the ropes and rigging are to manipulate the bucket (see Figs. 40 and 40A). The raising and lowering is a function of the hoist rope. The forward and rearward movements are controlled by the drag rope tension. Lateral control is maintained by

the swing motion. The bucket pitch attitude is controlled by the tension in the dump rope which is mainly influenced by the drag rope. However, to varying degrees, the hoist rope may affect attitude depending on the relative position of the bucket to the point sheave.

Front-End Components: The front-end components consist basically of the boom and its support structures (see Fig. 41 and 42). Some smaller draglines may have *live* booms that can be altered to any required angle by the operator. However, since the boom angle has such a pronounced effect on the load-carrying capability of the boom and the rotational inertia of the machine, the angle is usually fixed between 27° and 40° by design. Boom support structures vary greatly between manufacturers and models, but generally take on one of a number of configurations. The gantry (or A-frame) is typically found on smaller model machines (see Fig. 43). When the boom lengths exceed the 60 to 90 m (200 to 300 ft) range, it becomes necessary to alter the front-end geometry of the machine. In order to provide sufficient angle between the main boom support ropes and the boom to meet design load requirements for acceptable boom loading, the machine may be equipped with a mast-gantry (see Fig. 42A) or a tri-structure (see Fig. 42B). In addition to the main support ropes, some larger booms will utilize one or more sets of intermediate boom support ropes to aid in the distribution of boom support loading. Crawler and small walking draglines are similar; because of their size, the front end components are less complicated and live booms tend to be more common.

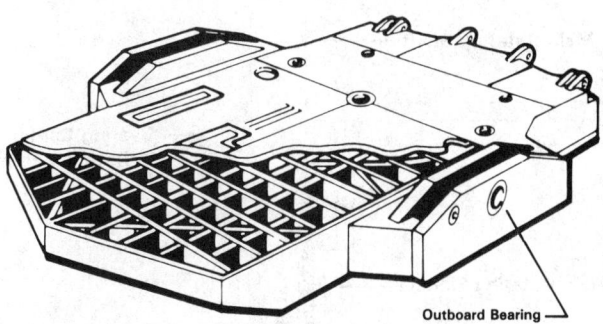

Fig. 34A. Upper frame (rotating or revolving frame).

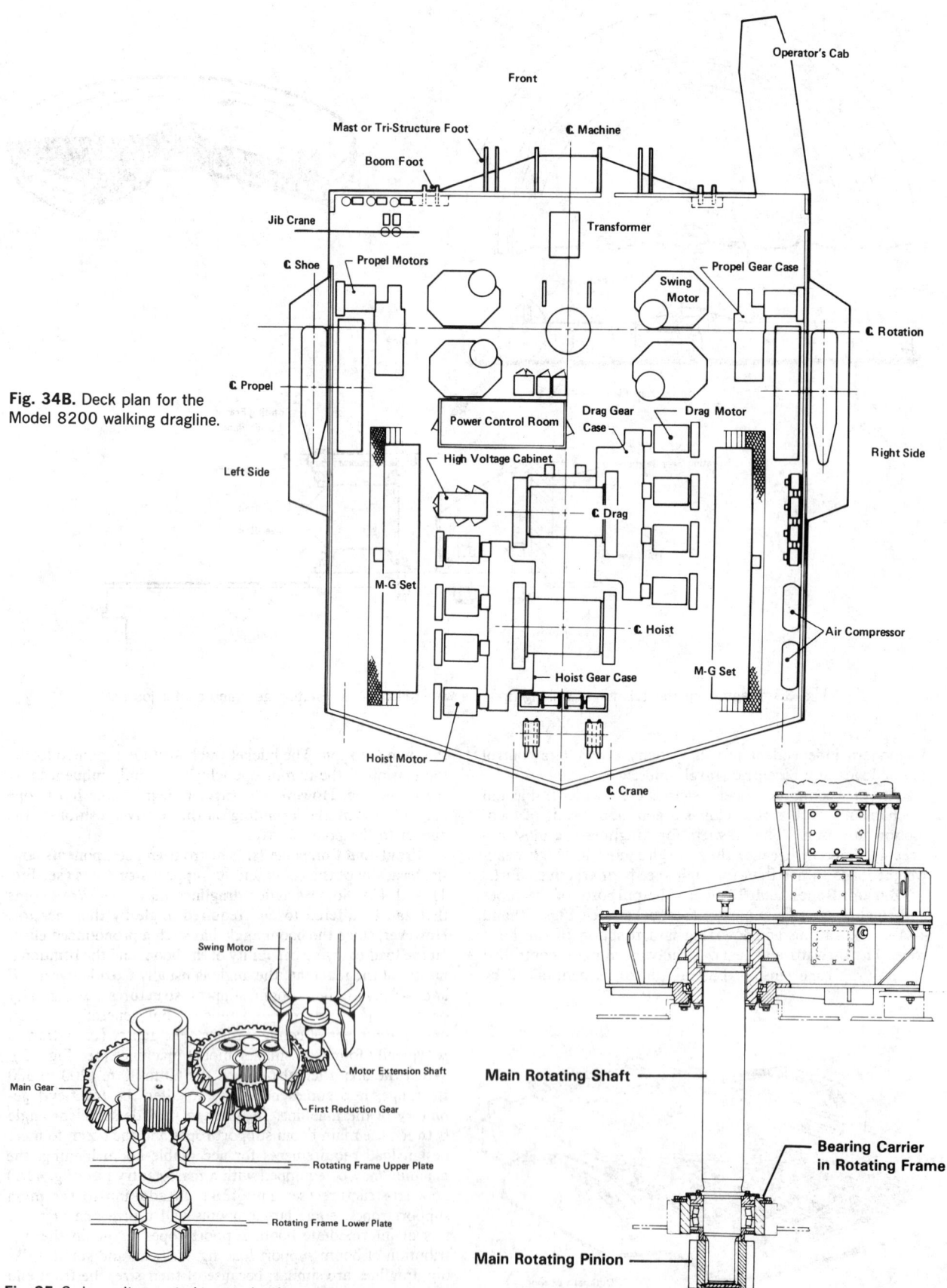

Fig. 34B. Deck plan for the Model 8200 walking dragline.

Fig. 35. Swing unit mounted to rotating frame. Note: gears above upper plate are enclosed in gear case, not shown.

Fig. 36. Swing gear case assembly.

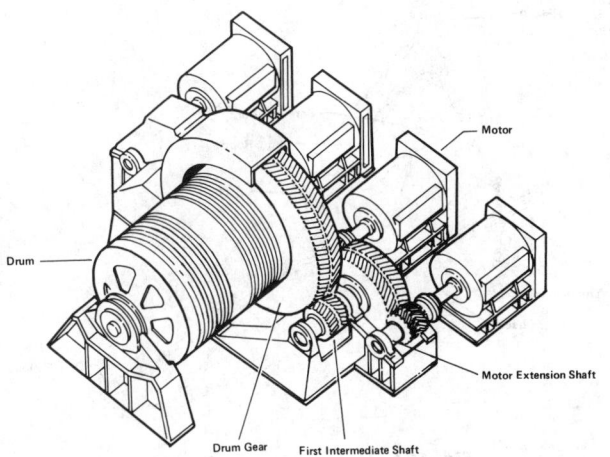

Fig. 37. Drum and motor arrangement (independent hoist and drag).

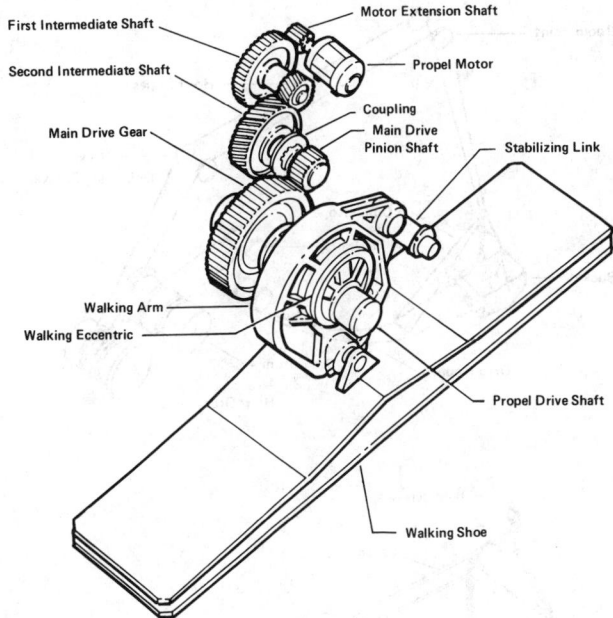

Fig. 39. Walking mechanism.

Machine Features—Electrical

The drive motors on the swing, drag, hoist, and propel, as mentioned previously, are normally d-c motors due to the requirement for variable speeds and for high torque at low speeds. A-c systems have recently been introduced for use in smaller excavators.

D-c voltage is provided by one of three methods: the first, and most common, is an a-c feed to the machine and a Ward Leonard system to convert to d-c voltage. The Ward Leonard system accepts an a-c feed at one of a number of preselected voltages, which in turn is transformed and used to drive an a-c synchronous motor(s). The a-c motor mechanically drives a set of d-c generators that provide the d-c voltage to the drive (hoist, drag, swing, and propel) motors.

The second method of a-c feed utilizes a static power conversion (SPC) system to convert the a-c voltage to d-c voltage. This system eliminates the rotating equipment, a-c

synchronous motors, and d-c generators (MGs) by the use of solid-state thyristors. The thyristors alter the incoming a-c voltage to d-c voltage. This system is fairly new and is currently offered only on smaller machines.

The third method is the diesel generator system in which the dragline has an on-board or independent skid-mounted diesel generator(s) that provides the necessary a-c voltage to either a Ward Leonard or a SPC system. The diesel generator systems are typically confined to the 15-m³ (20-cu yd) class and smaller machines.

The a-c system, referred to as *a-c variable frequency drive,* utilizes an a-c feed, converts the power to d-c voltage, and

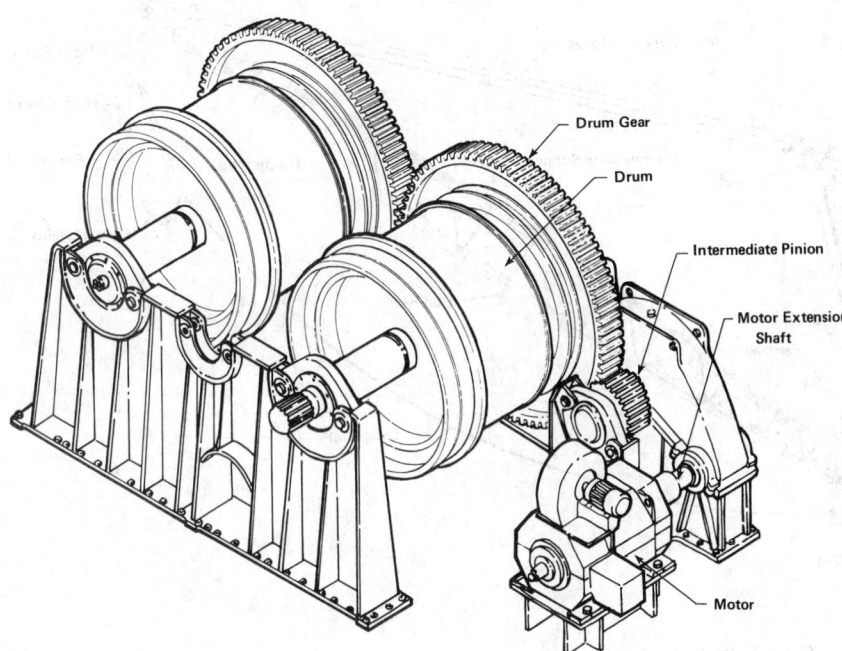

Fig. 38. Drum and motor arrangement (synchronous hoist and drag).

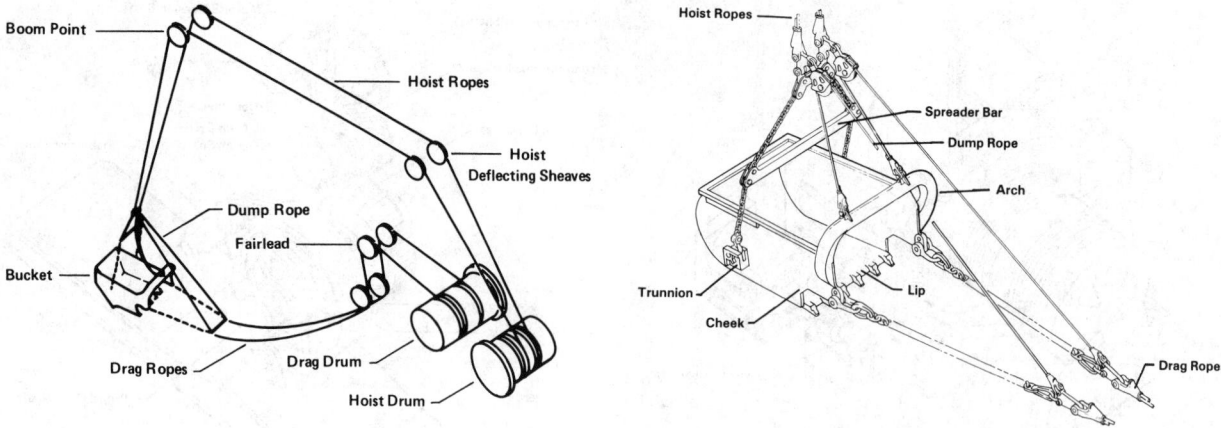

Fig. 40A. Twin dump rope bucket.

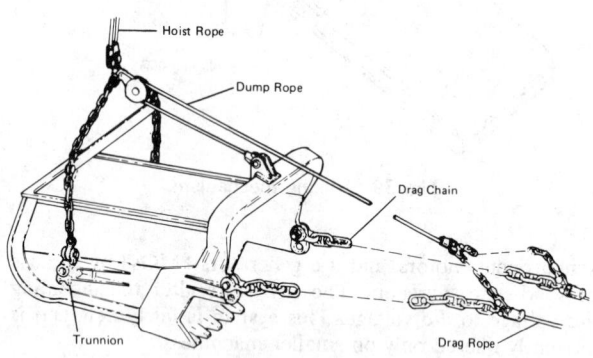

Fig. 40. Bucket rigging and ropes.

then back to a-c voltage at a different frequency. Both conversions are done by static power conversion equipment. This final a-c power is used to drive a-c motors. The frequency variation is in response to the operator's controls and allows the use of a-c motors as drive motors. This system has been only recently applied to draglines and is typically confined to the 15-m³ (20-cu yd) class and smaller machines.

Application

In the most common dragline operations, the dragline removes overburden material to uncover an ore (product) in a pit that is the most recent in a series of parallel adjacent pits. These pits are relatively long and narrow. The over-

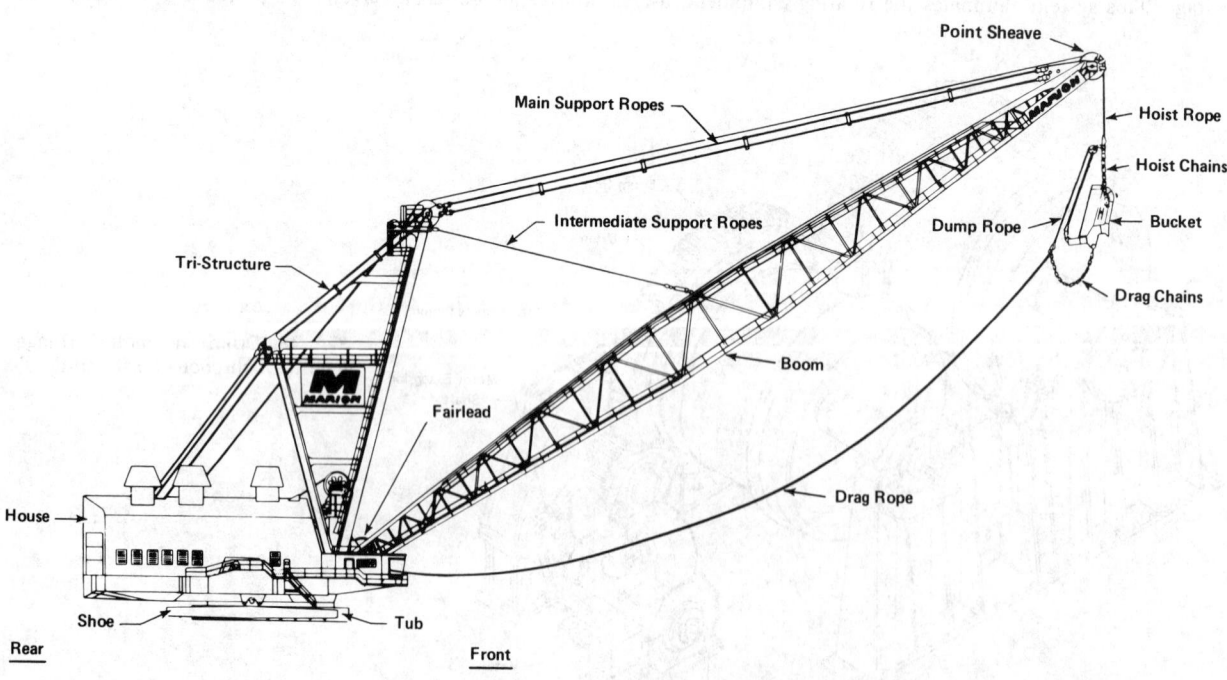

Fig. 41. Walking dragline nomenclature.

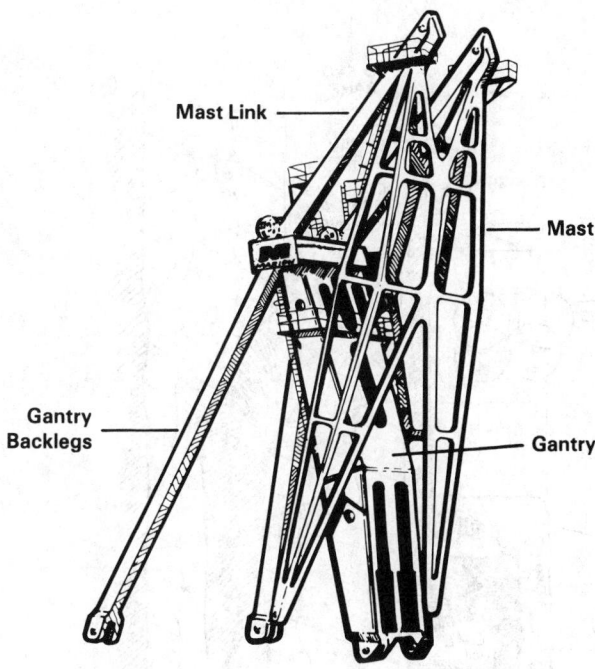

Fig. 42A. Mast and gantry.

Fig. 43. Gantry (A-Frame) style dragline, Marion Model 7620.

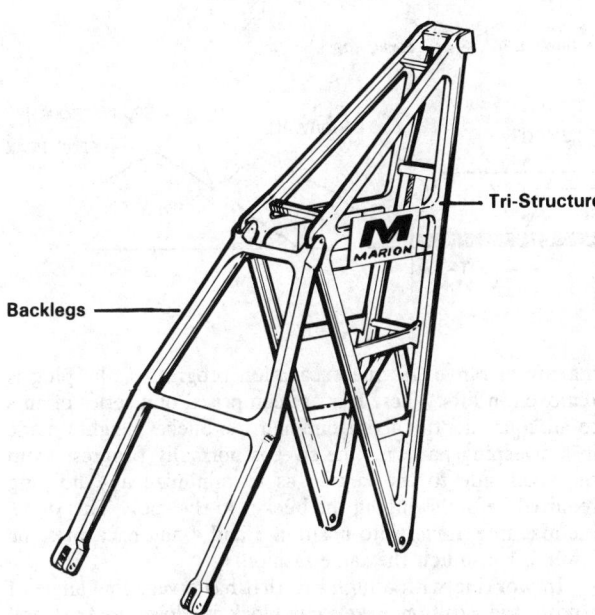

Fig. 42B. Tri-structure.

burden material from the current pit is placed in the previous adjacent pit from which the product has been removed by auxiliary equipment (see Fig. 44).

Dragline pits vary greatly in overall length due to the influence of geology, topography, and man-made obstacles. For larger machines, pits in the order of 300 to 3000 m (1,000 to 10,000 ft) long are common. In shorter pits, the sequencing of product removal is difficult and ramp construction becomes more frequent. In longer pits, power dis-

tribution systems become expensive and complex and dragline propel distances can be excessive.

Dragline pit widths commonly vary between 25 to 60 m (80 to 200 ft) variance from either extreme occurs to cope with local conditions and equipment. Pit width is influenced by the maneuverability of product removal equipment, depth of overburden, blasting pattern, material characteristics, dragline advance rate, and dragline dumping radius. In North America, 37 m (120 ft) is a common width in conventional operations.

Box Cuts: In order to initiate this series of parallel adjacent pits, there must obviously be an original pit. This pit is referred to as the *box cut* and has no previous pit available in which to spoil the material. There are a number of methods used to develop the box cut which vary depending on field conditions. Fig. 45 is a typical method.

Operating Methods: The standard dragline application is the *simple side casting* technique in which the dragline removes overburden from a specified length of the pit, referred to as a *set,* swings approximately 90°, and casts into an empty pit. Set lengths for larger machines are in the order of 30 m (100 ft) or about 16 steps for the dragline. To remove the overburden from a set, a dragline may use from two to four tub positions before retreating to start a new set. When the dragline removes the final set at the end of the pit, it may reverse direction and start a new pit. However, the dragline requires the product to be removed from the completed pit before starting the new pit to prevent burying product with spoil. Ideally, the product is being removed in conjunction with the dragline operation, and a constant interval is maintained between the dragline working face and the product face. The dragline must either wait until product material has been removed to provide a place to spoil or it must propel to a location where there is space for spoiling. This time may be utilized as maintenance time or the dragline may perform an auxiliary operation that can be spoiled outside of the working pit, e.g., ramp development. If the wait time is excessive, the dragline may propel to the opposite end of the pit where the product has been removed and initiate a new pit in the same direction as the previous one.

Long propel movements between working faces are referred to as *deadheads.* There are a number of variations in *deadheading* methods to best utilize machine time. Operations commonly operate from end to end as already described. Operating from the ends toward the middle (or vice versa) or to have two pits in close proximity are also common. This will depend on such factors as haulage ramp locations, product pit inventory requirements, terrain and geologic conditions, etc.

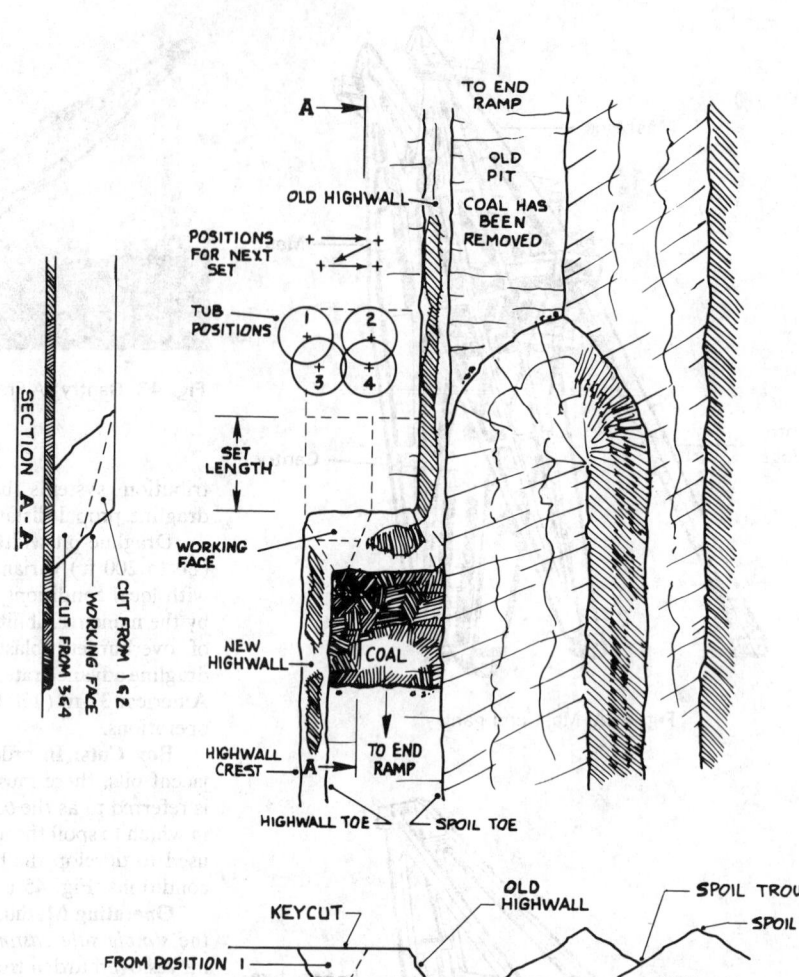

Fig. 44. Dragline working area plan
section and elevation.

Digging Positions: In fairly consolidated overburden, the four position set (see Fig. 44) is commonly used. The dragline progresses through the positions in numerical order. The first two rear positions are set back far enough to ensure that no material would be too close to the fairlead to be reached. In shallow digging, the first two positions may suffice to reach the product surface. In deeper pits, the digging may soon reach the point where the drag ropes are scouring through the shoulder of the digging face; consequently, the two forward positions are needed to clear the ropes. In the first two (rear) positions, the upper lift of the set is removed. In the last two positions, the machine has moved forward to the edge of the digging face to reach down for the lower lift of the set. The split between these lifts may be equal but usually varies to accommodate geological conditions.

Position 1 fixes the alignment and slope of the new highwall by a *key cut* along the full length of the set. This trenchlike cut, confined to a bottom width of only a single bucket as it works down, allows maximum lateral control of the bucket with a minimum of lateral strain on the boom. When the bottom of the first lift of the key has been reached (generally when the drag ropes begin to touch the crest of the working face), the machine moves laterally into position 2. This *plug* or *block* position allows the dragline to spoil at

maximum range. As the excavation progresses, the plug is removed in lifts. These lifts are comprised of a series of cuts to an equal depth (about one-half the bucket height) made in a sweeping pattern. The sweeps normally progress from the spoil side to the key so as to minimize any hoisting required before swinging the bucket to the spoil. This done, the machine moves into position 3 and 4 and excavates the lower lift in much the same fashion.

In working with a highwall that has a very low angle of repose, the position 3 key cut block is closer to the spoil than was the position 1 block. In the extreme of this situation, position 3 may move so much closer to the spoil as to overlap position 4. This situation then creates one type of three position set. The positions closest to the spoil (2 and 4) are positioned as close to the highwall as possible to make maximum use of the dump radius. Position 2 also removes the top of the plug so as to reduce the bucket clearance height that the second key position (position 3) must have before swinging.

This description, while typical, should be considered general. Set lengths and digging positions will vary, depending upon operating conditions and utilization of machine capabilities.

The side cast technique is commonly employed in op-

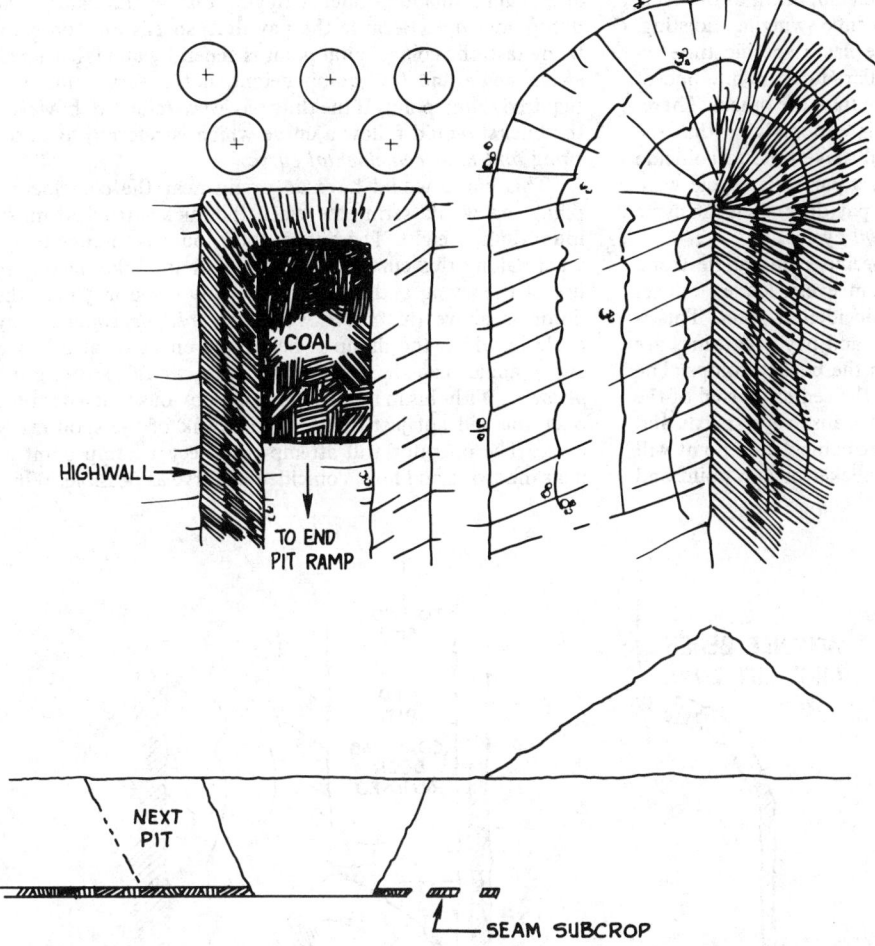

Fig. 45. Box cut.

erations that only want to uncover the mineral or product. Some operations, however, also want to load out the product with the same dragline. This is typically observed in phosphate operations in the southeastern United States. These operations remove the product (called matrix) from the pit with an identical digging method, but dump it into sumps or hoppers on the highwall where it is transported to a processing plant. These operations may not use key cuts because both the highwall depth and angle can be so low that they may only use a two position set (positions 2 and 4); nor do they need to keep the spoil out of the active pit since the matrix is constantly removed. This leads to some very small swing angles and low dump heights, which result in *pay drag dependent* cycles. This is something that does not often happen during normal side casting operations. Similar applications include operations that dump the product on the highwall to be reclaimed by other equipment, or smaller draglines loading directly into trucks or conveyor hoppers.

Digging Cycle: The digging cycle of a dragline is comprised of five components: (1) drag to fill, (2) hoist and swing, (3) dump, (4) return swing, and (5) positioning. The time required for each of these functions will vary depending on a number of factors: slope of the digging face, material characteristics, bank preparation, digging depth, hoisting height, swing angle, rope and swing speeds, as well as operator proficiency. Because of the diversity of these influences, even machines of the same model and design will have

different cycle times. Typical designed cycle time for larger machines is within the range of 50 to 60 sec for a 90° swing with a low dump. Smaller draglines are, as a rule, generally faster than larger draglines.

The dragline cycle begins with the bucket lowered in the pit and positioned to penetrate the bank. Dragging it into the digging face fills the bucket. Although buckets are designed for the teeth to have a good angle of attack in the relaxed position, sensitive handling of the hoist tension at this time can improve the penetration rate and reduce bucket fill time. In a good situation, the bucket will fill to overflowing in two to three times its length. Once filled, hoisting and drag pay out commence almost simultaneously, followed immediately by swinging as the bucket clears the trench. As the bucket swings and climbs, proper tension between hoist and drag holds the bucket in the carry position. As the dumping point is approached, the swing control is reversed (plugged) and the drag allowed to pay out until the bucket is unbalanced and the load dumps. Due to the swing inertia of the machine, the direction of swing will not change for several seconds after the controls are reversed, giving the bucket time to dump without delay. During the return swing, the hoist is payed out and the drag is reeved in so as to begin the positioning of the bucket for the next bite. The swing control is reversed to stop the swing motion, then neutralized as the bucket settles into position. The proficiency with which these five functions are carried out contributes significantly to the productivity of the machine.

Spoiling Techniques: The second component of the cycle is actually three independent movements: swinging, hoisting, and paying out drag. Each of these has a specific time requirement. For almost any particular dump point, one of these movements will take more time than the others. Therefore, two movements are retarded intentionally so that the slower, or *dependent,* movement will have time to coincide at the dump point. If, for example, a particular cycle were a long swing, short hoist, and short pay drag, then the cycle is considered to be a *swing dependent* cycle.

There is a point in space, a specific swing angle from, height above, and distance out from every loaded bucket pickup point, referred to as the coincidental point. This is the point at which the swing, hoist, and pay drag times are equal and as short as possible when the bucket dumps. The location of this point, referenced to the center journal of the machine, varies between machines because rope speeds and swing speeds are very individual. The coincidental point will also move with the location of the bucket pickup point and

any changes made in bucket rigging that would change the dump envelope. Because the pay drag speeds are comparatively fast, the coincidental point is generally at a fairly small swing angle and low dump height, not a very commonly required dump point. If the drag pay were retarded, however, the bucket would follow a curve which is referred to as the *swing and hoist coincidental curve.*

This curve would have an origin near the coincidental point and would terminate when the bucket reached maximum dump height. The bucket then could be dumped anywhere along this curve for the operator to make maximum use of the swing and hoist speeds. Any cycle in which the dump is above this curve would be *hoist dependent.* Any cycle in which the dump is below the curve or at a larger swing angle than the termination point would be *swing dependent.* With this in mind, it becomes obvious that a dragline operator will not just dump on the peak of the spoil every cycle. The operator will attempt to select a dump point as near the swing and hoist coincidental curve as possible. While

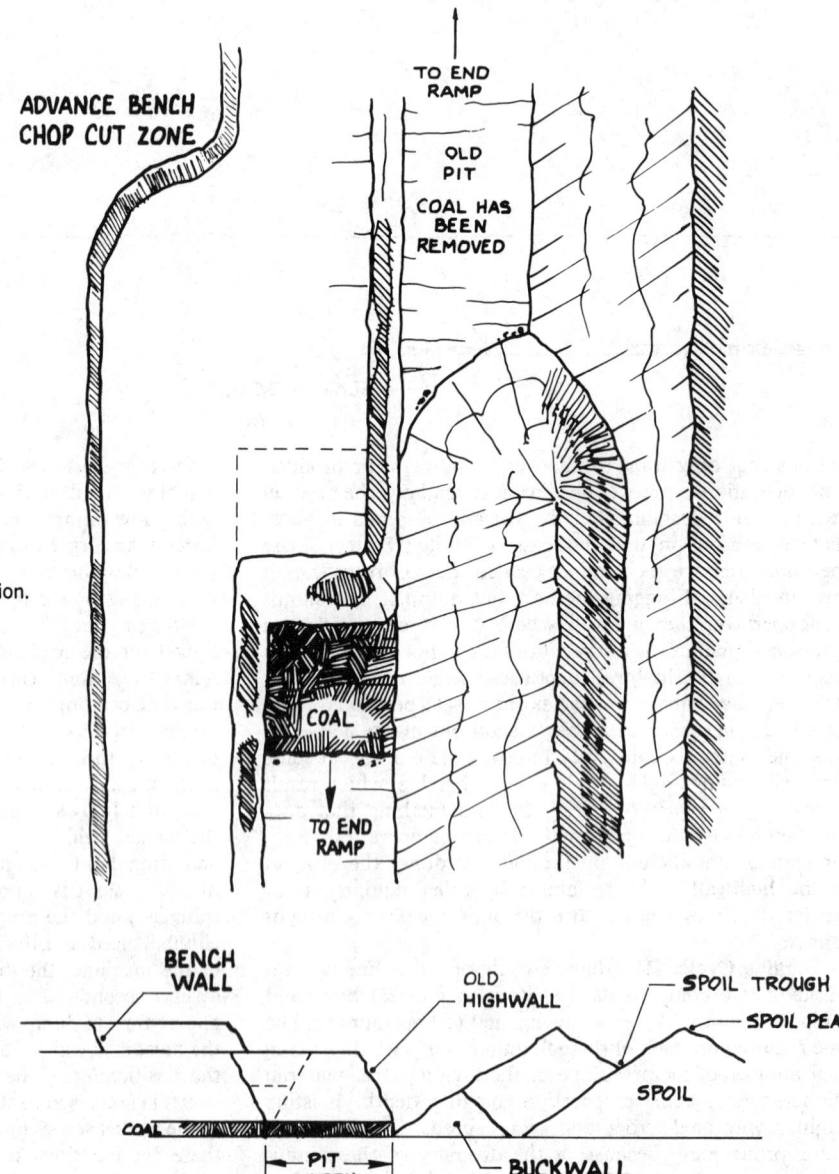

Fig. 46. Advance bench plan and section.

this is not practical for every cycle, the more coincidental cycles that occur the more efficient the operation will be.

Special Operating Techniques: One of the reasons the dragline is such a versatile tool is that it can be applied in a variety of manners. One common method is the *advance bench* (see Fig. 46). This method is very useful in areas with uneven terrain or in overburdens with a top layer of unconsolidated material. Here the set is split into an upper and lower bench. The lower bench is removed conventionally. The upper bench, however, requires a somewhat different technique called *chop cutting*. This is the technique that must be applied when a dragline excavates above its working level. Specifically, the bucket is pulled down a face rather than up or across it. The bucket is usually held in a dump position, teeth down, lowered on to the face, and dragged forward. Operating in a chop-down mode is hard on the rigging, ropes, and bucket, which leads to increased downtime, repair costs, and decreased productivity. Productivity is further decreased by the lower fill factor and increased drag to fill time that result. Advance benching generally requires a longer swing angle as well. The resulting efficiency varies; however, a minimum of 80 to 90% of a conventional rate should be used for initial estimates. It is most effective in unconsolidated overburdens with an advance bench height of no more than the fairlead height. There are operations that utilize a chop down in consolidated materials with bench heights considerably higher than the fairleads. *Chop cutting* is sometimes employed below the working level. Spoil side operations may use it to clean the highwall, rather than using a key cut.

If the material in the advance bench is extremely unconsolidated, it is sometimes convenient to utilize a *buckwall* (see Fig. 46) made of competent material removed from another area of the set. This dry competent material is placed at the toe of the spoil, thus forming a retaining wall. Unconsolidated material from the advance bench chop cut is then placed and contained behind it. This buckwall will add considerable stability to the spoil.

When a dragline does not have enough dump radius to place the spoil far enough away, it may be necessary to utilize one of a number of different spoiling techniques. This requirement for extra dump radius can occur on a localized basis around spoil side ramps, spoil or highwall slumps, high spots in the overburden, or inside curves. It may be required on an extended basis in operations that utilize a dragline with an inadequate dump radius for the overburden conditions. This may be because of limited machine selection at purchase time, capital cost savings for a smaller machine, relatively short period in deep overburden (e.g., 2 years of a 20-year mine life), etc. The advanced bench reduces the required dump radius by the horizontal component of the slope of the bench wall; a small amount, admittedly, but at times adequate. The most common technique utilized in reach limited situations is the *extended bench* (see Fig. 47). In this technique, the dragline places the driest, most competent material from the set against the old highwall. Enough material is placed there that, when leveled, a bench is formed that the dragline will move onto, so as to position itself nearer the spoil. As the excavating progresses, the extended bench is removed; unfortunately some material will have to be rehandled. Although efforts should always be made to avoid rehandling of any kind, in this situation it becomes a necessary function.

In two-seam operations, the placement of the spoil and the sequencing of equipment become more complex. Three methods are common, but not exclusive, in two-seam operations: (1) the two-bench, (2) the pullback, and (3) the extended lower bench. All three methods will move identical overburden combinations; however, they require different sizes of draglines and involve varied amounts of rehandle. The selection of the best method will depend on the relative costs and its feasibility in various scenarios. These three methods could also be applied to single-seam operations in deep overburden.

The most straightforward method is the two-bench method (see Fig. 48). In this method, the upper burden is removed first and spoiled into the bottom of the previous pit and the exposed upper coal removed. The lower burden is removed by simple side cast and spoiled on top of the first spoil. This can be accomplished with one dragline making two passes down the pit or two different draglines working in tandem.

The method has a few difficult points, primarily that the upper burden highwall must be setback (see Fig. 48, unit W_B) enough to provide swing clearance for the rear of the dragline on the lower bench during the key cut. Without this extra space, auxiliary equipment (dozers) must be used to make the key cut. Even though the use of dozers in the key is not uncommon, some operators consider working this close to the highwall unsafe in their conditions. A *setback* increases the required radius for the upper dragline. Also, the dragline on the lower bench needs more height, so any single machine in this application will probably be mismatched for one of the benches.

To help equalize the dragline size requirements, a spoil bench pullback method could be considered (see Fig. 49). This method removes the upper burden in the same manner; however, the spoil from this burden is then leveled and the second pass is made with the dragline on the spoil bench. Operations on the spoil bench may show productivities of 75 to 85% of conventional operations.

Another method would be to use an extended bench for the lower burden (see Fig. 50). In this method, some or all of the material from the upper burden is placed against the lower burden highwall forming the extended bench.

Productivity

Dragline productivity is based upon two primary factors: (1) the number of digging cycles occurring in a given period of time; and (2) the number of bank cubic yards in each bucket. Influencing each of these are several secondary factors, that all combine to govern the productivity rate of a machine. Contributing to the number of buckets are such things as scheduled time, availability, utilization, and cycle time. The yards per bucket are influenced by bucket design, bank preparation, swell factor, and operator skill. Each of these in turn also has its tertiary factors that are part of the foundation of good productivity.

To fully understand how to assess or forecast dragline productivity, clear definitions of some of the more ambiguous variables must be established. In particular, hours and yardage each have distinct categorical variations critical to dragline calculations (see Fig. 51).

1. *Calendar Hours (Hc):* This is the actual total hours in a given period of observation or prediction.

2. *Scheduled Hours (Hs):* This is the time during which the machine is expected to operate. This is the time that remains when hours for scheduled and uncontrollable delays (H_{DS}) are deducted from calendar hours. Scheduled delays are such periods as holidays, vacation, and nonscheduled shifts. Uncontrollable delays are such periods as strike time, adverse weather, power source outage, and other periods

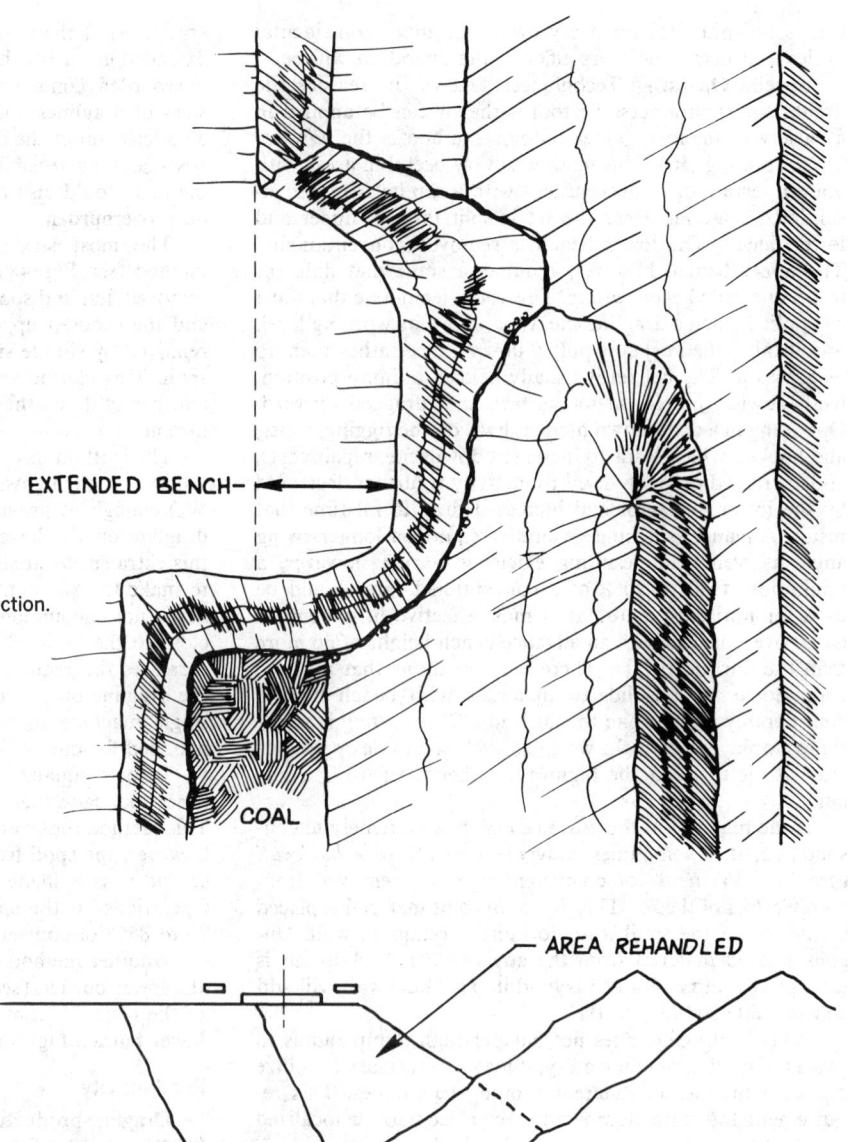

Fig. 47. Simple extended bench plan section.

EXTENDED BENCH-

COAL

- AREA REHANDLED

during which the machine or crews, either operating or support, are given no opportunity to perform. Large dragline operations are typically scheduled to operate around the clock all year long or seldom less than 8000 hr per year.

3. *Available Hours (HA):* This is that part of the scheduled hours *(Hs)* the machine is mechanically and electrically ready to be operated. This is the time that remains when hours for mechanical and electrical delays *(HDM)* are deducted from scheduled hours. These are any delays due to mechanical or electrical problems or repairs, $Hs - HDA = HA$.

4. *Utilized or Operating Hours (HU):* This is that part of the available hours *(Ha)* during which the dragline is actually operating at full potential. This is the time that remains when hours for operational delays (HDU) are deducted from available hours. These are any delays for losses of digging time such as moving cable or propelling and delays directly due to operational problems, e.g., supervision, labor, and support, $HA - HDM$ or $Hs - HDU - HDM = HU$.

There are basically two types of cubic yards: the *bank* cubic yard *(BCY)* and the *loose cubic yard (LCY)*. Material, when referred to as being in a bank state, regardless of whether or not it actually is, is considered to have a density equal to that of similar undisturbed (in situ) material. Material once excavated becomes loose and its density is decreased and volume increased. This is considered a material characteristic and varies with fragmentation, water content, etc.

The *swell factor (Fs)* is the ratio of these densities or, more generally, a ratio of volumes of equal weight, as in $LCY/BCY = Fs$; this is sometimes expressed as the inverse of this ratio. As expressed here, *Fs* for most overburdens varies from 1.1 to 1.6. In dragline operations, the material will undergo a number of swell changes. There is swell after blasting, swell in the bucket, and swell in the spoil. For the most part, swell in a dragline bucket is considered to be the same as swell in the spoil or about 1.2 to 1.4.

Productivity of a dragline is based on two primary factors: (1) the actual number of cycles that occur in a given time; and (2) the actual volume in BCY moved every cycle.

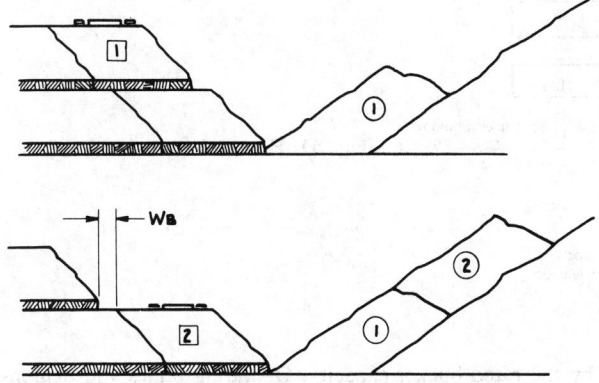

Fig. 48. Two-seam two-bench method.

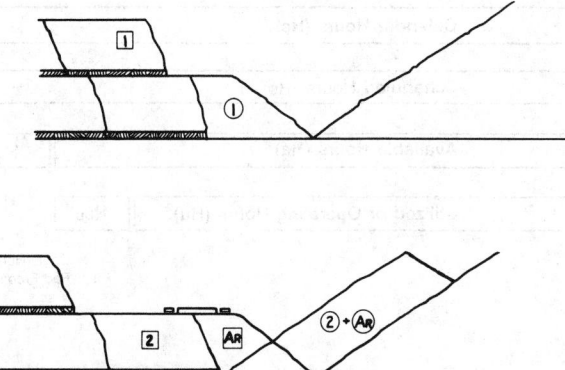

Fig. 50. Two-seam extended bench method.

The number of cycles per year is affected by four parameters:

1. *Observed Cycle Time (Tc):* The time required for the dragline to complete one entire cycle usually expressed in seconds.

2. *Scheduled Hours (Hs):* As previously defined.

3. *Availability (A):* This parameter reflects the decrease in scheduled time due to mechanical and electrical delays. A delay for a mechanical or electrical problem is scheduled time that the machine cannot operate because of a mechanical or electrical failure or repair; this includes wear part replacement, welding, etc., as well as major repair work. It does not, however, include delay time for mismanaged transportation of parts or repair labor. Availability varies with the age of the machine, the difficulty of the working conditions, the effectiveness of preventive maintenance the machine receives, the number of hours scheduled, etc. This availability percentage is not only critical to performance predictions but serves also as an indication of the efficiency of the maintenance program. The inclusion or exclusion of preventive maintenance as down hours varies between operators. Operations that exclude it will report higher availabilities. When attempting to correlate availability data, care must be taken to understand the calculating method.

If field data is obtainable, the availability may be calculated by dividing the hours available by the hours scheduled: $(HA)/(Hs)$ or $(Hs - HDM)/(Hs) = (A)$. Conversely, (A) may be estimated and multiplied by the scheduled hours to find available hours: $(A) \times (Hs) = HA$. Availability will vary from 0.95 to 0.70; 0.85 is considered typical.

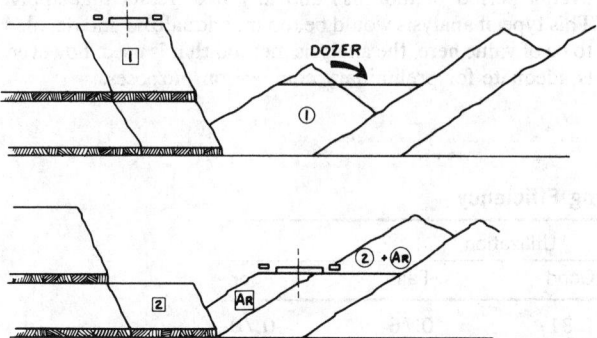

Fig. 49. Two-seam spoil bench pullback method.

4. *Utilization (U):* This factor reflects the amount of available time the machine is operated. A delay for an operational problem is available time lost that the machine is not operating because it is waiting on support, e.g., dozer work, cable handling, blasting delays, etc., or performing a nonproductive function, e.g., scaling down the highwall, propelling, preparing pad, or just generally being operated inefficiently. Note that utilization is calculated as a percentage of available hours, not scheduled hours. Utilization varies with the competency of the mine personnel, the efficiency of the mine plan, support equipment commitments, etc. This utilization percentage is used in performance predictions and is also an indication of machine management efficiency. If field data are obtainable, the utilization may be calculated by dividing the hours utilized by the hours available: $(HU)/(HA)$ or $(Hs - HDM - HDU)/(Hs - HDM) = (U)$. Conversely, (U) may be estimated and multiplied by the available hours to find utilized hours: $(U) \times (HA) = HU$ or $(U) \times (A) \times (Hs) = HU$. Utilization will vary from approximately 0.95 to 0.70; 0.85 is considered typical. *Note:* Both A and U may be combined into an *operating efficiency* (see Table 8).

The BCY moved every cycle is affected by three parameters:

1. *Rated Bucket Capacity (Bc):* Normally expressed in cubic yards; the units, however, are more accurately expressed as *loose cubic yards rated per cycle* (LCYR/cycle). This capacity is defined by SAE standard J67.

2. *Swell Factor (Fs):* As previously defined, expressed as *loose cubic yards per bank cubic yard* (LCY/BCY).

3. *Fill Factor (Ff):* This is the percentage of the bucket that actually fills with material, expressed as *loose cubic yards actual per loose cubic yard rated* (LCY/LCYR). Average fill factors normally vary from 0.80 to 1.00.

Fill factor and swell factor are often combined into a *bucket factor* (BCY/LCYR). Bucket factor is more practical to calculate from field data than swell and fill. It is surveyed yards (including rehandle) divided by the bucket (or cycle) count.

An estimate of dragline production can now be calculated that is sensitive to adjustments in these seven basic parameters. Not all mining operations will define these basic parameters in this same manner; some operations will use fewer and some more parameters. The equation that follows, however, applies the fundamentals in a straightforward reproducible manner that provides a basis for comparing machines and application methods.

$$(3600/Tc) \times (Hs/Year) \times A \times U \times (Bc/Fs) \times Ff$$

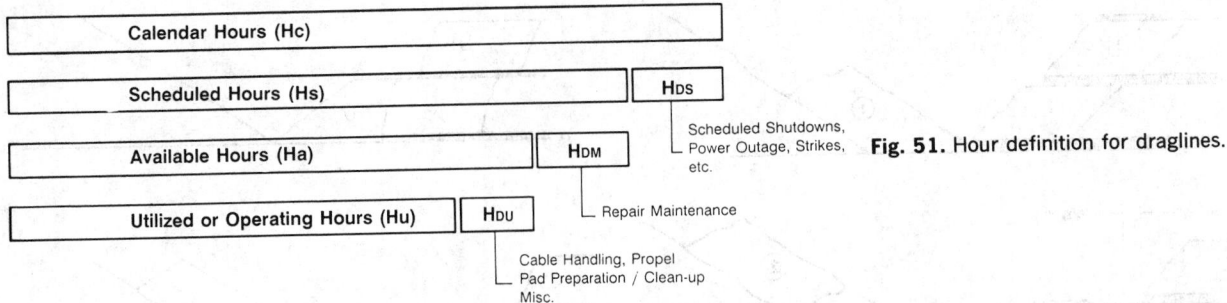

Fig. 51. Hour definition for draglines.

Example:

$$3600 = \text{seconds/utilized hour}$$
$$Tc = 57 \text{ seconds/cycle}$$
$$Hs = 8000 \text{ scheduled hours}$$
$$A = 0.85 \text{ available hours/scheduled hour}$$
$$U = 0.85 \text{ utilized hours/available hour}$$
$$Fs = 1.3 \text{ LCY/BCY}$$
$$Ff = 0.9 \text{ LCY/LCYr}$$
$$Bc = 1 \text{ LCYr/cycle}$$

The resulting production rate is approximately 250,000 BCY/year.

Since a one cubic yard bucket was applied, it can be said the production rate is 250,000 BCY/year per cubic yard of bucket. More appropriately, this *production factor* would be 250,000 (BCY/year) per (LCYr/cycle), and when multiplied by the rated bucket capacity per cycle (LCYr/cycle) will provide production rate per year. This, however, assumes that all the other parameters remain constant for all bucket sizes which is seldom true. Although for first round ball park estimating, this factor will prove easy and adequate. For any one mining method in a particular material at the same mine, the only parameters that will usually change between two different draglines are the cycle time and the rated bucket capacity. The production factor would remain the same in metric units, if a 1 m³ bucket were used, the resulting factor would be 250,000 (BCM/year) per (LCMr/cycle).

There is a common misconception that the dragline with the largest bucket is the largest producer; this is only true, of course, if all the other parameters remain constant. The dragline with the smaller bucket will, in fact, most likely have a faster cycle time because it may have less rotational inertia (it swings less load); in such an instance it will swing faster. If the dragline has a longer radius, less swing angle is required to achieve the same reach. This could also result in a faster swing time. Therefore, a 46-m³ (60-cu yd) dragline that cycles in 57 sec will actually produce less than a 45-m³ (59-cu yd) dragline that cycles in 55 sec.

The production factor may also be calculated from actual field production data by dividing the actual yearly production by the rated bucket capacity. Operating mines will demonstrate production factors from 200,000 to 300,000 (BCY/year)/(LCYr/cycle) quite commonly and values outside of this range do occur. The productivity factor also provides a means for measuring on an hourly basis efficiency of a mining method.

A caution should be pointed out here that some operations report production in *prime*, (also called *virgin* or *in situ*) BCY year labeled as just BCY/year. This is particularly true in countries other than the United States. *Prime* volume is the actual volume of overburden above the coal that was uncovered and does not include rehandle. When in fact most mines experience a minimum of 5 to 10% additional operational rehandle for ramps, bench fill, etc. Operations that utilize an extended bench or other rehandle methods will show even higher rehandle percentages. The production factor calculated using *prime* BCY will be smaller. Prime is adequate for comparing machines using the same mining method at the same mine; however, it is not a true measure of the productivity of the dragline. In situations where clarity is required, the use of the terminology *total* (including rehandle) bank cubic yards (TBCY) and *prime* (not including rehandle) back cubic yards (PBCY) is advisable.

Ownership and Operating Costs

When selecting a dragline, the owning and operating costs are always the bottom line. Theses costs are generally computed, for comparison purposes, on the basis of either cost per hour (usually scheduled hours), per BCY, or per ton of product. It is easiest, however, to compute the cost per year and divide it by the required base unit. Table 9 is an estimated ownership and operating cost for a typical 46-m³ (60-cu yd) dragline followed by a line-by-line explanation of the required computations. The ownership cost here is calculated by a simple method. A true ownership method would take into account the discounted cash flow over the payment schedule (draglines are not sold or paid for in a lump sum, but instead, over a period of months) and any tax credits that apply. This type of analysis would be too individualized and detailed to be of value here; the analysis method that is used, however, is adequate for preliminary comparison purposes.

Table 8. Operating Efficiency

Availability	Utilization			
	Excellent	Good	Fair	Poor
Excellent	0.84	0.81	0.76	0.70
Good	0.78	0.76	0.71	0.64
Fair	0.72	0.69	0.65	0.60
Poor	0.63	0.61	0.57	0.52

Table 9. 46-m³ (60-cu yd) Dragline—Ownership and Operating Costs Example

Ownership

1. Purchase price	US $16,330,000
2. Options and extras	817,000
3. Freight	226,000
4. Trail cable	50,000
a. Subtotal	$17,423,000
5. Ballast	90,000
6. Erection	2,958,800
7. Total price	US $20,471,800
8. Depreciation life 20 years	
9. Depreciation cost per year	US $ 1,023,600
$\frac{(Line\ 7)}{(Line\ 8)}$	
10. Average investment per year	10,747,700
$\frac{(Line\ 7)\ \times\ (Line\ 8\ +\ 1\ year)}{2\ \times\ (line\ 8)}$	
11. Interest, taxes, and insurance per year	
ITI% × (Line 10)	1,934,600
12. Total ownership costs per year	$ 2,958,200
(Line 9) + (Line 11)	

Operating

13. Maintenance and supply cost per year	US $1,219,600
7% × (Line 4a)	
14. Electric power cost per year	882,000
15. Labor costs	256,000
16. Total operating costs per year	US $ 2,357,600
17. Total ownership and operating cost per year	US $ 5,315,800

Total

18. Ownership and operating cost/ scheduled hour @ 8000 hr/year	US $ 664
19. Ownership and operating cost/BCY @ 15.0 × 10⁶ BCY/year	US $ 0.354

Line 1: Purchase price is normally f.o.b. factory; see manufacturer for estimate.

Line 2: 3 to 7% of line 1 is a good average; this will include one extra bucket and a standard selection of options.

Line 3: Varies with location, estimate here is for a 1600-km (1000-mile) ship distance at $60/ton. See manufacturer for specific shipping weights and freight estimates.

Line 4: This cost is estimated for 450 to 600 m (1500 to 2000 ft) of 350 MCM cable at about $30 per 0.3 m (ft). Conductor size varies with machine and length varies with pit plan.

Line 5: Cost estimated at $210/ton delivered for steel punchings. See manufacturer for specific requirement.

Line 6: Based on approximately $37/man-hour, the man-hours required are best received from manufacturer, however, 26 to 30 man-hr per ton of ship weight is a rough estimate.

Line 11: Estimated interest 14% + taxes 2.5% + insurance 1.5%, so total ITI is 18%. This will vary between locations.

Line 13: Includes costs for ropes, wear parts, lubricants, major parts, etc., as well as repair labor, (labor comprises 20 to 45% of costs). The 7% multiplier is based on actual experienced costs. It is, however, averaged for the life of the machine. Some years will have costs significantly above this percentage and some significantly below. For crawler-mounted diesel draglines, this maintenance factor is more in the order of 10%.

Line 14: This cost is based on 14.7 × 10⁶ kWh/yr at $0.06/kWh. Total power consumption is best obtained from manufacturer; however, consumption varies from 0.75 kWh/(BCY/yr) to 1.35 kWh/(BCY/yr) for small to large machines, respectively. For diesel machines, fuel consumption per hour is usually readily available from dragline or engine manufacturers. Be sure to understand whether the consumption is reported in operating hours or scheduled hours before converting to a per year cost.

Line 15: Includes wages and benefits for operator and oiler for every scheduled hour (8000 hr/year here).

Line 19: Production based on 250,000 (BCY/yr) (cu yd of bucket) total ownership and operating cost should be between $0.25 and $0.40/BCY for normal operations.

Dragline Selection

An important concept to keep in mind when sizing or selecting a dragline is to select the dragline for the mine plan, not the mine plan for the dragline. The two major parameters used to select the right dragline are, fundamentally, dump radius and allowable load. Under certain operating conditions other parameters, such as ground bearing pressure of the tub, the rear end swing clearance, dump height, etc., may also affect selection.

1. *Dump Radius (Rd):* This is the horizontal distance from the machine's center of rotation to the hoist rope when the bucket is in a vertically suspended position. Part of this dump radius is consumed by *stand-off (So)*. This is the distance from the center of rotation to the crest of the old highwall when the machine is in a digging position. The remaining dump radius is considered the *effective radius (Re)*; thus, $Rd - So = Re$. The stand-off distance will vary depending on machine size, operator preference, and overburden conditions. Lacking field data, a minimum stand-off distance of 50% of the width of the dragline from the outside of the shoes, or 75% of the tub diameter, are commonly used values for planning purposes.

2. *Allowable Load:* Sometimes called maximum suspended load, this is the maximum weight of bucket, rigging, and material for which the dragline is designed to provide optimum performance. Standard duty buckets, including rigging, weigh about 2000 lb per rated cubic yard of capacity, though they may vary from 1800 to 2300 lb/LCYr, depending upon the specific application. Some special purpose buckets for small draglines are as light as 1500 lb per rated cubic yard. Overburden weights are site specific; however, 3000 lb/LCY is a commonly used approximation. The combined weight then of a standard bucket and material load would be approximately 5000 lb per rated cubic yard. Therefore, an operation that requires a dragline with a 50-cu yd bucket actually requires a dragline with about a 250,000-lb allowable load. Allowable load should be selected for the 100% full bucket (peak bucket load) even though the average fill is less, e.g., 90%.

Calculating Radius Requirements: The effective radius (Re) required to place a spoil pile far enough away so that the spoil remains entirely in the previous pit is purely a function of the material. It is calculated by realizing that the cross-section area of the cut when swelled equals the cross-section area of the spoil; this method requires that the pits be of equal length. Knowing, then, the areas of both sides, the pit width(s), and the angles of repose, it is a minor problem in trigonometry to calculate standardized equations for a number of mining scenarios (see Figs. 52-57).

The effective radius (Re) of a dragline is the true indicator of its digging depth capacity; it is not boom length and not even dump radius. Although, for example, two draglines may have the same dump radius, if one were a smaller, lighter machine, it would most likely require less stand-off and therefore have a greater effective radius. Hence, the smaller dragline in the same overburden will be able to dig to a greater depth because it can actually place spoil farther from the crest of the highwall than can the larger machine.

Calculating Rated Bucket Requirements: It is not the objective here to explain how to calculate overburden removal requirements, so it will be assumed that the actual BCY/year requirement is known for each mining scenario selected. In addition it will be assumed that this number has already been corrected for coal losses in the pit and plant and for the overburden rehandle expected for the particular mining scenario.

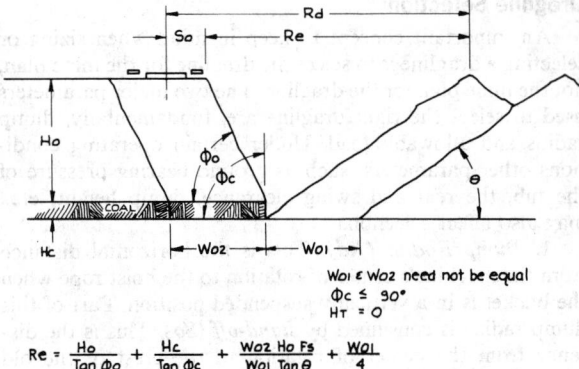

$$Re = \frac{Ho}{Tan\,\phi o} + \frac{Hc}{Tan\,\phi c} + \frac{Wo2\,Ho\,Fs}{Wo1\,Tan\,\theta} + \frac{Wo1}{4}$$

$$Ho = \frac{\left(Re - \frac{Wo1}{4} - \frac{Hc}{Tan\,\phi c}\right)}{\left(\frac{1}{Tan\,\phi o} + \frac{Wo2\,Fs}{Wo1\,Tan\,\theta}\right)}$$

Fig. 52. Simple side cast range diagram and equation.

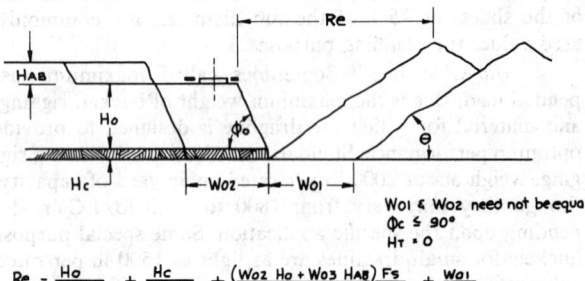

$$Re = \frac{Ho}{Tan\,\phi o} + \frac{Hc}{Tan\,\phi c} + \frac{(Wo2\,Ho + Wo3\,Hab)\,Fs}{Wo1\,Tan\,\theta} + \frac{Wo1}{4}$$

Fig. 53. Advance bench range diagram and equation.

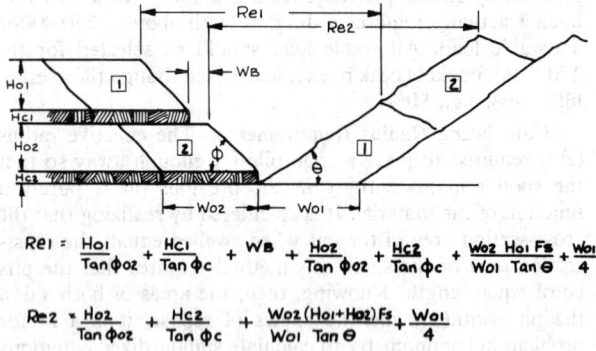

$$Re1 = \frac{Ho1}{Tan\,\phi o2} + \frac{Hc1}{Tan\,\phi c} + WB + \frac{Ho2}{Tan\,\phi o2} + \frac{Hc2}{Tan\,\phi c} + \frac{Wo2\,Ho1\,Fs}{Wo1\,Tan\,\theta} + \frac{Wo1}{4}$$

$$Re2 = \frac{Ho2}{Tan\,\phi o2} + \frac{Hc2}{Tan\,\phi c} + \frac{Wo2\,(Ho1 + Ho2)\,Fs}{Wo1\,Tan\,\theta} + \frac{Wo1}{4}$$

Fig. 54. Two-seam two-bench range diagram and equation.

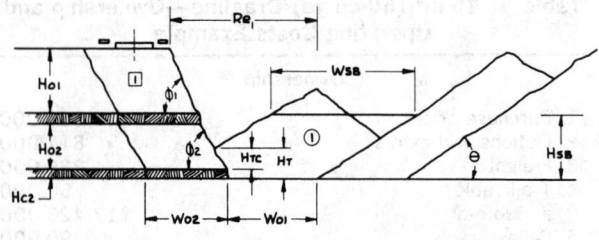

$$Re1 = \frac{Ho1}{Tan\,\phi 1} + \frac{Ho2 - Htc}{Tan\,\phi 2} + \frac{Hsb - Ht}{Tan\,\theta} + \frac{Wsb}{4}$$

Fig. 55. Two-seam pullback range diagram and equation.

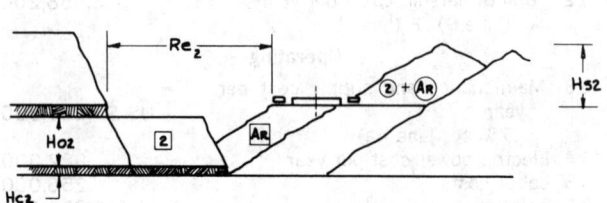

$$Re2 = \frac{Hsb - Ht}{Tan\,\theta} + \frac{Ho2 - Htc}{Tan\,\phi 2} + Wo2$$

$$Hsb = \frac{Wo2\,Ho1\,Fs}{Wo1} \frac{(Ht)^2 - (Hc2)^2}{2\,Tan\,\phi 2} + \frac{(Ht)^2}{2\,Tan\,\theta}}{Wo1 + \frac{Htc}{Tan\,\phi 2} + \frac{Ht}{Tan\,\theta}}$$

$$Wsb = Wo1 + \frac{Htc}{Tan\,\phi 2} + \frac{Ht}{Tan\,\theta}$$

$$Hs2 = \frac{Wo2\,(Ho1 + Ho2)\,Fs}{Wo1} + \frac{Wo1}{4}\,Tan\,\theta - Hsb$$

$$Ar = \frac{2\,Hsb\,Ht - (Ht)^2}{2\,Tan\,\theta} + \frac{2\,Htc\,Hsb - 2\,Htc\,Hc2 - (Htc)^2}{2\,Tan\,\phi 2}$$

Fig. 56. Support equations for two-seam pullback.

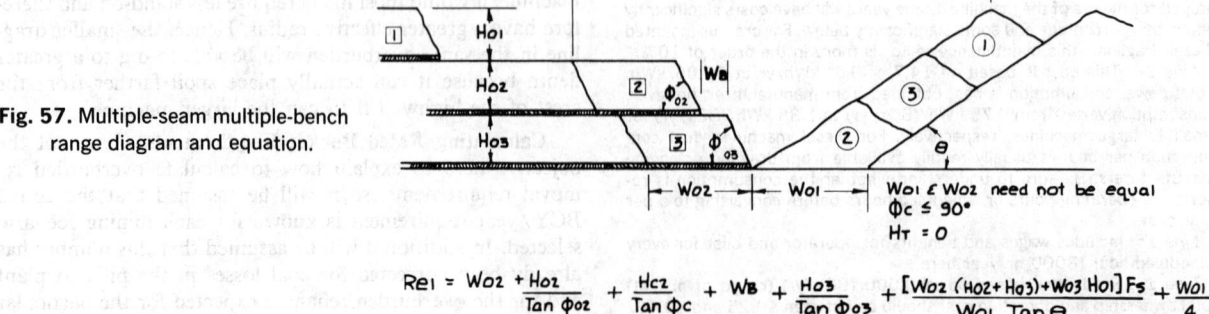

Fig. 57. Multiple-seam multiple-bench range diagram and equation.

$$Re1 = \frac{Wo2 + Ho2}{Tan\,\phi o2} + \frac{Hc2}{Tan\,\phi c} + WB + \frac{Ho3}{Tan\,\phi o3} + \frac{[Wo2\,(Ho2 + Ho3) + Wo3\,Ho1]\,Fs}{Wo1\,Tan\,\theta} + \frac{Wo1}{4}$$

Once the actual overburden requirements are ascertained in BCY/year, the first step is to approximate the bucket requirement by dividing the required BCY/year by the production factor (described earlier) of 250,000 (BCY/yr)/ (LCYr/cycle).

This ballpark bucket size with the previously calculated effective radius (*Re*) should narrow down the machine selection to a few models. The productivity equation should be fine-tuned then to include the individualized parameters for the operation, material, and machine cycle times. Manufacturers should provide precise swing times and rope speeds to estimate cycle times for particular models.

Selecting the Dragline: Once the models have been selected and the respective productivities sensitized, separate ownership and operating costs should be calculated and the costs per unit of product compared. This will provide a quantitiative comparison and show the least expensive scenario, *if all assumptions are correct.* At this point, check the sensitivity of the parameters and aspects of the plan that would be affected by the unknown, e.g., variances in spoil angle, weather, spoil slides, etc. Finally, assess the credibility of the scenarios; which one takes the most chances, which one is the most feasible? Often this type of qualitative comparison will eliminate the scenario that is favored economically.

DESIGNATION OF VARIABLES

Due to the complexity and variety of conditions that these dragline range equations encompass, it is advantageous to use variable names or designations that are easily and readily recognizable. For that reason, a system of *base* variables with standard designations has been established. These base variables are then subscripted with alpha characters to adapt them and/or subscripted with numeric characters to order them. The base variable set is as follows:

A	Areas
F	Factors
H	Heights
R	Radii
S	Separations
W	Widths
θ	Angles of repose of spoiled materials
ϕ	Angles of repose of in situ materials

The following is a list of the most common variations. Not all variations are listed; however, variable and subscript definitions remain constant and may be mixed as required.

VARIABLES USED IN RANGE EQUATIONS*

AB	Advance Bench (used only as subscript, e.g., H_{AB}).
AR	Area of rehandled material.
B	Bench (used only as subscript, e.g., W_B).
EB	Extended bench (used only as subscript, e.g., W_{EB}).
Fs	Swell factor (30% expressed as 1.3).
Hc	Height of coal; seams are numerically subscripted in order from top to bottom.
Ho	Height of overburden; cuts or interburdens are numerically subscripted in order from top to bottom.
Hs	Height of spoil numerically subscripted in order of placement.
HT	Height of toe of spoil on highwall measured from pit bottom.
HTC	Height of toe of spoil above the top of the coal.

Rd	Dump radius of designated dragline.
Re	Effective radius of designated dragline.
SB	Spoil bench (used only as subscript, e.g., H_{SB}).
So	Separation from highwall (machine center of rotation to crest of highwall).
Wo	Width of overburden (pit) numerically subscripted from old to new.
θ	Angle of repose of spoil.
ϕO	Angle of repose of highwall numerically subscripted from top to bottom.
ϕC	Angle of repose of coal (in situ) numerically subscripted from top to bottom.

MISCELLANEOUS EXCAVATING TOOLS FOR MINING

FRED R. SARGENT

Backhoes

Backhoes are generally considered secondary tools in the surface mining industry and are at times used for scraping off and cleaning up overlying material from surfaces. The primary use of the backhoe is for construction work such as trenching and excavation of footers for maintenance facilities and office buildings.

The introduction of the large 7.6 to 23-m³ (10 to 30-cu yd) capacity hydraulic backhoes now available to the mining industry has opened many possibilities to improve the handling of overburden or mineral, particularly in coal loading operations. The backhoe allows the choice of locating the truck on top of the overburden or in deep mineral seams, which can be a significant advantage in mines with wet pit floors. As a result, the hoe is easier to service; haul truck ramps are reduced or eliminated; and it may not be necessary to run electrical trail cables into the pit area.

The operating range of a typical hydraulic backhoe equipped with a 12.3-m³ (16-cu yd) capacity overburden bucket or a 17.6-m³ (23-cu yd) capacity coal loading bucket is illustrated in Fig. 58.

If the haul truck loading point can be located in front and slightly to either side of the backhoe, cycle time may be significantly reduced. This is particularly effective in deep seam coal loading when the mine enjoys dry pit floors.

While a dragline may be used similarly to the hydraulic backhoe, the rigid boom and arm of the hoe allow the operator more precise control of the bucket position, which provides more efficient dumping and greater clean coal recovery. See Fig. 59 for standard nomenclature for a hydraulic backhoe.

Clamshells

Clamshells are particularly useful in applications requiring vertical lifting of loose materials from one level to another. A number of operations may require removal of material from a level lower than where the machine sits, or it may be necessary to carry material above the machine.

The clamshell bucket is dumped by releasing the closing line and holding it with the lowering rope. A tag line is used to position the bucket as it is lowered to its digging position and to keep the bucket from spinning.

Some examples of different clamshell uses include opening new canals, cleaning out ditches or settlement ponds, loading from a stockpile into a truck, unloading barges, or loading into a hopper.

Output of Clamshells: The weight of the bucket and the weight of the maximum payload combined comprise the

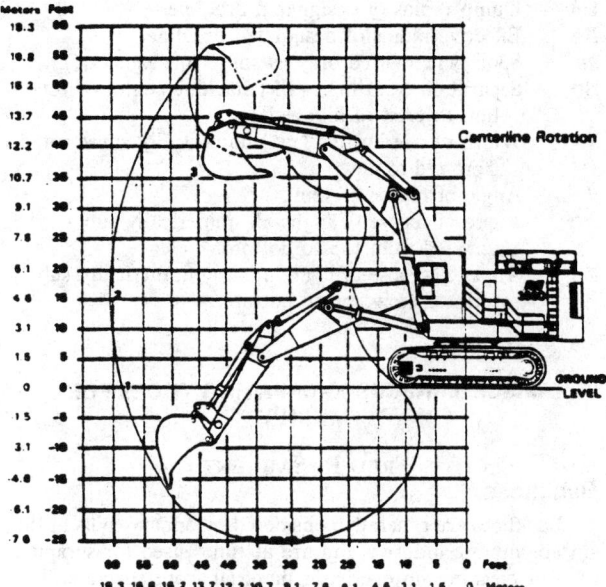

Fig. 58. Working range of a hydraulic backhoe.

allowable working loads at various radii of operation. These allowable working loads will vary with the percent of tipping designated.

Table 10 shows an allowable working load specification for a specific machine at the various radii of operation. Table

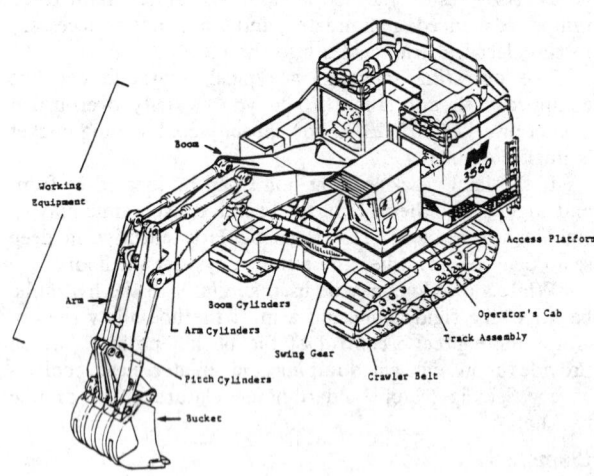

Fig. 59. Hydraulic backhoe nomenclature.

Table 10. Allowable Working Loads, (lb) 5.4-m³ (7-cu yd) (nominal) Clamshell Unit

Radius, ft	Boom Length		
	100 ft	120 ft	140 ft
80	36,000*	36,000*	36,000*
85	36,000*	36,000*	36,000*
90	36,000*	36,000*	36,000*
95	—	36,000*	36,000*
100	—	35,700	35,100
105	—	33,300	32,800
110	—	31,200	30,600
115	—	—	28,700
120	—	—	26,900
125	—	—	25,300

Combined weight of bucket and contents should not exceed values in the table. Values are based on 67½% of tipping, except for those with an asterisk (∗), which indicate capacities based on structural competence. Operation is not intended nor approved where allowable load is not shown. Metric equivalents: lb × 0.453 592 4 = kg; ft × 0.3048 = m.

11 gives approximate weights and dimensions of clamshell buckets.

Working Ranges of Clamshells: Fig. 60 shows the lettered dimensions of the working ranges which are generally provided by manufacturers of the various clamshell units.

Production Rate: Example: Assume that a 5.4-m³ (7-cu yd) standard rehandling clamshell bucket will be used to transfer wet sand from a canal into a continuous spoil pile.

Specifications for this calculation:

Boom length	30.5 m (100.0 ft)
Dumping radius	17.9 m (59.0 ft)
Dumping height	20.3 m (66.6 ft)
Digging depth	13.9 m (45.6 ft)
Allowable load, maximum	21 773 kg (48,000 lb)
Hoist speed	1.9 m/s (375 fpm)
Swing speed	3.2 rpm
Bucket size	5.4 m³ (7 cu yd)
Bucket weight, empty	6872 kg (15,150 lb)
Bucket weight, loaded	15 309 kg (33,750 lb)

The following average cycle time and production are based on:

Dumping height above water level	3 m (10.0 ft)
Average swing angle	90°
Bank yard factor	0.87
Operating hours per day	20

Cycle Time

Closing time to load	6 sec
Hoisting time	9 sec
Swing time	6 sec

Table 11. Approximate Weights and Dimensions of Clamshell Buckets

Size, cu yd	Height Open, ft, in.	Height Closed, ft, in.	Weight of Empty Bucket, lb
3½	11-11	10-2	8,300
4	12-3	10-4	8,855
5	13-8	11-7	11,290
6	14-11	12-4	14,750
7	15-3	12-8	15,150

Metric equivalents: ft × 0.3048 = m; in. × 25.4 = mm; lb × 0.453 592 4 = kg.

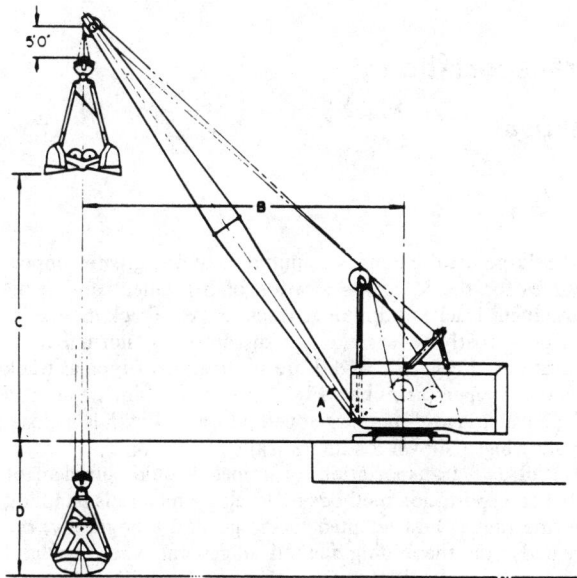

Production per Day

Wet sand, lb per bank cu yd	3,500
Bucket fillability	1.00
Size of bucket	5.4 m³(7 yd³)
Total cycle time	39 sec
Cycles/hour	92
Bank cu. yd/cycle	6
Tons/cycle	10.5
Bank cu yd at 100% operating efficiency	552
Operating efficiency-assume	85%
Bank cu yd/scheduled operating hour	469
Tons/scheduled operating hour	821
Bank cu yd/day	9,380
Tons/day	16,420

Fig. 60. Working ranges of a clamshell crane unit in a barge-mounted application. A, boom angle 40°; B, dumping radius 26m (85 ft); dumping height 15m (50 ft); digging depth 19m (62 ft).

Dumping time	5 sec
Return swing and lower bucket	8 sec
Position bucket	5 sec
Total cycle time	39 sec

REFERENCES

Anon., 1982, *SAE Handbook,* Society of Automotive Engineers, Warrendale, PA.

Anon., 1983, "Dragline Operator Training," Coal Extraction and Utilization Research Center, Southern Illinois University, Carbondale, IL, Jan.

Cummins, A.B., and Given, I.A., eds., 1973, *SME Mining Engineering Handbook,* AIME, New York.

Hirner, F.J., 1981, "Estimating Production and Costs for Electric Mine Shovels," *Mining Engineering,* Feb.

Hollingsworth, J., *History and Development of Strip Mining Machines,* Bucyrus Erie Co.

Meyer, G., "Large Hydraulic Excavators in Open Pit Mining Operations," Mannesmann Demag Ltd.

Pfleider, E.P., ed., 1968, *Surface Mining,* AIME, New York.

6.5 Haulage and Transportation

6.5.1 Rail Haulage

R. V. RAMANI

INTRODUCTION

Prior to World War II, rail haulage was the principal type of transportation in large pits. At the present time, in most surface mines, the overwhelming bulk of the mined material is transported from the mine to the crusher by trucks. However, railroad mining at the Morenci mine, Arizona, has been a successful operation since the start of open pit operations nearly 50 years ago (Hoppe, 1977). Here, each day over 136 kt (150,000 st) of ore and waste are hauled by railroad to the ore crusher and waste dumps. The average one-way haul to the crusher is 5.5 km (3.4 miles), and to the waste dumps, 9.3 km (5.8 miles). In the Bingham Canyon mine, Utah, railroad transport has played an important role in handling nearly 454 kt (500,000 st) of ore and waste per day (Jackson, 1980). A continuous blend of new developments with existing equipment has characterized the current materials transport system in Bingham. In 1963, truck haulage began augmenting the all-rail transportation system. In 1977, the conversion of the pit rail system from all electric to diesel-electric was started. As Bingham reaches deeper into the crust of the earth, in-pit crushers and conveyors are planned to replace the rail haulage.

Under favorable conditions, on a cost per ton transported basis, rail haulage is superior to other haulage methods for the transport of mined products to crushers and dumps. The advantage of rail haulage over other methods will dramatically increase with larger haul distances and greater production volumes. Typically, surface mines experience lengthening waste hauls as nearby disposal areas reach their capacities. An increase in length of haul, however, has only limited effect on rail haulage cost.

The most severe requirement for railroad application is the need to maintain a gentle grade for the track. Level grades are obviously the most desirable. Grades of 0.5 to 1% may be regarded as gentle. Relatively steeper grades may be employed within the open pit, a maximum of 3% for uphill hauls and 4% for downhill hauls. Track grades may go as high as 8% or 9% for some mines and industrial applications. The tracks outside the pit, leading to the concentrator or crusher, or the dump can be level or in favor of the load. If loaded trains can operate on level track or on descending grades, substantially lower haulage cost will be realized. However, the haulage cost will not reflect a reduction equal in magnitude to the drop in lift. There will be significant savings in the expense of maintaining diesel engines, wheels, generators, and traction motors. Partly offsetting these savings will be the higher costs of train brake systems maintenance. For the same lift, as the grade is lowered, length of haul increases. For example, a 3% ramp out of an open pit, only 91 m (300 ft) deep, must be almost 3.2 km (2 miles) long. For a 2% ramp, it will be nearly 4.8 km (3 miles). The greater the differences in elevation, the greater are the railroad track grade space requirements.

Capital investment in railroad transport can be considerable. Mines considering this method must have large reserves to support long-life and large production operations.

Also, large areal extent is required to afford greater opportunities for the judicious location of permanent and semi-permanent tracks to minimize track moves. Track moves are not only costly to execute but disruptive to normal mine operations. Figs. 1, 2, and 3 are illustrative of typical track layouts in open-pit mines. Fig. 4 presents a bird's eye view of the track layout in a large open pit mine. Fig. 5 is a close-up showing a shovel loading a train.

Railroad transportation in mines should supplement other transportation methods to develop a materials handling scheme that is best adapted to the general topography, the ore body, and the mining method. In general, where the haul from the mine to the location of processing facilities is over 4.8 or 6.4 km (3 or 4 miles), the topography is generally favorable so that massive earthmoving for railroading is not necessary, and sufficient minable reserves and market are in place for the mined product; the use of a field rail loading point and rail haulage to the processing facility is definitely feasible. In many cases, it is also a desirable alternative. For example, at the Muskinghum, OH, mine, a fully automated railroad transports coal 24 km (15 miles) from a truck-unloading field station near the mining area to an unloading and storage facility, which is the beginning of a 7.2 km (4½-mile) belt conveyor system to the Muskinghum river power plant (Anon., 1969).

Even in unfavorable topography, if the haul distance is long and the mine life quite high, the application of rail haulage should be explored. The effect of terrain will be severe on track grade construction costs; it is however, rather minimal on the operating costs of rail haulage systems. Specif-

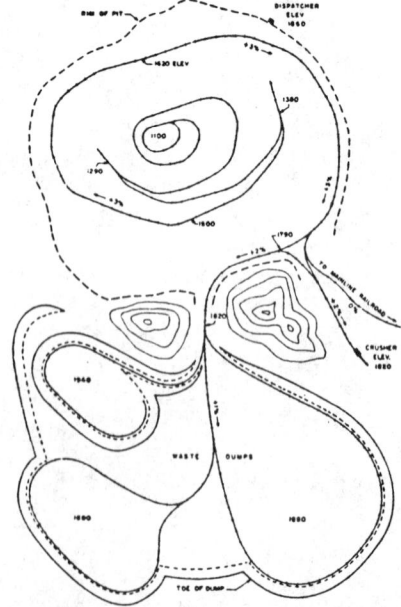

Fig. 1. New Cornelia mine main-line track layout.

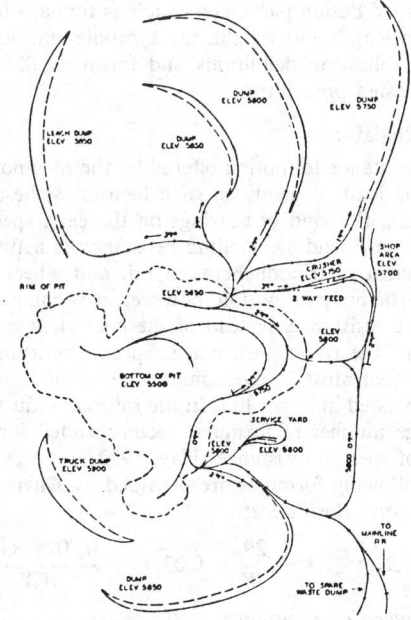

Fig. 2. Study of a track layout for a large open pit copper mine.

ically, in rolling hills or mountainous territories, the one-time cost of grade construction can be fairly well estimated at the start of the project. The normal service life of railroad equipment is higher than that of most other mine transport equipment. Operating and maintenance costs of tracks, locomotives, and railroad cars will not be significantly affected. Viewed in the context of the repetitive, uncertain, and es-

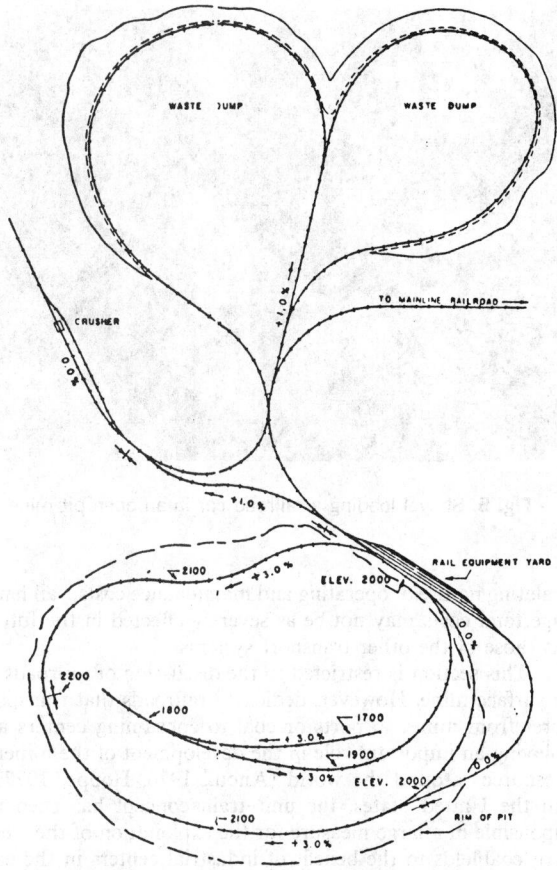

Fig. 3. Design features of a track plan for a large open pit mine.

Fig. 4. Bird's eye view of a open pit mine with rail haullage

Fig. 5. Shovel loading a railroad car in an open pit mine.

calating nature of operating and maintenance costs, rail haulage total costs may not be as severely affected in the future as those of the other transport systems.

This section is restricted to the discussion of railroads in a surface mine. However, dedicated railroads that transport ores from mines to ports or coal to consuming centers are playing an important role in the development of the mineral resources around the world (Anon., 1976; Hoppe, 1977a). In the United States, the unit train concept has been responsible in a large measure for the exploitation of the western coalfields to the benefit of industrial centers in the east and south.

BASIC DESIGN AND EQUIPMENT DATA

The relations of force, matter, and motion are dealt with in the branch of mechanics called kinematics. The understanding of these relationships is essential for the correct design of a rail haulage transportation system.

The accelerating and resisting forces in railroad design are shown in Fig. 6. The resistance to movement comes from many sources, the following being important: locomotive, trailing loads, grade, curve, and acceleration. Adhesion between the wheel tread of the locomotive and the steel rails is responsible for initiating the train movement and sustaining it. Adhesion is determined by rail and wheel material, foreign materials on the wheel tread, wheel diameter, axle weight, and rail condition. Adhesive force that can be developed by the locomotive depends on motor connection, truck characteristics, and wheel slip system characteristics. There are many different formulations to express the fundamental relationships between force, mass, acceleration, velocity, distance travelled, time, and horsepower as a function of the

mine railroad design parameters such as tonnage to be handled, train length and weight, track profile, and locomotive size. The following definitions and formulas illustrate the railroad design procedures.

Rolling Resistance

The resistance to motion offered by the locomotives and the trailing loads depends on such factors as the condition of the track, the kind of bearings on the cars, speed of the train, and rail bending. Rolling resistance is a function of the coefficient of friction between rail and wheel, and the weight of the body in motion; however, it is common to use the specific resistance, instead of the coefficient of friction, to calculate the rolling resistance. Specific resistance is the resistance that must be overcome to move one ton of load, and is expressed in lb per ton. In the railroad industry, there are a large number of formulas recommended for the calculation of specific resistance (Hay, 1982).

The following formulas are provided by Davis (1926):
Locomotive Resistance:

$$R_L = 1.3 + \frac{29}{W} + 0.03\ V + \frac{0.0024\ AV^2}{WN} \qquad (1)$$

Passenger Car Resistance:

$$R_P = 1.3 + \frac{29}{W} + 0.03\ V + \frac{0.00034\ AV^2}{WN} \qquad (2)$$

Freight Car Resistance:

$$R_F = 1.3 + \frac{29}{W} + 0.045\ V + \frac{0.0005\ AV^2}{WN} \qquad (3)$$

where:

W = average weight/axle, tons
N = number of axles
V = speed, mph
A = frontal area, ft^2

In general in mine railroads, the contribution from wind resistance may be negligible. The speed of modern-day trains, coupled with the direction of prevailing winds, can have a substantial influence on the total resistance.

Grade Resistance

When a train is moving against a slope, an additional resistance is encountered due to the weight of the train and the grade. Grade is defined as a percent and is the number of feet rise per 100 ft length of track.

Grade resistance in lb per ton of load is given by:

$$(1\ \text{ton})(2000\ \text{lb/ton})(\sin \alpha) = 2000 \sin \alpha$$

where α = the angle of the slope from the horizontal. Since the grade of railroad tracks is rather small, it is generally assumed that $\sin \alpha \approx \tan \alpha$. If grade is expressed in percent, then

Fig. 6. Accelerating and resisting forces in railroad design.

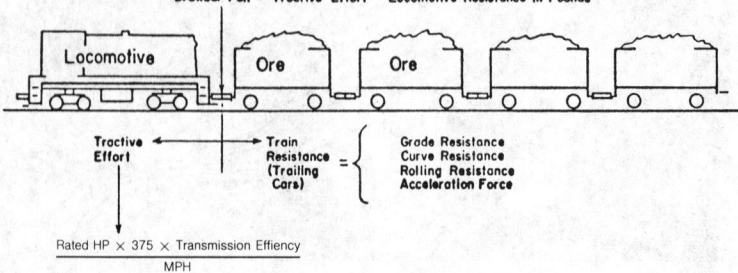

grade resistance in lb per ton load

$$= 2000 \times \tan \alpha = 2000 \, \frac{G}{100} = 20 \, G \quad (4)$$

where G = grade of the track, percent.

Curve Resistance

The resistance due to curves is caused by the slipping of the wheel tires on the rails in the axial direction, the friction between the flange of the wheels and the edge of the rails, and the different lengths of the inner and outer arcs of the rails. The curve resistance depends on many factors such as track gage on the curve, wheel base, wheel diameter, speed, radius of curve, and number of cars on the curve. The following equation from the Plymouth Locomotive Works (Anon., undated) brochure can be used to calculate the curve resistance:

$$\text{curve resistance, lb/ton} = \frac{225 \, (B + K)}{r} \quad (5)$$

where:

B = wheelbase of the wagons, ft
K = track gauge, ft
r = radius of the curve, ft.

A curve is also defined by its degree. A 1° curve is one in which a hundred feet of track is 1/360 of a complete circle. The radius of a 1° curve is 5730 ft. Degree of a curve is defined as

$$\text{degree of a curve} = \frac{5730 \text{ ft}}{\text{radius of the curve, ft}}$$

An approximate method for considering the resistance due to curves is to increase the total tractive resistance on level track by 20 to 30%. Another method is to use a resistance of 0.5 to 1 lb per ton per degree of curve. The recommended empirical equation is:

curve resistance = 0.8 lb per ton per degree of curve. (6)

Acceleration

A moving train consists of parts in linear motion and in rotation. The force required to overcome the inertia of a moving body, i.e., to accelerate or decelerate, can be calculated from the relationship: force = mass $\times$ acceleration, where mass is in lb-sec^2/ft and acceleration is in ft/sec^2. It is common in railroad calculations to express the weight in tons, the force in pounds, and the acceleration in miles per hour per second (mphps). The force required to impart a linear acceleration of 1 mphps to a weight of one ton is calculated as follows:

$$\text{mass} = \frac{(1 \text{ ton})(2000 \text{ lb/ton})}{(32.2 \text{ ft/sec}^2)} = 62.11 \text{ lb-sec}^2/\text{ft}$$

1 mphps acceleration

$$= \frac{(1 \text{ mile})(5280 \text{ ft/mile})}{(1 \text{ hour})(3600 \text{ sec/hour})(\text{sec})} = 1.47 \text{ ft/sec}^2$$

force = mass $\times$ acceleration = $62.11 \times 1.47 = 91.3$ lb

To allow for the acceleration of the mass in rotary motion, the calculated force must be increased between 5 to 10%. Thus, the total acceleration force required to impart a linear acceleration of 1 mphps and the needed rotational acceleration for the wheels, motors, and gears is approximately 100 lb per ton. Thus, if T is the weight of the train in tons and a is the linear acceleration in mphps, then the accelerating force (F_a) required is:

$$F_a = 100 \, aT \text{ lb} \quad (7)$$

Adhesion

Adhesion or coefficient of friction is defined as attachment between the wheel tires of the locomotive and the rail. The pulling force or the tractive effort a locomotive can deliver depends on the coefficient of friction between the wheel and the rail. It is a function of the material of the wheel tires and the rail; the conditions of the rail, whether wet, dry, or sanded; and to some extent on the springing and center of gravity of the locomotive. Adhesion is expressed as a percent and is defined as:

$$\% \text{ adhesion} = \frac{\text{available tractive force (lb)}}{\text{total weight on drivers (lb)}} \times 100 \quad (8)$$

$$\text{Available tractive force, lb} = \left(\% \, \frac{\text{adhesion}}{100} \right) \quad (9)$$
$$\text{(total weight on the drivers, lb)}$$

Adhesion values should be determined from a careful study of the conditions in the mine. Values for conditions usually encountered in mines are shown in Table 1 (Staley, 1949; Szklarski, Dudek, and Machowski, 1969). A starting adhesion of 25 to 30% can be attained. For a locomotive on the run, the values of adhesion are somewhat lower; running adhesions of 18 to 20% are normally attainable under average rail conditions. Higher running adhesions, up to 25%, are possible under excellent conditions. A locomotive speed-tractive effort curve provides the values of tractive effort that the locomotive can generate at different speeds. The curve for a 45-ton locomotive is shown in Fig. 7. Locomotive speed-tractive effort charts are available from the manufacturers.

Net Tractive Effort

This is defined as the difference between the available tractive effort and the various resistances to train movement. The net tractive effort is the force that creates acceleration.

Table 1. Adhesion Between Locomotive Wheels and Level Track (Percent)

Conditions	Wheels	
	Chilled Cast Iron	Steel Tires or Wheels
Dry rails, with sand	25	33
Dry rails, no sand	20	25
Moist rails	15	20
Wet rails	5-10	5-15

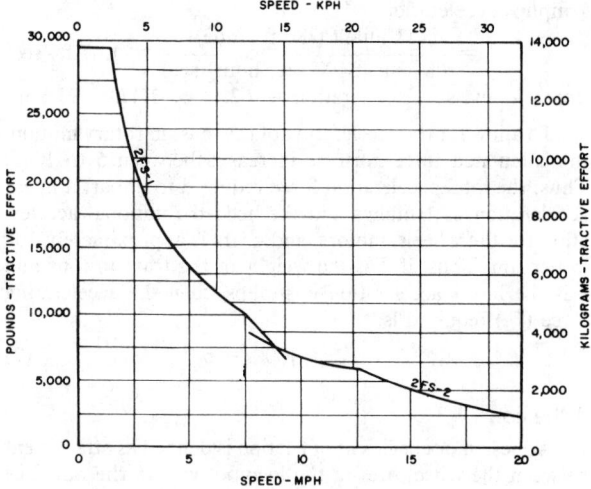

SPEED - KPH

Fig. 7. Locomotive speed-tractive effort chart.

Balancing Speed

The speed at which the net tractive effort is zero is called the balancing speed. At this speed, all the available tractive effort is used to overcome the various resistances to train movement, leaving no tractive effort for acceleration.

Useful Formulas

The following general transportation formulas are summarized from the earlier discussion.

Acceleration (mph/sec)
$$= \frac{\text{net tractive effort (lb)}}{100 \times \text{weight of trains (ton)}} \quad (10)$$

Time for which acceleration
is assumed constant (sec)
$$= \frac{\text{change in speed (mph)}}{\text{acceleration (mph/sec)}} \quad (11)$$

Distance travelled (ft)
$$= 1.467 \times \text{average speed (mph)} \times \text{time (sec)} \quad (12)$$

Distance travelled (miles)
$$= 0.000278 \times \text{average speed (mph)} \times \text{time (sec)} \quad (13)$$

$$\text{Time (minutes)} = \frac{\text{distance (miles)}}{\text{average speed (mph)}} \times 60 \quad (14)$$

$$\text{Trailing load (tons)} = \left[\frac{\text{tractive effort (lb)}}{\text{total resistance (lb/ton)}}\right] \\ - [\text{locomotive weight (tons)}] \quad (15)$$

Rail horsepower
$$= \frac{\text{tractive effort (lb)} \times \text{speed (mph)}}{375} \quad (16)$$

Horsepower input to generator
$$= \frac{\text{rail horsepower}}{\substack{\text{efficiency of the transmission} \\ \text{(generator + motor)}}} \quad (17)$$

Engine horsepower
= horsepower input to generator plus allowances for auxiliaries (radiator, fan, air compressor, etc.) (18)

$$\text{Gear ratio (G.R.)} = \frac{\text{Number of teeth on gear}}{\text{Number of teeth on pinion}} \quad (19)$$

$$\text{Motor rpm} = \frac{336 \times \text{speed (mph)}}{\text{wheel diameter (in.)}} \times \text{G.R.} \quad (20)$$

Braking

In addition to knowing the capacity of a locomotive for starting and hauling a train, its ability for stopping the train should be considered. When power is cut off to the locomotive, there is an accumulated reserve of kinetic energy in the train, which is used to overcome the resistance to motion. Under *favorable* conditions, the velocity of the train will drop and the train may come to a stop by coasting. The distance over which the train will come to a stop depends on the rate of drop of velocity (or deceleration). However, when taking load *down* a grade, there usually is no deceleration even after the power is cut off to the locomotive. In fact, the train may tend to accelerate rather than slow down. From the practical point of view of positive control over the train movement, braking requirements and performance must be investigated. The value for adhesion during braking is assumed to be 80 to 85% of that during hauling. Consequently, the braking effort is 80 to 85% of the tractive effort.

Stopping Distance and Time

The retarding force (Fr) is the sum of the braking effort, the train resistance, and the grade (combined with curve) resistance. An ascending grade aids retardation; a descending grade reduces retardation.

F_r = braking effort
$\quad$ + rolling resistance ± grade resistance (21)

Knowing the retarding force, the theoretical stopping distance, and the theoretical time to stop can be calculated as follows.

Kinetic energy of the train at the time of braking =
$$\tfrac{1}{2} mv^2 = \tfrac{1}{2} \frac{w}{g} v^2$$
where w = weight of the train, lb
$\quad v$ = velocity, ft/sec
$\quad g$ = acceleration due to gravity, ft/sec^2
$\quad\quad$ = 32.2 ft/sec^2

If S is the stopping distance (ft), and the retarding force is F_r (lb), then

$$\text{work done} = F_r \times S$$

and

$$F_r \times S = \tfrac{1}{2} \frac{w}{g} v^2$$

$$S = \tfrac{1}{2} \frac{w}{g} \frac{v^2}{F_r}$$

Expressing w and v in practical units of tons (W) and mph (V),

$$S = \tfrac{1}{2} \left(\frac{W \times 2000}{32.2}\right) \times \left(\frac{5280}{3600}\right)^2 V^2 = 66.8 \left(\frac{WV^2}{F_r}\right)$$

The rotary parts of the train also will have some kinetic energy. Assuming this to be approximately 5% of the linear component, the stopping distance can be increased by the same amount, i.e.,

$$S = \frac{70 \, WV^2}{F_r} \qquad (22)$$

The stopping distance S can also be expressed as the product of the average velocity during the stopping time, and the stopping time, t, i.e.,

$$S = \left(\frac{V}{2}\right)(t)$$

where t = stopping time, seconds.
Since

$$F_r \times S = \frac{1}{2} \frac{w}{g} v^2$$

$$F_r \times \frac{v}{2} \times t = \frac{1}{2} \frac{w}{g} w^2 \, v$$

$$t = \frac{wv}{gF_r}$$

As before, expressing the weight and velocity in practical units of tons and mph, respectively, and increasing the stopping time by 5% to account for the kinetic energy in the rotating components,

$$t = \frac{W(2000)}{(32.2)} \left(\frac{5280}{3600}\right) \left(\frac{V}{F_r}\right)(1.05) = \frac{95.6 \, WV}{F_r} \qquad (23)$$

In practice, the stopping distance and time vary due to many factors. The rate and time of propogation of brake application vary. The differences due to the human element are ever present. In fact, the above theoretical values should be increased by 25 to 30% for these and other safety reasons (Hay, 1982). Where the braking distances are critical, they should be established by tests.

The stopping distance and time formulas are generalized below for both acceleration and deceleration from one speed to the next.

If the train accelerates from a velocity V_1 to a velocity V_2

$$\text{Acceleration time, sec} = \frac{95.6}{NTE}(V_2 - V_1) \qquad (24)$$

$$\text{distance, ft} = \frac{70}{NTE}(V_2^2 - V_1^2) \qquad (25)$$

where NTE = net tractive effort available for acceleration, lb. If the train decelerates from a velocity V_2 to a lower velocity V_1,

$$\text{deceleration time, sec} = \frac{95.6}{F_r}(V_2 - V_1)$$

$$\text{distance, ft} = \frac{70}{F_r}(V_2^2 - V_1^2) \qquad (26)$$

where F_r = retarding force, lb.

Example 1: Calculate the size of a locomotive (weight in tons) to start and haul a load of 350 tons at 4 mph up a gradient of 1%. The track is 4 ft-$8\frac{1}{2}$ in. gage and has a minimum curve of 200 ft radius which coincides with the gradient. The wagons have a wheel base of 9 ft-6 in. The maximum acceleration specified is 0.1 mphps. Adhesion is 25%. Assume a train and locomotive rolling resistance of 20 lb per ton.

Solution: Let L be the weight of the locomotive in tons.

$$\text{Available tractive effort, lb} = \left(\frac{25}{100}\right)(L)(2000) = 500 \, L$$

Total resistance, lb
    = rolling resistance of the locomotive and the trailing loads
    + grade resistance of the locomotive and the trailing loads
    + curve resistance
    + resistance due to acceleration
    = $20[L + 350] + 20 \times 1 \times [L + 350]$
    $+ \dfrac{225 \, [4 \text{ ft } 8\frac{1}{2} \text{ in.} + 9 \text{ ft } 6 \text{ in.}]}{200} \times [L + 350]$
    $+ 100 \times 0.1 \times [L + 350]$
    = $65.98L + 7000 + 5593 + 7000 + 3500$
    = $65.98L + 23093$

$$500L = 65.98L + 23093$$
$$L = 53.21 \text{ tons} \approx \text{ select a 54-ton locomotive.}$$

The speed-tractive effort curve of the selected locomotive should be such that it will develop approximately 26,655 lb of tractive effort at 4 mph.

$$\text{Rail horsepower} = \frac{54 \times 0.25 \times 2000 \times 4}{375}$$
$$= 288.0$$

$$\begin{array}{l}\text{Hp input to generator} \\ \text{assuming 80\% efficiency}\end{array} = \frac{288.0}{0.8} = 360.0$$

$$\begin{array}{l}\text{Engine horsepower allowing} \\ \text{15\% for accessories}\end{array} = \frac{360.}{0.85} = 423.5$$

Example 2: Determine the number of 30-ton, four-axle cars that can be hauled by a 45-ton diesel electric locomotive at 10 mph up a 0.2% grade through an average track curve of 2°. The rolling resistance for the locomotive is 5 lb/ton and for the cars, 10 lb/ton. The speed-tractive effort curve of the 45-ton locomotive is shown in Fig. 7.

Solution: The tractive effort at 10 mph can be read from the curve in Fig. 7 as approximately 7500 lb.

Total resistance of the train = rolling resistance
                         + grade resistance
                         + curve resistance
Let N be the number of cars that can be hauled.
Total weight of the train = $(45 + 30 \, N)$
Rolling resistance = $(45 \times 5) + (30N)(10) = 225 + 300N$
Grade resistance = $(45 + 30N)(20)(0.2) = 180 + 120N$
Curve resistance = $(45 + 30N)(0.8)(2) = 72 + 48N$

$$\text{Total resistance} \qquad 477 + 468N$$

For balancing speed,
    Tractive effort = total resistance
$$7500 = 477 + 468N$$
Therefore, N = 15 cars.

Example 3: A locomotive weighing 45 tons is hauling 450 tons down a straight gradient of 1 in 100 at 4 mph. Percent adhesion during braking is 15. The rolling resistance is 12 lb/ton for the locomotive and for the trailing load. Calculate the stopping distance and the stopping time.

Solution: Retarding force, F_r = braking effort + rolling resistance ± grade resistance

The grade in this instance will assist the train movement, and decrease the retarding force.

$$Fr = \left[45 \times 2000 \times \frac{15}{100}\right] + [45 + 450][12]$$

$$- [45 + 450][2000]\left[\frac{1}{100}\right]$$

$$= 13500 + 5940 - 9900$$

$$= 9540 \text{ lb.}$$

$$S = \frac{70\,WV^2}{F_r} = \frac{70 \times 495 \times (4)^2}{9540} = 58.11 \text{ ft}$$

$$t = \frac{95.6\,WV}{F_r} = \frac{95.6 \times 495 \times 4}{9540} = 19.84 \text{ sec.}$$

When a train starts from rest, the tractive effort available is maximum. The net tractive effort is also maximum and contributes to accelerate the train. As the train accelerates and gains speed, the tractive effort decreases, and so does the net tractive effort; when net tractive effort is zero, balancing speed is achieved. The rate of change of acceleration is not uniform, and depends on the speed of the train and the resistance; therefore, the speed-time-distance calculations become important. A complete haul of a train may involve a series of acceleration and braking cycles. The following example illustrates the method of calculating the speed-time-distance data.

Example 4: Consider the case of a 65-ton locomotive whose speed-tractive effort chart is shown in Fig. 8. The train is powered by four traction motors. The ampere-tractive effort curve for one of the motors is shown in Fig. 9. The locomotive hauls eight 35 ton cars. The train has to haul up a grade of 0.5% on a straight track with no curves. The locomotive and train rolling resistances are 10 lb/ton. The train starts from rest and when a speed of 12 mph is reached, the power is cut off and the train is braked at a uniform rate of 1 mphps. Calculate the distance travelled, the elapsed time from start to stop, and the continuous current rating of the motor.

Solution:

Total load = (65 + 8 × 35) tons = 345 tons
Train resistance = 345 × 10 = 3450 lb
Grade resistance = 345 × 20 × 0.5 = 3450 lb
Curve resistance =   = 0 lb
Total resistance (excluding acceleration) = 6900 lb

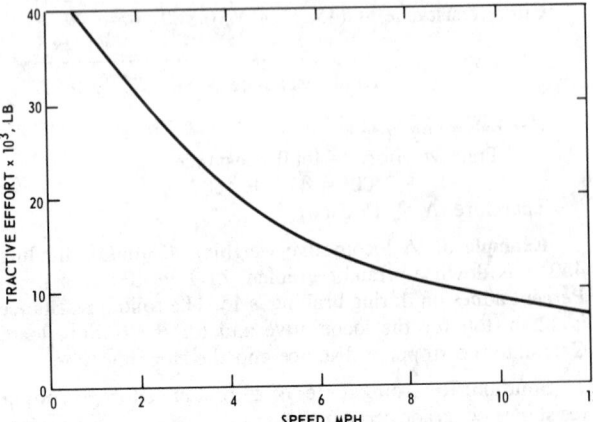

Fig. 8. Speed-tractive effort chart for a 59-t (65-ton) diesel-electric locomotive.

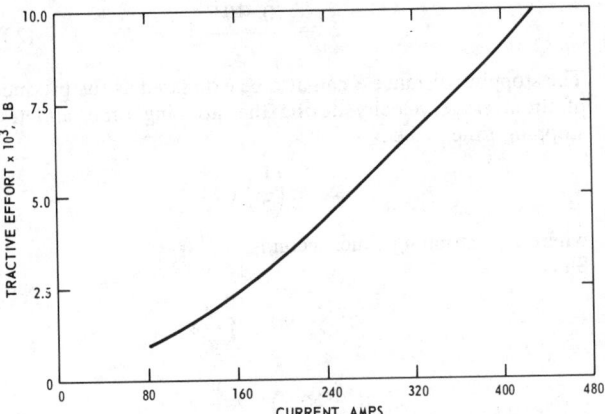

Fig. 9. Ampere-tractive effort curve for a tractive motor.

The speed-time-distance calculations for this problem are best done in a tabular form as shown in Table 2. It is convenient to increase the speed from rest in increments of 2 mph, though smaller or larger increments may be used for greater or lesser accuracy, respectively.

Since there are four motors on the locomotive, each motor contributes to one-fourth of the total tractive effort of the locomotive. The calculations for the continuous current rating are also best performed in a tabular manner as shown in Table 3. Entries in columns 1, 2, 3 and 6 of Table 3 are obtained from Table 2, columns 1, 2, 7, and 10, respectively. Entries in column 4 are obtained by dividing corresponding entries in column 3 by the number of motors, which is 4 in this case. Entries in column 5 are obtained from Fig. 9, using entries in column 4 as the entry points.

$$I^2 = \frac{I_1^2 d_{t_1} + I_2^2 d_{t_2} + I_3^2 d_{t_3} + \cdots}{T} \qquad (28)$$

where

I = the continuous current rating of the motor, amperes

d_{t_i} = the time required to traverse the track section at the average speed (see explanation for Column 10 in Table 2), sec

I_i = the current drawn by the motor for the tractive effort required, obtained from the ampere-tractive effort curve of the motor, amperes

T = the total elapsed time from the start and to the stop of the motor,

$$I^2 = \frac{3,304,038}{116.28} = 28414.50$$

$$I = 168.57 \text{ amperes}$$

In summary, from start to stop, the train took 116.28 seconds and travelled 1515.7 ft. The required current rating of each of the four motors should be greater than 169 amperes.

The above examples are meant to be illustrative of the application of the design formulas to planning and engineering problems. In practice, computerized calculation procedures are recommended for evaluating railroad designs. Railroad computer programs include not only speed-time-distance calculations but several other factors such as fuel consumption and continuous current rating calculations. There is no need to make simplifying assumptions for ease in calculations. Many alternatives to a railroad design problem can be evaluated. The design can be optimized with

Table 2. Speed-Time-Distance Calculations

1 Speed mph	2 Average speed mph	3 Train resistance lb	4 Grade resistance lb	5 Curve resistance lb	6 Total resistance lb	7 Locomotive tractive effort lb	8 Net tractive effort lb	9 Acceleration mphps	10 Time d_t sec	11 Time T sec	12 Distance d_d ft	13 Distance D ft
0										0		0
	1	3450	3450	0	6900	37500	30600	0.887	2.25		3.30	
2										2.25		3.3
	3	3450	3450	0	6900	25000	18100	0.525	3.81		16.8	
4										6.06		20.1
	5	3450	3450	0	6900	17500	10600	0.307	6.51		47.8	
6										12.57		67.9
	7	3450	3450	0	6900	12500	5600	0.162	12.35		126.8	
8										24.92		194.7
	9	3450	3450	0	6900	10000	3100	0.090	22.22		293.4	
10										47.14		488.1
	11	3450	3450	0	6900	8100	1200	0.035	57.14		922.0	
12										104.28		1410.1
	6	3450	3450	0	6900	—	—	−1.0	12.00		105.6	
0										116.28		1515.7

Column 1 - Speed in mph taken at 2 mph increments.
Column 2 - Average speed during speed change.
Column 3 - 4 and 5 are self explanatory.
Column 6 - The sum of columns 3, 4, and 5.
Column 7 - Locomotive tractive effort, obtained from the speed-tractive effort curve for the 65-ton locomotive at the speed shown in column 2.
Column 8 - Net tractive effort available for acceleration (column 7 minus column 6).
Column 9 - The acceleration obtained by dividing the net tractive effort available for acceleration by the product of 100 and the total weight of train with locomotive (see equation 10).

Column 10 - Time in seconds required to effect the change in speed shown in column 1, under the accelerating shown in column 9. The time is obtained by dividing the increment in speed in column 1 by the acceleration shown in column 9 (see equation 11).
Column 11 - The summation of time increments under column 10.
Column 12 - The distance in feet travelled in the time increment of column 10 at the average speed shown in column 2 (see equation 12).
Column 13 - The total distance covered, in feet.

Table 3. Continuous Current Rating Calculations

1 Speed, mph	2 Average speed, mph	3 Locomotive tractive effort, lb	4 Tractive effort per motor, lb	5 Current amperes, I	6 d_t, sec	7 $I^2 d_t$
0						
	1	37500	9375	410	2.25	378,225
2						
	3	25000	6250	300	3.81	342,900
4						
	5	17500	4375	240	6.51	374,976
6						
	7	12500	3125	190	12.35	445,835
8						
	9	10000	2500	170	22.22	642,158
10						
	11	8100	2025	140	57.14	1,119,944
12						
	6	—	—	—	12.0	0

$$T = 116.28$$
$$\Sigma I^2 d_t = 3,304,038$$

respect to any chosen parameter such as energy consumption and run time. Railroad simulation programs are available from several sources including manufacturers and consulting engineering companies.

EQUIPMENT AND TRACK

The pulling force to overcome the resistance to motion is provided by a locomotive. The pulling force or the tractive effort generated by a locomotive is a function of its horsepower and its speed:

$$TE = \frac{(375)(hP_e - hP_a)(\eta)}{(V)} \quad (29)$$

where

TE = tractive effort or pulling force, lb
hP_e = horsepower of the locomotive engine shaft
hP_a = horsepower to the auxiliaries
V = speed of the locomotive, mph
η = transmission efficiency, usually between 0.80 to 0.85

Using an η of 0.82,

$$TE = \frac{(308)(hp)}{V} \quad (30)$$

where hp is the manufacturer's rated horsepower for the locomotive.

Industrial locomotives may be for gages from 18 to 66 in., and railroad locomotives may be for gages of 24 to 66 in. The profile of the track, the maximum grade to be encountered, the number of cars and the weight of the cars will determine, as explained before, the size of the locomotive. There are many types of railroad cars—gondola cars, box cars, hopper cars, tank cars, etc. The cars can be bottom-dump or side-dump; about 80% of the cars in use today are bottom-dump hopper cars. Gondola cars are restricted to hauling ore to the crushers. Cars to haul ore and waste in open pit mining are built more substantially than regular cars and may require special mechanisms to facilitate dumping. Ample tractive effort must be available to pull the train over the grade without slipping, which occurs when the tractive effort at the wheels exceeds the available adhesion between the wheels of the locomotive and the rail. Once a locomotive has been selected, the train size will be dependent on operational factors such as the overall adhesion and the condition of the track.

There are many possible prime movers for the locomotive. The common prime movers are diesel, diesel-electric, and electric. Steam engines were widely used in the past. Efforts are continuing to develop new prime movers which are in experimental stages such as steam turbine, oil-gas, and coal-gas turbines. Discussion here will be limited to diesel-electric and electric locomotives.

A diesel-electric locomotive is essentially an electric locomotive with a self-contained power plant, the diesel engine-generator combination. Compared to diesel-electric, an electric locomotive has more horsepower per ton weight ratio and greater adhesion. Electric models (25 kV or 50 kV, 60 Hz) with over 6000 hp are available. They can be of four or six axle designs. Diesel electrics provide 3000 to 3600 hp per unit and there are a few 5500-hp experimental units. An electric locomotive is obviously limited to those tracks with a power supply, a third rail or a catenary. An electric locomotive costs considerably more than a diesel-electric but it has a much longer life, nearly twice that of the 15 to 20 years first life (before rebuilding) of the diesel engine. The

availability of the electrics is also higher, about 95%. Diesel-electrics availability will average about 85% because of servicing and engine repairs. Overall maintenance costs for electrics can be one-half to one-quarter of those for diesel electric. With electrics, there is no fuel storage and handling, and no water treatment or freeze-up problems. The electric operation is clean and nonpolluting.

There are a number of requirements with electrification. Direct current (D.C.) locomotives are not generally available and are usually of a special design. Rail bonds have to be increased in size and interference problems with other wayside systems must be studied. The interference factors with the power supply such as unbalanced voltage or current (because the locomotive is a single-phase load), power factor (varies with locomotive design and catenary impedance), harmonic currents (due to phase control system on the locomotive), and peak power demand must be studied (Henderson, 1983). The choice of diesel-electric or electric locomotive is very application dependent and cannot be generalized. Diesel-electrics have proven operating experience in industrial and mining applications. In several surface coal mines, however, the transport of coal from the mine to the power plant is achieved with dedicated high-speed electric railroads. Table 4 lists some salient statistics on rail haulage applications in open pit mines.

The railroad track is a complex structure put together with a few components. It supports the train load and guides the train movement. Safety, dependability, and performance of a railroad are determined by the condition of the track, which is subject to a variety of static and dynamic loads. It is a flexible load-distributing structure requiring great care during construction on such matters as rail size, ties, alignment, subgrade, ballast, drainage, track surface, track geometry, and rail joints. Reference has already been made to track grades. In mines, track grades may go as high as 8 to 9%. Track curves may be 35-ft in radius (degree of the curve = 5730/35 = 164°). One way to reduce the centrifugal force on a curve is to reduce the train speed on the curved track. Another way is to raise the outside rail as much as 4 in. over the inside rail around the curve. This height of the outside rail above the inside rail is called *superelevation*. The superelevation of the outside rail can be calculated from the formula,

$$h = \frac{av^2}{gR}$$

where

h = superelevation, ft
a = the rail gage, ft
v = velocity, ft/sec
g = acceleration due to gravity, ft/sec^2
R = radius of the curve, ft.

Expressing the velocity in mph (V), h in inches, and the curve in degrees $\left(D = \dfrac{5730}{R}\right)$,

$$h = \frac{a \times \left(\dfrac{5280}{3600}\right)^2 \times V^2}{32.2 \times \left(\dfrac{5730}{D}\right)} [12] = 0.00014aDV^2 \quad (31)$$

where

h = superelevation, in.
V = speed, mph
D = degree of the curve.

Table 4. Surface Mines Utilizing Rail Haulage, 1966

Name of mine	Name of Company	Location	Type of ore	Locomotive data type	hp	wt (tons)	Car sizes (st)	Cars per train	Average haul distance loaded (ft)	Continuous grade (%)
Chuquicamata	Chile Exploration Co.	Chuquicamata Chile, SA	Copper	Diesel-electric	1200 1750	120&190 (used as pushers)	68.5	14	10,000 (ore and waste)	+3
Erie	Erie Mining Co.	Hoyt Lakes, Minnesota, US	Tacopite	Diesel-electric	1200 1800 2000 2400	124 130 134 128	95	9	39,000	+2
Kolwezi Operations	Union Miniere du Haut Katanga	Republic of the Congo (now Zaire), Africa	Copper	Electric	640	70	44	6	16,000	?
Morenci	Phelps Dodge Corp.	Morenci, Arizona, US	Copper	Diesel-electric	1750 1800	130	83	11	21,000	+3 & −4
New Cornelia	Phelps Dodge Corp.	Ajo, Arizona, US	Copper	Diesel-electric	1750 2250	125&140	65&62 65&62	7 8	22,000 22,000	+3 +3
Bingham	Kennecott Copper Corp.	Bingham, Utah, US	Copper	Electric	1200 1600	85&125	86.5 (ore) 73 (waste)	14-21 (ore) 8 (waste)	35,000 (ore) 25,000 (waste)	−5 −5
Toquepala	Southern Peru Copper Corp.	Toquepala, Peru, SA	Copper	Diesel-electric	1800	135	83	10	17,000 (ore) 13,000 (waste)	−4
Kounrad		Kazakhstan, USSR	Copper	Electric	?	150	95&100	5	? (ore) 18,000 (waste)	+3

A combination of reduced speed and superelevation is used to counteract the centrifugal forces on curves.

MINE RAILROAD SAFETY

Mine railroads present unique operational and maintenance problems. Many commercial railroad practices are not directly applicable to mine railroads because of the conditions under which the tracks and equipment are used. The duty cycles for the mine locomotives and ore and waste cars are different. Within the mine, the track location is semipermanent. Frequent track shifts and additions and deletions are necessary. On the waste dumps, the track will be shifted to a new crest when dumping becomes increasingly difficult from existing locations. The subbase material for mine railroads may be weak, particularly on the waste dumps. The ballast material is usually crushed rock from the mine. Fortunately, mine trains operate at comparatively low speeds. However, the loss of safety and efficiency that can result from poor railroad systems requires that great care be exercised in the design, construction, inspection, and maintenance of mine railroads. Also, records of inspections and all track and train problems are essential for improvement of track safety. The discussion hereafter generally follows that by Brauns and Orr (1968).

Rolling Stock Maintenance

Adequate repair shops for locomotives and cars are essential. Except for its arrangement, a locomotive shop should be as well-equipped as a modern repair shop for off-highway trucks. Other maintenance work on locomotives logically divides itself into two horizons. That on brake systems, traction motors, etc. is most easily performed at track level. Work on the engines, generator, air compressor, and other components mounted atop the frame is best performed from the level of the frame or deck. Platforms at that level are a great convenience to mechanics. Minor repairs and overhauls are best accomplished in a shop building designed for the purpose. Automatic welding equipment for body work is very desirable. For any extensive body repair or rebuilding work, equipment for cutting and forming large steel plates is essential.

Track Maintenance

Track maintenance is vital to an efficient rail haulage system. Good alignment and surface on main line tracks is necessary if trains are to travel rapidly. Rough track not only slows the train but contributes greatly to the damage of rolling stock, crossties, and rail wear. This is normally the largest element of rail haulage cost, sometimes amounting to as much as 50% of that cost.

Guidelines for the construction, inspection, and maintenance of underground rail haulage track are presented in a research report to the US Bureau of Mines (Cunney, Rudd, and Hawkins, 1977). Many of these guidelines should be useful to the surface mine rail haulage tracks as well. A track maintenance program is usually offered by a manufacturer of rail products (Anon., 1980). The program includes a system for identifying existing and potential trouble spots, and setting timetable priorities for eliminating them.

A number of track maintenance machines have been developed for the common carrier railroads in recent years which are very useful in pit track work. Some of these are tie tampers, power jacks, ballast distributors and shapers, track liners, tie spacers, spike drivers, and bolt tighteners. Tampers and lining machines have recently been provided with controls capable of bringing track automatically to a present grade and line.

Rail Waste Dumps

These are more expensive to establish and maintain than are truck waste dumps. A grade must be prepared upon which track can be laid to initiate the dump. The initial track should be at least a train length long, and it is preferable that it extend for several train lengths. Dump height is largely dictated by the type of terrain available for waste disposal. Dumps in mountainous areas are likely to be high, while those in areas of little relief are generally low. High dumps require less frequent track throws, but they are more susceptible to settlement. Preventing the track from developing excessive grades as a result requires continual work. Rail dumps routinely develop a berm along their crests as the dumped material builds out from the track. When the berm becomes sufficiently high to seriously impede the action of muck sliding out of a dump car, the dump is said to be *plugged*. It is then necessary to blade off the berm and shift the track out to the new crest.

Waste trains are commonly pushed onto dumps rather than pulled. This is a precaution to save the locomotive from the likelihood of derailment and from the rare possibility that the dump crest may slough. More derailments occur on dumps than in any other part of a pit track system. Track quality is necessarily only fair because the track receives considerable abuse from frequent moves and derailments. Material covers the outside rail as the plug develops, frequently causing cars to jump the track.

Traffic Control Systems

For two or more trains to operate on a single track safely, some means of regulating traffic is needed. Simple manually operated signals suffice where only two trains are involved, or in the case of one train operating under conditions of restricted speed and clearance, as when track is being repaired. Automatic block-signal systems are a very effective means of regulation. Blocks are sections of track electrically bonded so as to form complete circuits, and these are connected to light signals. In single-track territory, the signal indications in the blocks both in front of and behind an occupied block will likely be stop. The signal in the block behind may display an approach aspect, meaning proceed at reduced speed and be prepared to stop at the next signal. Block signals on a single track can control traffic in one direction only. Two-way traffic can be accommodated if passing tracks are located at convenient intervals, and for most efficient operation two tracks are required. Many mine trains are equipped with two-way radios for communications between engineman, supervisors, and maintenance men. These radios also afford a means of traffic control.

Where train traffic is dense, a centralized traffic-control (CTC) system is superior to all other types of systems. CTC is unequalled in its ability to handle trains rapidly and safely over a given stretch of track. The operator, working from a control panel containing a diagram of the tracks in the system, selects the route of each train. Lights on the panel indicate the positions and movements of the trains as well as the aspects of the signals and positions of the switches. Signals either along the wayside, in the locomotive cab, or sometimes both, convey information to the locomotive engineer how a train is to proceed along the route.

Automated Operations of Trains

Within the past few years there have been revolutionary developments in the field of automatic train operation. Much experimental work is being conducted with a view to improving automation of commuter trains. Undoubtedly this will also benefit rail haulage in making available to mine managers a choice of several modes of automatic train control.

Three types of systems are remote control, supervisory control, and programmed control. At least two of the types have been applied to surface mine railroads. Radio remote-control systems are in use at the Morenci and New Cornelia mines. Wired remote control is more easily applied to electric locomotives. Supervisory control has train movement controlled by dispatcher who directs train movements between various points. The operator directs the train movements from point to point but the protection against conflicting moves, the speed of train movement, and train safety are built into the system, thus freeing the dispatcher for major operations only. The programmed-control system has no planned manual supervision. Some systems are in operation in which a sequence of events is performed during a complete cycle while in others, part of a cycle must be completed before the rest of the cycle can proceed. Provision is made for overriding control of the program so that an operator or maintenance man can correct a defect or stop operation at any time.

RAILROAD COSTS

The procedures for estimating the costs of a surface mine railroad are no different from those used for any other industrial activity. The basic requirements for getting dependable results are (1) developing accurate production data, (2) performing good engineering and design calculations, (3) estimating correctly the capital and operating costs, and (4) following a well-accepted costing procedure. Cost information from prior projects can be useful in the estimation process but too much reliance should not be placed on these. Railroads have a large initial investment in tracks, yards, structures, signals, and car and locomotive fleet. Significant cost changes from prior years can therefore occur due to changes in technology. Secondly, major railroad jobs like earthwork and trackwork are very site dependent. The influence of the site conditions, such as topography and slope, transcends both capital and operating costs. The construction costs per mile of track can vary from $100,000 to $1,000,000 or more.

The election of diesel-electric vs. electric locomotives has similar effects. The initial cost of a 65-ton to 144-ton, four axle diesel-electric switching locomotive with 600 to 1100 hp motor is between $500,000 to $700,000. Diesel-electric surface traction locomotives cost from $900,000 to $1,200,000 each. The lower maintenace, greater availability, and larger life of electric locomotives combine to lower the number of electrics required for a given production. However, electrics initial costs are higher. A 6000 hp, 190-ton electric locomotive costs nearly $2,000,000.

It is also important to develop appropriate units for evaluating costs of the various alternatives. Unit costs used by the US railroad companies include the cost per train mile, the cost per revenue ton mile, cost per locomotive mile, and cost per track mile maintained. For commercial railroads, accounting rules are prescribed by the Interstate Commerce Commission (ICC). ICC statistics may be useful for estimating some operating costs such as those for track maintenance (Poole, 1962).

In practice, good planning, engineering and design procedures should determine the various construction requirements, equipment specifications, personnel requirements, operating performances, and usage of principal supplies. For example, earthwork and trackwork must be calculated from the engineering drawings. Equipment requirements must be based on the production and mining plan. Fuel consumption should be obtained from the speed-time-distance calculations. Costs associated with the maintenance of rolling stock (locomotives and cars) and track account for a major fraction of the total operating costs, as much as 60 to 80%. Correct estimation of these is important. The reliability of the cost estimates can be greatly improved by working closely with manufacturers and suppliers. The following estimating procedure is useful to determine the cost of ownership and operation.

Cost of Ownership

The capital investment in a mine railroad will include cash outlays for such items as right-of-way acquisition, tracks, structures, signals, locomotives, and cars. The cost of railroad cars is between $40,000 to $50,000 per car. The cost is variable depending on the type of car and attachments put on it to facilitate loading, hauling, and dumping. The investment will consist of the labor, equipment, and materials associated with the following:

1. Land surveys and all construction engineering.
2. Land and land damages, including right-of-way acquisition.
3. Earthwork such as clearing and grubbing, earth and rock excavation.
4. Construction of culverts, bridges, and trestles.
5. Trackwork such as rails, ties, and switches.
6. Buildings such as repair shops, engine houses, and tool houses.
7. Communications and signalling.
8. Equipment, locomotives, cars, and any other specialized equipment such as those for track inspection.

The costs of ownership consists of two components, (1) depreciation of the investment due to wear and tear during use, and (2) cost of interest, taxes, and insurance.

Depreciation: The calculation of depreciation requires an estimation of the useful life of the investment item, i.e., an estimation of the life of the track, locomotives, cars, etc. In the United States the Internal Revenue Service provides depreciation guidelines and rules. For such preliminary cost estimation as discussed here, it is sufficient to use a straight line depreciation with no allowance for investment tax credit. Annual depreciation costs for each capital item will be calculated as follows:

$$D_i = \frac{I_i}{L_i} \tag{32}$$

$$D = \sum_{I=1}^{N} D_i \tag{33}$$

where

D_i = annual depreciation on capital item i, dollars
I_i = total investment in capital item i, dollars
L_i = life of item i, years

D = annual depreciation, dollars
N = number of capital items ($i = 1, 2, \ldots N$)

Interest, Taxes, and Insurance: The interest costs account for the cost of using capital to construct the railroad and to purchase equipment and facilities. As a result of ownership, there will also be costs of insurance and property taxes. Annual interest, taxes, and insurance costs are usually calculated as a percentage of the average investment in the equipment. The average investment is based on the number of years used for depreciation.

$$P = \text{Int} + \text{Tax} + \text{Ins}$$

$$T_i = \left(\frac{1}{2}\right)\left(\frac{L_i + 1}{L_i}\right)\left(\frac{P}{100}\right)(I_i) \tag{34}$$

$$T = \sum_{i=1}^{N} T_i \tag{35}$$

where

P = factor for interest, taxes, and insurance, percent
Int = interest rate, percent
Tax = property taxes, percent
Ins = insurance rate, percent
T_i = annual interest, taxes, and insurance costs for the capital item i, dollars
T = annual interest, taxes, and insurance costs, dollars.

Yearly cost of ownership is given by

$$R = D + T \tag{36}$$

where R = yearly ownership costs, dollars.

Operating Costs

According to an analysis of the 1978 US railroad operating costs, maintenance of equipment accounted for 26% of the total operating costs; maintenance of way and structures, 19%; transportation costs, 46%; and general and administrative expenses, 9%. Although these figures are not applicable to mine railroads, this breakdown provides a rational basis to categorize mine railroad operating costs.

1. Maintenance of way and structures includes all costs of maintaining the *fixed plant* such as tracks, terminals, and signals. Materials and labor constitute the major cost items.
2. Maintenance of equipment covers the repairs to locomotives, cars, and maintaining workshop facilities.
3. Transportation category includes all costs associated with running the trains, yards, terminals, signals, and other expenses involved in the the direct operation of the railroad. Labor and locomotive fuel, in that order, are the most important cost components in nonautomated railroads.
4. General and administrative costs are expenses associated with the administration of the railroad. In a mine railroad, this may not be an important cost center.

The major operating cost categories in a mine railroad in increasing order of importance are traffic control, power, locomotive repair, car maintenance, locomotive operation, and track maintenance. Each of these cost categories is both significant and identifiable. Costs records should also be broken down into the traditional operating cost categories of labor, supplies, power, and administrative and overhead costs for each category.

For example, data from an open pit mine (1984 costs) reveal that the rail haulage operating cost per ton hauled, calculated on the basis of nearly 40 million annual tons, was approximately $0.50. The average one-way haul distance was 4.5 miles. The trains consisted of 12 cars, each 45 yd^3 in capacity, hauled by a 2000 hp diesel electric locomotive. The cost per ton-mile averaged $0.12. Of the total cost of $0.50 per ton, the labor costs were approximately 20¢/ton, the supplies, 15¢/ton, and general and administrative costs, the remaining 15¢/ton. Another breakdown of this $0.50 was as follows: energy costs/ton, 15%; labor and supplies costs for maintenance of rolling stock, 15%; labor and supplies costs for maintenance of the way and structures, 33%; and cost of operating labor such as engineers, foreman, and supervisors, 37%.

An example will illustrate the application of the basic design formulas and costing procedure to the selection of a rail haulage system for an open pit mine. The example is contrived and simplified. It is presented only to outline the procedure.

Example 5: Given the following data, perform a job analysis for the rail haulage system, select the equipment, and calculate the ownership and operating costs.

General Data
ore reserve: 250 million tons
waste: 250 million tons
ultimate depth of pit: 800 ft
working days per year: 250
ore production per day: 30,000 tons
waste production per day: 30,000 tons
shift length: 8 hours (450 minutes working time)
number of shifts/day: 3

Average haul data for a trip
mining track: 1000 ft, 0%, average speed 5 mph
main line ramp: 26,640 ft, 3% against load, average
   curvature 10°, 800 ft lift, average speed 10 mph
rim of pit to dump: 8000 ft, 0%, average speed
   25 mph for the first 6000 ft, and 5 mph for the
   last 2000 ft
speed of an empty train in good track: 35 mph;
   maximum downhill speed is 20 mph.

Equipment Data
side dump cars: 40 yd^3 or 80 ton capacity
empty weight: 40 tons
rolling resistance of cars: 5 lb/ton
locomotive rolling resistance: 5 lb/ton
curve resistance: 0.8 lb/ton per degree of curve
train size: 7 cars
loading time per car including spotting time:
   5 min
dump time per car: 1 min
delay time per trip: 5 min
locomotive availability: 0.9
car availability: 0.83
cost of a locomotive: $500,000
cost of a car: $40,000

Seven miles of track grade construction work has to be done at a rate of $50,000 per mile. The track construction cost on the working levels and dumps is $120,000 per mile. It is estimated that 25 miles of track are to be laid on the dumps and the working levels. The mainline is a double-track (length = 26,640 ft + 6,000 ft = 32,640 ft), and construction costs average $200,000 per mile. Additional investment in depreciable equipment (signals, maintenance machines, etc.) is $2,000,000. It is proposed to use a 20-year useful life for all investment and 15% of the average annual investment for interest, taxes, and insurance. The cost of a gallon of fuel-oil is $1.00, and each gallon will produce 11,500 ft-tons of gross work. Assume that fuel cost represents 20% of the operating costs.

Solution: The most severe condition for the locomotive haulage must be identified. The locomotive will be selected to haul the train up a 3% grade, 10° curve, at 10 mph.
Locomotive Selection:
weight of a loaded car = 40 + 80 = 120 tons
total weight of cars = 7 × 120 = 840 tons
let W = weight of locomotive, tons
total resistance to be overcome = tractive effort
   required = $W(5 + 20 \times 3 + 0.8 \times 10)$ +
   $840(5 + 20 \times 3 + 0.8 \times 10) = (73W + 61320)$ lb.

Assuming that at a running adhesion of 25% a speed of 10 mph is possible,

$$(0.25)(2000)(W) = 73W + 61320$$
$$W = 143.61 \text{ tons}$$

Choose a 144-ton locomotive for each train.

Cycle Time of a Train: The cycle time is the sum of the loading, travel loaded, dump, travel empty times, and delay time per trip.

loading time, 7 cars @ 5 min/car	=	35.00 min
travel loaded time:		
1000 ft @ 5 mph = 2.27 min		
26,640 ft @ 10 mph = 30.27 min		
6000 ft @ 25 mph = 2.73 min		
2000 ft @ 5 mph = 4.55 min	=	39.82 min
dumping time @ 1 min per car	=	7.00 min
travel empty time:		
2000 ft @ 5 mph = 4.55 min		
6000 ft @ 35 mph = 1.95 min		
26,640 ft @ 20 mph = 15.14 min		
1000 ft @ 5 mph = 2.27 min	=	23.91 min
delay time per trip	=	5.00 min
cycle time per trip	=	110.73 min

Number of Locomotives and Cars Required:

$$\text{number of trips per train} = \frac{450 \text{ min/shift}}{110.73 \text{ min}} = 4.06 \text{ trips}$$

Assume that a train will make four trips per shift.
production per train per shift = 4 × 7 × 80 = 2240 tons

$$\text{number of trains required/shift} = \frac{20,000 \text{ tons/shift}}{2240}$$
$$= 8.93 \approx 9.$$

Since the locomotive availability is 0.9 and the car availability is 0.83,

$$\text{number of locomotives required} = \frac{9}{0.9} = 10.$$

$$\text{number of cars required} = \frac{9 \text{ trains} \times 7 \text{ cars/train}}{0.83} = 76$$

Fuel Consumption: The total work done in this haulage problem is calculated with the following assumptions:

1. the average lift for the load is 400 ft, since the final pit depth is given as 800 ft;

2. the average one-way haul to the dump is assumed to be 3 miles from the midpoint in the pit;

3. in the return trip, excepting for the first 8000 ft, the favorable gradient will power the trip; and

4. the rolling resistance for the train is assumed to be 10 lb/ton here.

The work done by a locomotive in a haul cycle is calculated as follows:

total load on a trip = weight locomotive + trailing loads
$= 144 + 7 \times 120 = 984$ tons

work done to lift the load, 400 ft $= 984 \times 400$ ft-tons
$= 393,600$ ft-tons

work done to overcome the rolling resistance for 3 miles

$$= \frac{984 \times 3 \times 5280 \times 10}{2000} = 77,933 \text{ ft-tons}$$

work done to overcome the rolling resistance of 800 ft of track in the empty trip (weight of empty trip)

$$= 424 \text{ tons}) = \frac{424 \times 8000 \times 10}{2000} = 16,960 \text{ ft-tons}$$

total work done in a trip $= 488,493$ ft-tons

fuel consumption for a trip

$$= \frac{488,493}{11,500} = 42.48 \text{ gallons}$$

fuel oil consumption per day (4 trips/locomotive, 9 locomotives/shift, 3 shifts/day)
$= 42.48 \times 4 \times 9 \times 3 = 4588$ gallons

Cost Analysis: The cost analysis will be performed on a cost/hour and cost/ton basis.

hourly production, tons $= \dfrac{\text{shift production}}{8 \text{ hr}} = \dfrac{20,000}{8}$

$= 2500$ tons

The capital costs are the sum of investments in track grade construction, track construction, locomotive, cars and other items.

track grade construction, 7 miles @ 50,000/mile	= $ 350,000
track construction, 25 miles @ 120,000/mile	= $ 3,000,000
double track construction (6.18 × 2) miles @ 200,000/mile	= $ 2,473,000
locomotive cost, 10 @ $500,000/each	= $ 5,000,000
cars costs, 76 @ $40,000/each	= $ 3,040,000
additional investment (depreciables only)	= $ 2,000,000
	$15,863,000

Cost of Ownership: This includes depreciation using the 20-year useful life, and the charges for interest, taxes, and insurance.

hours of work in 20 years assuming 8 hr/shift, 3 shifts/day, 5 days/week, 50 weeks/year = 120,000

depreciation cost/hour $= \dfrac{\$15,863,000}{120,000} = \$132.19$

interest, taxes and insurance/hour (assuming 6000 hour/year) $= \left(\dfrac{21}{2 \times 20}\right)(15,863,000)\left(\dfrac{15}{100}\right)\left(\dfrac{1}{6000}\right)$
$= \$208.20$

total cost of ownership/hour $= \$132.19 + \$208.20 = \$340.39 \approx \$341$.

total cost of ownership/ton $= \dfrac{\$341}{2500} = \$0.14$

If it is figured that taxes and insurance amount to 5%, then the $0.14 can be broken down as follows: depreciation, $0.05; taxes and insurance, $0.03; and return (or interest) on average investment, $0.06.

Operating Costs: The total operating costs in this example are calculated as a factor of the fuel-oil costs. Recapping the previously calculated figure, the fuel oil consumption per day was 4588 gallons.

fuel-oil cost/day, 4588 @ $1/gallon = $4588

total operating cost/day $= \dfrac{\$4588}{0.20} = \$22,940$

operating cost/hour $= \dfrac{\$22,940}{24} = \$955.83 = \$956$

operating cost/ton $= \dfrac{\$956}{2500} = \$0.38$

total ownership and operating cost/hour $= \$341 + \$956 = \$1297.00$

ownership and operating costs/ton

$$= \frac{1297.00}{2500.00} = \$0.52$$

It may be necessary now to perform sensitivity analysis on such factors as the size of the trip, the grade of the track, the size of the cars, and the speed of the train. The simple costing procedure outlined here can be very useful for such an analysis. More detailed cost analysis incorporating investment tax credit, accelerated depreciation schedules, and financing considerations will be necessary in the final stages of the design and selection of the rail haulage system.

REFERENCES

Anon., 1969, "Automated Railroad Carries Coal," *Coal Age,* Vol. 74, No. 12, pp. 62-67.

Anon., 1976, "The Application of Modern Transport Technology to Mineral Development in Developing Countries," Centre for Natural Resources, United Nations, New York, pp. 115-131.

Anon., 1980, "Industrial and Mine Track Maintenance," *Engineering and Mining Journal,* Vol. 181, No. 9, pp. 150-152.

Anon., undated, *What Size Diesel Shunter?* Plymouth Locomotive Works, Plymouth, Ohio, 15 pp.

Brauns, J. W. and Orr, D. H., Jr., 1968, "Railroad," Section 9.1, *Surface Mining,* E. P. Pfleider, ed., AIME, New York, pp. 531-552.

Cunney, E. G., Rudd, T. J., and Hawkins, S., 1977, "Construction, Inspection and Maintenance of Mine Haulage Track." Vols. I, II, and III, Final Report to US Bureau of Mines, Washington, DC, 185 pp.

Davis, W. J., Jr., 1926, "Traction Resistance of Electric Locomotives and Cars," *General Electric Review, 29,* pp. 685-708.

Hay, W. W., 1982, *Railroad Engineering,* John Wiley & Sons, New York, 758 pp.

Henderson, H., 1983, Personal communication, General Electric Co., Erie, PA.

Hoppe, R. W., 1977, "Open-pit Mining in Arizona," *Engineering and Mining Journal,* Vol. 178, No. 6, pp. 95-106.

Hoppe, R. W., 1977a, "Sangaredi—An African Plateau of Bauxite," *Engineering and Mining Journal,* Vol. 178, No. 8, pp. 83-90.

Jackson, D., 1980, "Keeping Equipment on Line and Available," *Engineering and Mining Journal,* Vol. 181, No. 9, pp. 70-75.

Poole, E. C., 1962, *Cost—A Tool for Railroad Management,* Simmons-Boardman Publishing Corp., New York.

Staley, W. W., 1949, *Mine Plant Design,* McGraw Hill, New York, pp. 420-457.

Szklarski, L., Dudek, W., and Machowski, J., 1969, *Underground Electric Haulage,* Pergamon Press, New York, 409 pp.

6.5.2. Trucks

Ronald M. Hays

GENERAL APPLICABILITY

Trucks designed exclusively for off-highway use are the principal equipment for material transport in surface mines. In the mid-1930s, off-highway trucks were introduced to the mining industry; the first trucks of approximately 13.6-t (15-ton) capacity replaced highway trucks modified for mine haulage. By the mid-1950s, 22.7- to 27.2-t (25- to 30-ton) off-highway trucks were in common usage and trucks were available with 54.4-t (60-ton) capacity. In the late 1950s, single rear-axle trucks were introduced. During the 1960s and 1970s, bigger and better off-highway trucks were developed, with a 318-t (350-ton) capacity truck introduced in the mid-1970s.

Large and efficient off-highway trucks make possible the development of large low-grade ore and coal deposits. These deposits, with high stripping ratios and large quantities of both waste and ore or coal, are minable because of the economies of scale achieved with large off-highway trucks. Due to greater truck capacity and the resultant increased capital and operating costs, it has become increasingly important to analyze haulage requirements.

OFF-HIGHWAY TRUCK DESCRIPTIONS

Off-highway trucks can be classified into three main types: (1) conventional rear dump; (2) tractor-trailer, bottom, side, and rear dump; and (3) integral bottom dump. These main truck types are illustrated in Fig. 1.

The conventional rear dump is the dominant off-highway truck for surface mines. These trucks, with rigid frame and heavy duty dump body, are suitable for hauling almost all materials. The dump body mounted on the truck chassis is raised by an integrally mounted hydraulic hoist system. Trucks are powered by onboard diesel engines with either mechanical or electric wheel drives. Mechanical drives are available for trucks up to 118-t (130-ton) capacity while electric wheel drives are available in the 77.1- to 318-t (85- to 350-ton) capacity range.

Most conventional rear dump trucks have two axles with dual tires on the rear-drive axle. A two-axle truck is manufactured with both front and rear axle drive and dual tires on both axles. Some trucks are manufactured with three axles. These may have single or dual tires on the rear tandem axles and may be driven by one or both rear axles. Some small-capacity, three-axle trucks have articulated frames with front drive, rear drive, and tag axles, all with single tires.

Tractor-trailer units have a separate tractor as prime mover and a trailer connected by an articulated hitch. The tractors used for conventional rear dump trucks can be modified by removing the dump body and replacing it with a frame-mounted hitch. Tractors usually are powered by diesel engines and have mechanical drives; however, units with electric wheel drives are available. These units commonly have three axles: two on the tractor and one on the trailer. The rear-drive axle on the tractor and the trailer axle usually have dual tires.

Smaller tractor-trailer units up to 54.4-t (60-ton) capacity are available in the *overhung* or two-axle design. A scraper tractor with a single axle is used as the prime mover. This unit has greater maneuverability than the three-axle unit.

Tractor-trailer units are designed for rear, side, or bottom dump. Rear dump units commonly are used for ore or rock haulage, but also are available for lighter materials such as coal. These units usually have an integral hydraulic hoist system with collapsible wheel base. Thus, when the unit is dumped, the drive and trailing axles collapse closer together as the axle with brakes locked remains stationary.

Side dump tractor-trailer units are used for ore or rock haulage. These units may have either spillover dumping or drop-side dumping gates which carry material away from the tires. The dump body is raised by either an integral hydraulic hoist system or an external hoist or skyhook. Units with internal hydraulic hoist systems have a greater net vehicle weight (NVW), but can discharge anywhere. With an external hoist, units can be dumped only at hoist locations such as at a crusher. Units needing an external hoist usually are not as versatile because of dumping limitations.

Bottom dumps are the most common tractor-trailer units. These units are limited to free flowing material with trailers designed to haul specific materials such as alluvial material

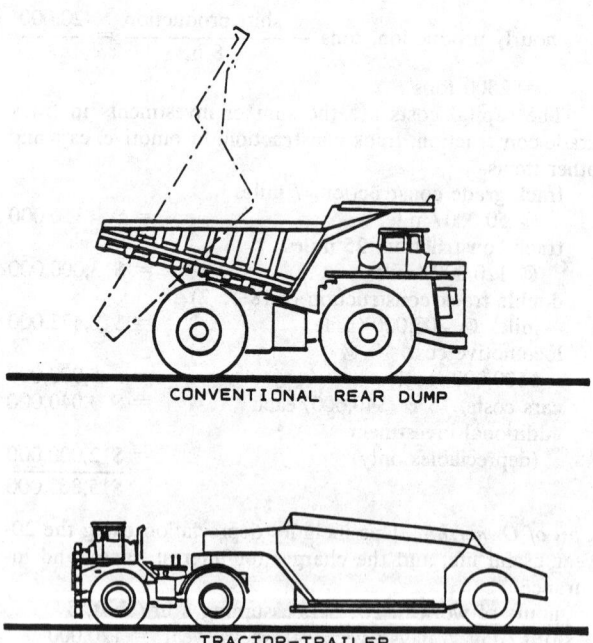

CONVENTIONAL REAR DUMP

TRACTOR-TRAILER

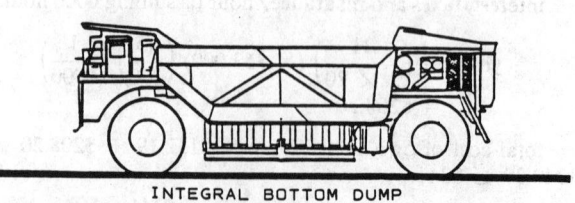

INTEGRAL BOTTOM DUMP

Fig. 1. Large off-highway truck types.

or coal. Material is dumped through hydraulically operated doors designed for controlled opening and closing and for maximum ground clearance when open.

A wagon (dolly) or second trailer sometimes can be used with side and bottom dump tractor-trailer units. These wagons have two axles with dual tires usually on both axles.

Integral bottom dump trucks with rigid frame and large volumetric capacity are available in the 136- to 154-t (150- to 170-ton) capacity range for hauling coal. These trucks are powered by diesel engines with either mechanical or electric drives. These trucks have two axles with single tires on both front and rear; single tires on front and dual tires on rear; or dual tires on both front and rear. Dump doors are hydraulically operated. This truck design eliminates the hitch, gooseneck, and one axle, usually providing a high payload to gross vehicle weight (GVW).

Advantages and disadvantages of the three main truck types are presented in Table 1.

Drive Components

Engines used in off-highway trucks are either high speed (1800- to 2100-rpm) or low speed (900- to 1000-rpm) diesel engines. High-speed engines with outputs up to 1342 kW (1800 hp) are used for trucks with capacities up to 200 t (220 ton). Low-speed engines with outputs in the 1230- to 2461-kW (1650- to 3300-hp) range have been used for very large trucks in the 181- to 318-t (200- to 350-ton) capacity range.

High-speed diesel engines are either two-cycle or four-cycle engines. In recent years, manufacturers have improved engine reliability, durability, performance, and efficiency. Fuel efficiency has been substantially improved, principally by turbocharging and after-cooling, which increase air input to the engine. With turbocharging, engines also can be operated at higher altitudes without derating.

Low-speed diesel engines are two-cycle engines similar

Table 1. Large Off-Highway Truck Types: Advantages and Disadvantages

Conventional Rear Dump Truck

Advantages	*Disadvantages*
Versatility, can haul wide variety of materials	Poor for long-distance, high-velocity haul because of relatively high tire loading
Good gradeability because of favorable kilowatt (horsepower) to weight ratio (182 to 243 kg/kW) (300 to 400 lb/hp)	Must stop, turn, and back up to dump load
Good traction because of favorable weight distribution (65% or more of GVW on drive wheels)	Low-rated payload to GVW (rated payload 55 to 60% of GVW).
Good performance under unfavorable road conditions	
Good maneuverability because of short wheel base and small turning circle	
Suitable for severe loading impact	
Good for dumping into restricted hoppers and over banks	
Provides maximum flexibility	
Up to 318-t (350-ton) capacity.	

Tractor-Trailer Truck

Advantages	*Disadvantages*
Versatility, can haul wide variety of materials depending on dump body type	Poor gradeability because of unfavorable kilowatt (horsepower) to weight ratio (274 to 365 kg/kW) (450 to 600 lb/hp)
Separate prime mover or tractor and trailer	Poor traction because of unfavorable weight distribution (35 to 45% of GVW on drive wheels)
Good for long-distance, high-velocity haul because of relatively low tire loading due to number of tires and to weight distribution	Poor under unfavorable road conditions
Good for dumping over hopper, building windrow, and dumping while moving	Poor maneuverability because of large turning circle and long overall tractor and trailer length
Favorable rated payload to GVW, especially for bottom dump coal haulers (rated payload 60 to 65% of GVW).	Material must be free flowing if bottom dump
	Not suitable for severe loading impact of bottom dump.

Integral Bottom Dump Truck

Advantages	*Disadvantages*
Medium gradeability (228 to 335 kg/kW) (375 to 550 lb/hp)	Material must be free flowing
Medium traction (50 to 65% of GVW on drive wheels)	Not suitable for severe loading impact.
Medium performance under unfavorable road conditions	
Medium maneuverability	
Good for long-distance haul because of favorable tire loading (not always the case)	
Good for dumping over hopper	
Favorable rated payload to GVW (rated payload 60 to 70% of GVW).	

to those used on railroad locomotives. Low-speed engines have greater weight per kilowatt (horsepower) than high-speed engines, which is a major disadvantage in haulage equipment.

Engine power is transmitted to truck wheels either through mechanical or electric wheel drive. Mechanical drive consists of a transmission and final drive. The transmission usually is a fully automated power shift transmission consisting of torque converter, clutch-operated planetary gearing, and hydraulic controls with automatic direct drive lockup in all ranges. The transmission usually incorporates a hydraulic retarder so that the transmission can be utilized for braking. The recent development of transmissions for 746- to 1007-kW (1000- to 1350-hp) engines has extended mechanical drives to rear dump trucks in the 109- to 118-t (120- to 130-ton) capacity range. The final drive consists of a drive shaft, differential, full floating axle shafts, and planetary gears within drive wheels.

Electric wheel drive usually consists of a d-c generator or alternator, an exciter, d-c electric wheel motor mounted within the rim of each drive wheel, and planetary drive assembly within each drive wheel. In some 154-t (170-ton) capacity and larger rear dump trucks, electric motors are mounted inboard of the drive wheels. Electric wheel drive permits the engine to operate at constant speed with a smooth power loading. The electric motor also can serve as a retarder with the motor converting to a generator and energy dissipated as heat by air-cooled resistors.

With electric wheel drive, the onboard diesel engine can be assisted by an external overhead electric power source. Under trolley assist, electric power is collected from a fixed overhead conductor system and supplied directly to the wheel motors.

Brakes

Off-highway trucks commonly are equipped with four braking systems: retarder, service brakes, emergency brakes, and parking brakes. Primary braking is performed by the retarder. Mechanical drive trucks have an hydraulic retarder which is integral with the transmission or rear wheel service brakes. The hydraulic retarder integral with the transmission is positioned between the torque converter and planetary assembly. The retarder consists of a bladed rotor turned by the converter output shaft, a fixed casing or stator with vanes, and an oil circulation system. Retarding is accomplished by filling the casing with oil, which causes fluid resistance against rotor movement. The retarder integral with the rear wheel service brakes consists of oil-cooled, multiple disk brakes and an oil circulation and cooling system.

Electric wheel trucks are equipped with dynamic retarders which consist of the wheel motor, resistor grid, and grid cooling system. With dynamic retarding, wheel motors become generators and generated energy is dissipated as heat by the resistor grid.

Service brakes provide secondary braking and should be used only for truck stopping. Service brakes are either internal expanding shoe-drum or disk brakes. These brakes may be air, air-over-hydraulic, or all-hydraulic activated. Shoe-drum brakes are air-cooled while disk brakes may be either air- or oil-cooled. Trucks sometimes are equipped with shoe-drum brakes on the front axle and disk brakes on the rear axle.

Emergency brakes usually are integral with service brakes and can be applied manually or automatically upon loss of operating pressure in the service brake system.

Parking brakes on mechanical drives usually are an in-ternal expanding shoe type on the transmission output shaft. For electric wheel drives, parking brakes may be integral with service brakes or mounted on the armature of the wheel motors.

Tires

Large truck tires are either bias ply or radial construction. The bias ply tire carcass is constructed of body plies that are 20 to 40 layers of rubber-cushioned nylon fabric with alternating plies of cord running from bead to bead and crossing the thread centerline at an approximate 30° angle (bias). In the tread area, body plies are overlaid with tread plies that improve carcass strength and protect the body plies. The radial tire carcass is constructed of a single heavy ply of steel cables running radially (90° bias angle) from bead to bead. The carcass is overlaid in the tread area by plies of steel cable or belts crossing the tread centerline at an angle with the angle reversed from the preceding belt. These belts provide the tire's radial strength and maintain the tire's shape under load. For both tires, the tread contacting the ground transmits truck weight to the ground and provides traction, flotation, cut resistance, and long wear.

Radial tires have the following advantages when compared with bias ply tires: (1) less heat generation and buildup; (2) lower fuel consumption; and (3) longer tread life. Heat generation due to tire flexing and internal friction among the various plies is reduced because of the single ply carcass. Radial tires also have less heat buildup because the steel ply conducts heat away from the tread and the thin carcass provides rapid heat dissipation. Heat buildup is especially important because the tire starts to weaken at about 93.3°C (200°F), making it more susceptible to impacts, cuts, and fatigue. At about 132.2°C (270°F), rubber *reversion* occurs and tire separation is imminent. Lower fuel consumption occurs because of less internal friction. Longer tread life results from less relative motion in the radial tire's footprint or *squirm*. However, the radial tire has the disadvantage of a thin sidewall, which makes the tire more susceptible to sidewall damage.

Tire size nomenclature designates tire cross section and rim diameter. As an example, a 30.00-51(46) tire is a bias ply standard base tire having an approximate width of 0.762 m (30 in.) between the outside of the sidewalls, a rim diameter of 1.275 m (51 in.), and a ply rating of 46. The same size radial tire is designated 30.00R51. Ply rating is an index of tire strength and does not necessarily mean the number of plies in the tire. The Tire and Rim Association (T&RA) has the following code identification for off-highway earthmover tires:

    E-1 Rib
    E-2 Traction
    E-3 Rock
    E-4 Rock Deep Tread
    E-7 Flotation.

The E-3 and E-4 tires are most common for mine haulage with the E-4 tire having approximately 50% more tread than the E-3 tire. Radial tire manufacturers usually use different tire identification codes.

Tire load-carrying capacity, based on T&RA ratings, is the maximum load the tire should carry. This carrying capacity, at times, may be exceeded by 10 to 15% due to variations in truck loading without drastically reducing tire life. This carrying capacity increases as tire ply rating and inflation pressure increase. Optimum standard base tire pressures (cold) are 483 kPa (70 psi) for bias ply nylon tires and 586 kPa (85 psi) for radial tires. It is important to maintain proper tire inflation because underinflation causes heat buildup and overinflation decreases impact and cut resistance.

Most off-highway tires have a tonne-kilometers per hour (t-km/h) (ton-miles per hour, tmph), rating, which is a tire work index related to heat buildup. The t-km/h (tmph) requirements are mean tire load times the hourly average velocity. Mean tire load is the average of the empty and loaded tire loads. Hourly average velocity is the maximum average velocity during any hour of operation. The t-km/h (tmph) requirement must not exceed the tire's t-km/h (tmph) rating to avoid excessive heat buildup.

In off-highway tire selection, factors to be considered in order of importance are: (1) t-km/hr (tmph) rating, (2) load-carrying capacity, and (3) tread thickness.

Dump Bodies

Rear dump bodies are available in several basic styles with the following most common:

Transverse V-shaped, flat bottom floor;
Transverse V-shaped, longitudinal V-shaped floor;
Horizontal, flat bottom floor;
Horizontal, longitudinal V-shaped floor.

For transverse V-shaped bodies, floor plates slope downward from rear to front, forming a transverse V with the front plate. For horizontal bodies, floor plate(s) are horizontal. A flat bottom floor has a flat floor plate. A longitudinal V-shaped floor has floor plates sloped downward from the side plates toward the center. All body types may have an upward sloped tail chute or ducktail at the rear for load retention. Bodies have either ribs or horizontal stiffeners to provide structural strength and rigidity.

Rear dump bodies usually are constructed of high strength alloy steels; however, lighter weight aluminum bodies also are available. Bodies heated by routing exhaust gases through structural supports prevents material from freezing to the body. Steel or rubber liners, plates, and bars may be installed to reduce wear. Top extensions (sideboards) and/ or tailgates can be installed to increase volumetric capacity. However, added weight due to liners, sideboards, and tailgates decreases tonnage capacity of the truck.

Bottom dump trailers for tractor-trailer units are constructed in a variety of styles. Trailers for coal haulage usually are longer, narrower, and lower than rear dump bodies. Trailer bodies usually are constructed of high strength alloy steel, but of lighter gage than that used in rear dump bodies.

Rear dump and trailer bodies are designed for hauling specific materials resulting in a range of volumetric capacities, construction materials and methods, and body protection for the same tonnage capacity. However, the body always is designed to facilitate dumping and minimize material sticking and hangup and to reduce weight consistent with acceptable strength and wear.

Specifications

Off-highway truck manufacturers provide specifications of importance to the truck operator. These specifications frequently are determined in accordance with procedures established by the Society of Automotive Engineers (SAE) or other organizations.

Design payload or tonnage capacity specified is the weight the truck is designed to carry with manufacturer's standard equipment. If optional equipment substantially increases truck weight, the design payload must be reduced by a corresponding weight. Actual payload is the actual weight of material loaded, which depends on material density, loading equipment, and operating conditions and varies from load to load. Truck performance and operating cost are based on actual payload rather than design payload. As a general rule, average actual payload should be as near as practical to and never exceed 110% of the design payload.

Volumetric capacity of the truck body is based on SAE struck and heaped 2:1 capacities. Neither of these capacities represents the volume of material that can be loaded into and hauled by the truck. Experience indicates that actual capacities are about 105 to 130% of SAE struck capacity with the highest percentage for the largest trucks with wide bodies and top extensions.

Engine kilowatt (horsepower) is based on rated brake and flywheel power. Rated brake power is engine kilowatt (horsepower) output at governed rpm and with manufacturer's approved fuel setting. Brake kilowatt (horsepower), torque, and fuel consumption curves usually conform to SAE standard J816b with conditions of 152.4-m (500-ft) altitude [0.7366 m (29.00 in.) Hg dry], 29.4°C (85°F) intake air temperature, and 0.965 cm (0.38 in.) Hg water vapor. Attached engine accessories usually include water pump, lubricating oil pump, fuel system, and air cleaner. Flywheel power is available engine kilowatt (horsepower) output or input power to the transmission for mechanical drives or to the generator or alternator for electric wheel drives. In addition to accessory power losses for brake kilowatt (horsepower), flywheel kilowatt (horsepower) usually includes losses for fan, air compressor, alternator, drive motor cooling fan, and hydraulic pump (no load). Actual engine kilowatt (horsepower) may vary ±5% from specified kilowatt (horsepower).

Net vehicle weight (NVW) is the weight of the empty truck with manufacturer's standard equipment. NVW may be specified as dry, wet, or operating weight. Wet or operating NVW includes lubricants, coolant, fuel, and operator. The NVW must be adjusted for installed optional equipment. NVW for a specific truck can be expected to vary ±5%. Gross vehicle weight (GVW) is NVW plus actual payload. In manufacturer's specifications, GVW is NVW with standard equipment plus design payload.

Manufacturers specify truck weight distribution on each axle for an empty and loaded truck on level ground. However, weight distribution is not solely determined by the truck, but also by load placement and haulage road grade. Weight distribution is important in determining tire loading and traction.

Truck loading height is important in matching truck and loading equipment. There must be adequate clearance so that the loading bucket or dipper does not collide with the truck dump body top rail.

Clearance turning circle indicates distance required for the truck to make a 180° turn without backing. This determines truck maneuverability and can influence mine plan and truck spotting procedures.

Overall truck dimensions are important in the design of loading and dumping areas, haulage roads, and maintenance facilities. These dimensions also influence mine layout and loading equipment selection.

Special Equipment

Off-highway trucks can be equipped with a variety of special equipment to reduce maintenance, protect equipment, improve communications, and increase productivity. Maintenance can be reduced by special air cleaners, replaceable radiator core tubes, automatic lubrication systems, and automatic tire inflation systems. Instrumentation and recorders such as the tachograph are available to monitor vital drive train data. Systems for truck protection include multifunction alarms, various interlocks, supplementary steering, ground level shutdown, and fire suppression. Special mirrors are available to improve operator visibility. Both voice and digital communication systems are available for truck monitoring and control. Truck productivity can be increased by fast fueling systems, cab climate control, and load indicators.

HAULAGE ROAD CONSIDERATIONS

Surface mine haulage roads must be located to minimize haulage cost while being consistent with other mine planning objectives. These roads must be constructed to insure efficient and safe truck travel.

Length

Haulage road length is a major consideration in truck selection, tire life, truck production, and haulage cost. Generally, road length should be minimized consistent with reasonable road grades.

Grade

Haulage road grades usually are determined as percent slope [(vertical rise/horizontal distance) × 100]. Uphill grades or vertical rises are adverse or unfavorable grades, expressed as a positive (+) percent. Downhill grades or vertical falls are favorable grades, expressed as a negative (−) percent. Optimum grades for conventional rear dump trucks usually are 7 to 10%, but must be evaluated for specific trucks. For short distances, grades may be as high as 15%. Tractor-trailer units operate on lower grades than conventional rear dump trucks. Surface mining regulations for coal limit overall grades to a maximum of 10%.

Maximum unfavorable grades are limited by truck drive train and traction and ability to arrest the truck's backward movement. Maximum favorable grades are limited principally by truck retarding capability and emergency stopping distance. However, the ability to arrest a possible runaway truck also must be evaluated. Small trucks can descend steeper grades at higher speeds than larger trucks. For example, on a −10% grade, a loaded 45.5-t (50-ton) truck might descend at a maximum speed of 34 km/h (21 mph) while for a 90.7-t (100-ton) truck, the maximum speed should be reduced to 24 km/h (15 mph).

Sight Distance

Haulage roads must be designed with adequate sight distance for the operator to stop a truck traveling at operating speed before reaching a hazard. Sight distances are limited on vertical curve crests due to road surface and on horizontal curves due to berms, steep rock cuts, trees, structures, etc.

Construction

The haulage road must have a stable road base that adequately supports the heavy weight of off-highway trucks. In many mines, natural strata provide a satisfactory base. The road base must be overlaid by surface material such as crushed stone or gravel, coarse tailings, or stabilized earth. Surface material should have a high adhesion or traction coefficient under operating conditions.

Haulage road width for one- and two-lane travel should be 2.0 and 3.5 times truck width, respectively. The road must be sloped to the sides, usually 2.08 cm/m (0.25 in. per ft), for proper drainage. Drainage ditches must be placed along roadsides to control runoff from the road and surrounding area to avoid road damage.

At curves, road surface elevation must be raised from the inside to the outside of the curve (superelevation). This superelevation is necessary to counteract the outward centrifugal force, reducing stress on tires, steering, and other vehicle components. Required superelevation is a function of friction factor, vehicle velocity, and curve radius. At sharp curves, it often is necessary to widen the road because of minimum turning circle and clearance.

Traffic signs should be used for posting velocity limits, stops, curve and intersection warnings, culvert crossings, controlling traffic, designating limited accesses, and indicating safety accesses. Berms must be placed parallel to the outside bank of elevated roads as a safety feature. Truck runaway escape lanes and/or collision berms are necessary to arrest runaway vehicles.

Resistance

Grade and rolling resistance provide resistance to truck movement. Grade resistance is the retarding or assisting force of gravity which is negative when traveling uphill and positive when traveling downhill. Grade resistance, the motion resistance due to the haulage road grade, is expressed as a percent grade or grade resistance factor. The grade resistance factor, *GRF*, is calculated by

$$GRF = 10 \text{ kg/t} \times \% \text{ grade}$$

or

$$GRF = 20 \text{ lb/ton} \times \% \text{ grade}$$

where 1% of vehicle weight is equivalent to 10 kg/t (20 lb per ton). The grade resistance, *GR*, is calculated by

$$GR = 10 \text{ kg/t} \times \text{vehicle weight in tonnes} \times \% \text{ grade}$$

or

$$GR = 20 \text{ lb/ton} \times \text{vehicle weight in tons} \times \% \text{ grade}$$

or

$$GR = GRF \times \text{vehicle weight in tonnes (tons)}$$

As an example, a truck weighing 136 078 kg (300,000 lb) traveling down a −8% grade has a grade resistance factor of −80 kg/t (−160 lb/ton) and a grade resistance of −10 886 kg (−24,000 lb).

These equations use percent grade or tangent of the grade angle which is satisfactory for low grades where tangent and sine of the grade angle are assumed equal. For steep grades, these equations should use sine of the grade angle.

Rolling resistance is motion resistance due to tire flexing under load, wheel bearing friction, tire penetration into the ground, and wind or air movement. Rolling resistance usually is expressed in kilogram per tonne (pound per ton) of vehicle weight, but also may be expressed as a percent of vehicle weight where 10 kg/t (20 lb/ton) equals 1%. Rolling resistance factor, *RRF*, in kilogram per tonne (pound per ton) is calculated by

$$RRF = R_f + R_s P + (KAV^2/W)$$

where R_f is the resistance due to tire flexing and wheel bearing friction, commonly 15 kg/t (30 lb/ton) for radial tires and 20 kg/t (40 lb/ton) for bias ply tires;

R_s is the resistance per centimeter (inch) of tire penetration, 5.9 kg/t (12 lb/ton);

P is tire penetration, cm (in.);

K is the coefficient of air resistance, about 0.0066 SI (0.0035 English) for the off-highway truck;

A is the truck frontal area, m² (sq ft);

V is the air velocity, km/h (mph); and

W is the truck vehicle weight, tonne (ton).

Resistance due to tire flexing and wheel-bearing friction is affected by tire type, tire inflation pressure, and wheel-bearing design. Tire penetration is affected principally by haulage road surface and construction, but also by tire type, tread pattern, and tire inflation pressure. Air resistance is usually not considered in determining rolling resistance.

There is evidence that rolling resistance is affected by vehicle velocity with the rolling resistance factor, RRF, being approximated by

$$RRF = 1.075\ RRF_n + 0.115V - 9.00 \qquad \text{(SI)}$$

or

$$RRF = 1.075\ RRF_n + 0.50V - 18.00 \qquad \text{(English)}$$

where RRF_n is the nominal rolling resistance factor in kg/t (lb/ton) and V is the vehicle velocity in km/h (mph).

Table 2 presents typical or nominal rolling resistance factors for bias ply tires. For radial tires, these factors should be reduced by 5 kg/t (10 lb/ton). These factors normally are used independent of the vehicle velocity, but can also be used when considering vehicle velocity. Rolling resistance, RR, is calculated by

$$RR = RRF \times \text{vehicle weight in tonnes (tons)}$$

As an example, a truck weighing 136 078 kg (300,000 lb) with radial tires and 3.81-cm (1.5-in.) tire penetration, has a rolling resistance factor, RRF, of 37.5 kg/t (75 lb/ton) and a rolling resistance, RR, of 5103 kg (11,250 lb) assuming no wind resistance. With wind resistance resulting from a frontal area of 20.4 m² (220 sq ft) and a velocity of 32 km/ h (20 mph), the RRF and RR are increased to 38.5 kg/t (77 lb/ton) and 5239 kg (11,550 lb), respectively. If vehicle velocity is considered and the vehicle has a velocity of 32 km/h (20 mph), RRF and RR are 36.3 kg/t (72.6 lb/ton) and 4940 kg (10,890 lb), respectively. However, if velocity is reduced to 16 km/h (10 mph), the RRF and RR decrease to 33.8 kg/t (67.6 lb/ton) and 4599 kg (10,140 lb), respectively.

Total resistance, TR, is the sum of grade and rolling resistance. Total resistance is the total resistance force to vehicle motion and usually is expressed as effective grade or percent of vehicle weight. As an example, consider a truck weighing 136 078 kg (300,000 lb) traveling down a −8% grade with a rolling resistance factor of 37.5 kg/t (75 lb/ton) (equivalent to 3.75%). The total resistance, TR, is calculated by

$$TR = GR + RR = -10\ 886\ \text{kg} + 5103\ \text{kg} = -5783\ \text{kg}$$

or

$$TR = GR + RR = -24{,}000\ \text{lb} + 11{,}250\ \text{lb} = -12{,}750\ \text{lb}$$

and the effective grade, EG, is

$$EG(\%) = GR(\%) + RR(\%) = -8\% + 3.75\% = -4.25\%$$

Traction

Traction is the usable driving force developed by the truck tire on the road surface. This usable driving force is limited by either drive train or road surface. The maximum force, F_M, which can be transferred from tire to road surface is a function of the coefficient of traction between tire and road surface, μ, and the normal force at the road surface, N. Thus, maximum force is

$$F_M = \mu N = \mu W \cos\theta$$

where W is weight on the tire and θ is grade angle. Coefficients of traction for various haul road surface materials are presented in Table 3.

As an example, consider a 77-t (85-ton) conventional rear dump truck traveling loaded up a 10% grade on a loose gravel haulage road with a rolling resistance factor of 100 kg/t (200 lb/ton) and a coefficient of traction of 0.35. As-

Table 2. Typical Rolling Resistance Factors Bias Ply Tires

Haulage Road Surface	kg/t (lb/ton)	Equivalent % grade
Cement, asphalt, or soil cement without tire penetration	20 (40)	2
Hard-packed gravel, cinders, or crushed rock	30 (60)	3
Firm packed earth or light surfacing	32.5 (65)	3.25
Moderately packed gravel, cinders, or crushed rock	50 (100)	5
Rutted or unmaintained earth	75 (150)	7.5
Loose sand and gravel	100 (200)	10
Soft, muddy, rutted, and unmaintained material	100-200 (200-400)	10-20

Table 3. Coefficient of Traction for Various Haulage Road Surface

Haulage Road Surface	Coefficient of Traction	
	Rubber Tires	Tracks
Concrete, new	0.80-1.00	0.45
Concrete, old	0.60-0.80	
Concrete, wet	0.45-0.80	
Asphalt, new	0.80-1.00	
Asphalt, old	0.60-0.80	
Asphalt, wet	0.30-0.80	
Gravel, packed and oiled	0.55-0.85	
Gravel, loose	0.35-0.70	0.50
Gravel, wet	0.35-0.80	
Rock, crushed	0.55-0.75	0.55
Rock, wet	0.55-0.75	
Cinders, packed	0.50-0.70	
Cinders, wet	0.65-0.75	
Earth, firm	0.55-0.70	0.90
Earth, loose	0.45	0.60
Sand, dry	0.20	0.30
Sand, wet	0.40	0.50-0.55
Snow, packed	0.20-0.55	0.25
Snow, loose	0.10-0.25	
Snow, wet	0.30-0.60	
Ice, smooth	0.10-0.25	0.12
Ice, wet	0.05-0.10	
Coal, stockpiled	0.45	0.60

suming the GVW is 136 078 kg (300,000 lb), the resistance to movement or required rimpull is 27 216 kg (60,000 lb). Assuming weight on rear drive tires is 90 718 kg (200,000 lb), the maximum force that can be transferred from drive tires to road surface is 31 751 kg (70,000 lb). Since maximum rimpull and force that can be transferred are greater than the resistance of 27 216 kg (60,000 lb), the truck can move up this grade. However, if the truck is empty with only 31 751 kg (70,000 lb) on the rear drive tires, resistance to movement is reduced to 11 793 kg (26,000 lb). This is greater than the 11 113 kg (24,500 lb) that can be transferred from drive tires to road. Thus, the empty truck cannot travel up this grade without improving the road surface to reduce the rolling resistance factor.

Velocity Limits

Truck velocity limits are necessary on favorable downhill grades, curves, loading and dumping areas, and congested traffic areas. On favorable grades, truck velocity depends primarily on retarding capability and emergency stopping distance. On curves, truck velocity must be consistent with the road superelevation. In loading and dumping areas, truck velocity usually is limited to 13 to 16 km/h (8 to 10 mph), depending on ground conditions and congestion.

Maintenance

Good haulage road maintenance is essential to avoid costly truck operating and maintenance problems. A motor grader should be used continually to fill and smooth potholes, ruts, depressions, and bumps; maintain cross slopes; and remove spillage, ice and snow. Haulage road dust should be controlled by watering and/or chemicals. However, excessive water should be avoided.

PERFORMANCE

The truck's performance characteristics determine its ability to haul material effectively under specific haulage conditions. Performance is determined by the interaction between truck and haulage environment with certain trucks designed to haul most effectively under specific haulage con-

ditions. Of particular importance are truck weight and its distribution, drive train, retarding and braking systems, and tires. For maximum utilization and availability, it is also important that the truck be designed and manufactured for reliability, maintainability, and driver acceptance. Specific haulage conditions affecting performance are haulage road distances, grades, construction, and maintenance; altitude and temperature; and operating procedures and practices.

Horsepower Utilization

The performance curve or rimpull vs. velocity curve characterizes drive train performance when the truck is operating against positive resistance to movement. Figs. 2 and 3 present typical performance curves for mechanical and electric wheel drive trucks, respectively. The vertical axis is the rimpull or force the drive train can produce at the drive tires-ground contact. The horizontal axis is the truck velocity. The performance chart shows the maximum rimpull the drive train can provide when the truck is traveling at the specified velocity. Because the truck must overcome resistance to movement, the performance chart indicates the maximum truck velocity when resistance force is equal to truck rimpull force.

The performance curve for the mechanical drive truck (Fig. 2) has a series of discontinuities representing gear changes. The lower portions of the curve occur when the power is transmitted by the torque converter; the upper portions occur when the transmission is locked up and slippage is eliminated. The performance chart for the electric wheel drive (Fig. 3) has three discontinuities due to voltage and current limitations and electric field shunting. At low velocities, rimpull is limited by maximum current. At high velocity, velocity is limited by maximum voltage.

As an example, consider a 109-t (120-ton) capacity truck with the performance curve shown in Fig. 2 and an assumed resistance to movement of 12 700 kg (28,000 lb). With this resistance to movement at equilibrium with the rimpull, the truck can maintain a velocity of 21 km/h (13 mph). If resistance to movement increases, the truck would decelerate until equilibrium between the higher resistance to movement

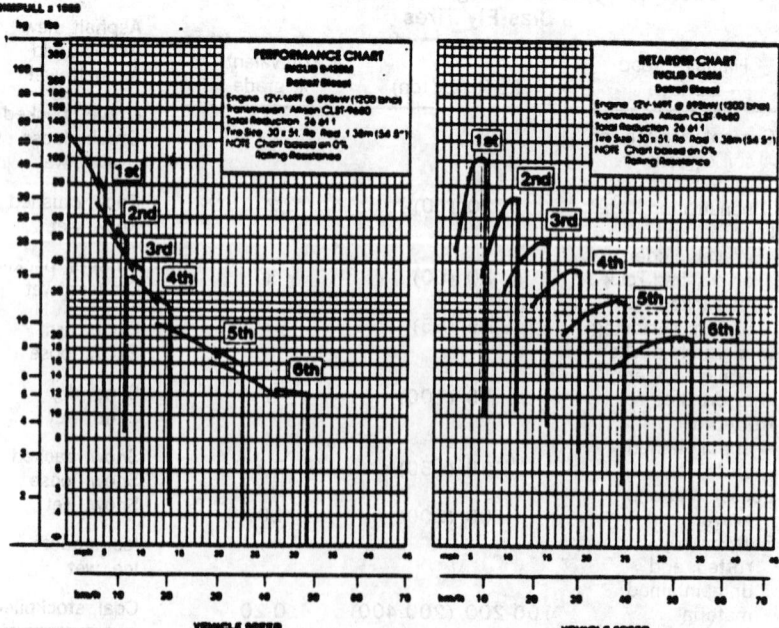

Fig. 2. Performance and retarding curves for mechanical drive trucks.

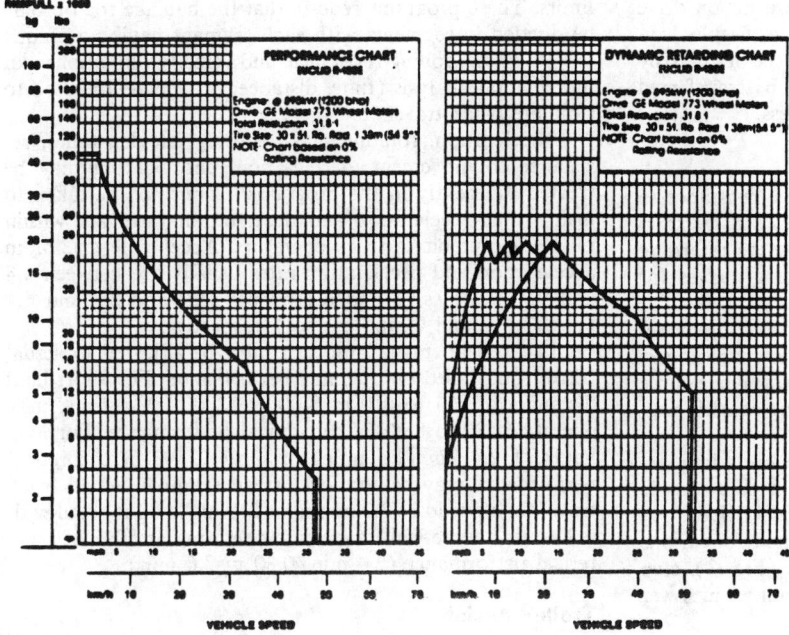

Fig. 3. Performance and retarding curves for electric wheel drive trucks.

and rimpull is attained at a lower speed. If resistance to movement decreases, the truck could accelerate until attaining equilibrium.

The rim kilowatt (horsepower), kW_R (hp_R), or power the drive train can deliver at the drive tires-ground contact is calculated by

$$kW_R = \frac{\text{rimpull (kg)} \times \text{velocity (km/h)}}{367.1}$$

or

$$hp_R = \frac{\text{rimpull (lb)} \times \text{velocity (mph)}}{375}$$

Drive train efficiency is the portion of engine kilowatt (horsepower) available as rim kilowatt (horsepower). Drive train efficiency usually is based on engine flywheel kilowatt (horsepower).

It is important that the drive train operate at a relatively high efficiency level, usually about 80 to 90%. Drive train energy losses occur as heat which must be dissipated to the environment. Mechanical drives are designed with heat exchangers connected to the cooling system to cool the torque converter, the principal heat source. Cooling capacity usually is adequate to dissipate 30% of the flywheel kilowatt (horsepower) on a continuous basis under normal ambient temperature and humidity. Thus, continuously operating the drive train at efficiencies below 70% efficiency may cause overheating and eventual component failure. Electric wheel motors are provided with forced air ventilation for cooling. Heat generation exceeding cooling capacity can cause deterioration of insulation and eventual motor damage. If the truck will be operated in the low efficiency range or under extreme ambient conditions, the manufacturer should be consulted to determine truck suitability and possible additional cooling capacity.

The manufacturer's performance curve must be adjusted for altitude, temperature, and tire size if different from those specified. As altitude increases or temperature rises, air density decreases, adversely affecting engine power output. En-

gine manufacturers provide information on performance derating for each specific engine. Naturally aspirated engines may have to be derated at altitudes greater than 304.8 m (1000 ft) above sea level while turbocharged and after-cooled engines may not need to be derated until altitudes of 1524 m (5000 ft) or more above sea level. Adjusted drive train performance can be obtained by derating the rimpull on the manufacturer's performance curve. As an example, consider an engine that is derated to 93% of rated kilowatt (horsepower) at an elevation of 3048 m (10,000 ft). Adjusted rimpull then is 0.93 times the rimpull specified by the manufacturer's performance curve.

The manufacturer's performance curve is for a specified tire size. Since rimpull is wheel or hub torque divided by effective loaded tire radius, and velocity is wheel rpm divided by tire revolutions per kilometer (mile), both rimpull and velocity must be adjusted for tire size. Both adjustments use tire revolutions per kilometer (mile) which is proportional to effective tire loaded radius. Rimpull, F_i, velocity, V_i, and tire revolutions per kilometer (mile), R_i, are adjusted as follows:

$$F_2 = F_1 \times (R_2/R_1)$$

and

$$V_2 = V_1 \times (R_1/R_2)$$

where i is 1 for manufacturer's performance curve and i is 2 for adjusted values with optional tires. This same approach also may be used to adjust the performance curve for different drive train gear ratios.

The rimpull specified by the manufacturer's performance curve may be derated to penalize the engine for extended operating hours or poor mechanical condition. However, operating experience indicates that the performance of a well maintained and properly adjusted engine does not substantially deteriorate during normal engine life.

For rimpull to be a usable driving force, this force must be transmitted from the drive tires to the road surface causing truck movement. Transfer of rimpull to the road surface may be limited because of poor road surface or weight on drive tires (see the earlier section on *Traction*). Thus, usable driving force (usable rimpull or traction) may be limited by either rimpull or maximum force that can be transferred from the tire to road surface, whichever is less.

Retarding

When the truck is operating with negative resistance to movement, retarding capabilities of the truck's drive train and related components are characterized by the retarding curve. Figs. 2 and 3 present typical retarding curves for mechanical and electric wheel drive trucks, respectively. The vertical axis is the retarding or rimpull force resisting movement at the drive tires-ground contact. The horizontal axis is truck velocity. The retarding curve and performance curve are similar except the rimpull on the retarding chart is negative, a force resisting movement.

The retarding chart shows retarding force, excluding service brakes, available when the truck is traveling at a specific velocity. The retarding curve for the mechanical drive truck (Fig. 2) has a series of discontinuities related to various transmission gears. For electric wheel drive (Fig. 3), step extended range retardation provides discontinuities in lower velocity ranges.

Accepted practice is to operate the truck at velocities lower than indicated by manufacturer's retarding curve because the curve is the maximum retarding capability. For mechanical drives, the truck usually is operated at least one gear lower than indicated by the retarding curve. As an example, consider a 109-t (120-ton) truck with the retarding curve in Fig. 2. If the required retarding force is 12 700 kg (28,000 lb), the truck could be operated at 40 km/h (25 mph) in fifth gear. However, a velocity of 30 km/h (19 mph) in fourth gear is preferred for safe operation. For both mechanical and electric wheel drives, there is a time lag between the operator activating the retarder control and activation of the retarding system. During this time lag, the truck may accelerate and gain velocity on a downhill or favorable grade. Thus, retarder control must be activated at a velocity lower than indicated by the retarding curve.

The manufacturer's retarding curve is for a specified tire size. If the tire size differs from the size specified, both retarding force and velocity of the retarding curve must be adjusted in the same manner as rimpull and velocity were adjusted for the performance curve.

Retarding systems are not capable of stopping a truck. Thus, service brakes are needed to bring the truck to a complete stop.

Gradeability

Gradeability is the ability of a truck to negotiate a given haulage road, taking into consideration total resistance to movement, including grade and rolling resistance (see the section on *Resistance*). Travel time required for a truck to travel from one location to another is based on the truck performance curve, retarder curve, net vehicle weight (NVW), and the actual payload; haulage road grade and rolling resistance; velocity limits and other operating constraints; and operator performance.

Computer simulation programs are available for calculating theoretical travel time. These programs provide quick, accurate, and inexpensive means for calculating travel time, permitting extensive analysis and comparison of trucks, haulage road designs, and operating practices. These computer programs use a variety of methods and procedures, but estimated travel time for most programs is within acceptable

limits. These programs require that the haulage road profile be divided in segments with each segment having a unique length, grade, rolling resistance, and velocity limitations. An incremental analysis (time, distance, or velocity) is used to simulate truck travel.

Acceleration, rotating mass, shifting and lag times, resistance to movement, deceleration, and braking may be treated differently in the programs. Care must be taken to insure that acceleration and deceleration rates are within ranges acceptable to truck operators. Acceptable maximum acceleration and deceleration rates usually are between 4.8 and 9.7 km/h/s (3 and 6 mph per sec) and 3.2 and 6.4 km/h/s (2 and 4 mph per sec), respectively.

Theoretical travel time then must be adjusted to actual travel time based on operator performance. For short haul distances of 305 to 610 m (1000 to 2000 ft), a truck may operate at 80 to 85% of theoretical performance. For long hauls of 2.4 km (1.5 mile) or more, 95% of theoretical performance may be attained. As an example, actual truck travel time would be 3.0 min for a 610-m (2000-ft) haul with a 2.4-min theoretical travel time operating at 80% of theoretical performance (2.4 min/0.80 = 3.0 min).

Trolley Assist

Trolley assist for electric wheel drive trucks uses an overhead electric power source to supply electric power for electric wheel motors. This external electric power supplements the power available from the onboard diesel engine and d-c generator or alternator.

An electric trolley assist system consists of an overhead power source along the haul road and a truck-mounted power collection and auxiliary control equipment. The overhead power source is primarily an a-c power distribution system feeding a rectified d-c power distribution system consisting of fixed wires or bus bars suspended above the haulage road. Transformer-rectifier substations step down and convert commercially available a-c power to d-c power. Substation size and location depend on specific mine requirements: truck traffic density, length of trolley haul, truck specific drive system, and frequency of overhead equipment relocation.

The fixed overhead dc transmission system can use either wire or bus bars. Wire has the advantage of lower initial cost, but the disadvantages of lower current and wire sag. Electric current requirements may require larger wire and multiple feeder points. Sag may be corrected by closer pole spacings, tensioning devices, and full caternary support. Bus bars are advantageous at mines with steep grades and high traffic density. Bus bars may have lower overall cost because of greater flexibility to meet changing traffic densities, less feeder points, and easier repair of sectionized bus bars.

Electric power is collected from the overhead d-c transmission system by either a trolley pole or pantograph mounted on the truck. The trolley pole can be used with overhead wire or bus bars, but the pantograph can be used only with overhead wire. The trolley pole can accommodate terrain changes and reasonable steering variation, but requires an entry guide assembly to engage the trolley pole to the overhead wire or bus bar. The pantograph can engage the overhead wire at any point along the wire, but has the disadvantages of being a larger structure and decreasing operator visibility.

Trolley electronic controls, logic relays, and power contractors are truck-mounted. Existing retarding resistors are also used as trolley accelerating step resistors. Protective features include ground leakage current detection, grid resistor protection, wheel motor overcurrent protection, line voltage protection, interlocking protection, and deadman control. Power collection and control equipment can be installed in new or in-service electric wheel drive trucks.

A trolley assist equipped electric wheel drive truck can be operated in either *diesel* or *trolley* operating mode. In *diesel* mode, the truck is powered by the onboard diesel engine. In *trolley* mode, the truck is powered principally by the external electric power source with the onboard diesel engine providing power for grid resistor and motor cooling air and all engine driven auxiliary systems. The truck operator can override automatic trolley controls to stop on grade, adjust speed, or switch to diesel operating mode.

Major advantages of using trolley assist are reduced diesel fuel consumption, increased productivity, improved deep pit capability, and decreased diesel engine maintenance. Major disadvantages are increased onboard truck equipment, electrical maintenance, tire wear, and electric wheel motor maintenance.

Trolley assist is most applicable at mines with high cost and limited availability of diesel fuel and with low cost and abundant electric energy. At these mines, lower cost commercial electric power can replace high cost diesel fuel. Increased productivity occurs because trolley assisted trucks usually can achieve higher velocities on grades. Power output of the truck diesel engine generally is less than the power output of the electric wheel motors. Thus, the trolley operating mode permits a truck to operate at the maximum power output of the electric wheel motors rather than the lower power output of the diesel engine.

Improved deep pit capability occurs because electric wheel motors are more efficient with trolley assists, reducing heat buildup on long grades. Diesel engine maintenance is reduced because of reduced engine duty cycle, but electric wheel motor maintenance increases because of increased motor output. Tire wear will increase because of increased rimpull and higher average speeds. However, the cost associated with increased wheel motor maintenance and tire wear may be offset by increased productivity.

Automatic Truck Control (ATC)

The ATC system is an electronic system that controls the haulage truck fleet by providing automatic control of all vehicle functions normally accomplished by the operator of each truck. Functions controlled include truck steering, direction, acceleration, velocity, retarding, braking, stopping, separation, dumping, etc. The system also monitors oil pressure, water temperature, tire pressure, and other critical truck functions.

The system consists of wayside and onboard control equipment. Wayside equipment consists of a guide wire buried in the haulage road in the truck's path, block control units, and a central controller which provides predetermined signals. Onboard antennas receive commands from wayside equipment and then equipment controls the truck's functions. Onboard sensors monitor critical truck functions and transmit the data to the wayside equipment. In the event of malfunction, onboard equipment brings the truck to a safe stop. For loading and dumping, the truck may be operated manually or by radio remote control.

ATC systems have the potential advantages of reducing operating labor, increasing truck and shovel utilization, decreasing truck maintenance and downtime, decreasing truck fuel consumption, assuring proper truck loading and dumping destinations, and safer operating conditions.

TRUCK CYCLE TIME

Truck cycle time is time required for the truck to complete a single cycle including spot and load, haul loaded, turn and dump, return empty, wait, and delays. The productivity of an operating truck is dependent on the average actual payload and average cycle time. Matching of loading equipment (front end loader, shovel, etc.) and truck are important to ensure appropriate payloads and loading times. The method of spotting and loading trucks influences spot and load times. Haulage road layout and design, including length, grade, rolling resistance, curves, and velocity limits affect travel times. Travel times also will be affected by driver skill and attitude, road maintenance, and truck maintenance. Velocity or travel time for the slowest truck in the fleet tends to set the fleet pattern, especially with a no overtaking policy. Space and ground conditions, as well as the necessary support equipment, influence turn and dump times. Climate and weather conditions can result in poor equipment performance and operating delays. Truck haulage, with many operating trucks, is capital intensive, requiring good engineering and supervision to ensure maximum productivity from available equipment.

Large-scale truck haulage systems are complex operations requiring synchronization of loading and hauling equipment. To analyze this complex system, it is necessary to break down and examine various detailed elements in the truck cycle time. This analysis is based on time studies and related operating experience. Using mine planning information to establish mining sequence and haulage road layout, truck cycle times and the resultant system production can be analyzed.

Complex haulage systems usually are analyzed by computerized simulation. Simulation may be either simple deterministic or sophisticated stochastic simulation. Deterministic simulation uses constant values for system parameters such as actual payload and parameters related to truck cycle time. Stochastic simulation uses probabilistic techniques, such as the Monte Carlo method, to vary values for system parameters. Stochastic simulation requires extensive field data on parameter variability. These data can best be collected with real time monitoring systems such as those developed to provide information for computerized truck dispatching systems. Data from previous stochastic simulation also may be incorporated into deterministic simulation, so that deterministic simulation may include probabilistic effects for certain haulage situations.

Spot and Load

Spot time is the time needed for the truck to maneuver into position for loading. Load time is the time required for the loading machine to make the required number of passes to load the truck. These times may interact because the loading machine may perform parts of its work cycle while the truck is being spotted.

Combined spot and load time depends on space and ground conditions, loading equipment, loading method, rock fragmentation, and match of loading equipment and trucks. Trucks can be loaded with a variety of equipment: wheeled or tracked front-end loaders, hydraulic or cable shovels, backhoes, draglines, bucket-wheel excavators and conveyer belt loaders, and hoppers or bins with controlled gates.

Wheeled front-end loaders are commonly used to load trucks from blasted rock faces and stockpiles. The loader is highly mobile and maneuverable, but requires good ground conditions because of high ground-bearing pressures at the tire-ground contact. Digging forces of the front-end loader generally are less than for cable or hydraulic shovels. Thus, a loader may require a dozer to push material to the loader, breaking fragment interlocking of hard digging material. The loader and truck must be properly matched, with the loader having adequate bucket capacity and reach.

Three principal loading methods using front-end loaders are:

1) Single truck loading where the truck is spotted by backing and stopping near the face and loaded from one side of the truck by a single loader.

2) Tandem truck loading where the truck is spotted by backing toward and stopping near the face and loaded from both sides of the truck by two loaders, one on each side of the truck.

3) Drive-by loading where the truck drives forward and stops near the loader and is loaded from one side of the truck by a single loader (common for tractor-trailer units).

Other more complex methods have been developed, such as staggered tandem and chain loading. The loading method depends on bench slope, available space, production requirements, ore grading constraints, and operator experience.

Spot time for large rear dump trucks usually is between 0.40 and 0.70 min. Spot time for tractor-trailer units may range from 0.15 to 1.00 min., depending on loading method and the need for backing.

Load time is the time needed for the loader to load the truck. Load time depends on the number of passes required to load the truck and the loader work cycle time. Number of passes, N_P, can be calculated by

$$N_P = \frac{C_t}{C_l \times F_f \times F_s \times \rho}$$

where C_t is the truck capacity, (ton); C_l is the loader rated capacity, m³ (cu yd); F_f is the loader bucket fill factor, decimal; F_s is the material swell factor, decimal; and ρ is the material bank bulk density, t/m³ (st per cu yd). Bucket fill factor is the ratio of volume of loose material in the bucket to rated volume capacity of the bucket. Swell factor is the ratio of the material's loose to bank bulk densities.

Truck capacity usually is based on truck design payload, but, in some cases, may be limited by truck volumetric capacity. As a general rule, the number of passes should be an integer so that the truck average actual payload does not exceed design payload. The most economical number of passes usually is four to six for conventional rear dump trucks. Less than four passes usually requires a very large loader with excessive capital cost while more than six passes increases truck requirements because of increased loading time.

Front-end loader work cycle time is the time required for a loader to make one complete cycle of dig from bank, retreat from face, advance to truck, dump bucket, retreat from truck, advance to face, and position at face. Table 4 presents nominal work cycle times for various capacity front-end loaders. These nominal times are based on the loader being properly operated and having adequate digging forces and appropriate bucket for the materials being excavated. Thus, it may be necessary to adjust these nominal times based on the specific loader, operator experience, material characteristics, and operating conditions. These nominal times also may increase because of excessive loader travel and/or maneuver time due to poor truck spotting. Long or variable work cycle times indicate inadequate loader capabilities or poor operating procedures. In some cases, inadequate loader digging forces can be overcome by blasting for

Table 4. Front-End Loader Work Cycle Times

Bucket Capacity m³ (cu yd)	Nominal Minutes
0-3.1 (0-4)	0.45
3.1-5.4 (4-7)	0.50
5.4-7.6 (7-10)	0.55
7.6-15.3 (10-20)	0.60

better fragmentation and/or by a dozer pushing to the loader.

If spot time is less than the loader work cycle time, combined spot and load time, t_{t1}, in minutes can be calculated by

$$t_{tl} = N_p t_{lc}$$

where t_{lc} is the loader work cycle time in minutes. If the spot time is greater than the loader work cycle time, combined spot and load time, t_{t1}, can be calculated by

$$t_{tl} = (N_p - 1)\, t_{lc} + t_s = N_p t_{lc} + t_s - t_{lc}$$

where t_s is the truck spot time in minutes.

Hydraulic and cable shovels commonly are used to excavate and load harder digging materials with cable shovels generating the greatest digging forces. Because shovels are mounted on a crawler track assembly, these machines can be designed to operate with low ground bearing pressures. Poor shovel mobility usually requires that a dozer perform cleanup around the shovel.

The four principal loading methods using shovels are:

1) Single backup loading where the truck is spotted by backing on the operator side of shovel and stopping near the face so that the shovel, with tracks aligned toward the face, has a maximum swing of 90°.

2) Double backup loading where trucks are spotted and loaded on both sides of the shovel by trucks alternating backing and stopping near the face on the operator side and then on the opposite side of the shovel with shovel tracks aligned toward the face and a maximum swing not exceeding 90°.

3) Drive-by loading where the shovel tracks are oriented parallel with the face and the truck drives forward and stops directly opposite the shovel, with a maximum shovel swing not exceeding 180°.

4) Modified drive-by loading where the shovel tracks are aligned parallel with the face and the truck drives forward under the shovel swing path with the dipper dumped before the truck stops, then the truck is stopped and spotted by backing and stopping near the face, with a maximum shovel swing angle of 120°.

For all loading methods, the trucks should be spotted so that the dipper swing path is along the major axis of the truck body, presenting a maximum target area and center of payload. The loading method selected depends on bench shape, available space, production requirements, ore grading constraints, and operator experience.

Spot times for trucks at shovels are similar to spot times for front-end loaders and depend on the loading method. As with front-end loaders, load times for shovels depend on the number of passes and shovel work cycle time. For shovels, the number of passes may be calculated in the same manner as described for front-end loaders, with the most economical number of passes usually ranging from three to six for conventional rear dump trucks. However, the number of passes sometimes may be calculated using dipper factor rather than dipper fill and material swell factors. The dipper factor is the ratio of volume of material in the dipper based on bank material or volume to rated volume capacity of the dipper.

These factors are related by

$$F_d = F_f \times F_s$$

where F_d is the shovel dipper factor, decimal; F_f is the shovel dipper fill factor, decimal; and F_s is the material swell factor, decimal.

The shovel work cycle time is the time required for the shovel to make one complete cycle of dig from and clear the bank, swing to dump position, dump dipper, swing back to bank lowering dipper, and position to dig. Table 5 presents cable shovel nominal work cycle times based on hard digging material and a 90° swing. These nominal times should be decreased 0.08 min. for easy digging and 0.04 min. for medium digging conditions and increased 0.04 min. for very hard digging conditions. These nominal times also must be adjusted by increasing or decreasing work cycle time 0.01 min. for each 5° increase or decrease in average swing.

The cable shovel nominal work cycle times in Table 5 are based on modern electric mining shovels with complete a-c drive systems using sophisticated electronic controls. The work cycle times will vary depending on specific drive system and digging conditions. Substantially longer times may occur for older shovels having Ward Leonard drive systems. For hydraulic shovels, nominal work cycle times can be reduced approximately 0.05 min. In addition to drive system and digging conditions, shovel work cycle time also may be influenced by shovel design, operator experience, bank height, and other operating conditions.

The combined spot and load time for shovels is calculated in the same manner as previously described for front-end loaders.

Travel

Travel time includes time to haul loaded to the the dump point and return empty to the loading equipment. Travel time depends on truck gradability, haulage road, operating constraints, and operator performance and can be calculated best by computer simulation programs (see the section on *Gradability*). Operating constraints such as velocity limits are required for downhill or favorable grades, curves, narrow bridges and tunnels, cross traffic areas, and congested traffic areas. Velocity limits may be further restricted because of adverse weather such as snow, ice, fog, and rain. Velocity limits also may be required to keep truck tires below their

tonne-kilometers per hour (t-km/h) (ton-miles per hour, tmph) rating. The velocity of the slowest truck in the fleet may restrict velocity of other trucks. Good haulage road maintenance will reduce travel time.

Turn and Dump

Turn and dump time depends on truck type and size, material characteristics, dump arrangements, space available, ground conditions, and operating practices. A rear dump truck enters the dump area, turns, back up, stops, raises body to dump payload, and lowers body. The time to hoist and lower the body of rear dump trucks ranges from about 0.30 to 0.70 min; time generally increases with payload. A bottom dump tractor-trailer unit drives forward through or over the dump area, sometimes dumping without even stopping.

Nominal turn and dump times are 1.00 min. for conventional rear dump trucks and 0.50 min. for tractor-trailer units. Nonfree-flowing, sticky, and/or frozen material may substantially increase these times and reduce actual payload. These nominal times may be less when dumping into a hopper and greater when dumping on a waste dump. At waste dumps, auxiliary equipment such as a dozer or grader usually is required for proper dump construction. While space usually is not limited on waste dumps and stockpiles, poor ground conditions may exist. Operating practices that maintain the dump area and minimize traffic congestion and safety hazards will decrease turn and dump times.

Wait

Truck wait time occurs when the truck must wait to be loaded or to dump because of nonsynchronization in the haulage system. Wait time occurs because of the following:

1) Over-trucking exists when the number of trucks in the system exceeds loading and/or dumping capabilities.

2) Bunching results when the spacing between trucks is reduced due to mixing faster and slower trucks or fleet start-up.

3) Mismatching of equipment exists when the system has variable size equipment with variable performance characteristics (i.e., small and large trucks in the same fleet) resulting in variable truck cycle times.

4) Weather conditions such as rain or ice which result in variable truck cycle times.

5) Operator performance causing variations in truck cycle times due to human participation.

Wait times can be reduced by improving truck allocation and supervision and by dispatching systems.

Wait time due to variable truck cycle times can be estimated based on the coefficient of variation of cycle times. Coefficient of variation, C_v, is calculated by

$$C_v = \frac{\sigma}{\bar{x}}$$

where σ is the standard deviation and $\bar{x}$ the mean. The coefficient of variation normally is from 0.10 to 0.20 for truck cycle times. A low coefficient of variation indicates a well synchronized system with a minimum of wait time.

Methods for estimating wait time include field studies, efficiency factors, and simulation. If practical, wait time should be estimated by simulation based on field studies rather than simple job efficiency factors.

Delays

Operational delays that reduce production can be classified as fixed or variable delays. Fixed delays are predictable

Table 5. Shovel Work Cycle Times Hard Digging Conditions

Dipper Capacity m³ (cu yd)	Nominal Minutes*
4.6(6)	0.47
6.9(9)	0.48
9.2(12)	0.49
11.5(15)	0.50
15.3(20)	0.52
19.1(25)	0.53

* Based on 90° swing.

as to time of occurrence and duration such as shift change, equipment inspection, breaks, refueling, blasting, etc. Fixed delays usually are not considered in truck cycle time. Variable delays are not predictable as to time of occurrence and duration. Variable delays include delays for haulage road maintenance, loading area cleanup, driver relief stop, loading oversize rock, etc., and are considered in the truck cycle time. Unscheduled mechanical delays are variable delays, but these are not considered in the truck cycle time.

Variable delays causing variations in the truck cycle time will result in increased truck wait time. These delays and associated wait time can be estimated by the same methods used for estimating wait time.

Total Cycle Time

Total truck cycle time is the summation of spot and load, haul loaded, turn and dump, haul empty, wait, and delay times. Total cycle time usually is expressed as an average time for specific loading equipment and trucks operating on a specific haulage road.

PRODUCTION AND FLEET REQUIREMENTS

Truck production and fleet requirements are affected by many factors: mine plan, haulage roads, mine production requirements, loading equipment, truck performance and cycle time, operating methods and practices, matching of loading equipment and trucks, and equipment availability and utilization. Methods used to estimate and evaluate truck production and fleet requirements vary from simple rules of thumb to complex computerized simulation.

Matching Trucks and Loading Equipment

Based on truck cycle and spot and load times, trucks and loading equipment must be matched to avoid excessive over- or under-trucking the loader or shovel. The theoretical number of trucks that can service a loader or shovel, N_T, can be derived by

$$N_T = \frac{t_{tc}}{t_{tl}}$$

where t_{tc} is the total theoretical truck cycle time (no wait time) in minutes and t_{tl} is the truck spot and load time in minutes. In calculating the theoretical number of trucks, N_T, the truck wait time, must be excluded from the truck cycle time. For loading methods where an arriving truck is spotted while another truck is loaded, truck spot and load time is replaced by loader or shovel loading time.

If the number of trucks is less than N_T, the loader or shovel is under-trucked. Truck wait times are less with under-trucking than with over-trucking; however, loader or

shovel wait times are higher with under-trucking. One purpose of truck dispatch systems is to control over- and under-trucking to achieve minimum combined loading and hauling cost.

Table 6 presents average truck cycle times for a fleet of one to six 109-t (120-ton) trucks. Because of interference among trucks which causes variation in actual truck cycle times, wait time increases as fleet size increases. In Table 6, the theoretical number of trucks is 4.94 (15.80/3.20) or five trucks.

Availability and Utilization

Availability and utilization do not have standardized definitions and, thus, these terms must be used with care. Irrespective of availability and utilization, percentage of use or actual operating time is needed to estimate truck production. Actual operating time can be defined as the time a truck operates in its designated operating mode such as load, haul, dump, return, wait, and variable delays. This time cannot be estimated using availability and utilization data as commonly defined, but can be estimated by developing detailed time distributions.

Tables 7 and 8 present truck scheduled and actual time distributions, respectively, for a truck fleet operating for different mine schedules and mechanical downtime requirements. The two schedules are for mines operating 5 or 20 shifts per week with mechanical maintenance available 10 or 21 shifts per week, respectively. The two mechanical downtime requirements are for ratios of actual total repair time to actual active operating time of 0.25 or 0.50. The ratio of operating shift actual unscheduled repair time to total actual repair time is 0.20. Because of unscheduled repair time, available operating time (Table 8) is less than scheduled operating time (Table 7). Truck fleet time can be obtained by multiplying truck scheduled and actual times by the number of trucks in the fleet.

The probability of a single truck being available at any specific time, P, can be determined by

$$P = \frac{\text{available operating time}}{\text{scheduled operating time}}$$

Using case C as an example, the scheduled and available operating times in Tables 7 and 8 are 144.00 and 138.33 hr, respectively, resulting in a 0.9606 or 96% probability of a single truck being available.

Assuming that availability of a specific truck is independent of the availability of any other truck, the probability of k trucks being available, P_k, can be determined by

$$P_k = p^k \times (1 - p)^{n-k} \times C^{-n}{}_k$$

where P is the probability of a single truck being available,

Table 6. Truck Cycle Times 109-t (120-ton) Actual Payload

Number of trucks in fleet	1	2	3	4	5	6
Truck cycle time, minutes						
Spot and load	3.20	3.20	3.20	3.20	3.20	3.20
Haul loaded	7.50	7.50	7.00	7.00	7.00	7.00
Turn and dump	0.60	0.60	0.60	0.60	0.60	0.60
Return empty	4.00	4.00	4.50	4.50	4.50	4.50
Wait	0.00	0.00	0.45	1.15	2.40	4.40
Delays, variable	0.50	0.50	0.50	0.50	0.50	0.50
Total	15.80	15.80	16.25	16.95	18.20	20.20

Table 7. Truck Scheduled Time Distributions No Scheduled Standby

Case	A,B	C,D
Scheduled, shifts/week		
Operating	5	20
Maintenance	10	21
Distribution, hours/week		
Operating Shifts		
Operating, scheduled		
Idle	1.50	6.00
Operating, active	29.50	118.00
Delays, fixed	5.00	20.00
Subtotal	36.00	144.00
PM, scheduled	1.50	6.00
Repair, scheduled	2.50	10.00
Repair, unscheduled	—	—
Standby	—	—
Total	40.00	160.00

n is the total trucks in fleet, k are the available trucks, and $C^n_k = n!/k!(n-k)!$

As an example, assuming case C with a probability of 0.9606 and which has a five truck fleet, the probability of four trucks being available is calculated by

$$P_4 = (0.9606)^4 \times (1 - 0.9606)^{5-4}$$
$$\times \frac{5 \times 4 \times 3 \times 2 \times 1}{4 \times 3 \times 2 \times 1 \times 1}$$
$$P_4 = 0.1677 = 16.77\%$$

Thus, four trucks will be available for operation 16.77% of the scheduled operating time.

Production

Truck production, assuming 100% availability and/or utilization can be calculated by

$$P_t = \frac{60 \times L_t}{t_{tc}}$$

where P_t is the truck production rate based on actual active-operating time, t/h (stph); L_t is the truck actual payload, t (ton); and t_{tc} is the truck total cycle time, minute.

Trips per hour made by a truck, T, can be calculated by

$$T = \frac{60}{t_{tc}}$$

Thus, truck production rate can be calculated by

$$P_t = T \times L_t$$

Truck fleet production, P_f, at a specific time is calculated by

$$P_f = k \times P_t$$

where k are the available trucks in the fleet at a specific time.

Table 9 presents estimated production for a fleet of five 109-t (120-ton) trucks with the truck cycle times in Table 6 and active operating times in Table 8. It should be noted that truck cycle time, truck trips per active operating hour, truck and fleet production per active operating hour, and fleet production per available operating hour are the same for all cases. However, fleet production per scheduled op-

erating hour and per shift are different because these are dependent on active operating hours per shift.

Truck fleet production in Table 9 has been calculated by two different procedures. In the first procedure, production is based on all five trucks being available. Production per available and scheduled operating hours are based on ratios of active to available and scheduled operating times. Production per shift is based on production per active operating hour and active operating hours per shift.

In the second procedure, production is based on the probability of trucks being available during scheduled operating time. Table 10 presents the procedure used to estimate production per scheduled operating hour for case C. For each case, production per available operating hour is based on truck cycle time in Table 6 and the ratio of active to available operating time. Production per shift is based on production per scheduled operating hours per shift.

Table 11 presents the actual time distributions when the truck fleet production requirements are limited to less than fleet production capabilities. This results in truck standby time. In certain situations, such as case D, time distributions and probability of a truck being available may have to be estimated by successive approximations.

It should be noted that the truck fleet production rates presented in Tables 9 and 11 assume that loading equipment is always available as required by trucks. Since this normally is not the situation, the production rate for the loading and haulage system will be less than the production rate for the truck fleet because of the probability of loading equipment not being available.

Table 8. Truck Actual Time Distributions No Scheduled Standby

Case	A	B	C	D
Scheduled, shifts/week				
Operating	5	5	20	20
Maintenance	10	10	21	21
Ratio*				
R_1	0.25	0.50	0.25	0.50
R_2	0.20	0.20	0.20	0.20
Distribution, hours/week				
Operating shifts				
Operating, available				
Idle	1.44	1.39	5.76	4.96
Operating, active	28.34	27.27	113.36	97.66
Delays, fixed	4.80	4.62	19.21	16.55
Subtotal	34.58	33.28	138.33	119.17
PM, scheduled	1.50	1.50	6.00	6.00
Repair, scheduled	2.50	2.50	10.00	10.00
Repair, unscheduled	1.42	2.72	5.67	24.83
Standby	—	—	—	—
Total	40.00	40.00	160.00	160.00
Maintenance				
Idle	38.33	33.09	1.33	—
Repair	1.67	6.91	6.67	8.00
Total	40.00	40.00	8.00	8.00
Available operating time	34.58	33.28	138.33	119.17
Scheduled operating time	36.00	36.00	144.00	144.00
Probability	0.9606	0.9244	0.9606	0.8276

* R_1 = Ratio of total repair time to active operating time.
  R_2 = Ratio of unscheduled repair time to total repair time.

Table 9. Truck Fleet Production

Case	A	B	C	D
Truck actual payload, tonne (ton)	109 (120)	109 (120)	109 (120)	109 (120)
Number of trucks	5	5	5	5
Scheduled, shifts/week				
Operating	5	5	20	20
Time operating, active				
Hours/week	28.34	27.27	113.36	97.66
Hours/shift	5.67	5.45	5.67	4.88
Truck cycle time, minute	18.20	18.20	18.20	18.20
Truck trips/active operating hour	3.30	3.30	3.30	3.30
Truck production				
Tonne (ton)/active operating hour	359 (396)	359 (396)	359 (396)	359 (396)
Truck fleet production				
Based on all trucks				
Tonne (ton)/active operating hour	1,795 (1,980)	1,795 (1,980)	1,795 (1,980)	1,795 (1,980)
Tonne (ton)/available operating hour	1,471 (1,623)	1,471 (1,623)	1,471 (1,623)	1,471 (1,623)
Tonne (ton)/scheduled operating hour	1,413 (1,558)	1,360 (1,500)	1,413 (1,558)	1,217 (1,343)
Tonne (ton)/shift	10,174 (11,227)	9,790 (10,791)	10,084 (11,227)	8,765 (9,662)
Truck fleet production				
Based on probability				
Tonne (ton)/scheduled operating hour	1,430 (1,576)	1,390 (1,532)	1,430 (1,576)	1,173 (1,403)
Tonne (ton)/shift	10,294 (11,347)	10,006 (11,030)	10,294 (11,347)	9,164 (10,102)

Truck Requirements

The number of trucks required for the fleet is determined by comparing various fleet production capabilities and costs with production requirements and selecting the lowest cost fleet with adequate production capability. Table 12 presents estimated production for loading and haulage systems using fleets of five and six 109-t (120-ton) trucks with one or two loaders or shovels having an 0.85 or 85% probability of being available.

Loading and haulage system production for five-truck fleets with one loader or shovel having a 1.00 probability is the same as presented in Table 9. With one loader or shovel having an 0.85 probability, production is 85% of production with 1.00 probability. If two loaders or shovels are available, but only one is used at any one time, the probability of at least one loader or shovel being available is 0.9775 or 97.75%. Thus, with two loaders or shovels, production is 97.75% of production with 1.00 probability. Production for the six-truck fleets was estimated by the same procedures as for the five-truck fleets.

If production of 9072 t (10,000 st) per shift is required for the loading and haulage system, this required production for cases A, B, and C can be obtained with (1) five trucks and two loaders or shovels or (2) six trucks and one loader or shovel. For case D, six trucks and two loaders or shovels are necessary for the required production. Loading and haulage costs using the various fleets must then be estimated to determine the lowest cost system.

The methods presented illustrate some of the complexities in estimating truck production and fleet requirements. In determining requirements, it is necessary to consider various truck cycle times, scheduled and actual time distributions, equipment probabilities of being available, and production requirements.

COSTS

Costs generally associated with trucks are ownership and operating costs. Costs for purchase and operation of a specific truck vary widely depending on delivered truck price including shipping cost, current finance charges, truck appli-

Table 10. Truck Fleet Production for Case C
Based on Probability of 0.9606

Trucks k	Probability of k Trucks Available P_k	Production Tonne (ton)/Available Operating Hour	Production Tonne (ton)/Scheduled Operating Hour
0	0.0000	0 (0)	0 (0)
1	0.0000	339 (374)	0 (0)
2	0.0006	678 (747)	0 (0)
3	0.0138	988 (1089)	14 (15)
4	0.1677	1263 (1393)	212 (234)
5	0.8179	1471 (1623)	1203 (1327)
Total	1.0000		1429 (1576)

Table 11. Truck Actual Time Distributions Truck Fleet Production of 9072 t (10,000 ton) per Shift

Case	A	B	C	D
Scheduled, shifts/week				
Operating	5	5	20	20
Maintenance	10	10	21	21
Ratio*				
R_1	0.25	0.50	0.25	0.50
R_2	0.20	0.20	0.20	0.20
Distribution/hours/week				
Operating shifts				
Operating, available				
Idle	1.27	1.26	5.08	4.90
Operating, active	24.98	24.72	99.90	96.56
Delays, fixed	4.23	4.19	16.93	16.36
Subtotal	30.48	30.17	121.91	117.82
PM, scheduled	1.50	1.50	6.00	6.00
Repair, scheduled	2.50	2.50	10.00	10.00
Repair, unscheduled	1.25	2.47	5.00	24.28
Standby	4.27	3.36	17.09	1.90
Total	40.00	40.00	160.00	160.00
Maintenance				
Idle	39.00	34.11	4.02	—
Repair	1.00	5.89	3.98	8.00
Total	40.00	40.00	8.00	8.00
Available operating time	30.48	30.17	121.91	117.82
Scheduled operating time	31.73	32.64	126.91	142.10
Probability	0.9606	0.9243	0.9606	0.8291

* R_1 = Ratio of total repair time to active operating time.
R_2 = Ratio of unscheduled repair time to total repair time.

cation and loading and haulage conditions, local fuel and lubricant prices, parts availability and prices, local labor availability, skills, and wages, etc. Reliable cost estimates must be based on an accurate assessment of mining conditions and current local data. Thus, methods presented for cost estimating are general and not intended for precise cost estimates.

Truck ownership and operating costs may be presented as annual or hourly costs. Hourly costs usually are based on truck actual operating time which does not include idle, delay, preventive maintenance, repair, and standby times.

Ownership and operating costs, in conjunction with truck fleet production, are used to determine haulage production cost. Haulage cost usually is a major portion of the total mining cost and cost per unit of material mined. However, haulage cost does not reflect the costs for drilling and blasting, loading, haulage road construction and maintenance, supervision, and other activities that may influence haulage cost. Thus, truck fleet selection decisions must be based on total mining cost rather than just haulage cost because the truck fleet interacts with other unit operations and equipment.

Ownership Costs

Ownership costs usually are fixed charges incurred by having equipment at the mine. For tax purposes, these costs may occur either as capital, financial, depreciation, or operating charges and may be determined both as annual and hourly costs.

Truck capital expenditure or depreciable value is delivered truck cost including basic truck and optional equipment,

freight, and erection at the mine, but excluding tires. Truck and optional equipment cost may vary from the manufacturer's list price, depending on the truck market at time of purchase. List price normally includes tires, but trucks frequently are purchased without tires because the mine may purchase tires directly at a lower cost.

Freight cost depends on the truck shipping weight which can be provided by the manufacturer. Freight cost can be substantial for export shipments which also may require import duty payments. Erection cost depends on the degree of truck breakdown for shipment and availability of erection equipment and personnel. Tire cost should be obtained from local tire distributors. Tire cost based on fleet discounts usually is 45 to 55% of tire list price plus tax.

Useful truck life varies widely depending on truck application and loading and haulage conditions, operating methods and procedures, and maintenance programs. Truck life usually is determined by operating hours. Small mechanical drive trucks have a useful life of 20,000 to 30,000 hr and large electric wheel drive trucks have a useful life of 30,000 to 40,000 hr. At some mines, trucks are reconditioned and rebuilt to extend useful life almost indefinitely. Truck operating hours per year depend on truck scheduling and may approach 6000 hr per year for mines operating 20 shifts per week.

Average annual investment can be calculated by

$$\text{average annual investment} = \text{depreciable value} \times \frac{N+1}{2N}$$

where N is useful life in years. Annual charges for interest, insurance, and taxes are often estimated as percentages of average annual investment.

Truck capital expenditures may be recovered by an annual or hourly depreciation charge. Depreciation charge may be calculated using the truck's depreciable value and useful life or some other method established by tax codes. Charges for interest, insurance, and taxes are often difficult to determine for specific equipment because these may be covered by corporate financing, insurance policies, and tax rates.

Table 13 shows capital expenditures for a 109-t (120-ton) conventional rear dump truck with electric wheel drive. Ownership costs are based on 30,000 hr useful life over a six-year period.

Operating Costs

Operating costs are variable charges directly applicable to truck operation. These normally include fuel, tires and tire repair, preventive maintenance, mechanical repairs, and operator wages. Operating costs normally do not include indirect mine costs associated with mine facilities and supervision. Operating costs can best be estimated from field studies and analysis of mine records.

Fuel cost depends on fuel consumption and local fuel price. Fuel consumption, C_f, in liters per hour (gallons per hour) can be calculated by

$$C_f = \frac{C_{sf} \times P}{\rho_f} \times F_l$$

where C_{sf} is the engine specific fuel consumption at full power, kg fuel/bkW-h (lb fuel per bhp-hr); P is the rated brake power, kW (hp); ρ_f is the fuel density, kg/L (lb/per gal); and F_l is the engine load factor, decimal.

Engine specific fuel consumption at full power ranges from 0.213 to 0.268 kg/bkW-r (0.35 to 0.44 lb per bhp-h) with lower fuel consumption for newer engines with turbo-

Table 12. Loading and Haulage System Production

Case	A	B	C	D
Truck actual payload, tonne (st)	109 (120)	109 (120)	109 (120)	109 (120)
Number of trucks	5	5	5	5
Loading equipment				
Number of loaders	1	1	1	1
Probability	1.00	1.00	1.00	1.00
System production				
Tonne (st)/shift	10,294 (11,347)	10,006 (11,030)	10,294 (11,347)	9,164 (10,102)
Loading equipment				
Number of loaders	1	1	1	1
Probability	0.85	0.85	0.85	0.85
System production				
Tonne (st)/shift	8,750 (9,645)	8,506 (9,376)	8,750 (9,645)	7,790 (8,587)
Loading equipment				
Number of loaders*	2	2	2	2
Probability	0.85	0.85	0.85	0.85
System production				
Tonne (st)/shift	10,062 (11,092)	9,781 (10,782)	10,062 (11,092)	8,958 (9,874)
Number of trucks	6	6	6	6
Loading equipment				
Number of loaders	1	1	1	1
Probability	1.00	1.00	1.00	1.00
System production				
Tonne (st)/shift	11,222 (12,370)	10,993 (12,118)	11,222 (12,370)	10,287 (11,340)
Loading equipment				
Number of loaders	1	1	1	1
Probability	0.85	0.85	0.85	0.85
System production				
Tonne (st)/shift	9,539 (10,515)	9,344 (10,300)	9,539 (10,515)	8,744 (9,639)
Loading equipment				
Number of loaders*	2	2	2	2
Probability	0.85	0.85	0.85	0.85
System production				
Tonne (st)/shift	10,970 (12,092)	10,746 (11,845)	10,970 (12,092)	10,056 (11,085)

* Only one loader used; other loader on standby.

charging and after-cooling. Engine manufacturers provide standard engine performance data, including brake kilowatt (horsepower) and specific fuel consumption, for the entire engine operating speed (rpm) range. Diesel fuel density ranges from 0.84 to 0.96 kg/L (7.0 to 8.0 lb per gal) with No. 2 diesel fuel having a normal density of about 0.85 kg/L (7.1 lb per gal). Fuel density data should be obtained from local suppliers.

Engine load factor is the portion of full power required by the truck. While idling and retarding, a truck engine operates at about 10% of full power with a load factor of 0.10. When moving, load factor is the ratio of rimpull required or used to available rimpull at the velocity at which the truck is moving. During acceleration, the engine usually operates at full power or maximum available rimpull with a load factor of 1.00. For operation at constant velocity, the rimpull required is equal to the resistance to movement. Table 14 presents load factors for the three main truck types when operating under light, average, and heavy haulage conditions. Load factors also can be estimated by computer simulation programs which estimate travel time.

As an example, fuel consumption for a 109-t (120-ton) truck with an 895-kV (1200-hp) engine operating with a 0.40 engine load factor is 97.3 L/h [(0.231 kg/bkW-h × 895 kW/0.85 kg/L) × 0.40] (25.7 gal per hr) [(0.38 lb per bhp-hr × 1200 hp / 7.1 lb per gal) × 0.40] based on a fuel density of 0.85 kg/L (7.1 lb per gal.)

Hourly fuel cost is fuel consumption in liters per hour (gallons per hour) times local fuel price.

Tire cost includes tire replacement and tire repair costs based on tire life. Tire life is affected by tire construction, type, and maintenance; truck application and velocity; haulage road construction, maintenance, grades, and curves; and tire load and position (front, drive, or trailing). Tire life is measured in kilometers (miles) or hours. Based on favorable conditions, E-3 (rock) and E-4 (rock deep tread) bias ply tires can be expected to have a tire life of 40 200 to 56 300 km (25,000 to 35,000 miles) and 56 300 to 72 400 km (35,000 to 45,000 miles), respectively. Radial tires with deep tread can be expected to have a tire life of 64 400 to 80 500 km (40,000 to 50,000 miles). Based on average truck velocity of 16.1 km/h (10 mph), these tire lives are equivalent to 2500 to 3500 hr for E-3 tires, 3500 to 4500 hr for E-4 tires, and 4000 to 5000 hr for deep tread radial tires. Tire life tends to be less for smaller trucks than for larger trucks and is reduced substantially at mines where frequent premature tire failures occur due to overloading, rock cuts, and punctures. Mine experience also has demonstrated that tire life can be increased substantially by very favorable operating conditions.

The hourly tire and tire repair cost is 115% of the hourly tire replacement cost when the tire replacement cost is tire cost divided by tire life in hours. This allows 15% of tire replacement cost for tire maintenance and repair. Tire cost

Table 13. Truck Ownership and Operating Costs 109-t (120-ton), 895-kW (1200-hp) Electric Wheel Drive

	$/hr	$/yr
Capital Cost		
Standard truck, fob factory		$675,000
Optional equipment, fob factory		32,000
Freight (193,000 lb at 5.7¢/lb)		11,000
Erection		6,000
Delivered cost		$724,000
Tires (30.00-51(46) E-4)		43,000
Depreciable value		$681,000
Truck life		
Years		6
Hours		30,000
Average annual investment		$397,250
Ownership costs		
Depreciation	22.70	113,500
Interest (15% of average yearly investment)	11.92	59,588
Insurance (2% of average yearly investment)	1.59	7,945
Taxes (2% of average yearly investment)	1.59	7,945
Total ownership costs	37.80	188,978
Operating costs		
Fuel (0.40 load factor)		
97.3 L/hr at $0.330/L		
(25.7 gal/hr at $1.25/gal)	32.13	160,625
Tires and tire repair (4000 hr)		
($43,000/4000 hr) × 1.10	11.83	59,125
Preventive maintenance		
(0.20 × $32.13)	6.43	32,125
Repair (0.40 repair factor)		
[($681,000/10,000) × 0.40 × 0.65 × 1.00]	17.71	88,530
Operator	18.00	90,000
Total operating costs	86.10	430,405
Total ownership and operating costs	123.90	619,383

often is reduced by recapping, with recapping prices obtained from local tire distributors.

Preventive maintenance cost includes lubricating oils and greases, filters, and miscellaneous parts and labor required for truck servicing. Preventive maintenance cost is about 20% of fuel cost. A more detailed estimate of both parts and labor can be based on truck service capacities; manufacturer's recommended service intervals; and local oil, grease, filter, and labor costs.

Repair cost includes all truck maintenance expenditures not considered preventive maintenance. This includes both parts and labor for truck overhauls and rebuilds. Two com-

mon methods for estimating repair costs are based on either depreciable value or fuel consumption.

Using depreciable value, repair cost, R_c, in dollars per hour can be estimated by

$$R_c = \frac{V_d}{10,000} \times F_r$$

where V_d is depreciable value and F_r is repair factor. It is important that depreciable value be a current value to reflect current repair cost. Table 15 presents truck repair factors for a range of operating conditions. These repair factors are for mechanical drive trucks and must be reduced for electric wheel drive trucks. Table 16 presents adjustment factors for adjusting the repair factor for truck age. As an example, the estimated average hourly repair cost in Table 13 for the 109-t (120-ton) electric wheel drive truck with a 0.40 repair factor and 30,000-hr life is $17.71/hr [($681,000/10,000) × 0.40 × 0.65 × 1.00]. The estimated repair cost while operating between 20,000 to 30,000 hr is $22.13/hr [($681,000/10,000) × 0.40 × 0.65 × 1.25].

Using fuel consumption to estimate repair cost relates repair cost to engine load factor or work performed by the truck. Using current (1983) parts and labor prices, repair cost for conventional rear dump trucks ranges from $0.24 to $0.42/L ($0.90 to $1.60 per gal) of fuel for mechanical drive trucks and from $0.16 to $0.28/L ($0.60 to $1.05 per gal) for electric wheel drive trucks. These repair costs are

Table 14. Truck Engine Load Factors

Truck Type	Load Factor*		
	Light	Average	Heavy
Conventional rear dump	0.25	0.35	0.50
Tractor-trailer	0.35	0.50	0.65
Integral bottom dump	0.25	0.35	0.50

* *Light*: Considerable idle, loaded hauls on favorable grades and good haulage roads.

Average: Normal idle, loaded hauls on adverse grades and good haulage roads.

Heavy: Minimum idle, loaded hauls on steep adverse grades.

Table 15. Truck Repair Factors

| | Repair Factor* | | | |
	Favorable	Average	Unfavorable	Very Unfavorable
Conventional rear dump	0.30	0.50	0.75	1.00
Tractor-trailer	0.25	0.40	0.60	0.80

* Repair factors for mechanical drive truck; multiply by 0.65 for electric wheel drive truck.

based on average cost for a 30,000-hr truck life. As an example, repair cost in Table 13 for the 109-t (120-ton) truck could range from $15.42 to $26.99/hr using this estimating method.

Operator cost is wages for the truck operator and normally includes base wage, fringe benefits, training, etc.

Production Costs

Haulage production costs usually are presented as cost per tonne (ton) or cost per tonne-kilometer (ton-mile). Cost per tonne (ton) is simply total truck cost divided by the total tonnes (tons) hauled by the truck. Cost per tonne-kilometer (ton-mile) considers the distance the material is hauled and is calculated by total truck cost divided by cumulative tonne-kilometers (ton-miles) which is cumulative weight times distance. As an example, a truck with a 109-t (120-ton) payload making 3.3 round trips per active operating hour with an operating cost of $86.10/hr (Table 13) has an operating cost of $0.2397/t ($0.2174/ton). If the one-way haulage distance is 2438 m (8000 ft), the operating cost is $0.0983/tonne-km ($0.1435/ton-mile).

REFERENCES

Alberte, T., 1973, "150 Ton Carrier Features Two-Axle Design, Engine Module, " *Diesel & Gas Turbine Progress,* North American ed., Dec.

Anon., 1982, "Better Haul Roads Speed Operations," *Coal Age,* Vol. 87, No. 3, Mar., pp. 96-99

Anon., 1978, "Braking System Problems, Off-Highway Vehicles, Review of Progress," Energy Resources Conservation Board, Calgary, Alberta, Canada, March 13.

Anon., 1978, "New Brake System Meets Tough Standards," *Engineering & Mining Journal,* Vol. 179, No. 8, Aug., pp. 80-81.

Anon., 1982, "Open-Pit Haul Trucks," *Engineering & Mining Journal,* Vol. 34, No. 6, June, pp. 134-136.

Barnes, R.J., King, M.S., and Johnson, T.B., 1979, "Probability Techniques for Analyzing Open Pit Production Systems," *16th Application of Computers and Operations Research in the Mineral Industry,* T.J. O'Neil, ed., SME-AIME, New York, pp. 462-276.

Beebower, A.G., and Nadzam, J.P., 1971, "Off-Highway Haulage Vehicle Diesel Electric Drive Performance," 710154, Society of Automotive Engineers, Inc., Jan., 14 pp.

Table 16. Truck Repair Factor Adjustment Factor for Truck Age

| | Adjustment Factor | |
Truck Life, hr	Incremental	Average
5,000	0.65	0.65
10,000	0.80	0.73
15,000	0.95	0.80
20,000	1.10	0.88
30,000	1.25	1.00
40,000	1.75	1.19

Butler, J.M., and Buerschinger, D.R., 1972, "WABCO's 200 Ton Truck Electro-mechanical Drive System," 720754, Society of Automotive Engineers, Inc., Sept., 7 pp.

Burton, A.K., 1975-1976, "Off-Highway Trucks in the Mining Industry (A five part series)," *Mining Engineering,* Aug.-Dec., Jan.

Burton, A.K., 1981, "The Effect of Oil Economics on Open Pit Mine Haulage Trends," *Open Pit Haulage Systems in the 80's,* 81-M&E-02, SME-AIME, Feb., pp. 1-5.

Carr, D.B., 1977, "Dart's 1000 hp Mechanical Drive Off-Highway Trucks," 770549, Society of Automotive Engineers, Inc., Apr., 12 pp.

Chironis, N.P., 1980, "Haulage Trucks Still Supreme," *Coal Age,* Vol. 85, No. 11, Nov., pp. 94-109.

Conger, H.M., 1972, "Economics Associated with New Designs and Applications of Large Equipment—Hauling," *Mining Congress Journal,* Vol. 58, No. 11, Nov., pp. 46-49.

Crosson, C.C., and Sumner, H.B., 1982, "Trolley Assisted Truck Haulage," *Engineering & Mining Journal,* Vol. 183, No. 6, June, pp. 88-98.

Cummings, B., 1977, "Matching Loaders and Trucks for Open-pit Mining," *Canadian Mining Journal,* Vol. 98, No. 10, Oct., pp. 52-53, 55.

Dawson, V.E., 1975, "Observations Concerning On-Site Brake Testing of Large Trucks in British Columbia," 750560, Society of Automotive Engineers, Inc., Apr., 33 pp.

Deshmukh, S.S., 1070, "Sixing of Fleets in Open Pits," *Mining Engineering,* Vol. 22, No. 12, Dec., pp. 41-45.

Felix, G., 1976, "Development of a 350-Ton Haulage Truck," 760408, Society of Automotive Engineers, Inc., Apr., 5 pp.

Goodbary, E.R., 1977, "Design Considerations for the Goodbary Bottom Dump Haulage Trucks," 770548, Society of Automotive Engineers, Inc., Apr., 6 pp.

Greshan, R.B., and Moray, R.D., 1974, "Electric-Drive Trucks at Canadian Johns-Manville's Jeffrey Mine," *Canadian Mining and Metallurgical Bulletin,* Vol. 67, No. 742, Feb., pp. 47-50.

Jarman, S.D., 1981, "Truck Evaluation Program," *Surface Coal Mining and Reclamation Symposium,* Coal Age Coal Conference & Expo VI, McGraw-Hill, Inc., pp. 190-205.

Kaufman, D., 1980, "Mechanical Versus Electric Drive, How the Availability of New Larger Mechanical Transmissions May Influence the Design of Off-Highway Haulage Trucks," *Engineering & Mining Journal,* Vol. 181, No. 8, Aug., pp. 533-538.

Kaufman, D., and Bowen, R., 1981, "Analyze Haul Truck Costs Wisely," *Coal Age,* Vol. 86, No. 3, Mar., pp. 70-77.

Kaufman, W.W., and Ault, J.C., 1977, "Design of Surface Mine Haulage Roads—A Manual," Information Circular 8758, US Bureau of Mines, 68 pp.

Kelley, D.M., 1972, "Euclid Turbine-Electric Rear Dump," 720375, Society of Automotive Engineers, Inc.

Lake, D.M., Brzezniak, W., and Hendry, M.C., 1981, "Truck Haulage Using Overhead Electrical Power to Conserve Diesel Fuel and Haulage Economics," GER-3278, General Electric Co., Erie, PA, Feb., 19 pp.

Lerick, G.E., 1979, "Use of Large Trucks in Taconite Operations," *Mining Congress Journal,* Mar., pp. 47-51.

Lyon, L.S., 1981, "New Developments in Large Haulage Trucks for the 1980's," *Open Pit Haulage Systems in the 80's,* 81-M&E-02, SME-AIME, Feb., pp. 67-82.

Martin, J.W., Martin, T.J., Bennett, T.P., and Martin, K.M., 1982, *Surface Mining Equipment,* 1st ed., Martin Consultants, Inc., Golden, Colorado, 453 pp.

Miller, J.E., 1981, "Development of Automatic Truck Control (ATC)," *Mining Engineering,* Vol. 182, No. 2, Feb., pp. 175-178.

Morgan, W.C., and Peterson, L.L., 1968, "Determining Shovel-Truck Productivity," *Mining Engineering,* Vol. 20, No. 12, Dec., pp. 76-80.

Nadzam, J.P., and Beebower, A.G., 1971, "In Growing Numbers, Electric Trucks Make Their Mark at Open-pit Mines," *Engineering & Mining Journal,* Vol. 172, No. 8, Aug., pp. 55-61.

Naft, M.H., 1977, "Integration of Component Design for a 170 Ton Off-Highway Truck," 770741, Society of Automotive Engineers, Inc., Sept., 12 pp.

Olson, E.K., Jr., 1974, "Truck Servicing and Maintenance Program at Kaiser's Eagle Mountain Mine," *Mining Congress Journal,* Vol. 60, No. 3, Mar., pp. 20-26.

Pugh, R. D., 1975, "Design Problems on Rear Dump Haulers," *Mining Magazine,* July, pp. 27, 29, 31.

Smiley, C.H., 1979, "Mechanical Drive Improves Mid-size Haulers," *Coal Mining & Processing,* Vol. 16, No. 4, Apr., pp. 78-80.

Sons, C., 1982, "Reducing Earthmover Tire Costs," *Mining Engineering,* Vol. 34, No. 1, Jan., pp. 27-30.

Sullivan, A.M., 1982, "Tires—They Make the Coal Move 'Round," *Coal Age,* Vol. 87, No. 4, Apr., pp. 138-153.

Thomas, P.R., and Till, R.H., 1974, " A Simplified Method for the Measurement of Vehicular Rolling Resistance," 740423, Society of Automotive Engineers, Inc., Apr., 12 pp.

Trindal, W.S., 1973, "Technical Analysis of Off-Road Tires," 730853, Society of Automotive Engineers, Inc., Sept., 15 pp.

Vanchina, J.M., 1982, "Trolley Assistance for Mine Trucks," *Mining Congress Journal,* Vol. 68, No. 4, Apr., pp. 61-63.

Walker, G., 1976, "Recommended Standards for the Service Brakes on Large Trucks in Mountain Mining Service," 760430, Society of Automotive Engineers, Inc., Apr., 8 pp.

White, J.W., Arnold, M.J., and Clevenger, J.G., 1982, "Automated Open-Pit Truck Dispatching at Tyrone," *Engineering & Mining Journal,* Vol. 183, No. 6, June, pp. 76-84.

6.5.3 Belt Conveyors

LARRY D. DUNCAN AND BRIAN J. LEVITT

In recent years the changes in the relative costs of equipment, labor, and fuels have made belt conveyors the most economical means of moving large quantities of bulk materials for distances up to several miles. The reliability, high-volume capability, and energy efficiency of belt conveyors have made them dominant for in-plant bulk handling for a long time. The recent costs of petroleum-based fuels and increased emphasis on environmental protection have created more usage of conveyors for overland conveyance of several miles. One of the more recent developments for long distance conveying has been the use of horizontal curves that adapt to terrain and eliminate the need for costly belt transfer structures and drives.

This section on belt conveyors is too limited in scope to serve as a design manual for conveyors. An excellent in-depth guide for detailed design of conveyors is *Belt Conveyors for Bulk Materials,* published by the Conveyor Equipment Manufacturers Association. This section is intended to serve as a guide in the selection of a type of belt conveyor, determination of size, energy requirements, and budget costs for the conventional type of conveyors. This information should enable a mining engineer to do preliminary sizing, system optimization, and capital and operating cost comparisons with different system configurations and alternative transport modes, such as trucks.

GENERAL APPLICABILITY

For surface mining applications, the use of belt conveyors can be divided into the following general applications: (1) in-pit conveyors; (2) overland conveyors; (3) in-plant conveyors; and (4) reclamation-landfill-conveyor systems.

In-Pit Conveyors

With the cost of petroleum-based fuels and increasing restraints on environmental pollution by particulate emissions, belt conveyor systems are being developed to replace mobile vehicles in several applications. Conventional belt conveyor construction, crawler-mounted conveyors, shiftable conveyors with bucket wheel excavators and extendable conveyors are all being used in conjuction with portable crushers and long slope belts.

Overland Conveyors

The advent of high-tension steel cable belting, cable-supported belts, wire rope idler support systems, and horizontal curve designs that reduce transfer stations are helping to reduce conveying costs. Conveying systems that adapt to the terrain are being used in single and multiple flights several miles long. Belt systems are operating in the US that are 24 km (15 miles) long. In Australia, an overland conveying system is in operation that is 50 km (31 miles) long. Cable-supported conveyors, where the cable acts as the tension element rather than the belt, are being used abroad for very long overland systems. The manufacturer claims higher tension capability, therefore a longer conveyor capability, with this system. The use of cable-supported idlers on conventional belt systems can save 10 to 20% of the initial cost of the conveyor structure. Horizontal curve conveyors have been used abroad and recently have found an application in the US.

Common usage for overland belt systems are for 900 to 1200 mm (36 to 48 in.) belts running 3.1 to 6.1 m/s (600 to 1200 fpm). These belts would typically transport 200 to 750 kg/s (800 to 3000 tph) of coal, for example. Energy costs are very low and particulate generation is nil for a covered conveyor system with adequate dust control equipment at the transfer points.

In-Plant Conveyors

In-plant application using conventional belt conveyor construction is the oldest and most common usage of conveyor belts. Conveyors using 20° and 35° troughed idlers are the most common. Belt speeds are typically lower in-plant than for overland conveyors to reduce dusting and belt and chute wear. Typical speeds range from 1.0 to 3.1 m/s (200 to 600 fpm).

Reclamation-Landfill-Conveyor Systems

Belt conveyor systems for landfill and solid waste disposal applications now are being used more often for the same cost and environmental reasons as in-pit and overland systems. Since the discharge point is constantly changing, special conveyors and discharge equipment have been developed for these applications. Some of these are crawler-mounted self-aligning conveyors, shiftable conveyors, extendable conveyors, and portable conveyors.

Normally these conveyors are used in conjunction with wheel or crawler-mounted stackers and trippers. The conveyor system usually advances over its own, self-created fill.

TYPICAL CONVEYOR CONSTRUCTIONS

Several types of belt conveyor designs are available for specific applications. Although more than one type of conveyor may solve a particular materials handling problem, the proper selection will optimize installation and operating costs.

Conventional Stringer Conveyors

Conventional conveyors are turnkey designed and engineered on an individual basis. The construction consists of parallel channel stringer members that support the troughing and occasionally return idlers (Fig. 1). Angle iron support legs are attached to the stringers, with the opposite ends being fixed to embedded anchor bolts, embedded plate, or a conveyor support truss.

A light gage deck plate is sometimes mounted on top of the stringers. The deck plate prevents conveyed material, which may fall off of the side of the belt, from landing on the return strand and becoming wedged between the belt and pulley, causing damage to the belt. V-plow belt cleaners are also usually mounted upstream from a vertical takeup, internal drive system, or tail pulley to keep material from causing belt damage.

Drive and terminal pulley shafts are normally mounted in pillow block bearings, mounted on heavy angle or wide-flange beams. The drive equipment varies depending on the design requirements and space available. Today soft-start devices, electrical or mechanical, are often used on conveyors with drives over 111 855 W (150 hp). They effectively reduce the belt tension and shock imparted by a high torque motor during start-up.

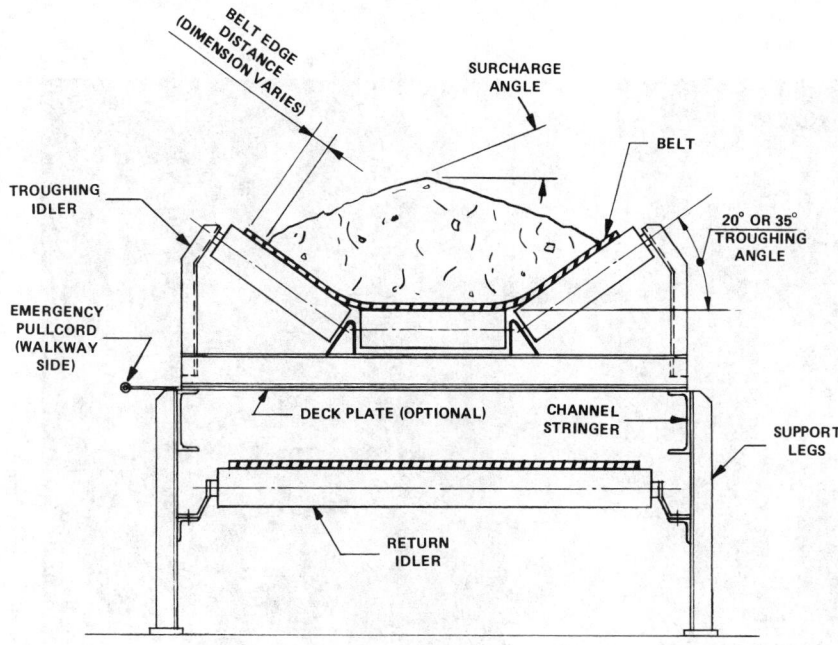

Fig. 1. Conventional stringer conveyor cross section.

Gravity takeups can be horizontally or vertically mounted either between the head and tail pulleys or at the tail pulley. Screw takeups are generally mounted at the tail pulley, but are limited to conveyor lengths of about 30 m (100 ft).

Conventional stringer conveyors are mounted on platforms and inclined ramps indoors and outdoors. They are rigid, easy to clean around, and provide good access for belt and idler maintenance. They are also used in "walk-through" galleries. Walk-through galleries are normally self-enclosing trusses, with roofing, siding and deck plate or floor grating. Galleries come in a wide range of designs, including concrete floors, corrugated arch enclosures, and steel tubes. Totally enclosed galleries must have provisions for heat vents, cleanup, and drains.

Conventional stringer conveyors are by far the most commonly used type today. They are used for in-plant, overland, mobile, and shiftable conveyors. Conventional belt conveyors are limited in their ability to transport up steep inclines. Many users limit slopes to 12° to 15° of incline from the horizontal. Some materials can be conveyed with conventional belts at steeper angles, up to 20°. Tables listing maximum conveyor inclines for many materials are available in the CEMA book *Belt Conveyors for Bulk Materials* mentioned earlier.

When selecting a conveyor slope it is important to consider the effect that frost or ice formation can have on letting material slide down a belt. This might lower a slope 5° to 10° from what otherwise could have been used.

Truss-Mounted Conveyors

Truss-mounted conveyors are commonly used for conveyors that are above grade or platform levels, in lieu of wall-through truss galleries. The troughing idlers are mounted directly to a box truss and the return idlers below or within the framework. This construction is more compact and economical than a walk-through truss, but does not include an enclosed walkway.

In outdoor applications a hood is normally used to protect the carrying belt from wind and rain. A walkway may or may not be added to the side of the truss for maintenance access. The truss-mounted conveyor is used extensively for portable/mobile conveyor construction, and in mild weather applications in place of walk-through galleries.

Wire Rope Conveyors

The stringers on these conveyors consist of two suspended wire ropes in lieu of channel members (Fig. 2). The wire ropes support the troughing idlers and belt. The return idlers are normally mounted on the support legs which also hold up the wire ropes. Components, head and tail terminals, and takeups are very similar to conventional conveyors.

Wire rope conveyors are used sometimes in overland applications to save capital and installation costs. They are also used extensively on underground applications because they are lightweight, easy to install and disassemble, and tend to be self-aligning.

CONVEYOR TYPES AND ADAPTATIONS

Shiftable Conveyors

These conveyors can be either channel or wire rope stringer supported (channel is most common) and contain the same basic design characteristics of conventional and wire rope conveyors. The difference with this conveyor is its ability to be repositioned with the use of mobile equipment (Fig. 3). Rails mounted on the conveyor support base, coupled with a tractor and mounting assembly, move the conveyor to the desired position. Shiftable conveyors are generally horizontal with the discharge end being elevated for material transfer. They are primarily used in surface mining and land reclamation operations.

Cable Conveyors

Two endless steel wire ropes form the support for the belt as well as carrying the tension forces for this type of conveyor (Fig. 4). The belt rests on the cables, while the tension required to move the belt is transmitted entirely through the endless steel wire ropes. The combination of grooved cable guides built into the belt and a sheave support

Fig.2. Wire rope conveyor (courtesy of Continental Conveyor and Equipment Company, Inc.).

Fig. 3. Shiftable conveyor (courtesy of Continental Conveyor and Equipment Company, Inc.).

system keeps the belt continuously supported and properly aligned.

This conveyor is most economical in lengths over 4.8 to 8.0 km (3 to 5 miles). The cables require a special drive with large sheaves to transmit the drive torque to the cable, which makes them uneconomical for short horizontal applications.

Air-Supported Conveyors

The belt for an air conveyor runs in a concave steel trough and is supported by a thin film of air (Fig. 5). Below the steel trough, a sealed plenum contains pressurized air supplied by a blower. The air passes up through the plenum and escapes from the sides of the belt. The air thus replaces the troughing idlers which would otherwise be necessary. The return belt is supported with conventional return idlers. The drive system and takeup is also similar to a conventional belt conveyor.

Cleat/Sidewall Conveyors

Cleats (or transverse walls) and sidewalls attached to a flat belt comprise this type of conveyor (Fig. 6). The sidewalls mounted to the base belt are pleated, thus providing an accordion effect around pulleys. The cleats mounted to the sidewalls and base belt enable this type of conveyor to elevate many materials vertically. On long installations, gravity-type takeups can be provided similar to conventional conveyors.

Crawler-Mounted Conveyors

This type of conveyor functions similar to a shiftable type conveyor in that the conveyor is designed to move primarily laterally (sideways). It is mounted on crawlers or tires which are self-transporting. It is made in sections to accommodate changes in slope and is usually equipped with devices to automatically keep it in line laterally. It can be made with movable load points or a movable tripper, which can discharge anywhere along its length to grade or another conveyor (stacker, for example).

This type of conveyor would be used where movement is more frequent than for a shiftable conveyor, where pit (or surface) conditions are not suitable for a shiftable conveyor, or where lengthwise movement must also be accomplished.

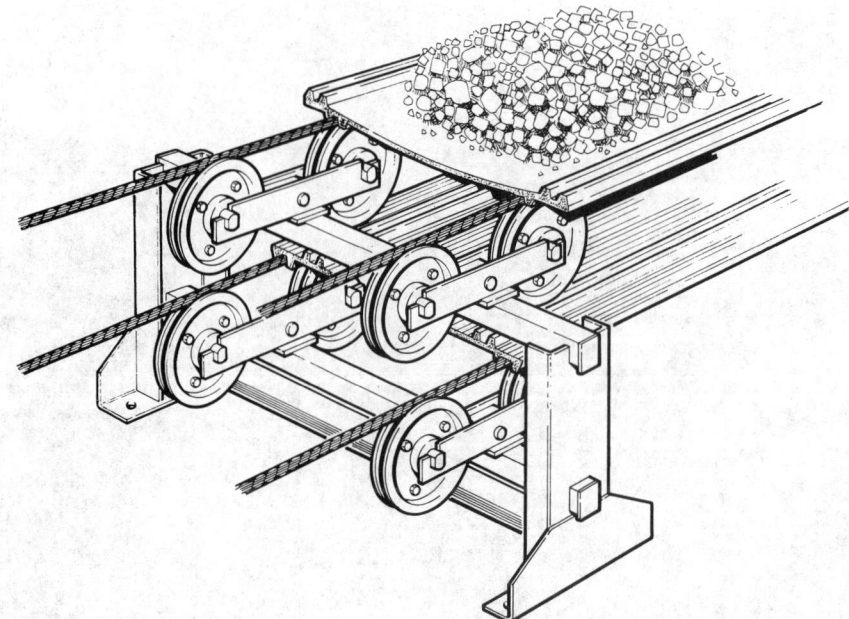

Fig. 4. Cable conveyor; typical eight-pulley line stand (courtesy of Cable Belt Conveyors, Inc.).

Mobile Conveyors

These come in a variety of lengths and configurations. Some are self-moving and others must be towed. They can be used in piggyback fashion where several can be added in a string to function as an extendable conveyor, or individually as with a radial stacker.

Pipe Conveyors

The pipe conveyor is a relatively new type of belt conveyor developed in Japan, but now marketed in this country (Fig. 7). It utilizes a special conveyor belt that is wrapped

into a circle or "pipe" shape after being loaded with material. The pipe enclosure is maintained by using idlers, which on a conventional conveyor are normally used to form a trough for the belt. For the pipe conveyor, five or six idlers are used to form a circle. At the discharge point the belt is unwrapped and the material is discharged over a pulley in the conventional way. The return run is also formed into a circle.

The manufacturer states that the pipe conveyor is very "clean" environmentally, since the load is totally enclosed and also the return run does not "dribble" like a conventional belt along the return run. The pipe conveyor is compact in

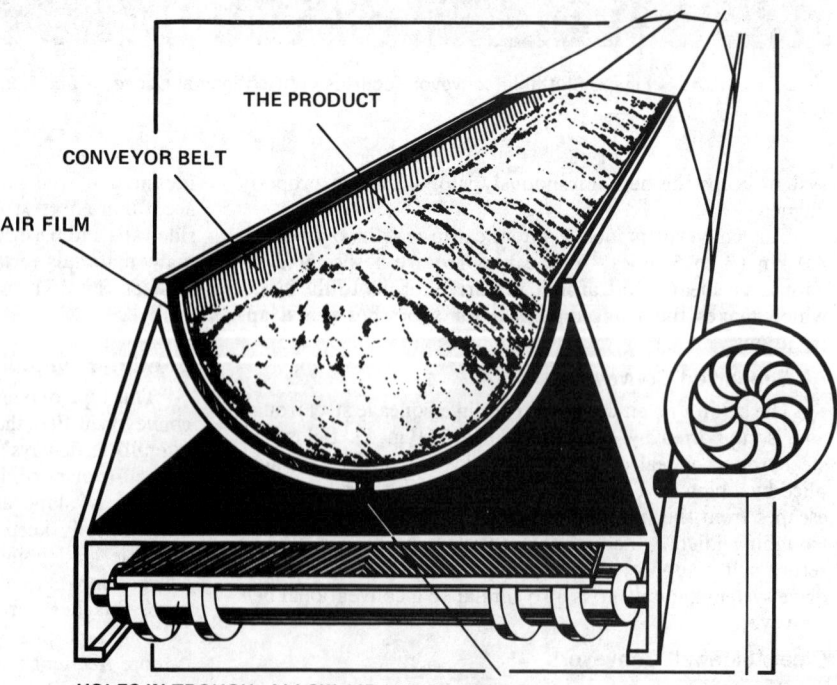

Fig. 5. Air conveyor (courtesy of Wolverine Corp.).

THE PRODUCT

CONVEYOR BELT

AIR FILM

HOLES IN TROUGH ALLOW AIR UNDER BELT TO
SUPPORT BELT AND MATERIAL ON A FILM OF AIR   AIR FAN

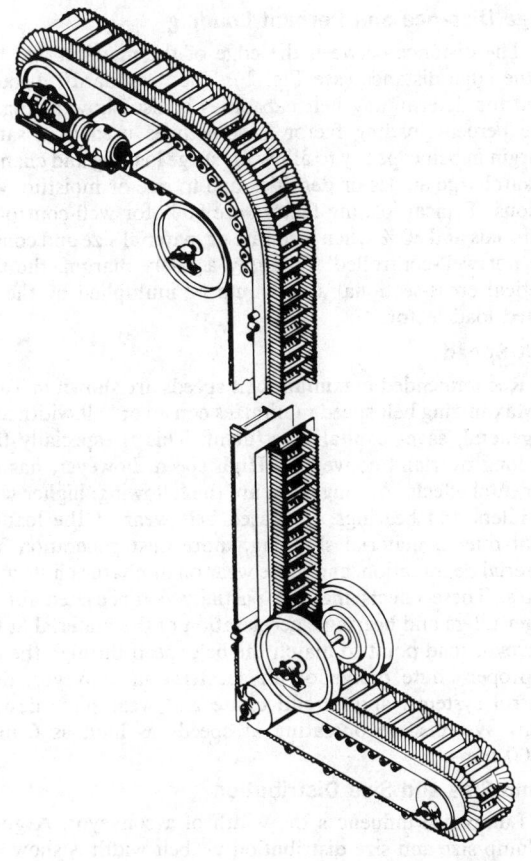

Fig. 6. Cleat/sidewall conveyor (courtesy of Scholtz-EFS Corp.).

cross section, normally would not require a hood to cover the belt, and can negotiate horizontal curves with a much shorter radius and more simply than a conventional belt.

Sandwich Conveyors

This conveyor adaptation consists of two conveyor belts, one lying on top of the other with the load "sandwiched" in between. It can be used on steep inclines and some designs can double back over themselves forming a *C* shape. They have found the most usage in self-unloading ships, bringing the bulk cargo from the hold to above deck. They are also utilized as high-angle conveyors using conventional belting and idlers.

Horizontal Curve Conveyors

A few of the preceding types of conveyors, including conventional, cable, and pipe conveyors, can be designed to "go around corners" as well as go up and down hills.

The advantage of this conveyor is the ability of the conveyor path to go around an obstacle such as a hill or building without having to utilize two or more separate conveyors with a transfer station in between each flight. This saves drives, pulleys, transfer structures, and dust emission points at the transfer.

Conventional and cable belt conveyors must have very large curve radii, generally over 500 m (1650 ft). A pipe conveyor type can utilize a much tighter curve.

The design of a conventional-type horizontal curve conveyor is complex and requires many detailed calculations, most often done by a computer. The inherent tension that pulls the belt and load has a component of its pull toward the center of the curve, off of the supporting idlers. The design of a conventional horizontal curve conveyor utilizes idlers (or cable sheaves, on a cable belt) that are tilted, with

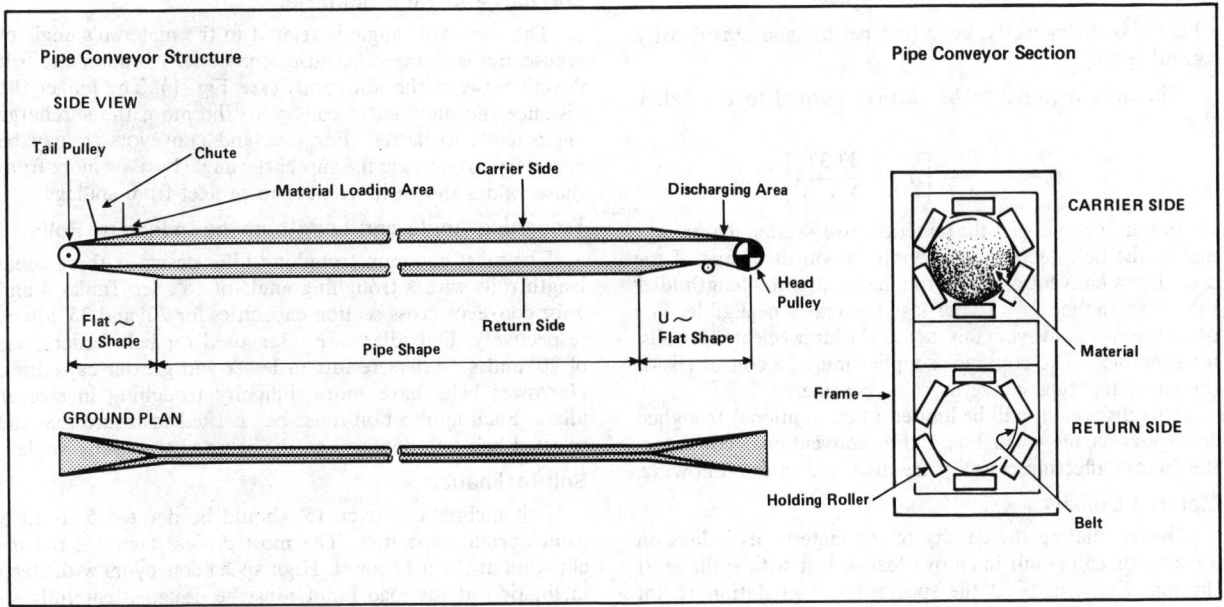

Fig. 7. Pipe conveyor (courtesy of Robins Engineers and Constructors).

the higher side toward the center of the curve. The idlers are also normally skewed, putting the idler toward the inside of the curve further ahead in the direction of belt travel than the idler on the outside of the curve.

The combination of tilt and skew utilizes gravity and friction between the belt and idler to counteract the belt pull toward the curve center. Since the belt tension is changing along the belt length, either the tilt or skew must vary to keep the belt near the center of the idler trough. During belt acceleration or deceleration the belt tensions can change radically and quickly. During weather changes (temperature or moisture) the friction between the belt and idlers can change considerably. These factors make the design of horizontal curves very complex, therefore system designs must be handled carefully.

As of this writing, less than 30 horizontal curves are known to exist worldwide. Their use is becoming more common as successful installations are being completed. Most are in mild climates, so the effects of large temperature variations and freezing conditions have not been thoroughly put to the test.

FACTORS IN SIZING A BELT CONVEYOR

Determining the size or width of a belt conveyor in order to handle a specified mass of material per unit time depends upon the following factors: (1) conveyor type, (2) material density, (3) edge distance and percent loading, (4) belt speed, (5) lump size and size distribution, (6) surcharge angle of material, (7) troughing angle and length of individual idler rolls, and (8) belt inclination.

The volume per unit of time that a conveyor handles is determined by the material's cross-sectional area normal to the belt and belt speed.

$$Q = A \times V \times 3600 \qquad (Q = A \times V \times 60$$

where Q is the capacity, m^3/hr (cu ft per hr); A is the cross-sectional area, m^2 (sq ft); and V is the belt speed, m/s (fpm).

The mass that the conveyor handles is:

$$T = Q \times D \left[T = \frac{Q \times D}{2000} \right]$$

where T is the capacity, kg/s (ton per hr) and D is density, kg/m^3 (pcf).

The area required to be carried, normal to the belt is then

$$A = \frac{T}{D \times V} \left[A = \frac{33.3T}{D \times V} \right]$$

For an inclined belt the material cross-sectional area normal to the belt decreases, depending upon the shape of the area. For a conventional 35° troughed, equal roll length idler set on an incline of 15°, the loss in area is negligible. For other types of conveyors this loss can be large (cleat sidewalls, for example). The conveyor supplier must be consulted for the particular type conveyor being considered.

This discussion will be limited to conventional troughed conveyors as shown in Fig. 1. For conventional conveyors the factors affecting capacity are discussed in the following.

Material Density

Overestimating the density of the material as it lays on a conveyor can result in an overloaded belt with spillage. If the material density at the specified size gradation is not known, it must be determined in order to accurately size the conveyor.

Edge Distance and Percent Loading

The distance between the edge of the material and belt is the edge distance (see Fig. 1). The "standard" distance used for determining belt capacities is also show in Fig. 1. The percent loading factor is sometimes used as a safety margin in belt capacity to allow for surge loading and changes in surcharge angles or densities due to size or moisture variations. Typical loading factors are 90% for well-controlled materials and 80% when feed rate or material size and consist are not well-controlled. To apply a safety margin, the theoretical cross-sectional area would be multiplied by the selected load factor.

Belt Speed

Recommended maximum belt speeds are shown in Table 1. Maximizing belt speed minimizes conveyor belt width and, in general, saves capital investment. This is especially true for long overland conveyors. High speed, however, has detrimental effects. Among them are the following: higher wear on idlers and bearings, increased belt wear at the loading point due to material skidding, more dust generation and material degradation, and more wear on discharge chute wear plates. These effects may be partially compensated for by larger idlers and bearings, acceleration of the material at the conveyor load point to match the belt speed through the use of proper chute design or an acceleration conveyor, dust control systems, and careful chute and wear plate design. Many systems are operating at speeds as high as 6 m/s (1200 fpm).

Lump Size and Size Distribution

Lump size influences the width of a conveyor. A guide to lump size and size distribution vs. belt width is shown in Table 2. Dangerous spillage can result from improper lump allowance and loading skirts. Chute size must also be taken into account so lumps do not bridge in the chutes. Most chute designers make chute widths a minimum of two and a half to three times the maximum lump size. Larger lumps or dropping distances must utilize impact idlers in order to save belt and idler damage at the load points. Belting and idler manufacturers provide guides for proper design.

Surcharge Angle of Material

The surcharge angle is related to the material's angle of repose, but is changed because of material vibration and belt flexing between the idler rolls (see Fig. 1). The longer the distance the material is conveyed, the more the surcharge angle tends to flatten. For overland conveyors it may be necessary to decrease the surcharge angle by 5° or more from those values shown in Table 3 to protect from spillage.

Troughing Angle and Length of the Individual Rolls

The most common troughing idler design is three equal length rolls with a troughing angle of 35°. See Tables 4 and 5 for conveyor cross-section capacities for 20° and 35° idlers, respectively. Flat idlers are often used for belt feeders; use of 20° and 45° idlers results in lesser and greater capacities. Narrower belts have more difficulty troughing in steeper idlers. Each application must be checked to insure that the selected belt will conform to the selected troughing angle.

Belt Inclination

Belt inclinations over 15° should be derated 5 to 10% from normal capacities. The most critical factor is the inclination at the load point. High speed conveyors with steep inclination at the load point must be designed carefully so the load settles quickly on the belt at the belt speed. Otherwise, back up will occur in the chute or spillage will result.

Table 1. Recommended Maximum Belt Speeds

MATERIAL BEING CONVEYED	BELT SPEEDS		BELT WIDTH	
	m/s	(fpm)	mm	(in)
GRAIN OR OTHER FREE-FLOWING, NONABRASIVE MATERIAL	2.5 3.6 4.1 5.1	(500) (700) (800) (1000)	450 600 - 750 900 - 1050 1200 - 2400	(18) (24 - 30) (36 - 42) (48 - 96)
COAL, DAMP CLAY, SOFT ORES, OVERBURDEN AND EARTH, FINE-CRUSHED STONE	2.0 3.0 4.1 5.1	(400) (600) (800) (1000)	450 600 - 900 1050 - 1500 1800 - 2400	(18) (24 - 36) (42 - 60) (72 - 96)
HEAVY, HARD, SHARP-EDGED ORE, COARSE-CRUSHED STONE	1.8 2.5 3.0	(350) (500) (600)	450 600 - 900 OVER 900	(18) (24 - 36) OVER (36)
FEEDER BELTS, FLAT OR TROUGHED, FOR FEEDING FINE, NONABRASIVE, OR MILDLY ABRASIVE MATERIALS FROM HOPPERS AND BINS	0.3 - 0.5 (50 - 100)		ANY WIDTH	

Adapted From: Belt Conveyors for Bulk Materials, Second Edition, Conveyor
Equipment Manufacturers Association, p. 54.

BELT CONVEYOR DESIGN EXAMPLE

The information presented herein provides a quick means of calculating the horsepower required to operate a conventional belt conveyor. This graphical approach is presented in the CEMA *Handbook*. To perform a detailed analysis of a belt conveyor and component selection, the CEMA *Handbook*, belt manufacturer catalogs, conveyor companies, and

engineering firms are suggested sources of additional information. The calculated belt tension and required horsepower may vary based on the friction factors used by the different sources.

The following is an example of how to size a belt conveyor and establish its horsepower requirement by use of a quick graphical method. It is most accurate when used for belt

Table 2. Belt Width Necessary for a Given Lump Size. Fines: No Greater
Than 1/10 Maximum Lump Size

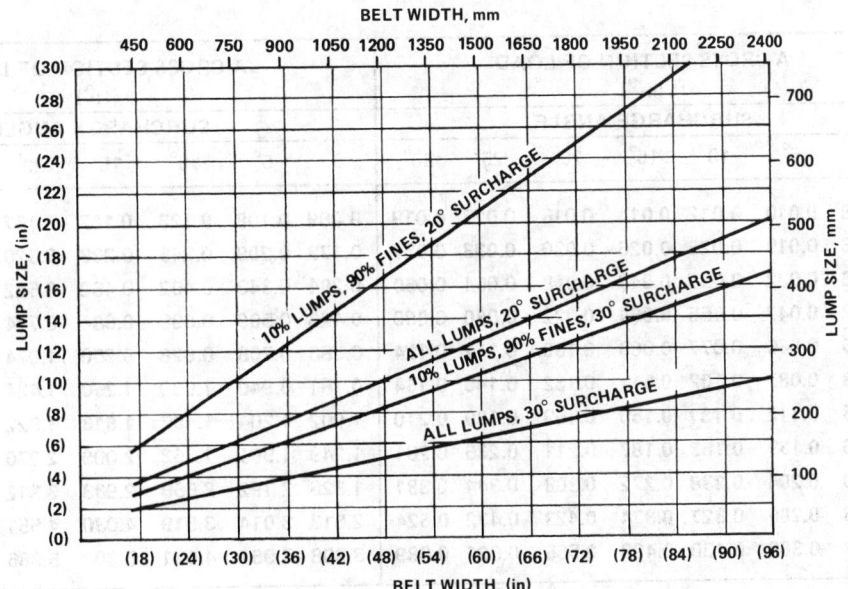

Adapted From: Belt Conveyors for Bulk Materials, Second Edition, Conveyor
Equipment Manufacturers Association, p. 53.

Table 3. Flowability - Angle of Surcharge - Angle of Repose

Very free flowing	Free flowing	Average flowing		Sluggish
5° Angle of surcharge	10° Angle of surcharge	20° Angle of surcharge	25° Angle of surcharge	30° Angle of surcharge
0° - 19° Angle of repose	20° - 29° Angle of repose	30° - 34° Angle of repose	35° - 39° Angle of repose	40° - up Angle of repose
MATERIAL CHARACTERISTICS				
Uniform size, very small rounded particle, either very wet or vert dry, such as dry silica sand, cement, wet concrete, etc.	Rounded, dry polished particles, of medium weight, such as whole grain and beans.	Irregular, granular or lumpy materials of medium weight, such as anthracite coal, cottonseed meal, clay, etc.	Typical common materials such as bituminous coal, stone, most ores, etc.	Irregular, stringy, fibrous, interlocking material, such as wood chips, bagasse, tempered foundry sand, etc.

Adapted From: Belt Conveyors for Bulk Materials, Second Edition, Conveyor Equipment Manufacturers Association, p. 39.

conveyors up to about 600 m (2000 ft) in length and horizontal or inclined. Declined, overland, and/or high capacity conveyors calculations should be directed to the previously mentioned sources for final determination of power requirements.

Conveyor Specifications

Material: Bituminous coal, run-of-mine, 0.15 × 0 m (6 × 0 in.)

Density: 800 kg/m³ (50 pcf)
Angle of repose: 38°
Capacity: 352.8 kg/s (1400 stph)
Troughing idler type: 35°
Length, center to center: 457.2 m (1500 ft)
Lift: 61 m (200 ft)

Step 1—Select belt speed.
With the assistance of Table 1, the material characteristics

Table 4. 20° Troughed Belt - Three Equal Rolls Standard Edge Distance = 0.055b + 0.9 Inch (b = belt width)

BELT WIDTH		A-CROSS SECTION OF LOAD (m²)							A-CROSS SECTION OF LOAD (ft²)						
		SURCHARGE ANGLE							SURCHARGE ANGLE						
mm	(in)	0°	5°	10°	15°	20°	25°	30°	0°	5°	10°	15°	20°	25°	30°
450	(18)	0.008	0.010	0.012	0.014	0.016	0.017	0.019	0.089	0.108	0.128	0.147	0.167	0.188	0.209
600	(24)	0.016	0.019	0.023	0.026	0.030	0.033	0.037	0.173	0.209	0.246	0.283	0.320	0.359	0.399
750	(30)	0.026	0.032	0.037	0.043	0.048	0.054	0.060	0.284	0.343	0.402	0.462	0.522	0.585	0.649
900	(36)	0.039	0.047	0.055	0.064	0.072	0.080	0.089	0.423	0.509	0.596	0.684	0.774	0.866	0.960
1050	(42)	0.055	0.066	0.077	0.088	0.100	0.112	0.124	0.588	0.708	0.828	0.950	1.074	1.201	1.332
1200	(48)	0.073	0.087	0.102	0.117	0.132	0.148	0.164	0.781	0.940	1.099	1.260	1.424	1.592	1.765
1350	(54)	0.093	0.112	0.131	0.150	0.169	0.189	0.210	1.002	1.204	1.407	1.613	1.822	2.037	2.258
1500	(60)	0.116	0.139	0.163	0.187	0.211	0.236	0.261	1.249	1.501	1.753	2.009	2.270	2.537	2.812
1800	(72)	0.170	0.204	0.238	0.272	0.308	0.344	0.381	1.826	2.192	2.560	2.933	3.312	3.701	4.102
2100	(84)	0.233	0.280	0.327	0.374	0.423	0.472	0.524	2.513	3.014	3.519	4.030	4.551	5.085	5.635
2400	(96)	0.307	0.369	0.430	0.493	0.556	0.621	0.689	3.308	3.967	4.631	5.302	5.986	6.687	7.411

Adapted From: Belt Conveyors for Bulk Materials, Second Edition, Conveyor Equipment Manufacturers Association, p. 58.

Table 5. 35° Troughed Belt - Three Equal Rolls Standard Edge Distance =
0.055b + 0.9 Inch (b = belt width)

BELT WIDTH		A-CROSS SECTION OF LOAD (m²)							A-CROSS SECTION OF LOAD (ft²)						
		SURCHARGE ANGLE							SURCHARGE ANGLE						
mm	(in)	0°	5°	10°	15°	20°	25°	30°	0°	5°	10°	15°	20°	25°	30°
450	(18)	0.013	0.015	0.016	0.018	0.020	0.021	0.023	0.144	0.160	0.177	0.194	0.212	0.230	0.248
600	(24)	0.026	0.029	0.032	0.035	0.038	0.041	0.044	0.278	0.309	0.341	0.373	0.406	0.440	0.474
750	(30)	0.042	0.047	0.052	0.057	0.062	0.067	0.072	0.455	0.506	0.557	0.609	0.662	0.716	0.772
900	(36)	0.063	0.070	0.077	0.084	0.091	0.098	0.106	0.676	0.751	0.826	0.903	0.980	1.060	1.142
1050	(42)	0.087	0.097	0.107	0.117	0.126	0.137	0.147	0.940	1.044	1.148	1.254	1.361	1.471	1.585
1200	(48)	0.116	0.129	0.141	0.154	0.168	0.181	0.195	1.248	1.385	1.523	1.662	1.804	1.949	2.099
1350	(54)	0.149	0.165	0.181	0.198	0.215	0.232	0.250	1.599	1.774	1.950	2.128	2.309	2.494	2.686
1500	(60)	0.185	0.205	0.226	0.246	0.267	0.289	0.311	1.994	2.211	2.429	2.651	2.876	3.107	3.345
1800	(72)	0.271	0.300	0.330	0.359	0.390	0.421	0.453	2.913	3.229	3.547	3.869	4.197	4.532	4.879
2100	(84)	0.372	0.412	0.453	0.494	0.536	0.578	0.623	4.007	4.440	4.876	5.317	5.766	6.226	6.701
2400	(96)	0.490	0.543	0.596	0.650	0.705	0.761	0.819	5.274	5.842	6.415	6.994	7.584	8.189	8.812

Adapted From: Belt Conveyors for Bulk Materials, Second Edition, Conveyor
Equipment Manufacturers Association, p. 59.

and environmental considerations, the following belt speed is initially selected: 2.79 m/s (550 fpm).

Step 2—Determine surcharge angle.
From Table 3, the surcharge angle is selected based on the known material characteristics: 25°.

Step 3—Calculate cross-sectional area of material on conveyor:

$$A = \frac{T}{D \times V} \quad \left[A = \frac{33.3\,T}{D \times V} \right]$$

where A is cross-sectional area, m² (sq ft); T is capacity, kg/s (stph); V is belt speed, m/s (fpm); and D is density, kg/m³ (pcf).

$$A = 0.145 \text{ m}^2 \text{ (1.7 sq ft)}$$

Step 4—Determine belt width.
Either Table 4 or Table 5 is selected based on the type of troughing idler used. From Table 5 for 35° idlers, 25° surcharge angle, and the calculated cross-sectional area, use a 1200 mm (48 in.) belt width. Note that at this point the maximum cross section listed in the table can be substituted back into step 3 to calculate a new belt speed with the conveyor 100% loaded.

Step 5—Determine weight per unit length of belt and revolving idler parts.

From Table 6: 61 kg/m (41 lb per ft)

Table 6. Weight Per Linear Foot of Belt and Revolving Idler Parts, kg (lb)

BELT WIDTH	MATERIAL DENSITY kg/m³ (lb/ft³)			
mm (in)	800 (50)	1600 (100)	2400 (150)	3200 (200)
450 (18)	17.9 (12)	20.8 (14)	25.3 (17)	25.3 (17)
600 (24)	23.8 (16)	28.3 (19)	34.2 (23)	34.2 (23)
750 (30)	29.8 (20)	35.7 (24)	43.2 (29)	43.2 (29)
900 (36)	41.7 (28)	52.1 (35)	61.0 (41)	71.4 (48)
1050 (42)	50.6 (34)	62.5 (42)	72.9 (49)	87.8 (59)
1200 (48)	61.0 (41)	75.9 (51)	102.7 (69)	114.6 (77)
1350 (54)	71.4 (48)	86.3 (58)	116.1 (78)	132.4 (89)
1500 (60)	89.3 (60)	104.2 (70)	129.5 (87)	147.3 (99)
1800 (72)	110.1 (74)	123.5 (83)	168.2 (113)	193.5 (130)
2100 (84)	151.8 (102)	189.0 (127)	221.7 (149)	245.5 (165)
2400 (96)	174.1 (117)	212.8 (143)	269.4 (181)	269.4 (181)

Adapted From: Belt Conveyors for Bulk Materials, Second Edition Conveyor
Equipment Manufacturers Association, p. 131.

Step 6—Calculate power required to drive an empty conveyor, PR_1.

From Table 7:

$$PR_1 = \frac{P_1 \times V}{0.5} \qquad \left[PR_1 = \frac{P_1 \times V}{100} \right]$$

where P_1 is power to drive empty belt for each 0.5 m/s (100 fpm) belt speed and PR_1 is 13 310 W (17.6 hp).

Step 7—Calculate power required to elevate material, PR_2.

From Table 8:

$$PR_2 = P_2 \times H$$

where P_2 is the power to elevate material per m (ft) of lift, H is lift m (ft), and PR_2 is 225 349 W (302 hp).

Step 8—Calculate power required to convey material horizontally, PR_3.

From Table 9:

$$PR_3 = \frac{P_3 \times T}{25} \qquad \left[PR_3 = \frac{P_3 \times T}{100} \right]$$

where P_3 is the power to convey material horizontally per 25 kg/s (100 stph) and PR_3 is 45 250 W (60.2 hp).

Adding steps 6, 7, and 8 (PR_1, PR_2, PR_3), the total power required, PR

$$PR = 283\,909 \text{ W } (379.8 \text{ hp})$$

Assuming a 5% power loss through drive components, the required motor power is:

$$PR = 298\,852 \text{ W } (399.8 \text{ hp})$$

ECONOMICS

When contemplating the installation of a belt conveyor system, engineering economics is a prime consideration. During the conceptual design stage, management may wish to determine any or all of the following: erected, operating, maintenance, taxes, and insurance costs. A present worth analysis is generally used to compare alternatives over the life of the operation. Several books are devoted entirely to the topic of engineering economy. This discussion will be

Table 7.   Horsepower Required to Drive Empty Conveyor

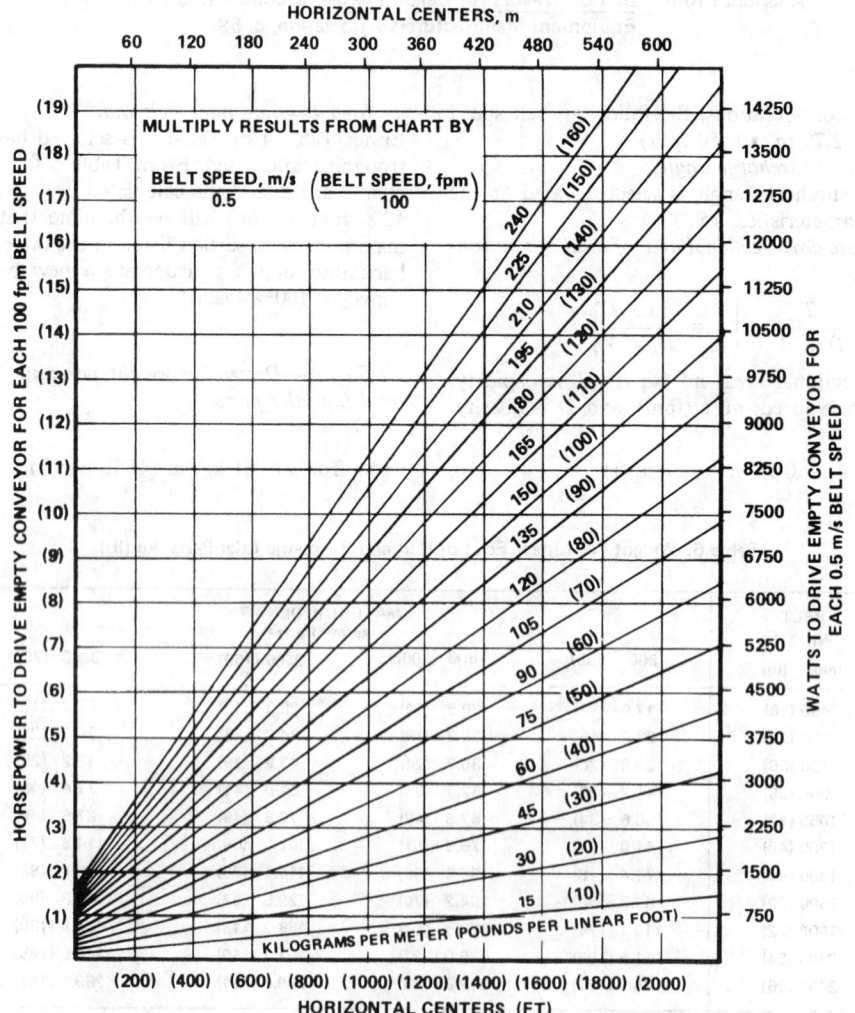

Adapted From: **Belt Conveyors for Bulk Materials**, Second Edition, Conveyor Equipment Manufacturers Association, p. 131.

Table 8.   Horsepower Required to Elevate Material

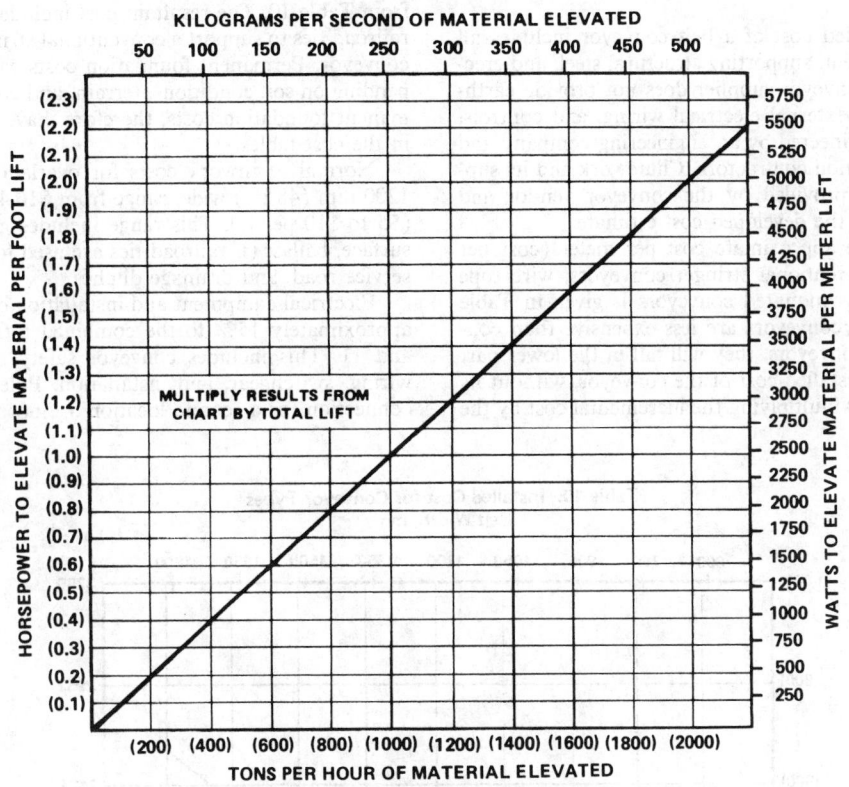

Adapted From: <u>Belt Conveyors for Bulk Materials</u>, Second Edition, Conveyor
Equipment Manufacturers Association, p. 132.

Table 9.   Horsepower Required to Convey Material Horizontally

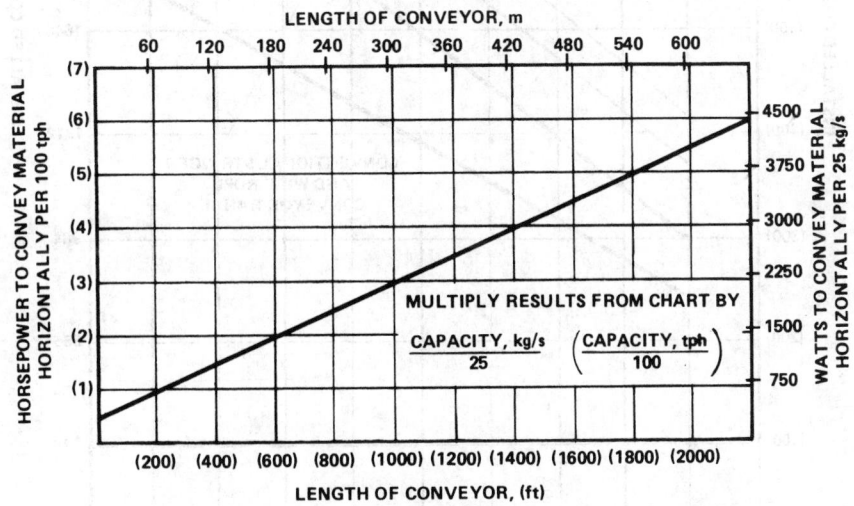

Adapted From: <u>Belt Conveyors for Bulk Materials</u>, Second Edition, Conveyor
Equipment Manufacturers Association, p. 133.

limited, therefore, to developing an estimate of installed, operating, and maintenance costs.

Installed Cost

The total installed cost of a belt conveyor includes all mechanical equipment, supporting structural steel, and erection. Typically a conveyor supplier does not provide earthwork, foundations, external electrical wiring, and controls. These are often engineered by an engineering company and erected by construction contractors. Chutework and its support are generally provided by the conveyor vendor and should be added to the developed cost estimate.

A range for the approximate cost per meter (cost per linear foot) for conventional stringer conveyors, wire rope conveyors, and truss mounted conveyors is given in Table 10. Since wire rope conveyors are less expensive than conventional stringer conveyors, they will fall in the lower part of the range. The installed cost of the conveyor, without its drive, is obtained by multiplying the incremental cost by the total length. From the power calculation presented previously, an erected drive assembly cost can be selected from Table 11 and added to the erected conveyor cost established from Table 10. The resultant cost includes an allowance for railroad ties to support a conventional stringer or a wire rope conveyor. Permanent foundation costs can vary greatly depending on soil conditions, terrain, and conveyor loads. Permanent foundation costs, therefore, have not been included in the cost tables.

Normal earthwork costs for overland conveyors up to 1200 mm (48 in.) wide, range from $16.40 to $32.80 per m ($5 to $10 per ft). This range includes grubbing, a graded surface, ballast (if railroad ties are used for support), gravel service road, and drainage ditches.

Electrical equipment and installation costs normally add approximately 15% to the combined total from Tables 10 and 11. This includes conveyor safety switches, controls, wiring, switchgear, and installation. Pole lines or a power connection for a remote location are not included.

Table 10. Installed Cost for Conveyor Types

Table 11. Installed Cost for Drive Equipment

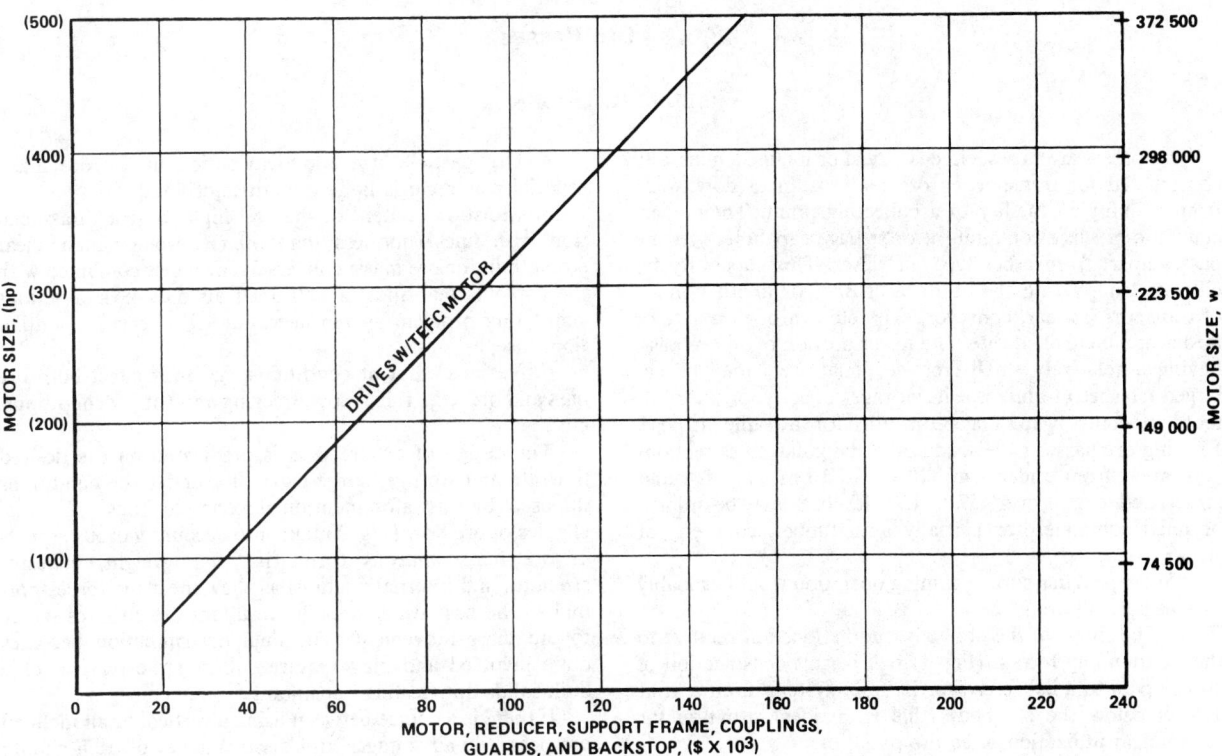

Many conventional stringer conveyors are suspended between supports and are in inclement climates and, therefore, are supported by totally enclosed gallery type trusses. For spans of 22.9 to 45.7 m (75 to 150 ft), the cost of the enclosed truss is approximately $2575 to $2950 per m ($785 to $900 per ft) respectively. This cost is for single conveyors up to 1200 mm (48 in.) wide and includes erection, but not the conveyor (see Tables 10 and 11).

Bents support gallery type trusses or truss-mounted conveyors. The cost of bents for a typical 30.5 m (100 ft) span truss can be interpolated through a height range from 1.5 to 48.8 m (5 to 160 ft) at $825 to $131,000, respectively, for each bent required.

Operating and Maintenance Cost

The operating cost of a conveyor system basically is comprised of two parts: labor and power. The cost for operating labor can vary greatly depending on the number of conveyors, their location, operating time per day, type of operating facility (mine, quarry, power plant, etc.), and the degree of automation. Depending on these factors, the number of operating personnel who periodically observe the system operation and the control room operator or local control operator(s) man-hours can be determined. This is best established by the operating company based on established policies, but can be approximated by the engineer.

The cost of power can also vary greatly. A good preliminary approximation is 5 to 6¢ per kW/h, based on actual power consumed. Declining conveyors, regenerative in nature, require special considerations since the conveyor will generate instead of consume power.

The annual cost of maintenance can be approximated using a rule of thumb of 1 to 2% of the installed conveyor cost developed from Tables 10 and 11.

All of the estimated costs presented herein are based on second quarter, 1985 dollars, and should only be used to establish an order-of-magnitude cost for a conveyor.

REFERENCES

Anon., 1980, "Curvoduc Goes Round Mountains and Over Hills, Not Through Them," *World Mining,* Sept.

The Conveyor Equipment Manufacturer's Association, 1979, *Belt Conveyors for Bulk Materials,* 2nd ed., CBI Publishing Co., Boston.

Diebold, W.H., 1984, "Steep Angle In-Pit Conveying," *Bulk Solids Handling,* Vol. 4, No. 1, Mar.

Dos Santos, J.A., 1984, "Sandwich Belt High Angle Conveyors," *Bulk Solids Handling,* Vol. 4, No. 1, Mar.

Grimmer, K.J., and Beumer, B., Jr., 1972, "Design and Operation of Curve-going Conveyors with Standard Belts," *Fordern und Heben,* No. 4, Mar. (in German)

Tagge, F.W., Snyder, H.J., and Duncan, L.D., 1983, "Modeling the Economics of Trucks and Overland Conveyors for Lignite Transportation," *Coal Technology '83,* Vol 6, pp. 139-171.

6.5.4 Ore Passes

WILLIAM R. GOODNER

Ore passes are raises, either vertical or inclined, generally constructed for transporting ore long distances downward from a dumping facility to a collecting point. Their characteristic of *surge* containment or storage capability sets ore passes apart from other types of raises. This capability facilitates transferring the ore to a crusher or another mode of transport (railcar, conveyor, skip, etc.) through a gate or feeder at a controlled rate. The lower portion of an ore pass, having a relatively small cross-sectional area, may be enlarged (slashed) to increase its storage capacity, or intermittent level control gates may be installed for the same purpose. Existing ore passes have round or rectangular cross sections with areas from under 1 m² (10 ft²) to 16 m² (172 ft²) and larger, and lengths over 457 m (1500 ft). They may be unlined or lined completely or partially with timber, concrete, or steel plate.

An ore pass in a surface-mining operation would probably be feasible only if:

1. The shape of the ore body and its location relative to the surrounding terrain (Fig. 1) will permit construction of an ore pass to a belt conveyor or train system in an adit at a level below the ore body (this is an ideal situation for economical utilization of an ore pass); or,

2. The depth of an ore body is such that at some mining level (Figs. 2 and 3), the cost of constructing and operating a conveyor and/or skip system and ore pass in conjunction with an existing pit haul system compares favorably with the cost of uphill hauling with the existing system (this is with the assumption that it is impractical to locate a conveyor and/or skip in the benched area of the pit); or

3. The shape and location of the ore body dictates surface mining preceding or concurrent with an underground mining operation (Fig. 4).

Characteristics of both the ore pass and the ore affect flowability. These characteristics may prevent efficient use of an ore pass and must also be considered. Generally included are:

1. Integrity of the ore-pass wall, i.e., geological conditions should be adequate (within acceptable erosion limits) to withstand the impact and abrasion of the falling ore; walls containing loose rock will require proper lining;

2. Degree of ore-pass wall roughness;

3. Dimensions of the ore pass, i.e., cross-sectional area, height of vertical free fall, length of an inclined pass, etc., related to hangups, and crushing and packing of the ore;

4. Ore particle size and distribution in the ore mass, especially as regards large quantities of fines;

5. Moisture content of the ore; high content in association with fines (but less than *mud*) develops high shear strength in the ore mass; conversely, dry fines combined with the copious quantities of entrained air may pose a serious ventilating problem—even necessitating a separate ventilation raise;

6. Various ambient conditions that may affect both ore pass and ore, e.g., rain, snow, freezing and/or icy conditions, etc.

The design of an ore pass is predicated on the desired flowrate and storage/surge capability under the conditions imposed by the aforementioned characteristics. The mechanics of ore flow (Fig. 5) take into account various aspects of ore mass analysis (primarily, unit weight, cohesive strength, and internal friction) as they affect the forces normal to the pass walls, and the resultant ore-pass resistance to ore mass movement. The final determination becomes quite involved and may require enlisting the services of a bulk-solids-flow testing consultant.

If there is a choice between using a vertical or an inclined ore pass, the advantages and disadvantages of each should be noted.

1. A vertical pass is shorter for the same elevation difference.

2. Generally, ore has less contact with the walls of a vertical pass; therefore, there is less wear, rough walls have less affect on ore flow, and there may be less need for lining.

3. There is more autogenic crushing in a vertical pass.

4. Vertical passes require no knuckle and have none of the problems inherent with knuckles.

5. Vertical passes are more easily slashed to increase storage capacity.

6. Mechanical boring of a vertical pass is cheaper, faster, and more accurate.

7. The horizontal component of an inclined pass replaces other more expensive horizontal transport methods.

8. Ore velocity is usually less in an inclined pass; therefore, there is less problem with entrained air, fragmentation, and dust.

9. For some conditions of ore and ore pass, arching is less likely in an inclined pass due to the imbalance of normal forces (see Fig. 5).

In the past, there have been designers who advocated

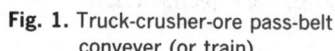

Fig. 1. Truck-crusher-ore pass-belt conveyer (or train).

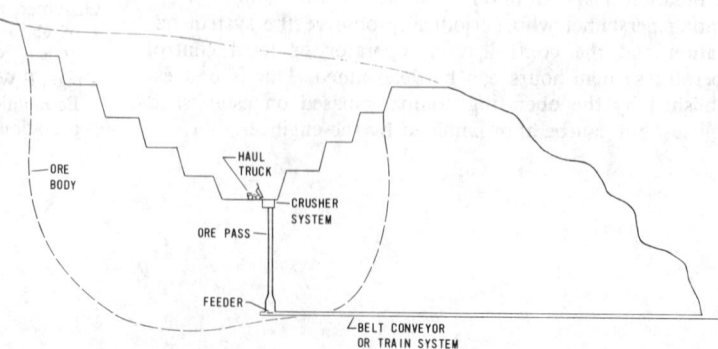

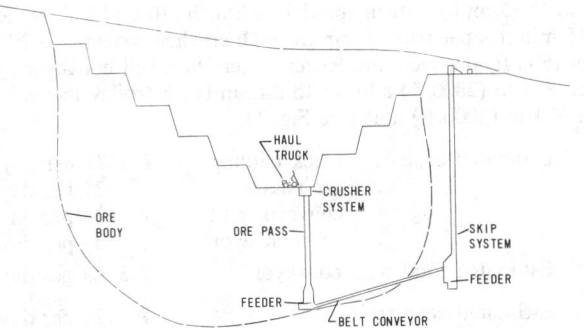

Fig. 2. Truck-crusher-ore pass-belt conveyor-skip hoist.

providing an inspection raise alongside the ore pass; however, the present consensus is that this is unnecessary if the ore pass is properly designed and constructed.

Ore-pass construction using state-of-the-art drill-and-blast techniques is gradually being replaced by mechanical raise-boring. This transition has been accelerated by the decreasing number of miners qualified for drill-and-blast

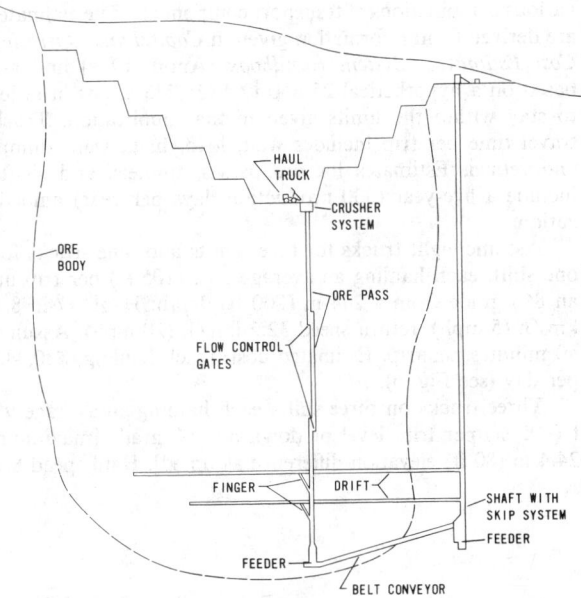

Fig. 4. Combined surface and underground mining transport systems.

methods. Raise-boring is also attractive because it is less hazardous for the miners, is faster and less costly, and produces a smoother wall than blasting. The advent of raise-boring contractors has made this method even more appealing; mining companies need not be saddled with the irretrievable costs of owning boring machinery which is used only occasionally. At the time of this writing (1982), raise-boring is rapidly coming into its own. Better materials, improved design of machinery, and advanced miner know-how are continuing to reduce costs and time requirements.

It is somewhat difficult to establish a cost of an ore pass to compare with other transport methods. Cost considerations should include amortization and operating (maintenance) costs along with personnel costs. The operating cost of an ore pass is generally very low and might be considered a tradeoff for capital cost. Following are estimated costs for

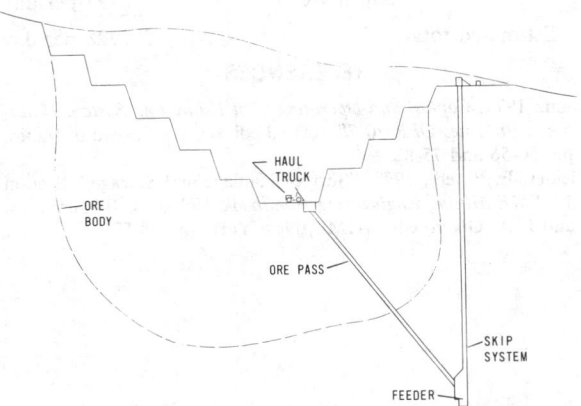

Fig. 3. Truck-crusher-ore pass-skip hoist.

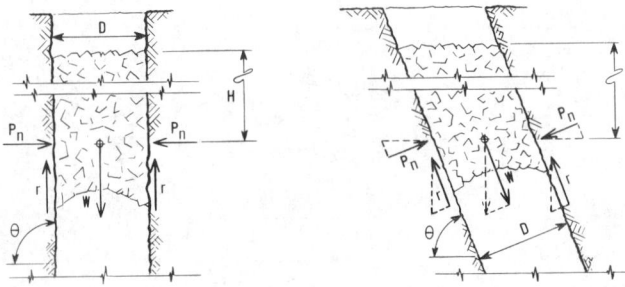

Fig. 5. Mechanics of ore flow in a circular ore pass.

P_n : Pressure normal to side-wall (Function of γ, ϕ, c, H, D and θ)

r : Resistance of wall to ore-mass movement (Function of γ, μ, ϕ, c, H, D and θ)

W : Pressure on arch (Function of γ, H and D)

γ : Weight of ore per unit volume; μ : friction coefficient of ore and wall; ϕ : angle of ore mass internal friction; c : cohesive strength of ore mass; H : depth of ore mass; D : ore pass diameter; θ : angle of ore pass above horizontal.

various combinations of transport components. The estimates are derived from information given in *Capital and Operating Cost Estimating System Handbook* (Anon, 1979) and are based on a hypothetical 25 650 t/d (28,275 st pd) in order to stay within the limits given in that publication. Truck travel time per trip includes wait, load, haul, wait, dump, and return. Estimates for ore passes, tunnels, and shafts include a five-year (330 production days per year) amortization.

Assume eight trucks for three shifts and nine trucks for one shift, each hauling an average 95 t (105 st) per trip up an 8% grade from a 244 m (800 ft) depth. Haul speed 8.1 km/h (5 mph), return speed 32.2 km/h (20 mph). Assume 40 minutes per trip. Estimated cost: truck hauling, $30,442 per day (see Fig. 6).

Three trucks on three shifts each hauling an average 95 t (105 st) per trip, level or down an 8% grade [maximum 24.4 m (80 ft) elevation difference assumed]. Haul speed 8.1

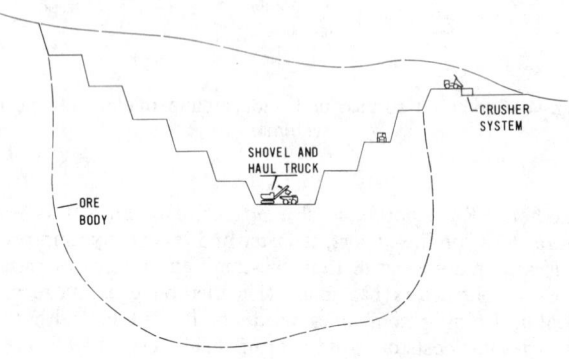

Fig. 6. Truck-crusher (at ground level).

km/h (5 mph), return speed 16.6 km/h (10 mph). Assume 15 minutes per trip. Dump through crusher system to a 209 m (600 ft) ore pass and feeder system to a belt conveyor in an 853 m (2800 ft) adit, or to a train (with trolley locos) in a 914 m (3000 ft) adit (see Fig. 1).

Estimated cost:	truck hauling	$ 2571 per day
	ore pass	21 per day
	conveyor adit	108 per day
	belt conveyor	369 per day
Estimated total with conveyor		$ 3069 per day
Estimated cost:	train adit	$ 174 per day
	train	6504 per day

Labor cost herein is based on the assumption that crew for underground rail system is one-third that for surface spur or main line.

| Estimated total with train | $ 9270 per day |

Trucks and ore pass are the same as shown in Fig. 1. Belt conveyor is 252 m (828 ft) long, inclined 15° upward to a 365.8 m (1200 ft) deep skip hoist system (see Fig. 2).

Estimated cost:	truck hauling	$ 2571 per day
	belt conveyor tunnel	32 per day
	belt conveyor	558 per day
	skip hoist shaft	1070 per day
	skip hoist	3497 per day
Estimated total		$ 7728 per day

REFERENCES

Anon., 1979, *Capital and Operating Cost Estimating System Handbook, BuMines OFR 10-78,* revised edition, US Bureau of Mines, pp. 50-56 and 75-82.

Pfleider, E. P., ed., 1973, "Surface Haulage and Storage," Section 18, *SME Mining Engineering Handbook,* Vol. 2, A. B. Cummins, and I. A. Given, eds., AIME, New York, pp. 18-59.

6.5.5 Scrapers

RONALD M. HAYS

TYPES AND APPLICABILITY

Tractor-scraper units have a track or wheeled (rubber-tired) tractor and a wheeled scraper connected by a hitch. Rubber-tired tractors and scrapers may have one or two axles. Currently, the most popular tractor-scrapers have a one-axle tractor with the engine *overhung* forward of the single drive axle and a one-axle scraper with the axle at the rear of the scraper. These scrapers operate well on adverse grades, can *duckwalk* through poor ground or mud, and have a small turning circle.

Tractor-scrapers are powered by a single diesel engine on the tractor or dual diesel engines with one on the tractor and the other on the rear end of the scraper. Dual-engine or tandem-powered machines with single-axle tractor and scraper offer greatest traction because all gross vehicle weight (GVW) is distributed on drive axles.

Tractor-scrapers are designed for either pull-, push-, or self-loading. Pull-loading scrapers use a track or wheeled tractor with adequate pulling force and traction to load the scraper. Push loading of conventional scrapers usually requires the assistance of a track or wheeled dozer (pusher) for efficient scraper loading. These scrapers may be either single engine or tandem powered. Two conventional dual engine scrapers may operate in a push-pull arrangement by

coupling the scrapers together. The front scraper is loaded first while being pushed by the rear scraper. The rear scraper then is loaded while being pulled by the loaded front scraper. Self-loading scrapers have an elevating mechanism or auger to raise material into the scraper. These various types of tractor-scrapers are illustrated in Fig. 1.

In surface mines, scrapers are used to excavate and haul a wide range of materials, including top soil, clay, silt, sand, gravel, well-broken rock, and ripped coal or rock. Scrapers have heaped capacities and rated payloads ranging up to 33.6 m³ (44 cu yd) and 47 t (52 ton) for general use and 42.8 m³ (56 cu yd) and 37 t (41 ton) for coal.

Conventional single engine scrapers load difficult-to-excavate materials when pushed by one or more track dozers. The pushing force usually ranges from 1 to 2 kg (lb) of push per kilogram (lb) of scraper payload. Table 3 (in Section 6.5.2) presents tractive coefficients which can be used to determine pushing capabilities of track and wheeled dozers. Conventional single engine scrapers are designed with good gradeability, having 182 to 243 kg/kW (300 to 400 lb per hp). These scrapers have fair traction with 50 to 55% of weight distributed on the drive axle.

These scrapers are the most efficient scrapers in terms of rated payload to gross vehicle weight (GVW) with payload

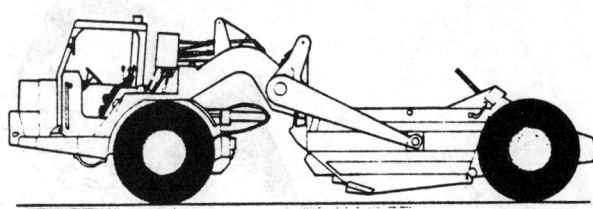
CONVENTIONAL SINGLE ENGINE

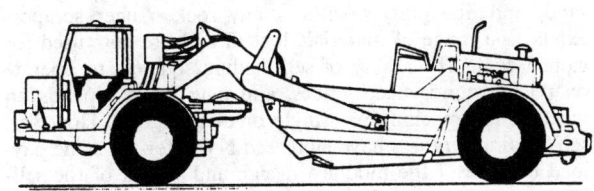
CONVENTIONAL DUAL ENGINE

PUSH-PULL

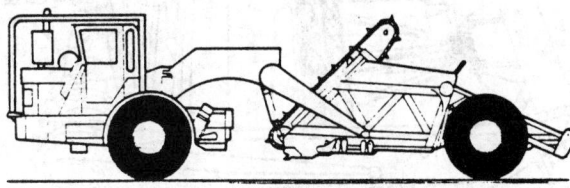
ELEVATING SINGLE ENGINE

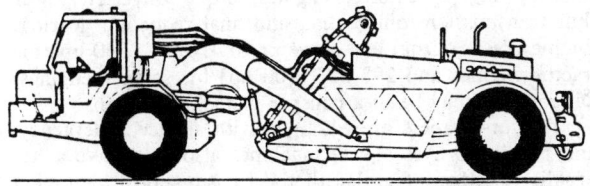
ELEVATING DUAL ENGINE

Fig. 1. Tractor-scraper types.

comprising about 45% of GVW. Principal advantages of conventional single engine scrapers are the wide range of materials which can be loaded, short load time, and favorable machine weight. Principal disadvantage is the need for pusher(s), requiring synchronization of scrapers and pusher(s). These scrapers perform best on relatively flat haulage roads over medium to long haul distances.

Conventional dual engine scrapers when pushed by track dozer(s) can load the same materials as single engine scrapers. However, when two scrapers operate as push-pull scrapers, these scrapers are limited to easily excavated materials. Conventional dual engine scrapers are designed with excellent gradeability, having 122 to 281 kg/kW (200 to 300 lb per hp). Because of this excellent gradeability, dual engine scrapers can operate at a higher velocity than single engine scrapers. Traction is the maximum possible because these scrapers have all the weight distributed on drive axles and have locking differentials to prevent any drive wheel from spinning free. Because of the additional drive train, tandem-powered scrapers have a higher net vehicle weight (NVW) for the same rated payload than single engine scrapers. Thus, the payload comprises a slightly lower percentage of GVW. Principal advantages of conventional dual engine scrapers are the wide range of materials that can be loaded when using dozer pusher(s), short load time, excellent gradeability, excellent traction, and ability to operate without dozer pusher(s) when operating in push-pull arrangement. These advantages can result in increased production. Principal disadvantage is increased machine weight resulting in higher ownership and operating costs. These scrapers perform best on poor ground conditions, steep grades, and short to medium hauls.

Self-loading scrapers only can load easily excavated materials. Elevating scrapers can load materials such as top soil, sand, and fine gravels without any rock. Auger scrapers extend the range of materials loaded and are now used for ripped coal. The ability of self-loading scrapers to operate on poor ground, steep grades, and long hauls depends on whether the machine has single or dual engines. However, self-loading scrapers have increased NVW for the same payload because of the modified design and weight of the self-loading mechanism. The principal advantage of self-loading scrapers is the ability to operate independently of other equipment such as a dozer or another scraper for assistance during loading. Principal disadvantages are limited range of materials that can be loaded, increased loading time, and increased machine weight.

DESIGN, SPECIFICATIONS, AND PERFORMANCE

Various types of tractor-scrapers are designed to perform best under specific loading and haulage conditions.

Design

Drive components providing power to the drive wheels consist of high speed diesel engines, torque converter, power shift transmission, differential, and final reduction gearing. Engine power output can range up to 410 kW (550 hp) for tractor engines and 298 kW (400 hp) for scraper engines. Differentials may be of a nonspin or locking design.

Tractor-scrapers are equipped with service, emergency, and parking brakes. These machines also may have a hydraulic retarder integral with the transmission for braking on favorable grades.

Tractor-scrapers use either bias ply or radial tires. These usually are wide base traction (E-2) or rock (E-3) tires. Tires

must have adequate load-carrying capacity and tonne-kilometers per hour (ton-miles per hour) rating.

The conventional scraper has three basic operating parts: bowl, apron, and ejector. The bowl forms a box with a rigid floor and sides. The floor is cut off forward of the center line and fitted with a cutting edge. The apron is the box front, which is opened to load and empty the bowl and closed during haulage. The ejector is the bowl rear which moves forward to force material out of the bowl. Fig. 2 shows the arrangement of bowl, apron, and ejector.

For the elevating scraper, the apron is replaced by an elevator consisting of two roller chains carrying crossbars or flights. The elevator bottom is near the bowl cutting edge with the elevator sloping back over the bowl. Elevator flights dig and lift material into the bowl. Ejection is usually accomplished by first moving the bowl backward against the

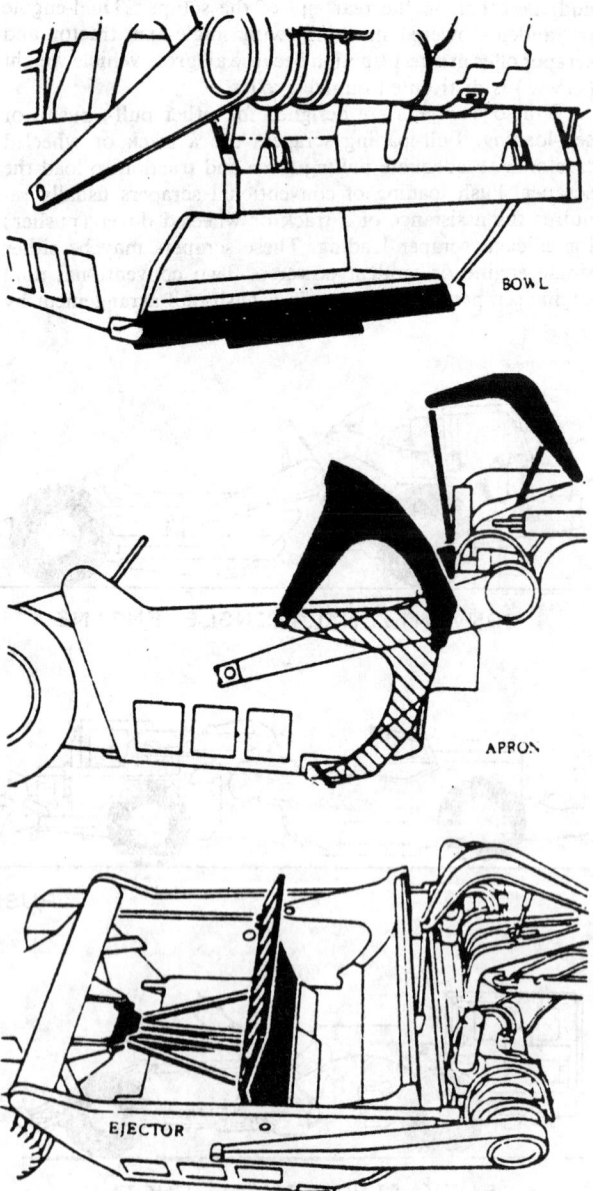

Fig. 2. Scraper bowl, apron, and ejector.

stationary ejector to open the space between the elevator and bowl cutting edge and then moving the ejector forward to complete unloading.

Specifications

Tractor-scraper manufacturers provide important machine specifications such as rated payload, heaped and struck volumetric capacity, engine kilowatt (horsepower), net vehicle weight (NVW), gross vehicle weight (GVW), weight distributions, clearance turning circle, width of cut, maximum depth of cut and spread, clearance under axles and bowl, and overall dimensions. Tractor-scrapers can be equipped with optional equipment to reduce maintenance, protect equipment, and increase productivity.

Performance

Drive train performance and retarding capabilities are characterized by performance and retarding curves, respectively. The drive train may have from 4 to 11 forward speeds and one or two reverse speeds. These curves are similar and used and adjusted in the same manner as performance and retarding curves for trucks. Travel time for the scraper to travel from one location to another is based on performance and retarding curves, net vehicle weight (NVW), actual payload, haulage road grade and rolling resistance, velocity limits and other operating constraints, and operator performance. Scraper theoretical travel time can be calculated using the same methods and computer simulation programs utilized to calculate truck travel time.

CYCLE TIMES

Two cycle times are important in estimating or evaluating scraper production: tractor-scraper and pusher cycle times.

Scraper Cycle Time

Tractor-scraper cycle time is the time required for the scraper to complete a single cycle including load, haul loaded, spread, return empty, wait, and delay. Scraper production is dependent on average actual payload and average cycle time. If pusher(s) are necessary, scraper fleet and pusher(s) must be matched to ensure adequate pushers with sufficient pushing force.

Load time depends on scraper type, material characteristics, pushing or pulling force, and operating conditions. Table 1 presents load times for various tractor-scrapers when operating under favorable, average, and unfavorable conditions. It is important that load time be minimized for economical scraper operation.

Travel time includes haul loaded and return empty times. These depend on the same factors and are calculated by the same computer programs as truck travel times. Manufac-

Table 1. Tractor-Scraper Load Times

	Load Time, Minutes		
	Favorable	Average	Unfavorable
Conventional, push loaded			
Single engine	0.40	0.70	1.00
Dual engines	0.35	0.60	0.90
Conventional, push pull			
Dual engines*	0.60	0.90	1.50
Elevating			
Single engine	0.60	0.90	1.30
Dual engines	0.45	0.70	1.00

* Time for loading both scrapers.

Table 2. Tractor-Scraper Spread Times

	Spread Time, Minutes		
	Favorable	Average	Unfavorable
Conventional			
Single engine	0.30	0.60	1.00
Dual engines	0.30	0.50	0.90
Elevating			
Single engine	0.40	0.70	1.10
Dual engines	0.30	0.60	1.00

turers provide distance vs. travel time graphs as a function of total resistance for various tractor-scrapers operating empty and loaded. These graphs can be used to estimate approximate travel times for haulage roads with simple profiles. Operating constraints such as velocity limits are required on downhill or favorable grades and curves. These limits depend on scraper design and load. Tractor-scrapers have a low center of gravity favoring stability, but the machine may *jackknife* under certain operating conditions.

Spread time is the time required for a scraper to maneuver and dump material. This time depends on scraper type, material characteristics, ground conditions, and operating conditions. Table 2 presents spread times for various tractor-scrapers. Material usually is spread in thin layers so that the scraper can maintain velocity and minimize equipment necessary to grade the dump.

Wait time occurs when the scraper must wait for a pusher at the loading area or for assistance at the dump. Wait time is caused by an inadequate number of pushers, inadequate pushing force causing long and variable load times, variable size scrapers with variable performance characteristics, poor weather conditions, and operator performance causing variation in cycle times.

Delay time includes variable delays which are not predictable as to time of occurrence and duration. These delays include haulage road maintenance, preparation of cut and dump area, driver relief stop, etc.

Total scraper cycle time is the summation of load, haul loaded, spread, return empty, wait, and delay times.

Pusher Cycle Time

Pusher cycle time is the time required for loading the scraper, boosting the scraper out of the cut, returning pusher for the next scraper, maneuvering the pusher, waiting, and delays. Pusher load time is the same as the scraper load time. Boost time usually ranges from 0.06 to 0.18 min. Pusher return and maneuver times depend on the pusher pattern.

Three commonly used pusher patterns are: (1) back-track loading, the simplest pattern, for short wide cut areas; (2) chain loading for long narrow cuts; and (3) shuttle loading for long cuts with loaders alternating direction. These patterns are illustrated in Fig. 3. Note that back-track loading is the only pattern that requires the pusher to operate in reverse and to return to the beginning of the cut.

Table 3 presents pusher cycle times without wait and delay times for various pusher patterns. Wait time occurs when the pusher must wait for a scraper. Wait time may be caused by an inadequate number of scrapers, variable size scrapers with variable performance characteristics, poor weather conditions, etc. Delay time includes variable delays such as preparation of the cut, driver relief stop, etc. In many cases, wait time can be used to prepare the cut and reduce delay time.

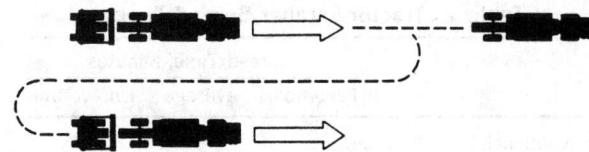

BACK TRACK LOADING

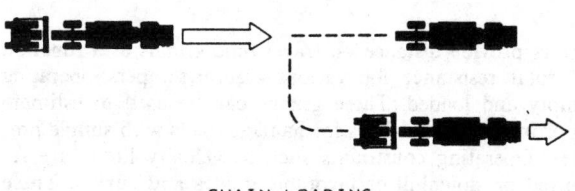

CHAIN LOADING

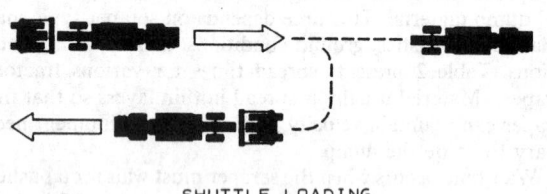

SHUTTLE LOADING

Fig. 3. Pusher patterns.

The theoretical number of scrapers that a pusher can service, N_p, can be derived by

$$N_p = \frac{t_{sc}}{t_{pc}}$$

where t_{sc} is the total theoretical scraper cycle time (no wait time) in minutes and t_{pc} is the total theoretical pusher cycle time (no wait time) in minutes.

PRODUCTION AND COSTS

Tractor-scraper production, fleet requirements, and costs are estimated and evaluated by the same general methods used for trucks. Actual operating time can be estimated by developing detailed scheduled and actual time distributions. Probability theory also can be used to estimate scraper availability.

Production

Scraper production, assuming 100% availability and/or utilization, can be calculated by

$$P_s = \frac{60 \times L_s}{t_{ts}}$$

where P_s is the scraper production rate based on actual active operating time, m³/h (cu yd per hr) or tonne per hour (stph); L_s is the scraper actual payload, m³ (cu yd) or tonne (ton); and t_{ts} is the scraper total cycle time in minutes.

Scraper production commonly is based on bank volume. Since scrapers usually are filled to heaped (1:1) capacity, bank volume of material in the scraper can be determined by

$$L_s = C_s \times F_s$$

where L_s is the scraper actual payload, bank m³ (cu yd); C_s is the scraper rated heaped (1:1) capacity, m³ (cu yd); and F_s is the material swell factor, decimal.

Trips per hour made by the scraper, T, can be calculated by

$$T = \frac{60}{t_{ts}}$$

Thus, scraper production also can be calculated by

$$P_s = T \times L_s$$

Fleet production, P_f, at a specific time is calculated by

$$P_f = k \times P_s$$

where k is number of available scrapers in the fleet. The number of scrapers required for the fleet is determined by comparing various fleet production capabilities and costs.

Costs

The ownership cost for scrapers is determined in the same manner as for other rubber-tired equipment, with depreciable value being delivered scraper cost less tires. Useful scraper life ranges from 12,000 to 16,000 hr.

Operating costs include fuel, tires and tire repair, preventive maintenance, mechanical repairs, cutting edge replacement, and operator wages. Hourly fuel consumption and cost can be calculated by the same method described for trucks. Table 4 presents tractor-scraper engine load factors when operating under light, average, and heavy loading and haulage conditions. Scraper tire life is greatly influenced by ground conditions, loading procedure, and operator experience as well as tire construction, type, loads, and maintenance; scraper velocity; and haulage road design, construction, and maintenance. As an example, tire life can be shortened greatly due to excessive wear and cuts by tire

Table 3. Pusher Cycle Times Without Wait and Delay Times

Pusher pattern	Pusher Cycle Time, minutes*		
	Favorable	Average	Unfavorable
Back track loading	0.75	1.25	1.80
Chain loading	0.60	0.95	1.40
Shuttle loading	0.60	0.95	1.40

* For tandem pushers, multiply pusher cycle time by 1.20.

Table 4. Tractor-Scraper Engine Load Factors

	Load Factor*		
	Light	Average	Heavy
All tractor-scrapers	0.40	0.55	0.75

* *Light*: Considerable idle, loaded hauls on favorable grades with low rolling resistance.
Average: Normal idle, load hauls on level grades with low rolling resistance.
Heavy: Minimum idle with steady cycling, loaded hauls on adverse grades with high rolling resistance.

Table 5. Tractor-Scraper Repair Factors

	Repair Factor*			
	Favorable	Average	Unfavorable	Very Unfavorable
Conventional, push loaded				
Single engine	0.30	0.65	1.00	1.30
Dual engines	0.35	0.70	1.05	1.40
Conventional Push pull				
Dual engines	0.35	0.72	1.08	1.45
Elevating				
Single engine	0.40	0.75	1.10	1.50
Dual engines	0.45	0.80	1.15	1.60

* Based on 0-16,000 hr tractor-scraper life.

spinning when loading ripped rock. Typical life of E-3 bias ply tires is 2000 to 3000 hr for drive wheel tires and 2500 to 3500 hr for tires on trailing axles. Hourly tire and repair cost is 115% of hourly tire replacement cost.

Preventive maintenance cost, including lubricating oils, greases, filters, miscellaneous parts, and labor, is about 15% of fuel cost. A more detailed estimate can be made based on specific machine data and local costs. Hourly repair cost, R_c, can be estimated by

$$R_c = \frac{V_d}{10,000} \times F_r$$

where V_d is the depreciable value and F_r is the repair factor. Table 5 presents tractor-scraper repair factors for a range of operating conditions. Cutting edge replacement cost varies widely depending on material loaded, loading procedure, and operator experience. This cost may be 2 to 10% of the repair cost. Operator wages normally include base wage, fringe benefits, training, and other operator related costs for operating the scraper.

Scraper loading and haulage costs usually are determined as cost per bank cubic meter (cubic yard). These include costs for both tractor-scraper and, if used, pusher.

Fleet Production and Cost

Table 6 presents estimated fleet production and costs as well as cost per bank cubic meter (cubic yard) for conventional single engine scrapers and pusher loading and excavating loose earth with a bank bulk density of 1780 kg/m³ (3000 lb per cu yd). At a loading velocity of 2.74 km/h (1.7 mph), the track dozer can provide 34 020 kg (75,000 lb) of push which is slightly more than required pushing force of 33 750 kg (74,400 lb). Thus, the track dozer can provide adequate pushing force to load the scraper. As the number of scrapers in the fleet increases, scraper load, travel, spread, and delay times remain constant, but scraper wait time increases because of nonsynchronization of scrapers and pusher. As an example, there is no wait time with three scrapers because there is excessive pusher capacity. As the scraper wait time increases, pusher wait time decreases.

Hourly scraper unit costs in Table 6 decrease as the number of scrapers increase because lighter work cycles result from increased wait times. However, hourly pusher unit costs increase as the number of scrapers increases because the pusher has a heavier work cycle. Based on scraper and pusher cycle times, the pusher theoretically can serve 5.7 (8.00 min/ 1.4 min) scrapers. However, the lowest cost per bank cubic meter (cubic yard) occurs with five scrapers.

Table 6. Tractor-Scraper Fleet Production and Costs

Material
Description: Loose earth
Bank density: 1780 kg/m³ (3000 lb/cu yd)
Loose density: 1424 kg/m³ (2400 lb/cu yd)
Swell factor: 0.80
Traction coefficient:
Rubber tires: 0.45
Track: 0.60

Time distribution
Scheduled: 10 shifts/week
Active operating time: 5.50 hr/shift

Scraper
Type: Conventional single engine, 336 kW (450 hp)
NVW: 43 090 kg (95,000 lb)
Capacity: 23.7 m³ (31 cu yd) heaped
Rated load: 34 020 kg (75,000 lb)
Weight distribution, drive wheels:
Empty: 70%
Loaded: 55%
Actual load: 23.7 m³ × 1780 kg/m³ × 0.80 = 33 750 kg (74,4000 lb)

Table 6.—Continued

Tractive force—traction limitation:
Empty	0.45 × 43 090 kg × 0.70 = 13 573 kg (29 925 lb)
Loaded	0.45 × 76 840 kg × 0.55 = 19 018 kg (41 927 lb)

Tractive force—rimpull limitation at 2.74 km/hr (1.7 mph): 29 485 kg (65,000 lb)

Cut dimensions:
Length	27.4 m (90 ft)
Width	3.50 m (11.5 ft)
Depth	19.8 cm (7.8 in.)

Push required: 33 750 kg (74,400 lb)

Pusher
Type	Track dozer, 343 kW (460 hp)
NVW:	58 740 kg (129,500 lb)

Tractive force—traction limitation: 0.60 × 58 740 kg = 35 244 kg (77,700 lb)
Tractive force—drawbar pull limitation at 2.74 km/h (1.7 mph): 34 020 kg (75,000 lb)
Pattern: Back-track loading

Number of scrapers	3	4	5	6	7
Scraper cycle time, minutes					
Load	0.70	0.70	0.70	0.70	0.70
Haul loaded	4.25	4.25	4.25	4.25	4.25
Spread	0.60	0.60	0.60	0.60	0.60
Return empty	2.10	2.10	2.10	2.10	2.10
Wait	0.00	0.18	0.48	1.13	2.05
Delays, variable	0.35	0.35	0.35	0.35	0.35
Total	8.00	8.18	8.48	9.13	10.05
Number of pushers	1	1	1	1	1
Pusher cycle time, minutes					
Load	0.70	0.70	0.70	0.70	0.70
Boost	0.10	0.10	0.10	0.10	0.10
Return	0.30	0.30	0.30	0.30	0.30
Maneuver	0.15	0.15	0.15	0.15	0.15
Wait	1.27	0.64	0.30	0.12	0.04
Delays, variable	0.15	0.15	0.15	0.15	0.15
Total	2.67	2.04	1.70	1.52	1.44
Scraper					
Trips/active operating hr	7.50	7.35	7.06	6.58	5.95
Scraper production					
Bank m³/active operating hr	142	139	134	125	113
(Bank cu yd/active operating hr)	(186)	(182)	(175)	(163)	(148)
Fleet production					
Bank m³/shift	2343	3067	3680	4116	4345
(Bank cu yd/shift)	(3065)	(4012)	(4814)	(5384)	(5683)
Unit cost					
Scraper, $/hr	107.50	106.40	104.70	101.40	97.45
Pusher, $/hr	89.22	100.05	105.85	111.30	114.14
Fleet cost					
Scrapers, $/shift	1773.75	2340.80	2879.25	3346.20	3751.83
Pusher, $/shift	490.71	550.28	582.18	612.15	627.77
Total, $/shift	2264.46	2891.08	3461.43	3958.35	4379.60
Production cost					
$/bank m³	0.9665	0.9426	0.9406	0.9617	1.0080
($/bank cu yd)	(0.7388)	(0.7206)	(0.7190)	(0.7352)	(0.7706)

6.5.6. Wheel Loaders

Ronald M. Hays

LOAD-AND-CARRY APPLICATIONS

Wheel front-end loaders commonly are used as load-and-carry units where the loader excavates, transports, and dumps or spreads material. Typical load-and-carry applications are:

1. Load at face and carry to hopper, crusher, stockpile, dump, etc.
2. Strip, carry, and stockpile topsoil.
3. Load from stockpile or surge piles and carry to hopper.
4. Load, carry, and spread material for construction and maintenance of roads, dikes, settling ponds, etc.

The wheel loader has excellent mobility and is widely used for small or intermittant load-and-carry applications. Larger wheel loaders also are used as primary haulage equipment for short transport distances of up to 183 m (600 ft) with level to gentle slopes.

DESIGN, SPECIFICATIONS, AND PERFORMANCE

Wheel front-end loaders used in mining applications usually are designed with a center-articulated design and a rear-mounted diesel engine. Power is transmitted to all four wheels by either a mechanical drive train or electric wheel motors.

Wheel loaders can be equipped with bias ply, radial, or beadless tires. The Tire and Rim Association (T&RA) code identification for loader tires is the same as for dozer tires. The beadless tire consists of a carcass covered by a replaceable steel-reinforced rubber mounting belt with anchor plates and mounting lugs and steel shoes bolted directly to the mounting belt. When used in load-and-carry applications, tires must have adequate tonne-kilometer per hour (ton-mile per hour) rating as well as adequate load carrying capacity. Since loader tire load limits often are based on a peak speed of 8.0 km/h (5 mph) and maximum one way distances of 76 m (250 ft), tire loads must be reduced for the higher velocities and longer transport distances often encountered in load-and-carry applications.

Wheel loader manufacturers provide important machine specifications such as rated payload, nominally heaped (2:1) and struck volumetric capacity, engine kilowatt (horsepower), operating weight, weight distribution, tipping loads, carry position, bucket width, clearance circle, and overall dimensions. For wheel loaders, the payload is only 15 to 25% of loaded operating weight, resulting in a relatively inefficient haulage unit.

Wheel loader drive train performance and retarding capabilities are characterized by performance and retarder curves, respectively. Wheel loaders with mechanical drives commonly have two to four forward and reverse gears. Reverse velocities are important in load-and-carry applications since the loader may travel the same distance in reverse as in forward. These curves are similar and used and adjusted in the same manner as performance and retarding curves for other mobile equipment.

LOADER CYCLE TIME AND PRODUCTION

When used in load-and-carry applications, loader cycle time consists of load, travel, dump, maneuver, and delay times. Load time is the time to dig into the bank and load the bucket. Load time depends on material characteristics, loader digging forces and bucket size, loading procedure, and operator skill. Travel time includes time for travel both loaded and return empty between the face or pile and the dump or spread point. During travel the bucket should be near the carry position to maintain a low center of gravity. Travel times can be calculated by the same computer programs used to calculate travel time for other mobile equipment. Manufacturers also provide distance vs. travel time graphs as a function of total resistance for loaded and empty wheel loaders. These graphs usually are satisfactory for short travel distances with simple haulage profiles. Dump time is the time to raise and dump the bucket or spread material at the dump or spread point. Dump time depends on the loader's hydraulic cycle times and dumping or spreading procedure. Maneuver time is the time required for the loader to maneuver and possibly turn at both the face or pile and dump or spread point. Delay time includes nonpredictable variable delays due to area maintenance and cleanup, driver relief stop, etc. Loader total cycle time for load-and-carry applications is the summation of load, travel loaded, dump, return empty, maneuver, and delay times.

Loader production, assuming 100% availability and/or utilization, can be determined by

$$P_l = \frac{60 \times C_l \times F_f \times F_s}{t_{tlc}}$$

where P_l is the loader production rate based on actual active operating time, bank m³/hr (bank cu yd per hr); C_l is the loader rated capacity, m³ (cu yd); F_f is the loader bucket fill factor, decimal; F_s is the material swell factor, decimal; and t_{tlc} is the loader total cycle time for load-and-carry in minutes.

Loader production for load-and-carry applications usually is expressed in bank volume.

6.5.7 Dozers

Ronald M. Hays

TYPES AND APPLICABILITY

Bulldozers, commonly called dozers, are a self-propelled tractor equipped with a front-mounted blade (dozer) or other assembly. Dozers are of two basic types: (1) track or crawler tractor and (2) rubber-tired or wheel tractor. Crawler dozers also may be equipped with a rear-mounted ripper for loosening unconsolidated soil or fragmenting rock. Both crawler and wheel dozers range up to 90 718 kg (200,000 lb) in operating weight. Dozers normally are powered by an onboard diesel engine with crawler and wheel dozers having flywheel power up to 522 kW (700 hp) and 611 kW (820 hp), respectively.

In surface mines, dozers are used for many applications, including:

Land clearing.
Construction and maintenance of access and haulage roads.
Benching and preparation of loading area.
Cleanup around excavating and/or loading equipment.
Relocate electric cable skids.
Maintain waste dump by spreading material, pushing material over dump slope, and grading dump area and slopes.
Cleanup around dump area.
Push material to hopper for loading.
Push overburden short distance as a primary method for stripping overburden.
Push spoil to contour spoil piles for reclamation.
Push load tractor-scrapers.
Assist disabled vehicles.
Relocate pumps, pipes, conveyors, etc.
Rip unconsolidated soil and rock.
Construction, maintenance, and cleanup of stockpiles.

Both crawler and wheel dozers are used for all these applications, except ripping which is accomplished by crawler dozers only.

Crawler dozers usually are used for more difficult operating and ground conditions because of high drawbar pull or pushing force provided by the power train; high coefficient of traction in loose earth, soil, and rock; low ground bearing pressure; and excellent stability. Wheel dozers are advantageous where conditions are more consistent such as dozing uniform material with a high coefficient of traction. With favorable ground conditions, wheel dozers generally provide greater pushing force when operating above speeds of 3.2 to 4.8 km/h (2 or 3 mph) than crawler dozers of the same class. Wheel dozers also are highly mobile and can be moved quickly from one location or application to another. Dozers are versatile and are essential at almost all surface mines.

CRAWLER DOZERS

Drive trains for crawler dozers use high speed diesel engines with power transmitted to tracks through mechanical drives. Drive train configuration varies depending on manufacturer and dozer size. Common drive train components are flywheel clutch to disengage engine and drive train when stopping or shifting gears; torque converter to provide shock-absorbing slippage between engine and track, provide torque multiplication and reduce number of transmission gear ratios, and provide for on-the-move power shifting; manual and power shift transmissions to provide for change in speed-

power ratio and reverse travel; power takeoff to power accessories and hydraulic systems; rear drive assembly to provide final drive gear reduction for power to drive wheels or sprockets and clutches and brakes for steering; drive sprockets to provide positive drive to the tracks; and tracks to support the dozer and provide traction. The dozer undercarriage consists of track frames, drive sprockets, front idlers, rollers, track adjustor, chain links and pins, and track shoes.

The two basic crawler dozer undercarriage designs are shown in Fig. 1. Conventional design has the drive at the rear of the track frames while the newer design has the drive sprockets elevated above and separated from the track frames. The dozer blade is controlled by lift, tilt, and angling hydraulic cylinders. The dozer ripper also is controlled by hydraulic cylinders.

Drive train performance is characterized by a performance curve of drawbar pull vs. velocity such as shown in Fig. 2. Drawbar pull is the horizontal force at the drawbar or the force the drive train can produce at the track-ground contact. Manufacturer's curves must be adjusted for altitude and temperature if different from those specified. The maximum force which can be transferred from the tracks to ground is the product of the coefficient of traction between tracks and ground (see Section 6.5.2, *Trucks,* Table 3) times the normal force on the ground due to the dozer operating weight. Thus, usable force is limited by either this maximum force or drawbar pull.

Fig. 1. Crawler dozer designs. Top: Conventional sprocket design. Bottom: Elevated sprocket design.

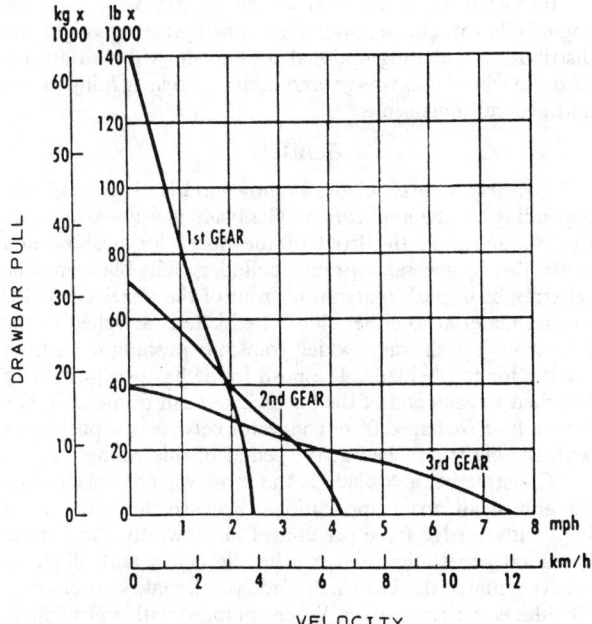

Fig. 2. Crawler dozer performance curve, 268-kW (360-hp) engine.

As an example, consider a 268 kW (360 hp) dozer with an operating weight of 41 227 kg (91,000 lb) that is pushing loose gravel with a 0.50 coefficient of traction. Based on the performance curve, the dozer drive train can provide a maximum force or drawbar pull of 63 500 kg (140,000 lb). However, the maximum force which can be transferred from the track to ground is 20 638 kg (45,500 lb).

Important dozer specifications are blade capacity, engine kilowatt (horsepower), machine operating weight including optional equipment, track width and ground contact area, and general dimensions. Dozer ground bearing pressure depends on machine operating weight and track ground contact area. Various width tracks are available for each dozer. A variety of track shoe types are available for special applications.

A pair of crawler tractors may be linked together and operated as a single machine with one operator. The two machines may be linked in-line with a front-to-rear swivel fastener or side-by-side with a rigid tube at the rear and a common blade at the front.

Crawler dozers may be equipped with a power and velocity sensing system which allow the operator to maximize effective dozer work. This system functions by measuring dozer true ground velocity and pushing force. The operator then can vary the blade cutting depth or loading, which alters the velocity and pushing force so that the dozer operates at full power and the most efficient portion of the dozer performance curve.

WHEEL DOZERS

Wheel dozers used in mining applications usually have a center-articulated design as shown in Fig. 3. The engine is at the rear, providing even weight distribution between front and rear axles. The operator's compartment usually is above the pivot point.

Drive trains for wheel or rubber-tired dozers use high speed diesel engines with power transmitted to all four wheels

by mechanical or electric wheel drives. Drive train varies depending upon manufacturer and dozer size with only very large dozers using electrical wheel drives. For mechanical drives, power is transmitted from the engine to all four wheels by a torque converter and transmission, drive shaft and necessary universal joints to front axle and drive shaft to rear axle, front and rear differentials, and final planetary drives. Electric wheel drive dozers use a diesel engine to drive an a-c generator that provides electric current to a solid-state electric power converter. The converter provides d-c current to each electric wheel motor which then drives the final planetary gearing. Because each wheel has its own wheel motor, wheel spin or slippage can be controlled by reducing power as the wheel loses traction.

Drive train performance is characterized by a performance curve of rimpull vs. velocity as shown in Fig. 4. This performance curve is for a mechanical drive dozer and is interpreted in the same manner as the truck performance curve. Manufacturers' performance curves must be adjusted for altitude, temperature, and tire size if different from those specified. These adjustments are the same as made for trucks. As with crawler dozers, usable force is limited by machine traction or rimpull.

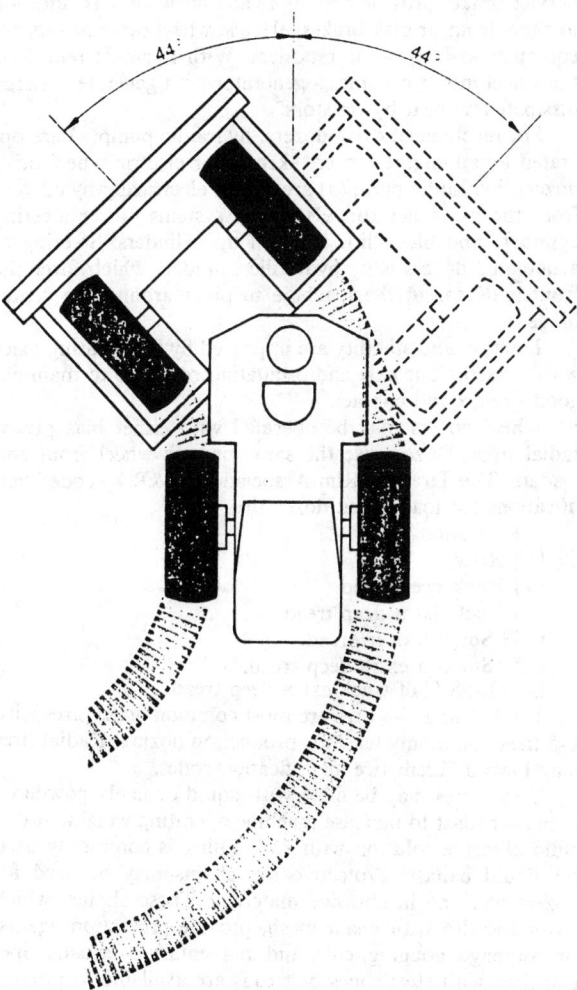

Fig. 3. Wheel dozer design.

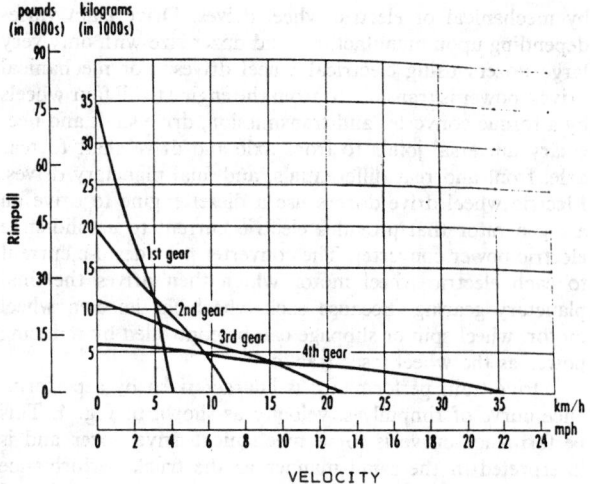

Fig. 4. Wheel dozer performance curve, 250-kW (335-hp) engine.

Wheel dozers are equipped with service, emergency, and parking brakes. Service brakes may be shoe-drum or disk brakes with air, air-over-hydraulic, or all hydraulic activated. Service brakes provide braking to all four wheels. In addition to shoe-drum or disk brakes, electric wheel drive dozers are equipped with dynamic retarders. With dynamic retarders, the wheel motor becomes a generator with generated energy dissipated as heat by resistors.

For mechanical drive dozers, hydraulic pump(s) are operated by an engine power takeoff. For electric wheel drive dozers, hydraulic pump(s) are driven electrically by current from the a-c generator. Hydraulic systems power steering cylinders and blade lift, tilt, and tip cylinders. Steering of articulated dozers is by hydraulic cylinders which force the front and rear of the machine to pivot around the center hinge.

Traction and stability are improved by free floating axles with a fixed front axle and oscillating rear axle to maintain good tire-ground contact.

Wheel dozers may be operated with either bias ply or radial tires. Dozers use the same tires as wheel front end loaders. The Tire and Rim Association (T&RA) code identifications for loader and dozer tires are:

L-2 Traction.
L-3 Rock.
L-4 Rock deep tread.
L-5 Rock extra deep tread.
L-4S Smooth deep tread.
L-5S Smooth extra deep tread.
L-5/L-5S Half track extra deep tread.

The L-4 and L-5 tires are most common dozer tires with L-5 tires commonly used for production dozing. Radial tires may have different tire identification codes.

Dozer tires may be filled with liquid or finely powdered mineral ballast to increase machine operating weight. A calcium chloride solution with 75% filling is commonly used for liquid ballast. Protective tire chains may be used for dozers working in abrasive materials. These chains, which cover the tire with chain mesh, provide protection against tire slippage, gouging, cuts, and momentary overloads. Special tires with steel shoes or treads are available to provide complete protection against abrasion and cuts in the tire tread or wear area.

Important wheel dozer specifications are blade capacity, engine kilowatt (horsepower), machine operating weight and distribution including optional equipment, width to outside of tires, wheelbase, tire-ground contact area, turning circle, and general dimensions.

BLADES

A variety of dozer blades as shown in Fig. 5 are available, depending on the application. The blade is a massive structure mounted on the front of the dozer for pushing and controlled by several hydraulic cylinders. The blade may be set straight to push material in front of the dozer or angled to cast material to either side. The blade is attached to the tractor by a push frame which transmits pushing force from the tractor to the blade. The push frame has two push arms attached to each end of the blade. The push frame also may have a V or U-shaped front connector between the push arms with the blade attached at the center of this connector.

The straight or S blade is the most versatile blade used for almost all dozer applications. Because the blade has a high cutting edge force per unit of blade width, the S blade has good penetration and can handle heavy and tough-to-excavate materials. The small blade width makes dozers with S blades easy to maneuver. When equipped with a push plate, the S blade is used for push loading scrapers.

The universal or U blade has large wings that increase blade capacity. The U blade is excellent for pushing big loads over long distances such as in reclamation, stockpile work, charging hoppers, and trapping rock for front-end loaders. This blade also is excellent for spreading fill and finish grading. The U blade has relatively poor penetration and is best for lighter and relatively easy to excavate or push materials. Large-capacity U blades are available for pushing light non-cohesive materials such as stockpiled coal.

The angling or A blade is essentially a straight blade of lower height, more moldboard curvature, and less capacity than the S blade. The A blade can be angled right or left for side-casting, backfilling, windrowing, benching, and ditching. This blade also can be positioned straight, but usually has less cutting edge force per unit of blade width than the S blade. Thus, in the straight position, the A blade operates most effectively in light and easy to excavate or push materials.

For relative comparisons, the capacity of straight, universal, and angle blades is calculated in accordance with SAE recommended practice. However, this blade capacity cannot be used satisfactorily to estimate actual blade loading or dozer production.

The cushion or C blade is a smaller blade used principally for push-loading scrapers. This blade has rubber cushions to absorb impact when contacting the scraper push block. The narrow blade width increases dozer maneuverability in congested scraper cuts and reduces the possibility of cutting scraper tires. The C blade also can be used for scraper cut maintenance and other ground dozing applications.

Many blades have been developed for special applications such as large angle and U blades pushed by two side-by-side tractors, narrower and deeper V blades, lightweight blades of various designs, bucket-type blades, and straight blades with high sidewalls. The dozer also may be equipped with a special application assembly such as grading bar, tree cutter, rake, etc.

DOZING

When dozers are used for continuous production dozing to push large volumes of material over short distances, pro-

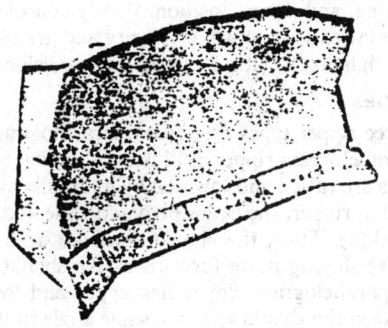

STRAIGHT 'S' BLADE

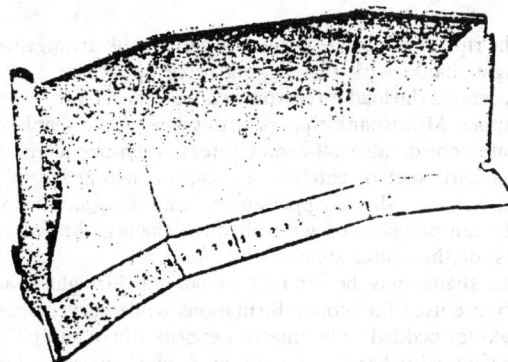

UNIVERSAL 'U' BLADE

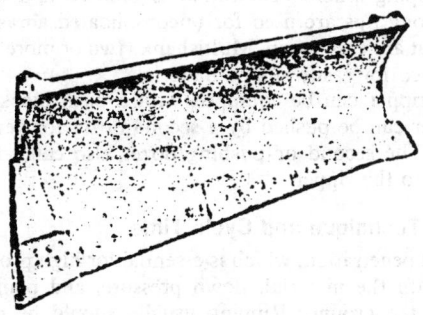

ANGLING 'A' BLADE

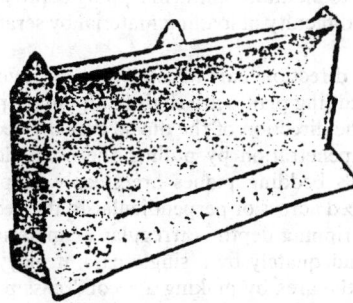

CUSHION 'C' BLADE

Fig. 5. Dozer blades.

duction depends on blade loading and dozer cycle time. Production dozing usually is accomplished with a straight or universal blade. When dozers are used for intermittent dozing such as cleanup, benching, haulage road construction and maintenance, etc., production usually is controlled by factors other than blade loading and cycle time.

Blade Loading

The capacity of straight, universal, and angle blades for crawler and wheel dozers are calculated in accordance with SAE recommended practice. Actual blade loading may deviate substantially from blade capacity because of blade design, drive train performance, operating weight, traction, material characteristics, terrain, operating procedure, operator technique, etc.

Three functions performed by the dozer blade are to excavate material using the cutting edge, transport material, and spread material. Dozer drawbar pull and/or operating weight and coefficient of traction between the track or tires and the ground determine the pushing force available for cutting and pushing material. It usually is desirable to operate the dozer near a velocity that maximizes power delivered at the track-ground contact by balancing the cutting depth and blade loading. Cutting depth and material characteristics determine the pushing force required with resistance increasing with increased cutting depth. Consolidated materials that are difficult to cut require a high pushing force in the cut. For single pass dozing of difficult to excavate materials, blade loading often can be maintained by reducing cutting depth to decrease cutting resistance and increasing cut length to increase blade loading. Multiple passes with shallow cutting depth also may be used for difficult to excavate materials. This reduces blade loading, making more pushing force available for cutting material.

Since the load slides or rolls along the ground during transport, friction (internal, blade to material, and material to ground), material density, and blade loading determine the pushing force required for transport. Pushing force requirements are based on experience or field testing. Some dry noncohesive (*dead*) and sticky materials are hard to push. When pushing material downhill, blade loading usually can be increased because gravity is assisting the dozer. Conversely, pushing material uphill decreases blade loading.

Operating procedure such as cutting a slot, operating two blades side by side, or pushing between windrows increase blade loading by reducing side spillage. Operator technique varies considerably. Because of material spillage off the sides of the blade as the dozer transports material, the operator must replace this spilled material to maintain full blade loading.

Blade loading is measured in volume of loose material. Thus, the material swell factor is used to convert from bank volume excavated or material bank bulk density to actual blade loading or material loose bulk density. Actual blade loading can be determined in the field by:

1) Counting number of dozer passes, surveying the cut to determine volume of material excavated, and then calculating average blade loading per pass.

2) Weighing actual blade load, determining material loose bulk density, and calculating blade loading.

3) Measuring actual blade load dimensions and calculating blade loading.

Dozer Cycle Time

Dozer cycle time is the time required for the dozer to complete a full pass including cut, transport, spread, return, fixed, maneuvering, and delay times.

Cut time is the time required to excavate the material in the cut and load the blade. Cut length ranges from 7.6 to 23 m (25 to 75 ft), depending on the dozer and excavating difficulty. Load times usually range from 0.15 to 0.45 min with the longer times for longer cuts with difficult-to-excavate material.

Transport time is the time required to push material from the end of the cut to the beginning of the spread or pile. This time depends on the pushing force required to transport the material. Based on a required pushing force of 13 608 kg (30,000 lb) and the performance curve in Fig. 2, this dozer could transport the load at a velocity of 4.0 km/h (2.6 mph) in second gear. If the transport distance is 61 m (200 ft), the transport time would be 0.87 min.

Spread time is the time required to spread or dump the blade load. Dozers usually spread material while moving forward and raising the blade, spreading a layer of material. Spread times usually range from 0.08 to 0.12 min.

Return time is the time required for the dozer to return backwards to the beginning of the cut. Crawler and wheel dozer return velocities usually are limited to about 8.0 km/h (5 mph) and 16.1 km/h (10 mph), respectively.

Fixed time is the time required for transmission shifting. Maneuvering time is the time required to position the dozer prior to beginning to excavate material. Delay time is due to variable delays such as cut maintenance, driver relief stop, etc.

Total cycle time is the summation of cut, transport, spread, return, fixed, maneuvering, and delay times. The dozer cycle time is greatly affected by operator experience and technique, visibility, terrain, nonuniform ground conditions, and material characteristics.

Dozer Production

Dozer production, assuming 100% availability and/or utilization, can be calculated by

$$P_d = \frac{60 \times L_d \times F_s}{t_{td}}$$

where P_d is the dozer production rate based on actual active operating time, bank m^3/hr (bank cu yd per hr); L_d is the blade loading, loose m^3 (loose cu yd); F_s is the material swell factor, decimal; and t_{td} is the dozer total cycle time in minutes.

Dozer production commonly is in bank volume of material excavated, but the blade loading and volume of material spread, dumped, or piled is in loose volume. Dozer manufacturers can provide graphs, information, and valuable assistance in estimating dozer production.

RIPPING

Crawler dozers can be equipped with a ripper mounted on the rear of the tractor. The ripper shank(s) is pulled through soil or rock to loosen or fragment the material. The material then can be pushed by a dozer or loaded by a front-end loader or scraper.

The rippability of the material is influenced by the material's compressive strength; bedding planes, joints, and fractures; brittleness; and softness and weakness caused by weathering. Glacial tills, shales, sandstones, conglomerates, schists, and coals usually are rippable while massive and homogeneous formations usually are nonrippable. Rippability can be estimated based on seismography testing, which determines the velocity of seismic waves through the material. The seismic wave velocity indicates the degree of material consolidation, including rock strength, stratification

and fracturing, and decomposition. Poorly consolidated materials have lower seismic wave velocity and are easier to rip than hard, tight rock with a higher seismic velocity.

Ripper Types

The three ripper types are radial, parallelogram, and adjustable parallelogram ripper as illustrated in Fig. 6. All three ripper types are raised and lowered by hydraulic cylinder(s). For the radial ripper, the beam pivots to raise and lower the ripper shank(s). Thus, the shank(s) moves in an arc with the shank(s) sloping more forward as the shank(s) is lowered. The parallelogram ripper has upper and lower hinge arms that keep the shank(s) at the same angle to the ground as the shank(s) is lowered. With the adjustable parallelogram ripper, hydraulic cylinders serve as the upper hinge arms. Thus, the shank(s) angle to the ground can be varied to the optimum angle for penetration, either before or during ripping.

The ripper may have a single or multishank arrangement. A single shank, with the digging force applied at a single point, centers the load forces and reduces strain on the tractor and ripper. Multishank rippers with two or more shanks can generate considerable off-center forces when one shank impacts a hard spot or catches or pushes a slab ahead of the shank. For multishank rippers with three shanks, the ripper usually can be operated with all three shanks, the two side shanks, or the center shank only.

The shank may be straight or curved. Straight shanks usually are used for blocky formations while curved shanks are used for bedded or laminated deposits where an uplifting action causes further fragmentation. Each shank is equipped with a detachable point or tip, which can be of various designs with shorter tips providing greater resistance to breakage.

A single center shank with a short tip is usually used for heavy ripping where penetration is difficult and shock is severe. Long tips are used for unconsolidated abrasive materials that are easy to rip. Multishank (two or more shanks) can be used for easily ripped materials.

The ripper can be equipped with a push block so that the ripper can be pushed by a second dozer for very heavy ripping. The second or pushing dozer also can add down pressure to the ripper.

Ripping Technique and Cycle Time

Good penetration, which is essential for high production, depends on the material, down pressure, and point or tip angle to the ground. Ripping usually should be near full depth since shallow ripping is irregular and causes high tip wear. However, ripping to partial depth may be desirable because of limited drawbar pull, traction, or penetration or when natural planes of weakness occur at partial depth. It is important to maintain uniform ripping depth and breakage to minimize difficulty in loading material by scrapers or other equipment.

Ripping direction can affect ripper production. If the material's bedding planes are slightly dipping, ripping should proceed in the direction of the dip so that the bedding planes assist point penetration by pulling the point downward. If the material's bedding planes are steeply dipping, ripping should proceed across or perpendicular to the bedding planes or strike. If ripping depth is irregular or material cannot be fragmented adequately by a single pass, it may be desirable to cross rip the area by making a second pass perpendicular to the first pass. If the surface is sloped, downhill ripping should be used whenever possible.

Ripper production is affected by the spacing between

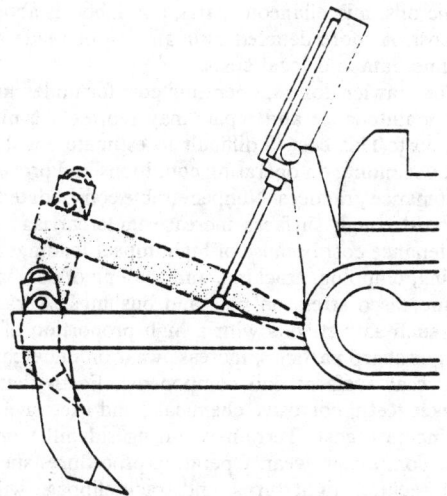

RADIAL RIPPER

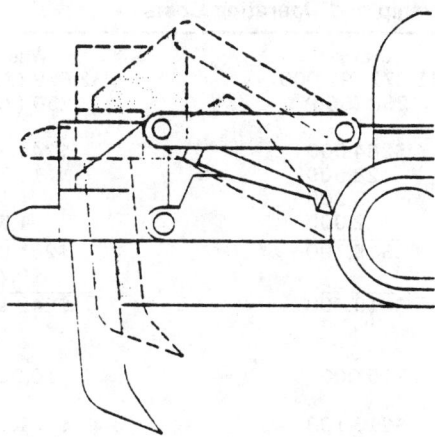

PARALLELOGRAM RIPPER

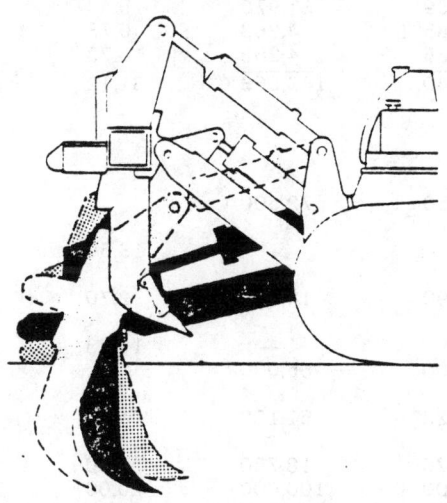

ADJUSTABLE PARALLELOGRAM RIPPER

Fig. 6. Ripper types.

ripper passes. Wide spacing increases production, but may result in irregular ripping depth or poor fragmentation. Thus, spacing must be close enough to insure a uniform ripping depth and fragmentation fine enough for loading, haulage, and possibly crushing without major difficulties.

When ripping, the tractor usually is operated in first gear at velocities of 1.6 to 2.4 km/h (1 to 1.5 mph), including slippage and stalls. Higher velocities tend to increase wear on both tractor undercarriage and ripper shank(s) and point(s).

The ripper cycle time consists of rip, turn, and delay times. Rip time is the time required to make a single pass while ripping and is based on the rip distance and average dozer speed. Turn time is time required to turn the tractor and maneuver to begin the next pass. Turn time usually is 0.20 to 0.35 min. Delay time is time for variable delays. Ripper total cycle time is the summation of rip, turn, and delay times.

Ripper Production

Ripper production depends on the material's rippability, the tractor and ripper, ripping technique and procedure, and operator experience and ability.

Since ripper production depends on many factors that cannot be quantified, reliable methods of estimating production depend on field studies. The two principal methods are to:

1. Determine average cycle time; measure average rip distance, rip spacing, and depth of penetration; calculate volume of material ripped; and then calculate production rate.

2. Record time spent ripping, remove and weigh ripped material or survey the area after removal of ripped material and calculate volume of ripped material, and then calculate production rate.

Ripper production rate is usually in bank volume or bank m^3/hr (bank cu yd per hr). Based on ripper cycle time, ripper production, assuming 100% utilization and/or availability, can be calculated by

$$P_r = \frac{60 \times L \times w \times p}{t_{tr}}$$

where P_r is the ripper production rate based on actual active operating time, bank m^3/hr (bank cu yd per hr); L is the rip distance (length of ripped area), m (yd); w is the rip spacing (distance between ripper passes), m (yd); p is the depth of penetration, m (yd); and t_{tr} is the ripper total cycle time in minutes.

As an example, consider a 58 967 kg (130,000 lb) dozer with a single shank ripper that rips at an average velocity of 1.6 km/h (1 mph), turn time of 0.30 min, delay time of 0.50 min, 152 m (167 yd or 500 ft) rip distance, 0.914 m (1 yd or 3 ft) rip spacing, and 0.610 m (0.67 yd or 24 in.) depth of penetration. Rip time then is 5.68 min for a total cycle time of 6.16 min. Ripper production rate is 851 bank m^3/h (1113 bank cu yd per hr). This production rate does not include any idle time, fixed delays, or downtime for preventive maintenance.

COSTS

Ownership costs for dozers and rippers are determined in the same manner as for other mobile equipment and include depreciation, interest, insurance, and taxes. Delivered cost includes factory cost of tractor; blade, ripper and any optional equipment; freight; and erection at the mine. For crawler dozers, depreciable value is the same as delivered cost. However, for rubber-tired dozers, the cost of tires is deducted from delivered cost to obtain depreciable value. For crawler tractors operating at mines, useful life ranges from about 8000 to 18,000 hr depending on the specific dozer, application, and operating conditions. Ripping consolidated rock is one of the severest applications for a crawler tractor. For wheel dozers, useful life ranges from about 8000 to 12,000 hr for mining applications. Table 1 shows ownership costs for both a crawler and a wheel dozer.

Operating costs for crawler dozers include fuel, preventive maintenance, undercarriage and track maintenance, mechanical repairs, blade cutting edge replacement, ripper shank and point replacement, and operator wages. Wheel dozers have operating costs for tires and tire repair but do not have costs for undercarriage and track maintenance.

Hourly fuel consumption and cost can be calculated by the same method used for other mobile equipment. Table 2 presents crawler and wheel dozer engine load factors for light, average, and heavy applications. Preventive maintenance cost, including lubricating oils, greases, filters, hy-

draulic oils, miscellaneous parts, and labor, is about 20% of fuel cost. A more detailed estimate can be made on specific machine data and local costs.

For crawler dozers, operating cost for undercarriage and track maintenance and repair may represent a major operating cost. This cost is difficult to estimate and highly variable, depending on operating conditions and procedures and maintenance practices. Nonpenetrable consolidated materials and hard rough surfaces increase undercarriage and track maintenance cost because of high impact loadings that cause bending, chipping, cracking, and spalling of contact surfaces and increased stress on pins and bushings. Abrasive materials, such as wet soils with a high proportion of hard, angular, or sharp particles, increase wear on undercarriage and track wear surfaces and components. Earth that packs in sprocket teeth, corrosive chemicals, and excessive soil moisture increase cost. Terrain requiring sidehill work will increase component wear. Operating procedures such as high-speed backing, tight turns, and track slippage will increase undercarriage and track maintenance cost. Maintenance

Table 1. Crawler and Wheel Dozer Ownership and Operating Costs

Dozer Type	Crawler		Wheel	
Shipping weight, kg (lb)	41,277 (91,000)		32,659 (72,000)	
Rated brake power, kW (hp)	268 (360)		250 (335)	
Capital cost				
Standard dozer, fob factory	$284,000		$244,500	
Optional equipment, fob factory	28,500		11,750	
Freight (5.7¢/lb)	5,200		4,100	
Erection	2,000		1,500	
Delivered cost	$319,700		$261,850	
Tires	—		19,700	
Depreciable value	$319,700		$242,150	
Dozer life				
Years	3		2	
Hours	15,000		10,000	
Average annual investment	$213,133		$181,613	
	$/hr	$/yr	$/hr	$/yr
Ownership costs				
Depreciation	21.31	106,567	24.22	121,075
Interest (15% of average annual investment)	6.39	31,970	5.44	27,242
Insurance (2% of average annual investment)	0.85	4,263	0.73	3,632
Taxes (2% of average annual investment)	0.85	4,263	0.73	3,632
Total ownership costs	29.40	147,063	31.12	155,581
Operating costs				
Fuel (0.60 load factor)				
43.9 L/hr at $0.330/L				
(11.6 gal/hr at $1.25/gal)	14.50	72,500		
40.9 L/hr at $0.330/L				
(10.8 gal/hr at $1.25/gal)			13.50	67,500
Preventive maintenance				
(0.20 × fuel cost)	2.90	14,500	2.70	13,500
Tires and tire repair (2000 hr)				
($19,700/2000 hr) × 1.15	—	—	11.33	56,640
Undercarriage and track maintenance	13.00	65,000	—	—
Repair				
($319,700/10,000) × 0.32	10.23	51,152		
($242,150/10,000) × 0.30			7.26	36,323
Wear items	3.75	18,750	2.80	14,000
Operator	20.00	100,000	20.00	100,000
Total operating costs	64.38	321,902	57.59	287,963
Total ownership and operating costs	93.78	468,965	88.71	443,544

Table 2. Dozer Engine Load Factor

Dozer Type	Load Factor*		
	Light	Average	Heavy
Crawler	0.45	0.60	0.75
Wheel	0.45	0.60	0.80

Light: Considerable idle or travel with no load.
Average: Normal idle, normal production dozing, back track push loading scrapers, steady shovel cleanup.
Heavy: Minimum idle and reverse travel, heavy production dozing, chain and shuttle push loading scrapers, steady ripping.

Table 3. Dozer Repair Factor

Dozer Type	Repair Factor*		
	Light	Average	Heavy
Crawler	0.16	0.32	0.50
Wheel	0.20	0.30	0.40

* Based on 0-18,000 hr crawler dozer life and 0-12,000 hr wheel dozer life.

practices with periodic wear measurements and servicing reduce undercarriage and track maintenance cost.

For wheel dozers, tire life is influenced by operating conditions and procedures, as well as tire construction, type, loads, and maintenance. Typical life of L-3 bias ply tires is 1500 to 2500 hr with tire life being extended by L-4 and L-5 tires. Tire life is shortened by tire spinning. When dozing rock, excessive wear and rock cuts, rips, and punctures can shorten tire life. Hourly tire and tire repair cost is 115% of hourly tire replacement cost.

Hourly repair cost can be calculated by the same method used for other mobile equipment. Table 3 presents repair factors for crawler and wheel dozers. The depreciable value used to determine the repair cost must be a current value to reflect current repair cost.

The cost of replacing and welding blade-cutting edges, ripper shank and points, and other high-wear items varies widely depending on materials, applications, and operator experience. Operator wages normally include base wage, fringe benefits, training, and other related costs.

Dozer and ripper production costs usually are presented as cost per bank cubic meter (cubic yard). For dozing, this production cost does not consider the distance the material is transported.

6.5.8 Haulage Systems Simulation Analysis

R. V. RAMANI

INTRODUCTION

In a mining operation, the haulage system is the intermediate component between the loading and the dumping systems. Together, the three form the materials handling system. Materials handling systems are characterized by a combination of inventory, waiting line, allocation, and replacement processes. Effective and efficient haulage systems can only be developed through a detailed consideration of these processes in a systems analysis framework. Otherwise, overloads and production bottlenecks will result at unexpected places in the mining system. The term *systems analysis* is used here to describe an integrated approach which views an entire system of components as an entity rather than as an assemblage of parts, i.e., a system in which each component is designed to fit properly with other components rather than to function by itself.

In surface mines, the transport of material from production faces to dumping sites is accomplished by rail, truck, belt conveyor, hydraulic transport or skips. In a systems analysis model of this haulage system, to provide both continuity and completeness, it is common to include the two most important interconnected systems—the loading system and the dumping system. Some of the earliest applications of computers in surface mining were in the materials handling system (Ware, 1955; Dunlop and Jacobs, 1955). Manufacturers of locomotives, trucks, and belt equipment use computer simulation programs to predict the performance of the haulage equipment over given haul profiles. Since shovel-truck systems are most common in open pit mining, many computer models have been developed to study these systems. In view of its importance and relevance to the analysis of haulage systems, this section on simulation covers broadly, with numerical examples, a range of topics in the areas of modeling, probability distributions, Monte Carlo sampling, deterministic and probabilistic simulation, and simulation models.

MATHEMATICAL MODELLING

In industrial and business settings, decision-making is constrained by many external factors, such as competition and the market, over which there may be little or no control. On the other hand, there is a high degree of selectivity with regard to internal factors such as methods of operation, equipment, and management. For example, in the selection of a haulage system for a surface mine, the location of the ore body, the ore grade and mineralization, though important, cannot be varied. However, there is considerable flexibility in the selection of the haulage method (trains, trucks, belts, pipelines, etc.), in the choice of equipment for the haulage method (small, large, bottom dump, side dump, etc.), and in the location of the mine facilities (dumps, ore pockets, bins, etc.). The objective of decision-making, in the systems analysis framework, is to select from the entire set of controllable and noncontrollable factors, those combinations of controllable factors which contribute most to the overall growth of the company. Symbolically, the problem can be represented as:

$$\text{Optimize } Z = F(X_i, Y_j) \ i = 1, 2, \ldots M; j = 1, 2 \ldots N$$

where

Z = Measure of effectiveness (cost, profit, production, etc.)
X_i = Controllable variables
Y_j = Noncontrollable variables
F = Functional operator.

This symbolic model outlines the essentials of the systems approach to decision-making. First, the controllable and uncontrollable variables of the system must be identified; second, the measure of effectiveness of the decision-maker must be defined, third, the scale for the measure of effectiveness must be established, and fourth, the relationships between the controllable and uncontrollable variables must be functionally defined. And finally, the solution (or the result) of the functional relationships must be integrated into a measurement on the effectiveness scale to serve as an indication of the value of the outcome of a particular decision.

The simplicity of the symbolic model belies the complexity of the real world decision-making environment. There are aspects of a decision-making environment, particularly, the sociohumanistic aspects, which may not be easily modeled. In practice, the measure of effectiveness is rarely single-valued. Problems are encountered in defining the boundaries of the system being analyzed. Often, it is not possible to solve the functional relationships using classical methods of mathematics.

SYSTEMS SIMULATION

Mathematical models have been classified as linear or nonlinear; stable or unstable; steady-state or transient; open or closed; and deterministic or stochastic (Forrester, 1961). Unstable, unconstrained models lead to explosive oscillations and are not representative of real-world systems. The model development may be based on theoretical knowledge of the system or on the analysis of data from the system. In the latter method, the values for many system variables are recorded under different operating conditions and empirical relationships between input and output variables are determined through the use of statistical techniques. The model development may also use various combinations of theoretical knowledge and empirical findings. There is an arsenal of tools and techniques and approaches to develop a model of a system.

Mathematical models of complex systems are difficult not only to develop but also to solve. For a particular haulage system study, assuming the loading and dumping systems are fixed along with the production rate, Z can be defined as the total haulage cost, X_i as the controllable variables of a haulage system (the type of haulage system, the size and number of haulage equipment, haul road profile and conditions, etc.) and Y_j, the uncontrollable variables (the desired production and the loading and dumping systems and their characteristics and locations). As stated here, the decision-maker is interested in minimizing the haulage cost for a defined production through the best choice of the type of haulage system and the size and number of equipment for the chosen system. Related objectives include minimizing waiting times and maximizing utilization of all equipment on a priority basis.

In surface mining the complex interdependence between the variables in the haulage system and the dynamic interactions between the haulage system and other related systems make a haulage system model too intricate to develop with classical tools and techniques. Specifically, in a shovel-truck haulage system, the loading time of a truck and the load (amount of material) on a truck are statistical variables. The performance of the truck on a haul road is a function of the truck speed-rimpull curve, the haul road profile and conditions, and the load on the truck. The truck travel times are also a function of the number and types of trucks on the haul road. Thus, a model of the shovel-truck system must have static, dynamic, kinematic, and statistical elements. In these instances, a systematic approach to improve on intuition and experience is to build a *simulation model* of the system. System simulation is defined as the technique of solving problems by following changes over time in a dynamic model of the system. Simulation models can be discrete, continuous, or a combination of both, in the dimensions of time and variables' value (Emshoff and Sisson, 1970; Gordon, 1969). The technique does not attempt to solve the equations of a model; rather it observes the way in which the controllable variables of the model change with time. While dynamic *physical* models can be useful for simulation studies, the model in most cases, is a computer program in which the theoretical knowledge and empirical findings about the system are incorporated. The inputs to the program consist of the values of the controllable and noncontrollable variables. The outputs of the program are measures of performance or cost. There is usually no explicit optimization in simulation. The best solution is found by conducting a number of experiments, the inputs to each of which are developed on the basis of the results of prior simulations, intuition, and experience. In essence, simulation is an experimental problem-solving technique.

The simulation approach is applicable to systems where classical analytical techniques are not feasible. There are no precise guidelines as to how to simulate a system. These two attributes—general applicability and lack of guidelines—have given simulation great flexibility and wide use, but because of these attributes, there also is a possibility for misapplication and abuse. Therefore, model builders must ensure that the simulation model of a system is in fact a valid representation of reality.

Three points are worth noting at this stage. First, the objective of modeling is to create abstractions of reality on which experiments can be performed to understand reality. Therefore, attempts to develop models that duplicate reality in *all* aspects may not be very useful or possible. In very detailed models, variations of outputs may not be tractable to changes in inputs, i.e., the cause and effect relationship may not be clearly established to enable decision-making. The objective should be to capture the *essence* of the real world system. Second, no model, however detailed and allegedly related to a system for which it is a model, can explain all observations. In the acceptance or rejection of a model for use, therefore, there should always be an element of judgment. Third, data is necessary for the development and application of any model. One should not, however, demand that data gathering be a prerequisite to model building. Some systems may require model building first and the data collection later. It is difficult to establish the sufficiency of data and, in this respect, the data needs of each system must be evaluated individually.

The concepts presented here are illustrated in a number of models—deterministic, statistical, and simulation—that are developed in subsequent sections.

CYCLE TIME MODEL

The cycle time for a haul unit can be expressed as (Suboleski, 1975):

$$LCT = STL + LT + TL + STD \\ + DT + TE + AD \qquad (1)$$

where:

LCT = cycle time of a haul unit, min
STL = spot time at the loader, min
LT = load time of the haul unit, min
TL = travel time (loaded), min
STD = spot time at the dump, min
DT = dump time, min
TE = travel time (empty), min
AD = average delay on the haul cycle, min (wait at dump, wait at the loader, or slow down on the haul road).

Also,

$$LT = \frac{TCP}{LR} \qquad (2)$$

where:

TCP = capacity of the haul unit, short tons*
LR = loading rate of the loader, st/min

With shovel-truck systems,

$$LT = \left[\frac{TCP}{BCP}\right]^* \times CTL \qquad (3)$$

where:

BCP = bucket capacity of the loader, st (adjusted for fill factor, etc.)
CTL = cycle time of the loader, min (adjusted for loading efficiency, etc.)

[ ]* implies that the value is rounded up to the next higher integer.

$$TL = \frac{HD}{SL} \qquad (4)$$

$$TE = \frac{HD}{SE} \qquad (5)$$

where:

HD = haul distance from the loader to the dump, ft
SL = speed of a loaded haul unit, ft/min.
SE = speed of an empty haul unit, ft/min.

The value of AD can be determined from time study data. It can also be calculated from an analysis of the system.

Assuming that there is no delay at the dump or the haul road, the number of haul units required to keep the loading unit busy is given by:

$$N = \left[\frac{LCT}{STL + LT}\right]^* \qquad (6)$$

where [ ]* implies that the value is rounded up to the next higher integer.

---

* Metric equivalents: st × 0.907 184 7 = t; ft × 0.3048 = m; mph × 1.609 = km/h.

Example 1

Consider the following case:

BCP = 5 st
CTL = 0.5 min
STL = 0.20 min
STD = 0.15 min
DT = 0.35 min
TCP = 27.5 st
HD = 5,000 ft
SL = 2,000 ft/min (22.73 mph)
SE = 2,500 ft/min (28.41 mph)

$$LT = \left[\frac{TCP}{BCP}\right]^* [CTL] = \left[\frac{27.5}{5}\right]^* [0.5] = 6 \times 0.5 = 3.0 \text{ min}$$

$$TL = \frac{5,000}{2,000} = 2.5 \text{ min}$$

$$TE = \frac{5,000}{2,500} = 2.0 \text{ min}$$

$$LCT = (0.20 + 3.0 + 2.5 + 0.15 + 0.35 + 2.0) \text{ min} = 8.20 \text{ min}.$$

In the above example, the haul unit is at the loader for 3.20 min ($STL + LT$), i.e., another haul unit cannot spot, till the loaded haul unit pulls away from the loader.

$$N = \left[\frac{LCT}{STL + LT}\right]^* = \left[\frac{8.20}{0.20 + 3.0}\right]^* = [2.56]^* = 3$$

Two haul units will keep the loader waiting for the haul unit to arrive from the dump. The average wait time of the loader, for every other haul unit, will be $(2.56 - 2)(3.2) = 1.80$ min. With three haul units, each truck will wait for $(3 - 2.56)(3.2) = 1.41$ min at the loader before the loader starts loading it. These are easily explained by drawing a simple Gantt chart of the status of the loading and hauling units as a function of time. Fig. 1 shows the Gantt chart for a loader with two and three haul units. Equating the loader and haul unit times for the interval I in Fig. 1 for the two haul units case, the waiting time (W) of a unit can be calculated:

$$2(STL + LT) + W = STL + LT + TL + STD + DT + TE \quad (7)$$
$$W = (TL + STD + DT + TE) - (STL + LT)$$

For the two haul unit cases, the total wait time (TW) for n haul cycles is

$$TW = \left(\frac{n-1}{2}\right)\{[TL + STD + DT + TE] - [STL + LT]\} \quad (8)$$

These equations can be generalized for N haul units as follows:

$$W = [TL + STD + DT + TE] - (N-1)[STL + LT] \quad (9)$$

$$TW = \left[\frac{n-1}{N}\right]\{[TL + STD + DT + TE] - (N-1)[STL + LT]\} \quad (10)$$

By increasing N, W can be decreased. If N is sufficiently large, then W can become negative. A negative W implies that the first haul unit completes its cycle before all other haul units have been loaded.

For the example under consideration, with only one haul unit, it is clear that the loader will wait for 5.20 min between trips. With two haul units,

$$W = [2.5 + 0.15 + 0.35 + 2.0] - [0.2 + 3.0] = 1.8 \text{ min}.$$

i.e., the loading unit will be waiting 1.8 min every other trip. This is shown in the Gantt chart. With three haul units,

$$W = [2.5 + 0.15 + 0.35 + 2.0] - 2[0.2 + 3.0] = 5.0 - 6.4 = -1.4 \text{ min}.$$

$W = -1.4$ implies that each truck waits for 1.4 min at the loader every trip. This is also clear from the Gantt chart.

Example 2

In the above problem, the one-way haul distance increases to a maximum of 4572 m (15,000 ft). Calculate the number of hauling units required as a function of the haul distance such that the shovel waiting time is zero.

Let N = number of haul units

$$W = [TL + STD + DT + TE] - [N-1][STL + LT]$$

$$W = \left[\frac{HD}{2000} + 0.15 + 0.35 + \frac{HD}{2500}\right] - [N-1][0.2 + 3.0]$$

$$= \left[\frac{2.5\ HD + 2\ HD}{5000} + 0.5\right] - 3.20\ N + 3.20$$

$$= \frac{4.5\ HD}{5000} - 3.20\ N + 3.7$$

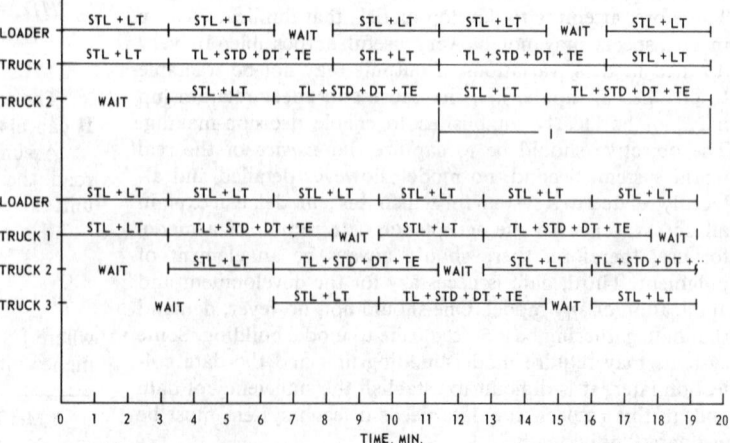

Fig. 1. Gantt chart.

Condition for loader wait time to be zero or lower is given by

$$W = \frac{4.5 \, HD}{5000} - 3.20 \, N + 3.7 \leq 0$$

or

$$N \geq \frac{4.5 \, HD}{16000} + 1.16$$

$$N \geq 0.00028 \, [HD] + 1.16$$

For: $HD = 5000$ ft $N > 2.56$ $N = 3$
       $HD = 7500$ ft $N > 3.26$ $N = 4$
       $HD = 10000$ ft $N > 3.96$ $N = 4$
       $HD = 12500$ ft $N > 4.66$ $N = 5$
       $HD = 15000$ ft $N > 5.36$ $N = 6$

The shift production is a complex function of many factors such as the number of loading units, the number of haul units, the number of dumps, availability of these units, the haul road length and hauling speeds. The total performance of the mining system is the result of the interaction between all these factors. Deterministic calculations as previously presented are useful for preliminary evaluation. On the other hand, the various elemental times in a cycle can be and often are random variables. Therefore, whenever data is available, application of statistical techniques provides better answers.

TRUCK FLEET SIZING

One of the important criteria in surface mine equipment selection is the determination of the truck fleet size. It is common to establish the total fleet size as:

$$\text{Total fleet size} = \frac{\text{number of trucks required in operation}}{\text{availability of a truck}} \quad (11)$$

$$\text{Availability} = \frac{\text{possible hours} - \text{downtime hours}}{\text{possible hours}} \quad (12)$$

For example, if 16 trucks are required to meet the production, and the availability of a truck is 0.8, then the total fleet size is calculated as $(16/0.8) = 20$. Twenty trucks in the fleet are inadequate to meet the operational requirements of 16 trucks. A better way to calculate the truck fleet size using a binomial distribution is presented by Connel (1969, 1973). Defining:

P_n = probability that exactly n units are available
P_a = probability that a single unit is available
P_{na} = probability that a single unit is not available = $(1 - P_a)$
N = total number of units in the system
Nc_n = combination of N things, taken n at a time ($n \leq N$).

$$P_n = Nc_n \, (P_a)^n \, (P_{na})^{N-n} \quad (13)$$

Also,

$$P_{1n} = \sum_{x=n}^{N} N_{c_x} \, (P_a)^x \, (P_{na})^{N-x} \quad (14)$$

where P_{1n} = probability at least n units will be available. As N and n increase, it is preferable to use a computer program to calculate these probabilities.

Example 3

Given that the availability of a single truck is 0.8, and that the truck fleet size is 20, calculate (1) the probability that exactly 16 trucks will be available, and (2) probability that at least 16 trucks will be available.

Note: $N = 20$, $n = 16$, $P_a = 0.8$, $P_{na} = 0.2$

$$P_{16} = 20_{c_{16}} \, (0.8)^{16} \, (0.2)^{20-16} = 0.2182$$

$$P_{1_{16}} = P\,(n \geq 16) = P_{16} + P_{17} + P_{18} + P_{19} + P_{20}$$
$$= 0.2182 + 0.2054 + 0.1369 + 0.0576 + 0.0115$$
$$= 0.6296.$$

Thus, with a fleet size of 20, there is only a 62.96% chance that at least 16 trucks will be available. In other words, there is a 37.04% chance that less than 16 trucks will be available. This is too high a probability for having a lower number of trucks than required for maintaining production. This probability should not normally exceed five to ten percent.

Example 4

For the data in Example 3, calculate the truck fleet size to ensure that there is at least a 95% probability of having 16 units available. Let N = truck fleet size. It is required that $P_{1_{16}} > 0.95$.

$$P_{1_{16}} = \sum_{x=16}^{N} N_{c_x} \, (P_a)^x \, (P_{na})^{N-x}$$

i.e., $\sum_{x=16}^{N} N_{c_x} \, (P_a)^x \, (P_{na})^{N-x} > 0.95$

In the previous inequality, solve for the value of N by trial and error. The values of $P_{1_{16}}$ for various values of N are as follows:

N	$P_{1_{16}}$
22	0.8670
23	0.9285
24	0.9638
25	0.9827
26	0.9921

A fleet size of 24 trucks will be inadequate.

The fleet size can be decreased by increasing the availability of the individual truck. This may be possible by such measures as buying better trucks, improving maintenance, and reducing avoidable delays. An alternate approach, though not recommended, is to allow a higher probability for not having the required number of trucks in operation.

Example 5

There are three types of trucks with availabilities of 0.8, 0.85, and 0.90, respectively. Calculate the fleet size required for each truck type to ensure that there are 90, 95, and 99% chances of having 16 trucks in operation.

Solution

There are nine cases to be evaluated: three truck types × three different probabilities for 16-truck operation.

Number of trucks needed in the fleet
for 16-truck operation

Truck availability	90% of the time	95% of the time	99% of the time
0.8	23	24	26
0.85	21	22	24
0.90	19	20	21

Thus, the fleet size is variable depending on the truck availability and the desired probability for having the required number of trucks. The justification to select the appropriate truck type and the fleet size can be made only after a thorough cost analysis.

MONTE CARLO SAMPLING

A useful application of probability and statistics involves sampling by the Monte Carlo method. For example, the loading time of a haul unit including spotting time may be exponentially distributed. The dumping time including spotting time, on the other hand, may be uniformly distributed. The travel times for the loaded and empty trip may be normally distributed. The truck cycle time, therefore, is the sum of four random variables. Through Monte Carlo simulation, truck cycle times can be generated. For applying the Monte Carlo technique, one proceeds as follows (Sasieni, Yaspan, and Freidman, 1959):

1. Calculate the cumulative probability function $F(x)$ of the variable x over its range (see Fig. 2):

$$y = F(x) = \int_{-\infty}^{x} f(x)d_x$$

where $f(x)$ is the frequency density function of x.

2. Choose a random number, r, between 0 and 1 from a table of random numbers (for example, see Table 1).

3. From the cumulative probability function $F(x)$, find the value of x corresponding to $y = r$.

This simulated value of x is distributed according to the frequency density function of the variable x. It is clear from the figure that the probability of having a simulated value between x_1 and $x_1 + d_x$ is proportional to $f(x_1)d_x$.

$$P(x_1 < \text{simulated value} < x_1 + d_x) = dy_1 = f(x_1)d_x$$

In the event the variable is discrete, then $F(x) = \sum_{u=0}^{x} f(u)$.

Example 6

The cycle time of a haul unit follows the density function:

$$f(t) = \frac{1}{20}(t + 3) \qquad 6 \leq t \leq 8$$
$$f(t) = 0 \qquad\qquad \text{elsewhere}$$

where t = cycle time of truck in minutes. Generate the average of ten cycle times by the Monte Carlo technique. Compare this average with the average of the distribution.

Solution

$$F(t) = \int_{6}^{t} f(t)d_t$$

$$= \frac{1}{20}\int_{6}^{t}(t + 3)d_t$$

$$= \frac{1}{20}\left[\frac{t^2}{2} + 3t - 36\right] \qquad 6 \leq t \leq 8$$

Calculating the value of $F(t)$ for values of $t = 6, 6.5, 7, 7.5,$ and 8, the following results are obtained: $F(6) = 0.0$, $F(6.5) = 0.23$, $F(7) = 0.48$, $F(7.5) = 0.73$, $F(8) = 1.0$.

The cumulative probability function is plotted in Fig. 3. Choosing ten random numbers from the random number table, the cumulative distribution is sampled with the following results (see box marked A in Table 1).

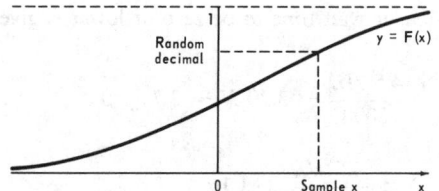

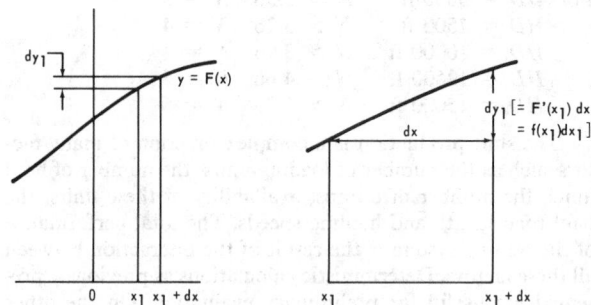

Fig. 2. Monte Carlo sampling procedure and justification.

Random number	Truck time, t
10097	6.20
37542	6.75
08422	6.18
99019	8.00
12807	6.25
66065	7.30
31060	6.60
85269	7.70
63573	7.25
73796	7.50
	69.73 average = 6.973 min

The theoretical average cycle time is obtained as follows:

$$\mu_t = \int_{6}^{8} t f(t)d_t = 7.03 \text{ min}$$

The variance of the cycle time can be calculated from:

$$\sigma_t^2 = \int_{6}^{8} (t - 7.03)^2 f(t)d_t$$

If the Monte Carlo sampling procedure were carried to a greater number of samples, then the average of the samples will approach the theoretical average.

The application of the Monte Carlo technique to distributions frequently used in mine simulation studies is summarized in Table 2 (after Taylor, et al., 1966). It is fairly easy to write computer programs to generate random numbers that are distributed according to some distributions. Tables of random numbers distributed normally with $\mu = 0$ and $\sigma^2 = 1.0$ simplify the generation of normally distributed random values:

$$X_N = \mu N + R_N \cdot \sigma_N$$

where X_N = normally distributed variable with mean, μ_N and variance, σ_N^2;

R_N = a number from a table of normalized random deviations (Table 3, see the box marked A).

Table 1. Table of Random Digits

10097	32533	76520	13586	34673	54876	80959	09117	39292	74945
37542	04805	64894	74296	24805	24037	20636	10402	00822	91655
08422	68953	19645	09303	23209	02560	15953	34764	35080	33606
99019	02529	09376	70715	38311	31165	88676	74397	04436	27659
12807	99970	80157	36147	64032	36653	98951	16877	12171	76833
A									
66065	74717	34072	76850	36697	36170	65813	39885	11199	29170
31060	10805	45571	82406	35303	42614	86799	07439	23403	09732
85269	77602	02051	65692	68665	74818	73053	85247	18623	88579
63573	32135	05325	47048	90553	57548	28468	28709	83491	25624
73796	45753	03529	64778	35808	34282	60935	20344	35273	88435
98520	17767	14905	68607	22109	40558	60970	93433	50500	73998
11805	05431	39808	27732	50725	68248	29405	24201	52775	67851
83452	99634	06288	98083	13746	70078	18475	40610	68711	77817
88685	40200	86507	58401	36766	67951	90364	76493	29609	11062
99594	67348	87517	64969	91826	08928	93785	61368	23478	34113
65481	17674	17468	50950	58047	76974	73039	57186	40218	16544
80124	35635	17727	08015	45318	22374	21115	78253	14385	53763
74350	99817	77402	77214	43236	00210	45521	64237	96286	02655
69916	26803	66252	29148	36936	87203	76621	13990	94400	56418
09893	20505	14225	86514	46427	56788	96297	78822	54382	14598
91499	14523	68479	27686	46162	83554	94750	89923	37089	20048
80336	94598	26940	36858	70297	34135	53140	33340	42050	82341
44104	81949	85157	47954	32979	26575	57600	40881	22222	06413
12550	73742	11100	02040	12860	74697	96644	89439	28707	25815
63606	49329	16505	34484	20419	52563	43651	77082	07207	31790
61196	90446	26457	47774	51924	33729	65394	59593	42582	60527
15474	45266	95270	79953	59367	83848	82396	10118	33211	59466
94557	28573	67897	54387	54622	44431	91190	42592	92927	45973
42481	16213	97344	08721	16868	48767	03071	12059	25701	46670
23523	78317	73208	89837	68935	91416	26252	29663	05522	82562
04493	52494	75248	33824	45862	51025	61962	79335	65337	12472
00549	97654	64051	88159	96119	63896	54692	82391	23287	29529
35963	15307	26898	09354	33351	35462	77974	50024	90103	39333
59808	08391	45427	26842	83609	49700	13021	24892	78565	20106
46058	85236	01390	92286	77281	44077	93910	83647	70617	42941
32179	00597	87379	25241	05567	07007	86743	17157	85394	11838
69234	61406	20117	45204	15956	60000	18743	92423	97118	96338
19565	41430	01758	75379	40419	21585	66674	36806	84962	85207
45155	14938	19476	07246	43667	94543	59047	90033	20826	69541
94864	31994	36168	10851	34888	81553	01540	35456	05014	51176
98086	24826	45240	28404	44999	08896	39094	73407	35441	31880
33185	16232	41941	50949	89435	48481	88695	41994	37548	73043
80951	00406	96382	70774	20151	23387	25016	25298	94624	61171
79752	49140	71961	28296	69861	02591	74852	20539	00387	59579
18633	32537	98145	06571	31010	24674	05455	61427	77938	91936
74029	43902	77557	32270	97790	17119	52527	58021	80814	51748
54178	45611	80993	37143	05335	12969	56127	19255	26040	90324
11664	49883	52079	84827	59381	71539	09973	33440	88461	23356
48324	77928	31249	64710	12295	36870	32307	57546	15020	09994
69074	94138	87637	91976	35584	04401	10518	21615	01848	76938

Example 7

The load on a 109-t (120-st) coal truck is normally distributed with a mean of 190 t (120 st) and a standard deviation of 9 t (10 st). The mine production on a particular day is reported as 20 trucks. Using the table of random numbers, estimate the tons produced.

The load on a truck $= 120 + R_{Ni} (5)$

$$\text{The load on 20 trucks} = \sum_{i=1}^{20} [120 + R_{Ni} (5)]$$

$$= 20 \times 120 + 5 \sum_{i=1}^{20} R_{Ni}$$

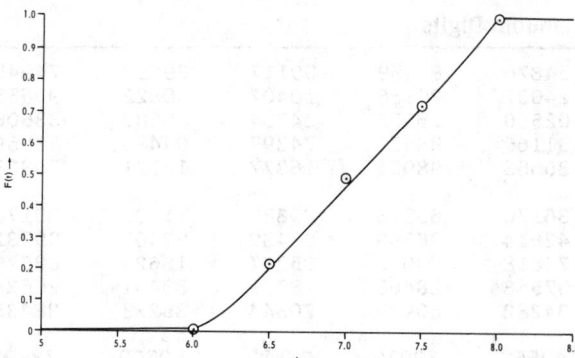

Fig. 3. Cumulative probability function for F(t) in example 6.

Choose 20 consecutive random numbers from the table of normalized random deviations (Table 3).

$$\text{The load on 20 trucks} = 20 \times 120 + 10\ [-7.625]$$
$$= 2400 - 76.3 = 2323.7 \text{ st}$$
$$\text{Average load per truck} = \frac{2323.7}{20} = 116.9 \text{ st}$$

Depending on the choice of the 20 random normal numbers, the total production and average production per truck will be different. For example, choosing the numbers in the box marked B in Table 3, the answers for the load on 20 trucks and average load per truck are 2476.50 and 123.8 st, respectively.

After data is collected on various variables of interest, it is necessary to analyze and fit the data, if appropriate, to some known distribution. Simulation of haulage systems can be considerably simplified if the time study data can be fitted to standard probability distributions.

A measure of the discrepancy between the observed and expected frequency is supplied by the χ^2 statistic defined as

$$\chi^2 = \sum_{i=1}^{n} \frac{(o_i - e_i)^2}{e_i}$$

where

o_i = observed number of observations in a cell
e_i = expected numbers of observations in a cell
n = number of cells with more than 5 expected observations

Table 2. Generation of Random Variables According to An Underlying Distribution

Distribution	Parameters	Mean	Variance	x is distributed according to the distribution in column 1. r = uniform random number	Comments
Uniform distribution					
$f(x) = \dfrac{1}{b-a}\ a, < x < b,$ $= 0$ elsewhere	a, b	$\dfrac{a+b}{2}$	$\dfrac{(b-a)^2}{12}$	$x = a + (b-a)\,r$	$0 \leq r \leq 1$
Exponential distribution					
$f(x) = \alpha e^{-\alpha t}$ $\alpha > 0, x \geq 0$	α	$\dfrac{1}{\alpha}$	$\dfrac{1}{\alpha^2}$	$x = -\dfrac{1}{\alpha} \ln r$	$0 \leq r \leq 1$
Normal distribution					
$f(x) =$ $\dfrac{1}{\sigma_x \sqrt{2\pi}} e^{-\frac{1}{2}\left(\frac{x-\mu_x}{\sigma_x}\right)^2}$ $-\infty < x < \infty$	μ_x, σ_x	μ_x	σ_x^2	$x = \mu_x + \sigma_x \left[\sum_{i=1}^{12} r_i - 6\right]$	See note 1 $0 \leq r_i \leq 1$
Binominal distribution					
$f(x) = n_{c_x} p x (1-p)^{n-x}$	n, p	np	$np(1-p)$	$r_i \leq p,\ x_i = x_{i-1} + 1$	See note 2 $0 \leq r_i \leq 1$
$x = 0,1,2 \ldots n$				$r_i \geq p,\ x_i = x_{i-1}$	
$P \geq 0, P \leq 1.0$				$x_0 = 0$	
(P = Probability of success)					
Poisson distribution					
$f(x) = \dfrac{e^{-\lambda} \lambda^x}{x!}$	λ	λ	λ	$t_i = -\ln r_i$	See note 3 $0 \leq r_i \leq 1$
$x = 0,1,2,\ldots$ $\lambda \geq 0$				$\sum_{i=0}^{x} t_i \leq \lambda \leq \sum_{i=0}^{x+1} t_i$	

Note 1: Twelve random numbers, r_i ($i = 1, 2, \ldots 12$) are generated and used to calculate the value of x.

Note 2: There are n trials conducted. x_o is initialized to 0. For each trial, a random value r_i is calculated as shown. At the end of n trials, the value of x_n is binomially distributed.

Note 3: A number of trials $i = 0, 1, 2, \ldots$ are conducted. For each trial, a random value r is generated and the value of t is calculated. When the stated relationship is satisfied at some $i = x$, then x is poisson distributed.

Table 3. Random Normal Numbers, $\mu = 0$ and $\sigma = 1$

01	02	03	04	05	06	07	08	09	10
0.464	0.137	2.455	−0.323	−0.068	0.296	−0.288	1.298	0.241	−0.957
0.060	−2.526	−0.531	−0.194	0.543	−1.558	0.187	−1.190	0.022	0.525
1.486	−0.354	−0.634	0.697	0.926	1.375	0.785	−0.963	−0.853	−1.865
1.022	−0.472	1.279	3.521	0.571	−1.851	0.194	1.192	−0.501	−0.273
1.394	−0.555	0.046	0.321	2.945	1.974	−0.258	0.412	0.439	−0.035
0.906	−0.513	−0.525	0.595	0.881	−0.934	1.579	0.161	−1.885	0.371
1.179	−1.055	0.007	0.769	0.971	0.712	1.090	−0.631	−0.255	−0.702
−1.501	−0.488	−0.162	−0.136	1.033	0.203	0.448	0.748	−0.423	−0.432
−0.690	0.756	−1.618	−0.345	−0.511	−2.051	−0.457	−0.218	0.857	−0.465
1.372	0.225	0.378	0.761	0.181	−0.736	0.960	−1.530	−0.260	0.120
−0.482	1.678	−0.057	−1.229	−0.486	0.856	−0.491	−1.983	−2.830	−0.238
−1.376	−0.150	1.356	−0.561	−0.256	−0.212	0.219	0.779	0.953	−0.869
−1.010	0.598	−0.918	1.598	0.065	0.415	−0.169	0.313	−0.973	−1.016
−0.005	−0.899	0.012	−0.725	1.147	−0.121	1.096	0.481	−1.691	0.417
1.393	−1.163	−0.911	1.231	−0.199	−0.246	1.239	−2.574	−0.558	0.056
−1.787	−0.261	1.237	1.046	−0.508	−1.630	−0.146	−0.392	−0.627	0.561
−0.105	−0.357	−1.384	0.360	−0.992	−0.116	−1.698	−2.832	−1.108	−2.357
−1.339	1.827	−0.959	0.424	0.969	−1.141	−1.041	0.362	−1.726	1.956
1.041	0.535	0.731	1.377	0.983	−1.330	1.620	−1.040	0.524	−0.281
0.279	−2.056	0.717	−0.873	−1.096	−1.396	1.047	0.089	−0.573	0.932
−1.805	−2.008	−1.633	0.542	0.250	−0.166	0.032	0.079	0.471	−1.029
−1.186	1.180	1.114	0.882	1.265	−0.202	0.151	−0.376	−0.310	0.479
0.658	−1.141	1.151	−1.210	−0.927	0.425	0.290	−0.902	0.610	2.709
−0.439	0.358	−1.939	0.891	−0.227	0.602	0.873	−0.437	−0.220	−0.057
−1.399	−0.230	0.385	−0.649	−0.577	0.237	−0.289	0.513	0.738	−0.300
0.199	0.208	−1.083	−0.219	−0.291	1.221	1.119	0.004	−2.015	−0.594
0.159	0.272	−0.313	0.084	−2.828	−0.439	−0.792	−1.275	−0.623	−1.047
2.273	0.606	0.606	−0.747	0.247	1.291	0.063	−1.793	−0.699	−1.347
0.041	−0.307	0.121	0.790	−0.584	0.541	0.484	−0.986	0.481	0.996
−1.132	−2.098	0.921	0.145	0.446	−1.661	1.045	−1.363	−0.586	−1.023
0.768	0.079	−1.473	0.034	−2.127	0.665	0.084	−0.880	−0.579	0.551
0.375	−1.658	−0.851	0.234	−0.656	0.340	−0.086	−0.158	−0.120	0.418
−0.513	−0.344	0.210	−0.736	1.041	0.008	0.427	−0.831	0.191	0.074
0.292	−0.521	1.266	−1.206	−0.899	0.110	−0.528	−0.813	0.071	0.524
1.026	2.990	−0.574	−0.491	−1.114	1.297	−1.433	−1.345	−3.001	0.479
−1.334	1.278	−0.568	−0.109	−0.515	−0.566	2.923	0.500	0.359	0.326
−0.287	−0.144	−0.254	0.574	−0.451	−1.181	−1.190	−0.318	−0.094	1.114
0.161	0.886	−0.921	−0.509	1.410	−0.518	0.192	−0.432	1.501	1.068
−1.346	0.193	−1.202	0.394	−1.045	0.843	0.942	1.045	0.031	0.772
1.250	−0.199	−0.288	1.810	1.378	0.584	1.216	0.733	0.402	0.226
0.630	−0.537	0.782	0.060	0.499	−0.431	1.705	1.164	0.884	−0.298
0.375	−1.941	0.247	−0.491	0.665	−0.135	−0.145	−0.498	0.457	1.064
−1.420	0.489	−1.711	−1.186	0.754	−0.732	−0.066	1.006	−0.798	0.162
−0.151	−0.243	−0.430	−0.762	0.298	1.049	1.810	2.885	−0.768	−0.129
−0.309	0.531	0.416	−1.541	1.456	2.040	−0.124	0.196	0.023	−1.204
0.424	−0.444	0.593	0.993	−0.106	0.116	0.484	−1.272	1.066	1.097
0.593	0.658	−1.127	−1.407	−1.579	−1.616	1.458	1.262	0.736	−0.916
0.862	−0.885	−0.142	−0.504	0.532	1.381	0.022	−0.281	−0.342	1.222
0.235	−0.628	−0.023	−0.463	−0.899	−0.394	−0.538	1.707	−0.188	−1.153
−0.853	0.402	0.777	0.833	0.410	−0.349	−1.094	0.580	1.395	1.298

A (boxed, column 05, rows 16–35) B

07 column boxed rows 26–45

Source: From tables of the RAND Corp., by permission.

The degrees of freedom of the test is given by $(n - 1 - k)$ where k is the number of parameters of the distribution estimated from the data.

Example 8

Time study data on the dumping time of a 29-t (32-st) truck is analyzed and grouped as follows (O'Neil, 1966):

Dumping time interval, min	Number of observations
0.545 — 0.745	27
0.745 — 0.945	11
0.945 — 1.145	4
1.145 — 1.345	1
1.345 — 1.545	1

Develop an appropriate frequency distribution for the dumping times.

Solution

The histogram of the observed values is plotted (Fig. 4).

Mean of the samples: 0.74
Standard deviation of the samples: 0.16.

From the histogram, it appears that the dumping times are distributed according to an exponential distribution. The exponential distribution is of the form

$$F(x) = \alpha e^{-\alpha x} \qquad \alpha > 0, x > 0$$

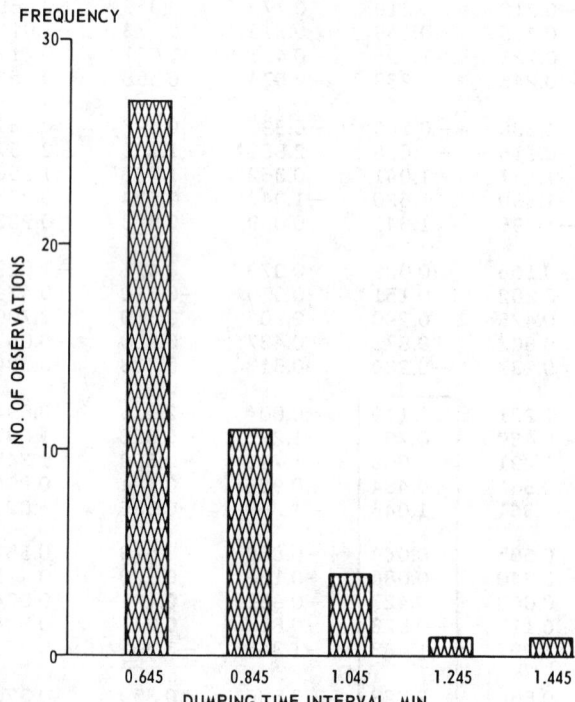

FREQUENCY

NO. OF OBSERVATIONS

DUMPING TIME INTERVAL, MIN.

Fig. 4. Histogram for data in example 8.

where $\dfrac{1}{\alpha}$ = mean of the exponential distribution, μ_x. To consider the lower limit of the distribution as zero, the data is translated by subtracting 0.545 min from each observation.

Mean of the translated data = 0.195

$$\alpha = \frac{1}{\mu_x} = \frac{1}{\bar{x}} = \frac{1}{0.195} = 5.13$$

The observed and theoretical frequencies are calculated in Table 4.

Chi-Square Goodness of Fit

$$\chi^2 \, (1 \, df) = \sum_{i=1}^{n} \frac{(o_i - e_i)^2}{e_i} = \frac{(27 - 28.6)^2}{28.6}$$
$$+ \frac{(11 - 10.12)^2}{10.12} + \frac{(6 - 5.28)^2}{5.28}$$
$$= 0.09 + 0.08 + 0.10 = 0.27$$

From tables $\chi^2_{0.95} \, (1 \, df) = 3.84 \qquad \chi^2_{0.05} \, (1 \, df) = 0.0039$

At 90% confidence level, it is concluded that the dumping times are exponentially distributed. In a simulation run the dumping time of a 29-t (32-st) truck will be calculated as:

$$DT = 0.545 - \frac{1}{0.195} \log r$$

where r = a random number in the interval (0, 1).

Example 9

Time study data on truck loading times of a 29-t (32-st) truck is analyzed and grouped as follows:

Load time interval, sec	Mid-point, sec x_i	Number of observations, f_i
60 — 62	61	5
63 — 65	64	18
66 — 68	67	42
69 — 71	70	27
72 — 74	73	8
		Σ 100

Develop an appropriate frequency distribution for the loading times.

Table 4. Exponential Distribution Fitting

Distribution time interval, min	Translated dumping time interval, min	Number of observations	Observed Relative frequency	Observed Cumulative frequency	Theoretical Relative frequency*	Theoretical Number of observations
0.545–0.745	0.0–0.2	27	0.62	0.62	0.65	28.60
0.745–0.945	0.2–0.4	11	0.25	0.87	0.23	10.12
0.945–1.145	0.4–0.6	4	0.09	0.96	0.08	3.52
1.145–1.345	0.6–0.8	1	0.02	0.98	0.03	1.32
1.345–1.545	0.8–1.0	1	0.02	1.00	0.01	0.44
		44	1.00		1.00	44

*Example of theoretical relative frequency calculation:

$$\text{Note } f(x) = 5.13 e^{-5.13x}, \qquad \alpha > 0, x > 0$$
$$F(x) = \int_0^x 5.13 e^{-5.13x} \, d_x = 1 - e^{-5.13x}$$

$$P_r(0.0 < x < 0.2) = 1 - e^{-5.13x} \Big|_0^{0.2} = 1 - 0.3584 = 0.6416 \sim 0.65$$

Solution

Plot the histogram of the data (Fig. 5). The histogram of the observed values suggests that the truck loading times may be normally distributed.

The mean and standard deviation of the observed data are:

$$\bar{x} = \frac{\Sigma f_i x_i}{\Sigma f_i} = \frac{6745}{100} = 67.45 \text{ sec}$$

$$s^2 = \frac{\Sigma f_i (x_i - \bar{x})^2}{\Sigma f_i} = \frac{852.75}{100} = 8.53 \text{ sec}$$

$$s = 2.92 \text{ sec}$$

The observed and theoretical frequencies are calculated in Table 5.

Chi-Square Goodness of Fit

$$\chi^2 \ (2 \ df) = \sum_{i=1}^{n} \frac{(o_i - e_i)^2}{e_i} = \frac{(5-4)^2}{4} + \frac{(18-21)^2}{21}$$

$$+ \frac{(42-39)^2}{39} + \frac{(8-8)^2}{8} = 0.833$$

$$X^2_{0.975} \ (2 \ df) = 7.38 \qquad X^2_{0.025} \ (2 \ df) = 0.51$$

At 95% confidence level, it is concluded that the truck loading times are normally distributed. In a simulation run, the loading time for a truck will be calculated as

$$LT = 67.45 + R_N * 2.92 \text{ sec}$$

where R_N = random normal number with $\mu = 0$, $\sigma^2 = 1.0$.

Example 10

A study on the number of tire failures per week in a surface mine revealed the following data for one year.

Number of tire failures/week x_i	Number of weeks f_i
0	23
1	17
2	7
3	2
4	1
5 or over	0

Examine the probability distribution of the data.

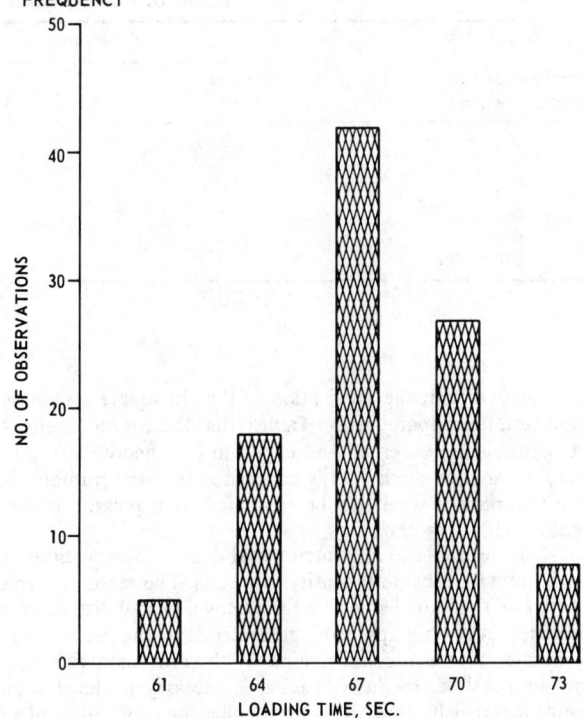

Fig. 5. Histogram for data in example 9.

Solution

Assuming that tire failures occurring during a week are independent, then the number of tire failures per week may be poisson distributed.

$$\bar{x} = \frac{\Sigma f_i x_i}{\Sigma f_i} = \frac{41}{50} = 0.82$$

The observed frequencies will be compared with the expected or theoretical frequencies of a poisson distribution with the parameter $\lambda = 0.82$.

Theoretical relative frequency:

$$f(x) = \frac{\lambda^x e^{-\lambda}}{x!} = \frac{(0.82)^x e^{-0.82}}{x!} = \frac{0.44(0.82)^x}{x!}$$

Table 5. Normal Distribution Fitting

Class boundary x_i	Class i	Z for class boundary $Z = \dfrac{x_i - \mu}{\sigma}$	Area under normal curve from $-\infty$ to Z P_i	Area for each class-A_i $P_{i+1} - P_i$	Expected frequency $N \times A_i$ e_i	Observed frequency f_i o_i
59.5		−2.72	0.0037			
	59.5–62.5			0.0409	4.09~4	5
62.5		−1.70	0.0446			
	62.5–65.5			0.2068	20.68~21	18
65.5		−0.67	0.2514			
	65.5–68.5			0.3892	38.92~39	42
68.5		0.36	0.6406			
	68.5–71.5			0.2771	27.71~28	27
71.5		1.39	0.9177			
	71.5–74.5			0.0753	7.53~8	8
74.5		2.41	0.9930			

Distribution Fitting: Note that $Z = \dfrac{x_i - \mu}{\sigma}$. Approximate μ by $\bar{x}$ and σ by S, i.e., $\mu = 67.45$ and $\sigma = 2.92$. $N = 100$ observations.

Table 6. Poisson Distribution Fitting, λ = 0.82

Number of tire failure/week	Observed		Theoretical	
	Number of weeks	Relative frequency	Relative frequency	Number of weeks
0	23	0.46	0.44	22
1	17	0.34	0.36	18
2	7	0.14	0.15	7.5
3	2	0.04	0.04	2.0
4	1	0.02	0.01	0.5
5 or over	0	0.00	0.00	0.0
	Σ = 50	Σ = 1.0	Σ = 1.0	Σ = 50

Even without the application of the chi-square goodness of fit test, it is apparent from Table 6 that there is an excellent fit between the observed and expected (or theoretical) data (i.e., number of weeks). It is concluded that the number of tire failures per week can be estimated by a poisson distribution with λ = 0.82.

It is desirable to fit experimental data or observations to well known probability density functions. The reason for this is that all the knowledge about the known probability density function is readily applicable to understand the process and generate statistical relationships. With some data this may not be possible. In these cases, it is necessry to develop an empirical density function. Empirically, the probability of an outcome is measured by the relative frequency of that outcome. Therefore, an empirical density function associates relative frequencies with outcomes (Maisel and Gnugnoli, 1972). An example of a data set that may not reasonably fit a known density function is the data on delay duration times. The time during which the equipment is not working, as collected during time studies, may be made up of time to service the delays associated with such things as the mechanic availability, the spares required, and the ready availability of the spares. In these cases, the data is the result of many processes interacting with each other (such as age of the equipment, maintenance policy, number of mechanics, delays other than equipment downtime, etc.). Therefore, it may not be possible to fit the data to a well-known density function. Use of an empirical density function in this case is appropriate.

Example 11

The following grouped data represents the load on a 68-t (75-st) truck.

Amount of ore st	Number of truck loads
69-70	2
70-71	4
71-72	2
72-73	4
73-74	8
74-75	12
75-76	10
76-77	4
77-78	10
78-79	8
79-80	8

Fit a density function to this data for use with simulation experiments.

Solution

From an examination of the histogram of the values (Fig. 6), it is concluded that there seems to be no particular known distribution to which the data may be fitted. Therefore, an empirical density function will be developed for the data (Table 7).

If it is assumed that the load on a truck is a *continuous* variable, in the range 63 through 73 t (69 through 80 st), then the empirical distribution will be continuous. If it is assumed that the load on the truck is *discrete,* then the empirical distribution will be discrete. These are shown in Fig. 7. For generating a load on a truck, Monte Carlo sampling will be employed. Depending on the type distribution fitted, whether continuous or discrete, there is a small difference in the values for the load on the truck. In practice, the manner in which a variable is treated, continuous or discrete, is fairly obvious.

A common problem that one encounters in materials handling system is matching the various pieces of equipment in order to minimize unnecessary delays. A typical problem is the number of trucks to be assigned to a loading shovel. If the number of trucks is too low, the shovel wait time may

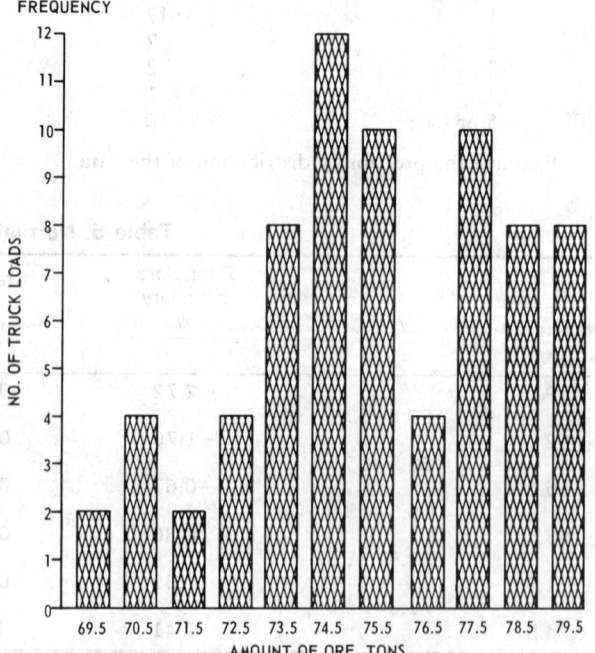

Fig. 6. Histogram for data in example 11.

Table 7. Empirical Distribution Fitting

Weight of the load, internal	Mid-point	Number of truck loads	Relative frequency	Cumulative frequency
69–70	69.5	2	0.0278	0.0278
70–71	70.5	4	0.0556	0.0833
71–72	71.5	2	0.0278	0.1111
72–73	72.5	4	0.0556	0.1667
73–74	73.5	8	0.1111	0.2778
74–75	74.5	12	0.1667	0.4444
75–76	75.5	10	0.1389	0.5833
76–77	76.5	4	0.0556	0.6389
77–78	77.5	10	0.1389	0.7778
78–79	78.5	8	0.1111	0.8889
79–80	79.5	8	0.1111	1.0000
		$\Sigma = 72$	$\Sigma = 1.0000$	

be prohibitively high. If the number of trucks is too high, then the truck wait time may be too costly. While the general objective is to minimize the shovel wait time, a more appropriate objective may be to minimize the total waiting costs. Defining M and N as the number of shovels and trucks in the system, and C_s as the cost of shovel waiting time ($/hr), C_t as the cost of truck waiting time ($/hr), W_m as the cumulative shovel wait time in a shift, and W_n as the cumulative truck wait time in a shift, it is clear that

$$W_m = F_1 (M,N)$$

$$W_n = F_2 (M,N)$$

$$C_T = W_m * C_s + W_n * C_t$$

where F_1 and F_2 are functional operators and C_T is the total cost of wait times in a shift.

It is clear that, except in very simple systems, W_m and W_t are not easily determined as a function of M and N. The next example illustrates the application of simulation to determine the values for W_m and W_n.

Example 12

Consider an operation where two shovels are engaged in loading six trucks. The shovels are close enough for the trucks to be loaded by the next available shovel.

The truck loading time is normally distributed with mean = 4 min and standard deviation = 1.0 min. The time each truck is away from the shovel (i.e., travel loaded, dump, and travel empty times and any delays on the haul road and the

dump) is also normally distributed with mean = 16 min and standard deviation = 2.0 min.

Set up a simulation experiment to evaluate (1) the shovel wait times for the trucks, and (2) the truck wait times for the shovel. Assume that all six trucks are at the shovels at the start of the simulation.

Solution

Note that all calculated times are rounded off to the nearest first decimal point.

Step 1

LT_i = Time to load a truck by the ith shovel
TA_j = Time away from the shovel for the jth truck

The following equations are established to calculate the various times:

$$LT_1 = 4 + RN_1 * 1 \qquad TA_3 = 16 + RN_5 * 2$$
$$LT_2 = 4 + RN_2 * 1 \qquad TA_4 = 16 + RN_6 * 2$$
$$TA_1 = 16 + RN_3 * 2 \qquad TA_5 = 16 + RN_7 * 2$$
$$TA_2 = 16 + RN_4 * 2 \qquad TA_6 = 16 + RN_8 * 2$$

where

RN_k is the kth random normal number stream. Note that the first eight columns in Table 3 has been chosen for the eight random number streams.

Step 2

Calculation of loading times for shovel 1 (LT_1 = 4.0 $+ RN_1 * 1.0$) and shovel 2 ($LT_2 = 4.0 + RN_2 * 1.0$)

Shovel 1		Shovel 2	
RN_1	LT_1	RN_2	LT_2
0.464	4.5	0.137	4.1
0.060	4.1	−2.526	1.5
1.486	5.5	−0.354	3.6
1.022	5.0	−0.472	3.5
1.394	5.4	−0.555	3.4
0.906	5.0	−0.513	3.5
1.179	5.2	−1.055	3.0
−1.501	2.5	−0.488	3.5
−0.690	3.3	0.756	4.8
1.372	5.4	0.225	4.2
−0.482	3.5	1.678	5.7
−1.376	2.6	0.150	3.8
−1.010	3.0	0.598	4.6
−0.005	4.0	−0.899	3.1
1.393	5.4	−1.163	2.8
−1.787	1.4	−0.261	3.7

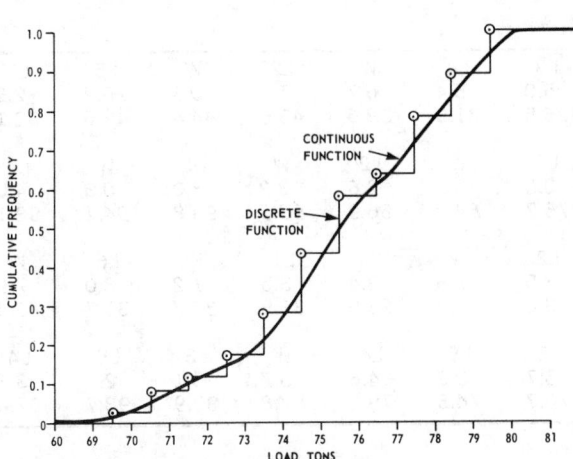

Fig. 7. Cumulative distribution function for data in example 11.

Step 3

Calculate a set of "time away from shovel (TA_i)" for each of the six trucks $(TA_i = 16.0 + RN_i * 2.0)$

Truck 1		Truck 2		Truck 3		Truck 4		Truck 5		Truck 6	
RN_3	TA_1	RN_4	TA_2	RN_5	TA_3	RN_6	TA_4	RN_7	TA_5	RN_8	TA_6
2.455	20.9	−0.323	15.4	−0.068	15.9	0.296	16.6	−0.288	15.4	1.298	18.6
−0.531	14.9	−0.194	15.6	0.543	17.1	−1.558	12.9	0.187	16.4	−1.190	13.6
−0.634	14.7	0.697	17.4	0.926	17.9	1.375	18.8	0.785	17.6	−0.963	14.1
1.279	18.6	3.521	23.0	0.571	17.1	−1.851	12.3	0.194	16.4	1.192	18.4
0.046	16.1	0.321	16.6	2.945	21.9	1.974	19.9	−0.258	15.5	0.412	16.8
−0.525	14.9	0.595	17.2	0.881	17.8	−0.934	14.1	1.579	19.2	0.161	16.3
0.007	16.0	0.769	17.5	0.971	17.9	0.712	17.4	1.090	18.2	−0.631	14.7
−0.162	15.7	−0.136	15.7	1.033	18.1	0.203	16.4	0.448	16.9	0.748	17.5
−1.618	13.8	−0.345	15.3	−0.511	15.0	−2.051	11.9	−0.457	15.1	−0.218	15.6
0.378	16.8	0.761	17.5	0.181	16.4	−0.736	14.5	0.960	17.9	−1.530	13.9

Step 4

Set up two registers, one for recording the status of the shovels and another for recording the status of the trucks (Tables 8 and 9).

Define:

$M =$	Number of shovels in the system
$N =$	Number of trucks in the system
$ESTS_i =$	Elapsed shift time of shovel i, $i = 1, 2, \ldots M$
$ESTT_j =$	Elapsed shift time of truck j, $j = 1, 2, \ldots N$
$ST =$	Time span for simulation (e.g., one shift, one month, etc.)

The procedure to simulate follows:

Step 4(a). Choose the truck with the smallest elapsed shift time *(EST)*.

Let it be the kth truck with $EST = ESTT_k$.

If $(ESTT_k > ST)$, go to Step 4(h).

Step 4(b). Choose the shovel with the smallest elapsed shift time *(EST)*.

Let it be the gth shovel with $EST = ESTS_g$.

Step 4(c). Let $(ESTS_g − ESTT_k) = w$

(1) if $w < 0$, go to Step 4(d)

(2) if $w > 0$, go to Step 4(e)

(3) if $w = 0$, the truck can be assigned to the shovel immediately. Go to step 4(f).

Step 4(d). Shovel g is waiting for a truck. The waiting time of the shovel, w_s

$$w_s = ESTT_k − ESTSg$$

Update the *EST* status of shovel g

$$ESTS_g = ESTS_g + w_s$$

Go to Step 4(f).

Step 4(e). Truck k is waiting for the shovel to be available. The waiting time of the truck, w_t, is

$$w_t = ESTS_g − ESTT_k$$

Update the *EST* status of truck k

$$ESTT_k = ESTT_k + w_t$$

Step 4(f) Generate a load time for the kth truck by the gth shovel.

Let this time be L_{gk}

Update the *EST* status of the shovel and the truck

$$ESTS_g = ESTS_g + L_{gk}$$
$$ESTT_k = \text{ESTT}_k + L_{gk}$$

Step 4(g). Generate a time away for the kth truck from the loading shovels. Let this time be, t_k

Update the *EST* status of the kth truck.

$$ESTT_k = ESTT_k + T_k$$

Go to Step 4(a).

Step 4(h) Simulation time has been exceeded. Summarize the results.

(1) Total production, truck loads or tons

(2) Number of trips/truck

Table 8.

	Activity											
Shovel 1	Activity	L1	L4	L6	W	L3	L4	W	L2	W	L5	L1
	Time duration	4.5	4.1	5.5	7.4	5.0	5.4	6.7	5.0	0.8	5.2	2.5
	EST	4.5	8.6	14.1	21.5	26.5	31.9	38.6	43.6	44.4	49.6	52.1
Shovel 1	Activity	W	L2	W	L1	L6	W	L2	W	L1	W	L6
	Time duration	8.9	3.3	2.5	5.4	3.5	8.6	2.6	3.9	3.0	0.3	4.0
	EST	61.0	64.3	66.8	72.2	75.7	84.3	86.9	90.8	93.8	94.1	98.1
Shovel 2	Activity	L2	L3	L5	W	L2	W	L5	L1	W	L6	W
	Time duration	4.1	1.5	3.6	10.3	3.5	1.6	3.4	3.5	1.2	3.0	7.9
	EST	4.1	5.6	9.2	19.5	23.0	24.6	28.0	31.5	32.7	35.7	43.6
Shovel 2	Activity	L3	L4	L6	W	L3	L5	L4	W	L3	L5	L4
	Time duration	3.5	4.8	4.2	8.9	5.7	3.8	4.6	8.7	3.1	2.8	3.7
	EST	47.1	51.9	56.1	65.0	70.7	74.5	79.1	87.8	90.9	93.7	97.4

Activity Guide:

L_i = Loading ith truck

W = Waiting for truck

EST = Elapsed shift time.

Table 9.

Truck 1	Activity	L1	TA	*W*	L2	TA	*W*	L1	TA	L1	TA	L1	TA
	Time duration	4.5	20.9	*2.6*	3.5	14.9	*3.2*	2.5	14.7	5.4	18.6	3.0	16.1
	EST	4.5	25.4	*28.0*	31.5	46.4	*49.6*	52.1	66.8	72.2	90.8	93.8	109.9
Truck 2	Activity	L2	TA	L2	TA	L1	TA	L1	TA	L1	TA		
	Time duration	4.1	15.4	3.5	15.6	5.0	17.4	3.3	23.0	2.6	16.6		
	EST	4.1	19.5	23.0	38.6	43.6	61.0	64.3	84.3	86.9	103.5		
Truck 3	Activity	*W*	L2	TA	L1	TA	L2	TA	L2	TA	L2	TA	
	Time duration	*4.1*	1.5	15.9	5.0	17.1	3.5	17.9	5.7	17.1	3.1	21.9	
	EST	*4.1*	5.6	21.5	26.5	43.6	47.1	65.0	70.7	87.8	90.9	112.8	

Truck 4	Activity	*W*	L1	TA	*W*	L1	TA	*W*	L2	TA	*W*	L2	TA	*W*	L2	TA
	Time duration	4.5	4.1	16.6	*1.3*	5.4	12.9	*2.3*	4.8	18.8	*3.8*	4.6	12.3	*2.3*	3.7	19.9
	EST	4.5	8.6	25.2	*26.5*	31.9	44.8	*47.1*	51.9	70.7	*74.5*	79.1	91.4	*93.7*	97.4	117.3

Truck 5	Activity	*W*	L2	TA	L2	TA	L1	TA	*W*	L2	TA	L2	TA
	Time duration	*5.6*	3.6	15.4	3.4	16.4	5.2	17.6	*3.5*	3.8	16.4	2.8	15.5
	EST	*5.6*	9.2	24.6	28.0	44.4	49.6	67.2	*70.7*	74.5	90.0	93.7	109.2

Truck 6	Activity	*W*	L1	TA	L2	TA	*W*	L2	TA	*W*	L1	TA	L1	TA
	Time duration	*8.6*	5.5	18.6	3.0	13.6	*2.6*	4.2	14.1	*2.0*	3.5	18.4	4.0	16.8
	EST	*8.6*	14.1	32.7	35.7	49.3	*51.9*	56.1	70.2	*72.2*	75.7	94.1	98.1	114.9

Activity Code: L_k = Loaded by k^{th} shovel;
TA = time away from shovel;
W = waiting for shovel.
EST = Elapsed shift time.

(3) Waiting time of each truck to shovel
(4) Waiting time shovel for trucks.

As an example of the procedure, consider the status of the various equipment in the system at about 15 min into the shift. From the shovel status table, it is seen that shovel 1 has loaded three trucks, specifically trucks 1, 4, and 6 from the time simulation began. Its elapsed shift time (EST) is 14.1 min. Also, shovel 2 has loaded three trucks, specifically trucks 2, 3, and 5 and its EST is 9.2 min.

Checking on the truck status table, it is seen that all the trucks are away from the shovel and that truck 1 will be back at shovels at an EST of 25.4 min, truck 2 at an EST of 19.5 min, truck 3 at 21.5 min, truck 4 at 25.2 min, truck 5 at 24.6 min, and truck 6 at 32.7 min.

Using the notation above,
Step 4(a): $k = 2$ $ESTT_2 = 19.5$
Step 4(b): $g = 2$ $ESTS_2 = 9.2$
and
Step 4(c): $ESTS_2 - ESTT_2 = 9.2 - 19.5 = -10.3$
Therefore, shovel 2 is waiting for trucks and this waiting time is 10.3 min.

Step 4(d): $ESTS_2 = 9.2 + 10.3 = 19.5$
At Step 4(f), truck 2 is loaded by shovel 2, and the loading time is 3.5 min (see Step 2, shovel 2).
Step 4(f): $ESTS_2 = 19.5 + 3.5 = 23.0$
$ESTT_2 = 19.5 + 3.5 = 23.0$
At Step 4(g) for truck 2, the time away from shovels (TA) is generated. This is 15.6 min (see Step 3, truck 2).
Step 4(g): $ESTT_2 = 23.0 + 15.6 = 38.6$
i.e., truck 2 will be back at the shovels at an EST of 38.6 min. The procedure is returned to Step 4(a), and it is seen that now $k = 3$, $ESST_3 = 19.5$, and Step 4(b), $g = 1$, $ESTS_1 = 14.1$. The status of shovel 1 and truck 3 will be updated using the above procedures and simulation will continue up to the time specified for the experiment. In this example, the simulation time was specified as at least 100 min of truck operation.

The results of this 100 min of simulation are:
Number of truck loads from shovel 1 = 14
Number of truck loads from shovel 2 = 16
         Total production = 30 truck loads
     Number of trips: Truck 1 = 5
                 Truck 2 = 5
                 Truck 3 = 5
                 Truck 4 = 5
                 Truck 5 = 5
                 Truck 6 = 5
     Total truck loads = 30
Shovel waiting times:
         shovel 1 for trucks = 39.1 min
         shovel 2 for trucks = 36.6 min
   Total shovel wait time = 75.7 min
   Truck wait times: truck 1 = 5.8 min
               truck 2 = 0 min
               truck 3 = 4.1 min
               truck 4 = 14.2 min
               truck 5 = 9.1 min
               truck 6 = 13.2 min
     Total truck wait time = 46.4 min

It is readily seen that by adding more trucks to the system, the shovel waiting time can be decreased. Note that, *on the average*, there can be five trucks assigned to each shovel:

Average truck cycle time = 16 + 4 = 20 min
(average TA + average LT)
     Average load time = 4 min
Number of trucks/shovel = 20/4 = 5

The above simulation exercise can now be continued with seven, eight, nine and ten trucks. The objective is to minimize the total cost of shovel waiting time and truck waiting time.

Obviously, it is very time-consuming and tedious to simulate this simple problem for over 100 min. However, the procedure can be easily programmed for the computer.

A generalized flow diagram of how such a computer program can be used to select the number of trucks assigned to a shovel is shown in Fig. 8.

TRUCK/RAIL MOVEMENT SIMULATION

The mathematical modeling of the movement of a haul unit can be attempted by both deterministic and probabilistic approaches. From basic mechanics, the continuous change of the position of a body can be established. Consider the following definitions.

f = force, (kg) lb
m = mass kg-s²/m (lb-sec²/ft)
a = acceleration, m/s² (ft/sec²)
V_t = velocity at time t, m/s (ft/sec)
S = distance traveled in the time interval (0, t), m (ft)
V_o = velocity at time 0, m/s (ft/sec).

By definition, $a = f/m$; acceleration will be constant only when for a constant mass, the force exerted does not change. Under constant acceleration conditions,

$$d_v = a * d_t \text{ and } d_s^2 = a * d_t^2$$

and, it will be a simple step to integrate these equations to find the velocity and distance traveled as a function of time. In fact, at time t,

$$V_t = V_o + a * t, \text{ and } S = V_o * t + (1/2) a * t^2$$

In mine haulage, however, the assumption of constant acceleration does not generally hold for all the trips. The force needed to create the acceleration is a function of the speed-rimpull characteristics of the equipment, the coefficient of adhesion between the vehicle wheels and the road surface, the load on the truck, and the frictional and gradient resistance of the haulage profile. The load on the truck may not be constant every trip. Consequently, the frictional and gradient resistances, the accelerating force and acceleration, all vary over the same haul road profile. In addition, as the number of trucks in the system increases, there may be mov-

ing queues. Two approaches have been used to capture this variation in truck/rail travel times, namely deterministic simulation and probabilistic simulation.

Deterministic Simulation

In the deterministic approach vehicle motion is simulated for increments of a small interval of time, Δt. During this small time interval, acceleration is assumed to be constant, and using the force-mass-acceleration relationship, and the motion formulas, the velocity and position of the vehicle are determined at the end of the time interval. The required data for this simulation include the speed-rimpull characteristic of the equipment, specifications of the haul road profiles such as length, gradient, and maximum speeds, and maximum accelerations. The simulation approach can be explained with reference to Figure 9, the speed-rimpull characteristics of a truck and the haul road profile, respectively. The total weight of the truck will be:

$$W_T = W_o + W_c$$

W_T = total weight of the trucks, st
W_o = tare weight of the truck, st
W_c = weight of ore (or waste) loaded, st.

The rolling resistance in lb/st and grade of the section X-Y in percent are K and G, respectively, and the acceleration due to gravity is g ft/sec². Assuming that this truck is at Junction X at time t with velocity V_1 mph, and that the acceleration of the truck is constant for the time increment Δt, the following can be developed:

Available rimpull at $X = R_1$ lb
Resistance to motion = $(G + K) * 20 * W_T$ lb
Acceleration force = $(R_1 - (G + K) * 20 * W_T)$ lb

a = acceleration

$$= \frac{(R_1 - (G + K) * 20 * W_T)}{W_T + 2000)/g} = \text{ft/sec}$$

Defining V_2 as the velocity at time $t + \Delta t$, and D as the distance traveled in time Δt, it is seen that

$$V_2 = V_1 + a * \Delta t$$
$$D = V_1 * \Delta t + (1/2)a * (\Delta t)^2$$

Now, the available rimpull at the location X_1 in the haul road section, as can be seen from the rimpull curve, is R_2. This will determine a new acceleration force and therefore, new acceleration, for the next time interval Δt, which in turn will be used to calculate the distance traveled and the velocity in that interval. These steps are repeated until the entire Section X-Y is traversed. At the end of Section X-Y, data is initialized for the next haul road segment and the simulation is continued until the haul unit reaches the dump.

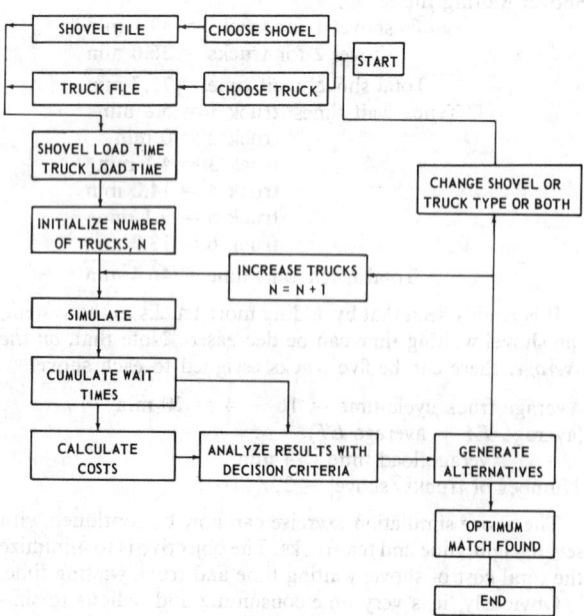

Fig. 8. Generalized flow diagram for truck selection.

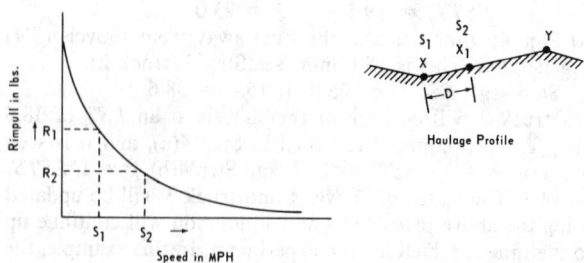

Fig. 9. Simulation of motion in a computer model.

Obviously, as the total weight of the truck (W) changes from one trip to another, the travel times also change.

Probabilistic Simulation

The basis for probabilistic simulation is time study data. Depending on the detail of the simulation, the haul road will be broken down into a number of distinct segments. The times to travel each segment under various operating conditions are collected. The data will be analyzed and grouped. Finally, the data will be described by a standard or empirical probability distribution function. The parameters of the distribution can change with different truck types, and the haul road segment. The travel time for a particular segment is obtained through Monte Carlo sampling of the distribution.

Deterministic vs. Probabilistic Simulation

There can be some significant differences in the results obtained from the two approaches. In the deterministic approach, the truck-travel times are generated as a function of the load on the truck. In probabilistic simulation, unless the travel time distributions are conditioned by the truck load, the generated travel times will be independent of the truck load; this is not true. On the other hand, the deterministic approach may give optimistic results for travel times as it does not include operator variability. Operator variability is automatically incorporated in the time-study generated travel-time distributions. When a new mine is being planned or major changes are planned in existing mines, time study data for these conditions is nonexistent. Historical time-study data has to be adjusted on the basis of experience to develop input for the changed conditions. Such limitations do not exist in the deterministic approach. Finally, the predicted performance of the deterministic method may serve as the standard for comparing operator performance. In practice, for shovel-truck systems, considerable data is to be collected for both the approaches. Depending on the purpose of the study, either approach or both can be advantageously used.

SHOVEL-TRUCK SIMULATION

The complexity of most open pit operations—a large number of different shovel types operating in various locations throughout the pit, different truck types assigned to each shovel, the number of waste and ore dumps, changing grades in haul road segments, truck-passing priorities, etc.—requires simulation models that will permit a complete analysis and evaluation of the shovel-truck system. The development of computer-oriented shovel-truck simulation models was well advanced by the mid-1960's and several models are currently available from consultants, manufacturers, and universities. A typical flow chart for a shovel-truck system is shown in Fig. 10.

Monte Carlo based shovel-truck models can be explained by reference to Fig. 11, where ore is moved from various shovel locations, along a network of haulage routes, to several dumping stations (Bauer and Calder, 1973). Through extensive time studies in the field, data is collected on the load times, the travel times, and dump times. Depending on the detail of the model, the haul road may be broken down into a number of distinct segments. The time to travel each segment is collected. The magnitude of the data collection effort depends on such factors as the number of different types of shovels and trucks employed, of the number of haul road segments in the haul route, the number of dump locations, and the operating procedures. Empirical or standard statistical distributions can be fitted to the observed load, travel, and dump times. These distributions permit the *random* se-

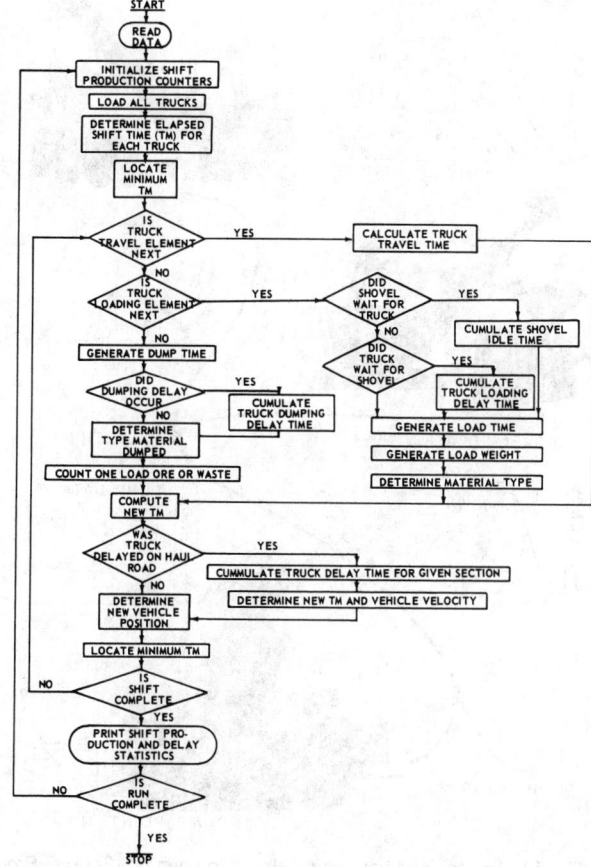

Fig. 10. Computer program flow diagram for a shovel-truck simulator.

lection of event times for the defined sequence of operations. To simulate a system, the operator must define the number, type and location of shovels, dumping stations to be used, the haul road segments from the shovel location to the dump station, and the number and type of truck assigned to each shovel. The distribution for the load, travel, and dump times associated with each shovel, truck, haul road segment, and dump stations are also defined. Other required data will include the shift length, the number of shifts to be simulated, the shovel, truck and dump availability, and the necessary delays for such activities as refueling, lunch, and start-up.

A hybrid simulation model employing probabilistic simulation for generating load and dump times, and deterministic simulation for moving trucks on the haul roads is also available (O'Neil and Manula, 1966). As opposed to the Monte Carlo model, the truck travel times are calculated as a function of the load and speed-rimpull characteristics. In most cases, all other input data for the two approaches are the same.

Essentially, a shovel-truck model, using a computer, cycles trucks between their assigned shovels and dumps over specified haul routes. Records are kept of all waiting times at the loading and dumping points, and at haul road intersections, and of the ore and waste production. A shovel truck simulation model can have a number of applications including equipment selection, equipment performance analysis, equipment assignment, equipment scheduling, production planning, mine layout planning, and haul road design.

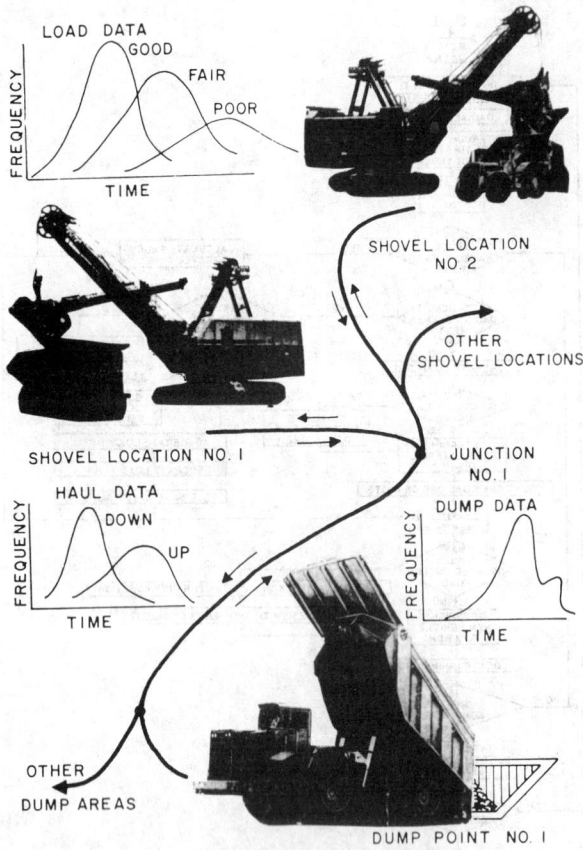

Fig. 11. Typical load-haul-dump circuit in open pit (Source: *Proceedings*, 10th International APCOM Symposium, Republic of South Africa, p. 274).

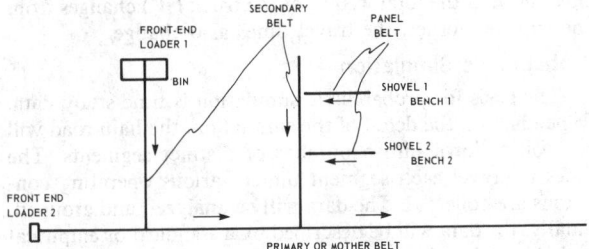

Fig. 12. Surface mine belt network.

The first realistic simulation model of mine belt networks, based on probabilistic simulation of arrival of loads on a belt was developed in 1965 by Sanford and Manula. Bucklen, et al. (1968), developed an improved belt haulage simulator based on the first model. In 1969, Sanford and Manula developed a time-increment belt simulator to integrate the belt network with other components of the production system to develop a more complete materials handling simulator. Tan Sizhe (1985) has developed a fairly complex belt simulation model which incorporates variable feeder and discharge locations and several modes for describing the load arrivals to the belt.

MATERIALS HANDLING SIMULATOR

The real potential of the simulation technique is not fully realized when it is used to develop a model of a single entity such as a truck, a rail, or a belt in a mine. Since all the haulage units in a system interact with each other, there is a need to consider all the units at one and the same time. The simplest relationship may be with the loading shovels and a crusher system. A more complex relationship may be similar to the one portrayed in Fig. 13, a sequence of materials handing activities from two mines. A fairly detailed Open Pit Materials Handling Simulator (OPMHS) is available for the planning and analysis of excavation and transportation activities in surface mining operations (Ramani and Manula, 1978). This simulator consists of a number of subassemblies which can be used in any order to model a surface mine (Fig. 14). The load subassembly incorporates bucketwheel excavators, stripping dragline, stripping shovel, and

BELT SIMULATION

Proper design of a conveyor belt system to handle the desired material flow rate involves many things, material characteristics such as density, lump size, percent fines; conveyor data such as width, length, angle of incline, lift (or drop), and speed; drive related information such as location and number of drives, type of motor, and horsepower; belt support data such as idlers, their spacings and angles; and loading conditions such as number of loading points and loading rates. In the design of a haulage system consisting of only one belt and one loading point, it may not be too difficult to establish an optimum flow rate for the belt. Even here one has to consider the adequacy of the surge capacity at the loading point for unexpected belt breakdowns. In typical mine application, however, a face belt may be loaded from several sources. In addition this belt may feed a submain or gathering belt, which in turn may feed a primary or mother belt (Fig. 12). In this belt network the problem of belt selection is compounded by many additional factors, such as (1) the arrival time of the loads on each belt, (2) the length of the loads on each belt, (3) the relationships between the various belt loading points and discharge points, and (4) the location and amount of surge capacity in the system. Unless the entire system is laid out with detailed consideration to each of these factors, peak loads of unknown magnitude and frequency can result. The need for the use of computer simulation techniques to design a belt network is obvious.

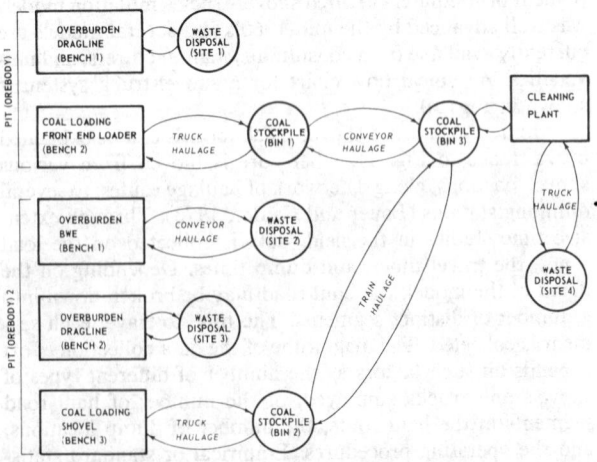

Fig. 13. Complex materials handling scheme in a surface coal mine in competing draglines, loaders, trucks, belts, and trains.

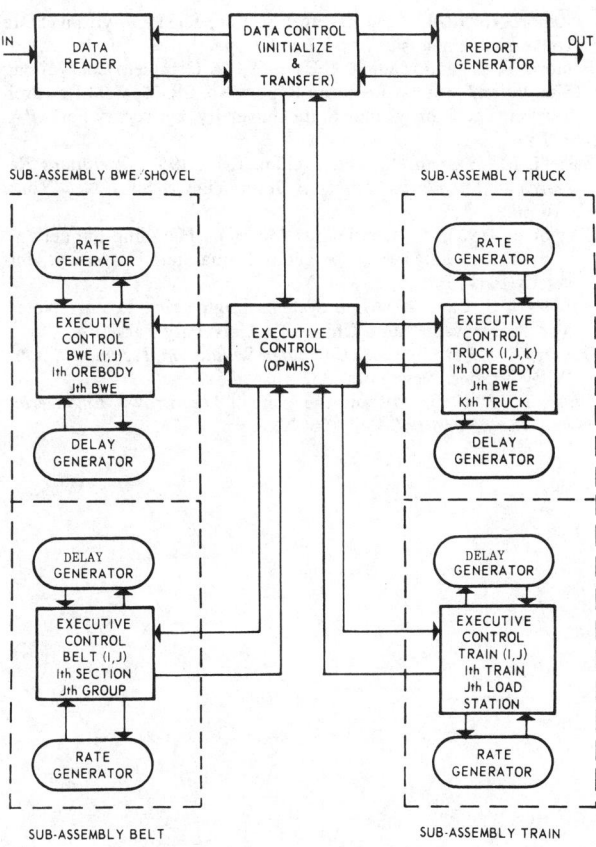

Fig. 14. Flow diagram of the open pit materials handling simulator.

loading shovel. The haulage assembly incorporates trucks, belts, and trains. There are provisions for surge bins and stockpiles. Using a combination of deterministic and probabilistic, and time-increment and event-oriented simulation, the equipment and materials are moved through the mining system. The output includes production and equipment performance and delay statistics. The simulator has been applied to a number of operations of varying complexities. A new version of this simulator incorporating geological and equipment characteristics for describing production has been recently released (Ramani and Albert, 1985).

SIMULATION LANGUAGES

Computerization of the mathematical model is an important step in a simulation study. The simulation model is, in fact, a computer program. The choice of the language for writing the computer program is a major decision. There are multipurpose languages, such as FORTRAN, ALGOL, and APL and simulation specific languages, such as GPSS, GASP, SIMSCRIPT, CSMP, DYNAMO, and SLAM. The applicability of the language will be determined by the scope and orientation of the simulation model. Where the model is extremely procedure-oriented, involving a number of complex decision-processes and logical relationships, such as the deterministic simulation of motion, the use of a multipurpose language will be the safest choice. On the other hand, where the model is generally process-oriented involving the flow of discrete items at discrete points in time, such as the Monte Carlo simulation model, a simulation language would be

natural to use. Simulation of inventory and waiting line processes can be ideally carried out in a simulation language. These broad generalizations are provided here only to indicate the importance of this subject rather than to serve as guidelines. Mine materials handling simulators have been developed in some of the simulation languages. However, the use of a multipurpose language, particularly FORTRAN, has been the general trend.

SUMMARY

There is little doubt that experience and judgement are important for decision-making in systems characterized by complex interactions of technical, economic, and social components, and by a large number of risky outcomes. However, quantitative analysis of many aspects of the system, as an aid to decision-making, is feasible. One such system is the materials handling system in surface mines. Simulation models of materials-handling systems such as shovel-truck and belt conveyor networks are available. These models provide an insight into the complex interactions that take place in real world systems, and enable the decision-maker to select alternatives for the system under evaluation to enhance desirable interactions and suppress undesirable consequences. Thus, simulation models are useful for planning new mines and evaluating operating mines. However, in the real world, fluctuations in performance will be encountered due to changes in weather, road conditions, age and wear and tear of the equipment. These conditions may go unsimulated. It is often necessary to adjust the simulation results with factors to approximate the expected average performance. Finally, simulation is an experimental technique. A simulation model's purpose is to help the decision-maker select the best values of the controllable variables to obtain the best system performance. Even with a good simulation model, efficient experimental design is required to decipher, at acceptable cost, the relationships between the controllable variables, the uncontrollable variables, and the measures of performance.

REFERENCES

Bauer, A., and Calder, P. N., 1973, "Planning Open Pit Mining Operations Using Simulation," *Proceedings of 10th International Symposium on the Application of Computer Methods in the Mineral Industry,* The South African Institute of Mining and Metallurgy, pp. 273-278.

Bucklen, E. P., Suboleski, S. C., Prelaz, L. J., and Lucas, J. R., 1968, *Computer Applications in Underground Mining Systems,* Research and Development Report No. 37, Interim Report No. 1, OCR Contract No. 14-01-0001-410, Office of Coal Research, US Dept. of the Interior, Washington, DC.

Connell, J. P., 1969, "Determination of Fleet Availability Using Probability Calculations," SME AIME Preprint No. 69-AR-46, 17 pp.

Connell, J. P., 1973, "Truck Haulage," (Section 18.2), *SME Mining Engineering Handbook,* AIME, New York, pp. 18-16 to 18-33.

Dunlop, J. W., and Jacobs, H. H, 1955, "How Operations Research Solved the Dragline Problem," *Engineering and Mining Journal,* Vol. 156, No. 8, pp. 79-83.

Emshoff, J. R., and Sisson, R. L., 1971, *Design and Use of Computer Simulation Models,* The MacMillan Company, New York, 302 pp.

Forrester, J. W., 1961, *Industrial Dynamics,* The M.I.T. Press, Cambridge, MA, 464 pp.

Gordon, G., 1969, *System Simulation,* Prentice-Hall, Inc., Englewood Cliffs, NJ, 303 pp.

Maisel, H., and Gnugnoli G., 1972, *Simulation of Discrete Stochastic Systems,* Science Research Associates, Chicago, IL, 465 pp.

O'Neil, T. J., 1966, "Computer Simulation of Materials Handling in Open Pit Mining," unpublished M.S. Thesis, The Pennsylvania State University, University Park, PA, 92 pp.

O'Neil, T. J., and Manula, C. B., 1966, "Computer Simulation of Materials Handling in Open Pit Mining," Special Research Report No. SR-56, Coal Research Section, The Pennsylvania State University, University Park, PA, 96 pp.

Ramani, R. V., and Albert, E. K., 1985, "A Computer Simulation Model for Surface Mine Planning, Volume 4: User's Manual for OPMHS/2," Final Report on Contract J0295005 to US Bureau of Mines, The Pennsylvania State University, University Park, PA.

Ramani, R. V., and Manula, C. B., 1978, "Application of a Total Systems Simulator to Coal Stripping," Volume 1 of the Final Report to the Bureau of Mines, US Department of the Interior, NTIS: PB 294 627/AS, 37 pp.

Sanford, R. L., and Manula, C. B., 1965, "A Simulation Model on the Optimal Design of Belt Conveyor Systems," Special Research Report No. SR-47, Coal Research Section, The Pennsylvania State University, University Park, PA, 82 pp.

Sanford, R. L., and Manula, C B., 1969, "A Complete Coal Mining Simulation," Special Research Report No. SR-75, Coal Research Section, The Pennsylvania State University, University Park, PA, 197 pp.

Sasieni, M., Yaspan, A., and Freidman, L., 1959, *Operations Research—Methods and Problems,* John Wiley & Sons, New York, 316 pp.

Tan Sizhe, 1985, "A Continuous Materials Handling Simulator," unpublished MS Thesis, The Pennsylvania State University, University Park, PA.

Suboleski, S. C., 1975, Mine Systems Engineering Lecture Notes, The Pennsylvania State University, University Park, PA.

Taylor, T. H., et al., 1966, *Computer Simulation Techniques,* John Wiley & Sons, New York, 352 pp.

Ware, T. M., 1955, "OR and the Mine of Tomorrow," *Engineering and Mining Journal,* Vol. 156, N. 8, pp. 75-78.

6.5.9 Haulage System Analysis: Queuing Theory

JORGEN ELBROND

INTRODUCTION

In an open pit operation the trucks move from the shovels to the dump-crusher and back. Sometimes they go to the repair shop or to the fueling station; they regularly go to the places where drivers are exchanged and where meals are taken. Occasionally the trucks have to wait at a shovel, at the dump-crusher, at the repair shop or at the fueling station when there already is a truck being loaded at the shovel or being fueled or when all positions at the dump-crusher or all bays at the repair shop are occupied. These situations are caused by variabilities of the loading, dumping, repair, and fueling times and of the time intervals between trucks arriving at these facilities.

These waiting times reduce the capacity of the operation. It is quite obvious that the waiting times increase when trucks are added to an existing system if no other changes are made to the system. The productivity per truck will thus decrease (while the productivity of the shovels will increase).

The estimation of these waiting times is an important task in the design and equipment selection for a new open pit operation or when changes in an existing operation are being considered. Also important is the estimation of the trucks' travel times, full and empty, on the pit roads, for example by a truck performance calculator, and the estimation of the loading, dumping and repair times.

The estimation of waiting times is the subject of operations research techniques such as simulation by random numbers or the queuing theory. These techniques together with the performance calculators of trucks and shovels are important tools in the design process, in equipment selection, and in the management of the daily operation.

The queuing theory offers an interesting approach to the estimation of waiting times because of its calculation speed and its relative simplicity compared with simulation by random numbers. Sometimes the queuing approach can completely substitute simulation; sometimes it offers an interesting supplement to simulation in that it can fill in missing points in a simulation study quickly and cheaply. In truck dispatching, where a forward estimate of waiting times is important information for the dispatcher, it offers the only way fast enough to provide that information.

WAITING TIME IN A CLOSED CIRCUIT

The waiting time of clients returning to a server in a closed circuit (Fig. 1) is a function of the service time (TS),
its standard deviation (SRT), the return time of clients (RT), its standard deviation (SRT), and the number of clients (N) in the circuit. Mathematical solutions of the waiting time are found in the following cases:

1. Service time, TS, of random length, STS-TS, and return time, RT, of random length, SRT=RT. In this solution, from Palm (1947), the waiting time is called WP. Posner and Bernholtz (1967) have shown that this solution is valid regardless of the standard deviation, SRT, of the return time, RT; a result which is used in the interpolation procedure described later.

2. Service time, TS, of constant length, STS=0, and return time, RT, of random length, SRT=RT. This solution, called WA, is from Ashcroft (1957).

3. Both service time, TS, and return time, RT, constant. STS=0, SRT=0. This is the trivial case. This waiting time is called WB.

The formulas are shown in Appendix I.

ADJUSTMENT TO INDUSTRIAL SITUATIONS

In an industrial circuit one would find neither random nor constant return times, but often would find service times which are less random but not constant. In repair shops service times tend to be random. Using the previous mathematical solutions supported by a series of simulation experiments, an interpolation procedure (Fig. 2) has been developed. The axis are the coefficients of variation of the service time, STS/TS, and of the return time, SRT/TR. The adjusted waiting time, WS, is a linear combination of WP, WA, and WB. The waiting time of trucks in shovel/dump circuits is mostly situated at STS/TS and SRT/RT both being about 0.3 (Elbrond, 1974).

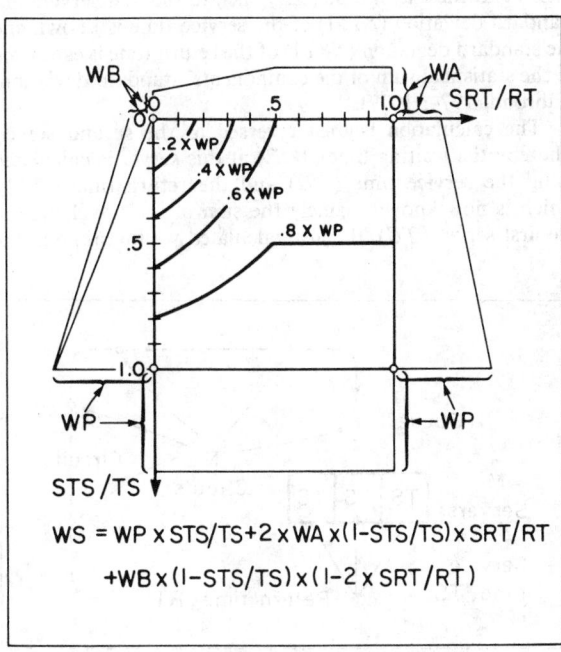

$$WS = WP \times STS/TS + 2 \times WA \times (1-STS/TS) \times SRT/RT$$
$$+ WB \times (1-STS/TS) \times (1-2 \times SRT/RT)$$

Fig. 2. Interpolation procedure.

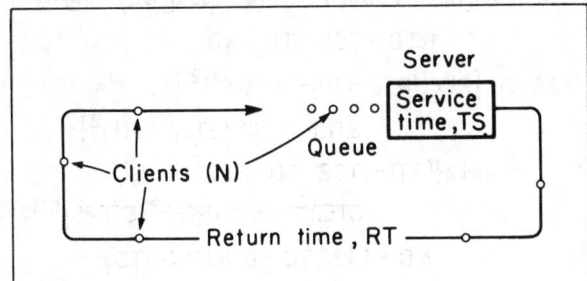

Fig. 1. Closed circuit with one server.

743

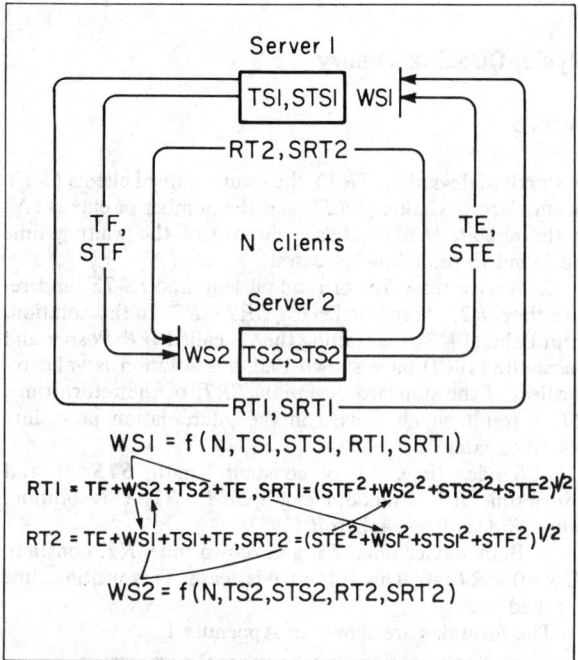

$$WSI = f(N, TSI, STSI, RTI, SRTI)$$

$$RTI = TF + WS2 + TS2 + TE, SRTI = (STF^2 + WS2^2 + STS2^2 + STE^2)^{1/2}$$

$$RT2 = TE + WSI + TSI + TF, SRT2 = (STE^2 + WSI^2 + STSI^2 + STF^2)^{1/2}$$

$$WS2 = f(N, TS2, STS2, RT2, SRT2)$$

Fig. 3. Circuit with two servers in series.

THE CIRCUIT WITH TWO SERVERS IN SERIES

The circuits in which the trucks in an open pit operation are moving have at least two servers, a shovel and a dump-crusher. As there is no mathematical solution to the waiting times at the two servers when neither service times nor return times are of random length, a special procedure (Fig. 3) has been developed, whereby the waiting time ($WS1$) at the first server is calculated using the service time (TS1) and the return time ($RT1$), which is known and which is the sum of the travel time (TF) to the second server, its service time ($TS2$) and the travel time (TE) back to the first server. The standard deviation ($STS1$) of the service time is known and the standard deviation ($SRT1$) of the return time is estimated as the statistical sum of the components' standard deviations (Elbrond, 1974; 1977).

The calculation is then reversed to the second server, whereby the waiting time ($WS2$) at this server is calculated using the service time ($TS2$) and the return time ($RT2$), which is now known, namely the sum of the travel time to the first server (TE), the just calculated waiting time at the

first server ($WS1$), its service time ($TS1$), and the travel time back (TF). The standard deviation of the service time ($STS2$) is known and the standard deviation of the return time ($SRT2$) is estimated as the statistical sum of the components' standard deviations, whereby the standard deviation of the estimated waiting time is assumed to be equal to the waiting time itself. The procedure then reverses to the first server to calculate a new waiting time using the newly calculated waiting time at the second server in the return time. The procedure converges rapidly. There is no mathematical proof for this procedure. It has, however, been possible to verify waiting time estimates by time studies. The convergence bears some resemblance to a real operation where waiting times also may oscillate.

CIRCUITS WITH PARALLEL SERVERS

A service station may consist of two or more servers working in parallel (Fig. 4). This would be the case of a dump-crusher with two dumping positions, provided that the crusher is capable of absorbing two truckloads at a time, and also the case of a repair shop for trucks with several repair bays. If the servers have equal random service times and are fully collaborating with a common queue, then the waiting time can be calculated by means of a procedure (Palm, 1947) which is similar to the one mentioned earlier but slightly more complex due to the number M of servers.

The formula for the calculation of the waiting time is given in Appendix II. This is the case of repair shops. If the service times of the M fully collaborating servers are not

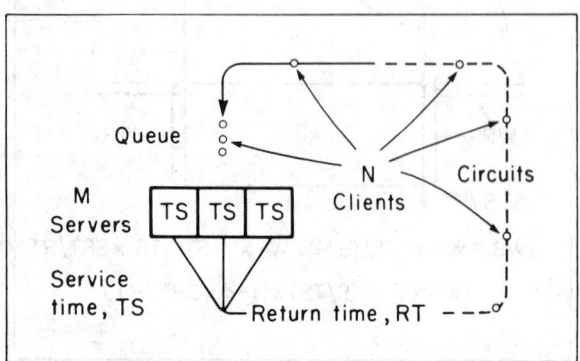

Fig. 4. Circuit with parallel servers.

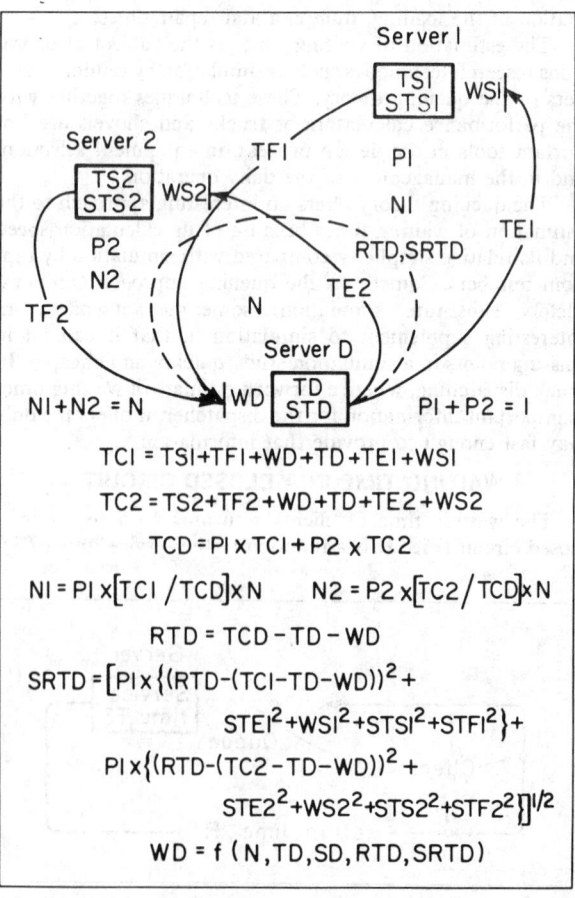

$$TCI = TSI + TFI + WD + TD + TEI + WSI$$

$$TC2 = TS2 + TF2 + WD + TD + TE2 + WS2$$

$$TCD = PI \times TCI + P2 \times TC2$$

$$NI = PI \times [TCI/TCD] \times N \qquad N2 = P2 \times [TC2/TCD] \times N$$

$$RTD = TCD - TD - WD$$

$$SRTD = [PI \times \{(RTD - (TCI - TD - WD))^2 + STEI^2 + WSI^2 + STSI^2 + STFI^2\} + PI \times \{(RTD - (TC2 - TD - WD))^2 + STE2^2 + WS2^2 + STS2^2 + STF2^2\}]^{1/2}$$

$$WD = f(N, TD, SD, RTD, SRTD)$$

Fig. 5. System with two circuits.

random even though of equal length, the necessary formula does not exist to establish an interpolation formula as the one shown in Fig. 2. In this case it is proposed to calculate the waiting time (WS) as was done in the case of one server and with a number of clients equal to N/M. This WS divided by WP, the waiting time for one server, becomes the correction factor to apply to the waiting time for M servers (Palm, 1947).

The procedure is described in Appendix III. This is the case of a double dump-crusher. In the total procedure described in this section, it is assumed that dispatching of trucks occurs at their departure from the dump-crusher. In this case the shovels are noncollaborating and have individual queues. If the dispatching is effectuated later, a group of shovels could become partly or wholly collaborating with a common queue. The estimation of the resulting waiting time can be made using an approach that finds a weighted average of the noncollaborating case and the totally collaborating case, taking into account the possible different loading times, production rates, and cycle times.

SEVERAL CIRCUITS

In an open pit operation several circuits join at the same dump-crusher. The single circuit concept has been expanded by using a weighted average return time to the common server.

The Two-Circuit System

In a two-circuit system (Fig. 5), the N trucks are first distributed between the two circuits ($N1$ and $N2$) according to the desired split ($P1$ and $P2$) of production between the two shovels and to the cycle times ($TC1$ and $TC2$) in the two circuits. These cycle times are not yet completely known since they must include the waiting times at the shovels ($WS1$ and $WS2$) as well as at the common dump-crusher (WD). The initial truck distribution is therefore made only according to the split and the net cycle times (without waiting times). The waiting time at the shovel in the first circuit is then calculated using the procedure developed in the preceding paragraphs, followed by the calculation of the waiting time at the shovel in the second circuit. Then the weighted return time (RTD) to the common dump-crusher is calculated from the weighted cycle time (TCD) minus the dumping time (TD) and minus the waiting time (WD) at the dump. The standard deviation ($SRTD$) of the return time to the dump-crusher is calculated knowing the production split between the two circuits ($P1$ and $P2$) and the individual return times' standard deviations. The waiting time (WD) at the dump-crusher can now be calculated using the weighted return time and its standard deviation.

The next step in the calculation is to make an adjustment to the distribution of trucks in the two circuits using the new cycle times, which now include the first estimates of the waiting times at the shovels and the dump-crusher. The procedure is then repeated until convergence, after a few iterations.

Noninteger Number of Trucks: At the truck distribution it will occur that noninteger numbers of trucks are to be distributed. Although the formulas for waiting time can cope with noninteger numbers of clients, except when smaller than 1, it is more appropriate to arrange the calculation so that the waiting time is calculated as the weighted average of the integer above and the integer below the noninteger number of clients. The weights are found from the two integer numbers' deviations from the noninteger. This is also more in line with reality.

Waiting in Line After Interruptions of Operations: Additional waiting time of trucks, prolonging their cycle times, occurs after an interruption of production at the shovel. For example after shift change or after a meal break, all the trucks in a shovel circuit depart from the same place at the same time to arrive in a group at the shovel. Another example is a short breakdown of the shovel, which does not result in rescheduling of the trucks. They will accumulate at the shovel. When the interruption is finished (Fig. 6), the first truck gets loaded immediately whereas the second waits during one loading time, the third waits during two loading times, etc. After some time, which can be estimated to be between ½ and 1 cycle time, the initial waiting time can be assumed to have dropped to the stationary waiting time, which is calculated by the queuing procedure Fig. 7 shows an example of interruptions during an 8-hr shift causing accumulation of trucks at the shovel.

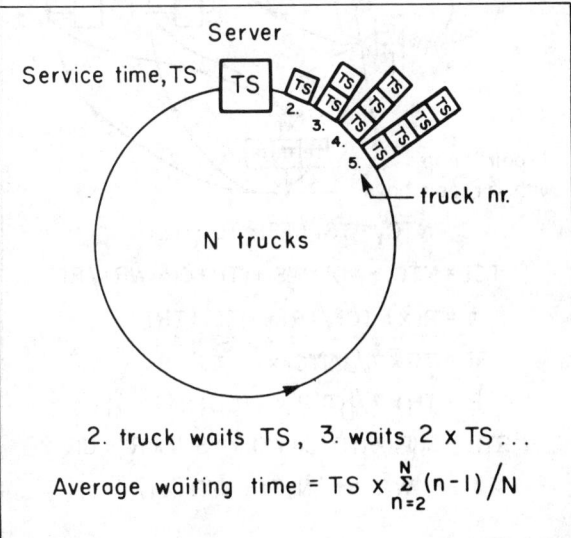

Fig. 6. Accumulation of trucks and initial waiting time.

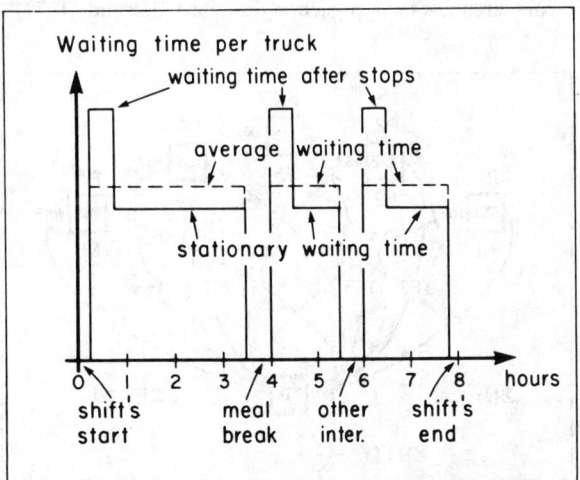

Fig. 7. Shift's composition with initial and stationary waiting times.

Several Circuits with a Common Dump

The expansion of the procedure for two circuits with a common dump-crusher to a procedure with several circuits is easy to conceive (Fig. 8). The return time (RTD) from the dump-crusher is now weighted from several circuits as is its standard deviation ($SRTD$).

Several Circuits in Waste and Ore

The procedure can also easily be expanded to include circuits between waste shovels and dumps (Fig. 9). The trucks are distributed according to the split between shovels' production, which now includes the waste/ore ratio, and the cycle times. The only difference between the ore circuits and the waste circuits is that the latter do not normally have waiting time at the dump.

AVAILABILITY

In the discussion so far 100% presence of shovels and trucks has been assumed. All machines are operated according to the desired distribution of production and waste/ore ratio except for the short breakdowns of shovels, which do not lead to the rescheduling of trucks and are not included in the nonavailability figure of the shovels. Their effect on the shovels is included in the available shift production time. The procedure must now be completed with due consideration of the effect of nonavailabilities of shovels and trucks.

The conventional definition of mechanical availability of trucks is the ratio between working hours and the sum of working hours and repair hours. Working hours include the waiting times at shovels and at the dump-crusher and the repair hours include the waiting time outside the repair shop. Since these waiting times are dependent on the system which is being studied, it seems appropriate to introduce a more basic availability measure, from which the conventional mechanical availability and the effective utilization can be worked out. Such a measure could be called the *internal availability* defined as the ratio between functioning hours and the sum of those hours and the effective repair hours. None of the hours should include waiting time at shovels, dump-crusher, or outside the repair shop. This measure could be obtained by monitoring devices installed on the trucks and would reflect the wear and tear of the truck due to its operation. From this internal availability, the effect of the repair shop with its repair bays can be included in the capacity estimation procedure by means of a superimposed repair circuit. The approach is described Elbrond (1979).

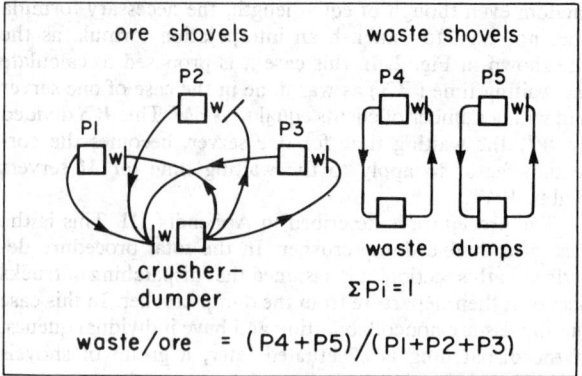

$$\text{waste/ore} = (P4+P5)/(P1+P2+P3)$$

Fig. 9. System with waste and ore circuits.

TRUCKS' REPAIR CYCLES

When trucks break down and need repair, they are taken to the repair shop where they might have to wait if all the repair bays are occupied. An extension of the cycle approach opens the possibility of including the trucks' nonavailability due to breakdowns and of calculating their waiting time ahead of the repair shop by superimposing a repair circuit on the production circuits. In this circuit, the trucks visit the repair facility, consisting of one or more repair bays, according to the internal availability, which is transformed into a number of production cycles between repair cycles by means of the average length of repair (Fig. 10). The simplest approach is made possible by the fact that the lengths of repair tend to be random. For each shovel circuit the number of truck cycles (Ri) between repairs is calculated from the internal availability (η), the net cycle time ($NTCi$) and the

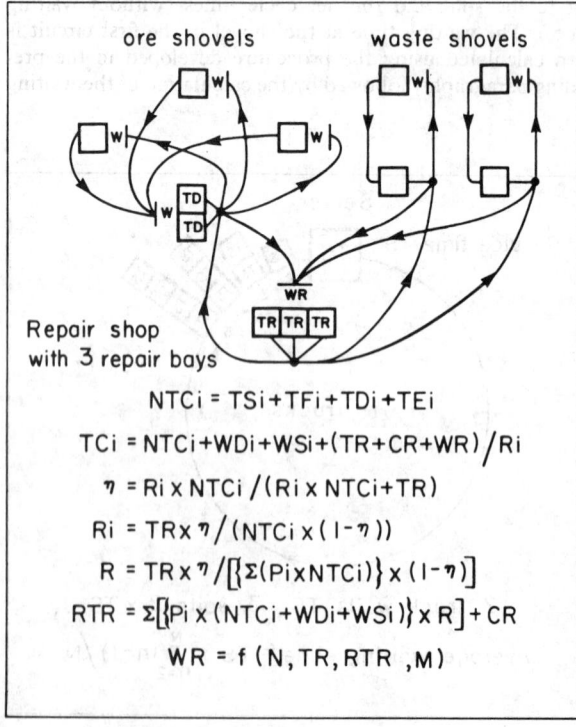

$$NTCi = TSi+TFi+TDi+TEi$$

$$TCi = NTCi+WDi+WSi+(TR+CR+WR)/Ri$$

$$\eta = Ri \times NTCi/(Ri \times NTCi+TR)$$

$$Ri = TR \times \eta/(NTCi \times (1-\eta))$$

$$R = TR \times \eta/[\{\Sigma(Pi \times NTCi)\} \times (1-\eta)]$$

$$RTR = \Sigma[\{Pi \times (NTCi+WDi+WSi)\} \times R] + CR$$

$$WR = f(N, TR, RTR, M)$$

Fig. 10. Superimposed repair circuit.

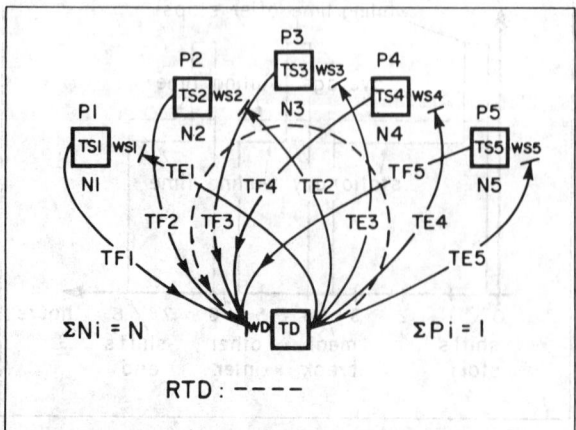

$$\Sigma Ni = N \qquad \Sigma Pi = 1$$

$$RTD : \text{----}$$

Fig. 8. System with several circuits.

average length of repair (TR). The total cycle in a shovel circuit (TCi) now includes a fraction of the repair time (TR), of the waiting time (WR) ahead of the repair facility yet to be calculated, and the time it takes to bring the truck to and from the mine (CR). The average number of truck cycles between repairs (R) and the return time to the repair shop (RTR) are found by weighting procedures. The waiting time at the repair shop (WR) is calculated by means of the simple formula for Palm's waiting time in the case of more than one server (Appendix II) from the total number of trucks (N), the repair time (TR), the return time (RTR), and the number of repair bays (M). The calculated waiting times in the shovel circuits and in the repair circuit gradually enter the total cycles, but the procedure converges rapidly.

SHOVELS' AVAILABILITY

When shovels break down, the repair is executed by a repair team at the shovel site. It has been found by observations that the frequencies of shovels' operational presence over a shift can be estimated reasonably well by means of the binomial law (Fig. 11), which expresses the probability $Pn(N)$, of having n shovels present out of N, when the internal availability (η) is known. It is also possible to express the probability of having a given set of shovels present. The shovels do not need to have an identical internal availability, which makes it possible to evaluate the effect of shovels of various internal availabilities.

For a given set of shovels, the trucks are distributed according to a modified split of production between the remaining shovels. The new distribution maintains the pro-

portions between the remaining shovels. This maintains the overall distribution considering all shovel presence combinations.

THE CAPACITY CALCULATION

The total capacity calculation can now be carried out using the models described in the previous paragraphs.

For each set of available shovels, the trucks are distributed according to the modified split of production between shovels. The cycle times in the various circuits are estimated by the adjusted iterative procedures and the capacity per shift of waste and ore, considering the available shift hours, is calculated. The total capacity is obtained by the integration of the probabilities of the various shovel sets' presence with their respective shift's capacity. Information on mechanical availability according to the conventional definition can be obtained as well as the effective utilization. The variation of shift capacity of waste and ore can be estimated.

CALCULATION EXAMPLE AND RESULTS

Five shovels, three in ore and two in waste, $W/O = \frac{2}{3}$ with an internal availability of 0.8, load 20 trucks with an internal availability of 0.8.

Shovel	Loading min.	Travel min.	Dumping min.	Pi	
1	5	15	2	0.25	ore
2	5	20	2	0.20	ore
3	5	15	2	0.15	ore
4	4	10	2	0.20	waste
5	4	10	2	0.20	waste

Four repair bays in the repair shop. Average repair time = 720 min. Average time to bring a truck to and from the pit = 120 min. Shift time = 360 min out of 480 min.
Results:
126 truckloads of ore per shift
86 truckloads of waste per shift
Effective utilization of trucks = 0.58.
Fig. 12 shows (for a different set of parameters) the suc-

Shovel number

I 2 • i • N-I N

Production split : $\sum_{i=1}^{N} Pi = 1$

PI P2 • Pi • P(N-I) PN

Shovel presence (I oui, O non)

I I I I I I

Probability of presence : $PN(N) = \binom{N}{N} \times \eta^n$

Shovel presence

O I I I I I I

New production split

$P'I = O, P'2 = P2 / (I - PI)$ etc.

Probability of set $= \eta^{(N-I)} \times (I - \eta)$

Probability of one shovel missing

$P_{(N-I)}(N) = \binom{N}{N-I} \times \eta^{(N-I)} \times (I - \eta)$

Probability of n shovels missing

$Pn(N) = \binom{N}{n} \times \eta^{(N-n)} \times (I - \eta)^n$

η : Shovels' internal availability

Fig. 11. Shovels' availabilities.

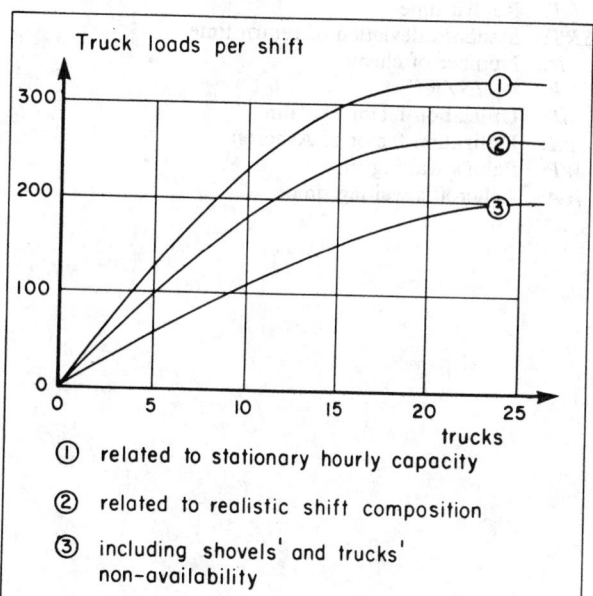

Truck loads per shift

① related to stationary hourly capacity

② related to realistic shift composition

③ including shovels' and trucks' non-availability

Fig. 12. Various capacity reductions.

cessive capacity reductions when including (1) the strict queue waiting time estimates, (2) adding the shift composition and initial waiting times after interruptions, and (3) the availabilities of shovels and trucks.

CONCLUSIONS

The calculation method described in this section provides a means of estimating the capacity of the loading and transportation of material in open pit mines utilizing trucks. It has been verified in practice (Elbrond, Piché, and Caines, 1979) and is easily adjustable. The computer programming is straightforward and computing time is short.

REFERENCES

Palm, C., 1947, "The Distribution of the Work Force for the Operation of Automatic Machines," *Industritidningen Norden,* Vol. 75, pp. 75-80, 90-94, and 119-123, in Swedish.

Ashcroft, H., 1957, "The Productivity of Several Machines under the Care of One Operator," *Journal of the Royal Statistical Society,* Series B, Vol. 12, pp. 13-27.

Posner, M., and Bernholtz, B., 1967, "Two-Stage Closed Queuing Systems with Time Lags," *Journal of Canadian Operations Research Society,* Vol. 5, pp. 82-99.

Elbrond, J., 1974, "A Procedure for the Calculation of Trucks' Waiting Time and Shovels' Efficiency," *Rapport Technique EP 74-R-13,* Ecole Polytechnique, Montréal, 14 pp.

Elbrond, J., 1977, "Calculation of an Open pit Operation's Capacity," SME-AIME Preprint 77-A0-357, 12 pp.

Elbrond, J., 1979, "Shovels, Trucks and Repair Bays: A Comprehensive Evaluation," SME-AIME Preprint 81-400, 5 pp.

Elbrond, J., Piché, A., Caines, R. E. G., 1979, "A New Procedure for the Calculation of an Open pit Operation's Capacity at the Carol Lake Operations of the Iron Ore Company of Canada," *Proceedings of "The 16th International Symposium on the Application of Computers and Operations Research in the Mineral Industries (APCOM),"* T. J. O'Neil, ed., AIME, New York, pp. 477-492.

APPENDIX I

Notations

- TS: Service time
- STS: Standard deviation of service time
- RT: Return time
- SRT: Standard deviation of return time
- N: Number of clients
- k: $= TS/RT$
- ρP: Utilization factor of Palm
- ρA: Utilization factor of Ashcroft
- WP: Palm's waiting time
- WA: Ashcroft's waiting time

WB: Waiting time in constant case
WS: Adjusted waiting time

TS, STS, RT, SRT, N
$\quad k = TS/RT$

$$\rho P = 1 - \left[1 + \sum_{n=1}^{N} \prod_{i=1}^{n} \{(N - i + 1) \times k\}\right]^{-1}$$

$$\rho A = \left[1 + \left\{N \times k \times \left(1 + \sum_{n=1}^{n-1} \prod_{i=1}^{n} (N - i) \times (e^{k \times i} - 1)/i\right)\right\}^{-1}\right]^{-1}$$

$WP = N \times TS/\rho P - (TS + RT)$
$WA = N \times TS/\rho A - (TS + RT)$
$WB = N \times TS - (TS + RT)$, if neg set $= 0$

$WS = WP \times STS/TS + 2 \times WA \times (1 - STS/TS)$
$\quad \times SRT/RT +$
$\quad WB \times (1 - STS/TS) \times (1 - 2 \times SRT/RT)$

APPENDIX II

Notations
Same as under Appendix I.
M: Number of parallel servers
H: Average number of occupied servers
ρP: H/M
TS, RT, N, M

$$H = \left[1 + \sum_{n=1}^{N} \prod_{i=1}^{n} \{(N - i + 1) \times k/m\}\right]^{-1} \times \left[\sum_{n=1}^{N} m \times \prod_{i=1}^{n} \{(N - i + 1) \times k/m\}\right]$$

when $1 \leq n \leq M - 1$ then $m = n$
when $M \leq n \leq N$ then $m = M$
$WP = N \times TS/(\rho P \times M) - (TS + RT)$

APPENDIX III

Notations same as under Appendix II.
1. Calculate by means of procedure under Appendix I.

$WP, WA, WB,$ and WS using a number of clients equal to N/M.

2. Calculate a correction factor.

$$CF = WS/WP$$

3. Apply this correction factor to the WP calculated using the procedure under Appendix II.

$$WS(N) = WP \times CF$$

6.6 Reclamation

6.6.1 Introduction

LEE W. SAPERSTEIN

By the nature of reserve exhaustion, surface mining is a temporal use of the land surface. The time span of mining can be blinkingly short as in a borrow pit for road building material or seemingly eternal as in some of the massive open pits for copper. Nonetheless, inherent in the nature of the mining process is the fact that there will be a time when mining ceases. Consequently, good mine design and good mining practice require that attention be paid to the post-mining use of the land formerly used by a mine. In this section, *reclamation* is defined as those operations that prepare mined land for postmining use. Reclamation also includes those steps that stabilize mined land in an environmental sense. Hence, reclamation is integral to a total plan for erosion and sediment control. Reclamation is not a single step supplementing mining, rather it is a series of integrated operations that begins with initial mine planning, continues through the extraction phase, and does not end until the new, postmining land use begins.

The emphasis of this section is on reclamation in active mining. The methods described are, of course, applicable to the reclamation of other disturbed lands such as refuse and waste piles or tailings dams. Many of the methods have been developed and tested during the reclamation of previously abandoned (derelict or orphan) mined lands. Although a rigid distinction is not made here between the reclamation of active and abandoned mine lands, note is made that others may do so. In British terminology (Down and Stocks), *restoration* means the return of newly mined land to postmining productivity, whereas *reclamation* means the recovery of derelict land (abandoned industrial land including that from mining) to usefulness. American usage of the word restoration has caused it to mean a strict replication of conditions existing before mining.

The operations of reclamation depend upon those applied scientific disciplines that make up the environmental sciences. They include soil mechanics for ground stability, soil science for soil productivity, hydrology and geohydrology for surface and ground water flows, agronomy for revegetation, ecology for baseline studies, and the social sciences for land-use planning. This section includes a review of those parts of the foregoing subjects necessary for good reclamation. Hydrology and geohydrology are treated in separate sections because they apply as well to considerations in mining other than reclamation.

Methods of reclamation have improved and there has been a substantial growth in their application in a relatively short period of time. Concurrent with this growth has been a substantial increase in reference literature. The bibliography in this section includes many general references that give a more extensive treatment of the subject than is possible here. The Office of Surface Mining (OSM) is charged with maintaining an information repository on surface mining. Information is retrieved from the repository by use of the SEAM system (USDI).

Barrett, J., et al., 1980, "Procedures Recommended for Overburden and Hydrologic Studies of Surface Mines," General Technical Report INT-71, Intermountain Forest and Range Experiment Station, Forest Service, US Dept. of Agriculture, 106 pp.

D'Appolonia Consulting Engineers, 1975, "Engineering and Design Manual: Coal Refuse Disposal Facilities," 1975 0-579-/601, Mining Enforcement and Safety Administration, US Government Printing Office, Washington, DC, 826 pp.

Down, C.G., and Stocks, J., *Environmental Impact of Mining,* John Wiley & Sons, New York.

Goldberg, E.F., and Power, G., 1972, "Legal Problems of Coal Mine Reclamation," 14010 F2U 03/72, US Environmental Protection Agency, March, 236 pp.

Grim, E.C., and Hill, R.D., 1974, "Environmental Protection in Surface Mining of Coal," EPA-670/2-74-093, US Environmental Protection Agency, October, 277 pp.

Johnson, W., and Paone, J., 1982, "Land Utilization and Reclamation in the Mining Industry, 1930-1980," Information Circular 8862, US Bureau of Mines, 22 pp.

Majumdar, S.K., Brenner, F.J., and Miller, E.W., eds., 1987, *Environmental Consequences of Energy Production: Problems and Prospects,* The Pennsylvania Academy of Science, Easton, PA, 531 pp.

Mills, T.R., and Clar, M.L., 1976, "Erosion and Sediment Control: Surface Mining in the Eastern U.S.," EPA-625/3-76-006, 2 Vols., US Environmental Protection Agency, October, 242 pp.

National Research Council, 1986, *Abandoned Mine Lands, A Mid-Course Review of the National Reclamation Program for Coal,* National Academy Press, Washington, DC, 221 pp.

National Research Council, 1981, *Surface Mining: Soil, Clay, and Society,* National Academy Press, Washington, DC, 233 pp.

National Research Council, 1981, *Coal Mining and Ground-Water Resources in the United States,* National Academy Press, Washington, DC, 197 pp.

National Research Council, 1975, *Mineral Resources and the Environment,* National Academy Press, Washington, DC, 348 pp.

National Research Council, 1974, *Rehabilitation Potential of Western Coal Lands,* Ballinger Publishing Co., Cambridge, MA, 198 pp.

Ramani, R.V., and Clar, M.L., *User's Manual for Premining Planning of Eastern Surface Coal Mining,* 6 Vols., US Environmental Protection Agency, Cincinnati, OH.

Schaller, F.W., and Sutton, P., eds., 1978, *Reclamation of Drastically Disturbed Lands,* American Society of Agronomy, Madison, WI, 742 pp.

US Dept. of the Interior, 1977-Present, *SEAMALERT* (Surface Environment and Mining Alert to Current Literature), Office of Surface Mining, Div. of Information and Records Management.

Veith, D.L., et al., 1985, "Literature on the Revegetation of Coal-Mined Lands: An Annotated Bibliography," Information Circular 9048, US Bureau of Mines, 296 pp., 805 entries.

Vogel, W.G., 1981, "A Guide to Revegetating Coal Minesoils in the Eastern United States," General Technical Report NE-68, Northeastern Forest Experiment Station, US Dept. of Agriculture, 190 pp.

6.6.2 Reclamation Planning

RAJA V. RAMANI, RICHARD J. SWEIGARD, AND MICHAEL L. CLAR

INTRODUCTION

Mining, in its various forms, has utilized 2.31 Mha (5.7 million acres) of land in the United States from 1930 through 1980. This represents 0.25% of the nation's total land area, far less than the amount of land devoted to any other use such as agriculture, urban development, national parks, wildlife refuges, forest service wilderness, or highways. Of the total area affected by mining, 47% has been reclaimed by the mining industry meaning that approximately 0.13% of the nation's land area is currently being used by the mineral industries, has been abandoned, or has been reclaimed by some organization other than the mine operator (Johnson and Paone, 1982).

Land disturbance by the mining industry can be classified in several ways. Two useful classifications involve (1) mineral commodities and (2) mining functions. From Fig. 1 (commodity distribution) it is apparent that coal mining is by far the largest land user in the mining industry. In Fig. 2 (functional distribution), the affected total land area is classified according to mining function. Surface mining is responsible for 85% of the total land utilized. Underground mining accounts for a relatively small percentage (5%). Mineral processing wastes occupy the remaining (10%) of the land (Johnson and Paone, 1982). Therefore, a large component of the efforts in reclamation planning, research, and regulatory controls is directed toward surface mining, particularly the surface mining of coal.

Since one of the primary goals of reclamation is to restore the land-use capability to disturbed land, reclamation planning is necessarily related to land-use planning. The scope of this section, therefore, goes beyond the strict sense of reclamation plan preparation as mandated by some federal and state legislative acts and addresses mineral resource utilization in conjunction with renewable resource conservation. The term *reclamation planning*, as used in this section, encompasses the consideration of natural and cultural factors that impact land use, economic and technical feasibility of reclamation plans, mining methods, and current planning processes employed by the mining industry and the public sector.

LEGAL BACKGROUND

Concern on government's part for effective reclamation of mined land has been evidenced through the years in both the passage of regulations and increased research and development efforts. Prior to 1965, only seven states had specific laws requiring postmining land reclamation. The first state surface mine reclamation act was adopted by West Virginia in 1939 followed by Indiana (1941), Illinois (1943), Pennsylvania (1945), and Ohio (1947). By 1977, when the Federal Surface Mining Control and Reclamation Act was passed, 31 of the 37 states with coal surface mining operations were exercising some degree of control over those operations. The first state laws basically addressed limited grading and revegetation of spoil piles. Over the years, more comprehensive state laws evolved through modifications and amendments to existing legislation. In the 1950s and 1960s, the laws were modified to include stricter grading requirements, soil con-

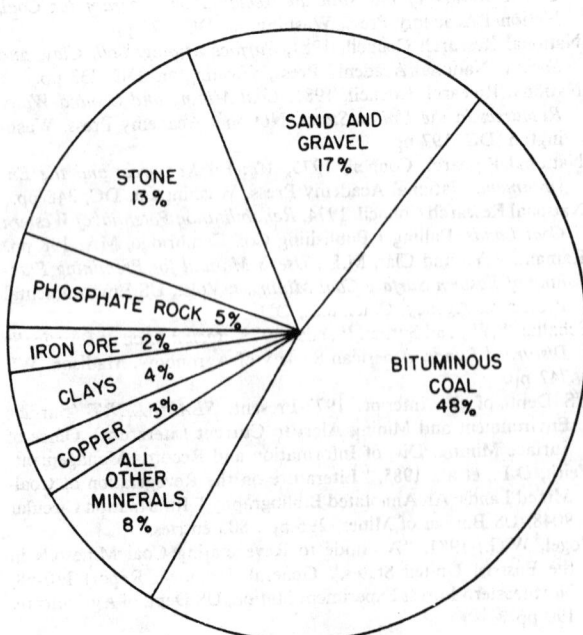

Fig. 1. Percentage of land used by selected commodity, 1930-1980 (Johnson and Paone, 1982).

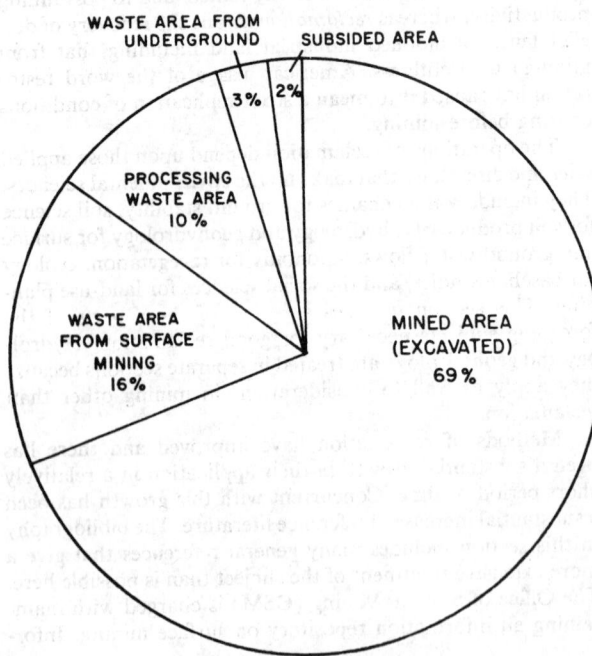

Fig. 2. Percentage of land used by mining function, 1930-1980 (Johnson and Paone, 1982).

servation measures, and water quality control. Modifications in the late 1960s and 1970s went beyond simple reclamation standards to the control of all major activities before, during, and after mining. Not all state programs addressed each of the major areas of concern. In addition, the levels of control on the same area varied from state to state. States with a long history of surface mine regulation had developed the most comprehensive and detailed regulatory programs.

By the early 1970s, it became clear that surface mining of coal would come under increasing scrutiny on environmental matters. The move was already underway to enact federal legislation (Imhoff, Friz, and LaFevers, 1976). In 1977, after nearly ten years of legislative efforts and three presidential vetoes, the federal government took a major step to standardize public policy with the passage of the Surface Mining Control and Reclamation Act (PL 95-87), hereafter referred to as the 1977 Act. Although the gradual evolution of various state laws and regulations had an impact on surface mine planning and design, PL 95-87 had immediate and far-reaching effects. With objectives of eliminating the detrimental effects that unequal standards had on interstate commerce and of providing adequate protection of all natural resources, the federal government implemented a plan that included, among other provisions, minimum performance standards and comprehensive data requirements for mining permit applications. This action has had many direct and indirect impacts on surface mine planning and design. It has also affected the way mining companies organize to perform these new or expanded functions.

The legal authority for regulating surface mine reclamation and for controlling land-use decisions originated and developed along different lines. The primary objective of surface mine reclamation is to preserve or restore the premining land-use capabilities. On the other hand, the general goal of land-use planning, as conducted by the public sector, is to protect the health, morality, and well being of the public. This is generally done through the formulation of plans that are intended to guide development, conserve resources, and prevent hazardous or conflicting uses of the land. Although the overall goals may differ, the relationship between surface mine reclamation and land-use planning becomes apparent when decisions must be made concerning the use of reclaimed surface-mined land. The jurisdiction over these matters can become confusing, particularly since it involves both public and private interests. A detailed review of the public policy or its impacts on reclamation planning is beyond the scope of this chapter. The following is abstracted from an expanded discussion by Ramani and Sweigard (1983).

The formulation of land-use plans and the imposition of land-use controls is considered to be a right of each state. Most states have, however, passed this authority to the local governments. Eleven states, however, have returned some form of control back to the central state government. This right does not apply to federal land. The Federal Land Policy and Management Act of 1976 required the Bureau of Land Management to prepare comprehensive plans for the management of all public lands. Attempts in the past to institute a national land-use policy, or even to pass legislation that might pressure states to require comprehensive planning affecting private land, have not been successful. Although the federal government has ostensibly avoided any land-use planning legislation that would interfere with privately owned land, the impact of several environmental bills on land-use planning is apparent. There exists a great amount of control on land development and management through a complex maze of federal, state, and local laws pertaining to air, water,

solid waste, wilderness preservation, wildlife protection, etc. (Office of Technology Assessment, 1979). As an example, in Fig. 3 the effect on the environment from each operation in coal mining and handling is illustrated. Federal legislation addressing each of these impacts is indicated in Fig. 4. A listing of agencies involved in the review and issuance of permits for surface coal mining operations is given in Table 1. Clearly, the move is toward some form of a common system of land-use planning and land development with participation from local, state, regional, and federal agencies.

The 1977 Act was concerned principally with the surface mining of coal. Surface mining of noncoal minerals is generally regulated under a variety of legislation on health and safety, environmental pollution control, and natural resource conservation at the federal and state level (COSMAR, 1979). Local authorities regulate surface mining through land-use controls that provide for public and professional participation, action, and enforcement (Curry and Fox, 1978). It is clear, however, that procedures mandated by the 1977 Act for coal will affect the attitudes of regulatory agencies toward the mining of noncoal minerals.

RECLAMATION PLAN

Several provisions of the 1977 Act and the regulations promulgated thereunder deal with the protection of the environment from surface mining damages. Two sections of the 1977 Act are important from reclamation and land-use planning aspects. Section 508—Reclamation Plan Requirements—is intended to ensure that there will be an approved postmining land use plan for the mined area. Section 522—Designating Areas Unsuitable for Surface Coal Mining—is to withdraw areas from surface mining. Specifically with regard to obtaining a mining permit, a reclamation plan must be submitted that shall be of such a detail to demonstrate that reclamation can be achieved as required by federal or state programs. The various components that are required by law in the reclamation and operations plan are listed in Table 2. Such a plan must cover, among other aspects, the following important points with regard to the uses of land:

1) The uses existing at the time of the application, and if the land has a history of previous mining, the uses that preceded any mining.

2) The capability of the land prior to any mining to support a variety of uses, giving consideration to soil and foundation characteristics, topography, and vegetative cover.

3) The use that is proposed to be made of the land following reclamation, including a discussion of the utility and capacity of the reclaimed land to support a variety of alternative uses and the relationship of such use to existing land-use policies and plans, and the comments of any owner of the surface, state, and local governments, or agencies thereof that would have to initiate, implement, approve, or authorize the proposed use of the land following reclamation.

4) A detailed description of how the proposed postmining land use is to be achieved and the necessary support activities that may be needed to achieve the proposed land use.

5) The consideration that has been given to making the surface mining and reclamation operations consistent with surface owner plans, and applicable state and local land-use plans and programs.

6) The consideration that has been given to developing the reclamation plan in a manner consistent with local physical, environmental, and climatological conditions.

The requirement that companies must submit a reclamation plan is not new. However, that surface mine planners

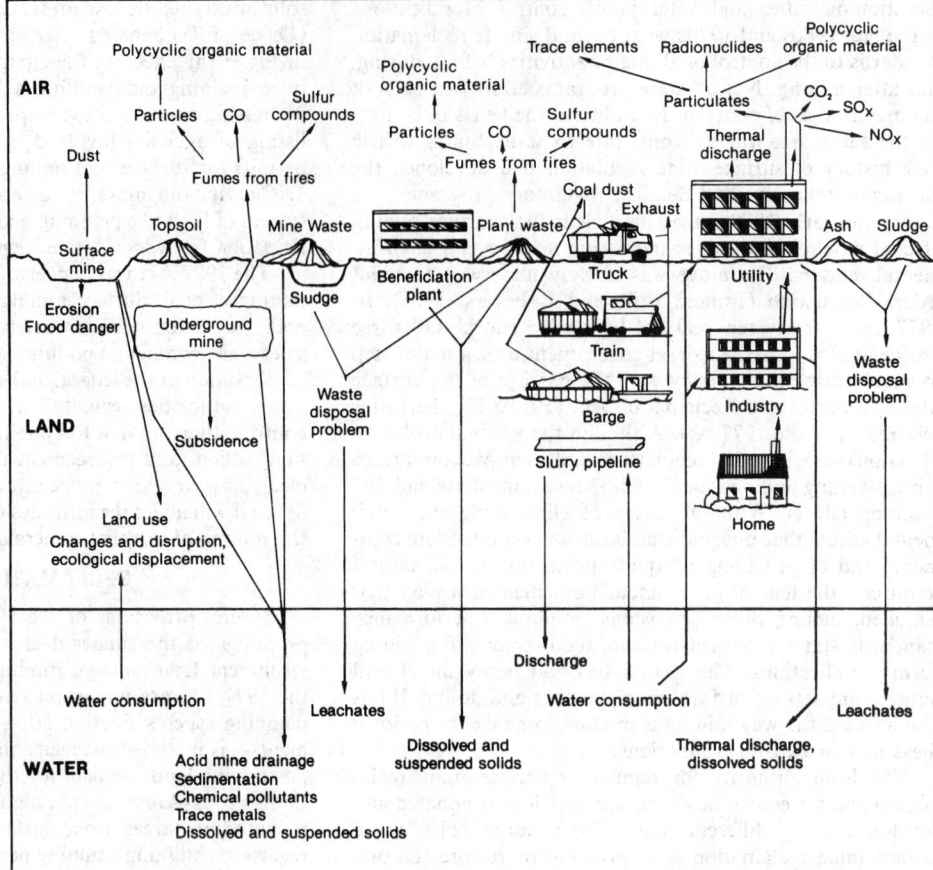

Fig. 3. Effects of coal-related activities on air, water, and land resources (Office of Technology Assessment, 1979).

must investigate "the utility and capacity of the reclaimed land to support a variety of alternative uses" [PL 95-87, Sec. 502(a)(8)] was not a requirement in most state laws. It is beyond the scope of traditional reclamation planning and enters into the area of environmental site planning or land-use planning. This provision of the law requires additional expertise in and the inclusion of many disciplines in the mine planning process.

In short, the planning of surface mines in general, and specifically that of reclamation and postmining uses of land has become very complex. Premining definition of long-range land-use plans for minable lands makes it possible to minimize any disturbance of the land that is not part of the normal mining or land-use plans and to limit reclamation activities to those required by the future-use plans. The integrated mining, reclamation, and land-use planning diagram, Fig. 5, defines information flow and interactions that should take place for the evaluation of mining, reclamation, and land-use plans before a mining plan is put into practice. The loops through which the mining and reclamation plans have to pass include not only federal, state, and local agencies, but special interest groups and the general public. The planning of surface mines thus is very involved and must be very thorough, with preserving or enhancing the long-term use of the land as a major objective. The application of an integrated mining, reclamation, and land-use planning concept requires an analysis of the interactions that must take place between the various levels of land-use planners. The development of procedures to evaluate land-use alternatives is also required (Ramani, 1978).

CURRENT RECLAMATION PRACTICES

Rehabilitation of surface-mined land is being accomplished through the systematic application of reclamation technology. Spoil banks that can be revegetated present minor problems and have great potential for development. Various reclamation programs that are being actively pursued include restoring the ground for agricultural and livestock farming, reforestation, recreation, and housing and industrial sites (Ramani, et al., 1977; Ramani and Grim, 1978). The possibilities for development under these conditions are limited only by cost-benefit consideration. Examples of reclaimed lands are shown in Figs. 6 through 10. There are, however, marginal and problem spoils (acid, toxic, etc.) requiring special attention and additional planning. Concern for good mining practices and effective land reclamation programs is evidenced in both the passage of regulations and increased research and development efforts. The various unit operations of reclamation are discussed elsewhere in this chapter; however, a time sequence of reclamation activities for surface coal mining is given in Table 3. This time sequence provides an orderly approach to grading, soil handling and stabilization, revegetation, and hydrologic control, all necessary for returning surface-mined land to productive uses.

PREMINING PLANNING

The planning problem with regard to surface mining is one of maximizing production and maintaining environmental quality, consistent with the constraints imposed by the regulatory agencies and the need to satisfy the profit objec-

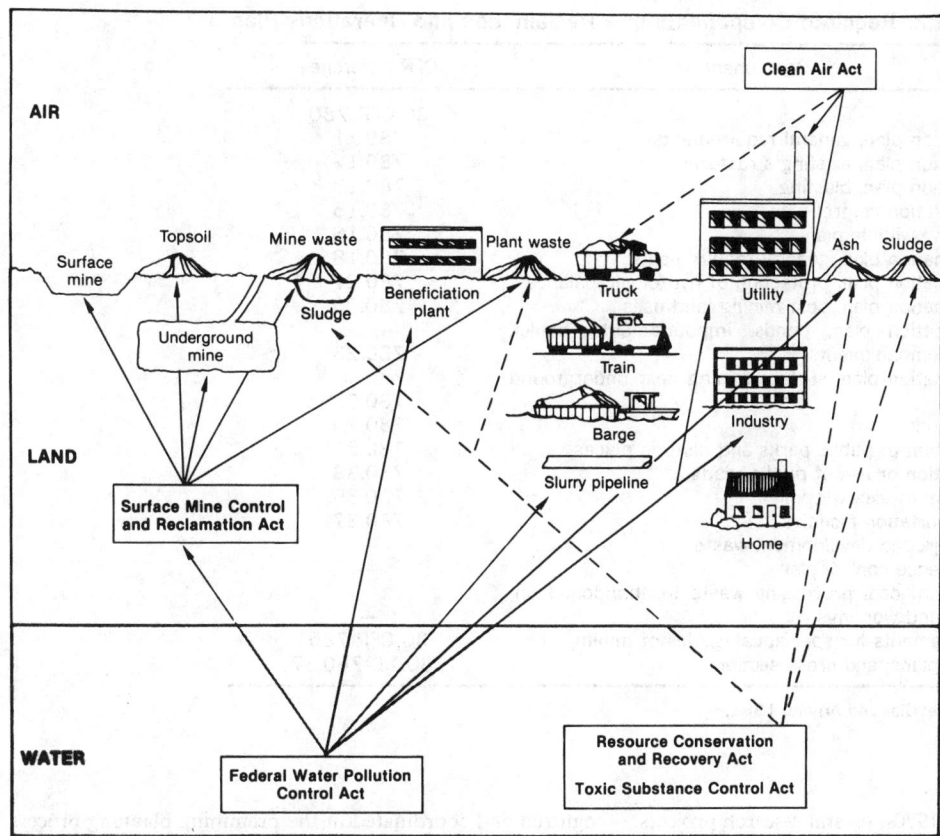

Fig. 4. Federal legislation to control the impacts on air, water, and land resources (Office of Technology Assessment, 1979).

Table 1. Listing of Agencies Involved in the Review and Issuance of Permits for Surface Coal Mining Operations.

Agency	Role	Regulatory Citation
1. Office of Surface Mining (OSM)	Regulatory authority lead agency	SMCRA (30 USC 1254)
2. State and Federal Fish and Wildlife Management Agency	Commenting agency	OSM Permanent Regulatory Program (30 CFR 779.20); Endangered Species Act of 1973 (16 USC 1531 et seq.); The Fish and Wildlife Coordination Act, as amended (16 USC 661, et seq.)
3. State Historic Preservation Officer and Advisory Council on Historic Preservation	Commenting agency	Historic and Archeological Data Preservation Act of 1974, as amended (16 USC 469, et seq.)
4. Environmental Protection Agency (EPA)	Commenting agency	Clean Water Act of 1977 (33 USC 1251, et seq.); Federal Clean Air Act of 1977, as amended (Classified to 42 USC 7401, et seq.); Resource Conservation and Recovery Act (42 USC 3251, et seq.)
5. Corps of Engineers	Commenting agency only if "navigable waters" are involved	Clean Water Act of 1977, Section 404 (33 USC 1251 et seq.)
6. Mining Safety and Health Administration (MSHA)	(Involvement is to be determined)	
7. State agencies (Note: the specific names of the state agencies will vary from state to state.)	The following state agencies could be involved in a commenting capacity: 1. Water Quality Control Agency 2. Water Resources Agency 3. Clean Air Protection Agency 4. Solid Waste Management Agency 5. Transportation (Highway) Agency 6. Dam Safety Agency 7. Hazardous Waste Management Agency 8. Others	
8. Local agencies	Local planning authority	

Source: Clar and Arnold, 1981.

Table 2. Required Components of a Reclamation and Operations Plan

Component	CFR Reference
	30 CFR 780
Operation plan, general requirements	780.11
Operation plan, existing structures	780.12
Operation plan, blasting	780.13
Air pollution control plan	780.15
Fish and wildlife plan	780.16
Reclamation plan, general requirements	780.18
Reclamation plan, protection of hydrologic balance	780.21
Reclamation plan, post mining land uses	780.23
Reclamation plan, ponds, impoundments, banks, dams, embankments	780.25
Reclamation plan, surface mining near underground mining	780.27
Diversions	780.29
Protection of public parks and historic places	780.31
Relocation or use of public roads	780.33
Disposal of excess spoil	780.35
Transportation facilities	780.37
Underground development waste	—
Subsidence control plan	—
Return of coal processing waste to abandoned underground workings	—
Requirements for special categories of mining	30 CFR 785
Maps, plans, and cross sections	780.11–780.37

Source: Clar and Arnold, 1981.

tives of the companies. In the 1970s, several research projects were initiated to aid in the premining planning of surface coal mining (Grim and Hill, 1974; Dames and Moore, 1976; Imhoff, et al., 1977; Ramani and Clar, 1978; Pugilese, et al., 1979). Mining actions use up some resources to produce minerals. The mining process, along with processes both upstream and downstream from mining, produce waste products; all these affect the environment. There are direct costs and revenues to the company and public from mining. Also, there are indirect costs and benefits. All these must be rec-

ognized and coordinated in the premining planning process. The necessary planning steps are to (1) make an inventory of the premining conditions; (2) evaluate and decide on the postmining requirements of the region, consistent with the needs and desires of the affected groups; (3) analyze alternative mining and reclamation schemes to achieve best the objectives; and (4) develop an acceptable mining, reclamation, and land use scheme that is the most suitable under the technical, social, and economic conditions (Ramani, et al., 1977; Riddle and Saperstein, 1978).

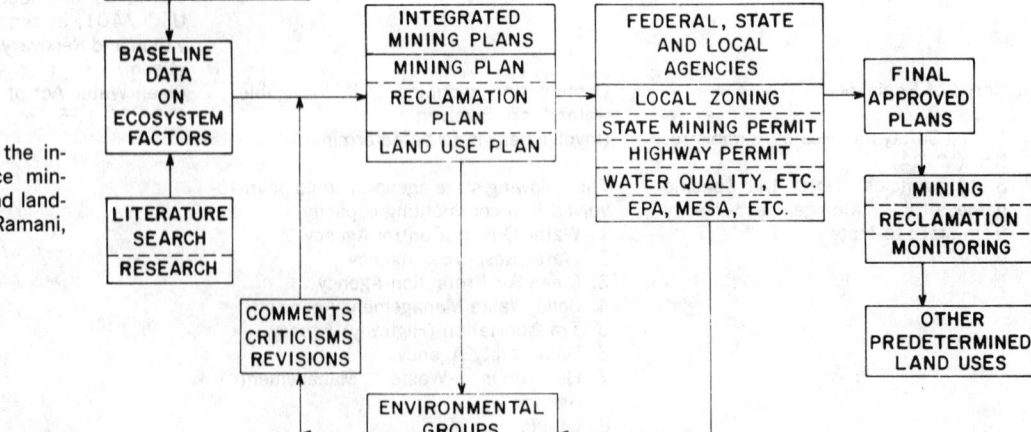

Fig. 5. Process of the integration of surface mining, reclamation, and land-use planning (Ramani, 1978).

Fig. 6. Modified area bituminous coal mine reclaimed for agricultural use (photo courtesy of Bradford Coal Co., Inc.).

Fig. 7. Bituminous area mine reclaimed for low-density residential use around final cut impoundment.

Fig. 8. Reclaimed sand and gravel pit used for high-density residential development.

Fig. 9. Public golf course as an example of recreational use of reclaimed modified area mine (photo courtesy of Hepburnia Coal Co.).

Information Needs

For planning reclamation, information is needed in several areas. The initial concern from the environmental standpoint was the ability to reclaim the land in a minimal manner; today, however, the environmental studies required prior to opening surface mines are much more detailed. Any attempt to extract mineral resources involves consideration of physical, biological, social, physiosocial, biosocial, and psychosocial factors of the mining environment. From the initial construction of access facilities for prospecting to the final completion of the mining and beneficiation operations, the ecosystem balance is continually being altered.

Since the ultimate goal of the reclamation plan is to ensure that the surface-mined land is returned to a productive use, it is appropriate to consider information needs for postmining land-use planning. Land-use factors can either be classified as natural or cultural. Important factors that must be considered are listed in Table 4. A brief discussion of the major factors follows. Detailed discussion of these factors can be found in Clar (1982), Clar and Ramani (in press), and Ramani and Sweigard (1983).

Natural Factors

Natural land-use factors include, at least, the geomorphic, climatic, hydrologic, and soil characteristics of a site. Although these characteristics can be altered by man, they were initially the result of nature.

Topography: Topographic relief is the difference in elevation between high and low points in a specific site or region. Relief and its relationships to climate, soils, hydrology, and plants are of concern to mine planners. The configuration or topographic relief of the land surface, the direction in trends of mountain chains, and the proximity to large water bodies must be considered together since these factors largely determine the direction of prevailing winds, humidity of the atmosphere, and amount of precipitation. Slope or gradient affects runoff and drainage, hence, water content of the soil (Toumey and Korstian, 1947). Topography influences the selection of the type of mining and mining equipment. Severe terrains also influence the methods of revegetation, and limit alternatives for future land use.

Climate: Extremes in climate affect the selection of equipment. Seasonal fluctuations determine desirable planting pe-

Fig. 10. Water-based recreational use of reclaimed area mine.

Table 3. Time Sequence for Reclamation Activities

I. During site preparation:
1. Install control measures (diversion, sediment traps and basins, etc.).
2. Clear and grub, marketing lumber if possible; stockpile brush for use as filters; run brush through woodchipper and use chips for mulch.
3. Stabilize areas around temporary facilities such as maintenance yard, power station, and supply area.

II. During overburden removal:
1. Divert water away from and around active mining areas.
2. Remove topsoil and store it if possible and/or necessary.
3. Selectively mine and place overburden strata if possible and/or necessary.

III. During coal removal:
1. Remove all coal insofar as possible.
2. For the purpose of controlling postmining ground-water flows, break, or conversely prevent damage to, the strata immediately below the coal seam as desired.

IV. Immediately after coal removal:
1. Seal the high wall if necessary.
2. Seal the low wall if necessary.
3. Backfill—bury toxic materials and boulders, dispose of waste, ensure compaction.

V. Shortly after coal removal:
1. Rough grade and contour, taking these factors into consideration:
   a. Time of grading—specific time limit; tied to advance of mining; seasonal considerations.
   b. Slope steepness.
   c. Length of uninterrupted slope.
   d. Compaction.
   e. Reconstruction of underground and surface drainage patterns.
2. If necessary, make mine spoil amendments (root zone), taking these factors into consideration:
   a. Type of amendment—fertilizers, limestone, fly ash, sewage sludge, or others.
   b. Depth of application.
   c. Top layer considerations—temperature (color), water retention (size consist, organics), mulching, and tacking.

VI. Immediately prior to first planting season:
1. Fine-grade and spread topsoil, taking seasonal fluctuations into consideration.
2. If necessary, manipulate the soil mechanically by ripping, furrowing, deep-chiseling or harrowing, or constructing dozer basins.
3. Mulch and tack.

VII. During the first planting season:
Seed and revegetate, considering time and methods of seeding, choice of grasses and legumes.

VIII. At regular, frequent intervals:
Monitor and control—slope stability; water quality, both chemical (Ph, etc.) and physical (sediment); vegetation growth.

Source: Ramani, 1978.

riods. In addition to the problem of permafrost in severely cold climates, operation of mobile equipment becomes difficult. Heavy rains can cause erosion, complicating maintenance of water quality control and establishment of vegetation.

Temperature acts as a constraint, particularly to agricultural and silvicultural land uses. Cyclic and seasonal changes in temperature determine the length of the growing season, potential evapotranspiration, precipitation occurring as snow, and frequency of freezing and thawing. These conditions are of prime concern for agricultural purposes, particularly, plant growth. A high freeze-thaw frequency accelerates weathering of overburden materials, increasing the infiltration characteristics of the materials.

Precipitation interacts with several other factors includ-ing atmospheric pressure, temperature, solar radiation, wind, humidity, evaporation, and the geology and geomorphology of an area. These interactions determine the potential for water resource availability, a major factor in determining the potential of an area for resource development. A water supply deficiency, such as that encountered in the arid regions, not only makes reclamation of mined lands difficult but may preclude any type of postmining development. In the humid regions, the situation is reversed. Here excessive rainfall can represent major design problems.

Altitude: Altitude, the height above sea level, has a pronounced effect on the climatic characteristics of a site. The atmosphere is less dense at higher altitudes and consequently is incapable of absorbing and retaining as much heat as is retained at lower altitudes. A rise of 91 m (300 ft) in altitude

Table 4. Information Needs for Reclamation and Postmining Land-Use Planning

I. Natural Factors
  A. Topography
    1. Relief
    2. Slope
  B. Climate
    1. Precipitation
    2. Wind—airflow patterns, intensity
    3. Humidity
    4. Temperature
    5. Climate type
    6. Growing season
    7. Microclimatic characteristics
  C. Altitude
  D. Exposure (aspect)
  E. Hydrology
    1. Surface hydrology
      a. watershed considerations
      b. flood plain delineations
      c. surface drainage patterns
      d. amount and quality of runoffs
    2. Ground water hydrology
      a. ground water table
      b. aquifers
      c. amount and quality of ground water flows
      d. recharge potential
  F. Geology
    1. Stratigraphy
    2. Structure
    3. Geomorphology
    4. Chemical nature of overburden
    5. Coal characterization
  G. Soils
    1. Agricultural characteristics
      a. texture
      b. structure
      c. organic matter content
      d. moisture content
      e. permeability
      f. pH
      g. depth to bedrock
      h. color
    2. Engineering characteristics
      a. shrink-swell potential
      b. wetness
      c. depth to bedrock
      d. erodibility
      e. slope
      f. bearing capacity
      g. organic layers

  H. Terrestrial ecology
    1. Natural vegetation, characterization, identification of survival needs
    2. Crops
    3. Game animals
    4. Resident and migratory birds
    5. Rare and endangered species
  I. Aquatic ecology
    1. Aquatic animals—fish; water birds, resident and migratory
    2. Aquatic plants
    3. Characterization, use, and survival needs of aquatic life system

II. Cultural Factors
  A. Location
  B. Accessibility
    1. Travel distance
    2. Travel time
    3. Transportation networks
  C. Size and shape of the site
  D. Surrounding land use
    1. Current
    2. Historical
    3. Land-use plans
    4. Zoning ordinances
  E. Land ownership
    1. Public
    2. Industry
    3. Private
  F. Type, intensity, and value of use
    1. Agricultural
    2. Forestry
    3. Recreational
    4. Residential
    5. Commercial
    6. Industrial
    7. Institutional
    8. Transportation/utilities
    9. Water
  G. Population characteristics
    1. Population
    2. Population shifts
    3. Density
    4. Age distribution
    5. Number of households
    6. Household size
    7. Average income
    8. Employment
    9. Educational levels

results in a fall of $-17°C$ (1°F) in temperature. This decrease in temperature largely accounts for the greater amount of precipitation on the windward side of mountains, ridges, or hills. The lowering of temperature, however, is greatly modified by the configuration of the land and by the air currents. Valleys, coves, and ravines may be more exposed to danger from frost than the adjacent slopes several hundred meters (feet) higher in elevation (Toumey and Korstian, 1947).

Exposure: Exposure or aspect refers to the direction of the slope of the land with respect to the points of a compass. The aspect of a slope interacts closely with altitude and slope angle to determine the amount of sunlight received by a site. Sunlight, in turn, modifies the moisture content and the temperature of the soil and air. A northern slope is considerably more moist and cooler than a southern slope. Exposure

influences plant growth chiefly through its impact upon temperature and soil water. Thus a slope exposed to the sun and wind often bears a different vegetation from one less exposed to either. Great differences may exist in temperature and atmospheric humidity on different exposures only short distances apart (Clar, 1982).

Hydrology: Hydrology, both surface and underground, is important for any reclamation program planning. A knowledge of water movement and quality in the mine area and its modification during mining is essential. In all cases erosion and sediment production are immediate concerns in the critical postmining, prevegetation time period, although their control during storms is a long-term problem that requires detailed planning.

Drainage systems are a complex component of the natural

environment and one that greatly influences the land-use suitability of a site. An analysis of an existing drainage system requires at least some understanding of stream course delineation, drainage basin delineation, the type of drainage pattern, drainage texture, slope, landform, and the ground water flow pattern. Drainage system analysis also adds to the understanding of basic bedrock structure and discontinuity patterns (Stranberg, 1967).

Surface mining operations create a significant impact on the drainage system. They necessitate the complete restructuring of the surficial and bedrock geology down to and slightly below the coal seam. Blasting and fragmentation of the consolidated strata and the spoiling sequence employed can result in substantial changes in numerous elements of the drainage systems. These generally include changes in the infiltration capacity of the surface soil and in the permeability, porosity, and conductivity of the subsoil and bedrock materials, and in the geochemical processes at work in these materials. Variations in the characteristics of these fundamental properties can result in marked changes in drainage patterns, watershed boundaries, and the availability of water supplies.

Geology: Geological aspects of the mineral occurrence (e.g., attitude, depth, thickness, stratigraphy, etc.) are the basis for establishing alternative mining and reclamation schemes, including slope stability. The nature of the overburden, especially those components such as sulfur that have potential environmental impacts, must be studied. The structural and stratigraphic evaluations affecting the hydrological conditions are critical for effective mining and reclamation. These conditions may impose limitations on potential land uses and may dictate that special reclamation practices be employed.

Soils: The identification of the material to be classified as topsoil is essential for their proper removal, storage, and postmining disposition. The characteristics of topsoil should be such that it ensures an effective growth of vegetative cover. Topsoil and subsoil, though not always the most ideal materials for supporting growth, do have certain advantages over other materials, including higher organic matter, better moisture-holding capacity, natural seed source, and ample soil microorganisms. The removal and storage of topsoil and

replacement of it on top of the graded spoil, in many instances, leads to better revegetation potential that, in turn, leads to better erosion control. In many older mining districts, spoil banks are only sparsely revegetated. This is due in part to the lack of nutrients in the spoil and the water flow characteristics of the restored strata. In cases where bedrock material, shale, sandstone, siltstone, and fireclay have been mixed with the original topsoil covering, inadequate fine-grained matrix material and organic matter may be available to provide nutrients or moisture-holding capacity. Too frequently, the fine-grained material sinks through the interstices in the coarse matrix. High infiltration capacity, high porosity, and high permeability are too commonly encountered when no attempt is made at top layer restoration and compaction. Color of the spoil has an effect on the surface temperatures of the spoil banks. Temperatures in excess of 66°C (150°F) have been reported on spoil surfaces composed of dark shales.

Summary: Natural land-use factors generally have a greater influence on low intensity land uses than on higher intensity uses. Natural factors are particularly important to agricultural, forestry, wildlife, and recreational uses. An estimate of the relative importance of each natural factor in determining land-use suitability is given in Table 5.

Cultural Factors

Cultural factors include all of those geographic, demographic, and economic characteristics that are the result of man's activities.

Location: Geographic location establishes the proximity to population centers and developed facilities, such as transportation systems, power supply, labor pool, manufacturing and supply services, and shops. All of these, if available, enhance the attractiveness of the site. The proximity to cities also increases the potential of the site for intensive postmining use. However, proximity to urban centers can also be an obstacle to new surface mine development due to adverse public pressure.

Accessibility: The value of a parcel of land is directly related to the ease or difficulty of access to roads and streets and the availability of transportation facilities as well as other utilities, including water and gas mains, sewers, and electrical

Table 5. Relative Importance of Natural Factors* as Determinants of Land Use Suitability

Natural Factors	Land Use Types						
	Forestry and Wildlife	Recreational	Agricultural	Residential	Institutional	Commercial	Industrial
Topographic relief	2	3	1	2	2	2	2
Slope	1	3	1	2	1	1	1
Altitude	2	3	2	3	3	3	3
Exposure	2	3	2	3	3	3	3
Drainage	1	3	1	1	1	1	1
Temperature	1	2	1	3	3	3	3
Precipitation	1	2	1	3	3	3	3
Consolidated overburden	2	2	1	2	3	3	3
Soils							
agricultural properties	2	2	1	3	3	3	3
engineering properties	3	3	2	1	1	1	1

Source: Clar and Ramani, in press.
*1 = Factor has high degree of influence on suitability of site for that particular land use.
 2 = Factor has moderate degree of influence.
 3 = Factor has low degree of influence.

and telephone lines. In addition, other services such as mail delivery; police and fire protection; and garbage, trash, and snow removal are significant (Weimer and Hoyt, 1966). The accessibility of a site is often measured by two parameters: travel distance and travel time.

Surface mining operations generally increase the accessibility of a given site. This increase is due to the addition of haul roads necessary for hauling the mined product to the markets or preparation plant for further processing. Sometimes rail lines are provided that may be restricted to onsite operations or that may provide a spur to a trunk line. It is also common to provide electric and telephone lines to the site. All of these positive aspects of the mining operations should be recognized (Clar and Ramani, in press).

Size and Shape of the Site: Size is an important factor in determining land use, and use in turn has a very direct bearing on income production. Size also determines effectiveness of use; a case in point is the trend toward larger farms as agriculture becomes more highly mechanized. Value, of course, is not directly proportional to size. Up to a point, additions to the size of a parcel of land tend to increase its income-producing capability and hence its value; thereafter, such additions tend to be of diminishing importance.

The shape of the land parcel may determine the possible uses to which it may be put and hence affect its income-producing capability. While lots or irregular shapes may be used to advantage for residential purposes, regularity is usually desirable for business, industrial, or agricultural uses.

Surface mine sites exhibit a great deal of variation in size and shape. They can vary in size from 4 to 6 ha (10 to 15 acres) of annual disturbance for a small operation to a range of from 120 to 200 ha (300 to 500 acres) of annual disturbance for large area mines. The shape of the site is also related to the type of mining system used. This is particularly true for contour mining operations that result in long, narrow strips of reclaimed land.

Surrounding Land Use: The status of the neighboring land, particularly with respect to type and intensity of use, and future trends has a strong influence on the suitability of the mine land to support various activities. In general, the use adopted at any specific site should be compatible with surrounding uses in terms of views, prevailing winds, protective belts of planting, open space, noise, and access routes.

Postmining land-use plans must comply with any existing local land-use plans. Compliance with such plans, if they exist, will generally insure compatibility with surrounding land uses.

Land Ownership: Most surface mined lands are privately owned. In 1973 ownership of these lands was distributed as follows (USDA, 1973):

Type of ownership	%
Public ownership:	
Federal	5
State/local	4
Subtotal	9
Private ownership:	
Mining industry	52
Farm	23
Other	16
Subtotal	91
Total	100

Ownership of the land and of mineral resources can often provide a source of significant conflict. In many instances, the surface and mineral ownerships have been severed. Often the right to extract the mineral resources is contingent upon providing a postmining land-use plan that is acceptable to the owner of the surface rights (Clar and Ramani, in press). It has also been observed that mining companies are more likely to propose an alternative land use for company-owned land than for leased land.

Type and Intensity of Use: A US Department of Agriculture (USDA) survey of mine lands showed that many were scattered small acreages best treated as part of the total conservation management of the farm and other areas with which they are intermingled (USDA, 1973). Nearly 80% of the sites were in forest, farm, or grassland or were reverting to forest at the time of the survey. These same uses were being made of land adjacent to 86% of the sites. Less than 2% of the acreage land had been set aside solely as outdoor recreation or wildlife areas; usually these are compatible with other uses of the land.

Surface mining operations are seldom permitted adjacent to highly urbanized areas. Typically the surrounding land will be undeveloped or at best used for agriculture, silviculture, and related activities. It must be kept in mind, however, that the land use selected may in turn restrict the development potential of adjoining property. Communities within the mining regions should carefully study and map their patterns of growth and expansion and seek to develop land use control mechanisms to reduce the loss of valuable mineral resources to urban expansion. Loss of sand and gravel and building stone deposits due to preemptive land development is a common occurrence and must be avoided.

Population Characteristics: Population characteristics of the surrounding area are key factors in determining land-use suitability. This information is provided decennially by the US Census Bureau and local planning agencies often prepare updated estimates between censuses. Although the total population is important, the percent change in population over recent years is a more useful parameter. Growth is the primary force behind land-use change. For an alternative postmining land use to be feasible, there must be some growth in the area, particularly for high intensity land uses, such as community development. Other demographic parameters such as population density, average household size, age distribution, and household incomes are useful in planning for postmining land use.

Summary: Cultural factors have a greater influence over land-use suitability for higher intensity uses such as residential, institutional, commercial, and industrial uses. These uses are already related to the proximity of population centers, particularly to the demand for services and products from these institutions. An estimate of the relative importance of each cultural factor in determining land use suitability is shown in Table 6.

LAND-USE ANALYSIS

The current land-use patterns may provide the best guidelines for use after reclamation. For example, forested hillside land adjoining a pasture may be more desirable as a pasture after mining if the need is for pasture lands. Similarly, orphaned mined land in a wilderness area may be more useful if restored as a forest. Future land-use plans contrary to premining use are unusual and not the common practice. Acceptance of mountaintop removal and valley fill mining methods that create new flat areas in regions where such areas are in short supply reflect the permanent positive po-

Table 6. Relative Importance of Cultural Factors* as Determinants of Land Use Suitability

Cultural Factors*	Forestry and Wildlife	Recreational	Agricultural	Residential	Institutional	Commercial	Industrial
Location	3	1	2	1	1	1	1
Accessibility	3	2	3	1	1	1	1
Size and shape of site	3	3	1	2	2	1	1
Surrounding land uses	3	2	3	1	1	1	3
Land ownership	3	2	3	2	2	2	2
Type and intensity of use	3	3	3	1	2	1	2
Population characteristics	3	2	2	1	2	1	2
Regulatory constraints	3	2	2	1	1	1	1
Company attitudes	2	2	2	1	1	1	1

Source: Clar and Ramani, in press.
*1 = Factor has high degree of influence on suitability of site for that particular land use.
2 = Factor has moderate degree of influence.
3 = Factor has low degree of influence.

tential of the temporary drastic disturbance of lands due to mining.

Identification of the current and probable future land uses can be summarized into the following five categories: (1) "wilderness" or unimproved use; (2) limited agriculture or recreation use with little development; (3) developed agriculture or recreation use such as for crop land, water sports, or vacation resorts; (4) suburban housing or light commercial and industrial use; and (5) urban housing or heavy commercial and industrial use.

Analysis of alternative long-term uses requires compliance with socioeconomic and public constraints as well as the economic goals of the mining company. Irrespective of the choice of short-term and long-term land use, the plans must be linked to the mining plan in space, time, and method. Definition of the long-range plans does achieve a purpose by minimizing disturbance unnecessary for the determined future use and limiting those reclamation activities that are not a part of the normal mining or land-use plans. Reference has already been made to the enactment of legislation at the federal and state levels pertaining to surface mining and reclamation. Mining companies today often have to meet

with nongovernmental, historical, wildlife, and other environmental groups to allay their concerns.

The integrated surface mining, reclamation, and land-use planning process illustrated in Fig. 5 requires delineation of the definitive tasks involved and identification of the locus of the responsibilities. In Fig. 11, two distinct land-use planning processes are recognized—comprehensive and site planning. The major tasks of each process are also identified. The respective roles of the public planners and the mining companies in these two processes are identified in Fig. 12. The site planning process itself is outlined in Fig. 13. In the site planning process, several feasible alternatives are to be evaluated through an iterative procedure applying social, economic, and environmental impact measurement techniques. This expanded portion of the detailed land-use planning process is illustrated in Fig. 14. Three case studies for reclamation and land-use planning for surface coal mines using the processes outlined in this section are reported by Ramani and Sweigard (1983). For the same mining area, there may be a number of feasible postmining land use alternatives. Also, since surface mining affords an opportunity to completely reshape the land forms, the potential for selectively

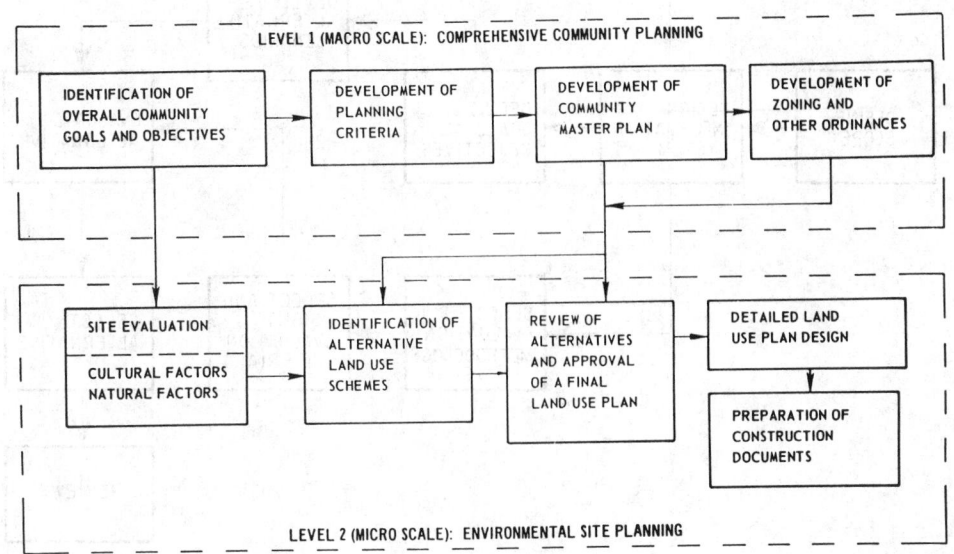

Fig. 11. Two levels of the land-use planning process.

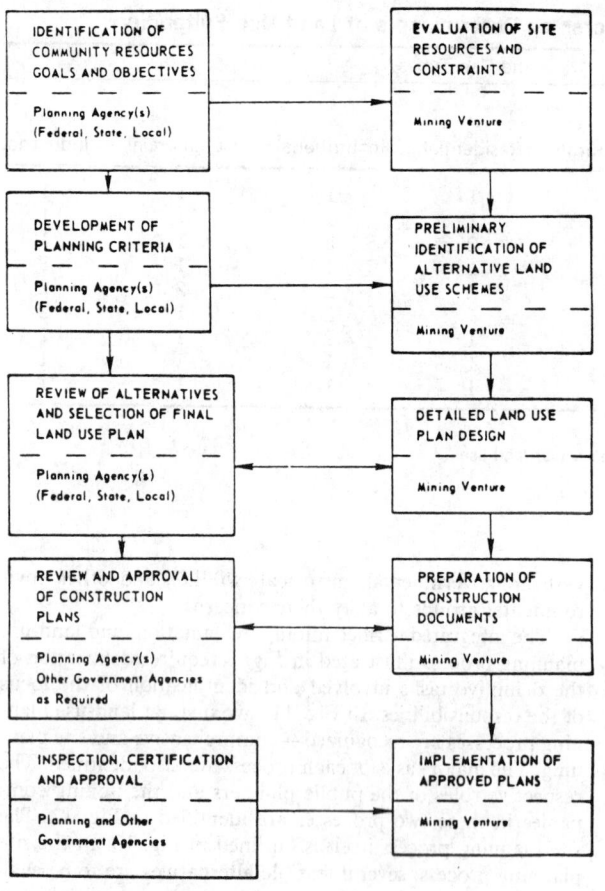

Fig. 12. Integrated public planning and mine planning.

reclaiming the land parcel for different end uses is great. Figs. 15, 16, and 17 show three different alternatives for reclamation and land-use development in a mining area. In practice, each plan will represent different amounts of earthmoving, grading, topsoiling, and vegetation planting for agricultural, wilderness, water and recreational, industrial, and residential uses.

The three case studies by Ramani and Sweigard (1983) indicate a general pattern of land-use planning and land management by the larger surface mine operators. In the formulation of postmining land-use plans, one of the most important factors is the postmining topography. Postmining topography is approximated by applying mine plan data, appropriate swell factors, and box and end cut locations to the premining topography. Another key factor is the location and thickness of high capability soils. This factor is particularly important in determining the location of agricultural lands. Location of property boundaries and public right-of-ways and limitations on certain equipment are also taken into consideration. For example, the grade limitation on mechanical tree planters may influence the location of forest land or the type of planting pattern that is employed. Premining land use, however, may play the most important role in determining the postmining land-use plan. Surface mine planners often endeavor to balance, as nearly as possible, the premining and postmining areas devoted to various land uses. Although the distribution of land among various uses may not change significantly, planning skills are required in designing a site plan that spatially orients the land uses in an efficient and aesthetically pleasing manner. Five primary postmining land-use designations are generally used: row crops, pasture/hay, forest, wildlife habitat, and water. Some additional land may be designated for public roads, etc. The planners may recognize a potential for higher intensity uses such as residential development or industrial uses; however,

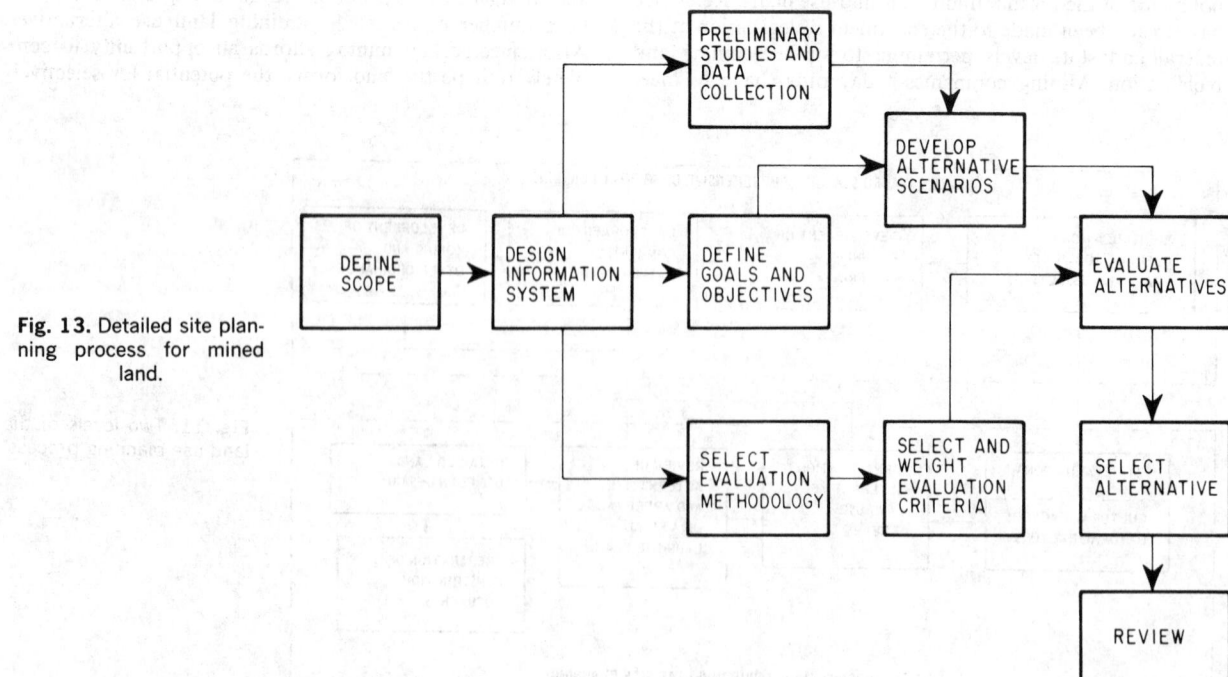

Fig. 13. Detailed site planning process for mined land.

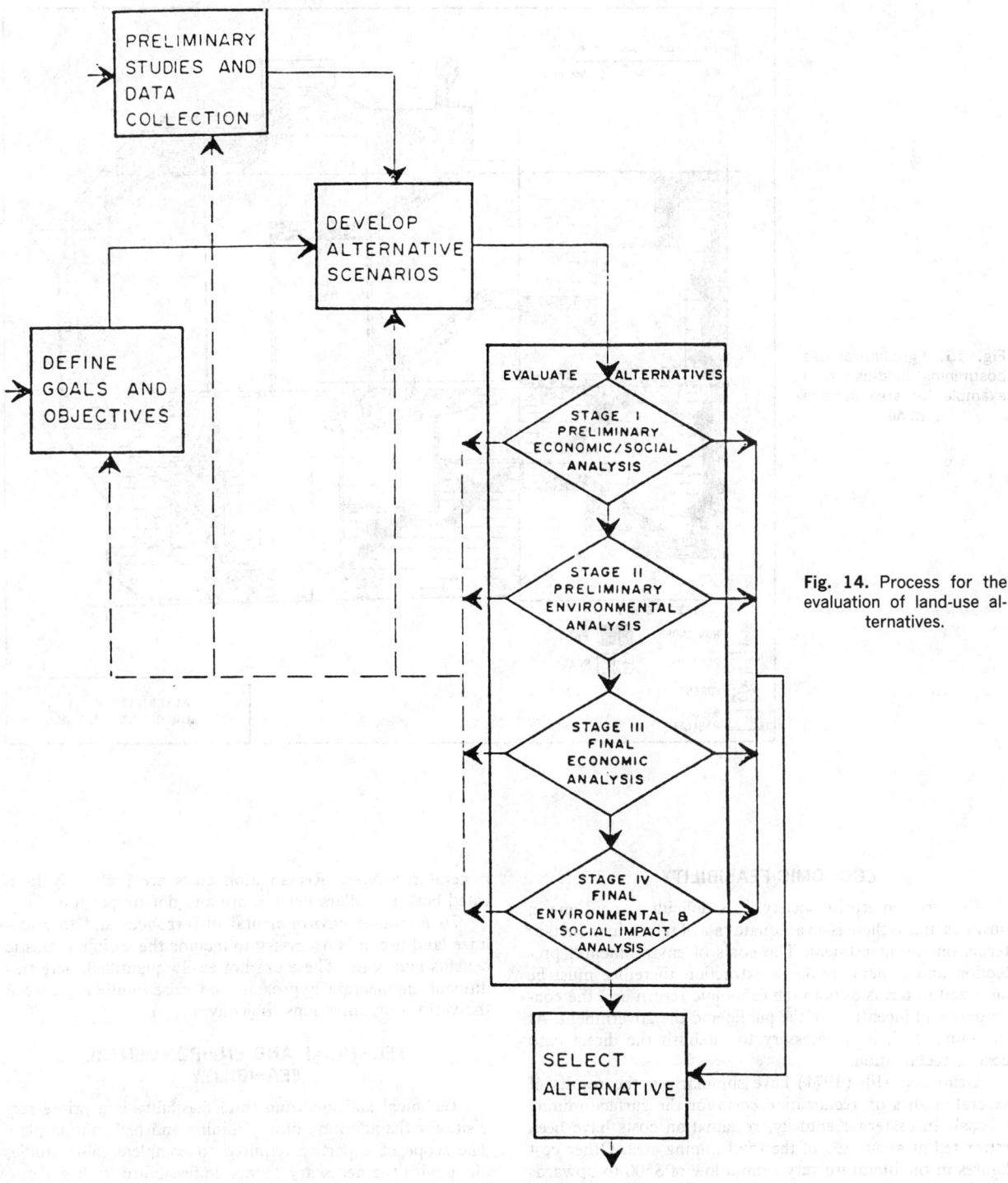

Fig. 14. Process for the evaluation of land-use alternatives.

these are not proposed as part of the required postmining land-use plan. Instead, the land is returned to one of the five primary uses until the bond has been released.

After all reclamation work has been completed, leased land is returned to the original owner. The company retains the right of access until the final bond release. Land that has been purchased by the company is generally retained for at least a few years. If possible, this land will be used for agricultural production. After a period of time, if it is de-

termined that the land is not needed for any future mining operations and is in fact a surplus, it may be considered for disposal. Disposal of surplus land is initiated by the land or real estate department. Before any land is sold or donated, input is solicited from the engineering department, the environmental engineering staff, the tax department, and the accounting department. After all of the necessary internal approvals have been obtained, the land is appraised and offered for sale.

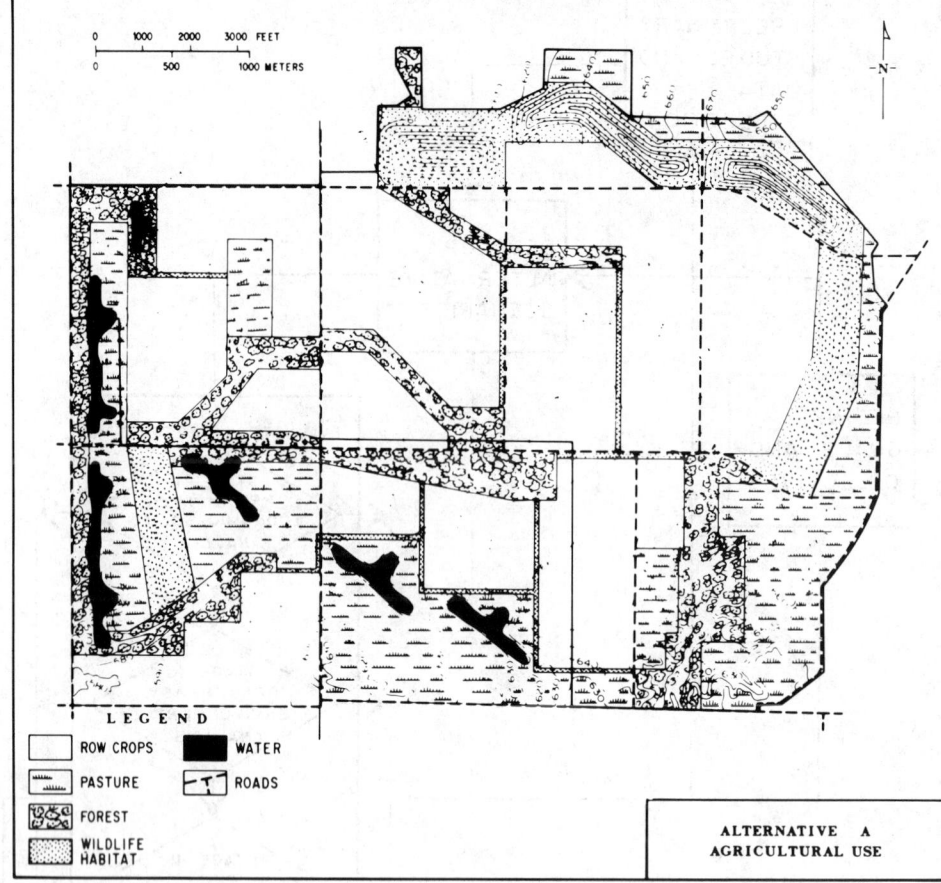

Fig. 15. Agricultural use postmining land-use plan example for area surface mine.

LEGEND

☐ ROW CROPS	■ WATER
PASTURE	⊢T⊣ ROADS
FOREST	
WILDLIFE HABITAT	

ALTERNATIVE A
AGRICULTURAL USE

ECONOMIC FEASIBILITY

In a free enterprise society, it is difficult to continue in business unless there is an adequate rate of return in monetary terms on the investment. The goals of environmental protection and mineral resource extraction therefore must be analyzed in terms of both the economic return and the constraints and incentives of the public and the government. At the mine level, it is necessary to establish the direct cash cost of reclamation.

Grim and Hill (1974) have summarized the results of several studies on reclamation costs for the surface mining of coal. In eastern Kentucky, reclamation costs have been estimated at about 8% of the total mining cost. Other cost figures in the literature vary from a low of $500 to upwards of $8000 per acre depending on the severity of the land effects and the reclamation requirements. Doyle and Chen (1973) have prepared a useful report on the analysis of pollution control costs. Their report addresses the costs involved in revegetation, acid mine drainage, stream diversion, strip mine spoil and waste, backfilling and regrading, etc.

Since reclamation costs are site-specific, the requirements of labor, equipment, and supplies for each site must be estimated. A convenient basis for estimation purposes may be identification of that volume of earth disturbed which must be contoured. The costs of reclamation are calculated for the disturbed land, and then assigned to the tons of coal or mineral recovered. Reclamation costs are preferably indicated both in dollars per hectare and dollars per ton.

To minimize environmental disturbances and to maximize land use, it is necessary to include the sociohumanistic benefits and costs. These are not easily quantified, defy traditional engineering approach, and raise conflicts between the various organizations' objectives.

TECHNICAL AND ENVIRONMENTAL FEASIBILITY

Technical and environmental feasibility is a prime requisite for the adoption of any mining and reclamation plan. The scope of expertise required to complete these studies along with the necessary economic feasibility studies is outlined in Table 7. This list illustrates the complexities that may be involved in the coordination of the overall premining planning effort.

It is preferable to closely coordinate the reclamation plans with the mining plans since nonconcurrent reclamation methods tend to be environmentally less optimal and more expensive. Reclamation efforts initiated long after the mining activity has ceased generally serve a limited land use. Concurrent reclamation with positive controls aids in the development of well-conceived future land uses with a long-term program to finish and maintain the reclamation plans. Concurrent methods allow total spoil manipulation and/or complete spoil profile construction (Stefanko, et al., 1973; Saperstein and Secor, 1973).

Mining Methods

Technical requirements of reclamation will vary depending upon the degree of restoration. In the standard surface mining approaches of today, selective replacement of material is relatively simple. Contour mining approaches such as block cut, box cut, haul back, mountaintop removal, and modified area mining are generally sufficiently flexible to allow segregation and selective replacement (Stefanko, et al., 1973; Saperstein and Secor, 1973; Grim and Hill, 1974; Ramani, et al., 1977). Larger scale area methods employing traditional overcasting are less flexible. However, selective placement of spoils in these cases can also be achieved by a combination of equipment and methods. For example, the requirement to save topsoil and spread it over the regraded land has led mining companies to use scrapers that remove the topsoil, run around the pit, and dump over formerly graded soil. Selection of shovel-truck systems, as opposed to conventional overcasting, also provides the needed flexibility in soil placement. In fact, the overburden conditions can be closely monitored, and the placement of the overburden achieved so as to ensure (1) burial of toxic spoils and less desirable material such as boulders or impermeable clays, (2) topsoil replacement, (3) desired stratigraphic sequence in the spoil, (4) flatter or other desired valley configurations, (5) desired quality of coal by blending in the pit, and (6) greater recovery of coal. The goal for premining planning and scheduling for resource extraction and reclamation is to develop such steps as an integral part of the mining method.

Backfilling

In backfilling, three major areas of long-term concern are (1) slope stability, (2) ground-water control, and (3) water quality maintenance (Phelps, et al., 1981). The primary source of slope failure is the prolonged action of water through erosion and mass wasting. The primary result of water's actions on slopes is erosion and sediment production. Long uninterrupted slopes are undesirable as compared to ones with terraces and diversions that decrease slope lengths and direct runoff water to safe outlets. Additionally, while backfilling, the permeability of the fill for water percolation can be modified. Such modifications are achieved by sealing or leaving the coal seam open, and by selective placement of the spoil. One of the concerns of water quality control in addition to the acid potential is the sediment. The interconnection between slope stability and water quality (sediment) is enhanced by apparently conflicting approaches to these problems. Gross stability is enhanced by establishing grade and cover to minimize water infiltration and permeation. Conversely, erosion and sediment control are achieved by encouraging infiltration and permeation to minimize runoff. The proper approach would entail, first, the establishment of as heavy a cover crop after backfilling as possible to encourage vegetative interception and evaportranspiration as opposed to surface or subsurface runoff. However, not all of the water can be handled in this manner. Provision must be

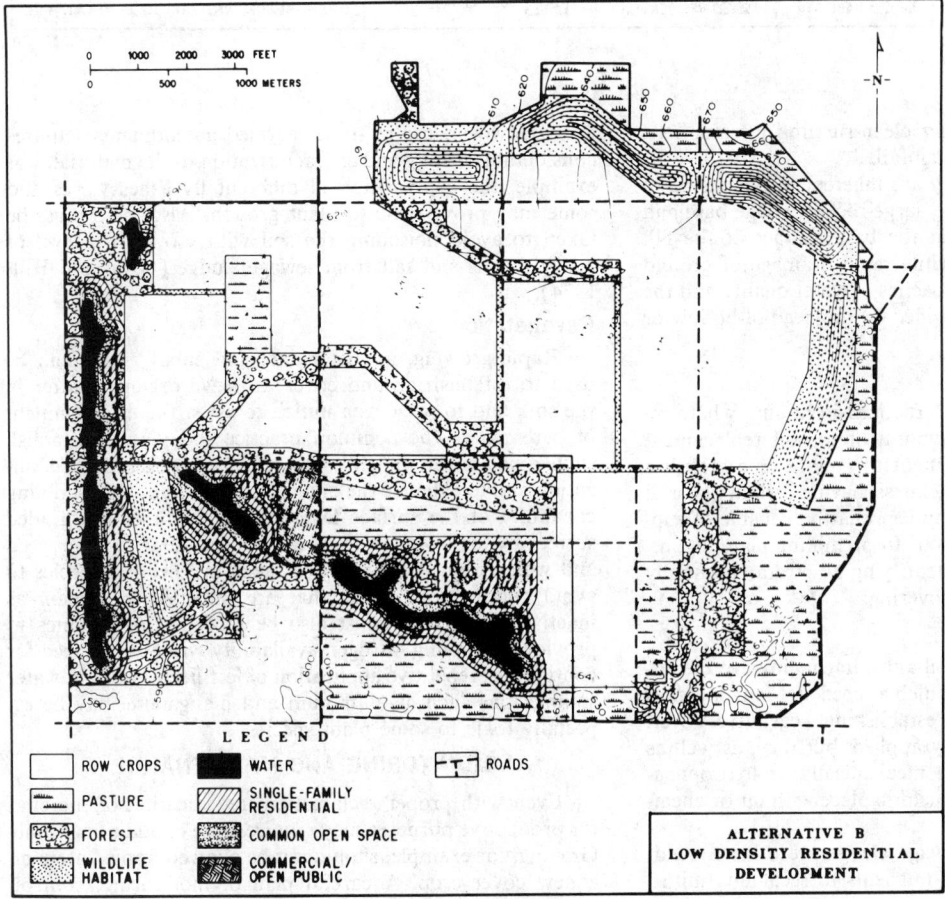

Fig. 16. Combined low-density residential use and agricultural use example for area mine.

LEGEND

- ROW CROPS
- PASTURE
- FOREST
- WILDLIFE HABITAT
- WATER
- SINGLE-FAMILY RESIDENTIAL
- COMMON OPEN SPACE
- COMMERCIAL / OPEN PUBLIC
- ROADS

ALTERNATIVE B
LOW DENSITY RESIDENTIAL
DEVELOPMENT

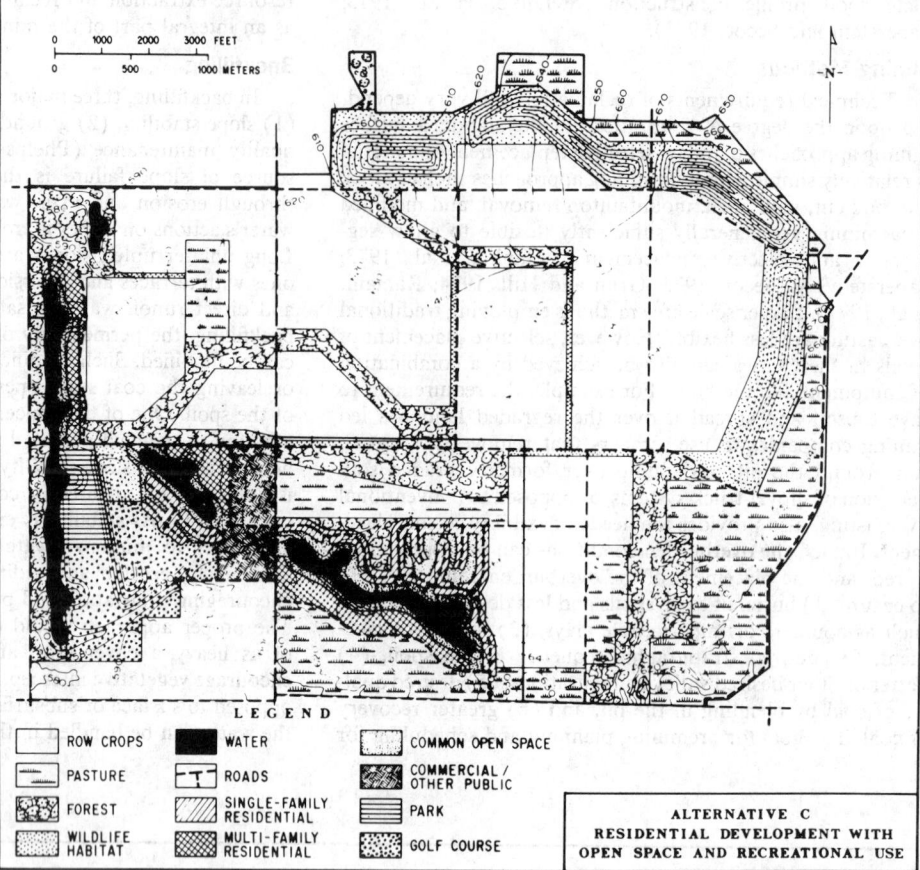

Fig. 17. Postmining land-use example showing residential development with open space and recreational use.

LEGEND

☐ ROW CROPS	■ WATER	COMMON OPEN SPACE
PASTURE	┬┬ ROADS	COMMERCIAL / OTHER PUBLIC
FOREST	SINGLE-FAMILY RESIDENTIAL	PARK
WILDLIFE HABITAT	MULTI-FAMILY RESIDENTIAL	GOLF COURSE

ALTERNATIVE C
RESIDENTIAL DEVELOPMENT WITH
OPEN SPACE AND RECREATIONAL USE

made for storms by designing clean-running interceptions and diversions for excessive rainfall.

Stability and water quality are inherent in the design of fills and backfills. If a suitable, large-rock drainage base and contact zone is established at the base contact of the fill, slippage can be minimized while rapid drainage of ground water is enhanced. Such approaches of water quality and the control of stability are exemplified by the head-of-hollow or valley-fill concepts

Topsoil

Replacement of topsoil is practiced in mining where required by regulations. The dictum of topsoil replacement could be viewed as a requirement to replace or establish a layer of material of specific thickness, having certain physical and chemical characteristics and, primarily, capable of supporting some specific vegetation. In premining planning, attention is to be directed at identifying all strata potentially adaptable as such a surface covering.

Soil Amendments

Common approaches to soil amendments involve the addition of lime, fertilizer, and mulch as needed. Many mulches have been used to assist in establishing vegetative cover. Straw or hay can be manually applied, but they, as well as wood chips, are often spread mechanically or hydropneumatically. Mulches should be held in place with tar or chemical emulsion.

Fly ash and stabilized sewage sludge are other amendments which can add needed nutrients to reclaimed mined land. However, use of these recycled amendments requires considerable testing and characterization of the material. For example, the composition of different fly ashes varies and some may prove toxic to plant growth. Also, care must be taken to avoid poisoning the soil with too high a level of heavy metals and salt from sewage sludge (Grim and Hill, 1974).

Revegetation

Rapid-growing, annual plants and small grains can be used to establish ground cover, to leave organic matter in the soil, and to serve as a mulch to assist the establishment of perennials. The common practice today is to establish such quick-cover mulch crops to assure vegetative stabilization and to enhance the organic content and water-holding capability of the surface layer. Scarification of the graded surface by furrowing or harrowing is also done to act as seed and water traps. Care must be taken in selecting species to avoid competition. Areas that are intended to remain as inactive grass cover must also be seeded with legumes to provide continuing nitrogen availability without the need for continuous, repetitive application of fertilizers. Certain water soluble ions such as aluminum and manganese may be especially toxic to some plant species.

MONITORING AND MAINTENANCE

Even with proper reclamation, rededication of the land for productive purposes may require lapse of additional time. Grazing, for example, should not be allowed immediately on a new cover crop. A careful plan of monitoring, control, maintenance, and development is required before the land is totally restored.

Monitoring slopes to assure stability and control and to maintain water quality is very important. To minimize slope stability and water quality problems initially and to prevent them from occurring in the future, at the time of grading it is necessary to keep all surfaces as minimally sloped as pos-sible. Additionally, through regular inspections, any failure or runoff must be corrected by refilling, compacting, and rapid establishment of cover. To control the erosion effects of flowing water, it is necessary to place interceptor and diversion channels to reduce down-slope flows.

Where coal mine drainage cannot be prevented, a long-term program to neutralize the discharge is necessary. In

Table 7. Areas of Technical Expertise Essential in Premining

Mine Planning Phase	Planning Activities	Areas of Specialization
Legal requirements analysis	Identification of regulatory constraints related to land use	*Land-use planner Attorney or paralegal specialist
Land and reserve acquisition	Prepare land use/land cover maps	Land-use planner Photogrammetrist/cartographer Plant biologist
	Prepare land ownership map	Photogrammetrist/cartographer Surveyor
Market development	Check market potential of site	Geographer Transportation engineer Land-use planner
Financial evaluation	Check if land development potential of the site will justify reclamation to a higher, more costly land use	Engineering economist Land-use planner Real estate specialist Fiscal planner
Coal beneficiation studies and plant design	Determine the impact of waste disposal on the postmining uses of land	Mineral processing engineer Environmental engineer Agronomist Geologist Hydrogeologist
Environmental impact studies	Evaluate the impact mining will have on the site with respect to capability and productivity	Mining engineer Environmental engineer Agronomist Geologist Hydrogeologist Terrestrial ecologist Plant biologist Agricultural engineer Archeologist Land-use planner Social scientist
Preliminary mine planning	Preliminary identification of postmining land uses	Mining engineer Land-use planner Agronomist Engineering economist
Permits acquisition	Land use information and postmining land use plan	Land-use planner Environmental engineer Agronomist
Administrative detail analysis	Submitted/approval of final land use plan	
Detailed mine planning	Detailed land use plan design	Land-use planner (specifically landscape architect) Mining engineer Environmental engineer Agricultural engineer Agronomist Hydrogeologist Plant biologist Engineering economist

*Refers to someone trained in regional planning, landscape architecture, or site design.
Source: Ramani and Sweigard, 1983.

most surface-mining situations, much of the necessary layout is present in sediment-control ponds. Limestone, lime, anhydrous ammonia (where no discharge to surface drainage is encountered), soda ash, sodium hydroxide, or other suitable neutralizing reagents are commonly used to treat acid mine drainage. However, isolation of toxic material and concurrent backfilling is probably the most desirable approach to minimization of acid formation problems.

One other aspect of maintenance involves the sediment and erosion control installations. With appropriate planning and no unusual storms, such maintenance of these features may be minimal. However, control structures may become clogged or filled, and should be cleared and cleaned when necessary.

SUMMARY

Reclamation, that primarily began as voluntary revegetation measures, has developed into a complex field encompassing economic, environmental, and social considerations and requiring significant planning effort. The evolution of surface mining regulations is partially responsible for this transition. Also, the recognition of surface mining as a temporary land use rather than a permanent land dedication has contributed to this development.

One of the most important steps in reclamation planning is data collection. Data collection by itself, however, is not enough. Rather, it is the careful analysis of the economic, technical, and environmental characteristics of the site and the mining operation that transform the data into useful information. The expansion of reclamation planning in scope and complexity has created a need for additional professionals in the mining industry to perform these necessary analyses.

Some efforts toward achieving the integration of mineral resource planning and land-use planning are underway (Ramani and Sweigard, 1982). A notable program is the US Geological Survey's Resource and Land Investigation (RALI) program. For example, the Missouri River Basin Commission along with the RALI program directed a project to help state and local planners develop data and methodology required to analyze problems associated with western coal and energy development (WCPAP, 1979). Another attempt at long-term planning with regard to mineral resources is the Model Mineral Reservation and Mine Zoning Ordinance prepared by the Wisconsin Geological and Natural History Survey (Preston, et al., 1974; Friz, 1975). This model code has dealt with problems of land-use planning but has not had wide acceptance because the resource characterization necessary to apply this type of code is generally lacking.

Demands on available lands will continue to increase due to the projected increase in population and industrial expansion. Though there is an upper limit on the available lands, the multi-use nature of the lands will make it possible to meet these demands. However, this will not happen without adequate planning. Land uses can be broadly categorized as nonrenewable and renewable. Lands with minable resources can be dedicated at some point in time to the extraction of resources. Such a dedication, however long-term, is only temporary because of the nonrenewable nature of the minable resources. Implied here is that the land will be available before and after mining for other uses. Planning for such lands must have the objective that, as far as practicable, no nonrenewable uses are preempted and no renewable uses are permanently destroyed.

REFERENCES

Clar, M. L., 1982, "An Analysis of Requirements and Guidelines for Surface Mine Land Planning," M.S. Thesis, The Pennsylvania State University, University Park, PA, pp. 30-105.

Clar, M. L., and Arnold, M. J., 1981, "Exploration and Permit Application Review/Approval Procedures for Coal Mining Operations Under a Federal Program," Final Report by Hittman Associates for US Office of Surface Mining Reclamation and Enforcement, Washington, DC.

Clar, M. L., and Ramani, R. V., User's Manual for Premining Planning of Eastern Surface Coal Mining, Volume 6: Mine Land Planning, EPA Grant No. R803882, University Park, PA (in press).

COSMAR, Committee on Surface Mining and Reclamation, 1979, "Surface Mining of Non-Coal Minerals—A Study of Mineral Mining from the Perspective of the Surface Mining Control and Reclamation Act of 1977," National Academy of Sciences, Washington, DC.

Curry, W. J., III, and Fox, C. A., Jr., 1978, "A Role for Local Governments in Controlling Strip Mining Activities," prepared for The Western Pennsylvania Conservancy, Pennsylvania Department of Community Affairs and Pennsylvania Department of Environmental Resources, Environmental Planning Information Series, Report No. 3.

Dames and Moore, 1976, "Development of Pre-mining and Reclamation Plan Rationale for Surface Coal Mines," 3 Vol. Final Report to US Bureau of Mines, Washington, DC.

Doyle, F. J., and Chen, C. Y., 1973, "Analysis of Pollution Control Costs," A Report to Appalachian Regional Commission by Michael Baker, Jr., Inc., Beaver, PA.

Friz, T. O., 1975, "Mineral Resources, Mining, and Land-Use Planning in Wisconsin," IC 26, Wisconsin Geological and Natural History Survey, Madison, WI.

Grim, E. C., and Hill, R. D., 1974, "Environmental Protection in the Surface Mining of Coal," EPA Publication No. 670-2-74-093.

Imhoff, E. A., Friz, T. O., and LaFevers, J. R., 1976, "A Guide to State Programs for the Reclamation of Surface Mined Areas," Circular No. 731, US Geological Survey, Reston, VA.

Imhoff, E. A., et al., 1977, "Integrated Mined-Area Reclamation and Land Use Planning," prepared for the Resource and Land Investigations (RALI) Program of the US Department of the Interior.

Johnson, W., and Paone, J., 1982, "Land Utilization and Reclamation in the Mining Industry, 1930-1980," Information Circular 8862, US Bureau of Mines, Washington, DC.

Office Technology Assessment, 1979, "The Direct Use of Coal," Congress of the United States, Washington, DC.

Phelps, L. B., et al., 1981, "Burial of Potentially Toxic Surface Mine Spoil," Final Report on Contract No. J0188136 to US Bureau of Mines, Washington, DC, 331 pp.

Preston, J., Strauss, E., and Friz, T., 1974, "Model Mineral Reservation and Mine Zoning Ordinance," IC 24, Wisconsin Geological and Natural History Survey, Madison, WI.

Pugliese, J. M., et al., 1979, "Quarrying Near Urban Areas: An Aid to Premine Planning," Information Circular 8804, US Bureau of Mines, Washington, DC.

Ramani, R. V., 1978, "Integration of Land Use Planning with Surface Mining," Earth and Mineral Sciences, The Pennsylvania State University, University Park, PA, Vol. 47, No. 5, pp. 33-39.

Ramani, R. V., and Clar, M. L., 1978, User's Manual for Premining Planning of Eastern Surface Coal Mining, Volume I—Executive Summary, EPA Pub. No. 600/7/78-180.

Ramani, R. V., and Grim, E. C., 1978, "Surface Mining—A Review of Practices and Progress in Land Disturbance Control," Chap. 14, Reclamation of Drastically Disturbed Lands, American Society of Agronomy, Madison, WI, pp. 241-270.

Ramani, R. V., et al., 1977, "Pre-mining Planning for Environmental Control in Surface Coal Mines," Preprint No. 77-F-387, Society of Mining Engineers of AIME, New York, 25 pp.

Ramani, R. V., and Sweigard, R. J., 1982, " Impacts of Land Use Planning on Mineral Resources," Preprint No. 82-418, Society of Mining Engineers of AIME, New York, 17 pp.

Ramani, R. V., and Sweigard, R. J., 1983, "Development of a Procedure for Land Use Potential Evaluation for Surface-Mined Land," 4-Vol., Final Report on Grant No. G1115428 to US Bureau of Mines, Washington, DC, 114 pp.

Riddle, J. M., and Saperstein, L. W., 1978, "Premining Planning to Maximize Effective Land Use and Reclamation," Chap. 13, *Reclamation of Drastically Disturbed Lands,* American Society of Agronomy, Madison, WI, pp. 223-240.

Saperstein, L. W., and Secor, E. S., 1973, "Improved Reclamation Potential with the Block Method of Contour Stripping," *Proceedings,* Research and Applied Technology Symposium on Mined-Land Reclamation, National Coal Association.

Stefanko, R., Ramani, R. V., and Ferko, M. R., 1973, "Analysis of Strip Mining Methods and Equipment Selection," Research and Development Report No. 61, Interim Report No. 7, Office of Coal Research, US Department of the Interior, Washington, DC.

Stranberg, C. H., 1967, *Aerial Discovery Manual,* John Wiley and Sons, Inc., New York.

Toumey, J. W., and Korstian, C. F., 1947, *Foundation of Silviculture Upon an Ecological Basis,* John Wiley and Sons, Inc., New York.

USDA, United States Department of Agriculture, 1973, "Restoring Surface-Mined Land," Miscellaneous Publication No. 1087, US Government Printing Office, Washington, DC.

WCPAP, 1979, "Western Coal Planning Assistance Project," 4 Vol., Final Report by Mountain West Research Inc. to Missouri River Basin Commission and US Geological Survey, Reston, VA.

Weimer, A. M., and Hoyt, H., 1966, *Real Estate,* 5th ed., The Ronald Press Co., New York.

ADDITIONAL REFERENCES NOT CITED

Adams, L. M., Capp, J. P., and Eisentrout, E., 1971, "Reclamation of Acidic Coal Mine Spoil with Fly Ash," Report of Investigation 7504, US Bureau of Mines.

American Law Institute, 1976, *A Model Land Development Code,* Washington, DC.

Avery, T. E., 1975, *Natural Resources Measurements,* 2nd ed., McGraw-Hill Book Co., New York.

Bondurant, D. M., 1971, *Proceedings,* Revegetation and Economic Use of Surface-Mined Land and Mine Refuse Symposium, Dec. 2-4, sponsored by School of Mines, College of Agriculture and Forestry, Appalachian Center, West Virginia University, Morgantown, WV.

Bosselman, F. P., and Callies, D. L., 1975, "The Quiet Revolution in Land Use Control," *Management & Control of Growth,* Vol. 1, The Urban Land Institute, Washington, DC.

Brady, N. C., 1974, *The Nature and Properties of Soil,* 8th ed., MacMillan Co., New York.

Bryson, R. A., and Hare, F. K., 1974, *World Survey of Climatology Volume II, Climates of North America,* Elsevier Scientific Publishing, New York.

Council of State Governments, 1975, "State Housing Actions: Programs and Alternatives," *Management & Control of Growth,* Vol. 3, The Land Institute, Washington, DC.

Dechiara, J., and Koppelman, L. E., 1978, *Site Planning Standards,* McGraw-Hill, New York.

Giles, R. H., Jr., ed., 1969, *Wildlife Management Techniques,* 3rd ed. revised, Wildlife Techniques Manual Committee, Wildlife Society, Washington, DC.

Hart, G., and Byrnes, W. R., 1960, "Trees for Strip-Mined Lands," Paper 136, 36, Northeastern Forestry Experiment Station.

Jensen, D. R., 1967, "Selecting Land Use for Sand and Gravel Sites," Report, Department of Landscape Architecture, University of Illinois, Urbana, IL.

Limstrom, G. A., 1960, "Forestation for Strip-Mined Land in the Central States," *U.S. Department of Agriculture, For. Ser., Agricultural Handbook No. 166,* 74, Forest Service, US Department of Agriculture.

Pennsylvania Research Committee on Coal Mine Spoil Revegetation, 1971, "A Guide for Revegetating Bituminous Strip Mine Spoils in Pennsylvania."

Pickels, G., 1970, "Realizing the Recreation Potential of Sand and Gravel Sites," Department of Landscape Architecture, University of Illinois under sponsorship of the National Sand and Gravel Association, Silver Spring, MD.

Rubenstein, H. M., 1968, *A Guide to Site and Environmental Planning,* John Wiley and Sons, New York.

Soil Conservation Service, 1969, "Standards and Specifications for Soil Erosion and Sediment Control in Developing Areas," Maryland Soil Conservation Service, US Department of Agriculture.

US Bureau of Mines, 1976, "Evaluation of Current Surface Coal Mining Overburden Handling Techniques and Reclamation Practices: Phase III: Eastern U.S.," Contract Report S0144081, July.

US Department of Agriculture, 1951, *Soil Survey Manual,* USDA Handbook No. 18, Soil Survey Staff, Bureau of Plant Industry, Soils and Agricultural Engineering, Agricultural Research Administration, US Government Printing Office.

US Department of Agriculture, 1973, "Restoring Surface-Mined Land," Miscellaneous Publication No. 1087, US Government Printing Office, Washington, DC.

US Geological Survey, 1976, *Topographic Maps, Tools for Planning,* US Government Printing Office.

Witmer, D. B., 1966, "Soils and Their Role in Planning a Suburban County," Chap. 3, *Soil Surveys and Land Use Planning,* L. J. Bartelli, et al., eds, Soil Science Society of America and American Society of Agronomy, Madison, WI.

6.6.3 Unit Operations of Reclamation

L. B. Phelps

INTRODUCTION

The unit operations of reclamation are generally different for open pit and quarry mining and for strip mining. In open pit and quarry mining, they usually are (1) pit slope reduction, (2) pit slope stabilization, (3) vision barriers, and (4) management of permanent rock and waste dumps outside the mining area. In contrast, strip mining reclamation unit operations consist of (1) backfilling the pit, (2) leveling the backfill, (3) providing stable contoured outside waste dumps, (4) topsoil handling, and (5) revegetation. Each of these unit operations will be dealt with in the context of appropriate mining type.

OPEN PIT MINING

Major reclamation planning for open pit mines is centered around the design and construction of outside rock dumps, waste piles, heap leaching piles, tailings ponds, and to a much lesser extent the pit itself. In the management of rock and waste dumps, it is necessary to plan for permanent structures. As such, they will remain in place when mining ceases at the pit. Therefore, major factors to be considered in design for reclamation are also related to their operation. The dumps should be constructed at the head of valleys in zones where there is, at most, intermittent stream flow. Construction of the dumps in such locations minimizes their visual impact on the area and provides the minimum surface area to be stabilized and revegetated. Where it is not possible to construct the dumps in such valley heads, the dumps should be designed and shaped to blend in with the topography. By such blending, it is more cost effective to establish the level of revegetation and other requirements to provide a reclaimed mine area.

Waste dumps or piles should be designed for the life of the reserve. Experience shows us too many piles that have had to be moved because they were located, in the first instance, in what became an awkward place. Typical practice is to start dumps close to the pit and then spread them away from the pit. This, over time, can lead to land-use conflicts with growing communities. Where the potential for land-use conflicts exists, it may be preferable to locate initial waste disposal sites at the ultimate boundaries of the mine property and have them grow back toward the mine and away from the community. Properly designed, stabilized, and reclaimed piles can, in this manner, be added to the stock of developable land rather than subtracting from it.

Reclamation of the pit is generally taken as providing stable slopes and allowing the pit to fill with water.

QUARRY MINING

A quarry is a specialized open pit operated to extract construction materials. As such, it is often found close to the community whose building needs its supplies. Therefore, reclamation of quarries must be concerned with general neighborliness—avoidance of undue noise, visual assault, dust, and risk to the populace—as well as environmental control.

Final reclamation of the quarry is generally limited to providing for stable slopes. A common way to refer to these is as maximum natural stable slopes when saturated. This refers to slopes that do not need the use of rock bolts, tensioned cables, or wire mesh to prevent rock falls or other slope failures. As an example, in Pennsylvania the height of each rock face is limited to a maximum of 15 m (50 ft) with a minimum 7.6-m (25-ft) bench. Careful mine scheduling will bring a large percentage of the quarry slopes into final configuration before exhaustion of the property. This minimizes unrecovered reclamation costs. A hypothetical example of mine scheduling to reduce final reclamation costs in quarrying follows:

Where water is to be impounded in the mined out quarry, the slope of the pit walls above the low water line should approximate the original contour or slope to a maximum of 35° measured from the horizontal. There should be a bench below the lowest water level that allows safe exit from the impoundment.

In addition to these requirements for the pit, visual barriers of mounded topsoil stored in the property line setback area or of permanently stored wastes should be used to restrict visibility of the pit from well-traveled areas. These mounds also reduce noise emanating from the quarry and its physical plant while the quarry is in operation. The mounds should be physically stable and planted with ground cover. Where there is sufficient space, and when the mounds are permanent, several rows of evergreen trees should be planted. The rows of trees should be staggered and mixed with a number of low-growing shrubs. To provide shielding in the winter and when the trees reach maturity, evergreens of a variety retaining limbs near the ground should be selected for the plantings.

STRIP MINING

The major difference between strip mining and open pits and quarries is that the strip mine tends to be developed promptly to its full depth and then progresses horizontally as the resource is recovered. Thus, in the progression of mining, considerable land surface is disturbed and large strip excavations result. These mined-out pits become ideal storage locations for the spoil being generated in the adjacent mining area. The characteristics have resulted in the current requirement that the mined-out pits be filled and returned to approximate original contour. Thus the reclamation process in strip mining is much more complex with several unit operations taking place simultaneously that directly impact on each other from a cost, operation, and planning standpoint. These unit operations are (1) backfilling, (2) leveling,

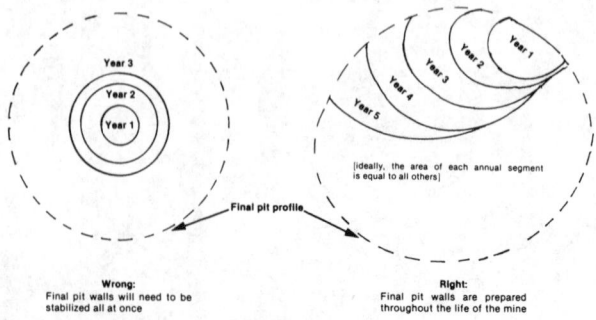

Wrong:
Final pit walls will need to be stabilized all at once

Right:
Final pit walls are prepared throughout the life of the mine

[ideally, the area of each annual segment is equal to all others]

Final pit profile

(3) development of stable outside spoil areas, (4) topsoil handling, and (5) seeding. A sixth, and continuing, operation is maintenance of reclaimed sites.

Backfilling.

In these bedded deposits, the newly removed overburden is placed as spoil in the adjacent mined-out area. This is the backfilling process. The more overburden that can be placed in the mined-out area, the smaller are the outside dumps and frequently the more cost effective the mining process.

In area strip mines, the percentage of total overburden that must be placed outside the mined area is usually small, sometimes 1% or less of the total overburden to be removed. The percentage that must be placed outside the mined area in mountaintop removal, in contrast, is substantial. Depending upon the steepness of the mountain slopes, the width of the mountain at the lowest seam, the number of seams mined, and the swell of the overburden, as much as 60% of the overburden must be placed outside of the mined areas, typically in valley fills (Cook and Kelly, 1976).

Current laws require that the mined-out areas of strip mining be returned to approximate original contour. However, there is a provision for modification of this requirement where a change in postmining land use is approved. In many such cases, this means larger outside dumps.

Where erosion control structures are to be used, they should be incorporated into the backfilling plan so that they can be constructed during the backfilling operation. These structures include erosion-control bench terraces and other structures such as down drains and reverse slopes.

In area strip mining, backfilling begins with the box cut being opened to the resource. Spoil from this cut must be placed on the surface or in another mined area to complete a fill. In the latter case, a permit must have been issued to allow the spoil to be placed there. Two methods of surface placement using dragline sidecasting are shown. In Fig. 1

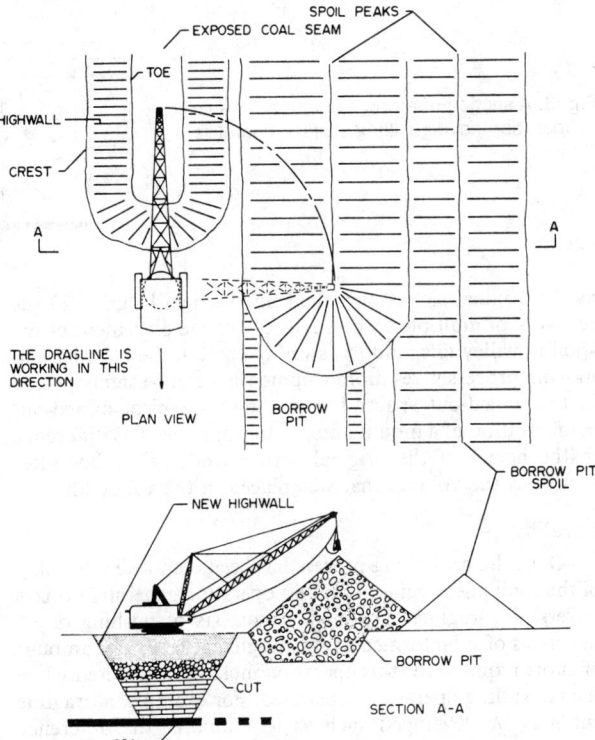

Fig. 2. Plan and cross-sectional view of a box cut being produced using the rehandle (borrow pit) method (Bucyrus-Erie Co.).

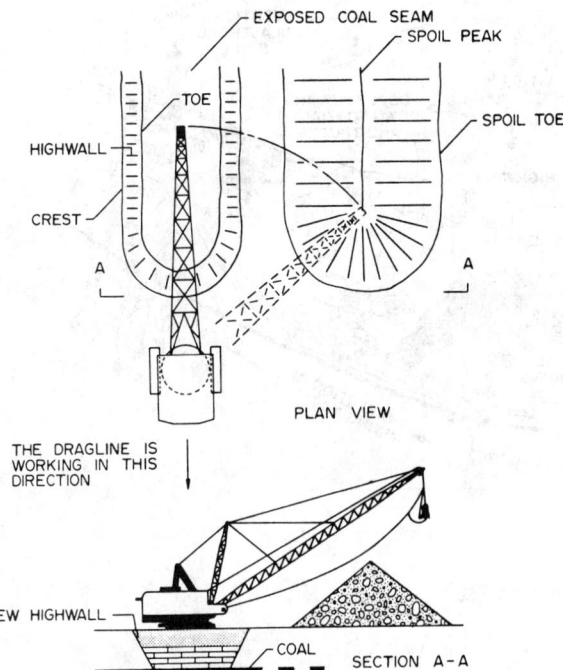

Fig. 1. Plan and cross-sectional view of a box cut, end cut method (Bucyrus-Erie Co.).

the overburden is not deep and can be placed directly on the surface. In Fig. 2, where the overburden is deep, the placement procedure is dictated by the type and size of equipment being used.

If a truck haulage system is being used to open the box cut, the spoil location becomes much more flexible. In such cases, consideration should be given to hauling the box cut overburden to a storage position next to the planned final cut. By such placement, it can be used to fill in the final cut and thus reduce or avoid a depression there while at the same time minimizing or avoiding having to reclaim outside box-cut spoil dumps.

The spoiling procedure for intermediate cuts is simply to cast the overburden in the mined-out area of the previous cut. The stacking ability of the spoil, the amount of spoil being placed in the mined-out cut, the space available in the mined-out cut, and the spoiling equipment used all are factors in the backfilled spoil profile. For dragline simple sidecasting, the mined-out spoil profile is determined using a range diagram such as shown in Fig. 3.

In contour mining, backfilling also takes place by continuously filling in the mined-out areas. The major concern in this type of strip mining is that filling in the mined zone to approximate original contour requires careful planning of spoil handling to allow sufficient material to be available for the final cuts. This aspect of contour mining is critical to the backfilling step if the best mining costs are to be achieved. Continuous backfilling is shown in Fig. 4. Intermittent backfilling, or block cut, is shown in Fig. 5.

In mountaintop removal mining, backfilling seldom achieves the approximate original contour. There are several reasons for this: (1) spoil slopes very seldom can be as steep

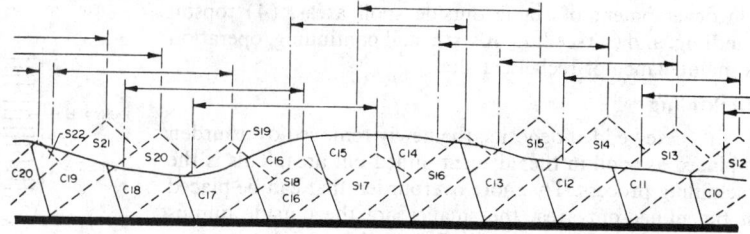

Fig. 3. A section of a range diagram for a large operation having rolling surface contours.

as the original slopes over the same vertical height, (2) the recovery of multiple seams requires outside placement of the spoil in valley fills, and (3) swell of spoil in the blasting and moving process gives more volume than can be safely placed in the mined-out space. Fig. 6 shows a typical mined-out configuration of a mountaintop removal mine. The difference in the height of the original surface and the surface after mining is the volume that was placed in the valley fill.

Leveling.

Once the backfilling process has been completed, leveling of the spoil piles begins. This unit operation generally utilizes dozers. In dragline cast spoil, it consists of pushing down the crests of spoil formed by the dragline, Fig. 7. The amount of effort required for this operation must be considered when the backfilling operation is planned. For example, a dragline cut may be developed such as to minimize the difference between spoil peaks and valleys. Fig. 8 illustrates this and how changes in the cut width, all other parameters remaining

the same, will materially affect the amount of material needed to be moved in the leveling operation.

In addition to cut width, another variable, the swing angle, needs to be optimized in dragline backfilling. Through variable, but planned, swing angles, the spoil peaks can be spread and reduced in height and thereby help to minimize the volume of spoil to be leveled.

Another piece of equipment having application in soil or soft rock overburden removal is the bucket wheel excavator. A closed or open connected machine can be used to advantage to level the peak/valley formation created by a dragline. The stacker for either type of machine can be positioned such that it fills in the valleys and thereby greatly reduces the effort required in the leveling unit operation. In addition, this machine may be used to keep soil and soft rock overburden on top of the zone being backfilled and leveled.

Finally a shovel/truck advance-bench operation stripping ahead of a dragline can also provide a very flexible spoiling operation. Selective placement of overburden can be

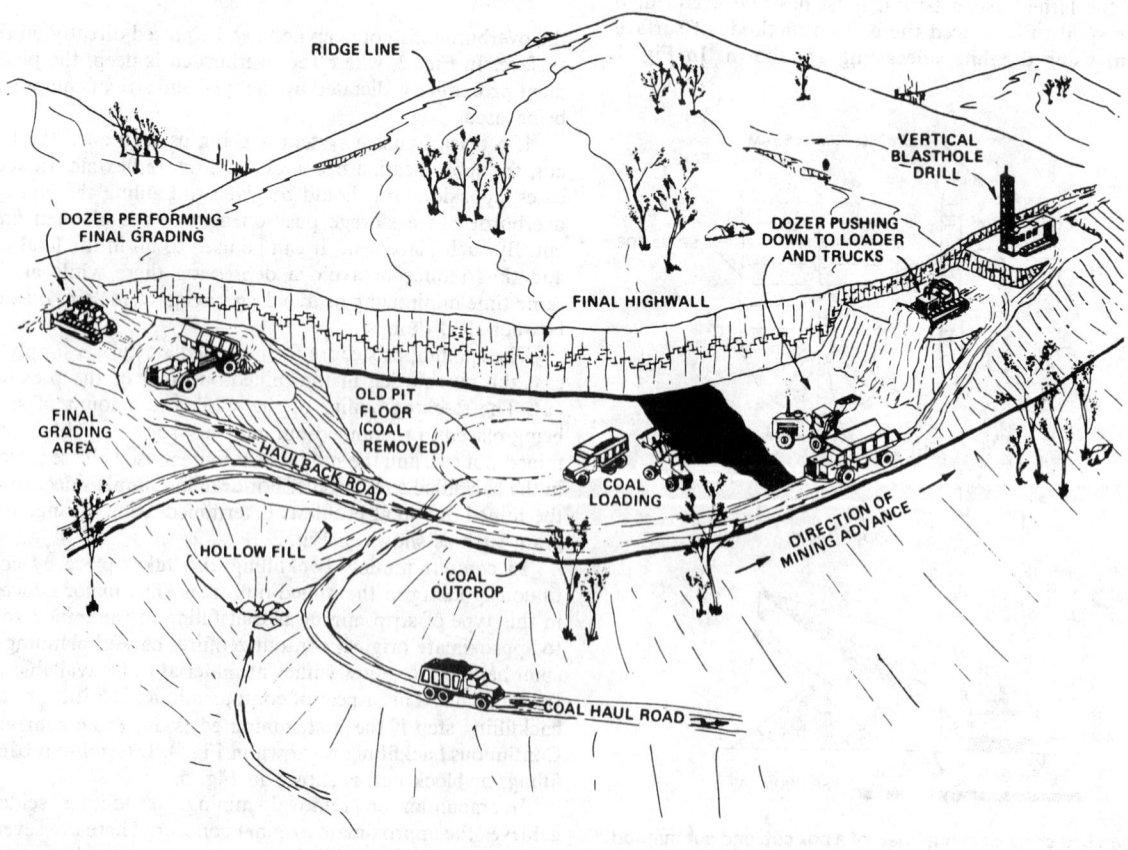

Fig. 4. Single seam, single cut haulback mining method (Cook and Kelly, 1976).

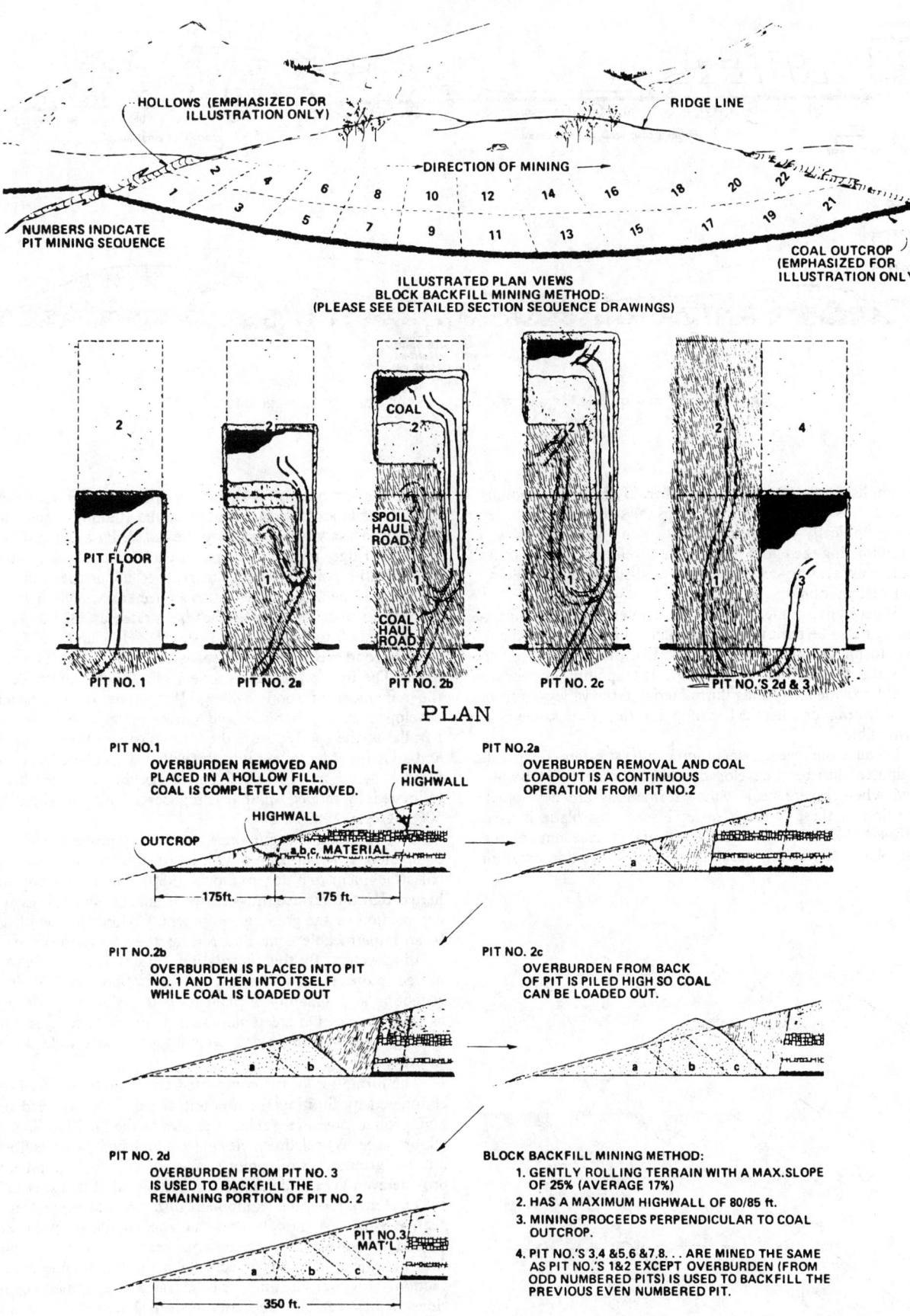

PLAN

SECTION

Fig. 5. Plan and section views of block backfill method (Cook and Kelly, 1976).

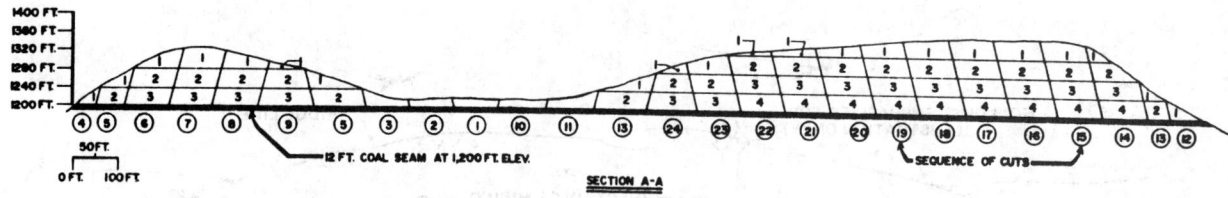

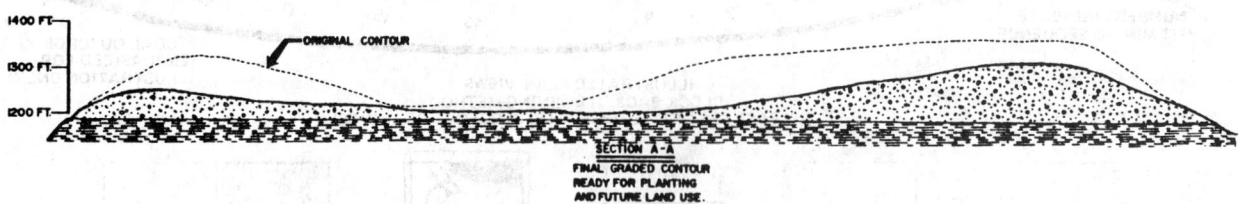

Fig. 6. Section view of cuts, original and final contour plan (Mathematica, Inc., 1977).

accomplished in an effective manner. If the truck spoiling procedure is designed in lifts, it keeps the stripped overburden stratigraphically on top in the backfilling sequence as well as minimizing the effort required for leveling. This and other tandem excavating schemes can be used effectively to achieve final reclamation.

Where environmentally possible, interburden and part-ings can also be removed with a truck/shovel operation and used to fill in valleys in the spoil. If a dragline is used to strip the interburden or partings, the spoiling procedure should provide for placing this material in the valleys, if from an operating or chemical standpoint this is possible and permissible.

Because the leveling operation can be substantial, special equipment has been developed. This equipment can be uti-lized where severe peak/valley formations are developed. The first of these is the V dozer blade. This blade is used with a 287-kW (385-hp) size dozer and is excellent for the first phase of leveling peaks. It is best suited for areas in

which a great deal of peak/valley spoil is generated each year. The blade is used to establish a flat running surface on the spoil peaks without backing the dozer. By moving down the spoil ridge, material is cast to each side. Leveling con-tinues to the point where it has reached about the width of the blade. Then, the blade is disconnected because it is no longer effective. The V blade has been rated at 3823 to 7646 m^3/hr (5,000 to 10,000 cu yd per hr) (Howland, 1975).

A second piece of equipment is the 14.6 m (48-ft) angle blade. The use of this blade is very effective for the bulk of the remainder of spoil leveling. It operates on the bench developed by the V blade and moves material by running parallel to the spoil crest and continuously cutting the spoil and allowing it to roll off into the valley. This blade has been rated at 4587 m^3/hr (6000 cu yd per hr) and has been estimated to reduce spoil leveling costs by approximately 50% (Goris, 1980).

Final grading is begun once the leveling process has been completed. It is performed with motor graders or a grading bar attachment to a dozer (Goris, 1980). The main criterion here is that the surface thus created is not subject to excessive compaction by the grading equipment. Compaction will lead to an impermeable zone that retards the infiltration of per-colating waters. By this retardation, topsoil can become sat-urated more rapidly, leading to vegetation survival rate reduction and difficulty of plant roots to penetrate as deeply as they should. On steep slopes, this compaction layer can also become a plane along which topsoil will slide when saturated.

The formation of the compacted layer can be reduced or eliminated by limiting the amount of travel by wheeled or high ground-pressure tracked vehicles as the final leveling is taking place. Wheeled vehicles place a high compactive effort into the ground. Two or more passes of wheeled equipment provide even greater compaction to the spoil (Phelps, et al., 1981). Track-mounted equipment also provide some com-pactive effort. Although they have a significantly lower ground pressure than wheeled equipment, their movement does provide a vibrating effect which produces some com-paction. However, in many cases, tracked equipment pro-vides the best vehicle, especially in steep slope regrading and leveling. If such a compacted layer has been formed, it should be broken up prior to spreading topsoil. This can be done

Fig. 7. Dozer grading of sidecast spoils (Cook and Kelly, 1976).

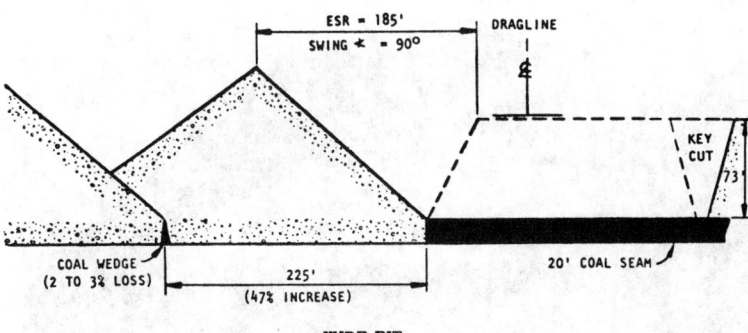

Fig. 8. Comparison of spoil bank spacing and height with wide and narrow pits (Mathematica, Inc., 1975)

through the use of disks, deep-chisel plows, or multishank rippers.

TOPSOILING

The topsoiling unit operation takes place after leveling. The operation places the growing medium for the plant cover over the rocky spoil in preparation for the final seeding phase. Topsoil is either brought from areas actively being cleared of soil in preparation for stripping or it is brought from topsoil stockpiles. The preferred method is to replace the topsoil directly on leveled backfill and not to stockpile. This procedure reduces rehandle and therefore the cost of topsoiling while at the same time minimizing the degradation of the soil as a plant medium.

Spreading topsoil on the regraded surface should be done in such a manner as not to recompact the surface upon which the topsoil is being placed. One method that can accomplish this is by dumping the soil from trucks and spreading with a dozer. In this manner, no tracking is done on the surface under the topsoil. Any compaction of the topsoil can be broken up by contour disking in preparation for planting. Another method for moving topsoil with a minimum of damage to the soil is to remove the entire profile with a front-end loader and tram it to its new location where it is replaced intact. This is akin to the establishment of lawns with turfs. Obviously, long tram distances would make this expensive. In small, or high-quality sites, this method bears consideration.

Where scrapers are used for relocating topsoil, care must be taken to minimize soil compaction (Holland and Phelps, 1986). This implies removal of the soil in as thick a lift as the machine can cut and replacement of the soil in a single lift of desired thickness. For example, one 305-mm (12-in.) lift would be preferred to two 152-mm (6-in.) lifts. Haul routes should avoid, as much as possible, crossing soil zones and should be kept, on the removal side, on the subsoil and, on the replacement side, on the graded mine spoil.

SEEDING

Seeding is done with standard farm seeding machinery. This consists of operating chiseling-tillage equipment along contour to prepare the surface, broadcast seeding, and mulching. Mulching is used to provide adequate protection of the seedbed from erosion prior to the establishment of cover. At least 0.23 t (0.25 st) of mulch per 0.4 ha (acre) should be used, depending upon the level of sediment control desired. If the area is to go back into crops, straw mulch should be used, otherwise a hay mulch may be used.

Seeding can be done with a conventional broadcast seeder or, where steep slopes prevent this, hydroseeding. Hydroseeding is performed with a tank truck or tank trailer filled with a mixture of seed, fertilizer, water, and sometimes mulch (Fig. 9). The mixture is sprayed over the slopes using a monitor mounted on the truck or trailer. The tank is equipped with an agitator to prevent settling of the components of the mixture. When large areas require seeding, or conditions are wet, aerial application methods are very effective.

REFERENCES

Bucyrus-Erie Co., 1976, "Surface Mining Supervisory Training Program," South Milwaukee, WI.

Cook, F., and Kelly, W., 1976, "Evaluation of Current Surface Coal Mining Overburden Handling Techniques and Reclamation Practice, Phase III: Eastern U.S.," Contract Report SO 144081, US Bureau of Mines, Washington, DC.

Goris, J.M., 1980, "Reducing Costs for Recontouring Mined Land," *Surface Coal Mining Reclamation Equipment and Techniques,* Information Circular 8823, US Bureau of Mines, Washington, DC.

Fig. 9. Reinco Model HG-30, 11.4-kL (3000-gal) Hydrograsser (Reinco).

Holland, L.J., and Phelps, L.B., 1986. "Topsoil Compaction During Reclamation: Field Studies," *Proceedings,* National Symposium on Mining, Hydrology, Sedimentology, and Reclamation, Lexington, KY, pp. 55-62.

Howland, J.W., 1975, "Three New Tools for Improving Land Reclamation Efficiency," *Mining Congress Journal,* March, pp. 20-21.

Mathematica, Inc., 1977, "Design and Evaluation of Cross-Ridge Mountaintop Mining," Vol. 1, Phase 1 Report on Project No. J0166112, US Bureau of Mines, Washington, DC.

Mathematica, Inc., 1975, "Evaluation of Current Surface Coal Mining Overburden Handling Techniques and Reclamation Practices, Phase I: Western US," Contract Report SO 144081, US Bureau of Mines, Washington, DC.

Phelps, L.B., Saperstein, L.W., Wells, W.B., and Yeung, D., 1981, "Burial of Potentially Toxic Surface Mine Spoil," Final Report on Project No. J0188136, USBM OFR 182-82, US Bureau of Mines, Washington, DC.

Reinco, 1980, "Hydrograssers and Power Mulchers," Reinco, Plainfield, NJ.

6.6.4 Mining and Reclamation Case Study

R. F. Goodrich and G. F. McKereghan

EDITOR'S INTRODUCTION

The following case study is included here because it gives insight into the complexity of mine planning for regions of well-developed land use. The state of Florida has enacted reclamation requirements, supplemented by county-enacted zoning ordinances, both of which require a high degree of planning before mining. We are grateful to IMC for sharing this interesting example.

RECLAMATION CASE STUDY: P-2 AREA

Land-Use Plans

The study area is located in Polk County, Florida, at IMC's Phosphoria phosphate mine and is designated P-2. The premining land uses were 20 ha (50 acres) of creek forest (Six-Mile Creek), 8 ha (20 acres) of citrus groves, 8 ha (20 acres) of railroad right-of-way, 72.8 ha (180 acres) of improved pasture, 145.7 ha (360 acres) of unimproved pasture, and 8 ha (20 acres) of power-line right-of-way and substation. The area covers about 283 ha (700 acres) of land in total.

Postmining land uses were planned to include 8 ha (20 acres) of railroad right-of-way, [4 ha (10 acres) relocated on reclaimed land], 39 ha (97 acres) of mine access corridors (dragline, power line, pipeline), 48 ha (118 acres) (three interconnected lakes) of water return and storage system, and 8 ha (20 acres) of power-line right-of-way/substation (never mined out). The mine access requirements were railroad/creek relocation (actual and potential) on the west, mine access (feed-line/power) on the north, and dragline/pipeline route on the east. The feed-line requirement on the north was justified by a reduction of 610 m (2000 ft) of pipe relocation from the Phosphoria washer to the Noralyn plant. Over the life of the Phosphoria washer, this would give a substantial reduction in power cost. Dragline access [61 m (200 ft)] on the east and north was required for IMC Noralyn mine draglines to move north and west to future mine areas. The entire P-2 area [283 ha (700 acres)] has been approved in accordance with state reclamation plans.

An important interim use of the P-2 area was for a settling area. About 180 ha (445 acres) of the mined-out area was being used, P-2 pond, for this purpose at the time the aerial photograph (Fig. 2) was taken. Pond size was determined by requirements set by the other area mines and limitations imposed by relocation requirements.

Mining and Processing

Strip mining was conducted with a walking electric dragline (Bucyrus-Erie 1250-B). Typical mining depth was 10.7 m (35 ft) with 5.5 m (18 ft) of sandy overburden and 5 m (17 ft) of matrix. Typically, the matrix consists of one-third phosphate ore, one-third silica sand (-16 to $+150$ mesh), and one-third clay/silt (-150 mesh).

The matrix is pumped by water slurry from the pit to the Phosphoria washer where the phosphate pebble ($-\frac{3}{4}$ in. $+11$ mesh) and feed (-1 mm to $+150$) are separated from the clay. Because Phosphoria has no beneficiation plant, the feed size particles are pumped to the Noralyn beneficiation plant about 5.6 km (3.5 miles) away. The clay size particles are disposed of in settling areas located in mined out areas with dams 6 to 12 m (20 to 40 ft) above grade. The bene-

ficiation plant recovers the phosphate concentrate (-16 to $+150$ mesh) from the silica sand by flotation. The tailings sand is then used as reclamation fill.

Reclamation

Reclamation fill was on-site cast overburden moved and recontoured with motor scrapers and bulldozers. Some overburden was pumped to the reclamation site from the mining dragline pit. In addition, some prestrip overburden was hauled to the reclamation zone to facilitate opening a new cut to avoid rehandling of overburden by the mining dragline. Two state road right-of-ways were mined against and immediately buttressed with shaped cast overburden. The three lakes were reclaimed to state and county standards. All reclaimed land areas were backfilled to approximate original contour. Premining coordination was accomplished between operations, reclamation waste disposal, and mine planning to identify zones requiring maximum overburden spoiling by the mining dragline to reduce earthmoving costs. Most of the remaining cast overburden in the area of the three lakes was removed with earthmoving equipment to build land areas, dams, and to maximize water storage volumes. The lakes were stocked with fish and will provide excellent recreational areas for Fin and Feather Club members.

Revegetation of the land areas was accomplished with 25 kg (55 lb) per acre of mixed seed consisting of 2.3 kg (5 lb) per acre of brown top millet (nurse crop), 15.9 kg (35 lb) per acre of Argentine bahia, and 6.8 kg (15 lb) per acre of Bermuda. Fertilizer application was 10N-10P-10K at 227 kg (500 lb) per acre. Mulch on slopes was 3.6 t (4 st) per acre of hay.

At this time the P-2 settling area is full and reclamation was started about 2½ years ago. The reclamation of the P-2 settling area is to dewater the clays by draining excess water and rim-ditching the area, plus using low ground pressure flotation tractors to form a stable surface for pasture and tree planting. The clay surface will consolidate and lower from final clay deposition elevation about 3 m (10 ft) above original ground to a final reclamation surface at approximate original grade. The exterior dams will be resloped to no steeper than 1.2 m (4 ft) high to 0.3 m (1 ft) vertical. Although not planned, tailings could have been used from the Noralyn beneficiation plant to cap the P-2 settling area clays and improve the final reclamation surface beyond agricultural usage to possible light industrial use.

The Six-Mile Creek recirculating water system was tied (inflow) to the three lakes which are interconnected by culverts. Water decanted from the P-2 pond via spillways has the capability of entering the system on the north by direct inflow and on the south by routing it through the water return system and relocated Six-Mile Creek.

Costs

The total outside services cost for the P-2 area reclamation [103 ha (255 acres)] was $1,409,000 or $5526 per acre. Total earthmoving reclamation cost was $1,343,000 and total contract services/supplies (culverts, spillway, grassing, mulching, and fertilizer) was $66,000. The reclamation costs were incurred from 1977 through 1979. The P-2 settling area was constructed from 1978 through 1979 at a cost of $1,929,000.

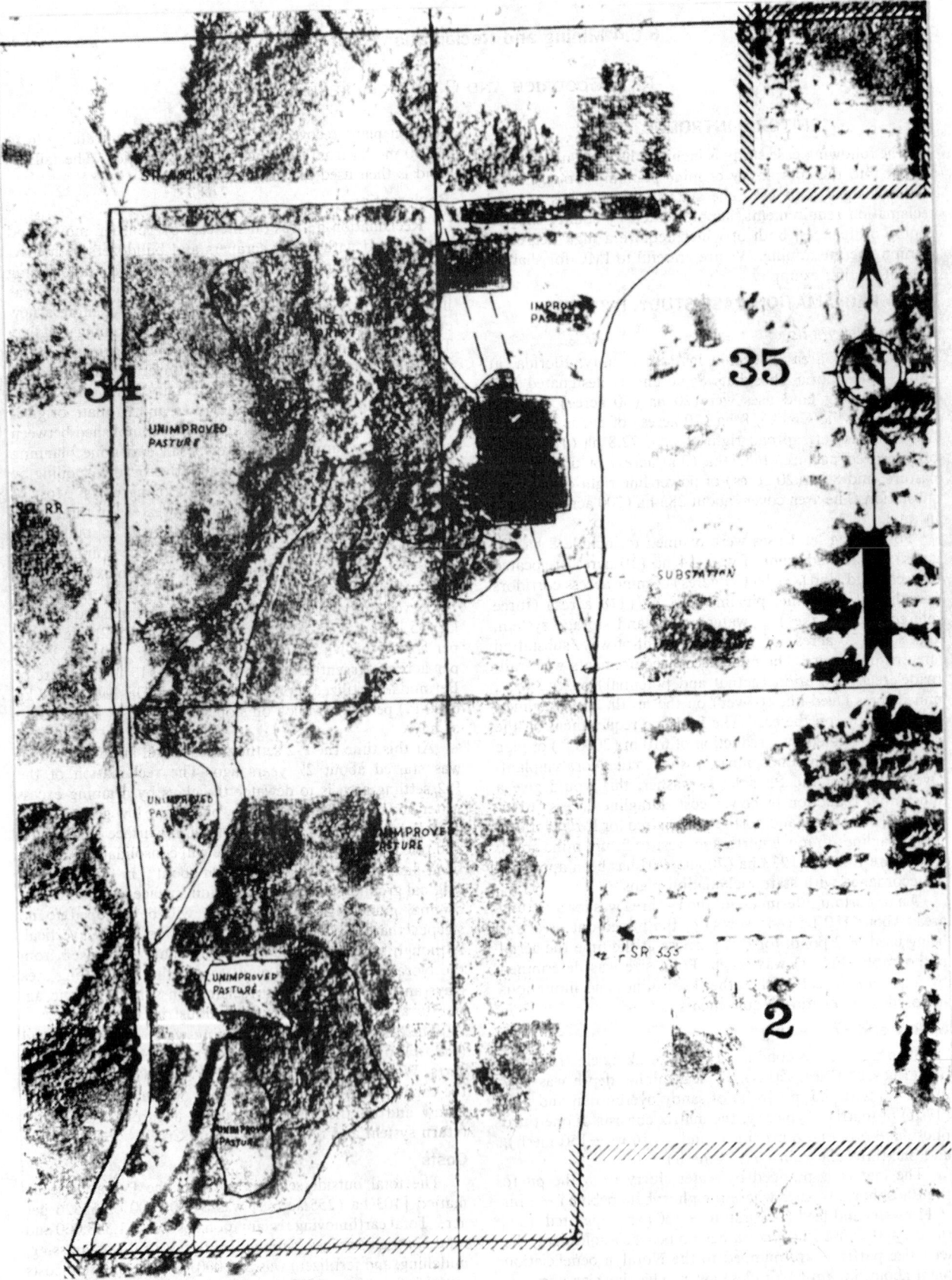

Fig. 1. A sketch of premining land uses in the IMC P-2 area.

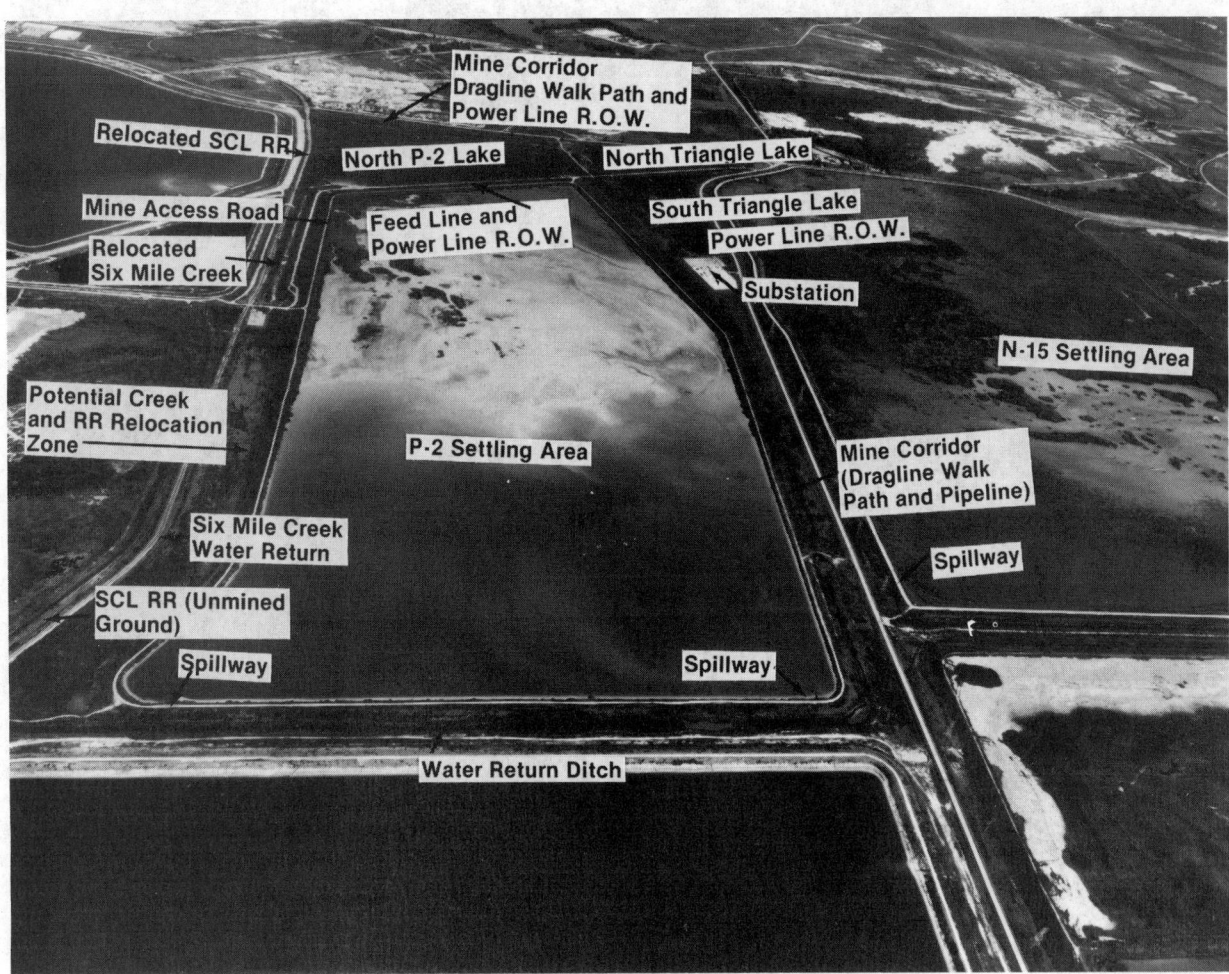

Fig. 2. An aerial photograph of the P-2 area showing the interim settling pond and reclamation zones. View is to the north.

Fig. 3 Looking south over the P-2 area with an unreclaimed zone in the foreground.

6.6.5 Topsoil Handling—A Biomass Productivity Approach

A.S. ROGOWSKI

B.E. WEINRICH

INTRODUCTION

The primary goal of any mining operation must remain the profitable recovery of a deposit, however, the need to reclaim land to previous or better use following mining cannot be overemphasized. Such reclamation constitutes a great challenge to ingenuity, because there may be an opportunity to create productive agricultural complexes where none existed before. Equally important is the preservation of existing agricultural land by restoring it to its original productivity. Novel approaches to reclamation have demonstrated that a mined area can be handsomely reclaimed, revegetated, and subsequently used for a variety of purposes, ranging from farming and ranching to recreation and wildlife habitat. Although initially the new "soil" will need to be supplemented with fertilizer and organic matter, eventually a full-fledged soil profile will form, perhaps not unlike the original profile it replaced.

The purpose of this section is to discuss different approaches to soil handling based on comparative biomass productivity criteria, subject to time and the existing climatic constraints. Although the reconstituted material is not immediately like the soil it replaces, it can be made into an adequate medium for plant growth through proper handling. With time, it should also develop its own particular layering (horizons) and characteristics. Consequently, the strict emphasis usually found in regulations on requirements for horizon and topsoil replacement is here modified to stress the need for a good rooting medium, and where applicable, to suggest possible measures to control excessive percolation, impeded drainage, and wind and water erosion potential.

An attempt is made to provide a broad overall framework of methodology for topsoil handling on a nationwide scale. It should be remembered, however, that this outline is no substitute for a detailed reclamation plan drawn with the help of the local Soil Conservation Service and University Extension specialists who are most knowledgeable about the details of soils, climate, and suitable vegetation in a particular area.

Scope and Extent

Although coal reserves constitute probably the largest areal deposits (Fig. 1) to be surface-mined, many other mineral resources produce large volumes of waste that are or will be in need of reclamation and revegetation. Some of the more important deposits are indicated in Fig. 1. Significant mining for uranium in Wyoming; copper in Utah, Arizona, and Montana; phosphate in Florida; and clay in Georgia result in areas that will need to be reclaimed in a manner similar to that now used by the coal industry. Examples of soil profiles from mining areas in many states will be used to illustrate potential problems and to describe an approach to topsoil management. The approach involves computation of a soil profile productivity index based on rather simple and readily available chemical and physical parameters. These parameters include soil properties such as pH, bulk density, texture, plant available water, aeration porosity, electrical conductivity, sodium adsorption ratio, and an estimate of root distribution. Computer values of the productivity index for a reclaimed spoil profile are then modified by a biomass productivity potential of a given area, which is a function of mean annual temperature and precipitation at a particular site.

SOIL AND TOPSOIL

The productivity controlling zone in the reclamation operation is the topmost layer of the mined profile, which consists of a relatively thin upper crust of a regolith formed by weathering of the underlying rock mantle or sedimentary deposits. By examining the composition and behavior of this topmost layer over time, it may be possible to reconstitute the premined properties and structure of soils as closely as possible following reclamation. The soils that existed before mining started were formed during a long period and were subject to climatic, vegetative, topographic, and parent-material constraints. The challenge of present-day reclamation techniques is the ability to reconstruct the premined soil conditions, or, preferably to improve the soil or the landscape that existed prior to mining. In the humid East, concern with coal mine drainage tends to dominate much of the reclamation effort with less emphasis placed on surface conditions; however, great areas of the West could be made more suitable for plant growth by modifying present soil surface conditions.

In the continental United States, soils have formed in response to varying climate, different parent materials, topography, and vegetative cover conditions. In some areas the nutrients have been leached out by abundant rainfall; in others, extensive decomposition of organic matter has created a nutrient-rich plant-growth medium. While some soils are far too dry to support much plant or animal life, others have developed impeding layers that obstruct water movement and root growth. Soil properties vary from field to field and from point to point. To consider topsoil handling in proper perspective, a brief review of background information is needed with particular emphasis on how these factors affect reconstructed profiles following reclamation.

Soil Properties

Definition of Soil: Soil can be defined as a system composed of partially or wholly weathered minerals; of animal, plant, and microbial residues in various stages of decay; and of living and metabolizing microbiota. This complex dynamic system in a loose state of aggregation also contains voids partially or completely filled with water, carbon dioxide, oxygen, and nitrogen and to a lesser extent with other gases. The mineral portion of the soil consists of unweathered rock fragments and weathered sands, silts, and clays. A vertical section through a soil showing different layers called *horizons* is known as a *soil profile*.

Soil Texture and Structure: Relative proportions of sand, silt, and clay (Fig. 2) determine the soil texture and have a significant effect on its chemical and physical properties. Mature soils have microscopic and macroscopic structure. The macroscopic structure, which can be seen with the naked eye, refers to the aggregation of primary particles into granular, blocky, platy, prismatic, or columnar clusters called *peds*. The micro-structure, which cannot be seen without the

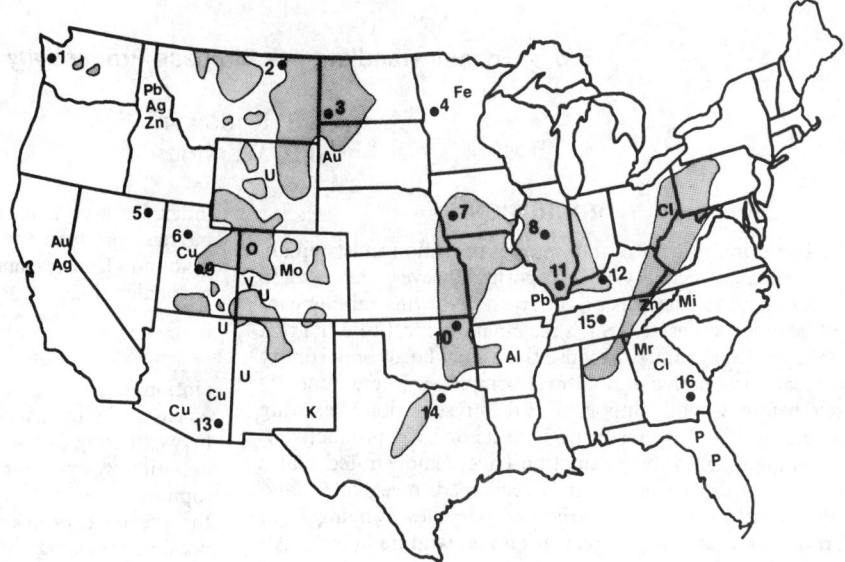

Fig. 1. Distribution of coal reserves, shaded areas; of metals: Ag-silver, Au-gold, Cu-copper, Fe-iron, Mo-molybdendum, Pb-lead, V-vanadium, U-uranium, and Zn-zinc; of minerals: Cl-clay, K-potash, Mi-mica, Mr-marble, and P-phosphate; and of oilshale-O. Location of typical profiles: 1. Grays Harbor Co., WA; 2. Blaine Co., MT; 3. Bowman Co., ND; 4. Otter Trail Co., MN; 5. Elko Co., NV; 6. Salt Lake Co., UT; 7. Shelby Co., IA; 8. Logan Co., IL; 9. Sanpete Co., UT; 10. Ottawa Co., OK; 11. Johnson Co., IL; 12. Jefferson Co., KY; 13. Cochise Co., AZ; 14. Lamar Co., TX; 15. Williamson Co., TN; and 16. Brantley Co., GA.

aid of a microscope, is the structure of soil fabric. It may be thought of as a distribution in space of clay domains, humus domains, and microbial microhabitats among, and often binding together, the organic debris and mineral constituents. It depends on the mineralogical composition, profile microbial count, and species diversity, as well as organic matter content and particle size distribution. Young soils exhibit little structure either on the macro- or micro-scale and their profiles can be subdivided into weakly defined A and C horizons only. Mature soils usually have a better defined structure and their profiles can be differentiated readily into A, B, and C horizons. Mine soils, following reclamation, usually have no distinct natural horizons, and are not likely to show much structure for many years.

Clay and Organic Matter: The topmost layer of soil usually contains the largest amounts of partially or fully decomposed organic matter called *humus*. Ordinarily, both the microbial population and humus decrease with depth and vary regionally. They are normally higher in the East and Midwest and lower in the dry West. In the continental United States they decrease from North to South because elevated temperatures lead to rapid decomposition of organic matter (National Research Council, 1981). The clay fraction, along with humus and microbiota, constitutes the active portion of soil fabric known as *the exchange complex*. The exchange complex is the primary seat of exchange reactions taking place in the soil and contributes to biochemical weathering of soil profile where soil clay minerals such as kaolinite, vermiculite, and montmorillonite are among some of the most important and interesting end products of the weathering cycle. Many Western soils having identifiable horizons are actually remnants of profiles that have developed under different climatic and vegetation regimes, whereas more recent Western soils developed in colluvium show little horizon differentiation.

The fertility level of Eastern and Midwestern topsoils is usually described by such properties as cation exchange capacity, base saturation, pH, total acidity, and carbon:nitrogen ratios (C:N). Fig. 3 shows a comparison of these properties between an original soil and some spoil materials in Pennsylvania. Western soils frequently contain more sodium, calcium, and magnesium salts in their exchange complex than their Eastern counterparts. If present in excess these can be damaging to plant growth. Measures such as electrical conductivity (*EC*) and sodium adsorption ratio (*SAR*) are generally used to describe the degree of the problem; the higher the *EC* and *SAR*, the bigger the problem.

Cation exchange capacity (CEC) in milliequivalents per 100 g (3.5 oz) of a soil is defined as a sum of exchangeable cations in a soil. It consists of calcium, magnesium, potassium, and sodium cations attached to the clay or organic matter constituents of the soil, plus such exchangeable hydrogen, aluminum, and traces of iron, manganese, and ammonium as may be present. The *CEC* generally reflects the soil humus content, pH, and the type and amount of clay

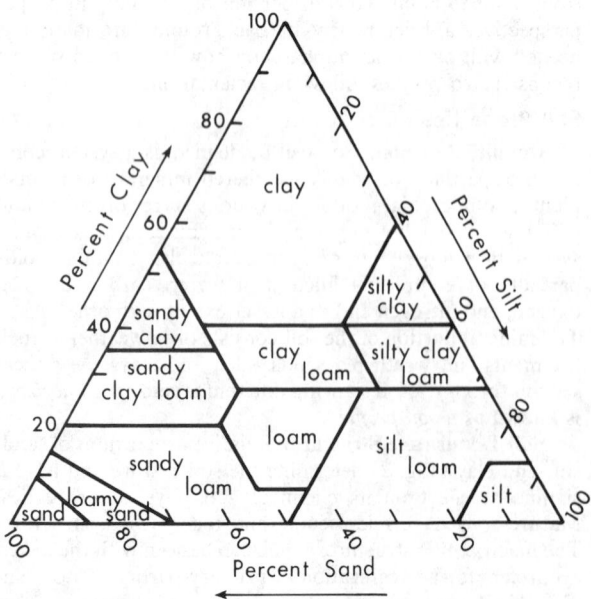

Fig. 2. Textural classification of soils.

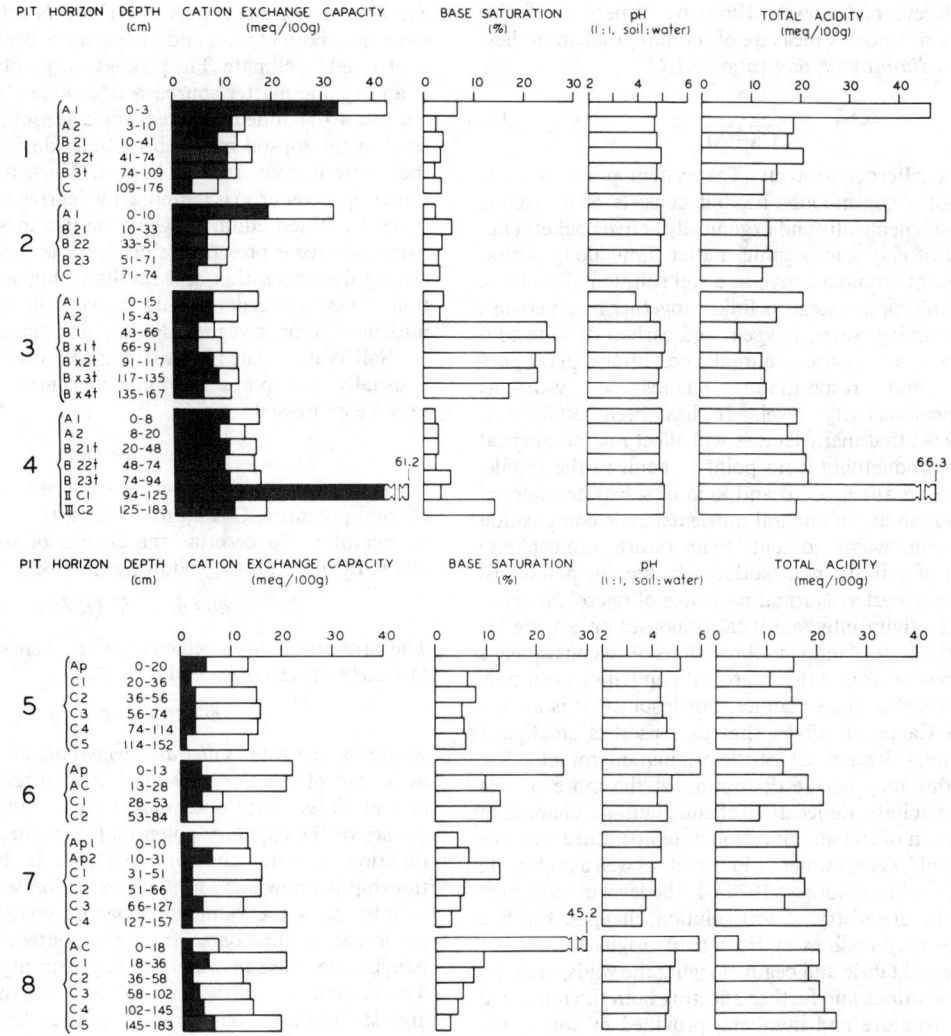

Fig. 3. Comparison of cation exchange capacity on the < 2 mm (0.08 in.) material and on the total volume of soil including coarse fragments (shaded), base saturation, pH, and total acidity in natural soils (1 to 4) and mine soils (5 to 6) in Pennsylvania (from Petersen, et al., 1978).

present. It varies with texture, ranging from 2- for sands to 60 meq/100 g for clays and clay loams. Variations within the same texture class often occur because of differences in organic matter content and pH level (Lyon, et al., 1952).

Base saturation (BS), usually defined as a fraction of *CEC* occupied by basic cations of calcium, magnesium, potassium and sodium describes how well the colloidal exchange complex of clay and humus is saturated with these cations. The difference, *(CEC-BS),* gives an estimate of *exchange acidity* of a given soil and represents acidity due to hydronium and aluminum ions. In soils of the humid regions there is usually a considerably higher level of exchange acidity and much lower pH values than in the soils of the arid regions. The *pH,* defined as the logarithm of the reciprocal of hydrogen ion concentration, reflects the concentration of hydrogen ions in soil solution. The resistance to change in pH for a given soil depends, however, on the amount of hydrogen and aluminum ions present on the exchange complex. When this is high, as in many Eastern soils, the soils are said to be highly *buffered* and may require much lime

to neutralize and make them suitable for plant growth. Consequently, horizons with low pH and high values of exchangeable acidity make poor candidates as topsoiling materials. In many Western soils the exchange complex is low in hydrogen but high in calcium, magnesium and sodium ions. As will be seen, this may lead to other problems.

When concentration of soluble salts of calcium and magnesium in soils increases because of water removal by evaporation or transpiration, excessive levels of salinity that will adversely affect plant growth may develop. The level of soil salinity is measured by the *electrical conductivity (EC)* of soil solution. Rhodes (1982) suggests that the only practical way to reduce excessive soluble salts in such soils is to leach them out of the root zone. Since many Western soils may contain such layers at the surface, it is particularly important not to use them for topsoiling during reclamation.

The second problem met with frequently in the West is posed by soils that contain excess sodium on their exchange complex. Such soils when leached with low-electrolyte concentration water, for example natural rain, lose their permeability (Reeve and Bower, 1960) by dispersion of clay and plugging of voids. A measure of sodium problem in these soils is the *sodium absorption ratio (SAR)*.

$$SAR = \frac{Na}{\sqrt{Ca + Mg/2}} \qquad (1)$$

Soil as a Microecosystem: The system previously described is not a continuous one, but consists of a discrete distribution of chemically and organically active pockets (microhabitats) of clay and organic matter films along with a population of microbiota active in a soil solution. This loose assembly of microecosystems is linked together by a network of pores containing water, oxygen, and carbon dioxide continua that vary in response to climatic conditions, plant photosynthesis, and respiration demands as well as microbiological activity level. It has been established (Stotzky, 1974) that many factors will affect microbiological activity and productivity from point to point in the profile. Among these are the amount and kind of substrate material (humus), availability of mineral nutrients, ionic composition of soil solution, water content, temperature, atmospheric composition of soil air, pH, oxidation-reduction potentials, and particulate matter. Normal measures of microbial activity and productivity integrate local responses on a scale far larger than the scale of individual reactions. As a consequence the predictions regarding the microbial populations can only be made in a rather gross manner. For instance, it is known (Miller and Cameron, 1976) that as topsoil is stockpiled during mining, it loses much of the original microbiota. The reason for this may be the disruption of the more or less stable microhabitats, desiccation of soil solution, changes in the composition of soil air, elevation of temperature, too low or too high a pH, excess water or lack of it, as well as substrate wastage and decline (Stotzky, 1974). If the level of biological activity or the chemistry of soil solution changes, the fine clay particles may wash away from their original stable positions in the soil fabric and begin clogging the voids, creating anaerobic conditions and further affecting both the microbial population structure and numbers, provided of course the climatic conditions are such that microbial population can exist. Consequently, the response in a stockpiled soil is a complex one and specific causes for the loss of productivity are hard to define. Additions of organic matter and immediate placement of removed topsoil over a reclaimed spoil will minimize the effects of disruption.

Carbon-Nitrogen Ratio: It is in this context that the carbon-nitrogen ratio should be considered. In soil layers, such as those recommended for topsoiling, a close relationship exists between the organic matter and nitrogen contents. The ratio of carbon to nitrogen in the organic matter on the average is about 10 or 12:1 (Lyon, et al., 1952). The ratio varies regionally; it is lower in arid and warmer regions, higher in humid and cooler ones. For any one region, however, the ratio tends to stay about the same, provided the management of soils is comparable. In general, C:N ratio in plant residues is high (greater than 20:1), while in the soil microbiota it is on the order of 6:1. Thus, upon addition of organic residues to the soil there is an increase in soil decay organisms, accompanied by an evolution of carbon dioxide and a decrease in C:N ratio. In the process of humification, carbon is lost from the soil in a form of gas CO_2, whereas nitrogen remains tied up in the microbial tissues. Eventually as the amount of available organic matter decreases, microbial activity slows down and some ammonia appears. With ensuing nitrification the product—nitrate nitrogen—is leached out of the profile by percolating water, and a quasi-

equilibrium condition prevails. The C:N ratio stays constant since now both carbon and nitrogen are being lost at the rate controlled by climate. The process has practical implications. If an organic matter source with a wide C:N ratio (such as oat straw) is added to a reclaimed topsoil, nitrate nitrogen level in the topsoil may substantially decrease until most of the straw is converted to humus. Thus, a growing or germinating cover crop is temporarily deprived of much nitrogen it needs. When adding organic matter to spoils, a high nitrogen source is preferred, since it depletes soil nitrogen least during decomposition and results in higher humus production. These concepts should be kept in mind when using mulches and or sewage sludge applications on disturbed land.

Soil Water: The movement of soil water within the voids is usually in response to profile concentration gradients and may be expressed as:

$$Q_N = C\nabla\phi \qquad (2)$$

where Q_N is the water flow velocity vector, ϕ is the vector of total potential, C is a proportionality coefficient, and ∇ is an operator. To describe the change of soil water content (θ) with time (t), the following may be written,

$$\partial\theta/\partial t = \nabla\cdot(K\nabla\phi) \qquad (3)$$

The proportionality coefficient, K, is then known as the soil *hydraulic conductivity* and

$$\phi = \psi + z + \rho \qquad (4)$$

where ϕ, the total *soil water potential*, may be represented as a sum of *capillary potential* ψ, *gravitational potential z*, and at times other potentials, such as *osmotic potential* ρ. Values of the capillary potential (tensiometer pressure) as a function of water content are shown in Fig. 4a; the relationship is known as *soil moisture characteristic*. Quite often in mine soils the values of moisture characteristic have to be corrected for coarse fragment content with an accompanying decrease in water-holding capacity of the new soil. The amount of soil water held between the tensions of 33 and 1500 kPa is considered to be the *plant available water* and it is assumed that the water at tensions less than 33 kPa drains down to the water table under the influence of gravity. This is not strictly correct since in many soils water is often held in the root zone at tensions less than 33 kPa, and is available to some plants at tensions greater than 1500 kPa. Yet the notion of available water content provides a useful practical guide for field personnel.

Soil hydraulic conductivity values (K) vary as a function of soil water (Fig. 4b) and tensiometer pressure (Fig. 4c). A value of K at or near 0 tension is called a *saturated conductivity*. Fig. 4 shows examples of moisture characteristic and hydraulic conductivity for some Pennsylvania spoils. The plot of hydraulic conductivity values as a function of water content (Fig. 4b) suggests that for these spoil materials small changes in water content will result in large changes of hydraulic conductivity. This means that, even when spoil water content varies little, substantial changes in the flow rate can occur.

Soil Air: Plant roots and most microorganisms respire, using oxygen (O_2) and giving out carbon dioxide (CO_2). Gaseous and dissolved oxygen and carbon dioxide frequently have markedly heterogeneous distributions within a soil fabric. This can lead to the occurrence of simultaneous anaerobic and aerobic reactions in close proximity of each other within the profile. Zones that impede diffusion, dead end pores, or sites where CO_2 production is high (for example, zones where organic matter is actively converted to humus by decay or-

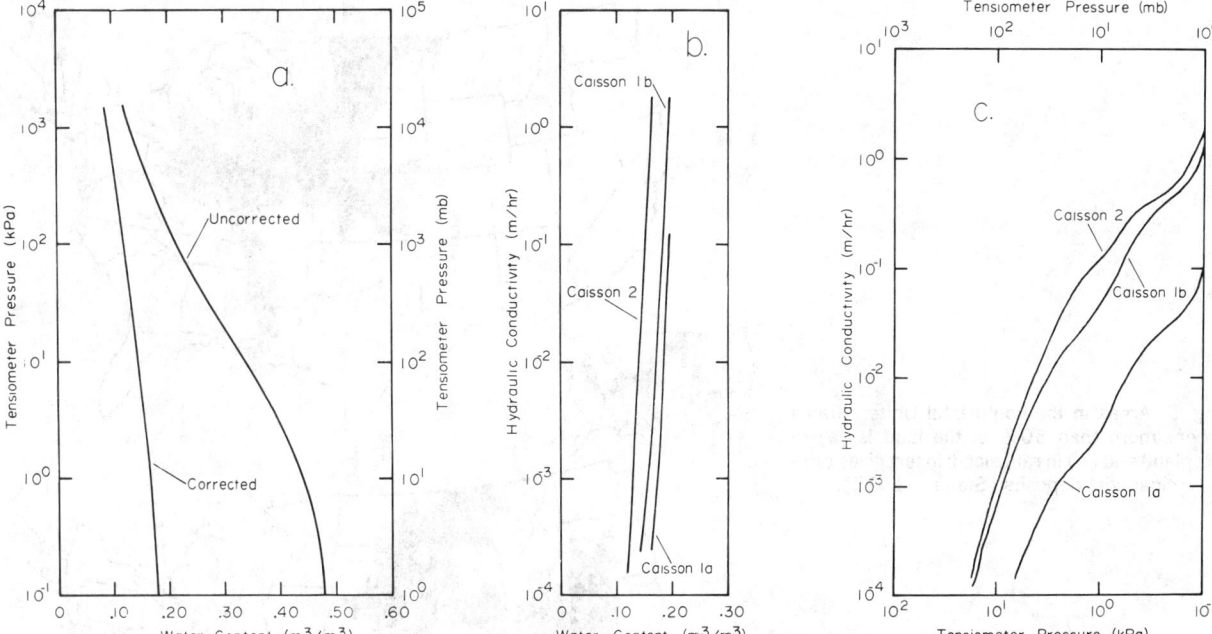

Fig. 4. For some Pennsylvania mine soils, (a) moisture characteristic corrected and uncorrected for coarse fragments, (b) hydraulic conductivity (corrected) as a function of water content, and (c) tensiometer pressure (from Rogowski and Jacoby, 1979).

ganisms) often result in substantial differences in concentration over short distances. The oxygen-in-soil solution stimulates nitrification, while high concentrations of CO_2 ($> 50\%$) may be toxic to soil fungi. Free CO_2 is generally present in soil solutions below pH 6.3; above that, carbonates and bicarbonates predominate (Levy and Toutain, 1982). A well-aerated soil may contain 18 to 20.5% of O_2 on the average. This can drop to 10% after a rain and can be as low as 2% around respiring roots. In general, soil CO_2 level increases at the expense of O_2; usually its concentration will range between 0.3 and 3%. However, following incorporation of large amounts of organic matter, CO_2 levels may reach 10% or more. In most natural soils, O_2 concentration decreases rapidly with depth. Since mine soils often contain more coarse fragments and larger pores than natural soil, high oxygen concentrations can sometimes be observed at depth. Under these conditions and in pyrite-containing spoils, carbonate neutralization of the acid, which is produced by pyrite oxidation, is the dominant source of CO_2. Levels of CO_2 as high as 18% have been recorded at 6 m (20 ft) in some Pennsylvania spoils (Jaynes, et al., 1983).

Soils and Land Use

This section will consider how the nature of soil itself may affect the manner in which it should be handled. Fig. 5 shows the areas where more than 50% of the land is in cropland (Fig. 5a) or rangeland (Fig. 5b), and Fig. 6 shows the location of major *soil orders* in relation to coal deposits in the continental United States. An order is a grouping of soils with similar taxonomic characteristics.

Appalachian coalfields are generally found on Inceptisols and Ultisols, those in the Midwest cropland are overlain by Alfisols and Mollisols, while on the Western range coal deposits underlie Entisols, Mollisols, and Aridisols. Table 1 shows the general properties of these soil orders primarily involved in surface mining for coal as well as other orders that may be important in areas where mineral deposits other than coal are exploited. Brief descriptions follow (Jenny, 1980).

Inceptisols and Ultisols found in Appalachia are most amenable to change and manipulation. Since Inceptisols are young shallow soils and Ultisols have been extensively leached of nutrients, opportunity exists when reclaiming these areas to produce a better growth medium than the one they replace. The reclaimed area should be structured to absorb extensive percolation and to minimize erosion hazard. It should have, if possible, an organic matter high in nitrogen added to it, while being fertilized and limed extensively to provide sufficient nutrients for plant growth. Good plant growth and an oxidizing topsoil environment will in turn go a long way to minimize acid mine drainage on these areas.

A different situation exists in the Midwest. Here Mollisols and Alfisols are generally highly productive agricultural soils. This cropland has to be reclaimed so as to, at least, maintain soil productivity. Particular attention must be paid to preserving rich topsoil layers and restoring the physical properties, such as bulk density and aeration of original horizons.

In the West the overriding problem is lack of moisture. Some soils may be highly productive (Mollisols) and should be reclaimed accordingly, if supplemental irrigation is available. Other soils often exhibit high sodium content, or horizons in which clay, gypsum, or carbonates have accumulated (Aridisols). Under these conditions different horizons should be selected for use in order to minimize detrimental effects on vegetation and to maximize moisture usage.

Broadly speaking, Midwestern soils should be reclaimed as croplands, those in the West as rangelands, and Appalachian soils as woodlands. Thus, the original soil and climate will largely dictate the agronomic potential of the area and how best it should be handled. An example of a taxonomic soil description illustrates how much information about a

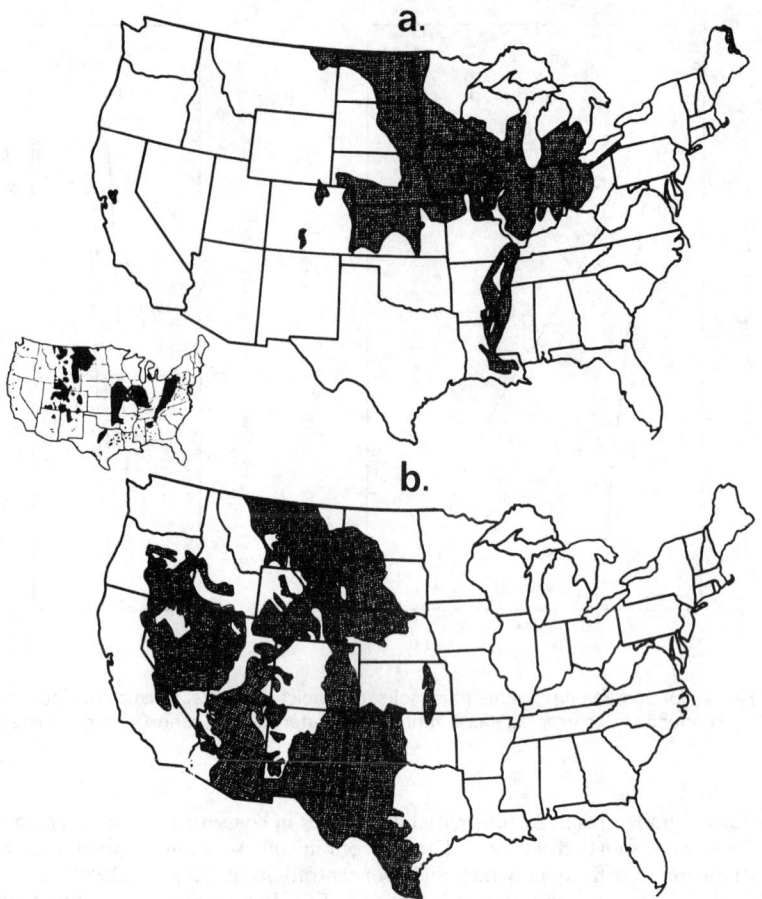

Fig. 5. Areas in the continental United States where more than 50% of the land is (a) in cropland and (b) in rangeland; insert gives principal coal deposits (Stewart, 1975).

given soil can be gained from its name alone (Soil Survey Staff, 1975). A profile in Grays Harbor County, Washington (No. 1 in Fig. 1), is classified as an Andic Dystrochrept. This soil belongs to an Inceptisol order. Thus, it is likely to be a relatively young soil, quite heterogeneous, with little clay translocation (Table 1). Since it is an Andic Dystrochrept, it is a moderately deep, freely drained, brownish or reddish soil commonly containing volcanic ash material (Soil Survey Staff, 1975, pp. 227, 228, 247, 249, 250). In a similar manner, taxonomic names of other soils not only describe the profile in reasonable detail, but also suggest where problems may arise.

Soil Variability

In attempting to predict a response of a soil, it should be remembered that soil properties are quite variable in space as well as in time. When general soil surveys are used for reclamation planning, it is well to recall (Beckett and Webster, 1971) that only about 50% of the sites chosen at random within a mapped soil series are likely to be occupied by profiles matching the definition of the profile for which the unit is named. Consequently, a large spatial variability is to be expected. Soils are dynamic: in time, moisture, temperature, and aeration statuses of a profile will change; materials will weather; topsoil will erode; plants will grow, die, and decay. The soil can never be adequately sampled to account for all the variability likely to be present in a soil profile. In general, Beckett and Webster (1971) concluded that soil variability increased with the size of the area sampled. Although Harradine (1949) found that young immature soils

showed more variability than developed mature profiles, young mine soils such as the ones discussed here may be less variable than soils from which they are derived. This, of course, should be the case when discrete soil layers become well mixed during handling, storage, and subsequent spreading during reclamation. If no overall mixing occurs, resulting mine soil would be quite variable over an area. This variability will be enhanced in time as spoils settle and erode and as plant cover changes.

CLIMATE AND SOIL MOISTURE REGIME

Some potential problems in topsoil handling that arise out of differences in soil moisture regimes will now be considered. These problems center primarily on shortage or excess of plant-available moisture, infiltration, runoff, water storage, and erosion. The status of soil moisture is only in part a function of climate. Many deep, permeable soils under high and evenly distributed rain have water available to plants most of the time. Even soils in arid climates are not necessarily dry. Their moisture status depends on their position in the landscape, since they can receive moisture from sources other than rain, such as deep seepage or snowmelt. However, since mining generally disturbs these features that have evolved over a considerable time period, moisture regimes of reconstructed profiles should be considered as being solely a result of climatic variables. The graphs in Figs. 7, 8, and 9 are based on average values of precipitation, temperature, and potential evapotranspiration (PE) and can provide a rather simple description of a profile's moisture regime. The area between the solid line (for example in Fig. 7),

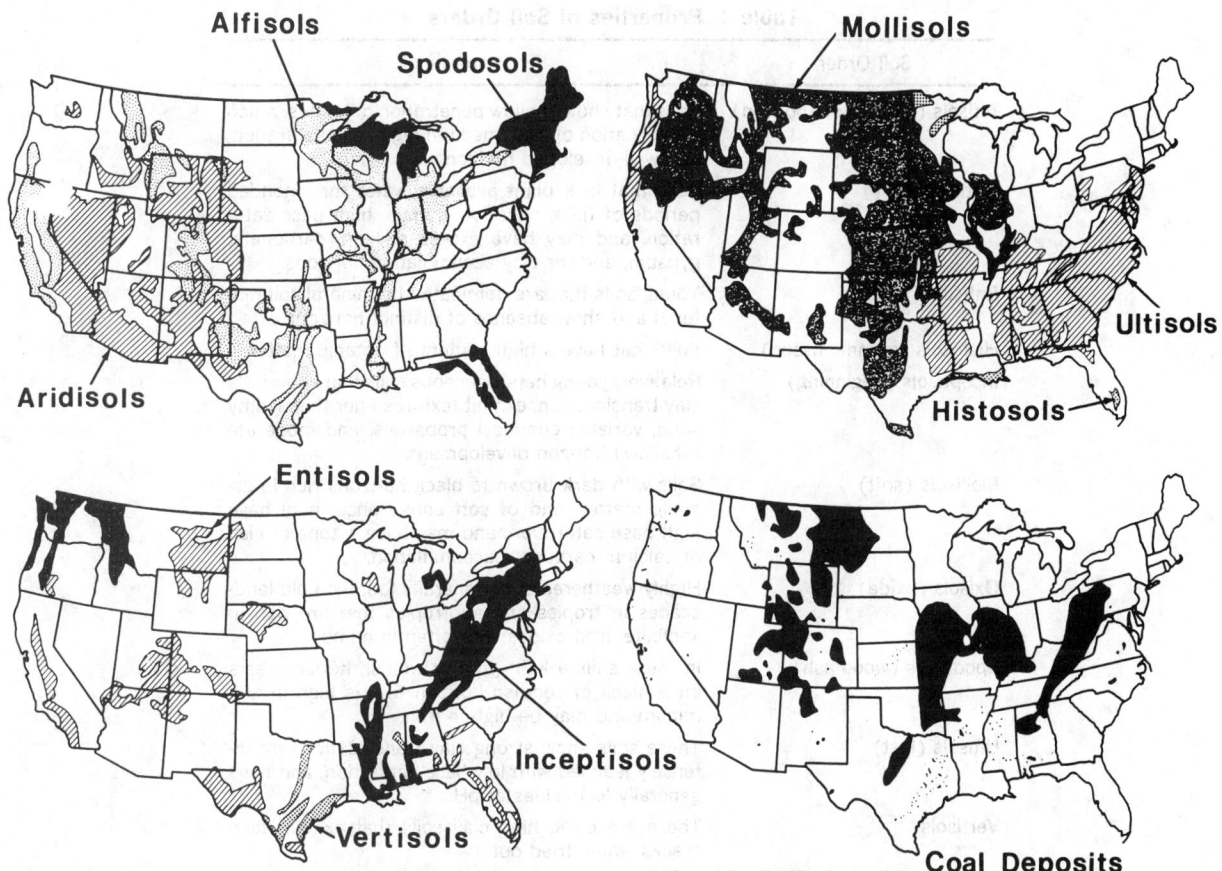

Fig. 6. Distribution of principal soil orders and coal deposits in the continental United States.

which joins all monthly precipitation normals, and the dashed line, which joins all PE normals, gives the status of soil moisture. Starting at the point where precipitation becomes greater than PE, the area to the right marked R (for recharge) shows the amount of water stored in the soil profile. Recharge usually begins sometime in the fall, extends to the extreme right of the diagram (early winter), and is continued on the extreme left (winter and summer). The amount of recharge is limited by the available water capacity (AWC) of the soil, in which case a vertical line is drawn when AWC becomes exhausted, and the area to the right of it is labeled S for surplus. Such surpluses may result in runoff, indicating a potential erosion problem in a given area. Recharge and surplus become limited when PE again exceed precipitation. The area to the right, marked U (for utilization of water by plants), shows the amount of PE necessary to remove water held by soil at less than 1500 kPa tensions. When all the available water is removed from the profile, excess PE, labeled D, is the soil-moisture deficit before recharge begins. Constructing one of these diagrams for an appropriate soil is relatively easy if sufficient data are available. In particular, such a diagram should be made for a reconstructed mine soil profile to develop a meaningful revegetation plan. The data needed are soil available-water capacity, which for practical purposes may be taken as the amount of water held by the soil between 10 to 33 kPa and 1500 kPa*, monthly mean

temperature, and precipitation. Using appropriate tables and formulas, potential evapotranspiration (Thornthwaite, 1948) and soil moisture regime status can be calculated.

Computation of Soil Moisture Regime

The following illustrates a procedure for computing an estimated soil moisture budget, given profile water-holding capacity, and average monthly values of temperature and precipitation. Suppose that a site of interest is located in central Pennsylvania near the village of Pine Glen, 25 km (15.5 miles) northeast of the FAA weather station in Philipsburg. Normal monthly temperature and precipitation based on the 1941–1970 period and extracted from NOAA (1978) are given in lines 1 and 2 of Table 2.

Profile water-holding capacity down to 100-cm (39.4 in.) depth was estimated for Gilpin channery silt loam and Wharton silt loam as the sum of differences between water content, θ_o, near saturation (0.9 θ_o), and water content, θ_{15}, at 1500 kPa times layer thickness. Thus, potential water-holding capacity for Gilpin soil was found to be 291 mm (11.5 in.), and for Wharton soil it was found to be 230 mm (9 in.) in

---

*For many soils these values are generally listed as 1/10, 1/3 and 15 bar in the County Soil Survey, or are available from the Soil Con-

servation Service office in different areas. Since they are usually given on a weight basis they need to be multiplied by bulk density to express them on the *per volume* basis. Soils-5 file, available at a nominal charge from Ames Statistical Laboratory, Ames, IA (c/o Harvey Terpstra), generally lists a range of available water for individual horizons of soils in inches/inch. These volumetric values are based on the difference between soil water held at 1500 and 33 kPa (15⅓ bar).

Table 1. Properties of Soil Orders

Soil Order	Properties
Alfisols (aluminum and iron)	Soils that show shallow penetration of humus, much translocation of clay, medium high base saturation, and well-developed horizons.
Aridisols (dry)	Soils that lack plant available water for extended periods of time, have low humus, high base saturation, and may have excess sodium, carbonate, gypsum, and/or clay accumulation horizons.
Entisols (recent)	Young soils that are dominated by mineral soil material and show absence of distinct horizons.
Histosols (organic tissue)	Soils that have a high content of organic matter.
Inceptisols (beginning)	Relatively young heterogeneous soils that show little clay translocation, exhibit textures finer than loamy sand, variable chemical properties, and moderate (shallow) horizon development.
Mollisols (soft)	Soils with dark brown to black horizons rich in organic matter, and of soft consistence, may have high base saturation, and may have a zone of clay or calcium carbonate accumulation.
Oxisols (oxide)	Highly weathered soils generally found on old landscapes in tropics and subtropics that are rich in kaolinite, iron oxides, and often in humus.
Spodosols (wood ash)	In these soils a light gray or whitish horizon rests on a black or reddish horizon that is high in aluminum and may be high in iron.
Ultisols (last)	These soils show strong clay translocation, are intensely leached with low base saturation, and have generally low values of pH.
Vertisols	These are dark, high clay soils that exhibit deep cracks when dried out.

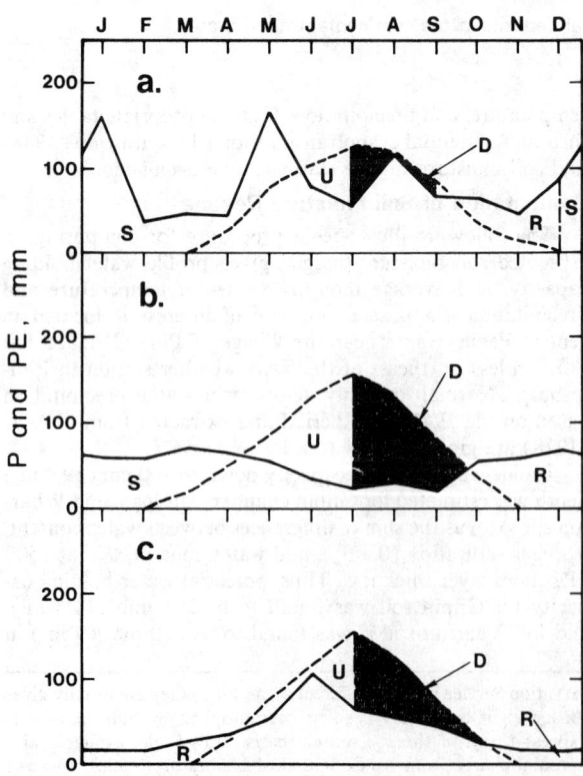

Fig. 7. Water budgets: (a) for Wharton soil, Centre Co., PA; (b) for a typical profile from Salt Lake Co., UT; and (c) for a typical profile from Bowman Co., ND; S = surplus, R = recharge, D = deficit, and U = utilization.

the top 100 cm (39.4 in.) of the profile. The monthly temperature and precipitation normals along with the profile water-holding capacity constitute the primary input into this moisture budgeting technique.

The first step is to compute potential evapotranspiration using Thornthwaite's (1948) method. The unadjusted (30 12-hr days) potential evapotranspiration (E) in millimeters is calculated as

$$E = 16(10\ T/I)^a \qquad (5)$$

where T is the mean monthly temperature in °C, I is the annual heat index computed as the sum of monthly heat indices (i),

$$i = (T/5)^{1.514} \qquad (6)$$

and a is an exponent approximated by

$$a = 0.000\ 000675I^3 - 0.0000771I^2$$
$$+ 0.01792I + 0.49239 \qquad (7)$$

Using Eqs. 5, 6, 7 and the monthly temperature data in Table 2, monthly heat indices i (Table 2, line 3) are computed and then the annual heat index $I = 36.48$ as a sum of monthly indices followed by exponent $a = 1.0763$ and unadjusted evapotranspiration E_v (Table 2, line 4). Values of unadjusted evapotranspiration E_v need to be adjusted for number of days in a month and approximate hours of sunlight. This is accomplished by extracting the appropriate correction factor from Table 3 (Table 2, line 5) and multiplying it by the unadjusted value of E_v. Table 3 was extracted from Table V of Thorthwaite (1948). For this example, a correction for latitude 41° is used.

Now water balance is computed for a reconstituted 100-cm (39.4-in.) profile at the site. Adjusted potential evapo-

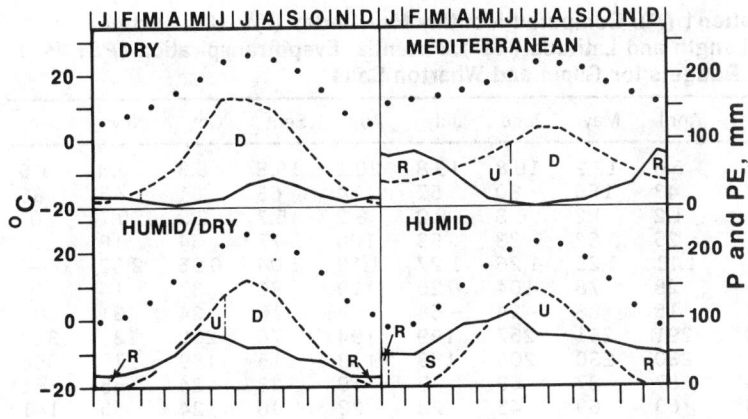

Fig. 8. Primary climatic and moisture regimes of the United States: hot/dry climate, mediterranean climate, humid climate, and humid/dry climate; S = surplus, D = deficit, R = recharge, U = utilization; ° = temperature (Soil Survey Staff, 1975).

transpiration (*PE*) in millimeters (Table 2, line 6) is obtained by multiplying E_v in line 4 by correction in line 5 of Table 2. These values are then subtracted from monthly precipitation to provide the P-PE index in line 7. At sites where precipitation exceeds potential evapotranspiration (annual value of P-PE > 0) the soil profiles likely will be near or at field capacity in the spring of the year, which is the case in Pennsylvania. Thus, water budgets for Gilpin, lines 8 and 10, and Wharton soils, lines 9 and 11 in Table 2, start with field capacity values in April. For subsequent months, values of (P-PE) in line 7 are subtracted from or added to the available water. If, as in lines 10 and 11, a water deficit develops, it is continued for as long as (P-PE) < 0. When (P-PE) in line 7 becomes positive, a new accumulation cycle begins (see values for October in lines 10 and 11 of Table 2). Fig. 7a graphically presents the water budget for Wharton

soil. The results show that under the assumption of 100 cm (39.4 in.) topsoiled profile, adequate moisture conditions for plant growth will prevail. However, should the operator reduce the thickness of topsoil to 30 cm (11.8 in.), moisture deficits (shaded area) can develop in July, August, and September. The water budgets in Table 2 are also predicated on normal values of precipitation. Should precipitation fall below normal, or should available water capacity of topsoiling materials be less than projected, moisture deficits can develop in other months also. It is worth noting that in January, February, and March excess precipitation may accumulate and be stored as snow, with potential for snowmelt, runoff, and erosion in early spring.

A somewhat different situation prevails when there is a slight overall annual moisture deficit, as for example, on a typic haploxerol soil (Table 4, Fig. 7b) in Salt Lake County,

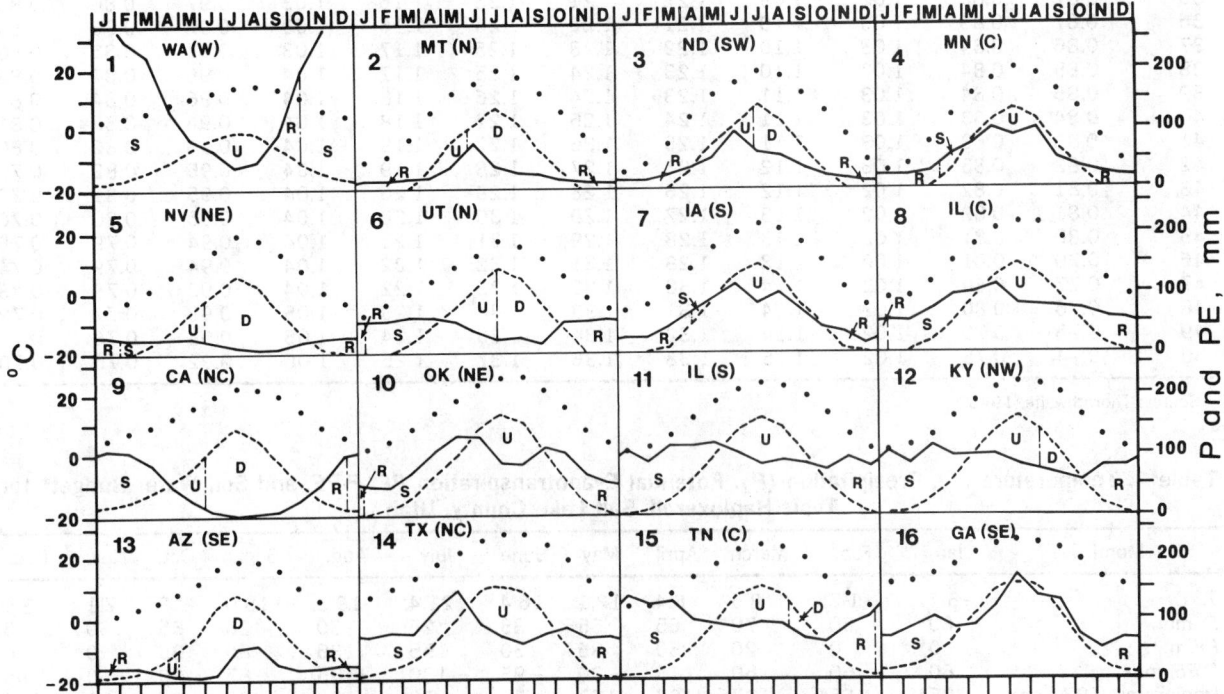

Fig. 9. Climatic data and soil moisture regimes for typical profiles: 1. Grays Harbor Co., WA; 2. Blaine Co., MT; 3. Bowman Co., ND; 4. Otter Trail Co., MN; 5. Elko Co., NV; 6. Salt Lake Co., UT; 7. Shelby Co., IA; 8. Logan Co., IL; 9. San Pete Co., UT; 10. Ottawa Co., OK; 11. Johnson Co., IL; 12. Jefferson Co., KY; 13. Cochise Co., AZ; 14. Collin Co., TX; 15. Williamson Co., TN; and 16. Brantley Co., GA; S = surplus, D = deficit, R = recharge, U = utilization; ° = temperature (Soil Survey Staff, 1975).

Table 2. Temperature (T), Precipitation† (P), Computed Monthly Heat Index (i)‡, Unadjusted Evapotranspiration (E$_v$), Correction for Day Length and Latitude (41°), Potential Evapotranspiration (PE), P-PE, and Soil Water Budgets for Gilpin and Wharton Soil§

Line	Month:	Jan.	Feb.	March	April	May	June	July	Aug.	Sept.	Oct.	Nov.	Dec.
1	T, °C	−7.2	−9.1	−0.6	5.6	12.9	16.8	19.8	20.1	15.8	8.3	4.0	−1.6
2	P, mm	159	33	47	43	166	80	57	114	63	61	45	86
3	i	0	0	0	1.2	4.2	6.3	8.0	8.2	5.7	2.2	0.7	0
4	E$_v$	0	0	0	25	62	83	99	100	77	34	18	0
5	Correction	—	—	—	1.11	1.25	1.26	1.27	1.19	1.04	0.96	2.82	—
6	PE, mm	0	0	0	28	78	104	125	119	81	37	14	0
7	P-PE mm	159	33	47	15	88	−24	−68	−5	−18	24	31	86
8	Gilpin, 291 mm	320*	353*	400*	291	291	257	199	194	176	200	231	317*
9	Wharton, 230 mm	415*	448*	495*	230	230	206	138	133	115	139	170	356*
10	Gilpin, 87 mm	300*	333*	380*	87	87	63	−5	−10	−28	24	55	141*
11	Wharton, 69 mm	300*	353*	400*	69	69	45	−23	−28	−46	24	55	141*

† For Philipsburg, PA, Lat. 40°54′, Long. 78°05′.
‡ $i = 0$ if $T \leq 0°C$.
§ Available water values are given in brackets for 100 cm profile (lines 8 and 9) and 30 cm profile (lines 10 and 11); starred values (*) designate months when water in excess of profile water holding capacity possibly can be stored as snow if $T \leq 0°C$.

Table 3. Mean Possible Duration of Sunlight in the Northern Hemispheres (Lat. 25-50), Expressed in Units of 30 Days of 12 Hr Each

N. Lat.	Jan.	Feb.	March	April	May	June	July	Aug.	Sept.	Oct.	Nov.	Dec.
25	0.93	0.89	1.03	1.06	1.15	1.14	1.17	1.12	1.02	0.99	0.91	0.91
26	0.92	0.88	1.03	1.06	1.15	1.15	1.17	1.12	1.02	0.99	0.91	0.91
27	0.92	0.88	1.03	1.07	1.16	1.15	1.18	1.13	1.02	0.99	0.90	0.90
28	0.91	0.88	1.03	1.07	1.16	1.16	1.18	1.13	1.02	0.98	0.90	0.90
29	0.91	0.87	1.03	1.07	1.17	1.16	1.19	1.13	1.03	0.98	0.90	0.89
30	0.90	0.87	1.03	1.08	1.18	1.17	1.20	1.14	1.03	0.98	0.89	0.88
31	0.90	0.87	1.03	1.08	1.18	1.18	1.20	1.14	1.03	0.98	0.89	0.88
32	0.89	0.86	1.03	1.08	1.19	1.19	1.21	1.15	1.03	0.98	0.88	0.87
33	0.88	0.86	1.03	1.09	1.19	1.20	1.22	1.15	1.03	0.97	0.88	0.86
34	0.88	0.85	1.03	1.09	1.20	1.20	1.22	1.16	1.03	0.97	0.87	0.86
35	0.87	0.85	1.03	1.09	1.21	1.21	1.23	1.16	1.03	0.97	0.86	0.85
36	0.87	0.85	1.03	1.10	1.21	1.22	1.24	1.16	1.03	0.97	0.86	0.84
37	0.86	0.84	1.03	1.10	1.22	1.23	1.25	1.17	1.03	0.97	0.85	0.83
38	0.85	0.84	1.03	1.10	1.23	1.24	1.25	1.17	1.04	0.96	0.84	0.83
39	0.85	0.84	1.03	1.11	1.23	1.24	1.26	1.18	1.04	0.96	0.84	0.82
40	0.84	0.83	1.03	1.11	1.24	1.25	1.27	1.18	1.04	0.96	0.83	0.81
41	0.83	0.83	1.03	1.11	1.25	1.26	1.27	1.19	1.04	0.96	0.82	0.80
42	0.82	0.83	1.03	1.12	1.26	1.27	1.28	1.19	1.04	0.95	0.82	0.79
43	0.81	0.82	1.02	1.12	1.26	1.28	1.29	1.20	1.04	0.95	0.81	0.77
44	0.81	0.82	1.02	1.13	1.27	1.29	1.30	1.20	1.04	0.95	0.80	0.76
45	0.80	0.81	1.02	1.13	1.28	1.29	1.31	1.21	1.04	0.94	0.79	0.75
46	0.79	0.81	1.02	1.13	1.29	1.31	1.32	1.22	1.04	0.94	0.79	0.74
47	0.77	0.80	1.02	1.14	1.30	1.32	1.33	1.22	1.04	0.93	0.78	0.73
48	0.76	0.80	1.02	1.14	1.31	1.33	1.34	1.23	1.05	0.93	0.77	0.72
49	0.75	0.79	1.02	1.14	1.32	1.34	1.35	1.24	1.05	0.93	0.76	0.71
50	0.74	0.78	1.02	1.15	1.33	1.36	1.37	1.25	1.06	0.92	0.76	0.70

Source: Thornthwaite, 1948.

Table 4. Temperature (T), Precipitation (P), Potential Evapotranspiration PE, P-PE, and Soil Water Budget* for Typic Haploxeroll, Salt Lake County, Utah

Month:	Jan.	Feb.	March	April	May	June	July	Aug.	Sept.	Oct.	Nov.	Dec.
T, °C	−5.7	−1.7	2.5	6.4	12.1	16.4	21.4	19.3	15.0	7.9	7.1	−3.5
P, mm	60	60	70	65	55	35	25	30	20	55	60	65
PE, mm	0	0	20	50	85	130	155	135	85	35	0	0
P-PE, mm	60	60	50	15	−30	−95	−130	−105	−65	20	60	65
Haploxeroll, 197 mm†	205*	265*	197	197	167	72	−58	−163	−228	20	80	145

Source: Soil Survey Staff, 1975.
*Starred values designate months when water in excess of profile holding capacity possibly can be stored as snow.
† Available water capacity.

Utah. In the spring, on a reclaimed 100-cm (39.4-in.) profile of this soil, potential water-holding capacity would probably be satisfied; however, large water deficits could develop later on in the season and create a serious problem with revegetation.

A quite different situation prevails on a Typic Natriboroll soil in Bowman County, North Dakota (Table 5, Fig. 7c). Here only 39% of the available water-holding capacity is satisfied in the spring, resulting in much lower plant available water.

The three water budgets for Centre County, Pennsylvania, Salt Lake County, Utah, and Bowman County, North Dakota, illustrate the procedure used, the relative magnitudes and time of occurrence of moisture deficits (D), and likely field status of soil moisture in a reclaimed profile. Such a balance will have a direct bearing on how the topsoil for a given area needs to be handled, how much topsoil is needed, how reclamation should proceed, and what might be the best adapted plant species to use at the site. Of particular interest is the comparison of Fig. 7c with Figs. 7a and 7b. The profiles in Centre County, Pennsylvania, and Salt Lake County, Utah, appear to have surplus (S) moisture in the spring, possibly giving rise to some runoff and erosion but in general providing adequate moisture for most plant species. Although Fig. 7c does not show the moisture deficit (D) until some of the profile water has been utilized (U), the fact that the profile is actively recharging in the spring and that no surplus exists suggests that spring moisture conditions may at times be limiting.

Climatic Considerations

Fig. 8 shows the four primary soil climatic and moisture regimes encountered in the continental United States. A hot-dry climate is characterized by an extensive water deficit through most of the year. A humid climate regime is common to soils that have adequate well-distributed rainfall and generally suffer no water deficits. Extensive surpluses, however, may lead at times to considerable runoff and erosion. A humid/dry climate regime is somewhat less adequate than the humid regime in supplying plant needs because in the summer and early fall large water deficits can develop. Lastly, there is the Mediterranean type of climate characterized by recharge during winter months and hot-dry summers. These soil climatic regimes will now be considered for some of the areas where mining and reclamation prevail in the hope of gaining some prior insight into necessary and appropriate operations for handling the topsoil.

Fig. 9 shows some representative climatic data and soil moisture regimes for a cross section of sites in the continental United States (Soil Survey Staff, 1975) listed in Fig. 1. The climatic data and annual profile water budgets range from an abundant rainfall site in Grays Harbor County, Washington, typifying a very wet humid climate, to a desert site

in Cochise County, Arizona, with a hot and dry climate. Whereas an overabundance of soil moisture prevails at all times in the former, the latter is dry and lacks adequate moisture for most of the year. Thus, topsoil handling operations in western Washington may have to contend with wet soggy conditions, while such operations in Arizona would require supplemental irrigation to establish and maintain vegetation. The soils of Montana, North Dakota, Utah, and Nevada in Fig. 7 have a humid/dry type of climate and are likely to show profile moisture deficits of varying length and severity, suggesting that, although topsoil handling operations should be carried out during these times of deficit, the lack of adequate moisture may limit revegetation on these soils. Moisture regimes of soils from Minnesota, Iowa, Illinois, northeastern Oklahoma, eastern Texas, and Georgia exemplify a humid climate and adequate soil moisture conditions for plant growth. The danger here, however, is the large potential for erosion. Much of Appalachia (Tennessee and Kentucky data), although generally classified as humid, often displays an incipient humid/dry soil moisture regime, largely because of shallow, draughty soils. Although precipitation in Appalachia is usually quite adequate—and at times excessive enough to cause erosion—short periods of moisture deficits in the late summer and early fall are common. Finally, the moisture budget and climatic data from California typify a Mediterranean climate with rainy winters, hot dry summers, and substantial soil moisture deficits in summer and early fall. Topsoil handling is limited to summer periods, whereas reclamation and revegetation are dependent upon moisture reserves generated in winter and spring.

Local and Regional Implications

Fig. 10 shows some very general regional climatic attributes of the continental United States that ought to be kept in mind when considering soil handling procedures. The shaded area in Fig. 10a shows areas where potential contribution of cropland to watershed sediment yield is high, while Fig. 10b represents annual potential percolation in excess of 180 mm (7 in.) (Stewart, 1975, pp. 15, 26). Topsoil handling in these areas should incorporate safeguards against excessive erosion and should endeavor to provide sufficiently deep and pervious reconstituted profiles to accommodate percolation in excess of 180 mm (7 in.). Data in Figs. 10 a and b appear to exclude much of Appalachia. However, a very steep precipitation gradient and the variable, usually rather shallow, soil conditions make broad generalizations for Appalachia difficult if not impossible. If anything, the threat of severe erosion in Appalachia is often compounded by the constraint of limited profile storage. The challenge of topsoil handling in this area is not only to maximize profile storage in order to reduce potential erosion but also to minimize deep percolation that may contribute to acid mine drainage.

Mine reclamation and revegetation operations should

Table 5. Temperature (T), Precipitation (P), Potential Evapotranspiration (PE), P-PE, and Soil Water Budget for Typic Natriboroll, in Bowman County, North Dakota, AWC = 216 mm

Month:	Jan.	Feb.	March	April	May	June	July	Aug.	Sept.	Oct.	Nov.	Dec.
T, °C	−14.2	−11.4	−4.3	6.0	14.3	18.5	22.9	21.4	15.7	8.6	0.0	−10.0
P, mm	14	14	25	36	54	104	64	57	43	21	14	11
PE, mm	0	0	0	29	79	114	154	129	79	29	0	0
P-PE, mm	14	14	25	7	−25	−10	−90	−72	−36	−8	14	11
Natriboroll, 216 mm*	39	53	78	85	60	50	−40	−112	−148	−156	14	25

* Available water capacity.

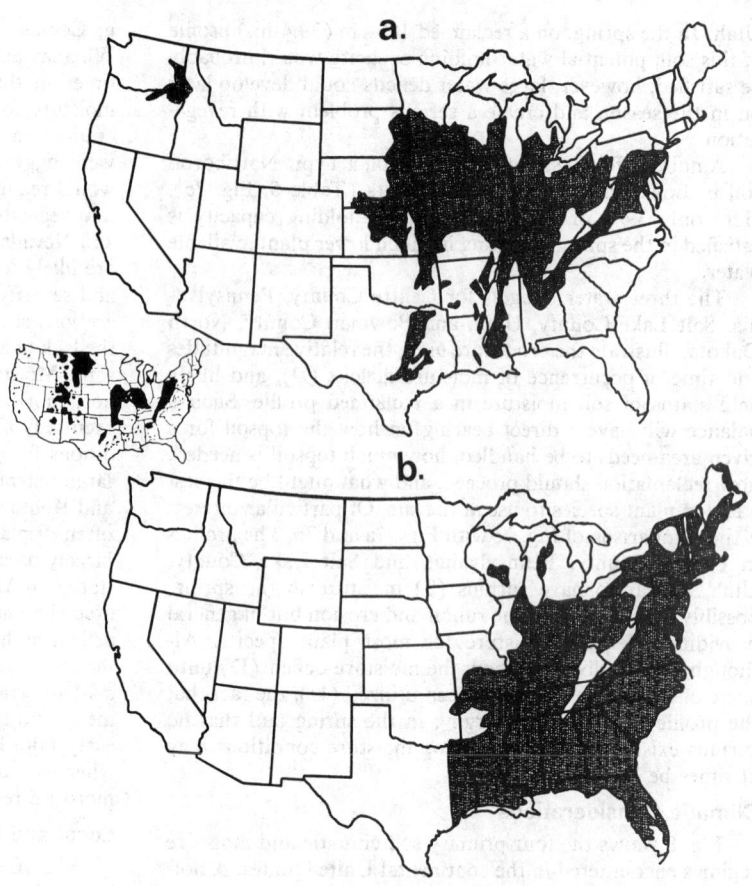

Fig. 10. Some general climatic attributes of the continental United States: (a) potential contribution of cropland to watershed sediment yield and (b) annual potential percolation in excess of 180 mm (7 in.) (from Stewart, 1975).

take into account not only the type of climate but also a projected spoil-profile water regime. The chances are that available water will be less, as much as 25 to 75% less (Pedersen, et al., 1980), in a reconstituted profile compared to natural soil. There are several reasons for this. In the East, large amounts of coarse fragments, mixed in with the spoil during reclamation, lower the water-holding capacity of the spoil, while in parts of the West high sodium and clay contents can substantially reduce infiltration and profile recharge during periods when moisture in the form of rain or snow is available. The amount and structure of topsoil material and the conditions at the soil-spoil interface can greatly affect both the total amount of water stored in the profile and water available for plant growth. In addition, compaction by heavy equipment can reduce aeration and increase amounts of water which, because it is held at higher tensions, is less available to plants. It should be remembered that natural soil profiles and layering result from long-time soil-forming processes that operate in response to climate, vegetation type, and relative amounts of available parent materials; on the other hand, man's knowledge of profile reconstruction is relatively new and short-lived.

ASSESSMENT OF PRODUCTIVITY POTENTIAL

A taxonomic soil description may contain much pertinent information about a soil, yet it says very little in a quantitative way of how productive a given soil is prior to mining and how productive it will be following mining. Consequently, it gives us little or no information on how the soil should be handled for optimum results.

Depending on whether the area mined is in the East, Midwest, West, or even South, it would be desirable to show objectively how mining affects soil productivity. It would also be desirable to propose two or more alternate ways of topsoil handling that attempt to minimize the impact of mining operations. To do this, we need to resort to modeling of soil productivity before and after a simulated mining operation.

Some currently available biomass productivity models approach the whole issue of productivity strictly from the perspective of climate. In Table 6, for example, biomass productivity (B) is calculated using mean annual values of temperature (Eq. 8) and precipitation (Eq. 9) in the so-called Miami model (Lieth, 1975, p. 246),

$$B_T = 3000/(1 + \exp(1.315 - 0.119T)) \qquad (8)$$

$$B_P = 3000(1 - \exp(-0.000664P)) \qquad (9)$$

$$\begin{aligned} B = B_T, \ B_T \leq B_P \\ B = B_P, \ B_P < B_T \end{aligned} \qquad (10)$$

where B, B_T, and B_P are biomass productivity in (g/m²/a), T is the mean annual temperature (°C), and P is the mean annual precipitation (mm).

The site productivity is taken as the lesser of temperature (B_T) or precipitation (B_P) dependent values (these are the starred values in Table 6). The major drawbacks of the model are that it does not take soil into account and that it computes productivity on a rather gross scale. Other models approach the productivity potential through soil effects alone. Kiniry, et al. (1982), assumed that the productivity index was a

Table 6. Biomass Productivity Potential as a Function of Mean Annual Temperature (T) or Precipitation (P) for Selected Areas in Fig. 1

County	State	Loc.	Temperature, T, °C	Precipitation, P, mm	Biomass Productivity	
					T, g/m²/a	P, g/m²/a
Grays Harbor	Washington	SW	10	1524	1406*	1909*
Blaine	Montana	NC	3	381	832	670*
Bowman	North Dakota	SW	6	381	1062	670*
Otter Trail	Minnesota	C	4	603	905*	990
Elko	Nevada	NE	10	178	1406	334*
Salt Lake	Utah	N	10	254	1406	466*
Shelby	Iowa	S	9	787	1318	1221*
Logan	Illinois	C	11	965	1496	1419*
Sanpete	Utah	C	10	254	1406	466*
Ottawa	Oklahoma	NE	15	1016	1846	1472*
Johnson	Illinois	S	14	1219	1961	1665*
Jefferson	Kentucky	NW	13	1143	1673	1596*
Cochise	Arizona	SE	16	178	1929	334*
Collin	Texas	W	18	1016	2087	1472*
Williamson	Tennessee	C	14	1321	1761	1752*
Brantley	Georgia	SE	20	1270	2231	1709*

* Starred values are taken as site biomass productivity.

function of certain physical and chemical soil properties which are or may become limiting. The primary soil properties of interest were bulk density, aeration porosity, available water, pH, and electrical conductivity. The model was weighted with depth by root distribution. When climatic variables of temperature and precipitation were included in the computations, Kiniry, et al. (1982), findings showed a higher correlation between the calculated index and field measured yields than when they were not included.

The Kiniry, et al. (1982), approach has been modified by the authors here, and combined with Lieth's (1975) Miami model for assessment of site productivity potential before and after mining. This combined model will guide the user in handling his topsoil. It will allow the user to choose which soil horizons he wishes to save and what soil properties he needs to maintain, improve, or amend in the reconstituted spoil. Coupled with the soil-moisture budgeting procedure, the model can make a realistic appraisal of the site's reclamation potential. Because the values used in the computations of moisture budgets and productivity factors are in fact distributions in space and time, spanning a range of values with different probabilities of occurrence, this model should be used as a relative rather than as an absolute predictor of productivity. In this sense it is particularly suitable for use in mining operations where the user generally wants to know how pre- and postmining conditions relate to one another.

Productivity Index

Relative productivity of a site (P) in grams per meter squared per year (ounces per square foot per year) can be written,

$$P = B \sum_{i=1}^{m} W \prod_{j=1}^{n} x_j \qquad (11)$$

where B is the biomass productivity, previously defined in Eq. 10; Σ is the summation operator over $i = 1, 2 \ldots m$ horizons, or layers; W is the relative root distribution weighting function to 1 m (3.3 ft) depth, dimensionless; Π is the product operator; $x_{j \ldots n}$ are productivity factors, $n = 6$; $x_{j=1}$ is the available water, by volume; $x_{j=2}$ is the bulk

density, g/cm³; $x_{j=3}$ is the aeration porosity, by volume; $x_{j=4}$ is pH; $x_{j=5}$ is EC, ds/m; and $x_{j=6}$ is other, such as sodium absorption ratio, topography, or nitrogen content.

The undisturbed field soil profile usually consists of several distinct horizons in a solum of different depths. Similarly, plant root distributions vary. Some plants are shallow rooted, others have roots going down to considerable depths. However, in reconstituted topsoil, root distribution, at least initially, should be quite similar and reasonably constant for most profiles. The model is simplified by considering a 100-cm (39.4-in.) deep soil profile, assuming that such a profile approximates an average realistic depth of reclaimed topsoiled profile available for plant growth. It is furthermore assumed that below this 1 m (3.3 ft) of topsoil, there is an unconsolidated spoil material compacted at the surface and consisting of rock fragments or unweathered and mixed alluvial, lacustrine, or aeolian deposits with little or no potential to support plant life. For comparison purposes, the 100-cm (39.4-in.) depth is used throughout all examples, realizing that some topsoiled profiles may be shallower and others deeper than the value discussed here. Taylor and Terrell (1982) compiled information on rooting depth of over 200 plant species. Their data show that, on the average, 1.5 ± 1 m (5 ± 3 ft) is the depth reached by many roots or their branches, to which the absorption of water and nutrients takes place.

The structure of the model given in Eq. 11 is fairly simple. Productivity factors based on soil data are computed individually for each horizon by considering the minima, or at times the maxima, required for plant growth. These values are then multiplied together to show where the problems may arise. The products are subsequently multiplied by an appropriate weighting factor for each horizon, approximating plant root distribution, summed, and multiplied by a site biomass productivity potential (Table 6), giving an overall site productivity index. Details of the model operation are available in Rogowski and Weinrich (1982, 1987) and Rogowski (1985); computer code can be obtained by writing to the senior author.

The factors, x_j, which go into Eq. 11 are themselves a major source of soil productivity. They should be understood both from the viewpoint of manipulating the model and reclaimed soil.

Weighting Factor (Roots): The roots distribution weighting factor, w, can be computed using the method proposed by Horn (1971) and adapted by Kiniry, et al. (1982). The principal assumption of this model is that the biomass production is a function of root growth that, in turn, depends primarily on available water, bulk density, pH, and where applicable, on conductivity, sodium adsorption ratio, and nutrient status of the soil. The weighting function is dependent upon total rooting depth, R, and the depth, r, of an individual layer in the soil profile. When integrated with depth and divided by total profile depletion, the method gives the root distribution weighting function (W) of Eq. 11 shown in Fig. 11. Others (Gardner, 1964) have described root distribution as decreasing logarithmically with depth.

Available Water Capacity: Available water capacity (AWC) in millimeters, $x_{j=1}$, should realistically describe the actual amount of water available for plant use during the growing season. This figure will vary depending on the climate, type of soil, and water use efficiency of a particular plant species. Table 7 shows available water capacity values (AWC) based on texture (Kiniry, et al., 1982; Peterson, et al., 1968). It has also been customary to express AWC as the difference between water contents of 33 and 1500 kPa.

Table 7. Potential Available Water Capacity Estimated from Soil Texture

Texture	cm/cm	mm*
Sand		
Coarse sand	0.016	16
Medium sand	0.030	30
Fine sand	0.066	66
Loamy sand	0.070	70
Loam		
Sandy loam	0.115	115
Fine sandy loam	0.130	130
Loam	0.180	180
Silt loam	0.190	190
Silt	0.200	200
Very fine sandy loam	0.200	200
Clay		
Clay	0.100	100
Sandy clay	0.110	110
Silty clay	0.115	115
Sandy clay loam	0.125	125
Clay loam	0.145	145
Silty clay loam	0.145	145

* In 100 cm profile; for textures where $0.15 \leq$ clay ≤ 0.40 estimated water-holding capacity was computed as the average of textural limits for a given texture class from AWC = 0.25-0.35 clay.

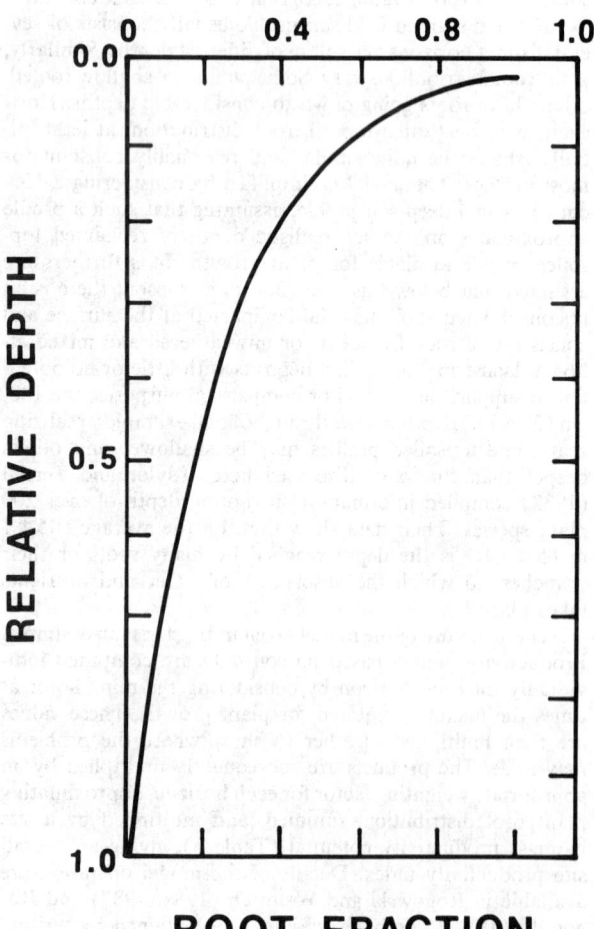

Fig. 11. Estimated distribution of roots with depth in a dimensionless profile.

These values may be satisfactory for use in estimation, provided they reflect actual soil conditions during a growing season from planting to harvest. In the humid East, where the annual percolation is usually in excess of 180 mm (7 in.) (Fig. 10b), this generally is the case; however, further west, climatic consideration may override textural properties and the precipitation simply may not be enough to fill the available storage capacity. It is therefore suggested that AWC be chosen as the lesser of the values in Table 7 (or a value computed from a desorption curve at 33 and 1500 kPa) and the actual measured or estimated field values. While a limiting AWC value of 0.20 is probably satisfactory in the East and Midwest, a higher value may need to be used in the West where high evaporative demand and low rainfall lead to soils with pronounced summer moisture deficits. Under such conditions, model computations may also suggest to what extent supplemental irrigation is necessary on the newly reclaimed spoils.

Bulk Density and Aeration Porosity: Both soil bulk density and aeration porosity are modified during topsoil handling. Assuming that reserves of moisture and nutrients are adequate, soil productivity depends largely on how well plant roots can access these reserves. It has long been established that compacted layers impede root growth by preventing root elongation, limiting respiration, and at times contributing to water logging. Soil bulk density (BD), $x_{j=2}$, is commonly used as an index of compaction. Its effects, however, should be evaluated relative to soil texture, moisture, and moisture content at the time the soil is handled. Soil moisture content is particularly relevant to consider on reclaimed profiles where dense layers can often be produced if the topsoil is handled at or near so-called *optimum moisture content*. It is a well-known fact in soil mechanics that com-

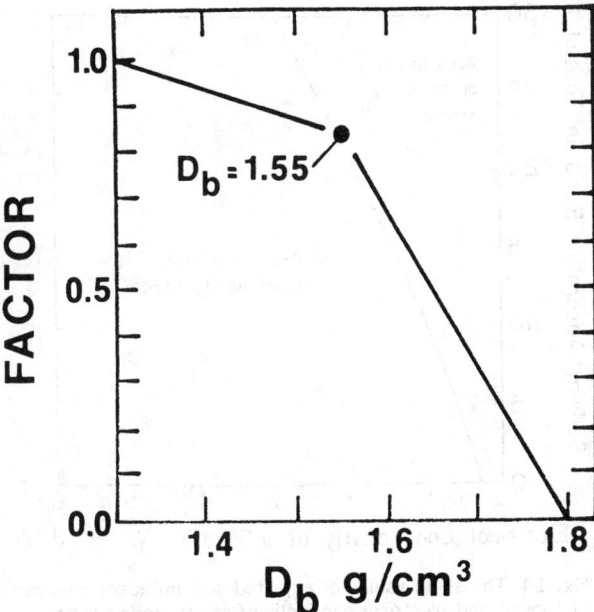

Fig. 12. Productivity factor bulk density.

paction at the optimum moisture content will result in maximum attainable values of bulk density for a given soil layer (Terzaghi and Peck, 1948). Such values of bulk density and of the associated moisture content are usually listed in SCS County Soil Survey publications in the engineering properties tables. On the other hand, agriculturalists know that if a seedbed is not firm enough to be in good contact with a planted seed, poor germination and excessive droughtiness of a soil may result. To account for these limitations, a bulk density productivity factor (D) suggested by Kiniry, et al. (1982), Fig. 12, is used in the model. In general, BD suitable for plant growth will range from 1.30 to 1.80 g/cm³ (0.047 to 0.065 lb per cu in.). A good discussion of the effects of bulk density is given by Pearson (1965) and again by Bowen (1981).

In the productivity model, most physical and hydraulic effects of bulk density are assumed to be incorporated into the factor D. Other effects, such as adequate aeration of root growth medium so necessary for proper respiration and functioning of the roots, are grouped under the productivity factor aeration porosity. The two are related through the equation,

$$P_a = 1.0 - (BD/2.65) - \theta \qquad (12)$$

where P_a, $x_{j\,=\,3}$, is the aeration porosity of a fully recharged profile, θ is the moisture content at field capacity, and 2.65 is the particle density of a mined soil. Field capacity will vary with soil texture, structure, and amount of organic matter present in the profile. In general, it should be somewhere between moisture contents measured at 10 and 33 kPa. Critical aeration porosity ($P_{crit} = 0.10$), when root growth may become restricted, ranges from 0.05 to 0.15 pore space by volume (Canell and Jackson, 1971; Pearson, 1965). Realistically aeration porosity effect should also include (but does not) a built-in dependence on time, a geometry factor to describe degree of continuity between air-filled pores and concentration level of CO_2. In some of the mine spoils where root respiration may compete for pore oxygen with oxidation of pyrite or iron in the profile, or when heavy additions of organic material (such as sludge) place an additional demand

on pore oxygen, it may well be advisable, particularly for deeper layers, to set P_{crit} higher than the recommended value. An operator should be concerned if at any time during the growing season, particularly when soil is wet, or following a heavy incorporation of organic matter, the soil oxygen concentration is likely to drop near critical level for even a short period of time (Canell and Jackson, 1981). In the event of water logging, for instance, aeration porosity may become quite critical.

At times, however, it may be desirable to minimize organic matter conversion or nutrient leaching in the topsoil that is stockpiled before being spread. Under these conditions, bulk density or aeration porosity may be manipulated so as to make the weighting factors for BD and P_a as small as possible.

pH: Soil reaction (pH), $x_{j\,=\,4}$, values appear particularly well suited to characterize the productivity response of reconstituted mine soil profiles in the East. Critical pH (Spurway, 1941) varies among soils and plant species and with time. The response to pH on acid soils may result from hydrogen, aluminum, and manganese toxicities and/or calcium or magnesium deficiencies. Thus, soils at the same value of pH could have limited yields for different reasons and the limiting factors would operate at different intensities in time and space (Adams, 1981; Pearson, 1965). Consequently, the model, which follows Neal's (1979) original formulation, should be used with caution and can be adjusted if sufficient information about a particular site or plant species is readily available. In the meantime, the model will provide sufficient guidance for the potential user in differentiating between the layers that may cause problems and those that will not.

The pH productivity factor (pH_F) is shown in Fig. 13, where pH denotes a measured value of pH in the 1:1 aqueous solution.

Electrical Conductivity (EC): An argument similar to that offered for pH applies as well to the effects of salinity on plant growth (Hoffman, 1981). Salinity effects may vary spatially with soil type, texture, and moisture status while different plant species will exhibit different tolerances. Salinity associated problems are almost certain to be present if reclamation is carried out in arid regions when original soils contain sufficient soluble salts derived either from marine

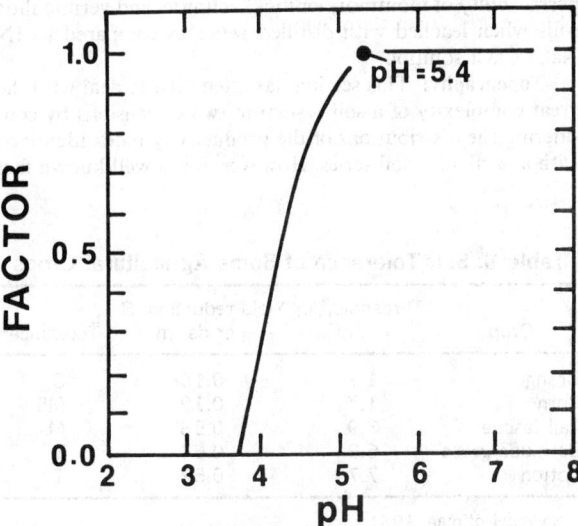

Fig. 13. Productivity factor pH.

deposits or from soil weathering. The problems can also arise in semiarid regions whenever rainfall is equal to evapotranspiration, as a result of upward artesian flow from aquifers, from overirrigation and from resulting saline seeps in adjacent areas, or if high water tables are present. Consequently, particular attention needs to be paid to the water regime and potential flow pathways in reconstituted profiles. Hoffman (1981) rates plants (his Table 9.3.1, p. 315) according to their salt tolerance, and suggests an appropriate form of the productivity factor, EC_F

$$EC_F = 1.0 - B(EC - A) \qquad (13)$$

where A is salinity threshold value, B is the yield reduction per unit of salinity increase, and EC is the electrical conductivity of soil saturation extract in ds/m. Selected values abstracted from the Hoffman (1981) table are shown in Table 8. Cursory inspection of Table 8 will show that EC_F can vary considerably depending on the crop used. Nevertheless, the model will alert the user that the salinity problem may exist if the profile is reclaimed in a similar form to original soil.

Sodium Saturation Ratio (SAR): Soils or soil horizons that have excess sodium on their exchange complex are known as sodic soils or sodic horizons. Such soils when leached with low electrolyte-content waters may show a marked decrease in permeability (Reeve and Bower, 1960; Frenkel, et al., 1978). Many studies (Rhoades, 1982) have dealt with reclamation of sodic soils and with evaluation of the irrigation water quality (Rhoades, 1972; Oster and Rhoades, 1976). Guidelines regarding the suitability of irrigation waters for agriculture, based on the type of predominant clay mineral, have been proposed (Ayers and Westcot, 1976) and questioned by more recent findings (Shainberg, et al., 1981; Frenkel, et al., 1978, and Suarez, et al., 1983). Currently the general consensus appears to be that the permeability of sodic soils depends on the electrolyte level a soil maintains—substantial decreases having been observed with low electrolyte contents (Shainberg, et al., 1981; Rhoades, 1982)—and may decrease with increasing pH (Suarez, et al., 1983). Such variations in permeability would have a significant effect on infiltration and subsequently on the amount of water available to plants for a given soil. Fig. 14 from Rhoades (1983) summarizes the present situation for some of the more sensitive arid land soils. On the average, Frenkel, et al. (1978) have observed an 83% reduction in permeability of montmorillonitic, kaolinitic, and vermiculitic soils when leached with distilled water as compared to 1N NaCl-CaCl solution.

Topography: This section has attempted to deal with the great complexity of a soil system in two dimensions by considering the distributions of the productivity index identified with a particular soil series. However, it is a well-known fact

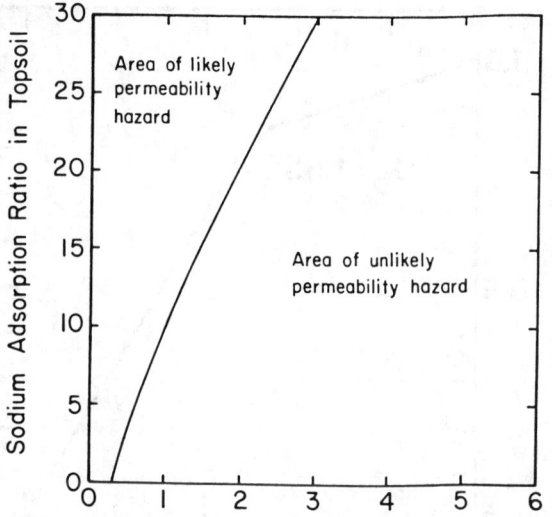

Fig. 14. Threshold values of adjusted sodium adsorption ratio of topsoil and electrical conductivity of infiltrating water (assumed to be in equilibrium with soil solution) for maintenance of soil permeability (from Rhoades, 1982).

in pedology (Jenny, 1980) that local relief, aspect, and drainage are among the most significant modifiers of the soil profile, particularly on a small scale. Thus, south-facing slopes may have a different vegetation, moisture, or temperature regime than north-facing ones. Soils developed on the hilltops are likely to be shallow compared to those developed near the base. Properties such as cation exchange capacity (CEC) or clay content may vary significantly with elevation (Fig. 15) and drainage and available moisture may range

Table 8. Salt Tolerance of Some Agricultural Crops

Crop	Threshold, A, ds/m	Yield reduction, B, per ds/m	Tolerance*
Orange	1.7	0.16	S
Corn	1.7	0.12	MS
Tall fescue	3.9	0.53	MT
Bermuda grass	6.9	0.64	T
Cotton	7.7	0.52	T

Source: Hoffman, 1981.

* S = sensitive, MS = moderately sensitive, MT = moderately tolerant, T = tolerant.

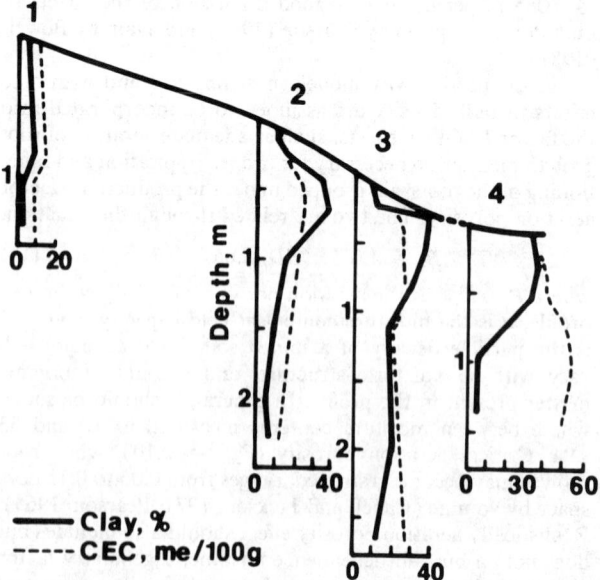

Fig. 15. Distribution of clay and cation exchange capacity in a semiarid toposequence (from Nettleton, et al., 1968); Nos. 1, 2, 3, and 4 denote successive soil series that have developed at different elevations.

from excessive to impaired depending on position relative to the slope and degree of profile anisotropy (Zaslavsky and Rogowski, 1969).

Considerations such as these should be incorporated into the productivity assessment primarily through a representative sampling of the area to be reclaimed through use of parameter values that adequately reflect the area's heterogeneity. This is of paramount importance when attempting to reclaim land in Appalachia and elsewhere where mountainous or rapidly changing conditions prevail. Locally, climatic regimes, such as those in Fig. 8, normally separated by many hundreds of kilometers (miles), can appear within a span of a few hundred meters (feet) or exist separately on the windward or leeward sides of the same hill.

TYPICAL PROFILES

Fig. 1 shows the approximate location of counties that contain the typical profiles discussed here; Table 6 gives the values of biomass productivity index based on climate alone, and Appendix 1 lists the input and output values obtained following the execution of the Comparative Biomass Productivity Model.

Fig. 16 illustrates a typical computer program output sheet for a soil, in this case profile #3, a Mollisol, from Bowman County, North Dakota. The lower table in Fig. 16 contains input parameters such as bulk density, available water content, pH, and electrical conductivity as well as the mean annual temperature, precipitation, and clay content in each horizon. In computing available soil water-holding capacity, values higher than 0.20 cm/cm were set equal to 0.20. The actual water-holding capacity for this and other profiles listed in Appendix 1 was taken as the customary difference between water content at 33 and 1500 kPa. When

a profile is or can be fully recharged in the spring, the foregoing procedure may be correct. As mentioned previously, soils can and do retain water at tensions lower than 33 kPa (⅓ bar), and some plants can utilize water at tensions higher than 1500 kPa (15 bar). Thus, a good practical estimate of soil water-holding capacity is the difference between water held at 0.9 saturation and 1500 kPa (15 bar). In areas with insufficient rainfall, such as the Bowman County, North Dakota, example in Fig. 16, a more nearly correct estimate is to subtract, as was done in Table 9, water content at 1500 kPa from an estimated profile water content in the spring. As a last resort, when desorption data are lacking but values for clay content are available, water capacity can be estimated from clay content and climatic data, as discussed earlier (Table 7).

Productivity factor values given in the upper table of Fig. 16 for profile No. 3 constitute the output pertinent to choosing an appropriate procedure for topsoil handling. The magnitude of the five productivity factors is listed by depth and horizon for this soil in columns 6 through 10. The factors range from 0 (critical value exceeded) to 1 (no soil limitations to root growth). Their product in column 11 shows the quality of each layer. Thus in Fig. 16, the bottom layer [> 97 cm (38 in.)], a 30 to 53-cm (12 to 21-in.) layer, and a 71 to 97-cm (28 to 38-in.) layer show low product values. In the bottom layer critical density is exceeded [1.84 g/cm³ (0.066 lb per cu in.)] while the 30 to 53-cm (12 to 21-in.) layer exhibits high density throughout. These layers, because of relatively higher clay contents, may compact too much during reclamation and should be handled with care. A more serious condition is that of the high electrical conductivity values that reduce the *EC* productivity factor in all layers below 30 cm (12 in.). The analysis suggests that at this site

PRODUCTIVITY POTENTIAL FOR MOLLISOL · · · · · · · · · · · BOWMAN COUNTY, NORTH DAKOTA

PRODUCTIVITY FACTOR / INDEX

SOIL	NAME	DEPTH(CM)	HORIZON	WATER	DENSITY	AERATION	PH	EC	PRODUCT	ROOTS	PRODUCTIVITY	BIOMASS(G/SQ M/YR)
3	P87	0 8	A	1.0000	1.0000	1.0000	1.0000	1.0000	1.0000	0.2653	0.6960	466.7
3	P87	8 18	A	1.0000	0.9076	0.9737	1.0000	1.0000	0.8996	0.2108	0.4307	288.8
3	P87	18 25	A	1.0000	0.9144	0.9605	1.0000	1.0000	0.8830	0.1104	0.2411	161.7
3	P87	25 30	A	0.9500	0.9212	0.9563	1.0000	1.0000	0.8383	0.0662	0.1436	96.3
3	P87	30 38	B	0.8000	0.1700	0.8807	1.0000	0.9685	0.1200	0.0886	0.0881	59.1
3	P87	38 53	B	1.0000	0.4688	0.8383	1.0000	0.3303	0.1327	0.1218	0.0775	51.9
3	P87	53 71	B	1.0000	0.8008	0.8343	1.0000	0.8378	0.5610	0.0889	0.0613	41.1
3	P87	71 97	B	0.9500	0.3692	0.8017	1.0000	0.8378	0.2410	0.0476	0.0115	7.7
3	P87	97 100	I	1.0000	0.0	0.7943	1.0000	0.2303	0.0	0.0005	0.0	0.0

INPUT AND PROGRAM PARAMETERS

SOIL	NAME	DEPTH(CM)	HORIZON	CLAY	DEN(G/CC)	WATER	PH	EC (MMHOS/CM)	RAIN(MM)	TEMP(C)
3	P87	0 8	A	0.0910	1.26	0.2000	6.40	0.0	381.0	6.0
3	P87	8 18	A	0.1050	1.43	0.2000	6.90	0.0	381.0	6.0
3	P87	18 25	A	0.0950	1.42	0.2000	7.70	0.0	381.0	6.0
3	P87	25 30	A	0.0610	1.41	0.1900	8.40	0.0	381.0	6.0
3	P87	30 38	B	0.1400	1.75	0.1600	9.00	3.40	381.0	6.0
3	P87	38 53	B	0.1950	1.66	0.2000	8.60	11.70	381.0	6.0
3	P87	53 71	B	0.0930	1.56	0.2000	8.50	5.10	381.0	6.0
3	P87	71 97	B	0.0890	1.69	0.1900	9.30	5.10	381.0	6.0
3	P87	97 100	I	0.1890	1.84	0.2000	8.90	13.00	381.0	6.0

Fig. 16. Computer printout containing input and output of the comparative biomass productivity model for a profile in Bowman Co, ND, without correction for SAR.

Table 9. Computations of Available Water Capacity

Depth, cm		Available water capacity, cm/cm*			
		1	2	3	4
0	0	0.32	0.20	0.12	0.60
8	18	0.29	0.20	0.11	0.55
18	25	0.21	0.20	0.08	0.40
25	30	0.19	0.19	0.07	0.36
30	38	0.16	0.16	0.06	0.38
38	53	0.20	0.20	0.08	0.40
53	71	0.21	0.20	0.08	0.40
71	97	0.19	0.19	0.07	0.37
97	100	0.24	0.20	0.09	0.45
Profile capacity, mm, 216				84	

*1. Computed as water content at 0.33 kPa (⅓ bar) less water content at 1500 kPa (15 bar).
2. Values listed in Fig. 16.
3. Computed as the difference between the amount of water held by the profile following recharge in spring prior to planting and water content of 1500 kPa. (See Table 5 and Fig. 7c.) Subsequently all values under (1) were reduced in the ratio of actual to potential water-holding capacity 84/216 = 0.39.
4. Column 3/column 2.

the 0 to 30-cm (0 to 12-in.) layer is best for plant growth; therefore it should be segregated and handled with care.

Column 12 in Fig. 16 gives the estimated profile root distribution, which when multiplied by the values in column 11 (product) and summed, gives the cumulative profile productivity index. The total profile productivity value of 0.6960 when multiplied by the biomass productivity (Table 6) gives the profile biomass productivity factor of 467 g/m²/a (1.53 oz per sq ft per year).

Because of limited recharge at this site (see Fig. 7c and Table 5), only part ($^{84}/_{216}$) of the available water capacity will be filled. Consequently, available water should be reduced by that amount (Table 9, column 3). The profile productivity will then be reduced in the ratio of column 3 to column 2 as given by column 4. When this correction is applied to

column 11 and subsequent steps are carried out as before, the profile biomass productivity drops to 236 g/m²/a (0.77 oz per sq ft per year).

The discussion of Fig. 16 illustrates the type of data the model will provide the user. With this kind of information, topsoil handling is guided so as to achieve optimum results, and individual results can be compared with results from other areas in the nation. Table 10 summarizes the output of the comparative biomass productivity model for some typical profiles. The input information for these profiles, chosen for soils from or near major mining areas, was extracted from a Soil Survey Staff (1975) publication, *Soil Taxonomy*. The inputs and outputs for all profiles are listed in Appendix 1. Here only selected ones will be discussed to illustrate a point.

Table 10. Biomass Productivity and Limitations for Selected Locations Near or in Mining Areas

No.	Location	Biomass,‡ g/m²/a	Limitations*	Order†
1	Grays Harbor Co., Washington	1169	pH	I
2	Blaine Co., Montana	186	AW, BD, A, SAR	M
3	Bowman Co., North Dakota	464	AW, BD, A, EC, SAR	M
5	Elko Co., Nevada	224	AW, BD, A	M
6	Salt Lake Co., Utah	464	C	M
7	Shelby Co., Iowa	837	AW, BD, A	M
8	Logan Co., Illinois	1016	AW, BD, A	M
9	Sanpete Co., Utah	447	AW, C	M
10	Ottawa Co., Oklahoma	862	BD, A	A
11	Johnson Co., Illinois	1340	BD, A, pH	A
12	Jefferson Co., Kentucky	949	AW, BD, A, pH	A
13	Cochise Co., Arizona	51	AW, BD, A	AR
14	Collin Co., Texas	3	BD, SAR	V
15	Williamson Co., Tennessee	852	AW, BD, A, pH	A
17	Greenbrier Co., West Virginia	589	BD, pH	I

* Limitations: limiting factors, AW = available water, BD = bulk density, A = aeration, pH = pH, EC = electrical conductivity, C = Climate, SAR = sodium adsorption ratio.
† Soil order: I = inceptisol, M = mollisol, A = alfisol, AR = aridisol, V = vertisol.
‡ 1 g/m²/a × 0.004467 = 1 ton per acre per year.

The biomass productivity values in Table 10 range from a low of 3 g/m²/a (0.0098 oz per sq ft per year) for a dense Vertisol in Texas (No. 14) and 51 g/m²/a (0.167 oz per sq ft per year) for a dry Aridisol in Arizona (No. 13) to highs of over 1000 g/m²/a (3.28 oz per sq ft per year) for an Inceptisol in Washington (No. 1) and a Mollisol in Illinois (No. 3), [these numbers translate to potential biomass production ranging from 1.4 kg (3 lb) to more than 13.4t/ha (6 tons per acre per year)]. Table 10 ranks these soils according to their productivity. Thus, most care should be taken in handling the really productive soil, like the profiles Nos. 1, 8, or 11, to restore them to nearly the same productivity values. Less productive soils, on the other hand, afford an opportunity for modifying the profile so as to increase productivity.

Profile No. 14, a Vertisol from Texas, has high sodium content, excessive bulk density throughout, and, associated with it, low aeration values in most horizons. If the area was managed properly so as to decrease the bulk density and increase aeration in the upper 56 cm (22 in.), this soil, despite adverse SAR, would have a potential biomass productivity in the neighborhood of 600 g/m²/a (1.97 oz per sq ft per year). Profile No. 13 (Arizona) lacks water, has high density, and low aeration values. Supplemental irrigation and proper management to decrease density and increase aeration could substantially increase productivity.

Even some of the better profiles can be improved. For instance, profile No. 1 from Washington, which is limited by pH, would benefit from liming. Profile No. 8 from Illinois, which is somewhat dry and in parts limited by bulk density and adequate aeration, could benefit from both an increase in water-holding capacity, perhaps by adding organic matter, and from an overall improvement in management, by decreasing density and increasing aeration. Here the user will have to make a choice. If the site is reclaimed as is, the projected profile productivity value would be 1016 g/m²/a (3.3 oz per sq ft per year). If the bulk density (BD) and aeration in the reconstituted profile were engineered so as to make their productivity factors equal to 1 (i.e., $BD \leq 1.30$), the biomass productivity value would increase to 1295 g/m²/a (4.24 oz per sq ft per year) for a productivity gain of about 27%. If on the other hand, water-holding capacity (AWC) was increased and other factors were kept constant, a somewhat lower increase to 1237 g/m²/a (4 oz per sq ft per year) (a 21% gain) would result. If both AWC and BD were improved, a biomass productivity increase to 1419 g/m²/a (4.65 oz per sq ft per year) (Table 6) could take place (a 40% gain). Under these circumstances, the profile would have no apparent limitations to plant growth and would be controlled by climate alone. A similar situation appears to exist in the two Utah profiles (Nos. 6 and 9) where climate plays the dominant role in determining the productivity potential. Under such circumstances reclamation to status quo and supplemental irrigation may be advised.

In much the same way, the other profiles listed can be analyzed. Their limitations to plant growth are expressed primarily in terms of available water (Nos. 2, 3, 5, 7, 8, 9, 12, 13, 15), bulk density and aeration (Nos. 2, 3, 5, 7, 8, 10, 11, 12, 13, 14, 16), pH (Nos. 1, 12, 15), EC (No. 3), SAR (Nos. 2, 3, 14), and climate (Nos. 6, 9). A different selection of soil profiles, with emphasis on Appalachia, would perhaps show a very strong influence of pH as, for example, in profile No. 17 (not shown in Fig. 1). Insufficient data other than climatic were found for profile Nos. 4 and 16; hence their biomass productivity values are not included.

RECOMMENDATIONS AND CONCLUSIONS

How best to handle topsoil on a mined area? The user should answer this question by first placing it in the context of the area's location. He will note that while abundant rain, erosion potential, and low pH values will create problems in the East, lack of water, salinity, excess sodium, and high bulk density will govern the use of reclamation techniques and topsoil handling in the West. The user should next examine the climate to see what kind of moisture budget he may expect in his area and simultaneously check the soils at the mine site for any possible problems that may arise because of their taxonomic classification. Having satisfied himself about the broad implications of location, climate, and soils, the user will need to get more specific. Execution of the comparative biomass model will point out the problem areas. Having identified his problem soils, the user may then want to select different options available, and to simulate possible outcomes, also by using the comparative biomass model. Having decided on the desired course of action, the user would want to consult with the soil conservation service or university extension personnel in this area to select specific techniques recommended to overcome particular problems on the area of interest.

It is hoped that the procedure outlined may streamline the solving of problems associated with topsoil handling, may identify areas that need closer attention, and may suggest techniques to improve productivity of a mined site compared to original soil.

REFERENCES

Adams, F., 1981, "Alleviating Chemical Toxicities," *Modifying the Root Environment to Reducing Stress.* G.F. Arkin and H.M. Taylor, eds., ASAE Monograph No. 4, American Society of Agricultural Engineers, St. Joseph, MO, p. 269.

Ayers, R.S., and Westcot, D.W., 1976, "Water Quality for Agriculture," Irrigation and Drainage Paper No. 29, Food and Agriculture Organization of the United Nations, Rome, Italy.

Beckett, P.H.T., and Webster, R., 1971, "Soil Variability: A Review," *Soils and Fertilizers,* Vol. 34, No. 1, pp. 1–14.

Bowen, H.D., 1981, "Alleviating Mechanical Impedance," *Modifying the Root Environment to Reducing Stress,* G.F. Arkin and H.M. Taylor, eds., ASAE Monograph No. 4, American Society of Agricultural Engineers, St. Joseph, MO, p. 2111.

Bower, C.A., Ogata, G., and Tucker, J.M., 1968, "Sodium Hazard of Irrigation Waters as Influenced by Leaching Fraction and by Precipitation or Solution of Calcium Carbonate," *Soil Science,* Vol. 106, pp. 29–34.

Canell, R.Q., and Jackson, M.B., 1981, "Alleviating Aeration Stress," *Modifying the Root Environment to Reducing Stress,* G.F. Arkin and H.M. Taylor, eds., ASAE Monograph No. 4., American Society of Agricultural Engineers, St. Joseph, MO, p. 191.

Chang, J.H., 1968, *Climate and Agriculture,* Aldine Publishing Co., Chicago.

Farazdaghi, H., Kaschani, A., and Asadi, N., 1982, "A Mathematical Model for Seasonal Dry Matter Production of Crops under Different Nitrogen and Irrigation Water Supplies," Paper No. NAR 82-303, ASAE 1982 North Atlantic Regional Meeting, Burlington, VT.

Feddes, R.A., 1981, "Water Use Models for Assessing Root Zone Modification," *Modifying the Root Environment to Reducing Stress,* G.F. Arkin and H.M. Taylor, eds., ASAE Monograph No. 4., American Society of Agricultural Engineers, St. Joseph, MO, p. 347.

Frenkel, H., Goertzen, J.O., and Rhoades, J.D., 1978, "Effects of Clay Type and Content, Exchangeable Sodium Percentage, and Electrolyte Concentration on Clay Dispersion and Soil Hydraulic Conductivity," *Soil Science Society of America Journal,* Vol. 42, No. 1, pp. 32–39.

Gardner, W.R., 1964, "Relation of Root Distribution to Water Uptake and Availability," *Agronomy Journal,* Vol. 56, pp. 41–45.

Harradine, F.F., 1949, "The Variability of Soil Properties in Relation to State of Profile Development," *Proceedings,* Soil Science of America, Vol. 13, pp. 302–311.

Hoffman, G.J., 1981, "Alleviating Salinity Stress," *Modifying the Root Environment to Reducing Stress,* G.F. Arkin and H.M. Taylor, eds., ASAE Monograph No. 4., American Society of Agricultural Engineers, St. Joseph, MO, p. 305–341.

Horn, F.W., 1971, "The Prediction of Amounts and Depth Distribution of Water in a Well-Drained Soil," MS Thesis, University of Missouri, Columbia.

Jaynes, D.B., Rogowski, A.S., Pionke, H.B., and Jacoby, E. L., Jr, 1983, "Spoil Atmosphere and Temperature in a Reclaimed Coal Stripmine," *Soil Science,* 136: 164–177.

Jenny, H., 1980, "The Soil Resource," *Ecological Studies 37,* Springer-Verlag, New York.

Kiniry, L.N., Scrivner, C.L., and Keener, M.E., 1983, "A Soil Productivity Index Based upon Predicted Water Depletion and Root Growth," Research Bulletin 1051, Univ. of Missouri, Columbia, MO.

Levy, G., and Toutain, F., 1982, "Aeration and Redox Phenomena in Soils," *Constituents and Properties of Soils,* M. Bonneau and B. Souchier, eds., Academic Press, New York.

Lieth, H., 1975, "Modeling the Primary Productivity of the World," *Primary Productivity of the Biosphere,* H. Lieth and R.H. Whitiaker, eds., Springer-Verlag, New York.

Lutwick, L.E., and Smith, A.D., 1977, "Yield and Composition of Alfalfa and Crested Wheat Grass, Grown Singly and in Mixture, as Affected by N and P Fertilizers," *Canadian Journal of Plant Science,* Vol. 57, pp. 1077–1083.

Lyon, T.L., Buckman, H.O., and Brady, N.C., 1952, *The Nature and Properties of Soils,* The Macmillan Co., New York.

Miller, R.M., and Cameron, R.E., 1976, "Some Effects on Soil Microbiota of Topsoil Storage during Surface Mining," 4th Symposium on Surface Mining and Reclamation, National Coal Association, pp. 131–139.

National Research Council, Committee on Soil as a Resource in Relation to Surface Mining, 1981, *Surface Mining: Soil, Coal, and Society,* National Academy Press, Washington, DC.

Neill, L.L., 1979, "An Evaluation of Soil Productivity Based on Root Growth and Water Depletion," unpublished MS Thesis, University of Missouri, Columbia.

Nielson, D.R., Biggar, J.W., and Erh, K.T., 1973, "Spatial Variability of Field Measured Soil Water Properties," *Hilgardia,* Vol. 42, No. 7, pp. 215–259.

NOAA, 1978, *Climatological Data, Annual Summary, Pennsylvania,* Vol. 83, No. 13, US Dept. of Commerce, National Climatic Center, Asheville, NC, pp. 1–15.

Oster, J.D., and Rhoades, J.D., 1976, "Various Indices for Evaluating the Effective Salinity and Sodicity of Irrigation Waters," *Proceedings,* International Salinity Conference, Texas Technology University, Lubbock, Texas, Aug., pp. 1–14.

Pearson, R.W., 1965, "Soil Environment and Root Development," *Plant Environment and Efficient Water Use,* N.H. Pierre, ed., American Society of Agronomy, Madison, WI, p. 9511.

Pedersen, T.A., Rogowski, A. S., and Pennock, R., Jr, 1978 "Comparison of Morphological and Chemical Characteristics of Some Soils and Minesoils," *Reclamation Review,* Vol. 1, pp. 143–156.

Pedersen, T.A., Rogowski, A.S., and Pennock, R., Jr, 1980, "Physical Characteristics of Some Minesoils," *Soil Science Society of America Journal,* Vol. 44, No. 2, pp. 321–328..

Peterson, G.W., Cunningham, R.L., and Matelski, R.P., 1968, "Moisture Characteristics of Pennsylvania Soils: Moisture Retention as Related to Texture," *Proceedings,* Soil Science Society of America, Vol. 32, pp. 271–275.

Prine, G.M., and Burton, G.W., 1956, "The Effects of Nitrogen Rate and Clipping Frequency upon the Yield, Protein Content and Certain Morphological Characteristics of Coastal Bermuda Grass," *Agronomy Journal,* Vol. 48, pp. 296–301.

Reeve, R.C., and Bower, C.A., 1960, "Use of High Salt Waters as a Flocculent and Source of Divalent Cations for Reclaiming Sodic Soils," *Soil Science,* Vol. 90, pp. 139–144.

Rhoades, J.D., 1972, "Quality of Water for Irrigation," *Soil Science,* Vol. 113, No. 4, pp. 277–284.

Rhoades, J.D., 1982, "Reclamation and Management of Salt-Deffected Soils after Drainage," Soil and Water Management Seminar, Nov. 29-Dec. 2, Lethbridge, Albe., Canada.

Riley, C.M., 1965, *Our Mineral Resources,* John Wiley and Sons, New York.

Rogowski, A.S., and Jacoby, E.L., Jr., 1979 "Monitoring Water Movement through Strip Mine Spoil Profiles," *Transactions ASAE,* Vol. 22, pp. 104–109, 114.

Rogowski, A.S., and Weinrich, B.E., 1982, "Effects of Erosion on Productivity: A Geostatistical Approach," Paper No. NAR82-204, Annual Meeting, North Atlantic Region, American Society of Agricultural Engineers, Burlington, VT.

Rogowski, A.S., and Weinrich, B.E., 1987, "Modeling the Effects of Mining and Erosion on Biomass Productivity," *Ecological Modeling 35,* pp. 85–112.

Rogowski, A.S., 1985, "Evaluation of Potential Topsoil Productivity," *Environmental Geochemistry and Health,* Vol. 7, No. 3, pp. 87–97.

Shainberg, I., Rhoades, J.D., Suarez, D.L., and Prather, R.J., 1981, "Effect of Mineral Weathering on Clay Dispersion and Hydraulic Conductivity of Sodic Soils," *Soil Science Society of America Journal,* Vol. 45, No. 2, pp. 287–291.

Soil Survey Staff, 1975, *Soil Taxonomy,* Agriculture Handbook No. 436, US Govt. Printing Office, Washington, DC.

Spurway, C.H., 1941, "Soil Reaction (pH) Preferences of Plants," Special Bulletin 306, Michigan Agricultural Experiment Station.

Stewart, B.A., 1975, *Control of Water Pollution from Cropland,* Vol. 1, Agricultural Research Service, US Govt. Printing Office, Washington, DC.

Stotzky, G., 1974, "Activity, Ecology, and Population Dynamics of Microorganisms in Soil," *Microbial Ecology,* A. Lasken and H. Lechevalier, eds., CRC Press, Cleveland, OH.

Suarez, D.L., Rhoades, J.D., Lavado, R., and Grieve, C.M., 1983, "Effect of pH on Saturated Hydraulic Conductivity and Soil Dispersion."

Taylor, H.M., and Terrell, E.E., 1982, "Rooting Pattern and Plant Productivity," *Handbook of Agricultural Productivity,* Miloslav Rechligl, Jr., ed., Vol. 1, CRC Press, Inc., Boca Raton, FL.

Terzaghi, V., and Peck, R.B., 1948, *Soil Mechanics in Engineering Practice,* John Wiley and Sons, New York, p. 5611.

Thornthwaite, C.W., 1948, "An Approach Towards a National Classification of Climate," *Geographical Review,* Vol. 38, pp. 55–94.

Thornthwaite, C.W., and Mather, J.R., 1955, "Publication in Climatology," Technical Report No. 5, Laboratory of Climatology, Centerton, NJ.

Unger, P.W., Eck, H.V., and Murie, J.T., 1981, "Alleviating Plant Water Stress," *Modifying the Root Environment to Reducing Stress,* G.F. Arkin and H.M. Taylor, eds., ASAE Monograph No. 4., American Society of Agricultural Engineers, St. Joseph, MO, p. 6111.

Visher, S.S., 1966, *Climatic Atlas of United States,* Harvard University Press, Cambridge, MA.

Zaslavsky, D. and Rogowski, A.S., 1969, "Hydrologic and Morphologic Implications of Anisotropy and Infiltration in Soil Profile Development," *Proceedings,* Soil Science Society of America, Vol. 33, No. 4, pp. 594–599.

APPENDIX 1
PRODUCTIVITY POTENTIAL FOR SELECTED PROFILES

Appendix Table 1. Productivity Potential for Inceptisol, Grays Harbor County, Washington

Soil	Name	Depth, cm	Horizon	Productivity Factor						Index		
				Water	Density	Aeration	pH	EC	Product	Prod.	Roots	Bio, g/m^2/a
1	P10	0	A	1.0000	1.0000	1.0000	0.7056	1.0000	0.7056	0.8313	0.4208	1169.2
1	P10	15	A	1.0000	1.0000	1.0000	0.9082	1.0000	0.9082	0.5344	0.3205	751.6
1	P10	38	B	1.0000	1.0000	1.0000	0.9407	1.0000	0.9407	0.2433	0.2264	342.2
1	P10	76	B	1.0000	1.0000	1.0000	0.9407	1.0000	0.9407	0.0303	0.0322	42.6

Input and Program Parameters

Soil	Name	Depth, cm	Horizon	Density, g/cm^3	Clay	Water	pH	EC, ds/m	Rain, mm	Temp., °C
1	P10	0	A	1.07	0.3960	0.2000	4.50	0.0	1524.0	10.0
1	P10	15	A	1.05	0.2550	0.2000	4.90	0.0	1524.0	10.0
1	P10	38	B	1.16	0.2590	0.2000	5.00	0.0	1524.0	10.0
1	P10	76	B	1.01	0.2920	0.2000	5.00	0.0	1524.0	10.0

Appendix Table 2. Productivity Potential for Aridisol, Blaine County, Montana

Soil	Name	Depth, cm		Horizon	Productivity Factor							Index		
					Water	Density	Aeration	pH	EC	SAR	Product	Roots	Prod.	Bio, g/m²/a
2	P23	0	8	A	0.6500	0.9552	0.9736	1.0000	1.0000	1.0000	0.6045	0.2653	0.2776	186.2
2	P23	8	13	B	0.6000	0.4688	0.8720	1.0000	1.0000	0.0	0.0	0.1153	0.1173	78.6
2	P23	13	18	B	0.5000	0.8008	0.8577	1.0000	1.0000	0.0422	0.0146	0.0955	0.1173	78.6
2	P23	18	23	B	0.6000	0.8008	0.8498	1.0000	1.0000	0.5933	0.2431	0.0812	0.1159	77.7
2	P23	23	43	C	0.6000	0.8008	0.8370	1.0000	1.0000	0.9889	0.4002	0.2304	0.0961	64.5
2	P23	43	71	C	0.6000	0.1036	0.7591	1.0000	0.8078	0.5109	0.0204	0.1643	0.0040	2.7
2	P23	71	100	C	0.8000	0.0704	0.7262	1.0000	0.6840	0.4399	0.0126	0.0480	0.0006	0.4

Input and Program Parameters

Soil	Name	Depth, cm		Horizon	Clay	Density, g/cm³	Water	pH	EC, ds/m	SAR	Rain, mm	Temp., °C	Ca	Mg	Na	HCO³, meq/L
2	P23	0	8	A	0.1380	1.36	0.1300	5.60	0.0	0.0	381.0	3.0	0.0	0.0	0.0	0.0
2	P23	8	13	B	0.3680	1.66	0.1200	7.30	0.90	0.0	381.0	3.0	1.1	1.5	6.2	0.0
2	P23	13	18	B	0.3570	1.56	0.1000	8.00	1.10	0.66	381.0	3.0	1.1	2.8	8.7	0.0
2	P23	18	23	B	0.2600	1.56	0.1200	8.30	2.90	17.21	381.0	3.0	8.5	10.4	18.6	7.8
2	P23	23	43	C	0.2180	1.56	0.1200	8.60	2.20	21.75	381.0	3.0	3.1	4.2	16.0	9.9
2	P23	43	71	C	0.2860	1.77	0.1200	8.30	5.49	28.05	381.0	3.0	11.5	19.5	37.6	7.3
2	P23	71	100	C	0.2930	1.78	0.1600	8.10	7.10	31.23	381.0	3.0	18.9	26.3	48.0	8.1

Appendix Table 3. Productivity Potential for Mollisol, Bowman County, North Dakota

Soil	Name	Depth, cm	Horizon	Productivity Factor							Roots	Index	
				Water	Density	Aeration	pH	EC	SAR	Product		Prod.	Bio, g/m²/a
3	P87	3–8	A	1.0000	1.0000	1.0000	1.0000	1.0000	1.0000	1.0000	0.2653	0.6919	464.0
3	P87	8–18	A	1.0000	0.9076	0.9737	1.0000	1.0000	1.0000	0.8996	0.2108	0.4266	286.1
3	P87	18–25	A	1.0000	0.9144	0.9605	1.0000	1.0000	1.0000	0.8830	0.1104	0.2370	158.9
3	P87	25–30	A	0.9500	0.9212	0.9563	1.0000	1.0000	1.0000	0.8383	0.0662	0.1395	93.5
3	P87	30–38	B	0.8000	0.1700	0.8807	1.0000	0.9685	1.0000	0.1200	0.0886	0.0840	56.3
3	P87	38–53	B	1.0000	0.4688	0.8383	1.0000	0.3303	0.7458	0.0989	0.1218	0.0734	49.2
3	P87	53–71	B	1.0000	0.8008	0.8343	1.0000	0.8378	1.0000	0.5610	0.0889	0.0613	41.1
3	P87	71–97	B	0.9500	0.3692	0.8017	1.0000	0.8378	1.0000	0.2410	0.0476	0.0115	7.7
3	P87	97–100	—	1.0000	0.0	0.7943	1.0000	0.2303	0.6254	0.0	0.0005	0.0	0.0

Input and Program Parameters

Soil	Name	Depth, cm	Horizon	Clay	Density, g/cm³	Water	pH	EC, ds/m	SAR	Rain, mm	Temp., °C	Ca	Mg	Na	HCO³, meq/L
3	P87	0–8	A	0.0910	1.26	0.2000	6.40	0.0	0.0	381.0	6.0	0.0	0.0	0.0	0.0
3	P87	8–18	A	0.1050	1.43	0.2000	6.90	0.0	0.0	381.0	6.0	0.0	0.0	0.0	0.0
3	P87	18–25	A	0.0950	1.42	0.2000	7.70	0.0	0.0	381.0	6.0	0.0	0.0	0.0	0.0
3	P87	25–30	A	0.0610	1.41	0.1900	8.40	0.0	0.0	381.0	6.0	0.0	0.0	0.0	0.0
3	P87	30–38	B	0.1400	1.75	0.1600	9.00	3.40	68.74	381.0	6.0	1.6	1.2	38.6	10.0
3	P87	38–53	B	0.1950	1.66	0.2000	8.60	11.70	87.26	381.0	6.0	18.9	21.5	132.0	9.0
3	P87	53–71	B	0.0930	1.56	0.2000	8.50	5.10	81.31	381.0	6.0	1.0	1.5	51.0	5.8
3	P87	71–97	B	0.0890	1.69	0.1900	9.30	5.10	81.31	381.0	6.0	1.0	1.5	51.0	5.8
3	P87	97–100	—	0.1890	1.84	0.2000	8.90	13.00	81.31	381.0	6.0	1.0	1.5	51.0	5.8

Appendix Table 5. Productivity Potential for Mollisol, Elko County, Nevada

Soil	Name	Depth, cm		Horizon	Productivity Factor						Roots	Index	
					Water	Density	Aeration	pH	EC	Product		Prod.	Bio. g/m²/a
5	P98	0	5	A	0.9985	0.9688	0.9887	1.0000	1.0000	0.9564	0.1821	0.6701	224.1
5	P98	5	10	A	0.9110	0.9756	0.9924	1.0000	1.0000	0.8810	0.1323	0.4960	165.9
5	P98	10	20	A	0.7125	0.9484	0.9791	1.0000	1.0000	0.6649	0.1958	0.3794	126.9
5	P98	20	30	B	0.7920	0.9824	0.9872	1.0000	1.0000	0.7655	0.1426	0.2493	83.4
5	P98	30	53	B	0.7820	0.3360	0.8484	1.0000	1.0000	0.2378	0.2104	0.1401	46.9
5	P98	53	63	I	1.0000	0.9076	0.8591	1.0000	1.0000	0.7754	0.0560	0.0901	30.1
5	P98	63	69	I	0.5580	0.8260	0.8565	1.0000	1.0000	0.3953	0.0256	0.0467	15.6
5	P98	69	100	I	0.8495	0.9008	0.8733	1.0000	1.0000	0.6606	0.0553	0.0366	12.2

Input and Program Parameters

Soil	Name	Depth, cm		Horizon	Clay	Density g/cm³	Water	pH	EC, ds/m	Rain, mm	Temp., °C
5	P98	0	5	A	0.1570	1.34	0.1997	6.60	0.0	178.0	10.0
5	P98	5	10	A	0.2210	1.33	0.1822	6.60	0.0	178.0	10.0
5	P98	10	20	A	0.2480	1.37	0.1425	6.40	0.0	178.0	10.0
5	P98	20	30	B	0.2800	1.32	0.1584	6.50	0.0	178.0	10.0
5	P98	30	53	B	0.6030	1.70	0.1564	6.60	0.0	178.0	10.0
5	P98	53	63	I	0.4120	1.43	0.2000	6.90	0.0	178.0	10.0
5	P98	63	69	I	0.0390	1.55	0.1116	8.10	0.0	178.0	10.0
5	P98	69	100	I	0.0440	1.44	0.1699	7.90	0.0	178.0	10.0

Appendix Table 6. Productivity Potential for Mollisol, Salt Lake County, Utah

Soil	Name	Depth, cm		Horizon	Productivity Factor						Roots	Index	
					Water	Density	Aeration	pH	EC	Product		Prod.	Bio, g/m²/a
6	P130	0	8	A	1.0000	1.0000	1.0000	1.0000	1.0000	1.0000	0.2653	0.9978	464.6
6	P130	8	23	A	1.0000	1.0000	1.0000	1.0000	1.0000	1.0000	0.2920	0.7325	341.1
6	P130	23	41	A	1.0000	1.0000	1.0000	1.0000	1.0000	1.0000	0.2126	0.4405	205.1
6	F130	41	84	B	1.0000	1.0000	1.0000	1.0000	1.0000	1.0000	0.2162	0.2279	106.1
6	P130	84	100	B	0.9035	0.9348	1.0000	1.0000	1.0000	0.8446	0.0139	0.0117	5.5

Input and Program Parameters

Soil	Name	Depth, cm		Horizon	Clay	Density, g/cm³	Water	pH	EC, ds/m	Rain, mm	Temp., °C
6	P130	0	8	A	0.2310	0.97	0.2000	8.00	0.61	254.0	10.0
6	P130	8	23	A	0.2330	0.94	0.2000	7.90	0.53	254.0	10.0
6	P130	23	41	A	0.2320	0.85	0.2000	7.90	0.43	254.0	10.0
6	P130	41	84	B	0.2270	1.05	0.2000	8.20	0.40	254.0	10.0
6	p130	84	100	B	0.2620	1.39	0.1807	8.40	0.37	254.0	10.0

Appendix Table 7. Productivity Potential for Mollisol, Shelby County, Iowa

Soil	Name	Depth, cm	Horizon	Productivity Factor						Index		
				Water	Density	Aeration	pH	EC	Product	Roots	Prod.	Bio, g/m²/a
7	P90	0 18	A	0.7500	0.9212	0.9358	1.0000	1.0000	0.6466	0.4761	0.6853	836.7
7	P90	18 33	A	0.8500	0.9484	0.9493	1.0000	1.0000	0.7609	0.2121	0.3774	460.9
7	P90	33 46	A	0.7500	0.9552	0.9561	1.0000	1.0000	0.6828	0.1243	0.2160	263.8
7	P90	46 69	B	0.7500	0.9688	0.9667	1.0000	1.0000	0.6988	0.1321	0.1311	160.1
7	P90	69 86	B	0.7500	0.9552	0.9680	1.0000	1.0000	0.6930	0.0448	0.0388	47.4
7	P90	86 100	B	0.8500	0.9008	0.9600	1.0000	1.0000	0.7388	0.0106	0.0078	9.5

Input and Program Parameters

Soil	Name	Depth, cm	Horizon	Density, g/cm³	Clay	Water	pH	EC, ds/m	Rain, mm	Temp., °C
7	P90	0 18	A	1.41	0.3040	0.1500	5.60	0.0	787.0	9.0
7	P90	18 33	A	1.37	0.3350	0.1700	5.70	0.0	787.0	9.0
7	P90	33 46	A	1.36	0.3280	0.1500	5.80	0.0	787.0	9.0
7	P90	46 69	B	1.34	0.3040	0.1500	5.80	0.0	787.0	9.0
7	P90	69 86	B	1.36	0.2820	0.1500	5.90	0.0	787.0	9.0
7	P90	86	B	1.44	0.2690	0.1700	5.90	0.0	787.0	9.0

Appendix Table 8. Productivity Potential for Mollisol, Logan County, Illinois

Soil	Name	Depth, cm		Horizon	Productivity Factor						Roots	Index	
					Water	Density	Aeration	pH	EC	Product		Prod.	Bio, g/m²/a
8	P02	0	18	A	0.8000	1.0000	1.0000	1.0000	1.0000	0.8000	0.4761	0.7157	1015.8
8	P02	18	38	A	0.9000	0.9008	0.9604	1.0000	1.0000	0.7958	0.2652	0.3348	475.2
8	P02	38	51	B	0.7500	0.8396	0.9282	1.0000	1.0000	0.5933	0.1084	0.1238	175.7
8	P02	51	66	B	0.7500	0.5020	0.8821	1.0000	1.0000	0.3400	0.0829	0.0595	84.4
8	P02	66	94	B	0.8000	0.6680	0.8533	1.0000	1.0000	0.4644	0.0656	0.0313	44.4
8	P02	94	100	B	0.8500	0.6348	0.8489	1.0000	1.0000	0.4594	0.0019	0.0009	1.2

Input and Program Parameters

Soil	Name	Depth, cm		Horizon	Clay	Density, g/cm³	Water	pH	EC, ds/m	Rain, mm	Temp., °C
8	P02	0	18	A	0.2520	1.30	0.1600	6.70	0.71	965.0	11.0
8	P02	18	38	A	0.2850	1.44	0.1800	5.60	0.71	965.0	11.0
8	P02	38	51	B	0.3640	1.53	0.1500	5.70	0.71	965.0	11.0
8	P02	51	66	B	0.4250	1.65	0.1500	5.80	0.71	965.0	11.0
8	P02	66	94	B	0.3710	1.60	0.1600	6.40	0.71	965.0	11.0
8	P02	94	100	B	0.2490	1.61	0.1700	7.50	0.71	965.0	11.0

Appendix Table 9. Productivity Potential for Mollisol, Sanpete County, Utah

Soil	Name	Depth, cm		Horizon	Productivity Factor						Roots	Index	
					Water	Density	Aeration	pH	EC	Product		Prod.	Bio, g/m²/a
9	P37	0	10	A	1.0000	1.0000	1.0000	1.0000	1.0000	1.0000	0.3143	0.9598	446.9
9	P37	10	18	A	1.0000	1.0000	1.0000	1.0000	1.0000	1.0000	0.1618	0.6455	300.5
9	P37	18	38	A	1.0000	1.0000	1.0000	1.0000	1.0000	1.0000	0.2652	0.4837	225.2
9	P37	38	53	C	0.7625	1.0000	1.0000	1.0000	1.0000	0.7625	0.1218	0.2185	101.7
9	P37	53	74	C	0.8860	1.0000	1.0000	1.0000	1.0000	0.8860	0.0988	0.1256	58.5
9	P37	74	100	C	1.0000	1.0000	1.0000	1.0000	1.0000	1.0000	0.0381	0.0381	17.8

Input and Program Parameters

Soil	Name	Depth, cm		Horizon	Clay	Density, g/cm³	Water	pH	EC, ds/m	Rain, mm	Temp., °C
9	P37	0	10	A	0.2440	0.98	0.2000	8.50	0.62	254.0	10.0
9	P37	10	18	A	0.2560	0.98	0.2000	8.40	0.55	254.0	10.0
9	P37	18	38	A	0.2900	0.90	0.2000	8.40	0.57	254.0	10.0
9	P37	38	53	C	0.2810	0.99	0.1525	8.60	0.59	254.0	10.0
9	P37	53	74	C	0.2780	0.99	0.1772	8.80	0.48	254.0	10.0
9	P37	74	100	C	0.2010	0.99	0.2000	9.00	0.48	254.0	10.0

Appendix Table 10. Productivity Potential for Mollisol, Ottawa County, Oklahoma

Soil	Name	Depth, cm		Horizon	Productivity Factor						Roots	Index	
					Water	Density	Aeration	pH	EC	Product		Prod.	Bio, g/m²/a
10	P91	0	15	A	1.0000	0.8328	0.8377	1.0000	1.0000	0.6977	0.4208	0.5858	862.2
10	P91	15	23	A	1.0000	0.8532	0.8455	1.0000	1.0000	0.7188	0.1365	0.2922	430.1
10	P91	23	46	B	0.8085	0.8328	0.8416	1.0000	1.0000	0.5677	0.2553	0.1941	285.6
10	P91	46	66	B	0.7300	0.5352	0.8159	1.0000	1.0000	0.3234	0.1200	0.0491	72.3
10	P91	66	100	B	0.5695	0.3360	0.7793	1.0000	1.0000	0.1532	0.0674	0.0103	15.2

Input and Program Parameters

Soil	Name	Depth, cm		Horizon	Clay	Density, g/cm³	Water	pH	EC, ds/m	Rain, mm	Temp., °C
10	P91	0	15	A	0.1540	1.54	0.2000	6.50	0.0	1016.0	15.0
10	P91	15	23	A	0.1640	1.51	0.2000	6.40	0.0	1016.0	15.0
10	P91	23	46	B	0.2430	1.54	0.1617	6.20	0.0	1016.0	15.0
10	P91	46	66	B	0.3190	1.64	0.1460	5.50	0.0	1016.0	15.0
10	P91	66	100	B	0.4220	1.70	0.1139	5.40	0.0	1016.0	15.0

Appendix Table 11. Productivity Potential for Alfisol, Johnson County, Illinois

Soil	Name	Depth, cm		Horizon	Productivity Factor						Roots	Index	
					Water	Density	Aeration	pH	EC	Product		Prod.	Bio, g/m²/a
11	P09	0	13	A	1.0000	0.9688	0.9887	1.0000	1.0000	0.9578	0.3806	0.8047	1339.6
11	P09	13	23	A	1.0000	0.9416	0.9753	1.0000	1.0000	0.9231	0.1767	0.4402	732.7
11	P09	23	33	B	0.9500	0.9008	0.9556	0.9943	1.0000	0.8200	0.1309	0.2770	461.2
11	P09	33	51	B	0.9500	0.7344	0.8975	0.8685	1.0000	0.5590	0.1615	0.1697	282.5
11	P09	51	74	B	0.9500	0.6680	0.8620	0.9082	1.0000	0.5065	0.1122	0.0794	132.2
11	P09	74	100	B	1.0000	0.8396	0.8576	0.8215	1.0000	0.5933	0.0381	0.0226	37.7

Input and Program Parameters

Soil	Name	Depth, cm		Horizon	Clay	Density, g/cm³	Water	pH	EC, ds/m	Rain, mm	Temp., °C
11	P09	0	13	A	0.1260	1.34	0.2000	5.90	0.0	1219.0	14.0
11	P09	13	23	A	0.1320	1.38	0.2000	5.80	0.0	1219.0	14.0
11	P09	23	33	B	0.2060	1.44	0.1900	5.30	0.0	1219.0	14.0
11	P09	33	51	B	0.2910	1.58	0.1900	4.80	0.0	1219.0	14.0
11	P09	51	74	B	0.3000	1.60	0.1900	4.90	0.0	1219.0	14.0
11	P09	74	100	B	0.2740	1.53	0.2000	4.70	0.0	1219.0	14.0

Appendix Table 12. Productivity Potential for Alfisol, Jefferson County, Kentucky

Soil	Name	Depth, cm		Horizon	Water	Density	Aeration	pH	EC	Product	Roots	Prod.	Bio, g/m²/a
						Productivity Factor						Index	
12	P34	0	15	A	0.8510	0.8736	0.8830	1.0000	1.0000	0.6565	0.4208	0.5946	948.7
12	P34	15	34	B	0.7275	0.8600	0.8756	1.0000	1.0000	0.5492	0.2786	0.3184	508.0
12	P34	34	61	B	0.9935	0.8940	0.8887	0.7056	1.0000	0.5530	0.2210	0.1653	263.8
12	P34	61	81	B	1.0000	0.8464	0.8795	0.6367	1.0000	0.4764	0.0709	0.0431	68.8
12	P34	81	94	A	0.9930	0.8872	0.8821	0.6367	1.0000	0.4940	0.0179	0.0093	14.8
12	P34	94	100	B	0.8800	0.4688	0.8726	0.7056	1.0000	0.2555	0.0019	0.0005	0.8

Input and Program Parameters

Soil	Name	Depth, cm		Horizon	Clay	Density, g/cm³	Water	pH	EC, ds/m	Rain, mm	Temp., °C
12	P34	0	15	A	0.1310	1.48	0.1702	6.60	0.0	1143.0	13.0
12	P34	15	34	B	0.2460	1.50	0.1455	5.50	0.0	1143.0	13.0
12	P34	34	61	B	0.2210	1.45	0.1987	4.50	0.0	1143.0	13.0
12	P34	61	81	B	0.1750	1.52	0.2000	4.40	0.0	1143.0	13.0
12	P34	81	94	A	0.1460	1.46	0.1986	4.40	0.0	1143.0	13.0
12	P34	94	100	B	0.1820	1.66	0.1760	4.50	0.0	1143.0	13.0

Appendix Table 13. Productivity Potential for Aridisol, Cochise County, Arizona

Soil	Name	Depth, cm		Horizon	Water	Density	Aeration	pH	EC	Product	Roots	Prod.	Bio, g/m²/a
						Productivity Factor						Index	
13	P56	0	5	A	0.4790	0.4024	0.7321	1.0000	1.0000	0.1411	0.1821	0.1530	51.2
13	P56	5	15	A	0.3865	0.4024	0.7321	1.0000	1.0000	0.1139	0.2387	0.1273	42.6
13	P56	15	23	A	0.4315	0.4688	0.7373	1.0000	1.0000	0.1487	0.1365	0.1002	33.5
13	P56	23	30	B	0.5175	0.4356	0.7378	1.0000	1.0000	0.1663	0.0954	0.0799	26.7
13	P56	30	43	B	0.8805	0.3028	0.7290	1.0000	1.0000	0.1954	0.1350	0.0640	21.4
13	P56	43	58	B	1.0000	0.2364	0.7197	1.0000	1.0000	0.1711	0.1056	0.0376	12.6
13	P56	58	66	B	0.8650	0.2364	0.7165	1.0000	1.0000	0.1468	0.0393	0.0196	6.5
13	P56	66	84	C	0.3200	0.8008	0.7369	1.0000	0.9455	0.1760	0.0535	0.0138	4.6
13	P56	84	97	C	0.5225	0.8008	0.7473	1.0000	1.0000	0.3103	0.0134	0.0044	1.5
13	P56	97	100	C	0.7900	0.7344	0.7490	1.0000	1.0000	0.4341	0.0005	0.0002	0.1

Input and Program Parameters

Soil	Name	Depth, cm		Horizon	Clay	Density, g/cm³	Water	pH	EC, ds/m	Rain, mm	Temp., °C
13	P56	0	5	A	0.0560	1.68	0.0958	6.30	0.0	178.0	16.0
13	P56	5	15	A	0.1030	1.68	0.0773	6.00	0.0	178.0	16.0
13	P56	15	23	A	0.1300	1.66	0.0863	6.60	0.0	178.0	16.0
13	P56	23	30	B	0.1720	1.67	0.1035	7.10	0.0	178.0	16.0
13	P56	30	43	B	0.5040	1.71	0.1761	7.40	0.0	178.0	16.0
13	P56	43	58	B	0.4770	1.73	0.2000	7.90	1.63	178.0	16.0
13	P56	58	66	B	0.2590	1.73	0.1730	7.90	2.84	178.0	16.0
13	P56	66	84	C	0.1450	1.56	0.0640	8.20	3.70	178.0	16.0
13	P56	84	97	C	0.1450	1.56	0.1045	8.40	2.68	178.0	16.0
13	P56	97	100	C	0.1450	1.58	0.1580	8.40	2.46	178.0	16.0

Appendix Table 14. Productivity Potential for Vertisol, Collin County, Texas

Soil	Name	Depth, cm		Horizon	Productivity Factor								Index	
					Water	Density	Aeration	pH	EC	SAR	Product	Roots	Prod.	Bio, g/m²/a
14	P127	0	15	A	0.7000	0.0372	0.6491	1.0000	1.0000	0.2953	0.0050	0.4208	0.0021	3.1
14	P127	15	56	A	0.6500	0.0	0.5802	1.0000	1.0000	0.8591	0.0	0.4610	0.0	0.0
14	P127	56	84	A	0.6500	0.0	0.5728	1.0000	1.0000	0.8404	0.0	0.1043	0.0	0.0
14	P127	84	100	A	0.7000	0.0	0.5692	1.0000	1.0000	0.8591	0.0	0.0139	0.0	0.0

Input and Program Parameters

Soil	Name	Depth, cm		Horizon	Clay	Density, g/cm³	Water	pH	EC, ds/m	SAR	Rain, mm	Temp., °C	Ca	Mg	Na	HCO³, meq/L
14	P127	0	15	A	0.3620	1.79	0.1400	6.40	0.34	1.43	1016.0	18.0	1.6	0.3	1.9	1.5
14	P127	15	56	A	0.4410	1.91	0.1300	6.30	0.44	5.40	1016.0	18.0	1.1	0.1	2.9	2.8
14	P127	56	84	A	0.4460	1.91	0.1300	6.60	0.65	7.80	1016.0	18.0	1.2	0.4	4.5	2.5
14	P127	84	100	A	0.4570	1.92	0.1400	7.70	1.59	13.66	1016.0	18.0	3.9	0.5	10.6	3.0

Appendix Table 15. Productivity Potential for Alfisol, Williamson County, Tennessee

Soil	Name	Depth, cm		Horizon	Productivity Factor							Index	
					Water	Density	Aeration	pH	EC	Product	Roots	Prod.	Bio, g/m²/a
15	P51	0	13	A	0.7695	0.8736	0.8830	0.9407	1.0000	0.5584	0.3806	0.4864	852.1
15	P51	13	36	B	0.5815	0.8396	0.8585	1.0000	1.0000	0.4235	0.3403	0.2738	479.8
15	P51	36	64	B	0.6870	0.8532	0.8593	0.9407	1.0000	0.4736	0.2028	0.1297	227.2
15	P51	64	91	B	0.6880	0.8736	0.8662	0.8685	1.0000	0.4503	0.0720	0.0337	59.0
15	P51	91	100	B	0.6160	0.6680	0.8590	0.8215	1.0000	0.2918	0.0043	0.0013	2.2

Input and Program Parameters

Soil	Name	Depth, cm		Horizon	Clay	Density, g/cm³	Water	pH	EC, ds/m	Rain, mm	Temp., °C
15	P51	0	13	A	0.2450	1.48	0.1539	5.00	0.0	1321.0	14.0
15	P51	13	36	B	0.6500	1.53	0.1163	5.50	0.0	1321.0	14.0
15	P51	36	64	B	0.6630	1.51	0.1374	5.00	0.0	1321.0	14.0
15	P51	64	91	B	0.6800	1.48	0.1376	4.80	0.0	1321.0	14.0
15	P51	91	100	B	0.6440	1.60	0.1232	4.70	0.0	1321.0	14.0

Appendix Table 17. Productivity Potential for Inceptisol, Greenbrier County, West Virginia

| Soil | Name | Depth, cm | Horizon | Productivity Factor | | | | | | | Index | |
				Water	Density	Aeration	pH	EC	Product	Roots	Prod.	Bio, g/m²/a
17	P29	0 5	A	1.0000	0.9212	0.9358	0.9659	1.0000	0.8327	0.1821	0.4190	589.3
17	P29	5 18	A	1.0000	0.9212	0.9358	0.9082	1.0000	0.7830	0.2940	0.2674	376.0
17	P29	19 33	B	1.0000	0.2696	0.8176	0.8215	1.0000	0.1907	0.1949	0.0372	52.3
17	P29	34 58	C	1.0000	0.0	0.7134	0.8215	1.0000	0.0	0.1939	0.0	0.0
17	P29	59 100	C	1.0000	0.0	0.6650	0.8215	1.0000	0.0	0.1012	0.0	0.0

Input and Program Parameters

Soil	Name	Depth, cm	Horizon	Clay	Density, g/cm³	Water	pH	EC, ds/m	Rain, mm	Temp., °C
17	P29	0 5	A	0.1600	1.41	0.2000	5.10	0.0	1168.0	10.0
17	P29	5 18	A	0.1820	1.41	0.2000	4.90	0.0	1168.0	10.0
17	P29	19 33	B	0.2320	1.72	0.2000	4.70	0.0	1168.0	10.0
17	P29	34 58	C	0.2020	1.85	0.2000	4.70	0.0	1168.0	10.0
17	P29	59 100	C	0.2020	1.85	0.2000	4.70	0.0	1168.0	10.0

6.6.6 Revegetation

Russell J. Hutnik

Guy W. McKee

INTRODUCTION

The benefits of a vegetative cover on land disturbed by mining are well known—erosion will be controlled, thereby lessening stream sedimentation; beauty will be restored to the landscape; and the land will become productive again.

As soon as the mining operation ceases, a natural process begins that, if left alone, eventually will culminate in a vegetative community in harmony with climatic and other environmental forces acting upon that particular site. This process, although inexorable, usually is too slow for today's society. And, the final community may not be the most desirable one. Emphasis therefore is placed on the rapid establishment of a vegetative cover composed of desirable species. Recently, laws have been passed to penalize operators who fail to achieve this goal.

Desirable species are those that can best achieve the goals of revegetation, as listed in the opening paragraph, both in the long-term as well as in the short-term. The vegetative cover, therefore, needs to be both rapidly achieved and persistent. As time goes on, the original community should develop into a natural community with little or no loss in its beneficial aspects at any time during this developmental period. Alternatively, the initial plant community can be managed so as to prevent natural succession, as in the case of agricultural land.

Natural communities usually are diverse. This diversity of species enables them to withstand disturbances caused by factors such as insects, diseases, and weather extremes. Diversity of plant species is highly desirable, even in the initial vegetation established on mined land.

ADVANCE PLANNING

Although all reclamation systems should be capable of achieving an immediate, permanent, and diverse vegetative cover, the operator, in developing a plan for a specific site, must consider three important points: (1) vegetation establishment is an integral part of the entire mining and reclamation operation; (2) methods will vary depending upon the objectives of revegetation and the plant species used; and (3) establishment methods will vary depending upon local differences in climate, geology, and soils.

The first point means that the operator should plan from the beginning how to achieve an adequate vegetative cover at a reasonable cost. In many cases, slight modification in the mining, grading, and resoiling operations can facilitate establishment of a vegetative cover. As far as possible, pyritic and other phytotoxic material should be kept out of the rooting zone. Compaction by heavy equipment should be kept to a minimum, and sharp differences in chemical or physical properties between the topsoil and underlying spoil layers should be avoided. Depending upon the plant species chosen and the method of seeding, some degree of surface irregularity can aid plant establishment. Examples are minor depressions from dozer cleats, disks and harrows, or cultipackers and similar tillage tools. If these goals are not accomplished before the revegetative phase, revegetation may be more expensive and less certain.

The second point means that the operator must have specific reclamation goals in mind. These could be restoration of vegetative conditions that existed prior to mining, development of a new and more productive plant community while maintaining the same land use, or development of a new land use pattern for the site. Usually, the operator will need to follow the dictates of laws and regulations currently governing mining and reclamation. Nevertheless, there is much to be gained by surpassing the minimum legal requirements; the operator will be better able to respond to future changes in the laws and regulations. In addition, the operator can reap the benefits of better public relations while improving the area disturbed by mining.

The third point means that environmental factors controlling plant establishment will vary from region to region and from site to site within a region. Adapted species and recommended practices will likewise vary between and within regions.

Consequently, there is no one "best" system of revegetation. Certain elements may be common for all sites but others will vary greatly from one site to another. This section will concentrate on the general aspects—those common to most revegetation systems. For systems applicable to a given region and land use, the reader should consult appropriate publications. Examples are "A Guide for Revegetating Coal Minesoils in the Eastern United States" (Vogel, 1981), a series of reports sponsored by the Fish and Wildlife Service, USDI (Rafaill and Vogel, 1978; Leedy, 1981; Nawrot, et al., 1982) and a series of user guides for mining and reclamation issued under the Surface Environment and Mining Program (SEAM) (USDA Forest Service, 1979a, 1979b, and 1982). Valuable information is contained in books such as *Reclamation of Drastically Disturbed Land* (Schaller and Sutton, 1978) and *The Restoration of Land* (Bradshaw and Chadwick, 1980). In addition, revegetation guides have been specifically developed for many states.

LIMITING SITE FACTORS

The selection of species and plant establishment techniques depends largely upon the nature of the site to be revegetated. An essential first step is to characterize the site, emphasizing those physical and chemical factors that are likely to limit or prevent plant establishment and growth. The reclamation specialist has a variety of options for revegetating sites with few, if any, limiting factors. But, where one or more factors seriously limit plant establishment, options may be greatly curtailed.

Much can be learned about the site to be revegetated by collecting data and samples during other phases of the mining operation. Especially important is the nature of the growing medium, both the topsoil (if it is to be replaced) and the subsoil. In properly planned operations, the spoil material best suited for plant growth should be placed near the surface, just beneath the surface topsoil layer. An analysis of drill cores and soil samples will help the operator identify overburden spoil materials and soil horizons best suited for plant growth. In some cases, the best spoil materials may be superior to the so-called "topsoil" as a medium for plant growth. Topsoil generally is considered to be fertile and a good medium for plant growth. However, much of the topsoil on strip-mined sites is acidic, infertile, and low in capacity to hold and supply water and essential mineral elements. Topsoil, or rather surface soil, by virtue of its position at the surface may or may not be a good medium for plant growth and should be carefully analyzed prior to use.

Topsoil and subsoil samples should be collected from a number of locations on the site to be revegetated. Separate soil tests should be carried out for each 2 to 4-ha (5 to 10-acre) segment. The sample for each separate soil test should be composed of 10 to 20 subsamples. Most laboratories routinely test soil samples for pH; buffer pH; the concentration of important nutrient elements such as potassium, potash, calcium, and magnesium; conductivity or the concentration of certain soluble salts; and cation exchange capacity. Based upon such tests, recommendations are made for lime and fertilizer treatment. Although nitrogen usually is deficient on surface-mined land, there is no good rapid laboratory test for available nitrogen. Additional tests may be desirable in those areas where other soil properties may be limiting. Examples of such special tests are textural analysis, concentration of exchangeable aluminum and manganese (especially in Appalachian localities), concentration of sodium (especially in arid and semiarid regions of the west), and bulk density (especially in the midwest).

Additional observations, such as slope steepness, slope length, slope position, aspect, and color of the surface material, may reveal potential problems from high surface temperatures, water availability, and erosion potential.

SPECIES SELECTION

Availability

In most regions of the US, reclamation specialists can borrow freely from a large reservoir of research and experience in agronomy, horticulture, range science, and forestry. Species are selected based on their performance in previous plantings throughout the region. Species suitability, at times, can also be gleaned from studies of plant succession for the region involved. In general, early successional species should be best suited for revegetation projects. For such native or naturalized species, however, seed or other propagules may not be readily available in quantity. Also the native and naturalized successional species may not be capable of achieving a rapid cover. In fact, little is known about the reproductive habits and establishment of many native and naturalized plant species.

Adaptation

Since conditions on agricultural fields, even worn out and abandoned ones, may differ greatly from those on mined land, the suitability of various plant species for revegetating strip-mined sites must be based upon field research and upon a detailed knowledge of plant species. Even if a species performs well in field trials, the reclamation specialist is faced with a dilemma if seed is not readily available. Should he encourage the development of a seed enhancement program, which often is costly and time-consuming and for which there is no guarantee of success, or should he be content with less suitable species for which seed is in plentiful supply? In the latter case, the site may require extensive treatment with lime and fertilizer or other amendments.

Actually, a plant species often is much too broad a category for determining the suitability of plant materials for use on disturbed sites. Considerable adaptive differences have been found among populations from different parts of a species range, i.e., ecotypes in the case of wild plants or cultivars in the case of cultivated plants. Ecotypic differentiation has been found to occur even in the same locality among plants growing on vastly different substrates. Thus, poor performance in field plantings may be due to a poor choice of ecotypes or cultivars rather than to a general lack of adaptation of the plant species. However, the range of adaptation within a given plant kind or plant species is limited. Thus, there may be statistically significant differences in aluminum tolerance between cultivars of birdsfoot trefoil (*Lotus corniculatus* L.), but the probability of finding a genotype of birdsfoot trefoil as tolerant to aluminum as that of serecia lespedeza [(*Lespedeza cuneata* (Dum.) G. Don)] seems remote.

Species, ecotypes, and cultivars used for reclamation should be adapted to local biotic, climatic, and edaphic conditions. Problems sometimes develop with exotics, i.e., non-native perennial species or ecotypes, which grow well so long as weather conditions are near normal but which are eventually damaged when extreme weather conditions occur. Therefore, selections of plant species and cultivars should be based on long-term field trials and observations.

The number of species suitable for mine-land reclamation is far less than the number of climatically suited species, however; environmental factors other than climate can greatly limit the number of suitable plant species. This is especially true for so-called *orphan* banks—those which have been mined in the past and, for one reason or another, never satisfactorily revegetated.

Soil conditions on mined land often are such that the species selected must be able to tolerate extremes of surface temperatures, low or high levels of soil moisture, and a very low level of fertility. In the northeast, tolerance to high levels of acidity and exchangeable aluminum is also a desirable trait, as is tolerance to high levels of alkalinity and exchangeable sodium in semi-arid and arid regions of the west.

In addition to their differential ability to tolerate environmental extremes, plant species usually possess other traits that make them either more or less desirable for reclamation purposes. Among these are their susceptibility to insects and diseases, their potential to spread and become pests on sites adjacent to the mined land, their potential for wildlife food and cover, their ability to fix atmospheric nitrogen, their ability to spread vegetatively especially on actively eroding sites, their ability to withstand heavy recreational use, and their potential for producing valuable products such as Christmas trees and timber products.

Because of the desirability of achieving a diverse vegetative cover, plant species often must also be able to compete with and persist with other planted or wild species. Birdsfoot trefoil, for example, a species often used to revegetate strip-mined sites, is relatively intolerant of competition from associated grasses and weeds. Therefore, establishment practices and management should be selected to favor the birdsfoot trefoil. This means that the growth forms and habits of associated species must be compatible with the birdsfoot trefoil. If shorter growing species are mixed with taller species, the shorter species should possess a certain degree of shade tolerance. Warm-season species, i.e., those that make most of their growth in the summer, often are at a disadvantage when planted with cool-season species. A special problem is the difficulty of establishing woody plants in a dense herbaceous cover.

It often is highly desirable that the herbaceous species used be able to spread rapidly into bare areas, either through vegetative means or through seed production and dissemination. Such characteristics also enable a species to maintain itself as individual plants die out. Some species that otherwise are suited, for example *serecia lespedeza* in Pennsylvania, are not recommended because the growing season is so short that few seeds are produced in most years. Such problems are especially serious on strip mines at high elevations or at high latitudes.

Species tolerances and traits can be found in such publications as "Plant Materials for Use on Surface-Mined Lands in Arid and Semiarid Regions" (Thornburg, 1982), "Plant Performance on Surface Coal Mine Spoil in Eastern United States" (Ruffner, 1978), and descriptions of individual species published by the US Forest Service and the US Soil Conservation Service. Information can also be obtained from the Plant Materials Centers and Plant Materials Specialists of the US Soil Conservation Service and from plant scientists at agricultural experimental stations located in the various coal-mining states.

ESTABLISHMENT OF HERBACEOUS COVER

The key to successful revegetation of mined land is to ameliorate those factors that limit plant establishment and to create favorable conditions for seed germination and seedling growth.

Control of pH and Fertility

Chemical properties, such as pH and fertility, can be enhanced through liming and fertilizing. Rates should be based on laboratory analyses of soil samples. Proper control of pH is essential, since many plant nutrients are available only within a fairly narrow range centered around neutrality. Furthermore, toxic materials, such as aluminum and manganese, are present in soluble form only at pH values below 5.5 or above 8.0. Although lime can be applied in many forms, agricultural-grade ground limestone generally is recommended.

Fertilizers can be applied in many different forms. In the most commonly used fertilizers, the nutrients are readily available and some nutrients, particularly nitrogen and potassium, leach readily. Thus, increases in nutrient concentration may be short-lived, especially in coarse-textured soils with low cation exchange capacities. It is therefore important that a plant community be rapidly established; once established, the growing plants will conserve nutrients through recycling. On coarse-textured soils it may be desirable to topdress with additional fertilizer, especially nitrogen, in the second year or use a slow-release nitrogen fertilizer when seeding. The need for subsequent nitrogen fertilization can be minimized by including a legume or other nitrogen-fixing species, such as one of the alders (*Alnus* spp.), in the mixture. If supplies are available at competitive costs, waste products high in one or more essential nutrients can be substituted for all or part of conventional inorganic nutrients. These waste products include animal manure and treated municipal wastes such as sludge, garbage, and compost. Because the nutrients in such wastes are largely in organic form, they are released slowly. However, supplemental fertilization may be needed to compensate for use of nitrogen by the microorganisms decomposing the organic wastes. In the case of treated municipal wastes, care must be taken that heavy metal concentrations are within acceptable levels. Many states have laws and regulations governing the use of sewage sludge and similar wastes on land. Generally, the sludge must be analyzed before application and the site monitored after application of sludge.

The lime, fertilizer, and waste material should be incorporated deeply into the soil. Ideally, this can be achieved by liming and fertilizing the subsoil and tilling or disking. Following respreading of the topsoil, the site is again limed, fertilized, and tilled or disked.

Adverse chemical problems associated with high levels of soluble salts or sodium are most difficult to correct. Leaching by excessive irrigation water is effective if the site is well drained. Chemical amendments also may be used. Two types of amendments are (1) soluble ammonium or calcium salts, such as ammonium sulfate, calcium chloride, or calcium sulfate (gypsum) applied individually or in combination; and (2) acids or acid formers such as sulfur, sulfuric acid, iron sulfate, or aluminum sulfate. Waste products, such as scrubber wastes from coal-burning power plants, may provide an economical source of calcium salts.

Seedbed

Physical problems related to soil texture are not easily overcome. If coal-burning electric generating stations are located nearby, the fly ash produced by the station often can be used to improve the physical and chemical conditions of the spoil or soil. The fly ash needs to be incorporated into the soil. The addition of organic matter, proper tillage, and the use of mulches can improve moisture relationships resulting from poor texture. Compaction is often a problem because of the use of heavy equipment in grading and resoiling operations. Ripping and disking, or deep harrowing with a heavy cultivator or harrow, is recommended both before and after respreading the topsoil. Moisture, aeration, and seedbed conditions will be improved and the often sharply contrasting interface between the spoil and the overspread soil will be reduced.

Prior to seeding and planting, the seedbed should be prepared. Ideally, the surface should be relatively loose and friable with small cracks and crevices within which the seed can lodge. Thus, seeds will not be washed out during rainstorms, and water and fine sediment will tend to collect around the seed, thereby improving conditions for germination and seedling growth. As was previously mentioned, modern grading and resoiling practices often tend to leave a smooth, hard-crusted or compacted surface—a very unfavorable seedbed.

A favorable seedbed can be created by several methods. The simplest is to do the final grading and resoiling with as little equipment use as possible, leaving the surface in a somewhat rough condition. A final pass by crawler tractor along the contour will create cleat tracks or depressions favorable to seed germination.

There are many types of harrows, disks, and chisel plows that can be used to scarify the surface; however, many sites are too rocky or too steep for use of conventional, lightweight farm equipment. Various types of rock-picker equipment are available but their use generally is of dubious economic value. If mine soils are extremely compacted, deep ripping may be required followed by disking or harrowing. The US Forest Service has an active project to develop equipment specifically for use in revegetation of strip mines and publishes annual reports entitled "Vegetative Rehabilitation and Equipment Workshop" (USDA Forest Service 1981, for example). If mine soils are extremely compacted, deep ripping may be required, followed by disking or harrowing.

In the arid and semi-arid regions of the west, and on rangelands, special equipment often is used to create a pitted or gouged surface. Snow and rain tend to collect in these depressions, thereby improving the growing condition for plants.

Seeding Techniques

Seed can be either broadcast on the surface or drilled into the soil to the proper depth. The latter is preferred since broadcast seeds tend to be washed off the surface. Broadcast seeds also are subjected to extremes of moisture and temperature. However, seed drilling cannot be used on steep or excessively rocky land. Rangeland drills may be better suited for mine soil use than are drills developed for cropland use.

Broadcast seeding often is successful if a suitable seedbed exists and if the soil is cultipacked or otherwise treated after the seed is broadcast to cover the seed with a thin layer of soil. Aerial seeding can be used for mined land too rough or too inaccessible for ground equipment. However, unless mulched, seeding failures are common on such areas because adequate seedbed preparation seldom is done on these sites. When seed is broadcast, attempts to cover by dragging a tree or an old dozer track over the seeded area are poor techniques. Some seeds will be covered too deeply and others not deep enough. Use of a mulch after broadcast seeding is excellent and should be used when possible.

One system of broadcast seeding that is highly successful is hydroseeding. It can be used on steep rocky slopes, providing an access road exists. In this system, lime, fertilizer, seeds, and legume inoculant are mixed with water and applied as a slurry through high-pressure pumps and directional nozzles. All areas hydroseeded should be mulched.

Mulches

Regardless of the method of seeding, a mulch application is recommended. Mulches protect the surface against erosion and prevent desiccation and excessively high temperatures. Mulches can mean the difference between revegetation success and failure. They are especially valuable where the seeds have been broadcast and not subsequently covered with soil. Although a cellulose or similar type mulch can be incorporated into the hydroseeder slurry, it may prevent the seed from coming into contact with the mineral soil. Better results usually are obtained if the mulch is applied in a separate operation following hydroseeding.

The most commonly used mulches are grass hay and cereal straw. Straw mulch usually does not decay as rapidly as hay mulch but, on steep slopes, hay may do a better job of controlling erosion. Hay mulch may contain weed seeds, but this can be beneficial on mine soils where species diversity is desired. It has been suggested that hay and volunteer plants with their contained seeds be cut from old fields for use as mulch on mine soils to increase the proportion of native plants in the resulting vegetation.

The hay or straw should be applied uniformly to a depth of about 2 to 2.5 cm (0.75 to 1 in.); 5.5 to 7 t per ha (2.5 to 3 st per acre) of mulch should be sufficient. The mulching material should be "tacked" or otherwise held in place. Examples of tacking materials are asphalt emulsion, cellulose fiber, and certain chemicals. Mulch also can be anchored in place by pressing the hay or straw stems into the soil using a heavy disk with the blades set straight.

Other commonly used mulches are wood residues such as bark and wood chips. For most sites 85 to 115 m^3 per ha (45 to 60 cu yd per acre) [1.0 to 1.3 cm (⅜ to ½ in.) deep] should be sufficient. However, bark or wood chip mulch, because of the large fragment size, may limit the establishment of grasses and legumes. Cellulose fiber is commonly used in hydroseeding. However, as previously mentioned, mulching as a separate operation is the preferred method.

Materials, such as nets and netting, plastic film, and even crushed rock or pebbles, can be used as mulch or to hold mulches in place. They may be expensive but necessary on erosive sites.

Time of Seeding

Seeding usually is most successful if done in the early spring as soon as the soil is workable. At that time, moisture conditions are optimum. However, state and federal laws may require that a vegetative cover be established as soon after the final grading and resoiling as practical. This means that seeding may have to be done at any time during the growing season.

A rapid cover can be achieved with perennial grasses and legumes if seeded in the spring. However, it is often desirable to include annuals, especially small grains, in the mixture to achieve quick cover while the longer-lived and more persistent grass and legume species are becoming established. Spring oats (*Avena sativa* L.) can be seeded through the late summer. They usually winter-kill but the dead plants help protect the soil until the desired grasses and legumes resume growth the following spring. Species such as winter wheat (*Triticum aestivum* L.) or winter rye (*Secale cereale* L.) can be seeded in late summer and early fall, thereby providing soil protection over the winter.

At many locations, grasses and legumes can be planted along with a small grain, even during the late summer or early fall. The small grain provides quick cover for soil stabilization and the grasses and legumes provide long-term cover. However, in the central and northern United States, legume seedlings may require ten or more weeks of growing weather past germination to achieve sufficient growth and organic reserves to be winter hardy. Otherwise, they may winter-kill. For that reason, late plantings of legumes should contain a high percentage (35% or more) of "hard seed," i.e., seed with a coat impervious to water. Such hard seeds usually do not germinate the year seeded but do germinate the following year or years. Thus, hard seeds are an insurance against loss of all legume seedlings. In fact, many times the legume component on a strip-mined site came from hard seeds.

Legume Inoculation

Legumes should be inoculated with the appropriate species and strains of *Rhizobia,* the bacteria that induce the formation of nodules on the roots and fix atmospheric nitrogen into a form usable by plants. Since the root-nodule bacteria are sensitive to heat and drying, the legume seeds should not be inoculated until just before planting. Likewise, the inoculant should be kept out of the sun and kept cool. If the sites are particularly adverse, or if the seed is to be applied by a hydroseeder, apply inoculant liberally, up to five times the rate recommended on the legume packet.

Seed Quality and Seeding Rate

Seed should be high in quality, i.e., high in germination and purity. All states require that all seed sold be labeled for content of pure seed, inert matter, and weed seed as well as the germination percent and the date when the seed was tested. Seed for revegetation purposes should have been tested within the past five months.

In order to determine seeding rates, particularly with chaffy seed, one should first calculate pure live seed (PLS). This is the product of percentage pure seed and germination percentage divided by 100. The percentage PLS is then divided into the recommended pounds of seed and this value multiplied by 100. Thus, if the recommended seeding rate for a certain grass is 9 kg/ha (8 lb per acre) and the purity is 96% with a germination of 90%, the adjusted planting rate would be 10.4 kg/ha (9.3 lb per acre). The calculations are as follows: 9 kg/ha, the recommended planting rate, divided by 86.40 PLS (96% × 90% ÷ 100 = 86.40) = 0.104 × 100 = 10.4 kg/ha.

Seed Mixes

There are two contrasting approaches with respect to formulating species mixtures. Some reclamation specialists prefer simple herbaceous mixtures, i.e., two to four species. A few specialists support complex mixtures of 6 to 12 or more plant species. In this latter case, the rationale is that there will be at least one species adapted to each microsite. However, complex mixtures often are very competitive, may be more difficult to seed, and the type of community established is less predictable. Often only a few species predominate, and some of the species in the mixture are so sparse that they have little influence on the nature of the community.

Species included in a mixture should be compatible rather than competitive during the establishment phase. Even with compatible species, one species seeded at an excessive rate can depress the establishment and growth of those species seeded with it. This is more likely to occur with small-seeded species, such as redtop (*Agrostis gigantea* Roth) or weeping lovegrass [(*Eragrostis curvula* (Schrad.) Nees)], each of which has many seeds per kilogram. Redtop may have 10 million seeds per kilogram and weeping lovegrass 3 million. Seeding rates on especially adverse sites or on sites that are broadcast seeded should be increased by 50 to 100% over rates recommended for general use on less adverse sites.

ESTABLISHMENT OF TREES AND SHRUBS

Planting Materials

A number of different types of planting materials can be used. Seed of woody species can be included in the herbaceous seed mixture. This has been successful with species that have rapid early growth, such as black locust (*Robinia pseudoacacia* L.). However, for many species the combination of slow growth during the first year and competition from grasses and legumes can result in seedlings too weak to survive the first winter. If a herbaceous cover is not part of the reclamation plan, pines can be successfully direct seeded, at least in the southern Appalachians. Seedlings of large seeded species, such as black walnut (*Juglans nigra* L.) and northern red oak (*Quercus rubra* L.) have been successfully established if the seed is buried and not destroyed by rodents.

Since there have been many failures with the direct seeding of woody plant species, most reclamation specialists prefer to plant seedlings. Seedlings are grown in tree nurseries for one or two years (sometimes longer), lifted, packaged in bundles, and shipped to the operator for planting in a bareroot condition. A recent development is the raising of seedlings in individual containers, usually in a greenhouse, and often for one year or less. The seedlings are then planted with the growing medium intact around the roots.

Studies comparing the two methods have shown mixed results. Apparently success is dependent more on quality of planting stock and care in planting than on the nature of the root at time of planting. Containerized seedlings do not require as much care during planting because the roots are not as likely to dry out. Also, the planting season can be extended later into the growing season with container-grown seedlings that have been kept refrigerated until the time of planting. Offsetting these advantages is the generally larger size and more robust nature of the two-year-old bareroot stock raised in conventional nurseries compared to containerized seedlings.

A few woody species, particularly willows (*Salix* spp.) and hybrid poplars (*Populus* spp.), reproduce readily by stem cuttings. These can be planted on mine soils in either a rooted or unrooted form. Usually better results are obtained with unrooted cuttings; transplant shock often is a problem with rooted cuttings. However, environmental conditions, especially soil moisture levels, must remain favorable for several weeks following the planting of unrooted cuttings, until the roots are sufficiently developed. For this reason, it is usually difficult to establish unrooted cuttings in an existing herbaceous vegetation where shading and competition for water can affect establishment of the cutting.

Other types of plant materials have not been widely used but may have merit for certain species and under special conditions. Where nursery grown stock is difficult to obtain, as in interior Alaska, wildings (individual plants transplanted from the wild to another site) and plugs (clumps of vegetation, which may contain several plants) are dug and replanted to another site. An extension of the latter is to replace the entire organic mat that had been removed prior to mining or other disturbance. This practice may be useful in especially sensitive sites, such as Alaska tundra regions. For a few species, root and rhizome cuttings seem feasible, although for the most part, techniques have not been fully developed. An example is sweetfern [(*Comptonia peregrina* (L.) J. M. Coult)].

In some cases no plant materials are added. The topsoil is limed, fertilized, and respread on the site. Vegetation develops from seed and root or rhizome fragments already in the topsoil and from seed blown into the area. This practice is especially appropriate if the goal is the restoration of the site to the former condition. However, it usually is not appropriate if a rapid cover is needed for erosion control or is mandated by law. Seed hay mulch also has been used. Seeds in the mulch are sufficient in some cases to establish a cover.

Quality of Planting Stock

Seedlings should have a good balance between root and top development and should have a high degree of robustness. The latter is best measured by the diameter of the stem at ground level, the larger the diameter, the more robust the seedling. However, seedlings beyond a certain size are difficult to plant and survival often drops because of improper planting. Recommended minimum and maximum diameters at the root collar vary with species, but often are given as 0.4 cm (0.15 in.) and 0.9 cm (0.35 in.), respectively.

Roots of older plants often are pruned to a length of 15 to 20 cm (6 to 8 in.) to facilitate planting. Tops of hardwood seedlings should also be pruned to achieve a good balance between the top and root systems. Top pruning is not recommended for conifer seedlings. Seedlings should have a healthy, vigorous appearance. Seedlings with poorly developed root systems and conifer seedlings with sparse or yellowish foliage should be discarded.

The quality of tree and shrub seedlings could be enhanced in many cases if they were inoculated with the proper mycorrhizal fungi, thereby leading to more efficient nutrient uptake. Considerable progress has been made in this area, and it seems likely that such inoculated seedlings will be commercially available in the near future. Many nursery-grown seedlings are naturally inoculated in the seedbed, although the fungi species present may not be the best suited for conditions prevailing on the mined sites.

Hybrid poplar cuttings should be 20 to 30 cm (8 to 12 in.) in length and 1 to 2 cm (⅜ to ¾ in.) in diameter. They should be taken from stems or branches of trees during the dormant season. Survival may be poor if the cuttings are taken just before the buds begin to expand.

Care of Planting Stock

Upon receipt, the unopened bundles of seedlings should be moistened and stored in a cool, shaded place. If they cannot be planted within a few days, they should either be placed in cold storage or "'heeled in." Heeling in consists of digging a trench in the soil in a moist shady location, breaking the bundles, and spreading the seedlings out evenly, filling in with soil and watering.

Care should be taken that the roots do not dry out either before or during planting. During planting, seedlings should be carried either in a canvas planting bag or bucket with moist peat in the bottom, or in a bucket with sufficient water to cover the roots. Root drying during planting can be retarded by dipping the roots in a kaolin clay slurry or in gel-like substances that are commercially available.

Time of Planting

Unlike establishing a herbaceous cover where planting can be done almost any time if care is used, planting of tree and shrub seedlings is best limited to the spring and, in some cases, the fall. Soil moisture, temperature, and light conditions are most favorable in the spring. Seedling survival is poor if the seedlings are planted after the initiation of growth. Containerized seedlings are an exception, but generally, even these have a lower survival when planted in the summer as compared to early spring.

Fall planting is recommended only in regions with mild winters. Otherwise, root growth is not sufficient to carry the seedling through the winter. Fall-planted seedlings are more susceptible to frost heaving than spring-planted ones.

Species Mixtures

As suggested for herbaceous cover, a mixture of woody species rather than a single one should be planted on any one site. Such mixtures provide a greater range of adaptation to the variation in microclimates and soils that often exists in close proximity on disturbed areas. Species mixtures can provide a variety of benefits in the form of site protection and wildlife food and habitat, and maintain these benefits more fully from one season to another and from year to year than a monoculture. There is also less likelihood that a diverse plant cover will be destroyed by environmental stresses such as drought, diseases, and insects. If hybrid poplars are selected for planting, a number of different clones should be used.

Tree and shrub mixtures can be planted by small blocks of each species, by strips (several rows of the same species), by individual rows, or by individual plants. Mixtures of species by small blocks are relatively easy to make and chances for success are usually better than for other types of mixtures. Mixtures by individual plants are not only difficult to design and carry out, but the seedlings are subject to interspecific competition. If one species is less competitive than the others in a mixture, the seedlings of that species tend to be suppressed or even eliminated from the stand.

Nevertheless, there are distinct benefits from mixtures by individual plants or individual rows. Inclusion of a nitrogen-fixing plant species in the mixture, either as individual plants or as individual rows, often will increase the growth of the associated species. This may not happen if the nitrogen-fixing species has rapid juvenile growth, and the associated species are slower growing and relatively shade intolerant. It is, therefore, important that the reclamation specialist take into account the growth habits and the relative tolerance of the various species when mixtures by individual plants or rows are planned.

Planting mixtures in strips is especially appropriate if the planned postmining land use is for wildlife purposes. A series of strips of species differing in growth habit will result in a large "edge" effect. Furthermore, all necessary habitat elements, i.e., food, escape cover, and reproductive cover, for a wildlife species can be found within a relatively small area.

Spacing

Tree seedlings should be spaced far enough apart so that merchantable products can be obtained in the first thinning, but close enough together so that the trees will develop a good form. Hardwood species, except hybrid poplars, should be at somewhat closer spacings than the conifers. Recommended spacings are 1.8 × 2.0 m (6 × 7 ft) or 2 × 2 m (7 × 7 ft) for hardwoods, 2.5 × 2.5 m (8 × 8 ft) for hybrid poplars [(or 2 × 2 m (7 × 7 ft) if planted in alternate rows with other species)], and 2.0 × 2.5 m (7 × 8 ft) or 2.5 × 2.5 m (8 × 8 ft) for conifers. Closer spacings [1.8 × 1.8 m (6 × 6 ft)] may be desirable for wildlife plantings, for critical sites such as steep slopes, or for adverse sites where substantial mortality may occur. Even closer spacing [1.5 × 1.5 m (5 × 5 ft)] is recommended for Christmas tree plantations. Shrubs, depending upon the species, may require closer spacings.

Planting Methods

The seedlings can be hand planted, or, if the area is not too rocky or steep, machine planted. The two most commonly used hand-planting tools are the dibble, or planting bar, and the mattock. These can create holes 15 to 20 cm (6 to 8 in.) or more in depth. Because the dibble is forced down with the foot, it may be a better choice on rocky or compacted mine soils than the mattock.

It is imperative that the roots of seedlings be placed properly within the hole. The seedlings should be planted to the same depth as they had been growing in the nursery bed. The roots should be well distributed within the hole. Roots that are restricted or bent (J shaped) should be avoided. Such trees seldom develop properly, and after several years usually die.

After the soil is replaced around the roots of the seedling, it should be tamped in with the foot. If air spaces are left around the roots, the roots may dry and die; even if the roots don't die, water and nutrient uptake may be impeded.

Container-grown seedlings have similar requirements as to proper planting depth and firming the soil around the root plug. Care should be taken to keep the root plug intact when removing the seedling from the container and placing it in the hole.

Unrooted cuttings can also be planted with a dibble or mattock, but preferably with a planting bar made from a sharpened steel rod about 2 cm (¾ in.) in diameter. They should be planted so that only about 5 cm (2 in.) of the cutting is above the ground level, the uppermost bud should be about 2.5 cm (1 in.) above the ground level. This will make it more likely that the base of the cutting is in contact with moist soil, that a single stem will develop rather than multiple shoots, and that the tender, developing shoot does not make contact with the hot surface of the mine soil. As with seedlings, the soil should be packed closely around the cuttings so that there is good contact with the soil.

Tractor-drawn tree planting machines are able to plant

seedlings at a rapid rate. However, the terrain must be suitable. They cannot be readily used on sites that have been gouged or pitted to collect precipitation. Planting machines should be operated on the contour since most machines leave a small furrow that can serve to channel runoff and eventually lead to gullies. It is desirable to firm the soil around the seedlings following planting.

WOODY-HERBACEOUS COMBINATIONS

Many tree and shrub species exhibit poor growth during the first year following planting. Because of this "transplanting shock," they are not highly competitive. Survival often will be low if they are planted in an already established dense herbaceous cover. Ironically, the better the job an operator does in achieving short-term site protection, the poorer are his chances to obtain long-term objectives. Tree or shrub survival and growth may be reduced, even when the herbaceous cover is less than complete or where woody seedlings are planted at the same time that the seed is sown.

Several steps can be taken to reduce this problem. As previously mentioned, woody species can be used that are especially competitive even during the first year following planting. However, this option will limit the tree or shrub species available for use by the operator. A second method, if permitted by reclamation laws and regulations, is to leave bare, i.e., unseeded strips, about 0.6 m (2 ft) in width on the contour. Woody plant seedlings are planted in the bare strip. This will permit the woody plant seedlings to become established during the year or two required for the herbaceous vegetation to spread into the unseeded area. Another possibility is to reduce the fertilization and/or herbaceous seeding rates so as to achieve a less dense, and hence less competitive, herbaceous cover. However, there is a risk that cover so obtained will not meet legal requirements or achieve the objectives of short-term site protection.

If possible, the tree or shrub seedlings should be planted at the time of seed sowing. However, this is not always possible. If planting must be done in a dense cover, some form of competition control is necessary. Herbiciding a strip about 0.6 m (2 ft) wide into which the seedlings are planted or herbiciding a spot around each seedling are promising methods. Mechanical control, either by plowing strips or by scalping around the seedlings, often is less successful because regrowth of forbs, grasses, and legumes into the denuded areas can be rapid.

POST-PLANTING OPERATIONS

Although desiccation is one of the major causes of revegetation failure on surface mines, irrigation is recommended only in special cases and then only for a short period of time. The community that is eventually established must be able to persist under the prevailing climatic conditions. Therefore, water-collecting measures, such as contour furrowing, gouging, and pitting, and water-conservation measures, such as mulching, are preferred over irrigation. Irrigation is a recommended practice in arid regions with less than 25 cm (10 in.) of annual precipitation, providing sufficient supplies of water are available. Only enough water should be applied to secure vegetation establishment. Irrigation should be reduced during the second growing season and discontinued in subsequent years. Two types of irrigation systems useful in revegetation efforts are drip and sprinkler.

Drip irrigation systems use less water but are more expensive and less flexible than sprinkler systems. Basin or flood irrigation systems generally are of little use in revegetating surface-mined areas.

In many areas it is also necessary to protect the site against livestock, wildlife, or human traffic. If the planted species are highly attractive to animals, it may be necessary to keep the animals out by fencing or by protecting individual woody plants by wrapping the stems in plastic or metal. Access routes can be barricaded to protect the site from vehicular traffic. However, it often is difficult to completely exclude off-road vehicles.

The site should be monitored annually, or even more often, for insect and disease outbreaks. If these become serious enough, it may necessary to apply recommended insecticides or fungicides.

COVER MAINTENANCE

Following the initial establishment period, the vegetation should be checked to determine if cover and seedling survival are sufficient. Soil samples should be collected from areas with insufficient vegetation and sent to a soil testing laboratory for analysis. Areas with inadequate cover should be limed and fertilized, if needed, and reseeded or replanted with tree or shrub seedlings. Even on areas with adequate vegetative cover, soils should be sampled every four to five years to identify and correct nutrient deficiencies before the cover begins to deteriorate.

GENERAL REFERENCES

Bradshaw, A.D., and Chadwick, M.J., 1980, *The Restoration of Land,* University of California Press, Berkeley, CA, 317 pp.

Leedy, D.L., 1981, "Coal Surface Mining Reclamation and Fish and Wildlife Relationships in the Eastern United States," Vol. I, FWS/OBS-80-24, USDI Fish and Wildlife Service, 75 pp.

Nawrot, J.R., Woolf, A., and Klimstra, W.D., 1982, "A Guide for Enhancement of Fish and Wildlife on Abandoned Mine Lands in the Eastern United States," FWS/OBS-80/67, USDI Fish and Wildlife Service, 101 pp.

Rafaill, B.L., and Vogel, W.G., 1978, "A Guide for Revegetating Surface-Mined Lands for Wildlife in Eastern Kentucky and West Virginia," FWS/OBS-78/84, USDI Fish and Wildlife Service, 89 pp.

Ruffner, J.D., 1978, "Plant Performance on Surface Coal Mine Spoil in Eastern United States," SCS-T5-155, USDA Soil Conservation Service, 76 pp.

Schaller, F.W., and Sutton, P., Eds., 1978, *Reclamation of Drastically Disturbed Lands,* American Society of Agronomy, Madison, WI, 742 pp.

Thornburg, A.A., 1982, "Plant Materials for Use on Surface-Mined Lands in Arid and Semiarid Regions," SCS-TP-157, USDA Soil Conservation Service, 88 pp.

USDA Forest Service, 1979a, "User Guide to Vegetation," General Technical Report INT-64, USDA Forest Service, 85 pp.

USDA Forest Service, 1979b, "User Guide to Soils," General Technical Report INT-68, USDA Forest Service, 80 pp.

USDA Forest Service, 1981, Vegetative Rehabilitation and Equipment Workshop, 35th Annual Report, Equipment Development Center, Missoula, MT, 84 pp.

USDA Forest Service, 1982, "Wildlife User Guide for Mining and Reclamation," General Technical Report INT-126, USDA Forest Service, 77 pp.

Vogel, W.G., 1981, "A Guide for Revegetating Coal Mine Soils in the Eastern United States," General Technical Report NE-68, USDA Forest Service, 190 pp.

6.6.7 Revegetation Case Study

ALTEN F. GRANDT

EDITOR'S INTRODUCTION

The following case study was provided by Alten F. Grandt, Director of Reclamation, Peabody Coal Co. The study was performed to determine if partially reclaimed mine land in Illinois could have its productivity raised by further reclamation treatment. After mining, the land was graded and seeded for use as pastureland. The experiment introduced topsoil onto the graded land two decades after mining. Yields on the experimental plots were virtually indistinguishable from those on reference plots.

The case study is included here for two reasons. The first is to demonstrate that, with time and care, mined agricultural land can be made to produce the highest yield of row crops. The second reason is to give readers a model of a demonstration that can be used by producing companies when their ability to reclaim agricultural land is in question.

(A) Purpose: To measure the effect of topsoil on graded, surface-mined land and to compare corn and soybean (row-crop) yields on original soil and graded surface-mined lands.

(B) Site Characteristics

   1. Location: Salem Township, Knox County, Illinois, 1974-1980.

   2. Mine type: Area surface mine, Illinois No. 5 coal; overburden consisted of loess, Illinoian glacial till, and unconsolidated shales; maximum depth, 21 m (70 ft).

   3. Operation: Mined 1954 and 1955 with Marion 30.6-m³ (40-cu yd) shovel (pre-law). Graded in 1955 with D-7 and D-8 dozers, no topsoil. Seeded in 1955 with disk, broadcast seeder, and harrow (one operation). Species seeded, alfalfa, 5.4 kg (12 lb) per acre, bromegrass, 2.3 kg (5 lb) per acre.

   4. Interim Use: Cut for hay and/or grazed by livestock (cattle).

   5. Study: Determine the value of replacing topsoil for row crops (corn, soybeans) on graded, surface-mined land.

   6. Treatment: 30 cm (12 in.) A horizon (Ipava silt loam) hauled with WABCO D-111A, self-loading scraper, placed on 20-year-old graded area. Graded area chisel plowed 23 cm (9 in.) prior to topsoil application.

(C) Soil test

	pH	P2(Bray)	K
		kg per ha	
1. Ipava silt loam, 1974	6.3	103	104
2. Topsoil on graded spoil, 1974	5.1	154	210
3. Graded spoil, 1954	7.3	176	201
1974	7.5	217	217

(D) Fertility treatment

   1. Corn   100-60-60

   2. Soybeans   None

   3. Herbicide and insecticide   Normal rates as on adjacent farm land.

                      Corn: Atrazine-Lasso   Beans: Treflan

(E) Corn Yield

	5-yr avg. (1975-1979)	Range of observations, Bu/Ac	
	Bu/Ac	Lowest	Highest
1. Ipava silt loam (ref. area)	139.8	104	168
2. Topsoil on graded spoil	125.5	101	164
3. Graded spoil only	84.6	67	119

(F) Soybean Yield

	3-yr avg. (1978-1980)	Range of observations, Bu/Ac	
	Bu/Ac	Lowest	Highest
1. Ipava silt loam (ref. area)	39.7	24.6	51.0
2. Topsoil on graded spoil	39.9	32.4	46.5
3. Graded spoil only	28.6	22.9	37.0

818

6.7 Water and Air Management

6.7.1 Water Management

F. A. MEEK, JR.

Introduction

The role of water management in the surface mining industry has changed dramatically in the past 25 years. Proper management and conservation of a mining area's water resources have always received some consideration in the design and implementation of surface mining operations. However, laws passed and regulations implemented in recent years have increased the need for more sophistication in the role of water management.

In the early years of surface mining, water management was primarily concerned with the operational aspects of mining. In the eastern mining regions, where annual precipitation is in excess of 88.9 cm (35 in.) per year, the primary concerns were to provide collection and conveying systems to prevent precipitation and ground water from flooding out or adversely affecting operations. In the western areas of the country, where mining is primarily conducted in arid regions, the primary water management was to conserve and store precipitation for future operational uses.

The operational uses of water at surface mining operations are diverse and dependent on the size of the operation, geologic conditions, and the characteristics of the material being mined. However, the major categories for water usage are as follows: (1) dust control; (2) product preparation; (3) potable and sanitation purposes.

Dust Control

In recent years, products have been made available that reduce the amounts of water required to provide adequate dust control for surface mine haulage ways. However, one of the primary operational requirements for water remains in the area of dust control. The control of dust in pits, haulage routes, preparation operation, and on material conveying systems is required for the safety and health of employees and the overall efficiency of the operation. The primary method for controlling dust in these situations is by the use of both mobile and stationary water spraying systems.

Product Preparation

A primary concern in the development of major operations is the availability of an adequate water supply to provide product preparation. In both coal mining in the East and hard rock mining of ore bodies in the West, the primary cleaning processes are normally by some form of heavy media separation or flotation. In heavy media separation, large amounts of water are mixed with a heavy media substance (magnetite, sand, etc.) and the resulting suspension is adjusted to have a specific gravity that falls between the product being produced and the waste material associated with the material being mined. In flotation, finely ground particles are suspended in quantities of water and selectively floated. Most of the newer coal and all ore preparation plants are designed to reuse the large amounts of water required for processing, but water loss due to surface moisture on and absorption in the products still requires a large makeup water supply. In some cases, the feasibility of mining an area is dependent on the availability of water for product preparation.

Potable and Sanitation

Most surface mining operations today provide bathing and sanitation facilities for employees. Health departments in most states require that water provided for drinking purposes comply with regulations set forth for noncommunity public water supplies. This requires that proposed water supplies must be tested for primary and secondary contaminate levels prior to approval for use. Depending on the source (ground or surface water), before water is considered potable, it must be flocculated, filtered, and chlorinated (with a minimum of 30 min of chlorine contact time). Since there are many variations in raw water sources and governing regulations throughout the country, it is recommended that the state and county health departments be contacted prior to the design of any mine-related drinking water supply.

In developing surface mining operations, the mine planning engineer must carefully evaluate the water requirement against the water availability of the proposed site. The method for evaluation is normally by development of a water budget, shown in Table 1.

The water budget shown in Table 1 indicates that under average conditions, the water supply is adequate for the needs of the operation. However, in addition to a yearly budget, monthly requirements and availability should also be assessed in the designing process. In most situations, water requirements are higher during periods of low precipitation. In addition, the drought frequency of an area should be considered. For areas with low water availability, storage and impoundment facilities can be utilized to distribute supplies over the low precipitation periods.

SURFACE MINE PLANNING AND PERMITTING

With the passage and implementation of the Clean Water Act and subsequent amendments and, more importantly, the Surface Mining Act of 1977, performance standards dealing with the protection of ground and surface water systems were implemented. The Clean Water Act primarily deals with the regulation and enforcement of water quality standards for mine discharges and receiving streams. The Surface Mining Act (PL 95-87) has resulted in regulations that cover virtually all phases of surface mining operations. The Act is primarily oriented to regulate the coal surface mining industry, but many standards set forth have been adopted by state regulatory agencies dealing with the surface mining of materials other than coal. The impact of this legislation on the role of water management in surface mining has been significant.

The portion of the Act dealing with water management is as follows:

Protection of the Hydrologic System: The permittee shall plan and conduct coal mining and reclamation operations to minimize disturbance to the prevailing hydrologic balance in order to prevent long-term adverse changes in the hydrologic balance that could result from surface coal mining and reclamation operations, both on- and off-site. Changes in water quality and quantity, in the depth to ground water, and in the location of surface water drainage channels shall be minimized such that the postmining land use of the disturbed land is not adversely affected and applicable federal and state statutes and regulations are not violated. The permittee shall conduct operations so as to minimize water pollution and shall, where necessary,

Table 1. Example Water Budget

Operational Requirements

Material Preparation
Total water requirement = 7000 gpm
Water recycle   = 98%
Water loss   = 2%
Preparation plant operating time = 14 hr per day; 240 days per year

Total usage rate = 7000 gpm $\times$ 14 hr per day $\times$ 60 min per hr
$\times$ 240 days per yr = 1.4×10^9 gal per year

Water recycle = 1.4×10^9 gal per year $\times$ 0.98 = 1.37×10^9 gal per year

Water loss = Water uptake requirement = 2.8×10^7 gal per year
1.17×10^5 gal per day

Dust Control
Haul road requirement* = 2 gal per linear ft per day
Haul road length   = 7800 ft
Haul road requirement = 2 gal per linear ft per day $\times$ 7800 linear ft
$\times$ 240 days per year = 3.7×10^6 gal per year

*Water requirements for haul roads are dependent on traffic conditions, weather, and road surface qualities.

Potable and Sanitation Requirements
Number of people employed = 226
Shower and drinking requirements = 62 gal per man per day
Sanitary facilities = 10 gal per man per day
Potable and sanitation requirements = 226 men per day $\times$ 72 gal per man per day
$\times$ 240 days per year = 3.9×10^6 gal per year

Total Water Requirements for Operation
Material preparation = 2.8×10^7 gal per year
Dust control   = 3.7×10^6 gal per year
Potable and sanitation = 3.9×10^6 gal per year
Total = 3.6×10^7 gal per year

Water Availability

Surface Water Sources
Source No. 1: A 3 sq mile drainage area
Source No. 2: A 1.2 sq mile drainage area
Average runoff = 6 in. of precipitation per year
Surface water available = 4.2 sq mile $\times$ 640 acre per sq mile
$\times$ 0.5 ft $\times$ 325829 gal per acre-ft
= 4.4×10^8 gal per year

Ground-Water Source
Source No. 1: A well with a yield of 3.5 gpm
Ground-water available = 3.5 gpm $\times$ 5.2×10^5 min per year
= 1.8×10^6 gal per year

---

use treatment methods to control water pollution. The permittee shall emphasize surface coal mining and reclamation practices that will prevent or minimize water pollution and changes in flows in preference to the use of water treatment facilities. Practices to control and minimize pollution include, but are not limited to, stabilizing disturbed areas through grading, diverting runoff, achieving quick growing stands of temporary vegetation, lining drainage channels with rock or vegetation, mulching, sealing acid-forming and toxic-forming materials, and selectively placing waste materials in backfill areas. If pollution can be controlled only by treatment, the permittee shall operate and maintain the necessary water-treatment facilities for as long as treatment is required. (Anon., 1977.)

In addition, the Act breaks down the requirements for the protection of the hydrologic system in the following categories and describes specific requirements and procedures:
(1) Water quality standards and effluent limitations
(2) Surface-water monitoring
(3) Diversion and conveyance of overland flow away from disturbed area
(4) Stream channel diversions
(5) Sediment control measures
(6) Discharge structures
(7) Acid and toxic materials
(8) Ground water
(9) Water rights and replacement
(10) Alluvial valley floors west of the 100th meridian west longitude
(11) Permanent impoundments
(12) Hydrologic impact of roads
(13) Hydrologic impact of other transport facilities
The Act has primarily been a guidance document, providing for standardization of individual state surface mining regulatory agency's procedures and programs. Therefore, there may be some variations between individual state regulatory programs, but these variations are normally not sig-

nificant. To summarize, this legislation requires that, prior to approval to conduct surface mining, the operator must first determine the "probable hydrologic consequences" of mining. To make this determination the following steps must be taken: (1) evaluate the surface and ground-water systems in both the areas being mined and adjacent areas in regard to quantity and quality; (2) evaluate overburden and spoil materials to be mined in regard to their acid, toxic, or alkaline forming characteristics; and (3) design and conduct surface mining operations so as to minimize pollution or quantity changes to both ground- and surface-water systems.

To implement and enforce the provisions of the Act (PL 95-87), the Office of Surface Mining was established within the Department of the Interior. In order to establish guidance and uniformity in the development of these hydrologic investigations, a document was prepared by the Office of Surface Mining entitled, Part 1 "The Determination of the Probable Hydrologic Consequences" and Part 2 "The Statement of the Results of Test Borings or Core Samplings." The document and Act do not specifically mandate the procedures for making the required investigations. However, the document does provide guidance.

In the early stages of mine development in the state in which the mining is to be conducted, the surface mine regulatory authorities should be called in to discuss the methodology for assessing the premining hydrologic conditions for the site. The success or failure of a reclamation plan in regard to the protection of the hydrologic system cannot be evaluated if the premining conditions of the area are not known. A hydrologic investigation and determination should be thoroughly planned and executed. Suggested methodology for conducting an investigation follows.

Area of Investigation

Laws require that hydrologic protection must be provided for both the area to be mined and any downstream adjacent areas that could be affected. It is therefore recommended that the premining hydrologic investigation cover not only the mining area, but also adjacent areas that may be affected.

Premine Surface Water Quantity/Quality Investigation

A surface water monitoring program should accurately define the premining quality and quantity condition of streams and drainways on the mining area and downstream to at least the first receiving stream. The locations of sampling points should be placed on reference maps with coordinates of all locations for future reference. Monitoring points should bracket the mining area with points both above and below the proposed mine area. The sampling frequency should be at a minimum of once per month for a period of six consecutive months. Quantity of water is an important part of the investigation. One full year of consecutive monthly samples would, therefore, more accurately reflect flow conditions during seasonal fluctuations.

Water sample collection should be conducted in accordance with ASTM or standard method procedures. The samples should be analyzed in the field for pH, temperature, and specific conductance. Laboratory analysis should include total dissolved solids, suspended solids, metals normally associated with mining being conducted, dissolved oxygen, sulfate, alkalinity, and acidity. Flow measurement at all sampling points should be taken at corresponding sampling times. Flow measurement should be conducted by an approved method such as weirs, flumes, stream cross sections, staff gages, etc.

The premining monitoring program for surface drainage should also include quality and quantity analysis for both peak and low flow conditions. During periods of low and high flow, the aquatic and plant life of streams receive the most stress due to elevated stream temperatures and increases in dissolved constituents and the deposition of sediments in streambeds.

Biota Sampling

A premining surface water investigation should include a sampling of biota for streams within and adjacent to the area to be mined. Sampling should be conducted during periods of low flow, if possible. The analysis should include identification of species, diversity, and density. A determination of stream biomass should be obtained for a unit area of the stream. The biota sampling should be conducted at a time corresponding to quantity and quality analysis. Physical conditions, such as stream width, depth, riffle/pool, and bottom type (gravel, boulder, etc.), should be recorded at the time of sampling.

Ground-water Monitoring Program

Premining ground-water conditions also must be evaluated in order to predict the probable hydrologic consequences of mining. The frequency of analysis and parameters analyzed for ground water should parallel those described under surface water investigation. In addition, a depth to water measurement should be taken when quality samples are pulled. At least twice during the monitoring period (preferably during low flow conditions), ground-water monitoring wells should be tested for yield and transmissivity.

The location and number of monitoring wells required to accurately determine ground-water conditions is dependent on the geologic conditions of the site. If only one aquifer is encountered in the mining area, the number of wells would be much less than if several aquifers are present. When the stratigraphy and ground-water location is not well known, it is often necessary to drill an observation hole to identify aquacludes, aquifers, stratigraphy, etc., before a ground-water monitoring program can be designed and implemented. In general, a ground-water monitoring program should include an observation well in the following locations: (1) in each major aquifer to be disturbed in the mining operation within the mined area, (2) in the first aquifer below the seam to be mined within the mined area, (3) in each major aquifer upgradient and adjacent to the mined area, and (4) in each major aquifer and the first aquifer below the seam to be mined down gradient and adjacent to the mined area. When the materials or seams proposed to be mined are located above drainage, springs may sometimes be used as observation points as opposed to wells. Physical information recorded for observation points should include location (coordinates), date drilled, driller, stratigraphy log, depth of hole, diameter of hole, depth to water, and casing details (type, diameter, perforations, etc.).

The surface and ground-water premining water quality investigation programs should be designed around the needs and conditions of the area being mined. In eastern areas of the country where the deposits primarily being mined are located above drainage, the aquifers encountered in the mining process are normally small in size and have limited usage. The topography of eastern mining areas and the high yearly precipitation make mining conditions such that surface water streams and drainways are normally in close proximity to areas being mined. These conditions would require that much more emphasis be placed on conducting an adequate surface water monitoring program rather than a ground-water monitoring investigation. In the western and midwestern areas of the country where mining is normally conducted in flat and arid or semiarid regions, perennial surface water drainways normally are not in close proximity to the mined areas. These regions, however, normally have large regional aquifers that are the primary sources of water for industrial, agricultural, and potable purposes. When these aquifers are encountered in the mining process, the need for a detailed ground-water hydrologic investigation is a necessity.

Statement of the Results of Test Borings or Core Samples

In order to predict and minimize the probable hydrologic consequences of surface mining, the physical and chemical nature and reactions upon weathering of the overburdens and spoils to be disturbed in the mining process must be evaluated. The earth's surface is in a continual state of weathering. Mountains are eroded by precipitation, wind, and other naturally occurring phenomena. Minerals are dissolved and transported by ground and surface water systems. This process is extremely slow and significant changes in both topography and surface and ground-water systems normally take millions of years to occur. Therefore, the topography, geology, surface and ground-water systems of a proposed surface mining area are in a relatively steady-state condition prior to mining. Ground-water courses and flow paths are well established and reactions that determine the water chemistry are occurring at a steady-state rate to produce ground-water quality that is relatively constant.

During the surface mining process, the rock and earth material (overburden) above the deposit being mined is broken up, excavated, and replaced back on the mine site. This results in new ground-water flow paths in the reconstructed backfills. When the overburden is fragmented, new reactive surfaces for rock/water contact in the backfill are created. It is therefore essential to know the physical and chemical nature of overburdens and other material encountered in the mining process. Because surface and ground-water systems are interconnected, this is essential in any area of the country.

Overburden sampling is normally accomplished by one of the following methods: (1) continuous cores, (2) highwall sampling, and (3) collection of rotary drill cuttings. The most effective sampling is, by far, the use of continuous cores. In this technique, the stratigraphy of the area is well defined. Unweathered samples, precise thickness, and accurate physical qualities of the overburden to be encountered can be obtained. In areas where previous mining has occurred and highwalls are not backfilled, highwall sampling sometimes can be used. In this technique, samples are chipped away and stored for analysis from the various strata above the deposit being mined. This technique can provide general overburden information for the area to be mined. However, because of the difficulty in collection of representative samples, the weathered nature of most highwalls, and the variability in overburdens from site to site, this method has limited applications. An overburden sampling technique that is better than highwall but less optimum than a continuous core is the use of rotary drill cuttings. In this technique, a production drill is set up on the proposed mining area at a predetermined location. The dust curtain on the drill is pulled up enough to enable a shovel to collect drill cuttings blown out the top of the hole. The drill steels are marked in 1.5 to 3-m (5 to 10-ft) increments, and samples are collected and identified according to their depth from surface. Care must be taken to clear the hole of all cuttings between sample increments. This technique provides unweathered samples for laboratory analysis, but the physical nature and stratig-

raphy of the overburden is difficult to determine using this method.

The ideal location and number of overburden test holes is dependent on the geology of the area and the nature of the deposit to be mined. In areas where previous mining has occurred with no adverse impacts on ground or surface water systems, one overburden test site may be adequate to confirm reaction characteristics. However, in areas where mining has occurred with adverse effects, no previous surface mining has occurred, or overburden characteristics are known to vary over short distances, more drilling and sampling will be needed. In these areas, three holes for every 40.5 ha (100 acres) to be mined will normally confirm overburden characteristics or define the need for more drilling.

Test holes should be drilled in locations that will allow collection of overburden samples for all materials encountered in the mining process. Spacing of holes should be such that the distribution of data is optimum.

Locations of test holes or other sampling points should be located by survey (x,y,z coordinates) and placed on maps for future reference. Suggested information to be recorded in the field while drilling is being conducted is as follows:

1) Site identification number
2) Date drilled
3) Driller name/company
4) Logged by
5) Weather conditions
6) Stratigraphy including height, general rock type, mineralology, grain type/size, matrix, reaction to Hcl, presence of any staining, and hardness
7) Location of ground water encountered during drilling

Overburden sampled by coring, highwall, or production drill methods should be stored in a manner that provides protection from weathering, easy identification, and compact storage.

Preparation of Samples For Analysis

Proper preparation of overburden samples is very important in obtaining accurate data on the acid/base reaction nature of material encountered in the surface mining process. The expense of site preparation, mobilization, and drilling will be wasted if the preparation of the samples obtained is not handled correctly. Continuous core samples should be prepared according to stratigraphy. Each individual strata of the column should be separated and the entire strata crushed to 9.5 mm ($\frac{3}{8}$ in.) × 0. When strata thickness exceeds 3 m (10 ft), the sample should be crushed in 3-m (10 ft) increments. If rock durability analysis is to be performed on the material, a 76.2 to 152.4-mm (3 to 6-in.) section of the strata uncrushed should be retained. Once the sample is crushed, it is split by *cone and quartering*, or other standard methods for sample preparation, into four approximately equal samples. One of the four samples is further pulverized to less than 75 μm (200 mesh) for acid/base accounting, petrographic examination, and other analytical procedures. The remaining samples are retained for simulated weathering tests and for a reserve. Highwall and production drill samples should be prepared for analysis in the same manner. However, the sample size and location are predetermined at the time of collection for these techniques.

Overburden Analytical Testing Procedures

In order to predict or determine what spoil/water reactions will occur on reclaimed surface mine sites, analytical procedures have been developed to help indicate the acid/

base properties of overburdens and mine spoils from core borings and other sampling techniques. The most commonly used procedure is termed *acid-base accounting* ("Field and Laboratory Procedures for Overburden and Mine Spoils," EPA publication). In this procedure, the samples are analyzed for their maximum acid-producing potential and the amount of neutralizing potential (alkalinity) present.

Acid-base accounting includes three analytical procedures performed on each sample tested: total sulfur, calcium carbonate equivalents, and paste pH. The total sulfur analysis is performed to give the maximum amount of pyrite (FeS_2 acid-producing) that will be present in the sample. The analysis is performed by heating a pulverized sample to temperatures hot enough to liberate SO_2. The SO_2 is then oxidized to H_2SO_4 and back titrated to give the sulfur contents of the sample. This is done with the same type of sulfur analyzer used in the analysis of sulfur in coal. The calcium carbonate equivalents are found by placing a pulverized overburden sample in hydrochloric acid and water and heating the sam-

ple until all alkaline materials present have reacted with the HCl. The sample is then back titrated and the amount of calcium carbonate ($CaCO_3$) equivalents present in the sample is calculated. The paste pH test is accomplished by making a paste of the overburden sample and distilled water and measuring the pH of the mixture. This procedure is intended to give information on the current acid-base state of the sample, rather than how the sample will react in the future.

In the acid-base accounting, the results of the acid-producing (*T*. Sulfur) and alkaline-producing ($CaCO_3$ equivalents) tests previously described are reported in $CaCO_3$ equivalents. The acid-producing nature is expressed in tons $CaCO_3$ needed per 1000 ton of material, and the alkaline-producing nature is expressed in tons $CaCO_3$ present per 1000 tons of material. By comparing these two values of an overburden sample, the overall alkaline or acidic nature of the material is determined.

An example acid-base accounting report form will help demonstrate the concept.

Example

Hole ID — DDH 514
Longitude —
Latitude —
Surface elevation — 617.5 m (2026 ft)

No.	Sample ID (ft)	Rock Type	Fizz	pH	*T*. sulfur	CaCO₃ Present	Max CaCO₃* Requirements	Amt. Need	Amt. Excess
						St CaCO₃ per 1000 St Material			
1	26.2–32.3	Brown shale	0	6.5	0.60%	5.4	18.75	13.35	
2	32.3–45.0	Gray sandstone	1	6.5	0.10%	24.3	3.125		21.18

*Maximum needed is obtained by multiplying the percent sulfur of the sample by 31.25.
Metric conversions: ft × 0.3048 = m; st × 0.907 184 7 = t.

The preceding example shows that sample No. 1 has more acid-producing capabilities than alkaline material present. For the sample, 13.35 tons of $CaCO_3$ (limestone) would be needed to make 1000 tons of the material neutral. Sample No. 2 has an excess above neutral of 21.18 tons of $CaCO_3$ per 1000 tons of material present. It has been suggested that a sample must have 5.0 tons $CaCO_3$ per 1000 tons material or more in the amount needed column before it should be considered acid-producing. However, this cannot be assumed to hold true in every case. Samples that are relatively neutral

(between 5 ton per 1000 tons needed and 5 ton per 1000 tons excess) by acid-base accounting procedures should be further evaluated by analyzing for forms of sulfur or simulated weathering tests.

When the acid-base accounting tests have been performed, this information should be integrated with the individual strata volumetrics of an area to be mined to see the overall acid-base materials balance. An example of this process is shown as follows:

Strata ID (ft)*	Volume	Density	Total Tons	Excess CaCO₃	Amt. Needed	Needed	Excess
				Tons CaCO₃/1000 Ton Materials		Total CaCO₃/Strata	
26.2–32.3	600 acre-ft	150 pcf	1960200		13.35	26168.7	
32.3–45.0	1000 acre-ft	146 pcf	3179880	21.18			67350

Metric equivalents: ft × 0.3048 = m; acre-ft × 1233.489 = m³; pcf × 16.018 46 = kg/m³; st × 0.907 184 7 = t.

By conducting the foregoing exercise for all materials encountered in the surface mining process, the overall acid-base nature of the overburden is assessed.

Simulated Weathering Tests

When acid-base accounting tests indicate that materials are relatively neutral, simulated weathering tests can be per-

formed to quantify the acidic or alkaline nature of the individual strata. Specific procedures are outlined in Sobek (1978). In general, the tests are designed to subject the materials tested to conditions that are similar to those of a surface mine backfill. Samples of the individual strata are crushed to 6.35 mm (¼ in.) × 0 size. Three to five hundred grams (10.6 to 17.6 oz) of sample is placed in a simulated weathering cell. The cell is normally a glass or plastic closed container, measuring approximately 203.2 × 203.2 × 101.6 mm (8 × 8 × 4 in.) in size, fitted with one hole near the top and one hole near the bottom of the container. The sample is initially wetted and both humid and dry air is passed over the sample (via a mechanical air pump) for three-day periods. On the seventh day, the air is turned off, the holes are plugged, and 200 to 500 mL (6.8 to 16.9 oz) of distilled water is poured into the container and allowed to stand for 1 hr. The water is then drained off and collected for laboratory analysis. The suggested analyses to be performed are volume, pH, total alkalinity, total acidity, total iron, total manganese, and specific conductance. Simulated weathering tests have been recommended to run from 7 to 15 weekly cycles.

In addition, to determine the acidic or basic nature of relatively neutral samples, this procedure reflects the rate at which acidity or alkalinity will be produced. Caution should be used, however, in using this information to predict specific rates of alkalinity or acidity that a mine site will produce in drainage. Due to the inconsistent size gradation of overburdens and mine spoils and the effects of adjacent overburden on an individual strata's reaction rates, predictions that deal with specific quantities or qualities are likely to be incorrect.

The analytical procedures previously described deal with identifying the acidic and alkaline characteristics of individual stratas. The surface mine backfill consists of a conglomeration of these individual stratas determined by the topography, geology, equipment, and method of operation used to mine the area. The effects of this strata mixing can alter the overall acid-base reactions of the individual strata from those reflected in the sampling and analysis procedures. Simulated weathering tests can be designed to account for this mixing effect if the sequence of overburden removal can be predetermined. By combining individual strata samples in the same proportion that they will exist on the mine area backfill, this combining effect can be assessed. In addition, the quality of water moving through a surface mine backfill can affect the acid-base reaction mechanisms.

Predicting the Hydrologic Consequences of Surface Mining

The investigations described earlier for both surface and ground water systems are intended to show the surface mine operator the quantity and quality of the systems present on the proposed mine area. The overburden testing procedures described indicate the presence or absence of acidic or alkaline producing compounds associated with the materials to be disturbed during the surface mining process.

In general, the areas to be addressed in predicting effects on ground water systems are: transmissivity, storage coefficient, aquifer yields, water quality, ground water/surface water hydrologic balance, and usage of ground water in the area to be mined. In the surface mining process, the ground water systems located on the mine site above the deposits being mined will be removed. After mining and reclamation is complete, the ground water systems will reestablish. The changes in ground water systems from premining to postmining are normally dependent on changes to fracturing of previously stratified rock; removal of aquacludes; increasing or decreasing infiltration rates; increasing or decreasing discharge rates to surface water systems; and increased surface area and reaction surfaces for water-rock contact.

In mining areas where ground water systems consist of multiaquifers that support commercial and/or residential use, investigations will normally require the aid of computer models that simulate ground water flows and quality according to conditions. In any event, the exercise will require expertise in geology, hydrology, chemistry, and mining engineering. In many cases, the effects of surface mining on ground water systems may be beneficial in the areas of well yield, storage capacity, and quality. The amount of investigation required to adequately predict the results of mining should correspond to the presence and usage of ground water systems on a proposed mine area.

Primary areas to address when predicting the effects of surface mining on surface water systems are: (1) changes in stream flow quantities, (2) channel stability, (3) erosion and sediment yields, (4) water quality, (5) aquatic biota, and (6) ground water/surface water hydrologic balance. Sedimentation and erosion controls that adequately protect streams located down gradient from surface mining operations have been in use for many years. Stream channel protections have been accomplished throughout the country by the use of riprap, levees, and vegetation. Surface mined areas normally have elevated infiltration and ground water storage coefficients, in comparison to premine conditions that generally help stream flow and hydrologic balance conditions by reducing peak flow conditions and increasing base flow conditions.

By far, the most difficult task in the area of hydrologic protection for surface mine operations is in the area of water quality protection. When surface mine overburdens do not contain alkaline or acidic-forming compounds, operational design can concentrate on efficient overburden removal, materials handling, and backfilling. However, when polluting materials are prominent in the overburdens, operations must be designed to minimize the production and transport of pollutants in the hydrologic regime. Unlike sediment control and stream stability measures, these techniques are site specific and dependent on the individual condition of the site. These preventative techniques are the result of a currently active research effort that will continue to expand the technology. The procedures should be considered state-of-the-art, but are not yet considered accepted design practices throughout the country. Because of the need of many operators throughout the country to know how to handle polluting overburdens, the primary techniques in use are discussed.

POLLUTION PREVENTATIVE MINING TECHNIQUES

Blending of Overburdens

In proposed mining areas where the ratio of potentially polluting to nonpolluting overburdens is low, blending is often used to avoid water quality problems. Studies have indicated that when polluting material constitutes 5% or less of the total overburden, dilution normally provides enough control to minimize water quality problems. Blending techniques coincide particularly well with surface operations utilizing draglines as the primary excavators. In operations using shovel/loaders and trucks, more design is needed to assure a thorough blending of materials.

In mine areas where the polluting materials present are disulfide minerals (acid-producing), blending can act as an acid preventative mechanism rather than just dilution. Alkalinity, if present in enough quantity, in a spoil system inhibits disulfide oxidation. When overburden acid/base analysis and stratigraphic volumes determine that the total overburden has in excess of 20 tons $CaCO_3$ per 1000 tons of material, blending can be an effective acid preventative technique. It is important to realize that the neutrality of the total overburden, in regard to acid-base and volume analysis, does not assure that blending will prevent acid mine drainage from occurring. The solubility of $CaCO_3$ is limited under atmospheric conditions and disulfide oxidation is enhanced when exposed to air and water (atmospheric conditions). This unbalanced reaction rate results in a system being acid producing when alkaline and acidic components are theoretically neutral. However, when alkaline components are significantly in excess (20 tons $CaCO_3$ per 1000 tons material), both neutralization and reduction in disulfide oxidation occurs if the materials are thoroughly blended. In areas where the alkaline components of overburden are more soluble than limestone, less excess alkalinity is needed. Simulated weathering tests should be conducted on overburden from the proposed mine site to assess alkaline/acid reaction rates on the blended materials. When the amounts of polluting material present in overburden is significant, blending is not an applicable preventative technique. In these cases, selective handling and placement of polluting materials may be required.

Selective Handling and Placement

A method used in the mining industry to minimize the transport of soluble salts from the oxidation and/or solubilization of polluting materials associated with fragmented overburden is the use of selective handling and placement. The technique involves:

1) Identification and location of polluting materials in the overburden or spoil to be disturbed in the mining process.

2) Designing the removal and placement of such material in a manner as to minimize its contact with ground or surface water flows. Designing the removal and placement of such material as to minimize oxidation of pyritic (acid forming) compounds in the mined material.

The identification of the polluting material was described earlier in the "Statement of the Results of Test Borings or Core Samplings." A graphic depiction of the results of such analysis is shown in Fig. 1.

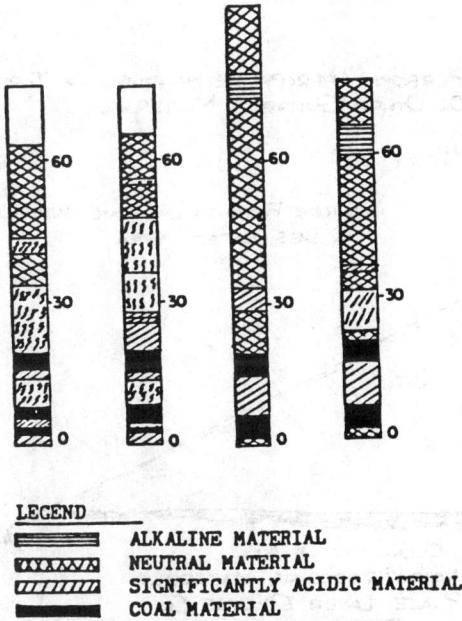

LEGEND
- ALKALINE MATERIAL
- NEUTRAL MATERIAL
- SIGNIFICANTLY ACIDIC MATERIAL
- COAL MATERIAL
- SLIGHTLY ACIDIC MATERIAL

Fig. 1. Stratigraphic column.

Once materials are identified, the task of designing removal and placement plans can be developed. When surface mine overburdens and spoils are drilled, shot, and excavated in the mining process, the surface area of the materials is greatly increased. Unweathered alkaline and acid-producing compounds associated with these freshly exposed surfaces are now capable of contacting air (oxidation) and water (transport). As a result, the weathering products can enter into the ground and surface water by stems. In a surface mine backfill, the porous nature of the backfill material makes it extremely difficult to exclude air from the system. Changes in barometric pressures can cause air movement in and out of what appears to be very compacted backfills. However, if polluting materials are selectively removed and placed in areas that will not come in direct contact with water, the pollutants are not transported into the hydrologic system. Figs. 2 through Fig. 5 are graphic examples of how selective handling and placement can be used on a contour surface coal mining operation. In this example, the relatively steep outslopes of the regraded will promote surface runoff as opposed to infiltration. With the placement of the coarse nontoxic material against the highwall and on the pit floor, ground water is directed away from the toxic material, thus preventing transport of weathering products.

In areas of the country where the materials being mined are located below the ground water table, another form of selective handling and placement is used. In order for pyritic (FeS_2) material to produce acid, it must first be oxidized.

Reaction Process

$$2Fe\,S_2 + 7O_2 + 2H_2O \rightarrow 2Fe^{++} + 4SO_4 + (4H^+)$$
$$Fe^{++} + 1/4\,O_2 + H^+ \rightarrow Fe^{+++} + 1/2\,H_2O$$
$$Fe^{+++} + 3H_2O \rightarrow Fe\,(OH)_3 + (3H^+)$$

Oxygen (O_2) is necessary for the oxidation of pyrite to occur. The maximum concentrations of oxygen in water, if not continually replaced, will limit pyrite oxidation. Fig. 6 shows material placement for this form of selective handling. It is very important that toxic materials be placed below seasonal and drought fluctuation in the water table. If toxic materials are temporarily stored out of these reducing conditions, or if the water table buildup over the toxic material takes considerable time, weathering (oxidation) of the toxic-forming compounds could occur. If this should happen, the soluble salt resulting from the oxidation will enter the ground water system.

Sealing Techniques

When topographic conditions of a mine site dictate a gentle or flat regraded slope, much of the yearly precipitation will infiltrate into the porous backfills. If the amounts of potentially polluting overburdens are large and spread throughout the mined area, significant amounts of infiltrating water will contact these overburdens, creating a potential for pollution to occur. By the use of selective handling and placement and subsequent sealing of the surface of potentially polluting materials, the hydrology of the backfill is manipulated to minimize the problem.

Fig. 7 shows a mountaintop situation where selective handling and placement and sealing techniques are used. After product removal, a blanket of porous nonpolluting material is placed on the pit floor. This is done to provide positive drainage for any artesian effects and to support the polluting material above any accumulated water on the pit

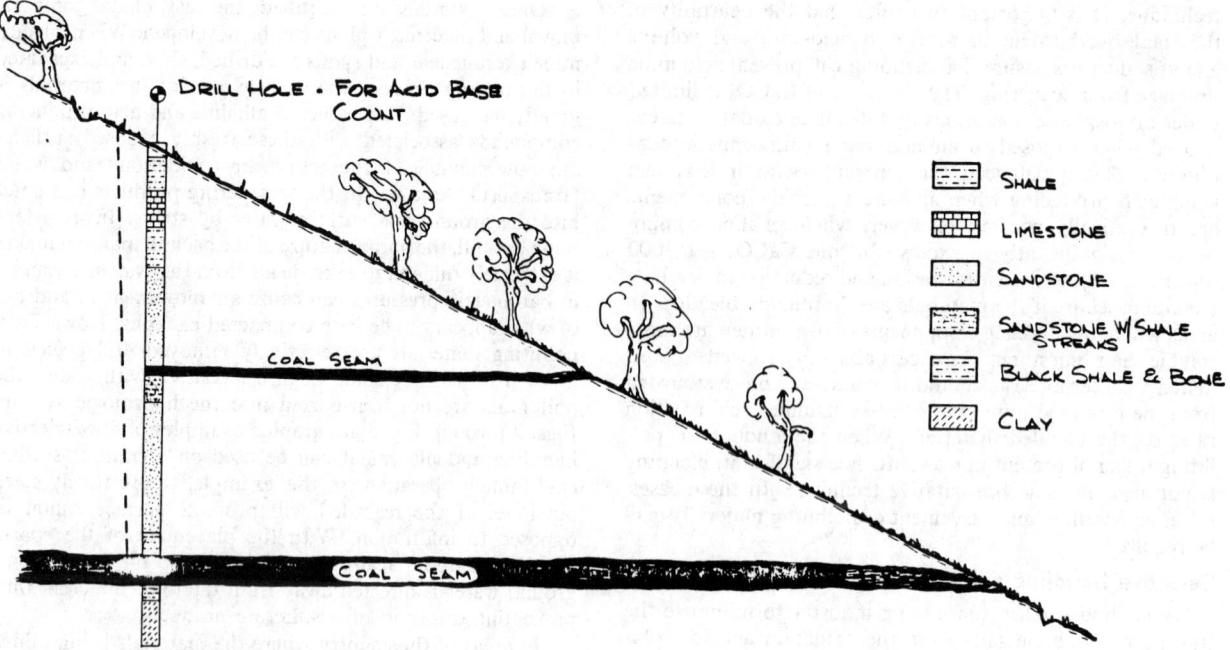

DRILL HOLE · FOR ACID BASE COUNT

COAL SEAM

COAL SEAM

SHALE

LIMESTONE

SANDSTONE

SANDSTONE W/SHALE STREAKS

BLACK SHALE & BONE

CLAY

Fig. 2. Selective handling and placement. Drainage system is to be installed before mining operations begin.

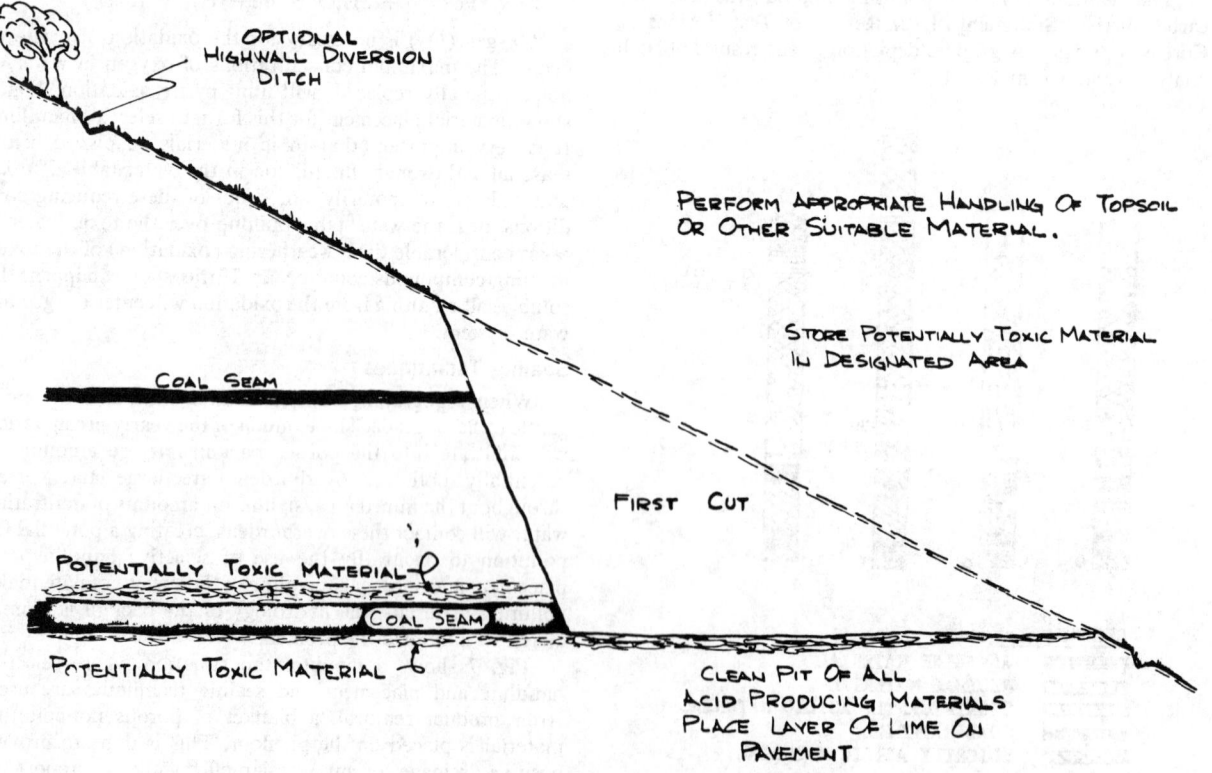

OPTIONAL HIGHWALL DIVERSION DITCH

PERFORM APPROPRIATE HANDLING OF TOPSOIL OR OTHER SUITABLE MATERIAL.

STORE POTENTIALLY TOXIC MATERIAL IN DESIGNATED AREA

COAL SEAM

FIRST CUT

POTENTIALLY TOXIC MATERIAL

COAL SEAM

POTENTIALLY TOXIC MATERIAL

CLEAN PIT OF ALL ACID PRODUCING MATERIALS PLACE LAYER OF LIME ON PAVEMENT

Fig. 3. Selective handling and placement. First cut.

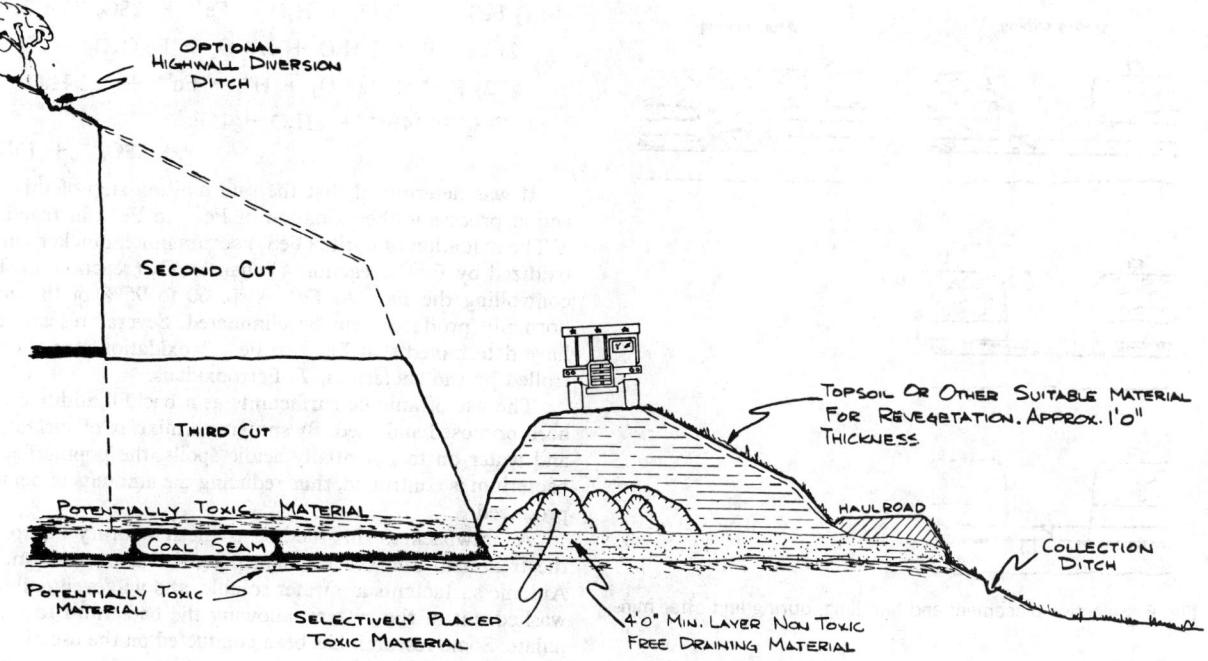

Fig. 4. Selective handling and placement. Second cut.

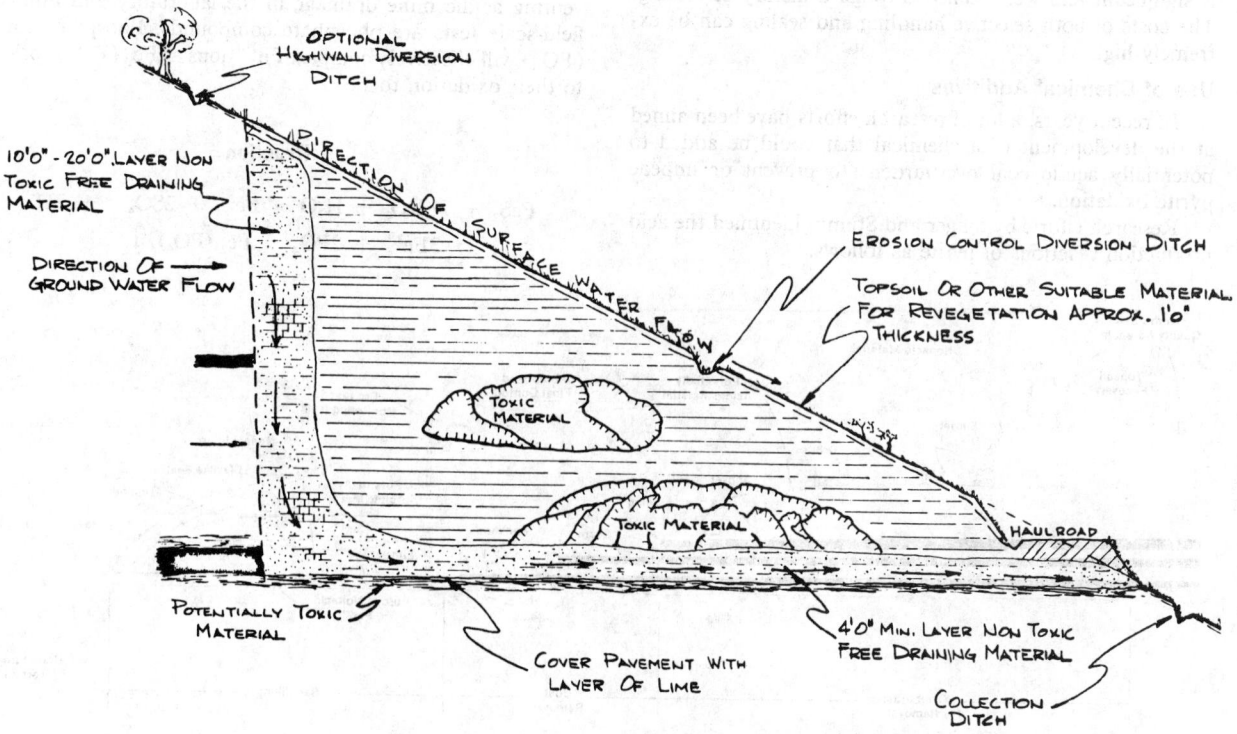

Fig. 5. Selective handling and placement. Backfill and regrading plan.

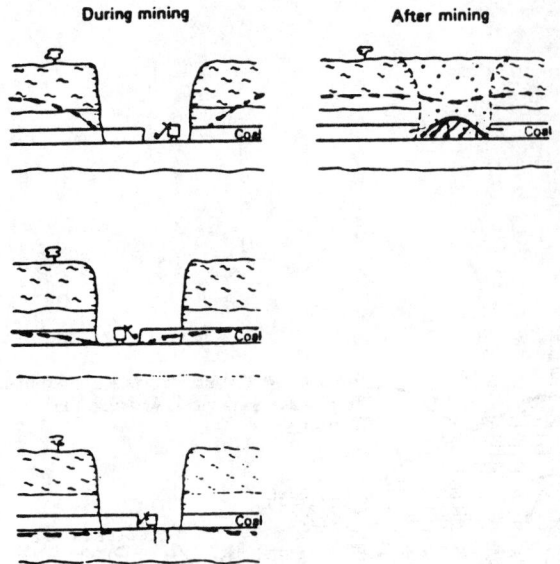

Fig. 6. Selective placement and handling, during and after mining.

floor. The potentially polluting material is then placed on the pad and brought to approximate final contour. In this case, the compacted clay seal is placed over the material. To provide positive drainage above the seal and to provide a rooting zone for revegetation, a cover of 1.2 to 1.8 m (4 to 6 ft) of nonpolluting material is placed on top of the seal.

In areas where suitable clay material is nonabundant, other sealing materials such as PVC or polyethelene have been used. It should be noted that this technique is normally only used in areas where the potential for pollution to occur is significant and well defined through a history of mining. The costs of both selective handling and sealing can be extremely high.

Use of Chemical Additives

In recent years, a lot of research efforts have been aimed at the development of a chemical that could be added to potentially acidic coal overburdens to prevent or impede pyrite oxidation.

Research efforts by Singer and Stumm identified the acid production reactions of pyrite as follows:

1) $FeS_2 + 7/2 \, O_2 + H_2O \rightarrow Fe^{2+} + 2SO_4^{-2} + 2H^+$

2) $Fe^{2+} + 5/2 \, H_2O + 1/4 \, O_2 \rightarrow Fe(OH)_3 + 2H^+$

3) $Fe^{2+} + 1/4 \, O_2 + H^+ \rightarrow Fe^{3+} + 1/2 \, H_2O$

4) $FeS_2 + 14F^{3+} + 8H_2O \rightarrow 15Fe^{3+} + 2SO_4^{2-} + 16H^+$

It was determined that the rate limiting step of this reaction process is the oxidation of Fe^{2+} to Fe^{3+} in reaction 3. The oxidation of pyrite (FeS_2) occurs much quicker when oxidized by Fe^{3+} (reaction 4) than by O_2 (reaction 1). By controlling the Fe^{+2} to Fe^{+3} step, 60 to 95% of the acid normally produced can be eliminated. Several researchers have determined that Fe^{++} to Fe^{+++} oxidation step is controlled by the bacterium, *T. Ferrooxidans*.

The use of anionic surfactants as a backfill additive has been proposed and used. By spraying a mixture of surfactant and water on to potentially acidic spoils, the population of bacterium is controlled, thus reducing the amounts of acidity produced.

A drawback to this method is the temporary nature of the treatment and the problem potential of overtreatment. Anionic surfactants are water soluble, and will eventually be washed out of the system, allowing the bacterium to repopulate. Some research has been conducted on the use of slow release pellets containing the controlling substance. Tests have indicated the treatment life of the slow release pellets is from two to five years.

The surfactant controls *T. Ferrooxidans* by breaking down their protective outer membrane. Concentrations as low as 25 ppm have been shown to adequately control the bacterium. Most suggested or required stream standards for surfactant concentrations are less than 1 ppm. Therefore, careful evaluation of the spoils adsorption and release capacities and potential adverse effects must be conducted prior to treatment.

Other chemical additives that have shown success in preventing acidic mine drainage in the laboratory and limited field-scale tests are phosphate compounds. Phosphate ions (PO_4) will effectively complex Fe^{++} ions as $Fe_3(PO_4)_2$ prior to their oxidation to Fe^{+++}.

Reaction

$$FeS_2 + 7/2 \, O_2 + H_2O \rightarrow Fe^{2+} + 2SO_4^{-2} + 2H$$
$$3Fe^{2+} + 2PO_4 \rightarrow Fe_2(PO_4)_3 \downarrow$$

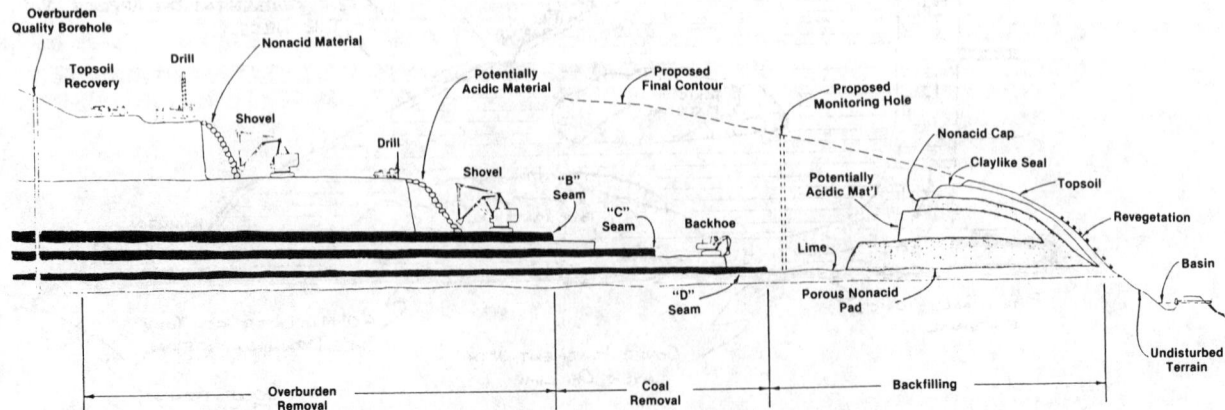

Fig. 7. Sealing techniques.

By complexing the ferrous ion (Fe^{++}), the phosphate ion has accomplished two significant steps in preventing acid mine drainage. They are (1) preventing Fe^{++} oxidation to Fe^{+++} (established above as the rate limiting step), and (2) preventing one-half of the potential acidity of pyrite from occurring.

The research shows that if the phosphate ions are present, acidities produced from pyrite oxidation are significantly reduced. Several phosphate compounds have been proposed for use as overburden additives. The primary ones are:

1) Diammonium phosphate (18-46-0, fertilizer)
2) Triple super phosphate (0-46-0, fertilizer)
3) Apatite rock

Both diammonium and triple super phosphate compounds are readily available and provide an immediate source of phosphate ions. The relatively high solubility of these compounds provides good mobilization of the phosphate ion to the acid-producing surfaces within a backfill, making the compound ideal for use on previously constructed fills. The high solubility consequently makes the treatment only temporary in nature, unless all reacted surfaces are complexed with the initial treatment. Recommended dosage rates of the compound are dependent on amount of acid-producing material present, adsorption capacity of spoil material treated, and hydrologic conditions of the site. Since these materials are water soluble, care must be taken to prevent PO_4 ions from entering ground or surface water systems. Most states limit discharge concentrations of phosphate because of its eutrification effects.

Apatite rock ($CA_5(X)(PO_4)_3$) ($X =$ F, Cl, OH) is a raw mineral mined and further refined to produce the two phosphate compounds discussed previously. The material is mined and washed to liberate clays from the matrix, resulting in an initial product that ranges from 30 to 45% PO_4, with a gradation similar to sand. Apatite rock is relatively insoluble except under acidic conditions. This provides control on the release of PO_4 ion into the system, allowing the compound to be inert until acid production occurs. Studies indicate that the optimum application rate is 0.3 tons of 64% BPL (40% PO_4) per 100 tons of acidic material. This is independent of the potential acidity of the material treated. An important factor in obtaining success from this treatment method is a thorough blending of the apatite rock with the material being treated.

A chemical additive that has had the most usage in surface mining for control of acidic drainage is lime ($CaCO_3$). The bacterium, *T. Ferrooxidans*, that are responsible for the rate-determining step in the overall acid production reaction process, thrive in pH conditions less than 4.5. If pH conditions within a backfill can be kept in a neutral or alkaline condition, the population of bacterium is controlled, thus controlling the acid production. Two basic schools of thought on how best to apply and use lime as an acidic control measure are: (1) topical application over the spoil area and (2) segregation of acidic material and blending of lime within the backfill. Under relatively mild acidic conditions, topical applications may provide enough alkalinity to prevent the acceleration of pyrite oxidation or neutralize what acidity is produced. However, $CaCO_3$ is relatively insoluble under ambient conditions and mobilization of alkalinity to the acid-producing reaction sites within a fill is limited. By segregating the potentially polluting materials and add mixing lime or other alkaline products to the spoil, more assurance is obtained that the alkaline material is near the acid-producing surfaces.

All of the acid preventative techniques discussed add significant costs to a surface mining operation. An operator must thoroughly evaluate the mining conditions and polluting potential of the areas to be mined. Current laws require that pollution resulting from mining operations must be treated to acceptable limits as long as the condition exists. Consideration must be given to the costs of long-term treatment as opposed to higher mining cost due to the implementation of an acid preventative technique for the life of the mine.

WATER TREATMENT

In surface mining operations, large quantities of rock and earth material are disturbed from their original consolidated state and exposed to the environment until final reclamation and revegetation of the area is achieved. During this period, the potential exists for large quantities of solids to be transported, via surface water runoff, into streams and waterways. Environmental regulations mandate that (1) water draining from all disturbed areas associated with surface mining must pass through a sediment control structure prior to entering a receiving stream, and (2) the suspended solid concentrations of surface mine discharges cannot exceed a 30 mg/L average or a 60 mg/L maximum concentration for any one-month period (some variations are given under extreme conditions). Regulating authorities require that sedimentation control structures be sized to provide 154.2 to 2467 m^3 0.125 to 2.0 acre-ft of storage capacity per acre of disturbed area (depending on individual state regulatory programs).

For most geologic conditions, this sediment storage volume is more than enough to meet suspended solids effluent limits. However, in areas with high concentrations of clays and other material present in the overburdens, it may be necessary to use chemical flocculants to achieve compliance quality discharges.

Sedimentation from surface mining operations can be controlled by using common sense mining methods and observations. Recommended guidelines and practices are as follows:

1) Schedule mining operations in such a manner as to minimize nonstabilized land areas.
2) Attempt to limit topsoil removal and replacement to one operation.
3) Construct and stabilize control measures in advance of mining operations.
4) With the use of diversions, prevent flowing water from entering highly erodable areas.
5) Use energy dissipaters to slow down flow velocities in ditches, diversions, etc.
6) Inspect and maintain controls after each significant storm event.
7) Completely integrate drainage, erosion, and sediment control into every stage of the surface mining operation.

Tables 2 and 3 list the various types of sediment control measures commonly used throughout the industry.

A common water treatment requirement for many surface mining operations is pH adjustment and metals removal. Regulations require that surface mine discharges be within the pH range of 6.0 to 9.0. Various metal ion concentrations for discharges are enforced, depending on the material to be mined. However, the most common requirements are for iron and manganese concentrations associated with acid or ferruginous mine drainage. Even in areas of the country where acid or iron producing overburden is not present, temporary treatment is often required for pit accumulation or product tailings runoff.

When evaluating the water treatments requirement for

Table 2. Drainage, Erosion, and Sediment Controls

Check dams	Energy dissipater	Sawdust
Chemical stabilization	Gabions	Shredded Bark
Compaction	Rock riprap	Straw
Culvert	Water pool	Wood cellulose
Open top	Erosion checks	Wood chips
Ditch relief	Filter	Wood fiber
Deep chiseling	Bales	Mulch anchoring
Diversion	Bales and filter cloth	Asphalt tacking
Bare soil	Berm	Chemical application
Bituminous	Brush	Matting
Excelsior	Brush and filter cloth	Netting
Fiberglass	Gravel or rock	Punching
Grass	Gabion	Pit drainage
Jute	Gouging	Pitting
Plastic sheeting	Grass cover	Riprap
Rock riprap	Grubbing	Road dips
Dozer basin	Incline tube settler	Roughened surface
Downdrain	Hydrocyclone	Scarifying
Chutes	Level spreader	Sediment basin
Excelsior mat	Matting and netting	Sediment trap
Fiberglass	Excelsior	Excavated
Flexible pipe	Fiberglass	Filter fabric
Flumes	Jute	Rock riprap
Half-round pipe	Paper yarn	Selective grading
Jute mesh	Plastic	Surface area exposure
Plastic sheets	Mulch	Swirl concentrator
Rigid pipe	Cellulose	Terracing
Rock lined	Fiberglass	Trenching
Sectional	Hay	Vegetative filter
	Hydromulch	Water bars

an operation in regard to pH and metals control, several factors must be assessed. These factors include longevity of treatment requirement, concentrations and quantities of water requiring treatment, and power and sludge handling requirements.

Longevity of Treatment Required

Surface mining operations normally are continually moving. Areas are disturbed and the products mined are exposed. The area is then backfilled and restabilized. pH and metals removal treatment requirements are often only for a short period of time. Preparation and refuse disposal facilities for operations, however, are normally at a fixed location for extended periods of time.

pH adjustments are achieved by the addition of alkaline reagents to the water being treated. Table 4 shows the most commonly used alkaline reagents for pH adjustments with their cost and neutralization capabilities. The information indicates the most cost-effective compounds are limestone, quicklime, and hydrated lime. However, the solubility of these compounds is much less than the solubility of either caustic soda or soda ash. This requires mechanical agitation of the compounds with the raw water being treated to obtain adequate treatment and reagent utilization. The solubility of the other two compounds shown allows the reagent to be added with no mechanical mixing to achieve proper treatment.

Figs. 8 and 9 show the types of systems most commonly used in conjunction with the reagents discussed. The high solubility of NaOH solutions allows for a simple flow regulation feeding system to be utilized. The system is initially calibrated and feed rates are automatically adjusted for all but extreme fluctuations in raw water flow. An example calculation of treatment cost for the two systems will best

reflect the importance of assessing the longevity of treatment requirements when determining treatment system needs.

Example

Water to be treated
Amount = 567.8 L/min (150 gpm)
Quality = 500 mg/L ($CaCO_3$ acidity)
Duration of treatment requirement = 2 years

Caustic soda system
Storage tank 11356.2 L (3000 gal) = $2,500
Feeder system = $1,000
Installation = $ 500

Initial cost of installation = $ 4,000
Operational cost
  Chemical cost = $70,042
  Power cost = -0-
  Operational labor and repairs = $ 1,000
Total O and O cost = $75,042

Hydrated Lime System
Lime storage bin 22.7 t (25 st) = $15,000
Mix tank; mixer = $ 3,000
Pumps = $ 4,000
Lime feeder, bin activator = $10,000
Installation = $30,000
Initial cost of installation = $62,000

Chemical cost = $ 5,819
Power cost = $ 2,000

Table 3. Control Measures for Mining Operations

Drainage, Erosion, and Sediment Control Measures	Access and Haul Road	Exploratory Drilling	Timber Stand Removal	Clearing and Grubbing	Topsoil Segregation	Topsoil Stockpile	Drill Bench Construction	Overburden Removal	Head-of-Hollow Fill	Mineral Extractions	Overburden Placement	Topsoil Placement	Reclaimed Area Stabilization	Sediment Basin Removal	Small Scale Control Removal	Road Closure
Surface Stabilization																
Selective grading and shaping	X	X		X	X	X	X		X		X	X		X	X	X
Mechanical preparation	X					X			X		X	X		X	X	X
Mulches	X					X			X				X	X	X	X
Mulch tack and anchoring	X					X			X				X	X	X	X
Matting and netting	X					X							X	X	X	X
Chemical addition	X					X			X				X	X	X	X
Grass cover	X					X			X				X	X	X	X
Runoff Interception and Conveyance																
Diversion ditch or channel	X	X	X	X	X		X	X	X	X	X	X	X	X		
Downdrain	X	X	X	X	X			X	X	X	X	X	X	X		X
Open-top culvert	X															X
Level spreader													X	X	X	
Ditch relief culvert	X															
Terracing									X			X	X			
Road dip	X															X
Water bars																X
Pit pumping										X						
Sedimentation																
Sediment basin								X	X	X	X	X	X			
Sediment trap	X	X	X	X	X	X	X	X	X	X	X	X	X	X	X	
Check dam	X	X	X	X	X	X	X	X	X	X	X	X	X	X		
Vegetative filter						X	X	X	X	X	X	X	X	X	X	
Brush barrier	X	X	X	X	X	X	X	X	X	X	X	X	X	X	X	
Filter fabric	X	X	X	X	X	X	X	X	X	X	X	X	X	X		
Swirl concentrator						X	X	X	X	X	X	X	X			

Table 4. Alkali Neutralization Capabilities

Reagent		Neutralizing* Equivalent	Cents/Lb	Cents per Neutralization Equivalent
Limestone	$CaCO_3$	8.36	0.23	1.92
Quicklime	CaO	4.67	1.05	4.91
Hydrated lime	$Ca(OH)_2$	6.18	1.19	7.35
Caustic soda	$NaOH$	6.67	13.40†	89.40
			3.90‡	26.02
Soda ash	Na_2CO_3	8.84	3.80	33.58

* Pounds of reagent to neutralize 1 mg/L acidity in 1 million gal of water.
† 50% solution in 55-gal drums.
‡ Flake or pellets in bulk.
Metric equivalents: gal × 3.785 412 = L; lb × 0.453 592 4 = kg.

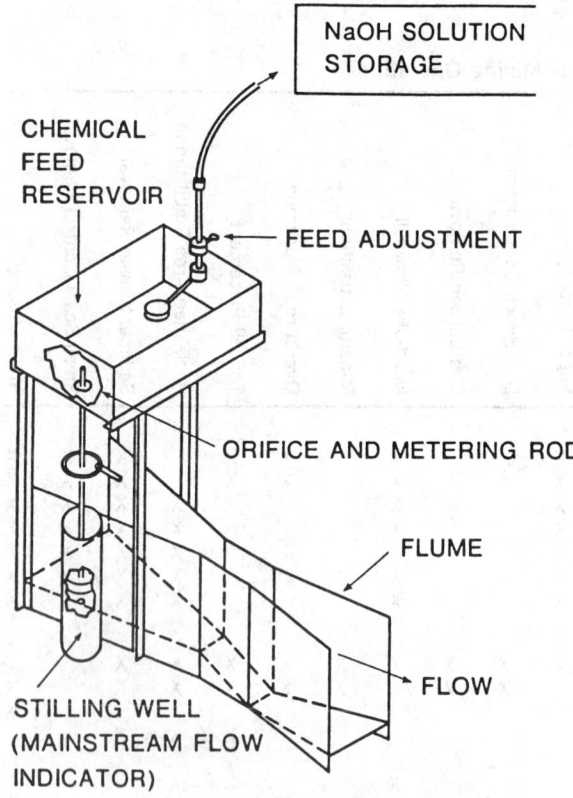

Fig. 8. Sodium hydroxide feed unit.

Operational labor and repairs   = $ 3,000
Total O and O cost   = $72,819

Calculation
Quantity to treat = 150 gpm × 525,600 min/year × 2
   year = 1.58 × 10^8 total gal treated
Reagent requirement = 1.58 × 10^8/1.0 × 10^6 × 500
   mg/L acidity × neutralizing equivalent

hydrated lime = 4.89 × 10^5 lb
caustic soda = 5.27 × 10^5 lb

Reagent cost
   hydrated lime = 4.89 × 10^5 × 1.19 = $5819.10
   caustic soda = 5.27 × 10^5 × 6.67 = $70,042.00

This example indicates that the hydrated lime treatment system will provide a slight cost advantage over the less sophisticated sodium hydroxide system. The salvage value of the system may add some additional cost saving. However, for most short duration treatment requirements, the less costly systems provide a better approach.

Quantity and Quality of Water to be Treated

In addition to the longevity of treatment requirements, the quantity and quality of the raw water to be treated must be thoroughly assessed prior to design of any treatment system. Typical acidic mine drainage has various concentrations of metals, in addition to low pH conditions. Table 5 lists various metal ion concentrations from mining operations.

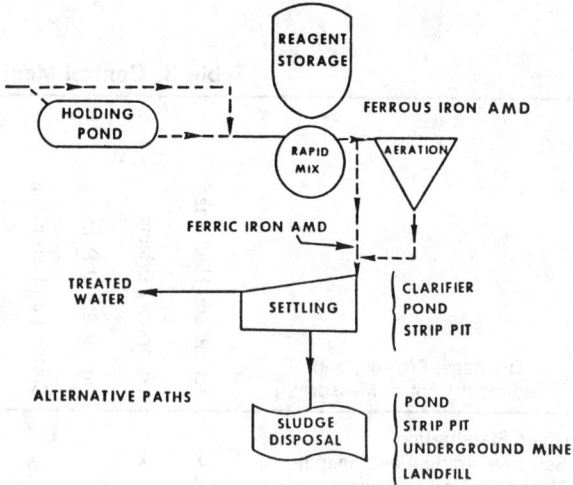

Fig. 9. Flowsheet for solid reagent AMD treatment system.

The most common method of metals removal is obtained by pH adjustment and metal sludge removal. Metal ions solubility is normally associated with the pH of the solution. Figs. 10 and 11 show the effects of neutralization on the primary and trace metal concentrations in solution. Typically, the metal ions are precipitated out as hydroxide compounds as the pH of the raw water is raised.

For example,

$$[\text{Metal Ion}^{++}] + 2\text{OH}^- = [\text{Metal}] \text{OH}_2 \downarrow.$$

As a result, the type and concentration of the metal ions in the raw water to be treated may determine the type of system and reagent to be used. For example, limestone is the most economical reagent available for neutralization purposes. However, Fig. 12 shows that the pH obtained by limestone treatment is not high enough to allow precipitation of metal ions associated with mine waters.

By far the most operationally cumbersome step in the mine water treatment process is the handling of precipitated metal sludges. Once the pH of raw water is treated to the point that metal hydroxide precipitation occurs, a separation of the precipitated material and the clarified water must be achieved. The simplest procedure is to provide settling capacity with the use of ponds or impoundment structures. When precipitation occurs, the metal hydroxide normally has several molecules of water tied up as water of hydration ($\text{Fe OH}_3 \cdot 6\text{H}_2\text{O}$). This results in a very light sludge product that will contain only 5 to 30% solids by weight. The sludges produced are often colloidal in size and may require chemical flocculation.

Prior to design of mine drainage treatment systems, it is necessary to predict or measure sludge production volumes so that settling and handling systems can be designed accordingly.

Sludge production and densities are dependent on type and concentration of metal ions present in the raw water, type of reagent and treatment system used, and the use of mechanical or chemical densifiers.

Fig. 13 shows the densification nature of an example of precipitated metal complex sludge. By running a bench scale neutralization test on the water to be treated (or water that is similar to what is expected) with the neutralization reagent to be used, the initial sludge production can be measured.

Table 5. Selected Mine Drainage Analyses, Mg/L

Pollutant Source	pH	Al	As	B	Be	Ca	Cd	Co	Cr	Cu	Fe	Hg	Mg	Mn	Na	Ni	P	Pb	Se	Sr	Te	Ti	Zn
Iron ore	5					260		2.0	0.01	1.0	180	2.0	120	18	15	0.1		0.1					8.0
Copper	3.5			2.0			1.3		0.4	90	2000	0.07		100		0.2	0.15	4.9	0.04	120	0.6		0.7
Lead/zinc	8.1						0.02			0.04	2.5			57			0.08	0.3					38
Lead/zinc	3.0						0.06			0.05								0.8					7.3
Gold	6		0.08	0.2		87	0.03				25		80	12	80	0.1		0.2	0.13	0.8	0.1		
Silver	8			0.1		45					2.0		32	6.3	12	0.09							0.8
Bauxite	2.8	88									64		45	7.7		0.3						1.1	0.03
U/Ra/V		0.5	0.03	0.01	0.01	120					15		26	0.3	140			0.01					0.03
Platinum						95					0.3							0.8					13
Coal mining-max		530									9300			92		5.6							
Coal mining-mean	2.4	43									350			7.3		0.7							1.5

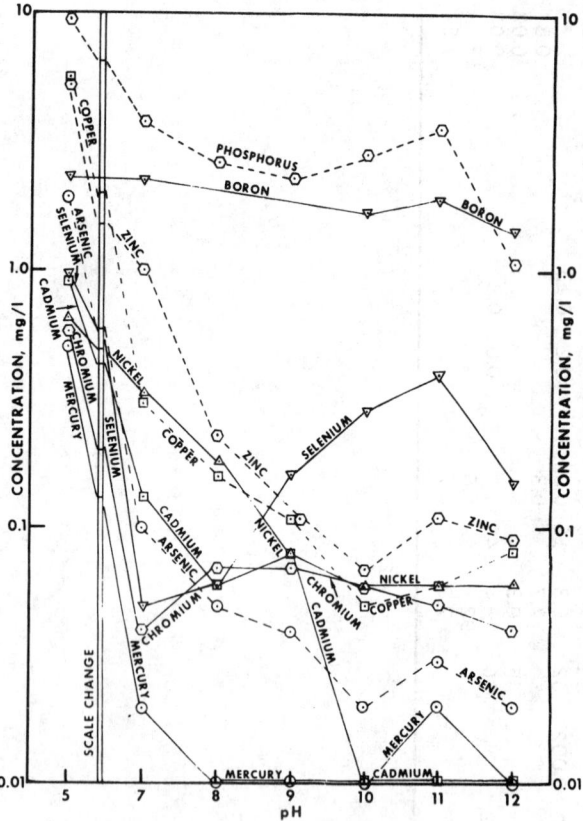

Fig. 10. Effect of pH on trace element concentration in the neutralization process effluent.

Example Calculation

Raw water quantity = 150 gpm
Sludge production = 100 mL sludge/liter
Initial sludge density = 5% solid (by weight)

Initial Sludge Production
3208.5 cu yd per month =

$$\frac{150 \text{ gpm} \times 3.785 \text{ L/gal} \times 100 \text{ mL/L} \times 43,200 \text{ min/month}}{764,418.6 \text{ mL/cu yd}}$$

Example Sludge Production Rate (for Six Month Period) with Densification Properties According to Fig. 13

		Month			
1	2	3	4	5	6
3208.5	2292	1123	943	844	844
	3208.5	2292	1123	943	844
		3208	2292	1123	943
			3208	2292	1123
				3208	2292
					3208

Total volume sludge accumulated	9254 cu yd

Metric equivalents: gal × 3.785 412 = L; cu yd × 0.764 554 9 = m^3.

This example calculation indicates that for the proposed quantity and quality of water tested, 7075 m^3 (9254 cu yd) of storage capacity is required to handle precipitated metal sludges for a six-month period of time. In addition, sizing must allow adequate sludge/clear-water freeboard and water velocities to properly separate the solids from suspension.

Handling and disposing of this precipitated sludge is generally accomplished by one of the following methods:

Cleaning Method
  1. Clam shell and trucks
  2. Endloader and trucks
  3. Pumping systems

Disposal Locations
  1. Strip pits
  2. Refuse disposal areas
  3. Abandon underground workings

Typical systems normally have two settling basins constructed for each system. While one basin is being cleaned, the other is in active use. Because of the nature of typical sludges and the problems encountered in handling this material, many installations utilize large impoundments for sludge removal. The structures are sized to handle several years of disposal with the sludge left in place and reclaimed as opposed to continual cleaning.

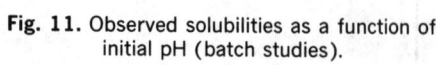

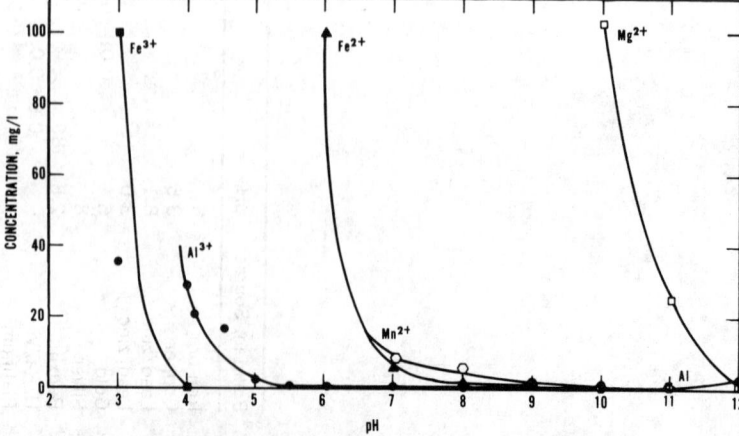

Fig. 11. Observed solubilities as a function of initial pH (batch studies).

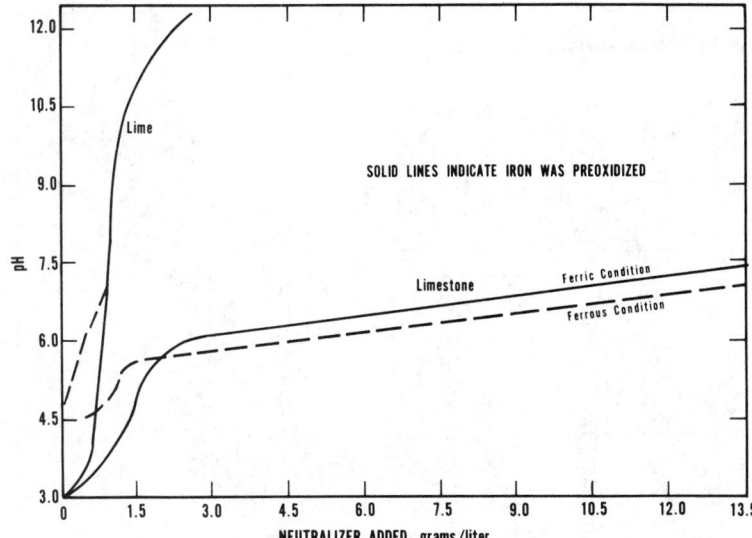

Fig. 12. Titration curve for lime and limestone.

WATER CONVEYANCE AND POND DESIGN

One of the primary functions of the surface mining engineer in the role of water management is the design and proper installation of water conveyance and sediment control structures. The proper sizing and erosion protection of ditch lines, sediment channels, and ponds is essential for water quality compliance and an efficient operation. Open channel conveyance structures and ponds that are undersized could result in overtops that can adversely affect operations (flood active pits) and transport large sediment loads off the mine area.

In the design of open channel conveyance structures, the primary areas of consideration are (1) maximum water flow that will enter the channel, (2) maximum capacity of the channel, and (3) erosion protection of the channel. The factors influencing the maximum flow that will enter an open channel are rainfall, slope of land, and runoff characteristics of the drainage area. Regulatory authorities require that open channels constructed on surface mines be designed to handle certain frequency storms. Permanent channels may be required to be designed to handle a 100-year frequency storm, while a temporary channel may only have to handle a 1-year frequency event. Fig. 14 shows the precipitation associated with a 5-year 24-hr storm event and the variation within the United States. The runoff characteristics of a drainage area are given by the relationship: (SCS method)

$$Q = \frac{(P - 0.2S)^2}{P + 0.8S}$$

where P is the accumulated precipitation, Q is the accumulated runoff volume (inches), and $S = (1000/CN) - CN$.

CN is known as the curve number and defines the absorption/infiltration/runoff characteristics of soils. The Soil Conservation Service (SCS) has classified over 4000 soils into four hydrologic soil groups. The curve numbers for the hydrologic soil groups with various land use descriptions are shown in Table 6. Some states, in cooperation with SCS, have assigned specific curve numbers for active and reclaimed surface mines. With the curve number and precipitation over time given, the runoff entering a particular open channel can be calculated.

Table 7 shows the effective runoff for a storm event using a curve number of 89. By combining the effective runoff with the area collected by an open channel and the lag time for the runoff to move to the channel, peak channel flow is given. The lag time (time of concentration) is given by the following formula:

$$tc = 0.928 \, (nL)^{.6}/ie^{0.4} \, S^{0.3}$$

where n is mannings, L is feet, ie is inches per hour (excess), S is the slope of land in foot per foot, and tc is minutes.

The result of these calculations is the development of a hydrograph (Fig. 15) that graphically shows the runoff to be handled for a precipitation event.

Because of the complexity of these calculations, most state regulatory authorities have developed technical handbooks that categorize typical drainage conditions to produce graphs that give peak discharge and runoff rates for a number of conditions. Fig. 16 gives peak discharge rates for small wa-

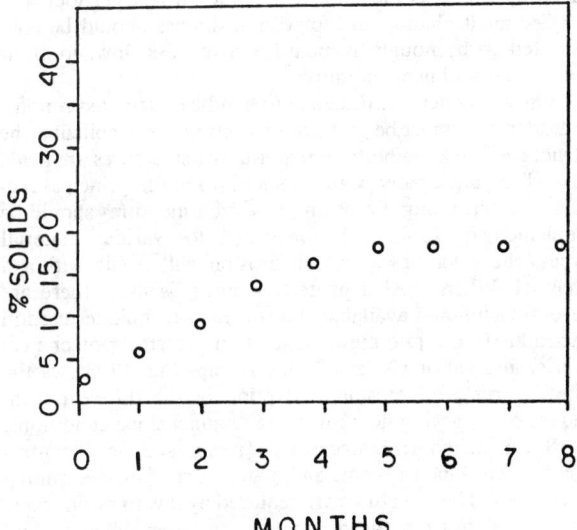

Fig. 13. Example sludge densification.

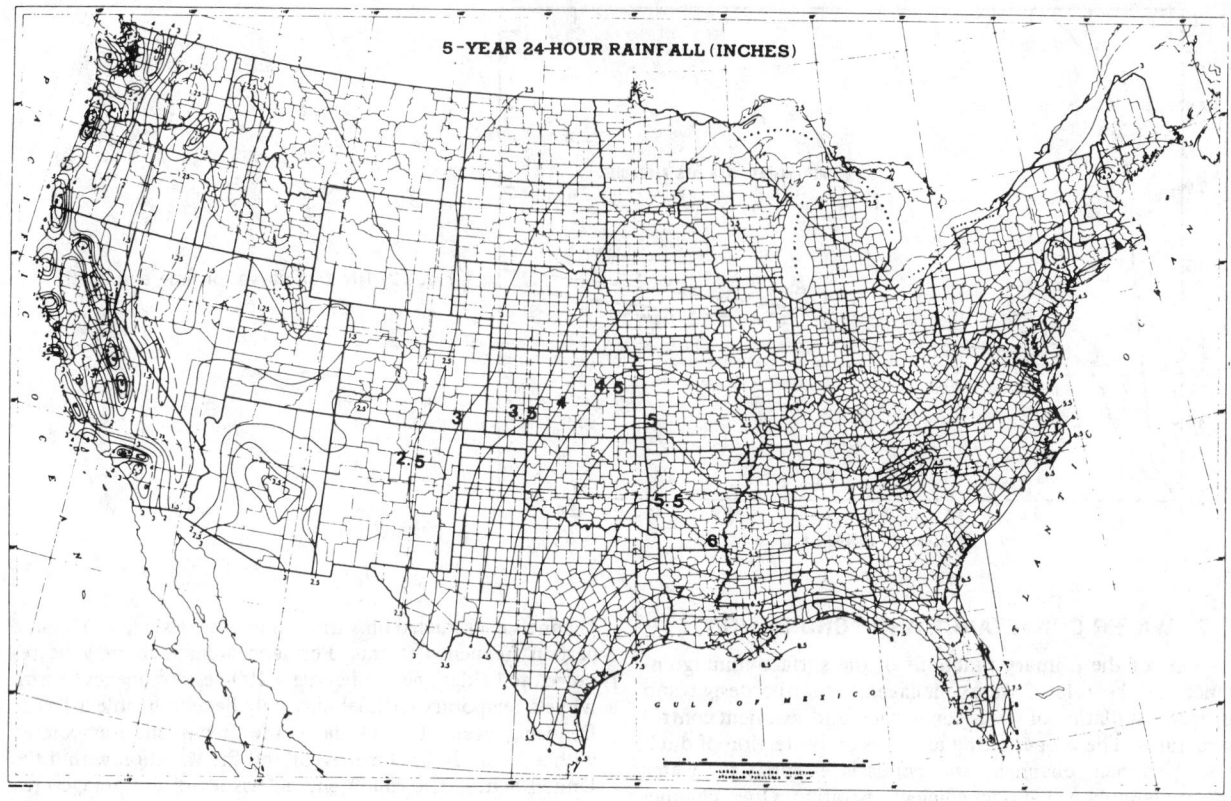

Fig. 14. Precipitation distribution of an event storm.

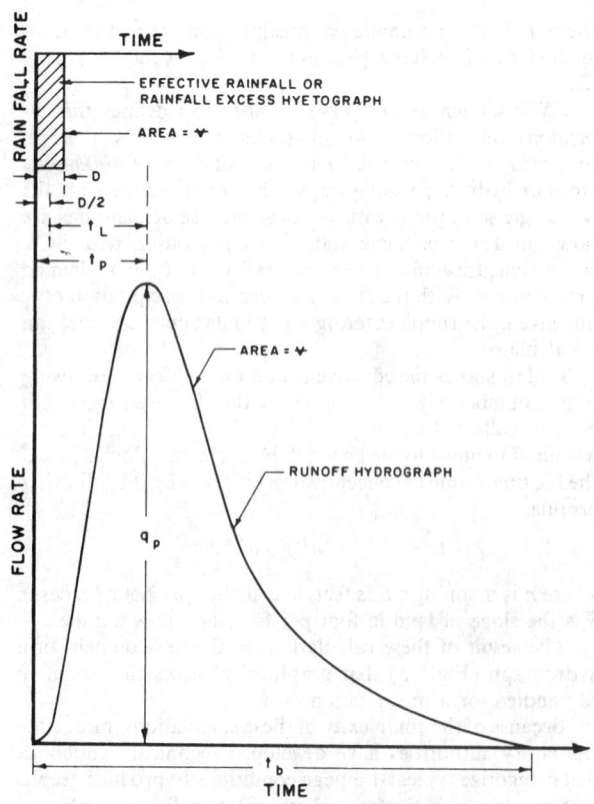

Fig. 15. Example runoff hydrograph.

tersheds with moderate slope conditions and a curve number of 85 for varying rainfall events.

The capacity of an open channel may be written as $Q = VA$, where Q is capacity in cfs, V is velocity in fps, and A is the cross-sectional area of the channel in square feet. Velocity is given by the following manning equation:

$$v = 1.49/N \, R^{2/3} \, S^{1/2}$$

where N is the manning number (see Table 8), R is feet (length), and S is the slope of channel in feet per foot.

Sediment channel and diversion ditches should be constructed with enough freeboard, above peak flow, to allow for partial sediment buildup.

Once channels are designed to handle anticipated runoffs, consideration must be given to protecting and stabilizing the structure. The erodibility of material the structures are made of and the anticipated water velocities are the principal factors in determining if erosion protection measures should be implemented. Table 8 demonstrates, for various material types, the velocities at which erosion will occur within a channel. When erosion protection must be used, there are several techniques available. The primary techniques used in the industry are jute mesh, vegetation growth, geotechnical fabric, and varying sizes of rock riprap. Fig. 17 shows the effective range of erosion protection of the different techniques over varying flow rates and channel slope conditions.

Sediment control structures (pond, sediment control channel, etc.) have become an integral part of surface mining operations. The structures are required by law to avoid transport of solids from the mining site to stream channels off the site. Proper protection against overtopping and failure of sediment structures is a necessity for safety and environ-

Table 6. Runoff Curve Numbers for Selected Agricultural, Suburban, and Urban Land Use (Antecedent Moisture Condition II)

Land Use Description		A	B	C	D
Cultivated land*: without conservation treatment		72	81	88	91
with conservation treatment		62	71	78	81
Pasture or range land: poor condition		68	79	86	89
good condition		39	61	74	80
Meadow: good condition		30	58	71	78
Wood or forest land: thin stand, poor cover, no mulch		45	66	77	83
good cover†		25	55	70	77
Open spaces, lawns, parks, golf courses, cemeteries, etc.					
good condition: grass cover on 75% or more of the area		39	61	74	80
fair condition: grass cover on 50% to 75% of the area		49	69	79	84
Commercial and business areas (85% impervious)		89	92	94	95
Industrial districts (72% impervious).		81	88	91	93
Residential:‡					
Average lot size	Average % Impervious§				
⅛ acre or less	65	77	85	90	92
¼ acre	38	61	75	83	87
⅓ acre	30	57	72	81	86
½ acre	25	54	70	80	85
1 acre	20	51	68	79	84
Paved parking lots, roofs, driveways, etc.#		98	98	98	98
Streets and roads:					
paved with curbs and storm sewers#		98	98	98	98
gravel		76	85	89	91
dirt		72	82	87	89

* For a more detailed description of agricultural land use curve numbers refer to *National Engineering Handbook*, Section 4, "Hydrology," Chap. 9, Aug. 1972.
† Good cover is protected from grazing and litter and brush cover soil.
‡ Curve numbers are computed assuming the runoff from the house and driveway is directed toward the street with a minimum of roof water directed to lawns where additional infiltration could occur.
§ The remaining pervious areas (lawn) are considered to be in good pasture condition for these curve numbers.
# In some warmer climates of the country a curve number of 95 may be used.
Metric equivalent: acre × 4046.856 = m², acre × 0.404 687 3 = ha.

Table 7. Effective Runoff for a Storm Event

Time, min	Incr. Depth, in.	Accum. Depth,* in.	Accum. Effective Rain,† in.	Incr. Effective Rain,‡ in.
0				
15	0.05	0.05	0.00	0.00
30	0.08	0.13	0.00	0.00
45	0.10	0.23	0.00	0.00
60	0.25	0.48	0.04	0.04
75	0.60	1.08	0.34	0.30
90	1.35	2.43	1.38	1.04
105	0.30	2.73	1.66	0.28
120	0.12	2.85	1.76	0.10
135	0.08	2.93	1.84	0.08
150	0.05	2.98	1.88	0.04
165	0.04	3.02	1.92	0.04
180	0.04	3.06	1.95	0.03
Total	3.06	3.06	1.95	1.95

* Accumulation of column 2.
† CN = 89.
‡ Incremental values from column 4.
Metric equivalent: in. × 25.4 = mm.

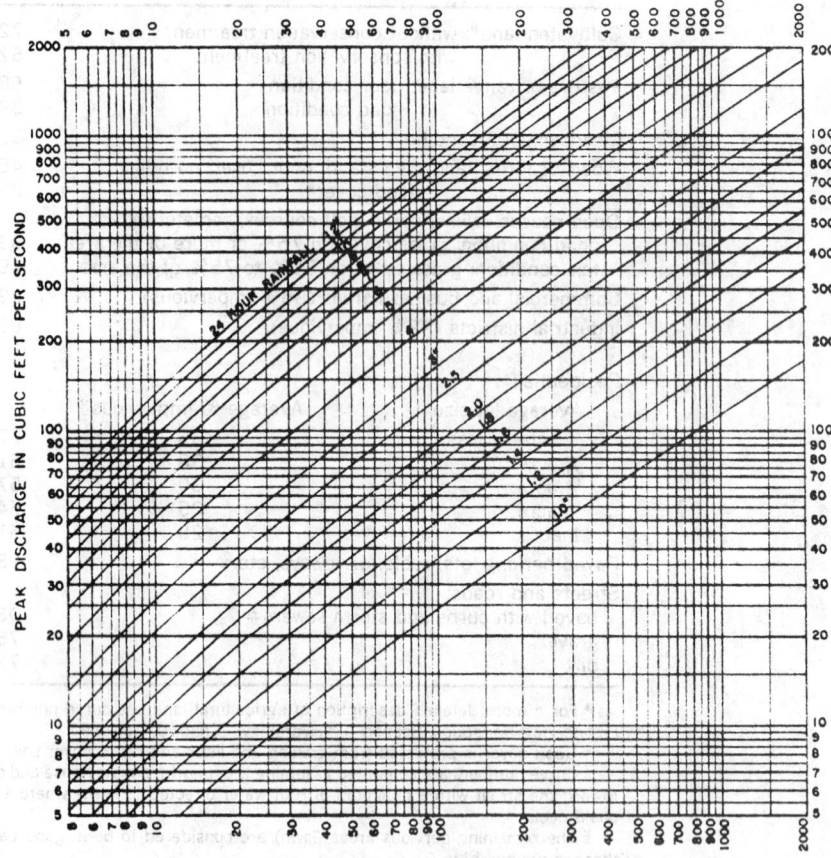

Fig. 16. Estimated peak flow quantities, discharge for small watersheds, type II storm distribution.

SLOPES— MODERATE
CURVE NUMBER— 90

Table 8. Limiting Velocities and Tractive Forces for Open Channels (Straight after Aging)

Material	n	For Clear Water		Water Transporting Colloidal Silts	
		Velocity, fps	Tractive Force, psf	Velocity, fps	Tractive Force, psf
Fine sand colloidal	0.020	1.50	0.027	2.50	0.075
Sandy loam noncolloidal	0.020	1.75	0.037	2.50	0.075
Silt loam noncolloidal	0.020	2.00	0.048	3.00	0.110
Alluvial silts noncolloidal	0.020	2.00	0.048	3.50	0.150
Ordinary firm loam	0.020	2.50	0.075	3.50	0.150
Volcanic ash	0.020	2.50	0.075	3.50	0.150
Stiff clay very colloidal	0.025	3.75	0.260	5.00	0.460
Alluvial silts colloidal	0.025	3.75	0.260	5.00	0.460
Shales and hardpans	0.025	6.00	0.670	6.00	0.670
Fine gravel	0.020	2.50	0.075	5.00	0.320
Graded loam to cobbles when noncolloidal	0.030	3.75	0.380	5.00	0.660
Graded silts to cobbles when colloidal	0.030	4.00	0.430	5.50	0.800
Coarse gravel noncolloidal	0.025	4.00	0.300	6.00	0.670
Cobbles and shingles	0.035	5.00	0.910	5.50	1.100

From Lane, 1955.
Metric equivalents: fps × 0.304 800 = m/s.

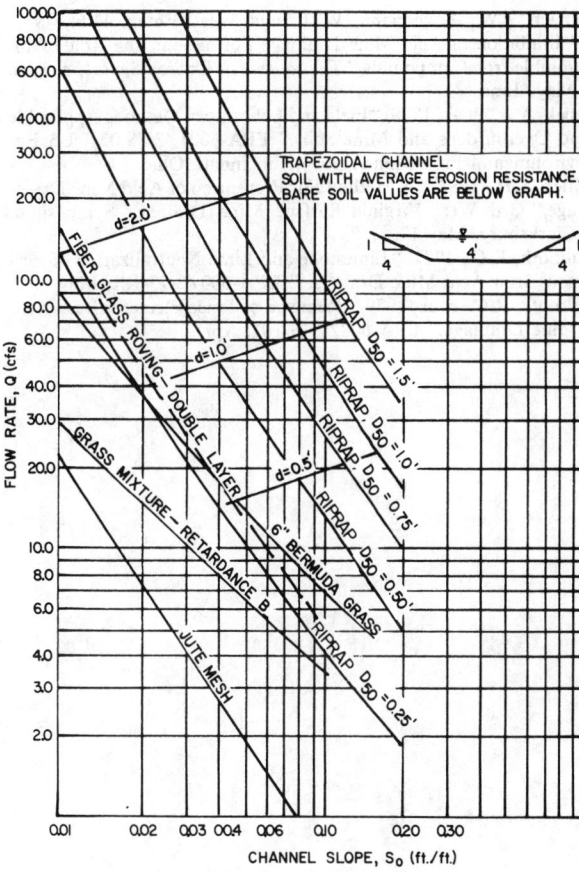

Fig. 17. Erosion protection provided if below line.

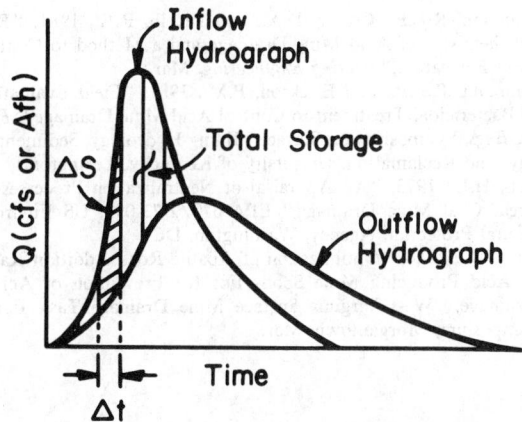

Fig. 19. Inflow vs. outflow hydrograph.

$$Q = \frac{a\,(2gH)^{1/2}}{1 + Ke + Kb + KcL^{1/2}}$$

where Q is cfs; a is the area of pipe in square feet; g is gravity, 32.2 ft per sec^2; H is the head in feet above the horizontal section; Ke is $\approx$ 1.0; Kb is $\approx$ 0.5; Kc is the friction loss of pipe; and L is the length of pipe in feet

A typical configuration is shown in Fig. 18.

A resulting hydrograph for a typical structure is shown in Fig. 19. Sediment structures should be built in accordance with a preplan that considers: (1) required capacities, (2) stability and erosion protection for the embankment, (3) permeabilities and required compaction for embankment materials, and (4) conservative engineering design.

mental protection. Flood routing and protective designs are required by several state and federal agencies. Design requirements are normally based on a certain frequency storm event, with storm frequency dependent on the size, how long the structure will be in existence, and the potential hazard if failure occurs.

Analysis of flood routing through a structure is normally accomplished by creating a hydrograph of the design storm to the inlet of the structure and a hydrograph of flow through the structure. The information required to develop the hydrograph of flow through the structure is stage storage information for the structure and discharge capacities of the outlet structures. Discharge capacities for structure with drop inlet and piped discharge are expressed by:

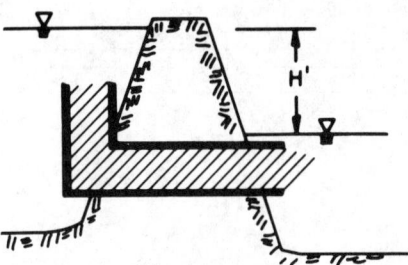

Fig. 18. Illustration of pipe flow control on a drop inlet spillway, outflow controlled.

SUMMARY

Water management and pollution control are, and will continue to be, an increasingly demanding discipline in the surface mining industry. Innovation in the application of existing technology and the development of new technology will be a necessity for effective utilization of this country's surface minable resources. The broad technical background needed in the field makes it both challenging and rewarding.

REFERENCES

Anon., 1977, "Surface Mining Reclamation and Enforcement Provisions," *Federal Register*, Part II, Dec. 13, Office of Surface Mining Reclamation and Enforcement, Dept. of the Interior.

Anon., 1979, "Suggested Guidelines for Method of Operation in Surface Mining of Areas with Potentially Acid-Producing Materials," West Virginia Surface Mine Drainage Task Force.

Anon., 1984, *Technical Handbook of Standards and Specifications for Erosion and Sediment Control, Excess Spoil Disposal, Haulage Ways for Mining Operations in West Virginia,* West Virginia Division of Natural Resources.

Barfield, B.J., Warner, R.C., and Haan, C.T., 1981, *Applied Hydrology and Sedimentology For Disturbed Areas,* Oklahoma Technical Press.

Flynn, J.P., 1969, "Treatment of Earth Surface and Subsurface for Prevention of Acidic Drainage from the Soil," US Patent 3,443,882, May 13.

Fung, R., 1981, "Surface Coal Mining Technology Engineering and Environmental Aspects," Penn. State, US Bureau of Mines, Noyes Data Corp., Park Ridge, NJ.

Grim, C., and Hill, R., 1974, "Environmental Protection in Surface Mining of Coal," EPA-6701/2-74-093, US Environemntal Protection Agency, Washington, DC.

Kleinmann, R.L.P., Crerar, D.A., and Pacelli, R.R., 1981, "Biogeochemistry of Acid Mine Drainage and a Method to Control Acid Formation," *Mining Engineering,* Mar.

Kleinmann, R.L.P., and Erickson, P.M., 1981, "Field Evaluation of Bactericidal Treatment to Control Acid Mine Drainage," *Proceedings,* Symposium on Surface Mining Hydrology, Sedimentology, and Reclamation, University of Kentucky, Lexington.

Lovell, H.L., 1973, "An Appraisal of Neutralization Processes to Treat Coal Mine Drainage," EPA-670/2-73-093, US Environmental Protection Agency, Washington, DC.

Meek, F.A., 1984, "Optimization of Apatite Rock Addition Rates to Acid Producing Mine Spoils For the Prevention of Acidic Drainage," West Virginia Surface Mine Drainage Task Force Symposium, Morgantown, Mar.

Nebgen, J.W., Engelmann, W.H., and Weatherman, D.F., 1981, "Inhibition of Acid Mine Drainage Formation: The Role of Insoluble Iron Compounds," *The Journal of Environmental Sciences,* May/June.

Sobek, A.A., et al., 1978, "Field and Laboratory Methods Applicable to Overburdens and Mine Soils," EPA-6001/2-78-054, US Environmental Protection Agency, Cincinnati, Ohio.

Stiller, A.H., 1982, "A Method For Prevention of Acid Mine Drainage," 3rd West Virginia Surface Mine Drainage Symposium, Clarksburg, May 17.

Wilmoth, R.C., 1977, "Limestone and Lime Neutralization of Ferrous Iron Acid Mine Drainage," EPA-600/2-77-101, May.

Wilmoth, R.C., et al., 1979, "Removal of Trace Elements from Acid Mine Drainage," EPA-600/7-79-101, Apr.

6.7.2 Air Quality Management

CLIFFORD F. COLE
RICHARD L. KERCH

INTRODUCTION AND OVERVIEW

Owners and operators of surface mines are concerned about the air quality impact of their mines for two reasons. First, they want to be good neighbors and provide a healthy environment for both the mine employees and for the residents (often the employees and their families) who live near the mine. Second, state and federal regulations require that surface mines comply with stringent rules designed to limit pollutant emissions and to maintain ambient air quality at a level that will protect the public health and welfare. To achieve these goals, permit systems have been established by federal, state, and local regulatory agencies.

Permit systems require that the owners and operators of facilities that are significant sources of air pollutants, including surface mines, obtain permits before construction and operation of the facility can begin. The permit requirements differ from state to state, and federal permit rules frequently differ from state rules, but generally applicants are required to submit detailed design and engineering plans of their proposed facility, including the location of the proposed air pollution sources, the control systems to be installed, estimates of the air pollutant emission rates, and a demonstration that all applicable air quality standards and regulations will be met.

The regulatory agency then compares the predicted impacts against standards or guidelines, and if no harmful impact is anticipated, the agency will issue a permit. Air quality permits granted at surface mines frequently require that the owners or operators install pollution monitoring instruments in the vicinity of the mine to ensure that once operation begins pollutant concentration standards will not be exceeded.

A large surface mine may have to secure more than one air quality permit before mining can commence, and a lead time of as much as three years may be needed to gather ambient air quality data, prepare the permit application, and go through the review process before obtaining these permits. Clearly the acquisition of the necessary air quality permits is a major environmental priority in opening and operating a surface mine.

There are presently six major air pollutants whose emission rates and ambient concentrations are regulated by the federal government through the United States Environmental Protection Agency (USEPA). These so-called criteria pollutants differ in their characteristics, sources, and health effects:

1) Particulate matter is used to mean any dispersed matter, solid or liquid, in which the individual particles are larger than single molecules but smaller than about 500 μm (microns). Particles in this size range can remain suspended in the atmosphere from a few seconds to several months. Particulate matter is undifferentiated with respect to size, chemical or biological composition, optical or surface properties, motion, or source. Particulate matter is defined according to its size. Total suspended particulate matter (TSP) is the term given to particulate matter collected by a high volume sampler, which captures particles in the size range from 0.1 to about 100 μm in diameter. TSP has long been used as a measure of airborne particulate matter. Particulate matter smaller than 10 μm in diameter is termed PM-10. EPA has adopted PM-10 as its measure of particulate matter, in light of evidence that only particles smaller than 10 μm are harmful to health and welfare. The use of TSP at the federal level will be phased out. However, many states and counties still use TSP as the sole measure of particulate matter.

At surface mines the sources of TSP are those areas or activities that cause dust to be entrained into the air—haul road traffic, ore or coal handling, wind erosion on exposed areas, overburden removal, etc.

2) Sulfur dioxide (SO_2) is one of the gaseous products of combustion emitted in the burning of sulfur-bearing fuels. It is also liberated during certain smelting operations and is produced in many industrial processes. Surface mines are a relatively minor source of SO_2. The major source of SO_2 at surface mines is the combustion of diesel fuel in mining vehicles and other diesel-powered equipment. Ancillary facilities such as coal-cleaning plants or smelters can emit hundreds of times more SO_2 than the mines themselves.

3) Carbon monoxide (CO) is a colorless, odorless, poisonous gas caused by incomplete combustion of fuels. At surface mines, vehicle exhaust is the main source of CO, although the amount of CO emitted by mining is insignificant compared to the huge quantities generated by urban traffic.

4) The term nitrogen oxides (NO_x) is used to represent the composite atmospheric concentrations of nitric oxide (NO) and nitrogen dioxide (NO_2). The standard is expressed in terms of nitrogen dioxide. Both of these compounds are formed during the combustion of fossil fuels. NO_x emissions are often further classified as thermal NO_x (formed from nitrogen in the air reacting with oxygen at high temperatures) and fuel NO_x (resulting from the oxidation of nitrogen contained in the fuel). The primary sources of NO_x at a surface mine are internal combustion engines used in diesel and gasoline powered vehicles and mining equipment.

5) Lead (Pb) particles come primarily from the combustion of leaded gasoline and from lead smelters. Most surface mines are insignificant sources of lead.

6) Ozone (O_3) is a secondary pollutant that is caused by complicated reactions of hydrocarbons and nitrogen oxides in the presence of sunlight. Surface mines are an insignificant source of ozone.

In addition to the criteria pollutants listed previously, EPA also regulates other air pollutants, including:

Asbestos	Vinyl Choloride	Radon 222
Beryllium	Fluorides	Benzene
Mercury	Sulfuric Acid Mist	Radionuclides
Arsenic	Hydrogen Sulfide	Total Reduced Sulfur

Most of the permitting effort, the pollution control effort, and the air monitoring work at surface mines is devoted to particulate matter. Environmental protection personnel at surface mines will generally find that this is their greatest air quality concern.

On a nationwide basis, surface mines account for only a small fraction of the particulate matter emitted in the United States. Combining estimates of TSP emission rates in the US

Table 1. Annual TSP Emissions by Source Category*

Source	Emissions, thousand st	Emissions, % of total
Roads†	280,813	47.2
Agriculture	268,667	45.1
Construction	21,547	3.6
Mining	9,589	1.6
Industrial processes	5,952	1.0
Combustion‡	5,291	0.9
Fires	1,939	0.3
Auto-truck exhaust	1,213	0.2
Incinerators	441	0.1

* From 1976 and 1977 data.
† Paved and unpaved roads combined.
‡ Power generation and space heating.
Metric equivalent: st $\times$ 0.907 184 7 = t.

grouped by source categories (Evans and Cooper, 1980) one can see that mining emits 1.6% of all TSP, as shown in Table 1. The major sources of TSP are unpaved roads and agriculture, which together contribute over 90% of all particulate in the US. However, on a local basis in a remote area, a surface mine can be the dominant source of TSP.

The remainder of this section explains the major air pollution regulations with which a surface mine must comply and outlines the steps required to obtain air quality permits. Additionally, the air pollution monitoring phase is discussed, as are the sources and controls of air pollution at surface mines. Finally, methods of estimating the pollutant impact at a surface mine are presented, and the section concludes with a brief illustrative example.

AIR POLLUTION REGULATIONS

The authority to regulate sources of air pollutants, such as surface mines, is granted to many different governmental agencies through various forms of enabling legislation. At the federal level, the Clean Air Act (CAA) empowers the federal Environmental Protection Agency (EPA) to set national ambient air quality standards for all designated pollutants and to require that individual states submit a state implementation plan (SIP) outlining how the state will achieve and maintain these standards. The CAA was amended in 1977 to include the Prevention of Significant Deterioration (PSD) program, whose purpose is to prevent areas where air is already cleaner than the national air quality standards from becoming significantly more polluted, and the visibility protection program. Visibility protection and maintenance is a special permit consideration that must be addressed in situations where the proposed facilities' emissions might impair the visibility in or around any special federal land use areas known as Class I areas (e.g., national parks and wilderness areas). The techniques for assessing visibility impacts of surface mines on Class I areas have not yet been adopted by EPA. Although visibility impacts of surface mines on Class I areas have not routinely been considered in the permit review process, this analysis is expected to become more important in the future. It is also worth noting that the CAA provides additional protection to Class I areas for other air quality related values such as effects on flora and fauna. All sources of air pollution are subject to these federal regulations unless specifically exempted. While other federal agencies may also include restrictions or provisions that control air pollutants in their regulations, none are as broad or as all-inclusive as the EPA regulations.

At the state level, all 50 states have their own air quality control agencies with authority generally granted by state legislation. For the most part, the state rules parallel EPA regulations. The use of different measures of particulate matter (PM-10 and TSP) mentioned previously is a notable exception. Mines located in states that use TSP as an indicator of particulate matter may have to comply with both state TSP regulations as well as federal EPA PM-10 regulations. The authority for some federal programs, such as PSD, has been delegated to those state agencies that have desired to have their own programs.

Finally, local government may in some cases administer air quality regulations at the county, township, or city level. It is the responsibility of mine managers to be familiar with all applicable regulations and to ensure that the mine complies with them. Specific questions or information requests can be directed to EPA regional offices or to state air pollution control agencies. A summary of EPA regional offices appears in Table 2.

Ambient Standards

EPA has established two kinds of National Ambient Air Quality Standards (NAAQS): primary standards to protect the public health and secondary standards to protect the general welfare, including such things as injury to plants, animals, and materials. These national standards (Anon., 1979), shown in Table 3, are measures of pollutant concentration which must not be exceeded in the ambient air. Various averaging time intervals are specified for all of the pollutants to account for effects associated with exposure to both acute and long term concentrations. *Ambient air* has not been well defined. Practically, national ambient air quality standards (NAAQS) are most often enforced for areas outside of the company property if public access to the property is prevented (Hawkins, 1978). Mine Safety and Health Administration (MSHA) standards, where they exist, apply to areas inside mine property.

Whether or not a region is in compliance with the NAAQS is usually determined by measuring pollutant concentrations within that region. Monitoring networks measure concentrations over various time periods to determine whether an area is currently attaining ambient air standards (*attainment areas*), or whether a region's pollutant concentrations exceed the standards (*nonattainment areas*). A region can be an attainment area for one pollutant but a nonattainment area for another.

Because the EPA's PM-10 measure of particulate matter

Table 2. Addresses of Federal EPA Regional Offices*

Region I—Connecticut, Maine, Massachusetts, New Hampshire, Rhode Island, Vermont
    John F. Kennedy Federal Building
    Boston, MA 02203

Region II—New Jersey, New York, Puerto Rico, Virgin Islands
    Jacob K. Javits Federal Building
    26 Federal Plaza
    New York, NY 10278

Region III—Delaware, District of Columbia, Maryland, Pennsylvania, Virginia, West Virginia
    841 Chestnut Building
    Philadelphia, PA 19106

Region IV—Alabama, Florida, Georgia, Kentucky, Mississippi, North Carolina, South Carolina, Tennessee
    345 Courtland, NE
    Atlanta, GA 30365

Region V—Illinois, Indiana, Michigan, Minnesota, Ohio, Wisconsin
    230 South Dearborn
    Chicago, IL 60604

Region VI—Arkansas, Louisiana, New Mexico, Oklahoma, Texas
    1201 Elm Street
    Dallas, TX 75270

Region VII—Iowa, Kansas, Missouri, Nebraska
    726 Minnesota Ave.
    Kansas City, MO 66101

Region VIII—Colorado, Montana, North Dakota, South Dakota, Utah, Wyoming
    999 18th Street, Denver Place
    Suite 500
    Denver, CO 80202

Region IX—Arizona, California, Hawaii, Nevada, Guam, American Samoa
    215 Fremont Street
    San Francisco, CA 94105

Region X—Washington, Oregon, Idaho, Alaska
    1200 Sixth Avenue
    Seattle, WA 98101

*(Anon., 1987)

Table 3. National Ambient Air Quality Standards

Pollutant	Time Interval*	Primary Standard, $\mu g/m^3$	Secondary Standard, $\mu g/m^3$
Particulate Matter			
PM-10	Annual average	50	50
	24-hr average	150	150
TSP‡	Annual average†	75	60
	24-hr average	260	160
Sulfur dioxide	Annual average	80	—
	24-hr average	365	—
	3-hr average	—	1300
Nitrogen dioxide	Annual average	100	100
Carbon monoxide	8-hr average	10,000	10,000
	1-hr average	40,000	40,000
Ozone	1-hr average	240	240
Lead	3-month	1.5	—

* All short-term standards may not be exceeded more than one time per year, except for ozone and particulate matter which are treated in a different manner.
    † Geometric mean for TSP; all other pollutants and averaging times are arithmetic means.
    ‡ Former NAAQS which may be retained by some states.

has only recently been adopted, there has not been sufficient time to measure PM-10 concentrations in order to determine which areas of the United States exceed particulate matter NAAQS. However, based on statistical analyses of existing TSP concentrations, and on limited PM-10 measurements, the EPA has identified 68 areas with at least a 95% probability of exceeding PM-10 NAAQS, and an additional 112 areas with greater than 20% probability of exceeding particulate matter NAAQS (Anon., 1987a). Many of these expected areas of exceedance are in the West and Southwest.

Individual states may have their own ambient air quality standards that must be met in addition to the national standards. These state standards are often equal to in numerical value, but occasionally more stringent than, the national ambient air quality standards. Some states may also have standards for additional pollutants not federally regulated.

State regulatory agencies ensure that new pollution sources, such as surface mines, will meet ambient air quality standards through a construction permit system. The permit process allows the agency to prohibit the construction and operation of new sources that might cause a violation of ambient air quality standards. When a proposed new source applies for a permit to construct, the agency requires as part of the permit review and approval process that the applicant demonstrate that the pollutant contribution from the proposed source, when numerically combined with preexisting pollutant concentrations in the air, will not exceed ambient air quality standards. This demonstration is most often made by computing the quantity of pollutants that will be emitted from a source, and then simulating the dispersion of these pollutants in the atmosphere by using an air quality model to calculate the total ambient air concentrations to which the public will be exposed in the vicinity of the source. The means of estimating air pollution impact from surface mines is discussed later in more detail.

Prevention of Significant Deterioration

The Prevention of Significant Deterioration (PSD) program, originally adopted by EPA in 1974, is designed to prevent attainment areas of the country from becoming significantly more polluted. The Clean Air Act was amended by Congress in 1977 to include the statutory authority and requirement for a PSD program. The PSD program applies only to attainment areas. Regulations implementing the PSD program were first adopted at the federal level and enforced by EPA. States have the option of promulgating their own PSD regulations, and if an individual state's regulations are approved by EPA, then the authority for the PSD program is delegated to that state. EPA retains the authority in states that choose not to have their own PSD program. Since most states' PSD rules mimic the federal regulations and since some states adopt the federal regulations by reference, the discussion in this section will focus on the federal PSD program.

The PSD program is made up of three major elements: an increment system for rationing the remaining clean air resources, a permitting system for implementing the program, and a control technology program requiring that the best available control technology (BACT) be installed and operated on all new sources subject to the regulations.

Under the PSD program (Anon., 1980a), the attainment areas of the US are divided into three categories or classes. Class I areas are those where the least amount of future air quality deterioration is to be allowed (up to 7% of the standard). These areas include many national parks, international parks, Wilderness Areas and national memorial parks. Class II areas, which at present include all of the remainder of the US, are allowed additional air quality deterioration (up to approximately 25% of the standard). Class III areas would be allowed still greater levels of pollution. There are no Class III areas, and none are anticipated since a rather complicated redesignation process must be followed to reclassify a Class II area to Class III status. The amount of increased pollution in Class I, II, or III areas is limited by the PSD program to maximum allowable increases, known as increments. In no case can the air quality be allowed to deteriorate to the point that the NAAQS are violated. The numerical values of these increments are shown in Table 4 for SO_2, TSP, and NO_2. These three pollutants are the only ones presently regulated under the federal PSD program, although there are statutory provisions to regulate the other criteria pollutants. Notice that the particulate matter PSD increment in Table 4 is expressed in terms of TSP concentration, whereas the National Ambient Air Quality Standards (Table 3) use PM-10 as a measure of particulate matter. This disparity results from the recent adoption of PM-10 as a measure of particulate matter. EPA intends to establish PM-10 increments in the future. Until that time, however, two different measures of particulate matter concentration persist in air quality regulations.

The allowable increments define the maximum allowable air quality deterioration over and above a baseline level. The baseline level is generally the prevailing ambient concentration on the day that the first pollutant source with significant pollutant emissions applies for a PSD permit. This day is called the baseline date. Some states, however, have adopted fixed baseline dates. The concepts of baselines date, baseline concentration, and baseline area are very poorly defined and often confusing. States have been given the authority to define these important terms; as such, no universal definition exists. All major sources of SO_2, TSP, and NO_2 which apply for PSD permits after the baseline date are considered consumers of PSD increment. Once a source consumes part of the increment, that portion is unavailable to other sources. No new source will be granted a PSD permit if its ambient concentration contributions exceed the remaining increment. There are some portions of the nation where the available PSD increment has already been totally consumed, and no new

Table 4. PSD Increments ($\mu g/m^3$)

Pollutant	Time Period	Class I	Class II	Class III
TSP	Annual average	5	19	37
	24-hr	10	37	75
SO_2	Annual average	2	20	40
	24-hr	5	91	182
	3-hr	25	512	700
NO_2	Annual average	2.5	25	50

Table 5. List of 27 Specific Source Types

1. Coal cleaning plants (with thermal dryers)
2. Kraft pulp mills
3. Portland cement plants
4. Primary zinc smelters
5. Iron and steel mills
6. Primary aluminum ore reduction plants
7. Primary copper smelters
8. Municipal incinerators capable of charging more than 227 t/d (250 stpd) of refuse
9. Hydrofluoric, sulfuric, or nitric acid plants
10. Petroleum refineries
11. Lime plants
12. Phosphate rock processing plants
13. Coke oven batteries
14. Sulfur recovery plants
15. Carbon black plants (furnace process)
16. Primary lead smelters
17. Fuel conversion plants
18. Sintering plants
19. Secondary metal production plants
20. Chemical process plants
21. Fossil-fuel boilers (or combination thereof) totalling more than 73 MW (250 million Btu/hr) heat input
22. Petroleum storage and transfer units with a total storage capacity exceeding 300,000 bbl
23. Taconite ore processing plants
24. Glass fiber processing plants
25. Charcoal production plants
26. Fossil fuel-fired steam electric plants of more than 73 MW (250 million Btu/hr) heat input
27. Any other stationary source category which, as of Aug. 7, 1980, is being regulated under Section 111 or 112 of the Act.

growth will be allowed without a suitable offset and perhaps not even then.

The PSD program is administered through a permitting process at the federal level, or if the state has been granted authority, at the state level. Not all air pollution sources are subject to PSD. Only sources above a certain size threshold (*major sources*) or source modifications above a certain size threshold (*major modifications*) are required to obtain PSD permits. For PSD purposes, a *major source* is one that has the potential to emit over 226 t (250 st) per year of any criteria pollutant or a source belonging to a list of 27 specific source types (shown in Table 5) that has the potential to emit over 90 t (100 st) per year of a criteria pollutant. Potential to emit means the maximum capacity of a stationary source to emit a pollutant under its physical and operational design. Any limitation on the capacity of the source to emit, including air pollution control equipment or restrictions of hours of operation or on the type or amount of material combusted, stored, or processed can be treated as part of its design if the limitation or effect it would have on emissions is generally enforceable, i.e., through the operating permit.

Special provisions apply to fugitive dust (the windblown particulate matter that could not pass through a stack or a vent) under PSD. These provisions are extremely important to surface mine operators since fugitive dust accounts for the vast majority of particulate matter emissions at surface mines. EPA provisions allow a fugitive dust source to escape PSD review if two conditions are met: (1) the source is major only because of the magnitude of fugitive emissions and (2) the source is not included in the list of sources in Table 5. The reason for this so-called fugitive dust exemption is that

there is ample evidence that natural dust with particles in the size range emitted at surface mines is not detrimental to health.

However, once the source triggers PSD review, e.g., the source is major for another pollutant, then fugitive dust is always included in calculating PSD increment consumption for particulate matter. It is also possible that a nonmajor stationary source with fugitive dust emissions in excess of the applicable PSD threshold may get entangled in increment consumption indirectly through EPA requirement on the states to *track* increment consumption for both major and nonmajor sources.

Legal challenges have been made to the present fugitive dust exemption by environmental groups. In particular, EPA has proposed to add surface coal mines to the list of 27 sources which would result in the loss of the fugitive dust exemption for that source category. There are many indications that if EPA regulations were changed to force surface mines to undergo full PSD review, many mines could not be permitted at an economic designed capacity because the magnitude of their fugitive dust emissions would exceed the available TSP increment near the mine (Anon., 1979b; Cabe, Hooper, and Schmidt, 1979; and Anon., 1982). The final resolution of this fugitive dust controversy may have a great impact on the surface mining industry.

There is one other important requirement under PSD which all new sources must meet—each source must apply Best Available Control Technology (BACT). Whether a piece of control equipment or a control method qualifies as BACT is to be judged on a case-by-case basis, taking into account energy, environmental, and economic impacts and other costs (Anon., 1978b). While the use of water spray to control dust on a haul road, for example, may qualify as BACT in one instance, elsewhere chemical dust suppressants may be required. Control technology used in similar installations frequently influences BACT determinations and these previous applications of BACT have been compiled in a reference book (Anon., 1985a).

If a source, such as a surface mine, does have to apply for a PSD permit, EPA requires that four major analyses or demonstrations be made before a permit can be granted:

1) The applicant must demonstrate that the source will not cause exceedances of the allowable PSD increments and will not exceed ambient air quality standards. These demonstrations must be made by predicting the air quality impact of new sources with EPA-approved dispersion models.

2) The applicant must demonstrate that the new source will apply BACT.

3) The applicant must analyze any air quality impacts projected for the area as a result of growth associated with the proposed source.

4) Using air pollution sampling instruments, the applicant may have to monitor the air quality concentrations in the vicinity of the source both before and after construction.

Nonattainment Regulations

It is not often that a proposed surface mine would be located in or near to a nonattainment area but the situation has occurred. In nonattainment areas, where pollutant standards are not being met, EPA is faced with addressing conflicting goals—allowing new industrial growth to add to the pollution levels already in violation of the law, while at the same time improving the quality of the air to meet ambient standards (Raffle, 1979). EPA has seemingly resolved the conflict by promulgating its *Offset Policy,* wherein new pollutant sources are allowed to locate in a nonattainment area only if they can reduce pollutant emissions elsewhere in the area so as to more than offset their emission rate increases. The net effect should be an overall improvement in air quality.

The nonattainment regulations encompass four separate provisions, each of which must be met before a source in a nonattainment area will be allowed to construct (Anon., 1980). First, the new source must agree to limit its pollutant emissions to the Lowest Achievable Emission Rate (LAER), meaning that pollution control equipment and operating practices will be such that pollutant emissions will be as low as those from any similar existing source. In practice, LAER determinations are often set by precedent established in pre-

vious permitting efforts. The second provision is that the applicant must certify that other pollutant sources in the state owned or operated by the applicant are complying with all applicable regulations. Third, an emissions offset must be provided for the same pollutant that the new source will emit in an amount at least as great as the proposed increase. Fourth, the effect of the offset in combination with the proposed increase must result in a net benefit in air quality in the affected area. This last provision essentially means that the offset must occur reasonably near the intended new pollutant source location—a decrease in TSP emissions 64 km (40 miles) from a proposed new mining operation, for example, would not achieve the desired net benefit in air quality and probably would not be approvable as an offset.

EPA did not intend that every pollutant source, regardless of size, abide by its nonattainment regulations. Clearly, to regulate sources which emit only insignificant quantities of pollutants would do little to improve air quality and would impose a needless burden on the applicant and EPA alike. These regulations are therefore limited to *major* new sources or *major modifications* to existing major sources. A major source is defined as one which emits, or has the potential to emit, 90 t (100 st) per year of any nonattainment pollutant. A major modification is a change in a major source that results in a net emissions increase that equals or exceeds the rates shown in Table 6.

Emissions Regulations

The next type of air pollution regulations to which surface mines are subject is emission limitations. Emission limitation regulations directly specify the maximum quantity of pollutants that a source may legally emit within a given time. Emission regulations are conceptually different than ambient air regulations or PSD, each of which seek to control the resulting ambient air concentrations downwind of a pollutant source, and only indirectly control the pollutant emission rate of that source.

Emission regulations take many different forms. At the federal level, New Source Performance Standards (NSPS) restrict the emissions of pollutants from over 50 different source types that EPA feels have a significant impact on air quality. Although surface mines by themselves do not appear on this list, there are new source performance standards for some facilities commonly found on mine property, e.g., coal cleaning plants and petroleum storage vessels. Most states have adopted the federal NSPS and made them a part of their state laws.

Table 6. Significant Net Emissions Increase Rates

Pollutant	Emission Rate, st per year	de minimis† concentration, μg/m³
Particulate matter		
TSP	25	10 (24-hr)
PM-10	15	10 (24-hr)
Sulfur dioxide	40	13 (24-hr)
Nitrogen dioxide	40	14 (annual)
Carbon monoxide	100	575 (8-hr)
Ozone (VOCs)	40*	not applicable
Lead	0.6	0.1 (3 month)

* A net emissions increase for ozone is considered significant if a source emits 36.3 t/a (40 stpy) or more of volatile organic compounds (VOCs).
† Ambient concentration and associated averaging time in parentheses.

Another kind of emission limitation that may be applicable is the process weight rate regulation, which restricts the amount of particulate matter from industrial processes as a function of the material throughput. These regulations would not apply to a surface mine per se, but they would apply to specific processes such as coal or ore crushing. Almost all states have their own process weight regulations. As an example, the maximum particulate emission rate from a new process source in Wyoming (Anon., 1982c) that has a throughput rate of less than 27 t (30 st) per hour is given by

$$E = 3.59 \, P^{0.62}$$

where E is the allowable emission rate in pounds of particulate per hour and P is the process weight rate (throughput) in tons per hour.

A third type of emission regulation is the fugitive dust regulation, commonly adopted by state regulatory agencies. This regulation specifies the level of control that is expected to be used at fugitive dust sources and may require such controls as water spraying on dirt roads, prompt revegetation of stripped areas, covering of haul trucks using public roads, etc. These regulations for fugitive dust are generally nonquantitative.

Other kinds of emission regulations sometimes encountered include volumetric in-stack gas concentration limits; mass per unit time regulations which limit emissions to a specific number of pounds per hour; and mass per unit of activity regulations.

AIR POLLUTION MONITORING

Air quality monitoring networks consist of a collection of instruments that measure air pollutant concentrations and meteorological parameters in the vicinity of a surface mine. Before construction of a mine begins, the regulatory agency often requires that between four months and one year of pollutant concentration data be gathered. The purpose of this effort is to determine what the existing background concentrations are so that when the additional pollution concentrations expected from the mine are calculated, they can be added to the existing background to check whether the mine will comply with air quality standards. The rules that dictate which pollutants must be sampled are complex (Anon., 1978a), but frequently a mine will be required to monitor only for particulate matter. Pollution sources that are large emitters of gaseous pollutants and that may be located adjacent to surface mines (smelters, coal-fired power plants, etc.) are often required to monitor SO_2, NO_x, CO, and ozone prior to construction.

The requirement to monitor background pollutant concentrations may be waived by regulatory agencies under certain circumstances. If a source's pollutant emission rates are expected to be less than the significant levels shown in Table 6, or if existing background concentrations or a source's expected concentration contributions are less than so called de minimis levels shown in Table 6, then the source may be granted a monitoring exemption. In all cases, mine management should check with the responsible regulatory agency before taking advantage of monitoring exemptions.

In areas distant from conventional weather stations, it may be necessary or desirable to gather meteorological data prior to construction of a surface mine. The parameters most often measured are wind speed, wind direction, standard deviation of wind direction, and temperature. Occasionally, relative humidity or dew point, solar insolation, evaporation rate, and precipitation are also measured. Some of these data will later be used as input to computerized pollutant dispersion models that predict the air quality impact of a proposed mine. The remaining data may be useful in planning reclamation.

Particulate Matter Monitoring

The most common ambient air quality monitoring device in the mining industry is the high volume sampler. Total suspended particulate (TSP) concentrations are measured with a hi-vol. The hi-vol consists of a filter holder and a blower motor housed in a shelter. The electric blower, identical to the kind often used in vacuum cleaners, draws air through a filter which captures particulate matter. By weighing the filter before and after exposing it, the mass of particulate matter can be determined. Additionally, the volume of air that passes through the filter is measured, so that the ratio of particulate mass to total air volume yields the average TSP concentration. The hi-vol is usually run for 24 hr at a time so that the resulting concentration is directly comparable to 24-hr ambient air quality standards. A hi-vol and shelter are shown in Fig. 1.

The accuracy with which a TSP concentration can be determined by the hi-vol method depends upon how accurately the hi-vol filter can be weighed and the volume flow measured. A typical hi-vol with flow controller costs about $3000 (1988 dollars). With the recent adoption of PM-10 as a measure of particulate matter concentration, special samplers that collect only particulate matter smaller than 10 μm are being used. The PM-10 samplers are very much like hi-vols, but with carefully designed inlets that admit only particles with mass median diameters less than 10 microns. These particles are collected on filters and subsequently weighed to compute concentration in the same way as with the hi-vol sampler. A PM-10 sampler costs about $5,000. The frequency with which PM-10 must be sampled, and the means of determining compliance with the PM-10 standards, are somewhat more complicated than for TSP.

Operation and maintenance of particulate matter samples, including the filter weighing activity and particulate concentration computations, can be performed by mine personnel. Otherwise, the filters can be sent to commercial laboratories, some of which offer complete turnkey services including filter analysis, data reporting, instrument calibration, and maintenance.

Fig. 1. Hi-vol sampler (courtesy of Flatiron Cos.; photograph by A. Tuggle).

The ratio of PM-10 to TSP concentrations has been measured at several places including mining and non-mining areas. Based upon these measurements, it is estimated that this ratio varies from about 25% to 60%.

Gaseous Pollutant Monitoring

In rare instances, such as when a surface mine has a thermal dryer or is to be located adjacent to a smelter, mine management may be required to sample gaseous pollutants like SO_2, NO_x, CO, or ozone. Instruments that measure these pollutants are much more costly, and require more attention than do particulate monitors. Usually gaseous pollutant monitors are housed in temperature-controlled sheds or trailers. Ambient air is drawn into the shed through glass manifolds and is introduced to the instruments which sense pollutant concentrations as low as a few parts per trillion. Real-time measures of pollutant concentrations detected by these instruments can be recorded on strip chart recorders, or can be digitized and recorded on magnetic tape cassettes, for subsequent analysis. Unlike the particulate samplers intermittent operation, gaseous pollutant instruments are run continuously. The capital cost of a complete gaseous pollutant sampling trailer with SO_2, NO_2, and ozone instruments and recorders can be as high as $50,000. The operating costs of such a sampling station, including routine maintenance and data reduction, would be roughly $30,000 per year.

Meteorological Monitoring

Before a surface mine is opened it may be necessary to collect meteorological data for subsequent use in pollution dispersion modeling studies. Meteorological measurements such as wind speed, wind direction, temperature, and precipitation, can be made continuously with self-contained, unmanned sampling stations, like the ones shown in Fig. 2.

These stations record data onto paper strip charts or magnetic cassettes. Meteorological data collection stations require a source of electrical power, which sometimes can be provided by solar cell and battery units, or by portable gas generators, where electrical utility service is unavailable. The capital cost for a sampling station with cassette recorders similar to the one shown in Fig. 2 is roughly $10,000, and the operating costs, including maintenance and data reduction, may be about $5000 per year.

SOURCES AND CONTROL OF AIR POLLUTION

Recognizing the sources of air pollution at a surface mine and knowing how to control them is at the heart of good air quality management. All actions that ultimately limit the emission or entrainment of pollutants into the air have a direct consequence on pollutant concentrations at the mine site, and if a mine exceeds an ambient air pollution standard, only reduction or cessation of pollutant emissions will help meet that standard.

Because particulate matter emission rates far exceed the emission rates of other pollutants at surface mines, the activities and sources that emit particulate matter deserve special attention. Particulate matter sources can be divided into two categories: mining activities that generate windblown dust (fugitive dust), and materials handling and processing activities whose particulate matter emissions are, or could be, emitted through a vent or a stack (nonfugitive dust).

The distinction between these two types of particulate matter sources is more than academic. As discussed previously, the magnitude of a mine's nonfugitive particulate matter emissions determines whether the mine is subject to PSD regulations. Furthermore, the way in which dust is controlled from these two source types differs. Fugitive dust source

Fig. 2. Remote meteorological sampling stations (courtesy of Flatiron Cos.).

emissions can be minimized somewhat by water spraying, reclamation, etc., but because of the exposed nature of these sources, it is impossible to totally eliminate fugitive dust emissions. Dust from enclosed, nonfugitive emission sources, on the other hand, can be vented to filters and other air cleaning devices that remove as much of the particulate matter as desired.

The remainder of this section discusses the various fugitive dust mining sources, the nonfugitive sources characteristic of materials handling and processing, and gaseous pollutant sources. In each case pollution control devices and practices are described.

Fugitive Dust Sources

Fugitive dust is generated by mining activities that disturb topsoil, overburden, or ore, causing dust to be emitted into the air. Fugitive dust that remains airborne is measured by the samplers discussed previously. Sources of fugitive dust at surface mines can be conveniently discussed in the same sequence that mining occurs (Anon., 1982b):

Construction: Construction of a mine involves clearing vegetation, site grading, building roads, and erecting permanent facilities. Operation of graders, dozers, and other heavy equipment "*kicks up*" fugitive dust into the air. These dust emissions can be controlled by restricting the area of disturbance and by spraying water on disturbed areas to prevent dusty conditions.

Topsoil Stripping: Scrapers used to strip topsoil can cause considerable dust emissions. Fugitive dust emissions caused by stripping can be reduced by watering, by restricting the speed of the scrapers, by minimizing the length of time that the stripped areas are exposed to wind erosion, and by limiting the total area stripped.

Drilling: Blasthole drilling, whether in overburden, ore bodies, or coal, is a source of fugitive dust. Careful planning can minimize the number of holes drilled, thereby saving money and reducing dust emissions. Additionally, dust from drilling can be controlled by water injection systems or by shielding the drilling area to prevent dust from being carried by the wind. When air circulation is used, dust at the top of the drill hole can be collected by a shroud or skirt and vented through a mechanical dust collector on the drilling rig.

Blasting: Blasting is a spectacular, but surprisingly insignificant, source of fugitive dust. Because blasting occurs infrequently, it does not generate as much dust over a period of time as more continuous fugitive dust sources such as topsoil stripping. Dust from blasting can be minimized by careful stemming and by avoidance of overshooting.

Overburden, Ore, and Coal Removal: Although several different pieces of equipment are used for material removal—draglines, shovels, scrapers, wheel excavators, dozers, and front-end loaders—operation of each is a significant source of fugitive dust at surface mines. In surface mining the fugitive dust is generated by shovels and draglines first when material is scooped into the dipper or bucket, and again when the material is dropped into the spoil site. Similarly, in open pit mining dust is created in scooping and dropping material onto trucks. In either case, dust can be controlled by minimizing the drop distance that material falls onto spoil piles or trucks. When scrapers and dozers are used, little can be done to control dust emissions other than to restrict scraper speed, minimize dozer push distances, or in extreme instances curtail operation during high winds.

Material Hauling: For most surface mines, haul truck traffic over unpaved roads is the predominant source of fugitive dust. Dust is picked up by the action of the trucks' tires, and then drawn into the aerodynamic wake behind the truck. The most commonly observed source of fugitive dust at surface mines is the plume of dust behind a large, fast-moving haul truck. The continuous nature of this source explains why haul truck traffic accounts for as much as 75% of all particulate emissions at some western mines. The magnitude of dust emissions from haul trucks increases with truck speed, truck weight, the physical size of the vehicle, the number of tires per truck, the silt content of the unpaved road, and of course, the total number of truck miles traveled. Haul road dust can be controlled most cost effectively by limiting the haul road length and minimizing the number of round trips that each truck must make. From the standpoint of dust control, the use of larger payload haul trucks is preferable to the use of smaller trucks which must make more circuits. Another means of limiting haul road fugitive dust emissions is to restrict truck speed. It should be noted that parameters such as haul length, truck speed, payload, and other haul cycle factors are generally chosen to optimize productivity rather than to minimize dust (Bishop, 1968).

A more direct means of reducing haul truck fugitive dust is to treat the road surface to minimize dust. This can be accomplished by routine watering, by removal of loose debris on roadways, and by application of dust suppressing chemicals. Watering is the most commonly used dust control and, of course, its effectiveness depends on the frequency of water spraying and on the climate at the mine. The correct application rate is one that eliminates visible dust plumes behind haul trucks but does not make the road muddy. Routine spraying requires available water and the appropriate equipment and labor. Haul roads cannot always be watered in freezing weather.

The application of chemicals to a haul road inhibits dust by forming a hard, relatively smooth surface. Most chemical surface treatments require special preparation of the road before application, and while chemical dust suppression is initially costly, savings may be realized by less frequent need to apply water. Table 7 compares features of several dust suppression chemicals.

Stockpiles: Topsoil, overburden, ore, and coal are sometimes stored in exposed stockpiles before being replaced or shipped off site. Stockpiles are sources of fugitive dust when material is added to or removed from the pile and when wind erosion blows material from the pile. Minimizing the size and number of stockpiles reduces the wind erosion dust, as does vegetation or water sprays. Windbreaks can be built on the predominant upwind side of the stockpile to reduce emissions.

Fugitive dust from coal stockpiles is often avoided by building facilities that enclose the load-in, reclaim, or storage areas altogether. A shroud attached to a conveyor stacker, for example, isolates coal load-in from wind action. Similarly, an underground reclaim conveyor eliminates windblown fugitive dust. The ultimate control is to use a fully enclosed coal storage barn or silo which obviates the exposed stockpile altogether.

Table 7. Costs for Chemical Road Dust Suppressants

Palliative	Costs per Gallon,* $	Application Rate, gal per sq yd	Dilution	Cost for 2000 sq yd Section, $	Maintenance Frequency	Maintenance Costs, Annual, $
Petroleum resin	1.22	1	1:4	496	every 6 wks. (1:9 dil.)	199 per occur. 725 per year
Emulsified asphalt	2.95	2.25	None	13,275	none	none
Calcium lignosulfunate†	0.184	1	1:1	418	every 6 wks. (1:3 dil.)	209 per occur. 1811 per year
Magnesium chloride	0.425	0.5	None	579	every 6 mos. (1:1 dil.)	289 per occur. 578 per year

Source: Larson, 1981.

* Does not include freight cost to mine.

† Application rate and costs are assumed.

Metric equivalents: gal $\times$ 3,785 412 = L; sq yd $\times$ 0.836 127 4 = m^2.

Wind Erosion: Wind erosion of unvegetated or exposed areas can be a source of fugitive dust. The best control measure is to minimize the area exposed to wind erosion and to limit the time over which areas are exposed. Prompt revegetation restricts wind erosion dust.

Nonfugitive Dust Sources

Nonfugitive dust sources are those whose particulate emissions can be contained by an enclosure of some sort and exhausted through a vent or stack. The reason for enclosing these dust sources is to prevent windblown dust emissions and to allow dust to be routed to air cleaning devices. Nonfugitive dust sources include most materials handling and processing facilities at surface mines, such as crushers and rail car loading stations. These nonfugitive dust sources are often termed *point sources* because their emissions come from a single, identifiable point such as a stack, in contrast to emissions from a fugitive dust source which emanate from a wide area.

Point source emissions are amenable to control by air-cleaning devices. Air-cleaning devices used to remove particulate matter from facilities at surface mines are generally of three distinct types—inertial separators, wet control devices, or fabric filters (Danielson, 1973). An inertial separator uses an abrupt change in the flow direction of a gas stream to separate out heavier particulate matter. Cyclone separators and baffles are common inertial separators which are rugged, simple, and inexpensive, but they are only capable of removing larger particulate matter. Wet collection devices, such as water sprays or scrubbers, introduce a water spray into a gas stream to impinge smaller particulate matter. Disposal of the waste water is sometimes a problem, and in freezing weather the water spray must be shut off or an anti-freezing solution must be used. Fabric filters (baghouses), the third type of control device, force air through a cloth filter to provide a very efficient, but expensive, means of removing particulate matter as small as 1.0 μm in diameter. Facilities at surface mines may use any one or a combination of these three devices.

There are a number of nonfugitive dust sources at surface mines:

Material Dumping: Ore and coal are dumped into a hopper after being transported from the surface mine and this activity creates dust emissions. These emissions can be controlled by minimizing the distance that the material falls, and by spraying the hopper openings and vents with water to intercept escaping dust. More sophisticated, and expensive, dust control results by enclosing the dump area and by exhausting the displaced air through a baghouse. When the material dumping occurs intermittently, as when haul trucks dump into a crusher, the spray or baghouse can be actuated automatically when the dumping occurs.

Crushing and Screening: Depending upon the material being processed, there can be significant dust emissions at primary and secondary crushers, grizzlies, and screens. Crushing is an especially dusty operation because it creates and exposes new material faces which are the sources of fines. Enclosing the crushers and screens is a first step towards controlling this dust, and water sprays and negative pressure dust pickups exhausted through a cyclone or baghouse can be utilized to further reduce dust emissions.

Conveyors and Transfers: Belt conveyors are almost always preceded by a crushing operation, so that unless fines have been removed or the material is wet, conveyors and transfers can be very dirty. Minimizing the fall distance from the feeder to the conveyor belt reduces dust emissions at transfers, as does the use of water sprays or negative pressure dust pickups vented to a baghouse.

Enclosing the conveyor on the top and sides prevents wind from blowing dust into the air. Total enclosure of the conveyor further reduces emissions, but interferes with maintenance. For very dusty materials with dry, powdery fines, it may be necessary to reduce belt speeds.

Storage: Fully enclosed silos or storage barns avoid the windblown dust problems associated with stockpiles, but there is still a need to ventilate these enclosures and to control dust displaced during loading. Forced air ventilation will often exhaust through a cyclone or baghouse.

Rail Car Loading: Dust is generated when product or ore is transferred into railcars. If the material is loaded by conveyor then a telescoping chute can be used to minimize the fall distance to the car and a fabric shroud can protect the material from wind. Water sprays are somewhat effective in controlling dust, but they increase the load mass and may cause freezing. Some coals are sprayed with oil which inhibits dust and also increases the heating value of the product.

Gaseous Pollutant Sources

Although the magnitude of gaseous pollutant emissions are dwarfed by dust emissions at most surface mines, gaseous emissions occasionally can be a problem. There are four sources of gaseous pollutants—sulfur dioxide, carbon monoxide, oxides of nitrogen, and hydrocarbons—at surface mines:

Vehicle Exhaust: Diesel and gasoline exhaust contains all four combustion product pollutants (TSP, SO_2, NO_x and CO) to some degree. The best way to control these pollutant emissions is to maintain the engine with regular service.

Industrial Boilers: Some surface mines employ small industrial boilers to provide steam for space heating, or less often, for electrical generation. These boilers can be major sources of nitrogen oxides and hydrocarbons. If high sulfur coal is burned, then sulfur dioxide may be emitted in large quantities. The pollution control measures that a mine manager has at his disposal include good operating practice and routine maintenance.

Thermal Drying: Thermal drying of coal or concentrate at a mine or mill can be a significant source of gaseous pollutant emissions. Scrubbing systems can be used to minimize sulfur dioxide emissions when coal is burned to produce heat. If natural gas is used as a fuel for thermal drying, SO_2 emissions will be negligible. NO_x and CO emissions must also be addressed.

Storage Tanks: Fuel storage tanks can be sources of hydrocarbon emissions as air is displaced during tank filling, and as natural heating and cooling of the tanks expels air.

QUANTIFYING AIR POLLUTANT EMISSIONS

Once the sources of air pollutants at a surface mine have been identified, the magnitude of the pollutant emissions from these sources can be estimated. The quantification, called an emissions inventory, serves three purposes. First, as explained earlier, it determines whether certain air pollution regulations are applicable to a particular mine or mine modification. Second, the emissions inventory is frequently required in air quality permit applications, and if a computer model simulation of a mine's impact on the surrounding air quality is needed, then the emissions inventory is used as an input to the model. Third, the emissions inventory provides the mine manager with a relative estimate of the magnitude of pollutant sources so that control efforts can be applied in a cost effective manner.

The fugitive emissions from a mining activity are most frequently determined from an emission factor, i.e., an equation that relates pollutant emission rates to mining or climatological parameters. For example, the particulate emissions caused by unloading run-of-mine coal into a crusher may be a function of the moisture content of the coal, the average wind speed at the hopper, and the number of tons of coal unloaded. As the coal surface moisture increases, the dust emissions would decrease; as the wind speed or the coal throughput increase, the emissions would increase. Additionally, the effects of various pollution control efforts can be taken into account in the equations. Emission factor equations reduce the emissions inventory task to a simple paper and pencil exercise, which can be initiated as soon as a detailed mine plan is available.

Emission factors for fugitive dust sources are derived from empirical tests made at operating mines. In these tests ambient air pollutant concentrations are measured downwind of a mining activity, and then using an air quality dispersion model or other techniques, the emission rate needed to account for the observed concentrations is determined. If the field tests are conducted over a comprehensive range of climatological and operating conditions, then the emission rate can be statistically related to these conditions. Because of the inherent errors in field measurements, the wide variation in pollutant emission rates and the large number of possible explanatory variables, fugitive dust emission factors are not very accurate. Even rigorously conducted field experiments yield emission factors that are accurate only to within a factor of three (Axetell and Cowherd, 1981); typical emission factors may be significantly less accurate (Cole, et al., 1985). Despite this inaccuracy, emission factors are widely used to estimate pollutant emission rates because there is at present no better quantitative method.

Emission Factor Parameters

Emission factor equations include a wide array of parameters upon which pollutant emissions depend. Some of the most common parameters, or variables, are described below (Curreri and Heinold, 1982):

Silt Content: Silt content reflects the amount of fines in an overburden, topsoil, ore, or mineral product, and determines how dusty the handling of these materials will be. Silt content is expressed in percent and is equal to the mass fraction of material that passes through a 75-μm (200-mesh) screen. Note that this definition of silt commonly used in emission factors may differ from that used in other disciplines.

Moisture Content: This is the mass fraction of water, expressed in percent, of a material. The higher the moisture content, the lower the particulate emissions.

Drop Height: In some instances dust emissions are proportional to the distance that ore or product drops while being transferred or loaded.

Vehicle Miles Traveled: Dust emissions from roads are generally proportional to the total vehicle miles traveled (VMT) on the road.

Wind Speed: Fugitive dust emissions frequently depend upon the average surface wind speed at a mine. As wind speeds increase, the aerodynamic drag force on particles also increases, thereby entraining more dust into the air.

Wet Days: Many emission factors use a term called *number of wet days,* which is merely the average number of days per year that precipitation exceeds 0.254 mm (0.01 in.) at a particular location. The number of wet days per year affects the surface moisture at a mine, which in turn affects the

emission rate of windblown dust. Fig. 3 shows the mean number of wet days per year for the United States.

Precipitation-Evaporation Index: The Thornthwaite precipitation-evaporation index (*P-E index*) is a more sophisticated measure of surface moisture. Values of the P-E index are displayed in Fig. 4.

Climate Factor: A climate factor, equal to

$$\frac{0.345u^3}{(P-E)^2}$$

where u is the wind speed in miles per hour and P-E is the Thornthwaite precipitation-evaporation index as used in computing wind erosion dust emissions. The climate factor can be computed directly if u and P-E are known, or it can be determined from a handbook (Jutze and Axetell, 1974).

Emission Factors and Control Efficiencies

There are many different emission factors for computing fugitive and nonfugitive pollutant emission rates at surface mines. The key to selecting the best factor to predict emissions from a particular mining activity is to choose one that best represents prevailing operational and meteorological conditions. Factors to predict coal dust emissions from crushing at a mine in Montana, for example, ideally would be ones derived from western coal mine data, where climatology and coal characteristics would be similar. Since selection of specific emission factors is a very important part of preparing an emissions inventory, state or federal regulatory agencies should be consulted. The most widely used, and commonly accepted, emission factor handbook is EPA's *Compilation of Air Pollutant Emission Factors* (Anon., 1985).

Tables 8 through 13 present a number of emission factors often used to prepare pollutant emission inventories. Many of the factors pertain to surface coal mines because much of the measurement work has focused on coal mining. The emission factors are grouped by mining activity.

Material Removal: Table 8 summarizes emission factors used to predict particulate emissions caused by material removal and load-in to haul trucks. Emissions differ according to the equipment used and the material handled.

Drilling and Blasting: Table 9 shows emission factors that apply to drilling and blasting operations. Particulate emissions are expressed in pounds per drill hole or pounds per blast.

Dumping, Crushing, and Conveying: In Table 10 factors are shown that predict particulate emissions from dumping, crushing, and conveying. Emissions caused by dumping into a hopper can be reduced by 50% by using water sprays, and by 95% by controlling emissions with a negative pressure enclosure exhausted to a baghouse (Anon., 1979a). Similarly, use of a fabric filter baghouse at conveyor transfer points and crushers reduces emissions predicted by the equations by 99%.

Stockpiles: Particulate emission factors associated with stockpiles are summarized in Table 11. Fugitive dust emissions are caused by load-in and load-out operations and by natural wind erosion. Active stockpiles which are disturbed by equipment emit more dust than inactive stockpiles in which the material forms a crust. Emissions during load-in operations can be reduced by 50% by water spray, or by as much as 90% by applying chemical dust suppressants at the load-in point. Wind erosion dust losses can be reduced by 90% on inactive stockpiles by chemical stabilization or by revegetation. The ultimate control efficiency at stockpiles is achieved by enclosing them, an effort which may be prohibitively expensive.

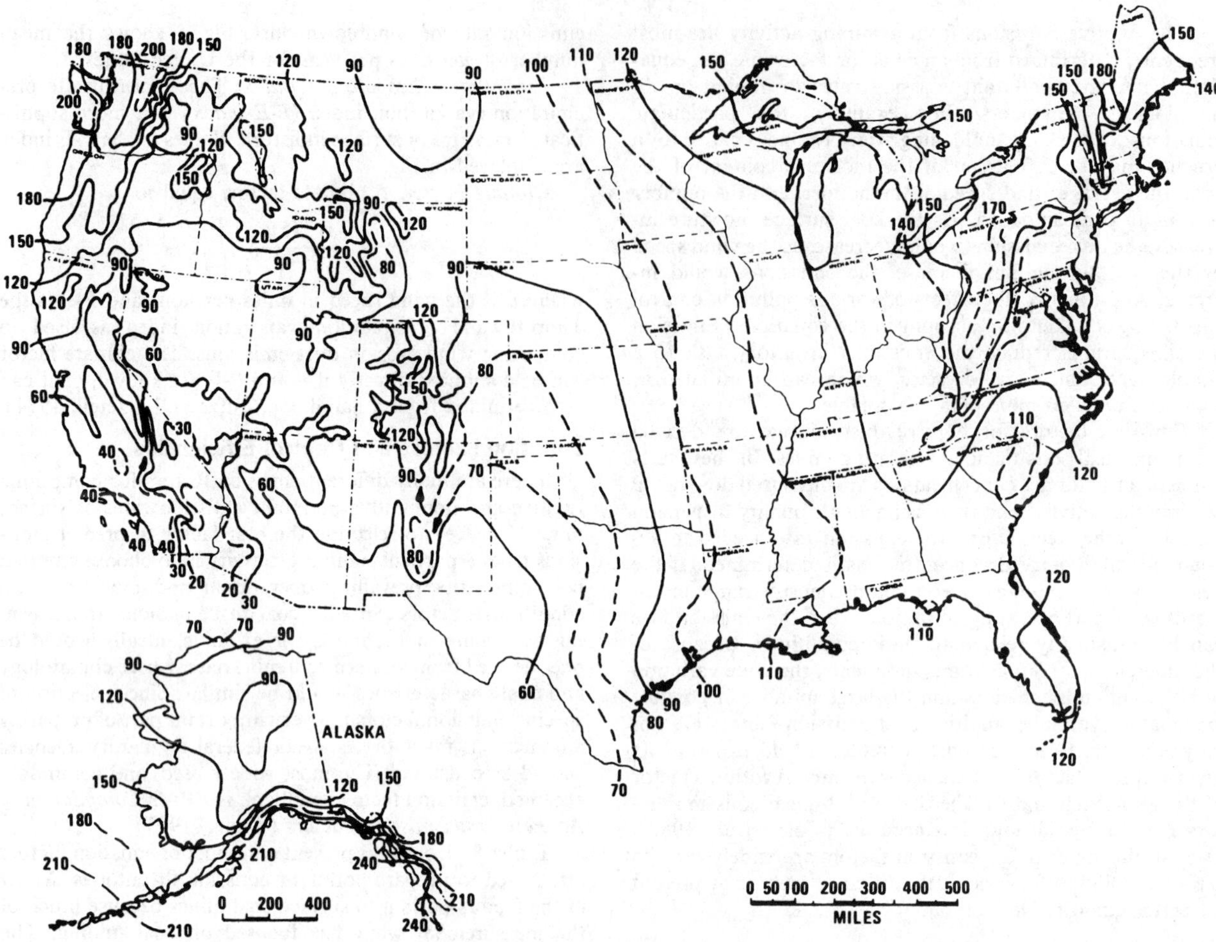

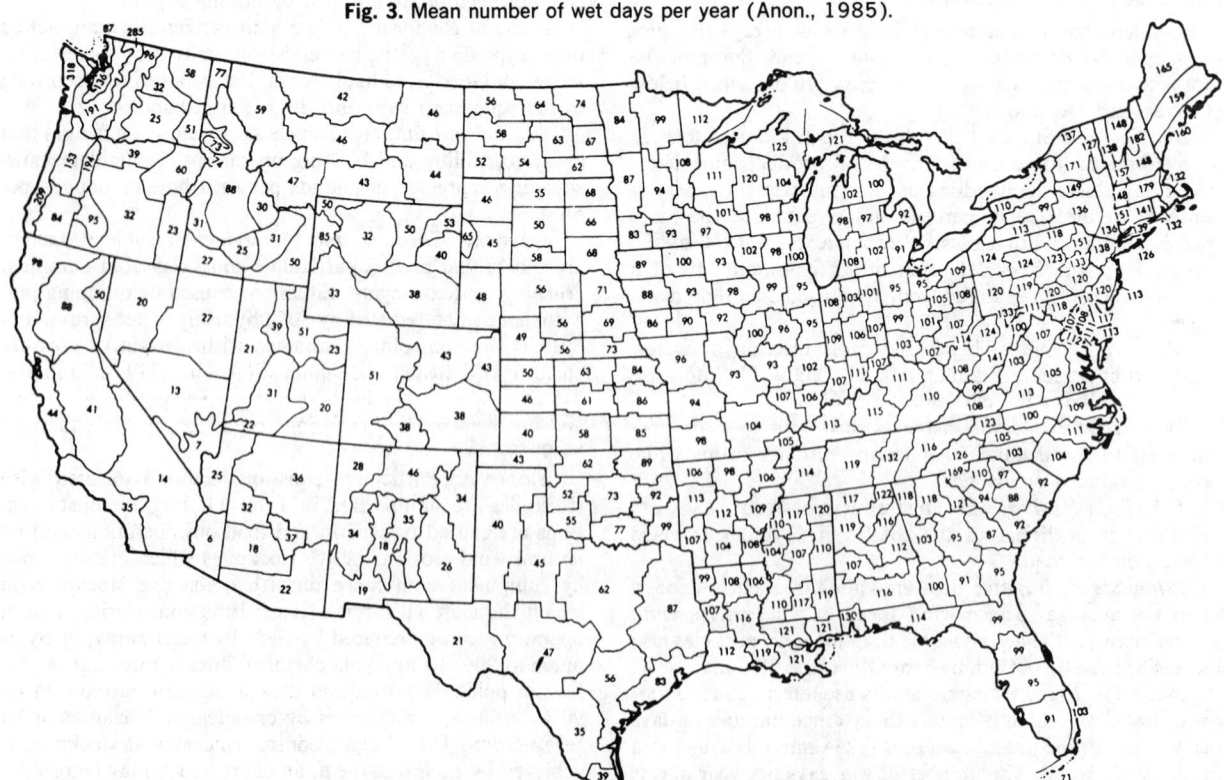

Fig. 3. Mean number of wet days per year (Anon., 1985).

Fig. 4. Values of Thornthwaite's precipitation-evaporation index (Anon., 1985).

Table 8. Particulate (TSP) Emission Factors for Material Removal and Loading

Activity	Factor	Reference	Comments
Topsoil removal	16 lb per scraper hr	Anon., 1976a	
	0.38 lb per cu yd	Anon., 1978	Scraping and dumping
	$(2.7 \times 10^{-5}) S^{1.3} W^{2.4}$ lb per VMT		Scraper Travel
Overburden removal	0.037 lb per st	Anon., 1979a	truck and shovel
	16 lb per scraper hr	Anon., 1979a	scraper
	0.053 lb per cu yd	Anon., 1978	Wyoming dragline
	$0.0021 \, d^{1.1} M^{-0.3}$ lb per cu yd where d = drop distance, ft; M = moisture content, %	Axetell and Cowherd, 1981	dragline
Product removal	0.12 lb per st	Anon., 1979a	coal; front-end loader
	0.05 lb per st	Anon., 1979a	uranium ore
	0.02 lb per st	Anon., 1976a	lignite; truck and shovel
	0.0035 to 0.014 lb/st	Anon., 1979a	coal; truck and shovel
	$1.16 \, M^{-1.2}$ lb per st where M = moisture content, %	Axetell and Cowherd, 1981	coal; truck and shovel

Metric equivalent: lb × 0.453 592 4 = kg; lb/st × 0.5 = kg/t; lb/yd³ × 0.593 276 4 = kg/m³; st × 0.907 184 7 = t; ft × 0.304 8 = m.

Haul Roads and Exposed Areas: Because haul roads are often the major source of dust at surface mines, a great deal of effort has been devoted to quantifying and controlling haul road emissions. Table 12 shows a widely used haul road emission equation, which indicates that truck speed is a factor in dust emission rates. Reducing truck speeds can significantly reduce dust, but at the expense of productivity. Haul road watering is generally thought to reduce emissions by 50%, while the use of chemical dust suppressants may reduce emissions by 85%, depending upon the frequency of application, the type of chemical used, the condition of the road, and other factors. Wind erosion dust from exposed areas can be controlled by minimizing the exposed or stripped areas, by frequent watering, and by prompt revegetation.

Equipment Exhaust: Table 13 lists emission factors for the gaseous pollutants emitted from heavy equipment operation (Anon, 1975). These pollutants are best controlled by properly maintaining the engines.

COMPUTING AIR POLLUTION IMPACT

A detailed emissions inventory will show which are the largest sources of pollutants at a surface mine, but it will not by itself indicate what the ambient air pollutant concentrations in the vicinity of the mine will be. To do this, air pollution dispersion models are required.

A dispersion model is a complex set of equations, most often solved with a digital computer, that predicts ambient pollutant concentrations given data describing source emission rates, source location and configuration, and meteorology. The model mathematically simulates the dispersion of air pollutants as they are transported away from the mine by prevailing winds.

The federal Clean Air Act Amendments of 1977, and many state regulations, explicitly require the use of dispersion models to assess the air quality impact of new or proposed pollutant sources. Dispersion models are used during the

Table 9. Particulate (TSP) Emission Factors for Drilling and Blasting

Activity	Factor	Reference	Comments
Drilling	0.22 lb per hole	Anon., 1979a	Wyoming coal
	1.5 lb per hole	Anon., 1979a	Overburden
	1.3 lb per hole	Axetell and Cowherd, 1981	Overburden
Blasting	14.2-85.3 lb per blast	Anon., 1979a	Overburden
	25.1-78.1 lb per blast	Anon., 1979a	Coal
	$961 \, A^{0.8} D^{-1.8} M^{-1.9}$ lb per blast where A = area blasted, sq ft; D = hole depth, ft; M = moisture content, %	Axetell and Cowherd, 1981	Coal and overburden

Metric equivalent: lb × 0.453 592 4 = kg; ft × 0.304 8 = m; ft² × 0.092 903 04 = m².

Table 10.   Particulate (TSP) Emission Factors for Dumping, Crushing, and Conveying

Activity	Factor	Reference	Comments
Dumping into hopper	0.005-0.027 lb per st 0.007 lb per st 0.04 lb per st	Anon., 1979a Anon., 1979a Anon., 1979a	Coal; bottom dump truck Coal; end dump truck Uranium; end dump truck
	$(1.05 \times 10^{-4})\, s\, u\, h\, M^{-2}\, Y^{-0.33}$ lb per st where s = silt content, %; u = wind speed, mph; h = drop height, ft; M = moisture content, %; Y = dumping device capacity, cu yd	Curreri and Heinold, 1982	Coal
Crushing and screening	0.02 lb per st 0.06 lb per st 0.10 lb per st	Anon., 1979a Anon., 1979a Anon., 1979a	Primary coal crushing Secondary coal crushing Coal screening
Conveyors and transfers	0.055 lb per st	Currier and Neal, 1979	Coal emissions for each conveyor section
	0.165 lb per st	Currier and Neal, 1979	Coal emissions for each transfer point
	$(5.8 \times 10^{-5})\, s\, u\, h\, M^{-2}$ lb per st where s = silt content, %; u = wind speed, mph; h = drop height, ft; M = moisture content, %	Curreri and Heinold, 1982	Coal emissions for each transfer point
	1.5 lb per st	Jutze et al., 1977	Dry phosphate rock; includes loading into rail cars
	2.0 lb per st	Jutze et al., 1977	Iron ore
	1.64-5.0 lb per st	Jutze et al., 1977	Lead ore

Metric equivalents: lb/st × 0.5 = kg/t; mph × 1.609 344 = km/h; ft × 0.304 8 = m; yd³ × 0.764 554 9 = m³.

Table 11. Particulate (TSP) Emission Factors for Stockpiles

Activity	Factor	Reference	Comments
Stacker load-in	$0.04\, k_1 \dfrac{(S/1.5)}{(P\text{-}E/100)^2}$ lb per st where S = silt content, %; $P\text{-}E$ = precipitation evaporation index; k_1 = material factor	Jutze et al., 1977	k_1 = 0.75 for coal and iron ore k_1 = 1.0 for ore bedding
	$0.12 \dfrac{0.33}{(P\text{-}E/100)^2}$ lb per st where P-E = precipitation evaporation index	Sterk, 1980	
Wind erosion	$(1.05 \times 10^{-7})\, sfd\, (365\text{-}w)$ lb per st where s = silt content, %; f = percentage of time wind speed exceeds 19 km/h (12 mph); w = wet days per year	Sterk, 1980	
	$1.2\, u \dfrac{(365\text{-}w)}{365}$ lb per acre-hr where u = wind speed, m/s; w = wet days per year	Collins, 1979	
Stockpile removal	$0.05\, k_3 \dfrac{(S/1.5)}{(P = E/100)^2}$ lb per st where S = silt content, %; $P\text{-}E$ = precipitation evaporation index; k_3 = material factor	Jutze et al, 1977	k_3 = 0.75 to 0.8 coal = 0.75 for iron ore
	$78.4\, S^{1.2}\, M^{-1.3}$ lb per hr where S = silt content, %; M = moisture content, %	Axetell and Cowherd, 1981	Coal unloading by dozer

Metric equivalents: lb/st × 0.5 = kg/t; lb/hr × 0.0001 259 979 = kg/s.

Table 12. Particulate (TSP) Emission Factors for Roads and Exposed Areas

Activity	Factor	Reference
Haul trucks	$0.0067\ W^{3.4}\ L^{0.2}$ lb per VMT where L = road surface silt loading, g/m^2 W = number of vehicle wheels	Anon., Axetell and Cowherd, 1981
Wind erosion from exposed areas	AIKCLV st per acre-yr where C = climate factor; K,L,V = 1.0 normally; A and I depend on soil type as shown:	Collins, 1979

Soil type	A	I
Rocky	0.025	38
Sandy	0.010	134
Fine	0.041	52
Clay loam	0.025	47

Metric equivalents: lb × 0.453 592 4 = kg; mph × 1.609 344 = km/h; st × 0.907 184 7 = t.

regulatory permitting process to determine whether mines will cause exceedances of ambient air standards or PSD increments.

Gaussian Models

There are many different air quality models, but the most frequently used models employ a Gaussian formula as the governing equation to predict pollutant concentrations. This equation imposes a Gaussian or normal distribution of pollutant concentration about the centerline of the pollution plume.

A simplified version of the Gaussian equation that predicts the ground level pollution concentrations at a location x meters downwind and y meters crosswind of an elevated pollution point source is (Turner, 1970):

$$C(x,\ y,\ 0,\ H) = \frac{Q}{\pi\ \sigma_y\ \sigma_z\ u}$$
$$\exp\left[-\frac{1}{2}\left(\frac{y}{\sigma_y}\right)^2\right] \exp\left[-\frac{1}{2}\left(\frac{H}{\sigma_z}\right)^2\right] \quad (1)$$

where C is the pollution concentration, g/m^3; Q is the pollutant emission rate, g/s; σ_y is the standard deviation of concentration in the crosswind direction, m; σ_z is the standard deviation of concentration in the vertical direction, m; u is the mean wind speed, m/s; y is the distance from the plume centerline to the point predicted concentration (measured perpendicular to the plume centerline), m; and H is the height of the plume above ground level, m.

A computerized Gaussian air quality model solves Eq. 1 repeatedly for each pollutant source and each *receptor* (the

location where an ambient pollution concentration is to be computed) that the model user requests. To compute the impact of a surface mine on ambient TSP concentrations, for example, one would input information about all particulate matter emitting activities and specify the location of receptors that were of interest. Using meteorological data (wind direction, wind speed, atmospheric stability, temperature) provided by the modeler, the model computer program would sum the computed concentration contribution from each source at all of the receptors. Very often model receptors are located in a uniform array in the vicinity of a mine so that the resulting grid of predicted pollutant concentrations can be used to draw contours of equal concentration.

Although it is not necessary to have a detailed understanding of the inner workings of Gaussian computer models, a familiarity with certain model characteristics is helpful in choosing models and in evaluating model results. The most important characteristics are (Anon., 1982a):

Dispersion Coefficient: Pollutants in the atmosphere are dispersed and diluted by the turbulent motion of the air. Highly turbulent or unstable atmospheres tend to mix pollutants with clean air rapidly so that pollutant concentrations decrease promptly with downwind distance from the source. In stable atmospheres less mixing occurs and pollutant concentrations along the plume centerline remain higher. The stability of the atmosphere, measured by wind direction fluctuations or inferred from vertical temperature data, is generally divided into six categories ranging from very unstable to very stable (class A to F). The Gaussian models take into account the varying degrees of turbulence by means of the σ_y and σ_z dispersion coefficients in Eq. (1). Values of σ_y

Table 13. Emission Factors for Heavy Duty Diesel Equipment Operation (Pounds of Pollutant Per Gallon of Fuel Consumed)*

Pollutant	Wheeled dozer	Scraper	Grader	Front-end loader	Haul Truck
Carbon monoxide	0.123	0.085	0.055	0.099	0.123
Hydrocarbons (VOC)	0.013	0.019	0.013	0.043	0.013
Nitrogen oxides	0.286	0.259	0.254	0.321	0.286
Sulfur dioxide	0.031	0.031	0.031	0.031	0.031
Particulate	0.015	0.027	0.022	0.029	0.018

* *Metric equivalent:* lb/gal × 119. 826 4 = kg/m^3.

and σ_z are derived from empirically determined equations for each of the six stability categories.

However, the values of the dispersion coefficients used in most models differ depending upon whether the coefficients were meant to apply in urban areas or rural areas. Obviously, for most surface mining applications, models that employ rural dispersion coefficients should be used.

Source Configuration: Eq. (1) yields pollutant concentration downwind of a source whose emissions come from a discrete, infinitesimally small point. This point source equation simulates dispersion from surface mining sources small enough to be reasonably idealized as a point, such as a thermal dryer stack or the exhaust vent from a baghouse. But many mining activities emit pollutants from a long linear configuration (e.g., a haul road), or from a large area (e.g., wind erosion from exposed areas). These sources, called line sources and area sources, are best modeled by variations of Eq. (1) that properly account for the area extent of the emitting source. Some dispersion models include these variations. One of the first tasks in modeling the ambient pollutant concentrations from a surface mine is to categorize each source as a point, a line, or an area idealization.

Time Interval: The Gaussian equation applies only to *steady-state* situations in which none of the important variables (wind direction, wind speed, etc.) change with time. It is also only strictly applicable for flat terrain. For modeling very short time intervals (on the order of 1 hr) it is often reasonable to assume that time-dependent variables remain constant. But to simulate pollution dispersion over longer periods of time, two methods of extending the time scale are used. The first method breaks up the desired time interval into individual, sequential 1-hr increments, models each hour separately, then averages the long-term pollutant concentration at each receptor. Using this hour-by-hour approach, an annual average concentration would be computed by separately calculating concentrations for each of 8760 hr in a year and then averaging them. The second means of extending the steady-state Gaussian equation to longer time intervals is to group each of the time-dependent meteorological variables (wind speed, wind direction, and stability class) into predetermined categories. Usually wind speed and stability class are subdivided into six categories each, and wind direction is subdivided into the 16 major compass points (N, NNE, NE, etc.), yielding $6 \times 6 \times 16 = 576$ combinations. Pollutant concentration at each receptor is then calculated for each of the 576 possible combinations. Next, the frequency with which each combination occurs is determined from meteorological data, and a joint frequency distribution (commonly called a STAR array) is derived. The STAR array is multiplied by the 576 concentrations to yield long-term average concentrations. Of these two methods, the hour-by-hour approach is attractive because in the process of calculating annual average concentration, it can also calculate 3- and 24-hr concentrations for comparison with air quality standards. The STAR data method is attractive because it uses far less computer time than the hour-by-hour approach.

Data Input: Two kinds of input data are required to run a dispersion model: pollutant source data and meteorological data. Source data include the pollutant emission rate of each source at the surface mine, often expressed in grams per second, and the location and extent of the source, usually expressed in a coordinate system consistent with the model. For point sources such as stacks, the gas exit temperature, exit velocity, stack diameter, and stack height are needed so that the model can compute the height of the plume above ground level. For line and area sources the plume rise is more difficult to simulate. The source data needed in an air quality dispersion model are derived largely from an emissions inventory.

The meteorological data needed for dispersion models differs depending upon whether the model is a sequential hour-by-hour type or a long-term STAR type. The sequential models require wind speed, wind direction, and stability class for each hour modeled. The STAR data models simply require a matrix of joint frequency distributions. In addition, both model types often require inversion height (or *mixing depth*) and ambient temperature from which point source plume rise is computed. The meteorological data are either derived from on-site measurements collected at the surface mine over a period of time, or are obtained from measurements taken at nearby weather stations or airports. The National Climatic Data Center (NCDC), a facility within the National Oceanic and Atmospheric Administration of the US Department of Commerce, can provide meteorological data over most of the United States at a modest cost. Because many air quality regulatory agencies have specific policies regarding the use of meteorological data in models, it is wise to consult with the appropriate agency if the model predictions are to be used in support of a permit application.

Deposition: Particulate matter emitted from surface mines is removed from the atmosphere by gravitational settling and deposition. These phenomena tend to reduce ambient air particulate concentrations below those which would be experienced in the absence of deposition and settling. The magnitude of the reduction depends upon the size of the particulate matter and the distance at which the concentrations are measured, but it is possible that 60% of the airborne TSP mass at a surface mine may be removed from the atmosphere by deposition and settling (Wurmbrand, 1981). If an air quality dispersion model that does not account for these phenomena is used to simulate TSP concentrations, then it should be recognized that the predicted concentrations may be unrealistically high. At the present time there are no EPA-approved models that do a good job of simulating deposition and settling.

Selection of Air Quality Models

There are many Gaussian air quality models in use, but only a handful include features that make them desirable for modeling pollutant dispersion from surface mines. Three of the most widely used of these are discussed:

CDM: The Climatological Dispersion model (CDM) is a STAR data-type model that predicts annual average or seasonal concentrations (Busse and Zimmerman, 1973). CDM admits point sources and area sources only; line sources such as haul roads must be simulated by a series of adjacent point sources. A rural version of CDM, called CDMW, is used in many western states (Dailey, 1979).

ISC: The Industrial Source Complex (ISC) model (Bowers, Bjorklund, Cheney, 1979) is comprised of a short-term and a long-term version. The short-term version is an hour-by-hour model, while the long-term version uses STAR data input. Both versions admit point and line source configurations, as well as a variation of an area source termed a volume source. The ISC model attempts to account for many meteorological phenomena that many models ignore, among them stack downwash, rural or urban dispersion coefficients, limited complex terrain adjustment, and particle settling (but not deposition). The inclusion of particle settling capability coupled with very flexible source idealization has made this a popular surface mining model. However, these features come at the expense of greater computer execution time than the other Gaussian models and are still not as good as they should be.

VALLEY: The VALLEY model (Burt, 1977) is a STAR data-type model that simulates annual average concentrations or, by creating an abbreviated STAR data input, can predict 24-hr concentrations. VALLEY is one of a few models that accepts receptor heights situated above the pollutant source, and for this reason it is often used to estimate pollutant concentrations in mountainous terrain. The method of handling high terrain causes the model to be notorious for predicting very high concentrations on elevated terrain when used in a *worst case* mode. VALLEY is a screening model that admits point and area source configurations, but does not account for particle settling or deposition.

EXAMPLE APPLICATION

In this section the individual steps leading to the acquisition of an air quality permit for a hypothetical surface coal mine are discussed. This example illustrates how air pollution regulations, ambient air monitoring efforts, emission inventories, and pollution modeling tie together.

The proposed mine in this hypothetical example is to be an open pit, truck and shovel operation with a life of 27 years. Maximum production of 14.5 Mt (15 million st) per year will be reached during the last 14 years of mine life. Run-of-mine coal will be transported from the pit by haul trucks to a two-stage crushing complex, and from there by conveyor to a storage silo. Coal will be shipped by rail to markets.

The only air quality permit needed to begin construction was that administered by the state air pollution regulatory agency. The state air pollution regulations embodied most of EPA PSD requirements and enforced ambient standards through a permitting process. New pollutant sources were required to secure state air quality permits, granted only after the applicant demonstrated that the proposed source would meet state ambient air quality standards, would apply BACT, and would meet applicable emission limitations. Additionally, new sources were required to submit one year of pre-construction monitoring data to the state agency to ensure that the sum of existing pollutant concentrations and anticipated future concentrations caused by the mine would not violate standards.

Because the requirement for one year of monitoring data was identified as the most time-consuming step in the permitting process, the mining company immediately set up the necessary monitoring equipment. Three PM-10 samplers (two collocated) were situated at two sites located along the prevailing wind direction on opposite boundaries of the proposed mine site. A 10-m (32.8-ft) tall meteorological tower was erected and instrumented to measure wind speed, wind direction, temperature, and wind direction fluctuation data.

An air pollution consulting firm was hired to design and operate the sampling system and reduce the data into quarterly reports for submission to the state regulatory agency.

Development of a comprehensive mining plan was begun shortly after initiation of the ambient monitoring effort. The mining company's engineers, working in conjunction with the company's environmental staff, specified coal handling equipment and pollutant emission control devices for competitive bid by vendors. The particulate matter control equipment was selected with two goals in mind: total point source emissions were to be kept below 227 t/y (250 stpy) so that the mine would not be subject to PSD regulations and equipment had to qualify as BACT. Performance guarantees from equipment vendors and a review of previous decisions by the regulatory agency as to what devices qualified as BACT ensured that these two goals were met. The truck dump, where bottom dump haul trucks unload into receiving hoppers, will be controlled by water sprays, and by negative pressure dust pickups vented to a baghouse. Similarly, the primary and secondary crushers will be vented to fabric filter baghouses. Above ground conveyors will be covered on the sides and top and vented to a baghouse. Prior to loading into railcars, coal will be stored in silos.

The mining company also made stringent commitments for the control of fugitive dust emissions. Prompt revegetation, minimization of disturbed area, and wind breaks were proposed as a means of controlling windblown dust. Dust from drilling is to be controlled by cyclone separators and shrouds and blasting is to be curtailed during high winds. The mining company planned to control dust from haul roads by use of chemical dust suppressants, by limiting haul truck speed to 64 km/h (40 mph), and by maintaining roads with a grader.

As soon as the mining plan was finished, and the dust control equipment and fugitive dust abatement practices were chosen, a complete emissions inventory was developed. The state regulatory agency asked that the mining company determine pollutant emissions and perform pollution modeling simulations only for the *worst-case* year—the year during which maximum particulate concentrations could be expected to occur off mine property. Initial calculations showed that the worst case year would be 2001, when TSP emission rates would be greatest because of long haul road lengths and high production levels. Equally important, that year is one in which the mine pit would be close to the plant boundary so that activities in the pit might induce high concentrations to which the public could be exposed.

Using the factors appearing in Tables 8 to 13 and other factors, an emissions inventory for the year 2001 was prepared. The point source emissions were computed to be well below the 227-t/y (250-stpy) limit, as illustrated in Table 14, and as a consequence the mine is clearly exempt from

Table 14. Particulate Nonfugitive Emission Rates from Mine Point Sources

Source	Emission Rate, st per year
Truck dump	19.2
Primary crusher	1.6
Secondary crusher	4.8
Transfers and conveyors	16.0
Silo load-in/load-out	0.1
Total	41.7

Metric equivalent: st × 0.907 184 7 = t.

Table 15. TSP Fugitive Dust Emission Rates

Source of Activity	Emission Rate, st per year
Topsoil removal and scraper travel	55.1
Overburden drilling	1.7
Overburden blasting	0.5
Overburden removal and replacement	181.5
Coal drilling	0.6
Coal blasting	0.8
Coal removal	144.6
Road construction and maintenance	168.0
Haul roads	566.3
Wind erosion	70.7
Total	1,189.8

Metric equivalent: st × 0.907 184 7 = t.

PSD regulations. The fugitive TSP emissions summarized in Table 15 were expected to be considerably larger than the point source emissions. The haul road emissions account for nearly one-half of TSP emissions at the mine.

The next step in the preparation of the permit application was to model the air quality impact of the proposed mine. The particulate matter emission rates were divided into point, area, and volume source configurations and the location of each was identified on a map of the mining area. The source emission rate and location information were coded into a format for use in the ISC dispersion model. Similarly, the meteorological data collected during the year-long field data gathering effort were coded for use with the computer models. The models were then run, and the annual average and 24-hr TSP concentrations predicted to occur at model receptors outside the mine property were tabulated.

Because specific emission factors for PM-10 were not available, the state agency asked that modeled TSP contributions be multiplied by a constant factor of 0.27 to estimate PM-10 concentration contributions. The multiplicative factor was based on previous studies that suggest that 27% of total suspended particulate matter in the vicinity of surface coal mines is less than 10 μm in size (Anon., 1982).

The peak annual average PM-10 contribution from the mine was found to be 13 $\mu g/m^3$, which when added to the measured background concentration of 11 $\mu g/m^3$, was in compliance with the 50 $\mu g/m^3$ ambient air standard. The peak 24-hr modeled concentration of 48 $\mu g/m^3$, when coupled with the highest measured daily PM-10 background of 16 $\mu g/m^3$, was similarly lower than the federal 24-hr secondary standard of 150 $\mu g/m^3$. Hence, the modeling study indicated that the proposed mine would meet applicable PM-10 standards. The emission rates of other pollutants—SO_2, CO, NO_x, and hydrocarbons—were so small that the state air pollution regulatory agency concluded that no modeling would be needed to show compliance with ambient standards for these pollutants.

With the emissions inventory and modeling effort completed, the permit application was nearly finished. Maps and drawings of the proposed mine, required by the state agency, were added, as were descriptions of pollution control equipment and anticipated annual coal production rates. Additionally, computations comparing expected emission rates with state emission limitations (New Source Performance Standard for coal preparation plants and process weight rate restrictions) were provided. The permit application was submitted to the state agency 15 months after the year-long field monitoring study was begun.

The state agency, following a statutory time frame for reviewing air quality permit applications, deemed the application complete 60 days after receipt and issued a preliminary determination of *conditional* approvability 90 days after judging the application complete. The preliminary approval was granted on the condition that the mining company agrees to continue monitoring particulate matter concentrations after mining begins.

REFERENCES

Anon., 1975, "Heavy Duty Construction Equipment," *Compilation of Air Pollutant Emission Factors,* 3rd ed., AP-42, US Environmental Protection Agency, National Technical Information Service, Washington, DC, pp. 3.2.7-2.

Anon., 1976, "Subpart Y—Standards of Performance for Coal Preparation Plants," *Federal Register,* 41 FR 2232, US Environmental Protection Agency, Jan. 15.

Anon., 1976a, "Evaluation of Fugitive Dust Emissions from Mining, Task 1 Report," PEDCo Environmental Specialists, Inc., US Environmental Protection Agency, Cincinnati, Apr.

Anon., 1978, "Survey of Fugitive Dust from Coal Mines," Project 3311, PEDCo Environmental Inc., US Environmental Protection Agency, Denver, Feb.

Anon., 1978a, "Ambient Monitoring Guidelines for Prevention of Significant Deterioration (PSD)," EPA-450/2-78-019, US Environmental Protection Agency, Research Triangle Park, NC.

Anon., 1978b, "Guidelines for Determining Best Available Control Technology (BACT)," US EPA Office of Air Quality Planning and Standards, Washington, DC, Dec.

Anon., 1979, *Cleaning the Air: EPA's Program for Air Pollution Control,* Office of Public Awareness, US Environmental Protection Agency, Washington, DC, pp. 10.

Anon., 1979a, "Compilation of Past Practices and Interpretations by EPA Region VIII on Air Quality—Mining," US EPA, Region VIII, Denver.

Anon., 1979b, "A Comparison of Alternative Approaches for Estimation of Particulate Concentrations Resulting from Coal Strip Mining Activities in Northeastern Wyoming," Environmental Research and Technology, Inc., US Dept. of Energy Document P-3545-106, Oct.

Anon., 1980, "Appendix S—Emission Offset Interpretative Ruling," *Federal Register,* 45 FR 52729, US Environmental Protection Agency, Aug. 7.

Anon., 1980a, "Requirements for Preparation, Adoption and Submittal of Implementation Plans: Approval and Promulgation of Implementation Plans," *Federal Register,* 40 CFR 51.24, 52.21, US Environmental Protection Agency, Aug. 7.

Anon., 1982, "Characterization of PM 10 and TSP Air Quality Around Western Surface Coal Mines," PN 3525-35, PEDCo Environmental, Inc. and TRC Environmental Consultants, Inc. US Environmental Protection Agency, Research Triangle Park, NC, June.

Anon., 1982a, "Air Quality Modeling Handbook for the Mining Industry (Draft)," Camp Dresser & McKee, Inc., American Mining Congress, Washington, DC, pp. 2.1–2.29.

Anon., 1982b, *BLM Air Quality Handbook for Surface Coal Mines (Draft)*, Morrison-Knudsen Co., Inc., US Bureau of Land Management, Denver, pp. 26–31.

Anon., 1982c, "Section 14—Control of Particulate Matter," *Wyoming Air Quality Standards and Regulations 1982*, Wyoming Division of Air Quality, Wyoming Department of Environmental Quality, Cheyenne, pp. 13–14.

Anon., 1985, "Fugitive Dust Sources," *Compilation of Air Pollutant Emission Factors*, 4th ed., AP-42, US Environmental Protection Agency, National Technical Information Service, Washington, DC.

Anon., 1985a, *BACT/LAER Clearinghouse—A Compilation of Control Technology Determinations, Vol I-IIC*, US Environmental Protection Agency, National Technical Information Service, Washington, DC., June 1985.

Anon., 1987, "1987 APCA Governmental Agencies Directory," *Journal of the Air Pollution Control Association*, Vol. 37, No. 4, April, pp. 438–439.

Anon., 1987a, "National Ambient Air Quality Standards for Particulate Matter," *Federal Register*, 52 FR 29382, US Environmental Protection Agency, Aug. 7.

Axetell, K., and Cowherd, C., 1981, "Improved Emission Factors for Fugitive Dust from Western Surface Coal Mining Sources, Volume II: Pre-Publication Copy," US Environmental Protection Agency, Cincinnati, Nov.

Bishop, T.S., 1968, "Trucks," *Surface Mining*, E.P. Pfleider, ed., AIME, New York, pp. 553–588.

Bowers, J.F., Bjorklund, J.R., and Cheney, C.S., 1979, "Industrial Source Complex (ISC) Dispersion Model User's Guide (Vols. 1 and 2)," publication No. EPA-450/4-79-0, US Environmental Protection Agency, Research Triangle Park, NC, Dec.

Burt, E.W., 1977, "VALLEY Model User's Guide," publication No. EPA-450/2-77-018, US Environmental Protection Agency, Research Triangle Park, NC, Sept.

Busse, A.D., and Zimmerman, J.R., 1973, "User's Guide for the Climatological Dispersion Model," EPA-RA-73-024, US Environmental Protection Agency, Research Triangle Park, NC, Dec.

Cabe, D.B., Hooper, M.W., and Schmidt, A.E., 1979, "Influence of Alternative Definitions of Exempt Fugitive Dust Sources on the Impact of PSD Regulations on Surface Coal Mines," 72nd Annual Meeting of the Air Pollution Control Association, Cincinnati, Ohio, June.

Cole, C.F., et al., 1985, "Quantification of Uncertainties in EPA's Fugitive Emissions and Modeling Methodology at Surface Coal Mines," TRC 2784-V12, TRC Environmental Consultants, Inc. Englewood, CO, February.

Collins, C.A., 1979, "Fugitive Dust Emission Factors Memorandum," Wyoming Air Quality Division, Cheyenne, Jan. 24.

Curreri, J.A., and Heinold, D.W., 1982, "Estimating Emissions and Dispersion Modeling of Fugitive Dust from Coal Handling and Storage Operations: A Review," Conference, Midwest Section of the Air Pollution Control Association, Kansas City, MO, April 27–30.

Currier, E.L., and Neal, B.D., 1979, "Fugitive Emission from Coal-Fired Power Plants," Conference, 72nd Annual Meeting of the Air Pollution Control Association, June 24–29, Cincinnati.

Dailey, B., 1979, Private communication.

Danielson, J.A., 1973, "Air Pollution Control Equipment for Particulate Matter," *Air Pollution Engineering Manual*, AP-40, US Environmental Protection Agency, Research Triangle Park, NC, pp. 91–168.

Evans, J.S., and Cooper, D.W., 1980, "An Inventory of Particulate Emissions from Open Sources," *Journal of the Air Pollution Control Association*, Vol. 30, No. 12, Dec., pp. 1298–1303.

Hawkins, D.G., 1978, "Environmental Protection Agency Memorandum on Meeting Ambient Standards, PSD Increments, over Company Property," *Environment Reporter Current Developments*, Jan., pp. 1564–1565.

Jutze, G., and Axetell, K., Jr., 1974, "Investigation of Fugitive Dust, Vol. I—Sources, Emissions, and Control," Contract No. 68-02-0044, US Environmental Protection Agency, Cincinnati.

Jutze, G.A., et al., 1977, "Technical Guidance for Control of Industrial Process Fugitive Particulate Emissions," PB-272288, US Environmental Protection Agency, Research Triangle Park, NC, Mar.

Larson, A.G., 1981, Private communication.

Raffle, B.I., 1979, "Prevention of Significant Deterioration and Nonattainment Under the Clean Air Act—A Comprehensive Review," Environment Reporter Monograph No. 27, Bureau of National Affairs, Inc., Washington, DC, May 4.

Sterk, L., 1980, "Current Available Fugitive Particulate Emission Factors Memorandum," Illinois Environmental Protection Agency, Springfield, IL, Apr. 28.

Turner, D.B., 1970, "Estimates of Atmospheric Dispersion," *Workbook of Atmospheric Dispersion Estimates*, US Environmental Protection Agency, Research Triangle Park, NC, pp. 5–6.

Wurmbrand, M.M., 1981, "Testimony Presented in Behalf of Western Energy Company," US Environmental Protection Agency Conference on Air Quality Modeling, Aug. 11.

6.8 Open Pit Rock Mechanics

RICHARD D. CALL

JAMES P. SAVELY

INTRODUCTION

Historically, the application of rock mechanics to surface mining has primarily been the analysis of the stability of slopes and was considered more theoretical than applied. In the first edition of *Surface Mining* (Pfleider, 1968), slope stability was included in the section on research and development. In the intervening years, slope design has become an integral part of mine planning rather than a research curiosity. This has come about with the development of improved data collection techniques (Call, et al., 1976), computer-aided stability analysis (Ross-Brown, 1979), and a reliability cost-benefit interfacing with mine planning (Kim, 1977). Since 1968, two comprehensive reference works on slope design have been published (Hoek and Bray, 1981; CANMET, 1976).

In addition to slope design, there are a number of other applications of rock mechanics to surface mining that utilize a common database. Table 1 lists the applications and the database. For purposes of this section, soil mechanics and rock mechanics are not differentiated. The underlying mechanics of stress strain and strength are the same, regardless of the classification of the material, as are the basic principles of determining the physical properties of a material and analyzing its response to imposed stress. Therefore, when reference is made to *rock,* it may be geologically a soil.

It has been the experience of the authors, both as mine staff geotechnical engineers and as consultants, that ground control difficulties more often have been the result of inadequacy in organization and implementation of a rock mechanics program than the lack of, or failure of, the technology to solve the problem. Therefore, in this section, the authors have placed the emphasis on what to do, when to do it, and how to integrate rock mechanics with other aspects of mining rather than on the theoretical aspects of mathematical analysis.

An effective rock mechanics program requires careful organization and planning. Data collection must precede analysis so that problems are anticipated; otherwise, there may be insufficient time to collect data and important information may be lost. For example, displacement and water level time trends cannot be measured retroactively, bench faces may become covered or inaccessible, and drill core ground up for assay. On the other hand, there is never the time, manpower, or budget to measure everything; so the analytical approach must be kept in mind during data collection to ensure the appropriate information is collected with the resources available.

Interpretation and analysis should be kept current with data collection; otherwise, data collection can become an end in itself. File cabinets full of raw data may give the appearance of productivity but do not, in and of themselves, result in an optimum mine plan.

In the subsequent section, the emphasis is on data collection first to provide the background for analysis and designs. Geologic data collection and presentation are discussed in more detail than other aspects of rock mechanics because of their importance and the number of times we have found the geologic database inadequate.

Design Approach

Mine design and operational decisions are primarily cost-benefit optimizations. The objective is to extract the mineral

Table 1. Rock Mechanics Applications and Data Requirements for Surface Mining

	Data										
	Geology			Material Properties		Site Conditions			Operational Factors		
Application	Structural Domains	Major Structure	Fabric	Substance	Fracture	Hydrology	Stress Field	Seismicity	Mining Plans	Equipment	Production Rates and Costs	Current Slope Geometry and Displacement
Slope design	1	1	1	1	1	1	2	1	1	3	2	2
Slope management		1				1			1	2	2	1
Diggability	1	2	1	1	1	2	3			1	2	
Blasting	1	2	1	1	2	2	2			1	1	
Trafficability	1			1		1			2	1		
Bearing capacity	1			1	1	1				1		
Crushing and grinding	1		1	1						1		
Leaching	1	2	1	2	2	1			2			

reserve at the lowest cost, or to accept or reject a mining option on the basis of benefits being greater or less than the cost. In this context, the role of rock mechanics is to predict the behavior of the rock in response to mining in a manner that costs and benefits can be assigned.

The prediction of rock behavior is by no means straightforward. To make a rational analysis, a conceptual model must be developed that is mathematically tractable and cost effective. The complexity of natural materials and processes precludes an exact modeling; thus, an analysis is only an approximation of the real world.

Even with the simplification of modeling, the present analytical capability exceeds the ability to obtain the prerequisite input data on material properties, geology, and site conditions for the following reasons:

1) Material properties vary from point to point and access is limited, so obtaining representative samples is difficult.

2) There are uncertainties in field measurements and testing.

3) The magnitude and time of occurrence of the phenomena that affect rock behavior, such as rainstorms or earthquakes, are governed by such a complex interrelationship of factors that it approaches a chance event.

This uncertainty in the input data and analysis precludes an exact prediction of rock behavior. For this reason, the probabilistic approach has gained widespread acceptance in open pit rock mechanics. By appropriate sampling and testing strategies, the distribution of rock properties can be quantitatively estimated. These distributions can be used in explicit mathematical models or in Monte Carlo simulations so that the results of an analysis can be computed as a probability distribution rather than a deterministic single value based on an average or assumed single input value. For example, the stability of a slope can be expressed as the probability of failure, which is more useful for an economic risk analysis than a safety factor.

GEOLOGY

Applied geology emphasizes an overall view of general geology of the mine, which includes the distribution of rock units and the spatial relationships of major structures. For specific rock mechanics studies, additional information on major structure and rock fabric is needed. Rock fabric is the orientations and characteristics of minor structure, such as joint sets and foliation.

Major structures are treated as individuals in slope design because their location and extent are known. Bench face mapping and geologic interpretation are used to define major structure. Rock fabric, however, is treated statistically in design because a single orientation measurement and property determination on a joint may or may not represent those of the joint set. The statistics of populations where orientations and characteristics are described, not by a single value but by a distribution of values, are needed because of the recognized variability in joint properties within the joint set. Cell mapping, set mapping, and detail line mapping are used to obtain data for statistical analyses.

Bench Face Mapping

The bench face mapping technique is described by Peters (1978). This is the basic method for pit mapping whereby major structures, such as faults and contacts (traceable for at least two benches), are described by the points where they intersect the toe and crest of the bench. True strike directions are seen only where structures cross bench levels. A map showing several bench levels should be carried in the mapping

to tie geology through to successive benches (Fig. 1). In addition to plotting the major structure, comments should also be made on rock type, alteration, mineralization, rock hardness, and fracture frequency.

An appropriate mapping scale is the key to successful pit mapping. Mapping should provide sufficient information for interpretation, and it must be provided in time to be useful. Thus, a mapping scale should be selected that will allow mapping to keep pace with the mining. Extensive detail in one small area is less important than providing general geologic knowledge for the entire mine. Ideally, mapping is kept current within 100 m (400 ft) of the mining advance. The following guidelines can be used for selecting an appropriate mapping scale:

Pit Diameter	Mapping Scale
<500 m (<1500 ft)	1:400 (1:600)
500-1500m (1500-5000 ft)	1:1000 (1:1200)
>1500 m (>5000 ft)	1:2000 (1:2400)

Data from field sheets should be posted onto mylars the same day the mapping is done. These mylars should contain only factual information, not interpretation.

Geology Interpretation and Presentation

Interpretations of geology should be made on section mylar or sepia copies from the factual mylars. Keeping factual information separate from the interpretations makes reinterpretation at a later date possible with a minimum of effort.

The overall geologic picture is provided by a complete set of interpretive geology level maps and cross sections at a scale that shows the complete view of the mine plan. Generally, these maps are at the same scale as that used for mine planning. If geologic information is merely collected and not interpreted, the geology will not be used effectively in ore reserve estimation, ore control, mine planning, or slope design. Mine geologists must extend interpretations, based on intuition if necessary, for a distance of one and a half pit depths beyond the final pit limit. Otherwise, mine planners and engineers must make the necessary geologic assumptions usually based on less knowledge than the geologists have. One assumption that can be made in error is that no useful geologic information is available.

From the basic level maps and cross sections, geology is easily transferred to plans and the current pit composite. Geology maps are made for every planned mining level. Cross sections are made to show views of geology in the vertical plane on intervals determined by the ability to project geologic information. Overlays are generally made on selected maps to show alteration, mineralization, grade, or other features of interest.

Level maps and cross sections contain the following basic information: (1) pre-mine topography, (2) current topography, (3) original surface geology, (4) drill holes that show geology and grade, (5) old underground workings, and (6) mining push backs and final pit limit. The process of developing geology plan maps is illustrated in Fig. 2. Cross-section maps are developed in a similar manner.

Rock Fabric Data Collection

To develop an understanding of minor geologic features, such as joint sets or foliation, mapping must provide a systematic and consistent measurement. This implies the use of coding forms suitable for data entry and compatible with computer processing (Fig. 3). Coding provides the desirable aspect of a checklist to ensure that all needed information

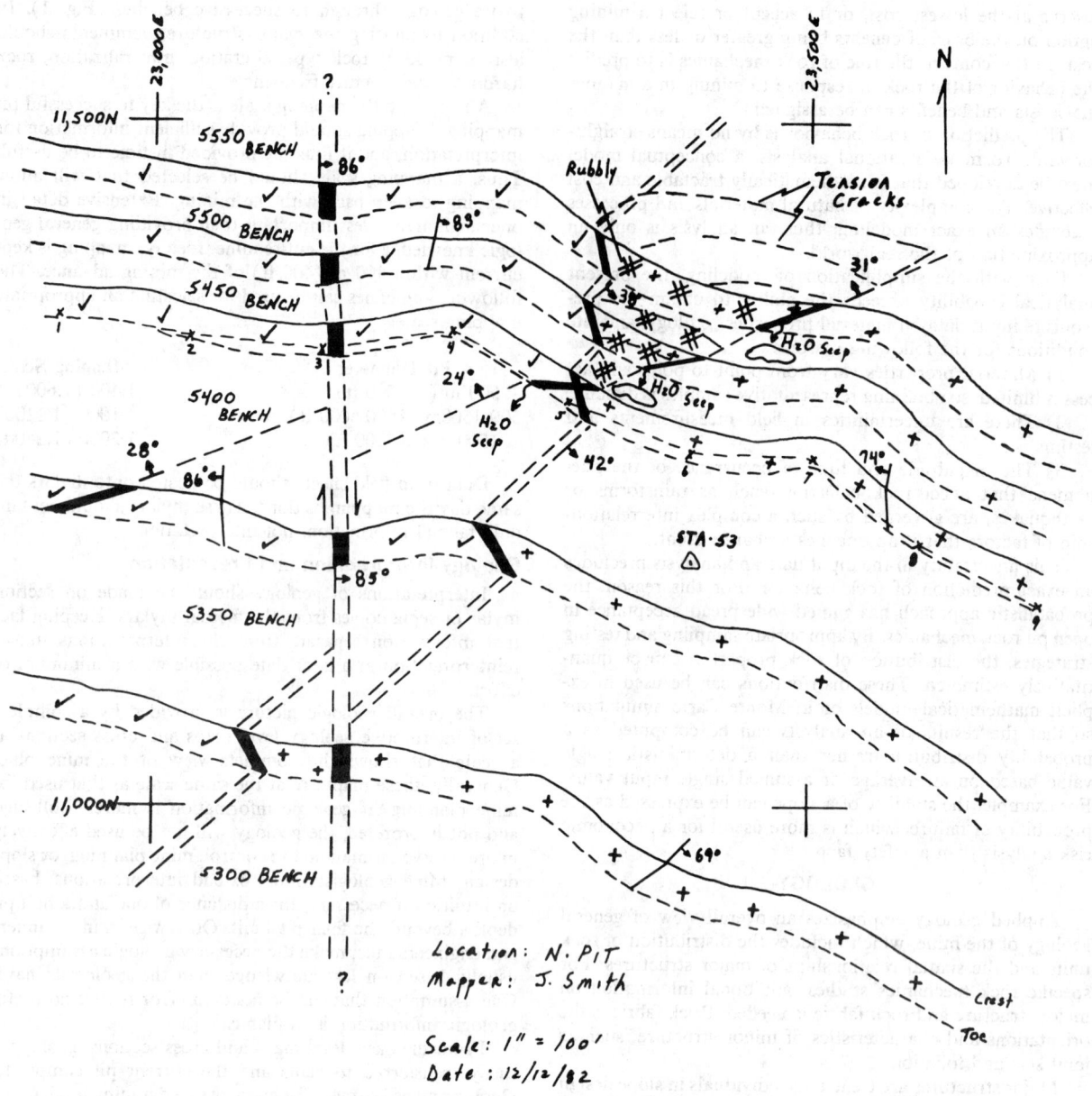

Fig. 1. Bench face mapping field sheet.

is recorded. Minor geologic features are analyzed statistically, which requires that an unbiased sample of the true distribution of values be obtained. All numbers should be recorded exactly, and not rounded, to provide continuous rather than discrete information.

There are three methods for obtaining rock fabric data: (1) cell mapping (Call, et al., 1976); (2) set mapping (Call, et al., 1976); and (3) detail line mapping (Piteau, 1970; Call, et al., 1976). Each method has a specific purpose, and each has advantages and disadvantages.

Cell Mapping: This method is used when there are large, extensive exposures of rock, such as along benches in an open pit or on large natural outcrops. Consecutive mapping cells are established along the strike of the exposure. Orientations are characteristics of the most significant structures are recorded. This mapping provides a continuous measurement

that gives an estimate of the recurrence of structure orientations and, thus, a probability of occurrence. Experience has shown 30 to 40 cells are needed in each structural domain to describe the statistical distributions needed in slope design.

Cell mapping is subjective because it relies on the observer's judgment as to which fracture sets are significant; therefore, an observer bias is built into the data. The advantages to this method are that fracture characteristics and their frequency of occurrence can be determined over large areas in relatively short time.

Set Mapping: This method is used in place of cell mapping when rock exposures are not suitable for establishing consecutive cells, or for reconnaissance-type mapping. It provides information on fracture orientations and characteristics, but not quantitative information on recurrence over a large area. It is a fast mapping method, but like cell mapping, the observer makes a judgment as to the significance of each fracture set, thus introducing an observer bias to the data.

Detail Line Mapping: This method has the least observer bias of the three methods for obtaining statistical information on the fracturing, but it also provides the least area coverage. Generally, this method is used to determine rock fabric when there is no previous knowledge of the fracture patterns or distribution forms of the fracture characteristics. This method is a spot sampling technique where sampling sites are chosen along the bench face or other available rock exposure.

At each mapping site, a measuring tape, which serves as a reference line, is stretched along the rock exposure. The orientation and characteristics of every fracture or its projection, which intersects the line, are recorded. Usually, an arbitrary length cutoff of about 0.3 m (1 ft) is used and fractures with lengths less than the cutoff are not recorded.

Major disadvantages to this method are that it is tedious and time-consuming, and only a small area is covered. Unless supplemental cell mapping or set mapping is done, significant fracture sets can be completely missed.

Rock Fabric Data Processing

Data from the coding forms are entered into the computer from which Schmidt plots (lower hemisphere, equal area projections) are produced (Billings, 1954; Hoek and Bray, 1970). These plots are used in determining structural domains and potential failure modes (Fig. 4).

Conventional methods for developing histograms and determining distribution forms are used to describe the fracture orientations and characteristics in preparation for stability analysis and slope design.

Structural Domains

A structural domain is an area usually bounded by a major structure, such as faults or contacts, where orientation patterns of the fractures and their characteristics can be considered similar. An example of structural domain definition is given in Fig. 4, and more discussions can be found in Chap. 2 of the *CANMET Pit Slope Manual* (1976).

Oriented Core

Oriented core is often used to supplement surface mapping data and to collect samples of natural fractures in areas behind proposed pit walls and at depth. Orientation of the entire core from the drill hole is not necessary for fracture studies. All that is needed is a statistical sampling of the fracture attitudes. For this reason, some of the simpler and less expensive core orientation methods, such as the clay imprinter are recommended (Call, et al., 1982). More sophisticated methods of core orientation are available from Christensen in Salt Lake City.

ROCK MECHANICS PROPERTIES AND ROCK MASS STRENGTH

It is important to differentiate between the rock substance which is intact rock, and the rock mass, which includes both intact rock and rock fabric.

Strength refers to the "maximum stress that a body can withstand without failing by rupture or continuous deformation." Application in analysis and design determines the loading conditions which define the strength value of interest.

In open pit slope design, compressive strength of the substance is important as a classification criteria. Also, block flow stability analysis, where crushing of the rock is consid-

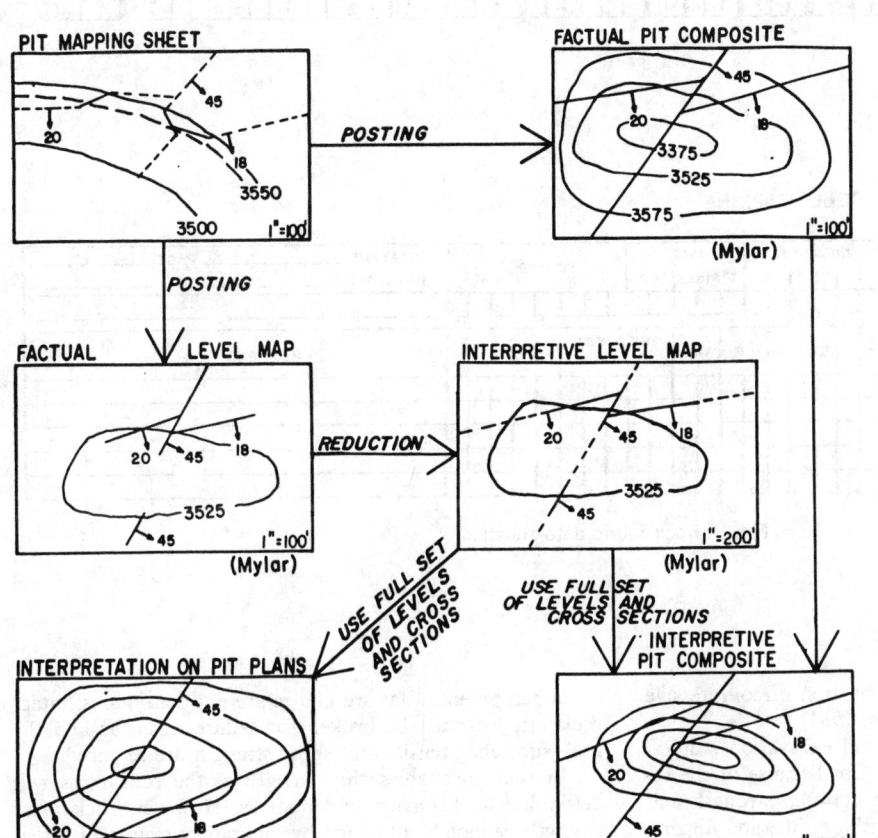

Fig. 2. Development of geologic maps.

DATA SHEET FOR STRUCTURE MAPPING

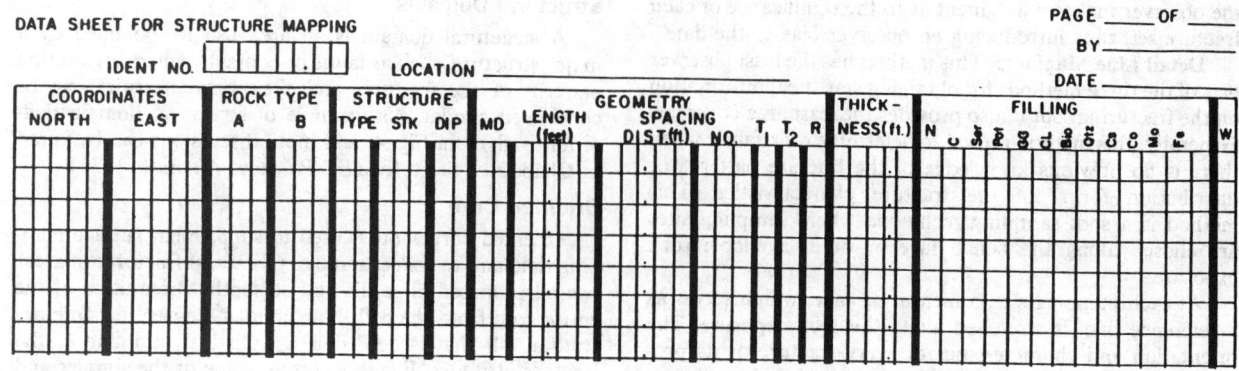

DATA SHEET FOR DETAIL LINE MAPPING

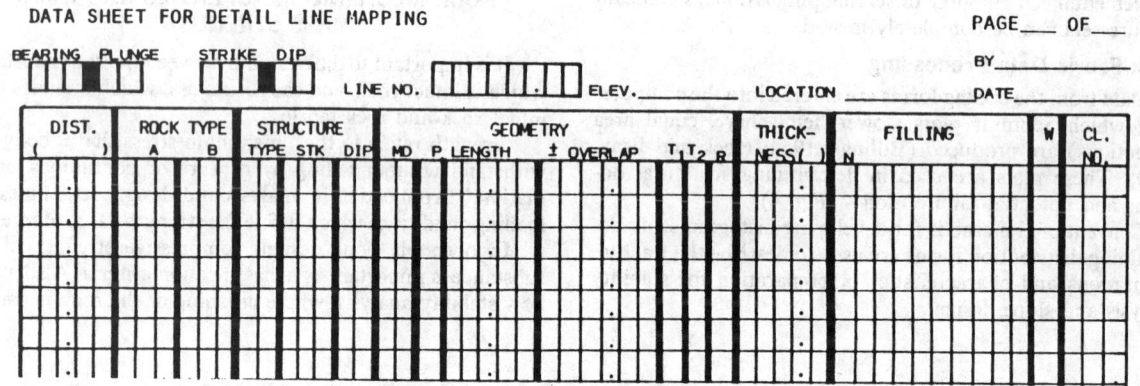

DATA SHEET FOR CELL MAPPING

Fig. 3. Rock fabric data sheets.

ered, uses the mean and standard deviation of compressive strength as input parameters (Coates, 1981).

In crushing, drill performance, and excavation studies, compressive strength becomes important because of the relationships that might be developed between strength and energy requirements, abrasiveness, drilling rate, and digging or ripping characteristics.

When potential failure planes are not continuous, intact rock bridges must be broken for failure to occur and the rock substance tensile and shear strength are required.

In rock mechanics the strength of the rock mass will determine its behavior under stress. However, rock mass strength cannot be obtained by laboratory testing or direct measurement; it must be inferred from the measurable components of the rock mass: rock substance strength, fracture strength, rock fabric, major structure, and block size. Determining rock fabric and major structure characteristics were discussed previously. The other parameters can be determined from laboratory tests, field methods, and estimation methods (Table 2).

Laboratory Tests

Uniaxial and triaxial compression tests are conducted in the laboratory on cylindrical samples, usually of drill core, to provide rock substance strength. Deformational properties, Poisson's ratio, and modulus of deformation are also determined from the uniaxial test.

Poisson's Ratio (μ) is the ratio of lateral strain (ϵ_{lat}) to longitudinal strain (ϵ_{long}) under normal uniaxial stress (σ_n) and is a measure of the directional variability in the deformability of the rock substance.

Modulus of Deformation (E) is a measure of the stiffness of the rock substance under normal uniaxial stress. It is the ratio of uniaxial stress to longitudinal strain.

Tensile Strength

Of the two most common tests for determining tensile strength, indirect tension (Brazilian) and direct tension, the Brazilian test is the least expensive, easiest, and most commonly used.

A Brazilian test consists of diametrically loading a disk of rock core until it fails. Theoretically, the diametrical loading induces a tensile stress in the center of the disk, and failure occurs parallel to the direction of loading.

The direct tension test pulls a cylindrical rock sample at both ends until the specimen fails.

Direct Shear

The direct shear test consists of taking two blocks of rock that are separated by a natural fracture, applying a load perpendicular to the fracture, then measuring the shear load required to displace the blocks relative to each other. Samples of fault gouge or low strength rock, where shear failure of the rock substance is expected, are tested in a similar manner by shearing through a single block or core of the intact

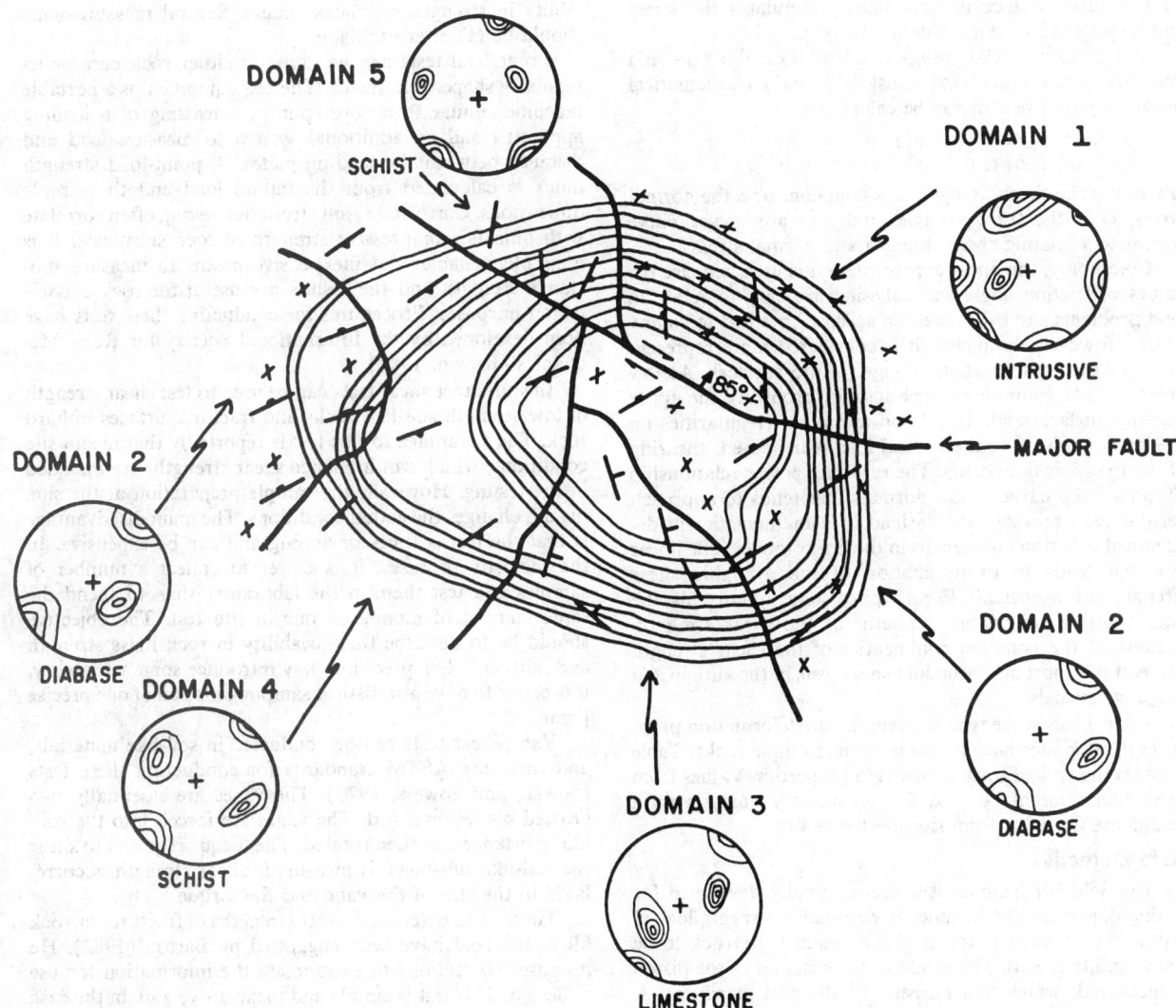

Fig. 4. Structural domains and rock fabric plots.

Table 2.   Methods of Determining Characteristics of the Components of the Rock Mass

Rock Mass Component	Laboratory Test	Field Method	Estimation Method
Substance compressive strength	Uniaxial, Triaxial	point load test, Schmidt rebound hammer	manual index tests
Substance tensile strength	Indirect (Brazilian), Direct	none	tensile strength = uniaxial compressive strength ÷ 10
Substance shear strength	Direct	in-situ direct, vane	shear strength = uniaxial compressive strength ÷ 2
Fracture shear strength	Direct	in-situ direct, tilt	Barton's classfication, back analysis of failure, geometries
Rock fabric	N/A	mapping	none
Major structure	N/A	mapping	none
Block size	N/A	RQD, core logging, screen tests, inspection of muck piles and dumps, volumetric joint count, block size index, photoanalysis	simulation of fracture geometry

material. Although the shear strength of intact rock can be obtained from triaxial tests, the direct shear test is preferred by the authors since it more closely simulates the stress conditions used in slope stability analysis.

The resulting relationship between normal stress and shear stress can be analyzed statistically, and a mathematical linear or power *best* fit can be calculated:

$$\tau = C + \sigma_n \tan\phi \quad \text{(linear)}$$
$$\tau = k\, \sigma_n^m \quad \text{(power)}$$

where τ is the shear strength, C is cohesion, σ_n is the normal stress, ϕ is the friction angle, and k, m are power curve parameters relating shear strength and normal stress.

Commonly, the linear approximation is used because the values of friction angle and cohesion are easy to relate to field problems and it has been an accepted method for many years. However, the power fit is considered more representative of the shear strength along fracture surfaces. At low normal loads, common to slope stability problems, the upper fracture surface tends to ride up and over irregularities on the lower surface. As the normal load is increased, shearing of the irregularities occurs. The resulting power relationship shows a steep curve at low normals that tends to approach zero at zero normal load, instead of a mathematically determined cohesion intercept as in the linear model. The linear intercept leads to overestimation of the available shear strength at low normals. For a limited range of normals, the linear fit is a reasonable estimator of the shear strength. Because of the potential nonlinearity of the shear strength curve, it is important to conduct shear tests in the anticipated range of normals.

Table 3 lists some typical strength and deformation properties of rock substance for some common mine rocks. Table 4 lists some typical fracture strength properties. Values from these tables might be used for preliminary analysis until results are available from site-specific testing.

Field Methods

The Schmidt hammer is a tool originally developed for testing concrete. The hammer is essentially a spring-loaded piston. The cocked piston is placed against the rock to be tested and triggered. The height of the rebound of the piston is measured, which is a measure of the rock hardness. A correction is made to standardize results based on the orientation of the hammer during testing. Schmidt hardness is

related to the compressive strength of the rock substance (Brown, 1981). The test is unreliable and considerable variablity in strength estimates occurs. Several measurements should be taken at each site.

Point load tests can be done on either rock core or irregularly shaped specimens. The test equipment is a portable machine similar to a core splitter, consisting of a loading apparatus and an additional system to measure load and distance between the loading plates. A point-load strength index is calculated from the failure load and the sample dimensions. Corrected results from this testing often correlate with uniaxial compressive strength of rock substance. It is a simple, reliable, and inexpensive means to measure substance strength and the results are useful for rock classification purposes. Procedures for conducting these tests have been developed by the International Society for Rock Mechanics (Brown, 1981).

In situ direct shear tests can be used to test shear strength of low strength soil-like rocks and fracture surfaces in hard rock. The advantage to this test is reportedly that in situ site conditions, which can influence shear strength, are included in the testing. However, the sample preparation at the site, in fact, changes the actual conditions. The main disadvantage is that the test is time-consuming and can be expensive. In the majority of cases, it is better to collect a number of samples and test them in the laboratory than to spend the same amount of money on one in situ test. The objective should be to describe the variability in rock mass strength and, although less precision may introduce some variability, it is better to have a statistical sampling instead of one precise point.

Vane shear tests can be conducted in soil-like materials, and there are ASTM standards for conducting these tests (Sowers and Sowers, 1970). The vanes are essentially two crossed blades on a rod. The vanes are forced into the soil-like substance and then rotated. The torque required to shear the soil-like substance is measured. Shear strength is correlated to the size of the vane and the torque.

Tilt tests to determine shear strengths of fractures or rock fill in the field have been suggested by Barton (1982). He proposes corrections to extrapolate the information for use in design. This test is simple and inexpensive and, in the case of fractures, simply involves a measure of the size of the blocks tested and measuring the tilt angle at which one block

Table 3. Rock Substance Properties

Material	Uniaxial Compressive Strength, psi*			Brazilian Disk Tensile Strength, psi						Cohesion, psi			Density, pfc
	Typical	High	Low	Typical	High	Low	Typical	High	Low	Typical	High	Low	Typical
Igneous Intrusive													
Fresh granite	30,000	50,000	20,000	2000	5000	1000	55	65	45	3000	4500	1500	167
Altered granite (porphyry copper)	12,000	20,000	6,000										165
Quartz monzonite	14,000	20,000	8,000	1200	2000	700	48	60	36	3200	4800	1600	164
Quartz diorite	27,500	40,000	16,500	3150	4750	1550							175
Diabase	40,000	57,000	20,000	3000	5200	800	60						185
Igneous Extrusive													
Rhyolite	21,500			1050									147
Dacite	19,000	31,500	6,500	800									162
Andesite	24,000	39,000	9,000	1050			46			2800			162
Basalt	20,000	50,000	14,000	1900	3000	900	49			5400			167
Welded tuff	10,000	18,000	3,000	300			36			900			
Metamorphic													
Gneiss (foliated)	19,500	35,000	10,500	1500	2100	900	50	70	45	3200	4700	1700	170
Schist (∥ to foliation)	6,800	8,300	5,300										172
Schist (⊥ to foliation)	13,500	18,000	9,300	1000	1200	800							172
Quartzite	35,000	55,000	15,000	2300	5000	1000	52	65	45	4300			170
Dolomite	25,000	45,500	4,500	1900	3400	400	50	60	40	1000			175
Slate	26,000	33,000	19,000	2450			45	60	40	4000	8000	2000	170
Sedimentary													
Siltstone (mid-US)	900	1,800	500	50	90	30							131
Sandstone (cemented)	12,000	32,000	8,000	900	1800	0	44	55	33	1600	3200	0	125
Limestone	16,000	26,000	8,000	1100	1800	50	46	57	35	2500	3700	1200	156
Clay shale (mid-US)	25,000	8,000	500	300	500	50	44	50	35	1150	1700	50	144
Conglomerate (southwest)	1,200	2,000	400	300	500	0	35	40	20	250	400	100	138
Miscellaneous													
Coal (subbituminous, lignite)	2,500	4,500	500	300	500	100	47	65	30	40	60	20	81
Fault gouge (clay with rock)													
Fault gouge (clay)													
Fault breccia													
Broken rock													120
Gravels (well graded)													120
Sand and silt													125
Clay													100

Metric equivalents: psi × 6.894 757 = kPa; pfc × 16.018 46 = kg/m³.

of rock slides on another. For rock-fill materials, Barton proposes the construction of a tilt box to contain the rock fill, but the test procedure is the same as for fractures.

Rock Mass Characteristics

Block size is an important characteristic of the rock mass for crushing and grinding, leaching, excavation, drilling, and blasting, as well as the effect it has on rock mass strength.

Block size index is one measure of block size (Brown, 1981). To determine the index, a bench face is selected that appears consistent in fracturing and rock type. Typical maximum, minimum, and mode block sizes are measured, and the number of different fracture sets bounding the measured block are noted. Information on the number of fracture sets

forming the block is used to adjust block volume calculations. Size distribution curves might be estimated from this data for each structural domain.

The volumetric joint count is the sum of the number of joints per meter for each joint set. A bench face is selected as in the block size index determination. For each joint set, average true spacings of the joints in each set are calculated from the number of joints in the set occurring over a specified distance measured normal to the joint set. The volumetric joint count is the sum of the number of joints per unit length for all sets. The following is an example.

Set 1: 6 joints in 20 m (65.6 ft)
Set 2: 2 joints in 10 m (32.8 ft)

Table 4. Rock Fracture Properties

Material	Peak Friction Angle			Peak Cohesion, psi			Residual Friction Angle			Residual Cohesion, psi		
	Typical	High	Low	Typical	High	Low	Typical	High	Low	Typical	High	Low
Igneous intrusive												
Fresh granite	34	40	30	24	40	7	30	35	28			
Altered granite (porphyry copper)							30					
Quartz monzonite	31	35	25	20	40	0	27	30	25	1.5	3	0
Quartz diorite	36						23	27	20			
Diabase	44											
Igneous extrusive												
Rhyolite												
Dacite												
Andesite	40	45	35	120	250	50	37	45	25			
Basalt	26						35	38	31			
Welded tuff	61	65	55	90	150	25	50	60	40	28		
Metamorphic												
Gneiss (foliated)	30	35	25				32	35	30			
Schist (∥ to foliation)												
Schist (⊥ to foliation)												
Quartzite	34						28	32	26			
Dolomite							31	36	26			
Slate							28	30	25			
Sedimentary												
Siltstone (mid-US)	40	50	30	50	65	40	27	34	18	5	15	0
Sandstone (cemented)	33	43	23	20	30	0	33	43	23	2	26	0
Limestone	37	47	27	50	100	0	37	47	27			
Clay shale (mid-US)	20	32	14	2.5	7	1	15	18	8	0.3	0.5	0
Conglomerate (southwest)	39	45	30	50	90	25	32	35	29	15	25	0
Miscellaneous												
Coal (Subbituminous, lignite)	26			12			12					
Fault gouge (clay with rock)	21	27	15	21	35	10	19	25	8			
Fault gouge (clay)												
Fault breccia												
Broken rock	37	41	34				37	41	34			
Gravels (well graded)	40	44	36				40	44	36			
Sand and silt	35	41	30				35	41	30			
Clay	21	30	10	10	20	5	14	16	11			

*Metric equivalent: psi $\times$ 6.894 757 = kPa.

Set 3: 20 joints in 10 m (32.8 ft)
Set 4: 20 joints in 5 m (16.4 ft)
Vol. __ Count = 6/20 + 2/10 + 20/10 + 20/5
       = 0.30 + 0.20 + 2.00 + 4.00
       = 6.5 joints/m^3

Block shape, number of joint sets, and joint lengths should be recorded to give a better description of the block size.

Screen tests could also be devised for size gradation determinations by constructing several size grizzlies and coarse screens. These are time-consuming tests and equipment construction can be expensive. Screening can be done only on blasted or loose rock and representative sampling is difficult.

Inspection of muck piles and dumps also gives an indication of block size and shape. The procedure is similar to the block size index where maximum, minimum, and mode block sizes are measured.

Usually, blasting tends to open up existing fractures and actual rock breakage is small. Therefore, the block sizes measured from blasted rock and dumps are often a reasonable estimate of in situ block size.

Rock quality designation (RQD) and core logging also provide information on the degree of fracturing, thus, the block size. In some instances, screening of the entire drill core will give a size distribution curve that is reasonably accurate for the lower range of block sizes. RQD, in the strictest sense, is a measure of all pieces of core greater than 10 cm (4.0 in.) in length expressed as a percentage of the total length in the drill run, and could be considered as a measure of +10 cm (+4.0 in.) rock blocks.

Estimation Methods

The Unified Soil Classification, which includes both quantitative and descriptive information, is a classic manual index test to determine characteristics of soils (Lambe and Whitman, 1969). Other soil indices can be found in Sowers and Sowers (1970). In geology and rock mechanics work, the rock hardness index proposed by Jennings and Robertson (1969) and Piteau (1970) are used (Table 5). Kirsten (1982) gives a field identification procedure for estimating compressive strength and vane shear strength.

Rule of thumb criteria for estimating strength has developed over the years. One is for tensile strength of rock. Tensile strength is one-tenth to one-twentieth the compressive strength. One-tenth is generally the closer estimate. A similar rule might be used for shear strength of intact rock substance. Maximum shear strength will not exceed half the uniaxial compressive strength.

Barton (1982) has proposed a classification scheme based on the roughness of joints and joint wall hardness. This classification is used to estimate peak shear strength. Caution in using this method is warranted because analysis of many slopes should not be done using peak shear strength values. Residual strengths may be a better estimate of actual conditions.

If slopes are available where failed geometries are present, back analysis can be done to determine strength values required for stability by assuming the failures are at limiting equilibrium. In many cases, these results are the "best estimates" for shear strength because they have the loading, geometry, and shear strength relationships of actual field conditions. However, one significant parameter that often cannot be reconstructed for the back analysis is the water condition at the time of failure. The water condition can be very critical to the stability calculation, and the assumption used will affect the results of the back analysis.

It should be noted that stable slopes are also useful indicators of strength because they give a lower bound, just as failed slopes give the upper bound of shear strength. If enough slopes in both failed and unfailed conditions can be observed, the actual strength values can be bracketed, and the ability to accurately estimate shear strength improves. McMahon (1976) reports on his study and use of back analysis in slope design.

Simulation to predict block size is becoming more common. The usual procedure is to randomly sample joint orientations and spacings and to make some assumption regarding joint lengths to model the fracturing. Numerical methods are generally used to calculate the size and number of blocks. Conceivably, this modeling could be used to develop size distribution curves for the rock blocks. There are, however, still major difficulties to overcome regarding appropriate length values before this method can be applied by people other than specialists. Additional work is being done using key block theory to describe rock block shapes and volumes (Goodman and Shi, 1985).

Table 5. Rock Hardness Index

Grade	Description	Field identification	Approximate range of uniaxial compressive strength, MPa
S1	Very soft clay	Easily penetrated several inches by fist	<0.025
S2	Soft clay	Easily penetrated several inches by thumb	0.025–0.05
S3	Firm clay	Can be penetrated several inches by thumb with moderate effort	0.05–0.10
S4	Stiff clay	Readily indented by thumb but penetrated only with great effort	0.10–0.25
S5	Very stiff clay	Readily indented by thumbnail	0.25–0.50
S6	Hard clay	Indented with difficulty by thumbnail	>0.50
R0	Extremely weak rock	Indented by thumbnail	0.25–1.0
R1	Very weak rock	Crumbles under firm blows with point of geological hammer, can be peeled by a pocket knife	1.0–5.0
R2	Weak rock	Can be peeled by a pocket knife with difficulty, shallow indentations made by firm blow with point of geological hammer	5.0–25
R3	Medium strong rock	Cannot be scraped or peeled with a pocket knife, specimen can be fractured with single firm blow of geological hammer	25–50
R4	Strong rock	Specimen requires more than one blow of geological hammer to fracture it	50–100
R5	Very strong rock	Specimen requires many blows of geological hammer to fracture it	100–250
R6	Extremely strong rock	Specimen can only be chipped with geological hammer	>250

Table 6.   Attributes Used in Some Developed Classification Schemes

Classification Methods	Rock Substance Strength	Drill Core Quality, RQD	Joint Spacing	Joint Orientation	Joint Strength	Rock Genesis	Ground-Water Conditions
CSIR rock mass rating (Bieniawski, 1974)	X	X	X	X			X
NGI tunneling index (Barton, et al., 1974)	X	X	X	X	X		X
Coates, 1981	X		X			X	
Deere, 1968	X		X			X	
Müller and Hofmann, 1970	X		X				
Zavodni and McCarter, 1977	X	X	X				X
GSL rock quality (Franklin, et al., 1971)	X		X			X	
Kirsten, 1982	X		X	X	X		

Plate tests and radial jacking tests can be used to estimate the deformation characteristics of the rock mass.

ROCK MASS CLASSIFICATION

Rock mass classification schemes are generally not sufficient design criteria for slopes. Their applicability has more merit for excavating, crushing, and leaching. Franklin, et al. (1971) discussed classification schemes based on fracture spacing and compressive strength of the rock substance to describe the rock mass. They then related it to the excavation characteristics of digging, ripping, and blasting.

Kirsten (1982) adapts classification to excavation and calculates an excavability index which relates to the equipment needed. He has developed his classification for both soil-like and hard rock applications.

A useful classification scheme provides comparisons of rock mass properties within a mine as well as a comparison between different mines. The classification should indicate some behavioral characteristic in a quantifiable manner. Then the classification is used to predict rock mass behavior in an area where the classification system is identifiable, which allows better planning and design before mining begins.

Available Classification Schemes

Classification schemes have been developed for a variety of purposes. Some mines may be able to adopt one of these classifications or modify it slightly to fit their needs. Others would have to develop their own classification based on the attributes of interest. Table 6 presents a list of some of the better known classifications and the attributes upon which they are based. Table 7 presents current classification schemes and their applications.

Developing a Specific Classification Scheme

Classification schemes are based on time, space, physical properties, and relationship between properties. An example of a time-related attribute would be the seasonal fluctuation in ground-water level or the time-dependent movement of a pit slope. Space-related attributes are most common and would include the variations in such attributes as rock hardness or fracture frequency over mine areas, which in themselves are examples of physical properties of the rock mass. Relationships between properties would be exemplified by RQD, a property of the rock mass that is dependent upon both rock hardness and fracture frequency (Deere, 1968). Coates (1981) gives further discussion on developing meaningful classifications.

The most effective method for developing a useful classification is to decide on the attributes of interest and to develop a set of overlay maps displaying the areal distribution. Varnes (1974) is an excellent reference on the method of attribute selection and map development.

SLOPE DESIGN

Slope design involves analysis of the three major components of a mine slope: bench configuration, interramp angle, and overall slope angle (Fig. 5). Bench configuration is

Table 7.   Information Provided by Some Developed Classification Schemes

Classification Method	Ground Stability	Block Size Distribution	Bearing Capacity	Diggability	Blasting Requirements	Crushing/Grinding Requirements
CSIR rock mass rating (Bieniawski, 1974)	X					
NGI tunneling index (Barton, et al., 1974)	X					
Coates, 1981	X					
Deere, 1968	X					
Müller and Hofmann, 1970	X	X	X			
Zavodni and McCarter, 1977	X					
GSL rock quality (Franklin, et al., 1971)	X		X	X	X	
Kirsten, 1982		X		X	X	

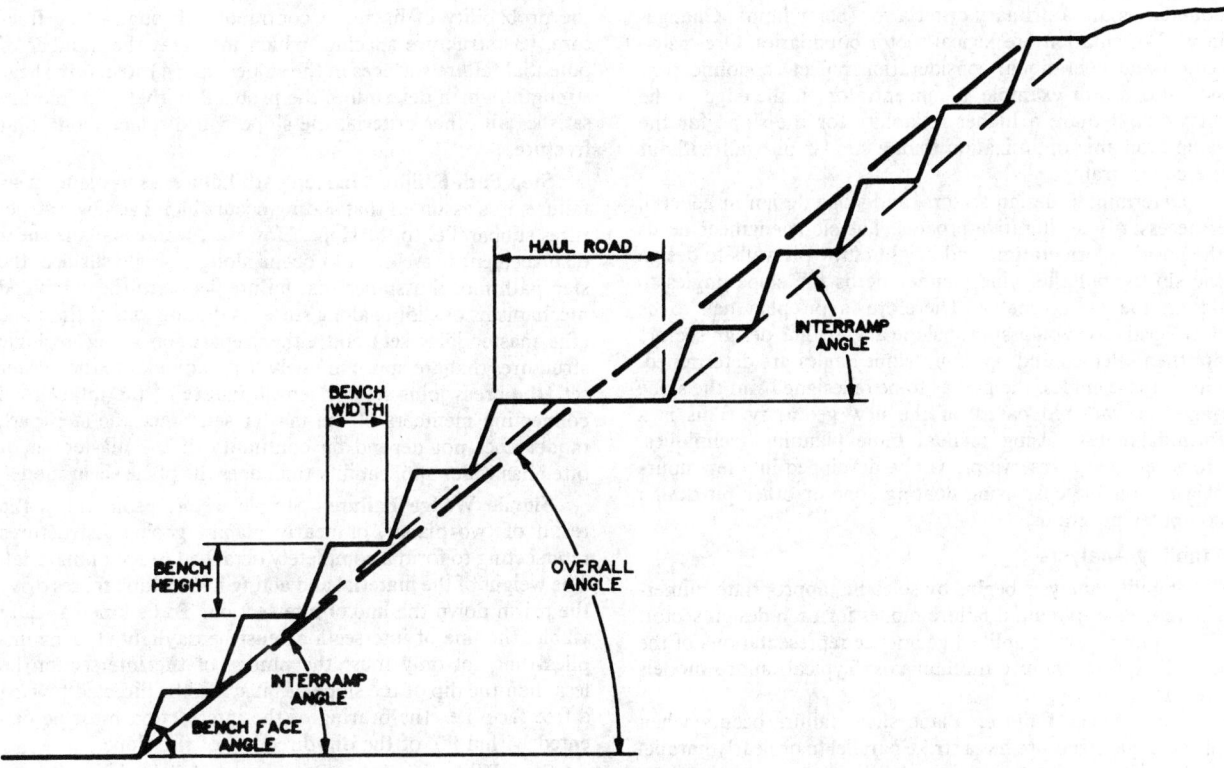

Fig. 5. Definition of bench face, interramp, and overall angles.

defined by bench height, width, and face angle; the interramp angle is defined by the bench configuration; and the overall slope angle is defined by interramp sections separated by haul roads or mining levels. If through-going structures do not produce the possibility of large-scale failure and joints lengths are short, all slope angles will depend on the bench configuration.

Fig. 6 is a flow chart of the slope design process from data collection to mine design. After data collection, the steps in the design process are:

1) Determine design sectors.
2) Conduct stability analysis to estimate probability of failure and expected failure tonnages for bench interramp and overall slopes.

3) Develop maximum interramp slopes based on catch bench criteria.
4) Determine optimum slopes with a cost benefit analysis.

Design Sectors

Design slope angles within an open pit are influenced by rock strength, geologic structure, hydrologic conditions, pit wall orientation, pit wall height, ore distribution, and operational conditions.

Since any or all of these parameters vary from place to place in an open pit, the pit must be divided into design sectors within which these parameters are similar or will have a similar impact on slope design. Structural domain

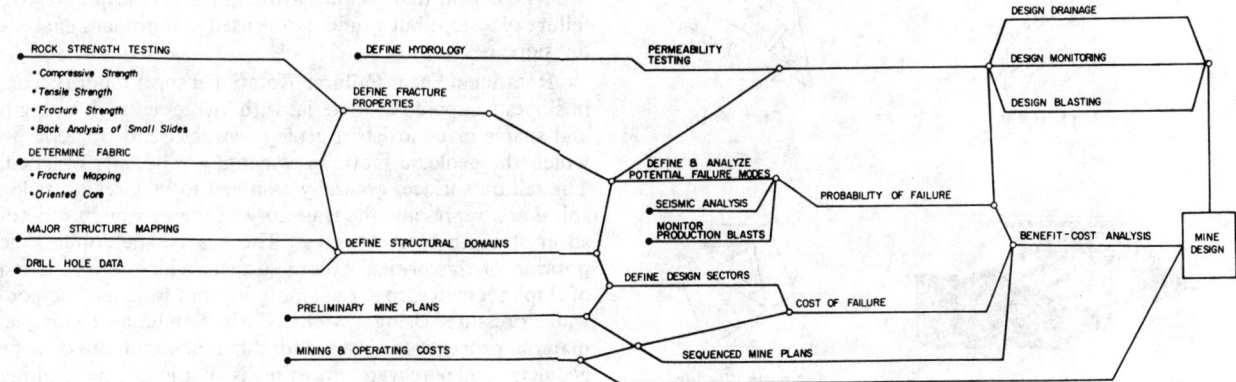

Fig. 6. Slope design flow chart.

boundaries are a primary criteria for sector limits. Changes in wall orientation are logical sector boundaries. Ore distribution and operational considerations affect economic considerations. For example, a concentrator on the edge of the pit would require a higher reliability for the slope for the same economic optimization than a similar pit wall without the concentrator.

Determining design sectors, and slope design in general, is necessarily an iterative process. The slope engineer needs the position, orientation, and height of the pit walls to design the slopes, but the mine planner needs the slope angles to design the pit geometry. Therefore, a pit plan has to be developed based on assumed slope angles. The design sectors are then selected and optimum slope angles are determined. Given these angles, the pit has to be redesigned and the slope angles reevaluated based on the new geometry. This is a formidable task using manual mine planning techniques. However, a new reserve pit can be developed in a few hours at a reasonable cost using floating cone or other pit design computer programs.

Stability Analysis

Stability analysis begins by selecting appropriate numerical models of potential failure modes for each design sector. These models are simplified geometric representations of the actual expected failure mechanisms. Typical failure models are shown in Fig. 7.

Plane Shear Failure: Plane shear failure occurs when the geologic structure has a strike parallel to or nearly parallel to the strike of the slope face and a dip flatter than the slope angle. Plane shear analysis determines the risk of sliding along structures of this type. Controlling factors in the analysis are (1) orientations of geologic structures, which determine whether the fractures have strikes (within about 20°) to the strike of the face and are daylighted (dip angles less than slope angle); (2) structure lengths, which determine

the probability of having a continuous through-going fracture; (3) structure spacing, which indicates the number of potential failure surfaces in the slope; and (4) structure shear strength, which determines the probability that, if a fracture satisfies all other criteria, the slope will displace along that fracture.

Step Path Failure: In step path failure, as in plane shear failure, it is assumed that sliding occurs along geologic structures subparallel to the slope. However, whereas plane shear displacement is assumed to occur along a single surface, the step path model assumes that failure is due to the combined mechanisms of sliding along surfaces dipping out of the slope (the master joint set) and either separation along geologic structures that are approximately perpendicular to the master set (the cross joint set) or tensile failure of the intact rock connecting members of the master set. Since the step path model does not depend on continuity of the master set, it often has wider applicability than does the plane shear model.

Simple Wedge Failure: Simple wedge geometry is the result of two planar, or nearly planar, geologic structures intersecting to form a completely detached prism of material. The weight of the material and acting hydrostatic forces drive the prism down the line of intersection. To be kinematically viable, the line of intersection must be daylighted. This implies that not only must the plunge of the intersection be less than the dip of the slope, it must also be directed toward a free face; i.e., the bearing of the intersection must be oriented within 90° of the dip direction of the slope.

Step Wedge Failure: Step wedge failure is similar to simple wedge, but in this case the structures that intersect to form the wedge do not need to be single, continuous features. Rather, as with step path, the combination of different structural sets forms the failure surfaces. A variation on this mode is a wedge formed on one side by a single planar surface and on the other by a step path geometry. There is a lack of sufficiently developed analysis for this failure mode, and simplifications are necessary if analysis is required.

Topping Failure: Some authors have proposed this failure mode as a primary failure mechanism (Goodman, 1980; Hoek and Bray, 1981; Brown, 1981). The topping failure mechanism relies on the development of thin slabs of rock that dip away from the slope face. In order to be a viable failure mode, the weight of the slabs must be directed outside their bases. Unless the slabs are very thin, sliding or crushing at the toe must occur before topping is initiated, and the analysis should concentrate on this toe area as the primary failure mechanism. Topping then would be a secondary failure mechanism that would have implications in progressive failure of a slope but would not be used as a primary analysis for slope design.

Rotational Shear Failure: Rotational shear failure occurs in slopes composed of material with low intact rock strength and sparse or nonexistent geologic structure, or material in which the geologic fabric is essentially randomly oriented. The failure surface, generally assumed to be circular or log spiral arc, represents the trajectory of the minimum ratio of shear strength to shear stress. The analysis determines the position of this critical failure surface, which is a function of slope geometry, material strengths, unit weights, and pore water pressure. Using a Monte Carlo simulation technique, material properties are varied for different conditions of slope geometry and pore water pressure distributions. The resulting distribution of the ratio of shear strength to shear stress provides an estimate of the probability of failure.

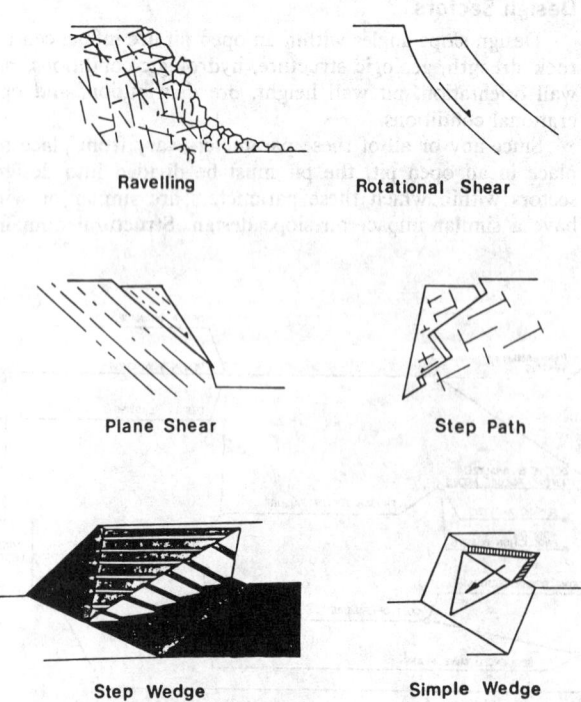

Fig. 7. Typical failure modes.

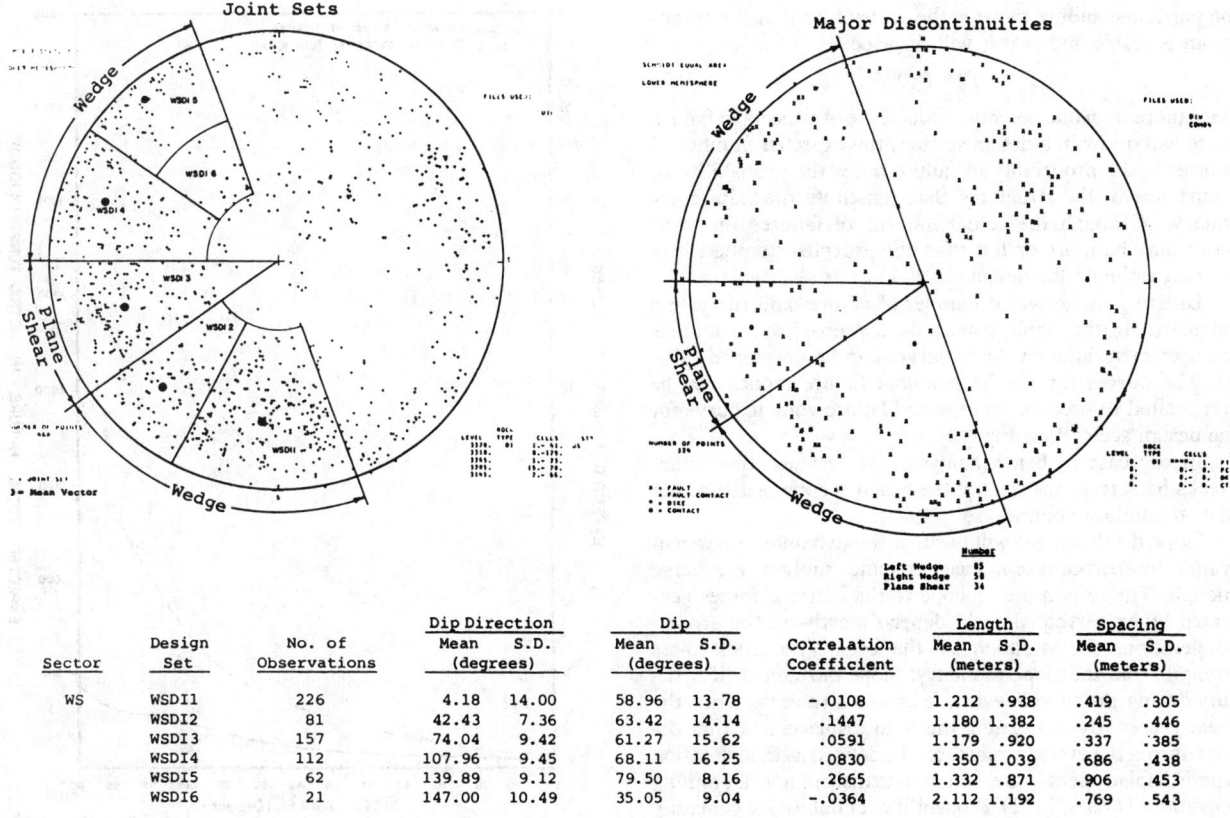

Sector	Design Set	No. of Observations	Dip Direction Mean (degrees)	Dip Direction S.D.	Dip Mean (degrees)	Dip S.D.	Correlation Coefficient	Length Mean (meters)	Length S.D.	Spacing Mean (meters)	Spacing S.D.
WS	WSDI1	226	4.18	14.00	58.96	13.78	-.0108	1.212	.938	.419	.305
	WSDI2	81	42.43	7.36	63.42	14.14	.1447	1.180	1.382	.245	.446
	WSDI3	157	74.04	9.42	61.27	15.62	.0591	1.179	.920	.337	.385
	WSDI4	112	107.96	9.45	68.11	16.25	.0830	1.350	1.039	.686	.438
	WSDI5	62	139.89	9.12	79.50	8.16	.2665	1.332	.871	.906	.453
	WSDI6	21	147.00	10.49	35.05	9.04	-.0364	2.112	1.192	.769	.543

Fig. 8. Design set determination.

Block Flow

For deep pits or where the rock substance strength is low, the stresses in the wall, particularly in the toe area, can exceed the compressive strength of the rock substance. This can lead to crushing and progressive deterioration of the slope. Coates (1981) presents a simplified analysis to check for the potential of block flow. If the simplified analysis indicates that a potential, more detailed investigation is warranted, use a finite element analysis to estimate the stress distribution and a comprehensive triaxial strength testing program conducted to estimate the distribution of rock strength.

Determination of Potential Failure Geometry

By plotting the pit wall orientation of a design sector on Schmidt plots of the rock fabric and major structures, the impact for stability analysis can be developed (Fig. 8). The fractures and major structures are sorted by the failure-type orientations and the attitude, distribution length, and spacing distributions computed. These design sets may not correspond to geologic sets although the orientation boundaries can be adjusted somewhat to avoid splitting a geologic set. We have found that defining sets by visual or mathematical analysis, while appropriate for geologic fabric analysis, is less satisfactory for slope design, and it is best to use the wall orientation for determining design sets.

Probability of Instability

Determination of the probability of slope instability depends on the ability to quantify the variable character of the geologic parameters. Since physical geologic conditions, such

as discontinuity lengths, orientations, spacings, and shear strength vary within the rock mass, statistical distributions are used to represent these geologic parameters. Estimation of these statistical distributions generally requires a representative statistical sample, which often consists of many observations.

In the structurally controlled failure models, the probability of failure (P_f) for a single occurrence of a specified failure mode has three parts:

1) the probability that the dip exists (P_d);

2) the probability that the structure is long enough (P_l); and

3) the probability of sliding (P_s).

The probability of dip (P_d) and the probability of length (P_l) are calculated from the statistical distributions of the geologic structures.

The probability of sliding (P_s) is determined by calculating the probability that the shear stress exceeds the shear strength along the failure surface. This probability is calculated from the distribution of safety factors generated either by a technique called Monte Carlo simulation, which involves an iterative process of randomly sampling strength values from the shear strength distribution and subsequently calculating the safety factor, or by the application of closed form mathematical modeling.

Using the calculated mean and standard deviation of the distribution of safety factors and assuming a standard normal distribution, the probability of sliding, or the percentage of the total area of the distribution less than 1.0, can be calculated.

The probability of failure (P_f) for a single occurrence of the particular failure mode is the probability that the mechanism is viable and that it will displace.

$$P_f = P_1 * P_d * P_s.$$

Since more than one potential occurrence of a specified failure mode can occur in a design section, the expected number of failures is the probability of failure times the probability of occurrence of the structures that constitute the failure geometry. Although the actual number of failures that will occur may be more or less than the expected number, it is the best estimate for design.

Utilizing the expected number of failures and the values calculated in the stability analysis, a probability of failures and expected failure volume curves can be developed (Fig. 9). The curves for all the potential failure modes can be composited to produce an expected failure volume curve for the design sector (see Fig. 10).

In the case of bench analysis, the distance the failure breaks back from the crest of the bench is composited rather than the failure volume.

Slope displacement will occur if the dynamic forces generated by earthquake-induced ground motion are large enough. The response of a slope to the external forces generated by an earthquake will depend mostly on the ground acceleration, the duration of the event, the rock mass strength, and the slope geometry. Slope movement, if it occurs during the seismic event, is assumed to cease when the event ceases. By calculating the total displacement that occurs during the event, a failure can be defined as that situation where displacement is great enough to disrupt normal mining operations (Glass, 1982). Probabilities of failure are generally increased, but often not significantly, when earthquake forces are included.

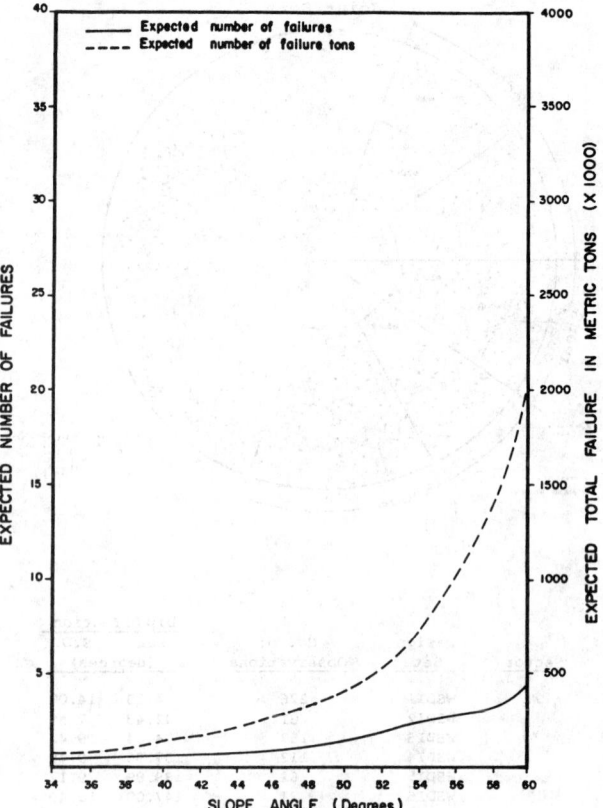

Fig. 10. Sector E, composite failure mode multibench, number of failures and failure tonnage.

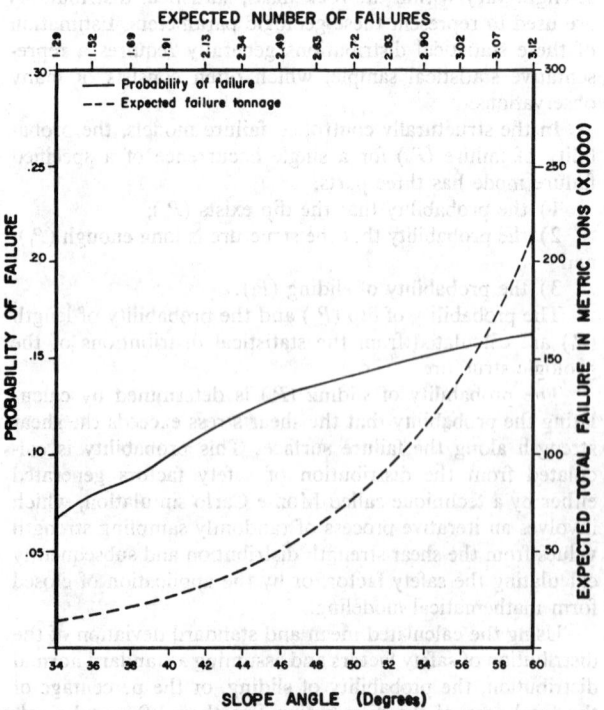

Fig. 9. Composited slope heights for multiple bench wedge analysis, wedge HL2R1, sector H.

Bench Design

Bench faces are normally mined as steeply as possible; as a result, rock falls and raveling are inevitable. Thus, it is customary, and in many cases mandated by mining regulations, that catch benches are left in the pit wall to retain rock falls and raveling.

Analyses of rock fall mechanics by Ritchie (1963) demonstrated that falling rocks impact relatively close to the toe of the slope, but because of horizontal momentum and spin, can roll considerable distances from the toe. Based on his analysis, Ritchie developed width and depth criteria for a ditch at the toe of a slope to protect highways from rock fall. The concept was that the rock would impact in the ditch and the side of the ditch would stop the horizontal roll.

It is not practical to excavate a ditch in an open pit catch bench, but the same effect can be achieved by casting up a berm (Fig. 11). Assuming the berm can be emplaced with slopes of 1.25 to 1, the modification of Ritchie's criteria presented in Fig. 11 is recommended for open pit catch benches. For a given bench height and corresponding design bench width, the upper limit of the interramp slope angle becomes a function of the bench face angle.

The bench face angle, however, is not a unique value because the variability of the rock fabric produces varying amounts of backbreak. Backbreak is defined as the distance from the design bench crest to the actual bench crest. Fig. 12 is an example of the cumulative frequency distribution of measured bench face angles and theoretical bench face angles. The theoretical bench face angle is obtained from stability analyses assuming a vertical bench face and is the upper limit

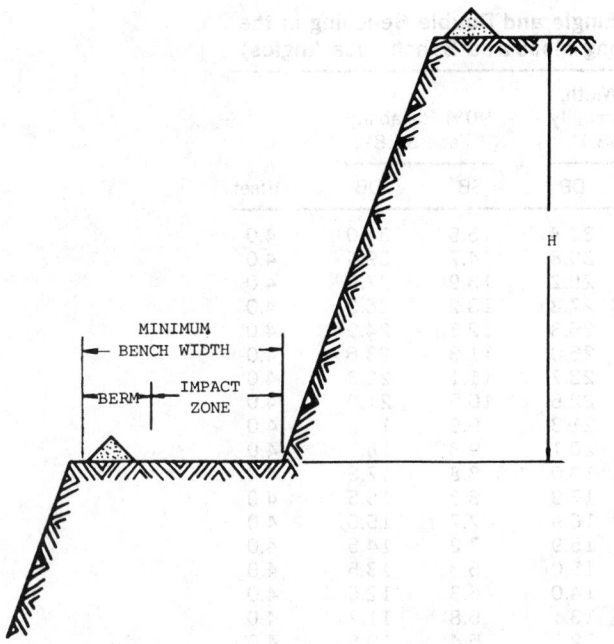

Fig. 11. Design catch bench geometry. Minimum bench width = 4.5m + 0.2H; berm height = 1m + 0.04H.

Bench Height, m	Impact Zone, m	Berm Height, m	Berm Width, m	Bench Width, m
7.5	1.5	0.8	2	3.5
15	5	1	3	8
30	7	1	3	10

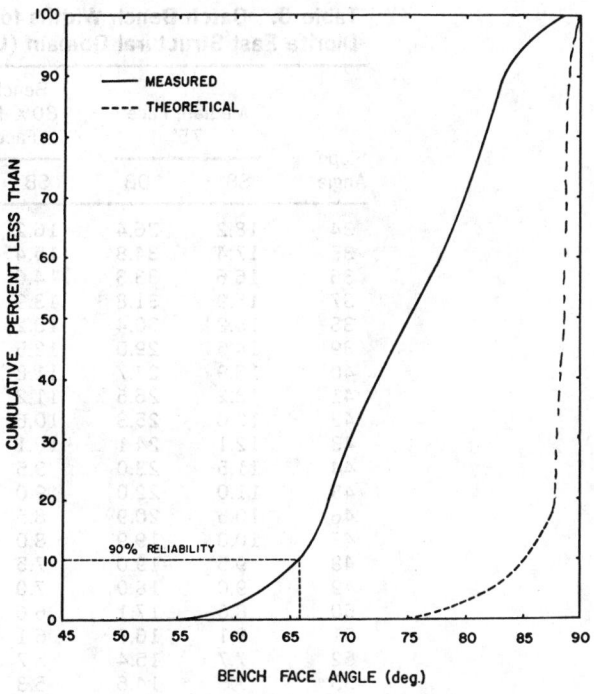

Fig. 12. Bench face angle distribution.

of possible bench face angles because it does not include the effect of blasting and digging. Comparison of measured and theoretical face angles at several properties gave a difference of 19° to 25°, except where the bench face was controlled by a strong geologic structure, such as bedding or foliation. In those cases, the measured and theoretical bench face angles were the same.

For an operating property, the measured bench face angles can be used for design. For a new property, the theoretical bench face angle adjusted for the effects of blasting must be used. Rather than choosing the mean bench face angle which would result in 50% of the catch benches less than the design width, or the minimum bench face angle which would result in unnecessarily flat slope angles, it is recommended that a desired catch bench reliability be chosen based on the potential for rock fall and the exposure of personnel. Catch benches in raveling ground being mined by front-end loaders should have a higher reliability than catch benches in massive ground mined with a large rope shovel. A bench face angle should then be chosen to give the desired reliability. For example, if a 90% reliability is desired, the bench face angle would be the angle where 90% of the bench faces will be steeper than the design angle shown in Fig. 12. Using the reliability criteria, the average bench width will be greater than the design width. In the sample shown in Table 8, a 52° interramp slope with a 90% reliability for a 10-m (32.8-ft) bench width has an average bench width of 15.4 m (50.5 ft).

Production bench heights are selected on the basis of equipment size and grade control requirements. There is considerable benefit in increasing the bench height on final

pit walls by leaving a catch bench every other mining level (double benching). A 5° to 8° increase in interramp slope may be achieved by double benching, assuming the same bench face angle. There is also the possibility of increasing the bench face angle with double benching because the majority of backbreak occurs as small failures of the crest. These small failures have less effect on the face angle of a high bench.

Controlled Blasting

The objective of production blasting is to produce the fragmentation required for excavation; however, the blast damage resulting from uncontrolled production blasting at the final wall reduces the bench face angles and, hence, the interramp slope angle. Therefore, controlled blasting in the vicinity of the final wall is desirable.

Measurement of blast damage indicates that a peak particle velocity of about 63.5 cm/s (25 ips) produces displacement on existing fractures and creation of new fractures. By monitoring blasts, the relationship between scaled distance and peak particle velocity can be established. Fig. 13 shows the results of such monitoring. Since the relationship is site-specific, it is preferable to monitor blasts at the property. The curves in Fig. 13 should be used only as a first estimate.

This scaled distance relationship can be transformed to a curve showing the maximum charge weight vs. distance from the face for 63.5 cm/s (25 ips) peak particle velocity at the face (Fig. 14). At distances where the maximum charge is greater than the charge per hole, blasting can be controlled by limiting the number of holes per delay. At closer ranges, the charge per hole must be reduced and the spacing adjusted on the holes *picked*.

The position of the blastholes relative to the catch benches should also be considered. By strategically placing the holes, damage to the catch benches can be reduced (Fig. 15). If blastholes are placed directly over the underlying bench crest, the subgrade damages the crest and reduces the bench width.

Table 8. Catch Bench Widths for Single and Double Benching in the Diorite East Structural Domain (Using Measured Bench Face Angles)

Slope Angle	Median, Face 75°		Bench Width, 80% Reliability Face 68.1°		90% Reliability Face 65.8°		Offset
	SB	DB	SB	DB	SB	DB	
34	18.2	36.4	16.2	32.4	15.5	31.0	4.0
35	17.4	34.8	15.4	30.8	14.7	29.4	4.0
36	16.6	33.3	14.6	29.2	13.9	27.8	4.0
37	15.9	31.8	13.9	27.8	13.2	26.3	4.0
38	15.2	30.4	13.2	26.3	12.5	24.9	4.0
39	14.5	29.0	12.5	25.0	11.8	23.6	4.0
40	13.9	27.7	11.8	23.7	11.1	22.3	4.0
41	13.2	26.5	11.2	22.5	10.5	21.0	4.0
42	12.6	25.3	10.6	21.3	9.9	19.8	4.0
43	12.1	24.1	10.1	20.1	9.3	18.7	4.0
44	11.5	23.0	9.5	19.0	8.8	17.6	4.0
45	11.0	22.0	9.0	17.9	8.3	16.5	4.0
46	10.5	20.9	8.5	16.9	7.7	15.5	4.0
47	10.0	19.9	8.0	15.9	7.2	14.5	4.0
48	9.5	19.0	7.5	15.0	6.8	13.5	4.0
49	9.0	18.0	7.0	14.0	6.3	12.6	4.0
50	8.6	17.1	6.6	13.1	5.8	11.7	4.0
51	8.1	16.3	6.1	12.2	5.4	10.8	4.0
52	7.7	15.4	5.7	11.4	5.0	10.0	4.0
53	7.3	14.6	5.3	10.5	4.6	9.1	4.0
54	6.9	13.8	4.9	9.7	4.2	8.3	4.0
55	6.5	13.0	4.5	8.9	3.8	7.5	4.0
56	6.1	12.2	4.1	8.2	3.4	6.8	4.0
57	5.7	11.4	3.7	7.4	3.0	6.0	4.0
58	5.4	10.7	3.3	6.7	2.6	5.3	4.0
59	5.0	10.0	3.0	6.0	2.3	4.5	4.0
60	4.6	9.3	2.6	5.3	1.9	3.8	4.0
61	4.3	8.6	2.3	4.6	1.6	3.1	4.0
62	4.0	7.9	1.9	3.9	1.2	2.5	4.0
63	3.6	7.2	1.6	3.2	0.9	1.8	4.0
64	3.3	6.6	1.3	2.6	0.6	1.1	4.0

Note: Single bench height = 15.0 m.

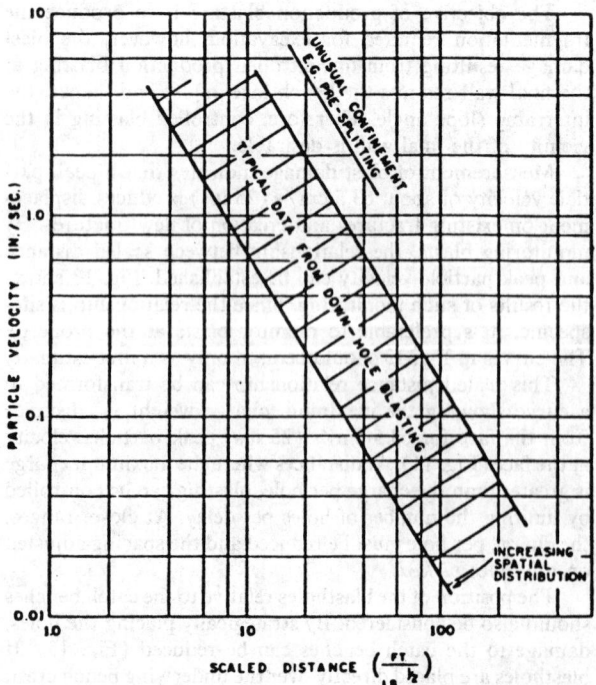

Fig. 13. Ground response to blasting (from Oriard, 1971).

Cost-Benefit Analysis

The simplest method for determining the optimum slope is to compute the incremental cost of slope failure and the incremental benefit per degree of slope increase (Fig. 16). If the incremental benefit is greater than the incremental cost, it is profitable to increase the slope angle. As the incremental cost exceeds the incremental benefit, it becomes unprofitable to increase the slope angle. In general, incremental benefit decreases while incremental cost increases with slope angle. The angle where the two curves cross is the economic optimum, as shown in the example in Fig. 16.

The cost of slope failure is determined by assigning cost models to expected failure volumes and possible mining responses. The benefit is the market value of the recoverable commodity minus the mining and processing costs. Kim (1977) discussed cost models and cost benefit analysis in more detail.

A more sophisticated analysis is to run a Monte Carlo simulation of the sequenced mine plans, applying the probability of failure schedule to include slope failure costs in the cash flow analysis. This way, the effect of interim slope failure and the time cost of money can be included. This type of analysis was developed for the CANMET *Pit Slope Manual* (Kim, 1977).

The cost-benefit approach, using probabilities of failure, provides a methodology by which the risks and costs of failure can be compared with the corresponding benefits for

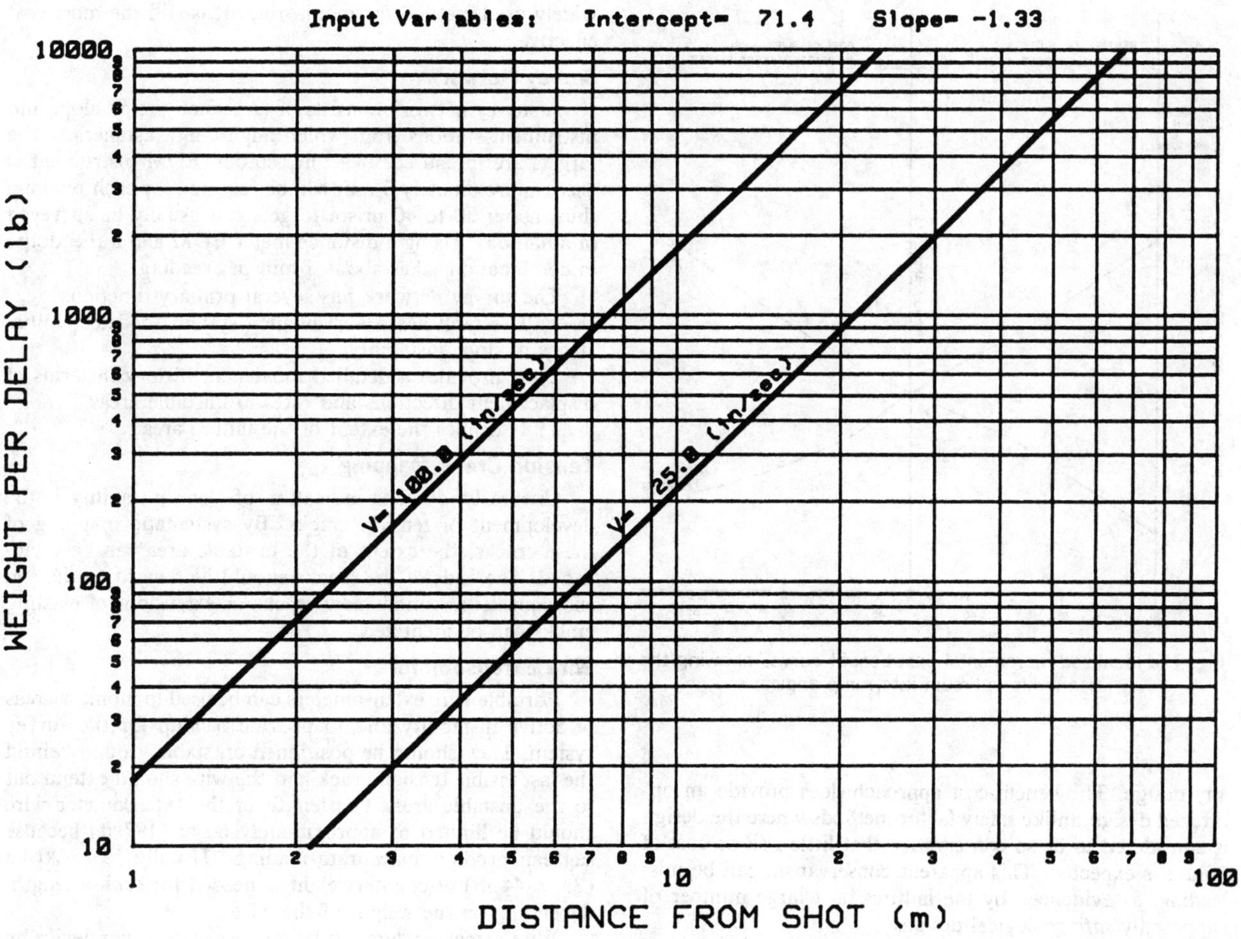

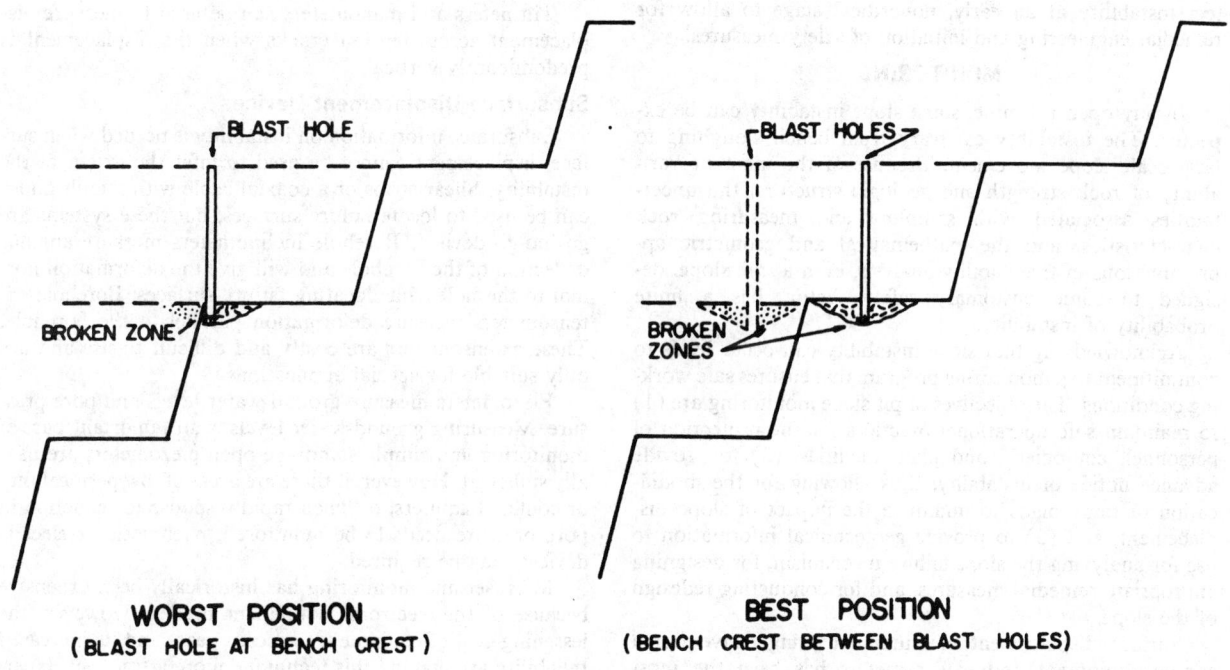

Input Variables: Intercept= 71.4   Slope= -1.33

Fig. 14. Controlled blasting criteria (diorite).

Fig. 15. Blasthole layout.

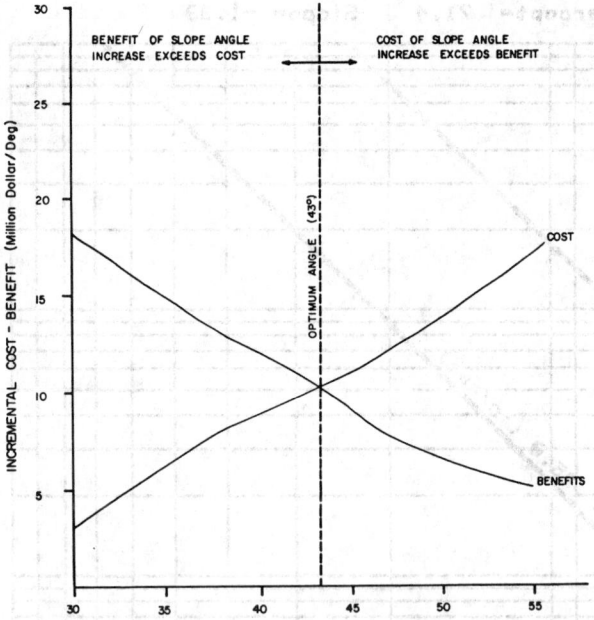

Fig. 16. Idealized incremental cost-benefit curves showing the economic optimum interramp angle.

any design. The benefit-cost approach does provide an optimized design unlike safety factor methods where the design is considered to be so conservative that little risk or cost of failure is expected. This apparent conservatism can be misleading, as evidenced by the failures of a large number of supposedly *safe* geological designs.

The fact that a slope is designed with some risk of failure must not be viewed as a disregard for safety concerns. Almost any economically viable option will have some probability of failure, and it is better to be aware of the level of that risk. Sufficient, suitable monitoring must be provided to detect instability at an early, noncritical stage to allow for remedial engineering and initiation of safety measures.

MONITORING

In any open pit mine, some slope instability can be expected. The instability can vary from bench sloughing to large-scale slope movement. Because of the inherent variability of rock strength and geologic structure, the uncertainties associated with sampling and measuring rock characteristics, and the mathematical and geometric approximations of the stability analysis, even a *safe* slope, designed to some customary safety factor, has a finite probability of instability.

Acknowledging that slope instability can occur leads to commitment to a monitoring program that ensures safe working conditions. The objectives of pit slope monitoring are (1) to maintain safe operational practices for the protection of personnel, equipment, and plant facilities; (2) to provide advance notice of instability, thus allowing for the modification of mine plans to minimize the impact of slope displacement; and (3) to provide geotechnical information to use for analyzing the slope failure mechanism, for designing appropriate remedial measures, and for conducting redesign of the slope.

Surface displacement measurement using conventional survey equipment and extensometers has been the most

widely used method for monitoring. It is still the most cost effective.

Survey Network

A survey network consists of targets on the pit slope and instrument stations from which angles and distances to the targets are measured. If a total station EDM instrument is used, approximately 3 min will be required for each reading; thus, about 30 to 40 prism targets can usually be surveyed in a half day. Using a distance meter EDM and a theodolite in combination takes about 5 min per reading.

The survey network has several primary functions.

1) It establishes a surveillance system to detect initial stages of slope instability.

2) It provides a detailed movement history in terms of displacement directions and rates in unstable areas.

3) It defines the extent of the failure area.

Tension Crack Mapping

One early, obvious indication of slope instability is the development of tension cracks. By systematic mapping of these cracks, the extent of the unstable area can be established. The ends of the cracks should be flagged so that on subsequent inspections new cracks or extensions of existing cracks can be identified.

Wire Extensometers

Portable wire extensometers can be used to monitor areas of active instability and to provide backup for the survey system. They should be positioned on stable ground behind the last visible tension crack and the wire should extend out to the unstable area. The length of the extensometer wire should be limited to approximately 60 m (197 ft) because sag can produce inaccurate readings. Usually 15 to 20 kg (33 to 44 lb) of counterweight is needed for such a length, depending on the weight of the wire.

Wire extensometers can be set up as a warning device by affixing a switch several centimeters (inches) above the counterweight. Significant displacement will trip the switch and activate a warning light or siren.

Other Surface Displacement Devices

Tiltmeters and manometers can be used to measure displacement across tension cracks when the displacement is predominantly vertical.

Subsurface Displacement Devices

Subsurface information on instability is needed when surface displacement cannot be used to infer the extent of the instability. Shear strips or a coaxial cable with a fault finder can be used to locate failure surfaces, but these systems are go/no-go devices. Borehole inclinometers measure angular deflection of the borehole and will give the deformation normal to the hole, thus locating failure surfaces. Borehole extensometers measure deformation parallel to the borehole. These extensometers are costly and difficult to use and are only suitable for special applications.

Piezometers measure ground-water levels and pore pressure. Measuring ground-water levels is an important part of monitoring and simple standpipe open piezometers are usually sufficient. However, if there are areas of low permeability or confined aquifers, or when rapid response to reduction in pore pressure needs to be monitored, pneumatic or electric devices may be required.

Microseismic monitoring has historically been expensive because of the electronic equipment needed. However, the lessening cost of equipment in recent years and its increased reliability are making this technique more attractive. Experiments with microseismic recordings have established that there is a correlation between rock noises and slope movement.

Guidelines for Monitoring

1) Measure obvious things first. Surface displacement is the most direct and most critical aspect of slope instability.

2) Simpler is better. The reliability of a series system is the product of the reliability of the individual components. A complex electronic or mechanical device with a telemetered output to a computer has significantly less chance of being in operation when needed than do two stakes and a tape measure.

3) Precision costs money. The cost of a measuring device is often a power function of the level of precision. Measuring to ± 1 cm (0.39 in.) is inexpensive compared to measuring to ± 0.0001 cm (0.000039 in.). A micrometer is unnecessary for monitoring slope movement that has a velocity of 5 cm/d (2 ips).

4) Redundancy is required. No single device or single technique tells the complete story. Backup devices are needed.

5) Timely reporting is essential. Data collection and analysis must be rapid enough to provide information in time to make decisions. Reducing last week's data and telling the mine superintendent that the slope was moving Thursday when a shovel was buried Sunday does not lead to pay raises.

Data Reduction and Reporting

The following measurements or calculations should be made for each survey reading:

1) Date of reading, incremental days between readings, and total number of days the survey point has been established.

2) Coordinates and elevation.

3) Magnitude and direction of horizontal displacement.

4) Magnitude and plunge of vertical displacement.

5) Magnitude, bearing, and plunge of resultant displacement vector.

6) Rates of horizontal, vertical, and resultant displacements.

Both incremental and cumulative displacement values should be determined. Calculating the cumulative displacement from initial values rather than from summing incremental displacements minimizes the effects of occasional survey aberrations.

Slope displacements are best understood and analyzed when the data are graphically displayed. For engineering purposes, the most useful plots are:

1) Horizontal position (northing vs. easting).

2) Vertical position (elevation vs. change in horizontal position, plotted on a section in the mean direction of horizontal displacement).

3) Displacement vectors (plotted on a plan map).

4) Cumulative total displacement vs. time.

5) Incremental total displacement rate (velocity) vs. time.

6) Schmidt plots of total displacement vectors.

Daily precipitation and the number of tons mined beneath the slope should be added to the graphs as histograms to compare these records with slope movement.

A monthly slope stability report should be prepared for mine management. This report serves the dual purpose of providing information to decision makers and providing the discipline to document slope behavior. Direct, informal communication also should be maintained with pit operations on a daily basis in the case of mining in an active slide area.

REMEDIAL MEASURES

With an economically optimized slope design, some degree of slope instability can be expected. Minimization of the adverse effects of slope instability must be accomplished through judicious mine planning and establishment of operational contingencies.

There are several principles of slope mechanics that should be kept in mind in dealing with slope instability.

Slope failures do not occur spontaneously. A rock mass does not move unless there is a change in the forces acting on it. The common changes that lead to instability in an open pit are removal of support by mining, increased pore pressure, and earthquakes.

Most slope failures tend toward equilibrium. It is an observed phenomenon that as a slide displaces, the toe pushes out and the crest recedes. Such displacement reduces the driving force and increases the resistance force so that the displacement rate is reduced until movement stops. When high pore pressures are involved, a similar balance is attained. Displacement causes dilation of the rock mass. As a result, pore pressures drop and the effective shear strength increases. This mechanism explains the stick slip movement of some slides, in which recharge increases the pore pressure in tension cracks, resulting in renewed displacement. There are exceptions to this generalization, but they are usually the result of reduction of shear strength due to shearing of asperities or changes in the forces acting on the rock mass.

A slope failure does not occur without warning. Prior to major movement, measurable deformation and other observable phenomena, such as development of tension cracks, occur. These phenomena occur from hours to years before major displacement. However, single bench sloughing directly associated with mining does occur rapidly. While a slope failure does not occur rapidly without warning, deformation and tension cracks can occur without major displacement.

Detection of Instability

The first step in slope management is the identification of potential failure areas such as faults, breccia dikes, and/or jointing with attitudes that would form a failure geometry. Data for this identification would come from geologic pit mapping. Areas of higher water levels are also potentially unstable and should be identified.

The second step is monitoring areas that are potentially unstable and/or show evidence of instability by displacement and tension cracks.

On the basis of monitoring and mapping, the geometry of a failure can be determined and predictions made of future behavior.

Slide Management

When instability occurs there are a number of response options:

1) Leave the unstable area alone.

2) Continue mining without changing the mine plan.

3) Unload the slide through additional stripping.

4) Leave a step out.

5) Partial cleanup.

6) Mine out the failure.

7) Support the unstable ground with cable bolts.

8) Dewater the unstable area.

The choice of options or combination of options depends on the nature of the instability and the operational impact. Each case should be evaluated individually and cost-benefit comparisons conducted. The following are guidelines on the choice of options.

1) When instability is in an abandoned or inactive area, it can be left alone.

2) If the displacement rate is low and predictable and the area must be mined, living with the displacement while continuing to mine may be the best action.

3) Even though unloading has been a common response, in general it has been unsuccessful. In fact, there are situations involving high water pressure where unloading actually decreases stability.

4) Step-outs have been used successfully in several mines. The choice between step-out and cleanup is determined by the tradeoff between the value of lost ore and the cost of cleanup.

5) Partial cleanup may be the best choice where a slide blocks a haul road or fails onto a working area. Only that material necessary to get back into operation need be cleaned up.

6) Where the failure is on a specific structure and there is competent rock behind the structure, mining out the failure may be the optimum choice.

7) Mechanical support may be the most cost-effective option when a crusher, conveyor, or haul road must be protected.

8) Where high water pressure exists, dewatering is an effective method of stabilization that may be used in conjunction with other options.

Contingency Planning

Mine planning should have the flexibility to respond to slope instability. Rather than an "after the fact" crisis response to forced deviation from a rigid mine plan, contin-

gency plans should be prepared in advance so that the response to slope instability is well thought out.

Operational flexibility should be built into the mining plan; for example:

1) Adequate ore should be exposed and accessible so that production is not dependent on a single location.

2) There should be more than one access road into the pit for service vehicles.

3) Whenever possible, double access to working benches should be maintained.

4) Production scheduling should have a provision for slide cleanup.

Excavation

Equipment selection and estimation of production rates and costs require a knowledge of the rock mass properties. A simple classification system based on rock substance compressive strength and fracture frequency can be used (Franklin, 1971) (Fig. 17). Kirsten (1982) has developed a classification that includes fracture and equipment characteristics.

Regardless of the classification system used, it is important to zone the excavation area rather than use average values. Although the average rock properties may indicate that blasting is not required, ignoring the 20% that does require blasting can have a devastating impact on mining costs.

CRUSHING AND GRINDING

Miners often attempt to improve fragmentation by altering blasting practices, but generally after experience and disappointment, will go to crushing because it is the only way to guarantee a uniform product size. In hard rock,

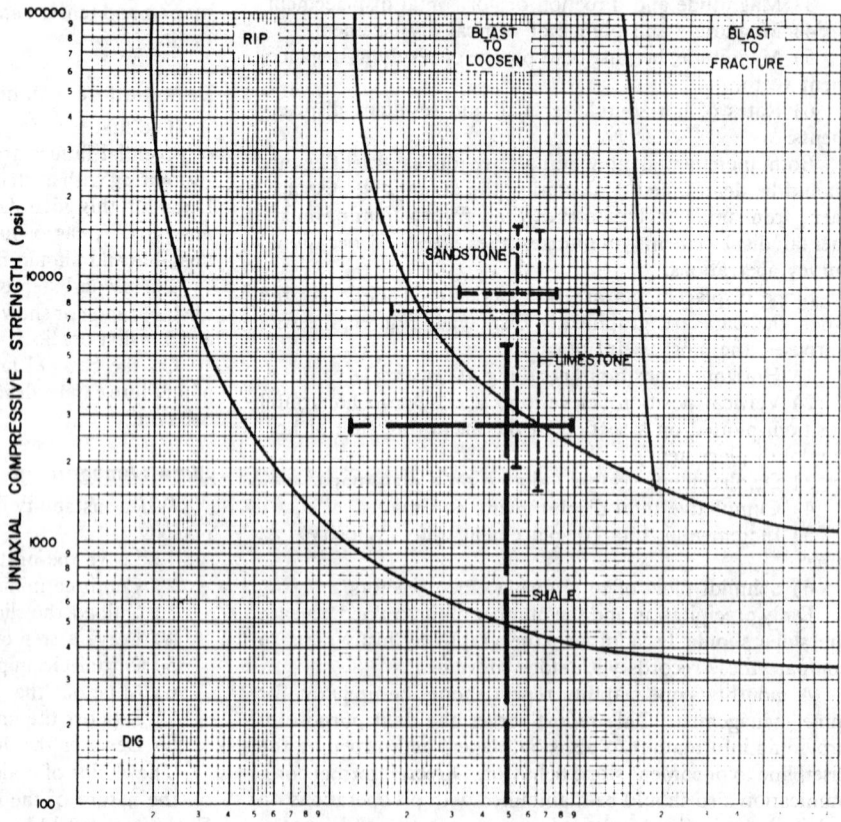

Fig. 17. Excavation classification (from Franklin, et al., 1971).

fragmentation can only be improved to a block size consistent with the fracture intensity. This is because the fractures are inherently weaker than the surrounding rock substance and the principles of fracture mechanics apply. These principles do not allow significant rock breakage, particularly with low velocity explosives such as ANFO, because less energy is required to open or extend existing fractures than to initiate new fracture and break rock substance. Some pulverization of the rock occurs immediately near the blasthole and, if costs allowed, high velocity explosives and a blasting pattern could be designed to pulverize the entire rock mass. Practically, however, the volume of pulverized rock in a blasthole is very small and probably insignificant when compared to the total volume of rock that is thrown into the muck pile.

Another consideration in attempting to improve fragmentation is that unless care is taken, mining costs could be significantly increased if the stability of final pit walls is decreased because of blast damage. To properly determine crushing requirements, the rock hardness, the size of the blocks to be crushed, and the tonnages needing crushing need to be known. These are parameters that will, in part, determine energy requirements, crusher size, grizzly size, abrasiveness, maintenance schedule, and equipment life. Siting of the crusher might be based on proximity to the material that requires the most crushing; thus, spatial distribution of large blocks must be known. Ultimately, conveyor belt size and other processing equipment will be selected based, in part, on the ability to crush materials to a specified size.

Rock Hardness

Compressive strength is the best indicator of rock hardness for crushing. Compressive strength can be estimated from laboratory tests on core from drill hole or rock blocks, from point load tests on irregular rock samples or drill core, or from a rock hardness classification. If rock hardness is plotted on level maps next to the drill hole locations or sample sites, the spatial distribution of hardness can be defined.

Block Size Prediction

Block size distribution can be determined by simulation and direct measurement. Both estimation methods require a knowledge of *fracture frequency,* i.e., a count of the number of fractures per meter (foot) without regard to orientation or the manner in which the fractures are separated. Rock Quality Designation, RQD, can also be used. A relationship between RQD and fracture frequency was reported by Priest and Hudson (1975). First it is necessary to define areas of similar fracture frequency on level maps. A good first approach is to contour fracture frequency on each level. Areas where data cannot be obtained from core holes can often be supplemented by mapping underground drifts and pit faces. Commonly, extrapolation is necessary, and this can lead to more sophisticated methods of estimation techniques, such as the geostatistics used in ore reserve estimation.

Simulation

After areas of similar fracture frequency have been defined, simulation of the fracturing in the rock mass can be done using Monte Carlo sampling of orientations and spacings. Block sizes can be calculated if all fracture lengths are assumed continuous, which is not a bad assumption because fractures are extended during the blasting. This type of simulation has been done to predict cavability for block caving (White, 1979). A simulation requires good input data, which includes nearly complete knowledge of the fracture orientations and the distribution of spacings. This data can be time-consuming to obtain and often requires special expertise in data reduction.

There are mathematical models that assume equidimensional fragment shape and that consider the probability of a drill hole or mapping line intersecting the maximum dimension of the fragment. This method results in an approximation of a size distribution curve based on fracture frequency. A curve would be estimated for each area defined on the level maps. Again, the mathematics are somewhat complex, and the results should be checked with direct measurement before extrapolation is done.

Direct Measurement

Most mining companies will not have the means or the desire to attempt simulation or mathematical modeling. Even when these methods are used, some direct measurement is still required to confirm predictions. Again, in direct measurement, areas of similar fracture frequency should be defined on level maps. Direct measurement is essentially a large-scale screening of as much tonnage from the muck pile as is feasible. A set of grizzlies or screens or some combination of screens and grizzlies would be constructed and several tests conducted to develop the size gradation curves for each area defined on the level maps.

Another less accurate method might be simply to estimate percentages from the muck pile. This would be a very crude approximation and many inaccuracies could result, but these approximations would still be better than no attempt to quantify the fragmentation and the crushing requirements. Consideration should be given to photogrammetry to assist in the estimation. Franklin, et al. (1988), report on encouraging results using photoanalysis to determine size distribution of blasted rock.

Both direct measurement methods require fracture frequency to be estimated prior to the blast to correlate fracture frequency with the resulting size distribution curve.

The Geotechnical Model

Once rock hardness and fracture frequency with the corresponding size distribution curves have been defined for each level, combining the data into a final rock classification for crushing requirements is done. A final level map is produced with areas defined according to rock hardness and block sizes. The result is a geotechnical model that can be used to estimate hardness, size, and tonnage in the same manner that tonnage and grade are estimated in ore reserve estimation. If a product of, say, less than 5 cm (2 in.) is desired, a tonnage by level which requires further crushing can be measured. It would be possible with this information to determine crushing requirements for each period in the mine life.

LEACHING

The geotechnical model is also useful for estimating the product size that ultimately reaches the leach dump. Recovery in a leaching operation is determined in large part by the amount of surface area on an individual rock fragment or block that is exposed to the leach solutions. If the tonnages of each size material are known, approximations can be made regarding the available surface area. Leach tests can be designed to give a more accurate prediction of expected recoveries. Thus, a more accurate estimation of recovery from the leach system can be made.

REFERENCES

Barton, N., 1982., "Shear Strength Investigations for Surface Mining," *Stability in Surface Mining, Vol. 3,* C.O. Brawner, ed., AIME, New York, pp. 171-196.

Barton, N., Lien, R., and Lunde, J., 1974, "Engineering Classification of Rock Masses for the Design of Tunnel Support, *Rock Mechanics*, Vol. 6, No. 4, pp. 189-236.

Bieniawski, Z.T., 1974, "Geomechanics Classification of Rock Masses and Its Application in Tunnelling," *Proceedings*, 3rd International Congress of Rock Mechanics, Vol. 11A, Denver, CO, pp. 27-32.

Brown, E.T., ed., 1981, *Rock Characterization and Monitoring, ISRM Suggested Methods*, Pergamon Press, New York, 211 pp.

Call, R.D., Savely, J.P., and Nicholas, D.E., 1976, "Estimation of Joint Set Characteristics from Surface Mapping Data," *Monograph on Rock Mechanics Applications in Mining*, W.S. Brown, S.S. Green, and W.A. Hustrulid, eds., AIME, New York, pp. 65-73.

Call, R.D., Savely, J.P., and Pakalnis, R., 1982, "A Simple Core Orientation Technique," *Stability in Surface Mining, Vol. 3*, C.O. Brawner, ed., AIME, New York, pp. 465-481.

CANMET, 1976, *Pit Slope Manual*, Canada Centre for Mineral and Energy Technology (CANMET), Ottawa, Ont., Canada, 10 Chaps. with supplements.

Coates, D.F., 1981, "Rock Mechanics Principles," Monograph 874 (rev. 1981), Dept. of Energy, Mines and Resources (CANMET), Ottawa, Ont., Canada.

Deere, D.U., 1968, *Geological Considerations in Rock Mechanics in Engineering Practice*, Staggy and Zienkiweicz, eds., John Wiley, New York.

Franklin, J.A., Broch, E., and Walton, G., 1971, "Logging the Mechanical Character of Rock," *Transactions*, Institution of Mining and Metallurgy, London, Vol. 80, pp. A1-A9.

Franklin, J.A., Maerz, N.H., and Bennett, C.P., 1988, "Rock Mass Characterization Using Photoanalysis," *International Journal of Mining and Geological Engineering*, No. 6, pp. 97-112.

Glass, C.E., 1982, "Influence of Earthquakes on Rock Slope Stability," *Stability in Surface Mining, Vol. 3*, C.O. Brawner, ed., AIME, New York, pp. 89-112.

Goodman, R.E., 1980, *Rock Mechanics*, John Wiley, New York.

Goodman, R.E., and Shi, G-h., 1985, *Block Theory and Its Application to Rock Engineering*, Prentice-Hall, Englewood Cliffs, NJ.

Hoek, E., and Bray, J., eds., 1981, *Rock Slope Engineering*, 3rd ed. rev., Institution of Mining and Metallurgy, London, 402 pp.

Jennings, J.E., and Robertson, A.M., 1969, "Procedures for the Prediction of the Stability of Slopes Cut in Natural Rock," *Proceedings*, 7th International Congress on Soil Mechanics, Mexico City, DF, Mexico.

Kim, Y.C., 1977, "Supplement 5-3," *Pit Slope Manual*, Report 77-6, Canada Centre for Mineral and Energy Technology (CANMET), Ottawa, Ont., Canada.

Kirsten, H., 1982, "A Classification System for Excavation in Natural Materials," *Civil Engineering in South Africa*, Vol. 24, No. 7.

Lambe, T.W., and Whitman, R.V., 1969, *Soil Mechanics*, John Wiley, New York.

McMahon, B., 1976, "Estimation of Upper Bounds to Rock Slopes by Analysis of Existing Slope Data," Report 76-14, Canada Centre for Mineral and Energy Technology (CANMET), 64 pp.

Müller, L., and Hofmann, H., 1970, "Selection, Compilation and Assessment of Geological Data for the Slope Problem, *Proceedings*, Open Pit Mining Symposium, Johannesburg, South Africa.

Oriard, L.L., 1971, "Blasting Effects and Their Control in Open Pit Mining," *Geotechnical Practice for Stability in Open Pit Mining*, C.O. Brawner and V. Milligan, eds., AIME, New York, pp. 197-222.

Peters, W.C., 1978, *Exploration and Mining Geology*, John Wiley, New York.

Piteau, D.R., 1970, "Geological Factors Significant to the Stability of Slopes Cut in Rock," *Proceedings*, Symposium on the Theoretical Background to the Planning of Open Pit Mines with Special Reference to Slope Stability, A.A. Balkema Publishers, Amsterdam, The Netherlands, pp. 55-71.

Priest, S.D., and Hudson, J.A., 1981, "Estimation of Discontinuity Spacing and Trace Length Using Scanline Surveys," *International Journal of Rock Mechanics and Mining Sciences and Geomechanics Abstracts*, Vol. 18, pp. 183-197.

Priest, S.D., and Hudson, J.A., 1976, "Discontinuity Spacings in Rock," *International Journal of Rock Mechanics and Mining Science and Geomechanic Abstracts*, Vol. 13, pp. 135-148.

Ritchie, A.M., 1963, "Evaluation of Rockfall and Its Control," *Highway Research Record*, Highway Research Board, Vol. 17, pp. 13-28.

Ross-Brown, D., 1979, "Analytical Design," *Open Pit Mine Planning and Design*, J.T. Crawford, III, and W.A. Hustrulid, eds., AIME, New York, pp. 161-183.

Sowers, G.B., and Sowers, G.F., 1970, *Introductory Soil Mechanics*, 3rd ed., McGraw Hill, New York.

Vanres, D.J., 1974, "The Logic of Geological Maps, with reference to their interpretation and use for engineering purposes," Professional Paper 837, US Geological Survey, 48 pp.

White, D., 1979, "Analysis of Rock Mass Properties for the Design of Block Caving Mines," SME Preprint 79-309, SME-AIME, Littleton, CO, 17 pp.

Zavodni, Z.M., and McCarter, M.K., 1977, "Main Slope Slide Zone, Utah Copper Division," *Monograph on Rock Mechanics Applications in Mining*, W.S. Brown, S.S. Green, and W.A. Hustrulid, eds., AIME, New York.

6.9 Waste Dumps

6.9.1 Mineral Leaching Technology: Heap—Dump—In Situ Leaching of Minerals

E.E. MALOUF

EXECUTIVE SUMMARY

1) Technically feasible and economically viable hydro-metallurgical processes have been developed permitting the recovery of values from low-grade materials that could not be processed economically by conventional processes. The fast-growing accumulation of technical data, from such diverse sources as the alteration of minerals in an ore body, bacteria-utilizing minerals (Malouf, 1968), the formation of minerals by the spewing of metallic compounds from vents in the seabed, followed by the rapid alteration of the minerals by bacteria (Matthews, 1981), and also the rapid alteration of the minerals in the crater of Mt. St. Helens (Matthew, 1981), all contribute to the development of even more efficient processes.

2) The minerals in ore bodies are altered in nature through a continuous ongoing process. The rate and degree of alteration is governed by the availability of oxygen and moisture to the indigenous bacteria that utilize the sulfides as the energy source for the bacterial life cycle (Malouf, 1968).

3) Major factors affecting the alteration of minerals are:

a) Once moisture enters the ore body a galvanic couple is established between dissimilar minerals which results in the galvanic corrosion of the minerals (Shreir, 1963).

b) Presence of oxygen from the atmosphere or dissolved in water accelerates the oxidation of sulfidic materials in the ore body and of ferrous ions to ferric ions by the indigenous bacteria.

c) The amount of fractures in the ore body along with the pores in the minerals control the rate of alteration.

4) Two general classifications of bacteria are involved in mineral alterations and hydrometallurgical processes: one is the autotrophs, both aerobic and anaerobic. These bacteria thrive on inorganic matter, with the aerobes requiring oxygen while the anaerobes can thrive in the absence of oxygen.

The other general classification is the heterotrophic bacteria, both aerobic and anaerobic. The heterotrophs differ from the autotrophs in that they also require organic matter in the development of energy for their life cycle.

The most common and most studied reactions of the autotrophs are the oxidation of sulfides to form metal sulfates and sulfuric acid, as well as the oxidation of ferrous ions to ferric ions, thereby providing a chemical oxidant for the oxidation of the sulfide minerals. Not as much investigation has been made on the heterotrophs as on the autotrophic bacteria, however one reaction that applies to the metallurgical processes is the ability of these microorganisms, in the absence of oxygen, to reduce sulfates to sulfides (Smith and Shumate, 1970; Postgate, 1965; Baas-Becking and Moore, 1961).

5) Oxidation of the sulfide minerals to sulfates, oxides, or hydroxides results in a severalfold increase in the molecular volume of the mineral, which in turn results in the expansion of the fractures and/or pores of the mineralization in the mineralized zone.

6) Mining of mineralized material increases exposure of the minerals to moisture and the oxygen of the atmosphere, thereby accelerating the oxidation of the sulfide minerals.

7) With the increased pollutants of sulfur and nitrogen compounds in the atmosphere, the bacterial oxidation of sulfide minerals embedded in limestone or dolomite host rocks has been accelerated to the extent that a normally alkaline or basic environment becomes acidic in a period of 30 years, whereas prior to the massive increase in air pollutants, the conversion from basic environment to an acid environment required 300 to 400 years.

8) The use of trucks for building mineral-containing dumps or heaps results in severe surface compaction as to restrict the convective flow of air through the pile of material and prevents the penetration of water or leach solutions into the interior of the dump or heap.

9) Presence of acid-consuming minerals, such as thaumasite, wollastonite, limestone, and/or dolomite, will retard the alteration and subsequent leaching of the minerals. Measurable amounts of metal ions in solution may not appear from beneath the tailings pile, dump, or heap until the environment or condition of the material has changed from a basic or alkaline state to an acidic state. This may require, under natural conditions, approximately 30 years.

10) If retardation of leaching of minerals is required, as the case may be in impounded concentrator tailings, one or more of the following steps could be used:

a) Compaction or sealing the surface of the mineral-containing material to prevent access of moisture and/or oxygen to the minerals.

b) Mix acid-consuming and clay minerals with the mineralized material. Store the mineralized material in an impervious clay basin.

c) Condition the mineralized material with a bactericide, such as iso-thiazolinone chloride, which is effective at very low concentrations.

d) Elimination of all sulfidic and/or organic matter from the material that has to be stored and not leached.

11) In the heap leaching of uranium, gold, or silver ores, the optimum height for the best recoveries of the values are in heaps up to 7 m (22 ft) high. Multiple *lifts* of the mineralized material diminishes recoveries of the contained values.

12) In the heap leaching of minerals containing values other than uranium, gold, or silver, the optimum height economic relationship are heaps up to 10 m (33 ft). Field experience data indicates that when sulfide or oxide mineral-bearing materials are leached in heaps up to 10 m (33 ft), a new *lift* can be placed on the *ripped* surface of the leached heap and leached as effectively as the first heap or *lift*. This technique has been used successfully in multiples of 12 *lifts*.

13) Microorganisms that alter sulfide minerals will, in a short period of time (30 days) once moisture and oxygen become available, evolve a species adapted to the environment created by the specific conditions of the ore body, heap, dump, or tailings area.

14) Laboratory optimum temperature for the autotrophic, aerobic microorganisms has been determined as 35°C

(Torma, 1977). However, species that carry on their life cycle at 85°C have been isolated from hot, humid, mine waste interiors and from hot sulfur-containing springs (Brierly, 1982).

15) Autotrophic, aerobic microorganisms have evolved that will survive acid solutions of pH 1.5 with copper sulfate concentrations of 15 gL copper (Malouf and Prater, 1961).

16) Biogenic reactions of significance in mineral alterations resulting in hydrometallurgical processes are:

a) Oxidation of sulfides to sulfuric acid and metal sulfates (Malouf, 1968).

b) Oxidation of ferrous ion to ferric ion to form a chemical oxidant that reacts with the sulfide minerals to form metal sulfates and sulfuric acid.

c) Reduction of sulfates to form sulfides (Postgate, 1965).

d) Microorganisms can function in a pH range of 1.5 to 11.

17) Microorganisms, such as *Thiobacillus thioparus* or *concretivorus*, can utilize the sulfur dioxide in the atmosphere to generate sulfuric acid and start the alteration of minerals that might be contained in acid-consuming host rock. These microorganisms function in a pH range of 7 to 11 (Posnjak and Merwin, 1969; Mayling, 1969).

18) Field process development data indicates that major improvement in the recovery of values from minerals can be achieved by preconditioning of the minerals with the lixiviant prior to leaching. The technology involves "wetting" the minerals, to be leached, with a suitable lixiviant of a greater concentration then would normally be used. The wetting of the minerals usually is made in transit as the mineral-bearing materials are moved to the heaps for leaching. The preconditioning increases the overall recovery by (Malouf, et al., 1958, 1962; Riggs, et al., 1978):

a) Starting the leaching reaction while the heap is constructed, thereby providing intimate contact of the lixiviant with the minerals prior to full-scale leaching.

b) Condition entrained solutions in the interstices of the mineralization, thereby accelerating the diffusion of the metallic ions into the leach solutions.

c) Eliminate the precipitated salts that might be in the pores or fractures of the mineralization blocking the entrance of leach solutions.

d) In the case of sulfide minerals, start the molecular volume change of the minerals, which increases the permeability of the mineralized zone to solutions.

e) Establishing the bacteria on the mineral surfaces to effect biogenic alteration of the minerals.

19) Field developments of in situ leaching of minerals have provided data (Malouf, 1971) indicating that:

a) Successful economic recovery of the values from the minerals contained in the block caved portion of an underground mining operation can be achieved (Fletcher, 1971).

b) The use of explosives to rubblize an ore body for subsequent leaching has not proven economically feasible. The formation of *fines* in the explosive-formed rubble *blinds* large volumes of the minerals from contact with the leach solutions.

c) Successful leaching operations have been developed by injecting the leach solutions into the mineralization by pressure. Sufficient pressure is used to dilate the pores of the formation without hydro-fracturing the rock (Malouf, 1971).

20) Pad construction for the recovery of leach solutions from under heaps of mineralized material involves the following factors:

a) In all cases, a minimum flow gradient of 5% from under the heap to the collection reservoir.

b) Removal of all vegetative or organic matter from the area to be overdumped to prevent the contamination of the leach solutions with organic compounds that might affect subsequent solvent extraction processes.

c) A compacted subsurface overlain with a minimum of 0.5 m (18 to 20 in.) of compacted clay. The clay pad is to be constructed in a minimum of three increments.

d) A compacted subsurface overlain with a minimum 15 cm (6 in.) asphalt-compacted surface. The asphalt pad must be protected with a sand or tailings impact cushion of material before construction of the heap.

e) A more recent concept in pad construction that is proving to be more effective than other types of construction is a chemical pad. The pad is constructed by placing 5 to 7 cm (2 to 3 in.) of compacted crushed limestone onto a compacted graded area that is to be used as the heap site. The compacted, crushed limerock is then sealed with a solution of ferric sulfate.

The particular advantage of this type of pad is its ability to self-heal any cracks or fissures that might develop from subsidence. This is the case if the leach solutions contain ferric ions.

21) Current research and development efforts on the elimination of cyanide anions from leach solutions to be discharged from gold and silver leaching operations have provided a new commercial approach. Data has been developed indicating that microorganisms, usually indigenous to the leaching operation, utilize the cyanide compounds as their energy source, producing nitrogen and carbon dioxide as the end product. One such microorganism is *Thiobacillus cyanoxidans.*

22) Biogenic mineral reactions can be used to process sulfide-bearing gold ores to alter and eliminate the sulfide minerals, leaving the gold available for recovery by conventional cyanide leach solutions.

INTRODUCTION

Hydrometallurgical processes for the recovery of values from minerals are controlled and accelerated reactions similar to the reactions found as continuous natural oxidation and leaching occurring in nature. Current studies provide data indicating that as soon as minerals are formed, bacteria are rapidly altering them, as has been observed around the vents in the seafloor and in the volcanic crater of Mt. St. Helens (Matthews, 1981). A common denominator, apparent through the study of various mineral deposits, indicates that all mineral deposits that have natural oxidation and leaching have the following zonal generalizations:

Leached capping zone.
Secondary enrichment zone.
Primary mineralization zone.
Roll-front accumulation of reduced metallic ion compounds that on oxidation become soluble compounds, such as the uranium mineralization found in sandstone formations. Some copper, molybdenum, and vanadium compounds are also found in such formations.

Data from the examination of leached capping rock specimens from various ore bodies indicate the presence of fossil imprints of minerals that have been leached. These observations have been made to a depth where the ancient water table had sealed the area to further oxidation. At the point where oxidation is prevented by the water table, a reduction zone results that precipitates the previously leached metallic ions into insoluble compounds, which combined with the minerals in the zone, result in secondary enrichment.

Once the mineralized sulfide material is placed into dumps, heaps, or tailing areas, the oxidation of the sulfides again occurs and the formation of water-soluble metal sulfates is accelerated by the availability of oxygen, water, and bacteria. The bacteria are usually indigenous to the ore body.

Data from examination of samples obtained from massive mine waste dumps and concentrator tailings ponds indicate that natural oxidation and leaching of the sulfide minerals occurs and the sulfuric acid generated by the oxidation of the sulfide minerals leaches the oxide minerals, even without the application of leach solutions. The moisture from rain and/or melted snow along with oxygen from the atmosphere is sufficient to establish the autotrophic, aerobic microorganisms that oxidize the sulfide minerals and form water-soluble metal sulfates.

Data obtained from samples from the interior of massive mine waste dumps indicate that areas deficient in oxygen, usually 50 m (150 to 160 ft) below the surface, show zones of secondary enrichment in the case of copper mineralization.

The following lists some of the reactions during the natural leaching of a copper sulfide ore body:

Leached Zone (Oxidation and Acid Generation)

$$2\ FeS_2 + 2\ H_2O + 7\ O_2 = 2\ FeSO_4 + 2\ H_2SO_4$$
$$2\ FeSO_4 + H_2SO_4 + \tfrac{1}{2}\ O_2 = Fe_2(SO_4)_3 + H_2O$$
$$3\ Fe_2(SO_4)_2 + 2\ K\ Al\ Si_3O_8 + 13\ H_2O = 2\ K\ Fe_3(SO_4)_2(OH)_6$$

feldspar   jarosite

$$Al_2Si_2O_5(OH)_4 + SiO_2 + H_2SO_4$$

kaolin   silica   acid

Leaching Zone (Oxidation and Solubilization of Copper Sulfides)

$$8\ Cu_2S + 2\ H_2SO_4 + O_2 = 8\ Cu_{1.75}S + 2\ CuSO_4 + 2\ H_2O$$
$$4\ Cu_{1.75}S + 3\ Fe_2(SO_4)_3 = 4\ CuS + 3\ CuSO_4 + 6\ FeSO_4$$
$$CuS + 2\ O_2 = CuSO_4$$

Enrichment Zone (Reduction and Neutralization)

$$4\ CuFeS_2 + 29\ CuSO_4 + 16\ H_2O = 13\ CuSO_4 + 4\ Cu_{1.75}S + 4\ FeSO_4 + 16\ H_2S$$
$$2\ CuFeS_2 + 3\ Cu_2SO_4 = 4\ Cu_{1.75}S + 2\ FeSO_4 + CuSO_4$$
$$FeS_2 + 3\ Cu_2SO_4 = 2\ Cu_2S + FeSO_4 + 2\ CuSO_4$$
$$2KAlSi_3O_8 + H_2SO_4 + H_2O = K_2SO_4 + Al_2Si_2O_5(OH)_4 + 4\ SiO_2$$

In the natural alteration and leaching of sulfide minerals, the following principal reactions involving the generation and/or consumption of acid occur. These reactions are of major importance in the design of hydrometallurgical processes.

Generation of Acid:
Oxidation of sulfides, particularly pyrite.
Hydrolysis and precipitation of ferric sulfate.
Consumption of Acid:
Oxidation of ferrous sulfate.
Acid attack or dissolution of gangue minerals (such as biotite, carbonates, and calc silicates).
Dissolution by acid of precipitated salts of iron, copper, aluminum, silica, and uranium.

Some typical chemical reactions showing the generation and consumption of sulfuric acid in both natural and applied hydrometallurgical processes are as follows:

1) *Oxidation of sulfides to generate sulfuric acid*
Chalcopyrite

$$CuFeS_2 + 5Fe_2(SO_4)_3 + 4H_2O = CuSO_4 + 11FeSO_4 + 4H_2SO_4 + S^0$$

b) Pyrite

$$FeS_2 + 8H_2O + 7Fe_2(SO_4)_3 = 15FeSO_4 + 8H_2SO_4$$

2) *Consumption of sulfuric acid*
a) Bacterial oxidation of ferrous ion

$$4FeSO_4 + O_2 + 2H_2SO_4 + bacteria = 2Fe_2(SO_4)_3 + 2H_2O$$

b) Gangue mineral reactions with sulfuric acid

$$K_2Fe_6[Si_2Al_2O_2](OH)_4 + 10H_2SO_4 = K_2SO_4 + 6FeSO_4 + Al_2(SO_4)_3 + 2SiO_2 + 2H_2O$$

(biotite)

3) *Hydrolysis and precipitation of ferric iron salts* will release sulfuric acid to the solution according to the following reaction:

$$3Fe_2(SO_4)_3 + 14H_2O = 2Fe_3(SO_4)_2(OH)_5 \cdot 2H_2O + 5H_2O$$

(basic iron sulfate)

The foregoing reactions are common to the various copper mine waste leaching operations and constitute the source of sulfuric acid that contributes to the leaching operation success.

HEAP OR DUMP DESIGN FOR ENHANCEMENT OF MINERAL LEACHING

Major factors affecting the leaching of minerals are:

Aeration is a most critical factor where the material to be leached contains sulfide minerals. Convected air supplying the required oxygen to the autotrophic aerobic bacteria will penetrate the face or sides of a dump or heap for a distance of approximately 65 to 70 m (200 ft). For the convection of air to occur it is necessary that the surface compaction of the dump or heap be removed. The compaction on the surface of a dump or heap occurs from the action of the truck tires on the material by the trucks bringing the mineralized material to the dump or heap. Compaction measurements made on a dump by 65-ton trucks indicate compaction occurs to a depth of 3 to 4 m (12 ft), with the upper 0.6 m (2 ft) constituting the most severe compaction. This compaction measurement gave approximately the same data as rock in place. Also the upper 0.6 m (2 ft) of compaction contained considerable *fines* formed by the grinding action of the truck traffic on the material.

Size of the heap or dump is a critical factor in the successful recovery of the values by leaching. Again, if the minerals to be leached are sulfides, the width of the heap or dump should be limited to 130 to 140 m (400 ft), as the convection of air supplying oxygen for the oxidation of the sulfide minerals will penetrate the sides of the heap or dump approximately 65 to 70 m (200 ft).

Field data also indicates that convection of air through a porous surface of a heap or dump will penetrate the pile approximately 15 to 20 m (50 ft); below this depth reducing conditions will become established.

A 12 to 15 m (35 to 50 ft) depth of material in a heap or dump should be leached before a new *lift* of material is placed on the heap or dump.

A third factor involves the composition of some of the barren rock being moved into the heap or dump during the mining operation. If this material contains acid-consuming minerals, such as thaumasite, wollastonite, or limestone, the solubilized metals will be precipitated and adsorbed by these minerals. These minerals should be selectively mined and disposed in areas outside of the heap or dump to be leached.

A fourth factor is the size of the material placed in the heap or dump. If the material is too coarse, a lower recovery can be expected as a longer leaching time will be required. Solution channels will be established that will prevent leach solution contact with all of the mineralized material for the time needed to achieve high recovery of the mineral values. During a mining operation of material to be leached, it may be possible to use *ripping* of the ore or extra explosives to fracture a greater amount of the ore.

In summary, mined ore should be classified into three general classifications:

Mineral-barren material, and/or material containing high acid-consuming minerals or clay; all of these materials should be kept out of a heap or dump that is to be leached.

Materials containing economic quantities of minerals that justify the cost of constructing proper heaps or dumps for the leaching of the minerals.

Materials containing uneconomic quantities of minerals should be disposed of in areas not suitable for leaching.

Heap or Dump Construction

Current technology in heap and/or dump construction successfully practiced for the recovery of such values as copper, gold, silver, and uranium is as follows. The same approach is applicable to all minerals to be leached; the processes differ in the leach solutions used. The common basis of construction is:

Construction of the heap and/or dump in a drainage basin that has a flow gradient or slope exceeding 5%.

An ideal location for construction is in an area where the bedrock is near the surface. This will result in good leach solution recovery at a minimum cost. If permeability data of the area on which the heap and/or dump is to be constructed indicates that leach solutions might be lost subsurface, the area must be made impervious to leach solutions. This can be accomplished by one of the following methods:

Remove all organic matter from the area. Grade and compact the area to obtain an uninterrupted flow gradient to a collection dam on bedrock.

If the area is overlain with 2 to 3 m (6.6 to 9.8 ft) of alluvium, no further action will need to be taken. Otherwise, the area must be covered with a compacted layer of 0.5 m (1.5 to 2 ft) of clay or river-bottom silt containing clay.

The compacted clay pad is then covered with 1 m (3.3 ft) of coarse rock to achieve good drainage of solutions from beneath the heap or dump and to provide an impact cushion of material to protect the pad during construction of the heap or dump.

A second approach is to use a 15 to 18 cm (6 in.) asphalt layer as a pad on the cleared, graded, compacted area. The same steps must be followed to protect the pad with a cushion of material during the construction of the heap or dump.

An asphalt pad, from field data, appears to lose some of the imperviousness (approximately 50%) after six to seven years of service.

A more recent development is the formation of an impervious base to a heap or dump by placing a 5 to 7 cm (2 to 2.8 in.) compacted layer of crushed limestone onto the cleared, graded, compacted area. The crushed limestone is then sprayed with a strong solution of ferric sulfate. The chemical coating forms a very impervious membrane.

If the leach solutions percolating through a heap or dump contain ferric ions, the pad under the heap or dump will continue to build up in accumulated ferric salts. In fact, fissures or cracks that might develop after a heap or dump is built would be *healed* in this manner.

Removal of all vegetation or organic matter from the area to be overdumped is necessary, because the organic products that would be picked up by the leach solutions would affect the solvent extraction operation.

The width of the heaps or dumps containing sulfide min-

erals should be limited to 120 m (400 ft), or less, thus allowing the maximum width that still supplies air by convection. The oxygen of the air is required for oxidation of the sulfide minerals.

If the heaps or dumps are constructed by moving the material by trucks, the upper 0.5 m (18 in.) should be removed. The *fines* generated by the action of the truck wheels on the material not only will affect the application of leach solutions but also will restrict the convective flow of air through the material.

After the removal of *fines* from the surface of the heap or dump, the surface must be *ripped* to a depth of 2 to 3 m (6.6 to 9.8 ft) in one direction, on 1-m (3.3-ft) centers, followed by *ripping* the surface again at right angles to the first series of rips. The action of ripping the surface of the heap or dump is to reestablish the pervious nature of the surface, which becomes compacted to a density of rock in place by the pounding action of the loaded trucks.

A technique that permits increasing recoveries of values from the leached minerals from a nominal 25 to 45% over a period of years to recoveries of 65 to 75% in a period of months involves *wetting* of the mineralized material with preconditioning solutions (Malouf, et al., 1958, 1962; Riggs, et al., 1978). In the case of leaching copper oxide minerals, the preconditioning solutions could be strong acid solutions containing 50 to 200 g/L of sulfuric acid. The amount and strength of the acid solutions is determined from laboratory and pilot plant test data. In the case of the leaching of silver- or gold-bearing materials, the preconditioning solutions could be varying concentrations of cyanide solutions or thiourea or thiosulfate solutions (Torma, 1977).

Heights of heaps or dumps containing copper minerals should not be greater than 10 to 15 m (35 to 50 ft). After leaching has been completed on the *lift* of material, additional *lifts* can be placed on the leached material. The multiple lift concept appears to function suitably in the field up to an overall height of 48 m (150 ft). In the heap or dump leaching of uranium, gold, or silver, field data indicates that loss of recovery appears after the height exceeds 6 to 7 m (20 ft).

Leaching Operation

Successful commercial leaching operations are being made in the recoveries of copper, uranium, gold, silver, potash, and sodium chloride.

The leaching of copper minerals usually is done with solutions made acidic with sulfuric acid. The effluent solutions from the leaching operation contain iron, aluminum, magnesium, and silica in concentrations ranging from 2 to 8 g/L. No effort as yet has been made to recover the aluminum or magnesium from the leach solutions, even though the concentration of each of these metals is in the range of 6 to 8 g/L.

The copper-bearing solutions usually are processed by concentration of the copper ions in a solvent extraction plant and electrowinning the copper from the concentrated solutions (Ashbrook, 1973). The use of the copper cementation process or the precipitation of copper on metallic iron is being used less and less because of the increasing cost of the metallic iron scrap for this purpose, the loss of the residual acid in the leach solutions, and the deleterious effects of the iron salts precipitating on the surface and in the interior of the heaps and dumps, thereby preventing solution contact with the minerals.

Leaching of heaps containing gold and/or silver values is accomplished with alkaline-cyanide solutions. The pregnant silver- or gold-bearing solutions are passed through columns of activated charcoal to absorb the values on the charcoal, followed by stripping the values from the carbon and recovery by electrolysis. In the case of strong silver-bearing solutions, the Merrill-Crowe process of precipitation with metallic zinc is used (McQuiston and Shoemaker, 1981).

No commercial attempts have been undertaken as yet to leach gold or silver values using thioureas or thiosulfate compounds (Torma, 1977).

Leach Solutions

The most commonly used leach solution in the leaching of copper minerals is made from the tailings solutions from a copper cementation operation or the raffinate from a solvent extraction operation, along with sufficient makeup water and sulfuric acid to adjust the pH of the leach solutions to the desired level. Makeup water is required to replace water or solutions lost to evaporation and/or seepage during leaching. The pH is controlled to buffer the various salts in the recycled solutions, to provide sufficient acidity to keep the leached copper salts in solution (pH 2.4 or lower), and to control the hydrolysis and precipitation of ferric ions in the pipelines, surface, and interior of the heap or dump. Occasionally, deposition of ferric iron salts in the interior of a heap or dump has developed to a degree that solutions are impounded in the interior and hydraulic rupturing of the side of the heap or dump occurs.

Controlling the pH of the leach solutions at pH 2.4 or lower will prevent the necessary microorganisms from becoming encapsulated in precipitated iron salts.

Methods of Leach Solution Distribution

Some of the more successful leaching operations control the distribution of the leach solutions to the surface of the heaps at the rate of 5 to 6 L/h/m² (0.1 to 0.2 gal/sq ft/hr).

The solutions may be distributed in one of several methods. One of the cheapest and most effective methods was developed by the Cyprus-Bagdad copper leaching operation at Bagdad, AZ. It consists of placing 0.5 m (18 in.) long pieces of surgical tubing 0.1 cm (⅜ in.) in diameter. These pieces of tubing are attached to nozzles from a 2.5-cm (1-in.) diam plastic pipe at 2.5 m (8 ft) intervals. Using solution pressures of 25 to 377 kPa (40 psig), a portion of the heap surface 3 to 4 m (10 to 12 ft) in diameter is *wetted* uniformly.

Plastic sprayers or *rainbirds* with stainless steel return springs are used for the effective distribution of leach solutions.

In arid climates, consideration should be given to using injection trickle tubes. The trickle tubes can be inserted in shallow drill holes that penetrate the compaction of the surface of the heap.

A natural segregation of the coarse and fine material occurs as a heap or dump is constructed. Furthermore, bulldozer leveling of the surface will cause additional segregation of the material. The overall effect is to form alternate layers of coarse and fine material in the heap or dump. The leach solutions follow a path of least resistance in passing through the material, leaving portions of the mineralization untouched by the leach solutions. Due to this, heaps and dumps should be built as uniformly as possible to assure maximum solution contact with the minerals to be leached, thereby assuring the highest recovery possible. If the material being placed on the heap or dump contains an excess of fines (+15%), agglomeration of the fines to the coarse material should be undertaken prior to placing the material in the heap or dump. The foregoing conditions apply to all types of materials that are leached, whether to recover copper, uranium, gold, or silver.

PROCEDURES FOR LABORATORY AND FIELD TESTING OF LEACHING MINERALS

Introduction

Laboratory and pilot plant data are required to determine the technical and economic feasibility of leaching values from mineralized rock, and in the case of disposal of mine or concentrator tailings material, the amount of possible pollutants that might need to be removed from the natural leaching solutions generated by rain or melted snow.

Design of the laboratory and/or pilot plant heap tests must be based on the intended results, namely:

1) Mineralized material to be leached for the maximum recovery of the contained values in the least possible time.

2) Mineralized material to be leached does not contain the values to sustain the cost required in No. 1, but leaching is to be accomplished at moderate cost.

3) Mineralized material to be leached at minimum cost, irrespective of recovery.

4) To establish the amount of pollutants that will be leached naturally from the mineralized material contained in the mine waste and/or concentrator tailings material.

The column size, heap size, and particle sizes would be applicable to any mineralized material to be leached. The changes from one mineral type to another would involve details of operation, such as lixiviant, whether *fines* should be removed or agglomerated, whether the minerals should be preconditioned with the lixiviant, and control of the salt content of the leach solutions.

Under the conditions outlined in No. 1, consideration has to be given to the possible added costs of crushing, agglomeration, and preconditioning, if required.

Under the conditions outlined in No. 2, leaching would be made without the costs of crushing, agglomeration, and possibly preconditioning.

In case No. 3, leaching would be made at a minimum cost. This would involve pumping the value-stripped solutions back onto the heap or dump, without the costs of special preparation of the mineralized material or the costs of solution adjustment of buffering the pH and salt content.

In case No. 4, accelerated laboratory column leach tests would need to provide data on the amount of polluting salts that might need to be processed to prevent pollution problems.

Successful extrapolation of the data from 1-t (0.9-st) column leach tests have been made in determining the recoveries and grade of solutions that might be expected from a commercial operation. This type of test involves obtaining 1 t (0.9 st) of representative ore that has been coarse crushed to −4 cm (−1½ in.). If the sample contains fines in excess of 15% of the total weight, agglomeration of the fines with the coarse fraction of the ore should be made before the sample is placed into the column. A column 0.5 m (20 in.) in diam by 3 m (10 ft) high is suitable to contain the sample.

Procedure for Test Column

The column is loaded with approximately 1 t (0.9 st) of the representative, thoroughly mixed crushed ore, without permitting segregation of the fines and coarse fractions. A perforated plastic plate in the bottom of the column can be used to support the sample. The surface of the column of ore is covered with an open weave nylon cloth to provide uniform distribution of the leach solutions over the surface of the material.

If the test is to be conducted to achieve the maximum recovery of the values in the least possible time, a preconditioning solution is applied to the material. The conditioning solution is usually of the same composition as the lixiviant that is to be used during the leaching, with the exception that the preconditioning solution is of a greater strength (four- to tenfold) than the normal leach solution.

The preconditioning solutions are allowed to penetrate the mineral pores, fractures, and interstices, thereby starting the reactions that result in the alteration of the minerals into water-soluble compounds.

A preconditioning-reaction period of 10 to 14 days is allowed before systematic leaching is started. The volume of preconditioning solution to be applied to the column of ore is equivalent to one pore volume displacement of the material in the column.

Leach solution application rate to the surface of the column of ore is 5 to 7 $L/h/m^2$ equivalent (0.1–0.2 gal/sq ft/hr). The solutions passing through the column of material for each 24-hr period are sampled. The sample of solution is analyzed as to pH, volume, temperature, and various salt and values content.

At the end of leaching, the column of mineralized material is washed with a displacement volume of leach solution, equivalent to 20% of the material weight.

The leached material from the column is then removed, dried, weighed, and crushed to ($-2mm$) (-10 mesh). Then it is sampled and analyzed for the remaining values to establish a metallurgical balance (McQuiston and Shoemaker, 1981).

IN SITU OR IN PLACE LEACHING OF MINERALS

Introduction

The development of in situ leaching of minerals has followed approaches of:

In place leaching, which involves the application of leach solutions to *caved* or *stoped* areas of an underground mine. The leach solutions are applied to the broken ore and recovered in the underground galleries of the mine.

In place leaching, wherein the ore to be leached has been broken by explosives, after which the leach solutions are applied to the broken ore.

In situ leaching, wherein the leach solutions are injected into the mineralized rock by controlled pressure. The leach solutions follow the strike, dip, and fracture set of the ore body to be recovered downdip in the ore body.

In situ leaching, wherein the ore is *hydrofractured* or *explosive-fractured* before injecting the leach solutions.

Commercial Applications

Several successful commercial in place leaching operations have been established in copper and uranium ore bodies. In these operations, *block-caving* mining techniques were used to obtain ore for conventional processing in a concentrator. After underground mining became unprofitable, the remaining *caved* ore in the mine was leached by percolating the leach solutions from the surface of the caved area down through the broken ore to the production galleries beneath the caved area. The value-bearing solutions are recovered on the impervious base of the collection galleries, pumped to the surface for processing to recover the values, and the barren solutions are adjusted as to the chemical content and recycled to the caved area.

A classical example of in place leaching is the copper leaching operation of the Miami Copper Co. at Miami, AZ (Fletcher, 1971). An abstract of this operation is: By 1955, the conventional mining by underground *block caving* to obtain ore for the concentrator had become unprofitable. Leaching of the caved ore was attempted and proved economically successful. The procedure consists of applying pH 1.9 acidified leach solutions to a portion of the surface depression caused by the caved material, allowing the leach solutions to percolate downward approximately 300 m (1000 ft) of mineralized material into the old established production galleries of the mining operation. When the copper content of the effluent solutions dropped to a predetermined level, the solutions on the surface of the caved area are moved to a new section. The previously leached area is allowed to oxidize approximately one year before leach solutions are again applied to the area. Approximately 680 kg (1.5 million lb) of copper per month have been produced in this manner since 1955.

Criteria for the successful implementation of inplace leaching are:

The mineralized material must be of a rock type that will readily *cave*.

A minimum of 25% of the ore body must be mined to provide sufficient void space to achieve proper caving of the mineralized material.

An impervious shale or a very dense rock base of uniform gradient should underlie the mass of caved material to achieve good solution recovery of the value-bearing solutions.

Recovery galleries or solution recovery well points must be below the water table.

If sulfide minerals are present in the caved material, oxygen from air flow through the material must be maintained by positive air pressure in the material.

Other successful in place commercial leaching operations involve leaching of uranium from the stopes and caved areas of underground uranium mines. Uranium is leached by spraying the underground workings with acidified ferric sulfate solutions that oxidize the insoluble four valent uranium ion into a water soluble hexavalent ion (Schlitt and Shock, 1979).

In situ leaching of uranium from sandstone formations is a continuing successful practice. This technique requires that the ore body be underlain with an impervious shale, clay, or dense rock formation beneath the water table, thereby permitting the recovery of the value-bearing solutions. Patterns of injection wells and recovery wells are designed for each ore body to achieve good solution contact with the mineralization as the solutions flow to the recovery wells.

In situ leaching of copper from the periphery of a mined-out open pit copper mine has been accomplished by pressure injection of the leach solutions into the mineralization. Successful operation on a semi-commercial scale was accomplished by taking advantage of the naturally highly fractured nature of the ore body, along with the strike and dip of the ore body to control the flow of the leach solutions to the recovery point (Malouf, 1971).

Rubblization with explosives of an ore body followed by leaching has proven to be uneconomic. Three different attempts to rubblize an oxide copper deposit with explosives and leach the copper from the rubblized material on a commercial scale have been made in the United States. In all instances, the amount of fines ($+20\%$) generated during the rubblization *blinded* a large proportion of the material from solution contact. Recoveries approximating 17% over a three-year period were achieved before further leaching was abandoned.

In summary, observations and studies made on in situ leaching of minerals indicate that the successful operation is controlled by:

The mineralization must be in a sandstone formation or a highly (naturally) fractured rock, with little or no clay minerals.

The ore body must be underlain with a uniform, dense shale, clay, or rock that would be impervious to leach solutions.

The solution recovery zone must be below the water table and have a uniform gradient or dip to a recovery point to minimize the cost of solution recovery while maximizing the recovery of values.

REFERENCES

Ashbrook, A.W., 1973, "A Review of the Use of Carboxylic Acids as Extractants for the Separation of Metals in Commercial Liquid-Liquid Extraction Operations," *Minerals Science and Engineering,* Vol. 5, No. 3, July, pp. 169–180.

Bass-Becking, L.G.M., and Moore, D., 1961, "Biogenic Sulfides," *Economic Geology,* Vol. 56, pp., 259–272.

Brierly, C., 1982, *Scientific American,* Sept.

Findley, R., 1981, *National Geographic,* Vol. 160, No. 6, Dec., pp. 732–733.

Fletcher, J.B., 1971, "In-Place Leaching—Miami Mine, Miami, Arizona," Preprint No. 71-AS-40, SME-AIME Annual Meeting, New York.

Malouf, E.E., 1968, "Dump Leaching," *Surface Mining,* E.P. Pfleider, ed., AIME, New York, pp. 762–770.

Malouf, E.E., and Prater, J.D., 1961, "Role of Bacteria in the Alteration of Sulfide Minerals," *Journal of Metals,* May.

Malouf, E.E., et al., 1958, "Cyclic Leaching of Metal Values Using Bacteria to Promote Conversion of Ferrous Iron to Ferric Iron," US Patent 2,829,964, Apr.

Malouf, E.E., et al., 1962, "Use of Ferric Sulfate and Mineral Acid in Leach Solutions at pH 1.9-2.8 for the Leaching of Metal Values, Also Using Continuous Bacterial Regeneration of Ferrous Ion to Ferric Ion," US Patent 3,260,593, Sept.

Malouf, E.E., et al., 1971, "Insitu Leaching of Copper," SME-AIME Annual Meeting, New York, Mar. 2.

Matthews, S.W., 1981, *National Geographic,* Vol. 160, No. 6, Dec., pp. 792–805.

Mayling, A.A., 1969, "Bacterial Leaching with an Alkaline Matrix," US Patent 3,455,679, July 15.

McQuiston, F.W., Jr., and Shoemaker, R.S., 1975, *Gold and Silver Cyanidation Plant Practice,* Vol. 1, AIME, New York, 187 pp.

McQuiston, F.W., Jr., and Shoemaker, R.S., 1981, *Gold and Silver Cyanidation Plant Practice,* Vol. 2, AIME, New York, 263 pp.

Posnjak, E., and Merwin, H.E., "The System, Ferric Oxide-Sulfur Trioxide-Water," Contribution from the Geophysical Laboratory of the Carnegie Institution of Washington.

Postgate, J.R., 1965, "Recent Advances in the Study of the Sulfate-Reducing Bacteria," *Bacteriological Reviews,* Dec., pp. 425–435.

Riggs, W., et al., 1978, "Leaching of Copper Using Strong Acid Solutions with Ferric Iron," US Patent 4,091,070, May.

Schlitt, W.J., and Shock, D.A., eds., 1979, *In Situ Uranium Mining & Ground Water Restoration,* AIME, New York, 137 pp.

Shreir, L.L., 1963, "The Microbiology of Corrosion," *Corrosion,* Vol. 1, *Corrosion of Metals and Alloys,* George Newnes Ltd., London.

Smith, E.E., and Shumate, K.S., 1970, "The Sulfide to Sulfate Reaction," FWPCA Grant No. 14010FPS, Feb.

Torma, A.E., 1977, "The Role of *Thiobaccillus ferrooxidans* in Hydrometallurgical Processes," *Advances in Biochemical Engineering,* Vol. 6, pp. 1–37.

6.9.2 Design and Operating Considerations for Mine Waste Embankments

M. K. McCARTER

INTRODUCTION

Waste embankments occupy a substantial portion of the area required for surface mining. Specific mines may have individual embankments in excess of 300 m (984 ft) high and volumes in excess of 500 Mm³ (653 million cu yd) of material (Campbell, 1981). Such structures represent a significant investment in both monetary and energy resources. Subsequent to construction, waste embankments usually have little if any practical use and may represent a long-term liability in the form of a potential source of pollution or stability hazard. Because of economic considerations and potential environmental impacts, waste embankments must be carefully engineered.

The word *dump* is commonly used throughout the mining industry in place of waste embankment and describes the mode of deposition rather than unsanitary or unsightly conditions. In coal operations, such structures may be referred to as tips, signifying the point where rail conveyances are tipped to discharge their contents. Both words should be considered as having similar connotations.

In the past, little consideration was given to selecting optimum dump sites (D'Applonia, 1977?). The rule of thumb was to dump waste occurring in the upper levels of a mine at high elevations and low waste at low elevations while minimizing the distance traveled. This procedure often minimizes immediate cost of waste disposal, but if used indiscriminately can result in expensive rehandling and adverse environmental impact. The Surface Mining Control and Reclamation Act of 1977 and subsequent state and federal legislation have added new dimensions to waste dump planning. At present, greater effort must be expended to properly site dumps to mitigate environmental effects and promote long-term reclamation. This section will briefly summarize essential concepts in siting, operating, and maintaining rock fills deposited by trucks, scrapers, rail cars, and conveyor belts.

WASTE DUMPS CLASSIFICATION

Waste dumps are broadly classified as water-impounding or nonimpounding. If the dump is situated across natural drainage so as to allow continuous or intermittent accumulation of water or saturated waste behind the structure, it is classified as impounding. If water cannot accumulate, it is classified as nonimpounding. Structures may retain interstitial water or impede percolating water and still maintain the nonimpounding status. Siting of impounding facilities requires consideration of the magnitude of future precipitation events, sedimentation control, and design of out-flow structures. In this respect, design of impounding structures requires more effort than the less complicated nonimpounding structures. The present discussion will focus only on nonimpounding dumps. Further information pertaining to impounding structures may be found elsewhere (D'Applonia, 1977?; Anon., 1971, 1974).

Rock dumps constructed in the process of surface mining commonly are composed of *durable rock* or *hard rock*. Hard rock is defined in 30 CFR 816.74 as "... rockfill consisting of at least 80 percent by volume of sandstone, limestone, or other rocks that do not slake in water." (See Brown, 1981 for a description of slaking tests.) Such materials are expected

to remain relatively inert and provide perpetual permeability of the gross structure. Materials that comply with slake durability as defined by the applicable regulatory authority can be deposited in single-lift structures. Coal waste that does not meet this criteria must be deposited in multiple compacted layers. Normally, regulations pertaining to coal are more stringent than those governing deposits of copper, gold, iron, molybdenum, etc. If a question exists concerning acceptable engineering practice, the engineer can usually be assured of a conservative design by adhering to the regulations for coal waste as defined in Title 30 Code of Federal Regulations.

In addition to impounding and nonimpounding status and durable vs. nondurable material, waste dumps can also be classified according to configuration, composition, and environmental sensitivity (Taylor and Greenwood, 1985).

Configuration

Natural topography provides several options for siting dumps. The dumps may be placed on relatively level ground, in valleys, on side hills, or along ridge tops. The location provides a corresponding adjective that is helpful in creating a mental picture of the structure.

Heaped: A heaped or fan dump is constructed on relatively flat terrain (Couzens, 1985). The final elevation of the crest of the dump may be achieved by compacting successive layers of fill with interconnecting ramps established between the original ground elevation and the current lift (Fig. 1). Alternatively, the structure can be completed by extending the top of a ramp laterally in a single lift (Fig. 2). If the lift is long relative to its width, it is referred to as a *finger dump*.

Valley Fill: Waste may be placed in a valley by continuous dumping on the downstream face from the planned final elevation. Valley fills may also be constructed from the bottom up using compacted lifts. The latter construction is more expensive and occasionally more hazardous in that steep downhill grades may be necessary to bring waste to the final elevation. The level surface of the dump is normally graded toward collection channels on the lateral boundaries or to a rock-core chimney drain established at the center of the dump (see 30 CFR 816.73).

A true valley fill is constructed below the ridge line separating the construction site from adjacent canyons (Fig. 3). If the valley fill is constructed to completely fill the disposal site to the elevation of the ridge line, the structure is more properly identified as a head-of-hollow fill (Fig. 4). If the dump extends from one side of a drainage channel to the opposite side in the shape of an earth dam, it is referred to as a cross-valley fill (Fig. 5). Such structures can be made temporarily nonimpounding by placing culverts in the drainage as part of the construction process. Maintaining a perpetual nonimpounding status for cross-valley fills is a difficult task in that metallic culverts and other open conduits are easily plugged by debris and may have a short life expectancy.

Side-Hill Fill: A very common practice in noncoal surface mines is to construct dumps over existing slopes (Fig. 6). Construction is normally accomplished by allowing the material to freeflow down the slope from a predetermined crest elevation. Side-hill fills may also constitute the first

Fig. 1. Heaped dump constructed in successive layers.

Fig. 2. Heaped dump constructed in a single lift.

Fig. 3. Valley fill.

Fig. 4. Head-of-hollow fill.

Fig. 5. Cross-valley fill.

Fig. 6. Side-hill fill.

construction phase for valley fills. Occasionally side-hill fills are referred to as wedge dumps (Couzens, 1985).

Ridge Fills: Embankments of this type are placed at a ridge crest so that material is deposited on both sides of the divide (Fig. 7). Ridge fills may be thought of as back-to-back side-hill fills.

Composition

Taylor and Greenwood (1985) identify four categories that can be used to identify the potential impact of mine wastes on biological systems and the environment in general. The categories are (1) nonhazardous/hazardous, (2) inert/reactive, (3) fugitive dust susceptible, and (4) leachate susceptible.

Nonhazardous/Hazardous: Section 1004(5) of the Resources Conservation Recovery Act classifies material that presents a substantial potential threat to human health or the environment as hazardous. Few, if any, overburden materials fit this classification unless they contain radioactive substances, soluble toxic compounds, or have been treated with chemicals such as cyanide or similar reagents.

Inert/Reactive: An inert material exhibits no chemical reactivity under normal environmental conditions. A reactive material, on the other hand, can undergo chemical change at normal temperature, moisture, and oxygen levels. Waste containing iron pyrite would be a good example of a reactive material.

Fugitive Dust Susceptible: This category is applied most frequently to mill tailings but can describe overburden waste that contains a high percentage of dust size particles. Such windborne material can restrict visibility and occasionally present a health hazard for individuals working in the immediate area.

Leachate Susceptible: Susceptibility to leachate action is closely related to processes involved in the inert/reactive category. Leachate susceptibility, however, may not involve chemical reactions. Substances may simply be soluble in aqueous solutions. The term leachate susceptible also implies a degree of permeability for the waste material. Not only must soluble materials be present, but the fill must allow infiltration and migration of solutions before adverse environmental consequences occur.

Environmental Sensitivity

The impact of a mine waste dump is directly proportional to the sensitivity of the environment in which the structure is located. Taylor and Greenwood (1985) offer a relative scale of sensitivity based on hydrologic, climatic, ecological, or social/political consequences. A dump located below a large watershed or in an area of high annual precipitation will be more vulnerable to flood flows and solution activity than one located in a dry climate. A dump located above a flood plain could impact water users downstream to a greater extent than a dump at a lower elevation. A dump located in the vicinity of abundant wildlife and vegetation would certainly have a higher environmental sensitivity than a similar structure located in a barren area. And finally, a dump located near communities or industrial facilities would have a higher sensitivity classification than one located in a remote area.

Taylor and Greenwood (1985) have proposed an alphanumeric classification for coal waste that incorporates factors describing impounding, configuration, composition, and sensitivity status. This scheme probably has much utility for regulatory purposes, but is of less value to engineers who need a more generic description. It is important to remember, however, that the foregoing factors need to be assessed in any complete description of a dump site.

Fig. 7. Ridge fill.

REGULATIONS

Siting of waste dumps in the United States is subject to regulatory control in the form of permits, design specifications, or design approvals. It is necessary for the engineer to become familiar with specific requirements by referring to the appropriate state and federal documents. Table 1 provides a listing of the various federal acts and the subjects with which they deal. Within the federal government, responsibility for approving mining plans rests with the agency assigned to manage the surface at the given location (e.g., Forest Service, Bureau of Land Management, Bureau of Indian Affairs, or Bureau of Mines-OSM). In addition, responsibility may be governed by the mineral commodity (e.g., locatables, leasables, or salables). Consequently, more than one federal agency may be involved in approving plans for a waste dump. In general, a federal, state, or local authority will approve or adopt the regulations of another legal jurisdiction provided the regulation is more stringent than its own. Where several agencies are involved, a joint review process may be implemented, allowing one agency to conduct approval processes for all (Vandre, 1985).

Requirements specified in the various regulations are designed to prevent adverse consequences and are generally conservative. The fact that a waste dump complies with all regulations, however, does not eliminate the possibility of future legal problems involving adjacent land owners. For this reason, the engineer should fully evaluate the risks involved in siting a specific dump and incorporate appropriate mitigating measures in the embankment design. For example, specific regulations require a safety factor of 1.5 for waste embankments associated with coal operations. If a particular dump is to be located uphill from a community (high environmental sensitivity), the engineer should carefully choose the "acceptable" technique for stability analysis and critically review how representative the laboratory and field data are. If uncertainty exists, it may be appropriate to raise the design safety factor above that required by the regulations.

A brief summary of general design requirements contained in the Surface Mining Control and Reclamation Act is provided in Table 2. These regulations are specifically for excess spoil associated with coal mining; however, obvious parallels do exist for dumps associated with other types of deposits.

SITE EVALUATION

Four steps are easily identified in selecting the optimum site for a waste dump:

1. Locate potential dump sites well beyond the surface intercept of the ultimate pit.
2. Rank available sites according to capacity and haulage cost per ton.
3. Evaluate geotechnical and hydrologic suitability of sites exhibiting acceptable capacity and minimum cost.
4. Estimate reclamation and mitigation costs if substantially different construction methods are to be used in the alternative sites.

The first step appears obvious but in reality is not easily accomplished. Generation of the ultimate pit surface is an ongoing engineering responsibility. As economic conditions change, the position of the ultimate intercept may also change. As a consequence, attempts to minimize immediate costs through short haulage profiles may be offset by the need to rehandle waste placed too close to the open pit. Some planning engineers establish a zone 100 m (328 ft) wide or more beyond the ultimate intercept in which no permanent facilities are placed.

Frequently the first two steps will clearly identify which site is optimum. Occasionally two or more alternatives will exhibit closely competing total hauling cost. In this case, environmental sensitivity and the cost of environmental impact mitigation and reclamation should be considered along with haulage cost.

Previous sections have described procedures for calculating haulage cost for a given profile. The profile for waste haulage begins at the centroid of the volume of material to be removed and ends at a point representing the average position of the dump crest for the planning period. The position of the crest as a function of time can be obtained by arbitrarily drawing a crest position within the dump limits and calculating the corresponding volume and tonnage. The time to bring the dump crest to the trial position is obtained from the average waste production assigned to that dump. Successive trial positions are assumed until the planning period is bracketed. The average crest position can then be estimated by eye.

The density of implaced waste is a function of initial swell of excavated material relative to the bank density and the recompaction of loose material in the embankment. Common earth materials will swell from 10 to 60% of their original volume when excavated (Gessel, 1981). In hard rock operations, the swell is typically between 30 to 45% (Bohnet, 1985). Compaction is a function of material type, size distribution, and placement method. Factors are variable and can range as high as 25% relative to bank conditions. Com-

Table 1. Summary of Legislation Affecting Mine Waste Dumps*

Subject	Act or Statute
Environmental analysis	National Environmental Policy Act of 1969 (NEPA) 42 USC 4321 et seq.
Water pollution	Federal Water Pollution Control Act Amendments of 1972, Clean Water Act of 1977, Clean Water Act Amendments 1978 33 USC 1251 et seq.
Control of solid waste	Resource Conservation and Recovery Act of 1976 (RCRA) 42 USC 6901 et seq.
Health and safety	Federal Coal Mine Health and Safety Act of 1977 30 USC 951 et seq.
Operation and reclamation	Surface Mining Control and Reclamation Act of 1977 (SMCRA) 30 USC 1201 et seq.
Hazardous waste	Toxic Substances Control Act (TSCA)

* Vandre, 1985; Whiting, 1985.

Table 2. Required Specifications for Excess Spoil

Specification	Reference
Leachate and surface runoff must not degrade surface or ground waters	30 CFR 816.74-1
Fill must be designed using recognized professional standards and certified by a registered professional engineer	30 CFR 816.74-3b
All vegetative and organic material must be removed from the disposal site prior to dumping	30 CFR 816.74-3c
Slope protection in the form of riprap or vegetation must be provided on all disturbed areas to minimize surface erosion	30 CFR 816.74-3e
Disposal areas must be located on the most moderately sloping naturally stable areas	30 CFR 816.74-3e
No depressions or impoundments will be allowed on completed fills (exception noted for head-of-hollow fills, 30 CFR 816.73)	30 CFR 816.74-3g
Fills are not to be constructed on natural terrain where the slope exceeds 36% unless special provisions are made to insure stability	30 CFR 816.74-3i
The fill must be inspected quarterly by a registered engineer or qualified professional specialist	30 CFR 816.74-3j
If the disposal area contains water courses, springs, or seeps, an underdrain consisting of durable rock with suitable filters must be provided	30 CFR 816.74-3l
Investigation and testing of foundation materials for stability analyses must be performed	30 CFR 816.74-3m
Fill placed against a highwall must have a static safety factor of 1.3	30 CFR 816.74-3o,3,iv
Valley fills must have a static safety factor of 1.5 and a safety factor of 1.1 under earthquake conditions	30 CFR 816.72-a
Surface water above a valley fill must be diverted away from the fill in channels designed to safely pass the runoff from a 100-year, 24-hr precipitation event or larger event if specified by the regulatory authority	30 CFR 816.72-d
Horizontal surfaces of valley fills must not slope greater than 5% (3% for head-of-hollow fills 816.73-b-3)	30 CFR 816.72-e
Vertical distance between terraces of valley fills must not exceed 15 m (50 ft)	30 CFR 816.72-e
Drainage must not be directed over the outslope of a valley fill	30 CFR 816.72-f
The outslope of a valley fill must not exceed 50%	30 CFR 816.72-g
The final configuration of the fill must be suitable for postmining land use	30 CFR 816.71-g

bining both swell and compaction, an average factor of about 1.9 t/m³ (120 pcf) is a good design density for single-lift truck dumps composed of porphyry or quartzite. Adjustments need to be made for uncommonly heavy material such as iron ore or unusually light material such as borax. The density should also be reduced for conveyor placement or increased for compacted lifts relative to end-dumping by truck.

Single-lift dump capacity is also a function of the angle of repose for the material. For most waste materials this angle ranges between 34 and 40°. Use of 37° is appropriate for volume estimates, but 34° should be used for estimating toe positions.

Hydrologic Considerations

Proper dump design requires a knowledge of existing drainage patterns that will be impacted by construction of a waste dump. A waste dump may interfere with these patterns by impeding flow from the upslope natural watershed or restricting the flow of springs and seeps which may be covered by the embankment.

Dump design must allow safe handling of the anticipated precipitation event specified by the regulatory authority. This design process requires establishing the following site conditions (Vandre, 1980): (1) area of watershed upstream from the embankment; (2) steepness of the slopes, soil type, vegetation cover, and infiltration rates for both natural and embankment areas; (3) runoff amounts and time distribution for the design event; (4) volume of runoff for the design event if runoff is to be impounded; and (5) peak flow if runoff is to be diverted. The foregoing reference also provides accepted procedures for using this information for surface water control. Consideration should also be given to the possible effect of surface runoff on the ground-water regime (Nelson and McWhorter, 1985; Whiting, 1985). Temporary impoundment can create excess pore pressure within the waste dump or within natural slopes downhill from diversion structures.

In addition to surface runoff, embankment stability can be adversely affected by burying springs or seeps. If it becomes necessary to cover such features, the location and seasonal flow rate should be established. This information will allow design of adequate underdrains if the fill material is not free draining (Cedegren, 1985).

Geotechnical Considerations

A well-designed geotechnical investigation provides necessary geological information and engineering properties of foundation and fill materials. The investigation should identify all factors affecting stability including significant topographic features, soil horizons and depths, rock types, depth of weathering, rock structure, alteration products, accumulation of organic material, and depth to ground water. Samples of soil and rock should be acquired for conventional laboratory testing and in situ field testing when appropriate. Standard techniques include surface geologic mapping and test pits for shallow reconnaissance and drill holes and geophysical surveys for deeper investigations (Welsh, 1985). Geologic maps should identify distribution of soil and rock types on a topographic base map. This map should also identify significant bedrock structural features such as bedding direction, faults, and fracture orientation. Surface features such as soil creep, landslide debris, and springs should be located.

Test pits, which can be excavated by hand, backhoe, or dozer, provide an excellent means of measuring depth of various soil horizons and extent of bedrock weathering. Such excavations provide access for removing samples for laboratory testing or field testing and direct observation of near-surface ground-water conditions. Test pits are usually limited to about 4 m (13 ft) in depth and should be logged immediately after excavation. Prompt attention will reduce the likelihood of premature sloughing, and prompt filling of the pit will help insure safe conditions.

For locations with deep soil accumulations or shallow ground water, drill holes may be required. Normally holes are limited to depths of about 30 m (98 ft), or 3 m (9.8 ft) into bedrock where the soil is relatively shallow. If the entire hole is in soil, auger drilling is usually preferred. When rock is encountered, rotary or core drilling may be necessary. The resulting holes provide the same information as test pits and may be used to subsequently monitor ground-water levels and leachate migration in areas adjacent to the completed waste dump.

Frequently geophysical techniques can be used in conjunction with test borings to determine continuity of subsurface horizons. The most applicable techniques include seismic refraction and resistivity surveys. Both methods can be applied with minimum site preparation and little disturbance to the environment. Proper interpretation of results, however, does require test pits or drill holes to confirm depth to contrasting horizons (Welsh, 1985).

In summary, a well-executed geotechnical investigation draws upon all existing data for a given site and concentrates on those factors most likely to affect the type of dump to be constructed. When properly planned, the geotechnical investigation provides cost-effective information required for stability analysis of the proposed structure.

STABILITY ANALYSIS

Evaluation of dump stability requires an appreciation of the various modes of failure. These modes include slumping, liquification, and foundation failure (Vandre, 1980; Pernichele and Kahle, 1971).

1. *Slumping*—This form of failure is most often confined to the crest area and upper reaches of the inclined surface of end-dumped embankments. End-dumping usually produces a natural segregation of material beginning with the fine fraction near the top and coarser material toward the bottom. Fine material tends to accumulate at a steeper angle than coarse material and periodically sloughs to maintain overall equilibrium. Deep-seated rotational failure is not as common. This form of failure is precipitated by excessive embankment height in cohesive material. Both embankment material and foundation soil can be involved if the soil is of significant depth.

2. *Liquification*—The most common form of liquification (shallow flow slide) is confined to near surface materials brought to saturation by excessive precipitation or snowmelt. The usual result is a mud or debris flow that can impact downstream areas for a considerable distance. A less common form of liquification is referred to as a *blowout*. This type of instability results from retention of water infiltrating from the surface. If sufficient water accumulates and a path is found to the inclined surface of the dump, the saturated debris can move outward allowing the crest to collapse.

3. *Foundation Failure*—Foundation failure can result from base translation or foundation spreading. Base translation most often occurs on steep slopes that are covered with a thin layer of cohesive soil or decaying organic material. Movement is characterized by translation of the entire embankment, or substantial portion of it, along the dump/soil interface. Foundation spreading is characterized by downhill translation of a portion of the toe of the embankment and subsequent squeezing of the foundation soil ahead of the toe. This form of instability is characteristic of rapidly advancing end-dumped embankments over saturated foundation soils. The rate of failure can be rapid with substantial impact to the crest and downslope areas.

A number of mathematical procedures are available for estimating the safety factor for proposed or existing waste dumps. Vandre (1980) presents a series of chart and table solutions covering most forms of failure. These charts are particularly useful for investigating safety factor sensitivity to assumed design parameters and material properties. More rigorous general solutions are available (Bishop, 1955; Morgenstern and Price, 1965; Janbu, 1973; Caldwell and Moss, 1985). Basic principles, assumptions, and differences among these various methods are discussed by Wright (1985).

The reader who wishes to pursue a detailed study of the various "accepted" analytical procedures will find that all methods generally will not yield the same safety factor for a given embankment. Experience and judgment, therefore, must play an important part in selecting suitable methods and input parameters. Vandre (1980) suggests the following factors in selecting appropriate analytical techniques: (1) consequence of instability, (2) confidence in investigation procedures and values for material properties, (3) reliability of design assumptions, (4) ability to predict adverse conditions, (5) control over construction practices, and (6) judgment based on past experience.

DUMP CONSTRUCTION

Specific construction practices are identified in the regulations for excess spoil dumps associated with coal. Little information, however, is available for waste embankments constructed by end-dumping in single lifts. The following list is presented to help identify major considerations in construction and operation of such structures (Bohnet, 1985):

1) New dumps established on steep hill sides are often subject to excessive settlement. The extent of settlement can be minimized by limiting the rate of growth by phasing in the new area gradually.

2) Side-hill fills should be initiated by a pioneer road cut along a contour slightly lower than the anticipated crest elevation. This road will provide a reference for maintaining proper elevation of the dump surface. The road may also provide a platform for light plants and drainage diversion.

3) The surface of the dump should be constructed with a 1 or 2% uphill grade toward the berm. This slope will provide adequate drainage and additional safety for trucks backing toward the berm. Maintaining this grade requires constant resurfacing of the dump to counteract settlement, which can amount to more than 0.3 m (1 ft) per day.

4) The width of the maintained surface of the dump should accommodate the turning radius of the trucks using the dump. For large vehicles, widths of 100 m (328 ft) are adequate with 30 m (98 ft) of crest for each vehicle assigned to the dump.

5) If the dump is to be reclaimed at a slope significantly less than the angle of repose, consideration should be given to benching the operating dump to limit the amount of dozer work required to produce the final slope.

6) The berm should be maintained at one-half of the diameter of the truck tire.

7) Light plants should be high enough to prevent blinding of drivers as they pull away from the berm at night.

8) Trucks that need assistance in dumping large boulders should not be permitted to dump directly over the berm. End-dumping under such conditions could cause the truck to tip over.

9) Stability of side-hill fills may be improved by establishing the operating crest perpendicular to the topographic contours and dumping in a direction roughly parallel to the contours (Campbell, 1985).

MONITORING

Once design decisions are made and construction begins, it is prudent to monitor the performance of waste dumps (McCarter, 1985). Monitoring through the use of instrumentation can improve reliability of performance data and help insure safe working conditions. Applicable instrumentation includes:

Conventional surveying—Nearly all mines have access to theodolites, levels, and EDM equipment. Such equipment provides an expedient way of measuring vertical settlement and total displacement of stations established on the surface of a dump. Adequate coverage, however, requires frequent measurements to establish short-term trends which are important in detecting impending instability.

Extensometers—This term is applied to any one of several devices that permit measuring the change in distance between two points. The two points may be located on the surface (Ko and McCarter, 1975) or below the surface in drill holes or buried pipes. Extensometers can be positioned so as to measure extension parallel to the surface or compaction in the vertical direction. Data may be collected on an intermittent basis or linked to recording devices to provide a continuous plot of movement as a function of time. In addition, extensometers can be automatically interrogated by a telemetry system or serve to activate warning devices when a preset rate of movement is exceeded.

Piezometers—Devices that are designed to measure pressure acting through the liquid phase in a saturated geologic material are referred to as piezometers. An open vertical hole that permits measurement of depth to standing water is perhaps the most common form of piezometer. An improvement over this observation well is provided by lowering a casing with a porous tip to a given depth in the hole. The casing serves to support the hole and provide a guide for sounding devices used to measure the water level. Multiple casings can also be installed in the same hole and isolated from each other by suitable packing materials (D'Applonia, 1977?). This procedure allows monitoring of pressures at various depths.

Observation wells are not suitable under all conditions. Special application may require more sophisticated electrical or pneumatic piezometers (McCarter, 1985). These devices provide an analog output proportional to the pressure at a specific point.

Piezometers can provide valuable information but they are usually difficult to maintain. Internal shear zones and high rates of compaction can crush or distort the casing or sever electrical or pneumatic lines. These problems are particularly common in deep holes. Shallow holes, on the other hand, generally are easier to maintain and can be used to probe for perched water which may contribute to blowout failures.

Inclinometers—Devices that sense a change in attitude with respect to the pull of gravity are commonly referred to as inclinometers. Inclinometers are usually placed in vertical holes and serve to measure deflection of the hole as differential movement occurs. Special adaptations allow inclinometers to be placed in inclined or horizontal holes (Anon., 1982).

Acoustic Emissions—Stressed geologic material, including fill, provides microseismic events that can be measured by suitable instrumentation (Leaird, 1981). The number of events per unit of time appears to be correlated with the magnitude of stress and/or movement. This technique offers promise for dump monitoring, but at this time has not been widely applied. The technology is certainly available, but additional research is needed on proper application and interpretation of data.

Instrumentation provides a convenient way of quantifying observations and allowing automatic collection of data. The importance of on-site, personal inspections should not, however, be overshadowed by the desire for technological sophistication. Experienced equipment operators, supervisory personnel, and field technicians often provide the most important source of information pertaining to the stability and performance of a waste dump. Sufficient care and attention should be given to establishing a mechanism for communicating observations to the individual responsible for operation of the dump. The following list provides a few gross manifestations of potential instability of end-dumped embankments: (1) difficulty in maintaining the berm or positive grade at the crest, (2) shallow steplike fractures near the crest, (3) a change in condensate patterns issuing from the surface in cold weather, (4) a sudden change in turbidity of discharge water, (5) convex upward curvature of the inclined surface of the dump, and (6) popping noise near the toe.

RECLAMATION

Reclamation of waste dumps involves shaping to acceptable configurations, applying of topsoil or appropriate soil conditioners, planting suitable vegetation, and ensuring successful growth for the required postmining period. The objective is to return the land to a condition as close as possible to the premining state or at least to leave the site in a condition compatible with the surrounding terrain. According to Richardson (1985), the major factors influencing the success of reclamation include: (1) climate; (2) chemical, hydrologic, and physical conditions of the fill material; (3) availability of suitable plant species; and (4) proper management of reclaimed sites.

The amount of precipitation, frequency, and the form (rain or snow) are obviously important factors in establishing and perpetuating vegetation. In addition, the amount of sunlight, temperature, and the number of freeze/thaw cycles significantly affect the length of the growing season and the water infiltration and retention characteristics for a given site.

The nature of the spoil and orientation of the embankment can also have a dramatic impact on revegetation efforts. South-facing slopes receive more direct sunlight and reach higher surface temperatures than north-facing slopes. Soil color influences surface temperatures, which can have an important impact on establishing initial growth. Physical characteristics, such as topsoil thickness, texture, particle size distribution, hardness, retention capacity, and hydrologic conductivity influence biologic suitability and should be determined by laboratory testing before planning reclamation efforts. Laboratory testing should also include determination of salinity; pH; nitrogen, phosphorous, and potassium content; and other chemical properties that will govern the productivity of the site.

Selection of appropriate species is critical to successful revegetation. The proper mix of grasses, shrubs, and forbs helps insure rapid coverage and establishment of environmental equilibrium. In addition, availability of seed and need for bare root or container-grown plants must be anticipated in order to minimize cost and time for completion of reclamation efforts.

Finally, proper management of the reclaimed land is essential. Careful control must be exercised over domestic livestock, and occasionally, native animal life must be excluded from the area until a stable ecology is established.

Reclamation of waste dumps requires application of known agronomic methods coupled with an appreciation of the special problems peculiar to specific mineral resources. Substantial research has been done in the field and detailed information is available (Richardson, 1985).

CONCLUSION

Waste management is an integral part of surface mine planning. Application of existing technology and careful engineering is necessary for economic optimization, safety, and long-term reclamation of mine waste dumps.

REFERENCES

Anon., 1971, *Tips—NCB (Production) Codes and Rules*, National Coal Board.
Anon., 1973, "Design of Small Dams," US Bureau of Reclamation, US Government Printing Office, Washington, DC.
Anon., 1982, Product Specification Brochure, SINCO (Slope Indicator Co.), Seattle, WA.
Bishop, A.W., 1955, "The Use of the Slip Circle in the Stability Analysis of Slopes," *Geotechnique*, Vol. 5, Mar., pp. 7–17.
Bohnet, E.L., 1985, "Optimum Dump Planning in Rugged Terrain," *Design of Non-Impounding Mine Waste Dumps*, M.K. McCarter, ed., AIME, New York, pp. 23–27.
Brown, E.T., 1981, "Suggested Methods for Determining Water Content, Porosity, Density, Absorption and Related Properties and Swelling and Slake-Durability Index Properties," *Rock Characterization Testing and Monitoring*, Pergamon Press, Oxford, England, pp. 89–94.
Caldwell, J.A., and Moss, A.S.E., 1985, "Simplified Stability Analysis," *Design of Non-Impounding Mine Waste Dumps*, M.K. McCarter, ed., AIME, New York, pp. 49–61.
Campbell, D.B., 1985, "Construction and Performance in Mountainous Terrain," *Design of Non-Impounding Mine Waste Dumps*, M.K. McCarter, ed., AIME, New York, pp. 145–149.
Cedergren, H.R., 1985, "Design of Drainage Systems for Embankments and Other Civil Engineering Works," *Design of Non-Impounding Mine Waste Dumps*, M.K. McCarter, ed., AIME, New York, pp. 109–119.
Couzens, T.R., 1985, "Planning Models: Operating and Environmental Implications," *Design of Non-Impounding Mine Waste Dumps*, M.K. McCarter, ed., AIME, New York, pp. 13–20.
D'Applonia Consulting Engineers, 1977?, *Engineering and Design Manual, Coal Refuse Disposal Facilities*, US Department of the Interior, Mining Enforcement and Safety Administration, US Government Printing Office, Catalog No. I 68.8: EN3.
Gessel, R.C., 1981, "The Basic Principles of Estimating," *Mineral Industry Costs*, J.R. Hoskins and W.R. Green, eds., Northwest Mining Assoc. Spokane, WA, pp. 92–93.
Janbu, N. 1973, "Slope Stability Computations," *Embankment-Dam Engineering, Casagrande Volume*, R.C. Hirschfield and S.J. Poulos, eds., John Wiley & Sons, New York, pp. 47–86.
Ko, K.C., and McCarter, M.K., 1975, "Dynamic Behavior of Pit Slopes in Response to Blasting and Precipitation," *Application of Rock Mechanics*, ASCE, pp. 363–383.
Leaird, J., 1981, "Geotechnical Waveguides," *Acoustic Emission Trends*, Vol. 2, No. 3, pp. 5.
McCarter, M.K., 1985, "Stability Monitoring," *Design of Non-Impounding Mine Waste Dumps*, M.K. McCarter, ed., AIME, New York, pp. 161–173.
Morgenstern, N.R., and Price, V.E., 1965, "The Analysis of the Stability of General Slip Surfaces," *Geotechnique*, Vol. 15, pp. 79–93.
Nelson, J.D., and McWhorter, D.B., 1985, "Water Movement," *Design of Non-Impounding Mine Waste Dumps*, M.K. McCarter, ed., AIME, New York, pp. 99–107.
Pernichele, A.D., and Kahle, M.B., 1971, "Stability of Waste Dumps at Kennecott's Bingham Canyon Mine," *Trans. SME-AIME*, Vol. 250, Dec., pp. 363–367.
Richardson, B.Z., 1985, "Reclamation in the Intermountain Rocky Mountain Range," *Design of Non-Impounding Mine Waste Dumps*, M.K. McCarter, ed., AIME, New York, pp. 177–192.
Taylor, M.J., and Greenwood, R.J., 1985, "Classification and Surface Water Controls," *Design of Non-Impounding Mine Waste Dumps*, M.K. McCarter, ed., AIME, New York, pp. 1–11.
Vandre, B.C., 1980, "Tentative Engineering Guide: Stability of Non-Water Impounding Mine Waste Embankments," USDA Forest Service, Intermountain Region, Ogden, UT.
Vandre, B.C., 1985, "Scoping Regulatory Requirements," *Design of Non-Impounding Mine Waste Dumps*, M.K. McCarter, ed., AIME, New York, pp. 79–88.
Welsh, J.D., 1985, "Geotechnical Site Investigation," *Design of Non-Impounding Mine Waste Dumps*, M.K. McCarter, ed., AIME, New York, pp. 31–34.
Whiting, D.L., 1985, "Surface and Groundwater Pollution Potential," *Design of Non-Impounding Mine Waste Dumps*, M.K. McCarter, ed., AIME, New York, pp. 89–98.
Wright, S.G., 1985, "Limit Equilibrium Slope Analysis Procedures," *Design of Non-Impounding Mine Waste Dumps*, M.K. McCarter, ed., AIME, New York, pp. 63–77.

6.10 Materials Handling

A.T. YU

SURFACE MINING LOADOUT SYSTEMS

It has been stated that once geology is set and rock is fractured, mining becomes essentially a materials handling proposition. In fact, a recent survey indicated ore and waste handling account for nearly 45% of total mining costs (Michaelson, 1974). The key to the success of a mining venture from that point on depends largely on how efficiently and economically one can move the materials.

Traditional Systems vs. Belt Systems

Over decades, ore or stripping has been loaded by cyclic machines such as shovels, front-end loaders, and draglines onto trucks or rail cars. As pit size and bench height increased, machines got bigger.

The traditional mode of operation is flexible, particularly where selective mining is required and equipment may be readily moved to desired locations as mining plans change. Breakdown of individual pieces of equipment is remedied by the rapid dispatch of spare units. Electrification of shovels and trucks greatly reduced mechanical wear and downtime with resultant significant increases in reliability. Recent introduction of hydraulic machines further improved equipment availability.

Continued escalation of labor costs coupled with the world energy debacle provided the impetus for the introduction and advances of *continuous* belt mining systems. If a belt system is properly applied, it can replace a fleet of trucks with only a few flights of belt conveyors and exchange similarly scores of drivers with one or two operators. Equally dramatic savings are evident in maintenance and repair as well as fuel.

A belt system is fed by a bucket wheel type loading machine if hard rock is not a problem, or traditional shovel and loader operation used in conjunction with in-pit crushing. Discharge of stripping is generally accomplished by a shiftable belt system, coupled with a crawler-mounted stacker (see the section on "Subsystems and Hardware, Belt Modules").

Storage/Reclaim Systems

To ensure continuous operation at the mine and uninterrupted loadout haulage, a *surge facility,* or a storage/reclaim system is required. These systems may be open or housed.

Open storage is cost effective. On the other hand, shelters may need to be built over the stockpiles, in some instances, to keep out weather and to contain dusting. Alternatively, large bins or silos may be used.

Stockpiling is generally accomplished by stackers (rail- or tire-mounted) or overhead devices such as trippers or shuttle belts. Stored material may be reclaimed by gravity underground or a wide variety of machines such as bucket wheel reclaimers, scrapers, or front-end loaders. Sizes of stockpiles can vary from a few thousand to several million tons, whereas bins or silos can be as big as 9 to 18 kt (10,000 to 20,000 st) [silos up to 90 kt (100,000 st) capacity have been planned in Japan].

Geotechnical consideration is often a major factor in the development, design, and management of an open storage/reclaim system. Poor soil can bring about ground failures that result in major losses of ore or concentrates as well as equipment. Where an alternate site is not available, relieve platforms supported on drilled piers can be a solution. Another approach is to build a berm around the stockpile to provide a counteracting stabilizing force. This is often accomplished by close monitoring with instrumentation coupled with a programmed stockpile reclaim strategy over a period of time until the soils are stabilized (Yu, 1982).

Large lumps or sticky material with clayish content impede steady flow and create a problem in reclaim operation. When reclaiming stockpiles by gravity, traveling plow feeders over a slotted tunnel roof have done wonders in promoting positive flow (Fig. 1). To ensure free flow, bins and silos should be designed with a *mass flow* principle. Otherwise, bin bottom activators such as a Vibrascrew type of pneumatic device should be used. In very difficult situations, pneumatic cannons may be installed at strategic locations to dislodge wedged-in material.

Conceptually a storage/reclaim system may be viewed as a necessary evil. Ore or concentrates in storage adds to inventory costs, requires double handling, occupies valuable space, and creates environmental problems, particularly in open storage. Consequently, meticulous fine-tuning is required to minimize the added cost. In this effort, one must recognize the fundamentally conflicting desires of engineers, operators, and owners. For example, the owner is primarily concerned with profit whereas the engineer may overemphasize state-of-the-art technology or overdesign the system for a limited throughput. The operator is inclined to add redundancy, larger-than-necessary storage, and easy maintenance at any cost.

Effective fine-tuning requires careful consideration of the needs and desires of all three parties concerned to optimize the system overall. The engineer must not overlook the operator. Both the owners and operators must keep abreast of developing new systems, their multitude of variations as well as their cost implications (Yu, 1982).

Mine and Plant Haulage Systems

Four principal haulage systems serve a mining operation: truck, rail, aerial tramway, and belt. Owing to their significant cost impact, numerous new systems are continuously being developed by manufacturers and the US Bureau of Mines.

By far the most commonly used piece of equipment in a mining operation is the haul truck. A truck can get into the most difficult area in a mine and it is generally regarded as having the greatest flexibility. Based on how the cargo is discharged, trucks are divided into side-dump, bottom-dump, and rear-dump types. For long distance haulage, the loads may be distributed into two separate carriers thus creating a truck-trailer configuration.

To attain the economies of scale, in the 1960s, payload capacities of haul trucks grew from 59 to 90 t (65 to 100 st). The introduction of electric drives increased reliability and reduced maintenance. The 1970s saw the development of the 154-t (170-st) diesel electric trucks. Over 1400 of these units were built in the 1970s, and are currently in use virtually everywhere in the world. Trucks over 272 t (300 st) in capacity are currently in active development for the 1980s.

900

Fig. 1. Traveling rotary plow feeder.

One of the major problems with trucking is the need for a large number of operators. Escalating fuel costs have added yet another problem. To cope with the high fuel costs, a trolley power-assist system has been developed. This system provides commercial electric power through an overhead line to power electric-drive haulage trucks in certain areas in the mine. Up to 75% of fuel savings was indicated in certain test installations.

To reduce labor, the ATC (automatic truck control systems) has been developed. The truck operation is directed from the remote control center. Although this system is now available for simple mining applications, additional development is required to achieve reliability and maneuverability in complex and large open pits (Lyon, 1981).

For high volume haulage over relatively difficult terrain, rail generally is preferred over trucking. The smooth rail tracks substantially reduce rolling friction. The multitude of rail cars in one string provides substantially greater capacity over individual trucks [6350 to 13 608 t (7,000 to over 15,000 st) per train load compared to 90 to 272 t (100 to 300 st) per truck]. Moreover, electrification is readily achievable with a railway operation.

Gradient and horizontal curve limitations make railroads costly to construct over rugged terrain [2% slope and 6° horizontal curve equivalent to a radius of 303 m (995 ft) have been normally regarded as a maximum]. In extreme cases, a 3% slope and a 10° curve equivalent to 175 m (573 ft) radius have been used. Braking under load can also be a safety problem in a downhill operation.

If the tonnage to be carried is not large [272 to 363 t/h (300 to 400 stph)], an aerial tramway could be a preferred choice of haulage in extreme rugged terrain. A bicable system can carry up to 4.5 t (5 st) per car running at 183 m/min (600 fpm) and up to 453.6 t (500 stph) (Anon, 1976). These systems can negotiate as high as a 30° gradient, compared to normally 15° to 18° for belt conveyors, up to 10° for truck roads, and 2° or 3° for rail tracks.

The need for reliable and economic means of haulage has resulted in the increased usage of belt conveyors. When properly designed, a belt conveyor system can traverse rugged terrain, carry substantial tonnage with little manpower and comparatively low power consumption (in a regenerative configuration, it can even generate power), and requires little maintenance (Anon, 1982). Recent development in the shiftable and extensive variety provides a great deal of flexibility to be applied in many underground and open pit mining operations (see the section on "Subsystems and Hardware, Belt Modules").

Compared to rail or trucks, one of the drawbacks of a belt conveyor system is its inability to carry excessively large lumps. To overcome this problem, in-pit crushers have been used. Today, belt conveyors can be as wide as 306 cm (120 in.), running at well over 305 m/min (1000 fpm), and can carry more than 36 kt/h (40,000 stph). The longest conveyor belt system ever installed was the 100-km (62-mile) system in Morocco carrying phosphates.

Principally responsible for the belt's increased length and capacities is the emergence of the stranded steel cable core. The steel cable core concept was first introduced in Europe. Today, the special brass coating insures strong bonding of the stranded wires to the rubber. The latest cable splicing techniques, when properly applied, enable the joints to have stronger tensile resistance than the cable itself. The allowable working tension of the steel cable core belt ranges from 44.6 kg/mm (2500 lb/in.) of belt width for 6.35 mm ($\frac{1}{4}$ in.) diam cable to 107 kg/mm (6000 lb/in.) for 9.5 mm ($\frac{3}{8}$ in.) diam cable.

A comparison of the advantages and disadvantages of several commonly used haulage systems is shown in Table 1.

Deep cleats made of rubber or metal have been used successfully to surpass the conventional 18° to 20° incline limit. To climb still steeper inclines, devices such as a wheel, a hugger belt, or a looped belt (Yu, 1971) have been successfully used. By banking idlers at key locations, a belt conveyor has been made to traverse a horizontal double S-curve of 610 m (2000 ft) radius. In the open pit mine at Garsdorf, Germany, a conveyor was able to round a horizontal curve of 305 m (1000 ft) radius. In New Caledonia, a single flight conveyor carries nickel ore 11 km (6.8 miles) by following the contour of the mountain, rounding four horizontal curves with a minimum radius of 1050 m (3445 ft) (Fig. 2). See Table 2.

In a conventional belt system, the cargo on the rubber belt is supported transversely on intermediate troughing idlers and longitudinally by the tension in the belt itself. The belt tension is in turn carried by its fabric carcass or the steel cable cores, as the case may be. To relieve the belt of high tensile stresses and to gain other benefits, the cable belt system was developed whereby the cargo on the belt is carried on the two supporting cables via transverse stiffness imbedded in the belt.

The cable belt system was first conceived late in the 1940s. The introduction of replaceable polyurethane-lined sheaves has been responsible for its recent success. Latest data indicate rope life varying from 18 months to 6 years. This system has scored success in long distance haulage of coal. An installation for crushed coal in West Virginia runs nearly 21 km (13 miles) with single flight length of 15 km (9.3 miles). In Australia two single flights of 30.4 km (18.9 miles) and 20 km (12.4 miles), have been used for bauxite (Anon, 1982).

The first cable belt application in a metal mining operation was built in 1976 in Arizona. It moves copper ore from Asarco's Mission Point 10 km (6.2 miles) to Anamax's Twin Buttes concentrator. The belt is 1067 mm (42 in.) wide, runs at 244 m/min (800 fpm) to handle 1814 t/h (2000 stph). The single flight negotiates a 60° bend at about 0.8 km (0.5 miles) from the primary crusher feed end. It has operated successfully in delivering its rated 4.5 Mt/a (5 million stpy) (see the case history on Twin Buttes).

Economics of each of the haulage systems is quite situation-dependent, particularly since operating costs over the life of the system can significantly overshadow the capital costs. Given the unpredictability of fuel cost fluctuations, haulage system economics must be analyzed on a case-by-case basis and updated by the changing worldwide and local economic scenarios. Rule of thumb estimates can be quite misleading.

One approach is to tabulate costs and cash flow analysis for each alternative considered over its entire service life. This would encompass the investment, fixed costs such as taxes and insurance, as well as operating costs. These numbers are then *present-valued* with an assumed rate to provide a total net present value of each of these alternatives.

A recent study (Benavides and Schuster, 1982) compares

Table 1. Comparison of Commonly Used Haulage Systems

Haulage System	Advantages	Disadvantages
Railroad	Rugged equipment-resistance to abuse Same installation may be used to haul equipment, supplies, and personnel No need for power transmission or large power plant if diesel equipment is used	High construction costs, especially in rugged terrain Inability to negotiate steep grades, consequently requiring much longer run Maintenance High accident rate in difficult terrain
Trucks	Low investment cost Highest flexibility Can carry equipment, supplies, and personnel	High maintenance, road, and equipment Require a team of operators and shop personnel High operating and fuel costs High accident rate
Belt conveyors	Low operating cost Comparatively lower investment cost for high tonnage and rugged terrain Requires least number of operators Reliability Low noise level	Normally cannot be used for hauling supplies, equipment, or personnel Rubber belt is vulnerable to damage unless all design precautions are exercised and operating restrictions are properly policed
Tramway	Ability to negotiate rugged terrain Environmentally favorable	Low capacity Relatively high maintenance

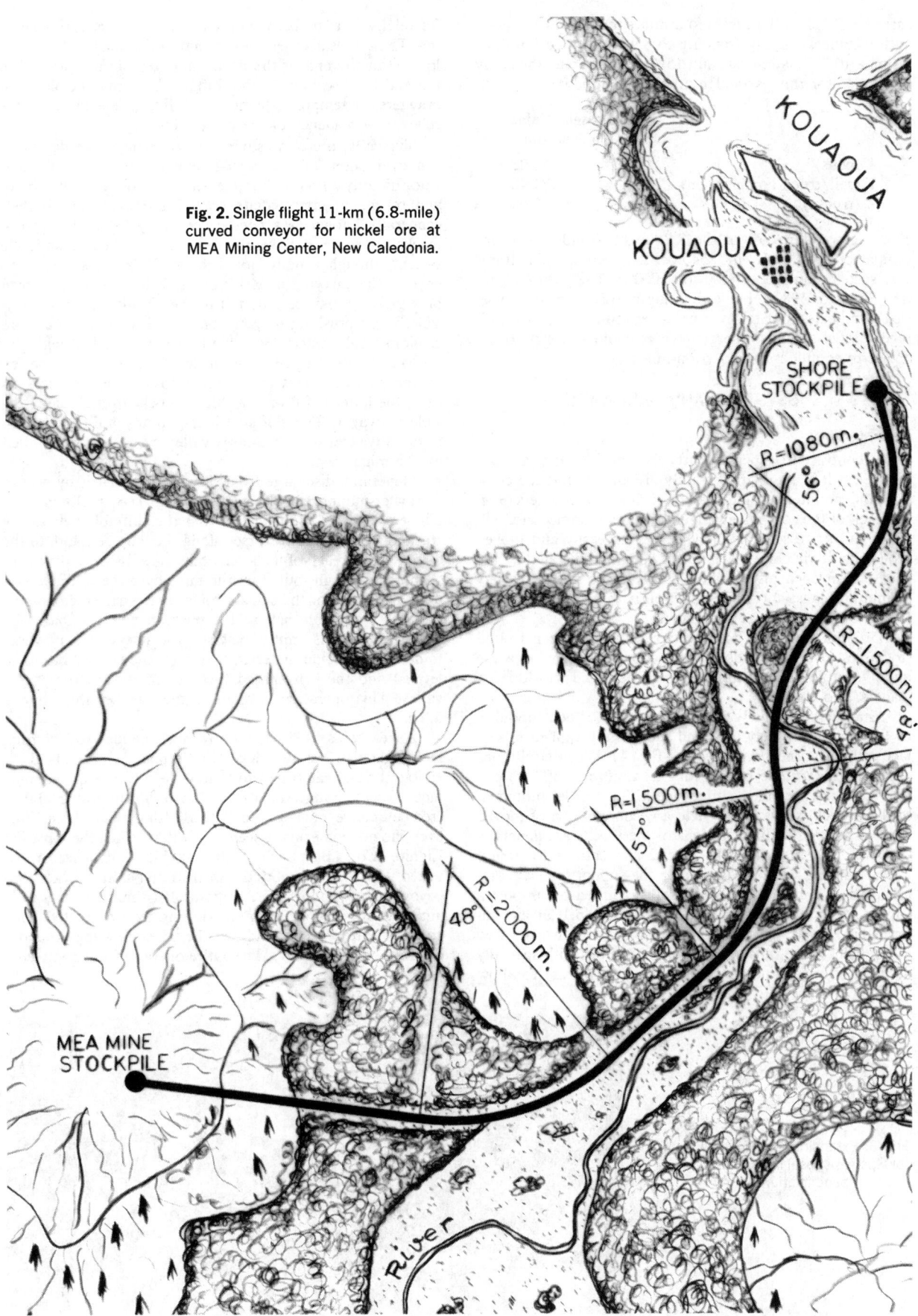

Fig. 2. Single flight 11-km (6.8-mile) curved conveyor for nickel ore at MEA Mining Center, New Caledonia.

three modes of haulage from four mines to a processing plant with distances varying from 1.6 to 8.9 km (1 to 5.5 miles). Using a 10% inflation rate and 15% discount rate, the study concluded for this particular situation as follows:

	Present Value (in $ million)
X (belts)	58.25
Y (rail/truck combination)	71.40
Z (trucks)	160.64

The construction cost of the 8.8-km (5.47-mile) belt was estimated to be $1771/m ($540 per ft). A comparable tramway system could cost $600 to $2000/m ($200 to 600 per ft) (Bonasso, 1983). For estimating purposes, recent experience indicates four to seven units per ton-mile. In downhill regenerative systems, energy savings could cause it to decrease to as much as 2¢ per ton-mile.

SUBSYSTEMS AND HARDWARE

Belt Modules

In contrast to a permanently fixed installation, a modularized belt is made up of readily dismantled unitized components, mounted on skids, rail ties, wheels, or crawlers (Fig. 3). Each of the components, including the intermediate, tail, head, and drive sections, may be steered easily and moved around to fit virtually any mode of operation dictated by the needs of an open pit operator, according to a wide variety of mining plans, and in conjunction with an assortment of excavating and haulage equipment.

Tire-mounted short portable conveyors have been around stockpiles for decades. Skid or tire-mounted dozer traps are quite versatile in cleanup in the pit when used with dozers, loaders, and trucks.

Among the virtually limitless varieties of belt modules, by far the most widely used and successfully applied in surface mining is the *shiftable* belt (Fig. 4). When conditions are right, a *shiftable* belt as long as 2286 m (7500 ft) can follow the advance of the mine face with a crawler-mounted, bucket-wheel excavator, to yield an economic ton of moved earth. The intermediate sections of these belts are generally supported on rail ties with special connections to permit lateral shifting without damage to the supporting structure (Fig. 5). A pair of rails is generally required to support a tripper for discharge of muck on the main belt at any intermediate point, or a traveling hopper to receive the feed to the belt. These rails are fastened to the ties with specially designed fittings (known as *Nebelung* sole plates) to allow

for relative motion between the rail and the ties during shifting. These rails also serve as a point of anchor for the rollers located at the end of the dozer-mounted shifting arm. The tail and head sections of the shiftable belts may be on skids, crawlers, or temporary concrete footings, depending on the relative permanency of the installation.

Typically, ore or waste is loaded through the discharge conveyor boom of a crawler-mounted, bucket-wheel excavator directly onto the shiftable belt. To add greater flexibility and reach, an intermediate crawler-mounted transfer belt (generally known as a *bandwagon*) may be used. Shovels, trucks, or scrapers have also been used to feed a shiftable system, though usually not directly. If large boulders are beyond the carrying capacity of the belt, a scalping screen or a grizzley may be required in the circuit. Wherever feasible, in-pit crushing may be considered to result in a more economic belt system by virtue of a narrower belt width.

Meanwhile, stripped overburden from a pit may be fed by scrapers or trucks in the pit onto a stationary belt. In turn, the latter is fed onto a shiftable belt through an extensible conveyor. The shiftable belt, initially built on a high ramp, advances toward a large valley eventually to be filled by the mine waste.

Generally, discharge from the shiftable belt is by way of a crawler-mounted boom stacker (Fig. 6). A track-mounted tripper serves as the transfer from the shiftable belt to the stacker. The stacker traverses along the long belt and, to the extent of the reach of its boom conveyor, fills the valley with strippings brought out from the mine over the shiftable system. The reach of the stacker boom is the extreme operable limit of a shiftable belt in the position where it has been shifted. Once this limit is reached, the stacker has nowhere to dump additional material; therefore, the ground has to be leveled and the belt shifted further out toward the empty valley. This operation is repeated until the entire valley is filled.

As early as 1890, the horizontal manual rail shifting technique was first introduced in Germany's lignite fields to enable the rail car tracks to follow the advance of the mine face (Goergen and Ambatiello, 1968). Shifting of modularized belt conveyors began in 1931. Shifting by machine, however, did not come on the scene until 1951, when the Bavarian Lignite Co. (Bayerischen Braunkohlen-Industrie, A.G.) tested what was probably the world's first shiftable belt conveyor in Schwandorf. The belt was 1016 mm (40 in.) wide and 518 m (1700 ft) long. In shifting it was pivoted horizontally with one end held and the far end swung around a distance of 25 m (82 ft). The entire operation was completed in 8 hr.

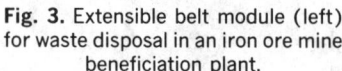

Fig. 3. Extensible belt module (left) for waste disposal in an iron ore mine beneficiation plant.

Fig. 4. Twin Buttes waste dump area when shiftable system began its operation, circa 1968.

The pioneer modules were mounted on a grillage of rails without rigid connections between longitudinal modules. A set of rails had to be laid alongside the belt to support the track-mounted shifting machine in Schwandorf. This was cumbersome and slow. Little wonder it soon gave way to the far more flexible *trackless* shifting system, featuring specially built shifting heads mounted on crawlers. The tripper rails on the belt modules were to be lifted slightly and dragged toward the shifting machine as the latter traversed forward, parallel to the new centerline of the belt. Though a great deal more sophistication and refinement have been introduced into today's shiftable belt modules, their fundamental concept and configuration has not departed materially from the forerunner of the early 1950s.

Outside Germany, Nchanga Consolidated Copper in Zambia's rich Copperbelt was probably one of the first users of the shiftable system (Anon, 1960). Search for an efficient and economic method to remove the large amount of lateritic overburden [projected at 9:1 waste-to-ore ratio at 244 m (800 ft) in depth] led to the bucket wheel and shiftable belt system. Commissioned in November 1958, the system features boom-mounted bucket wheel excavators on crawlers. Working with them are some 4267 m (14,000 ft) of belt

conveyors consisting of shiftables at the mine bench and the dump as well as transfer belts in between. The shiftable bench belts follow the advance of the mine face whereas the dump belt feeds a crawler-mounted stacker (Fig. 7).

Overland transfer belt F is ramped up to 21 m (70 ft) above grade to feed the extensible conveyor G, which may be extended as the valley is being filled. Conveyor G feeds the 914-m (3000-ft) long shiftable belt H built initially on a 21-m (70-ft) high levee. The crawler-mounted dump stacker has a 76.2-m (250-ft) boom conveyor. The valley is capable of receiving 76.5 million m^3 (100 million cu yd) of overburden.

Unlike the German lignite operation, where belt shifting is relatively infrequent, Nchanga's bench belts often have had to be shifted, shortened, or lengthened at three- to four-week intervals. Further, to minimize stripping and simultaneously to maximize ore exposure, the bench belts must be constantly shifted to follow the dip of the ore body. To accomplish this, three extensible shiftable bench belts are required.

The Nchanga belts are 1219 mm (48 in.) wide, each running at 244 m/min (800 fpm) to yield a maximum rate of 3 266 t/h (3600 stph). Each of the steel stringer modules

Fig. 5. A shiftable conveyor several thousand meters (feet) long can be shifted laterally into new positions by a special rig mounted on a bulldozer.

Fig. 6. Crawler-mounted boom stacker.

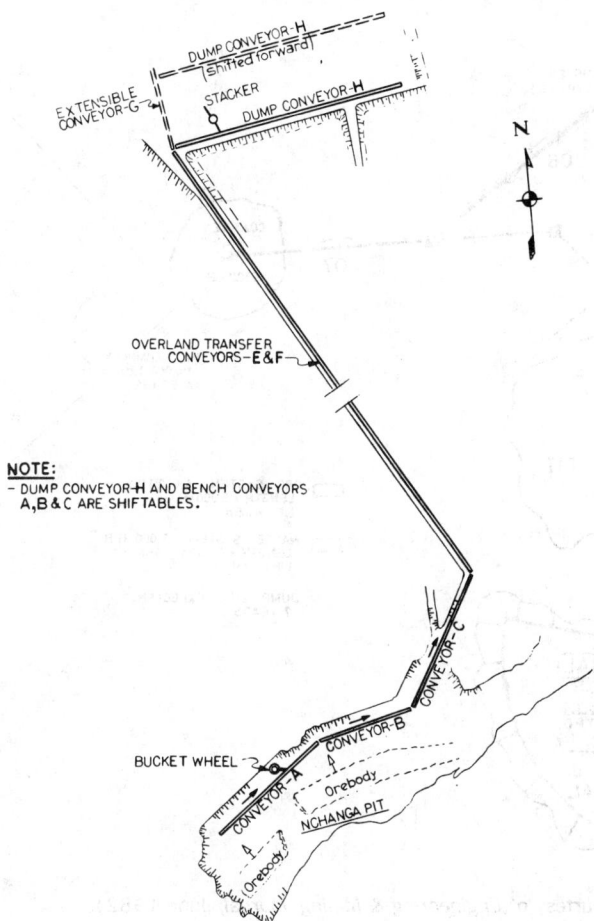

Fig. 7. Schematic diagram of the shiftable and extensible belt system at Nchanga pit.

carries 762-mm (30-in.) troughing idlers spaced at 1 m (3 ft 4 in.). The intermediate posts are mounted on 254 × 152 mm (10 in. × 6 in.) timber ties and the drive sections on skids. With the help of a D8 dozer equipped with a special roller arm, in 0.6 m (2 ft) *bites*, a 670.6-m (2200-ft) belt may be shifted a distance of 39.6 m (130 ft) in a matter of four 12-hr days. Moving an entire conveyor between benches, however, would normally take somewhere between two to three weeks.

Since Nchanga, the use of the shiftable belt has begun to spread. Some of the better known examples are the 3.2 Mt/ a (3.5 million stpy) Neyveli lignite operation in Madras, India; the Kursk magnetite mine in the USSR (Gartner, 1966); the Oroville Dam project in the US (Anon, 1966); the backfill project in Singapore (Von Campenhausen, 1967); the Washington Irrigation's Centralia coal project in the state of Washington; the Guyana Bauxite Co.'s mines in Mc-Kenzie; the Twin Buttes Copper operation in the US (see the case history on Twin Buttes); the Opencast Mine Most in Czechoslovakia; and Goonyella mine for coal in Australia.

A wealth of knowhow and operating experience in the shiftable belt modules has been amassed over the past three decades. Significant advances have been made in the methodology and hardware to make it a versatile and virtually foolproof system. The 3- or 5-roll Garland idler, having its individual rolls linked together by ball joints, provides greater impact resistance, promotes centering of the load, improves

alignment of the belt, and has the advantages of ease of installation and maintenance. Troughing angles have been increased to 45° and, in the case of 5-roll Garlands, to as much as 60°. V-type return idlers further help belt alignment.

In-Pit Crushing

DENNIS K. MORTENSEN

The major sources of all hard rock tonnages mined today come from either quarries or open pits. Ore is normally fractured within the pit by blasting and then truck-hauled to the pit rim for further crushing and transporting to the concentrator. Waste encountered in the mining process is generally truck-hauled out of the pit to designated dump areas. Cost estimates for this type of truck-haulage system range from 25% to 50% of the total mining costs of a typical open pit mining operation (McQuiston and Shoemaker, 1978). Fuel, lubricants, tire replacements, and operator's wages account for more than 80% of the cost of truck haulage and these costs probably will continue to increase at a steady rate (Kok, 1980).

Mining companies are continually looking for new ways to reduce these costs. Equipment replacement with larger units, as well as converting to conveyor haulage out of the pit, offer possibilities for lowering the cost per ton of material handled. The installation of belt conveyors as close to the mining face as possible would appear to be the logical solution to the cost problem. This would allow for maximum use of belt conveyors, which are much less labor-intensive and more energy-efficient than trucking systems. The fallacy of this solution to the problem is the impracticality of direct loading mine-run material onto a belt conveyor, mainly due to the material's excessive lump size. Some means of reducing the lump size to that which a belt conveying system can handle efficiently is necessary. An in-pit crusher, either fixed or portable and used in conjunction with a belt conveyor system, can offer one of the most effective available methods of reducing energy consumption and mining costs.

Although the potential for cost savings with in-pit crushing/conveying systems is recognized, many mining companies have considered implementing such systems, but few have done so. The need to mine both ore and waste at the same time makes it impractical to place crushing/conveying systems at each working face. It is therefore necessary to be able to use trucks to transport all material of the same classification to some convenient point from which it may be crushed and conveyed from the pit.

Many mines are required to simultaneously mine ore, leach rock, and waste. Such a mining operation would require the flexibility of truck haulage to change either rock type or destination at a moment's notice. Some mining companies have combined the flexibility of truck haulage systems with fixed in-pit crushing/conveying systems. In southern Arizona both the Anamax Mining Co. operation and the Duval Sierrita operation have done so.

At the Duval Sierrita* operation two fixed 1520 × 2260 mm (60 × 89 in.) gyratory crushers have been installed in the pit to crush ore (Fig. 8). The ore is conveyed out of the pit and transferred to a 1.37-m (54-in.) overland conveyor that extends to the concentrator. Duval's two original crushers, installed near the pit rim, were converted to handle waste by adding a waste conveying and stacking system. Table 3 shows the original operating components of Duval's system.

---

* Now Cyprus Sierrita.

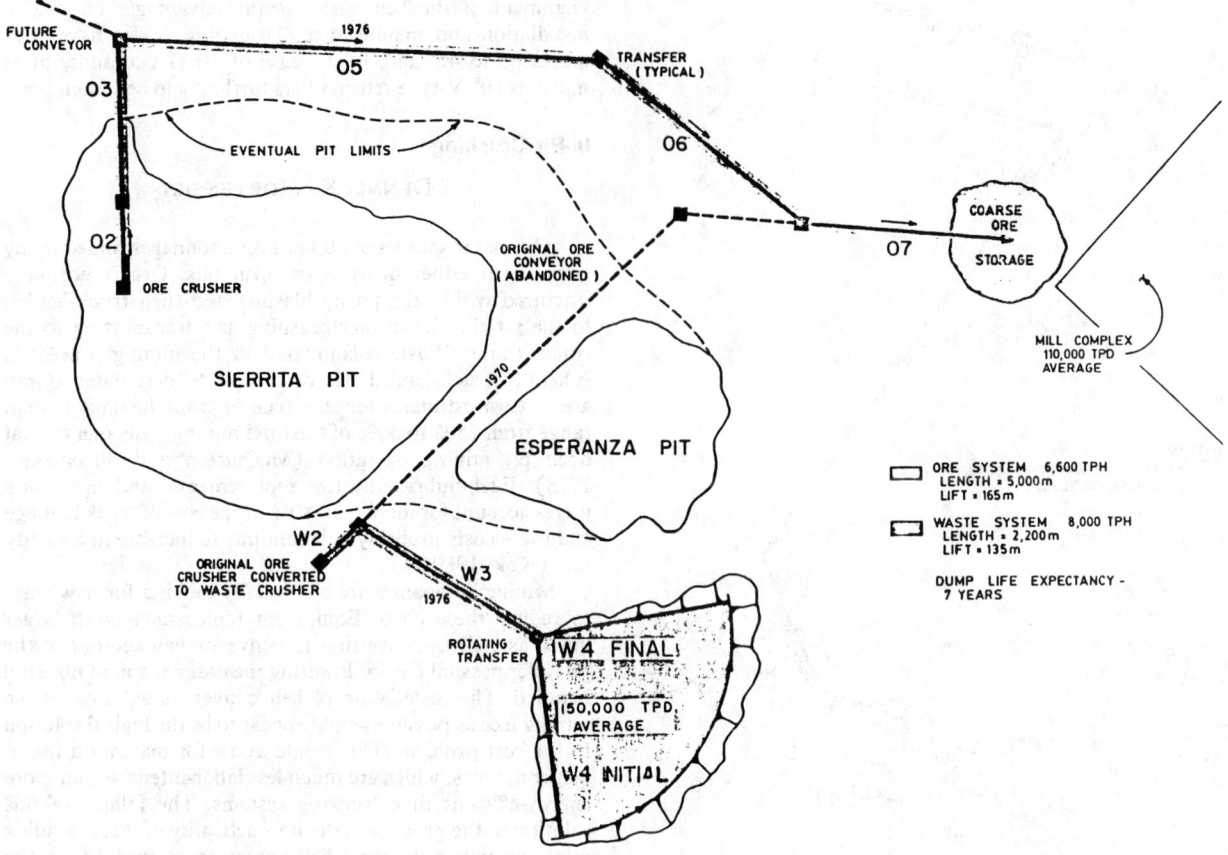

Fig. 8. Conveyor belt schematic of Duval Sierrita mine (courtesy of *Engineering & Mining Journal*, June 1982).

Table 2. Examples of Long Belt Conveyor Systems

Width, in.	System Length, ft	Max. Single Flight Length, ft	Speed, fpm	Capacity, stph	Material	Site	Year	Notes
47	12,140	12,140	492	1,760	Excavated earth	Kobe, Japan	1964	Move mountain to sea through culvert and tunnel
66	15,000	15,000	1100	12,000	Excavated earth	Portage Mtn. Dam Vancouver, BC	1965	Overland
42	54,048	10,591	492	1,200	Stone	Nagato, Japan	1964	Overland/tunnel
36/42	35,756	6,450	600	800/1200	Coal	Greene Cty., PA	1965	Overland/underground
36	50,300	16,400	590	2,240	Iron ore	Marcona, Peru	1967	Overland
42	33,377	10,950	500	1,300	Bauxite	Jamaica	1967	Overland
39	328,000	36,100	900	2,000	Phosphate	Morocco (Sahara)	1972	Overland
42	32,500	29,800	824	2,000	Copper ore	Twin Buttes, AZ	1978	Overland (Cable Belt)
31.5	36,600	36,600	0-710 Variable	560	Nickel ore	Mea mine, New Caledonia	1980	Overland with four curves (Curvoduc)

Metric equivalents: in. × 25.4 = mm; ft × 0.3048 = m; st × 0.907 184 7 = t.

At the Anamax Mining Co. operation, two fixed 1370 × 2030 mm (54 × 80 in.) gyratory crushers were installed in the pit and were able to crush either sulfide ore, oxide ore, or waste followed by conveying to the pit rim (Fig. 9). A transfer system at the pit rim could route sulfide ore, oxide ore, or waste to its final destination by 1.52-m (60-in.) conveying systems designed to handle each class of material. As the pit was deepened, an additional waste handling system was installed. The system consisted of an in-pit crusher and conveying and stacking system similar to the installation at Duval. Table 4 shows the original operating components of Anamax's system.

With regard to both waste and ore handling, the question of where to crush becomes very important from the standpoint of permanence and convenience. The crushing/conveying system should, ideally, be positioned so as to minimize any inconvenience it may create with regard to the future expansion of the mining operation. A location convenient to pit working faces, but not shielding ore that would otherwise be mined, is desirable. However, rarely is either a pit wall or an in-pit crushing/conveying system allowed to remain with any degree of permanency.

Semimobile crusher/conveyor installations address both questions of permanence and convenience and may be a better solution to the problem. These systems can be built as several subunits and placed in the main working area of the pit. Haulage distances to the crusher can be kept short and the system can be moved once the haulage distances become too long. Either minimal or no foundations are required for the equipment, thus resulting in a further cost savings. The systems can be designed so that each of the subunits, e.g., apron-type feed conveyor or crusher, can be transported as a single unit to its next location.

Past development of mobile primary crushers has centered mainly in Germany where soft iron ore and limestone deposits are readily accessible and adaptable to movable systems. Impact type crushers with large feed openings resulted in large reduction ratios but generally suffered from low capacities. Single-toggle jaw crushers and short-shaft gyratories were also used until manufacturers responded with larger crushers, feeders, and movable supports. Today mobile crushers have capacities of up to 3630 t/h (4000 stph).

One of the first large capacity, readily portable units suitable for large shovel operations was built in Mexico in 1975. An Allis-Chalmers 1370 × 1880 mm (54 × 74 in.) gyratory crusher, incorporated in a Weserhütte design, was fed by a 24.6 m long by 2.2 m wide (81 ft by 87 in.) apron feeder sloped at a 25° angle. The unit, operating in a limestone deposit, was fed by front-end loaders and originally operated one shift per day, six days per week. The unit was moved by three hydraulically operated feet, which walked at a speed of 0.02 m/s (4 fpm). The normal operating tonnage for the unit, which was equipped with a Hydroset unit for adjusting the crusher setting, was 680 t/h (750 stph). As of late 1982,

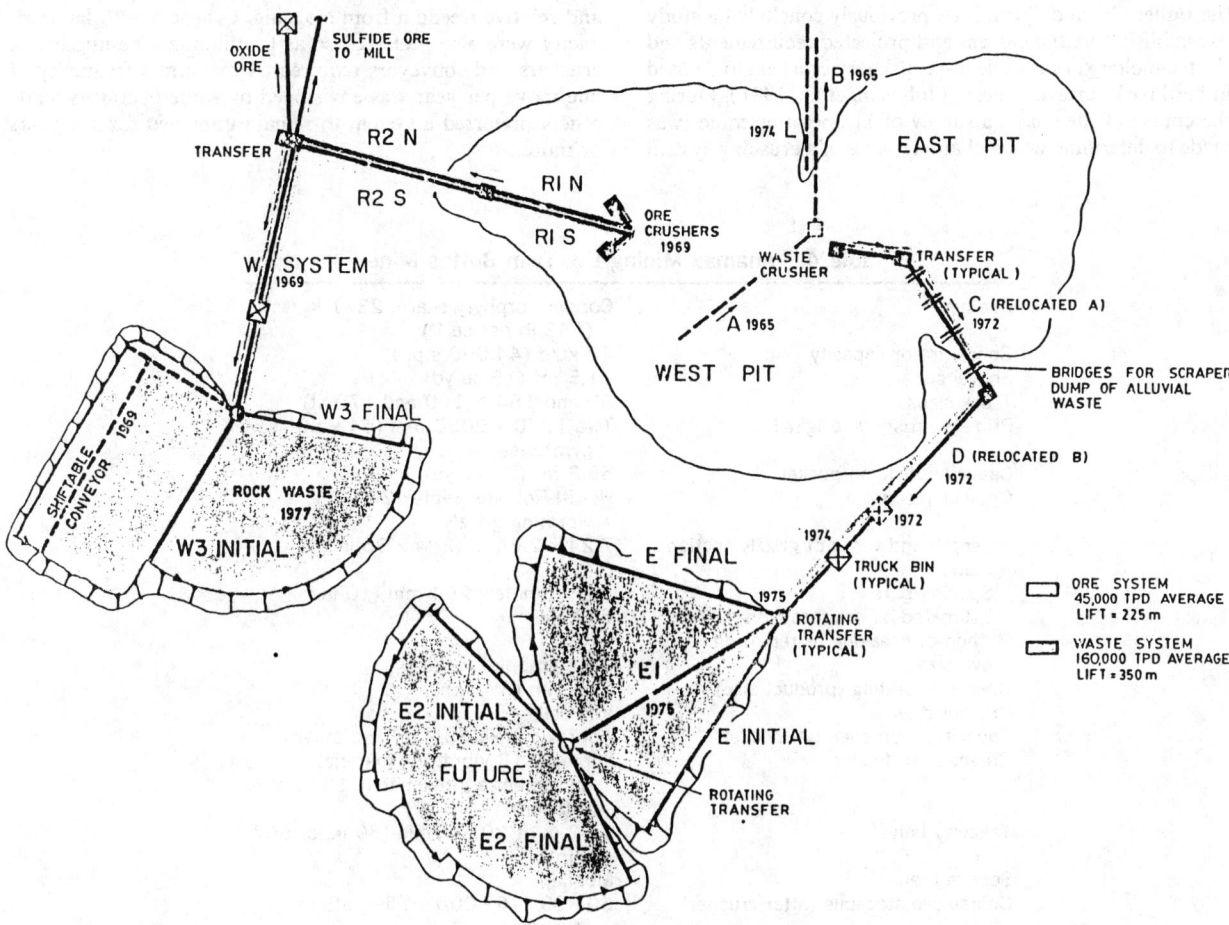

Fig. 9. Conveyor belt schematic of Twin Buttes mine (courtesy of *Engineering & Mining Journal*, June 1982).

Table 3. Duval Corp. Sierrita Mine

Ore type	Copper-porphyry 1442 kg/m³ (90 lb/cu ft)
Concentrator capacity	78 900 t/d (87,000 stpd)
Shovel size	11.5 m³ (15 cu yd)
Truck sizes	109 and 136 t (120 and 150 st)
Primary crusher, original	Two 1520 x 2260 mm (60 x 89 in.) gyratories
Capacity of dump pocket	± 225 t (± 250 st)
Crusher feeder	Not used
Grizzly	Not used
Method of breaking or removing oversize	Rock hook, grapple, sling, and crane
Open side setting (product size)	152 mm (6 in.)
Crusher drive	Direct
Capacity of crushed ore pocket	270 t live (300 st)
Crushed ore feeder	2.13-m apron, 7.9 to 14.9 m/min (84-in., 26 to 49 fpm)
Wear belt and takeaway belt	1.52 m, 192 m/min (60 in., 630 fpm)
Service crane	Mobile

Source: McQuiston and Shoemaker, 1978.

the crusher was still operating as a portable crusher. The crusher has been moved an average of once every two years since 1975.

The US Bureau of Mines (USBM), in conjunction with the Fuller Co. and Gard, Inc., previously concluded a study to establish "... the current and projected requirements and the technology of movable in-pit primary crushers to be used in hard rock surface mines." (Johnson, et al., 1981). During the course of the study, a survey of 11 operating mines was made to determine which characteristics of a crushing system were most desired by mine personnel. Table 5 shows the results of that survey. All of the mine personnel surveyed were interested in portable crushers. The gyratory-type primary crusher was preferred due to its reliability, durability, and relative freedom from clogging. Crushers with large capacity were also preferred so as to minimize the number of crushers and conveyors required. A maximum frequency of one move per year was envisioned by some operators while others preferred a system to remain unmoved for five years or more.

Table 4. Anamax Mining Co. Twin Buttes Mine

Ore type	Copper-porphyry skarn, 2371 kg/m³ (148 lb per cu ft)
Concentrator capacity	40 kt/d (44,000 stpd)
Shovel size	11.5 m³ (15 cu yd)
Truck sizes	90 and 154 t (100 and 170 st)
Primary crushers, original	Two 1370 x 2030 mm (54 x 80 in.) gyratories
Capacity of dump pocket	55.8 m³ (73 cu yd)
Crusher feeder	Hewitt-Robbins Eliptex 2-E-13 vibrating grizzly
Length and width of grizzly section	7.24 x 2.44 m (284 x 96 in.)
Slope	6°
Spacing bars	Approximately 254 mm (10 in.)
Estimated % ore through grizzly	65%
Method of breaking or removing oversize	Rock grapple
Open-side setting (product size)	216 mm (8½ in.)
Crusher drive	Vee belt
Capacity of crushed ore pocket	Approximately 38.2 m³ (50 cu yd)
Crushed ore feeder	Eliptex E-13 vibratory, material movement approximately 18 m/min (60 fpm)
Takaway belt	1.52 m at 203 m/min (60 in. at 667 fpm)
Service crane	Mobile
Coarse ore stockpile (after crusher)	60 800 t (67,000 st) live, after secondary

Source: McQuiston and Shoemaker, 1978.

Based on the USBM survey and general crusher principles, the following design criteria have been recommended for in-pit movable crushers:

1) The crusher should be able to reduce all run-of-mine material normally received.

2) The crushed product should be at the desired size for conveying.

3) The crusher should produce the desired throughput.

4) The system should be readily relocatable.

5) Operating costs should be comparable to a fixed installation.

6) The crushing/conveying system should be labor efficient.

7) Component wear life should be designed at a maximum.

8) An operating availability of not less than 85% should be designed into the crusher.

9) Maintenance should be rapid with efficient parts availability.

10) Site preparation should be minimized.

Currently, most portable in-pit crushing systems are finding use in the lower tonnage quarry stone industry.

Dutec, a division of Duval Corp., introduced a semiportable crushing system designed to reduce operating costs by incorporating the generally accepted design criteria for movable crushers (Anon, 1982). The equipment was assembled at the company's Sierrita operation in southern Arizona for start-up in 1982. The three-section system consisted of a semiportable feeder, crusher, and discharge conveyor. The subunits were designed so that each section could be moved with a crawler-type transporter up a maximum grade of 12% at a speed of 0.22 m/s (0.5 mph). The 895-kW (1200-hp) diesel-powered transporter had a lifting capacity of 1089 t (1200 st), 20% greater than needed to move the heaviest section. It was anticipated that the entire three-unit system could be moved a distance of 0.805 km (0.5 mile) in the pit to a newly prepared site and be ready to run in a 48-hr period. The new equipment was probably the largest crusher-feeder system in the world, as well as the first use of a semiportable gyratory crusher in an open pit copper mine. The crushing unit, a 1520 × 2260 mm (60 × 89 in.) gyratory, had a throughput capacity of 3630 t/h (4000 stph) and all units were serviced by a 360-t (400-st) maximum capacity rough terrain crane.

In open pit mining, the replacement of fixed crushers and truck haulage with semimobile crushers and conveyor haulage can result in an effective means of reducing both energy consumption and constantly mounting mining costs.

The systems now available or in the planning stages offer savings that appear to be readily achievable. However, each situation must be studied on its own merits since no two cases are sufficiently similar to permit a blanket statement of policy.

Bucket Wheel Machine

Much of the pioneering work for the wheel-on-boom excavator was done in Germany's brown coal (lignite) industry. This machine features a series of digging buckets along the perimeter of a steel wheel, supported at the head end of a steel boom (Fig. 10). The wheel is positioned at the face of an excavation bank or a stockpile by traversing the machine and moving the boom. As the wheel rotates, the buckets dig up the material and discharge it onto a belt conveyor supported by the boom. The boom is, in turn, supported by an A-frame. The entire machine is usually mounted on crawler tracks for maneuverability. These machines may also be rail-mounted. The boom is normally raised or lowered by a hoist winch. The boom conveyor conveys the excavated material onto other connecting conveyor systems.

Today the capacities of these machines can vary from 199 to 10 704 m³/h (260 to 14,000 cu yd per hr). In the extreme case, the wheel on the boom can have a vertical range of movement as much as several hundred meters. The largest wheel has a diameter of nearly 15 m (50 ft). The digging effort can go up to 1019 kg/cm²/cm (14,500 psi per in.) on an individual bucket tooth. These machines have been used successfully to excavate medium to hard digging materials such as coal, gypsum, phosphate rock, chalk, and soft limestone.

The wheel-on-boom excavators are now used worldwide in mining operations, especially for stripping of overburden. Smaller machines are used for stockpiling and reclaiming.

In an open pit operation, typically the height of a bench should be about half the diameter of the wheel. Thus a 16-m (52.5-ft) high face may be excavated in four 4-m (13-ft) benches by an 8-m (26-ft) diam bucket wheel. As the boom-mounted bucket wheel slews through the face, it makes a crescent-shaped cut (Fig. 11). The tangent of the crescent parallels the direction of machine travel. Both ends of the crescent are perpendicular to the direction of travel. The boom slewing speed should be varied as the secant of the boom slew angle from the direction of travel to compensate for the varying depths of the crescent. This increases the width of the cut to obtain a constant bite of depth times width. Cutting the extreme tapered ends of the crescent is inefficient. The depth becomes too shallow and boom slewing must slow, stop, and reverse direction. Consequently, the total slew angle for mobile boom-mounted bucket wheel reclaimers should be limited to about 160°.

Quality Control

Trucks are weighed by platform scales with the weighing mechanism set underground. Reading and recording are generally carried out in a control cab at the ground level. Computerized operation provides automatic waybill printout and weight totalizing. Rail cars are weighed in a similar manner.

Automated and accurate weighing can be accomplished at loading if material is metered into trucks or rail car via weigh hoppers. With the use of a programmable controller, waybills from each car in the long train are automatically printed out as loading proceeds (Yu, 1982).

Mechanical scales for weighing in transit have been in existence since the 1920s. With the emergence of electronic and nuclear scales in the 1950s and 1960s, respectively, there is now a choice of three varieties. Under ideal maintenance conditions, the accuracies of a mechanical or electronic scale can be expected to be within ¼ of 1%. Although it has less maximum accuracy, the nuclear scale works on the principle of mass indications and hence has the advantage of being less sensitive to fluctuation of belt tension. Consequently, it may be installed on such locations on a stacker belt where belt tension fluctuates widely and electrical or mechanical scales would not render any reasonable accuracy.

The counterpart of weighing in a quality control subsystem is sampling. The technology and methodology developed for bulk sampling can be quite complex, often involving high-powered statistics as the conceptual foundation. A practical approach, however, has been delineated (Yu, 1972).

As the quantity of raw materials to be handled has increased, so has the need for quality control. Accordingly, long strides have been made in the technique of automatic,

Fig. 10. Close-up of a boom-mounted wheel reclaimer.

mechanical sampling. Substantial progress has also been made in the development of equipment and sampling theories as well as codes and specifications on an international basis (Anon., 1982).

CASE HISTORIES

DENNIS K. MORTENSEN

Twin Buttes (Copper)

The Twin Buttes mine,* located approximately 40 km (25 miles) south of Tucson in the Pima Mining District of

_____
* Twin Buttes is now closed.

southern Arizona, was operated by Anamax Mining Co. (Fig. 12). Anamax Mining Co. was a partnership, formed in 1973, in which 50% interest was held by Atlantic Richfield Co. and its wholly owned subsidiary, Anaconda Arizona, Inc., and 50% was held by Amax Arizona, Inc., a wholly-owned subsidiary of Amax Inc. Mining activity in the district dates back to the 1870s when surface deposits of copper were first commercially mined.

In 1963 The Anaconda Co. entered into a long-term lease agreement with Banner Mining Co. to determine the feasibility of developing an open pit mine at the Twin Buttes site (Knaebel, 1970). Based on Anaconda's exploration data, the decision was made to develop the mine; thus, the largest known preproduction stripping operation in the history of mining began in mid-1965. It linked earthmovers and con-

Fig. 11. Crescent-shaped cuts in an open pit operation by crawler-mounted wheel excavators.

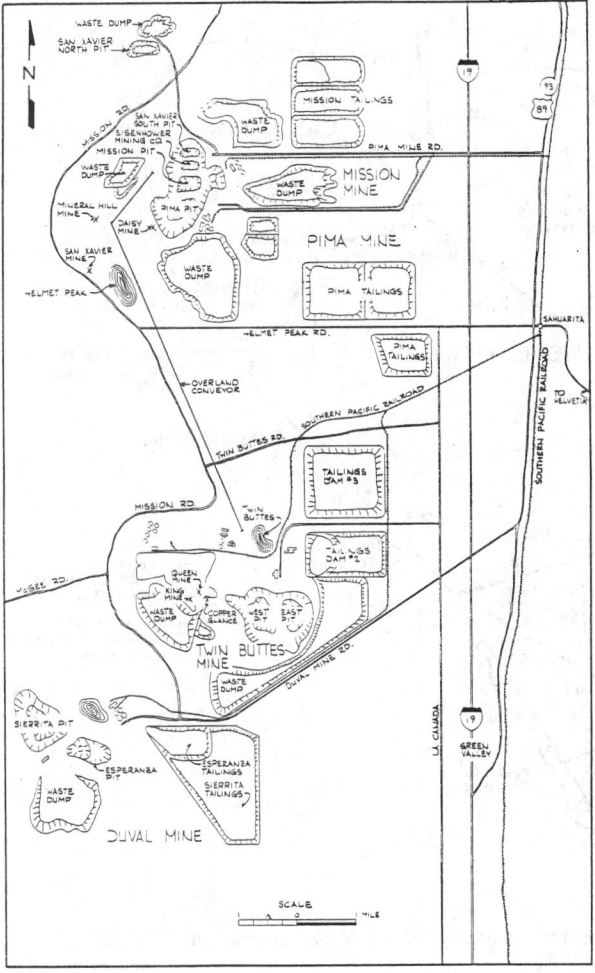

Fig. 12. Pima mining district.

veyor belts in an unusual approach to the development of a pit that became 2200 m long, 1700 m wide, and 365 m deep (7200 × 5600 × 1200 ft).

In order to reach the ore more than 227 Mt (250 million st) of alluvium and waste rock, at a depth of 152 to 244 m (500 to 800 ft), had to be removed.

Extensive studies to determine the most economical way of handling the material resulted in the decision to install an in-pit conveyor system. The system was operated in conjunction with a massive scraper and truck-shovel operation. The decision by Anaconda was significant to the mining industry since it marked the first installation of a high-capacity, high-tension belt conveyor system in a domestic hardrock, open-pit mine.

The first material handling system designated A and B consisted of two 1.52-m (60-in.) wide belt conveyors (Fig. 13) (Almond and Huss, 1982). These conveyors, with a capacity of 7260 t/h (8000 stph), transported the material out of the pit at a 25° slope and overland a total of about 3200 m (2 miles). The material was deposited into a 910-t (1000-st) loadout bin from which it was trucked to tailings pond sites for dam construction.

In 1969, after removing some 150 m (500 ft) of alluvium, waste rock capping was encountered. In anticipation of the waste rock, two individual 1370 × 2030 mm (54 in. × 80 in.) gyratory crushers had been installed in the pit wall,

designed to be fed from end-dump haul trucks (Figs. 14 and 15). Crushed ore or waste from these crushers fed at a rate of 4540 t/h (5000 stph) onto two parallel, 1.52-m (60-in.) belt conveyor systems designated R1-R2 North and R1-R2 South.

A transfer system near the pit rim allowed either of the R belts to feed material onto either of three other belt systems. One belt ran to the north and carried only sulfide ore to the secondary crushing plant. Another belt running to the north carried either oxide or sulfide ore to the truck bin. The third belt ran to the south and carried only waste to a dump where the material was distributed by a shiftable belt conveyor equipped with a track-mounted stacker conveyor. In 1968, the capital cost of each crusher installation was estimated at $1.1 million.

In 1972, the A and B conveyors, as well as the truck feed bin, were disassembled and reinstalled on the south side of the pit. Labeled C and D, they were 730 m (2400 ft) and 915 m (3000 ft) long, respectively. Truck bridges were built across the C conveyor at various bench elevations and material from bottom-dump haulage trucks was fed through a grizzly-feeder arrangement onto the conveyor. At the end of the D belt, bottom-dump trucks reclaimed the material from the truck hopper and hauled it to the waste dump areas.

In 1975, a third 1370 × 2030 mm (54 in. × 80 in.) crusher and conveyor system was installed lower in the pit for waste rock only. The new system also conveyed the crushed rock at a rate of 4540 t/h (5000 stph) to the tail end of the existing C conveyor for transport to the dump area. At the same time, the D conveyor was extended to feed onto a shiftable belt conveyor, E1, also equipped with a track-mounted stacker-conveyor.

Future expansion of the dump area anticipated the E1 belt to become a fixed installation. A new shiftable belt, E2, would be fed by E1. The existing stacker-conveyor would be moved to the new E2 belt for distribution of waste.

By the end of 1982, more than 1.22 Gt (1.34 billion st) of material had been removed from Twin Buttes at a stripping ratio of more than 6.6 to 1.

In 1976, Anamax and Asarco Inc. concluded several years of negotiations and signed the Eisenhower partnership agreement to form Eisenhower Mining Co. The partnership was formed to permit exploitation of Anamax's Palo Verde property and adjacent Asarco holdings as a single mining unit (Brost, 1979). In order to transport Anamax's portion of the ore to the Twin Buttes site, a decision was made to install a conveying system 10.3 km (6.4 mile) long (Fig. 12). The system, again fed by a 1370 × 2030 mm (54 in. × 80 in.) gyratory crusher, was supplied and installed by Cable Belt Ltd. England (Fig. 16). The unique feature of the system was the separation of the driving (tension) medium from the material carrying medium. In the Cable Belt system, the material was carried on a transversely stiffened belt, riding on two endless loops of wire rope that transmit the driving tension. Molded rubber grooves on the belt grip the wire ropes, which are supported at intervals by pulleys (Figs. 17 and 18). The wire ropes were driven by Koepe friction wheels at the drive unit, which incorporated a differential to allow for slight differences in rope movement. Because the ropes supply the driving tension, only enough tension was applied to prevent folding of the belt.

The 1.07-m wide (42-in.) belt was designed to carry 1815 t/h (2000 stph) of copper ore at a speed of 251 m/s (824 fpm). The system was commissioned in early 1979 and by year-end 1982 had transported approximately 16 Mt (18 million st) of ore. Capital cost of the system has been estimated at $18.6 million, including primary crusher installa-

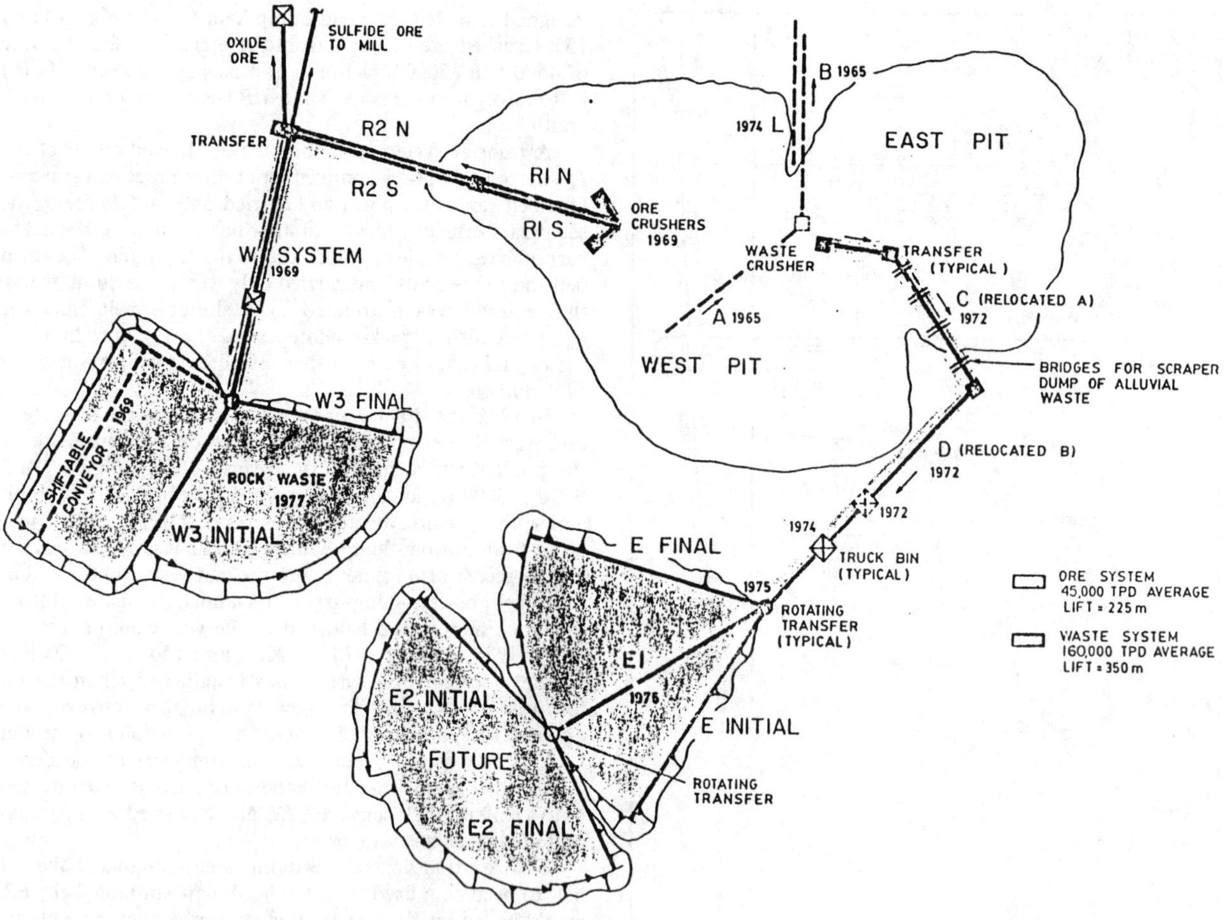

OXIDE ORE

SULFIDE ORE TO MILL

TRANSFER

R2 N

R2 S

W SYSTEM
1969

RI N

RI S

ORE CRUSHERS
1969

B 1965

1974 L

EAST PIT

WASTE CRUSHER

TRANSFER
(TYPICAL)

C (RELOCATED A)
1972

A 1965

WEST PIT

BRIDGES FOR SCRAPER
DUMP OF ALLUVIAL
WASTE

SHIFTABLE CONVEYOR 1969

W3 FINAL

ROCK WASTE
1977

W3 INITIAL

D (RELOCATED B)
1972

1974   1972

TRUCK BIN
(TYPICAL)

E FINAL

1975

ROTATING
TRANSFER
(TYPICAL)

E2 INITIAL

FUTURE

EI

1975

E2 FINAL

E INITIAL

ROTATING
TRANSFER

ORE SYSTEM
45,000 TPD AVERAGE
LIFT = 225 m

WASTE SYSTEM
160,000 TPD AVERAGE
LIFT = 350 m

Fig. 13. Conveyor belt schematic of Twin Buttes mine (courtesy of *Engineering & Mining Journal*, June 1982).

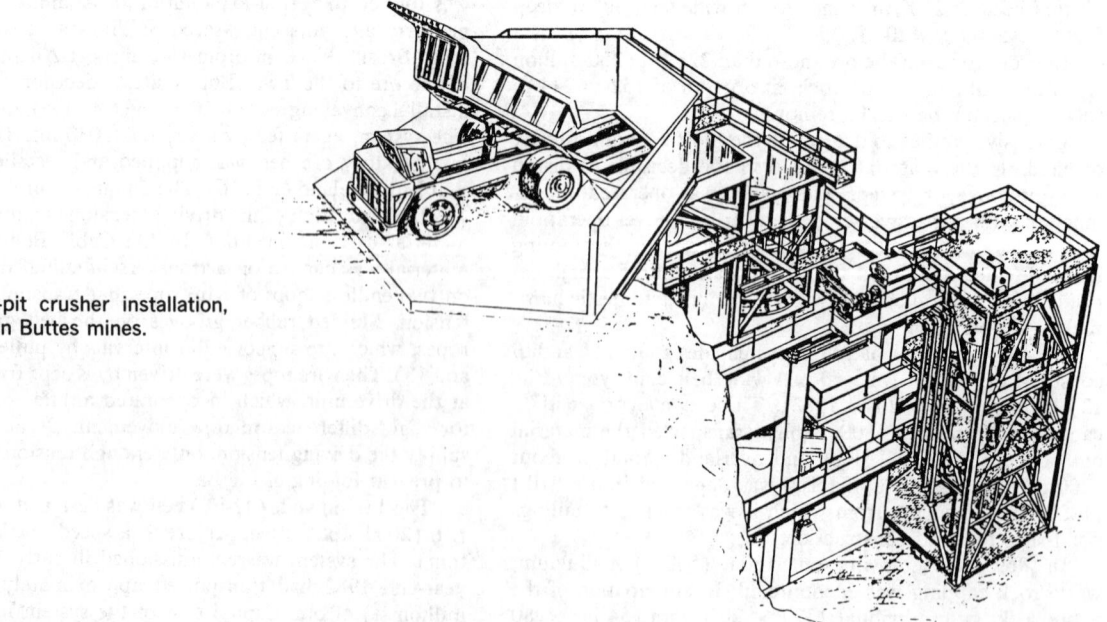

Fig. 14. In-pit crusher installation,
Twin Buttes mines.

Fig. 15. In-pit crusher installation, Twin Buttes mine.

tion, overland conveyor, and sampling plant located at the discharge end of the conveyor. In spite of problems encountered during the early months of operation, full production of 4.5 Mt (5.0 million st) per year was attained at favorable operating costs.

Bingham Mines of Kennecott Minerals (Copper)

Kennecott Mineral's (formerly a unit of Standard Oil of Ohio, then BP Minerals America, and now RTZ) Bingham mine started its operation in 1904. Since then, over 4.5 Gt (5,000,000,000 st) of copper ore, flux, and waste material have been removed. It is the first open pit mine in the copper industry, one of the world's largest mining operations, and has produced more copper than any individual mine in history.

Through 1981, ore milled plus flux amounted to over 1.4 Gt (1,500,000,000 st) with waste removed to dumps nearly 1.8 Gt (2 billion st) at a stripping ratio of 1.9 to 1. The total amount of copper produced in this period was over 11 Mt (12 million st).

The world's largest man-made excavation covers 7.28 km² (1900 acres). The pit measured 4 km (2.5 miles) wide at the top and it is over 0.8 km (0.5 mile) deep from the bottom to the top (Fig. 19).

Fifty-three 15.2-m (50-ft) benches in the mine vary in widths from 10.7 m (35 ft) to 38 m (125 ft). Kennecott's 1982 plan called for daily production rates as follows:

Material	Daily Rates
Ore	99 792 t (110,000 st)
Rail waste	40 824 t (45,000 st)
Truck materials	279 145 t (307, 700 st)
Total materials hauled per day	419 761 t (462,700 st)

On Feb. 11, 1981, a world record for hard rock materials blasted, loaded, and hauled during a 24-hr period was established. On that day, the mine trucks and trains hauled 579 382 t (638,649 st). In 1982, the stripping ratio increased to 3.1:1.0, with average copper content of ore being approximately 0.58%.

The Bingham mine features the most extensive rail haulage and loadout system in the world. In 1982 there were 153 km (95 miles) of standard gage railroad track all under direct centralized traffic control with 87 rail switches within the C.T.C. system and 36 under separate radio control. There are four entry points into the mine railroad proper: one switchback and three through tunnels. The rail cars are loaded directly by shovels at different strategic benches following the breaking of the rock (Fig. 20).

There were 39 electrical shovels varying from 4.6 to 20.6 m³ (6 to 27 yd³) capability and 58 locomotives (23 electric and 35 diesel-electric) in 1982. In addition, 98 haul trucks were used. These end-dump units varied from 59 to 136 t (65 to 150 st) in capacity.

Operating under the Utah Copper Div. of Kennecott, the Bingham pit is located 48 km (30 miles) southwest of Salt Lake City. The ore mined is shipped by rail into the concentration plant outside the pit 21 km (13 miles) north of the mine at the mouth of the Bingham Canyon.

Tailings from the concentrating plant are slurried to a nearby pond, where they have been stored since 1907. The pond now covers more than 2266 ha (5600 acres) and is 33 m (110 ft) high. More than 1.2 Gt (1.3 billion st) of material are impounded in the pond.

To further improve loadout operations, Kennecott embarked on an extensive study to explore the use of in-pit crushing and conveyor haulage. Until this could be imple-

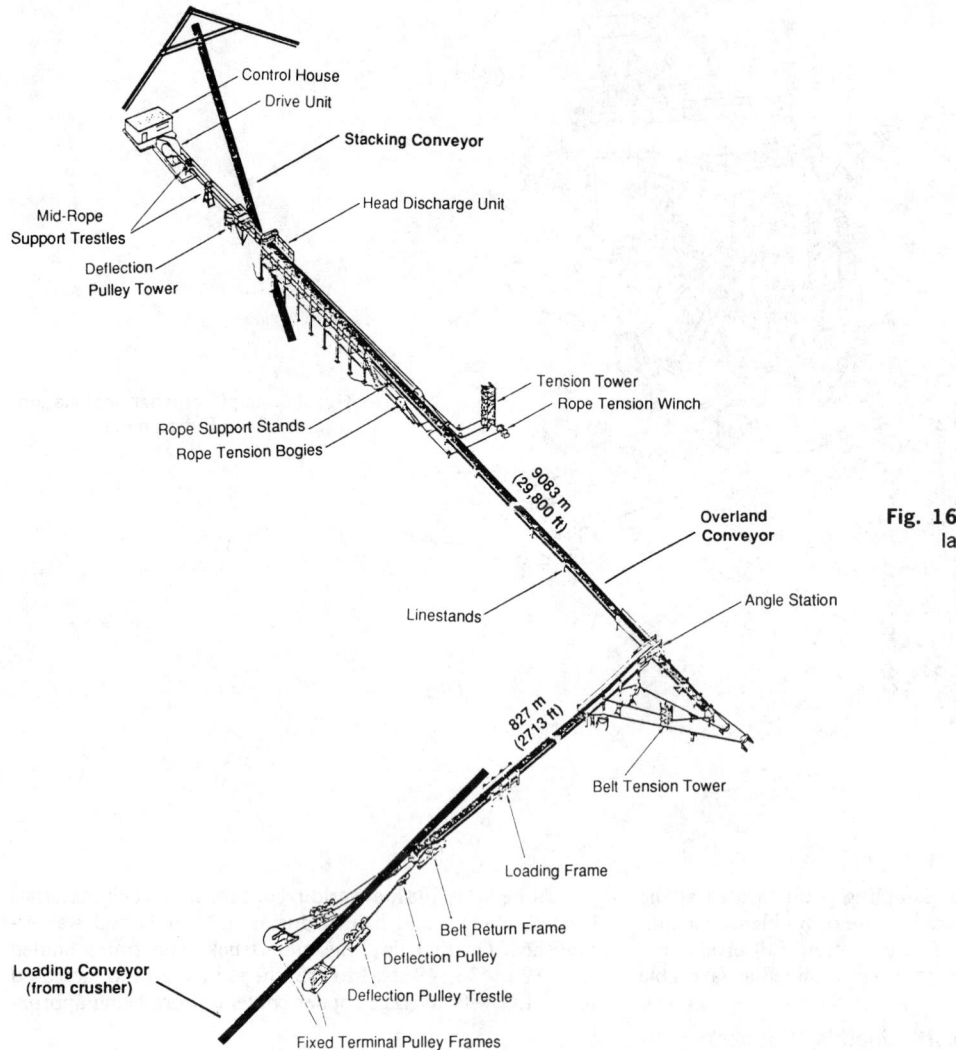

Control House
Drive Unit
Stacking Conveyor
Head Discharge Unit
Mid-Rope
Support Trestles
Deflection
Pulley Tower
Tension Tower
Rope Tension Winch
Rope Support Stands
Rope Tension Bogies
9083 m
(29,800 ft)
**Overland
Conveyor**
Linestands
Angle Station
827 m
(2713 ft)
Belt Tension Tower
Loading Frame
Belt Return Frame
Deflection Pulley
**Loading Conveyor
(from crusher)**
Deflection Pulley Trestle
Fixed Terminal Pulley Frames

Fig. 16. Cable belt conveyor installation, Twin Buttes mine.

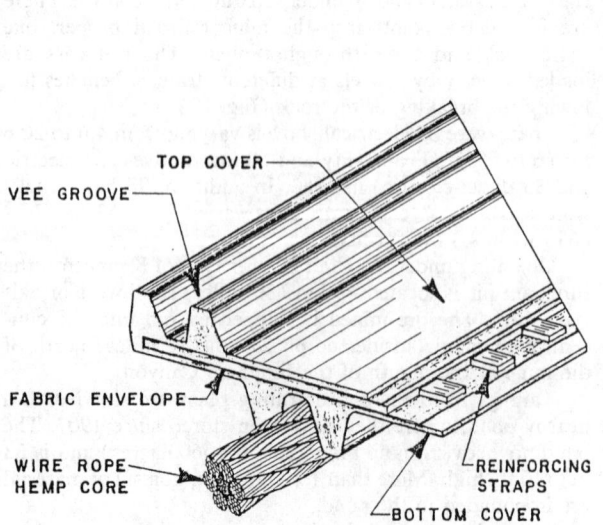

TOP COVER
VEE GROOVE
FABRIC ENVELOPE
WIRE ROPE
HEMP CORE
REINFORCING
STRAPS
BOTTOM COVER

Fig. 17. Cutaway view of cable belt and wire rope support, Twin Buttes mine.

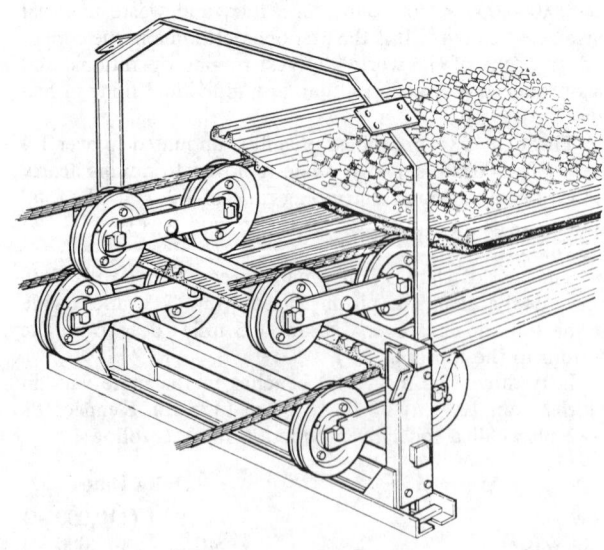

Fig. 18. Cutaway view of cable belt and wire rope support pulleys, Twin Buttes mine.

Table 5. Crusher System Characteristics Desired by Mine Personnel

Gyratory type
Large feedthrough capacity [3630 t/h (4000 stph)]
Freedom from clogging
Maximum frequency of moves: one/year
Average frequency of moves: once every two years
Large feed openings [1370 mm (54 in.) minimum]
Crush rock with compressive strength of up to 3520 kg/cm² (50,000 psi)
High reliability of system 85% +
Freedom from bridging
Low maintenance cost
All-weather operation (−40°C to 40°C) (rain, snow) (−40°F to 120°F)
22 hr per day operation, 350 days per yr
Moderate noise
Dust control at transfer points
Operate with 12 to 15-m (40 to 50-ft) bench heights
Operate with 45-m (150-ft) minimum bench widths
Move up 10% maximum grades [15 m (50 ft) wide]
Relocation within two weeks
Surge capacity of 360 t (400 st)
Ideally compatible with trucks or trains
Minimum of conveyors, but also redundancy where possible

mented, all in-pit operations were converted to truck, leaving the rail system to serve ore haulage from the mine to the concentrators only.

Kennecott's $400 million project at the Utah Copper Div. (UCD) includes in-pit ore crushing and the construction of new grinding and flotation facilities near the mine. The project incorporates some of the largest state-of-the-art crushing, conveying, grinding, and flotation equipment available in the industry (Fig. 21).

In the Bingham mine, a relocatable 1.5 × 2.7-m (60 × 109-in.) gyratory crusher (Allis-Chalmers) and conveyor system replaces the three rotary car dumpers and gyratory crushers at the Bonneville, Magna, and Arthur concentrators.

The existing 1000-car ore haulage railroad network is being replaced by an 8-km (5-mile) coarse ore conveyor system of 1.8-m (72-in.) wide belts which deliver ore from the mine to the new Copperton concentrator. The conveyor system has 12.9 MW (17,300 hp) connected, and a design capacity of 9.1 kt/h (10,000 tph). The crusher and conveyor systems were designed and supplied by PHB Weserhütte AG, West Germany.

The 70-kt/d (77,000-tpd) grinding facility includes three semiautogenous grinding (SAG) mills, 10.4 m diam by 4.6

Fig. 19. Bingham mine of Kennecott Minerals: One of the world's largest man-made excavations, covering 769 ha (1900) acres). Photo by Don Green.

Fig. 20. Rail cars being loaded directly by shovels at different strategic benches following breaking of the rock.

m long (34 ft diam by 15 ft long) and six ball mills, 5.5 m diam by 8.5 m long (18 ft diam by 28 ft long). These units are replacing 14 Symons crushers, 50 roll crushers, 5 rod mills, 50 primary ball mills, and 83 secondary ball mills.

The concentrator (flotation) contains 97 large flotation cells, including 33 85-m³ (3000-cu ft) WEMCO cells. This facility is replacing over 2000 small flotation cells.

Tailings from the flotation process are transported 21.4 km (13.3 miles) in a 1.2-m (48-in.) diam concrete-lined pipe to the existing UCD tailings pond near Magna, UT. The tailings pipeline, together with the copper concentrate and process water pipeline, are installed in a common corridor between the Copperton and Magna concentrator facilities.

At the existing Garfield smelter, a new pressure filtration plant receives concentrate from the 152.4-mm (6-in.) slurry pipeline, eliminating the need for rail transport of the concentrate.

Commissioning of the new facilities took place in early 1988, with completion in late 1988. Kennecott, formerly owned by BP Minerals America, is now owed by RTZ.

Mt. Newman Mining Co. of Western Australia (Iron)

Mt. Newman is one of the world's newest major iron ore operations. Started in April 1969, it is a joint venture between Amax Iron Ore Corp. (a subsidiary of Amax, Inc. of the United States) with 25% interest, Pilbara Iron Ltd. (a subsidiary of CSR Ltd., Australia) with 30%, BHP Minerals Ltd. (a subsidiary of The Broken Hill Proprietary Co., Ltd. of Australia) with 30%, Seltrust Mining Corp. Pty. Ltd. (a subsidiary of Selection Trust Ltd.) 5%, and Mitsui-C Itoh

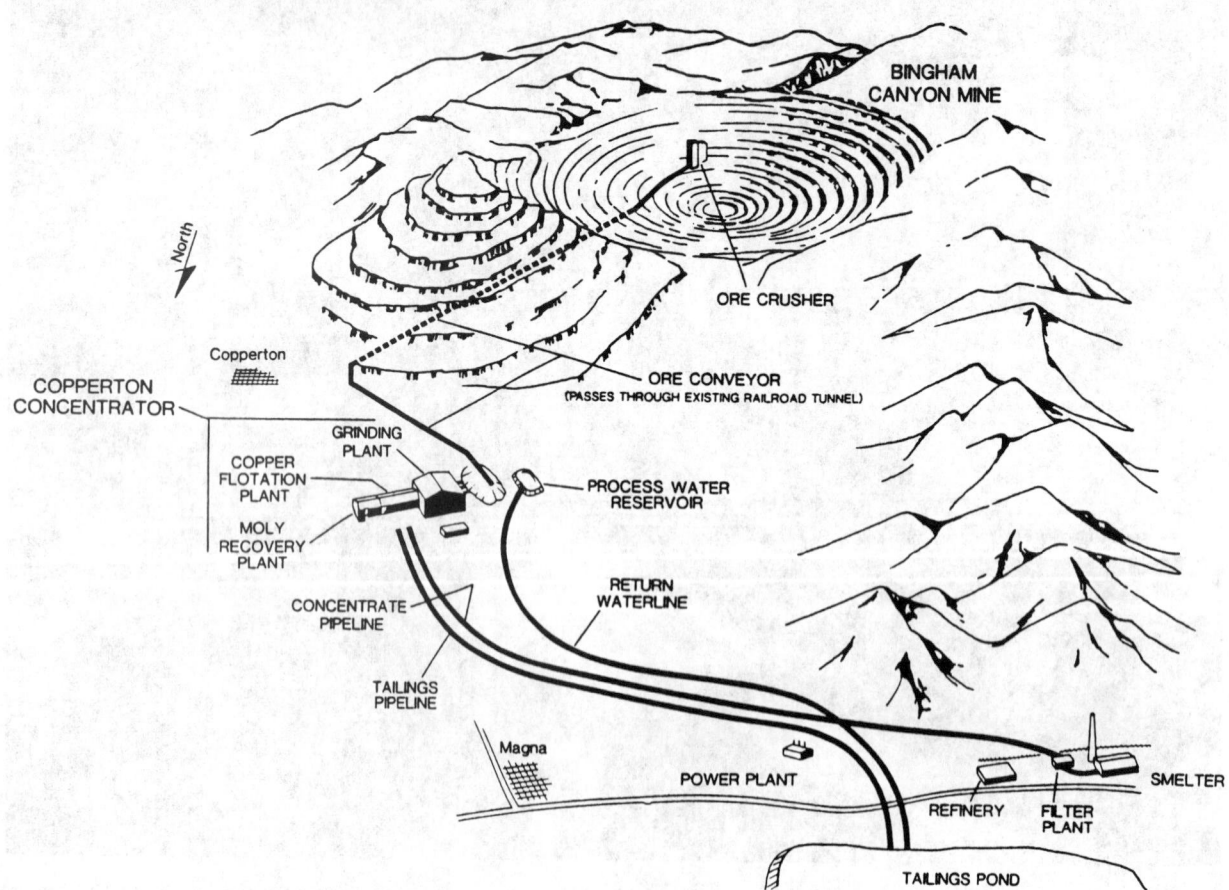

Fig. 21. Utah Copper moderization project.

Fig. 22. Mt. Whaleback at Newman, Western Australia.

Iron Pty. Ltd. (a joint subsidiary of Mitsui & Co. Ltd. and C Itoh and Co. Ltd. of Japan, respectively) with 10% interest.

The operation features the world's largest single open pit iron ore mine, Mt. Whaleback at Newman, Western Australia (Fig. 22). Ore is hauled over a 426-km (265-mile) railroad to Port Hedland where it is processed, stockpiled, and shipped (Fig. 23).

By Mar. 31, 1983, Mt. Newman Mining Co. had shipped a total of 326 Mt (360 million st) of iron ore to Australia's BHP steel works and Japanese, Asian, and European steel mills. With the exception of the first two years of start-up operation, Mt. Newman has shipped 20 to 30 Mt (22 to 33 million stpy) of ore with 32 Mt (35 million st) being a record established in 1979.

Mt. Whaleback was originally a 5.5-km (3.4-mile) long hill rising 185 m (607 ft) above the plains. The mine can produce annually more than 40 Mt (44 million st) of iron ore and about 80 Mt (88 million st) of waste material. Typical daily total material movement is between 200 and 320 kt (330,693 and 352,740 st), of which up to 140 kt (154,000 st) of crushed ore is produced and railed to Port Hedland.

The 15-m (50-ft) benches are drilled by Bucyrus-Erie 60Rs. Blasted ore is loaded by 20 electric shovels varying from 7.6 to 18.5-m³ (9.9 to 24.2-cu yd) capacity. The shovels load a fleet of diesel-electric rear dump trucks for haulage to the crushing plant (Fig. 24). The total fleet comprises 52 109-t (120-st) trucks and 22 189-t (208-st) trucks.

Crushing and Rail Loadout: There are two crushing plants, which together can crush 160 kt/d (176,000 stpd) of ore through primary and secondary crushers to a 200-mm (8-in.) product for loadout.

Ore trains are loaded in two concrete tunnels. Two stackers rated at 4 kt/h and 5.5 kt/h (4409 and 6067 stph), respectively, form surge piles above each tunnel.

Each tunnel is served by a rail loop and a complete train can be loaded at 13.7 kt/h (15,102 stph) in either tunnel. A train is reversed into a tunnel until the locomotive driver is instructed to halt by an operator supervising loading operations on a separate two-way radio channel from a control cab high on the side of the tunnel.

In each tunnel, 13 openings equipped with chutes are lowered to gravity-feed waiting cars. As the cars are filled, the chutes choke automatically and are raised by pneumatic cylinders to the tunnel roof, depositing approximately 100 t (90 st) of ore in each car. When one set of cars has been loaded, the train is moved on and the operation repeated until the entire train has been loaded.

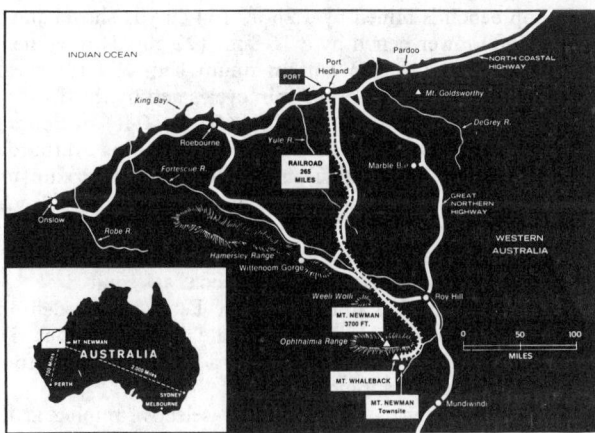

Fig. 23. Map of Mt. Newman and Port Hedland in the Pilbara region of Western Australia with railway connecting the mine and port.

Fig. 24. A large electric shovel dumps 50 t (55 st) of waste material into a 189-t (208-st) Haulpak on Mt. Whaleback, prime ore body of Mt. Newman Mining Co.

Fifty-two 2 680-kW (3593-hp) locomotives, 2000 110-t (121-st) capacity ore cars, and about 100 90-t (100-st) capacity cars are in use as of January 1983, to serve the railroad with a rated capacity over 48 Mt/a (53 million stpy) of ore. A typical train consists of 144 loaded cars drawn by three 2 680-kW (3593-hp) locomotives. Round trip to the port averages about 20 hr. Experiments are underway with 180 and 240 ore car trains using various locomotive combinations.

Port Operations: Port Hedland can accommodate ocean carriers varying in size from 5,000 dwt to 225,000 dwt, with maximum LOA of 315 m (1033 ft). Fully loaded draft at mean high water is 15.9 m (52 ft).

Rail cars are automatically positioned and rotary-dumped in one of the two car-dumping stations at Hedland at 104-sec cycle time [10 500 t/h (11,574 stph)] in the triple dumper, and 92.5 sec in the tandem dumper.

From the car dumper, ore is routed to a tertiary screening-crushing operation prior to stockpiling. Two screening-crushing plants feature 28 Allis-Chalmers double-deck screens and 12 Allis-Chalmers hydrocone crushers. Minus 30 mm +6 mm ore comprises the lump product, −6 mm ore comprises the fines product. Plus 30 mm is fed to the crushers and recirculated back with fresh feed to the screens.

A computer control system continuously monitors ore flow, bin levels and trends, equipment availability, etc. It also has the ability to automatically stop or start the equipment as required by programmed parameters of the throughput in quality control.

Stockpiling, Reclaiming, and Shiploading: Four rail-mounted boom stackers [two at 4 kt/h (4400 stph) and two at 5.1 kt/h (5622 stph)] are in use at Hedland. Each of the stackers is equipped with a 48-m (158-ft) boom with a reach from ground level to 12.5 m (41 ft). The booms are slewable to form stockpiles on either side of the tracks.

A transportable conveyor rated at 2 kt/h (2205 stph) serves a 2 Mt (2,205,000 st) circular fines stockpile. The total stockpiling capacity at Hedland is approximately 9 Mt (10 million st).

Reclaiming of the ore at Hedland is accomplished by bucket wheel reclaimers [two 3 kt/h (3,300 stph) crawler-mounted, and two 6 kt/h (6615 stph) rail-mounted]. The booms are 14 m (46 ft) long.

The crawler-mounted units have 26-m (85-ft) discharge booms at the rear to feed into a self-propelled hopper car on top of the reclaim conveyor with 770 m (2526 ft) length

along the stockpile. The rail-mounted machines can slew 360° to reclaim ore over either side of the track.

At the end of the reclaim conveyor, an automatic sampling system can take up to 1497 kg (3300 lb) primary increments, on a 6-min time interval basis. Two 8130 t/h (8962 stph) shiploaders travel along a 658-m (2159-ft) two-berth ore pier. The largest ship loaded has been a 225,000 dwt bulk carrier. Two 160,000 dwt vessels can be loaded simultaneously.

The Carter Mining Co.'s Rawhide Mine, Gillette, WY (Coal)

The Rawhide mine is located north of Gillette, WY. Having commenced shipping in August 1977, it is one of the newest open-pit coal mines in the United States. Two seams of coal are separated by several meters (feet) of clay parting. The coal is approximately 30.5 m (100 ft) thick and is in turn covered by an average of 30.5 m (100 ft) of overburden. Initial mining is at the rate of 6.8 Mt/a (7.5 million stpy) to be ultimately increased to 22 Mt/a (24 million stpy). The area to be mined covers some 2428 ha (6000 acres), with an average stripping ratio of 1.3:1.0 to 1.6:1.0.

The 4.6 m (15-ft) thick upper level Roland seam is mined by front-end loaders into trucks. The 3-m (10-ft) parting underneath is also removed the same way. The 27-m (90-ft) Smith seam below is mined in two benches. The 18-m (60-ft) top bench is mined by a 26-m³ (34-cu yd) shovel and 9-m (30-ft) lower bench by a 16.8-m³ (22-cu yd) dragline.

The 163-t (180-st) bottom dump Unit Rig trucks are loaded by the shovels at 3.5-min cycles and by the draglines in 5 to 6 min. Two 15 to 18-m (50 to 60-ft) overburden benches are each 45.7 m (150 ft) wide. Removing overburden above the 1.6-km (1-mile) mine face takes approximately three months and uncovers 1.8 Mt (2 million st) of coal. Open inventory varies between 1.8 and 3.6 Mt (2 and 4 million st). Stripping is hauled by 145-t (160-st) rear-dump trucks, which can also be used for coal as required.

Once the coal is mined out, the land (overburden and topsoil) is returned to its approximate original contour. The reclaimed area will be restored to a condition equal to or better than its premining use.

Coal blending is carried out by selective mining at the mine face and at the loadout silos by dispatching trucks to particular loading locations based on customer quality specifications.

Truck dump facilities are designed for both bottom-dump

and rear-dump trucks. Presently the round trip from mine face to the dump station takes 8 to 10 min, under 1.6 km (1 mile) in each direction.

Originally there were four silos; two silos and a second rail loop were added in 1981. Each silo will accommodate one unit train. There is no other storage since open storage is prohibited in Wyoming.

REFERENCES

Almond, R.M., and Huss, C., 1982, "Open Pit Crushing and Conveying Systems," *Engineering & Mining Journal,* June.

Anon., 1960, "Wheel and Belt Strip Nchanga Pit," *Engineering & Mining Journal,* Vol. 161, Jan.

Anon., 1966, "Material Haulage Systems at the Oroville Dam." Special Report on Modern Open Pit Trends, *Engineering & Mining Journal,* Nov.

Anon., 1976, "The Application of Modern Transport Technology to Mineral Development in Developing Countries," UN Publication No. ST-ESA/43, Centre for Natural Resources, Energy, and Transport, United Nations, NY.

Anon., 1982, "Gaseous Fuels; Coal and Coke; Atmospheric Analysis," Pt. 26, ASTM Standards. American Society for Testing & Materials, Philadelphia, PA.

Anon., 1982a, "Belt Conveyors—Highly Competitive in Distance Transport," *International Bulk Journal,* Feb.

Anon., 1982b, "Duval Corp. Develops New Portable Ore Crushing System," *Skillings Mining Review,* Nov. 6, pp. 12-14.

Anon., 1988, "BP Minerals completes $400 Million Modernization at Bingham Canyon," *Mining Engineering,* Nov., pp. 1017-1020.

Benavides, F.M., and Schuster, R.M., 1982, "Economic Comparison and Evaluation of an Overland Conveyor vs. Alternate Transportation Methods," *Mining Engineering,* Vol. 34, No. 2, Feb., pp. 176-181.

Bonasso, S.G., 1983, "Discussion: Economic Comparison and Evaluation of an Overland Conveyor vs. Alternate Transportation Methods," *Mining Engineering,* Vol. 35, No. 1, Jan., pp. 59-60.

Brost, F.B., 1979, "Construction and Operation of a Cable Belt Conveying System at Twin Buttes," *Mining Engineering,* Vol. 31, No. 12, Dec., pp. 1686-1692.

Gartner, K., 1966., "The Development of Ore Strip Mines in the U.S.S.R.," *Hebn-und Fordertechnik,* heft 1.

Goergen, H. and Ambatiello, P., 1968, "Development and State of Rail and Belt Shifting Methods in Open Cast Mining," *Fordern und Heben,* Vol. 18, No. 5, Apr.

Johnson, R.M., Frizzell, E.M., and Utley, R., 1981, "Movable In-Pit Primary Crushers," American Mining Congress Convention, Denver, Co., Sept. 17-30.

Knaebel, J.B. 1970, "Development of the Twin Buttes Mine for Production," SME-AIME Annual Meeting, Denver, Feb. 15-19.

Kok, H.G., 1980, "The Use of Mobile Crushers in the Minerals Industry," Preprint No. 80-91, AIME Annual Meeting, Las Vegas, Feb. 24-28.

Lyon, L.S., 1981, "New Developments in Large Haulage Trucks for the 1980's," SME-AIME Annual Meeting, Chicago, Feb.

McQuiston, F.W., Jr. and Shoemaker, R.S., eds., 1978, *Primary Crushing Plant Design,* AIME, New York, 246 pp.

Michaelson, S.D., 1974, "Wanted—New Systems for Surface Mining," Conference on Productivity in Mining, Rolla, MO, May 13-15.

Von Campenhausen, B., 1967, "High Capacity Bucket-Wheel Conveyor Belt Stripping Applicable to Mining," *World Mining,* June.

Yu, A.T., 1971, "Bulk Handling—Three-Quarter Century Survey & Outlook," *Skillings' Mining Review,* Apr. 17.

Yu, A.T., 1972, "Bulk Sampling—A Simplistic Approach," *Skillings' Mining Review,* Feb. 12.

Yu, A.T., 1982, "The Two Newest Transshipment Terminals in the U.S.," *Bulk Solids Handling,* North American Special, Pt. 2, Apr.

Yu, A.T., and James, T.M., 1981, "How an Engineer/Operator/Owner Evaluates Stockpiling and Reclaiming Alternatives," 4th International Coal Utilization Conference, Houston, Nov. 17-19.

Yu, A.T., and Schumann, R.A., 1982, "Toledo—The Newest Transshipment Terminal on the Great Lakes," *Skillings' Mining Review,* Feb.

6.11 Maintenance, Plant Facilities, and Utilities

6.11.1 Maintenance Systems

ROBERT N. McINDOO

INTRODUCTION

Maintenance systems use available materials, facilities, and the organization efficiently and effectively. Systems for maintenance such as preventive maintenance, planning, scheduling, and work sampling are not new; what is new is the computer to file and store data, produce schedules, and to provide timely information.

Some of the more important maintenance systems will be described in this section. Where applicable, the role of the computer will be defined. The systems can be made to function manually; however, the computer can greatly enhance the application in terms of man-hours expended and timeliness of information.

THE MAINTENANCE ORGANIZATION

The maintenance organization is at best a blueprint of how the department is to meet its objectives. The best organization plan can fail if the people are inadequate or their role is misunderstood. The most poorly conceived organization can succeed if the people expected to work within it fully understand implied responsibilities and are highly trained and motivated. Ideally, a maintenance organization is designed around the needs, the facilities, and the geography of the mine or plant, without regard to personnel. The ideal is seldom found. Generally, the ideal is conceived and then modified to meet the abilities, experience, and personalities of the personnel available.

Objectives

The primary objective of a maintenance department is to provide sufficient facilities and equipment in a safe operating condition, at the lowest possible cost, to meet the operating plan.

Maintenance may be called on to perform nonmaintenance work such as construction, plant modifications, equipment improvement, etc.; however, these activities must be considered secondary to the primary function when designing an organization. An exception to this would be the power distribution section of electrical maintenance. Pole-line construction and relocation is essential to the orderly progression of mining.

Functions

The maintenance organization is no different than any other organization in that it must perform certain management functions to be successful. The degree to which it performs these functions determines its ultimate success.

Planning—the activity that determines the goals of the department and the means to implement those goals. This would include planning for facilities, equipment, procedures, and most of all, personnel.

Organizing—the activity that distinctly identifies the work to be done and allocates it in the most efficient manner.

Leading—the activity that motivates the organization to meet departmental goals and to satisfy individual needs in concert with departmental needs.

Controlling—the activity that measures planned performance against actual performance. This requires detailed planning, maintenance standards, and monetary budgets.

Implicit in these managerial activities are personnel needs. This involves recognizing organizational needs and finding the proper people to fill those needs. Knowing the strengths and weaknesses of the individuals dictates the need for training.

Types of Organizations

Normally the maintenance organization has three levels of management:

First-Line Supervision—foremen who directly supervise the mechanics and craftsmen.

Middle Supervision—general foremen who supervise two or more first-line foremen. In large complex shops, there may be a lead foreman who directs shift foremen and reports to a general foreman.

Top-Level Supervision—maintenance managers or superintendents who direct the entire maintenance organization.

Generally there are maintenance planners, clerks, and maintenance engineers as staff to support the levels.

There are three basic types of organizations: (1) central maintenance organization, (2) area maintenance organization, and (3) central/area maintenance organization.

Central Maintenance Organization: This is a line organization generally built around craft skills and functions. See Fig. 1. This type of organization clearly divides work re-

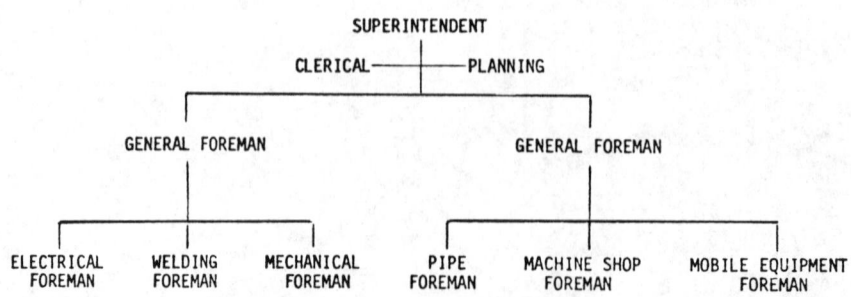

Fig. 1. Central maintenance organization.

922

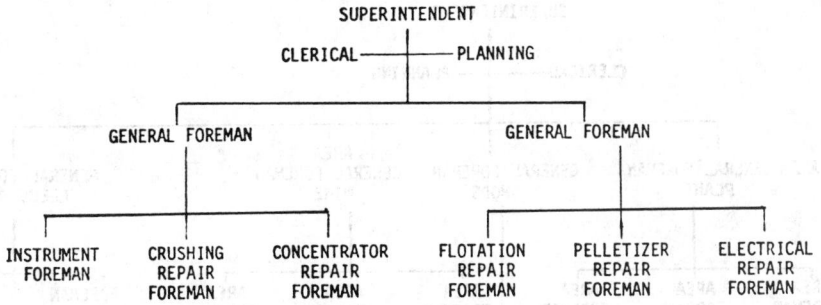

Fig. 2. Functional organization.

sponsibilities. It has the disadvantage of encouraging "inbreeding," that is, craftsmen tend to be promoted only within their group.

A variation of the centralized maintenance organization is the functional organization. See Fig. 2. This organization may or may not have some support from a central-shop group.

The pure area organization is seldom used. See Fig. 3. Generally, area maintenance is used with centrally controlled shop support.

Central/Area Maintenance Organization: This organization works well in a large operation which has defined plant functions and different mines or different ore bodies within a mine. See Fig. 4.

A variation of the preceding organizations appears where mobile equipment assumes a major responsibility for the operation. See Fig. 5.

In some surface-mining operations, all mobile-equipment maintenance and other field maintenance have been placed under the control of the mine superintendent. A separate central shops department performs support functions. This has the advantage of placing the responsibility for equipment condition on the user of the equipment. It has the serious disadvantage of possibly disregarding maintenance requirements in favor of short-term operational gains. The basic interests of the mine superintendent will inevitably favor operations to the long-range detriment of maintenance. Further, there is a conflict between the interests of the one maintenance group and the central-shops group. One may want improvements, modifications, or just reliability; the other's interests lie in the turnover of parts or overhauls.

Summary

The maintenance organization, like any organization, must be designed to meet goals and objectives. The form is less important than the definition of roles and the adequacy and motivation of the personnel available to fill those roles. The type of organization is dictated by the size and geography of the operation as well as the degree and kind of maintenance involved.

TRAINING

To increase productivity, maintain efficiency, and to assure qualifications for advancement, the maintenance organization must provide training for both its craft group and supervisors. Technology changes. Today's skills become obsolete. Update to meet change is essential for both the craft group and the supervisors. The skills a foreman brings to his job are generally technical in nature. Knowledge of human motivation, group psychology, report writing, communication, and all the other managerial requirements are hardly inherent. New supervisors particularly need training, but all potentially promotable supervisors should be prepared for advancement.

Craft Group Update

There are three avenues for updating craft skills:

1. *Manufacturer's Training Schools*—Most manufacturers of mining and plant equipment have training schools to provide the basic understanding of their equipment as well as the hands-on maintenance required. These are generally excellent, but have the disadvantage of high cost. The sponsoring company must provide travel, housing, food, and in some cases, tuition for the training. A good approach is to send select supervisors to these schools and then have these supervisors hand-tailor the material for presentation to the craftsmen.

2. *Vendor Training Schools*—Oftentimes the manufacturer's vendors are prepared to provide training on-site for the craftsmen. This has the advantage of being less expensive.

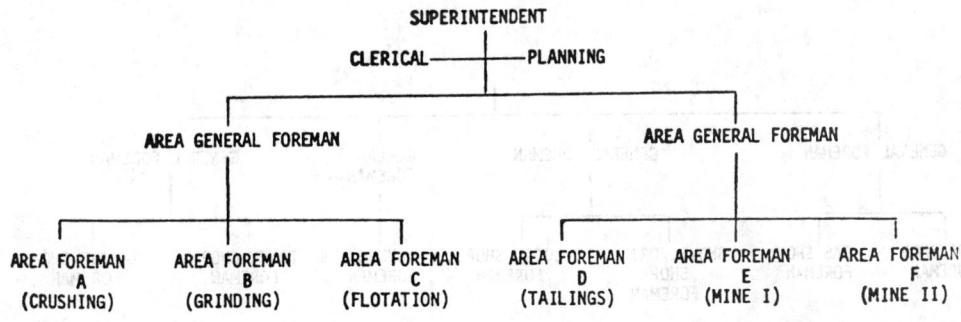

Fig. 3. Area maintenance.

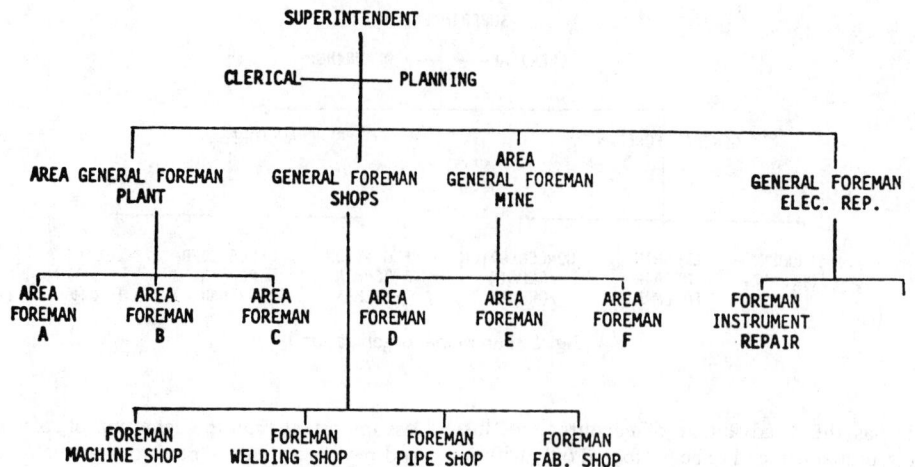

Fig. 4. Central/area maintenance.

However, this type of training should be previewed to assure conformance with company policy and capability.

3. *In-House Training*—In-house training programs, developed to suit particular needs, and presented by foremen or other supervisors, are the most successful and the least costly. Training aids, slides, film, and technical data can be obtained from vendors and original equipment manufacturers, but the program can be tailored to the needs and capabilities of the maintenance organization. Because the supervisor is, in effect, the expert, he gains in stature and credibility. Further, the followup on the job site then becomes ongoing training.

In order to prepare journeymen for training on new equipment, new techniques, and complex componentry, it is often necessary to provide training in basics:

Hydraulics—basic theory, valve operations, pumps, piping, and auxiliary components.

Bearings and Seals—types of bearings and seals, application, storage, and handling.

Electrical Systems and Controls, etc.

The areas of basic training should be tailored to individual organizational needs. Again, training in basics should be done in-house whenever possible.

Supervisory Training

Supervisory training falls into two categories: new supervisor orientation and preparation for promotion.

1. *New Supervisor Orientation*—The new maintenance supervisor is generally technically oriented and trained. However, he lacks managerial skills and needs help to bridge the transition from *doing* to getting others to do. Some general areas for training would be:

a) Motivation

b) Common Problem Solving—typical problems facing the supervisor such as discipline, the reluctant worker, absenteeism, drugs, etc., and how to deal with them.

c) Company Policies

d) Union Contracts

e) Management Skills—planning, leading, organizing, and controlling.

2. *Training for Promotion*—Annual evaluations, formal or informal, are a requisite to determine career goals for supervisors and to recognize the training to meet those goals. Promoting from within an organization develops *esprit de corps* and enhances morale. It is maintenance management's obligation to prepare potentially promotable supervisors to be qualified for advancement. Some areas where training may be helpful are:

a) Technical—computer technology, metallurgy, etc.

b) Accounting—cost-control techniques, budgets, cost analysis.

c) Communications—speech, report-writing, conducting meetings.

d) Industrial Relations—grievance-handling, arbitration, contract negotiation.

Training Methods

Training can be effective or a bore, depending on how it is done. Visual aids are all important. Movies, slides, and participative seminars are means to interest and stimulate

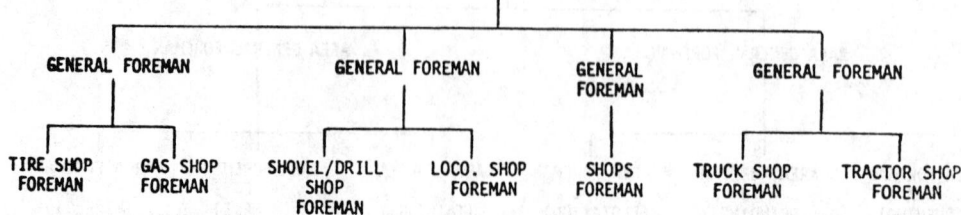

Fig. 5. Variation of maintenance organization when mobile equipment assumes a major responsibility.

thought. Under no circumstances should a script be read. Training is expensive. It behooves maintenance management to plan and execute training with the same attention to detail as for a plant overhaul. Training today is the key to tomorrow's success.

MANNING THE MAINTENANCE DEPARTMENT

No task facing a maintenance department is as difficult as justifying manpower. In operations, the manpower for an increase in production is rather straightforward: One drill operator, one serviceman, one shovel runner, five truck drivers, one crusher crew, etc., all from easily documented standards. But with the variety of crafts involved, maintenance has a much harder time. Then if production requirements increase, or worse, decrease, maintenance has an even harder time. Traditionally, maintenance has done a little guessing, or at best, used a certain rule-of-thumb to project crew sizes. But there are better ways.

Philosophy of Maintenance

The degree, or philosophy, of maintenance depends on a number of factors, not the least of which is distance from supporting vendors. The management of a mining operation must make a decision early on as to this degree: maximum, general, or minimum self-maintenance.

This decision will have a great effect on the maintenance workload, and hence the size of the department.

Methods for Manning

Backlog Method: In an established maintenance department, production increases or decreases are reflected in the shop backlog of work. With an effective work order system, this backlog is easily measured. A stabilized backlog of work equivalent to three to four weeks of man-hours is considered ideal. A rising backlog of 15 to 20% would indicate the need for more craftsmen; the number needed would equal the number of man-hours divided by hours per work week to restore the 3 to 4-week equilibrium. Similarly, a declining backlog would require consideration for reducing manpower.

The backlog is an effective tool for determining the number and kind of craftsmen needed for the work force. It has the advantage of being readily documented, but has the disadvantage of being a *responsive* measurement as opposed to an *anticipatory* measurement. It also presupposes the ability to increase or decrease manpower in the short run. Even if this were possible, it might not be desirable.

The backlog method is most effective for service shops where the work is somewhat discretionary as to timing.

Historical Data Method: The work of mobile equipment shops is for the most part nondiscretionary. Preventive maintenance inspections, servicing, and running repairs cannot be backlogged, although they can be preplanned (see the section on the "Work Order System—Planning and Scheduling"). Overhauls and component replacements can be planned and scheduled, but only with fairly large time-frame parameters. To provide the necessary manpower for these shops (and they include shovel repair, drill repair, truck and tractor shops, gas-vehicle shops, etc.), there has been developed another manning method using actual repair hour per operating hour statistics for each class of mobile equipment.

The computer accumulates the operating hours for each class of trucks, for example, and would also accumulate the repair hours charged to that class of trucks. It would then produce a repair hour per operating hour statistic for a period of time, usually a month, and a summary for six months and a year (see Figs. 6 and 7). The mining plan would indicate

COST ACCT NUMBER	UNIT DESCRIPTION	TOTAL OPER HOURS	RATED HOURS	RATED HR PER OPER HR	APPR. HOURS	TOTAL REPAIR HR PER OPER HR
150	Cat D9 Crawler Dozer	1366.00	299.00	.22	.00	.22
154	Cat 824 R Tire Dozer	4492.00	1282.25	.29	.00	.29
157	Cat 988 RT Loaders	273.00	85.00	.31	.00	.31
158	Cat Road Graders	2832.00	458.50	.16	.00	.16
163	Cat D8 Crawler Dozer	1127.00	299.50	.27	.00	.27
164	Cat 988 7yd Loader	926.00	183.00	.20	.00	.20
801	Unit Rig 85 Ton Ed Tk	343.00	814.00	2.37	33.00	2.47
802	Dart 120 Ton End Dump	301.00	164.00	.54	.00	.54
803	Dart 150 Ton End Dump	8.00	29.00	3.63	.00	3.63
804	80T Terex E. D. Truck	1967.00	976.00	.50	.00	.50
805	120 T Lectra Haul	1031.00	206.50	.20	.00	.20
806	Euclid 50 T ED Truck	1212.00	1213.25	.92	.00	.92
807	Unit Rig 100T ED Trk	1152.00	2008.00	1.74	.00	1.74
809	Unit Rig 100T ED Trk	2297.00	1155.00	.50	.00	.50
810	Euclid 100T End Dump	975.00	177.50	.18	.00	.18
815	Austin West Grader	26.00	13.00	.50	.00	.50
820	Mich 275A RT Loader	1204.00	221.50	.18	.00	.18
822	Mich 280 Dozer	44.00	4.00	.09	.00	.09
824	Cat R.T. Dozers	854.00	278.00	.33	.00	.33
831	Cat 834 RT Dozer	160.00	52.00	.33	.00	.33
870	90T Euclids 1970	981.00	1380.75	1.41	.00	1.41
871	90 Ton Dart SD Truck	684.00	567.25	.83	.00	.83
872	90 Ton Euc SD Truck	726.00	884.75	1.22	.00	1.22
873	1975 Dart 90 Ton SD	7399.00	3055.75	.41	8.00	.41
874	90 Ton Euc SD Truck	16.00	120.50	7.53	.00	7.53
869	90 Ton Trailer	9806.00	823.75	.08	.00	.08
881	Oxygen Tractors	864.00	1378.50	1.60	.00	1.60
883	Water Sprinkler Trks	1224.00	364.50	.30	.00	.30
884	Fuel Trucks	844.00	656.50	.78	.00	.78
Totals		45235.00	19151.25	.42	41.00	.42
Total Repair Hours	18190.25					
Total Labor Hours	22699.00					
Percent Repair Hours to Labor Hours	.80					

Fig. 6. Automotive mechanic, repair hour/operating hour.

COST ACCT NUMBER	UNIT DESCRIPTION	TOTAL OPER HOURS	RATED HOURS	RATED HR PER OPER HR	APPR. HOURS	TOTAL REPAIR HR PER OPER HR
150	Cat D9 Crawler Dozer	1366.00	205.00	.15	.00	.15
154	Cat 824 R Tire Dozer	4492.00	306.00	.07	.00	.07
157	Cat 988 RT Loaders	273.00	3.00	.01	.00	.01
158	Cat Road Graders	2832.00	71.00	.03	.00	.03
163	Cat D8 Crawler Dozer	1127.00	103.00	.09	.00	.09
164	Cat 988 7yd Loader	926.00	51.50	.06	.00	.06
801	Unit Rig 85 Ton Ed Tk	343.00	91.50	.27	17.50	.32
802	Dart 120 Ton End Dump	301.00	23.00	.08	.00	.08
803	Dart 150 Ton End Dump	8.00	2.00	.25	.00	.25
804	80T Terex E. D. Truck	1967.00	90.50	.05	2.50	.05
805	120 T Lectra Haul	1031.00	16.00	.02	.00	.02
806	Euclid 50 T ED Truck	1313.00	258.00	.20	3.00	.20
807	Unit Rig 100T ED Trk	1152.00	549.75	.48	8.00	.48
809	Unit Rig 100T ED Trk	1197.00	186.00	.08	.00	.08
810	Euclid 100T End Dump	975.00	42.00	.04	.00	.04
815	Austin West Grader	26.00	3.00	.12	.00	.12
820	Mich 275A RT Loader	1204.00	45.00	.04	.00	.04
822	Mich 280 Dozer	44.00	2.00	.05	.00	.05
824	Cat R.T. Dozers	854.00	33.25	.04	.00	.04
831	Cat 834 RT Dozer	160.00	1.00	.01	.00	.01
870	90T Euclids 1970	981.00	158.50	.16	3.50	.17
871	90 Ton Dart SD Truck	684.00	52.00	.08	.00	.08
872	90 Ton Euc SD Truck	726.00	49.50	.07	.00	.07
873	1975 Dart 90 Ton SD	7399.00	280.00	.04	15.00	.04
874	90 Ton Euc SD Truck	16.00	3.00	.19	.00	.19
869	90 Ton Trailer	9806.00	970.25	.10	2.00	.10
881	Oxygen Tractors	864.00	127.00	.15	.00	.15
883	Water Sprinkler Trks	1224.00	53.75	.04	.00	.04
884	Fuel Trucks	844.00	44.00	.05	.00	.05
Totals		45235.00	3820.50	.08	51.50	.09
Total Repair Hours	3756.75					
Total Labor Hours	12360.50					
Percent Repair Hours to Labor Hours	.30					

Fig. 7. Welder, repair hour/operating hour.

tonnages or yardage that can be translated into truck operating hours. If there is a mix of sizes and kinds of trucks, the production requirements would be assigned first to the fleet with the highest availability, or greatest possible use, then to the next fleet with the second highest availability, and so on until the production requirements are met.

The fleet operating hours are then multiplied by the repair hour per operating hour statistic for that fleet. Each fleet will then generate the repair hours required to produce the operating hours assigned to it. The total of all the fleets would then represent the total repair hours (in this case, automotive diesel mechanics; see Fig. 8) needed under this operating plan. The same approach can be used for welders, shovel, and drill mechanics. In the case of a shop which services a number of mobile equipment shops, such as a machine shop, the repair hours needed would be the sum of the hours generated by multiplying the machinist's repair hour per operating hour for shovels, drills, trucks, etc. by the operating hours of that equipment.

These repair hours are then divided by a statistic that is the percent of the labor hours of a shop that historically has been expended on the equipment in the mining plan. For a truck shop, approximately 80% of the labor hours of the shop are expended on the equipment maintained by the shop. (The computer can easily produce this figure along with the repair hour per operating hour.) The remaining 20% is expended on project work, safety meetings, shop tools, and unmeasured equipment.

The repair hour-total now represents the needed total to meet the plan, including labor hours for construction, equipment modifications, safety meetings, etc.

To this total must be added absentee hours and vacation obligation.

Summary

During times of economic uncertainty, the production requirements may fluctuate greatly. It is imperative that the maintenance organization have a reliable tool to predict manpower needs. The repair hour per operating hour method provides such a tool (Fig. 9). It is reliable, dynamic, and flexible. As productivity improves, the repair hour per operating hour reflects the improvements, reducing the statistic. As the equipment gets older, the statistic increases (see Fig. 10). Best of all, the logic of the approach is appealing to mine and plant management, eliminating the usual adversary posture toward maintenance manpower.

EQUIPMENT AND COMPONENT HISTORIES

Equipment and component records are essential to a maintenance organization. Records dictate the preventive maintenance program, control component replacement, in-

EQUIP. DESCRIPTION	NO.OF UNITS	REP.HR/ OP/HR	JAN - FEB OP/HR	REP/HRS	MAR - DEC OP/HR	REP/HR
Crawler Tractors	12	.31	700	217	720	224
Rubber Tired Dozers	21	.25	1520	380	1640	410
Front End Loaders	10	.52	180	95	280	146
Graders	14	.17	840	143	1080	184
Mobile Cranes	8	.64	40	26	40	26
85T Unit Rig E. D.	1	.60	100	60	80	48
120T Dart E. D.	1	.65	100	65	80	52
150T Dart E. D.	1	.80	100	80	80	64
85T Terex E. D.	6	.62	300	186	250	155
120T Unit Rig E. D.	2	.45	200	90	180	81
R50 Euclid E. D.	5	.50	500	250	500	250
100T Unit Rig E. D. (Old)	9	.94	240	226	220	207
100T Unit Rig E. D. (New)	6	.45	800	360	800	360
Euclid 100T End Dump	1	.30	100	30	100	30
Euclid Side Dump Tractors (Old)	11	.60	400	240	800	480
Dart Side Dump Tractors (Old)	7	.55	250	138	600	330
Euclid Side Dump Tractors (New)	4	.49	500	245	400	196
Dart Side Dump Tractors (New)	22	.28	2000	560	2320	650
Ore Haulage Trailers	41	.06	3150	189	4120	248
Oxygen/Water/Fuel Trucks	22	.38	880	335	1240	472
			12900	3915	15530	4613

JAN - FEB		MAR - DEC	
Total Repair Hrs/Week	= 3915	Total Repair Hrs/Week	= 4613
Repair Hours divided by .79	= 4956	Repair Hours divided by .79	= 5840
Absentee % is 6.1	= 303	Absentee % is 6.1	= 357
Total Hours Required	= 5259	Total Hours Required	= 6197
5259 divided by 40	= 132	6197 divided by 40	= 155
Vacation Replacements Required	= 15	Vacation Replacement Req'd	= 15
Total Mechanics Required	= 147	Total Mechanics Required	= 170
Present Force	= 157	Present Force	= 157
Additional Mechanics Required	= -10	Additional Mechanics req'd	= 13

1. Repair Hour/Operating Hours X Operating Hours = Repair Hours
2. Percentage of Total Department Hours charged to above Mobile Equipment=79%
3. Operating Hours Obtained from Mine Department Equipment Requirements.
4. Repair Hours are Automotive Diesel Mechanic Hours.

Fig. 8. Automotive maintenance manpower requirements.

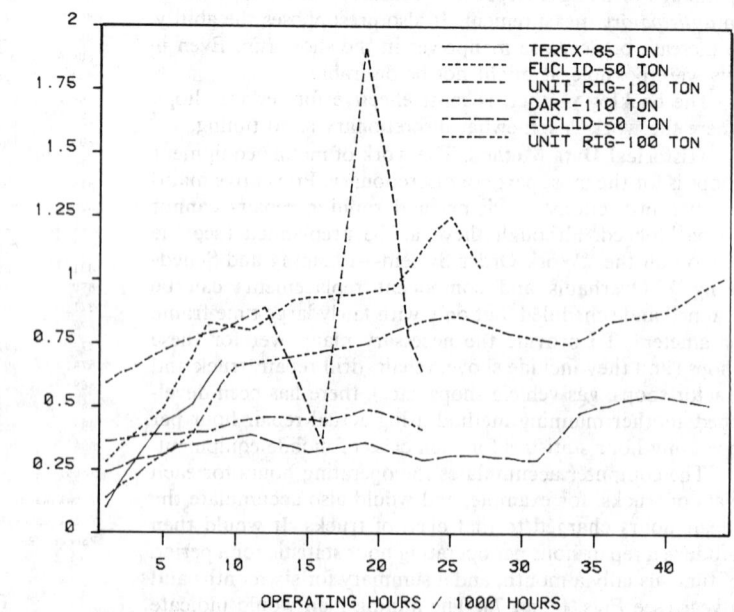

——	TEREX-85 TON
····	EUCLID-50 TON
– –	UNIT RIG-100 TON
— —	DART-110 TON
– –	EUCLID-50 TON
— —	UNIT RIG-100 TON

Fig. 9. Repair hour/operating hour for end dump trucks.

OPERATING HOURS / 1000 HOURS

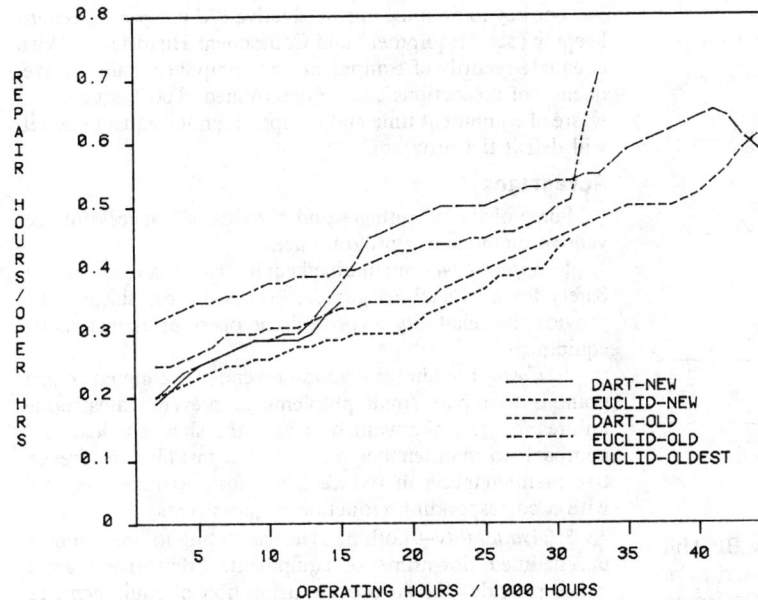

Fig. 10. Repair hour/operating hour for side dump trucks.

dicate the need for remedial investigation, and provide documented evidence for equipment replacement. Records can also be used to project major overhauls and thus maintenance budgets.

Good equipment and component histories can be maintained manually, but a manual system requires a prodigious amount of clerical time. The computer can eliminate that problem. With an appropriate software program, the computer is eminently qualified to perform record keeping with little or no paperwork.

The kind of information required will vary with the kind of equipment. The following data would be typical for mobile equipment, in this case, off-highway haulage trucks:

1) Unit number
2) Model and serial number
3) Basic specification: engine, transmission, wheel motors, etc.
4) Date in service
5) Total operating hours
6) Components
   a) Engine serial number, previous rebuilder
      1) Operating hours since last change
      2) Date of last change
      3) Serial numbers of previous engines in sequence
   b) Transmission serial number, previous rebuilder
      1) Operating hours since last change
      2) Date of last change
      3) Serial number of previous transmissions in sequence
   c) Wheel motors
      1) Operating hours since last change
      2) Date of last change
      3) Serial number of previous wheel motors in sequence
   d) Generator
      1) Operating hours since last change
      2) Date of last change
      3) Serial number of previous generators in sequence
   e) Differential
      1) Operating hours since last change
      2) Date of last change

f) Drivelines, front, rear, interaxle
   1) Operating hours since last change
   2) Date of last change
g) Brakes, front, rear
   1) Operating hours since last change
   2) Date of last change
   3) Standard or XX drums by location
h) Hydraulic system, pump, left and right hoist cylinder control valve, kickover valve, flow control valve
   1) Operating hours since last change
   2) Date of last change
i) Radiator
   1) Operating hours since last change
   2) Date of last change
j) Steering system, pump
   1) Operating hours since last change
   2) Date of last change
k) Suspension, left front, right front, left rear, right rear
   1) Operating hours since last change
   2) Date of last change
l) Front axle, left and right: king pins, spindle, inner and outer wheel bearings
   1) Operating hours since last change
   2) Date of last change
m) Rear axle, left and right: spindle, wheel bearings, planetaries, bull gear, axle shaft
   1) Operating hours since last change
   2) Date of last change
n) Dump body
   1) Operating hours since last change
   2) Date of last change

Component changes are entered into the program on a daily basis. With this information, a glance at the CRT will give the current status of any truck.

Individual component histories are important to determine trends, predict overhauls, and to find the problem areas. Typical component records for trucks would include the engine and transmission histories, as shown in Figs. 11 and 12. Other components such as wheel motors, generators, final drives, etc. can be handled in a similar manner.

Basic equipment histories for plant equipment have been

ENGINE COMPONENT HISTORY

```
Model Number              Horsepower              Serial Number

_____              _____              _____

1. In Unit number_____

2. Operating hours_____    Fleet Overhaul Ave. Hrs._____

3. Date installed_____

4. Total hours in service_____   Number of Overhauls_____

                                        Average Hrs. at Overhaul_____

5. Repair History

   a.  :

   n.  :

6. Date removed_____Oper. Hrs._____Cause of Failure_____

   Overhaul W.O.#_____Reinstalled in Unit #_____Spare_____

7. Total Cost This Repair_____Cost per Hour_____
```

Fig. 11. Engine component history.

put into a computer program by Kennecott (now BP Minerals America) using their *MAP* or *Maintenance Assist Program*. This program integrates a parts catalog cross-reference catalog, equipment spare parts, work order system, equipment specifications, preventive maintenance, and equipment histories into an integrated matrix.

Summary

Equipment and component histories are vital to a maintenance operation. By keeping track of equipment and component life and cost, the outputs will show which brands of equipment and components give best value. With this information the maintenance personnel may make the kind of decision that is required to keep the mining cost low. They can be maintained manually, but the manipulation of data is well suited to a computer program. The advantages of computer programs are many: saving clerical time, ease of input, ease of access, ability to recall information in different formats, and timeliness of information.

PREVENTIVE MAINTENANCE

Preventive maintenance (PM) is the predetermined inspection and subsequent repair of equipment and facilities to maintain high performance, prevent premature failure, and extend life.

Cleaning, inspecting, testing, adjusting, and lubricating are all functions of preventive maintenance.

TRANSMISSION COMPONENT HISTORY

```
Model Number          Assy. Number          Serial Number

_____          _____          _____

1. In Unit number_____

2. Operating hours_____   Fleet Overhaul Hours_____

3. Date installed_____   Number of Overhauls_____

4. Total hours in service_____   Ave. hours at overhaul_____

5. Repair history

   a.  :

   n.  :

6. Date removed_____Oper. Hours_____Cause of Failure_____

   Overhaul W.O. #_____Reinstalled in Unit #_____Spare_____

7. Total Cost This Repair_____Cost per Hour_____
```

Fig. 12. Transmission component history.

The key to establishing an effective PM program is record keeping (see "Equipment and Component Histories"). With adequate records of equipment and component failures, frequency of inspections can be determined. Too frequent is a waste of equipment time and manpower; not frequent enough will defeat the program.

Advantages

Some of the advantages and rewards of a successful preventive maintenance program are:

1. *Safety*—well-maintained equipment is safe equipment. Safety features such as guards, brakes, horns, alarms, etc., provide the reliability to protect the operator as well as the equipment.

2. *Cost*—the aim of a good preventive maintenance program is to repair small problems to prevent catastrophic failures, or "for the want of a nail, the shoe was lost . . ." approach to maintenance. After initial installation, preventive maintenance can reduce labor costs as much as 15% with a corresponding reduction of parts costs.

3. *Availability*—nothing is as frustrating to operations as unscheduled downtime of equipment. Preventive maintenance provides for an orderly inspection of equipment, resulting in timely scheduled repairs, thus minimizing failures in operation. Except for safety items, it is possible to defer repairs determined by inspection until a more propitious time, when manpower can be scheduled or the equipment can be released. This is particularly important in plants where down days are scheduled. Availability as a quantitative measure

$$\frac{\text{Total hours} - \text{downtime hours}}{\text{Total hours}}$$

can be increased as much as 20%. The value of this improvement can be easily measured by the cost of the units and operators not required to produce the required tonnages.

4. *Planned vs. Emergency*—planned and scheduled work is done on straight-time hours. Emergency work is often overtime work. Preventive maintenance reduces emergency work, hence the need for overtime.

5. *Productivity*—unscheduled downtime interrupts production and causes inefficiencies in the operation. PM will reduce unscheduled downtime and increase productivity. Further, equipment that is well maintained, adjusted, and lubricated will operate at peak efficiency, thus improving productivity. Operator acceptance alone will improve performance when there is confidence in and satisfaction with well-maintained equipment.

6. *Equipment Life*—a good PM program will extend the productive economic life of equipment.

Designing the System

The first step in designing a preventive maintenance program is to define the frequency of inspections. This can best be determined from records of equipment or component failures. Probably the best measurement is operating hours, although calendar time, tons, miles, and even gallons of fuel burned are sometimes used. Each basic kind of equipment may require a different frequency of inspection.

The checklist or PM form provides a guide for the inspector. It is a quantitative judgment of the equipment being inspected, and a directive to planning supervision as to what repairs are necessary. Inspections on the same equipment may have different degrees of inspection to be performed. For example, an off-highway truck may have a 100, a 500, and a 2000 hr inspection form, each varying as to the depth of the inspection. Calendar-time inspections work well if operating levels are constant, but can cause problems when mine plans fluctuate. The use of calendar time leads to a tendency to overinspect during periods of low operations and underinspect when operations are at a higher level.

When devising checklists, equipment manufacturers, vendors, and lubrication suppliers can be most helpful. A word of caution: Equipment manufacturers and lubrication specialists generally have a tendency to "guild the lily," suggesting overinspection, so take their advice and adjust to suit (see typical mobile equipment checklists: Figs. 13, 14, and 15).

Structuring the Preventive Maintenance

Because preventive maintenance includes servicing, the program should include two levels of skills. The serviceman or oiler can be very helpful in providing information, but the actual inspection should be performed by a craftsman. Ideally, these two skills would work together.

There are two schools of thought on the craftsman selected to perform PM inspections.

1) PM inspections are considered an integral part of the workload, and assignment of the inspections is made to any qualified craftsman from the daily pool not committed to other work.

2) The PM inspections are assigned to specific individuals in the craft group who have an interest in and are trained for PM inspections.

Both approaches can be successful. However, in a large operation the problem of training a large workforce suggests selecting and training those craftsmen needed who have an interest in PM. Many good craftsmen make poor inspectors if their driving interest is in normal trade work. Some craftsmen will overinspect or underinspect. The PM inspector must

---

DATE_____
UNIT NO._____

A. CAB
1. Windshield Wipers
2. Glass
3. Weatherstripping
4. Horn
5. Heater
6. Seat
7. Gauges
8. Wig-Wag
9. Headlights
10. Dashlights
11. Clearance Lights
12. Dome Light
13. Stop,Tail & Back-up Lights
14. Battery & Cables
15. Mirrors
16. A.I.D.

B. ENGINE
1. Throttle Linkage
2. Adjust All Belts
3. Fan & Fan Hub
4. Alternator Mounts
5. Starter Mounting
6. Air Hoses & Clamps
7. Engine Mounts
8. Radiator & Mounts
9. Rad. Shutter & Linkage
10. Blower Shutdown & Screen
11. Water Hoses & Clamps
12. Oil Lines & Clamps
13. Oil Pan Bolts
14. Comp. Mounting Bolts
15. Underhood Light
16. Engine P.T.O. Output Brg.

C. TRANSMISSION
1. Shift Linkage
2. Front & Rear Mounts
3. Output Shaft Brg. Ret. Bolts
4. Oil Lines & Clamps
5. Handbrake, Adjust

D. AIR SYSTEM
1. Air Lines & Clamps
2. Alcohol Injector
3. Bleed Air Tanks
4. Check & Adjust Brakes

E. DRIVELINES
1. Parellel
2. Drivelines & Joints

F. FRAME
1. Bumper & Tow Hitches
2. Rock Kickers
3. Fifth Wheel
4. Cab Mounts
5. Nose Cone Bushings

G. REAR END
1. Springs & Mounting
2. Torque Rods & Brackets
3. Differential Leaks

H. HOIST SYSTEM
1. Hoist Pump Pressure
2. Oil Pipes & Clamps
3. Hyd. Control Linkage
4. Hoist Cylinders & Mounts

I. FRONT AXLE
1. Repack Front Wheel Bearings
2. Dye-Check Spindles
3. Check Kingpin & Bushings
4. Check Draglinks & Ballstuds
5. Check Tie Rod
6. Springs & Bolts

J. STEERING SYSTEM
1. Steering Cylinders
2. Steering Pump Pressure
3. Oil Lines & Clamps

K. SAFETY
1. Steps
2. Grab Irons
3. Cab Canopy
4. Box
5. Box Target
6. Fire Extinguisher
7. Fenders & Decking
8. Side Panels & Clamps
9. Box Hinges & Pins
10. Fixed Fire Extinguisher Test

L. GENERAL
1. Service & Oil Samples
2. Road Test
3. Retorque Front Wheel Bearings

✓ GOOD REPAIR   X REPAIRED
O TO BE REPAIRED   NA NOT APPLICABLE   INSPECTED BY_____

FOREMAN_____

Fig. 14. For diesel trucks, 2000-hr preventive maintenance inspection.

DATE_____
UNIT NO._____
W. O. NO._____

A. CAB
1. Gauges
2. Windshield Wipers
3. Glass
4. Door Hardware
5. Air Horn
6. Heaters
7. Wig-Wag

B. ENGINE
1. Air Filter Hoses
2. Check All Belt Adjustments
3. Coolant Lines
4. Check All Oil Lines
5. Blower Shutdown
6. Shutters

C. TRANSMISSION
1. Bell Housing Bolts
2. Converter Oil Lines
3. Shift Linkage
4. Shift Cable
5. Parking Brake
6. Output Shaft Bearing Retainer Bolts
7. R-100 Eng. Output Shaft Brg.

D. FRONT AXLE
1. Tie Rod
2. Drag Link & Ball Studs

E. DRIVELINES
1. Drive Line
2. "U" Joints

F. REAR END
1. Springs
2. Spring Bolts
3. Differential Mounting Studs
4. Differential Pinion Seal

G. FRAME
1. Engine Mounts
2. Transmission Mounts
3. Rock Ejectors

H. AIR SYSTEM
1. Air Lines
2. Starter
3. Brake Fluid
4. Adjust All Brakes

I. HOIST
1. Hoist Lines
2. Hoist Caps
3. Power Take Off

J. ELECTRICAL SYSTEM
1. Headlights
2. Back-up Light
3. Tail Light
4. Stop Light
5. Dash Light
6. Dimmer Switch
7. Dome Light
8. Clearance Light
9. Under-Hood Light
10. Generator
11. Engine Alarm System
12. Battery and Cables

K. SAFETY
1. Steps
2. Grab Irons
3. Fenders
4. Mirrors
5. Box Target
6. Tire Lug Nuts
7. Firxed Fire Extinguisher Test

✓ GOOD REPAIR   X REPAIRED
O TO BE REPAIRED   NA NOT APPLICABLE   NAME_____

CHECK NO._____

FOREMAN_____

Fig. 13. For diesel trucks, 100-hr preventive maintenance inspection.

UNIT NO._____

A. GENERAL
1. Engine Oil Sample
2. Change Engine Oil & Filters
3. Clean/Change Fuel Filters
4. Change Trans. Oil & Filters
5. Clean Trans. Suction Screen
6. Service Air Cleaners
7. Condition of Hyd. Oil
8. Change Hyd.Filters/Clean Screens
9. Inspect Gear Cases
10. Grease
11. Other

B. ENGINE
1. Check/Adjust Belts
2. Fan & Hub
3. Coolant Hoses & Radiator Mounts
4. Engine Mounts
5. Oil Lines
6. Throttle Linkage
7. Clean Engine Breather
8. Other

C. TRANS. & CONVERTOR
1. Shift Linkage
2. Transmission Mounts & Guards
3. Clean Transmission Screen
4. Other

D. AXLES
1. Diff. Hsg. Mounting Bolts
2. Drivelines & Flanges
3. Wheel Brgs. & Seals
4. Steering Trunnions & Wheel Stops
5. Tie Rods & Draglinks
6. Steering Jacks & Brgs.
7. Steering Hyd. Lines
8. Other

E. FRAME
1. Hoist Mounts
2. Cradle Assembly
3. Axle Mounts
4. Draw Bar
5. Center Pin
6. Other

F. ELECTRICAL SYSTEM
1. Lights & Switches
2. Alternator & Mounting
3. Batteries & Cables
4. Windshield Wipers
5. Other

G. CAB
1. Cab Mounts
2. Gauges
3. Glass
4. Door Hardware
5. Heater & Weather Stripping
6. Seat & Floor Mat
7. Other

H. BRAKES
1. Air or Hyd. Lines
2. Compressor - Master Cyl - Hyd Pump
3. Brake Chambers
4. Treadle Valve
5. Brake Adjustment
6. Other

I. BLADE - BUCKET
1. Dozer Trunnions & Arms
2. Pins & Braces
3. Hoist & Tilt Jacks
4. Mold-Board & Cutting Edge
5. Bucket Lip & Teeth
6. "A" Frame & Circle
7. Other

J. SAFETY
1. Steps & Grab Irons
2. Fire Extinguisher
3. Horn & Back-Up Alarm
4. Parking Brake
5. Other

✓ GOOD REPAIR   X REPAIRED
O TO BE REPAIRED   NA NOT APPLICABLE   DATE_____

INSPECTED BY_____

Fig. 15. For dozer-loader-grader, 1000-hr preventive maintenance inspection.

be able to distinguish between *worn* and *worn out*. It is much simpler to train a select group to apply predetermined reasonable standards of performance.

Scheduling Preventive Maintenance

Once the program has been designed, the problem of scheduling becomes paramount. Records for each piece of equipment must be maintained, correlating inspection with operating hours, tons, or whatever appropriate measure is used. This can be done manually, but is a natural for the computer. Programs can be designed to list each piece of equipment, hours since last inspection, those which are due for inspection, and any other pertinent data required, such as filter sizes, part numbers, oil quantities, coolant, and gear lube capacities, etc. Hard copies can be produced from onsite printers, which in effect become schedules. Operating hours and completed inspections are the input. The program can produce summaries of fleet or section PM performance, labor expended, etc., to provide maintenance management with a measure of the overall PM program.

Introducing the Program

It is essential that a preventive maintenance program, or any other maintenance system, have not only the blessing of management, but its whole-hearted support. The advantages must be cogently presented, with realistic quantitative cost benefits outlined.

Once this support is obtained, it is equally essential to consult with the operating superintendents. They must be convinced that the program is to their advantage. They should be involved in the decision process as to what equipment should be inspected, the frequency of inspection, and what flexibility is allowable to maintain production. With their cooperation, the program can be successful; without it, the PM program becomes only a maintenance problem, rife with maintenance vs. operating confrontations.

WAREHOUSE COMPUTER-AIDED PARTS REQUISITIONING AND DELIVERY SYSTEM

A maintenance program is only as good as its skilled craftsmen and supervisory organization. But a corollary to that is: a good maintenance program is only as effective as the availability of parts.

Generally, in most mining operations, maintenance determines the kind and quantity of parts required (maximum-minimums, or order points, order quantities), but is not directly responsible for the warehouse function. Cooperation between these two responsibilities is essential to the health of any maintenance organization and to assure obsolete parts do not build up in the warehouse.

The usual arrangement is to have a central warehouse or a central warehouse with satellite warehouses throughout the mine and plant to supply operations and maintenance. Withdrawal cards signed by supervisors are presented to the warehouse counter by craftsmen or service truck drivers for parts. Part numbers are obtained by supervision or the craftsmen from equipment parts manuals, with or without superceding part numbers or cross-referencing. The warehouse usually supplies a parts location reference which may or may not be obsolete. The cards would be punched for computer processing or would be manually sorted to provide inventory control and cost data.

This standard scenario satisfies only the function of warehousing and inventory control. Maintenance suffers from the many inefficiences built into it.

1) Craftsmen travel to and from their place of work to obtain parts. Queuing at the warehouse is common. Work sampling suggests that 15% of a craftsman's time is wasted obtaining parts (see "Work Sampling").

2) Supervision will spend from 25% to 40% of their time looking up part numbers in parts books, checking superceding numbers, or trying to cross-reference parts.

3) Parts that are normally in the warehouse may be out-of-stock. The craftsmen make the trip and come back empty-handed.

It has long been the maintenance manager's dream to have a complete up-to-date catalog of parts at supervision's fingertips, to be able to order parts remote from the warehouse, and to have those parts delivered to the job site. With the advent of the computer with remote cathode ray tube (CRT) audio-video display units, the dream has become a reality.

The first step requires the cataloging of every part in the warehouse. Parts are entered into the computer program by fleet and cost center, by all known part numbers including cross-referencing to other equipment, and by description of part and warehouse location. The catalog information appears as a screen on the CRT as shown by the inventory control main menu, Fig. 16 and detail screens, Fig. 17.

With this program, information screens may be called up by part number, description, location, equipment number, or cost center to determine if the desired part or parts are available in the warehouse. If they are available, the ordering screen, Fig. 18 is called up and up to seven separate orders can be placed. After entering the appropriate data (see instruction screen, Fig. 19), the order key is depressed on the CRT (Fig. 20), and a printer in the warehouse prints out a three-part order, Fig. 21. A hard-copy printout can also be obtained from the shop printer. The warehouseman puts up the order and, depending on location for delivery and size of order, delivers to the designated location by electric truck [0.4t (0.5 st) golf cart], forklift, or pickup delivery truck, Fig. 22. One part of the order form is retained by the warehouse. The other two parts are signed by the receiving responsibility, with one part returned to the warehouse as evidence of receipt. Warehouse inventories are adjusted daily to account for material received and disbursed. Cost accounting information is distributed to the proper accounts.

CODE	FUNCTION	RESTRICTIONS
1	Order Parts From Warehouse	None
2	Ordering Parts General Info	None
3	Order File Procedures	Stores & Acctg. Only
PF4	Catalog File Browse	None
PF5	Class and Location Inquiry	None.
PF6	Part Number Or Desc. Browse	None
PF7	Class and Location Browse	None
PF8	Complete Description Browse	None
9	Receiv Items Into Stock	Stores Dept. Only

Employee No.   The number will not display as you
                 are keying it.

---------Instructions----------

1. Key in the code or hit tne appropriate PF key that you desire.
   If a code 1 is selected, key employee number also. For
   codes 1, 2, 3, 9 hit the enter key.

Fig. 16. Inventory control main menu on the CRT screen.

```
KEY CST-UNT-FAC-COM                    CATALOG FILE BROWSE

135  Pad,Cushioning-End   17892       11-019201   7 ea   25.88
135  Ring,Rub             948022      11-019432   3 ea   55.75
135  Screw,Cap 1/2X2-1/2  10033       11-319482   8 ea     .20
135  Screw,Cap 5/8X1-1/2  17877       11-218472  12 ea    1.00
135  Seal                 16949       11-019472   2 ea   41.25
135  Seal,Oil Frt Wheel   53X3646,Garlock 11-019424 6 ea 48.60
135  Stud,1 X 4-1/2 Front 947898      13-116384  24 ea    4.75
135  Washer,Thrust Frt Wh 26389,71796A 11-019683    ea   49.25

810  Euclid 100T End Dump      505 R100 Euclid
100  Cooling System

137  Belt                 4042188     63-011802   3 ea  177.93
137  Bolt,Brake Valve     0179839     63-029212   2 ea     .04
---------------Items To Be Ordered---------
     CL & LOC  QTY   U/M  BRIEF DESC
1.   0        00000  *    *
2.   0        00000  *    *
3.   0        00000  *    *
4.   0        00000  *    *
5.   0        00000  *    *
6.   0        00000  *    *
7.   0        00000  *    *
```

```
1. To see next group, hit enter.
2. To backup a page, hit PF9. If
   you backup, use PF10 to go
   forward.
3. For another browse key in
   CST-UNT-FAC-COM and hit the
   PF4 Key
4. PF5 - CLS/LOC     PF1-ORD FORM
   PF6 - PT NO BROWSE PF12-ABORT
   PF7 - CLS/LOC BROWSE
   PF8 - COMP DESC BROWSE
```

Fig. 17. Catalog file browse on the CRT screen.

DIRECT ORDERING OF PARTS GENERAL INFORMATION SCREEN

Here are some things to remember when ordering parts from the Warehouse.

1. Use the inquiry screens to find the items you wish to order. Key the class and location, quantity, unit of measure and a brief description of each item you wish to order.

2. If you do not need any inquiries to order any items you can proceed directly to the order screen and key in the items.

3. When filling in the cost coding on the items you wish to order, be very careful that you key it in properly. If you do not key any coding in for an item, it is assumed to be the same as the previous item.

4. When entering the quantity, remember to key in the leading zeros if you are ordering less than 10000.

5. One important thing must be remembered. No error checking is being done for you on any thing that you key in. If what you key in is not correct, your order will not be processed properly. So, in order to insure that your order is filled be sure to double check your order thoroughly before sending it to the Warehouse.

6. If you only have a 4 digit employee number, be sure to preceed it with a zero when keying it in.

7. Whenever using the PF keys you must hold down the alter key at the same time.

----------------HIT THE ENTER KEY TO RETRUN TO THE MAIN MENU----------

Fig. 19. General information for direct ordering of parts on the CRT screen.

The delivery system can be set up on a regular time-delivery schedule. However, the ideal is delivery on demand. In a central-shop complex experience has shown that in excess of 100 deliveries a shift are possible using the demand system. Distances to receiving areas would be the determining factor.

The inventory can be adjusted as each transaction occurs providing current inventory status. The transactions can also be held and the adjustments *batched* every 24 hr. The *real time* approach is best for maintenance. The batching method provides better inventory control. Parts receiving by the warehouse also becomes a part of the overall program.

Fig. 20. Ordering parts on a CRT in the planning office.

```
          STORES MATERIAL WITHDRAWAL          ORDER NO.
                                              0859203
CHARGE TO BE USED ON:                         DATE
DELIVERY POINT                                06/30/82
ORDERED BY HAUSCHILD MARK H      COUNTERMAN NO
RECEIVED BY
                                        BRIEF      QUANTITY
CST  UNT  FAC  W-O   CL & LOC  QTY  U/M DESCRIPTION DELIVERED
                              00000
                              00000
                              00000
                              00000
                              00000
                              00000
                              00000
-----------INSTRUCTIONS-------------FOLLOW EXACTLY AS STATED-----------
```

1. Add charge to be used on & delivery point to order. Add all cost coding to order. Sight verify complete order. Hit PF1 again to transmit to Warehouse.

2. If you wish to return to inquiries (not order yet) hit the PF2 key.

Fig. 18. Stores material withdrawal on the CRT screen.

Fig. 21. Receiving the order in the warehouse.

Fig. 22. Delivering the order.

Summary

The time lost procuring parts has a three-pronged effect; time lost by craftsmen obtaining the parts, the supervisor's time searching for part numbers, and the delay in completing the job itself. The computer-aided parts-requisitioning and delivery system will reduce labor by 15%, will reduce the time to perform the work a similar amount, and will give the foreman more time to supervise.

WORK-ORDER SYSTEM—PLANNING AND SCHEDULING

The most effective tool maintenance can have for the control of parts and labor is a functioning work-order planning and scheduling system.

An effective work-order system precedes planning and scheduling. The maintenance work order is an authorized document to perform a maintenance function or an improvement to equipment. Plant engineering work orders would be nonroutine capital or expensed work, such as construction, remodeling, or work entailing major expenditures and requiring detailed drawings. Both types of work orders can use the same format and procedures for handling.

Much has been written on the forms and procedures for a work-order planning and scheduling program. This section will detail in outline form one approach to the problem, utilizing the computer and CRTs instead of paper forms.

The advantages of a work-order planning and scheduling system are:

1) It provides preparation and planning to insure the effective use of labor and materials. With proper planning there are no surprises. Jobs are not started, then stopped for lack of materials. Work is scheduled and personnel are dedicated to the task. Planned and scheduled work is from 5% to 15% lower in cost.

2) It reduces downtime. Work is not interrupted. Crises do not develop. Equipment is returned to operations as scheduled.

3) Priorities are established. Essential work is done first. Each ordering responsibility knows when his work will be done. Flexibility of the system allows for changes as priority changes occur.

4) Supervision is relieved of details and expediting. Supervisors can depend on the system for detail planning, allowing them to function as supervisors.

These advantages of the basic system are much enhanced by utilizing the computer and CRTs to provide the memory filing, the work-order request, and the timely cost accounting. The computer can provide 48-hr cost data using the batch method. If programmed for on-line data update, the cost information can be available on a shift basis. The cost information is intended to provide maintenance with the means to quickly evaluate progress of a project against current expenditures of labor and materials. Timely information can be used to correct problems before they become serious.

The system as outlined in the following distinguishes between the traditional work order for a fabricating shop, machine shop, pipe shop, etc., where there is generally discretionary time for planning and scheduling, as opposed to mobile-equipment shops, such as truck shops, shovel repair, drill repair, etc., where work is generated by preventive maintenance and breakdowns and is repetitive in nature.

In the traditional shops, work can be repetitive, but oftentimes it is unique, *one-shot* type of work requiring detailed step-by-step planning. The planning often requires more than one craft. The Work-order planning form (screen) allows for step planning and subsequent accounting so that estimates for each step can be compared against the actual.

The Mobile equipment shop work order would be processed in the same manner as the traditional work order, but because the work is repetitive, the work has been preplanned, defining work procedures, safety hazards, and parts lists (as outlined under "Mobile Equipment Work Order Request"). The intention of preplanned jobs is to provide the one best approach to the job and to expedite parts ordering (see "Warehouse Computer-Aided Parts-Requisitioning and Delivery"). In this case, each job will use the same five-digit work-order number but will have a sequential suffix number to identify each time the job is done. The program is designed to give the labor and material for each job performance, with an average of all previous jobs, thus providing an historical standard.

WORK SAMPLING

Mine and plant management are most happy with direct measurement of performance. A shovel produces so many tons per hour; a truck hauls so many ton-miles per hour; a drill produces so many feet per hour; and the total work force produces so many tons per man-day. Maintenance cost per productive unit based on labor and supply costs if reported on a month-by-month basis shown for a 12-month rolling period shows trends and anomalies that are important to management and that should be explained where they deviate from a normal or predictable pattern. But maintenance defies such direct standard measurements. The need for measurement is not just to satisfy management, but also to provide a tool to examine problem areas and to measure the effect of change. The simplest, most reliable, and least expensive measurement is provided by the technique of work sampling.

Theory of Work Sampling

Work sampling is a statistical-measurement technique which observes and records an *activity* at a point in time. Based on the laws of probability and the binomial distribution curve, work sampling states that the percentage of observations of an activity reflects to a known degree the average percent of time spent on that activity. Work sampling can be applied to any nonrepetitive activity, such as machine operation or product quality, but as applied to maintenance, it is basically concerned with human activity.

Activities to be Measured

The basic human activities of maintenance are (1) work, (2) travel, and (3) idle. These activities are often modified to include preparation, a necessity in any maintenance work, and cleanup. Depending on circumstances, other activities may be added to further pinpoint problems.

Confidence Level

Reliability of work sampling is dependent on random observation and on a sufficient number of observations. Randomness is achieved by selecting the time of observation by random numbers or by throwing a die, and by varying the observer's route randomly.

The estimated number of observations required can be calculated after making some assumptions. A 95% confidence level with +5% tolerance limit is generally acceptable. The given activity, say *work,* is then estimated. The formula is then:

$$N = \frac{4(1 - P)}{E^2 P}$$

where N is the required number of observations, P is the percent of time activity occurs, and E is the element tolerance of error.

If we assume a craftsman's *work* activity at 60%, +5% element tolerance, we have:

$$N = \frac{4(1 - P)}{E^2 P} = \frac{4(1 - 0.60)}{(0.05)^2 \, 0.60} = \frac{1.6}{0.0015} = 1067$$

This means that 1067 observations would have to be made to achieve a 95% confidence level with +5% error.

Observers

An ongoing work sampling program would normally be conducted by the industrial engineering department. Their technicians would be the observers and would produce the reports. There is an advantage to this arrangement in that the observer would be unbiased and trained as an observer. However, using supervisors as observers has the advantage of imprinting the skill of objective observation on them that they will consciously or unconsciously use from that time forward.

Using the Data

Work sampling does not of itself provide answers, but it can point out problem areas requiring methods improvement. When changes are made, working sampling can measure the effect of that change.

The following is a comparison of typical work-sampling results:

	A	B
% Work	41.5	58.5
% Travel	16.3	9.6
% Idle	27.5	24.4
% Preparation	12.1	6.4
% Cleanup	2.5	1.1

The initial study *A* indicated that travel was high and preparation was high, adversely affecting the *work* activity. A warehouse computer-aided parts-requisitioning and delivery system was installed. The results are shown in study *B.* The idle remains high, indicating further efforts are needed.

It is not uncommon for first-time work sampling to reveal a 35 to 40% work activity. The shock to the maintenance management can produce amazing improvements. As method improvements are applied to trouble areas, the work activity will increase from 10 to 20%.

Work sampling is a tool to be used in conjunction with other management tools. It is not a panacea. It is a happy compromise between subjective observation and continuous certain time studies.

COMPUTER-AIDED DIAGNOSTICS

The advent of the microprocessor coupled with a computer can provide maintenance management with a means to diagnose the condition of engines, transmissions, hydraulic systems, or other componentry, quickly and surely. With appropriate sensing devices signaling the microprocessor (or directly to the computer), abnormal values are detected.

There are two approaches to providing the data to the computer: the static system and the dynamic warning and diagnostic system.

Static System

Hamilton Testing, Inc. has developed a static system for checking the condition of bus components and for component fault diagnoses. Pressure sensing transducers are external with a multiport disconnect to the bus. On-board sensors are brought to an electronic circuit box and interfaced with a computer device. Various tests have been programmed to provide high and low limits and actual test values. Unacceptable values are indicated on a printout with an asterisk. This system has not as yet been developed for off-highway mining equipment, although work is in progress.

The objection to this approach is that it does not measure the values under actual operating conditions. Too often problems that show up under operating conditions cannot be duplicated under no-load conditions in the shop. The main advantage is that all of the pressure-related sensors are separate from the unit, resulting in lower cost per unit and possibly greater reliability.

Dynamic Warning and Diagnostic System

A large taconite mining company and a communications manufacturing company have designed an automated Mine-Management System. The main purpose is to utilize computer software to optimize mine output and truck utilization while minimizing operating costs. Incorporated into the system is a microcomputer on-board each unit with appropriate sensors. These sensors signal on an exception basis, directed toward preventing catastrophic failures, or to signal a trend toward upper/lower limits. The signal would be sent to the sensor processor (microprocessor) and transmitted through a radio signal to the main computer and CRT display for appropriate action. The functions to be monitored are:

1) Engine speed	analog
2) Oil pressure	analog
3) Coolant temperature	analog
4) Transmission temperature	analog
5) Crankcase pressure (high)	binary
6) Coolant level (low)	binary
7) Oil level (low)	binary
8) Air restriction (high)	binary
9) Vehicle motion	binary

This system will be polled by the computer every 20 seconds to provide protection from catastrophic failures.

For diagnostic work, a "piggyback" microprocessor will be required. This will provide for additional functions to be measured and integrated against engine speed. For example, sensors will measure air restrictions after the air cleaner, ahead of the blower, and in the air box, thus pinpointing any air induction problems. The second microprocessor will also signal through the sensor processor whenever abnormal conditions occur. The piggyback processor will have a nonvolatile memory and will be polled by the main computer.

The epitome of a dynamic system would be to have a software program that would take each troubleshooting problem through a step-by-step examination of the data until the cause is located. This concept is not too far from reality.

Other Systems

Cummins Engine Co. has developed a static system for engine diagnosis called Compuchek. Because of the unique nature of their fuel pump, they feel the system very closely simulates actual operation.

Summary

On-board diagnostics for trucks will be a reality. With nonvolatile memory and an appropriate interface, any maintenance department with access to a computer could use the system. With troubleshooting programs, diagnosing problems can be performed by technicians with "best mechanic" results.

REFERENCES

Anon, 1981, "Computer Helps Build Machine Availability," *Construction Equipment*, Aug.

Anon, "Maintenance Assist Program," Kennecott Copper Corp.

Johnson, D., 1982, "Vehicle Monitoring with Microcomputers in a Mine Management Systems" *Proceedings*, American Mining Congress, Oct.

Jonkman, J., 1982, "Maintenance Control at the Kennecott Minerals Company," *17th Application of Computers and Operations Research in the Minerals Industry*, AIME, New York.

Mackie, J., 1982, "Development of the Computer-Aided Mine Management System at Reserve Mining Company," *Proceedings*, Minnesota Section, of AIME, Jan.

Syska and Hennessy, "A Guide to Improved Maintenance Management," Vol. 7, Engineering Management Division.

Tomlinson, P.D., 1980, "Maintenance Management for the Mining Industry," Paul D. Tomlinson Associates, Inc., Sept.

Winkel, T., 1981, "The Evolution of the Computerized Mine Maintenance Recordkeeping at KMC," *Proceedings*, American Institute of Industrial Engineers Fuel Conference, Dec.

6.11.2 MAINTENANCE EQUIPMENT AND FACILITIES

C. W. HOFFMAN

INTRODUCTION

Mine engineers or managers faced with the task of mine design and development have a unique opportunity to implement a long-term cost control program associated with the design and location of the mine maintenance facilities. It has long been recognized that as the size of mining equipment increases, so does cost per hour for downtime associated with the equipment. The type of equipment used will influence initial capital costs as well as long-term operating costs. The use of in-pit crushing stations and conveyors may reduce the need for diesel-powered equipment. The physical location of the maintenance facilities can either increase or decrease operating costs. The actual arrangement of the shop facilities can have a marked effect on the efficiency with which the shop may be operated. The choice of equipment used in the shop will also have an effect on the future operating costs of the mine.

Operating managers are now aware that maintenance is *not* a necessary evil. It is rather, an integral part of an operating facility and in many cases has a budget larger than that of the production group. The high initial cost of large, modern mining equipment and the growing cost to maintain this equipment require that careful consideration be given to maintenance facilities planning. The goal of management must be to construct these facilities at the lowest possible cost consistent with economical long-term operation.

This section will present fundamental considerations used to locate and design a surface maintenance facility that will enable maintenance to be carried out in a practical and efficient manner.

Major considerations in surface maintenance facility design are site selection, maintenance philosophy, mine equipment, building design and features, and shop equipment.

SITE SELECTION

Numerous factors will influence the location of surface mining maintenance facilities.

1. *Ultimate Pit Limits*—Obviously, it is undesirable to locate a large maintenance shop over the ore body and then have to move it at some future date. The maintenance shop should be located outside the ultimate pit boundary but not at such a distance that unnecessary travel time is a result.

2. *Mine Development*—The shop should be located in such an area that the access road from the mine to the shop will remain fixed for long periods of time. The projected development plan will have to be completed before the road system can be done. If possible, the shop should be located such that the access road is level or downhill from the mine.

3. *Natural Flow Point*—If a permanent crusher is to be used in or near the mine, it would be advantageous to locate the shop near the crusher. Caution about dust from dumping and crushing operations should be noted. If trucks or railcars are to haul directly to a milling facility, consideration will have to be given to locating the shop near or in the milling facility complex. Centralization of warehouse facilities and support shops such as welding, electrical, and machine shop could result in considerable cost savings. Whatever the case, the main service facility should be located near a natural flow point such as a crusher, mill, or dump.

4. *Ground Conditions*—An area that has good, natural drainage away from the mine should be considered. It is obviously expensive to do a lot of blasting and excavating if a large flat site has to be constructed on a rocky outcrop or ridge. This condition should be avoided if possible. Conversely, low or swampy areas should be rejected. Faulted areas where ground movement may occur should also be rejected.

5. *Climatic Conditions*—In northern climates, it is unwise to locate a shop facility on an exposed slope or ridge. If possible, roads should not be constructed on exposed ridges. In wet climates the opposite may be true and it would be unwise to locate the shop facilities in a drainage basin. Overall climatic conditions will influence building structure and will be discussed in another section.

6. *In-Pit Service Facility*—If an in-pit preventive maintenance (PM) service facility is planned, then the main shop can be located at a greater distance from the mine. If it is convenient, locate the main shop at the mill facility. Trucks, railcars, locomotives and other mobile equipment would only have to leave the pit for major service or repairs since PM and running repairs would be done in the pit. Equipment fueling would also be done at the in-pit service shop.

MAINTENANCE PHILOSOPHY

Corporate or personal maintenance philosophy will have an effect on the design of shop facilities. Factors such as the type of PM service, mechanical availability levels desired, budget considerations, use of outside vendors to perform maintenance services, and decisions based on mine life have an effect on maintenance philosophy developed at a particular property.

PM Service Philosophy

The type of PM service desired will have to be finalized prior to facility design. If a pit stop or assembly line-type PM service is to be used, a drive-through stall, ideally two bays long, will be needed. The total number of PM stalls needed will be determined by the size of the mobile equipment fleet. If PM is to be done in the mine, bays do not need to be provided in the main shop to perform PM service. In an effort to reduce travel to a minimum, the in-pit service center should be constructed with several *quick fix* or running repair bays. Again, fleet size will determine need.

Specific decisions as to where each type of equipment will be repaired need to be answered prior to facility design. More in-pit work means fewer bays will be needed in the main shop. If the majority of repairs will be done in the main shop, then space must be provided for those repairs.

Crew size and distribution will also affect facility design. Theoretically, maintenance forces distributed evenly on three shifts for a continuous operation should provide for the most economical utilization of plant facilities. In reality this does not work. Dislike of shift work and lack of management control are the most common reasons this approach does not work. The fact that bay space will not be utilized on the dark shifts should be taken into consideration when calculating necessary floor space.

935

Mechanical Availability

Estimates of mechanical availability need to be calculated in determining equipment fleet size as well as facility size. The size of the equipment fleet will determine the shop size. Generally speaking, each piece of equipment that is out of service for maintenance servicing or repairs should be allocated bay space. This includes such services as PM inspections, wash bay, tire shop (locomotive and car shops if there are railcars), fueling station, and load testing as well as running repairs and major repairs.

Equipment to be serviced in the mine maintenance facility will generally be all mobile equipment such as haul trucks, railcars, locomotives, loaders, dozers, water trucks, graders, cranes, and service vehicles. Availability calculations should be made for each vehicle type.

Budget Consideration

Foremost in the mind of the owner is cost and return on investment. This consideration will influence every decision to be made from feasibility studies to when to replace equipment and when to cease operations at the mine facility. Budget considerations will affect fleet size, equipment size, and type. Cost may dictate the equipment supplier. Since all mine equipment is not created equal, the selection of one equipment type over another may influence shop door width or crane height, which will in turn be reflected in facility changes and cost differences. It is probably best to design a *gold-plated* facility, then delete equipment and space, if necessary, at a later time when all budget figures have been assembled.

As much information as possible about operating costs and equipment availability should be gathered prior to making equipment selection. Operating mines are the best suppliers of this information.

Outside Services

The use of local vendors to provide various maintenance services will affect the size and cost of service facilities. The proximity to the site of these service vendors and the life of the mine are important factors. The cost of services is probably the chief factor. The decision is easily made if the mine is in a remote location and far from suppliers' services. The decision gets harder to make when the services are easy to obtain. Guaranteed availability by a supplier may change the fleet size, which may reduce necessary shop size. If the maintenance of specific pieces of equipment is contracted, it may be possible to reduce the number of service bays.

This is becoming more common when leasing equipment such as the gasoline-powered vehicles. If components such as engines and transmissions are to be rebuilt by the owner, a shop will have to be provided. If these components are to be sent out, this cost of construction can be avoided.

The amount of work done by outside contractors or vendors may also be a function of the availability of skilled maintenance employees in remote locations. It is possible that a large portion of maintenance work will, out of necessity, be done by vendor employees. This fact not only makes a difference in the facilities needed but also has a profound effect on the amount of control that management can exercise.

Mine Life

The economic life of the mine will play an important part in the design of maintenance facilities. Normally, long-life installations will be constructed with a view to performing maintenance services for the long haul. That is, the buildings will be heavier duty, more and better service equipment will

be installed, room for expansion will be provided, etc. Short pit life will dictate that everything will be built or purchased on a lowest cost basis even if cost to operate is higher. Generally, mines that have short-term pit life will have marginal or less than adequate facilities and most of the major maintenance and component rebuild will be done by outside equipment vendors. This is not necessarily wrong; management must decide if it is economically feasible to equip and train their maintenance forces to perform at a lower cost than that of vendor personnel.

MINE EQUIPMENT

The numbers and type of equipment used in the mine will have a direct impact on facility design. The amount of equipment will obviously be the largest factor in shop design.

In the recent past, the size of the haulage truck or rail fleet has dictated the size of the maintenance facility. This is true in most mines today, but in the near future, the ever-increasing alternate methods of materials transport will have a profound effect on haulage truck and rail usage. This reduction in haulage truck and rail will reduce the need for shop space and also reduce the number of mechanics needed to maintain the fleets.

When the mine plan is sufficiently developed, estimates of fleet size can be made. Initially, the number of material loading machines, usually shovels, is determined. This number is based on the overall tonnage proposed for the mine. In medium to large open pit iron ore or copper mines, shovels have been the most popular method of mining and loading ore and waste. Smaller open pit operators usually have a choice of front-end loaders, small cable shovels, or hydraulic shovels. Coal mines will probably use a combination of draglines, shovels, and front-end loaders. Tar sands, lignite, or other soft material mines may utilize bucket wheel reclaimers, draglines, shovels, and front-end loaders. Another choice of equipment may involve dozers with rippers and scrapers.

Generally, large, semi-mobile mining equipment is serviced at or near the working face and does not require a great deal of space in a central shop. A bay may be reserved for shovel and drill sideframe, boom, or stick repair, but that is mainly welding work and may be done in the welding shop. Since shovel and drill repairs are done in the mine, the maintenance facilities must be moved to the piece of equipment to be repaired. A mobile shop can be easily constructed by modifying one or more semi-trailers. One trailer could act as a mobile warehouse and would be stocked with commonly used spare parts such as nuts, bolts, pins, V-belts, filters, hydraulic hose, and pumps. Another trailer could serve as a mobile lunch room and would enable work to proceed without all the normal travel time associated with doing field maintenance work. Another trailer could be used as a welding shop and power center. Power would be taken from the mine electrical system and stepped down for tools, lighting, and welding equipment. For shovel overhauls, it is desirable to have a reinforced concrete pad with a pit located in the center so the gudgeon pin can be removed. Permanent jack pads and power outlets would also be a feature of this service pad.

After the mining equipment requirements have been determined, the method of ore and waste transport should be addressed (truck haulage, rail haulage, conveyors, or a combination of each). The most popular form is truck haulage with units ranging from 31.8 t (35 st) to 154.2 t (170 st). As a result of the present high cost of fuel, this mode of transport will undergo radical changes in the future. The use of in-pit crushing and conveying of ore and waste will reduce the number of trucks needed. The use of trolley assist to speed up haulage can reduce fleet size. Some portable in-pit crushing schemes eliminate trucks altogether because the shovel loads directly into a crusher which feeds a conveyor system.

However, in the near future, trucks will still be the most popular method of transporting ore and waste, although fleet size will be down as a result of the previously mentioned equipment. The size of the trucks will most generally be determined by the shovel or loader size. The number of trucks will be determined by factors such as rated production, mechanical availability, length of hauls, in-pit haul grades, etc. The economic advantage of size is well known in haulage truck selection. It still takes one driver to operate a 31.8 t (35 st) truck as well as a 154.2 t (170 st) truck. Labor costs for operations, as well as maintenance, can be greatly reduced by using fewer, larger haulage trucks.

On the other side of the coin, however, the impact on production is greater when a larger haulage truck is out of service than a smaller unit. This places increasing pressure on maintenance forces to provide more in the way of services to operations.

Maintenance forces must be given the tools to provide this increased service level. These tools include: a management philosophy committing maintenance personnel to improved productivity; improved employee training; maintenance management systems that work; preventive maintenance programs that stress failure prevention; effective repair procedures designed to correct known deficiencies; and effective tools and supplies.

BUILDING DESIGN AND FEATURES

In this section, facility design will generally be based on a worst case approach. That is, the general discussion will be based on the design of a hypothetical remote shop location where the mine needs to be almost self-supporting. Components will be rebuilt on site, personnel will be local, and vendor participation will be minimal. The operation is basically a shovel and truck haul system.

Mines located in warmer climates can eliminate all obvious cold weather details. It is obvious that more consideration be given to drainage in tropical climates than the dry southwestern United States. Lighting will have to be much better in cold weather buildings than those located in the South because skylights and translucent panels cannot be used. Mines where local, outside contractors can perform rebuilds and overhauls of equipment components economically can omit component rebuild facilities.

Individual components of this hypothetical facility will include buildings, truck shop, in-pit service shop, machine shop, welding shop, electrical shop, component rebuild shop, automotive (gasoline shop), central warehouse, change and lunch room facilities, and administrative office.

The central factor involved in the design of the maintenance facilities is efficiency. This should be the prime consideration with each function of the facility. The natural flow of personnel and equipment should also be a prime consideration. Unless built-in travel time for both personnel and equipment is kept to a minimum, an efficient and cost effective maintenance program cannot be realized.

All the various maintenance functions that will take place at the facility should be reviewed as a unit so that flow mistakes are not made. Building locations and the location of shops within those buildings should be designed to eliminate bottlenecks or backtracking.

In most cases it is advantageous to combine all maintenance services into one building except for in-pit PM services. In the far north this is done automatically to conserve heat but it does not always satisfy the travel time component. Experience shows that services buildings in warm climates tend to be scattered about with no apparent reason for their location.

Lost productivity resulting from mechanics' travel time can be as high as 2 or 3 hr per day, counting coffee breaks, lunch, washup, start of shift and end of shift time, and several trips to the tool crib or warehouse. Two or 3 hr is a realistic number. It is difficult to design everything around the mechanic's change room but that is the general idea. The largest group of mechanics should be located centrally to the change room and warehouse. Generally, a T-shaped building with the shops located in the horizontal bar of the T and offices located in the vertical bar will satisfy this criteria.

A variation of the T design is to produce a U or semi U-shaped shop configuration. This will give more shop space for cargo equipment fleets but still satisfy the criteria of keeping mechanic travel time to the minimum. A U-shaped shop with drive-through truck bays at one end will satisfy those individuals who want to avoid "trapping" haul trucks when another truck parks on the apron in front of an occupied bay. In a well-run shop there is little chance that this situation will occur.

Generally mechanics leave their work areas for breaks such as lunch and coffee, trips to the restrooms, consultations with their supervisor, and to draw parts and supplies from the warehouse. The proximity of the work stall to the lunchroom and change room will reduce travel time to these two areas. Travel time to the restroom can be reduced by placing minimum service units in the outer limits of the shop. Some travel to the warehouse can be eliminated by: utilizing phones and a parts runner; storing fast moving low cost items out in the shop; preplanning work and delivering parts prior to the need; or scheduling high parts usage work in bays close to warehouse issue windows. (For a computerized system, see Section 11.1).

Buildings

Generally, the mine maintenance buildings will be preengineered structural steel frame buildings. In severe winter climates the buildings will be clad with factory insulated wall panels composed of factory finished steel skins bonded to an insulated core. The roof will be built up of a smooth, single ply elastomeric membrane applied over a metal deck and sloped to aid in runoff removal.

A parapet roof over the large shops is recommended in snow and ice conditions to reduce chances of ice fall. Drains located inside the shop walls will insure that freeze ups are eliminated. Leaving the perimeter of the roof uninsulated for 0.9 to 1.2 m (3 to 4 ft) will promote melting and eliminate ice buildup behind the parapet.

The lower buildings in the complex will utilize a sloping roof with a large overhang to protect personnel from snow and ice falls. Doors should be protected with sloping ice shields.

Buildings constructed in warmer climates will have uninsulated ribbed prefinished metal siding and roofing. It is the practice in warm climates to leave 1.8 to 3 m (6 to 10 ft) at the bottom of shop walls without sheeting. This may be a good practice for processing plants but it is not good for shops. Dirt is a destructive agent and must be kept out of all shops.

In particularly hot climates, insulated walls and roofs may be necessary in reducing shop heat. Makeup air cooled by evaporative systems and contained by using rollup doors may lead to a more productive shop. Shops in warm climates may use translucent panels to provide additional lighting to the work area. Northern facilities cannot afford the heat loss through the uninsulated panels.

Shop lighting, especially in the high bays, is not as simple as it appears. Generally the truck bed and upper deck prevent direct overhead light from reaching the important lower frame components. This makes inspections and repairs more difficult. High bay lights should be 1000 watt high-pressure sodium for maximum illumination at the lowest cost. The light spacing should be closer than for a machine or welding shop. Additional lighting should be used in the PM service bays to aid in frame and component inspections. This lighting should be about 2.4 m (8 ft) off the shop floor. All internal walls should be finished in a gloss white or light beige to reflect as much light as possible. This will also prevent dust or exhaust smoke buildup. Finally, portable 250-watt mercury-vapor units on small stands should be provided on an as-needed basis.

Heating in the far north is usually an expensive proposition. Generally, natural gas will not be available so fuel oil will be the probable fuel source. Shop heating in the large shops can be done with a composite of forced air unit heaters and radiant heaters. Forced air ducted down to floor level between each door works well in controlling heat loss and delivering heated air to the equipment and personnel. Radiant heat panels on inside walls supply heat to areas in the center of the shop.

Heating in the in-pit service shop will, of necessity, be supplied by radiant heaters. This is because the doors will be cycling much more frequently than the doors of the main service shop and warm air will escape. Hot water heated floors may be a partial answer to the in-pit service shop heating problem.

Heating in offices and lunch and change rooms can easily be done by radiant means, either electric or hot water. If a hot water system is used, thought should be given to heating the shop floors.

Truck Shop

The haulage truck repair shop will consist of numerous bays or stalls reserved for truck repair. Historically, the bays have been side-by-side and only one bay deep. Many recently constructed truck shops are two bays deep and the trucks are parked facing each other. Space conservation, mechanic travel time in big shops, heat conservation in northern climates, and reduced "trapping" are all reasons for this design change. Obviously a fleet of less than 10 to 15 trucks can utilize the side-by-side arrangement. The drive-through double stalls are more suited for larger fleets.

Assembly line PM techniques utilize work stations located in a row with specific services performed at each station. The trucks are moved and the mechanic is stationary. This type of service can be done either at an in-pit service shop or in the central shop. Large mines may need several service lines to allow for either daily or weekly service. A drive-through shop two bays deep is ideal for this type of service concept.

Bay Size: The bays should be constructed for the largest truck deemed feasible for use during the foreseeable future regardless of initial design size. As ore grades diminish and

haul distance increases, it may become necessary to replace the original fleet with larger trucks. The bays should be 3.7 to 4.6 m (12 to 15 ft) wider than the truck to allow for service and movement around the truck. Parts that are removed and need to be replaced can be stored in the area beside each truck. The need to remove tires and wheel motors requires that sufficient lateral space be provided in each bay. Space for movement of service equipment such as forklifts or portable type cranes should be considered.

Bay length can be determined by adding truck length plus engine module length and 3 m (10 ft) for service work space. Some length should also be included for fleet upsizing or model changes. The length should be calculated with the bed in the lowered condition.

Column location should not interfere with component removal such as the engine module. The overhead crane should be able to work over the complete unit during repairs.

If drive-through bays are used, such as two bays where trucks would be parked facing each other, an aisle 3.7 to 6 m (12 to 20 ft) wide can provide access between the trucks and will not require crane support. This center lane can provide forklift access as well as access for foot travel. If this transportation aisle is provided, then the area directly in front of each truck can be used for part storage as well as a space to provide mechanic tool storage.

Work benches should not be placed between stalls. Most work benches become storage racks for new and discarded parts that eliminate space for work surface. Each mechanic should have a large roll-around tool box. These boxes can be modified to include a work surface where small components can be cleaned and repaired. Normally, components are removed and replaced in a repair bay. The component is then repaired in a component repair shop.

Bay Height and Crane Supports: The floor to crane hook distance of the truck bays is an important matter. The overhead crane should be able to clear a raised truck bed with the hook completely raised. Reducing the crane clearance height is an invitation to disaster, and should not be considered as a means to conserve construction funds.

The truck shop crane(s) should have sufficient capacity to remove large components such as engine modules and rock boxes. It is preferable that the shop cranes have a main hoist and a high speed auxiliary hoist for light lifts. In large shops, it is preferable to have two cranes in each crane aisle to prevent lost time waiting for crane support. The cranes should be either pendant or radio operated. There is little call for full-time crane operators.

Bay Doors: The door openings should be 1.5 to 1.8 m (5 to 6 ft) wider than the widest part of the truck. The edges of the opening should be protected by ground level steel or concrete wheel guides. These guides will prevent expensive accidents and should not be omitted. In northern climates, sectional doors seem to provide the best service. They are simple to maintain and are insulated to provide some heat retention. Large rollup doors seem to have guide problems and are less resistant to high wind forces. Door height should be 1.2 to 1.5 m (4 to 5 ft) higher than the maximum truck height with the bed down. Doors designed to open high enough to accept trucks with the bed up should be avoided.

Floors: All shop floors should be reinforced concrete, 203 mm (8 in.) thick. The top surface should have hardener additives to protect it from chipping or cracking. The surface finish should be smooth so as to allow for ease of cleaning. A rough surface promotes oil and grease buildup. All walkways and storage areas should be painted with an abrasive resistant epoxy surface. The area directly under the trucks may be left unpainted.

A floor drain should be located in the center of each bay to carry away water and fluid leaks. The main header for all the drains should be directly under the drains and should have no elbows or low points to allow for sand buildup. The floor drain should run to a sand sump outside the shop building where a front-end loader can remove accumulated sand and mud. A weir should divide the sump and an oil skimmer should be provided to contain oils and solvents. Oil from the skimmer should be piped to a buried sump which can be drained by vendor or plant forces.

Other Factors: The oil sump can be piped to a shop-wide oil recovery system. Each bay would have quick disconnect type fitting in the floor near the front one-third of the bay. A drain hose could be attached to the engine and the oil removed by air pump. A flush-mount quick disconnect cap would cover the fitting when not in use. A portable pan on wheels with a hose connection could be used if a tank or oil line is to be drained.

A sloped reinforced concrete apron with hardened surface should extend one and one half times the circumference of a truck tire [approximately 15 m (50 ft)] to provide for tire cleaning prior to the shop entrances. The apron should have embedded railroad rail runways to reduce concrete damage.

An area to park mine equipment must be provided near the truck entrance to the shop. This storage area should be sufficient in size to park from one-third to one-half of the fleet size.

To increase shop productivity, hose reels providing engine coolant, engine oil, hydraulic oil, grease, and air should be placed at the top end of every second bay. Coolant and lubricant supplies should be stored in bulk tanks buried in a convenient location at a building end. Grease should be supplied in bulk 1814 kg (4000 lb) containers.

Water outlets and welding outlets should be conveniently located between bays. Oxy/acetylene hose reels connected to a central system eliminate costly bottle handling.

The number of repair bays necessary to service the truck fleet will be proportional to the fleet size and estimated mechanical availability. At least one bay should be available for each truck that is out of service for maintenance. A 40-truck fleet with an estimated mechanical availability of 70% would ideally require 12 service bays. These bays would not all be main shop but would include other places where a truck could be serviced such as PM service bays, tire shop bay, wash rack, and engine load test area.

The average worldwide availability figures for haulage trucks are in the range of 65 to 70%. Using 65 to 70% availability would be conservative in designing a shop and would allow for bad weather such as fall and spring thaw conditions in the Northern Hemisphere and monsoon seasons in the Pacific areas. Bay space will have to be provided for support equipment such as water trucks, motor graders, front-end loaders, rubber-tired dozers, mobile cranes, track dozers, and any other large pieces of mobile equipment. The addition of these types of support equipment can double the size of the truck garage.

In-Pit Service Shop

A service shop located at the edge of the pit will increase productivity by reducing travel time for fuel, PM services, and small running repairs. The intent is to keep all mine equipment at the mine unless it needs maintenance that will require more than 4 hr or so to complete.

The size of the in-pit service shop will vary depending on the amount of equipment that is serviced on a daily basis. All rubber-tired equipment should be fueled and inspected on a daily basis. The complete fleet should then be scheduled for PM inspection on some repetitive cycle. One method is short duration frequent PM checks on a 125-hr interval (weekly). Another method is to see the equipment less frequently for a longer period of time. Either way, all parts and supplies that will be needed to perform the scheduled PM services must be stocked at the pit service shop. Generally those supplies are oil and air filters, V-belts, lights, mirrors, horns, electrical relays, and all consumable fluids such as oil, fuel, and coolant.

The service shop normally performs four functions: (1) daily fueling and lube checks; (2) weekly or bimonthly PM inspections; (3) running repairs (4 hr or less); and (4) tire service.

These functions should not interfere with one another. That is, lube and fuel service is done in specific bays, PM checks are done in another area, and running repairs in another. The tire service area will also be in a separate bay.

The bays in the service shop should be served by an overhead crane of 4.5-t (5-st) capacity. The minimum lifting height of a crane should accommodate a raised bed of a haul truck.

The fueling and lube service bays should be equipped with hose reels that supply liquids such as fuel, engine oil, hydraulic oil, coolant, grease, water, and air. Where possible, automatic filling systems should be used to speed service and to avoid contamination. Several systems are available and each has specific advantages. A boom-mounted hose cluster for fuel, hydraulic, and engine oil that can be attached to a similar connecting cluster of piping at a single point on a haul truck can speed service.

The PM service bay, which will be used for 125-, 250-, and 1000-hr inspections should not be used for running repairs. Equipment will be scheduled in at preset times and any delays in moving equipment out will delay more vehicles. Other bays are provided for running repairs. To support this pit-stop type of maintenance scheme, all spare parts or supplies which can be utilized in a short duration service of 1 to 2 hr should be stored where the mechanic can readily obtain them. To avoid lost time issuing parts, they must be in open stock.

The tire shop should be part of the in-pit service shop. A double bay large enough to allow a large forklift to change haulage truck tires is a requirement. A tire tear-down and assembly machine is necessary and can be located in the bay next to the one in which the equipment is parked. In warm weather, the tires can be changed on the apron to the tire shop.

Design of the truck wash bay should receive special attention. In warm climates it need only be a large slab of concrete sloping into a settling sump with a 1034 kPa (150-psig) water supply. It is also desirable to have a raised catwalk on each side of the wash bay. Cold weather design is a much more complex problem.

A large bay, one and one-half times the normal width, will be necessary for a wash bay, used in the far north. The extra width is necessary because the wash attendant needs to be able to stand away from the equipment while he is hosing it down. Medium pressure [1034 kPa (150 psig)] and high volume is necessary to remove dried mud. A high pressure, hot water cleaner will also be needed to clean areas such as the engine, hydraulic pumps, tanks, and other oily areas. High pressure hot water is also necessary for degreasing prior to painting.

The floor should drain to the center and then out the side of the building via a covered launder to a sand settling sump. The sump should be large enough to be cleaned by a front-end loader. An oil skimmer will also be needed in the sump.

The building should be of drive-through construction and the exit door should point toward the front of the in-pit service shop to reduce travel time. The building construction should be the same as the rest of the shop with a ceiling height higher than a raised truck bed. A catwalk should be constructed on each side of the shop to allow for cleaning of the upper parts of the haul trucks. The wash bay is sometimes used as a thawing shed so the normal heating equipment requirements need to be upgraded. Large quantities of heated air are required to dry a truck that is being prepared for service. Large unit heaters are the solution.

SHOPS AND EQUIPMENT

Machine Shop

When it comes to the machine shop, most mines over build. That is, they supply the shop with too many machines and then hire machinists to run them and then make work for the machinists.

The modern maintenance machine shop should only do maintenance work, not production work. A mine should purchase pins, shafts, bushings, etc. Maintenance work is removing and replacing worn parts, not making new parts. A place for a moderate sized machine shop may be provided, but do not fill all the open space with expensive, unused equipment.

A mine maintenance machine shop should have some basic machine tools and may include:
1 lathe with 610-mm (24-in.) swing
1 363-t (400-st) hydraulic press
1 large drill press
2 small drill presses
1 305-mm (12-in.) cap cutoff saw
1 large pedestal grinder
1 small pedestal grinder
1 key seater
1 small milling machine
1 portable boring bar
Normally, remotely located mines can have parts sent out for repair and returned in 90 days. Sufficient warehouse stores will reduce the need for expensive machine shop tools and operators.

All machine shops should be designed with material flow in mind. Dirty parts enter one end and clean parts leave the other end. A steam cleaner or high pressure, hot water cleaner should be located in the disassembly area. Sufficient floor space should be provided to work on several projects at once. The area where the machine tools are located should not be directly under the crane path to the assembly area.

Purchase only that equipment for which a future use can be determined and justified. Rely on the spare parts inventory and factory represented rebuild service. A machine shop is not a warehouse in which to store parts. Aisles should be painted around machines and shop. Because the weld shop produces abrasive dust from arc-air cutting, the machine and weld shop should be separated.

The machine shop should have high pressure sodium bay lighting and light colored, reflective walls as do the other shops. An overhead crane with a 13.6-t (15-st) main hoist and 4.5-t (5-st) auxiliary hoist will be sufficient for most cases. A forklift should be part of the shop tool list.

Weld Shop

The weld shop, as the machine shop, should not be considered a spare part production shop. It should be designed as a maintenance or repair shop and should be sized and equipped accordingly.

The weld shop should be large enough to accept a haulage truck rock box and several shovel dippers at one time. Generally the shop should have a large work area at one end, smaller work areas along a central aisle at the other end, and metal working machines along both sides of center in the middle of the shop.

This design provides for ease of access to all shop machines but gives some order to welders working on small projects in booths at one end, away from others using an arc-air gouging tool on a dipper bucket at the other end of the shop.

Shop equipment may include:
4 600-amp welding machines
6 400-amp welding machines (MIG)
1 3 m × 12.7-mm (10 ft × ½-in.) capacity press brake
1 3 m × 12.7-mm (10 ft × ½-in.) capacity shear
1 large pedestal grinder
1 small pedestal grinder
1 ironworker
1 305-mm (12-in.) rolls 19 mm by 1.2 m (¾ in. by 4 ft) capacity
1 12.7-mm (½-in.) drill press
2 welding rod heaters
The weld shop should be provided with lighting similar to that of the machine shop. The walls, however, should be clad in a light-colored, sound-absorbing material. As arc-air gouging is a source of objectionable sound, attempts should be made to reduce this source.

Crane service over the welding shop should be able to lift a haul truck box or a shovel dipper. The auxillary hook should be 4.5-t (5-st) capacity.

Electric Shop

Generally speaking, the electric shop of a moderately sized open pit mine does not need to be a large shop off by itself. Motor or generator problems that can be solved without rewinding are generally mechanical, such as replacing bearings, welding and turning shafts, and cleaning and inspecting. For this reason, the electric shop should be located in an area of the machine shop. Many motor components are repaired in a machine shop, then transported for reassembly back to the electrical shop so it is natural to make the electrical shop part of the machine shop.

The instrument repair portion of the electric shop should be an isolated room where benches can be set up so electronic components can be tested and repaired. Cleanliness is an important precondition for electronic work. The cost of many of the test instruments also requires that they be located in a semi-secure area.

Component Rebuild Shop

The decision to rebuild components vs. sending them out for repair should be a deliberate one. For a component rebuild shop to be cost effective, it must be able to compete with vendor and local shops. This is normally difficult to do if the mine in question does not have a sufficient volume of work to allow the component rebuild shop to be competitive with the vendor or contractor. This means if the number of pieces or components to be rebuilt is small, then a contractor or vendor can supply these components at a cost far less than a mine operation can rebuild the component for themselves.

If some components are to be rebuilt in-house, sufficient work benches and sufficient area should be provided for the miscellaneous components to be rebuilt. Support equipment for rebuilding components depends on the type of components to be rebuilt and should be installed accordingly.

Automotive (Gasoline Shop)

The general shop facility requirements for large production equipment also apply to gas vehicles. Some special consideration for these small units are:

1. A hydraulic hoist capable of raising a 9-t (10-st) truck bed should be considered.

2. A duct or trench system to remove exhaust gases to prevent the accumulation of carbon monoxide. A system of sealed trenches in the floor, connected to an exhaust fan with openings for each stall is most convenient. A metallic hose is inserted into the trench and attached to the vehicle exhaust pipe. An overhead duct system can be used if an overhead crane is to be excluded.

Central Warehousing

Storeroom or warehouse facilities should, if possible, be planned to adjoin the section of the maintenance shops that have the greatest need for spare parts. Sufficient spare parts must be kept on hand to allow for minimum shop service. Parts coverage must cover the time period for components to be shipped, repaired, and received.

The warehouse itself must be planned to facilitate handling of materials with the least amount of effort. High-use items, possible fasteners, mobile equipment repair parts, bearings, seals, and belts, should be stored where they are easily accessible—requiring a minimum of travel for disbursement. Access aisles must be provided for forklift trucks, and bin aisles must be of proper width to accommodate two-wheel or four-wheel hand trucks, a minimum of 0.9 m (3 ft).

Receiving areas should be provided with sufficient space to accommodate forklift unloading of trucks. A recessed truck platform-level ramp contained within the building will provide all-weather loading and unloading of trucks and is desirable in spite of the floor space it requires. In cases where rail deliveries are anticipated, it is desirable to include a boxcar-level ramp for unloading and loading with a forklift.

Material storage requirements must allow consideration for both size and weight of materials. Therefore, storage capacity must be determined with great care. Compartmental drawer-type cabinets should be provided for very small materials, and steel adjustable shelf cases should be planned to accommodate other varied sizes of materials. Heavy adjustable pallet racks with engineered safety factors for carrying capacity must be provided for very heavy materials.

Other considerations for storage must include special cabinets for gaskets; door-type cabinets for materials such as bearings, seals, and other items which should remain relatively dust or dirt free; and cabinets with lock doors to contain high value security items or items on which the issue must be strictly controlled.

Special racks must also be provided for cables, ropes, rolls of gasket materials, and other commodities that require special handling because of odd shapes, lengths, or safety in handling.

Consideration must also be given to storage of acids, lubricants, compressed gas cylinders, bulk of bagged materials, steel, and heavy wear or repair materials. Generally for safety reasons or because of their bulk, these items cannot be stored in the warehouse proper and will require outside storage or storage in buildings suited to the material. Heating such a building is unnecessary, but a heated building is desirable for lubricants and other fluid materials, some of which cannot stand varied temperatures. Again, in providing for such special storage, provision should be made for railroad access with boxcar level ramps or docks that will permit forklift unloading and a crane facility to provide for unloading heavy materials from gondola-type cars.

In addition to adequate storage facilities, stress should be placed on a cataloging system. A placement or commodity numbering system must be provided that will lend itself readily to easy determination and ability to find materials. This system must be determined with the knowledge that warehousemen are generally in-house trained, but often do not have an accounting background. Visiline, Visiguide, and other cataloging aids are very helpful in this area.

Lunch and Change Room Facilities

Amply designed and centrally located change rooms and washrooms can reduce lost time in the shop. Circular wash fountains are space savers and are easy to clean. In general, all materials used for washrooms should be selected with ease of cleaning and maintenance in mind. Ceramic tile or the less expensive glazed tile brick make excellent wall and floor materials.

Change room size would be determined by the maximum number of employees per shift. There should be one urinal and one water closet for every 10 to 15 employees. A good ventilation system should be provided. Change rooms and eating space (lunchrooms), are generally combined, but separated from washrooms. Expanded metal front lockers allow clothes to dry. Each maintenance worker should have one locker, and for those who work both inside and in the field, two lockers are needed.

Administrative Offices

Shop supervision requires the privacy of an office. Clerks with files, office equipment, etc., must have a convenient, comfortable place to work. If planned carefully with the shop, offices can be a valuable adjunct. Some points to keep in mind are:

1. Offices should be central to the shops.

2. The office of the shop foreman should be located adjacent to the shop and be large enough to accommodate small meetings.

3. An outside entrance is desirable.

4. Good lighting and many electrical outlets are needed.

5. Ceilings should be of acoustical material; walls and floors should have surfaces that can be easily cleaned.

SUMMARY

Maintenance facilities, as well as the overall mine development, should be given considerable planning and thought. Well-planned, well-designed shops which are custom built to an individual mine's needs will reduce the overall maintenance cost of the operation, which in turn, will reduce the overall operating cost of the mine, allowing a maximum profit for the operating company.

6.12 Health and Safety in Surface Mining

6.12.1 General Overview

Douglas W. Huber

DEFINITION OF HEALTH AND SAFETY

The subjects health and safety are closely related. Health, simply defined, is freedom from illness and disease. For surface mining the definition of health includes positive control of the miners' exposure to potentially harmful toxic dusts, chemicals, radiation, and noise. Such control is obtained through proper design and operation of the mine.

Safety is defined as being free of conditions that may lead to mental or physical bodily injury or the destruction of property. For surface mines the definition of safety includes all miners' activities ranging from slips and falls to protection from electric shock, machinery, eye injury, and drowning. Safety also deals with individual miner's impulsive behavior, attempts to shortcut safe work procedures, or acts of bravado.

In a broader sense safety includes training of miners also. Since any mining activity or work can be conducted in numerous ways, training is required to insure that objectives are met. Such training should emphasize doing the work safely along with doing the work well. A natural relationship exists between working safely and productively. Since miners, like all workmen, do not always instinctively work safely, a well executed training program yields multiple benefits.

MANAGEMENTS' ROLE IN HEALTH AND SAFETY

The foregoing definitions of health and safety include aspects of management involvement. Surface mines, similar to other industrial activities, are operated by management for profit. Management brings together and controls workmen, materials, labor-saving devices, and energy to produce a salable product with profit as a fundamental and necessary goal.

The conduct of activities which take place must be controlled within a framework of moral and legal obligations, at the lowest possible total cost. The overall cost of mining must remain low if the mine is to be competitive, to yield profit, and continue to stay in operation until its reserves are exhausted.

Any management decisions made will obviously affect this continuity in many ways. Managers, like miners, need training in order to make correct decisions. No decisions are more important than those that affect the welfare of the miners.

Managements' Function

Management controls the productive climate developed at any mine, or for that matter, any assemblage of workers in any industry. Part of any busy, producing activity is worker attitude toward health and safety. It can be said that miners' attitude is the key to safe, productive mines, and management controls the key. No adversary relationships can be allowed to exist which involve health and safety issues.

One difficulty in establishing and maintaining a health and safety program in surface mines is the appearance that hazards do not seem to be great. In fact, when compared to underground mines, surface mines are relatively safe places to work (see Table 1, Sec. 6.12.2). However, it is just such relative comparisons that can contribute to complacency regarding health and safety in surface mines. Any accident or occurrence that is *noticeable* must be treated with the same concern and action used for *actual* incidents of injury or property damage. Complacency and general acceptance of occurrences and "near misses" as being part of the job must be guarded against by management. All accidents are merely higher-order occurrences and near misses. Such accidents become definable because of injury (and human misery) and identifiable dollar value of property damage.

Management must monitor and control its own and the miners' notion that workers' accidents are part of the job. The mere fact that accidents just seem to happen does not justify acceptance of accidents as unavoidable. Such acceptance of accidents is probably *the* major problem in surface mining health and safety.

Managements' function is to *control* and thus meet those objectives established through formal planning. A safety policy, signed by top management and containing sufficient operating guidelines, is a basic mine management control function.

Evolution of Management Involvement

In the early days of mechanized surface mining, management consisted of two or three men who concentrated all of their management talents on excavating dirt and rock in order to expose the valuable material. Techniques and equipment evolved that yielded amazingly low unit mining costs. This "no frills" approach was dictated by declining deposit grade in some cases or competition for markets plagued by wide swings in market price. A conservative, and often too narrow, overall management style evolved due to continuous economic stress.

During this early stage of mechanized surface mining only two major functions were recognized: (1) the removal of waste rock and dirt, and (2) the loading and hauling of the valuable product to a place of processing and end use.

During the course of the foregoing surface mining development the so-called "rising expectations" of society (following the Korean War era) was also evolving. Broad concerns for human rights and man's effect on the environment focused attention on the need for changes in traditional surface mining systems. Concern for the welfare of miners and environmental protection has had considerable impact on surface mining.

As will be discussed in the following pages, legislation was enacted for improving conditions in underground coal mines. Later legislative attention was directed to metal and nonmetallic mines. Following an underground coal mine disaster, additional federal laws were enacted to improve health and safety in coal mines. Surface mines, although relatively safe, were carried along in the fervor to make all mines safer and healthier places to work. Additional legislation followed, covering training and other safety aspects of all mines. Legislative attention to environmental concerns has had a similar impact on surface mines.

Newer concepts of management were adopted by the mining industry to assist survival. As costs for mined products increased, prices obviously rose also. Predictably, domestic mines that were marginal (mainly due to low ore grade) closed in favor of higher grade and less costly international sources. Some mining localities suffered and the future for some domestic mineral sources is not promising.

From the simple two-function approach to surface mining of earlier times, a more complicated division is required. These are suggested to be:

1) Environmental studies and premining surface preparation.

2) Waste material removal and placement.

3) Mineral excavation, loading, hauling, and processing.

4) Environmental protection during and following mining, including surface restitution.

5) Planning and conducting health and safety functions, including training, during the mine operating cycles.

Managements' Current Position

It may be argued that surface mine management has always had a concern for health and safety, even though virtually nothing was written about it. Some managements undoubtedly included health and safety intuitively as part of the two basic functions mentioned earlier. However, modern surface mine management has learned that the effects of poor control of health and safety are too important to be so ill-defined. The health and safety of the work force bears as directly on the economic well-being of the mine as waste-to-ore or overburden-to-coal ratio does.

From a strictly legal viewpoint, a mine possessing reasonable economic feasibility becomes an economic liability if legally closed down for poor safety performance. Less dramatically, safe mines are productive mines. It is generally recognized in industry that most mines with high yields of product per man-day also have low accident rates. These mines are well-managed and examples can be found of such mines which operate profitably in all segments of the mining industry.

Training Responsibility: Decisions on how best to accomplish any task bear directly on the quality and quantity of work performed. Usually, work techniques acquired through experience are passed on from supervisors and senior workers to those newly hired. Occasionally new methods arise, are tested, and used, rejected, or modified. This informal training process is historic, and it works, but does not develop the full potential of the work force. Supervisors also need training, usually by outside trainers, such as provided by mining equipment manufacturers. Management must assist by providing time, training facilities, and such expert training assistance to supervisors.

A formal training program can be based on the mandatory federal regulations as stated in Part 48, Title 30 CFR. This training should be continuous and of high quality. It should capture the imagination of the trainees and use site-developed ideas and techniques. With proper control, methods for improving equipment operating techniques, based on job safety analyses, will always include safe procedures and easier ways to accomplish the task. Periodic evaluation (depending on type of task) of effectiveness of the training will indicate modifications required.

Results of such training are often noteworthy. The fact emerges that highly productive mines are safe mines, with low miner turnover and absentee rates.

Training Records: Training should be documented in as direct and simple a fashion as possible. Periodic reviews of training records, when coupled to records of safety performance (MSHA inspection results or accident statistics) or operation costs, will indicate areas needing improvement. Any improvement needed can be set as a goal and training emphasis shifted to accomplish this change required.

Training records tend to stand as evidence of care taken by management to improve health and safety and miners' environment. While some record is mandatory (MSHA regulation), management can use the information to improve and protect supervisors, the individual miners, and the mining system.

The Cost of Health and Safety

Studies that have been conducted into the cost of health and safety at surface mines understandably suffer from a small data base and lack of uniformity in accounting methods. This lack of data stems from the competitive nature of most mining and the obvious differences that exist in surface mining. However, one such study (Johansen, 1978) of a sampling of US surface coal mines indicates that accidents accounted for 47% of the total cost of safety. (Total safety costs are considered to be direct and indirect costs, including estimates of cost of lost production time following accidents.)

Although conditions will vary from mine to mine, health and safety costs, excluding special training and personnel-type costs, can be assumed to be in the range of 2 to 6 % of the total cost of production in surface mines. This percentage includes the cost of accidents, but excludes cost of major catastrophes and large compensation and property damage payments which are unusual in surface mine experience. However, whatever individual surface mine health and safety costs are, substantial dollar amounts can be saved by reducing accidents.

Cost saving through accident reduction is a reasonable management goal. However, it overlooks the improvement in productivity possible when employees recognize company concern is personal and directed toward them as individuals.

No figure for cost of health and safety can be estimated that covers the human misery that accompanies any accident. Both mental and physical injuries can occur which have no price tag; neither can a dollar asset figure be placed on safe mines. Such mines have an aura of well-being. They are highly productive, have a stable work force, and good relations with customers and neighbors.

In summary, top level management attention to health and safety is part of its overall responsibility. This attention yields positive results in the mining organization's stability, efficiency, and ultimate profits.

The following list states the basic conditions required for safety at surface mines:

1) Top management's commitment to health and safety as evidenced by a signed statement of policy.

2) Integration of health and safety into all management functions from evaluation of managers, right down to specific miners' work procedures.

3) Supervisors trained in cost implications of health and safety and provided with training assistance.

4) Inspection and recording of workplace conditions and work procedures by supervisors each work shift.

5) Reporting system to management of all *near-misses* and *accidents* with responsible *investigation* and *action suggestions.*

6) Internal health and safety inspection system, including supervisor and miner participation.

7) Scheduled evaluation of health and safety status.

LABOR ORGANIZATIONS' ROLE IN HEALTH AND SAFETY

From a historical perspective, concern for health and safety began in underground mining. Surface mining was limited to small deposits and specialty systems such as dredging that did not present the hazards of underground mining. The underground miners themselves recognized the dangers of their work and probably were the first source of stimulus for improving safety conditions.

No discussion of the subject would be complete without recognition of the contribution to health and safety by these historic members, their organizations, and their contemporary counterparts.

Most early mines had organizations that were developed and supported by the miners for their mutual benefit. With virtually no safety regulation in the early days, mines were operated with little understanding and, probably in some cases, regard for the needs of the miners. Hence, miners' organizations were formed to fill this need, which varied considerably from one mining camp to another.

Most of the many present mining labor unions developed from earlier miners' organizations. Wages, work rules, benefits, and safety provisions were issues that attracted new members. Some miners' unions developed into sophisticated organizations which today control large sums of money from retirement and health benefit funds. Such unions, composed of large numbers of individual miners, wield influence at the ballot box in both national and local elections. Politically, the labor vote, and this includes the miners' unions, has had a significant impact in elections and on health and safety legislation enacted.

Labor Union Impact

Mine health and safety laws and subsequent regulations contain the indelible imprint of union participation in the regulatory process. The many-sided lobbying efforts that yield the rules under which the mining industry operates is on-going and has produced dramatic results in the past 15 years. Certainly the miners' unions have substantially influenced this noteworthy achievement.

It would be unfair to single out specific unions since so many have participated over the years. Perhaps the coal mining unions could claim the major role since the regulations are most stringent for coal mining. However, this claim is also offset by recognition that underground coal mines have the highest disaster potential due to gas and dust ignition and a generally weaker rock environment.

Health and safety regulations reflect the differences between coal and noncoal mining by the existence of separate rules for each. Limited distinction was recognized for the differences between surface and underground mining, which may be surprising. Many of the surface coal mining regulations reflect underground mining concepts and conditions. This parallels the existence of strong underground coal mining unions and indicates the strength of the position of the unions during the rule-making process.

Labor organizations currently use health and safety issues as a major part of day-to-day operating procedure. In some instances tangible improvement in safe working technique is obtained. In others, a clouding of real issues, not totally based on safe procedures, takes place and true safety progress is hindered. In general, however, the miners' labor organizations have promoted health and safety to the benefit of the individual surface miner. This, of course, yielded benefits to the firm and industry also.

Labor Union Future Involvement

Improvement in surface mining health and safety statistics is the commendable goal of surface miners' unions. Experience in well-managed surface mines clearly points out that the well-being of the miners is best served when a nonadversary relationship exists between the union, the mine management, and the regulatory agency: MSHA (Mine Safety and Health Administration).

Until mutual trust and good communication exists between the three parties, real progress in advancing the cause of surface mine health and safety at any given mine cannot be made.

6.12.2 The Federal Role in Health and Safety

LEO MISAGI

INTRODUCTION

Federal legislation aimed at collecting and distributing statistical data and improving health and safety in the mining industry was enacted at the beginning of the 20th century. Before that, miners and the United States public seemed to accept the unsafe working conditions as natural, and risk-taking as part of the job. Each new mine disaster has prompted a new legislation related to mining. The progress of recent years in mine safety and health has done much to change the old fatalistic attitude toward hazards of mining.

HISTORY

The Organic Act of 1910 (Public Law 179, 61st Congress) had established the US Bureau of Mines to deal directly with the issues of mine safety. Further milestone in the same direction was the 1941 law that required inspection and investigation of health and safety conditions, accidents, and occupational diseases in coal mines (Public Law 49, 77th Congress).

Mine Safety Code for Bituminous Coal and Lignite Mines, issued in 1946, was the first formal set of federal mine safety regulations, still without requirements for enforcement. The 1952 law, aimed at prevention of major disasters in coal mines, empowered federal inspectors to write notices of violations and orders of withdrawal in cases of imminent danger (Public Law 552, 82nd Congress). The 1966 amendments to the latter law have extended mine safety and health regulations to small mines—any mine in which no more than 14 individuals are regularly employed underground (Public Law 376, 89th Congress).

The first legislation dealing directly with issues of safety and health in metal and nonmetallic mines was enacted in 1961. It authorized a study of the causes of accidents and the means of prevention of injuries and health hazards in this category of mining (Public Law 300, 87th Congress). It was followed by the 1966 law that required inspections of metal and nonmetallic mines, enforcement of standards, closure orders for imminent dangers, promulgation of new health and safety standards, establishment of a standard advisory committee, state plan provisions, and special emphasis on occupational health problems (Public Law 577, 89th Congress).

Penalties for violations, and fines and imprisonment for willful violations were provided in the Federal Coal Mine Health and Safety Act of 1969 (Public Law 91-173, 91st Congress). This law contained mandatory health and safety regulations, required the promulgation of standards, periodic inspection of mines, closure orders for imminent danger, and notices for violations of standards.

The most recent mining legislation—Public Law 95-164, 95th Congress, titled Federal Mine Safety and Health Amendments of 1977—amends the 1969 Federal Coal Mine Health and Safety Act (Public Law 91-173) and repeals the Federal Metal and Nonmetallic Mine Safety Act of 1966 (Public Law 577, 89th Congress). Briefly, the current law

• Eliminates the disparity of protection between coal and noncoal miners and between miners and other workers of the nation.

• Retains the health and safety standards set up by the previous acts for coal and noncoal mines.

• Strengthens the means for compliance with the law.

• Emphasizes protection of miners' health.

• Provides new procedures for assessing civil penalties for violating mine health and safety statutes and regulations at all mining operations.

• Calls for broad mandatory training of all US mine workers.

• Sets up new procedures for the rule-making process.

• Gives a greater role to miners or their representatives for improving their health and safety.

• Establishes an independent Mine Safety and Health Review Commission outside the Department of Labor to hear contested citations and withdrawal orders, thus eliminating the conflicts created by allowing the administering agency to review its own actions.

• Requires the operators to maintain records of exposure to toxic substances, of employee training, as well as other useful records.

• Prohibits discrimination against miners who complain about safety and health, take part in actions, or exercise their rights under the Act.

• Entitles miners to be rehired and reinstated to their former positions with back pay and interest.

• Holds liable any person discriminating against a miner for all expenses incurred by the miner in bringing a discrimination complaint.

• Retains the provisions in the 1969 Coal Mine Act for limited state participation in mine safety and health enforcement and expands its scope to include noncoal mines.

• Allows states to enforce standards that are not in conflict with federal standards.

• Transfers the mine safety and health authority from the Department of the Interior to the Department of Labor to eliminate the long-standing jurisdictional conflicts between the two agencies.

INSPECTION OF MINES

The Mine Safety and Health Administration (MSHA) of the US Department of Labor is the federal agency in charge of inspecting and investigating of health and safety conditions in the nation's mines. An MSHA inspector must have documents identifying him/her as representative of the US Secretary of Labor, who has the right to enter any mine to make any inspection or investigation. Like all employees of the United States government, MSHA mine inspectors are expected to maintain high standards of ethical, moral, and other conduct in enforcing the laws and regulations related to mine safety and health.

In cases of dispute between labor and management, an inspector is impartial; unless specifically directed by their supervisors, inspectors are not required to cross picket lines or make an inspection or investigation of the mine. Inspectors are not required to sign the forms designed to release the company from responsibilities related to safety and health at the mine. Inspectors are not to perform any work at a mine, or assist any employee or mine official in the performance of production-related jobs. Inspectors are there for inspection, investigation, and discussions directly related to occupational safety and health.

Mining activities and conditions at the time of inspection should represent the routine. Inspection of mines and mills on idle work shifts are limited to areas where conditions are practically the same as they are on the active shifts. Advance notice of an imminent inspection is prohibited by law when the purpose of inspection is to detect any imminent dangers that may exist at the mine and determine whether there is compliance with the mandatory health or safety standards or with any citation, order, or decision issued under requirements of the Act. Advance notice is allowed if the inspector intends to gather information about the mandatory health or safety standards, or obtain, utilize, and disseminate information relating to health and safety conditions, the causes of accidents, diseases, and physical impairments originating in such mines.

A representative of the operator and a person representing his fellow miners must be given the opportunity to take part in inspection of the mine and the conference related to it. A pre-inspection conference is held to inform the operator and the miners' representative what the inspector wants to do. A post-inspection conference is held to inform the parties involved about the inspector's findings.

The law prescribes two regular, complete inspections for every surface mine in the United States. Additional spot inspections are made to determine the status of citations, notices, and orders issued during a previous inspection, or to collect additional samples and monitor potentially hazardous conditions.

MSHA inspectors investigate any health and safety complaints brought to their attention prior to or during an inspection. They also conduct a special inspection of the mine if their office receives a written complaint about conditions of health and safety or an imminent danger threatening the lives of miners.

The US Code of Federal Regulations—Title 30, Mineral Resources—and the inspection and investigation manuals for federal mine inspectors contain detailed information on mine inspectors' duties. Copies of these can be obtained through the US Government Printing Office.

HEALTH AND SAFETY STATISTICS

The Safety and Health Technology Center of MSHA in Denver, CO, collects and publishes various statistics related to mining. The latest data on mine safety and health can be obtained by writing to Mining Information Systems Division, Denver Safety and Health Technology Center, Mine Safety and Health Administration, P.O. Box 25367, Denver Federal Center, CO 80225.

Tables 1 and 2 show the injury incidence rates in the United States and the number of injuries in surface mining, coal and noncoal, classified by industry and types of accidents. Both tables were adapted from *Mine Injuries and Worktime Quarterly* published by MSHA's Mining Information Systems Division.

Terms Used in Safety Statistics

Incidence Rate is the number of injuries per 200,000 employee-hours, rounded to two decimal places. *Fatal injuries* are those occurrences resulting in death. *NFDL* (nonfatal with days lost) injuries are nonfatal occurrences with lost workdays, that is, nonfatal injuries that result in days away from work, statutory days charged, or days of restricted work activity. *NDL* (no days lost) injuries are occurrences having no lost workdays, that is, nonfatal injury occurrences resulting only in temporary loss of consciousness or medical treatment other than first aid.

According to the generally accepted classification, accidents may be the result of contact with energy sources such as mobile equipment or moving parts of stationary equipment and machinery, electricity, explosives, compressed gases, caustic chemicals, flame, extremely hot or cold surfaces, toxic and noxious substances, injurious radiation (ultraviolet, ionizing); being struck by or against other objects; being caught in, on, or between other objects; being rubbed or abraded by friction, pressure, or vibration; falling to the same or other level; being engulfed by solid material, liquid, or gas (including drowning); suffering from overexertion; and bodily reaction to a voluntary or involuntary motion.

ORGANIZATION OF THE MINE SAFETY AND HEALTH ADMINISTRATION

The Assistant Secretary for Mine Safety and Health in the US Department of Labor has two administrators—one in charge of coal, and the other, metal and nonmetal mine safety and health—who enforce the law through a system of district, subdistrict, and field offices located in various states.

Coal Mine Safety and Health Districts' Areas of Jurisdiction and Office Locations

District 1: Maine, New Hampshire, Vermont, Massachusetts, Rhode Island, Connecticut, New York, New Jersey, Delaware, and eastern Pennsylvania. *District Office:* Wilkes-Barre, PA.

District 2: Western Pennsylvania. *District Office:* Pittsburgh, PA.

District 3: Maryland and northern West Virginia. *District Office:* Morgantown, WV.

District 4: Southern West Virginia. *District Office:* Mt. Hope, WV.

District 5: Virginia. *District Office:* Norton, VA.

District 6: Eastern Kentucky. *District Office:* Pikeville, KY.

District 7: North Carolina, South Carolina, Georgia, Florida, Alabama, Mississippi, Tennessee, Puerto Rico, Virgin Islands, and central Kentucky. *District Office:* Barbourville, KY

District 8: Ohio, Indiana, Illinois, Michigan, Wisconsin, and Minnesota. *District Office:* Vincennes, IN.

District 9: Iowa, Missouri, Arkansas, Louisiana, Texas, Oklahoma, Kansas, Nebraska, South Dakota, North Dakota, Montana, Wyoming, Colorado, New Mexico, Arizona, Utah, Idaho, Washington, Oregon, Nevada, California, Alaska, and Hawaii. *District Office:* Denver, CO.

District 10: Western Kentucky. *District Office:* Madisonville, KY.

Metal and Nonmetal Mine Safety and Health Districts' Areas of Jurisdiction and Office Locations:

Northeastern District: Maine, New Hampshire, Vermont, Massachusetts, Connecticut, Rhode Island, New York, New Jersey, Pennsylvania, West Virginia, Maryland, Delaware, Virginia, and the District of Columbia. *District Office:* Pittsburgh, PA.

Southeastern District: Kentucky, Tennessee, North Carolina, South Carolina, Georgia, Florida, Alabama, Mississippi, Puerto Rico, and the Virgin Islands. *District Office:* Birmingham, AL.

Table 1. Number of Injuries in Surface Mining, Injury-Incidence Rates per 200,000 Employee-Hours, Average Number of Workers, Employee-Hours, and Production, by Kind of Mineral Mined and Work Location, January-December, 1987*

	Bituminous Coal			
	Strip mines	Auger mines	Other surface mines	Total surface mines
Fatal	14	—	—	14
Fatal incidence rate	0.03	—	—	0.03
NFDL	1,540	14	5	1,559
NFDL incidence rate	3.27	4.39	2.90	3.27
NDL	929	4	1	934
NDL incidence rate	1.97	1.25	0.58	1.96
All occurrences	2,483	18	6	2,507
All incidence rate	5.27	5.64	3.47	5.27
Average number of workers	49,071	605	247	49,923
Employee-hours reported	94,223,419	637,787	345,352	95,206,558
Production reported (st)	520,624,820	3,174,302	806,215	524,605,337

	Pennsylvania Anthracite Coal			
	Strip mines	Culm bank	Dredge	Total surface mines
Fatal	—	—	—	—
Fatal incidence rate	—	—	—	—
NFDL	46	11	—	57
NFDL incedence rate	8.09	6.95	—	7.81
NDL	5	1	—	6
NDL incidence rate	0.88	0.63	—	0.82
All occurrences	51	12	—	63
All incidence rate	8.97	7.58	—	8.63
Average number of workers	759	184	12	955
Employee-hours reported	1,136,971	316,541	6,538	1,460,050
Production reported (st)	1,785,536	872,357	4,789	2,662,682

	Metal	Nonmetal	Stone	Subtotal surface mines
Fatal	4	2	15	21
Fatal incidence rate	0.03	0.03	0.05	0.04
NFDL	439	166	1,268	1,873
NFDL incidence rate	3.84	2.58	4.13	3.86
NDL	271	122	648	1,041
NDL incidence rate	2.37	1.89	2.11	2.14
All occurrences	714	290	1,931	2,935
All incidence rate	6.24	4.50	6.30	6.05
Average number of workers	12,691	7,058	31,346	51,095
Employee-hours reported	22,871,321	12,877,544	61,348,688	97,097,553

January-December, 1987—Continued

	Sand and gravel	Total surface mines
Fatal	17	38
Fatal incidence rate	0.06	0.05
NFDL	978	2,851
NFDL incidence rate	3.38	3.68
NDL	529	1,570
NDL incidence rate	1.83	2.03
All occurrences	1,524	4.459
All incidence rate	5.26	5.75
Average number of workers	35,066	86,161
Employee-hours reported	57,942,179	155,039,732

*Adapted from: *Mine Injuries and Worktime Quarterly, Jan.-Dec. 1987*, Mine Safety and Health Administration, US Dept. of Labor, 1987. Metric equivalent: st $\times$ 0.907 184 7 = t.

Table 2. Number of Injuries by Mineral Industry in Surface Mining, and Accident Classification, January-December, 1987*

Accident Classification	Coal			Metal		
	Fatal	NFDL	NDL	Fatal	NFDL	NDL
Electrical	—	10	3	—	2	3
Entrapment	—	—	—	—	—	—
Exploding vessels under pressure	—	5	5	—	2	1
Explosives and breaking agents	1	6	1	—	2	1
Falling, rolling, or sliding material	1	3	2	—	1	—
Fall of face, rib, side or highwall	1	9	3	—	—	1
Fire	—	9	3	—	—	1
Handling material	—	512	348	—	153	118
Hand tools	—	150	207	—	45	54
Nonpowered haulage	—	5	1	—	2	—
Powered haulage	4	216	76	1	58	17
Haulage trucks	4	124	33	1	38	9
Front-end loaders	—	30	8	—	4	5
All other powered haulage	—	62	35	—	16	3
Hoisting	—	1	—	—	—	—
Ignition or explosion of gas or dust	—	2	1	—	—	—
Impoundment	—	—	—	—	—	—
Inundation	—	—	—	—	—	—
Machinery	7	220	145	3	46	33
Dozer	2	48	16	—	6	3
Drill	2	9	3	—	3	2
All other machinery	3	163	126	3	37	28
Slips or falls of person	—	445	132	—	119	39
Stepping or kneeling on object	—	46	12	—	16	6
Striking or bumping	—	1	5	—	4	3
Other	—	30	38	—	10	10
Total	14	1,670	982	4	460	287

Table 2. Number of Injuries by Mineral Industry in Surface Mining, and Accident Classification, January-December, 1987—Continued

Accident Classification	Nonmetal			Stone		
	Fatal	NFDL	NDL	Fatal	NFDL	NDL
Electrical	1	2	1	1	12	1
Entrapment	—	—	—	—	—	—
Exploding vessels under pressure	—	—	—	—	3	2
Explosives and breaking agents	—	—	—	—	5	1
Falling, rolling or sliding material	—	1	2	1	8	3
Fall of face, rib, side or highwall	—	—	1	—	5	1
Fire	—	3	—	—	5	1
Handling material	—	45	38	—	423	242
Hand tools	—	17	25	—	153	149
Nonpowered haulage	—	—	—	—	3	1
Powered haulage	1	29	8	6	168	44
Haulage trucks	—	13	3	3	87	17
Front-end loaders	—	3	1	2	40	12
All other powered haulage	1	13	4	1	41	15
Hoisting	—	—	—	—	1	—
Ignition or explosion of gas or dust	—	—	1	—	3	—
Impoundment	—	—	—	—	—	—
Inundation	—	—	—	—	—	—
Machinery	—	18	15	7	159	92
Dozer	—	3	1	—	6	—
Drill	—	1	2	1	27	4
All other machinery	—	14	12	6	126	88
Slips or falls of person	—	45	20	—	271	88
Stepping or kneeling on object	—	5	3	—	28	9
Striking or bumping	—	—	2	—	6	4
Other	—	3	6	—	26	15
Total	2	168	122	15	1,279	653

North Central District: Minnesota, Wisconsin, Michigan, Illinois, Indiana, Iowa, and Ohio. *District Office:* Duluth, MN.

South Central District: Missouri, New Mexico, Oklahoma, Arkansas, Texas, and Louisiana. *District Office:* Dallas, TX.

Rocky Mountain District: Montana, North Dakota, South Dakota, Wyoming, Utah, Colorado, Kansas, and Nebraska. *District Office:* Denver, CO.

Western District: Washington, Oregon, Idaho, California, Nevada, Arizona, Alaska, and Hawaii. *District Office:* Alameda, CA.

Figs. 1 and 2 show the areas of jurisdiction of MSHA's regional offices for coal and noncoal mine health and safety. Current addresses and telephone numbers can be obtained from: Office of Information and Public Affairs, Mine Safety and Health Administration, US Department of Labor, 4015 Wilson Boulevard, Arlington, VA 22203.

Table 2. Number of Injuries by Mineral Industry in Surface Mining, and Accident Classification, January-December, 1987—Continued

Accident Classification	Sand and Gravel		
	Fatal	NFDL	NDL
Electrical	6	12	14
Entrapment	—	—	1
Exploding vessels under pressure	—	3	1
Explosives and breaking agents	—	2	—
Falling, rolling, or sliding material	—	1	—
Fall or face, rib, side or highwall	—	1	—
Fire	—	5	3
Handling material	—	333	184
Hand tools	—	117	101
Nonpowered haulage	1	3	—
Powered haulage	3	111	40
Haulage trucks	—	27	10
Front-end loaders	3	32	12
All other powered haulage	—	52	18
Hoisting	—	1	—
Ignition or explosion of gas or dust	—	1	—
Impoundment	—	—	—
Inundation	—	—	—
Machinery	4	117	89
Dozer	1	7	5
Drill	—	—	—
All other machinery	3	110	84
Slips or falls of person	3	225	80
Stepping or kneeling on object	—	17	7
Striking or bumping	—	4	3
Other	—	25	16
Total	17	978	529

*Adapted from: *Mine Injuries and Worktime Quarterly, Jan.-Dec.,* 1987, Mine Safety and Health Administration, US Dept. of Labor, 1987.

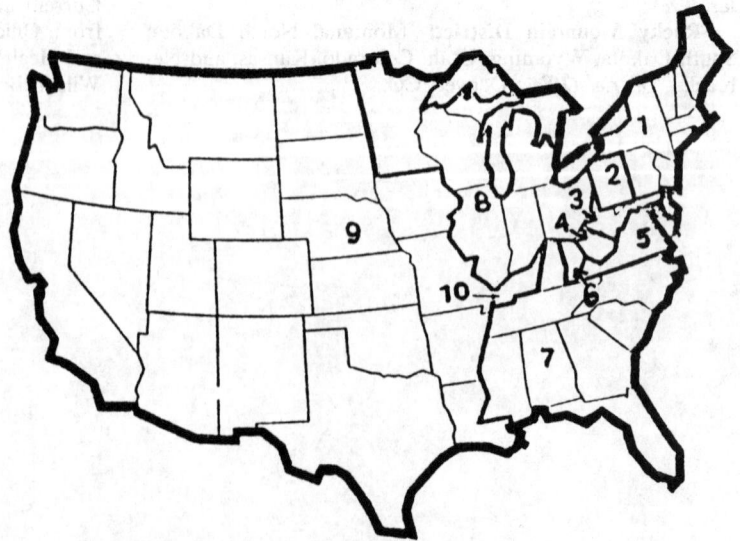

Fig. 1. MSHA Coal Mine Safety and Health Districts.

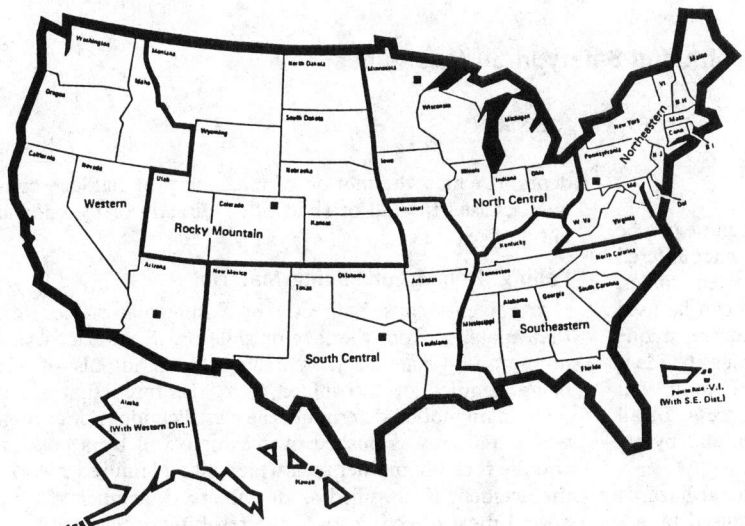

Fig. 2. MSHA Metal and Nonmetal Safety and Health Districts. Black squares indicate district offices.

6.12.3 Health and Safety in Surface Mines

WESLEY G. JOHNSON

SURFACE MINE HAZARDS

Health and safety programs for any surface mine can be planned on the basis of recognition of hazards encountered in each activity of the mining process. In problem mines, statistics of severity and frequency of accidents can be used for program planning. Conditions will arise that are unique to a particular surface mine. Such conditions must be identified and safe solutions found, which, with training, will become part of the overall safe production process. In all cases continuous training by formal on-the-job, and by-example methods must be developed and used.

One measure of the seriousness of occupational hazards in surface mining is the number of fatalities caused by accidents. Another is the magnitude of NFDL—nonfatal with days lost—injuries.

First, based on statistics shown in Table 2 of Sec. 6.12.2, are the most severe accidents, ranked from the highest to the lowest according to the number of fatalities.

Powered Haulage

Accidents related to motion of powered haulage equipment. Included are accidents involving conveyors, front-end loaders, forklifts, shuttle cars, load-haul-dump units, locomotives, railroad cars, haulage trucks, pickups, automobiles, and personnel carriers.

Machinery

Accidents related to motion of the machinery. Included are all electric and air-powered tools and mining machinery such as drills, slushers, winches, draglines, power shovels, and compressors.

Exploding Vessels under Pressure

Accidents involved with bursting of air hoses, air tanks, hydraulic lines, hydraulic hoses, stand pipes, etc., due to internal pressure.

Explosives and Breaking Agents

Accidents involving the deterioration of manufacture explosives. Also in this category should be premature detonations caused by lightning or stray currents with electric detonators.

Fall of Face, Rib, Pillar, Side or Highwall

Accidents in this classification include falls of material while scaling down or placing retaining support and also slides or slumps of in-face material. Not included are accidents in which the motion of machinery or haulage equipment caused the fall or slide either directly or by knocking out support.

Falling, Rolling, or Sliding Material

Accidents caused directly by falling material or other than material from the face or side; or, if material was set in motion by machinery, by haulage, by handtools, or while being handled or distributed, etc., the force that set the material in motion determines the classification. For example, where a rock was pushed over a high wall by a bulldozer and the rock hit another rock which hit and injured a worker, the accident is classified as machinery. Machinery (a bulldozer) most directly caused the resulting accident.

Next, based on available statistics (Table 2, Sec. 6.12.2) are the accidents that contribute most to the work days lost because of injuries.

Slips or Falls of Person

Accidents include slips or falls while getting on or off machinery and haulage equipment which is not moving, and slips while servicing or repairing equipment or machinery. Also included are slips and falls resulting from ice and snow.

Handling Material

Accidents related to handling packaged or loose material while lifting, pulling, pushing, or shoveling.

Machinery, Powered Haulage, and Hand Tools

Machinery and powered haulage are the same as defined earlier. Accidents classified in the hand tools category are those related to nonpowered tools.

Stepping or Kneeling on an Object

Accidents are classified in this category only when the object stepped or kneeled on contributed most directly to the accident.

Other

Accidents not classified elsewhere.

In mills, fatalities caused by the slip or fall of persons, electrical, handling material, powered haulage, and *other* accidents head the list of occupational hazards in surface milling operations of the United States.

Surface mine health and safety hazards listed in Table 1 and the notes following the table should be considered in planning and follow-up processes required for safe operation of any surface mine.

Table 1. Surface Mine Health and Safety Concerns

Function	Health and Safety Hazard or Concern	Detail No.
Mine Location		
Climate	Effects of cold, heat, or humidity	1
Light conditions	Visibility	2
Remoteness	Effects of boredom; stability of work force	3
Proximity to other workings	Effects of caving, slope failures; water inflow	4

Table 1. Surface Mine Health and Safety Concerns—Continued

Access roads and haulage ways	Separation from public roads; degree of curve, amount of grade, sight distance, condition of berms, markings	5
Water Control	Potential for subsurface and surface inflow; handling procedures, pumps, sumps, and pipelines	6
Ground Stability	Bank or highway design, potential for slope failure in waste or overburden coal or ore, tailing dumps, dams, and impondment	7
Drilling and Blasting	Mobile equipment operating and maintenance procedure	8
	Explosives transportation, storage, and use procedures	9
	Fire safety	10
Mobile Equipment Operations	Mobile equipment operating and maintenance procedure (specific to individual types and sizes of equipment)	8
	Fire safety	10
Fixed Equipment Operations	Fixed and portable equipment operating and maintenance procedures (includes pumps, generators, and air compressors)	11
	Fire safety	10
Shop Operations	Mobile, fixed, and portable equipment operation and maintenance procedures	8,11
	Crane and hoist procedures	12
	Welding and cutting (gas and electric procedures)	13
	Shop tool and machine operating procedures	14
	Fire safety	10
Other Maintenance Operations	Fueling and lubrication procedure	15
	Tire maintenance procedure	16
	Fire safety	10
Mill Operations	Provision for protective guards, walkways, lighting, dust, noise, and fume suppression	17
Warehouse Operations	Forklift operations, lifting and carrying, slings and chockers	18
	Fire safety	10
Electrical Use and Distribution Facilities	Protection from contact, grounding, and flashover; fault protection and maintenance procedure	19
	Fire safety	10
Change, Toilet, and Lunch Room Facilities	Protection from transmission of disease, slips and falls, adequate facilities	20 21
Employee Protection	Individual protective equipment including eyes	22
	Toxic dust and fume identification safety procedures	23
	Fire safety	10
	Slips and falls	21
	First aid, including emergency vehicle	24
	Alcohol and other drugs on and off the job	25
	Parking lots	26
	Driving to and from work	27

Detail 1—The hazard effects of cold, heat, and humidity on miners and equipment must be analyzed and prepared for. Personal protective equipment and special training may be required. At mines where large ranges occur, special facilities and equipment may be required. Mobile equipment must be outfitted with heaters and air conditioners. Operating conditions will change as freezing or thawing occurs. Inflows of water may vary widely, requiring special training for all personnel.

Detail 2—Light conditions may vary from long days with strong sunlight, requiring eye protection, to long black nights requiring glare-proof lighting.

Detail 3—For the individual, boredom affects safety; for society, increasing mine remoteness reduces hazards associated with contamination of water supplies by mining, flooding through failure of dams, etc., dust, and noise. (Individually, miners *must* be protected from adverse effects of dust, noise, and exposure to toxic materials.)

Detail 4—Slope stability must be maintained to protect miners and equipment. In addition to pit walls and

Table 1. Surface Mine Health and Safety Concerns—Continued

spoil banks, slope stability includes design and construction of drainage structures, truck dumps, roads and ramps, or any excavation or builtup earth structure with walls (slopes) of natural materials.

Design of slopes hinges on engineering, geologic, and hydrologic factors associated with the rock mass. Usually as steep a slope angle as possible is sought for economic reasons (reduction of waste to ore ratio, for instance). Pitted against the steep slope angle are consideration of safety and economic loss (injury and property damage). Economic considerations include equipment damage and possible disruption of mining sequences.

Detail 5—In order to eliminate hazards associated with public use of mine roads or vice versa, all mine access and haul roads should be totally closed to public use. (Public action and the attendant liability is uncontrollable; hence no access is required.)

Design of mine roads must match the needs of the largest vehicle using the road, including degree of curve, superelevation, sight distance, placement and condition of berms and guard rails (which are usually useless for haulage units), traffic plan, and marking and illumination.

Status of road maintenance, dust suppression activities, adherence to traffic rules, and vehicle condition are continuous safety concerns.

Detail 6—Review of plans for water control (both surface runoff and pit drainage) may reveal hazards. Work systems to carry out water control plans may reveal inadequate safety provisions for possible drowning; burial by slope failure; and injuries from installing and moving pumps, hoses, and pipelines. In low temperature areas special precautions are required. Pump crews require special training in prevention of electrical shock, burns, and fire during fueling operations. Proper lifting, carrying, and slips and falls protection must be taught.

Detail 7—Adequacy of slope or highwall design and adherence to design during mining should be monitored. Condition of such walls for dribble (rock and other material falling off walls), especially in areas where freeze-thaw occurs. Attitude of pit and highwall supervisory personnel regarding highwall safety may indicate training is required.

Provision for formal inspections of pit slope or highwall for indications of weakness are required. Waste or spoil areas are included. Impondment dams and other water control structures must be monitored for signs of failure, in some cases as required by law (see MSHA regulations).

Detail 8—Mobile equipment operating and maintenance procedure: involves task training to insure operator knowledge or operation of controls, warning devices, and safety features. Observation and grading of operation expertise by competent authority, with periodic update to insure highest quality of operation. Training includes understanding of operating principles of mechanical and electrical components and fire safety.

Such equipment includes: draglines, bucket wheel excavators, cross pit conveyors, shovels, special excavators, haulage trucks, tractor scrapers, truck tractors, wheel tractors, front-end loaders (articulated and conventional), motor graders, backhoes (excavators), blasthole and exploration drills, conventional trucks, and special equipment such as water distributors, cable movers and reel vehicles, tire changer vehicles, compactors, ripper attachments, cranes, hoists, and forklifts, all types of farm equipment, and pumps (gasoline, diesel, and electric).

Detail 9—Surface mines consume considerable quantities of explosives which are controlled vigorously by federal and state agencies through laws and regulations. The following federal agencies control explosives used in mining. *Dept. of Commerce:* Interstate transportation of explosives; *Dept. of Transportation:* Transport on interstate highways (most explosives are delivered by truck); *Dept. of Treasury:* Bureau of Alcohol, Tobacco, and Firearms (BAF) licenses manufacturers, agents, and users of explosives. A mine is a user and requires a license; *Federal Bureau of Investigation (FBI):* Investigates all thefts of explosive materials. Disappearance of explosives must be reported within 24 hr; *Dept. of Labor:* Mine Safety and Health Administration (MSHA). Controls safety in the storage, transportation, and use of explosives on mine property; *Dept. of the Interior:* Office of Surface Mining (OSM) controls protection of the environment from use of explosives at surface coal mines.

The best source of information regarding the subject of explosives is: Institute of Makers of Explosives (IME), 1575 I St. NW, Suite 950, Washington, DC 20005.

Detail 10—*Fire Safety:* Although widely acknowledged as important, fire protection often receives less than full support or attention. Continuous monitoring and training, including actual fire suppression drills, are worthwhile. Fire protection plans for facilities must be reviewed and adjusted to eliminate hazards. Fire suppression equipment, in addition to handheld extinguishers, may be warranted for installation on the expensive equipment at surface mines.

Modification of dust-suppression water trucks to include fire-fighting capability is worthwhile. Provisions for adequate storage and maintenance of water/fire trucks, plus training for operators, is required.

Safety inspections should include inquiry into fire protection adequacy for all parts of the mine.

A good source of information on fire protection is: National Fire Protection Association, 470 Atlantic, Boston, MA 02210.

Detail 11—Fixed equipment operating and maintenance procedure: involves task training and continuous update and monitoring of knowledge on operation, adjustment, repair, and fire safety of all fixed equipment such as cranes, fans, pumps, air compressors, fueling and lubrication devices, welders, electric motors, gear cases, crushers, feeders, and conveyors.

Detail 12—All mechanical equipment has individual operational characteristics which must be mastered to insure uniformly safe practice. Hoists and cranes require special care in that working conditions vary continually. Each load or *pick* is different, requiring experience for proper execution. Only rated or approved persons should operate cranes and hoists. Special training is required, especially in use of slings, boom angles, and outrigger (if used) placement. The Power Crane and Shovel Association is a good source of information (Milwaukee, WI).

Detail 13—Proper welding and cutting is an art that can be acquired only through training and practice. From a safety viewpoint, welding must be monitored to insure adequacy of safety training, including fire safety. Many fires are started by poor welding practice.

Detail 14—Formal training is required prior to operation of shop machine tools. A roster of qualified operators may be required to prevent injuries and damage to equipment. Ordinary tools require a common sense approach as do shop operations in general. The National Safety Council, Chicago, IL, has materials that may be of assistance for such training.

Detail 15—Fueling and lubrication procedure must be controlled to eliminate fire risks. Technical innovations such as the Wiggins Fast-Fuel System reduce dangers (and fuel loss). A safety check of the equipment can be made during the lubrication process.

Detail 16—Tire maintenance requires special training in the dangers involved, since extreme hazards exist. Inflation of mounted tires must follow a prescribed routine. Task training with periodic update is necessary, with observation of tire work monitored by supervisors and safety inspectors.

Detail 17—Mill or process plant operations contain potential hazards mainly associated with fixed and mobile machinery and equipment. Regulations address these hazards in general, such as machine guarding; walkways; handrails and toeboards; protection from noise, dust, and toxic fumes; and adequacy of lighting. Other hazards will exist, depending upon the process being used such as electrical, flame or heat, and explosive atmosphere dangers.

Conduct of repair work creates special hazards requiring foolproof lock-out procedures, adequate scaffolding and ladders, and lifting devices.

Supervisory and safety inspection attention, with continuous training, is necessary.

Detail 18—Warehouse operations contain specific hazards associated with fork lift equipment and falls of

Table 1. Surface Mine Health and Safety Concerns—Continued

stacked or piled materials. Lifting and carrying techniques must be monitored and updated through training. Lifting devices, including hoists and cranes, require specifically trained operators.

Dangers of fire and explosion may exist, requiring attention to housecleaning.

Detail 19—Electrical safety in surface mines requires constant attention. Because of the expanded use of high voltages (up to 25,000 V), much of the energy used is distributed by ground-laid cables which are exposed to damage in numerous ways (fly rock from blasting, equipment rock spills, and damage from being driven over). Moving these cables has become a daily activity which must be monitored carefully to insure safe practices. Proper equipment must be used in order that the cable is not damaged internally by too small bending radius or excessive tension loads when dragging it.

Detail 20—Protection from transmission of disease may vary to some extent according to climate. Precautions in regard to drinking water and foodstuff consumed at the mine are necessary. Availability of bathing facilities, clothes lockers, and toilet facilities is mandated by regulation. Housekeeping of such facilities must be provided for and monitored by supervisors and mine health and safety personnel.

Detail 21—Slips and falls account for many injuries in surface mines. The workforce attention to hazardous conditions in regard to slips and falls must be obtained by continuous training. Access to equipment by operators requires proper use of handholds and safety steps. Proper footwear and safety belts may be required. Continuous attention to walkways, ladders, scaffolds, and manlifts is required.

Detail 22—Certain personal protective equipment is required by regulation, such as protective footwear (steel-toe boots), hard hats, eye protection (depending upon hazards such as welding, grinding, and chipping), special gloves for electric cable handling, and filter masks. Many surface mines require, in addition, safety glasses (including sunglasses) welder protective clothing, work gloves (for some jobs), tight (nonbaggy or loose) clothing, no jewelry, and tied-up hair.

Ear protection in the form of muffs and plugs is required for certain jobs. Much of this equipment may be provided to each miner, including in some cases insulated lunch and water containers, protective clothing for mechanics, and hardhat liners for cold climates.

Detail 23—Use of toxic materials centers in processing mills. Continuous monitoring of chemical materials used will control hazards associated with miners' exposure to toxic materials. Particulate filter masks and positive ventilation of work areas will assist in resolving some of the hazards. Guidelines for control and use of toxic materials can be obtained from the National Institute of Safety and Health (NIOSH). A table of threshold limit values (TLV) will assist in checking the mine.

Detail 24—First aid training is required for all miners by regulation. Many mines have extended first aid provisions to include sponsorship of emergency medical technician (EMT) training for volunteer employees. EMT personnel can assist with annual first aid refresher training, are assigned as crew members for emergency vehicles, and contribute to the well being of the mine work force and surrounding community.

Detail 25—The use of alcohol and other drugs by employees is recognized as a source of potential hazards. Use on the job is not tolerated but carryover effects can produce disasters. Supervisory recognition of such effects and method of dealing with the situation requires careful handling to avoid a major personnel problem. Training for supervisors is suggested, with professional guidance in some cases.

Detail 26—Total commitment to mine safety requires concern for all aspects of employee safety. Parking lots can be unsafe areas which must be monitored. Mine equipment must not use employee parking facilities. Access and traffic flow are planned with safe sight distances and are well drained and lighted. In climates which warrant their use, electrical outlets are provided for engine heaters and snow removal is provided. Where conditions warrant, fences and control of entry are required, with no access to mine property by private vehicles.

Detail 27—Concern for mine safety provides a base from which miners will carry safety concepts over into their private lives. Training in defensive driving plus the daily practice of safe equipment operation will influence miners to become safer drivers on public highways. The awesome toll in so-called "accidents" on the highway can be improved through such training.

6.12.4 Health and Safety Organization

Archie M. Gilliss

As has been discussed in the preceding pages, health and safety in surface mining begins with management's commitment to make it happen. As part of the production process, health and safety must be part of any planning involving the mine work force. This commitment must be evident in all levels of management to insure that the safety policy becomes part of all mine activity. The most logical starting point is at the top level.

A written policy is required stating top management's feeling toward company safety efforts and it should be displayed for each miner to see on a daily basis.

Health and safety controls are based on form of organization and management objectives. The following lists three types of health and safety organizations:

1. A separate health and safety organization, reporting directly to management, that oversees mine safety activity (a staff function).

2. A combined function organization where either employee relations or personnel departments oversee health and safety, and report to management (two-hat concept).

3. A separate department that reports to both the mine manager and upper level management. (Details of procedure and responsibility are carefully defined, as this is a line and staff organization.)

While there is no *best* organization, the last type is very effective.

Depending upon requirements, the training function can be established as part of the health and safety department. Safety personnel will conduct or arrange training, developing a strong relationship through contact with mine personnel.

The need for training is often misunderstood and sometimes confused with the need for job performance evaluation, a production and safety department function. When performed, training must be done in a manner that reflects a positive safety attitude.

Statements like, "We've got a training problem because our workers aren't safety conscious," or "He really must change his attitude. . . ," are often listed as the problem when it is merely a symptom of the problem. The training function must work closely with the safety function if it is a separate department. The difference between what is being done and what is supposed to be done in the job can be termed as a "performance discrepancy." The method used to correct such a discrepancy may not be training; however, working with the safety and production departments, performance may be changed through positive and, in some instances, negative reinforcement.

HEALTH AND SAFETY PERSONNEL ACTIVITIES

The basic function is one of service to the mine operating department. This philosophy is carried out by inspecting and observing safety practices and assisting mine management in correcting deficiencies. Potential health hazards are detected and solutions suggested, but, in all cases, mine management is responsible for health and safety matters. Specifically, the following services are rendered:

1) Conducting periodic health and safety inspections to insure compliance with company and regulatory agencies, with formal reports.

2) Assisting in correction of deficiencies.

3) Organizing and conducting safety and other training; maintaining training records.

4) Investigating all accidents; issuing formal reports.

5) Maintaining statistical records of accidents and health records and periodically reporting status.

6) Interfacing with regulatory safety agencies as required.

7) Planning emergency procedures for fires, storms, and other catastrophes, and conducting drills.

8) Researching improvements for all hazardous procedures.

9) Functioning as a health and safety information source for mine personnel.

10) Extending influence to off-the-job health and safety.

11) Develop improved ways to reduce injury, accident, and work environment costs.

HEALTH ITEMS REQUIRING ATTENTION

Specific health items for surface miners must be monitored as required by law. These items involve the work environment of each miner and any potential injury, hygienic, or disease situation which might arise:

1) Exposure limit to noise: maximum allowable, 8 hr 90 dBA, 15 min 115 dBA.

2) Exposure limit to dust: 2 mg/m^3 (coal) [includes threshold limit values (TLV)] limits for a long list of toxic materials as set by the American Conference of Governmental Industrial Hygienists].

3) Skin irritations caused by chemicals, lubricants, cleaning solutions, and various fumes.

4) Adequate illumination.

5) Protection for eyesight.

6) Potable drinking water.

7) Adequate number and cleanliness of sanitary facilities (shower and toilet facilities).

8) Adequate ventilation, including positive fume removal from welding and cutting operations.

9) Monitoring of correct use and storage of carcinogenic and radioactive materials.

10) Monitoring of exposure to excessive heat or cold.

11) Continuous review of fire safety conditions.

PRINCIPLES USED: ACCIDENT PREVENTION EFFORTS

There are no quick remedies or gimmicks that can improve a safety record or performance. It is sticking to basic principles and using basic philosophy that has been developed and perfected throughout the years by professionals that will improve safety performance. General industry concepts are perfectly acceptable in coal mining and vice versa. In the long run, positive results will follow.

Basic principles are guidelines that broadly describe what has to be done to have an effective accident prevention program. The *how to* details are different in each application. Using ingenuity, imagination, and past history data, effective programs can be developed that are basic, but workable.

The foundation for accident prevention is the establishment of a cause or causes of accidents and eliminating same.

Accidents are caused by an interaction of people, equipment, material, and environment. Two very basic components of this are *unsafe practices or acts* and *unsafe conditions*. Of the two, unsafe conditions are the simplest to control. Changing unsafe practices is more of a challenge, as this portion deals with the human factor. The only way to prevent unsafe practices is to eliminate their causes. Going back to the basic foundation means finding the cause and eliminating that cause. This is the elimination of such causes as:

1) Inadequate job skills.
2) Lack of knowledge.
3) Lack of incentive to perform work safely.
4) Physical and mental limitations.
5) Motivations that conflict with working safely.
6) Lack of job procedures.

The basic way to eliminate unsafe conditions is to correct the condition and the cause of the condition.

M. U. Eninger (Eninger, 1981) lists 16 essential principles that outline a program of what needs to be done to prevent or eliminate unsafe practices and unsafe conditions. A listing of these 16 (ten associated with unsafe practices and six with unsafe conditions) follows:

Principles for eliminating unsafe practices:

1) Provide an overall safety orientation to new employees before they start to work.

2) Provide an occupation safety orientation to employees moving into new occupations.

3) Develop standard written job procedures for known hazardous jobs in each occupation.

4) Develop and publish general safety rules that cover the most probable and serious unsafe policies.

5) Provide thorough initial job safety training to employees who have moved into a new occupation, using job training guides when available.

6) Check job safety training with planned follow-up safety observations.

7) Conduct a program of regular, planned safety contacts to maintain employee awareness of hazards and precautions.

8) Always reinstruct employees observed working unsafely. Never walk away from an unsafe practice.

9) Provide prejob instructions to employees assigned to do infrequently done, high hazard jobs.

10) When possible, pair off inexperienced employees with experienced employees who can share their job safety know-how.

Principles for eliminating unsafe conditions are as follows:

1) Where supervision has authority to order correction, prompt correction of unsafe conditions should be ordered.

2) Where such authority is lacking, the unsafe condition should be reported to higher supervision, together with recommendations for corrective action.

3) Where supervision has authority to correct the source cause, the correction at the source should be ordered or taken.

4) Where such authority is lacking, the source cause(s) should be reported to higher supervision, together with recommendations for corrective actions.

5) When correction of an unsafe condition necessitates delay, supervision should take whatever temporary precautions are necessary to reduce the hazard.

6) Supervision should always follow up to be sure that correction ordered or requested has been applied.

A review of these 16 items or principles will show very little change in accident prevention concepts since the early 1960s.

Regardless of the industry or the business establishment, the role of management in any safety program takes precedence over any of the other elements. Where the responsibility for preventing accidents and providing healthful work environment is pushed off to the safety department or safety committee, any reduction in the accident rate is pure coincidence. Every member of the management team must have a role in the safety program. This thought is not new, but it is seldom employed.

A careful review of a successful safety program will show an involvement at all levels of management. Employee involvement in the safety program will also be very evident.

In today's environment, the same management principles and concepts that are applied to quality, cost, purchasing, and production must also be applied to safety. If most safety programs were examined, the results would show that the same management techniques being used to solve production problems are not being used to solve safety problems. Production problems are being left to line management, safety problems are handled by the safety department.

Management can hardly be expected to fulfill various safety functions that take even a little additional time unless they see a good reason or a need to do so. Past practices may have been to let the safety department handle safety, and they have gotten by. Hence, the number one job in mining is to teach or convince management that safety is a good practice of a good manager. As has been stated throughout this subsection, every successful safety program starts from the top of the organization and works down. The company president or mine manager passes the responsibility down through the ranks of management to the first line supervisor. However, in doing so, each level retains a participation part in the program. The first line supervisor now becomes the focal point. It is there that the program either dies or flourishes. Supervisors must be convinced that the sincere efforts they make to prevent any type of accident will cause their employees to rapidly accept safety leadership. Poor examples will destroy a program. In a sense, a good safety program that is built around the supervisor is a cornerstone of good employee relations.

There are many benefits other than a reduction in accident rates when a good safety program is achieved. Lower turnover, less absenteeism, steadier work, and improved morale are a few of these benefits.

Safety and health are just as much a part of a worker's job as productivity. The key is accountability. Once accident prevention procedures have been written and responsibilities defined, individual employees should be held accountable for successful implementation. Standards identifying responsibility areas for each management level need to be set. The responsibilities, as listed, then provide the managers with basic requirements on which they can develop their own specific responsibilities.

In safety it is not enough to have good intentions, but all employees must recognize the good work and efforts that are put forward.

WORK FORCE INVOLVEMENT

Health and safety input from the work force is basic in maintaining a safe mine. Formal recognition of this can be made through assignment of safety duties to miners in all parts of the mine. In some cases, safety committees are established through company efforts that suggest health and safety improvements. Committee members may accompany MSHA inspectors (as required by law) on formal inspections. Some companies prefer to have individual miners participate in all inspections, including both those conducted by the company safety person and MSHA inspectors.

Details of good safety practice are discussed openly during the course of inspections. Proposed changes in operations are reviewed to determine if such changes truly improve safety or if some other aspect needs to be altered. Possibly no facility change is required, but change in operator method or attitude is determined to be best. Such communication, later documented as a course of action with management approval, is fundamental in order to achieve a no-accident status at any mine.

Accident investigation data produced during the investigation process should be reviewed with all individuals involved or who possibly could be involved in the work process.

HEALTH AND SAFETY ORGANIZATIONS

The following organizations play important roles in surface mine health and safety. Due to the evolving nature of the subject, additional information may be available.

Society of Mining Engineers, Inc.

Mine Safety and Health Administration, US Dept. of Labor

National Safety Council

American Society of Safety Engineers

American Institute of Governmental Industrial Hygienists

Certified Safety Professional Organization

National Fire Protection Association

American National Standards Institute (National Electrical Safety Code)

Holmes Safety Association

REFERENCES AND BIBLIOGRAPHY

Anon., 1982, *Accident Facts*, National Safety Council, Chicago, IL.

Bacow, L.S., 1980, *Bargaining For Job Safety and Health*, MIT Press, Cambridge, MA.

Blake, R.R. and Moutan, J.S., 1981, *Productivity: The Human Side*, AMACOM, New York.

Boley, J.W., 1977, *A Guide to Effective Industrial Safety*, Gulf Publishing Co., Houston, TX.

Day, J.M., 1979, *The Federal Mine Safety and Health Act*, Practicing Law Institute, New York.

Denton, D.K., 1982, *Safety Management: Improving Performance*, McGraw-Hill, New York.

De Reamer, R., 1980, *Modern Safety and Health Technology*, John Wiley & Sons, New York.

Drucker, P.F., 1980, *Managing in Turbulent Times*, Harper and Row, New York.

Eninger, M.V., 1981, *Operation Zero—Accident Prevention Fundamentals for Managers and Supervisors*, Normax Publications, Inc., Pittsburgh, PA.

Ferry, T.S., 1981, *Modern Accident Investigation and Analysis, An Executive Guide*, John Wiley & Sons, New York.

Hammer, W., 1981, *Occupational Safety Management and Engineering*, 2nd ed., Prentice-Hall, Englewood Cliffs, NJ.

Johansen, T., and Reeder, R.T., 1978, "Cost of Safety," 4th Institute on Coal Mine Health and Safety, Colorado School of Mines, Golden, CO.

McGregor, D., 1960, *The Human Side of Enterprise*, McGraw-Hill, New York.

Petersen, D., and Goodale, J., 1980, *Readings in Industrial Accident Prevention*, McGraw-Hill, New York.

Petersen, D., 1982, *Human Error Reduction and Safety Management*, Garland Press, New York.

Petersen, D., 1980, *Analyzing Safety Performance*, Garland Press, New York.

Petersen, D., 1978, *Safety by Objectives*, Aloray, Inc., River Vale, NJ.

Petersen, D., 1978, *Techniques of Safety Management*, McGraw-Hill, New York.

Petersen, D., 1976, *Safety Supervision*, AMACOM, New York.

Petersen, D., 1975, *Safety Management, A Human Approach*, Aloray, Inc., Englewood, New Jersey.

ReVelle, J.B., 1980, *Safety Training Methods*, John Wiley & Sons, New York.

Tarrants, W.E., 1980, *The Measurement of Safety Performance*, Garland STPM Press, New York.

Wiant, W.F., "Major Federal Mining and Mineral Laws and Related Documents," Manuscript, National Mine Health and Safety Academy, Beckely, WV.

6.13 Communications and Controls

R. C. VOIGE

INTRODUCTION

Communication and control systems provide a vital link in a surface mine operation. The interaction of operating personnel through various types of communication instruments and the similar control of equipment through telemetry instruments is essential to a well-run operation. In recent years computers located on mobile machinery have also used radio communication to transfer digital data between the mobile units and fixed base stations.

The major use of communication and control systems at surface mines is productivity and safety-related. Telephones and mobile radios enable operating personnel to change plans instantly and have these plans carried out at remote locations. Also, the status of mining equipment can be monitored continuously and preventative maintenance initiated at the first sign of a malfunction. Computer-aided dispatch systems using a mobile radio can maximize the use of haulage vehicles. The safety of blasting operations is improved by the use of special warning signals over communication systems. Equipment operators who are performing a potentially hazardous task alone have mobile radios to contact rescue personnel in case of an emergency.

There are four general types of communication and control systems in use at surface mines today. They are:

Wired Voice Systems (telephone, public address, intercom)

Mobile Voice Systems (two-way radio, paging)

Wired Monitoring and Control Systems (telemetry, remote equipment activation, equipment and machine monitoring)

Mobile Monitoring and Control Systems (vehicular telemetry and location, shovel and dragline telemetry, remote equipment activation)

WIRED VOICE SYSTEMS

The most widely used wired voice system is the telephone. Telephones provide basic communication throughout the mine property and tie into public systems. High reliability and simple operation have been the cornerstone of telephone operation in the United States. Microprocessor technology has been incorporated into automatic switchboard design, producing features such as call waiting, call forwarding, automatic call back, and call transfer that make telephone communication more productive.

Wiring for telephone systems consists of lines to each instrument and trunk lines to the local public telephone exchange. The automatic or, in some cases, manual switchboard is the hub of the system. Each telephone instrument must have at least one pair of wires connected to it and instruments with push-button line selection require a pair of wires for each "line." A telephone wire plant (wire and cable system) for a mine will include an appropriate number of 6- and 12-pair cables from small groups of phones that feed into local junction boxes. Then larger 25-, 50-, and 100-pair cables complete the circuit path to the centrally located switchboard equipment. Terrain, climate, and other factors will determine whether aerial or buried cable is used to connect remote locations.

A second type of wired communication system is the public address system. Large open areas such as the warehouse and shop can be equipped with a high-powered amplifier (25 to 200 W) and multiple speakers to provide communication to roving personnel. Audio systems have one outstanding advantage over telephone systems in that, with a high-powered amplifier, a paging call can be heard over and above the high ambient noise conditions found at a surface mining operation. Telephones or intecom stations are used to respond to the audio page.

An intercom system usually consists of a number of stations that have low-power amplifiers (0.5 to 3 W) and built-in microphones. Remote stations are selected by use of a simple switch. A multi-pair cable to each station is required for this low-cost type of system.

MOBILE VOICE SYSTEMS

Mobile communication is now accepted as the best way to contact surface mining personnel who are usually found in vehicles or on foot throughout the mine property. The basic components of a mobile communication system are an antenna, transmitter-receiver, microphone, speaker, and power supply. The vehicular transceiver (transmitter-receiver) is usually mounted in some out-of-the-way place in a car, shovel, truck, or locomotive. A microphone and speaker are mounted inside the vehicle (or cab) for the convenient use of the operator or driver. The antenna will be mounted high on the vehicle or at a safe location to minimize breakage. Most surface mine systems will also use an a-c-powered base station radio with an antenna tower.

Mobile communication can help reduce a surface mine's capital requirements. Fewer road trucks, fewer switching locomotives, fewer loaders and material handling tractors, are needed to attain productivity goals when their operators can be directed by mobile communication. Fewer clock hours and man-hours are needed. Fewer people are needed to perform many tasks; mobile communication can be instrumental in the transformation of waiting and idle times into working time. When a man and a machine finish a task at one part of the mine, mobile communication can instantly direct them to another part of the mine for another task. Without such a tool, management would have to hire another man and buy another machine. Employees who are physically alone are never out of touch when their supervisors, subordinates, and peers can reach them by mobile communication.

Mobile communication is an exchange of information between two or more parties who are free to walk, run, or drive in separate directions as they communicate with each other. The medium is usually a two-way radio, so called because each instrument can transmit and receive. But in day-to-day business, many mobile communications turn out to be the multiway radio, in which several mobile parties transmit to each other. In certain two-way radios, the speaker is silent until a transmitted tone or digital signal opens the receiver to the sender's message. There are selective two-way systems that activate a small group of receivers out of a population of many. There are selective one-way systems, reaching a single pocket-size radio pager with an audible tone or a tone-and-voice message.

A radiotelephone uses a duplex system, that is, both parties can talk and/or listen at the same time. Most two-way mobile systems are simplex, with a push-to-talk button on the microphone. At each unit, pushing the button activates

---

960 SURFACE MINING

the microphone and transmitter while shutting off the receiver and speaker. Releasing the button reverses the condition; the operator can now listen. The push-to-talk arrangement conserves power, since the receiver's consumption is much less than that of the transmitter.

Communication can be in the form of a voice message or at times a data message between terminals.

The mobile communication of business (including mining) is carried out principally on frequency modulation (FM) radio, on frequencies which the Federal Communications Commission (FCC) limits to certain classes of users. For example, there are sets of frequencies for public safety, manufacturing, construction, agriculture, forestry, trucking, railroads, utilities, petroleum producers, and, of course, miners. The FCC must restrict not only the allocation of frequencies but also the radios' transmitting power to minimize interference and give access to as many legitimate users as possible.

Mobile communication is normally of limited range: within the plant, within the property, within the city or county. But the communication is not necessarily direct from radio to radio. In some systems, a two-way radio will transmit to a repeater, a radio which automatically re-transmits the message—at higher power—to other two-way radios in the system.

Mobile communication, then, increases the ability of management to manage its mobile people. It should be noted that mobile communication is not an electronic tether. It cannot prevent wandering of mind or body; it cannot prevent inefficiency. But, in the hands of motivated people and organizations, mobile communication is invaluable. It brings mobile minds together; it increases the value of each mind, the productivity of each person, and the return on the organization's investment in its people and property.

Heinrich Hertz produced radio waves in 1888. It occurred to Guglielmo Marconi that such waves could be used in communication. He demonstrated the mobile radio in 1898 by following the Kingstown Regatta in a tug and flashing the results to the office of a Dublin newspaper. A radio SOS brought aid to the crew of a sinking lightship in 1899, to the passengers of the disetressed steamship *Republic* in 1909, and to the passengers of the *Titanic* in 1912. But the idea of radio communication from a conveyance smaller than a ship was not immediately pursued.

The first broadcast-band automobile radios were installed in the late 1920s, followed by the first police department radios in the early 1930s. An FM radio pioneer, Daniel Noble, found that FM was more reliable than AM (amplitude modulation) in mobile communication, and he developed a two-way system for the Connecticut state police. During World War II, the walkie-talkie military two-way radio was said to have been the most essential piece of communication equipment in the Battle of Normandy and the subsequent invasion of Europe. After World War II, the way was clear for the production of FM two-way radios for police, taxi, and transportation users.

Subsequent developments, in the 1950s, included a car telephone, complete with a dial, to interconnect with telephone-company lines, the pocket paging receiver, and the widespread use of transistors.

In the 1960s, new manufacturing techniques improved the stability and reliability of the mobile equipment. By replacing the old vacuum tubes with solid-state devices (principally transistors), manufacturers were able to reduce the size and weight of their units. They were able to make compact multichannel radios and to develop a mobile teleprinter,

which provided hard-copy readout in a vehicle. Later printers also had keyboards for data entry.

In the 1970s, there was a sizable growth in the mobile communication of digital data and in the use of integrated circuits and microprocessors in such sophisticated products as the portable handheld radiotelephone.

Table 1 is a list of FCC classifications of service. There is wide-spread use of mobile communication in business and government. A fuller list of mobile communicators would, at least, include the federal and state governments, all utilities, the public transit system, and such industries as airline, marine transportation, telephone, mining, and construction. But while applicability would seem to be infinite, the portion of the radio spectrum available for mobile communication is finite.

Land mobile communication takes place at 25 to 50 MHz (low band), 132 to 174 MHz (high band, otherwise known as very high frequency or VHF), 403 to 430, and 450 to 512 MHz (ultrahigh frequency or UHF). During the 1970s, the band from 806 to 870 MHz was allocated to mobile radio users.

The low band is widely used by mining, county, and state law enforcement units, long-haul truck and bus lines, pipeline operators, electric and gas utilities, and other organizations that require long-range mobile communication. Low-band range may be as high as 80 km (50 miles) in flat terrain, less in hilly terrain. But it has a disadvantage in skip interference, in which a user's transmission skips far beyond its normal range. Low-band sytsems on opposite coasts of the United States have been known to cause mutual interference.

The high band is used by those whose operations are generally within a limited area. Range may vary from 8 to 64 km (5 to 40 miles), although greater ranges are often experienced.

The UHF band is ideal for short-range surface mining communication. It provides a clear two-way radio link.

The following is an overview of the capabilities of mobile communications equipment: the mobile and personal two-way radio, radio repeaters, radio pagers and paging encoders, monitors, control centers, and base stations.

Mobile Two-Way Radio

A unit designed for vehicular installation is called a mobile radio. The microphone, with its push-to-talk button, is shaped to fit the hand. A coiled cord connects the microphone to the control head, containing the on-off, volume, frequency selection, and squelch controls. A simple squelch circuit minimizes the reception of noise when no one is transmitting. A more sophisticated squelch circuit keeps the speaker silent

Table 1. FCC Service Classifications

Public Safety	Industrial Radio
Fire radio	Business*
Forestry conservation	Forest products
Highway maintenance	Manufacturers*
Local government	Petroleum
Police	Power
Special emergency	Special industrial*

Common Carrier	Land Transportation
Wireline car telephone	Railroad
Radio common carrier car telephone	Taxicab
	Other

*These services are available for use in mining activities.

until a brief sequence of inaudible tones opens the circuit and lets the message through. Typically, the user's transmitter begins each transmission with the tone sequence that opens the squelch circuits in that user's vehicles only. A more advanced system is digital squelch, which uses digital codes instead of tone codes. In the most economical FM two-way mobile radios, the control head, transmitter, receiver, and speaker are all in a single housing. In the more common configuration, the transmitter and receiver are in a separate, locked case, installed under a seat or in the trunk; the speaker also is separate from the control head. This allows room for the components that provide particularly high power (within FCC-mandated limits), high selectivity (reducing adjacent channel interference), low distortion, and clear audio output. It also allows room for the various optional functions. In addition to the squelch, they may include:

Busy Indicator: A light goes on when the frequency is being used by other people. Without a busy indicator, the user of a vehicular two-way radio has to monitor the frequency, that is, open the squelch circuit and listen, before transmitting, to be sure the frequency is not being used. When a driver transmits without monitoring his frequency, he may obliterate parts of another user's message. If the obliterated message concerns a life-or-death emergency, such carelessness may have a fatal consequence.

Transmit Timer: Another possible problem can be the inadvertent keying of the transmitter in the vehicle. The microphone may have been put into place carelessly, jamming the button and causing the errant transmitter to interfere with all two-way radios in the system. As a solution, the unit may have a timer, which shuts off the transmitter after a predetermined period.

Selective Signaling Codes: The "party-line" aspect of mobile communication has its disadvantages. Hearing every two-way communication, all day, may increase a driver's fatigue and make him less sensitive to his own messages. And, in many operations, one driver's job is of no concern to another.

The solution is selective signaling. It enables a base station operator to contact a single driver or a group of drivers. Various tone codes or digital codes act as keys that unlock certain receivers, either individually or in groups. The user thereby combines the mobility of radio with a degree of address exclusivity. And if a driver happens to be out of his vehicle when a selectively signaled call comes in, a "call received" light on the decoder's panel stays on until the driver removes his microphone from its hang-up box. Some selective signaling equipment will sound a vehicle's horn, briefly, or turn on the headlights as a reminder that a call has come in.

Scanning Monitor: This function sequentially samples each channel of a multi-channel system, and stops automatically when it detects the presence of a carrier or a message. One channel is designated as the priority receive channel, and if that channel receives a signal during a scan, the receiver automatically reverts to it.

Radio Paging: A radio pager is a compact radio receiver that emits an alerting signal when a unique tone-code activates the receiver's tone-sensing circuitry. A *tone-only pager* emits a signal only; a *tone-and-voice* pager allows a voice to follow the signal. Most pagers are about the size of a cigarette package and weigh 170 g (6 oz) or less.

Many two-way radio users have radio paging as part of their own communication systems. Pressing a series of buttons at the control center will signal a single pager. The operator can signal a group of pagers, or all the pagers, in systems that have group-call or fleet-call capability.

It may have a push-to-listen button; the user hears the voice message only if he pushes the button. It may be able to receive a page without sending an alert; later, pushing a button will retrieve the stored signal. Or the signal may come through immediately, not as an audible tone but as a vibration; this is useful when the user is in a conference or in a noisy area. Or the pager may be capable of a tone that is either pulsating (say, a signal to call the office) or steady (a signal to call home). From an array of paging products, the user can find the unit that fits his working style and environment.

A monitor is a pager that enables the listener to hear his own tone-and-voice message and, by flipping a switch, to monitor all conversation on the frequencies. They may be used by mines, for example, to carry tornado and blizzard warnings.

The earliest mobile two-way radios had vacuum tubes, which had to be mechanically cooled and which failed frequently. The latest units are 100% solid state, that is, free of moving parts, heated filaments, or vacuum gaps. Some are resistant to dust, shock, and vibration and are so weatherproof that they can be mounted outside the vehicle. The earliest units required crystals to establish their transmitting frequency. Many still do. But in the state-of-the-art mobile two-way radios, the crystal has been replaced by an electronic frequency synthesizer. To change or add frequencies, the user replaces the memory module of the radio's microprocessor control system.

For many years, 5 ppm was the standard of mobile transmitter stability. The synthesizer's stability is 2 ppm—a significant improvement in the transmission of voice and particularly in the transmission of data.

Portable Two-Way Radio

Today's radios bring mobile communication wherever a person can go. A person has complete freedom of movement—and yet he is never alone, never without direction, never without access to authority, to assistance, and to the full resources of his organization. To a user of portable two-way radio, weight is no problem: a multifrequency radio with battery and digital squelch may weigh less than 0.7 kg (1½ lb). Some recharge in as little as 1 hr; some are designed and approved by MSHA; some give their operators a choice of four frequency channels with the turn of a single knob; some allow them to communicate on as many as eight frequencies. Some are available with noise-canceling headsets, allowing the user to speak and listen with his portable in extremely noisy areas. At least one model slips into and out of a case mounted under a dashboard. When the user is in his vehicle, he has a true mobile radio, complete with a mobile microphone, speaker, antenna, and a connection to the automobile battery. When he is about to leave the vehicle, he takes out the portable, which instantly becomes a self-contained, two-way communication device.

Repeaters

Low-power mobiles, portables, and base stations can operate in a system that requires communication over a wide area, even an area containing hills and valleys. The system makes use of radio repeaters, sensitive receivers which pick up weak signals and feed them to transmitters, which repeat them at higher power.

In many areas, a repeater is at the top of a hill or mountain. It picks up signals from several users and retransmits them at the users' assigned frequencies. A user's base station may be connected to a rooftop dish antenna, which beams the signal at the repeater antenna many miles away. A rooftop antenna is advantageous because FM transmission takes a line-of-sight path; it does not follow the earth's curvature as AM transmission does.

Base Stations

These are the nonmobile transmitting/receiving radios through which management or its representatives communicate with people in the field. A base station may be self-contained in a desktop cabinet, complete with controls and a microphone connection. Or the base station may be a remote-control model; a control console and the microphones or desk sets may be elsewhere. Some base station systems have telephonelike desk sets with push-to-talk buttons instead of a microphone and speaker. The base station may be local/remote: operated during the day as a local-control station and during the night, as a remote-control station accessible by ordinary wireline telephone from a management person's home, which would also have a control console. Or, as a combination base station and repeater, it would transmit/receive for the base units and repeat for the field units.

Control Consoles

These control the more complex base station functions. While a simple control station, a desktop unit may determine the receive/transmit, squelch, and frequency selection; a floor-based control console contains an array of controls.

Most consoles have a modular design, which makes customization economical. Various switch panels may fit into various console bays and cabinets which, when assembled, comprise a control center. In this way, the center can grow with the organization. A fully equipped console may have vehicle status indicators, an illuminated map, a system of spotting lights (to spot the vehicles' positions on the map), a paging encoder, tape recorders, an intercom station, and more.

WIRED MONITORING AND CONTROL SYSTEMS

Surface mines have used hard-wired monitoring and control systems for a variety of applications including pumps and valves. Simple audio frequency (tone) systems and more complex digital systems utilizing FSK (frequency shift keying) are used to transmit information to and from remote locations.

An example of this use follows. The water level in a holding pond has risen to a critical level because of failure of the main pump. A monitoring system would pick up a signal from a water-level sensor and indicate this problem on an annunciator panel in the mine office. An auxiliary pump at the pond can then be manually or automatically started by a control system from the mine office.

In recent years, microcomputers have expanded the use of monitoring on draglines, shovels, front-end loaders, and other mobile mining equipment. A basic system would include a microprocessor to process data from various sensors and to drive an operator's display. Additional memory, display capability, sensors, and hard copy reports are optional features. In addition the microprocessor, an analog to the digital processor, monitors the voltage and current levels of the hoist, drag/crowd, and swing motors on a dragline or shovel monitor. Synchro encoders are used on the hoist and drag/crowd drums and swing gear to sense bucket position.

The microprocessor monitors the various machine parameters and gives the operator the option of displaying appropriate parameters during each machine cycle. Table 2 lists typical operator display parameters. Table 3 lists typical

Table 2. Operator Display Parameters

Totals	Averages	Last Cycle
Time of day	Yardage/tonnage	Yardage/tonnage
Cycles	Load distance	Load distance
Yardage/tonnage	Load time	Load time
Swing angle (absolute)	Load energy	Load energy
Bucket position	Swing angle	Swing angle
	Swing time	Swing time
	Swing energy	Swing energy
	Cycle time	Cycle time
	Cycle energy	Cycle energy
	Power demand	Steps
	Strip rate	Strip rate

Source: McDonnell Douglas Electronics Co.

operator warnings that are programmed into the microprocessor. This real-time performance information is very useful in improving dragline productivity.

On-board computers are also being used to monitor operating parameters of haulage and loading vehicles. Engine, drivetrain, hydraulic, and lubrication system failures are displayed for the operator. Performance data can be stored and then retrieved for periodic maintenance.

MOBILE MONITORING AND CONTROL SYSTEMS

Most of the monitoring and control functions described in the previous section on wired systems can also be performed using radios as the link. Truck operating and warning parameters can be combined with an automatic vehicle identification system.

Data communication gives the mobile user direct access to the memory of a distant computer. And it gives management an up-to-the-minute report of its mobile units' activity and productivity.

A maintenance man, for example, can enter the equipment model number and get a full alphanumeric display of information from the computer on parts, service notes, or other information to help in the maintenance of the equipment.

A typical mobile keyboard display terminal may provide a 240-character message display in a 6 × 40 format, for example, and may have an alphanumeric keyboard and several functional keys. It sends and receives through the vehicle's mobile radio and the system's base station. To and from the base station, the data flows through a mobile com-

Table 3. Operator Warnings

Parameter	Display	Source
Hoist limit	Hoist limit	Hoist length
Drag/crowd limit	Drag limit or crowd limit	Drag/crowd length
Tightline	Tightline	Bucket position
Hoist stall	Hoist stall XX	Hoist amperage
Crowd stall	Crowd stall XX	Crowd amperage
Multipass	Multipass	Cycle phase
Peak power demand	Power	15-min. average

Source: McDonnell Douglas Electronics Co.

munication processor, a central processing unit, and the system's control console. A mobile communication processor controls all the functions required to encode, decode, check, and correct messages in a mobile data terminal system. It transforms outbound messages from digital to analog and inbound from analog to digital.

In the vehicle, an alternative to the complete mobile terminal is a status-message terminal. It has several push-buttons, each of which sends an abbreviated routine message that gives the vehicle's status and helps the dispatcher manage the fleet. In a haulage system, for example, such messages might include *loading, enroute to site, waiting, discharging,* and *empty.* The system saves time for the driver and dispatcher; it gives management up-to-the-minute information on material flow; and it makes the briefest possible use of the channel.

Radio communication of data is also being used in haulage truck dispatching. When a truck passes a radio signpost—a roadside transmitter designed as part of the system—the signpost transmits its location digitally to the truck's mobile radio unit, which repeats the information (along with digital identification) in a message to the base station. The data then flows through a mobile communication processor to the central processing unit. With that information and a special software package, the computer is able to dispatch a haul truck fleet efficiently. Through the use of a computer making millions of calculations to determine the proper dispatching sequence, substantial savings can be obtained. It is estimated that a significant portion of the total cost in running a fleet of haul trucks can be saved.

The system will also monitor truck parameters such as low on fuel, low on oil pressure, high on engine temperature, and other readings, all with automatic vehicle identification.

Data communications can also be used to control and monitor remote location. It can be used to monitor a water tank and to turn the pump on when the tank gets low or it can be used to monitor a pond's level and sound an alarm if it is in danger of overflowing.

Data communications by radio can be used for communications between man and man, man and machine, man and computer, computer and machine.

SUMMARY

The use of electronic equipment for communication, monitoring, and control at surface mines is increasing each year. New applications for radio and computer equipment will continue to improve both productivity and safety.

6.14 Productivity

MICHAEL B. KAHLE

INTRODUCTION

Productivity is a simple idea that is often complex to measure and to apply in public or private policy—but it is vital to our performance. In the past century, rising productivity has been the source of higher living standards through wages and benefits that increased much faster than prices, the source of shorter hours of work, more leisure and longer life expectancy, and the source of enhanced public goods such as education and public health. Rapidly rising productivity acts as partial long-term antidote to inflation.

During the last 10,000 years, man has made dramatic improvements in his productivity. Usually this improvement has occurred through larger or more efficient equipment, along with improvements in labor relations and better resource evaluation. The last 50 years has seen tremendous growth in equipment size and technology. This growth started with the steam engine and moved through the eras of the internal combustion engine and the electric motor. Larger and larger horsepowered units provided higher-capacity excavating and haulage systems. Growth will continue in this area but probably not at the pace previously experienced.

The other element of productivity that has seen changes over the last five decades is the workforce. In earlier generations, workers were ruled by security needs. A good job was one that paid the bills and kept the family together, but times have changed. The notion of a full employment economy, increased education and sophistication, and richness of support programs for the unemployed have raised our workers to a new plateau. They now insist that work provide not only security but, equally important, heavy doses of recognition, meaning, and growth, all of which can affect worker productivity.

The third element of the productivity equation is the better understanding and utilization of our natural resources. It has not been until the last decade that this element has demonstrated its importance to productivity. Due to the greater capital need for mining and the escalating cost of mining operations, a greater emphasis will be placed on mine planning. This planning will include a greater use of geostatistics to evaluate reserves and plan mine production and more sophisticated computer applications in mine production monitoring and evaluation.

RESOURCE PRODUCTIVITY

Mine Planning

Planning for productivity is a continuing activity which begins with the initial pit design and continues throughout the mine life. Sound decisions made in the initial pit and plant layout will pay back dividends in terms of productivity for many years, while poor ones will come back to haunt the operations. The annual preparation of long-range forecasts offers the planner an opportunity to modify the operating plan to take into account changing economic conditions, new mining techniques, more productive equipment, etc. The periodic short-range forecasts direct the daily mine operations, making the most productive use of existing equipment and personnel while staying within the parameters of the long-range plan.

Pit Planning: The techniques used to develop the geologic model have been discussed in an earlier chapter. The end product of the resource modeling is a computer model made up of uniformly sized blocks of material. The block size is selected after considering the mining bench height, the structure and mineralization detail, the exploration drill hole spacing, the anticipated mining methods and equipment sizes, and the limitations imposed by practical data processing. Each block has a defined spatial location within the model and contains ore grades, mineralogy, and rock type required for resource evaluation. Estimates of the metal prices, metallurgical recoveries, and mining and milling costs provide data for the dollar evaluation of the resource. Sound engineering judgment of the following factors will greatly influence the economics of the final design and the productivity of the proposed pit:

1. Haul road location, widths, and grade.
2. Pit slopes, as determined by rock type structures, water, slope height, and road location.
3. Location of the ore and waste material within the pit.
4. Utility access.
5. Crusher and mill location.
6. Dump location.

Forecasting: Long-range forecasts, or schedules, are prepared annually to cover various periods of time or through mine life. The preparation of the forecast includes calculation of stripping ratios, blending of ores from different mine areas, estimation of tons remaining, and provides the basis for the preparation of long- and short-range budgets. The computerized mine plan together with a computer program that simulates the mining of the resource has greatly reduced the drudgery of this work and has allowed more time for the examination of different options, such as reading mill feeds, changing ore cutoffs, or blending tonnages from different pushbacks. The end result is a permanent record on maps or a computer list of the benches or partial benches that will be mined each year with the corresponding ore grades. This record can be compared to the actual mining at the end of the year and discrepancies between actual results and forecasts can be explained if need be. Upon completion of the long-range forecast, historical data on productivities and costs can be used to determine the equipment, manpower, and dollars needed to produce the forecast tons and the long-range budget is completed. Here again, computerized financial programs can be used to produce the budget along with sensitivities and probabilities of the plan.

Short-range forecasts are prepared as needed at the individual mine. They can be made as simple or as complex as needed and are primarily used to guide the operations for a given period of time to assure an orderly development of the long-range plan, a continuous supply of ore, and a uniform mill feed. A detailed four-week forecast would be laid out on copies of the bench maps showing the weekly areas to be mined by each shovel and/or loader. The size of the areas shown will be determined by the productivity of that particular machine and by the tons needed to produce a desired blend of ore. Upon completion, the forecast is presented to the operations and maintenance personnel and their input is encouraged. These discussions often result in modification of the forecast for one reason or another, but the end result is a smooth-running mine and an orderly development of the longer-range plan.

Operational Data: The need for accurate operational data becomes apparent in the preparation of the long-range and short-range forecasts and budgets. A few years ago, the amount of information that was accumulated was proportional to manpower available to record it. Today, computers have simplified this chore and accumulate, process, and deliver more data in the desired format.

Production data and the associated operating hours are recorded and accumulated by shift, day, month, and year for each loader, shovel, drill, and truck; tonnages mined are totaled by bench and pushback; ore tonnages hauled to various discard dumps are recorded also. Other information can be added when needed, such as haulage distances, tonnages into and out of stockpiles, and tonnages mined outside of plan. The computer output of this data can be expressed in many ways; however, the planner is most interested in the equipment productivities expressed in tons per shift.

The source of the data to be recorded is usually available from reports filed by the shift foremen or from time cards turned in by employees at the end of the shift. This data is only as reliable as the source and, in some cases such as truck haulage records, human errors can result in poor data. To prevent this and to expand the data available on pit haulage, haulage truck monitoring systems have been developed.

Computer dispatching of haulage trucks is currently being used in open pit mines around the world. The primary benefits are increasing effective truck utilizations between 10 and 15%, eliminating material going to the wrong destination, improvement of pit activity reporting, and providing a more consistent flow of material to the concentrator. Current trends are to improve the usage of dispatching to blend complex ores, and to couple dispatching systems with on-board monitoring of truck functions, such as engine compression, wheel motor amperage, and tire temperature. Computer dispatching coupled with trolly assisted truck haulage has the long-range potential of evolving into driverless trucks. The US Bureau of Mines is currently evaluating human monitoring systems, developed for the military, for continually testing driver alertness throughout the shift. This, too, would be coupled with dispatching systems.

Miniaturization and ruggedization of solid-state electronic components, together with advances in computer technology and radio communications hardware, now facilitates sophisticated on-board monitoring of mining and materials handling equipment. As industry's acceptance of on-board monitoring increases, the technology will probably evolve to automation and finally robotize. Hence, on-board computerized monitoring is a significant development which will ultimately change the way new mines are operated, staffed, and managed.

Mine operators and equipment manufacturers in Australia, South Africa, Germany, Sweden, Canada, and the United States are active in automating and computerizing processes within the mining/materials handling sequence. Materials transport, formerly dominated by truck haulage, is being automated with conveyor systems and computerized by truck dispatching. Productivity of conventional excavation equipment, such as shovels and draglines, is being increased by monitoring performance with microprocessors and providing a real time feedback to the operators.

In general, the on-board monitoring systems provide the following:
1. Log and summary report of key operating parameters.
2. Automatic detection of the excavation/haulage cycles so performance statistics can be presented on a per cycle basis.
3. Visual display of performance data.
4. Chronological log of operating times, delays, and alarms.
5. Shift reports.
6. Operator feedback display.
7. Maintenance monitoring.
8. Simple diagnostics.
9. Operator interaction with the microprocessor.

Some of the issues which should be considered in selecting and using on-board computer systems are:
1. The number of input data ports.
2. The ruggedness and simplicity of the hardware.
3. Redundancy of critical functions.
4. System flexibility and ease of field modifications.
5. Filtration and isolation of input signals from each other and from radio signals.
6. Compatibility with existing radio communications systems.
7. Standardization of hardware components.
8. Compatibility with mainframe computing facilities.
9. Provisions for routine maintenance.

The present trend toward automation and computerization of mining processes is expected to continue. Future manpower requirements for a mining operation are anticipated to emphasize the maintenance functions rather than the operating functions.

Maintenance Planning

Computers are used in a variety of maintenance applications from scheduling preventative maintenance (PM) inspections to reporting operating costs. PMs and cost analysis did not evolve from the computer, but rather the computer was applied to these areas because it easily and accurately handles the information manipulation necessary for such programs. This results in a more accurate, up-to-date data base for maintenance planners to work from.

The basis of a solid computerized maintenance system is the gathering and reporting of operating information; operating hours, repair hours, and labor and supply costs. Accurate up-to-date information insures accurate computerized programs.

Preventative Maintenance: Preventative maintenance programs aim to discover maintenance problems before they become serious enough to disrupt production or damage equipment. Once identified, the repairs can be scheduled to minimize production delays and maximize maintenance labor utilization. Instead of equipment breaking down during production and repair crews being summoned immediately on an emergency basis, the equipment can be scheduled down, its production capability replaced by other equipment, and men scheduled for the repair. This is a productive use of men and equipment.

The computer easily controls a PM program by using operating hours to calculate the time remaining on the PM cycle. For example, haul trucks can be called in every 200 hr, shovels every 300, and support equipment every 250 hr. The computer even identifies the specific routine to be followed. The payoff for any PM program is great, since studies have shown that increased PM activities result in increased equipment availability and a more productive maintenance force, hence more productivity through scheduling.

The computer can easily take PM scheduling one level further, by scheduling PM activities on components. The computer knows on which piece of equipment the component is located, and takes the proper operating hours when calculating the PM inspection cycle time remaining. Imagine tracking components by hand.

Predictive Maintenance: Predictive maintenance is a variation on the PM theme: predict the life of a component. A component's life is determined and the computer gathers operating hours against that expected life. As the remaining expected life shortens, the component is flagged for attention. The component's replacement can be scheduled to minimize production delays and maximize the maintenance workforce productivity. Engines and wheel motors are prime candidates for this, as their replacement involves days of downtime, many manhours, and thousands of hours.

On-Board Component Monitoring: Computers are being used (in conjunction with computerized truck dispatch systems) to monitor components on-board. Several components have benchmark values assigned to them in the computer and each time a truck is interrogated, the component values are reported to the computer. The computer then checks the value against the benchmark and examines the variance. An excessive variance can be flagged. Examples of components easy to monitor are engine oil temperature and pressure, wheel motor oil temperature, and wheel motor cooling air flow.

Quality and Performance Monitoring: Computers can increase productivity by the timely reporting of performance and quality. For example, fuel consumption indicates performance. Engine rebuild life expectancy vs. actual hours indicates quality. A computer can quickly and accurately track many such factors. Repair history can be computerized for fast query and retrieval to identify equipment failure trends that may indicate poor replacement parts or repair procedures. Equipment availability calculations and utilization figures indicate the performance of equipment and maintenance programs. Again, timely accurate information allows a more productive management of the resources.

Oil Analysis Program: Oil is usually sampled in conjunction with the PM inspection. The oil is analyzed by computerized equipment that translates the reading into meaningful values. These values are compared to acceptable values and tracked for either an unacceptably high value or a trend of differential, final drive, wheel motors, and hydraulic systems of all-haul glycol, fuel dilution, viscosity, and for certain metals, such as aluminum, copper, lead, iron, silicon, chromium, silver, tin, and nickel. If the presence of any of these exceed the manufacturer's guidelines, the machine is called into the garage and the problem corrected.

Electronic Analysis: For smaller equipment such as pickups and light trucks, electronic engine analyzers are used to diagnose mechanical problems and speed up repairs. For haul trucks, a load box center performs the same diagnostic service. With it, the electrical system of a diesel-electric haul truck can be tested by putting the system under simulated loads while the truck is stationary.

Tires: Tires are tracked by computer as to their location and life. Hours to failure and mode of failure are reported on and studied for trends that may mean tire manufacturing problems or operating procedure problems. Tire location is also tracked for tire rotation. New tires are mounted on the front, run there for approximately 200 hr, and then rotated to the rear. The computer does this tracking effortlessly. Tires

removed from a truck, but with tread remaining, are cataloged by the computer so that they can be matched with similar tires for mounting.

Cost Monitoring: Everything revolves around costs. If there is a system that accurately reports costs in a timely fashion, timely cost reduction decisions can be made. Some of the areas where computers enhance cost reporting are determining unit costs per piece of equipment and unit costs per component. This gives a measure of when the equipment is no longer economical. It can identify how much each component costs, allowing comparisons between pieces of equipment. If these reports are tied to trend analysis, problems can sometimes be identified before they become serious. Detailed costs analysis allows you to study the feasibility of sending repairs out, test new materials, identify areas for new procedures (different ways to line buckets), or just watch the results of cost reduction decisions.

EQUIPMENT PRODUCTIVITY

Over the past century, surface mining has changed from a predominately rail haulage system to its present truck haulage method. The needs of large scale open-pit development have been met with the larger capacity machines which in turn, need faster, larger, and heavier mobile equipment for materials transport. But ever-greater tonnages, increased energy costs, inflation, and longer hauling distances from larger and deeper pits are making truck haulage an economic question mark. Present and future surface mining operations are now contemplating a major increase in the use of conveyor belts. Conveyors are viewed as a way to slow rising fuel and labor costs at surface mines.

Excavators

Draglines: To improve productivity, US surface mines have relied on larger and larger stripping machines. This trend, however, has been tapering off as planning techniques become more sophisticated to get the most of the equipment on hand. As operators are forced to cope with increasing overburden depths, machine cycle times are longer, thus reducing machine productivity. Operating factors are being critically examined and simulated by operators and include maximum allowable suspended load, bucket selection to suit overburden conditions, strip profiles, boom angles and lengths to determine maximum operating dump radius, ballasting, and maintenance delays.

Several firms have developed dragline production simulator computer models. The model can simulate months or years of mining in minutes. Early results of the simulator tests on actual mining operations reportedly have achieved a 16.5% increase in production.

Shovels: Shovel productivity, tons per hour, is determined by dipper size, swing time, truck box size, and truck spotting conditions, as shown in the following equation:

$$\text{ton/hr} = \frac{\text{truck box m}^3 \text{ (cu yd)}}{\text{shovel dipper m}^3 \text{ (cu yd)}}$$
$$\times \frac{\text{shovel}}{\text{swing}} + \frac{\text{truck}}{\text{spot}} \times \frac{\text{truck}}{\text{box}}$$
$$\text{time} \quad \text{time} \quad \text{t (st)}$$

The shovel swing time is determined by the machine operating characteristics, material type, and swing arc to the truck. The spot time factor normally varies between truck-shovel configuration and single or double spot loading conditions. Table 1 presents representative cost and productivity data for a wide range of electric shovel sizes.

Table 1. Electric Shovel Data

Shovel size, m³ (cu yd)	Capital cost, $ *	Operating Maintenance cost, $/hr†	Metric tons (st) per hr
6 (8)	1,000,000	40-80	725-1100 (800-1200)
9 (12)	1,300,000	40-110	900-1600 (1000-1800)
11 (15)	1,500,000	40-140	900-2040 (1000-2250)
15 (20)	2,100,000	40-140	1100-2720 (1200-3000)
19 (25)	3,000,000	40-140	1100-3400 (1200-3750)

* Includes freight, sales tax, and erection charges.
† Includes direct labor, operating supplies, lubrication, electric power, teeth and adaptors, hoist cables, and maintenance costs.

The introduction of the larger 23 m³ (30 cu yd) class shovel has recently provided an improvement in excavator productivity. As the improved operating economies led to greater unit production, the 23 m³ (30 cu yd) electric mining shovel is used more often in today's shovel/truck operations. Mine planners find its larger size cost-effective because they can specify fewer benches—this larger shovel can handle bank heights from 12 to 15 m (40 to 50 ft) only. Moreover, operations management recognizes the 23 m³ (30 cu yd) shovel's inherently greater production capability—it can load in three to four passes the large 154-t (170-st) (and larger) trucks currently available.

Realizing the 23 m³ (30 cu yd) shovel's mine value, both because of its cost-effectiveness and larger capital investment, manufacturers have taken steps to optimize the productivity and reliability of this shovel size.

Among the steps taken were the following:

1. Incorporate proven and new design techniques which increase the excavator's reliability.

2. Achieve long life for machinery and structures.

3. Improve maneuverability for faster shovel positioning.

4. Reduce maintenance and improve performance of electrics.

5. Employ large-capacity, wide-front dippers for better fill factors.

Loaders: In the past, loaders have not had the power to challenge rope-operated shovels. Now they do. In addition, larger unit sizes [15 m³ (20 cu yd)] in recent years have allowed hydraulic loaders, with their improved mobility, high selectivity, and use in a wider range of applications, to achieve higher productivity. Two new methods for loading trucks faster, at reduced production costs and higher productivity, have been devised. The new methods, staggered tandem and chain loading, reduce truck loading time, not haulage time.

Using staggered tandem loading, one loader and possibly, a second, dumps into the truck before it backs into place against the loading face. This reduces the time for one truck to pull out and another to back in for loading, thus increasing truck productivity.

The second method, chain loading, requires a loader for each bucket load needed to fill the truck. The loaders line up along the loading face and move straight in and out from the face. The truck passes between the face and the loaders and stops underneath the raised bucket of each loader to receive a bucket load. Chain loading reportedly took 12 sec to load a 77-t (85-st) truck vs. 1 min to get a truckload with the conventional tandem method.

Trucks

Truck Effectiveness: Haulage truck productivity is generally assessed in terms of tons per hour, based on a payload carried per trip, times the trips per hour. Because of the variables involved, estimating truck productivity is probably the most complex portion of haulage truck evaluation. Computerized simulator programs have been developed to assist the miners in evaluating truck productivity. These programs contain the dipper and swing times for various excavators as well as performance and braking curves, dump, spot and delay times, box capacity, speed limits, and rolling resistance for haulage trucks. Table 2 shows typical output.

The key data are the simulated cycle time and the tons per hour. The potential effect of truck size on shovel productivity can be noted in the table and serves as a valuable operating management tool.

Trolley Trucks: Trucks with trolley assist remain an area of interest to US surface operators. Rapidly increasing diesel fuel costs can provide trolley assist as an economically attractive alternative. Reports indicate that using a 136-t (150-st) truck, predictions show a 9% increase in productivity gain and a 31% decrease in operating cost. Studies indicate that trolley assist performs best with grades greater than 5% and should be considered relatively permanent installation.

Sishen-Saldanha Irai Ore Project in South Africa is a good example of the success of the trolley truck method. When the conversion is complete, there will be 66, 150-wt trucks converted for use on eight overhead trolley assist lines. Results to date show a system that has reduced fuel consumption 30 to 40%, shortened haulage cycle time, and is expected to significantly reduce motor wear. When a fully loaded truck going up an 8% grade engages the overhead line, it can nearly double its speed, typically accelerating from 12 to 22 kW/hr.

Computer Dispatching: Computer-based truck dispatching has proved a production improvement to system over previous conventional voice radio based manual/visual dis-

Table 2. Haulage Truck and Shovel Effectiveness Report Shovel–Truck Assignment Sheet

Bank cond	Op. hr available shovels	Time trucks	Spot conditions	Truck fleet	Load capability	Cycle time	Per truck		Per shovel		Trucks to assign
							Trips per hr	t/h	Load time	t/h	
M	60.0	60.0	2	1	65	27.48	2.2	141	2.24	1566	11.0
M	60.0	60.0	2	2	67	27.94	2.1	143	2.24	1614	11.2
M	60.0	60.0	2	3	67	29.17	2.1	137	2.24	1614	11.7
M	60.0	60.0	2	4	100	29.94	2.0	200	3.36	1662	8.3
M	60.0	60.0	2	5	100	27.75	2.2	216	3.36	1662	7.7
M	60.0	60.0	2	7	100	30.24	2.0	198	3.36	1662	8.4
M	60.0	60.0	2	8	100	31.15	1.9	192	3.36	1662	8.6
M	60.0	60.0	2	9	100	28.40	2.1	211	3.36	1662	7.9
M	60.0	60.0	2	10	100	28.11	2.1	213	3.36	1662	7.8
M	60.0	60.0	2	18	120	30.40	2.0	236	3.92	1726	7.3
M	60.0	60.0	2	19	150	30.50	2.0	295	5.04	1701	5.8
M	60.0	60.0	2	20	120	30.57	2.0	235	3.92	1726	7.3
M	60.0	60.0	2	21	150	29.94	2.0	300	5.04	1701	5.7
M	60.0	60.0	2	22	150	29.83	2.0	301	5.04	1701	5.6
M	60.0	60.0	2	23	113	29.88	2.0	226	3.92	1625	7.2
M	60.0	60.0	2	24	113	29.98	2.0	226	3.92	1625	7.2
M	60.0	60.0	2	25	100	29.15	2.1	205	3.36	1662	8.1
M	60.0	60.0	2	26	100	26.88	2.2	223	3.36	1662	7.4

patch. Phelps Dodge Corp., Tyrone mine, has successfully used such a system. Productivity improvements associated with the dispatch system are a direct result of optimum route selection and a subsequent increase in productive operating time. The Tyrone mine has experienced a 14% improvement in productive operating time over the previously used radio dispatch method. In addition, secondary benefits included less load misrouting and a generation of timely and accurate production reports. The productivity gains translate into an increase in production and/or a reduction in equipment requirements, resulting in substantial operating savings. Fig. 1 compares the equipment requirements of a dispatch truck fleet vs. a dedicated truck fleet using a computer-based dispatching system.

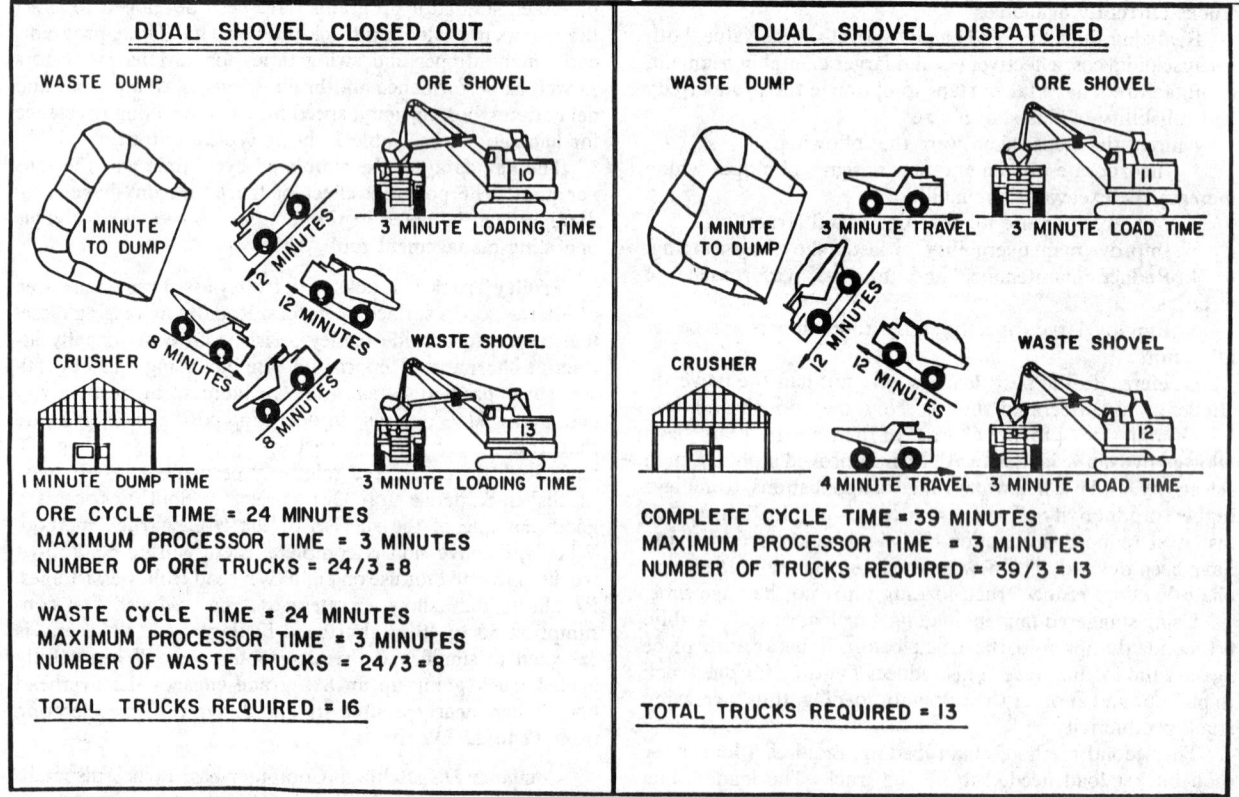

Fig. 1. Closed out (dedicated) vs. dispatched truck assignments.

Fig. 2. Movable concept with constant feed.

Conveyors and Mobile Crushers

As labor and fuel costs increase, and as open-pit and strip mines become larger, belt conveyors will become an important factor in holding down production costs and increasing productivity. There are generally three types of overland conveyor systems:

1. A movable source feeding a fixed terminal (bucket wheel excavator mines coal and conveys to plant).

2. Fixed source feeding fixed terminal (permanent crusher to plant).

3. Fixed source to movable terminal (permanent crusher to waste dump spreader).

Each of these systems has its problems. However, in every case, for distances up to several miles, a well designed and maintained belt conveyor system will prove to be more economical over a period of years and thus more productive than either truck or rail system. When conveyors are used in the pit, the long-range mining plan must be adapted to make the best use of the conveyor system's capabilities for increased productivity.

Usually, run-of-mine materials at most mines must be reduced in size before being transported by conveyors. Therefore, mobile or semi-mobile crushers offer one of the most effective available methods for in-pit conveyor systems. A mobile crusher is a completely self-contained unit mounted on a frame that is moved by means of a transport mechanism along the wall of the mine pit as mining progresses. The large primary crushers can be moved by setting the crusher

on a large frame which can be moved by tracks, wheels, or walking mechanisms. This portability promotes minimum haulage distances between shovels and the crusher by allowing the crusher to be placed at a location central to the shovels. The expense of the larger supporting structures and moving mechanisms is justified by their contribution to reduced fuel consumption and fewer haulage trucks and overall contribution to productivity levels. Fig. 2 shows an in-pit movable crusher-conveyor system.

LABOR PRODUCTIVITY

Productivity expresses a relationship between quantity of inputs—be it labor, materials, energy, capital, or some other aggregation—and the quantity of output. Output per man-hour, output per unit of capital, output per unit of energy, or output per unit of total input factors are all common measures; however, output per man-hour is the most frequently used or available measure. One of production's key components is a motivated, cohesive workforce that is dedicated to high standards of production and quality.

In earlier generations, workers were ruled by security needs; a good job was one that paid the bills and kept the family together. But times have changed; the notion of a full employment economy, increased education and sophistication, and a rich menu of support programs for the unemployed have raised workers to a new plateau. Workers now want not only security but, equally important, heavy doses of recognition, meaning, and growth.

In order that these expectations and demands be met and still achieve the required productivity, managers must recognize the legitimacy of the workers' search for self-esteem. Self-esteem has two interrelated aspects: self-confidence and self-respect. This is where good management becomes important. The modern manager must lead his workers into feelings of competence, involvement, and importance. A company can address such needs through better supervisor training and more direct worker involvement in the problems and activities of the workplace.

Management by Objectives

Management by objectives focuses on getting supervisors to commit themselves to meeting objectives. The objectives must be reasonable and attainable, but not necessarily easy. The objectives must be specifically documented and must be reviewed and updated routinely. The key word in this approach to better management planning is *accountability*. Employees will shirk responsibilities, but they will do what they are being held accountable for. In addition, employees need to know clearly their areas of accountability. This requires management to clearly communicate each employee's accountability.

Production Incentives

Production incentives are generally goal-oriented rather than oriented toward problem-solving. However, production incentives have proven useful in improving worker productivity. Generally, incentives are attractive once a company is pleased with its productivity and wants to push for a bit more. Incentives never should be used to attain an initial productivity goal.

Incentives can be a two-edged sword. If the goals are set too high, workers will become frustrated and can be a tremendous source of resentment. The royalty on each ton mined also cuts absenteeism, because workers do not get the bonus check when they miss work. However, management must be careful that the plan does not pay for what production should be achieving anyway.

Quality of Performance

An approach used to help a worker's personal efficacy and a sense of personal worth is that of quality circles. The quality circle techniques are familiar to approximately 1800 American companies and more recently have been introduced to the mining industry. The general purpose of the quality circle is to improve performance. Quality circles recognize that the people doing the work are most familiar with the problems that block their productivity on the job. The circles provide a mechanism for solving those problems that tend to demoralize and wear down employees.

A quality circle is usually made up of employees, both supervisor and workers, who volunteer to be in the circle. A foreman normally serves as the leader, but like other members, he has one vote in conducting circle business and decisions. The members of the circle are trained in brainstorming, cause-and-effect analysis, data collection techniques, graphs, and basic group dynamics. The circle group uses brainstorming to draw up a list of possible problems. Once a circle selects a problem, its first task is to gather data to discover the cost of the problem in terms of lost man-hours, wasted materials, downtime, or some other productivity measure. The circle group then uses its inherent knowledge to develop solutions and analyze the potential savings.

The circle will usually make a presentation to mine management proposing a solution. Each circle member participates in the presentation; the circle process is dynamic. Problems are routinely added by members, nonmembers, or management to the list of possible projects. Problems are often solved without formal circle action. Companies with the circles commonly find that problems will fall off the list because people are talking about problems and find that they can easily solve them themselves. The solutions are not presented to management because no expenditures or management action are involved.

Morale is a difficult attitude to measure. However, experience has shown that employees who are involved in solving workplace problems are more satisfied and generally more productive.

Chapter 7

Mine Capital and Operating Costs

Robert F. Winkle, Editor

7.1 Introduction

When a mining property has been discovered, the exploration work completed, the feasibility studies and other planning activities finished, and the market analyses compiled, the ultimate go-no go decision depends on the capital required and the operating costs anticipated. Mining ventures and projects are undertaken with a view of gaining some benefit: earning a return to the stockholders for use of their funds (Jones, 1968). Net profitability is the bottom line.

Mining, as earlier chapters in this book will attest, is a complex undertaking with many variables (some of which are not under the control of corporate management) to be considered. First, the ore body itself will be a major factor in final capital expenditure decisions. For example, with a greater capital expenditure for a large plant, the deposit could be worked out in a shorter time span. Conversely, in order to maximize total profits, there would be an initial lower capital investment and a longer operational life for the deposit before it is finally depleted (Jones, 1968). Two other factors are the increasingly stringent environmental requirements imposed by governmental entities and the taxation treatment of mining profits from the aspects of depletion, depreciation, and investment credits.

From the late 1960s on, air and water pollution became a national conscience issue in the United States, with a proliferation of legislation, administrative agencies, comprehensive federal and state regulations, and a good deal of bombast in political campaigns, from the media, and by environmental groups such as the Sierra Club and Friends of the Earth. Although environmental concerns have had a greater impact on mining in the developed countries, recently such concerns have been surfacing even in such remote areas as Papua New Guinea (Tinsley, et al., 1985). The continuing stress on environmental regulation and control, needless to say, has its impact on capital and operating costs. For example, new smelter construction expenditures must include funds for extensive emission controls.

Space herein precludes coverage of the detailed economic and evaluation studies entailed in a capital expenditure decision, but numerous publications are available (Gentry and O'Neil, 1984; Jones, 1968; Stermole, 1987; Tinsley, et al., 1985).

Capital and operating costs vary with the commodity to be mined and the terrain in which it is located.

Iron Ore

Most surface operations are large scale and enormous quantities of ore and rock must be moved. However, much of the ore to be moved is dense and abrasive. Large equipment is indicated, but abrasive and corrosive aspects must be considered when maintenance costs are calculated. The section on iron ore that follows summarizes the basic philosophy, responsibilities, cost control methods, and equipment selection procedures that are applied in surface iron ore mining operations.

Coal

Surface coal mine capital and operating costs have not only been affected by the environmental control of air and water pollutants, but also by the increasingly stringent requirements for restoration of the land surface after mining ceases.

The costs of mining coal vary, depending on whether the operation is in the eastern or western United States.

In a section that follows, the costs of mining in a typical eastern coal mining state are delineated. Cost coverage begins with the permitting process, proceeds to engineering and construction, employment, equipment, other operating expenses, and concludes with taxes and fees. A case study of a West Virginia dragline operation, in a previous section, shows all the equipment considerations and statistical data that impact capital decisions and subsequent operating costs.

Costs of the unique mining conditions in the Powder River Basin of Wyoming in the western US are highlighted. The equipment to be used is a major capital factor and the philosophy entailed and the specific considerations involved in selection are discussed. Many variables are involved in these operations, such as personnel deployment, efficiency considerations, whether to contract or use in-company capabilities, scheduling, and the support structure needed.

Base Metals

The fundamentals of methods used to estimate mine capital and operating cost for base metal surface mine feasibility studies are detailed in the final section of this chapter. Since many of these operations are large scale, equipment requirements, selection, and operating costs are major factors in the capital decision-making process. Statistical data are given for each type of equipment that might be required in an operation. Numerical tabulations of initial capital expenditures and operation, replacement, and personnel requirements are provided.

From the commodity-oriented sections in this chapter, it is apparent that the process involved in making a capital investment decision is indeed a complex one. The many variables that pertain to a surface mine operation also affect operating costs once the decision has been made to put a deposit/property into production. All of these analyses must be carefully and thoroughly done in order to achieve the final objective—a bottom-line net profit and return to the stockholders.

REFERENCES

Gentry, D.W., and O'Neil, T.J., 1984 *Mine Investment Analysis,* AIME, New York.

Jones, C., 1968, "Economic Analysis for Mining Ventures and Projects, *Surface Mining,* E.P. Pfleider, ed., AIME, New York.

Stermole, F.J., and Stermole, J.M., 1987, *Economic Evaluation and Investment Decision Methods,* 6th ed., Investment Evaluations Corp., Golden, CO.

Tinsley, C.R.; Emerson, M.E.; and Eppler, W.D., eds., 1985, *Finance for the Minerals Industry,* AIME, New York.

Vogely, W.A., ed., 1985, *Economics of the Mineral Industries,* 4th ed., AIME, New York.

7.2 Mine Capital and Operating Cost

7.2.1. Iron

KENNETH J. WEBER

BASIC PHILOSOPHY

The purpose of a mine is to profitably extract and market an essential product for the life of a particular mineral deposit by a method that conserves all resources. The conservation of an essential natural resource is synonomous with full use of the ore body and is only accomplished if the operation minimizes the expenditure of human, physical, time, and financial resources. Operations designed on any other basis are subject to failure during periods of economic stress with the resulting social problem of unemployment and the environmental problem of waste of reserves. Therefore, the basic rules of mine planning and operation are:

Move the Minimum Quantity of Material

Bank slopes are designed at the maximum safe angle that will protect the operators and comply with governmental regulations. Flatter slopes increase the stripping ratio and decrease the reserves of economical ore, thereby limiting the life of the mine or wasting the resource.

Surge stockpiles between operating sequences are minimized to provide only the predetermined quantity required to maintain scheduled production. Excess drilling, blasting, and surge stockpiles ahead of or within the crusher or plant circuit represent unnecessary expenditures.

Move It the Shortest Possible Distance

The distance between the ore body and the final product loadout facility is designed to eliminate all unnecessary movement and to use gravity assistance wherever possible. In the mine, for example, the crusher is located as close as safely possible to the ore body to utilize the cost advantage of conveying over rail or truck haulage.

Move It with the Minimum Amount of Equipment

Productivity, units of product per unit of time worked, is designed into the system by minimizing the equipment fleet. It is a fundamental principle of inefficiency that a use is always found for surplus equipment, thereby increasing the operating, maintenance, clerical, and supervisory work force as well as the facilities and supplies required to sustain them.

Move It with the Minimum Number of People

Improper manpower scheduling results in fluctuation in employment with the effect of demoralizing the work force and reducing productivity.

Move It in the Shortest Possible Time

Conservation of time adds the necessary sense of urgency to an operation, develops talent in the work force, and generates a pride in the attainment of objectives.

These rules seem simple and repetitive; however, an investigation of existing or past operations shows that basic design is often influenced by the technical specialty or opinion of the person in charge, resulting in a waste of resources because these basic rules have not been applied to decisions.

EQUIPMENT SELECTION

The selection process is best illustrated by a simplified example, Table 1. In this case, a typical taconite operation based on an average of seven Lake Superior mines is utilized.

Based on this data, the equipment options may be charted as shown in Table 2. Note that the *Number Required* at *Effective Actual Capacity* is based on the *Estimated Mechanical Availability* (operating hours/operating hours +

Table 1. Typical Taconite Operation Based on Average of Seven Lake Superior Mines

Operating Data		
Annual pellet production, dry lt		6,700,000
Percent weight recovery		31.6
Annual crude ore production, dry lt		21,200,000
Annual crude ore production, natural lt		21,900,000
Annual rock stripping, natural lt		14,000,000
Truck cycle time in minutes: crude ore		30
rock stripping		35
Required primary crusher product size in inches		5.5
Pit operating time: days per year		350
shifts per week		20
Drill penetration rate, feet per shift (120,000# pulldown): ore	15-in. bit	180
	12¼-in. bit	210
rock	15-in. bit	240
	12¼-in. bit	280
Bench height in feet		40
Drill pattern: ore	15-in. bit	36 × 36
	12¼-in. bit	32 × 32
rock	15-in. bit	38 × 38
	12¼-in. bit	34 × 34

Metric equivalents: lt × 1.016 047 = t; ft × 0.3048 = m; in. × 25.4 = mm.

Table 2. Equipment Selection Options

Unit	Type	Size	Rated Capacity	Estimated Mechanical Availability, %	Effective Actual Capacity	Unit Cost, $	Units Required At Rated Capacity	Units Required At Est. Act. Capacity
Primary crusher system	Gyratory	60 × 102	3000 ltph @ −6-in.	75	2250 ltph	8,750,000	0.9 (1)	1.2 (2)
		54 × 75	2500 ltph @ −5.5-in.	75	1875 ltph	8,750,000	1.1 (2)	1.5 (2)
			1500 ltph @ −6-in.	75	1125 ltph	4,375,000	1.8 (2)	2.4 (3)
			1200 ltph @ −5.5-in.	75	900 ltph	4,375,000	2.3 (3)	3.0 (3)
Trucks: Ore		200 st	179 lt	70	125 lt/load	1,200,000	7.6 (8)	11.0 (11)
		170 st	150 lt	80	120 lt/load	900,000	9.1 (10)	11.4 (12)
		130 st	116 lt	80	93 lt/load	700,000	11.8 (12)	14.7 (15)
		100 st	90 lt	80	72 lt/load	600,000	15.2 (16)	19.0 (19)
Rock		200 st	135	70	95 lt/load	1,200,000	7.6 (8)	10.8 (11)
		170 st	113	80	90 lt/load	900,000	9.0 (9)	11.4 (12)
		130 st	87	80	70 lt/load	700,000	11.7 (12)	14.6 (15)
		100 st	68	80	54 lt/load	600,000	15.0 (15)	18.9 (19)
Shovels: Ore		27 cu yd	2650 ltph	70	1850 ltph	3,900,000	1.0 (1)	1.5 (2)
		18 cu yd	1750 ltph	75	1310 ltph	3,000,000	1.6 (2)	2.1 (3)
		14 cu yd	1350 ltph	75	1010 ltph	2,300,000	2.0 (2)	2.7 (3)
Rock		27 cu yd	2000 ltph	70	1400 ltph	3,900,000	0.9 (1)	1.3 (2)
		18 cu yd	1310 ltph	75	985 ltph	3,000,000	1.3 (2)	1.8 (2)
		14 cu yd	1010 ltph	75	760 ltph	2,300,000	1.7 (2)	2.3 (3)
Drills, rotary: Ore		130,000 #	280 ft/shift 15-in. hole	75	210 ft/shift	1,250,000	0.9 (1)	1.2 (2)
		120,000 #	300 ft/shift 12-in. hole	80	240 ft/shift	1,250,000	1.1 (2)	1.4 (2)
		130,000 #	257 ft/shift 15-in. hole	70	180 ft/shift	1,100,000	1.0 (1)	1.4 (2)
		120,000 #	280 ft/shift 12-in. hole	75	210 ft/shift	1,100,000	1.2 (2)	1.6 (2)
Rock		130,000 #	360 ft/shift 15-in. hole	77	277 ft/shift	1,250,000	0.5 (1)	0.7 (1)
		120,000 #	400 ft/shift 12-in. hole	82	328 ft/shift	1,250,000	0.6 (1)	0.7 (1)
		130,000 #	330 ft/shift 15-in. hole	73	240 ft/shift	1,100,000	0.6 (1)	0.8 (1)
		120,000 #	365 ft/shift 12-in. hole	77	280 ft/shift	1,100,000	0.7 (1)	0.9 (1)

Metric equivalents: lt × 1.016 047 = t; st × 0.907 184 7 = t; in. × 25.4 = mm; cu yd × 0.764 554 9 = m³; ft × 0.3048 = m.

repair hours). There is a mathematical risk of lost production in basing the fleet size only on mechanical availability. For example, Table 2 shows that a fleet of 12 170-ton (154-t) trucks is needed to keep 9 operating on ore haulage. Referring to a random probability chart indicates that the probability of 9 of 12 operating 100% of the time is 96.64%. In other words, the risk of losing 3.36% of required production must be considered in determining whether the ore fleet will be 12 or 13 trucks.

Equipment Size and Type

Considering the large quantities of ore and rock involved, the mining equipment selected will be the largest available that will handle the tonnage, provide the required blend and operate at optimal cost. The cost is particularly important since to date the largest shovels, those above 13.8 m³ (18 cu yd), and trucks, those above 170 tons (154t) are unable to operate in the dense, abrasive iron ores at a reasonable maintenance cost. History indicates that these limits increase with experience.

The selection of a manufacturer to supply equipment requires objective evaluation. In general, all of the major suppliers can provide units which will perform adequately. The differences are in individual operating and maintenance items; therefore, the evaluation should be based on a comparison of common components and the selection based on the overall rating. A typical example for haulage trucks is presented in Table 3.

The system described is based on common practice and experience to provide a general guide to equipment selection. The planner should not restrict himself to the past—there is a great need for innovation in material handling. In any system, consideration must be given to technical advances in computerization of pit design, fleet dispatching, and engine diagnosis and maintenance programs. In addition, items such as semi-portable primary crushers and screening plants, conveyors, electric truck assist, and blasting procedures are undergoing changes which will improve the efficiency of future mines.

Equipment Purchase Schedule

The schedule is based on quantities to be excavated. Crude ore requirements are usually constant while stripping varies with the depth, dip, and internal contamination of the ore body. The purchase schedule follows the annual mining and stripping requirements as determined from the reserve (life of deposit) mining plan. The fleet size will be minimal during preproduction stripping, increase to a maximum needed to maintain production plus a stripped ore inventory, and then decrease finally to the size required for production and the removal of internal rock.

Planned *high grading* or *selective mining* is a completely acceptable procedure to delay expenditures; unplanned is not. The key is to maintain production and to avoid a panic stripping program.

Contracting Out

Equipment used for stripping and mining is usually the same; therefore, the contracting out of either is not considered. By definition, miners are expert earth movers which makes a decision to contract stripping an admission of incompetence.

Preproduction stripping by the mine operator is a valuable break-in period for the work force and equipment. It should be fully utilized to develop the work habits of the team before the routine of normal operations commences.

Table 3. Comparison of Types of Haulage Trucks

	Manufacturer			
Component	A	B	C	D
Engine				
Generator/alternator				
Wheel motor/traction motor				
Mechanical drive				
Electrical, general				
Frame				
Truck body				
Suspension				
Hydraulics				
Brakes				
Retarders				
Cabs				
Electrical maintenance accessibility				
General maintenance accessibility				
Estimated tire life				
Estimated fuel consumption				
Estimated mechanical availability				
Mechanical, general				
Payload to empty weight ratio				
Manufacturing facility				
Spare parts availability				
Cost selected spare parts and component replacements				
Size and location of service facilities				
Availability of warranteed rebuilt modular exchanges				
Operator acceptance				
Total rating*				

*Note: Rate components on a 1 to 5 scale. Poor = 1, average = 3, superior = 5.

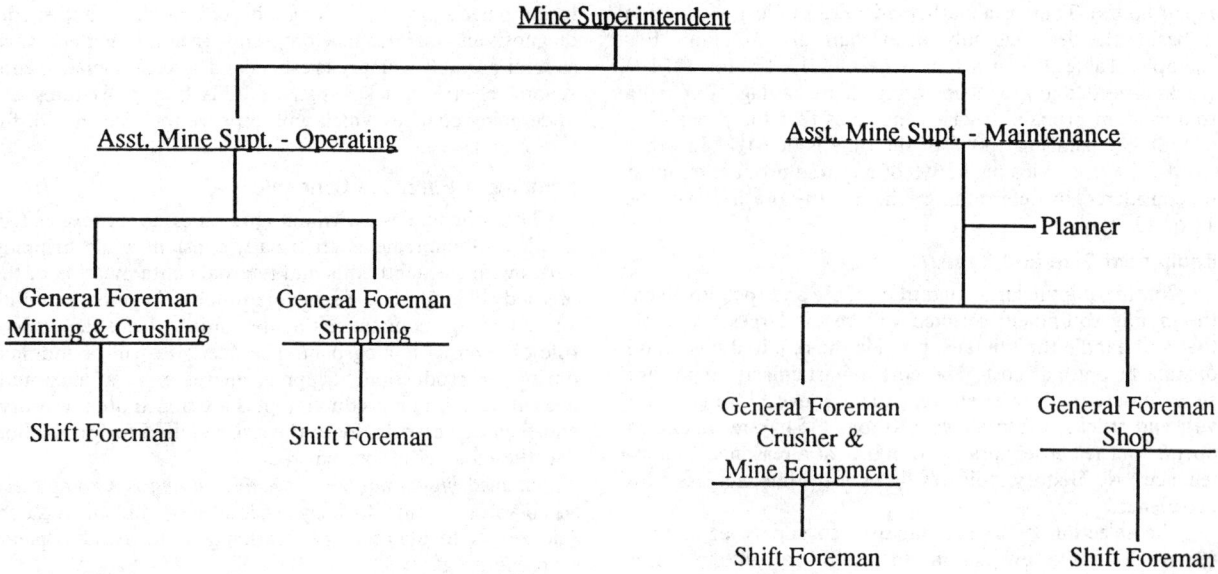

Fig. 1. Table of organization.

The contracting out of preproduction stripping should only be considered in these cases: (1) stripping by methods different from the mining method such as dredging, scrapers, or dragline; (2) stripping of quantities greater than will be required during normal operations.

TYPICAL OPERATING COSTS

Since this section deals with the mining portion of a total operation, the costs referred to are those under control of the mine superintendent. Comparing operations and operating costs is complicated because organizational responsibility and accounting methods vary greatly among mining companies. In this example, we assume that the mine superintendent is responsible for delivering crushed crude ore to the plant and that his table or organization is as shown in Fig. 1.

Budgetary responsibility follows organizational responsibility; therefore, the mine superintendent controls and is accountable for the items shown in Fig. 2. The superintendent is responsible for budgeting all items shown in Fig. 2. He is accountable for consumption of all supplies and labor and is entitled to a complete financial report.

This mining cost breakdown is not typical of a particular mining company; it is an example of the general areas of cost responsibilities. To pinpoint areas requiring attention, the mine superintendent requires detailed current and historical cost and performance data. With this information, he should be able to evaluate the relative influence of mining, stripping, and capital expenditures on his overall performance, and his overall performance to the profitability of the company. This management evaluation of all controllable costs is frequently neglected by:

Emphasizing production and production costs only.

Neglecting departmental influence on total performance, for instance, the potential increase in crushing and grinding costs by reduction of mining costs.

Cost control is a procedure used to generate profits; it should not restrict innovation or the utilization of changing technology.

The actual cost of producing a long ton of pellets in the Lake Superior-Canada areas varies from about $21.00 to $26.00 f.o.b. minesite excluding stripping, royalties, depreciation, and taxes. It is important to note that the cost of transportation to the steel mills adds significantly to this cost, thereby eliminating many rich ore bodies from consideration until sometime in the future.

The major reasons for these cost variations are:

Size—Large operations have lower costs because fixed costs per unit of production are less. It is doubtful whether a new operation of less than 5 Mt/a (5,000,000 ltpy) would be seriously considered.

Weight Recovery—A difference of 1% influences the cost per long ton of pellets from $0.45 to $0.60.

Stripping—The advantages of a low stripping ratio are obvious as illustrated by the *all materials ratio* which is:

$$\frac{\text{Tons crude ore } + \text{ tons rock stripping } + \text{ tons surface stripping}}{\text{Tons pellets}}$$

Equipment—Modern, large, energy sensitive and computer or processor controlled equipment gives newer operations a significant cost advantage.

Table 4 illustrates the proportionate cost percentages for an average of seven North American taconite operations.

LEASING VS. PURCHASING

The initial investment required to build an iron ore plant is extremely high, about $100 per annual ton; therefore, an operation is expected to last at least 25 years. The life of mining equipment such as drills, shovels, trucks, locomotives, rail cars, etc., has also increased due to the higher initial cost and to component replacement which enables the operator to maintain his fleet at much higher efficiencies than was possible in the past. In general, the only reason for changing equipment is to take advantage of new unit efficiency, technology, or to increase the unit size and this can be minimized by strict attention to the initial investment.

Leasing is a tool normally used to offset a current shortage of cash, or to obtain equipment for short term (10 years or less) projects. It should be remembered that the financial institutions which provide leasing services are also in business to make a profit. This adds to the fleet cost and usually

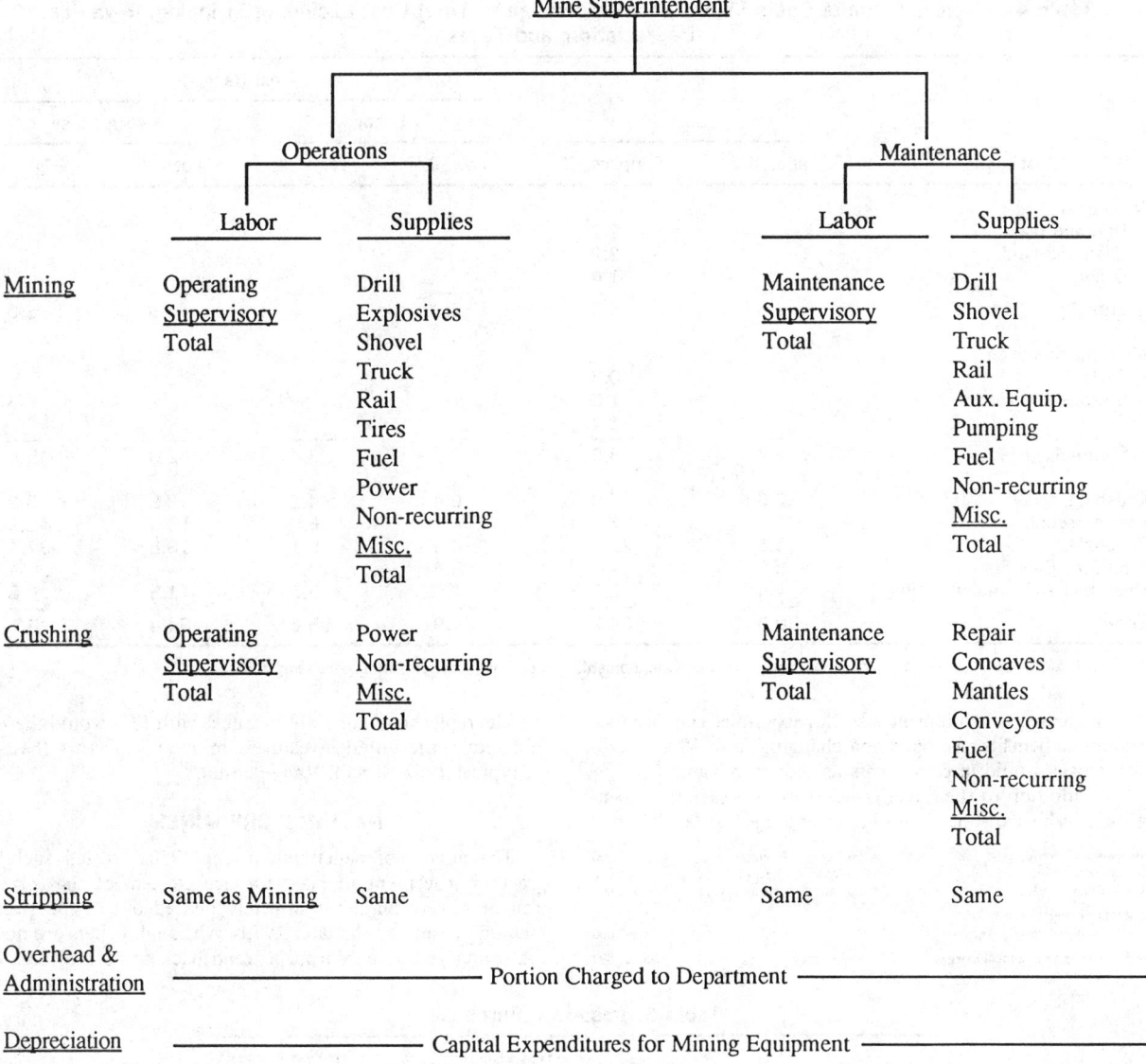

Fig. 2. Table of cost accountability.

results in a decision to purchase rather than lease, assuming cash is available. There may be advantageous exceptions to this depending on the tax laws prevalent at the time and in the area being considered. For instance, leasing can be attractive in a cash short situation when given the additional flexibility of either taking the investment tax credit (10%) or allowing the lessor (financial institution) to keep the tax credit. It is also important in evaluating a lease vs. purchase option to "present value" all lease costs at a consistent rate. However, recent history has proven that economic conditions and interest rates can vary significantly from time to time and, therefore, an evaluation in constant dollars must also be considered.

To illustrate the comparison of these various options, the following hypothetical example of the costs of obtaining a 170-ton (154-t) haulage truck is shown in Table 5. It should be noted that lease rates are subject to actual lending rates available at the time of equipment acquisition and there are many ways of approaching the leasing of capital equipment.

Balloon payments and guarantee residual values at termination used in this example are only two ways of demonstrating the use of the leasing tool that can reduce current operating costs and conserve cash flow. It is therefore essential that if leasing is a consideration, the purchasing and treasury departments be relied on to evaluate the lease proposals.

CAPITAL VS. OPERATING COSTS

Total operating costs cannot be significantly reduced by the expenditure of capital unless the expenditure is for a change in technology, a change in methods, increased equipment size, or improved efficiency. In other words, under given conditions (production, equipment size and quality, work force) the equipment hours required are the same regardless of the work schedule. If the schedule is reduced while maintaining the same production, the savings in supervisory, overhead, and shift differential costs cannot justify the capital cost of more equipment of equal size and quality.

Table 4.   Typical Taconite Costs Shown as a Percentage of Total Cost Excluding Stripping, Royalties, Depreciation, and Taxes

| Cost Item | Cost | | Cost Range* | | | |
| | Labor, % | Supplies, % | Labor | | Supplies | |
			Low, %	High, %	Low, %	High, %
Mine operating						
Drill and blast		2.7				
Load and haul		2.2				
Other		1.4				
Subtotal	2.6	6.3	2.0	3.6	4.5	8.0
Mine maintenance						
Drills		0.3				
Shovels and trucks		1.8				
Other		1.4				
Subtotal	2.7	3.5	2.2	3.4	2.0	5.6
Crushing	2.0	3.9	0.4	4.2	0.8	5.9
Concentrating	5.1	27.2	4.7	6.0	13.0	42.9
Pelletizing	5.5	25.6	4.5	9.1	18.8	37.4
Employee Benefits	8.5	—	3.4	10.2	—	—
Overhead and Administration	4.4	2.7	2.9	5.5	1.5	6.8
Total	30.8	69.2	24.9	35.6	64.4	74.6

* Cost variations due to size of operation, weight recovery, equipment, and crushing-concentrating pelletizing flowsheet.

A change in equipment size, for example, can produce economic benefits. An operation changing from 100- to 170-ton trucks would improve costs as shown in Table 6.

In addition to the direct costs, there is a savings in benefits, overheads, supervisory labor, and clerical labor:

Benefits and overheads	$ 9,500/year × 76 employees	$ 722,000
Supervisory labor	$48,000/year × 7 employees	336,000
Clerical labor	$38,000/year × 3 employees	114,000
Subtotal: indirect costs		$1,172,000
direct costs		3,849,000
Approximate total annual savings		$5,021,000

The replacement of 100-ton trucks with 170s would therefore return the initial investment in about 4.3 years (based on typical 1982 Mesabi Range costs).

NATURAL ORE MINES

The advent of pelletizing in the 1950s created such a great improvement in blast furnace efficiencies that other iron ores were almost completely phased out of the steelmaking process by the later 1970s. Although pellets are now the primary source of iron, a trend back to natural ores is

Table 5. Lease vs. Purchase

| Cost | Five-Year Lease | | | Ten-Year Lease | |
| | A | B | C | D | E |
	Purchase	Retain[1] Ownership	Release[2] Ownership	Retain[3] Ownership	Release[4] Ownership
Year 1	$900,000	$108,000	$144,000	$108,000	$144,000
2	—	198,000	144,000	198,000	144,000
3	—	198,000	144,000	198,000	144,000
4	—	198,000	144,000	198,000	144,000
5	—	198,000	144,000	198,000	144,000
6	—	612,000	?	198,000	144,000
7	—	—	—	198,000	144,000
8	—	—	—	198,000	144,000
9	—	—	—	198,000	144,000
10	—	—	—	288,000	144,000
Total cost, constant dollars	900,000	1,512,000	720,000	1,980,000	1,440,000
Total cost, present valued @ 10%	—	1,014,200	600,400	1,169,500	973,300

[1] Lease costs based on a rate of 5.5% of principal payable quarterly in arrears with "balloon" payment for ownership equal to 68% of original value (10% tax credit taken in first year).

[2] Lease costs based on a rate of 4.0% of principal payable quarterly in arrears, with guaranteed residual value equal to 68% of original cost (10% tax credit kept by lessor). If lessor must sell equipment for less than $612,000, lessee must pay the difference. Example: assume equipment is sold for 30% of original value ($270,000), then lessee must pay $342,200, increasing total lease cost to $1,062,000, or a present value of $758,200.

[3] Lease costs same as note 1 except no "balloon" payment. Purchase equipment at lease termination for fair market value, i.e., estimate 10% of original cost.

[4] Lease costs same as note 2 except no guaranteed residual value. Lessor sells to others at fair market value.

Table 6. Economic Benefits of a Change in Equipment Size

	100-Ton Trucks	170-Ton Trucks
Number required (see Table 2)	38	24
Personnel: drivers	124	76
mechanics	48	30
welders	12	8
tiremen	3	3
electricians	8	5
laborers	8	5
Total	203	127
Annual operating hours: ore	121,700	73,000
rock	120,100	72,300
total	241,800	145,300
Cost per operating hour	$70	$90
Total direct operating cost	$16,926,000	$13,077,000
Capital cost: each		$900,000
Total		$21,600,000
Saving, annual operating cost		$3,849,000

Metric equivalent: ton × 0.907 184 7 = t.

becoming evident primarily due to the huge reserves of high grade ore in Brazil and Australia.

Natural ore mines have the advantage of lower or no concentrating costs and the disadvantage that sintering or pelletizing is required to produce a high quality blast furnace feed.

Mine operating costs in North America are approximately one-half that of taconite operations and vary from $9.00 to $13.00 per long ton excluding the stripping cost. Operating costs in South America are approximately one-half that in North America primarily due to:

Continuous operations. North American mines are normally closed during winter months because of freezing.

Lower labor costs.

Table 7 illustrates the typical cost percentages for natural ore mines on the two continents.

SUMMARY

Basic philosophy, responsibilities, cost control methods, and equipment selection procedures are not significantly changed by time or location. The application of this material to a particular operation, however, must be updated to current inflation, interest rates, governmental regulations and taxes, and to prevailing labor, material, and equipment costs.

ACKNOWLEDGMENT

The assistance of Frank B. O'Connor in preparing the Leasing vs. Purchasing portion is appreciated.

Table 7. Typical Natural Ore Mine Costs Shown as a Percentage of Total Costs Excluding Stripping, Royalties, Depreciations, and Taxes

Cost Item	North America		South America	
	Labor, %	Supplies, %	Labor, %	Supplies, %
Mine operations				
Drilling and blasting	0.6	0.9	0.3	2.0
Loading and Hauling	3.1	3.7	2.1	6.3
Other	2.9	3.5	1.0	1.3
Subtotal	6.6	8.1	3.4	9.6
Mine maintenance				
Drills	0.2	0.3	0.1	0.6
Shovel-truck	3.6	3.9	1.9	5.9
Other	7.7	2.4	1.5	1.1
Subtotal	11.5	6.6	3.5	7.6
Crushing and screening or concentrating	10.1	8.6	12.8	35.8
Administration and general	8.1	9.4	8.0	19.3
Winter expense	13.3	17.7	—	—
Total	49.6	50.4	27.7	72.3

7.2.2. Mining Coal in West Virginia with a 72-Cubic Yard Dragline

KENNETH WOODRING
RUTH SULLIVAN

INTRODUCTION

Ashland Coal, Inc. founded in 1975, is based in Huntington, WV. The company markets nearly 4.5 Mt (5 million st) of coal annually to both domestic and foreign customers. Operational areas include eastern Kentucky and southern West Virginia. Total coal reserves are approximately 904 Mt (1 billion st). Hobet Mining, Inc., Ashland's West Virginia operating division, is headquartered at its No. 21 mine site in Boone County near Danville. Ashland's West Virginia holdings include about 182 kha (45,000 acres) representing reserves totaling 361 Mt (400 million st) of high quality steam coal. Mining is presently conducted in Boone and Logan Counties.

Ashland Coal acquired Hobet Mining in March 1977. At that time a preparation plant (Beth Station) with a capacity of 1.8 Mt (2 million st) was under construction on the Hobet 21 mine site. One of the first orders of business was to design a mine plan to provide 1.8 Mt/a (2 million stpy) of production for the preparation plant which has a 20-year life. The coal production would come from a more than adequate reserve base. The decision was made to develop the surface reserves first and maximize cash flow by potentially more economical surface mining methods.

Through the early 70s surface mining in mountainous southern West Virginia was generally done by contour mining and/or augering techniques. As market conditions changed and larger equipment became available, in the mid-70s, contouring at several mining locations gave way to mountaintop removal in order to maximize reserve recovery. The mountaintop method was explored for the development of the Hobet 21 mine property. This mining method involves the removal of all overburden and coal from the top of the mountain down to the lowest coal horizon to be mined in a particular mountain. Generally, the overburden is hauled to valley fills in hollows adjacent to the area being mined. Overburden can also be backstacked on the solid mountain area where the lowest seam has been removed. These particular coal seams are relatively flat lying. The overburden consists mostly of shale or sandstone and must be drilled and shot.

Generally, loaders and trucks or shovels and trucks have been used for this type mining. The contiguous acreages at Hobet's 21 mine site prompted the consideration of larger mining equipment, i.e., a dragline.

HOBERT MINING METHODS

A look at a typical cross section of the reserve is now appropriate. Fig. 1 shows the various available coal seams, the relative thicknesses of overburden separating them, and the original ground line. Average relief is 152 to 183 m (500 to 600 ft). Typical mountaintop removal involves removal of all overburden down to the middle Stockton seam and placement of the overburden in the adjacent valley with some backstack in the middle Stockton bench. Certainly, the dragline application on the original topography would be impossible because of the steep slopes, irregular shapes, and overall relief. A dragline requires a relatively flat surface area on which to operate.

Using some imagination, look again at the cross section and consider removing the overburden only to the 5-Block level, placing it in the adjacent valley fill on the right. The result is shown in Fig. 2. The flat area on which to work the dragline is now available. Placing a dragline on the 5-Block level and digging to the upper Stockton creates a rather conventional dragline operation. The adjacent valley fill then provides an area to cast the pit.

Looking at Fig. 3, notice the dragline has mined the first pit, placing the spoil on the adjacent fill area. Fig. 4 shows that some of the subsequent pits have been mined, and the dragline is about half complete with the mining area. Finally, Fig. 5 shows the regraded section after the mining is complete. Notice how the spoil piles have been leveled and the outslopes are graded and benched. The resulting topography is more gentle than the original, but the general trend of high and low areas has not changed.

Fig. 6 illustrates the plan view of a potential mining area. The seam outcrops are marked. Fig. 7 indicates the appearance of the same area after the overburden is removed down to the 5-Block seam and placed in the adjacent fills. Again, the 5-Block bench becomes the flat working area for the dragline. Notice the map shows additional mining was done in three areas. These additional cuts are to the Stockton level. They create casting areas for the dragline in the absence of valley fills. In fact these are relief cuts made for the dragline with the loader/truck mining method. All the overburden from above the 5-Block seam and the relief cuts was incorporated into valley fills and dragline access ramp. Fig. 8 demonstrates how the actual dragline pits are layed out. Pits are numbered consecutively; the arrows associated with the pit numbers indicate the direction of the dragline movement and casting.

EQUIPMENT SELECTION

We have talked only generally about equipment assignments. Let's take a closer look. Loader/truck or shovel/truck is appropriate for overburden from the top of the mountain down to the 5-Block. A loading shovel is most appropriate on the interval between the Middle Kittanning and 5-Block seams due to the larger bench widths and greater volume of overburden. The smaller irregular area above the Middle Kittanning is more appropriate for the loader/truck combination. The parting between the Upper and Middle Stockton seams is quite thin [approximately 2.4 m (8 ft)] and is another good loader/truck candidate. A shovel would not be desirable for use on the 2.4-m (8-ft) digging face and lacks the mobility to move quickly between pits. The dragline is best suited for the interval between the 5-Block and Upper Stockton seam.

Some assumptions were necessary for designing a mine plan as to the relative costs of overburden removal by different types of equipment. These assumptions were, of course, elaborated on with respect to the economic evaluation of the overall project. However, this is not a part of this discussion. Ashland's management and engineers assumed that the dragline would operate at 50%, or half the cost of loader/truck and the shovel/truck would operate at 75% the cost of the

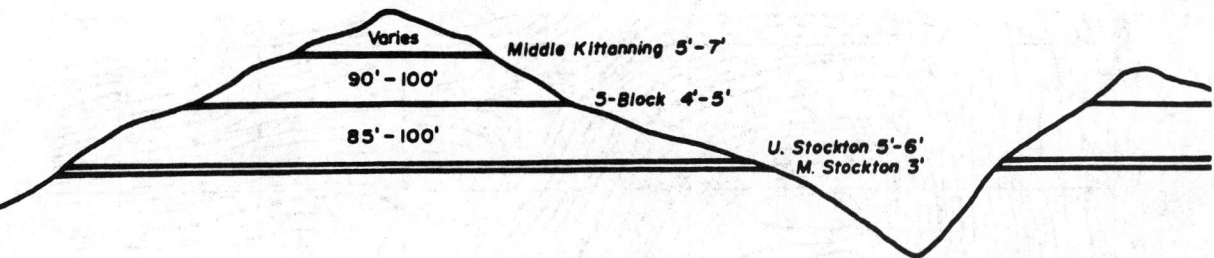

Fig. 1. Original section.

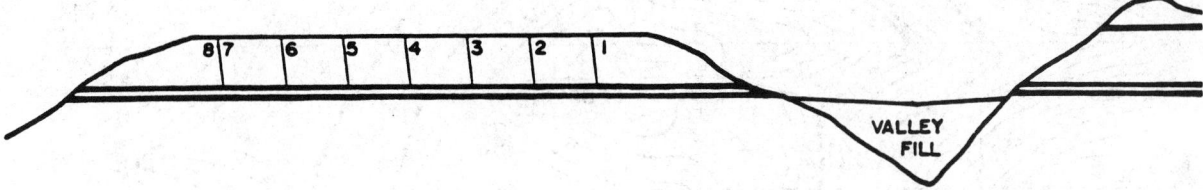

Fig. 2. Upper seams removed to 5-Block level; area ready for dragline operation.

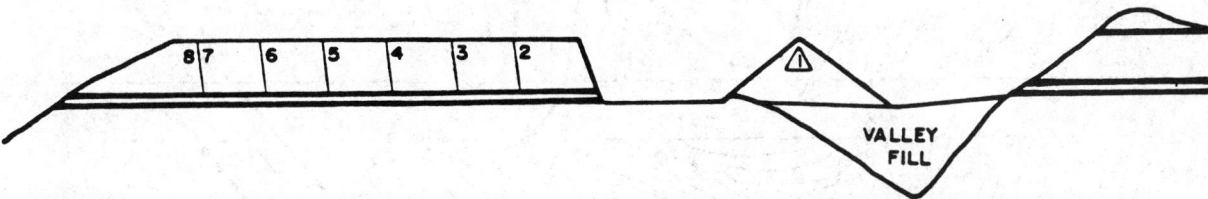

Fig. 3. Beginning dragline operation.

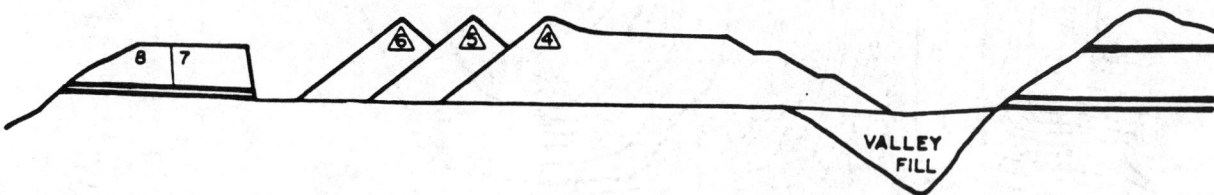

Fig. 4. Begin regrading; spoil from cuts 1, 2, and 3 required.

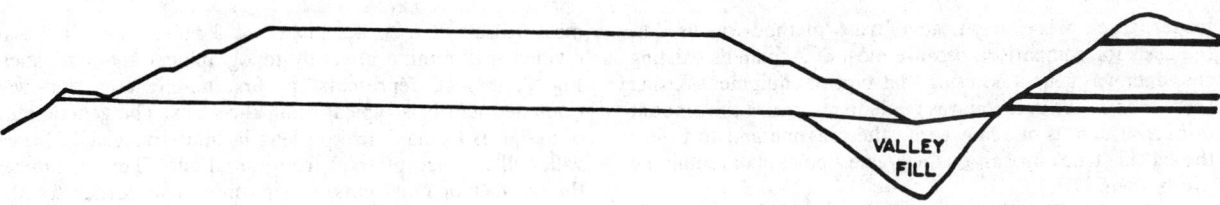

Fig. 5. Regraded section.

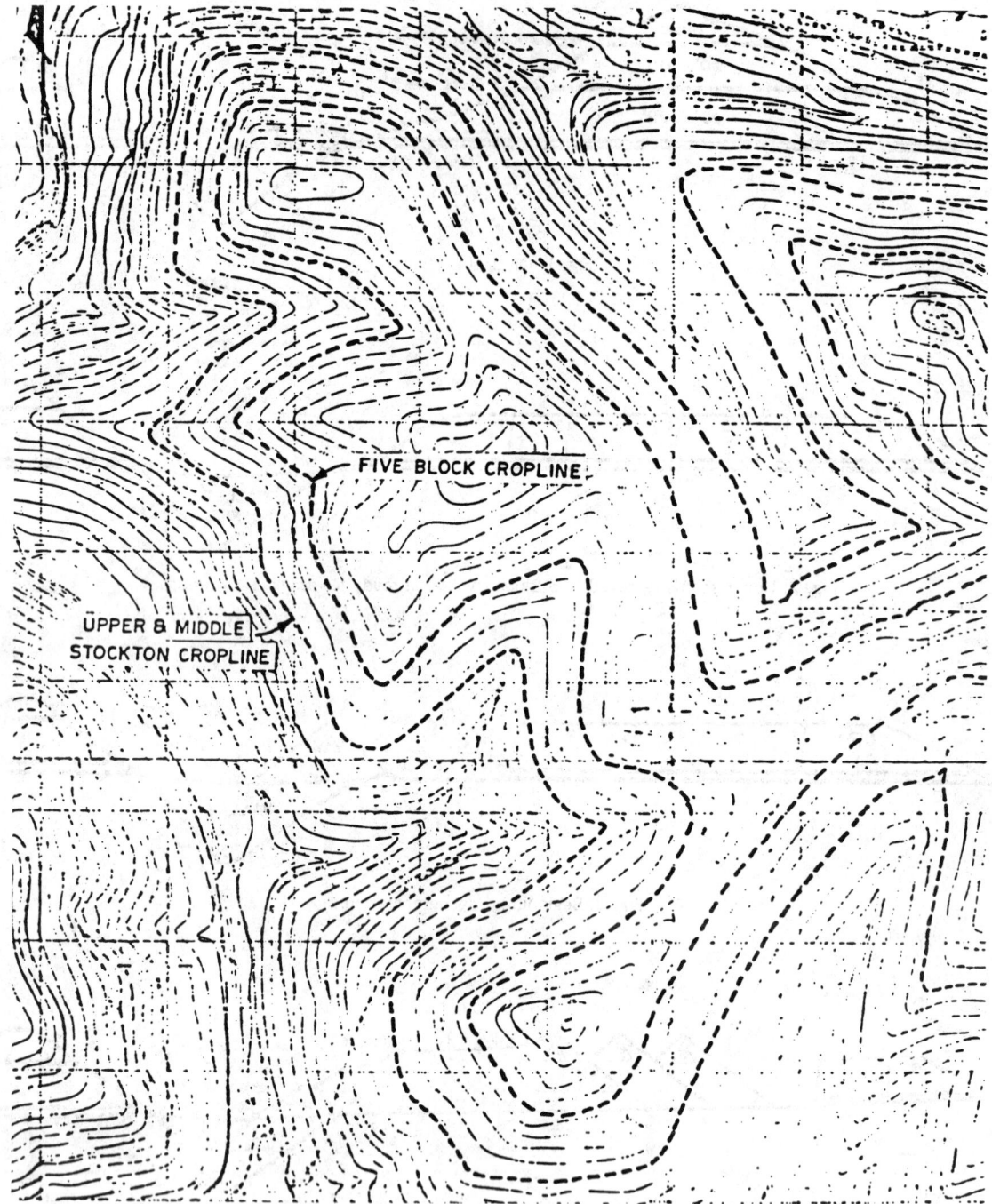

Fig. 6. Virgin mining area.

loader/truck procedure. Loader/truck method was used as the basis for comparison because most of Ashland's existing cost data was generated using that type of equipment. Using these criteria the decision was made to maximize the amount of overburden to be removed by the dragline and to utilize the shovel/truck process on the greatest portion of remaining overburden.

Approximately 4047 ha (10,000 acres) of coal-bearing land was required for the 20-year mine life, and that roughly measures about 8 × 4.8 km (5 × 3 miles). The area was divided into mining areas by topographical characteristics. Fig. 8, in fact, represents the first mining area that was examined and it is typical of all the areas. The general rule of design is to maximize pit lengths and strategically locate valley fills to accept spoil from initial cuts. This minimizes the number of relief cuts, leaving more overburden for the dragline. Notice that where pit lengths are short the machine is alternated between pits to allow adequate time for coal

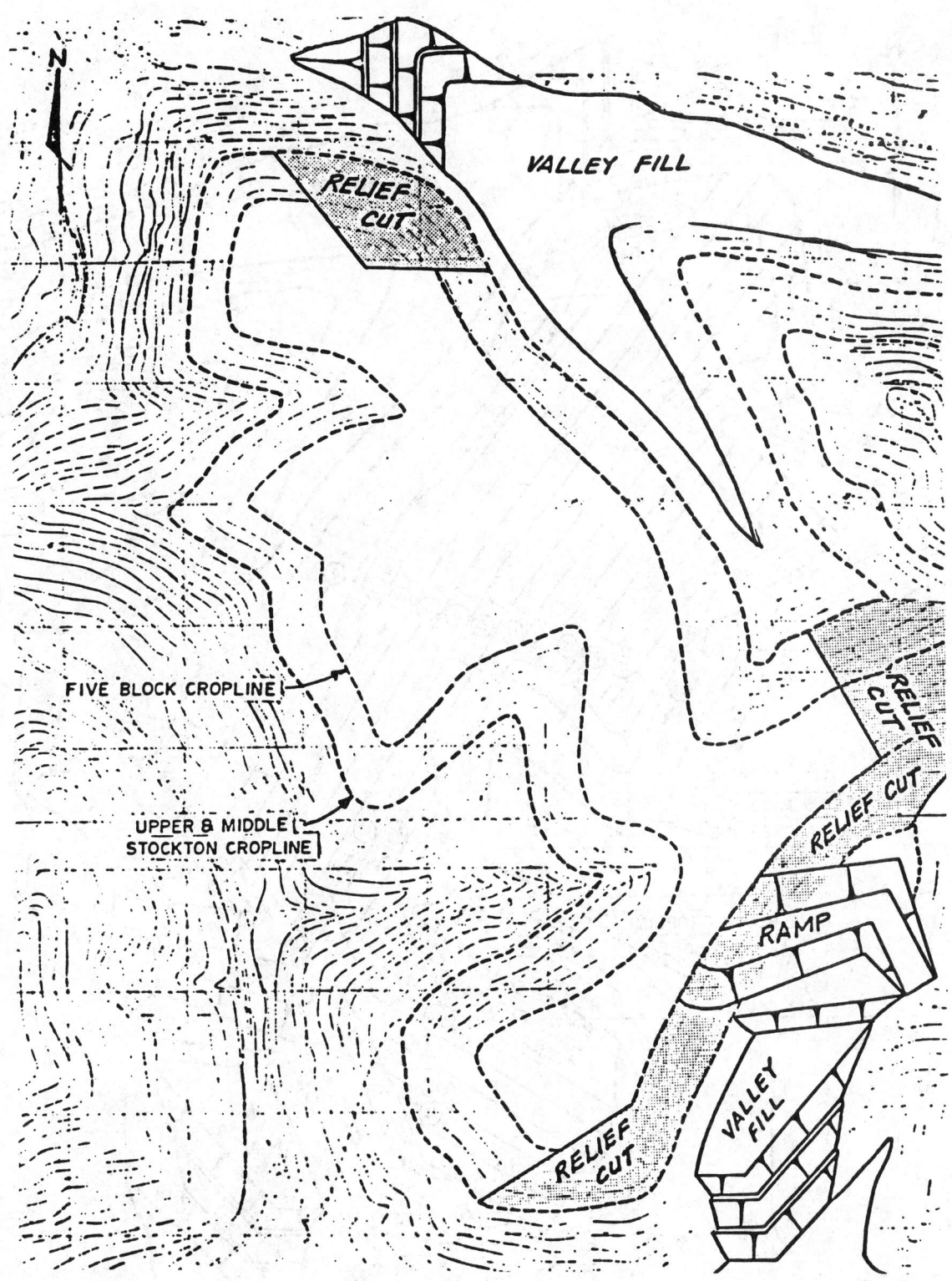

Fig. 7. Area developed for dragline mining.

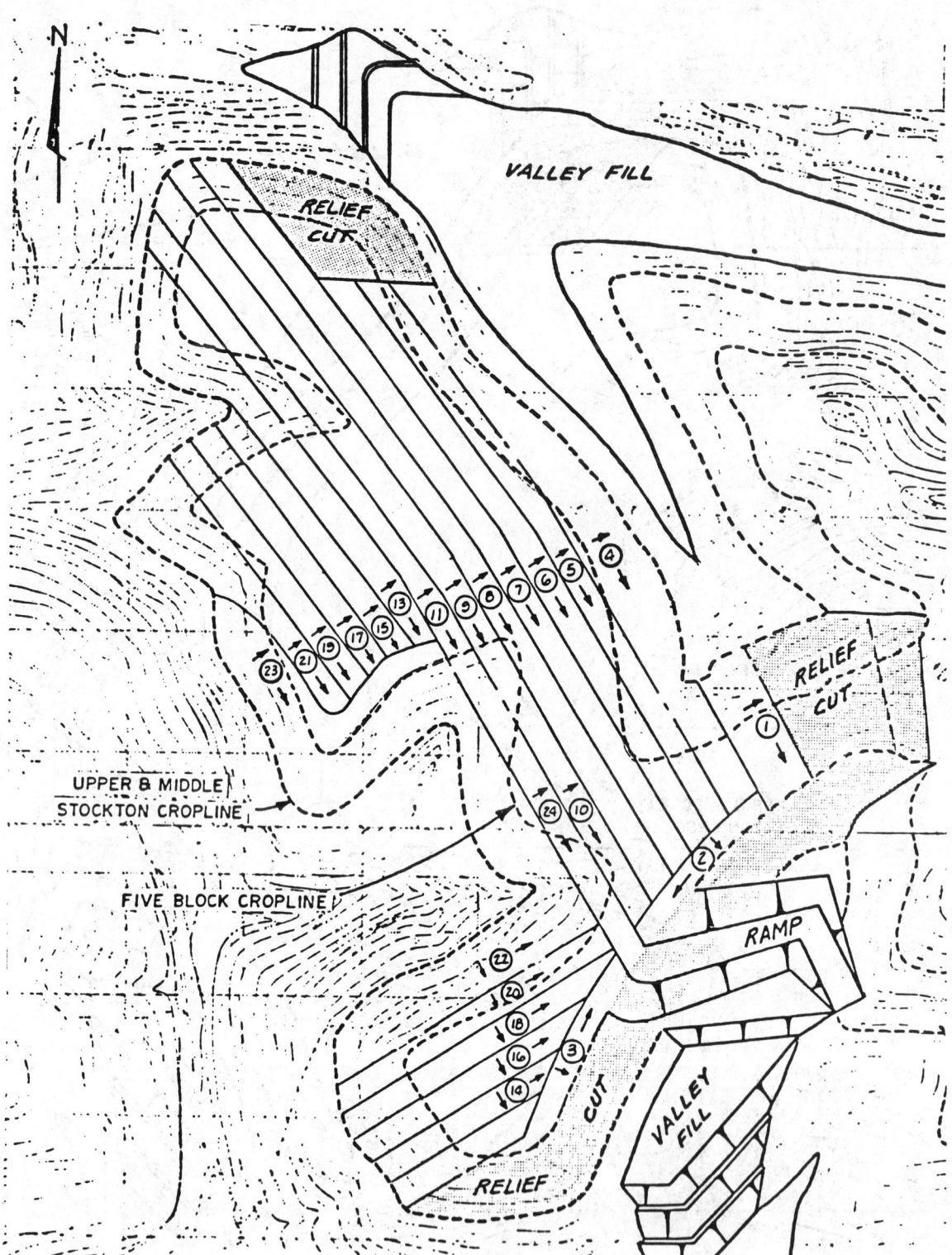

Fig. 8. Dragline pit outline.

Table 1. Reserve Analysis

Area	Raw tons	Overburden bank cu yd	Ratio	Dragline yd	%	Conventional yd	%	Dragline tons	Working Ratio	Conventional tons	Working Ratio
1	3,612,994	26,630,480	7.4:1	10,829,630	40.7	15,791,875	59.3	941,881	11.5:1	2,671,113	5.9:1
2	4,140,627	39,629,676	9.6:1	25,794,071	65.0	13,247,011	35.0	2,303,577	11.2:1	1,837,050	7.2:1
3	4,475,370	42,519,093	9.5:1	28,693,629	67.5	13,825,464	32.5	2,507,790	11.44:1	1,967,580	7.03:1
4	1,456,140	12,379,100	8.5:1	6,316,666	51.0	6,062,439	49.0	643,250	9.82:1	812,840	7.46:1
5	711,200	6,403,800	9.0:1	2,541,667	40.0	3,862,133	60.0	207,273	12.26:1	503,928	7.66:1
6	389,700	3,135,508	8.05:1	2,583,200	83.0	547,308	17.0	249,700	10.37:1	140,000	3.91:1
7	1,486,912	9,125,000	6.14:1	5,900,000	64.7	3,225,000	35.3	522,559	11.29:1	964,353	3.34:1
8	959,012	9,297,367	9.69:1	6,834,128	74.0	2,417,315	26.0	534,982	12.77:1	424,030	5.70:1
9	2,902,800	25,281,927	8.71:1	13,438,250	53.0	11,882,505	47.0	1,384,152	9.70:1	1,518,648	7.85:1
10	1,913,200	15,156,749	7.92:1	9,629,629	64.0	5,456,429	36.0	992,511	10.44:1	990,689	5.51:1
11	1,015,398	10,358,244	10.20:1	3,694,444	35.7	6,663,800	64.3	253,141	14.59:1	762,257	8.74:1
12	1,831,560	18,751,103	10.24:1	9,495,370	50.6	9,255,732	49.4	746,411	12.72:1	1,084,149	8.53:1
13	1,160,215	9,590,356	8.27:1	6,975,042	72.7	2,615,314	27.3	590,755	11.81:1	569,460	4.59:1
14	1,221,825	10,443,565	8.55:1	7,510,555	71.9	2,933,010	28.1	665,475	11.29:1	556,350	5.27:1
15	4,428,300	40,877,279	9.23:1	21,094,080	52.0	19,621,093	48.0	2,255,040	9.35:1	2,173,260	9.03:1
18	4,528,559	50,664,030	11.19:1	22,944,444	45.3	27,719,586	54.7	1,822,853	12.59:1	2,705,706	10.20:1
19	1,446,920	12,445,530	8.60:1	8,559,197	69.0	3,859,114	31.0	784,080	10.95:1	662,848	5.82:1
20	7,519,750	90,612,410	12.00:1	41,342,597	45.6	49,269,873	54.4	2,710,371	15.25:1	4,809,379	10.42:1
21	2,638,261	45,670,120	17.30:1	18,208,333	39.8	27,461,787	61.2	674,844	26.98:1	1,963,417	13.98:1
Final Total	47,838,743	478,971,337	10.01:1	252,418,733	53.0	225,116,528	47.0	20,720,640	12.18:1	27,118,099	8.30:1

Metric equivalents: ton × 0.904 184 7 = t; cu yd × 0.764 554 9 = m³; yd × 0.9144 = m.

Table 2.   General Equipment Requirement

Production requirement (clean tons)	2,000,000
Average plant recovery	79%
Raw tons required (2,000,000 ÷ 0.79)	2,531,600
Overburden ratio	10.01:1
Total overburden removal requirement	25,341,300
Maximum dragline percentage	53%
Maximum dragline overburden portion	13,430,900
Conventional percentage	47%
Conventional overburden portion	11,910,400

Metric equivalent: ton × 0.904 184 7 = t.

Table 3.   Dragline Operating Hour Calculations

Calendar time	8760
Vacation and holiday	− 600
Unscheduled time (PM, shift change, boom inspection)	− 825
Scheduled time	7335
Noncontrollable delays (strikes, fog)	− 272
Controllable time	7063
Electrical and mechanical delays, (15%)	−1059
Operational delays (20%)	−1201
Productive hours	4803

and parting removal. The ramp shown is set on a 6% grade and provides dragline access to the 5-Block elevation. Later it serves as the exit ramp when mining is completed. Grade changes such as these are necessary only three times in the mine life. After the third area is mined, the machine stays at the 5-Block level. The dragline has a 1.6-km (1-mile) walk between the first and second mining areas—the walk will require approximately 24 hr. All other mining areas contact each other.

After the mine layout was completed, each area's overburden was categorized into dragline work and conventional shovel/truck or loader/truck work. Layouts include cut and fill balances for conventional mining and pit layouts for the dragline. The results of this examination appears in Table 1.

An overall mining ratio of 10.01:1 is shown. The table also points out that 53% of the total overburden could be moved with a dragline. All parting between the Upper and Middle Stockton seams is included with the conventional equipment work.

For annual mine production of 1.8 Mt (2 million clean st) at 79% recovery through the preparation plant, the numbers in Table 2 apply. The dragline application could effect up to 10.3 Mm³ (13,430,900 bank cu yd) annually of the 19.4 Mm³ (25,341,300 bank cu yd) total requirement.

Dragline Sizing

Since the annual production requirement has been determined, we can now look at dragline sizing, actually, a range of sizes. To start, each manufacturer of walking draglines offers an appropriate model to accommodate Hobet's requirements. Each model is available with various bucket-boom configurations to suit different mining conditions. Therefore, once a model was selected, we were still working with a range of bucket-boom options. The options were:

Bucket Size, m³ (BCY)	Operating Radius, m³ (ft)
49.7 (65)	94.2 (309)
55 (72)	90.2 (296)
58.1 (76)	82 (269)

Annual production capabilities will be discussed first and then boom geometry. Table 3 illustrates annual operating hour calculations for a dragline operation. Table 4 summarizes dragline cycle calculations and calculates productive capability per operating hour per bucket yard. The appropriate annual production for the aforementioned options is shown in Table 5. By this analysis the 58.1-m³ (76-yd) option would seem most productive, but let us now consider boom geometry.

Boom geometry is a function of pit geometry. Pit geometry is, of course, a function of pit width and overburden

depth. The pit was assigned a 36.6-m (120-ft) width to provide adequate width for coal and parting removal. Overburden depths (distance from 5-Block to Upper Stockton) vary over the mining areas and are represented in Fig. 9. While the average depth is 28 m (92 ft), the dragline will dig as deep as 35 m (115 ft) and as shallow as 23 m (74 ft). These extremes must be taken into consideration as well as the average depths.

Another factor affecting boom geometry is the blasting technique. Two techniques that could be applied are open-face and buffered. Open-face blasting would involve shooting only the width of the pit the dragline was digging. It would provide a solid highwall to dig against and provide for some cast yardage from the blasting operation. Problems associated with the method include the loss of bench height during shooting and difficulties in sequencing blasting with the dragline movement and coal parting removal.

Buffered shooting would be completed many pits ahead of the dragline, therefore eliminating, to a great extent, the sequencing associated with open face blasting. This method maintains bench height and, from experience at the mine site, provides better fragmentation of material. In consideration of these factors it was decided to design the operation for buffered shooting. The highwall angles Hobet used are 1/4:1, the swell factor is 35%, and the spoil angles are 38°. The machine center pin is held 12 m (40 ft) from edge of the bench.

The three machine options, Fig. 10, are demonstrated on a pit cross section with the average digging depth of 28 m (92 ft). The maximum reach capability of each option is the working radius less the 12-m (40-ft) setback for the center

Table 4.   Dragline Cycle Calculations

Drag time	15 sec
Loaded swing	20 sec
Dump time	3 sec
Empty swing	18 sec
Preparation to drag	3 sec
	59 sec
Swell factor	0.74
Bucket fill	90%
Operator efficiency	90%

$$\text{Yd/hr/bucket-yd} = 60 \text{ sec/min} \times \frac{60 \text{ min/hr}}{59 \text{ sec/cycle}}$$
$$\times\ 0.74 \times 0.90 \times 0.90 = 36.57$$

Metric equivalents: yd × 0.9144 = m; cu yd × 0.764 554 9 = m³.

Table 5. Dragline Production Capabilities

Working Radius	Bucket Size	Operating hr/yr	Yd/hr bucket-yd	BCY/Yr
269	76	4803	36.57	13,349,100
296	72	4803	36.57	12,646,500
309	65	4803	36.57	11,417,000

Metric equivalents: yd × 0.9144 = m; cu yd × 0.764 554 9 = m³.

pin. The 49.7- and 55-m³ (65- and 72-cu yd) options reach the spoil peak easily digging 28 m (92 ft), but 58.1 m³ (76 cu yd) falls short and requires a bench extension. Any bench extension for this mining cross section requires a significant percentage of rehandle for the first increment as is demonstrated by the cross-hatched area. The total bench extension required here was 0.6 m (2 ft). The first increment required 25% rehandle, the remaining 0.6 m (2 ft) required an additional 2% rehandle. The total 27% rehandle is counterproductive and undesirable, thus making the 58.1-m³ (76-cu yd) option less productive than the 58-m (72-cu yd) unit.

Moving up toward the maximum 35-m (115-ft) vertical digging height the 58-m³ (72-yd) machine reaches its limitations at 31.5 m (103.5 ft). To go higher we would have to do a costly bench extension as described or do what is pictured in Fig. 11. Here the spoil line intersects the top of the Upper Stockton seam rather than the bottom of the Middle Stockton as shown in Fig. 10. The shifting of the spoil allows the 58-m (72-yd) machine to dig up to 35 m (115 ft). The 58.1-m³ (76-yd) still requires a bench extension, the 49.7-m³

(65-yd) is, of course, adequate. This spoil shifting creates some rehandle in the coal pit by loaders, but the amount is relatively small. Also, referring back to Fig. 9, only three of the mining areas would require a spoil shift for the 58-m³ (72-yd) machine, so the other 17 can be mined without any rehandle.

As a result of studies like these and consideration of the flexibility and conservatism, it was determined the best-suited machine option is the 58-m³ (72-cu yd) bucket with the 90.2-m (296-ft) working radius. The unit has an annual operating capability of 9.67 Mm³ (12,646,500 bank cu yd). Although this is less than the 10.3 Mm³ (13,430,900 bank cu yd) possible as shown in Table 2, the move up to the next larger model machine is great, in both size and cost. The result would be a machine far too large for the application.

Shovel/truck Portion

Considering the 19.4 Mm³ (25,341,300-bank cu yd) total requirement for the mine, the conventional portion becomes 9.71 Mm³ (12,694,800) as a result of the foregoing exercise.

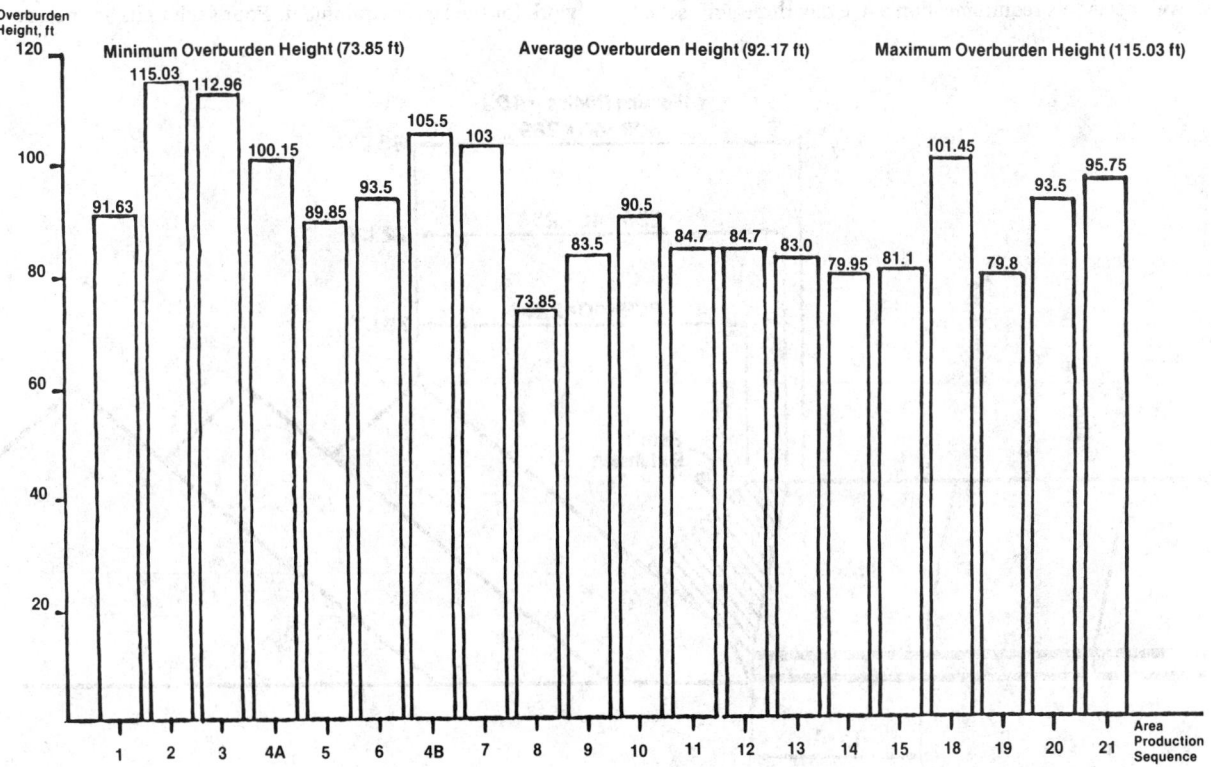

Fig. 9. Average overburden height for dragline areas. Metric equivalent: ft × 0.3048 = m.

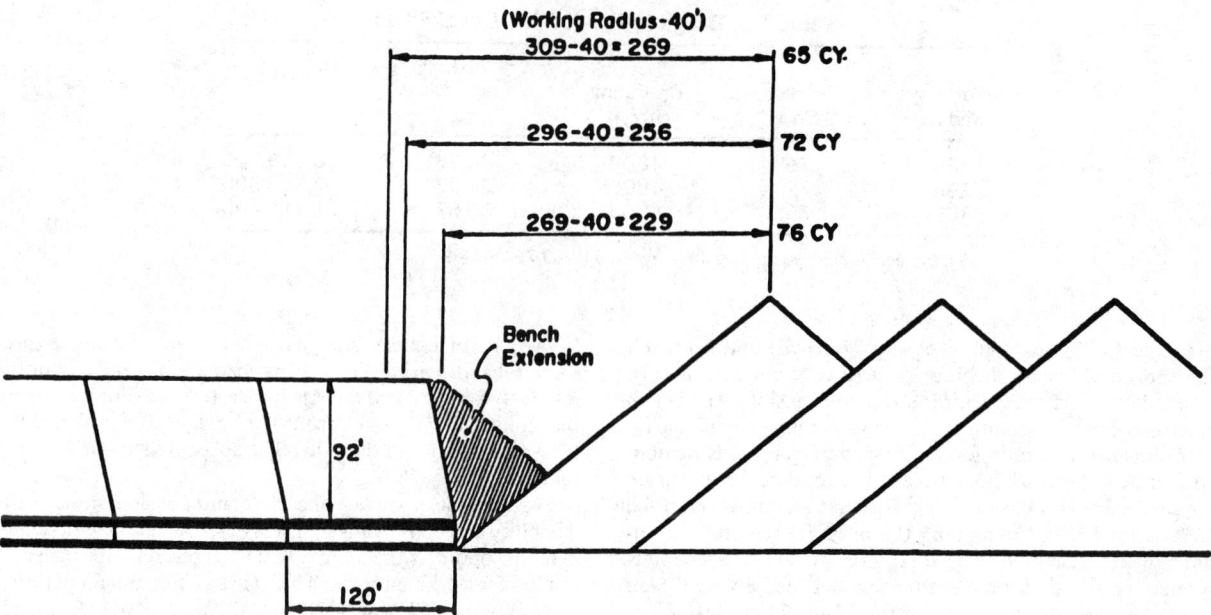

Fig. 10. Ranging diagram, average overburden.

As discussed before, the shovel/truck portion of this amount should be maximized to the fullest extent. The physical limitations are basically the amount of overburden between the 5-Block and Middle Kittanning seams and the amount of overburden in relief cuts on the Stockton which are large enough to work a shovel. This number for the 20-year mine life is 107 Mm³ (140 million bank cu yd) or 5.4 Mm³ (7 million bank cu yd) per year. A 20.6-m³ (27-cu yd) loading shovel meets this requirement on a five-day three-shift sched-

ule as indicated in Table 6. The machine is matched with 120-ton haulers.

Loaders

The remaining conventional yardage is left for 9.6-m³ (12½-cu yd) loaders teamed with 85-ton (108.5-t) trucks. These units provide the mobility and flexibility to make the small relief cuts, pull partings, and perform general support work for the larger equipment. Four such units are necessary

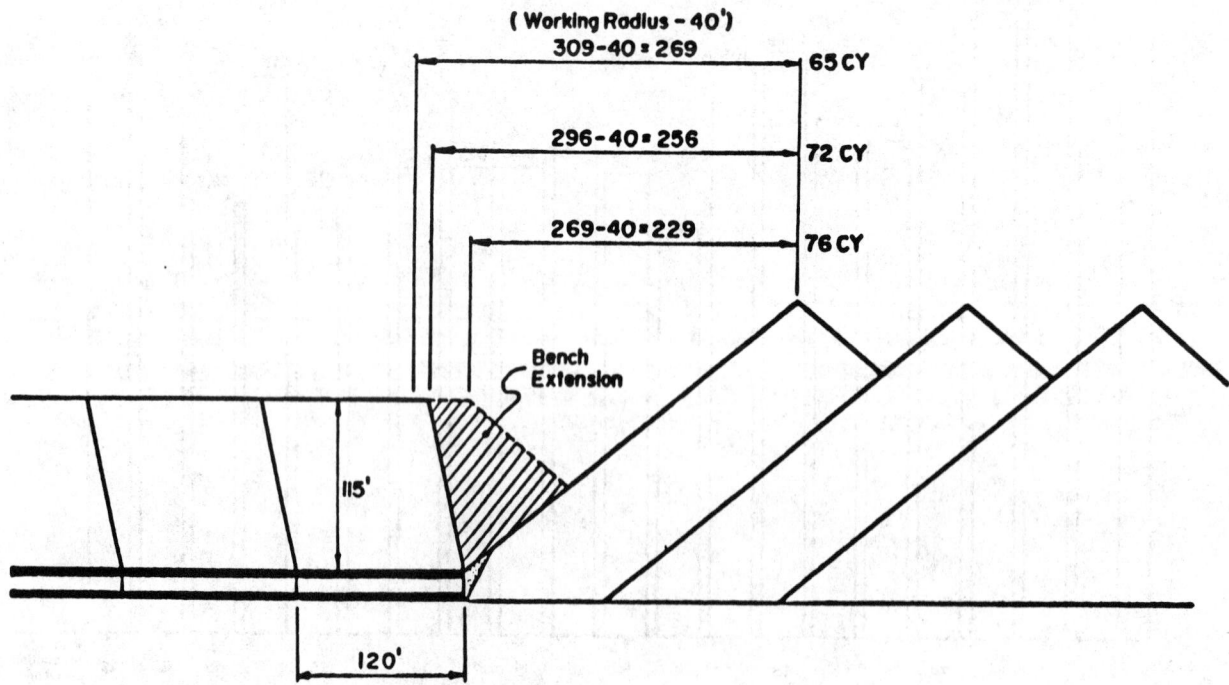

Fig. 11. Ranging diagram, maximum overburden.

Table 6. Shovel Operating Statistics

Calendar time	8760
Vacation and holiday	− 504
Unscheduled time (Saturday, Sunday, PM)	− 2696
Scheduled time	5560
Noncontrollable delays (strikes)	− 72
Controllable time	5488
Electrical and mechanical delays (15%)	− 823
Operational delays (20%)	− 929
Productive hours	3736
Bank cu yd/productive hour	1849*
Annual bank cubic yards	6,908,000

* Based on 1.75-min cycles per 120-ton truck, 59.9 BCY/load, 90% operator efficiency.

Metric equivalents: cu yd × 0.764 554 9 = m^3, ton × 0.904 184 7 = t.

on a five-day two-shift schedule to complete the overburden removal requirement. Annual production per loader unit is demonstrated in Table 7.

Table 8 summarizes the overburden removal equipment. The total capability of the selected equipment meets the annual mine requirement.

EQUIPMENT PHASE-IN

Dragline

The dragline, a Bucyrus-Erie 1570W, was purchased in February 1981. Erection was started in November 1981 and completed in March 1983. The machine walked off the erection pad March 26 and started digging on the 28th. The machine sits on a 20-m (66-ft) diam base, the walking shoes are 3.7 × 21.3 m (12 × 70 ft) each. The working weight of more than 3.6 Mg (8 million lb) includes 453.6 kg (1 million lb) of ballast. The boom is 99 m (325 ft) long on 35°. Hoist and drag ropes are 95.25-mm (3¾-in.) diam. Motor generator sets total 7.46 MW (10,000 hp). Twenty-two thousand nine hundred volts ac enters the machine and is reduced to 6900 volts ac in a 10-MVA transformer.

The dc motors are as follows:

Drag	2.98 × 969.8 kW	(4 × 1300 hp)
Hoist	4.48 × 969.8 kW	(6 × 1300 hp)
Swing	2.98 × 77.96 kW	(4 × 1045 hp)
Propel	0.746 × 373 kW	(1 × 500 hp) [per side]

Table 7. Wheel-Loader Operating Statistics

Calendar time	8760
Vacation and holiday	− 504
Unscheduled time (Saturday, Sunday, 1 shift/ weekday)	− 4416
Scheduled time	3840
Noncontrollable delays (strikes)	− 48
Controllable time	3792
Mechanical delays (10%)	− 379
Operational delays (20%)	− 662
Productive hours	2751
Bank cu yd/productive hour	575*
Annual bank cubic yards	1,582,000

* Based on 3.5-min cycle per 85-ton truck, 39.5 BCY/load, 85% operator efficiency.

Metric equivalents: cu yd × 0.764 554 9 = m^3, ton × 0.904 184 7 = t.

Table 8. Overburden Removal Equipment List

No.	Item	Size, yd	Annual BCY
1	B-E 1570W dragline	72	12,647,000
1	B-E 295 BII shovel	27	6,908,000
4	Cat 992C	12½	6,328,000
	Total		25,883,000

Metric equivalents: yd × 0.9144 = m; cu yd × 0.764 554 9 = m^3.

Hobet's machine has six hoist motors rather than the standard four and 77.96 kW (1045 hp) swing motors rather than the standard 596.8 kW (800 hp). This optional equipment was added to ensure optimum cycle times in all situations. More specifically, the amount of swing limiting and/or hoist limiting has been favorably reduced.

For the purpose of erection a 39.6-m (130-ft) long by 30.48-m (100-ft) wide by 12.2-m (40-ft) high steel building of rigid frame construction was built on wheels and rails over the site. The enclosure covered the base and revolving frame during welding, machining, and equipment alignment. The wheels and rails allowed the building to be moved away from time to time, permitting the 122-t (135-st) erection derrick to set the necessary parts in place. On completion of the revolving frame the building was moved away for the last time. The same structure now serves as a permanent four-bay shop building approximately 122 m (400 ft) from the erection pad. A 929-m^2 (10,000-sq ft) office and warehouse was attached to complete the mine's new service complex.

An on-board computer was added to the dragline to monitor productivity and to track any delays, providing a high degree of accuracy. The associated computer console, in the operator's cab, visually aids the operator with machine positioning and monitoring cycle times. As different digging techniques are explored, the system will provide rapid documentation for comparison purposes. Component life can also be studied with the system.

Loading Shovel

The 20.6-m^3 (27-cu yd) loading shovel is a Bucyrus-Erie 295-BII. Purchased in December 1981, the machine was placed in operation in July 1982 following a two-month erection period. Machine electrics are Bucyrus's new static a-c "Acutrol" system—it is the first such machine operating in the United States with General Electric components. The a-c motors are as follows:

Hoist	0.746 × 932.5 kW	(1 × 1250 hp)
Swing	1.49 × 167.9 kW	(2 × 225 hp)
Crowd	0.746 × 167.9 kW	(1 × 225 hp)
Propel	0.746 × 503.6 kW	(1 × 675 hp)

The trailing cable voltage is 7200 ac. The working weight is slightly less than a 680 kg (1½ million lb). The boom is 15.8 m (52 ft) high. Shovel benches are held to a vertical height of 13.7 m (45 ft).

Haul Trucks

Matched with the 295-BII shovel are WABCO 120D haul trucks. The trucks provide a three-pass loading match—the double back-in loading system is employed. The trucks were purchased with the 776 deep pit wheel option and extended range dynamic braking to aid in safely handling the downhill hauls from the 5-Block level to the fills.

Table 9.　Major Equipment List

72-yd dragline	1
27-yd shovel	1
10⅝-in. drill	1
9-in. drill	1
7⅞-in. drill	2
6¾-in. drill	2
120-ton end dumps (overburden)	4
85-ton end dumps (overburden)	15
7000-gal water trucks	3
90-ton coal haulers	8
12½-yd wheel-loaders	5
8-yd wheel-loaders	4
Motor graders	3
Scrapers	3
Bulldozers	14

Metric equivalents: yd × 0.9144 = m; in. × 25.4 = mm, ton × 0.904 184 7 = t, gal × 0.003 785 4 = m³.

Loaders

Wheel loaders for overburden removal are Cat 992C models. Equipped with 9.6-m³ (12½-yd) buckets, the units four-pass load 85-ton (76.9-t) Cat 777 end-dump trucks.

Dozers

The mine's dozer fleet is composed of Cat D10s, D9s, and D8s. Spoil leveling is generally done with the D10s. The D9s are normally used to feed material to end loaders and the D8s are used on fills and drill benches.

Drilling

Drilling varies in size, dependent upon the application. Dragline drilling is done with a Marion M-3 unit equipped with a 269.9-mm (10⅝-in.) rotary bit and 18.3-m (60-ft) drill pipe. Since maximum dragline overburden is 35 m (115 ft),

Table 10.　Manning Table Summary

UMWA Personnel	
Dragline	16
Shovel	6
Loaders (overburden)	8
Rock Trucks	39
Dozers	26
Drills	17
Shooters	9
Graders	6
Water trucks	6
Loaders (coal)	6
Coal trucks	20
Mechanics	17
Welders	10
Electricians	4
Utility	10
Total surface mine	200
Total preparation plant	29
Total UMWA	229
SALARY PERSONNEL	
Supervision	20
Clerical	12
Security	6
Total salary	38

Table 11.　Dragline Operating Statistics

Item	Original Budget	Current Budget
Optimum cycle time, sec	59	59
Operator efficiency, %	90	95
Effective cycle time, sec	65.6	62.1
Bucket fill factor	90	100
Swell factor	0.74	0.78
BCY/bucket	48	56
BCY/operating hour	2,633	3,250
Annual BCY	12,646,000	15,711,000

Metric equivalent: cu yd × 0.764 554 9 = m³.

the M-3 can drill all situations with only one steel change. The unit is equipped with Marion's automatic drill control system and operates on 7200 ac. Drill patterns are normally 8 × 8 m (27 × 27 ft).

A Robbins RR10 drill is used with the shovel. Equipped with a 22.9-cm (9-in.) rotary bit and 7.6-m (25-ft) drill pipe, the machine drills on a 5.5 × 5.5 m (18 × 18-ft) pattern. Driltech D40s and D50s round out the drill fleet and provide drilling for the loader units.

Loading

Coal loading is accomplished with both 998B and 992C Cat loaders. Coal shooting is unnecessary; the loaders break the coal from the seam quite easily because of the well established cleavage patterns. Haulage is in 65-ton (58.8-t) Cline and 85-ton (76.9-t) Dart haulers. Tandems have been used rather than bottom-dump trailers due to the mile-long grade at a 9% drop to the preparation plant.

Major equipment is listed in Table 9.

MANPOWER

Manning levels for the operation are summarized in Table 10. Production for the 200 UMWA surface mine employees will be approximately 36 t (40 tons) per man-day.

OPERATING EXPERIENCE

Since the first writing of this article about the time of the dragline startup in April 1983, Hobet has gained much information on the operation and performance of this machine. The original approach was very conservative due to the difficult application. Our experience since April 1983 has demonstrated the machine to be more productive and flexible than anticipated.

The operating statistics documented by the dragline's onboard computer readily indicate this and have been incorporated into the current budgeted operating performance, Table 11. The production rate of 12 Mm³ (15,711,000 BCY per year) reflects a higher BCY operating hour rate of 3250, mainly attributable to the higher operator efficiency and bucket fill factor.

Table 12.　Shovel Operating Statistics

Item	Original Budget	Current Performance
Mechanical/electrical delays, %	15	15
Operating delays, %	20	24
BCY/operating hour	1849	1812
Annual BCY	6908	6517

Metric equivalent: cu yd × 0.764 554 9 = m³.

Table 13. Loading Statistics

Equipment	Size, yd	Orig. Budget, Annual BCY	Current Budget, Annual BCY
BE 1570W dragline	72	12,647,000	15,711,000
BE 295 BII shovel	27	6,908,000	6,517,000
992C (three shifts)	12½	—	2,342,000
992C (two shifts)	12½	(×4) 6,328,000	1,562,000
Total		25,883,000	26,132,000

Metric equivalents: yd × 9144 =m, cu yd × 0.764 544 9 = m³.

The current budgeted operating statistics are intended to reflect the average performance for the life of the machine. The dragline's actual production statistics for fiscal 1984 indicate a better performance because the machine is still relatively new.

The increased production and flexibility of the dragline has permitted Hobet to revise the pit layout in some areas and cycle the dragline through a series of shorter pits in three or sometimes four adjacent areas. This more complex pit sequencing and design, while incurring more bench extension, rehandle, and walking, has two advantages: it provides additional time for coal removal from the pits and maximizes the percentage of overburden designated to the dragline.

The increased production, and maximization of the percentage dragline overburden, coupled with a maximum capacity at the preparation plant of 1.8 Mt/a (2 million stpy) permitted the mine to reach budgeted coal production while maintaining the previously established end loader output. It is envisioned, however, that the number of spreads will eventually increase as the ratios for the development work above the 5-Block seam increase in the future mining areas.

The shovel has not performed as expected due to higher than expected operational delays created by a shortage of trucks; however, with the recent addition of a fifth Wabco 120, the shovel performance is expected to match budget expectations, Table 12.

The endloaders have performed as expected. One spread operates three shifts per day primarily in inter burden between the Upper and Middle Stockton while the other operates two shifts a day on development work along with the shovel. The total yards handled per year is 20 Mm³ (26,132,000 bank cu yd), Table 13.

The approximate capital cost expenditures for major equipment is shown in Table 14.

Table 14. Approximate Capital Cost Expenditures for Major Equipment

Mine	
72-yd dragline and distribution equipment	$24,000,000
27-yd shovel and distribution equipment	4,000,000
120-ton rock trucks, 5	3,550,000
Rubber-tired dozer	250,000
12½-yd endloaders, 2	1,320,000
85-ton rock trucks, 6	3,420,000
10⅝-in. drills, 2	1,970,000
9-in. drills, 2	736,000
6-in. drill	280,000
Coal loaders, 2	650,000
Coal trucks, 6	3,060,000
Graders, 2	540,000
Water trucks, 2	840,000
Dozers, 10	3,392,000
	$48,008,000

Preparation Plant	
Preparation plant, storage, rail sidings	$24,000,000
Refuse haulers	840,000
Refuse dozer	310,000
	$25,150,000

Metric equivalents: cu yd × 0.764 554 9 = m³, ton × 0.904 184 7 = t, in. × 2.54 = cm, in. × 25.4 = mm.

7.2.3 Cost of Mining Eastern Coal*

How much does it cost to mine a ton of coal? This is the most complicated and critical question in the coal industry. It is crucial to know not only how much it costs to mine, but also what price the market will bear. A piece of coal with its own inherent overhead must be matched to the marketplace. This basic fact governs how, when, and if the coal will be mined.

There are some constants in mining expenses, but there are endless variables. To arrive at a reasonably accurate figure for mining a ton of coal, it is necessary to set forth certain conditions. In the following example the coal tract is 50.6 ha (125 acres) in West Virginia to be mined by the contour surface method. The seam is 101.6 cm (40 in.) thick. Hiring, equipment selection and procurement, and permit application may all be done simultaneously, but the permit must be secured before any mining or construction can take place.

PERMITTING AND BONDING COSTS

Permitting

When the permit application goes to the West Virginia Dept. of Natural Resources (DNR), it must be accompanied by a $500 filing fee. This is *not refundable* in the event the permit is denied. The NPDES permit, also under the jurisdiction of DNR, requires an additional fee of $50. Then there is the lands inquiry fee of $100 to determine if the proposed site contains any unique historical, environmental, or geological qualities. When a surface mine application (SMA) number is assigned, the prospective operator must purchase legal advertising to announce his intention to mine coal, to alert any potential opponents to the permit. This ad must run on four separate days, each a week apart, and, in this case, will cost a total of about $150. The legal ad process will be repeated three times, one for each phase of the bond release. Add in the cost of legal ads for intent to blast and the total advertising bill is $750, making the total face cost of the permit application of $1400.

Bonding

The bond is the amount of money committed by the operator to insure proper and complete reclamation. If reclamation standards are not met, the bond is forfeited and presumably used to contract reclamation for the area.

The bond is set by law at $1000 per acre ($10,000) minimum. For a permit of 101.6 ha (125 acres), the bond will be $125,000. It would be extremely difficult for a new business to qualify for bonding by insurance or related companies. Because a mining bond is more in the nature of collateral, as opposed to an actual cash outlay, the qualified operator can obtain bonding at a rate which averages about $12.50 per $1000.

Following successful reclamation, the bond is released in three phases. Phase 1 comes at the end of the mining operation when the entire area has been regraded to specifications. This amounts to about 60% of the original bond. Phase 2, an additional 25%, may be released two years after completion of reclamation when vegetative cover is well established. Phase 3, the final 15%, is held for an additional three years, a total of five years after completion of the

operation. In the case cited here this takes place eight years after the job was begun.

The cost on Phase 1 of the surety bond, $125,000 for three years, would be $4687.50. Phase 2, $50,000 for two more years, will cost $1250. Phase 3, $18,750 for three years, amounts to $703.12. Thus, the total bonding cost would be $6640.62.

ENGINEERING AND CONSTRUCTION

The permit application is quite voluminous and requires engineering expertise which is usually beyond the in-house capability of the small company. For an operation of the size under discussion, plans for haul roads, drainage and valley fills, surveying, maps, and hydrological studies will commonly cost about $25,000. Again, this is a front-end cost which is not recoverable in the event of permit denial.

The construction phase of the operation will entail two major projects. One is the drainage system, of which the primary component will be sediment ponds. Three ponds of average size, professionally designed to capture all runoff from a mining operation, will cost about $35,000. The haul road, like all other mine construction, must be carefully engineered. The typical haul road in West Virginia will cost about $100 per foot. Thus, a road of only one mile in length will run about $528,000. The total engineering and construction costs will be $588,000. These are all expenses which are incurred *before* any coal production is carried out.

EMPLOYMENT COSTS

A permit of 101.6 ha (125 acres) can be expected to yield about 30 ha (75 acres) of minable coal. At 2200 t/ha (6000 net st per acre), the projected total tonnage from the permit would be 408 kt (450,000 st) net.

Setting a production goal of 13.6 kt (15,000 st) per month and allowing for start-up, slow production as the permit plays out, and final reclamation, the projected life of the operation would be about 36 months. Given current productivity figures, this calls for a work force of about 20 miners, plus at least one foreman.

Wages

The average coal miner in West Virginia costs $585.69 per week (1983 figures). This amounts to a weekly payroll, including the foreman, of $12,463.80. Times 52, the annual payroll is $648,117.60. For the three-year life of the operation, and naively allowing for no wage increase over that period, the basic employment cost would be $1,944,352.80. This, as all employers know, is the tip of the iceberg.

Surcharges

Coal employers, even surface coal operators, pay 50¢ per ton into the federal black lung fund. This amounts to an additional $225,000 over three years. The state collects, for the same purpose, 2.4% of the payroll and collects it three years beyond the life of the operation. In this case, the cost would be $93,328.93.

Workers' compensation is paid to the state at varying rates. The rate for new companies (less than five years in business) in the coal industry is $6.50 per $100 of payroll. This would amount to $126,382.93 over three years.

Both the state and federal governments take a bite for unemployment compensation. The state gets an annual amount equivalent to 5.5% of the first $8000 per employee.

---

*Adapted from an article by Dan Miller in *Green Lands*. Costs cited are those of 1983.

This would be $9240 per year, $27,720 for three years. The federal government gets 0.2% of the first $7000 per employee per year, or $294 per year, $882 for three years.

FICA, the federal social security tax, is 7% of each employee's salary. There is a ceiling on this, but it is raised each year. For the 21 employees, the cost will be $45,368.23 yearly and times three is $136,104.69 for the life of the operation.

The cost of government is summarized in Table 1.

Other Fringe Benefits

There are other fringe benefits as well. Standard hospitalization, for instance, will cost about $375 per month per employee. In this case, the expense will be $7875 per month, $94,500 per year, $283,500 for three years. Practicality dictates that each employee undergo a thorough examination prior to employment to identify preexisting medical conditions. This would amount to an expense of about $4500.

The UMWA extracts $1.60 per ton from signatory companies for its pension fund. That is $720,000 on this permit. The Union also gets $1.07 per hr. Over three years that is 113,100 hr, or a total of $115,022.70. Costs of unionism are summarized in Table 2.

The Mine Safety and Health Administration (MSHA) also requires a bathhouse for employees or payment of a "bathhouse waiver." Small companies, particularly operating in a given area for only three years, will find it more convenient and economical to pay the waiver of $2 per day per employee. That adds another $31,200 to the employment costs.

There is also the uniform allowance. This amounts to $150 per year per employee, payable on the first payday for each employee. Even assuming no turnover in the work force, this cost would be $9000 for three years.

Training is required for all employees in surface mining procedures. Annual refresher training for this work force will cost $10,002. Training for the required emergency medical technician (EMT) will cost $1266.92.

Additionally, a labor bond must be posted with the Labor Commissioner of West Virginia to ensure payments of wages and benefits in the event of shutdown. The bond amounts to four weeks of payroll plus 15%. At current rates, the cost of the bond will be $6206.97.

Total Payroll

Thus, payroll nearly doubles, considering all the auxiliary costs of putting a miner to work. To be precise, the payroll is $1,944,352.80 and the extra costs amount to $1,790,117.14, for a total cost of employment of $3,734,469.94.

Table 1. Cost of Government

	Cost per Ton
Federal Black Lung payments	$0.50
State Black Lung payments	0.21
Workers' compensation	0.28
Unemployment (state and federal)	0.07
FICA (Social Security)	0.30
Permits and bonds	0.02
Safety training	0.03
B&O tax	1.48
Property and fuel tax	0.13
Abandoned Mine Lands (federal)	0.35
Special Reclamation (state)	0.01
Bathhouse waiver	0.07
	$3.45

Table 2. Cost of Unionism

	Cost per Ton
Pension fund (by ton)	$1.60
Pension fund (by hour)	0.25
Uniform allowance	0.02
	$1.87

EQUIPMENT

The variety of equipment "spreads" and methods of obtaining, financing, and disposing of them is endless. The average surface mining operation will involve bulldozers, loaders, off-highway trucks, drills, graders, and other trucks specialized for such functions as coal hauling, explosives storage and hauling, water pumping, water spreading, maintenance, seeding, welding, emergency transport, and others. Site, market, and company considerations dictate the particular combination of these pieces to be used on any particular operation. There is no "average."

However, using accepted industry figures for cost-per-ton of standard surface mining functions, and deleting from these employment costs covered previously, approximate estimates can be made on the cost of operating and maintaining the necessary equipment over the three-year life of the project.

In surface mining, by far the single biggest expense is that of exposing the coal, that is, removing and storing, or disposing of, overburden. In mining with a 10:1 overburden ratio, a good-to-average situation in West Virginia, this expense will amount to $7,593,750 over the life of the operation. Loading the coal for transport to the tipple will be $393,740. Haul road maintenance will be $196,875. Final regrading of the slopes and preparing for seeding will cost $393,750. These four basic functions will total $8,578,125, which represents equipment costs on this project.

OTHER OPERATING EXPENSES

Assuming the company does not own its own tipple, it will encounter a cost of $2 per raw ton for this function. The operator can expect a loss of at least 20% of raw tonnage during the cleaning process. To net 408 kt (450,000 tons) of clean coal, it will be necessary to haul and clean 526 kt (572,500 raw tons). Therefore the tipple cost will be $1,125,000.

The standard cost for hauling is $1 per ton for the first mile and 10¢ per ton for each additional mile. Assuming a 9.7-km (6-mile) haul, the cost would be $1.50 per raw ton, or $843,750.

Another common operating cost is "wheelage," that is, hauling coal over the property of an adjacent landowner. The standard fee is 10¢ per ton, a total of $56,250.

The actual revegetation of reclaimed land is another function likely to be contracted by the small mining company. At approximately $700 per acre, this will cost $93,750.

In many cases, the coal operator is mining on private property and therefore must pay a royalty to the mineral owner. This party gets either a price per ton rate or, more likely, a negotiated percentage of the sale price, typically 8%. Assuming a sale price of $38.50 per ton, the royalty will be $1,386,000.

TAXES AND FEES

Government, in addition to playing a leading role in employment costs, has several direct taxes for the coal op-

Table 3. Tons Mined Required to Meet Overhead Costs

Overhead Item	Cost	Tons
Permits	$ 1,400.00	36
Bonding	6,640.62	173
Engineering	25,000.00	649
Construction	588,000.00	15,273
Payroll	1,944,352.00	50,503
Government imposed payroll expenses	626,894.44	16,283
Union imposed payroll expenses	875,222.70	22,733
Hospitalization and medical	288,000.00	7,481
Overburden removal	7,593,750.00	197,241
Coal loading	393,750.00	10,227
Haul road maintenance	196,975.00	5,114
Slope preparation	393,750.00	10,227
Tipple	1,125,000.00	29,220
Hauling	843,750.00	21,915
Wheelage	56,250.00	1,461
Revegetation	93,750.00	2,435
Royalties	1,386,000.00	36,000
Property tax	550.80	14
Fuel tax	56,250.00	1,461
B&O tax	667,012.50	17,325
Abandoned Mine Lands Fund	157,500.00	4,091
Special Reclamation Fund	4,500.00	117

erator. Basic to these is the West Virginia Business & Occupation (B&O) tax, which squeezes coal at the rate of 3.85% of the sale price. On 408 t (450,000 tons) net at $38.50 per ton, the B&O tax comes to $667,012.50.

The coal property tax in the county where the permit is located is currently (1983) $255 per acre. For 30 ha (75 acres) of coal, the tax base would be $19,125, which is assessed at a rate of 60%, or $11,475. The Class III Levy is 2.4%, or $275.40, which would be the actual assessment for the first year. Assuming the coal is mined at the rate of one-third during each of the three years of operation, the actual assessment in the second year would be $183.60. The taxes for the third year will be $91.80, for a three-year total of $550.80.

Fuel tax is extracted at the rate of 4.85¢ per gal. It takes a little over 2 gal of fuel to mine a ton of raw coal. This means a tax of approximately 10¢ per ton, or $56,250 over three years.

The federal Office of Surface Mining takes 35¢ per net ton for the reclamation of abandoned mine lands. This amounts to $157,500 on this operation. West Virginia currently gets an additional 1¢ per ton for its own Special Reclamation Fund, a total of $4500.

The total direct taxes and fees on this operation will amount to $885,813.25 over three years, or $1.97 a ton. If that does not seem like much, consider that such government revenue is generated from one small operation. Multiplied by West Virginia's total 1983 tonnage, these figures amount to one quarter of a billion dollars for that year. These are only direct taxes and fees and do not include the various government-imposed payroll expenses.

VARIANCES

The hypothetical situation set forth here required numerous assumptions. It should be emphasized that this would be a relatively small operation. Even so, it is clear that any adjustments in costs ricochet through three years and 408 kt (450,000 tons) with a startling effect.

For example, this operation was calculated to need 20 employees plus one foreman. The addition of even one hourly worker adds $129,384.85 to the overhead in three years. A wage increase of only 50¢ per hour adds $67,045.83 to total employment costs.

A decade ago the West Virginia State Legislature tacked on a modest 0.35% surcharge to the B&O tax (for coal only). This increases the hypothetical mining costs here by $60,637.50. Suppose the financially troubled social security system is "transfused" with an additional 1% in FICA payments. That is another $19,443.53 in overhead.

Wheelage, the rate paid for hauling coal across someone else's property is generally 10¢ a ton or, if the route crosses the grounds of two property owners, there is an additional 10¢ per ton, another $56,250. Add five miles to the "haul," the distance from mine to tipple, you have increased the cost by 50¢ a ton, or $225,000 added to overhead.

Table 4. Overhead Cost per Net Ton

Permit and bonding	$ 0.02
Engineering	0.06
Construction	1.31
Basic payroll	4.32
Government imposed payroll expenses	1.39
Union imposed payroll expenses	1.94
Hospitalization and medical	0.64
Equipment	19.06
Tipple	2.50
Hauling	1.87
Wheelage	0.13
Revegetation	0.21
Royalties	3.08
Direct taxes and fees	1.97
	$38.50

Table 5. Net Profit After Expenses

Income for 450,000 tons at $38.50 per ton	$17,325,000.00
Equipment	−8,578,125.00
	$ 8,746,875.00
Tipple, hauling, wheelage, revegetation	−2,118,750.00
	$ 6,628,125.00
Payroll	−1,944,352.80
	$ 4,683,772.20
Other employment costs	−1,790,117.14
	$ 2,893,655.06
Royalties	−1,386,000.00
	$ 1,507,655.06
Taxes and fees	−885,813.30
	$ 621,841.76
Permits, bonding, engineering, construction	−621,040.62
Net Profit	$ 801.14

In regard to hauling, a recently imposed diesel fuel tax of 6¢ per gal for highway vehicles will certainly find its way back to basic haul rates. This would increase hauling costs of the operation's total tonnage by $1296.

Haul roads also vary greatly and the operator does not have a great deal to say about it. If this operator happened to need an extra mile of haul road, he would also happen to need an extra $724,875 to build and maintain it.

And how about royalties? The rate in this situation was 8%. If it were 9%, the overhead would increase by $173,250.

The most devastating change in operating costs is that of overburden ratio. In the operation described here, the ratio is 10:1. If that were 12:1, mining costs would go up by $2,461,618.

SUMMARY

Using all the figures cited for the initial permit case, the total overhead for this operation comes to $17,324,898.81.

Even this figure assumes no major problems in securing a permit, no prolonged work stoppages, average weather conditions, no unpleasant surprises with the coal seam, nor any other of the myriad problems which can plague an operation on a day-to-day basis.

With this overhead, the break-even price for mining a ton of coal is $38.50. Table 3 shows how many tons of coal must be mined to pay for various overhead items. Table 4 shows the cost of these items per net ton. This price, $38.50, does not include the charges for transport to the final destination, which could vary as widely as an in-state power plant to a Japanese steel mill. Selling the coal at the tipple will necessarily hold down the sale price.

Building in a standard profit margin for the operator will push the necessary sale price well past $40 and probably past the operator's ability to market his coal. In other words, the bottom line here is that this coal tract, with no outstanding adverse conditions, can probably not be profitably be mined by the small operator (Table 5).

7.2.4 Powder River Basin Open Pit Coal Mines

WILLIAM E. WALESKI

INTRODUCTION

In the Powder River Basin, which contains over half of Wyoming's coal resources, mining conditions are unique. That is principally because the basin itself is unique. At no other place on earth is there such a combination of vast reserves and diverse conditions. Geological and topographic features are variable. Extreme weather conditions are common. Temperatures range from below $-29°C$ ($-20°F$) in the winter to in excess of $38°C$ ($100°F$) in summer. Although the area is generally considered to be a desert, rain, snow, wind, and tornados are not uncommon. As new governmental regulations require innovative approaches, mining operations tend to be innovative. Some aspect of virtually every conceivable mining method can be found there. Even the workers in the mines are dissimilar. There are copper miners from Arizona, iron miners from Nevada and Minnesota, gold and silver miners from South Dakota, uranium miners from Wyoming, coal miners from North Dakota and Utah, and molybdenum miners from Colorado. There are farmers, ranchers, oilfield hands, teachers, lumberjacks, secretaries, and housewives. All this combines to create exceptional situations.

Basin Description

The Powder River Basin is a structural and topographic basin that covers more than 31 079 km² (12,000 sq miles) of northeastern Wyoming and southern Montana. It is bounded by the Black Hills to the east, the Bighorn Mountains to the west, the Hartville Uplift to the south, and the Missouri River breaks to the north. A more thorough description of the basin and its coal can be found in "Update on the Powder River Coal Basin" (Glass, 1976). The basin contains an estimated 544 to 635 Gt (600 to 700 billion st) of coal.

The geological conditions encountered in the Powder River Basin are quite variable. The coalbeds being mined range from 3 to 67 m (10 to 220 ft) thick. Some mines operate in a single seam while others mine multiple seams along with the intervening rocks. Often splits and partings are encountered. Dips vary from horizontal to 10° and intense local rolling is very common. Coal quality is wide ranging thus requiring diverse approaches to scalping and ultimately to mining itself.

Water inflow, both surface and subsurface, typically affects every mine in the basin to some degree. Powder River Basin mines tend to be of large areal extent, hence, they often intercept numerous local drainages and sometimes major stream tributaries. Also, the overburden, underburden, and the coal are usually all aquifers with differing hydraulic properties that are likely to change with both location and weather conditions. It should be noted that pit water not only affects blasting, excavation, and haulage, but also highwall stability.

Often the full range of physical conditions can be found in a single mine. Environmental and regulatory constraints and the sundry responses thereto combine to further compound the diversity. Most of the companies represented in the basin operate or plan to operate at least two mines and it is interesting that no company operates both its mines in the same manner.

Mining Methods

The manner in which coal mining has been approached in the Powder River Basin is as discordant as the physical conditions existing in the mines. The only common denominator in Powder River Basin coal mining is that the vast majority of the operators utilize a truck-shovel combination to some extent. Besides the standard truck-shovel combinations, draglines, hydraulic excavators—both shovel and backhoe configurations—loaders, scrapers, Easy Miners, and one bucket wheel excavator are in use. Typically, this non-shovel equipment is used in conjunction with haul trucks.

One operator uses a dragline to recover the lower portion of the coal seam with the dragline sitting on top of the coal loads directly into haul trucks. This arrangement avoids operating any equipment on the extremely soft underburden. This, in turn, lowers the operating costs by eliminating excessive shovel setup and cleanup costs, additional haulroad construction and maintenance, and considerable wear and tear on the haul fleet.

Another operator uses *Easy Miners* in conjunction with truck haulage. This mine has extremely shallow overburden and consequently uses only scrapers and dozers for overburden removal. The coal is excavated by *Easy Miners* and loaded directly into trucks. Front end loaders are used as backup coal excavators. The *Easy Miner* configuration generates the potential for a continuous mining operation with the only restrictions being storage capacity for mined coal and the availability of trucks.

One bucket wheel excavator is presently in operation in the Powder River Basin and has been used to mine both coal and overburden as well as scoria for mine road surfacing material. In all of these applications it was used to load haul trucks. To date, this particular machine has largely been a failure and may, at least in part, be responsible for the poor reception that bucket wheels have had in America. However, it appears that its failure may well have been a result of misapplication by using too light a machine for heavy-duty work. Recently, however, this same unit has been more than adequate in excavating unconsolidated sand overburden and well shot coal.

As can be seen from the foregoing descriptions, the non-shovel excavators are typically used as shovel substitutes to load haul trucks. This is not to imply that draglines are never used in their original design application—to side cast overburden. It simply indicates a tendency in the basin to pattern all operations after the standard truck-shovel operations.

EQUIPMENT SELECTION PHILOSOPHY

General

In selecting mining equipment, the total possible combinations of equipment are many. The engineer should try to arrive at the best combination, even though that combination might be less than perfect. The engineer will need to exercise his best judgment in deciding when to abandon the pursuit of perfection in favor of advancing the cause of the best workable compromise.

Bias is often a major factor in engineering analysis. Opinions of how specific equipment will function in the operation may or may not be based on fact. Occasionally, the information used to judge performance in the past may not be applicable in the present review. What worked under one set of conditions may not work under different conditions and vice versa. The engineer should strive to eliminate bias from the analysis—his own, his coworkers', his supervisor's, the manufacturer's—to produce a totally objective result. The best analysis will be the most accurate and honest one.

In a similar vein, even the most casual observer will notice that a piece of equipment liked by its operator will be more productive than a piece of equipment that is not liked. The actuality of this statement exists regardless of what the verifying numbers might indicate. No one has developed a simulation method that accounts for operator bias. If the operator does not like it, the unit will not work, and may experience an inordinate amount of downtime. Often, an operator's dislike for the equipment to which he is assigned stems from nothing more than the fact that the new unit differs from what he is used to. Here the engineer should include in his alternatives statements comparing the selection of lesser equipment to take advantage of operator preference with the selection of the optimum equipment, combined with a *selling* of the optimum to the operator.

There are occasions when management will choose something different from what the engineer's analysis indicated as the optimum selection. Management may have had access to additional or better information. They might even be making what seems to be an error. The engineer must remember that he is part of the management team, therefore, he might have to compromise his personal belief of what the best selection is in order to give management what that team feels it wants and needs. Although the engineer has a duty to provide his management team with the best advice and counsel, he also has a duty to the team and its decisions. At that point, he should make his feelings known—stating his position, all the available facts, and the reasons for his recommendation without belaboring any of them—and then support the team's decision. This can be a tremendous blow to one's idealism, but it is a necessary part of an engineer's development. Every engineer, at some point in his career, experiences this frustration. He may have produced his best analysis ever. He may even have produced the best analysis that has ever been done by anyone. Yet he finds his recommendations not being followed. It is a bitter pill to swallow. The engineer should remember three things above all others:

1) He is not the first person that this has happened to nor will he likely be the last. He is not alone in his frustration.

2) The fact that the results of an engineer's best work are not followed is in no way a reflection on him or his abilities.

3) Even the best study can be outweighed by someone else's 20 or more years of experience.

Keeping these things in mind will help the engineer to avoid bad feelings that can last for years and, ultimately, will aid him in his career advancement. Future experience will prove that teamwork is more important than an individual effort.

Generic Considerations

Before someone can select equipment, he will need to know what equipment is available and who the suppliers are. Likewise, in order to apply the proper equipment to given conditions, some understanding of those conditions and actual mining procedures must be garnered. Therefore, the mining engineer should make every effort to stay abreast of the developments in his profession. Professional journals, conventions, and seminars provide excellent base data. Contacts with manufacturers and their representatives will generate more specialized information. The purchasing department can be a good source of information. Purchasing departments should circulate brochures on new equipment and even set up meetings with manufacturers for presentations on new products and developments.

The best information is, of course, hands-on experience. Enlightened companies provide ample opportunities for engineers to serve as front-line supervisors. As such, the engineer will be able to try some of his own ideas in the field and experience first hand the problems that are inherent in production. If the opportunity for hands-on experience is not available, the engineer should try to glean all the information he can as an observer. He should visit as many different operations as possible, both in his company and elsewhere. This is an easily accomplished task in the Powder River Basin where there are 20 mines within a 161-km (100-mile) radius of Gillette, WY. Interviews with supervisors, operators, and maintenance personnel will give insight into good points and bad points, likes and dislikes, and specific uses and limitations of various equipment types. Here again, the engineer should be aware of possible biases. As an observer, the engineer would have flexibility to determine the reasons for delays and downtime. He should time-study every aspect of the individual operations, paying special attention to ways to streamline procedures, methods for eliminating bottlenecks and delays, and to cost-cutting potential. As the engineer observes new and different techniques, he should remember that being different does not make something either good or bad. Furthermore, something that he observes as being less than ideal may be an outstanding producer when used in other combinations or other applications. Lastly, the engineer should try to determine if conditions exist at other mines in the immediate area that require special mining techniques. Those same conditions could exist at his mine and may well make an otherwise ideal piece of equipment useless.

Specific Considerations

The selection of equipment for a particular operation can be broken down into five steps:
1) Look at the property.
2) Look at the available equipment.
3) Generate a mining plan matching equipment to the property.
4) Develop alternative plans.
5) Pick the optimum alternative.

Looking at the property should include consideration of the size and configuration, reserves, production levels, material properties, regulatory constraints, and the possibility of expanded operations in the future. A general rule of thumb in the Powder River Basin has been that big production requires big equipment. Therefore, mines there have tended to try to field the largest equipment available. The results of this effort have largely been negative. Haul road materials lack the bearing capacity to carry the increased wheel loads. This results in bumpy roads, lower speeds, increased maintenance, lost production, and increased costs. Slightly smaller equipment with wheel loads that do not exceed the bearing capacity of the road materials may well alleviate problems while actually boosting production and profits. On the other hand, building better roads could also be a possible alternative. Another important aspect of equipment selection in the Powder River Basin is the ability to handle materials selectively. Overburden of specific sand content is often selectively placed in the backfill in order to re-establish aquifers and reconstruct stream channels. Likewise, boulders are stockpiled for later use as riprap or rabbit habitat. This special handling sometimes requires special equipment.

Generating a plan that matches equipment to the property should include matching equipment to equipment. There has been a tendency of late to mismatch equipment. A good example of this is trying to load big trucks with small front end loaders. It can be done, but it is not very efficient. Another consideration is the problem of scale with regard to maintenance. Big equipment requires big tools to work on it and big facilities to house it. Furthermore, when big equipment breaks down in the mine, it must be repaired where it sits. This is an item of major concern in the Powder River Basin where winters can be very long and harsh.

The engineer will find that his attempts to reconcile the possible equipment combinations with all the possible conditions that can be encountered to be a monumental task. The best way to handle all the possible alternatives is through the development of production parameters. Production parameters are a quantitative tool for describing equipment capabilities. Typically, they are based on distance, time, speed, and capacity. Each make and model is so described. The interaction of various types of equipment is then studied to determine an optimum mining rate that agrees with the physical conditions at the mine. This is usually done through computer simulations. Units are matched that maintain the rate. Costs are then applied to further optimize the selection.

Each of us tends to see things in a specific manner. Our own biases get in the way of our creativity. The development of alternative plans forces one to look at things with a different point of view and with differing perspectives. Whereas we have a tendency to be protective of our ideas, it causes us to view other approaches more open mindedly. Quite often, a good plan can be made even better through the incorporation of aspects that by themselves seem less than adequate. Furthermore, working through an alternative solution will make a much more vivid impression than just casually thinking about it. Compromises will appear much more attractive.

It is important to bear in mind that tradeoffs often exist between capital and operating costs. A classic example of this exists in the Powder River Basin—the capital-intensive operating philosophy vs. the labor-intensive operating philosophy. Labor-intensive operations are characterized by comparatively larger workforces, higher equipment utilization, and lower capital costs. There is only one piece of equipment for each operator with only enough spare units to account for availability. Typically, the labor-intensive operations use some type of multiple-use truck to haul both coal and overburden as needed. Trucks can switch from dirt to coal, or vice versa without missing a load, thus providing flexibility. The capital-intensive operations are characterized by smaller workforces, lower equipment utilization, and higher capital costs. There are usually two or more pieces of equipment for each operator. Here flexibility is sacrificed in order to optimize specific functions of the operation. The capital-intensive operations typically run two distinct truck fleets, one for coal and one for overburden. The coal trucks are usually a large volume, bottom-dump truck featuring rapid discharge. This gives a larger payload and faster cycle times. Switching from coal to overburden necessitates changing trucks. Using a dirt truck on the coal haul sacrifices the larger payload and, hence, efficiency. Again, the choice between capital-intensive and labor-intensive operating philos-

ophies is one between optimizing specific mining functions and flexibility.

CAPITAL COSTS

Capital costs are those arising from the purchase of goods that are in turn used for production: buildings, plants, machinery, heavy equipment, etc. The only way to avoid capital costs is to not buy anything. No matter how small an operation might be, it will incur some capital expenditures. The objective, then, is to maximize the return on the investment. This return is a function of the interplay between both capital and operating costs. There are ways to reduce a mine's capitalization. However, they all will involve some sort of compromise.

Big Equipment vs. Small

The purchase of comparatively smaller equipment does effect a savings in capital cost. Small equipment, by virtue of its size, uses less iron and, hence, costs less. Bigger equipment is heavier, requires a larger engine, and burns more fuel. It also requires larger tires, maintenance facilities, and tools. On an hourly basis smaller equipment costs less to own and less to run. However, when considered on a production basis, the analysis becomes much more complex. Large equipment carries a bigger payload. Small equipment can cycle faster and thus carry more loads per shift. At this point, an analysis comparing specific equipment is necessary. Considerations should include:

1) What is the specific application?
2) Can the unit in question accomplish the task?
3) Are qualified operators available to operate the unit? Could some be trained? What would this training cost?
4) What is the total cost—both capital and operating—on a production (per ton) basis?
5) How will the unit be utilized in slack periods?
6) Which unit will be most beneficial in all respects?

An interesting analysis along these lines is presented by Kaufman and Bowen (1981).

Delayed Purchase

The concept of delayed purchase is based on the theory of time value of money, whereby a unit of money is seen to have more value today than it will in the future. Placing a property into production requires the expenditure of capital. Does this mean that all new projects should be delayed to effect a savings of capital? No, because money earned today has more value than money earned in the future when viewed in a similar light. The object, then, is to spend capital in such a manner as to maximize the return on the investment.

One method of achieving this goal is to mine on an increasing ratio strategy. Stripping ratio is defined as the ratio of the specific yards of overburden needed to be removed to the specific tons of coal production facilitated by the removal of the overburden. Now, by mining areas of low overburden first and gradually progressing into areas of thicker overburden, the stripping ratio will gradually increase, provided the thickness of the coal remains constant. This allows a minimum acquisition of equipment to take place initially. Furthermore, additional equipment is required bit by bit over a period of time. Ideally, a minimum of capital is being expended while production generates income at an optimum rate.

Another approach is to build part of a facility or develop part of a mine and then expand it later. Modern mining equipment is becoming increasingly larger and more complex. Extended periods of time are required for its assembly at the mine site. Experienced operators are unavailable; therefore, extensive training programs are required. Starting a project on a small scale provides several benefits. First, there is no temptation to attempt to assemble a large fleet of equipment all at the same time. Second, it facilitates training by making it possible to deal with new operators in small groups and on an individual basis. Third, it delays some of the capital spending until some time in the future while freeing up capital for investment in other projects now. This approach provides the opportunity for diversification while still furnishing cash flow. Furthermore, it allows, at least to a limited extent, a sort of trial use of equipment before commitment of the entire investment. Lastly, it makes problems more manageable— it permits them to be examined and solved while they are small and easy to handle rather than after the project grows to full size and complexity.

Elimination of Peak Equipment Requirements

Quite often, variable overburden thicknesses will require large volumes of overburden to be moved for relatively short durations of time. Similarly, periods of reduced overburden movement will also occur. This happens in the Powder River Basin even though coal production remains constant as the result of undulating surface topography. The typical mining plan analysis responds to these production changes by calling for adding and deleting equipment and personnel as production demands change. If equipment were to be acquired on the basis of this type of plan, periods of extremely high capital expenditure would be encountered. Likewise, after the production peak has passed, the excess equipment and personnel needed to handle the peak would be idled or operate at a reduced rate. Neither of these conditions is desirable. A better way to deal with variable production rates is to design so that a production level is maintained that averages the peaks and valleys. The overburden benches advance at a relatively higher rate prior to approaching areas of thicker overburden and the bench configuration expands. Upon entering the area of thicker overburden, the bench advance slows down and the benches begin to stack. As the area is passed, once again the rate of bench advance increases and the configuration expands. Acquiring the necessary equipment to attain the required production plateau may cause capital to be expended slightly sooner than otherwise desired. However, this avoids having to purchase excess equipment and then idling it later. Furthermore, adequate planning will insure that the early capital acquisition will be much less than that required for peak conditions.

Work Scheduling

The schedule upon which a mine operates can have a great impact on its capital equipment requirements. Some mines in the Powder River Basin opt to provide their workforces with weekends off. This five-days-a-week plan is a benefit to employees and also keeps overtime to a minimum. However, it does not make full utilization of equipment. Other mines schedule complete operations on seven days a week to take advantage of the total capacity of their equipment. Still other mines follow schedules somewhere in between. If two identical mines with the same size equipment are producing the same tonnage and one operates five days a week while the other operates seven days a week, the seven-days-a-week mine will be able to meet its production requirements with less equipment.

Operating schedules can be variable, from 5 days worked and 2 days off on straight shifts to 7 and 2, 7 and 2, 7 and 3 on rotating shifts even to 28 and 2. Also, they can be flexible on a daily basis. An overlapping system where the shifts actually change on the equipment instead of at the change room can be very beneficial. Not only would this eliminate some of the lost production time associated with shift changes, but it would also improve turnovers with respect to the peculiarities of the individual equipment and the exchange of information on conditions in the mine. Tradeoffs exist involving employee attitudes and productivity and between capital and operating costs. The critical aspect is to get the maximum benefit for the money spent.

Contracting the Work

It may be possible to eliminate capital equipment by contracting various phases of the work. Where allowable under local labor agreements, contracting can remove the need for entire fleets of equipment. Furthermore, it excludes having to provide operators, maintenance, parts, and supervision for that part of the operation. In addition, payment is only for work that is actually done. There is no charge for idle or slack time. Finally, all the work is guaranteed— there is no payment until the work is right. An added bonus is the potential bargains that competitive bidding can produce. In the recent past, one Powder River Basin operator was having his topsoil handling done under contract for 40¢ less per yard than another operator could do it with his own equipment.

Warehousing

The items sitting on mine warehouse shelves can be a tremendous sink of expended capital. Parts for equipment no longer in service can sit there forgotten. Thus they create a demand for additional storage space that is not really needed. Parts for existing equipment that will never be needed could collect dust forever. Each represents money spent. No effort is made to delete unnecessary items from inventory and return them for credit. It should be noted that, in most cases, vendors charge a restocking fee of between 10 to 25% if parts are returned for credit. Here again a tradeoff exists.

Often, when a company purchases an altogether new piece of equipment, they will request a "list of recommended spares" from the manufacturer. All too often, every item on the list is stocked in the company's warehouse. These lists are meant to serve as a guide. Typically they show each individual component item as well as assemblies. Stocking the entire list, then, results in duplication of supplies. Supplier's warehouses have helped alleviate the problem somewhat. Prior to making an equipment purchase, it may be possible for the buyer to negotiate with the probable vendors to have the vendor agree to stock spare parts and supplies and specific quantities thereof. This would decrease capital spending for both supplies and warehouse space. However, breakdowns on off-shifts and the delays caused by having to wait for parts still generate a need for on-site warehousing. What is needed is a thorough analysis of each specific type of machine's parts consumption. While some parts are consumed on an almost daily basis, others are virtually never needed. A probability analysis would indicate which parts and how many thereof would be kept on site. Less frequently used items would be left to the supplier's warehouse. The warehouse stocking policy should be modified as needed on the basis of experience and the ease of obtaining nonstocked parts. The deciding factor should be economic expedience, that which generates the least ultimate cost.

Specific Mine Costs

As indicated previously, the Powder River Basin mines are typically truck-shovel operations. Total capitalization on a per mine basis ranges from $50 to $200 million. The shovels most commonly found are P&Hs 2300 and 2800 and Bucyrus

Erie's 295B. The most common trucks are Unit Rig's M-120-17 and MK-36 and WABCO's 120C and 170C Haulpacs. Approximate capital costs for shovels have been ranging from $3 to $5 million each and are about $700,000 each for trucks. Capitalization for a Powder River Basin mine's physical plant can be quite variable depending upon office architecture and decor and the size of the equipment maintenance facilities. Capital costs for coal handling facilities vary with the type of storage and length of the rail loop. Silos cost around $335 per ton of storage and slot storage systems are about $230 per ton of storage. Unit train railroad loops cost about $90 0.3m (per ft) of length while weigh-in-motion track scale systems are around $900,000.

An additional source of capital expense is the mine permitting process. A new mine in Wyoming would probably need 150 individual authorizations including: an industrial siting permit; a land quality permit; an environmental impact statement; state engineer's authorizations; water quality permits; a solid waste permit; an air quality permit; and a permit from the Federal Bureau of Alcohol, Tobacco, and Firearms. Each individual water well, whether it be for mine water usage or just for monitoring purposes, requires a separate permit both from the state engineer's office and the Water Quality Div. of the Dept. of Environmental Quality. The air quality permit alone can take two years to acquire data and model potential dust emissions from the proposed mine prior to initial construction.

The most exhaustive of these permits is the land quality permit. It is divided into four parts as follows:

1. *The Adjudication Section*—A description of who owns what and copies of surface owner consents for areas both in the actual mine and adjacent thereto that will be overstripped or otherwise affected by adjacent mining.

2. *The Land Description Section*—A complete description of the land in its premining condition, its past uses, its present uses, and the current condition of all resources (soils, vegetation, wildlife, etc.) of the area to be mined. The heart of Wyoming's Environmental Quality Act is the requirement that all land disturbed by mining be reclaimed to a use equal to or greater than its premining use. Therefore information must be presented to answer every conceivable potential question.

3. *The Mine Plan Section*—A description of how the operation will affect resources both on and off the site. It details all activities associated with the extraction of coal and the movement of overburden and all activities that will be undertaken to protect, preserve, and minimize the impact on the environment.

4. *The Reclamation Plan Section*—An explanation of how the environmental resources of the site will be restored, reconstructed, or enhanced. This is done by identifying the post-mining land use and detailing its condition and how that condition will be achieved.

Now, what does it take to satisfy these requirements? Estimates for Wyoming's Powder River Basin mines indicate that a complete permit application requires the full-time efforts of six environmental scientists, two mining engineers, one attorney, two draftsmen, and two secretaries. In addition, up to four part-time temporary employees may be needed for proofreading, editing, folding, assembling, etc., of the physical data. This translates to in excess of 100,000 hr of labor and can range in cost from $300 to $800 per acre of land affected by mining. It should be noted that the permit application for the first major mine operating in Powder River Basin contained 2700 pages of text and 150 maps. The full-time contributors are still engaged in monitoring activities, updating and amending the permit, reporting on activities, and insuring that all aspects of the operation comply with the requirements of the permit. Also, the Land Quality permit must be renewed every five years. Therefore, almost as soon as work is completed on one, it must start again for the next submittal.

OPERATIONAL COSTS

Operational costs are those that vary with the rate of production and other conditions. They include the cost of labor, raw materials, fuel, power, and overhead. They are the costs most easily controlled by eliminating waste and inefficiency. In the Powder River Basin, these costs range from $2 to $5 per ton of coal produced. Where an individual mine's operating costs actually fall in this range will depend upon its production level and how the costs are viewed. That is, one mine may view a specific cost as a part of its capital expense while another mine will view it as an operating expense.

Cost Functions

Operational costs are usually broken down into or tracked on the basis of the functions actually being performed. The entire mining operation is divided up into separate distinct functions. The costs associated therewith are grouped and subtotaled in a similar fashion. The functional breakdown can be made as simple or complex as desired. Additional complexity maybe desirable as a means of tracking cost pass-throughs and as an aid in achieving cost effectiveness. Generally speaking, complexity beyond that which provides a direct benefit should be avoided. Increased complexity is typically achieved by a further breakdown of a more simple system. For example, a Powder River Basin mining operation could be divided on the basis of environmental, coal, overburden, and, where applicable, interburden functions. Each of these functions can be further divided, as shown in Table 1. Charges based on individual operating units can fall under headings in the table. That is, the cost of operating labor on a specific coal shovel would be subtotaled under III.B.1.a. or the cost of repair supplies for a specific dirt truck would be subtotaled under II.C.2.b.

Typically, the data collecting and storing process has been computerized. The shift foreman's daily time sheets or employees' time cards are the initial source of data for tracking functional costs. It would also be possible to use equipment cards or reports for this purpose. In fact, equipment cards could prove to be the best source of information as they also indicate time lost due to delays. Fig. 1 shows an explanation of one method of encoding the data for computer input. Again, the information being entered can be made more or less complex as desired. The first five columns provide a general description of an employee's daily work. The last four columns specify exact activities and the accounts to which they should be charged. Each activity or change requires an entire line to describe it. The total of the hours column for all the lines should equal the employee's total time for the day. Fig. 2 shows an equipment list with the respective codes to be entered in column 8 in Fig. 1. Fig. 2 also shows codes for miscellaneous labor not associated with specific pieces of equipment and codes for special accounts. It should be noted that a special account code replaces all of columns 6, 7, 8, and 9 without any portion of it being assigned to any specific column. In this respect it acts as a flag. Under this coding plan a spotter on an overburden dump would be shown as AC FA AAC.JJ B. A truck driver hauling overburden in truck 237 would be shown as AC AA

Table 1. Breakdown of Operational Costs by Function

I. Environmental
  A. Premining
    1. Access Roads
    2. Sediment Control
    3. Drainage
    4. Topsoiling
      a. Excavating
      b. Haulage
      c. Stockpiling
  B. Post Mining
    1. Access Roads
    2. Sediment Control
    3. Drainage
    4. Topsoiling
      a. Recovery
      b. Haulage
      c. Spreading and Contouring
    5. Seeding
    6. Fencing
    7. Monuments
II. Overburden
  A. Preparation
    1. Drilling
      a. Labor
        (1) Operating
        (2) Repair
      b. Supplies
        (1) Operating
        (2) Repair
      c. Power
    2. Blasting
      a. Labor
        (1) Operating
        (2) Repair
      b. Supplies
        (1) Operating
        (2) Repair
  B. Loading (excavating)
    1. Labor
      a. Operating
      b. Repair
    2. Supplies
      a. Operating
      b. Repair
    3. Power
  C. Haulage
    1. Labor
      a. Operating
      b. Repair
    2. Supplies
      a. Operating
      b. Repair
  D. Roads
    1. Labor
    2. Supplies
III. Coal
  A. Preparation
    1. Drilling
      a. Labor
        (1) Operating
        (2) Repair

      b. Supplies
        (1) Operating
        (2) Repair
      c. Power
    2. Blasting
      a. Labor
        (1) Operating
        (2) Repair
      b. Supplies
        (1) Operating
        (2) Repair
  B. Loading (excavating)
    1. Labor
      a. Operating
      b. Repair
    2. Supplies
      a. Operating
      b. Repair
    3. Power
  C. Haulage
    1. Labor
      a. Operating
      b. Repair
    2. Supplies
      a. Operating
      b. Repair
  D. Roads
    1. Labor
    2. Supplies
  E. Receiving
    1. Labor
      a. Operating
      b. Repair
    2. Supplies
      a. Operating
      b. Repair
    3. Power
  F. Processing
    1. Labor
      a. Operating
      b. Repair
    2. Supplies
      a. Operating
      b. Repair
    3. Power
  G. Storage
    1. Labor
      a. Operating
      b. Repair
    2. Supplies
      a. Operating
      b. Repair
    3. Power
  H. Loading (of unit trains)
    1. Labor
      a. Operating
      b. Repair
    2. Supplies
      a. Operating
      b. Repair
    3. Power

IV. Interburden
  A. Preparation
    1. Drilling
      a. Labor
        (1) Operating
        (2) Repair
      b. Supplies
        (1) Operating
        (2) Repair
      c. Power
    2. Blasting
      a. Labor
        (1) Operating
        (2) Repair
      b. Supplies
        (1) Operating
        (2) Repair
  B. Loading (excavating)
    1. Labor
      a. Operating
      b. Repair
    2. Supplies
      a. Operating
      b. Repair
    3. Power
  C. Haulage
    1. Labor
      a. Operating
      b. Repair
    2. Supplies
      a. Operating
      b. Repair
  D. Roads
    1. Labor
    2. Supplies
V. Miscellaneous
  A. Lighting
    1. Labor
      a. Operating
      b. Repair
    2. Supplies
      a. Operating
      b. Repair
    3. Power
  B. Pumping
    1. Labor
      a. Operating
      b. Repair
    2. Supplies
      a. Operating
      b. Repair
    3. Power
  G. General
    1. Labor
      a. Operating
      b. Repair
    2. Supplies
      a. Operating
      b. Repair
    3. Power

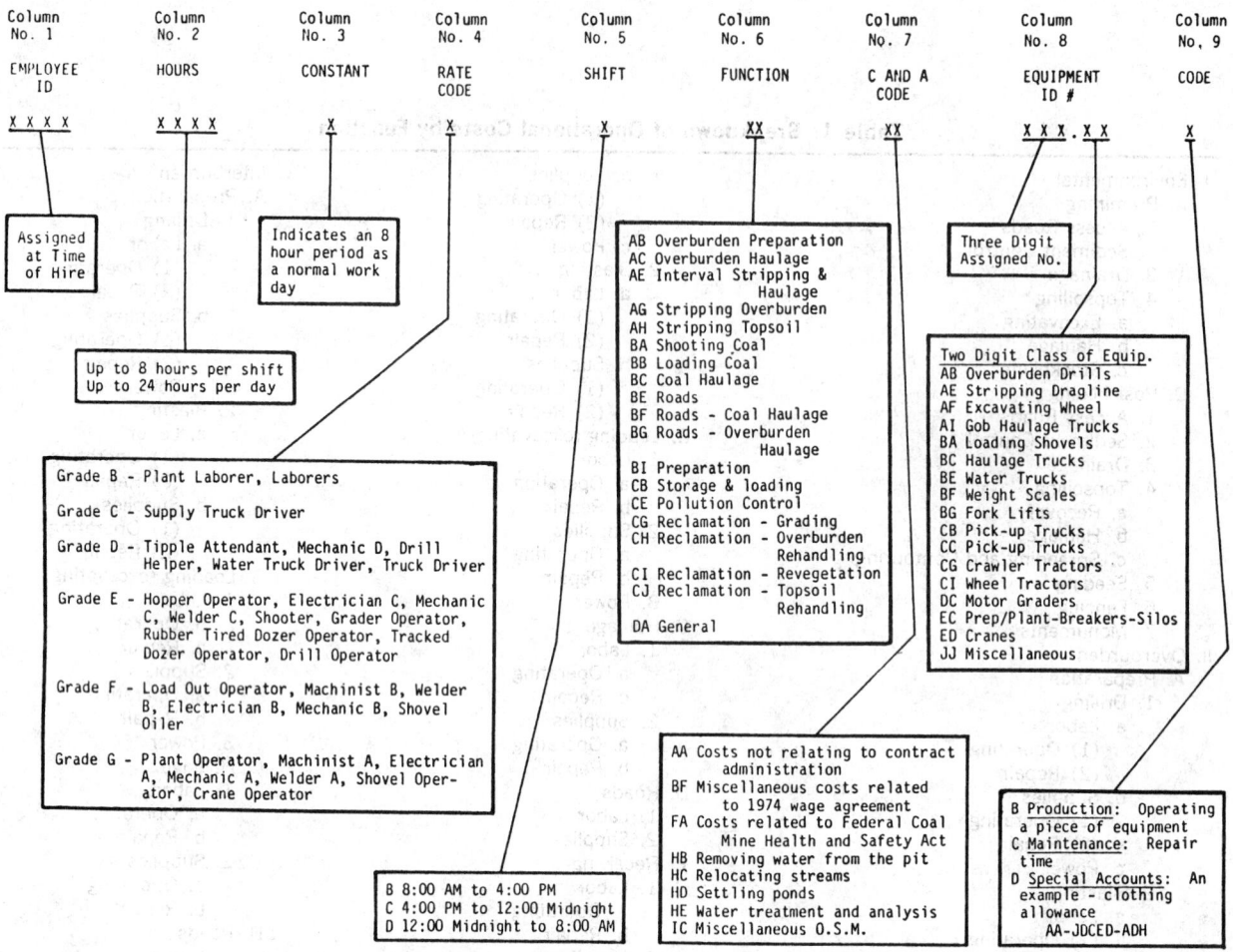

Column No. 1	Column No. 2	Column No. 3	Column No. 4	Column No. 5	Column No. 6	Column No. 7	Column No. 8	Column No. 9
EMPLOYEE ID	HOURS	CONSTANT	RATE CODE	SHIFT	FUNCTION	C AND A CODE	EQUIPMENT ID #	CODE
X X X X	X X X X	X	X	X	X X	X X	X X X . X X	X

Assigned at Time of Hire

Indicates an 8 hour period as a normal work day

Up to 8 hours per shift
Up to 24 hours per day

Grade B - Plant Laborer, Laborers

Grade C - Supply Truck Driver

Grade D - Tipple Attendant, Mechanic D, Drill Helper, Water Truck Driver, Truck Driver

Grade E - Hopper Operator, Electrician C, Mechanic C, Welder C, Shooter, Grader Operator, Rubber Tired Dozer Operator, Tracked Dozer Operator, Drill Operator

Grade F - Load Out Operator, Machinist B, Welder B, Electrician B, Mechanic B, Shovel Oiler

Grade G - Plant Operator, Machinist A, Electrician A, Mechanic A, Welder A, Shovel Operator, Crane Operator

AB Overburden Preparation
AC Overburden Haulage
AE Interval Stripping & Haulage
AG Stripping Overburden
AH Stripping Topsoil
BA Shooting Coal
BB Loading Coal
BC Coal Haulage
BE Roads
BF Roads - Coal Haulage
BG Roads - Overburden Haulage
BI Preparation
CB Storage & loading
CE Pollution Control
CG Reclamation - Grading
CH Reclamation - Overburden Rehandling
CI Reclamation - Revegetation
CJ Reclamation - Topsoil Rehandling
DA General

Three Digit Assigned No.

Two Digit Class of Equip.
AB Overburden Drills
AE Stripping Dragline
AF Excavating Wheel
AI Gob Haulage Trucks
BA Loading Shovels
BC Haulage Trucks
BE Water Trucks
BF Weight Scales
BG Fork Lifts
CB Pick-up Trucks
CC Pick-up Trucks
CG Crawler Tractors
CI Wheel Tractors
DC Motor Graders
EC Prep/Plant-Breakers-Silos
ED Cranes
JJ Miscellaneous

AA Costs not relating to contract administration
BF Miscellaneous costs related to 1974 wage agreement
FA Costs related to Federal Coal Mine Health and Safety Act
HB Removing water from the pit
HC Relocating streams
HD Settling ponds
HE Water treatment and analysis
IC Miscellaneous O.S.M.

B Production: Operating a piece of equipment
C Maintenance: Repair time
D Special Accounts: An example - clothing allowance
AA-JDCED-ADH

B 8:00 AM to 4:00 PM
C 4:00 PM to 12:00 Midnight
D 12:00 Midnight to 8:00 AM

Fig. 1. Explanation of computer coding for functional costs.

CDH.BC B. Portions of the miscellaneous labor code numbers can also be combined to provide very specific descriptions of the work being done. For instance, a member of the pump crew who is repairing a pump used to remove water from an area of an overburden dump could be shown as AC HB ACE.JJ C. In these examples dumps are viewed as being part of the overburden haulage function. However, it would also be possible to add a separate item to the list in Fig. 2 specifically to designate work done on dumps.

With the system outlined previously, it is possible to examine and, hopefully, improve the operating cost picture for every aspect of a mining operation. The functional cost breakdown combined with data from operators' and maintenance reports (load count, delay times, type of delay, mileage, repair time, item being repaired or replaced, etc.) facilitate a very thorough analysis of operating costs. For example, the maintenance reports for a particular shift showing several broken haul truck hoist cylinders would be a definite indication of overloading. This type of cost analysis, combined with an enthusiastic, progressive approach to mining, allows a company to cut costs, improve benefits, and, at the same time, increase its profitability.

Efficiency Considerations

Good supervision can have a tremendous effect on decreasing operational costs. One of the most effective ways to reduce operating costs is by improved shovel staffing. Fig. 3 gives an indication of how a shovel's productivity is affected as trucks are added to the haul. Each additional truck increases the shovel's production rate until the maximum rate for the shovel is reached. In this example the maximum rate is reached when the sixth truck is added. After that, adding more trucks does not increase production. Each additional truck displaces another truck in the cycle and only adds to the queue. Theoretically, shovel productivity then decreases as more trucks increase the congestion around the shovel and interfere with the cleanup machine. Fig 4 shows the effect on truck productivity of adding trucks to a single shovel. Here the productivity per truck remains constant until the capacity of the shovel is exceeded and then decreases with each additional truck. This is where good supervision comes into play. If the work force is of sufficient size so as to permit the start-up of an additional shovel through judicious use of personnel, the supervisor can effect an increase in production while actually decreasing the operating cost.

Fig. 5 again shows how shovel productivity varies with increasing numbers of trucks. In this case a second shovel is started after the sixth truck is assigned to the first shovel. The corresponding plot of truck productivity is given on Fig. 6. Fig. 7 shows the cost comparison on a unit basis for the two cases. Here the direct financial benefit of adding a shovel is readily apparent. Another way to look at this is on the basis of how production increases at a very minimal increase in total cost. Fig. 8 shows a comparison of how total cost

Fig. 2

O. B. Drill	AAB.AB	Welding Truck	BCC.BI
O. B. Drill	AAC.AB	Welding Truck	BCD.BI
Coal Drill	AAD.AB	Welding Winch Truck	ACE.BI
Pump Truck	AAB.AD	Electrician's Truck	BCE.BI
Pump Truck	AAC.AD	Electrician's Truck	BCF.BI
BWE	AAB.AF	Powder Truck	ABC.BJ
Small Backhoe	AAB.AI	Powder Truck	ABD.BJ
Large Backhoe	AAC.AI	Powder Truck	CFC.BJ
Shovel #1	AAB.BA	Anfo Truck	CGH.BJ
Shovel #2	AAC.BA	Plant Winch Truck	CFD.CC
Shovel #3	AAD.BA	Supply Truck	JCD.CC
Shovel #4	AAE.BA	Supply Truck	BAD.CD
Shovel #5	AAF.BA	Crew Van	AGA.CD
Shovel #6	AAG.BA	Crew Van	AGB.CD
Shovel #7	AAH.BA	Crew Van	AGC.CD
Hydraulic Shovel	AAI.BA	Crew Van	DBB.CD
Dirt Truck	BHB.BC	Crew Van	DBC.CD
Dirt Truck	BHC.BC	Track Dozer	DAJ.CG
Dirt Truck	BHD.BC	Track Dozer	DJH.CG
Combination Truck	BHE.BC	Track Dozer	ECG.CG
Dirt Truck	BHI.BC	Track Dozer	EGE.CG
Dirt Truck	BHJ.BC	Hopper Machine	AIH.CI
Dirt Truck	BIA.BC	RT Dozer	AHE.CI
Dirt Truck	BIB.BC	RT Dozer	AII.CI
Dirt Truck	BID.BC	RT Loader	AIJ.CI
Dirt Truck	BIE.BC	RT Dozer	AJA.CI
Dirt Truck	BIF.BC	RT Dozer	AJH.CI
Dirt Truck	BIG.BC	RT Dozer	AJI.CI
Dirt Truck	BIJ.BC	Skid Steer Loader	BAC.CI
Dirt Truck	BJA.BC	Small RT Loader	BAJ.CI
Dirt Truck	BJB.BC	RT Dozer	BBC.CI
Dirt Truck	BJC.BC	RT Loader	ADE.DA
Coal Truck	CCE.BC	Motor Grader	ADC.DC
Coal Truck	CCF.BC	Motor Grader	ADI.DC
Coal Truck	CCG.BC	Motor Grader	ADJ.DC
Coal Truck	CCH.BC	Motor Grader	AGB.DC
Dirt Truck	CDB.BC	Plant #1 Scales	CB AA AAH.BF
Dirt Truck	CDC.BC	Plant #2 Scales	CB AA AAI.BF
Dirt Truck	CDD.BC	Old Scales	CB AA AAJ.BF
Dirt Truck	CDE.BC	New Light Scales	CB AA ABA.BF
Dirt Truck	CDG.BC	New Heavy Scales	CB AA ABB.BF
Dirt Truck	CDH.BC	Preparation Plant #1	BJ AA ABC.EC
Coal Truck	CDI.BC	Preparation Plant #2	BJ AA ABJ.EC
Coal Truck	CDJ.BC	Load Outs #1 & #2	CB AA ABG.EC
Coal Truck	CEA.BC	Load Outs #3 & #4	CB AA ABH.EC
Coal Truck	CEB.BC	Misc. O.B. Prep.	AB AA AAB.JJ
Dirt Truck	CED.BC	Misc. O.B. Haulage	AC AA AAC.JJ
Dirt Truck	CEE.BC	Misc. O.B. Stripping	AG AA AAG.JJ
Dirt Truck	CEF.BC	Misc. Topsoil Stripping	AH AA AAH.JJ
Combination Truck	CEG.BC	Misc. Coal Prep.	BA AA ABA.JJ
Combination Truck	CEH.BC	Misc. Coal Loading	BB AA ABB.JJ
Combination Truck	CEI.BC	Misc. Coal Haulage	BC AA ABC.JJ
Coal Truck	CGA.BC	Misc. Roads (Others)	BE AA ABE.JJ
Coal Truck	CGB.BC	Misc. Roads (Coal)	BF AA ABF.JJ
Coal Truck	CGC.BC	Light Plants	BF FA ABF.JJ
Coal Truck	CGD.BC	Misc. Roads (O.B.)	BG AA ABG.JJ
Water Truck	ADA.BE	Misc. Coal Processing	BJ AA ABJ.JJ
Water Truck	ADF.BE	Misc. Storage & Loading	CB AA ACB.JJ
Water Truck	ADG.BE	Misc. Pollution Control	CE AA ACE.JJ
Water Truck	AFA.BE	Pit Pumping	CE HB ACE.JJ
Water Truck	AFB.BE	Relocating Streams	CE HC ACE.JJ
Fork Lift	ACE.BG	Settling Ponds	CE HD ACE.JJ
Fork Lift	ADB.BG	Water Treatment	CE AA ACE.JJ
Man Lift	AEE.BG	Misc. Reclamation—Grading	CG AA ACG.JJ
Cherry Picker	AEI.BG	Misc. Rec. O.B. Rehdl.	CH AA ACH.JJ
Fuel Truck	AAB.BI	Misc. Rec. Reveg.	CI AA ACI.JJ
Fuel Truck	ACD.BI	Misc. General	
Lube Truck	AAC.BI	Special Accounts	DA AA ADA.JJ
Lube Truck	ABH.BI	Vacation Pay	GGCAB-AAA
Mechanic's Truck	BCA.BI	Holiday Pay	GGCAC-AAA
Mechanic's Truck	BCB.BI	Personal or Sick Pay	GGCAE-AAA

Bereavement Pay	JDCDF-ADH	Orientation—New	
Jury Duty	JDCDH-ADH	Employees	JDCGD-ADH
Call Out Pay	JDCDJ-ADH	Orientation—Old Employees	JDCGF-ADH
Clothing Allowance	JDCED-ADH	Guards' Equipment	JFEDA-ADH
Safety Meetings	JDCFG-ADH	In-Town Vehicles	JGBAE-ADH

Fig. 2. Computer codes for equipment, miscellaneous labor, and special accounts.

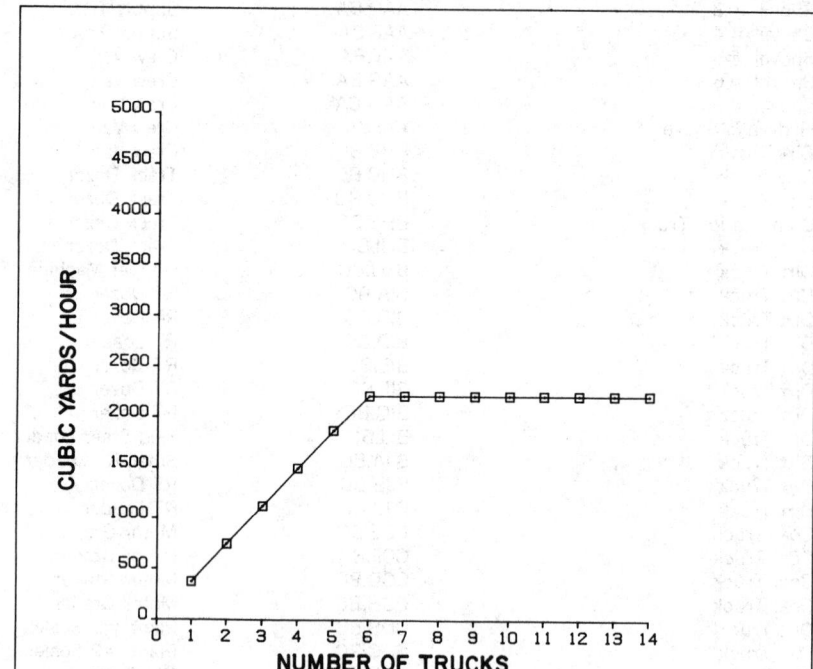

Fig. 3. Shovel productivity variation with increasing number of trucks on a single shovel.

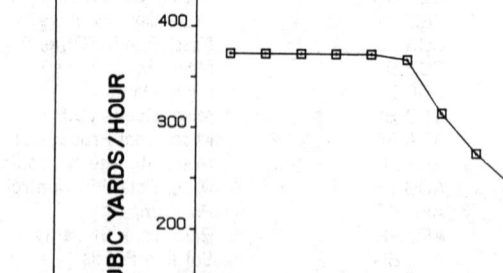

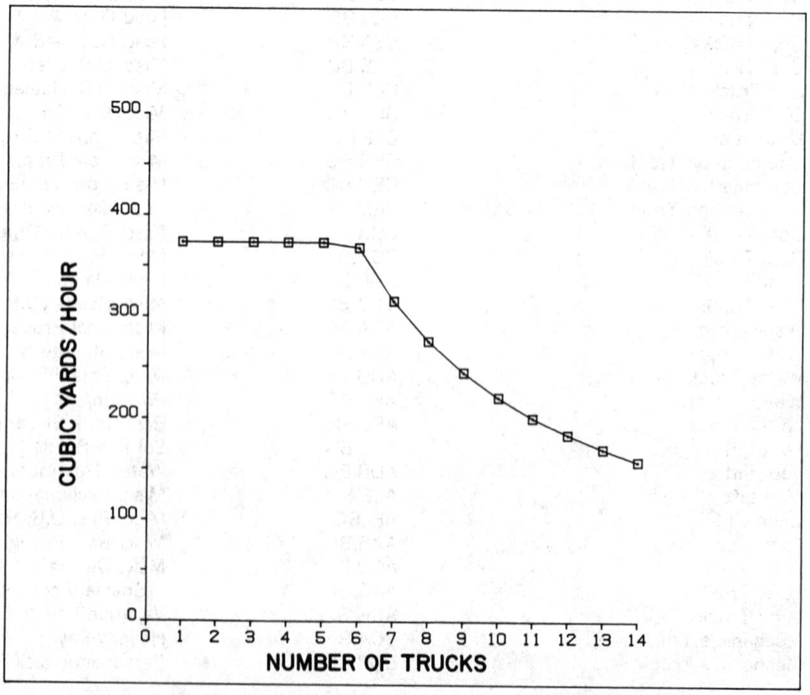

Fig. 4. Truck productivity variation with increasing number of trucks on a single shovel.

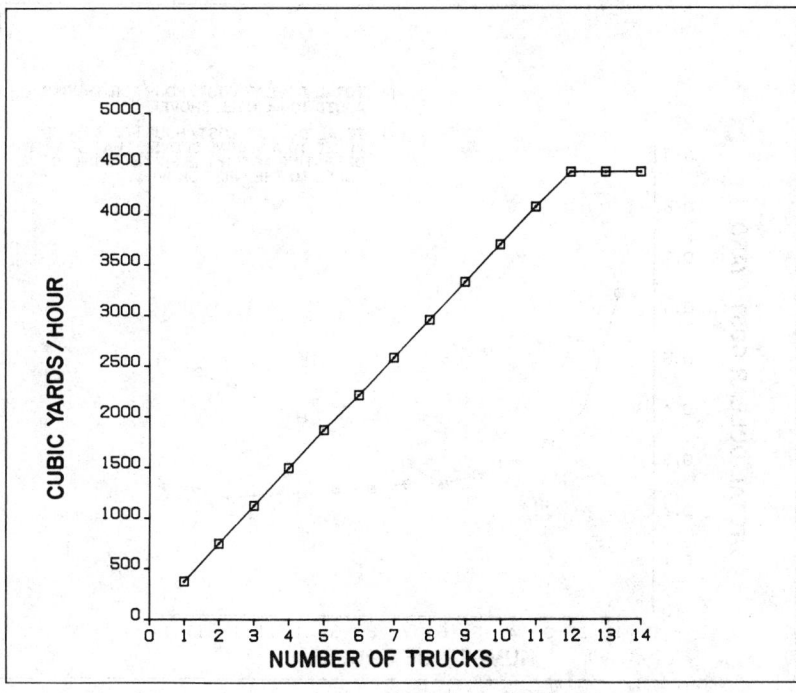

Fig. 5. Shovel productivity variation with increasing number of trucks. A second shovel is started after the sixth truck is assigned to the first shovel.

per hour varies with increasing numbers of trucks for one and two shovels. The additional increase in cost that occurs when the second shovel is started is the result of additional electrical consumption and the "upgrading" of hourly personnel. This cost is only about $30 per hr in this example and is based upon actual time studies and operating costs. At the same time, production is being increased by about 283 m³/hr (370 cu yd per hr) for each truck added to the second shovel up until the capacity of the shovel is exceeded. This is predicated on the idea that both equipment and personnel are presently available at the mine site. Ownership costs for the additional shovel are charged against the scheduled production whether the additional shovel runs or not. Nothing is added. The supervisor just makes more judicious use of what he already has at his disposal. People are upgraded to run the second shovel. Cleanup machines and graders do double duty. Trucks from different shovels might haul to the same dump. Idle time and queues are eliminated. Everything meshes. The end result is that, on an incremental basis, the seventh truck hauls muck for less than 9¢ per yard,

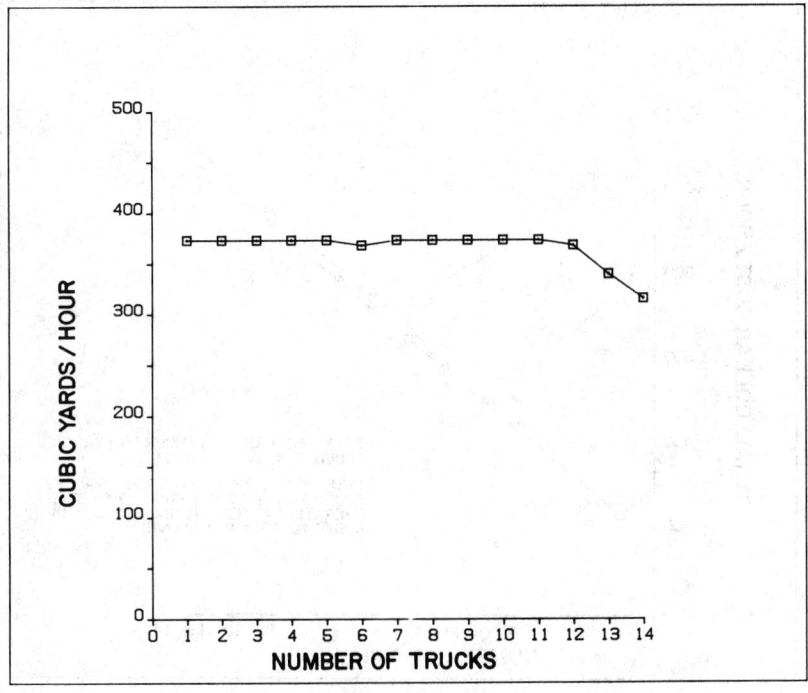

Fig. 6. Truck productivity variation with increasing number of trucks. A second shovel is started after the sixth truck is assigned to the first shovel.

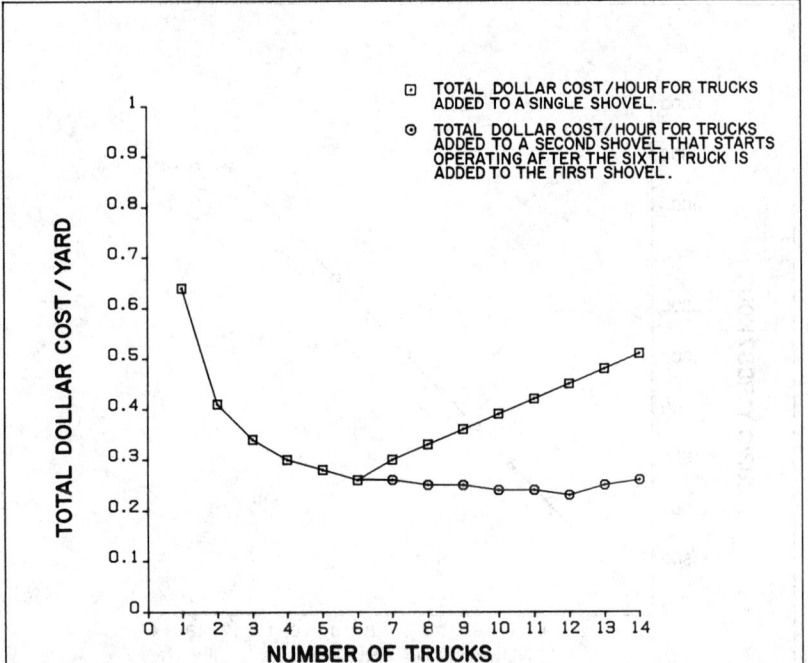

Fig. 7. Comparison of unit cost variations with increasing number of trucks on one and two shovels.

the eighth for less that 4½¢ per yard, the ninth for less that 3¢ per yard, etc. It's like mining for free.

Please note that Figs. 3 through 7 also point to an alternative use of personnel. For instance, if there is only one spare person available on the shift, i.e., there are seven operating trucks on the first shovel described by the figures and no other hauls can be shortened to free up additional trucks, the operator of the seventh truck might be used for a special project. He could be used to man a dozer and build a ramp that ultimately would shorten hauls and allow the operation of the additional shovel at a later time. He could be used to

build a sedimentation pond. He could be used to set up an additional shovel for use on the next shift. Two or more trucks might make a longer haul that would equate to a standard cycle single truck haul. The possibilities are many. The better supervisor will find them and take full advantage of them to benefit the company.

Final Analysis

Mining, perhaps more than any other field of endeavor, presents opportunities for compromise. Different conditions, situations, and needs present different tradeoffs. The engineer

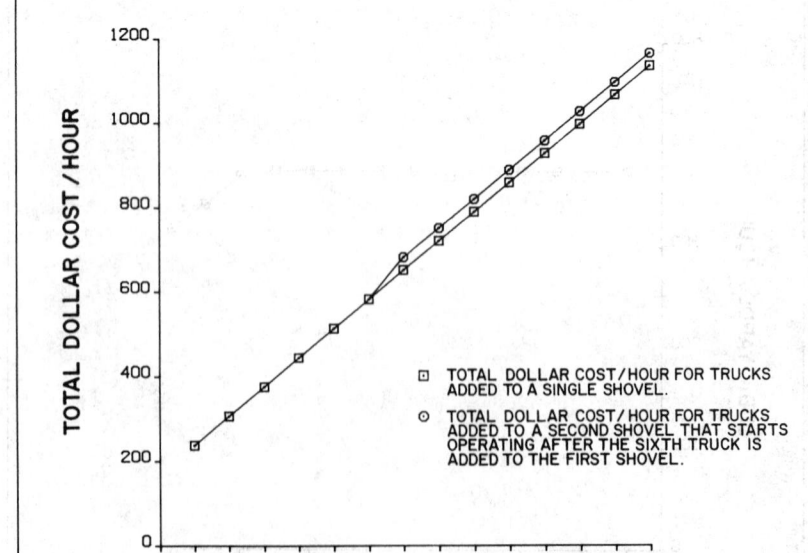

Fig. 8. Comparison of total cost per hour variations with increasing number of trucks on one and two shovels.

may find himself occupying this middle ground at any time. It might be trying to advance the cause of the best workable compromise or yielding to a team decision with respect to equipment selection. It might occur while trying to balance operator bias with the potential for increased production from new equipment. It could be in matching specific equipment to specific conditions in the course of optimizing some less than perfect plan. The balancing of capital and operating costs again offers the likelihood of reconciling alternatives. This includes the choices involved in big equipment vs. little equipment analyses, attempts to eliminate peak equipment requirements, and development of a warehouse stocking policy. Consideration of delaying a purchase or of different work schedules also displays inherent compromises. Finally, attempts to improve efficiency present a myriad of potential compromises. Should a longer haul be made now in order to facilitate shorter hauls later? Should extra employees be used on special projects or to boost production? Will avoiding a long haul now necessitate that it be done at some later time when trucks may not be available? Throughout it all, the engineer should strive to choose the alternative that will generate the maximum benefit to his company.

BIBLIOGRAPHY

Anon., 1982, *Application for a Permit to Mine, Belle Ayr Mine, Campbell County, Wyoming*, AMAX, Inc., submitted to the Wyoming Dept. of Environmental Quality, Cheyenne, WY, pp. 2.4-1 to 2.5-33.

Calahan, G.E., 1982, Private communication.

Chironis, N.P., 1980, "Haulage Trucks Still Supreme," *Coal Age*, Vol. 85, No. 11, Nov., pp. 94-109.

Connell, J.P., 1973, "Truck Haulage," *SME Mining Engineering Handbook*, Vol. 2, A.B. Cummino and I.A. Given, eds., AIME, New York, Sect. 18, pp 18-19-18-28.

Dinsmoor, P.C., 1983, Private communication.

Glass, G.B., 1976, "Update on the Powder River Coal Basin," *Geology and Energy Resources of the Powder River*, Guidebook, R.B. Laudon, W.H. Curry III, and J. S. Runge, eds., Wyoming Geological Association, Casper, pp. 209-220.

Glass, G.B., 1977, "Wyoming Coal Deposits," *Proceedings*, Colorado Geological Survey, Resource Series 1, pp. 74-75.

Gralla, L.J., 1983, Private communication.

Ham, D.J., 1983, Private communication.

Kaufman, D., and Bowen, R., 1981, "Analyze Haul Truck Costs Wisely," *Coal Age*, Vol. 86, No. 3, Mar., pp. 70-77.

Lien, T.J., 1983, Private communication.

Rentz, R.F., 1983, Private communication.

Sandlin, C.H., 1976, Private communication.

Waleski, W.E., and Force, S.J., 1979, "New Look at Effective Use of Trucks," *Surface Coal Mining and Reclamation Symposium*, Coal Conference and Expo V, McGraw-Hill, New York, pp. 141-145.

Welch, T.J., 1982, Private communication.

Woodward, C.J., 1980, *Graphs*, Aug., AMAX Coal Co., Indianapolis, IN, 59 pp.

7.2.5 Base Metal Open Pit Mining

G. S. ZIMMER

INTRODUCTION

This section covers the fundamentals of methods used to estimate mine capital and operating costs for base metal open pit feasibility studies. Methodology is based on the combining of historical data with engineering and practical judgments to yield a best estimate of costs expected for a given mining plan.

A prerequisite for cost development is the generation of a mining plan and a resultant production schedule, which sequentially describes the flow of waste and ore material during selected time intervals. It should be noted that, while mine planning and cost development are sequential operations, they are also interactive as it is possible in many cases to reduce or to defer costs by revising the mine plan.

Following the development of a production schedule, the basic costing sequence consists of the following:
1) Determining production parameters.
2) Selecting mining equipment.
3) Developing equipment productivities.
4) Derivation of equipment requirements.
5) Developing a mine equipment capital and replacement schedule.
6) Calculating mine equipment capital costs.
7) Estimating personnel requirements.
8) Estimating mine operating costs.
9) Calculating mine capital costs including mine equipment, mine facilities, and preproduction operating costs.

Mine operating costs are developed by using equipment productivities to derive the number of shifts required to move a scheduled volume of material within a mining period. An operating cost per shift, in terms of maintenance labor, repair parts, and consumables, is developed for each type of equipment and is multiplied by the required shifts to obtain a total cost per mining period. Operating and maintenance labor requirements are based on providing sufficient manpower to yield the required equipment shifts plus related supplementary functions and these numbers of personnel are used to generate total labor costs per mining period. Repair parts and operating consumables costs are developed by utilizing the pertinent portions of the equipment operating cost per shift along with the respective number of required shifts to calculate a period cost. Mine overhead costs, consisting of salaried labor, salaried fringes, and hourly fringes, are similarly developed on a mining period basis. All period costs are divided by units of total material (ore plus waste) mined during the period to obtain a total unit cost and by units of ore mined to obtain a cost per ore unit that reflects the operating waste-to-ore ratio.

Operating unit costs are summarized by mining function (e.g., drilling, blasting, loading, hauling, roads and dumps, crushing/conveying, general mine, general maintenance, and general administration) with each function consisting of operating labor, maintenance labor, parts and consumables, and mine overhead.

A flowchart summarizing the development of mine capital and operating costs is shown in Fig. 1.

EQUIPMENT SELECTION

The equipment selection process consists of first determining the type of equipment required followed by determining the optimum sizes of equipment and the number of units required for a given production schedule.

The type of equipment required is a function of mine life, ore and pit geometry, site location, topography, and total material movement. A shallow ore body with a long life at a high tonnage rate may be most suited to a dragline stripping operation while a similar geometry with a short mine life may favor a scraper operation. A deep mine with a medium-to-long mine life and a high tonnage rate would probably favor a shovel operation, while a similar, but short-lived, operation would probably be more economical with loaders.

Once the type of equipment has been determined, the size and number required for each category must be determined. Equipment is usually sized as large as possible to comfortably meet production targets. The large equipment will take advantage of economies of scale and consequently will minimize unit costs, maintain the fleet at a minimum size, and will minimize labor requirements. Factors affecting or modifying equipment size include bench height, material density, ore selectivity, operating geometry, work schedule, and crushing capacity.

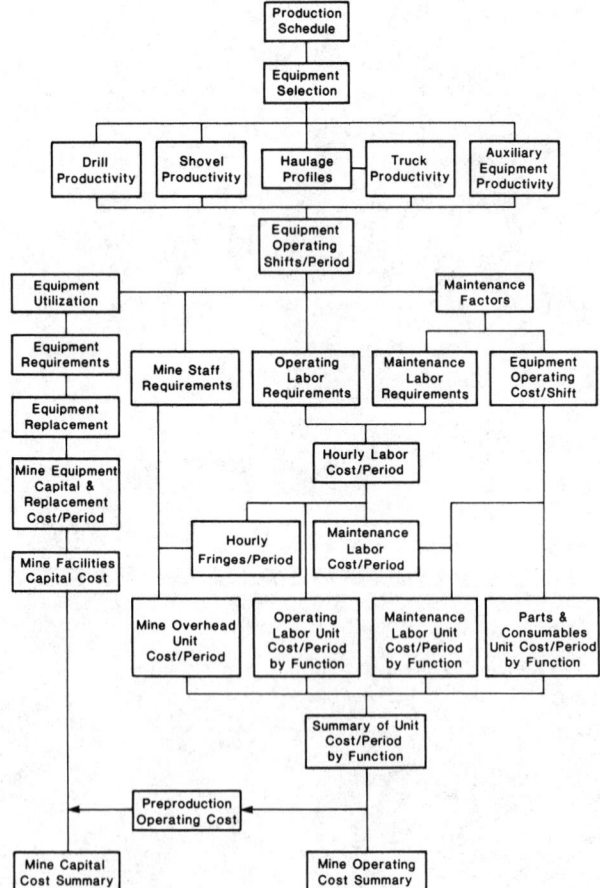

Fig. 1. Mine capital and operating cost flowsheet.

EQUIPMENT REQUIREMENTS

After the type and size of all major equipment has been determined, the numbers of units required for each period of the mining schedule are calculated by first developing a productivity per operating shift for each primary equipment category (drills, shovels, and trucks). Annual or period volumes are divided by these shift productivities to obtain the number of operating shifts required to move the material. The required operating shifts are divided by the number of

Table 1. Drilling Productivity Estimate

Drill type	12.25-in. electric
Material to drill	Ore and waste
In situ density of material	12.50 cu ft/ton
Blasting Parameters	
Bench height	50 ft
Subgrade drilling	10 ft
Hole depth	60 ft
Hole diameter	12.25 in.
Explosives specific gravity (ANFO)	0.82 g/cm^3
Powder factor	0.35 lb/ton
Powder rise	30 ft
Stemming height	30 ft
Column load	41.88 lb/ft
Burden	30 ft
Spacing	30 ft
Tons drilled/hole	3,600 tons
Drilling Productivity	
Tons/ft drilled	60.00 tons/ft
Penetration rate	70 ft/hr
Down-the-hole drilling, time/shift	5.5 hr
Ft drilled/operating shift	385 ft
Tons drilled/operating shift	23,100 tons
Scheduled days/year	360 days
Shifts/day	3 shifts
Total scheduled shifts/year	1,080 shifts
Maximum drill utlization	60%
Maximum operating shifts/year	648 shifts
Maximum tons × 1000/drill year	14,969 kt
Average tons drilled/scheduled shift	13,860 tons

Metric equivalents: ton × 0.907 184 7 = t, ft × 0.3048 = m, in. × 25.4 = mm, cu ft × 0.028 317 = m^3, lb × 0.453 592 = kg, lb/ft × 1.488 164 = kg/m.

Table 2. Shovel Productivity Estimate

Type of Equipment	
Shovel	25 cu yd
Truck	170-ton
Material to Load	Ore and waste
In situ density	12.50 cu ft/ton
Swell factor	1.33
Loose density	16.63 cu ft/ton
Bench height	50 ft
Productivity/Load	
Shovel bucket size	25 cu yd
Bucket fill factor	90%
Truck size	170-ton
Theoretical passes to load	4.65 passes
Actual average passes to load	5 passes
Average swing cycle time	32 sec
Spot time between loads	15 sec
Total time/load	2.92 min
Productivity/Shift	
Effective minutes/shift	350 min
Average truck loads/shift	119 loads
Truck load/factor	95%
Average truck load	161.5 tons
Average shovel production/operating shift	19,218 tons
Productivity/Year	
Scheduled days/year	360 days
Shifts/day	3 shifts
Total scheduled shifts/year	1,080 shifts
Maximum shovel utilization	75%
Maximum operating shifts/year	810 shifts
Maximum tons × 1000/shovel year	15,567 kt

Metric equivalents: cu yd × 0.764 555 = m^3, cu ft × 0.028 317 = m^3, ft × 0.3048 = m, ton × 0.907 184 7 = t.

scheduled shifts in the period, which yields the number of operating shifts per scheduled shift or the average number of units operating during each scheduled shift. The operating shifts per scheduled shift is divided by the proper utilization for each equipment category to obtain the number of units in the fleet, rounded up to the nearest whole number. Tables 1 through 5 illustrate the development of drill, shovel, and truck requirements.

Table 3. 12.25-in. Drill Requirements

Period	Total (tons × 1000)	Tons/ Shift	Reqrd Shifts	Sched Shifts	Units Required		
					Optg/ Shift	% Util	Total Fleet
PP*	91,500	20,790	4,401	2,160	2.04	60	4
1	59,000	23,100	2,554	1,080	2.36	60	4
2	58,750	23,100	2,543	1,080	2.35	60	4
3	58,750	23,100	2,543	1,080	2.35	60	4
4	58,200	23,100	2,519	1,080	2.33	60	4
5	53,500	23,100	2,316	1,080	2.14	60	4
6–7	85,200	23,100	3,688	2,160	1.71	60	3
8–10	120,825	23,100	5,231	3,240	1.61	60	3
11–15	170,000	23,100	7,359	5,400	1.36	60	3
16–20	111,250	23,100	4,816	5,400	0.89	60	2

* Preproduction = 2.0 years @ 90% of drill capacity.
Metric equivalents: ton × 0.907 184 7 = t, in. × 2.54 = cm.

Table 4. 25-yd Shovel Requirements

Period	Total (tons × 1000)	Tons/ Shift	Reqrd Shifts	Sched Shifts	Units Required		
					Optg/ Shift	% Util	Total Fleet
PP*	91,500	17,639	5,268	2,160	2.44	75	4
1	59,000	19,299	3,057	1,080	2.83	75	4
2	58,750	19,299	3,044	1,080	2.82	75	4
3	58,750	19,299	3,044	1,080	2.82	75	4
4	58,200	19,299	3,016	1,080	2.79	75	4
5	53,500	19,299	2,772	1,080	2.57	75	4
6–7	85,200	19,299	4,415	2,160	2.04	75	3
8–10	120,825	19,299	6,261	3,240	1.93	75	3
11–15	170,000	19,299	8,809	5,400	1.63	75	3
16–20	111,250	19,299	5,765	5,400	1.07	75	2

* Preproduction = 2.0 years @ 90% of shovel capacity.
Metric equivalents: ton × 0.907 184 7 = t, cu yd × 0.764 554 9 = m³.

Productivity Per Operating Shift

Drilling productivity is a function of rock density, drilling pattern, and penetration rate. Rock density is either measured or estimated from specific gravity tables. Drilling pattern may be set by operating experience or estimated by various formulas based on hole diameter, bench height, subgrade depth, powder factor, and rock density. Penetration rate is either based on operating experience or on drillability tests. Average down-the-hole drilling time per 8-hr shift is normally 5.5 hr. However, this time is sensitive to bench height and penetration rate and a low bench height and/or soft rock will result in a lower drilling time per shift as more time is spent moving the drill from hole to hole. An example of drill productivity is shown in Table 1.

Loading (shovel) productivity is a function of rock density, swell factor, bucket size, bucket fill factor, truck size, truck load factor, swing cycle time, and truck spotting time. Material swell factor may be determined by actual measurement, estimated by comparison to other operations, or estimated from available tables. Bucket size is determined by machine size, material loose density, and range available from manufacturer. Bucket fill factor is a function of bench height and digging conditions and should fall within a range of 85 to 90% for well-fragmented material. Truck size is usually stated as a carrying weight in short or metric tons and trucks should be sized such as to be loaded in three to five passes from the loading equipment. A truck factor of 90 to 95% of rated capacity is normally used to prevent overloading and to exclude moisture, as the aforementioned densities are usually dry weight. Swing cycle time usually increases with equipment size and is a function of material density, digging conditions, swing angle, truck height, and operator proficiency. Swing cycle time may be obtained from operating conditions or estimated from manufacturer's tables or graphs. Truck spotting time is the time required for a haulage unit to be positioned under the shovel. Note that this time does not include truck queuing time or time waiting for trucks in an under-trucked operation. Spotting time may be minimized by using double-side loading. Table 2 is an example of a typical shovel productivity calculation. In this table, a shift is based on 350 effective min, which consists of a full 8-hr shift (480 min) less 60 min for lunch and start/end of shift delays less 10 min per hr for the remaining 7 hr to allow for operational delays. The number of truck loads possible per shift, assuming the loading unit is fully trucked, is the effective minutes per shift divided by the total time required

to load a single haulage unit. The resulting number of loads should be rounded down to the nearest full load.

Haulage unit productivity is a function of haul profile and truck speeds. Since haul profiles may vary considerably over mine life, truck productivity should be calculated for each mining period specified in the production schedule. To determine an average profile for a mining period, measurements are made from the centroid of each bench mined to the proper destination (crusher, stockpile, waste dump, etc.). Bench measurements include total horizontal distance, cumulative lift, and cumulative fall. To obtain an average annual or period profile, these bench measurements are weighted by the respective volumes mined. Maximum truck speeds for both loaded and empty conditions are determined from manufacturers' charts and are modified to suit project conditions and safety requirements. As most mines do not have passing lanes, truck productivity is effectively limited by the slowest unit and average speeds are usually less than the maximum allowed. Consequently, speeds in the vicinity of 90% of maximum allowed should be used. These average speeds are combined with the various segments of the haul profile to determine total travel time (loaded plus return) per unit, and a fixed time (queuing time plus spot and load time plus dumping time) is added to obtain total cycle time per load. Number of loads per truck operating shift is obtained by dividing 350 min per shift by the cycle time per load and rounding down to the nearest load. The number of loads per shift times the truck capacity times the truck load factor yields truck productivity per operating shift.

Equipment Operating Shifts and Units Required

Primary equipment operating shifts for each mining period are obtained by dividing the total volume of material mined in the period by the equipment productivities previously developed for drills, trucks, and shovels. Note that the foregoing calculations assume that the various mining operations (ore, internal waste, and stripping) required in each period are positioned such that equipment may be exchanged and substituted without a significant loss in productivity. In many cases, particularly when utilizing a pushback or phase mining system, operating geometry separates ore mining from waste stripping to the extent that these areas must be treated as separate entities for drilling and loading purposes, which will result in additional equipment. The utilization of the more mobile haulage units may be significantly increased to cover these situations by means of a computer truck dis-

Table 5. 170-ton Truck Requirements

Period	Total (tons × 1000)	Haul Profile (Feet)			Cycle Time (min.)			Shift Productivity‡			Reqrd Shifts	Sched Shifts	Units Required		
		Up	Down	Flat	Haul*	Fixed†	Total	Loads/Shift	Tons/Load	Tons/Shift			Optg/Shift	% Util	Total Fleet
PP*	91,500	400	2,050	2,550	6.41	4.92	11.33	30.5	161.5	4,926	18,575	2,160	8.60	70	13
1	59,000	425	850	4,325	6.46	4.92	11.38	30.5	161.5	4,926	11,977	1,080	11.09	70	16
2	58,750	925	50	3,800	6.22	4.92	11.14	31.0	161.5	5,007	11,734	1,080	10.86	70	16
3	58,750	375	850	4,200	6.64	4.92	11.56	30.0	161.5	4,845	12,126	1,080	11.23	70	17
4	58,200	775	950	4,700	8.14	4.92	13.06	26.5	161.5	4,280	13,598	1,080	12.59	70	18
5	53,500	225	575	4,925	6.81	4.92	11.73	29.5	161.5	4,764	11,230	1,080	10.40	70	15
6–7	85,200	1,250	550	5,050	8.96	4.92	13.88	25.0	161.5	4,038	21,100	2,160	9.77	70	14
8–10	120,825	525	1,325	4,600	8.03	4.92	12.95	27.0	161.5	4,361	27,706	3,240	8.55	70	13
11–15	170,000	1,575	850	5,175	10.14	4.92	15.06	23.0	161.5	3,715	45,760	5,400	8.47	70	13
16–20	111,250	6,725	0	2,225	15.90	4.92	20.82	16.5	161.5	2,665	41,745	5,400	7.73	70	12

* Based on average truck speeds:

	Loaded		Empty
Up @ 8%	=	8 mph	20 mph
Down @ 8%	=	15 mph	20 mph
Flat @ 0%	=	20 mph	20 mph

† Fixed time:

Queueing at shovel	=	1.00 min
Spot and load	=	2.92 min
Dump	=	1.00 min
Total	=	4.92 min

‡ Based on 350 effective minutes/operating shift

Metric equivalents: ton × 0.907 184 7 = t, mph × 2.239 936 = m/sec.

patching system and to a lesser extent by radio dispatching. Examples of primary equipment calculations are shown in Tables 3, 4, and 5.

Mine auxiliary equipment consists of secondary drills (secondary drilling and wall control), front-end loaders (road maintenance and shovel backup), rubber-tired dozers (shovel cleanup), medium track dozers (dump maintenance), large track dozers (rip hard toes, road construction, and dump maintenance), road graders, and water/sanding trucks. Op-

erating shifts for auxiliary equipment are determined subjectively and are influenced by climatic conditions, length of road systems, number and location of working faces, rock type, and mining rate.

At this time, equipment requirements should be reviewed and the reasons for peaks identified. It is possible, due to adverse haulage configurations, for truck requirements to jump by several units without a corresponding increase in volume moved, which would not have been evident in the

Table 6. Equipment List
(in first quarter 1982 US dollars × 1000)

Item	Repl Life	Unit cost			
		FOB Factory	Land Freight	Erection	Total
Drilling					
12.25-in. drill	N/R	992.0	7.2	35.0	1,034.2
9-in. diesel drill	7	500.0	2.9	9.0	511.9
Secondary drill	N/R	180.0	1.0	0.0	181.0
Blasting					
Prill truck	3	65.0	0.4	0.0	65.4
Explosives truck	3	22.0	0.2	0.0	22.2
Loading					
25-yd shovel	N/R	3,250.0	89.3	100.0	3,439.3
15-yd loader	5	678.8	9.3	0.0	588.1
Hauling					
170-ton truck	7	820.0	12.8	28.0	860.8
Auxiliary					
7-yd loader	5	330.5	4.6	0.0	335.1
R.T. dozer	5	233.0	3.3	0.0	236.1
Track dozer (400 hp)	5	537.9	7.2	0.0	545.1
Track dozer (300 hp)	5	295.0	4.4	0.0	299.4
Motor grader (250 hp)	5	257.0	2.0	0.0	259.0
35-ton truck*	5	277.8	3.2	0.0	281.0
Motivator	N/R	250.0	2.0	0.0	252.0
Lube/fuel truck	3	100.0	0.4	0.0	100.4
Cable reel truck	3	125.0	0.7	0.0	125.7
Backhoe	5	160.0	0.7	0.0	160.7
Dewatering truck	5	30.0	0.4	0.0	30.4
Crushing plant	N/R	600.0	2.0	0.0	602.0
S/D cable (1000 ft)	3	16.0	0.0	0.0	16.0
Cable couplers	3	0.8	0.0	0.0	0.8
Mobile light plant	5	7.8	0.1	0.0	7.9
Man buses	3	12.0	0.4	0.0	12.4
Flatbed truck	3	50.0	0.5	0.0	50.5
Engineering equipment	N/R	100.0	0.0	0.0	100.0
Mobile radios	5	1.5	0.1	0.0	1.6
Base station radio	10	2.5	0.5	0.0	3.0

* Includes 30 283-L (8000-gal) water unit.

Note: N/R indicates equipment not replaced during 22-year production life (2 years preproduction and 20-year mine life).

Metric equivalents: in. × 25.4 = mm, ft × 0.3048 = m, cu yd × 0.765 554 9 = m³, ton × 0.907 184 7 = t, hp × 0.745 699 = kW.

Table 7. Equipment Replacement Schedule for 170-Ton Truck
(Replacement Life = 7 Years)

Period	Reqd	On Hand	Add	No. of Units at Indicated Age							
				1	2	3	4	5	6	7	Retire
PP-1	13	0	13	13	0	0	0	0	0	0	0
PP-2	13	13	0	0	13	0	0	0	0	0	0
1	16	13	3	3	0	13	0	0	0	0	0
2	16	16	0	0	3	0	13	0	0	0	0
3	17	16	1	1	0	3	0	13	0	0	0
4	18	17	1	1	1	0	3	0	13	0	0
5	15	18	0	0	1	1	0	3	3	10	10
6	14	8	6	6	0	1	1	0	3	3	3
7	14	11	3	3	6	0	1	1	0	3	3
8	13	11	2	2	3	6	0	1	1	0	0
9	13	13	0	0	2	3	6	0	1	1	1
10	13	12	1	1	1	0	2	3	6	0	0
11	13	13	0	0	1	1	0	2	3	6	6
12	13	13	0	0	1	1	0	2	3	6	6
13	13	7	6	6	0	1	1	0	2	3	3
14	13	10	3	3	6	0	1	1	0	2	2
15	13	11	2	2	3	6	0	1	1	0	0
16	12	13	0	0	2	3	6	0	2	0	0
17	12	13	0	0	0	2	3	6	1	1	1
18	12	12	0	0	0	0	2	3	6	1	0*
19	12	12	0	0	0	0	0	0	2	10	0*
20	12	12	0	0	0	0	0	0	2	10	0*

Note: When spare units occur, youngest units used first.

* Assuming a 20-year mine life, trucks should not be purchased during the last two years, if at all possible. At end of 20 years, the above truck fleet will consist of the following: two units, 6 years old; three units, 7 years old; six units, 8 years old; and one unit, 9 years old.

Metric equivalent: ton × 0.907 184 7 = t.

mine plan. Should this situation occur, the mine plan should be reviewed to determine if advance stripping would alleviate the equipment peak.

MINE EQUIPMENT CAPITAL AND REPLACEMENT COSTS

To develop capital and replacement costs for mine equipment, the list of primary and auxiliary equipment is expanded to include support, service, and maintenance equipment. All items are assigned a replacement life that is a function of operating conditions, usage, and mine life. Table 6 shows typical replacement lives and example purchase costs for major mine items, excluding site-specific maintenance and electrical equipment.

Based on a given replacement life, a replacement schedule is developed for each item. These schedules are usually very simple to develop, but truck replacements can be complex when changes in fleet requirements do not coincide with replacement life. Table 7 is an example of a truck replacement schedule using the truck requirements developed in Table 5.

Initial equipment capital costs are calculated and a contingency of 10 to 20% is generally added to provide for a spare parts inventory. The costs of equipment additions and replacements are calculated for each mining year throughout the mine life. Note that equipment is usually purchased in the year prior to actual use to insure that units are available on January first.

In addition to mine equipment costs, capital costs must be developed for mine facilities including maintenance shops and equipment, crushing and conveying systems, explosives storage, fuel storage, electrical distribution systems, warehousing, offices, and mine access.

At this point, mine capital costs should be complete except for preproduction operating costs, which are developed as part of operating costs.

OPERATING COSTS

As previously noted, operating costs are developed by using equipment productivities to determine the number of operating shifts required to move a given volume of material, followed by providing sufficient manpower, replacement parts, and consumables to cover the required operating shifts.

Maintenance Factors

As an integral part of operating cost development, factors have been developed that relate maintenance labor and repair parts costs per operating hour to delivered equipment capital costs. These factors have been developed by manufacturers and operators and are averages that will yield parts and labor costs within reasonable ranges. Factors representing average conditions for surface mining operations in the western United States are shown in Table 8. These factors are based on medium-to-heavy equipment usage in medium-to-extreme operating conditions. For use in other areas, particularly foreign operations, these factors must be adjusted to consider the effects of additional shipping charges, import duties, climatic conditions, maintenance skill levels, and maintenance labor rates.

Personnel Requirements

Mine personnel are usually subdivided into four main groups: (1) administration, (2) mine operations, (3) mine maintenance, and (4) engineering/geology. Salaried personnel in these groups are subjectively determined and, while providing adequate supervisory coverage, both job categories and numbers of personnel within categories usually reflect corporate policy and may vary considerably from company to company.

Table 8. Western US Maintenance Factors

Category	Class/Size	Maintenance Factors*	
		Parts	Labor
Drills	Diesel to 6.75-in.	0.0390	0.0252
	Electric 9-12.25-in.	0.0310	0.0170
Shovels	Electric 8-30 yd	0.0300	0.0140
Loaders†	To 6 yd	0.0360	0.0240
	7-25 yd	0.0540	0.0360
Haul trucks†	Mechanical drive	0.0360	0.0240
	Electrical drive	0.0280	0.0120
Track dozers	200-400 hp	0.0800	0.0470
	400+ hp	0.0826	0.0444
R.T. dozers†	150-350 hp	0.0510	0.0320
Road graders†	150-250 hp	0.0580	0.0350
Scrapers/water units†	225-550 hp	0.0540	0.0360

* Used with delivered cost in thousands excluding erection cost.
† Factors applied to delivered cost less tires.
Metric equivalents: in. $\times$ 25.4 = mm, cu yd $\times$ 0.764 554 9 = m³, hp $\times$ 0.745 699 = kW.

Hourly paid operating personnel must be sufficient to man the required number of equipment operating shifts per scheduled shift developed earlier. Using year 1 in Table 3 as an example, an average of 23.6 drill shifts per scheduled shift is required. To cover periods where more than the average of 2.36 units is available, a high average of 3.00 operating shifts per shift is assumed, which results in a total of three operators per shift times four shifts, equaling 12 drillers required. To provide for vacation, sick leave, and absenteeism (VSA), an allowance of 10 to 15% is added, bringing the total number of drillers for this period to 14. Operators are similarly calculated for shovels, trucks, and auxiliary equipment for each mining period, followed by a subjective estimate of support personnel (blasting crew, crusher operators, laborers, etc.). A more detailed method of determining major equipment operating personnel is based on the use of a binomial distribution to estimate the attainable equipment operating shifts per shift based on a given availability applied to a specific fleet size.

Primary maintenance personnel requirements (mechanics, welders, and electricians) are based on the number of equipment operating shifts per scheduled shift for major mine equipment and are derived for four maintenance areas (drills, shovels, trucks, and auxiliary equipment) by utilizing the maintenance labor factors in Table 8 to calculate a required number of maintenance shifts per operating shift. These maintenance shifts per operating shift are comprised of a proportion of mechanics, welders, and electricians that varies with equipment type and is based on historical distributions. Average maintenance shifts required per equipment shift and the distribution of mechanics, welders, and electricians for the southwestern United States is summarized in Table 9. Using a 154-t (170-st) truck as an example, the derivation of required maintenance shifts per operating shift is as follows:

Truck cost (delivered cost less tires and erection)	=	$ 759.0
Maintenance labor factor	=	$\times$ 0.0120
Maintenance labor, $ per operating hour	=	$ 9.11

(Operating hours/shift)/(maintenance hour/ shift)	=	$\times$ 0.875
Average tradesman pay rate ($/hr)	=	$\div$ 12.00
Required maintenance shifts/operating shift	=	0.66

For each maintenance area, the numbers of mechanics, welders, and electricians are calculated as in the following example:

Shifts/man/year (42 hr/wk $\div$ 8 hr/day $\times$ 52 wks/yr − 5 holidays − 10 days vacation)	=	258
Drill operating shifts/year	=	2,554
Drill maintenance shifts/operating shift	=	$\times$ 1.25
Total maintenance shifts required	=	3,193
Mechanic shifts @ 60%	=	1,916
Welder shifts @ 10%	=	319
Electrician shifts @ 30%	=	958
Total mechanics	= 1,916/258 =	7.43 or 8
Total welders	= 319/258 =	1.24 or 2
Total electricians	= 958/258 =	3.71 or 4

In order to treat auxiliary equipment as a single maintenance area, an average number of maintenance shifts per operating shift is obtained by weighting the individual maintenance shifts per operating shift by the number of operating shifts in a typical year for each type of equipment. The average maintenance shifts per operating shift shown in Table 9 are based on operating these units as primary equipment. When these units are used as auxiliary equipment, the required MS/OS may be reduced by at least half, due to lower maintenance priorities, easier usage, and overlapping capabilities.

Hourly manpower requirements and the resulting cost per period for mine operations and mine maintenance are calculated for each mining period as shown in Table 10.

Operating Labor Cost/Ton

Operating labor costs for each period are accumulated by function and divided by the total tonnage mined during

Table 9. Maintenance Labor Factors and Distribution

Equipment	Maint Shift/ Op Shift	Distribution (%)		
		Mech.	Welders	Elec.
Drills: 6.75-in.	0.55	60	10	30
9.00-in.	0.86	60	10	30
11.00-in.	0.96	60	10	30
12.75-in.	1.25	60	10	30
Shovels: 12-yd	1.50	40	15	45
15-yd	1.79	40	15	45
20-yd	2.36	40	15	45
30-yd	3.26	40	15	45
Loaders: 5.25-yd	0.32	40	15	45
7.00-yd	0.71	40	15	45
12.50-yd	1.40	40	15	45
Trucks: 35-ton	0.42	70	15	15
50-ton	0.60	70	15	15
85-ton	0.84	70	15	15
120-ton	0.48	70	15	15
170-ton	0.65	70	15	15
R.T. dozers				
814 (Class)	0.28	60	30	10
824 (Class)	0.45	60	30	10
Track dozers				
D-7G (Class)	0.57	60	30	10
D-8K (Class)	0.78	60	30	10
D-9H (Class)	1.02	60	30	10
D-9L (Class)	1.13	60	30	10
D-10 (Class)	1.87	60	30	10
Graders: 12G (Class)	0.27	60	30	10
14G (Class)	0.39	60	30	10
16G (Class)	0.55	60	30	10
Water Trucks				
30 283 L (8000 gal)	0.37	60	30	10
Scrapers (Standard)				
20-yd	0.52	60	30	10
31-yd	0.82	60	30	10
38-yd	0.99	60	30	10
44-yd	1.04	60	30	10
Scrapers (Tandem)				
20-yd	0.65	60	30	10
31-yd	1.05	60	30	10
44-yd	2.37	60	30	10
Scrapers (Elevating)				
22-yd	0.73	60	30	10
34-yd	0.93	60	30	10
34-yd	1.15	60	30	10

Metric equivalent: in. $\times$ 25.4 = mm, cu yd $\times$ 0.764 554 9 = m³, ton $\times$ 0.907 184 7 = t.

Table 10. Operating and Maintenance Personnel Requirements, Years 6 and 7
(Two Years in Period)

Job Title	No. Reqd	Pay Rate, $/hr	Yearly Cost, $/man†	Total Cost, $ × 1000
Mine Operations				
Driller	9	13.00	29,250	526.5
Drill helper	5	11.00	24,750	247.5
Blaster	2	12.50	28,125	112.5
Powderman	6	11.00	24,750	297.0
Shovel operator	10	13.50	30,375	607.5
Shovel, helper	5	11.00	24,750	247.5
Haul truck driver	46	12.50	28,125	2,587.5
R.T. dozer operator	8	12.00	27,000	432.0
Track dozer operator	10	12.00	27,000	540.0
Grader operator	6	12.00	27,000	324.0
Water truck driver	5	12.00	27,000	270.0
Utilityman	6	11.00	24,750	297.0
Trainee	2	11.00	24,750	99.0
Laborer	2	10.50	23,625	94.5
Subtotal	122			6,682.5
Mine Maintenance				
Heavy eqpt mech	20	13.50	30,375	1,215.0
Mechanic helper	10	11.50	25,875	517.5
Welder	13	12.50	28,125	731.3
Electrician	19	14.00	31,500	1,197.0
Lineman	2	12.50	28,125	112.5
Auto mechanic	3	12.00	27,000	162.0
Lubeman	8	11.50	25,875	414.0
Tireman	8	11.50	25,875	414.0
Machinist	3	13.00	29,250	175.5
Plumber/pipefitter	2	12.50	28,125	112.5
Carpenter/printer	2	12.00	27,000	108.0
Trainee	4	11.00	24,750	198.0
Laborer	4	10.50	23,625	189.0
Subtotal	98			5,546.3
Total	220			12,228.8
Fringes @ 35%				4,280.1
Grand Total	220			16,508.9

* Includes 15% vacation, sick leave, and absenteeism (VSA) allowance.

† Annual rate based on 1928 hr at straight time + 215 hr of overtime to give equivalent time of 2250 hr per man per yr.

the respective period. Using Table 10 as a source, Table 11 summarizes the operating labor cost per ton.

Maintenance Labor Cost per Ton—Parts and Consumables Cost per Ton

The maintenance factors for parts and labor shown in Table 8 are used to develop an operating cost per shift (exclusive of operating labor) for each type of major operating equipment. In addition to these factors, relevant fuel, power, tire, bit, pipe, and stabilizer costs are also used. These calculations for drills, shovels, and haul trucks are shown in Tables 12, 13, and 14.

For each mining period, required operating shifts have previously been derived and are utilized in conjunction with the cost per shift and period tonnage to obtain the maintenance labor cost per ton for each function or equipment classification. In order to provide for unspecified maintenance (small vehicles, plumbers, carpenters, trainees, laborers, and miscellaneous), a general maintenance cost is developed by subtracting the cost per period for each equipment type from the total maintenance labor cost shown in Table 10. Table

15 is an example of maintenance labor cost per ton calculations.

The parts and consumables cost per ton is similarly developed for each mining period using equipment shifts and the respective cost per shift. In addition, the blasting consumables cost per ton must be derived, an allowance for mine utilities should be included under general mine, an allowance for gasoline and small vehicle maintenance should be included under general maintenance, and an allowance for general supplies should be included under general administration. Table 16 shows parts and consumables cost per ton calculations for the same mining period as the previous examples.

Mine Overhead Cost/Ton

Mine overhead generally consists of salaried and supervisory costs directly concerned with mining, the fringe benefits associated with these salaried labor costs, and the hourly fringes previously developed and shown in Table 10. Using the same mine period as in previous examples, assuming a salaried labor cost of $4,318 \times 10^3$ and 35% fringes, mine

Table 11. Operating Labor Cost per Ton, Years 6 and 7
(2.0 Years)
Total Material Mined = 85 200 kt

Function	Total Cost, $ × 1000	Cost/Ton, $
Drilling		
12.25-in. drills	774.0	0.0091
Blasting	409.5	0.0048
Loading		
24-yd shovels	855.0	0.0100
Hauling		
170-st trucks	2587.5	0.0304
Roads and dumps		
R.T. dozers	432.0	0.0051
Track dozers	540.0	0.0063
Graders	324.0	0.0038
Water trucks	270.0	0.0032
General mine*	490.5	0.0058
General maintenance	0.0	—
General administration	0.0	—
Total	6682.5	0.0785

* Includes utilitymen, trainees, and laborers.
Metric equivalents: in. × 2.54 = cm, ton × 0.907 184 7 = t, cu yd × 0.764 554 9 = m³.

Table 12. Equipment Operating Cost per Shift

Drill size	12.25 in.
Delivered cost, $ × 1000	$999.2
Power consumption kWh/hr	350
Bit cost	$4800
Bit life, hr	145
Cost Breakdown ($/hr)	
Bit cost	$33.10
Drill pipe and stabilizer @ ¼ bit cost	$8.28
Repair parts	
(0.0310 × delivered cost in $ × 1000)	$30.98
Maintenance labor	
(0.0170 × delivered cost in $ × 1000)	$16.99
Power cost @ $0.050/kWh	$17.50
Operating hr/shift	6.5
Drilling hr/shift	5.5
Cost Breakdown ($/shift)	
Bit cost @ 5.5 hr/shift	$182.05
Drill pipe and stabilizer @ 5.5 hr/shift	$45.54
Repair parts @ 6.5 hr/shift	$201.37
Maintenance labor @ 6.5 hr/shift	$110.44
Power cost @ 6.5 hr/shift	$113.75
Total cost/shift	$653.15

Metric equivalent: in. × 2.54 = cm.

Table 13. Equipment Operating Cost per Shift

Shovel size	25.00 yd
Delivered cost, $ × 1000	$3339.3
Power consumption, kWh/hr	850
Cost Breakdown, $/hr	
Power cost @ $0.050/kWh	$42.50
Repair parts	$100.18
(0.0300 × delivered cost in $ × 1000)	
Maintenance labor	$46.75
(0.0140 × delivered cost in $ × 1000)	
Total operating cost/hr	$189.43
Operating hr/shift	6.5
Cost Breakdown, $/shift	
Maintenance labor	$303.88
Parts and consumables	$927.42
Total operating cost/shift	$1231.30

Metric equivalent: cu yd × 0.764 554 9 = m³.

Table 14. Equipment Operating Cost per Shift

Truck size	170 ton
Delivered cost, $ × 1000	$832.8
Delivered cost less tires	$759.0
Fuel consumption, gal/hr	30
Cost per tire	$123.00
Number of tires	6
Tire life (hr)	4000
Cost Breakdown, $/hr	
Fuel cost @ $1.100/gal	$33.00
Repair parts	
(0.0280 × delivered cost in $ × 1000 − tires)	$21.25
Maintenance labor	
(0.0120 × delivered cost in $ × 1000 − tires)	$9.11
Tire cost	$18.45
Total operating cost/hr	$81.81
Operating hr/shift	7.0
Cost Breakdown, $/shift	
Maintenance labor	$63.77
Parts and consumables	$508.90
Total operating cost/shift	$572.67

Metric equivalents: ton × 0.907 184 7 = t, gal × 0.003 785 4 = m³.

Table 15. Maintenance Labor Cost per Ton, Years 6 and 7
(2.0 Years)
(Total Material Mined = 85,200 kt)

Function	Oprtg Shifts	Maint Labor, Cost/Shift	Total Cost, $ × 1000	Cost/Ton
Drilling				
12.25-in. drills	3,688	$110.44	$407.3	$0.0048
Blasting	N/A	N/A	—	—
Loading				
25-yd shovels	4,415	303.88	1341.6	0.0157
Hauling				
170-ton trucks	21,100	63.77	1,341.6	0.0158
Roads and dumps				
R.T. dozers	2,208	45.18	99.8	0.0012
Track dozers	3,112	102.31	318.4	0.0037
Graders	2,135	54.80	117.0	0.0014
Water trucks	1,425	37.24	53.1	0.0006
General mine	N/A	N/A	—	
General maintenance	N/A	N/A	1863.6	0.0219
General administration	N/A	N/A	—	—
Total			$5546.3	$0.0651

Metric equivalents: in × 2.54 = cm, cu yd × 0.764 554 9 = m³, ton × 0.907 184 7 = t.

Table 16. Parts and Consumables Cost per Ton, Years 6 and 7
(2.0 Years)
(Total Material Mined = 85,200 kt)

Function	Oprtg Shifts	P&C Cost/Shift	Total Cost ($ × 1000)	Cost/Ton
Drilling				
12.25-in. drills	3,688	$542.71	$2,001.5	$0.0235
Blasting	N/A	N/A	4,260.0	0.0500
Loading				
25-yd shovels	4,415	927.42	4094.6	0.0481
Hauling				
170-ton trucks	21,100	508.90	10,737.8	0.1260
Roads and dumps				
R.T. dozers	2,208	174.01	384.2	0.0045
Track dozers	3,112	297.64	926.3	0.0109
Graders	2,135	110.64	236.2	0.0028
Water trucks	1,425	158.90	226.4	0.0027
General mine*	N/A	N/A	1,704.0	0.0200
General maintenance†	N/A	N/A	3,408.0	0.0400
General administration‡	N/A	N/A	480.0	0.0056
Total			$28,459.4	$0.3341

* Based on allowance of $0.02/ton for mine utilities.
† Based on allowance of $0.04/ton for gasoline and small vehicle repairs/maintenance.
‡ Based on general supplies at $20,000 per month.
Metric equivalents: in. × 2.54 = cm, cu yd × 0.764 554 9 = m³, ton × 0.907 184 7 = t.

overhead cost per ton for years 6 and 7 are calculated as follows:

Salaried labor cost	= $	4,318.0
Salaried fringes	=	1,511.3
Hourly fringes	=	4,280.1
Total ($ × 1000)	= $	10,109.4
Total kt mined	= ÷	85,200
Cost/ton	= $	0.1187

Operating Cost Summaries

A detailed summary may be prepared for each mining period in which each functional cost is subdivided into operating labor, maintenance labor, parts and consumables, and mine overhead. Table 17 summarizes the mining period of years 6 and 7 in this format. In Table 18, functional costs per total ton are summarized for each mining period and also expressed as total cost and cost per ton of ore.

Table 17. Cost per Total Ton by Function, Years 6 and 7
(2.0 Years)
(Total Material Mined = 85,200 kt)

Function	Oprtg Labor	Maint Labor	P&C	Mine Overhead	Total
Drilling	0.0091	0.0048	0.0235	—	0.0374
Blasting	0.0048	—	0.0500	—	0.0548
Loading	0.0100	0.0157	0.0481	—	0.0738
Hauling	0.0304	0.0158	0.1260	—	0.1722
Roads and dumps	0.0184	0.0069	0.0209	—	0.0462
General mine	0.0058	—	0.0200	—	0.0258
General maintenance	—	0.0219	0.0400	—	0.0619
General administration	—	—	0.0056	0.1187	0.1243
Total	0.0785	0.0651	0.3341	0.1187	0.5964

Table 18. Operating Cost Summary

Period	Total, Tons × 1000	Ore, Tons × 1000	Unit Cost ($/Ton of Total Material)									Total Cost, $ × 1000	$/Ton Ore
			Drill	Blast	Load	Haul	Roads and Dumps	Gen. Mine	Gen. Maint.	Gen. Admin.	Total		
pp	91,500	—	0.0401	0.0611	0.0713	0.1793	0.0587	0.0273	0.0639	0.1287	0.6404	$ 58,597	—
1	59,000	17,500	0.0346	0.0535	0.0723	0.1847	0.0435	0.0245	C.0587	0.0984	0.5702	33,642	1.92
2	61,200	17,500	0.0348	0.0541	0.0726	0.1759	0.0446	0.0251	0.0592	0.1017	0.5680	34,762	1.99
3	60,400	17,500	0.0347	0.0540	0.0725	0.1742	0.0448	0.0255	0.0599	0.1052	0.5708	34,476	1.97
4	58,200	17,500	0.0350	0.543	0.0729	0.1686	0.0453	0.0255	0.0602	0.1092	0.5710	33,232	1.90
5	53,500	17,500	0.0362	0.0545	0.0735	0.1698	0.0459	0.0258	0.0610	0.1163	0.5830	31,191	1.78
6&7	85,200	35,000	0.0374	0.0548	0.0738	0.1722	0.0462	0.0258	0.0619	0.1243	0.5964	50,813	1.45
8-10	120,825	52,500	0.0381	0.0550	0.0740	0.2695	0.0487	0.0261	0.0624	0.1357	0.7095	85,725	1.63
11-15	170,000	87,500	0.0385	0.0551	0.0744	0.3412	0.0548	0.0278	0.0637	0.1402	0.7957	135,269	1.55
16-20	111,250	97,500	0.0467	0.0572	0.0766	0.4007	0.0625	0.0301	0.0658	0.2351	0.9747	108,435	1.24
Total	871,075	350,000									0.6965	606,142	1.73

Metric equivalent: ton × 0.907 184 7 = t

Chapter 8

Management and Organization

Donald O. Rausch, Editor

8.1 Introduction

DONALD O. RAUSCH

An integral part of any mining entity is management of the mineral, financial, and personal resources of the firm. Although management is not unique to this industry, managing the complexities of an efficient surface mining operation is certainly a challenging experience. This chapter gives the reader an overview into many of the primary considerations faced by various levels of mine management.

One of the paradoxes that everyone faces at one time or another is that corporate and middle management can often be at odds with each other. Theoretically all levels should be pulling together for the common good of the corporation, and yet major internal conflicts can tend to be the rule rather than the exception. These conflicts generally exist because each management level has different perspectives than do other management levels. For example, corporate management tends toward objectives that are longer-term, broad, idealistic and perhaps, even, somewhat ill-defined; their concern is for maintenance and growth of the corporation. Conversely, mine and other middle level management have objectives that are more precise and exacting, tending toward the shorter term; their concern is to keep the corporation solvent through production and sales, allowing for a future by keeping present business circumstances healthy.

More specifically, there has been a movement afoot since publication of the first edition of *Surface Mining* toward management involvement in *strategic planning*. Strategic planning impacts all levels of management, although the effort is most often initiated at the corporate level. This type of planning can be loosely defined as "where the enterprise is going and how it will get there."

Strategic planning is an especially useful tool in mining because the industry is very capital-intensive, experiences long paybacks, extreme cyclical swings, and has depleting assets. Strategic planning enables the mining firm to analyze its strengths and weaknesses to optimally develop its business. However, strategic planning is only a tool and is not a substitute for good management.

Good managers recognize their jobs as not just developing and applying policies and procedures but also as working with people—people with varying abilities and aspirations and with unique feelings and thoughts. Thus a major consideration in mining enterprises is management of human resources. Human resource management not only deals with such things as recruiting, selection, and compensation, but also with training and development, performance measurement, communication, and union/management relations.

While applying policies and procedures managers must

always realize one of their major functions is control. A well-managed enterprise is an active (not a reactive) entity; effective action can only be achieved through a well-maintained system of control. Critical to good financial control is effective reporting. Generally, the smallest responsibility area of a mining company's concern is the cost center. Cost centers are usually the operating, maintenence service, and administrative functions. Accurate and timely cost reports are imperative if these functions are to manage effectively; i.e., the cost control cycle must provide for an effective analysis of costs as they occur. Other important considerations are effective budgeting, procedures, and production reporting. All of these different facets should be integrated allowing for effective management analysis of the mining concern.

Too often individuals in mining tend to overlook, or at least, not to recognize the full importance of the sales and marketing of mined products. This is unfortunate because efforts expended toward marketing mined products directly impact the mining enterprise; i.e., sales generate revenues and without revenues no company survives for very long. Marketing is, however, more than just selling—it actually is the complicated interaction of four different functions: product, distribution, promotion, and pricing. This interaction touches on an array of ancillary specialties as well, and ultimately affects directly or indirectly everyone in the company.

In today's society the concern of permitting has become increasingly important for the mine operator. One piece of legislation which has had a major impact for many in the industry is the Surface Mining Control and Reclamation Act of 1977. Interpretation, compliance, and internal enforcement of this act usually comes under the realm of the corporate legal department. Close coordination between the legal staff and the mine operator can greatly reduce, or even eliminate, penalties as well as mitigate possible confrontations with federal inspectors.

Since management seems to be constantly faced with limited cash flows and high interest rates, it is even more important today that cost effective purchasing and inventory procedures be practiced. Besides these more traditional concerns, today's materials manager is now, more than ever, involved with such diverse things as the sizing and location of warehouse facilities to the complexities of information and data processing. Thus purchasing decisions, inventory management, storage, and maintaining spare parts can have a significant impact on the corporate bottom line.

A final major involvement by management is in government and public affairs. Everyday more and more corporations are coming under close public scrutiny; presentation of a positive corporate image should be a top priority of management. Compliance with the myriad of relatively recent Congressional Acts is a necessity for any mining concern. Furthermore, it is imperative for mining firms to communicate to respective legislators their views on both existing and possible regulations; no one can better express the views of mining companies than the companies themselves.

The following sections will give the reader better insight into problems and solutions confronting all levels of mining corporate management.

8.2 Management Philosophies

CHARLES W. BERRY

DAVID E. FLETCHER

In addressing the issue of managerial philosophies and comparing ideas to determine if there are differences between corporate management and mine management, it can be stated that the first and most obvious difference is that corporate management is *goal-oriented* while management at the mine level is *objective-oriented*. The major differences between goals and objectives are that goals are broad, idealistic, not necessarily achievable phenomena, and relatively difficult to measure. An example of a corporate goal might be stated, "It is a goal of this firm to be one of the leaders in the industry serving the consumer with a high quality product, providing safe, meaningful, and secure kinds of work for its employees, while at the same time being a strong, secure, and high return investment for stockholders."

Goals are stipulated by policy statements and originate from the representatives of the ownership of the firm and the top levels of management. Goals are the things toward which activities are directed, but by their very nature are not specific enough to lend themselves to exact measurement. In the preceding example, as desirable as all of those things are that are suggested, there is very little that can be done to actually determine if any of those goals have been met.

Objectives on the other hand are specific, precise, exact statements of what is to be accomplished by the various elements of the firm. Being specific and precise, objectives can be measured in some commonly accepted and understood manner, such as, quantity produced, time to produce, or meeting deadlines and scheduled times. Unit costs should be measured in terms of final product from the operation rather than in terms of intermediate products, e.g., cost per pound of refined copper is much more meaningful than cost per ton of copper ore.

Objectives also must be supportive of the goals of the organization. One method of visualizing the relationship of goals and objectives is to view goals as parabolic in form and covering the entire range of activities of the firm, while objectives are the pillars or supports that maintain or hold up the parabola (see Fig. 1). Another major difference can be derived from this illustration; goals are representative of the totality of corporate activities while objectives are the ac-

complishments of individual parts of the organization. For example, every industrial organization has a need for some kind of production activity and a product or service. There is a need for some kind of marketing and distribution objective, and there is a need for finance, i.e., the development of financial reserves and resources and some method of allocating those assets. These are the traditional line activities found in any organization, regardless of size or complexity. If the firm is to achieve any degree of success it must achieve planned, predetermined objectives in each of these areas.

The well-managed firm will also have established methods and procedures for evaluating the achievements of the various parts of the organization. Included in this evaluation are the positions, functions, and personnel hierarchy related to the objectives supporting the goals of the firm (see Fig. 2). Concomitant with the evaluation process to determine the consistency of what has been accomplished with what was planned, a determination should be made with regard to the integration of objectives of various parts of the firm. The objectives developed by each part of the organization must not only be supportive of corporate goals, but must be integrated with and mutually supportive of the objectives of every other part of the firm. Extending this idea one step further, it can be seen that cooperation among the various parts of the organization is necessary if this approach to management is to be successful. Much has been said about the viciousness and cruelty of intercorporate battles over scarce resources. Whether these admonitions and caveats are true is not the issue, but what can be said is that the recognition of the objectives of various parts of the organization and the interdependence of these objectives is a viable way to develop the cooperation necessary for the success of the total corporate structure; Fig. 3 illustrates this interdependence. As can be seen, the structure can possibly survive the nonachievement of one objective or perhaps even two, but if others fail as well, the total structure will collapse.

The remarks in the preceding paragraphs are neither new nor startling; rather they are a brief discussion of the management style or philosophy identified as *management by objectives*. To summarize these ideas in the context of cor-

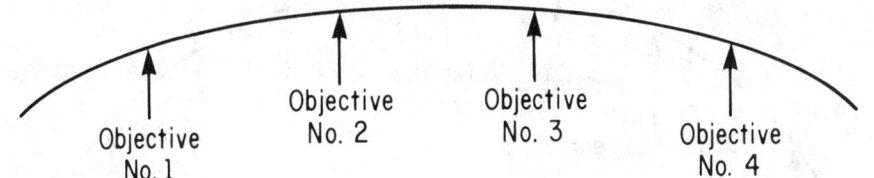

Corporate Goal
Parabolic in nature, covering the entire spectrum of organizational activities

Objective No. 1   Objective No. 2   Objective No. 3   Objective No. 4

Objectives
Specific, measurable, attainable, interdependent, and supportive of corporate goals

Fig. 1. Goal-objectives relationship.

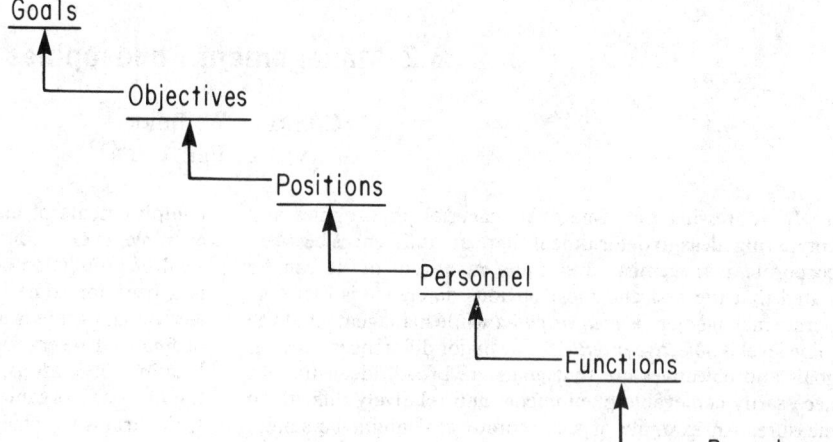

Fig. 2. Procedures-to-goals hierarchy.

porate goals and mine management, it can be seen that the corporate goals are supported by the objectives of mine management. In the mining industry, the production and operations aspect of the firm can be associated with specific objectives such as quantities of ore taken from the ore body, the grade of ore being recovered, the number of tons of ore delivered, at what cost, etc. The idea of management by objectives, then, seems reasonable for the mining industry. Further it can be observed that it is necessary to integrate the objectives of the operational aspects of the mining venture with the objectives of the finance element of the organization and the marketing objectives of the firm.

Anyone who is looking at corporate goals and the objectives of the various elements that make up the corporation can see that there is ample opportunity for conflict among and between departmental objectives, as each department has its own ideas about priorities of function and need for resources and rewards. A classic example in the mining industry is this: A corporate goal may be to provide the public with a needed natural resource without destroying the environment; on the other hand, a corporate goal may also be to provide a high return to the stockholders of the firm. Whatever procedure is selected to extract the minerals from the earth will have to be done recognizing the inherent conflict in these goals. Obviously there are methods of mining that are more destructive to the environment than others and these methods may be the most cost effective in terms of production. Other methods that may be less obvious and intrusive to the terrain may have significantly greater costs

associated with them with the attendant result of reduced earnings.

These policy decisions with regard to goals of the corporation will have a limiting effect on the objectives of various components of the organization. Mining operations will be limited in the techniques that can be used, the finance people will have to allocate resources of the firm based on whatever constraints the goals provide, and the marketing function will have additional limitations on the price that they can obtain.

Within the organization there are other potential conflicts. The production element may come in conflict with the finance and marketing components over such basic issues as who gets what share of resources in terms of money, manpower, equipment, as well as over issues such as authority, power, responsibility, and accountability. Consider the potential for different priorities in a situation where marketing wants materials available in certain quantities, at certain grades, and at a very specific time in order to meet the needs of a customer. To meet these demands, production management will probably have a case for soliciting more manpower, equipment, and funds from the financial managers. Finance will then find that to meet its objectives and to respond to the needs of the operating unit, it will have to require a higher price for the firm's product, which obviously again falls back on the marketing managers. Add to this the desire of operations managers to have the option of using what they perceive to be the most efficient and cost effective methods. They would like and need to have operations that are

Fig. 3. Interdependence of objectives to support goals. There is no significance to the location of the various functions in this model. Marketing, for example, is not at the top for any particular reason. The placement has been random, although an individual organization may wish to use this design to show priority of function.

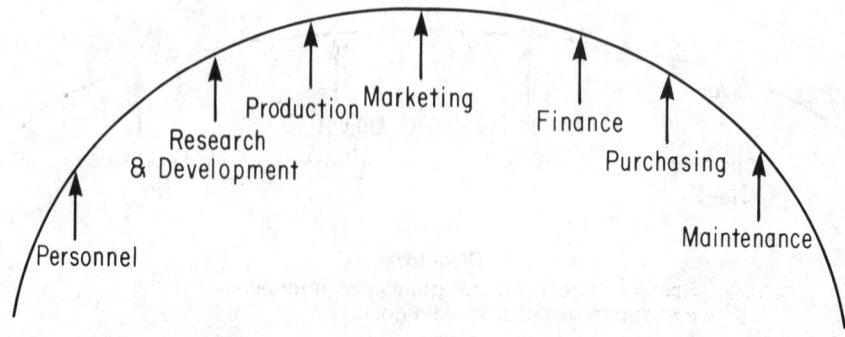

safe and satisfying for their workers. These desires and aspirations may, although not necessarily, come in conflict with the finance managers who would like to invest as little as possible to obtain the highest possible earnings to investment ratios.

This discussion could continue for a much greater period than is devoted in these pages. The solution to these issues is to have an increased understanding of the problems and objectives of all the parts of the organization. It is difficult, if not impossible, for a manager of any phase of the firm to fully appreciate the position of another element of the organization if there is not a comprehensive understanding, not only of what other parts of the organization are trying to achieve, but also how the accomplishment of those objectives integrate with his or her own departmental objectives.

A final comment about the potential for intraorganizational conflict is that managers of the various elements of the organization, whether in the line functions of production, finance, and marketing, or in the traditional staff positions of personnel, research and development, purchasing, or whatever other staff functions may exist, must recognize that as much as they may wish to achieve their departmental objectives, they cannot do so solely at the expense of some other element of the organization. The result of this conflicting strategy is clear; the corporate goals cannot be jointly supported because some of the supportive objectives cannot be achieved and eventually the whole organization will fail. Good, clear communications among departments is essential for sound management.

Management in the mining industry has significant dif-

ferences from management in other industries. These differences are listed in Table 1.

Mining is very capital-intensive with investment in plant, property, and equipment normally single purpose; thus sound capital investment planning is essential. Market for product is derived demand rather than direct demand, e.g., the consumer purchases an automobile whose beginning raw materials are iron ore, coal, limestone, lead, zinc, copper, etc., rather than purchasing those mineral commodities directly. The remoteness of most mines is important in considering labor infrastructure and transportation. With mines, the ore reserve is finite; replacement ore bodies must be explored for and developed. Each mine is rather unique and each normally requires special engineering; in manufacturing and merchandising, many factories and stores are duplicates of each other.

The authors believe that it is important to highlight the differences between mining and petroleum. The commodity prices for most mining products are more changeable than petroleum products. In the exploration stage, petroleum normally requires large expenditures relative to total investment. However, in mining, most investment capital is in the development-construction phase rather than in exploration. In petroleum, once a reservoir is discovered, production lead time can be less than a year. Mining, on the other hand, requires several years for development and construction after an ore body is discovered. Because of the long pre-production period and changeable commodity prices in mining, timing of the development-construction-production sequence is financially critical.

Table 1. Differences Between Mining and Other Industries

	Mineral	Other
Investment	Most single purpose Highly capital-intensive	Many multipurpose Less capital-intensive
Demand	Mostly derived Little product differentiation	Both direct and derived Product differentiation
Location	Cannot move mine Infrastructure critical Labor problems in remote areas	Can select site Infrastructure minimized by site selection Can go to good labor
Raw Materials	Ore Reserve Time and money to find Difficulty in estimating Depletable	Purchase Multiple suppliers
Production	Each mine unique Variable raw material Extraction important	Similarity between operations Uniform raw materials

Differences Between Mining and Petroleum

Mining	Petroleum
Exploration investment low relative to total investment	Exploration investment high relative to total investment
Once reserve is found, long time (6 years) for feasibility, permitting, and construction	Once reservoir is found, short time (6 months) for construction (except some offshore)
Relatively large construction investment	Relatively low construction investment

8.3 Strategic Planning

MILTON H. WARD

INTRODUCTION

Eric Hoffer states, "The only way to predict the future is to have the power to shape it." How to create this power is the subject of this section.

Business, or strategic, planning is the determination of where the enterprise wants to go and how it expects to get there. It involves the critical long-range thinking that assures fulfillment of the most important management responsibilities, survival and continued profitable growth of the business. Over the past few decades, a quasi-standardized method of strategic planning has been developed, generally accepted procedures formulated, and the practices of many companies reported in great detail. Specific programs and results of practitioners are well documented, and information on the subject has been reported in a number of excellent volumes and business publications. An intense interest in strategic planning developed during the 1970s; in the early part of the current decade (1980s), it was a highly marketable and profitable product for management consultants. Strategic planning now has its own special techniques, nuances, and lexicon, and its use in some form is considered a necessity for the progressive company.

Space limitations do not permit an in-depth review of all facets of this subject, but a brief description of the formal steps and techniques is required. Equally important are observations and opinions as to which steps are critical, and how these can be applied to the mineral industry. Both the techniques and the philosophy will be addressed, and in addition, a number of highly regarded books, texts, and resource materials are listed in the bibliography for those who desire to pursue this subject further. It should be noted early on that there are varying opinions on the merits of strategic planning; especially questioned is the value of the complex techniques of strategy development. This is a legitimate challenge, and while a host of practices are reported herein, all are not advocated. The wise planner will, as with any decision, review the options and select only those that fit his particular business and level of planning sophistication.

JUSTIFICATION

It has been said that few industries have a greater need for strategic planning than mining. This need results from the very nature of the industry: It is capital-intensive, involves long payback, is based on depleting assets (ore bodies), and usually deals in products that are inevitably battered by ever-repeating business cycles. Additionally, many of the parameters of the assets are unknown until the deposits are completely extracted. Former President Herbert Hoover, a renowned mining engineer and highly successful businessman, said of the business "... We must plunge in, learn, and repent. Not only is the useful life of our mining works indeterminate, but their very character is uncertain in advance."

Good business planning can do little to alter the size or shape of the ore body, but it can instill flexibility and prepare the company for change. Change will come as deposits are extracted. A mine experiences a life cycle similar to that of many products, which in the parlance of strategic planning leads to a "planning gap" (Fig. 1). This term describes the shortfall between desired sales and earnings performance of the enterprise, and what is more likely to occur if enough new earning sources (new mines) are not added to the company's asset portfolio. This gap develops because most products, industries, or mines follow a life cycle that results in a tapering off of earnings in the latter stages of their lives. Various terms have been coined to describe these stages as illustrated in Fig. 2. Development, growth, maturity, and aging are typical.

In mining, the *Development* stage is the period when the company is completing a mine and just beginning to produce. It is often characterized by low earnings, high debt, and small dividends. The next stage, *Growth* or *Expansion,* has good earnings, increasing output, improving efficiencies, and allows payment of dividends. This phase is followed by *Maturity,* when the design or high levels of production are achieved, operating costs are low, earnings are at a peak, and dividends are increased. The final stage, *Aging,* encompasses the period when earnings are decreasing and the product or mine moves into the twilight of its years. Mine operations in this stage are characterized by lower grade, poor efficiencies, and high costs, and if growth of the company is to be sustained, a new replacement operation should already be developed and production under way. Anticipating the need for additional growth, new mines, and a satisfactory mineral inventory, plus development of a plan for acquiring these, are tasks of the strategic planner. An example of how asset additions may be made to sustain growth is illustrated in Fig. 3.

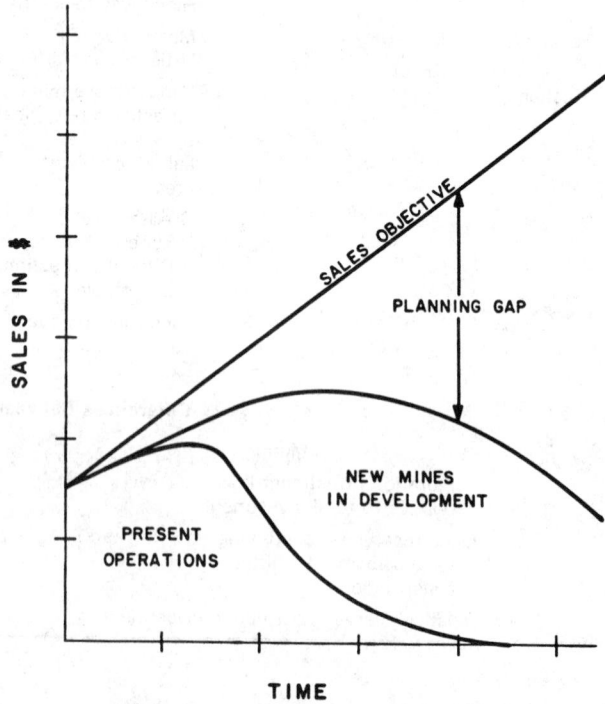

Fig. 1. The planning gap.

1026

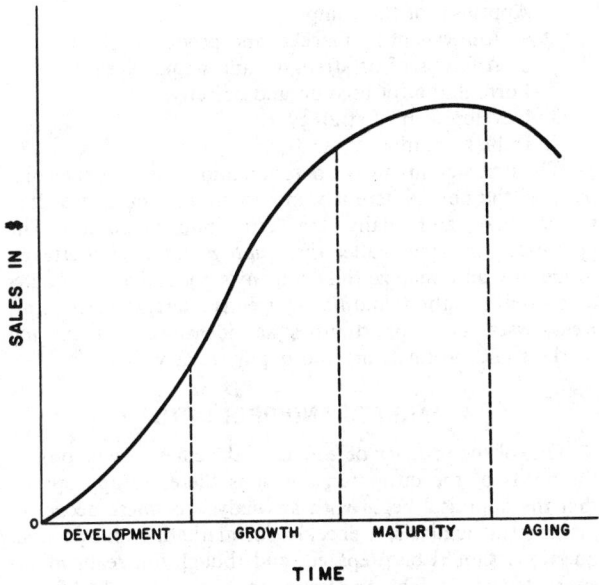

Fig. 2. Product or mine life cycle.

Few companies, however, can show such an attractive growth profile, and even if a company is successful in finding and developing new mining properties, few can produce a solid record of increasing earnings. The upward trend is frequently thwarted by problems and delays in mine development and disruptions caused by the low phase of the business cycle. Independent analysis shows that the majority of new operations experience an extended time delay in attaining their design level of production, and that a high percentage of these have significant cost overruns. Such startup problems, coupled with more frequent and deeper recessions and the incresingly difficult task of finding low-cost deposits, clearly challenge the objectives of increasing growth and earnings.

The history of free enterprise nations tracks a cyclical pattern, with expanding Gross National Product (GNP)

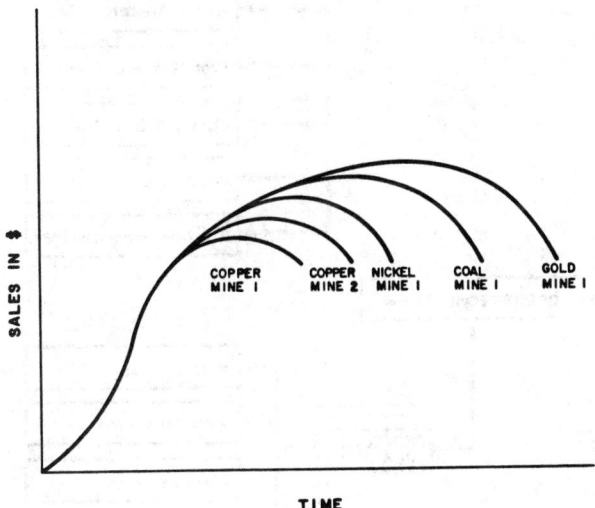

Fig. 3. Extending earnings life.

phases followed by periods of contraction (recessions). Producers of mineral and metal commodities are all too familiar with the problems that accompany recessions: increasing inventories and declining unit prices. Traditionally, the level of inventories has been a critical leading indicator of future activity in any business. Increasing stocks often portend a decline in prices, while a depletion of inventories forecasts a pick up in activity and higher prices. Unfortunately, producers are slow to accept the fact that a change is under way and that inventories are truly rising, particularly if the buildup follows an active, extended period of growth. When times are good, the tendency is to believe that the recessionary phase can be postponed, and when business is horrid, businessmen tend to think that the depressed state of affairs will never end. This feeling of either euphoria or overwhelming depression is understandable since commodity prices generally reflect either the nadir or peak of the cycle. But the cycle does repeat; at least it has twenty times since 1900 and eight times since World War II. In each instance, fortunately, the recession was followed by a period of recovery. Proper strategic planning should allow the enterprise to better prepare for, and react to, these changes. Proper planning requires certain conservative action to ameliorate the severity of the recession and wise investment and positive steps to capitalize on the higher-price periods that accompany or closely trail the boom phase of the cycle.

How well a company adapts to change, formulates its action plans, and develops new strategies for future growth will determine whether it survives and prospers. A good strategic plan can be an invaluable tool for guiding management, tempering optimistic projections, and reducing *seat-of-the-pants* decision-making. Granted, planning cannot eliminate the business cycle or place mineralization in a barren horizon, but it can indicate the trend of things to come and suggest methods for coping with the future. The case for formal planning is based on its ability to:

1. Encourage systematic management thinking.
2. Lead to better coordination of company efforts.
3. Foster controls and standards.
4. Force an identification of objectives.
5. Better prepare for unexpected developments.
6. Better define management's responsibility.

Strategic planning cannot solve all problems, but it will prevent many, and it can reduce the number of poor decisions.

TYPES OF PLANNING

Recorded history shows an evolution of planning over several millenia, ranging from the accomplishments of the Egyptians 5000 years ago and Alexander's exploits almost 2500 years ago down to the much-touted production and marketing achievements of the Japanese today. In recent decades, the strategic planning function has received an exceptionally large amount of attention, with a variety of classifications being formulated. The purpose of planning determines its classification and terminology, and often different management writers use different terms to describe the same type of planning. Also, purposes may blend or overlap one another, but practically all systems attempt to differentiate between strategic and operational planning.

Strategic Planning

This broadest or highest level of planning is concerned with the objectives and goals of the firm, where it wants to go, and how it expects to get there. It is under the aegis of top management and is principally concerned with the economic, financial, and technical aspects of the business environment, with particular attention given to the future of the enterprise.

Operational Planning

This form of planning is involved with short-term aspects of the business, and unfortunately, it often consumes most if not practically all of management's planning attention. It is concerned with maximization of return and problems of budgeting, scheduling, monitoring, and controlling the daily or short-term (one to three years) activities of the firm.

Demarcation of strategic and operational planning is not easy or clear-cut, and classification is further complicated by the use of additional terms and definitions. Tactical (means and goals for attaining objectives), year-to-date (budgetary or shorter-term), and administrative (combination of staff and operational) planning are terms that further define and circumscribe the types of planning.

There is also a diversity of planning concepts or systems that incorporate the total range of planning, and a good example is shown in Fig. 4 (Stanford Research Institute). This formidable-looking flowsheet incorporates most of the plans required by a major business entity.

STEPS FOR DEVELOPING PLAN

Development of the strategic plan follows certain steps that are common to all types of business. These steps may be separated into four major categories:

1. Appraisal of the company:
   a. Analysis of its markets and products.
   b. Analysis of its strengths and weaknesses.
2. Formulation of mission and objectives.
3. Development of strategy.
4. Implementation of strategy.

The first step involves a determination of where the company is; the next, where it wants to go; the third, how it is to get there; and finally, the actual implementation. The appraisal, sometimes called the *situation audit,* is undertaken to identify and analyze the company's performance to date, its current situation, and its key trends. Here, the strengths, weaknesses, and opportunities are identified and existing markets and products are thoroughly reviewed.

MARKETS AND PRODUCTS

One of the primary objectives of the enterprise is to meet the needs of the customer, and it is therefore appropriate that the appraisal begin with an analysis of these needs. In making this analysis, a checklist of marketing and product questions should be prepared, and thoughtful, realistic answers developed. The questions should be tailored to fit the business under review—obviously, a consumer products company is different from the metal miner—but surprisingly the questions can be similar. The following might be typical:

Who are our customers?

How are the customers geographically distributed?

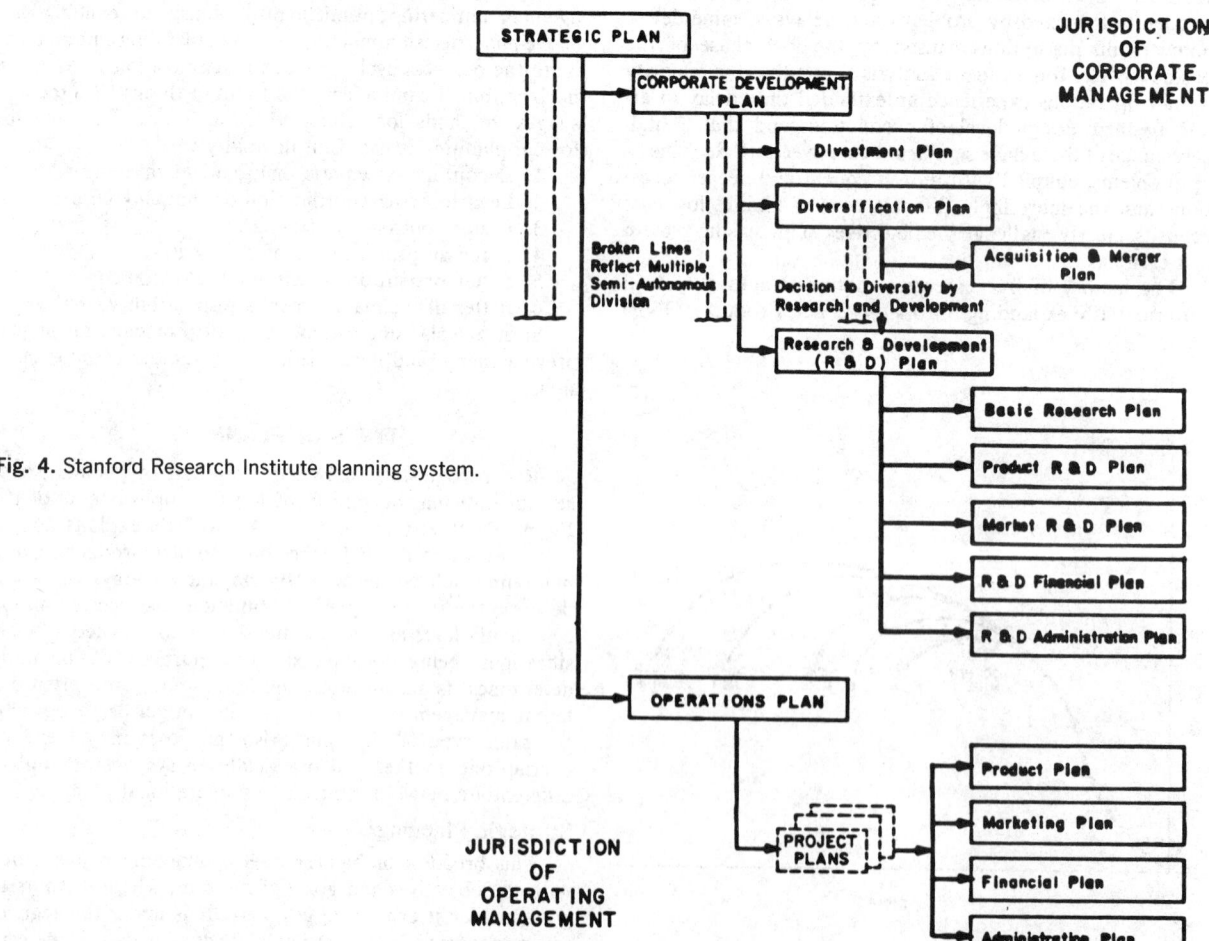

Fig. 4. Stanford Research Institute planning system.

Is the market developing, growing, maturing, or aging?

What is our market potential?

What percent of the market do we now have?

What is our cost position relative to other producers?

Is the market highly seasonal? Highly cyclical?

Is the market dominated by a few companies?

How difficult is entry?

What are the key factors with regard to competition; price, quality, or service?

What is the role of foreign competition?

Is there unrestricted entry to our market by foreign competition?

Is the need for our commodity expected to change in the near term, next decade or two?

Undertaking a thorough analysis of markets and the competition could be a new experience for the typical mining company. Indeed, observers often criticize the mining industry for not giving sufficient attention nor applying creativity to this important facet of the business. The appraisal should consider the questions tabulated above and the answers to these, plus a rigorous assessment of other marketing factors, all of which could well lead to new strategies. Creating a demand for a specific commodity, such as copper, is no mean task, since by the definition of commodity, one pound of copper is identical to a pound produced by the competition. The innovative marketer realizes that he must sell more than just copper. The sales story must contain a host of benefits, including the company's ability to provide:

A reliable supply of the metal.

Timely transportation and delivery.

An acceptable sales price.

Pricing flexibility that may involve discounts, premiums, back-pricing, or other methods that meet the needs of the customer.

Similarly, the customer should be made aware of the company's technical competency, high productivity, large ore reserves and ability to find more ore, and its capacity to fund the needs of the operation. The company's expertise in the futures market and how this special knowledge can work to the benefit of the customer should also be explained.

This tabulation of marketing factors will be helpful in both identifying other opportunities and in focusing on problems. The list is not complete, but answers to these questions can be quite sobering. A US copper miner making this appraisal prior to the 1981-84 recession might have seen a mature industry, easy entry, a true product commodity that can be readily produced by many other companies, and one susceptible to substitution. And maybe he would have been surprised to find that his cost position was dangerously high—in the upper 25th percentile of world producers. Analysis would also have shown that lower-cost foreign competition can easily enter and capture his market, that copper is the basic foreign exchange generator and employer of foreign competitors, and that this competition will maintain its production even when copper prices drop precipitously. Last, and perhaps most important, the US copper miner would have noted that copper is an exceptionally cyclical commodity with a price greatly affected by the business cycle. This appraisal should call for action; possibly action to make the company more competitive or steps to lead away from copper and into more recession-proof businesses.

ANALYSIS OF STRENGTHS AND WEAKNESSES

"He is a fool who tries to match his strength with the stronger," warned Hesiod. Determination of when and where to make the match depends on knowledge; this is why a review of the firm's strengths and weaknesses is critical. The review stimulates thinking and influences components of the plan. Needless to say, an honest, objective appraisal, like any self-analysis, is difficult to make. Often we magnify our strengths but must search for our weaknesses.

Strengths

The purpose of ascertaining the strengths of the enterprise is to identify areas in which protective action plans can be built. This review should cover financial, operating, personnel, and other assets, including special talents, patents, and facilities of the company. A listing of strengths for a diversified mining entity might include:

International operating experience.

Worldwide shipping knowledge.

Experience in the formation and operation of joint ventures.

Skills in managing low-cost mineral and metal operations.

Extensive reserves.

Marketable, workable patents for recovering byproducts.

Excellent, well-staffed research and development facilities.

A high quality, well-established mineral exploration team.

Exceptionally competent financial and legal groups.

A young, experienced management team.

A good cash flow situation.

Additional borrowing capacity.

Weaknesses

A French maxim notes that "One always knocks oneself on the sore place." It is amazing how weaknesses often manifest themselves at an inopportune time. A thorough search can identify sore spots and allow development of plans for overcoming or minimizing these. Shortcomings of the enterprise depend on the particular company, but a tabulation might show:

Diversity of products and markets is lacking.

High percentage of revenues flow from commodities.

Participation in only cyclical businesses.

Little influence over sale price of products.

Significant foreign investment exposure.

Poor record of projecting metal and mineral prices.

Limited marketing expertise in specialty minerals and metals.

No experience in making acquisitions.

Limited experience outside natural resource field.

Stock value unrecognized in the marketplace.

A thorough ongoing appraisal and critique of strengths and weaknesses usually expands these lists and often turns up unexpected additions—many favorable but some that merit corrective action. This phase of the planning process may justify the retention of an outside consultant or independent third party to assure objectivity.

CORPORATE MISSION

One of Webster's definitions for the word *mission* is "...a ministry commissioned by a religious organization to propagate its faith." This definition closely parallels that of *corporate mission,* which is the idea or statement that sets forth the direction and faith of the enterprise. Simply stated, a company's mission denotes the nature of its activities or the lines of business in which it will engage. Formulation of this statement requires profound thinking since it is the most important decision that senior management will make. "What is our business and what should it be?" must be asked at the inception of the enterprise and throughout its existence;

answering this is the primary responsibility of top management. The highly regarded management consultant, Peter Drucker, states that every great business builder from the Medici family and the founders of the Bank of England down to IBM's Thomas Watson had a definite idea and a clear theory of the business that was to be pursued.

Once determined, the mission can be conveyed to the organization and used as its guiding beacon. Some companies distribute their written charter to their constituents—management, workers, customers, and the public. Setting forth the mission in writing is recommended, and the practice is becoming more common, but many corporations still choose to guard this information from all except those who have a need to know. Other firms have no written mission—this is cloaked in the mind of the CEO, and he only metes out this information as required. He believes that secrecy restricts the flow of data to the competition and thwarts criticism or discontent from those who might hold different views about the future direction of the firm.

However, on balance, most would support articulation and distribution of the mission statement, if not externally, then at least within the organization. The advantages are several, including motivating top management to analyze the business and to actually think through and formulate the mission. It also provides a consistent, coordinated guideline for internal planning and actions, and informs the shareholder of the company's business so a more knowledgeable investment can be made. The mission statement can be brief or lengthy; it may speak only to the business with which the organization is charged, or it can be a meld of business purposes, aims, and philosophies of operation. Examples follow.

Actual Case—A Major Technical Corporation: "Texas Instruments exists to create, make, and market useful products and services to satisfy the needs of our customers throughout the world. Because economic wealth is essential to the development of our society, we measure ourselves by the extent to which we contribute to the economic wealth—as expressed by sales growth and asset return. We believe our effectiveness in serving our customers and contributing to the economic wealth of society will be determined by our innovative skills."

Actual Case—A Major Natural Resource Company: The following are excerpts from the company's annual report which outlines a portion of this entity's mission and businesses:

"M.I.M. Holdings Limited (MIM) is a mining and mineral processing company . . . its principal activities are: . . . Mining of copper and silver-lead-zinc ores . . . Refining of Mt. Isa's copper . . . Refining of lead and silver from Mt. Isa . . . Mining of coal . . . Transporting and stevedoring . . . Trading in nonferrous metals . . . Recycling of scrap lead . . . Exploration for base metals, coal, iron ore, gold, and uranium . . . Exploration for oil and gas . . ."

The public document from which this information was taken also notes the goals and major operating strategies of the enterprise.

Hypothetical Example—A Mining Company: "The Red Metal Mining Company is in the business of mining and milling copper ores."

Every business has its mission, whether it is stated in lofty inspiring words or only lightly etched in the mind of the CEO, and the entity that gives thoughtful attention to this is more likely to be satisfied with the results of its labors than the company uncertain of its reason for existence.

Objectives

Writings of strategic planning often include the words *goals* and *objectives* as though they were synonymous. In some companies they are, but in others *goals* are considered to be more short-term and of less importance than *objectives*. Generally, objectives last for years and relate to the broad issues of the enterprise, such as growth, profitability, stability, and image, while goals are more specific, focus internally, and are often more limited in duration.

The objectives which the organization seeks in order to fulfill its mission must be consistent with the fundamental purpose of the enterprise. Each objective should contribute to the mission, aid in decision-making, and give direction to operation of the firm. Objectives, which should be written, must be operational and serve as action guides for all critical areas of the company, and they must be conveyed clearly to those charged with achieving them. They should be challenging but attainable. The likelihood of fulfilling objectives should be reasonable, and in actual practice they are often designed with a 50% probability of success.

The following is an example of objectives that were developed several years ago by a major chemical and consumer-products company:

Maintain a minimum annual growth in earnings of 10%.
Raise dividends in line with earnings.
Reduce interest-bearing debt to 45% of total capital.
Increase the return on beginning-shareholders' equity to 18%.

Periodically, management of the company has reported its performance and success in achieving these goals, and to date exceptionally fine results have been communicated. Obviously publicly advertising objectives can lead to embarrassment if severe shortfalls occur. However, to be effective, objectives must be developed, discussed, and distributed internally, and many believe that giving public attention spurs performance and is well worth any loss of face that might accompany a shortfall in performance.

The objectives listed above relate to financial activities, and this sector of the business is critical, especially since survival requires profit and maintenance of a solid financial base. Therefore, it is appropriate that dollar-defined objectives be developed and pursued. However, formulation of goals should not cease at this point. Attention should also be given to social objectives and every facet of the business important enough to be the subject of planning.

Strategy

The action plans and programs undertaken to achieve corporate objectives are termed *strategies*. This is one of many similar definitions of strategy, and a few others are: 1) Strategy is the method whereby corporate objectives and goals are to be attained by the allocation of resources to activities; 2) Strategies consist of major action programs used to achieve missions and goals, and simply stated, 3) Strategies are how we get what we want. Objectives can be achieved in a variety of ways, but each method or path consumes resources of the company, and the path selected constitutes its strategy. The development of strategy requires a tabulation of many possible activities that may fulfill or meet the company's objectives, and then selection of those that can be afforded and which are most likely to lead to success.

It is apparent from earlier comments that the language of business planning is not without dispute. One writer's *mission* may be another's *objective* and one's *objectives* may be another's *goals*. A similar conflict exists between *strategies* and *tactics*, and in some writings the two words are synonymous. In others they are distinctly different, with *strategies* being developed first and *tactics* being the action taken to implement the strategies. As an example, a firm may have a strategy to enter the gold mining business by acquiring a smaller company that holds undeveloped gold properties. Its tactic could be to purchase the smaller company by the use of stock rather than cash. Understanding the specific definitions of planning is important, but grasping the concept and having the ability to formulate and implement it is critical.

The first step in setting strategy is identification of activities that the company should consider undertaking, followed by selection of actions that are more likely to facilitate achievement of company objectives, allocation of resources to the selected activities, assignment of responsibilities, and fixing a time schedule for action and completion, and finally, identification of measures to be monitored in assessing progress. Strategies, like objectives, should be developed for every key area of the business. The selection process involves an assessment and decision as to whether to formulate new strategies or to maintain or drop old ones.

Methods for developing strategies are numerous and diverse, and generation of new concepts is an ongoing activity in this field. Management consultants find this to be an attractive area for marketing their services, and many have created sophisticated models for identifying opportunities. Among these are portfolio strategy, product-market matrix, research and development possibilities, and a host of other methods.

Portfolio Strategy—This method utilizes special categorized information to guide strategic growth decisions. The concept, which is based on the relationship of market share to growth for different businesses, was developed by The Boston Consulting Group. It uses a matrix that consists of four quadrants as shown in Fig. 5.

Here, consideration is given to the product's many characteristics, including its profits, cash flow, and market po-

sition. Based on this information, the product is assigned to one of the four quadrants. The Boston Consulting Group termed the quadrants (and the products in them) *Opportunities?, Stars, Cash Cows,* and *Dogs. Opportunities?* is the title given to Quadrant 1 because it supposedly has potential to capture a large share of a rapidly growing market. The second quadrant, *Stars,* is so named because it encompasses the best performers which "have a large market share in an increasing market." *Cash Cows* in Quadrant 3 reflects the divisions or products that supply funds for Quadrants 1 and 2. *Dogs* is the term given to divisions and products that are lodged in Quadrant 4. These are the worst performers that have no obvious chance for improvement. In addition to the quadrants, the matrix contains circles located in the various quadrants, and the relative size of the circles represents the product's assets or its annual volume of output or level of sales. Matrices of all or a selected group of these elements are assessed in determining the prospects of a particular product, commodity, or business.

The portfolio terminology and method of analysis has been in vogue for several years, and can be helpful in analyzing a particular strategy. As an example, a company may choose to divest itself of the *Dogs* (perhaps a nickel operation) and use these funds for additional investment in *Stars* (maybe gold operations) and *Opportunities?* (possibly silver operations). This method of appraisal also forces a review of the relative importance of divisions, and if considered over a period of time, it indicates progression, or lack thereof. Is the *Opportunity* division really moving toward a *Star* or is it stagnant? Are the *Dogs* beginning to cost too much to retain, and are their markets declining? These and other important strategic questions can be partially appraised by reviewing the company's portfolio, and such analysis may be helpful in the development of a more attractive strategy. The method, however, is not without drawbacks, and many of its shortcomings will be discussed later.

Research and Development—Recently, the advantages of R&D have been the subject of many technical conferences, seminars, and discussions in industry and government. This activity can indeed generate new products and suggest strategies for expansion, and it is the primary growth base for many companies, especially those in the chemical, pharmaceutical, and electronics industries. Similarly, the R&D effort of mining companies can lead to growth opportunities. New processes can advance the viability of low-grade mineral resources, increase the recovery of values from tailings, and allow extraction of additional elements from a given ore. Examples include pelletizing lower-grade iron deposits to create a source of iron ore, development of new reagents that lead to higher recovery in flotation, and piloting of more sophisticated processes that allow recovery of uranium from phosphoric acid streams. Additionally, R&D generates new opportunities by allowing the company to market and license new technology.

Monitor the Competition—Observing activities of the competition is a worthwhile practice that should be followed by all organizations, especially the smaller company. An ongoing review of others can provide information on marketing, R&D, and operating methods and techniques. This approach is especially helpful when considering entry into a new business. Many corporations have made the decision not to try to be Number One (or maybe not even Number Two) in pioneering a new business. They prefer to participate later, after the entry risk has been reduced. The decision of a company to be the initial participant in an industry is greatly influenced by its attitude toward risk-taking. Some

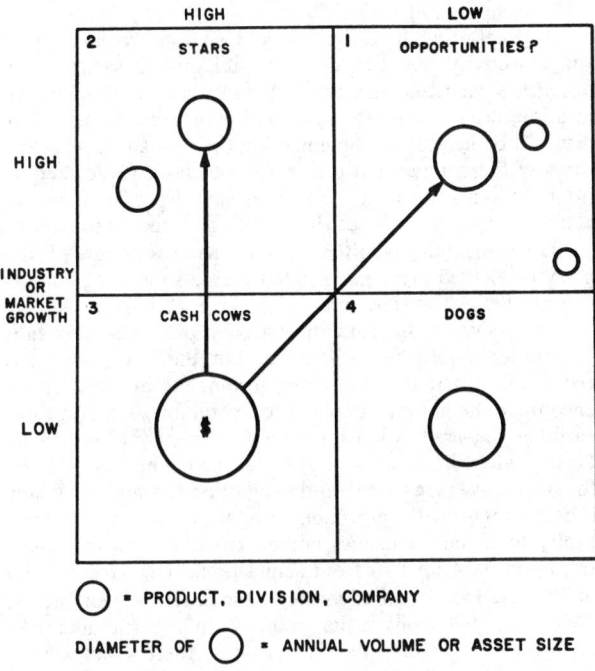

Fig. 5. Industry growth—market share.

mining companies would place seabed mining in this high-risk category. They believe the initial cost and risk level is too high, but they stand ready to try to carve out a niche if the competition is successful in its endeavors. This approach, which requires close monitoring of the competition, is also commonly used by explorationists. Here, the high cost of drilling a new risky area is avoided, but if the pioneering company is successful, the alert geologist moves in to stake around or adjacent to the area being drilled. This delayed-action strategy has advantages, but obviously it can also result in lost opportunities.

Soliciting Suggestions—The quest for new growth strategies can be extended out past top management and the formal planning group to include all members of the organization. This practice involves an explanation of the need for new ideas, followed by the use of the simple suggestion box. It is reported that this is a creative mainstay of Japanese industry, yielding one million suggestions for Toyota over a two-year period and resulting in savings of $250 million. Solicitation of employee ideas is still used extensively in the US, although in recent years the practice has waned in the mining industry.

Opportunistic Approach—Some companies prefer to wait alertly for Lady Luck to bring opportunities. The follower of this approach argues that the cost, risk, and effort required to force its way into a particular market at a specific time are too great. It chooses to maintain its financial and technical competency and be poised to act when the appropriate opportunity arrives. Disraeli's words that, "Everything comes if a man will only wait," are indeed appropriate for this tack. As an example, such a mining company would not identify a product or commodity for development, but instead would just wait for an attractive property to come to its attention. In instances of this type, the copper miner that desired diversification would make no decision to enter a particular business, but would consider gold, coal, or any favorable prospect that became available. This has been a rewarding strategy of the well-financed company, allowing it to purchase underpriced assets and operations during the heart of recessions. Also, this opportunistic approach has led to acquisitions by *white knights* that came to the rescue and acquired or merged with companies being pursued by unwanted suitors. Obviously there are disadvantages to this practice, the primary one being the inability to dictate the time and manner in which action and growth are undertaken.

Intuition of Top Management—Intuition is the power or faculty to attain knowledge without rational thought and inference, or simply stated, a quick and ready insight. This is a method, or at least appears to be a method, frequently utilized by CEOs in arriving at growth strategies. In many instances this practice, which is about as simple as possible, yields excellent results, leads to major new businesses, and has been the foundation for growth for a number of major corporations. However, in other instances this practice has led to decisions and investments that have been disastrous.

Strategy Selection and Action Programs

Huxley advises that, "The great end of life is not knowledge but action," and if results are to be attained, action must be taken. In industry, implementation of action plans can commence only after the strategy is fixed, and this should involve rigorous decision-making and detailed analytical appraisal. Reviewing current strategy is a good starting point, and in doing this, the planner should ask:

What is our present strategy?
What trends suggest future change?

Which portions of our strategy are still appropriate?
Which activities should be dropped?
Which should be altered? When? How?

It is possible that existing strategies should be abandoned if, for example, it is apparent that the program will fail for technical reasons (ore reserves are found to be lower grade or the new revolutionary process will not work), financial reasons (capital or operating costs for the business are going to be exorbitant), or marketing reasons (operations by others have provided a glut, prices have dropped, and the proposed operation will not be competitive).

It is easier to stay with an existing well-known activity, but if projected growth, R&D, or competitive analysis along with other factors indicate a bright outlook for the gold business, a medium attractiveness for copper, and a dismal future for nickel, all things being equal, the company should consider directing its new investment to the favorable sector. This decision, however, should be scrutinized and challenged by asking if the new strategy is consistent with the existing and future business environments, the company's resources, and its management experience and capabilities. And, it should be determined whether the company is truly prepared to accept the risks of pursuing a new strategy and seeing it through. This analysis requires thoughtful answers to a myriad of questions, of which the following are only a small sampling:

Is the strategy likely to initiate antitrust opposition?
Does the strategy bring the enterprise in direct competition with a powerful adversary?
Does the strategy utilize the company's strengths and avoid its weaknesses?
Does the company have managers to implement the strategy?
Is sufficient capital available or can it be obtained to see the strategy through, even if a recession develops at a critical point during the development period?
Does the strategy take the company too far from its current market and expertise?
Are the basic decision data and assumptions reasonably accurate and realistic?

A penetrating investigation and analysis requires time and objectivity, but can be a valuable aid in selecting or rejecting a particular strategy. If the strategy stands up under scrutiny, often the investigative effort provides information that will be helpful in implementing the new strategy. Also, study of the competition may turn up candidates for acquisition or joint venturing, properties that are in need of exploration funding, or areas that justify grassroot exploration.

Diversification is often selected as a strategy of the growth-oriented firm, and a noted business planning author, George Steiner, states that the managerial imperatives are ". . . diversify or die, and diversify by plan." Some might take exception to this statement, but the low prices and reduction in demand that often accompany business cycles encourage the mining company to search for countercyclical earnings. Diversification means a movement into new markets, product lines, processes, or services. The justifications for such moves are varied, and in addition to stability include a desire for growth, a broader product line, and an opportunity to reduce costs. As noted earlier, growth can come from internal efforts or from acquisition. Peter Grace, CEO of W. R. Grace, is an advocate of the latter. He comments: "My basic philosophy is that you have to be big to take risks and you have to take risks to grow. If there's one way to get big without buying companies, I haven't found it yet." On the other hand, H. Barclay Morley, CEO of Stauffer

Chemical, takes the opposite view, espousing a strategy of internal development. Both companies have been exceptionally successful, and both methods of growth should be considered, but the successful strategy will depend on the philosophy of top management and the particular strengths within the enterprise. Peter Grace has done a fine job of restructuring his company and managing growth by acquisition, but many others have been less fortunate.

Volumes have been written on acquisitions, mergers, and diversification, and while indepth coverage cannot be given here, several key points should be made. First, statistics and analysis do not support the thesis that mergers and acquisitions improve corporate performance and shareholder investment. And secondly, analysis does indicate that diversification and growth are most effective if made in related fields and businesses. Furthermore, studies have shown that most acquisitions, especially acquisitions of huge companies, do not work out. Obviously participants do not proudly advertise their failures, but some experts say that seven out of ten acquisitions fall short of their initial objectives. A noted specialist in corporate mergers states that "... of the 50,000 mergers undertaken over the past 30 years, over half of the companies acquired by larger companies have been weakened, damaged, or totally destroyed." This is a strong indictment of mergers, and the acquisition-minded strategist should be aware of the record and the odds faced in pursuing such growth.

Similar studies show that firms that diversify by developing or acquiring units related to their existing activities are more profitable than those that pursue strategies of unrelated diversification. Further analysis indicates that the most important dimension of relatedness is that of management competence and attitude. This idea is also supported by a comprehensive study done by UCLA's Richard Rumelt who found that "... related-constrained (controlled) diversification strategies unquestionably gave the best performance." Rumelt also states that "... these companies have strategies of entering only those businesses that build on, draw strength from, and enlarge some central strength or competence." While such firms frequently develop new products and enter new businesses, they are loath to invest in areas that are unfamiliar to management. The central point of Rumelt's study is that companies that branch out somewhat, yet still stay close to their primary skill, outperform all others. Few mining companies have been successful in diversification well outside mining, and even fewer can show excellent results.

An obvious reason for staying close to familiar fields is to use existing capability. If the goal of the enterprise is to be the premium producer and marketer of gold in the US and its strategic plan gives no consideration to personnel having the capabilities to accomplish these tasks, it may be wishful thinking to assume the existing staff is competent and able to perform the required tasks. The firm that succeeds is the one that is willing to develop the clear, superior capabilities that are decisive to success. Skills can be learned, and if the need exists, the wise company will attempt to gain the required experience before entering the business.

IMPLEMENTATION

Strategy implementation requires decisions about organization, communication methods, control procedures, and performance measurements. Achievement of these tasks requires practices and attitudes for performing most major functions. Of paramount importance is a keen interest on the part of top management and a planned program of implementation.

Organization

Responsibility for implementing the strategy must be assigned, and the timing, methods, financial needs, and potential markets determined. Proper coordination will require the attention of those directly and indirectly involved, but also the support and monitoring of top management. It is also likely that some form of new organizational structure will be required. If so, there should be a willingness to alter the structure since the existing organization may hinder new efforts and act as a roadblock to progress. Alfred Chandler, the noted business historian, coined the phrase, "Structure follows strategy," in his 1962 landmark study, *Strategy and Structure,* and this concept is still embraced by management specialists. It means simply that the organizational structure should be tailored to the strategy to be pursued.

An alteration in strategy can indeed lead to a realignment of the organizational structure as some activities may be expanded while others are curtailed. If the new strategy and organization are not in tune, it is likely that the desired objective will not be achieved. Ideally, the staff and management that formulate the plan would also be involved in its implementation. It is understandable that those who helped shape the program will more readily appreciate and accept its objectives and strategy, and in addition should be more qualified to successfully shepherd it to fruition. Also, most organizations prefer to assign implementation of the new tasks to existing managers, especially if the new activity is related to an ongoing business. Lastly, the structure of the organization should be as simple or complex as required, but the simpler the structure, the better.

Communication

As noted above, utilizing the planning team in the implementation phase is invaluable, and nowhere is this more apparent than in communicating and disseminating the program. Proper action can only follow if all involved understand the purpose of proposed strategies, and few are more knowledgeable about this than the draftsmen. The chief communicator should be the firm's CEO since his opinions are highly valued and his enthusiastic support illustrates a strong commitment to the organization. His best communication device is the written plan itself. If properly stated and distributed, it will give the basic strategy, proposed action, and methods for attaining the desired objectives. Whether special written instructions should be utilized depends on the number of individuals involved in implementation, the amount of time that will pass before the work is completed, the degree of complexity or detailed information involved, and if the matter is of such importance that special written steps are warranted to eliminate misunderstandings.

Controls and Performance Measurement

The primary objective of controls is to steer action toward the desired goals. The process starts with the manager agreeing with his superior about the results that he is expected to achieve within a certain period of time. This agreement should be based on an understanding about the organization, the desired goals, how performance will be measured, and the freedom and restrictions of the manager. At the end of the period, performance is appraised, shortcomings and better-than-expected results are discussed, and objectives for the future are established. Motivation to perform is based on a variety of rewards, including financial gains, recognition, social status, promotional opportunities, and a personal sense of achievement. The key to any incentive program is establishing a belief that appraisals will be undertaken and that they will be objective, with rewards given fairly.

PROBLEMS AND CAVEATS OF PLANNING

While many think that strategic planning, if performed *properly*, assures success, the cold light of reality often shows this not to be the case. Some of the original assumptions may be invalid, underestimated, or just plain wrong, and if so, and a continuation of the strategy strongly indicates the likelihood of failure, a prompt change of direction or a complete halt to implementation is justified. Any plan or program should be flexible, as it would be foolhardy to follow a strategy when new facts or better knowledge show some key assumptions to be wrong.

Similarly, if the programs that formed the basis for the strategy change, modifications should be initiated. As an example, if funding for entry into a new business is dependent upon cash flow from ongoing operations and a severe recession develops, prudence would call for an alteration of plans. Additionally, if trends and events turn out dramatically different from estimate, the company should be prepared to act accordingly. Frequently projections and assumptions are wrong. Competent planners and managers will attempt to foresee pitfalls and fix the limits and deviations that may be tolerated from the original plan, and they should be prepared to respond (stop, go slow, sell off) if the actual situation deviates markedly from the forecast. Well-thought-out sensitivity, *worst case* and *what if* (negatively inclined), analyses enable the enterprise to act in a rational, panic-free manner if assumptions prove inaccurate. Fear, uncertainty, and over-concern for disaster can cause indecision and inaction, but failure to challenge new corporate directions can be devastating.

Obviously the sooner an error-based strategy is detected, the less problems and damage it will cause. In many instances this does not mean abandonment of the original strategy, but it may call for significant modification of plans or a deferral of action. As an example, if construction of a major new silver operation for a well-established copper company is based on revenue from its copper business, consideration should be given to selling forward a portion of the projected copper production or arranging financing, even though it may appear that borrowing is unnecessary. Likewise, loan commitment fees that give security and provide funds for completion of a project may be a small price to pay for allowing construction to proceed on plan. Similarly, contingency plans can be developed for cost overruns, changes in markets, technical errors, and other critical assumptions, and such plans are best considered in the early stages of planning rather than when disaster strikes.

There are innumerable pitfalls in developing and executing the strategic plan, and recognition of these can circumvent problems and reduce frustrating and time-consuming deliberation. Improper management of the planning function itself can ring the death knell. Slow implementation decisions, too much reliance on committees, too many planners, poor communication of plans, and lack of support at the top are commonly listed as reasons for failure. Chief among these is disinterest or lack of CEO support. Commitment of the CEO is required of many management programs and functions, but it is critical to strategic planning because this activity is often viewed as superfluous. While it can be complex and confusing to the uninitiated, and is time-consuming, it should not be perceived as interfering with placing rock in the box. The firm hand of the top man is required, but more important is the fact that the strategic plan should reflect the thinking and goals of the CEO. It is impossible for his desires and judgments to be incorporated without his direct involvement.

Planning Is No Substitute for Management

Strategic planning is no substitute for good management. If management fails to mind the store, the store may go broke while the sophisticated plan is being assembled. Maintaining production, low costs, high productivity, a good balance sheet, and control of the enterprise are still necessary functions and will remain the key responsibilities of management. Several years ago the planning of a particular international natural resource company was held up as the model for our industry. It had large market research, forecasting, and strategic planning groups, and its bold acquisition and expansion activities were supposed to give a diversity that could withstand the cycles of our industry. Time has passed, and in the 1981-84 recession, the company had more than its share of difficulties, and it is doubtful that it will ever recover to its prerecession strength and lofty industry position. A number of other resource companies had similar experiences, and this was attributed to a disregard for basic business principles (poor marketing, elevated debt level, high costs, and excessive overhead) and a belief that higher prices would persist indefinitely. These comments are not an indictment of strategic planning, but again prove that proper management is a multifaceted activity requiring constant attention.

In the ever-changing field of management science, and with the challenges created by the severe swings of the business cycle, few planning concepts withstand criticism indefinitely. Currently it is questioned whether strategic planning is the primary force that moves a corporation forward, or whether planning and its required structure are just two of several key factors. Many management consultants are now marketing an ability to assist business in developing skills and programs for *all* the critical factors. The McKinsey 7-S Framework (Fig. 6) has been a highly touted concept within the last year or two. The 7-S approach calls for the *Strategy* and *Structure* of Strategic Planning, but these must be accompanied by *Staff* (the people of the enterprise), *Skills* (the company's know-how), *Style* (manner in which management

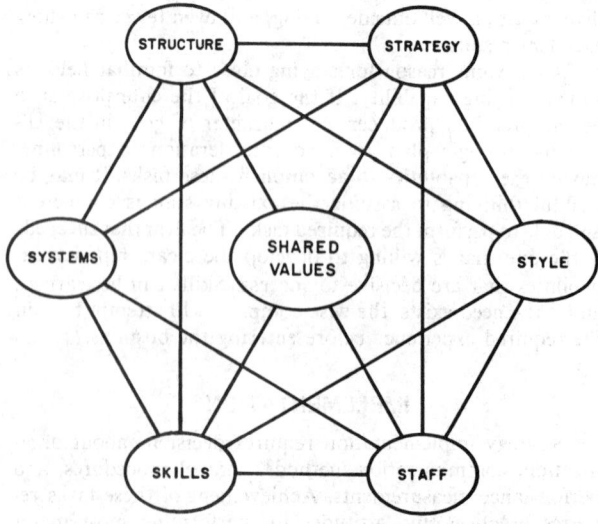

Fig. 6. McKinsey 7-S framework.

and employees conduct themselves), *Systems* (enterprise's methods of communication), and *Shared Values* (the so-called culture of the organization). These 7-Ss are further explained by Waterman of McKinsey and Co.:

Strategy—A coherent set of actions aimed at gaining a sustainable advantage over competition, improving position vis-a-vis customers, or allocating resources.

Structure—The organization chart and accompanying baggage that show who reports to whom and how tasks are both divided up and integrated.

Staff—The people in an organization. Here it is very useful to think not about individual personalities but about corporate demographics.

Style—Tangible evidence of what management considers important by the way it collectively spends time and attention and uses symbolic behavior. It is not what management says that is important, but the way management behaves.

Systems—The processes and flows that show how an organization gets things done from day to day (information systems, capital budgeting systems, manufacturing processes, quality control systems, and performance measurement systems all would be good examples).

Shared Values—(or superordinate goals). The values that go beyond, but might well include, simple goal statements in determining corporate destiny. To fit the concept, these values must be shared by most people in an organization.

Skills—A derivative of the rest. Skills are those capabilities that are possessed by an organization as a whole as opposed to the people in it. The concept of corporate skill as something different from the summation of the people in it seems difficult for many to grasp; however, some organizations that hire only the best and the brightest cannot get seemingly simple things done while others perform extraordinary feats with ordinary people.

Proponents of this concept claim that success is based on the proper interaction of the factors with one another and with the corporate environment, and for strategy to work, all the factors must fit. They must each support the strategy, and if not, changes must be made to accommodate the anomalous variables, or the strategy should be modified or abandoned.

Over the years other multivariable concepts have been proffered, including the *Management Diamond* (task, structure, people, information and control, and environment) by Harold Leavitt, and *Six Factors* (same as 7-S less *Skills*) by Harvard's Anthony Athos and Richard Pascale. It is apparent that there are many opinions on the necessary components of good management, and most would agree that strategic planning is just one of them.

Responsibility for Strategic Planning

All key elements (internal and external) of the business must be considered in formulating a strategic plan, and this requires inputs and analysis from practically every manager of the firm. Likewise, the development and execution of strategies involves the contribution, support, and action of operating managers as well as staff personnel. Normally the greater the degree of participation, the more effective the plan. Therefore, there are benefits in assigning certain planning responsibilities to practically all managers. Clearly the CEO is the prime mover, and his participation and commitment must be conveyed to all concerned. He should have the broadest perspective of the firm's business today and its potential tomorrow, and therefore he must be viewed as the chief planner. Granted, assistance can be drawn from the entire organization, and a small planning staff may be jus-

tified, but the CEO must make the key planning decisions. Many companies have no separately responsible staff for planning; others have only one or two individuals; and as would be expected, some have large, well-supported departments. Usually the size and complexity of planning staffs depend on the size and complexity of the organization, the style and preferences of top management, and the existing organizational structure of the firm.

Others are also involved in planning. The CEO is the key architect, but his decisions should be critiqued by the board of directors. This group has the responsibility for seeing that the company has an effective strategy, and obviously this requires that they understand what constitutes an effective strategy and know the strategy is pursued. The CEO should keep the board informed by at least reporting the highlights of the planning process, and in some companies, all major policy and strategy decisions are discussed by the board prior to taking action. The latter approach allows better feedback and is more likely to give management the benefit of the board's thinking and judgment. In summary, many are involved and have responsibilities for strategic planning, managers, senior officers, the CEO, and the board of directors, and all can contribute, but the CEO is the linchpin.

Now that the justification for strategic planning has been considered, its formal procedures scanned, and some of its shortcomings noted, the question to ask is: Is it really worth the effort? This author believes the answer is a resounding, but qualified, *Yes.* The qualifications relate primarily to the company's existing level and degree of planning and to the procedures that should be followed.

If in the past the company has done little strategic planning, it should move forward with deliberation. Action should generally follow the flowsheet in Fig. 7, but in small steps. After gaining experience and confidence in the method, more intricate and comprehensive planning can be considered. One should resist the temptation to launch a full-scale campaign until the knowledge and acceptance level of the organization have evolved to the point where these new ideas can be assimilated.

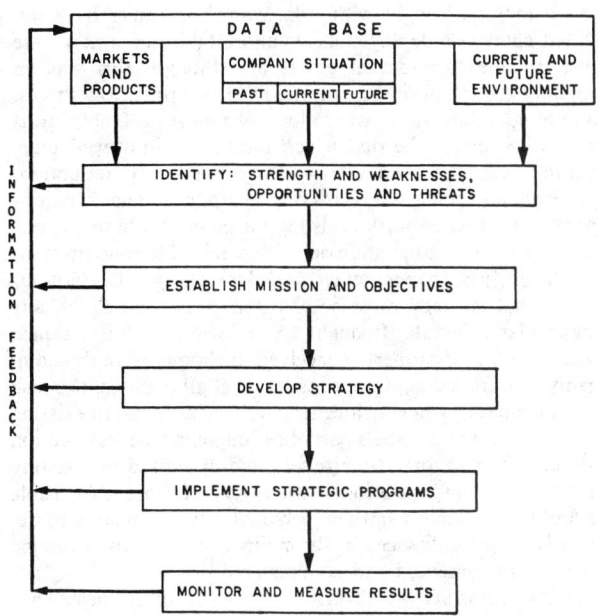

Fig. 7. Strategic planning—action flow.

Planning can be initiated in a gathering of the top corps of management during which the merits and need for strategic planning are discussed, questions are asked and answered about what the firm should look like five years in the future, and preliminary decisions are reached about how to guide the enterprise along a path for fulfilling this new ambition. To be sure, all these decisions cannot be made in one session, but the preliminary discussion can be used to obtain the commitment of senior management and set the stage for future planning. The steps of Fig. 7 are recommended, but they should be tempered as follows:

Strengths and Weaknesses—The study of the company's strengths, weaknesses, markets, and competition is invaluable. This analysis points out the threats and opportunities of the enterprise, and serves as the basis for all other planning—therefore it should be done well, incorporating inputs and ideas from all levels of the organization.

Missions and Objectives—The business of the company (its mission) and goals (objectives) that it must pursue to fulfill its mission should be understood by all employees. The initial list of objectives should be limited and relate only to those key activities that may make or break the business. The list can include minimum financial objectives (targets of sales, net income, return on investment), objectives that will maintain competitiveness of the company (productivity, cost control, methods improvement, diverse markets), and objectives that focus on the people and spirit of the organization. In the beginning, objectives should not be overly ambitious. Give the system a chance to evolve and be accepted.

Strategy Development—Keep it simple. Use the ideas indicated by the appraisal, new concepts of top management, quality approaches suggested by R&D, and worthwhile practices of the competition. Do not be reluctant to wait for the opportunities that will come to the competent, well-managed company. Go slow in relying on the complex fashionable techniques of strategy development. Portfolio theory, product-market matrices, and other fashionable, sophisticated methods may have value, but they can be time-consuming, frustrating, and difficult to apply to the minerals industry. A strategy that leads to dramatic restructuring, challenging acquisitions, and new, *exciting growth* business has appeal, but it may be fraught with risk. Many companies have initiated change, redeployed assets and established a new base just in time to find that others also thought this was an *exciting growth* business. Inevitably, oversupply, low prices, and tough competition will follow. The most profitable strategy will probably be that which focuses on improved marketing practices, productivity increases, cost reduction, incremental expansion, recovering byproducts, and other opportunities related to the existing business. While this is not as glamorous as other approaches, it can lead to solid growth.

Strategy Implementation and Monitoring—Do not assume that strategy must be changed every year or on any particular schedule. It ought to be reviewed and critiqued annually, but if properly conceived, it should serve the company well for years, if not decades. If at all possible, the task of implementing new strategies should be assigned to existing management and made a part of the ongoing business. Action should be monitored, controlled, and appraised by existing methods and organizational structure, and those responsible should be aware that their personal advancement and rewards will be influenced by the manner in which the strategic plan is implemented and administered.

Diversification—A diversity of products or businesses has value, particularly if the products follow a countercyclical pricing trend; that is, if the bottom of the copper market does not occur at the same time as a slump in phosphoric acid (fertilizer) sales. Recognize that diversification can be accomplished without going too far afield. The impact of metal cycles may be mitigated to a large extent by developing or producing a variety of metals, certain industrial minerals, specialty chemicals, or energy fuels. In a deep, severe world recession, however, few if any businesses will be fully immune from decreased demand and low prices.

The value of broad scale diversification can and has been challenged. Statistics support the conclusion that nondiversified companies, on average, achieve faster growth and superior returns on investment. This is the reason that restructuring, a return to the core business, and divestment of unrelated activities are now in vogue. Relatedly, the manager that can create efficiency, reduce costs, and increase output is now in demand. How so? Why and where did the broad gage strategist and portfolio manager go? Well, times change, philosophy and management practices shift, and the needs of the enterprise are different. The sensitive manager responds. The natural resource company must seek growth, especially since its resources (ore bodies) are depleting, but many would say this should only be in related areas—areas in which management has most of the skills and knowledge required for success. If a move to a completely different business, such as consumer goods or service sectors, seems justified because of unusual circumstances or an exceptional opportunity, this can best be accomplished with partners that bring experience and knowledge to the venture.

In placing strategic planning in perspective, one should keep in mind the primary responsibilities of management—to assure survival and profitable growth of the enterprise—and realize that this is accomplished by performing a few tasks and performing them well. The 7-S, 6-S, or X-S key functions of management discussed earlier are worthwhile and should be understood and possibly applied, but none should overwhelm the others. Strategy development is an important and necessary activity, but the key strategy should be to first manage the fundamentals of the business, then seek the new.

Productivity and performance studies have shown that the average business fails to realize half its potential, and if this is the case, the guiding principle of Theodore Roosevelt, ". . . to do what you can, with what you have, where you are," should be the first step in strategic planning.

REFERENCES

Ackoff, R.L., 1981, *Creating the Corporate Future*, John Wiley and Sons, Inc., New York, NY.

Ansoff, H.I., 1965, *Corporate Strategy*, McGraw Hill, New York, NY.

Cannon, J.T., 1968, *Business Strategy and Policy*, Harcourt, Brace and World, New York, NY.

Chandler, A.D., 1962, *Strategy and Structure*, MIT Press, Cambridge, MA.

Davidson, K.M., 1981, "Looking At The Strategic Impact of Mergers," *The Journal of Business Strategy*, Vol. 2, No. 1, Summer, Warran, Gorham and Lamont, Inc., Boston, MA, pp. 13-22.

Drucker, P.F., 1973, *Management Tasks, Responsibilities, Practices*, Harper and Row, New York, NY.

Drucker, P.F., 1980, *Managing In Turbulent Times*, Harper and Row, New York, NY.

Ellis, D.J., Pekar, P.P., Jr., 1980, *Planning for Non Planners*, AMACOM, New York, NY.

Gluck, F.W., Kaufman, S.P., Walleck, A.S., 1980, "Strategic Management For Competitive Advantage," *Harvard Business Review*, Vol. 58, No. 4, July-August, pp. 154-161.

Gup, B.E., 1980, *Guide to Strategic Planning*, McGraw Hill, New York, NY.

Haspelagh, P., 1982, *Portfolio Planning: Uses and Limits,* The Boston Consulting Group.

Hodgetts, R.M., 1979, *Management: Theory, Process and Practice,* W.B. Saunders Co., New York, NY.

Hoover, H.C., 1909, *Principles of Mining,* McGraw Hill, New York, NY.

Kiechel, Walter, III, 1982, "Corporate Strategist Under Fire," *Fortune,* Vol. 106, No. 13, Dec. 27, pp. 33-39.

Lavenstein, M.C. and Skinner, W., 1980, "Formulating A Strategy of Superior Resources," *The Journal of Business Strategy,* Vol. 1, No. 1, Summer, Warren, Gorham and Lamont, Inc., Boston, MA, pp. 4-10.

Moyer, R., 1982, "Strategic Planning In Coal," *Mining Congress Journal,* June, American Mining Congress, Washington, D.C.

Navin, T.R., 1978, *Copper Mining and Management,* University of Arizona Press, Tucson, AZ.

O'Connor, R., 1980, *Preparing Managers for Planning,* The Conference Board, New York, NY.

Peters, T.J., and Waterman, R.H., Jr., 1982, *In Search of Excellence,* Harper and Row, New York, NY.

Porter, M.E., 1980, *Competitive Strategy,* Free Press, Collier McMillan, London.

Rothschield, W.E., 1976, *Putting It All Together, A Guide to Strategic Thinking,* AMACOM, New York, NY.

St. Thomas, C.E., 1965, *Practical Business Planning,* American Management Association, New York, NY.

Sherman, P.M., 1982, *Strategic Planning for Technological Industries,* Addison-Wesley Publishing Co., Reading, MA.

Steiner, G.A., 1969, *Top Management Planning,* McMillian Company, London.

Tregoe, B.B., and Zimmerman, J.W., 1980, *Top Management Strategy,* Simon and Schuster, New York, NY.

Ward, M.H., 1979, "Selection of Target Metals and Minerals—Management's Key Decision," *Mining Congress Journal,* Vol. 69, No. 2, Feb., American Mining Congress, Washington, D.C.

Waterman, R.H., 1982, "The Seven Elements of Strategic Fit," *The Journal of Business Strategy,* Vol. 2, No. 3, Winter, Warren, Gorham and Lamont, Inc., Boston, MA, pp. 69-73.

8.4 Human Resource Management

ANNETTE DUPREE

INTRODUCTION

A company has available many resources to ensure success and profitability. Among these, the human resource is a most valuable one. After all, what is an organization made up of but people at work.

The administration, maintenance, and management of people in a company is called *human resource management*. In the last decade, human resource management has evolved from a narrow concern to a general management awareness of the important role employees play in the success of a company.

Earlier human resource administration concentrated on maintenance aspects like hiring, firing, and paperwork. Today's human resource responsibility is much broader and is a basic management responsibility, from the President or Chief Executive Officer on down. Personnel administration permeates all levels and types of managers and every manager needs to be an effective human resource administrator.

The personnel specialists in a company aids managers in getting more effective results from their people. Specialists provide advice, counsel, and procedures for consistent administration of personnel policies. The human resource specialists play a strategic role, linking their activities with the overall business objectives of the company.

Human resource management is not simply a matter of developing and applying policies and procedures, but a matter of working with people who have unique feelings and thoughts. It is concerned with the worker as an individual, worker satisfaction, communication, motivation, and productivity.

> The decade of the eighties brings many new challenges to the field of Human Resource Management in the mining industry. Technology will continue to expand as it has since the late fifties but at a quicker pace. Just as capital expansions, the emergence of mining and mineral conglomerates and labor confrontations dominated the first half of this century, the future of mining Human Resource Management will center around the development of employees, the elimination of meaningless work and the mutual success of the employee and the organization working together for their mutual benefit (Scobel, 1982).

Human resource management has several functions:
- Recruiting and selection.
- Compensation.
- Training and development.
- Performance measurement.
- Communication.
- Union-management relations.

These functions are all interrelated and they do not work independently of each other.

RECRUITING AND SELECTION

Having a sound selection system is one way to insure a solid foundation of human resources. The success of a company, in many ways, is dependent upon its ability to recruit, select, place, and retain qualified people.

Recruiting is anything done to fill a vacancy. Recruiting may be directed internally by use of inter or intradepartmental promotion or transfer, or externally where people are sought from outside the company.

To a great extent, the success achieved in hiring competent people is a direct result of recruiting. A large enough pool of applicants must be attracted so no compromise is made on the selection standards.

What method is chosen to recruit depends on the nature of the job, the immediacy of need, and the cost. Sources available for recruiting include college placement offices, trade or business associations, newspaper advertisements, employee referrals, public employment services, private employment agencies, and advertisements in trade journals. Effective recruiting should be done on a continual basis, not only when a job opening occurs.

The objective of the employment interview is to gather enough information about the applicant concerning skills and interests to make a hiring decision. Very simply, the objective of the employment interview is to match the applicant to the job.

Employment interviewing can be more successful by planning and preparing for the interview. Most interviews are started with little preparation. The steps in preparation are:
1. Determining the job content.
2. Developing questions.
3. Examining the application blank.

Determining the content involves breaking down the job into the duties, skills, experience, and education needed to perform the job. It is most important that the requirements be specifically related to the job and are minimum qualifications. The job content can come from a prepared job description, the manager, or a person who has performed the job.

The primary purpose of questioning is to get information from the applicant as a basis for making a sound hiring decision. Effective questioning stimulates the applicant to think and motivates him to discuss. Of the two types of questions, directive and nondirective, the latter is usually most appropriate. Nondirective, or open, questions are ones that do not lead the applicant to a yes or no answer. Open questions, like "How do you feel about your last job?" encourage applicants to think and share their ideas and observations. Directive questions are best when a specific answer is needed. Otherwise, directive questions lead the applicant to the desired answer.

The application form contains a lot of information and should be reviewed carefully. It gives the interviewer information about the skill and experience level of the applicant. Be thorough in reviewing the application, considering the way in which the form was filled out. Is it neatly done? Is it complete, signed, and dated?

Choose a private place to interview the applicant, one where there will be no interruptions or phone calls. At first, spend a few minutes talking to help the applicant feel at ease. Outline the interview to let the applicant know in advance what is going to happen during the interview. The more the interviewer can create a pleasant environment and put the applicant at ease, the more productive and effective the interview will be.

During the interview obtain information from the applicant relating to training, work experience, job progress, earning progress, relationships with supervisors, and reasons for leaving previous jobs. Also, the interviewer should give information concerning the specifics of the job, the company, the benefits, and other things the applicant needs to know.

After the information is obtained, it must be analyzed and evaluated. A proper match of the individual to the job must be made. It is important to take the guesswork, or gut feelings, out of the decision-making process and base hiring decisions on job related factors. The interviewer must have a clear understanding of the job, what he or she is looking for and if it was found in the interview.

COMPENSATION

The basis of the employee-employer relationship is that employees are hired to do work for which they are compensated by the employer.

Compensation addresses two issues. One is direct compensation, such as wages and salaries. The other is indirect compensation or benefits. Establishing compensation objectives and developing plans and programs that will help achieve these objectives is very important.

Objectives, once developed, should be communicated to all employees. Each employee at a mine, for example, should understand that levels of pay are determined by job skills and responsibilities, what other mines are paying for comparable jobs, how well the employee performs his job, and how long the employee has been in the job.

Generally accepted principles for effective wage and salary administration have been identified through the years as relating to one or more of the following:

1. A fair day's pay for a fair day's work.
2. Attract, retain, and motivate high performing people.
3. Encourage and increase productivity.
4. Maintain internal and external equity among jobs within the company and realistically compare jobs to similar ones outside the company.
5. Budget and control.
6. Lessen complaints and antagonisms.

For the employer, the cost of compensating employees is not limited to wages and salaries. A large percentage of payroll costs is in the form of indirect compensation or benefits.

Indirect compensation should provide employees with a choice of benefits that meet their needs and wants. Benefits can be group medical, life, accident, and dental insurance or in the form of paid time off such as holidays, vacations, and leaves of absence. Capital accumulation programs in the form of savings and investment plans, profit sharing, and employee stock ownership plans are becoming more important as the employee sees the value of the dollar declining through inflation, the instability of the Social Security System, and the high cost of education, housing, etc.

TRAINING AND DEVELOPMENT

Whether or not a company has a formal training program, training is going on all the time. Employees, through time, pick up skills, habits, and attitudes of others in the workplace.

Training is an investment in a person so he can make greater contributions to the company's future. Training can be on-the-job, classroom instruction, apprenticeships, or self-designed programs.

An effective training program can:
1. Improve employee performance, quality, and quantity.

2. Prepare the employee to meet new challenges in technology, product, or environment.
3. Promote and transfer employees.
4. Improve employee morale, reduce absenteeism and turnover.
5. Lead to fewer on-the-job accidents.

Training needs should first be assessed. Who will benefit? What type of training is needed? A system should also be set up to evaluate the effectiveness of the training. Training charts can be helpful in assessing and meeting training needs. See Fig. 1.

All new employees, regardless of experience and education, need to be introduced to the new work environment and taught the specifics of the new job. The initial induction, or orientation, is where training and development begins.

It is important that employees start off in the right direction, feel welcome, and know the policies under which they will be working. Orientation should include an introduction to the company (history and future plans), the compensation package, the company product or service, how the employee fits into the overall picture, and the basic work rules and policies. See Fig. 2, Checklist for Employee Orientation, for an outline of topics to be covered. In many ways the success of the new employee depends on the amount of time spent and the amount of information covered in the orientation process.

PERFORMANCE MEASUREMENT

Increasing productivity is a concern of many companies today. Saying a "200% increase in productivity is needed" has little meaning to an employee. It must be translated into the specific contribution the employee can and should make.

Of the variety of plans and approaches available to increase productivity, one thread that seems to run through all of the approaches is meaningful and timely feedback on employee performance. Performance measurement and evaluation is a tool to better communicate work expectations.

Objective performance standards are the key to a successful performance appraisal program. A performance standard is the performance condition that will exist when a job is being done according to the expectation of the company and the employee. Performance standards are written from job-related criteria that are measurable, attainable, and observable.

Measurability refers to how well the job is being done, i.e., loads per day, how fast loaded. Measurability of standards looks at quality, quantity, time and possibly, cost factors.

Of course, standards should be written so the employee can attain them. Performance standards should be practical and realistic.

Observability is the ability to actually see the standard of performance being met. Can the manager and employee see the levels of measurability being met? Can the manager and the employee see progress or lack of progress in job performance?

Without standards, a performance system might be based on personality traits, like attitude, initiative, or cooperation. This results in judging the employee as a person. The method of measurement is unclear and is based on the subjectivity or opinion of the rater. Systems wihout well communicated and defined standards of performance can promote defensiveness in the employee being evaluated and result in disagreements in performance rating. The employee does not know what is expected and may find out when it is too late.

Performance appraisal is not an every-six-month or once-

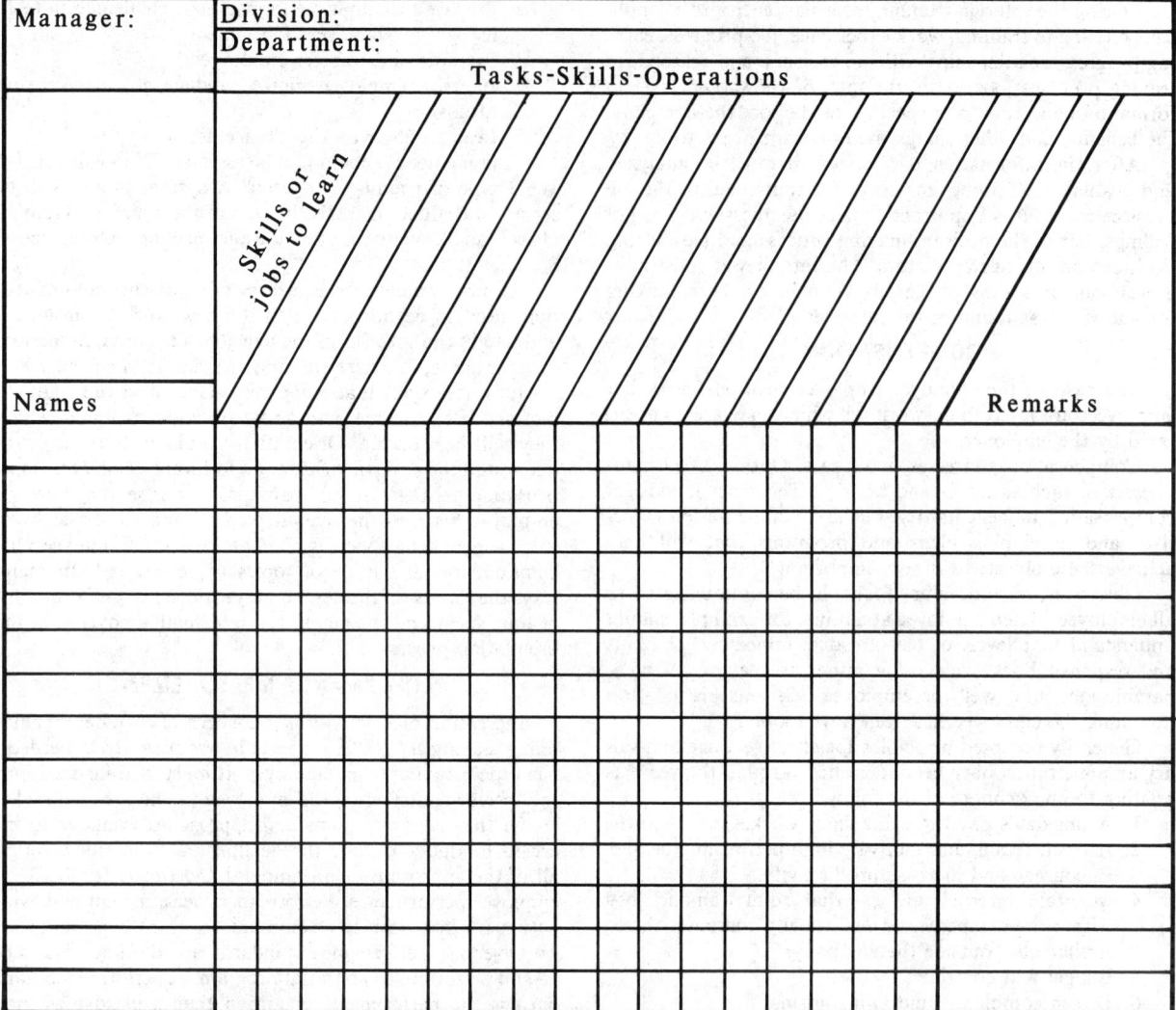

Fig. 1. Training Chart

a-year occurrence. It is daily feedback and daily activity for the manager and the employee.

A *formal* performance interview is putting in writing what has been collected throughout the performance period. There should be no surprises in store for the employee.

Performance measurement or management systems must also look at mutual goal setting and development to be successful and effective. Employees have a stronger commitment to goals if they participate in the development.

COMMUNICATION

A communication program is a vital part of the human resource effort. Communication systems can greatly assist work flow and create a climate of understanding in a company. In fact, communication is such an essential part that any discussion of human resource management deals with communication.

Recruiting and selection, matching the applicant to the job, is done through communication. Interviewing is a conversation with the purpose of making a selection decision.

Acceptance by employees of a wage and salary plan is achieved through communication. The more the employee knows about the way a company pays, the greater the acceptance.

Training and performance appraisal is daily communication and interaction between management and employees.

The effectiveness of a company's policies and procedures will depend on the communication system. An employee handbook is an excellent device to communicate to the employee and his family, the benefits and policies of the company. A formal policies and procedures manual helps managers communicate and apply company policies and work rules in a fair and consistent manner. The communication process is especially important when new practices are introduced or changes are made in existing policies.

Other communication devices are suggestion systems, bulletin boards, and opinion surveys.

Communication is a two-way design, keeping employees informed and giving them an opportunity to express their ideas, questions, complaints, and concerns to management.

_____ 1. Location and use of time clock, time cards, or time sheets
_____ 2. Exact hours of work; lunch period; break periods
_____ 3. Location of dressing and rest rooms
_____ 4. Smoking rules
_____ 5. Parking facilities
_____ 6. Safety requirements and fire regulations
_____ 7. First aid facilities
_____ 8. Rate of pay, pay week, incentives, and overtime
_____ 9. Benefits: holidays, insurance, retirement, etc.
_____10. Introduce to coworkers; introduce to work area
_____11. Copy of employee handbook
_____12. Job assignment and instructions (prepare, tell, show, practice, and check)
_____13. Location and use of bulletin boards
_____14. Location and issue of tools and supplies
_____15. Use of telephone
_____16. Leaving department during working hours
_____17. Reporting absence or lateness
_____18. Discuss employee's relation to the department and the company (function of the department, how production is scheduled, how the department fits into the overall production operation, importance of job, etc.)
_____19. Tour of plant or office
_____20. Work standards—quality and quantity, improvement will come with experience, etc.
_____21. Reporting change of address, telephone number, marital status, emergency notification information, etc.
_____22. Importance of good housekeeping and time for clean-up—work station, department, plant and grounds, etc.
_____23. Recreation facilities, programs, social clubs, etc.
_____24. Performance appraisal, promotion determinations
_____25. Grievance or problem solving procedure

_____   _____
Employee Signature   Date

Fig. 2. Checklist for Employee Orientation

A necessary part of the communication effort is letting people know where their jobs fit in the company. The orderly arrangement of jobs and duties is of little value to employees unless they know what they are supposed to do and what their role is in meeting the overall company objectives.

Organizational structure is concerned with (1) role and task definition, (2) organization charts, and (3) communication and information flow.

Organization charts show the role and task definition and provide the grouping of activities into a logical structure. It provides the identification and relationships of functions of the company and is a graphic picture of the lines of authority of the company, showing by job title who reports to whom. Organization charts also help provide answers to questions such as:

1. Who am I; what is my position in the company?
2. What do I do?
3. To whom am I accountable?
4. Who is accountable to me?

UNION-MANAGEMENT RELATIONS

Ineffective human resource administration can possibly bring a new role to management, union-employee relations. If a company has arbitrary policies and procedures, wages and benefits that are not consistent and competitive, and has poor lines of communication, union intervention is likely.

Even with bargaining expanding to areas of benefits, work hours, and work rules, union organizers use the issues of wages and working conditions as big selling points. The employer who is perceived by the employees as not competitive in the area of wages or unfair in the treatment of employees could be in a position that looks attractive to union organizers.

Organizing can also be instigated by employees within the company. The National Labor Relations Act, a federal law, refers to such acts as "protected concerted activity," which is described as two or more employees or one, if the employee speaks for several others, who protest a working condition or who attempt to seek union representation.

No matter if the organizing attempt comes from an outside union organizer or the disgruntled employee, the employer has to acknowledge the activities and their protection under the law.

One of a union's first steps in an organizing drive is to obtain authorization cards from employees. See Fig. 3. When this occurs, there are various responses management may make. The following guidelines will assist management in answering questions that frequently arise. First, however, some general precautions:

1. Do NOT examine any documents or cards a union representative offers to show management.

2. Do NOT discuss any contract proposals or company policies with a union representative.

3. If management receives a letter from the union asserting that it represents a majority of employees, respond with these words and NOTHING ELSE: "I have a good faith doubt your labor organization does in fact represent an uncoerced majority of my employees in an appropriate bargaining unit, and I suggest you may wish to avail yourself of the orderly procedures available at the National Labor Relations Board to resolve any question of representation."

4. Bear in mind that the rules regarding campaign conduct are extremely complex and that union organizers are trained professionals. Lacking competent advice, an employer can very easily become entangled with the National Labor Relations Board.

DESIRING TO BE REPRESENTED BY _____UNION, I HEREBY AUTHORIZE SUCH ORGANIZATION TO BE MY EXCLUSIVE COLLECTIVE BARGAINING REPRESENTATIVE.

NAME _____ S. S. NO. # _____

ADDRESS _____ TELEPHONE _____

CITY _____ STATE _____ ZIP _____

NAME OF COMPANY _____

OCCUPATION _____ DATE HIRED _____ SALARY _____

DATE _____ SIGNATURE _____

Fig. 3. Authorization Card

CONCLUSIONS

People are the core element in any company. The management of people, human resources management, will no doubt play a bigger role in the future. With changes in the workforce, and pressures from technology and the economy, greater emphasis must be placed on the management of this most valuable resource.

REFERENCES

Burack, E.H. and Smith, R.D., 1977, *Personnel Management: A Human Resource Systems Approach,* St. Paul, p. 376.

Anon., *Business Week,* 1981, "The New Industrial Relations," May 11.

Cullen, D.E. and Greenbaum, M.L., 1966, *Management Rights and Collective Bargaining: Can Both Survive? Bulletin 58,* New York State School of Industrial and Labor Relations, Cornell University, Ithaca, New York.

Dale, E., 1965, *Management: Theory and Practice,* New York.

Drucker, P., 1974, *Management: Tasks, Responsibilities, Practices,* New York.

Famularo, J., 1979, *Organization Planning,* revised ed., New York.

Hicks, H.G., 1972, *The Management of Organizations: A Systems and Human Resources Approach,* 2nd ed., New York.

Johnson, R.G., 1979, *The Appraisal Interview Guide,* New York.

Lopez, F.M., 1975, *Personnel Interviewing: Theory and Practice,* New York.

Miner, M.G. and Miner, J.B., 1973, *A Guide to Personnel Management,* Washington, D.C.

Scobel, D.N., 1982, "Business and Labor—From Adversaries to Allies," *Harvard Business Review,* Nov.-Dec., p. 129.

Stanton, E.S., 1977, *Successful Personnel Recruiting and Selection,* New York.

8.5 Reporting

Darryl Marshall and L. S. Heyborne
October 23, 1987

RESPONSIBILITY COST ACCOUNTING AND BUDGETING

Introduction

The objective of responsibility accounting is to provide reports of the costs for which each manager is responsible. It provides a method for budget development, and timely comparisons of actual expenditures with budgets. This encourages each manager to deal frequently with budgets and costs. Managers act to correct unfavorable trends in costs versus budgets. They realize the effect of action taken, and the importance of maintaining control over costs.

It is important that managers understand the overall cost picture of the larger group, whether this is the department or the company. The costs associated with the manager's responsibility must be carefully delineated. A manager's responsibility report should include only those costs over which he has direct control. For example, managers should not receive information in their cost reports about equipment depreciation if they have no control over the purchase of new equipment. Nor should they be responsible for costs distributed from overhead or administrative cost centers.

It is not the objective of responsibility accounting to present an accurate product cost or financial picture to the responsible manager. The objective is to make him aware of the effect of his actions on the costs in his responsibility area.

This section describes the principles and operation of a responsibility accounting system. It discusses the breakdown of departments into cost centers, and the generation of budgets for them. It also discusses responsibility reports comparing budgets to expenditures. Finally, it includes a description of a budget and cost control cycle.

Cost Centers and Cost Elements

Cost Centers are the smallest division into responsibility areas. Examples of typical cost centers within a mining department are administration, production drilling, production blasting, and hauling. A manager is responsible for the costs within each cost center.

Cost centers contain *cost elements,* such as labor, drill bits, and cleaning supplies. Careful assignment of elements of costs to the cost centers is important. It is typical in a mining department to have separate cost centers for haul truck maintenance and for material hauling. A clear definition of which materials and supplies apply to which of these cost centers is important. Wear steel is typically assigned to the operations cost centers. Mechanical spares are typically assigned to the maintenance cost centers.

There are commonly three types of cost centers within an operation. The first is an operating cost center. Examples of operating cost centers are the production drilling, production blasting, and hauling cost centers mentioned previously.

Second are service cost centers. Examples of service cost centers are the maintenance cost centers, such as haul truck repair and electrical power distribution. The reporting system distributes costs from the service cost centers to the operating cost centers. These costs appear in the responsibility reports twice. First they appear in the service cost center as a direct cost under the responsibility of the service cost center head. They also appear under the operating cost center as an indirect cost or distributed cost. The responsibility cost report for a department must show a credit for internally billed costs to avoid double accounting.

The third type of cost center is administrative. These costs are overhead. The system does not distribute them to operating or service cost centers.

There are two types of elements within cost centers: direct and indirect. Direct cost elements include labor. They may include a number of different types of labor such as managerial, clerical, and wage. Other direct elements include operating supplies, maintenance supplies, and services and utilities purchased from outside the company. Indirect cost elements distribute charges from within the company for shops and other services. These are the distributed costs. Fig. 1 is a typical cost center report for an operating cost center. Fig. 2 is a report for a maintenance cost center, and Fig. 3 is a report for an administrative cost center.

Budgets and Estimates

The components of an operating budget system are:
- Prices.
- Basic standards.
- Dollar standards.
- Determinants.
- Determinant quantities, budgeted and actual.

Prices for direct cost elements are the standard unit price of the item. This may be the price per pound of blasting agent, or per hour for labor. Both the actual price charged and the price used in the budget development should be a constant standard price. This is because the cost center head cannot affect the price of these items. The standard price is adjusted from time to time.

Prices for indirect cost elements are the standard charging rate for the service cost center. The system develops the rate by dividing the total dollar budget for the service cost center by the budgeted units sold.

Basic standards are performance goals. They are consumption or usage rates, such as kilograms of frother per ton of concentrate. Kilograms of frother is the basic standard. Tons of concentrate is the determinant.

Dollar standards are either developed or are stand-alone standards. If there is a basic standard associated with a cost element, the system develops a dollar standard. It is the product of the basic standard and the price. If there is no basic standard, an analyst enters the dollar standard. An example of the latter is a standard for miscellaneous consumables of $500 per month. $500 is the dollar standard. Month is the determinant.

Determinants are descriptions of factors. Dollar standards are dollars per unit of a determinant. Examples of determinants are months, tons of ore produced, or truck operating hours.

Determinant quantities are the number of associated months, tons or hours. Budgeted determinant quantities are estimates of the factors. Multiplying these by the dollar standards before the year yields an annual estimate of expendi-

1310 — LOADING   **C. BRAVO**   **ASSISTANT SUPERINTENDENT**

FOR THE MONTH ENDED DECEMBER 31, 19   PAGE 20

CURRENT PERIOD				ELEMENT	DESCRIPTION	YEAR TO DATE			
ACTUAL	BUDGET	VARIANCE	PERCENT			ACTUAL	BUDGET	VARIANCE	PERCENT
17,170	17,825	655		111	HOURLY—REGULAR PAY	211,474	211,731	257	
	1,233	1,233		114	HOURLY—DOUBLE TIME	8,166	10,744	2,578	
17,170	19,058	1,888		100	LABOR	219,640	222,475	2,835	
				239	OPER MATL—RECAPPING	12		12*	
15	99	84		271	OPER SUP—CLEANING & SANITARY	525	1,090	565	
255	760	505		200	OPER MATERIALS AND SUPPLIES	4,964	8,420	3,456	
				313	MTCE MATL—PLANT SPARE PARTS	25	750	725	
				300	MAINTENANCE MATERIALS	25	750	725	
17,425	19,818	2,393	113.7		TOTAL INCURRED COSTS	224,629	231,645	7,016	103.1

* DENOTES UNFAVORABLE VARIANCE

Fig. 1. Operating cost center responsibility report.

1631 — PRODUCTION TRUCK REPAIR CREW   **G. LATORRE**   **GENERAL FOREMAN**

FOR THE MONTH ENDED DECEMBER 31, 19   PAGE 38

CURRENT PERIOD				ELEMENT	DESCRIPTION	YEAR TO DATE			
ACTUAL	BUDGET	VARIANCE	PERCENT			ACTUAL	BUDGET	VARIANCE	PERCENT
7,404	4,484	2,920*		111	REGULAR PAY	81,522	80,493	1,029*	
294	320	26		114	DOUBLE TIME	2,862	4,502	1,640	
7,698	4,804	2,894*		100	LABOR	84,384	84,995	611	
				222	OPER MATL—GASOLINE	5		5*	
				253	HEAVY DUTY TIRES & TUBES	675		675*	
4	55	51		244	OPER MATL—OTHER CHEMICALS	127	610	483	
24		24*		244	OPER SUP—STATIONERY & OFFICE	223		223*	
173	771	598		271	OPER SUP—CLEANING & SANITARY	2,578	8,942	6,364	
201	826	625		200	OPER MATL & SUP	3,608	9,552	5,944	
7,899	5,630	2,269*	71.3		TOTAL INCURRED COST	87,992	94,547	6,555	107.4

* DENOTES UNFAVORABLE VARIANCE

Fig. 2. Maintenance cost center responsibility report.

1130 — MINE MAINTENANCE ADMIN.   R. ALVARADO   SUPERINTENDENT

FOR THE MONTH ENDED DECEMBER 31, 19   PAGE 4

CURRENT PERIOD				ELEMENT	DESCRIPTION	YEAR TO DATE			
ACTUAL	BUDGET	VARIANCE	PERCENT			ACTUAL	BUDGET	VARIANCE	PERCENT
272	909	637		111	HOURLY—REGULAR PAY	3,495	1,782	1,713*	
17		17*		114	HOURLY—DOUBLE TIME	95		95*	
13,909	13,537	372*		121	SALARIED—REGULAR PAY	198,784	154,044	44,740*	
694	1,754	1,060		124	SALARIED—DOUBLE TIME	14,268	20,009	5,741	
14,892	16,200	1,308		100	LABOR	216,642	175,835	40,807*	
133	220	87		264	OPER SUP—STATIONERY & OFFICE	4,443	3,440	1,003*	
89	61	28*		271	OPER SUP—CLEANING & SANITARY	672	672	0	
				279	OPER SUP—LABORATORY	3	3	0	
222	281	59		200	OPER MATL & SUPPLIES	5,118	4,115	1,003*	
290	350	60		621	CONTRACT TRUCKING	3,698	4,200	502	
290	350	60		600	OUTSIDE CHARGES	3,698	4,200	502	
				789	COSTS BILLED TO OTHERS	28		28*	
				791	TRAVEL AND EXPENSE ACCOUNTS	274		274*	
				700	OTHER DIRECT COST AND CREDIT	302		302*	
15,404	16,831	1,427	109.3		TOTAL INCURRED COSTS	225,760	184,150	41,610*	81.6

* DENOTES UNFAVORABLE VARIANCE

Fig. 3. Administrative cost center responsibility report.

tures. Actual determinant quantities are used at the end of an accounting period. The system multiplies them by dollar standards to yield responsibility budgets.

For variable costs, most systems use production determinants, such as tons hauled. For fixed costs, most systems use time determinants, such as calendar month. Variable budgets are functions of actual production levels. If actual production is zero, the total operating budget for the variable cost elements is zero. Budgets increase in direct proportion to the production level. An examination of the budget calculation detail report in Fig. 4 will give a clearer picture of the relationship between the different components described.

The selection of determinates is critical to the development of a budgeting system. A typical determinate for hauling is per 100 ton of material hauled. The budget for diesel fuel and wear steel for haulage trucks is a function of this determinate. In other words, the cost center head earns a budget of so many gallons (liters) of diesel fuel per 100 tons hauled and so many pounds (kilograms) of wear steel per 100 tons hauled.

Charges for the trucks themselves are distributed from the maintenance cost center to the haulage cost center at a standard rate. The standard rate per truck hour reflects the total cost of maintaining the truck. In other words, it is the business of the hauling cost center to haul materials with trucks. To do this, the hauling cost center rents the trucks from the maintenance cost center. It adds wear steel, diesel fuel, and other operating supplies.

Truck operating hours generate a budget in the maintenance cost center. It is the business of the maintenance cost center to provide haul trucks to the haulage cost center. Therefore, for every hour of actual truck usage, the maintenance cost center earns a budget for maintenance labor and supplies.

Responsibility Reports

Responsibility reports compare actual expenditures with budgets for each supervisor and manager. There are two types of responsibility reports. The first is the same form as a cost center report, illustrated in Figs. 1, 2, and 3. Consolidation of cost centers to higher levels follows the same format. Three cost center heads may report to one superintendent. The superintendent receives a report that combines the information for those three cost center heads. Fig. 5 illustrates such a report. This is an upper level statement for mine operations. Note that the report includes the name of the superintendent of mine operations which personalizes it. The format is the same as for cost center reports, with costs segregated by cost elements.

Fig. 6 illustrates a second format for presentation of upper level responsibility reports. This report lists the constituent segments within the superintendent's responsibility. In this case, they are mine operations administration, drilling and blasting, and mine production. The reports show credit for charges from one cost center to another. Note that the net responsibility costs for the two reports is the same.

Fig. 4. Budget Calculation Detailed Report
(Annual)

381—Grinding

Determinant Number	Description	Element Description	Determinant Quantity	Price	Basic Standard	Dollar Standard	Budget
1	Work day	Wage labor	240	48.00	21		241,920
1	Work day	Wage fringes	240	14.40	21		72,576
2	Month	Salary labor	12	3200.00	2		76,800
2	Month	Salary fringes	12	960.00	2		23,040
3	kt ore	Steel liners	3,500	4.09	27.0		386,505
3	kt ore	Rubber liners	3,500	4.50	3.0		47,250
3	kt ore	3.5-in.* rods	3,500	0.628	170.0		373,660
3	kt ore	2-in. balls	3,500	0.743	560.0		1,456,280
3	kt ore	1-in. balls	3,500	0.837	16.0		46,872
4	Year	Misc. equipment	1			30,000	30,000
Net Grinding							2,754,903

*Metric equivalent: in. $\times$ 25.4 = mm.

Fig. 5 Upper Level Consolidated Cost Center Report

MINE OPERATIONS   **R. W. GREEN**   **SUPERINTENDENT**

REPORT NO. 201   **FOR THE MONTH ENDED DECEMBER 31, 19**   Page 1

| Current Period | | | | | | Year to Date | | | |
Actual	Budget	Variance	Percent	Element	Description	Actual	Budget	Variance	Percent
101,279	100,210	1,069 *		111	Hourly-Regular Pay	1,219,798	1,215,268	4,530 *	
1,082	8,161	7,079		114	Hourly-Double Time	45,673	81,485	35,812	
4,721	4,215	506 *		121	Salaried-Regular Pay	53,917	50,672	3,245 *	
103	322	219		124	Salaried-Double Time	2,940	3,439	699	
108,534	114,780	6,246	105.8	100		1,343,467	1,377,829	34,362	102.6
6,420 CR	993	7,413		202	Drilling Natl-Percussion Bit	1,384 CR	15,654	17,238	
6,370 CR	2,907	9,277		205	Drilling Natl-Other	35,028	44,406	9,378	
651	707	56		211	Blasting Natl-Dynamite	10,562	11,551	989	
				214	Blasting Natl-Aluminium	1 CR		1	
				217	Blasting Natl-Slurry Product	236,370	146,070	90,300 *	
9,281	2,582	6,699 *		219	Blasting Natl-Other	107,621	41,632	65,989 *	
5	100	95		222	Oper Natl-Gasoline	824	1,200	376	
				233	Heavy Duty Tires & Tubes	319		319 *	
				246	Oper Natl-Other Chemicals	22		22 *	
91	55	36 *		264	Oper Sup-Stationery & Office	1,301	610	691 *	
168	894	726		271	Oper Sup-Cleaning & Sanitary	4,225	9,936	5,711	
				281	Oper Natl-Milk Liners	34		34 *	
11,792	6,605	5,187 *		283	Oper Natl-Wear Steel & Matl	63,933	85,692	21,759	
39,593	20,069	19,524 *		285	Oper Natl-Conveyor Belting	124,332	260,816	76,484	
246,423	203,967	42,456 *	82.8	200	Oper Materials and Supplies	2,969,925	3,088,383	118,458	104.0
380	133	247 *		311	MTCE Natl-Mobile Equip Parts	7,432	6,429	1,003 *	
586	110	476 *		319	MTCE Natl-Other Spare Parts	2,918	1,220	1,698 *	
1,088	408	680 *	37.5	300	Maintenance Materials	16,096	10,829	5,867 *	63.5
				745	Insurance	141		141 *	
589,248	605,366	16,118 *	102.7		Net Responsibility Cost	7,334,072	7,679,127	345,055 *	104.7

Fig. 6. Upper Level Responsibility Segment

MINE OPERATIONS   **R. E. GREEN**   **SUPERINTENDENT**

REPORT NO. 201   **FOR THE MONTH ENDED DECEMBER 31, 19**   **PAGE NO. 22**

| Current Period | | | | | | Year to Date | | | |
Actual	Budget	Variance	Percent	Element	Description	Actual	Budget	Variance	Percent
11,635	13,506	1,871	116.1		Mine Operations Admin	140,365	161,223	20,858	114.9
203,448	221,731	18,283	109.0		Drilling & Blasting	3,201,137	3,276,364	75,227	102.4
374,629	370,593	4,036 *	93.9		Mine Production	4,001,290	4,250,260	248,970	106.2
					Less				
464 CR	464 CR				Intra-Responsibility Charges	8,720 CR	8,720 CR		
589,248	605,366	16,118	102.7		Net Responsibility Cost	7,334,072	7,679,127	345,055	104.7

The highest level report in the system is for the top level mine manager or company manager. It is similar in form to those illustrated in Figs. 5 and 6. Fig. 7 is an illustration of the use of the responsibility reports.

The reporting system deals with the last part of the objective stated previously. The highest level report, which goes to the vice president and general manager, shows there is a problem in the current month. There is a large variance in operations. Step 2 shows that the mining department accounts for the majority of the variance. In step 3, examining the responsibility reports within mining shows that mine shop maintenance and maintenance planning area has a variance accounting for most of the problem. Step 4 is an examination of the responsibility reports for mine maintenance. This shows an even larger variance for mine mechanical maintenance, truck and tractor. Step 5 is an examination of the report for mine mechanical maintenance, truck and tractor. Note that the responsibility report illustrated under step 5 shows costs both by element and by section. The total dollars on both reports is the same. An examination of both reports shows that the problem area is in the production truck maintenance department and in the maintenance materials element. Step 6 is an examination of the cost center report for production truck maintenance. The problem is located in the maintenance materials, mechanical mobile equipment. Finally, in step 7, a detailed costs transaction listing shows the makeup of the actual charges. They are located in the maintenance material, mechanical mobile equipment element. This report shows the use of eight repaired engines and four magnet frames. Thus, the report provides information for all levels of management to locate the source of cost overruns. Managers can take corrective action.

The system illustrated provides each manager with the information necessary to control costs. The managers are accountable for the costs within their responsibility. However, no system can reduce costs. To do this, people must change their behavior. One method of providing for this is through a budget and cost control cycle.

Budget and Cost Control Cycle

The cost control cycle must provide for an effective analysis of costs as they occur. It must provide awareness of costs to the manager and awareness of the effect of their management decisions on costs. This awareness must be in a time frame that will allow effective changes or modifications in decisions affecting costs. In other words, it cannot all be after the fact. An examination of last month's costs is meaningless unless it has an impact on spending this month.

The first step in the cycle is to estimate the cost. This is the mid-month estimate. A cost analyst prepares it with input on the major segments or elements of cost from the operating managers. This exercise does not produce a detailed report. It produces a report by element grouping of the total costs for the operations. Its purpose is to make the operations managers at all levels aware of any unusual expenditures that are taking place during the current month. The managers take corrective action if problems develop.

The second step is a comparison of the mid-month estimate with actual costs. Analysts compare them immediately after the close of the month. The comparison must be timely and not in great detail. Its value diminishes if it is not timely. It is available the third working day after the close of the month. By one day later, analysts present it to top level management with explanations for major variances. The responsible managers provide the explanations.

Within the first week after the close of the month, the system produces cost center reports and other responsibility statements. Cost center heads hold a series of meetings. They analyze major variances between budgets and actual costs. They then prepare explanations of the variances for the superintendents, general superintendents, and higher levels. Top management holds a final meeting by the middle of the following month. It is critical that the meetings do not degenerate into a series of alibi sessions. The objective is to analyze variances between actual expenditures and budgets. People must change behavior in order to reduce variances. Managers must also analyze favorable variances. They may show that planned maintenance is not taking place.

The monthly cycle develops a high level of cost awareness. Operating managers develop annual budgets with staff assistance from the cost analysts. It is not advisable to change the annual operating budget unless warranted by a change in operating conditions, such as addition or deletion of machinery. The variable budgeting system automatically provides for changes in the production and operating levels.

It is important to remember that the purpose of the system is to control spending. Reducing variances by increasing budgets can be contrary to this. It is even more critical to remember that to control spending requires behavior modification. Reduction of budgets does not control spending. It can, however, serve to delude management. It can cause them to ignore real spending problems until it is too late for effective control.

PRODUCTION REPORTING

Introduction

Production reporting serves the primary purpose of advising managers of production performance. Production reports may express performance in either a quantitative or a qualitative manner or a combination of the two.

Production reports provide each responsibility head with the necessary information to assess production performance. The production performance relates to equipment, to people, or to both.

The system tailors production reports to fit the needs of each manager. They should contain only the information required, and no more.

Types of Reports

There are several types of production reports. The first consists of those prepared by the operator on a daily basis. An example is the shift report prepared by the operator of a power shovel. The report includes such data as equipment ID number, date, shift, location, and number of loads by category of material loaded. It also includes time of operation related to each product, delay hours or nonproduction time, and the reasons for these. The system prepares similar reports for trucks, drills, front-end loaders, draglines, and other pieces of mobile or semi-mobile equipment.

There are similar reports for other equipment such as tractors, graders, or cranes that provide service or support functions.

Plant operators prepare reports that express productivity in crude ore in and product out. It is also common to report on repairs made or required.

All reports mentioned have ingredients in common. They include an accounting of time, a description of the work done, and a report of equipment condition. Reports relating to some equipment yield production data that is quantifiable. They are expressed in units of work or production per unit

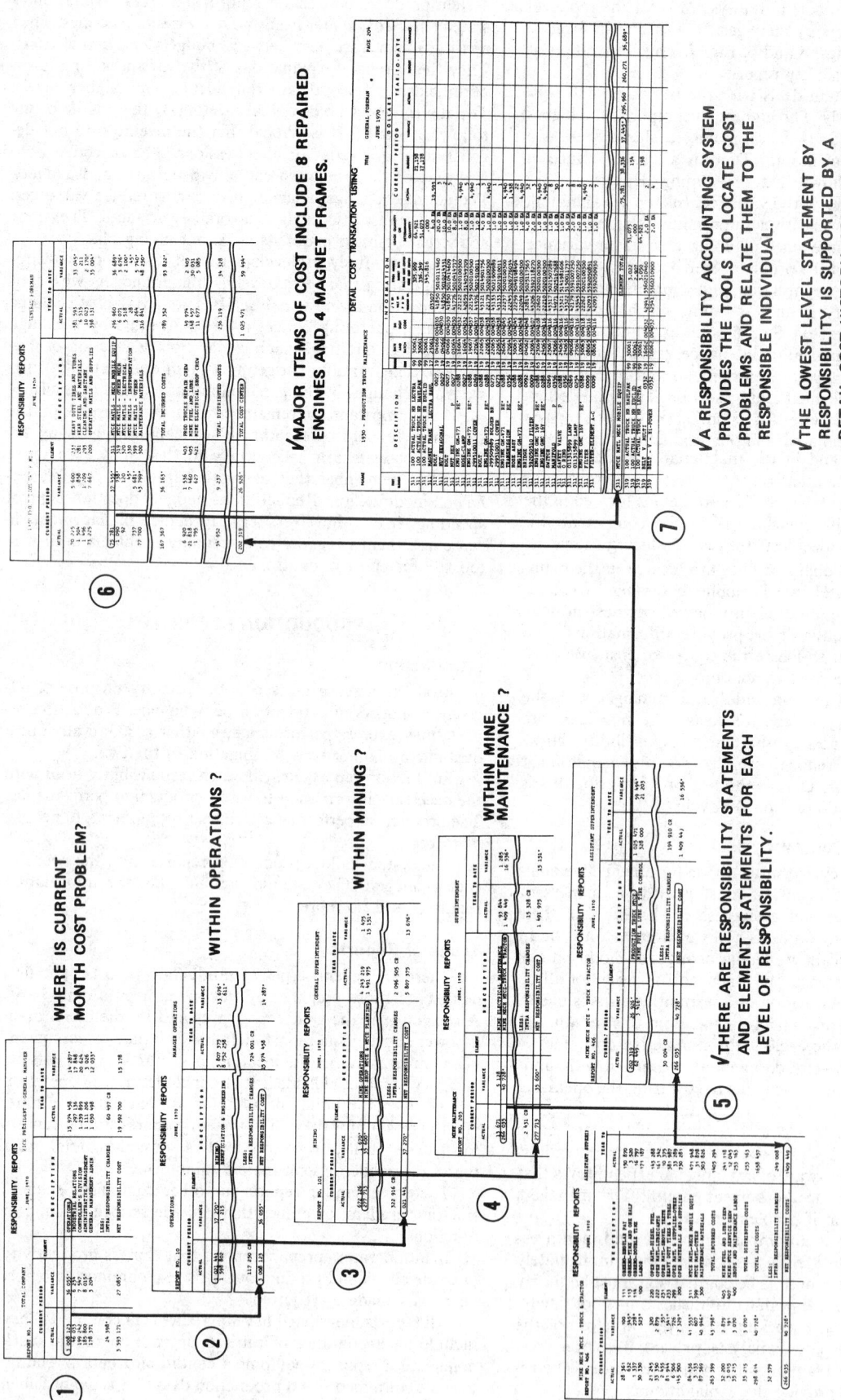

Fig. 7. Responsibility accounting cost statements. *Source:* Wraith III, W., 1973, "Section 30.8, Management Cost Control Systems—A Case Study," *SME Mining Engineering Handbook,* Vol. 2, I.A. Given, eds., AIME, New York pp. 30-64 and 30-65.

of time. The purpose of this type of production report is to provide raw data for input into more refined reports.

The second type of production reports are summaries. Data are compiled to provide production data on a time-related basis. In the summarization and compilation steps, it is common to convert the production units from a shift report to a more desirable unit of measure. Excavators and haulage trucks commonly report production in loads. In the summary process, it is common practice to apply a volume and/or a weight factor to convert loads to volume or weight. These units are more meaningful than loads in quantifying production. Similar steps relate meters (feet) of blast hole drilled to volume or weight. They may express explosive consumption in kilograms (pound) per unit of weight or unit of volume.

Reports of this second type show whether or not production goals are being met. Simple raw production quantities are readily available at all operations. They serve as one of the early indicators of operating performance. They are only an indicator, however, for in some cases high production levels are attained by sacrificing cost.

A third type of production report uses data from the two preceding report types. It expresses the productivity of equipment or of people. This information is one of the primary ingredients for repair or replacement evaluation. Time rate of production data is also valuable in comparing productivity of similar pieces of equipment of different makes.

Similarly, a comparison of the output of different people operating similar pieces of equipment serves as a basis for identifying substandard performers.

Production reports sometimes express productivity in quality rather than quantity. A common report sometimes incorporates quality and quantity data to present a composite basis for comparison.

Summary

The primary objective of every mining operation is economic mining and processing of a mineral. This should return a profit commensurate with the risk involved. A carefully conceived and carried-out system of production reporting is a valuable tool in the attainment of this objective.

The reporting system should yield timely and succinct information to the operator. The rapid advances in the field of electronic data processing promotes this process. Plant operations commonly use electronic means of controlling the operation. Computers produce reports that cover a wide range of information relative to production and general operating and maintenance data. Regardless of the methods of data collection and correlation, the reports are a valuable tool to the operator.

The effective operator is one who quickly reviews the information contained in the various forms of shift or monthly production reports. He makes maximum use of the data. The wise use of production report data results in greater productivity, improved quality, and lower product cost.

8.6 Sales and Marketing

Steven H. Grundstedt

Jerald S. Morris

WHAT IS MARKETING?

Frequently the marketing of mined products is narrowly depicted as the selling of run-of-mine ore to a custom mill, or the selling of concentrates to a smelter. These factors are certainly important, especially to metallic mines, however, marketing in mining is often a much broader field and can encompass a wide range of interrelated disciplines.

When marketing is discussed in mining, one should be cognizant of the variety of commodities produced from the world's mines. One convenient categorization of mine-run products breaks them down into four general areas: (1) minerals from which metals are extracted; (2) minerals used for their chemical and fertilizer properties; (3) minerals and materials used for their special or physical properties (e.g., gemstones and aggregates); and (4) minerals and materials used as sources of energy (Smith, 1982).

Since mining, and surface mining in particular, produces such a wide range of commodities, the marketing of these commodities is very diverse. Such marketing efforts will be simple or complex, dependent upon the commodities involved and upon trade requirements relative to grade and usable form (Parks, 1957).

Simply defined, marketing involves those activities that transfer goods and services from producer to consumer. Marketing is usually broken down into four main segments referred to as the *marketing mix:* (1) product, (2) distribution, (3) communications or promotion, and (4) price (Rewoldt, 1977). The marketing mix is applicable not only to commodities produced from surface mines, but also for all other products sold from any industry.

Product is that which is being offered for sale. This can be anything from run-of-mine metallic ore to different segregated sizes of limestone aggregate. It can be precious metal concentrates, ore, or refined bullion, depending on the final mine product. In the case of a completely vertically integrated copper company, the final product being offered for sale may be various grades and sizes of copper wire. In marketing, the term *product* may also refer to any of several different ancillary functions, such as the research and development of new (or modification of old) products; the decision to add or drop a product; and decisions to *brand* (national, private, or secondary) a product.

Distribution involves the movement of product from the place of production (mine, mill, refinery, manufacturing plant, packaging plant, etc.) to the place of sale. This not only includes modes of physical transportation but also the channels of product distribution as well; for example, direct sales, wholesalers vs. retailers, discounters, trade channels, etc.

Promotion is the actual selling of the product. Sales is often considered to be interchangeable with marketing, but, as has been shown, this is not correct (albeit sales is a very important component of marketing). Promotion, then, includes such things as personal selling, advertising, dealer incentives, budgeting, and media and message selection.

Price is involved with all of those actions and decisions concerning the dollar amount received for selling the product.

At what price level should the product be offered? Should the product ever be offered for less than this base price (e.g., volume discounts or geographic discounts)? Should a price schedule be changed, and why? These are all questions that are answered and subsequently implemented under the price segment of marketing.

The four segments of the marketing mix are not all mutually exclusive. Quite the contrary, they should be considered as interrelated, synergistic parts of the dynamic marketing process. A good example of the synergistic effect of the parts of the marketing mix is product service. Service is of major concern to the mine or production department since they have to produce a product that meets customer specifications; they are concerned with serving the customer a quality product. But service is also a major concern to the sales (promotion) people who directly interface with the customer. If a product is delivered and doesn't meet specifications, it is a problem that must be rectified by the production people; however, it is a problem as well for the sales people who must keep the customer satisfied through good service in order to generate future sales.

PLANNING, STRATEGY, AND RESEARCH

The first consideration of any marketing planning effort is to establish the marketing objectives. Actually, since marketing deals with the very basics of a business enterprise, the marketing objectives are dictated by, and must be consistent with, company objectives. Consequently, the marketing effort is limited by the financial resources of the firm. Thus, marketing objectives of expected sales volume and expected market position have to be able to achieve company forecasted gross margins and profit within company budgetary constraints.

During the planning stage, all major marketing policies must be formulated. These policies should comprehensively cover all major activities of the marketing department, including sales, price, distribution, advertising, and customer service as well as personnel policies. Formulation of policy, especially pricing policy, should take into consideration all regulatory guidelines. These guidelines can be obscure and complex and inadvertent violations of law must be carefully avoided.

The other important aspect having its inception in the planning stage is the assimilation of all of the facts, ideas, and support available from research and engineering, production personnel, finance, law, etc. (Rewoldt, 1977). The objective of this assimilation process could include any or all of the following: identification of existing or new markets; conceptualization and/or creation of new markets; the decision to enter (not enter) existing or new markets; the decision to expand (not to expand) present market position; maintenance of present market position. The assimilation process sets the stage for marketing strategy and any warranted subsequent action.

This assimilation process can be very simple, as would be the case of a small open-pit mine marketing its high-grade ore to a custom mill or smelter. This process could be quite extensive and complex as would be the case for a large industrial minerals producer trying to establish a new, high-volume use for its product.

One of the severe constraints that all mining companies must contend with is that their mine's ore body is limited in both volume and quality. Hence, there is an ultimate limit on the amount (and on the specification) of product that can be delivered from a mine site. Practically, however, some mines have ore body lives that may exceed many decades (as is the case, for example, with some industrial minerals and coal producers, as well as a few foreign metalliferous producers).

When a mine has a limited life (i.e., the ore body will be exhausted in the short- to medium-term), the primary constraint on marketing efforts is generally the productive capacity of the mine. This is often the case for many smaller metal mines whereby a market can be found for more metal than the mine is able to produce (albeit this market is usually for variable and unpredictable prices). If funding is available, increased capacity is only warranted when there are adequate ore reserves and if an adequate return can be achieved on the investment necessary for expansion.

For very long-lived ore bodies, the limiting factor is not mine production (assuming funding would be available for desired expansion), it is the market. An example of this might be a Wyoming bentonite mine with very large reserves. Expansion of production is not warranted because larger penetration into the very competitive bentonite markets has not been successfully achieved.

Marketing people must be cognizant of the productive limitations of the mine and of the company. They must also be cognizant of the processing capabilities and other skills of the firm. They must not sell product that the mining company is not capable of producing or that is not feasible to produce.

The utilization of marketing research introduces the scientific method to mine marketing. As such, the more dependent a mining company becomes on marketing research, the less subject it is to the inefficiency and waste incurred by marketing efforts based wholly on past experience, intuition, and pure change (Nystrom, 1951). However, marketing research, when warranted by the size of the marketing effort, should be given well-defined goals within a well-defined budget. Much time and money can be inefficiently spent on prospective products for which there will be no adequate market or on markets that do not meet corporate objectives. Close coordination with all phases of the marketing mix is necessary in order to derive the most benefit from research.

There are many phases of marketing research that would be appropriate for some mining companies but not for others. Depending on the particular mineral or metal commodity (or refined product), as well as particular corporate objectives, the research effort should concentrate on different analyses. The following comprehensive list better defines the specific types of marketing research that can be undertaken (after Nystrom, 1951):

1. *Product Analysis.* This field embraces all applications of marketing research techniques designed to develop new products or adapt old ones so that they will have maximum acceptability to the user.

2. *Brand Position Analysis.* Current or periodic study of competitive volume of the different brands in a product field.

3. *Consumer Surveys.* Designed to determine the users of product.

4. *Sales Organization and Operation Research.* Time and duty analyses of sales activities; cost accounting analyses of operations of individual salesmen; studies of methods of com-

pensating sales personnel; and studies of sales training methods and procedures.

5. *Wholesale and Retail Distribution Analysis.* Studies of the various channels of distribution.

6. *Sales Record Analysis.* Internal analyses using the accounting records of sales as a foundation for solving a number of specific marketing problems.

7. *Distribution Cost Analysis.* Studies aimed directly at reducing the cost of distribution, which averages approximately 60% of the final cost of commodities.

8. *Quantitative Market Analysis.* Determines the amount of a commodity that a given market can be expected to absorb.

9. *Attitude and Opinion Research.* Used chiefly as a basis for public relations and advertising activities.

10. *Advertising and Sales Promotion Research.* Applying and developing marketing research procedures as a basis for developing and executing advertising campaigns.

11. *Price Analysis.* Measuring market demand at various price levels.

12. *Market Trend Analysis.* Observe and interpret changing conditions and forecast future market conditions.

Some of the above mentioned types of marketing research are better suited to manufacturers of consumer goods while others are better suited for producers of industrial commodities. However, all have been included because at one time or another, a comprehensive mine marketing research program would probably include at least parts of all the techniques listed.

PRODUCT

Products extracted from surface mines are varied in composition, grade, and possible coproducts or byproducts. This variability includes large mines producing high volumes of low-grade material, usually sold at relatively small profit margins. Examples may be a large molybdenum porphyry pit or a large iron taconite pit. The other end of the spectrum would be small surface mines producing low volumes of high-grade material that is usually sold at relatively high profit margins. Examples may include a small foreign mine extracting high-grade manganese ore from an outcropping vein to a small domestic mine extracting high-grade uranium ore from a sandstone outcrop.

After mine-run ore is extracted, it may be directly marketable or further upgrading and/or treatment may be necessary. Chrome ore and manganese ore are marketable directly from the mine as long as shipping grades exceed 44% Cr_2O_3 for chrome ore and 48% manganese (low impurities) for the latter (Anon., 1982). Steelmaking and chemical industries use these ores as raw materials. Certain lower grades of hematite, containing free silica, were beneficiated to merchantable grades by washing, although this is not common practice in the United States currently. Coal is cleaned and often graded to size in preparation for market. Note that the physical character of the material is not changed by such beneficiation process. Some types of treatment for market are more specialized, as with asbestos, which is cleaned, expanded, and graded carefully (Parks, 1957).

Many ores require transformation treatment between mine and the final market. Although most small, some intermediate, and a few large mines will sell their ore or concentrates directly to a custom smelter, the fact remains that the final market usually depends on a refined metallic product. Copper, lead, and zinc ores are typical of mine-run products that are milled, smelted, and ultimately refined (Parks, 1957).

Industry standard specifications for marketable metal, ore, alloys, and nonmetallics can be found in publications like *Metals Week* and *American Metal Market*.

Most ores and concentrates sold in the United States have standardized marketable production units in short tons (2000 lb). Notable exceptions, however, are iron ore, phosphate rock, and sulfur, which all have marketable units in long tons (2240 lb). Most foreign ores are sold on the metric tonne (2204.6 lb) basis. Refined metal, in the United States, is generally sold in avoirdupois pound units; precious metals (gold, silver, and the platinum group metals) are sold in troy ounces; and mercury is sold in 76-lb flasks.

It is very seldom that a mining company would be involved in the marketing of a totally *new* product. More than likely, this new product would be the modification of an already mined (either by the company in question or other mining companies) product. Examples might be a new standard length or a new standard diameter of refined copper wire, or the marketing (by a mining company) of minted silver coins (silver produced from their mines), or the marketing of a new size of crushed aggregate. However, discovering a market for a *new* product does occasionally happen. For example, a use may be discovered for a previously uneconomic mineral. (Natural zeolites, a relatively new industrial mineral with many unique marketable properties, is a good case-in-point.)

In any event, management should take the proper marketing evaluation steps (whether these be formal or informal) to ascertain economic viability of introducing the new or modified product. This product analysis should include some or all of the following: engineering development; pretesting; further engineering development; test marketing; go/no-go decision; and marketing plan. During this product analysis, cost estimates, price estimates, and life cycle estimates should be ascertained as accurately as possible. In this way, a proper economic analysis can be accomplished to establish the profitability of the new product as well as the expected return on invested capital to get the new product into production. The economic analysis is usually the culmination of all the preproduction efforts and is the mechanism by which management makes the final go/no-go decision for the new product (Rewoldt, 1977).

Ancillary decisions that often follow the introduction of a new/modified product are such things as packaging, warranty policy, and product service. Although often more appropriate for consumer products, the mining firm may want to brand and/or patent their new/modified product.

DISTRIBUTION

Once a sale is made, the product must be made available to the customer to be put to its end use. The overall distribution strategy should allow dependable product deliveries by the most efficient means, at reasonable cost. Distribution is often an area where seller/buyer problems, if any, materialize (Converse, 1964).

Generally speaking, the producer/seller must identify the distribution channels most effective for his select markets. These channels could involve direct sales to the customer or indirect sales by utilizing the services of middlemen. If distributors, wholesalers, retailers, or brokers are to be used, the appropriate ones must be recruited and incorporated in the overall distribution scheme.

Surface mined products of a high-margin, low-volume nature, will tend more toward the use of middlemen, at least more than the distribution of low-margin, high-volume products. This is due primarily to the fact that the former commodities are much less sensitive to transportation factors and hence are distributed over a much broader geographic area. High tonnage sales of products such as aggregates, trona, sulfur, potash, coal, and lime tend to be to more localized customers, or at least to a narrower portfolio of customers, which allows the more direct sales approach, without middlemen. There are exceptions to all cases, certainly, and the surface mine producers must be flexible enough to utilize the distribution channels most efficient for their own needs.

The aspect of physically transporting the product to a buyer is a very important one and can involve any of the basic transport methods via truck, rail, water, or air. The selection of the transportation mode is dependent on many factors, including product bulk, geographic location, distance to buyer, quantity to be moved, availability and accessibility to types of transportation, availability of loading and unloading facilities, freight rates, time frame allowed by buyer, nature of product, and many more.

Often the choice of transportation methods is very limited and may even exclude the benefit of competition (i.e., service available from only one truck or rail line). Fortunate is the surface mine producer who has his choice of several transportation lines, allowing a better chance for improvement on transportation costs. This also usually results in better service and more willingness by the entity providing the transportation to resolve any problems.

Then there is the age-old question of whether to utilize outside transportation services or provide them internally, such as buying and operating a fleet of trucks. Again, there are so many variables to consider for each site specific case but, generally, producers will not provide their own transportation unless careful studies clearly show economic advantage, especially considering amortization of investment. Chances are, however, unless the producer has no choice, normal available transportation services will be utilized.

Inventories can get out of hand, however, either by poor management or poor planning and can result in a rather large working capital requirement. Over the long run, inventory levels should be maintained just enough so that interruptions of dependable product delivery to the buyer are minimized or eliminated.

Assuming product specifications and quality control are maintained, proper implementation of product distribution is possibly the most important aspect of fostering repeat customers. But even under the best of conditions, problems can arise so policies need to be developed to resolve these situations.

PROMOTION

Buyer Behavior

Most surface mined products, whether they be high-margin, low-volume or low-margin, high-volume, are typically sold to such industries as manufacturing, construction, building, and fabrication. These sales transactions often involve middlemen such as brokers and distributors. If product sales are made directly to the buyer, this is usually done through departments of buyers and purchasing agents. The customer purchasing control during the buying process can be centralized or decentralized. It is important, then, to recognize that industrial buyer behavior is governed by stages of internal decision-making, the nature of which has to be understood in order to devise proper product promotion approaches. The number of decision-making stages, of course, varies from company to company and depends on the situation.

A customer's willingness to buy depends on the amount of product information he needs, the extent of availability of competition, the extent of available substitutions that will accomplish the same end use, and the thoroughness with which the buyer intends to research the possible purchase (Rewoldt, 1977).

The final product sale, then, can be as routine as taking orders over the telephone or it can be a very complex and involved approach, depending on buyer behavior during the purchase decision process.

Buyer behavior can be affected by the nature of the salable product. For instance, the final decision maker in a construction company that needs to purchase aggregate (low-margin, high-volume commodity) for a road building job may choose to purchase from an aggregate producer that did not have the lowest price bid. His reasons may simply be because that particular producer has an established presence and the credibility to produce the aggregate. This situation is one in which a dependable aggregate supply as an integral part of the overall construction program is absolutely essential for the construction job to be successful. On the other hand, a purchasing agent buying copper for a fabrication plant may make his purchase decision based on price alone due to his circumstances and indifference to the kinds of concerns that bothered the construction company. Recognizing these different characteristic commodity effects on buyer behavior will help the seller to promote his product. The aggregate producer should promote his past track record and emphasize his ability to perform dependably. The copper company wishing to sell to the fabricator should know that the sale depends almost entirely on offering the lowest price.

There is a tremendous amount of logistics involved in coordinating the product distribution cycle and, depending on the size of operation, a lot of information has to be processed, stored, and scrutinized. The computer can greatly facilitate the handling of this information. If no in-house computer is available, then outside services (such as time-sharing using a remote terminal to access a computer owned and maintained by a service company) can be used. Production, inventories, costs, customer lists, etc. can be accommodated nicely by computer and, properly implemented, can keep track of enough variables and consolidate information in such a way as to allow better chances of improving overall efficiency. Even small producers can take advantage of today's relatively low cost computer packages available for standard business needs. Again, site specific circumstances will dictate whether data processing services would be cost effective.

One of the major factors in the product distribution chain is the maintenance and management of product inventory. Ideally, product inventories would be kept at zero, with deliveries scheduled directly from production. Obviously, this is impossible because the real world is not ideal and, in fact, is subject to a myriad of practical considerations and problems of getting a product from the production department to the end user. A *buffer* is needed to accommodate this inability to perfectly match production rate to customer demands, and hence inventories are a necessary part of the equation.

Promotional Strategy

Unless a surface mine operation is producing a commodity that is in constant demand and has a shortage of supply, some effort is going to have to be expended to promote the product to increase chances of a sale. Sales promotion, then, is the presentation of product knowledge to prospective buyers to whatever degree of detail is needed. It can also, and often does, involve a propaganda approach in an attempt to predispose the buyer to a favorable consideration of a producer and his product(s) (McGarry, 1964).

Propaganda to persuade customers to buy products can occur via advertising and/or personal selling. Advertising depends heavily on communications media and often on psychology. The psychological ploy appeals to either reasoning or emotion or both.

Most advertising for surface mined products is done through mining publications and brochures and tends to be directed at the psychology of technical reasoning, such as promoting products of good quality control, dependable delivery, and suitable specifications. Even the personal selling approach emphasizes the technical reasoning aspects because decisions to buy mining commodities usually depend on the relative economics of selecting a commodity from one competitor or another. This economic equation for the buying decision often involves much more than just price.

The aggressiveness with which products are promoted and the budget appropriated for this program are dependent on several things including the seller's philosophy, type of product, and current market conditions. One producer may treat product promotion as an investment and another may consider product promotion expenditures to be cost of product sold on a unit basis.

Depending on the nature of surface mined products to be sold, a fair amount of promotional attention is frequently given to product identity. Once a product identity is established and can be related to a good track record of reliable performance, then subsequent sales promotions can utilize that product name or identity to enhance future sales and to draw comparisons to promote other products from the same producer.

If a mine producer's products are of a nature that can be utilized by distributors or manufacturers, often the established presence of these entities and their product identities indirectly benefit their raw material suppliers (in this case the surface mine producer). This is because their raw materials are promoted for them, perhaps in a remanufactured or blended form, but promoted nevertheless.

PRICING

Exchange Prices

The subject of pricing and prices is exceedingly complex, especially when dealing with the intricacies of mining commodities. Many of the commodities are very cyclical in nature and the profits of many mines are intimately tied to the ups and downs of the economy. Some commodities have very extreme price swings during a business cycle, especially base and precious metals. Thus, these industries may experience a period of one or two years of very high profits only to weather an extended period of low profits or even excessive losses.

Some commodities, such as coal and uranium, for the most part are sold under medium- to long-term contractual arrangements. In this way the producer is guaranteed a stable market and prices while the buyer is guaranteed a stable source of supply.

In the world metal markets, most metals are sold through exchanges. In London (and there are counterparts in other large European cities) metals are usually sold through the London Metal Exchange (LME). This exchange has many member firms specializing in the purchase and sale of metals in all of their forms, as well as various other mineral commodities. In general, the functions of this exchange are very similar to those of the United States' grain exchanges. That is, there are traders seeking the best advantages and protection against price fluctuations (Anon., 1968).

Although spot (immediately deliverable) metal is available, the mechanism of sale on the LME is usually the futures contract. In this contract, a buyer purchases a warrant for so many metric tonnes of a metal (which are not necessarily available at the moment) for delivery at some time in the future, usually three months hence. The price offered for the metal will be different, usually not lower, from the prices offered for spot delivery (prices are quoted in pounds sterling per metric tonne). Such contracts are negotiable in the same way as spot metals. By trading these futures contracts, traders again seek protection against unfavorable price swings by hedging against current sales. Suppliers, such as smelters, also use the device to help smooth out production schedules, at prices that have near-term assurance (Anon., 1968).

The London Metal Exchange often sets the current prices for Europe and much of the world. Its publicly quoted prices represent the actual purchases recorded from day to day.

On the North American continent there is no centralized metal exchange in the same sense, though the Metals Section of the New York Commodity Exchange (COMEX) does handle some volumes of metals entering commerce in the United States. The COMEX does trade commodities futures contracts for copper, gold, and silver. (Other exchanges trading in metals are the New York Mercantile Exchange which trades in platinum futures, and the Chicago Board of Trade, which trades in silver futures.) Some primary producers of copper base their cathode price on formulas incorporating specific COMEX copper futures contract prices (Anon., 1982).

The commodity futures markets in the United States are similar to futures markets in London and elsewhere. In US markets, people and/or firms purchase contracts that guarantee the delivery of a standardized amount of a metal at a certain price on a certain future date (contracts can be purchased up to two years forward). People participating in the futures market are generally classified as either speculators or hedgers. Speculators buy contracts *long* (anticipating a rise in price) or sell contracts *short* (anticipating a decline in price). If the speculator has correctly anticipated the subsequent rise or fall in price (i.e., subsequent to his *long* purchase or *short* sale of a particular metal futures contract), he will profit; if the opposite is the case, however, then he will experience a loss. Hedgers, unlike speculators, buy and sell futures contracts in order to avoid the risks associated with price fluctuations. In this way hedgers, being actual producers and consumers of specific commodities, ensure they will be able to deliver or to receive a specified quantity of a commodity guaranteed at a predictable price (Aronson, 1979).

Contract Prices

Most metal sales in this country are, however, generally effected through private contracts, though through much the same pattern of brokers as exists at the LME, many of whom are highly specialized metal traders. Producers may refer to *Metals Week* for price quotes (a subsidiary publication of *Engineering and Mining Journal*), which is functionally the US counterpart to the LME's price quotes. This service gathers data on virtually all major minerals commodities sales made in the United States, and is thus able to publish authoritative average prices in daily, weekly, and monthly terms. In some metals, *Metals Week* collaborates with specialized trading companies, such as Handy and Harman, Inc.

who predominate in dealings in silver and other precious metals (Anon., 1968). *Metals Week* is an excellent source of price quotations for base, precious, and miscellaneous metals as well as for different ores and concentrates, ferroalloys, and many industrial and nonmetallic minerals. Another major source for price quotes is the authoritative publication *American Metal Market*.

New Pricing

When trying to establish a price for a new product (i.e., a product for which there is no established price) such as a new specification of a particular industrial mineral, cost-based pricing has many advantages. This type of pricing is explainable, understandable, justifiable, predictable, politically acceptable, and widely used. Properly used, it can be salable and profitable (Bailey, 1978).

Other factors that should enter into the pricing decision of a new product are market conditions, prospective competitive pricing, anticipated growth rate of the market, anticipated market share, and the market's expectations of prices and its ability to pay. Pricing policies for the new product have to be considered. Will pricing discounts (quantity, functional, trade) or advertising allowances be offered? Will the new price take into account the trade channels for distribution of the new product?

Legal Considerations

When establishing any pricing, a mining company has to be aware of all of the possible legal pitfalls. Antitrust has always been one of the most important legal considerations affecting major industrial pricing strategy. Officials of the Department of Justice have become increasingly more critical (through enforcement by criminal prosecution) of antitrust violations such as price-fixing and *predatory economic conduct*. Prudent management should give greater attention to avoiding involvement in antitrust proceedings. One such area is price leadership. When increasing the price of a product, it is the leader (usually a company with a large market share) which takes the risk of severe governmental scrutiny. The price leader should be prepared to explain its actions to a Congressional committee or some other investigating body.

Another trend in antitrust enforcement involves what can be described as *price signalling* or *price-fixing*. For many years, it was understood that a price-fixing conspiracy required an element of common understanding among competitors. However, more recently, a meeting in which one party expressed his intention to increase prices while another party sat quietly by and then later increased his own price has been found by juries to constitute this necessary element of common understanding. For this reason, lawyers frequently advise clients to avoid, or at least to document, meetings with competitors and to make a dramatic exit from a properly scheduled meeting that veers into forbidden waters (Bailey, 1978).

Another legal area that should concern mining company management is price discrimination. Although price discrimination is an area of law more frequently applicable to consumer products cases, it also has some application to the pricing of industrial products. Price discrimination is restricted under the Robinson-Patman Act and has been criticized in recent years for fostering undue price rigidity. Despite this apparent pricing rigidity, the statute still enjoys considerable support in Congress, and may be enforced by injured parties even if government enforcement is limited. Thus, violations (or alleged violations) of the Act may be subject to civil lawsuits even if there is no challenge by the government (Bailey, 1978).

REFERENCES

Anon., 1968, *Mining Explained,* Northern Miner Press Ltd., Toronto, pp. 175-176.

Anon., 1982, "E & MJ Markets," *Engineering and Mining Journal,* Vol. 183, No. 10, Oct., pp. 19-21.

Aronson, R.J., 1979, *The Scorecard: A Guide to the Real World of Economics and Business,* W. B. Saunders, Philadelphia, London, Toronto, p. 12.

Bailey, E.L., ed., 1978, *Pricing Practices and Strategies,* Conference Board Report No. 751, The Conference Board, Inc., New York, pp. 40-56.

Converse, P.D., 1964, "The Other Half of Marketing," *The Environment of Marketing Behavior, Selections from the Literature,* R.J. Holloway and R.S. Hancock, eds., Wiley and Sons, New York, p. 261.

McGarry, E.D., 1964,"The Propaganda Function in Marketing," *The Environment of Marketing Behavior, Selections from the Literature,* R.J. Holloway and R.S. Hancock, eds., Wiley and Sons, New York, p. 244.

Nystrom, P.H., ed., 1951, *Marketing Handbook,* Ronald Press, New York, pp. 82-86.

Parks, R.D., 1957, *Examination and Valuation of Mineral Property,* 4th ed., Addison-Wesley, Advanced Book Program, Reading, MA, pp. 264-265.

Rewoldt, S.H., et al., 1977, *Introduction to Marketing Management, Text and Cases,* 3rd ed., Richard D. Irwin, Inc., Homewood, IL, pp. 8-16, 269-274.

Smith, V.K., and Krutilla, J.V. eds., 1982, *Explorations in Natural Resource Economics,* Resources for the Future, Inc., Johns Hopkins University Press, Baltimore, p. 354.

8.7. Surface Mining Control and Reclamation Act of 1977

DAVID S. HEMENWAY

INTRODUCTION

This section presents a discussion of legal counsel's role in the government regulatory area dealing with mine land reclamation with specific reference to coal mining. The Surface Mining Control and Reclamation Act of 1977 (Act) has had a dramatic impact on the coal mining industry. It is not the intention to present here an indepth analysis of the legal implications of the Act, but to present an overview of the lawyer's role in assisting coal mine operators in complying with the requirements of the Act and regulations.

It seems helpful to understand the role of the other disciplines in an undertaking such as compliance with the Act. Hopefully the nonlawyer reader will have a better understanding of the assistance and advice that legal counsel can provide. The lawyer reader will be struck by the superficial treatment given the subject. In that vein, it was sought to direct counsel to sources of knowledge and not to provide that knowledge here which could well fill a legal treatise.

COMPLYING WITH THE ACT'S REGULATIONS

Permitting

Early participation by counsel in the permitting application process, not only in the legal facets described subsequently, but also in identifying the studies and the like which may be necessary is imperative. For example, extensive environmental studies may be required in the early stages in order to secure a sufficient baseline data.

Counsel should understand the corporate structure of the operator in undertaking the legal facets of the permitting function. Sec. 778.13 (all references to final EIS, Vol. III, "Draft Final Regulations," supra.) provides a detailed informational reporting requirement concerning corporate structures including names and addresses of officers, directors, and principal shareholders. It further requires legal ownership of lands contiguous to the permit area be identified and reported.

Sec. 778.14(a) provides for a disclosure of expense relating to suspension of permits and revocation of bonds for a five-year period preceding the permitting application. If such has occurred a detailed report is required (778.18(b)). A listing and status report is required for all violations issued by any government agency of reclamation or air or water environmental protections by the operator, its subsidiary, affiliates, or persons controlling the operator (778.14(c)).

Documentation of the legal right to conduct mining operations on the permit area must be submitted and where the mineral estate is severed from the surface a consent form executed by the surface owner must be provided if a surface mining operation is planned (778.15). In addition, the Act describes certain land areas as unsuitable for mining and provides that other land areas may be unsuitable if certain criteria are met (521 Act). If such unsuitability proceedings or impacts are present, (Parts 761, 762, 764, and 769) a status description plus compliance with the respective requirements must be submitted. (See also Sec. 510(c) of the Act.)

One other significant consideration which needs early attention is the method the operator intends to use in financing the mine operation. Sec. 506(b) of the Act provides that the maximum permit term of five years *may* be extended by showing an extended period of time is necessary to obtain financing "for equipment and opening of the operation." Documentation from the source of the funds is necessary to support an extended permit term (778.17). Bonding and insurance requirements must be met when submitting the permit application (778.18, see Part 860 as amended and Sec. 509 of the Act).

To obtain data on the informational requirements of the Act counsel should review Sec. 507 which contains informational requirements and Sec. 508 which sets forth the reclamation plan requirements. A checklist can be formulated to make sure that in the initial stages no subject is overlooked. This process can be extremely helpful as the state application form may mask the specific identity of the subject matter. The performance standards of the Sec. 515, surface, and Sec. 516, underground, and Parts 816 and 817 of the regulations should be reviewed in order that the reclamation plan addresses those requirements. Finally, each state's regulatory program needs a subject matter view for purpose of identifying additional or unique treatment of subjects.

Due to the fact that most states have implemented permanent programs, the initial permanent permitting process is now underway. The operator has two months from the date of the state program approval to submit an application. A permit is required within six months to continue operation. There are procedures to avoid the harsh result of being forced to shut down due to the state's inability to process a timely filed "complete" application. These include permitting additional acreages during the interim program and a two-tier review process. (See also part 773.) However, one development arising out of the identification of the problem may enable the operator to achieve a savings in cost and time. The state agency and other permittees may have submitted environmental studies which may provide beneficial baseline and other data which may be subject to incorporation into the studies the operator needs to perform. This may also assist in the identification and selection of consultants to perform studies not within the inhouse capabilities of the operator.

After review and planning has been accomplished, counsel may propose a meeting with the state regulatory authority. At this meeting, a description and a general outline of the studies and other aspects of the mine project can be given for reaction. Counsel also can achieve a clarification of any issues concerning legitimacy of the information requested. In this way, the operator can ascertain any areas where emphasis or additional studies need to be undertaken. These officials are the people who will review the application and issue the permit; therefore it is desirable to open lines of communication at an early date. The operator's personnel, including counsel, may seek advice from his counterpart representing the agency. This action may help avoid pitfalls and unfortunate conditions being placed on the permit.

To this point our discussion has involved no unique problems such as whether the operation in question is subject to jurisdiction and regulation. This question most frequently arises with a tipple or coal processing facility which is situated at a considerable distance from the mine operation and which is not an adjunct to a given mine (785.21). Other unique areas which should be given careful consideration include steep slopes (Appalachia 785.15), prime farmland grandfather (Midwest 785.15) and alluvial valley floors (West 785.19). These matters have all been subject to the litigation process (supra.) and it is recommended that careful scrutiny of those matters and administrative determinations by the Office of Hearings and Appeals be pursued.

During the preparation of the application and underlying supporting documentation, it is desirable to review progress in those areas. Legal questions which are bound to arise can be identified and a response made. In addition, counsel should make certain all legal records and documentation have been reviewed for accuracy. Contact should be maintained with the operator to ascertain any changes in the violation history, the operator's structure, and officers and directors who must be identified in the permit application.

Finally, when the permit application package is complete, a thorough review should be made by counsel to determine that all subjects have been appropriately covered and that there are no inconsistencies within the application. The agency review process including the public notification and participation features should be scrutinized during this evaluation (773.13, et seq.).

Upon issuance of the permit counsel should immediately review with the operator all conditions and requirements imposed by the regulatory authority. The conditions or requirements of the permit are subject to administrative and judicial review. The operator may seek to have conditions of the permit invalidated (775.11, et seq.). (See also 43 CFR, Part 4.) This means that unlike MSHA and similar to the NPDES permitting, a permit holder may seek judicial review of a permit condition which is deemed not lawful or inappropriate under the circumstances.

Compliance

When a viable permit is secured, the next area of concern is educational. Participation by operations personnel in the permitting project varies greatly. In most instances, that participation is minimal. It is imperative that operations personnel fully understand the requirements of the permit and the regulations governing the mine operations at the earliest possible date. Hence, a presentation to these personnel on the permit's conditions and terms and the regulations governing the operations must be undertaken. Counsel should be prepared to participate in the area of compliance and to respond to questions involving interpretations of the permit requirements.

The need for early understanding by operator personnel cannot be overemphasized. Remember the new permit holder occupies somewhat the position of the "new kid on the block." The regulatory authority may make early and frequent inspections to insure permit compliance. Initial compliance failures may cause the regulatory authority to conduct more aggressive and frequent inspections.

Operator personnel should be requested to avoid the temptation to make legal interpretations of the permit conditions or regulations. Rather they should be encouraged to consult with counsel immediately upon identifying any question of interpretation. Counsel by the same token needs to respond to those inquiries in an expeditious manner with short reasoned answers. Counsel must understand operations

personnel have no need for lengthy and learned discussions—precise and to the point responses are needed.

Another area of compliance assistance which counsel might propose is the use of an internal compliance team. This team would be composed of nonmine personnel having disciplines in the major permit areas, e.g., environmental (water), reclamation, and legal. This team would routinely visit the mine and perform an inspection. A conference can then be held with operations personnel. Here detailed corrective measures for problem areas can be discussed and an understanding reached as to problem areas and proper corrective course of action. Counsel can provide not only legal guidance, but may act as a mediator if disagreements arise.

There are disadvantages to formal procedures. Informal consultations between the disciplines and operating personnel may prove more effective. The temptation of nonmine personnel to impose their precepts of mining procedure on operations personnel is lessened through an informal structure. Finally, the necessity of maintaining open communications between these personnel is lessened in the formal procedure as the critiques which necessarily will arise may not be received with great enthusiasm by the mine. Counsel should advise the operator as to the pros and cons of such a procedure. The results of government enforcement however may dictate the advisable procedure to implement.

Enforcement

In the area of enforcement, counsel's participation is probably most evident. The responsibility here is somewhat negative in nature. It is either to reduce by degree and/or tenor a sanction or void the sanction. It is negative in that it does nothing in a positive fashion to enhance operator production or profit. The adoption of defined procedures and identity of responsibilities will help to alleviate the negative impact of enforcement.

Counsel should be familiar with the sanctions provided by the Act, the federal and state regulations. The federal and state regulations will not be dissimilar; hence all references are to the Act and the federal regulations. The permanent regulations dealing with enforcement were not a part of the overall repromulgation of permanent regulations. Applicable federal requirements are found in the Act at Secs. 517, 518, 521, and 525 and in the regulations at 30 CFR Parts 840, 842, 843, and 845. (843.11 and 843.12 were amended after publication of the CFR revised 1981 edition.) OSM inspectors and state regulatory inspectors have a right of entry on the mine property without a search warrant (517(b)(3)). Citizens who have filed a complaint may accompany an inspector but only to visit those areas which were the subject of the complaint (517(h)).

The Act provides four enforcement sanctions which may be imposed by the state regulatory authority. They are:

Notice of Violation: A notice of violation requires abatement of the violation within a set time period. A civil penalty of up to $5000 may be imposed if corrective action is not taken by the operator. The amount is determined on the basis of the negligence of the operator, gravity of the violation, history of violations at the mine, and the good faith exercised in abatement.

Cessation Order: A cessation order requires cessation of mining activities on the area affected until abatement of the violation has occurred (521(a)2,3). A cessation order is issued if a violation has not been abated within the time set in a notice of violation, if the condition or practice creates an imminent danger to health or safety of the public, or is or can be reasonably expected to cause significant imminent environmental harm (regulatory definition of terms found at 701.5). A penalty must be assessed if a cessation order is issued based on the above criterion.

Suspension or Revocation of the Permit: If a series of violations caused by the unwarrantable failure of the operator is found by the regulatory authority to exist, then a show/cause order is issued as to whether a permit should be suspended or revoked.

Injunctive Relief: Injunctive relief may be sought by the regulatory authority for interfering with inspections or for failure to comply with any order of the regulatory authority (521(c)).

In the event a cessation order is issued, mine personnel should be informed of their right to an informal conference at or near the mine site (521(e)5). A judgment should be made as to whether such a conference is desirable under the circumstances.

It is to be noted that if an informal conference is not held, the cessation order expires after 30 days of issuance. However, it should be borne in mind that nothing would prevent an inspector from reissuing a new cessation order at the time the old one expires.

Counsel should advise the operator when an inspector enters the mine property (usually he reports to the mine office); a qualified individual previously designated by the operator should accompany the inspector on his rounds. This individual should be versed in the mine operation, the permit terms and conditions, and should be able to explain in detail various reclamation activities. Frequently an apparent violation is reversed once an explanation of the mining or reclamation operation is made. In this way, the cost and expense of legally contesting the notice of violation or cessation order is avoided. The designation of key individuals to accompany the inspector should instill a cooperative and creditable relationship as distinguished from the adversary relationship which could arise.

A procedure should be adopted whereby the mine notifies counsel immediately upon receipt of a notice of violation, a cessation order, and if they are served with an order to show cause as to why the permit should not be suspended or revoked. A report form completed by the person accompanying the inspector serves this purpose. The circumstances surrounding the violation and the penalty assessment elements of negligence, gravity, history, and good faith abatement should be described on the form.

Upon a receipt of the notice or order and the report form, counsel can begin an investigation of the conditions and circumstances surrounding the incident. Counsel should ascertain whether the cessation order meets the statutory test of immediate danger to public health or safety or is of significant imminent environmental harm. Appropriate recourse through an expedited hearing with the office of hearings and appeals is available (525(c), 43 CFR 4.1180, et seq.).

In most instances, seeking of expedited relief may not prove productive. The next step is to await receipt of the penalty assessment. At that time, the operator has a right to request an informal conference with the regulatory authority. Exercise of this right stays the 30-day time limit on requesting a formal hearing. Informal conferences may serve as a productive means of achieving either a mitigation of the penalty or completely avoiding it. Preparation for conference should be thorough. A knowledgeable person from the mine should accompany counsel to the conference. As penalty assessments are levied on four elements—negligence, gravity, history, and good faith abatement—it is desirable to present mitigating evidence on the elements in the categories.

If the conference results do not prove successful, then the next step would be to request a formal hearing. At this point it would be treated as any other administrative appeal or court case in that normal trial preparation and conduct is available. The only exception to this is that the administrative law judge must follow the regulatory authority's point system in assessing penalties (43 CFR 4.1157). In essence so many points are charged for each element and then totaled. The total points are then converted to dollars of assessment (e.g., see 845.14). Also at the time notice of contest is submitted and a formal hearing requested the penalty must be paid. A refund will be made if the assessment is lowered (518(c)).

CONCLUSION

In conclusion, there are three areas where counsel participation could prove helpful. These are:

Permitting: includes the compilation of compliance, operator's structure, and land ownership information and oversight, guidance, and in remote cases, advocacy.

Compliance: is primarily to avoid enforcement problems through education.

Enforcement: the primary role is as an advocate and negotiator on behalf of the coal mine operator.

In performing these responsibilities, counsel must bring with him an understanding of litigation and the regulatory activities which have transpired. Particular emphasis should be given to the litigation of the regulatory programs promulgated by the Office of Surface Mining and to the ever-changing content of the regulations themselves. In this fashion, appropriate and able advocacy may be more readily available.

The nonlawyer reader may have a better understanding of the reasons for counsel's advice and the fact that some advice may have to be of a qualified nature. Hopefully, the reader has been convinced of the need for an open dialogue between the operator and counsel at an early date. Finally, the need for early and thorough preparation by counsel and the operator should be emphasized.

8.8 Purchasing and Inventories

Daniel P. Plute

MATERIALS MANAGEMENT

Many mining and processing companies find that material and spare part costs account for 40 to 60% of their operating and maintenance budgets. They also find that the capital invested in inventory to support the mining venture runs into the millions of dollars—money that may be nonproductive and often wasted if not properly managed.

Maintaining higher than required inventory balances ties up capital; keeping a low balance increases the risk of prolonging equipment downtime or production delays. The key is defining the proper mix, the correct balance, planning purchases to meet production and maintenance schedules, and locating assured sources of supply with reasonable prices and acceptable service.

Managing the purchasing, inventory, and warehousing functions requires that each area be accountable and responsible for its activities. The following responsibility statements work well in industry:

1. *Purchasing*—The role of Purchasing is to provide materials requested by Operations and Maintenance and Inventory Management at the required time, at the lowest possible cost. Cost consideration must include the price of the materials, the cost of transporting it to its destination, and the cost of the clerical and paperwork required to order, receive, store, issue, and pay for the materials. Purchasing often determines which vendor is to be used and may switch vendors to obtain a better price.

Purchasing does not have the authority to substitute a different item or different brand to obtain a better price or delivery time without the approval of the user. It is absolutely necessary that Purchasing work closely with the user so that the user can make appropriate decisions relative to the tradeoff between purchase price, delivery time, item specifications, and quality.

2. *Inventory Management*—The role of Inventory Management is to request material for stock in a timely manner and in quantities that provide the best level of support to users at the lowest cost. Inventory Management establishes the rules that determine when an item is to be ordered and in what quantity. They work directly with Maintenance and Operations in determining what items are to be stocked. Maintenance provides them with projected usage information for normal stock items and minimum stock levels for critical and insurance items.

3. *Warehousing*—Warehousing's responsibility is to receive, store, issue, and deliver materials to users efficiently and to prepare accurate transaction documents for these functions. It is the responsibility of the warehouse to develop and adhere to adequate procedures for the receiving, issuing, storing, and securing materials. The warehouse is responsible for keeping accurate records on all material within its purview.

4. *Operations and Maintenance Requirements for Spare Parts*—The responsibility of operations and maintenance is to communicate in an accurate and timely manner to Inventory Management and to Purchasing what their requirements are for spare parts. For parts used on an ongoing basis, they will communicate to Inventory Management any expected changes from normal usage patterns and any new

items which should be maintained in warehouse stock. They also work closely with Inventory Management in the analysis of slow and nonmoving items to determine what items can be sold as surplus or declared obsolete. In case of an emergency requiring exceptional purchasing activity and followup, they work directly with Purchasing providing them with information to help them obtain the material required.

Operations and maintenance personnel deal directly with vendor representatives in determining an item's specification, application requirements, and what items can be substituted for standard items.

DETERMINING SPARE PARTS AND MATERIAL REQUIREMENTS AND MANNING LEVELS FOR A MINING VENTURE

The amount of capital required for the inventory of spare parts and operating supplies will vary with the mine type, the distance from the supply centers, production and operating plans, the rules developed for reordering and the quantity to be purchased, and the maintenance program used at the property. Normally, 4 to 7% of the capital cost of the fixed plant and equipment is allocated for inventory investment.

The manning level required will also vary with the mine and ore type, processes included, production plan, operating schedule, and type of information processing system. Fig. 1 shows Materials Management Statistics for several North American mining companies. It includes information by mine and ore type, production levels, purchasing and warehousing statistics, manning levels, and performance of employees.

Selecting the Initial Complement of Spare Parts and Operating Supplies for New Properties or Property Modification

The initial selection of spare parts and operating supplies should begin after the Request for Expenditure (RFE) has been approved. A work plan and critical path schedule should be prepared to monitor project activities and track key completion dates.

The responsibility for selecting the inventory item mix should be assigned to key maintenance, operations, and inventory management personnel. Final selection approval should come from the mine manager, senior financial personnel, and the materials manager.

Factors to consider in the spare part and operating supply selection should include:

Equipment availability and operating goals.

Capital availability.

Equipment manufacturer's spare part recommendations.

Previous experience of Maintenance and Operating Managers.

Experience of other mining companies with similar equipment and conditions.

Management review of Process and Instrument drawings to identify critical backup requirements.

Consumption estimates for chemicals, reagents, and operating supplies.

Wear-life estimates for major material such as liners, refractories, grinding media, and belting, etc.

Fig. 1. North American Mining Companies Materials Management Performance and Statistics

Product	Mine Type	Production (tpy)			Inv. Statistics		Purchasing Stats. (per month)		Employment (number of employees)							Purchasing†				Whse,‡ Issues/ Day	No. of Inv. Items/ Employee	No. of Issues/Day Per Employee
		Ore (millions)	Waste	Total	No. of Items	Issues Per Month	No. of PO's	No. of Line Items	Total	Whse	%	Inv. Mgt.	%	Purch.	%	POs/ Day	Lines/ Day	POs-Day/ Employee	Lines-Day/ Employee			
1. Copper	UG	23	3	26	60,000	17,650	2,550	8,380	78	56	72	8	10	14	18	121	399	9	29	588	1,072	11
2. Copper	UG			38	N.A.	7,000	760	2,376	44	12	75	2	12.5	2	12.5	62	310	31	155	367	667	31
3. Coal	UG	4.1	1.7	5.8	8,000	11,000	1,300	6,500	16	15	68	2	9	5	23	52	186	10	37	194	N.A.	13
4. Uranium	Surface	0.131	0.003	0.134	N.A.	5,830	1,089	3,912	22	17	63	5	18.5	5	18.5	32	101	7	20	240	N.A.	14
5. Uranium*	UG (1 UG, 1 Under Const.)	0.120	0.003	0.123	N.A.	7,200	668	2,119	27													
6. Copper	Concent.	0.764	—	0.764	N.A.	4,875	879	1,856	15	9	60	1	5	6	40	42	88	7	15	163	N.A.	18
7. Copper	UG			11	N.A.	N.A.	1,197	N.A.	19	13	68	4	12	5	27	57	N.A.	4	N.A.	83	N.A.	4
8. Phosphate	Surface	47.5	34.4	81.9	N.A.	2,500	2,600	5,700	34	22	65	1	6	8	23	124	271	16	34	9	N.A.	N.A.
9. Coal	Surface			0.799	N.A.	281	93	418	9							5	20	N.A.	N.A.			
10. Coal	Surface			2,667	8,000	N.A.	N.A.	N.A.	16	14	78	1	25	3	16	39	174	13	58	133	571	10
11. Coal	Surface	4.8	39	43.8	6,550	4,000	810	3,650	18	2	50			1	25	34	N.A.	34	N.A.	67	1,638	34
12. Copper	UG	0.90	0.01	0.91	N.A.	2,000	710	N.A.	4													
13. Aluminum	Smelter			0.299	13,000	5,424	1,238	5,424	29	13	45	3	10	13	45	59	258	5	20	181	1,000	14
14. Iron Ore	Surface	13.8	12.9	26.7	150,000	37,500	5,560	15,400	173	93	54	61	35	19	11	265	733	14	39	1,250	1,613	17
	Concentrator			19.5																		
15. Copper, Silver, Gold Smelter, Ref., etc.	Metal Proc. Plant			0.80	8,317	3,400	1,067	N.A.	20	8	40	3	15	9	45	52	N.A.	12	N.A.	113	1,040	14
16. Uranium	Mine	0.31	1.9	2.21	8,791	N.A.	N.A.	N.A.	13	10	77	2	15	1	8	N.A.	N.A.	N.A.	N.A.	N.A.	879	N.A.
	Mill			0.31																		
Averages					32,832	9,055	1,466	5,067	34	22		8		7		80	358	14	45	282	1,060	16

* Under development.
† 21-day month.
‡ 30-day month.

Operating supply and spare part availability and vendor lead time projections.

Availability of storage space.

The inventory item identification process must include activities to determine the initial purchase order quantity and delivery date, to estimate annual requirements, and to define application or minimum quantity requirements.

Defining Stock Level Requirements for Inventory Items

The Inventory Management supervisor is responsible for determining when to order and the quantity to purchase for all inventory items.

The Inventory Department should process the approved inventory item requests and determine when to review the item for reordering (the reorder point); calculate the suggested order quantity (the economic order quantity); determine the items' category (i.e., bearings, haul truck, shovel, crusher, mill, etc.); and the catalogue and ordering description for the item.

Estimated usage figures and vendor lead time projections, included on the new stock item request, can be used for the calculations. After six months usage history has been captured or when production or maintenance plans change, the safety stock, reorder point, and economic order quantity levels should be adjusted.

Production and maintenance plans should be used to determine actual quantity and delivery date requirements for items consumed in the production process and for spare parts used in planned maintenance programs. Quantity and delivery date projections should be modified when plans change.

The Inventory Department should also calculate the average and maximum stock levels, and prepare budgets for the inventory investment and new stock item requests. Monthly, actual, and budget comparisons should be made to identify variances from plan.

The calculations for average and maximum value levels are:

Average inventory value level = safety stock quantity + one-half of the economic order quantity × unit price

Maximum inventory level = safety stock quantity + one economic order quantity × unit price

Determining Warehouse Size Requirements and Location

The warehouse facility should be attached to, or located near, the maintenance shop. The Maintenance Department is the largest user of warehouse service; therefore, locating the warehouse near the maintenance shop area will reduce idle and spare parts transportation time.

The amount of space required for inside storage will be influenced by the mine production plan, weather conditions, and the number of items planned for the inventory mix. Heated and cold storage areas may be required.

For estimates on inside and outside space requirements, allow 0.2 m² (2.2 sq ft) per line item for inside storage and 0.4 m² (4.7 sq ft) per line item for outside storage; pallet rack storage, allow 1.2 m² (13.1 sq ft) per pallet rack.

Aisleways and material handling equipment routes should be spaced so that equipment can be moved easily and safely. For large equipment, i.e., 1814-kg (4000-lb) forklifts, a 1.8-m (6-ft) aisle will be required while a 0.9-m (3-ft) aisle should be planned for walkways. A large aisleway 1.8-m (6-ft) should be provided for moving material handling equipment between the receiving and issue areas.

Storage Bins, Storage Location Codes, and Warehouse Security

A mixture of storage drawer modules, storage bins, and pallet racks will be required. A ratio of 50% modules, 30% bins, and 20% pallet racks can be used for estimating.

The modular cabinets should be located near the issue window and used for storing smaller fast-moving items. The bins should be situated next to the modular cabinets. The pallet rack area should be situated near wall exposures and near the receiving station.

The modular cabinet and storage bin aisleways should face the issue window for easy access to storage areas.

Storage location codes and row content labels should be placed at the end of rows. The row content labels should include a brief description of row contents, i.e., Row 1—Unit Rig M100 and the number of bin or cabinet sections.

Storage location codes should include designations for warehouse area, row, section, shelf, drawer number, and bin number. Number codes should be used.

Bin labels should be affixed to bin areas and should contain information on item descriptions, part number, equipment type, where used, and the computer number.

Warehouse storage areas (inside and out) should be secured and have controlled entry. Items subject to pilferage, i.e., tools, automotive tuneup parts, etc., should be stored in a locked cabinet.

Adequate lighting is essential to good housekeeping and identification of items.

MANAGING THE CAPITAL INVESTED IN SPARE PARTS AND OPERATING SUPPLIES

Millions of dollars are committed to purchase and maintain spare parts, components, and operating supplies. Often, the dollars are mismanaged. Capital is invested in overstocked supplies and spare parts; on the other hand, investment levels that are too low increase the risk of production losses and may contribute to underutilization of maintenance labor.

An optimum inventory investment level can be reached by planning material requirements using maintenance schedules and production plans and by analyzing historical usage and cost information to project needs.

The programs that manage the inventory investment and material and supply costs must be flexible. They must contain features that allow rapid adjustments commensurate with changes in mining plans and increases or decreases in production levels. Materials management must adjust its strategy as economic and operating conditions change.

To achieve its goal, materials management must have the correct set of *tools* that will allow it to make timely adjustments to its business plan, minimize the capital required to purchase inventory spare parts and operating supplies, and to purchase the required material on the required date at the lowest possible cost. The following programs are recommended:

Material Requirements Planning

Mine, maintenance, and processing plans are used to determine spare part and operating supply requirements for the short term. Adjustments to purchase commitments are made as the plans change—delivery dates and order quantities are modified to meet current requirements. An economic evaluation is made on the feasibility of cancelling the purchase order or reducing stock levels.

Successful use of material requirements planning requires a strong communication link between operations, maintenance, and materials management. Purchasing must relay information on delayed deliveries to maintenance so that planners and schedulers can adjust their maintenance plan, and operations must advise inventory management on changes to production plans so that inventory requirements and inventory levels can be adjusted.

Inventory Stratification

Inventory items are stratified into categories and classes to improve inventory management capabilities. The categories are active and inactive—the active classes are *A, B,* and C; the inactive classes are *X, Y,* and *Z.* The active classes are determined by the annual usage dollar amount, the inactive items by the dollar value on hand. (Active items are items with issue activity during the last 12 months; inactive items have no issue activity during the last 12 months.)

The active and inactive item categories and their dollar parameters are:

Active Item Class	Annual Usage Dollar Amount
A	Exceeds $1500
B	Between $500 and $1500
C	Less than $500

Inactive Item Class	On-Hand Dollar Value
X	Exceeds $500
Y	Between $300 and $500
Z	Less than $300

(The usage value and on-hand value parameters may vary.)

A items represent approximately 5 to 8% of the active, high-usage-value items; comprise 40 to 60% of the active item inventory investment; and account for 70 to 90% of the dollars spent annually on material.

B items are the intermediate-usage-value items representing 10 to 20% of the active inventory investment and account for 10 to 20% of the dollars spent annually on material.

C items represent 75 to 80% of the active items which have less individual effect on inventory investment. As a group, they represent approximately 15 to 25% of the active inventory investment and account for approximately 5 to 8% of the dollars spent annually on material.

X items represent 10 to 15% of the high-inventory-value inactive items and account for 70 to 80% of the inactive inventory investment. These items offer the greatest opportunity for recovery of investment through internal use or other disposal.

Y items are of secondary importance and represent 15 to 25% of the inactive inventory investment, accounting for 25 to 35% of the inactive items.

Z items represent approximately 60 to 70% of the relatively unimportant low-inventory value items and offer the least opportunity for investment recovery because of the substantial clerical cost required to analyze so many items of low individual value. *Z* items account for approximately 3 to 5% of the value in the inactive inventory.

Once stratified, different controls and programs should be used to manage the categories and classes.

Active Item Management

Economic Order Quantity—The economic order quantity balances the costs of preparing purchase orders and carrying inventory. The economic order quantity is a suggested order quantity and should be adjusted for vendor packaging, for actual quantity requirements, weight required for C/L or T/L shipments and for quantity discounts in some situations, etc.

One formula for calculating the economic order quantity is:

$$\text{EOQ in dollars} = K \sqrt{\text{annual usage dollars}}$$

$$\text{EOQ in units} = \text{EOQ\$} \div \text{unit price}$$

$$K = \sqrt{\frac{2B}{I}}$$

where

B = cost of preparing purchase orders (line item)
I = cost of carrying inventory (decimal)

Costs to consider in determining purchase order and inventory carrying costs include:

Purchase Order Costs (line item)
Purchasing forms
Salaries
Telephones
Telex
Computer costs
Envelopes
Mailing
Purchasing department operating costs

Inventory Carrying Costs (decimal)
Cost of capital
Storage costs
Deterioration and obsolescence
Insurance
Warehouse salaries

In the mining industry, a K factor between 3.5 and 6 is considered normal. For new operations and operations converting to the economic order quantity system, an initial K factor of 6 is recommended. After six months of use, the K factor should be reduced gradually to a factor of 4 or to the factor calculated by the formula (Figs. 2 and 3).

Reorder Point and Safety Stock Levels: Historical or estimated usage information (new items), vendor lead time, the items' class, and safety stock factors are used to determine the stock level at which the item should be reviewed for order action.

The formulas for calculating safety stock and reorder point levels are:

Safety Stock

Safety Stock = Safety Stock Factor × Lead Time Usage

Safety stock is that portion of the inventory maintained as a buffer against delayed deliveries or an erratic demand pattern. It can be expressed in months, weeks, or days of usage and should vary according to the lead time and classification of the item.

Safety Stock Parameters

If lead time is	Items	Safety stock factor (of lead time use)
Over 10 weeks	*A* and *B*	0.60
	C	0.80
6 to 10 weeks	*A* and *B*	0.50
	C	0.70

Fig. 2. Inventory Model Current Practice
(Order Items Six Times per Year)

Item Class	Number of Items	Annual Usage Value	Line Item Orders	Order Value	Average* Inventory Value
A	200	$500,000	1200	$417.00	$41,700
B	300	60,000	1800	33.00	4,950
C	1200	27,000	7200	3.75	2,250
	1700	$587,000	10200		$48,900

Application of economic order quantity concepts: order A items, 10 times year; B items, 5 times year; and C items, 3 times year.

A	200	$500,000	2000	$250.00	$25,000
B	300	60,000	1500	40.00	6,000
C	1200	27,000	3600	7.50	4,500
	1700	$587,000	7100		$35,500

Results:
 Number of line item orders reduced 30%
 Average inventory value reduced 27%

* Average inventory value = item count × order value × 0.5

2 to 5 weeks	A and B	0.40
	C	0.60
Less than 2 weeks	A and B	0.30
	C	0.50

Average weekly use = annual usage quantity ÷ 52
Average daily use = annual usage quantity ÷ number of operating days/yr
Lead time use = lead time × average weekly usage

Reorder Point

ROP = safety stock quantity + lead time usage quantity

Reorder Point (ROP) is set at usage over the lead time plus safety stock. Thus, for an A item with a lead time of 3 weeks, and an average weekly usage of 12 units, the reorder point would be 50 units, a safety stock of 14 plus lead time use of 36.

Technical Review: The object of this program is to defer, reduce, or cancel reorders for high unit price, and selected A items. Actual quantity requirements are determined by reviewing maintenance and operating plans. Consistent ap-

plication of this concept will result in significant deferral, or cancellation, of cash outlays and will reduce inventory levels.

Since it is impractical to review all reorders, the Technical Review Program requires a detailed review of the reorders that cover the most important reorder decisions affecting materials and purchases; are for items of a high unit price; or are for items designated as critical.

Overstocked Item Analysis: An inventory item with a current balance of more than the safety stock plus one EOQ is considered to be overstocked. The following action should be taken to reduce overstocks:

1. Review quantity on-order to determine if it should be reduced, cancelled, or deferred.

2. Reduce order point and EOQ to reflect current or projected usage trend.

3. Use overstock as a substitute for another inventory item.

4. Initiate disposal or return-to-vendor action.

Overcommitted Item Analysis: An overcommitted inventory item is one having an on-hand and on-order quantity in excess of the sum of the reorder point quantity and one

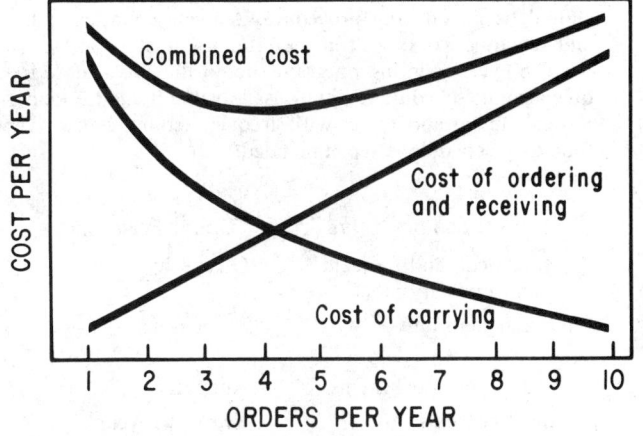

Fig. 3. Economic order quantity. Annual cost of carrying stock vs. annual cost of ordering under varying order quantity.

economic order quantity. The following action should be taken to reduce overcommitted items:

1. Cancel outstanding purchase order quantities to reduce the on-order quantity.

2. Defer delivery of the on-order quantity until the stock on hand level returns to normal.

3. Review the usage and lead time projections and assure that order point and order quantity are consistent with the current operating plan.

A Item Control Program: *A* items tie up the majority of inventory investment and represent 70 to 90% of dollars spent on materials and spare parts. Inventory and cost reduction benefits can be realized by carrying out an organized analytical program for each item in the *A* category. The following are recommended:

1. Competitive bidding should be used effectively and consistently to reduce prices.

2. Annual or longer contracts should be negotiated with scheduled deliveries to minimize stock levels.

3. Salvage and reuse of items should be investigated.

4. Usage rate should be analyzed to determine if redesign, quality control, or improved operating or maintenance practices will lower the usage rate.

Low Value Program: Frequently used fastener products, fittings, small electrical terminals, grease zerks, pipe products, and low unit price items commonly used in an area should be stored in open bins near maintenance work stations. Backup stock for low value items should not be carried in inventory.

The following are characteristics of low value items:

1. Consumables and common hardware items with a low unit value.

2. Specialized items used in one area.

3. Not subject to pilferage or higher dollar loss.

4. Available from local suppliers.

5. Lead time less than 14 days.

A low value overhead account should be used to track costs to each low value station. Warehouse employees should inspect bin areas for low balances and replenish them. Maintenance employees are assigned housekeeping duties.

Inactive Item Management Program

The amount of capital misappropriated to inactive spare parts and operating supplies will vary in mining operations. Often, the investment in inactive items represents between 25 and 45% of the total inventory investment.

The following inactive item control program should be used to manage the inactive portion of the inventory investment:

Inactive Item Review—The inactive items should be sorted into the *X, Y,* and *Z* classes, by category, where used, and reviewed annually.

The inventory management department should obtain information on the item or category of items, i.e., purchase order number, date added to inventory, approved by and issue history, and request assistance from operation and maintenance department employees in determining the current status of the items.

The following status categories should be used:

1. *Obsolete*—The item is no longer required.

2. *Critical or Insurance*—The item must be kept in inventory to reduce the risk of curtailing or stopping a major portion of the production flow. Manager or supervisor level approval should be required for this classification. A critical/insurance item code and cost center account number, where used, should be added to the item's record.

3. *Inactive—Retain in Inventory*—Maintenance department employees may be uncertain as to the immediate need for an inactive item. Research may be required to define the item's status. Management authorization should be required to retain an inactive item in the inventory mix and the retention time period clearly defined. A followup review should be performed after three months. If the item's status has not been determined, either return to vendor, sale, or disposal action should be started.

The inactive item investment must be controlled and monitored to reduce capital requirements for spare parts and materials. It is an area where new capital can be recovered from nonproductive assets. Maintenance and operating employees must realize that it costs the average mining company 30 to 35% of the item's value to keep it in inventory. Employees must realize that with sound planning programs, items can be ordered and delivered, just-in-time for their needs.

Inventory Investment Controls

New Stock Item Budget—A new stock item budget and management controls will help regulate inventory investment increases. Each year, materials management, operations, and maintenance department supervisors analyze the current mine plan, the approved capital appropriation budget, and the maintenance department's major maintenance and rebuild plans, and subsequently, project new requirements for the inventory. A budget is prepared for each mine process and the maintenance department and submitted for approval by the mine manager.

Requests for additions to the inventory mix should be submitted on new item requisitions and approved by the department manager. Once a department is over budget on new item additions, senior management approvals should be required for the request.

Monthly, materials management should prepare a report on the new item additions, comparing budget and actual dollar amounts for the current period and year-to-date.

Verifying Investment Value and Quantity Levels—Accurate quantity and value levels are important for the financial, materials management, operations, and maintenance departments. Checking a record to confirm an on-hand balance and then finding the bin empty creates maintenance and operation scheduling problems and financial reporting errors. In addition, reordering an item before its balance reaches the reorder point level may create an overstock situation; reordering at a level lower than the reorder point increases the risk of stock-out.

Record accuracy checks and financial record verification should be an ongoing procedure. A cycle counting policy and schedule works well in the mining industry.

The cycle counting program should include a more frequent count schedule for high usage value items (*A* items), critical items, and items with frequent count errors. The following schedule is recommended:

Item Class or Status	Count Frequency
A, critical, and frequent count error items	Quarterly
B and high unit price items	Semiannual
C and new stock items	Annually
Inactive items	Every two years

Inventory record checks can also be performed when an item is at a zero balance level. The warehouse clerk flags the issue requisition as having a zero balance and the inventory control clerk compares the physical and record balances. If a discrepancy exists, the clerk adjusts the record or requests a second count. (Expediting action should be initiated if the item is on order and the item's scheduled count date adjusted.)

A management report should be prepared on cycle counting activities. The report could include:

1. The number of items counted this month and year-to-date.

2. The value of adjustments (debit and credit) for the current period and year-to-date.

3. The number of items to be counted the next period and for the year.

4. Highlights on adjustments and problems.

5. Performance statistics (the number of counts per employee per day, etc.).

Items with frequent count errors should be evaluated and controls improved. The following controls may reduce count errors: (1) improved warehouse security; (2) locked storage cabinets; (3) issued by supervisor only; (4) controlled access to warehouse storage areas.

Active Item Budget—A budget should be prepared annually for the active investment. The active item budget should include financial requirements for each category based on the mine production and processing plans; the Maintenance Department's schedule for planned maintenance and major overhaul activities; and an analysis of past performance and future requirements when parameters are changed. An active item simulation program (Fig. 4) could be used to define future needs based on changes in market and mine conditions or the system's reorder decision rules. The budget should be adjusted monthly for changes in production and maintenance plans, usage, lead time, and unit price changes.

Inactive Item Reduction Goal—A goal should be prepared annually for reducing the amount of capital misappropriated to slow moving and no longer needed spare parts and material. The goal should be prepared based on a review of inactive item classes, a profile of the inactive item investment (Fig. 5), past performance, and a review of the production and maintenance plan and equipment utilization estimates. The goal should include monthly capital reduction estimates by material category and it should be adjusted for mine and market condition changes.

PURCHASING

Introduction

The Purchasing Department plays an important role in a mining venture. They must wisely and safely spend millions of dollars for material, spare parts, equipment, services, and equipment leases. In spending the operating capital, they must insure that the vendor can perform the service or deliver the requested material at a competitive price, on time, and as requested. Mishandling a purchase order could result in lost revenues, downtime, and idle labor.

In addition to its purchasing function, the department should have responsibility for the sale of obsolete items, returning material to vendors, contract and blanket order negotiations, vendor selection, expediting, routing of incoming material, evaluating vendor performance, and adjusting purchase order commitments for changes in operating plans and market conditions.

The following programs are recommended:

Quotation Policy

The department should request price and delivery quotations for multiple source items when a requisition exceeds $2000. Quotation replies should be analyzed and reviewed with the requesting department managers for nonstock requisitions, capital item purchases, and service and performance contracts. The department should use good business judgment in determining if a quotation should be requested for requisitions under $2000.

Expediting Policy

Expediting action should be initiated for late delivery items, inventory items with a low or zero balance on hand, and items that require improved delivery dates and priority code *AA* purchase orders.

The department should contact the vendor and request status information on the scheduled shipping date, route of shipment, transportation agent, and bill of lading number.

New or revised delivery date information should be conveyed to the requisitioning department so they can update their schedule and plan the work or project.

Trial of Commodity and Value Analysis Programs

For stock items and commonly used direct purchase items, standard brand and vendors should be established. A program to evaluate the use of different brands or vendors or substitute items by performing a trial of the new item is called a *trial of commodity* program.

A trial of commodity may be initiated by Maintenance, Operations, or Purchasing. A trial of commodity form is completed and a copy attached to a purchase requisition for the material to be tried. The user is responsible for executing the trial and reporting on the results. If, as a result of a trial, a new item is selected as a substitute item, the order description and vendor's record should be updated.

In the value analysis program, spare parts and materials are evaluated for price, wear life, cost, and end use application. The analysis includes a determination of ultimate cost per unit of production or operating hour, maintenance cost for part replacement, and the cost of carrying the inventory of spare parts. Financial results are reviewed and approved by management before the new product or spare part is approved for purchase.

Priority Requisition Processing and Procedures

A priority requisition processing policy and procedure is needed to define priority codes and conditions, control rush and emergency requisitions, establish procedures for processing priority requisitions, and to define approval level required for priority codes.

Priority codes are assigned to requisitions according to the severity and effect on production or cost. A service delay or production loss caused by lack of an item should justify priority handling and air or special transportation.

Four codes and processing conditions should be assigned. They are:

Code	Situation	Processing Requirement
AA	Critical	Immediate
A	Important	Within two hours
B	Urgent	Day after requisition received
C	When requested	Within two days

For critical and important condition (*AA* and *A* codes) the department manager or vice president should be notified

Fig. 4. Ace Mining Company Active Item Simulation Report

COMM CODE / R-O-P	DESCRIPTION	PRESENT CONDITIONS: SAFETY STOCK QTY / VALUE	R-O-Q	MO OH	TYP I CL STR	VALUE AVG STK	EST. PO'S	PRICE	PROPOSED: SAFETY STOCK QTY / VALUE,$	S-O-H	MO OH	R-O-P	R-O-Q / USAGE MONTH	TOTAL VALUE AVG STK	A-M-C EST PO'S	MAX-A-M-C L/T INS QTY / DIFFERENCE VALUE AVG STK	EST. PO'S
141500450 / 4	WIRE, AUTO: "AYCLIFFE" 4MM	7 / 55		2	05 N B	128	2	18.3139	1 / 18	9	0	165 / 2	0 / 4	19 / 55	1 / 5	21 / 0 -73	0 / 3
141500523 / 0	WIRE: COPPER BARE 6mm HARD	213 / 62	50	4	06 N B	450	0	0.8015	39 / 31	1000	2	802 / 99	0 / 63	200 / 57	16 / 3	90 / 0	0 / 3
141500604 / 383	WIRE, BATTERY: STRANDED COPPER	229 / 228	367	1	05 N A	608	8	1.6093	75 / 121	400	0	644 / 262	0 / 152	2325 / 243	193 / 15	24 / 0 -393	0 / 7
141500612 / 6	WIRE, AUTO: "AYCLIFFE" 1.6MM	11 / 4 / 59		1	05 N B	181	3	14.6537	2 / 29	15	0	220 / 5	2 / 7	41 / 81	3 / 6	21 / 0 -365	0 / 3
141500620 / 4	WIRE, AUTO: "AYCLIFFE" 1.8MM	5 / 3 / 41		5	06 N C	77	2	13.6331	2 / 14	9	1	123 / 2	0 / 3	8 / 34	0 / 3	30 / 0 -100	0 / 1
141500638 / 8	WIRE, AUTO: "AYCLIFFE" 1.8MM	11 / 4 / 62		1	05 N B	141	5	13.6211	3 / 47	6	0	94 / 8	3 / 7	47 / 102	3 / 7	30 / 0 -43	0 / 2
141500808 / 31	WIRE, EARTH: COPPER BARE	33 / 24 / 86		3	05 N B	130	4	3.5643	5 / 18	50	0	178 / 13	0 / 22	105 / 57	8 / 5	24 / 0 -73	0 / 1
141500816 / 5	WIRE: PVC INSULATED 20-SWG	10 / 19 / 32		4	05 N C	32	3	4.8662	1 / 5	8	0	39 / 2	0 / 7	13 / 22	1 / 2	30 / 0 -10	0 / -1

SUMMARY

TOTAL SAFETY STOCK:	3199	6235	1162	2780	-3455
TOTAL AVG STK LEVEL:	7358				
TOTAL EST. P.O.'S:	138		208	70	
MONTHS STOCK ON HAND:	3			70	

Record Count	Items	Value		Total Difference:	
ACTIVE	40	7358		Percent	56.94 %
				% TOTAL STOCK VALUE	-55.41 / 26.74 / -46.96
				% TOTAL STOCK VALUE:	66.34

INACTIVE	23	5564	43.06	43.48
TOTAL	63	12922	100.00	

VALUE IN SAFETY STOCK:

Recap of Active Stock	NO OF ITEMS	VALUE	%	ANNUAL ISS VALUE	%
A Items	5	2091	28.28	13035	54.40
B Items	19	3496	47.51	8899	37.14
C Items	16	1721	24.20	2027	8.46
Totals	40	7358	100.00	23961	100.00

Notes

Line 1 (left to right) Computer number, description stock type, insurance code (Y = yes; No = no), item class (A, B, C), strategic item (Y = yes; No = no), price, SOH = stock on hand, value, usage (current month; annual total), A-M-C = average monthly consumption, MAX A-M-C (peak demand quantity), L/T = lead time (days), insurance quantity.

Line 2 (left to right) present conditions. R-O-P = Reorder point, R-O-Q = order quantity, safety stock (quantity and value), MO OH = months on hand (estimated number of months supply on hand); Value avg. STK = value of average stock level (safety stock quantity + 0.5 of ROQ × unit price). EST PO's = estimated number of POs year (annual usage ÷ ROQ or number prior year).

Proposed conditions (modifications to safety stock and ROQ calculation. Safety stock, quantity and value, MO OH = months on hand, ROP, ROQ and value avg. STK (same as present conditions), EST P.O.s = estimated number of P.O.s year (annual usage ÷ ROQ). Difference (present to proposed—average stock level and estimated number of POs).

Summary (recap for group)
Value of safety stock, present, proposed and difference
Value of Average stock, present, proposed and difference
Estimated POs, present, proposed and difference
Value of stock on hand, present, proposed and difference
Months of stock on hand (value on hand ÷ average monthly issue value)
% difference in average stock level (3455 ÷ 6235)
% reduction in total stock level (3455 ÷ 12,922)
% reduction from active item value (3455 ÷ 7358)
% value in safety stock (3199 ÷ 7358)

Record count and recap information. Record count is a profile of inventory items in the category and percentages. Recap information is a profile of the active item class, the value and percent on hand and annual issue value and percent.

Fig. 5. Ace Mining Company Inactive Item Profile

COMM CODE LST ISS DTE LST ISS QTY MTH LST ISS	DESCRIPTION	ST T.O.D.	I SC S SHELF	LIFE	STD PRICE EXPECTED S/O DATE	SOH	VALUE	R.O.P.	R.O.G.	SAFETY STOCK QTY	VALUE	INS QTY	L/T
692001098 NA 0	BAND, WEAR: BEARING CAP	09 730723	N C Y	0	13.5400 NA	9	122	0	0	0	0	0	180
692001585 810519 6 20	BOLT: 7/16"−14 X 7/8" LG HEX	05 740330	N C Y	0	0.1063 970401	51	5	0	0	0	0	0	180
692001624 800407 8 34	BOLT: 1/2"−13 X 3" LG HEX HEAD	09 740330	N C Y	0	0.3146 001101	50	16	0	0	0	0	0	180
692001713 NA 0	BOLT: 1/2"−13 X 3 1/2" LG HEX	03 740330	N C Y	0	0.4400 NA	20	9	0	0	0	0	0	180
692001852 770808 3 66	BOLT: 3/8"−16 X 1" HEX HEAD	09 740330	N C Y	0	0.0300 140401	17	1	0	0	0	0	0	180
692001878 800312 20 35	BOLT: 1/2"−13 X 3 3/4" LG HEX	09 740330	N C Y	0	0.6700 970901	100	67	0	0	0	0	0	180
692002086 800221 1 36	CAP: (GE-41B531703P1)	05 690101	Y A Y	0	192.2797 520201	23	4422	0	11	0	0	0	180
692002206 820105 4 13	CLAMP: (H2 WHEEL)	05 690101	N C Y	0	2.7649 030601	75	207	0	31	0	0	0	150
692002557 740214 0 108	CONTACTOR: (GE 17CM5J5)	09 690101	N C Y	0	73.0300 830201	3	219	0	0	0	0	0	180
692002743 810122 3 24	CLAMP, BEARING: C/W PIN	09 690101	N C Y	0	24.6300 031001	31	764	0	0	0	0	0	180
692107624 820622 32 7	RING: (GE 1X7010) (772K-4)	01 690101	N C N	0	1.0000 830901	32	32	0	0	0	0	0	180
692107690 NA 0	RING, RETAINING:	05 800428	N C Y	0	4.2568 NA	54	230	0	0	0	0	0	60

SUMMARY

		RECAP OF INACTIVE STOCKS	
TOTAL NO OF INACTIVE ITEMS:	18	LAST ISSUE DATE > 1 < 2 YRS:	6
VALUE:	11391	VALUE:	4912
VALUE OF SAFETY STOCK:	0	% OF TOTAL:	42.24
% OF TOTAL:	0.00	LAST ISSUE DATE > 2 < 3 YRS:	4
		VALUE:	4655
TOTAL NO OF STRATEGIC ITEMS:	17	% OF TOTAL:	40.87
		LAST ISSUE DATE > 3 < 4 YRS:	2
TOTAL NO OF INSURANCE ITEMS:	2	VALUE:	1342
VALUE:	0	% OF TOTAL:	11.78
% OF TOTAL	0.00	LAST ISSUE DATA > 4 < 5 YRS:	0
		VALUE:	0
		% OF TOTAL:	0.00
		LAST ISSUE DATE > 5 YRS:	6
		VALUE:	522
		% OF TOTAL:	5.11

Notes
Line 1. Includes item status information on the inactive item (see active item simulation legend for abbreviations).
Line 2. LST ISS DTe = date last issued; LST ISS Qty = quantity last issue; MTH LST ISS = number of months since last issue; TOD = date added to file; shelf life = shelf life for item; expected stock-out date = estimate date stock on hand will be at a zero balance.
Summary. Total number and value of inactive items and value and percentages for safety stock, strategic, and insurance items.
Recap. A profile of inactive stocks (number, value and percent of total) by last issue date.

and he should approve the coding. The conditions for critical and important situations are:

1. *Critical*—Department production goals or operating objectives are affected and cannot be made up.

2. *Important*—Department production goals or operating objectives are affected but can be made up at additional cost.

Evaluating Quantity Discounts

Maintaining stable prices or reducing unit prices in an inflationary environment requires that buyers search the market for price, quality, and availability. It also requires that the buyer evaluate the cost of carrying an increased quantity in stock if a larger quantity is purchased. A quantity discount evaluation form (Fig. 6) should be prepared before the larger quantity is purchased. The cost savings from a lower purchase price must be greater than the cost of carrying a larger quantity in inventory.

Source of Supply/Spare Part Cross-Reference Program

Single purchase source spare parts and materials offer opportunities for unit price reductions and lower material and spare part costs. Product cross-reference information is available from many fastener, oil filter, bearing, seal, and V-belt vendors.

In addition to vendor cross-reference catalogues, warehouse employees can obtain spare part source information

from labels or identification tags attached to the spare part or from maintenance employees.

Purchasing original equipment spare parts from alternate vendors generally reduces unit prices between 30 to 60%. Product specification information, including mounting position, amp, volts, psi pressures, dimensions, and product hardness, etc., must be provided to alternative vendors. Product warranty and maintenance compatibility information should be obtained before committing future requirements to the new source.

Contracts and Blanket Orders

Contracts and blanket orders should be negotiated for the purchase of major operating supplies, essential spare parts, high usage dollar products, and frequent purchase materials. Simplified order release and scheduled delivery date techniques should be used to order and release material.

Negotiating favorable contract terms will reduce or stabilize material and supply costs and provide an assured source of supply. The following terms and conditions should be negotiated:

1. Price.
2. Invoice payment.
3. Product warranty.
4. Backup stocking.
5. Product returns.
6. Cancellation.

Fig. 6. Quantity Discount Evaluation

Date: 1-5-83

Part Number: 14236	Description: Terminal/Electrical		U.M. EA, Stock # 16-24-073	
Item	Consideration	E.O.Q. Order	1st Discount Level Order	2nd Discount Level Order
1	Order Quantity	400	1,000	10,000
2	Unit Price Delivered	$ 1.20	$ 1.10	$ 1.00
3	Savings Per Unit		$ 0.10	$ 0.20
4	Savings Per Year		$ 670.00	$1,340.00
5	Annual Usage (Units)	6,700	6,700	6,700
6	Safety Stock (Units)	125	125	125
7	Average Inventory	325	625	5,125
8	Average Inventory Value	$ 390.00	$ 687.50	$ 5,125.00
9	Value of Inventory Increase		$ 297.50	$ 4,735.00
10	30% Cost to Carry Inventory		$ 89.25	$1,420.50
11	Cost Advantage (Disadvantage)		$ 580.75	$ (80.50)

Quantity Ordered: _____ Notified: _____
Purchasing Comments: Scheduled 400 for delivery on 1/27/83; 200 on 2/18/83;
200 on 3/1/83 and 200 on 3/17/83. Advised to ship via Fast Freight prepaid.

_____

_____

_____

BUYER: James Price

7. Contract extension.

8. Quantity discounts for noncontract items.

9. Vendor services, etc.

Surplus and Obsolete Item Disposal

Material, spare parts, and equipment that will no longer be required because of a change in the mine plan or production goal should be sold or transferred to other properties. The sale of nonproductive assets will improve the company's cash position. When the mine's operating plan changes, or when the production goal has been decreased, the purchasing department must review its order commitments and stock levels. Purchase order cancellation, delivery date deferrals, or return to vendor action should be started immediately.

INFORMATION PROCESSING

Large volumes of transactions are prepared and processed by the purchasing, inventory management, and warehousing departments. Records must be retained for thousands of items. They must include accurate and timely information on an item's status, description, value on hand, incoming shipments, etc.

In addition to inventory records, Purchasing must retain information on open and outstanding orders, vendors, contracts, prices, receipts, contract expiration dates, purchase order values, and accounting information, etc.

Fast retrieval of accurate information is essential so that managers can make reasonable financial and operating decisions on the amount of operating capital required for the purchase of material and spare parts; on scheduling a major overhaul; on retaining or returning a spare part; on placing an order to replenish stock levels, and decisions on material and spare part costs for a piece of mining equipment, processing plant, or mine area. Purchasing and warehousing must contribute to the data base.

An on-line computer system that interacts with accounting, maintenance, and operations should be used for purchasing, inventory management, and warehousing. The design of the system must respond to the needs of users, the operating environment of the property, and must support the objectives of materials management. These objectives should include:

1. A specific level of support to the operation minimizing total ordering and inventory carrying costs.

2. Efficient purchase order releasing and order status reporting.

3. Highlighting problems through exception and summary reporting.

4. Providing appropriate vendor and buyer evaluations.

5. Reducing routing clerical activities.

6. Timely input and processing of system information.

7. Predicting spare part and material requirements accurately, using current operating and maintenance plans and historical usage information.

8. Minimizing the capital required for inventory investment.

CRT screens should be located in work areas in the purchasing, inventory management, and warehousing departments. The screens should be used to inquire on file information and for data entry. CRT screen displays should include a screen menu on files that can be accessed and self-help messages for data entry. The terminals should have controls for accessing file information, updating files, and data entry. Sign-on and password controls are also necessary for system security.

The following reports or screen displays are recommended:

1. *Status Reports*—provide historical and current information on inventory items and purchase orders.

2. *Exception Reports*—provide historical and current information on items or conditions outside of normal operating parameters.

3. *Performance Reports*—provide information on operating conditions and dollar levels.

4. *Summary Reports*—highlight material activities, i.e., investment levels by category, item class and location, by property, process and storage area.

5. *On-Request Reports*—one-time or infrequent use reports on special situations or conditions. The report should provide information on a single item or group of items.

The system must be capable of responding to changes in operating plans, maintenance schedules, and market conditions.

MATERIALS MANAGEMENT SUMMARY REPORT

The materials management summary report, Fig. 7, should measure department performance relative to goals or objectives and provide status information on inventory management, purchasing and warehousing activities, and manpower. The report should include current month, prior month, and year-to-date information, performance and goal statistics, and variance analysis. Detailed reports should provide backup information.

Fig. 7. Materials Management Summary Report
Ace Mining Company
March 1987

	Current Month Number Value	Prior Month Number Value	Year-to-Date Number Value
1. Purchasing			
A. Order activity			
Stock orders	____ ____	____ ____	____ ____
Direct orders	____ ____	____ ____	____ ____
Capital orders	____ ____	____ ____	____ ____
Total	____ ____	____ ____	____ ____
B. Expediting and priority processing			
Past due orders	____	____	____
Rush orders	____	____	____
Priority *AA* orders	____	____	____
C. Contracts negotiated	____	____	____
D. Number of air shipments	____	____	____
2. Inventory Management			
A. Active item investment			
A Items	____ ____	____ ____	____ ____
B Items	____ ____	____ ____	____ ____
C Items	____ ____	____ ____	____ ____
Total	____ ____	____ ____	____ ____

	Current Month Number Value	Prior Month Number Value
B. Investment by major category		
Fuels and lubes	____ ____	____ ____
Chemicals and reagents	____ ____	____ ____
Explosives	____ ____	____ ____
Grinding media	____ ____	____ ____
Tires	____ ____	____ ____
Crusher	____ ____	____ ____
Refractories	____ ____	____ ____
Mining and processing	____ ____	____ ____
Equipment spare parts	____ ____	____ ____
Insurance items	____ ____	____ ____
Total	____ ____	____ ____
C. Inactive item investment		
X Items	____ ____	____ ____
Y Items	____ ____	____ ____
Z Items	____ ____	____ ____
D. Overstocked items	____ ____	____ ____
E. Overcommitted items	____ ____	____ ____

	Goal	Current Month Number Value	Variance Number Value	Year-to-Date Number Value
F. Cycle counting				
Items counted	____	____	____	____
Items adjusted	____	____	____	____
Count accuracy	____	____	____	____
Adjustments				
Debit	____	____	____	____
Credit	____	____	____	____
G. Investment statistics				
Months of stock on hand	____	____	____	____
Turnover rate	____	____	____	____
Value of issues	____	____	____	____
Value of receipts	____	____	____	____
H. Investment Goals				
Active items	____	____ ____	____ ____	____ ____
Inactive items	____	____ ____	____ ____	____ ____
Insurance items	____	____ ____	____ ____	____ ____
Overstocked item	____	____ ____	____ ____	____ ____
Reduction	____	____ ____	____ ____	____ ____
Obsolete item reduction	____	____ ____	____ ____	____ ____

Fig. 7. Materials Management Summary Report—Continued

	Current Month		Prior Month		Year-to-Date	
	Number	Value	Number	Value	Number	Value
Sales						
New stock items						
3. Warehousing						
Issues						
Backorders						
Service level						
Returns from operations						
Returns from maintenance						
Shipments received						
Shipments to vendors						
Safety violations						
Lost time accidents						

	Goal	Current Month	Variance
4. Manning Levels			
Purchasing			
Inventory management			
Warehousing			
Overtime			

8.9 Government and Public Affairs

CHARLES S. BURNS

Over the years a mining company's relations with government, particularly the federal government, have become increasingly important to the point that they are now a vital management function.

Based in Washington, D. C., the American Mining Congress is the national trade association which represents before the federal government the producers of most of America's metals, industrial and agricultural minerals; manufacturers of mining and mineral processing machinery, equipment and supplies; and engineering and consulting firms and financial institutions that serve the mining industry. Together with the National Coal Association, it also represents the coal industry. In 1963, the AMC had a staff of 14 professionals and 24 support personnel for a total of 38 employees. Twenty years later, in 1983, its staff had grown to a total of 90, including 47 professionals and 43 secretarial and clerical employees.

The major steel companies, which have captive coal operations, have long maintained their own government affairs offices in Washington. However, with the exception of the steel industry, in 1963 there was only one independent coal company with a Washington office. In 1983 there were at least ten such offices. Only one copper mining company had a Washington office in 1963. Twenty years later almost every major company has Washington representation.

Why has the mining industry dramatically expanded its presence in Washington over the last twenty years? The answer can best be given in a recitation of federal government actions. In 1963 there was:

1. No *Federal Metal and Nonmetallic Mine Safety Act* (enacted in 1966 and incorporated with the Coal Safety Act into one statute in 1977).

2. No *Wilderness Preservation Act* (a wilderness system composed of 9.2 million acres was enacted in 1964).

3. No *Solid Waste Disposal Act* (enacted in 1966).

4. No *Resource Conservation and Recovery Act* (federal and state controls over hazardous wastes were enacted in 1976).

5. No *Wild and Scenic River System* (enacted in 1968).

6. No strong federal *Water Pollution Control Act* (federal guidelines for states to require pollution controls enacted in 1972).

7. No stringent *Clean Air Act* (legislation establishing national ambient air standards for pollutants enacted in 1970).

8. No *National Environmental Policy Act* (1970).

9. No *Alaska Native Claims Act* (1971).

10. No *Alaska National Interest Lands Conservation Act* (1980).

11. No *Noise Control Act* (1972).

12. No *Endangered Species Act* (1973).

13. No *Safe Drinking Water Act* (1974).

14. No *Toxic Substances Control Act* (1976).

15. No *Federal Land Management and Policy Act* (1976).

16. No *National Forest Management Act* (1976).

17. No *Surface Mining Control and Reclamation Act* (1977).

18. No *Uranium Mill Tailings Radiation Control Act* (1978).

This is only a partial listing of federal statutes that have affected mineral exploration, development, and production. And, of course, with each federal statute there comes a federal regulatory scheme.

What has happened in twenty years to so change the landscape? In the early 1960s the mining industry's principal concerns on the federal level were taxation, the management of government stockpiles of strategic minerals, and access to the public lands for prospecting for and mining of mineral resources. The pertinent committees of Congress concerned with public land management and mining interests generally were the House and Senate Interior committees (the latter is now known as the Committee on Energy and Natural Resources).

These committees were dominated, if not controlled, by members of Congress from the Western states who understood the problems of the mining industry and who, in large part, sought to encourage mineral development. Twenty years later these committees, though still containing Western members, have been deliberately filled with members from other parts of the country who represent a strong environmental bias and are, in many cases, openly hostile to mining.

The major cause of these changes in the Washington political landscape is, of course, the emergence of the environmental movement as a powerful political force. In the early sixties, the leaders of the old line, respected conservationist organizations sought federal legislation to impose environmental requirements on mineral development designed to protect the environment while, at the same time, permit mining to continue. By the early 1970s, however, the environmental movement had been radicalized. At most of the environmental organizations, positions of staff leadership were assumed by ultraleft professional activists. Expanding their efforts beyond Congress to the courts and, during the Carter Administration, to the Executive Branch itself, these militants sought not just to control mining, but to prohibit it. Exploiting legitimate environmental concerns as a means to their end, they have sought to establish a no-growth policy for the nation in an effort to change fundamentally the nation's economic system.

Clearly a mining company attempting to operate in such a climate cannot long succeed without devoting major attention to government relations and public affairs. That is the reason for the growing number of mining representatives in Washington.

USE OF TRADE ASSOCIATIONS

Companies maintaining Washington offices make extensive use of their industry trade association as well as the general business associations such as the Chamber of Commerce of the United States and the National Association of Manufacturers. Certainly companies without Washington representation should look to their association for help.

Such help is a two-way street and a company must be willing to devote expertise and man-hours to the enterprise. For example, the American Mining Congress has 29 standing committees and numerous subcommittees, their membership drawn from the ranks of its member companies. Each year a company is asked to recommend individuals to serve on one or more of these committees. Athough the AMC staff guides the work of the committees and performs a secretariat role, policy is molded and substantive work is done by the members themselves. A list of the standing committees reflects the myriad of problems and issues emanating from actions taken or considered by the federal government:

Accounting
Capital Formation
Cement Advisory Council
Coal Leasing
Coal Mine Safety
Coal Policy Council
Coal Preparation and
    Waste Disposal
Coal Surface Mining
Communications
Diesel
Energy Resources Policy
Environmental Matters
Export Council
Financial Advisory Council
Industrial Relations
Manufacturer Technical
Minerals Availability
Mining and Minerals
    Education
Noncoal Mine Safety
Noncoal Surface Mining
Occupational Health
Product Liability
Public Lands
Reclamation Technology
Silver
Synthetic Fuels
Tax
Undersea Minerals
    Resources
Uranium

Clearly a company cannot have representation on every committee. However, it should attempt to obtain membership on those committees dealing with issues which are of prime importance to it or, at the very least, be placed on the mailing list to receive information.

Specialists should be chosen to serve on committees that require technical expertise, such as, tax, mine safety, or reclamation technology. Otherwise, it is usually best to place a generalist in charge of coordinating the company's overall government relations activities.

Mining companies also should maintain membership in the national associations that represent business in general, such as the US Chamber of Commerce and the National Association of Manufacturers. Both of these organizations have Natural Resources committees composed of representatives of member companies. In testimony before Congress and in policy statements directed to the Executive Branch and the public, their positions are generally similar to those of the mining associations, but they bring to those positions support from a broader national constituency.

State mining associations have been established in most states in which surface mining is conducted. These state associations monitor state legislative and administrative developments that impact upon the industry and assist the national associations in their efforts on the federal level. State associations generally have small staffs (usually a director,

sometimes called a *secretary,* with one or two assistants). In order for them to be effective, corporate participation is essential.

THE VALUE AND FUNCTION OF A WASHINGTON OFFICE

Despite their efforts on a variety of issues, trade associations can only assist the industry in collective endeavors. As a result there are many times when a company requires its own Washington representation. Frequently, a company will have a problem with the federal government that is uniquely its own, whether it be negotiation of a lease or a regulatory or legislative issue that is site specific. Policy differences in a particular segment of the industry may prevent the trade association from achieving a consensus and thus participating in the effort. For example, the uranium industry was divided over legislation to impose restrictions on the importation of foreign uranium. Domestic producers favored it. US companies with extensive foreign production opposed it. As a result, the Mining Congress could not take a position on the subject.

Although there are some exceptions, most corporate Washington offices are small, consisting of one or two government relations specialists and support staff. In general, public relations staff is maintained at corporate headquarters.

It is the responsibility of the Washington representative to monitor those activities of the federal government that have a direct impact upon the company and report such developments to senior management. The Washington representative must present policy positions of the company to members of Congress and their staffs and officials of the Executive Branch. In doing this, the representative is frequently accompanied by company personnel who are specialists on the particular issue being discussed.

The Washington representative's *box* on the table of organization varies from company to company. Sometimes it is a function falling under the purview of the general counsel. In other companies the Washington representative reports to a senior vice president for corporate and legal affairs, or, in rare instances, directly to the chairman or president.

The exact positioning in the chain of command is not important as long as the Washington representative has ready access to the highest policy-making levels of the company. Many times a corporate policy decision on an impending government action must be made swiftly and such decisions may be influenced by political facts and nuances that can be inadvertently misstated if reported through intermediate parties. That is why it is vital for the Washington representative to have direct access to those in the company who must make the decision.

GRASSROOTS NETWORK

Most companies with Washington offices have established *constituency building* or *grassroots lobbying* programs which are developed and implemented by the Washington office. The purpose of such a program is to facilitate and supplement corporate activities designed to influence public policy through both the electoral and legislative processes.

Once established, such a program can become remarkably effective. For example, one company with facilities located in 15 states appoints an employee at each location to be the *Government Affairs Executive* for that area. Such an assignment is in addition to his regular duties. The executive (known as a GAE) is expected to become personally acquainted with the congressman representing his district and the two US senators from that state. Plant tours by these officials are encouraged.

The Washington office, through publication of a newsletter and other communications, keeps the network of GAEs informed of federal issues of importance to the company. At the appropriate time, the GAEs are asked to contact their assigned members of Congress (by letter, telegram, or, if urgent enough, by telephone) expressing the company position on a given issue. Once each year all the GAEs are brought to Washington for several days of meetings, briefings, and visits with their senators and congressmen on Capitol Hill.

The value of such a program is considerable. It is one thing for the Washington representative (or the chairman, for that matter) of a company operating in a congressman's district to contact him on a legislative issue. It is quite another thing—and far more meaningful—for that contact to be made by the plant manager who works, lives, and *votes* in the congressman's district.

POLITICAL ACTION COMMITTEES

Recent years have seen a steady rise in the number of companies which have formed political action committees. Such committees have been utilized by labor unions and other special interest groups for decades. A change in federal law in 1974 clarified a corporation's legal right to operate such a committee, known as a PAC.

The PAC permits employees to pool their individual political contributions and thereby render meaningful support to candidates for Congress who recognize the problems of industry and who will work to establish an economic climate that is conducive to business growth. Employees are solicited by the PAC, generally by letter, and this is followed by discussion of the program at informal employee meetings. Employee contributions to the PAC are voluntary and strictly confidential.

Many PACs are managed by the firm's Washington representative who either sits on the committee that decides which candidates will receive contributions or serves as an advisor to the committee.

PUBLIC RELATIONS

The management of corporate relations with the media, the public, and the shareholders is a function best kept separate from government relations. For obvious reasons, the company lobbyist is not the ideal spokesman to explain a plant closing to the press and public nor should he be exposed to questions on legitimate, but perhaps delicate, negotiations being held with legislators or other government officials.

Nevertheless the roles are somewhat similar and are best supervised by an officer in charge of corporate, legal, and public affairs. Volumes have been written on effective public relations techniques and such information will not be repeated here. Suffice it to say that honesty and accessibility— a willingness on the part of company leaders to meet the press and public—are the overriding principles to follow.

CHAPTER 9

Case Studies

Bruce A. Kennedy, Editor

1. Introduction

Bruce A. Kennedy

The first eight chapters of this book have dealt with the details of the numerous and diverse parts of surface mining. In the case of a mining project, all these components must be pulled together to result in an efficient and profitable operation. Each mining project is unique and consequently the final form of the eventual operation will be influenced by many factors. These will include, but certainly not be limited to, commodity, geographic location, geology, environmental considerations, scale, available labor skills, capital and operating costs, etc., etc.

This chapter contains seven case histories of open pit mining different commodities in various locations in the world. The mines examined are Hambach, West Germany—coal; Palabora, South Africa—copper; Metcalf/Morenci, USA—copper; Cuajone, Peru—copper; Chuquicamata, Chile—copper; Shirley Basin area, USA—uranium; and Island Copper, Canada—copper. The only criterion in the selection of these particular mining operations was to demonstrate the wide diversity of situations and problems that can be encountered in surface mining and the solutions that can be found to deal with them.

9.2. The Hambach Open Pit Mine

DIETER HENNING
HANS WEISE

INTRODUCTION

The Federal Republic of Germany has a total of some 56 Gt (62 billion st) of lignite resources, of which approximately 55 Gt (61 billion st) are contained within a triangular area bounded by the cities of Cologne, Aachen, and Dusseldorf (Fig. 1). Based on today's energy prices, about 35 Gt (39 billion st) of this total is economically minable. This so-called Rhenish mining area (*Rheinisches Revier*) thus contains Europe's largest continuous lignite deposits.

Presently, Rheinische Braunkohlenwerke AG (Rheinbraun), Cologne, is operating five open pit mines in the Rhenish mining area: Frimmersdorf, Fortuna-Garsdorf, Frechen, Ville, and Zukunft. The combined lignite production from these mines approaches 120 Mt/a (132 million stpy). With the exception of the Frimmersdorf mine, reserves in all mines will have been exhausted by the early 1990s.

The output of mine extensions and/or new developments (at Inden, Bergheim, and Hambach) will serve to maintain present production levels. Beyond the year 2000, however, three large-scale open pit mines will still be in operation, with lignite production as follows:

Frimmersdorf/Garzweiler 45 to 50 Mt/a (50 to 55 million stpy)

Hambach 45 to 50 Mt/a (50 to 55 million stpy)

Inden 20 to 25 Mt/a (22 to 28 million stpy)

THE HAMBACH MINING AREA: HYDROLOGY AND DEVELOPMENT HISTORY

The Hambach deposit, which was initially explored in the 1940s, is located in the so-called Erft block, situated near the geographic center of the Rhenish mining area. This 85-sq-km (33 sq-mile) area (Fig. 2) holds lignite reserves of some 2.5 Gt (2.8 billion st), the mining of which involves removal of 15.4 km³ (20 billion yd³) of overburden, yielding an overall stripping ratio of 6.2:1 m³/t (199 ft³/st).

The western boundary of the deposit is formed by the so-called *Rur peripheral fault* (Fig. 3). Here, the seams split up initially into six segments, which dip 3 to 4% to the northeast. In the southeastern area of the mining field, the seams coalesce into a continuous unit with a thickness of up to 70 m (230 ft). In the remaining sections, the individual seams are separated by interburden of varying thicknesses. The lignite has a mean heating value of 9800 kJ/kg (4213 Btu/lb), with an average water content of 50% and ash content of about 7%.

The rock overlying the lignite seams has a thickness varying between approximately 150 m (492 ft) in the west and 400 m (1312 ft) in the east. The overburden, like the intermediate rock, consists of nonindurated sediments: 70% sand and gravel, 30% cohesive clays and silts. Unlike other deep open pit mines presently in operation in the Rhineland, this mine is traversed from the southeast to the northwest by only a few small faults, with vertical displacements seldom more than 15 m (50 ft).

Due to the presence of several large interbedded aquifers,

it is nearly impossible to mine the deep seams of the Hambach mine without first providing extensive dewatering. Satisfactory in-pit working conditions are assured only by proper ground water collection procedures (Leuschner, 1972).

From the hydrological point of view, the Erft Basin, developed for the first time by the Hambach mine, is to a great extent shielded against adjacent water-bearing areas by the existing fault system. Thus, the lowering of the ground water table for the Hambach mine will mainly be confined to the Erft Basin, which covers some 850 sq km (328 sq miles). Additional relief is provided by the Erft Basin drainage system, which has been in operation for more than two decades. This drainage system was initially constructed to insure sufficient stability for slopes in neighboring mines which extended into the Erft Basin.

In the next few years, the quantity of ground water raised from the Erft Basin will need to be increased only slightly and should decline in the late 1980s. When the mine reserves are exhausted, another 100 dewatering wells will have been drilled in the mining field at its rim to be added to the present array of 260 filter wells being operated in the Erft Basin.

The pump wells reach depths of 520 m (1706 ft); the water is raised by submersible pumps with capacities of 15 to 30 m³/min (530 to 1060 ft³/min). These quantities are either discharged into natural water courses or are collected to meet the demand of local public and industrial users. So far, lowering the ground water level has caused only limited damage to wild and cultivated plant life since the fertile loess and forest soils covering the surface have a high water storage capacity, allowing plants to meet their water needs from precipitation.

The lignite deposits in the area of Erft Basin are minable beyond the boundaries now fixed for the Hambach mine; however, primarily population density and landscape planning criteria have defined the boundaries of the mining area and its division into Hambach I and Hambach II (Fig. 2). The long-term planning scheme provides, first of all, for the Hambach I field to be worked leaving the decision on Hambach II to be taken in the mid-1990s. Extensions going beyond these boundaries in the next century are quite conceivable, if necessitated by future energy requirements.

The mine plan selection process for the Hambach deposit involved studies of more than 20 variants in an attempt to satisfy two basic criteria: the smallest possible outside dump volumes, and an expeditious startup schedule (Leuschner, 1972). The selected mine plan incorporated a slewing operation with a bench length of approximately 5 km (3.1 miles) and the slewing point located in the southwestern part of the field. With a clockwise slewing pattern, the overall deposit can be mined from one slew point with a nearly constant bench length (Fig. 2).

In 1978 development work started in the flattest part of the mining field with a box cut in parallel operations. Due to its length [some 5 km (3.1 miles)], this box cut was not excavated separately, but widened by the first bucket wheel excavators and gradually deepened. Lignite was reached and mined for the first time in 1983. By then the cut will be approximately 5 km (3.1 miles) long, 1.8 km (1.2 miles) wide

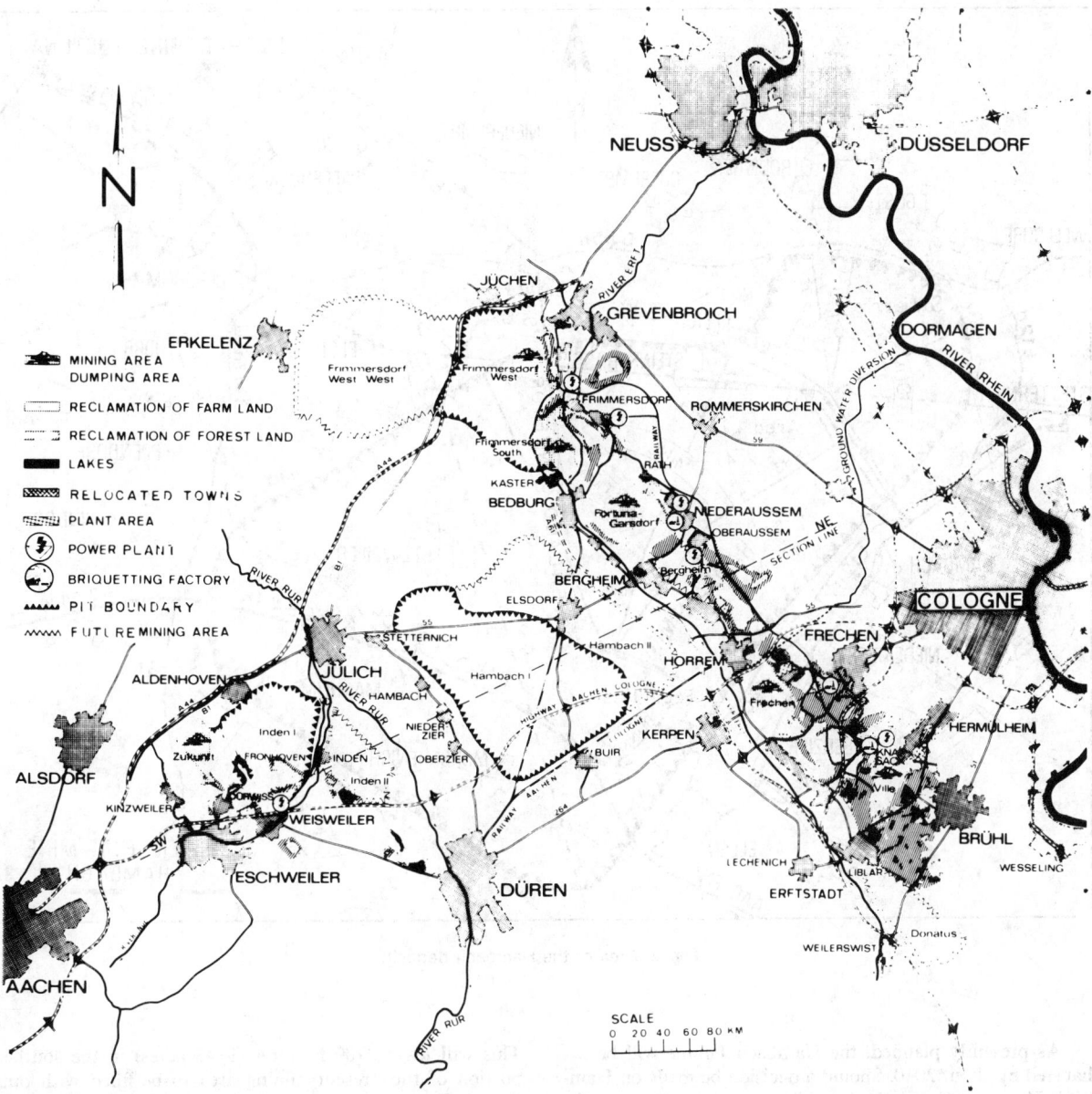

Fig. 1. The Rhenish open pit mining area.

and about 160 m (525 ft) deep. Subsequently, slewing operations will commence.

The following concept was worked out to handle the outside dump volume of 2.7 km³ (3.5 billion yd³). In the first five years, the overburden volumes were disposed of solely in the *Sophienhoehe* outside dump, located northwest of the box cut. The *Sophienhoehe* will eventually be comprised of nearly 1 km³ (1.3 billion yd³) of overburden. The site chosen has the advantage that the *Sophienhoehe* can later be continued directly to the future inner dump. Another 1.7 km³ (2.2 billion yd³) of outside dump volumes will be transported to the Fortuna-Garsdorf, Frechen, and Bergheim mines to fill the final mine voids. Two conveyors, some 14 km (8.7 miles) in length each, will move more than 0.10 km³ (130 million yd³) of overburden volume per year from Hambach to these mines (Fig. 4).

In 1982 three equipment groups were in operation (Fig. 5a), with the fourth bench being developed and another excavator starting operations. Overburden transport to the Fortuna mine will commence when construction is completed on the first long-distance conveyor. Later, it will be supplemented by a second parallel conveyor.

To build up the first inside dump, the first spreader will transport overburden from the *Sophienhoehe* to the mine. Via a ramp system, the three dumping benches will alternately be worked with two equipment systems (Fig. 5). By that time, a volume of nearly 0.80 km³ (1046 million yd³) of outer dump material will have been moved. Late in 1990, the *Sophienhoehe* outer dump will be completed. The mine will attain its full operating capacity of 50 Mt/a (55 million stpy) in the mid-1990s, with eight excavators planned for operation during the final stages (Fig. 5c).

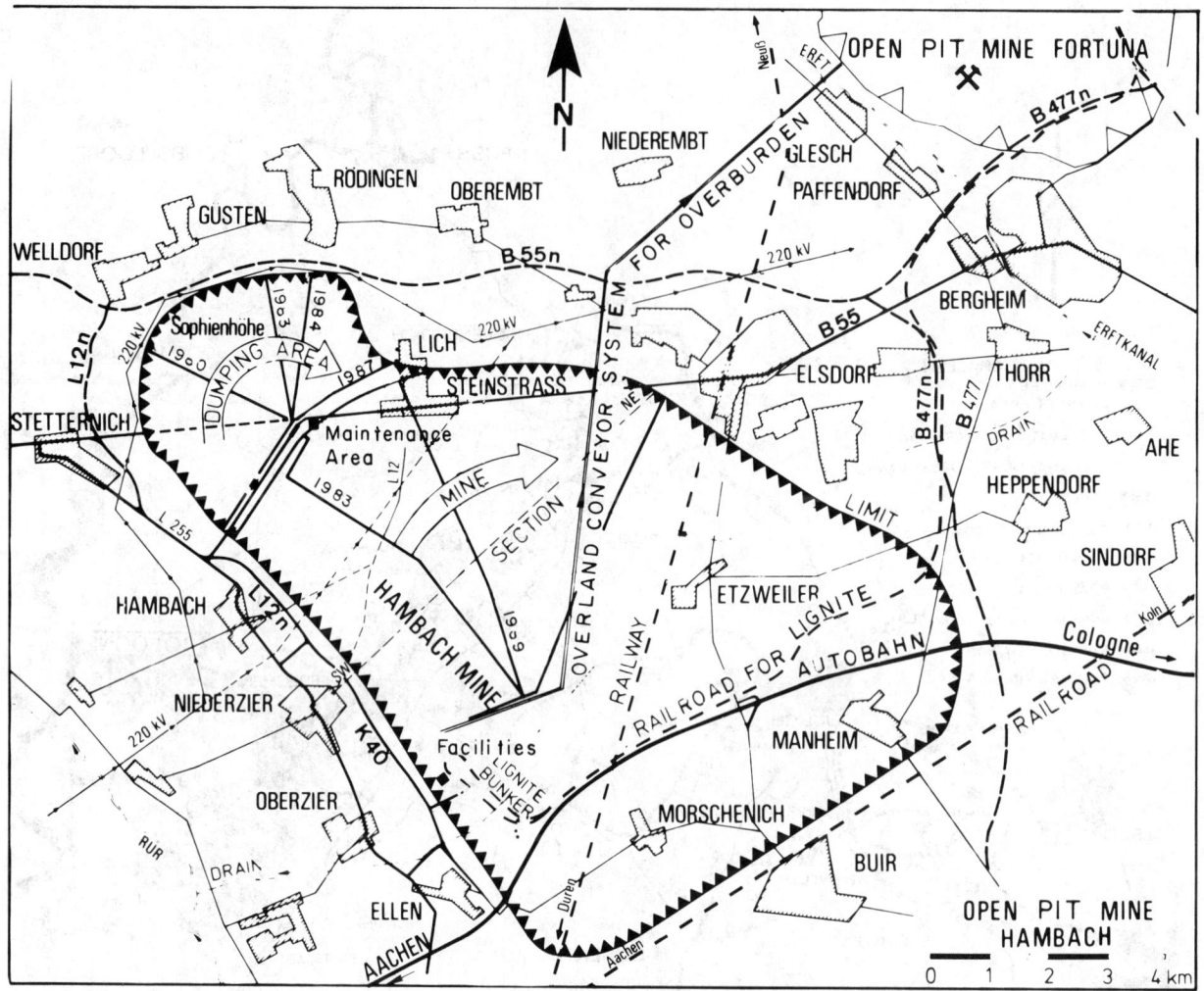

Fig. 2. Area of the Hambach deposit.

As presently planned, the Hambach I mine will be exhausted by about 2040. Should a decision be made on Hambach II or on mining the large lignite reserves beyond the southern boundary, higher output will be possible by employing additional equipment groups without any change in basic operating concepts.

RECLAMATION AND POPULATION RESETTLEMENT

The Hambach mining area consists mainly of farmland (48%) and forest (47%). To replace the large wooded area disturbed by mining operations, the *Sophienhoehe* outer dump is to be reclaimed as forest, thus providing a quiet recreation area. Planning for the high dump, therefore, necessitates special monitoring of the continuously updated landscaping data. By mid-1982 some 230 hectares (568 acres) had been reclaimed on the peripheral slopes and on the plateau, with some areas opened to the public (Fig. 6).

Filling the final voids at the Fortuna, Frechen, and Bergheim mines will produce 2600 hectares (6425 acres) of reclaimed land, mostly agricultural. In total, 3800 hectares (9390 acres) in the Hambach area will be restored for forestry, and 1000 hectares (2471 acres) for agricultural purposes.

This will leave 3700 hectares (9143 acres) in the southern portion of the present mining area to be filled with outer dump volumes from mine extensions, part of which will possibly be laid out as residual lakes.

Compared with the population density of the Rhenish area [420 persons per sq km (0.39 sq mile)], the population density in the Hambach lignite field [about 60 persons per sq km (0.39 sq mile)] is very low. In addition to some farms that were already acquired in the initial phase or that are presently being resettled, only the township of Lich-Steinstrass will be relocated to Juelich. This relocation will be carried out in accordance with the wishes of the inhabitants, and will be completed by the year 2000 (Figs. 2 and 7). The last community to be reached by mining operations (in about 50 years) will be Morschenich.

As a part of mine development at Hambach, substitute roads were constructed for Highway L12 between Steinstrass and Niederzier and for a section of Federal Highway 55 (Fig. 2). Thus, the connecting road between Federal Highway B55 and Highway L12, the so-called *mine road* open to public transport, branches off between the mine and the Sophienhoehe. A 200 m (656 ft) section of the road runs through a tunnel over which the conveyors to the outer dump are

CROSS SECTION SW-NE

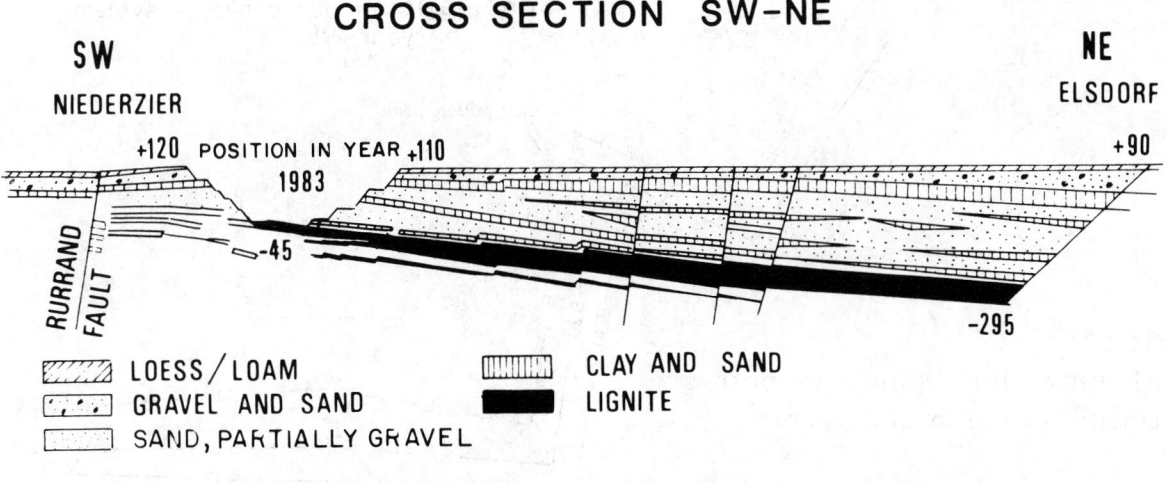

Fig. 3. Cross section SW-NE of the Hambach mine. All elevations are shown in meters. Datum is mean sea level.

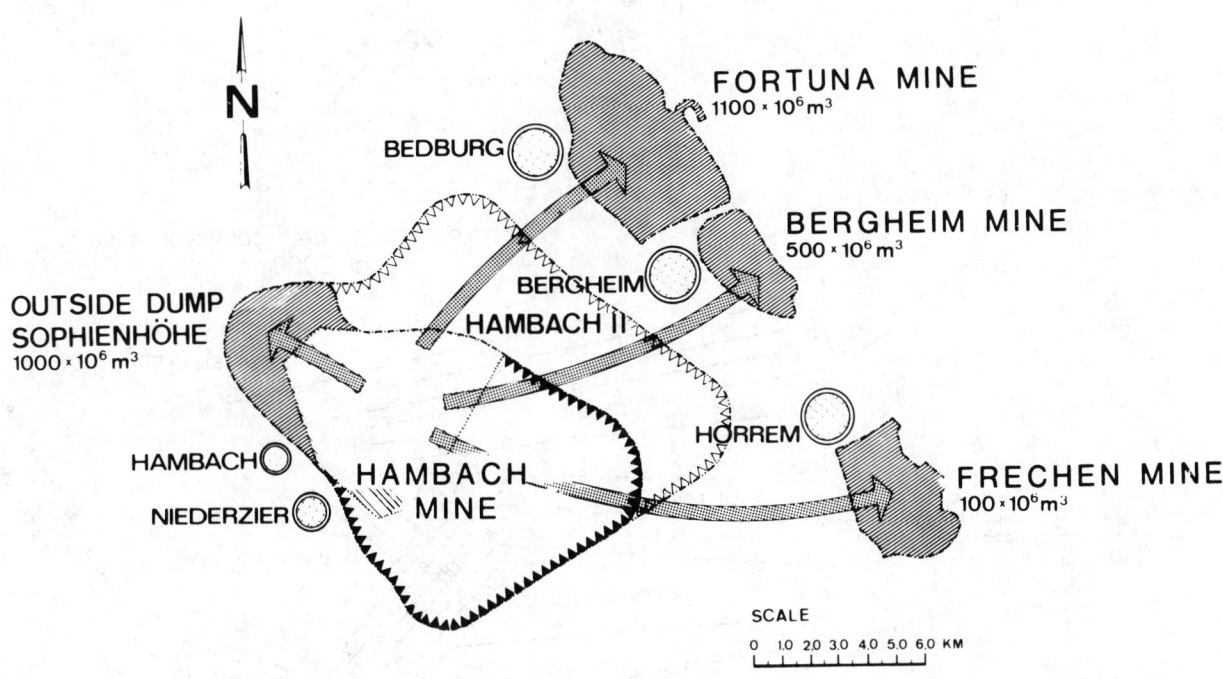

Fig. 4. Distribution of overburden from the Hambach mine.

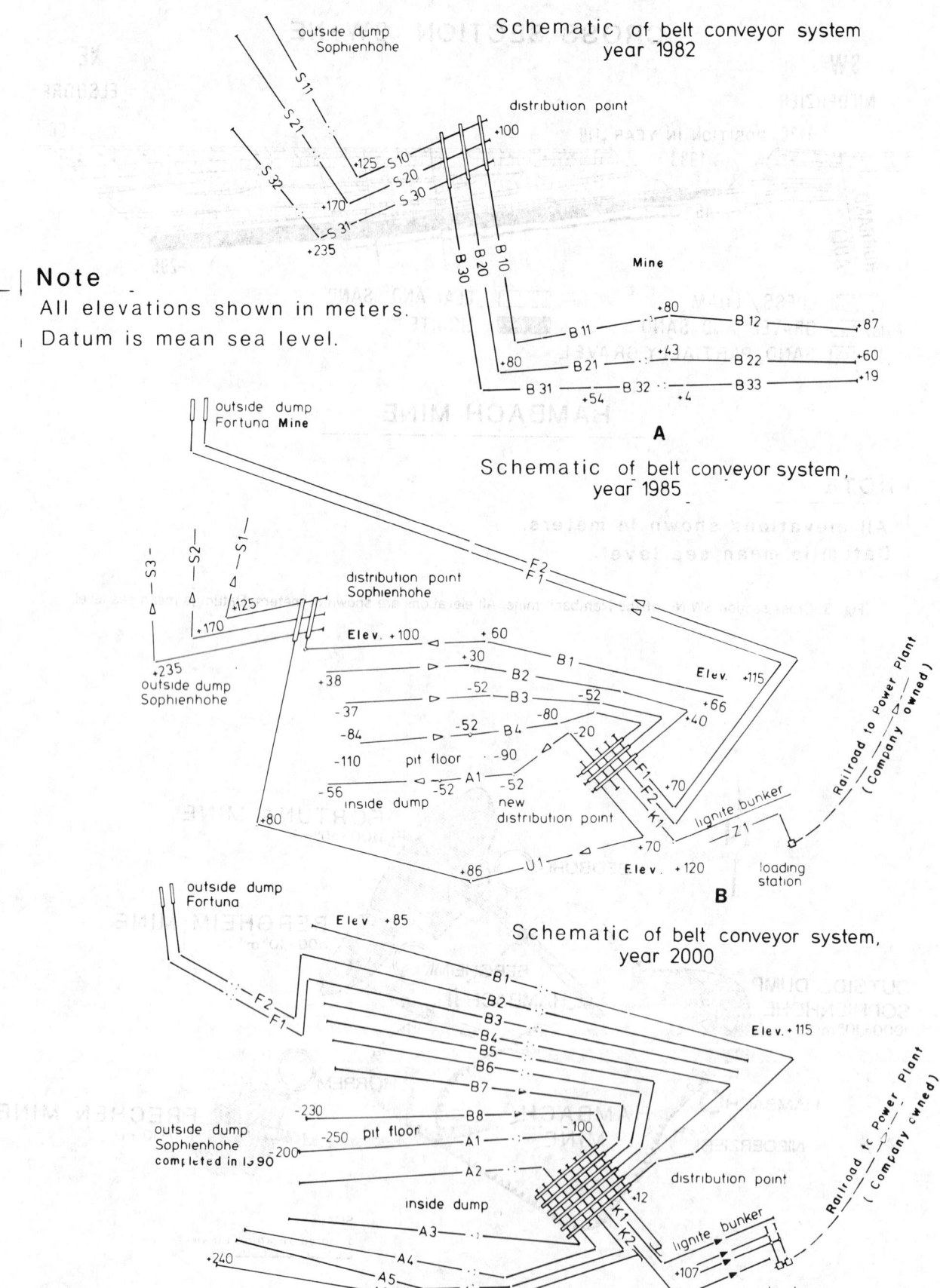

Fig. 5. Schematics of the belt conveyor system in 1982, 1985, and 2000. All elevations are shown in meters. Datum is mean sea level.

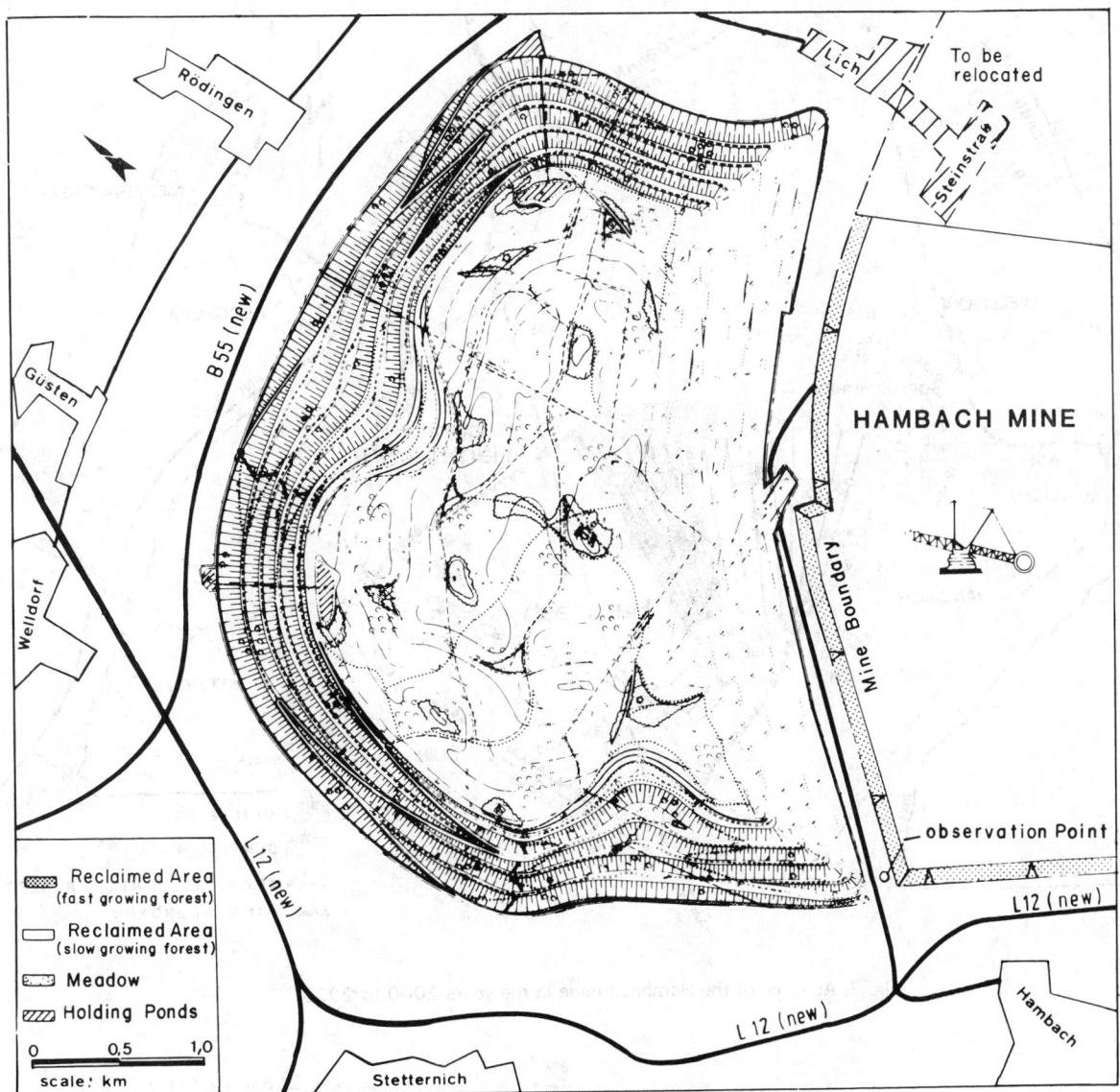

Fig. 6. Final configuration of outside dump *Sophienhöhe*.

routed. The final substitute road for Highway B55 will be built on the northern rim of the Sophienhohe in the late 1980s.

After the year 2000 a section of Autobahn A4 and the federal railroad line from Dueren to Neuss will have to be relocated parallel to the present railroad line from Cologne to Aachen, in addition to other sections of federal and state highways.

EQUIPMENT SELECTION AND SITE PREPARATION

Equipment selection philosophy for the Hambach mine was based on improving technology and increasing machine capacity. These advances were stimulated by the need to offset a portion of the cost increase resulting from mining under stripping ratios less favorable than those found in other mines in the Rhenish mining area. The design and operational geometry of the excavators and spreaders are comparable to time-tested equipment with a capacity of 110 000 m³/d (143,875 yd³/d). The large-scale equipment and conveyors used at Hambach, however, have a daily capacity of 240 000 m³(313,908 yd³). Increasing capacity not only requires greater driving power (especially in the area of the bucket wheel and conveying routes) but also involves increased service weights and higher belt speeds (Figs. 8 and 9).

The conveyor systems, consisting of 2800-mm (110-in.) wide belts of ST 4500 quality, have been designed for a belt speed of 7.5 m/s (24.6 ft./s) and driving power of as much as 6 × 2000 kW. The design and steel construction of the belt conveyor driving stations and supporting structures were based on experience gained with units operating at the For-

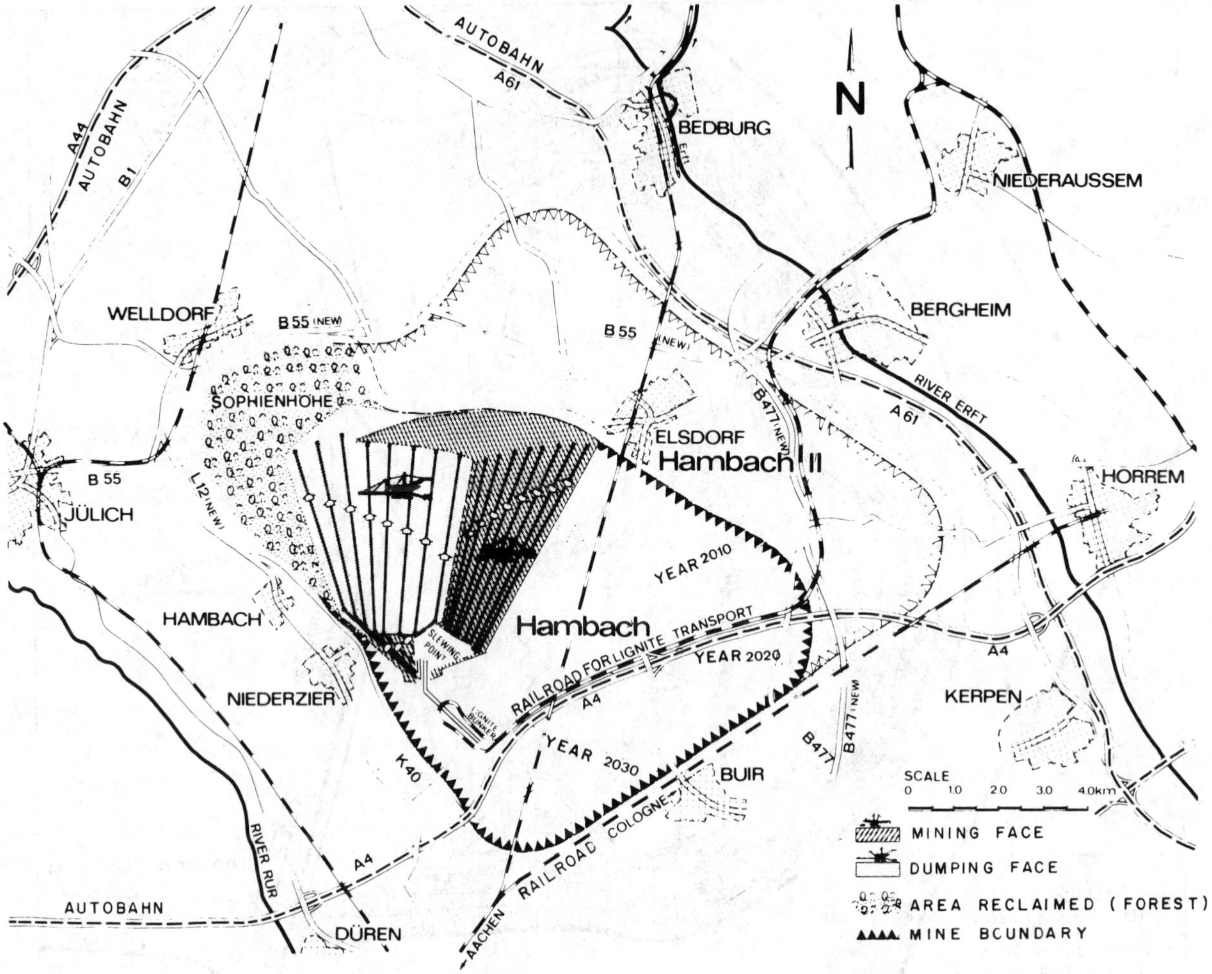

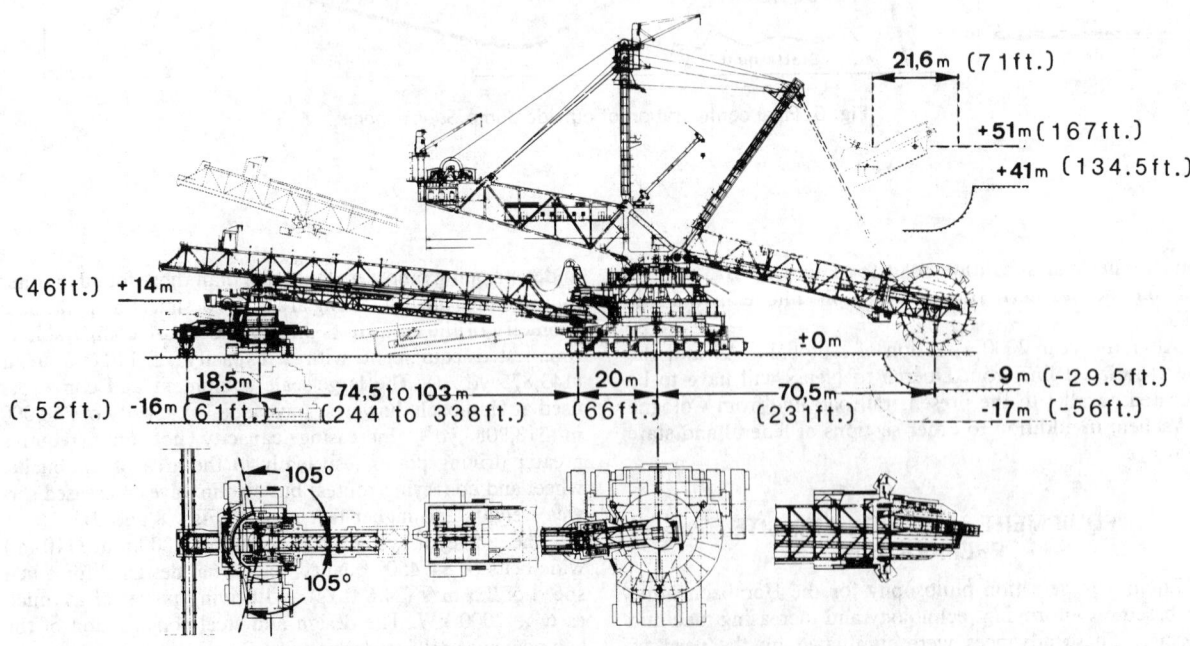

Fig. 7. Advance of the Hambach mine in the years 2000 to 2030.

Fig. 8. Bucket wheel excavator with a daily capacity of 240 000 bank m³.

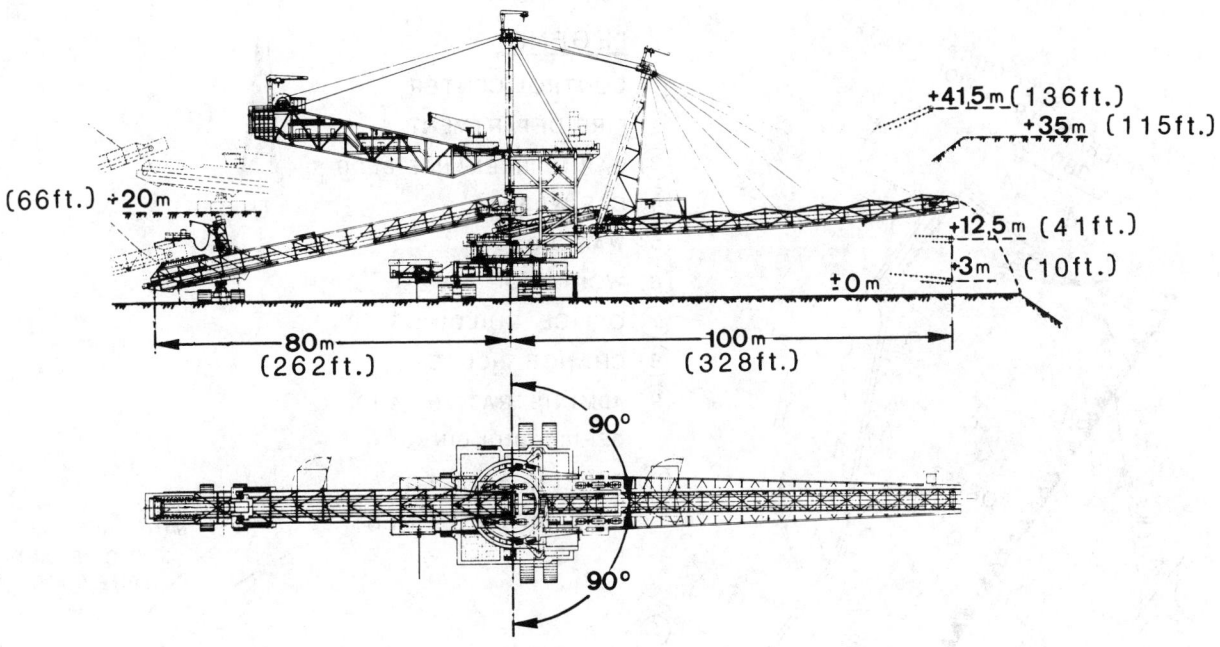

Fig. 9. Spreader with a daily capacity of 240 000 bank m³.

tuna mine, but feature two major improvements: (1) the spacing of the three-part garlands of the upper belt was increased from 1.25 m to 1.875 m (4.10 to 6.15 ft), thus reducing the number of garlands per support, from 6 to 4, and (2) the tail stations have been completely redesigned and can be transported by mobile crawlers. The mine's auxiliary systems and vehicles are largely identical in type with those used at other Rheinbraun mines.

In mid-1982, the Hambach mine had an equipment inventory consisting of:

20 motor graders, including 5 bulldozers on wheels
12 loaders and cleaning vehicles
 4 pipe layers
11 auxiliary excavators and
28 other items, such as crawlers, cranes, low-loaders, rolling mill engines, among other equipment,

and a vehicle fleet consisting of

45 trucks, personnel carriers, and tractor trucks
42 passenger cars and
70 trailer wagons.

Power is supplied to the mine via a two-system, 380-kV overhead line by RWE (Rheinisch Westfaelisches Elektrizitaetswerk, West Germany's major electricity supplier) and three 200-MVA transformers that feed 30 kV into the mine's central switchgear plant. The equipment in the mine itself is connected radially to the switchgear. Some 350 km (217 miles) of 30-kV service cables and supply lines for equipment, conveyors, and the drainage system were laid by the end of 1980.

In order to assemble the first three large-scale equipment groups (excavator, spreader, and tripper car), the necessary sites were prepared at the slew point of the Hambach mine late in 1975, with machine assembly commencing in 1976. Machinery components were transported by suppliers to these sites via highway only. Assembly work proceeded on

schedule, with erection of the belt conveyor head and tail stations beginning early in 1978 (Fig. 10).

Starting in early 1977, Hambach's surface installations were built in a large complex near the assembly sites for the large-scale equipment. They included social, operational, and office buildings, workshops, stores, fire house, central control station, central switchgear plant, and numerous facilities for the mine's services and waste disposal systems (Fig. 10). A small base with social and operational buildngs as well as a workshop was installed for the work force allocated to the dump site at the slew point of the Sophienhoehe outside dump. Here, too, is the Sophienhoehe belt distribution point where shiftable transfer stations (shunting heads) handle the transfer of material from the excavators to the spreaders (Fig. 11).

In 1976 the first wells draining westward toward the Ruhr river were put into operation near the slew point area of the mine. The drainage system was designed to handle a maximum water quantity of 2 m³/s (71 ft³/s). Early 1978 saw the completion of the first extensive intercepting scheme that routed water eastward to the Erft river. This scheme consisted of an 11-km (6.8-mile) long steel pipe system and the 8-km (5-mile) long Wiebach conduit system. The conduit system is comprised of concrete pipes as much as 2 m (6.6 ft) in diameter and having a maximum flow rate of 6 m³/s (219 ft³/s).

Late in 1980, the second drainage system to the Erft, namely the Finkelbach conduit system in the northeast, started operating, likewise with a capacity of up to 6 m³/s (219 ft³/s). All pipes were laid underground with the disturbed land restored on completion of the work.

In addition to the present 142 deep wells and 13 surface wells [maximum 25 m (82 ft) deep], approximately 290 exploration and 149 piezometric boreholes have been drilled since 1974 to explore and examine the geological, geomechanical, and hydrological conditions of the lignite and overlying strata. Overall, the mining field so far developed has a density of approximately 25 boreholes per sq km (0.39 sq mile).

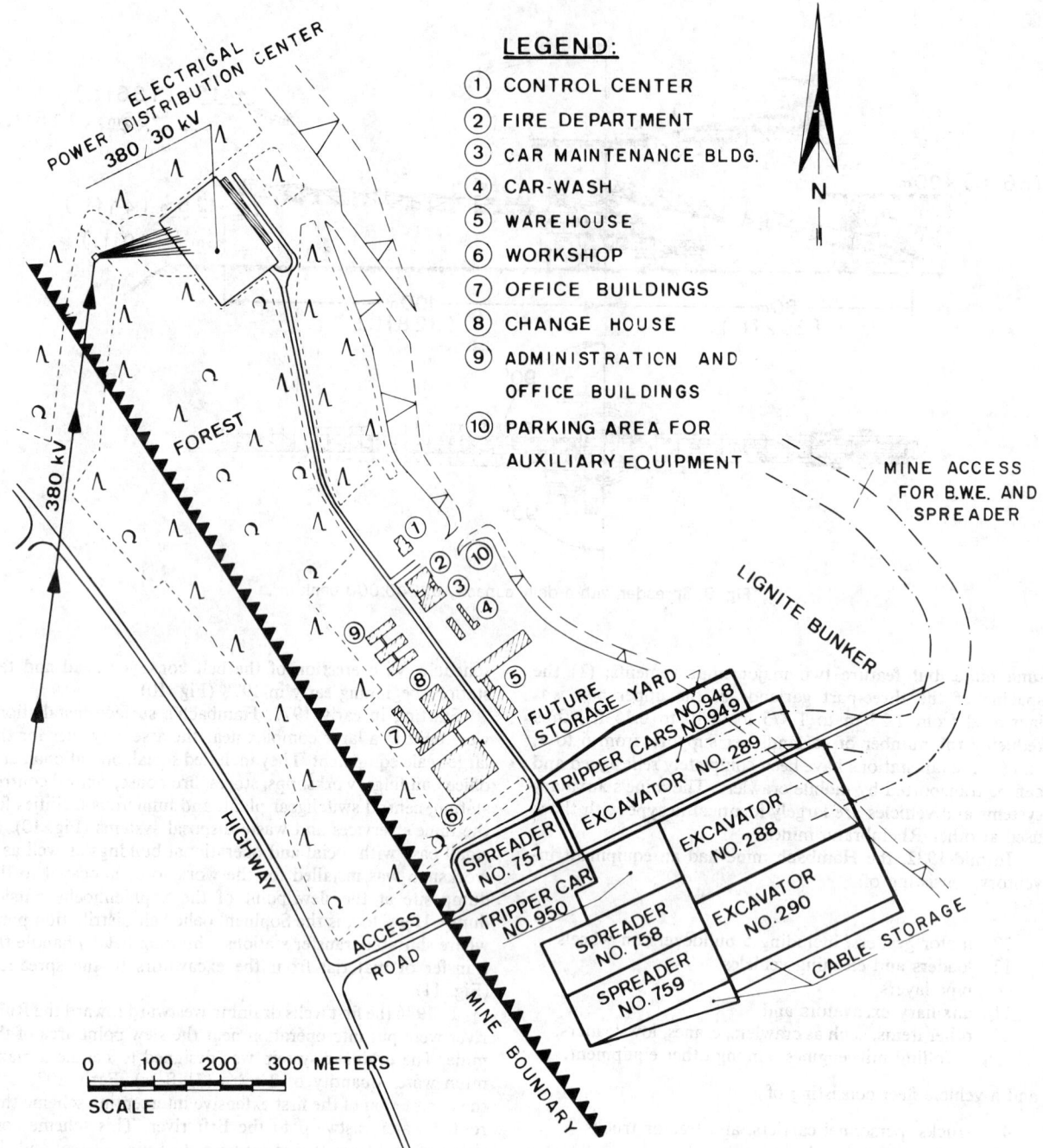

Fig. 10. Site plan of the facilities and erection area in 1978.

In mid-1977, work was started on the various jobs associated with stripping the surface. Wooded areas in the path of the excavators and stackers were deforested, grubbed and, after an intensive search for buried shells from World War II, made available for archeological investigations. Both the routes for the large-scale equipment, belt conveyor driving stations, and conveyors were all given a thick top gravel layer. The supports for the belt conveyors were assembled on the spot, with assembly and conversion work planned and supervised by means of general and detailed network plans. The critical factor in the construction of the conveyors, which did not involve any major problems, was the vulcanizing and electrical installation work along the conveyor routes.

DEVELOPMENT PLANNING AND OPERATIONS START-UP

The first equipment group was transported on schedule from the assembly site to its point of operation on Sept. 1, 1978 (Fig. 12). Experience previously gained in starting up the large-scale 200 000-m³ (261,590 yd³) equipment in the Fortuna mine was used to insure that transport, gauging, and adjustment of the equipment caused no major problems. On Sept. 22, overburden removal started at the northwestern rim of the box cut. Three weeks later the second excavator was put into operation on the same bench, but excavating different material.

The equipment operated initially in a somewhat confined

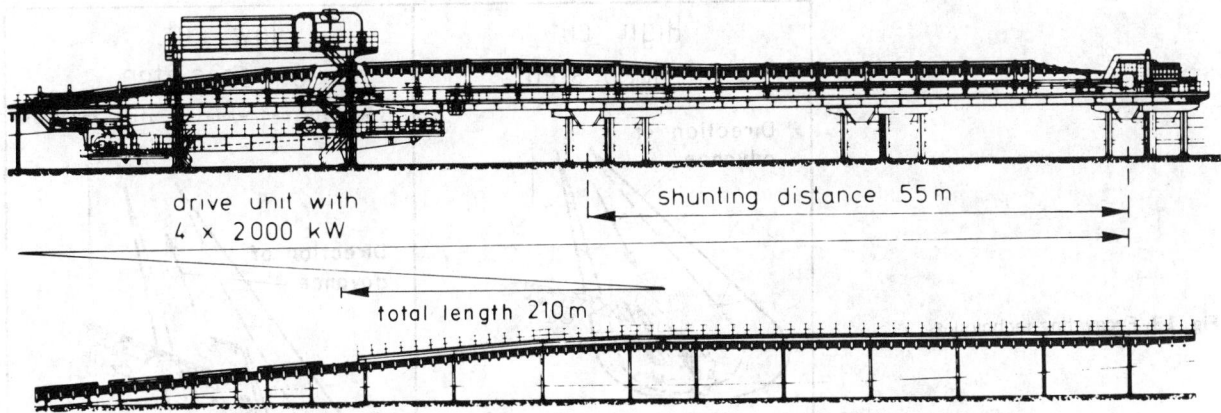

Fig. 11. Conveyor shunting head unit with a capacity of 37 500 t/hr.

space. Pollution control called for a minimum distance of 350 m (1148 ft) to be observed between the western working rim and the conveyors installed on the site. Thanks to its design, the excavator can cut 14 m (46 ft) deep from one conveyor position in the deep step mode and as much as 17 m (58ft) below the traveling level of the excavator in deep cut mode (Fig. 13). The conveyors are moved down into the trenches thus mined, or, if time or space do not permit the provision of space for maneuvering, relocated in the new cuts.

In the first year of development, only a relatively small dumping area was available, clearly bounded by final slopes and Federal Highway 55 to the north. Starting out from the typical cross-section for final slopes of the dump (Fig. 14), comprehensive planning was necessary to schedule the sequence of mining blocks and select the quality of the material required at the dump. To insure a stable final slope, first a 30-m (98-ft) wide drainage ditch was excavated in the center

of the safety embankment. At the bottom of the final slope, cohesive material was removed to guarantee a perfect bond between the safety embankment, forest gravel layer, and the original subsoil.

In the first year of development, problems arose primarily in using equipment in wet loose-rock layers. Throughout the development area a layer of *Reuver* clay below a 10 to 15-m (32.8 to 49.2-ft) thick alluvial gravel bed was encountered. The Reuver clay consists of alternating bands of varyingly cohesive strata with water-saturated fine-sand stringers below 1 to 2 m (3.3 to 6.6 ft) of residual water. The stratification of these layers ruled out prior drainage by wells. Despite extensive mine drainage measures preceding and accompanying the excavator work, overburden movement was subject to serious disruptions and performance limitations. Only after the Reuver clay had been completely worked through in the development area was the residual water able to percolate through to the underlying permeable strata.

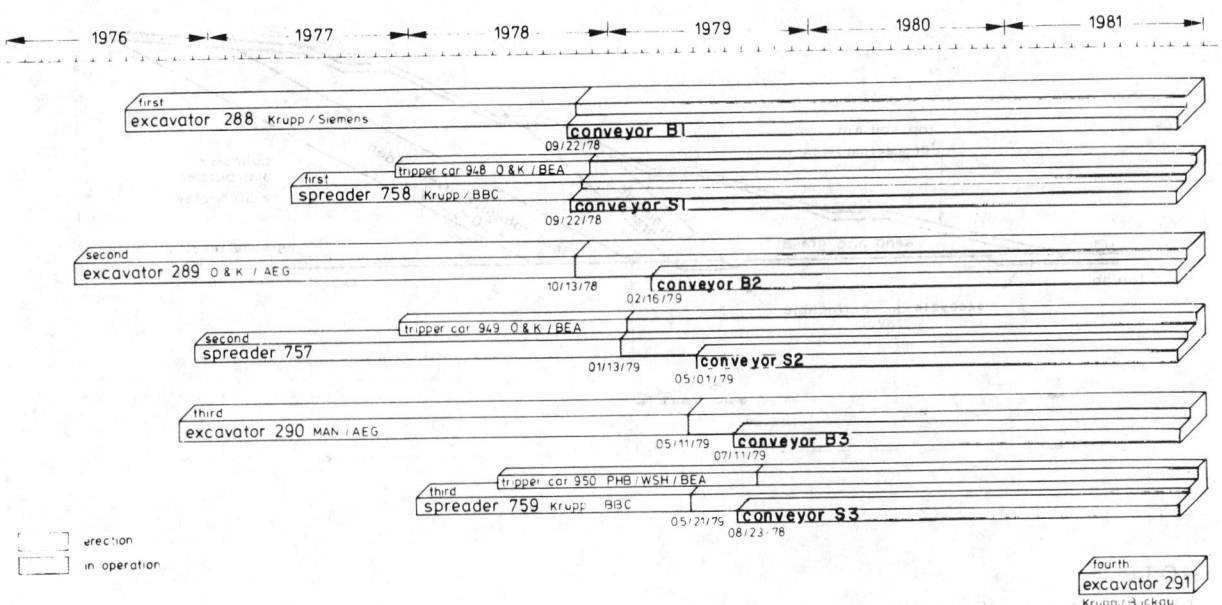

Fig. 12. Equipment erection and start-up schedule for the Hambach mine.

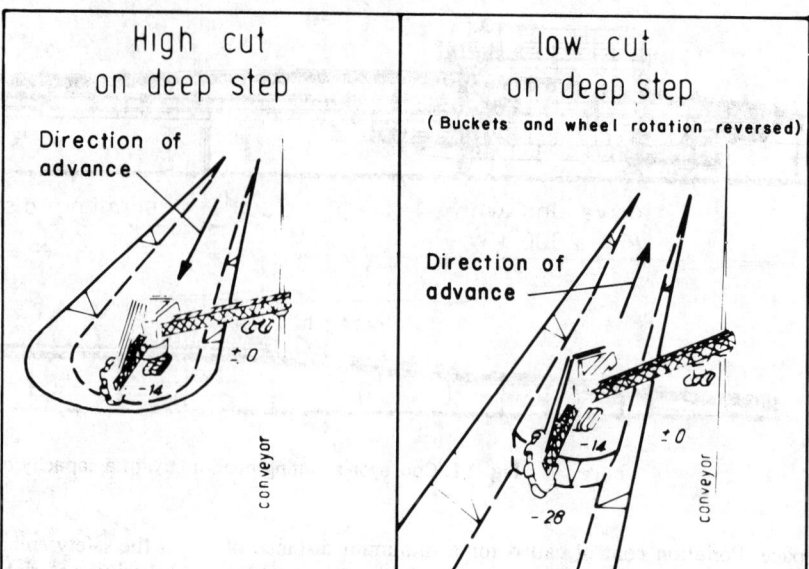

Fig. 13. Excavating techniques.

Over one year of operations at the Hambach mine dealing with water-saturated, cohesive overburden masses not only involved various problems in mining and conveying work due to breakdowns and limits to performance, but also led to a great number of technical and organizational improvements in materials handling. The overburden, normally still lumpy when extracted, changed its consistency to become a viscous sludge, even after short distance conveying. The avoidance of transfer point plugging, belt track control, and scraper system efficiency were improved greatly by design modifications. Selective use of plastic wear parts in the side chutes of the transfer points and the installation of rubber

aprons on the baffle plates reduced material caking, and in particular, spillage.

The dump structure required extensive investigations in planning and geomechanics in order to safely and effectively dispose of the wet, cohesive mixed soil. This was especially true with regard to the confined dump space and stringent conditions governing the quality of material required to build up the final slopes. The most important requirement was the development of suitable methods to insure proper material distribution. These methods included limiting slope height and gradient, and safeguarding the stability both of the individual slopes and of the entire dump body through flexi-

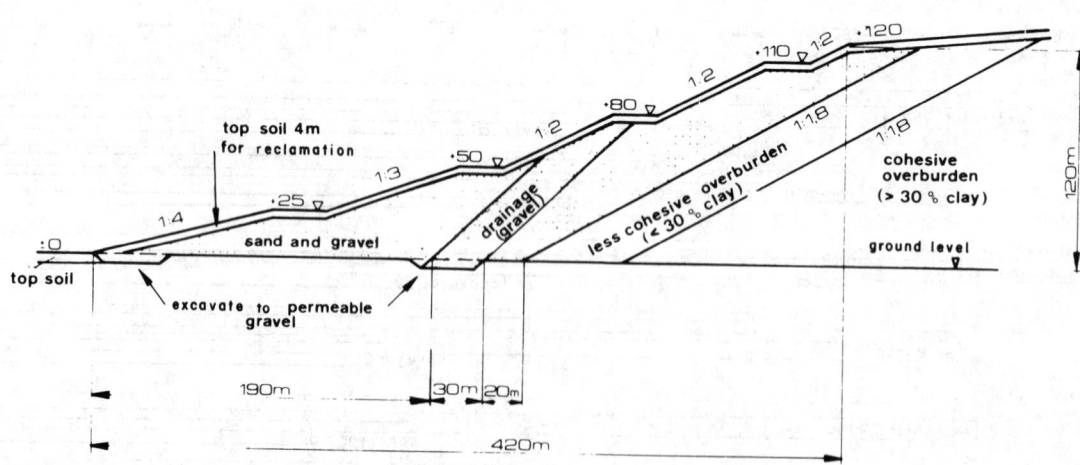

NOTE

All elevations shown in meters.

Fig. 14. Typical cross section of final slope configuration for outside dump. All elevations shown in meters.

bility in equipment utilization. Examples can be found in Fig. 15.

In the first six months of 1979, it was not possible to complete the safety embankment on the lowermost dump due to the lack of sufficient volumes of stable gravel and sand. Only in the boundary area between mixed soils and safety embankment and the area parallel to the face conveyor system was it possible to create stable polder dams in order to obtain the maximum dump space for the excessive quantities of wet mixed soils (Figs. 16 and 17). The final slope was completed from materials from the upper sections of the dump that were built up later.

Mixed soils were disposed of in well-planned, stable, polder dams which used minimum volumes of noncohesive overburden. Restoration of the load-bearing capacity of the masses dumped into the polders was effected through lengthy periods of consolidation. These geotechnical considerations, however, needed to be balanced with optimization of dumping space and the demands on system performance. From these practical considerations, an efficient method was finally devised that involved building up the dump in slices. This method, produced stable polders, maximum mixed soil dump space with individual slope heights of between 12 and 15 m (39.3 and 49.2 ft) in both high and deep dumping and extremely low slope gradients. Bench block was the operation mode for the spreaders (Fig. 16) due to the various material types encountered (Henning, 1980a). This method insured the controlled dumping of masses of low consistency and at the same time allowed greater scope for decision-making on material handling and spreader movement.

Especially problematic was the restoration of the load-bearing capacity of the cohesive mixed soils, which was done to insure equipment mobility and dump mass stability. The originally established thickness of 5 m (16 ft) for grade dumping proved to be too low. It was discovered that during dumping such a grade thickness could not always be observed, since the movement of the material in the impact area of the bulk masses falling from the spreader produced a change in the thickness and composition of the grade. The

crawlers of spreaders and tripper cars sank dangerously into the unstable soils, which in turn led to movement in the dump slopes. An increase in the firm grade to a thickness of 7 m (23 ft) and strict adherence to an adjusted dumping scheme aimed at reducing the excess pore water pressure in the polders eventually proved effective (Henning, 1980b). A typical dump-slope cross-section is shown in Fig. 17, which illustrates that on each bench permeable and/or only slightly cohesive material is required to build up the 7-m (23-ft) thick grade and the polder dams. High dumping of the upper dump bench ends with an 8-m (26-ft) thick permeable layer with the top 2 m (6.6 ft) preferably being constructed with Quaternary gravel as a substructure for loess fill. A total of 50% of the material being mined is classified as permeable material, as opposed to 40% of the material dumped at the Sophienhoehe. The 10% surplus quantity is required to insure adequate and consistent material supplies (Loeper, 1982).

In 1978, only 0.006 km³ (8.1 million yd³) of overburden was moved, i.e., less than planned. In addition to the difficulties already explained, the primary cause of the lower output was the fact that in the first four months of operation there was no night shift due to a noise pollution control order issued by the mining regulatory authorities.

From the outset, the development phases were arranged to insure optimal emission control. Prior to start-up of operations, a 7-m (23-ft) high, 3.2-km (2-mile) long protective barrier was erected on the western rim of the mine. During the initial assembly of the conveyors in the field, minimum distances of 350 m (1148 ft) between work site and houses were observed. Bench planning insured that the conveyors were relocated in the cuts made by the excavators after the shortest possible period in the field.

Nevertheless, it was not possible to avoid serious problems with the neighboring residential communities, particularly during the start-up phase. The erection of additional noise-abatement barriers and walls at the stationary conveyors near the belt distribution point at the Sophienhoehe, the replacement of all upper belt idlers with specially designed

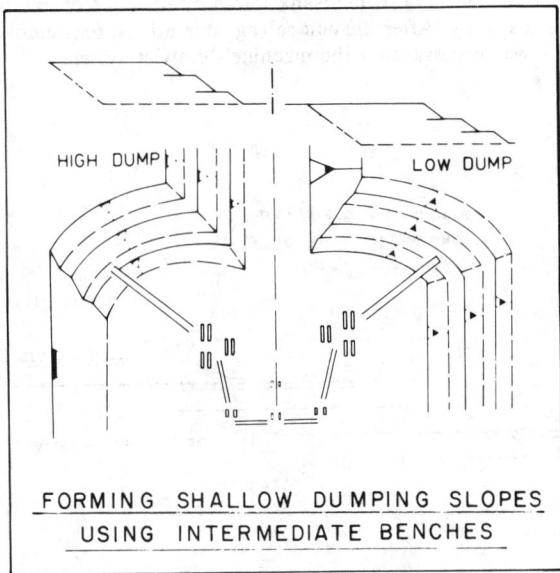

FORMING SHALLOW DUMPING SLOPES
USING INTERMEDIATE BENCHES

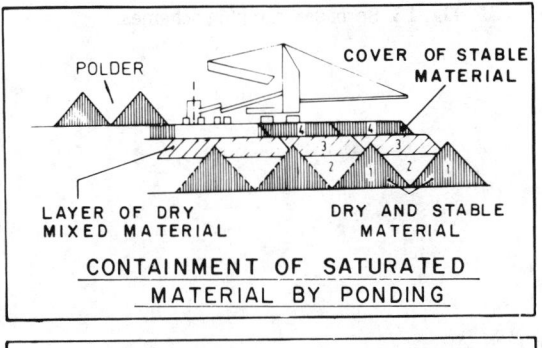

CONTAINMENT OF SATURATED
MATERIAL BY PONDING

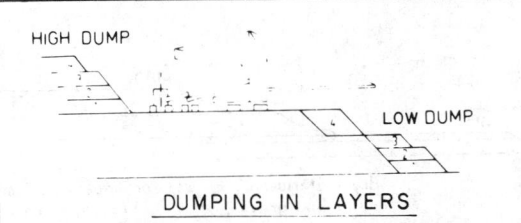

DUMPING IN LAYERS

Fig. 15. Typical dumping schemes for slope stability.

Spreader dumping schemes

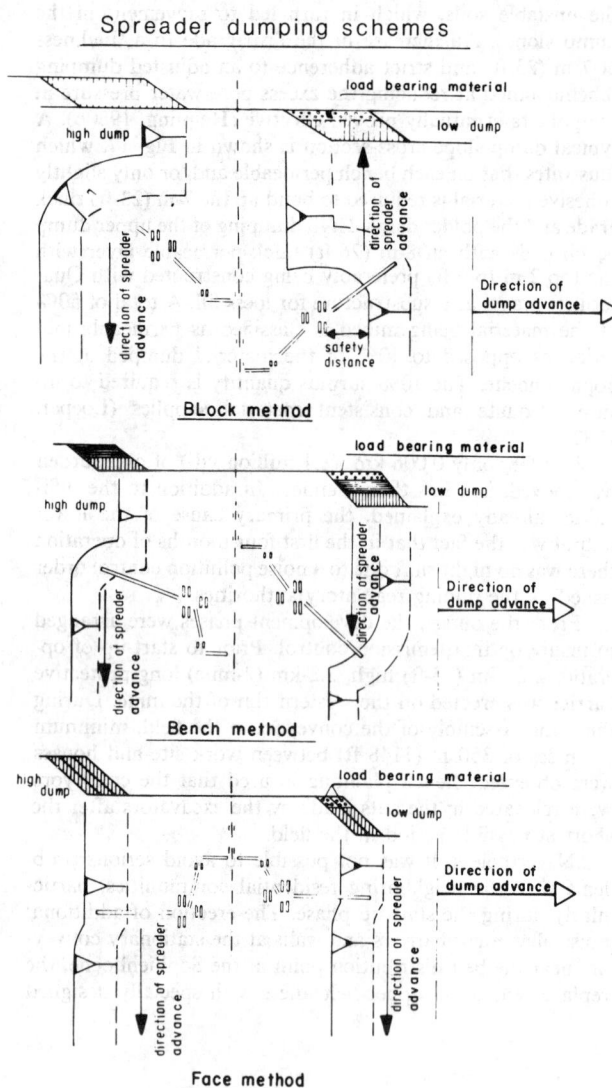

Fig. 16. Spreader dumping schemes.

machine-balanced idlers having an average reduced noise level of 7 dB (A), along with organizational measures in the deployment of equipment, all helped to attain an effective reduction in noise pollution caused by mining operations.

Later, a great number of innovative technical and organizational noise-control measures were tested and implemented at the Hambach mine. The new equipment and installations incorporate the latest technological developments and meet the technical and noise emission requirements of the mining regulatory authorities. The current developments are aimed at achieving an effective noise reduction at the sources on mobile equipment and conveyors. In addition, possibilities of reducing noise levels by shielding and covering drive units are presently under study, with special consideration given to prevailing conditions in open pit mines.

To reduce dust pollution, experiments are underway on sprinkling devices that later on will be used primarily in areas of exposed lignite. Partial reforestation of the upper mine slopes and greening of the western and southern final slope system by spraying a mixture of grass and clover seeds, granulated fertilizer, and a plastic adhesive will inhibit dust formation and at the same time provide protection from surface water erosion.

FULL-PERFORMANCE OPERATIONS

In mid-1979, the preconditions existed for the transition to full-performance operations. On the mining side, dry, sandy-gravel layers, exposed below the *Reuver* clay, were urgently required to build up the needed safety embankment on the western rim of the Sophienhohe (Fig. 18a). By contrast, the mixed-soil dump space in the center of the dump was already fully utilized. By late July 1979, 0.036 km³ (47 million yd³) of overburden had been moved under the difficult conditions of the start-up phase already described.

In August 1979, the third conveyor began operating on the outside dump. Together with the improved quality of the material, this addition led to a considerable increase in overburden output (Fig. 19b). What also became clear is how much output is dependent on the occurrence of sand and gravel masses necessary to lay out the final slopes, polder dams, and dump slope. The decrease in output in April 1980, as shown in Fig. 19b, was due to a prolonged breakdown of spreader 759, which sank into a dump slope of low bearing capacity. After difficult salvage operations, reassembly work was necessary on the machine's crawler system.

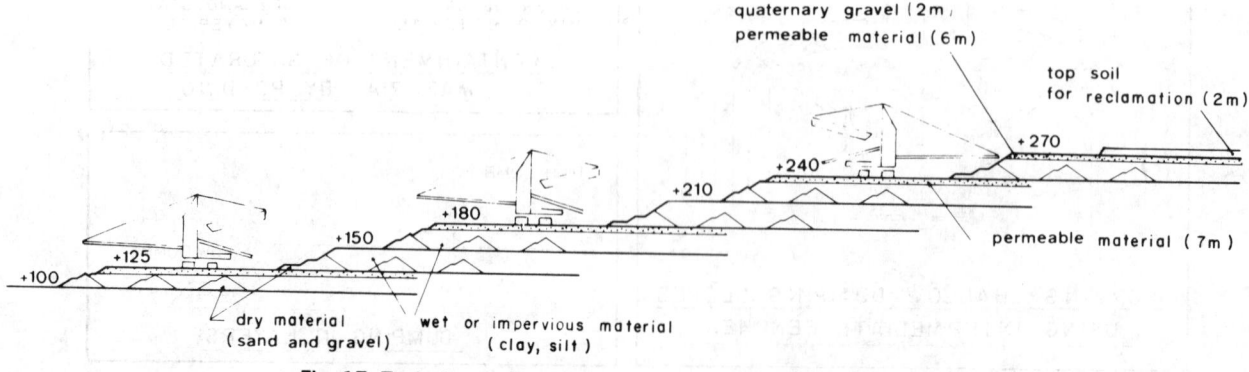

Fig. 17. Typical cross section of working benches for outside dump.

In 1979, total overburden of 0.075 km³ (98 million yd³) was moved at the Hambach mine. With the same equipment, overburden transport volume rose from 0.125 km³ (163.4 million yd³) in 1980 to 0.140 km³ (183.5 million yd³) in 1981. The overburden quantity of 0.079 km³ (103 million yd;3) removed in the first six months of 1982 suggests further gains in machine capacity.

Fig.18d show diagrams of the development work on the mine and dump including overburden movement, land used, and land reclaimed by the end of 1981. The section-wise development of the mining operations into greater depths depended very much on the overburden quality required for the outer dump, in particular, the demand for sand/gravel masses and reclaimable forest gravel for the completion of the surface in the final slope area.

The situation at the end of 1980 shows the operations after total overburden movement of 0.206 km³ (270 million yd³). The box cut as shown is approximately 5 km (3 miles) long, 1 km (0.6 miles) wide and as much as 70 m (230 ft) deep. The area of the future lignite bunker has been cut out by one of the 240 000 m³ (313,908 yd³) excavators. The contours of the final slope system of the Sophienhoehe are quite prominent.

In 1981, the mine developed primarily in depth. The goal was the earliest possible excavation of the section for the new belt distribution point in the southwestern part of the box cut. Later in 1981, this cut was about 5.5 km (3.4 miles) long, 1.4 m (4.6 ft) wide and as much as 110 m (361 ft) deep. During that year, the outer dump, Sophienhoehe, grew to a height of 167 m (548 ft) (Fig. 18d). By the fall of 1981, forestry reclamation was confined to the berms and inclines of the final slope system, with a separate unit operation being set up to spread topsoil. The 1 to 2-m (3.3 to 6.6-ft) thick layer of loess in front of the dump is extracted by a small bucket wheel excavator and, after being transported over a 1000-mm (39-in.) wide conveyor, distributed by a spreader in 2-m (6.6-ft) lifts on the dump surface. The total area of 185 hectares (457 acres) reclaimed by the end of 1981 was increased by another 72 hectares (178 acres) to 257 hectares (635 acres) by the end of the planting period in April 1982. In mid-1982, the first sections of the reclaimed areas were opened to the public for recreation.

The utilization of the mining equipment's capacity has been increased almost continuously since startup of operations in 1978 (Figs. 19a and b). Crucial for the load figures obtained was the nature and quality of the overburden as well as restraints on output, which were partially influenced by the arrangement of the conveying and dumping systems. The steady improvement in working time is due to the decrease in operational disruptions during service and to the successful implementation of technical and organizational measures.

The output of the individual excavators in terms of daily, monthly, and annual overburden movement varies considerably. Over longer periods, all items of equipment have reached a rated capacity of 240 000 m³ (313,908 yd³) based on 19.2 operating hours per day (i.e., 80% of 24 hours). The comparatively great differences in average and/or maximum daily output are due to the already mentioned factors limiting output and operating hours, namely, material quality, conveying, and dumping, the impact of which was particularly felt in the development stage of the mine. The maximum output in overburden movement reached at Hambach [365 700 m³/d (478,318 yd³) for one bucket wheel excavator and 933 480 m³/d (1,220,946 yd³/d) for three equipment groups] furnishes convincing proof of the operating efficiency of the selected equipment.

Mining system downtime had a considerable influence on machinery utilization. This downtime consisted of standstills attributable to the lack of receiving capacity in the spreaders caused by breakdowns and adverse material properties (Fig. 20). After the initial difficulties in the conveyor system had been overcome, system downtime fell to mid-1979 levels that were considered normal. The various causes of downtime in the conveying system were nonspecific, thus eliminating evidence for fundamental technical or organizational troublespots. Only at those conveyors with separate feed points some distance from the tail station, was downtime above average. This downtime was caused by belt damage and operational breakdowns attributable to defective belt tracking. In total, disruptions in conveyors on the dumping side were considerably higher than those in the conveyors on the mining side. These higher values correspond directly to deterioration of the wet, cohesive material during transport as well as to the difficult grade conditions under which the tripper cars (on the dumping side) operated.

The operational control center (BU) of the Hambach mine is where all operations information is collected. A single room houses the operational control center, breakdown center, radio center, and belt control station. Central recording, identification, and evaluation of downtime and performance data from the mining and conveying systems is carried out by a process computer. Intersystem communication is an important precondition for the control and monitoring of operations and subsequent feedback to mine planning. Further development of computerized decision aids for mine operations control at Hambach will be an important task in the years to come. These developments include modified processes of operational control which, for example, will allow a faster retrieval of survey data than was previously possible.

Due to the increase in volumes transported, a higher probability of damage occurring and the greater risk of consequential damage, more attention is being paid to the continuous local supervision of equipment and conveyors. Automatic conveyor monitoring instrumentation, having widespread use in the Rhenish mining area (for spillage, belt tension, slippage, excess load, excess torque, and positioning of transfer points on shunting heads, etc.) is employed, as well as a mobile staff of monitoring personnel. During the start-up phases, which were affected by temperature extremes and adverse material properties, further stationary monitoring personnel were temporarily required at critical points, such as belt distribution points, separate transfer points, and cleaning facilities. The use of automated control for belt tracking, caking, and damage detection (in particular, longitudinal cracks, still not adequately solved) will mean the development of improved equipment able to meet tough operational requirements.

The equipment at the Hambach mine has proved to be successful in full operation. Technical and organizational developments in the mine's maintenance sector are being carried out with priority given to improving operational safety and economics. In the mechanical sector, current efforts are mainly centered on belt transfer points, pulley linings, equipment traveling gears, vulcanizing techniques, and pollution control.

Modifications at the belt transfer points have improved operating conditions along with a decreased incidence of damage and tracking difficulties. These modifications include use of raised separate impact sections; installation of top chutes with modified longitudinal sealing; change in type of garland-type impact idlers; and insertion of rubber aprons in front of the baffle walls.

Results of the initial experiments with automatic belt-track controls have proven promising. The wear of the pulley

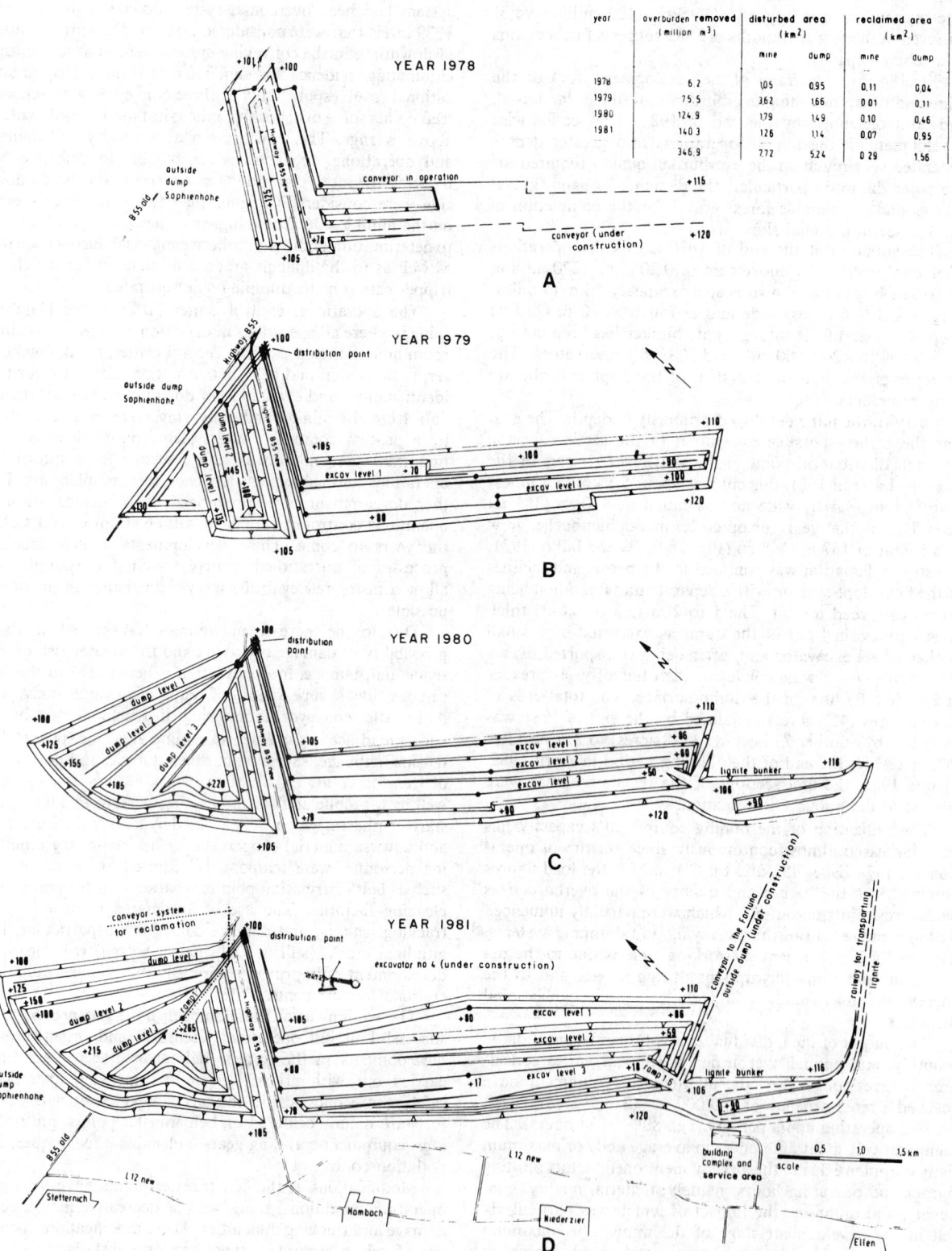

year	overburden removed (million m³)	disturbed area (km²)		reclaimed area (km²)	
		mine	dump	mine	dump
1978	6.2	1.05	0.95	0.11	0.04
1979	75.5	3.62	1.66	0.01	0.11
1980	124.9	1.79	1.49	0.10	0.46
1981	140.3	1.26	1.14	0.07	0.95
	346.9	7.72	5.24	0.29	1.56

Fig. 18. Development of the Hambach mine: 1978-1981.

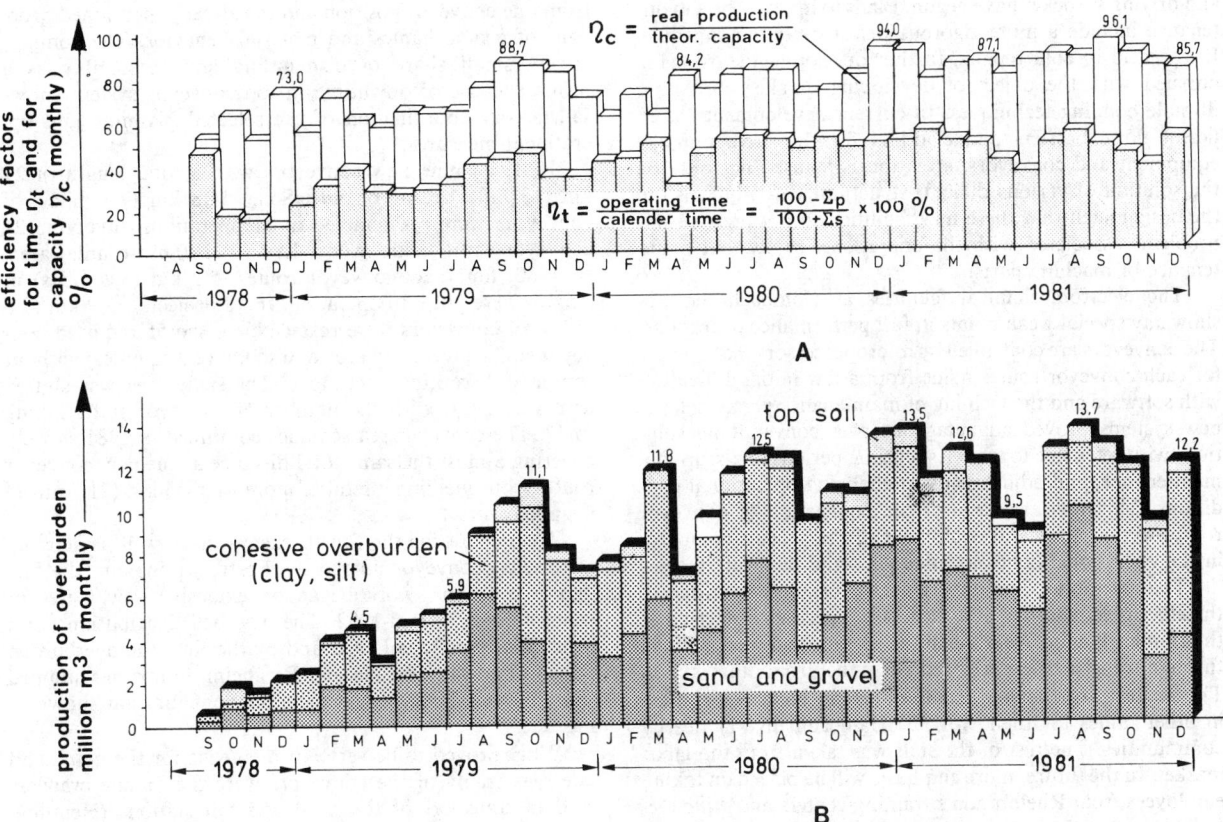

Fig. 19. Development of overburden production and efficiency factors for time (η_t) and for capacity (η_c) of the excavators, 1978-1981, monthly.

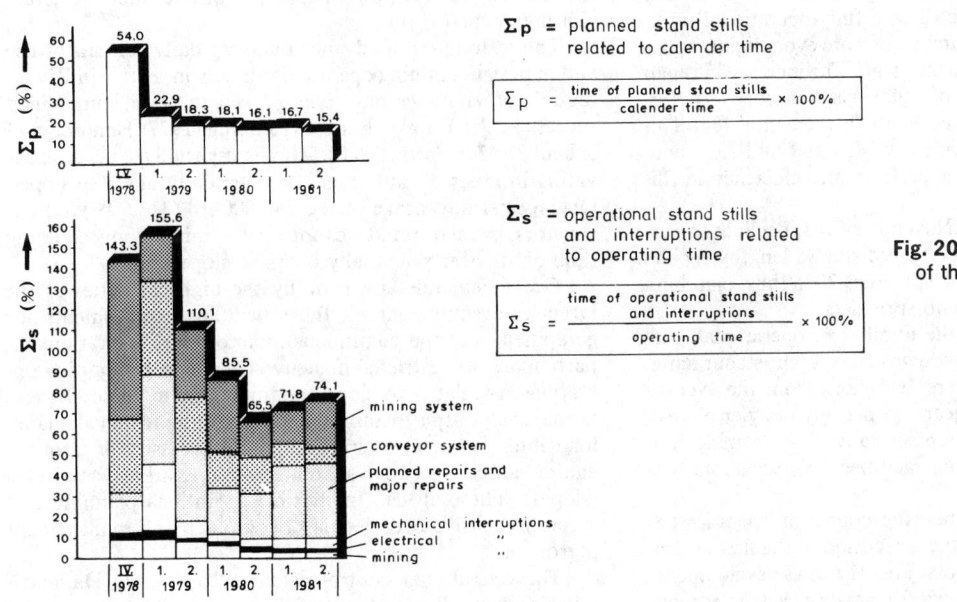

Fig. 20. Standstills and interruptions of the excavators, 1978-1981.

linings, mainly due to material return and belt contamination, will be reduced through improved belt tracking and by the use of scrapers and more stable linings. In order to prolong service life, structural improvements on the traveling gear and driving sprocket have begun. Plans to improve belt maintenance include a more rigorous defect detection schedule. In the field of cold repairs, further developments are being pursued with the object of developing quicker and more durable bonding techniques. In current developments of efficient noise barriers and cladding for the drive units of equipment and conveyors, attention is focused not only on the solution of various difficult structural tasks, but also in the design such that these new facilities do not present new problems associated with the operational safety and maintenance of machine parts.

The electromechanical facilities at Hambach did not show any special weak points in full performance operations. The conveyors are controlled by microprocessors that operate for each conveyor route. Aside from a few initial difficulties with software and the training of maintenance personnel, the new systems proved advantageous over conventional solutions with reference to troubleshooting, periods of disruption, maintenance, and adjustments to changing operational conditions. This was also true of the connection of conveyor routes to the peripheral equipment of the process computer in the operational control center via serial data interfaces.

The assembly of an experienced workforce was one of the chief preconditions of a successful operations start-up at the Hambach mine. Nearly half the present staff was obtained through transfer of experienced personnel from other mines. The majority of the staff, some of whom had been trained in other mines for their work at Hambach, live in nearby communities. The rest of the staff was taken from the labor market. In the future, more emphasis will be placed on taking employees from Rheinbraun's training centers and *Jungwerker* (young worker) departments. The training of new staff members and young personnel has made great demands on the experienced personnel, particularly on the supervisory staff. Prospects of promotion in a new mine have greatly enhanced staff involvement in the difficult start-up phase.

The downtime for scheduled maintenance and scheduled repairs is considered high, reflecting the problems of maintenance organization in an open pit mine under development. Greater time is spent on performing necessary wear-related repairs than would be expected in a fully operational mine, due mainly to the small number of conveyor flights and a comparatively small maintenance staff. The increase in repair time from the second half of 1980 was caused by the replacement of the bucket wheel shaft in excavator 288. This repair took 10 weeks between early May and mid-July, contributing to the decline in the performance efficiency in this period.

Between September and November 1981, declines in performance efficiency were experienced due to lengthy repairs on the bridge suspension of spreader 757, thus matching three excavators with only two spreaders.

The trends in reducing the number of operational, mechanical, and electrical breakdowns have been encouraging. In individual cases, there were deviations from the average values shown for traveling gears, (especially crawler plates), as an interaction between equipment and difficult grade conditions, and also for cleaning facilities on excavators and spreaders.

In new mine developments, the course of operations is very much affected by planning, execution of the first assembly, and relocation of conveyors. Fig. 21 shows developments in conveyors with a breakdown for newly constructed conveyors, operating conveyors, and relocated and reassembled

conveyors. The dumping section in particular frequently required changes at short notice in the belt routing and in some cases a disproportionate amount of reassembly work, resulting from comparatively small effective dumping space from one conveyor position and the already mentioned problems of geomechanics and material behavior. The comparatively small share of dismantling and reassembly as a primary cause of downtime in the conveyor system shows satisfactory coordination of the selected planning and operational measures.

In 1979, nine new conveyors with a total length of 21 km (13 miles) came on line. Some 11.8 km (7.3 miles) of conveyors were reassembled in 31 operations, involving 21 conveyor shifts made over a 1.8-sq km (0.69 sq-mile) area. In 1980, four new conveyor routes, 5.1 km (3.2 miles) in length, were put into operation. In 18 projects, 13.9 km (8.6 miles) of conveyors were reassembled; the 55 required conveyor shifts covered an area of 6 sq km (2.3 sq miles) without seriously disrupting operations. The same area was shifted across in 1981, with the number of conveyor shifts falling to 47. The conveyor reassemblies continued in 1981 as well, covering almost the same total distance as the new conveyor routes, thus yielding a total of more than 35 km (21.7 miles) installed so far.

To prepare for the fourth excavator as well as start-up of the first conveyor link to the Fortuna mine and the first lignite conveyor, work began on extensive new conveyor constructions in mid-1982. The new belt distribution point is being gradually enlarged and by the time the overburden transport to the Fortuna and Bergheim dumps has stopped after the year 2000, it will have the configuration shown in Fig. 22.

What proved to be very advantageous for the quick and safe reassembly of the conveyors were the mobile crawlers used in transport of the head and tail stations (Henning, 1977). The smaller crawlers used for the first time at Hambach for powered tail stations of 210 t (231 st) service weight have proven to be especially useful. Even shifts into cuts with slope gradients of 1:5 have not met with any particular problems during shifting or maintenance work. The system of 30/6-kV transformers and power switches mounted on the head and tail stations chosen at Hambach has the advantage that conveyor shifts no longer require special preparation time for the power supply, prior and after the actual shifting is carried out.

The techniques of advance planning daily equipment use and organizing mining operations already in service in Rheinbraun's other mines have also proven to be an unqualified success at the Hambach mine (Henning, 1977; Henning and Schenk, 1977; Krug, 1980). This is especially true of excavating in areas of fault zones or potentially unstable slopes. One special advantage of the 240 000-m³ (313,908-yd³) excavators is their rapid advance rates minimizing standing time on or near potentially unstable slopes.

Great demands are made by the high capacities of the large-scale equipment on the scheduling requirements for equipment and the continuous monitoring of operations; in particular: the restricted maneuvering space in dumping operations as related to slope conditions; permissible changes in material; output planning limits resulting from calculated load intervals in mass spreading; and necessity of frequent equipment changes to avoid dangerous operational situations (slopes). These all call for flexible operations planning and a comprehensive overview of operations by all management personnel.

The operational control center (*BU*) of the Hambach mine is where all operations information is collected. A single room houses the operational control center, breakdown cen-

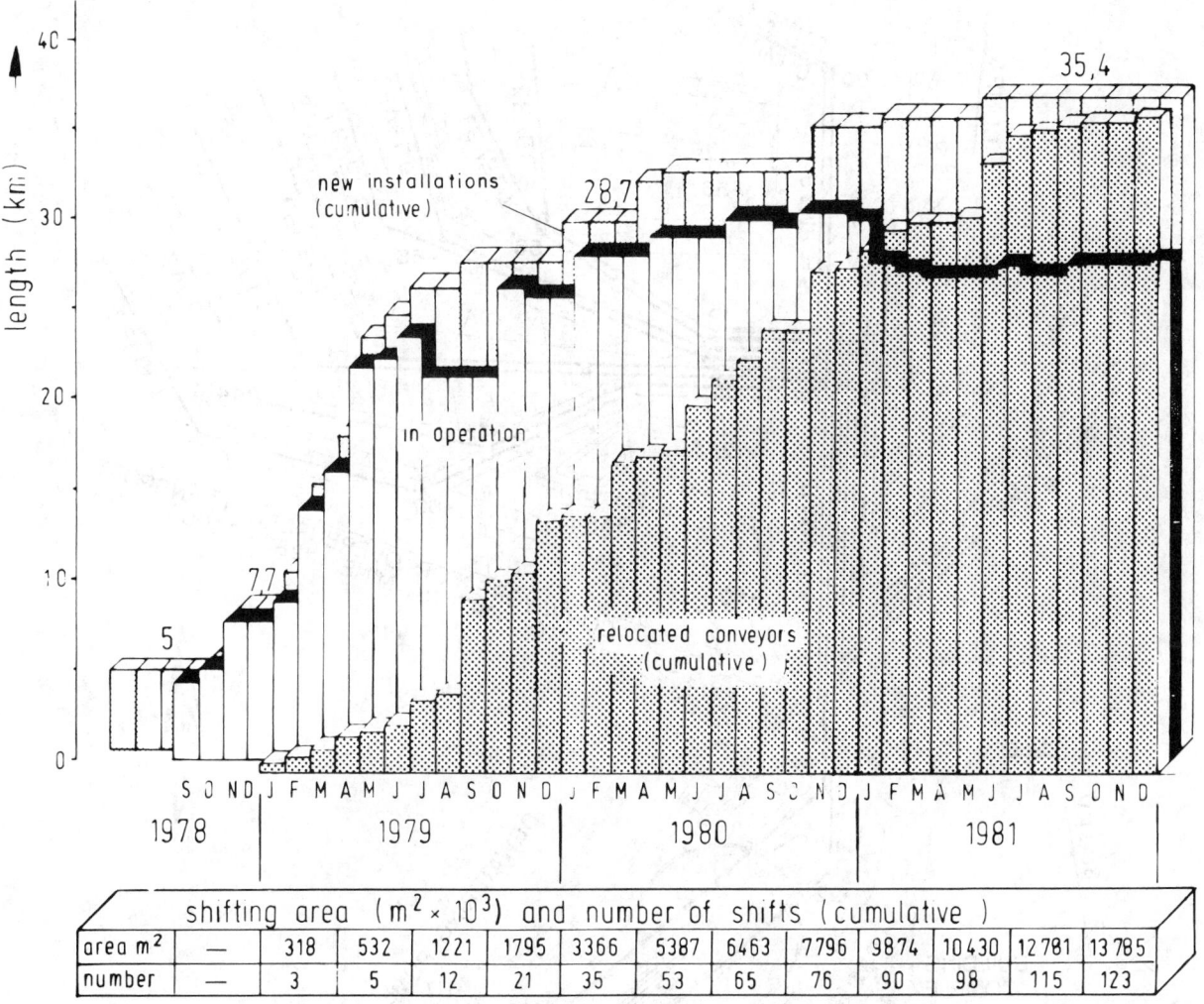

Fig. 21. Development of the belt conveyor system, 1978-1981.

shifting area (m² × 10³) and number of shifts (cumulative)													
area m²	—	318	532	1221	1795	3366	5387	6463	7796	9874	10 430	12 781	13 785
number	—	3	5	12	21	35	53	65	76	90	98	115	123

ter, radio center, and belt control station. Central recording, identification, and evaluation of downtime and performance data from the mining and conveying systems is carried out by a process computer. Intersystem communication is an important precondition for the control and monitoring of operations and subsequent feedback to mine planning. Further development of computerized decision aids for mine operations control at Hambach will be an important task in the years to come. These developments include modified processes of operational control which, for example, will allow a faster retrieval of survey data than was previously possible.

Due to the increase in volumes transported, a higher probability of damage occurring and the greater risk of consequential damage, more attention is being paid to the continuous local supervision of equipment and conveyors. Automatic conveyor monitoring instrumentation, having widespread use in the Rhenish mining area (for spillage, belt tension, slippage, excess load, excess torque, and positioning of transfer points on shunting heads, etc.) is employed, as well as a mobile staff of monitoring personnel. During the start-up phases, which were affected by temperature extremes and adverse material properties, further stationary monitor-

ing personnel were temporarily required at critical points, such as belt distribution points, separate transfer points, and cleaning facilities. The use of automated control for belt tracking, caking, and damage detection (in particular, longitudinal cracks, still not adequately solved) will mean the development of improved equipment able to meet tough operational requirements.

The equipment at the Hambach mine has proved to be successful in full operation. Technical and organizational developments in the mine's maintenance sector are being carried out with priority given to improving operational safety and economics. In the mechanical sector, current efforts are mainly centered on belt transfer points, pulley linings, equipment traveling gears, vulcanizing techniques, and pollution control.

Modifications at the belt transfer points have improved operating conditions along with a decreased incidence of damage and tracking difficulties. These modifications include use of raised separate impact sections; installation of top chutes with modified longitudinal sealing; change in type of garland-type impact idlers; and insertion of rubber aprons in front of the baffle walls.

Results of the initial experiments with automatic belt-

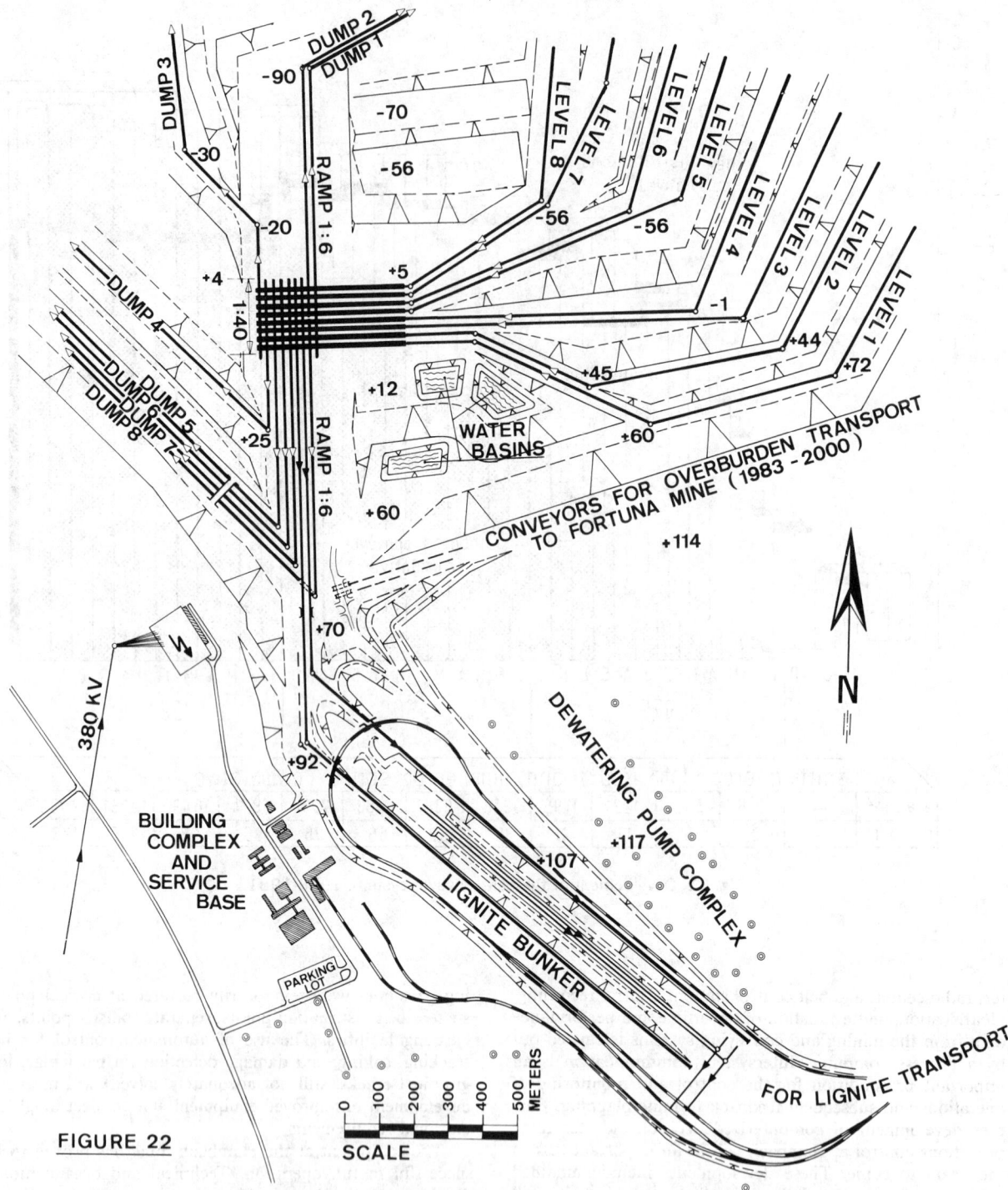

Fig. 22. Distribution point with connection of eight excavating and eight dumping units.

track controls have proven promising. The wear of the pulley linings, mainly due to material return and belt contamination, will be reduced through improved belt tracking and by the use of scrapers and more stable linings. In order to prolong service life, structural improvements on the traveling gear and driving sprocket have begun. Plans to improve belt main-

tenance include a more rigorous defect detection schedule. In the field of cold repairs, further developments are being pursued with the object of developing quicker and more durable bonding techniques. In current developments of efficient noise barriers and cladding for the drive units of equipment and conveyors, attention is focused not only on the solution of various difficult structural tasks, but also in the design such that these new facilities do not present new problems associated with the operational safety and maintenance of machine parts.

The electromechanical facilities at Hambach did not show any special weak points in full performance operations. The conveyors are controlled by microprocessors that operate for each conveyor route. Aside from a few initial difficulties with software and the training of maintenance personnel, the new systems proved advantageous over conventional solutions with reference to troubleshooting, periods of disruption, maintenance, and adjustments to changing operational conditions. This was also true of the connection of conveyor routes to the peripheral equipment of the process computer in the operational control center via serial data interfaces.

The assembly of an experienced workforce was one of the chief preconditions of a successful operations start-up at the Hambach mine. Nearly half the present staff was obtained through transfer of experienced personnel from other mines. The majority of the staff, some of whom had been trained in other mines for their work at Hambach, live in nearby communities. The rest of the staff was taken from the labor market. In the future, more emphasis will be placed on taking employees from Rheinbraun's training centers and *Jungwerker* (young worker) departments. The training of new staff members and young personnel has made great demands on the experienced personnel, particularly on the supervisory staff. Prospects of promotion in a new mine have greatly enhanced staff involvement in the difficult start-up phase.

Early in 1978, the Hambach mine, with some 170 employees, became an operational unit within Rheinbraun's western group. Further staff developments for the mining, maintenance, and technical administration sectors are shown in Fig. 23. Late in 1981, some 1100 persons were employed at the Hambach mine. To this number must be added 150 employees working for Hambach in the mining group's commercial administration. As more equipment and conveyors are installed and operated, staff size will also increase. In the mid-1990s, when eight large-scale equipment groups and conveyors of about 100 km (62 miles) in length will be operational, the Hambach mine will have a projected workforce numbering nearly 2500.

Hambach's plant and equipment has been extended according to plan. The fourth excavator, the assembly of which began in May 1981, and the first overburden conveyor to the Fortuna mine, became operational in 1983. The first lignite was mined in that year. Lignite bunkers, train loading systems, and the Hambach railroad users have been completed. So has the extension of the stationary belt distribution line into the mine's slewing point and the installation of the second conveyor to Fortuna. The build-up of the first inner dump and the start-up of the fifth excavator will soon be completed. Present progress made in planning and implementing these projects suggests that they can be carried out on schedule.

Capital expenditures on mine development, equipment, and other facilities, including projected inflation, will total some 5 billion DM (Deutsche Mark) by 1990. By the end of 1981, some 1.2 billion DM had been invested in equipment for the Hambach mine. Including development costs of about 750 million DM incurred in the same period, the Hambach mining project, having cost some 2 billion DM so far, has become Rheinbraun's single most important capital spending item.

SUMMARY

Development of the Hambach open pit lignite mine began in the western sector of the 85-sq km (32.8-sq mile) large

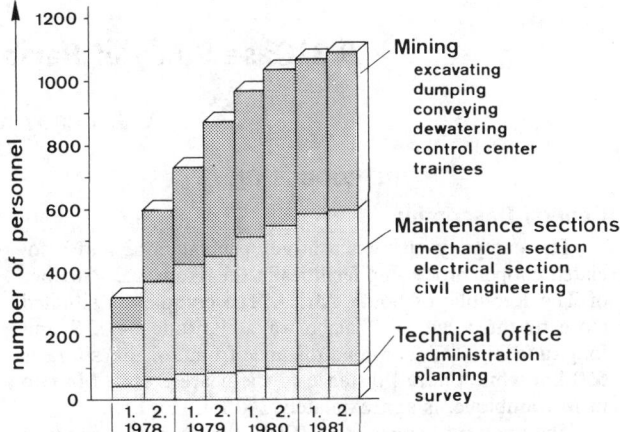

Fig. 23. Personnel development, 1978-1981.

mining area (the so called *Erft Block*) that has lignite reserves of 2.5 Gt (2.8 billion st) and an overburden-to-lignite ratio of 6.2:1. Starting in September 1978, three large-scale equipment groups with rated capacities of 240 000 m³/d (313,908 yd³/d) each were successively put into operation. In the first year of the development phase, numerous equipment disruptions occurred due to the wet, cohesive overburden being excavated and difficulties in maintaining stable pit slopes. Full operations were started in August 1979. By the end of 1981, overburden volume totalling some 0.347 km³ (454 million yd³) had been moved. Valuable operational experience was gained in the following areas: planning and performance of extensive assembly and reassembly jobs, solving difficult problems in geomechanics, noise pollution abatement, and personnel training. On the whole, development work at the Hambach mine is being carried out according to plan.

REFERENCES

Henning, D., 1977, "Applications Engineering and Start-up of 200,000-cu-m Equipment and 3-m-wide Conveyors in the Fortuna Open Pit Mine," *Braunkohle*, Vol. 29, Nos. 1 and 2, pp. 7-14, in German.

Henning, D., 1980a, "Problems in Rock Mechanics in the Development of the Hambach Open Pit Mine," *Tunnel-und Stollenbau*, pp. 10-22, in German.

Henning, D. 1980b, "Stability Determinations of Dump Slopes," *Braunkohle*, Vol. 32, No. 6, pp. 161-168, in German.

Henning, D., and H. Schenck, 1977, "Organization and Monitoring of Operations at the Fortuna Open Pit Mine," *Braunkohle*, Vol. 29, No. 9, pp. 363-369, in German.

Krug, M., 1980, "Applied Planning Methods in the Development of the Hambach Open Pit Mine, 1978/1979," *Braunkohle*, Vol. 32, No. 3, pp. 71-81, in German.

Leuschner, H.J., 1972, "Planning Criteria for the Development of the Hambach Open Pit Lignite Mine," *Braunkohle*, Vol. 24, No. 2, pp. 41-50, in German.

Leuschner, H.J., 1976, "The Hambach Open Pit Lignite Mine—A Synthesis of Raw Material Extraction and Landscape Architecture," *Braunkohle*, Vol. 28, pp. 111-123, in German.

Loeper, K., 1982, "Use of Spreaders and Material Planning on the Sophienhoehe Dump," *Braunkohle*, Vol. 34, in German.

Stahl, H. 1976, "Preliminary Mine Surveying Work for the Hambach Open Pit Mine," *Braunkohle*, Vol. 28, No. 8, pp. 289-297, in German.

Thiede, H.J., 1981, "Planning and Development of the Hambach Open Pit Mine," *Erzmetall 1981*, pp. 140-146, in German.

9.3 Case Study of Palabora Mining Company Ltd.

A. J. LEROY AND J. W. LILL

INTRODUCTION

General Description

The company operates a large open pit mine and associated copper processing facilities in the Transvaal Province of The Republic of South Africa. The complex is situated close to the town of Phalaborwa at latitude 24°00′S and longitude 31°07′E. The road distance from Johannesburg is 550 km while the rail distance to the nearest port, Maputo in Mozambique, is approximately 340 km (Fig. 1.).

The open pit copper mine (Fig. 2) is geared to produce 307 kt/d of material on a six-day week basis to feed a continuous milling and flotation operation at a rate of 80 kt/d of ore. Other works include a magnetic separation plant for the removal of magnetite, a heavy minerals plant with its associated chemical treatment sections, a conventional coal-fired smelter with its attendant sulfuric acid and gas scrubbing plants, an electrolytic refining tankhouse with a capacity of 142 kt/a, and a modern continuous copper rod casting plant which serves the needs of South Africa's copper wire and cable industries. A vermiculite open pit mining operation and beneficiation plant form part of the integrated complex which is manned by some 3900 employees. High quality refined copper is marketed as cathode and continuous cast rod. Byproducts from the copper operation are uranium calcine, zirconium compounds, reground magnetite for use in coal preparation plants, sulfuric acid, refined nickel sulfate, and refinery anode slimes. The latter contain gold, silver, platinum, and palladium. The vermiculite operation produces concentrates in six commercial size fractions.

DISCOVERY OF THE COPPER DEPOSIT

Archaeological Investigations

It has been shown that the ancients recovered copper from the ore body on Loolekop Hill over 1000 years ago. They dug and transported oxide copper ore, predominantly as malachite, to syenite hills close by. The copper was smelted in primitive but effective clay furnaces.

Modern Exploration

A prospector, Carl Mauch, was the first to record the copper occurrence while carrying out reconnaissance mapping of a portion of the Transvaal Province and of Zimbabwe during the period 1868-1871. More detailed exploration was not made until 1912 when Hans Merensky initiated an investigation aimed primarily at the occurrences of vermiculite and apatite which were found at Loolekop and in the immediate vicinity. Merensky continued to feature prominently in developments in the Phalaborwa area until 1951.

Phosphate mining was commenced in 1932 in a valley below Loolekop Hill but the operation failed after the extraction of 4500 t. In August 1939, the Transvaal Ore Co. Ltd. was formed for the purpose of exploiting the known vermiculite occurrence. This company was acquired from the Hans Merensky Trustee Co. by Palabora Mining Co. Ltd. and became a subsidiary of the latter on May 10, 1963.

Merensky continued his investigations to prove the economic viability of recovering phosphates and during the latter 1940s and early 1950s made extended efforts to dispose of his claims. Ultimately during 1951, the Industrial Devel-

opment Corp. of South Africa, at the request of the government, financed a new company, The Phosphate Development Corp. Ltd. (FOSKOR).

FOSKOR was created with the objective of exploiting the Phalaborwa phosphates in order to produce a concentrate which would satisfy the fertilizer requirements of South Africa. This objective has been successfully achieved. During 1952, the Geological Unit of the Energy Board discovered the presence of uranothorianite in the carbonatite on Loolekop. During the ensuing few years the Dept. of Mines and the Geological Unit conducted a prospecting program which included some underground development and surface diamond drilling. It was established that the radioactive minerals were of no economic significance but that copper mineralization could render the deposit valuable.

Local mining companies showed only mild interest in this low-grade copper deposit where the values were 1% or less. The techniques of exploiting large tonnages of low-grade copper ore at shallow depths were very different from the deep level underground operations at which South African engineers were adept. Elsewhere in the world, notably in the Americas, mining companies had acquired wide experience in operating low-grade open pit copper mines with their related financial and technical problems. At that time some argued that copper was already overproduced; seemingly to confirm this, the London Metal Exchange fixings plunged from over R800 per long ton to R320 between 1955 and 1958. It was against this background that two prominent overseas mining groups, the Rio Tinto-Zinc Corp. of London (RTZ) and the Newmont Mining Corp. of New York, became interested in a closer study of the Loolekop copper occurrence. Palabora Mining Co. Ltd. (PMC) was registered in August 1956 to carry out this task.

GEOLOGY

Geology of the Palabora Igneous Complex

The Palabora igneous complex (Fig. 3) located in the Archean Shield of northeastern Transvaal is unique among

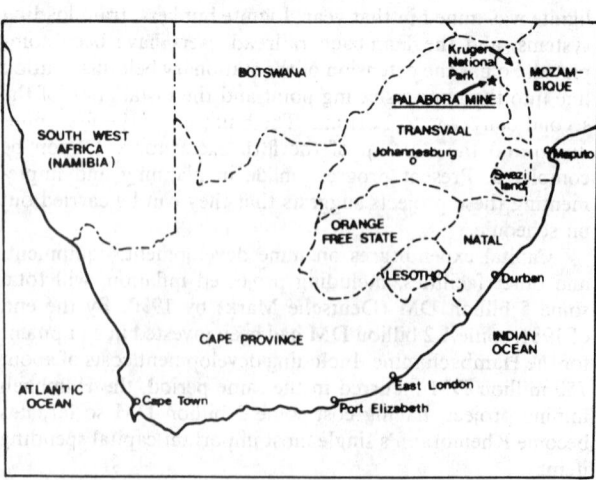

Fig. 1. Location of Palabora copper mine in southern Africa.

Fig. 2. Aerial view of open pit.

the many described African alkaline complexes in that its carbonatite member is the site of an economic deposit of copper ore. Magnetite, uranothorianite, baddeleyite, gold, silver, and nickel sulfate are subsidiary products of the copper mining venture, while the ultramafic rocks of the complex are host to economic deposits of apatite and vermiculite.

The complex resulted from an alkaline intrusive cycle which emplaced, in successive stages, pyroxenite, syenite, and ultrabasic pegmatoids. The age of the complex is generally assumed to be greater than 2060 million years. Pyroxenite intruded first in a north-south elongated, kidney-shaped stock which covers an area of approximately 1655 ha, i.e., roughly 6.4 km N-S by 2.6 km E-W. A corona of feldspathic pyroxenite, formed by interaction with the Archean gneiss country rock, is peripherally distributed. Syenite plugs were forcibly injected into the gneiss surrounding the main pyroxenite mass and this phase was followed by an extended period of non-violent and partly metasomatic activity which formed irregular, vertically disposed pegmatoids at three centers in the pyroxenite pipe and caused some fenitization of the Archean gneiss in contact with the pyroxenite. During this period the foskorite and banded carbonatite in the centrally located pegmatoid body were emplaced.

Subsequent fracturing of the consolidated infilling of this latter subsidiary pipe and renewed igneous activity led to the intrusion of a dikelike body of transgressive carbonatite at the intersection of two prominent fracture zones and a stockwork of transgressive carbonatite veinlets crosscutting the older rocks along preferred trends. Intensive post-carbonatite fracturing along preexisting zones of rupture provided channels for copper-bearing mineralizing solutions to permeate the carbonatite-foskorite pipe infilling with veinlets of copper sulfide and other subsidiary minerals.

The ultrabasic pegmatoid bodies consist of two main rock types with constituent minerals in varying proportions, i.e., an olivine-phlogopite-diopside rock and a diopside-phlogopite-apatite rock. In places the constituent minerals of these rocks attain pegmatoid dimensions. Hydration of the phlogopite in the weathered zone of the pegmatoid bodies has formed commercial deposits of vermiculite. Apatite is an important constituent of pyroxenite, foskorite, and some of the pegmatoid rocks, attaining economic concentrations over large areas.

The complex was subsequently invaded by a number of northeast trending dolerite dikes of Karoo age.

Structure and Lithology of the Loolekop Copper Ore Body

The ore body is an elliptically shaped vertical pipe intrusive into the central core of the Palabora igneous complex (Figs. 4 and 5).

Unlike the central core, the Loolekop pipe is elongated in an east-west direction and is 1.44 km long by 0.80 km wide. It is a composite intrusion with the following age sequence: foskorite, banded carbonatite, and transgressive carbonatite. The pipe has a concentric interbanded structure.

Foskorite is composed of olivine, magnetite, apatite, and phlogopite. Magnetite makes up on average about 30% of the mass, while apatite with a small fluorine content constitutes about 15%. The phlogopite is normally present in subordinate amounts. Apart from carbonatite veining, the foskorite contains numerous irregularly shaped patches of calcite. The constituent minerals are very variable in proportion, the magnetite content, for instance, rises to well over 50% in places. In other areas, considerable masses are composed almost entirely of olivine.

Banded carbonatite is essentially composed of magnetite-rich sövite with minor amounts of apatite, dolomite, chondrodite, olivine, phlogopite, and biotite. The magnetite comprises some 20% of the rock. This carbonatite is characterized by a crude banding caused by alignment of the magnetite and accessory minerals in rudimentary layers concordant with the concentric banding of the overall lithological structure.

Transgressive carbonatite is mineralogically very similar to the banded variety but lacks the crude mineral banding. Its diagnostic feature is the crosscutting, clearly intrusive relationship to the foskorite and banded carbonatite.

The highest concentration of copper sulfide minerals occurs in the transgressive carbonatite. In descending order of abundance, the primary copper sulfide minerals are chalcopyrite, bornite, cubanite, valleriite, and chalcocite.

Titanium is present in solid solution with the magnetite and also in the form of titanium minerals. The magnetite which is associated with the transgressive carbonatite has an average titania content of less than 0.5% whereas the titania content of magnetite in foskorite may exceed 5.0%. Weathering of the copper sulfide minerals was generally limited in depth to a few meters. This, coupled with the fact that high grade ore outcropped on Loolekop, greatly reduced the preproduction mining tonnage required before start-up.

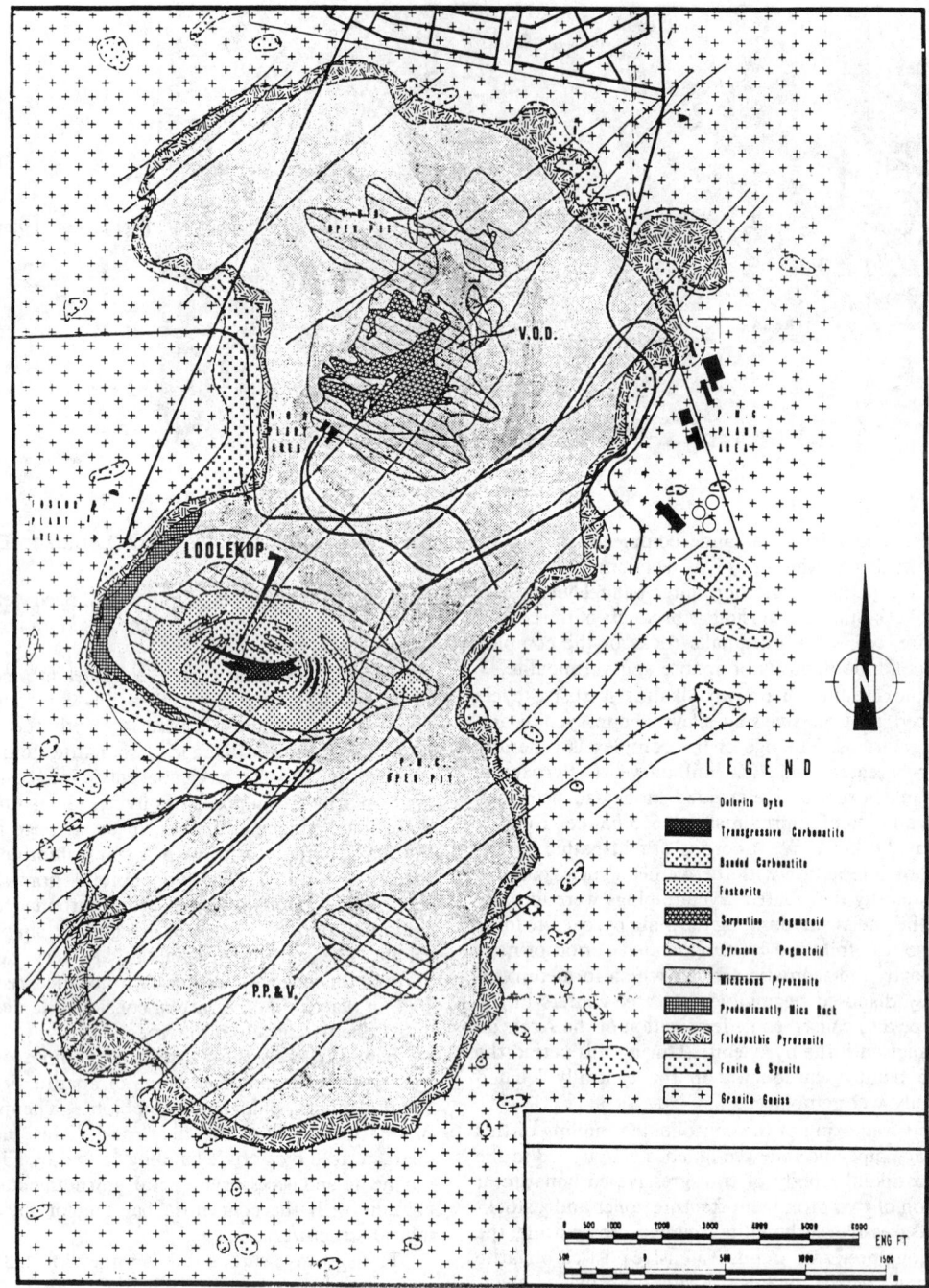

Fig. 3. Plan of geology of the Palabora Igneous Complex.

GEOLOGICAL EXPLORATION AND METALLURGICAL EVALUATION OF THE ORE BODY

General

The initial investigations commenced in June 1957 and were completed in September 1962. These investigations included surface diamond drilling, underground diamond drilling and bulk sampling, laboratory-scale metallurgical testing, and the operation of a pilot plant. These investigations proved the existence of a large reserve of copper ore which could be processed by conventional crushing, grinding, and flotation techniques to provide a concentrate suitable for smelting.

Surface Diamond Drilling

Initially 108 inclined surface diamond drill holes, with a total length of 373 km, were drilled on sections at intervals of 76.2 m on the long axis of the ore body. These holes gave coverage to a vertical depth of approximately 366 m below the crest of Loolekop Hill or roughly 305 m below general ground surface.

A further three deep holes, which intersected the ore body at a vertical depth of 914 m and indicated that the ore body had a remarkable vertical continuity in respect of lithology, mineralogy, and mineral grades, were drilled in late 1962.

This remarkable vertical continuity was confirmed by later diamond-drilling campaigns. In 1970, a further 20 inclined holes with a combined length of 17.3 km were completed. These holes provided detailed ore body coverage to a vertical depth of 610 m below surface. During 1976, six additional holes intersected the ore body at a vertical depth of 1.2 km below surface.

Early Bulk Sampling and Metallurgical Investigation

Numerous laboratory-scale metallurgical tests carried out on core drill samples, from various locations and depths in the ore body, indicated a relatively high recoverability of copper sulfide minerals.

An underground sampling campaign became necessary to obtain bulk samples for treatment in a pilot plant in order to confirm the laboratory tests and suggested metallurgical processes. The campaign served two other essential purposes. First, it made it possible to check the accuracy of underground borehole samples against the corresponding bulk sample grades and, second, the validity of the grade projections from the surface boreholes on which the ore reserve computations were based could be verified. Mineralogical

and grade distribution studies showed that a complete horizontal section through the ore body at any elevation was representative of the ore body as a whole, because of the strong vertical continuity.

The elevation of the bulk sampling level was 122 m below the crest of Loolekop which protruded some 60 m above the average surface level. A shaft, 76 m deep, was sunk at the base of the hill and from this 665 m of access drives and 1.8 km of bulk sampling crosscuts were driven. Of the 7.1 km of horizontal diamond drilling carried out, 1868 m were in advance of bulk sampling crosscuts and the remainder were beyond and between the crosscuts.

Pilot-plant operation confirmed the results of laboratory metallurgical testing while the grade comparisons noted previously proved to be most satisfactory.

In May 1962, Palabora Mining Co. Ltd. obtained the services of two firms of international repute, Bechtel and Western Knapp Engineering, to perform an independent analysis of the work carried out to date. The Joint Venture report, received in January 1963, concluded that the investigations carried out by the company had been done in accordance with accepted and sound procedures and that the pilot-plant results could be reasonably reproduced in full-scale operation.

The Joint Venture companies were subsequently appointed to complete the detailed engineering design and to undertake the construction of the copper processing plant

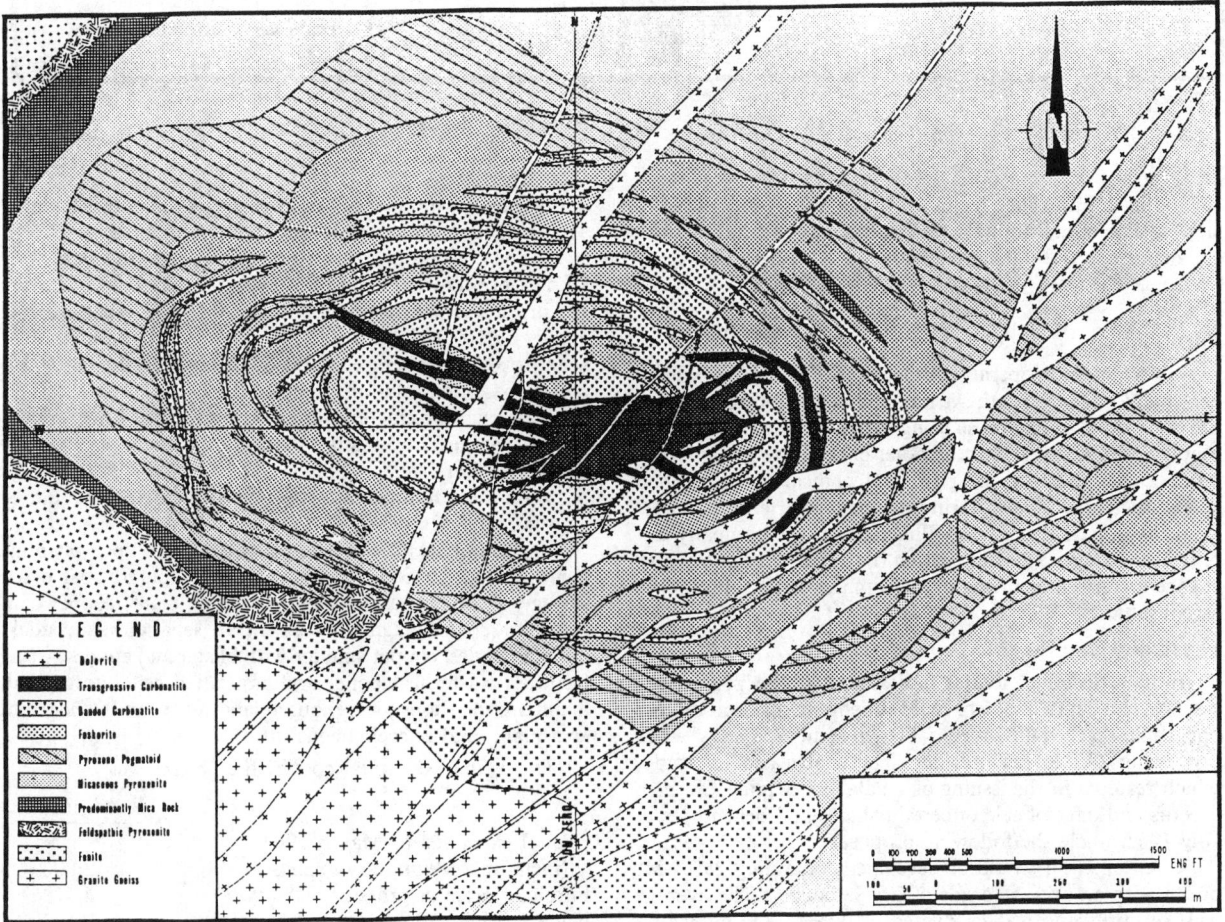

Fig. 4. Lithology of the Loolekop deposit.

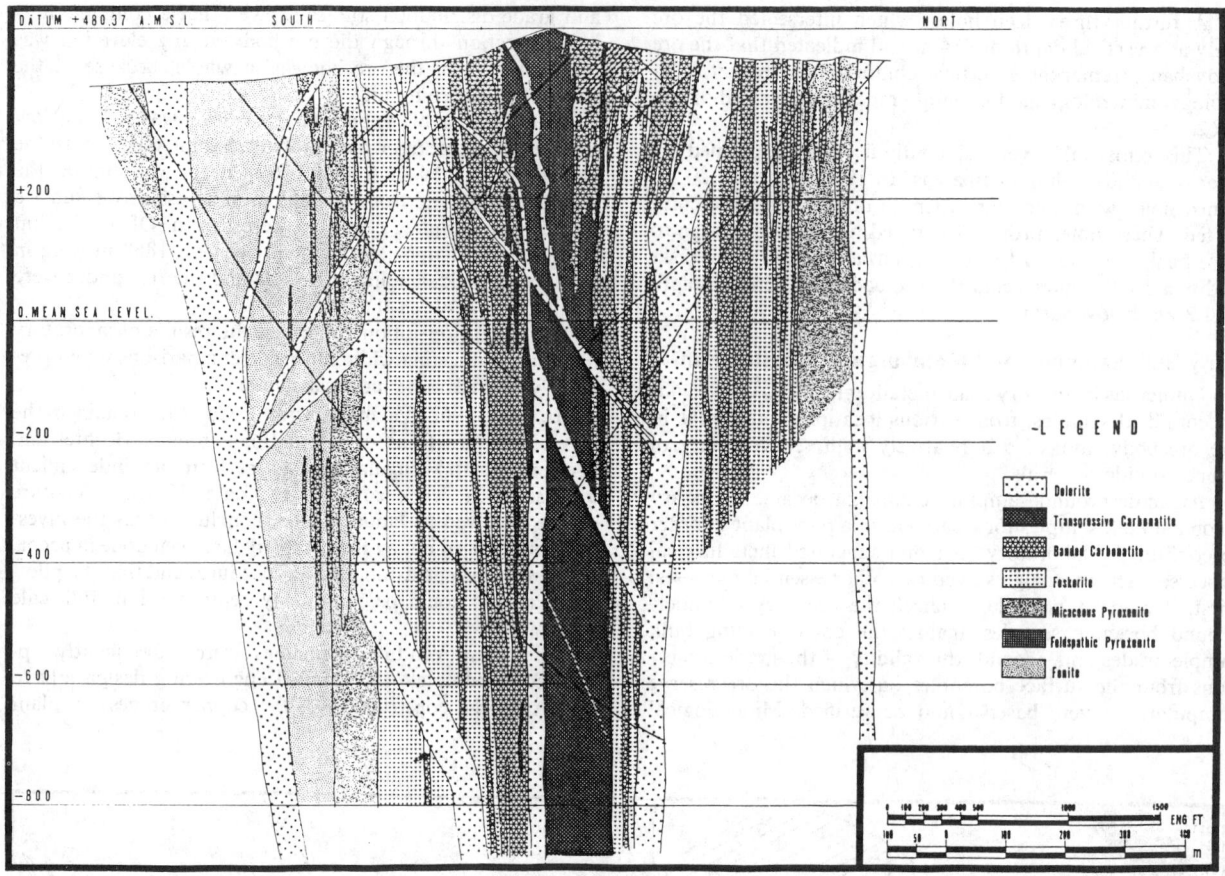

Fig. 5. Section through Loolekop ore deposit.

(including all auxiliary facilities) under the overall supervision of Palabora Mining's staff and senior engineers from the major sponsoring shareholders.

FEASIBILITY STUDIES

Introduction

Given that any operation mounted at Palabora would be extremely large by South African standards and in a copper ore body quite unlike those being exploited elsewhere, it was highly desirable to reassure all other prospective major shareholders that most of the possible technical problems had been anticipated. Thus, during the geological and metallurgical investigation period at Phalaborwa, the company not only used expertise existing within the parent groups but also engaged many eminent engineers from the various disciplines as consultants.

Feasibility Studies

Initial feasibility studies indicated that the proven ore reserves to a depth of 366 m could support an open-pit mining operation that would have a life of some 25 years when processing 27.2 kt per mill day. The final feasibility study, which resulted in the issuing of a Palabora Mining Co. prospectus and offer of sale of certain shares and debentures in July 1963, made the following proposals:

• An open pit would be developed containing 285.8 Mt of ore with an average grade of 0.68% copper. The average waste to ore ratio was calculated at 0.89:1. The pit would have a final depth of 305 m below general ground level, while

its surface measurements would be approximately 1.5 km by 915 m. Overall final pit slopes were approximately 45°.

• The concentrator would be capable of processing 29.9 kt/d of ore. During the initial five-year period of operation the ore grade was estimated to be 0.79% copper using a copper cutoff grade of 0.40%. In the following nominal ten-year period the average grade of the ore would be 0.65% copper based on a cutoff grade of 0.30%.

• Annual production of anode and/or blister copper would average 72.6 kt/a for years one through five and 61.7 kt/a thereafter.

• At these milling rates the life of the project would be approximately 26 years.

The major facilities to be provided were: open-pit mine facilities and services, primary crushing plant, coarse ore storage, secondary and tertiary crushing plant, copper concentrator, coarse magnetite concentrate separation and stockpiling facilities, copper smelter with blister and anode-casting facilities, waste heat boiler and associated generating plant, maintenance shops, warehouse, laboratory and offices, an acid plant, and miscellaneous utilities.

The estimated capital cost of the project was:

	Rand, millions
Capital expended to March 1963 (including preliminary expenses)	3,806
Purchase price of property	4,475
Underwriting commission	0,170
Preproduction mine development	2,104

**Table 1. Comparison of Copper and Byproduct Output from Palabora
Ore in Year of First Production and 1986**

	Year of first production	Annual production		Period years
		First year, t	1986, t	
Copper				
Copper ore treated	1966	10 900 000	29 412 000	21
Anode copper	1966	61 900	107 553	21
Cathode copper	1968	37 200	104 800	19
Copper casting	1968	27 900	67 400	19
Byproducts				
Sulfuric acid (100%)	1966	46 300	147 100	21
Magnetite concentrate	1966	2 300 000	8 450 000	21
Anode slimes, precious-metal content	1968	5.1	17.0	19
U308 Calcine	1971	60.8	185.4	16
Baddeleyite processed to chemicals	1974	3 100	12 000	13
P205 in tailings	1975	365 700	580 000	12

Mine plant and equipment	7,933	
Process plant and equipment	39,722	
Overhead and administration costs during construction	2,150	
European housing and amenities	1,600	
Loan interest during the construction period	3,150	
Cash required for initial production	8,320	
Expenses of raising capital funds, transfer, and stamp duty	0,280	
Grand total	73,710	

In connection with the foregoing it should be noted that:

• The total did not include R2 400 000 to be raised by way of first mortgage loans from building societies on 445 white employee houses in the town of Phalaborwa together with bachelor quarters and other amenities.

• Housing for black employees was to be constructed by the government at no cost to the company in the town of Namakgale and leased by the government to the employees.

• A railway spur line from Hoedspruit to Phalaborwa had been completed by the South African Railways system in late 1962. The company had made suitable arrangements with the South African Electricity Supply Commission (ESCOM) for the necessary power supply to the project. Other arrangements had been made to secure adequate water supplies from a barrage to be erected on the Olifants River.

• The construction period was expected to last for approximately three years.

PROJECT FINANCE

Capital Sources

It was proposed to raise the necessary capital as follows:

	Rand, millions
Ordinary share capital including share premium (28 220 000 shares)	29,128
Sale of 6.5% mortgage debentures (126 250 units of R100 each)	12,625
Loan from Kreditanstalt fuer Wiederaufbau (108 million Deutsche Marks)	19,250
Equipment Loans	4,207
Overdraft facilities to be provided by Barclays Bank for working capital	8,500
Total	73,710

The authorized share capital of the company was 28.5 million shares of R1.00 each. Of the total number, 24.72 million were held by the sponsoring shareholders, predominantly RTZ and Newmont. A further 2.50 million shares would be taken up in equal amounts by The South African Industrial Development Corp. (IDC), S.A. Mutual, and Sanlam. The latter two concerns were mutual life insurers.

One million shares were placed on the open market on the condition that each subscriber was obliged to subscribe simultaneously for four debentures and 100 shares or any multiple thereof.

A further 86 250 debentures would be taken up by the IDC, S.A. Mutual, and Sanlam in equal amounts.

The shares placed on the market and those taken up by the three concerns noted commanded a premium of 25¢ a share. Kept in reserve, and at the Board of Directors disposal, were a further 280 000 shares.

If required, the Board was authorized to raise additional capital by the issue of three million 7% cumulative redeemable preference shares of R1.00 each at par. In the event that yet more capital was required to bring the project into production, the sponsoring shareholders undertook to provide loans bearing interest of not more than 7.5%.

The large loan from the German Corporation, Kreditanstalt Fuer Wiederaufbau was secured, inter alia, by a notarial bond by pothecating the company's interest in certain moveable assets and for the pledge of the entire share capital of the Transvaal Ore Co. (the vermiculite operation) authorized at approximately four million Rand. An agreement was signed in May 1963 between the company and Norddeutsche Affinerie in which the company agreed to sell to the latter 572 kt of blister copper. In the following month, a further agreement was entered into between the company, Kreditanstalt Fuer Wiederaufbau and Norddeutsche Affinerie in terms of which the company assigned its rights under the contract with Norddeutsche Affinerie as security for payment of amounts due to Kreditanstalt Fuer Wiederaufbau in terms of the provisions of the agreement with the latter.

CONSTRUCTION, START-UP, AND SUBSEQUENT HISTORY

Construction and Start-up

Financing and construction activities proceeded smoothly to enable the commencement of mining activities in 1964 and the commissioning of the milling facilities at the end of 1965. Anode copper was first produced in early 1966 which was the first year of full production on the project.

Subsequent History

The design capacity of the mill was attained after only a few months of operation and, due to metallurgical improvements, the milling rate had reached 43.5 kt/d by mid-1968. A mill expansion was completed in 1969, involving the addition of a sixth milling section, designed to raise milling capacity to 49 kt/d. In the ensuing years this capacity was invariably exceeded by margins of well over 1 Mt/a.

In 1977 a further mill expansion was completed which involved the installation of a third primary crusher and two autogenous grinding mills. With this expansion, planned milling capacity was increased to 75 kt/d. After initial design problems in the autogenous mills had been corrected, the design capacity was again exceeded and current capacity stands at 80 kt/d.

The foregoing history of mill expansions could not have taken place without corresponding far-reaching changes in the open-pit operations. These changes are discussed in the following section.

OPEN PIT DESIGN AND MINE PLANNING

Introduction

The original open-pit design envisaged a pit life of 26 years, ore production of 30 kt/d and an overall mining rate of 75 kt/d for the first years of operation. A series of mine expansions associated with the increased milling rates described and/or the need to extend the life of the operation has led to a new termination date in the year 2000, repre-

senting a total pit life span of 35 years. The peak mining rate for the mine plan presently in effect occurred in 1980 and was 350 kt/d. A uniform ore requirement of 80 kt/d of ore is achieved through the life of the plan. At present the cutoff grade is 0.15% copper and the average mill head grade is 0.50% copper.

In addition to copper minerals in the ore, iron in the form of magnetite and phosphate in the form of apatite are present in significant quantities. The amount of titania in the magnetite has an effect on its salability and consequently the ore is subdivided into two streams with ore containing magnetite with a titania content of less than 1.0% reporting to one stream and the balance to another. This separation is initiated at the mining stage and is maintained through the mill. Fortunately, this division coincides with the low- and high-phosphate-bearing ores, the economic portion being associated with the high titania stream.

Palabora's mining lease does not permit the company to recover phosphorus minerals. Under various agreements, which in certain cases allows the mining of claims belonging to FOSKOR, phosphate-bearing tailings from the concentrator and rock from the open pit are delivered to FOSKOR's facilities situated close to the open pit.

Because of the pipelike formation of the ore body, successive pit expansions have had to carry the burden of ever increasing quantities of peripheral waste that necessitate high initial waste: ore stripping ratios. These ratios and, therefore, the peak mining rates will diminish in the latter portion of a mining plan's life, as mining activity enters the ore body fully with depth.

In 1978, based on outlook for copper, it appeared that the then pit design marked the economic limit of open-pit mining. A further expansion (a mean of 116 m radially) was made possible under a new agreement with FOSKOR whereby FOSKOR agreed to make certain financial contributions in exchange for greatly increased deliveries of high-phosphate-bearing material resultant from the expansion.

Mining of this expanded pit (pit 5, Fig. 6), which is still the design in effect, commenced in 1980. The current plan

Fig. 6. Northeastern corner of the pit showing the 1980 extension cut with plant in background.

Table 2. Key Parameters Associated With the Various Major Pit Plans

Pit No.	Designation	Date Commenced	Daily mining rate, kt/d		Pit tonnage, millions of t			Ultimate pit depth, m	Completion year
			Ore	Peak	Ore	Waste	Total		
1	Feasibility Pit	1964	30 000	75 000	286	254	540	366	1991
2	BD-2AM.36	1970	52 500	205 000	471	478	949	467	1990
3	BD 13D	1974	74 000	300 000	655	861	1516	668	1994
4	BD 13E	1978	75 000	290 000	612	780	1392	668	1992
5	100F	1980	80 000	350 000	837	1240	2077	836	2000

requires an average mining rate in excess of 325 kt/d over the period 1980-1987 with sharply reducing mining rates in subsequent years once the major portion of the waste mining has been completed.

It is interesting to note that the original "feasibility pit" of 1963 contained a total of 535 Mt of material. The most recent plan contains over 2.1 Gt or just slightly less than four times the original.

Salient information on the various pit designs are given in Table 2. Graphical representations of achieved and projected pit production levels and limits are as shown in Figs. 7, 8, and 9.

8.2 Mine Planning

The high production rates demanded from a pit of relatively small surface area has made mine planning of a high order absolutely essential to the success of the mining program.

The original ore reserves used for the early mine planning were of a polygonal type and were extremely ponderous to use. In the late 1960s, a cubic block matrix model, eventually to comprise some 650 000 blocks, was adopted to represent the ore body. Each of the ore reserve blocks (15 × 15 m × bench height) may be assigned as many as 31 fields of information. Included in the present model is such information as rock density, copper mineral inventory, copper and accessory mineral grades, blasting restrictions applicable to the particular rock type, and details of the property rights within the mining lease. Coordinate information is also assigned to each block.

The block model is stored on hard disk and is ideally suited to access by computer. Sophisticated evaluation programs are used to determine optimum pit shape and the economics of successively larger incremental expansions to the base pit design. Having established the overall pit design, another suite of specially designed computer programs, which have been upgraded to include interactive graphics features, are used to access the ore reserve data for planning and evaluation of various mining options. This system allows for the development of sequential mine plans as well as providing for updating of mine plans, the rapid evaluation of different cutoff grades, rates of ore exposure, and changes to pit geometry for geotechnical reasons. For example, it is not uncommon for the cutoff grade of the ore to vary for given differing economic conditions and availability of downstream plant throughput capacity. Similarly long-term plans can be quickly adjusted to cater for changes in mining rate to smooth equipment requirements.

Typically, sequential long-term mine plans are developed at quarterly intervals through to end of mine life. These plans then form the framework within which more detailed short-term mine plans and shovel, drill, and blast schedules are produced by pit operations engineers.

As part of the short-term planning process appropriately detailed six-month, monthly, two weekly, and daily plans are developed by the pit planning engineer. A key to the success of short-term planning is the close and interactive relationship which exists between pit planning, production, drill and blast, geology, and survey personnel.

8.3 Pit Slope Design

After a number of years of testing presplit and perimeter blasting techniques, a successful means of controlled blasting was developed in 1978. The method was amenable to a rapid rate of final wall exposure where it is not unusual to require the development of over 2 km of final wall in a single month. It also minimized the amount of blast shatter on final faces and reduced back break on bench crests. As a direct result of the use of this controlled blasting technique, it became possible to introduce double benching, effectively further increasing the available catchment area of the safety berms and thereby making steep wall angles a practical possibility.

Subsequently it was decided to undertake a detailed geotechnical study to determine optimum safe slope angles for the pit, incorporating double benching (Fig. 10). Structural mapping of pit walls was carried out routinely since 1974; however, with the new emphasis on slope stability these efforts were intensified. By 1980 over 32 km of bench face had been mapped in detail and more than 3000 rock samples had been tested for mechanical strength. The results of the study indicated that the competence of the rock mass was such that the possibility of deep-seated failure was remote. Rather, the factor controlling safe slope angles lay in the provision for an adequate catchment area to provide for the localized wedge failures which would occur over the life of the pit. To optimize the pit design variable, slope angles ranging from 40° to 58° were incorporated in the design, dependent on the nature of the wedges present in the wall and the wall orientation. The use of artificial support measures for areas of local weakness and the provision for adequate dewatering of final walls were included as part of the design philosophy.

All of these developments have been incorporated into the latest pit expansion which was commenced in 1980.

The design of the current final pit shell and projection of the various slope angles used in its design are shown in Figs. 11 and 12, respectively.

MINING OPERATION AND ITS EVOLUTION

Original Equipment

The geometry of the deposit and the need to deliver material to multiple destinations combined to make Palabora a conventional hard rock truck and shovel operation. Because Palabora was distant from the sources of supply for its major items of equipment, which were mainly from North America, it was logical to select items which had been well proven in the field.

1966 through 2000

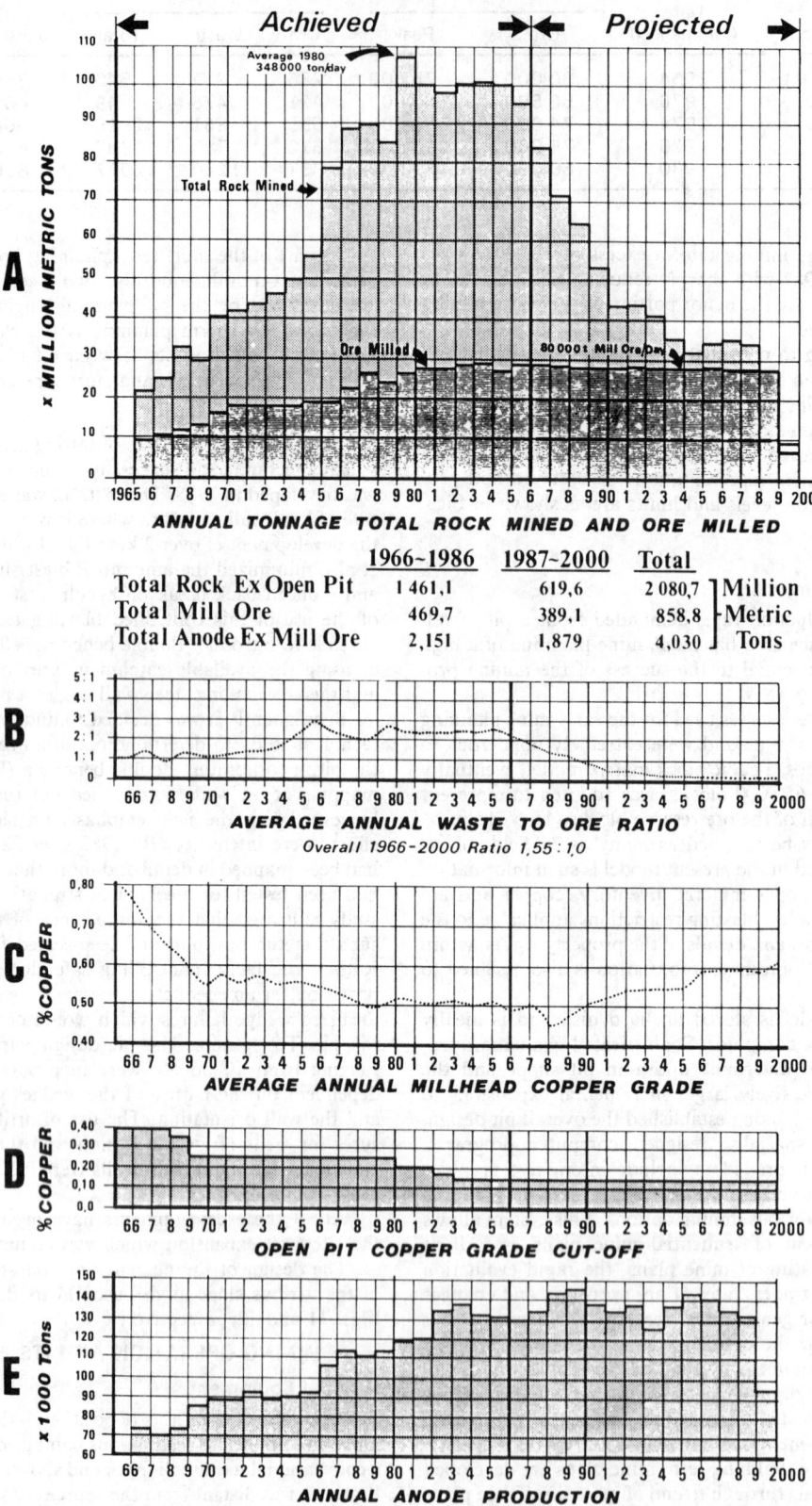

	1966~1986	1987~2000	Total	
Total Rock Ex Open Pit	1 461,1	619,6	2 080,7	Million
Total Mill Ore	469,7	389,1	858,8	Metric
Total Anode Ex Mill Ore	2,151	1,879	4,030	Tons

Fig. 7. Achieved and projected mining production levels and limits for life of open pit, showing (A) annual tonnage of total rock mined and ore milled; (B) average annual waste:ore ratio (1.50:1.00 overall, 1966–1999); (C) average annual mill-head copper grade in percent; (D) open-pit copper cutoff grade in percent; and (E) annual tonnage of anode production from mill ore.

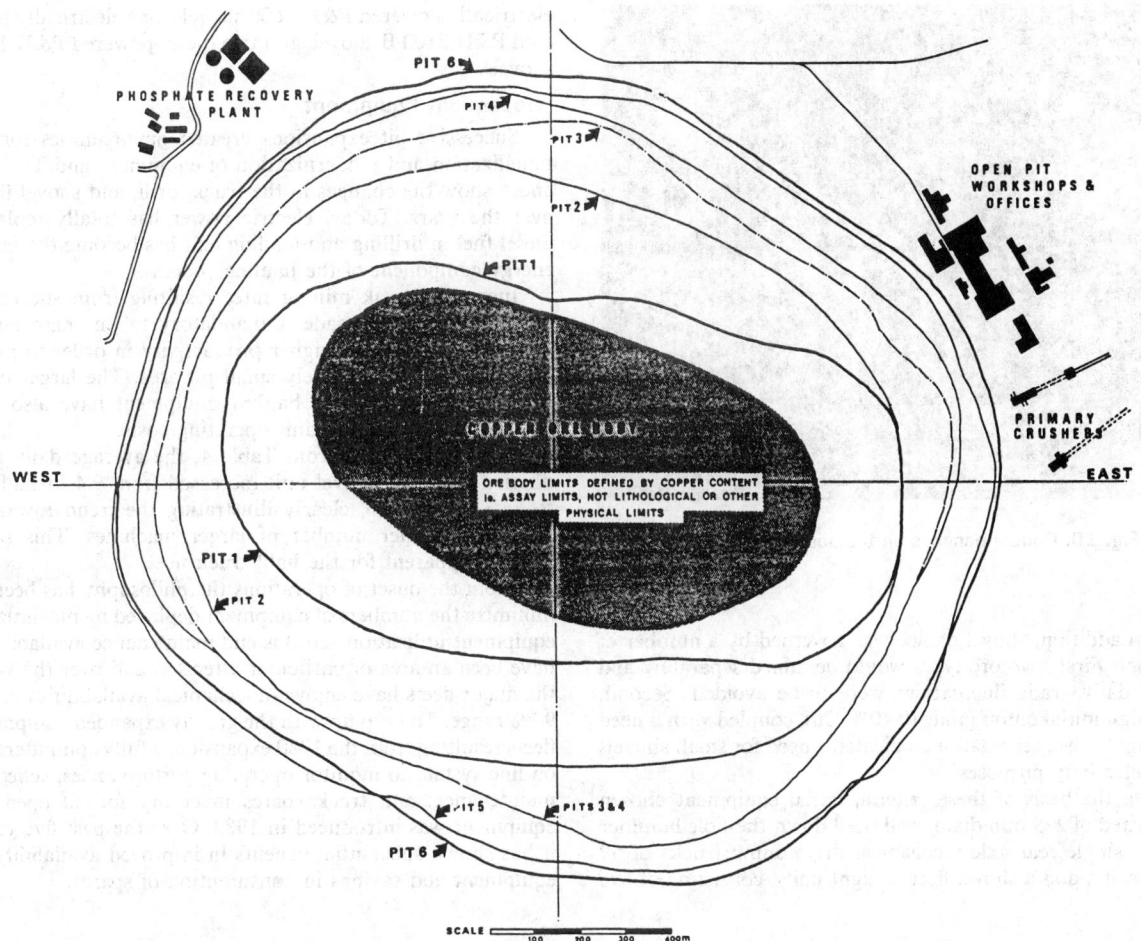

Fig. 8. Surface pit limits of major mining plans implemented since 1966. Long axis: pit 1, 1.54 km; pit 5, 1.79 km. Short axis: pit 1, 0.76 km. Ore body limits defined by copper content (assay limits, not lithological or other physical limits).

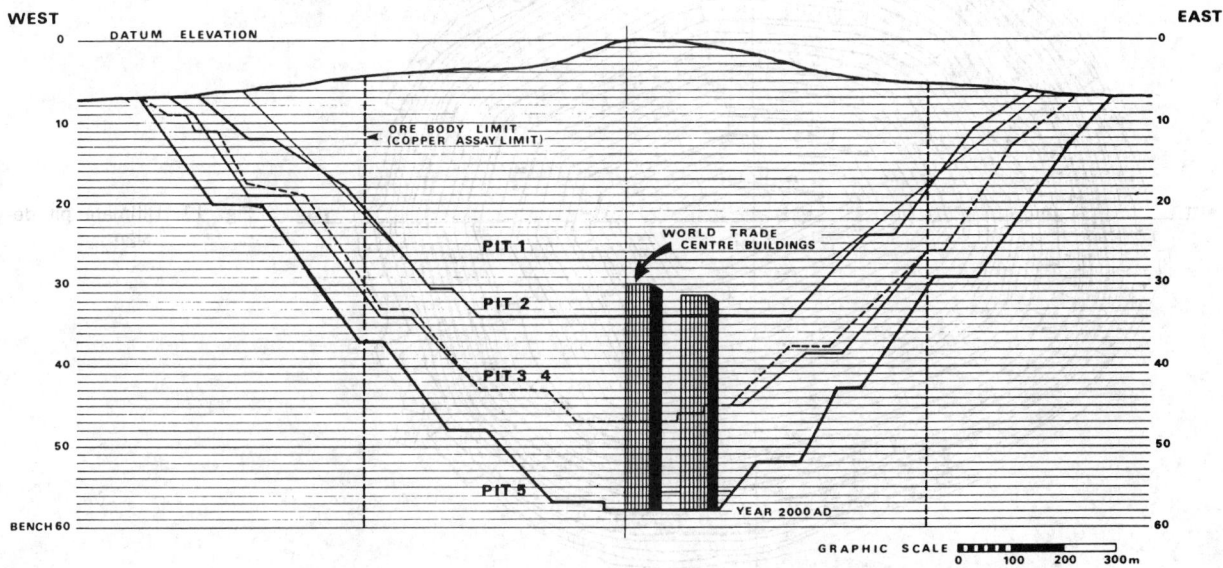

Fig. 9. East-west cross-sectional view of major pit mining plans with superimposed World Trade Center buildings. Depth below datum elevation: pit 1, 366 m (27 benches); pit 5, 805 m (56 benches).

Fig. 10. Double benches on the south side of the pit.

In addition, shovel choice was governed by a number of factors. First, two ore types would be mined separately and wide daily grade fluctuations were to be avoided. Second, the high initial cutoff grade (0.40% Cu), coupled with a need for rigid waste segregation, indicated a need for small shovels for selectivity purposes.

On the basis of these criteria, initial equipment chosen consisted of 229 mm diam, well tried down-the-hole hammer drills, single rear axle mechanical drive dump trucks of 59 t capacity, and a shovel fleet of eight units, consisting of five electrically powered P&H 1600 shovels, one electrically powered P&H 2100 B shovel, and two diesel-powered P&H 1400 shovels.

Subsequent Equipment

Successive pit expansions created opportunities for rationalization and modernization of equipment and Tables 3 and 4 show the changes in the truck, drill, and shovel fleets over the years. Today, electric power has totally replaced diesel fuel in drilling and loading and has become the major energy component of the hauling process.

Increasing peak mining rates resulting from successive pit expansions has made it mandatory to purchase larger equipment capable of higher productivity in order to avoid congestion in the relatively small pit area. The larger units of drilling, loading, and hauling equipment have also had the effect of containing unit operating costs.

As may be noted from Table 4, the average daily production level per shovel unit increased from 9.4 kt in 1966 to 34.3 kt in 1986, clearly illustrating the trend toward a relatively smaller number of larger machines. This same trend is apparent for the haul truck fleet.

From the onset of operations the philosophy has been to minimize the numbers of equipment deployed by maximizing equipment utilization. To this end maintenance availabilities have been an area of particular attention and over the years the major fleets have enjoyed mechanical availabilities in the 90% range. To keep up with the greatly expanded equipment fleets resulting from the 1980 expansion, a fully computerized on-line system to monitor operating performances, schedule maintenance, and track spares inventory for all open pit equipment was introduced in 1982. Over the past five years it has shown substantial benefits in improved availability of equipment and savings in consumption of spares.

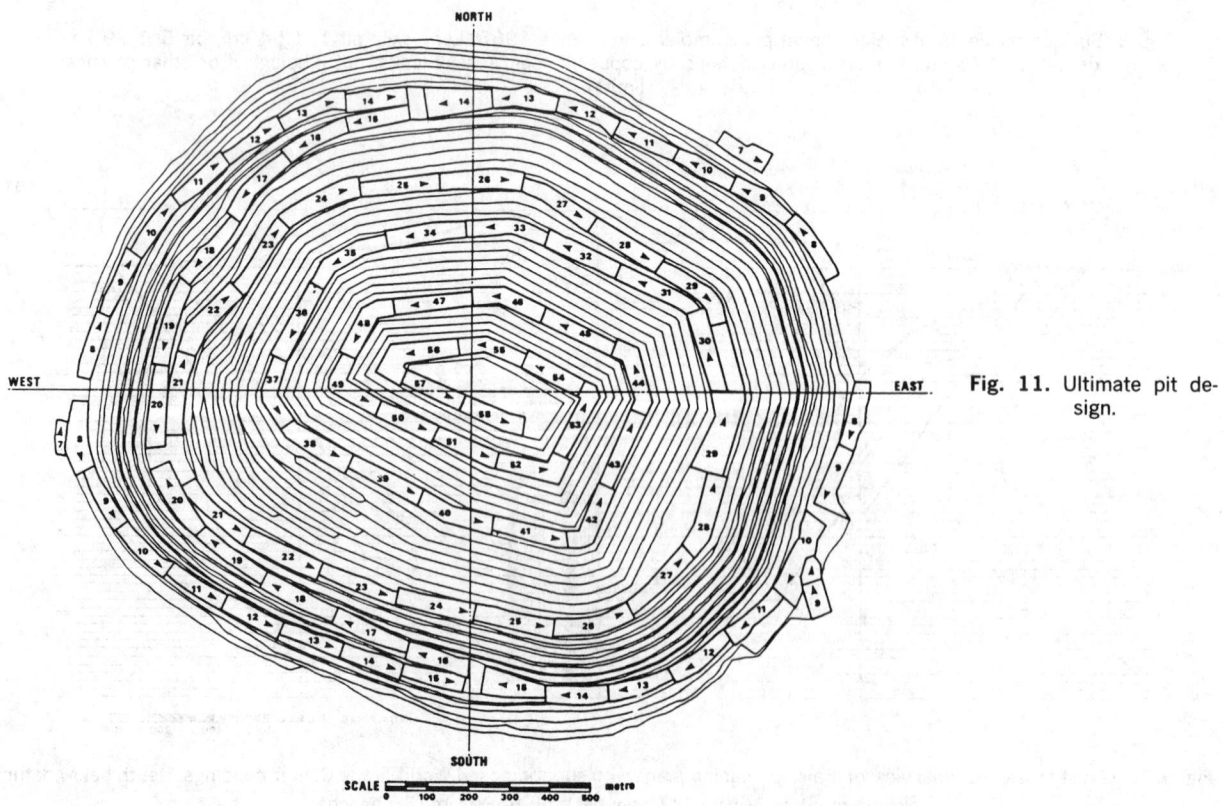

Fig. 11. Ultimate pit design.

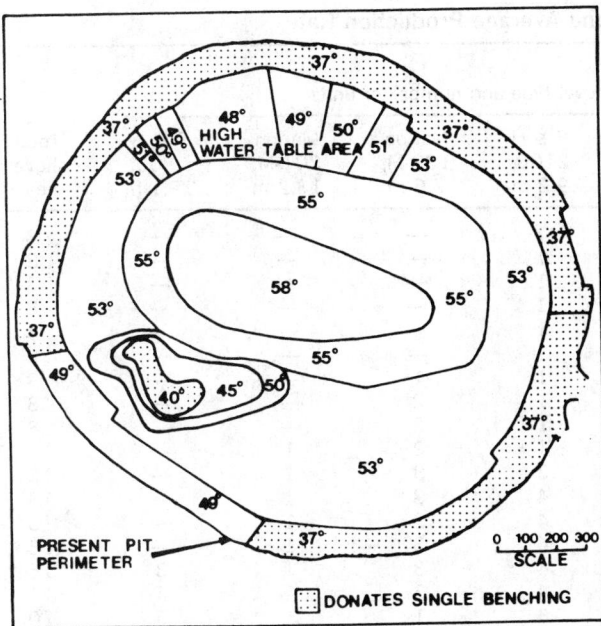

Fig. 12. A plan view showing the inter-ramp slope angles used in the Palabora pit slope design.

Drilling and Blasting

Drilling is accomplished by a fleet of eight Gardner Denver model 120 rotary blasthole drills. These units are equipped with 28-m masts, capable of single pass drilling on Palabora's 15.2-m bench height. Currently a 311-mm hole size is used for all blastholes. Four distinct bit types are used for drilling the various formations. These are "sulfide," "fenite," "dike," and "sulfide dike contact."

Drill patterns are varied depending on the particular area of the pit and material being blasted. Typical drill patterns range from 6.7 × 7.6 m to 7.0 × 8.1 m. Hole patterns are staggered and normally 2.4 m of subdrill is used for each hole.

In the central core of transgressive carbonatite, grade values showed little fluctuation and five row shots are allowed. Elsewhere in the ore body, grade variations are much more pronounced and three row shots are the norm with the objective of minimizing dilution. In pure waste areas no limits are placed on the numbers of rows shot.

Primary blasting is carried out using emulsion explosives made under a license agreement with Ireco. Three grades of emulsion with varying aluminum content can be produced. A typical blasthole will be loaded with 1000 kg of emulsion consisting of 350 kg of aluminized emulsion bottom load and the balance being made up of zero percent aluminum content emulsion. Each hole is stemmed with approximately 5.8 m of drill cuttings. Patterns are shot on the long axis with 40 millisecond delays between rows.

Trolley Assistance

Rock haulage has become the single most expensive process in the entire mine and plant operation. It accounted for only 33% of the total mine and plant operating cost in 1975, but was made more expensive mainly by the energy crisis of the 1970s; by 1981 it was responsible for 57% of the operating cost; In the years 1978 to 1980 alone the price of diesel fuel in South Africa increased by over 320%. This, combined with the increase in mining rate as a result of the 1980 pit expansion, was a pressing reason for finding an alternative or less expensive mode of rock haulage. The consumer cost of electrical power in South Africa is among the lowest in the world and it was accepted that its' escalation in real terms would be appreciably less than that of diesel fuel.

Studies undertaken to investigate alternative or complementary means of material handling indicated that the use of electric trolley assistance for the haul truck fleet would offer the fastest and most attractive return to the company. In 1981, after less than two years of development work, Palabora introduced a system of trolley assistance for the

Table 3. Truck and Drill Fleet Changes

Year	Total truck fleet capacity, t	Dart KW, 59 t	Unit Rig M 100, 91 t	Unit Rig MK 36, 154 t	Euclid R 170, 154 t	Unit Rig MK 36, 172 t	Total truck fleet	Quarry Master QM 500	Bucyrus Erie 45R	Bucyrus Erie 60R	Gardner Denver GD 120	Total drill fleet
1966		26	—	—	—	—	26	6	—	1	—	6
1967	1.77	30	—	—	—	—	30	6	—	1	—	7
1968	1.95	30	2	—	—	—	32	6	—	1	—	7
1969	2.04	30	2	—	—	—	32	6	1	1	—	8
1970	2.13	30	4	—	—	—	34	6	1	1	—	8
1971	2.56	28	10	—	—	—	38	3	1	2	—	6
1972	2.83	28	13	—	—	—	41	1	1	3	1	6
1973	2.87	24	16	—	—	—	40	—	1	3	1	5
1974	4.16	23	19	7	—	—	49	—	1	3	2	6
1975	5.72	13	19	21	—	—	53	—	1	3	3	7
1976	7.25	5	19	34	—	—	58	—	1	3	7	11
1977	7.35	4	19	35	—	—	58	—	1	3	7	11
1978	8.01	—	22	39	—	—	61	—	—	3	7	10
1979	9.20	—	25	45	—	—	70	—	—	3	8	11
1980	12.30	—	27	64	—	—	91	—	—	3	9	12
1981	12.46	—	10	75	—	—	85	—	—	—	12	12
1982	12.92	—	10	78	—	—	88	—	—	—	12	12
1983	13.42	—	7	74	9	—	90	—	—	—	—	—
1984	13.42	—	7	74	9	—	90	—	—	—	—	—
1985	12.34	—	7	67	9	—	83	—	—	—	—	—
1986	11.65	—	6	62	11	3	82	—	—	—	—	—

Table 4. Shovel Fleet Changes and Average Production Rate

Year	Average daily production rate, t	Average daily production rate per shovel unit, t	Shovel type and number of units						Total shovel fleet
			P & H 1400, 3.1 m	P & H 1600, 4.6 m	P & H 2100, 9.2 m	Marion 182-M, 7.6 m	Marion 201-M, 13.2 m	P & H 2800, 19.1 m	
1966	75 295	9 412	2	5	1	—	—	—	8
1967	82 325	10 291	2	5	1	—	—	—	8
1968	106 623	13 328	2	5	1	—	—	—	8
1969	96 679	12 085	2	5	1	—	—	—	8
1970	133 436	16 680	2	5	1	—	—	—	8
1971	138 528	17 316	2	5	1	—	—	—	8
1972	146 255	16 251	2	5	1	1	—	—	9
1973	152 734	19 092	—	5	1	2	—	—	8
1974	161 881	20 235	—	5	1	2	—	—	8
1975	183 548	20 394	—	5	1	2	1	—	9
1976	255 763	21 314	—	5	3	3	1	—	12
1977	289 579	22 275	—	5	4	3	1	—	13
1978	292 305	22 485	—	5	4	3	1	—	13
1979	279 020	21 463	—	5	4	3	1	—	13
1980	348 195	23 213	—	5	4	3	1	3	16
1981	326 810	29 710	—	—	4	3	1	4	11
1982	321 325	32 133	—	—	3	1	1	5	10
1983	326 773	33 175	—	—	3	—	—	6	9
1984	327 928	35 298	—	—	3	—	—	6	9
1985	324 370	36 555	—	—	3	—	—	6	9
1986	308 708	34 301	—	—	3	—	—	6	9

haul truck fleet. Initially a total of some 2.7 km of adverse 8% haul ramp was equipped for trolley assist.

Under trolley assist the haul truck collects electric power from the fixed overhead conductor system. The power is supplied directly to the electric motors of the haul truck located in the rear wheel hubs. This eliminates the need for the truck engine to drive the truck alternator to generate its own electricity. Fuel consumption for a 154-t capacity truck hauling up an 8% adverse grade is 24 L/km diesel fuel. That same truck on trolley assist uses approximately 0.5 L diesel fuel per kilometer of travel. In 1982, the first full year of trolley-assisted operation, diesel fuel savings amounted to 17 mL. By 1986 these savings increased to 37 mL (Fig. 13). Energy cost savings from the use of lower cost electrical energy as a result of trolley assistance amounted to R12.6 million in the same year.

As a result of increased vehicle speed while on trolley, significant productivity improvements have been achieved. In 1986 this higher speed and consequently improved truck cycle time equated to a reduction in required truck fleet of seven 154-t capacity units.

Because the truck engine does not work as hard while on trolley, the engine lasts longer. Engine life has over the past six years more than doubled as a direct result of trolley assistance. Wheel motor armature life has also improved as a result of the shorter time spent on grade. On the basis of calculation, it is estimated that armature life is presently 160% greater than would have been the case without trolley assistance. Although an item of initial concern, experience has shown that there is no net reduction in tire life resulting for the use of trolley assist.

Currently, 7.7 km of trolley-assisted ramps (Fig. 14) are in use at the Palabora pit. With the significant energy cost savings realized by the system and the other benefits which more than double the savings, there is every incentive to advance successive planned phases of the sytem as the pit deepens.

Haul Truck Control

With 14 different material classifications and a corresponding complex haul route system, the potential to improve efficiency by computer-aided control of trucks and related production equipment was recognized early and Palabora installed the first automatic computer-aided truck dispatch-

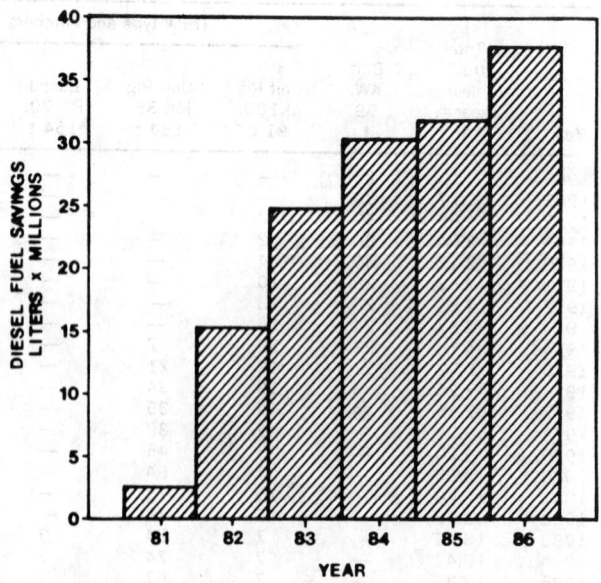

Fig. 13. Diesel savings as a result of trolley assist.

Fig. 14. Hauling using trolley assist.

ing system in 1972. This system, although simple, was effective and achieved an immediate increase in fleet utilization. The system worked well and enjoyed a high operating availability.

Recently, however, technology had developed sufficiently that it became feasible to introduce a more sophisticated truck dispatching system, which integrated all aspects of open pit production. In 1985 the old dispatching system was replaced by a customized Modular Mining dispatch system. With this system each piece of production equipment, shovel, trucks, crushers, is equipped with its' own microprocessor (Fig. 15) which communicates with a central VAX 780 computer by radio. On the basis of information received, the central computer is able to determine all vital parameters such as shovel loading rates, crusher digestion rates, and

hauling rates, etc. By a network of key points along haul routes, the system is able to track haul truck traffic throughout the haul cycle. In addition, equipment operators provide further information through purpose built terminals located conveniently in each operator's cab. The central computer is able to use linear programming techniques to optimize equipment assignments and consequently improve productivity.

Within a few months of commissioning the new system had proven itself to the extent that it was possible to reduce the truck fleet by five units. The productivity improvement was found to be 7% greater than the previous arrangement. In addition, the system offers numerous other advantages, including the ability to evaluate production performance, more accurate production records, and a complete record of activities of all major mining equipment.

In-Pit Crushing and Conveying

Trolley-assisted haulage was chosen in preference to alternative materials handling methods in the early 1980s. However, the possibility of crushing and conveying remained under study and in 1985 a concept involving a simple crusher and conveying system with the crusher positioned on the final wall of the pit emerged as a cost-effective solution.

The in-pit crushing system was looked at in a context of long-term ore haulage requirements. The study showed that the in-pit conveying system would be economically and practically more attractive if it could be extended periodically to follow pit depth advances so as to eventually cater for a changeover to underground mining.

The current in-pit crushing and conveying concept achieves these goals in three distinct phases, each of which will be required to stand successively on its own economic merits.

Phase 1 of the system is currently under construction and consists of a 60 × 89-in. (152 × 226-cm) gyratory crusher conventionally installed below a flat area at bench 28 (Fig. 16), some 300 vertical m below the pit rim. Ore from the production benches will be trucked to the crusher via the trolley-assisted haul ramps. The crushed ore will be fed to a 1.1-km long 1800-mm wide conveyor belt up an inclined (15.5°) tunnel. After exiting the tunnel at surface,

Fig. 15. Dispatch panel mounted in a haul truck cab.

Fig. 16. In-pit crusher excavation on bench 28.

the material will be conveyed by a series of shorter belts and deposited on the existing coarse ore stockpiles. Two of the three existing surface crushers will remain functional and on standby as backup to the new in-pit unit and also to process low-grade surface stockpiles as required by the mine plan.

Phase 1 will be operational early in 1988 and could remain in use for the duration of the pit life but is currently designed to be in operation for 6.5 years during which time an extension conveyor tunnel will be developed from the phase 1 crusher chamber behind the pit wall to the phase 2 station at bench 44, a further 250 m vertical lower down.

The phase 3 crusher will in turn be located 60 m below the conventional pit bottom at bench 58 and will be housed in an underground chamber behind the pit walls. It will connect to the phase 1 transfer station by a dogleg conveyor tunnel that bypasses the lower leg of the phase 2 conveyor, which becomes redundant. The total vertical lift to surface would be around 780 m through the three conveyor sections.

The operating cost of the phase 1 crusher conveyor system will be significantly below the present trolley-assisted haulage cost. In 1988 terms, the cost of conveying the material by a conveyor from bench 28 to the surface is estimated at R0.19t as compared to the cost of hauling which would amount to

R0.82t for the same lift. The phase 1 conveyor will allow the truck fleet to be reduced by over 14 units.

A diagram of the three phases of the crushing and conveying system are shown in Fig. 18.

Support Equipment

As is usual in this type of operation, there is a very comprehensive fleet of support equipment at Palabora. The major items are listed in Table 5.

Equipment Performance

Before quoting performance figures it is necessary to understand the basis for calculating the statistics. The central formulae in use are: mechanical availability (MA) percent is defined as:

$$\frac{(\text{used time} + \text{operational delay time}) \times 100}{(\text{Total time} - \text{standby})}$$

or alternatively

$$\frac{\text{time available for operation} - \text{standby} \times 100}{(\text{Total time} - \text{standby})}$$

where used time is the time equipment is operated, operational delay time is time that equipment is not used due to being delayed for operational reasons, total time is total calendar time when mining operations are being carried out, and standby is time when equipment is not scheduled to operate but mining operations are in progress.

Mining Organization and Manpower

Of the present employee strength of approximately 3900, one-quarter is of European descent. In the commissioning and the early operational years of Palabora, an exodus of mining personnel from the Zambian Copper Belt provided Palabora with a significant portion of its total staff complement. They were skilled in almost every facet of copper mining and metallurgical operations. The balance of the skilled white labor force was obtained from the long-established mining industry of South Africa. Heavy reliance was placed on overseas specialist assistance, especially from North America, and although this has now diminished, Palabora still utilizes overseas expertise in the form of visiting consultants.

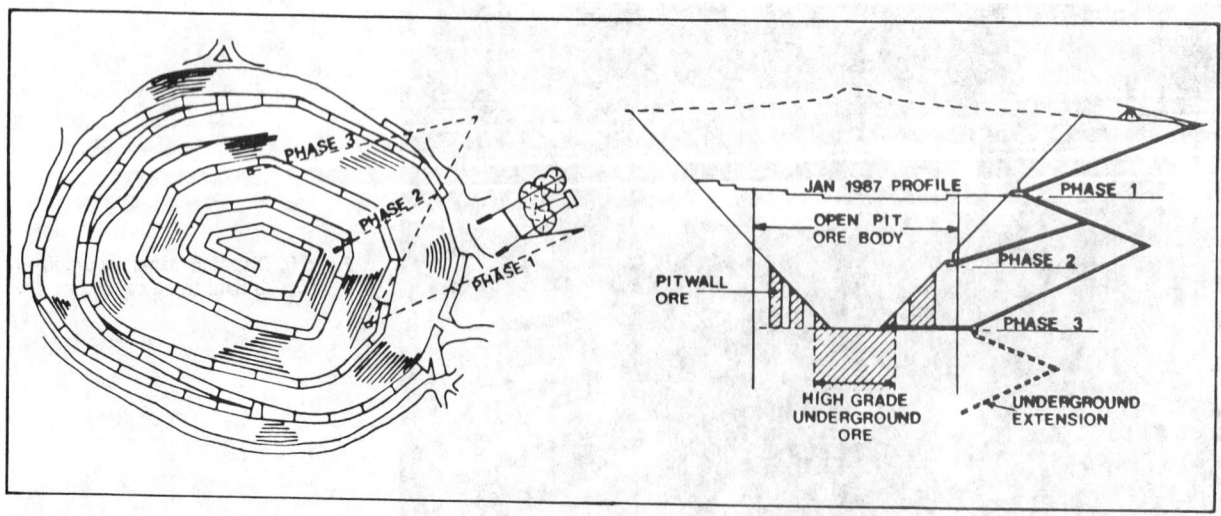

Fig. 17. Diagrammatic plan and section view of the open pit showing the three-phase in-pit crusher development.

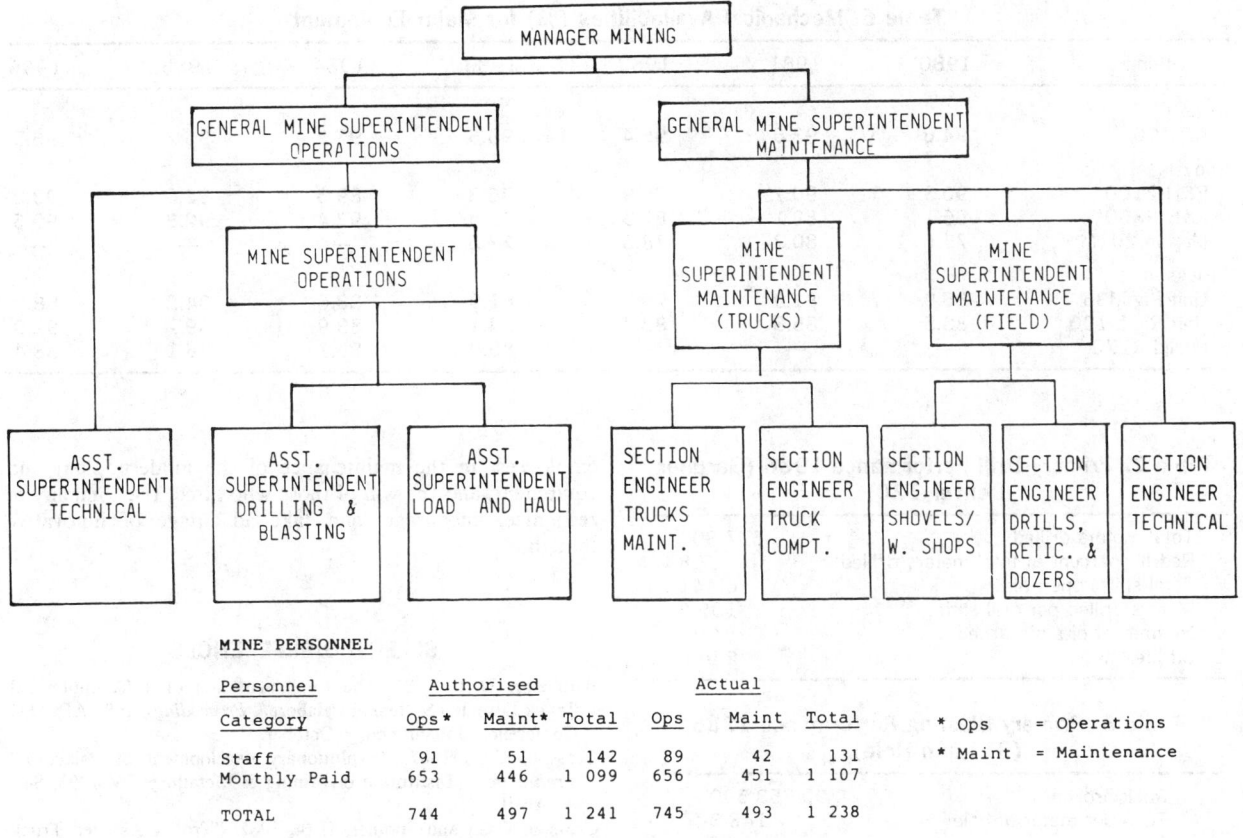

MINE PERSONNEL

Personnel Category	Authorised			Actual			
	Ops*	Maint*	Total	Ops	Maint	Total	
Staff	91	51	142	89	42	131	* Ops = Operations
Monthly Paid	653	446	1 099	656	451	1 107	* Maint = Maintenance
TOTAL	744	497	1 241	745	493	1 238	

Fig. 18. Organization of the mining department.

Palabora managed to establish a relatively efficient and stable labor force from the very onset of operations. Locally recruited people have adapted to the modern industrial environment through sustained training in the classroom and "on-the-job." Turnover has been minimized by the adoption of a policy, unique at the time in South Africa, of accommodating all black employees with their families as close to the mine as possible. Supporting social services, such as

Table 5. Support Equipment Complement (1986)

Item	No. of Units
Rubber tire dozers, Caterpillar 824C	19
Track dozers, Caterpillar D9L	8
Graders, Caterpillar 16G	9
Front-End Loaders, Caterpillar 992C	2
Bobcat 631-D	3
Drop Ball Cranes, P&H 670 TC	2
Slurry Pump Trucks	5
Water Tank Trucks, Converted M100 trucks	6
Diesel Tank Trucks, Converted M100 trucks	3
Personnel Carriers, 40-seat	2

Five voice radio channels are in use as follows:

1. Operations—all supervisory staff and selected equipment.
2. Maintenance and drill and blast sections.
3. Other divisions on the mine, e.g., concentrator and smelter.
4. Truck dispatch to shovels.
5. Truck dispatch to graders and rubber-tired dozers.

schooling, recreational facilities, hospitals, and a technical training school, have been provided progressively over the years. As a result, a remarkably stable black labor force has been achieved, in which levels of technical and operational proficiency have steadily increased. Judged by any standards, an annual turnover maintained at less than 6% of those employed must be considered exemplary.

The company boasts of a fully integrated salary structure and equal conditions of employment for all employees. Employment and promotional opportunities are provided on a basis of merit and experience only.

Fig. 19 shows the organizational structure of the mining division and 1986 authorized manning levels. In 1986 an open pit productivity of 257.9 t per man shift was achieved. This figure takes into account shifts operated by nonpit personnel such as surveyors, geologists, and other members of the mine engineering department.

CONCLUSION

The past 20 years have seen Palabora evolve to take advantage of new technology and the economies of greater throughput to maintain its position as one of the world's low-cost copper producers. A measure of this is the fact that despite the ruling depressed price levels for copper in 1986, Palabora managed to achieve its highest level of profitability.

Much obviously depends on the future price of copper. However, Palabora is in a comparatively strong position whatever situation arises. Benefits from the many cost-savings measures introduced over the years will continue to accrue to the company but Palabora's real strength in the future continues to lie in the training and organization of its

Table 6. Mechanical Availabilities (%) for Major Equipment

Item:	1980	1981	1982	1983	1984	1985	1986
Drills							
GD 120	94.6	93.6	95.4	95.6	97.2	96.2	96.5
Shovels							
P&H 2100	90.3	90.7	86.9	86.3	89.3	92.0	93.2
P&H 2800	86.1	88.1	87.5	89.5	92.4	92.3	90.6
Marion 201M	75.6	80.3	78.5	77.2	—	—	—
Trucks							
Unit Rig M36	86.9	81.9	79.8	81.9	83.6	84.2	88.3
Unit Rig M100	88.5	85.8	83.4	84.1	89.9	89.7	92.0
Euclid R170	—	—	—	85.9	85.1	86.1	88.4

Table 7. Primary Drill Performance 1986 (Gardner Denver 120)

Total meters drilled	677 809
Redrill (percent of total meters drilled)	7.8
Total shifts operated	6 441
Meters drilled per drill shift	105.2
Number of bits discarded	256
Bit life, m	2 948

Table 8. Primary Blasting Performance 1986 (311-mm Hole)

Rock broken, t	90 753 500
Tons per meter of hole	143.3
Powder factor t/kg	3.04

Table 9. Loading Performance, 1986

Rock loaded, t	24 896 181	70 494 492
Operating hours	17 077	33 185
Loading rate, t/h	1458	2124

Table 10. Haul Truck Performance, 1986

Rock hauled, t	95 390 673
Load factor, t	144.1
Truck operating hours	494 593.6
Average loads per truck operating hour	1.34
Total km recorded, round trip	7 160 600
Average one-way haul, km	5.40
Average km per truck operating hour, one way	7.22
T-km per truck operating hour	1 043.1

employees, in the maintenance of the modern plant and equipment, and the will of those who direct the company to recognize, encourage, and take advantage of innovative thought.

SELECTED REFERENCES

Batchelor, D.H., 1987, The Implementation of a Computerized Truck Dispatch System at Palabora," *Proceedings,* 1987 APCOM Conference, Johannesburg, October.

Crosson, C.C., 1987, "Evolutionary development of Palabora," *Transactions,* Institution of Mining & Metallurgy, Vol. 93, Sec. A, April.

Crosson, C.C., and Sumner, H.B., 1982, "Trolley-Assisted Truck Haulage," *Engineering and Mining Journal,* June.

Fauquier, G.P., 1982, "The Development of Controlled Perimeter Blasting with Double-Bench Mining at Palabora Open Pit," 88th Annual Northwest Mining Assn. Convention, Spokane, WA, December.

Lill, J.W., "Operating Features and Control Aspects of the Palabora Copper Open Pit," *The Planning and Operation of Open-Pit and Strip Mines,* Symposium Series S7, South African Institute of Mining and Metallurgy, pp. 161-174.

Lill, J.W., Gliddon, J.P., and Wade, G.C., 1988, "Palabora—Changing to meet the challenges of the 80's," *Mining Engineering,* August.

Lombard, A.F., Ward-Able, N.M., and Bruce, R.W., 1964, "The Exploration and Main Geological Features of the Copper Deposit in Carbonatite at Loolekop, Palabora Complex," *The Geology of Some Ore Deposits of South Africa,* Vol. 11, Transactions of Geological Society of South Africa.

Martin, D.C., Steenkamp, N.S.L., and Lill, J.W., "Application of a Statistical Analysis Technique for Design of High Rock Slopes at Palabora Mine, South Africa."

9.4 Morenci/Metcalf

John L. Bolles
Gary A. Loving
James L. Madson

INTRODUCTION

One of the large copper properties in the world today is The Phelps Dodge Morenci, Inc. operation located near the confluence of the San Francisco and Gila rivers in eastern Arizona.

The Morenci operation encompasses two operating open pit mines, two concentrators, extensive dump leaching, two copper precipitation plants, a large solvent extraction/electrowinning facility including three separate solvent extraction circuits, two power plants, shops and maintenance facilities necessary to support those operations, and a townsite where most of the employees live. It is the largest of Phelps Dodge Corp.'s copper mining and processing complexes, normally producing about 50% of the company's new copper output.

Early History

First discovery of mineralization in the district was recorded in 1864. The earliest production of any consequence occurred in 1872 when copper oxide ore of very high grade was mined and processed by direct smelting methods. Some of that early copper production was hauled overland by mule- and oxen-train to Kansas City, Missouri.

By the 1880s, a number of companies were operating in the district. These included the Detroit Copper Mining Co., the Arizona Copper Co., the Longfellow Copper Co., the Shannon Copper Co., and several smaller companies.

Phelps Dodge's entry into the region, and in fact into copper mining, occurred in 1881 when it acquired partial ownership of Detroit Copper Mining Co.

Ownership in the district was gradually consolidated until only the Detroit Copper Mining Co. and the Arizona Copper Co. remained. After 1897 Phelps Dodge assumed full ownership of the Detroit Copper Mining Co. In 1921, Phelps Dodge purchased the Arizona Copper Co. to become sole operator in the district.

With the exception of some surface *glory holing,* all mining prior to 1932 was by underground methods. Both high-grade copper oxide and copper sulfide veins were mined. As mining progressed, exploratory drilling revealed a very large low-grade deposit of disseminated chalcocite, called the Clay ore body, in porphyry host rock.

Initiation of the Morenci Open Pit Copper Mine

Pioneer exploratory work on the Clay ore body by Arizona Copper Co. and later by Phelps Dodge Corp., revealed that the porphyry copper deposit lay beneath 60 to 150 m (200 to 500 ft) of leached capping. Phelps Dodge began a systematic drilling program on the Clay ore body in 1928 to confirm and augment information gained earlier by Arizona Copper and Detroit Copper companies. The drilling, which continued until November 1930, confirmed the existence of more than 181 Mt (200 million st) of material averaging slightly more than 1% copper.

One of the key questions to be resolved was whether to mine the ore by underground block caving or by the open pit method. In considering the choice the following items were studied:

Minimum dilution of the ore, maximum recovery of the ore, low maintenance costs, favorable production costs, production flexibility, uniform ore production and stripping rates, availability of space to locate waste and leach dumps, good physical layout for haulage systems, minimum pre-ore stripping, 209 Mt (230 million st) of ore at 1.06% Cu, and a waste to ore ratio of 1:1.

After a year of investigation, it was determined that the open pit mining method could recover about 10% more copper, reduce final average mining costs by more than 30%, and could ultimately yield from 20 to 25% greater profits.

Development of the mine, however, was only one phase of the large and costly enterprise. The old No. 6 Concentrator and its various units at Morenci were inadequate to process the huge tonnage that the new mine was designed to produce and the old facilities were not advantageously located. Further evaluation indicated that it would not be economically feasible to enlarge and modernize these existing facilities.

A decision was made to construct a new infrastructure—crushing plant, mill, smelter, power plant, and auxiliary facilities necessary for a major mining operation.

The site selected for the crushing plant, mill, smelter, and other units would lie almost halfway between the towns of Morenci and Clifton and would allow the ore to be delivered by favorable haul from the mine over a significant portion of the mine's life. By the time the preliminary study was completed, the United States had entered the Great Depression and development was suspended pending improvement in the economic climate.

In 1937, encouraged by the economic outlook, Phelps Dodge reactivated the Morenci Project. In June, the company made a public offering of $20,285,000 of 15-year convertible 3-½% debentures. They soon were sold and work on the mine and related facilities began. The balance of the total estimated cost of $35 million came from corporate earnings as the work progressed.

The size, quantity, and cost of the equipment and facilities required for a task of such magnitude taxed the imagination. The new mine at Morenci was to become a huge, complex, and thoroughly mechanized industrial operation.

To provide for the great increase in personnel that the new operation required, Phelps Dodge laid out a new townsite. The company built more than 300 housing units that would be rented to employees at nominal rates. The project also included the expansion of recreation and school facilities and the construction of a new hospital with the most modern equipment.

In 1937, the pre-ore stripping of 44 Mt (49 million st) of overburden commenced at the Morenci mine with shovels and trucks developing 15-m (50-ft) high mining benches and railroad access. The trains were added later to handle the movement of large tonnages of material over long distances including ore deliveries to the Morenci primary crusher. The physical features of the topography permitted the installation of the main line haulage tracks outside final pit limits and also permitted the favorable downhill movement for an estimated 85% of the material to be handled.

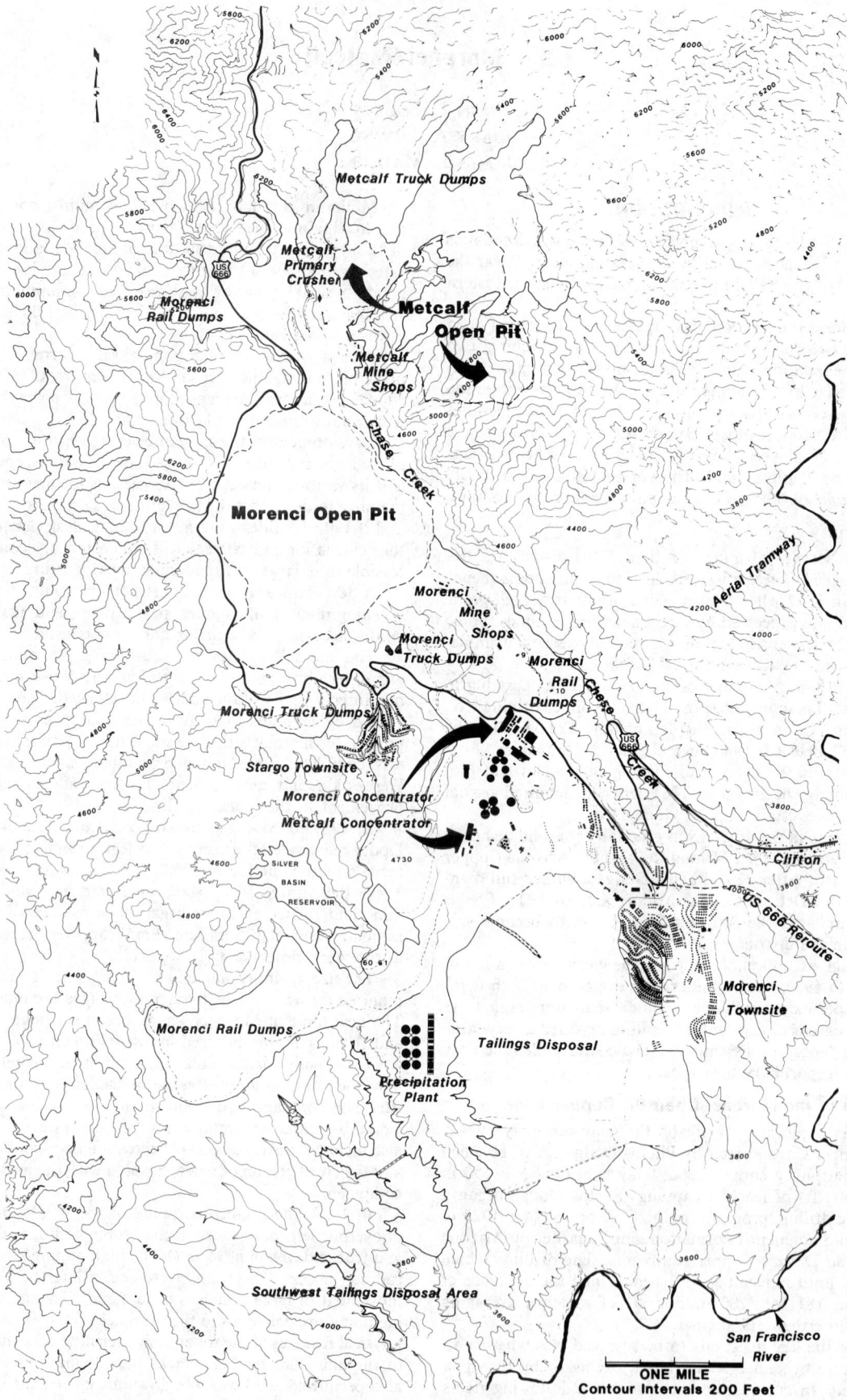

Fig. 1. Map of the Morenci district.

Copper concentration studies were resumed in 1938 following a six-year hiatus to test various types of equipment and to develop a final concentrator flowsheet. Ground was broken for the new concentrator and smelter in 1939. The first copper anode was poured on April 26, 1942.

Wartime Expansion

The Morenci operation was designed to process 22.7 kt (25,000 st) of copper ore per day. However, in 1941 the nation was facing international war, and the federal government urgently requested Phelps Dodge to expand the Morenci Project by 80%, to a capacity of 41 kt (45,000 st) of copper ore per day. Responding to this request, the company designed, constructed, and completed a major expansion project in slightly more than two years.

By the close of World War II, Phelps Dodge had spent approximately $42 million on the Morenci project and the federal government's Defense Plant Corp. had provided an additional $26 million. The facilities provided by the US government were sold to Phelps Dodge after the war ended.

The Postwar Years

Additions and improvements raised the capacity of the Morenci concentrator to its present level 59 kt (65,000 st) of copper ore per day with a corresponding increase in mining rates.

Major improvements in recent years have included construction of the largest solvent extraction/electrowinning complex in the United States to recover copper leached from mine dumps. The mine dumps contain copper-bearing material with grades too low for processing through the concentrator.

Since the initiation of the Morenci mine in 1937, about 2.0 Gt (2.2 billion st) of rock have been moved, of which about 690 Mt (760 million st) were ore averaging 0.86% copper. Approximately 530 Mt (584 million st) of material remain to be mined under current plans. The present Morenci pit covers some 595 ha (1,470 acres), measuring about 2.9 km (1.8 miles) north to south and 2.4 km (1.5 miles) east to west. Mining, which began at 1691 m (5550 ft) above sea level, has continued down to the 1219 m (4000 ft) level. Current mine plans indicate that the final bottom of the pit will be at the 1021 m (3350 ft) elevation.

Metcalf, A Second Open Pit in the Morenci Ore Body

In 1970, development of the nearby Metcalf mine was initiated to replace, insofar as overall company production is concerned, the production lost when the open pit and underground mining operations ceased at Bisbee, Arizona, in 1974 and 1975. The Metcalf development, operated as part of the Morenci Branch, also required construction of a new concentrator built southwest of the Morenci concentrator.

The Metcalf copper deposit is an upfaulted portion of the Morenci ore body and is located a short distance north and on the opposite side of Chase Creek from the Morenci open pit mine. Development of the Metcalf area began in 1870 when Robert Metcalf struck a rich copper oxide vein upstream from Clifton. Metcalf returned to the area about two years later and located the Longfellow claim and mine. The mine was acquired by the Lezinsky brothers, then by the Arizona Copper Co., and finally by Phelps Dodge in 1921 as part of the acquisition of the Clay ore body.

Exploration drilling at Metcalf began in the 1960s and in 1969 the company announced plans to develop the Metcalf mine. By December, 1974, the Metcalf project was substantially completed—at a cost of almost $200 million.

Due to the extremely rugged terrain at the mine site, it was necessary to locate the concentrator for Metcalf ore adjacent to the Morenci concentrating and smelting complex. However, the primary crusher was located near the mine. After the Metcalf ore is crushed using a 1.52 m (60 in.) gyratory crusher at the mine, the ore was hauled approximately 8 km (5 miles) by rail to the concentrator. The facility was designed to handle 27 kt (30,000 st) of ore per day, although it has a proven capacity of 41 kt (45,000 st) of ore per day.

Pre-ore stripping resulted in the removal of 68.6 Mt (75.6 million st) of overburden. To maintain ore development, a stripping ratio of 2.5:1 was required. However, that ratio was not achieved for several years following startup due to economic constraints. Since production began in 1974, 39 Mt (43 million st) of copper ore averaging 0.76% copper have been mined. All material mined since stripping began totals about 220 Mt (242 million st).

During the years 1977 through 1980, a very substantial portion of the Metcalf concentrator ore supply was provided from the Morenci open pit. Early in 1981 the Metcalf open pit was shut down, the entire 100 kt/d (110,000 stpd) of ore production for the two concentrators (Morenci and Metcalf) was shifted to the Morenci mine along with the mining equipment to take advantage of the lower waste-to-ore ratios and slightly higher ore grade. There remains about 0.9 Gt (1.0 billion st) of material to be mined at Metcalf.

Consolidation of Potential Mining Areas at Morenci

Phelps Dodge completed the consolidation of all potentially minable copper properties in the Morenci district in 1981. This was accomplished through purchase of the Western Copper property for $10 million. The property, acquired from Hanna Mining Co., consists of 81 patented claims covering 754 ha (1862 acres). The property is located on the east side of Chase Creek at the eastern limit of the Morenci mine and at the southern limit of the Metcalf mine. It is part of the Morenci copper mineralization, as is the Metcalf deposit. Portions of the Western Copper property are included in the final pit limits of both the Morenci and Metcalf mines.

The relatively low elevation of the Western Copper property makes it ideally suited for use as a waste disposal area for both the Metcalf and Morenci mines. Substantial reductions in haulage costs are anticipated because of shorter distances and favorable grade compared to disposal of Metcalf waste on dumps located at higher elevations north and east of the Metcalf mine.

MINES DEVELOPMENT

As previously indicated, the construction of a major copper mining complex in a remote setting requires a great amount of auxiliary development. In addition to direct mine and plant structures and equipment, a significant amount of pre-mine fact gathering, planning, and design, as well as support facility logistics development are required.

Water Supply Development

Because of the increased copper production brought on by the expansion of the Morenci mine and concentrator in the 1940s, there also was a corresponding need for additional water. In order to meet this water demand, innovative arrangements were developed to import the majority of the required water from various sources outside the Morenci district. To accomplish this, several dams, pumping plants, pipelines, powerlines, and support facilities were constructed, as shown in Fig. 2.

Fig. 2. Map of Arizona showing the location of Morenci's water development system.

In 1947 Horseshoe Dam was the first constructed retention structure for the collection of surface runoff, followed by Jacques Dam in 1954 and Blue Ridge Dam in 1965. Then, through various agreements and water exchanges, Phelps Dodge was permitted to divert water from the Black River, 43 km (27 miles) northwest of Morenci, and transfer it into Eagle Creek. The water diverted from the Black River flows south to the lower reaches of Eagle Creek where it is pumped 9.6 km (6 miles) to Morenci. A well field in the upper Eagle Creek area also has been developed to augment surface water sources with groundwater as necessary.

Water use by the entire Morenci Branch operations consists of 30.837 Mm³ (25,000 acre-ft) per year of which 617 000 m³ (500 acre-ft) per year are used by the mines for dust control, cleaning of equipment, change room services, and other minor uses. Water is pumped to head tanks located near the top of the mine and then gravity fed to the place of use. Rubber hose of 38 mm (1.5 in) diameter is used to furnish water to the open pit mining faces. This provides a flexible, inexpensive transport system.

Utilities

Electrical power is distributed to the Morenci mine through two 46 kV power transmission lines. This power line feeds a 15 MVA substation in the bottom of the Morenci pit. Voltage is transformed down from 46 kV to 4160 V and is distributed from the substation through radial power lines placed perpendicular to the mining faces. Switchhouses are located at the bottom of selected wooden power poles in active mining areas from which insulated electrical cable provides power to the mining equipment.

One of the 46 kV power lines extends past the Morenci substation to the Metcalf mine. This voltage is then transformed at two 5 MVA substations to feed the shops, the Metcalf solvent extraction plant, and primary crusher. The total connected load in the Morenci and Metcalf mines is about 25 megawatts.

Natural gas is provided to the Reduction Works area through two main pipelines: a 152 mm (6 in.) and 203 mm (8 in.) line, both at 4100 kPa (600 psi). The natural gas then is redistributed to the Morenci mine shops through a 102 mm (4 in.) feeder line at 621 kPa (90 psi). The annual mine consumption is about 708 m³ (25,000 ft³). Propane, delivered by truck, is used at the Metcalf offices and shop area, since an active mining area is located between the Morenci mine shops and the Metcalf mine shops.

Compressed air at about 621 kPa (90 psi) is provided to the Morenci mine shop area from the Morenci power plant through a 152 mm (6 in.) pipeline. Compressed air at about 758 kPa (110 psi) is provided to the Metcalf shop area by two stationary 10 m³/min (355 cfm) air compressors located at the Metcalf shop area and distributed through a 102 mm (4 in.) pipeline.

Fire protection for the mine complex is provided through the use of fire hydrants located near principal mine buildings, fire extinguishers on mobile equipment, and water trucks equipped with fire hoses or water cannons. A volunteer fire department trained to fight industrial fires is also available.

Mine Planning and Mine Engineering

Before development of Morenci/Metcalf mines could begin, it was essential to determine the location of the copper ore body, thickness of the overburden, grade of the ore, type and character of the rock to be milled, and location of dumping areas. This was accomplished by drilling a portion of the 1450 vertical diamond drill and churn drill holes now in the data base, as well as extensive sampling and some diamond drilling from the underground workings.

For the surface drilling, 122-m (400-ft) grid with a center hole in each 122-m (400-ft) square initially was used. Later, this was modified to eliminate the center holes. The ore body at the Morenci/Metcalf mines is a large porphyry copper deposit. The dominant rock types that host ore are monzonite porphyry, granite, and granite porphyry, with minor amounts of limestone, shale, and quartzite.

Sulfide ore mineralization is typified by supergene chalcocite and covellite replacing pyrite, chalcopyrite, and minor amounts of bornite. This sulfide mineralization occurs in veins, veinlets, microveinlets, and disseminations.

Oxide ore minerals present within the mining limits include azurite, malachite, tenorite, brochantite, chrysocolla, and cuprite. Minor amounts of native copper are encountered locally. Minor amounts of gold, silver, and molybdenite also are included in the ore and are collected with the copper sulfides in the concentrating process.

The material in the minable reserve is classified into three separate materials: ore, leach, and waste. Material classifications by contained copper are as follows:

Ore	$\geq 0.40\%$ Cu
Leach	$\geq 0.20\%$ Cu $\leq 0.39\%$ Cu
Waste	$\leq 0.19\%$ Cu

Material that is above 0.40% copper, but exhibits more than half of the contained copper as oxides, is reclassified as leach and handled accordingly.

In evaluating the drilling results, the area of influence of each drill hole was determined by using the polygonal technique. Drill hole assays averaged in 15-m (50-ft) vertical increments corresponding to the mining levels were used in assigning polygonal block grades for each drill hole. An average density of 0.354 m³/t (12.5 ft³/st) was used to calculate tonnages included within the polygonal blocks. In the absence of geological information, drill holes that ended in ore were not projected downward. Drill holes that ended in non-ore material were projected downward as waste to determine a boundary for adjacent holes.

Final copper reserve mining slopes are set at 51° based on geological evidence, rock strength characteristics, shape of the banks, and length of the slope. Initially, the Morenci pit mining limits were projected to be on a 45° overall slope, but were changed to a 37° overall slope, and later to a 51° overall slope.

The final reserve slopes are calculated using an economic model based on the value of the copper recovered from a block of material and the amount of overburden removal that the specified value can support.

A mine planning staff is located at Morenci to develop annual and long-term mine plans, as well as to review district copper ore reserves, and to prepare feasibility studies for improved mining methods. Presently the polygon method of reserve calculation is under review and efforts are being made to reevaluate the district's mining reserves through computer-generated block modeling and geostatistics. Other feasibility studies focus on improving mining efficiencies and alternatives to the conventional haulage methods.

The inter ramp slope for final mining limits is 51° with a uniform bench height of 15 m (50 ft) and a working bench width of 30 to 61 m (150 to 200 ft) in most areas. Because of the topography at the Metcalf mine, only truck haulage is used, while the Morenci mine uses both truck and rail haulage. All mine trackage is standard 1435-mm (56.5-in.) gage. The maximum available degree of curvature for main line track is 12°. The main line is constructed of 60-kg (133-lb) rail and normally is located outside pit limits or placed in a semipermanent location. It is a heavily traveled route and therefore consists of double track: one for the loaded trains and the other for empty trains.

The main truck haulage system is designed so that the maximum ramp grade is 10% and ramp width is 30 m (100 ft). This 30-m (100-ft) width allows for large berms on each side of the ramp as well as a drainage ditch. Runaway truck ramps and crash berms built of reverberatory furnace slag are provided for downhill hauls. The minimum radius of curvature for the main haulage road is 61 m (200 ft).

Internal truck ramps are used where necessary and can be narrower than main ramps. Dump loads are designed to be on an adverse grade of 2 to 3% to allow for settlement and to limit the possibility of a truck rolling backwards through the dump berm.

Equipment Selection

The decision regarding type of haulage equipment was based on the following factors: Topography of the area, tonnage to be removed, speed and distance that material must be moved, and overall economics.

The selection of rail haulage equipment was based on the layout of the final track haulage system. To determine the final layout, the maximum safe grade and the size of the train first had to be ascertained.

The train size also had to be flexible enough to meet the varying operating conditions with the maximum car size. Maximum grade was determined to be 4% with a most efficient train size being eight 30.6-m³ (40-yd³) self-dumping side dump cars. Because of the interchangeability and the need to facilitate train dispatching, the same type and size cars were used both for copper ore deliveries to the primary crusher and for waste and leach material deliveries to the mine dumps.

In the original selection of the locomotive power to be used, serious consideration was given to total electrification of the rail haulage system, but because of frequent changing of the bench track as well as narrow mining benches and heavy blasting, this concept was abandoned. The decision thus was made to use an overhead trolley electrification system only on the main line track system in combination with diesel-electric and battery-electric locomotives for bench and dump track.

Plans for a track extension in 1955 called for the electrification of an additional 8 km (5 miles) of main track. However, due to the high cost of such a venture and improvements over the previous two decades in diesel-electric locomotives, all trolley electric locomotives were replaced with 1306-kw (1750-hp) diesel-electric locomotives. These larger locomotives also were capable of improving efficiency by pulling longer trains.

The present fleet of locomotives consists of 3 895-kW (1200-hp), 13 1306-kW (1750-hp), 2 1343-kW (1800-hp) and 12 1492-kW (2000-hp) diesel-electric locomotives. The present railroad car fleet consists of 200 32.9-m³ (43-yd³) side dump cars, and 53 90-t (100-st) bottom dump cars used only for ore deliveries from the Metcalf primary crusher to the Metcalf concentrator.

Although the Morenci open pit has historically been a rail haulage mine, truck haulage has become increasingly important over the years. The decision to use trucks in the preliminary stripping operations was based on the need to prepare the mine for rail haulage and the success of truck haulage in other large-scale excavations.

Initially, a fleet of 18 17.2-m³ (22.5-yd³) trucks was used to open new levels, to remove overburden from the upper benches and to build railroad access. The fleet in 1970 was upgraded to 47 90-t (100-st) trucks and in the 80s, the old truck fleet was phased out and replaced with 28 154-t (170-st) trucks. Trucks in the Morenci mine now are used for all direct production from the active mining faces while rail is used to transfer ore to the Morenci and Metcalf concentrators.

The selection of loading equipment was based on the following factors: loading rate, size and type of material being loaded, haulage unit being loaded, and life of the operation.

Initially 3.4-m³ (4.5-yd³) electric shovels were purchased for mine development. These shovels were much more efficient and dependable than earlier models and could load the large tonnages required on a daily basis. They also were compatible with the rail and truck haulage units. Shovel equipment life coincided with the expected mine life of 30 years. As the mine developed, shovels were upgraded to larger units. The shovel fleet currently consists of six 11.5-m³ (15-yd³), and six 16.8-m³ (22-yd³) electric shovels. In addition, two 9.2-m³ (12-yd³) rubber-tired front-end loaders are being used as backup production units or for special projects.

Initially, the selection of drilling equipment was based on the following factors: tonnage to be broken per day, size and depth of hole to be drilled, and type of material being drilled.

The large electric churn drill was selected at Morenci because its drilling rate and hole size would allow breakage of a maximum amount of material.

In 1940, all the primary drilling was performed by 12 electric churn drills using 229-mm (9-in.) diameter bits. Blast holes were drilled from 2.4 to 3 m (8 to 10 ft) below grade.

Drill bits were reconditioned by a mechanical shaper and hardened for reuse. In 1956, a portion of the primary drilling was performed by large rotary drills capable of developing 311-mm (12-¼-in.) diameter blast holes. The rotary drill used tricone bits and 10-m (33-ft) long steel sections. Two steel changes were required per hole. The present primary drill fleet consists of 10 rotary drills: 6 electric and 4 diesel-electric. The diesel-electric drills were used for the early development of the Metcalf mine, two of which have been converted to electric drills. Single pass drilling now is available with the larger drills as well as variable hole sizes ranging from 229 to 381 mm (9 to 15 in.) in diameter.

DESCRIPTION OF OPERATIONS—MORENCI MINE

Breaking Ground

In the Morenci pit, there are certain operating conditions that determine the pattern of blast holes and the powder loading ratios. Drilling access must be maintained and material must be available at each shovel on a continuous basis. To accommodate these constraints, hard toes and high bottoms resulting from unbroken rock, must be avoided.

Drilling and blasting practices in the Morenci open pit have undergone a slow evolution. Initially, 15-m (50-ft) benches were established using 229-mm (9-in.) churn drilled blast holes averaging 17.6 to 18.2 m (58 to 60 ft) in depth. Also, there was a tendency to lighten the burden on the holes in hard ground by decreasing the distance between holes and by shortening the length of the toe. This practice broke the toe well, although it resulted in poor fragmentation in the upper part of the bank. It also yielded a low tonnage per vertical foot of blast hole.

Later, it was found that the heavier the burden without overloading, the better the resultant fracturing. Currently, most blast hole drilling uses 311-mm (12¼-in.) diameter rotary bits.

A small amount of secondary drilling is accomplished with air drills equipped with 64-mm (2½-in.) button carbide bits.

During 1987, rotary drill blasts accounted for 70.7 Mt (77.9 million st) of broken ground in the Morenci pit and involved 402 km (1.32 million ft) of drilling. This resulted in 176 t of broken rock per meter (59 st per ft) of blasthole.

Throughout the years as the types of available explosives changed, the patterns and loading ratios also have varied. In the early days of the mine, gelatin dynamite was used. This was followed by bagged ANFO beginning in 1960. In the early 1970s, bulk, prilled ammonium nitrate was used and in 1980, a mixture of ammonium nitrate and aluminum was introduced.

Currently, ammonium nitrate and fuel oil account for about 50% of the explosive utilized in primary bank blasts, although aluminized ANFO in a 10% mixture occasionally is used. An increasing number of wet holes have been encountered that cannot be dewatered, sleeved, and blasted with ANFO. Consequently, such holes are loaded with a slurry form of ANFO.

Trucks outfitted with special compartments are used for loading the ANFO into blast holes. Ammonium nitrate prills, fuel oil, and flaked or granulated aluminum are augered or pumped separately from compartments to a nozzle where they are mixed with the other ingredients while being injected into the hole. Each hole is single primed with an in-the-hole delay cap in a 0.45-kg (¾-lb) cast TNT booster hung on a 20-grain downline. Individual holes are tied to a 25-grain

trunkline, which is delayed on the surface between rows when a multiple row pattern is used. Slag, gravel, or drill hole cuttings scraped in with a backhoe serve as the stemming material. All blasts are initiated with an electric blasting cap by the blasting foreman. Typically, six benches are guarded against entry during a blast: one above the drill hole collars, the bench being blasted, and four below.

Typical powder factors vary among the different rock types encountered in the Morenci pit; however, most average approximately 5.0 t broken per kg of explosive (2.5 st broken per lb, etc.).

Although the production work schedule is a 24-hr-per-day continuous operation, the drilling and blasting crews work a 40-hr week, seven days per week schedule. Drilling is conducted on a three-shift rotation and all blasting is done on day shift. Each drill shift will result in approximately 90 kt (100,000 st) of material when blasted.

Loading

Loading of haulage units is accomplished with 12 electric shovels varying in dipper sizes from 11.5 to 15 m³ (9 to 22 yd³). All shovels are powered by 4160 V via trailing power cables. Two 9.2-m³ (12-yd³) front-end loaders are used for cleanup operations and low bank loading and also serve as backup units to the shovels.

At Morenci, loading consists of production loading of all material at the mining face. Nonore material is delivered by truck to the dump while ore is delivered to several transfer locations. Once the ore has been delivered to the transfer site, it is reloaded onto trains and is transported to the concentrators.

Reloading of the ore is accomplished using three methods which include conventional 11.5 m³ (15 yd³) shovels, hydraulically operated drop chutes, or a mechanized feeder hopper.

Approximately 40% of ore is transferred onto rail by the drop chute method, 27% by the pan feeder and 33% conventionally. The pan feeder is the newest method of transfer loading incorporated at Morenci and consists of a large bin structure with a reciprocating pan feeder attachment at the base. The locomotive engineer operates the feeder and spots the train simultaneously resulting in a quicker, safer and more uniform loading process.

Construction of an in-pit crush convey ore handling system is scheduled for completion in 1989 which will replace all ore transfer at Morenci.

Shovel operators work three-shift-rotation on a continuous schedule. There are four shovel crews, each of which works seven days with two or three days off between shift changes. This schedule allows a continuous staffing with each employee working only one overtime day per month. In addition, a relief system for production personnel is utilized whereby employees are relieved at their equipment rather than being transported to the mine office or change room for relief.

Haulage

Haulage of material out of the Morenci pit is accomplished with trains and trucks. Presently all ore grade material is transferred onto trains for delivery to the concentrators. Trains are 15 33-m³ (43-yd³) capacity side dump cars mobilized by either 1306- or 1492-kW (1750- or 2000-hp) diesel-electric locomotives. The caboose is a side dump car that has been outfitted with an enclosed riding platform containing a seat, heater, two-way radio, and air gauges to indicate the pressures in the air brake systems. To

Fig. 3. Aerial view of the Phelps Dodge Corp.'s Morenci mine with the Metcalf mine in the background.

provide a headlight for the caboose, an air-powered generator is used, operated by compressed air from the dump line on the train. Each train is operated by one employee via radio remote control. The remote-controlled locomotive system at Morenci is the largest of its kind in the world. The locomotive engineer carries a packset on his or her shoulders and a control head on the belt that enables the engineer to transmit signals to the throttle, all brake systems, wheel sanders, horn, and the headlights when he or she is located at either end of the train or on the ground in the near vicity of the train.

Until the advent of remote control operation, two-man train crews, consisting of a locomotive engineer and a brakeman, were utilized. One-man operation of trains has significantly reduced the cost of train operation. Fail-safe features are built into the equipment, including automatic air application to stop the train whenever a coded tone radio command is not received clearly or whenever the locomotive engineer's body assumes a position more than 45° off vertical. Remote control enables the driver to have full control of the train from the leading end—whichever direction he is traveling.

Computer-controlled dispatching of rail and truck traffic has been installed in the Morenci pit, similar to the system for dispatching haulage trucks in operation at Phelps Dodge's Tyrone mine at Tyrone, NM. The system has increased effective operating time by reducing queuing at shovels and crushers. It incorporates on-board radio telemetry along with roadway sensors to measure and report train or truck location. With the addition of computerized dispatching, the productivity increases have offset the purchase cost of additional equipment to maintain production.

At the concentrators each train of ore is dumped pneumatically by the locomotive engineer. Each car is spotted at the dump pocket and a lever is pulled to open the valve that activates the dumping air cylinders.

Each train completes three to four cycles per 8-hr shift. The haulage operation, just as the loading operation, is continuous, operating 24 hours per day, seven days per week. A relief system for equipment operators identical to the loading operation relief plan is used.

The Morenci mine has a fleet of 28 haulage trucks rated at 154-t (170-st) pay-load capacity. Each of the trucks is powered by a 16-cylinder (1600- or 1800-hp) diesel engine that drives an a-c traction alternator that, in turn provides power to two d-c electric wheel motors that drive the rear dual wheels of the trucks. The trucks primarily are loaded by large electric shovels equipped with 16.8-m³ (22-yd³) dippers.

Historically, truck haulage in the Morenci mine accounted for less than 10% of the material moved. Recently, however, the accelerated development on the lower levels of the pit has increased truck haulage to 100% of total daily production while trains are utilized for ore transfer. Truck haulage will continue to be used in the Morenci pit in conjunction with other haulage methods. The in-pit crush and convey system that is scheduled to come on line by 1989 will replace rail haulage at Morenci.

REFERENCES

Cleland, R.G., 1952, *A History of Phelps Dodge, 1834-1950,* Knopf, New York, pp. 244-260.

Epler, W.C., 1975, "Production Begins at Phelps Dodge's New $200 Million Metcalf Operation," *Pay Dirt,* No. 430, Apr. 28, pp. 1-6.

Epler, W.C., 1981, "Morenci, First Investment Is Now Mainstay; Metcalf, Second Openpit on Morenci Orebody," *Phelps Dodge, A Copper Centennial, 1881-1981, Arizona Paydirt,* pp. 36-62.

Fenzi, W.E., 1942, *Some Aspects of Breaking Ground in the Morenci Open Pit Mine,* private publication by Phelps Dodge Corp.

Fenzi, W.E. and Ormsby, L., 1956, "Rail Haulage at the Morenci Open Pit Mine," *Mining Congress Journal,* Nov., pp. 86-90.

Hardwick, W.R., 1959, "Open-Pit Copper Mining Methods, Morenci Branch, Phelps Dodge Corporation, Greenlee County, Arizona," *Information Circular 7911,* US Bureau of Mines.

Hoppe, R., 1977, "Open Pit Copper Mining in Arizona," *Engineering and Mining Journal,* June, pp. 95-100.

Langton, J.M., 1972, "Ore Genesis in the Morenci-Metcalf District," paper presented at AIME Annual Meeting.

Larsen, J.U., 1947, "Some Phases of Drilling and Blasting Practices in the Morenci Open Pit," paper presented at AIME Open Pit Subdivision Meeting, March 21-22, pp. 1-4.

Lawson, W.C., 1938, "Preliminary Stripping of the Morenci Open Pit, Arizona," *Technical Publication No. 980,* AIME, Sept., pp. 1-15.

Lawson, W.C., 1940, "Preliminary Engineering at Morenci Open Pit," *Civil Engineering,* Vol. 10, No. 8, Aug., pp. 510-513.

Lawson, W.C., 1947, "Laying Panel Track at the Morenci Open Pit," *Technical Publication No. 2189,* AIME, July, pp. 1-13.

Orr, D.H., Jr., 1960, "Maintenance of Main Line Haulage Track in the Morenci Mine," paper presented at Annual Arizona Section AIME Meeting, pp. 1-9.

Orr, D.H., Jr. and Berra, F.G., 1965, "One Man Remote Control Rail Haulage," *Mining Engineering,* Apr., pp. 75-79.

Parsons, A.B., 1957, *The Porphyry Coppers in 1956,* AIME, New York, pp. 49-66.

Skillings, D.N., Jr., 1975, "Phelps Dodge Corp.'s Metcalf Porphyry-Type Project," *Skillings Mining Review,* Vol. 64, No. 25, June 21, pp. 12-17.

Train, A., Jr., 1941, *Ajo-Bisbee-Morenci,* private publication by Phelps Dodge Corp., pp. 85-103.

9.5 Case Study: Cuajone, Peru

Daniel Rodriguez Hoyle
Respectfully dedicated to my friend Frank W. Archibald,
formerly Chairman and President, Southern Peru Copper Corporation

INTRODUCTION

In December 1969, an agreement (the Cuajone Bilateral Agreement) was reached between the Peruvian government and Southern Peru Copper Corp. (SPCC) on a specific program to develop and operate the Cuajone ore body. The Agreement provided a maximum time limit of 6½ years for the placing in operation of the Cuajone project, with all financing, construction, and operation being the responsibility of SPCC. The government undertook to grant SPCC the benefits, terms, and guarantees contained in the then existing Peruvian mining legislation.

Cuajone is the second large investment made by SPCC in Peruvian mining. The first was the development of the Toquepala deposit, located 24 km southeast from Cuajone. Toquepala started production in January 1960 and, with subsequent expansions, amounted to a total investment of US $318 million by the end of 1982. Until 1976, when Cuajone started productive operations, Toquepala was the largest Peruvian copper producer. A summary of Peru's production covering the last three decades is given in Table 1.

Quellaveco, the third large ore deposit partially developed by SPCC, reverted to the state in 1971. At the time of its reversion, the reserves of this deposit were in the order of 191 Mt with an average content of 0.94% copper. In January 1972, the government assigned Quellaveco to its state company Empresa Minera del Perú (Minero Peru). The deposit is located 16 km north of the Toquepala mine and 11 km southeast of the Cuajone mine.

The Cuajone copper complex is owned 88.5% by SPCC and 11.5% by Billiton BV, a wholly owned subsidiary of the Royal Dutch/Shell Group. SPCC is owned 52.3% by Asarco Inc., 20.7% by the Marmon Group Inc., 16.3% by Phelps Dodge Overseas Capital Corp., and 10.7% by Newmont Mining Corp.

LOCATION

The Cuajone ore body is located in the Andes mountain range at 3500 to 3800 m elevation, 71 km from the Pacific Ocean in the Dept. of Moquegua in southern Peru. The deposit is sited 30 km northeast of the town of Moquegua. Fig. 1 shows the approximate location of the Cuajone and Toquepala mines, the Ilo copper smelter, the nearby Quellaveco ore body, and the nearest major cities.

DISCOVERY AND EXPLORATION

Copper mining in southwestern Peru was carried on sporadically since the end of the 19th century, and there are brief references in geographical literature of the time to copper occurrences in the general area of Toquepala and Cuajone. Narrow oxide and enriched sulfide veinlets were exploited on a very limited scale, but the desert nature and difficult accessibility of the area discouraged continued mining activities. Soon after the settlement of the border conflict between Peru and Chile in 1929, interest was renewed in the area and local residents started staking out mining claims. Prominent among these were Juan Oviedo Villegas (Toquepala) and Julio E. Gianella (Cuajone).

In 1937 the Toquepala and Cuajone prospects were recognized as potential porphyry-type copper deposits of major importance by geologist A. C. Schmedeman during an exploration mission for Cerro de Pasco Corp. Merit is due to Schmedeman for his geological insight, as the evidence of widespread copper mineralization in the two deposits was obscured by a deep leached capping mantle and/or thick post-ore volcanic rock cover.

Cerro de Pasco Corp. acquired mining rights to Cuajone in 1943 and explored the deposit during the period 1943 to 1945. The exploration program, which included the drilling of 40 diamond drill holes, was resumed by Cerro and New-

Table 1. Peru—Copper Production (Short Tons)

Period or Year	In Ore and Concentrates	In Blister and Mixed Bars	Refined	Total (Fine Content)
1950–1954*	11,801	1,143	25,148	38,092
1955–1959*	14,914	10,104	28,803	53,811
1960–1964*	22,502	138,056	38,603	199,161
1965–1969*	35,182	140,734	41,097	217,013
1970–1974*	42,393	150,596	40,980	233,969
1975	26,731	113,514	59,271	199,516
1976	17,637	64,795	154,450	236,882
1977	18,739	147,479	206,481	372,699
1978	52,488	150,056	201,450	403,994
1979	37,465	154,928	254,449	446,842
1980	34,185	137,898	246,351	418,434
1981	29,724	112,363	220,901	362,988
1982	41,224	104,258	224,549	370,031
1983	28,524	99,567	177,932	306,023
1984	33,293	120,876	216,907	371,076
1985	52,470	133,202	216,157	401,829

* Yearly average.
Metric equivalent: 1 st × 0.907 = 1 t.

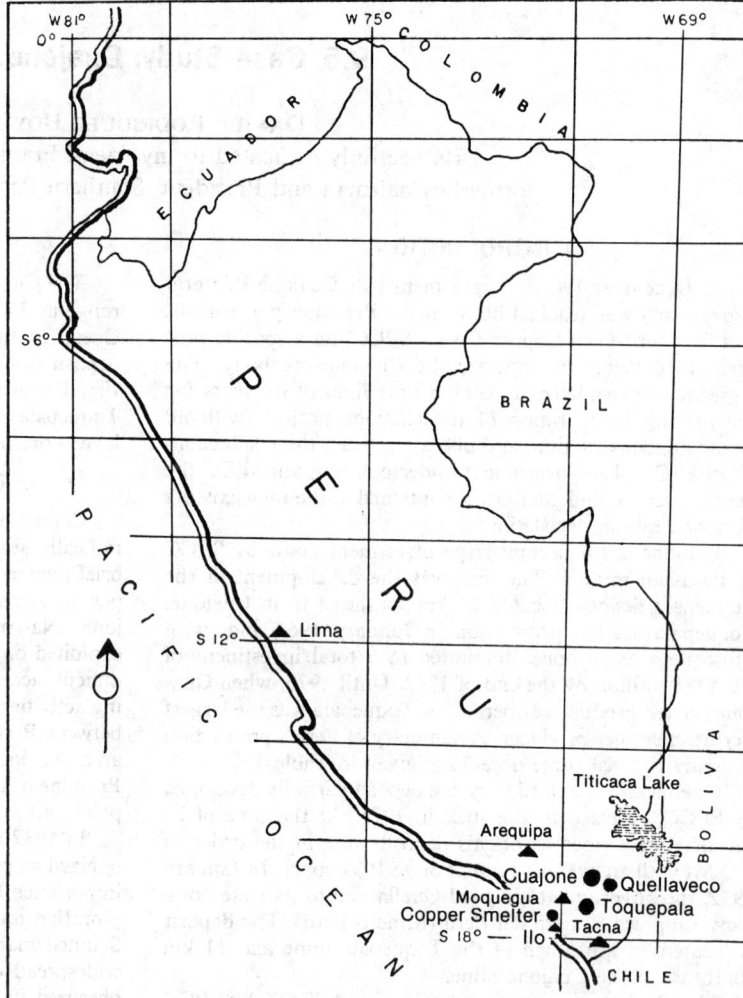

Fig 1. Cuajone location map.

mont Mining between 1952 and 1954 when an additional 88 churn boreholes were drilled, guided in part by the findings of a self-potential and resistivity survey conducted in the areas covered by post-ore volcanics.

During the period 1965-1966, SPCC completed the exploration of Cuajone with an additional 83 diamond boreholes and established the grade-tonnage potential of the deposit for mining by surface methods. A number of shafts and adits were driven and approximately 9 kt of sulfide and oxide ore samples were treated at a pilot plant in Toquepala to investigate the milling characteristics of the ores and to define an adequate flowsheet.

Using a 37° final pit slope, a minable reserve of approximately 425 Mt of sulfide ore averaging 1% copper at a 0.45% cutoff and containing 0.025% molybdenite; 23 Mt of oxide copper ore at 1.29% Cu; 109 Mt of low-grade sulfide ore averaging 0.32% Cu; and 1052 Mt of waste material were dimensioned. The ratio waste:ore averages 2.8:1.0.

GEOLOGY

Areal Geology

The mineral district is characterized by a Jurassic(?)-Cretaceous-Tertiary basement of homoclinal volcanic flow rocks, tilted gently to the west that extend in a northwesterly trend along the western flank of the Andes and are contin-

uous units between Toquepala and Cuajone. Along the northeast they are unconformably overlain by Plio-Pleistocene volcanic flows, pyroclastic units, and tuff beds. Along the southwest they are covered by continental deposits of the Moquegua formation of Pliocene age.

The basement rocks have been faulted and intruded along a northwesterly belt by large masses of the Andean diorite batholith. Two diorite apophysis occur in the Cuajone area, one east and the other west of the mineralized zone.

An intrusive complex composed of quartz monzonite, quartz latite porphyry, and andesite stocks intrudes the basement volcanic rocks within the Cuajone mineralized area.

Basement Rocks

The basement volcanic rocks present in the mineralized zone are Cuajone basaltic andesite, Quellaveco quartz porphyry, and Toquepala dolerite. These rocks make up the southwestern, southern, and eastern sides of the ore body.

Intrusive Complex

The central, western, northern, and northeastern portions of the ore body are constituted by intrusive stocks. The largest intrusive is a quartz latite porphyry body 600 to 800 m wide that trends northwest and extends 4 km to the northwest of the deposit. The southernmost portion of the latite body is located within the ore zone and the northern extension is unaltered and unmineralized.

The latite body has intruded an older monzonite rock, and roof pendant remnants of strongly altered, mineralized monzonite are found in the central and southern part of the mineralized zone.

The youngest intrusive rock is a fine-grained andesite that is thoroughly altered to quartz-sericite-chlorite and, in two separate bodies, cuts through the older intrusive rocks and basement volcanics along brecciated contacts. The bodies are nearly parallel, trend northeast, and are located in the northeastern and southwestern sides of the mineralized zone. A few narrow postmineral latite porphyry dikes cut basement rocks in the district.

Postmineral Volcanics

Two distinct groups of postmineral volcanics overlie the mineralized zone. The oldest volcanics, trachyte flows and tuffs, are found along the south rim of the mineralized area. A younger set of volcanic ignimbrite rock covers the northern half of the ore body.

Structures

The only major fault structure associated with the deposit cuts across the southwestern area of the ore body. Strike is N50°W with a steep dip to the southwest. The fault system or shear zone, 10 to 130 m wide, is filled with parallel tabular bodies of unmineralized postore angular breccia and barren latite dikes. Displacement of the basement rock sequences on both sides of the breccia body indicates a major premineral fault structure.

Widespread quartz veining lacework and intense rock jointing indicate a very strong preore-stockwork shattering action affecting the intrusive complex and surrounding rocks. Numerous small and steep fault systems with a predominant northwesten trend are common throughout the ore body.

Ore Body

The Cuajone ore body is composed of an upper, discontinuous, and roughly flat-lying layer of copper oxide mineralization that overlies a tabular layer of enriched sulfide. Directly beneath is the bulk of the ore body, composed of primary sulfides.

A capping of leached rock and postmineral volcanics covered the ore body almost entirely. This capping is a few meters thick along Quebrada Chuntacala in the central part of the ore body and up to 150 m thick on the southern and northern parts.

In plan view at its maximum areal extent, the ore body is oval in shape with the long axis oriented in a northwestern direction and measuring 1200 × 900 m approximately. The maximum horizontal extent of the ore body is found at the top of the enriched sulfide zone.

The ore body is funnel-shaped in depth; however, most of the lateral limits of ore are inclined inward at steep angles.

Copper Oxide Zone

The copper oxide zone is tabular and approximately horizontal, dipping gently to the west. Average thickness is 15 m. A total of 23 Mt of oxide material at 1.29% total copper exists within mining limits.

The predominant oxide copper minerals are malachite and chrysocolla. There are moderate amounts of tenorite (variety melaconite) and minor or trace amounts of brochantite, azurite, cuprite, and native copper. Some copper is also present in a brown to black amorphous form, probably a mixture of copper oxides and limonites. The oxide mineral assemblage is intermixed with substantial amounts of chalcocite, minor covellite and pyrite, and trace chalcopyrite.

Enriched Sulfide Zone

The enriched or supergene sulfide zone underlies oxide copper mineralization or leached capping and overlies primary or hypogene sulfide mineralization. In areal extent it is only slightly smaller than the maximum extent of the primary zone.

The zone has a semi-horizontal tabular configuration and is inclined to the west. Maximum thickness is 78 m but the average is close to 20 m. There are approximately 57 Mt of enriched sulfide ore at 1.57% copper grade.

The aphanitic basic rocks, intrusive andesite, and Cuajone basaltic andesite have yielded the finely textured alteration minerals, chlorite and biotite, that were replaced by large amounts of fine-grained sulfides. The intermediate acidic intrusives have coarser textures and less replaceable alteration minerals. Therefore somewhat better copper grade is found where intrusive andesite and basic volcanics predominate.

Pyrite content is low over most of the ore body. The propylitic alteration halo contains large amounts of vein pyrite. Pyrite to chalcopyrite ratios prevailing over the main central ore-body zone are between 1:1 and 2:1. Total sulfide content of the primary zone ranges from 4 to 9% by weight.

Molybdenite occurs as small platy clusters and, to a minor degree, as fine disseminations. The average molybdenite content of the ore is 0.025%. Areas in the intrusive andesite bodies contain up to 0.04% molybdenite. Silver and gold occur in very minor trace quantities throughout the ore body.

In addition to the widespread hydrothermal alteration minerals—quartz, sericite, biotite, chlorite, and clays—profuse quartz veining pervades the ore-body host rocks.

Chalcocite is the predominant supergene mineral replacing chalcopyrite, bornite, and pyrite. Minor covellite and digenite are found as intergrowths with chalcocite.

Primary Sulfide Zone

The Cuajone ore body is essentially a hypogene copper deposit. Over 86% of the sulfide ore reserves tonnage estimated for the deposit is in the primary zone.

Mineral assemblage consists of major pyrite and chalcopyrite and minor bornite and molybdenite. Trace enargite, sphalerite, and galena occur in vugs within quartz veins and breccia dikes. The most abundant sulfide is pyrite and the principal copper sulfide is chalcopyrite.

Chalcopyrite dissemination takes two principal forms: as discrete and moderate to very fine grains dispersed among the quartz-sericite-biotite alteration minerals, and as irregular clusters replacing laths and aggregates. Veinlet chalcopyrite is not common.

Pyrite is found largely in narrow to thick veinlets along with quartz. A moderate amount is dispersed as grains or clusters.

Hydrothermal Alteration

The halo of hydrothermal alteration surrounding the Cuajone ore body is from 3 to 4 km in diam. From the center of the ore body outward, four main zones or stages with diagnostic alteration mineral assemblages have been established: a central zone, entirely within the ore body that contains strong quartz-sericite, moderate clay, and minor potash feldspar; an outer zone within the ore body in Cuajone basaltic andesite that shows moderate to strong clay and biotite alteration; a zone of propylitic alteration at the limit of the ore body and beyond; and a zone at the limit of the ore body and beyond, only within Quellaveco quartz porphyry, that has pervasive silicification. The quartz latite dike carries interstitial carbonate. No significant tourmaline or anhydrite has been found.

FEASIBILITY STUDIES

Coincidental with the final Cuajone exploration phase (1965-1966), SPCC started feasibility studies through metallurgical testing of drill cores. Enriched and primary ores, as well as the various host rock types, were treated in this manner. During 1965, a 36 t/d pilot plant was erected at Toquepala to test bulk samples of Cuajone ores. Aside from conventional grinding and flotation, the material was also tested for autogenous grinding, but the results were not sufficiently conclusive to incorporate autogenous mills in the final plant design. The ore posed no serious metallurgical problems, except for some high clay ore types with slime-producing characteristics.

Column and heap leaching of Cuajone oxide ores, followed by solvent extraction and electrowinning, were also extensively tested at the pilot plant with very satisfactory results.

At the same time, a stripping and mining plan was developed, which included 136 Mt of premine stripping for an initial 27 kt/d ore production rate. The pit was to be mined using 15-m high benches, 7.6-m³ electric shovels, and truck and train haulage.

In 1973, the mining plan was modified to support a production rate of 40.8 kt/d of ore. That resulted in an increase to 236 Mt of premine stripping and the redimensioning of all equipment capacity.

In 1970, Fluor Utah Inc. was awarded the project's general contract for engineering and construction as well as management of subcontract work with local firms. However, SPCC maintained control in specific development areas such as basic mill and railroad design, mine and strip planning, ground-water exploration and development, exploration for aggregates, and all foundation engineering geological investigations. Engineering design was performed at San Mateo, CA, and at Lima and Toquepala in Peru.

The most critical factors in the overall Cuajone project design were the concentrator site location, the choice of concentrate conveyance routes, the disposal of mill tailings, and the procurement of an industrial and domestic water supply.

The Cuajone concentrator site finally selected is located at Botiflaca, 8 km southeast of the mine, the closest area to offer topographic construction advantages and expansion possibilities and at the desired elevation for the primary crusher to correspond with the designed pit exit level.

Rail transport offered the advantage of concentrate haulage from Cuajone to the existing Toquepala-Ilo-Industrial railway and of economical transportation of supplies from the port area and was considered a logical extension of the accessways serving Toquepala. But the rough terrain separating Cuajone from Toquepala, consisting of parallel drainageways that have dissected a high volcanic plateau and carved numerous deep NE-SW canyons in the region, ruled out the convenience of building a surface railroad. Tunnel alternatives were, therefore, investigated in great detail. Thirty-one alignments were analyzed and their geologic conditions studied in order to arrive at the final route consisting of 30 km of railroad, including five tunnels with an aggregate length of 27 km.

Because of the rugged topography, steep-gradient canyons, occasional flash flood conditions, and seismic risk, it was decided from the start to dispose of mill tailings into the sea through, essentially, the same outlet taken by the Toquepala mill tailings.

Water resources in southwestern Peru are very scarce and the few agricultural valleys and settlements have first priority on its usage. A short—two- to three-month—yearly precipitation regime and the lack of natural storage possibilities result in cyclical water shortages. Water supply for the project, estimated at a maximum rate of 1500 L/s, either had to be imported from distant Andean water sheds, or to be developed from highland ground-water sources. US consultants in the fields of hydrogeology and geophysics and company geologists were successful in proving and developing an extensive aquifer—the Capillune Formation—50 km northeast of Cuajone, at 4450- to 4550-m elevation, that satisfied the project's requirements.

Considerable studies for the expansion of SPCC's Ilo copper smelter and power plant on the sea coast were made, in order to handle the new Cuajone concentrate production and to satisfy the energy demands of the project.

Two new modern townsites were designed to be built to house the staff and workers and their families, necessary for the operation of the Cuajone mine and concentrator. Likewise, the expansion of the Ilo Camp was planned.

The construction of offices and other service buildings and the equipping of service shops and similar permanent and temporary installations were programmed in detail.

Paramount in the design of the Cuajone complex components was the need to integrate them smoothly, efficiently, and economically into the Toquepala infrastructure.

CUAJONE BILATERAL AGREEMENT

The Cuajone Bilateral Agreement is a case study in international mining transactions: it represented an important investment; was negotiated and carried out within circumstances of high political and economic risk; and was innovative in its financial and legal conception.

Antecedents

By Public Deeds of November 1954 and April 1955, the government of Peru and SPCC signed contracts for the exploitation, as independent units, of the Toquepala, Cuajone, and Quellaveco mineral deposits.

The company initiated its operations with the exploitation of Toquepala in 1960. This was effected under a bilateral contract which, in compliance with Law 16732, ended on Jan. 1, 1968. Said Law stated that new conditions would be negotiated for the exploitation of both the Cuajone and the Quellaveco mines.

The second clause of the Cuajone Bilateral Agreement, signed on Dec. 19, 1969, states that the Quellaveco ore body is the subject matter of a separate contract. However, as stated before, Quellaveco reverted to the state in January 1971.

Cuajone Agreement—Company's Obligations

Under the terms of the agreement, the company undertook the obligation to build the necessary installations, invest and obtain the corresponding financing, and operate the mining complex at a minimum ore production rate of 27.2 kt/d (all of the equipment and installations, including water and power supply systems, the smelter expansion, etc., to match this goal). The necessary investment was estimated to be US $355 million and the maximum construction period was fixed at 6½ years.

Modification in the Scope of the Project

In fact, as will be seen later in this section, it became necessary during the project's construction to substantially enlarge the size of the principal and many auxiliary installations. This increase in productive capacity was in the order of 50% in excess of what the bilateral agreement specified, and was the result of the decision made by the company to maintain a reasonable investment-per-ton-of-installed-capacity ratio. The initial estimated ratio was approximately US $3000 per 0.9 t/a of blister copper [US $355 million per 108.8 Mt/a of copper]. The decision to increase the scope of the project was triggered by the worldwide inflationary trend brought about by the abrupt change in the price of oil by OPEC in 1973. The resulting unit investment was US $4150 per 0.9 t/a. This relatively low ratio was met because the company had already carried out much of the infrastructural work and had firm purchase orders for an important proportion of the equipment and raw materials required during the construction phase of the project.

This case study does not describe the installations that the company undertook to carry out in compliance with its obligations under the bilateral agreement. Due to SPCC's decision to increase the scope of the project, the description and size of the various project components are shown here as they were actually built and the corresponding investments also as those effectively made.

Should SPCC have failed to comply with its contractual obligations, the penalty would have been the voidance of the concessions in the following cases:

• That SPCC should not invest or undertake duly verified expenditure commitments of a minimum of US $25 million, during the first 18 months of a work schedule to begin on Apr. 1, 1970.

• That within an 18-month period, counted from the date of signature of the agreement, SPCC should not assure the financing necessary for the continuation of the investment plan, submitting the corresponding scheme and/or assuming responsibility for executing said plan by its own means, accrediting said facts, as the case may be, with the pertinent documents. If, during any year, the company would not have invested at least 60% of the amounts corresponding to the annual investment programs, these conditions would have been considered as not having been met.

• The nonconclusion of the construction work within the period of 6½ years would terminate the project and obtain the minimum production.

The time corresponding to these penalties would not be counted in case of fortuitous or *force majeure* cases, such as strikes, wars, or any other that may have impeded the start or continuation of the works, despite the desire of the company to carry them out on schedule.

The company agreed not to transfer, adjudicate, convey, or, by any other means, deliver the agreement to third parties without the prior consent of the government.

The agreement specifies that it may be modified only by means of public deed, once the contracting parties have agreed upon such modification.

Finally, the company declared its express subjection to the laws and law courts of Peru, and forewent recourse to diplomatic protection regarding this Agreement.

Government's Obligations

In consideration for the investments to be made by the company and the execution of the works subject matter of the project, the government agreed to grant to the company the concessions, authorizations, permits, easements, rights, water rights, rights of passage, right-of-way, and other facilities that may have been requested by the company. The government further agreed to grant the support that may have been pertinent and legally possible for the negotiation and successful conclusion of the loan agreements that the company may have requested from foreign government institutions or from international financial institutions, without the government assuming any economic obligation whatsoever, nor affecting the interests of the state.

This did not imply, either directly or indirectly, that the government would have assumed or assumed the obligation to grant any security or collateral as a guarantee for the obligations that the company may have incurred.

The company agreed to deliver to the Central Reserve Bank of Peru, directly or by means of deposit in the name of the Central Reserve Bank in a corresponding bank abroad, the total amount of the foreign currency that it would receive as the f.o.b sales price of its export products. In exchange, the Central Reserve Bank of Peru undertook to furnish to the company immediately, for a value equivalent to the amount delivered to it, foreign exchange certificates issued nominatively and transferable by endorsement that may be necessary to meet payment of SPCC's invested funds; the interest on loans that may have accrued provided it had been taken into account in fixing the amount of the investment; and such justified services that the company would meet in the country or abroad to remunerate technicians engaged by the company, up to a certain specified limit.

In fulfilment of the guarantee for disposability granted under the preceeding paragraph, and for the purposes described therein, the Central Reserve Bank of Peru agreed to deliver to the company the necessary foreign exchange, against submittal of the corresponding certificates in the amounts necessary to cover the "Investment Recovery Program" (see further on in this section).

It was agreed by the Central Reserve Bank of Peru to grant to SPCC the guarantee of disposability of foreign exchange during the period of recovery of the capital invested.

The company agreed to register with the Central Reserve Bank of Peru—during the construction period—the documents certifying its imports and, in general, all receipts considered in the investment plan and in the investment recovery program.

The delivery of foreign exchange to the Central Reserve Bank of Peru would not be applicable in the cases of capital from loans obtained abroad that were earmarked, also abroad, for the purchase and importation of equipment, materials or other capital goods, or for other expenses and services included in the investment plan, but these investments had to be registered with the Central Reserve Bank of Peru.

The government granted free disposability for the export and sale by the company of the products and byproducts of the Cuajone mine. Therefore, no measures could be applied that would:

• Restrict the right to sell to any destination, once the needs of the local national consumption were met in the manner established by law.

• Suspend or postpone the said sales and/or exports.

• Require the sale in any foreign market at prices lower than may have been obtained in the international markets where the major part of the company's exports would be placed.

• Require the payment for the said products on the basis of barter or in currencies not valid for international payments.

The company was authorized to invest, after expiration of the agreement, for a period not to exceed five years, up to 50% of the net profit of each fiscal period in the establishment of refining plants and/or plants for the conversion of metallurgical products.

The government agreed to grant to the company authorizations and facilities for temporary imports, as well as for the sale or reexportation of such equipment or materials imported and specifically used during the construction period of the project. It also granted partial exemption from payment of certain customs duties on the imports of specialized equipment.

On the other hand, the company was required to acquire materials of national production, provided they should be available, in sufficient quantities at the time they would be required; of quality that would conform to the specific needs of the mining industry; be functionally adequate, and have competitive prices, the latter limit to be considered when the value of the national product did not exceed by more than 25% the CIF value of the similar foreign product.

Investment, Recoverable Investment, Investment Recovery Period

It was agreed that the taxation system, including the tax stability regime, would be applicable during the period required for the recovery of the investment, a period not to exceed 10 years, counted as from the date of start of mining exploitation operations at Cuajone (the 10-year period could be extended up to a maximum of 20 years under exceptional conditions, such as unforeseen very extended price depression periods).

For the purpose of determining the actual period for recovery, "Investment" was defined as the *cumulative* disbursement made in the purchase of equipment and machinery, the expenses of preparation and installation, the expenses of construction, and the interest on the loans, that is to say, all expenses incurred during the period comprised between the dates of initiation and completion of the project, up to the start of mining exploitation operations, including working capital and the losses due to exchange differences arising from credits, insurance, expenses, commissions, or premiums obtained abroad.

This all-encompassing investment figure would, however, be reduced, for contractual purposes, by the amount that the company could reinvest in the project from depletion reserves originating in other mining operations in the country (such as Toquepala). The resulting *cumulative* figure would be defined, under the agreement, as "recoverable investment."

The concept of recoverable investment had no other effect but to *help measure the time period* during which the taxation system, including the tax stability period, would be contractually applicable.

Once having started mining exploitation operations, the company would retain from its Cuajone profit and loss statements, prepared under Peruvian tax legislation, the following items of cash flow:

• The total profits, less the taxes thereon applied.

• The depreciation of the fixed assets, less all government-authorized investments made by the company subsequent to the initiation of operations.

• The depletion reserves originating from the Cuajone unit operations.

• The amortization of the mine preparation expenses.

• The interest accrued on the investment of capital obtained by loan, provided it had been considered in fixing the amount of the investment.

• The total of any ordinary or extraordinary reserves the company may make, with the exception of those reserves or provisions intended to cover the social benefits of employees.

The sums thus retained would constitute the "recovered investment" for the period in question. These recovered investments would be added from year to year and constitute the *cumulative* recovered investment.

The investment recovery period would be determined by the time elapsed to accumulate recovered investments into such a sum as to equal the recoverable investment.

At this point in time, the taxation system applicable to Cuajone would change rates (see further in this section) and remain at the new rate for a period (tax stability period) of exactly six years, after which the Cuajone operations would be subject to any tax rates then generally applicable in Peru.

The foregoing Cuajone agreement tax formula, which followed Peruvian legislation to the letter, provides a sophisticated solution to the complex problem of determining the contractual time period during which a given tax structure would apply. The formula allows for the precise determination of a *variable* contract period. Should, for instance, the operating company enjoy exceptionally high prices for its products, the period during which it would pay preferential tax rates would be correspondingly shortened, and vice versa. Consequently, neither the host country nor the investor company would suffer unduly for having agreed to certain terms for a *fixed period of time* during which *luck,* good or bad, for any party in the contract, would result in detriment or in windfall benefit to the other party.

Taxation System

The parties agreed on the following taxation system:

• Apply to the taxable income resulting from the operation of Cuajone, *during the investment recovery period,* a total of 47.5%. This tax rate would be raised to 54.5% during the *tax stability period,* comprising the six years immediately following the investment recovery period. This taxation system contractually exempted the company during the two aforementioned periods, from all other tax created or to be created, whether national, regional, or local, as well as from any other taxation upon the concession or upon the products obtained therefrom. The exemption covers the company from liens or obligations that might represent a reduction of its cash resources, such as forced investments, forced loans, and tax advances, except the 4% of the value of exports *on account* of the income tax paid by all mining concessionaires in Peru. The company has the right to deduct from any year's net income the losses it may have suffered during the preceding five years.

• Depreciate the fixed assets on a global weighted rate between 3% and 12%, to be determined annually by the company, with the objective of eliminating, by means of this system, the payment of taxes abroad.

• Calculate the depletion reserves originating from the Cuajone unit operations in accordance with the applicable mining and tax laws existing on the date of signature of the agreement. These were limited to the smaller of two values for any one year: 15% of the gross value of the mineral production or 50% of the profits, after having deducted the depletion allowance itself.

• Amortize the mine preparation expenses, which include the cost of prospecting, preparation, and development of the mine and removal of the overburden during the preoperative stage, at a rate of 10% per year.

• Amortize the interest accrued by the investment of capital obtained from loans, provided it had been considered

Table 2. Application of Funds by Work Centers

	Million US$
Roads	6.5
Energy and Communications	16.3
Water Supply System	31.3
Urban Centers	77.9
Shops and Warehouses	31.1
Concentrator	122.4
Railway and Tunnels	68.1
Transportation Equipment, including Railroads	17.4
Smelter Expansion	65.1
Power Plant Expansion	18.5
Mining and other Equipment	52.7
Premine Stripping and Preproduction Expenses	76.8
Interest Accrued During Construction Period	83.2
Mineral Concession	2.7
Working Capital	68.8
Exchange Adjustments	(0.4)
Investment Adjustments for Programs Completion	8.5
Total Investment: Application of Funds	746.9

in fixing the amount of the investment, at an annual rate of 5%.

• Retain any ordinary or extraordinary reserves the company would make, with the exception of those intended to cover the social benefits of employees, which must be kept inviolate.

For the purpose of determining the investment and its recovery, the company was authorized to carry an auxiliary book with a special account in US currency.

The company was also authorized to revalue annually, tax-free, the fixed assets and the mine preparation expenses, and to apply amortization and depreciation on the readjusted value of said assets, with the exception of repair parts and accessories, when there may occur fluctuations in the value of the national currency in relation to the US currency in excess of 5% per year.

It was agreed that the accounting systems of Toquepala and Cuajone would be carried out independently, with detailed provision being made in the agreement for attaining this objective.

CUAJONE INVESTMENT, FINANCING, AND RECOVERABLE INVESTMENT

The investment made in Cuajone and the financing of the required funds are summarized as follows:

Total Investment

The total investment made in Cuajone was US $746.9 million. The application of funds by work centers is given in Table 2. Table 3 shows the application of funds in chronological order.

Financing

Source of Funds: Cuajone was financed through equity participation, direct cash credits, equipment suppliers' credits, and copper purchasers' credits, as shown in the following table:

	Million US$
SPCC equity, Including Reinvestment of Depletion Reserves Originating from Toquepala	310.4
Billiton BV-Capital Association Participation	24.8
Direct Credits	215.7
Equipment Suppliers' Credits	142.0
Copper Purchasers' Credits	54.0
Total Investments: Source of Funds	746.9

Recoverable Investment

For the purpose of application of the Cuajone Bilateral Agreement (see pertinent section), the project's recoverable investment was determined as follows:

	Million US$
Total Investment	746.9
Toquepala Depletion invested in Cuajone	(74.6)
Working Capital difference	(19.6)
Mining Community—Workers Participation System	(22.5)
Cuajone Recoverable Investment	630.2

Recovered Investment

Table 4 shows the contractual recovered investment figures for the period 1977-1981.

PRODUCT SALES DISTRIBUTION

The Cuajone copper production is allocated in accordance with sales contracts subscribed on Sep. 19, 1974, as follows:

Cuajone Copper Sales Contracts

Contracts for First 90.7 kt/a of Blister: Duration of the contracts is 15 years after production start-up.

Table 3. Chronological Application of Funds

Period, Years	Million US$
1955–1969	14.0
1970	14.0
1971	19.7
1972	41.7
1973	95.2
1974	161.5
1975	213.7
1976	178.6
1977	8.5
Total Investment (Chronological)	746.9

Table 4. Recovered Investment, 1977–1981, Millions US$

	Dec. 31, 1977	Dec. 31, 1978	Dec. 31, 1979	Dec. 31, 1980	Dec. 31, 1981
Recoverable Investment	630.2	630.2	630.2	630.2	630.2
Cumulative Recovered Investment	—	(38.1)	(130.4)	(307.6)	(450.5)
Exchange Readjustment	—	0.2	4.4	12.0	16.3
Investment Pending Recovery	630.2	592.3	504.2	334.6	196.0
1) Profit (Loss)	(50.7)	19.1	57.8	30.3	(5.5)
2) Depreciation	79.4	65.0	63.8	85.5	68.7
Less: New Investments	—	(2.3)	(2.9)	(11.5)	(12.6)
3) Cuajone Unit Depletion			47.1	29.3	—
4) Amortization Mine Preparation Expenses	5.8	7.3	8.0	5.8	4.8
5) Amortization Accrued Interest	3.6	3.2	3.4	3.5	3.3
6) Other Reserves	—	—	—	—	—
Yearly Recovered Investment	38.1	92.3	177.2	142.9	58.7
Cumulative Recovered Investment	38.1	130.4	307.6	450.5	509.2

	Max.	
	%	dkt/a
1) British Insulated Callender's Cables Ltd.	15	13.6
2) Enfield Rolling Mills Ltd.	10	9.1
3) Billiton Metallurgie, B.V.	30	27.2
4) Imperial Metals Industries Ltd.	15	13.6
5) Dowa Mining Company Ltd. and Others (The Japanese Copper Purchasers)	30	27.2
	100	90.7

Excess Production above 90.7 kt/a of Blister: Duration of the contractual "investment recovery period."

	%
1) American Smelting and Refining Co. (Asarco)	51.50
2) Newmont Mining Corp.	10.25
3) Cerro Corp. (The Marmon Corp.)	22.25
4) Phelps Dodge Refining Corp.	16.00
	100.00

CUAJONE MINE—OPERATIONS

Premine Stripping

To prepare the pit for an initial uniform production of 27 kt/d of ore, it was originally planned to remove about 136 Mt of waste material during the construction phase of the project. Actually, when the premine period was concluded and mine production started on Oct. 21, 1976, at an ore rate of 40.8 kt/d, SPCC had removed 239 Mt of waste, including approximately 90 Mt of material that was used in the construction of a dam across Quebrada Chuntacala to connect the north and south portions of the pit and also prevent possible flooding by sporadic local rainstorms.

Mining Operations

Cuajone, like Toquepala, is mined by open pit methods. Both operate on a six-day-per-week schedule.

Most of the waste material is directly transported by trucks to the dump areas, while the minor tonnage is hauled by rail-truck combination through loading docks. Approximately 50% of the ore is hauled by trucks to the pit's loading dock for transfer to the trains; the other 50% is loaded directly on trains, also for delivery to the primary crusher. The oxide and low-grade sulfide ores are transported by trucks to selected dumps.

Normally, the 11.5-m³ shovels are worked in waste while the 6- and 6.9 m³ shovels are operated in ore. The 109-t trucks generally haul stripping waste only, while the 90- and 45-t trucks are employed either in ore or in nonore material (waste, oxides, and low-grade sulfides). The haul roads average 25 m wide and have maximum grades of 8%. Each train consists of 12 38-m³ cars drawn by a 1.7 MW GE locomotive.

ANFO is the normal blasting agent used, but Slurrex is employed for wet holes. The drilling pattern varies widely but averages 8 × 6 m. Tricone bits of 311 and 270 mm are currently used in primary drilling, and 76- and 63.5-mm bits are normally used in secondary blasting.

Major Mining Equipment

Initially, Cuajone had available the following major mining equipment capable of removing about 63.5 Mt/a of rock. Equipment consisted of 10 1.7-MW Diesel locomotives and 115 railroad cars; 7 11.5-m³ PH-2100 shovels; 3 6.9-m³ PH-1800 shovels; 2 6-m³ M-152 Marion shovels; 9 Bucyrus Erie rotary drills; 22 90-t M-100 Lectra Haul trucks; 20 109-t Wabco trucks; 8 Haulpack 45-t trucks; and front-end loaders, dozers, medium and light trucks, heavy graders, compressors, drills, cranes, and mobile equipment.

Mine Operating Summary

An operating summary of the mine is given in Table 5. The general mining operating summary to Dec. 31, 1981, is given in Table 6.

CUAJONE CONCENTRATOR OPERATIONS

Location and Capacity

The site at Botiflaca, about 7 km west of the Cuajone mine at an average elevation of 3400 m above sea level, was selected and the concentrator, with a capacity to treat 40.8 kt/d, was built. The design allows for an economical expansion of processing capacity to 49.9 kt/d.

Topographic conditions required the excavation and removal of over 9 Mt of rock at the concentrator site. In addition, over 3.6 Mt of material were excavated and removed in foundation work done for the concentrator and in constructing ancillary installations.

The Cuajone mine and Ilo industrial railroads interconnect in the concentrator area.

Description of the Process

The ore is transported from the mine to the 1524-mm primary crusher by a 9-km long railroad system in 91-t side

Table 5. Cuajone Mine Operations Summary for the Year 1981

	Millions st*
Production	
Waste Mined, Pit	39.49
Ore Mined, Pit	15.85
Ore Mined to Stockpile, Pit	—
Leach Mined, Pit	0.45
LGS to Dumps Stock	0.86
Total Mined Inside Pit	55.65
Total Handled Outside Pit	0.25
Total Mined	55.90
Ore Rehandled, Pit	—
Material Rehandled, Pit	—
Ore from Stockpile to Concentrator	0.02
Flux to Smelter	0.16
Total Mined and Rehandled	56.08
Total Ore to Concentrator	15.87
Waste to Ore Ratio	2.5:1
Grade of Ore, % Cu	1.068
Grade of Flux, % Cu	1.953
Average Daily Mine Production:	
Total Material, St*	219,307
Ore Delivered to Crusher, St*	62,828
Number of Days Worked	254.9†
Drilling	
Total Ft Drilled, Primary	939,942
Total Ft Drilled, Secondary	208,050
Total Shifts Drilled	3,354.7
Ft Drilled per Shift, Primary	280.94
Total Bits Used	91
Blasting	
Footage Blasted	832,374
Total Material Broken	55.11
Tons Broken per Lb Explosive, Primary	4.40
Tons Broken per Ft Drilled and Blasted	66.21
Tons Mined per Lb Explosive Used	4.39
Explosives Used, Millions Lb	12.72
Loading-Electric Shovels	
Tons Loaded by Operations	
Direct Rail Haulage	9.79
Dock Transfer	8.19
Truck, Hopper Transfer	0.19
Truck, Direct Dumping	37.75
Total Tons Loaded	55.92
Truck Haulage	
Total Tons Hauled	46.13
Total per Truck Shift, St	
120-Ton	2,016
100-Ton	1,016
50-Ton	744
All Trucks, Avg	1,744
Haulage Distance Miles, Avg	1.420
Train Haulage	
Total Hauled	18.16
Tons per Locomotive Shift, St	
Direct	2,393
Dock	2,486
Hopper	2,790
Avg	2,437
Haulage Distance Miles, Avg	5.055
Miscellaneous	
Equipment Availability, %	
Drills	75.4
Shovels	87.5
Locomotives	97.5

Table 5. Cuajone Mine Operations Summary for the Year 1981—Continued

	Millions st*
Trucks	84.9
Other Equipment	67.3
Manpower, Mine Division:	
Workers, daily	969
Employees, monthly	136
Staff‡ monthly	57
Total	1,162

  * Metric equivalents: st × 0.907 = t; ft × 0.305 = m; lb × 0.4536 = kg, mile × 1.609 = km.
  † 53.0 days lost due to strikes, work stoppage, etc.
  ‡ Peruvian and Foreign.

dump cars. Undersize and 190.5-mm crushed ore are conveyed to the intermediate ore storage from which it is fed to two 2-m standard and six 2-m short head secondary and tertiary crushers, in closed circuit with ten 1.8 × 4.9-m vibratory screens. The −12.7-mm product is transported to the fine ore storage from which the eight grinding sections will be fed. All material movements are effected by belt conveyor systems.

Each grinding section consists of a 5 × 6-m ball mill, driven by a 2238-MW motor, operating in closed circuit with eight 508-mm cyclone classifiers with corresponding pumps.

The cyclone overflow pulp from all sections passes a sand-slime separation circuit—a very important result of the massive pilot plant tests—consisting of 48 254-mm cyclones with pumps. Sand rougher flotation is effected by 132 2.3-m³ cells; the slime is treated by 84 8.5-m³ rougher cells. The tailings are discharged into three 131-m diam center caisson-type thickeners from which water is reclaimed and pumped back to the mill circuits.

After regrinding in four 3.2 × 5.2-m ball mills, the concentrates are cleaned, recleaned, and scavenged in 124 2.8-m³ flotation cells. The scavenger tailings are conveyed to the tailings thickener; the concentrate joins the feed to two 61-m diam middling thickeners for recycling in the regrind circuit.

The recleaner concentrate is treated in the molybdenite recovery section or sent directly to a 48.8-m diam thickener and fed into four 3.6 × 5.5-m drum filters and later into two 3 × 18.3-m stainless steel rotary dryers.

Water recovered from the concentrate and middling thickeners joins reclaimed water from the tailings thickeners.

Products

The copper concentrate is transported to the Ilo smelter by railroad cars through the five-tunnel system. The moly concentrate is also transported by railroad to the Ilo port for shipment.

The Cuajone tailings flow by gravity through a pipeline system, installed in railroad tunnels Nos. 1 to 4, to join the Toquepala tailings at Quebrada Simarrona and from there flow to the ocean.

Molybdenite Recovery Section

The molybdenite recovery section of the Cuajone concentrator has an installed capacity of 2.3 kt/d of copper-molybdenum concentrate.

This section incorporates feed aging tanks, attrition-conditioning equipment, rougher flotation cells, two stages of regrinding, and eight cleaning flotation stages.

ASMOL copper depressants are applied in the rougher and first cleaner steps and sodium cyanide in subsequent cleaning steps. The thickened final moly concentrate is leached with cyanide, filtered and washed in two stages, and dried and packaged in metal drums. The molydenite recovery section started production in June 1980.

Flux and Low-Grade Stockpile

Higher grade Cuajone mixed oxide-sulfide ores are crushed, screened, and transported to the Ilo smelter, at the rate of about 900 t/d for use as siliceous flux.

Other oxide ore and low grade sulfide [combined reserves in the order of 132 Mt] are being stockpiled at Cuajone for leaching at a future date.

Table 6. General Mine Operating Summary Up to Dec. 31, 1981

	dst*
Material in Dumps (Waste, including pre-mine stripping + low-grade sulfide + oxides)	525,766,200
Sulfide Ore Mined	85,589,200
Oxide Ore (Flux) to Smelter	1,194,200
Total Material Mined	612,549,600
Waste/Ore Ratio	6.1/1.0
Avg Ore Grade, Cu	1.28%

  Metric equivalent: st × 0.907 = t.

Major Concentrator Equipment

A list of major Cuajone concentrator equipment is given in Table 7.

Concentrator Operating Summary

An operating summary of the concentrator is given in Table 8.

Cuajone Molybdenite Recovery

A summary of the Cuajone molybdenite recovery section is given in Table 9.

General Concentrator Operations

The general concentrator operating summary up to Dec. 31, 1981, is given in Table 10.

ILO SMELTER EXPANSION, OPERATIONS

Location and Capacity

The SPCC Ilo smelter, located on the seacoast, 17 km north of Ilo port, has processed the Toquepala mine copper concentrates since 1960. The production capacity of this smelter [127 kt/a of blister copper] was expanded—within the Cuajone project—to a total of 290 kt/a of blister to process the combined productions of both mines. This capacity allows for the normal fluctuations in concentrate grade and tonnage of the ore treated in the two concentrators.

New Major Installations

A modern materials handling system was constructed to replace the old clamshell bucket installations. The concentrate and flux ore materials are now discharged by a rotary car dumper and transported by conveyor to feed the reverberatory furnaces.

Two new 11-m wide by 35-m long side-charging reverberatory furnaces have been installed to treat 1 kt/d of concentrates, flux, and flue dust. The gases generated by these furnaces are processed through four waste-heat boilers and a new Cottrell precipitator system before they are discharged through a new 5-m diam by 107-m high acid-brick-lined concrete stack.

On the converter side, three new 4-m diam by 11-m long furnaces have been installed. In addition, the capacity of one of the four original converters was enlarged. Waste gases are emitted through a new 5-m diam by 107-m high acid-brick-lined concrete stack after being processed by a new Cottrell precipitator system.

The product from the converters is transferred to two new 4-m diam by 11-m long holding furnaces or to the two original holding furnaces. Blister copper is cast in a new 9-m diam 26-mold casting wheel and in the original 24-mold casting wheel.

Finished blister copper is transported by rail to the Peruvian government Ilo refinery or to the Ilo port to be shipped abroad through SPCC's Ilo industrial pier.

The increased consumption of heavy fuel oil by the smelter and the power plant has required erection of additional fuel oil storage tanks.

A second Calciner (hearth-type) kiln capable of producing an additional 113 t/d of lime from coquina shells has been built to satisfy the increased demand. The coquina heavy-media separation plant, located 10 km south of Ilo, was improved and expanded accordingly.

The repair shop and warehouse capacities were enlarged. This required the construction of new buildings and the expansion or relocation of existing ones.

Slag, matte, flux, and other material handling systems were enlarged to satisfy the new plant requirements.

Ilo Smelter Major Equipment

A list of Ilo smelter major equipment is given in Table 11.

Smelter Operations

An operating summary of the Ilo smelter in 1981 is given in Table 12. A general smelter operating summary up to Dec. 31, 1981, is given in Table 13.

POWER GENERATION, TRANSMISSION, AND COMMUNICATIONS

Power Generation

To satisfy the Cuajone electrical requirements, the existing 110-MW nominal (77-MW net) SPCC Ilo power plant was expanded with the installation of a new 66-MW steam turbine generator. The steam requirements of all generating units are supplied by the four waste-heat boilers connected to the two original reverberatory furnaces in the smelter, by the three already installed direct-fired boilers, and by the four 43 092-kg/h waste-heat boilers that were installed in connection with the two new reverberatory furnaces required for the Ilo smelter expansion.

To provide adequate cooling water for the new turbine, the original seawater intake system was improved and expanded. This required the construction of new pumping, desanding, and piping installations.

The Cuajone water supply system included the installation of two 4.5-MW hydroelectric generating plants. Although the principal object of these plants is to provide power for operating the pumps in the water supply system, they are fully integrated into the company's electrical generation and transmission network.

All of these installations are connected to the government's Aricota hydroelectric system and operate under power-interchange contracts by which the company supplies energy to the cities of Ilo and Moquegua and is, in turn, supplied with the same amount of power from Aricota.

Power Transmission

The power generated at Ilo is transmitted to Toquepala and Cuajone through a Ilo-Toquepala-Cuajone-Ilo power loop, using the original Ilo-Toquepala line and the new 138-kV Toquepala-Cuajone (32-km) and Cuajone-Ilo (84.9-km) lines. For this purpose, it was necessary to expand the Ilo and Toquepala substations and to build the Cuajone substation. An additional substation was installed in Tala in the Toquepala-Cuajone line to supply energy, on a temporary basis, for the Cuajone tunnel system construction.

In addition to the main transmission system described, it became necessary to install transmission lines from the Cuajone substation to other substations in the mine (69 kV, 6.5 km) and the Cuajone hospital and townsite areas (13.8 kV, 6.5 km).

A 52-km long 69-kV transmission line from Cuajone substation to the pumping station at Lake Suche was installed to serve the water supply system needs. This line connects to the two 4.5-MW hydroelectric generating plants.

To supply power to the construction sites, shops, and provisional camps, 24 km of temporary transmission lines and eight substations were installed during the construction phase of the project.

Communications

The communications requirements between the different areas of Cuajone and between these areas and the installations in Toquepala and Ilo were satisfied by the installation of an

Table 7. Major Cuajone Concentrator Equipment

Plant Capacity, Ore Milled, tpd	45-50,000
Processing Equipment:	
Unloading Ore	Dumping from Rail Cars
Primary Crushing Plant	
Location	Mill
Number	1
Size/Manufacturer	60 × 89-in. Allis Chalmers
Designation	Gyratory Crusher
Motor Rating (hp)	500
Open Side Setting	7½ in.
Coarse Ore Feeders	
Number of Feeders	2
Size/Manufacturer	84-in. × 31-ft GEC Elliot
Designation	Pan Feeder
Motor Rating, (hp)	125
Coarse Ore Conveying System	
Number of Stages	2
Belt Size	60 in./72 in.
Belt Length, total	2620 ft
Motor(s) Rating, Total hp	2-800/250 = 1850
Coarse Ore Storage	
Capacity, Total, tons	220,000
Capacity, Live, tons	40,000
Fine Crushing Plant	
Number Secondary Crushers	2
Size/Manufacturer	7-ft Symons Standard Cone
Designation	Extra Heavy Duty Standard Cone Crushers
Motor Rating, hp	350
Number Tertiary Crushers	6
Size/Manufacturer	7-ft Symons Shorthead Cone
Designation	Extra Heavy Duty Cone Crusher
Motor Rating	350
Screening Equipment	
Location	Secondary Crushing
Number of Screens	4
Size/Manufacturer	6 × 16-ft Double Deck
Designation	Ty-Rock Vibrating Screens
Location	Tertiary Crushing
Number of Screens	6
Size/Manufacturer	6 × 16-ft Single Deck
Designation	Ty-Rock Vibrating Screens
Circuit	Closed Circuit
Fine Ore Storage/Feeders	
Capacity, Total, tons	86,400
Capacity, Live, tons	48,000
Number of Feeders/Section	3
Grinding Plant	
Number of Sections	2
Size/Manufacturer	16½ × 20-ft Allis Chalmers
Designation	Overflow Ball Mills
Motor Rating, hp	3000
Classification Equipment	
Number of Pumps	8 (1 per Mill)
Size/Manufacturer	16 × 14-in. Denver Hard
Designation	Hard Metal End
Motor Rating, hp	8-150 hp
No. of Classifiers/Section	4/Mill
Type of Classifier	Krebs Cyclone Classifiers
Size	D26B
Flotation Plant	
Number Rows/Cells Per Row	Sands* Slimes*
	3/22 3/14
	Total 132 Total 84
Size/Manufacturer	Sands: 80-cu-ft Galigher
	Slimes: 300-cu-ft Wemco
Designation	N-60 Galigher 300-cu-ft Wemco
	Agitairs Sand Slime Rougher
	Rougher 25 40
Motor Rating, hp	Sands: 25 hp Slime: 40 hp

Number Rows/Cell per Row Cleaner Flotation	Cleaners: 2/10/Each Section Recleaners: 1/10 Each Section Scavengers: 2/16 Each Section All Denver DR 100-cu-ft
Number of Stages	2
Number of Regrind Mills	2 per section
Size/Manufacturer	10-½ × 17-ft Allis Chalmers
Designation	Overflow Mills
Motor Rating, hp	800 hp each

Metric equivalents: ton × 0.907 = t, in. × 25.4 = mm, ft × 0.305 = m, cu ft × 0.028 = m³, hp × 0.746 = kW.
* Number per Section; the plant has two sections.

automatic telephone system with exchanges at the mine, the main office, the concentrator, the hospital, and the townsites and by a radio system that supplies communication with Lima. A telephone system, connected to the government's national telephone network, has been incorporated.

A portable-unit radio communication system has been installed in the vehicles of the operation, maintenance, and supervisory personnel. This system is controlled from a fixed central station and was extensively used during the construction phase.

WATER SUPPLY SYSTEM

Studies

A water evaluation study started in 1967 in the Vizcachas River-Pasto Grande watershed, which had been reserved by the government for the Cuajone project as per terms of the Bilateral Agreement, proved a firm supply of only 1000 L/s, not enough to fully satisfy the needs of the project.

Consequently, investigations centering on the development of underground water sources were initiated by SPCC. These studies were successful in discovering a large aquifer with sufficient capacity to satisfy the project requirements.

New Sources

A water supply system was installed and operates for industrial and domestic purposes of the concentrator and mine. Two small hydroelectric power generating plants, metallurgical and leaching systems, townsites, tailings disposal, etc., were designed and built to the necessary supply capacity to serve the installations that compose the Cuajone unit.

SPCC has developed up to 1500 L/s that can be pumped

Table 8. Cuajone Copper Plant Operation Summary for the Year 1981

	Dst* Millions	
Crushed, Primary	15.87	
Crushed, Secondary	15.88	
Milled	15.87	
Milled per Day, dst		52,136
% Moisture		3.36
Concentrate Produced, dst		454,061
Operating Days		304.34†
Plant Capacity, tpd		52,136
Feed:		
% Total Copper		1.068
% MoS₂		0.028
Recoveries:		
Total Copper, %		86.58
MoS₂, %		70.59
Concentrate Grades		
% Cu		32.33
% MoS₂		0.6923
Assay of Tailings, % Total Copper		0.147
Ratio of Concentration		34.94
Power Requirement		
kWh/ton milled		15.32
Grinding Ball Consumption, lb per ton Milled		1.61
Flotation Reagents, lb per ton Milled:		
Primary Collectors		0.0266
Frothers		0.0368
Lime, lb CaO		3.74
General Mill Grinds, % +100 Mesh		18.3

* Metric Equivalents: st × 0.907 = t, lb × 0.4536 = kg, lb per st × 0.5 = kg/t.
† 53.72 days lost due to strikes, work stoppage, etc.

Table 9. Cuajone Molybdenite Recovery Section Summary

Available Feed, dst	454,061
Available Feed Treated, dst	453,511
Molybdenite Concentrate Produced, Dry lb	6,082,057
Plant Capacity, tpd	2,500
Assay of Feed, % MoS$_2$	0.6923
Assay of MoS$_2$ Concentrate:	
% MoS$_2$	94.07
% Copper (after leaching)	0.85
% Insoluble	2.02
% Iron	1.76
Assay of Tailings, % MoS$_2$	0.0499
Ratio of Concentration	143.86
Actual Recoveries:	
Copper Plant, MoS$_2$ %	70.59
Molybdenite Plant, MoS$_2$ %	91.12
Overall MoS$_2$, %	64.26
Flotation Reagents, lb per ton feed:	
Sodium Sulfide	8.29
Arsenic Trioxide	2.71
ASMOL	11.00
Sodium Cyanide	2.07
Exfoam 636	0.53
DWT-287	0.10
Total Antifoam	0.63
Fuel Oil (Flotation)	0.04
Nitrogen Plant:	
Monoethanolamine	0.04
Fuel Oil	0.41 gal per ton

Metric equivalents: st × 0.907 = t, lb × 0.4536 = kg, lb per ton × 0.5 = kg/t, gal per ton × 4.173 = L/t.

from aquifer beds in the Capillune formation located in the Titijones, Huaitire-Gentilar, and Vizcachas Lake basins. An additional underground reserve capacity of 300 L/s has also been found at Vizcachas Lake basin for use in the future expansion of the Toquepala operations.

Water Requirements

The Toquepala and Cuajone operations requirements are as follows:

	L/s
Initial Phase	
Toquepala current needs	600
Cuajone initial needs	750
	1350

Final Phase	
Toquepala, including expansion	1000
Cuajone final needs	1500
	2500

Supply System

The Cuajone project main water supply system consists of a floating pumping station on Lake Suche and a high pressure booster pumping station located on its shore. This installation lifts the water approximately 220 m over a distance of 7 km through a 965-mm diam steel pipeline. From this point, the water flows by gravity through a 864-mm diam, 44-km long resin-reinforced fiberglass pipeline to the Cuajone concentrator area. This pipeline also conveys water

Table 10. General Concentrator Operating Summary Up to Dec. 31, 1981

	Dst
Ore Received	85,726,400
Avg Ore Grade, Cu	1.28%
Ore Processed	85,431,800
Copper Contained in Ore Processed	1,093,500
Copper Concentrates Produced	2,539,700
Avg. Grade of Concentrates, Cu	36.59%
Contained Copper	929,312
Moly Concentrates Produced	5,862
Avg. Grade of Concentrates MoS$_2$	91.25%

Metric equivalent: st × 0.907 = t.

Table 11. Ilo Smelter Major Equipment

Smelting Reverbatory Furnaces		
Type	Sprung SiO$_2$, Combined Dietrick/Suspension	Floating, Panelize Basic Arch
Number	2	2
Size	30×113 ft	36×116 ft
Fuel	Residual Fuel Oil	Residual Fuel Oil
Fuel Ration Mm Btu per Ton Solid	6.389	6.389
Smelting Rate, tpd	700	1150
Converter, Pierce Smith		
Size	3, 13×30 ft, and 4, 13×35 ft	
No. Tuyeres and Size	42, 2 in. and 48, 2 in.	
Blowing Rate, avg.	27,400 cfm	
Matte Grade, % Cu	35.73	
Blister Casting	4 Cylindrical Vessels, 13×35 ft	
Type	1, 24-Mold Casting Wheel 1, 26-Mold Casting Wheel	

Metric equivalents: ft × 0.305 = m, Btu × 0.252 = kcal, ton × 0.907 = t, in. × 25.4 = mm, cfm × 0.0283 = m³/min.

collected through a feed line from the Titijones basin wells field.

In the future, when the Cuajone and Toquepala units are expanded and their operations need more water, additional flow will be pumped from wells located in the Huaitire-Gentilar and Vizcachas Lake basins.

A 700,000−m³ reservoir was built at Viña Blanca, a site located between the two 4.5-MW hydroelectric generating plants near Cuajone. This volume can satisfy the fresh water requirements of the mine-concentrator complex during emergencies.

New Ilo Seawater Desalination Plant

The expansion of the Ilo smelter and other installations at Ilo, required an additional volume of water. To satisfy this need, SPCC built a 42-L/s seawater desalination plant, similar to the one already serving the then current Ilo smelter-Toquepala operations, which has a capacity of 32 L/s.

TUNNEL SYSTEM, INDUSTRIAL RAILROAD, AND TAILINGS DISPOSAL

Introduction

The original 206-km Ilo-Toquepala industrial railroad was extended to the Cuajone area, through a 30-km line of which 27 km were through tunnels. These tunnels allow for the installation of a standard gage railroad track and a pipeline system to convey the tailings by gravity from the Cuajone concentrator to Quebrada Simarrona, to mix with the Toquepala concentrator tailings, and from there flow down 93 km of dry creeks until reaching the Pacific Ocean at Ite Bay.

Preliminary Studies

Geologic studies of the areas to be crossed by the tunnels, including the drilling of approximately 4800 m of exploratory holes, were made during several years. Based on this geological exploration, tunnel support studies were made and the conclusion reached that probably 51% of the total tunnel length would not require support, 29% would require bolt and wire mesh support, and 20% would need steel and shotcrete. These predictions were supported by fact when the tunnel system was completed.

Load Characteristics of the Industrial Railroad

The industrial railroad has the following daily average load characteristics (include both the Toquepala and Cuajone operations): concentrates and blister copper, 4.5 kt; supplies, 2.3 kt; and silicious flux and other materials 907 t.

Tunnels

The system consists of five tunnels with a total length of 27,076 m and a net cross section of approximately 36 m². Tunnel data are given in Table 14.

Pipeline for Tailings Transportation

A system of 762-mm resin-reinforced fiberglass plastic pipelines was originally installed in four of the tunnels for transporting the Cuajone tailings. The R-5 tunnel, located south of Quebrada Simarrona, that is, past the point where the tailings from both concentrators mix, does not contain pipelines.

Although extensive and successful tests using Toquepala tailings had been carried out to determine the abrasion resistance characteristics of the plastic pipeline, it became necessary to replace the Cuajone tailings pipeline system. A steel-reinforced concrete flume was used to replace the plastic pipeline after the second year of operation when it became evident that the abrasion/flow characteristics of the Cuajone tailings exceeded the wear resistance of the plastic pipeline.

MAIN ROAD SYSTEM

The roads consist of permanent highways and access roads to the different areas.

Permanent Highways

Permanent highways have been built or improved for connecting the centers of operations, camps, etc. These permanent highways permit access to the Cuajone unit area from the Toquepala mine, from the city of Moquegua, and from the Andean highlands. A total of 123 km of permanent roads were built and 68 km of permanent highways were improved.

Access Roads

An extensive trail and road network was built during the construction phase of the project to serve the various sites and facilitate the supply of equipment and materials for

Table 12. Ilo Copper Smelter Operating Summary for the Year 1981

Reverberatory Department	Dst*
Furnace Days	895.67†
Concentrate Smelted from Toquepala	419,628
Concentrate Smelted from Cuajone	459,494
Total Concentrate Smelted	879,122
Reverb Silica Smelted	33,951
Secondaries Smelted	4,548
Coquina Smelted	64,652
Total Solids Smelted	982,273
Concentrate per Furnace Day	982
Total Solids per Furnace Day	1,097
Mm/Btu per ton Concentrate Smelted, Furnace	6.389
Mm/Btu per ton Concentrate, Air Preheater	0.828
Mm/Btu per ton Concentrate Smelted, total	7.217
Mm/Btu per ton Solids Smelted	6.460
Matte Produced	788,552
Lb Steam Produced per ton of Concentrate Smelted	3,493
Converter Slag Returned	565,191
Converter Slag per Furnace Day	631
% Cu in Toquepala Concentrates Smelted	24.40
% Cu Cuajone Concentrates Smelted	32.26
% Cu in Concentrates Smelted	28.51
% Cu in Matte	35.73
% Cu in Reverb Slag	0.63
Total Dust Recovered in Cottrell	3,202
% of Dust Recovery	91.62
% of Copper Recovery	96.38
Converter Department	Dst*
Matte Charged	788,552
Silicious Flux, Toquepala and Cuajone Mine Ore	200,107
Secondaries, Minero Peru Refinery	116
Secondaries, Smelter	18,290
Total Material Treated	1,007,065
Total Converter Days	1,662.27
Tons Matte per Converter Day	474
Matte Grade, % Cu	35.73
Blister Produced	252,654
Blister Converter Day, avg.	152
% Cu in Blister Produced	99.242
% SiO_2 in Converter Slag, avg.	24.20
Converter Efficiency, %	68.94
Total Dust Recovered in Cottrell	6,616
% of Dust Recovery	94.43
% of Copper Recovery	97.05
Manpower, Ilo Area	
Workers, daily	1,472
Employees, monthly	308
Staff‡ monthly	182

* Metric equivalents: st × 0.907 = t, Btu × 0.252 = kcal, lb × 0.454 = kg
† 52 days lost due to strikes, work stoppages, etc.
‡ Peruvian and foreign.

construction. Many of these roads have been kept in operating condition to provide access for inspection and maintenance purposes to pipeline, transmission lines, tunnel systems, and other installations.

TOWNSITES AND GENERAL SERVICE INSTALLATIONS

Permanent Townsites

Two permanent townsites have been built and a third substantially expanded to provide housing and other community services and facilities for the Cuajone operating personnel and their families. These new towns were constructed in accordance with standards and requirements of the Peruvian mining code relating to housing, educational, and medical facilities. The permanent townsites are: (1) Villa Cuajone to house management and supervisory personnel, (2) Villa Botiflaca to house employees and workers, and (3) expansion of Ilo camp.

Villa Cuajone faces the Torata Valley and is located 18 km west of the Cuajone mine, 10 km from the Cuajone concentrator, and 30 km from the city of Moquegua (road distances), at an average elevation of 2700 m above sea level. Three- and four-bedroom houses eight-unit apartment buildings, and other housing units have been constructed for mar-

Table 13. General Smelter Operating Summary Up to Dec. 31, 1981

	Cuajone Production, Dst
Copper Concentrates Received	2,539,700
Avg. Grade of Concentrates, Cu	36.59%
Concentrates Smelted	2,487,900
Avg. Grade of Concentrates, Cu	36.63%
Copper Contained in Concentrates	911,279
Concentrates Sold	37,000
Concentrates Stock	14,800
Blister Copper Produced	902,363
Avg. Grade of Blister, Cu	99.20%
Copper Contained in Blister	895,144

Metric equivalent: st × 0.907 = t.

ried and single status personnel and for visitors. Villa Cuajone has adequate educational, social, and recreational facilities; a shopping center; police and fire stations; and ancillary community services.

Villa Botiflaca is located 8 km west of the Cuajone mine, 2 km from the Cuajone concentrator and 38 km from the city of Moquegua (road distances), at an average elevation of 3325 m above sea level. Eight- and 12-unit three-bedroom apartment family buildings and single status residence halls and dormitories have been constructed. Villa Botiflaca has adequate social and recreational facilities, a church building, movie house, shopping center, police and fire stations, bakery, laundry and dry cleaning plants, and other community services.

Ilo Townsite, which originally housed the personnel operating the Ilo smelter, the power plant, the seawater desalination plant, port, and the industrial railroad and other installations related to the Toquepala operations, has been expanded to house and provide adequate services for the new personnel required in the area as a result of the incorporation of the Cuajone unit.

Temporary Camps

Ten temporary camps and ancillary installations were built to house and provide necessary services to contractors' personnel during the construction phase of the project. A work force of about 6000 men used these facilities.

Table 14. Cuajone Tunnel Data

Tunnel	Length, m	Gradient, %	Orientation
R-5	2326	−0.9	South-North
R-4*†	14 722	+1.0	South-North
R-3*	990	+1.2	Curved
R-2*	5448	+1.5	Southeast-Northwest
R-1*	3590	+1.5	East-West

* Tunnel R-1 to R-4 conduct Cuajone tailings.
† Tunnel R-4 has a 207-m long, 1.8-m diam ventilation shaft.
English equivalent: meter × 3.281 = foot.

Cuajone Hospital

A 69-bed hospital located near the Botiflaca-Villa Cuajone road junction was constructed and it has been provided with the most modern medical equipment and comfort. The original Ilo hospital has been expanded.

General Service Installations at Cuajone

These comprise facilities that provide services to all operations in the Cuajone area, rather than to specific areas or operational functions. They consist, essentially, of the general office, operations office, industrial warehouse, time office, flammable material storage building, and the Cuajone airstrip.

SHOPS AND SUPPORT INSTALLATIONS

Mine Shops

An ample mine heavy duty shop has been built and equipped to repair and service large-size mining machinery and equipment. This shop consists essentially of machine welding, shovel, repair, tractor, and tire and truck sections.

Other support installations in the area are the explosive magazine, electrical shop, engine rebuild shop, and tank farms.

Concentrator Area Shops

The light duty and paint shops for the repair of automotive vehicles, built for use during the construction phase, continues in operation. Additionally, the concentrator area includes a small warehouse, a carpenter shop, and a light mechanical shop.

Ilo Area Shops

The Ilo port and smelter shops were enlarged to take care of the requirements for the Cuajone project in that area.

Industrial Railroad Shops

These comprise the expansion of the Toquepala industrial railroad maintenance shop and the construction of a mine locomotive shop at Cuajone.

Camp Maintenance Shops

Camp maintenance shops were built at Villa Cuajone and Villa Botiflaca and the Ilo camp maintenance shop was expanded.

Principal Temporary Installations

To satisfy the Cuajone project's construction phase needs, several temporary shops and ancillary installations were built in different areas of the project. The principal installations are the Ilo steel fabrication shop, tunnel shops, concrete batch plant, block plant, aggregate plant, carpenter shop, reinforced plastic fiberglass mortar pipe plant, and a large diameter steel pipe fabricating shop.

CONSTRUCTION CONTRACTORS

Fluor Utah Inc. was the general contractor for engineering and construction management. The project employed over 30 independent contractors—mostly Peruvian construction companies, some of which associated, under joint venture contracts, with internationally recognized building contractors. During its construction phase, Cuajone employed about 6000 workers.

9.6 The Chuquicamata Complex

ENRIQUE MOREL DONOSO

INTRODUCTION

Of the four divisions of the National Copper Corp. of Chile (CODELCO-Chile), the Chuquicamata complex is the largest mineral resource. It is located in the province of Loa, II Region, Antofagasta, Chile, 1600 km (994 miles) north of Santiago, the nation's capital. It is also 240 km (149 miles) northeast of the port of Antofagasta and 150 km (93 miles) east of the port of Tocopilla, between 2500 and 3000 m (8202 and 9843 ft) above sea level (Fig. 1). The most important populated zones in the area are Calama with 90,000 inhabitants and Chuquicamata with 17,000.

The climate in the region corresponds to marginal high altitude conditions, is extremely dry and rainless, except during the *Bolivian winter* which occasionally produces torrential rains between December and March. The average annual temperature is 22.8 °C, subject to seasonal and daily variations.

The complex is based upon a porphyry copper deposit, 14 km (9 miles) long from north to south, with an average width of 1 km (0.6 miles) from east to west, in which are to be found the areas known as *Chuqui Norte,* still unexploited, Chuquicamata, which is the principal area under exploitation, and inmediately to the south, *Mina Sur,* also being exploited. The latter orebody is formed by the deposition of oxidized exogenous ore that gave origin to an exotic copper deposit 6 km (3.7 miles) in length from north to south.

The power required by the complex is obtained from two thermoelectric generating plants owned by the company. The most important, and the one that produces approximately 90% of the power used, is the Tocopilla Thermoelectric Plant located at the port. The remaining 10% is obtained from the Chuquicamata Thermoelectric Plant which generates low-cost energy by means of waste heat boilers that use the excess heat of the gases from the reverberatory furnaces in the smelter.

The water used by the complex comes from the Andes mountains, by means of two water supply pipelines for industrial water, 60 km (37 miles) and 70 km (43 miles) in length, and three supply pipelines for potable water 99 km (61.5 miles), 92 km (57 miles), and 108 km (67 miles) long, respectively. Through different retreatment and recovery systems, approximately 70% of the fresh water received is recirculated. The main systems are the sewer water treatment plant, tailings dam water recovery system; thickeners systems in the concentrator, recirculating system in the smelter; thermoelectric plant and substations.

Considering the history of the Chuquicamata complex, the magnitude of the deposit, the many technological changes, and the resulting projects developed during its exploitation, this is a very special mining project. Accordingly, the following points will be described: the history and geology of the deposit; project execution; description of operations; operations costs; and production and technology in the future.

THE HISTORY AND GEOLOGY OF THE DEPOSIT

History

Before the Incas conquered the region, it was inhabited by a small tribe of Indians who had descended from the Aimarás and the Quechuas, who were known as the *Chucos,* hence the name *Chuquicamata.*

Archaeologists and historians think that aborigines of the area of Atacama were the first to use the copper from this region in the making of domestic utensils and weapons. It is believed that Diego de Almagro, on his return to El Cuzco after the invasion of Chile, used copper from Chuquicamata to reshoe his horses.

At the close of the last century, some of the copper veins near Chuquicamata were being exploited, in particular the *Zaragoza, Balmaceda, Angélica, Poderosa, Lérida,* and *San Antonio* veins. In 1910, Albert C. Burrage, a lawyer and financier from Boston, obtained information on the extent of the Chuquicamata deposit. After a thorough study, he then interested the Guggenheim Brothers in the project, who in turn founded the Chile Exploration Company in Chile on January 11, 1912. The evaluation of the deposit was then undertaken and very soon the figures justified accelerated mining development.

Mining operations began in 1915 with the extraction of oxidized ore, principally atacamite, antlerite, brochantite, and kroehnkite with approximately a grade of 1.75% Cu. Since 1952, copper sulfides have been mined, principally chalcocite, covellite, and chalcopyrite.

During its 67 years of operation, the mine has lived the history of the development of mining equipment, from steam shovels, rail mounted shovels, chain shovels, wooden hammer and rope drills, steam locomotives, etc., to the modern equipment presently in use such as electromechanical shovels with a capacity of up to 21.43 m³ (28 yd³), electromechanical trucks 231 t (255 st) and modern rotary drills. As the production of the mine reached certain sectors of the oxide/sulfide transition zone, the need arose for construction of a plant to treat the ore by flotation and fire refining methods. The new plant went into service in 1952 and gradually became increasingly important in the total production of fine copper, while production based on oxidized ore declined. By 1971, the exploitable reserves of the oxide minerals in Chuquicamata had been exhausted. To replace this type of ore and to take advantage of the existing facilities in the leaching plant, in 1970 the first ore from the Exotica mine (today known as Mina Sur) was processed as part of the production of the Chuquicamata mining complex.

In all these years of operation in the main pit of Chuquicamata, until June of 1982, the following quantities of ore have been extracted:

Material	Millions of st
Oxidized	505
Sulfides	518
Mixed	49
Waste	938
Total	2,010

The exploitation has given the excavation the form of a large elliptic amphitheater whose present dimensions are 3.5 km (2 miles) in length, 2.0 km (1.2 miles) in width at the top, and a depth of 476 m (1562 ft) from the highest point. At the present time, the copper sulfides exploited are chalcocite, covellite, chalcopyrite, etc., and only a small reserve

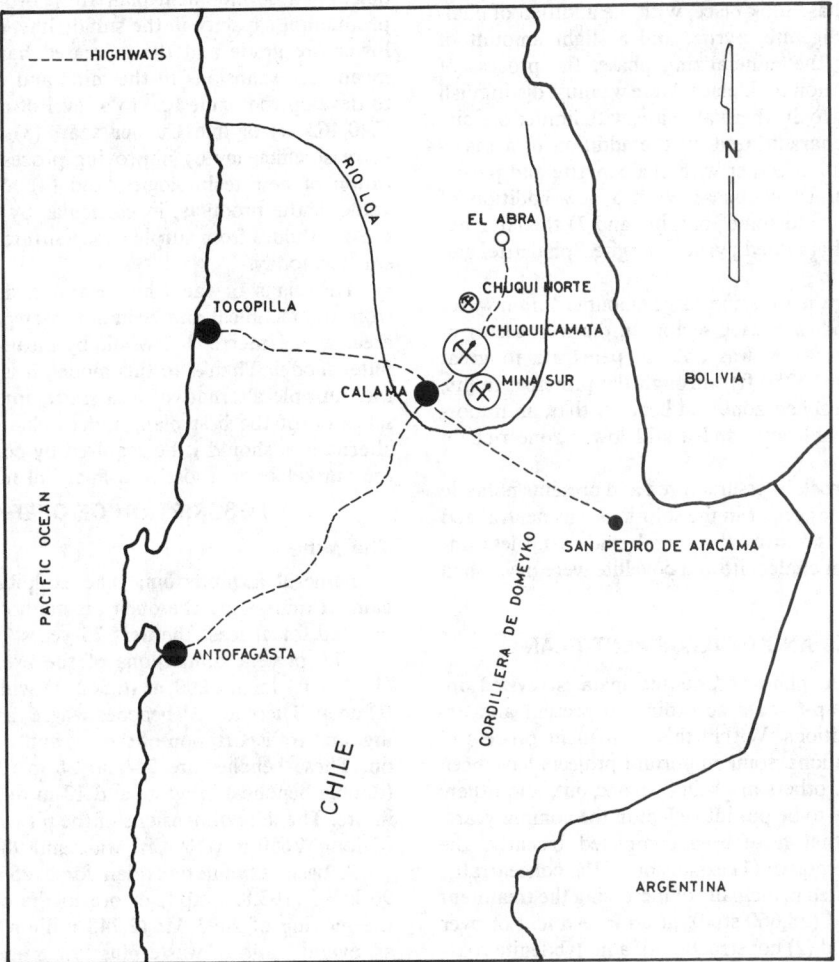

Fig. 1. The Chuquicamata complex and its location, National Copper Corporation of Chile (Codelco-Chile).

of oxidized ore remains. It is also worth noting that the deposit contains other useful elements such as molybendum in the form of sulfide (molybdenite), gold and silver, the recovery of which, as byproducts, is economically profitable.

In 1952, for the purpose of finding the northern limit of Chuquicamata, and because of the need to look for new reserves of oxidized ore, a drilling program using churn drills was carried out on an east-west profile every 500 m (1640 ft) up to 10 km (6.2 miles) north of the Chuquicamata pit. This led to the discovery of the mineralized zone called *Chuqui Norte*.

In 1963, drilling was carried out on the spent oxide dumps to the south of Chuquicamata to investigate the flow of solutions of in situ leaching tests. These drill holes discovered the exogenous deposit now exploited and known as *Mina Sur*.

Geology

The Chuqui porphyry copper is a porphyritic body of granodiorite vein type, 14 km (8.7 miles) long with a mean width of 1 km (0.6 miles) completely mineralized, but with a nucleus of greater mineralization in Chuquicamata and Chuqui Norte. The estimated age is between 32 and 34 million years.

The deposit is formed by magmatic porphyry that probably began to develop during the Lower Paleozoic period and during the latest stage of development in the Middle Tertiary, together with a process of potassic alteration, the first mineralizing phase took place, with the addition of chalcopyrite, bornite, digenite, pyrite, and a slight amount of molybdenite. After the mineralizing phase, the process of hydrothermal alteration took place. Here we must distinguish between 1) the early hydrothermal phase, with limited sericite quartz alteration, characterized by the addition of a maximum of molybdenite, together with chalcopyrite and pyrite, 2) the main hydrothermal phase, with a new addition of pyrite, chalcopyrite, and some enargite, and 3) the later hydrothermal phase that added pyrite, enargite, sphalerite, galena, and tetrahedrite.

Later on, during the Pliocene, large amounts of rain water formed enriched bodies in each sector. In the west sector of Chuquicamata, the water was able to penetrate to great depths [over 1000 m (3280 ft)] through the principal faults, forming an upper leached zone and beneath that, an important zone rich in chalcocite and a still lower zone rich in covellite.

To the east, the rock is less fractured and presents potassic alteration, so that the copper in the solutions was neutralized near the surface in the form of mineral oxides and less important zones rich in chalcocite and covellite were developed below.

PROJECTS AND DEVELOPMENT PLAN

The development plant at Chuquicamata is revised annually to bring it up-to-date according to present and expected future conditions. Within this permanent process of planning and operations, some important projects have been concluded recently, others are being carried out, and others are still under study to be put into effect in the coming years.

Among those that have been completed recently, the following should be noted: (1) expansion of the concentrator complex that consisted principally of increasing the treatment capacity by 26 kt/d (28,660 stpd), at an investment of over US$ 100 million, and (2) construction of a molybdenite roaster plant, with a capacity of 5.5 million kg/a (12 million lb per yr), costing approximately US$ 20 million.

Among the projects currently being executed, the most important are (1) expansion of the thermoelectric plant at Tocopilla by the addition of a 72 MW unit (N°12) at a cost of US$ 70 million; (2) a new primary crushing station with the capacity to treat up to 9600 t/h (10,582 stph), at a cost of US$ 100 million; and (3) development of the concentrate smelter (first stage), consisting of a group of related subprojects, including expansion of dryer capacity, enlargement of converter bay, installation of a modified converter of the Teniente type, a gas recovery system, an oxygen plant, and improved storage and handling of concentrate. The investment in this phase amounts to US$ 70 million.

The following projects have been approved for the next few years: (1) expansion and rebuilding of the refinery to increase production capacity from 370 kt/a (407,855 stpy) to 630 kt/a (694,456 stpy); (2) oxide tailings *ripios*, dump leaching and solvent extraction plant, to achieve a production rate of 200 t/d (220 stpd) of fine copper by 1988; (3) heap leaching to replace vat leaching, to achieve production of 140 t/d (154 stpd) of fine Cu. This technological change will permit significant reduction in production costs in the oxide line. This project also includes a solvent extraction plant with a capacity to treat 2000 m^3/h (1529 yd^3 per hr) of solution. Expansion of the Tocopilla thermoelectric plant with two more units (13 and 14) of 72 and 38 MW, respectively, is also planned.

The basic objectives of the projects included in the Chuquicamata development plan are (1) to maintain the copper production capacity in the sulfide line, compensating for the lower ore grade and the increased hardness of the ore by means of expansions in the mine and the concentrator; (2) to develop the oxide line to a level of approximately 200 kt (220,462 st) of fine Cu per year; (3) to reduce costs and increase efficiency by improving processes and the incorporation of new technologies; and (4) to increase the added value of the products, in particular by producing electrorefined cathodes from surplus concentrates that are sold or toll smelted today.

The plans of the Chuquicamata division are analyzed from the technical and economic viewpoints in the different areas as an interrelated whole by means of a complex computer model. Thanks to this model, it is possible to simulate the multiple alternatives in a short time, which permits the selection of the best plan, and to change some of the subalternatives should it be required by conditions imposed by the market or by those of a financial nature.

DESCRIPTION OF OPERATIONS

The Mine

Mineral Exploitation: The exploitation of the Chuquicamata mine is by the open pit method, which system will be used for at least the next 25 years.

The present dimensions of the excavation are 3500 m (11,483 ft) long, 2000 m (6,562 ft) wide, and 476 m (1562 ft) deep. There are 21 benches where drilling, blasting, loading, and transportation of the ore and waste rock are carried on. These benches are 24 and 26 m (78.7 and 85 ft) high (double benches) in waste and 13 m (43 ft) (single benches) in ore. The final dimensions of the pit will be 4100 m (13,451 ft) long, 2650 m (8,694 ft) wide and 986 m (3,235 ft) deep.

A basic exploitation plan for a 25-year period, feeding 96 kt/d (105,822 stpd) of ore to the concentrator, implies the moving of 2493 Mt (2,748 million st) of material with an overall ratio of waste plus low grade ore to ore of 1.89. The fine copper production level decreases from 1300 t/d (1433 stpd) at present to 800 t/d (882 stpd) by the end of the period, due to the reduction in ore grade as a result of the change from the present secondary enriched zone to the primary zones of the deposit. Chalcopyrite increases progressively while chalcocite and covellite decrease. The same grade change occurs with the molybdenum in the ore.

The mine next to Chuquicamata, Mina Sur, is exploited with benches 13 m (43 ft) high. The present dimensions of the pit are: 1500 m (4,921 ft) long, 1200 m (3937 ft) wide and 195 m (640 ft) deep. The oxidized ores are leached in vats which were used for the processing of oxides from the Chuquicamata pit. The basic production level of fine copper, 95 t/d (105 stpd) requires the handling in the next 22 to 24 years of 450 Mt (496 million st) in order to achieve the final pit.

The Chuquicamata Pit: Usually work goes on six days a week, in three 8-hr shifts per day, and Sunday is assigned to maintenance and special repairs to roads and access to the dumps, special earth moving jobs, and the relocation of heavy equipment and machinery. Also the preventive maintenance and major repairs on equipment and the primary belts in the crushing system are done on Sundays in order not to interrupt productive work during the week.

The Chuquicamata mine has a crusher-belt system located in the pit for handling ore. The main piece of equipment is an Allis-Chalmers gyratory primary crusher 1.37 × 1.88 m (54 × 74 in.) with a capacity of 5 kt/h (5512 stph) and a system of conveyor belts in two sections of 1524 mm (60 in.) wide and a total of 1830 m (6004 ft) in length, which transport the ore through a tunnel and which discharge the storage pile (120 kt capacity) at the concentrator.

Drilling—The first step of exploitation is the drilling of blast holes to a depth equal to the height of the bench plus 2 to 3 m (6.6 to 9.8 ft) subdrilling. The drilling is done with electromechanical rotary drills, eight Bucyrus Erie units, model 61-R, and one Gardner Denver model GD-60, powered by 5,000 V obtained from an aerial loop installed in the mine. Tricone bits of 279 and 311 mm (11 and 12¼ in.) diameter, with tungsten carbide inserts of different brands are used. The drilling pattern is designed with vertical holes laid out in rectangles or similar pattern, spaced at an average of 7 × 9 m (23 × 29.5 ft).

The average length of drill hole achieved daily is 1480 m (4856 ft), or the equivalent of 217 m/worked shift/machine, with approximately 37 400 m (122,703 ft) drilled per month. The cost of drilling is approximately 5.4% of the mine's total operating cost. Since the drill bits represent the greatest share of this cost, strict statistical control of yield and cost per bit per meter drilled is maintained.

Blasting—Blasting is carried out according to a plan determined by the planning and operations engineers and represents 16.4% of the mine's total operating cost.

The average explosive load for the mine is approximately 0.215 kg (0.474 lb) of explosive per ton blasted; 60% of the explosives are nitrocarbonitrates and the rest aqua-gels.

Nearly 3700 t/d (4079 stpd) of rock requires secondary blasting, for which three boom drills and one hydraulic rock-breaking hammer are used.

Loading—The total daily production of the mine is 300 kt (331,000 st) with a ratio of waste (including low-grade mineral) to ore of nearly 1.8. The loading is done by fourteen electromechanical shovels with a bucket capacity that varies between 9.17 (12 yd³) and 21.41 m³ (28 yd³).

The shovels used are Bucyrus Erie 190-B (1) and 280-B (2), Komatsu-Bucyrus 280-KB (6) and P&H 1,800 (5). To support the loading operation, 7.65 m³ (10 yd³) and 9.17 m³ (12 yd³) capacity Caterpillar 992 and Clark Michigan 475

front end loaders are used. The loading phase represents approximately 13.6% of the mine's total operating cost. The average shovel production is 12 420 t/worked shift/machine.

Transportation—At the present time, the transportation of material from the mine, whether to the crushing stations, storage piles, or waste dumps, is carried out solely by trucks. The fleet includes 90 units whose capacity varies between 109 t (120 st) and 231 t (225 st). The makes of truck used include Wabco 120-B and C, 170-C and D and 3200-B, Terex 33-15-B and Euclid R-170s.

The performance of the 154 t (170 st) units (57% of the total fleet) is approximately 1250 t-km/working hour, for an average haul distance of 4.5 km (3 miles).

It should be noted that since 1979, a computerized dispatch system controls the trucks and this has permitted substantial improvement in the operation, an increase in the effective usage of the equipment and a cost decrease. It also permits continuous control and homogenization of the ore blend sent to the plant through the control of the copper grade, molybdenum, and arsenic content. The cost of transportation is approximately 48% of the total operating cost of the mine.

Auxiliary Equipment—For production support jobs, there are 25 Caterpillar tractors, 17 rubber tired tractors, 9 bulldozers, and 6 front loaders of different brands and types. There are also 8 40 m³ (52 yd³) water trucks for sprinkling the roads, fuel trucks, compressors and personnel transport vehicles. The operation of the auxiliary equipment represents about 13.6% of the total operating cost.

The Mina Sur Pit: At present, Mina Sur moves 40 kt/d (44,092 stpd) of material, with a waste to ore ratio of 2.7. The equipment used includes one Bucyrus Erie 60-R drill, four Bucyrus Erie 190-B (2) and 280-B(2) electromechanical shovels, a fleet of 16 offroad trucks—8 Lectra-Haul M-100 and 8 Wabco 120-Bs—and 12 pieces of auxiliary equipment.

The ore is transported by truck to a primary crushing station outside the pit. Here an Allis-Chalmers 1.37 × 1.88 m (54 × 74 in.) gyratory crusher is used. The ore is then transported on a 2.6 km (1.6 miles) long, 1.22 m (48 in.) wide conveyor belt to the storage piles at Chuquicamata. The capacity of the system is 2500 t/h (2756 stph).

Chuqui Norte: This ore body is the north extension of the Chuquicamata deposit and contains low-grade copper oxides and sulfides covered by 80 m (262 ft) of over burden. The exploitation of this area in the future is under study.

Specific Projects

Bank K-1 Crushing Station—At the beginning of 1984, a new crusher/conveyor belt system was put into operation in the Chuquicamata pit to replace the old one.

It's advantages are shorter distances for truck transport in the mineralized section and an increase in crushing capacity up to 156 kt/d (171,961 stpd). The exploitation of the zone where the old system was will provide excellent quality ore and a low waste to ore ratio.

This new system includes two Allis Chalmers 1.52 × 2.77 m (60 × 109 in.) gyratory crushers and two parallel conveyor belts 1.83 m (72 in.) wide and approximately 2.5 km (1.6 miles) long. In 1981, construction began on the conveyor tunnel out of the pit. The tunnel is 5.7 × 9.1 m (18.7 × 30 ft) in section by 2340 m (7677 ft) long, with a 15.8% gradient for approximately 1500 m (4921 ft) and a 1.7% gradient for the remainder.

The Plants

Concentration: In the concentration plant, the flotation of sulfide ore from the Chuquicamata pit produces copper concentrate that is then sent to be smelted at the smelter. There is also some market for copper concentrates of approximately 38 to 40% Cu.

On the other hand, molybdenum concentrate (54% Mo) is obtained as a byproduct. This is packed in drums and sent for toll roasting, sale, or to the division's Roaster Plant.

Ore dressing is carried out in this area in various plants, through which the ore is submitted to size reduction stages and collective and selective flotation to obtain copper and molybdenum concentrates as final products.

The following is a description and flow chart of the five most important stages.

Crushing Plant—The size reduction stage begins in an Allis-Chalmers 1.37 m (54 in.) gyratory crusher at the mine, which delivers material of less than 0.20 m (8 in.) to a storage pile.

The material extracted from the upper benches of the mine is sent directly to the plant's primary crusher [Traylor 1.5 m (60 in.)]; the product is merged with the ore from the storage pile and together is sent to the coarse ore storage hopper [under 0.2 m (8 in.) ore] which has a capacity of 35 kt (38,581 st).

The secondary and tertiary crushing installations have a treatment capacity of 102 kt/d (112,436 stpd) and are divided into five sections. In each section the material from the coarse ore storage feeds a double deck Tyler F-900 vibrating screen. The ore under 12.7 mm (0.5 in.) is added to the final product of the plant while material coarser than 38 mm (1.5 in.) feeds to a standard Symons 2.1 m (7 ft.) crusher. The output from this crusher, together with the intermediate product of the Tyler F-900 screen feeds to four Tyler F-600 vibratory screens.

The material over 12.7 mm (0.5 in.) from these screens is reduced in two short head Symons 2.1 m (7 ft.) crushers and is added to the material under 12.7 mm (0.5 in.) forming the final fine product 8% + 12.7 mm (0.5 in.), which is sent to the fine ore storage bins of the concentrate. To assure a continuous ore supply, there is a fine ore reserve pile, with a 33 kt (36,376 st) capacity. Fig. 2 illustrates the crusher plant flow.

Concentrator Plant—This plant has twelve primary wet grinding mill sections, each of which is fed by two adjustable-speed belts. In each section, the fine ore is fed to a Marcy 3.1 × 4.3 m (10 × 14 ft) rod mill that in turn feeds two Marcy 3.1 × 3.7 m (10 × 12 ft) ball mills, each of which operates in closed circuit with a battery of Krebs D-20 hydrocyclones.

The overflow from the cyclones, at 28% + 212 μm (65 mesh), is fed to primary flotation. There are 12 flotation sections, seven of which are made up of Agitair N°120 cells. Section N°3 consists of a row of six Agitair N°120-A cells and a row of seven Denver DR-300 cells. Wemco 14.16 m³ (18.5 yd³) cells have been installed in section N°7. The original Agitair N°48 cells in the rest of the sections are being replaced by Agitair N°120 cells.

The primary tailings flow by gravity to six 91 m diameter (300 ft) thickeners, whose underflow of 52 to 55% solids is sent to the tailings drain. Previously, in two tailings retreatment plants, some of the sulfide tailings have been recovered. The water recovered from the overflow of the thickeners is recirculated to the concentrator. The primary concentrate from each of the primary flotation sections is thickened to 45% solids in a Dorr-Oliver 24 m (80 ft) diameter thickener.

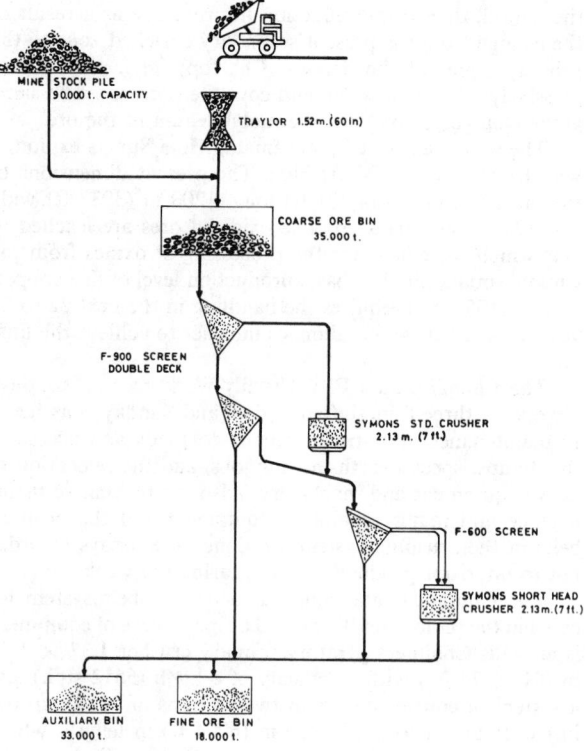

Fig. 2. Crushing plant, 102 kt/d capacity.

The underflow is reground in two Marcy 1.8 × 3.7 m (6 × 12 ft) regrind ball mills that operate in closed circuit with a battery of four Krebs D-10 cyclones. The overflow of the cyclones, at 15 to 17% solid and 80 to 90% −45 μm (325 mesh), is fed to two rows of 14 Agitair N°60 cells for cleaning. The concentrate from here feeds the recleaning circuit that consists of 12 rows of 20 Agitair N°48 cells.

The collective Cu-Mo concentrate is obtained from the first six cells of each row. The concentrate collected in the following four cells is recirculated to the feed end of the circuit. The tailings of these four cells and the tailings of the cleaning flotation feed the cleaning scavenger flotation cells (10 cells) from which a concentrate is obtained. After thickening in an Eimco 30.5 m (100 ft) diameter thickener, this concentrate is reground in two Marcy 1.8 × 3.7 m (6 × 12 ft) ball mills that operate in closed circuit with one battery of Krebs D-6 hydrocyclones each. The product is recirculated to the second bank of four cells. When the scavenger tailings reach over 9.6% Fe content, they are sent to a small pyrite recovery plant. The scavenger cells tailings are added to the primary tailings. The collective Cu-Mo concentrate is then sent to two Eimco 45.8 m (150 ft) diameter thickeners operating in parallel. The underflow of these thickeners, after dilution with fresh water, goes to a third thickener similar to the previous ones. The 60% solids underflow of this last one feeds the molybdenum recovery plant. Fig. 3 illustrates a flow diagram of the concentrator plant.

Expansion of the Concentration Plant—The expansion of the concentration plant was completed at the beginning of 1982, with an increase in ore treatment capacity to 96 kt/d (105,822 stpd). This plant has three wet grinding lines. Each line consists of one Marcy 4.1 × 5.5 m (13.5 × 18 ft) rod mill and one Marcy 5 × 6.4 m (16.5 × 21 ft) ball

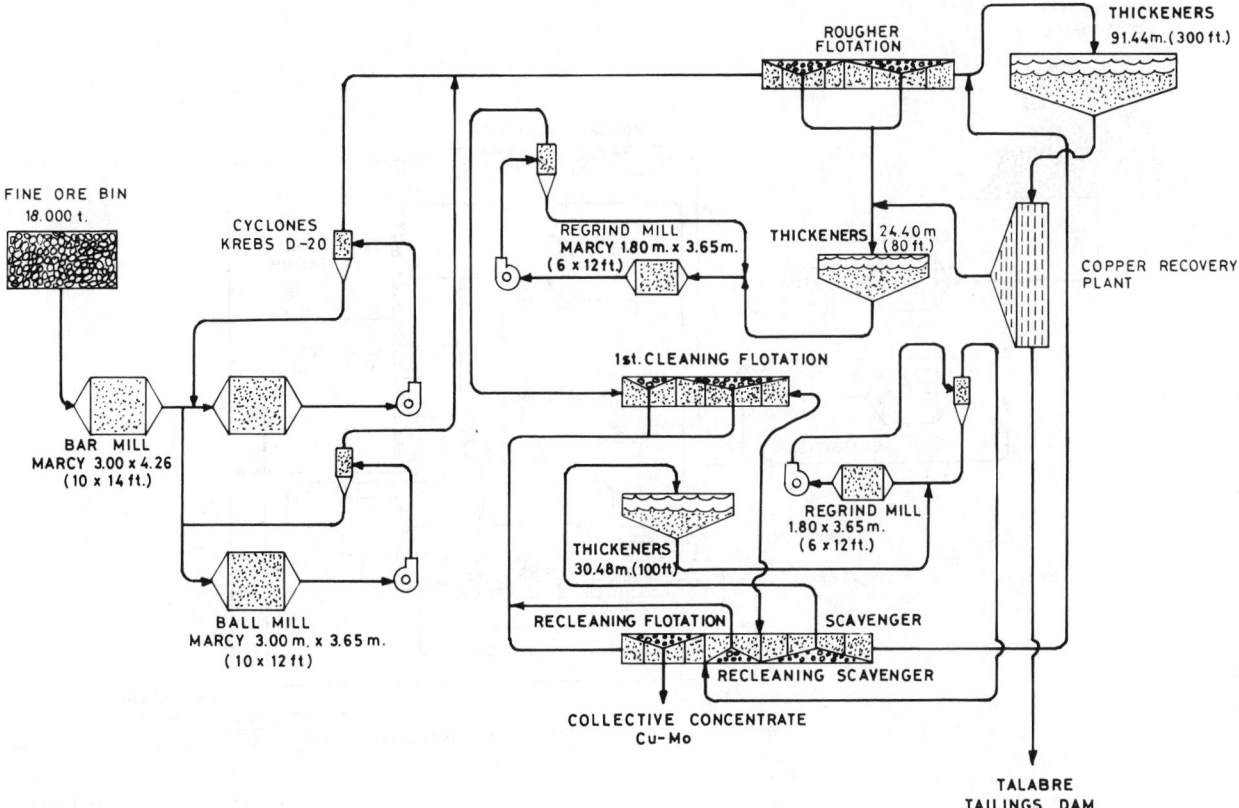

Fig. 3. Concentration plant, 70 kt/d capacity.

mill that operate with a battery of six Krebs D-26 hydrocyclones.

The overflow of the cyclones from the three lines is sent to a distributor that feeds three rows of 15 Denver 14.16 m³ (500 ft³) primary or rougher flotation cells. The primary concentrate produced is dumped to the battery of hydrocyclones of the regrinding circuit (Krebs D-15). The underflow is distributed to two Marcy 3.1 × 3.7 m (10 × 12 ft) ball mills. The overflow is sent to the cleaning circuit, formed by two rows of 13 Denver DR-300 [8.5 m³ (300 ft³)] cells. The cleaner concentrate is obtained from the first five cells, while the concentrate from the three middle cells is added wholly or partially to the cleaner concentrate or to the scavenger concentrate, obtained from the last five cells. The primary and scavenger tailings are sent to three tailings thickeners 99 m (325 ft) in diameter.

The cleaner concentrate feeds the recleaning circuit that is formed by a row of 13 Denver DR-300 [8.5 m³ (300 ft³)] cells. The collective concentrate is obtained from the first five cells and is added to the collective concentrate produced in the old plant. The concentrate from the three middle cells may be added totally or partially to the collective Cu-Mo concentrate or to the recleaning-scavenger obtained from the last five cells. These are also fed by the scavenger concentrate, after regrinding and classification in a closed circuit formed by a Marcy 1.8 m × 3.7 m (6 × 12 ft) ball mill and a battery of Krebs D-6 cyclones. The recleaning-scavenger refeeds the recleaning flotation and the tailings refeeds the scavenger flotation. The flow chart of the expansion of the concentration plant is shown in Fig. 4.

Molybdenum Recovery Plant—The collective Cu-Mo concentrate, after the two thickening stages described previously, is diluted to a solids content of 38 to 40% by weight and conditioned to a predetermined pH before being distributed to the four primary flotation rows. Each primary flotation row is made up of 20 Agitair N°48 cells, arranged in five banks of four cells each. The depressor reagent is added at the head of each row.

The primary tailings or general tailings of the plant are sent to the copper concentrate filter plant. The primary concentrate at 10 to 12% solids and a molybdenum grade that fluctuates between 7 and 15% feeds the cleaner flotation circuit. The cleaner flotation is carried out in two rows of 12 Agitair N°48 cells, arranged in three banks of four cells each. The tailings of this stage are sent to the collective concentrate thickeners and thus return to the circuit.

The cleaner concentrate is sent to a Dorr Oliver thickener 15.2 m (50 ft) in diameter. The underflow of this thickener, at 50 to 55% solids, is sent to two 250 m³ (8829 ft³) storage tanks, where the pulp is diluted to 25% solids before being fed to the second cleaner flotation.

The second cleaner flotation consists of two rows of 18 Agitair N°48 cells, arranged in a first bank of four cells, a second bank of six cells, and a third bank of eight cells. The depressor reagent is added at the head of each row and to each second bank. The tailings of the second cleaning are added to the feed of the first cleaner flotation.

The second cleaner concentrate feeds to a row of 18 Agitair N°48 cells, arranged in three banks of four, six, and eight cells each, where the third cleaning takes place. The

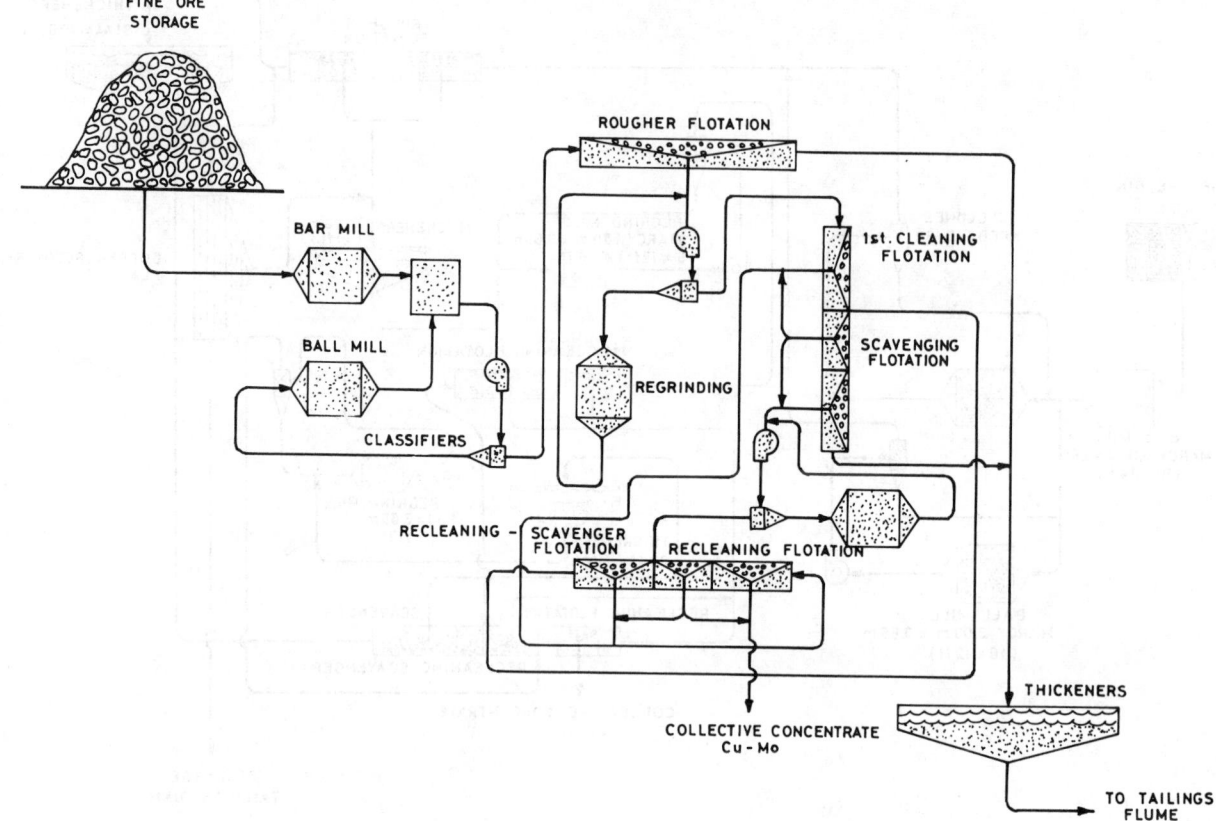

Fig. 4. Concentration plant expansion, 26 kt/d capacity.

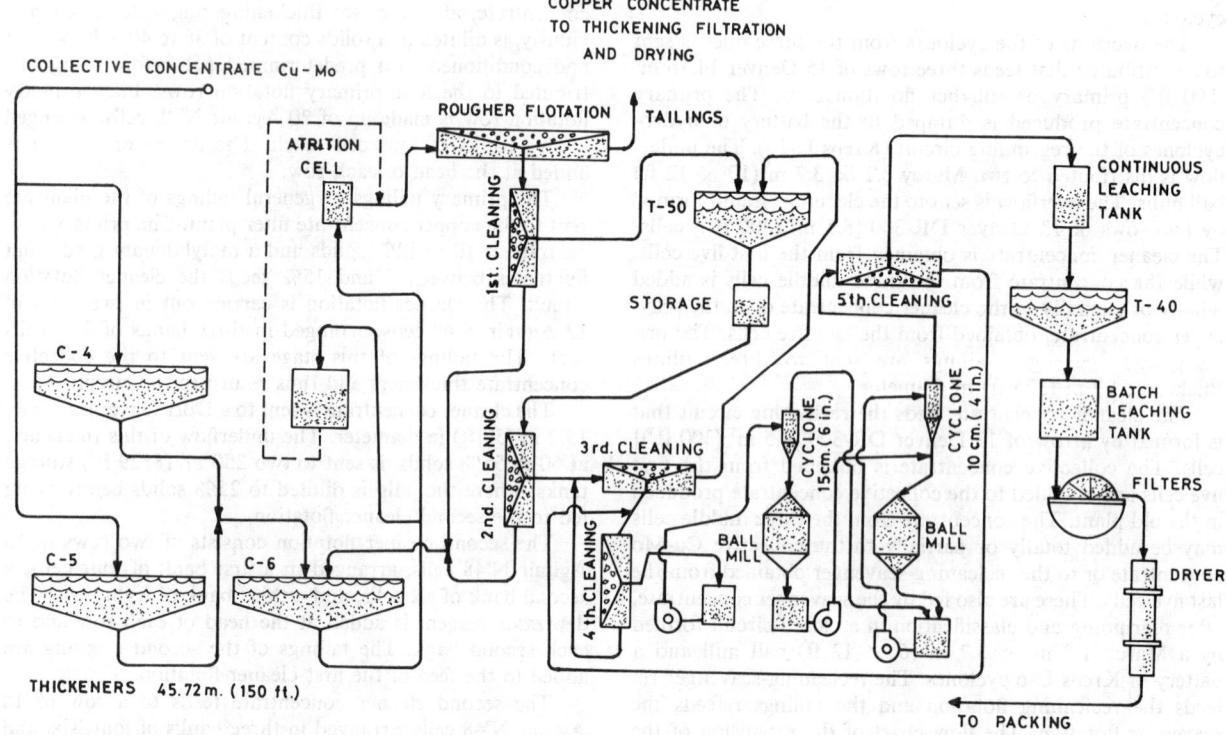

Fig. 5. Molybdenum recovery plant.

tailings of this stage, at approximately 10 to 15% solids, are added to the first cleaner concentrate. The concentrate, at 14 to 16% solids, is sent to a fourth cleaner flotation.

The fourth cleaner flotation is carried out in a row of 12 Agitair N°48 cells, arranged in two banks of four and eight cells each. The tailings are added to the second cleaner concentrate. The concentrate, at 16 to 20% solids, is sent for regrinding and classification in four Marcy 1.2 × 2.4 m (4 × 8 ft) ball mills, and the classifiers overflow goes to a distributor that feeds the fifth cleaner flotation.

The fifth cleaner flotation is performed in two banks of six Denver N°24 — 1.4 m³ (50 ft³) cells. The tailings of this stage are added to the third cleaner concentrate. The fifth cleaner concentrate, at 16 to 20% solids, is deposited in a box that feeds two 9 700-L (2562-gal) tanks arranged in series where continuous leaching takes place.

The pulp remains in the set of tanks for 1.5 hours and is then sent to a Dorr Oliver 12.2 m (40 ft) diameter thickener. The underflow of this thickener at 50 to 55% solids, is distributed to one of four batch leaching tanks, 4.9 m in diameter × 6.1 m high (16 × 20 ft) where the copper content is lowered to less than 0.15%. The underflow of each tank is diluted to 45% solids, and sent to an Eimco 8-disc 1.2 m (4 ft) diameter filter. The cake obtained from this filter is repulped with the addition of water and steam, and then sent to a second Dorr Oliver 1.8 m (6 ft) diameter filter. A final cake (25% humidity) is obtained and sent to one of the two drying tubes by means of helical conveyors. The drying tubes operate on superheated steam. Their capacity is 63.5 t/h (70 stph) and 45.3 t/h (50 stph), respectively.

The concentrate obtained at the discharge of each drying tube, with humidity between 0.5 and 2%, is stored in a bin. Part of the concentrate is packed in drums for shipment and the rest is transferred to the molybdenite roasting plant. Fig. 5 illustrates the flow of the molybdenum recovery plant.

The Molybdenite Roasting Plant—This plant has been designed to roast 5 443 109 kg (11,999,799 lb) of molybdenum concentrate per year to produce Molybdic oxide technical grade.

The process begins with the storage of molybdenum concentrate in two bins, which can either feed the area where concentrate is packed in drums, or to a 6.6 t/h (7.3 stph) capacity pneumatic conveyor system to the roasting plant. The transport of concentrate is performed in the dense phase and air is delivered by a 93.21 kW compressor. The concentrate is received in two 500-t (551-st) silos.

The roasting furnace is fed with Mo concentrate by means of a screw conveyor and a bucket elevator system. The bucket elevator is also fed by the return of dust from the gas treatment system. The roaster is a 6.6 m diameter (21.5 ft), 12-story Nichols oven, with a treatment capacity of 16.3 t/d (18 stpd) of molybdenum concentrate.

The roaster is equipped with kerosene combustion burners on the different floors and LPG is used for initiating firing. The gas from the roaster feeds to a cleaning system, which consists of an indirect cooler, a secondary cooler, two multicyclones in series, a multicyclone blower, a dry electrostatic precipitator, and a blower. The dust recovered in this system is returned to the roaster by a screw conveyor.

The molybdenum oxide, at a temperature of 482 °C (900 °F) at the roaster discharge, is fed to a size reducer to eliminate any clinker produced in the roasting process. Once a uniform size of material is obtained, it is fed to a cooler where the temperature of the oxide drops to 93 °C (200 °F). The cooled material is transferred to a bucket elevator that feeds a classifying screen. The oversize material is sent to a size reducer and returned to the screen. The undersize material is sampled, weighed, and transported by screw conveyor to one of nine storage bins. The packing facilities provide for packing the product in 10-kg (22-lb) cans or 115-kg (254-lb) drums. Fig. 6 illustrates the flow of the molybdenite roasting plant.

Concentrate Smelting: The smelting of copper concentrate, because of the characteristics and processes used, is defined as traditional smelting. The most important equipment and installations used at the present time are described in the following sections.

Concentrate Drying Ovens—The filtered copper concentrate coming from the filter plant reaches the drying units in the smelter by conveyor belts. These units (13) are cylindrical rotating furnaces, with a total capacity of 203 t/h (224 stph).

Once the humidity of the concentrate has been reduced to adequate limits (6 to 8%), it is carried by conveyor belts to storage and the charge preparation facilities that consist of three beds with a charge capacity of 8500 t (9370 st) each.

Reverberatory Furnaces—These units, four in total, have the following characteristics:

		Reverberatory Furnace Nos. 1, 2, 3	Reverberatory Furnace No. 4
Length	(m)	39.62	36.58
Width	(m)	9.14	10.67
Height	(m)	3.73	4.16
Bed area	(m²)	235.25	245.10

Here the prepared charge or *green charge* (concentrate + flux + circulating charge) is smelted, producing three products: matte, slag, and gases.

Matte is the rich copper product that continues to be processed. The slag, since it is a material with a Cu content of less than 1% is discarded and is transported to dumps by railroad in 6.37 m³ (225 ft³) ladle cars. The gases, from which the heat and dust are recovered, are eliminated through a 91.44 m (300 ft) stack that can evacuate approximately 450 000 Nm³/h.

In order to take advantage of the heat from the gases, each of the reverberatory furnaces has two boilers that can generate about 20 t/h (31 stph) of superheated steam used in the Chuquicamata thermoelectric plant for power generation, representing a production of 12,600 J (3.5 MW).

The dust recovered is stored for future recirculation. Smelting is achieved by a combination of air-oil burners or regular horizontal burners (AFB) and vertical oxy-fuel burners (OFB) that use No. 6 oil (Bunker C) as fuel. This last type is controlled by a computerized system in a centralized control room or console where the corresponding flows are regulated and verified.

Converter Furnaces—The matte, rich in copper, as it comes from the smelting units, is processed in conversion furnaces (7 Peirce Smith) that have the following characteristics:

	Converters Nos. 2-7	Converter No. 1
Length (m)	10.671	12.195
Diameter (m)	3.963	3.963
No. of nozzles	52	55
Air Feed:		
a) pressure (kPa)	138	138
b) flow (m³/min)	650-800	650-800
Production capacity (t/d)	200-250	200-250
Nozzles punding	mechanical	mechanical

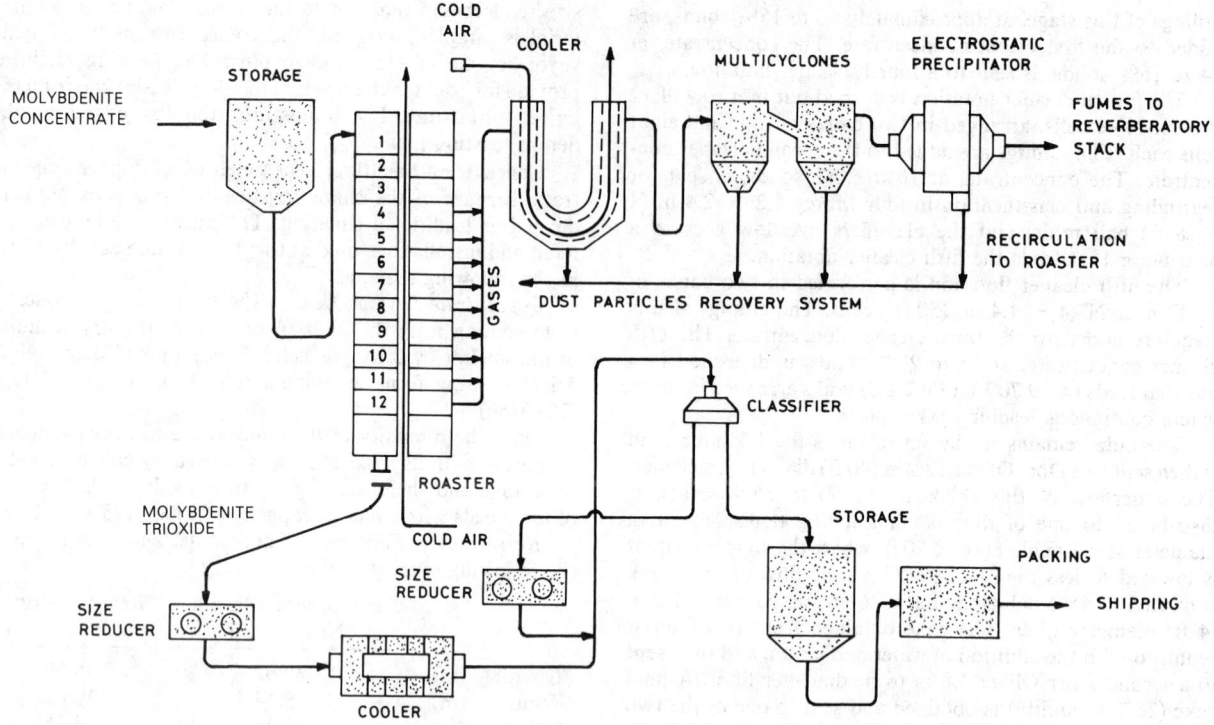

Fig. 6. Molybdenite roasting plant.

In these units, by the blowing of air through the nozzles and silica as flux, iron and sulphur are eliminated from the matte, converting it into blister copper after 6.5 to 7 hours of processing.

Modified Teniente Type Converter—This is a smelting conversion unit that is fed with concentrate, flux, and matte. Conversion takes place to white metal by blowing oxygen enriched air.

The gases generated are used in the production of sulfuric acid and the slag obtained in the process is sent to the treatment plant to recover the copper it contains.

The principal characteristics of the equipment are as follows:

Length (m)	18.29
Diameter (m)	4.04
No. of nozzles	67
Air Feed:	
a) pressure (kPa)	138
b) flow (m³/min)	650-800
c) oxygen enrichment (%)	30
Concentrate feed, t/d (dry)	700
Matte feed, t/d	655
White metal produced, t/d	750

Overhead Cranes—The handling of liquids and secondary materials in the conversion bay is performed by five cranes that have a central hook with a capacity of 92 t (10 st) and two auxiliary hooks with a 23-t (25-st) capacity each. This equipment handles ladles of 3.5 to 7.2 and 9.2 m³ (325 ft³) and metal bins for the transport of copper scrap.

Refining and Casting—The blister copper obtained from the converters is transported in a liquid state to the refining furnaces (6), which have the following chracteristics:

Length (m)	7.62
Diameter (m)	3.96
No. of nozzles	6
Capacity (t Cu/charge)	200-220
Casting system	automatic weighing

In these units, to obtain a product of 99.6 to 99.8% Cu content, the copper is submitted to a process of fire refining, consisting of two stages: (1) oxidation by air under pressure; and (2) reduction using kerosene atomized by steam or by eucalyptus logs as the reduction agent.

Casting Wheels—This equipment (3), one unit serving two refining furnaces, has the following characteristics:

—Diameter (m)	12.58
—No. anode molds	26
—Casting speed (t Cu/h)	40

The copper is casted in the form of anodes that weigh 365 kg (805 lb) (commercial anodes) and 385 kg (849 lb) (initial plate anodes). After the elimination of possible burrs, the anodes are sent to railroad cars to the electrolytic refineries. Fig. 7 shows the flow of the concentrate smelter.

Refineries: The electrometallurgical processes in this area are concerned with the electrorefining and electrowinning of copper and the treatment of anode slimes. These processes are carried out in the following installations: refinery No. 1 (E.R. No. 1); refinery No. 2 (E.R. No. 2) and the noble metals plant.

The electrolytic refineries in Chuquicamata have a production capacity up to 32 kt/month (35,274 st per month) of copper in the form of electro-refined cathodes (15,500 electro refinery E.R. No. 1 and 16,700 electro refinery E.R. No. 2) and up to 4 kt/month (4409 st per month) of electrowinning copper. Ore metal production is about 80-100 t/a (88-110 stpy).

Electrolytic Refinery No. 1—This refinery was initially designed for electrowinning. With the depletion of oxidized minerals and the increase of sulfide ore, the electrowinning circuits have been modified to electrorefining ones.

The electrorefining area includes circuits Nos. 3, 4, 5, 6A and 6B with a total of 786 cells. The electrowinning area includes circuits Nos. 1, 2, and 7 with a total of 318 cells.

Electrolytic Refinery No. 2—This refinery has all the technical advances known at the time of its construction in 1969. It has been adapted and modernized to maintain the quality and cost of cathode production. Originally designed to produce 13,500 t/month (14,881 st per month), now it produces 16,500 t (18,188 st) monthly.

Refinery No. 2 is divided into three bays, two lateral bays for production, and one center service bay that houses the entire hydraulic and thermal system (pumps, tanks, exchangers, etc.) designed for the circulation of the electrolyte, for heating it, and for the evaluation of anode slimes. Circuits 10, 11, 12, and 13 with a total of 960 cells, correspond to this refinery.

Circuits 10 and 11, with 160 cells, are for the production of the initial plates, and circuits 12 and 13, with 800 electrolytic cells are for the commercial production of electrorefined cathodes.

Technological Advances and Modernization—Permanent work to maintain the quality and competitiveness of the cathodes from the Chuquicamata refinery includes the following innovations: conditioning of the electrolyte, the liberation plant, protection and covering of cells, replacement of copper starting blanks by titanium starting blanks, and change of the riveting system in the bar-sheet joint by the welding system. The licence has been obtained for the use of initial cathode press, etc.

Technological Advances—The replacement of the copper blanks by titanium blanks was another innovation. For this purpose, in 1975 the Chuquicamata division bought 50 titanium plates. By means of metalographic analysis, the fundamental differences in grain structure in sheets obtained with titanium and copper blanks were assessed. Based on the results obtained, all the copper blanks were replaced by titanium ones (6000). Titanium has an electrical resistivity of 48.2 cm (19 in.), highly superior to that of copper, so that a greater consumption of power should be expected when titanium blanks are used.

Technical innovations introduced by the titanium blanks have led to the replacement of the present method of riveting in the sheet-bar joint (mechanical) by welding, which contributes to a decrease in power consumption and the elimination of corrosion from the effects of electrolyte penetration in the cavities in that zone.

Titanium's high resistance to corrosion permits operation with better temperatures (60°C) and concentrations of 185 g/L (10,807 gr/gal) acid, conditions that improve electrolytic conductivity, thereby lowering energy consumption.

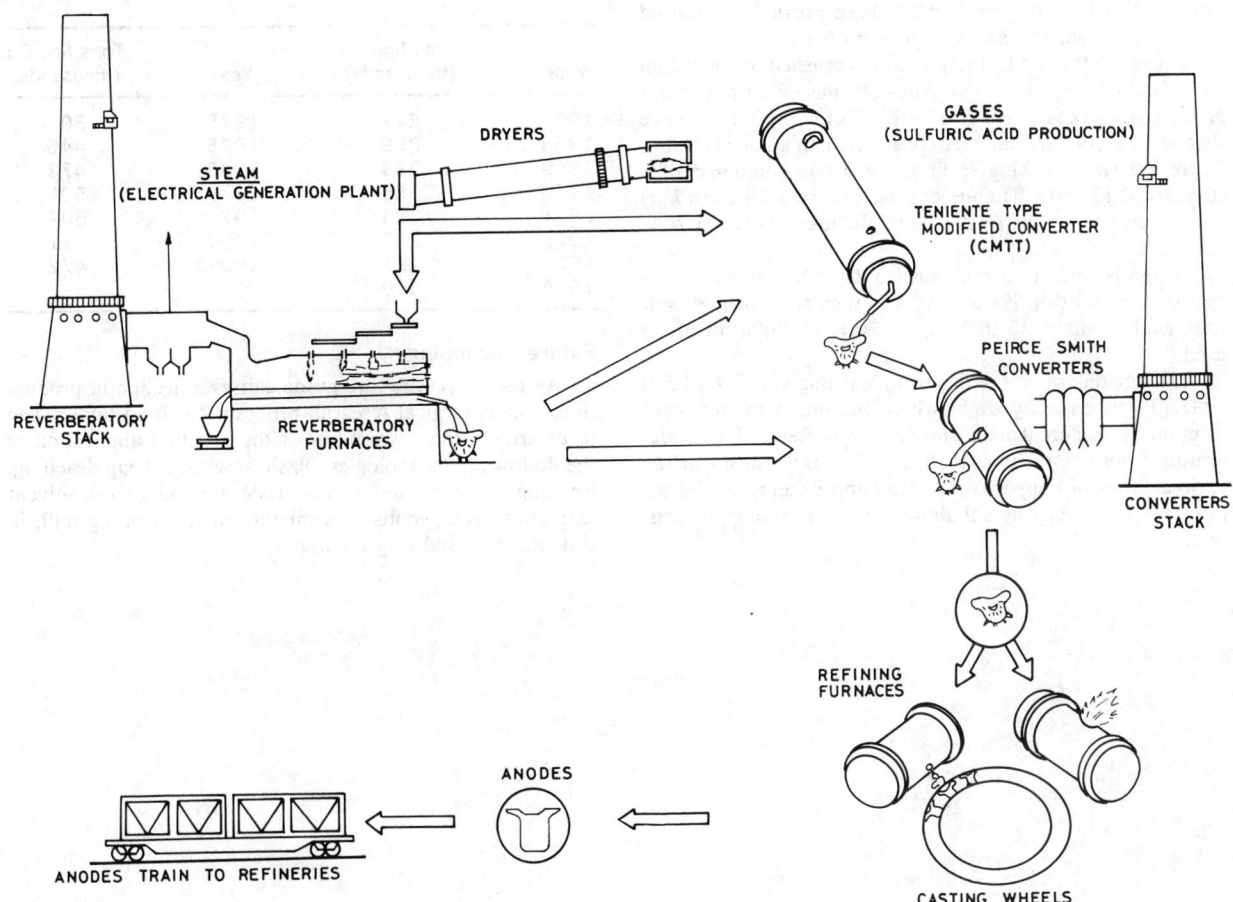

Fig. 7. Concentration smelter flow diagram.

The deposits obtained with titanium blanks are finer grained and free of protuberances. Since the useful life of titanium plates is far superior to that of copper plates, their acquisition has been highly advantageous.

Extraction: Under this heading the following functions related to hydrometallurgical treatment of ore and other copper oxides are grouped:

1. The recovery of copper from oxide ores, at the present time mainly from ore brought from Mina Sur, by leaching with sulfuric acid, to obtain a concentrated solution of copper sulfate.

2. The purification of the copper sulfate solutions, to eliminate impurities that affect the electrowinning of copper.

3. To supply solutions of the correct chemical quality and copper concentration to the electrowinning circuits in the refinery.

4. To produce the sulfuric acid necessary for carrying out the production and research programs, and at the same time, supply the needs of the other production units.

5. To carry out metallurgical research to determine future exploitation of copper oxide resources, as well as the hydrometallurgical treatment of low-grade mineral sulfides. This research is oriented toward the use of the greater amounts of sulfuric acid that will be generated when the new gas recovery projects in the concentrate smelter are implemented.

DESCRIPTION OF THE PROCESS

The extraction of copper from oxidized ores in Chuquicamata and *Mina Sur* is performed by a hydrometallurgical process, in which the ore that has been previously reduced in size is leached by sulfuric acid solutions.

The ore extracted from *Mina Sur* is crushed to a nominal size of 0.18 m (−7 in.) in an Allis-Chalmers 1.37 m (54 in.) crusher, and transported to a 60 kt (66,139 st) bin. The ore then feeds secondary and tertiary crushing in an open circuit. There are two crushing sections, each one composed of a standard 2.13 m (7 ft) cone crusher and two 2.13 m (7 ft) short head cones. The final product fluctuates between 20% and 40% + 9.5 m (⅜ in.).

Water is added to the crushed ore until it reaches a moisture content of 5% to 7%, and then concentrated sulfuric acid is added; 15 to 24 kg/t (30 to 48 lb/st) of ore is used.

The agglomerated ore is sent to leaching vats [14 12 500 t (13,779 st) capacity each] where the ore is treated with solutions of different acid and copper contents, for predetermined periods of time, that solubilize and transform the various types of copper oxide into copper sulfates. The solutions are replaced by solutions of decreasing copper content.

The leaching cycle consists of three essential stages: in the first, the extraction of primary solutions that alter purification (dechlorization-reduction-decanting), are sent to electrowinning; the second is the regeneration of secondary solutions, required for the processing of new batches, and a third stage of eight washings in decreasing copper and acid concentrations, the last being in water. The leached tailings (ripios) are extracted from the vats by means of four dredge bridges, two 6 ft (6.6 st) Mead Morrison and two 12 t (13 st) Wellman-Seaver-Morgan, and transported to dumps in 85 t (94 st) Lectra-Haul trucks.

OPERATING COSTS

Using 1981 as a base, operating costs are distributed as follows:

Wages	32%
Materials and fuel	42%
Services	7%
Transport to port	2%
Other income and expenditure	17%

PRODUCTION AND FUTURE TECHNOLOGY

Production

Since 1971, the Chuquicamata complex has been managed and operated by Chilean professionals and personnel. The following production figures show the performance from 1967 to 1981.

Year	Tons fine Cu (thousands)	Year	Tons fine Cu (thousands)
1967	277	1975	305
1968	279	1976	446
1969	283	1977	478
1970	265	1978	501
1971	285	1979	507
1972	265	1980	511
1973	297	1981	472
1974	389		

Future Technology

As the decreasing ore grade causes a decline in production, a metallurgical research program has been undertaken to determine the possibilities of the eventual application of the following technologies: flash smelting, heap leaching, leaching in waste dumps, low-grade ore and gravel, solvent extraction, autogenous or semiautogenous grinding mill, in situ leaching, and slag flotation.

9.7 Shirley Basin Mine

M. I. Ritchie

INTRODUCTION

The mines and mills of Pathfinder Mines Corporation have been actively producing uranium concentrates in Wyoming since early 1958 when the Lucky Mc facilities in the Gas Hills district, Fremont County, first achieved operating status. The company acquired claims in the Shirley Basin district, Carbon County, Wyoming, and shortly thereafter, in 1960, produced uranium ore from these holdings by underground mining methods. During the underground mine life from 1960 to 1963, approximately 99.8 kt (110,000 st) of ore was mined containing 544 311 kg (1.2 million lb) U_3O_8. The ore was transported to the Lucky Mc mill for processing to yellowcake. After gaining valuable experience regarding geologic and hydrologic conditions and studying various alternatives, patented solution mining techniques were introduced and successfully implemented from 1963 to 1970. Concentrate extracted from solution mining was some 680 388 kg (1.5 million lb) of U_3O_8. In November of 1968, the company announced that it would develop the uranium reserves by open-pit mining methods and would construct a processing facility capable of producing 1 133 981 kg (2.5 million lb) of U_3O_8 per year. For the period from 1968 to 1982, approximately 5.7 Mt (6.3 million st) of ore containing 9 525 440 kg (21 million lb) of U_3O_8 has been mined by surface mining methods. The location of the area is shown in Fig. 1.

In December, 1976, Pathfinder Mines Corporation (known as Lucky Mc Uranium Corporation) was incorporated as a wholly owned subsidiary of Utah International Inc. in connection with the merger of Utah International Inc. and the General Electric Co.

In March 1982, 80% of the capital stock of Pathfinder was acquired by COGEMA Inc., a United States subsidiary of Compagnie Generale Des Matieres Nucleaires (COGEMA), a French corporation. COGEMA, wholly owned by the French Atomic Energy Commission, is involved in uranium operations around the world.

CLAIM ACQUISITION AND EARLY EXPLORATION

After the initial exploration and development phases of the Lucky Mc mine were completed and the project was turned over to the mine management staff in the spring of 1955, the company's exploration group was charged with finding additional uranium reserves that could contribute ore for the new mill under construction. Initially, the area of study was within an 80-km (50-mile) radius of the Gas Hills district. Background studies and preliminary geologic criteria developed from other Tertiary basins in Wyoming yielded evidence that considerable potential existed for major uranium deposits in the Shirley Basin area. In response to this information, an exploration effort was set into motion and the restrictive 80-km (50-mile) radius around the Lucky Mc mill was extended to a 161-km (100-mile) radius.

By July 25, 1957, the activity in progress included compilation of land status, and maps were obtained from the Bureau of Land Management in Cheyenne, WY, by the Land Department. Preliminary geologic and field reconnaissance of Shirley Basin was assigned. The field camp was being assembled by closing the facilities at Copper Mountain northeast of Shoshoni, WY, and from surplus equipment in the Gas Hills.

On July 27, a convoy of 17 vehicles including one drill rig and trucks loaded with 102×102-mm (4×4-in.) claim stakes and assorted camping equipment left the Gas Hills and arrived in Shirley Basin.

The first priority was to stake claims of projected potential and to cut off the northward advance of competitor survey crews. The projected potential area was determined from surface geology maps, land status maps, and the known locations of surface mineralization. The recently acquired experience in the Gas Hills exploration program was most beneficial in this analysis. In the targeted area, the surface was covered with thin remnants of the White River formation. The subsurface Wind River formation was composed of arkosic sandstones and shales that had been deposited on an erosional surface of deformed Cretaceous sediments.

On July 28 the first claims were located and by July 30 over 250 claims were staked and filed in the Carbon County Recorder's office. On Aug. 17, 1957, it was reported that three of the first 11 drill holes had intersected significant mineralization, with the best intercept of 0.3 m (1 ft) of 0.80% e U_3O_8, in a zone 76 to 91 m (250 to 300 ft) deep in the eastern portion of the claim block. This area later was designated area 1. Progress during the summer and early fall was summarized in a Sept. 13 report:

To date, approximately 412 claims have been staked and validated. . . . In addition, 1080-acre state leases have been obtained in the area of greatest interest and negotiations have commenced for other favorable ground. A campsite has been located, water well drilled, and summer camp installed. The winterized camp from the Gas Hills will be moved to this site next week. Forty-four deep holes have been completed, twenty-four on a widespread net for geology, and twenty discovery holes have intersected high-grade mineralization. An estimated 150,000 st at 0.75% e U_3O_8 of potential ore is indicated by the drilling to date. . . . The geologic potential probably is greater than 300,000 st.

By December 1958, drilling had expanded the reserve to an estimated 1 496 855 kg (3.3 million lb) of U_3O_8 at an average grade of 0.69%. On Oct. 28, 1957, the US Atomic Energy Commission (AEC) had announced that future domestic allocation would be limited to ore reserves developed by the announcement date. This production limit was later modified by an announcement on Nov. 24, 1958, that provided for some expansion of production in Wyoming. The Shirley Basin annual allocation of 116 120 kg (256,000 lb) through 1962 was increased to 158 757 kg (350,000 lb) to the end of the contract period in 1966. With the identification of ample reserves to fill the allocation allowed by the AEC, Shirley Basin moved from an exploration program to mine development.

UNDERGROUND MINE DEVELOPMENT

The corporate decision to proceed with underground mine development and to plan deliveries of mined high-grade ore to the Lucky Mc mill for processing became a reality when construction commenced in the spring of 1959.

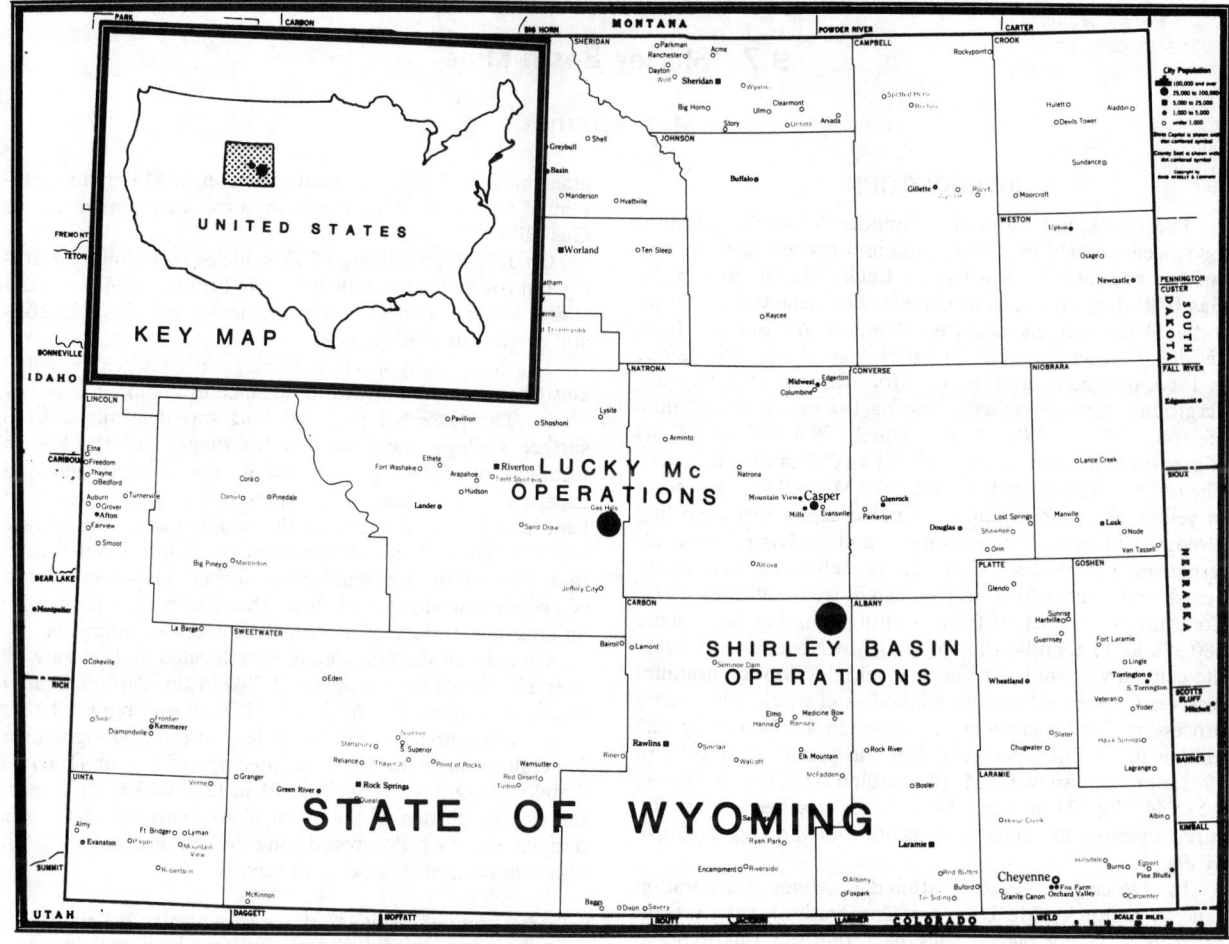

Fig. 1. Location of the Shirley Basin mine, Carbon County, Wyoming.

Surface Plant

The surface plant consisted of two main steel-frame, corrugated iron buildings. A 7.6 × 30.5-m (25 × 100-ft) building housed the 187 kW (250 hp) double drum hoist, two 0.850-m³/s (1800-cfm) compressors, and a standby power-plant consisting of three 300-kW diesel generators. The second building, connected to the headframe house, contained mine offices, warehouse, shop, change room, radiometric assay laboratory, and first aid room. Timber storage and framing facilities were included in the mine yard.

A mine townsite constructed near the shaft site included 84 trailer spaces, a bunkhouse, mess hall, community building, and school facilities for grades one through eight.

Mine Development

The mine access was through a 3.7-m (12-ft) diam, circular, concrete-lined, four-compartment vertical shaft 111 m (365 ft) deep. Men, supplies, ore, and waste were hoisted through two compartments in combination cage-skip car dumpers. The third compartment contained a manway and electric cables. The service compartment contained a 762-mm (30-in.) ventilation duct, a 152-mm (6-in.) pump column, a 152-mm (6-in.) compressed air line, and a 51-mm (2-in.) potable water line. The shaft was sunk in 1959 by the construction division of the company.

In general, development was by haulage drifts on the 2070-m (6790-ft) level at a depth of 99 m (324 ft), driven south and southwesterly along the trends of the 2067-m (6780-ft) and 2079-m (6820-ft) roll front systems. Drifting through the unconsolidated, saturated sediments required back and side spiling with excavation by pneumatic spaders. Drift support was achieved with timber sets that included bridge caps.

Haulage drifts had 762-mm (30-in.) gage track, 152-mm (6-in.) or 102-mm (4-in.) compressed air line, 25-mm (1-in.) drill-water line, power cables, 152-mm (6-in.) pump-discharge line, and vent tubing when needed. Track grade was maintained at 0.5%. Haulage in the track drifts was done with 1.8-t (2-st) ridge box cars. Trains were pulled with two 2.7-t (3-st) and two 0.9-t (1-st) storage-battery locomotives on the 2070-m (6790-ft) level. The 2079-m (6820-ft) level ore zones were developed through two-compartment timbered raises, driven from the haulage drifts.

The underhand stopes in the 2067-m (6780-ft) ore horizon were developed by square set *scrams* driven on a decline from the haulage drifts to the bottom of the ore and advanced through the ore to a grade times thickness cutoff of 1.20 with a minimum thickness of 1.2 m (4 ft).

Mining Methods

Mining the unconsolidated ore bodies required overhead spiling, square set support, and excavation with pneumatic spaders. The broken ore was removed with slushers, feeding into mine cars or chutes. Stoping advanced along the roll fronts and retreated towards the haulage drift or development raise.

Sand filling of completed stopes along the haulage drift achieved stabilization of the drift pillars. Drift maintenance and ground support repairs were a major program in those areas where the stopes had caved to the drift pillar line. The repair program involved up to 20% of the underground crew working on a nonproduction shift. Replacing the drift caps with 45.4 kg-(100-lb) *H* beams provided only temporary support. The installation of yieldable steel arches through this drift section provided stable ground support. Table 1 provides summary data on mining quantities for the years the underground mine was operated.

Ventilation

During the shaft sinking phase and early stages of development, ventilation was provided by a 18.7-kW (25-hp) fan that forced air through a 762-mm (30-in.) casing installed in the service compartment of the shaft. Later, connections were made to the two 762-mm (30-in.) boreholes drilled from the surface. Each borehole was equipped with a 29.8-kW (40-hp) fan, a 2.1-MJ (2-million-Btu) propane-fired heater, and emergency escape facilities. The 29.8-kW (40-hp) fans delivered 11.3 m^3/s (24,000 cfm) of heated, downcast air into the haulage level. Two additional 305-mm (12-in.) boreholes, equipped with 18.7-kW (25-hp) fans, increased the total volume of air upcast through the shaft to 26.9 m^3/s (57,000 cfm). Auxiliary ventilation to the working faces was provided with 11.2-kW (15-hp) fans through vent tubing.

Mine Drainage

The Wind River formation, composed of unconsolidated and uncemented sand, gravel, and clay, was an artesian aquifer completely saturated with water to within 30.5 m (100 ft) of the surface. A mine-dewatering well field, consisting of eight 305-mm (12-in.), gravel-packed wells was completed before shaft sinking commenced. At the time the station was cut, the water table had been lowered from the 30.5 m (100 ft) depth to an average depth of 104.2 m (342 ft). The pumping rate during this period was 0.28 m^3/s (4500 gpm). Eleven additional wells were installed as mine workings extended beyond the well field drawdown area. An average pumping rate of 0.32 m^3/s (5000 gpm) was required to hold the water table at a depth of 111 m (365 ft).

The pump station was constructed on the 2070-m (6790-ft) level at a depth of 98.8 m (324 ft) and two 37.3-kW (50-hp) pumps discharged 0.03 m^3/s (500 gpm). The settling sump ahead of the pumping sump was cleaned periodically with a sludge pump discharging into mine cars. Standby emergency pump capacity was achieved by the installation of two 56.0-kW (75-hp) submersible pumps in the shaft. The underhand stopes required continuous pumping, with submersible pumps discharging into a 152-mm (6-in.) line in the haulage drift.

Although longhole drilling for dewatering was employed to provide drainage ahead of and over the haulage drifts, on several occasions erosion through the longholes induced caving ahead of and back over the drift section. Another method of dewatering ahead of drifting was to discontinue the full-face advance and drive a square *scram* ahead of the drift. Drifting would be suspended until the area was drained.

Sampling and Grade Control

Ore control procedures consisted of channel sampling of drift and stope walls, muckpile grab, and sampling of muck in mine cars. Grade was determined with a beta-gamma scaler. Mined ore was stockpiled in four ore blending piles with a grade ranging from 0.30% to +0.90% U_3O_8. Division between piles was at increments of 0.20% U_3O_8. The submarginal ore blending pile included ores with a grade range of 0.10% to 0.29% U_3O_8.

In January 1963 the decision was made to phase out the conventional underground mine operation. The factors contributing to this decision were the poor mining conditions and the costly ground support required, revision of the AEC program allowing ore substitution without loss of allocation, and the promising results being obtained from solution min-

Table 1. Underground Production Summaries

Year	Mine Ore				Marginal Ore			
	Dry Tons*	Grade (%)	kg U_3O_8	lb U_3O_8	Dry Tons	Grade, %	kg U_3O_8	lb U_3O_8
1960	16,340	0.540	80,104	176,600	1,052	0.210	2,003	4,416
1961	40,710	0.591	218,269	481,200	4,805	0.232	10,126	22,325
1962	33,530	0.532	161,932	357,000	10,489	0.218	20,713	45,665
1963	13,910	0.495	62,460	137,700	2,666	0.208	5,021	11,070
1964†	1,050	0.016	136	300	977	0.947	8,372	18,456
Total	105,540	0.546	522,901	1,152,800	19,989	0.255	46,235	101,932

Year	Total Ore				Submarginal Ore			
1960	17,390	0.520	82,085	180,966	6,376	0.051	2,968	6,544
1961	45,510	0.553	228,404	503,545	9,050	0.051	4,187	9,231
1962	44,020	0.457	182,659	402,694	3,052	0.067	1,842	4,060
1963	16,572	0.449	67,485	148,779	1,383	0.075	938	2,067
1964	2,031	0.462	8,521	18,786	—	—	—	—
Total	125,523	0.500	569,154	1,254,770	19,861	0.055	9,935	21,902

* Short tons. Metric equivalent: st $\times$ 0.907 184 7 = t.
† Difference between stockpiles depleted and quantities milled.

ing tests. After all salvageable material was removed, the mine was allowed to flood in July 1964.

SOLUTION MINING

Following consultation with reservoir engineers, a pilot program was initiated in the spring of 1961 to test the feasibility of in-place leaching of uranium, or solution mining. Dewatering the mineralized zone was not surmised to be necessary as the natural ground water would act as a containment shell for the leaching solutions. Thus, an area was selected for the initial experiment as far away from the shaft and underground operation as property boundary and proven mineral reserve would allow. See Fig. 2 for the location of the various mining phases. The patented well design and development, pattern size and shape, leaching chemistry, and uranium recovery from production well solution has been adequately documented (Ritchie and Gardner, 1967; Ritchie and Anderson, 1968) and following the pilot progam, 680 386 kg (1.5 million lb) U_3O_8 were produced on a commercial basis to 1970. All material produced was finished to convertor specifications at the Lucky Mc mill.

FEASIBILITY OF OPEN PIT DEVELOPMENT

The first serious feasibility study to examine the economics of open-pit mining was submitted to management in August 1967. At the time, proven, indicated, and inferred reserves totaled 5.9 Mt (6.5 million st), 0.22% grade, con-

taining 12 927 383 kg (28.5 million lb) U_3O_8. Forty-two percent of this reserve total was selected, based on the most favorable strip ratio, for design into the preliminary pit study. The volume of ore, coupled with its relatively low grade, eliminated consideration of hauling to the Lucky Mc mill. Thus, an onsite concentrator was part of the economic feasibility study.

Following the favorable economic indications contained in the first study, a development drilling program of 76 200 m (250,000 ft) was authorized. The results expanded the proven reserve in area 2, the claim group previously exploited by underground mining and still producing from solution mining, to 8 255 381 kg (18.2 million lb) U_3O_8. The development drilling program successfully increased the total reserve by 14.4%, but more significantly, increased the proven category of reserves by 55.9%.

Six contiguous pit phases were engineered, requiring an estimated 65.8 Mm^3 (86 million cu yd) of overburden removal for the extraction of 4 082 331 kg (9 million lb) U_3O_8. The economies of this pit development and an appropriately sized mill formed the basis for a final feasibility study.

OPEN PIT DEVELOPMENT

On Nov. 1, 1968, the announcement was made by the company that the Shirley Basin mine would be converted to a large scale open-pit operation, and that a new 1089-t/d

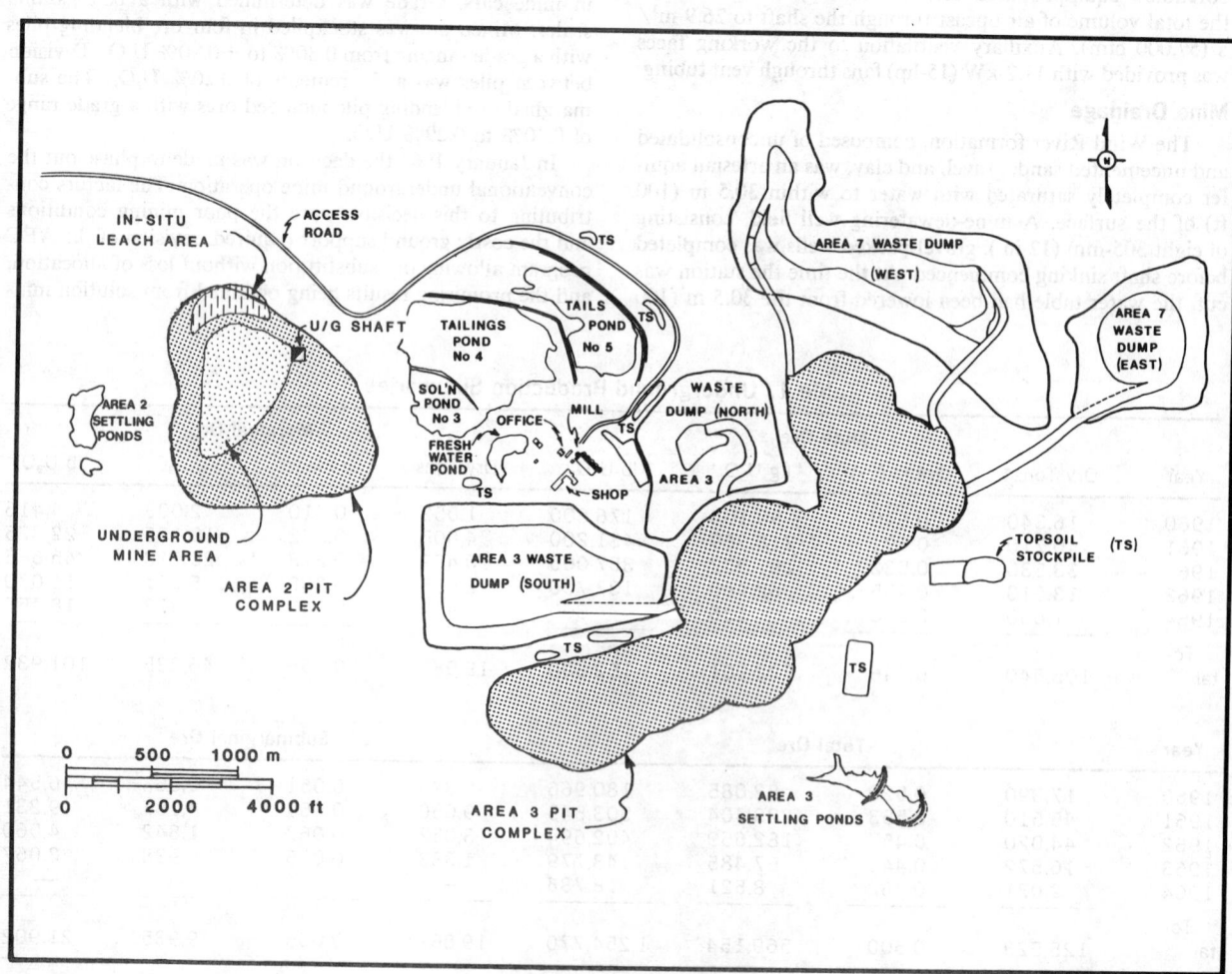

Fig. 2. Mine map, Shirley Basin mine.

(1200-stpd) mill would be constructed to accommodate the resulting ore production.

In the following year, a new warehouse and equipment maintenance facility were constructed, a new tailings basin and dam were completed, and an engineer-contractor was engaged to begin construction of the new mill. In addition, the first of the large-scale stripping equipment was delivered to begin the excavating of the initial pit.

The equipment purchased, after a study of various stripping options, was large-scale, three-axle scrapers, 24.5-m³ (32-cu-yd), 72.6 t (80-ton) capacity scrapers, and dual crawler tractors for push loading. Mid-sized, self-loading scrapers were acquired to load and haul the waste material between ore pockets in the ore zones, and hydraulic backhoes and 22.7-t (25-ton) trucks were used for the actual ore mining.

The method of excavating waste from around ore pods within the ore body and then mining the ore areas with backhoes had been pioneered and proven at the Lucky Mc mine. This method consisted of defining ore pods by the use of portable Geiger counters, very carefully excavating the adjacent waste, and then mining with a relatively small mining unit in order to minimize dilution and yet insure total excavation of any available ore. By the end of 1969, the bulk of the stripping equipment had been delivered and slightly over 7 Mm³ (9 million cu yd) had been removed from the initial pit.

Ground Control

In March 1970, the pit experienced the first of many major slope failures that were to plague the early years of this project. The original pit designs had been made using slopes of around 45°, but after suffering numerous failures even before reaching ore depth approximately 122 m (400 ft below surface), it became apparent that nothing steeper than 33° would remain in place long enough to complete ore extraction.

Another problem noted during this phase of operation was the poor performance of the scrapers selected for stripping. As the pit became deeper, water was encountered and a pumping system was installed to move this water to the surface for disposal. The area of active removal, however, became very wet at times and this, combined with high clay content material, made the loading of the three-axle scrapers very difficult. Near the end of 1970, a study was instituted to reevaluate the stripping methods being used and recommend alternatives. As a result of this study, it was decided to move the bulk of the waste material with a standard truck-shovel combination.

To proceed as soon as possible with this recommendation, the truck-shovel units were ordered and began arriving in mid-1971. By the end of that year, the bulk of waste material was being moved by the new truck-shovel combination, 2 11.5-m³ (15-cu-yd) electric shovels and 12 108.9-t (120-ton) trucks. With just a few startup and operator training problems the new stripping method won rapid acceptance and proved much more economical than the scraper method.

It was recognized that even though the shovels were the preferred method of overburden removal, they could not handle the task singlehandedly. Because of the sometimes rapidly changing conditions due to slope failures, some scrapers were necessary to cope with unexpected problems. Scrapers also were necessary to do the final stripping just above ore so that no ore would be inadvertently removed as overburden.

Pit wall slope failures continued through 1971 and this period saw many man-hours devoted to solving this problem.

Near the end of the year, a monitor net had been installed around the pit perimeter, a leading geotechnical engineering firm had been retained to do a slope stability analysis, and some experimental drilling designs to drain the slopes had begun.

All of these actions eventually proved fruitful. The monitoring net, while not preventive in nature, at least provided early warning about impending failures and provided data for future designs.

The geotechnical engineering firm provided a report in 1972 that analyzed the strength characteristics of the soils around the pit areas and gave various safety factors for different slopes configurations. A compound slope configuration was eventually utilized; this was essentially a 33° slope on the top section and 27° on the bottom section This was not a fail-proof slope, but did provide some measure of stability and decreased the magnitude of the slope failures.

Pit Dewatering

Various combinations of drilled holes were tried to accomplish the draining of pit walls. The most practical of these was a combination of vertical drains and horizontal drains. This method consisted of a series of vertical holes drilled from the surface near the pit perimeter. These holes were completed as wells with a slotted PVC pipe and gravel-packed for the entire length of the hole. At the same time, holes slightly elevated from the horizontal were drilled from the pit bottom in the general direction of the vertical drains. This combination of holes allowed any perched water or elevated water table to drain down the vertical holes and out the near-horizontal holes into the pit. From here, the water was collected in the normal pit dewatering sump and pumped to the surface. Another dewatering technique, useful in some areas, was to drill and complete a well into a zone known to provide water to a slope. This well would be pumped at a rate high enough to create a drawdown which resulted in drying of the slope. These methods of dewatering slopes aided stability considerably and still are being used today.

Mine Production

Slides continued to be a problem through the early years until 1974 when the first pit was excavated and mined without any slides of major proportions. Instability still continues to be a problem, but is dealt with by using techniques learned in the earlier years.

Stripping and mining quantities generally continued to increase from the beginning in 1969 through 1976. Table 2 shows stripping and mining quantities for the period 1969 through 1986. Also, during this period, three-axle scrapers were replaced with large two-axle, push-pull scrapers. These units were difficult to maintain but were far superior in the sometimes difficult stripping situations encountered.

During this period all stripping and mining were being carried out in one general location (area 2). Each new *pit* was actually an extension of the previous excavation and resulted in one large pit. As pit extensions were mined out, they provided areas for dumping the waste material from the pit then being excavated. This practice was perfected to the point where almost no material was being hauled to the surface but rather "backfilled" into previously mined-out pits.

During 1976 and early 1977, a large-scale study of a new mining area was intensified. This new area (area 3) had long been known to contain good mineralization and the mill was actually positioned in a location to take advantage of this new area should it ever be mined (the mill location being

Table 2. Open Pit Production Summary

Year	Stripping, m³	Stripping cu yd	Mining			
			Dry Tons*	Grade, %	kg U₃O₈	lb U₃O₈
1969	7,074,000	9,252,000	—	—	—	—
1970	6,727,000	8,799,000	—	—	—	—
1971†	10,542,000	13,788,000	250,000	0.158	358,800	791,100
1972	16,329,000	21,358,000	587,260	0.239	1,272,000	2,804,200
1973	13,076,000	17,103,000	357,900	0.188	611,200	1,347,400
1974‡	16,527,000	21,616,000	433,900	0.197	776,600	1,712,100
1975	13,751,000	17,986,000	452,700	0.180	737,100	1,625,100
1976	15,336,000	20,059,000	526,070	0.194	923,200	2,035,300
1977	18,088,000	23,658,000	628,700	0.177	1,010,600	2,227,900
1978	18,708,000	24,469,000	879,200	0.110	875,900	1,931,100
1979	19,341,000	25,297,000	674,810	0.140	859,600	1,895,200
1980	20,449,000	26,746,000	631,200	0.173	988,900	2,180,200
1981	12,218,000	15,980,000	467,750	0.136	578,200	1,274,700
1982	7,539,000	9,861,000	275,780	0.150	378,800	835,200
1983	6,587,000	8,616,000	208,900	0.246	466,500	1,028,500
1984	4,168,000	5,452,000	73,620	0.222	148,000	326,300
1985	2,463,000	3,222,000	166,620	0.180	271,800	599,300
1986	3,504,000	4,584,000	38,400	0.083	29,400	64,900

* Short tons. Metric equivalent: st × 0.907 184 7 = t.
† Contracted 2.4 Mm³ (3.2 million cu yd)—stripping.
‡ Contracted 1.9 Mm³ (2.5 million cu yd)—stripping.

about midway between the original mining area and the new one).

Pit designs were drafted and studied and a complete analysis of stripping methods was done for this new area. The analysis indicated that a truck-shovel combination would be the most economical method of stripping. In mid-1977, a new truck-shovel fleet, one 20.6-m³ (27-cu-yd) electric shovel and six 154.2-t (170-ton) trucks, was purchased and began stripping in this new area. The new area produced its first ore in September of 1979.

This period showed, again, a generally slow but steady increase in the volume of material stripped and tons of ore mined. This trend continued until September 1980 when the first of several cutbacks was deemed necessary. These cutbacks in production were, indirectly, adjustments to the decrease in uranium prices which had begun their decline in 1979.

The operation, at present (1983), continues to be mainly a truck-shovel stripping operation with aid from a scraper fleet. All mining activity in the original pit area (area 2) has been completed, and all stripping and mining activity is now carried on in the area started in 1977 (area 3). Large areas of the original pits, pit slopes, and dumps have been regraded and reclaimed in accordance with the state-issued mining permit.

The manpower growth kept pace with the operation, starting out with 128 employees in 1969 and peaking in mid-1980 at 620 employees. Presently, at the reduced rates of production, employment has decreased to 135 employees.

MILL CIRCUIT DEVELOPMENT AND PERFORMANCE

Contrary to traditional uranium mill design at the time, several unique features were introduced into the engineering of the Shirley Basin mill, setting a precedent that has since been emulated by many uranium processors. Fig. 3 presents the flow diagram for the mill process. The Shirley Basin design eliminated all crushing and conveying steps. Run-of-mine ore is introduced into the 1633-t/d (1800-stpd) mill from an exterior blending pile pad by a front-end loader through an inclined bar grizzly to the loading hopper. The material is fed at a rate of approximately 181.4 t/h (200 stph) on a pan feeder into a 5.5-m (18-ft) diam by 1.8-m (6-ft) long semiautogenous grinding mill; little grinding media is required in this mill. The material is ground, discharged from the mill, and pumped to a bank of six DSM-type (Dutch State Mines) screens. The sized product, all −850 μm (−20 mesh), is then pumped to pulp storage tanks and the oversize is recirculated back to the mill for regrind. Operating this circuit for 8 to 9 hr per day produces sufficient feed for the 24-hr continuous leaching operations. The pulp, at 60% solids, is stored in two 7.6-m (25-ft) diam by 15.2-m (50-ft) high, flat-bottomed, 345-kPa (50-psi) air-agitation tanks; no problems have been encountered with keeping coarse-grained particles in suspension once the inverse cone deadbed was formed. The pulp is metered from the storage tanks to leaching on a continuous basis. The leaching vessels consist of four conical-bottom, rubber-lined, air-agitation Pachuca tanks 6.9-m (22.5-ft) in diam by 14.2-m (46.5-ft) high. Fresh feed to leach is sampled and monitored in a mass flow station. Sulfuric acid and sodium chlorate are also added at this juncture; steam is also added in winter. The retention time in leach is approximately 18 hr and is sufficient to extract up to 98% of the uranium into solution. Leach discharge is classified in a hydrocyclone into two size fractions, coarse and fine. Each fraction is treated separately; the coarse material has the uranium in solution washed from it in a series of six countercurrent sand-washing hydrocyclones and the fine-sized fraction is washed through a series of six 30.5-m (100-ft) diam countercurrent decantation thickeners. Pulp streams in both circuits are washed by recycle solution and after optimum recovery, the uranium-depleted, coarse-sand underflow from the last cyclone and the thickened fines underflow from the last thickener are combined and pumped to the tailings ponds where all pulp and associated liquor are retained.

The pregnant solution overflowing the No. 1 thickener is pumped through a clarifying bed of granular activated

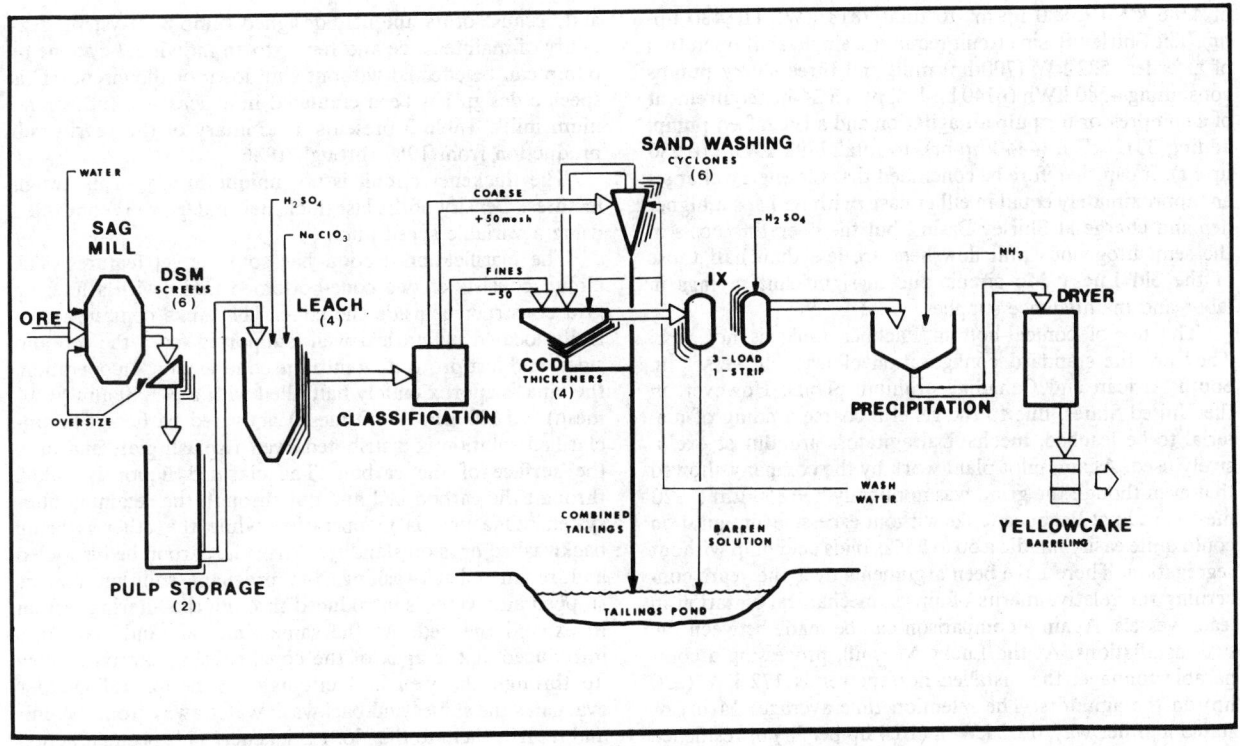

Fig. 3. Shirley Basin flow diagram for the mill process.

carbon and any extraneous slimes are removed. The completely clear liquor containing the extracted uranium is then fed to a set of fixed-bed ion exchange columns. The columns each contain a 8.5-m³ (300-cu-ft) bed of a weak-base tertiary amine resin. Each set of columns consists of four beds and in general, only three are employed for uranium absorption while the fourth is being eluted. The strip liquor, or eluate, has a chemical composition of 2 M sodium chloride and 1 M ammonium sulfate. Approximately 6½ bed volumes of strip liquor are sufficient to strip all of the uranium from the resin bed. Due to the exchange characteristics of the resin, the first 1½ bed volumes of liquor are high in uranium, high in sulfate, low in chloride, and have a high sodium ion concentration. This portion of the liquor is designated as the No. 1 concentrated eluate and is guided to a separate tank. The ensuing 2½ bed volumes, also high in uranium, contain high values of chloride. This liquor is designated as the No. 2 concentrated eluate and it too goes into a separate circuit tank. The final 2½ bed volumes are designated as recycle and are stored separately.

Because the stripping solution is neutral, the acidity of the Nos. 1 and 2 concentrated eluates is very low, and these solutions are pumped separately to respective precipitation tanks where the uranium product is precipitated by neutralization with ammonia gas. The precipitated products in each case flow to separate yellowcake thickening vessels where, in the case of the No. 1 circuit, the overflow material is polished and, being low in chloride, is discarded to tailings. The material overflowing the second yellowcake thickener is guided to eluate makeup tanks where the chloride concentration is replenished with fresh saturated brine before being recirculated to fresh eluate storage.

The settled uranium precipitate in the Nos. 1 and 2 circuits is combined after initial thickening. The solids so pro-

duced are fed to a third yellowcake thickener designated as the washing thickener. The overflow from this vessel is polished and discarded to waste as it contains no salvageable chemical constituents. The thickened solids are then dewatered in a drum filter prior to introduction to a multi-hearth, roaster-type dryer. The completely dried solids, discharging from the dryer, accumulate in an enclosed hopper and, on a batch basis, are automatically packaged in 0.21-m³ (55-gal) drums after being sized through a small pulverizer.

Some of the differences between traditional flowsheets of the period and the Shirley Basin facility are in the ore handling section where, with no conveyors or crushing, direct conversion of the run-of-mine ore to a pulp is achieved. Significant capital cost savings and a substantial reduction in operating costs were achieved. This direct conversion also eliminates the requirement for costly dust-control equipment normally associated with conveyors and crushing plants.

The decision to employ a semiautogenous mill and establishment of the dimensions resulted from pilot-plant test work conducted on several hundred tonnes of typical material run through a mill, 1.8 m (6 ft) diam by 0.61 m (2 ft) long, installed at one of the company's other mines.

DSM-type screens were favored, as opposed to cyclone classification, due to successful utilization of these units at the Lucky Mc mill and the fact that a closer control of grain size can be achieved, facilitating storage of the pulp in air-agitation vessels without segregation.

It is particularly interesting to analyze the power requirements in the ore handling section, comparing one flow scheme with the other. In Lucky Mc mill's old circuit, the single shift operation of crushing, conveying, partial drying, dust collection, and sampling required 3432 kWh (4600 hp-hr). Added to that, there is the conveying (from ore bins to mill), grinding, and pump feeding to leach on a 24-hr basis of 4386 kWh (5880 hp-hr) to total 7818 kWh (10,480 hp-hr). The Shirley Basin circuit requires a single-shift operation of a feeder, 522-kW (700-hp) mill, and three slurry pumps consuming 4580 kWh (6140 hp-hr), plus a 24-hr requirement of a compressor for pulp air agitation and a leach feed pump, adding 3312 kWh (4440 hp-hr), to total 7892 kWh (10,580 hp-hr). It can therefore be concluded that the energy charges are approximately equal in either case (with perhaps a higher demand charge at Shirley Basin), but the operating costs in the semiautogenous mill flowsheet are less than half those of the old Lucky Mc circuit due to significant savings in labor and maintenance supplies.

The use of conical-bottom Pachuca tanks is not new. They are the standard configuration of leaching vessels in South African and Canadian uranium plants. However, in the United States, due to the general coarse grading of material to be leached, mechanical agitators are almost exclusively used. Again, pilot plant work by the company showed that even though the grind was nominally a $-850~\mu m$ (-20 mesh) grading, Pachuca tanks without excessive air agitation could quite easily handle a 50 to 55% solids acid pulp without segregation. There have been arguments over the years concerning the relative merits of air vs. mechanical agitation in leach vessels. Again a comparison can be made between the two installations. At the Lucky Mc mill, processing a comparable tonnage, the installed horsepower is 172 kW (230 hp) on the agitators. The retention time averages 14 hr, or stated another way, 12.2 kW/h (16.4 hp per hr) of residence time. The Shirley Basin mill requires the maximum service of one 149-kW (200-hp) compressor, but the retention time is 18 hr, or 8.28 kW/h (11.1 hp per hr) of residence time. Although this comparison is not absolutely conclusive, the absence in a Pachuca of any moving parts in the acid and abrasive environment contributes to lower maintenance costs.

The sand-washing hydrocyclone circuit was introduced approximately four years after the initial commissioning of the plant. This circuit enabled plant throughput to be increased from 1089 t/d (1200 stpd) to the presently licensed 1633 t/d (1800 stpd). The circuit has functioned very well,

and because of its uniquely designed sump box system, flexibility of maintenance and repair to an individual cyclone or pump can be effected without shutdown of the circuit. This specific design has been emulated in at least two other uranium mills. Table 3 presents a summary of the yearly mill production from 1969 through 1986.

The thickener circuit is not unique in any of its design features. Density of the last thickener underflow is controlled using a variable speed pump.

The clarification section has some novel features. The circuit consists of two cone-bottomed tanks with a filtering grid constructed inside the cone. The tanks contain a centrally located cylindrical well, supported from the top and sides, and just protruding into the cone section. In operation, the tank is approximately half filled with a -1.40 mm (-14 mesh) $+600~\mu m$ ($+30$ mesh) activated carbon. The unclarified solution is distributed over a splash plate and onto the surface of the carbon. The clarified liquor is pulled through the carbon bed and out through the retaining filter system. One unit is in operation while the other is being backwashed or is on standby. When the carbon bed is fouled and requires backwashing, the pregnant solution flow is stopped and water is introduced through the filtering section to expand the bed. At the same time, air and water are introduced at the apex of the cone, causing carbon to flow up through the well and circulate. A peripheral launder evacuates the slimes and backwash water away from the unit and returns them to the No. 1 thickener. This peculiar action of backwashing not only effectively cleans the bed, but prevents gypsum from forming large conglomerate pieces with the carbon.

The Shirley Basin mill was the first commercial plant to use a weak-base resin. It was selected because it has capital cost advantages over an eluex system and a straight solvent extraction system for this volume of solution and also because it results in a more economical operation than either of the two aforementioned processes and will produce a premium grade of product. The weak-base resin does not exhibit as high a loading for uranium as does a strong base anionic

Table 3. Mill Production Summary

Year	Mill Dry Tons*	Grade, %	kg U₃O₈	lb U₃O₈	kg Produced	lb Produced
1969	—	—		—		
1970	—	—		—		
1971	196,870	0.155	276,800	610,200	260,300	573,800
1972	452,930	0.226	928,100	2,046,200	861,600	1,899,400
1973	444,900	0.225	907,000	1,999,700	838,600	1,848,900
1974	344,900	0.175	546,400	1,204,700	542,600	1,196,200
1975†	489,700	0.195	867,800	1,913,200	766,300	1,689,500
1976‡	512,100	0.199	923,400	2,035,800	860,000	1,895,900
1977	631,400	0.183	1,047,700	2,309,700	991,000	2,184,800
1978	615,500	0.152	850,100	1,874,200	810,600	1,787,100
1979	657,650	0.139	827,700	1,824,700	764,000	1,684,300
1980	657,820	0.170	1,018,500	2,245,400	985,100	2,171,700
1981	466,350	0.127	538,300	1,186,800	521,000	1,148,700
1982	296,080	0.130	336,200	741,300	341,200	752,300
1983	296,300	0.143	386,600	852,200	373,400	823,300
1984	279,100	0.111	280,300	618,000	273,300	602,500
1985	240,960	0.128	280,300	617,900	272,400	600,600
1986	288,200	0.092	241,500	532,400	227,000	500,400

* Short tons. Metric equivalent: st × 0.907 184 7 = t.
† Cyclone circuit added in mill.
‡ Boiler added in mill.

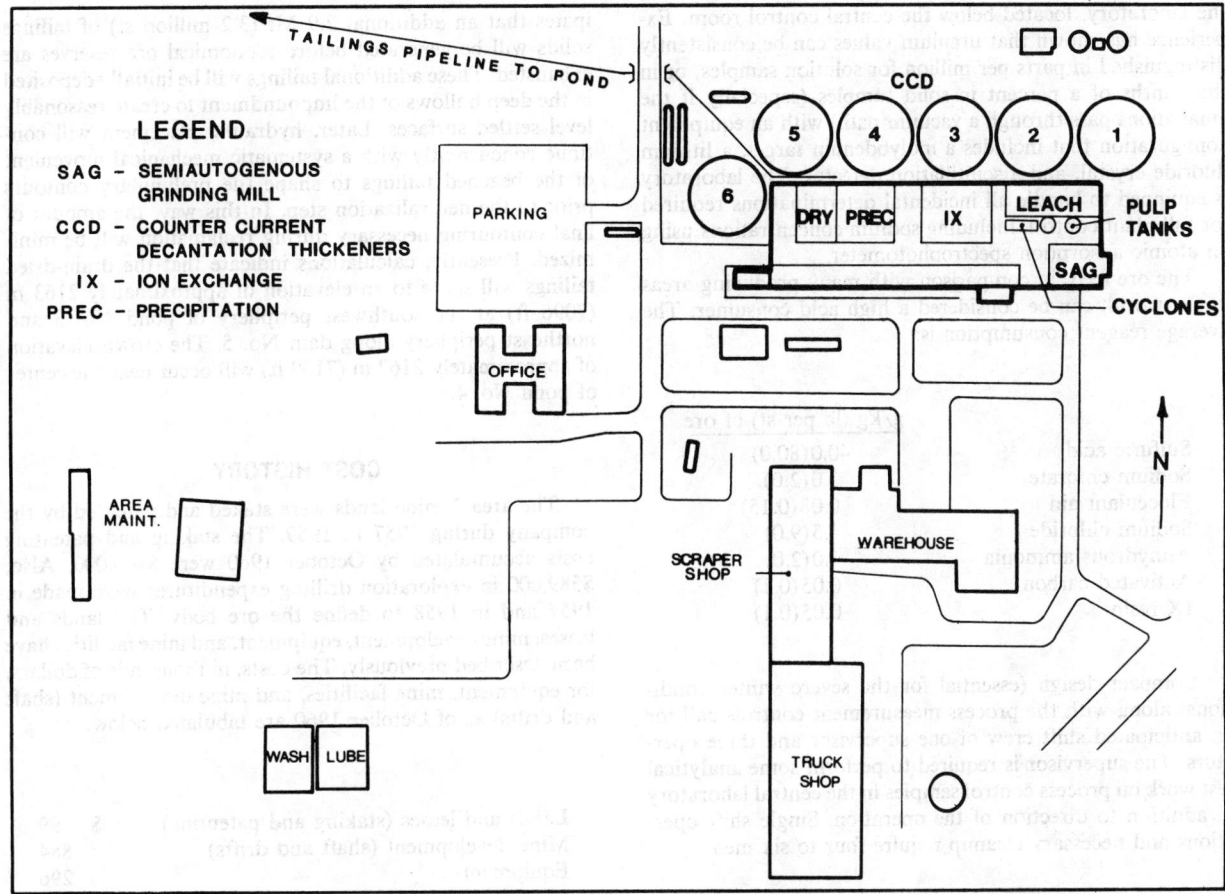

Fig. 4. Site plan of the surface facilities at the Shirley Basin mine.

resin. However, it is more selective and due to the ease with which it elutes with a neutral and inexpensive reagent, this lower capacity can be tolerated. The predominance of ammonium sulfate in the liquor fed to precipitation results in a product precipitate that is very low in sodium ion.

The reason for the dual or separated precipitation circuits is to effect a bleed of sodium ion from the circuit and at the same time conserve, by recycle, the chloride ion necessary for uranium stripping. The subsequent flowsheet in the product recovery circuit is designed around maximum washing of the precipitated product in order to reduce the halogen content in the final product.

The use of roaster-type dryers is not new; the application has been widely used in the uranium industry. The entire drying and packaging circuit is completely enclosed, with vapors and sublimed salts being vented through a scrubber prior to discharge to the atmosphere. All precautionary measures have been incorporated to avoid the exposure of employees to radioactive-bearing particles.

Except for the six 30.5-m (100-ft) thickener vessels, all other pieces of equipment are compact enough to be contained within one building structure, 131 m (430 ft) long and 57.9 m (190 ft) wide. Fig. 4 presents a site plan of the mine's surface facilities, including the mill building and some of its major circuits. This compactness of design facilitates control from a central instrumentation panel located in the heart of the mill building. Instrumentation in the mill follows the basic principle that instruments are most effective when they

accurately record process variables but do not control corrections. In almost all instances, correction must be made manually.

The control panel is divided into three segments:

1) At the top of the panel is a semi-graphic representation of the entire process, complete with machinery running lights.

2) All of the recording and indicating instruments are located in the center portion. Included in this package are 3 density-recording gages, 14 magnetic flowmeter readouts, oxidation-reduction potential and conductivity recorder showing the oxidizing level and acid concentration in the leach circuit, and numerous tank level indicators.

3) On the console portion of the panel, a full ion exchange graphic is depicted, including column pressures and valve operating switches. The entire ion exchange section is manually, but remotely, operated from this central point. The circuit design is sized so that only one cell in either exchange set can be stripped of its uranium at a time. This necessitates careful scheduling of the loading cycles, but at the same time, lessens the chances of error during elution. The circuit has a theoretical surge capacity of 6350.3 kg (14,000 lb) of U_3O_8 per 24-hr period, or double the nameplate capacity of the mill. This was a deliberate design consideration due to the expected fluctuations in the grade of ore from the open-pit operation.

Chemical and analytical control of the circuit is largely dependent upon an X-ray emission spectrograph installed in the laboratory, located below the central control room. Experience has shown that uranium values can be consistently distinguished in parts per million for solution samples, or in thousanths of a percent in solid samples (especially if the emanations pass through a vacuum path) with an equipment configuration that includes a molybdenum target, a lithium fluoride crystal, and a scintillation detector. The laboratory is equipped to handle all incidental determinations required for full plant control, including sodium concentrations using an atomic adsorption spectrophotometer.

The ore feed, in comparison with many producing areas in the world, can be considered a high acid consumer. The average reagent consumption is:

	g/kg (lb per st) of ore
Sulfuric acid	40.0(80.0)
Sodium chlorate	1.0(2.0)
Flocculant aid	0.08(0.15)
Sodium chloride	4.5(9.0)
Anhydrous ammonia	1.0(2.0)
Activated carbon	0.05(0.1)
IX resin	0.05(0.1)

Compact design (essential for the severe winter conditions) along with the process measurement controls call for an anticipated shift crew of one supervisor and three operators. The supervisor is required to perform some analytical test work on process control samples in the central laboratory in addition to direction of the operation. Single shift operations and necessary cleanup require four to six men.

TAILINGS MANAGEMENT

The present tailings impoundment occupies approximately 148 ha (365 acres) in a dry wash that originally drained northeastward toward Spring Creek. The impoundment is contained on three sides by natural landforms and on the fourth side by dams. The impoundment consists of three ponds. Fig. 5 shows a plan view of the tailings impoundment. Mining operations prior to 1970 created a series of mine water settling ponds (ponds 1, 2, and 3). The material in ponds 1 and 2 was excavated and removed during the development of the area 2 pit complex.

Pond 3 presently contains approximately 535 331 m³ (434 acre-ft) of solution, retained by a compacted earthfill dam (dam No. 3). Dam No. 4 was constructed to accommodate tailings generated by the mill, which began production in 1971; dam No. 4 is a compacted earthfill dam with a crest elevation of 2167 m (7110 ft) and an impound volume of approximately 4.9 Mm³ (4000 acre-ft).

Construction of dam No. 5 began in 1975. The dam is an engineered earthfill dam that was subject to continuous engineering inspection during construction. The dam has a crest elevation of 2167 m (7110 ft) with a maximum height of 19.8 m (65 ft) and a length of approximately 1859 m (6100 ft). The impound volume is 4 441 000 m³ (3600 acre-ft). Dam No. 5 consists of three zones: (1) a low permeability upstream embankment constructed of compacted mine overburden; (2) a chimney drain constructed of cycloned tailings underflow; and (3) a downstream cover of compacted mine overburden.

As of December 1986, ponds 4 and 5 contained approximately 6.2 Mt (6.8 million st) of tailings solids. There have been no solid tailings placed in pond 3. The company antic-

ipates that an additional 2.9 Mt (3.2 million st) of tailings solids will be generated before economical ore reserves are exhausted. These additional tailings will be initially deposited in the deep hollows of the impoundment to create reasonably level settled surfaces. Later, hydraulic placement will continue concurrently with a systematic mechanical movement of the beached tailings to shape the preliminary contours prior to the neutralization step. In this way, the amount of final contouring necessary during reclamation will be minimized. Presently, calculations indicate that the drain-dried tailings will settle to an elevation of approximately 2163 m (7096 ft) at the southwest periphery of pond No. 3 and northeast periphery along dam No. 5. The crown elevation of approximately 2167 m (7109 ft) will occur near the center of pond No. 4.

COST HISTORY

The area 2 mine lands were staked and patented by the company during 1957 to 1959. The staking and patenting costs accumulated by October 1960 were $100,000. Also, $589,000 in exploration drilling expenditures were made in 1957 and in 1958 to define the ore body. The lands and leases, mine development, equipment, and mine facilities have been described previously. The costs, in thousands of dollars, for equipment, mine facilities, and mine development (shaft and drifts) as of October 1960 are tabulated below:

Lands and leases (staking and patenting)	$ 99
Mine development (shaft and drifts)	884
Equipment	296
Mine facilities	568
Total	$1,847

In-situ leach facilities were set up in 1962. The initial capital expenditures by 1962 were $48,000. During 1962 to 1967 capital expenditures for the in-situ leach facilities accumulated to $129,000.

For the period 1958 to 1970, concentrate production was sold pursuant to an AEC contract at approximately $8 per lb except in 1969 and 1970. The commercial market began in the late sixties. Except for the start-up period from 1958 to 1963, the mines were profitable through 1970. Concentrate costs per pound were in the range of $4 to $7 per lb.

The feasibility study of the area 2 open pit mine and a concentrator at Shirley Basin was begun in 1968, while the feasibility study of the area 3 open pit mine was begun in 1976. Additional mineral rights were acquired in 1961, 1963, and 1978 for the area 3 open pit mine. The capital costs at year end, in thousands of dollars, are tabulated below for 1971, 1977, and 1981:

Capital Costs			
Type	1971	1977	1981
Lands and leases	$ 1,517	$ 1,584	$ 1,584
Development	12,923	14,701	16,518
Equipment	9,014	22,256	35,043
Mine facilities	1,714	3,216	6,052
Mill facilities	8,061	9,123	9,999
Townsite	—	643	638
Total	$33,229	$51,523	$69,834

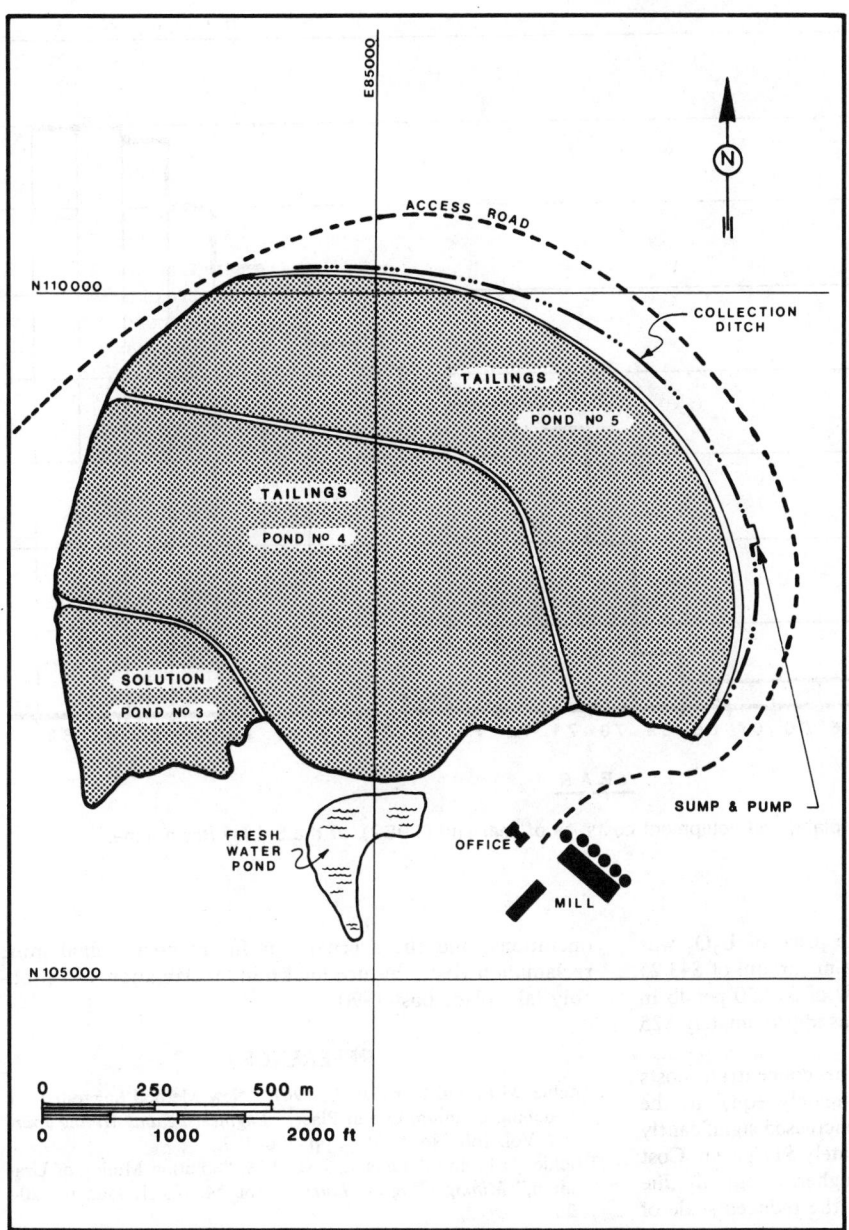

Fig. 5. Plan view of the tailings impoundment, Shirley Basin mine.

The costs of property, plant, and equipment from the inception of Shirley Basin to the end of 1982 are shown in Fig. 6.

Major capital expenditures made after 1963 are listed below:

1) In 1968, 12 72.6-t (80-st) scrapers were ordered for mine development operations of the area 2 surface mine.

2) In 1968, design work was authorized for a 1088.6-t/d (1200-stpd) mill.

3) In 1970, mine development advanced, but development was delayed by pit slope failures and scraper production performance. To compensate, 2 11.5-m³ (15-cu-yd) shovels and 12 108.9-t (120-ton) trucks were ordered.

4) In 1971, mill construction was completed and two shovels were placed in operation.

5) In 1972, a new scraper fleet replaced the old.

6) In 1977, a 20.6-m³ (27-cu-yd) shovel and six 154.2-t (170-ton) trucks were added to the equipment fleet.

7) In 1975, a cyclone circuit was added to the mill that increased mill capacity from 1088.6 t/d (1200 stpd) to 1633 t/d (1800 stpd).

Following 1977, additional capital expenditures were made:

1. In 1978, a truck shop extension was added to the shop.

2. In 1978, a new multiple-hearth dryer was constructed in the mill.

3. In 1978, Spring Creek was diverted to permit area 3 stripping.

4. In 1979, a hydraulic shovel and three 108.9-t (120-ton) trucks were acquired.

5. In 1980, six 108.9-t (120-ton) trucks were rebuilt.

By 1981, the scale of operations had been reduced, a

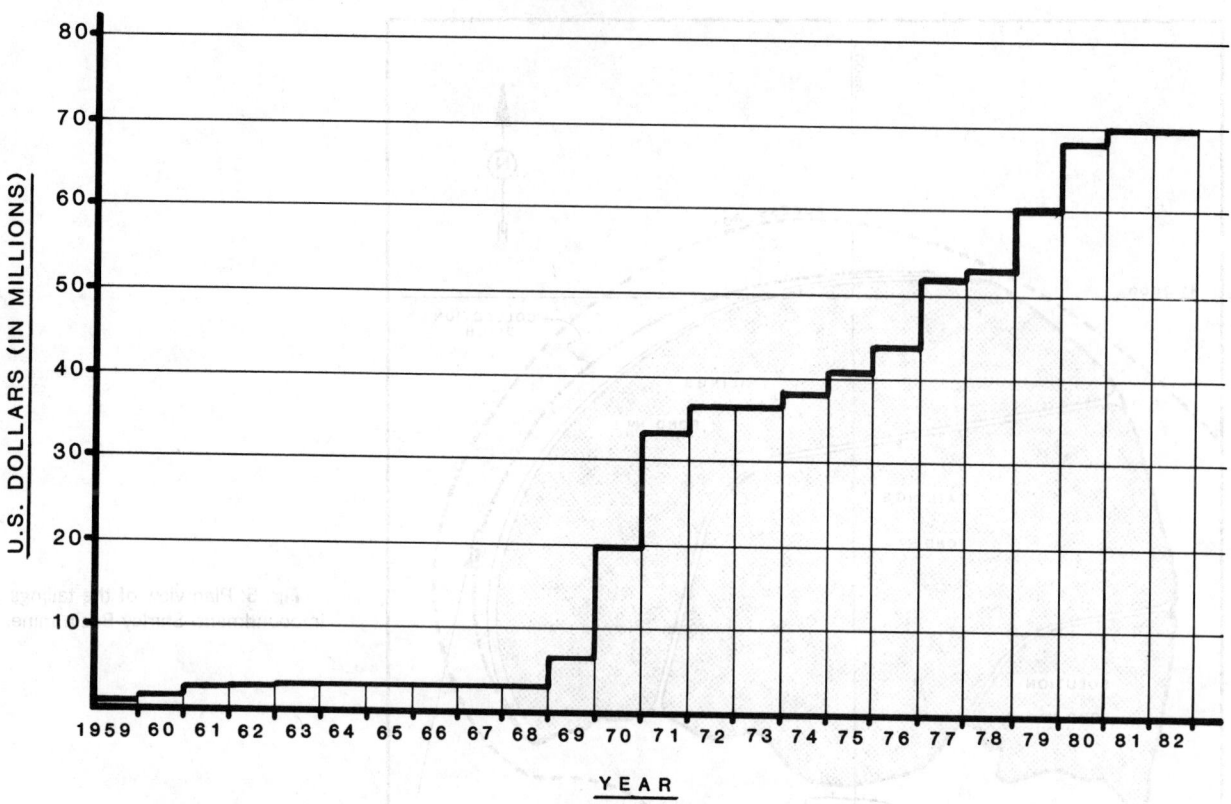

Fig. 6. Balance of property, plant, and equipment costs, as of year-end (1982), at the Shirley Basin mine.

reflection of the uranium market. The price of U_3O_8 was $6.05 per lb in 1971 and increased to a maximum of $43.25 per lb in 1978. On its descent to a low of $17.50 per lb in 1982, the price of U_3O_8 in mid-1981 was approximately $25 per lb.

At the start of area 2 in 1971, the concentrate costs including noncash costs were approximately equal to the market price. By 1978, the margin had increased significantly with the concentrate costs at approximately $17 per lb. Cost of concentrate sales in the 1980s was higher, essentially due to the higher strip ratio of area 3 pits, the reduced scale of

operations, and the accruals for future area 3 final mine reclamation costs. Future final mine reclamation will probably take place post-1990.

REFERENCES

Ritchie, M.I., and Gardner, J., 1967, "New Method Suggested for Leaching Uranium Ore in Place," *Engineering and Mining Journal,* Vol. 168, No. 5, May, pp. 106–107.

Ritchie, M.I., and Anderson, J.S., 1968, "Solution Mining of Uranium," *Mining Congress Journal,* Vol. 54, No. 1, Jan., pp. 20–26.

9.8 Island Copper Mine

Morton E. Pratt

Introduction

The Island Copper mine is located adjacent to the Rupert Inlet arm of Quatsino Sound on the north end of Vancouver Island in the Province of British Columbia, Canada (Fig. 1). The residential base for the mine's employees is the town of Port Hardy about 19 km (12 miles) from the mine on the Canadian-Alaskan Inland Waterway some 418 km (260 miles) northwest of Vancouver, British Columbia. The area is served by commercial airlines, bus lines, and trucking lines. The coastal freighter and barge systems that helped build and service the island communities beginning in the 1800s gave way to the trucking industry in 1979 when the national highway system was extended along the island making Port Hardy its western terminus (Figs. 2 and 3).

DISCOVERY AND EXPLORATION

In 1963 the Canadian government released an airborne magnetometer map of the north end of Vancouver Island. The anomalies shown on the map combined with his knowledge of the area prompted Gordon Milbourne, a well-known prospector, to investigate the area. He staked the initial claims in 1965 after finding small quantities of native copper and chalcopyrite in outcrops of volcanic rocks north of Francis Lake (sometimes called Bay Lake). He later exposed high-grade chalcopyrite in bedrock in two shallow pits in the same area about 1.5 km (0.9 mile) west of the present Island Copper pit. Initially Milbourne, like other explorers in the area, was searching for iron ore, as the island had a record of iron ore deposits. He found no commercial iron but he

Fig. 1. Aerial view of the Island Copper mine looking south down Rupert Inlet
(*Source:* North Island Studios, Port Hardy, BC, Canada).

1163

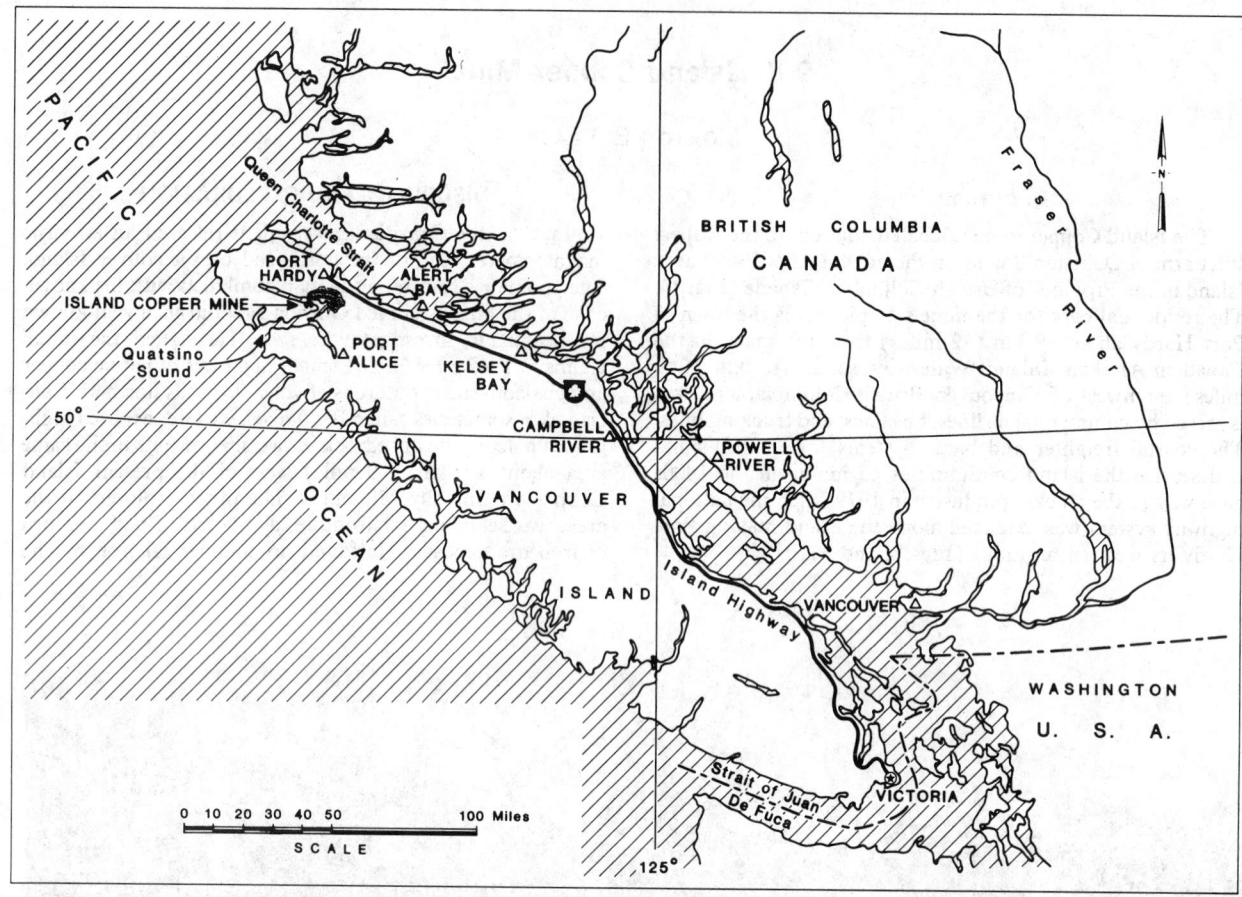

Fig. 2. Location of the Island Copper mine, Port Hardy, Vancouver Island, British Columbia.

had discovered traces of copper and courageously increased his holdings to 112 judiciously placed claims, strategically located over an anomaly that represented the disseminated copper deposit.

In 1966, Milbourne and Utah Construction and Mining Co. signed an option agreement permitting Utah to explore these claims. After initially disappointing diamond-drill results, Utah in 1966 moved away from the original Milbourne discovery site to a second area on which a soil geochemical survey had delineated a large copper anomaly. Ground magnetic surveys and induced polarization surveys were also completed and provided additional useful information in the anomalous area. This anomaly was tested with four shallow, small-diameter diamond drill holes. The results were encouraging and larger diamond drilling equipment was put on the project. In February 1967, a BQ size drill hole, No. 83, intersected 88 m (290 ft) of commercial-grade mineralization. With this discovery the exploration intensity increased. Additional diamond drills were committed to the project and by 1969 the drilling, guided by the soil geochemistry made effective because of the shallow cover over the ore, had delineated 254 Mt (280 million st) of ore at 0.52% copper and 0.017% molybdenum. The ore body was located within an oval-shaped configuration 2438 m (8000 ft) long, 1067 m (3500 ft) wide, and 366 m (1200 ft) thick.

GEOLOGY

The north end of Vancouver Island, north of Rupert and Holberg Inlets, is composed chiefly of Middle Triassic to Lower Jurassic volcanic and sedimentary rocks of the Vancouver group (Muller, et al., 1974). The group consists mainly of a thick pile of basalt flows, pillows, and breccias overlain by units of limestone; shale and siltstone; argillite; and an upper sequence of bedded and massive tuffs, formational breccias, flows, and minor sediments. These rocks are intruded by Jurassic granitic rocks occurring in a northwest trending zone and are overlain by Cretaceous conglomerates, siltstones, sandstones, and graywackes.

The structure of the area is characterized by block faults and gently dipping, generally undeformed strata offset by complex sets of steep to vertical faults. Regional metamorphism in the area is low- to very low-grade. Some contact-metamorphic aureoles occur where granitic intrusive rocks intersect carbonate sediments.

The Island Copper deposit is in the shape of an inverted *U* draped over a quartz-feldspar porphyry dike (QFP) intruding pyroclastic volcanic rocks. The dike parallels the regional strike at N70°W, dips 50° to the northeast, and is bowed up longitudinally in the middle while plunging and digitating to the east and west along strike.

The north or hanging wall ore limb occurs in dark to medium green-gray andesite tuffs and QFP, with andesite breccias marginal to the dike. The crest of the ore body extends over the QFP dike and related breccias to the footwall ore limb that consists predominantly of intermixed QFP and altered andesite. Rocks in the ore zone are generally highly altered andesite. Additionally, the ore is fractured and moderately to highly silicified.

In the early years of production at Island Copper, ore came predominantly from the andesite on the north or hanging-wall side of the porphyry dike. The ore character will change to greater amounts of porphyritic ore as the pit deepens and more ore comes from the south or footwall side of the dike. It is estimated that as much as 80% of the ore is of the andesite type.

Of the ore minerals, chalcopyrite and molybdenite, 70% are located in the fracture and cleavage planes of the fine-grained highly fractured andesite. The remaining 30% of the sulfide ore mineralization is finely disseminated within the andesite or the porphyry.

Sulfide minerals in order of abundance are pyrite, chalcopyrite, and molybdenite with very minor amounts of bornite, galena, and sphalerite. Gold, silver, and rhenium are present in the ore. Silicates, magnetite, calcite, clays, and minor amounts of the hydrocarbon gilsonite make up the remainder of the host rock.

PROJECT DEVELOPMENT

Engineering and Feasibility

In 1967 Utah Construction and Mining Co.'s San Francisco-based Technical Services Group was assigned to the project to complement Exploration Group activities. The two groups working within their specific expertise proceeded with the evaluation and development of the project. Elements of the evaluation process were:

1) Exploration drilling continued.

2) A shaft was sunk and drifts crosscutting the ore body were driven to obtain bulk ore samples used for metallurgical testing. The metallurgical test work was conducted in Utah's Palo Alto laboratory and in a Utah pilot plant in Cedar City, UT.

3) Contacts with government agencies, both federal and provincial, were developed in order to determine the conditions and regulations that would be applied to a mining project.

4) The socioeconomic aspects, such as employee housing, transportation, schools, stores, and other people-oriented aspects of the development were investigated both for a construction program and for a mining operation.

5) During this period of investigation and before the final financial feasibility was completed, specific technical choices for all of the components of a mining and concentrating operation were studied, tentative choices were made, and the cost of each was estimated. These included: timber removal from the property, mining methods, economic concentrator size, concentrator location, port location, access roads, water supply, electric power supply, waste dump location, and tailings disposal.

6) Product markets for the copper concentrate and for the potential molybdenum concentrate were investigated.

7) The Utah financial policy was to evaluate for project financing. The project would stand on its own financially as a producer of copper concentrate and finally would be totally financed by Utah.

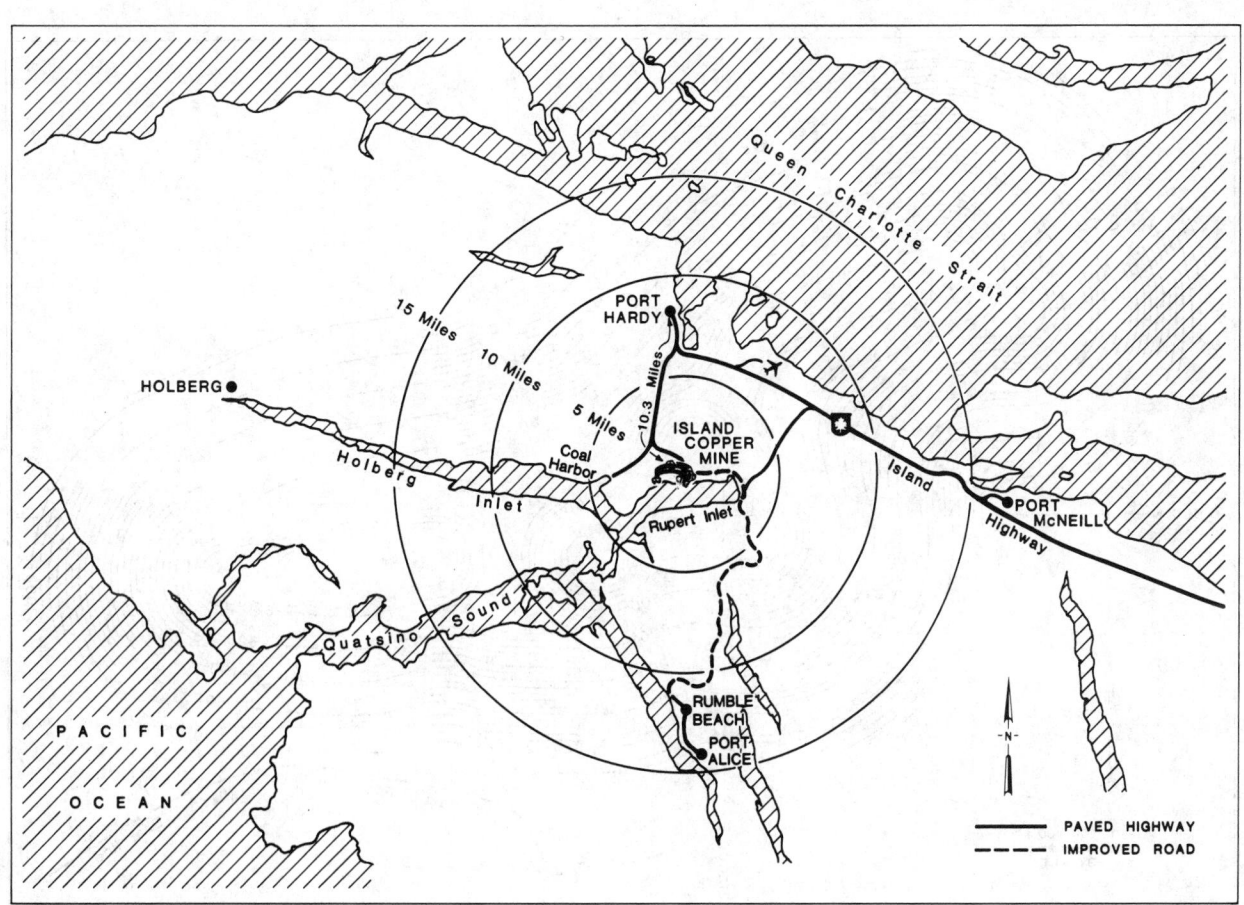

Fig. 3. Port Hardy: Island Copper mine area.

By early 1969 Utah had completed a series of feasibility studies. These evaluations were based on the discounted cash flow method using a 15% after-tax return on investment as the investment guideline. The objective of each evaluation was to:

1) Determine if a project based on a given set of assumptions satisfied the DCF investment criteria.

2) Determine how sensitive the project's profitability was to changes in various operating conditions and costs.

3) Compute the actual dollar cash flows of the project so that the effects on the company's funding arrangements could be ascertained.

4) Estimate the project's effects on the company's annual after-tax profits.

Based on the series of evaluations it was determined that:

The copper deposit was economic and not solely a mineral reserve.

A concentrator with a feed capacity of 29.9 kt/d (33,000 stpd) presented the best balanced size choice both financially and operationally.

Following the concentrator size selection, the remainder of the facility would be sized to serve the concentrator.

Mine Planning

At the completion of these assessments the selected design, designated as pit D for identification purposes, was used for final feasibility (Fig. 4). The basic tenets of this design were overall pit slopes of 45° on straight sections, about 34° on convex sections, and 37° for the glacial till overburden.

Ramps at 8% and 30.5 m (100 ft) wide, and 12-m (40-ft) working benches with 10.7-m (35-ft) safety berms spaced at 24-m (80-ft) intervals, completed the major design tenets.

Reserves were 256.7 Mt (283 million st) at 0.522% Cu and 0.029% MoS_2 with an overall stripping ratio of 2.23 t of stripping per tonne of ore. The following reserve Tables 1 and 2 show the distribution of ore and waste in pit D by grades and by bench elevations as estimated early in 1969.

Discussion with the Department of Transport had indicated that the government would approve a well designed waste dump in Rupert Inlet, and haul cycles were estimated for various mine operating periods based upon dumping half of the waste material on land and half into the water of the inlet. A marine dump was financially attractive as it would reduce the amount of lift required for much of the waste. Additionally, Rupert Inlet was up to 76.2 m (250 ft) deep, and the marine dump adjoining the pit would provide for a close large dump. Finally, the dump would increase the size of the rock barrier to be left between the pit and the inlet. Glacial till and swamp mud were excluded as not suitable for the marine dump due to instability and water discoloration by the mud and would be dumped on land.

A drilling and sampling program to determine the vulnerability to seepage of seawater through the barrier area was undertaken and the results indicated that seepage would not become a problem. As a safety measure, the Provincial Inspector of Mines requested a temporary slope on the pit side adjoining the inlet of less than 45% design slope until the pit was below sea level and the barrier wall more thor-

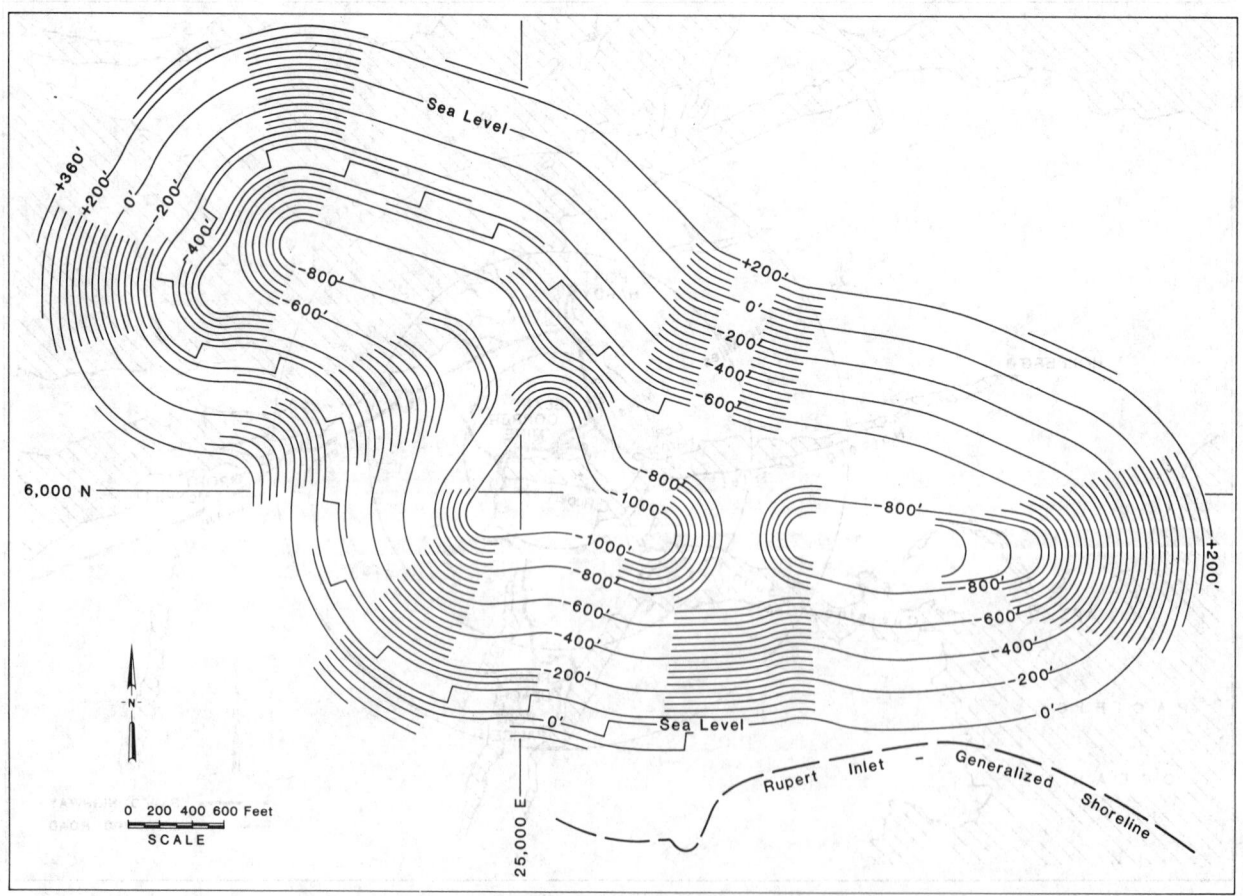

Fig. 4. Island Copper mine, pit D.

Table 1. Pit D Reserves by Grade Categories

Category	1000 St*	Average % Cu
Waste, unconsolidated overburden	82,095	—
Waste rock, 0 to 0.19% Cu	482,469	—
Marginal ore, 0.20 to 0.29% Cu	67,697	0.248
Ore, 0.30 to 0.39% Cu	78,686	0.347
Ore, 0.40 to 0.49% Cu	65,905	0.448
Ore, 0.50% Cu and over	138,762	0.656

* Metric equivalent: st × 0.907 184 7 = t.

oughly tested and evaluated. This restriction posed no initial problem and will be discussed later in this chapter.

The 29.9 kt/d (33,000-stpd) program called for 9.1 Mt (10 million st) of pre-mine stripping followed by 50.8 Mt (56 million st) of stripping during the first five years of operation distributed as shown in Table 3. Table 4 lists the initial (1969) fleet of mining equipment. Tables 5 and 6 list the current fleet of mining equipment.

CONSTRUCTION

The directors of Utah Construction and Mining Co. approved the expenditure and authorized the start-up of the project in May of 1969.

Construction was started during the summer of 1969 and the first copper concentrates were produced 28 months later in September of 1971. The rain forest was cleared and the

Table 2. Pit D Reserves

Mining Bench	St* of Ore over 0.3% Cu Equivalent	% Cu	St Overburden	St Waste Rock	Total St Stripping	Stripping Ratio Waste/Ore
360			106		106	—
320			1,082		1,082	—
280			6,059		6,059	—
240			10,995	3,021	14,016	—
200			15,307	6,708	22,015	—
160			19,946	11,134	31,080	—
120	1,344	0.53	12,897	30,013	42,910	31.93
80	4,007	0.476	6,572	39,873	46,445	11.59
40	5,025	0.52	6,354	42,692	49,046	9.76
0	7,509	0.49	2,777	44,705	47,482	6.32
−40	8,372	0.51		44,959	44,959	5.37
−80	8,251	0.54		40,708	40,708	4.93
−120	11,923	0.50		36,065	36,065	3.02
−160	12,931	0.49		32,859	32,859	2.54
−200	13,934	0.53		29,046	29,046	2.08
−240	12,913	0.50		27,781	27,781	2.15
−280	14,629	0.54		24,544	24,544	1.68
−320	18,081	0.51		18,855	18,855	1.04
−360	16,648	0.50		18,155	18,155	1.09
−400	16,601	0.52		15,952	15,952	0.96
−440	13,265	0.53		15,120	15,120	1.14
−480	15,809	0.51		12,236	12,236	0.77
−520	13,172	0.58		12,425	12,425	0.94
−560	12,987	0.55		9,965	9,965	0.77
−600	13,211	0.53		8,504	8,504	0.64
−640	12,155	0.49		7,252	7,252	0.60
−680	11,916	0.52		6,035	6,035	0.51
−720	7,899	0.56		4,429	4,429	0.56
−760	7,485	0.54		3,197	3,197	0.43
−800	7,153	0.55		1,893	1,893	0.26
−840	4,760	0.57		1,103	1,103	0.23
−880	4,044	0.57		593	593	0.15
−920	2,868	0.55		224	224	0.08
−960	2,501	0.54		40	40	0.02
−1000	1,960	0.51		80	80	0.04
Total	283,353	0.52	82,095	550,166	632,261	2.23

* All quantities in thousands of short tons. Metric equivalent: st × 0.907 184 7 = t.

Table 3. Island Copper Mining Schedule for 33,000 Stpd

| Period | Production during Period | | | | Waste Ratio | Material Available at End of Period | | | | |
| | Ore | | Waste, St | | | Ore Uncovered | | Ore Exposed in Bank | | |
	St*	Grade	Marginal	OB & Waste		St	Grade	Linear ft	Days of Ore	Grade
Pre-mine	—	—	476	9,524		1,738	0.44	2,615	11	0.49
Year 1	9,900	0.52	1,837	21,923	2.40:1	3,634	0.56	1,710	7	0.50
Years 2-5	46,200	0.52	4,569	106,311	2.40:1	5,598	0.50	5,820	24	0.51
Total Pre-mine through year 5	56,100	0.52	6,882	137,758	2.58:1†	5,598	0.50	5,820	24	0.51

* Thousands of short tons. Metric equivalents: st × 0.907 184 7 = t; ft × 0.3048 = m.
† Includes pre-mine stripping on 10,000,000 St.

area transformed into an operating open pit mine equipped to feed the newly constructed 29.9 kt/d (33,000-stpd) concentrator. A port was built on Rupert Inlet complete with barge and ship docks. A water system with pumping station on Lake Alice and 16 km (10 miles) of water pipeline to the concentrator was installed. Access roads were built and a housing subdivision was constructed in Port Hardy. The British Columbia Hydro and Power Authority constructed 193 km (120 miles) of 138 kV power transmission line from the Strathcona generating station near Campbell River on Vancouver Island to furnish the mine with electricity. The diesel-powered generating station at Port Hardy did not have the capacity to provide sufficient electrical energy to supply the plant demand of approximately 44 760 kW (60,000 hp).

Because a 64-km (40-mile) gap in the public highway along the island existed between Kelsy Bay and Beaver Cove on Vancouver Island, truck transport for construction material and operating supplies was not practical; therefore, the construction materials and supplies were assembled in a marshalling yard in Vancouver and barged up the Inland Waterway around the north end of Vancouver Island into Quatsino Sound and offloaded on a beachhead at the Rupert Inlet mill site.

The overall design of the project was directed and controlled by Utah's Technical Service Group under C. K. McArthur with detailed design and construction by Fluor-Utah Engineers and Contractors Inc. of Menlo Park, CA. Detailed engineering in Canada was contracted by the H. A. Simmons Co. of Vancouver, BC, operating under Fluor-Utah's supervision.

At the start of construction and to complement the efforts of the San Francisco-based management and design groups, the mine operating group (Utah Mines Ltd.) was established

in Vancouver and individuals moved as required to Port Hardy during the two-year construction period. The group started with a manager, mine superintendent, administrative manager, mill superintendent, chief mine engineer, owners' construction representatives, and office manager, and grew to a full operating crew at the site 28 months later when operation started.

During the two-year period of plant construction, the mine was being prepared for production. The rain forest was logged and stumps removed and burned, streams were relocated, swamps were drained, and roads were built to the areas initially required for stripping and mining. The pre-mine waste stripping program of 9.1 Mt (10 million st) was relatively minor. This was fortunate as the use of the 11.5-m³ (15-cu-yd) electric shovels selected for the project could await the completion of the electric power transmission line from down island, rather than having to install temporary power for a longer pre-mine program. The transmission line project, initially planned for completion in November 1970, was completed and the pit electrified in May 1971, four months before the scheduled start-up. When it became apparent that the initial schedule for pit electrification could not be met, the scope of work for the contractor retained to build the access roads was increased to include the stripping of waste and starting the development of mining faces. After the electricity arrived, the production equipment took over this development program. Also, an unforseen provincially-imposed requirement for the construction of an emergency mill tailings basin prior to the start of production forced the unscheduled quarrying of approximately 0.9 Mt (1 million st) of waste. Because of tidal action, the quarried waste had to be physically coarse and blocky as it was placed in marine dams used to cordon off a section of the inlet. Unfortunately,

Table 4. Mining Equipment

Earth Moving Equipment Included in the Original (1969) Start-up Program*	
3	P&H 2100B mining shovels
9	Unit Rig M-120 end dump trucks
2	Caterpillar Model 14 graders
1	Caterpillar Model 12 graders
1	Bucyrus-Erie 452 diesel-powered rotary drill
2	Bucyrus-Erie 60R electric-powered rotary drills
1	Caterpillar D-9 tractors
3	Caterpillar D-8 tractors
1	Caterpillar 988 loader
1	Caterpillar 824 wheel tractor

* As the concentrator's production rate increased, the size of the mining fleet was increased.

Table 5. Production* Per Working Day 1972-1982

| Period | Ore to Mill | | | | Waste to Dumps | | | | |
	Ore, 0.30 %	Marginal, 0.20 to 0.30 % Cu	St Total Ore	Waste Rock & Stockpiled Marginal Ore	Overburden Till	Total Waste	Total Production	Stripping Ratio, Tons Waste:Tons Ore
1972	22.5	0	22.5	33.5	37.3	70.8	93.3	3.14:1
1973	30.7	1.0	31.7	45.3	33.6	78.9	110.6	2.49:1
1974	35.8	2.9	38.7	53.9	21.9	75.8	114.5	1.96:1
1975	35.7	1.3	37.0	89.0	19.2	108.2	145.2	2.92:1
1976	36.5	1.3	37.8	92.8	15.1	107.9	145.7	2.85:1
1977	31.1	3.4	34.5	100.5	21.4	121.9	156.4	3.54:1
1978	30.1	10.4	40.5	122.3	8.3	130.6	171.1	3.23:1
1979	32.7	8.5	41.2	125.7	4.0	129.7	170.9	3.15:1
1980	36.9	5.6	42.5	116.8	7.4	124.2	166.7	2.92:1
1981	35.4	7.7	43.1	109.0	7.4	116.4	159.5	2.70:1
1982	42.2	4.9	47.7	118.4	4.4	122.8	170.5	2.59:1
1 yr. average	33.4	4.3	37.7	92.0	16.4	108.4	146.1	2.93:1†

* In hundreds of short tons. Metric equivalent: st $\times$ 0.907 184 7 = t.
† Includes pre-1972 production.

Table 6. Summary of Equipment Performance, Major Pieces of Earth Moving Equipment Currently in Use

No. of Units	Class of Equipment	Equipment Age,* Hours of Operation			Performance		
		Minimum	Maximum	Average	Mech. Eff., %	Utilization, %	Mech. Util., %
2	Marion 191-M mining shovels	16,616	17,915	16,766	74.3	65.2	87.8
3	P&H 2100-BL mining shovels	40,913	59,562	50,657	79.2	68.6	86.4
1	P&H 2100-B mining shovel, modified with a GB Corbody	—	—	50,307	80.4	68.9	85.7
11	Euclid R-170 end dump trucks, 154.2-t (170-ton) capacity	5,896	22,097	15,035	67.6	65.9	97.5
17	Unit Rig Mark 36 end dump trucks 154.2-t (170-ton) capacity	31,179	43,587	37,782	66.7	65.1	97.6
14	Unit Rig M-120 end dump trucks, 108.9-t (120-ton) capacity	46,608	55,185	52,130	59.2	36.9	62.2
5	Caterpillar Model 769B+C end dump trucks, 31.8-t (35-ton) capacity, one converted to water truck, 4 in use	800	20,940	10,560	68.8	22.0	32.0
1	Mack 35T end dump truck, converted to water truck	—	—	9,082	75.2	11.8	15.7
5	Caterpillar Model 16 graders	3,265	23,909	17,620	73.3	34.5	47.0
1	Caterpillar Model 14 graders	—	—	22,491	75.6	27.1	35.9
1	Caterpillar 988 wheel loader	—	—	10,434	84.3	31.3	37.1
2	Bucyrus-Erie 60R rotary drills, Series 2, EDE53 + 54	57,576	59,949	58,762	79.6	41.8	52.5
2	Bucyrus-Erie 60R rotary drills, EDE59 + 70	28,625	40,530	34,578	78.8	43.9	55.7
2	Bucyrus-Erie 45R rotary drill†	—	—	40,237	82.8	50.3	60.7
4	Caterpillar 824B wheel tractors, 3 in use	6,786	23,110	12,963	83.5	45.6	54.6
3	Fiat-Allis Model 41B tractors	3,831	7,856	5,844	71.6	52.4	73.2
9	Caterpillar D9H tractors, 8 in use	7,644	24,269	16,110	74.1	41.4	55.8
1	Caterpillar D8K tractor	—	—	12,139	80.5	19.9	24.7
1	Gradall	—	—	12,001	66.3	3.9	5.9
1	Backhoe, new	—	—	—	—	—	—

* Ten months average, 1982.
† Purchased as a diesel/electric in 1969 and subsequently converted to full electric.

the waste from the scheduled pre-mine stripping was fine and was not suitable for marine dams. Because of this requirement for coarse materials, a part of the truck and shovel hours scheduled for the pre-mine stripping program was diverted to build this tailings basin and the tonnage lost had to be made up after production started. The mining operation was able to cope with these unforseen requirements and to deliver ore to meet the mill's starting date.

MINING OPERATIONS

General

The Island Copper ore body is mined by open-pit methods. The final pit will be 2438 m (8000 ft) long, 1067 m (3500 ft) wide, and 488 m (1600 ft) deep, extending 378 m (1240 ft) below sea level. It is known that copper mineralization continues below the design depth of the pit and the future economics and technology of copper mining will determine the economic life of the ore body beyond current planning. Table 4 is a list of the equipment purchased to start the mining operation.

The production plan selected and adhered to for the past years has been modestly flexible and capable of producing at the stripping ratio required to keep the mill in ore while providing a reasonable quantity of immediately available ore to be used in the event of a catastrophic failure. Due to the physical configuration of the ore body and the land form over the ore body this planning has resulted in:

1) A minor pre-mine stripping program as pointed out previously.

2) A buildup to a higher than average stripping ratio in the early years of operation.

3) A tapering off in the latter years to a less-than-average stripping ratio.

The development of the pit over the past ten years has resulted in only minor changes in ore locations, minor pit slope changes, and some access design change. The total quantities remain essentially unchanged from the pit D design (Fig. 4).

The Island Copper ore reserves initially were calculated by the polygon method. Over the years, the tons and average grade mined as measured by the blasthole assays have compared closely to the tons and grade predicted by polygons generated by the diamond drill holes and the assays. Ore tons were underestimated by less than 2.5%, the grade overestimated by less than 3.5%, and the contained copper overestimated by about 1%. In 1979 in order to refine this measurement system, a regression of the mean actual polygon grade, using the blasthole grade, for classes of predicted polygon grades showed that the high-grade drill-hole assay assigned to polygons overestimated the actual polygon grade and that assigned low-grade assays underestimate the actual grade. To improve the grade predictions within individual polygons, correction factors derived from the regression curve have been used for the past three years in ore reserve calculations. To date, geostatistical reserve estimation methods such as kriging have been tested at Island Copper but have not been used to generate reserves. The continuity of ore in the deposit is such that reserve estimation errors from using polygons have not been a problem.

In 1978 the design of the road grades was increased from 8 to 10%. This design modification resulted in a minor reduction in the overall amount of stripping required.

After 11 years of operation and experimentation in design and blasting practice, the overall pit slopes vary from 50%

on the north, andesite, and final wall to 40% in the porphyries and the porphyritic mixtures. The steep to vertical faults mentioned in the geologic part of this section exercise degrees of local control over the pit slopes.

The barrier wall between the pit and Rupert Inlet is slightly flatter than the 45% design. An alluvium or till area below sea level was removed and refilled to proper compaction standards. There are no indications of salt water seeping into the pit.

Mining in this rain-forested basin provides unique challenges. First, the ore body and waste rock were overlain by glacial till that ranged in thickness from 4.6 m (15 ft) in the center of the pit to 76.2 m (250 ft) on the east and west ends. Benches are mined to a height of 15.2 m (50 ft) in glacial till and are mined to a 12-m (40-ft) height in rock. The removal of glacial till was efficient and at a high production rate only during the dry season (normally July and August) and occasionally during the winter, if the ground was frozen to a depth that would support the haul trucks. Unfortunately, the normal procedure in till was costly, as the wet till turned to mush under the weight of the trucks or shovels and required a pad of 0.9 to 1.8 m (3 to 6 ft) of shot rock on the mining bench in order to support the shovel and the trucks. The second requirement was to ditch left or right, close to and along the face, to drain the till, and to confine the water to the ditch in order to keep the bench as dry as possible. The extra work required reduced the production rates and forced the rehandling of the shot rock.

Second, the swamps that were scattered over the area were removed with the mining shovels, usually by mining into the swamp from a lower elevation with the shovel operating on a thick rock foundation with water diverted into ditches away from the shovel. Occasionally the mud flow feeding the shovel would be at a rate where the shovel stayed at one spot and dug for hours. More commonly the mud either came too fast or too slowly, which forced the shovel to retreat from the face or caused a delay while the rock pad was advanced. Either event reduced the rate of production.

The andesite used for the shovel pads had to be mined in conjuction with the till or swamp mud as this shot rock was the cheapest shovel pad and surfacing material available. Andesite, the best of the poor road surfacing material available, pulverizes under the heavy weight of the loaded truck's wheels, which results in mud on the roads, at the mining faces, and in the drainage ditches most of the year.

Mine operations, including supporting maintenance work, are continuous and on a schedule of three shifts a day, for seven days a week. Operations in the pit are controlled by a shift supervisor and two principal foremen, one supervises production, and one is assigned to pit services, including pit dewatering, road maintenance, and dump maintenance. Additionally, routine traffic in the pit is supervised from a dispatch office located on high ground at the edge of the pit. This traffic supervisor in the observation post can see all vehicles and equipment in the pit, and dispatch the trucks to the shovels by radio. Additionally, this dispatcher/supervisor acts as a second set of eyes for the shift foremen in preventing accidents and equipment breakdowns as well as the primary assignment of allocating trucks to ensure the most efficient balance of shovels and trucks. Banks of floodlights provide adequate visibility during hours of darkness.

Drilling and Blasting

Drilling is by electric rotary drills. The holes are 251 mm (9⅞ in.) in diam drilled to grade in glacial till and to 1.2 m (4 ft) below grade in rock. The glacial till could be dug unshot; however, shovel productivity increased materially when the till was drilled on wide spacing and shot lightly.

The rotary drills used are the Bucyrus-Erie 60Rs and one 45R. The 45R was purchased equipped as a diesel-electric and subsequently converted to full electric. The purpose of the 45R diesel was to provide drilling capacity for the pioneering work with a machine lighter than the 60Rs and with diesel power to permit it to operate in advance of the principal mining areas.

Drill patterns vary according to local rock conditions with detailed engineering and production monitoring to ensure efficiency. The current practice is generally as follows:

Material	Spacing	Powder Factor
Andesite	8.2 x 8.2 m (27 x 27 ft)	0.47 kg/m³ (0.8 lb/yd³)
Porphyry	7.6 x 7.6 m (25 x 25 ft)	0.59 kg/m³ (1.0 lb/yd³)
Till	9.1 x 9.1 m (30 x 30 ft)	0.30 kg/m³ (0.5 lb/yd³)

Initially explosives used were of three types, (1) ANFO in bulk with wet blasthole charges kept dry with plastic waterproof hole liners, (2) packaged water repellant types of ANFO products, and (3) packaged slurries for very wet holes and holes that could not be pumped dry. Since 1978 the only explosive used has been waterproof slurry manufactured by a contractor in a truck-mounted plant at the blasthole and delivered to the blasthole by hose.

Loading

Broken material from the pit is loaded into trucks by P&H or Marion electric shovels equipped with 11.5-m³ (15-cu-yd) buckets. Normally four or six shovels operate to provide the tonnages required.

The original loading shovels were P&H 2100 B mining shovels and now are P&H 2100 BLs and Marion 191M mining shovels with one 2100B left for emergencies and stockpiles. The 11.5-m³ (15-cu-yd) bucket is the standard and normally one shovel will operate with four to six trucks.

Hauling

The ore and waste haul trucks are diesel-powered with final drive by electric motors in each rear wheel. The initial haul truck fleet at the mine utilized the Unit Rig M120 truck, each of 108.9-t (120-ton) nominal capacity equipped with 59.6-m³ (78-cu-yd) capacity boxes. The effective payload for this truck averages 104.3 t (115 st) of ore or 99.8 t (110 st) of waste. In 1974 the fleet was augmented with Unit Rig Mark 36 trucks at a rated capacity of 154.2 t (170 ton) with 83.3 m³ (109-cu-yd) struck box capacity. Since 1979, truck additions have been the Euclid Model R-170 rated at 154.2 t (170-ton) capacity and carrying a 145-t (160-st) payload. The engines used are the General Motors V-12-149-T in the 108.9-t (120-ton) trucks and the General Motors V-16-149-T1 in the 154.2-t (170-st) class trucks. The electric motors in the wheels are the General Electric 772 series in the 108.9 t (120-ton) and the General Electric 776 series in the 154.2-t (170-ton) trucks.

Ore is hauled from the pit to the crusher over a distance of 1.6 to 2.4 km (1 to 1.5 miles), involving a maximum climb of up to 204 m (670 ft) from the bottom operating level 146 m (480 ft) below sea level to the crusher at 58 m (190 ft) above sea level. Haul roads are 36.6 m (120 ft) wide and ramp grades in the pit are 10%. Waste is dumped on land to the north of the pit and into a low-level marine fill to the south of the pit.

High rolling resistance due to the wet and muddy conditions is a serious trucking cost factor at this property. The rolling resistance on an average haul road during an average wet day at Island Copper is in the order of 2 to 3% higher for a given grade than it would be on a dry desert road or on a frozen road.

Productivity

The 1982 average rates of productivity are as follows:

Mining Rates

Ore	43.3 kt/d	(47,700 stpd)
Waste	111.4 kt/d	(122,800 stpd)
Total	154.7 kt/d	(170,500 stpd)

Loading Rate

Shovels	1.5 kt/operating h	(1692 st per operating hr)

Hauling Rates

Trucks—170T trucks or 170T truck equivalents	276 t/operating h	(304 st per operating hr)

Drilling Rate

Drills	22 m of penetration per operating hour	(73 ft of penetration per operating hour)

The designed capacity of the plant was a nominal 29.9 kt/d (33,000 stpd) which, considering the estimated 95.76% onstream factor, calculates to 28.7 kt/d (31,600 stpd) on an annual basis. The capacity was increased to 32.7 kt/d (36,000 stpd) in 1973 with the addition of three secondary ball mills. As the techniques of operating the six semiautogenous mills that are the prime grinding tools improved, it became possible to increase the plant throughput and the mine capacity increased accordingly.

Additionally and in response to copper price changes, it became economically attractive to process material in the low-grade category, ores containing between 0.20 or 0.30% copper. With the additional grinding capacity, the practice has been to treat low-grade ore during a copper market recession and higher-grade ore when the market was high. This system, while not the common operating pattern, has best fulfilled Utah's needs. Table 5 gives the average production rates and stripping ratios for the years 1972 through November of 1982.

Production from the pit, originally designed at 101.6 kt/d (112,000 stpd) of ore and waste, was increased to as high as 155.1 kt/d (171,000 stpd) in 1978. At the time of writing, production is at 43.3 kt/d (47,700 stpd) ore and low-grade ore and 111.4 kt/d (122,800 stpd) waste for a total of 154.7 kt/d (170,500 stpd).

The stripping ratio over the 11 year period since start-up has been 2.93 t waste to 1.00 t ore. This ratio is scheduled to decrease in future years to give a final overall ratio of the pit as initially designed of 2.1:1.

Auxiliary Activities

The wet climate and the requirements of pioneering new benches amidst the stumps of a logged-off rain forest have made it necessary to have a larger fleet of auxiliary vehicles than normally would be required for a mine of this size.

The auxiliary fleet includes five graders, three rubber-tired dozers, and large-sized track laying dozers. A front-end loader and 31.8-t (35-ton) class truck loading and hauling group is used for miscellaneous auxiliary work, including mud cleanup and the initial development of benches suitable for the large equipment. For ditching and ditch cleanup a

Gradall and a backhoe are used with the aforementioned auxiliary equipment.

Electrical power is used for the mining shovels, rotary drills, pumps, and pit lights. The pit is supplied from the main transformer station located adjacent to the mill by a conventional power transmission line fed by two 13.2 kV feeders. This transmission line circles the pit and short overhead lines, with the poles on movable bases, feed down the pit walls where needed to 2000-kVA mobile substations located on lowboy trailers within the pit. The circle or ring configuration of the main feed line provides for maximum flexibility in the location of substations and in balancing electrical loads. The mobile transformer stations in the pit convert 13.2-kV main supply to 4160 V for major equipment and to 575 V for auxiliary services. The transformers are moved to serve new working areas and are located near the shovel and drill sites. Electrical cables from the transformers to the mining machines are equipped at both ends with male couplings. The cables are connected through portable junction boxes. This configuration precludes error during cable stringing and the elevated junction boxes keep the couplings out of the mud. The cables are protected in working areas by markers and in road crossovers placed in ground-level rubber crossover channels or on overhead gantries.

Water is collected in the pit by gravity and by pumping diverted to a main sump normally located a level or two above the bottom level of the pit. The main sump, or sumps, have settling areas for mud removal and are equipped with parallel sets of main pumps. The first set is electrically driven and the backup set is powered by diesel engines. The diesel units provide pumping capacity for peak demands, as well as the capability to pump water from the pit during an electric power outage.

The current maximum pump capacity is 0.757 m^3/s (12,000 gpm) and is scheduled to be increased to 0.946 m^3/s (15,000 gpm). At the present time the water is pumped from the 680 bench sump, 97.5 m (320 ft) below sea level to 3 m (10 ft) above sea level. Water pumped from the pit is initially discharged into a silt settlement basin adjoining Rupert Inlet and from there percolates into the salt water of Rupert Inlet.

Maintenance

The Island Copper maintenance crews are on the schedule of the pit operating crews, i.e., three shifts per day seven days per week for the shift electricians and for the truck maintenance crews. The specialty shops, such as auxiliary equipment, lubrication, tires, field preventive maintenance, small vehicle, component rebuild, welding, and electrical, work two shifts per day seven days per week.

At Coal Harbor the Detroit diesel distributor installed a diesel engine rebuild shop for GM engines in 1971. This local service fulfilled the need for the engine rebuild shop that might be expected in this semi-remote area considering the number of truck diesel engines being operated by the mine. The Caterpillar Tractor Co. dealer established a general maintenance shop in Port Hardy in the late 1970s. Engines are transported to Vancouver for major rebuilding.

Electric wheel motors are rebuilt by the General Electric agent W. A. Stevens Co. in Vancouver or by Island Copper's component rebuild group.

Table 6 delineates the age of the equipment in use and the machine performance. The performance efficiencies were calculated on the following basis and using the following formulae:

1) The reporting of hours is on a 24-hr day basis.

2) Mechanical Efficiency:

$$\frac{\text{Hours Operated}}{\text{Hours Operated} + \text{Downtime Hours}}$$

3) Utilization:

$$\frac{\text{Hours Operated}}{\substack{\text{Elapsed Time} \\ \text{(Hours per Month or Year)}}}$$

4) Mechanical Utilization:

$$\frac{\text{Hours Operated} + \text{Downtime Hours}}{\text{Elapsed Time}}$$

Table 7 is a summary giving the distribution of equipment operating costs for a six-month period during 1982. These site- and time-specific costs reflect the use of new equipment on the most arduous work and as such are not a brand endorsement. The practice is to schedule new trucks for the haul from the bottom levels of the pit and the old trucks for the more nearly level hauls. The effect of this preferential scheduling is immediately apparent in operating costs and is also shown in the relatively high mechanical efficiency achieved by old equipment [59.2% for 50,000-hr 108.9-t (120-ton) trucks vs. 67.6% for 15,000-hr 154.2-t (170-ton) trucks as shown in Table 5].

The older equipment when scheduled for fewer hours maintains relatively low costs as shown in Table 6. These older machines are maintained during lulls in the maintenance schedule and components are run longer than could be expected for mainline production equipment.

Table 8 summarizes the capital investment made to bring the Island Copper mine into production. The author has used his judgment in distributing the costs of common facilities between the mine and the concentrator. The components of the mixture of expenditures, such as $2.5 million for housing when a similar deposit in another location might require no expenditures for housing or more than the Island Copper expenditure, illustrate the variability in capital costs for specific developments.

Table 9 lists the Canadian dollar operating costs for an eight-month period in 1982. In this period of rapidly changing costs they reflect only a point in time and cannot be evenly factored to give future costs. With time, the pit will deepen, the quantity of water to be pumped will increase, and fuel prices change, to mention just three of the many cost elements that will make up future mining costs at the Island Copper Mine.

REFERENCES

Brown, C.M., 1974, "Island Copper Mine, Milling for Copper and Molybdenum," reprint from *Western Miner*, Dec.

Brown, C.M., 1982, "The Selection of Instrumentation and Control Systems for Semi-Autogenous Grinding Circuits," *Design and Installation of Comminution Circuits*, A.L. Mular and G.V. Fergensen, II, eds., AIME, New York.

Evans, J.B., Ellis, D.V., and Pelletier, C.A., 1972, "The Establishment and Implementation of a Monitoring Program for Underwater Tailing Disposal in Rupert Inlet, Vancouver Island, British Columbia," *Tailing Disposal Today*, C.O. Aplin, and G.O. Argall, Jr., eds., Miller Freeman Publications, San Francisco, pp. 512–552.

Cargill, D.G., et al., 1976, *Island Copper*, Canadian Institute of Mining & Metallurgy Special Vol. No. 15, pp. 206–218.

Table 7. Summary of Equipment Costs, Major Pieces of Earth Moving Equipment Currently in Use

Class of Equipment	Operating Hrs Per Month 6 Mons. 1982 Group/Unit	Maint. Labor	Repair Parts	Fuel and Lube	Tires or Tracks	Wire Rope	Edges/Bits	Power	Outside Labor	Engine or Bucket	Power Train w/Motor	Total $ Canadian 1982 per Op. Hr
1 – P&H 2100B	518/518	31.63	23.30	4.32	-	8.44	2.26	23.16	-	1.34	1.63	96.08
2 – Marion 191-M mining shovels	982/491	38.40	25.51	4.97	0.28	9.18	10.21	23.19	0.04	15.26	-	127.04
3 – P&H 2100BL mining shovels	1493/498	39.19	27.65	5.36	4.46	10.59	9.35	23.14	0.01	11.79	3.33	134.87
11 – Euclid R-170 end dump trucks	5223/475	14.04	20.69	43.44	25.75	-	-	-	0.03	8.96	6.99	120.85
17 – Unit Rig Mark 36 end dump trucks	8183/481	12.08	15.94	44.33	26.44	-	-	-	0.02	6.12	5.02	105.95
14 – Unit Rig M-120 end dump trucks	3461/247	14.60	18.26	28.47	14.36	-	-	-	0.02	0.62	4.38	80.71
5 – Caterpillar 769 end dump trucks	778/156	12.67	5.48	8.59	3.82	-	-	-	1.12	0.29	2.19	34.16
1 – Mack 35T converted to water truck	121/121	14.39	4.81	12.13	6.69	-	-	-	-	-	0.90	38.92
5 – Caterpillar Model 16 graders	1018/204	4.32	2.49	7.60	2.85	0.54	3.98	-	0.11	0.01	0.80	22.16
1 – Caterpillar Model 14 grader	159/159	2.31	0.94	1.04	0.61	-	0.28	-	-	-	-	5.18
1 – Caterpillar Model 988 loader	200/200	4.90	2.33	11.07	-	-	1.34	-	-	-	-	19.64
2 – Bucyrus-Erie 60R-2 rotary drills	619/310	14.79	10.74	2.81	0.01	0.03	37.19	9.64	-	0.02	-	75.23
2 – Bucyrus-Erie 60R-3 rotary drills	705/353	14.88	17.00	3.24	3.33	0.03	31.87	9.59	-	-	1.25	81.19
1 – Bucyrus-Erie 45R rotary drill	378/378	10.90	17.86	2.95	-	0.17	21.87	9.77	-	-	-	63.52
4 – Caterpillar 824-B wheel tractor	1328/332	3.74	4.44	6.73	1.57	-	1.87	-	0.22	0.05	0.43	19.05
3 – Fiat-Allis Model 41B tractors	622/207	9.62	4.92	16.93	1.06	-	1.71	-	0.01	0.13	0.63	35.01
9 – Caterpillar D9-H tractors	2771/308	7.61	3.33	11.72	3.94	-	1.35	-	0.96	0.15	1.38	30.44
1 – Fiat-Allis Model 31 tractor	22/22	7.97	6.25	16.46	-	-	-	-	-	0.29	-	30.97
1 – Caterpillar D-8K tractor	146/146	6.78	4.48	9.37	4.44	-	1.04	-	0.47	0.84	0.35	27.77

Table 8. Mine Capital Costs 1969–1970

	Project US$ 1000s	Mine Dept., US$ 1000s
Pre-mine development*	15,883	15,883
Townsite, land and houses (half to mine)	2,509	1,255
Mining equipment	11,130	11,130
Camp buildings and facilities (half to mine)*	568	284
Subtotal		28,552
Buildings		
Assay laboratory (half to mine)*	388	194
Maintenance shop	1,158	1,158
Grease building	442	442
Plant warehouse (half to mine)*	263	132
Dock warehouse (half to mine)*	132	66
Office building (half to mine)*	166	83
Change house (75% to mine)*	257	,193
Subtotal		2,268
General facilities		
Water supply (10% to mine)*	4,184	418
Electrical system (15% to mine)*	6,710	557
Sanitation and auxiliary equipment (50% to mine)*	258	129
Subtotal		1,104
Grand Total		31,924

* Based on the author's estimate of use by the mine department.

Note: At the start of construction the Canadian dollar was pegged at 93¢ US. When freed to float in 1969, the exchange moved up to around 98¢ for the remainder of the construction period.

Hollister, V.F., 1978, *Geology of Porphyry Copper Deposits of the Western Hemisphere*, AIME, New York.

Lenton, W., et al., 1974, "Panel Discussion of Primary Autogenous Grinding in the Copper Industry," *Trans. SME-AIME*.

Pelletier, C.A., et al., 1982, "Round Table Seminar on Submarine and Lake Disposal of Mill Tailing," *Proceedings*, 14th International Mineral Processing Congress, Canadian Institute of Mining & Metallurgy, Montreal.

Waldichuk, M., and Buchanan, R.J., 1980, "Significance of Environmental Changes Due to Mine Waste Disposal into Rupert Inlet," Fisheries and Oceans Canada, British Columbia Ministry of Environment, Nov. 15.

Witherly, K.E., 1979, "Geophysical and Geochemical Methods Used in the Discovery of the Island Copper Deposit, Vancouver Island, British Columbia," Economic Geology Report 31, Geological Survey of Canada, pp. 685–696.

Young, M.J., and Rugg, E.S., 1971, "Geology and Mineralization of the Island Copper Deposit," *Western Miner*, Vol. 44, No. 2, Feb., pp. 31–40.

Table 9. Mine Operating Costs
Dollars Canadian per tonne of Ore or Waste, 1982

	$/t Canadian	$/St Canadian	$/t US	$/St US
Direct costs				
Drilling	0.04	0.04	0.04	
Blasting	0.11	0.10	0.09	
Loading	0.11	0.10	0.09	
Hauling	0.55	0.50	0.44	
Subtotal	0.81	0.74	0.66	
Indirect costs				
Supervision	0.03	0.03	0.03	
Engineering and geology	0.02	0.02	0.02	
Pit road maintenance	0.04	0.04	0.04	
Pit dewatering and electrical	0.02	0.02	0.02	
Shops	0.11	0.10	0.09	
Subtotal	0.26	0.21	0.20	
Total mining costs	1.07	0.95	0.86	
Conversion to US dollars @ 0.8:1.				0.76

Index

R
M $ 58.50
SM $48.50
LP $83.50